Smithells Metals Reference Book

Smithells Metals Reference Book

Eighth Edition

Edited by
W. F. Gale *PhD*
Professor, Auburn University,
Materials Research and Education Center, Auburn, AL, USA

T. C. Totemeier *PhD*
Staff Scientist, Idaho National Engineering and Environmental Laboratory,
Idaho Falls, ID, USA

A Co-Publication with

ASM INTERNATIONAL

The Materials Information Society

ELSEVIER
BUTTERWORTH
HEINEMANN

AMSTERDAM • BOSTON • HEIDELBERG • LONDON • NEW YORK • OXFORD
PARIS • SAN DIEGO • SAN FRANCISCO • SINGAPORE • SYDNEY • TOKYO

Elsevier Butterworth-Heinemann
The Boulevard, Langford Lane, Kidlington, Oxford, OX5 1GB
200 Wheeler Road, Burlington, MA 01803, USA

First published 1949
Second edition 1955
Third edition 1962
Fourth edition 1967
Fifth edition 1976
Reprinted 1978
Sixth edition 1983
Seventh edition 1992
Paperback edition (with corrections) 1998
Eighth edition 2004

British Library Cataloguing in Publication Data
A catalogue record for this book is available from the British Library

Library of Congress Cataloguing in Publication Data
A catalogue record for this book is available from the Library of Congress

ISBN 0 7506 7509 8

For information on all Elsevier Butterworth-Heinemann publications
visit our website at http://books.elsevier.com

Typeset by Charon Tec Pvt. Ltd, Chennai, India
Printed and bound in The Netherlands

Contents

Preface   xv

Acknowledgements   xvii

Disclaimer   xviii

List of contributors   xix

1   **Related designations**   1–1

Related designations for steels – Related designations for aluminium
alloys – Related designations for copper alloys – Related designations for
magnesium alloys

2   **Introductory tables and mathematical information**   2–1

2.1   Conversion factors   2–1
      SI units – Conversion to and from SI units – Temperature conversions,
      IPTS-48 to IPTS-68 – Corrosion conversion factors – Sieve Nos to
      aperture size – Temperature scale conversions
2.2   Mathematical formulae and statistical principles   2–13
      Algebra – Series and progressions – Trigonometry – Mensuration –
      Co-ordinate geometry – Calculus – Introduction to statistics

3   **General physical and chemical constants**   3–1

      Atomic weight and atomic numbers – General physical constants –
      Moments of inertia – Periodic system
3.1   Radioactive isotopes and radiation sources   3–5
      Positron emitters – Beta energies and half-lives – Gamma energies
      and half-lives – Nuclides for alpha, beta, gamma and neutron sources

4   **X-ray analysis of metallic materials**   4–1

4.1   Introduction and cross references   4–1
4.2   Excitation of X-rays   4–1
      X-ray wavelengths
4.3   X-ray techniques   4–11
      X-ray diffraction – Specific applications – Crystal geometry
4.4   X-ray results   4–39
      Metal working – Crystal structure – Atomic and ionic radii
4.5   X-ray fluorescence   4–49

5 Crystallography   5–1

5.1   The structure of crystals   5–1
      Translation groups – Symmetry elements – The point group –
      The space group
5.2   The Schoenflies system of point- and space-group notation   5–3
5.3   The Hermann–Mauguin system of point- and space-group notation   5–3
      Notes on the space-group tables

6 Crystal chemistry   6–1

6.1   Structures of metals, metalloids and their compounds   6–1
      Structural details – Comparison of Strukturbericht and
      Pearson nomenclature

7 Metallurgically important minerals   7–1

Minerals, sources, and uses

8 Thermochemical data   8–1

8.1   Symbols   8–1
8.2   Changes of phase   8–2
      Elements – Intermetallic compounds – Metallurgically important
      compounds
8.3   Heat, entropy and free energy of formation   8–8
      Elements – Intermetallic compounds – Selenides and tellurides –
      Intermetallic phases
8.4   Metallic systems of unlimited mutual solubility   8–15
      Liquid binary metallic systems
8.5   Metallurgically important compounds   8–20
      Borides – Carbides – Nitrides – Silicides – Oxides – Sulphides – Halides –
      Silicates and carbonates – Compound (double) oxides – Phosphides –
      Phosphides dissociation pressures – Sulphides dissociation pressures
8.6   Molar heat capacities and specific heats   8–39
      Elements – Alloy phases and intermetallic compounds – Borides –
      Carbides – Nitrides – Silicides – Oxides – Sulphides, selenides
      and tellurides – Halides
8.7   Vapour pressures   8–51
      Elements – Halides and oxides

9 Physical properties of molten salts   9–1

Density of pure molten salts – Densities of molten salt systems – Density of
some solid inorganic compounds at room temperature – Electrical conductivity of
pure molten salts – Electrical conductivity of molten salt systems –
Surface tension of pure molten salts – Surface tension of binary molten salt systems –
Viscosity of pure molten salts – Viscosity of molten binary salt systems

10 Metallography   10–1

10.1   Macroscopic examination   10–2
       Etching reagents for macroscopic examination
10.2   Microscopic examination   10–7
       Plastics for mounting – Attack polishing – Electrolytic polishing
       solutions – Reagents for chemical polishing – Etching – Color etching –
       Etching for dislocations

10.3 Metallographic methods for specific metals 10–29
Aluminium – Antimony and bismuth – Beryllium – Cadmium – Chromium –
Cobalt – Copper – Gold – Indium – Iron and steel – Lead – Magnesium –
Molybdenum – Nickel – Niobium – Platinum group metals – Silicon –
Silver – Tantalum – Tin – Titanium – Tungsten – Uranium – Zinc –
Zirconium – Bearing metals – Cemented carbides and other hard alloys –
Powdered and sintered metals
10.4 Electron metallography and surface analysis techniques 10–74
Transmission electron microscopy – Scanning electron microscopy –
Electron spectroscopy and surface analytical techniques
10.5 Quantitative image analysis 10–81
10.6 Scanning acoustic microscopy 10–82
10.7 Applications in failure analysis 10–83

11 Equilibrium diagrams 11–1
11.1 Index of binary diagrams 11–1
11.2 Equilibrium diagrams 11–7
11.3 Acknowledgements 11–524
11.4 Ternary systems and higher systems 11–533

12 Gas–metal systems 12–1
12.1 The solution of gases in metals 12–1
Dilute solutions of diatomic gases – Complex gas–metal systems –
Solutions of hydrogen – Solutions of nitrogen – Solutions of oxygen –
Solutions of the noble gases – Theoretical and practical aspects of
gas–metal equilibria

13 Diffusion in metals 13–1
13.1 Introduction 13–1
13.2 Methods of measuring D 13–4
Steady-state methods – Non-steady-state methods – Indirect methods,
not based on Fick's laws
13.3 Mechanisms of diffusion 13–8
Self-diffusion in solid elements – Tracer impurity diffusion coefficients –
Diffusion in homogeneous alloys – Chemical diffusion coefficient
measurements – Self-diffusion in liquid metals

14 General physical properties 14–1
14.1 The physical properties of pure metals 14–1
Physical properties of pure metals at normal temperatures –
Physical properties of pure metals at elevated temperatures
14.2 The thermophysical properties of liquid metals 14–9
Density and thermal expansion coefficient – Surface tension – Viscosity –
Heat capacity – Electrical resistivity – Thermal conductivity
14.3 The physical properties of aluminium and aluminium alloys 14–16
14.4 The physical properties of copper and copper alloys 14–19
14.5 The physical properties of magnesium and magnesium alloys 14–22
14.6 The physical properties of nickel and nickel alloys 14–25
14.7 The physical properties of titanium and titanium alloys 14–28
14.8 The physical properties of zinc and zinc alloys 14–29
14.9 The physical properties of zirconium alloys 14–29
14.10 The physical properties of pure tin 14–29
14.11 The physical properties of steels 14–30

15 Elastic properties, damping capacity and shape memory alloys   15–1

15.1   Elastic properties   15–1
     Elastic constants of polycrystalline metals – Elastic compliances
     and elastic stiffnesses of single crystals – Principal elastic compliances
     and elastic stiffnesses at room temperature – Cubic systems (3 constants) –
     Hexagonal systems (5 constants) – Trigonal systems (6 constants) –
     Tetragonal systems (6 constants) – Orthorhombic systems (9 constants)
15.2   Damping capacity   15–8
     Specific damping capacity of commercial alloys – Anelastic damping
15.3   Shape memory alloys   15–37
     Mechanical properties of shape memory alloys – Compositions and
     transformation temperatures – Titanium–nickel shape
     memory alloy properties

16 Temperature measurement and thermoelectric properties   16–1

16.1   Temperature measurement   16–1
     Fixed points of ITS-90
16.2   Thermocouple reference tables   16–4
16.3   Thermoelectric materials   16–10
     Introduction – Survey of materials – Preparation methods

17 Radiative properties of metals   17–1

Spectral normal emittance of metals – Spectral emittance in the infra-red metals –
Spectral emittance of oxidised metals

18 Electron emission   18–1

18.1   Thermionic emission   18–1
     Element-adsorbed layers – Refractory metal compounds –
     Practical cathodes
18.2   Photoelectric emission   18–4
     Photoelectric work function – Emitting surface
18.3   Secondary emission   18–5
     Emission coefficients
18.4   Auger emission   18–6
     Oxidised alloys – Insulating metal compounds
18.5   Electron emission under positive ion bombardment   18–8
18.6   Field emission   18–8
     Second Townsend coefficient γ electrons released per positive arriving

19 Electrical properties   19–1

19.1   Resistivity   19–1
     Pure metals – Alloys – Copper alloys – EC Aluminium
19.2   Superconductivity   19–7
     Transition temperatures and critical fields of elements –
     Critical temperatures of superconducting compounds

20 Magnetic materials and their properties   20–1

20.1   Magnetic materials   20–1
20.2   Permanent magnet materials   20–2
     Alnico alloys – Ferrite – Rare earth cobalt alloys –
     Neodymium iron boron – Bonded materials – Other materials –
     Properties, names and applications

20.3   Magnetically soft materials   20–9
       Silicon–iron alloys – Ferrites and garnets – Typical properties of
       silicon steels – Typical properties of some magnetically soft ferrites –
       Garnet material – Nickel–iron alloys – Amorphous alloy material
20.4   High-saturation and constant-permeability alloys   20–16
20.5   Magnetic powder core materials   20–17
20.6   Magnetic temperature-compensating materials   20–17
20.7   Non-magnetic steels and cast irons   20–18

21   Mechanical testing

                                                        21–1

21.1   Hardness testing   21–1
       Brinell hardness – Rockwell hardness – Rockwell superficial hardness –
       Vickers – Micro-hardness – Hardness conversion tables
21.2   Tensile testing   21–8
       Standard test pieces
21.3   Impact testing of notched bars   21–9
       Izod test – Charpy test
21.4   Fracture toughness testing   21–12
       Linear-elastic (K_{Ic}) – K–R curve – Elastic-Plastic (J_{Ic}, CTOD)
21.5   Fatigue testing   21–16
       Load-controlled smooth specimen tests – Strain-controlled smooth
       specimen tests – Fatigue crack growth testing
21.6   Creep testing   21–19
21.7   Non-destructive testing and evaluation   21–20
       Ultrasonic – Radiography – Electrical and magnetic methods –
       Acoustic emission testing – Thermal wave imaging

22   Mechanical properties of metals and alloys

                                                        22–1

22.1   Mechanical properties of aluminium and aluminium alloys   22–1
       Alloy designation system for wrought aluminium – Temper
       designation system for aluminium alloys
22.2   Mechanical properties of copper and copper alloys   22–26
22.3   Mechanical properties of lead and lead alloys   22–46
22.4   Mechanical properties of magnesium and magnesium alloys   22–49
22.5   Mechanical properties of nickel and nickel alloys   22–61
       Directionally solidified and single crystal cast superalloys
22.6   Mechanical properties of titanium and titanium alloys   22–82
22.7   Mechanical properties of zinc and zinc alloys   22–93
22.8   Mechanical properties of zirconium and zirconium alloys   22–94
22.9   Tin and its alloys   22–95
22.10 Steels   22–98
22.11 Other metals of industrial importance   22–157
22.12 Bearing alloys   22–160

23   Sintered materials

                                                        23–1

23.1   The PM process   23–1
23.2   The products   23–1
23.3   Manufacture and properties of powders   23–2
       Powder manufacture – Properties of metal powders and
       how they are measured
23.4   Properties of powder compacts   23–4
       Comparison tables of standard sieves – Properties of PM
       grade sponge iron powder
23.5   Sintering   23–8
23.6   Ferrous components   23–8
23.7   Copper-based components   23–9

23.8   Aluminium components   23–13
23.9   Determination of the mechanical properties of sintered components 23–13
23.10   Heat treatment and hardenability of sintered steels   23–15
23.11   Case hardening of sintered steels   23–15
23.12   Steam treatment   23–15
23.13   Wrought PM materials   23–15
        Refractory metals – Superalloys – Copper – Lead – Aluminium –
        Ferrous alloys – Aluminium matrix composites
23.14   Spray forming   23–28
23.15   Injection moulding   23–29
23.16   Hardmetals and related hard metals   23–31
        ISO classification of carbides according to use
23.17   Novel and emerging PM materials   23–36

24 Lubricants   24–1

24.1   Introduction   24–1
        Main regimes of lubrication
24.2   Lubrication condition, friction and wear   24–1
24.3   Characteristics of lubricating oils   24–2
        Viscosity – Boundary lubrication properties – Chemical stability –
        Physical properties
24.4   Mineral oils   24–3
24.5   Emulsions   24–6
24.6   Water-based lubricants   24–7
24.7   Synthetic oils   24–7
        Diesters – Neopentyl polyol esters – Triaryl phosphate esters –
        Fluorocarbons – Polyglycols
24.8   Greases   24–8
        Composition – Properties
24.9   Oil additives   24–11
        Machinery lubricants – Cutting oils – Lubricants for chipless-forming –
        Rolling oils
24.10   Solid lubricants   24–13

25 Friction and wear   25–1

25.1   Friction   25–1
        Friction of unlubricated surfaces – Friction of unlubricated materials –
        Friction of lubricated surfaces – Boundary lubrication –
        Extreme pressure (EP) lubricants
25.2   Wear   25–12
        Abrasive wear – Adhesive wear – Erosive wear – Fretting wear –
        Corrosive wear

26 Casting alloys and foundry data   26–1

26.1   Casting techniques   26–1
26.2   Patterns—crucibles—fluxing   26–10
        Pattern materials – Crucibles and melting vessels –
        Iron and steel crucibles—fluxing
26.3   Aluminium casting alloys   26–20
26.4   Copper base casting alloys   26–39
26.5   Nickel-base casting alloys   26–52
26.6   Magnesium alloys   26–56
26.7   Zinc base casting alloys   26–68
26.8   Steel castings   26–70
        Casting characteristics – Heat treatment

26.9   Cast irons   26–84
        Classification of cast irons – General purpose cast irons –
        Compacted graphite irons – Applications of special purpose cast irons
26.10   Acknowledgements   26–100

27 Engineering ceramics and refractory materials   27–1

27.1   Physical and mechanical properties of engineering ceramics   27–1
27.2   Prepared but unshaped refractory materials   27–7
27.3   Aluminous cements   27–7
27.4   Castable materials   27–8
27.5   Mouldable materials   27–14
27.6   Ramming material   27–14
27.7   Gunning material   27–14
27.8   Design of refractory linings   27–14

28 Fuels   28–1

28.1   Coal   28–1
        Analysis and testing of coal – Classification – Physical properties of coal
28.2   Metallurgical cokes   28–8
        Analysis and testing of coke – Properties of metallurgical coke
28.3   Gaseous fuels, liquid fuels and energy requirements   28–13
        Liquid fuels – Gaseous fuels – Energy use data for various
        metallurgical processes

29 Heat treatment   29–1

29.1   General introduction and cross references   29–1
29.2   Heat treatment of steel   29–1
        Introduction – Transformations in steels – Hardenability –
        Hardenability measurement – Austenitisation – Annealing –
        Quenching – Tempering – Austempering – Martempering –
        Carburising – Carbonitriding
29.3   Heat treatment of aluminium alloys   29–66
        Introduction to aluminium heat treating – A brief description of
        aluminium physical metallurgy – Defects associated with heat treatment –
        Solution treatment – Quenching – Ageing (natural and artificial) –
        Appendix: Quenchants

30 Metal cutting and forming   30–1

30.1   Introduction and cross-references   30–1
30.2   Metal cutting operations   30–1
        Turning – Boring – Drilling – Reaming – Milling
30.3   Abrasive machining processes   30–4
        Surface grinding – Cylindrical grinding – Centreless grinding –
        Plunge grinding – Creep feed grinding – Honing – Microsising –
        Belt grinding – Disc grinding
30.4   Deburring   30–5
30.5   Metal forming operations   30–5
        Introduction – Massive forming operations – Sheet-metal
        forming operations – Superplasticity
30.6   Machinability and formability of materials   30–8
        Definitions – Formability
30.7   Coolants and lubricants   30–9
        Liquid metal working fluids – Gaseous fluids and gaseous-liquid mixtures

30.8	Non-traditional machining techniques	30–10
	Electrical Discharge Machining (EDM) – Fast Hole EDM drilling –	
	Waterjet and abrasive waterjet machining – Plasma cutting –	
	Photochemical machining – Electrochemical machining –	
	Ultrasonic machining – Lasers	
30.9	Occupational safety issues	30–12
	Machine guarding – Hazardous materials – Noise exposure – Ergonomics	

31 Corrosion   31–1

31.1	Introduction	31–1
31.2	Types of corrosion	31–1
31.3	Uniform corrosion	31–1
	Galvanic corrosion – Erosion, cavitation, and fretting corrosion	
31.4	Localised forms of corrosion	31–4
	Crevice corrosion – Dealloying corrosion –	
	Environmental cracking—stress corrosion cracking and corrosion	
	fatigue cracking – Hydrogen damage – Intergranular corrosion –	
	Pitting corrosion	
31.5	Biocorrosion	31–10

32 Electroplating and metal finishing   32–1

32.1	Polishing compositions	32–1
32.2	Cleaning and pickling processes	32–2
32.3	Anodising and plating processes	32–6
32.4	Electroplating process	32–8
32.5	Plating processes for magnesium alloys	32–16
32.6	Electroplating process parameters	32–17
32.7	Miscellaneous coating processes	32–18
32.8	Plating formulae for non-conducting surfaces	32–19
32.9	Methods of stripping electroplated coatings	32–20
32.10	Conversion coating processes	32–21
32.11	Glossary of trade names for coating processes	32–23
	Wet processes – Dry processes	

33 Welding   33–1

33.1	Introduction and cross-reference	33–1
33.2	Glossary of welding terms	33–1
33.3	Resistance welding	33–6
	The influence of metallurgical properties on resistance weldability –	
	The resistance welding of various metals and alloys	
33.4	Solid-state welding	33–10
	Friction welding – Ultrasonic welding – Diffusion welding	
33.5	Fusion welding	33–15
	The fusion welding of metals and alloys—ferrous metals –	
	Non-ferrous metals – Copper and copper alloys – Lead and lead alloys –	
	Magnesium alloys – Nickel and nickel alloys – Noble metals –	
	Refractory metals – Zinc and zinc alloys – Dissimilar metals	
33.6	Major standards relating to welding, brazing and soldering	33–41

34 Soldering and brazing   34–1

34.1	Introduction and cross-reference	34–1
34.2	Quality assurance	34–1
34.3	Soldering	34–2
	General considerations – Choice of flux – Control of corrosion –	
	Solder formulations – Cleaning – Product assurance	

34.4   Brazing   34–9
       General design consideration – Joint design – Precleaning and
       surface preparation – Positioning of filler metal – Heating methods –
       Brazeability of materials and braze alloy compositions
34.5   Diffusion soldering or brazing   34–13

35   Vapour deposited coatings and thermal spraying   35–1

35.1   Physical vapour deposition   35–1
       Evaporation – Sputtering – Ion cleaning
35.2   Chemical vapour deposition   35–2
35.3   Thermal spraying   35–13
       Combustion wire spraying – Combustion powder spraying –
       Electric wire arc spraying – High velocity oxy-fuel spraying (HVOF) –
       Plasma spraying

36   Superplasticity   36–1

       Investigations on superplasticity of metal alloys – Summary of research
       on internal stress superplasticity

37   Metal-matrix composites   37–1

       Properties of reinforcing fibres at room temperature – Typical
       interactions in some fibre-matrix systems – Properties of aluminium
       alloy composites – Properties of magnesium alloy composites –
       Properties of titanium alloy composites – Properties of zinc
       alloy composites – Properties of co-deformed copper composites

38   Non-conventional and emerging metallic materials   38–1

38.1   Introduction   38–1
38.2   Cross references   38–1
38.3   Structural intermetallic compounds   38–1
       Sources of information on structural intermetallic compounds –
       Focus of the section – The nature of ordered intermetallics –
       The effect of ordering on the properties of intermetallics – Overview of
       aluminide intermetallics – Nickel aluminides – Titanium aluminides –
       Dispersion strengthened intermetallics and intermetallic matrix
       composites – Processing and fabrication of structural intermetallics –
       Current and potential applications of intermetallics
38.4   Metallic foams   38–18
38.5   Metallic glasses   38–20
       Metallic glasses requiring rapid quenching – Bulk metallic glasses
38.6   Mechanical behaviour of micro and nanoscale materials   38–23
       Mechanics of scale – Thin films – Nanomaterials – Nanostructures

39   Modelling and simulation   39–1

39.1   Introduction   39–1
39.2   Electron theory   39–1
39.3   Thermodynamics and equilibrium phase diagrams   39–3
39.4   Thermodynamics of irreversible processes   39–4
39.5   Kinetics   39–5
39.6   Monte Carlo simulations   39–6
39.7   Phase field method   39–7
39.8   Finite difference method   39–9
39.9   Finite element method   39–9
39.10 Empirical modelling: neural networks   39–10

40 Supporting technologies for the processing of metals and alloys   40–1

    40.1   Introduction and cross-references   40–1
    40.2   Furnace design   40–1
          Introduction – Types of furnaces – Heat calculations –
          Refractory design – Vacuum furnaces – Cooling
    40.3   Vacuum technology   40–8
          Introduction – Pressure units and vacuum regions –
          Pressure measurement – Pumping technologies – Vacuum systems –
          Residual gas analysis – Safety – Selective bibliography
    40.4   Metallurgical process control   40–16
          Metals production and processing – Modelling and control of
          metallurgical processes – Process control techniques – Instrumentation for
          process control – Process control examples in metals production –
          Keywords in process control

41 Bibliography of some sources of metallurgical information   41–1

Index   I–1

Preface

Over ten years have elapsed since the last edition of Smithells was published. Hence, a key objective of the editors for the present edition was to update the existing content. Thus, as can be seen from the table below, extensive changes have been made. In addition, the editors wished to expand the coverage of Smithells, both with respect to topics that have been overlooked in previous editions (such as metal working and machining) and to include aspects of metallurgy that have grown in importance, since the last edition was published (for example modelling).

Chapter	Title	Changes
1	Related designations	Rewritten completely, to reflect current designations.
2	Introductory tables and mathematical information	Major new section added on the statistics needed for process and quality control, etc.
3	General physical and chemical constants	No major changes.
4	X-ray analysis of metallic materials	Numerous edits to text, references updated and obsolete information removed.
5	Crystallography	Numerous edits to text.
6	Crystal chemistry	Updated comprehensively to bring all crystal structure and lattice parameter data into accord with commonly accepted values.
7	Metallurgically important minerals	Data table updated; introductory paragraphs rewritten.
8	Thermochemical data	No major changes.
9	Physical properties of molten salts	No changes.
10	Metallography	Rewritten completely to reflect modern specimen preparation techniques for light microscopy and contemporary practice in both electron microscopy and analytical techniques. New section added on applications in failure analysis. Numerous new references.
11	Equilibrium diagrams	Updated comprehensively to bring all phase diagrams into accord with commonly accepted data.
12	Gas–metal systems	Section on solutions of hydrogen updated completely and re-written.
13	Diffusion in metals	Numerous corrections, plus some new data.
14	General physical properties	All data on solid pure metals has been updated to match commonly accepted values. The section on liquid metals has been rewritten and now includes a more detailed narrative introduction.
15	Elastic properties, damping capacity and shape memory alloys	Minor corrections and updating of references.
16	Temperature measurement and thermoelectric properties	New section on thermoelectric materials, with narrative and data. Additional thermocouple data.

(continued)

Chapter	Title	Changes
17	Radiating properties of metals	Text rewritten completely.
18	Electron emission	Minor corrections.
19	Electrical properties	New references added on superconducting materials.
20	Magnetic materials and their properties	Text updated. Includes new references on contemporary magnetic materials.
21	Mechanical testing	Existing text edited extensively. Sections on tensile testing and fracture toughness testing re-written completely. New sections on fatigue testing, creep testing and non-destructive evaluation added.
22	Mechanical properties of metals and alloys	Minor editing of existing material; notes added concerning the applicability of the standards cited. New section added to provide data for cast Ni-base alloys.
23	Sintered materials	Edited extensively and updated.
24	Lubricants	Extensive revisions; new data tables.
25	Friction and wear	Extensive revisions; new data tables.
26	Casting alloys and foundry data	Extensive revisions, in particular extension and updating of the compositional and property data for foundry alloys.
27	Engineering ceramics and refractory materials	Extensive revisions, including new data on the properties of ceramics and standards for these materials.
28	Fuels	Extensive revisions, in particular to energy use data and standards.
29	Heat treatment	Rewritten entirely to reflect contemporary practice in steel and aluminium alloy heat-treatment. Provides a much greater level of detail than the previous edition.
30	Metal cutting and forming	New chapter providing a comprehensive overview of metal cutting and forming by both traditional and non-traditional techniques. Replaces a chapter that was focused narrowly on laser metal working.
31	Guide to corrosion control	Rewritten completely. Includes a new section on biocorrosion.
32	Electroplating and metal finishing	No changes.
33	Welding	Extensive revisions, in particular with respect to EN and ISO standards.
34	Soldering and brazing	Extensive revisions, principally to expand coverage with respect to processes and to include lead-free solders.
35	Vapour deposited coatings and thermal spraying	Coverage expanded to include a new section on thermal spraying. Some revision of content on vapour deposition.
36	Superplasticity	Rewritten completely to provide narrative and extensive tabular information on both microstructural and internal-stress superplastic systems.
37	Metal-matrix composites	Property data expanded and updated.
38	Non-conventional and emerging metallic materials	New chapter providing narrative and tabular information on structural intermetallic compounds, metal foams, metallic glasses and micro/nanoscale materials. Includes numerous references to both overviews and original research.
39	Modelling and simulation	New chapter providing an overview of the methods used to model metallic materials.
40	Supporting technologies for the processing of metals and alloys	New chapter containing information on furnace design, vacuum technology and metallurgical process control.
41	Bibliography of some sources of metallurgical information	New chapter comprised of a guide to major works of reference, texts, journals, conferences, databases and specialist search tools.

Acknowledgements

'Data! Data! Data!' he cried impatiently. 'I can't make bricks without clay.'

Sherlock Holmes,
in *The Adventure of the Copper Beeches*
by Sir Arthur Conan Doyle

The editors wish to thank Mr Daniel Butts and Miss Dina Taarea of the Materials Research and Education Center (MREC) at Auburn University (AU), for their extensive, tireless and immensely valuable assistance with many aspects of the preparation of this work. Additional help from Mr. Venu Gopal Krishnardula and other members of the Physical Metallurgy and Materials Joining research group at the AU–MREC is acknowledged with thanks. WFG is also very grateful to his wife, Dr Hyacinth Gale of the NSF Center for Advanced Vehicle Electronics (CAVE) at AU, for her unstinting help with this work. The editors would like to thank the many contributors to this edition, without whom this work would not of course have been possible. A debt to both the contributors and editors of previous editions of this work should also be recorded.

The editors would also like to remark that it is a little over half a century since the first edition of Smithells was published. Although many of the topics in this book (bulk metallic glasses, nanomaterials, etc.) lay far in the future in 1949, none of these would have been recorded here if it were not for the initial efforts of C. J. Smithells.

The following bodies kindly provided permission to reproduce copyrighted material for this edition, as acknowledged at the relevant location(s) in the text:

The Aluminum Association
The American Foundry Society (AFS)
ASM International
Association of Iron and Steel Engineers (AISE)
ASTM International
Cambridge University Press
Copper Development Association
CRC Press
Marcel Dekker
Metallurgical and Materials Transactions
National Center for Manufacturing Sciences (NCMS)
North American Die Casting Association (NADCA)
The Timken Company

In addition to the above, the following allowed use of copyrighted material in the previous edition:

British Standards Institute
Genium Publication Corporation
Institute of Gas Engineers
International Atomic Energy Agency
McGraw-Hill Book Company

W. F. Gale, Auburn, AL
T. C. Totemeier, Idaho Falls, ID

Disclaimer

Although great care was exercised in the preparation of this volume, the contributors, editors and publisher provide no warranty of any kind (explicit or implied) as to the accuracy, reliability or applicability of the data and information contained herein. All data in this volume are provided for general guidance only and **cannot** be used as evidence of: merchantability; suitability for engineering design; fitness for a particular purpose; safety in storage, transportation or use; or other similar or related purposes.

Information provided in this volume is derived from the relevant professional literature, or other sources believed to be reputable by the contributors and editors, as specified in the detailed references provided. Where copyrighted material has been employed, this is used with permission of the copyright owner. Copyrighted material, used with permission, is indicated at the relevant location in the text. Mention of trademarked product or corporate names is purely for the purposes of identification and these remain the property of their owners and there is no intent to infringe upon ownership rights therein. Neither the presence, nor the absence of discussions of materials, processes, products or other items provides any indication as to the extent to which such items actually exist, are available commercially, or constitute original inventions.

All information within this volume is used entirely at the reader's own risk. Since the circumstances in which the contents of this volume might be employed will differ widely, the contributors, editors and publisher can not guarantee favourable results and expressly deny any and all liability connected with the use (and consequences of the use) of the information and data contained herein.

This volume is intended for readers with suitable professional qualifications, supplemented by sufficient experience in the field, to be capable of making their own professional judgement as to the reliability and appropriateness of the information and data contained within. Before using any of this information or data, a suitably qualified professional/chartered engineer must be consulted, possible risks analysed/managed and rigorous testing conducted under the actual conditions to be experienced.

Under no circumstances, whether covered in the disclaimer above or not, shall the liability of the contributors, editors or publisher of this work exceed the original purchase price of this volume and this shall constitute the sole remedy available to the purchaser (and subsequent readers) of this book. No liability is accepted for consequential or implied damages of any kind resulting from use of this volume.

Contributors

*Contributors to this edition**	*Chapters/Sections*
J. R. Alcock *Advanced Materials Department, Cranfield University,* *Cranfield, Bedfordshire, UK*	20
P. N. Anyalebechi *Padnos School of Engineering, Grand Valley State University,* *Grand Rapids, MI, USA*	12
S. I. Bakhtiyarov *Solidification Design Center, Auburn University, Auburn, AL, USA*	14.2
H. K. D. H. Bhadeshia *Department of Materials Science and Metallurgy,* *University of Cambridge, Cambridge, UK*	39
JT. Black *Department of Industrial and Systems Engineering,* *Auburn University, Auburn, AL, USA*	30
D. A. Butts *Materials Research and Education Center, Auburn University,* *Auburn, AL, USA*	6, 11, 19.2, 20-Appendix
V. Dayal *Department of Aerospace Engineering and Engineering Mechanics,* *Iowa State University, Ames, IA, USA*	21.7
J. C. Earthman *Department of Chemical and Biochemical Engineering,* *University of California at Irvine, Irvine, CA, USA*	31.5
J. W. Fergus *Materials Research and Education Center, Auburn University,* *Auburn, AL, USA*	41
H. S. Gale *NSF Center for Advanced Vehicle Electronics, Department of Electrical* *and Computer Engineering, Auburn University, Auburn, AL, USA*	41
W. F. Gale *Materials Research and Education Center, Auburn* *University, Auburn, AL, USA*	6, 10.5, 11, 28.3.3, 38.1–38.5, 41
R. M. German *Department of Engineering Science and Mechanics,* *Pennsylvania State University, University Park, PA, USA*	23
W. Gestwa *Poznan University of Technology, Poznan, Poland*	29.2
C. Hammond *School of Materials, University of Leeds, Leeds, UK*	4, 5

---

*Where material was retained from the Seventh Edition, this was not the work of the individuals shown above.

Contributors to this edition*	*Chapters*
T. Hornig *Materials Science Institute, RWTH (Aachen University of Technology),* *Aachen, Germany*	35.1, 35.2
W. E. Lee *Department of Engineering Materials, The University of Sheffield,* *Sheffield, South Yorkshire, UK*	27
D. Y. Li *Department of Chemical and Materials Engineering, University of Alberta,* *Edmonton, Alberta, Canada*	24, 25
E. Lugscheider *Materials Science Institute, RWTH (Aachen University of Technology),* *Aachen, Germany*	35
D. S. MacKenzie *Houghton International, Valley Forge, PA, USA*	29.3
S. Maghsoodloo *Department of Industrial and Systems Engineering, Auburn University,* *Auburn, AL, USA*	2.2.7
M. F. Modest *Department of Mechanical Engineering, Pennsylvania State University,* *University Park, PA, USA*	17
S. B. Newcomb *Sonsam Ltd, Newport, Co. Tipperary, Ireland*	10.4
T. G. Nieh *Lawrence Livermore National Laboratory, Livermore, CA, USA*	36
R. A. Overfelt *Solidification Design Center, Auburn University, Auburn, AL, USA*	26
G. Ozdemir *Department of Industrial and Systems Engineering, Auburn University,* *Auburn, AL, USA*	2.2.7
L. N. Payton *Department of Industrial and Systems Engineering (now with Materials* *Research and Education Center), Auburn University, Auburn, AL, USA*	30
P. J. Pinhero *Idaho National Engineering and Environmental Laboratory,* *Idaho Falls, ID, USA*	33
D. L. Porter *Argonne National Laboratory - West, Idaho Falls, ID, USA*	22
B. C. Prorok *Materials Research and Education Center, Auburn University,* *Auburn, AL, USA*	38.6
M. Przylecka *Poznan University of Technology, Poznan, Poland*	29.2
D. Pye *Pye Metallurgical Consulting Inc., Meadville, PA, USA*	29.2, 40.2
R. J. Reid *CCLRC Daresbury Laboratory, Daresbury, Warrington, Cheshire, UK*	40.3
M. A. Reuter *Resource Engineering Section, Department of Applied Earth Sciences,* *TU Delft, Delft, The Netherlands*	40.4

---

*Where material was retained from the Seventh Edition, this was not the work of the individuals shown above.

*Contributors to this edition**	*Chapters*
K. Seemann *Materials Science Institute, RWHT (Aachen University of Technology),* *Aachen Germany*	35.3
M. R. Scheinfein *Department of Physics and Astronomy, Arizona State University,* *Tempe, AZ, USA* *(Now at Department of Physics, Simon Fraser University, Burnaby,* *British Columbia, Canada)*	18
C. A. Schuh *Department of Materials Science and Engineering, Massachusetts Institute* *of Technology, Cambridge, MA, USA*	36
C. Shannon *Department of Chemistry, Auburn University, Auburn, AL, USA*	16.3
Y. Sohn *Advanced Materials Processing and Analysis Center,* *University of Central Florida, Orlando, FL, USA*	13
D. Taarea *Materials Research and Education Center,* *Auburn University, Auburn, AL, USA*	14.1, 19.2, 20-Appendix, Tables 26.28, 26.34, 26.46, 26.47
T. C. Totemeier *Idaho National Engineering and Environmental Laboratory,* *Idaho Falls, ID, USA*	1, 10.7, 21, 22
G. E. Totten *G.E. Totten & Associates, LLC, Seattle, WA, USA*	29.2
H. Tsai *Department of Mechanical Engineering, Michigan State University,* *East Lansing, MI, USA*	15
G. F. Vander Voort *Buehler Ltd., Lake Bluff, IL, USA*	10.1–10.3
A. Williams *Department of Fuel and Energy, University of Leeds, Leeds, UK*	28
P. J. Withers *Manchester Materials Science Centre, University of Manchester/UMIST,* *Manchester, UK*	37
Y. Yang *Resource Engineering Section, Department of Applied Earth Sciences,* *TU Delft, The Netherlands*	40.4
B. Yaryar *Department of Mining Engineering, Colorado School of Mines,* *Golden, CO, USA*	7
Y. Zhou *Department of Mechanical Engineering, University of Waterloo,* *Waterloo, ON, Canada*	33, 34

---

*Where material was retained from the Seventh Edition, this was not the work of the individuals shown above.

1 Related designations

The following tables of related designations are intended as a guide to alloy correspondence on the basis of chemical composition. The equivalents should not be taken as exact, and in all cases of doubt the relevant national specification or standard should be consulted. The tables do not represent an exhaustive list of all alloys available; the references listed at the end of the chapter are more complete sources.

In the case of the United Kingdom, France, and Germany, the tables refer to designations and standards that have recently been superseded by European (EN) and international (ISO) standards. The older standards have been referenced because the alloy designations are still in common use, and because the new standards use in some cases several different designations for chemically identical alloys, depending on the product form. A list of all designations is beyond the scope of this book; the references should be consulted for further details.

Table 1.1 lists designations for steels, with subsections for steels of different types. Designations for wrought aluminium alloys are listed in Table 1.2; cast aluminium alloys are shown in Table 22.2. Table 1.3 gives related copper alloy designations, subdivided into coppers, brasses, bronzes, and nickel silvers. Magnesium alloys cast and wrought are in Table 1.4. Related designations for nickel and titanium alloys are listed in Sections 22.5 and 22.6, respectively.

Unified number designations—UNS designations are five-digit numbers prefixed by a letter that characterises the alloy system as shown below.

UNS Letter Designation[1]

A   Aluminium and aluminium alloys
B   Copper and copper alloys
D   Specified mechanical properties steels
E   Rare earth and rare earth like metals and alloys
F   Cast irons and cast steels
G   AISI and SAE carbon and alloy steels
H   AISI H-steels
J   Cast steels (except tool steels)
K   Miscellaneous steels and ferrous alloys
L   Low melting metals and alloys
M   Miscellaneous nonferrous metals and alloys
N   Nickel and nickel alloys
P   Precious metals and alloys
R   Reactive and refractory metals and alloys
S   Heat and corrosion resistant (stainless) steels
T   Tool steels
W   Welding filler metals

Table 1.1   RELATED DESIGNATIONS FOR STEELS

Nominal composition	USA AISI/SAE (UNS)	UK BS 970 (En)	Germany DIN (Wk. No.)	Japan JIS	Russia GOST	Sweden SIS	France AFNOR	China GB
1.1.1 Carbon steels								
C 0.06 Mn 0.35	1006 (G10060)	030A04	Ck7 (1.1009)	—	08Fkp	14 1147	XC6FF	05F
C 0.08/0.13 Mn 0.3/0.6	1010 (G10100)	045M10 (32A)	Ck10 (1.1121)	S9Ck	10kp	14 1311	CC10, AF34	10F
C 0.13/0.18 Mn 0.3/0.6	1015 (G10150)	050A15	Ck15 (1.1141)	S15Ck	15kp	14 1370	XC15	15F
C 0.17/0.23 Mn 0.3/0.6	1020 (G10200)	050A20 (2C, 2D)	Ck22 (1.1151)	S20C	20kp	14 1450	XC18, C20	20F
C 0.22/0.28 Mn 0.3/0.6	1025 (G10250)	080A25	Ck25 (1.1158)	S25C	M26	—	XC25	1025
C 0.27/0.34 Mn 0.6/0.9	1030 (G10300)	080A30 (5B)	Ck30 (1.1178)	S30C	30G	—	XC32	30
C 0.31/0.38 Mn 0.6/0.9	1035 (G10350)	080A35 (8A)	Ck34 (1.1173)	S35C	35	14 1572	XC35, C35	35
C 0.36/0.44 Mn 0.6/0.9	1040 (G10400)	080A40 (8C)	Ck40 (1.1186)	S40C	40	—	AF60, C40	40
C 0.42/0.50 Mn 0.6/0.9	1045 (G10450)	080M46	Ck45 (1.1191)	S45C	45	14 1672	XC45	45
C 0.47/0.55 Mn 0.6/0.9	1050 (G10500)	080M50	Ck50 (1.1206)	S50C	50	14 1674	XC50	50
C 0.55/0.65 Mn 0.6/0.9	1060 (G10600)	080A62 (43D)	Ck60 (1.1221)	S58C	60, 60G	14 1678	XC60	60
C 0.65/0.76 Mn 0.6/0.9	1070 (G10700)	080A72	Ck67 (1.1231)	—	70, 70G	14 1770	XC70	70
C 0.74/0.88 Mn 0.6/0.9	1080 (G10800)	080A83	80Mn4 (1.1259)	—	80	14 1774	XC80	80
C 0.90/1.04 Mn 0.3/0.5	1095 (G10950)	060A99	Ck101 (1.1274)	SUP 4	—	14 1870	XC100	—

1.1.2 Carbon-manganese and free-cutting steels

Composition								
C 0.10/0.16 Mn 1.1/1.4	1513 (G15130)	130M15 (201)	12Mn6 (1.0496)	—	—	—	12M5	—
C 0.15/0.21 Mn 1.1/1.4	1518 (G15180)	120M19	21Mn4 (1.0469)	SMnC420	20GLS	14 2135	20M5	18MnSi
C 0.18/0.24 Mn 1.1/1.4	1522 (G15220)	150M19 (14A, 14B)	20Mn6 (1.1169)	SMn21	20G2	14 2165	20M5	18MnVB
C 0.30/0.38 Mn 1.2/1.55	1536 (G15360)	120M36 (15B)	36Mn5 (1.1167)	SMn1	35GL	14 3562	32M5	35Mn2
C 0.36/0.45 Mn 1.35/1.65	1541 (G15410)	150M40	36Mn6 (1.1127)	SMn21	40G2	14 2120	40M5	40Mn2
C 0.35/0.43 Mn 1.35/1.65 S 0.13/0.20	1139 (G11390)	216M36 (15AM)	35S20 (1.0726)	SUM41	A40G	14 1957	35MF4	Y40Mn
C 0.40/0.48 Mn 1.35/1.65 S 0.24/0.33	1144 (G11440)	226M44	45S20 (1.0727)	SUM43	A40	14 1973	45MF6	—
C 0.12 Mn 0.7/1.0 S 0.16/0.23	1212 (G12120)	—	9S20 (1.0711)	SUM21	A11	—	12MF4	—
C 0.13 Mn 0.7/1.0 S 0.24/0.33	1213 (G12130)	230M07	9SMn28 (1.0715)	SUM22	—	14 1912	S250	—

1.1.3 Alloy steels

Composition								
C 0.1/0.2 Mn 0.3/0.6 Cr 0.6/0.95 Ni 2.75/3.25	3415	655M13 (36A)	14NiCr14 (1.5752)	SNC22H	12ChHN3A	—	12NC15	12CrNi3
C 0.23/0.28 Mn 0.7/0.9 Cr 0.4/0.6 Mo 0.2/0.3	4125	708A25	25CrMo4 (1.7218)	SCCrM1	25ChGM	14 2225	25CD45	—

(continued)

Table 1.1   RELATED DESIGNATIONS FOR STEELS—*continued*

Nominal composition	USA AISI/SAE (UNS)	UK BS 970 (En)	Japan JIS	Germany DIN (Wk. No.)	Russia GOST	Sweden SIS	France AFNOR	China GB
C 0.35/0.40 Mn 0.7/0.9 Mo 0.2/0.3	4037	605H37	SCCrM1	GS-40MnMo43	25ChGM	14 2225	25CD45	—
C 0.28/0.33 Mn 0.4/0.6 Cr 0.8/1.0 Mo 0.15/0.25	4130 (G41300)	708A30	SCM2 SCM430	—	30ChMA	14 2233	30CD4	30CrMo
C 0.33/0.35 Mn 0.7/0.9 Cr 0.6/1.10 Mo 0.15/0.25	4135 (G41350)	708A37 (19B)	SCM3H	CS-34CrMo4 (1.7220)	35ChM	14 2234	35CD4	35CrMo
C 0.38/0.43 Mn 0.75/1.00 Cr 0.8/1.10 Mo 0.15/0.25	4140 (G41400)	708A40 (19A)	SCM4	42CrMo4 (1.7225)	38ChM	14 2244	42CD4	42CrMo
C 0.38/0.43 Mn 0.6/0.8 Cr 0.7/0.9 Ni 1.65/2.00 Mo 0.2/0.3	4340 (G43400)	816M40 (110)	SNCM8	40NiCrMo6 (1.6565)	40ChMA	14 2541	35NCD6	—
C 0.18/0.23 Mn 0.45/0.65 Mo 0.45/0.60	4419 (G44190)	—	SCPH11	GS-22Mo4 (1.5419)	—	—	18MD4.05	—
C 0.13/0.18 Mn 0.7/0.9 Cr 0.7/0.9	5115 (G51150)	527A17	SCr21	16MnCr5 (1.7131)	15Ch	14 2127	16MC5	15Cr
C 0.17/0.22 Mn 0.7/0.9 Cr 0.7/0.9	5120 (G51200)	—	SCr22	20CrMnS3 (1.7121)	20Ch	—	20MC5	20Cr
C 0.28/0.33 Mn 0.7/0.9 Cr 0.8/1.1	5130 (G51300)	530A30 (18A)	SCr2H	30MnCrTi4 (1.8401)	27ChGR	—	28C4	30Cr

Composition								
C 0.33/0.38 Mn 0.6/0.8 Cr 0.8/1.05	5135 (G51350)	530A36 (18C)	38Cr4 (1.7043)	SCr3H	35Ch	—	38C4	35Cr
C 0.38/0.43 Mn 0.7/0.9 Cr 0.7/0.9	5140 (G51400)	530A40 (18D)	42Cr4 (1.7045)	SCr4H	40Ch	14 2245	42C4	40Cr
C 0.98/1.10 Mn 0.25/0.45 Cr 1.3/1.6	E52100 (G52986)	535A99 (31)	105Cr5 (1.2060)	SCr5	SchCh15	14 2258	100C6	—
C 0.48/0.53 Mn 0.7/0.9 Cr 0.8/1.1 V 0.15	6150 (G61500)	735A50 (47)	GS-50CrV4 (1.8159)	SUP10	50ChF	14 2230	50CV4	50CrVA
C 0.18/0.23 Mn 0.7/0.9 Cr 0.4/0.6 Ni 0.4/0.7 Mo 0.15/0.25	8620 (G86200)	805A20	21NiCrMo2 (1.6523)	SNCM21	AS20ChGNM	14 2506-03	20NCD2	20CrNiMo
C 0.38/0.43 Mn 0.75/1.0 Ni 0.4/0.6 Ni 0.4/0.7 Mo 0.15/0.25	8640 (G86400)	945A40 (10C)	40NiCrMo2 (1.6546)	SNCM6	38ChGNM	—	40NCD2	—

1.1.4 Stainless steels

Composition								
C 0.9 Cr 16/18 Ni 6.5/7.75 Al 0.75/1.5	17-7 PH (S17700)	301S81	X7CrNiAl17 7 (1.4564)	SUS631	—	—	177F00	1Cr17Ni7Al

(continued)

Table 1.1   RELATED DESIGNATIONS FOR STEELS—*continued*

Nominal composition	USA AISI/SAE (UNS)	UK BS 970 (En)	Germany DIN (Wk. No.)	Japan JIS	Russia GOST	Sweden SIS	France AFNOR	China GB
C 0.15 Mn 7.5/10.0 Cr 17/19 Ni 4/6 N 0.25	202 (S20200)	284S16	X8CrMnNi18 8 (1.3965)	SUS202	1Ch17N3G8AE	14 2357	—	1Cr18Mn8Ni5N
C 0.15 Mn 2.0 Cr 17/18 Ni 8/10	302 (S30200)	302S25 (58A)	X12CrNi18 8 (1.4300)	SUS302	Ch18N9	14 2331	302F00 Z10CN18.9	1Cr18Ni9
C 0.15 Mn 2.0 Cr 18/20 Ni 8/10.5	304 (S30400)	304S15 (58E)	X5CrNi18 10 (1.4301)	SUS304	0Ch18N10	14 2332	304F01 Z5CN18.09	0Cr18Ni9
C 0.03 max Mn 2.0 Cr 18/20 Ni 8/12	304L (S30403)	304S11 (58E)	X2CrNi19 11 (1.4306)	SUS304L	03Ch18N11	14 2352	304F11 Z2CN18.10	00Cr18Ni10
C 0.25 Mn 2.0 Cr 24/26 Ni 19/22	310 (S31000)	310S24	X12CrNi25 21 (1.4845)	SUSY310	Ch25N20	—	310F00 Z12CN25.20	2Cr25Ni20
C 0.8 Mn 2.0 Cr 16/18 Ni 10/14 Mo 2/3	316 (S31600)	316S16 (58J)	X15CrNiMo17 13 3 (1.4436)	SUS316	SW-04Ch19Ni1M3	14 2343	316F00	0Cr17Ni20Mo2
C 0.03 max Mn 2.0 Cr 16/18 Ni 10/14 Mo 2/3	316L (S31603)	316	X2CrNiMo18 14 3 (1.4435)	SUSY316L	03Ch16N15M3	14 2348	22CND17.12	00Cr17Ni14Mo2

Composition								
C 0.08 Mn 2.0 Cr 16/18 Ni 10/14 Mo 2/3 Ti 5x(C+N)/0.7	316Ti (S31635)	320S17 (58J)	X6CrNiMoTi17 12 2 (1.4571)	—	08Ch17N13M2T	14 2350	Z6CNDT17.12	0Cr18Ni12Mo2Ti
C 0.08 Mn 2.0 Cr 17/19 Ni 10/14 Mo 2/3 Nb 10xC/1.10	316Cb (S31640)	318C17	X6CrNiMoNb17 12 2 (1.4580)	SCS22	0Ch18N9MB	—	Z4CNDNb18.12M	—
C 0.08 Mn 2.0 Cr 17/19 Ni 9/12 Ti 5xC min	321 (S32100)	321S20 (58B/C)	X6CrNiTi18 10 (1.4541)	SUS321	08Ch18N10T	14 2337	Z6CNT18.10	0Cr18Ni11Ti
C 0.08 Mn 2.0 Cr 17/19 Ni 9/13 Nb+Ta 10xC min	347 (S34700)	347S31	X5CrNiNb18 10 (1.4546)	SUS347	Ch18N11B	14 2338	Z4CNNb19.10M	0Cr18Ni11Nb
C 0.15 Mn 1.0 Cr 11.5/13.0	403 (S40300)	410S21 (56A)	X10Cr12 (1.4006)	SUS403	15Ch13L	14 2301	Z15C13	1Cr12
C 0.15 Mn 1.0 Cr 11.5/13.5 Si 1.0	410 (S41000)	—	X15Cr13 (1.4024)	SUS416	12Ch13	14 2302	Z12C13M	1Cr13
C 0.15 min Mn 1.0 Cr 12/14 Si 1.0	420 (S42000)	420S37 (56C)	X20Cr13 (1.4021)	SUS420J2	20Ch13	14 2304	Z20C13N	2Cr13

(continued)

Table 1.1   RELATED DESIGNATIONS FOR STEELS—*continued*

Nominal composition	USA AISI/SAE (UNS)	UK BS 970 (En)	Germany DIN (Wk. No.)	Japan JIS	Russia GOST	Sweden SIS	France AFNOR	China GB
C 0.12 Mn 1.0 Cr 16/18	430 (S43000)	430S15 (60)	X8Cr17 (1.4016)	SUS430	12Ch17	14 2320	Z8C17	1Cr17
C 0.48/0.58 Mn 8/10 Cr 20/22 Ni 3.25/4.5 N 0.28/0.5	21-4N (S63008)	349S52	X5CrMnNiN21 9 (1.4871)	SUH36	5Ch20N4AGN	—	Z52CMN21.09	Y5Cr21Mn9Ni4N
1.1.5 Tool steels								
C 0.8/1.0 Cr 3.75/4.5 Mo 4.5/5.5 W 5.5/6.75 V 1.75/2.2	M2 (T11302rc)	BM2	S6-5-2 (1.3343)	SKH51	R6AM5	14 2722	6-5-2	W6Mo5Cr4V2
C 0.65/0.8 Cr 3.75/4.5 W 17.25/18.75 V 0.9/1.3	T1 (T12001)	BT1	S18-0-1 (1.3355)	SKH2	R18	—	18-0-1	W18Cr4V
C 1.5/1.6 Cr 3.75/5.0 Mo 1.0 max W 11.75/13.0 V 4.5/5.25 Co 4.75/5.25	T15 (T12015)	BT15	S12-1-4-5 (1.3202)	SKH10	R10K5F5	—	12-1-5-5	—
C 0.35/0.45 Si 0.8/1.2 Cr 3.0/3.75 Mo 2.0/3.0 V 0.25/0.75	H10 (T20810)	VH10	X32CrMoV3 3 (1.2365)	SKD7	3Ch3M3F	—	32DCV28	—

C 0.33/0.43 Si 0.8/1.2 Cr 4.75/5.0 Mo 1.1/1.6 V 0.3/0.6	H11 (T20811)	BH11	X41CrMoV5 1 (1.7783)	SKD6	4Ch5MF5	—	FZ38CDV5	—
C 0.32/0.45 Si 0.8/1.2 Cr 4.75/5.5 Mo 1.1/1.75 V 0.8/1.2	H13 (T20813)	BH13	X40CrMoV5 1 (1.2344)	SKD61	4Ch5MF15	14 2242	Z40CDV5	—
C 0.26/0.36 Cr 3.0/3.75 W 8.5/10.0 V 0.3/0.6	H21 (T20821)	BH21	X30WCrV9 3 (1.2581)	SKD5	3Ch2W8F	14 2730	Z30WCV9	—
C 1.4/1.6 Cr 11.0/13.0 Mo 0.7/1.2 V 1.1 max	D2 (T30402)	BD2	X155CrMo12 1 (1.2379)	SKD11	—	14 2310	Z200C12	3-2 Cr12MoV
C 2.0/2.25 Cr 11.0/13.5 W 1.0 max V 1.0 max	D3 (T30403)	BD3	X210CrW12 (1.2436)	SKD1	Ch12	—	Z200C12	3-1 Cr12
C 0.4/0.55 Si 0.10/1.2 Cr 1.0/1.8 W 1.5/3.0 V 0.15/0.30	S1 (T41901)	BS1	45WCrV7 (1.2436)	SKS41	4ChW2S	14 2710	55WC20	2-2 5CrW2Si
C 0.7/1.5	W1 (T72301)	BW1B	C125W1 (1.1663)	SK2	U12	14 1880	Y(2) 120	3 85
C 0.8/1.5 V 0.15/0.35	W2 (T72302)	BW2	100V1 (1.2833)	SKS43	—	—	Y105V	1-8V

Table 1.2   RELATED DESIGNATIONS FOR WROUGHT ALUMINIUM ALLOYS

UNS	ISO—Nominal composition	USA/Japan	UK Former BS	Germany DIN (Wk. No.)	Russia GOST	Sweden SIS	France Former NF	China GB
A91050	Al99.5	1050	1B	Al99.5 (3.0255)	A1	4007	A5	L4
A91070	Al99.7	1070		Al99.7 (3.0275)	A7	4005	A7	L2
A91080	Al99.8	1080	1A	Al99.8 (3.0285)		4004	A8	L1
A91100	Al99.0Cu	1100			A2		A45	
A91145		1145			AE, AT			
A91200	Al99.0	1200	1C	Al99.0 (3.0205)	A2	4010	A4	
A91350		1350	1E, E57S	EAl99.5 (3.0257)	A00	4008		
A92011	Al-Cu6BiPb	2011	FC1	AlCuBiPb (3.1655)		4355	A-U5PbBi	
A92014	Al-Cu4SiMg	2014	H15	AlCuSiMn (3.1255)	1380, AK8	4338	A-U4SG	
A92017	Al-Cu4MgSi	2017		AlCuMg1 (3.1325)	AK10, D17		A-U4G	LD9, LD10
A92024	Al-Cu4Mg1	2024	2L97, 2L98, L109, L110	AlCuMg2 (3.1355)	1160, D16		A-U4G1	LY6, LY9, LY12
A92031	Al-Cu2NiMgFeSi	2031	H12				A-U2N	
A92117	Al-Cu2Mg	2117	3L86	AlCu2.5Mg0.5 (3.1305)	1180, D18		A-U2G	
A92219		2219			1201			
A92618	Al-Cu2Mg1.5Fe1Ni1	2618	H16		1141, AK4-1			
A93003	Al-Mn1Cu	3003	N3	AlMnCu (3.0517)	1400, AMts	4054	A-M1	
A93004		3004		AlMn1Mg1 (3.0528)	AMts2		A-M1G	
A93103	Al-Mn1	3103	N3	AlCu2.5Mg0.5 (3.1305)	D18			
A93105	Al-MnMg	3105	N31, E4S	AlMn0.5Mg0.5 (3.0505)				
A94032		4032	38S					
A94043		4043	N21					
A94047	Al-Si12	4047	N21	S-AlSi5 (3.2245)		4225	A-S12	LT1
A95005	Al-Mg1	5005	N41	AlMg1 (3.3315)	1510, AMg1	4106	A-G0.6	

A95050	Al-Mg1.5	3L44	AlMg1.5 (3.3318)	—	—	A-G1.5	LF2
A95052	Al-Mg2.5	2L55	AlMg2.5 (3.3523)	1520, AMg2	4120	A-G2.5C	LF10, LF5-1, LF51
A95056	Al-Mg5	N6, A56S	AlMg5 (3.3555)	AMg5	4146	—	LF4
A95083	Al-Mg4.5Mn	N8	AlMg4.5Mn (3.3547)	—	4140	A-G4.5MC	—
A95086	Al-Mg4	—	AlMg4.5 (3.3345)	1540, AMg4	—	A-G4MC	—
A95154	Al-Mg3.5	N5	—	1530, AMg3	—	AG3	—
A95251	Al-Mg2	N4	AlMg2Mn0.3 (3.3525)	—	—	A-G2M	—
A95454	Al-Mg3Mn	N51	AlMg2.7Mn (3.3537)	—	—	A-G2.5MC	—
A95456	—	—	—	45Mg2, AMg6	4133	A-G5MC	—
A95754	Al-Mg3	—	AlMg3 (3.3525)	1530, AMg3	4103	A-G3M	—
A96060	—	—	AlMgSi0.5 (3.3206)	—	—	A-GS	—
A96061	Al-Mg1SiCu	H20, E91E	AlMgSiCu (3.3211)	1330, AD33	4104	A-GSUC	—
A96063	Al-MgSi	H9	—	1310, AD31	4212	—	—
A96082	Al-Si1MgMn	H30	AlMgSi1 (3.2315)	—	4102	A-SGM0.7	—
A96101	—	91E	E-AlMgSi0.5 (3.3207)	1310, AD31	—	A-GS/L	—
A96463	—	BTR6	Al99.85MgSi (3.2307)	1910, m ATsM	4212	A85-GS	—
A97005	Al-Zn4.5Mg	H17	AlZnMgCu0.5 (3.4345) (3.4335)	—	4425	A-Z5G	—
A97020	—	—	—	AtsPl	—	—	—
A97072	Al-Zn1	—	AlZn1 (3.4415)	—	—	—	LB1
A97075	Al-Zn6MgCu	2L95, L160, L161, L162, C77S	AlZnMgCu1.5 (3.4365)	1950, V95	—	A-Z5GU	LC4, LC9

Table 1.3   RELATED DESIGNATIONS FOR COPPER ALLOYS

1.3.1 Coppers

UNS	ISO—Nominal composition	USA	UK Former BS	Germany DIN (Wk. No.)	Japan JIS	Russia GOST	Sweden SIS	France Former NF	China GB
C10200	Cu-OF	OF	C103	SE-Cu (2.0070)	C1020	V3	5011	Cu/c1, Cu/c2	Cu-1, Cu-2
C11000	Cu-ETP, Cu-FRHC	ETP	C101	E-Cu57 (2.0060)	C1100	—	5010	Cu/a1, Cu/a2	T2
C11600	CuAg0.1	STP	—	CuAg0.1 (2.1203)	—	—	5030	—	TAg0.1
C12200	Cu-DHP	DHP	C106	SF-Cu (2.0090)	C1220	—	5105	Cu/a3	—
C12500	Cu-FRTP	FRTP	C104	(2.008)	—	—	5013	—	—
C14200	CuAs(P)	DPA	C107	CuAs(P) (2.0150)	—	—	—	—	—
C14500	Cu-Te, Cu-Te(P)	DPTE	C109	CuTeP (2.1646)	—	—	—	—	—
C16200	CuCd1	CDA 162	C108	CuCd1 (2.1266)	—	MTsBB	—	—	QCd1
C17000	CuBe1.7	CDA 170	CB101	CuBe1.7 (2.1245)	C1700	BHT 1.7	5055	CuBe1.7	QBe1.7
C17200	CuBe2	CDA 172	—	CuBe2 (2.1247)	C1720	BHT 1.9, M2, M3	—	—	QBe2
C17500	CuCo2Be	CDA 175	A 3/1	CuCo2Be (2.1285)	—	—	—	—	—
C18200	CuCr1	CDA 182	A 2/1	CuCr (2.1291)	—	Br.Kh, Br.Kh.5, Br.Kh.7, Br.Kh.8	—	—	—

1.3.2 Brasses

UNS	ISO—Nominal composition	USA	UK Former BS	Germany DIN (Wk. No.)	Japan JIS	Russia GOST	Sweden SIS	France Former NF	China GB
C21000	CuZn5	CDA 210	CZ125	CuZn5 (2.0220)	C2100	L96, LT96	—	U-Z5	H96
C22000	CuZn10	CDA 220	CZ101	CuZn10 (2.0230)	C2200	—	—	U-Z10	H90
C23000	CuZn15	CDA 230	CZ102	CuZn15 (2.0240)	C2300	L85, L87	5112	U-Z15	H85
C24000	CuZn20	CDA 240	CZ103	CuZn20 (2.0250)	C2400	L80	5114	U-Z20	H80
C26000	CuZn30	CDA 260	CZ106	CuZn30 (2.0285)	C2600	L69, L70, LMSH68-.06	5122	U-Z30	H70
C26800	CuZn33	CDA 268	CZ107	CuZn33 (2.0280)	C2680	L66, L68	—	U-Z33	—
C27400	CuZn37	CDA 274	CZ108	CuZn37 (2.0320)	C2740	L63	5150	U-Z37	H62, H63
C28000	CuZn40	CDA 280	CZ109	CuZn40 (2.0360)	C2800	L58, L60	5163	U-Z40	H59
C35300	CuZn35Pb2	CDA 353	CZ119	CuZn36Pb1.5 (2.0331)	—	LS63-2	5165	—	—
C36000	CuZn36Pb3	CDA 360	CZ124	CuZn36Pb3 (2.0375)	C3603	LS64-2	—	U-Z36Pb3	HPb63-3
C37700	CuZn38Pb2	CDA 377	CZ120	CuZn38Pb2 (2.0380)	C3561	LS59-1, LS59-1L, LS59-1V	5168	—	—
C38000	CuZn43Pb2	CDA 380	CZ122	CuZn43Pb2 (2.0402)	—	—	5168	U-Z39Pb2	—
C38500	CuZn39Pb3	CDA 385	CZ121	CuZn39Pb3 (2.0401)	—	—	5170	—	—
C44300	CuZn28Sn	CDA 443	CZ111	CuZn28Sn (2.0470)	—	LO70-1, LOAs70-1-005	—	—	HSn70-1
C46400	CuZn38Sn1	CDA 464	CZ112	CuZn38Sn (2.0530)	C4641	LO60-1, LO62-1	5220	U-Z29E1	HSn62-1

1.3.3 Phosphor bronzes

C51000	CuSn4	CDA 510	PB102		C5101	Br.OF5.5-.3	—	—	QSn4-0.3
C51100	CuSn4	CDA 511	PB101		C5101	Br.OF4-25	—	—	QSn6.5-0.1
C51900	CuSn6	CDA 519	PB103	CuSn6 (2.1020)	C5191	Br.OF6.5-15	5428	U-E5P	QSn6.5-0.4
C52100	CuSn8	CDA 521	PB104	CuSn8 (2.1030)	—	Br.OF7-2 Br.OF7-4	—	U-E9P	QSn7-0.2
C52400	CuSn10	CDA 524	—		—	Br.O10-1 Br.OF10-1	—	U-E9P	—
C54400	CuSn4Pb4Zn3	CDA 544	—		C5441	Br.OSTs4-4-.25 Br.OSTs4-4-2.5	—	—	QSn4-3 QSn4-4-2.5 QSn4-4-4

1.3.4 Aluminium-silicon bronzes

C60800	CuAl5	CDA 608	CA101	CuAl5As (2.0918)	—	BRA5	—	U-A6	—
C62300	CuAl10Fe3	CDA 623	CA103	CuAl10Fe (2.0938)	C6161	Br.VAZh	—	—	QAl9-4
C63000	CuAl10Fe5Ni5	CDA 630	CA105	CuAl11Fe6Ni5 (2.0978)	C6031	Br.AZhN10-4-4L Br.AZhN9-4-4	—	—	QAl10-4-4 QA/11-6-6
C65500	CuSi3Mn1	CDA 655	CS101	CuSi3Mn (2.1525)	—	Br.KMTs3-1	—	—	QSi3-1
C68700	CuZn20Al2	CDA 687	CZ110	CuZn20Al (2.0460)	—	LA77-2	—	U-Z22A2	—

1.3.5 Copper-nickel and nickel silvers

C70600	CuNi10Fe1Mn	CDA 706	CN102	CuNi10Fe1 (2.0872)	C7060	MNZhMTs10-1.5-1	5667	U-N10	BFe10-1-1
C71000	CuNi20Mn1Fe	CDA 710	CN104, NS108, NS109	CuNi20Fe (2.0878)	C7100	—	—	—	BFe30-1-1
C71500	CuNi30Mn1Fe	CDA 715	CN106, CN107	CuNi30Fe (2.0882)	C7150	HM70, MH33	5682	U-N30	—
C75200	CuNi18Zn20	CDA 752	NS106	CuNi18Zn (2.0740)	C7521	MNTs	5246	U-Z22N18	—
C75700	CuNi12Zn24	CDA 757	NS104	CuNi12Zn24 (2.0730)	—	—	5243	—	—
C77000	CuNi18Zn27	CDA 770	NS107	—	C7701	—	—	U-Z27N18	—

Table 1.4   RELATED DESIGNATIONS FOR MAGNESIUM ALLOYS

UNS	ISO—Nominal composition	USA ASTM	UK BS 2970	Germany DIN (Wk. No.)	Japan JIS	Russia GOST	Europe AECMA	France AFNOR
1.4.1 Cast alloys								
M11101	Mg-Al9Zn	AZ101A	MAG 3	—	—	MGS6, ML6	—	—
M11810	Mg-Al8Zn	AZ81A	MAG 1	G-MgAl8Zn1 (3.5812)	MB2, MS2	ML5	MG-C-61	G-A9
M11910	Mg-Al9Zn	AZ91A	—	GD-MgAl9Zn (3.5912)	MDC1A	—	—	—
M11914	—	AZ91C	MAG 3	G-MgAl9Zn (3.5912)	MC2	—	—	G-A9Z1
M12330	Mg-RE3Zn2Zr	EZ33A	MAG 6	G-MgSE3Zn2Zr1 (3.5103)	MC8	—	MG-C-91	G-TR3Z2
M13320	Mg-Th3Zn2Zr	HZ32A	MAG 8	MgTh3Zn2Zr1 (3.5105)	—	ML14	MG-C-81	G-Th3Z2
M16410	Mg-Zn4REZr	ZE41A	MAG 5	MgZn4SE1Zr1 (3.5101)	—	—	MG-C-43	G-Z4TR
M16510	Mg-Zn5Zr	ZK51A	—	—	MC6	ML12, VM65	—	—
M16610	Mg-Zn6Zr	ZK61A	—	—	MC7	VM65-3	—	—
M19995	Mg99.95	9995A	—	H-Mg99.95 (3.5002)	—	—	—	—
1.4.2 Wrought alloys								
M11311	Mg-Al3Zn1	AZ31B	MAG S-1110	MgAl3Zn (3.5312)	MT1, MB1, MS1	—	Mg-P-62	G-A3Z1
M11600	Mg-Al6Zn1	AZ61A	—	MgAl6Zn (3.5612)	MT2, MB2, MS2	MA3	Mg-P-63	G-A6Z1
M11800	Mg-Al8Zn	AZ80A	MAG 7	MgAl8Zn (3.5812)	MB3, MS3	MA5	Mg-P-61	G-A8Z1

REFERENCES

1. 'Metals and Alloys in the Unified Numbering System', 8th edition, Society of Automotive Engineers, Warrendale, PA, 1999.
2. W. C. Mack, ed., 'Worldwide Guide to Equivalent Irons and Steels', 4th edition, ASM International, Materials Park, OH, 2000.
3. W. C. Mack, ed., 'Worldwide Guide to Equivalent Nonferrous Metals and Alloys', 3rd edition, ASM International, Materials Park, OH, 1996.
4. D. L. Potts and J. G. Gensure, 'International Metallic Materials Cross-Reference', 3rd edition, Glenium Publishing, Schenectady, NY, 1988.
5. R. B. Ross, 'Metallic Materials Specification Handbook', 4th edition, Chapman and Hall, London, 1992.

2 Introductory tables and mathematical information

2.1 Conversion factors

Conversion factors into and from SI units are given in Table 2.5. The table can also be used to convert from one traditional unit to another. Convenient multiples or sub-multiples of SI units can be derived by the application of the prefix multipliers given in Table 2.4. Table 2.6 gives commonly required conversions.

The majority of the conversion factors are based upon equivalents given in BS 350:Part 1:1983 'Conversion Factors and Tables'.

Throughout the conversions the acceleration due to gravity (g) has been taken as the standard acceleration $9.806\,65\,\text{m s}^{-1}$. Units containing the word force like 'pounds force' are converted to SI units using this value of g.

The B.t.u. conversions are based on the definition accepted by the 5th International Conference on Properties of Steam, London, 1956, that 1 B.t.u. $\text{lb}^{-1} = 2.326\,\text{J g}^{-1}$ exactly. Conversions to joules are given for three calories; calories (IC) is the 'international table calorie' redefined by the 1956 conference referred to above as $4.186\,8\,\text{J}$. Calories ($15°\text{C}$) refers to the calorie defined by raising the temperature of water at $15°\text{C}$ by $1°\text{C}$ and calories (US thermochemical) is the 'defined' calorie used in some USA work and is defined at $4.184\,\text{J}$ exactly.

The conversions are grouped in alphabetical order of the physical property to which they relate but are not alphabetical within the groups.

2.1.1 SI units

In this edition quantities are expressed in SI (Système International) units. Where c.g.s. units have been used previously only SI units are given. However, familiar units in general technical use have been retained where they bear a simple power of ten relation to the strict SI unit. For instance density is given as g cm^{-3} and not as kg m^{-3}. Where Imperial units have been used (e.g. in Mechanical Properties, etc.) data are given in both SI units and Imperial units.

The basic units of the SI system are given in Table 2.1, derived units with special names and symbols in Table 2.2 and derived units without special names in Table 2.3.

Multiples and sub-multiples of SI units are formed by prefixes to the name of the unit. The prefixes are shown in Table 2.4. The prefixed unit is written without a hyphen—for instance a thousand million newtons is written giganewton—symbol GN. The name of the unit is written with a small letter even when the symbol has a capital letter, e.g. ampere, symbol A. In the case of the kilogram, the multiple or sub-multiple is applied to the gram—for instance a thousand kilograms is written Mg.

In this edition stress is expressed in Pascals (Pa). A pascal (Pa) is identical to a newton per square metre (N m^{-2}) and a megapascal (MPa) is identical to a newton per square millimetre (N mm^{-2}).

PRINTED FORM OF UNITS AND NUMBERS

The symbol for a unit is in upright type and unaltered by the plural. It is not followed by a full stop unless it is at the end of a sentence. Only symbols of units derived from proper names are in the upper case.

When units are multiplied they will be printed with a space between them. Negative indices are used for units expressed as a quotient. Thus newtons per square metre will be $N\,m^{-2}$ and metres per second will be $m\,s^{-1}$.

The prefix to a unit symbol is written before the unit symbol without a space between and a power index applies to both the symbols. Thus square centimetres is cm^2 and not $(cm)^2$.

Numbers are printed with the decimal point as a full stop. For long numbers, a space and not a comma is given between every three digits. For example $\pi = 3.141\,592\,653$. When a number is entirely decimal it will begin with a zero, e.g. 0.5461. If two numbers are multiplied, a $\times$ sign is used as the operator.

HEADING OF COLUMNS IN TABLES AND LABELLING OF GROUPS

The rule adopted in this edition is that the quantity is obtained by multiplying the unit and its multiple given at the column head by the number in the table.

For example when tabulating a stress of 2×10^5 Pa the heading is stress, below which appears 10^5 Pa, with 2.0 appearing in the table. If no units are given in the column heading, the values given are numbers only. In graphs the power of ten and units by which the point on the graph must be multiplied are given on the axis label.

TEMPERATURES

The temperature scale IPTS–68 has been replaced by the International Temperature Scale of 1990 (ITS–90). For details of this see chapter 16, where Table 16.1 gives the differences between ITS–90 and EPT–76 and between ITS–76 and between ITS–90 and IPTS–68. Figure 16.1 gives differences $(t_{90} - t_{68})$ between ITS–90 and IPTS–68 in the range $-260°$C to $1064°$C. Table 2.7 gives conversions between the old IPTS–68 and the old IPTS–48.

Table 2.1   BASIC SI UNITS

Quantity	Name of unit	Unit symbol
Length	metre	m
Mass	kilogram	kg
Time	second	s
Electric current	ampere	A
Thermodynamic temperature	kelvin	K
Luminous intensity	candela	cd
Amount of substance	mole	mol
Plane angle	radian	rad
Solid angle	steradian	sr

From 'Quantities, Units and Symbols', Royal Society, 1981.

Table 2.2   DERIVED SI UNITS WITH SPECIAL NAMES

Quantity	Name of unit	Symbol	Equivalent	Definition
Activity (radioactivity)	becquerel	Bq	s^{-1}	
Absorbed dose (of radiation)	gray	Gy	$J\,kg^{-1}$	
Dose equivalent (of radiation)	sievert	Sv	$J\,kg^{-1}$	
Energy	joule	J	$N\,m$	$m^2\,kg\,s^{-2}$
Force	newton	N	$J\,m^{-1}$	$m\,kg\,s^{-2}$
Stress or pressure	pascal	Pa	$N\,m^{-2}$	$m^{-1}\,kg\,s^{-2}$
Power	watt	W	$J\,s^{-1}$	$m^2\,kg\,s^{-3}$
Electric charge	coulomb	C	$A\,s$	$s\,A$
Electric potential	volt	V	$W\,A^{-1}$	$m^2\,kg\,s^{-3}\,A^{-1}$

(continued)

Table 2.2 DERIVED SI UNITS WITH SPECIAL NAMES—*continued*

Quantity	Name of unit	Symbol	Equivalent	Definition
Electric resistance	ohm	Ω	$V\,A^{-1}$	$m^2\,kg\,s^{-3}\,A^{-2}$
Electric capacitance	farad	F	$C\,V^{-1}$	$m^{-2}\,kg^{-1}\,s^4\,A^2$
Electric conductance	siemens	S	$A\,V^{-1}$	$m^{-2}\,kg^{-1}\,s^3\,A^2$
Magnetic flux	weber	Wb	$V\,s$	$m^2\,kg\,s^{-2}\,A^{-1}$
Inductance	henry	H	$V\,s\,A^{-1}$	$m^2\,kg\,s^{-2}\,A^{-2}$
Magnetic flux density	tesla	T	$Wb\,m^{-2}$	$kg\,s^{-2}\,A^{-1}$
Luminous flux	lumen	lm	cd sr	cd sr
Illumination	lux	lx	$cd\,sr\,m^{-2}$	m^{-2} cd sr
Frequency	hertz	Hz	s^{-1}	s^{-1}

From 'Quantities, Units and Symbols', Royal Society, 1981.
Note: Symbols derived from proper names begin with a capital letter.
In the definition the steradian (sr) is treated as a base unit.

Table 2.3 SOME DERIVED SI UNITS WITHOUT SPECIAL NAMES

Quantity	SI unit	Symbol
Area	square metre	m^2
Acceleration	metre/second squared	$m\,s^{-2}$
Angular velocity	radian/second	$rad\,s^{-1}$
Calorific value	joule/kilogram	$J\,kg^{-1}$
Concentration	mole/cubic metre	$mol\,m^{-3}$
Current density	ampere/square metre	$A\,m^{-2}$
Density	kilogram/cubic metre	$kg\,m^{-3}$
Diffusion coefficient	square metre/second	$m^2\,s^{-1}$
Electrical conductivity	siemens/metre	$S\,m^{-1}$
Electric field strength	volt/metre	$V\,m^{-1}$
Electrical resistivity	ohm metre	$\Omega\,m$
Entropy	joule/kelvin	$J\,K^{-1}$
Exposure (to radiation)	coulomb/kilogram	$C\,kg^{-1}$
Heat capacity	joule/kelvin	$J\,K^{-1}$
Heat flux density	watt/square metre	$W\,m^{-2}$
Latent heat	joule/kilogram	$J\,kg^{-1}$
Luminance	candela/square metre	$cd\,m^{-2}$
Magnetic field strength	ampere/metre	$A\,m^{-1}$
Magnetic moment	joule/tesla	$J\,T^{-1}$
Molar volume	cubic metre/mole	$m^3\,mol^{-1}$
Moment of inertia	kilogram/square metre	$kg\,m^{-2}$
Moment of force	newton metre	Nm
Molar heat capacity	joule/kelvin mole	$J\,K^{-1}\,mol$
Permittivity	farad/metre	$F\,m^{-1}$
Permeability	henry/metre	$H\,m^{-1}$
Radioactivity	1/second	s^{-1}
Speed (velocity)	metre/second	$m\,s^{-1}$
Specific volume	cubic metre/kilogram	$m^3\,kg^{-1}$
Specific heat-mass	joule/kilogram kelvin	$J\,kg^{-1}\,K^{-1}$
Specific heat-volume	joule/cubic metre kelvin	$J\,m^{-3}\,K^{-1}$
Surface tension	newton/metre	$N\,m^{-1}$
Thermal conductivity	watt/metre kelvin	$W\,m^{-1}\,K^{-1}$

(*continued*)

Table 2.3 SOME DERIVED SI UNITS WITHOUT SPECIAL NAMES—*continued*

Quantity	SI unit	Symbol
Thermoelectric power	volt/kelvin	$V\ K^{-1}$
Viscosity–kinematic	square metre/second	$m^2\ s^{-1}$
Viscosity–dynamic	pascal second	Pa s
Volume	cubic metre	m^3
Wave number	1/metre	m^{-1}

Table 2.4 PREFIXES FOR MULTIPLES AND SUB-MULTIPLES USED IN THE SI SYSTEM OF UNITS

Sub-multiple	Prefix	Symbol	Multiple	Prefix	Symbol
10^{-1}	deci	d	10	deca	da
10^{-2}	centi	c	10^2	hecto	h
10^{-3}	milli	m	10^3	kilo	k
10^{-6}	micro	μ	10^6	mega	M
10^{-9}	nano	n	10^9	giga	G
10^{-12}	pico	p	10^{12}	tera	T
10^{-15}	femto	f	10^{15}	peta	P
10^{-18}	atto	a	10^{18}	exa	E

From 'Quantities, Units and Symbols', Royal Society, 1981.

Table 2.5 CONVERSION FACTORS

To convert B to A multiply by	A	B	To convert A to B multiply by
	Acceleration		
10^2	centimetres/second squared	metres/second squared	10^{-2}
$3.937\,008 \times 10$	inches/second squared	metres/second squared	2.54×10^{-2}
$3.280\,84$	feet/second squared	metres/second squared	3.048×10^{-1}
$1.019\,716 \times 10^{-1}$	standard acceleration due to gravity	metres/second squared	$9.806\,65$
	Angle—plane		
$2.062\,65 \times 10^5$	seconds	radians	$4.848\,14 \times 10^{-6}$
$3.437\,75 \times 10^3$	minutes	radians	$2.908\,88 \times 10^{-4}$
$5.729\,58 \times 10$	degrees	radians	$1.745\,33 \times 10^{-2}$
$1.591\,55 \times 10^{-1}$	revolutions	radians	$6.283\,19$
$6.366\,20 \times 10$	grades	radians	$1.570\,80 \times 10^{-2}$
	Angular velocity		
$5.729\,58 \times 10$	degrees/second	radians/second	$1.745\,33 \times 10^{-2}$
$1.591\,55 \times 10^{-1}$	revolutions/second	radians/second	$6.283\,19$
$3.437\,75 \times 10^3$	degrees/minute	radians/second	$2.908\,88 \times 10^{-4}$
$9.549\,27$	revolutions/minute	radians/second	$1.047\,20 \times 10^{-1}$
	Area		
10^{28}	barn	square metres	10^{-28}
$1.550\,003 \times 10^3$	square inches	square metres	$6.451\,6 \times 10^{-4}$

(*continued*)

Table 2.5   CONVERSION FACTORS—*continued*

To convert B to A multiply by	A	B	To convert A to B multiply by
$1.076\,391 \times 10$	square feet	square metres	$9.290\,3 \times 10^{-2}$
$1.195\,990$	square yards	square metres	$8.361\,27 \times 10^{-1}$
$3.861\,02 \times 10^{-7}$	square miles	square metres	$2.589\,99 \times 10^{6}$
$2.471\,052 \times 10^{-4}$	acres	square metres	$4.046\,86 \times 10^{3}$
10^{-4}	hectares	square metres	10^{4}
$2.471\,052$	acres	hectares	$4.046\,86 \times 10^{-1}$
2.5×10^{-1}	acres	roods	4
$1.562\,5 \times 10^{-3}$	square miles	acres	6.40×10^{2}

Calorific value—volume basis

$2.683\,92 \times 10^{-2}$	British thermal units/cubic foot	joules/cubic metre	$3.725\,89 \times 10$
$4.308\,86 \times 10^{-11}$	therms/UK gallon	joules/cubic metre	$2.320\,80 \times 10^{10}$
$2.388\,46 \times 10^{-4}$	kilocalories/cubic metre	joules/cubic metre	$4.186\,8 \times 10^{3}$

Calorific value—mass basis

4.299×10^{-4}	British thermal units/pound	joules/kilogram	2.326×10^{3}
$2.388\,46 \times 10^{-4}$	International kilocalories/kilogram	joules/kilogram	$4.186\,8 \times 10^{3}$
$2.390\,06 \times 10^{-4}$	thermochemical kilocalories/kilogram	joules/kilogram	4.184×10^{3}

Compressibility

10^{-1}	square centimetres/dyne	metres/newton	10

Density

10^{-3}	grams/cubic centimetre	kilograms/cubic metre	10^{3}
$1.603\,59 \times 10^{-1}$	ounces/gallon (UK)	kilograms/cubic metre	$6.236\,03$
$6.242\,80 \times 10^{-2}$	pounds/cubic foot	kilograms/cubic metre	$1.601\,85 \times 10$
$3.612\,73 \times 10^{-5}$	pounds/cubic inch	kilograms/cubic metre	$2.767\,99 \times 10^{4}$
$1.002\,241 \times 10^{-2}$	pounds/gallon (UK)	kilograms/cubic metre	$9.977\,64 \times 10$
$8.345\,434 \times 10^{-3}$	pounds/gallon (US)	kilograms/cubic metre	$1.198\,26 \times 10^{2}$
$7.015\,673 \times 10$	grains/gallon (UK)	kilograms/cubic metre	$1.425\,38 \times 10^{-2}$

Diffusion coefficient

10^{4}	square centimetres/second	square metres/second	10^{-4}

Electric charge

$2.997\,93 \times 10^{9}$	electrostatic units	coulombs	$3.335\,64 \times 10^{-10}$
10^{-1}	electromagnetic units	coulombs	10

Electric current

$2.997\,93 \times 10^{9}$	electrostatic units	amperes	$3.335\,64 \times 10^{-10}$
10^{-1}	electromagnetic units	amperes	10

Electric current density

$2.997\,93 \times 10^{5}$	electrostatic units	amperes/square metre	$3.335\,64 \times 10^{-6}$
10^{-5}	electromagnetic units	amperes/square metre	10^{5}
10^{-4}	amperes/square centimetre	amperes/square metre	10^{4}
6.452×10^{-4}	amperes/square inch	amperes/square metre	1.55×10^{3}
$9.290\,2 \times 10^{-2}$	amperes/square foot	amperes/square metre	$1.076\,4 \times 10$

Energy—work—heat

$2.777\,778 \times 10^{-7}$	kilowatt hours	joules	3.6×10^{6}
$1.019\,72 \times 10^{-1}$	kilogram force metres	joules	$9.806\,65$
$2.373\,04 \times 10$	foot poundals	joules	$4.214\,01 \times 10^{-2}$
$7.375\,62 \times 10^{-1}$	foot pounds force	joules	$1.355\,82$
$3.725\,06 \times 10^{-7}$	horsepower hours	joules	$2.684\,52 \times 10^{6}$
$9.869\,23 \times 10^{-3}$	litre (dm^3) atmospheres	joules	$1.013\,25 \times 10^{2}$

(*continued*)

Table 2.5   CONVERSION FACTORS—*continued*

To convert B to A multiply by	A	B	To convert A to B multiply by
$2.388\,46 \times 10^{-4}$	kilocalories (IC)	joules	$4.186\,8 \times 10^{3}$
$8.850\,34$	inch pounds force	joules	$1.129\,9 \times 10^{-1}$
$9.478\,17 \times 10^{-4}$	British thermal units	joules	$1.055\,06 \times 10^{3}$
10^{7}	ergs	joules	10^{-7}
$6.241\,808 \times 10^{18}$	electron volts	joules	$1.602\,1 \times 10^{-19}$
$9.478\,13 \times 10^{-9}$	therms (Btu)	joules	$1.055\,06 \times 10^{8}$
$2.388\,46 \times 10^{-1}$	calories (IC)	joules	$4.186\,8$
$2.389\,201 \times 10^{-1}$	calories (15°C)	joules	$4.185\,5$
$2.390\,057 \times 10^{-1}$	calories (US thermochemical)	joules	4.184

Entropy

$2.388\,46 \times 10^{-1}$	calories (IC)/degree centigrade	joules/kelvin	$4.186\,8$
$5.265\,62 \times 10^{-4}$	British thermal unit/degree Fahrenheit	joules/kelvin	$1.899\,11 \times 10^{3}$

Force

10^{5}	dynes	newtons	10^{-5}
$3.596\,94$	ounces force	newtons	$0.278\,014$
$1.019\,72 \times 10^{2}$	grams force	newtons	$9.806\,65 \times 10^{-3}$
$2.248\,09 \times 10^{-1}$	pounds force	newtons	$4.448\,22$
$7.233\,01$	poundals	newtons	$0.138\,255$
$1.003\,61 \times 10^{-4}$	UK tons force	newtons	$9.964\,02 \times 10^{3}$
$1.124\,047 \times 10^{-4}$	US tons force	newtons	$8.896\,422 \times 10^{3}$
$1.019\,72 \times 10^{-1}$	kilograms force	newtons	$9.806\,65$

Fracture toughness

$1.019\,72 \times 10^{-4}$	(kilograms force/square centimetre)$\sqrt{}$(centimetre)	newton/$\sqrt{}$(metre^3)	$9.806\,55 \times 10^{3}$
$9.100\,42 \times 10^{-7}$	(kilopounds force/square inch) $\sqrt{}$(inch)	newton/$\sqrt{}$(metre^3)	$1.098\,85 \times 10^{6}$
$4.062\,73 \times 10^{-7}$	(tons force/square inch)$\sqrt{}$(inch)	newtons/$\sqrt{}$(metre^3)	$2.461\,4 \times 10^{6}$
$3.162\,26 \times 10^{-6}$	hectobars$\sqrt{}$(millimetre)	newtons/$\sqrt{}$(metre^3)	$3.162\,3 \times 10^{5}$

Heat—see Energy
Heat flow rate—see Power
Latent heat

$4.299\,23 \times 10^{-4}$	British thermal units/pound	joules/kilogram	2.326×10^{3}
$2.388\,46 \times 10^{-4}$	calories (IC)/gram	joules/kilogram	$4.186\,8 \times 10^{3}$
$3.345\,52 \times 10^{-1}$	foot pounds force/pound	joules/kilogram	$2.989\,07$
$1.019\,72 \times 10^{-1}$	kilogram force metres/kilogram	joules/kilogram	$9.806\,65$

Leak rate

$7.500\,64 \times 10^{3}$	lusec (micron Hg litre/second)	joules/second	$1.333\,22 \times 10^{-4}$

Length

10^{10}	angstroms (Å)	metres	10^{-10}
10^{6}	microns (μ)	metres	10^{-6}
$9.979\,84 \times 10^{9}$	kx units	metres	$1.002\,02 \times 10^{-10}$
$3.937\,01 \times 10$	inches	metres	2.54×10^{-2}
$3.280\,84$	feet	metres	3.048×10^{-1}
$1.093\,61$	yards	metres	9.144×10^{-1}
$6.213\,71 \times 10^{-4}$	miles	metres	$1.609\,344 \times 10^{3}$
$5.396\,118 \times 10^{-4}$	miles (naut UK)	metres	$1.853\,184 \times 10^{3}$
$5.399\,568 \times 10^{-4}$	miles (naut Int)	metres	1.852×10^{3}
1.81×10^{-1}	rods, poles or perches	yards	5.5
2.5×10^{-1}	chains	rods, poles, etc.	4.0
10^{-1}	furlongs	chains	10.0
1.25×10^{-1}	miles (UK)	furlongs	8.0

(continued)

Table 2.5　CONVERSION FACTORS—*continued*

To convert B to A multiply by	A	B	To convert A to B multiply by
1.66×10^{-1}	fathoms	feet	6.0
8.33×10^{-3}	cable lengths	fathoms	1.2×10^2
1.6447×10^{-4}	nautical miles	feet	6.080×10^3
1.8939×10^{-4}	miles (UK)	feet	5.280×10^3

Magnetic conversions—see Magnetic units and conversion factors, Chapter 20

Moment of force—see Energy

Moment of inertia

10^7	grams centimetre squared	kilograms metre squared	10^{-7}
3.41717×10^3	pounds inch squared	kilograms metre squared	2.92640×10^{-4}
2.37304×10	pounds foot squared	kilograms metre squared	4.21401×10^{-2}
5.46747×10^4	ounces inch squared	kilograms metre squared	1.82900×10^{-5}

Momentum

7.23301	foot pounds/second	kilogram metres/second	1.38255×10^{-1}
10^5	gram centimetres/second	kilogram metres/second	10^{-5}

Mass

5.643819×10^2	drams (Av)	kilograms	1.77185×10^{-3}
3.527399×10	ounces (Av)	kilograms	2.83495×10^{-2}
2.2046226	pounds (Av)	kilograms	4.5359237×10^{-1}
1.574731×10^{-1}	stones (Av)	kilograms	6.350293
7.873650×10^{-2}	quarters (Av)	kilograms	1.270059×10
1.968415×10^{-2}	hundredweights (Av)	kilograms	5.08023×10
9.842035×10^{-4}	tons (Av)	kilograms	1.01605×10^3
1.543236×10^4	grains or minims (Apoth)	kilograms	6.47989×10^{-5}
7.716180×10^2	scruples (Apoth)	kilograms	1.295978×10^{-3}
2.572063×10^2	drams (Apoth)	kilograms	3.88793×10^{-3}
3.215072×10	ounces (Apoth or Troy)	kilograms	3.11035×10^{-2}
1.543237×10^4	grains (Troy)	kilograms	6.479885×10^{-5}
10^{-3}	tonnes (metric)	kilograms	10^3
1.102311×10^{-3}	tons (short 2000 lb)	kilograms	9.07185×10^2
5.0×10^3	metric carats (CM)	kilograms	2.0×10^{-4}

Mass per unit area

8.92180×10^3	pounds/acre	kilograms/square metre	1.12085×10^{-4}
1.843348	pounds/square yard	kilograms/square metre	5.424912×10^{-1}
2.04816×10^{-1}	pounds/square foot	kilograms/square metre	4.882432
2.949357×10	ounces/square yard	kilograms/square metre	3.39057×10^{-2}
3.227055	ounces/square foot	kilograms/square metre	3.05152×10^{-1}

Mass per unit length

5.59973×10^{-2}	pounds/inch	kilograms/metre	1.78580×10
6.71971×10^{-1}	pounds/foot	kilograms/metre	1.48816
2.01591	pounds/yard	kilograms/metre	4.96055×10^{-1}

Power—Heat flow rate

10^7	ergs/second	watts	10^{-7}
3.41214	British thermal units/hour	watts	2.93071×10^{-1}
8.59845×10^{-1}	kilocalories (IC)/hour	watts	1.163
7.37561×10^{-1}	foot pounds force/second	watts	1.35582
2.38846×10^{-1}	calories (IC)/second	watts	4.1868
1.341022×10^{-3}	horsepower	watts	7.457×10^2
1.35962×10^{-3}	metric horsepower (CV) (PS)	watts	7.35499×10^2

Pressure—see Stress

Radioactivity

2.7×10^{-11}	curie	becquerel	3.7×10^{10}

(continued)

Table 2.5   CONVERSION FACTORS—*continued*

To convert B to A multiply by	A	B	To convert A to B multiply by
	Radiation–absorbed dose		
10^2	rem	sievert	10^{-2}
	Radiation exposure		
3.876×10^3	roentgen	coulomb/kilogram	2.58×10^{-4}
	Specific heat capacity—mass basis		
10^{-3}	joules/gram degree centigrade	joules/kilogram kelvin	10^3
$2.388\,46 \times 10^{-4}$	calories*/gram degree centigrade	joules/kilogram kelvin	$4.186\,8 \times 10^3$
$2.388\,46 \times 10^{-4}$	British thermal units/ pound degree Fahrenheit	joules/kilogram kelvin	$4.186\,8 \times 10^3$
$1.858\,63 \times 10^{-1}$	foot pounds force/ pound degree Fahrenheit	joules/kilogram kelvin	$5.380\,32$
$1.019\,72 \times 10^{-1}$	kilogram force metres/ kilogram degree centigrade	joules/kilogram kelvin	$9.806\,65$
	Specific heat—volume basis		
10^{-6}	joules/cubic centimetre degree centigrade	joules/cubic metre kelvin	10^6
$2.388\,459 \times 10^{-4}$	kilocalories*/cubic metre degree centigrade	joules/cubic metre kelvin	$4.186\,8 \times 10^3$
$1.491\,066 \times 10^{-5}$	British thermal units/cubic foot degree Fahrenheit	joules/cubic metre kelvin	$6.706\,61 \times 10^4$
	Stress		
$1.450\,377 \times 10^{-4}$	pounds force/square inch	newtons/square metre	$6.894\,76 \times 10^3$
$6.474\,881 \times 10^{-8}$	UK tons force/square inch	newtons/square metre	$1.544\,43 \times 10^7$
10	dynes/square centimetre	newtons/square metre	10^{-1}
10^{-5}	bars	newtons/square metre	10^5
10^{-7}	hectobars	newtons/square metre	10^7
$1.019\,716 \times 10^{-7}$	kilograms force/square millimetre	newtons/square metre	$9.806\,65 \times 10^6$
$7.500\,638 \times 10^{-3}$	torrs	newtons/square metre	$1.333\,22 \times 10^2$
$7.500\,638 \times 10^{-3}$	millimetres of mercury	newtons/square metre	$1.333\,22 \times 10^2$
$7.500\,638$	micron of mercury	newtons/square metre	$1.333\,22 \times 10^{-1}$
$9.869\,233 \times 10^{-6}$	atmospheres	newtons/square metre	$1.013\,250 \times 10^5$
1	pascals	newtons/square metre	1
$6.474\,8807 \times 10^{-2}$	UK tons force/square inch	megapascals	$1.544\,43 \times 10$
	Surface tension		
10^3	dynes/centimetre	newtons/metre	10^{-3}
$6.852\,178 \times 10^{-2}$	pounds force/foot	newtons/metre	$1.459\,39 \times 10$
$5.710\,148 \times 10^{-3}$	pounds force/inch	newtons/metre	$1.751\,268 \times 10^2$
	Temperature interval		
1	degrees Celsius (centigrade)	kelvins	1
1.8	degrees Fahrenheit	kelvins	5.55×10^{-1}
1.8	degrees Rankine	kelvins	5.55×10^{-1}
	Thermal conductivity		
10^{-2}	watts/centimetre degree centigrade	watts/metre kelvin	10^2
$2.388\,46 \times 10^{-3}$	calories/centimetre second degree centigrade	watts/metre kelvin	$4.186\,8 \times 10^2$
$8.598\,45 \times 10^{-1}$	kilocalories/metre hour degree centigrade	watts/metre kelvin	1.163
$5.777\,91 \times 10^{-1}$	British thermal unit/foot hour degree Fahrenheit	watts/metre kelvin	$1.730\,73$
$6.933\,47$	British thermal unit inch/square foot hour degree Fahrenheit	watts/metre kelvin	$1.442\,28 \times 10^{-1}$

(continued)

Table 2.5 CONVERSION FACTORS—*continued*

To convert B to A multiply by	A	B	To convert A to B multiply by
		Time	
1.66×10^{-2}	minutes	seconds	6.0×10
2.77×10^{-4}	hours	seconds	3.600×10^3
$1.157\,41 \times 10^{-5}$	days	seconds	8.64×10^4
$1.653\,44 \times 10^{-6}$	weeks	seconds	6.048×10^5
$3.170\,98 \times 10^{-8}$	years	seconds	$3.153\,6 \times 10^7$
$1.141\,552\,5 \times 10^{-4}$	years	hours	8.760×10^3
		Torque—see Energy	
		Velocity	
$3.280\,84$	feet/second	metres/second	$0.304\,8$
$1.968\,504 \times 10^2$	feet/minute	metres/second	5.08×10^{-3}
3.6	kilometres/hour	metres/second	$0.277\,778$
$3.728\,227 \times 10^{-2}$	miles/minute	metres/second	$2.682\,40 \times 10$
$2.236\,94$	miles/hour	metres/second	$0.447\,04$
$1.942\,60$	UK knots	metres/second	$0.514\,773$
$1.943\,85$	International knots	metres/second	$0.514\,444$
1.136×10^{-2}	UK miles/minute	feet/second	8.8×10
		*Viscosity—dynamic**	
10	poise	newton seconds/square metre	10^{-1}
$1.019\,72 \times 10^{-1}$	kilogram force seconds/square metre	newton seconds/square metre	$9.806\,65$
$6.719\,71 \times 10^{-1}$	poundal seconds/square foot	newton seconds/square metre	$1.488\,16$
$2.088\,542 \times 10^{-2}$	pound force seconds/square foot	newton seconds/square metre	$4.788\,03 \times 10$
		Viscosity—kinematic	
10^4	stokes	square metres/second	10^{-4}
$1.550\,03 \times 10^3$	square inches/second	square metres/second	6.4516×10^{-4}
$1.076\,392 \times 10$	square feet/second	square metres/second	$9.290\,3 \times 10^{-2}$
$5.580\,011 \times 10^6$	square inches/hour	square metres/second	$1.792\,111 \times 10^{-7}$
$3.875\,009 \times 10^4$	square feet/hour	square metres/second	$2.580\,640 \times 10^{-5}$
3.6×10^3	square metres/hour	square metres/second	2.77×10^{-4}
		Volume	
$6.102\,37 \times 10^4$	cubic inches	cubic metres	$1.638\,71 \times 10^{-5}$
$3.531\,473 \times 10$	cubic feet	cubic metres	$2.831\,68 \times 10^{-2}$
$1.307\,95$	cubic yards	cubic metres	$7.645\,55 \times 10^{-1}$
10^3	litres	cubic metres	10^{-3}
$2.199\,69 \times 10^2$	gallons (UK)	cubic metres	$4.546\,09 \times 10^{-3}$
$2.641\,72 \times 10^2$	gallons (US)	cubic metres	$3.785\,41 \times 10^{-3}$
$1.759\,755 \times 10^{-3}$	pints (UK)	cubic centimetres	$5.682\,61 \times 10^2$
$2.199\,69 \times 10^{-1}$	gallons (UK)	litres	$4.549\,09$
$3.519\,508 \times 10$	fluid ounces (UK)	litres	$2.841\,306 \times 10^{-2}$
		Work—see Energy	

* newton seconds/square metre $(Ns/m^2) \equiv$ pascal seconds (Pa s).

Table 2.6   COMMONLY REQUIRED CONVERSIONS

Acceleration	$g = 32$ feet/second squared	=	9.806 65 metres/second squared	$m\,s^{-2}$
Angle	1 radian	=	57.295 8 degrees	°
Area	1 acre	=	4 046.86 square metres	m^2
	1 hectare	=	10 000 square metres	m^2
Density	1 gram/cubic centimetre	=	1 000 kilograms/cubic metre	$kg\,m^{-3}$
Energy	1 calorie (IC)	=	4.186 8 joules	J
	1 kilowatt hour	=	3.6 megajoules	MJ
	1 British thermal unit	=	1 055.06 joules	J
	1 erg	=	10^{-7} joules	J
	1 therm	=	105.506 megajoules	MJ
	1 horsepower hour	=	2.684 52 megajoules	MJ
Force	1 dyne	=	10^{-5} newtons	N
	1 pound force	=	4.448 22 newtons	N
	1 UK ton force	=	9 964.02 newtons	N
	1 kilogram force	=	9.806 65 newtons	N
Length	1 angstrom unit (Å)	=	10^{-10} metres	m
	1 micron (μm)	=	10^{-6} metres	m
	1 micron (μm)	=	$0.039\,37 \times 10^{-3}$ inches	in
	1 thousandth of an inch	=	25.4 micrometres	μm
	1 inch	=	2.54 centimetres	cm
	1 foot	=	30.48 centimetres	cm
	1 yard	=	91.44 centimetres	cm
	1 mile	=	1.609 344 kilometres	km
Mass	1 ounce (Av)	=	28.349 5 grams	g
	1 ounce (Troy)	=	31.103 5 grams	g
	1 pound (Av)	=	453.592 grams	g
	1 hundredweight	=	50.802 3 kilograms	kg
	1 UK ton (Av)	=	1 016.05 kilograms	kg
	1 short ton (2 000 lbs)	=	907.185 kilograms	kg
	1 carat (metric)	=	0.2 grams	g
Power	1 horsepower	=	745.7 watts	W
Stress	1 pound force/square inch (p.s.i.)	=	6.894 76 kilopascals	kPa
	1 UK ton force/square inch	=	15.444 3 megapascals	MPa
	1 bar	=	100 kilopascals	kPa
	1 hectobar	=	10 megapascals	Mpa
	1 kilogarm force/square centimetre	=	98.006 5 kilopascals	kPa
	1 kilogram force/square millimetre	=	9.806 65 megapascals	MPa
	1 torr = 1 millimetre of mercury	=	133.322 pascals	Pa
	1 atmosphere	=	101.325 kilopascals	kPa
	1 pascal	=	1 newton/square metre	$N\,m^{-2}$
Surface tension	1 dyne/centimetre	=	1 millinewton/metre	$mN\,m^{-1}$
Velocity	1 foot/second	=	1.097 28 kilometres/hour	$km\,h^{-1}$
	1 mile/hour	=	1.609 344 kilometres/hour	$km\,h^{-1}$
Volume	1 cubic inch	=	16.387 1 cubic centimetres	cm^3
	1 cubic foot	=	28.316 8 cubic decimetres	dm^3
	1 cubic yard	=	0.764 555 cubic metres	m^3
	1 litre	=	1 cubic decimetre	dm^3
	1 litre	=	1.759 75 UK pints	pint
	1 UK gallon	=	4.546 09 cubic decimetres	dm^3
	1 UK gallon	=	0.160 544 cubic feet	ft^3

Table 2.7 CORRECTIONS TO TEMPERATURE VALUES IPTS-48 TO IMPLEMENT IPTS-68
(IPTS–68)–(IPTS–48) IN °C

t_{68}	0	−10	−20	−30	−40	−50	−60	−70	−80	−90	−100	$t_{68}°C$
−100	0.022	0.013	0.003	−0.006	−0.013	0.013	−0.005	0.007	0.012	(0.008 at O$_2$ point) 0.029	0.022	−100
−0	0.000	0.006	0.012	0.018	0.024	0.029	0.032	0.034	0.033	0.029		−0

$t_{68}°C$	0	10	20	30	40	50	60	70	80	90	100	$t_{68}°C$
0	0.000	−0.004	−0.007	−0.009	−0.010	−0.010	−0.010	−0.008	−0.006	−0.003	0.000	0
100	0.000	0.004	0.007	0.012	0.016	0.020	0.025	0.029	0.034	0.038	0.043	100
200	0.043	0.047	0.051	0.054	0.058	0.061	0.064	0.067	0.069	0.071	0.073	200
300	0.073	0.074	0.075	0.076	0.077	0.077	0.077	0.077	0.077	0.076	0.076	300
400	0.076	0.075	0.075	0.075	0.074	0.074	0.074	0.075	0.076	0.077	0.079	400
500	0.079	0.082	0.085	0.089	0.094	0.100	0.108	0.116	0.126	0.137	0.150	500
600	0.150	0.165	0.182	0.200	0.23	0.25	0.28	0.31	0.34	0.36	0.39	600
700	0.39	0.42	0.45	0.47	0.50	0.53	0.56	0.58	0.61	0.64	0.67	700
800	0.67	0.70	0.72	0.75	0.78	0.81	0.84	0.87	0.89	0.92	0.95	800
900	0.95	0.98	1.01	1.04	1.07	1.10	1.12	1.15	1.18	1.21	1.24	900
1 000	1.24	1.27	1.30	1.33	1.36	1.39	1.42	1.44	—	—	—	1 000

$t_{68}°C$	0	100	200	300	400	500	600	700	800	900	1 000	$t_{68}°C$
1 000	—	1.5	1.7	1.8	2.0	2.2	2.4	2.6	2.8	3.0	3.2	1 000
2 000	3.2	3.5	3.7	4.0	4.2	4.5	4.8	5.0	5.3	5.6	5.9	2 000
3 000	5.9	6.2	6.5	6.9	7.2	7.5	7.9	8.2	8.6	9.0	9.3	3 000

From BS 1826:1952 Amendment No. 1, 2 February 1970. Example: 1 000°C according to IPTS–48 would be corrected to 1001.24°C to conform to IPTS–68. For conversions from IPTS–68 to ITS–90 see Table 16.1.

Table 2.8   CORROSION CONVERSION FACTORS

The following conversion factors relating loss in weight and depth of penetration are useful in the assessment of corrosion.

$$\text{density of metal in grams/cubic centimetre} = d$$
$$\text{density of metal in kilograms/cubic metre} = 10^3 d$$

To convert B to A multiply by	A	B	To convert A to B multiply by
$10^{-3}\,d^{-1}$	millimetres	grams/square metre	$10^3 d$
$3.65 \times 10^{-1}\,d^{-1}$	millimetres/year	grams/square metre per day	$2.74\,d$
$8.76\,d^{-1}$	millimetres/year	grams/square metre per hour	$1.14 \times 10^{-1}\,d$
$3.937 \times 10^{-2}\,d^{-1}$	thousandths of an inch (mils)	grams/square metre	$2.54 \times 10\,d$
$1.44 \times 10\,d^{-1}$	mils/year	grams/square metre per day	$6.96 \times 10^{-2}d$
$3.45 \times 10^2\,d^{-1}$	mils/year	grams/square metre per hour	$2.90 \times 10^{-3}d$
$1.201 \times 10\,d^{-1}$	mils	ounces/square foot	$8.326 \times 10^{-2}d$

Table 2.9   TEST SIEVE MESH NUMBERS CONVERTED TO NOMINAL APERTURE SIZE FROM BS 410:1969

Wire cloth test sieves were designated by the mesh count or number. This method, widely used until 1962, was laid down in previous British Standards—BS 410. Sieves are now designated by aperture size: *see* BS 410:1969, for full details.

   The table gives the previously used mesh numbers with the corresponding nominal aperture sizes, the preferred average wire diameters in the test sieves and the tolerances.

			Aperture tolerances		
Mesh No.	Nominal aperture size mm	Preferred average wire diameter in test sieve mm	Max. tolerance for size of an individual aperture mm +	Tolerance for average aperture size mm ±	Intermediate tolerance mm +
3	5.60	1.60	0.50	0.17	0.34
$3\frac{1}{2}$	4.75	1.60	0.43	0.14	0.29
4	4.00	1.40	0.40	0.12	0.28
5	3.35	1.25	0.34	0.10	0.23
6	2.80	1.12	0.31	0.084	0.20
7	2.36	1.00	0.26	0.071	0.17
8	2.00	0.90	0.24	0.060	0.16
10	1.70	0.80	0.20	0.051	0.14
12	1.40	0.71	0.18	0.042	0.11
14	1.18	0.63	0.17	0.035	0.11
16	1.00	0.56	0.15	0.030	0.09
	μm	μm	μm +	μm ±	μm +
18	850	500	128	30	79
22	710	450	114	28	71
25	600	400	102	24	66
30	500	315	90	20	55
36	425	280	81	17	51
44	355	224	71	14	43
52	300	200	64	15	40
60	250	160	58	13	36
72	212	140	53	12	33
85	180	125	51	11	31
100	150	100	48	9.4	29
120	125	90	46	8.1	27

(continued)

Table 2.9　TEST SIEVE MESH NUMBERS—*continued*

Mesh No.	Nominal aperture size mm	Preferred average wire diameter in test sieve μm	Aperture tolerances		
			Max. tolerance for size of an individual aperture μm +	Tolerance for average aperture size μm ±	Intermediate tolerance μm +
150	106	71	43	7.4	25
170	90	63	43	6.6	25
200	75	50	41	6.1	24
240	63	45	41	5.3	23
300	53	36	38	4.8	21
350	45	32	38	4.8	21
400	38	30	36	4.0	20

Notes:
(1) No aperture size shall exceed the nominal by more than the maximum tolerance.
(2) The average aperture size shall not be greater or smaller than the nominal by more than the average tolerance size.
(3) Not more than 6% of the apertures shall be above the nominal size by more than the intermediate tolerance.

For perforated plate sieve sizes with square or round holes—*see* BS 410:1969. Other national standards for test sieves may be found for France in NF X11–501, for Germany in DIN 4188 and for USA in ASTM E11–61.

2.1.2　Temperature scale conversions

The absolute unit of temperature, symbol K, is the kelvin which is 1/273.16 of the thermodynamic temperature of the triple point of water—*see* Section 16. Practical temperature scales are Celsius (previously Centigrade) symbol °C, and Fahrenheit, symbol °F. An absolute scale based on Fahrenheit is the Rankine, symbol °R.

Where K, C, F and R, represent the same temperature on the Kelvin, Celsius, Fahrenheit and Rankine scales, conversion formulae are:

$$K = C + 273.15 \qquad C = \frac{5}{9}(F - 32)$$

$$F = \frac{9}{5}C + 32 \qquad R = F + 459.67$$

Rapid approximate conversions between Celsius and Fahrenheit scales can be obtained from Figure 2.1.

2.2　Mathematical formulae and statistical principles

2.2.1　Algebra

IDENTITIES

$$a^2 - b^2 = (a - b)(a + b)$$
$$a^2 + b^2 = (a - ib)(a + ib) \quad \text{where } i = \sqrt{-1}$$
$$a^3 - b^3 = (a - b)(a^2 + ab + b^2)$$
$$a^3 + b^3 = (a + b)(a^2 - ab + b^2)$$
$$a^n - b^n = (a - b)(a^{n-1} - a^{n-2}b + \cdots + b^{n-1})$$
$$a^n + b^n = (a + b)(a^{n-1} - a^{n-2}b + \cdots + b^{n-1}) \quad \text{when } n \text{ is odd}$$
$$(a \pm b)^2 = a^2 \pm 2ab + b^2$$
$$(a \pm b)^3 = a^3 \pm 3a^2b + 3ab^2 \pm b^3$$

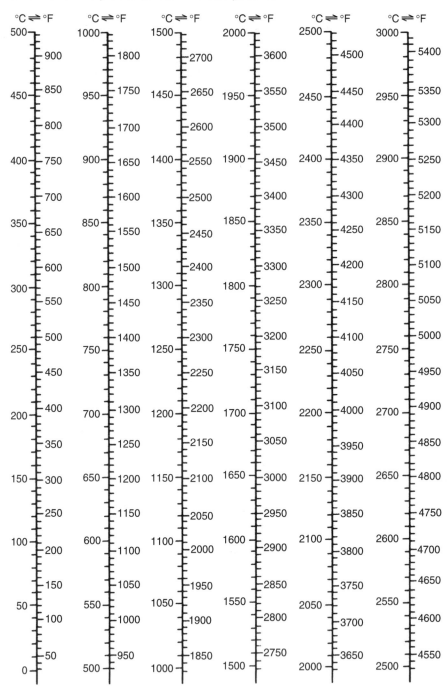

Figure 2.1   *Nomogram for approximate interconversion between Celsius and Fahrenheit temperature scales*

$$(a+b)^n = a^n + na^{n-1}b + \frac{n(n-1)}{2!}a^{n-2}b^2$$

$$+ \cdots + \frac{n(n-1)(n-2)\cdots(n-r+1)}{r!}a^{n-r}b^r + \cdots + b^n = \sum_{x=0}^{n}\binom{n}{x}a^{n-x}b^x$$

RATIO AND PROPORTION

If

$$\frac{a}{b} = \frac{c}{d}$$

then

$$\frac{a+b}{b} = \frac{c+d}{d}$$

and

$$\frac{a-b}{b} = \frac{c-d}{d}$$

In general,

$$\frac{a}{b} = \frac{c}{d} = \frac{e}{f} = \cdots = \left(\frac{pa^n + qc^n + re^n + \cdots}{pb^n + qd^n + rf^n + \cdots}\right)^{1/n}$$

where p, q, r and n are any quantities whatever.

LOGARITHMS

If

$$a^x = N, \quad \text{then } x = \log_a N$$

$$\log_a MN = \log_a M + \log_a N$$

$$\log_a \frac{M}{N} = \log_a M - \log_a N$$

$$\log_a(M^p) = p\log_a M$$

$$\log_a(M^{1/r}) = \frac{1}{r}\log_a M$$

$$\log_b N = \frac{1}{\log_a b} \times \log_a N$$

In particular,

$$\log_e N = 2.302\,58509 \times \log_{10} N \qquad e = 2.718\,28$$

$$\log_{10} N = 0.434\,294\,48 \times \log_e N$$

$$\log_e N \equiv \ln N \qquad \log_{10} N \equiv \lg N$$

THE QUADRATIC EQUATION

The general quadratic equation may be written

$$ax^2 + bx + c = 0$$

Solution $x = \dfrac{-b \pm \sqrt{b^2 - 4ac}}{2a}$

If

$$\Delta = (b^2 - 4ac)$$

the roots are real and equal if $\Delta = 0$
the roots are imaginary and unequal if $\Delta < 0$
the roots are real and unequal if $\Delta > 0$

Also, if the roots are α and β,

$$\alpha + \beta = -\frac{b}{a}$$

$$\alpha\beta = \frac{c}{a}$$

THE CUBIC EQUATION

The general cubic equation may be written

$$y^3 + a_1 y^2 + a_2 y + a_3 = 0$$

If we put $y = x - \frac{1}{3}a_1$ the equation reduces to $x^3 + ax + b = 0$, where

$$a = a_2 - \frac{1}{3}a_1^2 \quad \text{and} \quad b = \frac{2}{27}a_1^3 - \frac{1}{3}a_1 a_2 + a_3$$

Solution $x = z + v$ or $-\dfrac{z+v}{2} \pm i\sqrt{3}\left(\dfrac{z-v}{2}\right),$

where

$$z = \sqrt[3]{-\frac{1}{2}b + \sqrt{\frac{b^2}{4} + \frac{a^3}{27}}}$$

and

$$v = \sqrt[3]{-\frac{1}{2}b - \sqrt{\frac{b^2}{4} + \frac{a^3}{27}}}$$

Alternatively,*

$$x = 2\sqrt{\left(\frac{-a}{3}\right)}\cos\frac{\theta + 2k\pi}{3} \quad (k = 0, 1, 2)$$

where

$$\cos\theta = -\frac{b}{2}\left(\frac{-a^3}{27}\right)^{-1/2}$$

If

$$\Delta = \frac{b^2}{4} + \frac{a^3}{27}$$

there are two equal and one unequal root if $\Delta = 0$
three real roots if $\Delta < 0$
one real and two complex roots if $\Delta > 0$

---

* The second form of the solution is particularly useful when $\Delta < 0$ (i.e. in the case of three real roots).

2.2.2   Series and progressions

NUMERICAL SERIES

$$1 + 2 + 3 + \cdots + n = \frac{n}{2}(n+1)$$

$$1^2 + 2^2 + 3^2 + \cdots + n^2 = \frac{n}{6}(n+1)(2n+1)$$

$$1^3 + 2^3 + 3^3 + \cdots + n^3 = \frac{n^2}{4}(n+1)^2$$

ARITHMETIC PROGRESSION

$$a, a+d, a+2d, \ldots, a+(n-1)d$$

$$S_n = \frac{n}{2}[2a + (n-1)d]$$

where S_n denotes the sum to n terms.

GEOMETRIC PROGRESSION

$$a, ar, ar^2, \ldots, ar^{n-1}$$

$$S_n = \frac{a(r^n - 1)}{(r-1)} = \frac{a(1 - r^n)}{(1-r)}$$

where S_n denotes the sum to n terms.

If $r^2 < 1$ and $n \to \infty$,

$$S_\infty = \frac{a}{(1-r)}$$

TAYLOR'S SERIES

$$f(x) = f(a) + (x-a)f'(a) + \frac{(x-a)^2}{2!}f''(a) + \frac{(x-a)^3}{3!}f'''(a) + \cdots$$

MACLAURIN'S SERIES

$$f(x) = f(0) + xf'(0) + \frac{x^2}{2!}f''(0) + \frac{x^3}{3!}f'''(0) + \cdots$$

In the following series, the region of convergence is indicated in parentheses. If no region is shown, the series is convergent for all values of x.

BINOMIAL SERIES

$$(1 \pm x)^n = 1 \pm nx + \frac{n(n-1)}{2!}x^2 \pm \frac{n(n-1)(n-2)}{3!}x^3 + \cdots \quad (x^2 < 1)$$

$$(1 \pm x)^{-n} = 1 \mp nx + \frac{n(n+1)}{2!}x^2 \mp \frac{n(n+1)(n+2)}{3!}x^3 + \cdots \quad (x^2 < 1)$$

LOGARITHMIC SERIES

$$\log_e(1 \pm x) = \pm x - \frac{x^2}{2} \pm \frac{x^3}{3} - \frac{x^4}{4} \pm \cdots \quad (x^2 < 1)$$

$$\log_e \frac{(1+x)}{(1-x)} = 2\left(x + \frac{x^3}{3} + \frac{x^5}{5} + \cdots\right) \quad (x^2 < 1)$$

$$\log_e x = 2\left[\frac{x-1}{x+1} + \frac{1}{3}\left(\frac{x-1}{x+1}\right)^3 + \frac{1}{5}\left(\frac{x-1}{x+1}\right)^5 + \cdots\right] \quad (x > 0)$$

$$\log_e(a+x) = \log_e a + 2\left[\frac{x}{2a+x} + \frac{1}{3}\left(\frac{x}{2a+x}\right)^3 + \frac{1}{5}\left(\frac{x}{2a+x}\right)^5 + \cdots\right]$$

$$(a > 0, \ x + a > 0)$$

EXPONENTIAL SERIES

$$e = 1 + 1 + \frac{1}{2!} + \frac{1}{3!} + \frac{1}{4!} + \cdots$$

$$e^x = 1 + x + \frac{x^2}{2!} + \frac{x^3}{3!} + \frac{x^4}{4!} + \cdots$$

$$a^x = 1 + x\log_e a + \frac{(x\log_e a)^2}{2!} + \frac{(x\log_e a)^3}{3!} + \cdots$$

TRIGONOMETRIC SERIES

$$\sin x = x - \frac{x^3}{3!} + \frac{x^5}{5!} - \frac{x^7}{7!} + \cdots$$

$$\cos x = 1 - \frac{x^2}{2!} + \frac{x^4}{4!} - \frac{x^6}{6!} + \cdots$$

$$\tan x = x + \frac{x^3}{3} + \frac{2x^5}{15} + \frac{17x^7}{315} + \frac{62x^9}{2835} + \cdots \quad \left(-\frac{\pi}{2} < x < \frac{\pi}{2}\right)$$

$$\sin^{-1} x = x + \frac{1}{2}\cdot\frac{x^3}{3} + \frac{1}{2}\cdot\frac{3}{4}\cdot\frac{x^5}{5} + \frac{1}{2}\cdot\frac{3}{4}\cdot\frac{5}{6}\cdot\frac{x^7}{7} + \cdots \quad (-1 \le x \le 1)$$

$$\cos^{-1} x = \frac{\pi}{2} - \sin^{-1} x$$

$$\tan^{-1} x = x - \frac{x^3}{3} + \frac{x^5}{5} - \frac{x^7}{7} + \cdots \quad (-1 \le x \le 1)$$

If

$$x = 1, \ \frac{\pi}{4} = 1 - \frac{1}{3} + \frac{1}{5} - \frac{1}{7} + \cdots$$

$$\cot^{-1} x = \frac{\pi}{2} - \tan^{-1} x$$

SERIES FOR HYPERBOLIC FUNCTIONS

$$\sinh x = x + \frac{x^3}{3!} + \frac{x^5}{5!} + \frac{x^7}{7!} + \cdots$$

$$\cosh x = 1 + \frac{x^2}{2!} + \frac{x^4}{4!} + \frac{x^6}{6!} + \cdots$$

$$\tanh x = x - \frac{x^3}{3} + \frac{2x^5}{5} - \frac{17x^7}{315} + \frac{62x^9}{2835} + \cdots \quad \left(-\frac{\pi}{2} < x < \frac{\pi}{2}\right)$$

$$\sinh^{-1} x = x - \frac{1}{2} \cdot \frac{x^3}{3} + \frac{1}{2} \cdot \frac{3}{4} \cdot \frac{x^5}{5} - \frac{1}{2} \cdot \frac{3}{4} \cdot \frac{5}{6} \cdot \frac{x^7}{7} + \cdots \quad (-1 < x < 1)$$

$$\tanh^{-1} x = x + \frac{x^3}{3} + \frac{x^5}{5} + \frac{x^7}{7} + \cdots \quad (-1 < x < 1)$$

$$\coth^{-1} x = \frac{1}{x} + \frac{1}{3x^3} + \frac{1}{5x^5} + \frac{1}{7x^7} + \cdots \quad (-1 < x < 1)$$

2.2.3   Trigonometry

DEFINITIONS AND SIMPLE RELATIONSHIPS

$$\sin A = \frac{a}{b}$$

$$\cos A = \frac{c}{b}$$

$$\tan A = \frac{a}{c}$$

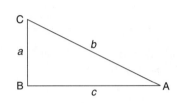

$$\operatorname{cosec} A = \csc A = \frac{b}{a} = \frac{1}{\sin A}; \quad \sec A = \frac{b}{c} = \frac{1}{\cos A}; \quad \cot A = \frac{c}{a} = \frac{1}{\tan A}$$

$$\operatorname{versin} A = \operatorname{vers} A = 1 - \cos A$$

$$\operatorname{coversin} A = \operatorname{covers} A = 1 - \sin A$$

$$\operatorname{haversin} A = \operatorname{hav} A = \frac{1}{2} \operatorname{versin} A$$

$$\sin^2 A + \cos^2 A = 1$$

$$\tan A = \frac{\sin A}{\cos A}; \quad \cot A = \frac{\cos A}{\sin A}$$

$$1 + \tan^2 A = \sec^2 A$$

$$1 + \cot^2 A = \operatorname{cosec}^2 A$$

RADIAN MEASURE

$$\pi \text{ radians} = 180°, \quad \pi = 3.141\,59\ldots$$

COMPOUND ANGLES

$$\sin(A + B) = \sin A \cos B + \cos A \sin B$$

$$\cos(A + B) = \cos A \cos B - \sin A \sin B$$

$$\sin(A - B) = \sin A \cos B - \cos A \sin B$$

$$\cos(A - B) = \cos A \cos B + \sin A \sin B$$

Table 2.10   SIGN AND VALUE OF THE FUNCTIONS BETWEEN 0° AND 360°

Degrees	0°		90°		180°		270°		360°	30°	45°	60°
Quadrant		1		2		3		4				
sin	0	+	1	+	0	−	−1	−	0	1/2	$1/\sqrt{2}$	$\sqrt{3}/2$
cos	1	+	0	−	−1	−	0	+	1	$\sqrt{3}/2$	$1/\sqrt{2}$	1/2
tan	0	+	∞	−	0	+	∞	−	0	$1/\sqrt{3}$	1	$\sqrt{3}$
cot	∞	+	0	−	∞	+	0	−	∞	$\sqrt{3}$	1	$1/\sqrt{3}$

Table 2.11   SUPPLEMENTARY AND COMPLEMENTARY ANGLES

$x =$	$90 \pm \alpha$	$180 \pm \alpha$	$270 \pm \alpha$	$n360 \pm \alpha$ (or $\pm\alpha$)
$\sin x$	$\pm\cos\alpha$	$\mp\sin\alpha$	$-\cos\alpha$	$\pm\sin\alpha$
$\cos x$	$\mp\sin\alpha$	$-\cos\alpha$	$\pm\sin\alpha$	$+\cos\alpha$
$\tan x$	$\mp\cot\alpha$	$\pm\tan\alpha$	$\mp\cot\alpha$	$\pm\tan\alpha$
$\cot x$	$\mp\tan\alpha$	$\pm\cot\alpha$	$\mp\tan\alpha$	$\pm\cot\alpha$

$$\tan(A + B) = \frac{\tan A + \tan B}{1 - \tan A \tan B}$$

$$\tan(A - B) = \frac{\tan A - \tan B}{1 + \tan A \tan B}$$

$$\sin 2A = 2 \sin A \cos A$$

$$\cos 2A = \cos^2 A - \sin^2 A = 1 - 2 \sin^2 A = \cos^2 A - 1$$

$$\tan 2A = \frac{2 \tan A}{1 - \tan^2 A}$$

$$\sin 3A = 3 \sin A - 4 \sin^3 A$$

$$\cos 3A = 4 \cos^3 A - 3 \cos A$$

$$\left.\begin{aligned}\sin A &= \frac{2t}{1 + t^2} \\[2mm] \cos A &= \frac{1 - t^2}{1 + t^2}\end{aligned}\right\} \quad \text{where } t = \tan\frac{A}{2}$$

$$\sin A + \sin B = 2 \sin\frac{A + B}{2} \cos\frac{A - B}{2}$$

$$\sin A - \sin B = 2 \cos\frac{A + B}{2} \sin\frac{A - B}{2}$$

$$\cos A + \cos B = 2 \cos\frac{A + B}{2} \cos\frac{A - B}{2}$$

$$\cos B - \cos A = 2 \sin\frac{A + B}{2} \sin\frac{A - B}{2}$$

PROPERTIES OF TRIANGLES

$$\frac{a}{\sin A} = \frac{b}{\sin B} = \frac{c}{\sin C} = 2R, \quad \text{where } R = \text{radius of circumcircle}$$

$$a^2 = b^2 + c^2 - 2bc \cos A$$

$$\tan \frac{B - C}{2} = \frac{b - c}{b + c} \cot \frac{A}{2}$$

$$a = b \cos C + c \cos B$$

Area of triangle $= \Delta = \frac{1}{2}ab \sin C = \sqrt{s(s-a)(s-b)(s-c)}$, where $s = \frac{1}{2}(a+b+c)$

$$\sin \frac{A}{2} = \sqrt{\frac{(s-b)(s-c)}{bc}}$$

$$\cos \frac{A}{2} = \sqrt{\frac{s(s-a)}{bc}}$$

$$\tan \frac{A}{2} = \sqrt{\frac{(s-b)(s-c)}{s(s-a)}}$$

$$A + B + C = 180°$$

$$\sin A + \sin B + \sin C = 4 \cos \frac{A}{2} \cos \frac{B}{2} \cos \frac{C}{2}$$

$$\cos A + \cos B + \cos C = 4 \sin \frac{A}{2} \sin \frac{B}{2} \sin \frac{C}{2} + 1$$

$$\tan A + \tan B + \tan C = \tan A \tan B \tan C$$

$$\cot \frac{A}{2} + \cot \frac{B}{2} + \cot \frac{C}{2} = \cot \frac{A}{2} \cot \frac{B}{2} \cot \frac{C}{2}$$

HYPERBOLIC FUNCTIONS

$$\sinh x = \frac{1}{2}(e^x - e^{-x})$$

$$\cosh x = \frac{1}{2}(e^x + e^{-x})$$

$$\tanh x = \frac{\sinh x}{\cosh x} = \frac{e^x - e^{-x}}{e^x + e^{-x}}$$

$$\text{sech } x = \frac{1}{\cosh x} = \frac{2}{e^x + e^{-x}}$$

$$\text{cosech } x = \frac{1}{\sinh x} = \frac{2}{e^x - e^{-x}}$$

$$\cosh^2 x - \sinh^2 x = 1$$

$$\sinh(-x) = -\sinh x$$

$$\cosh(-x) = \cosh x$$

$$\tanh(-x) = -\tanh x$$

$$\sinh(x + y) = \sinh x \cosh y + \cosh x \sinh y$$

$$\cosh(x + y) = \cosh x \cosh y + \sinh x \sinh y$$

(*See also* 'Series and progressions'.)

2.2.4 Mensuration

PLANE FIGURES

Parallelogram. Area = base × altitude

Triangle. Area = $\frac{1}{2}$ × base × altitude

Trapezium. Area = $\frac{1}{2}$(sum of parallel sides) × altitude

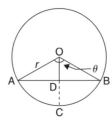

$\theta = \text{A}\hat{\text{O}}\text{B}$ (*measured in radians*) OD = x

Circle. Radius = r

Area = πr^2; circumference = $2\pi r$

Length of arc ACB = $r\theta$

Area of sector OACB = $\frac{1}{2}r^2\theta$

Length of chord AB = $2r\sin\dfrac{\theta}{2}$

Area of segment ACB = $\frac{1}{2}r^2(\theta - \sin\theta) = \dfrac{\pi r^2}{2} - \left[x\sqrt{r^2 - x^2} + r^2\sin^{-1}\left(\dfrac{x}{r}\right)\right]$

Ellipse. Area = πab, where a and b are the semi-axes

Perimeter = $2\pi\sqrt{\dfrac{a^2 + b^2}{2}}$ (approximate)

SOLID FIGURES

Rectangular prism. Sides a, b and c

Surface area = $2(ab + bc + ca)$

Volume = abc

Diagonal = $\sqrt{a^2 + b^2 + c^2}$

Sphere. Radius = r

Surface area = $4\pi r^2$; volume = $\frac{4}{3}\pi r^3$

Curved area of spherical segment, height $h = 2\pi rh$

Volume of spherical segment, height $h = \frac{1}{3}\pi h^2(3r - h)$

Curved area of spherical zone between two parallel planes, distance l apart = $2\pi rl$

Spherical shell. Internal radius = r, external radius = R

Volume of shell = $\frac{4}{3}\pi(R^3 - r^3)$

$= \frac{4}{3}\pi t^3 + 4\pi trR$

$= 4\pi r'^2 t + \dfrac{\pi}{3}t^3$

where t = thickness = $R - r$

and r' = mean radius = $\frac{1}{2}(R + r)$

Right circular cylinder.*   Height = h, radius = r

Area of curved surface = $2\pi rh$

Total surface area = $2\pi r(h + r)$

Volume = $\pi r^2 h$

Right circular cone.†   Radius of base = r, height = h

Area of curved surface = $\pi r\sqrt{r^2 + h^2}$

Volume = $\dfrac{\pi}{3}r^2 h$

Area of curved surface of frustrum (radii R and r) = $\pi(R + r)\sqrt{h^2 + (R - r)^2}$

Volume of frustrum (radii R and r) = $\dfrac{\pi h}{3}(R^2 + Rr + r^2)$

2.2.5   Co-ordinate geometry (two dimensions, rectangular axes)

The length d of the straight line joining the points A(x_1, y_1) and B(x_2, y_2) is given by

$$d = \pm\sqrt{(x_2 - x_1)^2 + (y_2 - y_1)^2}$$

If P(x, y) is a point on AB such that $\dfrac{AP}{PB} = \dfrac{m}{n}$,

$$x = \frac{mx_2 + nx_1}{m + n}; \quad y = \frac{my_2 + ny_1}{m + n}$$

STRAIGHT LINE

The general equation to a straight line is,

$$ax + by + c = 0$$

Slope = $-\dfrac{a}{b}$. Intercept on y-axis = $-\dfrac{c}{b}$

Alternative forms are

1. $\dfrac{y_2 - y_1}{x_2 - x_1} = \dfrac{y - y_1}{x - x_1}$ for the line joining the points (x_1, y_1) and (x_2, y_2)

2. $y = mx + c$ for the line of slope, m, cutting y-axis at point $(0, c)$.

3. $y - y_1 = m(x - x_1)$ for the line through the point (x_1, y_1), slope m.

4. $\dfrac{x}{a} + \dfrac{y}{b} = 1$ for the line making intercepts of a and b on the axes of x and y, respectively.

5. $x\cos\alpha + y\sin\alpha = p$, where p is the length of the perpendicular from the origin to the line and α is the angle between the perpendicular and the positive direction of the x-axis.

The perpendicular distance d from the point (x_1, y_1) to the line $ax + by + c = 0$ is

$$d = \pm\frac{ax_1 + by_1 + c}{\sqrt{a^2 + b^2}}$$

---

* Volume of *any* cylinder = area of base × vertical height.

† Volume of *any* cone or pyramid = $\frac{1}{3}$[area of base × perpendicular distance from vertex to base].

The angle θ between two lines of slopes m_1 and m_2 is given by:

$$\tan\theta = \frac{m_1 - m_2}{1 + m_1 m_2}$$

If the two lines are parallel, $m_1 = m_2$
If the two lines are perpendicular, $m_1 m_2 = -1$

TRIANGLE

The area of a triangle, vertices (x_1, y_1), (x_2, y_2) and (x_3, y_3) is

$$\Delta = \frac{1}{2}(x_1 y_2 - x_2 y_1 + x_2 y_3 - x_3 y_2 + x_3 y_1 - x_1 y_3)$$

CIRCLE

The general equation to a circle is

$$x^2 + y^2 + 2gx + 2fy + c = 0$$
$$\text{centre}(-g, -f); \quad \text{radius} = \sqrt{g^2 + f^2 - c}$$

Equation to tangent at the point (x_1, y_1) on the circle is

$$xx_1 + yy_1 + g(x + x_1) + f(y + y_1) + c = 0$$

An alternative form of the equation to the circle is

$$(x - a)^2 + (y - b)^2 = R^2,$$
$$\text{centre}(a, b); \quad \text{radius} = R$$

ELLIPSE

$$\frac{x^2}{a^2} + \frac{y^2}{b^2} = 1. \text{ Centre at origin and semi-axes } a \text{ and } b$$

Equation to tangent at point (x_1, y_1) is $\dfrac{xx_1}{a^2} + \dfrac{yy_1}{b^2} = 1$

PARABOLA

$$y^2 = 4ax. \text{ Vertex at origin, } \textit{latus rectum} = 2a$$

Equation to tangent at point (x_1, y_1) is $yy_1 = 2a(x + x_1)$

HYPERBOLA

$$\frac{x^2}{a^2} - \frac{y^2}{b^2} = 1. \text{ Centre at origin and semi-axes } a \text{ and } b$$

Equation to tangent at point (x_1, y_1) is $\dfrac{xx_1}{a^2} - \dfrac{yy_1}{b^2} = 1$

The asymptotes are $\dfrac{x}{a} \pm \dfrac{y}{b} = 0$

2.2.6   Calculus*

DIFFERENTIALS[†]

$\mathrm{d}ax = a\,\mathrm{d}x$

$\mathrm{d}(u+v) = \mathrm{d}u + \mathrm{d}v$

$\mathrm{d}uv = v\,\mathrm{d}v + v\,\mathrm{d}u$

$\mathrm{d}\dfrac{u}{v} = \dfrac{v\,\mathrm{d}u - u\,\mathrm{d}v}{v^2}$

$\mathrm{d}x^n = nx^{n-1}\,\mathrm{d}x$

$\mathrm{d}x^y = yx^{y-1}\mathrm{d}x + x^y\log_e x\,\mathrm{d}y$

$\mathrm{d}\sin x = \cos x\,\mathrm{d}x$

$\mathrm{d}\cos x = -\sin x\,\mathrm{d}x$

$\mathrm{d}\tan x = \sec^2 x\,\mathrm{d}x$

$\mathrm{d}\cot x = -\mathrm{cosec}^2 x\,\mathrm{d}x$

$\mathrm{d}\sec x = \tan x\,\sec x\,\mathrm{d}x$

$\mathrm{d}\mathrm{cosec}\,x = -\cot x\,\mathrm{cosec}\,x\,\mathrm{d}x$

$\mathrm{d}\,\mathrm{vers}\,x = \sin x\,\mathrm{d}x$

$\mathrm{d}\sin^{-1}x = (1-x^2)^{-1/2}\,\mathrm{d}x$

$\mathrm{d}\cos^{-1}x = -(1-x^2)^{-1/2}\,\mathrm{d}x$

$\mathrm{d}\tan^{-1}x = (1+x^2)^{-1}\,\mathrm{d}x$

$\mathrm{d}\cot^{-1}x = -(1+x^2)^{-1}\,\mathrm{d}x$

$\mathrm{d}\sec^{-1}x = x^{-1}(x^2-1)^{-1/2}\,\mathrm{d}x$

$\mathrm{d}\mathrm{cosec}^{-1}x = -x^{-1}(x^2-1)^{-1/2}\,\mathrm{d}x$

$\mathrm{d}e^x = e^x\,\mathrm{d}x$

$\mathrm{d}e^{ax} = ae^{ax}\,\mathrm{d}x$

$\mathrm{d}a^x = a^x\log_e a\,\mathrm{d}x$

$\mathrm{d}\log_e x = x^{-1}\,\mathrm{d}x$

$\mathrm{d}\log_a x = x^{-1}\log_a e\,\mathrm{d}x$

$\mathrm{d}x^x = x^x(1+\log_e x)\,\mathrm{d}x$

$\mathrm{d}\,\mathrm{vers}^{-1}x = (2x-x^2)^{-1/2}\,\mathrm{d}x$

$\mathrm{d}\sinh x = \cosh x\,\mathrm{d}x$

$\mathrm{d}\cosh x = \sinh x\,\mathrm{d}x$

$\mathrm{d}\tanh x = \mathrm{sech}^2 x\,\mathrm{d}x$

$\mathrm{d}\coth x = -\mathrm{cosech}^2 x\,\mathrm{d}x$

$\mathrm{d}\,\mathrm{sech}\,x = -\mathrm{sech}\,x\,\tanh x\,\mathrm{d}x$

$\mathrm{d}\,\mathrm{cosech}\,x = -\mathrm{cosech}\,x\,\coth x\,\mathrm{d}x$

$\mathrm{d}\sinh^{-1}x = (x^2+1)^{-1/2}\,\mathrm{d}x$

$\mathrm{d}\cosh^{-1}x = (x^2-1)^{-1/2}\,\mathrm{d}x$

$\mathrm{d}\tanh^{-1}x = (1-x^2)^{-1}\,\mathrm{d}x$

$\mathrm{d}\coth^{-1}x = -(x^2-1)^{-1}\,\mathrm{d}x$

$\mathrm{d}\,\mathrm{sech}^{-1}x = -x^{-1}(1-x^2)^{-1/2}\,\mathrm{d}x$

$\mathrm{d}\,\mathrm{cosech}^{-1}x = -x^{-1}(x^2+1)^{-1/2}\,\mathrm{d}x$

INTEGRALS

Elementary forms

1. $\displaystyle\int a\,\mathrm{d}x = ax$

2. $\displaystyle\int a\cdot f(x)\,\mathrm{d}x = a\int f(x)\,\mathrm{d}x$

3. $\displaystyle\int \phi(y)\,\mathrm{d}x = \int \frac{\phi(y)}{y'}\,\mathrm{d}y,\quad$ where $y' = \dfrac{\mathrm{d}y}{\mathrm{d}x}$

4. $\displaystyle\int (u+v)\,\mathrm{d}x = \int u\,\mathrm{d}x + \int v\,\mathrm{d}x,\quad$ where u and v are any functions of x

5. $\displaystyle\int u\,\mathrm{d}v = uv - \int v\,\mathrm{d}u$

6. $\displaystyle\int u\frac{\mathrm{d}v}{\mathrm{d}x}\,\mathrm{d}x = uv - \int v\frac{\mathrm{d}v}{\mathrm{d}x}\,\mathrm{d}x$

7. $\displaystyle\int x^n\,\mathrm{d}x = \frac{x^{n+1}}{n+1},\quad (n\neq -1)$

---

* From *Handbook of Chemistry and Physics*, Cleveland, Ohio, 1945.

[†] Differentials have been written as above for ease in use, e.g. $\mathrm{d}ax = a\mathrm{d}x$ instead of $(\mathrm{d}/\mathrm{d}x)ax = a$, which is the mathematically correct.

8. $\int \dfrac{f'(x)\,dx}{f(x)} = \log f(x), \quad [d\,f(x) = f'(x)\,dx]$

9. $\int \dfrac{dx}{x} = \log x, \quad \text{or} \quad \log(-x)$

10. $\int \dfrac{f'(x)\,dx}{2\sqrt{f(x)}} = \sqrt{f(x)} \quad [d\,f(x) = f'(x)\,dx]$

11. $\int e^x\,dx = e^x$

12. $\int e^{ax}\,dx = \dfrac{e^{ax}}{a}$

13. $\int b^{ax}\,dx = \dfrac{b^{ax}}{a\log b}$

14. $\int \log x\,dx = x\log x - x$

15. $\int a^x \log a\,dx = a^x$

16. $\int \dfrac{dx}{a^2 + x^2} = \dfrac{1}{a}\tan^{-1}\left(\dfrac{x}{a}\right), \quad \text{or} \quad -\dfrac{1}{a}\cot^{-1}\left(\dfrac{x}{a}\right)$

17. $\int \dfrac{dx}{a^2 - x^2} = \dfrac{1}{a}\tanh^{-1}\left(\dfrac{x}{a}\right), \quad \text{or} \quad \dfrac{1}{2a}\log\dfrac{a+x}{a-x}$

18. $\int \dfrac{dx}{x^2 - a^2} = -\dfrac{1}{a}\coth^{-1}\left(\dfrac{x}{a}\right), \quad \text{or} \quad \dfrac{1}{2a}\log\dfrac{x-a}{x+a}$

19. $\int \dfrac{dx}{\sqrt{a^2 - x^2}} = \sin^{-1}\left(\dfrac{x}{a}\right), \quad \text{or} \quad -\cos^{-1}\left(\dfrac{x}{a}\right)$

20. $\int \dfrac{dx}{\sqrt{x^2 \pm a^2}} = \log\left[x + \sqrt{x^2 \pm a^2}\right]$

21. $\int \dfrac{dx}{x\sqrt{x^2 - a^2}} = \dfrac{1}{a}\cos^{-1}\left(\dfrac{a}{x}\right)$

22. $\int \dfrac{dx}{x\sqrt{a^2 \pm x^2}} = -\dfrac{1}{a}\log\left[\dfrac{a + \sqrt{a^2 \pm x^2}}{x}\right]$

23. $\int \dfrac{dx}{x\sqrt{a + bx}} = \dfrac{2}{\sqrt{-a}}\tan^{-1}\sqrt{\dfrac{a+bx}{-a}}, \quad \text{or} \quad \dfrac{-2}{\sqrt{a}}\tanh^{-1}\sqrt{\dfrac{a+bx}{a}}$

More complex integrals are to be found in *Handbook of Chemistry and Physics*, Cleveland, Ohio, 1945.

DIFFERENTIAL EQUATIONS

Equations of the first order

1. $\dfrac{dy}{dx} = f(x)\phi(y) \quad$ (Variable separable type)

    Solution $\quad \int \dfrac{dy}{\phi(y)} = \int f(x)\,dx + A$

2a. $\dfrac{dy}{dx} = f\left(\dfrac{y}{x}\right) \quad$ (Homogeneous equation)

Let $v = \dfrac{y}{x}$

Solution $\displaystyle\int \frac{dv}{f(v) - v} = \log x + A$

2b. $\dfrac{dy}{dx} = \dfrac{ax + by + c}{a'x + b'y + c'}$ (Reducible to homogeneous form if $a'b - ab' \neq 0$)

Let $x = X + h;\ y = Y + k$, where h and k are the values of x and y which satisfy equations

$$\left.\begin{array}{c} ax + by + c = 0 \\ a'x + b'y + c' = 0 \end{array}\right\}$$

Solution   obtained as in (2a) by putting $Y = vX$

3. $M\ dx + N\ dy = 0$ where M and N are functions of x and y and

$$\frac{\partial M}{\partial y} = \frac{\partial N}{\partial x} \quad \text{(Exact equation)}$$

Here, a function $u(x, y)$ exists such that

$$du = M\ dx + N\ dy$$

Solution $\displaystyle u = \int M\ dx\ (y\ \text{constant}) + \int (\text{terms in } N \text{ independent of } x)\ dy = \text{constant}$

4. $\dfrac{dy}{dx} + Py = Q$ (Linear equation)

where P and Q are functions of x only (or constants).

Solution $\displaystyle y\, e^{\int P\,dx} = \int Q e^{\int P\,dx} dx + A$ (Integrating factor $= e^{\int P\,dx}$)

5. $\dfrac{dy}{dx} + Py = Qy^n$ (Bernouilli's equation)

where P and Q are functions of x only (or constants).
   Multiplying both sides of the equation by $(1 - n)y^{-n}$, and putting $z = y^{1-n}$ the equation becomes

$$\frac{dz}{dx} + (1 - n)Pz = (1 - n)Q$$

This equation is linear.
Solution   obtained as in 4. [Integrating factor $= e^{\int (1-n)P\,dx}$]

6. $y = x\dfrac{dy}{dx} + f\!\left(\dfrac{dy}{dx}\right)$ (Clairaut's form)

Let $\dfrac{dy}{dx} = p$ and differentiate the equation with respect to x,

$$\frac{dp}{dx}[x + f'(p)] = 0$$

Solution $\dfrac{dp}{dx} = 0$ or $x + f'(p) = 0$

i.e. $p = \text{constant} = A$ or $y = Ax + B$

This solution is obtained by eliminating p from this equation and the original equation. The solution contains no arbitrary constant and is known as the *singular solution*.

Equations of the second order

7. $\dfrac{d^2 y}{dx^2} = f(x)$

Solution $\displaystyle y = \int F(x)\ dx + Ax + B$,   where $\displaystyle F(x) = \int f(x)\ dx$

8. $\dfrac{d^2y}{dx^2} = f(y)$

Let $\dfrac{dy}{dx} = p;$ $\dfrac{1}{2}p^2 = \displaystyle\int f(y)\,dy + A$

Integration of this equation of the first order may then lead to the required solution.

9.* $a\dfrac{d^2y}{dx^2} + b\dfrac{dy}{dx} + cy = 0$ (Linear equation, constant coefficients, RHS $= 0$)

Auxilliary equation. $am^2 + bm + c = 0$
Solution

(a) If $b^2 > 4ac$ (roots m_1 and m_2), $Ae^{m_1x} + Be^{m_2x}$
(b) If $b^2 = 4ac$ (two equal roots, m_1), $e^{m_1x}(Ax + B)$
(c) If $b^2 < 4ac$ (roots $m_1 \pm im_2$), e^{m_1x} $(A \sin m_2x + B \cos m_2x)$
      or $Ce^{m_1x} \sin (m_2^x + \alpha)$
      or $Ce^{m_1x} \cos (m_2^x - \alpha)$

10.* $a\dfrac{d^2y}{dx^2} + b\dfrac{dy}{dx} + cy = P$ (Linear equation, constant coefficients, RHS a function of x only)

Solution $y =$ complementary function + particular integral.
The *complementary function* is the solution of the equation when $P = 0$ [*see* (9)].
The *particular integral* is *any* solution of the equation which involves no arbitrary constants.
The following examples of values of the particular integral may be useful:

(a) $P =$ constant $= m$

Particular solution $y = \dfrac{m}{c}$

(b) $P = p + qx + rx^2 + \cdots$ $(p, q, r \ldots$ are constants)
Particular solution $y = A + Bx + Cx^2 + \cdots$
    If $C \neq 0, P$ and y are of the same degree.
    If $C = 0, P$ and $\dfrac{dy}{dx}$ are of the same degree.

(c) $P = k\,e^{mx}$
Particular solution $y = A\,e^{mx}$
    If e^{mx} is a term of the complementary function, try $y = Ax\,e^{mx}$
    If $x\,e^{mx}$ is a term of the complementary function, try $y = Ax^2\,e^{mx}$.

(d) $P = l \sin nx + m \cos nx$ (l and m constants, or zero)
Particular solution $y = A \sin nx + B \cos nx$
    If $C \sin nx + D \cos nx$ is a term of the complementary function try
    $Ax \sin nx + Bx \cos$

Note: The constants A, B, C, D, etc. in the particular solutions are evaluated by substitution in the original equation.

2.2.7 Introduction to statistics

2.2.7.1 *Introduction and cross-references*

This section provides a brief introduction to statistics (Sections 2.2.7.2 and 2.2.7.3) and statistical process control theory (Section 2.2.7.4). Related material includes Sections 10.5 (Quantitative Image Analysis) and especially 40.4 (Metallurgical Process Control).
    The field of statistics is the science of collecting and summarising data and drawing inferences from the resulting data summary. The field is divided into two branches: descriptive statistics (section 2.2.7.2) and inductive (or inferencial) statistics (section 2.2.7.3).

---

* Reproduced with permission from G. W. Caunt, *Introduction to the Infinitesimal Calculus*, Oxford University Press, 1928.

2.2.7.2   *Descriptive statistics*

Descriptive statistics comprises of all methods that summarise collected data and is subdivided into 2 categories: (i) Numerical (or quantitative) measures of the mean (or the centre), and of variability. (ii) Pictorial and tabular measures: which consist of histograms and boxplots.

MEASURES OF CENTRE

Mean (or arithmetic average), median, geometric mean, harmonic mean, and the mode.

Example 1

The following data reports the oxidation-induction time (in minutes), X, for various commercial oils (data from Lubrication Engr., 1984: 75–83).

87, 103, 130, 160, 180, 195, 132, 145, 211, 105, 145, 153, 152, 138, 87, 99, 93, 119, 129 minutes.

Since there are 19 sample values, then the sample size is n = 19.

The value of the 1st observation is $x_1 = 87$, the 2nd observation is $x_2 = 103, \ldots$, and the nth sample value is $x_n = 129$.

The sample (arithmetic) mean is defined as

$$\bar{x} = \frac{1}{n}\sum_{i=1}^{n} x_i$$

For the above data $\sum_{i=1}^{19} x_i = 2\,563$, and hence the value of the sample mean is $\bar{x} = 2\,563/19 = 134.894\,7$ minutes.

In order to obtain the median, the data has to be put in order statistic format (i.e. rearranged from the smallest observation to the largest). The order statistics for the above data are $x_{(1)} = 87$, $x_{(2)} = 87$, $x_{(3)} = 93$, $x_{(4)} = 99$, $x_{(5)} = 103$, $x_{(6)} = 105$, $x_{(7)} = 119$, $x_{(8)} = 129$, $x_{(9)} = 130$, $x_{(10)} = 132$, $x_{(11)} = 138$, $x_{(12)} = 145$, $x_{(13)} = 145$, $x_{(14)} = 152$, $x_{(15)} = 153$, $x_{(16)} = 160$, $x_{(17)} = 180$, $x_{(18)} = 195$, $x_{(n)} = 211$. If the value of n is odd, like in this example, then the median or the 50th percentile is simply the value of the middle order statistic $x_{((n+1)/2)}$.

For our example, the median is $\hat{x}_{0.50} = \tilde{x} = x_{(10)} = 132$. However, if the data contain an even number of observations such as n = 10, then the sample median is given by $\hat{x}_{0.50} = \tilde{x} = (x_{(n/2)} + x_{(1+(n/2))})/2$. As an example, the lifetimes of 10 incandescent lamps (in ascending order) are 2\,120, 2\,250, 2\,690, 4\,450, 4\,800, 4\,950, 5\,700, 5\,840, 5\,888, 6\,990 hours. The median for this data is the average of the 5th and the 6th order statistics, i.e. $\hat{x}_{0.50} = \tilde{x} = (4\,800 + 4\,950)/2 = 4\,875$ hours. Note that, unless the data contain outliers or is very skewed (i.e. asymmetrical), the values of the sample mean and median should generally be fairly close. The mean time to failure (MTTF) of the ten lamps is $\bar{x} = 4\,567.800$ hours.

The Geometric mean is defined as $\bar{x}_g = (x_1, x_2, \ldots, x_n)^{1/n}$, i.e. $\bar{x}_g$ is the nth root of the product of all n sample items, $\prod_{i=1}^{n} x_i$, only if all x_i's > 0 for all i, and in general $\bar{x}_g \leq \bar{x}$. For the data of Example 1, $\bar{x}_g = (\prod_{i=1}^{19} x_i)^{1/19} = 130.564\,812 < 134.894\,7 = \bar{x}$. Geometric mean has applications in DOE (design of experiments) where at least 2 responses from each experimental unit are measured.

The Harmonic mean is defined as $\bar{x}_h = \left[\sum_{i=1}^{n} (1/x_i)/n\right]^{-1} = (n/\Sigma(1/x_i))$, $x_i \neq 0$ for all i, i.e. $\bar{x}_h$ is the inverse of the average inverses of x_i's. For the data of Example 1:

$$\sum_{i=1}^{19} (1/x_i) = (1/87) + (1/103) + (1/130) + \cdots + (1/129) = 0.150\,325\,6 \longrightarrow$$

$$\bar{x}_h = \frac{19}{0.150\,325\,588\,522\,53} = 126.392\,3 < \bar{x}_g.$$

The Harmonic mean has applications in ANOVA (analysis of variance) when the design is unbalanced and gives the average sample size over all levels of a factor, and in general $\bar{x}_h \leq \bar{x}_g \leq \bar{x}$.

The mode is the observation with the highest frequency. For the data of Example 1, MO1 = 87 and MO2 = 145 because both observations 87 and 145 have a frequency f = 2 (i.e. the data is bimodal). Most populations that should not be stratified for the purpose of sampling generally have a single mode. If a manufacturing product dimension has more than one mode, then there are quality problems with the process that manufactures the product. This is due to the fact that the ideal situation occurs when there is a single mode with a very high frequency at the ideal target of the product dimension.

MEASURES OF VARIABILITY

We present two measures of variability: the 1st measure, defined by R/d_2, is used in statistical process control when the size of the sample $n \leq 12$. The random variable R represents the sample range defined as $R = x_{(n)} - x_{(1)}$, where $x_{(n)}$ is the nth order statistic (i.e. the sample maximum) and $x_{(1)}$ is the sample minimum (i.e. the 1st order statistic). The parameter $d_2 = E(R/\sigma)$; the quantity R/σ, where σ is the process standard deviation, is called the relative range in the field of statistics and E represents the expected value operator. For a normal universe, the values of d_2 are given in Table 2.12 below.

For $n > 15$, the sampling distribution of sample range, R, becomes unstable and hence it is inadvisable to use R as a measure of variation regardless of the underlying parent population.

The 2nd measure of variability, and the most common, is the standard deviation (stdev) $= S$. In order to compute the stdev, S, we must always compute the variance first.

The sample variance, SV, is the average of deviations of x_i ($i = 1, 2, \ldots, n$) from their-own-mean squared, i.e. $SV = \sum_{i=1}^{n} (x_i - \bar{x})^2/n$, but SV generally underestimates the population variance σ^2 because $E(SV) = ((n-1)/n)\sigma^2$, where E stands for the expected-value operator. This implies that SV is a biased estimator of σ^2. In order to get rid of this bias, the quantity $S^2 = (1/(n-1)) \sum_{i=1}^{n} (x_i - \bar{x})^2$ is used as a measure of variability. Another reason for dividing the corrected sum of squares $CSS = \sum_{i=1}^{n} (x_i - \bar{x})^2$ by $(n-1)$, instead of n, in order to obtain a measure of variation, is the fact that the $CSS = \sum_{i=1}^{n} (x_i - \bar{x})^2$ has only $(n-1)$ degrees of freedom because of the constraint $\sum_{i=1}^{n} (x_i - \bar{x}) \equiv 0$. For the above two stated reasons, S^2 is loosely referred to as the sample variance. The square root of S^2 provides the stdev, S, and dividing S by $\sqrt{n}$ gives the (estimated) standard error of the mean $\hat{\sigma}_{\bar{x}} = S/\sqrt{n}$. A simple binomial expansion of $(x_i - \bar{x})^2$ shows that the $CSS = \sum_{i=1}^{n} (x_i - \bar{x})^2 = \sum_{i=1}^{n} x_i^2 - \left(\sum_{i=1}^{n} x_i\right)^2/(n) = USS - CF$, where $USS = \sum_{i=1}^{n} x_i^2$ is called the uncorrected sum of squares and $CF = \left(\sum_{i=1}^{n} x_i\right)^2/n = n(\bar{x})^2$ is called the correction factor.

For the data of Example 1, the value of the correction factor $CF = 2\,563^2/19 = 345\,735.210\,526\,316$, the $USS = 87^2 + 103^2 + 130^2 + \cdots + 129^2 = 368\,501$, and thus the $CSS = 368\,501 - 345\,735.210\,53 = 227\,65.789\,473\,7$. Since there are 19 sample values, then there are 18 degrees of freedom and as a result $S^2 = 22\,765.789\,473\,7/18 = 1\,264.766\,1$ minutes.[2] Clearly, the units of variance is that of the data's squared and hence the true measure of variation is the positive square root of variance, namely $S = \sqrt{S^2} = (1\,264.766\,1)^{1/2} = 35.563\,55$ minutes. Although in Example 1 the sample size $n = 19$ is a bit too large to use $\hat{\sigma} = R/d_2$ as a measure of variation, we will compute this statistic for comparative purposes. The values of d_2 for $n = 16, 17, 18, 19$, and 20 are 3.532, 3.588, 3.640, 3.689, and 3.735, respectively. The sample range for the data of the example is $R = 211 - 87 = 124$, and hence $\hat{\sigma} = R/d_2 = 124/3.689 = 33.613\,445\,4$. For majority of samples encountered in practice, the value of $\hat{\sigma} \leq S$. Note that $\hat{\sigma}$ without a subscript represents the stdev of individual elements in the sample while $\hat{\sigma}_{\bar{x}} = S/\sqrt{n}$ gives the estimate of the standard error of $\bar{x}$ given by $se(\bar{x}) = \sigma/\sqrt{n}$.

As yet another example, for the $n = 10$ time to failures of incandescent lamps (2 120, 2 250, 2 690, 4 450, 4 800, 4 950, 5 700, 5 840, 5 888, 6 990 hours), $S = 1\,687.018\,264$ while $\hat{\sigma}_x = R/d_2 = 1\,582.196\,231\,32$, and $\hat{\sigma}_{\bar{x}} = 1\,687.018\,264/\sqrt{10} = 533.482\,02$.

HISTOGRAM

Grouping a data set in the form of a histogram is appropriate only for large samples, say $n > 40$ units. In order to illustrate the construction of a histogram for a large data set, we will go through an example in a stepwise fashion.

Example 2

The data set given below consists of $n = 58$ observations on shear strength of spot welds made on a certain type of sheet (arbitrary units).

Table 2.12   VALUES OF d_2 FOR A NORMAL UNIVERSE

n	2	3	4	5	6	7	8	9	10	11	12	13	14	15
d_2	1.128	1.693	2.059	2.326	2.534	2.704	2.847	2.970	3.078	3.173	3.258	3.336	3.407	3.472

5 434, 5 112, 4 820, 5 378, 5 027, 4 848, 4 755, 5 207, 5 049, 4 740, 5 248, 5 227, 4 931, 5 364, 5 189, 4 948, 5 015, 5 043, 5 260, 5 008, 5 089, 4 925, 5 621, 4 974, 5 173, 5 245, 5 555, 4 493, 5 640, 4 986, 4 521, 4 659, 4 886, 5 055, 4 609, 5 518, 5 001, 4 918, 4 592, 4 568, 4 723, 5 388, 5 309, 5 069, 4 570, 4 806, 4 599, 5 828, 4 772, 5 333, 4 803, 5 138, 4 173, 5 653, 5 275, 5 498, 5 582, 5 188.

The 1st task in developing a histogram is to decide on the number of subgroups, denoted by C. The two guidelines for C are $C \cong \sqrt{n}$ and Sturges' practical guideline $C \cong 1 + 3.3 \log_{10}(n)$. It can be verified that for $n \geq 40$, $\sqrt{n} \geq 1 + 3.3 \log_{10}(n)$, and thus we generally select $1 + 3.3 \log_{10}(n) \leq C \leq \sqrt{n}$ with the option that both $1 + 3.3 \log_{10}(n)$ and $\sqrt{n}$ can be either rounded down or up to the next positive integer in order to obtain the value of C. For our shear strength data, $\sqrt{58} = 7.62$ and $1 + 3.3 \log_{10}(58) = 6.82$, and hence $6 \leq C \leq 8$. We settle on the value of $C = 6$ subgroups (or classes) for our histogram.

The 2nd task is to determine the length, Δ, of a subgroup, where it is mandatory to have equal class interval length. This is given by $\Delta = R/C$, where it is necessary to always round up to the same number of decimals as the data. For our shear strength data, $R = 5\,828 - 4\,173 = 1\,655$ and hence the length of each subgroup is equal to $\Delta = 1\,655/6 = 276$. Note that if the value of R/C is not rounded up in order to obtain Δ, then in most cases the C subgroups will not span over the entire data range.

The 3rd step in developing a histogram is to determine the class limits and boundaries for each subgroup, bearing in mind that class limits must have the same number of decimals as the original data while boundaries must have one more decimal than the original data. Table 2.13 gives the frequency distribution for the shear strength data, which clearly shows that the upper limit of the 1st subgroup is 4 448 and the lower class limit of the 2nd subgroup is 4 449. However, the upper boundary of the 1st class is $UB_1 = 4\,448.5$ which is equal to the lower boundary of the 2nd subgroup LB_2, i.e. $UB_1 = LB_2 = 4\,448.5$. In a similar manner, $UB_2 = LB_3 = 4\,724.5$, $UB_3 = LB_4 = 5\,000.5$, $UB_4 = LB_5 = 5\,276.5$, and $UB_5 = LB_6 = 5\,552.5$. Further, $\Delta_j = UB_j - LB_j = \Delta = 276$ for all j, and because there cannot be any overlapping of successive subgroups, then the $\sum_{j=1}^{C} f_j$ must always add to n ($= 58$ in this case). The midpoint of the jth subgroup is given by $m_j = (UB_j + LB_j)/2$, or the average of upper and lower class limits of the same subgroup. Table 2.13 shows that $m_1 = (4\,448 + 4\,173)/2 = 4\,310.5$, $m_2 = (4\,724.5 + 4\,448.5)/2 = 4\,586.5, \ldots$, $m_6 = 5\,690.5$; note that Δ must equal to $m_j - m_{j-1}$ for $j = 2, 3, 4, \ldots, C$, i.e. $\Delta = m_j - m_{j-1}$. The histogram from Minitab software is provided in Figure 2.2.

Table 2.13   FREQUENCY DISTRIBUTION FOR DATA IN EXAMPLE 2

Subgroup Limits	4 173–4 448	4 449–4 724	4 725–5 000
f_j	1	9	14
Class Limits	5 001–5 276	5 277–5 552	5 553–5 828
f_j	20	8	6

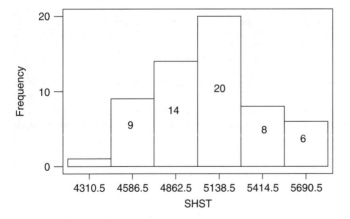

Figure 2.2   *Histogram of Shear Strength (SHST) Data*

If an experimenter does not possess the raw data but only the histogram is available, then the computing formulas for the mean and standard deviation for grouped data may be used as given below:

$$\bar{x}_g = \frac{1}{n} \sum_{j=1}^{C} (m_j \times f_j),$$

where the subscript g stands for grouped, and

$$S_g^2 = \frac{1}{n-1} \sum_{j=1}^{C} (m_j - \bar{g})^2 \times f_j = \frac{1}{n-1} \left[\sum_{j=1}^{C} m_j^2 \times f_j - \frac{(\sum_{j=1}^{C} m_j \times f_j)^2}{n} \right].$$

The raw mean and standard deviation for the shear strength data are $\bar{x} = 5\,057.552$ and $S = 344.680\,9$, while those of the histogram in Figure 2.2 are $\bar{x}_g = 5\,067.1207$ and $S_g = 341.218\,3$.

THE BOXPLOT

A boxplot is a graphical measure of variability, and it can also assess if the data contain mild or extreme outliers. For the sake of illustration, we will obtain the boxplot for the data of Example 2 in a stepwise manner.

Step 1. Draw a vertical line through the median $\tilde{x} = x_{0.50} = 5\,046$.
Step 2. Draw vertical lines through the 25th percentile $x_{0.25} = Q1$ and the 75th percentile $x_{0.75} = Q3$; then connect at the bottom and the top to make a (rectangular) box. For the data of Example 2, the box is shown below.

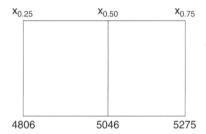

Step 3. Compute the values of interquartile range, $IQR = Q3 - Q1$, $1.5 \times IQR$, and $3 \times IQR$. For the Example 2, $IQR = 5\,275 - 4\,806 = 469$, $1.5 \times IQR = 703.50$ and $3 \times IQR = 1\,407.00$. Then all data points less than $Q1 - 3 \times IQR = 4\,806 - 1\,407 = 3\,399$ or larger than $Q3 + 3 \times IQR = 5\,275 + 1\,407 = 6\,682$ are extreme outliers. Example 2 data has no extreme outliers because no data point lies outside the interval $[3\,399, 6\,682]$. Next determine if there are any mild outliers by obtaining the interval $[x_{0.25} - 1.5 \times IQR, x_{0.75} + 1.5 \times IQR] = [4\,102.5, 5\,978.5]$. Since no data point falls outside the interval $[4\,102.5, 5\,978.5]$, then Example 2 has no mild outliers.
Step 4. Draw whiskers from $Q1 = x_{0.25}$ and $Q3 = x_{0.75}$ to the smallest and largest order statistics that are not outliers.

Note that only outliers for which assignable causes are found, and the corresponding corrective actions taken, should be removed from the data for further analysis, or else the outlier must remain as a part of the data.

LAWS OF PROBABILITIES

Definition. The occurrence probability (pr) of an event A is defined as the function

$$P(A) = \frac{N(A)}{N(U)}$$

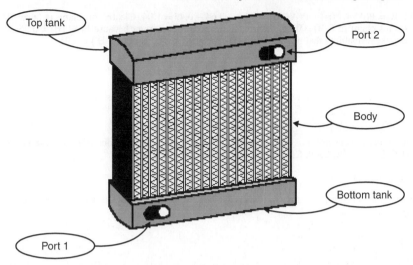

Figure 2.3   *Heat exchanger geometry for use with example 3*

iff (if and only if) all the N(U) elementary outcomes in the universe U are equally likely, and N(A) are those elementary outcomes that are favourable to the occurrence of the event A.

Example 3

An aluminium-alloy heat exchanger is furnace brazed so that the tubes, fins, and side rails, plus the headers used to attach the top and bottom tanks come out of the furnace as a single unit that makes up the body of the heat exchanger. After brazing, the top and bottom tanks are crimped into place. The orientation of the top and bottom tanks will determine the relative position of the ports through which coolant enters and leaves the heat exchanger. Thus, for the heat exchanger shown is in Figure 2.3, there are 4 possible outcomes because of the body symmetry. The two tanks can be assembled correctly as shown or vice versa on the other side of the body. The two unacceptable assemblies occur when port 1 (at the left of bottom tank) and port 2 (at the right of top tank) are on different sides. Letting F represent a port on the front and B a port on the back, the set of all possible outcomes for the assembly is $U = \{(F_1, F_2), (F_1, B_2), (B_1, B_2), (B_1, F_2)\}$. The conforming units are those with either (F_1, F_2) or (B_1, B_2) outcomes.

The function P(A) must satisfy the following axioms:

(1) $0 \leq P(A) \leq 1$ because $0 \leq N(A) \leq N(U)$.
(2) $P(U) \equiv 1$, and $P(\phi) \equiv 0$, i.e. the occurrence pr of the null set ϕ is identically zero.
(3) If A and B are mutually exclusive (MUEX), then $P(A \cap B) = P(\phi) = 0$, and $P(A \cup B) = P(A) + P(B)$; further, if $A_1, A_2, A_3, \ldots$ is a countably infinite jointly MUEX events, then

$$P\left(U_{i=1}^{\infty} A_i\right) = \sum_{i=1}^{\infty} P(A_i)$$

The event $A \cap B$ occurs iff both A and B occur simultaneously, and $A \cup B$ occurs iff at least one of the two events occur. The event A' occurs iff A does not occur, i.e. the event A' is the complement of the event A. Since A and A' are MUEX events, then $A \cup A' = U$ and as a result $P(A \cup A') = 1 = P(A) + P(A') \rightarrow P(A') = 1 - P(A)$.

Let events A and B belong to the same sample space U. Then in general $P(A \cup B) = P(A) + P(B) - P(A \cap B)$ for all events A and B in the same universe U.

Example 4

A repairman claims that the pr that an air conditioning compressor is 'all right' is $P(A) = 0.85$, and $P(B) = 0.64$ that its fan motor is all right, and 0.45 that both the compressor and fan belt are all right. Can his claim be true?

$P(A \cup B) = P(A) + P(B) - P(A \cap B) = 0.85 + 0.64 - 0.45 = 1.04$. Clearly, his claim is false! If he revises his claim to $P(B) = 0.55$, then his claim would be correct because $P(A \cup B) = 0.95$ in which case the probability that neither A nor B occur is $P[(A \cup B)'] = P(A' \cap B') = 0.05$.

PERMUTATIONS

The total number of different ways that n distinct objects can be permuted is given by

$$_nP_n = n!$$

As an example, consider the 3 distinct objects A, B, C. The permutations of these objects are ABC, ACB, BAC, BCA, CAB, CBA, of which there are six. Therefore,

$$_3P_3 = 3! = 6$$

The total number of ways that $r(\leq n)$ objects can be selected from n objects and then permuted is given by

$$_nP_r = \frac{n!}{(n-r)!}$$

For example, consider 4 objects A, B, C and D. Then the elements of $_4P_2$ are: AB, BA, AC, CA, AD, DA, BC, CB, BD, DB, CD, DC, of which there are $12 = 4!/(4-2)!$, i.e. the permutations of 4 objects taken two at a time is equal to $_4P_2 = 12$.

COMBINATIONS

The total number of different ways that r distinct objects can be selected from n $(r \leq n)$ is given by

$$_nC_r = \binom{n}{r} = {}_nP_r/r! = \frac{n!}{(n-r)!r!}.$$

For example, consider the combinations of the 4 objects A, B, C and D, taken 2 at a time:

AB, AC, DA, BC, DB, CD,

of which there are $6 = {}_4C_2 = 4!/(2! \times 2!)$ of them. Note that AB and BA are the same element of $_4C_2$, while they are two distinct elements of $_4P_2$.

Example 5

A production lot contains 80 units of which only 6 are nonconforming (NC) to customer specifications. A random sample of size $n = 10$ is drawn from the lot without replacement and the number of NC units in the sample is counted. What is the pr that the sample contains exactly 2 NC units? What is the pr that the sample contains at least one NC unit? Let the variable X represent the number of NC units in the random sample of size $n = 10$. Then,

$$P(X=2) = \frac{(_6C_2)(_{74}C_8)}{_{80}C_{10}} = \frac{\binom{6}{2}\binom{74}{8}}{\binom{80}{10}} = 0.137\,305\,316\,27,$$

and

$$P(X \geq 1) = 1 - P(X=0) = 1 - \frac{_{74}C_{10}}{_{80}C_{10}} = 0.563\,674\,217\,2.$$

Note that the procedure outlined in the above example is generally referred to as sampling inspection in the field of quality control (QC). Its objective often is to rectify an outgoing lot by first deciding whether to accept or reject a lot, and in case the lot is rejected, then it is subjected to 100% screening in order to remove all defective units before shipment to customer.

RANDOM VARIABLES AND THEIR FREQUENCY FUNCTIONS

A random variable (rv) is a real-valued function defined on the universe U. If the range space of a rv is discrete, then its frequency function is generally referred to as a pr mass function (pmf), and if its range space is continuous, then its frequency function is called a pr density function (pdf). For example, the rv of the Example 5 is discrete because its range space $R_x = \{0, 1, 2, 3, 4, 5, 6\}$ and its pmf is given by

$$p(x) = \frac{\binom{6}{x}\binom{74}{10 - x}}{\binom{80}{10}}, \quad x = 0, 1, 2, 3, 4, 5, 6$$

In fact the above pmf is called the hypergeometric in the field of statistics. Clearly, if p(x) is a pmf, then it is absolutely necessary that $\sum_{R_x} p(x) \equiv 1$, and similarly, if the rv, X, is continuous with pdf f(x), then it is paramount that $\int_{R_x} f(x)dx \equiv 1$ (or 100%). The population mean, μ, and population variance, σ^2, of a rv are given by (the symbol E used below represents the expected-value operator).

$$\mu = E(X) = \begin{cases} \sum_{R_x} xp(x), & \text{if X is discrete} \\ \int_{R_x} xf(x)\,dx, & \text{if X is continuous} \end{cases}$$

and

$$\sigma^2 = V(X) = E[(X - \mu)^2] = \begin{cases} \sum_{R_x} (x - \mu)^2\, p(x), & \text{if X is discrete} \\ \int_{R_x} (x - \mu)^2\, f(x)\,dx, & \text{if X is continuous} \end{cases}$$

The most two important pmfs in the field of statistics are the binomial and Poisson.

THE BINOMIAL PMF

Consider $n > 1$ independent (or Bernoulli) trials in successions, where the experimental result can be classified as either 'success' or 'failure'. The binomial rv, X, is defined as X = 'The number of generic successes observed in n trials', where $R_x = \{0, 1, 2, 3, \ldots, n-1, n\}$, and p = pr of a success at each single trial. Then, the binomial pmf is given by

$$b(x; n, p) = {}_nC_x p^x (1 - p)^{n-x}$$

The PDF (pr distribution function) in this equation is called the binomial with parameters n and p and is denoted by Bin(n, p). Its cumulative distribution function (cdf) is given by $F(x) = P(X \leq x) = B(x; n, p) = \sum_{i=0}^{x} b(i; n, p) = \sum_{i=0}^{x} {}_nC_i \times p^i \times q^{n-i}$, where $q = 1 - p$ represents the failure pr at each trial. It can be verified that the binomial pmf, b(x; n, p), has a mean of E(X) = np and a variance $\sigma^2 = npq$.

Example 6

A manufacturing process produces parts which are, on the average, 1% NC to customer specifications. A random sample of n = 30 items is drawn from a conveyor belt. Compute the pr that the sample contains exactly x = 2 NC units.

$$b(2; 30, 0.01) = {}_{30}C_2 (0.01)^2 (0.99)^{28} = 0.032\,83.$$

The pr that the sample contains at most two NC units is given by $P(X \leq 2) = B(2; 30, 0.01) = 0.996\,682\,3$, and the pr that the sample contains at least two NC units is given by

$P(X \geq 2) = 1 - P(X \leq 1) = 1 - B(1; 30, 0.01) = 0.036\,148$. The expected number of NC units in the sample is given by $\mu = n \times p = 0.30$ and the variance of X is given by $V(X) = \sigma^2 = npq = 0.297$.

Note that in the context of the Example 6 the statistic $\hat{p} = X/n$ is called the sample fraction NC and is used as a point estimate of process fraction NC, p. The binomial expansion of $(a + b)^n$ is given by

$$(a + b)^n = \sum_{i=0}^{n} {}_nC_{n-i} a^{n-i} b^i$$

which shows that $\sum_{x=0}^{n} b(x; n, p) = \sum_{x=0}^{n} {}_nC_x p^x (1 - p)^{n-x} = (p + q)^n = 1^n = 1$.

THE POISSON DISTRIBUTION

Consider many Bernoulli trials (i.e. $n \to \infty$) such that occurrence pr of success at each trial is small (say $p < 0.15$) and average number of successes per time unit, $\mu = n \times p = \lambda$, is a constant. Let the rv, X, denote the number of Poisson events (or generic successes) that occur during one time interval of unit length. Then the PDF (or pr distribution function) of X is given by

$$p(x; \lambda) = \frac{\lambda^x}{x!} e^{-\lambda}, \quad x = 0, 1, 2, 3, 4, \ldots \tag{2.1}$$

It can be shown that the Poisson pmf in (2.1) is simply the limiting distribution for a Bin(n, p) as $n \to \infty$ and simultaneously as $p \to 0$ but the product $n \times p$ stays fixed at a rate equal to λ, and is generally required that $n \times p \leq 20$. For $n \times p > 20$, the process becomes practically Gaussian (a pdf that will be defined later).

Example 7

The number of accidents per day in a certain city is Poisson distributed at an average rate of 5 accidents/day. The pr of no accidents occurring in the city during the next day is given by $p(0; 5) = (5^0/0!) e^{-5} = 0.006\,738$. The pr of at most 4 accidents occurring during the next day is $P(X \leq 4) = F_x(4; 5) = \sum_{x=0}^{4} (5^x/x!) e^{-5} = 0.440\,493\,3$.

The Poisson distribution can be used to compute probabilities over intervals of length t ($t \neq 1$ unit of time). Let Y = number of Poisson events occurring during an interval of length t ($t \neq 1$) where the average number of Poisson events per unit of time ($t = 1$) is $\lambda = \mu = np$. Then the average number of Poisson events per interval of length t is $E(Y) = \lambda t$. As a result the pmf for the rv Y is given by

$$p(y; \lambda t) = \frac{(\lambda t)^y}{y!} e^{-\lambda t}, \quad y = 0, 1, 2, 3, 4, \ldots$$

Consider the Poisson process of Example 7. We wish to compute the pr of exactly 6 accidents occurring during the next 2 days. Then, $\lambda = np = 5$ accidents/day, $t = 2$ days $\to \lambda \times t = 10$ accidents/two days $\to P(Y = 6) = (10^6/6!) e^{-10} = 0.144\,458\,2$, i.e. Y is $p(y; 10)$ read as Poisson distributed at an average rate of 10 accidents per two days. The pr that there will be exactly 11 accidents in the city during the next 3 days is $P(Y = 11) = (15^{11}/11!) e^{-15} = 0.066\,287\,4$. The pr that there will be at least 11 accidents in the next 3 days is given by $P(Y \geq 11) = 1 - P(Y \leq 10) = 1 - F_Y(10; 15) = 1 - \sum_{x=0}^{10} (15^x/x!) e^{-15} = 0.881\,536$.

It can be verified that the long-term (or weighted) average of X, as expected, is given by $\mu = E(X) = \lambda$ and $V(X) = \lambda = \mu$. As a result, all Poisson processes have a CV (coefficient of variation $= \sigma/\mu$) equal to $(100/\sqrt{\lambda})\%$.

CONTINUOUS FREQUENCY FUNCTIONS

If the range space of a rv is continuous, then the occurrence pr of a single point in the range space is zero, but the pr over a real interval [a, b] is given by $\int_a^b f(x)\,dx$, where f(x) is the probability density function (pdf) of the rv X. There are numerous pdfs in the field of statistics, each of which has specific applications.

The three most commonly encountered in practice are the uniform, normal (or Gaussian), and the exponential. The normal distribution is used to approximately model the distribution of almost any manufactured dimension, and the exponential has extensive applications in life testing and in Markov processes (specifically in queuing theory). The uniform distribution has extensive applications in statistical simulations because of the fact that all continuous cumulative distribution functions in the universe are uniformly distributed over the interval [0, 1].

THE UNIFORM pdf

A continuous rv, X, is said to be uniformly distributed over the real interval [a, b] iff its pdf is given by

$$f(x) = \begin{cases} \dfrac{1}{b-a}, & a \le x \le b \\ 0, & \text{elsewhere.} \end{cases}$$

This is expressed as $X \sim U(a, b)$. The population mean of a rv $X \sim U(a, b)$ is given by $\mu = E(X) = (a+b)/2$ and its population variance is given by $V(X) = \sigma^2 = (b-a)^2/12$.

THE NORMAL (OR GAUSSIAN) DISTRIBUTION $N(\mu, \sigma^2)$

A continuous rv, X, is said to be normally distributed with mean μ and variance σ^2 iff its pdf is given by

$$f(x; \mu, \sigma) = C\,e^{-((1/2)((x-\mu)/\sigma)^2)} = C\,\text{Exp}[-(x-\mu)^2/(2\sigma^2)], \quad -\infty < x < \infty$$

where C is a normalising constant and must be evaluated in such a manner that the total area under the pdf is exactly equal to 1 (or 100% pr), i.e. we must require that

$$\int_{-\infty}^{\infty} C\,e^{-((1/2)((x-\mu)/\sigma)^2)}dx = 1,$$

which leads to $C^2 = 1/(2\pi\sigma^2)$. Therefore, the only unique value of C that makes the value of the above improper integral equal to 100% is $C = 1/(\sigma\sqrt{2\pi})$. These developments show that a continuous rv, X, is $N(\mu, \sigma^2)$ [read as normally distributed or Gaussian with mean μ and variance σ^2] only if its pdf is given by

$$f(x; \mu, \sigma) = \frac{1}{\sigma\sqrt{2\pi}}e^{-((1/2)((x-\mu)/\sigma)^2)}, \quad -\infty < x < \infty.$$

The graph of the normal pdf in the equation above is symmetrical and is provided in Figure 2.4.

It can easily be verified that the modal point of a $N(\mu, \sigma^2)$ distribution, as shown in Figure 2.4, occurs at $x = \mu$. Secondly, its points of inflection occur at $x = \mu \pm \sigma$. Therefore, one invariant property of all normal distributions is the fact that the distance between the line $x = \mu$ and the point of inflection is exactly one standard deviation σ. Because of the enormous importance of this distribution, we list its first 4 moments, where the kth moment about a real constant C is given by $E[(X-C)^k]$.

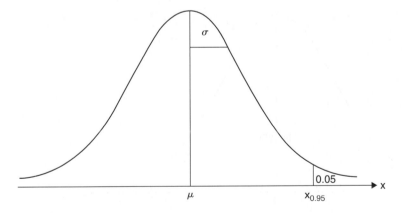

Figure 2.4 *The modal point of a $N(\mu, \sigma^2)$ distribution*

It can be verified that the 1st four moments of any Gaussian distribution are $\mu'_1 = \mu$, $V(X) = \mu_2 = \sigma^2$, $\alpha_3 \equiv \mu_3/\sigma^3 = 0$, and $\alpha_4 = \mu^4/\sigma^4 = 3.00$, where $\mu_4 = E[(X-\mu)^4]$ is called the 4th central moment of X, and μ'_k represents the kth origin moment.

It must be emphasised that the normal distribution has unlimited applications in all facets of human life (for example, in all sciences, in all fields of engineering, in agriculture, pharmacy and medicine, etc.). In fact the distribution of every dimension that is manufactured can be modelled closely by a normal curve! This implies that over 75% of all parent populations in the universe are assumed normally distributed with some mean μ and some variance σ^2. Application examples will now follow.

Example 8

The design tolerances (or consumers' specifications) on length of steel pipes is 30 ± 0.25 cm. The dimension or length of a pipe, X, is N(30, 0.006 4 cm^2). We wish to compute the pr that a randomly selected pipe has length below 30.10 cm?

$$P(X < 30.10 \, \text{cm}) = \int_{-\infty}^{30.10} \frac{1}{\sigma\sqrt{2\pi}} e^{-((1/2)((x-\mu)/\sigma)^2)} dx,$$

where $\mu = 30.00$ cm and $\sigma = 0.08$ cm. To simplify the integrand, let $(x-\mu)/\sigma = z$; then, $dz/dx = 1/\sigma$ or $dz = dx/\sigma$.

Substituting the transformation $z = (x-\mu)/\sigma$ into the above integral results in:

$$P(X < 30.10) = \int_{-\infty}^{1.25} \frac{1}{\sqrt{2\pi}} e^{-z^2/2} dz = \int_{-\infty}^{1.25} \phi(z) dz = \Phi(1.25).$$

The pdf $\phi(z) = (1/\sqrt{2\pi})e^{-z^2/2}$ is called the standard normal density function because its mean is zero and its variance is equal to 1. Unfortunately, $\phi(z)$ has no closed-form simple antiderivative, and therefore its cdf has been tabulated for different values of z (using numerical integration) in tables in almost all statistical texts. For example, see pages 722–723 of J. L. Devore 'Probability and Statistics for Scientists and Engineers', Duxbury Press, Belmont, CA, 2000. The cdf of the standard normal density N(0, 1) is universally denoted by $\Phi(z)$. For example, from Table A.3, p. 723 of Devore, $\Phi(1.5) = 0.933\,19$, $\Phi(-1.5) = 0.066\,81$, $\Phi(1.96) = 0.975$, $\Phi(-1.96) = 0.025$, $\Phi(3) = 0.998\,65$, $\Phi(-3) = 0.001\,35$ [Note that $\Phi(-z) = 1 - \Phi(z)$]. Thus, $P(X < 30.10) = F_X(30.10) = P(Z \le 1.25) = \Phi(1.25) = 0.894\,35$. The graph of $\phi(z)$ is shown in Figure 2.5. Microsoft Excel will provide both normal prs and the normal inverse. The reader can open Excel, click on Insert, Function, Statistical, scroll down to NORMSDIST, insert the desired value of z, then Excel will provide the corresponding value of cdf of z, that is denoted by $\Phi(z)$. Figure 2.5 graphs the standard normal density $\Phi(z)$ versus z.

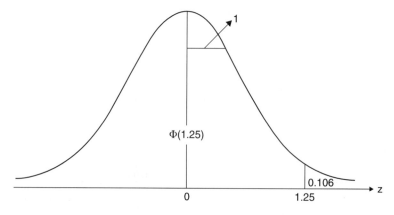

Figure 2.5   *Standard normal density $\Phi(z)$ versus z*

Suppose a pipe is selected at random. We wish to compute the pr that its length falls within 2σ of the mean $\mu = 30$ cm.

$$P(|X - \mu| \le 2\sigma) = P(29.84 \le X \le 30.16)$$

$$= P\left(\frac{29.84 - 30}{0.08} \le \frac{X - \mu}{\sigma} \le \frac{30.16 - 30}{0.08}\right)$$

$$= P(-2 \le Z \le 2) = \Phi(2) - \Phi(-2) = 0.977\,25 - 0.022\,75 = 0.954\,50.$$

Further, the pr that the length of a pipe is within 3 sigma of μ is given by $P(|X - \mu| \le 3\sigma) = P(-3 \le Z \le 3) = \Phi(3) - \Phi(-3) = 0.998\,65 - 0.001\,35 = 0.997\,30$. Finally, the $P(\mu - 4\sigma \le X \le \mu + 4\sigma) = \Phi(4) - \Phi(-4) = 0.999\,968\,33 - 0.000\,031\,671 = 0.999\,936\,66$.

Next, the fraction of the pipes that are not conforming to the lower specification limit $LSL = 29.75$ cm and upper specification limit $USL = 30.25$ cm is called the fraction NC (FNC) of the process, denoted by p, and is computed below.

$$p = FNC = P(X < 29.75) + P(X > 30.25) = P(Z \le -3.125) + P(Z \ge 3.125)$$

$$= \Phi(-3.125) + [1 - \Phi(3.125)] = 2 \times \Phi(-3.125) = 2 \times 0.000\,889\,025 = 0.001\,778\,051.$$

Suppose that the process in the Example 8 is not centred at $\mu = 30.00$ cm. Then, the process fraction NC under the condition that $\mu \ne 30.00$ cm will always exceed $p = 0.001\,778$ (at the same value of σ). For example, if $\mu = 30.05$ cm, then $Z_L = -3.75$, $p_L = 0.000\,088\,42$, $Z_U = 2.50$, $p_U = 0.006\,21$, and $p = 0.006\,298\,1$ at $\sigma = 0.08$ cm.

THE PERCENTILES OF A NORMAL DISTRIBUTION

For the standard normal density, $N(0, 1)$, the 95th percentile is denoted by $z_{0.05}$, and is defined such that the $P(Z \le z_{0.05}) = 0.95$. Similarly, the 90th percentile of Z is defined as $P(Z \le z_{0.10}) = 0.90$. This definition implies that for the standard normal density function (STNDF), $\Phi(z_{0.05}) = 0.95$ and $\Phi(z_{0.10}) = 0.90$. Table A.3 on page 723 of Devore shows that $z_{0.05} = 1.645$ while $z_{0.10} = 1.282$. By symmetry of the STNDF, it follows that the 5th percentile of Z is equal to $z_{0.95} = -1.645$ and the 10th percentile $z_{0.90} = -1.282$, i.e. the $P(Z \le -1.645) = \Phi(-1.645) = 0.05$ and the $P(Z \le -1.282) = \Phi(-1.282) = 0.10$. It should be noted that $z_{0.05} = 1.645$ is also called the 5 percentage point of the STNDF and $z_{0.10} = 1.282$ is called the 10 percentage point of a STNDF. It turns out that in the field of statistics, a percentage point, such as z_α, is the distance from the origin beyond which the right tail pr is exactly equal to α. Further, for the STNDF, it should be clear that $z_\alpha = -z_{1-\alpha}$ for all $0 < \alpha < 1$. Figure 2.5 shows that $z_{0.105\,65} = 1.250$ and hence $z_{0.894\,35} = -1.250$.

If the rv, X, is normally distributed but either $\mu \ne 0$, or $\sigma \ne 1$ (or both), the 95th percentile of X is defined as $x_{0.95}$ such that the $P(X \le x_{0.95}) = 0.95$, and similarly the 90th percentile of X is defined such that $P(X \le x_{0.90}) = 0.90$. Furthermore, in general $x_{0.95} \ne - x_{0.05}$ or $x_p \ne x_{1-p}$.

THE REPRODUCTIVE PROPERTY OF THE NORMAL DISTRIBUTION

Suppose $X_1, X_2, \ldots, X_n$ are n normally distributed rvs with means μ_i and variances σ_i^2 and correlation coefficients $\rho_{ij} = \sigma_{ij}/\sigma_i\sigma_j$, where $\sigma_{ij} = E[(X_i - \mu_i) \times (X_j - \mu_j)]$ is called the covariance between the two random variables X_i and X_j ($i \ne j$). Further, $c_1, c_2, \ldots, c_n$ are known or given constants. Then it has been proven in the theory of statistics that the linear combination (LC).

$$Y = c_1 X_1 + c_2 X_2 + \cdots + c_n X_n = \sum_{i=1}^{n} c_i X_i$$

is also normally distributed with mean

$$E(Y) = \sum_{i=1}^{n} c_i \mu_i$$

and variance

$$V(Y) = \sum_{i=1}^{n} c_i^2 \sigma_i^2 + 2 \sum_{i=1}^{n-1} \sum_{j>i}^{n} c_i c_j \sigma_{ij}, \quad \text{i.e., } Y \sim N \left(\sum_{i=1}^{n} c_i \mu_i, \sum_{i=1}^{n} c_i^2 \sigma_i^2 + 2 \sum_{i=1}^{n-1} \sum_{j>i}^{n} c_i c_j \sigma_{ij} \right).$$

If the dimensions X_i and X_j are mutually independent for all $i \neq j$, then $\sigma_{ij} = 0$, and the $V(Y)$ reduces to $\sigma_y^2 = \sum_{i=1}^{n} c_i^2 \sigma_i^2$. Note that zero covariance does not always imply independence.

Example 8 (continued)

Three steel pipes are selected at random from the $N(30, 0.0064)$ process and placed end-to-end. Compute the pr that the total length of the assembly exceeds 90.24 cm. [$n = 3$, each $c_i = 1$, and each $X_i \sim N(30.00, 0.0064)$]. $\rightarrow Y = X_1 + X_2 + X_3 \rightarrow E(Y) = 30 + 30 + 30 = 90.00$, $V(Y) = 0.0064 + 0.0064 + 0.0064 = 0.0048 \rightarrow Y \sim N(90, 0.0192)$, and hence

$$P(Y > 90.24 \text{ cm}) = P\left(\frac{y - \mu_y}{\sigma_y} > \frac{90.24 - 90.00}{0.138\,564} \right) = P(Z > 1.732\,051)$$

$$= 1 - \Phi(1.732\,051) = 1 - 0.958\,368 = 0.041\,632\,3.$$

THE CENTRAL LIMIT THEOREM (CLT)

If $X_1, X_2, \ldots, X_n$ are independent, but not necessarily normal, rvs with $E(X_i) = \mu_i$ and $V(X_i) = \sigma_i^2$, then the distribution of $Y = X_1 + X_2 + \cdots + X_n$, as $n \rightarrow \infty$, approaches normality with $E(Y) = \sum_{i=1}^{n} \mu_i$ and $V(Y) = \sum_{i=1}^{n} \sigma_i^2$. Note that the more skewed the frequency functions of the individual X_i's in the sum $\sum_{i=1}^{n} X_i$ are, the more slowly (in terms of n) the distribution of Y approaches normality. For example, if the X_i's are uniformly distributed, then for $n > 8$, the distribution of the sum, $Y = \sum_{i=1}^{n} X_i$, is approximately normal, while if X_i's are exponentially distributed, then a value of $n > 200$ is needed. The uniform skewness is $\alpha_3 = E\left[(x - \mu/\sigma)^3 \right] = 0$, while that of exponential distribution is $\alpha_3 = 2.00$.

In case X_i's have the same mean and variance, then the distribution of Y approaches normality with $E(Y) = n\mu$ and $V(Y) = n\sigma^2$. A general conservative rule of thumb is that a value of $n > 25\alpha_3^2$ and simultaneously $n > 5\alpha_4$ (and in most applications at least $n > 30$) is needed in order for the distribution of the sum $Y = \sum_{i=1}^{n} X_i$ to become approximately Gaussian. Again, the quantity $\alpha_3 = E[(X - \mu)^3/\sigma^3]$ is called the coefficient of skewness, or simply the skewness of X. The closer $\alpha_3(X)$ is to zero, the smaller n is needed for an adequate normal approximation to the distribution of the sum $\sum_{i=1}^{n} X_i$.

Example 9

A large freight elevator can transport a maximum of 10 000 kg of load. A load of cargo containing 50 boxes of a certain product is transported on the elevator every hour. The weights of the boxes are independent (not normal but identically distributed) each with $E(X_i) = 190$ kg and $V(X_i) = 900$ kg^2. Estimate (using the CLT) the pr that the maximum limit is exceeded on a given hour and disrupts the transporting process?

Solution: $Y = a$ Cargo Load $= \sum_{i=1}^{50} X_i \rightarrow E(Y) = 50 \times 190 = 9\,500$ and $V(Y) = 50 \times 900 = 45\,000.00$. $P(Y > 10\,000 \text{ kgs}) = P(Z > (10\,000 - 9\,500/212.132)) = P(Z > 2.357\,02) = 1 - \Phi(2.357\,023) = 1 - 0.990\,789 = 0.009\,211\,1.$

THE NORMAL APPROXIMATION TO THE BINOMIAL

Since a binomial rv, X_{Bin}, is the sum of n independent Bernoulli rvs, i.e. then the binomial rv $X_{Bin} = \sum_{i=1}^{n} X_i$, where each X_i is Bernoulli (i.e. X_i has a binomial distribution with the parameter $n \equiv 1$) with $E(X_i) = p$ and $V(X_i) = pq$. As a result of the CLT, as $n \rightarrow \infty$, the binomial pmf can be approximated by a normal pdf with $E(X) = np$ and $V(X) = npq$. However, because a binomial rv has a discrete range space, a correction for continuity has to be applied as will be illustrated in the following example. The approximation is adequate when n is large enough such that the product $np > 15$ and $0.10 < p < 0.90$. If $np < 15$, we would recommend the Poisson approximation to the Binomial. The general rule of thumb was that n must be large enough such that $n > 25\alpha_3^2$ and simultaneously $n > 5\alpha_4$, where for a single Bernoulli trial $\alpha_3^2 = (q - p)^2/(p \times q)$, and $\alpha_4 = (1 - 3p + 3p^2)/(p \times q) = (1 - 3\,pq)/$

$(p \times q)$. It can be shown that for a Bin(n, p) distribution the value of skewness is $\alpha_3 = (q - p)/\sqrt{npq}$ and the binomial $\alpha_4 = E[((X - \mu)/\sigma)^4] = E(Z^4)$ is given by $\alpha_4 = (1 + 3(n - 2)pq/npq)$.

Example 10

Consider a binomial distribution with $n = 50$ trials and $p = 0.30$ (so that $\mu = np = 15$). Note that we would like to have $np > 15$, but $25\alpha_3^2 = 19.05$ implies that the requirement on skewness $n = 50 > 25\alpha_3^2$ is easily satisfied and similarly, $5\alpha_4 = 14.876\,2$ shows that the α_4 requirement is also satisfied because $n > 14.876\,2$. Then the range space of the discrete rv, X_D, is $R_x = \{0, 1, 2, 3, \ldots, 50\}$. The exact pr of attaining exactly 12 successes in 50 trials is given by $b(12; 50, 0.30) = P(X_D = 12) = {}_{50}C_{12}(0.30)^{12}(0.70)^{38} = 0.083\,83$.

To apply the normal distribution (with $\mu = 15$ and $\sigma^2 = npq = 10.50$) in order to approximate the $P(X_D = 12)$, we 1st have to select an interval on a continuous scale, X_C, to represent $X_D = 12$. Clearly this continuous interval on X_C has to be $(11.5, 12.5)$, i.e. we will have a correction of 0.50 for continuity. In short, $X_D = 12 \cong (11.5, 12.5)_C$. Thus,

$$P(X_D = 12) \cong P(11.5 \leq X_C \leq 12.5) = P\left(\frac{11.5 - 15}{3.240\,4} \leq Z \leq \frac{12.5 - 15}{3.240\,4}\right)$$

$$= P(-1.080\,1 \leq Z \leq -0.771\,52) = \Phi(-0.771\,52) - \Phi(-1.080\,1)$$

$$= 0.220\,200 - 0.140\,044 = 0.080\,157.$$

Since the exact pr was 0.083 83, the percent relative error in the above approximation is

$$\% \text{ Relative Error} = \left(\frac{0.083\,83 - 0.080\,157}{0.083\,83}\right) \times 100 = 4.382\%.$$

Next we compute the exact pr that X exceeds 12, i.e. $P(X_D > 12) = 1 - B(12; 50, 0.30) = 1 - 0.222\,865\,8 = 0.777\,134$. The normal approximation to this binomial pr is

$$P(X_C \geq 12.5) = P(Z \geq -0.77\,152) = 0.779\,80 \cong P(X_D > 12).$$

The % error in this approximation is 0.343%.

As stated earlier, the normal approximation to the binomial should improve as n increases and as $p \to 0.50$. To illustrate this fact, consider a binomial distribution with parameters $n = 65$ and $p = 0.40$. Then, $\mu = np = 26, \sigma = (npq)^{1/2} = 3.949\,7$, and $P(X_D = 22) = {}_{65}C_{22}(0.40)^{22}(0.60)^{43} = 0.061\,7$. The normal approximation to this b(22; 65, 0.40) is $P(21.5 \leq X_C \leq 22.5) = P(-1.139\,33 \leq Z \leq -0.886\,15) = 0.187\,77 - 0.127\,28 = 0.060\,5$ with relative % error of 1.935%.

Further, $P(X_D < 23) = B(22; 65, 0.40) = \sum_{x=0}^{22} {}_{65}C_x(0.40)^x(0.60)^{65-x} = 0.188\,327$. The normal approximation to this binomial cdf corrected for continuity (cfc) is $P(X_C \leq 22.5) = P(Z \leq -0.886\,15) = \Phi(-0.886\,15) = 0.187\,769\,2$ with an error of 0.296 2%.

THE EXPONENTIAL pdf

A continuous rv, X, with pdf $f(x; \lambda) = \lambda e^{-\lambda x}$ is said to be exponentially distributed at the average rate of $\lambda > 0$. The cdf of the exponential density is given by

$$F(x) = \int_0^x \lambda e^{-\lambda t} dt = 1 - e^{-\lambda x}.$$

The exponential pdf has widespread and enormous applications in reliability engineering (or life testing) and in stochastic processes (specifically queuing theory). The majority of its applications occur because of its relation to the Poisson distribution as shown below.

Consider a Poisson event that occurs at a rate of λ per unit of time. The average number of occurrences during an interval of length t is λt. Let X(t) represent the number of Poisson events occurring during an interval of length t, i.e. $R_{X(t)} = \{0, 1, 2, 3, \ldots\}$. Then from the Poisson pmf

$$P[X(t) = 0] = \frac{(\lambda t)^0}{0!} e^{-\lambda t} = e^{-\lambda t}.$$

Next, define a continuous rv, T, as follows: $T =$ the time between the occurrences of 2 successive Poisson events (or $T =$ intercurrence or intervening time of 2 successive Poisson events). Clearly, the following two events are equivalent.

$$[X(t) = 0] = [T > t].$$

Thus,

$$P(T > t) = P[X(t) = 0] = e^{-\lambda t} = 1 - P(T \le t) = 1 - F(t) \longrightarrow F(t) = 1 - e^{-\lambda t}$$
$$\longrightarrow \quad f(t) = dF(t)/dt = \lambda e^{-\lambda t}$$

The above developments show that if the number of occurrences of an event is Poisson distributed, then the occurrence time, T, to the next Poisson event measured from the last occurrence is exponentially distributed.

Example 11

The number of downtimes of a computer network is Poisson distributed with an average of 0.20 downtime/week. Compute the pr of exactly 2 failures next week.

$$p(2; 0.20) = \frac{(0.20)^2}{2!} \times e^{-0.20} = 0.0164.$$

Next, the pr that the time between 2 future successive downtimes exceeds one week is given by $P(T > 1) = P[X(1) = 0] = e^{-0.20} = 0.81873$. The pr that exactly 2 failures occur in the next 3 weeks is computed as follows.

$$Y = \text{number of failures}/(3 \text{ weeks}) \longrightarrow \mu = \lambda t = (0.20) \times 3 = 0.60$$

$$P(Y = 2) = \frac{(0.60)^2}{2!} e^{-0.60} = 0.09879.$$

The pr that the time between 2 successive downtimes (or failures) exceeds 3 weeks is $P(T > 3) = P[X(3) = Y = 0] = e^{-0.60} = 0.548812$. The pr that the time between next 2 failures will be shorter than 2 weeks is given by $P(T < 2) = P[X(2) \ge 1] = 1 - P[X(2) = 0] = 1 - e^{-0.40} = 0.3297$. This pr can also be computed by the direct integration of the exponential pdf as $P(T < 2) = \int_0^2 0.20\, e^{-0.20t} dt = 1 - e^{-0.20 \times 2} = 0.3297$.

Finally, the pr that the time (since the last system failure) to the next downtime will be less than 4 weeks is $P(T < 4) = 1 - e^{-0.80} = 0.5507$.

In this last example, the $P(T < 4)$ also is referred to as the unreliability at time $t = 4$ weeks, and therefore, the reliability at time $t = 4$ weeks is defined as $R(4) = P(T > 4) = 1 - P(T \le 4) = 0.4493$. Note that by reliability at 4 units of time we mean the pr that a system survives beyond $t = 4$, i.e. in general $R(t) = \Pr(\text{survival} > t) = P(T > t) = \int_t^\infty f(x)\, dx$, where $f(t)$ represents the mortality (or failure) density function of the rv T.

2.2.7.3   *Statistical inference*

By statistical inference (SI) we mean estimation and/or test of hypothesis. There are two types of estimates: point and (confidence) interval estimate. A point estimate, or estimator,* is a random variable (or statistic) before the sample is drawn, but it becomes a numerical value after the sample is drawn estimating a population parameter (such as μ and σ). For example, any one of the statistics, the sample mean $\bar{x}$, the median $\tilde{x}$, the sample mode MO, etc., can be considered as point estimators of the population mean μ. If the parameter under consideration is the population variance σ^2, then we may use any one of the statistics

$$S^2 = \frac{1}{n-1} \sum_{i=1}^{n} (x_i - \bar{x})^2, \quad \hat{\sigma}^2 = \frac{1}{n} \sum_{i=1}^{n} (x_i - \bar{x})^2,$$

---

* The term estimator, $\hat{\theta}$, is applied to the random variable, and the word estimate is reserved for the numerical value of $\hat{\theta}$ taken on by the random variable after the sample is drawn and the experimental data have been inserted.

or $(R/d_2)^2$ as estimators of σ^2, where R is the sample range, $d_2 = E(R/\sigma)$, and the rv $w = R/\sigma$ is called the relative range in the field of QC. Note that any symbol with a hat atop it implies an estimator. The question is then how one chooses among several available estimators of a given process parameter? The obvious answer is to select the estimator that is the most accurate, i.e. whose value comes closest to the true value of the parameter being estimated in the long run.

Since a point estimator, $\hat{\theta}$, is a random variable (i.e. it changes from sample to sample), it has a pr density function, and for $\hat{\theta}$ to be an 'accurate' estimator, the pdf of $\hat{\theta}$ should be closely concentrated about the population parameter θ. The criterion used to decide which one of the two estimators $(\hat{\theta}_1$ or $\hat{\theta}_2)$ of the parameter θ is better now follows.

The statistic $\hat{\theta}_1$ is said to be a more accurate estimator of θ than $\hat{\theta}_2$ iff $E[(\hat{\theta}_1 - \theta)^2] < E[(\hat{\theta}_2 - \theta)^2]$. Since $E[(\hat{\theta} - \theta)^2]$ is defined to be the mean square error of $\hat{\theta}$, i.e. $MSE(\hat{\theta}) = E[(\hat{\theta} - \theta)^2]$, then $\hat{\theta}_1$ is a more accurate estimator than $\hat{\theta}_2$ iff MSE $(\hat{\theta}_1) <$ MSE $(\hat{\theta}_2)$.

PROPERTIES OF POINT ESTIMATORS

An estimator is said to be consistent iff the limit $(n \to \infty)$ of $\hat{\theta} = \theta$. If the population is finite of size N, then $\hat{\theta}$ is consistent iff limit of $\hat{\theta} = \theta$ as $n \to N$. For example, $\bar{x}$ is a consistent estimator of μ for both finite and infinite populations because limit of $\bar{x} = \mu$ as $n \to N$. However, S^2 is not a consistent estimator of σ^2 for finite populations but S^2 becomes consistent as $N \to \infty$.

One of the most important property of a point estimator is the amount of bias in the estimator. The amount of bias in a point estimator is defined as

$$B(\hat{\theta}) = E(\hat{\theta}) - \theta = E(\hat{\theta} - \theta)$$

and, therefore, $\hat{\theta}$ is an unbiased estimator iff $E(\hat{\theta}) = \theta$ (i.e. $B = 0$). Users make the common mistake that an unbiased estimator is one whose value is equal to the parameter it is estimating. This is completely false!

We now give the relationship between the accuracy of an estimator $\hat{\theta}$, measured by its MSE, and the amount of bias in $\hat{\theta}$. It can be shown that

$$MSE(\hat{\theta}) = V(\hat{\theta}) + B^2 = [E(\hat{\theta}^2) - E^2(\hat{\theta})] + B^2$$

This equation clearly shows that MSE $(\hat{\theta}) = V(\hat{\theta})$ iff $\hat{\theta}$ is an unbiased estimator of θ.

Example 12

Consider an infinite population (not necessarily a normal population) with parameters μ and σ^2, and a simple random sample of size of $n > 1$ is drawn. Then $E(\bar{x}) = \mu$, i.e. $\bar{x}$ is an unbiased estimator of the population mean μ. The identity

$$\sum_{i=1}^{n}(x_i - \bar{x})^2 \equiv \sum_{i=1}^{n}[(x_i - \mu) - (\bar{x} - \mu)]^2 \equiv \cdots \equiv \sum(x_i - \mu)^2 - n(\bar{x} - \mu)^2,$$

can be used to further prove that $E(S^2) = \sigma^2$ only for infinite populations. That is, S^2 is an unbiased estimator of process variance σ^2 for any infinite population.

The relative efficiency (REL-EFF) of $\hat{\theta}_1$ to $\hat{\theta}_2$ is defined as $MSE(\hat{\theta}_2)/MSE(\hat{\theta}_1)$. As an example, consider a random sample of size $n = 6$ from a population with unknown mean μ. Let $\hat{\theta}_1 = (x_1 + x_2 + x_3 + x_4 + x_5 + x_6)/6 = \bar{x}$, and $\hat{\theta}_2 = (2x_1 + x_4 - x_6)/2$. Then, $E(\hat{\theta}_1) = E(\hat{\theta}_2) = \mu$, i.e. both $\hat{\theta}_1$ and $\hat{\theta}_2$ are unbiased estimates of μ but the MSE $(\hat{\theta}_1) = \sigma^2/6$, $MSE(\hat{\theta}_2) = 1.5\sigma^2$ so that the REL-EFF of $\hat{\theta}_1$ to $\hat{\theta}_2 = (1.5\sigma^2)/(\sigma^2/6) = 900\%$.

SAMPLING DISTRIBUTIONS OF STATISTICS WITH UNDERLYING NORMAL PARENT POPULATIONS

In carrying out statistical inference, we will assume that the underlying distribution (or the parent population) is Gaussian (or approximately so) in which case the sampling distribution of sample elements $x_1, x_2, \ldots, x_n$ is normal and mutually independent of each other. A summary of the most commonly occurring normally-related statistics that are used to conduct SI on different population parameters will now follow.

(i) Suppose $X \sim N(\mu, \sigma^2)$ and a random sample of size n has been drawn with sample values $x_1, x_2, \ldots, x_n$. Since the statistic $\bar{x} = \left(\sum_{i=1}^{n} x_i\right)/n$ is a linear combination of normally and independently distributed (NID) rvs, then $\bar{x}$ is also $N(\mu, \sigma^2/n)$. This implies that the sampling distribution (SMD) of $Z = (\bar{x} - \mu)/(\sigma/\sqrt{n}) = ((\bar{x} - \mu)\sqrt{n})/\sigma$ is that of the standardised normal distribution $N(0, 1)$ so that Z can be used for SI on μ if the value of σ is known.

(ii) If σ is unknown and has to be estimated from the sample statistic S, then the SMD of $((\bar{x} - \mu)\sqrt{n})/S = (\bar{x} - \mu)/\hat{\sigma}_{\bar{x}}$ is that of a Student's t with $(n - 1)$ degrees of freedom.

(iii) To compare the means of two normal populations, we make use of the fact that the SMD of $\bar{x}_1 - \bar{x}_2$ is $N(\mu_1 - \mu_2, (\sigma_1^2/n_1) + (\sigma_2^2/n_2))$. In case the two population variances are unknown but equal, i.e. $\sigma_1^2 = \sigma_2^2 = \sigma^2$, then rv $[(\bar{x}_1 - \bar{x}_2) - (\mu_1 - \mu_2)]/[S_p\sqrt{(1/n_1) + (1/n_2)}]$ has a t sampling distribution with $\nu = n_1 + n_2 - 2$ df, where

$$S_p^2 = \frac{(n_1 - 1)S_1^2 + (n_2 - 1)S_2^2}{n_1 + n_2 - 2} = \frac{CSS_1 + CSS_2}{n_1 + n_2 - 2}.$$

(iv) It can be shown that if $Z \sim N(0, 1)$, then the rv Z^2 has a chi-square (χ^2) distribution with 1 degree of freedom (df). Using this fact, then it can be shown that the SMD of $S^2(n - 1)/\sigma^2 = \sum_{i=1}^{n} (x_i - \bar{x})^2/\sigma^2$ follows a χ^2 with $(n - 1)$ df (not n degrees of freedom).

(v) If S_x^2 represents the sample variance of machine X from a random sample of size n_x and S_y^2 represents that of machine Y, then from statistical theory the sampling distribution of $S_x^2/\sigma_x^2)/(S_y^2/\sigma_y^2)$ follows (Sir R.A.) Fisher's F with numerator degrees of freedom $\nu_1 = n_x - 1$, and denominator degrees of freedom $\nu_2 = n_y - 1$, where n_y is the number of items sampled at random from machine Y.

CONFIDENCE INTERVALS FOR ONE PARAMETER OF A NORMAL UNIVERSE

Let $(1 - \alpha)$ be the confidence interval (CI) coefficient (or confidence level) for the population mean μ. In the case of known σ for the normal underlying distribution, the two-sided confidence interval for μ is given by

$$\bar{x} - Z_{\alpha/2}\sigma/\sqrt{n} \leq \mu \leq \bar{x} + Z_{\alpha/2}\sigma/\sqrt{n}, \tag{2.2a}$$

where $Z_{\alpha/2}$ is the $(\alpha/2) \times 100$ percentage point of a standardised normal distribution. For example, $Z_{0.025} = 1.96$ is the same as the 97.5 percentile (or the 2&1/2 percentage point) of a $N(0, 1)$ distribution. If the quality characteristic of interest, X, is of smaller-the-better (STB) type (such as loudness of a compressor, or radial force harmonic of a tire), then from an engineering standpoint it is best to obtain an upper one-sided confidence interval for μ given by

$$0 < \mu \leq \bar{x} + Z_{\alpha}\sigma/\sqrt{n}. \tag{2.2b}$$

Conversely, if the quality characteristic, X, is of larger-the-better (LTB) type (such as tensile strength, efficiency, etc) then it is best to provide a lower one-sided confidence interval for μ, which is given by

$$\bar{x} - Z_{\alpha}\sigma/\sqrt{n} \leq \mu < \infty. \tag{2.2c}$$

If the population standard deviation, σ, of the normal universe is unknown and has to be estimated by sample statistic $S = \sqrt{(1/(n - 1)) \sum_{i=1}^{n} (x_i - \bar{x})^2}$, then in equations (2.2) for the CI on μ, replace σ by its estimator S and replace Z_{α} with $t_{\alpha;n-1}$ where $t_{\alpha;n-1}$ is the $\alpha \times 100$ percentage point of a Student's t distribution with $(n - 1)$ degrees of freedom.

Example 13

A normal process manufactures wires with a strength lower specification limit $LSL = 8.28$ units and variance $\sigma^2 = 0.029\,93$ units^2 ('units' is used to indicate some measure of strength). A random sample of size $n = 25$ items gave $\sum_{i=1}^{25} X_i = 217.250$ units. Our objective is to obtain the point and 95% interval estimators $(1 - \alpha = 0.95)$ for the process mean μ.

The point estimator of the process mean strength, μ, is

$$\hat{\mu} = \overline{x} = \frac{1}{25} \sum_{i=1}^{25} x_i = 8.690 \text{ units.}$$

Since μ is an LTB type parameter, then we develop a 95% lower CI for μ (note that there are no concerns about strength being too high from the consumers' standpoint). The lower 95% confidence limit for μ, from (2.2c), is given by $\mu_L = \overline{X} - Z_{0.05} \times \sigma/\sqrt{n} = 8.690 - 1.645 \times 0.173/\sqrt{25} = 8.690 - 0.0569 = 8.633\ 1$ units. Thus, we are 95% confident that μ lies in the interval $8.633\ 1 \leq \mu < \infty$. The number interval [8.633 1 units, ∞] no longer has a pr of 0.95 to contain the true value of μ because that pr is either 0 or 1. However, the random interval $\overline{x} - 1.645\sigma/\sqrt{n} \leq \mu < \infty$ has a pr of 95% to contain μ before the sample is drawn because $\overline{x}$ is a rv before the random drawing of the n sample values. This implies that if we wish to test the null hypothesis $H_0 : \mu = 8.621$ versus the one-sided alternative $H_1 : \mu > 8.621$ at the level of significance of $\alpha = 0.05$, then our 95% CI: $8.6331 \leq \mu < \infty$ dictates that we must reject H_0 in favour of H_1 at the 5% level of significance because the hypothesised value of process mean $\mu_0 \equiv 8.621$ is outside the 95% CI, or $\mu_0 \equiv 8.621 < \mu_L = 8.633\ 1$ units. On the other hand, our 95% CI does not allow us to conclude that $\mu > 8.650$ units at the 5% level because 8.650 is inside the 95% CI $8.633\ 1 \leq \mu < \infty$.

With the above example we have demonstrated that all confidence intervals in the universe are tests of hypotheses in disguise because a lower $(1 - \alpha)100\%$ one-sided CI for μ provides all possible right-tailed tests at the level of significance α for the parameter μ. In the above example if the process variance σ^2 were unknown and had to be estimated from the sample, then the sample values must be used to compute the USS $= \sum_{i=1}^{n} x_i^2$, say that for the above example the uncorrected sum of squares were computed to be $\sum_{i=1}^{n} x_i^2 = 1\,888.790$ units^2. Then, the CSS $= 1\,888.790 - 217.25^2/25 = 0.887\,50$, $S^2 = 0.887\,5/24 = 0.037\,0$, and $S = 0.192\,3$ units would be the point estimate of the unknown parameter σ. The corresponding 95% lower confidence limit for μ is given by $\mu_L = \overline{x} - t_{0.05;24} \times S/\sqrt{n} = 8.690 - 1.711 \times 0.192\,3/\sqrt{25} = 8.624$ units, which results in the confidence interval $8.624 \leq \mu < \infty$.

CONFIDENCE INTERVALS FOR PARAMETERS OF TWO NORMAL UNIVERSES

The Student's t distribution must be used (except when the two process variances are known) in order to compare two population means while the Fisher F distribution is used to compare two process variances. We provide an example of how to compare two normal population means.

Example 14

For the sake of illustration consider the Experiment reported in the Journal of Waste and Hazardous Materials (Vol. 6, 1989), where $X_1 =$ weight percent calcium in standard cement, while $X_2 =$ weight percent calcium in cement doped with lead. Reduced levels of calcium would cause the hydration mechanism to become blocked and allow water to attack various locations of the cement structure. Ten samples of standard cement gave $\overline{x}_1 = 90.0\%$ with $S_1 = 5.0$ while $n_2 = 15$ samples of lead-doped cement resulted in $\overline{x}_2 = 87.0$ with $S_2 = 4.0$. Assuming that $X_1 \sim N(\mu_1, \sigma^2)$ and $X_2 \sim N(\mu_2, \sigma^2)$, then the rv $[(\overline{x}_1 - \overline{x}_2) - (\mu_1 - \mu_2)]/[S_p\sqrt{(1/n_1) + (1/n_2)}]$ has a t sampling distribution with $\nu = n_1 + n_2 - 2$df. The expression for the two-sided 95% CI for $\mu_1 - \mu_2$ is given by $\overline{x}_1 - \overline{x}_2 - t_{0.025;23} \times S_p\sqrt{(1/n_1) + (1/n_2)} \leq \mu_1 - \mu_2 \leq \overline{x}_1 - \overline{x}_2 + t_{0.025;23} \times S_p\sqrt{(1/n_1) + (1/n_2)}$, where $t_{0.025;23} = 2.069$ and $S_p^2 = [9(25) + 14(16)]/(23) = 19.521\,74$. Substitution of sample results into the above expression yields $(\mu_1 - \mu_2)_L = 3 - 2.069 \times 1.8038 = -0.732$ and $(\mu_1 - \mu_2)_U = 3 + 3.732 = 6.732$. Since this 95% CI, $-0.732 \leq \mu_1 - \mu_2 \leq 6.732$, includes zero, then there does not exist a significant difference between the two population parameters μ_1 and μ_2 at the 5% level, i.e. the null hypothesis $H_0 : \mu_1 - \mu_2 = 0$ cannot be rejected at the 5% level of significance. This implies that doping cement with lead does not significantly (at the 5% level) alter water hydration mechanism. Note that the null hypothesis $H_0 : \mu_1 - \mu_2 = 7$ must be rejected at the 5% level of significance because 7 is outside the 95% CI.

When the variances of the two independent normal populations, σ_1^2 and σ_2^2, are unequal and unknown, then the CI for $\mu_1 - \mu_2$ must be obtained from the fact that the SMD of the rv

$$\frac{(\overline{x}_1 - \overline{x}_2) - (\mu_1 - \mu_2)}{\sqrt{\dfrac{S_1^2}{n_1} + \dfrac{S_2^2}{n_2}}}$$

follows a Student's t distribution with df ν given by the following equation.

$$\nu = \frac{(S_1^2/n_1 + S_2^2/n_2)^2}{\dfrac{(S_1^2/n_1)^2}{n_1 - 1} + \dfrac{(S_2^2/n_2)^2}{n_2 - 1}} \qquad (2.3)$$

where $\text{Max}(n_1 - 1, n_2 - 1) < \nu < n_1 + n_2 - 2$. As a general rule of thumb, we would recommend against using the pooled t procedure outlined in Example 14 if $S_1^2 > 2S_2^2$, or vice a versa. From the above discussions we conclude that in case $S_1^2 > 2S_2^2$, then the t statistic for testing the null hypothesis $H_0 : \mu_1 - \mu_2 = \delta$ is given by

$$t_0 = \frac{(\bar{x}_1 - \bar{x}_2) - \delta}{\sqrt{\dfrac{S_1^2}{n_1} + \dfrac{S_2^2}{n_2}}} .$$

If the test is 2-sided, then the acceptance interval for testing $H_0 : \mu_1 - \mu_2 = \delta$ at a level of significance 0.05 is given by $(-t_{0.025;\nu}, t_{0.025;\nu})$, where ν is given in equation (2.3).

2.2.7.4 *Statistical process control (SPC)*

The objective of SPC is to test the null hypothesis that the value of a process parameter is either at a desired specified value, or at a value that has been established from a long-term data. This objective is generally carried out through constructing a Shewhart control chart from $m > 1$ subgroups of data. Further, it is assumed that the underlying distribution is approximately Gaussian, and for moderately large sample sizes, it is also assumed that the SMD of the statistic used to construct the Shewhart chart is also Gaussian. When a sample point goes out of control limits, the process must be stopped in order to look for assignable causes, and if one is found, then corrective action must be taken and the corresponding point should be removed from the chart. In case no assignable (or special) causes are found, then the control chart has led to a false alarm (or a type I error). Since false alarms are very expensive and disruptive to a manufacturing process, all Shewhart charts are designed in such a manner that the pr of committing a type I error, α, is very small. The standard level of significance, α, of all charts are set at $\alpha = 0.002\,7$.

When departures from the underlying assumptions are not grossly violated, then a Shewhart control chart will generally lead an experimenter to 27 false alarms in $10\,000$ random samples. Moreover, setting the value of α at $0.002\,7$, will correspond to three-sigma control limits for a control chart as long as the normality assumption is tenable. We will discuss only two types of charts: (1) Charts for continuous variables, and (2) Charts for attributes, where the measurement system merely classifies a unit either as conforming to customer specifications or nonconforming to specifications (i.e. success/failure, 0/1, defective/effective, etc.).

SHEWHART CONTROL CHARTS FOR VARIABLES

As an example, suppose we wish to control the dimension of piston ring inside diameters, X, with design specifications X: 74.00 ± 0.05 mm. The rings are manufactured through a forging process. Since the random variable X is continuous, then we need two charts; one to control variability (or σ), and a second chart for controlling the process mean μ. If subgroup sample sizes, n_i, are identical and lie within $2 \le n \le 15$, then an R-chart (i.e. range-chart) should be used to monitor variability, but for $n > 15$, an S-chart should be used for control of variation. This is due to the fact that the SMD of sample range, R, becomes unstable for moderate to large sample sizes. Samples of sizes n_i $(i = 1, 2, \ldots, m)$ are taken from the process, generally at equal intervals of time, (where hourly or daily samples, or samples taken at different shifts, are the most common), and the number of subgroups m should generally lie within $20 \le m \le 50$. Samples should be taken in such a manner as to minimise the variability within samples and maximise the variability among samples, a concept that is consistent with design of experiments (DOE). Such samples are generally referred to as rational subgroups.

R- AND $\overline{X}$-CHARTS (FOR $2 \le n \le 15$)

In practice the R-chart should be constructed first in order to bring variability in a state of statistical control, followed by developing the $\overline{x}$-chart for the purpose of monitoring the process mean. In order to use the R-chart for monitoring process variation, the subgroup sample sizes n_i $(i = 1, 2, \ldots, m)$ must be the same, i.e. $n_i = n$ for all i, or else an R-chart cannot be constructed. All univariate control charts consist of a central line, CNTL, a lower control limit LCL, and an upper control limit UCL. Further, nearly in all cases, LCL = CNTL $- 3 \times$ se (sample statistic) and UCL = CNTL $+ 3 \times$ se (sample statistic), where in the case of the R-chart the sample statistic will be the sample range R, while for the $\overline{x}$-chart the sample statistic will be the sample mean $\overline{x}$. The pertinent formulas are provided below.

$$\text{CNTL}_R = \overline{R} = \frac{1}{m} \sum_{i=1}^{m} R_i$$

Note that we are taking the liberty to use the terminology standard error, se, as the estimate of the Stdev of the sample statistic.

The $\text{se}(R) = \hat{\sigma}_R = d_3 \overline{R}/d_2$, the values of d_2 are given in Table 2.12 for $n = 2, 3, \ldots, 15$, $d_3^2 = V(R/\sigma)$, and the values of d_3 for a normal universe are given in Table 2.14. Since the most common of all sample sizes for constructing an R- and $\overline{x}$-Chart is $n = 5$, we compute $\hat{\sigma}_R$ only for $n = 5$. From Tables 2.12 & 2.14, $\text{se}(R) = 0.864 \times \overline{R}/2.326 = 0.371\,45 \times \overline{R}$. Then for $n = 5$, the $\text{LCL}_R = \overline{R} - 3 \times 0.371\,45 \times \overline{R} = \overline{R}(1 - 1.1144) \to \text{LCL}_R = 0$, and $\text{UCL}_R = 2.114\,36 \times \overline{R}$. In fact, it can be shown that the $\text{LCL}_R = 0$ for all sample sizes in the range $2 \le n \le 6$, but $\text{LCL}_R > 0$ for all $n > 6$.

The central line for an $\overline{x}$-Chart is given by $\text{CNTL}_{\overline{x}} = 1/m \sum_{i=1}^{m} \overline{x}_i = \overline{\overline{X}}$ and the $\text{se}(\overline{x}) = \sigma/\sqrt{n}$. Since a point estimate of σ is given by $\hat{\sigma}_x = \overline{R}/d_2$, then $\hat{\sigma}_{\overline{x}} = \overline{R}/d_2\sqrt{n}$, as a result the $\text{LCL}_{\overline{x}} = \overline{\overline{X}} - 3 \times \text{se}(\overline{x}) = \overline{\overline{X}} - 3\overline{R}/d_2\sqrt{n}$, and $\text{UCL}_{\overline{x}} = \overline{\overline{X}} + 3\overline{R}/d_2\sqrt{n}$. For samples of size $n = 5$, these last two control limits reduce to

$$\text{LCL}_{\overline{x}} = \overline{\overline{X}} - 0.5768\overline{R}, \quad \text{and} \quad \text{UCL}_{\overline{x}} = \overline{\overline{X}} + 0.5768\overline{R}.$$

S- AND $\overline{x}$-CHARTS

If subgroup sample sizes differ and/or $n > 12$, then process variation must be monitored by an S-chart. The most common occurrence of an S-chart is when the sampling design is not balanced, i.e. n_i's $(i = 1, 2, \ldots, m)$ are not the same, then the experimenter has no option but to use an S-chart for the control of variation. The central line is given by

$$\text{CNTL}_S = S_p = \left[\frac{\sum_{i=1}^{m} (n_i - 1)S_i^2}{\sum_{i=m}^{m} (n_i - 1)} \right]^{1/2} = \left[\sum_{i=1}^{m} (n_i - 1)S_i^2/(N - m) \right]^{1/2}$$

where $N = \sum_{i=1}^{i=m} n_i$, and the quantity $(n_i - 1)S_i^2 = \sum_{j=1}^{j=n_i} (x_{ij} - \overline{x}_i)^2$ is called the corrected sum of squares within the ith subgroup. It can be shown (using the properties of χ^2) that for a normal universe the $E(S) = c_4\sigma$, where the constant $c_4 = \sqrt{2/(n-1)} \times \Gamma(n/2)/\Gamma[(n-1)/2]$ lies in the interval $(0.797\,884\,5, 1)$ for all $n \ge 2$ and its limit as $n \to \infty$ is equal to 1. Further, the authors have shown that for $n \ge 20$ the value of c_4 can be approximated to 5 decimals by $c_4 \cong 4n^2 - 8n + 3.875/(4n - 3)(n - 1)$. These discussions imply that, in the long-run, the statistic S underestimates the population standard deviation σ, and hence an unbiased estimator of σ for a normal universe is given by $\hat{\sigma}_x = S/c_4$ (this is due to the fact that $E(S) = c_4\sigma$).

Table 2.14   THE VALUES OF d_3 FOR A NORMAL UNIVERSE

n	2	3	4	5	6	7	8	9	10	11	12	13	14	15
d_3	0.853	0.888	0.880	0.864	0.848	0.833	0.820	0.808	0.797	0.787	0.778	0.770	0.763	0.756

To compute the se(S), we make use of the fact that $V(S) = E(S^2) - [E(S)]^2 = \sigma^2 - (c_4\sigma)^2 = (1-c_4^2)\sigma^2$. This development implies that $se(S) = (1 - c_4^2)^{1/2}\hat{\sigma}_x = (1 - c_4^2)^{1/2}(S/c_4) = S\sqrt{(c_4)^{-2} - 1}$. Thus the control limits are given by

$$LCL_i(S) = S_p - 3\,S_p\sqrt{(c_4)^{-2} - 1} = [1 - 3\sqrt{(c_4)^{-2} - 1}]S_p$$

$$UCL_i(S) = S_p + 3\,S_p\sqrt{(c_4)^{-2} - 1} = [1 + 3\sqrt{(c_4)^{-2} - 1}]S_p.$$

Note that the $LCL_S = 0$ when $2 \le n_i \le 5$, but $LCL_S > 0$ when $n > 5$. If the sampling design is balanced, i.e. $n_i = n > 12$ for all i, then replace S_p in the previous two equations with $\bar{S} = (1/m)\sum_{i=1}^{m} S_i$.

Once variability is in a state of statistical control (i.e. all sample S_i's lie within their own control limits), then an $\bar{x}$-chart is developed to monitor the process mean. The central line for an $\bar{x}$-chart is given by

$$CNTL_{\bar{x}} = \bar{\bar{x}} = \sum_{i=1}^{m} n_i\bar{x}_i/N$$

where $N = \sum_{i=1}^{i=m} n_i$. Since the $V(\bar{x}_i) = \sigma^2/n_i$, then $se(\bar{x}_i) = \hat{\sigma}_x/\sqrt{n_i} = (S_p/c_4\sqrt{n_i})$, and as a result

$$LCL_i(\bar{x}) = \bar{\bar{x}} - 3\frac{S_p}{c_4\sqrt{n_i}}, \quad \text{and} \quad UCL_i(\bar{x}) = \bar{\bar{x}} + 3\frac{S_p}{c_4\sqrt{n_i}}. \tag{2.4}$$

Note that if the sampling design is unbalanced, then all points on the S- and $\bar{x}$-charts have the same CTLNs but every point on both charts has its own control limits due to differing sample sizes. When the sampling scheme is balanced (i.e. $n_i = n$ for all i), then in equation (2.4) replace S_p with $\bar{S} = (1/m)\sum_{i=1}^{m} S_i$.

SHEWHART CONTROL CHART FOR FRACTION NONCONFORMING (THE P-CHART)

As an example, consider an injection moulding process that produces instrument panels for an automobile. The occurrence of splay, voids, or short shots will make the panel defective. Thus we have a binomial process where each panel is classified as defective (or 1) or as conforming (or 0). The binomial rv, X, represents the number of nonconforming panels in a random sample of size n_i, where it is best to have at least $m > 20$ subgroups in order to construct the p-chart, where p is the FNC of the process. The sample FNC is given by $\hat{p} = X/n$, and if $n > 30$ and np and nq > 10, then the SMD of $\hat{p}$ is approximately normal with mean p and $se(\hat{p}) = \sqrt{pq/n}$, where $q = (1 - p)$ is the process fraction conforming (or process yield). The central line is given by

$$CNTL_p = \frac{\sum_{i=1}^{m} X_i}{\sum_{i=1}^{m} n_i} = \frac{\sum_{i=1}^{m} n_i\hat{p}_i}{N} = \bar{p}$$

where $N = \sum_{i=1}^{i=m} n_i$ is the total number of units inspected by attributes in all m samples, X_i represents the number of NC units in the ith subgroup, and $\hat{p}_i = X_i/n_i$ is the sample FNC of the ith subgroup. Since the estimate of the $se(\hat{p}_i) = \sqrt{\bar{p}(1 - \bar{p})/n_i}$, then the control limits for the ith subgroup is given by $LCL_i(p) = \bar{p} - 3\sqrt{\bar{p}(1 - \bar{p})/n_i}$, and $UCL_i(p) = \bar{p} + 3\sqrt{\bar{p}(1 - \bar{p})/n_i}$. Note that, when subgroup sizes differ on a Shewhart p-chart, then every sample FNC, $\hat{p}_i$, has its own control limit. If the difference between maximum and minimum sample sizes do not exceed 10 units, then a p-chart based on average sample size should be constructed for monitoring process FNC. In all cases the central line stays the same, but the average control limits simplify to

$$\overline{LCL}(p) = \bar{p} - 3\sqrt{\frac{\bar{p}(1 - \bar{p})}{\bar{n}}}, \quad \text{and} \quad \overline{UCL}(p) = \bar{p} + 3\sqrt{\frac{\bar{p}(1 - \bar{p})}{\bar{n}}},$$

where $\bar{n} = (1/m)\sum_{i=1}^{m} n_i$. The reader is cautioned to the fact that if a p-chart based on an average sample size is used to monitor process FNC, then all points (i.e. all sample FNCs) that are close to their average control limits (whether in or out of control) must be checked against their own limits to determine their control nature.

Table 2.15 DATA FOR TEXTILE PROCESS EXAMPLE

Sample number (i)	1	2	3	4	5	6	7	8	9	10
Square metres	180	150	120	90	150	160	120	140	130	175
c_i	9	15	6	5	16	10	4	12	14	9
n_i	1.8	1.5	1.2	0.90	1.5	1.6	1.2	1.4	1.3	1.75
u_i	5.00	10.00	5.000	5.556	10.667	6.25	3.333	8.571	10.769	5.143
LCL_i	1.123	0.555	0.00	0.00	0.555	0.762	0.00	0.327	0.072 4	1.038
UCL_i	13.012	13.579	14.348	15.474	13.579	13.372	14.348	13.807	14.062	13.096

SHEWHART CONTROL CHART FOR NUMBER OF NONCONFORMITIES PER UNIT (THE u-CHART)

Since the construction of the u-chart is not as straight forward as the others discussed thus far, we will describe the methodology through an example. In practice, it is best to have at least $m = 20$ subgroups, but herein for simplicity we will use $m = 10$ samples of differing sizes. Consider a textile process that produces oilcloth in lots of differing sizes, measured in square metres. An inspector selects $m = 10$ lots at random and counts the number of defects, c_i, in each lot (or sample). The data is displayed in Table 2.15.

Note that in Table 2.15, because of different square metres, we have arbitrarily let 100 square metres equal to one unit, although 50, or, 10, or any other convenient square metres would work just as well. Further, $u_i = c_i/n_i$ represents the number of defects per unit. Note that because of differing sample sizes, it would be erroneous to compute the average number of defects per unit from $(1/m) \sum_{i=1}^{m} u_i = 7.028\,9$, where this last formula would work only if all n_i's were identical. The correct formula for the central line is given by

$$CNTL_u = \bar{u} = \frac{\sum_{i=1}^{m} c_i}{\sum_{i=1}^{m} n_i} = \frac{\sum_{i=1}^{m} n_i u_i}{\sum_{i=1}^{i} n_i} = \frac{100}{14.15} = 7.067\,1$$

It is well known that the number of defects per unit, c, follows a Poisson process and hence its variance is also given by E(c), i.e. V(c) can be approximated by $\bar{u}$. Unfortunately, the terminology and notation for a u-chart has been somewhat confusing in statistical and QC literature and we anticipate no change. Therefore, herein we attempt to remedy the notational problem to some extent. First of all, the fifth row of Table 2.15 actually provides the average number of defects per unit for the ith sample, and hence the proper notation for the fifth row should be $\bar{u}_i$ (not u_i as is used in QC literature) because a bar is generally placed on averages in the field of statistics. This implies that a u-chart is actually a $\bar{u}$-chart because it is the average number of defects per unit that is plotted on this chart. Secondly, the central line should be called $\bar{\bar{u}}$ because the CNTL gives the grand average of all average number of defects per unit. These discussions lead to the fact that firstly $V(\bar{u}) = V(c)/n$, and secondly $V(\bar{u})$ can be estimated by $\bar{u}/n$. Since we do not wish to deviate from QC literature terminology, we will stay with the existing notation and let u_i represent the average number of defects per unit with CNTL as $\bar{u}$ and the $se(u) = \sqrt{\bar{u}/n_i}$. Thus, the $LCL_i(u) = \bar{u} - 3\sqrt{\bar{u}/n_i}$, and $UCL_i(u) = \bar{u} + 3\sqrt{\bar{u}/n_i}$. The values of control limits for all the $m = 10$ samples are provided in the last two rows of Table 2.15. Table 2.15 clearly shows that each u_i is well within its own control limits, implying that the process is in a state of excellent statistical control. Note that in all cases when the value of LCL became negative, a zero LCL was assigned in row 6 of Table 2.15.

This example provides a good illustration of a process that is in an excellent state of statistical control, but one that is in all pr not capable of meeting customer specifications due to the fact that $\bar{u} = 7.067\,1$ is too large and customers in today's global market will generally demand lower average number defects per unit. If this manufacturer does not improve its process capability through quality improvement (QI) methods, it may not survive very long in global competition.

Further, SPC is not a QI method but simply an on-line procedure to monitor process quality and to identify where the quality problems lie. After problems are identified, then off-line methods (DOE or Taguchi methods) can be applied to fine-tune a process.

3 General physical and chemical constants

Table 3.1 ATOMIC WEIGHTS AND ATOMIC NUMBERS OF THE ELEMENTS

Name	Symbol	Atomic number %	International atomic weights* 1971 $^{12}C = 12$	Name	Symbol	Atomic number %	International atomic weights* 1971 $^{12}C = 12$
Actinium	Ac	89	(227)	Iridium	Ir	77	192.2_2
Aluminium	Al	13	26.981 54	Iron	Fe	26	55.84_7
Americium	Am	95	(243)	Krypton	Kr	36	83.80
Antimony	Sb	51	121.7_5	Lanthanum	La	57	138.905_5
Argon	Ar	18	39.94_8	Lawrencium	Lr	103	(260)
Arsenic	As	33	74.921 6	Lead	Pb	82	207.2
Astatine	At	85	(210)	Lithium	Li	3	6.94_1
Barium	Ba	56	137.3_4	Lutetium	Lu	71	174.97
Berkelium	Bk	97	(247)	Magnesium	Mg	12	24.305
Beryllium	Be	4	9.012 18	Manganese	Mn	25	54.938 0
Bismuth	Bi	83	208.980 4	Mendelevium	Md	101	(258)
Boron	B	5	10.81	Mercury	Hg	80	200.5_9
Bromine	Br	35	79.904	Molybdenum	Mo	42	95.9_4
Cadmium	Cd	48	112.40	Neodymium	Nd	60	144.2_4
Caesium	Cs	55	132.905 4	Neon	Ne	10	20.17_9
Calcium	Ca	20	40.08	Neptunium	Np	93	237.048 2
Californium	Cf	98	(251)	Nickel	Ni	28	58.69
Carbon	C	6	12.011	Niobium	Nb	41	92.906 4
Cerium	Ce	58	140.12	(Columbium)			
Chlorine	Cl	17	35.453	Nitrogen	N	7	14.006 7
Chromium	Cr	24	51.996	Nobelium	No	102	(259)
Cobalt	Co	27	58.933 2	Osmium	Os	76	190.2
Copper	Cu	29	63.54_6	Oxygen	O	8	15.999_4
Curium	Cm	96	(247)	Palladium	Pd	46	106.42
Dysprosium	Dy	66	162.5_0	Phosphorus	P	15	30.973 8
Einsteinium	Es	99	(254)	Platinum	Pt	78	195.0_9
Erbium	Er	68	167.2_6	Plutonium	Pu	94	(244)
Europium	Eu	63	151.96	Polonium	Po	84	(209)
Fermium	Fm	100	(257)	Potassium	K	19	39.09_8
Fluorine	F	9	18.998 40	Praseodymium	Pr	59	140.908
Francium	Fr	87	(223)	Promethium	Pm	61	(145)
Gadolinium	Gd	64	157.2_5	Protactinium	Pa	91	231.035 9
Gallium	Ga	31	69.72	Radium	Ra	88	226.025
Germanium	Ge	32	72.5_9	Radon	Rn	86	(222)
Gold	Au	79	196.966 5	Rhenium	Re	75	186.2
Hafnium	Hf	72	178.4_9	Rhodium	Rh	45	102.905 5
Helium	He	2	4.002 60	Rubidium	Rb	37	85.467_8
Holmium	Ho	67	164.930 4	Ruthenium	Ru	44	101.0_7
Hydrogen	H	1	1.007 9	Samarium	Sm	62	150.36
Indium	In	49	114.82	Scandium	Sc	21	44.955 9
Iodine	I	53	126.904 5	Selenium	Se	34	78.9_6

(*continued*)

Table 3.1   ATOMIC WEIGHTS AND ATOMIC NUMBERS OF THE ELEMENTS—*continued*

Name	Symbol	Atomic number %	International atomic weights* 1971 $^{12}C = 12$	Name	Symbol	Atomic number %	International atomic weights* 1971 $^{12}C = 12$
Silicon	Si	14	28.085 5	Thulium	Tm	69	168.934
Silver	Ag	47	107.868	Tin	Sn	50	118.71
Sodium	Na	11	22.989 8	Titanium	Ti	22	47.88
Strontium	Sr	38	87.62	Tungsten	W	74	183.8₅
Sulphur	S	16	32.06	Uranium	U	92	238.029
Tantalum	Ta	73	180.947₉	Vanadium	V	23	50.941₄
Technetium	Tc	43	(98)	Xenon	Xe	54	131.30
Tellurium	Te	52	127.6₀	Ytterbium	Yb	70	173.0₄
Terbium	Tb	65	158.925	Yttrium	Y	39	88.905 9
Thallium	Tl	81	204.383	Zinc	Zn	30	65.39
Thorium	Th	90	232.038	Zirconium	Zr	40	91.224

*Atomic Weights of the Elements 1981, *Pure and Applied Chemistry* 1983, **55** (7), 1101–1136. A value given in brackets denotes the mass number of the isotope of longest known half-life. Because of natural variation in the relative abundance of the isotopes of some elements, their atomic weights may vary. Apart from this they are considered reliable to ±1 in the last digit, or ±3 if that digit is subscript.

Table 3.2   GENERAL PHYSICAL CONSTANTS

The probable errors of the various quantities may be obtained from the reference given at the end of the table.

Quantity	Symbol	Value	Units
Acceleration due to gravity (standard)	g_n	9.806 65	$m\,s^{-2}$
Atmospheric pressure (standard)	A_0	$1.013\,25 \times 10^5$	Pa
Atomic mass unit	m_u	$1.660\,565\,5 \times 10^{-27}$	kg
Atomic weight of electron	$N_A m_c$	$5.486\,802\,6 \times 10^{-4}$	u^*
Avogadro number	N_A	$6.022\,045 \times 10^{23}$	mol^{-1}
Boltzmann's constant	k	$1.380\,662 \times 10^{-23}$	$J\,K^{-1}$
Bohr radius	a_0	$5.291\,770\,6 \times 10^{-11}$	m
Bohr magneton	μ_B	$9.274\,078 \times 10^{-24}$	$J\,T^{-1}$
Charge in electrolysis of 1 g hydrogen	F/H	$9.572\,378 \times 10^4$	C
Compton wavelength of electron	λ_C	$2.426\,308\,9 \times 10^{-12}$	m
Classical electron radius	r_e	$2.817\,938\,0 \times 10^{-15}$	m
Compton wavelength of proton	$\lambda_{c,p}$	$1.321\,409\,9 \times 10^{-15}$	m
Compton wavelength of neutron	$\lambda_{c,n}$	$1.319\,590\,9 \times 10^{-15}$	m
Density of the earth (average)	δ	5.518×10^3	$kg\,m^{-3}$
(core)		1.072×10^4	$kg\,m^{-3}$
Density of mercury (0°C, A_0)	D_0	$1.359\,508 \times 10^4$	$kg\,m^{-3}$
Density of water (max)	$\delta_m(H_2O)$	$9.999\,72 \times 10^2$	$kg\,m^{-3}$
Electronic charge	e	$1.602\,189\,2 \times 10^{-19}$	C
Electron rest mass	m_e	$9.109\,534 \times 10^{-31}$	kg
Electron volt energy	E_p	$1.602\,192 \times 10^{-19}$	J
Electron magnetic moment	μ_e	$9.284\,832 \times 10^{-24}$	$J\,T^{-1}$
Faraday constant	F	$9.648\,456 \times 10^4$	$C\,mol^{-1}$
Fine structure constant	α	$7.297\,350\,6 \times 10^{-3}$	—
Gas constant	R_0	8.314 41	$J\,K^{-1}\,mol^{-1}$
Gravitational constant	G	$6.672\,0 \times 10^{-11}$	$Nm^2\,kg^{-2}$
Ice point (absolute value)	T_0	$2.731\,5 \times 10^2$	K
Litre (12th CGPM 1964)	1	1.0 exactly $\times 10^{-3}$	m^3
(1963 weights and measures)		$1.000\,028 \times 10^{-3}$	m^3
Neutron rest mass	m_n	$1.674\,954\,3 \times 10^{-27}$	kg
Planck's constant	h	$6.626\,176 \times 10^{-34}$	Js
Proton rest mass	m_p	$1.672\,648\,5 \times 10^{-27}$	kg

(continued)

Table 3.2   GENERAL PHYSICAL CONSTANTS—*continued*

Quantity	Symbol	Value	Units
Radiation constant—first	$8\pi hc$	$4.992\,563 \times 10^{-24}$	Jm
Radiation constant—second	$c_2 = hc/k$	$1.438\,786 \times 10^{-2}$	mK
Rydberg's constant	R_∞	$1.097\,373\,177 \times 10^7$	m^{-1}
Stefan–Boltzmann constant	σ	$5.670\,32 \times 10^{-8}$	$Wm^{-2}\,K^{-4}$
Velocity of light	c	$2.997\,924\,580 \times 10^8$	$m\,s^{-1}$
Volume of ideal gas ($0°C$, A_0)	V_0	$2.241\,36 \times 10^{-2}$	$m^3\,mol^{-1}$
Wien's constant	$\lambda_m T$	$2.897\,8 \times 10^{-3}$	mK
Zeeman displacement		$4.668\,58 \times 10$	$m\,Wb^{-1}$

$^*u =$ unified atomic mass unit (amu) based on $u = \frac{1}{12}$ of the mass of ^{12}C.

REFERENCES

'Quantities, Units and Symbols', The Royal Society, 1981.

Table 3.3   MOMENTS OF INERTIA
Moment of inertia $= Mk^2$ where M is mass and k radius of gyration.

Body	Dimensions	Axis*	k^2
Uniform thin rod	length l	Through centre, perpendicular to rod	$l^2/12$
		Through end, perpendicular to rod	$l^2/3$
Rectangular lamina	sides a and b	Through centre, perpendicular to plane of lamina	$(a^2 + b^2)/12$
		Through centre, parallel to side b	$a^2/12$
Circular lamina	radius r	Through centre, perpendicular to plane of lamina	$r^2/2$
		Through any diameter	$r^2/4$
Annular lamina	radii r_1 and r_2	Through centre perpendicular to plane of lamina	$(r_1^2 + r_2^2)/2$
		Through any diameter	$(r_1^2 + r_2^2)/4$
Rectangular solid	sides a, b and c	Through centre, perpendicular to face ab	$(a^2 + b^2)/12$
Sphere	radius r	Through any diameter	$2r^2/5$
Spherical shell	radii r_1 and r_2	Through any diameter	$2(r_1^5 - r_2^5)/5(r_1^3 - r_2^3)$
Thin spherical shell	radius r	Through any diameter	$2r^2/3$
Right circular cylinder	radius r	Longitudinal axis through centre	$r^2/2$
	length l	Through centre perpendicular to longitudinal axis	$r^2/4 + l^2/12$
Hollow circular cylinder	radii r_1 and r_2	Longitudinal axis through centre	$(r_1^2 + r_2^2)/2$
	length l	Through centre perpendicular to longitudinal axis	$(r_1^2 + r_2^2)/4 + l^2/12$
Thin hollow circular cylinder	radius r	Longitudinal axis through centre	r^2
	length l	Through centre perpendicular to longitudinal axis	$r^2/2 + l^2/12$
Right circular cone	height h	Longitudinal axis through apex	$3r^2/10$
	base radius r	Through centre of gravity perpendicular to longitudinal axis	$3(r^2 + h^2/4)/20$
Ellipsoid	semi-axis a, n and c	Through centre along axis a	$(b^2 + c^2)/5$

*If the moment of inertia I_q about an axis through the centre of gravity is known, then the moment, I, about any other parallel axis may be obtained from

$$I = I_q + Mh^2$$

where h is the distance from the centre of gravity to the parallel axis.

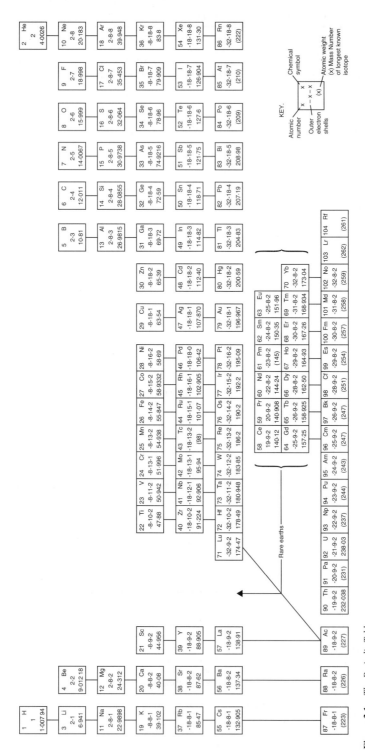

Figure 3.1 *The Periodic Table*

3.1   Radioactive isotopes and radiation sources

Tables 3.4, 3.5 and 3.6 are so arranged as to assist in selecting an isotope with a given half-life and decay radiation energy, for positron, beta and gamma emitters.

Tables 3.7 to 3.12 list the most commonly used commercially available alpha, beta and neutron sources.

Table 3.4   POSITRON EMITTERS (USEFUL NUCLIDES)

Isotope	Half-life	β^+ energies MeV %
Oxygen-15	2.0 min	1.73–99.9
Nitrogen-13	10.0 min	1.20–99.8
Bromine-80	17.4 min	0.85–2.6
Carbon-11	20.4 min	0.96–99.8
Manganese-52 m	21.1 min	2.63–96.4
Gallium-68	68.0 min	0.82–1.1
		1.90–87.9
Fluorine-18	109.8 min	0.63–100
Titanium-45	3.1 h	1.04–84.8
Scandium-44	3.9 h	1.48–94.4
Iron-52 (Daughter: Mn-52 m)	8.3 h	0.80–56.0
Zinc-62	9.3 h	0.61–7.6
Gallium-66	9.4 h	0.36–1.0
		0.77–0.7
		0.92–4.1
		1.78–0.4
		4.15–49.6
Copper-64	12.7 h	0.65–17.9
Niobium-90	14.6 h	1.50–53
Cobalt-55	17.5 h	0.44–0.27
		0.48–0.03
		0.65–0.46
Arsenic-72	26.0 h	0.81–0.5
		1.87–6.0
		2.50–64.0
		2.64–0.1
		3.33–16.5
Nickel-57	36.1 h	0.30–0.41
		0.46–0.87
		0.72–5.7
		0.84–33.1
Gold-194	39.5 h	1.16–0.64
		1.49–1.0
Bromine-77	56 h	0.34–0.73
Yttrium-87	80.3 h	0.45–0.16
Iodine-124	4.2 d	0.81–0.29
		1.53–11.2
		2.14–11.2
Manganese-52	5.6 d	0.58–29.6
Caesium-132	6.5 d	0.6–0.3
Iodine-126	13.0 d	0.47–0.2
		1.13–0.8
Vanadium-48	16.2 d	0.56–0.1
		0.70–50.0
		1.99–0.09
		0.94–25.7
Arsenic-74	17.8 d	1.54–3.4
Rubidium-84	32.8 d	0.78–14.4
		1.66–11.9
Cobalt-58	70.8 d	0.48–15.0
Cobalt-56	78.8 d	0.42–1.1
		1.46–18.8
Yttrium-88	106.6 d	0.76–0.22
Zinc-65	243.9 d	0.33–1.5
Sodium-22	2.60 yr	0.55–89.8
		1.89–0.06
Aluminium-26	7.2×10^5 yr	1.17–81.8

Table 3.5   BETA ENERGIES AND HALF-LIVES

Half-life β^--*energy* MeV	*<1 hour*			*1–10 hours*			*10 hours–1 day*		
0–0.3	^{101}Mo,	^{233}Th		^{31}Si, ^{127}Te, ^{177}Yb	^{105}Ru, ^{165}Dy	^{117}Cd, ^{171}Er,	^{28}Mg, ^{188}Re	^{77}Ge,	^{130}I,
0.3–0.5	^{88}Rb, ^{233}Th	^{101}Mo,	^{116m}In,	^{56}Mn, ^{97}Nb, ^{132}I, ^{177}Yb	^{61}Co, ^{105}Ru, ^{149}Nd,	^{65}Ni, ^{117}Cd, ^{171}Er,	^{28}Mg, ^{77}Ge, ^{187}W, ^{197}Pt	^{43}K, ^{97}Zr, ^{188}Re,	^{72}Ga, ^{130}I, ^{194}Ir,
0.5–0.7	^{70}Ga, ^{101}Mo, ^{199}Pt,	^{80}Br, ^{101}Tc, ^{233}Th	^{88}Rb, ^{116m}In,	^{65}Ni, ^{105}Ru, ^{132}I,	^{75}Ge, ^{117}Cd, ^{152m}Eu,	^{97}Nb, ^{127}Te, ^{171}Er	^{64}Cu, ^{97}Zr, ^{159}Gd, ^{194}Ir,	^{72}Ga, ^{130}I, ^{187}W, ^{197}Pt	^{77}Ge, ^{142}Pr, ^{188}Re,
0.7–1.0	^{49}Ca, ^{81}Se, ^{101}Tc, ^{144}Pr, ^{233}Th	^{69}Zn, ^{88}Rb, ^{116m}In, ^{155}Sm,	^{80}Br, ^{101}Mo, ^{117}In, ^{199}Pt,	^{41}Ar, ^{85m}Kr, ^{117}Cd, ^{152m}Eu,	^{56}Mn, ^{97}Nb, ^{132}I, ^{165}Dy,	^{75}Ge, ^{105}Ru, ^{139}Ba, ^{171}Er	^{28}Mg, ^{77}Ge, ^{159}Gd, ^{197}Pt	^{43}K, ^{97}Zr, ^{187}W	^{72}Ga, ^{130}I, ^{194}Ir,
1.0–1.5	^{38}Cl, ^{80}Br, ^{101}Tc, ^{155}In, ^{155}Sm,	^{49}Ca, ^{81}Se, ^{108}Ag, ^{123m}Sn, ^{199}Pt,	^{52}V, ^{101}Mo, ^{116m}In, ^{128}I, ^{233}Th	^{31}Si, ^{61}Co, ^{97}Nb, ^{132}I, ^{171}Er,	^{41}Ar, ^{65}Ni, ^{105}Ru, ^{149}Nd, ^{176m}Lu, ^{177}Yb,	^{56}Mn, ^{75}Ge, ^{117}Cd, ^{165}Dy,	^{24}Na, ^{77}Ge, ^{130}I, ^{194}Ir	^{43}K, ^{97}Zr, ^{187}W	^{72}Ga, ^{109}Pd, ^{188}Re,
1.5–2.0	^{27}Mg, ^{66}Cu, ^{81}Se, ^{128}I,	^{49}Ca, ^{70}Ga, ^{101}Mo, ^{155}Sm,	^{51}Ti, ^{80}Br, ^{108}Ag, ^{199}Pt	^{105}Ru, ^{149}Nd	^{117}Cd, ^{152m}Eu	^{132}I,	^{42}K, ^{77}Ge, ^{188}Re,	^{43}K, ^{97}Zr, ^{194}Ir	^{72}Ga, ^{130}I,
2.0–3.0	^{28}Al, ^{49}Sc, ^{66}Cu, ^{128}I,	^{38}Cl, ^{51}Ti, ^{88}Rb, ^{144}Pr	^{49}Ca, ^{52}V, ^{101}Mo,	^{41}Ar, ^{117}Cd,	^{56}Mn, ^{132}I,	^{65}Ni, ^{139}Ba,	^{28}Mg/^{28}Al, ^{77}Ge, ^{194}Ir	^{142}Pr,	^{72}Ga, ^{188}Re,
>3.0	^{38}Cl,	^{88}Rb					^{42}K,		^{72}Ga

1–10 days	10–100 days	100 days–1 year	>1 year
^{67}Cu, ^{76}As, ^{77}As, ^{82}Br, ^{99}Mo, ^{105}Rh, ^{131}I, ^{132}Te, ^{133}Xe, ^{143}Ce, ^{149}Pm, ^{151}Pm, ^{166}Ho, ^{169}Er, ^{175}Yb, ^{177}Lu, ^{198}Au, ^{199}Au, ^{222}Rn + D.P.	^{33}P, ^{35}S, ^{59}Fe, ^{95}Nb, ^{103}Ru, $^{129m/129}$Te, ^{124}Sb, ^{147}Nd, ^{160}Tb, ^{192}Ir, ^{203}Hg	^{45}Ca, ^{110m}Ag, ^{144}Ce, ^{115m}Cd, ^{182}Ta, ^{191}Os, ^{233}Pa	^{3}H, ^{14}C, ^{63}Ni, ^{85}Kr, ^{93}Zr, ^{106}Ru, ^{125}Sb, ^{129}I, ^{134}Cs, ^{147}Pm, ^{152}Eu, ^{154}Eu, ^{155}Eu, ^{210}Pb, ^{227}Ac, ^{228}Ra, ^{228}Th + D.P., ^{235}U + D.P., ^{238}U/^{234}Th
^{47}Sc, ^{67}Cu, ^{76}As, ^{77}As, ^{82}Br, ^{99}Mo, ^{121}Sn, ^{131}I, ^{133}Xe, ^{151}Pm, ^{161}Tb, ^{166}Ho, ^{169}Er, ^{175}Yb, ^{177}Lu, ^{199}Au, ^{222}Rn + D.P.	^{46}Sc, ^{59}Fe, ^{95}Zr, ^{103}Ru, ^{124}Sb, ^{126}I, $^{129m/129}$Te, ^{141}Ce, ^{147}Nd, ^{181}Hf, ^{185}W	^{91}Y, ^{110m}Ag, ^{123}Sn, ^{144}Ce, ^{115m}Cd, ^{182}Ta, ^{140}Ba, ^{160}Tb	^{60}Co, ^{87}Rb, ^{99}Tc, ^{125}Sb, ^{134}Cs, ^{152}Eu, ^{154}Eu, ^{226}Ra + D.P., ^{227}Ac + D.P., ^{228}Ra + D.P., ^{228}Th + D.P., ^{235}U + D.P.
^{47}Ca, ^{47}Sc, ^{67}Cu, ^{76}As, ^{77}As, ^{105}Rh, ^{111}Ag, ^{115}Cd, ^{131}I, ^{143}Ce, ^{151}Pm, ^{153}Sm, ^{161}Tb, ^{193}Os, ^{222}Rn + D.P.	^{86}Rb, $^{114m/114}$In, ^{115m}Cd, ^{124}Sb, $^{129m/129}$Te, ^{141}Ce, ^{160}Tb, ^{233}Pa	^{110m}Ag, ^{182}Ta, ^{140}Ba, ^{192}Ir	^{85}Kr, ^{90}Sr, ^{94}Nb, ^{125}Sb, ^{134}Cs, ^{137}Cs, ^{152}Eu, ^{154}Eu, ^{226}Ra + D.P., ^{228}Ra/^{228}Ac, ^{228}Th + D.P.
^{99}Mo, ^{111}Ag, ^{115}Cd, ^{122}Sb, ^{131}I, ^{143}Ce, ^{149}Pm, ^{151}Pm, ^{153}Sm, ^{186}Re, ^{193}Os, ^{198}Au, ^{222}Rn + D.P.	^{74}As, ^{84}Rb, ^{103}Ru, ^{124}Sb, $^{129m/129}$Te, ^{143}Pr, ^{147}Nd, ^{192}Ir	^{95}Zr, ^{127m}Te, ^{126}I, ^{170}Tm, ^{140}Ba, ^{160}Tb, ^{144}Ce/^{144}Pr	^{36}Cl, ^{152}Eu, ^{154}Eu, ^{204}Tl, ^{226}Ra + D.P., ^{227}Ac + D.P., ^{228}Ra/^{228}Ac, ^{228}Th + D.P.
^{47}Ca, ^{76}As, ^{99}Mo, ^{111}Ag, ^{115}Cd, ^{122}Sb, ^{140}La, ^{143}Ce, ^{149}Pm, ^{151}Pm, ^{186}Re, ^{193}Os, ^{210}Bi, ^{222}Rn + D.P.	^{74}As, ^{89}Sr, ^{126}I, $^{129m/129}$Te, ^{140}Ba/^{140}La	^{95}Zr, ^{123}Sn, ^{182}Ta	^{40}K, ^{137}Cs, ^{152}Eu, ^{154}Eu, ^{226}Ra + D.P., ^{227}Ac + D.P., ^{228}Ra/^{228}Ac, ^{228}Th + D.P., ^{238}U/^{234m}Pa
^{47}Ca, ^{76}As, ^{122}Sb, ^{140}La, ^{166}Ho, ^{222}Rn + D.P.	^{32}P, ^{59}Fe, ^{86}Rb, ^{91}Y, $^{114m/114}$In, ^{115m}Cd, ^{124}Sb, $^{129m/129}$Te, ^{140}Ba/^{140}La, ^{160}Tb	^{182}Ta	^{60}Co, ^{106}Ru/^{106}Rh, ^{154}Eu, ^{226}Ra + D.P., ^{228}Ra/^{228}Ac, ^{228}Th + D.P.
^{76}As, ^{90}Y, ^{140}La, ^{222}Rn + D.P.	^{124}Sb, ^{140}Ba/^{140}La	$^{110m/110}$Ag, ^{144}Ce/^{144}Pr	^{90}Sr/^{90}Y, ^{106}Ru/^{106}Rh, ^{226}Ra + D.P., ^{228}Th + D.P., ^{228}Ra/^{228}Ac, ^{238}U/^{234m}Pa
^{222}Rn + D.P.			^{106}Ru/^{106}Rh, ^{226}Ra + D.P.

Table 3.6   GAMMA ENERGIES AND HALF-LIVES

Half-life γ-energy MeV	<1 hour	1–10 hours	10 hours–1 day
0–0.3	^{27}Mg, ^{70}Ga, ^{81m}Se, ^{81}Se, ^{94m}Nb, ^{101}Mo, ^{101}Tc, ^{116m}In, ^{117}In, ^{155}Sm, ^{199}Pt	^{52}Fe, ^{61}Co, ^{75}Ge, ^{80m}Kr, ^{85m}Kr, ^{99m}Tc, ^{105}Ru, ^{117}Cd, ^{127}Te, ^{132}I, ^{134m}Cs, ^{139}Ba, ^{149}Nd, ^{152m}Eu, ^{165}Dy, ^{171}Er, ^{177}Yb, ^{180m}Hf	^{28}Mg, ^{43}K, ^{55}Co, ^{77}Ge, ^{97}Zr, ^{109}Pd, ^{123}I, ^{159}Gd, ^{187}W, ^{194}Ir, ^{197m}Hg, ^{197}Pt
0.3–0.5	^{51}Ti, ^{101}Mo, ^{101}Tc, ^{116m}In, ^{128}I, ^{155}Sm, ^{199}Pt	^{65}Ni, ^{75}Ge, ^{85m}Kr, ^{87m}Sr, ^{97}Nb, ^{105}Ru, ^{117}Cd, ^{127}Te, ^{132}I, ^{149}Nd, ^{152m}Eu, ^{165}Dy, ^{171}Er, ^{180m}Hf	^{28}Mg, ^{42}K, ^{43}K, ^{55}Co, ^{69m}Zn, ^{77}Ge, ^{97}Zr, ^{109}Pd, ^{123}I, ^{130}I, ^{159}Gd, ^{187}W, ^{194}Ir
0.5–0.7	^{11}C, ^{13}N, ^{15}O, ^{51}Ti, ^{80}Br, ^{81}Se, ^{101}Mo, ^{101}Tc, ^{104}Rh, ^{108}Ag, ^{117}In, ^{128}I, ^{144}Pr, ^{155}Sm, ^{199}Pt	^{18}F, ^{44}Sc, ^{45}Ti, ^{43}K, ^{52}Fe, ^{65}Ni, ^{66}Ga, ^{72}Ga, ^{68}Ga, ^{75}Ge, ^{97}Nb, ^{109}Pd, ^{105}Ru, ^{117}Cd, ^{132}I, ^{142}Pr, ^{149}Nd, ^{152m}Eu, ^{165}Dy, ^{194}Ir, ^{171}Er, ^{180m}Hf	^{55}Co, ^{64}Cu, ^{77}Ge, ^{97}Zr, ^{123}I, ^{130}I, ^{159}Gd, ^{187}W

(Including isotopes giving 0.51 MeV gamma rays from annihilation of positrons)

Half-life γ-energy MeV	<1 hour	1–10 hours	10 hours–1 day
0.7–1.0	^{27}Mg, ^{51}Ti, ^{66}Cu, ^{80}Br, ^{81}Se, ^{88}Rb, ^{101}Mo, ^{108}Ag, ^{116m}In, ^{128}I, ^{155}Sm, ^{199}Pt	^{45}Ti, ^{56}Mn, ^{61}Co, ^{65}Ni, ^{66}Ga, ^{105}Ru, ^{117}Cd, ^{132}I, ^{149}Nd, ^{152m}Eu, ^{165}Dy, ^{171}Er, ^{177}Yb	^{28}Mg, ^{42}K, ^{43}K, ^{55}Co, ^{72}Ga, ^{77}Ge, ^{97}Zr/^{97m}Nb, ^{123}I, ^{130}I, ^{109}Pd, ^{187}W, ^{194}Ir
1.0–1.5	^{27}Mg, ^{49}Ca, ^{52}V, ^{66}Cu, ^{70}Ga, ^{80}Br, ^{88}Rb, ^{101}Mo, ^{108}Ag, ^{116m}In, ^{128}I, ^{144}Pr, ^{155}Sm, ^{199}Pt	^{31}Si, ^{41}Ar, ^{44}Sc, ^{45}Ti, ^{65}Ni, ^{66}Ga, ^{68}Ga, ^{97}Nb, ^{105}Ru, ^{117}Cd, ^{132}I, ^{139}Ba, ^{149}Nd, ^{152m}Eu, ^{165}Dy, ^{177}Yb	^{24}Na, ^{28}Mg, ^{43}K, ^{55}Co, ^{64}Cu, ^{72}Ga, ^{77}Ge, ^{97}Zr, ^{130}I, ^{194}Ir
1.5–2.0	^{28}Al, ^{38}C, ^{52}V, ^{88}Rb, ^{101}Mo, ^{116m}In	^{41}Ar, ^{56}Mn, ^{65}Ni, ^{66}Ga, ^{68}Ga, ^{97}Nb, ^{105}Ru, ^{117}Cd, ^{132}I, ^{139}Ba	^{28}Mg/^{28}Al, ^{72}Ga, ^{142}Pr, ^{77}Ge, ^{194}Ir, ^{42}K, ^{97}Zr
2.0–3.0	^{38}Cl, ^{49}Ca, ^{88}Rb, ^{101}Mo, ^{116m}In, ^{144}Pr	^{44}Sc, ^{56}Mn, ^{66}Ga, ^{117}Cd, ^{132}I	^{24}Na, ^{72}Ga, ^{42}K, ^{77}Ge, ^{55}Co, ^{194}Ir
>3.0	^{49}Ca, ^{88}Rb	^{56}Mn, ^{66}Ga	

1–10 days	*10–100 days*	*100 days–1 year*	*>1 year*
^{47}Sc, ^{67}Cu, ^{67}Ga, ^{77}As, ^{77}Br, ^{82}Br, ^{97}Ru, ^{99}Mo, ^{105}Rh, ^{111}Ag, ^{111}In, ^{115}Co, ^{131}I, ^{132}Te, ^{133m}Ba, ^{133m}Xe, ^{133}Xe, ^{140}La, ^{143}Ce, ^{149}Pm, ^{151}Pm, ^{153}Sm, ^{161}Tb, ^{166}Ho, ^{174}Yb, ^{177}Lu, ^{186}Re, ^{193}Os, ^{197}Hg, ^{199}Au, ^{200}Bi, ^{222}Rn + D.P., ^{224}Ra, ^{225}Ac	^{59}Fe, ^{73}As, ^{83}Rb, ^{103}Pd/^{103m}Rh, ^{103}Ru/^{103m}Rh, ^{114m}In, ^{125}I, ^{105}Ag, ^{129m}/^{129}Te, ^{131m}Xe, ^{140}Ba, ^{131}Ba, ^{147}Nd, ^{160}Tb, ^{141}Ce, ^{175}Hf, ^{181}Hf, ^{169}Yb, ^{185}Os, ^{191}Os, ^{183}Re, ^{203}Hg, ^{223}Ra, ^{192}Ir, ^{233}Pa	^{57}Co, ^{75}Se, ^{110m}Ag, ^{113}Sn, ^{119m}Sn, ^{127m}Te, ^{139}Ce, ^{144}Ce, ^{153}Gd, ^{170}Tm, ^{181}W, ^{182}Ta, ^{195}Au	^{44}Ti, ^{109}Cd, ^{125}Sb, ^{129}I, ^{133}Ba, ^{152}Eu, ^{154}Eu, ^{155}Eu, ^{208}Po, ^{210}Pb, ^{226}Ra + D.P., ^{227}Ac + D.P., ^{228}Ra/^{228}Ac, ^{228}Th + D.P., ^{230}Th, ^{231}Pa, ^{232}U, ^{233}U, ^{235}U + D.P., ^{237}Np, ^{238}U/^{234}Th, ^{239}Pu, ^{241}Am
^{47}Ca, ^{67}Cu, ^{67}Ga, ^{77}Br, ^{87}Y, ^{97}Ru, ^{99}Mo, ^{105}Rh, ^{111}Ag, ^{115}Cd, ^{131}I, ^{132}Cs, ^{140}La, ^{143}Ce, ^{151}Pm, ^{153}Sm, ^{161}Tb, ^{175}Yb, ^{177}Lu, ^{193}Os, ^{198}Au, ^{206}Bi, ^{222}Rn + D.P., ^{224}Ra, ^{225}Ac	^{7}Be, ^{51}Cr, ^{85}Sr, ^{103}Pd, ^{103}Ru, ^{105}Ag, ^{115}Cd, ^{124}Sb, ^{126}I, ^{129m}Te/^{129}Te, ^{140}Ba/^{140}La, ^{131}Ba, ^{160}Tb, ^{169}Yb, ^{147}Nd, ^{181}Hf, ^{183}Re, ^{175}Hf, ^{223}Ra, ^{233}Pa, ^{192}Ir	^{75}Se, ^{106}Ru/^{106}Rh, ^{110m}Ag, ^{113}Sn/^{113m}In	^{125}Sb, ^{133}Ba, ^{134}Cs, ^{152}Eu, ^{154}Eu, ^{207}Bi, ^{226}Ra + D.P., ^{227}Ac + D.P., ^{228}Ra/^{228}Ac, ^{228}Th + D.P., ^{231}Pa, ^{232}U, ^{233}U, ^{235}U + D.P., ^{239}Pu
^{52}Mn, ^{72}As, ^{76}As, ^{77}As, ^{77}Br, ^{82}Br, ^{87}Y, ^{97}Ru, ^{99}Mo, ^{111}Ag, ^{115}Cd, ^{122}Sb, ^{124}I, ^{131}I, ^{132}Cs, ^{143}Ce, ^{149}Pm, ^{151}Pm, ^{153}Sm, ^{161}Tb, ^{166}Ha, ^{186}Re, ^{193}Os, ^{198}Au, ^{206}Bi, ^{222}Rn + D.P., ^{224}Ra, ^{225}Ac	^{48}V, ^{56}Co, ^{58}Co, ^{74}As, ^{83}Rb, ^{84}Rb, ^{85}Sr, ^{103}Ru, ^{105}Ag, ^{114m}In, ^{124}Sb, ^{126}I, ^{129m}Te/^{129}Te, ^{131}Ba, ^{140}Ba, ^{147}Nd, ^{160}Tb, ^{181}Hf, ^{185}Os, ^{192}Ir, ^{223}Ra	^{57}Co, ^{65}Zn, ^{88}Y, ^{106}Ru/^{106}Rh, ^{110m}Ag, ^{127m}Te, ^{144}Ce/^{144}Pr	^{22}Na, ^{26}Al, ^{85}Kr, ^{94}Nb, ^{125}Sb, ^{134}Cs, ^{137}Cs/^{137m}Ba, ^{152}Eu, ^{154}Eu, ^{207}Bi, ^{208}Po, ^{226}Ra + D.P., ^{227}Ac + D.P., ^{228}Ra/^{228}Ac, ^{228}Th + D.P., ^{238}U/^{234m}Pa
^{47}Ca, ^{52}Mn, ^{67}Ga, ^{72}As, ^{76}As, ^{77}Br, ^{82}Br, ^{97}Ru, ^{99}Mo, ^{111}Ag, ^{124}I, ^{131}I, ^{140}La, ^{143}Ce, ^{149}Pm, ^{151}Pm, ^{166}Ho, ^{186}Re, ^{193}Os, ^{206}Bi, ^{222}Rb + D.P.	^{46}Sc, ^{48}V, ^{56}Co, ^{58}Co, ^{83}Rb, ^{84}Rb, ^{89}Sr, ^{95}Nb, ^{95}Zr, ^{105}Ag, ^{114m}In, ^{115m}Cd, ^{124}Sb, ^{126}I, ^{129m}Te/^{129}Te, ^{131}Ba, ^{140}Ba/^{140}La, ^{160}Tb, ^{185}Os, ^{192}Ir	^{54}Mn, ^{88}Y, ^{106}Ru/^{106}Rh, ^{110m}Ag, ^{182}Ta, ^{210}Po	^{94}Nb, ^{134}Cs, ^{152}Eu, ^{154}Eu, ^{207}Bi, ^{226}Ra + D.P., ^{227}Ac + D.P., ^{228}Ra/^{228}Ac, ^{228}Th + D.P., ^{238}U/^{234m}Pa
^{47}Ca, ^{52}Mn, ^{72}As, ^{76}As, ^{77}Br, ^{82}Br, ^{122}Sb, ^{124}I, ^{132}Cs, ^{143}Ce, ^{166}Ho, ^{198}Au, ^{206}Bi, ^{222}Rn + D.P.	^{46}Sc, ^{48}V, ^{56}Co, ^{59}Fe, ^{74}As, ^{84}Rb, ^{86}Rb, ^{91}Y, ^{105}Ag, ^{115m}Cd, ^{124}Sb, ^{126}I, ^{129m}Te/^{129}Te, ^{131}Ba, ^{160}Tb	^{65}Zn, ^{106}Ru/^{106}Rh, ^{110m}Ag, ^{123}Sn, ^{144}Ce/^{144}Pr, ^{182}Ta	^{22}Na, ^{26}Al, ^{40}K, ^{60}Co, ^{134}Cs, ^{152}Eu, ^{154}Eu, ^{207}Bi, ^{226}Ra + D.P., ^{227}Ac + D.P., ^{228}Ra/^{228}Ac, ^{228}Th + D.P., ^{238}U/^{234m}Pa
^{72}As, ^{76}As, ^{82}Br, ^{124}I, ^{132}Cs, ^{140}La, ^{166}Ho, ^{206}Bi, ^{222}Rn + D.P.	^{56}Co, ^{58}Co, ^{84}Rb, ^{124}Sb, ^{140}Ba/^{140}La	^{88}Y, ^{106}Ru/^{106}Rh, ^{110m}Ag	^{26}Al, ^{152}Eu, ^{154}Eu, ^{207}Bi, ^{226}Ra + D.P., ^{228}Ra/^{228}Ac, ^{228}Th + D.P., ^{238}U/^{234m}Pa
^{72}As, ^{76}As, ^{124}I, ^{140}La, ^{206}Bi, ^{222}Rn + D.P., ^{72}As	^{48}V, ^{56}Co, ^{124}Sb, ^{140}Ba/^{140}La, ^{56}Co	^{88}Y, ^{106}Ru/^{106}Rh, ^{144}Ce/^{144}Pr	^{26}Al, ^{152}Eu, ^{154}Eu, ^{226}Ra + D.P., ^{228}Th + D.P.

Table 3.7   NUCLIDES FOR ALPHA SOURCES

Nuclide	Half-life	α-energies MeV	Associated β and γ radiation MeV
Americium-241	432.2 yr	5.44, 5.48	γ_{max} 0.060
Lead-210 (+daughters)	22.3 yr	5.305	β_{max} 1.17
			γ_{max} 0.80 (very weak)
Plutonium-238	87.74 yr	5.352, 5.452, 5.495	γ 0.096 (very weak)
Plutonium-239	2.41×10^4 yr	5.096, 5.134, 5.147	γ_{max} 0.451 (weak)
Polonium-210	138.4 d	5.305	γ 0.80 (very weak)
Radium-226 (+daughters)	1 600 yr	4.589–7.68	β_{max} 3.26
			γ_{max} 2.43

Table 3.8   NUCLIDES FOR BETA SOURCES

Nuclide	Half-life	β_{max} MeV	Associated α and γ radiation MeV
Carbon-14	5 730 yr	0.159	—
Cerium-144 + praseodymium-144	284.3 d	2.98	γ 0.034–2.19
Iron-55	2.7 yr	0.0052	γ 0.0059–0.0065
Krypton-85	10.72 yr	0.67	γ 0.51
Lead-210 + bismuth-210	22.3 yr	1.17	α 5.30
			γ_{max} 0.80 (very weak)
Nickel-63	96 yr	0.066	—
Promethium-147	2.62 yr	0.225	—
Ruthenium-106 + rhodium-106	368.2 d	3.6	γ 0.43–2.41
Strontium-90 + yttrium-90	29.12 yr	2.27	—
Thallium-204	3.78 yr	0.77	—
Tritium	12.35 yr	0.018	—
Yttrium-90	64.0 h	2.27	—

Table 3.9   NUCLIDES FOR NEUTRON SOURCES—POLONIUM-210 (ALPHA, N) SOURCES WITH VARIOUS TARGETS

Target	Neutrons s^{-1} TBq^{-1}	Neutron energy (MeV) Mean	Neutron energy (MeV) Maximum
Aluminium	5.4×10^5	—	2.7
Beryllium	6.8×10^7	4.3	10.8
Boron	5.4×10^6	—	5.0
Fluorine-19	2.7×10^6	1.4	2.8
Lithium	1.4×10^6	0.48	1.32
Magnesium	8.1×10^5	—	—
Oxygen-18	2.7×10^7	—	4.3
Sodium	1.1×10^6	—	—
Mock fission	1.1×10^6	1.6	10.8

Table 3.10   NUCLIDES FOR NEUTRON SOURCES—(GAMMA, N) SOURCES

Nuclide	Half-life	Target	Observed neutron s^{-1} TBq^{-1}	Observed neutron energy keV
Antimony-124	60.20 d	Beryllium	4.3×10^7	24.8
Radium-226 (+daughters)	1 600 yr	Beryllium	3.5×10^7	700 (max)
Radium-226 (+daughters)	1 600 yr	Deuterium (heavy water)	—	120
Thorium-228 (+daughters)	1.91 yr	Beryllium	—	827
Thorium-228 (+daughters)	1.91 yr	Deuterium (heavy water)	3.2×10^7	197

Table 3.11   NUCLIDES FOR NEUTRON SOURCES—(ALPHA, N) SOURCES WITH BERYLLIUM TARGETS

Nuclide	Half-life	Neutrons s^{-1} TBq^{-1}	Gamma emission μGy h^{-1} *at* 1 m *from* 10^6 neutrons s^{-1}
Actinium-227 (+daughters)	21.77 yr	4.9×10^8	80
Americium-241	432.2 yr	5.9×10^7	11.4
Lead-210 (+daughters)	22.3 yr	6.2×10^7	88
Plutonium-239	2.41×10^4 yr	3.8×10^7	17
Polonium-210	138.4 d	6.8×10^7	0.4
Radium-226 (+daughters)	1 600 yr	4.2×10^8	600
Thorium-228 (+daughters)	1.91 yr	6.8×10^8	300

Table 3.12   SPONTANEOUS FISSION NEUTRON SOURCE

Nuclide	Half-life	Neutrons s^{-1} mg^{-1}	Gamma emission μGy h^{-1} *at* 1 m *from* 10^6 neutrons s^{-1}
Californium-252	2.64 yr (effective)	2.3×10^9	0.7

REFERENCES

'Radionuclide Transformations, Energy and Intensity of Emissions', ICRP Publication 38, Pergamon, Oxford, 1983.
'Radioactive Decay Data Tables', DOE/TIC-11026, Technical Information Center, U.S. Department of Energy, 1981.

4 X-ray analysis of metallic materials

4.1 Introduction and cross references

X-rays are very short wavelength electromagnetic waves. Their range encompasses the interatomic distances in crystalline materials, typically 0.5 to 2.5 Å, which permit the analysis of crystalline and to a lesser degree amorphous materials by X-ray diffraction (XRD). When atoms are irradiated with radiation of sufficient energy, they emit characteristic X-ray spectra, fluorescence, which are analysed in X-ray fluorescent analysis (XRF) to give element analysis. Individual electrons are emitted from an irradiated surface with energies related to the electron levels in the atom. The energies of these electrons are analysed in X-ray photoelectron spectroscopy (XPS) to provide an elemental analysis of surface atoms together with information on the chemical bonding between them.

This chapter reviews the applications of XRD to the investigation of metallic materials. The subject is well covered in detail in several standard textbooks including 'X-ray Metallography' by A. Taylor;[1] 'Structure of Metals' by C. S. Barrett and T. B. Massalski;[2] 'X-ray Diffraction Procedures' by H. P. Klug and L. E. Alexander.[3] 'Elements of X-ray Crystallography' by L. V. Azaroff,[4] 'Structure Determination by X-ray Crystallography' by M. F. C. Ladd and R. A. Palmer,[5] 'An Introduction to X-ray Crystallography' by M. M. Woolfson,[6] 'Elements of X-ray Diffraction' by B. D. Cullity and S. R. Stock,[7] and 'The Basics of Crystallography and Diffraction' by C. Hammond.[8]

Chapter 5 discusses crystallography and crystal structure data can be found in Chapter 6.

4.2 Excitation of X-rays

X-rays are produced (excited) when an electron experiences a sudden acceleration or (more commonly in practice) deceleration as, for example, when a beam of fast-moving electrons strikes an atom. The frequency v or wavelength λ of the radiation produced as a result of energy loss E is given by the equation

$$E = hv = \frac{hc}{\lambda}$$

The maximum value of E, E_{max}, and corresponding smallest value of λ occurs when all the electron energy is lost on impact, i.e.

$$E_{max} = Ve = \frac{hc}{\lambda_{swl}}$$

where $V =$ accelerating voltage and $\lambda_{swl} =$ short wavelength limit or cut-off wavelength.
Substitution of numerical values gives

$$\lambda_{swl} = \frac{12.34}{V}$$

where λ_{swl} is expressed in Ångstrom units and V in kilovolts.

However, most electrons undergo repeated collisions of varying energy losses until finally stopped giving rise to a whole range of λ values – the so-called 'white' or 'Brehmstrallung' radiation.

If the intensity of the radiation is plotted against the wavelength, for relatively low applied voltages a curve is obtained rising from zero to a more or less pronounced maximum at a wavelength of about 4/3 λ_{swl}, falling again to a low value with increasing wavelength. It has been found that for this curve of so-called 'white radiation' both the total radiation and the intensity of the peak are closely proportional to the atomic number of the target material. Thus if white radiation is required a target of one of the heavy metals should be selected, and since the total intensity (the area under the curve) is proportional to the square of the applied voltage the latter should be as high as possible.

On increasing the voltage beyond a certain limit, which is different for different targets, a line spectrum begins to appear, superimposed upon the continuous or white spectrum, the wavelengths of the lines being constant and characteristic of the target. The lowest critical voltage at which these lines appear is that which endows the bombarding electrons with just sufficient energy to eject electrons from one of the inner shells in the target atoms. Each vacancy as it occurs is filled by an electron jumping in from one of the outer shells, the jump being accompanied by an emission of energy of frequency v given by the Bohr relationship

$$hv = W_1 - W_2$$

where $W_1 - W_2$ represents the change in potential energy of the system and h is Planck's constant.

The K series of lines is excited when electrons are ejected from the K shell and their places taken by electrons from the L, M or N shells. It consists principally of five lines $K\alpha_1$, $K\alpha_2$, $K\beta_1$, $K\beta_3$ and $K\beta_2$, the α lines being associated with jumps from the L shell, β_1 and β_3 with those from the M shell, and β_2 with those for the N shell. Other series of lines are emitted when the bombarding electrons eject electrons from the L, M or N shells. The K emissions which are the most intense are generally used in XRD while all the spectra are employed in XRF.

If V_0 is the critical voltage at which the K spectra first appear and e is the electron charge, then the energy needed to remove an electron from the K shell is V_0e which is equivalent to radiation of frequency v_k and wavelength λ_k given by

$$V_0e = hv_k = \frac{hc}{\lambda_k}$$

where c is the velocity of light. This critical wavelength λ_k, termed the K absorption edge of the material, together with slightly shorter wavelengths are heavily absorbed in the material. In XRD, to avoid fluorescence (high background) and low penetration into the sample a radiation is chosen which is on the long wavelength side of any absorption edge of the sample.

The empirical expression

$$I = K(V - V_0)^{1.7}$$

where K is a constant, relates the intensity, I, of the characteristic radiation to the difference between the tube voltage V and the critical voltage V_0 of the target material. It is found that the relative intensity of the characteristic radiation to that of the continuous wavelength (white) radiation is greatest for tube voltages of about four times the critical voltage. A method of achieving partial monochromatisation of the X-ray beam is to reduce the intensity of the K_β radiation with respect to that of the K_α by using a filter of an element with K absorption edge between the K_β and K_α wavelengths. The thickness of the filter is usually chosen to reduce the intensity of K_β to about 1/600 of the transmitted K_α. Generally, the element preceding the target element in the periodic table will act as a K_β filter. Filters cannot separate the α_1 and α_2 components and where they are not resolved in a diffraction pattern, d values are calculated using a weighted average wavelength where α_1, the stronger of the two, is given twice the weight of α_2. Full monochromatisation can be achieved with a single crystal monochromator oriented such that only $K\alpha_1$ component is diffracted. Monochromators in the form of a single crystal lamina bent to produce focusing are widely used in diffractometry to remove all unwanted radiation. Construction of focusing monochromators is described in 'International Tables for X-ray Crystallography'.[9]

Table 4.1 lists wavelengths, minimum excitation potentials and K_β filters for targets frequently used in XRD. Table 4.2 gives minimum excitation potentials for characteristic K, L, M, and N spectra while Tables 4.3 and 4.4 give K and L emission spectra and L_{III} absorption edges for all elements. Crystals suitable for monochromators are listed with comments on their performance in Table 4.5.

4.2.1   X-ray wavelengths

The X-unit defined by Siegbahn and very nearly equal to 10^{-3} Å was based on the atomic spacing in calcite as determined from the expression:

$$2d = \left(\frac{4M}{\rho N V}\right)^{1/3}$$

where M = molecular weight of calcite, ρ = density of calcite, N = Avogadro's number, V = volume of calcite rhombohedron with unit distance between the faces considered.

kX units were widely used up to about 1945 but were then gradually replaced by Ångström units as more precise methods for determining X-ray wavelengths were introduced. The nanometer (10^{-9} m) is now declared to be the standard unit for X-ray studies. However, in view of the much greater familiarity with Å, and the very simple conversion factor, in this edition the Ångström unit is used unless the contrary is stated.

1 kX unit = 1.002 02 Ångström units
1 Ångström unit, Å = 10^{-10} m or 10^{-1} nanometers

Emission lines are frequently quoted in keV

$$1\,\text{keV} = \frac{12.3975}{\text{Wavelength in Å}}$$

Table 4.1   WAVELENGTHS, EXCITATION POTENTIALS, AND β FILTERS FREQUENTLY USED IN CRYSTALLOGRAPHIC WORK

		Target				β-filter		Material content	
Element	Atomic No.	Line	Wavelength* Å	Excitation potential† kV	Element	Absorption edge	Mass absorption coefficient μ/ρ	g cm^{-2}	Thickness mm
Cr	24	$K\alpha_1$	2.289 62	5.98	V	2.267 5	77.3	0.009	0.016
		$K\alpha_2$	2.293 51						
		$K\alpha$	2.290 9						
		$K\beta_1$	2.084 80						
Mn	25	$K\alpha_1$	2.101 75	6.54	Cr	2.070 1	71	0.011	0.016
		$K\alpha_2$	2.105 69						
		$K\alpha$	2.103 1						
		$K\beta_1$	1.910 15						
Fe	26	$K\alpha_1$	1.935 97	7.1	Mn	1.895 4	61.9	0.012	0.016
		$K\alpha_2$	1.939 91						
		$K\alpha$	1.937 3						
		$K\beta_1$	1.756 53						
Co	27	$K\alpha_1$	1.788 92	7.71	Fe	1.742 9	58.6	0.014	0.018
		$K\alpha_2$	1.792 78						
		$K\alpha$	1.790 2						
		$K\beta_1$	1.620 75						
Ni	28	$K\alpha_1$	1.657 84	8.29	Co	1.607 2	51.6	0.015	0.018
		$K\alpha_2$	1.661 69						
		$K\alpha$	1.659 1						
		$K\beta_1$	1.500 10						
Cu	29	$K\alpha_1$	1.540 51	8.86	Ni	1.486 9	48.0	0.019	0.021
		$K\alpha_2$	1.544 33						
		$K\alpha$	1.541 8						
		$K\beta_1$	1.392 17						
W	74	$L\alpha_1$	1.476 35	12.1	Cu	1.380 2	42.0	0.019	0.021
		$L\alpha_2$	1.487 42						
Zn	30	$K\alpha_1$	1.435 11	9.65	Cu	1.380 2	42.0	0.019	0.021
		$K\alpha_2$	1.438 94						
		$K\alpha$	1.436 4						
		$K\beta_1$	1.295 22						
Au	79	$L\alpha_1$	1.276 39	14.4	Ga	1.195 7	37.0	0.028	0.047
		$L\alpha_2$	1.287 77						
		$K\alpha_2$	0.785 88						
		$K\alpha_2$	0.790 10						
		$K\alpha$	0.787 29	80.5	Sr	0.769 69	18.1	0.053	0.210

(continued)

Table 4.1 WAVELENGTHS, EXCITATION POTENTIALS, AND β FILTERS FREQUENTLY USED IN CRYSTALLOGRAPHIC WORK—*continued*

	Target					β-filter			
							Mass absorption coefficient μ/ρ	Material content	
Element	Atomic No.	Line	Wavelength* Å	Excitation potential[†] kV	Element	Absorption edge		g cm^{-2}	Thickness mm
Zr	40	Kβ	0.701 70						
Mo	42	Kα_1	0.709 26	20	Zr	0.688 8	17.2	0.069	0.108
		Kα_2	0.713 54						
		Kα	0.710 7						
		Kβ_1	0.632 25						
Rh	45	Kα_1	0.613 25	23.2	Ru	0.559 5	15.4	0.077	0.064
		Kα_2	0.617 61						
		Kα	0.614 7						
		Kβ_1	0.545 59						
Pd	46	Kα_1	0.585 42	24.4	Rh	0.534 1	14.6	0.091	0.073
		Kα_2	0.589 80		or				
		Kα	0.586 9		Ru	0.559 5			
		Kβ_1	0.520 52						
Ag	47	Kα_1	0.559 36	25.5	Pd	0.509 0	13.1	0.096	0.079
		Kα_2	0.563 78		or				
		Kα	0.560 9		Rh	0.534 1			
		Kβ_1	0.497 01						

* $\lambda K\alpha$ is here defined as $(2\lambda K\alpha_1 + \lambda K\alpha_2)/3$.
[†] The optimum voltage for operating a tube with raw alternating currents is approximately 5 times the excitation potential, and 4 times the excitation potential with fully smoothed direct current, but normally 80 kV cannot be exceeded owing to the danger of electrical breakdown.

Table 4.2 EXCITATION POTENTIALS IN kV FOR CHARACTERISTIC X-RAY SPECTRA

Atomic No. and element	K	L	M	N	Atomic No. and element	K	L	M	N
11 Na	1.07	—	—	—	49 In	27.80	4.21	0.83	0.12
12 Mg	1.30	—	—	—	50 Sn	29.06	4.42	0.88	0.13
13 Al	1.55	—	—	—	51 Sb	30.35	4.69	0.94	0.15
14 Si	1.83	—	—	—	52 Te	31.66	4.93	1.01	0.17
15 P	2.13	—	—	—	53 I	33.01	5.16	1.08	0.19
16 S	2.46	—	—	—	55 Cs	35.80	5.68	1.21	0.23
17 Cl	2.81	0.24	—	—	56 Ba	37.24	5.92	1.29	0.25
19 K	3.69	0.34	—	—	57 La	38.75	6.24	1.36	0.27
20 Ca	4.02	0.40	—	—	58 Ce	40.27	6.53	1.43	0.29
21 Sc	4.48	0.46	—	—	59 Pr	41.81	6.81	1.51	0.30
22 Ti	4.94	0.53	—	—	60 Nd	43.37	7.10	1.58	0.32
23 V	5.44	0.60	—	—	62 Sm	46.63	7.71	1.72	0.35
24 Cr	5.96	0.68	—	—	63 Eu	48.29	8.02	1.80	0.36
25 Mn	6.51	0.76	—	—	64 Gd	50.00	8.35	1.88	0.38
26 Fe	7.08	0.85	—	—	65 Tb	51.76	8.67	1.96	0.40
27 Co	7.67	0.92	—	—	66 Dy	53.55	9.01	2.04	0.42
28 Ni	8.29	1.01	—	—	67 Ho	55.36	9.35	2.13	0.43
29 Cu	8.94	1.10	—	—	68 Er	57.22	9.71	2.22	0.45
30 Zn	9.62	1.19	—	—	69 Tm	59.07	10.06	2.31	0.47
31 Ga	10.32	1.29	—	—	70 Yb	61.02	10.45	2.41	0.50
32 Ge	11.05	1.41	—	—	71 Lu	63.01	10.82	2.50	0.51
33 As	11.81	1.52	—	—	72 Hf	65.01	11.23	2.60	0.54
34 Se	12.59	1.65	—	—	73 Ta	67.09	11.63	2.69	0.57
35 Br	13.41	1.78	—	—	74 W	69.18	12.04	2.80	0.59
37 Rb	15.13	2.05	—	—	75 Re	71.28	12.46	2.91	—

(continued)

Table 4.2 EXCITATION POTENTIALS IN kV FOR CHARACTERISTIC X-RAY SPECTRA—*continued*

Atomic No. and element	K	L	M	N	Atomic No. and element	K	L	M	N
38 Sr	16.03	2.20	—	—	76 Os	73.54	12.91	3.03	0.64
39 Y	16.96	2.38	—	—	77 Ir	75.77	13.35	3.15	0.67
40 Zr	17.92	2.52	0.43	0.05	78 Pt	78.02	13.80	3.28	0.71
41 Nb	18.90	2.69	0.48	0.05	79 Au	80.42	14.29	3.41	0.79
42 Mo	19.91	2.86	0.51	0.06	80 Hg	82.69	14.77	3.55	0.82
44 Ru	22.02	3.21	0.59	0.06	81 Tl	85.28	15.27	3.69	0.86
45 Rh	23.12	3.39	0.62	0.07	82 Pb	87.66	15.79	3.84	0.89
46 Pd	24.23	3.60	0.67	0.08	83 Bi	90.03	16.31	3.99	0.96
47 Ag	25.40	3.81	0.72	0.10	90 Th	109.27	20.36	5.17	1.33
48 Cd	26.59	4.00	0.77	0.11	92 U	115.54	21.66	5.54	1.44

Table 4.3 K EMISSION LINES AND K ABSORPTION EDGES IN Å

Line transition intensity rel. to $K\alpha_1$	α_2 KL_{11} 50	α_1 KL_{111} 100	β_3 KM_{11} 15	β_1 KM_{111} 15	β_2 $KN_{11,111}$ 5	K Absorption edge
3 Li	228.0		—	—	—	226.5
4 Be	114.0		—	—	—	—
5 B	67.6		—	—	—	—
6 C	44.7		—	—	—	43.7
7 N	31.6		—	—	—	31.0
8 O	23.62		—	—	—	23.3
9 F	18.32		—	—	—	—
10 Ne	14.61		—	14.452	—	—
11 Na	11.910 1		—	11.575	—	—
12 Mg	9.890 0		—	9.521	—	9.511 7
13 Al	8.341 73	8.339 34	—	7.960	—	7.951 1
14 Si	7.127 91	7.125 42	—	6.753	—	6.744 6
15 P	6.160	6.157	—	5.796	—	5.786 6
16 S	5.374 96	5.372 16	—	5.032	—	5.018 2
17 Cl	4.730 7	4.727 8	—	4.403 4	—	4.396 9
18 A	4.194 74	4.191 80	—	3.886 0	—	3.870 7
19 K	3.744 5	3.741 4	—	3.453 9	—	3.436 45
20 Ca	3.361 66	3.358 39	—	3.089 7	—	3.070 16
21 Sc	3.034 2	3.030 9	—	2.779 6	—	2.757 3
22 Ti	2.752 16	2.748 51	—	2.513 91	—	2.497 3
23 V	2.507 38	2.503 56	—	2.284 40	—	2.269 0
24 Cr	2.293 61	2.289 70	—	2.084 87	—	2.070 1
25 Mn	2.105 78	2.101 82	—	1.910 21	—	1.896 4
26 Fe	1.939 98	1.936 04	—	1.756 61	1.744 2	1.743 3
27 Co	1.792 85	1.788 97	—	1.620 79	1.608 9	1.608 1
28 Ni	1.661 75	1.657 91	—	1.500 14	1.488 6	1.488 0
29 Cu	1.544 39	1.540 56	1.392 6	1.392 22	1.381 09	1.380 4
30 Zn	1.439 00	1.435 16	—	1.295 25	1.283 72	1.283 3
31 Ga	1.343 99	1.340 08	1.208 35	1.207 89	1.196 00	1.195 7
32 Ge	1.258 01	1.254 05	1.129 36	1.128 94	1.116 86	1.116 5
33 As	1.179 87	1.175 88	1.057 83	1.057 30	1.045 00	1.045 0
34 Se	1.108 82	1.104 77	0.992 68	0.992 18	0.979 92	0.979 78
35 Br	1.043 82	1.039 74	0.933 27	0.932 79	0.920 46	0.919 95
36 Kr	0.984 1	0.980 1	0.879 0	0.878 5	0.866 1	0.865 47
37 Rb	0.929 69	0.925 553	0.829 21	0.828 68	0.816 45	0.815 49
38 Sr	0.879 43	0.875 26	0.783 45	0.782 92	0.770 81	0.769 69
39 Y	0.833 05	0.828 84	0.741 26	0.740 72	0.728 64	0.727 62
40 Zr	0.790 15	0.785 93	0.702 28	0.701 73	0.689 93	0.688 77
41 Nb	0.750 44	0.746 20	0.666 34	0.665 76	0.654 16	0.652 91
42 Mo	0.713 59	0.709 30	0.632 87	0.632 29	0.620 99	0.619 77
43 Tc	0.679 32$^+$	0.675 02$^+$	0.601 88$^+$	0.601 30$^+$	0.590 24$^+$	(0.589 1)
44 Ru	0.647 41	0.643 08	0.573 07	0.572 48	0.561 66	0.560 05

(continued)

Table 4.3 K EMISSION LINES AND K ABSORPTION EDGES IN Å—*continued*

Line transition intensity rel. to Kα_1	α_2 KL$_{11}$ 50	α_1 KL$_{111}$ 100	β_3 KM$_{11}$ 15	β_1 KM$_{111}$ 15	β_2 KN$_{11,111}$ 5	K Absorption edge
45 Rh	0.617 63	0.613 28	0.546 20	0.545 61	0.535 03	0.533 8
46 Pd	0.589 82	0.585 45	0.521 12	0.520 52	0.510 23	0.509 2
47 Ag	0.563 80	0.559 41	0.497 69	0.487 07	0.487 03	0.485 8
48 Cd	0.539 42	0.535 01	0.475 73	0.475 10	0.465 33	0.464 09
49 In	0.516 54	0.512 11	0.455 18	0.454 54	0.445 00	0.443 88
50 Sn	0.435 24	0.425 92	0.424 67	0.431 84	0.431 75	0.424 68
51 Sb	0.474 83	0.470 35	0.417 74	0.417 09	0.407 97	0.406 63
52 Te	0.455 78	0.451 30	0.400 66	0.399 99	0.391 10	0.389 72
53 I	0.437 83	0.433 32	0.384 56	0.383 91	0.375 23$^+$	0.373 79
54 Xe	0.420 87$^+$	0.416 34$^+$	0.369 41$^+$	0.368 72$^+$	0.360 26$^+$	0.358 49
55 Cs	0.404 84	0.400 29	0.355 05	0.354 36	0.346 11	0.344 74
56 Ba	0.389 67	0.385 11	0.341 51	0.340 81	0.332 77	0.331 37
57 La	0.375 31	0.370 74	0.328 69	0.327 98	0.320 12	0.318 42
58 Ce	0.361 68	0.357 09	0.316 52	0.315 82	0.308 16	0.306 47
59 Pr	0.348 75	0.344 14	0.304 98	0.304 26	0.296 79	0.295 16
60 Nd	0.336 47	0.331 85	0.294 03	0.293 30	0.286 1$^+$	0.284 51
61 Pm	0.324 80	0.320 16	0.283 63$^+$	0.282 90$^+$	0.275 9$^+$	0.274 3
62 Sm	0.313 70	0.309 04	0.273 76	0.273 01	0.266 2	0.264 62
63 Eu	0.303 12	0.298 45	0.264 33	0.263 58	0.256 92	0.255 52
64 Gd	0.293 04	0.288 35	0.255 34	0.254 60	0.248 16	0.246 80
65 Tb	0.283 42	0.278 72	0.246 83	0.246 08	0.239 7$^+$	0.238 40
66 Dy	0.274 25	0.269 53	0.238 62	0.237 88	0.231 7$^+$	0.230 46
67 Ho	0.265 486	0.260 76	0.230 83	0.230 12	0.224 1$^+$	0.222 90
68 Er	0.257 11	0.252 37	0.223 41	0.222 66	0.216 7$^+$	0.215 66
69 Tm	0.249 10	0.244 34	0.216 36	0.215 56	0.209 8$^+$	0.208 9
70 Yb	0.241 42	0.236 66	0.209 6$^+$	0.208 84	0.203 3$^+$	0.202 23
71 Lu	0.234 08	0.229 30	0.203 09$^+$	0.202 31$^+$	0.196 9$^+$	0.195 84
72 Hf	0.227 02	0.222 22	0.196 86$^+$	0.196 07	0.190 8$^+$	0.189 81
73 Ta	0.220 31	0.215 50	0.190 89	0.190 09	0.185 10	0.183 93
74 W	0.213 83	0.209 01	0.185 18	0.184 37	0.179 51	0.178 37
75 Re	0.207 61	0.202 78	0.179 70	0.178 88	0.174 15	0.173 11
76 Os	0.201 64	0.196 79	0.174 43	0.173 61	0.169 90	0.167 80
77 Ir	0.195 90	0.191 05	0.169 37	0.168 54	0.164 05	0.162 86
78 Pt	0.190 38	0.185 51	0.164 50	0.163 68	0.159 29	0.158 16
79 Au	0.185 08	0.180 20	0.159 81	0.158 98	0.154 72	0.153 44
80 Hg	0.179 96	0.175 07	0.155 32	0.154 49	0.150 30	0.149 23
81 Tl	0.175 04	0.170 14	0.150 98	0.150 14	0.146 04	0.144 70
82 Pb	0.170 29	0.165 38	0.146 81	0.145 97	0.142 01	0.140 77
83 Bi	0.165 72	0.160 79	0.142 78	0.141 95	0.138 07	0.137 06
84 Po	0.161 30$^+$	0.156 36$^+$	0.138 92$^+$	0.138 07$^+$	0.134 28$^+$	—
85 At	0.157 05$^+$	0.152 10$^+$	0.135 17$^+$	0.134 32$^+$	0.130 67$^+$	—
86 Rn	0.152 94$^+$	0.147 98$^+$	0.131 55$^+$	0.130 69$^+$	0.127 08$^+$	—
87 Fr	0.148 96$^+$	0.143 99$^+$	0.128 07$^+$	0.127 19$^+$	0.123 68	—
88 Ra	0.145 12$^+$	0.140 14$^+$	0.124 69$^+$	0.123 82$^+$	0.120 39	—
89 Ac	0.141 41$^+$	0.136 42$^+$	0.121 43$^+$	0.120 55$^+$	0.117 21$^+$	—
90 Th	0.137 83	0.132 81	0.118 27	0.117 40	0.114 15$^+$	0.112 93
91 Pa	0.134 34$^+$	0.129 33$^+$	0.115 23$^+$	0.114 35$^+$	0.111 18$^+$	—
92 U	0.130 97	0.125 95	0.112 30	0.111 39	0.108 27$^+$	0.106 80

$^+$ Interpolated values.

Table 4.4 L EMISSION SPECTRA AND ABSORPTION EDGES IN Å

Line transition intensity rel. to α1	*l* Liii Mi 30	η Lii Mi 10	α2 Liii Miv 10	α1 Liii Mv 100	β1 Lii Miv 80	β4 Li Mii 20	β3 Li Miii 30	β2 Lii Nv 60	β5 Lii Oiv,v 60	γ1 Lii Niv 40	L Absorption edges Li	L Absorption edges Lii	L Absorption edges Liii
17 Cl	67.90	67.33	—	—	—	—	—	—	—	—	52.084	61.366	61.672
18 A	56.30+	55.9+	—	—	—	—	—	—	—	—	43.192	50.390	50.803
19 K	47.74	47.24	—	—	—	—	—	—	—	—	36.352	42.020	42.452
20 Ca	40.96	40.46		36.33	—	—	—	—	—	—	31.068	35.417	35.827
21 Sc	35.59	35.13		31.35	31.02	—	—	—	—	—	26.831	30.161	30.457
22 Ti	31.36	30.89		27.42	27.05	—	—	—	—	—	23.389	26.831	27.184
23 V	27.77	27.34		24.25	23.88	—	—	—	—	—	20.523	23.702	24.070
24 Cr	24.78	24.30		21.64	21.27	—	—	—	—	—	18.256	21.226	21.596
25 Mn	22.29	21.85		19.45	19.11	—	—	—	—	—	16.268	18.896	19.248
26 Fe	20.15	19.75		17.59	17.26	—	—	—	—	—	14.601	17.169	17.484
27 Co	18.292	17.87		15.972	15.666	—	—	—	—	—	13.343	15.534	15.831
28 Ni	16.693	16.27		14.561	14.271	—	—	—	—	—	12.267	14.135	14.448
29 Cu	15.286	14.90		13.336	13.053	—	—	—	—	—	11.269	12.994	13.258
30 Zn	14.02	13.68		12.254	11.983	—	—	—	—	—	10.330	11.840	12.106
31 Ga	12.953	12.597		11.292	11.023	—	—	—	—	—	9.535	10.613	10.855
32 Ge	11.965	11.609		10.4361	10.175	—	—	—	—	—	8.729	9.965	10.228
33 As	11.072	10.734		9.6709	9.4141	—	—	—	—	—	8.107	9.1281	9.3767
34 Se	10.294	9.962		8.9900	8.7358	—	—	—	—	—	7.467	8.4212	8.6624
35 Br	9.585	9.255		8.3746	8.1251	—	—	—	—	—	6.925	7.7523	7.9871
36 Kr	—	—		7.817+	7.576+	—	—	—	—	—	6.456	7.1653	7.4227
37 Rb	8.3636	8.0415	7.3251	7.3183	7.0759	6.8207	6.7876	—	—	—	6.006	6.6538	6.8752
38 Sr	7.8362	7.5171	6.8697	6.8628	6.6239	6.4026	6.3672	—	—	—	5.604	6.1856	6.3996
39 Y	7.3563	7.0406	6.4558	6.4488	6.2120	6.0186	5.9832	—	—	—	5.1931	5.7098	5.9141
40 Zr	6.9185	6.6069	6.0778	6.0705	5.8360	5.6681	5.6330	5.5863	—	5.3843	4.8938	5.3709	5.5737
41 Nb	6.5176	6.2109	5.7319	5.7243	5.4923	5.3455	5.3102	5.2379	—	5.0361	4.5911	5.0247	5.2260
42 Mo	6.1508	5.8475	5.4144	5.4066	5.1771	5.0488	5.0133	4.9232	—	4.7258	4.3207	4.7133	4.9093
43 Tc	—	—	—	5.1148+	4.8873+	—	—	—	—	—	4.0643	4.4271	4.6254
44 Ru	5.5035	5.2050	4.8538	4.8458	4.6206	4.5230	4.4866	4.3718	—	4.1822	3.8413	4.1765	4.3663
45 Rh	5.2169	4.9217	4.6055	4.5974	4.3741	4.2888	4.2522	4.1310	—	3.9437	3.6416	3.9490	4.1389
46 Pd	4.9525	4.6605	4.3759	4.3677	3.9902	4.0711	4.0346	3.9089	—	3.7246	3.4300	3.7136	3.8969

(continued)

Table 4.4   L EMISSION SPECTRA AND ABSORPTION EDGES IN Å—*continued*

Line transition →	l	η	α_2	α_1	β_1	β_4	β_3	β_2	β_5	γ_1	\multicolumn{3}{c}{L Absorption edges}		
	Liii Mi	Lii Mi	Liii Miv	Liii Mv	Lii Miv	Li Mii	Li Miii	Liii Nv	Lii Oiv,v	Lii Niv	Li	Lii	Liii
intensity rel. to α_1	30	10	10	100	80	20	30	60	60	40			
47 Ag	4.7076	4.4183	4.1629	4.1544	3.9347	3.8702	3.8331	3.7034	—	3.5226	3.2382	3.4948	3.6729
48 Cd	4.4801	4.1932	3.9650	3.9564	3.7382	3.6820	3.6450	3.5141	—	3.3356	3.0843	3.3224	3.5007
49 In	4.2687	3.9833	3.7807	3.7719	3.5553	3.5070	3.4698	3.3384	—	3.1621	2.9333	3.1550	3.3215
50 Sn	4.0717	3.7888	3.6089	3.5999	3.3849	3.3434	3.3059	3.1751	—	3.0012	2.7888	2.9949	3.1695
51 Sb	3.8883	3.6077	3.4484	3.4394	3.2257	3.1901	3.1526	3.0234	—	2.8516	2.6630	2.8230	2.9964
52 Te	3.7170	3.4383	3.2985	3.2892	3.0768	3.0466	3.0089	2.8822	—	2.7124	2.5027	2.6825	2.8516
53 I	3.5575	3.2798	3.1579	3.1486	2.9374	2.9121	2.8743	2.7505	—	2.5824	2.3898	2.5532	2.7190
54 Xe	—	—	—	3.0166+	—	—	—	—	—	—	2.2745	2.4306	2.5933
55 Cs	3.2670	2.9932	2.9020	2.8924	2.6837	2.6666	2.6285	2.5118	—	2.3480	2.1725	2.3127	2.4723
56 Ba	3.1355	2.8627	2.7855	2.7760	2.5682	2.5553	2.5164	2.4044	—	2.2415	2.0834	2.2022	2.3611
57 La	3.006	2.740	2.6753	2.6657	2.4589	2.4493	2.4105	2.3030	1.8470	2.1418	1.9789	2.1003	2.2579
58 Ce	2.8917	2.6203	2.5706	2.5615	2.3561	2.3497	2.3109	2.2087	1.7772	2.0487	1.8908	2.0094	2.1641
59 Pr	2.7841	2.512	2.4729	2.4630	2.2588	2.2550	2.2172	2.1194	1.7130	1.9611	1.813	1.9231	2.0771
60 Nd	2.6760	2.4094	2.3807	2.3704	2.1669	2.1669	2.1268	2.0360	1.6510	1.8779	1.7376	1.8424	1.9945
61 Pm	—	—	2.2926	2.2822	2.0797	—	2.0421	1.9559	1.5884	1.7989	1.6684	1.7658	1.9189
62 Sm	2.4823	2.2182	2.2106	2.1998	1.9981	2.0010	1.9624	1.8822	1.5378	1.7272	1.6011	1.6944	1.8446
63 Eu	2.3948	2.1315	2.1315	2.1209	1.9203	1.9255	1.8867	1.8118	1.4848	1.6574	1.5382	1.6259	1.7749
64 Gd	2.3122	2.0494	2.0578	2.0468	1.8468	1.8540	1.8150	1.7455	1.4349	1.5924	1.4787	1.5608	1.7096
65 Tb	2.2352	1.9730	1.9875	1.9765	1.7768	1.7864	1.7472	1.6830	1.3870	1.5303	1.4227	1.5011	1.6484
66 Dy	2.1589	1.8974	1.9199	1.9088	1.7106	1.7210	1.6822	1.6237	1.3418	1.4727	1.3693	1.4436	1.5903
67 Ho	2.0860	1.8264	1.8561	1.8450	1.6475	1.6595	1.6203	1.5671	1.2976	1.4174	1.3194	1.3900	1.5353
68 Er	2.015	1.7566	1.7955	1.7843	1.5873	1.6007	1.5616	1.5140	1.2555	1.3641	1.2709	1.3372	1.4824
69 Tm	1.9550	1.6963	1.7381	1.7268 8+	1.5304	1.5448	1.5063	1.4640	1.2155	1.3153	1.2271	1.2886	1.4314
70 Yb	1.8942	1.6356	1.6829	1.6719	1.4757	1.4914	1.4523	1.4155	1.1772	1.2677	1.1814	1.2415	1.3852
71 Lu	1.8360	1.5779	1.6303	1.6195	1.4236	1.4406	1.4014	1.3701	1.1405	1.2223	1.1401	1.1975	1.3404
72 Hf	1.7815	1.5233	1.5805	1.5696	1.3741	1.3922	1.3530	1.3264		1.1790	1.0986	1.1531	1.2957
73 Ta	1.7284	1.4711	1.5329	1.5220	1.3270	1.3458	1.3068	1.2845		1.1379	1.0608	1.1126	1.2543
74 W	1.6782	1.4211	1.4874	1.4764	1.2818	1.3016	1.2627	1.2446		1.0986	1.0250	1.0744	1.2153
75 Re	1.6306	1.3734	1.4440	1.4329	1.2386	1.2592	1.2203	1.2066		1.0610	0.9901	1.0365	1.1772
76 Os	1.5850	1.3279	1.4023	1.3912	1.1973	1.2184	1.1796	1.1698		1.0250	0.9557	1.0013	1.1414

77 Ir	1.5409	1.2845	1.3625	1.3513	1.1578	1.1796	1.1409	1.1353	1.1059	0.9909	0.92425	0.96700	1.1060
78 Pt	1.4995	1.2429	1.3243	1.3130	1.1199	1.1422	1.1039	1.1020	1.0724	0.9580	0.89405	0.93484	1.0731
79 Au	1.4596	1.2027	1.2877	1.2764	1.0835	1.1065	1.0679	1.0702	1.0404	0.9265	0.86378	0.90277	1.0403
80 Hg	1.4216	1.164	1.2526	1.2412	1.0487	1.0722	1.0336	1.0398	1.0099	0.8965	0.83531	0.87790	1.0094
81 Tl	1.3848	1.1277	1.2188	1.2074	1.0151	1.0392	1.0006	1.0103	0.9806	0.8675	0.80787	0.84355	0.97968
82 Pb	1.3499	1.0924	1.1865	1.1750	0.9829	1.0075	0.9691	0.9822	0.9526	0.8397	0.78153	0.81552	0.95112
83 Bi	1.3161	1.0586	1.1554	1.1439	0.9520	0.9769	0.9386	0.9552	0.9256	0.8131	0.75649	0.78910	0.92459
84 Po	1.2829	—	1.1255⁺	1.1139	0.9220	0.9475	0.9091	0.9294	0.8996	0.7875	0.73219	0.76377	0.89761
85 At	—	—	1.0967⁺	1.0850⁺	0.8935⁺	—	0.8814⁺	—	—	0.7629	0.70915	0.73873	0.87234
86 Rn	—	—	1.0690⁺	1.0572⁺	0.8661⁺	—	0.8544⁺	—	—	0.7393⁺	0.68675	0.71529	0.84845
87 Fr	1.1672	—	1.0423	1.0305	0.8394	—	0.8279⁺	0.858	—	0.7165⁺	0.66537	0.69290	0.82529
88 Ra	—	0.9074	1.0166	1.0047	0.8138	0.8407	0.8027	0.8354	0.8063	0.6946	0.64461	0.67114	0.80284
89 Ac	—	—	0.9918⁺	0.9799⁺	0.7890⁺	—	0.7782⁺	—	—	0.6735⁺	0.6248	0.6500	0.7816
90 Th	1.1151	0.8545	0.9679	0.9560	0.7652	0.7926	0.7548	0.7935	0.7647	0.6531	0.6061	0.6301	0.7615
91 Pa	1.0908	0.8295	0.9448⁺	0.9328	0.7423	0.7699	0.7323	0.7737	0.7452	0.6336⁺	0.5875	0.6106	0.7414
92 U	1.0671	0.8051	0.9226	0.9106	0.7200	0.7480	0.7103	0.7547	0.7263	0.6148	0.5697	0.5919	0.7223
93 Np	1.0428	0.7809	0.9011	0.8891	0.6985	0.7267	0.6892	0.7362	0.7081	0.5965	0.5531	0.5742	0.7039
94 Pu	1.0226	0.7591	0.8803	0.8683	0.6777	0.7062	0.6687	0.7185	0.6907	0.5789	0.5366	0.5571	0.6864

⁺ Interpolated values.

Table 4.5   CRYSTALS FOR PRODUCING MONOCHROMATIC X-RAYS

Crystal	Reflection	Spacing Å	Properties of reflection		Properties of crystal			Special uses
			Peak intensity	Breadth	Crystal imperfection	Stability	Mechanical properties	
β alumina	0 0 0 2 0 0 0 4	11.24 5.62	Weak Weak–medium	Moderate	Great	Perfect	Hard, brittle	For long wavelengths, but usable crystals hard to obtain
Mica	0 0 1 0 0 4	10.1 2.53	Weak	Small	Negligible for selected specimens	Fair	Flexible, easily cleaved	For point-focusing devices; exhibits irradiation effects
Gypsum	0 2 0	7.60	Medium–strong	Very small	Good specimens hard to find	Poor	Soft, flexible	For small-angle scattering; focusing long wavelengths
Pentaerythritol	0 0 2	4.40	Very strong	Moderate	Great	Poor	Soft, easily deformed	General purposes; exhibits irradiation effects
Quartz	1 0 1̄ 1	3.35	Weak–medium	Very small	Negligible	Perfect	Can be elastically bent	For small-angle scattering; focusing
Potassium bromide	2 0 0	3.29	Medium–strong	Moderate	Negligible	Slightly deliquescent	—	—
Fluorite	1 1 1 2 2 0	3.16 1.94	Medium–strong Very strong	Moderate	Small	Perfect	Moderately hard	For eliminating harmonics; general purposes; short wavelengths
Urea nitrate	0 0 2	3.14	Strong	Very large	Very great	Very poor	Very easily deformed	For large specimens; soon decays
Calcite	2 0 0	3.04	Medium	Small	Negligible	Perfect	Moderately soft	For small-angle scattering; isolation of α_1 or α_2
Rock salt	2 0 0	2.82	Medium–strong	Large	Great	Slightly deliquescent	Can be plastically bent in warm supersaturated saline	For focusing
Aluminium	1 1 1	2.33	Very strong	Moderate to large	—	Good	Soft, can be seeded and grown to shape, then plastically shaped at room temperature	For focusing; diffuse scattering
Diamond	1 1 1	2.05	Weak	Very small	Negligible	Perfect	Very hard	For eliminating harmonics
Lithium fluoride	2 0 0	2.01	Very strong	Small–moderate	Negligible	Perfect	Hard, can be plastically bent at high temperature	For focusing; diffuse scattering; general purposes
Graphite	0 0 2	3.35	Very strong	Small	Negligible	Perfect	Easily shaped	Widely used for monochromators on diffractometers

4.3   X-ray techniques

The techniques, involving X-rays, which are commonly employed in the investigation of metallic materials are X-radiography, X-ray diffraction analysis, X-ray small angle scattering, X-ray fluorescence, and X-ray photoelectron spectroscopy (XPS). The basis of these techniques and their metallurgical applications are given in Table 4.6.

4.3.1   X-ray diffraction

Diffraction effects are produced when a beam of X-rays passes through the three-dimensional array of atoms which constitutes a crystal. Each atom scatters a fraction of the incident beams, and if certain conditions are fulfilled then the scattered waves reinforce to give a diffracted beam. These conditions are expressed by the Bragg equation:

$$n\lambda = 2d_{hkl} \sin \theta$$

where n is an integer, λ the wavelength of the incident beam, d_{hkl} the interplanar spacing of the *hkl* planes, and θ the angle of incidence on the *hkl* planes.

A randomly oriented single crystal irradiated with monochromatic radiation is unlikely to have any planes at the correct orientation to satisfy the Bragg equation. Hence when single crystals or more usually individual grains are examined they are either rotated in the X-ray beam or continuous wavelength (white) radiation is used as in the Laue technique.

In the case of powder or fine grained material sufficient crystals are irradiated for some to have planes oriented to satisfy the Bragg equation. The number is increased by rotating or translating the specimen. A series of concentric diffraction cones with semiapex angle 2θ are produced which intersect a narrow film placed around the specimen in a series of arcs, which is the diffraction pattern. This is either recorded on film or scanned with an electronic radiation detector.

4.3.1.1   *Experimental methods*

The experimental methods of X-ray diffraction and their common metallurgical applications are reviewed in Table 4.7. Film methods are included (only to illustrate particular applications) as they have virtually been replaced by computer automated diffractometer methods. Fully automated diffractometer systems are now commercially available for all XRD applications, ranging from routine rapid qualitative and quantitative analysis to detailed single crystal structure analysis.

With these systems, diffraction data are collected and processed automatically, peaks are identified and 2θ, d, peak intensity and width and integrated intensities are measured.

The data are stored in preparation for further evaluation specific to the applications, for example phase analysis and quantification, crystallite size and strain determination from peak breadths and crystallographic evaluations. These are obtained from the determination of unit cell, indexing of peaks and lattice parameter measurements.

The speed of data collection can be increased some 50 to 100 fold by replacing the scintillation counter previously supplied as the standard detector on conventional diffractometers with a Position Sensitive Detector (PSD). With this conversion a complete diffraction pattern can be obtained in a few minutes. The increase in speed is achieved by simultaneously recording diffractions over a 2θ range of about 12° as compared with only, typically, 0.05° with a conventional detector. There are two different modes of operation, stationary and scanning. In the stationary mode diffractions over a 2θ range of about 12° are simultaneously recorded, similar to, but much faster than recording on film. In the scanning mode, the detector is moved round the 2θ circle like a conventional detector. The scanning mode is often called Continuous Position Sensitive Proportional Counter (CPSC).[10]

Particular applications of each mode are:

Stationary mode—small angle scattering
         phase transition
         residual stress determination
Scanning mode—to produce a rapid throughput of specimens
         phase transition where a complete diffraction pattern is needed
         use with a high temperature chamber

Table 4.6 METALLURGICAL INVESTIGATION TECHNIQUES INVOLVING X-RAYS

Technique	Principle of technique	Metallurgical applications
X-radiography	Sample placed between divergent X-ray beam and detector (usually a photographic film). Dark areas on negative show regions of low X-ray absorption (i.e. low density in specimen such as cracks) and vice versa	Crack and other defect detection in castings, welded fabrications, etc. Very widely used inspection and quality assurance technique
X-ray diffraction	Beam of X-rays of specific wavelength (λ) diffracted at certain angles (θ) by crystal planes of appropriate spacing (d) which satisfies the Bragg equation, i.e. $n\lambda = 2d \sin \theta$ Angular position of diffraction peaks, shape of peaks, and intensity peaks give information on crystal structure and physical state	Identification and quantitative analysis of crystalline chemical compounds (phase analysis and quantification, e.g. retained austenite determination) Crystal structure determination Residual macro- and micro-stress analysis and crystallite size Texture (preferred grain orientation) measurement Detection of cold work and imperfections (by stacking faults, etc.) N.B. All results obtained from thin (typically 50 μm) surface layer
Small angle X-ray scattering	Large regions (e.g. 10 → 10 000 Å) of inhomogeneous electron density distributions (e.g. vacancy clusters or Guinier–Preston zones) cause scattered radiation pattern near main beam	Determination of size, shape and composition (in terms of electron density) of Guinier–Preston zones, vacancy clusters, etc. in metallic transmission technique therefore limited to thin samples—typically 10–50 μm
X-ray fluorescence	Electrons ejected from inner shells of atoms cause emission of X-rays (fluorescence) characteristic of atomic species—exciting radiation may be photons (γ and X-rays), positive ions or electrons. Fluorescent X-rays analysed by wavelength or energy dispersive spectrometers.	Elemental analyses Na to U routine. C, N, O, F require specifically designed equipment. Hence used for routine analysis of composition of ores, semi-finished and finished metals and their alloys. Quantitative analysis normally employs calibrated standards. N.B. Results are obtained from surface layer of approx. 2 μm thickness
X-ray photo-electron spectroscopy (XPS)	Electrons from inner shells of atoms are ejected by X-radiation of specific wavelength. Energy of ejected electron measured by spectrometer gives information on binding energy of electron shells	Qualitative and quantitative analysis of surface layers, XPS signals typically come from depths of less than 50 Å. Elemental analysis He to U including, C, N, O, F. Information on the state of bonding of analysed elements

Table 4.7   X-RAY DIFFRACTION TECHNIQUES

Experimental technique	Description	Metallurgical applications
Single crystal		
Single-crystal camera Weissenberg camera Precession camera Automated single crystal diffractometer	Small single crystal or metallic grains oriented on a two-axis goniometer with a prominent crystallographic axis along the goniometer axis. Crystal rotated in a monochromatic or filtered X-ray beam. Produces a spot diffraction pattern related to the symmetry of the crystal.	Determination of unit cell dimensions Crystal structure analysis (i.e. determination of atom positions and thermal vibrations) etc.
Back reflection or transmission Laue camera	Stationary sample, single crystal or polycrystalline irradiated with white, continuous wavelength radiation. The back scattered, (back reflection) and forward scattered transmission diffraction patterns are recorded on flat films perpendicular to the incident beam. The symmetry of the diffraction pattern is related to the symmetry of the crystal in the beam direction. The transmission pattern can be electronically recorded on large-area detectors.	Texture studies on deformed (worked) metallic materials Confirmation of crystal symmetry Determination of crystal orientation
Powder or fine-grained sample		
Debye–Scherrer camera	Cylindrical compact of powder or wire sample (approx. 5 mm long and 0.5–1 mm in diameter) rotated (to avoid texture effects) in monochromatic X-ray beam. Diffraction cones intersect narrow cylindrical film (coaxial with sample) to give line spectrum.	Phase identification Quantitative analysis of mixtures of chemical compounds usually with aid of calibrated standards Composition of alloy phases by correlation of lattice parameters with varying constituent elements
Glancing angle camera	Diffraction spectra obtained by irradiating surface or edge sample at glancing angle and recording one half of total diffraction spectra (other half absorbed by sample) on photographic film.	Examination of bulk samples, simultaneous detection and analysis of surface film (e.g. oxides) and parent metal substrate
Guinier camera	Flat sample positioned on the circumference of the camera is irradiated by a monochromatic or filtered beam of X-rays diverging from a film around the circumference.	Improved resolution and inherently low background aids identification and comparison of specimens with small differences of structure. Diffractions at low Bragg angles (high *d* values) can be studied

Diffractometry	Flat sample irradiated by a diverging beam of monochromatic or filtered X-rays. Detector rotated at twice the angular speed of the specimen to maintain the Bragg-Brentano focusing conditions.	Phase identification and quantification (usually with calibrated standards)
		Studies of crystal imperfections, e.g. stacking faults, microstrain, etc. by detailed measurement of spectra profiles
		With appropriate attachments the following are possible:
		  Residual stress measurement
		  Texture (preferred orientation) determination
		  Thermal expansion parameter measurement
		  Phase transition monitoring

4.3.1.2   *Accessory attachments for diffractometers*

The range of applications of diffractometers is increased by accessory attachments for particular applications. These include:

- Specimen holder with rotation and translation scanning for analysing coarse grained solid and powder samples.
- Specimen changers for the sequential analysis of up to 40 specimens.
- Chambers for high and low temperature measurement. Temperature control and data collection can be automated and with a CPSC, rapid data collection system, a 3D display of 2θ and intensity against temperature can be obtained in a few hours.
- Preferred orientation is measured with a Texture Attachment. Intensity data are automatically collected and results displayed as either a conventional pole figure or as orientation distribution functions (ODF).[11]
- Attachments for quantitatively measuring residual stress.
- Diffraction patterns can be obtained from thin surface films by irradiating the surface at a shallow angle to increase the effective thickness of the film.

4.3.2   Specific applications

4.3.2.1   *Phase identification and quantitative measurements*

Phase identification and quantification depend on the accurate measurement of the interplanar (d) spacings and relative intensities of the diffractions in a diffraction pattern. For routine phase identification the observed d spacings and relative intensities are compared with standard X-ray data listed in the X-ray Powder Diffraction File (PDF) published by the International Center for Diffraction Data (ICDD), formerly the Joint Committee on Powder Diffraction Standards (JCPDS).[12] The file contains data for over 90 000 materials and is regularly updated. It can be obtained on computer files for computer search and match procedures which have virtually replaced the previously used manual procedures.

Interactive graphic search programmes are commonly employed. After entering any chemical information, the PDF files are automatically searched and possible matches listed. These are subsequently subtracted from the unknown pattern and the residual displayed for further analysis. With a mixture, the intensities of the file patterns can be varied to produce the best fit.

In practice, the success in analysing a mixture often depends on the availability of additional information. For example chemical analysis of individual phases or particles carried out by EDAX in a SEM, or a manual separation of phases for separate XRD analysis. The patterns can then be subtracted from the unknown pattern.

Quantitive determinations are carried out by comparing integrated intensities of selected diffractions. In order to allow for absorption, powders are usually analysed by adding a known fraction of standard calibrating powder and comparing the intensities of a diffraction from the component to be analysed with the intensity of a diffraction from the internal standard. Methods are thoroughly discussed in references 3, 9, 10 and 11.

X-ray diffraction analysis is used for the identification of atmospheric pollutants. Examples include silica[13] and asbestos particles collected on multipore filters. The method requires the taking of an X-ray scan directly from the filter and determining the amount of pollutant by comparison with calibration curves prepared from similar filters containing known amounts of the pollutants. As little as $2\,\mu\mathrm{g\,cm}^{-2}$ of silica can be detected by this method.

Most metallurgical samples of interest, however, are solid and cannot be analysed by the above methods. In this case either a calibration curve is constructed from well characterised standards, showing, for precisely defined diffraction conditions, the variation in intensity of a particular diffraction with percentage of the phase to be analysed. Alternatively, theoretical intensities are calculated for diffractions from each phase to be analysed and the ratios of their amounts determined from the observed intensity ratios. The method is illustrated below for the determination of retained austenite in steels but can be extended to other systems. X-ray data for calculating diffraction intensities are listed in Tables 4.8, 4.9, 4.10 and 4.11.

Table 4.8   ANGLES BETWEEN CRYSTALLOGRAPHIC PLANES IN CRYSTALS OF THE CUBIC SYSTEM

(HKL)	(hkl)	Values of α, the angle between (HKL) and (hkl)						
100	100	0°	90°					
	110	45°	90°					
	111	54° 44′						
	210	26° 34′	63° 26′	90°				
	211	35° 16′	65° 54′					
	221	48° 11′	70° 32′					
	310	18° 26′	71° 34′	90°				
	311	25° 14′	72° 27′					
	320	33° 41′	56° 19′	90°				
	321	36° 43′	57° 42′	74° 30′				
110	110	0°	60°	90°				
	111	35° 16′	90°					
	210	18° 26′	50° 46′	71° 34′				
	211	30°	54° 44′	73° 13′	90°			
	221	19° 28′	45°	76° 22′	90°			
	310	26° 34′	47° 52′	63° 26′	77° 5′			
	311	31° 29′	64° 46′	90°				
	320	11° 19′	53° 58′	66° 54′	78°41′			
	321	19° 6′	40° 54′	55° 28′	67°48′	79° 6′		
111	111	0°	70° 32′					
	210	39° 14′	75° 2′					
	211	19° 28′	61° 52′	90°				
	221	15° 48′	54° 44′	78° 54′				
	310	43° 5′	68° 35′					
	311	29° 30′	58° 31′	79° 58′				
	320	61° 17′	71° 19′					
	321	22° 12′	51° 53′	72° 1′	90°			
210	210	0°	36° 52′	53° 8′	66° 25′	78° 28′	90°	
	211	24° 6′	43° 5′	56° 47′	79° 29′	90°		
	221	26° 34′	41° 49′	53° 24′	63° 26′	72° 39′	90°	
	310	8° 8′	58° 3′	45°	64° 54′	73° 34′		
	311	19° 17′	47° 36′	66° 8′	82° 15′			
	320	7° 7′	29° 45′	41° 55′	60° 15′	68° 9′	75° 38′	82° 53′
	321	17° 1′	33° 13′	53° 18′	61° 26′	70° 13′	83° 8′	90°
211	211	0°	33° 33′	48° 11′	60°	70° 32′	80° 24′	
	221	17° 43′	35° 16′	47° 7′	65° 54′	74° 12′	82° 12′	
	310	25° 21′	49° 48′	58° 55′	75° 2′	82° 35′		
	311	19° 8′	42° 24′	60° 30′	75° 45′	90°		
	320	25° 9′	37° 37′	55° 33′	63° 5′	83° 30′		
	321	10° 54′	29° 12′	40° 12′	49° 6′	56° 56′		
		70° 54′	77° 24′	83° 44′	90°			
221	221	0°	27° 16′	38° 57′	63° 37′	83° 37′	90°	
	310	32° 31′	42° 27′	58° 12′	65° 4′	83° 57′		
	311	25° 14′	45° 17′	59° 50′	72° 27′	84° 14′		
	320	22° 24′	42° 18′	49° 40′	68° 18′	79° 21′	84° 42′	
	321	11° 29′	27° 1′	36° 42′	57° 41′	63° 33′	74° 30′	
		79° 44′	84° 53′					
310	310	0°	25° 51′	36° 52′	53° 8′	72° 33′	84° 16′	
	311	17° 33′	40° 17′	55° 6′	67° 35′	79° 1′	90°	
	320	15° 15′	37° 52′	52° 8′	74° 45′	84° 58′		
	321	21° 37′	32° 19′	40° 29′	47° 28′	53° 44′	59° 32′	
		65°	75° 19′	85° 9′	90°			
311	311	0°	35° 6′	50° 29′	62° 58′	84° 47′		
	320	23° 6′	41° 11′	54° 10′	65° 17′	75° 28′	85° 12′	
	321	14° 46′	36° 19′	49° 52′	61° 5′	71° 12′	82° 44′	
320	320	0°	22° 37′	46° 11′	62° 31′	67° 23′	72° 5′	90°
	321	15° 30′	27° 11′	35° 23′	48° 9′	53° 37′	58° 45′	63° 36′
		72° 45′	77° 9′	85° 45′	90°			
321	321	0°	21° 47′	31°	38° 13′	44° 25′	50°	60°
		64° 37′	69° 4′	73° 24′	81° 47′	85° 54′		

Table 4.9 SYMMETRY INTERPRETATIONS OF EXTINCTIONS*

Class of reflection	Condition for non-extinction (n = an integer)	Interpretation of extinction	Symbol of symmetry element
hkl	$h + k + 1 = 2n$	Body centred lattice	I
	$h + k = 2n$	C-centred lattice	C
	$h + l = 2n$	B-centred lattice	B
	$k + l = 2n$	A-centred lattice	A
	$\left.\begin{array}{l} h + k = 2n \\ h + l = 2n \\ k + l = 2n \end{array}\right\}$	Face centred lattice	F
	$\Leftrightarrow h, k, l,$ all even or all odd		
	$-h + k + l = 3n$	Rhombohedral lattice indexed on hexagonal reference system	R
	$h + k + l = 3n$	Hexagonal lattice indexed on rhombohedral reference system	H

* From M. J. Buerger, 'X-ray Crystallography', John Wiley & Sons, New York, 1942.

Table 4.10 MULTIPLICITY FACTORS FOR POWDER PHOTOGRAPHS

Cubic system

Laue or point-group symmetry	hkl	hhl	$0kl$	$0kk$	hhh	$00l$
$\bar{4}3m$, 43, $m3m$	48	24	24	12	8	6
23, $m3$	2×24	24	2×12	12	8	6

Hexagonal and rhombohedral systems

Laue or point-group symmetry	$hkil$	$hh\bar{2}hl$	$0k\bar{k}l$	$hki0$	$hh\bar{2}h0$	$0k\bar{k}0$	$000l$
$\bar{6}2m$, $6mm$, 62, $6/mmm$	24	12	12	12	6	6	2
6, $\bar{6}$, $6/m$	2×12	12	12	2×6	6	6	2
$3m$, 32, $\bar{3}m$	2×12	12	2×6	12	6	6	2
$3, \bar{3}$	4×6	2×6	2×6	2×6	6	6	2

Tetragonal system

Laue or point-group symmetry	hkl	hhl	$0kl$	$hk0$	$hh0$	$0k0$	$00l$
$\bar{4}2m$, $4mm$, 42, $4/mmm$	16	8	8	8	4	4	2
4, $\bar{4}$, $4/m$	2×8	8	8	2×4	4	4	2

Orthorhombic system

Laue or point-group symmetry	hkl	$0kl$	$h0l$	$hk0$	$h00$	$0k0$	$00l$
mm, 222, mmm	8	4	4	4	2	2	2

Monoclinic system

Laue or point-group symmetry	hkl	$h0l$	$0k0$
m, 2, $2/m$	4	2	2

Triclinic system

Laue or point-group symmetry	hkl
1, $\bar{1}$	2

Where the multiplicity is given, for example, as 2×6, this indicates two sets of reflections at the same angle but having different intensities.

Table 4.11 ANGULAR FACTORS

$\theta°$	$\sin^2\theta$	$\dfrac{1+\cos^2 2\theta}{\sin 2\theta}$	$\dfrac{1+\cos^2 2\theta}{\sin^2\theta\cos\theta}$	$\theta°$	$\sin^2\theta$	$\dfrac{1+\cos^2 2\theta}{\sin 2\theta}$	$\dfrac{1+\cos^2 2\theta}{\sin^2\theta\cos\theta}$
0	0.0000	∞	∞	45	0.500	1.000	2.828
1	0.0003	57.272	6563	47½	0.5436	1.011	2.744
½	0.0006	38.162	2916				
2	0.0011	28.601	1639.1	50	0.5868	1.046	2.731
2½	0.0019	22.860	1048				
3	0.0027	19.029	727.2	52½	0.6294	1.105	2.785
3½	0.0037	16.289	533.6	55	0.6710	1.189	2.902
4	0.0049	14.231	408.0	57½	0.7113	1.300	3.084
4½	0.0061	12.628	321.9				
5	0.0076	11.344	260.3	60	0.7500	1.443	3.333
6	0.0109	9.411	180.06				
7	0.0149	8.025	131.70	62½	0.7868	1.622	3.658
8	0.0193	6.980	100.31	65	0.8214	1.845	4.071
9	0.0243	6.163	78.80	67½	0.8536	2.121	4.592
10	0.0302	5.506	63.41	70	0.8830	2.469	5.255
12	0.0432	4.510	43.39	72	0.9045	2.815	5.920
14	0.0581	3.791	31.34	74	0.9240	3.244	6.749
16	0.0762	3.244	23.54	76	0.9415	3.791	7.814
18	0.0955	2.815	18.22	78	0.9568	4.510	9.221
20	0.1170	2.469	14.44	80	0.9698	5.506	11.182
22½	0.1465	2.121	11.086	81	0.9755	6.163	12.480
25	0.1786	1.845	8.730	82	0.9806	6.980	14.097
27½	0.2133	1.622	7.027	83	0.9851	8.025	16.17
				84	0.9891	9.411	18.93
30	0.2500	1.443	5.774	85	0.9924	11.344	22.78
				85½	0.9938	12.628	25.34
32½	0.2887	1.300	4.841	86	0.9951	14.231	28.53
35	0.3290	1.189	4.123	86½	0.9963	16.289	32.64
37½	0.3706	1.105	3.629	87	0.9973	19.029	38.11
				87½	0.9981	22.860	45.76
40	0.4131	1.046	3.255	88	0.9988	28.601	57.24
				88½	0.9993	38.162	76.35
42½	0.4564	1.011	2.994	89	0.9997	57.272	114.56
45	0.5000	1.000	2.828	90	1.000	∞	∞

4.3.2.2 *Determination of retained austenite in steel*

An important example of quantitative phase analysis by X-ray diffraction is the determination of retained austenite in steels. The method is based on the comparison of the integrated diffracted X-ray intensities of selected (*hkl*) reflections of the martensite and austenite phases. The necessary formulae and reference data are given below; for more details of the experimental methods the definitive paper by Durnin and Ridal[14] should be consulted.

The integrated intensity of a diffraction line is given by the equation:

$$I_{(hkl)} = n^2\, Vm\,(LP)\,e^{-2m}\,(F)^2 \tag{4.1}$$

in which $I_{(hkl)}$ = integrated intensity for a special (*hkl*) reflection; n = number of cells in cm³; V = volume exposed to the X-ray beam; (LP) = Lorentz-Polarisation factor; m = multiplicity of (*hkl*); e^{-2m} = Debye–Waller temperature factor; F = structure factor (which includes f, the atomic scattering factor.)

For n^2V we may substitute V/v^2, in which v is the volume of the unit cell.

If the ratio between the integrated intensities of martensite and austenite is denoted by P:

$$P = \frac{I\ \text{martensite }(\alpha)}{I\ \text{austenite }(\gamma)} = \frac{V_\alpha v_\gamma^2 m_\alpha (LP)_\alpha e_\alpha^{-2m}(F_\alpha)^2}{V_\gamma v_\alpha^2 m_\gamma (LP)_\gamma e_\gamma^{-2m}(F_\gamma)^2} \tag{4.2}$$

Each factor is determined from the International Tables[9] and depends on the reflection used. A factor G is then determined for each combination of α and γ peaks used; hence:

$$P = \frac{V_\alpha}{V_\gamma} \times \frac{1}{G} \tag{4.3}$$

If α and γ are the only phases present:

$$V_\gamma = \frac{1}{1 + GP} \tag{4.4}$$

Hence, measurement of the ratio (P) of two diffraction peaks and calculation of the factor G will give the volume fraction of austenite V_γ.

The factors involved in the calculation of G for two steels—16.8%Ni–Fe(0.35%C) and NCMV (a Ni–Cr–Mo–V steel with composition wt % 0.43C; 0.31Si; 0.57Mn; 0.009S; 0.005P; 1.69Ni; 1.36Cr; 1.08Mo; 0.24V; 0.11Cu), Mo, Co and Cr radiation and a selection of hkl peaks have been extracted from the 'International Tables for X-Ray Crystallography'[9] by Durnin and Ridal[14] and are presented in Table 4.12.

These factors may be used to calculate G for different radiations and peaks. The results are presented in Table 4.13.

When the alloy compositions are being investigated the factors which make up G must be determined from the International Tables.[9]

Accuracy obtainable using a diffractometer is in the region of 0.5% for the range 1.5–38 volume percentage of austenite. X-ray diffraction determination accuracy thus compares favourably with other techniques such as metallography, dilatometry and saturation magnetisation intensity methods which are all inaccurate below 10% austenite content.

The main source of error in X-ray determination of retained austenite comes from overlapping carbide peaks. The carbides and their diffraction peaks most likely to cause problems are summarised in Table 4.14.

4.3.2.3   X-ray residual stress measurements

Residual stresses can be divided into two general categories, macrostresses where the strain is uniform over relatively large distances and microstresses produced by non-uniform strain over short distances, typically a few hundred Å. Both types of stress can be measured by X-ray diffraction techniques.

The basis of stress measurement by X-ray diffraction is the accurate measurement of changes in interplanar d spacing caused by the residual stress. When macrostresses are present the lattice plane spacing in the crystals (grains) change from their stress-free values to new values corresponding with the residual stress and the elastic constants of the material. This produces a shift in the position of the corresponding diffraction, i.e. a change in Bragg angle θ. Microstresses however give rise to non-uniform variations in interplanar spacing which broaden the diffractions rather than cause a shift in their position. Small crystal size also give rise to broadening.

Measurement of Macro-residual stress[1,2,3,7,15]

The working equation used in most X-ray stress analysis is

$$\sigma\phi = \frac{d_\psi d_\perp}{d_\perp} \times \frac{E}{1 + v} \times \frac{1}{\sin^2 \psi} \tag{1}$$

where σ_ψ is the surface stress lying in a direction common to the surface and the plane defined by the surface normal and the incident X-ray beam. E and v are Young's modulus and Poisson's ratio respectively, d_ψ and $d_\perp$ are the interplanar spacings of planes with normals parallel to the surface normal and at an angle ψ to the surface normal. These angles are related to the direction of the incident X-ray beam, as shown in Fig. 4.1. Thus σ can be determined from two exposures, one with the incident beam inclined at θ to the surface to measure $d_\perp$ and the other at an angle ψ the first to measure d_ψ. This technique is known as the two-exposure technique.

Alternatively equation (1) can be rewritten as

$$d_\psi = d_\perp \left(\frac{1 + v}{E}\right) \sigma\phi \sin^2 \psi + d_\perp \tag{2}$$

Table 4.12   INTENSITY FACTORS FOR DIFFERENT RADIATIONS AND PEAKS[14]

Material	Radiation	Factor	Peak				
			α200	α211	γ200	γ220	γ311
16.8%Ni–Fe	Mo	Bragg angle, θ	14.41	17.73	11.49	16.32	19.21
NCMV	Mo		14.40	17.70	11.49	16.32	19.21
16.8%Ni–Fe	Co		38.67	49.89	29.99	44.92	55.85
16.8%Ni–Fe	Cr		53.15	78.05	39.74	64.65	—
Both compositions	All	Multiplicity, m	6	24	6	12	24
Both	Mo	Lorentz and polarisation (LP)	29.46	18.84	47.56	22.55	15.78
	Co		3.44	2.73	5.79	2.83	2.96
	Cr		2.81	9.26	3.29	4.01	—
16.8%Ni–Fe	Mo	Debye–Waller temp. e^{-2m}	0.910	0.869	0.943	0.889	0.847
NCMV	Mo		0.910	0.869	0.943	0.889	0.848
16.8%Ni–Fe	Co		0.908	0.869	0.941	0.889	0.852
16.8%Ni–Fe	Cr		0.912	0.869	0.943	0.889	
16.8%Ni–Fe	Mo	Atomic scattering, f_0	15.1	13.4	17.0	14.0	12.8
NCMV	Mo		14.7	13.1	16.6	13.7	12.5
16.8%Ni–Fe	Co		10.78	9.14	12.69	9.84	8.60
16.8%Ni–Fe	Cr		13.19	11.54	15.09	12.24	
Both	All	Structure factor, F	2	2	4	4	4
Both	All	$1/V^2$ (V is volume of unit cell)	1.79×10^{-3} kx units		4.68×10^{-4} kx units		

Table 4.13   AUSTENITE DETERMINATION FACTOR G FOR DIFFERENT RADIATIONS AND PEAK COMBINATIONS[14]

Material	Radiation	Peak combination					
		$\alpha200-\gamma200$	$\alpha200-\gamma220$	$\alpha200-\gamma311$	$\alpha211-\gamma200$	$\alpha211-\gamma220$	$\alpha211-\gamma311$
16.8%Ni–Fe	Mo	2.22	1.36	1.50	1.16	0.71	0.78
NCMV	Mo	2.23	1.38	1.51	1.15	0.72	0.78
16.8%Ni–Fe	Co	2.52	1.40	2.15	1.09	0.61	0.93
16.8%Ni–Fe	Cr	1.66	2.50	—	0.17	0.26	—

Table 4.14   INTERFERENCE OF ALLOY CARBIDE LINES WITH AUSTENITE AND MARTENSITE LINES[14]

Austenite and martensite 'd' spacings	Fe_3C	M_6C	V_4C_3	Mo_2C or W_2C	WC	$Cr_{23}C_6$	Cr_7C_3
(200)γ 1.80 Å	Clear	Clear	Clear	Clear	Clear	Strong overlap	Strong overlap
(200)α 1.43 Å	Clear	Strong overlap	Clear	Clear	Weak overlap	Weak overlap	Weak overlap
(220)γ 1.27 Å	Weak overlap	Strong overlap	Weak overlap	Strong overlap	Weak overlap	Medium overlap	Clear
(211)α 1.17 Å	Strong overlap	Clear	Clear	Weak overlap	Weak overlap	Medium overlap	Medium overlap
(311)γ 1.08 Å	Weak overlap	Medium overlap	Clear	Weak overlap	Clear	Strong overlap	Clear

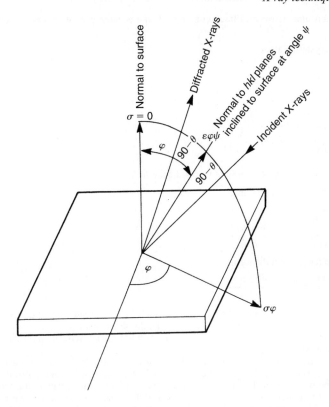

Figure 4.1

This shows that d_ψ is a linear function of $\sin^2 \psi$. The intercept on the d axis gives $d_\perp$ and the slope $d_\perp((1+\nu)E)\sigma\phi$. A positive slope corresponds to a tensile stress and a negative slope to compression. This technique takes advantage of measurements involving a number of d_ψ values.

The determination of interplanar spacings depends on the accurate measurement of the corresponding Bragg angle θ where

$$d_{hkl} = \frac{\lambda}{2 \sin \theta_{hkl}}$$

Back reflection diffractions are used as the highest sensitivity to changes in interplanar spacing are obtained as θ tends to 90°. In practice the specimen is rotated through ψ between exposures (or the tube and detector together through the same angle on portable systems for measuring large samples).

Due to the limited penetration of commonly used X-ray wavelengths, typically 4 μm for chromium radiation to 11 μm for molybdenum in steel, only surface stresses are measured. Stresses at lower depths in the sample are determined by repeating measurements after removing (electropolishing) layers of known thickness. The measured values are subsequently corrected for changes in stress resulting from the removal of upper layers.

Residual stress in small components can be measured on a converted diffractometer by adding a specimen support table with three orthogonal adjustments to permit the surface of the component to be brought into the beam. Fully automated portable systems are available for determining residual stress in large components.

Typical accuracy for steel is ± 1.5–3.0×10^7 Pa (± 1–2 ton in^{-2}).

MEASUREMENT OF MICROSTRESSES[2,16,17]

A worked surface gives rise to broadened diffractions due to a combination of microstresses and small crystallite size. An approximate method to separate the two effects is to assume that the breadth is the sum of the separate broadening from each effect.

The relationships between diffraction breadth β and average crystallite size ε and mean stress $\bar{\sigma}$ are:

1. Small crystallite alone

$$\beta_{\text{s.c.}} \simeq \frac{k\lambda}{\varepsilon_{hkl} \cos \Theta}$$

where k is a constant $\simeq 1$ and ε_{hkl} is the linear dimension perpendicular to the measured hkl plane.

2. Microstresses alone

$$\beta_{\text{M.S.}} \simeq \frac{4\bar{\sigma} \tan \Theta}{E_{hkl}}$$

where E_{hkl} is the elastic constant perpendicular to hkl.

When both effects are present, then

$$\beta = \beta_{\text{s.c.}} + \beta_{\text{M.S.}}$$
$$= \frac{\lambda}{\varepsilon_{hkl} \cos \Theta} + \frac{4\bar{\sigma} \tan \Theta}{E_{hkl}}$$

which can be rewritten in the form:

$$\frac{\beta \cos \Theta}{\lambda} = \frac{1}{\varepsilon_{hkl}} + \frac{4\bar{\sigma} \sin \Theta}{\lambda E_{hkl}}$$

If only these defects are present, then a plot of $\beta \cos \Theta / \lambda$ against $\sin \Theta$ should be a straight line, with intercept on the $\beta \cos \Theta / \lambda$ axis giving $1/\varepsilon_{hkl}$ and the slope $4\bar{\sigma}/\lambda E_{hkl}$.

For non-cubic materials, for example tetragonal and hexagonal, separate plots should be made for the hk0 and 001 results.

In the above formulae, diffraction breadth is measured as either the breadth in radians at half the peak height (HPHW) or as the integral breadth which is the integrated intensity of the diffraction divided by the peak height. It corresponds to the width of a rectangle having the same area and height as the diffraction. The latter is particularly useful in analysing peaks with partially resolved $\alpha_1\alpha_2$ doublets. The measured breadths include the instrumental breadth which is independent of any crystal defects. This can be measured for subsequent subtraction from the measured breadths by running a scan from a defect-free specimen.

4.3.2.4   *Preferred orientation*

Preferred orientation can be represented in two ways, either as a convential pole figure or as an inverse pole figure. A conventional pole figure shows the distribution of a low-index pole—normal to a crystallographic plane, over the whole specimen. With a cubic metal, these are generally constructed for $\{100\}$, $\{110\}$ and $\{111\}$ planes while for hexagonal metals usually the basal (0001) planes $\{10\bar{1}0\}$ or $\{10\bar{1}1\}$ planes are selected. A high density of poles shows the preferred direction of the pole with respect to the sample.

An inverse pole figure on the other hand shows how the grains are distributed with respect to a particular direction in the sample. Inverse pole figures are usually constructed for the principal directions of the sample, for example the extrusion, radial, and tangential directions in extruded material. The method is rapid, and data for a single direction can be determined from a conventional diffractometer scan taken from a surface which is perpendicular to the required direction. The method is based on the fact that all diffractions recorded on a conventional scan come from planes which are parallel to the surface and their intensities are related to the number of grains (volume of material) which has this orientation. The diffraction intensities are compared to those from a sample having random orientation. These can be either theoretical calculated values or values measured directly from a corresponding scan taken from a sample with random orientation. These relative intensities are called texture coefficients (TC) and are expressed mathematically as

$$\text{TC}_{(hkl)} = \frac{I_{(hkl)}/I^0_{(hkl)}}{\frac{1}{n}\sum_0^n I_{(hkl)}/i^0_{(hkl)}}$$

where

I = measured integrated intensity of a given hkl diffraction
I^0 = corresponding intensity for the same hkl diffraction from a random sample
n = total number of diffractions measured.

The TC_{hkl} values are proportional to the number of grains (volume of sample) which are oriented with an hkl plane parallel to the sample surface.

The values can be plotted on a partial sterographic projection and contour lines drawn through the plotted points to produce an 'Inverse Pole Figure' for that particular specimen direction. High values show the preferred grain orientation in the specimen direction.

4.3.2.5   *Specimen preparation*

Methods for preparing standard diffractometer specimens are discussed in references 1–3. The most common method is to pack loose powder into a flat cavity in an aluminium specimen holder, taking care not to introduce preferred orientation. Another method is to mix the powder into a slurry and smear some over a glass cover slip. Coarse powders which tend to slide out of the cavity during a scan can be mixed with petroleum jelly or gum tragacanth which themselves give negligible diffraction.

A problem when analysing a small amount of material is diffraction and scatter from the specimen holder. These effects are reduced by using a single crystal holder of a low element material which has been polished so the surface is just off a Bragg plane.

Most laboratories have developed procedures and specimen holders for non-standard applications. These have been surveyed by D. K. Smith and C. S. Barrett and the results published in 'Advances in X-ray Analysis'.[18] One application is the analysis of air-moisture reactive powders. A simple solution is to fill and seal the powder in thin walled glass quills inside a dry box. After removal from the glove box the quills are mounted in raft fashion across a recessed specimen holder. Sealed cells are also used to avoid handling delicate quills inside a glove box. One such method for lithium compounds was to load the powder inside a glove box into a recessed specimen holder and cover the powder with a flanged 0.001 in. thick aluminium foil dome. Petroleum jelly was lightly smeared round the flange to hold the dome on to the specimen holder and make a moisture seal.

Metal samples submitted for XRD analysis are generally metallurgically mounted and polished samples. Due to the limited penetration of the X-ray beam into the sample it is always advisable to first electropolish the surface. Useful electropolishes are listed in Chapter 10, Table 10.4. An electropolish frequently used in preparing specimens for retained austenite determinations is

    7 vol% perchloric acid
    69 vol% ethanol
    10 vol% glycerol
    14 vol% water

At a voltage of 23 V. Electrolyte maintained at temperature below 12°C.

EXTRACTION TECHNIQUES

Precipitates that are present at very low concentrations can be concentrated by dissolving away the matrix and collecting the residue. Typical solutions used for extraction are as follows.[19]

  (i) Carbides and certain intermetallic compounds
     (a) Immersion overnight in 5 or 10% bromine in methanol, or
     (b) Electrolytically in 10% hydrochloric acid in methanol at a current density of 0.07 A cm^{-2}. Extraction for 4 h provides sufficient powder residue.
     The intermetallic phases which can be extracted using these solutions are sigma phase, certain of the Laves phases, Fe_3Mo_2 etc.
 (ii) Gamma prime, eta phase, M(CN)
     (a) Electrolytically in 10% phosphoric acid in water at a current density of 10 A dm^{-2}. Time of extraction 4 h. It is sometimes necessary to add a small amount of tartaric acid in order to prevent the formation of tantalum and tungsten hydroxides,
     or
     (b) Electrolytically in 1% citric acid plus 1% ammonium sulphate in water. Current density 2 A dm^{-2}, duration 4 h. For quantitative work this solution is preferred to the phosphoric acid electrolyte.
     Certain of the Laves phases may also be extracted using either of these solutions.
(iii) Precipitates and intermetallic compounds in chromium
     Immersion overnight in 10% hydrochloric acid. This has been used to extract carbides and borides.

Precipitates can also be concentrated by heavily etching a surface to leave them proud of the surface as used to produce replicas in electron microscopy.[20]

Table 4.15 CRYSTAL GEOMETRY

System	d_{hkl} = Interplanar spacing	V = Vol. of unit cell
Cubic	$\dfrac{1}{d^2} = \dfrac{h^2 + k^2 + l^2}{a^2}$	$V = a^3$
Tetragonal	$\dfrac{1}{d^2} = \dfrac{h^2}{a^2} + \dfrac{k^2}{a^2} + \dfrac{l^2}{c^2}$	$V = a^2 c$
Orthorhombic	$\dfrac{1}{d^2} = \dfrac{h^2}{a^2} + \dfrac{k^2}{b^2} + \dfrac{l^2}{c^2}$	$V = abc$
Rhombohedral*	$\dfrac{1}{d^2} = \dfrac{(h^2 + k^2 + l^2)\sin^2\alpha + 2(hk + kl + hl)(\cos^2\alpha - \cos\alpha)}{a^2(1 - 3\cos^2\alpha + 2\cos^3\alpha)}$	$V = a^3\sqrt{(1 - 3\cos^2\alpha + 2\cos^3\alpha)}$
Hexagonal†	$\dfrac{1}{d^2} = \dfrac{4}{3}\left(\dfrac{h^2 + hk + k^2}{a^2}\right) + \dfrac{l^2}{c^2}$	$V = \dfrac{\sqrt{3}}{2}a^2 c = 0.866\,a^2 c$
Monoclinic	$\dfrac{1}{d^2} = \dfrac{h^2}{a^2\sin^2\beta} + \dfrac{k^2}{b^2} + \dfrac{l^2}{c^2\sin^2\beta} - \dfrac{2hl\cos\beta}{ac\sin^2\beta}$	$V = abc\sin\beta$
Triclinic	$\dfrac{1}{d^2} = \dfrac{1}{V^2}(s_{11}h^2 + s_{22}k^2 + s_{33}l^2 + 2s_{12}hk + 2s_{23}kl + 2s_{13}hl)$	$V = abc\sqrt{(1 - \cos^2\alpha - \cos^2\beta - \cos^2\gamma + 2\cos\alpha\cos\beta\cos\gamma)}$

System	ϕ = angle between planes $h_1 k_1 l_1$ and $h_2 k_2 l_2$
Cubic	$\cos\phi = \dfrac{h_1 h_2 + k_1 k_2 + l_1 l_2}{\sqrt{[(h_1^2 + k_1^2 + l_1^2)(h_2^2 + k_2^2 + l_2^2)]}}$
Tetragonal	$\cos\phi = \dfrac{(h_1 h_2/a^2) + (k_1 k_2/a^2) + (l_1 l_2/c^2)}{\sqrt{\{[(h_1^2/a^2) + (k_1^2/a^2) + (l_1^2/c^2)][(h_2^2/a^2) + (k_2^2/a^2) + (l_2^2/c^2)]\}}}$
Orthorhombic	$\cos\phi = \dfrac{(h_1 h_2/a^2) + (k_1 k_2/b^2) + (l_1 l_2/c^2)}{\sqrt{\{[(h_1^2/a^2) + (k_1^2/b^2) + (l_1^2/c^2)][(h_2^2/a^2) + (k_2^2/b^2) + (l_2^2/c^2)]\}}}$
Rhombohedral*	$\cos\phi = \dfrac{(h_1 h_2 + k_1 k_2 + l_1 l_2)\sin^2\alpha + (k_1 l_2 + k_2 l_1 + l_1 h_2 + l_2 h_1 + h_1 k_2 + h_2 k_1)(\cos^2\alpha - \cos\alpha)}{\sqrt{[(h_1^2 + k_1^2 + l_1^2)\sin^2\alpha + 2(h_1 k_1 + k_1 l_1 + h_1 l_1)(\cos^2\alpha - \cos\alpha)][(h_2^2 + k_2^2 + l_2^2)\sin^2\alpha + 2(h_2 k_2 + k_2 l_2 + h_2 l_2)(\cos^2\alpha - \cos\alpha)]}}$
Hexagonal[†]	$\cos\phi = \dfrac{h_1 h_2 + k_1 k_2 + \frac{1}{2}(h_1 k_2 + h_2 k_1) + \frac{3}{4}(a^2/c^2)\cdot l_1 l_2}{\sqrt{\{[h_1^2 + k_1^2 + h_1 k_1 + \frac{3}{4}(a^2/c^2)\cdot l_1^2][h_2^2 + k_2^2 + h_2 k_2 + \frac{3}{4}(a^2/c^2)\cdot l_2^2]\}}}$
Monoclinic	$\cos\phi = \dfrac{(h_1 h_2/a^2) + k_1 k_2 \sin^2\beta/b^2) + (l_1 l_2/c^2) - [(l_1 h_2 + l_2 h_1)\cos\beta/ac]}{\sqrt{\{[(h_1^2/a^2) + (k_1^2 \sin^2\beta/b^2) + (l_1^2/c^2) - (2h_1 l_1 \cos\beta/ac)][(h_2^2/a^2) + (k_2^2 \sin^2\beta/b^2) + (l_2^2/c^2) - (2h_2 l_2 \cos\beta/ac)]\}}}$
Triclinic	$\cos\phi = \dfrac{d_{h_1 k_1 l_1} \cdot d_{h_2 k_2 l_2}}{V^2}[s_{11}h_1 h_2 + s_{22}k_1 k_2 + s_{33}l_1 l_2 + s_{23}(k_1 l_2 + k_2 l_1) + s_{13}(l_1 h_2 + l_2 h_1) + s_{12}(h_1 k_2 + h_2 k_1)]$

where
$$s_{11} = b^2 c^2 \sin^2\alpha \qquad s_{12} = abc^2(\cos\alpha\cos\beta - \cos\gamma)$$
$$s_{22} = a^2 c^2 \sin^2\beta \qquad s_{23} = a^2 bc(\cos\beta\cos\gamma - \cos\alpha)$$
$$s_{33} = a^2 b^2 \sin^2\gamma \qquad s_{13} = ab^2 c(\cos\gamma\cos\alpha - \cos\beta)$$

* Rhombohedral axes.
[†] Hexagonal axes, co-ordinates $hkil$ where $i = -(h + k)$.

4.3.2.6   *Formulae and crystallographic data*

Formulae for calculating interplanar d_{hkl} spacings from lattice parameters and data for calculating intensities together with other useful information on crystal symmetry are given in Tables 4.8, 4.9, 4.10, 4.11 and 4.15.

INTENSITIES

The relative intensities of diffractions recorded on a diffractometer scan are given by the formula

$$1 \propto \left(\frac{1 + \cos^2 2\theta}{\sin^2 \cos \theta} \right) |F|^2 \cdot T \cdot p \cdot A$$

where $\left(\dfrac{1 + \cos^2 \theta}{\sin^2 \theta \cos \theta} \right) |F|^2 \cdot T \cdot p \cdot A = $ Combined polarisation Lorentz factor (Table 4.11)

$F = $ structure factor involving the summation of scattering from all atoms in the unit cell
$T = $ temperature factor, $e^{-(B \sin^2 \theta / \lambda^2)}$
$p = $ multiplicity factor, Table 4.10
$A = $ an absorption factor.

For a diffractometer specimen of effectively infinite thickness, $A = K/\mu$ where K is a constant and μ the linear absorption coefficient of the sample. A is therefore independent of θ. A criterion for this condition is that the thickness of the specimens is $> 3.2/\mu$.

Tables 4.16 and 4.17 give values of Mean Atomic Scattering Factors and Mass Absorption Coefficients, respectively.

Table 4.16 MEAN ATOMIC SCATTERING FACTORS*

$\frac{\sin\theta}{\lambda}$ Å⁻¹	0.0	0.1	0.2	0.3	0.4	0.5	0.6	0.7	0.8	0.9	1.0	1.1	1.2
H	1.000	0.811	0.481	0.251	0.130	0.071	0.040	0.024	0.015	0.010	0.007	0.005	0.0035
H⁻¹	2.000	1.064	0.519	0.255	0.130	0.070	0.040	0.024	0.015	0.010	0.007	0.005	0.0035
He	2.000	1.832	1.452	1.058	0.742	0.515	0.358	0.251	0.179	0.129	0.095	0.071	0.054
Li	3.000	2.215	1.741	1.512	1.269	1.032	0.823	0.650	0.513	0.404	0.320	0.255	0.205
Li⁺¹	2.000	1.935	1.760	1.521	1.265	1.025	0.818	0.647	0.510	0.403	0.319	0.254	0.203
Li⁻¹	4.000	2.176	1.743	1.514	1.269	1.033	0.826	0.654	0.516	0.408	0.323	0.257	0.205
Be	4.000	3.067	2.067	1.705	1.531	1.367	1.201	1.031	0.878	0.738	0.620	0.519	0.432
Be⁺¹	3.000	2.583	2.017	1.721	1.535	1.362	1.188	1.022	0.870	0.735	0.618	0.520	0.436
Be⁺²	2.000	1.966	1.869	1.724	1.550	1.363	1.180	1.009	0.855	0.721	0.606	0.508	0.427
B	5.000	4.066	2.711	1.993	1.692	1.534	1.406	1.276	1.147	1.016	0.895	0.783	0.682
B⁺¹	4.000	3.471	2.551	1.962	1.688	1.536	1.410	1.283	1.154	1.028	0.908	0.798	0.698
B⁺²	3.000	2.757	2.290	1.928	1.707	1.552	1.414	1.278	1.144	1.016	0.896	0.786	0.687
B⁺³	2.000	1.979	1.919	1.824	1.703	1.566	1.420	1.274	1.132	0.999	0.877	0.767	0.669
C	6.000	5.126	3.581	2.502	1.950	1.685	1.536	1.426	1.322	1.218	1.114	1.012	
C⁺²	4.000	3.686	2.992	2.338	1.910	1.672	1.533	1.429	1.332	1.233	1.131	1.030	
C⁺³	3.000	2.842	2.487	2.133	1.874	1.697	1.564	1.447	1.335	1.225	1.116	1.012	0.913
C⁺⁴	2.000	1.986	1.945	1.880	1.794	1.692	1.579	1.459	1.338	1.219	1.104	0.994	0.893
N	7.000	6.203	4.600	3.241	2.397	1.944	1.698	1.550	1.444	1.350	1.263	1.175	1.083
N⁺³	4.000	3.772	3.227	2.635	2.172	1.869	1.682	1.558	1.461	1.373	1.287	1.199	1.112
N⁺⁴	3.000	2.890	2.619	2.306	2.038	1.837	1.690	1.573	1.472	1.375	1.281	1.188	1.097
N⁻¹	8.000	6.688	4.631	3.186	2.364	1.929	1.694	1.551	1.446	1.352	1.263	1.170	
O	8.000	7.250	5.634	4.094	3.010	2.338	1.944	1.714	1.566	1.462	1.374	1.296	1.220
O⁺¹	7.000	6.493	5.298	4.017	3.016	2.356	1.956	1.717	1.567	1.461	1.374	1.296	
O⁺²	6.000	5.647	4.776	3.771	2.924	2.327	1.948	1.716	1.568	1.463	1.378	1.301	
O⁺³	5.000	4.760	4.151	3.410	2.745	2.246	1.913	1.701	1.562	1.463	1.382	1.308	
O⁻¹	9.000	7.836	5.756	4.068	2.968	2.313	1.934	1.710	1.566	1.46	1.373	1.294	
F	9.000	8.293	6.691	5.044	3.760	2.878	2.312	1.958	1.735	1.587	1.481	1.496	1.322
F⁻¹	10.00	9.108	7.126	5.188	3.786	2.885	2.323	1.972	1.747	1.596	1.486	1.399	1.419
Ne	10.00	9.363	7.824	6.987	4.617	3.536	2.794	2.300	1.976	1.760	1.612	1.504	1.52
Na	11.00	9.76	8.34	6.89	5.47	4.29	3.40	2.76	2.31	2.00	1.78	1.63	
Na⁺	10.00	9.551	8.390	6.925	5.510	4.328	3.424	2.771	2.314	2.001	1.785	1.634	1.524
Mg	12.00	10.50	8.75	7.46	6.20	5.01	4.06	3.30	2.72	2.30	2.01	1.81	1.65
Mg⁺²	10.00	9.66	8.75	7.51	6.20	4.99	4.03	3.28	2.71	2.30	2.01	1.81	1.65
Al	13.00	11.23	9.16	7.88	6.77	5.69	4.71	3.88	3.21	2.71	2.32	2.05	1.65
Al⁺¹	12.00	10.94	9.22	7.90	6.77	5.70	4.71	3.88	3.22	2.70	2.32	2.04	1.83

(continued)

Table 4.16 MEAN ATOMIC SCATTERING FACTORS*—continued

$\frac{\sin\theta}{\lambda}$ Å^{-1}	0.0	0.1	0.2	0.3	0.4	0.5	0.6	0.7	0.8	0.9	1.0	1.1	1.2
Al^{+2}	11.00	10.40	9.17	7.65	6.79	5.70	4.71	3.88	3.22	2.71	2.33	2.05	1.84
Al^{+3}	10.00	9.74	9.01	7.98	6.82	5.69	4.69	3.86	3.20	2.70	2.32	2.04	1.84
Si	14.00	12.16	9.67	8.22	7.20	6.24	5.31	4.47	3.75	3.16	2.69	2.35	2.07
Si^{-3}	11.00	10.53	9.48	8.34	7.27	6.25	5.30	4.44	3.73	3.14	2.67	2.34	
Si^{-4}	10.00	9.79	9.20	8.33	7.31	6.26	5.28	4.42	3.71	3.13	2.68	2.33	2.06
P	15.00	13.17	10.34	8.59	7.54	6.67	5.83	5.02	4.28	3.64	3.11	2.69	2.35
S	16.00	14.33	11.21	8.99	7.83	7.05	6.31	5.56	4.82	4.15	3.56	3.07	2.66
S^{-1}	17.00	15.00	11.36	8.95	7.79	7.05	6.32	5.57	4.83	4.16	3.57	3.08	2.67
S^{-2}	18.00	(15.16)	(10.74)	(8.66)	(7.89)	(7.22)	(6.47)	(5.69)	(4.93)	(4.23)	(3.62)	(3.13)	(2.71)
Cl	17.00	15.33	12.00	9.44	8.07	7.29	6.64	5.96	5.27	4.60	4.00	3.47	3.02
Cl^{-1}	18.00	16.02	12.22	9.40	8.03	7.28	6.64	5.97	5.27	4.61	4.00	3.47	3.03
A	18.00	16.30	12.93	10.20	8.54	7.56	6.86	6.23	5.61	5.01	4.43	3.90	3.43
K	19.00	16.73	13.73	10.97	9.05	7.87	7.11	6.51	5.95	5.39	4.84	4.32	3.83
K^{+1}	18.00	16.68	13.76	10.96	9.04	7.86	7.11	6.51	5.94				
Ca	20.00	17.33	14.32	11.71	9.64	8.26	7.38	6.75	6.21	5.70	5.19	4.69	4.21
Ca^{+1}	19.00	17.21	14.35	11.70	9.63	8.26	7.38	6.75	6.21	5.70	5.19	4.68	
Ca^{+2}	18.00	16.93	14.40	11.70	9.61	8.25	7.38	6.75	6.22	5.96	5.18	4.68	
Sc	21.00	18.72	15.38	12.39	10.12	8.60	7.64	6.98	6.45	5.96	5.48	5.00	4.53
Sc^{+1}	20.00	18.50	15.43	12.43	10.13	8.61	7.64	6.98	6.45	5.96	5.48	5.00	5.53
Sc^{+2}	19.00	17.88	15.27	12.44	10.18	8.64	7.65	6.98	6.45	5.96	5.48	5.01	4.54
Sc^{+3}	18.00	17.11	14.92	12.38	10.22	8.68	7.76	6.98	6.44	5.95	5.49	5.02	4.56
Ti	22.00	19.96	16.41	13.68	11.53	9.88	8.57	7.52	6.65		5.36	4.86	4.43
Ti^{+1}	21.00	19.61	16.55	13.64	11.53	9.98	8.56	7.52					
Ti^{+2}	20.00	18.99	16.52	13.75	11.50	9.86	8.58	7.52					
Ti^{+3}	19.00	18.24	16.27	13.82	11.58	9.84	8.55	7.53					
V	23.00	20.90	17.23	14.39	12.15	10.43	9.05	7.95	7.05	6.31	5.69	5.15	4.70
V^{+1}	22.00	20.56	17.37	14.36	12.15	10.44	9.05	7.95					
V^{+2}	21.00	19.94	17.35	14.46	12.12	10.41	9.07	7.96					
V^{+3}	20.00	19.19	17.11	14.54	12.19	10.38	9.04	7.97					
Cr	24.00	21.84	18.05	15.11	12.78	10.98	9.55	8.39	7.44	6.67	6.01	5.45	4.97
Cr^{+1}	23.00	21.50	18.20	15.07	12.78	10.99	9.54	8.40					
Cr^{+2}	22.00	20.89	18.18	15.18	12.75	10.97	9.56	8.40					
Cr^{+3}	21.00	20.15	17.96	15.26	12.82	10.94	9.53	8.41					
Mn	25.00	22.77	18.88	15.84	13.41	11.54	10.04	8.84	7.85	7.03	6.34	5.75	5.25
Mn^{+1}	24.00	22.44	19.02	15.79	13.42	11.55	10.04	8.84					

Mn^{+2}	23.00	21.84	19.01	15.90	13.38	11.53	10.06	8.84					
Mn^{+3}	22.00	21.10	18.80	15.99	13.45	11.50	10.03	8.85					
Mn^{+4}	21.00	20.30	18.42	15.97	13.54	11.53	10.00	8.82					
Fe	26.00	23.71	19.71	16.56	14.05	12.11	10.54	9.29	8.25	7.39	6.67	6.06	5.53
Fe^{+1}	25.00	23.39	19.85	16.52	14.05	12.12	10.54	9.29					
Fe^{+2}	24.00	22.79	19.85	16.62	14.02	12.09	10.56	9.29					
Fe^{+3}	23.00	22.06	19.65	16.71	14.08	12.06	10.54	9.30					
Fe^{+4}	22.00	21.26	19.28	16.71	14.18	12.09	10.50	9.28					
Co	27.00	24.65	20.54	17.29	14.69	12.67	11.05	9.74	8.66	7.77	7.01	6.37	5.82
Co^{+1}	26.00	24.33	20.68	17.25	14.70	12.68	11.04	9.74					
Co^{+2}	25.00	23.74	20.69	17.35	14.66	12.66	11.07	9.74					
Co^{+3}	24.00	23.01	20.50	17.44	14.72	12.63	11.04	9.76					
Ni	28.00	25.60	21.37	18.03	15.34	13.25	11.56	10.20	9.08	8.14	7.35	6.68	6.11
Ni^{+1}	27.00	25.28	21.52	17.98	15.34	13.25	11.55	10.20					
Ni^{+2}	26.00	24.69	21.53	18.08	15.30	13.24	11.58	10.19					
Ni^{+3}	25.00	23.97	21.35	18.18	15.36	13.20	11.56	10.21					
Cu	29.00	26.54	22.21	18.76	15.98	13.82	12.07	10.66	9.46	8.52	7.70	7.00	6.40
Cu^{+1}	28.00	26.22	22.35	18.71	15.99	13.83	12.07	10.66					
Cu^{+2}	27.00	25.64	22.37	18.81	15.95	13.81	12.09	10.65					
Cu^{+3}	26.00	24.93	22.20	18.91	16.00	13.77	12.07	10.68					
Zn	30.00	27.48	23.05	19.50	16.64	14.40	12.59	11.12	9.91	8.90	8.05	7.32	6.70
Zn^{+2}	28.00	26.59	23.32	19.55	16.60	14.40	12.61	11.12					
Ga	31.00	28.43	23.89	20.35	17.29	14.98	13.11	11.59	10.33	9.29	8.40	7.64	6.99
Ga^{+1}	30.00	28.12	24.03	20.19	17.30	14.99	13.10	11.60					
Ga^{+2}	28.00	26.84	23.90	20.39	17.30	14.93	13.11	11.61					
Ge	32.00	29.37	24.73	20.99	17.95	15.57	13.63	12.06	10.76	9.68	8.76	7.97	7.29
Ge^{+2}	30.00	28.50	24.91	21.03	17.92	15.57	13.65	12.06					
Ge^{+4}	28.00	27.02	24.45	21.18	18.05	15.53	13.60	12.07					
As	33.00	30.32	25.58	21.74	18.61	16.16	14.16	12.54	11.19	10.07	9.11	8.30	7.60
As^{+1}	32.00	30.02	25.72	21.68	18.63	16.16	14.16	12.54					
As^{+2}	31.00	29.45	25.76	21.77	18.58	16.16	14.18	12.53	11.62	10.46	9.47	8.63	7.91
As^{+3}	30.00	28.75	25.62	21.88	18.61	16.11	14.17	12.53	12.06	10.86	9.84	8.97	8.21
Se	34.00	31.26	26.42	22.49	19.28	16.75	14.69	13.02	12.50	11.26	10.21	9.31	8.53
Br	35.00	32.21	27.27	23.24	19.95	17.35	15.22	13.50					
Kr	36.00	33.16	28.12	24.00	20.62	17.95	15.76	13.98					

(continued)

Table 4.16   MEAN ATOMIC SCATTERING FACTORS*—continued

$\frac{\sin\theta}{\lambda}$ Å⁻¹	0.0	0.1	0.2	0.3	0.4	0.5	0.6	0.7	0.8	0.9	1.0	1.1	1.2
Rb	37.00	34.11	28.97	24.75	21.29	18.55	16.30	14.47	12.94	11.66	10.58	9.65	8.84
Rb⁺¹	36.00	33.82	29.11	24.70	21.31	18.55	16.30	14.48					
Sr	38.00	35.06	29.83	25.51	21.96	19.15	16.84	14.96	13.39	12.07	10.95	9.99	9.16
Y	39.00	36.01	30.68	26.28	22.64	19.76	17.39	15.46	13.84	12.48	11.32	10.34	9.48
Zr	40.00	36.96	31.54	27.04	23.32	20.37	17.94	15.95	14.29	12.89	11.70	10.68	9.80
Zr⁺⁴	36.00	34.72	31.39	27.25	23.39	20.31	17.92	15.97					
Nb	41.00	37.91	32.40	27.81	24.01	20.98	18.49	16.45	14.74	13.31	12.08	11.04	10.13
Mo	42.00	38.86	33.25	28.57	24.69	21.60	19.04	16.95	15.20	13.73	12.46	11.39	10.45
Mo⁺¹	41.00	38.59	33.39	28.51	24.72	21.59	19.04	16.96					
Tc	43.00	39.81	34.12	29.34	25.38	22.21	19.60	17.46	15.65	14.15	12.85	11.74	10.78
Ru	44.00	40.76	34.98	30.12	26.07	22.83	20.16	17.96	16.12	14.57	13.24	12.10	11.11
Rh	45.00	41.72	35.84	30.89	26.76	23.46	20.72	18.47	16.58	14.99	13.63	12.46	11.45
Pd	46.00	42.67	36.70	31.67	27.46	24.08	21.28	18.98	17.05	15.42	14.02	12.82	11.78
Ag	47.00	43.63	37.57	32.44	28.16	24.71	21.85	19.50	17.52	15.85	14.42	13.19	12.12
Ag⁺¹	46.00	43.37	37.71	32.38	28.18	24.70	21.85	19.50					
Cd	48.00	44.58	38.44	33.22	28.85	25.34	22.42	20.02	17.99	16.28	14.81	13.56	12.46
In	49.00	45.53	39.31	34.00	29.56	25.97	22.99	20.53	18.46	16.71	15.21	13.93	12.80
Sn	50.00	46.49	40.17	34.78	30.26	26.60	23.56	21.05	18.93	17.15	15.61	14.30	13.15
Sb	51.00	47.45	41.05	35.57	30.96	27.74	24.14	21.58	19.41	17.59	16.02	14.67	13.49
Te	52.00	48.40	41.92	36.35	31.67	27.87	24.71	22.10	19.89	18.03	16.42	15.05	13.84
I	53.00	49.36	42.79	37.14	32.38	28.51	25.29	22.63	20.37	18.47	16.83	15.42	14.19
Xe	54.00	50.32	43.66	37.93	33.09	29.16	25.87	23.16	20.86	18.92	17.24	15.80	14.54
Cs	55.00	51.27	44.54	38.72	33.80	29.80	26.46	23.69	21.34	19.36	17.65	16.18	14.90
Ba	56.00	52.23	45.41	39.51	34.51	30.44	27.04	24.22	21.83	19.81	18.07	16.57	15.25
La	57.00	53.19	46.29	40.30	35.23	31.09	27.63	24.76	22.32	20.26	18.48	16.95	15.61
Ce	58.00	54.15	47.16	41.09	35.94	31.74	28.22	25.30	22.81	20.71	18.90	17.34	15.97
Pr	59.00	55.11	48.04	41.89	36.66	32.39	28.81	25.84	23.31	21.17	19.32	17.72	16.33
Nd	60.00	56.07	48.92	42.69	37.38	33.04	29.40	26.38	23.80	21.62	19.74	18.11	16.69
Pm	61.00	57.02	49.80	43.48	38.10	33.69	29.99	26.92	24.30	22.08	20.16	18.51	17.05
Sm	62.00	57.98	50.68	44.28	38.82	34.35	30.59	27.46	24.80	22.54	20.58	18.90	17.42
Eu	63.00	58.94	51.56	45.08	39.55	35.01	31.19	28.01	25.30	23.00	21.01	19.29	17.79
Gd	64.00	59.91	52.45	45.88	40.27	35.66	31.79	28.56	25.80	23.46	21.44	19.69	18.16
Tb	65.00	60.87	53.33	46.68	41.00	36.33	32.39	29.11	26.31	23.93	21.87	20.09	18.53
Dy	66.00	61.83	54.21	47.49	41.73	36.99	32.99	29.66	26.81	24.39	22.30	20.49	18.90
Ho	67.00	62.79	55.10	48.29	42.46	37.65	33.59	30.21	27.32	24.86	22.73	20.89	19.27

Element													
Er	68.00	63.75	55.98	49.10	43.19	38.31	34.20	30.76	27.83	25.33	23.17	21.29	19.65
Tm	69.00	64.71	56.87	49.90	43.92	38.98	34.81	31.32	28.34	25.80	23.60	21.70	20.03
Yb	70.00	65.67	57.75	50.71	44.66	39.65	35.42	31.88	28.85	26.28	24.04	22.11	20.40
Lu	71.00	66.64	58.64	51.52	45.39	40.32	36.03	32.44	29.37	26.75	24.48	22.51	20.78
Hf	72.00	67.60	59.53	52.33	46.13	40.99	36.64	33.00	29.88	27.23	24.92	22.92	21.17
Ta	73.00	68.56	60.42	53.14	46.86	41.66	37.25	33.56	30.40	27.70	25.36	23.33	21.55
W	74.00	69.52	61.31	53.95	47.60	42.33	37.87	34.12	30.92	28.18	25.80	23.74	21.93
Re	75.00	70.49	62.20	54.76	48.34	43.01	38.48	34.69	31.44	28.66	26.25	24.16	22.32
Os	76.00	71.45	63.09	55.58	49.08	43.68	39.10	35.26	31.96	29.14	26.70	24.57	22.70
Ir	77.00	72.42	63.98	56.39	49.83	44.36	39.72	35.82	32.48	29.63	27.14	24.99	23.09
Pt	78.00	73.38	64.87	57.21	50.57	45.04	40.34	36.39	33.01	30.11	27.59	25.41	23.48
Au	79.00	74.35	65.77	58.02	51.31	45.72	40.96	36.96	33.53	30.60	28.04	25.83	23.87
Au^{+1}	78.00	74.14	65.88	57.96	51.35	45.70	40.97	36.97					
Hg	80.00	75.31	66.66	58.84	52.06	46.40	41.59	37.54	34.06	31.08	28.50	26.25	24.27
Hg^{+2}	78.00	74.65	66.90	58.79	52.05	46.41	41.59	37.53					
Tl	81.00	76.27	67.55	59.66	52.81	47.08	42.22	38.12	34.60	31.59	28.68	26.68	24.67
Tl^{+1}	80.00	76.07	67.67	59.59	52.84	47.06	42.23	38.12					
Tl^{+3}	78.00	75.03	67.82	59.71	52.76	47.08	42.24	38.10					
Pb	82.00	77.24	68.45	60.48	53.56	47.77	42.85	38.69	35.13	32.08	29.42	27.11	25.07
Pb^{+3}	79.00	76.00	68.71	60.53	53.50	47.46	42.87	38.68					
Bi	83.00	78.20	69.34	61.30	54.30	48.45	43.47	39.27	35.66	32.57	29.87	27.53	25.46
Po	84.00	79.17	70.24	62.12	55.05	49.14	44.10	39.85	36.19	33.06	30.33	27.96	25.86
At	85.00	80.13	71.13	62.94	55.80	49.82	44.73	40.43	36.73	33.55	30.79	28.38	26.26
Rn	86.00	81.10	72.03	63.76	56.56	50.51	45.36	41.01	37.26	34.05	31.25	28.81	26.66
Fr	87.00	82.07	72.93	64.58	57.31	51.20	45.99	41.59	37.80	34.55	31.71	29.24	27.06
Ra	88.00	83.03	73.82	65.41	58.06	51.89	46.63	42.17	38.34	35.04	32.17	29.67	27.46
Ac	89.00	84.00	74.72	66.23	58.82	52.58	47.26	42.75	38.88	35.54	32.64	30.10	27.87
Th	90.00	84.97	75.62	67.06	59.57	53.27	47.90	43.34	39.42	36.05	33.10	30.54	28.27
Pa	91.00	85.93	76.52	67.88	60.33	53.97	48.53	43.93	39.96	36.55	33.57	30.97	28.68
U	92.00	86.90	77.42	68.71	61.09	54.66	49.17	44.51	40.50	37.05	34.04	31.41	29.09

* Condensed from 'International Tables of X-ray Crystallography', Kynoch Press, Birmingham, England 1962.
Note: For elements of atomic number 22 or more factors are from Thomas–Fermi–Dirac Model.

Table 4.17   MASS ABSORPTION COEFFICIENT μ/ρ, CORRECTED FOR SCATTERING*

Absorber		Line	Ag 0.5609 / 0.4970	Pd 0.5869 / 0.5205	Rh 0.6147 / 0.5456	Mo 0.7107 / 0.6323	Zn 1.4364 / 1.2952	Cu 1.5418 / 1.3922	Ni 1.6591 / 1.5001	Co 1.7902 / 1.6207	Fe 1.9373 / 1.7565	Mn 2.1031 / 1.9102	Cr 2.2909 / 2.0848	Ti 2.7496 / 2.5138
H	1	$K\alpha$	0.371	0.373	0.375	0.381	0.425	0.434	0.446	0.462	0.482	0.509	0.545	0.658
		$K\beta$	0.366	0.368	0.370	0.376	0.414	0.421	0.430	0.442	0.458	0.479	0.506	0.595
He	2	$K\alpha$	0.195	0.197	0.199	0.207	0.348	0.384	0.430	0.490	0.569	0.673	0.813	1.261
		$K\beta$	0.190	0.192	0.194	0.200	0.306	0.334	0.369	0.414	0.474	0.554	0.661	1.010
Li	3	$K\alpha$	0.187	0.191	0.196	0.217	0.611	0.716	0.850	1.025	1.254	1.556	1.960	3.260
		$K\beta$	0.178	0.181	0.185	0.200	0.493	0.572	0.673	0.804	0.978	1.209	1.520	2.533
Be	4	$K\alpha$	0.229	0.239	0.251	0.307	1.245	1.498	1.823	2.245	2.796	3.526	4.474	7.636
		$K\beta$	0.208	0.215	0.224	0.258	0.958	1.149	1.393	1.711	2.130	2.688	3.439	5.884
B	5	$K\alpha$	0.279	0.295	0.314	0.393	1.965	2.386	2.927	3.628	4.545	5.757	7.378	12.58
		$K\beta$	0.245	0.257	0.270	0.327	1.490	1.806	2.213	2.742	3.438	4.365	5.614	9.674
C	6	$K\alpha$	0.400	0.432	0.464	0.625	3.764	4.603	5.680	7.075	8.899	11.31	13.53	24.83
		$K\beta$	0.333	0.356	0.383	0.495	2.815	3.446	4.257	5.310	6.697	8.542	11.02	19.08
N	7	$K\alpha$	0.544	0.597	0.658	0.916	6.132	7.524	9.311	11.62	14.64	18.63	23.95	40.96
		$K\beta$	0.433	0.471	0.515	0.700	4.557	5.603	6.950	8.698	11.00	14.05	18.16	31.47
O	8	$K\alpha$	0.740	0.821	0.916	1.313	9.337	11.48	14.22	17.76	22.39	28.48	36.61	62.52
		$K\beta$	0.570	0.628	0.696	0.980	6.917	8.53	10.60	13.28	16.80	21.48	27.76	48.07
F	9	$K\alpha$	0.943	1.056	1.187	1.737	12.79	15.72	19.48	24.34	30.67	39.01	50.10	85.38
		$K\beta$	0.710	0.790	0.884	1.276	9.459	11.67	14.51	18.19	23.02	29.43	38.02	65.72
Ne	10	$K\alpha$	1.308	1.471	1.662	2.463	18.53	22.79	28.23	35.26	44.41	56.43	72.39	123.0
		$K\beta$	0.965	1.082	1.218	1.792	13.96	16.91	21.03	26.37	33.36	42.62	55.01	94.82
Na	11	$K\alpha$	1.644	1.857	2.106	3.147	23.96	29.45	36.46	45.50	57.24	72.65	93.07	157.5
		$K\beta$	1.198	1.350	1.528	2.275	18.05	21.87	27.19	34.05	43.05	54.95	70.83	121.7

Radiation

Element	Z												
Mg	12	2.171 / 1.568	2.458 / 1.774	2.794 / 2.014	4.202 / 3.024	32.15 / 24.26	39.49 / 29.35	48.86 / 36.47	69.92 / 45.65	76.55 / 57.65	97.01 / 73.50	124.1 / 94.60	209.0 / 161.9
Al	13	2.671 / 1.917	3.030 / 2.174	3.450 / 2.474	5.208 / 3.736	39.90 / 30.14	48.98 / 36.55	60.53 / 45.24	75.37 / 56.57	94.57 / 71.35	119.6 / 90.82	152.7 / 116.7	255.9 / 198.7
Si	14	3.330 / 2.379	3.783 / 2.703	4.312 / 3.082	6.525 / 4.672	50.06 / 37.84	61.42 / 45.73	75.87 / 56.75	94.41 / 70.92	118.4 / 89.40	149.4 / 113.7	190.8 / 146.0	318.8 / 247.9
P	15	3.992 / 2.842	4.539 / 3.234	5.179 / 3.692	7.852 / 5.614	60.04 / 45.47	73.58 / 54.88	90.78 / 68.02	112.8 / 84.90	141.1 / 106.8	178.0 / 135.6	226.4 / 173.7	375.9 / 293.3
S	16	5.014 / 3.561	5.705 / 4.056	6.513 / 4.824	10.43 / 7.403	84.24 / 57.03	92.00 / 68.74	113.3 / 85.08	140.5 / 106.0	175.5 / 133.2	220.8 / 168.7	280.0 / 215.5	461.5 / 361.5
Cl	17	5.796 / 4.109	6.599 / 4.684	7.536 / 5.356	11.44 / 8.172	86.38 / 65.68	105.6 / 79.03	129.8 / 97.70	160.6 / 121.5	200.1 / 152.3	251.1 / 192.5	317.4 / 245.2	518.5 / 408.1
A	18	6.486 / 4.592	7.385 / 5.238	8.435 / 5.992	12.81 / 9.148	95.83 / 73.06	116.9 / 87.75	143.5 / 108.3	177.2 / 134.4	220.2 / 168.1	275.3 / 211.8	346.7 / 268.9	560.9 / 443.7
K	19	8.238 / 5.830	9.382 / 6.651	10.72 / 7.611	16.26 / 11.62	120.4 / 92.09	146.6 / 110.3	179.6 / 135.9	221.3 / 168.3	274.0 / 210.0	341.5 / 263.8	428.1 / 333.6	684.6 / 545.0
Ca	20	9.872 / 6.984	11.24 / 7.970	12.83 / 9.120	19.48 / 13.93	142.5 / 109.3	173.1 / 130.7	211.5 / 160.6	259.7 / 198.4	320.6 / 246.7	398.0 / 308.9	496.6 / 389.0	783.4 / 628.2
Sc	21	10.66 / 7.543	12.14 / 8.608	13.86 / 9.851	21.00 / 15.03	151.7 / 116.7	183.8 / 139.2	223.9 / 170.7	274.1 / 210.3	337.1 / 260.6	416.5 / 325.0	516.8 / 407.3	802.4 / 649.0
Ti	22	12.08 / 8.548	13.75 / 9.754	15.70 / 11.16	23.65 / 17.02	169.0 / 130.6	204.3 / 155.3	248.0 / 189.9	302.5 / 233.2	370.5 / 288.0	455.3 / 357.4	561.2 / 445.5	98.45 / 75.84
V	23	13.54 / 9.585	15.41 / 10.94	17.58 / 12.51	26.56 / 19.06	185.8 / 144.2	224.0 / 171.0	271.0 / 208.4	329.2 / 255.1	401.0 / 313.6	489.9 / 387.3	68.38 / 479.7(k)	116.2 / 89.57
Cr	24	15.67 / 11.11	17.83 / 12.67	20.34 / 14.49	30.67 / 22.03	210.7 / 164.3	253.1 / 194.2	305.0 / 235.9	368.7 / 287.5	446.7 / 351.8	55.67 / 431.9	70.08 / 53.30	118.8 / 91.70
Mn	25	17.39 / 12.34	19.78 / 14.08	22.54 / 16.08	33.94 / 24.42	228.3 / 179.1	273.3 / 210.7	327.8 / 255.0	394.1 / 309.5	57.20 / 376.6(k)	72.59 / 54.90	93.00 / 70.78	157.4 / 121.6

(*continued*)

Table 4.17 MASS ABSORPTION COEFFICIENT μ/ρ, CORRECTED FOR SCATTERING*—*continued*

Absorber		Ag Kα=0.5609 Kβ=0.4970	Pd 0.5869 0.5205	Rh 0.6147 0.5456	Mo 0.7107 0.6323	Zn 1.4364 1.2952	Cu 1.5418 1.3922	Ni 1.6591 1.5001	Co 1.7902 1.6207	Fe 1.9373 1.7565	Mn 2.1031 1.9102	Cr 2.2909 2.0848	Ti 2.7496 2.5138
Fe	26	19.91 / 14.14	22.63 / 16.11	25.79 / 18.41	38.74 / 27.92	254.9 / 200.2	303.7 / 235.6	362.5 / 283.9	53.48 / 342.8(k)	67.25 / 50.61	85.29 / 64.56	109.2 / 83.16	184.4 / 142.6
Co	27	21.83 / 15.52	24.79 / 17.68	28.23 / 20.19	42.31 / 30.55	271.5 / 215.7	322.0 / 251.5	48.18 / 301.6(k)	60.17 / 45.02	75.48 / 56.85	95.66 / 72.47	122.4 / 93.29	206.1 / 159.6
Ni	28	25.18 / 17.93	28.59 / 20.42	32.53 / 23.30	48.62 / 35.19	303.5 / 243.0	45.99 / 281.7(k)	56.15 / 42.49	69.88 / 53.17	89.03 / 67.17	112.7 / 85.49	144.1 / 110.0	242.1 / 187.7
Cu	29	26.55 / 18.93	30.12 / 21.54	34.24 / 24.58	51.05 / 37.03	40.41 / 30.53	49.61 / 36.91	61.32 / 45.82	76.35 / 57.31	95.81 / 72.28	121.2 / 92.01	154.7 / 118.2	259.3 / 201.3
Zn	30	29.30 / 20.92	33.22 / 23.80	37.74 / 27.13	56.09 / 40.78	45.73 / 34.57	56.10 / 41.77	69.30 / 51.83	86.23 / 64.78	108.1 / 81.65	136.7 / 103.9	174.2 / 133.3	291.1 / 226.4
Ga	31	31.04 / 22.20	35.16 / 25.24	39.90 / 28.76	59.12 / 43.11	49.49 / 37.44	60.69 / 45.22	74.92 / 56.08	93.15 / 70.05	116.7 / 88.22	147.3 / 112.1	187.6 / 143.8	312.5 / 243.5
Ge	32	33.49 / 24.00	37.90 / 27.27	42.98 / 31.04	63.43 / 46.39	54.70 / 41.34	66.91 / 49.90	82.55 / 61.85	102.6 / 77.20	128.3 / 97.14	161.9 / 123.3	205.9 / 158.0	341.8 / 266.7
As	33	36.29 / 26.07	41.04 / 29.60	46.49 / 33.66	68.36 / 50.14	60.44 / 45.81	74.03 / 55.26	91.25 / 68.44	113.3 / 85.36	141.6 / 107.3	178.4 / 136.1	226.6 / 174.1	374.8 / 293.0
Se	34	38.31 / 27.59	43.28 / 31.29	48.97 / 35.56	71.69 / 52.78	65.14 / 49.52	79.72 / 59.57	98.19 / 73.73	121.8 / 91.88	152.1 / 115.4	191.3 / 146.2	242.6 / 186.8	399.8 / 313.3
Br	35	41.95 / 30.29	47.33 / 34.32	53.50 / 38.96	77.97 / 57.61	72.75 / 55.26	88.97 / 66.55	109.5 / 82.31	135.7 / 102.5	169.2 / 128.6	212.7 / 162.7	269.2 / 207.6	441.8 / 346.9
Kr	36	44.18 / 31.98	49.79 / 36.20	56.19 / 41.05	81.49 / 60.46	78.20 / 59.47	95.57 / 71.56	117.5 / 88.44	145.4 / 110.0	181.2 / 137.9	227.4 / 174.3	287.4 / 222.0	469.4 / 369.5
Rb	37	47.51 / 34.49	53.47 / 39.01	60.26 / 44.18	83.32 / 64.78	85.88 / 65.39	104.9 / 78.61	128.8 / 97.1	159.2 / 120.6	198.2 / 151.0	248.2 / 190.6	313.2 / 242.4	509.2 / 401.8
Sr	38	50.74 / 36.95	57.03 / 41.74	64.18 / 47.22	92.04 / 68.92	93.51 / 71.30	114.1 / 85.63	140.0 / 105.6	172.9 / 131.2	214.8 / 164.0	268.7 / 206.7	338.3 / 262.4	547.3 / 433.0
Y	39	54.50 / 39.83	61.16 / 44.93	68.71 / 50.76	97.92 / 73.72	102.5 / 78.24	124.9 / 93.87	153.1 / 115.7	188.8 / 143.5	234.3 / 179.2	292.5 / 225.5	367.6 / 285.8	591.3 / 469.2
Zr	40	57.66 / 42.29	64.61 / 47.66	72.45 / 53.76	14.94 / 77.63(k)	110.7 / 84.64	134.8 / 101.5	165.1 / 124.9	203.3 / 154.8	251.9 / 193.0	313.9 / 242.5	393.6 / 306.7	629.3 / 501.0

Radiation

	Z													
Nb	41	61.23	68.49	76.65	16.23	119.5	145.4	177.8	218.7	270.5	336.5	420.9	668.6
		45.09	50.73	57.14	82.02(k)	97.71	109.6	134.8	166.9	207.7	260.5	328.8	534.1
Mo	42	63.87	71.30	79.63	19.90	145.9	177.2	216.5	265.8	328.3	407.5	508.4	802.1
		47.23	53.07	59.67	14.22	111.9	133.8	164.4	203.1	252.6	316.2	398.3	643.2
Tc	43	66.43	84.01	14.22	21.57	157.0	190.6	232.5	285.1	351.3	435.1	541.5	847.8
		49.36	55.37	62.14(k)	15.42	120.7	144.1	176.9	218.2	271.0	338.5	425.4	682.6
Ru	44	69.16	76.88	15.07	22.85	165.2	200.2	243.9	298.6	367.2	453.7	563.0	874.3
		51.65	57.83	64.78(k)	16.35	127.2	151.7	185.9	229.1	283.9	354.0	443.7	707.0
Rh	45	72.82	14.34	16.37	24.81	178.0	215.5	262.1	320.3	393.1	484.5	599.2	922.2
		54.68	61.10(k)	11.63	17.75	137.3	163.5	200.2	246.3	304.7	379.1	473.9	749.4
Pd	46	13.32	15.17	17.32	26.22	186.7	225.7	284.0	334.2	409.2	502.9	619.9	944.7
		56.28(k)	10.75	12.31	18.78	144.3	171.6	209.8	257.6	318.1	394.8	492.1	777.1
Ag	47	14.42	16.41	18.74	28.35	200.2	241.6	292.9	356.5	435.5	533.7	655.4	988.1
		10.20	11.64	13.32	20.31	155.0	184.1	224.7	275.5	339.5	420.4	522.4	812.1
Cd	48	15.11	17.20	19.63	29.68	207.8	250.5	303.1	368.1	448.4	547.8	670.1	998.4
		10.70	12.21	13.97	21.28	161.3	191.2	233.1	285.2	350.7	433.1	536.4	825.9
In	49	16.14	18.36	20.95	31.65	219.6	264.2	319.1	386.7	469.9	572.0	696.8	102.5
		11.42	13.03	14.91	22.71	170.9	202.2	246.1	300.6	368.7	454.0	560.4	853.9
Sn	50	16.96	19.29	22.01	33.21	228.2	274.0	330.2	399.2	483.5	586.3	710.6	103.0
		12.00	13.69	15.66	23.85	178.0	210.2	255.3	311.2	380.9	467.5	574.6	864.9
Sb	51	17.94	20.41	23.29	35.10	238.9	286.4	344.5	415.3	501.5	605.9	730.8	975.4
		12.71	14.50	26.58	25.23	186.8	220.3	267.1	324.9	396.5	485.1	594.0	883.5(l)
Te	52	18.54	21.09	24.04	36.21	243.8	291.8	350.1	420.9	506.5	609.2	730.6	513.1(l_{II})
		13.14	14.99	17.14	26.04	191.2	225.0	272.3	330.5	401.2	490.3	597.6	727.3(l_{I})
I	53	20.15	22.91	26.12	39.30	261.7	312.5	374.0	448.3	537.4	643.4	766.8	223.0(l_{III})
		14.29	16.29	18.63	28.29	205.9	241.7	291.9	353.4	428.7	520.6	631.5	447.8(l_{II})
Xe	54	21.03	23.91	27.24	40.94	269.5	321.2	383.3	458.0	546.8	651.2	642.2(l_{I})	243.9
		14.93	17.01	19.44	29.50	212.7	249.1	300.2	362.5	438.3	530.2	639.6(L)	470.1(l_{II})

(continued)

Table 4.17 MASS ABSORPTION COEFFICIENT μ/ρ, CORRECTED FOR SCATTERING*—*continued*

Radiation

Absorber		Ag Kα=0.5609 Kβ=0.4970	Pd 0.5869 0.5205	Rh 0.6147 0.5456	Mo 0.7107 0.6323	Zn 1.4364 1.2952	Cu 1.5418 1.3922	Ni 1.6591 1.5001	Co 1.7902 1.6207	Fe 1.9373 1.7565	Mn 2.1031 1.9102	Cr 2.2909 2.0848	Ti 2.7496 2.5138
Cs	55	22.40 / 15.90	25.46 / 18.13	29.00 / 20.71	43.54 / 31.39	283.1 / 224.2	336.6 / 261.9	400.6 / 314.9	476.9 / 379.1	567.0 / 456.9	671.3 / 550.1	669.7(I_1) / 659.8(L)	271.1 / 212.4
Ba	56	23.31 / 16.56	26.48 / 18.87	30.16 / 21.56	45.21 / 33.64	290.2 / 230.7	344.3 / 268.9	408.5 / 322.5	484.6 / 387.1	573.3 / 464.7	565.9(I_1) / 556.8(L)	417.0(I_{II}) / 555.1(I_1)	293.3 / 230.0
La	57	24.76 / 17.60	28.12 / 20.06	32.01 / 22.93	47.94 / 34.64	303.7 / 242.2	359.2 / 281.6	424.8 / 336.8	501.8 / 402.9	590.7 / 481.8	359.0(I_{II}) / 574.2(L)	197.7 / 576.6(I_1)	322.9 / 254.1
Ce	58	26.34 / 18.74	29.91 / 21.35	34.03 / 24.37	50.89 / 36.82	317.9 / 254.5	375.0 / 295.1	442.0 / 352.0	519.8 / 419.7	512.8(I_{II}) / 499.6(L)	383.1(I_{II}) / 497.1(I_1)	218.0 / 374.9(I_{II})	354.5 / 279.7
Pr	59	28.04 / 19.96	31.82 / 22.73	36.21 / 25.95	54.07 / 39.16	332.7 / 267.5	391.1 / 309.1	459.3 / 367.6	537.5 / 436.4	333.4(I_{II}) / 517.3(L)	190.6 / 518.3(I_1)	240.1 / 186.2	388.3 / 307.2
Nd	60	29.29 / 20.87	33.24 / 23.76	37.80 / 27.11	56.37 / 40.87	341.4 / 275.7	400.1 / 317.7	467.9 / 376.6	461.4(I_1) / 445.5(L)	349.4(I_{II}) / 442.4(i_1)	205.7 / 337.4(i_{II})	258.4 / 200.9	415.7 / 329.9
Pm	61	30.91 / 22.04	35.05 / 25.09	39.85 / 28.61	59.34 / 43.08	353.4 / 286.8	412.6 / 329.3	480.4 / 388.9	304.1(i_{II}) / 458.0(L)	169.4 / 458.5(i_1)	211.3 / 357.0(i_{II})	265.2 / 206.4	425.3 / 338.1
Sm	62	31.99 / 22.84	36.28 / 25.98	41.22 / 29.63	61.28 / 44.54	358.6 / 292.5	417.2 / 334.6	411.6(I_1) / 393.8(L)	314.9(I_{II}) / 390.9(i_1)	174.2 / 300.5(i_{II})	217.1 / 167.7	272.2 / 212.1	435.2 / 346.4
Eu	63	33.72 / 24.09	38.22 / 27.41	43.41 / 31.24	64.43 / 46.90	370.1 / 303.6	326.5(i_I) / 345.9(L)	276.0(I_{II}) / 340.7(i_1)	153.1 / 404.7(I_1)	189.4 / 317.7(i_{II})	235.6 / 182.4	294.7 / 230.2	468.1 / 374.0
Gd	64	34.66 / 24.79	39.26 / 28.18	44.57 / 32.11	66.04 / 48.14	371.8 / 306.9	367.1(i_{II}) / 348.1(L)	284.1(i_{II}) / 345.1(i_{II})	162.3 / 268.1(i_{II})	200.4 / 154.2	248.7 / 193.0	310.3 / 243.1	489.6 / 392.6
Tb	65	36.38 / 26.04	41.20 / 29.60	46.75 / 33.72	69.14 / 50.48	324.1(I_1) / 316.6(L)	249.3(I_{II}) / 357.6(L)	142.9 / 356.7(I_1)	175.1 / 281.9(I_{II})	215.8 / 166.5	267.3 / 208.0	332.6 / 261.3	520.7 / 518.1
Dy	66	37.83 / 27.11	42.82 / 30.80	48.56 / 35.06	71.69 / 52.41	217.7(I_{II}) / 323.2(L)	259.4(i_{II}) / 309.8(i_1)	152.7 / 242.5(i_{II})	186.9 / 143.4	229.9 / 177.8	284.1 / 221.7	352.5 / 277.8	547.4 / 442.6
Ho	67	39.83 / 28.57	45.06 / 32.45	51.07 / 36.93	75.26 / 55.11	230.1(I_{II}) / 333.8(L)	135.7 / 212.9(i_{II})	165.0 / 256.0(i_{II})	201.7 / 155.1	247.5 / 191.9	305.0 / 238.7	377.3 / 298.4	580.7 / 471.9
Er	68	41.73 / 30.03	47.19 / 34.02	53.46 / 38.70	78.62 / 57.66	241.4(I_1) / 292.2(I_1)	139.0 / 223.6(i_{II})	168.9 / 129.1	206.1 / 158.7	252.7 / 196.1	311.0 / 243.7	383.9 / 304.2	587.9 / 479.1

	Z												
Tm	69	43.04	48.63	55.06	80.81	117.6	142.2	172.7	210.6	257.8	316.9	390.6	595.2
		30.93	35.11	39.91	59.37	198.6(i$_{II}$)	230.9(l$_{II}$)	132.2	162.3	200.4	248.7	310.1	486.3
Yb	70	44.49	50.25	56.87	83.28	120.6	145.6	176.6	215.2	263.2	323.0	397.3	602.2
		32.02	36.33	41.28	61.30	206.2(l$_{II}$)	110.8	135.4	166.1	204.9	254.0	316.1	493.5
Lu	71	46.41	52.40	59.26	86.60	123.5	149.0	180.6	219.8	268.5	329.1	404.1	609.2
		33.45	37.93	43.08	63.85	215.4(l$_{II}$)	113.5	138.6	169.9	209.4	259.2	322.1	500.7
Hf	72	53.58	60.48	68.34	99.63	124.9	151.0	183.3	223.6	273.8	336.5	414.7	632.0
		38.66	43.82	49.75	73.60	302.4(l$_{II}$)	114.8	140.3	172.4	212.8	264.1	329.2	516.3
Ta	73	55.53	62.63	70.74	102.9	133.3	160.9	195.1	237.4	290.0	355.4	436.4	658.0
		40.13	45.46	51.58	76.16	103.3	122.6	149.7	183.5	226.1	279.9	347.9	540.7
W	74	57.57	64.88	73.24	106.3	142.1	171.3	207.3	251.7	306.7	374.6	458.3	682.9
		41.66	47.16	53.49	78.81	110.3	130.8	159.4	195.1	239.9	296.2	366.9	564.8
Re	75	59.37	66.87	75.42	109.1	151.4	182.0	219.8	266.3	323.6	394.0	479.9	706.0
		43.03	48.69	55.19	81.12	117.7	139.3	169.5	207.0	253.9	312.7	385.7	588.1
Os	76	60.97	68.63	77.36	111.6	158.5	190.5	229.6	277.7	336.5	408.4	495.4	719.6
		44.25	50.05	56.69	83.17	123.6	146.1	177.5	216.4	264.9	325.3	400.2	603.7
Ir	77	63.40	71.32	80.33	115.6	168.4	201.9	242.9	292.8	353.7	427.3	515.6	736.1
		46.09	52.10	58.98	86.32	131.7	155.3	188.3	229.1	279.6	342.1	419.0	623.5
Pt	78	65.68	73.93	83.09	119.2	178.5	213.7	256.3	308.2	370.9	451.9	535.0	749.6
		47.82	54.02	61.12	89.24	140.0	164.8	199.4	242.0	294.5	359.0	437.6	641.7
Au	79	67.84	76.19	85.67	122.5	188.3	224.9	269.2	322.7	386.9	463.2	552.3	758.7
		49.48	55.87	63.16	91.96	148.1	173.9	210.0	254.3	308.6	374.8	454.7	656.8
Hg	80	69.90	78.46	88.15	125.7	198.8	237.1	283.2	338.6	404.8	482.7	572.6	773.1
		51.07	57.62	65.11	94.57	156.8	183.8	221.6	267.7	324.0	392.3	474.0	675.9

(continued)

Table 4.17   MASS ABSORPTION COEFFICIENT μ/ρ, CORRECTED FOR SCATTERING*—*continued*

Radiation

Absorber		Ag Kα=0.5609 / Kβ=0.4970	Pd 0.5869 / 0.5205	Rh 0.6147 / 0.5456	Mo 0.7107 / 0.6323	Zn 1.4364 / 1.2952	Cu 1.5418 / 1.3922	Ni 1.6591 / 1.5001	Co 1.7902 / 1.6207	Fe 1.9373 / 1.7565	Mn 2.1031 / 1.9102	Cr 2.2909 / 2.0848	Ti 2.7496 / 2.5138
Tl	81	71.77 / 52.52	80.49 / 59.23	90.37 / 66.87	128.4 / 96.90	206.7 / 163.6	245.7 / 191.2	292.5 / 229.9	348.2 / 276.8	414.0 / 333.6	490.3 / 401.7	576.1 / 481.9	
Pb	82	73.74 / 54.07	82.63 / 60.93	92.67 / 68.75	131.2 / 99.31	215.4 / 171.1	255.6 / 199.6	303.1 / 239.2	359.5 / 287.2	425.6 / 344.7	500.5 / 413.0	583.1 / 492.2	
Bi	83	76.60 / 56.26	85.78 / 63.37	96.13 / 71.44	135.7 / 103.0	226.8 / 180.0	268.3 / 210.3	317.4 / 251.5	375.0 / 301.0	441.5 / 359.9	516.2 / 429.2	596.0 / 508.1	
Po	84	79.42 / 58.44	88.86 / 65.77	99.50 / 74.11	139.9 / 106.5	237.7 / 190.3	280.3 / 220.6	330.3 / 263.1	388.5 / 313.7	454.6 / 373.4	527.3 / 442.5	602.0 / 519.5	
At	85	81.40 / 60.05	90.97 / 67.53	101.7 / 76.00	122.1(l_1) / 108.8(L)	249.2 / 200.2	293.1 / 231.6	344.4 / 275.5	403.4 / 327.4	469.6 / 388.1	540.7 / 457.5	611.0 / 533.2	
Rn	86	82.38 / 60.91	91.97 / 68.44	102.7 / 76.96	122.5(l_1) / 109.8(L)	249.0 / 201.1	291.9 / 231.7	341.5 / 274.7	397.8 / 325.1	459.8 / 383.4	524.3 / 448.6		
Fr	87	83.36 / 61.77	92.98 / 69.35	103.7 / 77.92	91.18(l_{11}) / 110.8(L)	258.8 / 210.1	302.2 / 241.1	351.8 / 284.8	407.2 / 335.5	466.7 / 393.1	526.0 / 456.2		
Ra	88	85.43 / 63.47	95.17 / 71.20	106.0 / 79.91	95.03(l_{11}) / 113.1(L)	266.6 / 217.8	309.9 / 248.9	358.7 / 292.7	412.0 / 342.7	467.3 / 398.6			
Ac	89	86.25 / 64.30	95.95 / 72.04	106.7 / 80.75	97.50(l_{11}) / 102.1(l_1)	273.6 / 224.7	314.6 / 255.8	364.3 / 299.5	415.4 / 348.8				
Th	90	88.04 / 65.81	97.82 / 73.66	97.75(l_1) / 82.48(L)	100.3(l_{11}) / 75.07(l_{11})	278.6 / 230.6	320.5 / 261.1	366.1 / 304.0					
Pa	91	90.63 / 67.91	88.60(l_1) / 75.95(L)	72.19(l_{11}) / 84.95(L)	103.2(l_{11}) / 77.4(l_{11})	290.1 / 240.2	332.8 / 272.2	378.6 / 316.0					
U	92	93.18 / 69.94	89.92(l_1) / 78.17(L)	74.20(l_{11}) / 87.38(L)	105.9(l_{11}) / 79.70(l_{11})	294.9 / 246.7	336.7 / 275.2	380.7 / 320.3					

* Reproduced by permission from the *International Tables for X-ray Crystallography.*

4.4   X-ray results

4.4.1   Metal working

Table 4.18   GLIDE ELEMENTS OF METAL CRYSTALS

Structure	Metal	Low temperatures Glide plane	Glide direction	Elevated temperatures Glide plane	Glide direction	Most closely packed Lattice plane	Lattice direction
Cubic, face centred	Al, Cu, Ag, Au, Pb, Ni	(111)	[10$\bar{1}$]	(100)	[101]	1(111) 2(100) 3(110)	1[10$\bar{1}$] 2[100] 3[112]
	Cu-Au, α-Cu-Zn, α-Cu-Al, Al-Cu, Al-Zn, Au-Ag			—	—	—	—
Cubic, body centred	α-Fe	(101)	[11$\bar{1}$]	—	—	1(101) 2(100) 3(111)	1[111] 2[100] 3[110]
		(112)	[11$\bar{1}$]	—	—		
		(123)	[11$\bar{1}$]	—	—		
	W	(112)	[11$\bar{1}$]	—	—		
	Mo	(112)	[11$\bar{1}$]	(110)	[1$\bar{1}$1]	—	—
	K	(123)	[11$\bar{1}$]	(123)	[11$\bar{1}$]	—	—
	Na	(112)	[11$\bar{1}$]	(110)	[11$\bar{1}$]	—	—
		—	—	(123)	[11$\bar{1}$]	—	—
	β-Cu-Zn	(110)	[11$\bar{1}$]	—	—	—	—
		(112)(?)	—	—	—	—	—
	α-Fe-Si, 5% Si	(110)	—	—	—	—	—
Hexagonal, close packed	Mg, Zn, Cd, Be	(0001) or (10$\bar{1}$0) (10$\bar{1}$0)	[11$\bar{2}$0]	(10$\bar{1}$1) or (10$\bar{1}$2)	[11$\bar{2}$0]	(10$\bar{1}$0) (0001) (0001) (0001)	[11$\bar{2}$0]
	Ti, Zi	or (0001)				(10$\bar{1}$0) (10$\bar{1}$0)	
Tetragonal	β-Sn (white)	(110)	[001]	(110)	[$\bar{1}$11]	1(100) 2(110) 3(101)	1[001] 2[111] 3[100] 4[101]
		(100)	[001]	—	—		
		(101)	—	—	—		
		(121)	[10$\bar{1}$]	—	—		
Rhombohedral	As	—	—	—	—	1($\bar{1}$10) 2(11$\bar{1}$)	1[$\bar{1}$01]
	Sb	(111)	[10$\bar{1}$] and	—	—	—	—
	Bi	(11$\bar{1}$)	[101]	—	—	—	—
	Hg	(100) and complex	—	—	—	—	—

4.4.2   Crystal structure

Crystal structural data for free elements are given in Table 4.25. The coordination number, that is the number of nearest neighbours in contact with an atom, is listed in column 4 and the distances in column 5. In complex structures, such as α Mn where the coordination is not exact, no symbol is used and the range of distances between near neighbours is given.

A co-ordination symbol x in column 4 indicates that each atom has x equidistant nearest neighbours, at a distance from it (in kX-units) specified in column 5. The symbol x, y indicates that a

Table 4.19 PRINCIPAL TWINNING ELEMENTS FOR METALS

Crystal structure	Twinning plane, K_1	Twinning direction, η_1	Second undistorted plane, K_2	Direction, η_2	Shear
B.c.c.	{112}	$\langle 11\bar{1}\rangle$	{11$\bar{2}$}	$\langle 111\rangle$	0.707
F.c.c.	{111}	$\langle 11\bar{2}\rangle$	{11$\bar{1}$}	$\langle 112\rangle$	0.707
Rhombohedral (As, Sb, Bi)	{110}	$\langle 00\bar{1}\rangle$	{001}	$\langle 110\rangle$	
All c.p.h.	{10$\bar{1}$2}	$\langle\bar{1}011\rangle$	{$\bar{1}$012}	$\langle 10\bar{1}1\rangle$	
	{11$\bar{2}$1}	$\langle 11\bar{2}6\rangle$	{0001}	$\langle 11\bar{2}0\rangle$	
Some c.p.h.	{11$\bar{2}$2}	$\langle 11\bar{2}3\rangle$	{11$\bar{2}$4}	$\langle 22\bar{4}3\rangle$	
	{11$\bar{2}$3}	—	—	—	
Hg	{135}	$\langle\bar{1}21\rangle$	{$\bar{1}$11}	$\langle 0\bar{1}1\rangle$	
Diamond cubic (Ge)	{111}	$\langle 11\bar{2}\rangle$	{11$\bar{1}$}	$\langle 112\rangle$	0.707
Tetragonal (In, β-Sn)	{301}	$\langle\bar{1}03\rangle$	{$\bar{1}$01}	$\langle 101\rangle$	
	{101}	$\langle 10\bar{1}\rangle$	{$\bar{1}$01}	$\langle 101\rangle$	
	{130}	$\langle 3\bar{1}0\rangle$	{11$\bar{0}$}	$\langle 110\rangle$	0.299
	{172}	$\langle 312\rangle$	{112}		0.228
Orthorhombic α-U	{112}			$\langle 312\rangle$	0.228
	{121}			$\langle 311\rangle$	0.329
	{176}	$\langle 512\rangle$	{1$\bar{1}$1}	$\langle 123\rangle$	0.214

From C. S. Barrett and T. B. Massalski, 'Structure of Metals', 1980.[2]

given atom has x equidistant nearest neighbours, and y equidistant neighbours lying a small distance further away. These distances are given in column 5. In complex structures, such as α-Mn, where the co-ordination is not exact, no symbol is used, and the range of distances between near neighbours is given in column 5.

The Goldschmidt atomic radii given in column 6 are the radii appropriate to 12-fold co-ordination. In the case of the f.c.c. and c.p.h. metals the radius given is one-half of the measured interatomic distance, or of the mean of the two distances for the hexagonal packing. In the case of the b.c.c. metals, where the measured interatomic distances are for 8-fold co-ordination, a numerical correction has been applied. In some cases, where the pure element crystallises in a structure having a low degree of co-ordination, or where the co-ordination is not exact, it is possible to find some compound or solid solution in which the element exists in 12-fold co-ordination, and hence to calculate its appropriate radius. In a few cases no correction for co-ordination has been attempted, and here the figures, given in parentheses, are one-half of the smallest interatomic distances. It should be emphasised that the Goldschmidt radii must not be regarded as constants subject only to correction for co-ordination and applicable to all alloy systems: they may vary with the solvent or with the degree of ionisation, and they depend to some extent on the filling of the Brillouin zones.

Ionic radii vary largely with the valency, and to a smaller extent with co-ordination. The values given in column 8 are appropriate to 6-fold co-ordination, and have been derived either by direct measurement or by methods similar to those outlined for the atomic radii. All are based, ultimately, on the value of 1.32 Å obtained for O^{2+} ions by Wasastjerna,[21] using refractivity measurements. Ionic radii are also affected by the charge on neighbouring ions: thus in CaF_2 the fluorine ion is 3% smaller than in KF, where the metal carries a greater charge. It is not possible to give a simple correction factor, applicable to all ions: the effect is specific and is especially marked in structures of low co-ordination. Figures in arbitrary units indicating the power of one ion to bring about distortion in a neighbour (its 'polarising power'), and indicating the susceptibility of an ion to such distortion (its 'polarisability') are given in columns 9 and 10, respectively.

The crystal structures of alloys and compounds are listed in Chapter 6, Table 6.1. Other sources of data are references 12 and Pearson[22] which is particularly valuable as the variation of lattice parameters with composition as well as structure is given. Structures are generally referred to standard types which are listed in Pearson and in Table 6.2 in Chapter 6. Further information on crystallography can be obtained from International Tables for X-ray Crystallography.[9]

Table 4.20 ROLLING TEXTURES IN METALS AND ALLOYS

Metal or alloy	Texture		
	1	2	3
Face-centred cubic			
Cu	(110)/[11̄2]		
Cu	(123)/[12̄1]		
Cu*	(123)/[12̄1]		
Cu 70%–Zn 30%	(110)/[112̄]		
Cu 70%–Zn 30%*	(110)/[11̄2]		
Cu + 12 at. % Al	(123)/[12̄1]	(110)/[1̄12]	
Cu + 1.5 at. % Al	(110)/[11̄2]	(110)/[1̄12]	
Cu + 3 at. % Au	(123)/[12̄1]	(110)/[1̄12]	
Cu + 29.6 at. % Ni	(123)/[12̄1]	(110)/[1̄12]	
Cu + 49 at. % Ni	(123)/[12̄1]		
Ni	(123)/[12̄1]		
Au	(123)/[12̄1]		
Au + 10 at. % Cu	(110)/[1̄12]		
Al	(123)/[12̄1]		
Al	(110)/[1̄12]		
Al + 2 at. % Cu	(110)/[11̄2]		
Al + 1.25 at. % Si	(123)/[12̄1]		
Al + 0.7 at. % Mg	(110)/[11̄2]	(123)/[12̄1]	
Ag	(110)/[1̄12]		
Pb + 2 wt. % Sb	(110)/[011]		
Body-centred cubic			
α-Fe	(100)/[011]		
α-Fe	(100)/[011]	(112)/[11̄0]	(111)/[112̄]
Mo	(100)/[011]		
W	(100)/[011]		
V	(100)/[011]	(112)/[11̄0]	(111)/[112̄]
Fe + 4.16 wt. % Si	(100)/[011]	Scatter increases with Si Content	

Metal or alloy	Texture		
	1	2	3
Body-centred cubic (continued)			
Fe + 35 wt % Co	(100)/[011]		
Fe + 35 wt % Ni	(100)/[011]		
β-Brass	(100)/[011]	(112)/[11̄0]	(111)/[112̄]; (111)/[11̄0]
Hexagonal close-packed			
Be, $\frac{c}{a}=1.5847$	(0001)/[101̄0]		
Ti, $\frac{c}{a}=1.5873$	(0001) tilted approx. 25–30° round RD out of rolling plane; [101̄0] parallel RD		
Zr, $\frac{c}{a}=1.5893$	(0001)/[112̄0] (0001) tilted 30–40° round RD out of rolling plane; [101̄0] parallel Rd		
Mg, $\frac{c}{a}=1.6235$	(0001) parallel rolling plane		
β-Co, $\frac{c}{a}=1.623$	(0001) parallel rolling plane		
Zn, $\frac{c}{a}=1.8563$	(0001) tilted 20° round transverse direction out of rolling plane		
Cd, $\frac{c}{a}=1.8859$	(0001)/[101̄0] (0001) tilted ∼15° out of rolling plane round transverse direction		
Mg + Al (<4% by wt)			
Mg + 2% Mn			
Mg + 0.4% Co			
Rhombohedral			
α-U	(102)/[010]	(012)/[02̄1]	

* Straight-reverse rolling treatment.
From A. Taylor, 'X-ray Metallography', John Wiley and Sons Inc.

Table 4.21   FIBRE TEXTURES OF DRAWN AND EXTRUDED WIRE

| Metal | Parallel to drawing direction | | Parallel to extrusion direction |
	1	*2*	
Face-centred cubic			
Al, Cu			[110]
Ni, Pd, Ag, Au, Pb, Cu + 0.47% Ag, Cu + 0.45% Sb, Cu + 1.0% As, Cu + 0.009% Bi	[111]	[100]	
Cu-Ni (<32% Ni), Cu-Al (<2.16% Al), Cu-Zn (<2.35% Zn)			[110], [113] and [110]
Cu-Al (>4.4% Al), Cu-Zn (>4.8% Zn)	[111]		
α-Brass, α-Bronze, Ni + 20% Cr, Ni-Fe, austenite, 18/8 and 12/12 Cr-Ni steel	[111]	[100]	
Body-centred cubic			
α-Fe, β-Brass			[111] and [100]
Na, α-Fe, W, V, Nb, Ta	[110]		
Hexagonal close-packed			
Mg, $\frac{c}{a} = 1.6235$	[0001]$\perp$ [11$\bar{2}$0] [10$\bar{1}$0]		[0001]
Zn, $\frac{c}{a} = 1.8563$	[0001] approx. 72° to drawing direction		
Ti, $\frac{c}{a} = 1.5873$	[10$\bar{1}$0]	[10$\bar{1}$1]	
Zr, $\frac{c}{a} = 1.5893$	[0001]$\perp$		
Se, $\frac{c}{a} = 1.131$			[11$\bar{2}$0]

From A. Taylor, 'X-ray Metallography', John Wiley and Sons Inc. 1961

Table 4.22   TEXTURES IN ELECTRODEPOSITS*

Metal	Fibre textures
Ni	[100]; [100] + [110]; [112]
Cu	[110]; [100]
Ag	[111] + [100]; [111]; [110]
Pb	[112]
Au	[110]
Fe	[111]; [112]
Co	[110]
Cr	[100] + [111] (f.c.c.); [0001] (hexagonal)
Sn	[111]; [001]
Cd	[11$\bar{2}$2]
Bi	[211]; [100]

* From C. S. Barrett and T.B. Massalski, 'Structure of Metals', McGraw-Hill, New York, 1980.

Table 4.23 TEXTURES IN EVAPORATED AND SPUTTERED FILMS*

	Metal deposited	Texture	Technique
Face centred cubic	Ag	[111]; [100]; [110]	Evaporated
	Al	[111]; [100]; [110]	Evaporated
	Au	[110]; [111]	Evaporated
	Pt	[100]; [111]	Sputtered
	Pd, Cu, Ni	[111]	Evaporated
Body centred cubic	Fe	[111]	Evaporated
	Mo	[110]	Evaporated
Hexagonal	Cd, Zn	[0001]	Evaporated
Rhombohedral	Bi	[111]; [110]	Evaporated

*From C. S. Barrett and T. B. Massalski, 'Structure of Metals', McGraw-Hill, New York, 1980.

Table 4.24 TEXTURES OF CAST METALS*

Structure	Metal	Normal to cold surface
Body centred cubic	Fe–Si (4.3% Si)	[100]
	β-Brass	[100]
Face centred cubic	Al Cu Ag Au Pb α-Brass	[100]
Hexagonal close packet[†]	Cd($c/a = 1.885$)	Columnar grains, [001]; (100) ∥ to surface Chilled surface, [001]
	Zn($c/a = 1.856$)	Columnar grains, [001]; (100) ∥ to surface Chilled surface, [001]
	Mg($c/a = 1.624$)	Columnar grains, [100]; (205) ∥ to surface and (001) 37° from it
Rhombohedral	Bi	[111]
Tetragonal	β-Sn	[110]

* From C. S. Barrett and T. B. Massalski, 'Structure of Metals', McGraw-Hill, New York, 1980.
[†] Three indices system; equivalent indices in four indices systems are as follows: (001) = (0001) = basal plane; [100] = [2$\bar{1}\bar{1}$0] = diagonal axis of type I = close packed row of atoms in basal plane; [100] normal to surface = (120) parallel to surface.

The density of a material is calculated from crystallographic data with the relation

$$p_c = \frac{n\bar{A}}{VN}$$

where n is the number of atoms contained in the unit cell of volume V, A is Avogadro's constant and $\bar{A}$ is the mean atomic weight of the atoms. $\bar{A}$ is computed from the atomic percentages p_1, p_2, etc, of the elements forming the alloy and their atomic weights A_1, A_2, etc, using the formula

$$\bar{A} = \frac{p_1A}{100} + \frac{p_2A_2}{100} + \cdots$$

Table 4.25   ATOMIC AND IONIC RADII

1	2	3	4	5	6	7	8	9	10
		As element				In ionic crystals			
Atomic number	Symbol	Type of structure	Co-ordination No.	Inter-atomic distances	Gold-schmidt at. radii	State of ionisation	Gold-schmidt ionic radii	Polarising power	Polarisability
1	H	c.p.h	6, 6	—	0.46	H	1.54	0.62	—
2	He	—	—	—	—		—		—
3	Li	b.c.c.	8	3.03	1.57	Li^+	0.78	1.64	0.075
4	Be	c.p.h.	6, 6	2.22; 2.28	1.13	Be^{2+}	0.34	17.30	0.028
5	B	—	—	—	0.97	B^{3+}	0.2		0.014
6	C	d. / hex.	4 / 3	1.54 / 1.42	0.77	C^{4+}	<0.2	—	—
7	N	cub.	—	—	0.71	N^{5+}	0.1–0.2	1.15	3.1
8	O	orthorh.	—	—	0.60	O^{2-}	1.32	0.57	0.99
9	F	—	—	—		F^-	1.33		
10	Ne	f.c.c.	12	3.20	1.60	—		—	
11	Na	b.c.c.	8	3.71	1.92	Na^+	0.98	1.04	0.21
12	Mg	c.p.h.	6, 6	3.19; 3.20	1.60	Mg^{2+}	0.78	3.29	0.12
13	Al	f.c.c.	12	2.86	1.43	Al^{3+}	0.57	9.23	0.065
14	Si	d.	4	2.35	[1.17]	Si^{4-} / Si^{4+}	1.98 / 0.39	— / 26.30	— / 0.043
15	P	orthorh.	3	2.18	[1.09]	P^{5+}	0.3–0.4	—	—
16	S	f.c. orthorh.	—	2.12	[1.04]	S^{2-} / S^{6+}	1.74 / 0.34	0.66 / 51.90	7.25 / —
17	Cl	orthorh.	1	2.14	[1.07]	Cl^-	1.81	0.30	3.05
18	A	f.c.c.	12	3.84	1.92	—		—	—
19	K	b.c.c.	8	4.62	2.38	K^+	1.33	0.57	0.85
20	Ca	f.c.c. / c.p.h.	12 / 6, 6	3.93 / 3.98; 3.99	1.97 / 2.00	Ca^{2+}	1.06	1.78	0.57
21	Sc	f.c.c. / c.p.h.	12 / 6, 6	3.20 / 3.23; 3.30	1.60 / 1.64	Sc^{3+}	0.83	4.35	0.38
22	Ti	c.p.h.	6, 6	2.91; 2.95	1.47	Ti^{2+} / Ti^{3+} / Ti^{4+}	0.76 / 0.69 / 0.64	— / — / 9.76	— / — / 0.27

No.	Element	Structure	CN	Distance	Value	Ion	Radius		
23	V	b.c.c.	8	2.63	1.36	V^{3+}	0.65	—	—
						V^{4+}	0.61	—	—
						V^{5+}	~0.4	—	—
24	Cr	b.c.c. (α)	8	2.49	1.28	Cr^{3+}	0.64	—	—
		c.p.h. (β)	6, 6	2.71; 2.72	1.36	Cr^{6+}	0.3–0.4	—	—
25	Mn	cub. (α)	—	2.24–2.96	[1.12]	Mn^{2+}	0.91	—	—
		cub. (β)	—	2.36–2.68	[1.18]	Mn^{3+}	0.70	—	—
		f.c.t. (γ)	8, 4	2.58; 2.67	~1.37	Mn^{4+}	0.52	—	—
26	Fe	b.c.c. (α)	8	2.48	1.28	Fe^{2+}	0.87	—	—
		f.c.c. (γ)	12	2.52	1.26	Fe^{3+}	0.67	—	—
27	Co	c.p.h. (α)	6, 6	2.49; 2.51	1.25	Co^{2+}	0.82	—	—
		f.c.c. (β)	12	2.51	1.26	Co^{3+}	0.65	—	—
28	Ni	c.p.h. (α)	6, 6	2.49; 2.49	1.25	Ni^{2+}	0.78	—	—
		f.c.c. (β)	12	2.49	1.25				
29	Cu	f.c.c.	12	2.55	1.28	Cu^{+}	0.96	—	—
30	Zn	c.p.h.	6, 6	2.66; 2.91	1.37	Zn^{2+}	0.83	2.90	—
31	Ga	orthorh.	—	2.43–2.79	1.35	Ga^{2+}	0.62	7.80	—
32	Ge	d.	4	2.44	1.39	Ge^{4+}	0.44	20.66	—
33	As	r.	3, 3	2.51; 3.15	[1.25]	As^{3+}	0.69	—	—
						As^{5+}	~0.4	—	—
34	Se	hex.	2, 4	2.32; 3.46	[1.16]	Se^{2-}	1.91	0.55	6.4
						Se^{6+}	0.3–0.4	—	—
35	Br	orthorh.	1	2.38	[1.19]	Br^{-}	1.96	0.26	4.17
36	Kr	f.c.c.	12	3.94	1.97		—	—	—
37	Rb	b.c.c.	8	4.87	2.51	Rb^{+}	1.49	0.45	1.81
38	Sr	f.c.c.	12	4.30	2.15	Sr^{2+}	1.27	1.24	1.42
39	Y	c.p.h.	6, 6	3.59; 3.66	1.81	Y^{3+}	1.06	2.67	1.04
40	Zr	c.p.h. (α)	6, 6	3.16; 3.22	1.60	Zr^{4+}	0.87	5.28	—
		b.c.c. (β)	8	3.12	1.61				

(continued)

Table 4.25   ATOMIC AND IONIC RADII—*continued*

		As element				In ionic crystals			
Atomic number	Symbol	Type of structure	Co-ordination No.	Inter-atomic distances	Gold-schmidt at. radii	State of ionisation	Gold-schmidt ionic radii	Polarising power	Polarisability
41	Nb	b.c.c.	8	2.85	1.47	Nb^{4+}, Nb^{5+}	0.69, 0.69	10.50	—, —
42	Mo	b.c.c.	8	2.72	1.40	Mo^{4+}, Mo^{6+}	0.68, 0.65	—	—, —
43	Tc	—	—	—	—	—	—	—	—
44	Ru	c.p.h.	6, 6	2.64; 2.70	1.34	Ru^{4+}	0.65	—	—
45	Rh	f.c.c.	12	2.68	1.34	Rh^{3+}, Rh^{4+}	0.68, 0.65	—	—, —
46	Pd	f.c.c.	12	2.75	1.37	Pd^{2+}	0.50	0.78	—
47	Ag	f.c.c.	12	2.88	1.44	Ag^{+}	1.13	1.88	—
48	Cd	c.p.h.	6, 6	2.97; 3.29	1.52	Cd^{2+}	1.03	3.54	—
49	In	f.c.t.	4, 8	3.24; 3.37	1.57	In^{3+}	0.92	7.30	—
50	Sn	d. / tetra.	4 / 4, 2	2.80 / 3.02; 3.18	1.58	Sn^{4-}, Sn^{4+}	2.15, 0.74		—, —
51	Sb	r.	3, 3	2.90; 3.36	1.61	Sb^{3+}	0.90	0.45	—
52	Te	hex.	2, 4	2.86; 3.46	[1.43]	Te^{2-}, Te^{4+}	2.11, 0.89	0.21	9.6
53	I	orthorh	—	2.70	[1.36]	I^{-}, I^{5+}	2.20, 0.94		6.28
54	Xe	f.c.c.	12	4.36	2.18	—	—	—	—
55	Cs	b.c.c.	8	5.24	2.70	Cs^{+}	1.65	0.37	2.79
56	Ba	b.c.c.	8	4.34	2.24	Ba^{2+}	1.43	0.98	2.08
57	La	c.p.h. / f.c.c.	6, 6 / 12	3.72; 3.75 / 3.75	1.87, 1.87	La^{3+}	1.22	2.01	1.56
58	Ce	c.p.h. / f.c.c.	6, 6 / 12	3.63; 3.65 / 3.63	1.82, 1.82	Ce^{3+}, Ce^{4+}	1.18, 1.02	3.84	1.20
59	Pr	hex. / f.c.c.	6, 6 / 12	3.63; 3.66 / 3.64	1.83, 1.82	Pr^{3+}, Pr^{4+}	1.16, 1.00	—	—, —
60	Nd	hex.	6, 6	3.62; 3.65	1.82	Nd^{3+}	1.15	—	—

61	—	—	—	—	—	—	—	—	—
62	Sm	b.c.c.	8	3.96	2.04	Sm^{3+}	1.13	—	—
63	Eu	c.p.h.	6, 6	3.55; 3.62	1.80	Eu^{3+}	1.13	—	—
64	Gd	c.p.h.	6, 6	3.51; 3.59	1.77	Gd^{3+}	1.11	—	—
65	Tb	c.p.h.	6, 6	3.50; 3.58	1.77	Tb^{3+} / Tb^{4+}	1.09 / 0.89	—	—
66	Dy	c.p.h.	6, 6	3.48; 3.56	1.76	Dy^{3+}	1.07	—	—
67	Ho	c.p.h.	6, 6	3.46; 3.53	1.75	Ho^{3+}	1.05	—	—
68	Er	c.p.h.	6, 6	3.45; 3.52	1.74	Er^{3+}	1.04	—	—
69	Tm	c.p.h.	6, 6	3.45; 3.52	1.74	Tm^{3+}	1.04	—	—
70	Yb	f.c.c.	12	3.87	1.93	Yb^{3+}	1.00	—	—
71	Lu	c.p.h.	6, 6	3.44; 3.51	1.73	Lu^{3+}	0.99	—	—
72	Hf	c.p.h.	6, 6	3.13; 3.20	1.59	Hf^{4+}	0.84	—	—
73	Ta	b.c.c.	8	2.85	1.47	Ta^{5+}	0.68	—	—
74	W	b.c.c. (α) / cub. (β)	8 / 12; 2, 4	2.74 / 2.82; 2.52, 2.82	1.41 / 1.41	W^{4+} / W^{6+}	0.68 / 0.65	—	—
75	Re	c.p.h.	6, 6	2.73; 2.76	1.38	—	—	—	—
76	Os	c.p.h.	6, 6	2.67; 2.73	1.35	Os^{4+}	0.67	—	—
77	Ir	f.c.c.	12	2.71	1.35	Ir^{4+}	0.66	—	—
78	Pt	f.c.c.	12	2.77	1.38	Pt^{2+} / Pt^{4+}	0.52 / 0.55	—	—
79	Au	f.c.c.	12	2.88	1.44	Au^{+}	1.37	—	—
80	Hg	r.	6	3.00	1.55	Hg^{2+}	1.12	1.59	—
81	Tl	c.p.h. / b.c.c.	6, 6 / 8	3.40; 3.45 / 3.36	1.71 / 1.73	Tl^{+} / Tl^{3+}	1.49 / 1.06	2.72	—
82	Pb	f.c.c.	12	3.49	1.75	Pb^{4-} / Pb^{2+} / Pb^{4+}	2.15 / 1.32 / 0.84	5.67	—

(continued)

Table 4.25   ATOMIC AND IONIC RADII—*continued*

				As element		In ionic crystals			
1	*2*	*3*	*4*	*5*	*6*	*7*	*8*	*9*	*10*
Atomic number	*Symbol*	*Type of structure*	*Co-ordination No.*	*Inter-atomic distances*	*Gold-schmidt at. radii*	*State of ionisation*	*Gold-schmidt ionic radii*	*Polarising power*	*Polarisability*
83	Bi	r.	3, 3	3.11; 3.47	1.82	Bi^{3+}	1.20	—	—
84	Po	monocl.	—	2.81	[1.40]	—	—	—	—
85	At	—	—	—	—	—	—	—	—
86	Rn	—	—	—	—	—	—	—	—
87	Fr	—	—	—	—	—	—	—	—
88	Ra	—	—	—	—	Ra^+	1.52	—	—
89	Ac	—	—	—	—	—	—	—	—
90	Th	f.c.c.	12	3.60	1.80	Th^{4+}	1.10	3.31	—
91	Pa	—	—	—	—	—	—	—	—
92	U	orthorh.	—	2.76	[1.38]	U^{4+}	1.05	—	—
93	Np	—	—	—	—	—	—	—	—
94	Pu	—	—	—	—	—	—	—	—
95	Am	—	—	—	—	—	—	—	—
96	Cm	—	—	—	—	—	—	—	—

4.5 X-ray fluorescence

X-ray fluorescence occurs after an electron has been ejected from a shell surrounding the nucleus of an atom. The X-radiation is characteristic of the atom from which the electron has been ejected, and hence provides a means of identifying the atomic species. The ejection of an electron may be induced by irradiating the sample with photons (X or γ-rays) electrons, protons, charged particles or, indeed, any radiation capable of creating vacancies in the inner shells of the atoms of interest in the sample. The relative merits of each technique are given in Table 4.26. A further comparison of X-ray or radio-isotope sources for X-ray fluorescent spectroscopy is given in Table 4.27. Details of suitable available isotope sources are given in Table 4.28.[23]

Analysis of fluorescent X-rays is achieved by wavelength dispersion using crystal analyser (or several in a multichannel instrument), or by energy dispersion with solid-state detectors. Wavelength dispersion offers more accurate quantitative analysis, especially for the detection of small concentrations of elements where X-ray spectra from several elements overlap. Energy dispersion is preferred when rapid or quantitative analysis is required of an unknown sample.

Examples of the detection limits for X-ray excited samples are given in Tables 4.29 and 4.30, and for ion excited samples in Table 4.31.

Accuracy levels for elemental analysis are typically:

for X-ray excitation   better than 1%.
for electron and ion excitation   1–2%.

These values can be improved with very carefully calibrated standards, but are frequently much worse, especially when the specimen surface is rough. Unlike X-ray diffraction, powdered samples are the most difficult sample form to analyse.

Table 4.26 COMPARISON OF X-RAY FLUORESCENCE TECHNIQUES

Exciting radiation	Advantages	Limitations
Electrons	High-intensity energy-regulated sources easily produced Can be focused into submicron spot size Low cost Good light element detection	Specimen must be in vacuum with source Signal-to-background ratio relatively poor
Positive ions	Better signal-to-background ratio than electrons Can be focused Very sensitive to low concentrations	Specimens must be in vacuum with source Expensive equipment
Photons X-rays or γ-rays	Convenient—specimen need not be in vacuum Widely used Wavelength can be chosen for maximum sensitivity for element of interest	Cannot be focused Not as sensitive to small samples as positive ion excited methods Light elements (<Mg difficult)

Table 4.27 COMPARISON OF X-RAY AND RADIOACTIVE SOURCES FOR X-RAY FLUORESCENT SPECTROSCOPY

Radiation source	Advantages	Limitations
X-rays	Controllable high-intensity source which can be switched off when not required Intensity can be 10^2–10^4 times that of radioisotope source	Bulky, expensive equipment—requires high voltage generator
Radio isotopes	Cheaper, portable, and smaller than X-ray systems Can be built into process plant for local on-stream analysis	Permanent radioactive hazard Low intensity means long exposure times and/or larger samples Relatively small number of available isotopes (*see* Table 4.28)

Table 4.28 RADIO-ISOTOPE SOURCES FOR X-RAY FLUORESCENCE

Nuclide	Half-life	Emission energies keV
^{55}Fe	2.7 years	5.9
^{109}Cd	453 days	22.1, 87.7
^{125}I	60 days	27
^{241}Am	458 years	12.17, 60
^{57}Co	270 days	6.4, 1.22, 144
^{238}Pu	86.4 years	12–17

Source: Jaklevic and Goulding, in ref. 23, p. 33.

Table 4.29 3σ DETECTION LIMITS FOR X-RAY EXCITED SAMPLES

	Experimental conditions			
	Wavelength dispersion Cr tube 100 s *analysis time*	2 500 W	*Energy dispersive Ag tube* 10 min *analysis time*	
			Ag *filter* 1.8 W	Ag + *W filter* 22 W
Element	3σ *Detection limits in* 10^{-9} gcm^{-2}			
Mg	2		80	
Al	3		40	
Si	3		20	
S	9		12	
Cl	9		16	
Ca	2		28	
Fe	18			140 Fe impurity in Be window
Zn	7			30
Br	28			20
Pb	30			50

Source: J. V. Gilfrich, in ref. 23, p. 405.

Table 4.30 3σ DETECTION LIMITS FOR BULK SAMPLES

	Experimental conditions: Wavelength dispersion; All measurements in p.p.m.	
Element	*Iron and steel sample W target,* 2 240 W 10 min *analysis time*	*Fe and* Ni *base alloys W target* 2 025 W 100 s
Si	170	4
P		35
S		8
Ti	1.0	
V	1.9	
Cr	4.0	1
Mn	1.4	5
Ni	5.4	
Ca	8.5	12
As	6.8	
Zr	4.6	
Mo	4.5	22
Sn	3.9	

Source: J. V. Gilfrich, in ref. 23, p. 408.

Table 4.31 EXAMPLES OF DETECTION LIMITS FOR ION EXCITATION

Sample mount	Ions	Energy MeV	Current or charge	Time	Detection limit	Detection limit criterion
1 mg cm^{-2} VYNS	α	50	1 nA	400 s	Cu 1.9×10^{-12} g Sn 3.2×10^{-12} g Pb 5.5×10^{-12} g	P/B=0.1
10–20 μg cm^{-2} carbon or nitrocellulose	p^+	5	5 μC	100–200 s	K 1×10^{-9} g cm^{-2} Cu 2×10^{-9} g cm^{-2} Br 1×10^{-9} g cm^{-2} Au 5×10^{-9} g cm^{-2}	3σ Bgd
	α	5	5 μC	100–200 s	K 1×10^{-9} g cm^{-2} Br 10×10^{-9} g cm^{-2} Au 20×10^{-9} g cm^{-2}	3σ Bgd
40 μg cm^{-2} Carbon	p^+	1.5	5 μA	30 min	Ca 0.3×10^{-12} g Cu 1×10^{-12} g Ba 20×10^{-12} g Pb 10×10^{-12} g	100 counts above Bgd
4 μm Mylar	p^+	1	10 μC	500 s	Ca 7×10^{-9} g cm^{-2} Zn 18×10^{-9} g cm^{-2} Zr 300×10^{-9} g cm^{-2} Pb 90×10^{-9} g cm^{-2}	3σ Bgd
		0.3	10 μC	500 s	Ca 3×10^{-9} g cm^{-2} Zn 5×10^{-9} g cm^{-2} Zr 30×10^{-9} g cm^{-2} Pb 23×10^{-9} g cm^{-2}	3σ Bgd

Source: J. V. Gilfrich, in ref. 23, p. 406.

REFERENCES

1. A. Taylor, 'X-ray Metallography', John Wiley and Sons Inc., 1961.
2. C. S. Barrett and T. B. Massalski, 'Structure of Metals Crystallographic Methods, Principles and Data', Pergamon Press, Oxford, 1980 (new edition of C. S. Barrett, 'Structure of Metals', McGraw-Hill, New York, 1943).
3. H. P. Klug and L. E. Alexander, 'X-ray Diffraction Procedures for Polycrystalline and Amorphous Materials', John Wiley and Sons, 1974.
4. L. V. Azaroff, 'Elements of X-ray Crystallography'. McGraw Hill, New York, 1968.
5. M. F. C. Ladd and R. A. Palmer, 'Structure Determination by X-ray Crystallography', Plenum Press, New York, 1977.
6. M. M. Woolfson, 'An Introduction to X-ray Crystallography', Second Edition. Cambridge University Press, Cambridge, 1997.
7. B. D. Cullity and S. R. Stock, 'Elements of X-ray Diffraction', Third Edition. Prentice-Hall, New Jersey, 2001.
8. C. Hammond, 'The Basics of Crystallography and Diffraction'. Oxford University Press/International Union of Crystallography, 2001.
9. International Tables for X-ray Crystallography Volume I (1952), Volume II (1959), Volume III (1962), Volume IV (1974). Published for the International Union of Crystallography by the Kynoch Press, Birmingham and subsequently re-issued as the International Tables for Crystallography Volume A (1983), Volume B (1993), Volume C (1992) by Kluwer Academic Publishers, Dordrecht, Boston and London.
10. H. F. Gobel, *Advances in X-ray Analysis*, Vol. 22, p255, 1978, Plenum Press, New York; *Advances in X-ray Analysis*, Vol. 25, pp273, 315, 1981, Plenum Press, New York.
11. H. J. Bunge, 'Texture Analysis in Materials Science Mathematical Methods', Butterworths, 1982.
12. Joint Committee on Powder Diffraction Standards—International Centre for Diffraction Data, 12 Campus Boulevard, Newtown Square, PA 19073–3273.
13. K. J. Pickard, R. F. Walker and N. G. West, *Ann. Occup. Hyg.*, Vol. 29, No. 2, pp149–167, 1985.
14. J. Durnin and K. A. Ridal, *Journal of the Iron and Steel Institute*, January 1988, p60.
15. M. R. James and J. B. Cohen, *Advances in X-ray Analysis*, Vol. 22, p241, 1978.
16. A. J. C. Wilson, 'Mathematical Theory of X-ray Powder Diffractometry', Philips Technical Library, 1963.

17. A. R. Stokes, 'Line Profiles' X-ray Diffraction by Polycrystalline Materials, H. S. Peiser, H. P. Rooksby, A. J. C. Wilson. (Eds) The Institute of Physics, London, 1955.
18. D. K. Smith and C. S. Barrett, *Advances in X-ray Analysis*, Vol. 22, p1, 1978, Plenum Press, New York.
19. C. H. White, 'The Nimonic alloys', W. Betteridge and J. Heslop (Eds), p63, 2nd ed., 1974, Edward Arnold (Publishers) Ltd.
20. I. S. Brammar and M. A. P. Dewey, 'Specimen Preparation for Electron Metallography', Blackwell Scientific Publications, Oxford, 1966.
21. J. A. Wasastjerna, *Coment. Phys. Math.*, Helsinaf, 1923, **1**, 38.
22. W. P. Pearson, 'A Handbook of Lattice Spacings and Structure of Metals and Alloys', Pergamon Press, New York.
23. H. K. Herglotz and L. S. Birks (eds), 'X-ray Spectrometry', Marcel Dekker, New York, 1978.

5 Crystallography

5.1 The structure of crystals*

5.1.1 Translation groups

Metals and alloys, like most condensed matter, are aggregates of crystals; they are built up of units, consisting of small groups of atoms regularly and indefinitely repeated throughout the body by parallel translations. If the co-ordinates of the atoms within such a group are given, then three independent translations represented by vectors a, b, c, which are not all parallel to the same plane, suffice to specify the position of any other atom in the crystal. Let the vector from an arbitrary origin to an atom be

$$r = xa + yb + zc$$

then atoms of the same kind will be found at all points

$$r_n = (n_1 + x)a + (n_2 + y)b + (n_3 + z)c$$

where n_1, n_2, n_3 may be any positive or negative integers. Such a succession of regularly arranged points in space constitutes a space lattice.

The lattice may be regarded either as a system of translations relating identical points in a structure, or as a system of points arranged in parallel and equidistant nets, each net consisting of series of parallel and equidistant rows in which the points are spaced at equal distances. The points of such an array can be arbitrarily arranged in an infinite number of ways in parallel equidistant linear rows or planar nets; they can, in other words, be referred to an infinite number of systems of three primitive vectors, but investigation has shown that any structure possessing the symmetry observed in crystals can be referred to one of 14 lattices, defined by its primitive vectors and by the character of its unit cell, the latter being the parallelepiped formed by the three translations selected as units. In general, unit translations are selected so as to give the simplest cell having edges as short as possible, but there are several cases in which a more complex cell is chosen so as to display the symmetry of the lattice, or its relation to other lattices, to greater advantage.

The system of three vectors a, b, c, is described by their lengths a, b, c and by the angles between them: $(bc) = \alpha, (ca) = \beta, (ab) = \gamma$. The face of the unit cell which is parallel to the plane of the (a) and (b) axes, and which therefore intersects the (c) axis at distance c from the origin is termed the c face. Similarly, the face parallel to the b–c axial plane is the a face, and that parallel to the a–c axial plane the b face.

The simplest cell, having points only at its corners, is termed 'primitive' and is given the symbol Γ (Schoenflies) or P (Hermann). Other cells, termed 'face centred', have points at the corners and at the centres of two or more of their faces, and are given symbols indicating the faces carrying these additional points. Thus A, B, C, F represent centring on the a, b, c and all faces respectively. Finally there is the 'body centred' cell, having points at its corners and one additional point at the intersection of the body diagonals. This is given the symbol I (Hermann). Centred cells are indicated by Schoenflies by dashes. The 14 space lattices are listed in Table 5.1.

---

* This chapter can be cross-referenced with chapters 4 (X-ray diffraction) and 6 (crystal chemistry), plus section 10.4 (electron microscopy).

Table 5.1   THE FOURTEEN SPACE LATTICES*

System	Axes	Angles	Unit cells	Symbols	
				Schoenflies	*Hermann*
Triclinic	$a \neq b \neq c$	$\alpha \neq \beta \neq \gamma$	Primitive	Γtr	P
Monoclinic	$a \neq b \neq c$	$\alpha = \gamma = 90°$	*1* Primitive	Γm	P
		β obtuse	*2 c* face centred	$\Gamma m'$	C
Orthorhombic	$a \neq b \neq c$	$\alpha = \beta = \gamma = 90°$	*1* Primitive	Γo	P
			2 c Face centred	$\Gamma o'(c)$	C
			3 All face centred	$\Gamma o''$	F
			4 Body centred	F'''	I
Tetragonal	$a = b \neq c$	$\alpha = \beta = \gamma = 90°$	*1* Primitive	Γt	P
			2 Body centred	$\Gamma t'$	I
Cubic	$a = b = c$	$\alpha = \beta = \gamma = 90°$	*1* Primitive	Γc	P
			2 All face centred	$\Gamma c'$	F
			3 Body centred	$\Gamma c''$	I
Rhombohedral	$a = b = c$	$\alpha = \beta = \gamma \neq 90°$	Primitive	Γrh	R
Hexagonal	$a = b \neq c$	$\alpha = \beta = 90°$ $\gamma = 120°$	Primitive	Γh	C

* The 14 space lattices are also referred to as Bravais lattices (after Auguste Bravais 1811–63). The centred cells are chosen to display most clearly the symmetry of the lattice, but all space lattices may be represented by primitive cells. For example, the all-face and body-centred cubic lattices may be represented by rhombohedral unit cells having $\alpha = \beta = \gamma = 60°$ and $109.48°$ respectively. These are referred to as alternative settings of the unit cell of the lattice.

5.1.2   Symmetry elements

Symmetry elements may be classified as axes, planes, and centres. A body has an axis of symmetry when rotation through a definite angle about some line through it (the axis of rotation) causes it to assume its original aspect. Crystals can only have axes of 2-, 3-, 4- and 6-fold rotation, involving coincidence after rotation through 180°, 120°, 90° and 60° respectively. If a plane passes through a body such that every point on one side of the plane stands in mirror-image relationship to a corresponding point on the other side, the plane is said to be a reflecting plane or mirror plane of symmetry. A point within a body is a centre of symmetry or centre of inversion if a line drawn from any point of the body to the centre and extended to an equal distance beyond it encounters a corresponding point. Other symmetry operations are:

Rotoinversion (or inversion) axes of symmetry, involving rotation through 180°, 120°, 90° and 60°, combined with an inversion through a centre of symmetry.

Screw axes of rotation, combining rotation about a 2-, 3-, 4- or 6-fold axis with a translation of a specified length in the direction of the axis.

Glide planes, combining reflection with a translation parallel to the plane of the mirror.

In the case of screw axes, the amount of the shift must be a rational fraction of the translation along the same axis, the denominator of the fraction being the multiplicity of the rotation. Thus for a 6-fold axis, the shift may be 1/6, 2/6, 3/6 … of the translation. In the case of the glide plane, the shift must be one half of some translation in that plane. Thus it may be $a/2$ or $b/2$ parallel to the a and b axes, or of half the face diagonal in the direction parallel to that diagonal. If the cell is centred on that particular face, the shift may be one half of the distance to the centre, i.e. of the face diagonal.

5.1.3   The point group

The point group may be defined as a group of symmetry elements distributed about a point in space, and may be conveniently visualised as an assembly of points generated by the operation of the symmetry elements in question upon a single point having co-ordinates xyz referred to specified axes, the symmetry elements passing through the origin. Thus a symmetry plane, passing through the origin and containing the x and y axes, will generate, from the point xyz, an equivalent point of which the co-ordinates are $xy\bar{z}$. These two points serve to characterise the point group.

The 32-point groups define all the ways in which axes, planes and centres of symmetry can be distributed so as to intersect in a point in space, and correspond to the 32 classes of morphological crystallography.

5.1.4 The space group

The space group may be defined as an extended network of symmetry elements distributed about the points of a space lattice, and may be visualised as an assembly of points generated by the operation of symmetry elements on a series of points situated identically in each cell of the lattice. Whereas in the point group the repeated operation of any symmetry element must ultimately bring each point back to its original position, in the space group an operation need only bring the point to an analogous position in the same or in another cell of the lattice. Thus in the space group the more complex symmetry operations of screw axes and glide planes are possible, combining translation with rotation and reflection respectively.

Point groups, placed at the points of space lattices belonging to the same system of symmetry, give rise to the simplest of the 230 space groups. The remainder are generated by replacing the simple planes and axes of the point group by glide planes and screw axes.

5.2 The Schoenflies system of point- and space-group notation

The symmetry elements chosen by Schoenflies are axes of n-fold rotation, reflection planes, rotore-flection (rather than rotoinversion) axes of symmetry and centres of inversion. The symbols assigned to the various point groups are as follows:

C_n = groups having a single n-fold axis;
C_n^h = groups having a single vertical n-fold axis, together with a horizontal reflection plane;
C_n^v = groups having a single vertical n-fold axis, together with n vertical reflection planes;
D_n = groups having an n-fold axis and n two-fold axes at right angles to it;
V = a symbol frequently used as an alternative to D_2;
O = the two cubic groups which possess the maximum possible number of rotation axes, namely three 4-fold axes parallel to the cube edges, four 3-fold axes parallel to the cube diagonals and six 2-fold axes parallel to the face diagonals;
T = the three remaining groups of the cubic system;
S_n = groups having an n-fold axis of rotary reflection.

The suffix i signifies a centre of inversion.
The suffix s signifies a single plane of symmetry.
The suffix d signifies a diagonal reflection plane, bisecting the angle between two horizontal axes.

The symbols for the space groups are simple modifications of those for the point groups: the index and subscript of the point group are combined to give the subscript of the space group symbol, and an index is added representing the order in which Schoenflies deduced the symmetry of the group. Thus C_{nh}^m represents the mth group derived from the point group C_n^h.

Table 5.2 gives the point groups, their symbols and elements of symmetry, the crystal classes with which they correspond, and the co-ordinates of equivalent points.

5.3 The Hermann–Mauguin system of point- and space-group notation

The symbols used by Hermann and Mauguin to indicate the various symmetry operations are as follows:

Rotation axes: the number 2, 3, 4 or 6 denoting the multiplicity.
Screw axes: the symbol denoting the multiplicity of rotation, with a subscript indicating the magnitude of the shift. The complete set of screw axes is 2_1; 3_1, 3_2; 4_1, 4_2, 4_3; 6_1, 6_2, 6_3, 6_4, 6_5.
Rotoinversion (or inversion) axes: $\bar{2}, \bar{3}, \bar{4}, \bar{6}$, the numeral indicating the multiplicity.
Centre of inversion: $\bar{1}$.
Reflection plane: m.
Glide plane: with shift in the a direction: a
with shift in the b direction: b
with shift of 1/2 the face diagonal: n
with shift of 1/2 the centring translation: d.

The full space group symbol consists of the translation (or lattice) symbol followed by symbols of the symmetry elements associated with specified crystallographic directions in a specified order.

Table 5.2   POINT-GROUP NOTATION (SCHOENFLIES)

System	Class no.	Schoenflies symbol	Crystal class		
			Schoenflies	Dana	Miers
Triclinic	1	C_1	Hemihedry	Asymmetric	Asymmetric
	2	C_i	Holohedry	Normal	Central
Monoclinic	3	$C_s = C_1^h$	Hemihedry	Clinohedral	Planar
	4	C_2	Hemimorphic hemihedry	Hemimorphic	Digonal polar
	5	C_2^h	Holohedry	Normal	Digonal equatorial
Orthorhombic	6	C_2^v	Hemimorphic hemihedry	Hemimorphic	Didigonal polar
	7	$V = D_2$	Enantiomorphic hemihedry	Sphenoidal	Digonal holoaxial
	8	$V^h = D_2^i$	Holohedry	Normal	Didigonal equatorial
Tetragonal	9	S_4	Tetartohedry of 2nd sort	Tetartohedral	Tetragonal alternating
	10	$V^4 = D_2^d$	Hemihedry of 2nd sort	Sphenoidal	Ditetragonal alternating
	11	C_4	Tetartohedry	Pyramidal hemimorphic	Tetragonal polar
	12	C_4^h	Paramorphic hemihedry	Pyramidal	Tetragonal equatorial
	13	C_4^v	Hemimorphic hemihedry	Hemimorphic	Ditetragonal polar
	14	D_4	Enantiomorphic hemihedry	Trapezohedral	Tetragonal holoaxial
	15	D_4^h	Holohedry	Normal	Ditetragonal equatorial
Cubic	16	T	Tetartohedry	Tetartohedral	Tesseral polar
	17	T^h	Paramorphic hemihedry	Pyritohedral	Tesseral central
	18	T^d	Hemimorphic hemihedry	Tetrahedral	Ditesseral polar
	19	O	Enantiomorphic hemihedry	Plagihedral	Tesseral holoaxial
	20	O^h	Holohedry	Normal	Ditesseral central
Rhombohedral	21	C_3	Tetartohedry	Not named	Trigonal polar
	22	$C_3^i = S_5$	Hexagonal tetartohedry of 2nd sort	Trirhombohedral	Hexagonal alternating
	23	C_3^v	Hemimorphic hemihedry	Ditrigonal pyramidal	Ditrigonal polar
	24	D_3	Enantiomorphic hemihedry	Trapezohedral	Trigonal holoaxial
	25	D_3^d	Holohedry	Rhombohedral	Dihexagonal alternating
Hexagonal	26	C_3^n	Trigonal paramorphic hemihedry	Not named	Trigonal equatorial
	27	D_3^h	Trigonal holohedry	Trigonotype	Ditrigonal equatorial
	28	C_6	Tetartohedry	Pyramidal hemimorphic	Hexagonal polar
	29	C_6^h	Paramorphic hemihedry	Pyramidal	Hexagonal equatorial
	30	C_6^v	Hemimorphic hemihedry	Hemimorphic	Dihexagonal polar
	31	D_6	Enantiomorphic hemihedry	Trapezohedral	Hexagonal holoaxial
	32	D_6^h	Holohedry	Normal	Dihexagonal equatorial

Typical example	*Symmetry elements*	*Co-ordinates of equivalent points*
—	None	xyz
Copper sulphate	Centre of inversion	$xyz,\ \bar{x}\bar{y}\bar{z}$
—	Horizontal reflecting plane	$xyz,\ xy\bar{z}$
Tartaric acid	2-fold axis	$xyz,\ \bar{x}\bar{y}z$
Gypsum	2-fold axis and horizontal plane	$xyz,\ \bar{x}\bar{y}z,\ xy\bar{z},\ \bar{x}\bar{y}\bar{z}$
Topaz	2-fold axis and vertical plane	$xyz,\ \bar{x}\bar{y}z,\ \bar{x}yz,\ x\bar{y}z$
Sulphur	Three 2-fold axes	$xyz,\ x\bar{y}\bar{z},\ \bar{x}y\bar{z},\ \bar{x}\bar{y}z$
Barytes	Three 2-fold axes and horizontal plane	$xyz,\ x\bar{y}\bar{z},\ \bar{x}y\bar{z},\ \bar{x}\bar{y}z,\ xy\bar{z},\ x\bar{y}z,\ \bar{x}yz,\ \bar{x}\bar{y}\bar{z}$
—	4-fold rotary reflection	$xyz,\ \bar{y}x\bar{z},\ \bar{x}\bar{y}z,\ y\bar{x}\bar{z}$
Chalcopyrite	4-fold axis, two 2-fold axes and diagonal vertical plane	$xyz,\ x\bar{y}\bar{z},\ \bar{x}y\bar{z},\ \bar{x}\bar{y}z,\ yxz,\ \bar{y}x\bar{z},\ y\bar{x}\bar{z},\ y\bar{x}\bar{z},\ \bar{y}\bar{x}z$
Wulfenite	4-fold axis	$xyz,\ \bar{y}xz,\ \bar{x}\bar{y}z,\ y\bar{x}z$
Scheelite	4-fold axis, horizontal plane	$xyz,\ \bar{y}xz,\ \bar{x}\bar{y}z,\ y\bar{x}z,\ xy\bar{z},\ yx\bar{z},\ \bar{x}\bar{y}\bar{z},\ y\bar{x}\bar{z}$
—	4-fold axis, vertical plane	$xyz,\ \bar{y}xz,\ \bar{x}\bar{y}z,\ y\bar{x}z,\ \bar{x}yz,\ yxz,\ x\bar{y}z,\ \bar{y}\bar{x}z$
—	4-fold axis and four 2-fold axes	$xyz,\ \bar{y}xz,\ \bar{x}\bar{y}z,\ y\bar{x}z,\ x\bar{y}\bar{z},\ \bar{y}\bar{x}\bar{z},\ \bar{x}y\bar{z},\ yx\bar{z}$
Zircon	4-fold axis, four 2-fold axes, horizontal plane	$xyz,\ \bar{y}xz,\ \bar{x}\bar{y}z,\ y\bar{x}z,\ x\bar{y}\bar{z},\ \bar{y}\bar{x}\bar{z},\ \bar{x}y\bar{z},\ \bar{y}x\bar{z}$ $xy\bar{z},\ \bar{y}x\bar{z},\ \bar{x}\bar{y}\bar{z},\ y\bar{x}\bar{z},\ x\bar{y}z,\ \bar{y}\bar{x}z,\ \bar{x}yz,\ yxz$
Ullmanite	{ Three 2-fold axes coincident with cube axes, *and* Four 3-fold axes coincident with cube diagonals	{ $xyz,\ x\bar{y}\bar{z},\ \bar{x}y\bar{z},\ \bar{x}\bar{y}z$ $zxy,\ \bar{z}x\bar{y},\ \bar{z}\bar{x}y,\ z\bar{x}\bar{y}$ $yzx,\ \bar{y}\bar{z}x,\ y\bar{z}\bar{x},\ \bar{y}z\bar{x}$
Pyrites	As for *T* plus a horizontal plane	Those of *T* plus { $xy\bar{z},\ x\bar{y}z,\ \bar{x}yz,\ \bar{x}\bar{y}\bar{z}$ $zx\bar{y},\ zxy,\ \bar{z}\bar{x}\bar{y},\ z\bar{x}y$ $yz\bar{x},\ \bar{y}\bar{z}\bar{x},\ y\bar{z}x,\ \bar{y}zx$
Blende	As for *T* plus a diagonal vertical plane	Those of *T* plus { $yxz,\ \bar{y}x\bar{z},\ y\bar{x}\bar{z},\ \bar{y}\bar{x}z$ $xzy,\ x\bar{z}\bar{y},\ \bar{x}\bar{z}y,\ \bar{x}z\bar{y}$ $zyx,\ \bar{z}\bar{y}x,\ \bar{z}y\bar{x},\ z\bar{y}\bar{x}$
Cuprite	As for *T* plus six 2-fold axes coincident with face diagonals thereby converting the three original 2-fold axes into 4-folds	Those of *T* plus { $\bar{y}x\bar{z},\ y\bar{x}\bar{z},\ \bar{y}\bar{x}z,\ yxz$ $x\bar{z}\bar{y},\ \bar{x}zy,\ xz\bar{y},\ \bar{x}\bar{z}y$ $\bar{z}\bar{y}x,\ zyx,\ z\bar{y}\bar{x},\ \bar{z}yx$
Galena	As for *O* plus a horizontal plane	Those of all the four preceding classes
—	3-fold axis	$xyz,\ zxy,\ yzx$ referred to rhombohedral axes
Dioptase	3-fold axis and centre of inversion	$xyz,\ zxy,\ yzx,\ \bar{x}\bar{y}\bar{z},\ \bar{z}\bar{x}\bar{y},\ \bar{y}\bar{z}\bar{x}$ (rhombohedral axes)
Tourmaline	3-fold axis and vertical plane	$xyz,\ zxy,\ yzx,\ yxz,\ xzy,\ zyx$ (rhombohedral axes)
Quartz	3-fold axis and three 2-fold axes	$xyz,\ zxy,\ yzx,\ \bar{y}\bar{x}\bar{z},\ \bar{x}\bar{z}\bar{y},\ \bar{z}\bar{y}\bar{x}$ (rhombohedral axes)
Calcite	3-fold axis, three 2-fold axes and diagonal vertical planes	$xyz,\ zxy,\ yzx,\ \bar{y}\bar{x}\bar{z},\ \bar{x}\bar{z}\bar{y},\ \bar{z}\bar{y}\bar{x}$ $\bar{x}\bar{y}\bar{z},\ \bar{z}\bar{x}\bar{y},\ \bar{y}\bar{z}\bar{x},\ yxz,\ xzy,\ zyx$ (rhombohedral axes)
—	3-fold axis and horizontal plane	$xyz,\ (y-x)\bar{x}z,\ \bar{y}(x-y)z,\ xy\bar{z},\ (y-x)\bar{x}\bar{z},$ $\bar{y}(x-y)\bar{z}$ (hexagonal axes)
—	3-fold axis, three 2-fold axes and horizontal plane	$xyz,\ (y-x)\bar{x}z,\ \bar{y}(x-y)z,\ (x-y)\bar{y}\bar{z},\ yx\bar{z},$ $\bar{x}(y-x)\bar{z}$ $xy\bar{z},\ (y-x)\bar{x}\bar{z},\ \bar{y}(x-y)z,\ (x-y)\bar{y}z,\ yxz,$ $\bar{x}(y-x)z$ (hexagonal axes)
Nepheline	6-fold axes	$xyz,\ y(y-x)z,\ (y-x)\bar{x}z,\ \bar{x}\bar{y}z,\ \bar{y}(x-y)z,$ $(x-y)xz$ (hexagonal axes)
Apatite	6-fold axis and horizontal plane	Those of C_6 plus $xy\bar{z},\ y(y-x)\bar{z},\ (y-x)\bar{x}\bar{z},\ \bar{x}\bar{y}\bar{z},$ $\bar{y}(x-y)\bar{z},\ (x-y)x\bar{z}$ (hexagonal axes)
Greenockite	6-fold axis and vertical plane	Those of C_6 plus $\bar{x}(y-x)z,\ (y-x)yz,\ yxz,$ $x(x-y)z,\ (x-y)\bar{y}z,\ \bar{y}\bar{x}z$ (hexagonal axes)
—	6-fold axis and six 2-fold axes	Those of C_6 plus $\bar{x}(y-x)z,\ (y-x)\bar{y}z,$ $xy\bar{z},\ x(x-y)\bar{z},\ (x-y)\bar{y}z,\ \bar{y}\bar{x}z$ (hexagonal axes)
Beryl	6-fold axis, six 2-fold axes and horizontal plane	Those of all the four preceding classes

The direction associated with a reflection or glide plane is that of its normal: no direction can be specified for a centre of inversion.

The specified directions are:

Triclinic system: none
Monoclinic system: the *b* direction, i.e. the *b* axis.
Orthorhombic system: the *a*, *b*, *c* directions, in that order.

Tetragonal
Hexagonal   } *systems*:
Rhombohedral

the *a*, *b*, and $(a - b)$ directions, in that order. The direction represented by the vector difference is one of the diagonals of the *c*-face.

Cubic system: the directions *c* $(a + b + c)$, and $(a - b)$, in that order, i.e. the *c*-axis the cube diagonal and a diagonal of the *c*-face.

If a symmetry axis has a symmetry plane normal to it, the two symbols are combined in the form of a fraction, thus $\frac{2}{m}, \frac{4_1}{d}$, alternatively written $2/m$, $4_1/d$.

If one of the specified crystallographic directions has no symmetry element associated with it, this is indicated by inserting the symbol 1 in the appropriate position in the space group symbol. The 1 may be omitted without risk of misunderstanding if it occurs at the end of the space group symbol. Symmetry symbols may also be omitted if they can be derived from those already indicated. These abbreviated symbols are termed 'short'.

As already explained, the symmetry of a space group can be derived from that of a point group by placing the latter at the points of the various lattices appropriate to the crystal system, and by using glide planes and screw axes as well as reflection planes and rotation axes. Thus the point group symbol will contain no symbol for the lattice; its symmetry planes will be indicated by *m* and its axes by numbers specifying the multiplicity and without subscripts. Thus the point group $2/m$ will be associated with the space groups $P(2/m)$; $P(2_1/m)$; $P(2/c)$; $P(2_1/c)$; $C(2/m)$; and $C(2/c)$.

5.3.1   Notes on the space-group tables

For a full description of the space groups, reference should be made to the *International Tables for Crystallography*, Vol. A (1983) published for the International Union of Crystallography by Kluwer Academic Publishers.

If there are *n* symmetry elements associated with any space group, their operation upon any single point having co-ordinates *xyz* will give rise to a total of *n* points which may be termed geometrically equivalent. If, however, the co-ordinates *xyz* are such that the point lies, say, on an axis or plane of symmetry, then the number of equivalent positions will be reduced, while if it lies at the intersection of two elements the number will be reduced still further. A knowledge of these so-called special positions is of importance, because experience has shown that they are the positions which are frequently occupied by the atoms or ions in an actual crystal. In sodium chloride, for example, the four sodium ions are situated in one set of four equivalent positions, those having co-ordinates 000, $\frac{1}{2}\frac{1}{2}0, \frac{1}{2}0\frac{1}{2}, 0\frac{1}{2}\frac{1}{2}$, whilst the four chlorine ions are situated in another set of four points, having co-ordinates $\frac{1}{2}\frac{1}{2}\frac{1}{2}, 00\frac{1}{2}, 0\frac{1}{2}0, \frac{1}{2}00$. The co-ordinates of all the special positions for each space group were given by R. W. G. Wyckoff in *The Analytical Expression of the Results of the Theory of Space Groups* (Washington Carnegie Institution, 1930) and are also listed in the *International Tables*, Vol. A.

The last column of Table 5.3 gives the missing x-ray reflections characteristic of each space group. If the unit cell is centred on one or more faces, or is body centred, certain reflections will be absent, because in directions corresponding to the missing reflections the waves scattered by the atoms at the face or body centres will be exactly out of phase with those scattered by the atoms at the cell corners. In other words, the spacings of certain planes are halved, and odd-order reflections from these planes are destroyed. Thus with the body-centred lattice all reflections are absent for which $(h + k + l)$ is odd. Again, a glide plane halves the spacings in the direction of glide, and a 2-fold screw axis halves those along the axis. Consequently, odd order reflections are missing in these directions. Similarly with a 3-fold axis; the only reflections occurring in the direction of the axis are those for which (l) is a multiple of three.

In Table 5.3, x-ray reflections of the type indicated do not occur unless indices which are underlined are even, or unless the sum of indices joined together by brackets is even. Thus:

$00\underline{l}$ means that reflections will not occur unless *l* is even.
$0\underline{kl}$ means that reflections will not occur unless $(k + l)$ is even.
$\underline{hk}0$ means that reflections will not occur unless both *h* and *k* are even.
$\underline{hkl}$ means that reflections will not occur unless $(h + k + l)$ is even.

$\underset{\smile}{hkl}$ means that reflections will not occur unless the sums of any two indices are even.

$hh\underline{l}$ means that reflections will not occur if the first two indices are equal, unless the third index, l, is even.

A subscript 3, 4 or 6 means that the marked index, or the sum of the marked indices, must be a multiple of that number for reflections to occur. Thus

$\underset{\smile}{0kl}_4$ means that reflections will not occur unless $(k + l)$ is a multiple of 4.

$\underset{\smile}{hkl}_3$ means that reflections will not occur unless $h + 2k + l$ is a multiple of 3.

Table 5.3   THE HERMANN–MAUGUIN SYSTEM OF POINT- AND SPACE-GROUP NOTATION

Space group Hermann–Mauguin Full	Short	Schoenflies	Missing spectra	Space group Hermann–Mauguin Full	Short	Schoenflies	Missing spectra
Triclinic system				Class mm2 (short mm)—C_{2v} continued			
Class 1		C_1		Pma2	Pma	C_{2v}^4	$h0\underline{l}$
	P1	C_1^1	—	Pmn2$_1$	Pmn	C_{2v}^7	$h0\underline{l}$
Class $\bar{1}$		C_i		Pcc2	Pcc	C_{2v}^3	$0k\underline{l}, h0\underline{l}$
	P$\bar{1}$	C_i^1	—	Pca2$_1$	Pca	C_{2v}^5	$0k\underline{l}, h0\underline{l}$
Monoclinic system				Pcn2	Pcn	C_{2v}^6	$0k\underline{l}, h0\underline{l}$
Class m		C_s		Pba2	Pba	C_{2v}^8	$0k\underline{l}, h0\underline{l}$
	Pm	C_s^1	—	Pbn2$_1$	Pbn	C_{2v}^9	$0k\underline{l}, h0\underline{l}$
	Pc	C_s^2	$h0\underline{l}$	Pnn2	Pnn	C_{2v}^{10}	$0k\underline{l}, h0\underline{l}$
	Cm	C_s^3	$\underset{\smile}{hkl}$	Cmm2	Cmm	C_{2v}^{11}	$\underset{\smile}{hkl}$
	Cc	C_s^4	$\underset{\smile}{hkl}, h0\underline{l}$	Cmc2$_1$	Cmc	C_{2v}^{12}	$\underset{\smile}{hkl}, h0\underline{l}$
Class 2		C_2		Ccc2	Ccc	C_{2v}^{13}	$\underset{\smile}{hkl}, 0k\underline{l}, h0\underline{l}$
	P2	C_2^1	—	Amm2	Amm	C_{2v}^{14}	$\underset{\smile}{hkl}$
	P2$_1$	C_2^2	$0\underline{k}0$	Ama2	Ama	C_{2v}^{16}	$\underset{\smile}{hkl}, h0\underline{l}$
	C2	C_2^3	$\underset{\smile}{hkl}$	Abm2	Abm	C_{2v}^{15}	$\underset{\smile}{hkl}, 0k\underline{l}$
Class $\dfrac{2}{m}$		C_{2h}		Aba2	Aba	C_{2v}^{17}	$\underset{\smile}{hkl}, 0k\underline{l}, h0\underline{l}$
				Fmm2	Fmm	C_{2v}^{18}	$\underset{\smile}{hkl}$
$P\dfrac{2}{m}$		C_{2h}^1	—	Fdd2	Fdd	C_{2v}^{19}	$\underset{\smile}{hkl}, \underset{\smile}{0kl}_4, \underset{\smile}{h0l}_4$
$P\dfrac{2_1}{m}$		C_{2h}^2	$0\underline{k}0$	Imm2	Imm	C_{2v}^{20}	$\underset{\smile}{hkl}$
$P\dfrac{2}{c}$		C_{2h}^4	$h0\underline{l}$	Ima2	Ima	C_{2v}^{22}	$\underset{\smile}{hkl}, h0\underline{l}$
$P\dfrac{2_1}{c}$		C_{2h}^5	$h0\underline{l}, 0\underline{k}0$	Iba2	Iba	C_{2v}^{21}	$\underset{\smile}{hkl}, 0k\underline{l}, h0\underline{l}$
$C\dfrac{2}{m}$		C_{2h}^3	$\underset{\smile}{hkl}$				
$C\dfrac{2}{c}$		C_{2h}^6	$\underset{\smile}{hkl}, h0\underline{l}$	Class 222 (short 22)—$D_2(V)$			
				P222		$D_2^1(V^1)$	—
Orthorhombic system				P222$_1$		$D_2^2(V^2)$	$00\underline{l}$
				P2$_1$2$_1$2		$D_2^3(V^3)$	$h00, 0k0$
Class mm2 (short mm) C_{2v}				P2$_1$2$_1$2$_1$		$D_2^4(V^4)$	$h00, 0k0, 00l$
				C222		$D_2^6(V^5)$	$\underset{\smile}{hkl}$
Pmm2	Pmm	C_{2v}^1	—	C222$_1$		$D_2^5(V^6)$	$\underset{\smile}{hkl}, 00l$
Pmc2$_1$	Pmc	C_{2v}^2	$h0\underline{l}$	F222		$D_2^7(V^7)$	$\underset{\smile}{hkl}$
				I222		$D_2^8(V^8)$	$\underset{\smile}{hkl}$
				I2$_1$2$_1$2$_1$		$D_2^9(V^9)$	$\underset{\smile}{hkl}$

Table 5.3   THE HERMANN–MAUGUIN SYSTEM OF POINT- AND SPACE-GROUP NOTATION—*continued*

Class $\dfrac{2}{m}\dfrac{2}{m}\dfrac{2}{m}$ (short *mmm*)—$D_{2h}(V_h)$

Space group Hermann–Mauguin (Full)	Short	Schoenflies	Missing spectra
$P\dfrac{2}{m}\dfrac{2}{m}\dfrac{2}{m}$	Pmmm	$D_{2h}^1(V_h^1)$	—
$P\dfrac{2_1}{m}\dfrac{2_1}{m}\dfrac{2}{n}$	Pmmn	$D_{2h}^{13}(V_h^{13})$	hk0
$P\dfrac{2_1}{n}\dfrac{2_1}{n}\dfrac{2}{m}$	Pnnm	$D_{2h}^{12}(V_h^{12})$	0kl, h0l
$P\dfrac{2}{n}\dfrac{2}{n}\dfrac{2}{n}$	Pnnn	$D_{2h}^2(V_h^2)$	0kl, h0l, hk0
$P\dfrac{2_1}{m}\dfrac{2}{m}\dfrac{2}{a}$	Pmma	$D_{2h}^5(V_h^5)$	hk0
$P\dfrac{2_1}{n}\dfrac{2}{n}\dfrac{2}{a}$	Pnna	$D_{2h}^6(V_h^6)$	0kl, h0l, hk0
$P\dfrac{2}{m}\dfrac{2}{n}\dfrac{2_1}{a}$	Pmna	$D_{2h}^7(V_h^7)$	h0l, hk0
$P\dfrac{2_1}{n}\dfrac{2_1}{m}\dfrac{2_1}{a}$	Pnma	$D_{2h}^{16}(V_h^{16})$	0kl, hk0
$P\dfrac{2_1}{b}\dfrac{2_1}{a}\dfrac{2}{m}$	Pbam	$D_{2h}^9(V_h^9)$	0kl, h0l
$P\dfrac{2}{b}\dfrac{2}{a}\dfrac{2}{n}$	Pban	$D_{2h}^4(V_h^4)$	0kl, h0l, hk0
$P\dfrac{2}{c}\dfrac{2}{c}\dfrac{2}{m}$	Pccm	$D_{2h}^3(V_h^3)$	0kl, h0l
$P\dfrac{2_1}{c}\dfrac{2_1}{c}\dfrac{2}{n}$	Pccn	$D_{2h}^{10}(V_h^{10})$	0kl, h0l, hk0
$P\dfrac{2_1}{b}\dfrac{2_1}{c}\dfrac{2_1}{m}$	Pbcm	$D_{2h}^{11}(V_h^{11})$	0kl, h0l
$P\dfrac{2_1}{b}\dfrac{2}{c}\dfrac{2_1}{n}$	Pbcn	$D_{2h}^{14}(V_h^{14})$	0kl, h0l, hk0
$P\dfrac{2_1}{b}\dfrac{2_1}{c}\dfrac{2_1}{a}$	Pbca	$D_{2h}^{15}(V_h^{15})$	0kl, h0l, hk0
$P\dfrac{2_1}{c}\dfrac{2}{c}\dfrac{2}{a}$	Pcca	$D_{2h}^8(V_h^8)$	0kl, h0l, hk0
$C\dfrac{2}{m}\dfrac{2}{m}\dfrac{2}{m}$	Cmmm	$D_{2h}^{19}(V_h^{19})$	hkl
$C\dfrac{2}{m}\dfrac{2}{m}\dfrac{2}{a}$	Cmma	$D_{2h}^{21}(V_h^{21})$	hkl, hk0
$C\dfrac{2}{c}\dfrac{2}{c}\dfrac{2}{m}$	Cccm	$D_{2h}^{20}(V_h^{20})$	hkl, 0kl, h0l
$C\dfrac{2}{c}\dfrac{2}{c}\dfrac{2}{a}$	Ccca	$D_{2h}^{22}(V_h^{22})$	hkl, 0kl, h0l, hk0
$C\dfrac{2}{m}\dfrac{2}{c}\dfrac{2_1}{m}$	Cmcm	$D_{2h}^{17}(V_h^{17})$	hkl, h0l

Class $\dfrac{2}{m}\dfrac{2}{m}\dfrac{2}{m}$ (short *mmm*)—$D_{2h}(V_h)$—*continued*

Space group Hermann–Mauguin (Full)	Short	Schoenflies	Missing spectra
$C\dfrac{2}{m}\dfrac{2}{c}\dfrac{2_1}{a}$	Cmca	$D_{2h}^{18}(V_h^{18})$	hkl, h0l, hk0
$F\dfrac{2}{m}\dfrac{2}{m}\dfrac{2}{m}$	Fmmm	$D_{2h}^{23}(V_h^{23})$	hkl
$F\dfrac{2}{d}\dfrac{2}{d}\dfrac{2}{d}$	Fddd	$D_{2h}^{24}(V_h^{24})$	hkl, 0kl$_4$, h0l$_4$, hk0$_4$
$I\dfrac{2}{m}\dfrac{2}{m}\dfrac{2}{m}$	Immm	$D_{2h}^{25}(V_h^{25})$	hkl
$I\dfrac{2}{b}\dfrac{2}{a}\dfrac{2}{m}$	Ibam	$D_{2h}^{26}(V_h^{26})$	hkl, 0kl, h0l
$I\dfrac{2_1}{m}\dfrac{2_1}{m}\dfrac{2_1}{a}$	Imma	$D_{2h}^{28}(V_h^{28})$	hkl, hk0
$I\dfrac{2_1}{b}\dfrac{2_1}{c}\dfrac{2_1}{a}$	Ibca	$D_{2h}^{27}(V_h^{27})$	hkl, 0kl, h0l, hk0

Tetragonal system

Class $\bar{4}$		S_4	
$P\bar{4}$		S_4^1	—
$I\bar{4}$		S_4^2	hkl

Class 4		C_4	
$P4$		C_4^1	—
$P4_2$		C_4^3	00l
$P4_1$, $P4_3$		C_4^2, C_4^4	00l$_4$
$I4$		C_4^5	hkl
$I4_1$		C_4^6	hkl, 00l$_4$

Class $\dfrac{4}{m}$		C_{4h}	
$P\dfrac{4}{m}$		C_{4h}^1	—
$P\dfrac{4}{n}$		C_{4h}^3	hk0
$P\dfrac{4_2}{m}$		C_{4h}^2	00l
$P\dfrac{4_2}{n}$		C_{4h}^4	hk0, 00l
$I\dfrac{4}{m}$		C_{4h}^5	hkl
$I\dfrac{4_1}{a}$		C_{4h}^6	hkl, hk0 00l$_4$

Class $\bar{4}2m$ (or, in other orientation, $\bar{4}m2$)—$D_{2d}(V_d)$

	Schoenflies	Missing spectra
$P\bar{4}2m$	$D_{2d}^1(V_d^1)$	—

Table 5.3   THE HERMANN–MAUGUIN SYSTEM OF POINT- AND SPACE-GROUP NOTATION—*continued*

Class $\bar{4}2m$ (or, in other orientation, $4m2$) $D_{2d}(V_d)$—*continued*:

Hermann–Mauguin Full	Short	Schoenflies	Missing spectra
$P\bar{4}2c$		$D_{2d}^2(V_d^2)$	hhl
$P\bar{4}2_1m$		$D_{2d}^3(V_d^3)$	h00
$P\bar{4}2_1c$		$D_{2d}^4(V_d^4)$	hhl, h00
$P\bar{4}m2$		$D_{2d}^5(V_d^5)$	—
$P\bar{4}c2$		$D_{2d}^6(V_d^6)$	0kl
$P\bar{4}b2$		$D_{2d}^7(V_d^7)$	0kl
$P\bar{4}n2$		$D_{2d}^8(V_d^8)$	0kl
$I\bar{4}2m$		$D_{2d}^{11}(V_d^{11})$	hkl
$I\bar{4}2d$		$D_{2d}^{12}(V_d^{12})$	hkl, hhl$_4$
$I\bar{4}m2$		$D_{2d}^9(V_d^9)$	hkl
$I\bar{4}c2$		$D_{2d}^{10}(V_d^{10})$	hkl, 0kl

Class $4mm$ C_{4v}

Full	Short	Schoenflies	Missing spectra
P4mm	P4mm	C_{4v}^1	—
P4$_2$mc	P4mc	C_{4v}^7	hhl
P4$_2$cm	P4cm	C_{4v}^3	0kl
P4cc	P4cc	C_{4v}^5	0kl, hhl
P4bm	P4bm	C_{4v}^2	0kl
P4$_2$bc	P4bc	C_{4v}^8	0kl, hhl
P4$_2$nm	P4nm	C_{4v}^4	0kl
P4nc	P4nc	C_{4v}^6	0kl, hhl
I4mm	I4mm	C_{4v}^9	hkl
I4cm	I4cm	C_{4v}^{10}	hkl, 0kl
I4$_1$md	I4md	C_{4v}^{11}	hkl, hhl$_4$
I4$_1$cd	I4cd	C_{4v}^{12}	hkl, 0kl hhl$_4$

Class 422 (short 42) D_4

Full	Short	Schoenflies	Missing spectra
P422	P42	D_4^1	—
P4$_4$22	P4$_2$2	D_4^5	00l
P4$_1$22	P4$_1$2	D_4^3	} 00l$_4$
P4$_3$22	P4$_3$2	D_4^7	}
P42$_1$2	P42$_1$	D_4^2	h00
P4$_2$2$_1$2	P4$_2$2$_1$	D_4^6	00l, h00
P4$_1$2$_1$2	P4$_1$2$_1$	D_4^4	} 00l$_4$, h00
P4$_3$2$_1$2	P4$_3$2$_1$	D_4^8	}
I422	I422	D_4^9	hkl
I4$_1$22	I4$_1$2	D_4^{10}	hkl, 00l$_4$

Class $\frac{4}{m}\frac{2}{m}\frac{2}{m}$ (short $4/mmm$)—D_{4h}

Full	Short	Schoenflies	Missing spectra
$P\frac{4}{m}\frac{2}{m}\frac{2}{m}$	P4/mmm	D_{4h}^1	—

Class $\frac{4}{m}\frac{2}{m}\frac{2}{m}$ (short $4/mmm$)—D_{4h}—*continued*

Full	Short	Schoenflies	Missing spectra
$P\frac{4_2}{m}\frac{2}{m}\frac{2}{c}$	P4/mmc	D_{4h}^9	hhl
$P\frac{4_2}{m}\frac{2}{c}\frac{2}{m}$	P4/mcm	D_{4h}^{10}	0kl
$P\frac{4}{m}\frac{2}{c}\frac{2}{c}$	P4/mcc	D_{4h}^2	0kl, hhl
$P\frac{4}{m}\frac{2_1}{b}\frac{2}{m}$	P4/mbm	D_{4h}^5	0kl
$P\frac{4_2}{m}\frac{2_1}{b}\frac{2}{c}$	P4/mbc	D_{4h}^{13}	0kl, hhl
$P\frac{4_2}{m}\frac{2_1}{n}\frac{2}{m}$	P4/mnm	D_{4h}^{14}	0kl
$P\frac{4_2}{m}\frac{2_1}{n}\frac{2}{c}$	P4/mnc	D_{4h}^6	0kl, hhl
$P\frac{4_2}{n}\frac{2_1}{m}\frac{2}{m}$	P4/nmm	D_{4h}^7	hk0
$P\frac{4_2}{n}\frac{2_1}{m}\frac{2}{c}$	P4/nmc	D_{4h}^{15}	hk0, hhl
$P\frac{4_2}{n}\frac{2_1}{c}\frac{2}{m}$	P4/ncm	D_{4h}^{16}	hk0, 0kl
$P\frac{4}{n}\frac{2_1}{c}\frac{2}{c}$	P4/ncc	D_{4h}^8	hk0, 0kl, hhl
$P\frac{4}{n}\frac{2}{b}\frac{2}{m}$	P4/nbm	D_{4h}^3	hk0, 0kl
$P\frac{4_2}{n}\frac{2}{b}\frac{2}{c}$	P4/nbc	D_{4h}^{11}	hk0, 0kl, hhl
$P\frac{4_2}{n}\frac{2}{n}\frac{2}{m}$	P4/nnm	D_{4h}^{12}	hk0, 0kl
$P\frac{4}{n}\frac{2}{n}\frac{2}{c}$	P4/nnc	D_{4h}^4	hk0, 0kl, hhl
$I\frac{4}{m}\frac{2}{m}\frac{2}{m}$	I4/mmm	D_{4h}^{17}	hkl
$I\frac{4}{m}\frac{2}{c}\frac{2}{m}$	I4/mcm	D_{4h}^{18}	hkl, 0kl
$I\frac{4_1}{a}\frac{2}{m}\frac{2}{d}$	I4/amd	D_{4h}^{19}	hkl, hk0, hhl$_4$
$I\frac{4_1}{a}\frac{2}{c}\frac{2}{d}$	I4/acd	D_{4h}^{20}	hkl, hk0, 0kl, hhl$_4$

Cubic system

Class 23 T

Full	Short	Schoenflies	Missing spectra
	P23	T^1	—
	P2$_1$3	T^4	h00

Table 5.3   THE HERMANN–MAUGUIN SYSTEM OF POINT- AND SPACE-GROUP NOTATION—*continued*

Space group Hermann–Mauguin Full	Short	Schoenflies	Missing spectra
Class 23		*T—continued*	
$F23$		T^2	*hkl*
$I23$		T^3	*hkl*
$I2_13$		T^5	*hkl*
Class $\frac{2}{m}3$ (short $m3$)		T_h	
$P\frac{2}{m}3$	$Pm3$	T_h^1	—
$P\frac{2}{n}3$	$Pn3$	T_h^2	*hk0*
$P\frac{2_1}{a}3$	$Pa3$	T_h^6	*hk0*
$F\frac{2}{m}3$	$Fm3$	T_h^3	*hkl*
$F\frac{2}{d}3$	$Fd3$	T_h^4	*hkl, hk0*$_4$
$I\frac{2}{m}3$	$Im3$	T_h^5	*hkl*
$I\frac{2_1}{a}3$	$Ia3$	T_h^7	*hkl, hk0*
Class $\bar{4}3m$		T_d	
	$P\bar{4}3m$	T_d^1	—
	$P\bar{4}3n$	T_d^4	*hh*
	$F\bar{4}3m$	T_d^2	*hkl*
	$F\bar{4}3c$	T_d^5	*hkl, khl*
	$I\bar{4}3m$	T_d^3	*hkl*
	$I\bar{4}3d$	T_d^6	*hkl, hhl*$_4$
Class 432		O	
$P432$	$P43$	O^1	—
$P4_232$	$P4_23$	O^2	*h00*
$P4_132$	$P4_13$	O^7 ⎫	
$P4_332$	$P4_33$	O^6 ⎭	*h*$_4$*00*
$F432$	$F43$	O^3	*hkl*
$F4_132$	$F4_13$	O^4	*hkl, h*$_4$*00*
$I432$	$I43$	O^5	*hkl*
$I4_132$	$I4_13$	O^8	*hkl, h*$_4$*00*
Class $\frac{4}{m}\bar{3}\frac{2}{m}$ (short $m3m$)—O_h			
$P\frac{4}{m}\bar{3}\frac{2}{m}$	$Pm3m$	O_h^1	—

Space group Hermann–Mauguin Full	Short	Schoenflies	Missing spectra
Class $\frac{4}{m}\bar{3}\frac{2}{m}$ (short $m3m$)—O_h—continued			
$P\frac{4_2}{m}\bar{3}\frac{2}{n}$	$Pm3n$	O_h^3	*hhl*
$P\frac{4_2}{n}\bar{3}\frac{2}{m}$	$Pn3m$	O_h^4	*hk0*
$P\frac{4}{n}\bar{3}\frac{2}{n}$	$Pn3n$	O_h^2	*hk0, hhl*
$F\frac{4}{m}\bar{3}\frac{2}{m}$	$Fm3m$	O_h^5	*hkl*
$F\frac{4}{m}\bar{3}\frac{2}{c}$	$Fm3c$	O_h^6	*hkl, hhl*
$F\frac{4_1}{d}\bar{3}\frac{2}{m}$	$Fd3m$	O_h^7	*hkl, hk0*$_4$
$F\frac{4_1}{d}\bar{3}\frac{2}{c}$	$Fd3c$	O_h^8	*hkl, hk0*$_4$*, hhl*
$I\frac{4}{m}\bar{3}\frac{2}{m}$	$Im3m$	O_h^9	*hkl*
$I\frac{4_1}{a}\bar{3}\frac{2}{d}$	$Ia3d$	O_h^{10}	*hkl, hk0, hhl*$_4$

Rhombohedral system (all indices and multiplicities referred to hexagonal axes)

Space group Hermann–Mauguin	Short	Schoenflies	Missing spectra
Class 3		C_3	
	$C3$	C_3^1	—
	$C3_1, C3_2$	C_3^2, C_3^3	*00l*
	$R3$	C_3^4	*hkl*$_3$
Class $\bar{3}$		C_{3i}	
	$C\bar{3}$	C_{3i}^1	—
	$R\bar{3}$	C_{3i}^3	*hkl*$_3$
Class $3m$ (to indicate orientation distinguish $3m1$ and $31m$)		C_{3v}	
	$C3m1$	C_{3v}^1	—
	$C3c1$	C_{3v}^3	*h0l*
	$C31m$	C_{3v}^2	—
	$C31c$	C_{3v}^4	*hhl*
	$R3m$	C_{3v}^5	*hkl*$_3$
	$R3c$	C_{3v}^6	*hkl*$_3$*, h0l*

Table 5.3   THE HERMANN–MAUGUIN SYSTEM OF POINT- AND SPACE-GROUP NOTATION—*continued*

Class 32 (to indicate orientation distinguish 321 and 312) $D3$

Space group Hermann–Mauguin Full	Short	Schoenflies	Missing spectra
	$C321$	D_3^2	—
	$C3_121$	D_3^4 }	
	$C3_221$	D_3^6 }	$00l$
	$C312$	D_3^1	—
	$C3_112$	D_3^3 }	
	$C3_212$	D_3^5 }	$00l$
	$R32$	D_3^7	$\underline{hkl}_3$

Class $3\dfrac{2}{m}$ **(short $3m$) (to indicate orientation distinguish $3ml$ and $3lm$)** — D_{3d}

Full	Short	Schoenflies	Missing spectra
$C3\dfrac{2}{m}\bar{1}$	$C3m1$	D_{3d}^3	—
$C3\dfrac{2}{c}\bar{1}$	$C3c1$	D_{3d}^4	$h0\underline{l}$
$C3\bar{1}\dfrac{2}{m}$	$C31m$	D_{3d}^1	—
$C3\bar{1}\dfrac{2}{c}$	$C31c$	D_{3d}^2	$hh\underline{l}$
$R\bar{3}\dfrac{2}{m}$	$R\bar{3}m$	D_{3d}^5	$\underline{hkl}_3$
$R\bar{3}\dfrac{2}{c}$	$R\bar{3}c$	D_{3d}^6	$\underline{hkl}_3, h0\underline{l}$

Hexagonal system

Class $\bar{6}$ C_{3h}

	Short	Schoenflies	Missing spectra
	$C\bar{6}$	C_{3h}^1	—

Class 6 C_6

	Short	Schoenflies	Missing spectra
	$C6$	C_6^1	—
	$C6_3$	C_6^6	$00\underline{l}$
	$C6_2, C6_4$	C_6^4, C_6^5	$00\underline{l}$
	$C6_1, C6_5$	C_6^2, C_6^3	$00\underline{l}$

Class $\dfrac{6}{m}$ C_{6h}

Full	Short	Schoenflies	Missing spectra
$C\dfrac{6}{m}$		C_{6h}^1	—
$C\dfrac{6_3}{m}$		C_{6h}^2	$00\underline{l}$

Class $\bar{6}2m$ (in other orientation $\bar{6}m2$) — D_{3h}

Space group Hermann–Mauguin	Schoenflies	Missing spectra
$C\bar{6}2m$	D_{3h}^3	—
$C\bar{6}2c$	D_{3h}^4	$hh\underline{l}$
$C\bar{6}m2$	D_{3h}^1	—
$C\bar{6}c2$	D_{3h}^2	$h0\underline{l}$

Class $6mm$ C_{6v}

Full	Short	Schoenflies	Missing spectra
$C6mm$	$C6mm$	C_{6v}^1	—
$C6_3mc$	$C6mc$	C_{6v}^4	$hh\underline{l}$
$C6_3cm$	$C6cm$	C_{6v}^3	$h0\underline{l}$
$C6cc$	$C6cc$	C_{6v}^2	$h0\underline{l}, hh\underline{l}$

Class 622 D_6

Full	Short	Schoenflies	Missing spectra
$C622$	$C62$	D_6^1	—
$C6_322$	$C6_32$	D_6^6	$00\underline{l}$
$C6_222$	$C6_22$	D_6^4 }	
$C6_422$	$C6_42$	D_6^5 }	$00\underline{l}$
$C6_122$	$C6_12$	D_6^2 }	
$C6_522$	$C6_52$	D_6^3 }	$00\underline{l}$

Class $\dfrac{6\ 2\ 2}{m\,m\,m}$ (short $6/mmm$) — D_{6h}

Full	Short	Schoenflies	Missing spectra
$C\dfrac{6}{m}\dfrac{2}{m}\dfrac{2}{m}$	$C6/mmm$	D_{6h}^1	—
$C\dfrac{6_3}{m}\dfrac{2}{m}\dfrac{2}{c}$	$C6/mmc$	D_{6h}^4	$hh\underline{l}$
$C\dfrac{6_3}{m}\dfrac{2}{c}\dfrac{2}{m}$	$C6/mcm$	D_{6h}^3	$h0\underline{l}$
$C\dfrac{6}{m}\dfrac{2}{c}\dfrac{2}{c}$	$C6/mcc$	D_{6h}^2	$h0\underline{l}, hh\underline{l}$

6 Crystal chemistry

6.1 Structures of metals, metalloids and their compounds[†]

The elements have been arranged in the following order:

1. Group Ia—Li, Na, K, Rb, Cs
2. Group IIa—Be, Mg, Ca, Sr, Ba
3. Group IIIa—Sc, Y, La, Ce, Pr, Nd, Eu, Gd, Tb, Dy, Ho, Er, Tm, Yb, Lu, Ac
4. Group IVa—Ti, Zr, Hf, Th, Pa, U, Np, Pu, Am, Cm
5. Group Va—V, Nb, Ta
6. Group VIa—Cr, Mo, W
7. Group VIIa—Mn, Tc, Re
8. Group VIII—Fe, Co, Ni; Ru, Rh, Pd, Os, Ir, Pt
9. Group Ib—Cu, Ag, Au
10. Group IIb—Zn, Cd, Hg
11. Group IIIb—Al, Ga, In, Tl
12. Group IVb—Si, Ge, Sn, Pb
13. Group Vb—As, Sb, Bi
14. Metalloids etc.—B, C, P, N, Te, Se, S

A compound or solid solution composed of the elements ABCD . . . is placed under that element, A, B, C, D, . . . which occurs *last* in the above list. If there are several compounds containing this particular element, they are arranged in the following order:

(a) Compounds containing the same elements—in the order of increasing content of the second element.
(b) Compounds of different elements—in the order in which the other elements occur in the above list.

For example, Na_2K is described under K, because K comes after Na in the list; MgSr and Mg_2Sr are both entered under Sr, and are described in that order, in accordance with (a); Li_2Ca and Mg_2Ca are entered under Ca, and described in that order, in accordance with (b).

The second column of Table 6.1 gives the symbol for the structural type to which the element or compound is assigned, the notation used being that of *Strukturbericht*.[*] Detailed descriptions of the structural types are given in Table 6.2.

The third column of Table 6.1 gives the lattice constants of the various elements and compounds. For cubic crystals the single parameter a is given, in Å units; for tetragonal and hexagonal crystals a and c, in that order; for orthorhombic, a, b and c; for rhombohedral a, α; for monoclinic a, b, c, β.

Temperatures given in parentheses are those at which allotropic or polymorphic modifications are stable; figures such as $A = 28, M = 4$, also given in parentheses, refer to the number of atoms (A) or molecules (M) included in the unit cell; alternative structures are given where the available evidence is insufficient to permit of a decision being reached between them, and in such cases the authorities are quoted. Finally, it should be noted that in the case of the sulphides, only those having relatively simple structures have been included.

[†] For information on crystallography, see Chapter 5. Techniques that enable determination of crystal structure can be found in Chapter 4 and Section 10.4.

[*] Strukturbericht of Z. *Krystallographie*, Leipzig.

Although this system of classification is not used in any other compilation of intermetallic compounds, it has been retained as it has its own logic in keeping together compounds of similar systems in a Group; this would be lacking in a strictly alphabetical classification.

Since this compilation was first made, data on a very large number of intermetallic compounds have been published, far more than can be included in this list without extending it to over 1000 pages. Table 6.1 refers to the more frequently encountered compounds. Other compilations should be consulted for compounds not found in Table 6.1.

A comprehensive database is: 'Pearson's Handbook of Crystallographic Data for Intermetallic Phases' (4 volumes) by P. Villars and L. D. Calvert, published by A.S.M., Materials Park, OH, 1991. For an annually updated database, see the 'Powder Diffraction File', published by the International Center for Diffraction Data, Newtown Square, PA.

Pearson has introduced a new system of symbols based on the crystallography of the classes of isostructural compounds. This cannot be deduced readily from the Structurbericht Classification used in Table 6.1 and described in Table 6.2. Hence a comparison of the two nomenclatures is provided in Table 6.3.

The Pearson symbols take the form aBn, where:

a denotes the crystal system, i.e. cubic, hexagonal, etc.

B denotes the space lattice of the unit cells using the Hermann–Mauguin symbols, i.e. P, C, F, I, etc.

n denotes the number of atoms per unit cell.

The crystal systems are:

c = Cubic	o = Orthorhombic
h = Hexagonal	t = Tetragonal
m = Monoclinic	

The space lattices of the unit cells are:

P = Simple	C = One face-centred*
F = All Face-centred+	R = Simple (rhombohedral)
I = Body-centred	

*Monoclinic and Orthorhombic
+Cubic and Orthorhombic
e.g. A1(Cu) is cF4 and C14($MgZn_2$) is hP12.

As will be evident from Table 6.3, a Pearson symbol can be the same for several different structural types or prototypes. See note at end of references for sources of data.

Table 6.1   STRUCTURES OF METALS, METALLOIDS AND THEIR COMPOUNDS

Element or compound	Structure type	Lattice constants, remarks	Additional Refs.
Group Ia: Li, Na, K, Rb, Cs			
Li	A2	3.509 3	1
	A3	3.111; 5.093 (spontaneous transformation at −195°C)	
	A1	4.355 (high pressure phase)	
Na	A2	4.282 0(5)	
	A3	3.767; 6.154 (low temperature phase below −237°C)	
K	A2	5.330 17(34)	
Na_2K	C14	7.48; 12.27	
Rb	A2	5.585	
Cs	A2	6.141(7)	
Group IIa: Be, Mg, Ca, Sr, Ba			
Be	A3	2.282 6; 3.583 6	
$Be_{13}Mg$	D2₃	10.166	2
Ca (α)	A1	5.588 4 (<∼443°C)	
(γ)	A3	4.00(2); 6.50(2) (>∼443°C)	
Li_2Ca	C14	6.248(8); 10.23(2)	
$Be_{13}Ca$	D2₃	10.312	2
Mg_2Ca	C14	6.225; 10.18	

(continued)

Table 6.1   STRUCTURES OF METALS, METALLOIDS AND THEIR COMPOUNDS—*continued*

Element or compound	Structure type	Lattice constants, remarks	Additional Refs.
Group IIa: Be, Mg, Ca, Sr, Ba—*continued*			
Sr	$A1$	6.084 9(5)	
MgSr	$B2$	3.900	
Mg_2Sr	$C14$	6.426(8); 10.473(8)	
Ba	$A2$	5.013(5)	3
Mg_2Ba	$C14$	6.636(8); 10.655(8)	
Group IIIa: Sc, Y and Rare Earths			
Sc (α)	$A3$	3.310(1); 5.274(2)	
Y	$A3$	3.651 5(2); 5.747 4(4)	
$Be_{13}Y$	$D2_3$	10.24	
La (α)	$A3$	3.770; 12.159 (room temp.)	4
(β)	$A1$	5.31 (above 310°C)	4
MgLa	$B2$	3.97	
Mg_2La	$C15$	8.774	
Mg_3La	$D0_3$	7.478	
Ce (α)	$A1$	4.832(4)	
(β)	$A3$	3.16;	4
$Be_{13}Ce$	$D2_3$	10.376	
MgCe	$B2$	3.90	
Mg_2Ce	$C15$	8.733	
Mg_3Ce	$D0_3$	7.42	
Pr	$A1$	5.183	
	$A3$	3.671 5; 11.830	4
MgPr	$B2$	3.88	
Mg_3Pr	$D0_2$	7.42	
Nd	$A3$	3.656; 11.798	207
Eu	$A2$	4.582 7	208
Gd		3.633 6; 5.781	209
Tb		3.605 5; 5.696 6	
Dy		3.591 5; 5.650 1	
Ho	$A3$	3.576 4; 5.613 6	
Er		3.561 0(2); 5.592 9(10)	210
Tm		3.537 5; 5.554	
Yb	$A1$	5.484 7	
Lu	$A3$	3.503 1(4); 5.550 9(4)	
Ac	$A1$	5.670	6
Group IVa: Ti, Zr, Hf, Th and Pa, U, Np, Pu, Am, Cm			
Ti (α)	$A3$	2.950 3(6); 4.681 0(2) (<~882°C)	
(β)	$A2$	3.311 2 (>~882°C)	
$BeLi_4$	hex.	10.92; 8.94 ($M = 6$)	211
Be_2Ti	$C15$	6.453 2(4)	
$Be_{12}Ti$	$D2a$	29.44; 7.33 (may be considered as $D2b$ 7.35; 4.19)	
Ti_2Be_{17}	Th_2Ni_{17}	7.36; 7.30	
Ti_2Be_{17}	Nb_2Be_{17}	7.34; 10.73	
Zr (α)	$A3$	3.231 78; 5.148 31 (<~863°C)	7
(β)	$A2$	3.568(5) (>~863°C)	7
Be_5Zr	$D2d$	4.564(2); 3.485(2)	
$Be_{17}Zr_2$	$Be_{17}Nb_2$	7.538 1(1); 11.015(1)	
Be_2Zr	$C32$	3.814; 3.230	8
$Be_{13}Zr$	$D2_3$	10.044	
Hf (α)	$A3$	3.198; 5.061 < 1743°C	9, 10
(β)	$A2$	3.615	
Hf_2Be_{17}	Nb_2Be_{17}	7.494; 10.93	213
$HfBe_{13}$	$D2_3$	10.005	
$HfBe_5$	$D2d$	4.519; 3.465	
Th	$A1$	5.084 2	
$B_{13}Th$	$D2_3$	10.41	
Mg_2Th	$C36$	6.086; 19.64 (at and below 700°C)	12
	$C15$	8.570 (at and above 800°C)	12
Pa	$A6$	3.940; 3.244	13

(continued)

Table 6.1   STRUCTURES OF METALS, METALLOIDS AND THEIR COMPOUNDS—*continued*

Element or compound	Structure type	Lattice constants, remarks	Additional Refs.
Group IVa: Ti, Zr, Hf, Th and Pa, U, Np, Pu, Am, Cm—*continued*			
U (α)	$A20$	2.853 7; 5.869 5; 4.954 8	214
(β)	$D8_6$	10.758 9; 5.653 1 (range 668–776°C)	14
(γ)	$A2$	3.533 5 (>776°C)	
UH_3	$A15$	4.147(3)	
$Be_{13}U$	$D2_3$	10.248 87(2)	
TiU_2	$C32$	4.828; 2.847	16
Np (α)	Ac	(<278°C) 6.663; 4.723; 4.887	
(β)	Ad	4.897(2); 3.388(2) (range 278–540°C)	
$B_{13}Np$	$D2_3$	10.266 to 10.256	19
Pu (α)	monocl.	6.183; 4.822; 10.963; $\beta = 101.79$	20
(β)	monocl.	11.830; 10.449; 9.227; $\beta = 138.65$	215
(δ)	$A1$	4.634 7	21
(δ')	$A6$	3.339(3); 4.446(7)	21
(ε)	$A2$	3.637 5	21
Am	$P6_3/mmc$	3.468 1(2); 11.241(3)	216
Group Va: V, Nb, Ta			
V	$A2$	3.027	22
Be_2V	$C14$	4.385; 7.130	
$Be_{12}V$	$D2b$	7.35; 4.24	
Be_2Nb_3	$D5a$	6.49(1); 3.35(1)	217
$Be_{17}Nb_2$	$R\bar{3}m$	7.409(2); 10.84(1)	218
$Be_{12}Nb$	$D2b$	7.376(5); 4.258(5)	
Nb	$A2$	3.300	
Ta	$A2$	3.302 56(5)	
$BeTa_2$	$C16$	6.010(4); 4.890(4)	
$Be_{17}Ta_2$	$R\bar{3}m$	7.388; 10.74	
Be_2Ta	$C15$	6.507	
ZrV_3	$C14$	5.28; 8.65	
TiV	$A2$	3.159	219
Group VIa: Cr, Mo, W			
Cr	$A2$	2.884 4	
CrH	$B1$	3.860 5(5)	
Be_2Cr	$C14$	4.260; 6.988	
$Be_{12}Cr$	$D2b$	7.238(4); 4.174(2)	
TiCr	$A2$	3.12	219
$TiCr_2$	$C15$	6.932(4)	
$ZrCr_2$ (1)	$C14$	5.102; 8.273	24
(2)	$C15$	7.205	24
$NbCr_2$	$C15$	6.991	
$TaCr_2$ (1)	$C15$	6.985	
(2)	$C14$	4.950; 8.245	
Mo	$A2$	3.147 0	
Be_2Mo	$C14$	4.448; 7.310	
$Be_{12}Mo$	$D2b$	7.251(2); 4.234(1)	25
$Be_{13}Mo$	tetr.	10.27; 4.29 ($M = 4$)	
$Be_{22}Mo$	cubic f.	11.634(2)	
$BeMo_3$	$A15$	4.89	220
$ZrMo_2$	$C15$	7.587 5	26
U_2Mo (γ)	$C11b$	3.427; 9.834	27
W (α)	$A2$	3.163 79	
Be_2W	$C14$	4.559; 7.333	
WBe_{22}	O_h^7	11.631(2)	220
$Be_{12}W$	$D2b$	7.362(5); 4.216(5)	
ZrW_2	$C15$	7.613(2)	
Group VIIa: Mn, Tc, Re			
Mn (α)	$A12$	8.912 5 (<727°C)	
(β)	$A13$	6.315 2 (727–1100°C)	28
(γ)	$A1$	3.860 (1100–1138°C)	
(δ)	$A2$	3.080 (>1138°C)	30

(continued)

Table 6.1   STRUCTURES OF METALS, METALLOIDS AND THEIR COMPOUNDS—*continued*

Element or compound	Structure type	Lattice constants, remarks	Additional Refs.
Group VIIa: Mn, Tc, Re—*continued*			
YMn_2	$C15$	7.675 ($M = 8$)	221
YMn_4	bc.tetr.	8.808; 12.521 ($M = 12$)	222
YMn_{12}	$D2b$	8.541; 4.785	
Be_2Mn	$C14$	4.231; 6.909	
$Be_{12}Mn$	$D2b$	7.276(5); 4.256(5)	
$GdMn_2$	$C15$	7.74	
$TiMn_2$	$C14$	4.832; 7.930	31
$TiMn$	$D8b$	8.880; 4.542	
$ZrMn_2$	$C14$	5.026; 8.258	
$ThMn_{12}$	$D2b$	8.74(1); 4.95(1)	
Th_6Mn_{23}	$D8a$	12.443	
$ThMn_2$	$C14$	5.485; 8.896	
UMn_2	$C15$	7.162 8	
U_6Mn	$D2c$	10.312(5); 5.255(4)	
VMn_4	$D8b$	8.92; 4.61	
$NbMn_2$	$C14$	4.872; 7.975	
$TaMn_2$	$C14$	4.864; 7.943	
$CrMn_3$	$D8b$	8.86; 4.59	
Tc	$A3$	2.740 7(1); 4.398 0(1)	
Re	$A3$	2.761 5; 4.456 6	
Be_2Re	$C14$	4.345; 7.087	
$Be_{22}Re$	O_h^7	11.560(1)	
$ZrRe_2$	$C14$	5.259; 8.615	
$HfRe_2$	$C14$	5.254(1); 8.600(2)	223
URe_2	ortho.	<180°. 5.600; 9.180; 8.460	
	$C14$	>180°. 5.405; 8.683	224
Group VIII: Fe, Co, Ni; Ru, Rh, Pd; Os, Ir, Pt			
Fe (α)	$A2$	2.867 0; (<912°C and >1394°C)	
(γ)	$A1$	3.654 4 (912–1394°C)	
Be_2Fe	$C14$	4.215; 6.853	
Be_5Fe	$C15$	5.875 (stable only > ~1000°)	
Be_7Fe	hex.	4.137(2); 10.720(5)	
$Be_{12}Fe$	$D2b$	7.253(5); 4.232(5)	
$CeFe_2$	$C15$	7.300	
Gd_2Fe_3	cubic	8.25 ($M = 6$)	
$GdFe_3$	rhomb.	5.157; 24.70; 120°	
Gd_2Fe_7	ortho.	5.71; 6.78; 7.15 ($M = 2$)	
$GdFe_4$	hex.	5.15; 6.64 ($M = 2$)	225
$GdFe_5$	ch	4.92; 4.11 ($M = 1$)	
Gd_2Fe_{17}	hex.	8.496; 8.345	
$GdFe_2$	$C15$	7.39	
YFe_2	$C15$	7.37	221
$TiFe_2$	$C14$	4.79; 7.81	
$TiFe$	$B2$	2.976	
$TiFe_3O$	$E9_3$	11.15	
Ti_4Fe_2O	$E9_3$	11.28	32
Ti_2Fe	$E9_2$	11.305	33
$ZrFe_2$	$C15$	7.065	
$ZrFe_2$	$C36$	4.988; 16.32	
$ThFe_5$	$D2d$	5.111 3(4); 4.052 2(4)	
U_6Fe	$D2c$	10.302 2; 5.238 6(1)	
UFe_2	$C15$	7.058	226
VFe	$D8b$	8.965; 4.633	
$NbFe_2$	$C14$	4.845; 7.893	
$TaFe_2$	$C14$	4.825; 7.875	
$CrFe$ (σ)	$D8b$	8.796 6(6); 4.558 2(3)	35
$MoFe$ (σ)	$D8b$	9.190; 4.814	35
Mo_6Fe_7	$D8_5$	4.754 6(5); 25.716(3); 120°	
WFe_2	$C14$	4.737(3); 7.694(4)	
W_6Fe_7	$D8_5$	4.764(3); 25.85(2); 120°	

(continued)

Table 6.1   STRUCTURES OF METALS, METALLOIDS AND THEIR COMPOUNDS—*continued*

Element or compound	Structure type	Lattice constants, remarks	Additional Refs.
Group VIII: Fe, Co, Ni; Ru, Rh, Pd; Os, Ir, Pt—*continued*			
$MnFe_4$	$A3$	2.530; 4.079 (not an equilibrium phase)	
Co (α)	$A3$	2.5071(3); 4.0695(5)	227
(β)	$A1$	3.5688 (>422°C)	
BeCo	$B2$	2.624	
$Be_{12}Co$	$D2b$	7.237(5); 4.249(5)	
Be_5Co	$C15b$	5.852	
YCo_2	$C15$	7.22	221
La_3Co	DO_{11}	7.277(9); 10.02(10); 6.575(8)	228
$Co_{13}La$	$D2_3$	11.344	
$CeCo_5$	$D2d$	4.933; 4.011	
Co_2Tb	$D2d$	5.090; 12.52	
Co_5Ho	$D2d$	4.9138; 3.9750	
Gd_3Co	DO_{11}	7.027; 9.496; 6.296	235
GdCo	Bf	3.90; 4.87; 4.22 ($M = 2$)	
Gd_2Co_3	cubic	7.98 ($M = 6$)	
$GdCo_2$	$C15$	7.359	
$GdCo_3$	rhomb.	5.039; 24.522; 120°	
$GdCo_4$	hex.	5.47; 6.02 ($M = 2$)	
$GdCo_5$	$D2d$	4.960; 3.989	
$CeCo_2$	$C15$	7.15	
$TiCo_2$	$C36$	4.73; 15.41 (Co-rich)	
	$C15$	6.692	
TiCo	$B2$	2.993	
Ti_4Co_2O	$E9_3$	11.30 (Rostoker[32])	
Ti_2Co	cubic f.	11.283	
$ZrCo_2$	$C15$	6.950	
$ThCo_3$	rhomb.	5.03; 24.54; 120°	
UCo_2	$C15$	6.9783(4)	
UCo	Ba	6.3525	
U_6Co	$D2c$	10.36; 5.21	
VCo_3	DO_{21}	5.037; 12.29	229
V_3Co	$A15$	4.676	
$NbCo_2$	$C15$	6.760	
Co_2Ta	$C36$	4.747; 15.47	
Co_3Ta (α)	$C15$	6.788	
(β)	$C36$	4.728; 15.45	230
Co_2Ta (α)	$C14$	4.838; 7.835	
(β)	$C15$	6.752	
(γ)	$C36$	4.747; 15.47	
CrCo (σ)	$D8b$	8.84(1); 4.60(1)	36
$MoCo_3$	DO_{19}	5.194; 4.212	
Mo_6Co_7	$D8_5$	4.761(2); 25.58(2); 120°	
Mo_3Co_2	$D8b$	9.279; 4.871	
WCo_3	DO_{19}	5.120; 4.120	
W_6Co_7	$D8_5$	4.7398(17); 25.542(11); 120°	
FeCo	$B2$	2.8504 ⎫ transf. temp. ~730°C; constants at room temp.	
	$A2$	2.8550 ⎭	
Ni	$A1$	3.5350	
BeNi	$B2$	2.625	
$Be_{21}Ni_5$	cubic	7.619	
$MgNi_2$	$C36$	4.807; 15.77	
Mg_2Ni	Ca	5.205(1); 13.236(2)	
$CaNi_5$	$D2d$	4.952; 3.937	
YNi_2	$C15$	7.182	221
YNi_3	hex.	4.977(4); 24.44(2) ⎫	
YNi	ortho.	7.156(3); 4.124(1); 5.515(2) ⎭	231
YNi_5	$D2d$	4.867; 3.973	
$LaNi_5$	$D2d$	5.016; 3.983	
$LaNi_2$	$C15$	7.340	
$CeNi_5$	$D2d$	4.879; 3.994	
$CeNi_2$	$C15$	7.208(5)	

(continued)

Table 6.1   STRUCTURES OF METALS, METALLOIDS AND THEIR COMPOUNDS—*continued*

Element or compound	Structure type	Lattice constants, remarks	Additional Refs.
Group VIII: Fe, Co, Ni; Ru, Rh, Pd; Os, Ir, Pt—*continued*			
$CeNi_3$	$D0_{21}$	4.945(5); 16.48(1)	233
$PrNi_5$	$D2d$	4.967 5; 3.972 5	
$PrNi_2$	$C15$	7.287	
Gd_3Ni	$D0_{11}$	6.95; 9.68; 6.36	234
Gd_3Ni_2	tetr.	7.28; 8.61 ($M = 4$)	
$GdNi$	$B27$	5.428(2); 4.353(2); 6.931(2)	
$GdNi_2$	$C15$	7.213	
Gd_2Ni_7	hex.	4.953; 24.21	
$GdNi_4$	hex.	5.35; 5.83 ($M = 2$)	
$GdNi_5$	$D2d$	4.909; 3.965	
Gd_2Ni_{17}	Th_2Ni_{17}	8.43; 8.04	
Ni_5Sm	$D2d$	4.925; 3.960	
Ni_5Tb	$D2d$	4.894(5); 3.966(5)	232
Ni_5Ho	$D2d$	4.867; 3.966	
Ni_5Yb	$D2d$	4.842; 3.959	
Ti_2Ni	O_h^7–$Fd3m$	11.324	236
$TiNi_3$	$D0_{24}$	5.102 8; 8.271 9	
$TiNi$	$B2$	3.007	
Ti_2Ni	$E9_3$	11.29	
Ti_4Ni_2O	$E9_3$	11.30	
$ZrNi_5$	$C15b$	6.702(4)	38
Zr_2Ni	$C16$	6.483(3); 5.267(5)	238
$ZrNi_2$	$C15$	6.915 5	
$ZrNi_3$	$D0_{19}$	5.215(8); 4.244 2(6)	
$ZrNi$	ortho.	3.268; 9.937; 4.101	
Hf_2Ni	$C16$	6.405; 5.252	
$HfNi$	Bf	3.218; 9.788; 4.117	
$ThNi_5$	$D2d$	4.941; 4.000	
$ThNi_2$	$C32$	3.960(3); 3.844(4)	
UNi_5	$C15$	6.785	
UNi_2	$C14$	4.963 0(5); 8.235(1)	
U_6Ni	$D2c$	10.37; 5.21	
$PuNi_4$	monocl. 62/m	4.87; 8.46; 10.27; $\beta = 100°$ ($M = 6$)	239
$PuNi_3$	rhomb.	5.00; 24.35; 120°	240
$PuNi$	Bf	3.59; 10.21; 4.22	241
VNi_3	$D0_{22}$	3.54; 7.22	
VNi_2	orh.	2.559; 7.641; 3.549	
$V_{19}Ni_{11}$	$D8b$	8.97; 4.64	
$NbNi_3$	$D0_{22}$	3.62; 7.41	
$TaNi_3$	$D0a$	5.122; 4.235; 4.522	
Cr_2Ni	$A2$(tetr. deformed)	($c/a = 1.09$; metastable intermediate by quenching)	
$MoNi_4$	$D1a$	5.683; 3.592	
$MoNi_3$	$D0a$	5.064; 4.223; 4.449(2)	242
WNi_4	$D1a$	5.730(1); 3.553(1)	
$MnNi_3$	$L1_2$	3.593 (<510°C)	
$MnNi$	$L1_0$	3.69; 3.49 (<~700°C)	
	$A2$	2.974 (range ~700–900°C; constant at 745°)	
$FeNi_3$	$L1_2$	3.5550 (<586°C)	
Fe_5Cr_5Ni	cubic b.c.	8.88 (powder diagram similar $A12$-type)	
Ru	$A3$	2.705 8; 4.281 6; 2.705 3; 4.282 0	
Ru_2Be_3	$D5_3$	11.42	252
$RuBe_2$	$C14$	5.96; 9.18	
Ru_3Be_{10}	$D8_2$	11.03	
Ru_2Ce	$C15$	7.539	244
$TiRu$	$B2$	3.07	40
$ZrRu_2$	$C14$	5.131; 8.490	
URu_3	$L1_2$	3.977(3)	39
RuU_2	monocl.	13.106; 3.343; 5.202; 96.17°($M = 4$)	245
Rh	$A1$	3.803 0	
$RhMg$	$B2$	3.099(2)	246
$RhBe$	$B2$	2.739 7(5)	

(continued)

Table 6.1   STRUCTURES OF METALS, METALLOIDS AND THEIR COMPOUNDS—*continued*

Element or compound	Structure type	Lattice constants, remarks	Additional Refs.
Group VIII: Fe, Co, Ni; Ru, Rh, Pd; Os, Ir, Pt—*continued*			
Rh_2Sr	$C15$	7.695	
Rh_3Ti	$L1_2$	3.823 ⎫	248
Rh_3Hf	$L1_2$	3.911 ⎭	
Rh_3Nb	$L1_2$	3.865	
Rh_3Zr	$L1_2$	3.887(1)	
Rh_3V	$L1_2$	3.795	
Rh_3Ta	$L1_2$	3.86	
Pd	$A1$	3.887 4(1)	
BePd	$B2$	2.813	
Be_5Pd	$C15b$	5.982 (ordered as Be_4Be, Pd)	
	$C15b$	5.982 [disordered as Be_4 $(Be, Pd)_2$]	
$Be_{12}Pd$	$D2b$	7.271(5); 4.251(5)	
Pd_2Ca	$C15$	7.652 ⎫	
Pd_2Sr	$C15$	7.800 ⎬	247
Pd_2Ba	$C15$	7.967 ⎭	
$TiPd_3$	$D0_{24}$	5.489; 8.964	
Pd_3Zr	$D0_{24}$	5.611 9(2); 9.231 6(5) ⎫	248
Pd_3Hf	$D0_{24}$	5.593 5(2); 9.192 8(5) ⎭	
Pd_4Th	$L1_2$	4.113(3)	249
$PdTi_2$	$C11b$	3.09; 10.04	342
$PdZr_2$	$C11b$	3.299(1); 10.88(1)	342
$PdHf_2$	$C11b$	3.251; 11.061	342
PdMg	$B2$	3.17(1)	341
UPd_3'	$D0_{24}$	5.770; 9.652	39
Pd_2V	ortho.	2.750; 8.250; 3.751	250
Pd_3V	$D0_{22}$	3.847; 7.753	
$FePd_3$	$L1_2$	3.848	
FePd	$L1_0$	3.855(2); 3.714(2)	
Os	$A3$	2.733 8; 4.319 5	
$Be_{12}O_5$	monocl.	10.628(6); 8.480 3(3); 11.305(3); 96.55(1)°	
TiOs	$B2$	3.081	40
$ZrOs_2$	$C14$	5.20; 8.53	
UOs_2	$C15$	7.514(4)	39
TaOs (σ)	$D8b$	9.91(6); 5.10(8)	41
OsW_2 (σ)	$D8b$	9.63; 4.98	41
Ir	$A1$	3.839 0	
Ir_2Sr	$C15$	7.849	247
Ir_3Ti	$L1_2$	3.847	250
Ir_3Hf	$L1_2$	3.935	
Ir_3Nb	$L1_2$	3.890(2)	
Ir_3Zr	$L1_2$	3.94	
Ir_3V	$L1_2$	3.812	
$ZrIr_2$	$C15$	7.36	
UIr_2	$C15$	7.495 4	39
IrW	$B19$	4.452; 2.760; 4.811	
Pt	$A1$	3.923 3	
Pt_2Ca	$C14$	5.373; 9.311	
Pt_2Sr	$C15$	7.742	
Pt_2Ba	$C15$	7.947	
PtMg	$B20$	4.863	341
Pt_3Mg	$L6_0$	3.88; 3.72	341
$CePt_2$	$C15$	7.725	
$TiPt_3$	$D0_{24}$	5.520; 9.019	
Ti_3Pt	cubic.	5.030	43
$ZrPt_3$	$D0_{24}$	5.624; 9.213	
Pt_3Hf	$D0_{24}$	5.636; 9.208	248
Pt_3U	$D0_{19}$	5.748; 4.898	39
Pt_2U	ortho.	5.60; 9.68; 4.12	251
$FePt_3$	$L1_2$	3.87(2) (<700°C)	
FePt	Aa	3.905; 3.735	

(continued)

Table 6.1 STRUCTURES OF METALS, METALLOIDS AND THEIR COMPOUNDS—*continued*

Element or compound	Structure type	Lattice constants, remarks	Additional Refs.
Group VIII: Fe, Co, Ni; Ru, Rh, Pd; Os, Ir, Pt—*continued*			
Fe_3Pt	$L1_2$	3.730 (<850°C)	
$CoPt_3$	$L1_2$	3.831 (constant at 700°C; transformation to disordered state ~750°C; constant at 800°C = 3.829)	
CoPt	$L1_0$	3.80; 3.70 (constant at 700°C; transformation temp. to disordered $A1$-type ≈ 825°C)	
NiPt	$L1_0$	3.821; 3.591 (<600°C)	
Group Ib: Cu, Ag, Au			
Cu	$A1$	3.613	
BeCu (γ)	$B2$	2.702(3)	
Be_2Cu	$C15$	5.977 to 5.97	
Be_3Cu	hex.	4.179(3); 2.557(3)	
ZrIr	$B2$	3.318	
Zr_2Ir	$C16$	6.51; 5.62	
$MgCu_2$	$C15$	7.048	
Mg_2Cu	Cb	9.07; 5.284; 18.25	
$CaCu_5$	$D2d$	5.092; 4.086	
$LaCu_5$	$D2d$	5.186; 4.110	
CeCu	$B27$	7.19 (14); 4.30(8); 6.23(12)	253
$CeCu_2$	ortho. $1mma$	4.433; 7.064; 7.472	254
$CeCu_6$	ortho. $Pnma$	8.109; 5.098; 10.172	255
$CeCu_5$	$D2d$	5.104; 4.112	
Cu_5Tb	$D2d$	5.030; 4.090 (also $C15b$ 7.041)	256
Cu_5Sm	$D2d$	5.07; 4.104	256
Ti_3Cu	$L1a$	4.158; 3.594	
$CuTi_2$	$C11b$	2.953; 10.734	
Ti_2Cu	$D2_3$	11.24	
Ti_4Cu_2O	$E9_3$	11.47	
TiCu (δ)	$L1_0$	3.140; 2.856	
(γ)	$B11$	3.107; 5.919	
$TiCu_3$ (β)	$A20$	2.585; 4.527; 4.351 (>~600°C)	
(β)	$L6_0$	5.169; 4.342; 4.531 (<~ 600°C)	
$CuZr_2$	$C11B$	3.220 4; 11.183 2	342
$CuHf_2$	$C11B$	3.164 3; 11.153	342
$ThCu_2$	$C32$	4.383(4); 3.496(3)	47
Th_2Cu	$C16$	7.324(5); 5.816(5)	47
UCu_5	$C15b$	7.03	
$PdCu_3$	$L1_2$	3.722 (also A1 3.701)	
	tetr.	3.710(3); 25.655(21)	
PdCu	$A1$	3.766	
PtCu	$A1$	3.776 9(5)	
Ag	$A1$	4.086	
LiAg	$B2$	3.168	
Be_2Ag	$C15$	6.287	
MgAg	$B2$	3.311 4	
Mg_3Ag	cubic	17.622(7)	
Ca_5Ag_3	tetr.	8.039; 15.011	258
LaAg	$B2$	3.814	
CeAg	$B2$	3.785(16)	
PrAg	$B2$	3.723	
TiAg	$L1_0$	4.104; 4.077 (also B11 2.90; 8.14)	
ZrAg	$B11$	3.471; 6.603	
$AgZr_2$	$C11B$	3.253; 7.952	342
Th_2Ag	$C16$	7.581; 5.854	48
Pt_3Ag	$L1_2$	3.900	
Au	$A1$	4.07894(5)	
$NaAu_2$	$C15$	7.85(1)	
Na_2Au	$C16$	7.415(5); 5.522(5)	
BeAu	$B20$	4.668(1)	
Be_5Au	$C15b$	6.100	
MgAu	$B2$	3.266	

(continued)

Table 6.1   STRUCTURES OF METALS, METALLOIDS AND THEIR COMPOUNDS—*continued*

Element or compound	Structure type	Lattice constants, remarks	Additional Refs.
Group Ib: Cu, Ag, Au—*continued*			
Mg_3Au	$D0_{18}$	4.64; 8.46	
$TiAu_4$	$D1a$	6.460(7); 3.975(8)	
$TiAu_2$	$C11b$	3.419; 8.514	
Ti_3Au	$L1_2$	4.096	
Ti_3Au	$A15$	5.088 9(6)	43
Th_2Au	$C16$	7.462; 5.989	48
V_3Au	$A15$	4.880	49
Nb_3Au	$A15$	5.205	49
MnAu	ortho.	3.146; 3.182; 3.291	
Au_2Mn	$C11b$	3.359; 8.746	259
Au_4Mn	$D1a$	6.45; 4.03	260
CuAu	$L1_0$	3.968; 3.662	
	$A1$	3.872	
	ortho.	3.676(2); 3.956(3); 39.72(2)	
Cu_3Au	$L1_2$	3.747	
Group IIb: Zn, Cd, Hg			
Zn	$A3$	2.664 74(4); 4.946 9(1)	
LiZn	$B32$	6.234	
$NaZn_{13}$	$D2_3$	12.283 6(3)	
KZn_{13}	$D2_3$	12.335(5)	
Mg_2Zn_{11}	$D8c$	8.552(5)	
$MgZn_2$	$C14$	5.223(1); 8.566(3)	
$CaZn_{13}$	$D2_3$	12.185(5)	
$CaZn_5$	$D2d$	5.371; 4.242	
$SrZn_5$	orthorh. *Pnma*	13.147(7); 5.312(2); 6.707(3)	50
$SrZn_5$	$D2d$	5.549; 4.283	
$SrZn_{13}$	$D2_3$	12.242(2)	
$BaZn_5$	orthorh. *Cmcm*	10.788(8); 8.441(5); 5.318(3)	50
$BaZn_{13}$	$D2_3$	12.359(5)	
$LaZn_5$	$D2d$	5.416; 4.217	
LaZn	$B2$	3.76(1)	
CeZn	$B2$	3.70(1)	
PrZn	$B2$	3.692	
$TiZn_3$	$L1_2$	3.932 2(3)	51
$TiZn_2$	$C14$	5.064; 8.210	51
Th_2Zn	$C16$	7.614(4); 5.658(4)	47, 52
$ThZn_2$	$C32$	4.20; 4.17	261
$ThZn_4$	$D13$	4.273; 10.359	261
$ThZn_9$	$D2d$	5.237; 4.442	52
$ZrZn_{22}$	f.c.c.*Fd/3m*	14.103(1)	262
$ZrZn_6$	tetr.	12.7; 8.68	262
$NbZn_2$	$C36$	5.05; 16.32	264
$MnZn_{13}$ (ξ)	monocl. c.	13.483(5); 7.662 6(10); 5.134(3); 127.78(4)°	
$MnZn_3$ (α)	$L1_2$	3.86 (<320°C)	
~MnZn (β)	$B2$	3.070	
(β_1)	$A3$	2.733 9; 4.455 0	
$FeZn_{13}(\xi)$	monocl.	13.424(5); 7.608 0(10); 5.061(3); 127.30(4)°	
$CoZn_{13}(\xi)$	monocl.	13.306(33); 7.535(2); 4.992(12); 126.78°	
CoZn	$A13$	6.319	
NiZn (β_1)	$L1_0$	3.895; 3.214	
(β)	$B2$	2.908 3	
Pd_2Zn	$C23$	5.35; 7.65; 4.14	341
Pd_2Zn_3 (β_1)	$B2$	3.049(>~600°C)	
PdZn (β)	$L1_0$	4.100; 3.346	53
Pd_2Zn	$B2$	3.055 (>~600°C)	
PtZn (ξ)	$L1_0$	4.05; 3.51	
Pt_3Zn	$L1_2$	3.893 (<~800°C)	
$CuZn_3$ (δ)	$A2$	3.001(8) (stable between 550–700°C)	

(continued)

Table 6.1   STRUCTURES OF METALS, METALLOIDS AND THEIR COMPOUNDS—*continued*

Element or compound	Structure type	Lattice constants, remarks	Additional Refs.
Group IIb: Zn, Cd, Hg—*continued*			
Cu_5Zn_8 (γ)	$D8_2$	8.859(3)	
CuZn (β)	$A2$	(>450°C) 2.95	
(β')	$B2$	(<~450°C) 2.959	
Cu_2TiZn	$CsCl$ (B2) type	2.96	265
$AgZn_3$	$A3$	2.823 1(2); 4.440 7(3)	
Ag_5Zn_8	$D8_2$	9.351(3)	54
AgZn (β')	$B2$	3.088(3)	
(ξ)	Bb	7.636 0; 2.819 7	
$(Ag_{0.10}Zn_{0.90})_2$ Mg	hex.	5.21; 90.3 (long period structure)	
$AuZr_2$	$C11B$	3.28; 11.6	342
$AuHf_2$	$C11B$	3.230 9; 11.605 7	342
$AuZn_3$	cub $Pm\bar{3}n$	7.903	266
Au_5Zn_3	ortho.	6.345; 8.971; 4.486	
AuZn	$B2$	3.14	
Au_3Zn	tetr.	5.586(5); 33.4(1)	
	ortho.	16.650; 5.585; 5.594	
Cd	$A3$	2.979; 5.617	
$LiCd_3$	$A3$	3.083; 4.889	
LiCd	$B32$	6.715(2)	
Li_3Cd	$A1$	4.250	
$RbCd_{13}$	$D2_3$	13.88	
$CsCd_{13}$	$D2_3$	13.89	
$MgCd_3$	$D0_{19}$	6.233 5; 5.044 9	
MgCd	$B19$	5.0051(3); 3.221 7(2); 5.270 0(3)	
Mg_3Cd	$D0_{19}$	6.310; 5.080	
$CaCd_2$	$C14$	5.996(2); 9.654(2)	
SrCd	$B2$	4.003	55
BaCd	$B2$	4.213(2)	55
$BaCd_{11}$	tetr.	12.040(5); 7.740(5)	56
LaCd	$B2$	3.910	
CeCd	$B2$	3.855	
PrCd	$B2$	3.827	
TiCd	$B11$	2.904; 8.954 ($M = 4$)	267
Ti_2Cd	$C11b$	2.865(3); 6.71(1)	267
$ZrCd_2$	cubic f.c.	4.376 8	51
ZrCd	$B11$	3.124; 8.75	51
$NaCd_2$	$Fd\bar{3}m$	30.56	268
PdCd	$L1_0$	4.281 7(1); 3.631 0(1)	53
~$PtCd_5$ (γ)	b.c.c	19.804 2(9) (also f.c.c. 9.9200(3))	
PtCd	$L1_0$	4.183 5; 3.816 3	53
Cd_3Cu_4	tetr.	13.701; 9.944 (also f.c.c. 25.871(2))	
$AgCd_3$	$A3$	3.071; 4.816 2	
AgCd	$A3$	2.975 0; 4.825 5	269
AgCd (β)	$A2$	(>440°C) 3.382	
(β_1)	$A3$	(230–440°C) 3.007; 4.852	
(β')	$B2$	(<210°C) 3.332	
$AuCd_3$	$D0_{21}$	8.147(2); 8.511(2)	
Au_2Cd	$A3$	2.919 8; 4.804 5	
AuCd (β_1)	$B2$	3.323 2(5) (>60–80°C)	
(β')	$B19$	4.764 5(5); 3.154 0(5); 4.864 4(5) (<~60°C)	
Au_3Cd (α_1)	$L1_0$	4.111; 4.133 (<~400°C)	
Hg	$A10$	3.458; 6.684; 120°	59
$LiHg_3$	$D0_{19}$	6.240; 4.794	
LiHg	$B2$	3.286(1)	
Li_3Hg	$D0_3$	6.548	
NaHg	ortho.	7.19; 10.79; 5.21 ($M = 8$)	60
Na_3Hg_2	tetr.	8.52; 7.80 ($M = 4$)	60
KHg	triclinic	6.59; 6.76; 7.06 ($\alpha = 106°5'$, $\beta = 101°52'$; $\gamma = 92°47'$)	61, 62

(continued)

Table 6.1   STRUCTURES OF METALS, METALLOIDS AND THEIR COMPOUNDS—*continued*

Element or compound	Structure type	Lattice constants, remarks	Additional Refs.
Group IIb: Zn, Cd, Hg—*continued*			
K_5Hg_7	ortho. *Pbcm*	10.06; 19.45; 8.34 ($M = 4$)	270
KHg_2	ortho.	8.10; 5.16; 8.77 ($M = 4$)	61, 62
$MgHg_2$	$C11_b$	3.83; 8.781	
$MgHg$	$B2$	3.448 0(1)	
Mg_5Hg_3	$D8_8$	8.259; 5.931	
Mg_3Hg	$D0_{18}$	4.83; 8.62	
$SrHg$	$B2$	3.955	55
$BaHg$	$B2$	4.131	55
$BaHg_{11}$	$D2e$	9.586	
$LaHg_4$	cubic b.c.	10.968	
$LaHg_3$	$D0_{19}$	6.816; 4.971	
$LaHg_2$	$C32$	4.956; 3.637	
$LaHg$	$B2$	3.864	
$CeHg$	$B2$	3.808	
$PrHg$	$B2$	3.791	
$TiHg$	$L1_0$	4.26; 4.04	51
Ti_3Hg	$A15$	5.187(1)	51
	$L1_2$	4.165 4(4)	51
Hg_3Th	$A3$	3.361; 4.905	271
$HgTh$	$A1$	4.80 (high temp. phase $>500°C$)	
$ZrHg$	$L1_0$	4.45(1); 4.17(1)	51
$ZrHg_3$	$L1_2$	4.368	51
Zr_3Hg	$A15$	5.560(2)	51
$ThHg_3$	$A3$	3.361; 4.905	47
$ThAg$	f.c.c.	4.80	272
$NdHg$	$B2$	3.772	
UHg_3	$D0_{19}$	6.654(9); 4.888(5)	
UHg_2	$C32$	4.976; 3.218	
$NiHg_4$	b.c.c.	6.016	
$PdHg$	$L1_0$	4.271 6; 3.706 7	
Pt_3Hg	$A1$	3.979 5(5)	341
Au_3Hg	$A3$	2.918 0; 4.811 3	
$CdHg$	$A6$	3.932; 2.872	
$CdHg_2$	$C11_b$	3.965; 8.607	63
Group IIIb; Al, Ga, In, Tl			
Al	$A1$	4.049 50(12)	
$LiAl$	$B32$	6.370 3(5)	
Mg_2Al_3 (β)	$C14$	5.73; 9.54 (several other structures exist)	
$Al_{18}Mg_3Cr_2$	f.c.c.	14.53(1)	273
Y_3Al_2	tetr.	8.258; 7.702 5	
YAl	$B33$	3.870 7; 11.599; 4.396 0	274
Yal	$B2$	3.754(12)	
$CaAl_4$	$D1_3$	4.353; 11.07	
$CaAl_2$	$C15$	8.040	
$SrAl$	cubic	12.76	
$SrAl_4$	$D1_3$	4.460; 11.20	
$BaAl_4$	$D1_3$	4.566; 11.250	66
$LaAl_4$	$D1_3$	4.48; 10.42	
$LaAl_2$	$C15$	8.125	
$CeAl$	ortho.	9.27; 7.68; 5.76	67
$CeAl_4$	$D1_3$	4.380; 10.030	
$CeAl_2$	$C15$	8.060	
$NdAl$	ortho.	5.940; 11.728; 5.729	
$TiAl_3$	$D0_{22}$	3.853 7(3); 8.583 9(13)	
Ti_3Al (α_2)	$D0_{19}$	5.780; 4.647	60
$TiAl$	$L1_0$	4.001; 4.071	
$ZrAl_3$	$D0_{23}$	4.007 4(6); 17.286(4)	
Zr_3Al	$L1_2$	4.373	82
Zr_2Al	$C7$	4.882; 5.918	277

(continued)

Table 6.1   STRUCTURES OF METALS, METALLOIDS AND THEIR COMPOUNDS—*continued*

Element or compound	Structure type	Lattice constants, remarks	Additional Refs.
Group IIIb; Al, Ga, In, Tl—*continued*			
Zr_2Al_3	ortho. *Fdd*2	9.617(3); 13.934(3); 5.584(2)	278
Zr_4Al_3	hex.	5.424(1); 5.405(2)	279
Zr_3Al_2	tetr.	7.633 3(8); 6.996(2)	280
HfAl	*B*33	3.240; 10.803; 4.278	281
Hf_2Al_3	ortho.	9.474; 13.737; 5.501	282
$ThAl_3$	DO_{19}	6.499; 4.626	48
$ThAl_2$	*C*32	4.412(13); 4.173(13)	48
Th_3Al_2	tetr. *P*4/*mbm*	8.134; 4.221	48
Th_2Al	*C*16	7.614; 5.857	48
UAl_4	*D*1*b*	4.42; 6.30	
UAl_3	$L1_2$	4.262; 13.63	
UAl_2	*C*15	7.78	
$NpAl_2$	*C*15	7.79	70
$NpAl_3$	$L1_2$	4.260(1)	70
$NpAl_4$	*D*1*b*	4.42; 6.26; 13.71	70
$PuAl_3$	hex.	6.082; 14.433	81
VAl_3	DO_{22}	3.780; 8.322	
$NbAl_3$	DO_{22}	3.841(1); 8.609(2)	
Nb_2Al	tetr.	9.945; 5.171	283
Nb_3Al	cubic	5.186	337
$TaAl_3$	DO_{22}	3.851 3; 8.558 3	
Cr_2Al	*C*11*b*	3.005; 8.649	
$Cr_2Mg_3Al_{18}$ (E phase)	$Fd\bar{3}m$	14.53(1)	76
$MoAl_{12}$	b.c.c.	7.573	
$MoAl_7$	monocl.	5.12; 13.0; 13.5($\beta = 95°$)	284
WAl_4	monocl.	5.272; 17.771; 5.218; 100° 12′ (a/cell = 30)	285
Mo_3Al	*A*15	4.950(1)	
WAl_5	hex. *P*6_3	4.902 0; 8.857(1)	78
WAl_{12}	cubic. b.c., $m\bar{3}$	7.580 3(5)	77
$MnAl_6$	ortho. c	7.551(4); 6.499 4(3); 8.872 4(17)	79
$MnAl_4$	ortho. p	6.795; 9.343; 13.897	
MnAl	$L1_0$	3.92; 3.54	
	tetr. p	3.92; 3.57	
Mn_3Al_{10}	hex. p	7.543; 7.898	287
Mn_4Al_{11}	tricl.	5.095(4); 8.879(8); 5.051(4); α 89.35(7)°; β 100.47(5)°; γ 105.08(6)°	286
$MnAl_3$	ortho. *Pnma*	14.79; 12.42; 12.59 (*M* = 36)	288
(Mn, Cr)Al_{12}	cubic b.c., $m\bar{3}$	7.509(1)	77
Tc_2Al_3	$D5_{13}$	4.16; 5.13	
$TcAl_{12}$	b.c.c. (WAl_{12})	7.528(1)	289
$TcAl_4$	monocl.c	5.1(1); 17.0(1); 5.1(1); $\beta = 100(1)°$	
$ReAl_6$	isomorph with $MnAl_6$	7.609 1(9); 6.611 7(6); 9.023(10)	275
$ReAl_{12}$	WAl_{12}	7.5270(5)	
Fe_2Al_5	ortho. c	7.644; 6.411; 4.220	80
$FeAl_2$ (ν)	tricl.	4.88; 6.46; 8.80; $\alpha = 91.70°$; $\beta = 73.3°$; $\gamma = 96.90°$	
$FeAl_3$	monocl. c	15.509(3); 8.066(2); 12.469(2); $\beta = 107.72(2)°$	
FeAl (β_2)	*B*2	2.907 6	
Fe_3Al (B_1)	DO_3	5.791 (disordered A2 = 2.911)	
Co_2Al_9	*D*8*d*	10.149(6); 6.290(5); 8.556 5(5); $\beta = 142.40°$	
Co_2Al_5	$D8_{11}$	7.656 0; 7.593 2	
CoAl	*B*2	2.863	
$NiAl_3$	DO_{20}	6.598; 7.352; 4.802	
Ni_2Al_3	$D5_{13}$	4.036(1); 4.897(1)	
NiAl	*B*2	2.848(14)	
Ni_3Al	$L1_2$	3.571 8(2)	
Ni_2TiAl	$L2_1$	5.865	
$NiFe_2Al$ (β')	$L2_1$	5.758	
RuAl	*B*2	3.036	290
$RuAl_2$	*C*54	7.99; 4.71; 8.76	

(*continued*)

Table 6.1   STRUCTURES OF METALS, METALLOIDS AND THEIR COMPOUNDS—*continued*

Element or compound	Structure type	Lattice constants, remarks	Additional Refs.
Group IIIb: Al, Ga, In, T1—*continued*			
$RuAl_3$	$D0_{24}$	4.81; 7.84	
Pd_2Al_3	$D5_{13}$	4.219(1); 5.161(1)	
PdAl	hex.	15.659(4); 5.251(2)	83
	$B2$	3.0532 (560–700°C)	
OsAl	$B2$	3.001(1)	83
IrAl	$B2$	2.985	83
PrAl	ortho.	5.964(5); 11.777(10); 5.745(5) (also E1a type high-temp. phase 9.146; 7.625; 5.698)	83
$PtAl_2$	$C1$	5.910	
Pt_5Al_3	ortho.	5.413(1); 10.73(1); 3.950(1)	338
RhAl	$B2$	2.9806(5)	
Rh_2Al_5	$D8_{11}$	7.893; 7.854	
Rh_2Al_9	$D8d$	8.721; 6.428; 6.352; 94° 48′	
Pt_3Al	$L1_2$	3.877 5	
$CuAl_2$	$C16$	6.067(1); 4.877(1)	
CuAl (η)	monocl.	12.066; 4.105; 6.913; 55.04°	
Cu_3Al_2 (γ_2)	$C7$	4.146(1); 5.063(3)	
Cu_9Al_4 (γ)	$D8_3$	8.7068(3)	
$Mg_2Cu_6Al_5$	$D8c$	8.311(3)	
$MgCuAl_2$	$E1a$	4.00; 9.23; 7.14	
MgCuAl	$C36$	5.102; 16.76 (also other hex. structures)	
$MnCu_2Al$ (β')	$D0_3$	5.968	
$FeCu_2Al_7$	$E9a$	6.336(1); 14.870(2)	
Ag_2Al (γ)	$A3$	2.877 9; 4.622 5	
Ag_3Al (β')	$A13$	6.942	
(β)	$A2$	3.24(3) (hightemp. phase)	
$AuAl_2$	$C1$	5.995(1)	
Au_4Al	$A13$	6.92	
Ga	$A11$	4.524 2; 7.661 8; 4.519 5	
LiGa	$B32$	6.150(7)	
Mg_5Ga_2	$D8_9$	13.708; 7.017; 6.020	
Mg_2Ga	hex.	7.794(2); 6.893(1)	
$CaGa_2$	$C32$	4.325; 4.327	
$LaGa_2$	$C32$	4.320; 4.427	
$CeGa_2$	$C32$	4.321; 4.320	
$PrGa_2$	$C32$	4.281 7; 4.289 8	
α_2GaTi_3	$D0_{19}$	5.750(4); 4.636(3)	291
TiGa	$L1_0$	3.968; 3.970	85
Ti_2Ga	$B8_2$	4.51; 5.50	86
$TiGa_3$	$D0_{22}$	3.789; 8.734	
$ZrGa_3$	$D0_{22}$	3.963(4); 8.712(4)	
Zr_5Ga_3	$D8_8$	8.075; 5.738	291
Nb_3Ga	$A15$	5.176	337
V_3Ga	$A15$	4.819 4(5)	337
Mo_3Ga	$A15$	4.944(15)	
CoGa	$B2$	2.875	83
$NiGa_4$	$D8_2$	8.424(3)	
Ni_2Ga_3	$D5_{13}$	4.054(1); 4.882(2)	
NiGa	$B2$	2.886(1)	
Ni_2Ga	$A6$	3.75; 4.78	
RhGa	$B2$	3.006 3(8)	
PdGa	$B20$	4.965	
$PtGa_2$	$C1$	5.911	
Pt_2Ga_3	$D5_{13}$	4.22; 5.17	
PtGa	$B20$	4.910	
Pt_3Ga	$L1_2$	3.892 5 (several other structures exist)	
Pt_5Ga_3	ortho.	8.031; 7.440; 3.948	
$CuGa_2$ (θ)	tetr.	2.830; 5.839	
Cu_9Ga_4	$D8_3$	8.729 5	
	cubic	8.747(2)	
Cu_4Ga	$A1$	3.665	
Ag_3Ga	$A3$	2.886 9; 4.675 3	

(continued)

Table 6.1   STRUCTURES OF METALS, METALLOIDS AND THEIR COMPOUNDS—*continued*

Element or compound	Structure type	Lattice constants, remarks	Additional Refs.
Group IIIb: Al, Ga, In, Tl—*continued*			
AuGa$_2$	C1	6.075 8(10)	
AuGa	B31	6.262(2); 3.463(2); 6.397 1(5)	
In	A6	3.252 2; 4.950 7	
LiIn	B32	6.792 0(6)	
NaIn	B32	7.332	
MgIn$_2$	cubic	4.60	
MgIn (β'')	L1$_0$	4.606; 4.421 (A6 phase also exists)	
Mg$_2$In	C22	8.27; 3.42	
α_2Ti$_3$In	D0$_{19}$	5.92; 4.78	291
Mn$_3$In	D8$_3$	9.453(4)	
ZrIn	A1	4.418	291
ThIn$_3$	L1$_2$	4.694	292
Ni$_2$In$_3$ (δ')	D5$_{13}$	4.387(3); 5.295(4)	
NiIn (δ)	B2	3.060 (>760°C)	
(ε)	B35	4.536(5); 4.434(5)	
Ni$_2$In	C7	4.188 9; 5.123 0	
Ni$_3$In (γ)	D0$_{19}$	5.332(2); 4.234(2)	
PdIn$_3$	D8$_3$	9.433 4(5) (also D8$_2$ 9.42)	
Pd$_2$In$_3$	D5$_{13}$	4.535 2(2); 5.511 7(2)	
PdIn	B2	3.249 5(2)	
Pd$_3$In	A6	4.06; 5.36 (also L1$_0$ 4.0729(2); 3.791 8(2))	
Pt$_3$In$_7$	D8f	9.435	
PtIn$_2$	C1	6.366	
Pt$_3$In	L1$_2$	3.992(1)	342
Pt$_2$In$_3$	D5$_{13}$	4.53; 5.51	
Cu$_7$In$_3$ (γ)	tricl.	10.071(5); 9.126(5); 6.724(4); $\alpha = 90.22°$; $\beta = 82.84°$; $\gamma = 106.81°$	
Cu$_2$MnIn	D0$_3$	6.206	
AuIn	tricl.	(4.30; 10.59; 3.56; $\alpha = 90.54°$; $\beta = 90.00°$; $\gamma = 90.17°$)	80
Au$_3$In	D0a	5.857 2; 4.735 2; 5.150 4	
Tl	A3	3.458; 5.518 (<230°C)	
	A2	3.871(2) (>230°C)	
LiTl	B2	3.438(3)	
NaTl	B32	7.448 0(4)	
MgTl	B2	3.628	
Mg$_2$Tl	C22	8.082 9(4); 3.679 6(4)	
Mg$_5$Tl$_2$	D8g	14.285(3); 7.328(2); 6.197(2)	
CaTl$_3$	L1$_2$	4.796	
CaTl	B2	3.852 2(6)	
SrTl	B2	4.038	
LaTl$_3$	L1$_2$	4.803	
LaTl	B2	3.936(3)	
ThTl$_3$	L1$_2$	4.751	292
PrTl	B2	3.895	
Pd$_2$Tl	C23	5.719; 4.228; 8.363	
Pd$_3$Tl	D0$_{23}$	4.117; 15.274 (D0$_{22}$ high temp. phase also exists)	342
PtTl	B35	5.605; 4.639	
HgLi$_2$Tl	B2	3.352(2)	
Group IVb: Si, Ge, Sn, Pb			
Si	A4	5.430 6	
Mg$_2$Si	C1	6.35	
CaSi$_2$	C12	3.855(5); 30.6(1); $\gamma = 120°$	
CaSi	B33	4.545(3); 10.728(10); 3.890(3)	102
BaSi	B33	5.028(5); 11.929(6); 4.131(3)	294
LaSi$_2$	Cc	4.318; 13.84	
CeSi$_2$	Cc	4.191; 13.949	
PrSi$_2$	Cc	4.29; 13.76 (ortho. phase also exists)	
NdSi$_2$	Cc	4.142(6); 13.65(2) (ortho. phase also exists)	
SmSi$_2$	Cc	4.08; 13.51 (ortho phase also exists)	
YSi	Bf	4.25; 10.50; 3.82 ($M = 4$)	295

(continued)

Table 6.1   STRUCTURES OF METALS, METALLOIDS AND THEIR COMPOUNDS—*continued*

Element or compound	Structure type	Lattice constants, remarks	Additional Refs.
Group IVb: Si, Ge, Sn, Pb—*continued*			
YSi_2	ortho.	4.05; 3.95; 13.40 (several other structures exist)	
$TiSi$	$B27$	6.551(6); 3.633(3); 4.983(5)	296
Ti_5Si_3	$D8_8$	7.461 0(3); 5.150 8(1)	
Zr_2Si	$C16$	6.609(3); 5.298(3)	95, 96
Zr_5Si_3	$C16$	7.903; 5.581	95
$ZrSi_2$	$C49$	3.73; 14.60; 3.66	94
$HfSi_2$	$C49$	3.672(4); 14.57(1); 3.641(4)	94
$ThSi_2$	Cc	4.107; 14.07	
USi_2 (α)	Cc	3.99; 13.15 ⎫ Zachariasen[97]	
(β)	$C32$	3.839; 4.72 ⎭	
USi	$B27$	7.585; 3.903; 4.663	
U_3Si_2	$D5a$	7.329 9(4); 3900 5	
U_3Si	$D0c$	6.033; 8.690	
$NpSi_2$	Cc	3.97(1); 13.70(3)	
$PuSi_2$	Cc	3.98(1); 13.58(5)	99
VSi_2	$C40$	4.58; 6.37	
V_3Si	$A15$	4.7272 (tetr. phase also exists)	
V_5Si_3	$D8_8$	7.135; 4.842	100
$NbSi_2$	$C40$	4.819(2); 6.592(2)	
Nb_5Si_3	$D8_8$	7.52; 5.24 (also $D8_m$ and $D8_1$ phases)	100
$TaSi_2$	$C40$	4.782; 6.565	
$CrSi_2$	$C40$	4.45; 6.529	
$CrSi$	$B20$	4.622(1)	
Cr_3Si	$A15$	4.556(4)	
$MoSi_2$	$C11_b$	3.206; 7.846	101
Mo_3Si	$A15$	4.898	
Mo_5Si_3	$D8_m$	9.642 5(5); 4.909 6(3)	100
WSi_2	$C11_b$	3.213; 7.887	101
Mn_4Si_7	tetr.	5.525(1); 17.463(3)	
$MnSi$	$B20$	4.558	
Mn_5Si_3	$D8_8$	6.912(5); 4.812(5)	
Mn_3Si	$D0_3$	5.720	
$ReSi_2$	$C11_b$	3.138; 7.666	
$FeSi_2$	tetr.	2.690 1; 5.134 (also ortho. phase)	
$FeSi$	$B20$	4.517(5)	
Fe_5Si_3	$D8_8$	6.759(5); 4.720(5)	
Fe_3Si	$D0_3$	5.644	
$CoSi_2$	$C1$	5.356(1)	
$CoSi$	$B20$	4.450	
Co_2Si	$C37$	4.919; 3.725; 7.104	
$NiSi_2$	$C1$	5.406	
$NiSi$	$B31$	5.190(1); 3.330(1); 5.628(2)	106
Ni_2Si (θ)	hex.	3.836(1); 4.948(1) (>981°C)	106
(δ)	$C23$	5.022(1); 3.741(1); 7.088(1) (<981°C)	106
Ni_3Si (β_1)	$L1_2$	3.505 (<1040°C)	
Ru_2Si_3	tetr.	5.52; 4.46 (also ortho. phase)	
$RuSi$ (1)	$B2$	2.909(2)	
(2)	$B20$	4.707 5	
$RhSi$	$B31$	5.630; 3.061; 6.311	107
Rh_2Si	$C23$	5.370; 3.935; 7.360	
Rh_5Si_3	ortho.	5.317; 10.131; 3.895	
Rh_3Si_2	$B8_1$	3.949; 5.047 (also $B8_2$ with same lattice parameters reported)	293
$RhSi$	$B31$	5.630; 3.061; 6.311	
$PdSi$	$B31$	5.617(3); 3.386(2); 6.149(4)	
Pd_2Si	$C22$	6.496(5); 3.433(4)	108
Os_2Si_3	ortho.	11.157(2); 8.964(1); 5.580(1)	
$OsSi$	$B2$	2.960(1) (also $B20$ phase)	297
Ir_3Si	$D0c$	5.222; 7.954	
Ir_2Si	$C23$	5.284; 3.989; 7.615	
Ir_3Si_2	$B8_2$	3.963; 5.126 (also $D8_2$ with same lattice parameters reported)	293

(continued)

Table 6.1 STRUCTURES OF METALS, METALLOIDS AND THEIR COMPOUNDS—*continued*

Element or compound	Structure type	Lattice constants, remarks	Additional Refs.
Group IVb: Si, Ge, Sn, Pb—*continued*			
$IrSi_3$	$D0_{18}$	4.350; 6.630 (several other structures exist)	
$PtSi$	$B31$	5.595; 3.603; 5.932	
Pt_2Si	$C22$	6.525; 3.603 (various tetr. structures also reported)	110
$BeZrSi$	$B8_2$	3.722; 7.232	8
Pd_4Al_3Si	$B20$	4.840	83
$Cu_{15}Si_4 (\varepsilon)$	$D8_6$	6.915	
Cu_5Si	tetr.	8.815; 7.903	
Cu_7Si	$A3$	2.560 5; 4.184 6	
$Cu_{16}Mg_6Si_7$	$D8a$	11.65(2) ($A = 116$)	111
Cu_3SiMg_2	$C14$	5.004(5); 7.872(8)	
$Al_{13}Cr_4Si_4$	cubic F	10.917(1)	112
Al_9Mn_3Si	$E9c\ Al_{10}Mn_3$	7.42; 7.72	
Al_3FeSi_2	tetr. Ga_5Pd	6.07; 9.50	
$AlNi_2Si$	$B20$	4.537	83
Ge	$A4$	5.657 7	
Mg_2Ge	$C1$	6.380 ($C15$ phase also reported)	
$CaGe_2$	$C12$	3.949; 30.72; 120°	114
Y_5Ge_3	$D8_8$	8.46(1); 6.36(1)	298
$CeGe_2$	Cc	4.26; 14.22 $\Big\}$ (ortho. phases also reported)	299
$PrGe_2$	Cc	4.43(5); 14.00(5)	
$TiGe_2$	$C54$	8.588(8); 5.032(3); 8.862(6)	
Ti_5Ge_3	$D8_8$	7.563(2); 5.228(2)	
$ZrGe_2$	$C49$	3.810(3); 15.07(2); 3.769(2)	
Zr_5Ge_3	$D8_8$	8.029(2); 5.594(2)	115
Hf_5Ge_3	$D8_8$	7.941(1); 5.542(2)	298
$\alpha ThGe_2$	Cc	4.133; 14.202	
$ThGe_2$	$C49$	4.223(2); 16.911(6); 4.052(2) (other ortho structures also proposed)	301
$ThGe_3$	cubic	11.72	
V_3Ge	$A15$	4.763	
$NbGe_2$	$C40$	4.977(2); 6.809(3)	
$TaGe_2$	$C40$	4.938(2); 6.740(2)	
Ta_5Ge_3	$D8_8$	7.581; 5.15	115
Cr_3Ge	$A15$	4.625 2(8)	337
$CrGe$	$B20$	4.800(5)	
Cr_5Ge_3	$D8m$	9.427; 4.786	115
Mo_3Ge	$A15$	4.932	
Mn_5Ge_3	$D8_8$	7.197; 5.038	
$FeGe_2$	$C16$	5.908(3); 4.957(3)	
Fe_2Ge	$B8_2$	4.058; 5.038	
$CoGe$	monocl.	11.65; 3.807; 4.945; 101.1°	293
$CoGe_2$	ortho.	5.670(3); 5.670(3); 10.796(3) (pseudo tetragonal)	
$NiGe$	$B31$	5.381; 3.428; 5.811	
Ni_2Ge	$B8_1$	3.934; 5.056 (also $C23\ B8_2$ structures)	
Ni_3Ge	$L1_2$	3.570 (also $B32$)	
$GeRu$	$B20$	4.846	302
Ge_3Ru_2	ortho.	11.436(5); 9.238(4); 5.716(2) (tetr. at high T)	
$PdGe$	$B31$	5.782; 3.481; 6.259	
Pd_2Ge	$C22$	6.712(1); 3.408(1)	108
$OsGe_2$	monocl.	8.995(6); 3.094(4); 7.685(2); 119.17°	
Ir_4Ge_5	tetr.	5.615; 18.308	293
$IrGe_4$	hex.	6.215; 7.784	
Ir_3Ge_7	$D8f$	8.735(1)	
$IrGe$	$B31$	5.600(1); 3.493(1); 6.290(1)	
$PtGe$	$B31$	5.719(2); 3.697(1); 6.084(2)	
Pt_2Ge	$C22$	6.68; 3.53	108
Pt_3Ge	monocl.	7.930; 7.767; 5.520; $\beta = 135.28°$	293
Pt_3Ge_2	ortho.	12.240; 7.549; 6.854	
Pt_2Ge_3	ortho.	16.441 1(21); 3.377 1(3); 6.201 7(7)	
$PtGe_2$	$C18$	6.185; 5.767; 2.908	

(continued)

Table 6.1   STRUCTURES OF METALS, METALLOIDS AND THEIR COMPOUNDS—*continued*

Element or compound	Structure type	Lattice constants, remarks	Additional Refs.
Group IVb: Si, Ge, Sn, Pb—*continued*			
Cu_3Ge (1)	$D0_{18}$	4.169; 7.499 (>636°C)	
(2)	$D0a$	5.28; 4.22; 4.54; 2.64; 5.45; 4.19 (<636°C)	
(3)	hex.	4.17; 4.92 (570–750°C)	
Cu_5Ge	$A3$	2.612; 4.231	
Co_4GaGe_3	$B20$	4.64	83
Pd_5Al_4Ge	$B20$	4.87	83
Rh_5GaGe_4	$B20$	4.832	83
Sn grey	$A4$	6.489 2(1) (<13.2°C)	
metallic	$A5$	5.830 8; 3.181 0 (>13.2°C)	
$Na_{15}Sn_4$	$D8_6$	13.14(2)	
Mg_2Sn	$C1$	6.77 ($C15$ type phase also documented)	
$CaSn_3$	$L1_2$	4.732(3)	
$CeSn_3$	$L1_2$	4.722	
$PrSn_3$	$L1_2$	4.716(3)	
Ti_5Sn_3	$D8_8$	8.04; 5.45	
Ti_3Sn	$D0_{19}$	5.921(3); 4.762(2)	
Hf_5Sn_3	$D8_8$	8.36; 5.710	304
$ThSn_3$	$L1_2$	4.719	292
USn_3	$L1_2$	4.608 9(23)	
$MnSn_2$	$C16$	6.659(3); 5.447(3)	
Mn_2Sn	$C7$	4.404; 5.529	
$FeSn_2$	$C16$	6.542; 5.326	
$FeSn$	$B35$	5.307; 4.445	
Fe_3Sn_2	hex.	5.344(5); 19.845(5)	
Fe_3Sn	$D0_{19}$	5.44; 4.36 (oxygen stabilised phase)	
$CoSn_2$	$C16$	6.363(3); 5.456(3)	
$CoSn$	$B35$	5.278; 4.258	
Co_3Sn_2	hex.	(high temp.) 4.162; 5.233	
Co_3Sn_2	hex.	16.445; 5.179	
Co_2MnSn	$D0_3$	5.985	116
Ni_3Sn_4	$D7a$	12.214(6); 4.060(2); 5.219(3); 105.0(1)°	
Ni_3Sn_2	ortho.	7.061 5(8); 5.153 0(4); 8.090 3(12)	
	hex.	4.146; 5.253	
Ni_3Sn	$A3$	(>900°C)	
	$D0_{19}$	(<900°C), 5.305; 4.254	
Ni_2MgSn	$D0_3$	6.097(4)	
Ni_2MnSn	$D0_3$	6.059	116
$RhSn$	$B20$	5.122	
Rh_3Sn_2	$C7$	4.331; 5.542	
$RhSn_2$ (1)	$C16$	6.410(3); 5.656(3); (>500°C)	
(2)	$D2_b$	4.487(1); 17.717(5); (<500°C)	117
	ortho.	6.319(5); 6.319(5); 11.97(1) (pseudo tetragonal)	118
$PdSn_4$	$D1_c$	6.388 8(1); 6.441 5(1); 11.446 2(2)	
$PdSn_2$ (1)	ortho.	6.478(3); 6.478(3); 12.155(3) (pseudo tetragonal)	117
(2)	tetr.	6.490(1); 24.378(4) (low temp. mod.)	118
$PdSn$	$B27$	6.31; 3.86; 6.13	
Pd_3Sn_2	$C7$	4.389; 5.703	
Pd_3Sn	$L1_2$	3.976(1) (Disorders to $A1$ 3.970)	
$IrSn$	$B8_1$	3.98; 5.63	
$PtSn_4$	$D1_c$	6.382 3(1); 6.419 0(1); 11.366 6(2)	
$PtSn_2$	$C1$	6.433 1	
Pt_2Sn_3	$D5_b$	4.334 5; 12.960 6	
$PtSn$	$B8_1$	4.104(2); 5.436(2)	
Pt_3Sn	$L1_2$	4.000(1)	
Cu_6Sn_5 (η)	$B8_1$	4.192(2); 5.037(2)	119
Cu_3Sn (ε)	$D0_a$	5.49; 4.32; 4.74	119
	ortho.	5.529(8); 47.75(6); 4.323(5)	120
Cu_3Sn	$D0_a$	5.49; 4.32; 4.74	121
Cu_3Sn	$C15_6$	6.1176(10) (high temp. phase)	122
$Cu_{10}Sn_3$	hex.	7.330(4); 7.864(5)	122
$Cu_{41}Sn_{11}$	cubic	17.980(7)	
$Cu_{17}Sn_3$ (β)	$A2$	3.0261 (stable >600°C)	
Cu_2MnSn	$D0_3$	6.166	

(continued)

Table 6.1   STRUCTURES OF METALS, METALLOIDS AND THEIR COMPOUNDS—*continued*

Element or compound	Structure type	Lattice constants, remarks	Additional Refs.
Group IVb: Si, Ge, Sn, Pb—*continued*			
Ag_3Sn	$D0a$	5.968(9); 4.780 2(4); 5.184 3(9)	
$AuSn_4$	$D1c$	6.512 4(1); 6.516 2(1); 11.706 5(1)	
$AuSn_2$	ortho.	6.909; 7.037; 11.789	80
$AuSn$	$B8_1$	4.322 2; 5.522 2	
$InSn_4$	Af	3.205; 2.995	63
In_3Sn	$A6$	3.456; 4.391	
Pb	$A1$	4.950 8(4)	
$LiPb$	$B2$	3.561	
Li_3Pb	$D0_3$	6.687(3)	124
Li_7Pb_2	hex.	4.751(2); 8.589(4)	124
Li_8Pb_3	monocl. C2/m	8.240; 4.757; 11.03; $\beta = 104°\ 25'\ (M = 2)$	125
$Li_{10}Pb_3$	$D8_3$	10.082	
$Li_{22}Pb_5$	f.c.c.	20.08	305
$NaPb$	tetr.	10.580(5); 17.746(15)	126
$Na_{15}Pb_4$	$D8_6$	13.02	
KPb_2	$C14$	6.66; 10.76	127
Mg_2Pb	$C1$	6.836	
$CaPb_3$	$L1_2$	4.910	
$SrPb_3$	$L1_0$	4.965; 5.024	
$LaPb_3$	$L1_2$	4.902(2)	
$CePb_3$	$L1_2$	4.876(2)	
$PrPb_3$	$L1_2$	4.859 6(3)	
Ti_4Pb	$D0_{19}$	5.962; 4.814	128
UPb_3	$L1_2$	4.791	
$ThPb_3$	$L1_2$	4.853	292
$RhPb_2$	$C16$	6.674(3); 5.831(3)	
$PdPb_2$	$C16$	6.865(3); 5.844(3)	
Pd_3Pb_2	$C7$	4.465 0; 5.709 0	
Pd_3Pb	$L1_2$	4.035(1)	
$IrPb$	$B8_1$	3.993(5); 5.566(5)	
$PtPb_4$	$D1_d$	6.666(10); 5.978(10)	
$PtPb$	$B8_1$	4.26; 5.47	
Pt_3Pb	$L1_2$	4.058(1)	
Ag_4Pb	$A3$	2.92; 4.76	129
$AuPb_2$	$C16$	7.338(4); 5.658(4)	
Au_2Pb	$C15$	7.911(2)	
Group Vb: As, Sb, Bi			
As	$A7$	3.7597(1); 10.4412(2); 120°	
Li_3As	$D0_{18}$	4.450; 7.880	
$LiAs$	monocl. $P2_1/C$	5.79(1); 5.24(1); 10.70(2); $\beta = 117.4°$	306
Na_3As	$D0_{18}$	5.088; 8.982	
K_3As	$D0_{18}$	5.782; 10.222	
Mg_3As_2	$D5_3$	12.355	
$MgLiAs$	$C1_b$	6.19	
$CeAs$	$B1$	6.0815	
$NdAs$	$B1$	5.971	
$ZrAs$	$B1$	5.4335(6)	307
$TiAs$	$B8_1$	3.645; 6.109 (high temp. phase)	130
	B_i	3.64; 12.050	130
$CrAs$	$B31$	5.591; 3.573; 6.128 ($B8_1$ at high temp.)	
Cr_2As	$C38$	3.618(1); 6.350(1)	
$\sim MnAs$ (1)	$B8_1$	3.722; 5.702 (up to 27°C)	131
(2)	$B31$	5.72; 3.676; 6.379 (40 to 125°C)	132
Mn_2As	$C38$	3.785(5); 6.265(5)	
Mn_3As	$D0d$	3.788; 3.788; 16.29 (pseudo tetr.)	
$FeAs_2$	$C18$	5.300 1(5); 5.983 8(5); 2.882 1(4)	
$FeAs$	$B31$	5.440 1(5); 3.371 2(4); 6.025 9(5)	
Fe_2As	$C38$	3.63; 5.98	
$CoAs_3$	$D0_2$	8.208	
$CoAs_2$	$C18$	5.051(2); 5.873(2); 3.127(1) (monocl. phase also reported)	

(continued)

Table 6.1 STRUCTURES OF METALS, METALLOIDS AND THEIR COMPOUNDS—*continued*

Element or compound	Structure type	Lattice constants, remarks	Additional Refs.
Group Vb: As, Sb, Bi—*continued*			
CoAs	$B31$	5.263 1(3); 3.456 9(1); 5.801(1)	
NiAs$_2$	$C18$	4.782; 5.83; 3.545 (Rammelsbergite)	
	ortho.	5.771 8(5); 5.834 2(6); 11.421 4(12)	
NiAs	$B8_1$	3.619; 5.044	
Ni$_{11}$As$_8$	tetr.	6.852(7); 21.81(3)	
Ni$_5$As$_2$	hex.	6.815(1); 12.506(2)	
PdAs$_2$	$C2$	5.984 5(10)	
PtAs$_2$	$C2$	5.968 1(1)	
Cu$_3$As	cubic	9.619(1) ($M = 16$) (Domeykite)	
	hex.	7.102(20); 7.246(20)	
CuMgAs	$C38$	3.953; 6.225	
AgMgAs	$C1_b$	6.21	
ZnAs$_2$	monocl.	9.287(1); 7.691(2); 8.010(1); 102.47°	133
Zn$_3$As$_2$	hex.	7.27; 12.08	
ZnLiAs	$C1$	5.940(1)	
ZnNaAs	$C38$	4.176; 7.088	134
ZnAgAs	$C1$	5.912	134
Cd$_3$As$_2$	$D5_9$	8.96; 12.66	
AlAs	$B3$	5.661 0	
AlLi$_3$As$_2$	ortho.	11.87; 11.98; 12.11 [orthorh., deformed (D_{2h}^{27}) superstructure]	
GeAs$_2$	ortho. $Pbam$	14.76; 10.16; 3.728 ($M = 8$)	308
GaAs	$B3$	5.654	309
InAs	$B3$	6.062	309
SnAs	$B1$	5.716	
Sb	$A7$	4.300 7; 11.222; 120°	
Li$_3$Sb (α)	$D0_{18}$	4.701; 8.309	
(β)	$D0_3$	6.559	
NaSb	monocl.	6.80(2); 6.34(2); 12.48(4); 117.6(2)°	306
Na$_3$Sb	$D0_{18}$	5.355; 9.496	
K$_3$Sb	$D0_{18}$	6.025; 10.693 (also $C16$ phase reported)	
Cs$_3$Sb	$B32$	9.148 4 ($D0_3$ phase also reported)	135
Mg$_3$Sb$_2$	$D5_2$	4.573; 7.229	
CeSb	$B1$	6.388 (tetragonal phase at 9.5 Kelvin also reported)	
PrSb	$B1$	6.38	
NdSb	$B1$	6.321	
TiSb$_2$	$C16$	6.654(3); 5.806(3)	
TiSb	$B8_1$	4.115; 6.264	
Ti$_3$Sb	$A15$	5.222 8	310
	$D8m$	10.465; 5.263 9($M = 32$)	
ThSb	$B1$	6.318(1)	136
Th$_3$Sb$_4$	$D7_3$	9.3840(5)	136
ThSb$_2$	$C38$	4.34; 9.15	136
VSb$_2$	$C16$	6.554 1(4); 5.638 5(5)	131
V$_3$Sb	$A15$	4.939(1) ($M = 2$)	337
Nb$_3$Sb	$A15$	5.264 3($M = 2$)	337
CrSb$_2$	$C18$	6.018 3(4); 6.873 6(2); 3.270 4(2)	
CrSb	$B8_1$	4.115; 5.493	
MnSb	$B8_1$	4.130; 5.786	
Mn$_2$Sb	$C38$	4.080; 6.550	137
FeSb$_2$	$C18$	5.85; 6.575; 3.240	
FeSb	$B8_1$	4.124; 5.173	
CoSb$_2$	$C18$	5.600; 6.380; 3.370 (monocl. phase also reported)	
CoSb	$B8_1$	3.875 7; 5.180 2	
CoMnSb	$C1$	5.899 (other structures also reported)	
NiSb$_2$	$C18$	5.215; 6.358; 3.855	
NiSb	$B8_1$	3.932 5; 5.135 1	
Ni$_3$Sb	$D0a$	5.320 7(5); 4.280 8(3); 4.514 7(4)	
NiMgSb	$C1_b$	6.045	
Ni$_2$MgSb	$D0_3$	6.050(4)	
NiMnSb	$C1_b$	5.929	
Ni$_2$MnSb	$D0_3$	6.027	

(continued)

Table 6.1 STRUCTURES OF METALS, METALLOIDS AND THEIR COMPOUNDS—*continued*

Element or compound	Structure type	Lattice constants, remarks	Additional Refs.
Group Vb: As, Sb, Bi—*continued*			
RhSb	$B31$	5.971 8(7); 3.862 1(7); 6.324(9)	
PdSb$_2$	$C2$	6.460(1)	
PdSb	$B8_1$	4.078; 5.592	
IrSb$_2$	$E1_b$	5.63; 6.573; 3.575; 90.8°	80
PtSb$_2$	$C2$	6.440 0(4)	13
PtSb	$B8_1$	4.126; 5.481	
Cu$_2$Sb (γ)	$C38$	4.001 4; 6.104 4	
Cu$_3$Sb	$D0_3$	(>400°C) 6.00 (D0$_a$ structure also reported)	
CuMgSb	$C1_b$	6.152	
CuMnSb	$C1_b$	6.100	
Ag$_3$Sb	$D0_a$	5.99; 4.85; 5.24 (other structures also reported)	
AuSb$_2$	$C2$	6.658 3(5)	
ZnSb	Be	6.23; 7.76; 8.14	
ZnMg$_2$Sb$_2$	$D5_2$	4.428(4); 7.185(7)	
CdSb	Be	6.469(1); 8.251(2); 8.522(2)	
Cd$_3$Sb$_2$	ortho.	8.40; 7.89; 12.00	
CdCuSb	$C16$	6.262 5	
AlSb	$B3$	6.135 0	309
GaSb	$B3$	6.095 1	309
InSb	$B3$	6.47	309
Tl$_7$Sb$_2$	$L2_2$	11.618(6)	
SnSb	$B1$	5.880(4)	
Bi	$A7$	4.533 0(5); 11.797(1); 120°	
LiBi	$L1_0$	4.753; 4.247	
Li$_3$Bi	$D0_3$	6.708	
NaBi	$L1_0$	4.89; 4.80	
Na$_3$Bi	$D0_{18}$	5.448; 9.655	
KBi$_2$	$C15$	9.501(5)	
K$_3$Bi	$D0_{18}$	6.178; 10.933	
Mg$_3$Bi$_2$	$D5_2$	4.666; 7.401	
MgLiBi	$C1_b$	6.74	
LaBi	$B1$	6.61	
CeBi	$B1$	6.505(2)	
PrBi	$B1$	6.47	
ZrBi$_2$	ortho.	15.49; 10.18; 3.97	312
Th$_3$Bi$_4$	$D7_3$	9.562 ($M=4$)	311
ThBi$_2$	$C38$	4.495; 9.308	141
UBi	$B1$	6.356	141
MnBi	$B8_1$	4.28; 6.11 (ortho. phase also reported)	
NiBi	$B8_1$	4.07; 5.33	
NiMgBi	$C1_b$	6.166	
RhBi	$B8_1$	4.089 4; 5.664 2	220
Pd$_5$Bi$_3$	$B8_1$	4.48(5); 5.86(5)	
PtBi$_2$	$C2$	6.683 (several other structures also reported)	80
CuMgBi	$C1_b$	6.256	
Au$_2$Bi	$C15$	7.942(2)	
InBi	$B10$	5.015; 4.771	142
In$_2$Bi	$C7$	5.496; 6.579	
Pb$_3$Bi	$A1$	4.9714(3)	
Metalloids, *etc.*: B, C, P, N, Po, Te, Se, S			
B	Ag	8.75(2); 5.075(1)	
	hex.	10.96; 23.89 (105 atoms per cell)	144
	hex.	10.926 5(4); 23.809 6(13) (111 atoms per cell)	
	hex.	4.91; 12.57 (12 atoms per cell)	145
	hex.	10.932(2); 23.819(5) (141 atoms per cell)	
	tetr.	10.14(1); 14.17(1) (196 atoms per cell)	
Be$_2$B	$C1$	4.663	314
BeB$_6$	tetr.	10.16; 14.28	
MgB$_2$	$C32$	3.085; 3.523	146
CaB$_6$	cubic	4.151 2	

(continued)

Table 6.1   STRUCTURES OF METALS, METALLOIDS AND THEIR COMPOUNDS—*continued*

Element or compound	Structure type	Lattice constants, remarks	Additional Refs.
Metalloids, *etc.*: B, C, P, N, Po, Te, Se, S–*continued*			
SrB$_6$	cubic	4.199	
BaB$_6$	cubic	4.252(2)	
YB$_6$	cubic	4.102	
LaB$_6$	cubic	4.156 6	
EuB$_6$	cubic	4.19(1)	315
TbB$_6$	cubic	4.102	316
TbB$_4$	*D1e*	7.120(3); 4.042(2)	
CeB$_6$	cubic	4.139 6(4)	
CeB$_4$	*D1e*	7.198; 4.085	
PrB$_6$	cubic	4.148	
NdB$_6$	cubic	4.128(1)	
GdB$_6$	cubic	4.111 1(2)	
ErB$_6$	cubic	4.100	
YbB$_6$	cubic	4.102	
TiB$_2$	*C32*	3.03; 3.22	
TiB	*B27*	6.105; 3.048; 4.551 (B1 structure also reported)	149
ZrB$_{12}$	*D2f*	7.388(3)	151
ZrB$_2$	*C32*	3.169 4(2); 3.530 7(4)	151
ZrB	*B1*	4.65	148 (149, 150)
ThB$_6$	cubic	4.100	
ThB$_4$	*D1e*	7.122 6(5); 4.024 2(3)	
UB$_{12}$	*D2f*	7.475	
UB$_4$	*D1e*	7.077; 3.983	
PuB	*B1*	4.905(4)	317
PuB$_2$	*C32*	3.186 2(2); 3.949(2)	
PuB$_4$	*D1e*	7.101 8(3); 4.002 8(1)	
PuB$_6$	cubic	4.113 4(3)	
VB$_2$	*C32*	2.998; 3.055	
V$_3$B$_4$	*D7$_b$*	3.052; 13.35; 2.98	152
VB	*B$_f$*	3.060(3); 8.048(3); 2.972(1)	
NbB$_2$	*C32*	3.110; 3.270	
Nb$_3$B$_4$	*D7$_b$*	3.342; 14.12; 3.143	
NbB	*B$_f$*	3.298; 8.722; 3.164	
TaB$_2$	*C32*	3.098; 3.225	
Ta$_3$B$_4$	*D7$_b$*	3.036; 12.846; 2.958	
TaB	*B$_f$*	3.280(1); 8.670(3); 3.155(2)	
Ta$_2$B	*C16*	5.783(3); 4.866(3)	
CrB$_2$	*C32*	2.969; 3.066	
Cr$_3$B$_4$	*D7$_b$*	3.000 4(8); 13.018(3); 2.951 6(8)	318
CrB	*B$_f$* (also B33)	2.978 2(7); 7.870(1); 2.934 6(7)	
Cr$_5$B$_3$	*D8$_1$*	5.480; 10.01	153
Cr$_2$B	*D1$_f$*	14.56; 7.320; 4.208	153
Mo$_2$B$_5$	*D8$_i$*	3.007, 20.91, 120°	
MoB$_2$	*C32*	3.039; 3.055 (*C12* structure also reported)	
MoB	*Bg*	3.103; 16.95	
Mo$_2$B	*C16*	5.547(3); 4.739(3)	
W$_2$B$_5$	*D8$_i$*	3.011(3); 20.93(1)	
WB$_2$	*C32*	3.020(2); 3.050(2)	
WB (1)	*B$_f$*	3.19; 8.40; 3.07 (>1 850°C)	
(2)	*B$_g$*	3.0973(2); 16.956 7(25)	
W$_2$B	*C16*	5.568(3); 4.744(3)	
Mn$_3$B$_4$	*D7$_b$*	3.035; 12.838; 2.960	
MnB	*B27*	5.562; 2.976; 4.146	
Mn$_2$B	*C16*	5.148 4(6); 4.208 1(6)	
Mn$_4$B	*D1$_f$*	14.53; 7.293; 4.209	
ReB$_2$	hex.	2.900(1); 7.478(2)	319
FeB	*B27*	5.502; 2.947; 4.058	
Fe$_2$B	*C16*	5.110(3); 4.249(3)	
CoB	*B27*	5.260; 3.042; 3.954	
Co$_2$B	*C16*	5.012; 4.220	

(continued)

Table 6.1 STRUCTURES OF METALS, METALLOIDS AND THEIR COMPOUNDS—*continued*

Element or compound	Structure type	Lattice constants, remarks	Additional Refs.
Metalloids, *etc.*: B, C, P, N, Po, Te, Se, S—*continued*			
Co_3B	$D0_{11}$	5.232; 6.636; 4.418	154
Ni_2B	$C16$	4.991(3); 4.247(3)	
Ni_3B	$D0_{11}$	5.215 73(31); 6.618 41(46); 4.391 59(21)	154
RuB	cubic	7.02	320
RuB_2	orthorh.	2.865; 4.045; 4.645	
Rh_2B	orthorh.	5.42(1); 3.98(2); 7.44(3)	155
Pd_3B_2	hex.	6.48; 3.42	
Pd_5B_2	monocl. $C2/c$	12.786; 4.955; 5.472 ($\beta = 97°$ 2′; $M = 4$)	321
Pd_3B	$D0_{11}$	5.463; 7.567; 4.852	
OsB	cubic	7.03	
Pt_3B	tetr.	2.63; 3.83	
AlB_2	$C32$	3.005; 3.257	
AlB_{10}	orthorh.	8.881; 9.100; 5.690 ($M = 2$)	339
AlB_{12}	ortho.	16.623(5); 17.540(5); 10.180(5) (384 atoms per unit cell)	
		16.573(4); 17.510(3); 10.144(1) (396 atoms per unit cell)	
	tetr.	10.161(7); 14.283(8)	
$Al_6Mo_7B_7$	ortho.	7.03; 6.34; 5.76	
B_3Si	hex. $R\bar{3}m$	6.319; 12.713 (tetr. structure also reported)	322
C	$A4$	3.566 986(3) (diamond)	
	$A9$	2.464(2); 6.711(4)	
RbC_8	ortho.	4.926(4); 8.532(6); 22.472(10)	
CsC_8	hex.	4.945; 17.76	
Be_2C	$C1$	4.340(2)	
MgC_2	tetr.	3.92(2); 5.02(2)	
Mg_2C_3	hex.	7.43(3); 10.59(5)	
CaC_2	$C11_a$	3.89(1); 6.38(1)	
SrC_2	$C11_a$	4.11; 6.68	
BaC_2	$C11_a$	4.40; 7.06	
LaC_2	$C11_a$	4.000; 6.58	323
La_2C_3	$D5c$	8.816 2(5)	
CeC_2	$C11_a$	3.881(2); 6.487(3)	
PrC_2	$C11_a$	3.847(1); 6.428(2)	
NdC_2	$C11_a$	3.822; 6.386	
TiC	$B1$	4.327(1)	
ZrC	$B1$	4.694 1	
HfC	$B1$	4.628	
ThC_2	$C11_a$	4.14; 5.28	156
	Cg	6.684(2); 4.220(1); 6.735(2); 103.91°	157
ThC	$B1$	5.322 1(5)	
UC_2	$C11_a$	3.522(1); 5.988(1)	161
U_2C_3	$D5c$	8.074	162
UC	$B1$	4.960 6(5)	
Pu_2C_3	$D5c$	8.132	
PuC	$B1$	4.972 5	
VC	$B1$	4.172(3)	
V_4C_3	hex.	2.919(2); 27.79(3)	
NbC	$B1$	4.470	163
Nb_2C	$B8_1$	5.416 9(2); 4.971 9(6)	163
TaC	$B1$	4.4562(2)	
Ta_2C	$B8_1$	3.105; 4.940	
C_3Ta_4	hex.	3.116(5); 30.00(5)	164
Cr_7C_3	ortho.	4.526(5); 7.010(5); 12.142(5)	
Cr_7C_3	hex.	14.01; 4.53	
Cr_3C_2	$D5_{10}$	5.532 9(5); 2.829 0(2); 11.471 9(7)	
$Cr_{23}C_6$	$D8_4$	10.63	
MoC	B_h	2.903; 2.820 (several other hex. structures proposed).	
MoC	$B1$	4.26	
Mo_2C	$B8_1$	2.994 9(5); 4.373 24(6)	
WC	B_h	2.906; 2.837	
W_2C	$B8_1$	2.992; 4.725 (other hex. structures reported)	

(continued)

Table 6.1   STRUCTURES OF METALS, METALLOIDS AND THEIR COMPOUNDS—*continued*

Element or compound	Structure type	Lattice constants, remarks	Additional Refs.
Metalloids, *etc.*: B, C, P, N, Po, Te, Se, S—*continued*			
Mn_7C_3	ortho.	4.545; 6.962; 11.984 (isom. with Cr_7C_3)	
$Mn_{23}C_6$	$D8_4$	10.598	
$Fe_{20}C_9$	ortho.	9.06(3); 15.69(3); 7.94(1)	165
Fe_2C	ortho.	4.704(16); 4.318(5); 2.830(6)	166
Fe_3C	$D0_{11}$	5.078 7; 6.729 7; 4.514 4 (Cementite)	
('ε')	hex.	4.767; 4.354	167
Fe (+0.25%C)	$L'2$ (deformed tetr.)	2.842; 3.008 ⎫ Martensite at low temp.	
(+0.75%C)	$L'2$ (deformed tetr.)	2.850; 2.939 ⎭	
Fe (+C)	$A1$	3.654 4 (Austenite at high temp.)	
	$A2$	2.935 (Ferrite)	
Fe_3Mo_3C	$E9_3$	11.10	
$Fe_{21}Mo_2C_6$	$D8_4$ ordered	10.546	
Fe_3W_3C	$E9_3$	11.10	
$Fe_{21}W_2C_6$	$D8_4$ ordered	10.533	
Co_2C	$C18$	2.910; 4.469; 4.426	326
Co_3C	$D0_{11}$	5.07; 6.70; 4.53	
Co_3W_3C	$E9_3$	11.08	
$Co_3W_9C_4$	hex.	7.826; 7.826	168
$Co_3W_{10}C_4$	hex.	7.848; 7.848 (isom. Fe, Ni-compounds)	
Ni_3C	hex.	4.553; 12.92	169
CNi_3W_3	$E9_3$	11.15	170
Ni_5W_6C (η)	cubic	10.873	170
$Ni_3W_{16}C_6$	hex.	7.818 3; 7.818 0	170
Al_4C_3	$D7_1$	3.338; 25.117	
$AlMn_3C$	$E2_1$	3.872 2(1)	
$\sim AlFe_3C$	$E2_1$	3.78	
SiC	$B3$	4.358 1(5) ⎫ there are several other	
	$B5$	3.076; 5.048 ⎭ hexagonal modifications	
B_4C	$D1g$	5.60; 12.12	
P (red)	cubic	11.31 ($A = 66$) (also claimed to be 7.331)	
(white)	$I432$	18.51 ($A = 224$; other possible space groups; 1 m3 m and I43 m)	
Li_3P	$D0_{18}$	4.273; 7.594	
Na_3P	$D0_{18}$	4.980; 8.797	
Be_3P_2	$D5_3$	10.15(3)	
Mg_3P_2	$D5_3$	12.01(1)	
MgLiP	$C1$	6.023 ($C1_b$ structure also claimed)	
AlP	$B3$	5.500	
GaP	$B3$	5.449 9(7)	309
InP	$B1$	5.423	309
LaP	$B1$	6.034 6(9)	
CeP	$B1$	5.927	
PrP	$B1$	5.903	
NdP	$B1$	5.682	
Th_3P_4	$D7_3$	8.618	
U_3P_4	$D7_3$	8.203 9	
UP	$B1$	5.584 4(5)	
V_3P	tetr.	9.400; 4.750	
CrP	$B31$	5.346(6); 3.107(4); 5.999(6)	
Cr_3P	$D0_e$	9.188 7(1); 4.559 3(1)	
WP	$B31$	5.731(4); 3.248(2); 6.227(4)	
MnP	$B31$	5.240 9(3); 3.180 6(2); 5.902 7(4)	
Mn_2P	$C22$	6.059; 3.440	
Mn_3P	$D0_e$	9.178(3); 4.608(3) (isom. Cr_3P)	
FeP_2	$C18$	4.972 9(7); 5.656 8(8); 2.723 0(4)	
FeP	$B31$	5.208; 3.16; 5.812	
Fe_2P	$C22$	5.864 4; 3.456 0	
Fe_3P	$D0_e$	9.110; 4.458 (isom. Cr_3P)	
CoP	$B31$	5.06; 3.27; 5.58	
Co_2P	$C23$	5.63; 3.52; 6.60	

(continued)

Table 6.1   STRUCTURES OF METALS, METALLOIDS AND THEIR COMPOUNDS—*continued*

Element or compound	Structure type	Lattice constants, remarks	Additional Refs.
Metalloids, *etc.*: B, C, P, N, Po, Te, Se, S—*continued*			
Ni_2P	$C22$	5.88; 3.36	
Ni_3P	$D0_e$	8.812; 4.373 (isom. Cr_3P)	
Rh_2P	$C1$	5.502 1(3)	
Ir_2P	$C1$	5.543	
PtP_2	$C2$	5.695 0(5)	
Cu_3P	$D0_{21}$	6.88; 7.18	
ZnP_2	tetr.	5.073 6; 18.561 (several other structures claimed)	
Zn_3P_2	$D5_9$	8.088 9; 11.406 9	
ZnLiP	$C1$	5.779	
CdP_2	tetr.	5.285 2(7); 19.787(3) (isom. with ZnP_2)	
Cd_3P_2	$D5_9$	8.753 7; 12.266 9	
AlP	$B3$	5.500	
$AlLi_3P_2$	ortho.	11.47; 11.61; 11.73	
GaP	$B3$	5.449 9(7)	
InP	$B3$	5.847	
Li_3N	hex.	3.648(1); 3.875(1) (several other hex. structures claimed)	
Be_3N_2	$D5_3$	8.134(10)	
Mg_3N_2	$D5_3$	9.96(1)	
MgLiN	$C1$	4.970(2)	
Ca_3N_2	$D5_3$	11.473(1)	
(β)	$D5_2$	3.553; 5.812 (ortho. phase also reported)	
ScN	$B1$	4.505(7)	
LaN	$B1$	5.305(2)	
CeN	$B1$	5.044	
PrN	$B1$	5.163	
NdN	$B1$	5.132(2)	
EuN	$B1$	5.017(3)	172
GdN	$B1$	4.988 0	
SmN	$B1$	5.046(2)	172
YbN	$B1$	4.786(2)	172
TiN	$B1$	4.239(1)	
$TiLi_5N_3$	$E9_d$	9.700(2)	173
ZrN	$B1$	4.585(2)	
Th_2N_3	$D5_2$	3.883(2); 6.187(4)	
UN_2	$C1$	5.299	
U_2N_3	$D5_2$	3.700(2); 5.825(3) (also 10.68, $D5_3$)	
UN	$B1$	4.887(3)	
NpN	$B1$	4.898 7(5)	
PuN	$B1$	4.904 6	
VN	$B1$	4.134 70(2)	
NbN (1)	$B1$	4.377(1)	
(2)	$B8_1$	2.967; 5.538	
(3)	B_i	2.959; 11.271	
Nb_2N	B_i	3.058(1); 4.961(1)	
TaN	B_h	2.936; 2.885	174
Ta_2N	$B8_1$	3.048; 4.918	
CrN	$B1$	4.148	
Cr_2N	hex.	4.796; 4.470	
MoN	hex.	5.725; 5.608	
Mo_2N	tetr.	4.200; 8.010	
W_2N (β)	hex.	2.89; 22.85	
Mn_3N_2	tetr.	2.974(1); 12.126(4)	
Mn_4N	$E2_1$	3.871	
Fe_2N	hex.	4.787; 4.418	
Fe_3N (ε)	$B8_1$	2.705; 4.376	
Fe_4N (ε)	$E2_1$	3.896(2)	
Fe_4N (γ')	$B1$	3.790(1)	
Fe_8N (α'')	tetr.	5.720; 6.292	
Co_2N	$C18$	4.605 6(10); 4.344 3(10); 2.853 5(5)	
Co_3N	hex.	2.66; 4.35	
Ni_3N	hex.	4.616(1); 4.298(3)	

(continued)

Table 6.1 STRUCTURES OF METALS, METALLOIDS AND THEIR COMPOUNDS—*continued*

Element or compound	*Structure type*	*Lattice constants, remarks*	*Additional Refs.*
Metalloids, *etc.*: B, C, P, N, Po, Te, Se, S—*continued*			
Cu_3N	*D*0$_9$	3.813	
Zn_3N_2	*D*5$_3$	9.743(5)	
ZnLiN	*C*1$_b$	4.877(2) (ordered)	
Cd_3N_2	*D*5$_3$	10.79(2)	
AlN	*B*4	3.12; 4.97	
$AlLi_3N_2$	*E*9*d*	9.461(3)	
GaN	*B*4	3.190(1); 5.189(1)	
$GaLi_3N_2$	*E*9*d*	9.594(3)	
InN	*B*4	3.533(4); 5.693(4)	
α Si_3N_4	hex. D_d^2–P31c	7.818(3); 5.591(4)	327
β Si_3N_4	hex. C_{6h}^2–P6/3*m*	7.595(1); 2.902 3(6)	
$SiLi_5N_3$	*E*9*d*	9.436(2)	
Ge_3N_4	hex. *P*3/*c*	8.196 0(8); 5.930 1(5) (also P6/3*m* structure reported)	173
$GeLi_5N_3$	*E*9*d*	9.614(2)	173
BN	*B*3	3.615 3(4) (several other structures also reported)	
Al_5C_3N	*E*9$_4$	3.285; 21.65	
Po (α)	*Ah*	3.359(1)	
(β)	*A*10	5.093(2); 4.927(2); 120°	
PbPo	*B*1	6.590(3)	
Te	*A*8	4.312(1); 5.957(2)	175
Li_2Te	*C*1	6.504	
Na_2Te	*C*1	7.294(1)	
K_2Te	*C*1	8.148	
BeTe	*B*3	5.626 9(5)	
MgTe	*B*4	4.53(1); 7.38(2)	
CaTe	*B*1	6.356(1)	
SrTe	*B*1	6.659(6)	
BaTe	*B*1	7.005(10)	
EuTe	*B*1	6.594	
YbTe	*B*1	6.361(1)	
$TiTe_2$	*C*6	3.777(3); 6.498(6)	
TiTe	*B*8$_1$	3.834; 6.390	
Ti_5Te_4	tetr.	10.164; 3.772 0	328
UTe	*B*1	6.155	176
U_3Te_4	*D*7$_3$	9.31	176
UTe_2	*C*38	4.243(5); 8.946(5)	
VTe	*B*8$_1$	3.983; 6.133	
CrTe	*B*8$_1$	4.004 7(10); 6.241 8(13)	
$MoTe_2$	*C*7	3.510; 14.00	329
WTe_2	ortho.	3.496; 6.282; 14.073 (C7 phase also reported)	177
$MnTe_2$	*C*2	6.953 2(12)	
MnTe	*B*8$_1$	4.149 7; 6.76 (B1 phase also reported)	
$FeTe_2$	*C*18	5.260; 6.268; 3.876 (several other structures also reported)	
FeTe	*C*38	3.829(2); 6.288(4) } C2 structure also reported	
	*C*6	3.804; 5.405 } C2 structure also reported	
$CoTe_2$	*C*18	5.328; 6.310; 3.875 } C2 structure also reported	
CoTe	*B*8$_1$	3.890; 5.373	
$NiTe_2$	*C*6	3.854 2; 5.260 4	
NiTe	*B*8$_1$	3.965; 5.358	
$RuTe_2$	*C*2	6.391 0(2)	
$PdTe_2$	*C*6	4.036 5; 5.126 2	
PdTe	*B*8$_1$	4.20; 5.79	
$OsTe_2$	*C*2	6.425	
$PtTe_2$	*C*6	4.025 9; 5.220 9	
Cu_2Te	*Ch*	4.188; 7.251	
Cu_2Te	cubic f.c.	6.03	80
CuTe	ortho.	3.10; 4.02; 6.86	80
Ag_2Te	cubic	6.6 (high temp. mod.)	
	monocl.	5.09; 4.48; 6.96; 123.4°	
ZnTe	*B*3	6.079(5) (B1 structure also reported)	
CdTe	*B*3	6.482 (B4 structure also reported)	
HgTe	*B*3	6.453; 6.429 (Coloradoite)	175

(*continued*)

Table 6.1   STRUCTURES OF METALS, METALLOIDS AND THEIR COMPOUNDS—*continued*

Element or compound	Structure type	Lattice constants, remarks	Additional Refs.
Metalloids, *etc.*: B, C, P, N, Po, Te, Se, S—*continued*			
Ga$_2$Te$_3$	B3	5.892(3)	
InTe	B37	8.444; 7.136	63,80
In$_2$Te$_3$	B3	6.135	
TeTl	b.c.tetr.	12.95(1); 6.18(1)	
Te$_3$Tl$_2$	monocl.	17.413; 6.552; 7.910; 133.16°	331
SnTe	B1	6.315(1)	
PbTe	B1	6.438 4(4) (Altaite)	
As$_2$Te$_3$	monocl.	14.357 3(9); 4.019 9(2); 9.899 0(7); 95.107(5)°	178
Sb$_2$Te$_3$	C33	4.264(1); 30.458(7)	179
Te$_2$SbTl	F5$_1$	4.439; 23.20; 120°	332
Te$_2$BiTl	F5$_1$	4.509; 23.13; 120°	
Bi$_2$Te$_3$	C33	4.385 0(2); 30.487(2); 120°	
Se (γ)	A8	3.910(1); 5.080(1)	
(α)	Ak	15.018(1); 14.713(1); 8.789(1); 93.61°	
(β)	A1	12.85; 8.07; 9.31; 93° 8′	
Li$_2$Se	C1	6.001 4(4)	
Na$_2$Se	C1	6.799	
K$_2$Se	C1	7.675	
BeSe	B3	5.147 7(5)	
MgSe	hex.	4.145(5); 6.723(8)	
CaSe	B1	5.932(2)	
SrSe	B1	6.243 2(12)	
BaSe	B1	6.593(16)	
EuSe	B1	6.192	
YbSe	B1	5.933	
TiSe$_2$	C6	3.540(1); 6.008(3)	
TiSe	B8$_1$	3.571; 6.301 (B31 and rhomb. phases also reported)	
ZrSe$_3$	monocl.	5.41; 3.77; 9.45 ($\beta = 97.5°$)	340
ZrSe$_2$	C6	3.772(2); 6.125(5)	
U$_3$Se$_4$	D7$_3$	8.820(1)	334
U$_2$Se$_3$	D5$_8$	11.34; 4.057; 10.92	334
ThSe$_2$	C23	7.629; 4.435; 9.085	
Th$_7$Se$_{12}$	D8k	11.570(6); 4.23(1)	
Th$_2$Se$_3$	D5$_8$	11.57(5); 4.27(1); 11.34(5)	
ThSe	B1	5.889 6(2)	
USe$_3$	monocl.	5.652; 4.056; 10.469; 115.03°	180
VSe$_2$	C6	3.356; 6.150	
VSe	B8$_1$	3.660; 5.988	
Cr$_2$Se$_3$	hex.	6.28; 11.64 (several other structures reported)	
CrSe	B8$_1$	3.699 1(4); 6.072 1(8)	
CrNaSe$_2$	F5$_1$	3.732 3(2); 20.396(3); 120°	
WSe$_2$	C7	3.282(1); 12.96(1)	
MnSe$_2$	C2	6.430	
MnSe (α)	B1	5.46(6)	
(β)	B3	5.82	
(γ)	B4	4.12; 6.72	
FeSe$_2$	C18	4.804(2); 5.784(3); 3.586(2) (C2 phase also reported)	
FeSe	B8$_1$	3.632; 5.910	
	B10	3.775; 5.527 (L1$_0$ structure also reported)	
CoSe$_2$	C2	5.8593(5)	
CoSe	B8$_1$	3.613; 5.300	
NiSe$_2$	C2	5.973	
NiSe	B8$_1$	3.652; 5.347	
	B13	10.007; 3.333; 120°	
RuSe$_2$	C2	5.930	
RhSe$_2$	C2	5.963 9(6)	181
OsSe$_2$	C2	5.946(1)	
PtSe$_2$	C6	3.727 8; 5.081 3	
CuSe	B18	3.976; 17.243 (several other hex. and ortho. phases reported)	
Cu$_2$Se	C1	5.759(1) (Berzelianite) (numerous other structures also reported)	

(continued)

Table 6.1 STRUCTURES OF METALS, METALLOIDS AND THEIR COMPOUNDS—*continued*

Element or compound	Structure type	Lattice constants, remarks	Additional Refs.
Metalloids, *etc.*: B, C, P, N, Po, Te, Se, S—*continued*			
Ag_2Se	cubic	5.006(1) (ortho. phase also reported)	
ZnSe	B3	5.663 (B4 structure also reported)	
CdSe	B3	6.04(3)	
	B4	4.299 5; 7.012 0	
$CdCr_2Se_4$	$H1_1$	10.742 ('normal' $H1_1$-type)	182
HgSe	B3	5.65 (Tiemannite)	175
GaSe	hex.	3.754(1); 15.945(1)	63,
	hex.	several phases reported	184, 185
	hex.	3.755 3; 31.990(10)	63,
Ga_2Se_3	B3	5.446 (several other structures reported)	184
In_2Se_3	hex.	16.06; 19.28 (numerous other structures reported)	186
TlSe	B37	8.024; 7.18 (cubic structure also reported)	
$GeSe_2$	ortho.	6.953(1); 12.221(5); 23.04(3) (several other phases also reported)	
GeSe	B16	10.825; 3.833(4); 4.388(4)	336
SnSe	B16	11.62(1); 4.282(5); 4.334(4)	187
PbSe	B1	6.121 3(8)	
Sb_2Se_3	$D5_8$	11.793 8(9); 3.985 8(6); 11.647(7)	
Bi_2Se_3	C33	4.129; 28.60	80
S (orthrh.)	A16	10.464 6(1); 12.866 0(1); 24.468 0(3)	
(monocl.)	C_2^5h	11.02; 10.96; 10.96; 96.7°	
(rhomb.)	rhomb.	10.9; 4.27	
Li_2S	C1	5.708	
Na_2S	C1	6.523	
K_2S	C1	7.393	
Rb_2S	C1	7.65	
BeS	B3	4.863(5)	
MgS	B1	5.199 7(5)	
CaS	B1	5.700 (Oldhamite)	
SrS	B1	6.024(6)	
BaS	B1	6.387(3)	
La_2S_3	$D7_3$	8.692(3) (several other structures reported)	
Ce_2S_3	$D7_3$	8.608 4(18) (several other structures reported)	
Ce_3S_4	$D7_3$	8.625 0(5)	
CeS	B1	5.776	
DyS	B1	5.489(6)	
Ac_2S_3	$D7_3$	8.99(1)	
TiS_2	C6	3.402(1); 5.716(1)	
ThS_2	C23	7.267(3); 4.273(2); 8.615(4)	
Th_2S_3	$D5_8$	10.99(5); 3.96(3); 10.85(5)	
ThS	B1	5.685 1(3)	
ZrS_2	C6	3.663(2); 5.827(6)	
U_2S_3	$D5_8$	10.62; 3.86; 10.39	
US	B1	5.483	
Np_2S_3	$D5_8$	10.32(1); 3.86(5); 10.62(1)	
Pu_2S_3	$D7_3$	8.454 6(8)	
PuS	B1	5.53	
Am_2S_3	$D7_3$	8.434 4(3)	
VS	$B8_1$	3.360; 5.813 (several other structures reported)	
TaS_2	C6	3.365(2); 5.853(2) (numerous other phases reported)	
CrS	$B8_1$	3.419; 5.550	
	monocl.	3.826; 5.913; 6.089; 101.60°	
$CrNaS_2$	$F5_1$	3.554; 19.52; 120°	
$CrKS_2$	$F5_1$	3.602(6) 21.15(5); 120°	
MoS_2	C7	3.161; 12.295 (Molybdenite)	
WS_2	C7	3.154; 12.36 (Tungstenite)	
MnS_2	C2	6.101 6	
MnS	B1	5.222 6 (Alabandite)	
	B3	5.59	
	B4	3.988; 6.433	
$MnCr_2S_4$	$H1_1$	10.111	

(continued)

Table 6.1   STRUCTURES OF METALS, METALLOIDS AND THEIR COMPOUNDS—*continued*

Element or compound	Structure type	Lattice constants, remarks	Additional Refs.
Metalloids, *etc.*: B, C, P, N, Po, Te, Se, S—*continued*			
FeS_2	$C2$	5.293(2) (Pyrites)	
	$C18$	4.464; 5.44; 3.39 (Marcasite)	
FeS	$B8_1$	3.445; 5.763 (numerous minor variants)	183, 189
Fe_7S_8	hex.	6.867 3(9); 17.062(2) (several other structures proposed)	
$FeKS_2$	$F5a$	7.079(1); 11.304(2); 5.398(1); 113.20(1)°	
$FeCr_2S_4$	$H1_1$	9.998	193
CoS_2	$C2$	5.427(2)	
Co_3S_4	$H1_1$	9.404(1) (Linnaeite)	
CoS	$B8_1$	3.44; 5.79	
Co_9S_8	$D8_9$	9.930(2)	
$CoCr_2S_4$	$H1_1$	9.923	
NiS_2	$C2$	5.676 5(1)	
Ni_3S_4	$H1_1$	9.489	
NiS	$B8_1$	3.437; 5.350	
NiS	$B13$	9.607(1); 3.143(1) (Millerite)	
Ni_6S_5	ortho.	3.254; 11.334; 16.43	
Ni_3S_2	rhomb.	5.738(2); 7.126(2); 120°	
Ni_2FeS_4	$H1_1$	9.465	
RuS_2	$C2$	5.607(3)	
RhS_2	$C2$	5.607(4)	
PdS	$B34$	6.429(2); 6.611(2)	
OsS_2	$C2$	5.620(2)	
PtS_2	$C6$	3.543 2; 5.038 8	
PtS	$B17$	3.470 1(3); 6.109 2(8) (Cooperite)	
CuS	$B18$	3.760(1); 16.20(4) (Covellite)	
Cu_2S	$C1_b$	5.564 (several other structures also reported)	
Cu_3VS_4	$H2_4$	5.391(5) (Sulvanite)	
$CuFeS_2$	$E1_1$	5.286 4(8); 10.410 2(8) (Chalcopyrite)	
$CuFe_2S_3$	ortho.	6.231(2); 11.117(6); 6.467(1) (Cubanite)	
Cu_5FeS_4	tetr.	10.94; 21.88 (several other structures also reported)	
$CuCo_2S_4$	$H1_1$	9.472(1) (Carrollite)	
Ag_2S	monocl.	4.231; 6.930; 9.526; 125.48° (several other phases also reported)	
$AgFeS_2$	$E1_1$	5.66; 10.30	
ZnS	$B3$	5.382 9 (Zincblende) (low temp. mod.)	
	$B4$	3.822 6(3); 6.260 5(15) (Würtzite) (high temp. mod.)	
$ZnAl_2S_4$	$H1_1$	10.009 3(10) (<~1 000°C)	
	$B4$	3.743; 6.089; >~1 000°C	
CdS	$B4$	4.100; 6.650 (Greenockite)	
	$B1$	5.32	
$CdCr_2S_4$	$H1_1$	10.242	
HgS	$B3$	5.851 4(3) (Metacinnabarite)	
	$B9$	4.147 9(7); 9.496 0(15) (Cinnabar)	
$HgCr_2S_4$	$H1_1$	10.239(1)	
Al_2S_3	$B4$	3.579(4); 5.829(7)	
Ga_2S_3 (α)	$B3$	5.153; < 550°C	
(β)	$B4$	3.682(1); 6.031(1)	
InS	ortho.	4.447(1); 10.648(2); 3.944(1)	198
In_2S_3	$H1_1$	10.728 (various other structures also reported)	
In_2CaS_4	$H1_1$	10.774(9)	
In_2MgS_4	$H1_1$	10.719(6)	
In_2MnS_4	$H1_1$	10.739 6(5)	
In_2FeS_4	$H1_1$	10.618(3)	
In_2CoS_4	$H1_1$	10.646(5)	
In_2NiS_4	$H1_1$	10.518(1)	
In_2HgS_4	$H1_1$	10.812(7)	
TlS	$B37$	7.77(1); 6.79(1)	
Tl_2S	rhomb.	12.20; 18.20; 120°	
SiS_2	$C42$	9.545(3); 5.564(2); 5.552(2)	
GeS_2	$C44$	11.691(8); 22.41(3); 6.68(1)	

(continued)

Table 6.1   STRUCTURES OF METALS, METALLOIDS AND THEIR COMPOUNDS—*continued*

Element or compound	Structure type	Lattice constants, remarks	Additional Refs.
Metalloids, *etc.*: B, C, P, N, Po, Te, Se, S—*continued*			
GeS	$B16$	10.449(9); 3.653(4); 4.306(4)	
$(Fe, Ge)Cu_2S_4$	$E1_1$	5.332(1); 10.531(1)	
SnS_2	$C6$	3.66; 5.96	
SnS	$B16$	11.32(1); 4.050(4); 4.242(4) (several other structures also reported)	
$SnCu_2FeS_4$	tetr.	5.449(2); 10.757(3)	
PbS	$B1$	5.914 3(4) (Galena)	
$PbSnS_2$	$B16$	11.35; 4.05; 4.29 (Teallite)	
AsS	$B1$	7.153(4); 9.994(6); 12.986(6); 120.6° (Realgar)	
As_2S_3	$D5f$	4.256(2); 9.577(4); 12.191(5); 109.76(8)° (Orpiment)	
CoAsS	$C2$	5.576(2) (Cobaltite)	
NiAsS	$F0_1$	5.690 (Gersdorffite)	
CuAsS	ortho.	11.35; 5.456; 3.749	
Sb_2S_3	$D5_8$	11.380(5); 3.829(2); 11.189(5) (Stibnite)	
FeSbS	monocl.	6.02; 5.93; 6.02; 67.87° (Gudmundite)	
NiSbS	$F0_1$	5.933(3) (Ullmannite)	
$CuSbS_2$	$F5_6$	6.008(10); 3.784(10); 14.456(30) (Wolfsbergite)	
Bi_2S_3	$D5_8$	11.31; 3.98; 11.13 (Bismuthite)	
$NaBiS_2$	$B1$	5.75(2)	
$KBiS_2$	$B1$	6.00(1)	
$CuBiS_2$	$F5_6$	6.142 6(3); 3.918 9(4); 14.528 2(7) (Emplecite)	
$AgBiS_2$	$B1$	5.648	
	hex.	4.07(2); 19.06(5)	
Bi_2Te_2S	rhomb.	4.14; 29.45; 120°	

Table 6.2   STRUCTURAL DETAILS

See note at end of references for sources of these data.

*A*1 (Cu type)

Cubic: O_h^5–$Fm3m$; $a = 3.614\,91$; $A = 4$
Co-ordinates: $4Cu(O_h)$: $000, \frac{1}{2}\frac{1}{2}0$ ♪

*A*2 (W type)

Cubic: O_h^9–$Im3m$; $a = 3.163\,79$; $A = 2$
Co-ordinates: $2W(O_h)$: 000; $\frac{1}{2}\frac{1}{2}\frac{1}{2}$

*A*3 (Mg type)

Hexagonal: D_{6h}^4–$P6_3/mmc$; $a = 3.209\,44$, $c = 5.210\,76$; $A = 2$
Co-ordinates: $2Mg(D_{3h})$: $\frac{2}{3}\frac{1}{3}0$; $\frac{1}{3}\frac{2}{3}\frac{1}{2}$

*A*4 (Diamond type)

Cubic: O_h^7–$Fd3m$; $a = 3.566986(3)$; $A = 8$
Co-ordinates: $8C(T_d)$: 000; $\frac{1}{2}\frac{1}{2}0$♪; $\frac{1}{4}\frac{1}{4}\frac{1}{4}$; $\frac{3}{4}\frac{3}{4}\frac{1}{4}$♪

*A*5 (Tin type)

Tetragonal: D_{4h}^{19}–$I4/amd$; $a = 5.830\,8$, $c = 3.181\,0$; $A = 4$
Co-ordinates: $4Sn(D_{2d})$: 000; $\frac{1}{2}\frac{1}{2}\frac{1}{2}$; $\frac{1}{2}0\frac{1}{4}$; $0\frac{1}{2}\frac{3}{4}$

*A*6 (In type)

Tetragonal: D_{4h}^{17}–$F4/mmm$; $a = 3.252\,2$, $c = 4.950\,7$; $A = 4$
Co-ordinates: $4In(D_{4h})$: 000; $\frac{1}{2}\frac{1}{2}0$; ♪

*A*7 (As type)

Rhombohedral: D_{3d}^5–$R\bar{3}m$
Co-ordinates: Rhombohedral   (I), $2As(C_{3v})$: $\pm(xxx)$
                Rhombohedral   (II), $8As(C_{3v})$: $(000; \frac{1}{2}\frac{1}{2}0; ♪)\pm(xxx)$
                Hexagonal   (III), $6As(C_{3v})$: $(000; \frac{2}{3}\frac{1}{3}\frac{1}{3}; \frac{1}{3}\frac{2}{3}\frac{2}{3})\pm(00x)$

(*continued*)

Table 6.2   STRUCTURAL DETAILS—*continued*

	Rhomb. I A = 2		Rhomb II A = 8		Hexagonal III A = 6			
	a	α	a	α	a	c	x	c/a
As	4.12	54° 10′	5.57	84° 38′	3.759 7(1)	10.441 2(2)	0.226	2.78
Sb	4.50	57° 06′	6.20	87° 24′	4.300 7	11.222	0.233	2.61
Bi	4.74	57° 14′	6.57	87° 32′	4.535 0(5)	11.814(1)	0.237	2.61
Simple cub.	—	60°	—	90°	—	—	—	2.45

A8 (Se type)

Hexagonal: D_3^4–$P3_121$ (and D_3^6–$P3_221$)
Co-ordinates: $3Se(C_2)$: $x00; \bar{x}\bar{x}\frac{1}{3}; 0x\frac{2}{3}$ (and $x00; \bar{x}\bar{x}\frac{2}{3}; 0x\frac{1}{3}$)

	a	c	c/a	x
Se	3.956(1)	5.069(2)	1.28	0.22
Te	4.312(1)	5.957(2)	1.38	0.27
Simple cub.	—	—	1.23	0.33

A9 (Graphite type)

(α) Hexagonal: D_{6h}^4–$P6_3/mmc$ (if $z = 0$); or
    C_{6v}^4–$P6_3/mc$ (if $z \neq 0$); $a = 2.47(1), c = 6.93(1); A = 4$
    Co-ordinates: $2C(D_{3h}$ or $C_{3v})$: $000; 00\frac{1}{2}$
        $2C(D_{3h}$ or $C_{3v})$: $\frac{1}{3}\frac{2}{3}z; \frac{2}{3}\frac{1}{3}(\frac{1}{2} + z); z \approx 0$ (or very probably $z = 0$)
(β) Rhombohedral: D_{3d}^5–$R\bar{3}m; a = 2.46, c = 10.1; A = 6$
    Co-ordinates: $6C(C_{3v})$: $(000; \frac{2}{3}\frac{1}{3}\frac{1}{3}; \frac{1}{3}\frac{2}{3}\frac{2}{3}) \pm (00x)$, with $x \approx \frac{1}{6}$ (or very probably $x = \frac{1}{6}$)

A10 (Hg type)

Rhombohedral: D_{3d}^5–$R\bar{3}m$
Co-ordinates: Rhombohedral   (I): $1Hg(D_{3d})$: 000
        Rhombohedral   (II): $2Hg(D_{3d})$: $000; \frac{1}{2}\frac{1}{2}\frac{1}{2}$
        Rhombohedral   (III): $4Hg(D_{3d})$: $000; \frac{1}{2}\frac{1}{2}0; ♪$
        Hexagonal   (IV): $3Hg(D_{3d})$: $000; \frac{1}{3}\frac{2}{3}\frac{1}{3}; \frac{2}{3}\frac{1}{3}\frac{2}{3}$

	Rhombohedral			Hexagonal	Ideal cubic			
	I	*II*	*III*	*IV*	*I*	*II*	*III*	*Hex.*
a	3.00	4.90	4.38	3.46	$\frac{a}{2}\sqrt{2}$	$\frac{a}{2}\sqrt{6}$	a	$\frac{a}{2}\sqrt{2}$
α	70° 32′	41° 25′	98° 15′	—	60°	33° 33′	90°	—
c	—	—	—	6.71	—	—	—	$a\sqrt{3}$
A	1	2	4	3	1	2	4	3
c/a	—	—	—	1.94	—	—	—	2.45

A11 (Ga type)

Orthorhombic: D_{2h}^{18}–$Abma; a = 4.524\,2, b = 4.519\,5, c = 7.661\,8; A = 8$
Co-ordinates: $8Ga(C_s)$: $(000; 0\frac{1}{2}\frac{1}{2}) \pm (x0z; \frac{1}{2} + x, \frac{1}{2}, z)$ with $x = 0.079; z = 0.153$

A12 (α-Mn type)

Cubic: T_d^3–$I\bar{4}3m; a = 8.912\,5; A = 58$
Co-ordinates: $(000; \frac{1}{2}\frac{1}{2}\frac{1}{2}) + 2Mn(T_d)$: 000
        $+8Mn(C_{3v})$: $xxx; \bar{x}\bar{x}x; ♪$; with $x = 0.32$
        $+24Mn(C_s)$: $xxz; ♪; \bar{x}\bar{x}z; ♪; \bar{x}x\bar{z}; ♪; x\bar{x}\bar{z}; ♪$; with $x = 0.36; z = 0.04$
        $+24Mn(C_s)$ with similar co-ordinates but with $x = 0.09; z = 0.28$

(continued)

Table 6.2 STRUCTURAL DETAILS—*continued*

---

*A*13 (β-Mn type)

Cubic: O^6–$P4_33$ and O^7–$P4_13$; $a = 6.315(2); A = 20$
Co-ordinates: $8\text{Mn}(C_3)$: $xxx; (\frac{1}{2}+x)(\frac{1}{2}-x)\bar{x}; \text{⌡}; (\frac{3}{4}-x)(\frac{3}{4}-x)(\frac{3}{4}-x); (\frac{1}{4}-x)(\frac{3}{4}+x)(\frac{1}{4}+x); \text{⌡};$
        with $x = 0.061$
        $12\text{Mn}(C_2)$: $\frac{3}{8}\bar{x}(\frac{1}{4}+x); \text{⌡}; \frac{7}{8}(\frac{1}{2}+x)(\frac{1}{4}-x); \text{⌡}; \frac{1}{8}x(\frac{1}{4}+x); \text{⌡}; \frac{5}{8}(\frac{1}{2}-x)(\frac{3}{4}-x); \text{⌡};$
        with $x = 0.206$
    An alternative structure has been proposed (Wilson[9]) with space group O_h^6–$Fm3c$; $a = 12.58; A = 160$

*A*15 (Cr₃Si type)

Cubic: O_h^3–$Pm3n$; $a = 4.556(4); A = 8$
Co-ordinates: $2\text{Si}(T_h)$: $000; \frac{1}{2}\frac{1}{2}\frac{1}{2}$
        $6\text{Cr}(D_{2d})$: $\frac{1}{2}0\frac{1}{4}; \text{⌡}; \frac{1}{2}0\frac{3}{4}; \text{⌡}$

*A*16 (orh. S type)

Orthorhombic: D_{2h}^{24}–$Fddd$; $a = 10.464\,6(1), b = 12.866\,0(1), c = 24.486\,0(3); A = 128$
Co-ordinates: 4 times $32\text{S}(C_1)$ in $(000; \frac{1}{2}\frac{1}{2}0; \text{⌡}) + xyz; \bar{x}y\bar{z}; (\frac{1}{4}-x)(\frac{1}{4}-y)(\frac{1}{4}-z); (\frac{1}{4}+x)(\frac{1}{4}-y)(\frac{1}{4}+z);$
        $x\bar{y}\bar{z}; \bar{x}\bar{y}z; (\frac{1}{4}-x)(\frac{1}{4}+y)(\frac{1}{4}+z); (\frac{1}{4}+x)(\frac{1}{4}+y)(\frac{1}{4}-z)$

	x	y	z
S I	−0.017	0.083	0.072
S II	−0.094	0.161	0.200
S III	−0.167	0.105	0.125
S IV	−0.094	0.028	0.250

*A*20 (α-U type)

Orthorhombic: D_{2h}^{17}–$Cmcm$; $a = 2.844\,4, b = 5.868\,9, c = 4.931\,6; A = 4$
Co-ordinates: $4\text{U}(C_{2v})$: $(000; \frac{1}{2}\frac{1}{2}0) + 0y\frac{1}{4}; 0\bar{y}\frac{3}{4}; y = 0, 105$

A$_a$ (Pa type)

Tetragonal: D_{4h}^{17}–$I4/mmm$; $a = 3.940, c = 3.244; A = 2$
Co-ordinates: $2\text{Pa}(D_{4h})$: $000; \frac{1}{2}\frac{1}{2}\frac{1}{2}$

A$_b$ (β-U type)

Tetragonal: C_{4v}^4–$P4nm$; $a = 10.758\,9, c = 5.653\,1; A = 30$
Co-ordinates: $2\text{U}(C_{2v})$: $00z; \frac{1}{2}\frac{1}{2}(\frac{1}{2}+z); z = 0.66$
        $4\text{U}(C_s)$: $xxz; \bar{x}\bar{x}z; (\frac{1}{2}+x)(\frac{1}{2}-x)(\frac{1}{2}+z); (\frac{1}{2}-x)(\frac{1}{2}+x)(\frac{1}{2}+z); x = 0.11; z = 0.23$
        $4\text{U}(C_s)$ in similar position with $x = 0.32; z = 0.00$
        $4\text{U}(C_s)$ in similar position with $x = 0.68; z = 0.50$
        $8\text{U}(C_1)$: $xyz; \bar{x}y\bar{z}; (\frac{1}{2}+x)(\frac{1}{2}-y)(\frac{1}{2}+z); (\frac{1}{2}-x)(\frac{1}{2}+y)(\frac{1}{2}+z); xyz; \bar{y}\bar{x}z; (\frac{1}{2}+y)$
        $(\frac{1}{2}-x)(\frac{1}{2}+z); (\frac{1}{2}-y)(\frac{1}{2}+x)(\frac{1}{2}+z);$ with $x = 0.56; y = 0.24; z = 0.25$
        $8\text{U}(C_1)$ in similar position with $x = 0.38; y = 0.04; z = 0.20$
    Thewlis[14] compared the lattice constants at 720°C of pure β-U and Cr containing β-U.

	a	c
Pure β-U at 720°C	10.759	5.656
1.4 at % Cr-U alloy at 720°C	10.763	5.652
1.4 at % Cr-U alloy at 20°C	10.590	5.634

A$_c$ (α-Np)

Orthorhombic: D_{2h}^{16}–$Pmcn$; $a = 6.663(3), b = 4.723(1), c = 4.887(2); A = 8$
Co-ordinates: $4\text{Np}(C_s)$: $\pm(\frac{1}{4}yz); \pm(\frac{1}{4}, \frac{1}{2}-y, \frac{1}{2}+z); y = 0.208; z = 0.036$
        $4\text{Np}(C_s)$ in similar positions with $y = 0.842; z = 0.319$

A$_d$ (β-Np)

Tetragonal: D_4^2–$P42_1$; $a = 4.897(2), c = 3.388(2); A = 4$
Co-ordinates: $2\text{Np}(D_2)$: $000, \frac{1}{2}\frac{1}{2}0$
        $2\text{Np}(C_4)$: $\frac{1}{2}0z; 0\frac{1}{2}\bar{z}; z = 0.38$

(*continued*)

Table 6.2 STRUCTURAL DETAILS—*continued*

A_e (β'-TiCu$_3$ type)

Orthorhombic: D_{2h}^{17}–$Cmcm$; $a = 2.585$, $b = 4.527$, $c = 4.351$; $A = 4$
Co-ordinates: 4Ti or Cu(C_{2v}): $(000; \frac{1}{2}\frac{1}{2}0) + 0y\frac{1}{4}; 0\bar{y}\frac{3}{4}; y = 0.345$

A_f (HgSn$_{6-13}$ type)

Hexagonal: D_{6h}^1–$P6/mmm$; $a = 3.205$, $c = 2.984$; $A = 1$ (data for HgSn$_6$)
Co-ordinates: 1Hg or Sn (D_{6h}): 000

A_g (B type)

Tetragonal: $P4_2/nnm$; $a = 8.75(2)$, $c = 5.075(1)$; $A = 50$
Co-ordinates: 2B(S_4): $00\frac{1}{2}; \frac{1}{2}\frac{1}{2}0$
6 times 8B (C_1): $xyz; (\frac{1}{2}-x)(\frac{1}{2}+y)(\frac{1}{2}+z)$
$\bar{x}\bar{y}z; (\frac{1}{2}+x)(\frac{1}{2}-y)(\frac{1}{2}+z)$
$\bar{y}x\bar{z}; (\frac{1}{2}+y)(\frac{1}{2}+x)(\frac{1}{2}-z)$
$y\bar{x}\bar{z}; (\frac{1}{2}-y)(\frac{1}{2}-x)(\frac{1}{2}-z)$

	B I	B II	B III	B IV	B V	B VI
x	0.328	0.095	0.223	0.078	0.127	0.250
y	0.095	0.328	0.078	0.223	0.127	0.250
z	0.395	0.395	0.105	0.105	0.395	−0.078

A_h (α-Po type)

Cubic: O_h^1–$Pm3m$; $a = 3.359(1)$; $A = 1$
Co-ordinates: 1Po(O_h): 000

A_i (β-Po)

Rhombohedral: D_{3d}^5–$R\bar{3}m$; $a = 3.37$; $\alpha = 98°13'$; $A = 1$
Co-ordinates: 1Po(D_{3d}^5): 000

A_k (α-Se)

Monoclinic: C_{2h}^5–$P2_1/n$; $a = 9.31$; $b = 8.07$; $c = 12.85$; $\beta = 93.13°$; $A = 32$
Co-ordinates: 8 times 4Se(C_1): $\pm(xyz) \pm (\frac{1}{2}+x, \frac{1}{2}-y, \frac{1}{2}+z)$

	Se I	Se II	Se III	Se IV	Se V	Se VI	Se VII	Se VIII
x	0.321	0.427	0.317	0.134	−0.081	−0.156	−0.084	0.131
y	0.486	0.664	0.637	0.820	0.686	0.733	0.520	0.597
z	0.237	0.357	0.535	0.556	0.521	0.328	0.229	0.134

A_l (β-Se)

Monoclinic: C_{2h}^5–$P2_1/a$; $a = 12.85$, $b = 8.07$, $c = 9.31$; $\beta = 93° 08'$; $A = 32$
Co-ordinates: 8 times 4Se(C_1): $\pm(xyz) \pm (\frac{1}{2}+x, \frac{1}{2}-y, z)$

	Se I	Se II	Se III	Se IV	Se V	Se VI	Se VII	Se VIII
x	0.334	0.227	0.080	0.102	0.159	0.340	0.409	0.459
y	0.182	0.221	0.397	0.578	0.832	0.832	0.763	0.476
z	0.436	0.245	0.238	0.050	0.157	0.141	0.366	0.336

$B1$ (NaCl type)

Cubic: O_h^5–$Fm3m$; $a = 5.640$; $A = 8$
Co-ordinates: $(000; \frac{1}{2}\frac{1}{2}0; ♪) + 4$Na($O_h$): 000
$+ 4$Cl(O_h): $\frac{1}{2}\frac{1}{2}\frac{1}{2}$

$B2$ (CsCl type)

Cubic: O_h^1–$Pm3m$; $a = 4.123$; $A = 2$
Co-ordinates: Cs(O_h): 000; Cl(O_h): $\frac{1}{2}\frac{1}{2}\frac{1}{2}$

(*continued*)

Table 6.2 STRUCTURAL DETAILS—*continued*

$B3$ [Sphalerite (ZnS) type]

Cubic: $T_d^2 - F\bar{4}3\,m$; $a = 5.382\,9$; $A = 8$
Co-ordinates: $(000; \frac{1}{2}\frac{1}{2}0; \mathcal{J}) + 4\mathrm{Zn}(T_d)$: 000
$\qquad\qquad + 4\mathrm{S}(T_d)$: $\frac{1}{4}\frac{1}{4}\frac{1}{4}$

$B4$ [Wurtzite(ZnS) type]

Hexagonal: $C_{6v}^4 - P6_3mc$; $a = 3.822\,7(1)$, $c = 6.260\,7(1)$; $A = 4$
Co-ordinates: $2\mathrm{Zn}(C_{3v})$: $\frac{1}{3}\frac{2}{3}0$; $\frac{2}{3}\frac{1}{3}\frac{1}{2}$
$\qquad\qquad 2\mathrm{S}(C_{3v})$: $\frac{1}{3}\frac{2}{3}z$; $\frac{2}{3}\frac{1}{3}(\frac{1}{2}+z)$; $z \approx \frac{3}{8}$

$B8$ (α-NiAs type; β-Ni$_2$ In type)

Between the main types (α) and (β) there exist a number of intermediate arrangements due to the variation of the stoichiometric formulae. The axial ratio c/a may change from the value 1.75 (in type α) to 1.22 (in type β). Similarly, there is virtually a continuous change from the B8 type to the C6 type.

α-NiAs type (B8$_1$)

Hexagonal: $D_{6h}^4 - P6_3/mmc$; $a = 3.619$, $c = 5.044$, $c/a = 1.39$; $A = 4$
Co-ordinates: $2\mathrm{Ni}(D_{3d})$: 000; $00\frac{1}{2}$
$\qquad\qquad 2\mathrm{As}(D_{3h})$: $\frac{1}{3}\frac{2}{3}\frac{1}{4}$; $\frac{2}{3}\frac{1}{3}\frac{3}{4}$

β-Ni$_2$ In type (B8$_2$)

Hexagonal: $D_{6h}^4 - P6_3/mmc$; $a = 4.188\,9$, $c = 5.123\,0$, $c/a = 1.22$; $A = 6$
$\qquad\qquad 2\mathrm{Ni}(D_{3d})$: 000; $00\frac{1}{2}$; $2\mathrm{Ni}(D_{3h})$: $\frac{1}{3}\frac{2}{3}\frac{3}{4}$; $\frac{2}{3}\frac{1}{3}\frac{1}{4}$
$\qquad\qquad 2\mathrm{In}(D_{3h})$: $\frac{1}{3}\frac{2}{3}\frac{1}{4}$; $\frac{2}{3}\frac{1}{3}\frac{3}{4}$

$B9$ [Cinnabar (HgS) type]

Hexagonal: $D_3^4 - P3_121$ and $D_3^6 - P3_221$; $a = 4.147\,9(7)$, $c = 9.499\,60(15)$; $A = 6$
Co-ordinates: $3\mathrm{Hg}(C_2)$: $x00$; $\bar{x}\bar{x}\frac{1}{3}$; $0x\frac{2}{3}$; $x = 0.33$
$\qquad\qquad 3\mathrm{S}(C_2)$: $x0\frac{1}{2}$; $\bar{x}\bar{x}\frac{5}{6}$; $0x\frac{1}{6}$; $x = 0.21$

$B10$ (LiOH type)

Tetragonal: $D_{4h}^7 - P4/nmm$; $a = 3.552$, $c = 4.347$; $A = 4$
Co-ordinates: $2\mathrm{Li}(D_{2d})$: 000; $\frac{1}{2}\frac{1}{2}0$
$\qquad\qquad 2\mathrm{OH}(C_{4v})$: $0\frac{1}{2}z$; $\frac{1}{2}0\bar{z}$; $z = 0.20$
For FeSe: $z = 0.26$

$B11$ (PbO type)

Tetragonal: $D_{4h}^7 - P4/nmm$; $a = 3.974\,4(5)$, $c = 5.022\,0(5)$; $A = 4$
Co-ordinates: $2\mathrm{Pb}(C_{4v})$: $0\frac{1}{2}z$; $\frac{1}{2}0\bar{z}$; $z = 0.24$
$\qquad\qquad 2\mathrm{O}(C_{4v})$: the same with $z = 0.74$
For γ-TiCu: $z(\mathrm{Ti}) = 0.65$; $z(\mathrm{Cu}) = 0.10$

$B13$ [Millerite (NiS) type]

Rhombohedral: $C_{3v}^5 - R3m$; $a = 5.64$; $\alpha = 116°\ 35'$; $A = 6$
Co-ordinates: $3\mathrm{Ni}(C_s)$: xxz; $\mathcal{J}$; $x = 0$; $z = 0.264$
$\qquad\qquad 3\mathrm{S}(C_s)$: the same with $x = 0.714$; $z = 0.361$

$B16$ (GeS type)

Orthorhombic: $D_{2h}^{16} - Pbnm$; $a = 4.306(4)$, $b = 10.449(9)$, $c = 3.653(4)$; $A = 8$
Co-ordinates: $4\mathrm{Ge}(C_s)$: $\pm(xy\frac{1}{4})$; $\pm[(\frac{1}{2}-x)(\frac{1}{2}+x)\frac{1}{4}]$; $x = 0.167$; $y = 0.375$
$\qquad\qquad 4\mathrm{S}(C_s)$: the same with $x = 0.111$; $y = 0.139$

$B17$ [Cooperite (PtS) type]

Tetragonal: $D_{4h}^9 - P4_2/mmc$; $a = 3.470\,1(3)$, $c = 6.109\,2(8)$; $A = 4$
Co-ordinates: $2\mathrm{Pt}\,(D_{2h})$: $0\frac{1}{2}0$; $\frac{1}{2}0\frac{1}{2}$, $2\mathrm{S}(D_{2d})$: $00\frac{1}{4}$; $00\frac{3}{4}$

(*continued*)

Table 6.2   STRUCTURAL DETAILS—*continued*

---

*B*18 [Covellite (CuS) type]

Hexagonal: D_{6h}^4–$P6_3/mmc$; $a = 3.768(3)$, $c = 16.27(6)$; $A = 12$
Co-ordinates: $2\text{Cu}(D_{3h})$: $\pm(\frac{2}{3}\frac{1}{3}\frac{1}{4})$
  $4\text{Cu}(C_{3v})$: $\pm(\frac{1}{3}\frac{2}{3}z)$; $\pm(\frac{1}{3}, \frac{2}{3}, \frac{1}{2} - z)$; $z = 0.107$
  $2\text{S}(D_{3h})$: $\pm(\frac{1}{3}\frac{2}{3}\frac{1}{4})$
  $4\text{S}(C_{3v})$: $\pm(00z)$; $\pm(0, 0, \frac{1}{2} - z)$; $z = 0.063$

---

*B*19 (AuCd type)

Orthorhombic: D_{2h}^5–$Pmcm$; $a = 3.143\,7$, $b = 4.871\,7$, $c = 4.769\,0$; $A = 4$
Co-ordinates: $2\text{Au}(C_{2v})$: $\pm(0y\frac{1}{4})$; $y = 0.805$
  $2\text{Cd}(C_{2v})$: $\pm(\frac{1}{2}y\frac{1}{4})$; $y = 0.315$
For MgCd: $y(\text{Mg}) = 0.818$; $y(\text{Cd}) = 0.323$

---

*B*20 (FeSi type)

Cubic: T^4–$P2_13$; $a = 4.483$; $A = 8$
Co-ordinates: $4\text{Fe}(C_3)$: xxx; $(\frac{1}{2} + x)(\frac{1}{2} - x)\bar{x}$; $\therefore$; $x = 0.137$
  $4\text{Si}(C_3)$:   the same with $x = -0.158$
For BeAu: $x(\text{Be}) = -0.156$; $x(\text{Au}) = 0.150$
For RhSn: $x(\text{Rh}) = 0.142$; $x(\text{Sn}) = 0.159$

---

*B*27 (FeB type)

Orthorhombic: D_{2h}^{16}–$Pbnm$; $a = 4.064$, $b = 5.503$, $c = 2.946$; $A = 8$
Co-ordinates: $4\text{Fe}(C_s)$: $\pm(xy\frac{1}{4})$; $\pm(\frac{1}{2} - x, \frac{1}{2} + y, \frac{1}{4})$; $x = 0.125$; $y = 0.180$
  $4\text{B}(C_s)$:   the same with $x = 0.61$; $y = 0.036$
For MnB: $x(\text{Mn}) = 0.125$; $y(\text{Mn}) = 0.180$; $x(\text{B}) = 0.614$; $y(\text{B}) = 0.031$
For USi: $x(\text{U}) = 0.125$; $y(\text{U}) = 0.180$; $x(\text{Si}) = 0.611$; $y(\text{Si}) = 0.028$
For TiB: $x(\text{Ti}) = 0.123$; $y(\text{Ti}) = 0.177$; $x(\text{B}) = 0.603$; $y\text{B}) = 0.029$

---

*B*29 (SnS type)

Orthorhombic: D_{2h}^{16}–$Pmcn$; $a = 4.050(4)$, $b = 4.242(4)$, $c = 11.32(1)$; $A = 8$
Co-ordinates: $4\text{Sn}(C_s)$: $\pm(\frac{1}{4}yz)$; $\pm(\frac{1}{4}, \frac{1}{2} - y, \frac{1}{2} + z)$; $y = 0.115$; $z = 0.118$
  $4\text{S}(C_s)$:   the same with $y = 0.478$; $z = 0.850$
If this description, given in Strukturbericht vol. **3**, p. 14, is transformed to the following, it is virtually identical with the B16 (GeS type).
Orthorhombic: D_{2h}^{16}–$Pbnm$; $a = 4.33$, $b = 11.18$, $c = 3.98$; $A = 8$
Co-ordinates: $4\text{Sn}(C_s) \pm (xy\frac{1}{4})$; $\pm(\frac{1}{2} - x, \frac{1}{2} + y, \frac{1}{4})$; $x = 0.115$; $y = 0.382$
  $4\text{S}(C_s)$:   the same with $x = 0.022$; $y = 0.150$

---

*B*31 (MnP type)

Orthorhombic: D_{2h}^{16}–$Pcmn$; $a = 5.902\,7(4)$, $b = 3.180\,6(2)$, $c = 5.240\,9(3)$; $A = 8$
Co-ordinates: $4\text{Mn}(C_s)$: $\pm(x\frac{1}{4}z)$; $\pm(\frac{1}{2} - x, \frac{1}{4}, \frac{1}{2} + z)$; $x = 0.20$; $z = 0.005$
  $4\text{P}(C_s)$:   the same with $x = 0.57$; $z = 0.19$
For AuGa: $x(\text{Au}) = 0.184$; $z(\text{Au}) = 0.010$; $x(\text{Ga}) = 0.590$; $z(\text{Ga}) = 0.195$
For PdSi: $x(\text{Pd}) = 0.190$; $z(\text{Pd}) = 0.070$; $x(\text{Si}) = 0.570$; $z(\text{Si}) = 0.190$
For PtSi: $x(\text{Pt}) = 0.195$; $z(\text{Pt}) = 0.010$; $x(\text{Si}) = 0.590$; $z(\text{Si}) = 0.195$
For NiGe: $x(\text{Ni}) = 0.190$; $z(\text{Ni}) = 0.005$; $x(\text{Ge}) = 0.583$; $z(\text{Ge}) = 0.188$
For PdGe: $x(\text{Pd}) = 0.188$; $z(\text{Pd}) = 0.005$; $x(\text{Ge}) = 0.595$; $z(\text{Ge}) = 0.190$
For IrGe: $x(\text{Ir}) = 0.192$; $z(\text{Ir}) = 0.010$; $x(\text{Ge}) = 0.590$; $z(\text{Ge}) = 0.185$
For PtGe: $x(\text{Pt}) = 0.195$; $z(\text{Pt}) = 0.010$; $x(\text{Ge}) = 0.590$; $z(\text{Ge}) = 0.195$
For PdSn: $x(\text{Pd}) = 0.182$; $z(\text{Pd}) = 0.007$; $x(\text{Sn}) = 0.590$; $z(\text{Sn}) = 0.182$
For RhSb: $x(\text{Rh}) = 0.192$; $z(\text{Rh}) = 0.010$; $x(\text{Sb}) = 0.590$; $z(\text{Sb}) = 0.195$
For NiSi: $x(\text{Ni}) = 0.184$; $z(\text{Ni}) = 0.006$; $x(\text{Si}) = 0.580$; $z(\text{Si}) = 0.170$

---

*B*32 (NaTl type)

Cubic: O_h^7–$Fd3m$; $a = 7.480(4)$; $A = 16$
Co-ordinates: $(000; \frac{1}{2}\frac{1}{2}0; \therefore) + 8\text{Na}(T_d)$: $000; \frac{1}{4}\frac{1}{4}\frac{1}{4}$
  $+ 8\text{Tl}(T_d)$: $\frac{1}{2}\frac{1}{2}\frac{1}{2}; \frac{3}{4}\frac{3}{4}\frac{3}{4}$

(*continued*)

Table 6.2 STRUCTURAL DETAILS—*continued*

B34 (PdS type)

Tetragonal: $C_{4h}^2-P4_2/m; a = 6.429(2), c = 6.611(2); A = 16$
Co-ordinates: $2Pd(S_4):$ $00\frac{1}{4}; 00\frac{3}{4}; 2Pd(C_{2h}):$ $0\frac{1}{2}0; \frac{1}{2}0\frac{1}{2}$
    $4Pd(C_s):$ $\pm(xy0); \pm(\bar{x}y\frac{1}{2}); x = 0.475; y = 0.250$
    $8S(C_1):$ $\pm(xyz); \pm(xy\bar{z}); \pm(y, \bar{x}, \frac{1}{2} + z); \pm(y, \bar{x}, \frac{1}{2} - z)$
    with $x = 0.20; y = 0.32; z = 0.22$

B35 (CoSn type)

Hexagonal: $D_{6h}^1-P6/mmm; a = 5.278, c = 4.258; A = 6$
Co-ordinates: $1Sn(D_{6h}):$ $000; 2Sn(D_{3h}):$ $\frac{1}{3}\frac{2}{3}\frac{1}{2}; \frac{2}{3}\frac{1}{3}\frac{1}{2}$
    $3Co(D_{2h}):$ $\frac{1}{2}00; 0\frac{1}{2}0; \frac{1}{2}\frac{1}{2}0$

B37 (TlSe type)

Tetragonal: $D_{4h}^{18}-I4/mcm; a = 8.024, c = 7.18; A = 16$
Co-ordinates: $(000; \frac{1}{2}\frac{1}{2}\frac{1}{2}) + 4Tl(D_4):$ $00\frac{1}{4}; 00\frac{3}{4}; + 4Tl(D_{2d}):$ $\frac{1}{2}0\frac{1}{4}; \frac{1}{2}0\frac{3}{4}$
    $+ 8Se(C_{2v}):$ $\pm(x, \frac{1}{2} + x, 0); \pm(\frac{1}{2} + x, \bar{x}, 0); x = 0.179$

B_a (UCo type)

Cubic: $T^5-I2_13; a = 6.352\ 5; A = 16$
Co-ordinates: $(000; \frac{1}{2}\frac{1}{2}\frac{1}{2}) + 8U(C_3):$ $xxx; (\frac{1}{2} + x)(\frac{1}{2} - x)\bar{x}; \mathcal{J}; x = 0.035$
    $+ 8Co(C_3):$ the same with $x = 0.294$

B_b (ζ-AgZn type)

Hexagonal: $C_{3i}^1-P\bar{3}; a = 7.636\ 0; c = 2.819\ 7; A = 9$
Co-ordinates: $1Zn(C_{3i}):$ $000; + 2Zn(C_3):$ $\frac{1}{3}\frac{2}{3}z; \frac{2}{3}\frac{1}{3}z; z \approx \frac{3}{4}$
    $(1.5Zn + 4.5Ag)(C_1):$ $\pm(xyz);$ $\pm(\bar{y}, x - y, z); \pm(y - x, \bar{x}, z)$ with $x = 0.350;$
    $y = 0.032; z = 0.750$

B_c (CaSi type)

Orthorhombic: $D_{2h}^{17}-Cmmc; a = 3.90(3), b = 4.545(3), c = 10.728(10); A = 8$
Co-ordinates: $(000; 0\frac{1}{2}\frac{1}{2}) + 4Ca(C_{2v}):$ $\pm(\frac{1}{4}0z); z = 0.36$
    $+ 4Si(C_{2v}):$ the same with $z = 0.07$
By choosing different axes and origin from those given in the original paper, this type becomes virtually identical with the *Bf* (CrB type):
    $D_{2h}^{17}-Cmcm; a = 4.59, b = 10.80, c = 3.91; y(Ca) = 0.14; y(Si) = 0.43$

B_d (η-NiSi)

Orthorhombic: $D_{2h}^{16}-Pbnm; a = 5.628(2), b = 5.190(1), c = 3.330(1); A = 8$
Co-ordinates: $4Ni(C_s):$ $xy0; \bar{x}\bar{y}\frac{1}{2}; (\frac{1}{2} - x)(\frac{1}{2} + y)0; (\frac{1}{2} + x)(\frac{1}{2} - y)\frac{1}{2}; x = 0.184; y = 0.006$
    $4Si(C_s):$ the same with $x = 0.080; y = 0.330$
By choosing different axes and origin from those given in the original paper, this type becomes identical with the *B31* (MnP type):
    $D_{2h}^{16}-Pcmn; A = 5.62, b = 3.34, c = 5.18; x(Ni) = 0.184; z(Ni) = 0.006; x(Si) = 0.580;$
    $z(Si) = 0.170$

B_e (CdSb type)

Orthorhombic: $D_{2h}^{15}-Pbca; a = 6.476\ 9(1), b = 8.251(2), c = 8.522(2); A = 16$
Co-ordinates: $8Sb(C_1):$ $\pm(xyz); \pm(\frac{1}{2} + x, \frac{1}{2} - y, \bar{z}); \pm(\bar{x}, \frac{1}{2} + y, \frac{1}{2} - z); \pm(\frac{1}{2} - x, \bar{y}, \frac{1}{2} + z)$
    $8Cd(C_1):$ the same

	Sb			Cd *or* Zn		
	x	y	z	x	y	z
CdSb	0.136	0.072	0.108	0.456	0.119	−0.128
ZnSb	0.142	0.081	0.111	0.461	0.103	−0.122

(continued)

Table 6.2   STRUCTURAL DETAILS—*continued*

B_f (CrB type)

Orthorhombic: D_{2h}^{17}–$Cmcm$; $a = 2.970$, $b = 7.865$, $c = 2.936$; $A = 8$
Co-ordinates: $(000; \frac{1}{2}\frac{1}{2}0) + 4Cr(C_{2v})$: $\pm(0y\frac{1}{4})$; $y = 0.146$
$\qquad\qquad\qquad\qquad + 4B(C_{2v})$:   the same with $y = 0.440$
For NbB: $y(Nb) = 0.146$; $y(B) = 0.444$
For CaSi: $y(Ca) = 0.14$; $y(Si) = 0.43$ (cf. Bc type)

B_g (MoB type)

Tetragonal: D_{4h}^{19}–$I4/amd$; $a = 3.103$, $c = 16.95$; $A = 16$
Co-ordinates: $(000, \frac{1}{2}\frac{1}{2}\frac{1}{2}) + 8Mo(C_{2v})$: $\pm(00z); \pm(0, \frac{1}{2}, \frac{1}{4} + z)$; $z = 0.197$
$\qquad\qquad\qquad\qquad + 8B(C_{2v})$:   the same with $z = 0.35$

B_h (WC type)

Hexagonal: D_{6h}^{1}–$P\bar{6}m2 = 2.906$, $c = 2.837$; $A = 2$
Co-ordinates: $1W(D_{6h})$: 000; $1C(D_{3h})$: $\frac{1}{3}\frac{2}{3}\frac{1}{2}; \frac{2}{3}\frac{1}{3}\frac{1}{2}$

B_i (γ'-MoC type)

Hexagonal: D_{6h}^{4}–$P6_3/mmc$; $a = 2.932$, $c = 10.97$; $A = 8$
Co-ordinates: $4Mo(C_{3h})$: $\frac{1}{3}\frac{2}{3}z; \frac{1}{3}\frac{2}{3}(\frac{1}{2} - z); \frac{2}{3}\frac{1}{3}(\frac{1}{2} + z); \frac{2}{3}\frac{1}{3}\bar{z}; z \approx \frac{1}{8}$
$\qquad\qquad\quad 4C$:   in holes

B_l [Realgar (AsS) type]

Monoclinic: C_{2h}^{5}–$P2_1n$; $a = 9.994(6)$, $b = 12.986(6)$, $c = 7.153(4)$; $\beta = 120.6°$; $A = 32$
Co-ordinates: 4 times $4As(C_1)$ and 4 times $4S(C_1)$ in:
$\qquad\qquad\quad \pm(xyz); \pm(\frac{1}{2} + x, \frac{1}{2} - x, \frac{1}{2} + z)$

	As *I*	As *II*	As *III*	As *IV*	S *I*	S *II*	S *III*	S *IV*
x	0.118	0.425	0.318	0.038	0.346	0.213	0.245	0.115
y	0.024	−0.140	−0.127	−0.161	0.008	0.024	−0.225	−0.215
z	−0.241	−0.142	0.181	−0.290	−0.295	0.120	−0.363	0.048

B_m (TiB type)

Orthorhombic: D_{2h}^{16}–$Pnma$; $a = 6.105$, $b = 3.048$, $c = 4.551$, $A = 8$
Co-ordinates: $4Ti(C_s)$: $\pm(x\frac{1}{4}z); \pm(\frac{1}{2} - x, \frac{3}{4}, \frac{1}{2} + z)$; $x = 0.177$; $z = 0.123$
$\qquad\qquad\quad 4B(C_s)$:   the same with $x = 0.029$; $z = 0.603$
If the axes are changed from those of the original paper, this type becomes identical with the *B27* (FeB) type:

$$D_{2h}^{16}\text{–}Pbnm; a = 4.56, b = 6.12, c = 3.06; x(\text{Ti}) = 0.123; y(\text{Ti}) = 0.177;$$
$$x(\text{B}) = 0.603; y(\text{B}) = 0.029$$

C1 (CaF$_2$ type—MgAgAs type)

(α) Cubic: O_h^{5}–$Fm3m$; $a = 5.462\ 6$; $A = 12$
$\qquad$ Co-ordinates: $(000; \frac{1}{2}\frac{1}{2}0; \mathcal{J}) + 4Ca(O_h)$: 000
$\qquad\qquad\qquad\qquad\qquad + 8F(T_d)$: $\pm(\frac{1}{4}\frac{1}{4}\frac{1}{4})$
In those cases in which the *F*-position is occupied by two components in an ordered fashion—for example
in As(MgAg)—the space group is changed to
(β) Cubic: T_d^{2}–$F\bar{4}3m$; $a = 6.24$; $A = 12$
$\qquad$ Co-ordinates: $(000; \frac{1}{2}\frac{1}{2}0; \mathcal{J}) + 4As(T_d)$: 000
$\qquad\qquad\qquad\qquad\qquad + 4Ag(T_d)$: $\frac{1}{4}\frac{1}{4}\frac{1}{4}$
$\qquad\qquad\qquad\qquad\qquad + 4Mg(T_d)$: $\frac{3}{4}\frac{3}{4}\frac{3}{4}$

C2 [Pyrites (FeS$_2$) type]

Cubic: T_h^{6}–$Pa3$; $a = 5.293(2)$; $A = 12$
Co-ordinates: $4Fe(S_{3i})$: 000; $\frac{1}{2}\frac{1}{2}0; \mathcal{J}$
$\qquad\qquad\quad 8S(C_3)$: $\pm(xxx); \pm(\frac{1}{2} + x, \frac{1}{2} - x, \bar{x}; \mathcal{J})$; $x = 0.386$
For MnS$_2$, $x = 0.401$; for CoAsS and NiAsS (random distribution of As and S) $x = 0.385$; for PtBi$_2$: $x = 0.38$.

(continued)

Table 6.2   STRUCTURAL DETAILS—*continued*

---

*C*6 (CdI$_2$) type)

Hexagonal: D_{3d}^3–$P\bar{3}m1$; $a = 4.244$; $c = 6.859$; $A = 3$
Co-ordinates: 1Cd(D_{3d}):   000; 2I(c_{3v}): $\frac{1}{3}\frac{2}{3}z$; $\frac{2}{3}\frac{1}{3}\bar{z}$; $z \approx \frac{1}{4}$
There is virtually a continuous change from this type to the $B8$ type.

*C*7 (MoS$_2$ type)

Hexagonal: D_{6h}^4–$P6_3/mmc$; $a = 3.159(2)$, $c = 12.307(2)$; $A = 6$
Co-ordinates: 2Mo(D_{3h}): $\frac{1}{3}\frac{2}{3}\frac{1}{4}$; $\frac{2}{3}\frac{1}{3}\frac{1}{4}$.
   4S(C_{3v}): $\frac{1}{3}\frac{2}{3}z$; $\frac{2}{3}\frac{1}{3}\bar{z}$; $\frac{1}{3}\frac{2}{3}(\frac{1}{2} - z)$; $\frac{2}{3}\frac{1}{3}(\frac{1}{2} + z)$; $z = 0.62$

*C*11*a* (CaC$_2$ type)

Tetragonal: D_{4h}^{17}–$I4/mmm$; $a = 3.89(1)$, $c = 6.38(1)$, $c/a = 1.64$; $A = 6$
Co-ordinates: $(000; \frac{1}{2}\frac{1}{2}\frac{1}{2}) + 2$Ca($D_{4h}$):   000
   $+ 4$C(C_{4v}): $\pm(00z)$; $z = 0.38$

*C*11*b* (MoSi$_2$ type)

Tetragonal: D_{4h}^{17}–I/mmm; $a = 3.208$, $c = 7.900$, $c/a = 2.46$; $A = 6$
Co-ordinates: $(000; \frac{1}{2}\frac{1}{2}\frac{1}{2}) + 2$Mo($D_{4h}$):   000
   $+ 4$Si(C_{4v}): $\pm(00z)$; $z \approx \frac{1}{3}$
This type is a superstructure of the Aα(Pa) type.

*C*12 (CaSi$_2$ type)

Rhombohedral: D_{3d}^5–$R\bar{3}m$; $a = 10.4$; $\alpha = 21°30'$; $A = 6$
Co-ordinates: 2Ca(C_{3v}): $\pm(xxx)$; $x = 0.083$
   2Si(C_{3v}):   the same with $x = 0.185$
   2Si(C_{3v}):   the same with $x = 0.352$
Hexagonal axes: $a = 3.88$, $c = 30.4$; $A = 18$
Co-ordinates: $(000; \frac{2}{3}\frac{1}{3}\frac{1}{3}; \frac{1}{3}\frac{2}{3}\frac{2}{3}) + 6$Ca($C_{3v}$): $\pm(00x)$; $x = 0.083$
   $+ 6$Si(C_{3v}):   the same with $x = 0.185$
   $+ 6$Si(C_{3v}):   the same with $x = 0.352$

*C*14 (MgZn$_2$ type)

Hexagonal: D_{6h}^4–$P6_3/mmc$; $c = 5.223(1)$, $c = 8.566(3)$; $A = 12$
Co-ordinates: 4Mg(C_{3v}): $\pm(\frac{1}{3}\frac{2}{3}z; \frac{1}{3}, \frac{2}{3}, \frac{1}{2} - z)$; $z \approx \frac{1}{16} = 0.062$
   2Zn(D_{3d}):   000; $00\frac{1}{2}$
   6Zn(C_{2v}): $\pm(x, 2x, \frac{1}{4}; \bar{2}\bar{x}, \bar{x}, \frac{1}{4}; x\bar{x}\frac{1}{4})$; $x \approx -\frac{1}{6} = -0.170$

*C*15 (MgCu$_2$ type)

Cubic: O_h^7–$Fd3m$; $a = 7.048$; $A = 24$
Co-ordinates: $(000; \frac{1}{2}\frac{1}{2}0; ⅃) + 8$Mg($T_d$):   000; $\frac{1}{4}\frac{1}{4}\frac{1}{4}$
   $+ 16$Cu(D_{3d}): $\frac{5}{8}\frac{5}{8}\frac{5}{8}$; $\frac{7}{8}\frac{7}{8}\frac{5}{8}$;

*C*16 (CuAl$_2$ type)

Tetragonal: D_{4h}^{18}–$I4/mcm$; $a = 6.063(3)$, $c = 4.872(3)$; $A = 12$
Co-ordinates: $(000; \frac{1}{2}\frac{1}{2}\frac{1}{2}) + 4$Cu($D_4$): $\pm(00\frac{1}{4})$
   $+ 8$Al(C_{2v}): $\pm(x, \frac{1}{2} + x, 0; \frac{1}{2} + x, \bar{x}, 0)$; $x = 0.158$

	AuNa$_2$	MnSn$_2$	FeSn$_2$	CoSn$_2$	RhSn$_2$	TiSb$_2$	VSb$_2$	Ta$_2$B	Mo$_2$B	W$_2$B	Mn$_2$B
x	0.160	0.159	0.160	0.116	0.161	0.158	0.158	0.167	0.170	0.170	0.163

For the compounds FeGe$_2$, (Rh,Pd,Au)Pb$_2$ no deviation from $x = 0.158$ has been reported.

*C*18 [Marcasite (FeS$_2$) type]

Orthorhombic: D_{2h}^{12}–$Pnnm$; $a = 4.441$, $b = 5.425$, $c = 3.387$; $A = 6$
Co-ordinates: 2Fe(C_{2h}):   000; $\frac{1}{2}\frac{1}{2}\frac{1}{2}$
   4S(C_s): $\pm(xy0)$; $\pm(\frac{1}{2} + x, \frac{1}{2} - y, \frac{1}{2})$; $x = 0.20$; $y = 0.38$
For FeAs$_2$: $x = 0.18$; $y = 0.36$
For NiAs$_2$: $x = 0.22$; $y = 0.37$

(continued)

Table 6.2 STRUCTURAL DETAILS—*continued*

For SeSb$_2$: $x = 0.18$; $y = 0.36$
For FeP$_2$: $x = 0.16$; $y = 0.37$
For CoTe$_2$: $x = 0.22$; $y = 0.36$
For FeTe$_2$: $x = 0.22$; $y = 0.36$
For FeSe$_2$: $x = 0.21$; $y = 0.37$

C22 (Fe$_2$P type)

Hexagonal: $P\bar{6}2m$; $a = 5.8644$, $c = 3.4560$; $A = 9$
Co-ordinates: $3\text{Fe}(C_2)$: $\bar{x}00; 0\bar{x}0; xx0; x = 0.26$
                $3\text{Fe}(C_2)$: $x0\frac{1}{2}; 0x\frac{1}{2}; \bar{x}x\frac{1}{2}; x = 0.40$
                $1\text{P}(D_3)$: $00\frac{1}{2}$
                $2\text{P}(C_3)$: $\pm(\frac{1}{3}\frac{2}{3}z); z \approx \frac{1}{8} = 0.125$

C23 (PbCl$_2$ type)

Orthorhombic: D_{2h}^{16}–$Pmnb$; $a = 4.534$, $b = 7.622$, $c = 9.044$; $A = 12$
Co-ordinates: $4\text{Pb}(C_s)$: $\pm(\frac{1}{4}yz); \pm(\frac{1}{4})\frac{1}{2} + y, \frac{1}{2} - z); y = 0.246; z = 0.905$
                $4\text{Cl}(C_s)$:   the same with $y = 0.85$; $z = 0.93$
                $4\text{Cl}(C_s)$:   the same with $y = 0.95$; $z = 0.33$

For Co$_2$P, Ni$_2$Si* and ThS$_2$ the parameters are:

	P	Co	Co	Th	S	S	Si	Ni	Ni
y	0.250	0.862	0.970	0.250	0.850	0.965	0.236	0.825	0.958
z	0.900	0.930	0333	0.875	0.942	0.320	0.886	0.937	0.297

C32 (AlB$_2$ type)

Hexagonal: D_{6h}^1–$P6/mmm$; $a = 3.005$, $c = 3.257$; $A = 3$
Co-ordinates: $1\text{Al}(D_{6h})$: 000
                $2\text{B}(D_{3h})$: $\frac{1}{3}\frac{2}{3}\frac{1}{3}; \frac{2}{3}\frac{1}{3}\frac{1}{2}$

C33 (Bi$_2$Te$_2$S type)

Rhombohedral: D_{3d}^5–$R\bar{3}m$; $a = 10.31$; $\alpha = 24°10'$; $A = 5$
Co-ordinates: $2\text{Bi}(C_{3v})$: $\pm(xxx); x = 0.392$
                $2\text{Te}(C_{3v})$:   the same with $x = 0.788$
                $1\text{S}(D_{3d})$: 000

C34 [Calaverite (AuTe$_2$) type]

Monoclinic: C_{2h}^3–$C2/m$; $a = 7.1947(4)$, $b = 4.4146(2)$, $c = 5.0703(3)$; $\beta = 90.038(4)$; $A = 6$
Co-ordinates: $(000; \frac{1}{2}\frac{1}{2}0) + 2\text{Au}(C_{2h})$: 000
                           $+4\text{Te}(C_s)$: $\pm(x0z); x = 0.689; z = 0.280$

C36 (MgNi$_2$ type)

Hexagonal: D_{6h}^4–$P6_3/mmc$; $a = 4.807$, $c = 15.77$; $A = 24$
Co-ordinates: $4\,\text{Mg}(C3v)$: $+(\frac{12}{33}z); \pm(\frac{2}{3}, \frac{1}{3}, \frac{1}{2} + z); z \approx \frac{27}{32}$
                $4\text{Mg}(C_{3v})$: $\pm(00z); \pm(0, 0, \frac{1}{2} + z); z \approx \frac{3}{32}$
                $6\text{Ni}(C_{2h})$: $\frac{1}{2}00; 0\frac{1}{2}0; \frac{1}{2}\frac{1}{2}0; \frac{1}{2}0\frac{1}{2}; 0\frac{1}{2}\frac{1}{2}; \frac{1}{2}\frac{1}{2}\frac{1}{2}$
                $6\text{Ni}(C_{2v})$: $\pm(x, 2x, \frac{1}{4}; 2\bar{x}, \bar{x}, \frac{1}{4}; x, \bar{x}, \frac{1}{4}); x \approx \frac{1}{6}$
                $4\text{Ni}(C_{3v})$: $\pm(\frac{1}{3}\frac{2}{3}z); \pm(\frac{2}{3}, \frac{1}{3}, \frac{1}{2} + z); z \approx \frac{1}{8}$

C37 (Co$_2$Si type)

Orthorhombic: D_{2h}^{16}–$Pbnm$; $a = 7.104$, $b = 4.919$, $c = 3.725$; $A = 12$
Co-ordinates: $4\text{Si}(C_s)$: $\pm(xy\frac{1}{4}); \pm(\frac{1}{2} - x, \frac{1}{2} + y, \frac{1}{4}); x = 0.440; y = 0.070$
                $4\text{Co}(C_s)$:   the same with $x = 0.103$; $y = 0.090$
                $4\text{Co}(C_s)$:   the same with $x = 0.772$; $y = 0.193$
With a different choice of axes and origin, a similarity to the C23 (PbCl$_2$ type) becomes apparent:
Orthorhombic: D_{2h}^{16}–$Pmnb$; $a = 3.73$, $b = 4.91$, $c = 7.10$; $A = 12$

---

* If Toman's[199] description of δ-Ni$_2$Si is charged from *Pbnm* to *Pmnb*, and if the origin is chosen differently, this compound belongs to the C23 type.

(*continued*)

Table 6.2 STRUCTURAL DETAILS—*continued*

Co-ordinates: $4Si(C_s)$: $\pm(\frac{1}{4}yz); \pm(\frac{1}{4}, \frac{1}{2}+y, \frac{1}{2}-z); y = 0.07; z = 0.94$
 $4Co(C_s)$: the same with $y = 0.59; z = 0.897$
 $4Co(C_s)$: the same with $y = 0.693; z = 0.228$

With this orientation the lattice constants are virtually identical with those of δ-Ni_2Si ($a = 3.73, b = 4.99$, $c = 7.03$) which has the C23 type. A redetermination of the Co_2Si type might lead to a still closer similarity than that appearing in this description.

C38 (Cu_2Sb type)

Tetragonal: D_{4h}^7–$P4/nmm; a = 4.0014, c = 6.1044; A = 6$
Co-ordinates: $2Cu(D_{2d})$: $000, \frac{1}{2}\frac{1}{2}0$
 $2Cu(C_{4v})$: $0\frac{1}{2}z; \frac{1}{2}0\bar{z}; z = 0.27$
 $2Sb(C_{4v})$: the same with $z = 0.70$

For $Cu_{2-x}Te$: $z(Cu) = 0.27; z(Te) = 0.715$, with vacant sites in the $(0\frac{1}{2}z)$-position.
For Fe_2As and Cr_2As: $z(As) = 0.735; z(Fe$ or $Cr) = 0.33$. For $CuMgAs$: $z(As) = 0.75; z(Cu, Mg) = 0.335$.
For $AlNaSi_4$: $z(Si) = 0.79; z(Na, Al) = 0.37$.
 A very similar arrangement is shown by $CuGa_2$ with $a = 2.83, c = 5.84$. The 3 atoms per cell take the position $000; \frac{1}{2}\frac{1}{2}x; \frac{1}{2}\frac{1}{2}y$ with $x = 0.70; y = 0.27$. For comparison a larger cell with $a' = a\sqrt{2}$ can be chosen:

Tetragonal: C_{4v}^1–$P4mm; a' = 4.03, c = 5.84; A = 6$
Co-ordinates: $2Ga(C_{4v})$: $000; \frac{1}{2}\frac{1}{2}0$
 $2Ga(C_{4v})$: $0\frac{1}{2}z; \frac{1}{2}0z; z = 0.70$
 $2Cu(C_{4v})$: $0\frac{1}{2}z; \frac{1}{2}0z; z = 0.27$

C40 ($CrSi_2$ type)

Hexagonal: D_6^4(or D_6^5)–$P6_222$ (or $P6_422$); $a = 4.430(2), 6.365(5); A = 9$
Co-ordinates: $3Cr(D_2)$: $\frac{1}{2}00; \frac{1}{2}\frac{1}{2}\frac{1}{3}; 0\frac{1}{2}\frac{2}{3}$
 $6Si(C_2)$: $\pm(x, 2x, 0); x\bar{x}\frac{1}{3}; \bar{x}x\frac{1}{3}; 2x, x, \frac{2}{3}; 2\bar{x}, \bar{x}\frac{2}{3}; x \approx \frac{1}{6}$

C42 (SiS_2 type)

Orthorhombic: D_{2h}^{26}–$Icma; a = 5.564(2); b = 5.552(2); c = 9.545(3); A = 12$
Co-ordinates: $(000, \frac{1}{2}\frac{1}{2}\frac{1}{2}) + 4Si(D_2)$: $\pm(0\frac{1}{4}0)$
 $+ 8S(C_s)$: $\pm(x0z); \pm(x\frac{1}{2}\bar{z}); x = 0.208; z = 0.119$

C44 (GeS_2 type)

Orthorhombic: C_{2v}^{19}–$Fdd2; a = 11.691(8), b = 22.341(3); c = 6.68(1); A = 72$
Co-ordinates: $(000; \frac{1}{2}\frac{1}{2}0) + 8Ge(C_2)$: $000; \frac{1}{4}\frac{1}{4}\frac{1}{4}$
 $+ 16Ge(C_1)$: $xyz; x\bar{y}z; (\frac{1}{4}-x)(\frac{1}{4}+y)(\frac{1}{4}+z); (\frac{1}{4}+x)(\frac{1}{4}-y)(\frac{1}{4}+z)$ with $x = 0.125; y = 0.139; z = 0.00$
 $+ 16S(C_1)$: the same with $x = 0.022; y = 0.081; z = 0.183$
 $+ 16S(C_1)$: the same with $x = 0.153; y = -0.014; z = -0.183$
 $+ 16S(C_1)$: the same with $x = 0.063; y = 0.125; z = -0.278$

C46 [Krennerite ($AuTe_2$) type]

Orthorhombic: C_{2v}^4–$Pma2; a = 16.51(3), b = 8.80(3), c = 4.45(3); A = 24$
Co-ordinates: $2Au(C_2)$: $00z; \frac{1}{2}0z; z = 0$
 $2Au(C_s)$: $\frac{1}{4}yz; \frac{3}{4}\bar{z}; y = 0.319; z = 0.014$
 $4Au(C_1)$: $xyz; \bar{x}\bar{y}z; (\frac{1}{2}-; x)yz; (\frac{1}{2}+x)\bar{y}z; x = 0.124; y = 0.666; z = 0.500$
 $2Te(C_s)$: $\frac{1}{4}yz; \frac{3}{4}\bar{y}z; y = 0.018; z = 0.042$
 $2Te(C_s)$: the same with $y = 0.617; z = 0.042$
 $4Te(C_1)$: as $4Au (C_1)$ with $x = 0.003; y = 0.699; z = 0.042$
 $4Te(C_1)$: the same with $x = 0.132; y = 0.364; z = 0.500$
 $4Te(C_1)$: the same with $x = 0.119; y = 0.964; z = 0.500$

C49 ($ZrSi_2$ type)

Orthorhombic: D_{2h}^{17}–$Cmcm; a = 3.725, b = 14.774, c = 3.664; A = 12$
Co-ordinates: $(000; \frac{1}{2}\frac{1}{2}0) + 4Zr(C_{2v})$: $\pm(0y\frac{1}{4}); y = 0.106$
 $+ 4Si(C_{2v})$: the same with $y = 0.750$
 $+ 4Si(C_{2v})$: the same with $y = 0.355$

(continued)

Table 6.2   STRUCTURAL DETAILS—*continued*

$C54$ (TiSi$_2$ type)

Orthorhombic: D_{2h}^{24}–$Fddd$; $a = 8.2671(9), b = 4.800(5), c = 8.5505(11); A = 24$
Co-ordinates: $(000, \frac{1}{2}\frac{1}{2}0; \text{♪})$ $+ 8\text{Ti}(D_2)$: $000; \frac{1}{4}\frac{1}{4}\frac{1}{4}$
                $+ 16\text{Si}(C_2)$: $\pm(x00); (\frac{1}{4}+x)\frac{1}{4}\frac{1}{4}; (\frac{1}{4}-x)\frac{1}{4}\frac{1}{4}; x \approx \frac{1}{3}$

C_a(Mg$_2$Ni type)

Hexagonal: $D_6^4(D_6^5$–$P6_222(P6_422); a = 5.205(1), c = 13.236(2); A = 18$
Co-ordinates: $3\text{Ni}(D_2)$: $00\frac{1}{6}; 00\frac{3}{6}; 00\frac{5}{6}$
                $3\text{Ni}(D_2)$: $0\frac{1}{2}\frac{1}{6}; \frac{1}{2}0\frac{3}{6}; \frac{1}{2}\frac{1}{2}\frac{5}{6}$
                $6\text{Mg}(C_2)$: $\pm(\frac{1}{2}0z), \frac{1}{2}\frac{1}{2}(\frac{1}{3}+z); \frac{1}{2}\frac{1}{2}(\frac{1}{3}-z); 0\frac{1}{2}(\frac{2}{3}+z); 0\frac{1}{2}(\frac{2}{3}-z); z=\frac{1}{9}$
                $6\text{Mg}(C_2)$: $\pm(x, 2x, 0); x\bar{x}\frac{1}{3}; \bar{x}x\frac{1}{3}; 2x, x, \frac{2}{3}; 2\bar{x}, \bar{x}\frac{2}{3}; x=\frac{1}{6}$

C_b(Mg$_2$Cu type)

Orthorhombic: D_{2h}^{24}–$Fddd$; $a = 5.284, b = 9.07, c = 18.25; A = 48$
Co-ordinates: $(000; \frac{1}{2}\frac{1}{2}0; \text{♪})$ $+16\text{Cu}(C_2)$: $\pm(00z); \frac{1}{4}\frac{1}{4}(\frac{1}{4}+z); \frac{1}{4}\frac{1}{4}(\frac{1}{4}-z); z=0.128$
                $+16\,\text{Mg}(C_2)$:   the same with $z=0.411$
                $+16\,\text{Mg}(C_2)$: $\pm(0y0; \frac{1}{4}(\frac{1}{4}+y)\frac{1}{4}; \frac{1}{4}(\frac{1}{4}-y)\frac{1}{4}; y=0.161$

C_c(ThSi$_2$ type)

Tetragonal: D_{4h}^{19}–$I4_1/amd$; $a = 4.107, c = 14.07; A = 12$
Co-ordinates: $(000; \frac{1}{2}\frac{1}{2}\frac{1}{2})$ $+4\text{Th}(D_{2d})$: $000; 0\frac{1}{2}\frac{1}{4}$
                $+8\text{Si}(C_{2v})$: $\pm(00z); 0\frac{1}{2}(\frac{1}{4}+z); 0\frac{1}{2}(\frac{1}{4}-z); z=0.417$

C_e(CoGe$_2$ type)

Orthorhombic: C_{2v}^{17}–$Aba; a \approx b = 5.670, c = 10.796(3); A = 24$
Co-ordinates: $(000, 0\frac{1}{2}\frac{1}{2})$ $+4\text{Co}(C_2)$: $00z; \frac{1}{2}\frac{1}{2}z; z=-0.012$
                $+4\text{Co}(C_2)$:   the same with $z=-0.238$
                $+8\text{Ge}(C_1)$: $xyz; \bar{x}\bar{y}z; (\frac{1}{2}-x)(\frac{1}{2}+y)z; (\frac{1}{2}+x)(\frac{1}{2}-y)z$; with
                                      $x=0.342; y=0.158; z=-\frac{1}{8}$
                $+8\text{Ge}(C_1)$:   the same with $x=y=\frac{1}{4}; z=\frac{1}{8}$

The real composition is Co$_{0.9}$Ge$_2$.

C_f

Tetragonal: a series of types composed of alternating sheets of the Fluorite—and the $C16(\text{Al}_2\text{Cu})$—type.
Whereas the length of the a-axis is approximately constant, the length of the c-axis varies due to the
different possibilities of the sheet sequences.

	RhSn$_2$	PdSn$_2$
a	6.38	6.55
c	17.88	24.57

C_g(ThC$_2$ type)

Monoclinic: C_{2h}^6–$C2/c; a = 6.684(2), b = 4.220(1), c = 6.73(5), \beta = 103.91(1)°; A = 12$
Co-ordinates: $(000; \frac{1}{2}\frac{1}{2}0)$ $+4\text{Th}(C_2)$: $\pm(0y\frac{1}{4}); y = 0.202$
                $+8\text{C}(C_1)$: $\pm(xyz); \pm(x, \bar{y}, \frac{1}{2}+z)$; with $x = 0.29; y = 0.13; z = 0.08$

C_h (Cu$_2$Te type)

Hexagonal: D_{6h}^1–$P6/mmm; a = 4.188, c = 7.251; A = 6$
Co-ordinates: $2\text{Te}(C_{6v})$: $\pm(00z); z = 0.306$
                $4\text{Cu}(C_{3v})$: $\pm(\frac{1}{3}\frac{2}{3}z); \pm(\frac{1}{3}\frac{2}{3}\bar{z}); z = 0.160$

$D0_2$ (CoAs$_3$ type)

Cubic: T_h^5–$Im3; a = 8.208; A = 32$
Co-ordinates: $(000; \frac{1}{2}\frac{1}{2}\frac{1}{2})$ $+8\text{Co}(C_{3i})$: $\frac{1}{4}\frac{1}{4}\frac{1}{4}; \frac{3}{4}\frac{3}{4}\frac{1}{4}; \text{♪}$
                $+24\text{As}(C_s)$: $\pm(xy0; \text{♪}) \pm(x\bar{y}0; \text{♪}); x = 0.35; y = 0.15$

(continued)

Table 6.2   STRUCTURAL DETAILS—*continued*

DO_3 (BiF$_3$ or BiLi$_3$ type)

Cubic: 0_h^5–$Fm3m$; for BiLi$_3$: $a = 6.708$; $A = 16$ (for BiF$_3$ $a = 5.853(6)$)
Co-ordinates: $(000; \frac{1}{2}\frac{1}{2}0; \mathcal{J})$ $+4\text{Bi}(O_h)$: 000
$+4\text{Li}(O_h)$: $\frac{1}{2}\frac{1}{2}\frac{1}{2}$
$+8\text{Li}(T_d)$: $\pm(\frac{1}{4}\frac{1}{4}\frac{1}{4})$

DO_9 (ReO$_3$ or Cu$_3$N type)

Cubic: O_h^1–$Pm3m$; for Cu$_3$N; $a = 3.813$; $A = 4$ (for ReO$_3$ $a = 3.74774(4)$)
Co-ordinates: $\text{IN}(O_h)$: 000; $3\text{Cu}(D_{4h})$: $\frac{1}{2}00; \mathcal{J}$

DO_{11} (Fe$_3$C type)

Orthorhombic: D_{2h}^{16}–$Pbnm$; $a = 4.523$, $b = 5.090$, $c = 6.748$; $A = 16$
Co-ordinates: $4\text{Fe}(C_s)$: $\pm(xy\frac{1}{4})$; $\pm(\frac{1}{2} - x, \frac{1}{2} + y, \frac{1}{4})$; $x = 0.833$; $y = 0.040$
$8\text{Fe}(C_1)$: $\pm(xyz)$; $x, y, \frac{1}{2} - z$; $\pm(\frac{1}{2} - x, \frac{1}{2} + y, z; \frac{1}{2} - x, \frac{1}{2} + y, \frac{1}{2} - z)$;
$x = 0.333$; $y = 0.183$; $z = 0.065$
$4\text{C}(C_s)$: $\pm(xy\frac{1}{4})$; $\pm(\frac{1}{2} - x, \frac{1}{2} + y, \frac{1}{4})$; $x = 0.47$; $y = 0.86$

DO_{18} (Na$_3$As type)

Hexagonal: D_{6h}^4–$P6_3/mmc$; $a = 5.088$, $c = 8.982$; $A = 8$
Co-ordinates: $2\text{As}(D_{3h})$: $\pm(\frac{1}{3}\frac{2}{3}\frac{1}{4})$
$2\text{Na}(D_{3h})$: $\pm(00\frac{1}{4})$
$4\text{Na}(C_{3v})$: $\pm(\frac{1}{3}\frac{2}{3}z; \frac{2}{3}, \frac{1}{3}, \frac{1}{2} + z)$; $z = 0.583$

DO_{19} (Mg$_3$Cd type)

Hexagonal: D_{6h}^4–$P6_3/mmc$; $a = 6.310$, $c = 5.080$; $A = 8$
Co-ordinates: $2\text{Cd}(D_{3h})$: $\pm(\frac{1}{3}\frac{2}{3}\frac{1}{4})$
$6\,\text{Mg}(C_{2v})$: $\pm(2x, x, \frac{1}{4}; \bar{x}x\frac{1}{4}; \bar{x}, 2x, \frac{1}{4}; x \approx \frac{1}{6}$

DO_{20} (NiAl$_3$ type)

Orthorhombic: D_{2h}^{16}–$Pnma$; $a = 6.618(1)$, $b = 7.368(1)$, $c = 4.814(1)$; $A = 16$
Co-ordinates: $4\text{Ni}(C_s)$: $\pm(x\frac{1}{4}z)$; $\pm(\frac{1}{2} + x, \frac{1}{4}, \frac{1}{2} - z)$; $x = -0.731$; $z = -0.055$
$4\text{Al}(C_s)$:   the same with $x = 0.011$; $z = 0.415$
$8\text{Al}(C_1)$: $\pm(xyz)$; $\pm(\frac{1}{2} + x, \frac{1}{2} - y, \frac{1}{2} - z)$; $\pm(x, \frac{1}{2} - y, z)$; $\pm(\frac{1}{2} + x, y, \frac{1}{2} - z)$
with $x = 0.174$; $y = 0.053$; $z = 0.856$
Following the choice of co-ordinates and of the origin in the original paper, NiAl$_3$ was described in the Strukturbericht as a separate type. However, another choice of co-ordinates and of origin shows that the structure is very similar to the DO_{11} (Fe$_3$C) type:
Orthorhombic: D_{2h}^{16}–$Pbnm$; $a = 4.814(1)$, $b = 6.618(1)$, $c = 7.368(1)$, $A = 16$
Co-ordinates: $4\text{Al}(C_s)$: $\pm(xy\frac{1}{4})$; $\pm(\frac{1}{2} - x, \frac{1}{2} + y, \frac{1}{4})$; $x = 0.915$; $y = 0.011$
$8\text{Al}(C_1)$: $\pm(xyz, xy\frac{1}{2} - z)$; $\pm(\frac{1}{2} - x, \frac{1}{2} + y, z; \frac{1}{2} - x, \frac{1}{2} + y, \frac{1}{2} - z)$; $x = 0.356$;
$y = 0.174$; $z = 0.053$
$4\text{Ni}(C_s)$: $\pm(xy\frac{1}{4})$; $\pm(\frac{1}{2} - x, \frac{1}{2} + y, \frac{1}{4})$; $x = 0.445$; $y = 0.869$

DO_{21} (Cu$_3$P type)

Hexagonal: D_{3d}^4–$P6_3$ cm; $a = 6.88$, $c = 7.18$; $A = 24$
Co-ordinates: $6\text{P}(C_2)$: $\pm(x0\frac{1}{4})$; $\pm(0x\frac{1}{4})$; $\pm(\bar{x}\bar{x}\frac{1}{4})$; $x = 0.38$
$2\text{Cu}(C_{3i})$: $000, 00\frac{1}{2}$
$4\text{Cu}(C_3)$: $\pm(\frac{1}{3}\frac{2}{3}z)$; $\pm(\frac{1}{3}, \frac{2}{3}, \frac{1}{2} + z)$; $z = 0.17$
$12\text{Cu}(C_1)$: $\pm[xyz; \bar{y}(x - y)z; (y - x)\bar{x}z; \bar{y}\bar{z}(\frac{1}{2} + z); x(x - y)(\frac{1}{2} + z); (x - y)y(\frac{1}{2} + x)]$;
$x = 0.69$; $y = 0.07$; $z = 0.08$

See remarks by Haraldsen[200]

DO_{22} (TiAl$_3$ type)

Tetragonal: D_{4h}^{17}–$I4/mmm$; $a = 3.846$, $c = 8.594$; $A = 8$
Co-ordinates: $(000; \frac{1}{2}\frac{1}{2}\frac{1}{2})$ $+2\text{Ti}(D_{4h})$: 000
$+2\text{Al}(D_{4h})$: $00\frac{1}{2}$
$+\text{Al}(D_{2d})$: $0\frac{1}{2}\frac{1}{4}; \frac{1}{2}0\frac{1}{4}$

(continued)

Table 6.2   STRUCTURAL DETAILS—*continued*

$D0_{23}$(ZrAl$_3$ type)

Tetragonal: D_{4h}^{17}–$I4/mmm$; $a = 4.0074(6)$, $c = 17.286(4)$; $A = 16$

Co-ordinates: (000; $\frac{1}{2}\frac{1}{2}\frac{1}{2}$)   +4Zr($C_{4v}$): $\pm(00z)$; $z = 0.122 \approx \frac{1}{8}$
+4Al(D_{2h}): $0\frac{1}{2}0$; $\frac{1}{2}00$
+4Al(D_{2d}): $0\frac{1}{2}\frac{1}{4}$; $\frac{7}{2}0\frac{1}{4}$
+4Al(C_{4v}): $\pm(00z)$; $z = 0.361 \approx \frac{3}{8}$

$D0_{24}$(TiNi$_3$ type)

Hexagonal: D_{6h}^{4}–$P6_3/mmc$; $a = 5.1028$, $c = 8.2719$; $A = 16$

Co-ordinates: 2Ti(D_{3d}):   000; $00\frac{1}{2}$; 2Ti(D_{3h}): $\frac{1}{3}\frac{2}{3}\frac{1}{4}$; $\frac{2}{3}\frac{1}{3}\frac{3}{4}$
6Ni(C_{2h}): $\frac{1}{2}00$; $0\frac{1}{2}0$; $\frac{1}{2}\frac{1}{2}0$; $\frac{1}{2}0\frac{1}{2}$; $0\frac{1}{2}\frac{1}{2}$; $\frac{1}{2}\frac{1}{2}\frac{1}{2}$
6Ni(C_{2v}): $\pm(x, 2x\frac{1}{4}; 2\bar{x}, \bar{x}, \frac{1}{4}; x x\frac{1}{4})$; $x = \approx -\frac{1}{6}$

$D0_a$(β-TiCu$_3$ type)

Orthorhombic: D_{2h}^{13}–$Pmmn$; $a = 5.169$, $b = 4.342$, $c = 4.531$; $A = 8$

Co-ordinates: 2Ti(C_{2v}): $00z$; $\frac{1}{2}\frac{1}{2}\bar{z}$; $z = 0.655$
2Cu(C_{2v}): $0\frac{1}{2}z$; $\frac{1}{2}0\bar{z}$; $z = 0.345$
4Cu(C_s): $x0z$; $x0\bar{z}$; $(\frac{1}{2} + x)\frac{1}{2}\bar{z}$; $(\frac{1}{2} - x)\frac{1}{2}\bar{z}$; $x = \frac{1}{4}$; $z = 0.155$

$D0_c$ (U$_3$Si type)

Tetragonal: D_{4h}^{18}–$I4/mcm$; $a = 6.0328(4)$, $8.6907(6)$; $A = 16$

Co-ordinates: (000; $\frac{1}{2}\frac{1}{2}\frac{1}{2}$ +4U(D_4): $\pm(00\frac{1}{4})$
+8U(C_{2v}): $\pm(x, \frac{1}{2} + x, 0)$; $\pm(\frac{1}{2} + x, \bar{x}, 0)$; $x = 0.231$
+4Si(D_{2d}): $0\frac{1}{2}\frac{1}{4}$; $\frac{1}{2}0\frac{1}{4}$

$D0_d$ (Mn$_3$As type)

Orthorhombic: D_{2h}^{13}–$Pmmn$; $a \approx b = 3.788$, $c = 16.29$; $A = 16$

Co-ordinates: 3 times 2Mn(C_{2v}): $00z$; $\frac{1}{2}\frac{1}{2}z$;
2As(C_{2v}):   the same
3 times 2Mn(C_{2v}): $0\frac{1}{2}z$; $\frac{1}{2}0\bar{z}$;
2As (C_{2v}):   the same

	($00z$)-*position*				($0\frac{1}{2}z$)-*position*			
	Mn *I*	Mn *II*	Mn *III*	As *I*	Mn *IV*	Mn *V*	Mn *VI*	As *II*
z	0.194	−0.194	−0.435	0.409	0.307	−0.307	−0.066	0.091

$D1_3$ (BaAl$_4$ type)

Tetragonal: D_{4h}^{17}–$I4/mmm$; $a = 4.566$, $c = 11.250$; $A = 10$

Co-ordinates: (000; $\frac{1}{2}\frac{1}{2}\frac{1}{2}$)   +2Ba($D_{4h}$):   000
+4Al(C_{4v}): $\pm(00z)$; $z = 0.380$
+Al(D_{2d}): $0\frac{1}{2}\frac{1}{4}$; $\frac{1}{2}0\frac{1}{4}$

$D1_a$ (MoNi$_4$ type)

Tetragonal: C_{4h}^{5}–$I4/m$; $a = 5.683$, $c = 3.592$; $A = 10$

Co-ordinates: (000, $\frac{1}{2}\frac{1}{2}\frac{1}{2}$)+2Mo($C_{4h}$):   000
+8Ni(C_s): $\pm(xy0)$; $\pm(y\bar{x}0)$; $x = 0.400$; $y = 0.200$

$D1_b$ (UAl$_4$ type)

Orthorhombic: C_{2v}^{20} (or D_{2h}^{28}–$I2ma$ (or $Imma$); $a = 4.42$, $b = 6.30$, $c = 13.63$; $A = 20$

Co-ordinates: (000; $\frac{1}{2}\frac{1}{2}\frac{1}{2}$)+4U($C_{2v}$): $\pm(0\frac{1}{4}z)$; $z = 0.111$
+4Al(C_{2v}):   the same with $z = -0.111$
+4Al(C_{2v}):   000, $0\frac{1}{2}\frac{1}{2}$
+8Al(C_s): $\pm(0yz)$; $\pm(0, \frac{1}{2} - y, z)$; $y = -0.033$; $z = 0.314$

(*continued*)

Table 6.2 STRUCTURAL DETAILS—*continued*

$D1_c$ (PtSn$_4$ type)

Orthorhombic: C_{2v}^{17}–$Aba2$; $a = 6.382\ 3(1)$, $b = 6.419\ 0(1)$, $c = 11.366\ 6(2)$; $A = 20$
Co-ordinates: $(000;\ 0\frac{1}{2}\frac{1}{2}) + 4Pt(C_2)$: $00z$; $\frac{1}{2}\frac{1}{2}z$; $z = 0$
$\qquad\qquad\quad + 2$ times $8Sn(C_1)$: xyz; $\bar{x}\bar{y}z$; $(\frac{1}{2}+x)(\frac{1}{2}-y)z$; $(\frac{1}{2}-x)(\frac{1}{2}+y)z$

	x	y	z
Sn I	0.173	0.327	0.125
Sn II	0.33	0.17	−0.13

$D1_d$ (PtPb$_4$ type)

Tetragonal: D_{4h}^3–$P4/nbm$; $a = 6.666(10)$, $c = 5.978(10)$; $A = 10$
Co-ordinates: $2Pt(D_4)$: 000; $\frac{1}{2}\frac{1}{2}0$
$\qquad\qquad\quad 8Pb(C_s)$: $x(\frac{1}{2}+x)z$; $\bar{x}(\frac{1}{2}-x)z$; $(\frac{1}{2}+x)\bar{x}z$; $(\frac{1}{2}-x)xz$; $x(\frac{1}{2}-x)\bar{z}$; $\bar{x}(\frac{1}{2}+x)\bar{z}$; $(\frac{1}{2}+x)x\bar{z}$; $(\frac{1}{2}-x)\bar{x}\bar{z}$; $x = 0.175$; $z = 0.255$

$D1_e$ (UB$_4$ type)

Tetragonal: D_{4h}^5–$P4/mbm$; $a = 7.077$, $c = 3.983$; $A = 20$
Co-ordinates: $4U(C_{2v})$: $\pm(x, \frac{1}{2}+x, 0; \frac{1}{2}+x, \bar{x}, 0)$; $x = 0.31$
$\qquad\qquad\quad 4B(C_4)$: $\pm(00z; \frac{1}{2}\frac{1}{2}z)$; $z = 0.2$
$\qquad\qquad\quad 4B(C_{2v})$: $\pm(x, \frac{1}{2}+x, \frac{1}{2}; \frac{1}{2}+x, \bar{x}, \frac{1}{2})$; $x = 0.1$
$\qquad\qquad\quad 4B(C_s)$: $\pm\left[xy\frac{1}{2}; y\bar{x}\frac{1}{2}; (\frac{1}{2}+x)(\frac{1}{2}-y)\frac{1}{2}; (\frac{1}{2}+y)(\frac{1}{2}+x)\frac{1}{2}\right]$ $x = 0.2$; $y = 0.04$

$D1_f$ (Mn$_4$B type)

Orthorhombic: D_{2h}^{24}–$Fddd$; $a = 14.53$, $b = 7.293$, $c = 4.209$; $A = 40$
Co-ordinates: $(000;\ \frac{1}{2}\frac{1}{2}0;\ \mathcal{J}) + 16Mn(C_2)$: $\pm(0y0)$; $\frac{1}{4}(\frac{1}{4}+y)\frac{1}{4}$; $\frac{1}{4}(\frac{1}{4}-y)\frac{1}{4}$; $y = 0.333$
$\qquad\qquad\quad + 16Mn(C_2)$: $\pm(x00)$; $(\frac{1}{4}+x)\frac{1}{4}\frac{1}{4}$; $(\frac{1}{4}-x)\frac{1}{4}\frac{1}{4}$; $x = 0.083$
$\qquad\qquad\quad + 8B(C_2)$: at random in the same position with $x = 0.375$

$D1_g$ (B$_4$C type)

Rhombohedral: D_{3d}^5–$R\bar{3}m$; $a = 5.19$, $\alpha = 66°18'$; $A = 15$
(Hexagonal setting: $a = 5.60$, $c = 12.12$; $A = 45$)
Co-ordinates: $(000;\ \frac{1}{3}\frac{2}{3}\frac{1}{3};\ \frac{2}{3}\frac{1}{3}\frac{2}{3}) + 3C(D_{3d})$: $00\frac{1}{2} + 6C(C_{3v})$: $\pm(00z)$; $z = 0.385$
$\qquad\qquad\quad + 18B(C_1)$: $\pm(x\bar{x}z)$; $\pm(x, 2x, z)$; $\pm(2\bar{x}, \bar{x}, z)$; $x = 0.106$; $z = 0.113$
$\qquad\qquad\quad + 18B(C_1)$: the same with $x = \frac{1}{6}$; $z = 0.360$

$D2_1$ (CaB$_6$ type)

Cubic: O_h^1–$Pm3m$; $a = 4.151\ 2$; $A = 7$
Co-ordinates: $1Ca(O_h)$: 000
$\qquad\qquad\quad 6B(C_{4v})$: $\pm(\frac{1}{2}\frac{1}{2}x;\ \mathcal{J})$; $x = 0.20$

$D2_3$ (NaZn$_{13}$ type)

Cubic: O_h^6–$FM\bar{3}c$; $a = 12.283\ 6(3)$; $A = 112$
Co-ordinates: $(000;\ \frac{1}{2}\frac{1}{2}0;\ \mathcal{J}) + 8Na(O)$: $\pm(\frac{1}{4}\frac{1}{4}\frac{1}{4})$
$\qquad\qquad\quad + 8Zn(T_h)$: 000; $\frac{1}{2}\frac{1}{2}\frac{1}{2}$
$\qquad\qquad\quad + 96Zn(C_s)$: $\pm(0yz;\ \mathcal{J})$; $\frac{1}{2}zy$; $\mathcal{J}$; $0y\bar{z}$; $\mathcal{J}$; $\frac{1}{2}\bar{z}y$; $\mathcal{J}$; $y = 0.180\ 6$; $z = 0.119\ 2$

For Be$_{13}$U, neutron diffraction gave: $y = 0.178$; $z = 0.112$

$D2_b$ (ThMn$_{12}$ type)

Tetragonal: D_{4h}^{17}–$I4/mmm$; $a = 8.74(1), 4.95(1)$; $A = 26$
Co-ordinates: $(000;\ \frac{1}{2}\frac{1}{2}\frac{1}{2}) + 2Th(D_{4h})$: 000
$\qquad\qquad\quad + 8Mn(C_{2h})$: $\frac{1}{4}\frac{1}{4}\frac{1}{4}, \frac{3}{4}\frac{3}{4}\frac{1}{4};\ \mathcal{J}$
$\qquad\qquad\quad + 8Mn(C_{2v})$: $\pm(x00)$; $\pm(0x0)$; $x = 0.361$
$\qquad\qquad\quad + 8Mn(C_{2v})$: $\pm(x\frac{1}{2}0)$; $\pm(\frac{1}{2}x0)$; $x = 0.277$

(*continued*)

Table 6.2 STRUCTURAL DETAILS—*continued*

$D2_c$ (U_6 Mn type)

Tetragonal: D_{4h}^{18}–$I4/mcm$ (or subgroup); $a = 10.312(5)$, $b = 5.255(4)$; $A = 28$
Co-ordinates: $(000; \frac{1}{2}\frac{1}{2}\frac{1}{2}) + 4\text{Mn}(D_4)$: $\pm(00\frac{1}{4})$
$+ 8\text{U}(C_{2v})$: $\pm(x, \frac{1}{2} + x, 0); \pm(\frac{1}{2} + x, \bar{x}, 0); x = 0.405$
$+ 16\text{U}(C_s)$: $\pm(xy0; y\bar{x}0; x\bar{y}\frac{1}{2}; yx\frac{1}{2}); x = 0.213; y = 0.103$

$D2_d$ ($CaCu_5$ type)

Hexagonal: D_{6h}^1–$P6/mmm$; $a = 5.092$, $c = 4.086$; $A = 6$
Co-ordinates: $1\text{Ca}(D_{6h})$: 000
$2\text{Cu}(D_{3h})$: $\frac{1}{3}\frac{2}{3}0; \frac{2}{3}\frac{1}{3}0$; $3\text{Cu}(D_{2h})$: $\frac{1}{2}0\frac{1}{2}; 0\frac{1}{2}\frac{1}{2}; \frac{1}{2}\frac{1}{2}\frac{1}{2}$

$D2_e$ ($BaHg_{11}$ type)

Cubic: O_h^1–$Pm3m$; $a = 9.586$; $A = 36$
Co-ordinates: $3\text{Ba}(D_{4h})$: $\frac{1}{2}00; \natural$
$1\text{Hg}(O_h)$: 000; $8\text{Hg}(C_{3v})$: $\pm(xxx; xx\bar{x}; \natural); x = 0.155$
$12\text{Hg}(C_{2v})$: $\pm(xx0; \natural); \pm(x\bar{x}0; \natural); x = 0.345$
$12\text{Hg}(C_{2v})$: $\pm(xx\frac{1}{2}; \natural); \pm(x\bar{x}\frac{1}{2}; \natural); x = 0.275$

$D2_f$ (UB_{12} type)

Cubic: O_h^5–$Fm3m$; $a = 7.475$; $A = 52$
Co-ordinates: $(000; \frac{1}{2}\frac{1}{2}0;) \natural) + 4\text{U}(O_h)$: 000
$+ 48\text{B}(C_{2v})$: $\pm(\frac{1}{2}xx; \natural); \pm(\frac{1}{2}x\bar{x}; \natural); x = \frac{1}{6}$

$D2_g$ (Fe_8N type)

Tetragonal: D_{4h}^{17}–$I4/mmm$; $a = 5.720$, $c = 6.292$; $A = 18$
Co-ordinates: $(000; \frac{1}{2}\frac{1}{2}\frac{1}{2}) + 2\text{N}(D_{4h})$: 000
$+ 4\text{Fe}(D_{2d})$: $\frac{1}{2}0\frac{1}{4}; 0\frac{1}{2}\frac{1}{4}$
$+ 4\text{Fe}(C_{2v})$: $\pm(00z); z = 0.56$
$+ 8\text{Fe}(C_{2v})$: $\pm(xx0); \pm(x\bar{x}0); x = \frac{1}{4}$

$D5_2$ (La_2O_3 type)

Hexagonal: D_{3d}^3–$P\bar{3}m1$; $a = 3.937$, $c = 6.130$, $c/a = 1.56$; $A = 5$
Co-ordinates: $2\text{La}(C_{3v})$: $\frac{1}{3}\frac{2}{3}z; \frac{2}{3}\frac{1}{3}z; z \approx 0.23$
$1\text{O}(D_{3d})$: 000; $2\text{O}(C_{3v})$: $\frac{1}{3}\frac{2}{3}z; \frac{2}{3}\frac{1}{3}\bar{z}; z \approx 0.63$

Apart from the different c/a-values and small differences in the z-values this type is similar to the $D5_{13}$ (Ni_2Al_3) type.

$D5_3$ (Mn_2O_3 type)

Cubic: T_h^7–$Ia3$; $a = 9.414$; $A = 80$
Co-ordinates: $(000; \frac{1}{2}\frac{1}{2}\frac{1}{2}) + 8\text{Mn}(C_{3i})$: $\frac{1}{4}\frac{1}{4}\frac{1}{4}; \frac{1}{4}\frac{3}{4}\frac{3}{4}; \natural$;
$+ 24\text{Mn}(C_2)$: $\pm(x0\frac{1}{4}; \natural); \pm(x\frac{1}{2}\frac{3}{4}; \natural); x = -0.030$
$+ 48\text{O}(C_1)$: $\pm(xyz; \natural); \pm(x, \bar{y}, \frac{1}{2} - z; \natural); \pm(\frac{1}{2} + x, \bar{y}, z; \natural);$
$\pm(x, \frac{1}{2} + y, \bar{z}; \natural); x = 0.39; y = 0.15; z = 0.38$

$D5_8$ (Sb_2S_3 type)

Orthorhombic: D_{2h}^{16}–$Pbnm$; $a = 11.189(5)$, $b = 11.380(5)$, $c = 3.829(2)$, $A = 20$
Co-ordinates: $4\text{Sb}(C_s)$: $\pm(xy\frac{1}{4}); \pm(\frac{1}{2} - x, \frac{1}{2} + y, \frac{1}{4}); x = 0.33; y = 0.03$
$4\text{Sb}(C_s)$: the same with $x = -0.04; y = -0.15$
$4\text{S}(C_s)$: the same with $x = 0.88; y = 0.05$
$4\text{S}(C_s)$: the same with $x = -0.44; y = -0.13$
$4\text{S}(C_s)$: the same with $x = 0.19; y = 0.21$

For other substances:

	M I	M II	S I	S II	S III
U_2S_3, x	0.311	−0.008	0.878	−0.439	0.206
y	−0.014	−0.195	0.053	−0.129	0.230
Th_2S_3, x	0.314	−0.019	0.878	−0.439	0.206
y	−0.022	−0.200	0.053	−0.129	0.230

(continued)

Table 6.2   STRUCTURAL DETAILS—*continued*

$D5_9$ (Zn_3P_2 type)

Tetragonal: D_{4h}^{15}–$P4/nmc$; $a = 8.10, c = 11.45; A = 40$
Co-ordinates: $4P(C_{2v})$: $\pm(00z); \pm(\frac{1}{2}, \frac{1}{2}, \frac{1}{2}+z); z = 0.25$
    $4P(C_{2v})$: $0\frac{1}{2}z; \frac{1}{2}0\bar{z}; 0\frac{1}{2}(\frac{1}{2}+z); \frac{1}{2}0(\frac{1}{2}-z); z = 0.24$
    $8P(C_s)$: $\pm(xx0; \bar{x}x0; \frac{1}{2}+x, \frac{1}{2}+x, \frac{1}{2}; \frac{1}{2}-x, \frac{1}{2}+x, \frac{1}{2}); x = 0.26$
    $8Zn(C_s)$: $0xz; 0\bar{x}z; x0\bar{z}; \bar{x}0\bar{z}; \frac{1}{2}(\frac{1}{2}+x)(\frac{1}{2}-z); \frac{1}{2}(\frac{1}{2}-x)(\frac{1}{2}-z); (\frac{1}{2}+x)\frac{1}{2}(\frac{1}{2}+z);$
    $(\frac{1}{2}-x)\frac{1}{2}(\frac{1}{2}+z); x = 0.22; z = 0.10$
    $8\bar{Z}n(C_s)$:   the same with $x = 0.28; z = 0.39$
    $8Zn(C_s)$:   the same with $x = 0.26; z = 0.65$

$D5_{10}$(Cr_3C_2 type)

Orthorhombic: D_{2h}^{16}–$Pbnm$; $a = 11.471\ 9(7), b = 5.532\ 9(5), c = 2.829\ 0(2); A = 20$
Co-ordinates: $4C(C_s)$: $\pm(xy\frac{1}{4}); \pm(\frac{1}{2}-x, \frac{1}{2}+y, \frac{1}{4}); x = 0.11; y = -0.10$
    $4C(C_s)$:   the same with $x = -0.06; y = 0.22$
    $4Cr(C_s)$:   the same with $x = 4.406; y = 0.03$
    $4Cr(C_s)$:   the same with $x = -0.230; y = 0.175$
    $4Cr(C_s)$:   the same with $x = -0.070; y = -0.150$

$D5_{13}$ (Ni_2Al_3 type)

Hexagonal: $D_{3d}^3 - P\bar{3}m1; a = 4.036(1), c = 4.897(1), c/a = 1.21; A = 5$
Co-ordinates: $2Ni(C_{3v})$: $\frac{1}{3}\frac{2}{3}z; \frac{2}{3}\frac{1}{3}z; z = 0.149$
    $1Al(D_{3d})$: $000; 2Al(C_{3v})$: $\frac{1}{3}\frac{2}{3}z; \frac{2}{3}\frac{1}{3}z; z = 0.648$

Apart from the different c/a-values and small differences in the z-values, this type is similar to the $D5_2$ (La_2O_3) type.

For Ni_2Ga_3: $z(Ni) = 0.138; z(Ga) = 0.625$
For Ni_2In_3: $z(Ni) = 0.135; z(In) = 0.641$

$D5_a$ (U_3Si_2 type)

Tetragonal: D_{4h}^5–$P4/mbm; a = 7.329\ 9(4), c = 3.900\ 4(5); A = 10$
Co-ordinates: $2U(C_{4h})$: $000; \frac{1}{2}\frac{1}{2}0$
    $4U(C_{2v})$: $\pm(x, \frac{1}{2}+x, \frac{1}{2}); \pm(\frac{1}{2}+x, \bar{x}, \frac{1}{2}); x = 0.181$
    $4Si(C_{2v})$: $\pm(x, \frac{1}{2}+x, 0); \pm(\frac{1}{2}+x, \bar{x}, \frac{1}{2}); x = 0.389$

$D5_b$ (Pt_2Sn_3 type)

Hexagonal: D_{6h}^4–$P6_3/mmc; a = 4.334\ 5, c = 12.960\ 6; A = 10$
Co-ordinates: $4Pt(C_{3v})$: $\pm(\frac{1}{3}\frac{2}{3}z); \pm(\frac{1}{3}, \frac{2}{3}, \frac{1}{2}-z); z = 0.14$
    $4Sn(C_{3v})$:   the same with $z = -0.07$
    $2Sn(D_{3h})$: $\pm(00\frac{1}{4})$

$D5_c$ (Pu_2C_3 type)

Cubic: T_d^6–$I\bar{4}3d; a = 8.132; A = 40$
Co-ordinates: $(000; \frac{1}{2}\frac{1}{2}\frac{1}{2}) + 16Pu(C_3)$: $xxx; (x+\frac{1}{4})(x+\frac{1}{4})(x+\frac{1}{4}); (\frac{1}{2}+x, \frac{1}{2}-x, \bar{x}); \mathcal{J};$
    $(\frac{3}{4}+x)(\frac{1}{4}-x)(\frac{3}{4}-x); \mathcal{J}; x = 0.050$
    $+24C(C_2)$: $y0\frac{1}{4}; \mathcal{J}; (\frac{1}{2}-y)0\frac{3}{4}; \mathcal{J}; (\frac{3}{4}+y)0\frac{3}{4}; \mathcal{J}; (\frac{3}{4}-y)0\frac{1}{4}; \mathcal{J}; y = 0.28$

$D5_e$ (Ni_3S_2 type)

Rhombohedral: D_3^7–$R32; a = 4.08; \alpha = 89°25'; A = 5$
Co-oridinates: $3Ni(C_2)$: $\frac{1}{2}x\bar{x}; \mathcal{J}; x = \frac{1}{4}$
    $2S(C_3)$: $\pm(xxx); x = \frac{1}{4}$

$D5_f$ (As_2S_3 type)

Monoclinic: C_{5h}^2–$P2_1/n; a = 12.191(5), b = 9.577(4), c = 4.256(2); \beta = 109.76(8)°; A = 20$
Co-oridiantes: $4As(C_1)$: $\pm(xyz); \pm(\frac{1}{2}-x, \frac{1}{2}+y, \frac{1}{2}-z)$
All the other atoms in the same position.

	As *I*	As *II*	S *I*	S *II*	S *III*
x	0.268	0.482	0.410	0.340	0.125
y	0.187	0.313	0.120	0.380	0.305
z	0.161	−0.339	0.454	−0.046	0.455

(continued)

Table 6.2   STRUCTURAL DETAILS—*continued*

$D7_1$ (Al$_4$C$_3$ type)

Rhombohedral: D_{3d}^5–$R\bar{3}m$; $a = 8.53$; $\alpha = 22°28'$; $A = 7$
(Hexagonal setting: $a = 3.32$, $c = 24.95$; $A = 21$)
Co-ordinates: 1C(D_{3d}):   000
$\qquad\qquad\qquad$ 2C(C_{3v}): $\pm(xxx)$; $x = 0.217$
$\qquad\qquad\qquad$ 2Al(C_{3v}):   the same with $x = 0.293$
$\qquad\qquad\qquad$ 2Al(C_{3v}):   the same with $x = 0.128$

$D7_2$ (Co$_3$S$_4$ type)

Cubic: O_h^7–$Fd\bar{3}m$; $a = 9.404(1)$; $A = 56$
Co-ordiantes: $(000; \frac{1}{2}\frac{1}{2}0; \textit{J}) + 8\mathrm{Co}(T_d)$: $000; \frac{1}{4}\frac{1}{4}\frac{1}{4}$
$\qquad\qquad\qquad +16\mathrm{Co}(D_{3d})$: $\frac{5}{8}\frac{5}{8}\frac{5}{8}; \frac{7}{8}\frac{7}{8}\frac{7}{8}; \textit{J}$
$\qquad\qquad\qquad +32\mathrm{S}(C_{3v})$: $xxx; x\bar{x}\bar{x}; \textit{J}; (\frac{1}{4}-x)(\frac{1}{4}-x)(\frac{1}{4}-x); (\frac{1}{4}-x)(\frac{1}{4}-x)(\frac{1}{4}+x);$
$\qquad\qquad\qquad\qquad \textit{J}; x = -0.135$

An ordered variety of this type is described as $H1$ (Spinel) type.

$D7_3$ (Th$_3$P$_4$ type)

Cubic: T_d^6–$I\bar{4}3d$; $a = 8.618$; $A = 28$
Co-ordinates: $(000; \frac{1}{2}\frac{1}{2}\frac{1}{2}) + 12\mathrm{Th}(S_4)$: $\frac{1}{4}\frac{3}{8}0; \textit{J}; \frac{3}{4}\frac{1}{8}0; \textit{J}$
$\qquad\qquad\qquad +16\mathrm{P}(C_3)$: $xxx; (\frac{1}{4}+x)(\frac{1}{4}+x)(\frac{1}{4}+x); (\frac{1}{2}+x)(\frac{1}{2}+x)\bar{x}; \textit{J}; (\frac{3}{4}+x)(\frac{1}{4}-x)$
$\qquad\qquad\qquad\qquad (\frac{3}{4}-x); \textit{J}; x = \approx \frac{1}{12} = 0.083$

In the compounds (La, Ce, Ac, Pu, Am)$_2$S$_3$, $10\frac{2}{3}$ metal ions occupy the Th positions at random.

$D7_a$ (Ni$_3$Sn$_4$ type)

Monoclinic: C_{2h}^3–$C2/m$; $a = 12.214(6)$, $b = 4.060(2)$, $c = 5.219(3)$; $\beta = 105.0(1)°$; $A = 14$
Co-ordinates: $(000; \frac{1}{2}\frac{1}{2}0) + 2\mathrm{Ni}(C_{2h})$:   000
$\qquad\qquad\qquad +4\mathrm{Ni}(C_s)$: $\pm(x0z); x = 0.220; z = 0.350$
$\qquad\qquad\qquad +4\mathrm{Sn}(C_s)$:   the same with $x = 0.428; z = 0.675$
$\qquad\qquad\qquad +4\mathrm{Sn}(C_s)$:   the same with $x = 0.180, z = 0.800$

$D7_b$ (Ta$_3$B$_4$ type)

Orthorhombic: D_{2h}^{25}–$Immm$; $a = 3.036$, $b = 12.846$, $c = 2.958$; $A = 14$
Co-ordinates: $(000; \frac{1}{2}\frac{1}{2}\frac{1}{2}) + 2\mathrm{Ta}(D_{2h})$: $000\frac{1}{2}$
$\qquad\qquad\qquad +4\mathrm{Ta}(C_{2v})$: $\pm(0\bar{y}0); y = 0.180$
$\qquad\qquad\qquad +4\mathrm{B}(C_{2v})$:   the same with $y = 0.375$
$\qquad\qquad\qquad +4\mathrm{B}(C_{2v})$: $\pm(0y\frac{1}{2}); y = 0.444$

$D8_1$ (Fe$_3$Zn$_{10}$ type)

Cubic: O_h^9–$Im3m$; $a = 9.018(3)$; $A = 52$
Co-ordinates: $(000; \frac{1}{2}\frac{1}{2}\frac{1}{2}) + 12\mathrm{Fe}\ (C_{4v})$: $\pm(x00; \textit{J}); x \approx \frac{1}{3}$
$\qquad\qquad\qquad +16\mathrm{Zn}(C_{3v})$: $\pm(xxx; x\bar{x}\bar{x}; \textit{J}); x \approx \frac{1}{6}$
$\qquad\qquad\qquad +24\mathrm{Zn}(C_{2v})$: $\pm(xx0; \textit{J}; x\bar{x}0; \textit{J}); x \approx \frac{1}{3}$

$D8_2$ (Cu$_5$Zn$_8$ type)

Cubic: T_d^3–$I\bar{4}3m$; $a = 8.859(3)$; $A = 52$
Co-ordiantes: $(000, \frac{1}{2}\frac{1}{2}\frac{1}{2}) + 12\mathrm{Cu}(C_{2v})$: $\pm(x00; \textit{J}); x = 0.355 \approx \frac{1}{3}$
$\qquad\qquad\qquad +8\mathrm{Cu}(C_{3v})$: $xxx; x\bar{x}\bar{x}; \textit{J}; x = -0.172 \approx -\frac{1}{6}$
$\qquad\qquad\qquad +8\mathrm{Zn}(C_{3v})$:   the same with $x = 0.110 \approx +\frac{1}{6}$
$\qquad\qquad\qquad +24\mathrm{Zn}(C_s)$: $xxz; \textit{J}; x\bar{x}\bar{z}; \textit{J}; \bar{x}\bar{x}z; \textit{J}; \bar{x}x\bar{z}; \textit{J}; x = 0.313 \approx \frac{1}{3};$
$\qquad\qquad\qquad\qquad z = 0.036 \approx 0$

$D8_3$ (Cu$_9$Al$_4$ type)

Cubic: T_d^1–$P\bar{4}3m$; $a = 8.7023(5)$; $A = 52$
Co-ordinates: $6\mathrm{Cu}(C_{2v})$: $\pm(x00; \textit{J}); x = 0.356 \approx \frac{1}{3}$
$\qquad\qquad\qquad 6\mathrm{Cu}(C_{2v})$: $+(x\frac{1}{2}\frac{1}{2}; \textit{J}); x = 0.856 \approx \frac{1}{2} + \frac{1}{3}$
$\qquad\qquad\qquad 4\mathrm{Cu}(C_{3v})$: $xxx; x\bar{x}\bar{x}; \textit{J}; x = -0.172 \approx -\frac{1}{6}$
$\qquad\qquad\qquad 4\mathrm{Cu}(C_{3v})$:   the same with $x = 0.331 \approx \frac{1}{3} - \frac{1}{6}$
$\qquad\qquad\qquad 4\mathrm{Cu}(C_{3v})$:   the same with $x = 0.601 \approx \frac{1}{2} + \frac{1}{6}$
$\qquad\qquad\qquad 4\mathrm{Al}(C_{3v})$:   the same with $x = 0.112 \approx \frac{1}{6}$
$\qquad\qquad\qquad 12\mathrm{Al}(C_s)$: $xxz; \textit{J}; x\bar{x}\bar{z}; \textit{J}; \bar{x}\bar{x}z; \textit{J}; \bar{x}x\bar{z}; x = 0.812 \approx \frac{1}{2} + \frac{1}{3}; z = 0.536 \approx \frac{1}{2}$
$\qquad\qquad\qquad 12\mathrm{Cu}(C_s)$:   the same with $x = 0.312 \approx \frac{1}{3}; z = 0.036 \approx 0$

(continued)

Table 6.2 STRUCTURAL DETAILS—*continued*

$D8_4$ ($Cr_{23}C_6$ type)

Cubic: O_h^5–$Fm3m$; $a = 10.650(2)$; $A = 116$
Co-oridiantes: $(000; \frac{1}{2}\frac{1}{2}0; ♪) + 4Cr(O_h)$:   000
$\qquad\qquad\qquad +8Cr(T_d)$: $\pm(\frac{1}{4}\frac{1}{4}\frac{1}{4})$
$\qquad\qquad\qquad +32Cr(C_{3v})$: $\pm(xxx; x\bar{x}\bar{x}; ♪); x = 0.385$
$\qquad\qquad\qquad +48Cr(C_{2v})$: $\pm(xx0; ♪; x\bar{x}0; ♪); x = 0.165$
$\qquad\qquad\qquad +24C(C_{4v})$: $\pm(x00; ♪); x = 0.275$

$D8_5$ (Fe_7W_6 type)

Rhombohedral: D_{3d}^5–$R\bar{3}m$; $a = 9.02$; $\alpha = 30°31'$; $A = 13$
(Hexagonal setting: $a = 4.74, c = 25.75; A = 39$)
Co-ordiantes: $1Fe(D_{3d})$: 000
$\qquad\qquad 6Fe(C_s)$: $\pm(xxz; ♪); x = 0.09; z = 0.59$
$\qquad\qquad 2W(C_s)$: $\pm(xxx); x = \frac{1}{6} = 0.167$
$\qquad\qquad 2W(C_{3v})$: the same with $x = 0.346$
$\qquad\qquad 2W(C_{3v})$: the same with $x = 0.448$

$D8_6$ ($Cu_{15}Si_4$ type)

Cubic: T_d^6–$I\bar{4}3d$; $a = 9.615$; $A = 76$
Co-ordinates: $(000; \frac{1}{2}\frac{1}{2}\frac{1}{2}) + 12Cu(S_4)$: $0\frac{1}{4}\frac{3}{8}; ♪; 0\frac{3}{4}\frac{1}{8}; ♪;$
$\qquad +48Cu(C_1)$: $xyz; ♪; x\bar{y}(\frac{1}{2}-z); ♪; (\frac{1}{2}-x)y\bar{z}; ♪; \bar{x}(\frac{1}{2}-y)z; ♪; (\frac{1}{4}+y)(\frac{1}{4}+x)$
$\qquad (\frac{1}{4}+z); ♪; (\frac{1}{4}-y)(\frac{1}{4}+x)(\frac{3}{4}-z); ♪; (\frac{1}{4}+y)(\frac{3}{4}-x)(\frac{1}{4}-z); ♪; (\frac{3}{4}-y)(\frac{1}{4}-x); (\frac{1}{4}+z); ♪;$
$\qquad x = 0.12; y = 0.16; z = 0.04$
$\qquad +16Si(C_3)$: $xxx; x\bar{x}(\frac{1}{2}-x); ♪; (\frac{1}{4}+x)(\frac{1}{4}+x)(\frac{1}{4}+x); (\frac{1}{4}-x)(\frac{1}{4}+x)(\frac{3}{4}-x); ♪; x = 0.208$

$D8_8$ (Mn_5Si_3 type)

Hexagonal: D_{6h}^3–$P6_3/mcm$; $a = 6.912(5), c = 4.812(5); A = 16$
Co-ordinates: $4Mn(D_3)$: $\frac{1}{3}\frac{2}{3}0; \frac{2}{3}\frac{1}{3}0; \frac{1}{3}\frac{2}{3}\frac{1}{2}; \frac{2}{3}\frac{1}{3}\frac{1}{2}$
$\qquad\qquad 6Mn(C_{2v})$: $\pm(x0\frac{1}{4}; 0x\frac{1}{4}; \bar{x}\bar{x}\frac{1}{4}); x = 0.23$
$\qquad\qquad 6Si(C_{2v})$: the same with $x = 0.60$
For Mg_5Hg_3: $x(Mg) = 0.25; x(Hg) = 0.615$

$D8_9$ (Co_9S_8 type)

Cubic: O_h^5–$Fm3m$; $a = 9.930(2); A = 68$
Co-ordinates: $(000; \frac{1}{2}\frac{1}{2}0; ♪) + 4Co(O_h)$: $\frac{1}{2}\frac{1}{2}\frac{1}{2}$
$\qquad\qquad\qquad +32Co(C_{3v})$: $\pm(xxx; x\bar{x}\bar{x}; ♪); x \approx \frac{1}{8}$
$\qquad\qquad\qquad +8S(T_d)$: $\pm(\frac{1}{4}\frac{1}{4}\frac{1}{4})$
$\qquad\qquad\qquad +24S(C_{4v})$: $\pm(x00; ♪); x \approx \frac{1}{4}$

$D8_{10}$ (Cr_5Al_8 type)

Rhombohedral: C_{3v}^5–$R3m$; $a = 7.79$; $\alpha = 109°8'$; $A = 26$
(Hexagonal setting: $a = 12.70, c = 7.90; A = 78$)
Co-ordinates: $1Cr(C_{3v})$: $xxx; x = 0.097$
$\qquad\qquad 3Cr(C_s)$: $xxz; ♪; x = -0.103; z = 0.106$
$\qquad\qquad 3CR(C_s)$: the same with $x = 0.170; z = -0.172$
$\qquad\qquad 3Cr(C_s)$: the same with $x = 0.003; z = 0.352$
$\qquad\qquad 1Al(C_{3v})$: $xxx; x = -0.164$
$\qquad\qquad 3Al(C_s)$: $xxz; ♪; x = 0.006; z = -0.352$
$\qquad\qquad 3Al(C_s)$: the same with $x = 0.291; z = 0.058$
$\qquad\qquad 3Al(C_3)$: the same with $x = -0.322; z = 0.044$
$\qquad\qquad 6Al(C_1)$: $xyz; ♪; xzy; ♪; x = 0.330; y = -0.297; z = -0.042$

$D8_{11}$ (Co_2Al_5 type)

Hexagonal: D_{6h}^4–$P6_3/mmc$; $a = 7.656\,0, c = 7.593\,2; A = 28$
Co-ordinates: $2Co(D_{3h})$: $\frac{2}{3}\frac{1}{3}\frac{1}{4}; \frac{1}{3}\frac{2}{3}\frac{3}{4}$
$\qquad\qquad 6Co(C_{2v})$: $\pm(x, 2x, \frac{1}{4}; 2\bar{x}, \bar{x}, \frac{1}{4}; x\bar{x}\frac{1}{4}); x = 0.128$
$\qquad\qquad 2Al(D_{3d})$: $000; 00\frac{1}{2}$
$\qquad\qquad 6Al(C_{2v})$: $\pm(x, 2x, \frac{1}{4}; 2\bar{x}, \bar{x}, \frac{1}{4}; x\bar{x}\frac{1}{4}); x = 0.467$
$\qquad\qquad 12Al(C_s)$: $\pm(2x, x, z; \bar{x}, 2\bar{x}, z; \bar{x}xz; 2\bar{x}, \bar{x}, \frac{1}{2}+z; x, 2x, \frac{1}{2}+z; x, \bar{x}, \frac{1}{2}+z); x = 0.196;$
$\qquad z = 0.061$

(continued)

Table 6.2   STRUCTURAL DETAILS—*continued*

$D8_a$ (Th$_6$MN$_{23}$ type)

Cubic: O_h^5–$Fm3m$; $a = 12.443$; $A = 116$

Co-ordinates: $(000; \frac{1}{2}\frac{1}{2}0; ♪) + 24\text{Th}(C_{4v})$: $\pm(x00; ♪)$; $x = 0.207$

$+4\text{Mn}(O_h)$: $\frac{1}{2}\frac{1}{2}\frac{1}{2}$

$+24\text{Mn}(D_{2h})$: $\frac{1}{4}\frac{1}{4}0;♪$; $\frac{1}{4}\frac{3}{4}0; ♪$

$+32\text{Mn}(C_{3v})$: $\pm(xxx; x\bar{x}\bar{x}; ♪)$; $x = 0.378$

$+32\text{Mn}(C_{3v})$:   the same with $x = 0.178$

$D8_b$ (σ type, as for example: V$_3$Ni$_2$)

(a) According to Bergman and Shoemaker[202]:

Tetragonal: D_{4h}^{14}–$P4_2/mnm$; for σ-FeCr: $a = 8.7968(5)$, $c = 4.5585(3)$; $A = 30$

Co-ordinates: 2 atoms $A(D_{2h})$: 000; $\frac{1}{2}\frac{1}{2}\frac{1}{2}$

4 atoms $B(C_{2v})$: $\pm(xx0)$; $\pm(\frac{1}{2}+x, \frac{1}{2}-x, \frac{1}{2})$; $x = \frac{2}{5} = 0.400$

8 atoms $C(C_s)$: $\pm(xy0)$; $\pm(yx0)$; $\pm(\frac{1}{2}+x, \frac{1}{2}-y, \frac{1}{2})$; $\pm(\frac{1}{2}+y, \frac{1}{2}-x, \frac{1}{2})$;

$x = \frac{7}{15} = 0.468$; $y = \frac{2}{15} = 0.134$

8 atoms $D(C_s)$:   the same with $x = \frac{11}{15} = 0.735$; $y = \frac{1}{15} = 0.067$

8 atoms $E(C_s)$: $\pm(xxz)$; $\pm(xx\bar{z})$; $\pm(\frac{1}{2}+x, \frac{1}{2}-x, \frac{1}{2}+z)$; $\pm(\frac{1}{2}+x, \frac{1}{2}-x, \frac{1}{2}-z)$;

$x = \frac{11}{16} = 0.183$; $z = \frac{1}{4} = 0.250$

The space groups C_v^4–$P4nm$ and D_{2d}^8–$P\bar{4}n2$ could not be ruled out.

(b) According to Kasper, Decker and Belanger,[203] who investigated σ-CoCr: the same space group; $a = 8.84(1)$, $c = 4.60(1)$; $A = 30$.

Co-ordinates: 2 atoms $A(D_{2h})$: $00\frac{1}{2}$; $\frac{1}{2}\frac{1}{2}0$

4 atoms B as under (a) but $x = 0.100$

8 atoms C as under (a) but $x = 0.373$; $y = 0.027$

8 atoms D as under (a) but $x = 0.573$; $y = 0.227$

8 atoms E as under (a) but $x = 0.300$; $z = \frac{1}{4}$

$D8_c$ (Mg$_2$Cu$_6$Al$_5$ type)

Cubic: T_h^1–$Pm3$; $a = 8.311(3)$; $A = 39$

Co-ordinates: $6\text{Mg}(C_{2v})$: $\pm(x0\frac{1}{2}; ♪)$; $x = 0.32$

$6\text{Cu(or Zn I)} (C_{2v})$: $♪ \pm(x00; ♪)$; $x = 0.225$

$12\text{Cu(or Zn II)}(C_s)$: $\pm(\frac{1}{2}yz; ♪)$; $\pm(\frac{1}{2}y\bar{z}; ♪)$; $y = 0.243$; $z = 0.336$

$1\text{Al(or Zn III)}(T_h)$: $\frac{1}{2}\frac{1}{2}\frac{1}{2}$

$6\text{Al(or Zn IV)}(C_{2v})$: $\pm(x\frac{1}{2}0; ♪)$; $x = 0.16$

$8\text{Al(or Zn V)} (C_3)$: $\pm(xxx; x\bar{x}\bar{x}♪)$; $x = 0.215$

For Mg$_2$Zn$_{11}$: $x(\text{Zn I}) = 0.235$; $y(\text{Zn II}) = 0.243$; $z(\text{Zn II}) = 0.343$; $x(\text{Zn IV}) = 0.160$; $x(\text{Zn V}) = 0.222$.

$D8_d$(Co$_2$Al$_9$)

Monoclinic: C_{2h}^5–$P2_1/c = 10.149(6)$, $b = 6.290(5)$, $c = 8.5565(5)$; $\beta = 142.40°$; $A = 22$

Co-ordinates: 2 Al I$(C_i) = 000$; $0\frac{1}{2}\frac{1}{2}$

4 times Al, II, Al III, Al IV, Al V and Co in: $\pm(yyz)$; $\pm(\frac{1}{2}+x, \frac{1}{2}-y, z)$

	Co	Al *II*	Al *III*	Al *IV*	Al *V*
x	0.3335	0.2682	0.2309	0.9986	0.0417
y	0.6149	0.9619	0.2899	0.1931	0.6148
z	0.2646	0.4044	0.0889	0.3891	0.2159

$D8_e$ (Mg$_{32}$X$_{49}$ type)

Cubic: T_h^5–$Im3$; $a = 14.16$; $A = 162$

Co-ordinates: $(000; \frac{1}{2}\frac{1}{2}\frac{1}{2}) + 12\text{MG}(C_{2v})$: $\pm(x0\frac{1}{2}; ♪)$; $x = 0.605$

$+12\,\text{Mg}(C_{2v})$: $\pm$ the same with $x = 0.185$

$+16\,\text{Mg}(C_3)$: $\pm(xxx; x\bar{x}\bar{x}; ♪)$; $x = 0.185$

$+24\,\text{Mg}(C_s)$: $\pm(0yz; ♪; 0y\bar{z}; ♪)$; $y = 0.300$; $z = 0.115$

$+2\,\text{Al}(T_h)$: 000

$+24$ atoms (83% Zn; 17% Al)(C_s): $\pm(0yz; ♪; 0y\bar{z}; ♪)$; $y = 0.097$; $z = 0.157$

$+24$ atoms (44% Zn; 56% Al)(C_s):   the same with $y = 0.195$; $z = 0.310$

$+48$ atoms (48% Zn; 52% Al)(C_s): $\pm(xyz; ♪; x\bar{y}z; ♪; \bar{x}y\bar{z};♪; \bar{x}\bar{y}z;♪)$;

$x = 0.160$; $y = 0.190$; $z = 0.400$

(continued)

Table 6.2 STRUCTURAL DETAILS—*continued*

$D8_f$ (Ir$_3$Ge$_7$ type)

Cubic: O_h^9-Im3m; $a = 8.735(1)$; $A = 40$
Co-ordinates: $(000; \frac{1}{2}\frac{1}{2}\frac{1}{2}) + 12\,Ir(C_{4v})$: $\pm(x00; \,\natural)$; $x = 0.342$
$\qquad\qquad\qquad +12\,Ge(D_{2d})$: $\pm(\frac{1}{4}0\frac{1}{2}; \,\natural)$
$\qquad\qquad\qquad +16\,Ge(C_{3v})$: $(xxx; x\bar{x}\bar{x}; \,\natural)$; $x = 0.156$

$D8_g$ (Mg$_5$Ga$_2$ type)

Orthorhombic: $D_{2h}^{26}-Ibam$; $a = 13.708$, $b = 7.017$, $c = 6.020$; $A = 28$
Co-ordinates: $(000; \frac{1}{2}\frac{1}{2}\frac{1}{2}) + 8\,Ga(C_s)$: $\pm(xy0); \pm(xy\frac{1}{2})$; $x = 0.122$; $y = 0.262$
$\qquad\qquad\qquad +8\,Mg(C_s)$: the same with $x = 0.080$; $y = 0.660$
$\qquad\qquad\qquad +8\,Mg(C_2)$: $\pm(x0\frac{1}{4}); \pm(\bar{x}0\frac{1}{4})$; $x = 0.242$
$\qquad\qquad\qquad +4\,Mg(D_2)$: $\pm(00\frac{1}{4})$

[Unpublished work by E, Hellner; for Mg$_5$Tl$_2$: x(Tl) $\approx \frac{1}{8}$; y(Tl) $\approx \frac{1}{4}$]

$D8_h$ (W$_2$B$_5$ type)

Hexagonal: $D_{6h}^4-P6_3/mmc$; $a = 2.98$, $c = 13.87$; $A = 14$
Co-ordinates: $4W(C_{3v})$: $\pm(\frac{1}{3}\frac{2}{3}z); \pm(\frac{1}{3}, \frac{2}{3}, \frac{1}{2} - z)$; $z = 0.139$
$\qquad\qquad\quad 2B(D_{3h})$: $\pm(00\frac{1}{4})$; $+2B(D_{3h})$: $\pm(\frac{1}{3}\frac{2}{3}\frac{3}{4})$
$\qquad\qquad\quad 2B(D_{3d})$: $000; 00\frac{1}{2}$
$\qquad\qquad\quad 4B(C_{3v})$: $\pm(\frac{1}{3}\frac{2}{3}z); \pm(\frac{1}{3}, \frac{2}{3}, \frac{1}{2} - z)$; $z = -0.028$

$D8_i$ (Mo$_2$B$_5$ type)

Rhombohedral: $D_{3d}^5-R\bar{3}m$; $a = 7.19$; $\alpha = 24°10'$; $A = 7$
(Hexagonal setting: $a = 3.01$, $c = 20.93$; $A = 21$)
Co-ordinates (for hex. setting): $(000; \frac{1}{3}\frac{2}{3}\frac{1}{3}; \frac{2}{3}\frac{1}{3}\frac{1}{3}) + 6\,Mo(C_{3v})$: $\pm(00z)$; $z = 0.075$
$\qquad\qquad\qquad\qquad\qquad\qquad\qquad +6B(C_{3v})$: the same with $z = \frac{1}{3}$
$\qquad\qquad\qquad\qquad\qquad\qquad\qquad +6B(C_{3v})$: the same with $z = 0.186$
$\qquad\qquad\qquad\qquad\qquad\qquad\qquad +3B(C_{3d})$: $00\frac{1}{2}$

$D8_k$ (Th$_7$S$_{12}$ type)

Hexagonal: $C_{6h}^5-P6_3/m$; $a = 11.04$, $c = 3.98$; $A = 19$
Co-ordinates: $1Th(S_6)$: $\pm(00\frac{1}{4})$
$\qquad\qquad\quad 6Th(C_s)$: $\pm(xy\frac{1}{4}); \pm(\bar{y}, x-y, \frac{1}{4}); \pm(y-x, \bar{x}, \frac{1}{4})$; $x = 0.153$; $y = -0.283$
$\qquad\qquad\quad 6S(C_s)$: the same with $x = 0.514$; $y = 0.375$
$\qquad\qquad\quad 6S(C_s)$: the same with $x = 0.235$; $y = 0 \pm 0.010$

$E1_1$ (CuFeS$_2$ type)

Tetragonal: $D_{2d}^{12}-I\bar{4}2d$; $a = 5.2864(8)$, $c = 10.4102(8)$; $A = 16$
Co-ordinates: $(000; \frac{1}{2}\frac{1}{2}\frac{1}{2}) + 4\,Cu(S_4)$: $000; \frac{1}{2}0\frac{1}{4}$
$\qquad\qquad\qquad +4\,Fe(S_4)$: $00\frac{1}{2}; \frac{1}{2}0\frac{3}{4}$
$\qquad\qquad\qquad +8S(C_2)$: $\frac{1}{4}y\frac{1}{8}; \frac{3}{4}\bar{y}\frac{1}{8}; y\frac{37}{48}; \bar{y}\frac{17}{48}$; $y = 0.27$

$E1_a$ (MgCuAl$_2$ type)

Orthorhombic: $D_{2h}^{17}-Cmcm$; $a = 4.00$, $b = 9.23$, $c = 7.14$; $A = 16$
Co-ordinates: $(000; \frac{1}{2}\frac{1}{2}0) + 4\,Mg(C_{2v})$: $\pm(0y\frac{1}{4})$; $y = 0.072$
$\qquad\qquad\qquad +4\,Cu(C_{2v})$: $\pm(0y\frac{1}{4})$; $y = -0.222$
$\qquad\qquad\qquad +8\,Al(C_s)$: $\pm(0yz); \pm(0, y, \frac{1}{2} - z)$; $y = 0.356$; $z = 0.056$

$E1_b$ [AuAgTe$_4$ (Sylvanite) type]

Monoclinic: C_{2h}^4-P2/c; $a = 8.96(2)$, $b = 4.49(2)$, $c = 14.62(2)$; $\beta = 145°26'$; $A = 12$
Co-ordinates: $2Au(C_i)$: $000; 00\frac{1}{2}$
$\qquad\qquad\quad 2Ag(C_2)$: $\pm(0y\frac{1}{4}$; $y = 0.433$
$\qquad\qquad\quad 4Te(C_1)$: $\pm(xyz); \pm(x, \bar{y}, \frac{1}{2} + z)$; $x = 0.298$; $y = 0.031$; $z = 0.999$
$\qquad\qquad\quad 4Te(C_1)$: the same with $x = 0.277$; $y = 0.425$; $z = 0.235$

(continued)

Table 6.2   STRUCTURAL DETAILS—*continued*

$E9_3$ (Fe$_3$W$_3$C type)

Cubic: O_h^7–$Fd3m$; $a = 11.10$; $A = 112$

Co-ordinates: $(000; \frac{1}{2}\frac{1}{2}0; \wr) + 16\mathrm{Fe}(D_{3d})$: $\frac{5}{8}\frac{5}{8}\frac{5}{8}; \frac{5}{8}\frac{7}{8}\frac{7}{8}; \wr$

$+32\mathrm{Fe}(C_{3v})$: $xxx; x\bar{x}\bar{x}; \wr; (\frac{1}{4}-x)(\frac{1}{4}-x)(\frac{1}{4}-x); (\frac{1}{4}-x)(\frac{1}{4}+x)(\frac{1}{4}+x);$
$\wr; x = 1.175$

$+48\mathrm{W}(C_{2v})$: $\pm(x00; \wr); (\frac{1}{4}+x)\frac{1}{4}\frac{1}{4}; \wr; (\frac{1}{4}-x)\frac{1}{4}\frac{1}{4}; x = 0.195$

$+16\mathrm{C}(C_{3d})$: $\frac{1}{8}\frac{1}{8}\frac{1}{8}; \frac{3}{8}\frac{3}{8}\frac{3}{8}; \wr$

$E9_4$ (Al$_5$C$_3$N type)

Hexagonal: C_{6v}^4–$P6_3mc$; $a = 3.285$; $c = 21.65$; $A = 18$

Co-ordinates: $2\mathrm{Al}(C_{3v})$: $00z; 00(\frac{1}{2}+z); z = 0.150$

$2\mathrm{Al}(C_{3v})$: the same with $z = 0.345$

$2\mathrm{Al}(C_{3v})$: $\frac{1}{3}\frac{2}{3}z; \frac{2}{3}\frac{1}{3}(\frac{1}{2}+z); z = 0.045$

$2\mathrm{Al}(C_{3v})$: the same with $z = 0.456$

$2\mathrm{Al}(C_{3v})$: the same with $z = 0.240$

$2\mathrm{C}(C_{3v})$ the same with $z = 0.133$

$2\mathrm{C}(C_{3v})$: the same with $z = 0.369$

$2\mathrm{C}(C_{3v})$: $00z; 00(\frac{1}{2}+z); z = 0.001$

$2\mathrm{N}(C_{3v})$: the same with $z = 0.250$

$E9_a$ (FeCu$_2$Al$_7$ type)

Tetragonal: D_{4h}^6–$P4/mnc$; $a = 6.336(1), c = 14.870(2); A = 40$

Co-ordinates: $4\mathrm{Fe}(C_4)$: $\pm(00z); \pm(\frac{1}{2}, \frac{1}{2}, \frac{1}{2} + z); z = 0.300$

$8\mathrm{Cu}(C_s)$: $\pm(xy0); \pm(\frac{1}{2} + x, \frac{1}{2} - y, \frac{1}{2}); \pm(\bar{y}x0); \pm(\frac{1}{2} + y, \frac{1}{2} + x, \frac{1}{2}); x = 0.278; y = 0.092$

$4\mathrm{Al}(C_4)$: $\pm(00z); \pm(\frac{1}{2}, \frac{1}{2}, \frac{1}{2} + z); z = 0.122$

$8\mathrm{Al}(C_2)$: $\pm(x, \frac{1}{2} + x, \frac{1}{4}; \bar{x}, \frac{1}{2} - x, \frac{1}{4}; \frac{1}{2} - x, x, \frac{1}{4}); x = 0.167$

$16\mathrm{Al}(C_1)$: $\pm(xyz; \bar{x}\bar{y}z; \bar{y}xz; y\bar{x}z; \frac{1}{2} + x, \frac{1}{2} - y, \frac{1}{2} + z; \frac{1}{2} - x, \frac{1}{2} + y, \frac{1}{2} + z;$
$\frac{1}{2} + y, \frac{1}{2} + x, \frac{1}{2} + z; \frac{1}{2} - y, \frac{1}{2} - x, \frac{1}{2} + z); x = 0.203; y = 0.414; z = 0.100$

$E9_b$ (FeMg$_3$Al$_8$Si$_6$ type)

Hexagonal: D_{3h}^3–$P\bar{6}2m$; $a = 6.63, c = 7.94; A = 18$

Co-ordinates: $1\mathrm{Fe}(D_{3h})$: 000

$3\mathrm{Mg}(C_{2v})$: $x0\frac{1}{2}; 0x\frac{1}{2}; \bar{x}\bar{x}\frac{1}{2}; x = 0.445$

$1\mathrm{Al}(D_{3h})$: $00\frac{1}{2}$

$3\mathrm{Al}(C_{2v})$: $x00; 0x0; \bar{x}\bar{x}0; x = 0.403$

$4\mathrm{Al}(C_3)$: $\pm(\frac{1}{3}\frac{2}{3}z); \pm(\frac{1}{3}\frac{2}{3}\bar{z}); z = 0.231$

$6\mathrm{Si}(C_s)$: $x0z; x0\bar{z}; 0xz; 0x\bar{z}; \bar{x}\bar{x}z; \bar{x}\bar{x}\bar{z}; x = 0.750; z = 0.223$

$E9_c$ (Mn$_3$Al$_9$Si type)

Hexagonal: D_{4h}^6–$P6_3/mmc$; $a = 7.42, c = 7.72; A = 26$

Co-ordinates: $6\mathrm{Mn}(C_{2v})$: $\pm(x, 2x, \frac{1}{4}); \pm(2x, x, \frac{3}{4}); \pm(x\bar{x}\frac{1}{4}); x = 0.120$

$6\mathrm{Al}(C_{2v})$: the same with $x = 0.458$

$12\mathrm{Al}(C_s)$: $\pm(x, 2x, z; 2x, x, \bar{z}; x\bar{x}z; x, 2x, \frac{1}{2} - z; 2x, x, \frac{1}{2} + z; x, \bar{x}, \frac{1}{2} - z); x = 0.201;$
$z = -0.067$

$2\mathrm{Si}(D_{3d})$: $000; 00\frac{1}{2}$

$E9_d$ (AlLi$_3$N$_2$ type)

Cubic: T_h^7–$Ia3$; $a = 9.48; A = 96$

Co-ordinates: $(000; \frac{1}{2}\frac{1}{2}\frac{1}{2}) + 16\mathrm{Al}(C_3)$: $\pm(xxx); \pm(\frac{1}{2} + x, \frac{1}{2} - x, \bar{x}; \wr); x = 0.115$

$+48\mathrm{Li}(C_1)$: $\pm(xyz; \wr); \pm(x, \bar{y}, \frac{1}{2} - z; \wr); \pm(\frac{1}{2} - x, y, \bar{z}; \wr);$
$\pm(\bar{x}, \frac{1}{2} - y, z; \wr); x = 0.160; y = 0.382; z = 0.110$

$+8\mathrm{N}(C_{3i})$: $000; \frac{1}{2}\frac{1}{2}0; \wr$

$+24\mathrm{N}(C_2)$: $\pm(x0\frac{1}{4}; \wr); \pm(\bar{x}\frac{1}{2}\frac{1}{4}; \wr); x = 0.205$

For GaLi$_3$N$_2$: $x(\mathrm{Ga}) = 0.117; x(\mathrm{Li}) = 0.152; y(\mathrm{Li}) = 0.38\bar{1}; z(\mathrm{Li}) = 0.114; x(\mathrm{N}) = 0.215$

(*continued*)

Table 6.2 STRUCTURAL DETAILS—*continued*

$E9_e$[CuFe$_2$S$_3$ (Cubanite) type]

Orthorhombic: D_{2h}^{16}–$Pnma$; $a = 6.231(2)$, $b = 11.117(6)$, $c = 6.467(1)$; $A = 24$
Co-ordinates: $4Cu(C_s)$: $\pm(x\frac{1}{4}z)$; $\pm(\frac{1}{2} - x, \frac{3}{4}, \frac{1}{2} + z)$; $x = \frac{1}{8}$; $z = \frac{7}{12}$
  $8Fe(C_1)$: $\pm(xyz); \frac{1}{2} + x, \frac{1}{2} - y, \frac{1}{2} - z; \bar{x}, \frac{1}{2} + y, \bar{z}; \frac{1}{2} - x, \bar{y}, \frac{1}{2} + z)$; $x = \frac{1}{8}$; $y = \frac{1}{12}$; $z = \frac{1}{12}$
  $8S(C_1)$: the same with $x = \frac{1}{4}$; $y = \frac{1}{12}$; $z = \frac{5}{12}$
  $4S(C_s)$: $\pm(x\frac{1}{4}z)$; $\pm(\frac{1}{2} - x, \frac{3}{4}, \frac{1}{2} + z)$; $x = \frac{1}{4}$; $z = \frac{11}{12}$

$F0_1$ [NiSbS (Ullmannite) type]

Cubic: T^4–$P2_13$; $a = 5.933(3)$; $A = 12$
Co-ordinates: $4Ni(C_3)$: $xxx; (\frac{1}{2} + x)(\frac{1}{2} - x)\bar{x}; \rightarrow; x \approx 0$
  $4Sb(C_3)$: the same with $x \approx 0.385$
  $4Sb(C_3)$: the same with $x \approx 0.615$

$F5_1$ (NaHF$_2$ type)

Rhombohedral: D_{3d}^5–$R\bar{3}m$; $a = 5.05$; $\alpha = 40°2'$; $A = 4$
(Hexagonal setting: $a = 3.45$, $c = 13.90$, $c/a = 4.03$; $A = 12$)
Co-ordinates: $1Na(D_{3d})$: 000; $1H(D_{3d}) = \frac{1}{2}\frac{1}{2}\frac{1}{2}$
  $2F(C_{3v})$: $\pm(xxx)$; $x = 0.410$
For CaCN$_2$: $\alpha = 43°50'$; $c/a = 3.63$; $x = 0.37$
For NaCrS$_2$: $\alpha = 29°48'$; $c/a = 5.59$; $x = 0.236$
For NaCrSe$_2$: $\alpha = 30°18'$; $c/a = 5.49$; $x = 0.235$
For RbCrSe$_2$: $\alpha = 21°33'$; $c/a = 7.85$

$F5_6$ (CuSbS$_2$ type)

Orthorhombic: D_{2h}^{16}–$Pnma$; $a = 6.008(10)$, $b = 3.784(10)$, $c = 14.456(30)$; $A = 16$
Co-ordinates: $4Cu(C_s)$: $\pm(x\frac{1}{4}z)$; $\pm(\frac{1}{2} + x, \frac{1}{4}, \frac{1}{2} - z)$; $x = 0.25$; $z = 0.83$
  $4Sb(C_s)$: the same with $x = 0.23$; $z = 0.06$
  $4S(C_s)$: the same with $x = 0.63$; $z = 0.10$
  $4S(C_s)$: the same with $x = 0.88$; $z = 0.83$

$F5_a$ (KFeS$_2$ type)

Monoclinic: C_{2h}^6–$C2/c$; $a = 7.079(1)$, $b = 11.304(2)$, $c = 5.398(1)$; $\beta = 113.20(1)$; $A = 16$
Co-ordinates: $(000; \frac{1}{2}\frac{1}{2}0) + 4K(C_2)$: $\pm(0y\frac{1}{4})$; $y = 0.355$
  $+Fe(C_2)$: the same with $y = -0.008$
  $+8S(C_1)$: $\pm(xyz); \pm(x, \bar{y}, \frac{1}{2} + z)$; $x = 0.195$; $y = 0.111$; $z = 0.10$

$H1_1$ [Spinel (Al$_2$MgO$_4$) type]

Cubic: O_h^7–$Fd3m$; $a = 8.075$; $A = 56$
Co-ordinates: $(000; \frac{1}{2}\frac{1}{2}0; \rightarrow) + 8Mg(I_d)$: $000; \frac{1}{4}\frac{1}{4}\frac{1}{4}$
  $+16Al(D_{3d})$: $\frac{5}{8}\frac{5}{8}\frac{5}{8}; \frac{5}{8}\frac{7}{8}\frac{7}{8}; \rightarrow$
  $+32O(C_{3v})$: $xxx; x\bar{x}\bar{x}; \rightarrow; (\frac{1}{4} - x)(\frac{1}{4} - x)(\frac{1}{4} - x); (\frac{1}{4} - x, \frac{1}{4} + x, \frac{1}{4} + x)$;
  $\rightarrow; x \approx -\frac{1}{8}$

In some compounds, better agreement with observed intensities is obtained by assuming that the metal atoms are distributed at random among the 24 available sites, or that the trivalent element occupies all the 8-equivalent sites and half of the 16-equivalent ones. In some cases lattice sites may be vacant, e.g., γ-Al$_2$O$_3$ or In$_2$S$_3$.

$H2_4$[Cu$_3$VS$_4$ (Sulvanite) type]

Cubic: T_d^1–$P\bar{4}3m$; $a = 5.391(5)$; $A = 8$
Co-ordinates: $3Cu(D_{2d})$: $\frac{1}{2}00; 0\frac{1}{2}0; 00\frac{1}{2}$
  $1V(T_d)$: 000
  $4S(C_{3v})$: $xxx; x\bar{x}\bar{x}; \rightarrow; x = 0.235$

$H2_6$ [Stannite (FeCu$_2$SnS$_4$) type]

Tetragonal: D_{2d}^{11}–$I\bar{4}2m$; $a = 5.449(2)$; $c = 10.757(3)$; $A = 16$
Co-ordinates: $(000; \frac{1}{2}\frac{1}{2}\frac{1}{2}) + 2Fe(D_{2d})$: 000
  $+2Sn(D_{2d})$: $00\frac{1}{2}$
  $+4Cu(S_4)$: $0\frac{1}{2}\frac{1}{4}; \frac{1}{2}0\frac{1}{4}$
  $+8S(C_s)$: $xxz; \bar{x}\bar{x}z; x\bar{x}\bar{z}; \bar{x}x\bar{z}; x = 0.245; z = 0.132$

(*continued*)

Table 6.2   STRUCTURAL DETAILS—*continued*

$L1_0$ (CuAu type)

Tetragonal: D_{4h}^1–$P4/mmm$; $a = 3.964$; $c = 3.672$; $A = 4$
Co-ordinates: $(000; \frac{1}{2}\frac{1}{2}0) + 2\mathrm{Cu}(D_{4h})$:   000
$\qquad\qquad\qquad\qquad\qquad\ +2\mathrm{Au}(D_{4h})$: $\frac{1}{2}0\frac{1}{2}$
Superstructure of the $A1$ (Cu) type

$L1_2$ (Cu$_3$Au type)

Cubic: O_h^1–$Pm3m$; $a = 3.747$; $A = 4$
Co-ordinates: $3\mathrm{Cu}(D_{4h})$: $\frac{1}{2}\frac{1}{2}0$; ♪
$\qquad\qquad\quad\ 1\mathrm{Au}(O_h)$:   000
Superstructure of $A1$ (Cu) type

$L2_1$ (Cu$_2$MnAl type)

Cubic: O_h^5–$Fm3m$; $a = 5.968$; $A = 16$
Co-ordinates: $(000; \frac{1}{2}\frac{1}{2}0; ♪) + 4\mathrm{Al}(O_h)$:   000
$\qquad\qquad\qquad\qquad\qquad\quad +8\mathrm{Cu}(I_d)$: $\frac{1}{4}\frac{1}{4}\frac{1}{4}$; $\frac{3}{4}\frac{3}{4}\frac{3}{4}$
$\qquad\qquad\qquad\qquad\qquad\quad +4\mathrm{Mn}(O_h)$: $\frac{1}{2}\frac{1}{2}\frac{1}{2}$

Superstructure of the $A2$ (W) type; this type is virtually identical with the D_{o3} (BiF$_3$ or BiLi$_3$) type

$L2_2$ (Tl$_7$Sb$_2$ type)

Cubic: O_h^9–$Im3m$; $a = 11.618(6)$; $A = 54$
Co-ordinates: $(000; \frac{1}{2}\frac{1}{2}\frac{1}{2}) + 2\mathrm{Tl}(O_h)$:   000
$\qquad\qquad\qquad\qquad\qquad\quad +16\mathrm{Tl}(C_{3v})$: $\pm(xxx, x\bar{x}\bar{x}; ♪)$; $x = 0.17 \approx \frac{1}{6}$
$\qquad\qquad\qquad\qquad\qquad\quad +24\mathrm{Tl}(C_{2v})$: $\pm(xx0; ♪; x\bar{x}0; ♪)$; $x = 0.35 \approx \frac{1}{3}$
$\qquad\qquad\qquad\qquad\qquad\quad +12\mathrm{Sb}(C_{4v})$: $\pm(x00; ♪)$; $x = 0.29 \approx \frac{1}{3}$

L_1' (Fe$_4$N type)

Cubic: O_h^1–$Pm3m$ or O_h^5–$Fm3m$ (depending on the distribution of the N atoms); $a = 3.896(2)$; $A = 5$
Co-ordinates: $4\mathrm{Fe}$ at 000; $\frac{1}{2}\frac{1}{2}0$; ♪
$\qquad\qquad\quad 1\mathrm{N}$ at $\frac{1}{4}\frac{1}{4}\frac{1}{4}$; or at $\frac{1}{2}\frac{1}{2}\frac{1}{2}$, probably the latter; or at random at $\frac{1}{4}\frac{1}{4}\frac{1}{4}$; $\frac{3}{4}\frac{3}{4}\frac{3}{4}$; ♪; and (or) at
$\qquad\qquad\quad \frac{3}{4}\frac{3}{4}\frac{3}{4}$; $\frac{1}{4}\frac{1}{4}\frac{3}{4}$; ♪; and (or) at $\frac{1}{2}\frac{1}{2}\frac{1}{2}$; $\frac{1}{2}00$; ♪

$L'1_2$ ($\sim$AlFe$_3$C type)

Cubic: O_h^1–$Pm3m$; $a = 3.78$; $A \approx 5$
Co-ordinates: $1\mathrm{Al}(O_h)$:   000
$\qquad\qquad\quad 3\mathrm{Fe}(D_{4h})$: $\frac{1}{2}00$; ♪
$\qquad\qquad\quad 0.6$ to $0.9\mathrm{C}(O_h)$ at $\frac{1}{2}\frac{1}{2}\frac{1}{2}$

$L'2$ (Martensite type)

Tetragonal: D_{4h}^{17}–$I4/mmm$; $a = 2.84$, $c = 2.97$; $A = 2\mathrm{Fe} +$ (up to) $0.12\mathrm{C}$
Co-ordinates: $2\mathrm{Fe}(D_{4h})$ at 000; $\frac{1}{2}\frac{1}{2}\frac{1}{2}$
The C atoms at random: $\frac{1}{2}\frac{1}{2}0$ and (or) $00\frac{1}{2}$

$L'3$ (Interstitial $A3$ type)

Hexagonal: D_{6h}^4–$P6_3/mmc$ or D_{6h}^1–$P6/mmm$
Co-ordinates: 2 metal atoms (D_{3h}): $\frac{2}{3}\frac{1}{3}0$; $\frac{1}{3}\frac{2}{3}\frac{1}{2}$
$\qquad\qquad\quad$ C or N(C_{3v}): $\frac{1}{3}\frac{2}{3}z$; $\frac{2}{3}\frac{1}{3}(\frac{1}{2} + z)$; $\frac{1}{3}\frac{2}{3}\bar{z}$; $\frac{2}{3}\frac{1}{3}(\frac{1}{2} - z)$; $z \approx \frac{3}{8}$; or $00\frac{1}{4}$; $00\frac{3}{4}$

$L6_0$ (Ti$_3$Cu type)

Tetragonal: D_{4h}^1–$P4/mmm$; $a = 4.158$, $c = 3.594$; $A = 4$
Co-ordinates: $1\mathrm{Cu}(D_{4h})$:   000
$\qquad\qquad\quad 1\mathrm{Ti}(D_{4h})$: $\frac{1}{2}\frac{1}{2}0$
$\qquad\qquad\quad 2\mathrm{Ti}(D_{2h})$: $0\frac{1}{2}\frac{1}{2}$; $\frac{1}{2}0\frac{1}{2}$
Tetragonal deformed $L1_2$ (Cu$_3$Au) type; superstructure of $A6$ (In) type

$L'6$ (Interstitial $A6$ type)

Tetragonal: D_{4h}^{17}–$F4/mmm$ (or D_{4h}^1–$P4/mmm$, depending on the distribution of the N atoms)
Co-ordinates: 4 metal atoms (D_{4h}):   000; $\frac{1}{2}\frac{1}{2}0$; ♪;
$\qquad\qquad\quad$ N atoms in the holes: $\frac{1}{4}\frac{1}{4}\frac{1}{4}$; $\frac{1}{4}\frac{3}{4}\frac{3}{4}$; ♪; $\frac{3}{4}\frac{3}{4}\frac{3}{4}$; $\frac{3}{4}\frac{1}{4}\frac{1}{4}$; ♪; or $\frac{1}{2}\frac{1}{2}\frac{1}{2}$; $\frac{1}{2}00$; ♪

Table 6.3   COMPARISON OF STRUKTURBERICHT AND PEARSON NOMENCLATURE

Strukturbericht structural type	Typical compound or element	Pearson symbol
$A1$	Cu	cF4
$A2$	W	cI2
$A3$	Mg	hP2
$A4$	C (Diamond)	cF8
$A5$	Sn (Beta)	tI4
$A6$	In	tI2
$A7$	As	hR2
$A8$	Gamma-Se	hP3
$A9$	C (Graphite)	hP4
$A10$	Hg	hR1
$A11$	Ga	oC8
$A12$	Alpha-Mn	cI58
$A13$	Beta-Mn	cP20
$A15$	Cr_3Si	cP8
$A16$	Alpha-S	oF128
$A20$	Alpha-U	oC4
A_a	Pa	tI2
A_b	Beta-U	tP30
A_c	Alpha-Np	oP8
A_d	Beta-Np	tP4
A_e	Beta-$TiCu_3$	oF4
A_f	$HgSn_{10}$	hP1
A_g	B	tP50
A_h	Alpha-Po	cP1
A_i	Beta-Po	hR1
A_k	Alpha-Se	mP32
A_l	Beta-Se	mP32
$B1$	NaCl	cF8
$B2$	CsCl	cP2
$B3$	ZnS (Sphalerite)	cP2
$B4$	ZnS (Würtzite)	hP4
$B8_1$	NiAs	hP4
$B8_2$	Ni_2In	hP6
$B9$	HgS	hP6
$B10$	LiOH; PbO	tp4
$B13$	NiS	hR6
$B16$	GeS	oP8
$B17$	PtS	tP4
$B18$	CuS	hP12
$B19$	AuCd	oP4
$B20$	FeSi	cP8
$B27$	FeB	oP8
$B29$	SnS	oP8
$B31$	MnP	oP8
$B32$	NaTl	cF16
$B34$	PdS	tP16
$B35$	CoSn	hP6
$B37$	TlSe	tI16
B_a	UCo	cI16
B_b	AgZn	hP9
B_c	CaSi	oP16
B_d	NiSi	oP8
B_e	CdSb	oP16
B_f	CrB	oC8
B_g	MoB	tI16
B_h	WC	hP2
B_i	MoC	hP8
B_l	AsS	mP32
B_m	TiB	oP8
$C1$	CaF_2	cF12
$C2$	FeS_2	cP12
$C6$	CdI_2	hP3
$C7$	MoS_2	hP6

(*continued*)

Table 6.3 COMPARISON OF STRUKTURBERICHT AND PEARSON NOMENCLATURE—*continued*

Strukturbericht structural type	*Typical compound or element*	*Pearson symbol*
$C11_a$	CaC_2	tI6
$C11_b$	$MoSi_2$	tI6
$C12$	$CaSi_2$	hR6
$C14$	$MgZn_2$	hP12
$C15$	$MgCu_2$	cF24
$C15_b$	$AuBe_5$	cF24
$C16$	$CuAl_2$	tI12
$C18$	FeS_2 (Marcasite)	oP6
$C22$	Fe_2P	hP9
$C23$	$PbCl_2$	oP12
$C32$	AlB_2	hP3
$C33$	Bi_2Te_2S	hR5
$C34$	$AuTe_2$ (Calaverite)	mC6
$C36$	$MgNi_2$	hP24
$C37$	Co_2Si	oP12
$C38$	Cu_2Sb	tP6
$C40$	$CrSi_2$	hP9
$C42$	SiS_2	oI12
$C44$	GeS_2	oF72
$C46$	$AuTe_2$ (Krennerite)	oP24
$C49$	$ZrSi_2$	oC12
$C54$	$TiSi_2$	oF24
C_a	Mg_2Ni	hP18
C_b	Mg_2Cu	oF48
C_c	$ThSi_2$	tI12
C_e	$CoGe_2$	oC24
C_g	ThC_2	mC12
C_h	$CuTe_2$	hP6
$D0_2$	$CoAs_3$	cI32
$D0_3$	BiF_3	cF16
$D0_9$	ReO_3/Cu_3N	cF4
$D0_{11}$	Fe_3C	oP16
$D0_{18}$	Na_3As	hP8
$D0_{19}$	Mg_3Cd	hP8
$D0_{20}$	$NiAl_3$	oP16
$D0_{21}$	Cu_3P	hP24
$D0_{22}$	$TiAl_3$	oP16
$D0_{23}$	$ZrAl_3$	tI16
$D0_{24}$	$TiNi_3$	hP16
$D0_a$	$TiCu_3$	oP8
$D0_c$	U_3Si	tI16
$D0_d$	Mn_3As	oP16
$D1_3$	$BaAl_4$	tI10
$D1_a$	$MoNi_4$	tI10
$D1_b$	UAl_4	oI20
$D1_c$	$PtSn_4$	oC20
$D1_d$	$PtPb_4$	tP10
$D1_e$	UB_4	tP20
$D1_f$	MnB_4	oF40
$D1_g$	B_4C	hR15
$D2_1$	CaB_6	cP7
$D2_3$	$NaZn_{13}$	cF112
$D2_b$	$ThMn_{12}$	tI26
$D2_c$	U_6Mn	tI26
$D2_d$	$CaCu_5$	hP6
$D2_e$	$BaHg_{11}$	cP36
$D2_f$	UBe_{12}	cF52
$D2_g$	Fe_8N	tI18
$D5_2$	La_2O_3	hP5
$D5_3$	Mn_2O_3	cI80
$D5_8$	Sb_2S_3	oP20
$D5_9$	Zn_3P_2	tP40

(*continued*)

Table 6.3   COMPARISON OF STRUKTURBERICHT AND PEARSON NOMENCLATURE—*continued*

Strukturbericht structural type	Typical compound or element	Pearson symbol
$D5_{10}$	Cr_3C_2	oP20
$D5_{13}$	Ni_2Al_3	hP5
$D5_a$	U_3Si_2	tP10
$D5_b$	Pt_2Sn_3	hP10
$D5_c$	Pu_2C_3	cI40
$D5_e$	Ni_3S_2	hR5
$D5_f$	As_2S_3	mP20
$D7_1$	Al_4C_3	hP7
$D7_2$	Co_3S_4	cF56
$D7_3$	Th_3P_4	cI28
$D7_a$	Ni_3Sn_4	mC14
$D7_b$	Ta_3B_4	oI14
$D8_1$	Fe_3Zn_{10}	cI52
$D8_2$	Cu_5Zn_8	cI52
$D8_3$	Cu_9Al_4	cP52
$D8_4$	$Cr_{23}C_6$	cF116
$D8_5$	Fe_7W_6	hR13
$D8_6$	$Cu_{15}Si_4$	cI76
$D8_8$	Mn_5Si_3	hP16
$D8_9$	Co_9S_8	cF68
$D8_{10}$	Cr_5Al_8	hR26
$D8_{11}$	Co_2Al_5	hP28
$D8_a$	Th_6Mn_{23}	cF116
$D8_b$	Sigma (FeCr)	tP30
$D8_c$	$Mg_2Cu_6Al_5$ (Mg_2Zn_{11})	cP39
$D8_d$	Co_2Al_9	mP22
$D8_e$	$Mg_{32}X_{49}$ (X = Al, Zn)	cI162
$D8_f$	Ir_3Ge_7	cI40
$D8_g$	Mg_5Ga_2	oI28
$D8_h$	W_2B_5	hP14
$D8_i$	Mo_2B_5	hR7
$D8_k$	Th_7S_{12}	hP19
$E1_1$	$CuFeS_2$	tI16
$E1_a$	$MgCuAl_2$	oC16
$E1_b$	$AuAgTe_4$	mP12
$E9_3$	Fe_3W_3C	cF112
$E9_4$	Al_5C_3N	hP18
$E9_a$	$FeCu_2Al_7$	tP40
$E9_b$	$FeMg_3Al_8Si_6$	hP18
$E9_c$	Mn_3Al_9Si	hP26
$E9_d$	$AlLi_3N_2$	cI96
$E9_e$	$CuFe_2S_3$	oP24
$F0_1$	$NiSbS$	cP12
$F5_1$	$NaHF_2$	hR4
$F5_6$	$CuSbS_2$	oP16
$F5_a$	$KFeS_2$	mC16
$H1_1$	$MgAl_2O_4$	cF56
$H2_4$	Cu_3VS_4	cP8
$H2_6$	$FeCu_2SnS_4$	tP16
$L1_0$	$CuAu$	tP2
$L1_2$	Cu_3Au	cP4
$L2_1$	Cu_2MnAl	cF16
$L2_2$	Tl_7Sb_2	cI54
L'_1	Fe_4N	cF5
$L'1_2$	$AlFe_3C$	cP5
$L'2$	Fe-C (Martensite)	tI2
$L'3$	Fe_2N	hP3
$L6_0$	$CuTi_3$	tP4

REFERENCES

1. E. A. Owen and G. I. Williams, *Proc. Phys. Soc.*, 1954 (A), **67**, 895.
2. T. W. Baker and J. Williams, *Acta Cryst.*, 1955, **8**, 519.
3. C. S. Barrett, *J. Chem. Physics*, 1956, **25**, 1123.
4. F. H. Spedding, A. H. Daane and K. W. Herrmann, *Trans. Amer. Inst. Min. (Metall.) Engrs.*, 1957, **209**, 895.
5. F. H. Ellinger, *US Atomic Energy Comm. Publn.*, LADC–1460.
6. J. D. Farr, A. L. Giorgi and M. G. Bowman, *US Atomic Energy Comm. Publn.*, LA–1545, 1953.
7. K. Gordon, B. Skinner and H. L. Johnston, *US Atomic Energy Comm. Publn.*, NP–4737, 1953.
8. J. W. Nielsen and N. C. Baenziger, *Acta Cryst.*, 1954, **7**, 132.
9. P. Duwez, *J. Appl. Physics*, 1951, **22**, 1174.
10. J. D. Fast, *J. Appl. Physics*, 1952, **23**, 350.
11. D. M. Poole, G. K. Williamson and J. A. C. Marples, *J. Inst. Metals*, 1957–58, **80**, 172.
12. D. T. Peterson, P. F. Djilak and C. L. Vold, *Acta Cryst.*, 1956, **9**, 1936.
13. P. A. Sellers, S. Fried, R. E. Elson and W. H. Zachariasen, *J. Amer. Chem Soc.*, 1954, **76**, 5935.
14. J. Thewlis, *Acta Cryst.*, 1952, **5**, 790.
15. C. W. Tucker, *Acta Cryst.*, 1951, **4**, 425; 1952, **5**, 395.
16. A. G. Knapton, *Acta Cryst.*, 1954, **7**, 457.
17. A. M. Holden and W. E. Seymour, *Trans. Amer. Inst. Min. (Metall.) Engrs.*, 1956, **206**, 1312.
18. W. H. Zachariasen, *Acta Cryst.*, 1952, **5**, 664.
19. O. J. C. Runnalls, *Acta Cryst.*, 1954, **7**, 222.
20. W. H. Zachariasen and F. Ellinger, *J. Chem. Physics*, 1957, **27**, 811.
21. F. H. Ellinger, *Trans. Amer. Inst. Min. (Metall.) Engrs*, 1956, **206**, 1256.
22. M. A. Gurevich and B. F. Ormont, *Fizika Metall.*, 1957, **4**, 112.
23. A. E. Austin and J. R. Doig, *Trans. Amer. Inst. Min. (Metall.) Engrs.*, 1957, **209**, 27.
24. R. F. Domogala and D. J. McPherson, *US Atomic Energy Comm. Publn.*, COO–100, 1952.
25. R. F. Raeuchle and F. W. von Batchelor, *Acta Cryst.*, 1955, **8**, 691.
26. P. Duwez and C. B. Jordan, *J. Am. Chem. Soc.*, 1951, **73**, 5509.
27. E. K. Halteman, *Acta Cryst.*, 1957, **10**, 166.
28. E. Oehman, *Metallwirtschaft*, 1930, **9**, 825.
29. A. J. C. Wilson, *Bull. Am. Phys. Soc.*, 1934, **9**, No. 16.
30. Z. S. Basinski and J. W. Christian, *J. Inst. Metals*, 1952, **80**, 659.
31. H. Margolin and Elmars Ence, *Trans. Amer. Inst. Min. (Metall.) Engrs.*, 1954, **200**, 1267.
32. W. Rostoker, *J. Metals*, 1952, **4**, 209.
33. P. Duwez and J. L. Taylor, *J. Metals*, 1950, **2**, 1173.
34. E. T. Hayes, A. H. Robertson and W. L. O'Brien, *Trans. Am. Soc. Metals*, 1951, **43**, 888.
35. D. Bergman and D. P. Shoemaker, *Acta Cryst.*, 1954, **7**, 857.
36. G. J. Dickins, Abs. Dissert. Univ. Cambridge, 1954–55, 1957, 244.
37. G. J. Dickins, Audrey M. B. Douglas and W. H. Taylor, *Acta Cryst.*, 1956, **9**, 297.
38. Emma Smith and R. W. Guard, *Trans. Amer. Inst. Min. (Metall.) Engrs.*, 1957, **209**, 1189.
39. T. J. Heal and G. I. Williams, *Acta Cryst.*, 1955, **8**, 494.
40. C. B. Jordan, *Trans. Amer. Inst. Min. (Metall.) Engrs.*, 1955, **203**, 832.
41. M. V. Nevitt and J. W. Downey, *Trans. Amer. Inst. Min. (Metall.) Engrs.*, 1957, **209**, 1072.
42. R. M. Waterstrat and J. S. Kasper, *Trans. Amer. Inst. Min. (Metall.) Engrs.*, 1957, **209**, 872.
43. P. Duwez and C. B. Jordan, *Acta Cryst.*, 1952, **5**, 213.
44. L. Misch, *Z. Physikal Chem.*, 1935, (B), **29**, 42.
45. G. F. Kossolapow and A. K. Trapesnikow, *Metallwirtschaft*, 1935, **14**, 45.
46. A. Bryström, P. Kierkegaard and O. Knop, *Acta Chem. Scand.*, 1956, **6**, 709.
47. N. C. Baenziger, R. E. Rundle and A. I. Snow, *Acta Cryst.*, 1956, **9**, 93.
48. J. R. Murray, *J. Inst. Metals*, 1955–56, **84**, 91.
49. E. A. Wood and B. T. Mattheas, *Acta Cryst.*, 1956, **9**, 534.
50. N. C. Baenziger and J. W. Conant, *Acta Cryst.*, 1956, **9**, 361.
51. P. Pietrokowsky, *Trans. Amer. Inst. Min. (Metall.) Engrs.*, 1954, **200**, 219.
52. E. S. Makarov and L. S. Gudhov, *Krystallografiya*, 1956, **1**, 650.
53. H. Nowotny, E. Bauer and A. Stempfl, *Monatsh.*, 1950, **81**, 1164.
54. R. E. Marsh, *Acta Cryst.*, 1954, **7**, 379.
55. R. Ferro, *Acta Cryst.*, 1954, **7**, 781.
56. M. J. Sanderson and N. C. Baenziger, *Acta Cryst.*, 1953, **6**, 627.
57. E. A. Owen and E. A. O'D. Roberts, *J. Inst. Metals*, 1940, **66**, 389.
58. A. Byström and K. E. Almin, *Acta Chem. Scand.*, 1948, **1**, 76.
59. C. S. Barrett, *Acta Cryst.*, 1957, **10**, 58.
60. J. W. Nielsen and N. C. Baenziger, *Acta Cryst.*, 1954, **7**, 277.
61. E. J. Duwell and N. C. Baenziger, *Acta Cryst.*, 1955, **8**, 705.
62. N. C. Baenziger, E. J. Duwell and J. W. Conant, *US Atomic Energy Comm. Publn.*, COO–127, 1954.
63. K. Schubert, U. Rösler, W. Mahler, E. Dorre and W. Schütt, *Z. Metallk.*, 1954, **45**, 643.
64. G. V. Raynor and J. A. Lee, *Acta Met.*, 1954, **2**, 616.
65. E. S. Makarov, *Dokl. Akad. Nauk. USSR*, 1950, **74**, 935.
66. D. K. Das and D. T. Pitman, *Trans. Amer. Inst. Min. (Metall.) Engrs.*, 1957, **209**, 1175.
67. J. H. N. van Vucht, *Z. Metallk.*, 1957, **48**, 253.

68. E. Ence and H. Margoli, *Trans. Amer. Inst. Min.* (*Metall.*) *Engrs.*, 1957, **209**, 484.
69. W. Köster and A. Sampaio, *Z. Metallk.*, 1957, **48**, 331.
70. O. J. C. Runnals, *J. Metals*, 1953, **5**, 1460.
71. J. F. Smith and E. A. Ray, *Acta Cryst.*, 1957, **10**, 169.
72. A. J. Bradley and S. S. Lu, *Z. Krist.*, 1937, **96**, 20.
73. W. Hofmann and H. Wiehr, *Z. Metallk.*, 1941, **33**, 369.
74. K. Little, J. N. Pratt and G. V. Raynor, *J. Inst. Metals*, 1951, **80**, 456.
75. K. Little, *J. Inst. Metals*, 1953–54, **82**, 463.
76. S. Samson, *Nature*, 1954, **173**, 1185.
77. J. Adam and J. B. Rich, *Acta Cryst.*, 1954, **7**, 813.
78. *idem*, *Acta Cryst.*, 1955, **8**, 349.
79. A. D. I. Nicol, *Acta Cryst.*, 1953, **6**, 285.
80. K. Schubert, U. Rösler, M. Kluge, K. Anderko and L. Härle, *Naturwiss.*, 1953, **40**, 269, 437.
81. A. C. Larsen, D. T. Cromer and C. N. Stambaugh, *Acta Cryst.*, 1957, **10**, 443.
82. J. H. Keeler, *US Atomic Energy Comm. Publn.*, SO–2515, 1954.
83. P. Esslinger and K. Schubert, *Z. Metallk.*, 1957, **48**, 126.
84. M. G. Bown, *Acta Cryst.*, 1956, **9**, 70.
85. K. Anderko and U. Zwick, *Naturwiss.*, 1957, **44**, 510.
86. K. Anderko, *Naturwiss.*, 1957, **44**, 88.
87. M. B. Waldron, *J. Inst. Metals*, 1951, **79**, 103.
88. K. Robinson, *Acta Cryst.*, 1952, **5**, 401.
89. G. Phragmen, *J. Inst. Metals*, 1950, **77**, 489.
90. E. Hellner, *Z. Metallk.*, 1950, **41**, 401.
91. E. S. Makarov, *Izvest. Akad. Nauk S.S.S.R.*, 1943, (Khim), 264.
92. J. Reynolds, W. A. Wiseman and W. Hume-Rothery, *J. Inst. Metals*, 1952, **80**, 637.
93. E. Hellner and F. Laves, *Z. Naturforsch.*, 1947 (A), **2**, 180.
94. P. G. Cotter, J. A. Kohn and R. A. Potter, *J. Amer. Ceram. Soc.*, 1956, **39**, 11.
95. H. Schachner, H. Nowotny and R. Machenschalk, *Monatsh.*, 1953, **84**, 677.
96. P. Pietrokowsky, *Acta Cryst.*, 1954, **7**, 435.
97. W. H. Zachariasen, *Acta Cryst.*, 1949, **2**, 94.
98. G. Brauer and H. Haag, *Z. Anorg. Chem.*, 1949, **259**, 197.
99. O. J. C. Runnals and R. R. Boucher, *Acta Cryst.*, 1955, **8**, 592.
100. H. Schachner, E. Cerwenka and H. Nowotny, *Monatsh.*, 1954, **85**, 245.
101. H. Nowotny, R. Kieffer and H. Schachner, *Monatsh.*, 1952, **83**, 1243.
102. P. Eckerlin and E. Wölfel, *Z. Anorg. Chem.*, 1955, **280**, 321.
103. K. Schubert and H. Pfisterer, *Z. Metallk.*, 1950, **41**, 438.
104. A. Osawa and M. Okamoto, *Sci. Rep. Tohoku Imp. Univ.*, 1939, (i), **27**, 326.
105. B. Boren, *Arkiv. Kemi Mineral. Geol.*, 1933, **11A**, (10), 1.
106. K. Toman, *Acta Cryst.*, 1951, **4**, 462.
107. S. Geller and E. A. Wood, *Acta Cryst.*, 1954, **7**, 441.
108. K. Anderko and K. Schubert, *Z. Metallk.*, 1953, **44**, 307.
109. J. H. Buddery and A. J. E. Welch, *Nature*, 1951, **167**, 362.
110. K. Schubert, *Naturwiss.*, 1952, **39**, 351.
111. G. Bergman and J. L. T. Waugh, *Acta Cryst.*, 1953, **6**, 93.
112. K. Robinson, *Acta Cryst.*, 1953, **6**, 854.
113. K. Robinson and P. J. Black, *Phil. Mag.*, 1953, **44**, 1392.
114. P. Eckerlin, H. J. Meyer and E. Wölfel, *Z. Anorg. Chem.*, 1955, **281**, 322.
115. E. Parthé and J. T. Norton, *Acta Cryst.*, 1958, **11**, 14.
116. P. I. Kripyakevich, E. I. Gladyshevskv and O. S. Zarechnyuk, *Dokl. Akad. Nauk. S.S.S.R.*, 1954, **95**, 525.
117. K. Schubert and H. Pfisterer, *Z. Metallk.*, 1951, **41**, 433.
118. E. Hellner, *Fortschr. Mineral, Krist. Petrogr.*, 1951, **29–30**, 59.
119. A. Westgren and G. Phragmen, *Z. Anorg. Chem.*, 1928, **175**, 80.
120. O. Carlssohn and G. Hägg, *Z. Krist.*, 1932, **83**, 308.
121. S. T. Knobejewski and W. P. Tarassova, *Zh. Fiz. Khim.* (*J. Phys. Chem.*), 1937, **9**, 681.
122. H. Knödler, *Acta Cryst.*, 1957, **10**, 86.
123. H. Nowotny and H. Schachner, *Monatsh.*, 1953, **84**, 169.
124. A. Zalkin and W. J. Ramsey, *J. Phys. Chem.*, 1956, **60**, 234.
125. *idem*, *J. Phys. Chem.*, 1956, **60**, 1275.
126. R. E. Marsh and D. P. Shoemaker, *Acta Cryst.*, 1953, **6**, 197.
127. D. Gilde, *Z. Anorg. Chem.*, 1956, **284**, 142.
128. P. Farrar and H. Margolin, *Trans. Amer. Inst. Min.* (*Metall.*) *Engrs.*, 1955, **203**, 101.
129. R. D. Heidenreich, *Acta Met.*, 1955, **3**, 79.
130. K. Bachmayer, H. Nowotny and A. Kohl, *Monatsh.*, 1955, **86**, 39.
131. H. Nowotny, R. Funk and J. Pesl, *Monatsh.*, 1951, **82**, 513.
132. K. E. Fylking, *Arkiv. Kem. Min. Geol.*, 1935, (B), **11**, (48), 1.
133. H. Cole, F. W. Chambers and H. M. Dunn, *Acta Cryst.*, 1956, **9**, 685.
134. H. Nowotny and B. Glatzl, *Monatsh.*, 1951, **82**, 720.
135. K. H. Jack and M. M. Wachtel, *Proc. Roy. Soc.*, 1957, *A*, **239**, 46.
136. R. Ferro, *Acta Cryst.*, 1956, **9**, 817.

137. Le Roy Heaton and N. S. Gingrich, *Acta Cryst.*, 1955, **8**, 207.
138. R. N. Kuzmin, G. S. Khdanov and N. N. Zhuravlev, *Kristallografiya*, 1957, **2**, 48.
139. A. Osawa and N. Shibata, *Sci. Rep. Tohoku Imp. Univ.*, 1939, (i), **28**, 1, 197.
140. W. Hofmann, *Z. Metallk.*, 1941, **33**, 61.
141. R. Ferro, *Acta Cryst.*, 1957, **10**, 476.
142. W. P. Binnie, *Acta Cryst.*, 1956, **9**, 686.
143. J. L. Hoard, S. Geller and R. E. Hughes, *J. Am. Chem. Soc.*, 1951, **73**, 1892.
144. A. W. Laubengayer, D. T. Hurd, A. E. Newkirk and J. L. Hoard, *J. Am. Chem. Soc.*, 1943, **65**, 1924.
145. St. v. N. Szabo and C. W. Tobias, *J. Am. Chem. Soc.*, 1949, **71**, 1882.
146. V. Russell, R. Hirst, F. A. Kanda and A. J. King, *Acta Cryst.*, 1953, **6**, 870.
147. B. Post and F. W. Glaser, *J. Chem. Phys.*, 1952, **20**, 1050.
148. B. F. Decker and J. S. Kosher, *Acta Cryst.*, 1954, **7**, 77.
149. P. Ehrlich, *Z. Anorg. Chem.*, 1949, **259**, 1.
150. R. Kiessling, *Acta Chem. Scand.*, 1950, **4**, 164.
151. F. W. Glaser and B. Post, *J. Metals*, 1953, **5**, 1117.
152. D. Moscowitz, *Trans. Amer. Inst. Min. (Metall.) Engrs.*, 1956, **206**, 1325.
153. F. Bertant and P. Blum, *Compt. Rend.*, 1953, **236**, 1055.
154. Stig Rundqviet, *Nature*, 1958, **181**, 259.
155. R. W. Mooney and A. J. E. Welch, *Acta Cryst.*, 1954, **7**, 49.
156. M. von Stackelberg, *Z. Physical Chem.*, 1930, **B**, **9**, 437.
157. E. B. Hunt and R. E. Rundle, *J. Am. Chem. Soc.*, 1951, **73**, 4777.
158. H. A. Wilhelm and P. Chiotti, *Trans. Amer. Soc. Metals*, 1950, **42**, 1295.
159. R. E. Rundle, N. C. Baenziger, A. S. Wilson and R. A. McDonald, *J. Am. Chem. Soc.*, 1948, **70**, 99.
160. U. Esch and A. Schnieder, *Z. Anorg. Chem.*, 1948, **257**, 254.
161. L. M. Litz, A. B. Garrett and F. C. Croxton, *J. Am. Chem. Soc.*, 1948, **70**, 1718.
162. M. W. Mallett, A. F. Gerds and D. A. Vaughan, *J. Electrochem. Soc.*, 1951, **98**, 505.
163. G. Brauer, H. Renner and J. Wernet, *Z. Anorg. Chem.*, 1954, **277**, 249.
164. V. I. Smirnova and B. F. Ormont, *Dokl. Akad. Nauk. SSSR*, 1954, **96**, 557.
165. K. H. Jack, *Proc. Roy. Soc.*, 1948, **195**, 56.
166. L. J. E. Hofer, E. M. Cohen and W. C. Peebles, *J. Am. Chem. Soc.*, 1949, **71**, 189.
167. K. H. Jack, *Acta Cryst.*, 1950, **3**, 392.
168. N. Schönberg, *Acta Met.*, 1954, **2**, 837.
169. S. Nagakura, *J. Phys. Soc. Japan*, 1957, **12**, 482.
170. K. Whitehead and L. D. Brownlee, *Planseebar, Pulvermet*, 1956, **4**, 62.
171. K. H. Jack, *Acta Cryst.*, 1952, **5**, 404.
172. H. A. Eick, N. C. Baenziger and L. Eyring, *J. Am. Chem. Soc.*, 1956, **78**, 5987.
173. R. Jusa, H. H. Weber and C. Meyer-Simon, *Z. Anorg. Chem.*, 1953, **273**, 48.
174. G. Brauer and K. H. Zapp, *Naturwiss*, 1953, **40**, 604.
175. U. Zorll, *Z. Physik*, 1954, **138**, 167.
176. R. Ferro, *Z. Anorg. Chem.*, 1954, **275**, 320.
177. O. Knop and H. Haraldsen, *Canad. J. Chem.*, 1956, **34**, 1142.
178. J. Singer and C. W. Spencer, *Trans. Amer. Inst. Min. (metall.) Engrs.*, 1955, **203**, 144.
179. S. A. Semiletov, *Kristallografiya*, 1956, **1**, 403.
180. P. Khodadad and J. Flahaut, *Compte Rendu.*, 1957, **244**, 462.
181. S. Geller and B. B. Cetlin, *Acta Cryst.*, 1955, **8**, 272.
182. H. Hahn and K. F. Schröder, *Z. Anorg. Chem.*, 1952, **269**, 135.
183. H. Haraldsen, *Z. Anorg. Chem.*, 1941, **246**, 169.
184. K. Schubert, E. Dörre and M. Kluge, *Z. Metallk.*, 1955, **46**, 216.
185. L. I. Tatarinova, Yu K. Auleitner and Z. G. Pinsker, *Kristallografiya*, 1956, **1**, 537.
186. H. Miyazawa and S. Sugaike, *J. Phys. Soc. Japan*, 1957, **12**, 312.
187. A. Okazaki and I. Ueda, *J. Phys. Soc. Japan*, 1956, **11**, 470.
188. K. Schubert and H. Frieke, *Z. Metallk.*, 1953, **44**, 457.
189. Structure Reports, 1951, **11**, 246.
190. A. Bryström, *Arkiv Kemi. Mineral. Geol.*, 1945, **19B**, No. 8.
191. A. R. Graham, *Amer. Mineralogist*, 1949, **34**, 462.
192. F. Bertaut, *Compt. Rend.*, 1952, **234**, 1295.
193. H. Hahn, *Z. Anorg. Chem.*, 1951, **264**, 184.
194. D. Lundqvist, *Arkiv Kemi. Mineral. Geol.*, 1943, **17B**, No. 12.
195. N. V. Belov and V. P. Butuzov, *Dokl. Akad. Nauk. SSSR*, 1946, **54**, 717.
196. R. Ueda, *J. Phys. Soc. Japan*, 1949, **4**, 287.
197. Structure Reports, 1952, **12**, 156.
198. K. Schubert, E. Dörre and E. Günzel, *Naturwiss.*, 1954, **41**, 448.
199. K. Toman, *Acta Cryst.*, 1952, **5**, 329.
200. H. Haraldsen, *Z. Anorg. Chem.*, 1939, **240**, 337.
201. R. F. Raeuchle and R. E. Rundle, *Acta Cryst.*, 1952, **5**, 85.
202. B. G. Bergman and D. P. Shoemaker, *J. Chem. Physics*, 1951, **19**, 515.
203. J. S. Kasper, B. F. Decker and J. R. Belanger, *J. Appl. Physics*, 1951, **22**, 361.
204. W. B. Pearson and J. W. Christian, *Acta Cryst.*, 1952, **5**, 157.
205. Y.-C. Tang, *Acta Cryst.*, 1951, **4**, 377.

206. A. Schneider and U. Esch. *Z. Elektrochem.*, 1944, **50**, 290.
207. F. H. Spedding, A. H. Deane, G. Wakefield and B. H. Dennison, *Trans. Metall. Soc. A.I.M.E.*, 1960, **218**, 608.
208. F. H. Spedding, A. H. Deane and J. J. Hanah, *Trans. Metall. Soc. A.I.M.E.*, 1959, **212**, 179.
209. E. M. Savitsky, V. F. Terekhova and I. V. Burov, *Tsvet. Metally.*, 1960, **33**, 59.
210. E. M. Savitsky, V. F. Terekhova and O. P. Naumking, *Tsvet. Metally.*, 1960, **33**, 43.
211. D. V. Keller, F. A. Kanda and A. J. King, *J. Phys. Chem.*, 1958, **62**, 732.
212. A. Zalkin, R. G. Bedford and D. E. Sands, *Acta Cryst.*, 1959, **12**, 701.
213. R. M. Paine and J. A. Carrabine, *Acta Cryst.*, 1960, **13**, 680.
214. E. F. Sturken and B. Post, *Acta Cryst.*, 1960, **13**, 852.
215. W. H. Zadareisen and F. H. Ellinger, *Acta Cryst.*, 1959, **12**, 175.
216. D. B. McWhan, *US Atomic Energy Comm. Publn.* UCRL 9695, 1964.
217. A. Zalkin, D. E. Sands and A. H. Krikorian, *Acta Cryst.*, 1960, **13**, 160.
218. *idem*, *Acta Cryst.*, 1960, **13**, 713.
219. S. A. Spakner, *Trans. Metall. Soc. A.I.M.E.*, 1958, **212**, 57.
220. R. M. Paine and J. A. Carrabine, *Acta Cryst.*, 1960, **13**, 680.
221. B. J. Baudry, J. F. Haufling and A. H. Daane, *Acta Cryst.*, 1960, **13**, 743.
222. R. I. Myklebist and A. H. Deane, *Trans. Metall. Soc. A.I.M.E.*, 1962, **224**, 354.
223. N. H. Krikorian, W. G. Witteman and M. S. Cowme, *J. Phys. Chem.*, 1960, **64**, 1517.
224. B. A. Hatt, *Acta Cryst.*, 1961, **14**, 119.
225. V. F. Novy, R. C. Vickery and E. V. Kleben, *Trans. Metall. Soc. A.I.M.E.*, 1961, **221**, 585.
226. G. Katz and A. J. Jacobs, *J. Nucl. Mater.*, 1962, **5**, 338.
227. F. R. Morral, *J. Metals N.Y.*, 1958, **10**, 662.
228. D. J. Cromer and A. C. Larsen, *Acta Cryst.*, 1961, **14**, 1226.
229. S. Saito, *Acta Cryst.*, 1959, **12**, 500.
230. M. Korchynsky and R. W. Fountain, *Trans. Metall. Soc. A.I.M.E.*, 1959, **215**, 1053.
231. B. J. Baudry and A. H. Daane, *Trans. Metall. Soc. A.I.M.E.*, 1960, **218**, 854.
232. S. E. Haszko, *Trans. Metall. Soc. A.I.M.E.*, 1960, **218**, 763.
233. D. T. Cromer and G. E. Olsen, *Acta Cryst.*, 1959, **12**, 689.
234. V. F. Novy, R. C. Vickery and E. V. Kleben, *Trans. Metall. Soc. A.I.M.E.*, 1961, **221**, 585.
235. *idem*, *Trans. Metall. Soc. A.I.M.E.*, 1961, **221**, 588.
236. G. A. Yurks, J. W. Barton and J. S. Parr, *Acta Cryst.*, 1959, **12**, 909.
237. D. Kramer, *Trans. Metall. Soc. A.I.M.E.*, 1959, **215**, 256.
238. J. F. Smith and W. L. Laram, *Acta Cryst.*, 1962, **15**, 252.
239. D. T. Cromer and A. C. Larsen, *Acta Cryst.*, 1950, **13**, 909.
240. D. T. Cromer and C. E. Olsen, *Acta Cryst.*, 1959, **12**, 689.
241. D. T. Cromer and R. B. Roof, *Acta Cryst.*, 1959, **12**, 942.
242. S. Saito and P. A. Beck, *Trans. Metall. Soc. A.I.M.E.*, 1959, **215**, 938.
243. E. Rudy, B. Kieffer and H. Fröhlich, *Z. Metallk.*, 1962, **53**, 90.
244. W. Obrowski, *Z. Metallk.*, 1962, **53**, 715.
245. A. F. Berndt, *Acta Cryst.*, 1961, **14**, 1301.
246. V. B. Crompton, *Acta Cryst.*, 1958, **11**, 446.
247. E. A. Wood and V. B. Compton, *Acta Cryst.*, 1958, **11**, 429.
248. A. E. Dwight and P. A. Beck, *Trans. Metall. Soc. A.I.M.E.*, 1959, **215**, 976.
249. J. R. Thomson, *Proc. 11th Amer. Conf. on X-Ray*, 1963.
250. W. Köster and W. D. Hackl, *Z. Metallk.*, 1958–59, **12**, 647.
251. B. A. Hatt and G. I. Williams, *Acta Cryst.*, 1959, **12**, 685.
252. W. Obrowski, *Metall.*, 1963, **17**, 108.
253. A. C. Larsen and D. T. Cromer, *Acta Cryst.*, 1961, **14**, 514.
254. *idem*, *Acta Cryst.*, 1961, **14**, 73.
255. D. T. Cromer, A. C. Larsen and R. B. Roof, *Acta Cryst.*, 1960, **13**, 913.
256. S. C. Haszko, *Trans. Metall. Soc. A.I.M.E.*, 1960, **218**, 4, 763.
257. E. Ence and M. Mayoln, *Trans. Metall. Soc. A.I.M.E.*, 1961, **221**, 370.
258. R. P. Rand and L. D. Calvert, *Canad. J. Chem.*, 1962, **40**, 705.
259. E. O. Hall and J. Royan, *Acta Cryst.*, 1959, **12**, 607.
260. D. Watanabe, *J. Phys. Soc. Japan*, 1960, **15**, 1251.
261. P. Chiotti and K. J. Gill, *Trans. Metall. Soc. A.I.M.E.*, 1961, **221**, 573.
262. *idem*, *Trans. Metall. Soc. A.I.M.E.*, 1959, **215**, 892.
263. D. R. Petersen and H. W. Rinn, *Acta Cryst.*, 1961, **14**, 328.
264. C. L. Vold, *Acta Cryst.*, 1961, **14**, 1289.
265. W. Heine and U. Zwicker, *Z. Metall.*, 1962, **53**, 386.
266. E. Günzel and K. Schubert, *Z. Metall.*, 1958, **49**, 234.
267. R. V. Schablaski, B. S. Tani and M. G. Chesanov, *Trans. Metall. Soc. A.I.M.E.*, 1962, **224**, 867.
268. S. Samson, *Nature, Lond.*, 1962, **195**, 259.
269. D. V. Masson and C. S. Barrett, *Trans. Metall. Soc. A.I.M.E.*, 1958, **212**, 260.
270. E. I. Duwell and N. C. Baenziger, *Acta Cryst.*, 1960, **13**, 476.
271. W. Rostowker, *Trans. Metall. Soc. A.I.M.E.*, 1958, **212**, 393.
272. R. F. Domagala, R. P. Elliott and W. Rostotier, *Trans. Metall. Soc. A.I.M.E.*, 1958, **212**, 393.

273. S. Sansom, *Acta Cryst.*, 1958, **11**, 857.
274. T. Dagenham, *Acta Chem. Scand.*, 1963, **17**, 267.
275. L. M. d'Alte da Veiges, *Phil. Mag.*, 1962, **7**, 1247.
276. A. J. Goldat and J. G. Pair, *Trans. Metall. Soc. A.I.M.E.*, 1961, **221**, 639.
277. C. J. Wilson and D. Sand, *Acta Cryst.*, 1961, **14**, 72.
278. T. J. Rensuf and C. A. Beevers, *Acta Cryst.*, 1961, 14, 469.
279. C. G. Wilson, D. K. Thomas and F. J. Spooner, *Acta Cryst.*, 1960, **13**, 56.
280. C. G. Wilson and F. J. Spooner, *Acta Cryst.*, 1960, **13**, 4, 358.
281. L. E. Edshammer, *Acta Chem. Scand.*, 1961, **15**, 403.
282. idem, *Acta Chem. Scand.*, 1960, **14**, 2248.
283. C. R. McKinsey and G. M. Faubring, *Acta Cryst.*, 1959, **12**, 701.
284. J. W. H. Clare, *J. Inst. Mab.*, 1960–61, **89**, 232.
285. J. A. Bland and D. Clarke, *Acta Cryst.*, 1958, **11**, 231.
286. J. H. Bland, *Acta Cryst.*, 1958, **11**, 236.
287. M. A. Taylor, *Acta Cryst.*, 1959, **12**, 393.
288. idem, *Acta Cryst.*, 1961, **14**, 84.
289. L. M. d'Alte de Veiges and L. K. Walford, *Phil. Mag.*, 1963, **8**, 349.
290. W. Obrowski, *Metall.*, 1963, **17**, 108.
291. K. Andutro, *Z. Metallk.*, 1958, **49**, 165.
292. R. Ferro, *Acta Cryst.*, 1958, **11**, 737.
293. S. Bhan and K. Schubert, *Z. Metallk.*, 1960, **51**, 327.
294. E. I. Gladishevsky, *Dopov., Akad. Nauk ukr. R.S.R.*, 1959, **3**, 294.
295. E. Parthé, *Acta Cryst.*, 1959, **12**, 559.
296. B. J. Baudry and A. H. Daane, *M.A.*, 1962, 443.
297. L. M. Finme, *J. less common Metals*, 1962, **4**, 24.
298. E. Parthé, *Acta Cryst.*, 1960, **13**, 968.
299. E. I. Gladishevsky, *Dopov. Akad. Nauk ukr. R.S.R.*, 1959, **3**, 294.
300. N. Ageev and V. Samsonov, *Doklady Akad. Nauk S.S.S.R.*, 1957, **112**, 853.
301. A. G. Tharp, A. W. Searcy and H. Novotkny, *J. Electrochem. Soc.*, 1958, **105**, 473.
302. E. Raub and W. Fuzsche, *Z. Metallk.*, 1962, **53**, 779.
303. E. Weitz, L. Born and E. Hellness, *Z. Metallk.*, 1960, **51**, 228.
304. D. M. Bailey and J. F. Smith, *Acta Cryst.*, 1961, **14**, 57.
305. A. Zalkin and W. J. Ramsey, *J. Phys. Chem.*, 1958, **62**, 689.
306. D. T. Cromer, *Acta Cryst.*, 1959, **12**, 36.
307. W. Trzebratowski, S. Weglowski and K. Lukasgewag, *Roczniki Chem.*, 1958, **32**, 189.
308. J. H. Bryden, *Acta Cryst.*, 1962, **15**, 167.
309. C. Giesecki and H. Pfister, *Acta Cryst.*, 1958, **11**, 369.
310. A. Kockus, F. Gronvolde and J. Thorbioin, *Acta Chem. Scand.*, 1962, **16**, 1493.
311. V. N. Bykoff and V. V. Kazarnikov, *Kristallografiya*, 1959, **4**, 924.
312. I. Obinata, Y. Takechi and S. Saikewa, *Trans. Amer. Soc. Metals*, 1959, **52**, 156.
313. J. L. Hoard, R. E. Hughes and D. E. Sands, *J. Am. Chem. Soc.*, 1958, **80**, 4507.
314. D. E. Sands, C. F. Cline, A. Zalkin and C. L. Hoenig, *Acta Cryst.*, 1961, **14**, 309.
315. G. V. Samonov, V. P. Dzeganovsky and I. A. Simashko, *Kristallografiya*, 1959, **4**, 119.
316. Y. B. Paderno, T. I. Serebaykova and G. V. Samsonov, *Dokl. Akad. Nauk S.S.S.R.*, 1959, **125**, 317.
317. B. J. MacDonald and W. I. Stuart, *Acta Cryst.*, 1960, **13**, 447.
318. M. Elfstrom, *Acta Chem. Scand.*, 1961, **15**, 1178.
319. S. Laplace and B. Poste, *Acta Cryst.*, 1962, **15**, 97.
320. W. Obrowski, *Metall.*, 1963, **17**, 108.
321. E. Steinberg, *Acta Chem. Scand.*, 1961, **15**, 861.
322. B. Magnussen and C. Brossit, *Acta Chem. Scand.*, 1962, **16**, 449.
323. M. Atoji, K. G. Schneider, A. H. Waane, R. E. Rundle and F. H. Spedding, *J. Am. Chem. Soc.*, 1958, **80**, 1804.
324. G. Brauer and K. Lesser, *Z. Metallk.*, 1959, **50**, 8.
325. S. Nagakura, *J. Phys. Soc. Japan*, 1959, **14**, 186.
326. idem, *J. Phys. Soc. Japan*, 1961, **16**, 1213.
327. W. D. Forgang and B. F. Decker, *Trans. Metall. Soc. A.I.M.E.*, 1958, **212**, 343.
328. F. Gronwold, A. Kjetshus and F. Raun, *Acta Cryst.*, 1961, **14**, 93.
329. D. Pustiner and R. E. Newnham, *Acta Cryst.*, 1961, **14**, 691.
330. M. S. Mirgalowskaya and E. V. Skudnova, *Izv. Akad. Nauk. S.S.S.R.*, 1959, **4**, 148.
331. A. Stechen and P. Eckerlin, *Z. Metallk.*, 1960, **51**, 295.
332. E. F. Hockiup and J. C. White, *Acta Cryst.*, 1961, **14**, 328.
333. E. Parthé, *Acta Cryst.*, 1960, **13**, 865.
334. P. Khodad, *C. R. Akad. Sci., Paris*, 1960, **250**, 3998.
335. idem, *C. R. Akad. Sci., Paris*, 1959, **249**, 694.
336. A. Okasaki, *J. Phys. Soc. Japan*, 1958, **13**, 1151.
337. E. A. Wood, V. B. Compton, B. T. Matthias and E. Carengurt, *Acta Cryst.*, 1958, **11**, 604.
338. W. Klemm, F. Darn and R. Huck, *Naturwiss.*, 1958, **45**, 490.
339. J. A. John, G. Katz and A. A. Giardini, *Z. Krist.*, 1958, **111**, 52.

340. W. Krönert and K. Plieth, *Naturwiss.*, 1958, **45**, 416.
341. H. H. Stadelmeier and W. K. Hardy, *Z. Metallk.*, 1961, **52**, 391.
342. M. V. Nevitt and J. W. Downey, *Trans. Metall. Soc. A.I.M.E.*, 1962, **224**, 195.

Data updated in this edition of Smithells is from 'Pearson's Handbook of Crystallographic Data for Intermetallic Phases' (4 volumes) by P. Villars and L. D. Calvert, published by ASM, International Materials Park OH, 1991.

Supplemented by checking of selected phases not present in the above (NaCl, CsCl, LiOH, $PbCl_2$, P) using the Powder Diffraction File, International Center for Diffraction Data, Newton Square, PA, 2001.

7 Metallurgically important minerals

Table 7.1 gives data on the minerals from which the more widely used metallic and non-metallic elements and their compounds are extracted. The elements in this table are given in alphabetical order for easy reference. A 'mineral' is a naturally occurring, homogenous, inorganic solid material that has a definite, but not always fixed, chemical composition. It may be crystalline or amorphous and may have formed, in nature, by inorganic or biological processes.

Most elements form more minerals than those listed; minerals of primary importance are shown in bold type in column 2. Not all elements cited are metals although they have widespread utility in metallurgical and materials processing applications. For example silicon (Si) is a non-metal used in semiconductor (computer chip) applications, while chlorine and bromine are utilised in metal-extraction technologies. Some elements are hardly ever used as such but rather in compound forms. For example, boron (B) compounds have hundreds of uses in the form of B_2O_3, H_3BO_3, or $Na_2B_4O_7$ etc. Many elements, such as germanium (Ge)—an important component of semiconductor devices, are normally extracted as by-product of sphalerite (ZnS) or bauxite processing residues.

Extraction technologies often incorporate a number of process steps including concentration (flotation), smelting (pyro-metallurgy), and leaching (hydro-metallurgy). Parentheses indicate that the element or its compounds are recovered as by-products from the extraction residues of other primary materials. The mineral-producing countries are listed in order of decreasing production volume in column 5 while major end product sources are given in column 6. When data are not available or not certain, US import sources are cited in column 6. The figure for abundance given in column 1 is the concentration of the element (in p.p.m., parts per million) in the crustal rocks of the earth. The figure for metal production in column 7 refers to recovered primary element production except where indicated as contained quantities in produced ores or concentrates or as secondary refining production. *Salines* is used to include all salt waters, bitter lakes and the sea.

Abbreviations used in the table:

p.p.m. : Parts per million (or gram per metric ton)
m.t.   : Metric ton = 1 000 kg = tonne
N/A   : Not available or not disclosed due to strategic or proprietary reasons
Equiv. : Equivalent
Est.   : Estimated

BIBLIOGRAPHY

Mineralogical and economic data

C. S. Hurlbut Jr. and C. Klein, 'Manual of Mineralogy', 19th edition, John Wiley and Sons, New York, 1971.
R. V. Gaines, *et al.*, 'Dana's New Mineralogy', 8th Edition, John Wiley and Sons, New York, 1977.
C. Palache, H. Berman and C. Frondel, 'The System of Mineralogy of J. D. Dana', 7th edition, Volumes I and II, John Wiley and Sons, New York, 1951.
H. H. Read, Rutley's Elements of Mineralogy, 24th edition, Thomas Murby and Co., London, 1957.
'Minerals Yearbook; vol. 1, Mineral Industry Surveys', U.S. Geological Survey, Minerals Information Publications Services, Reston, 2001.

Table 7.1   MINERALS, SOURCES, AND USES

Element: symbol and abundance (p.p.m.)	Minerals	Formulae	Density (g cm^{-3})	Major mineral sources	Major sources	1999/2000 World production (m. t.)	Primary uses and applications
1	2	3	4	5	6	7	8
Aluminium Al 81 300	**Bauxite** **Gibbsite** Kyanite	(ore with Fe, Ti, clay impurities) $Al_2(OH)_6$ Al_2SiO_5	2–2.55 2.35 3.55–3.66	Australia, Guinea, Brazil Jamaica	USA, Russia, China, Canada	23.1×10^6	Containers, construction, electrical, machinery, equipment, aerospace
Antimony Sb 0.2–1	Senarmotite Valentinite Kermesite **Stibnite**	Sb_2O_3 Sb_2O_3 Sb_2S_2O Sb_2S_3	5.3 5.6 4.6 4.5–4.6	China S. Africa Russia Bolivia	China Mexico Australia	118×10^3	Flame retardant, lead-alloys, plastics filler, ceramics, glass, corrosion resistance
Arsenic As 1.8–5	**Arsenopyrite** Orpiment **Realgar**	FeAsS As_2S_3 AsS	6.07 3.49 3.48	China, Chile, Ghana, Mexico	China, Chile, Ghana, Mexico	In the form of As_2O_3 36.9×10^3	Wood preservation, glass additive, pesticides, Cu-Alloy
Barium Ba 4 000	**Barite** (or barites) Witherite	$BaSO_4$ $BaCO_3$	4.5 4.3	China, India, USA, Morocco	China, India, USA, Morocco	In the form of $BaSO_4$ 6.2×10^6	Chemicals, filler/extender, glass, well drilling
Beryllium Be 2–6	**Bertrandite** **Beryl**	$B_4Si_2O_7(OH)_2$ $B_3Al_2Si_6O_{18}$	2.59–.66 2.7	USA, Russia, Kazakhstan	USA	Metal Equiv.: 226	Electronics, alloy with Cu, defence applications
Bismuth Bi 0.2	**Native** (in PbS) **Bismuthinite**	Bi Bi_2S_3	9.7–9.8 6.4–6.5	China, Mexico, Peru	China, Mexico, Peru, Belgium	4.37×10^3	Alloys, solders, pharmaceuticals, glazes chemicals
Boron B 10	**Colemanite** **Tincal (Borax)** **Ulexite**	$Ca_2B_6O_{11} \cdot 5H_2O$ $Na_2B_4O_7 \cdot 10H_2O$ $NaCaB_5O_9 \cdot 8H_2O$	2.42 1.7 1.65	USA, Turkey, Russia	USA, Turkey, Russia	B_2O_3 Equiv.: 4.47×10^6	Glasses, insulators, soaps/detergents, chemicals, refining flux
Bromine Br 2.5	Bromyrite Embolite (also **NaBr, KBr in salines**)	AgBr Ag(Cl, Br)	6.5 5.3–5.4	USA, Israel, United Kingdom, China	USA, Israel, United Kingdom, China	542×10^3	Fire retardants, drilling, pesticides, water-treatment, photography, rubber

Element	Mineral	Formula	Density	Occurrence	Major producers	Production (tonnes)	Uses
Cadmium Cd 0.17	Greenockite (also from **ZnS** processing)	CdS	5.00	China, Australia, USA, Mexico	Japan, China, USA, Belgium	19.7×10^3	Batteries, pigments, coatings, plastics
Calcium Ca 36300	**Calcite (limestone)** **Aragonite** Gypsum	$CaCO_3$ (Hex.) $CaCO_3$ (Orth.) $CaSO_4 \cdot 2H_2O$	2.71 2.94 2.3	All countries	Limestone, cement and dolomite: USA, Canada, China, Australia	Extensive; US-usage calcite: 998×10^6 dolomite: 101×10^6	Minerals used in cement making, roads, construction-aggregates, agriculture, chemicals
Carbon C 200	**Graphite** Diamond	C (Hex.) C (Cub.)	2–2.3 3.52	China, India, Brazil, N. Korea	China, Brazil, Mexico, USA	602×10^3	Batteries, lubricants, electrodes, refractories
Cerium Ce 60–78	**Bastnazite** **Monazite**	$(Ce,La,Di)(CO_3)F$ $(Ce,La,Y,Th)PO_4$	4.9–5.2 4.6–5.4	China, Brazil, India, USA,	China, France, Japan, USA	Est. compounds: 6500	Lighters, lanterns, catalysts, glass
Caesium Cs 3	Pollucite	$CsAlSi_2O_6 \cdot H_2O$	2.9	Canada, Namibia, Zimbabwe	Canada, Germany, UK	N/A Est: 20	Electronics, photoelectric, medical
Chlorine Cl 130	**Halite** Sylvite Carnalite	NaCl KCl $KCl.MgCl_2 \cdot 6H_2O$	2.2 2.0 1.6	USA, China, Germany, France, Canada	USA, European Union, Canada, China	Est.: 50×10^6	Polymer production, pesticides, pulp/paper bleaching, water treatment
Chromium Cr 167	**Chromite**	$FeCr_2O_4$	4.5–4.8	S. Africa, Kazakhstan, India, Zimbabwe	S. Africa, Kazakhstan, India, Zimbabwe	Ores: Cr-Equiv.: content 4.32×10^6	Stainless steel, refractories, metal plating, chemicals
Cobalt Co 25	**Smaltite** **Linnaeite** **Erythrite** **Cobaltite**	$CoAs_2$ Co_3S_4 $Co_3As_2O_8 \cdot 8H_2O$ CoAsS	6.4 4.85 2.95 6–6.3	As byproduct of Cu and Ni: Australia, Canada, Cuba, Congo	Congo, Finland, Norway, Zambia, S. Africa	35×10^3	Superalloys, magnets, corrosion resistant alloys, tools, catalysts

(continued)

Table 7.1 MINERALS, SOURCES, AND USES —continued

Element: symbol and abundance (p.p.m.)	Minerals	Formulae	Density (g cm⁻³)	Major mineral sources	Major sources	1999/2000 World production (m. t.)	Primary uses and applications
1	2	3	4	5	6	7	8
Copper Cu 55–70	**Native Cu** **Chalcopyrite** **Malachite** Azurite **Covellite** Chrysocolla	Cu CuFeS$_2$ Cu$_2$(OH)$_2$ Cu$_3$(OH)$_2$(CO$_3$)$_2$ CuS CuSiO$_3 \cdot$ 2H$_2$O	8.95 4.1–4.3 4.05 3.77 4.6–4.76 2–2.2	Chile, USA, Indonesia, Australia	Chile, Japan, USA, China	11.4 × 10^6	Conductor (wires), ingots/rods, brasses, bronzes
Fluorine F 625	**Fluorite** Cryolite	CaF$_2$ Na$_3$AlF$_6$	3.0–3.25 2.97	China, S. Africa, Mexico, Brazil	China, S. Africa, Mexico, Brazil	As CaF$_2$ 4.52 × 10^6	Mainly as CaF$_2$ in HF production, ceramics, enamels, glass, steel-flux
Gallium Ga 15	Gallite (in **Bauxite, sphalerite, and fly-ash**)	CuGaS$_2$	4.2	Australia, Germany, Japan, Kazakhstan	France, Kazakhstan, Russia	Est.: 40–60	Photo-electric devices, electronics
Germanium Ge 1.5–7	Renierite Germanite (in **bauxite and fly-ash**)	(Cu,Zn)$_{11}$(Ge,As)$_2$ Fe$_4$S$_{16}$ Cu$_{26}$Fe$_4$Ge$_4$S$_{32}$	4.3 4.4–4.6	China, USA, Belgium, Russia	China, USA, Belgium, Russia	Est.: 70	Fibre optics, infrared optics, polymerisation, electrical/solar
Gold Au 0.005	**Native** Calaverite Krennerite Petzite	Au Au$_2$Te$_4$ (Mon.) Au$_8$Te$_{16}$ (Orth.) Ag$_3$AuTe$_2$	19.3 9.2–9.3 8.62 8.7–9.0	S. Africa, USA, Australia, China, Canada	S. Africa, USA, Australia, China, Canada	2 550	Jewellery, conductor in electronics, finance
Hafnium Hf 3–4.5	In **Zircon** at a concentration of (Zr:Hf) = 50:1	Zircon: ZrSiO$_4$	See Zircon	France, Germany, Japan, Canada	France, Germany, Belgium	Est.: 12	Neutron absorber, reactor-rods, lanterns, refractories
Indium In 0.1	Mainly with **sphalerite** at p.p.m. levels	Sphalerite: ZnS	See Sphalerite	China, Canada, France	Belgium, Canada, China, France	220	Low melting point solders, alkaline batteries

Element	Mineral	Formula	Density	Sources	Sources	Production	Uses
Iodine I 0.5	Iodyrite Lautarite Marshite (Iodine: from **caliche, salines, oil wells**)	AgI $Ca(IO_3)_2$ CuI	5.69 4.59 5.68	Chile, Japan, USA, China	Chile, Japan, USA, China	18×10^3	Pharmaceuticals, sanitation, animal feed, catalysts, photography
Iridium Ir 0.001	Native: see Platinum group metals (**PGM**)	Ir	22.65	See PGM	See PGM	Est.: 4	See PGM
Iron Fe 50 000	**Haematite** Magnetite **Pyrite** Siderite Limonite	Fe_2O_3 Fe_3O_4 FeS_2 $FeCO_3$ $2Fe_2O_3 \cdot 3H_2O$	4.9–5.3 5.18 4.8–5.1 3.7–3.9 3.6–4.0	China, Brazil, Australia, India	China, Brazil, Australia, India, USA	Fe Equiv.: 635×10^6; ore 92×10^6	Cast iron, steels, pigments, chemicals
Lanthanum La 30	**Monazite** **Bastnazite** Xenotime	$(Ce,La,Y,Th)PO_4$ $(Ce,La,Di)(CO_3)F$ YPO_4 (also has La)	5.27 4.9–5.2 4.46–5.1	China, USA, India, Russia	China, France, Japan, India, USA,	N/A Est.: less than 1000	Catalytic converters, ceramics, magnets, catalysts, TV screens
Lead Pb 13–16	**Galena** **Cerussite** Anglesite	PbS $PbCO_3$ $PbSO_4$	7.4–7.6 6.55 6.38	Australia, China, USA, Peru	USA, Australia, Canada, China	Mine output: 310×10^3	Acid batteries, solders
Lithium Li 20–65	**Spodumene** **Amblygonite** **Petalite**	$LiAlSi_2O_6$ $Li(F,OH)AlPO_4$ $Li(AlSi_4)O_{10}$	3.13–3.20 3–3.1 2.4	Chile, USA, Argentina, Australia	Chile, USA, Argentina	Minerals and brines: 37 000	Ceramics, glass, batteries
Magnesium Mg 20 900	**Magnesite** **Dolomite** Epsomite (also in **salines**)	$MgCO_3$ $CaMg(CO_3)_2$ $MgSO_4 \cdot 7H_2O$	2.8–3.0 2.8–2.9 1.67	Russia, China, USA, Slovakia	China, Australia, Canada	Compounds MgO Equiv.: 11.1×10^6	Compounds as refractories, flue-gas treatment, and as metal in aircraft
Manganese Mn 1 000	**Pyrolusite** **Rhodochrosite** **Manganite** **Psilomelane**	MnO_2 $MnCO_3$ $MnO(OH)$ $BaMn_{10}O_{16}(OH)_4$	4.4–5.0 3.7 4.3 4.7	S. Africa, Ukraine, Brazil, Gabon	China, S. Africa, Ukraine, France, Gabon	Ores: metal Equiv.: 7.28×10^6	Steel, batteries, chemicals

(continued)

Table 7.1 MINERALS, SOURCES, AND USES—*continued*

Element: symbol and abundance (p.p.m.)	Minerals	Formulae	Density (g cm^{-3})	Major mineral sources	Major sources	1999/2000 World production (m. t.)	Primary uses and applications
1	2	3	4	5	6	7	8
Mercury Hg 0.05–0.5	**Cinnabar**	HgS	8.09	Kyrgyzstan, Spain, Algeria, China	India, Singapore, Spain, UK	Mine output: minimum 1 640	Electrolytic Cl_2 and NaOH production, dentistry, gold amalgamation
Molybdenum Mo 1–15	**Molybdenite** Wulfenite Molybdite	MoS_2 $PbMoO_4$ $Fe_2O_3 \cdot 3MoO_3 \cdot 8H_2O$	4.7–4.8 6.3–7.0 4.5	USA, Chile, China, Mexico	USA, Chile, China, Mexico	129×10^3	Stainless steel, catalysts, lubricants
Nickel Ni 78	**Pentlandite** **Garnierite** **Millerite**	(Fe,Ni)S (Ni,Mg)SiO$_4$ NiS	5.0 2.2–2.8 5.3–5.6	Russia, Canada, Australia, New Caledonia	Russia, Japan, Canada, Australia	1.12×10^6	Steels, corrosion resistant coatings, Cu-alloys, magnets
Niobium (Nb) or Columbium 23	**Columbite** **Pyrochlore** also in **tin slags**	(Fe,Mn)(Nb,Ta)$_2$O$_6$ NaCb$_2$O$_6$F	5.2–6.4 4.45	Brazil, Canada, Australia	Brazil, Canada, Germany	32.6×10^3	Steel, jet engines, carbides
Osmium Os 0.005	Native in **Pt or Ir alloys**	Os	22.61	See PGM	See PGM	Est.: 22	See Platinum group metals (PGM)
Palladium Pd 0.01	Native in **Pt alloys and Cu-Ni ores**	Pd	12.02	Russia, S. Africa, USA, Canada	Russia, S. Africa, USA, Canada	174	See Platinum group metals (PGM)
Phosphorus P 1000	Apatite with F, Cl and/or OH: **Fluorapatite** **Hydroxylapatite**	(F,Cl,OH) Ca$_5$(PO4)$_3$	3.1–3.2	USA, Morocco, China, Russia,	USA, Morocco, China, Russia,	P$_2$O$_5$ Equiv.: 42×10^6; ore: 133×10^6	Fertilisers, phosphoric acid, chemicals
Platinum Pt 0.01	**Native Pt** **Sperrylite** **Braggite** **Cooperite**	Pt PtAs$_2$ PtS and (Pt,Pd,Ni)S PtS	21.45 10.6 10.0 9.5–10.2	S. Africa, Russia, Canada, USA	S. Africa, Russia, Canada, USA	155	See Platinum group metals (PGM)

Element	Minerals	Formula	Specific gravity	Producing countries	Producing countries	Production	Uses
Platinum group metals (PGM) 0.03	**Native and in alloys with Pt or PtS/PtAs$_2$, in Chromite (FeCr$_2$O$_4$)**	Pt, Pd, Rh, Ir, Os, Ru	see individual elements	S. Africa, Russia, Canada, USA	S. Africa, Russia, Canada, USA	365	Catalytic converters, petroleum refining catalysts, jewellery, ceramics, electronics
Potassium K 25 900	**Sylvite** **Carnalite** Nitre	KCl KCl · MgCl$_2$ · 6H$_2$O KNO$_3$	2 1.6 2.1	Canada, Russia, Germany, Belarus	Canada, Russia, Belarus, Germany	K$_2$O Equiv.: 25 × 10^6	Fertilisers, chemicals
Rhenium Re 0.001	Rheinite (in **Cu, Mo-ores**)	ReS$_2$	7.5	USA, Peru, Kazakhstan, Chile	USA, Chile, Germany, Russia	Mine output: 28.4	Catalysts, Mo-Ni-and W—Alloys, heat-sensors, electronics
Selenium Se 0.08	Clausthalite Berzelianite Naumannite (from **Cu-anode slimes**)	PbSe Cu$_2$Se Ag$_2$Se	7.8–8.08 6.71–7.23 7.0–8.0	Japan, Canada, Belgium Germany	Canada, Philippines, Korea, Australia,	1 460	Glass, metallurgy, electronics, agriculture
Silicon Si 277 200	**Quartz** Opal and numerous silicates	SiO$_2$ SiO$_2$ · nH$_2$O	2.65 2–2.25	China, Russia, Ukraine, USA, Brazil	China, Russia, Ukraine, USA, Brazil	Ferro-silicon: 4.3 × 10^6; Si: 7.2 × 10^6	Steels (as ferrosilicon additive) and computer chips as silicon
Silver Ag 0.085	**Native Ag** **Argentite** Pyrargyrite Hessite (also **gold recovery byproduct**)	Ag Ag$_2$S Ag$_3$SbS$_3$ Ag$_2$Te	10.5 7.19–7.36 5.7–5.9 8.4	Mexico, Peru, Australia, USA	Mexico, Peru, Australia, USA	18.3 × 10^3	Photography, batteries, electrical, electronics, catalysts
Sodium Na 28 300	**Halite** Mirabilite (Glauber's salt) **Trona** (also from **salines as NaCl**)	NaCl Na$_2$SO$_4$·10H$_2$O Na$_2$CO$_3$ · NaHCO$_3$ · 2H$_2$O	2.2 1.48 2.13	USA, China, Mexico, Germany, Russia, India	USA, China, Mexico, Germany, Russia, India	Na Equiv.: 100.23 × 10^6	Glass, soaps and detergents, textiles, household, pulp and paper. Use pattern: NaCl: 83.9% Na$_2$CO$_3$: 14.9% Na$_2$SO$_4$: 1.29%

(continued)

Table 7.1 MINERALS, SOURCES, AND USES—*continued*

Element: symbol and abundance (p.p.m.)	Minerals	Formulae	Density (g cm⁻³)	Major mineral sources	Major sources	1999/2000 World production (m. t.)	Primary uses and applications
1	2	3	4	5	6	7	8
Strontium Sr 375	**Celestine (or Celestite)** **Strontianite**	$SrSO_4$ $SrCO_3$	3.96 3.6–3.7	Mexico, Spain, Turkey, China, Russia	Mexico, Germany, China, France	Sr Equiv.: 139×10^3 China and Russia: N/A	TV tubes, magnets, fire-works
Sulphur S 260	**Native Sulphur** **Pyrite** **Chalcopyrite** **Galena** **Sphalerite (also from crude oil)**	S FeS_2 $CuFeS_2$ PbS ZnS	2.07 4.8–5.1 4.1–4.3 7.4–7.6 3.9–4.2	USA, Canada, China, Russia	USA, Russia, Saudi Arabia, Poland, UAE	57.2×10^6	Sulphuric acid, (phosphate fertilisers, oil refining, copper ore leaching)
Tantalum Ta 2.05	**Tantalite**	$(Fe, Mn)Ta_2O_6$	5.7–7.3	Australia, Brazil, Congo, Canada	Australia, Nigeria, Japan, China	836	Electronics, aerospace, carbides, heat- exchangers
Tellurium Te 0.06	**Tellurite** **Hessite** **Sylvanite** **Calaverite**	TeO_2 Ag_2Te $AuAgTe_2$ Au_2Te_4 (Mon.)	5.9 8.4 8–8.2 9.2–9.3	Japan, Peru, Canada	UK, Philippines, Belgium, Canada	Est. 140 (USA numbers N/A)	Steel additive, catalysts, chemicals, photoelectric devices
Thallium Tl 0.7	Carlinite Avicennite Fangite Lorandite (byproduct of **Pb, Zn, Cu ores**)	Tl_2S Tl_2O_3 Tl_3AsS_4 $TlAsS_2$	8.1 9.57 6.18 5.53	Canada, Belgium, Mexico	Belgium, Germany, Mexico, UK	Est.: 0.5	Electronics, radiation detection equipment, catalysts, medical
Thorium Th 10.2	**Monazite** Thorianite Thorite	$(Ce,La,Y,Th)PO_4$ ThO_2 $ThSiO_4$	5.27 9.7 5.3	Australia, India, Norway, Brazil,	France, Switzerland	6.2×10^3 (monazite concentrate)	Welding, catalysts, refractories, radar, microwave ovens
Tin Sn 2–40	**Cassiterite** Stannite	SnO_2 Cu_2SnFeS_4	6.8–71 4.4	China, Indonesia, Peru, Brazil, Bolivia, Australia	Peru, China, Bolivia, Brazil	238×10^3	Electrical, bronzes, solders, cans-containers, chemicals

Element	Mineral	Formula	SG	Occurrence	Major producers	Production	Uses
Titanium Ti 4 400	**Rutile** **Ilmenite** Sphene Leucoxene	TiO_2 $FeTiO_3$ $CaTiSiO_5$ Variant of Ilmenite	4.2 4.5–5.0	Australia, Brazil, India, China, Norway	Australia, Canada, Norway, S. Africa	Ti Equiv.: Est.: 1.493×10^6 as (ilmenite: 88.8%, rutile:11.2%)	Pigments (for paper, coatings, plastics), aerospace, armour, medical tools
Tungsten W 1.5–69	**Scheelite** **Wolframite** Ferberite Huebnerite	$CaWO_4$ $(Fe,Mn)WO_4$ $FeWO_4$ $MnWO_4$	5.9–6.1 7.1–7.9 7–7.5 7.0	Russia, China, Austria, Portugal	Russia, China, Austria, Portugal	37.4×10^3	Carbides, tools, armaments, steels, catalysts, high temperature applications
Uranium U 1.8–4	**Pitchblende** (or Uraninite) Torbernite **Carnotite**	$2UO_3 \cdot UO_2$ $Cu(UO_2)_2P_2O_8 \cdot 12H_2O$ $K_2O \cdot 2U_2O_3 \cdot$ $V_2O_5 \cdot 2H_2O$	6.4 3.5 4.7–5.0	Canada, Australia, Niger, Namibia	Canada, Australia, USA Niger, Namibia	Mine output: 31 172	Nuclear fuels, weapons, medicine
Vanadium V 145	Patronite Vanadinite Carnotite (also Rascoelite **U-ore processing byproduct)**	VS_4 $Pb_4Cl(VO_4)_3$ $K_2O \cdot 2U_2O_3 \cdot$ $V_2O_5 \cdot 2H_2O$ $3V_2O_5 \cdot 10SiO_2 \cdot 4H_2O$	2.82 6.9 4.7–5.0 2.97	S. Africa, China, Russia, Kazakhstan	China, Russia, S. Africa	43×10^3	Steels, catalysts, ceramics
Yttrium Y 33	**Monazite** **Xenotime**	$(Ce,La,Y,Th)PO_4$ YPO_4	5.27 4.46–5.1	China, India, Canada	China, France, UK, Japan	Mine output Y_2O_3 Equiv.: 2 400	TV monitors, fluorescence, lasers, jet engines, ceramics
Zinc Zn 50–132	**Sphalerite** **Smithsonite** Hemimorphite Willemite	ZnS $ZnCO_3$ $Zn_4Si_2O_7(OH)_2 \cdot H_2O$ $ZnSiO_4$	3.9–4.2 4–4.5 3.45 4.0–4.1	China, Australia, Peru, USA	China, Australia, Canada, Kazakhstan	Mine output: 8.73×10^6	Galvanising, alloys, brasses and bronzes
Zirconium Zr 165–220	**Zircon** **Baddeleyite**	$ZrSiO_4$ ZrO_2	4.68 5.4–6.02	Australia, S. Africa, USA,	France, Germany, Japan USA	760×10^3 (mineral concentrate) excludes USA	Nuclear reactors, heat exchangers,

8　Thermochemical data

Except where otherwise indicated, the data given in these tables have been selected from three main sources: 'Selected Values of the Thermodynamic Properties of the Elements', by R. Hultgren, P. O. Desai, D. T. Hawkins, M. Gleiser and K. K. Kelley, and by the same authors, 'Selected Values of the Thermodynamic Properties of Binary Alloys'; also 'Metallurgical Thermochemistry', 5th edn., by O. Kubaschewski and C. B. Alcock.* These works represent authoritative compilations of critically assessed data presently available, to which reference should be made for original sources or details of assessment.

See also the following: 'NIST-JANAF Thermochemical Tables', 4th edn., by M. W. Chase, Jr.

8.1　Symbols

θ_m = melting point in °C
θ_e = boiling or sublimation point in °C at 760 mmHg
θ_t = transition temperature in °C
θ_s = sublimation point in °C
T = absolute temperature in K
L_m = latent heat of fusion
L_e = latent heat of vaporisation $\left.\begin{array}{l}\\\\\\\\\end{array}\right\}$ in kJ mol^{-1} (or kJ g-atom^{-1}) or in Jg^{-1}
L_t = latent heat of transition
L_s = latent heat of sublimation
ΔV_m = volume change during melting $[(V_{liq} - V_{solid})/V_{solid}]\%$
ΔH_{298} = heat of formation at 298 K (25°C) in kJ mol^{-1} (or kJ g-atom^{-1}) or in Jg^{-1}.
　　　(The value of a heat evolved during a reaction is taken to be negative)
ΔG = maximum work (change of free energy) in kJ mol^{-1} (or kJ g-atom^{-1})
S_{298} = standard entropy at 298 K (25°C) in J K^{-1} mol^{-1} (or J g-atom^{-1})

p (mmHg) = vapour or dissociation pressure in mmHg
N_1 = mol fraction of the first component
N_2 = mol fraction of the second component
C_p = specific heat in J K^{-1} mol^{-1}
c_p = specific heat in J g^{-1} k^{-1}

---

* Now called 'Materials Thermochemistry', 6th edn., with P. J. Spencer added as an author.

8.2   Changes of phase (phase diagrams may be found in Chapter 11)

Table 8.1   ELEMENTS

Latent heats and temperatures of fusion, vaporisation and transition, and change in volume on melting.

Element	Melting point θ_m °C	Boiling point θ_e °C	θ_t °C	L_m at m.p. kJ g-atom^{-1} or mol^{-1}	L, kJ g-atom^{-1} or mol^{-1} L_s at 25°C	L_e at b.p.	L_t kJ g-atom^{-1} or mol^{-1}	ΔV_m %
Ag	960.8	2 200	—	11.09	284.2	257.8	—	(3.8)
Al	660.1	2 520	—	10.47	321.9	290.9	—	6.5
Am	—	2 600	—	—	—	238.6	—	—
As	817	603	—	As$_4$ = 4As + 118.1 kJ			—	10
Au	1 063	2 860	—	12.78	378.9	342.4	—	5.1
B	2 180	3 800	—	22.6	577.8	—	—	—
Ba	729	1 700	370	7.66	(192)	177.1	0.59	—
Be	1 287	2 470	1 254	12.22	324.4	292.6	2.55	—
Bi	271	1 564	—	10.89	207.2	179.2	—	−3.35
Br$_2$	−7.3	58	—	10.55	—	30.56	—	—
C (graph)	(3 800)	(5 000)	—	—	712	—	1.90 diam → graph	—
Ca	843	1 484	464	8.36	176.2	150.7	0.25	—
Cd	320.9	767	—	6.41	112.2	99.6	—	4.0
Ce	798	3 430	726	5.23	(407)	376.0	(2.9)	—
Cl$_2$	−101.0	−34.1	—	6.41	—	20.423	—	—
Co	1 495	2 930	{440, 1 120}	(15.5)	425	—	0.25, 0.92	3.5
Cr	1 857	2 672	—	(20.9)	397	342.1	—	—
Cs	29.8	700	—	2.09	78.7	66.6	—	2.6
Cu	1 083.4	2 560	—	13.02	341.2	304.8	—	4.2
Dy	1 409	2 560	1 384	—	—	—	—	—
Er	1 522	2 860	1 470	—	—	—	—	—
Eu	826	1 490	—	—	—	—	—	4.8
F$_2$	−219.6	−188.0	—	1.595	—	6.531	—	—
Fe	1 536	2 860	{914, 1 391}	15.2	398.6	340.4	{β5.11*, γ0.563*, $\gamma \to \delta$0.84}	3.5
Ga	29.7	2 420	—	5.594	285.0	270.5	—	−3.2
Gd	1 312	3 290	1 260	—	—	—	—	—
Ge	937	2 830	—	36.8	383.8	327.8	—	−5.1
H$_2$	−259.2	−252.5	—	0.117	—	0.909	—	(12.3)
Hf	2 227	4 600	1 940	24.07	611.3	571.1	(6.91)	—
Hg	−38.87	357	—	2.324	—	61.1	—	3.7
I$_2$	113.6	183	—	15.78	62.4	41.9	—	21.6
In	156.4	2 070	—	3.27	242.8	232.4	—	2.0
Ir	2 443	4 430	—	(26)	669.9	612.5	—	—
K	63.2	779	—	2.39	90.0	79.5	—	2.55
La	920	(3.420)	868	(8.37)	422.9	402.4	(2.9)	—
Li	181	1 324	—	2.93	161.6	147.8	—	(1.65)
Mg	649	1 090	—	8.79	146.5	127.7	—	4.12
Mn	1 244	2 060	{710, 1 090, 1 136}	(14.7)	291.0	231.1	{2.22, 2.22, 1.80}	(1.7)
Mo	2 620	4 610	—	35.6	664.5	590.3	—	—
N$_2$	−210.0	−195.8	−237.5	0.720	—	5.581	0.229 0	7.3
Na	97.8	883	—	2.64	108.9	98.0	—	2.5
Nb	2 467	4 740	—	29.3	722.2	683.7	—	—
Nd	1 016	3 070	862	7.14	323.6	—	2.98	—
Ni	1 453	2 910	358	17.16	429.6	374.3	0.58	4.5
Np	637	—	280, 577	—	—	—	—	—

* Only together with the tabulated heat capacities of iron: Table 8.10.   (*continued*)

Table 8.1 ELEMENTS—*continued*

Element	Melting point θ_m °C	Boiling point θ_e °C	θ_t °C	L_m at m.p. kJ g-atom^{-1} or mol^{-1}	L, kJ g-atom^{-1} or mol^{-1} L_s at 25°C	L_e at b.p.	L_t kJ g-atom^{-1} or mol^{-1}	ΔV_m %
O_2	−218.8	−183.0	$\begin{Bmatrix}-249.5\\-229.4\end{Bmatrix}$	0.445	—	6.8	$\begin{Bmatrix}0.093\,8\\0.743\,6\end{Bmatrix}$	7.4
Os	3 030	5 030	—	—	791	—	—	—
P (yellow)	44.1	280	—	2.64	$\begin{Bmatrix}140.7(P_2)\\58.8(P_4)\end{Bmatrix}$	$\begin{Bmatrix}51.9(P_4)\end{Bmatrix}$	17.6 yell. → red	3.5
Pb	327.4	1 750	—	4.81	196.4	178.8	—	3.5
Pd	1 552	2 940	—	(16.7)	377.2	361.7	—	—
Po	246	965	(100)	—	—	100.9	—	—
Pr	932	3 510	798	(11.3)	—	—	—	—
Pt	1 769	4 100	—	(19.7)	545.0	469.2	—	—
Pu	640	3 420	122, 205, 318, 452, 476	2.9	343.7	352.0	3.39, 0.59, 0.54, 0.08, 1.84	−2.5
Ra	700	1 500	—	—	—	—	—	—
Rb	38.8	688	—	2.198	87.5	75.8	—	2.5
Re	3 180	5 690	—	33.5	779.2	(712)	—	—
Rh	1 966	3 700	(1 030)	(22.6)	556.0	(494)	—	—
Ru	2 250	4 250	—	—	—	—	—	—
S (rhomb.)	112.8	444.5	95.5	1.235	—	*	0.38	(5.1)
S (mon.)	119.0	—	—	—	—	—	—	—
Sb	630.5	1 590	—	19.89	—	(167)(Sb$_2$)	—	0.8
Sc	1 538	(2 870)	1 334	—	376.0	—	—	—
Se (met.)	220.5	685	—	6.28	—	95.5(Se$_2$)	—	15.8
Si	1 412	3 270	—	50.66	450.1	384.8	—	(−10)
Sm	1 072	1 803	917	8.92	207.2	165.0	3.10	—
Sn	231.9	2 625	13	7.08	302.3	296.4	2.22	2.3
Sr	770	1 375	235, 540	(8.4)	177.1	154.5	—	—
Ta	3 015	5 370	—	(24.7)	782.5	—	—	—
Tb	1 360	3 220	1 290	—	—	—	—	—
Te	450	988	—	17.6	171.6(Te$_2$)	104.7(Te$_2$)	—	4.9
Th	1 750	4 790	1 325	—	576.1	(511)	—	—
Ti	1 667	3 285	882	(17.5)	469.3	425.8	3.34	—
Tl	304	1 473	234	4.3	180.9	166.2	0.38	2.2
U	1 132	4 400	662, 770	12.5	482.2	417.4	2.85, 4.878	—
V	1 902	3 410	—	16.74	510.2	457.2	—	—
W	3 400	5 555	—	35.2	847.8	(737)	—	—
Y	1 530	3 300	1 485	11.43	424.9	367.6	5.0	—
Yb	824	1 194	760	—	—	—	—	—
Zn	419.5	907	—	7.28	129.3	114.3	—	4.7
Zr	1 852	4 400	852	(19.3)	612.1	579.9	3.85	—

* L_e at b.p.: S_2, 106.4 (625°C); S_4, 96.0 (625°C); S_6, 66.2 (527°C); S_8, 63.1 (490°C).
ΔV_m, ref. 4.

Table 8.2a INTERMETALLIC COMPOUNDS

Latent heats and temperatures of fusion.

If an intermetallic phase is completely disordered, the entropy of fusion (L_m/T_m) can generally be calculated additively from the entropies of fusion of the components. If it is completely ordered, $-19.146\,(N_1 \log N_1 + N_2 \log N_2)$ is as a rule to be added to the calculated entropy of fusion.

Phase	N_2 10^{-2}	θ_m °C	L_m kJ g-atom^{-1}	L_m Jg^{-1}
δ-Ag-Cd	67.5	592	8.46 ± 0.42	76.2
γ-Ag-Zn	61.8	664	7.79 ± 0.33	95.5
δ-Ag-Zn	72.1	632	8.75 ± 0.42	131.5

(*continued*)

Table 8.2a   INTERMETALLIC COMPOUNDS—*continued*

Phase	N_2 10^{-2}	θ_m °C	L_m kJ g-atom^{-1}	L_m Jg^{-1}
Al$_2$Cu	33.3	590–605	12.6 ± 0.8	320.3
γ-Al-Mg	57.2	455	8.8 ± 0.8	346.7
AuCd	50.0	627	8.96 ± 0.50	57.8
AuPb$_2$	66.7	254–300	8.00 ± 0.75	39.4
AuSn	50.0	418	12.81 ± 0.33	81.2
β-Au-Zn	50.0	760	12.31 ± 0.54	93.8
ε-Au-Zn	88.9	490	7.45 ± 0.38	93.4
Bi-In	50.0	110	—	7.33
Bi-In$_2$	66.7	—	—	4.81
δ-Bi-Tl	40.0	214	7.24 ± 0.25	35.2
δ-Cd-Cu	38.4	555	9.71 ± 0.21	103.8
δ-Cd-Cu	40.0	562	9.55 ± 0.21	103.0
Cd–Mg	25.0	349–368	5.86	56.5
	50.0	415–430	6.05	87.9
	75.0	489–515	6.91	152.4
Cd$_2$Na	33.3	385	7.87 ± 0.63	95.5
CdSb	50.0	456	16.0 ± 0.33	136.9
GaLi	50.0	700–760	16.7 ± 0.8	435.4
GaSb	50.0	703	25.1 ± 1.7	131.0
Hg$_2$Na	33.3	350	8.8 ± 0.8	63
γ-Hg–Tl	28.6	14.5	2.01 ± 0.13	10.0
InLi	50.0	630	13.4 ± 0.8	219.8
InSb	50.0	425	24.7 ± 0.8	205.2
KNa$_2$	66.7	7	2.9 ± 0.1	103.4
Mg$_2$Pb	33.3	550	13.4 ± 1.3	155
MgZn$_2$	66.7	590	14.3 ± 0.8	260
Na$_5$Pb$_2$	28.5	400	7.18 ± 0.5	95
NaPb	50.0	368	8.4 ± 0.4	71
β-Na–Pb	71.5	320	7.1 ± 0.4	46
NaTl	50.0	250–305	8.4 ± 0.8	73.7
γ-Pb–Tl	87.5	329–339	5.23 ± 0.17	25.5
γ-Pb–Tl	63.0	379	5.65 ± 0.17	27.6

Ref. 3.

Table 8.2b   INTERMETALLIC COMPOUNDS

Latent heats and temperatures of transition.

The method of measurement is subject to error. Most of the reported values are probably too low.

Phase	N_2 10^{-2}	Transition	θ_t °C	L_t Jg-atom^{-1}
β-Ag–Cd	50.0	$\beta'-\beta$	211	712
AgZn	50.0	order–disorder	258	2 449
AuCu	50.0	order–disorder	408	1 779
AuCu$_3$	75.0	order–disorder	390	1 214
AuSb$_2$	66.7	$\beta-\gamma$	355	335
Cd$_3$Mg	27.0	order–disorder	95	963
CdMg	50.0	order–disorder	50–260	2 638
CdMg$_3$	75.3	order–disorder	80–165	1 256
CrFe	52.0	$\sigma-\alpha$	805	1 633
CuPt	50.0	order–disorder	800	3 810
Cu$_3$Pt	20.0	order–disorder	610	1 968
β-Cu–Zn	50.0	$\beta'-\beta$	470	2 219
FeNi$_3$	74.3	order–disorder	506	2 721
MnNi$_3$	75	order–disorder	(500)	3 119
Pd$_3$Sb	25	$\beta-\beta'$	950	10 300
Zn$_3$Sb$_2$	40.0	—	409–455	6.071

Ref. 3.

Table 8.3   OTHER METALLURGICALLY IMPORTANT COMPOUNDS

Latent heats and temperatures of fusion, vaporisation and transition.
(If not stated otherwise, the values of the latent heats are for the temperatures of transition, fusion, evaporation or sublimation, respectively. Boiling and sublimation points are for 1 atm pressure of the undissociated molecules).

Compound	$\theta_m, \theta_e, \theta_s$ or θ_t °C	$L_m, L_e, L_s,$ or L_t kJ mol^{-1}	Compound	$\theta_m, \theta_e, \theta_s$ or θ_t °C	$L_m, L_e, L_s,$ or L_t kJ mol^{-1}
	Carbides			*Halides—continued*	
CH$_4$	θ_m, −182.5	L_m, 0.938	CaCl$_2$	θ_m, 772	L_m, 28.5
	θ_e, −161.4	L_e, 8.323	CaBr$_2$	θ_m, 741	L_m, 28.9
Fe$_3$C	θ_t, 190	L_t, 0.75	CaI$_2$	θ_m, 779	L_m, (42)
	θ_m, 1 227	L_m, 51.62	CdF$_2$	θ_m, 1 072	L_m, 22.6
Mn$_3$C	θ_t, 1 037	L_t, 15.1		θ_e, 1 750	L_e, 225.2
			CdCl$_2$	θ_m, 568	L_m, 30.1
	Halides			θ_e, 961	L_e, 125.2
AgCl	θ_m, 455	L_m, 13.0	CdBr$_2$	θ_m, 565	L_m, 33.5
	θ_e, 1 564	L_e, 177.9		θ_e, 863	L_e, 113.0
AgBr	θ_m, 430	L_m, 9.2	CdI$_2$	θ_m, 390	L_m, 20.9
	θ_e, 1 560	L_e, 192.2		θ_e, 796	L_e, 106.3
AgI	θ_m, 557	L_m, 9.42	CeCl$_3$	θ_m, 817	L_m, 53.6
	θ_e, 1 506	L_e, 144.24	CeBr$_3$	θ_m, 732	L_m, 51.9
	Hex. → cub.		CeI$_3$	θ_m, 760	L_m, 51.1
	θ_t, 147	L_t, 6.07	CoCl$_2$	θ_m, 740	L_m, 59.0
AlF$_3$	θ_s, 1 280	L_s, 280.5		θ_e, 1 025	L_e, 157.4
Al$_2$Cl$_6$	θ_m, 193	L_m, 71.2	CrCl$_2$	θ_m, 815	L_m, 31.4
	θ_s, 160	L_s, 111.8		θ_e, 1 304	L_e, 198.9
Al$_2$Br$_6$	θ_m, 97	L_m, 22.6	CrCl$_3$	θ_s, 945	L_s, 237.8
	θ_e, 255	L_e, 45.6	CsF	θ_m, 703	L_m, 21.8
AlI$_3$	θ_m, 191	L_m, 16.3		θ_e, 1 210	L_e, 155.7
	θ_e, 385	L_e, 64.5	CsCl	θ_m, 646	L_m, 20.5
AsF$_3$	θ_m, −6	L_m, 10.38		θ_e, 1 300	L_e, 159.9
	θ_e, 58	L_e, 29.7		θ_t, 469	L_t, 2.5
AsF$_5$	θ_m, −80	L_m, 11.47	CsBr	θ_m, 636	L_m, 23.66
	θ_e, −53	L_e, 20.9		θ_e, 1 300	L_e, 150.7
AsCl$_3$	θ_m, −16	L_m, 10.13	CsI	θ_m, 621	L_m, 23.9
	θ_e, 130	L_e, 31.4		θ_e, 1 280	L_e, 150.3
AsBr$_3$	θ_m, 31	L_m, 11.7	CuCl	θ_m, 430	L_m, 7.5
	θ_e, 221	L_e, 47.7	CuBr	θ_m, 488	L_m, (9.6)
AsI$_3$	θ_m, 142	L_m, 9.2	CuI	θ_m, 588	L_m, (10.9)
	θ_e, 424	L_e, 59.5	FeCl$_2$	θ_m, 677	L_m, 43.1
BaF$_2$	θ_m, 1 290	L_m, 28.5		θ_e, 1 026	L_e, 126.4
	θ_e, 2 382	L_e, 270	FeCl$_3$	θ_m, 304	L_m, (43.1)
BaCl$_2$	θ_m, 962	L_m, 16.7		θ_e, 315	L_e, 60.7 (Fe$_2$Cl$_6$)
	θ_t, 922	L_t, 17.2	FeI$_2$	θ_t, 375	L_t, 0.83
BaBr$_2$	θ_m, 854	L_m, 31.4		θ_m, 590	L_m, 44.8
BaI$_2$	θ_m, 711	L_m, 26.4		θ_e, 935	L_e, 111.8
BeCl$_2$	θ_m, 415	L_m, 8.7	GaCl$_3$	θ_m, 78	L_m, 11.5
	θ_t, 403	L_t, 16.8		θ_e, 302	L_e, 62.8
	θ_e, 532	L_e, 104.7	Ga$_2$Cl$_6$	θ_e, 201	L_e, 44.0
BeBr$_2$	θ_m, 488	L_m, 18.9	GaBr$_3$	θ_m, 122	L_m, 11.7
	θ_e, 511	L_e, 100.0		θ_e, 314	L_e, 58.6
	θ_s, 473	L_s, 125.6	Ga$_2$Br$_6$	θ_e, 292	L_e, 50.2
BeI$_2$	θ_m, 480	L_m, 20.9	GaI$_3$	θ_m, 212	L_m, 16.3
	θ_e, 482	L_e, 96.3		θ_e, 349	L_e, 67.8
	θ_s, 488	L_s, (79.5)	Ga$_2$I$_6$	θ_e, 462	L_e, 82.5
BiCl$_3$	θ_m, 230	L_m, 23.9	GeCl$_4$	θ_e, 84	L_e, 29.4
	θ_e, 441	L_e, 72.4	GeBr$_4$	θ_e, 189	L_e, 36.0
BiBr$_3$	θ_m, 218	L_m, 21.8	HF	θ_m, −83	L_m, 3.94
	θ_e, 461	L_e, 75.4	HCl	θ_m, −114.2	L_m, 1.99
CF$_4$	θ_m, −183.6	L_m, 0.701		θ_e, −85.1	L_e, 16.161
	θ_e, −151	L_e, 13.063	HBr	θ_m, −86.9	L_m, 2.407
CBr$_4$	θ_m, 90	L_m, 3.957		θ_e, −67	L_e, 17.626
	θ_t, 47	L_t, 5.95	HI	θ_m, −50.8	L_m, 2.872
	θ_e, 190	L_e, 44.4		θ_e, −35.4	L_e, 19.778
CCl$_4$	θ_m, −23	L_m, 2.51	HfCl$_4$	θ_s, 316	L_s, 99.65
	θ_e, 77	L_e, 30.6	HgCl$_2$	θ_m, 278	L_m, 19.47
CaF$_2$	θ_m, 1 418	L_m, 29.7		θ_e, 304	L_e, 59.0
	θ_e, 2 510	L_e, 312.2	HgBr$_2$	θ_m, 238	L_m, 18.0
	θ_t, 1 151	L_t, 4.77		θ_e, 319	L_e, 59.0

(continued)

Table 8.3   OTHER METALLURGICALLY IMPORTANT COMPOUNDS—*continued*

Compound	$\theta_m, \theta_e, \theta_s$ or θ_t °C	$L_m, L_e, L_s,$ or L_t kJ mol^{-1}	Compound	$\theta_m, \theta_e, \theta_s$ or θ_t °C	$L_m, L_e, L_s,$ or L_t kJ mol^{-1}
Halides—continued			*Halides—continued*		
HgI$_2$ (yellow)	θ_m, 250	L_m, 18.8	OsF$_6$	θ_m, 33	L_m, 7.5
	yellow → red			θ_e, 47.5	L_e, 28.1
	θ_t, 127	L_t, 2.72	PCl$_3$	θ_m, −92	L_m, 4.5
HgI$_2$ (red)	θ_m, 256	L_m, 19.3		θ_e, 74	L_e, 30.6
	θ_e, 354	L_m, 59.9	PCl$_5$	θ_s, 163	L_s, 64.9
InCl	θ_m, 225	L_m, 9.2	PBr$_3$	θ_e, 174	L_e, 39.8
	θ_e, 608	L_e, 88.4	PbF$_2$	θ_m, 818	L_m, 17.4
InCl$_3$	θ_s, 498	L_s, 158.3		θ_e, 1 293	L_e, 160.4
InBr	θ_m, 275	L_m, 24.3	PbCl$_2$	θ_m, 498	L_m, 24.3
	θ_e, 660	L_e, 95.0		θ_e, 954	L_e, 123.9
InBr$_2$	θ_e, 630	L_e, 82.5	PbBr$_2$	θ_m, 370	L_m, 18.0
InBr$_3$	θ_s, 371	L_s, 108.4		θ_e, 914	L_e, 116.0
InI	θ_m, 365	L_m, 22.4	PbI$_2$	θ_m, 410	L_m, 16.2
	θ_e, 770	L_e, 96.7		θ_e, 872	L_e, 103.8
IrF$_6$	θ_m, 44	L_m, 5.0	PdCl$_2$	θ_m, 678	L_m, 18.4
	θ_e, 53	L_e, 27.2	PrCl$_3$	θ_m, 786	L_m, 50.7
KF	θ_m, 857	L_m, 28.26	PrBr$_3$	θ_m, 693	L_m, 47.3
	θ_e, 1 510	L_e, 186.7	PrI$_3$	θ_m, 735	L_m, 53.2
KCl	θ_m, 772	L_m, 26.6	PuF$_3$	θ_m, 1 426	L_m, (59.9)
	θ_e, 1 407	L_e, 162.4		θ_e, (2 120)	L_m, (321)
KBr	θ_m, 740	L_m, 25.6	PuF$_6$	θ_m, 52	L_m, 17.6
	θ_e, 1 383	L_e, 155.3		θ_e, 62	L_e, 30.1
KI	θ_m, 685	L_m, 24.07	PuCl$_3$	θ_m, 760	L_m, 63.6
	θ_e, 1 330	L_e, 145.3		θ_e, 1 790	L_e, 186
LiF	θ_m, 848	L_m, 26.8	PuBr$_3$	θ_m, 681	L_m, 51.5
	θ_e, 1 681	L_e, 214		θ_e, 1 460	L_e, 193
LiCl	θ_m, 610	L_m, 19.89	RbF	θ_m, 775	L_m, 23.0
	θ_e, 1 382	L_e, 150.7		θ_e, 1 390	L_e, 177.9
LiBr	θ_m, 550	L_m, 17.6	RbCl	θ_m, 717	L_m, (21)
	θ_e, 1 310	L_e, 148.2		θ_e, 1 381	L_e, 165.8
LiI	θ_m, 469	L_m, 14.7	RbBr	θ_m, 682	L_m, (18.4)
	θ_e, 1 171	L_e, 170.8		θ_e, 1 352	L_e, 154.9
MgF$_2$	θ_m, 1 263	L_m, 58.2	RbI	θ_m, 641	—
	θ_e, 2 332	L_e, 292.2		θ_e, 1 304	L_e, 150.7
MgCl$_2$	θ_m, 714	L_m, 43.1	ReF$_5$	θ_e, 221	L_e, 58.2
	θ_e, 1 418	L_e, 136.9	ReF$_6$	θ_m, 19	L_m, 4.2
MnCl$_2$	θ_m, 650	L_m, 37.7		θ_e, 34	L_e, 28.5
	θ_e, 1 190	L_e, 123.9	ReF$_7$	θ_e, 48	L_m, 7.5
MoF$_5$	θ_e, 214	L_e, 51.9		θ_e, 74	L_e, 36.0
MoF$_6$	θ_m, 17	L_m, 4.2	SF$_6$	θ_m, −51	L_m, 5.9
	θ_e, 34	L_e, 28.1		θ_s, −64	—
MoCl$_5$	θ_m, 194	L_m, 34	S$_2$Cl$_2$	θ_m, −80	—
	θ_e, 268	L_e, 68.8		θ_e, 138	L_e, 36.4
NaF	θ_m, 992	L_m, 32.7	SbCl$_3$	θ_m, 73	L_m, 12.6
	θ_e, 1 710	L_e, 216.9		θ_e, 219	L_e, 43.5
NaCl	θ_m, 801	L_m, 28.1	SbCl$_5$	θ_m, 2	L_m, (10.5)
	θ_e, 1 465	L_e, 170.4	SbBr$_3$	θ_m, 97	L_m, 14.7
NaBr	θ_m, 750	L_m, 26.2		θ_e, 280	L_e, (59.0)
	θ_e, 1 392	L_e, 38.0	SbI$_3$	θ_m, 170	L_m, 17.6
NaI	θ_m, 660	L_m, 23.7		θ_e, 401	L_e, 68.7
	θ_e, 1 304	L_e, 159.5	ScCl$_3$	θ_m, 966	L_m, 67.4
NbF$_5$	θ_m, 77	L_m, 12.1		θ_e, 967	L_s, 272.1
	θ_e, 233	L_e, 52.3	SeF$_6$	θ_m, −34	L_m, (8.4)
NbCl$_5$	θ_m, 205	L_m, 28.9		θ_s, −46	L_s, 26.8
	θ_e, 250	L_e, 54.8	SiF$_4$	θ_s, −95	L_s, 25.8
NdCl$_3$	θ_m, 760	L_m, 50.2	SiCl$_4$	θ_m, −70	L_m, 7.75
NdBr$_3$	θ_m, 684	L_m, 45.2		θ_e, 57	L_e, 28.9
NH$_4$Cl	θ_t, 184	L_t, 4.3	SiBr$_4$	θ_e, 153	L_e, 38.1
NH$_4$Br	θ_t, 138	L_t, 3.2	SiI$_4$	θ_e, 301	L_e, 50.2
NiCl$_2$	θ_m, 1 030	L_m, 77.5	SnCl$_2$	θ_m, 247	L_m, 12.77
NiCl$_2$	θ_s, 970	L_s, 225.3		θ_e, 652	L_e, 82.9
NiBr$_2$	θ_s, 919	L_s, 224.8	SnCl$_4$	θ_m, −33	L_m, 9.2
NpF$_6$	θ_m, 54	L_m, 17.6		θ_e, 115	L_e, 33.9
	θ_e, 55	L_e, 30.1	SnBr$_2$	θ_m, 232	L_e, 98.4
OsF$_5$	θ_e, 226	L_e, 65.7		θ_e, 639	L_m, (7.5)

(continued)

Table 8.3 OTHER METALLURGICALLY IMPORTANT COMPOUNDS—*continued*

Compound	$\theta_m, \theta_e, \theta_s$ or θ_t °C	$L_m, L_e, L_s,$ or L_t kJ mol^{-1}	Compound	$\theta_m, \theta_e, \theta_s$ or θ_t °C	$L_m, L_e, L_s,$ or L_t kJ mol^{-1}
Halides—continued					
SnBr$_4$	θ_m, 30	L_m, 12.6	ZnBr$_2$	θ_m, 402	L_m, 15.7
	θ_e, 205	L_e, 41.0		θ_e, 650	L_e, 98.4
SnI$_2$	θ_m, 320	—	ZnI$_2$	θ_m, 446	L_m, (18.8)
	θ_m, 714	L_e, 99.6		θ_e, 727	L_e, 96
SnI$_4$	θ_e, 145	L_m, 19.3	ZrF$_4$	θ_s, 908	L_s, 232.4
	θ_e, 348	L_e, 50.2	ZrCl$_4$	θ_s, 334	L_s, 103.8
SrF$_2$	θ_m, 1 477	L_m, 18.4	ZrBr$_4$	θ_s, 356	L_s, 108.0
	θ_e, 2 480	L_e, 297.2	ZrI$_4$	θ_s, 431	L_s, 121.4
SrCl$_2$	θ_m, 874	L_m, 15.9			
SrCl$_2$	θ_t, 730	L_t, 5.0	*Nitrides*		
SrBr$_2$	θ_m, 657	L_m, 10.5	Mg$_3$N$_2$	θ_t, 550	L_t, 0.46
	θ_t, 645	L_t, 12.1		θ_t, 788	L_t, 0.92
SrI$_2$	θ_m, 538	L_m, 19.7			
TaCl$_5$	θ_m, 217	L_m, 36.8	*Oxides*		
	θ_e, 234	L_e, 50.2	As$_4$O$_6$	θ_t, −33	L_t, 5.7
TaBr$_5$	θ_m, 269	L_m, 45.6		θ_m, 309	L_m, 36.8
	θ_e, 347	L_e, 62.4		θ_e, 459	L_e, 59.9
TaI$_5$	θ_m, 497	—	BeO	θ_m, 2 580	80.8
	θ_e, 543	L_e, 75.8		θ_e, 4 120	L_e, 471.0
TeF$_4$	θ_m, 130	L_m, 26.8	CdO	θ_s, 1 559	L_s, 251
TeCl$_2$	θ_e, 322	L_e, 64.1	Fe$_{0.95}$O	θ_m, 1 378	L_m, 31.0
TeCl$_4$	θ_m, 224	L_m, 18.8	Fe$_3$O$_4$	θ_m, 1 597	L_m, 138.2
	θ_e, 392	L_e, 70.3	H$_2$O	θ_m, 0	L_m, 6.016
ThCl$_4$	θ_m, 770	—		θ_e, 100	L_e, 41.11
	θ_e, 921	L_e, 152.8	MgO	θ_m, 2 825	L_m, 77.0
ThBr$_4$	θ_m, 679	L_m, 54.4	MnO	θ_m, 1 875	L_m, 54
	θ_e, 857	L_e, 144.4	Mn$_3$O$_4$	θ_t, 1 172	L_t, 20.3
ThI$_4$	θ_m, 566	L_m, 48.1	MoO$_3$	θ_m, 795	L_m, 48.4
	θ_e, 837	L_e, 131.9		θ_e, 1 100	L_e, 192.6
TiF$_4$	θ_s, 283	L_s, 90.0	NbO$_2$	θ_t, 817	L_e, 2.9
TiCl$_4$	θ_m, −25	L_m, 9.37	OsO$_4$ (white)	θ_m, 42	L_m, 9.6
	θ_e, 136	L_e, 36.22		θ_e, 130	L_e, 39.57
TiBr$_4$	θ_m, 39	L_m, 13.0	P$_4$O$_6$	θ_e, 175	L_e, 43.5
	θ_e, 233	L_e, 44.4	PbO (yellow)	θ_m, 886	L_m, 29.3
TiI$_2$	θ_s, 1 170	L_s, 222.7		θ_e, 1 470	L_e, (214.8)
TiI$_4$	θ_m, 154	L_m, 19.8	Re$_2$O$_7$	θ_m, 298	L_m, 64.1
	θ_e, 377	L_e, 56.1		θ_e, 363	L_e, 75.4
TlF	θ_e, 700	L_e, 116.0	SO$_2$	θ_m, −75.5	L_m, 7.5
TlCl	θ_m, 430	L_m, 15.9		θ_e, −10	L_e, 24.95
	θ_e, 816	L_e, 103.6	SO$_3$	θ_m, 17 (α)	L_m, 2.1
TlBr	θ_m, 460	L_m, 16.3		θ_s, 52 (γ)	L_s, 66.6
	θ_e, 825	L_e, 103.4	Sb$_4$O$_6$	θ_t, 570	L_t, (14.2)
TlI	θ_e, 845	L_e, 103.8		θ_m, 656	L_m, 109.9
	θ_m, 440	L_m, 14.8		θ_e, 1 425	L_s, 214.4 (m.p.)
UF$_4$	θ_m, 1 036	L_m, 42.7	SeO$_2$	θ_s, 316	L_s, (102.6)
	θ_e, 1 457	L_e, 221.9	SiO$_2$	θ_t, 250	L_t, 1.3
UF$_6$	θ_m, 64	L_m, 19.3	(cristobalite)	θ_m, 1 713	L_m, 10.9
	θ_s, 57	L_s, 48.1	SnO$_2$	θ_t, 410	L_t, 1.88
UCl$_4$	θ_m, 590	L_m, 44.8		θ_t, 540	L_t, 1.26
	θ_e, 789	L_e, 141.5	SrO	θ_m, 2 460	L_s, 523 (25°C)
UBr$_4$	θ_m, 519	L_m, 55.3	Tc$_2$O$_7$	θ_m, 119	L_m, 48.1
	θ_e, 777	L_e, 113.5		θ_e, 312	L_e, 58.6
VF$_5$	θ_e, 48.3	L_e, 46.5	TeO$_2$	θ_m, 733	L_s, 237.8 (m.p.)
VCl$_4$	θ_e, 160	L_e, 33.1	TiO	θ_t, 991	L_t, 3.43
WF$_6$	θ_m, 0	L_m, 1.76	Ti$_3$O$_5$	θ_t, 177	L_t, 9.38
	θ_e, 17	L_e, 26.6	VO$_2$	θ_m, 1 550	L_m, 56.9
WCl$_5$	θ_e, 298	L_e, 52.8		θ_t, 67	L_t, 4.29
	θ_m, 240	L_m, (17.6)	V$_2$O$_5$	θ_m, 670	L_m, 65.3
WCl$_6$	θ_m, 282	L_m, 6.7	WO$_3$	θ_m, 1 473	L_s, 456 (m.p.)
	θ_e, 337	L_e, 58.2	ZrO$_2$	θ_t, 1 175	L_t, 5.9
ZnF$_2$	θ_m, 875	L_m, 41.8			
	θ_e, 1 500	L_e, 184.2	*Sulphides*		
ZnCl$_2$	θ_m, 318	L_m, 10.5	Ag$_2$S	θ_t, 176; 586	L_t, 5.9; —
	θ_e, 732	L_e, 119.3		θ_m, 830	L_m, 11.3

(continued)

Table 8.3 OTHER METALLURGICALLY IMPORTANT COMPOUNDS—*continued*

Compound	$\theta_m, \theta_e, \theta_s$ or θ_t °C	$L_m, L_e, L_s,$ or L_t kJ mol^{-1}	Compound	$\theta_m, \theta_e, \theta_s$ or θ_t °C	$L_m, L_e, L_s,$ or L_t kJ mol^{-1}
	Sulphides—continued			*Sulphides*—continued	
CS_2	θ_e, 45.2	L_e, 27.2	HgS	θ_t, 345	L_t, 4.2
Cu_2S	θ_t, 103	L_t, 3.85	H_2S	θ_m, −85.3	L_m, 2.43
	θ_t, 350	L_t, 0.8		θ_e, −60.2	L_e, 18.686
	θ_m, 1 130	L_m, 10.9	MnS	θ_m, 1 530	L_m, 26.8
FeS	θ_t, 138	L_t, 2.39			
	θ_t, 325	L_t, 0.50	Na_2S	θ_m, 978	L_m, (6.91)
	θ_m, 1 195	L_m, 32.36	Sb_4S_6	θ_m, 550	L_m, 126.3 (25°C)
GeS	θ_m, 615	L_m, 21.4	—		L_s, 214.4 (m.p.)
	θ_s, 760	L_s, 145.3	ZnS	θ_t, 1 020	L_t, (13.4)

8.3 Heat, entropy and free energy of formation

Table 8.4 ELEMENTS

Standard entropies.

Element	S_{298} J K^{-1}	Element	S_{298} J K^{-1}	Element	S_{298} J K^{-1}
Ag	42.7	Hf	44.0	Rh	31.8
Al	28.34	Hg	76.20	Ru	28.5
As	35.2	I	58.6	S (rhomb.)	31.90
Au	47.39	In	58.2		
B (cryst.)	5.9			S (monocl.)	32.57
		Ir	35.6	S_2 (gas)	227.8
Ba	(67.8)	K	63.6	S_4 (gas)	306.1
Be	9.50	La	56.9	S_6 (gas)	376.0
Bi	56.9	Li	29.1	S_8 (gas)	471.4
Br_2 (liq.)	152.4	Mg	32.5		
C (graph.)	5.69			Sb	45.6
		Mn	31.82	Sc	34.3
Ca	41.66	Mo	28.6	Se (met.)	42.3
Cd	51.5	N_2	191.63	Si	18.8
Ce	64.1	Na	51.29	Sm	69.5
Cl_2	223.24	Nb	36.55		
Co	30.06			Sn	51.5
		Nd	73.3	Ta	41.4
Cr	23.9	Ni	29.81	Te	49.61
Cs	82.9	O_2	205.24	Th	53.42
Cu	33.37	Os	32.7	Ti	30.31
Dy	74.9	P (yellow)	44.17		
Er	73.3			Tl	64.27
		Pb	64.9	U	50.2
F_2	203.1	Pd	37.89	V	29.3
Fe	27.2	Pr	73.3	W	33.5
Ga	41.0	Pt	41.9	Y	44.4
Gd	66.2	Pu	51.5		
Ge	31.19			Zn	41.7
		Rb	76.6	Zr	38.9
H_2	130.75	Re	37.3		

Table 8.5a   INTERMETALLIC COMPOUNDS

Heats of formation in kJ and standard entropies.

Phase or compound	$-\Delta H$		S_{298}
	kJ g-atom^{-1}	kJ mol^{-1}	J K^{-1} mol^{-1}
Ag–Au	*See* Table 8.5c		—
Ag–Cd	*See* Table 8.5c		—
Ag–Zn	*See* Table 8.5c		—
Al_2Au (α)	42.1	126.4	—
AlAu (β)	38.7	77.4	—
$AlAu_2$ (γ)	34.9	104.7	—
Al_4Ca	—	218.5	138.1
Al_2Ca	—	216.8	85.4
Al_4Co (γ)	32.2	—	—
Al_5Co_2 (ε)	42.0	—	—
AlCo (ξ)	55.2	—	—
Al_7Cr (θ)	13.4	—	—
$Al_{11}Cr_2$ (η)	15.1	—	—
Al_4Cr (ε)	17.2	—	—
Al_9Cr_4 (γ_4)	15.9	—	—
Al_8Cr_5 (γ_2)	15.1	—	—
$AlCr_2$ (β)	10.9	—	—
Al_2Cu (θ)	13.4	—	—
AlCu (η_2)	20.0	—	—
$AlCu_2$ (γ)	23.0	—	—
Al-Fe	*See* Table 8.5c		—
Al_4La (β')	35.2	—	—
Al_2La	—	150.7	98.8
AlAs	—	122.6	60.3
AlSb	—	50.2	65.0
Al_3Ni (ε)	37.7	150.7	110.7
Al_3Ni_2 (δ)	56.5	282.6	136.5
AlNi (β)	59.2	118.5	54.1
$AlNi_3$ (α')	37.7	150.8	113.8
Al_4Pu	36.2	181.0	—
Al_3Pu	45.2	180.8	—
Al_2Pu	47.3	141.9	—
Al_3Ti (γ)	35.6	142.3	94.6
AlTi (δ)	37.0	—	52.3
$AlTi_3$ (ε)	27.6	—	—
Al_4U	26.0	129.7	163.3
Al_3U	28.6	114.3	136.0
Al_2U	32.9	98.8	106.7
Al_3V (ξ)	27.6	—	—
AsIn	—	57.8	74.7
An–Cd	*See* Table 8.5c		—
Au–Cu	*See* Table 8.5c		—
Au–Ni	*See* Table 8.5c		—
Au–Pb	*See* Table 8.5c		—
$AuSb_2$	—	19.5	119.3
AuSn	*See* Table 8.5c		—
Au_3Zn (α)	(18.0)	—	—
AuZn (β)	(26.0)	—	—
Ba_3Bl_2	—	670	—
$BaMg_2$ (β)	2.1	6.3	—
Ba_2Pb	97.6	293.0	—
BaPb	75.0	150.0	—
$BaPb_3$	44.0	176.0	—
Ba_3Sb_2	146.5	733	—
Ba_2Sn	—	377.0	126.8
$BaSn_3$	—	194.6	188.0
Bi_2Ca_3	—	528.0	117.9

(continued)

Table 8.5a   INTERMETALLIC COMPOUNDS—*continued*

Phase or compound	$-\Delta H$		S_{298}
	kJ g-atom^{-1}	kJ mol^{-1}	J K^{-1} mol^{-1}
BiK$_3$	—	226.5	198.0
BiMn	9.8	19.7	—
BiNa	32.7	65.3	109.6
BiNa$_3$	47.6	190.5	160.0
BiNi	3.9	7.8	88.3
Bi–Tl	*See* Table 8.5c		
CaCd$_3$	31.4	126.0	—
CaMg$_2$	—	39.3	—
Ca$_2$Pb	—	215.6	105.5
CaPb	—	119.7	80.8
Ca$_3$Sb$_2$	—	728.4	157.4
Ca$_2$Su	—	314	100.5
CaSn	—	159.0	70.7
CaSn$_3$	—	180.0	—
CaTl	81.6	163.2	—
CaZn (δ)	36.6	73.2	66.6
CaZn$_2$	31.4	94.2	101.7
CaZn$_5$ (ξ)	23.0	138.1	—
Cd–Hg	*See* Table 8.5c		—
Cd–Mg	*See* Table 8.5c		—
Cd–Sb	*See* Table 8.5c		—
Cd$_3$As$_2$	—	38.1	207.2
Cd$_6$Na (β)	7.5	—	—
Cd$_2$Na (γ)	11.7	—	—
CeMg (β)	8.0	—	—
Co$_5$As$_2$	—	111.3	223.1
Co$_2$As	18.8	56.5	96.3
Co$_3$As$_2$	—	113.8	160.3
CoAs	—	56.9	64.5
Co$_2$As$_3$	—	144.0	164.1
CoAs$_2$	—	92.1	92.9
CoSb (γ)	20.9	41.8	70.7
CoSb$_x$ ($x = 0.74 \rightarrow 0.96$)			
CoSn	14.7	29.3	71.6
Cr–Ni	*See* Table 8.5c		—
Cr$_2$Ta	9.0	27.0	88.1
Cu$_7$In$_3$ (δ)	8.5	—	—
Cu$_2$In (η)	8.0	—	—
Cu–Mg	*See* Table 8.5c		—
Cu–Pt	*See* Table 8.5c		—
Cu$_{11}$Sb$_2$ (η)	0.26	—	—
Cu$_9$Sb$_2$ (ϵ)	0.54	—	—
Cu$_2$Sb (γ)	4.2	—	—
Cu$_{31}$Sn$_8$ (δ)	5.36	—	—
Cu$_3$Sn (ε)	7.5	—	—
Fe–Ni	*See* Table 8.5c		—
Fe$_2$Pu (ζ)	9.1	27.3	—
FeSb$_2$ (ζ)	(9.6)	—	—
FeTi (ε)	20.3	40.6	—
Fe$_2$U (ε)	10.9	32.2	—
GaAs	—	81.6	64.2
GaSb	—	41.9	77.4
GeU	*See* Table 8.5c		—
Hg$_8$K	11.6	104.6	—
Hg$_4$K	18.0	90.0	—
Hg$_3$K	20.9	83.6	—
Hg$_2$K	38.7	77.4	—

(*continued*)

Table 8.5a   INTERMETALLIC COMPOUNDS—*continued*

Phase or compound	$-\Delta H$		S_{298}
	kJ g-atom^{-1}	kJ mol^{-1}	J K^{-1} mol^{-1}
HgK	28.1	56.1	—
Hg$_3$Li	28.2	113.0	—
Hg$_2$Li	34.8	104.4	—
HgLi	44.0	88.0	—
Hg$_4$Na	16.7	83.5	—
HgNa	21.4	42.7	—
Hg$_2$Na$_3$	18.8	94.4	—
Hg$_2$Na$_5$	13.4	93.8	—
HgNa$_3$	11.7	46.9	—
Hg$_5$Tl$_2$	*See* Table 8.5c		—
InSb	—	31.1	87.7
KNa$_2$	*See* Table 8.5c		—
K$_3$Bi	56.6	226.4	—
K$_3$Sb	47.1	188.4	—
KTl	28.3	56.5	—
LaMg(ζ)	9.0	18.0	—
Li$_4$Pb(γ)	35.2	—	—
LiPb(ζ)	30.6	—	—
Li$_3$Bi	—	232.3	—
Li$_3$Sb	—	180.0	—
LiTl	—	153.6	—
βLi–Sn	39.3	—	—
Li$_5$Sn$_2$(δ)	40.2	—	—
LiSn(ζ)	35.1	—	—
Mg$_2$Ge	—	115.2	73.0
Mg$_2$Sn	—	80.6	101.5
Mg$_2$Pb	—	48.1	119.3
Mg$_3$Sb$_2$	—	300.1	136.7
Mg$_3$Bi$_2$	25.3	126.8	191.9
Mg$_2$Ni	17.3	51.9	95.0
MgNi$_2$	18.8	56.5	88.7
MgZn(γ)	10.5	—	—
MgZn$_2$(ε)	10.9	—	—
Mg$_2$Zn$_{11}$(ζ)	10.0	—	—
MnAs	—	57.3	—
Mn$_2$Sb	—	32.6	136.9
Na$_3$As	—	217.7	—
Na$_3$Sb	—	197.6	176.2
NaSb	—	66.2	99.8
Na$_3$Bi	—	190.5	(160)
NaBi	—	65.3	(110)
Na$_4$Sn	11.7	58.6	—
Na$_2$Sn	20.1	60.3	—
NaSn	25.1	50.2	—
Na–Pb	*See* Table 8.5c		—
NaTl	*See* Table 8.5c		—
NbCr$_2$	—	20.9	83.7
NbFe$_2$	—	61.5	75.3
NbNi(γ)	22.6	—	—
NbNi$_3$(δ)	31.8	—	—
Ni$_2$Ge	—	110.1	90.8
NiAs	—	72.0	51.9
Ni$_3$Sb(δ)	18.8	—	—
Ni$_5$Sb$_2$(β')	21.8	—	—
NiSb(γ)	33.1	66.2	78.3
NiSb$_2$(ε)	24.7	—	—
Ni$_3$Sn(β')	25.2	103.0	131.4

(*continued*)

Table 8.5a   INTERMETALLIC COMPOUNDS—*continued*

	$-\Delta H$		S_{298}
Phase or compound	kJ g-atom^{-1}	kJ mol^{-1}	J K^{-1} mol^{-1}
$Ni_3Sn_2(\gamma)$	38.5	192.3	173.7
$Ni_3Sn_4(\delta)$	33.7	235.7	257.9
Ni_3Ti	35.1	140.3	104.6
NiTi	33.3	66.6	53.2
$NiTi_2$	27.9	83.7	83.6
Pb–Tl	*See* Table 8.5c		—
Pb–U	*See* Table 8.5c		—
Sb–Zn	*See* Table 8.5c		—
$TaFe_2$	—	57.8	106.7
$ThRe_2$	—	174.1	123.7
$ThMg_2$	—	31.4	92.5
UCd_{11}	—	45.6	583.5
UFe_2	—	32.2	104.7
URh_3	—	259.5	148.2
URu_3	—	217.6	108.4

Table 8.5b   SELENIDES AND TELLURIDES

Heats of formation in kJ and standard entropies.

Compound	$-\Delta H$ kJ mol^{-1}	S_{298} J K^{-1}	*Accuracy* kJ
Ag_2Se	43.5	150.3	1
Ag_2Te	36.0	153.6	0.5
Al_2Se_3	540.0	157.0	15
Al_2Te_3	319.0	188.4	4
As_2Se_3	102.6	194.6	18
As_2Te_3	37.7	226.5	7
$AuTe_2$	18.6	141.8	3
BaSe	393.5	89.6	40
BaTe	269.6	99.6	30
Bi_2Se_3	140.2	240.0	4
Bi_2Te	78.3	261.2	4
CaSe	368.4	69.1	25
CaTe	272.1	80.8	33
CdSe	144.8	83.3	14
CdTe	101.8	93.1	1
Cu_2Se	65.3	129.8	7
CuSe	41.9	78.3	5
Cu_2Te	41.9	134.8	11
$FeSe_{0.96}$	67.0	69.2	7
$FeSe_{1.14}$	66.1	87.7	4
Fe_3Se_4	212.2	280.0	3
$FeTe_{0.9}$	23.0	80.2	4
$FeTe_2$	72.0	100.3	5
GaSe	161.2	70.3	11
Ga_2Se_3	406.0	192.1	17
GaTe	123.5	85.4	11
Ga_2Te_3	272.1	222.7	13
GeSe	69.1	87.5	13
$GeSe_2$	113.0	112.6	21
GeTe	32.7	89.2	13
$H_2Se_{(g)}$	−29.3	218.9	1.3
$H_2Te_{(g)}$	−99.6	229.0	0.9

(continued)

Table 8.5b SELENIDES AND TELLURIDES—*continued*

Compound	$-\Delta H$ kJ mol^{-1}	S_{298} J K^{-1}	Accuracy kJ
HgSe	43.3	100.9	15
HgTe	31.8	113.0	4.5
InSe	118.0	81.6	13
In$_2$Se$_2$	326.5	201.3	17
In$_2$Te	79.5	157.0	2
InTe	72.0	105.7	2
In$_2$Te$_3$	191.7	238.6	3
In$_2$Te$_5$	191.7	—	2.5
K$_2$Se	372.6	—	2.5
La$_2$Se$_3$	933.5	202.3	21
La$_2$Te$_3$	784.9	223.4	25
Li$_2$Se	401.8	71.2	40
Li$_2$Te	355.8	77.4	—
MgSe	272.9	62.8	17
MgTe	209.3	74.5	21
MnSe	154.9	90.8	11
MnTe	111.3	93.8	9
MnTe$_2$	125.6	145.0	40
Na$_2$Se	343.2	—	13
NaSe	194.2	62.8	21
Na$_2$Te	343.2	94.6	25
NaTe	173.3	83.7	11
NaTe$_3$	210.6	145.3	21
NiSe$_{1.05}$	74.9	75.2	4
NiSe$_{1.143}$	79.7	77.2	2.5
NiTe	35.7	80.1	2
PbSe	99.6	102.6	4
PbTe	69.1	110.1	1
Pt$_5$Se$_4$	265.4	327.3	25
Re$_2$Te$_5$	122.6	253.3	25
Sb$_2$Se$_3$	127.7	211.4	5
Sb$_2$Te$_3$	56.5	246.1	2
SiSe$_2$	146.5	94.2	40
SnSe	88.7	86.2	7
SnSe$_2$	124.7	118.0	5
SnTe	60.7	98.8	1.5
SrTe	397.7	80.8	38
SrTe	259.5	—	25
Tl$_2$Se	94.2	173.7	2.5
TlSe	61.3	102.6	1.5
Tl$_2$Te	80.4	174.1	12
Use	275.9	96.6	21
ZnSe	159.1	70.3	9
ZnTe	119.3	78.2	1

The above values of the heat of formation ΔH were measured calorimetrically at room temperature or at about 600°C, or have been calculated from measurements of vapour pressure or electromotive force at different temperatures. They are probably correct to within ±10%. As the molar heats of alloys are obtained nearly additively from the atomic heats of the components (Neumann and Kopp's rule) all the heats of formation (even if measured at higher temperatures) are probably valid also at room temperature within the limits of error mentioned above. The formulae of the compounds are given only to indicate composition, independent of whether the phases form a broad or narrow homogeneous field, or are ordered or disordered.

Table 8.5c INTERMETALLIC PHASES

Heats, entropies and free energies of formation.

Phase	N_2	Temp. °C	$-\Delta H$ kJ g-atom^{-1}	$-\Delta G$ kJ g-atom^{-1}	ΔS J K^{-1} g-atom^{-1}	Remarks
Ag–Au, s.s.	0.55	600	4.02	8.08	5.02	
α Ag–Cd	0.40	400	7.16	—	4.15	
ξ Ag–Cd	0.564	400	6.78	—	4.52	
β Ag–Cd	0.494	450	6.62	—	—	
γ' Ag–Cd	0.60	400	8.37	—	1.88	Cd solid
ε Ag–Cd	0.70	400	6.45	—	2.85	
AgMg	0.50	500	19.47	17.08	1.38	
Ag–Pd, s.s.	0.40	727	5.65	—	−2.09	
β Ag–Zn	0.50	600	3.14	10.34	8.25	Zn solid
ζ Ag–Zn	0.50	51	6.45	—	—	
γ Ag–Zn	0.612	600	4.48	10.09	6.41	Zn solid
θ Al–Fe	0.26	900	28.1	—	−3.94	
η Al–Fe	0.30	900	28.26	—	−3.56	Al solid
ζ Al–Fe	0.33	900	27.2	—	−2.34	
AlFe	0.50	900	27.2	—	0.13	
α Al–Zn	0.20	380	−4.02	—	7.45	
α Au–Cd	0.35	427	17.38	15.78	−2.26	
β Au–Cd	0.50	427	21.35	18.42	−4.19	Cd liquid
δ' Au–Cd	0.62	427	19.05	16.20	−3.48	
ε Au–Cd	0.70	427	19.26	15.83	−4.90	
Au–Cu, s.s.	0.58	500	5.32	9.67	5.65	
Au$_3$Cu	0.26	25	4.02	4.90	2.97	
AuCu I	0.50	25	8.96	8.96	0.00	
AuCu II	0.50	400	6.03	8.79	4.10	
AuCu$_3$	0.75	25	6.87	7.24	1.26	
AuNi, s.s.	0.53	877	−7.5	—	8.71	
Au$_2$Pb	0.33	227	−0.96	1.34	4.61	
AuPb$_2$	0.67	227	2.09	1.80	−0.72	
AuSn	0.50	25	15.24	—	−0.23	
Bi–Tl, δ	0.47	150	2.81	5.28	5.86	
Bi–Tl, γ	0.80	150	3.18	5.15	4.69	
Bi$_2$U	0.333	25	34.2	—	−3.93	
Bi$_4$U$_3$	0.43	25	38.1	—	−5.0	
BiU	0.50	25	36.8	—	−4.96	
β Cd–Hg	0.50	25	4.35	4.06	−0.92	Hg liquid
Cd–Mg, s.s.	0.50	270	5.61	7.96	4.35	
Cd$_3$Mg	0.25	25	5.19	5.19	−0.04	
CdMg	0.50	25	8.37	8.08	−1.05	
CdMg$_3$	0.75	25	5.65	5.61	−0.17	
CdSb	0.50	25	6.67	6.42	−0.86	
CdTe	0.50	20	50.9	49.7	−4.12	
CeHg$_4$	0.80	342	66.6	7.54	−96.13	Hg gas
α Co–Fe, s.s.	0.50	870	6.6	—	—	
α Cr–Fe, s.s.	0.50	1 280	−5.9	−7.63	8.42	
σ Cr–Fe	0.55	750	−4.73	—	6.91	
Cr–Mo, s.s.	0.47	1 400	−7.24	—	7.70	
Cr–Ni, f.c.c.	0.50	1 265	−6.70	—	9.21	
Cu$_2$Mg	0.333	25	11.18	—	−8.4	
CuMg$_2$	0.667	25	9.6	—	—	
Cu–Ni, s.s.	0.65	700	−1.84	—	4.40	
Cu–Pd, s.s.	0.40	640	12.6	—	−5.32	
Cu–Pt, s.s.	0.50	640	15.5	—	−4.15	
Cu$_2$Sb	0.33	25	4.2	—	−3.56	
α Cu–Zn	0.39	25	9.38	10.22	2.81	
β Cu–Zn	0.50	—	9.17	—	4.19	

(continued)

Table 8.5c INTERMETALLIC PHASES—*continued*

Phase	N_2	Temp. °C	$-\Delta H$ kJ g-atom^{-1}	$-\Delta G$ kJ g-atom^{-1}	ΔS J K^{-1} g-atom^{-1}	Remarks
β' Cu–Zn	0.50	25	11.72	12.14	1.47	
γ Cu–Zn	0.615	—	12.35	—	0.54	
ε Cu–Zn	0.79	25	7.20	7.49	1.00	
γ Fe–Mn	0.5	1 127	5.0	—	—	
γ Fe–Ni, s.s.	0.65	1 000	4.35	—	3.60	
FeSb$_2$	0.67	560	3.4	—	−0.8	
Ge$_3$U	0.25	1 100	26.8	—	−1.47	
Ge$_2$U	0.33	1 100	29.3	—	−1.3	
Ge$_5$U$_3$	0.375	1 100	30.2	—	−1.3	
GeU	0.50	1 100	30.77	—	−1.09	
Ge$_3$U$_5$	0.625	1 050	29.48	—	−0.38	
γ Hg–Tl	0.286	−59	−0.29	1.05	6.45	
β In–Sn	0.42	100	−2.1	—	—	
KNa$_2$	0.667	25	−0.75	0.63	4.61	Na$_2$K, Na, K: liquid
	0.667	25	−0.04	0.29	1.13	Na$_2$K, Na, K: solid
Na$_5$Pb$_2$	0.286	25	20.5	—	−6.7	
Na$_9$Pb$_4$	0.31	25	20.9	—	−6.3	
NaPb	0.50	25	23.0	—	−8.8	
NaPb$_3$	0.714	25	13.8	—	−2.1	
NaTl	0.50	25	16.54	—	−3.98	
α Pb–Tl	0.50	250	1.80	4.19	4.56	
β-Pb–Tl	0.85	270	2.81	2.81	0.0	
Pb$_3$U	0.25	700	19.7	—	−0.4	} Pb solid
PbU	0.50	700	18.8	—	−1.3	
SbZn	0.50	25	9.52	8.85	−2.26	
Sn$_3$U	0.25	950	23.9	—	−0.63	} Sn solid
Sn$_5$U$_3$	0.375	950	27.2	—	−1.17	
Sn$_2$U$_3$	0.60	1 000	26.8	—	−1.05	
Th$_2$Zn	0.333	700	22.2	15.5	−7.75	
ThZn$_2$	0.667	700	40.2	25.5	−15.1	
ThZn$_4$	0.80	700	38.5	23.0	−15.1	
Th$_2$Zn$_{17}$	0.895	700	32.7	13.0	−16.96	
U$_2$Zn$_{17}$	0.895	700	24.3	7.5	−17.2	
ZrZn	0.50	500	63.6	32.2	−40.6	
ZrZn$_2$	0.667	500	56.1	30.1	−33.5	
ZrZn$_3$	0.75	500	54.22	28.5	−33.29	
ZrZn$_6$	0.86	500	40.6	18.8	−28.26	
ZrZn$_{14}$	0.933	500	28.1	9.42	−24.07	

ΔS = entropy of formation. N_2 = mol fraction of second component.   s.s. = solid solution.

8.4   Metallic systems of unlimited mutual solubility

While mutual solubility in the solid state is usually limited, that in the liquid state is frequently unlimited. Thus, the curves of concentration against integral heat, free energy and entropy of formation are convex with a maximum or a minimum. The thermochemical values at the concentrations corresponding to these maxima and minima are given in Table 8.6.

According to van Laar the form of the heat of mixing curve may be represented by

$$\Delta H = \frac{aN_2(1 - N_2)}{1 + b \cdot N_2}$$

where N_2 is the mol fraction of the second component and a and b are constants for each binary system.

For a system, completely disordered, the entropy of formation is often represented by this equation:

$$\Delta S = -19.155(N_1 \log N_1 + N_2 \log N_2) \, \text{J K}^{-1} \, \text{g-atom}^{-1}$$

which has a maximum value of $5.78 \, \text{J K}^{-1}$ at $N_1 = N_2 = 0.5$. Some deviations from this value may be due to experimental errors, but others are real.

For more detailed discussions *see* refs 1–3.

Table 8.6   LIQUID BINARY METALLIC SYSTEMS

Heats and entropies of formation.

		Heat of formation			Entropy of formation	
			ΔH_{max}			ΔS_{max}
System	°C	N_2	J g-atom^{-1}		N_2	J K^{-1} g-atom^{-1}
Ag–Al	1 000	0.30	−6 410		0.60	7.58
Ag–Au	1 077	0.50	−5 150		0.50	3.81
Ag–Bi	727*	0.75	1 700		0.50	6.17
Ag–Cu	1 150	0.45	4 260		0.50	6.27
Ag–In	827*	0.32	−5 680		0.70	3.82
Ag–Mg	1 050	0.50	1 215		—	—
Ag–Pb	1 000	0.53	3 720		0.47	7.62
Ag–Sb	977	0.20	−3 120		0.52	8.37
Ag–Sn	977	0.19	−3 220		0.56	6.68
Ag–Tl	702*	0.56	2 510		0.50	6.32
Al–Bi	900	0.40	7 340		—	—
Al–Cu	1 100	0.60	−9 435		0.60	9.25
Al–Ga	750	0.40	670		0.50	6.10
Al–Ge	927	0.53	−3 915		0.50	6.20
Al–In	900	0.45	5 735		0.50	6.36
Al–Mg	800	0.45	−3 390		0.50	4.94
Al–Sn	700	0.45	4 100		0.50	6.99
Al–Zn	727	0.51	2 575		0.50	6.71
Au–Bi	700*	0.50	625		0.50	6.97
Au–Cu	1 277	0.53	−4 400		0.53	6.90
Au–Pb	927*	0.40 / 0.90	−795 / 260		0.50	7.54
Au–Sn	550*	0.45	−11 595		0.42	7.91
Au–Tl	700*	0.50	140		0.50	7.86
Au–Zn	807*	0.52	−22 810		0.25	4.50
Bi–Cd	500	0.65	880		0.52	7.25
Bi–Cu	927†	0.60	5 420		0.50	7.59
Bi–Hg	321	0.70	555		0.50	5.55
Bi–In	627	0.55	−185		0.50	5.87
Bi–Pb	427	0.50	−1 100		0.50	5.94
Bi–Sb	927	0.50	560		0.50	8.15
Bi–Sn	327	0.50	105		0.50	5.48
Bi–Tl	477	0.62	−4 580		0.50	6.67
Bi–Zn	600	0.58	4 685		0.50	7.85
Cd–Ga	427	0.50	2 665		0.50	5.73
Cd–Hg	327	0.50	−2 625		0.43	5.28
Cd–In	527	0.46	1 435		0.50	6.32
Cd–Mg	650	0.50	−5 615		0.50	4.72
Cd–Pb	500	0.45	2 665		0.45	6.50
Cd–Sb	500†	0.47	−2 035		0.50	6.92
Cd–Sn	500	0.43	1 815		0.50	6.91

(*continued*)

Table 8.6 LIQUID BINARY METALLIC SYSTEMS—*continued*

System	°C	Heat of formation N_2	ΔH_{max} J g-atom^{-1}	Entropy of formation N_2	ΔS_{max} J K^{-1} g-atom^{-1}
Cd–Tl	477	0.40	2 300	0.50	6.78
Cd–Zn	527	0.50	2 090	0.50	5.86
Co–Fe	1 590	0.45	−2 610	0.58	3.80
Cs–K	111	0.59	126	—	—
Cs–Na	111	0.65	1 020	—	—
Cu–In	800*	0.23 / 0.82	−3 390 / 305	0.55	4.83
Cu–Mg	827*	0.44	10 445	0.50	2.53
Cu–Pb	1 200	0.45	6 800	0.49	6.85
Cu–Sb	917*	0.25 / 0.90	−5 660 / 345	0.55	8.20
Cu–Sn	1 127	0.25 / 0.90	−4 150 / 218	0.54	8.16
Cu–Tl	1 300	0.50	8 580	0.52	7.75
Fe–Si	1 600	0.48	−37 950	0.05 / 0.54 / 0.95	1.05 / −1.66 / 0.65
Ga–Zn	477	0.58	1 610	0.50	6.80
Hg–In	25[†]	0.48	−2 260	0.45	5.25
Hg–Na	400	0.40	−20 570	(0.04) / 0.39 / 0.12	(0.4) / −4.57 / 0.71
Hg–Pb	327	0.20 / 0.83	505 / −160	0.50	4.60
Hg–Sn	177[†]	0.40	905	0.50	5.17
Hg–Tl	310	0.28	−1 130	—	—
Hg–Zn	300[†]	0.33	440	0.42	4.89
In–Mg	700	0.60	−7 300	—	—
In–Pb	400	0.48	963	0.50	6.29
In–Sb	627	0.45	3 300	0.50	6.51
In–Sn	427	0.45	−200	0.48	6.76
In–Tl	500	0.42	575	0.50	5.48
In–Zn	427	0.58	3 235	0.53	6.99
K–Na	111	0.40	737	0.50	5.66
Mg–Pb	700	0.40	−10 050	0.53	5.21
Mg–Sn	500	0.40	−14 650	—	—
Mg–Tl	650	0.40	−7 040	0.57	6.66
Mg–Zn	700	0.60	−6 490	—	—
Na–Rb	111	0.42	1 260	—	—
Na–Pb	427	0.40	−17 100	0.05 / 0.20 / 0.80	0.34 / −2.85 / 2.26
Ni–Pd	1 600	0.50	1 200	0.50	5.41
Pb–Sb	632	0.55	−75	0.50	6.21
Pb–Sn	777	0.50	1 370	0.55	4.81
Pb–Tl	500	0.55	−1 090	0.50	5.22
Sb–Sn	632	0.50	−1 390	0.50	6.08
Sn–Tl	500	0.45	724	0.47	5.17
Sn–Zn	477	0.57	3 220	0.54	7.93

$N_2 =$ mol fraction of second component.

* or [†] indicates that the standard state for the first or second component, respectively, is the hypothetical supercooled liquid.

Table 8.7a   LIQUID BINARY METALLIC SYSTEMS AND SOME SOLID SOLUTIONS

Partial molar free energies $(-\overline{\Delta G})$ in kJ.

For the solution of 1 g-atom of metal C in a theoretically infinite amount of alloy of concentration N_C (mole fraction of the dissolved metal), $-\overline{\Delta G}$ is given in kJ.

System	Metal C	Temp. °C	N_C: 0.1	0.2	0.3	0.5	0.8
Ag–Al (sol.)	Al	449	37.36	s + s	20.33	s + s	s + s
Ag–Al (liq.)	Al	1 000	51.81	36.57	23.40	8.73	2.26
Ag–Au (sol.)	Ag	527	25.87	19.46	15.03	8.54	2.20
Ag–Au (liq.)	Ag	1 077	32.99	24.16	18.51	10.63	3.01
Ag–Cd (sol.)	Cd*	400*	34.87α	26.41α	19.01α	10.47ζ	1 + s
Ag–Cd (liq.)	Cd	950	42.41	29.53	20.91	10.03	2.40
Ag–Cu (liq.)	Cu	1 150	15.94	10.20	7.41	4.58	2.06
Ag–Hg (sol.)	Hg*	227*	19.28α	12.32α	6.24α	1 + s	1 + s
Ag–Mg (sol.)	Mg	500	63.56α	51.33α	s + s	32.57β'	1
Ag–Pb (liq.)	Pb	1 000	20.99	13.56	9.51	5.50	1.96
Ag–Pd (sol.)	Ag	927	25.85	21.00	18.16	13.52	3.49
Ag–Sn (liq.)	Sn	977	25.34	21.02	17.68	11.50	3.50
Ag–Tl (liq.)	Tl	702	s	1 + s	(5.31)	2.77	1.20
Ag–Zn (sol.)	Zn*	600*	30.96α	26.04α	19.69α	11.57β	s + 1
Al–Mg (liq.)	Mg	800	31.26	21.21	14.97	7.42	2.01
Al–Zn (sol.)	Zn	380	3.34	1.99	1.64	1.41	s + s
Al–Zn (liq.)	Zn	727	13.52	8.88	6.78	4.42	1.45
Au–Bi (liq.)	Bi	700	1 + s	1 + s	10.45	5.11	1.94
Au–Cd (sol.)	Cd*	427*	51.92α	44.12α	35.16α_2	18.84β	1
Au–Cd (liq.)	Cd	727	s	s	(38.0)	19.05	3.33
Au–Cu (sol.)	Cu	527	28.90	23.41	19.24	11.88	2.17
Au–Cu (liq.)	Cu	1 277	49.19	36.15	27.32	14.95	3.84
Au–Fe (sol.)	Fe	850	17.27	8.95	4.44	0.44	s + s
Au–Hg (sol.)	Hg*	227*	9.41α	3.20β	1 + s	1 + s	1 + s
Au–Pb (liq.)	Pb	927	1 + s	23.71	17.34	9.15	2.83
Au–Sn (liq.)	Sn	550	1 + s	44.03	30.87	13.51	2.28
Au–Tl (liq.)	Tl	700	1 + s	1 + s	(13)	8.00	2.19
Au–Zn (liq.)	Zn	807	s	68.04	51.84	26.32	3.86
Bi–Cd (liq.)	Cd	500	15.28	11.10	8.61	5.07	1.34
Bi–Hg (liq.)	Hg	321	9.77	6.52	4.69	2.52	0.74
Bi–K (liq.)	K	575	87.44	78.57	70.33	48.10	1 + s
Bi–Mg (liq.)	Mg	702	60.04	54.44	50.17	42.48	3.73
Bi–Pb (liq.)	Pb	427	17.24	12.52	9.48	5.27	1.43
Bi–Sn (liq.)	Sn	327	10.76	7.36	5.43	3.11	2.01
Bi–Tl (liq.)	Tl	477	25.15	21.10	17.73	10.90	2.50
Bi–Zn (liq.)	Zn	600	9.96	5.61	3.35	1.01	0.21
Cd–Cu (liq.)	Cu*	600*	16.01	12.34	10.22	6.60	1 + s
Cd–Ga (liq.)	Ga	427	4.71	2.92	1.96	1.28	0.77
Cd–In (liq.)	In	527	11.37	8.15	6.35	3.84	1.34
Cd–Mg (sol.)	Mg	270	24.45	19.38	15.57	7.87	1.42
Cd–Na (liq.)	Na	400	21.30	11.42	6.21	2.02	0.76
Cd–Pb (liq.)	Pb	500	7.30	4.91	3.83	2.62	1.15
Cd–Sb (liq.)	Sb*	500*	24.62	18.97	14.19	6.29	1 + s
Cd–Sn (liq.)	Sn	500	11.62	7.96	5.97	3.62	1.33
Cd–Zn (liq.)	Zn	527	8.88	5.62	4.18	2.53	1.04
Co–Fe (liq.)	Fe	1 590	30.29	21.09	16.05	10.21	3.65
Co–Fe (sol.)	Co, Fe	both solutes approx. ideal in γ phase					
Co–Pt (sol.)	Pt	1 000	64.59	48.82	37.08	19.75	4.35
Cr–Fe (solα)	Cr	1 327	23.75	16.60	12.82	8.10	2.70
Cr–Mo (sol.)	Cr	1 198	15.15	9.14	6.25	3.64	1.59
Cr–Ni (sol.)	Cr	1 277	35.76β	21.39β	12.24β	5.79β	α
Cr–V (sol.)	Cr	1 277	37.96	28.35	22.66	14.17	3.88
Cu–Fe (liq.)	Cu	1 600	8.86	4.81	3.78	2.85	1.66
Cu–Ni (sol.)	Ni	700	10.34	6.28	4.12	2.21	1.08
Cu–Pb (liq.)	Cu	1 200	12.18	6.74	4.53	2.88	1.96

(continued)

Table 8.7a   LIQUID BINARY METALLIC SYSTEMS AND SOME SOLID SOLUTIONS—*continued*

System	Metal C	Temp. °C	N_c: 0.1	0.2	0.3	0.5	0.8
Cu–Pt (sol.)	Cu	1 077	54.27	41.52	32.98	20.38	5.44
Cu–Sn (liq.)	Sn	1 127	57.37	30.57	18.93	8.87	2.84
Cu–Zn (sol.)	Zn*	500*	38.91α	29.67α	22.05α	(14.4β)	2.28ε
Cu–Zn (liq.)	Zn	1 060	42.58	31.28	23.66	12.94	—
Fe–Mn (liq.)	Mn	1 590	32.24	22.19	16.58	9.71	3.26
Fe–Ni (sol.)	Fe	927	—	—	13.02γ	7.45γ	2.30γ
Fe–Ni (liq.)	Ni	1 600	41.87	30.82	24.39	15.38	4.36
Fe–Ni (liq.)	Fe	1 600	48.15	33.54	23.75	11.74	3.60
Fe–Si (liq.)	Si	1 600	126.53	97.62	68.68	23.35	4.64
Fe–V (sol.)	V	1 327	59.24α	41.78α	30.21α	15.43α	3.91α
Ga–Mg (liq.)	Mg	650	43.89	35.68	28.86	17.50	3.04
Ga–Zn (liq.)	Zn	477	12.03	8.11	5.93	3.38	1.15
Hg–K (liq.)	K	327	69.98	44.39	25.56	6.33	1.09
Hg–Na (liq.)	Na	400	64.02	46.64	32.22	12.25	1.62
Hg–Pb (liq.)	Pb	327	68.61	56.51	47.76	33.45	1.22
Hg–Tl (liq.)	Tl*	25*	8.46	5.43	3.64	1 + s	1 + s
Hq–Zn (liq.)	Zn*	300*	8.12	5.55	4.08	2.26	1 + s
In–Sb (liq.)	Sb	627	31.33	23.07	16.56	8.08	1.88
In–Zn (liq.)	Zn	427	6.41	3.53	2.05	1.21	0.51
K–Pb (liq.)	Pb	575	49.77	42.69	35.17	17.78	2.24
K–Tl (liq.)	Tl	525	27.17	24.20	21.65	14.58	3.22
Mg–Pb (liq.)	Mg	700	38.47	33.45	28.97	19.10	4.84
Mg–Sn (liq.)	Mg	800	55.98	47.77	40.53	25.86	6.26
Mg–Zn (liq.)	Mg	650	32.26	22.51	16.22	8.37	2.24
Mn–Ni (sol.)	Mn	777	61.76γ	43.10γ	33.35γ	17.22η	(2.1γ)
Na–Pb (liq.)	Na	427	45.92	38.70	30.82	19.05	3.14
Na–Sn (liq.)	Na	500	48.76	41.44	34.95	s	2.68
Na–Tl (liq.)	Na	400	41.06	32.05	24.07	10.47	1.47
Pb–Sb (liq.)	Pb	632	18.86	13.32	9.99	5.69	1.75
Pb–Sn (liq.)	Sn	777	9.27	7.37	6.57	4.79	1.80
Pb–Tl (liq.)	Tl	500	16.32	11.90	9.16	5.43	1.41
Sb–Sn (liq.)	Sn	632	22.73	16.40	12.34	6.89	1.97
Sb–Zn (liq.)	Zn	677	1 + s	18.82	15.12	8.10	0.81
Sn–Tl (liq.)	Tl	450	9.93	6.71	5.06	3.15	1.20
Sn–Zn (liq.)	Zn	477	10.49	6.54	4.46	2.22	0.84

* Note standard state for metal C.

Table 8.7b   LIQUID BINARY METALLIC SYSTEMS AND SOME SOLID SOLUTIONS

Partial molar heats of solution ($\overline{\Delta H}$) in kJ.

System	Metal C	Temp. °C	N_c: 0.1	0.2	0.3	0.5	0.8
Ag–Al (liq.)	Al	1 000	−27.43	−18.71	−8.98	+5.53	+1.31
Ag–Au (sol.)	Ag	527	−14.28	−11.72	−9.29	−5.07	−0.88
Ag–Cd (sol.)	Cd*	400*	−23.42	−18.96	−15.42	—	—
Ag–Cu (liq.)	Cu	1 150	15.70	10.76	7.48	3.77	0.63
Ag–Mg (sol.)	Mg	500	−38.45	−31.18	(26.3)	—	—
Ag–Pb (liq.)	Ag	1 000	11.05	9.56	7.76	3.73	0.17
Ag–Zn (sol.)	Zn*	600*	−14.22	−11.08	−9.41	−7.89β	1 + s
Al–Mg (liq.)	Mg	800	−11.86	−9.14	−6.61	−2.68	−0.21
Al–Zn (sol.)	Zn	380	11.93	8.99	6.30	3.79	s + s
Al–Zn (liq.)	Zn	727	8.35	6.51	4.94	2.64	0.46
Au–Bi (liq.)	Bi	700	1 + s	1 + s	1.77	0.70	0.04
Au–Cd (sol.)	Cd*	427*	−55.26	−49.42	−39.31α₂	−20.32β	ε + 1
Au–Cu (sol.)	Cu	527	−12.37	−12.00	−10.92	−7.39	−0.37

(*continued*)

Table 8.7b   LIQUID BINARY METALLIC SYSTEMS AND SOME SOLID SOLUTIONS—*continued*

System	Metal C	Temp. °C	N_C: 0.1	0.2	0.3	0.5	0.8
Au–Fe (sol.)	Fe	850	18.48	14.14	12.02	9.67	s + s
Au–Pb (liq.)	Pb	927	−3.05	−2.18	−1.14	−0.09	+0.60
Au–Sn (liq.)	Sn	550	1 + s	−32.83	−22.70	−9.31	−0.48
Au–Zn (liq.)	Zn	807	s	−61.04	−48.12	−25.20	−1.93
Bi–Cd (liq.)	Cd	500	2.29	1.81	1.41	1.13	0.63
Bi–In (liq.)	In	627	−5.02	−4.80	−3.64	−2.35	−0.19
Bi–Na (liq.)	Na	500	−54.93	−66.36	−68.34	(−54)	1 + s
Bi–Pb (liq.)	Pb	427	−3.35	−2.90	−2.29	−1.07	−0.09
Bi–Sn (liq.)	Sn	327	0.34	0.27	0.20	0.10	0.02
Bi–Tl (liq.)	Tl	477	−10.96	−9.33	−7.89	−7.18	−1.12
Bi–Zn (liq.)	Zn	600	12.22	10.81	9.11	6.20	1.59
Cd–Ga (liq.)	Cd	427	9.19	6.60	4.89	2.67	0.84
Cd–Hg (liq.)	Cd	327	−7.85	−7.09	−5.66	−2.47	0.31
Cd–Mg (sol.)	Mg	270	−14.88	−14.39	−12.29	−5.92	−0.34
Cd–In (liq.)	Cd	527	14.15	3.54	2.89	1.70	0.39
Cd–Pb (liq.)	Cd	500	7.89	6.54	5.28	3.14	0.84
Cd–Sb (liq.)	Cd	500	1 + s	1 + s	(−4.5)	−3.24	+0.45
Cd–Sn (liq.)	Cd	500	5.45	4.40	3.52	2.16	0.52
Cd–Tl (liq.)	Cd	477	6.49	5.56	4.65	2.80	0.65
Cd–Zn (liq.)	Zn	527	6.82	5.37	4.04	2.12	0.40
Cr–V (sol.)	Cr	1 277	−7.48	−2.62	+4.43	−7.70	−1.49
Cu–Ni (sol.)	Ni	700	3.52	3.96	3.94	2.93	0.69
Cu–Pb (liq.)	Cu	1 200	21.70	16.80	13.06	6.99	1.96
Cu–Pt (sol.)	Cu	1 077	−28.65	−27.41	−25.30	−13.10	−3.33
Cu–Zn (sol.)	Zn*	500*	−28.69	−24.65	−17.36	(−15.7β)	−2.40ε
Fe–Mn (sol.)	Mn	1 177	−12.36	−10.95	−9.59	−6.61	−1.65
Fe–Ni (sol.)	Ni	927	−4.31	—	−8.83	−5.99	—
Fe–Ni (liq.)	Ni	1 600	−9.77	—	−9.72	−7.22	—
Fe–Si (liq.)	Si	1 600	−125.15	−109.73	−81.20	−27.80	−1.82
Fe–V (sol.)	Fe	1 327	−9.29	−5.98	−0.57	+14.17	+10.71
Ga–Zn (liq.)	Zn	477	4.52	3.73	3.08	2.01	0.51
Hg–Na (liq.)	Na	400	−72.24	−61.52	−40.18	−10.88	−0.35
Hg–Pb (liq.)	Hg	327	−0.73	+0.16	+0.65	+0.96	+0.45
Hg–Sn (liq.)	Sn*	177*	3.61	2.23	1.41	0.62	1 + s
Hg–Tl (liq.)	Tl*	25*	−4.37	−2.78	−2.06	1 + s	1 + s
In–Pb (liq.)	In	400	2.98	2.40	1.88	1.00	0.17
In–Sn (liq.)	In	427	−0.59	−0.46	−0.38	−0.21	−0.04
In–Tl (liq.)	Tl	350	2.12	1.51	1.05	0.42	0.04
In–Zn (liq.)	Zn	427	9.50	7.98	6.50	3.78	0.91
Na–Pb (liq.)	Na	427	−39.05	−37.55	−34.32	−23.86	−3.43
Na–Tl (liq.)	Na	400	−33.33	−30.22	−24.70	−10.98	−0.67
Pb–Sb (liq.)	Pb	732	−0.14	−0.20	−0.19	−0.04	+0.03
Pb–Sn (liq.)	Sn	777	4.69	3.48	2.59	1.34	0.25
Pb–Tl (sol.)	Tl	250	−4.50	−4.20	−3.98	−2.93	−1.08
Pb–Tl (liq.)	Tl	500	−3.11	−2.67	−2.18	−1.15	−0.15
Sb–Sn (liq.)	Sn	632	−4.02	−3.56	−2.91	−1.50	−0.15
Sb–Zn (liq.)	Zn	577	1 + s	−4.73	−5.39	−2.70	+0.96
Sn–Tl (liq.)	Tl	450	2.70	1.94	1.28	0.41	0.08
Sn–Zn (liq.)	Zn	477	8.22	7.47	6.43	4.13	1.35

* Note standard state for metal C.

8.5   Metallurgically important compounds

In Tables 8.8a to 8.8j values of the heats and free energies of formation from the constituent elements are given throughout in kJ mol^{-1} of compound and standard entropies in J K^{-1} mol^{-1} of compound. It is to be noted that the minus signs appear in the captions, and therefore positive numerical values denote exothermic formation.

The standard state of a reactant element is the most stable form at the temperature indicated unless otherwise stated in the headings; the ideal diatomic gases were used as the standard states for sulphur, bromine and iodine in computing the free energy values at room temperature. The standard state of the compounds is the ideal stoichiometric proportion, and the state of aggregation where necessary is indicated by s = solid, m = liquid and g = gaseous. Many compounds are stable with slight deviations from stoichiometric proportion. Compounds which may exist over a range of composition are indicated by an asterisk (*). For such compounds thermochemical values must be used with caution; dissociation pressures calculated from the free energy values may show considerable deviations from measured dissociation pressures.

Free energy values are given at five temperatures; further values can be obtained by linear interpolation provided that no change in the state of aggregation occurs in the interval. Even then, a linear interpolation will generally give a value within the accuracy of the data. The limits of accuracy are given in the last column of the tables; they apply to the values of the heat and free energy of formation below, say, 1 500 K. The errors may be greater at 2 000 K.

Dissociation pressures of sulphides and phosphides have only been determined for parts of many systems, the data for the lowest oxidation steps often being absent. No total thermochemical values for the formation of the compounds from the elements can therefore be given in Tables 8.8f and 8.8j but dissociation pressures of the compounds rich in sulphur or phosphorus appear in Tables 8.9a and 8.9b, respectively.

The thermochemical values for the metal borides and carbides suggested below can only be used as a guide. They are not sufficiently reliable for accurate calculations—mainly for two reasons: (1) transition metals do not form stoichiometric compounds with boron and carbon but rather extensively homogeneity ranges of which the investigators have taken too little heed. Thus the data below are oversimplifications. (2) it is very doubtful whether the experimental zero-point entropies of borides and carbides are generally zero. More likely, 'frozen-in disorder' of the substances studied accounted for substantial zero-point entropies, so that the values in column 3 when obtained from low-temperature heat capacities are not truly 'standard entropies'.

Table 8.8a BORIDES

Heats and free energies of formation in kJ and standard entropies.

Compound	$-\Delta H_{298}$ 25°C	S_{298} 25°C	$-\Delta G_{300}$ 27°C	$-\Delta G_{500}$ 227°C	$-\Delta G_{1000}$ 727°C	$-\Delta G_{1500}$ 1 227°C	Accuracy ±kJ
AlB_2	67.0	—	—	—	—	—	12
AlB_{12}	201.0	—	—	—	—	—	15
CeB_6	351.6	74.1	344.0	338.9	326.2	313.5	100
Co_2B	125.6	59.9	123.8	122.5	119.5	116.4	30
CoB	94.2	30.6	92.6	91.1	89.8	86.2	25
CrB	75.3	24.1	73.6	72.5	69.6	66.8	25
CrB_2	94.2	26.2	91.4	89.5	84.7	80.0	25
Fe_2B	71.2	56.7	70.1	69.3	67.4	65.5	20
FeB	71.2	27.7	69.6	68.5	65.8	63.1	20
HfB_2	336.1	42.7	332.2	329.5	323.0	316.5	10
MgB_2	92.1	36.0	89.6	88.0	83.0	60.0	25
MgB_4	105.1	51.9	103.8	103.0	100.4	77.5	25
MnB	75.3	32.4	73.7	72.6	70.0	67.3	30
MnB_2	94.2	34.5	91.4	89.6	85.0	80.4	30
NbB_2	251.2	37.7	248.0	245.9	240.6	235.2	25
Ni_4B_3	311.9	114.7	305.2	300.7	289.7	278.6	60
NiB	100.5	30.1	98.8	97.7	94.9	92.1	25
TaB_2	209.3	44.4	206.7	204.9	200.5	196.1	40
TiB_2	342.0	28.5	319.9	317.2	310.4	303.6	5
UB_2	164.5	55.1	162.1	160.6	156.6	152.7	25
UB_4	245.7	71.2	244.9	244.4	243.1	241.8	40
UB_{12}	433.2	139.8	438.8	442.6	452.0	461.4	40
ZrB_2	324.0	36.0	319.6	316.7	309.3	302.0	10

Table 8.8b   CARBIDES

Heats and free energies of formation in kJ and standard entropies.

Compound	$-\Delta H_{298}$ 25°C	S_{298} 25°C	$-\Delta G_{300}$ 27°C	$-\Delta G_{500}$ 227°C	$-\Delta G_{1000}$ 727°C	$-\Delta G_{1500}$ 1 227°C	$-\Delta G_{2000}$ 1 727°C	Accuracy ±kJ
Al_4C_3	215.8	88.7	203.6	—	174.1	153.3	122.2	8
B_4C	71.6	27.09	71.2	70.8	69.8	68.7	67.4	19
Be_2C	117.2	16.3	114.8	—	—	100.9	78.8	5
CaC_2	59.0	70.3	64.5(α)	69.5(α)	85.0(β)	103.0(β)	121.0(β)	13
Ce_2C_3	176.6	173.7	—	—	—	—	—	7
CeC_2	97.1	90.4	—	—	—	—	—	7
Co_2C	−16.7	74.5	−19.3	−12.1	−8.0	—	—	21
Cr_4C	98.4	135.6	100.0	72.0	104.2	118.3	—	9
Cr_7C_3	228.1	200.9	223.4	237.1	245.9	254.3	—	13
Cr_3C_2	109.7	85.4	108.6	109.8	115.3	122.4	—	13
Fe_3C	−25.1	104.7	−19.0	−14.2	−1.93	+4.90	—	4
CH_4	74.90(g)	186.3(g)	50.7(g)	32.7(g)	−19.3(g)	—	—	4
Mg_2C_3	−79.5	−62.8	—	—	—	—	—	33
MgC_2	−87.9	—	−84	—	—	—	—	21
Mn_7C_3	112.2	239.0	111.4	112.6	118.9	113.9	—	13
Nb_2C	186.3	64.0	—	—	—	—	—	13
NbC	138.1	35.1	—	—	—	—	—	13
SiC	67.0	16.54	—	58.2	55.3	53.2	41.9	9
Ta_2C	203.0	83.6	—	—	—	—	—	17
TaC	143.6	42.2	142.2	—	141.0	137.8	—	8
ThC	125.6	59	—	—	—	—	—	13
ThC_2	117.2	70.3	—	124.3	117.2	—	—	21
TiC	183.8	24.3	180.0	177.9	172.9	167.1	160.4	13
UC	90.9	59.5	92.1	93.4	96.7	95.5	92.5	13
U_2C_3	205.1	138.4	205.2	205.2	206.4	199.3	187.1	17
UC_2	96.3	67.8	98.4	100.0	101.7	99.6	93.8	13
VC	100.9	27.6	—	—	94.6	91.3	80.4	17
WC	37.7	41.8	37.3	36.4	35.6	34.8	33.5	13
W_2C	26.4	—	—	—	—	—	—	7
ZrC	202.0	33.1	198.3	196.2	190.7	184.4	—	21

Table 8.8c   NITRIDES

Heats and free energies of formation in kJ and standard entropies.

Compound	$-\Delta H_{298}$ 25°C	S_{298} 25°C	$-\Delta G_{300}$ 27°C	$-\Delta G_{500}$ 227°C	$-\Delta G_{1000}$ 727°C	$-\Delta G_{1500}$ 1 227°C	$-\Delta G_{2000}$ 1 727°C	Accuracy ±kJ
AlN	318.6	20.2	287.1	266.1	—	—	—	5
BN	254.1	14.8	—	—	—	48.1	27.6	5
Ba_3N_2	341.1	(152.4)	292.2	244.1	123.9	—	—	38
Be_3N_2	589.9	34.3	534.1	—	—	—	—	5
Ca_3N_2	439.6	108.0	377	335	230	—	—	23
Cd_3N_2	−161.6	—	—	—	—	—	—	13
CeN	326.6	—	295.2	274.2	222	—	—	38
Co_3N	−8.4	98.8	—	—	—	—	—	21
CrN	123.1	—	96.7	78.7	40.6	—	—	5
Cr_2N	114.7	—	93.8	79.5	51.1	28.0	—	5
Cu_3N	−74.5	—	—	—	—	—	—	13
Fe_4N	10.9	(156.2)	−4.2	−12.77	−41.0	—	—	17
GaN	109.7	29.7	—	—	—	—	—	10
Ge_3N_4	65.3	(167)	—	—	—	—	—	9
HfN	369.3	50.6	—	—	—	—	—	2.5
NH_3	46.1	192.47	16.7	−4.6	−61.5	—	—	2.1
InN	138.1	43.5	—	—	—	—	—	13
LaN	299.4	44.4	270.5	249.5	197.2	—	—	38

(continued)

Table 8.8c   NITRIDES—*continued*

Compound	$-\Delta H_{298}$ 25°C	S_{298} 25°C	$-\Delta G_{300}$ 27°C	$-\Delta G_{500}$ 227°C	$-\Delta G_{1000}$ 727°C	$-\Delta G_{1500}$ 1227°C	$-\Delta G_{2000}$ 1727°C	Accuracy ±kJ
Li_3N	196.8	—	154.1	125.6	—	—	—	2.1
Mg_3N_2	461.8	93.7	399.4	359.6	258.3	—	—	13
Mn_4N	126.9	—	—	—	—	—	—	4
Mn_5N_2	201.8	—	156.2	125.6	49.4	—	—	4
Mo_2N	69.5	(87.9)	70.8	37.3	−8.8	—	—	17
NbN	234	—	—	—	—	—	—	11
Nb_2N	248.6	67.0	—	—	—	—	—	4
Ni_3N	−0.8	—	−27.2	—	—	—	—	4
Si_3N_4	745.1	113.0	622.2	571.9	410.3	251.2	64.5	17
Sr_3N_2	391.0	123.5	323.6	278.4	—	—	—	23
Ta_2N	270.9	—	—	—	—	—	—	13
TaN	252.4	42.7	214.8	202.2	166.6	134.0	103.8	13
Th_3N_4	1 298.0	—	1 185	1 114	925	737	548	84
TiN	336.6	30.1	308.1	289.7	243.3	195.9	148.2	8
UN	294.7	62.7	260.0	242.0	196.8	152.0	106.7	42
U_2N_2	708.5	128.9	—	—	—	—	—	17
VN	217.3	37.3	—	—	—	—	—	17
V_2N	264.5	53.4	—	—	—	—	—	17
Zn_3N_2	22.2	140.2	—	—	—	—	—	8
ZrN	365.5	38.9	336.2	317.8	270.9	223.2	—	7

Table 8.8d   SILICIDES

Heats and free energies of formation in kJ and standard entropies.

Compound	$-\Delta H_{298}$ 25°C	S_{298} 25°C	$-\Delta G_{300}$ 27°C	$-\Delta G_{500}$ 227°C	$-\Delta G_{1000}$ 727°C	$-\Delta G_{1500}$ 1227°C	Accuracy ±kJ
Ca_2Si	209.3	—	—	—	—	—	13
$CaSi$	150.7	—	—	—	—	—	9
$CaSi_2$	150.7	—	—	—	—	—	13
Co_2Si	117.2	—	—	—	—	—	10
$CoSi$	95.0	42.7	93.2	91.9	88.9	85.8	9
$CoSi_2$	98.8	64.0	97.7	97.0	95.1	95.5	12
Cr_3Si	92.1	86.3	90.8	90.0	87.9	85.8	18
Cr_5Si_3	211.4	16.91	209.4	280.0	204.6	201.2	35
$CrSi$	53.2	43.9	53.6	53.8	54.4	55.0	10
$CrSi_2$	80.0	58.6	78.5	77.5	75.1	72.7	13
$FeSi$	76.9	41.8	75.6	74.8	72.7	70.9	5
Mg_2Si	79.1	63.8	73.1	69.1	55.0	—	5
Mn_2Si	79.5	103.8	76.4	74.3	69.0	63.8	15
Mn_5Si_3	200.9	238.6	207.8	212.5	224.0	235.6	40
$MnSi$	60.7	46.5	59.5	58.6	56.6	54.5	13
Mn_4Si_7	308.5	228.1	299.3	293.1	277.7	262.3	65
Mo_3Si	116.4	106.5	117.0	117.4	118.3	119.3	7.5
Mo_5Si_3	310.2	208.0	312.8	314.5	318.8	323.1	25
$MoSi_2$	131.9	65.1	131.6	131.4	130.8	130.3	10
$NbSi_2$	138.1	69.9	136.8	136.0	133.9	131.7	45
Ni_2Si	142.7	—	—	—	—	—	11
Ni_3Si_2	232.3	—	—	—	—	—	20
$NiSi$	89.6	44.4	88.4	87.5	85.4	83.3	7.5
$NiSi_2$	94.2	65.3	93.6	93.1	92.1	91.0	13
Re_5Si_3	157.4	256.0	161.3	164.0	170.5	177.1	70
$ReSi$	52.7	55.4	52.5	52.4	52.0	51.7	21
$ReSi_2$	90.4	74.1	90.2	90.0	89.6	89.2	30
Ta_2Si	125.6	105.5	126.7	127.4	129.2	131.0	20

(*continued*)

Table 8.8d SILICIDES—*continued*

Compound	$-\Delta H_{298}$ 25°C	S_{298} 25°C	$-\Delta G_{300}$ 27°C	$-\Delta G_{500}$ 227°C	$-\Delta G_{1000}$ 727°C	$-\Delta G_{1500}$ 1 227°C	Accuracy ±kJ
Ta$_5$Si$_3$	334.9	280.9	340.2	343.7	352.4	361.2	35
TaSi$_2$	119.3	75.3	118.2	117.5	115.6	113.8	13
Th$_3$Si$_2$	279.6	166.4	270.2	263.9	248.1	232.4	50
ThSi	126.0	62.8	123.2	121.3	116.6	111.9	17
Th$_3$Si$_5$	477.6	231.9	470.9	466.4	455.2	444.1	85
ThSi$_2$	170.8	89.2	170.3	169.9	169.0	168.1	30
Ti$_5$Si$_3$	579.8	—	—	—	—	—	65
TiSi	129.8	49.0	49.0	48.95	48.9	48.8	15
TiSi$_2$	134.4	61.1	132.0	131.0	127.6	124.2	21
U$_3$Si	92.1	167.4	91.5	91.1	90.1	89.1	13
U$_3$Si$_2$	170.8	197.6	173.6	175.5	180.2	184.9	11
USi	84.6	66.6	83.9	83.4	82.2	81.0	5
U$_3$Si$_5$	354.6	231.5	353.7	353.1	351.5	350.0	17
USi$_2$	129.8	82.0	127.1	126.9	124.0	121.1	3
USi$_3$	130.6	106.3	130.5	130.45	130.3	130.1	3

Table 8.8e OXIDES

Heats and free energies of formation in kJ and standard entropies.

Compound	$-\Delta H_{298}$ 25°C	S_{298} 25°C	$-\Delta G_{300}$ 27°C	$-\Delta G_{500}$ 227°C	$-\Delta G_{1000}$ 727°C	$-\Delta G_{1500}$ 1 227°C	$-\Delta G_{2000}$ 1 727°C	Accuracy ±kJ
Ag$_2$O	30.6	121.8	10.68	—	—	—	—	2.1
Al$_2$O$_3$	1 678.2	51.1	1 584.0	1 520.8	1 362.4	1 146.9	—	17
As$_2$O$_3$	653.77	122.80	577.4	526.3	—	—	—	3.3
As$_2$O$_5$	914.8	105.5	771.6	675.7	—	—	—	6.3
B$_2$O$_3$	1 272.5	54.0	1 192.9	1 139.8	—	—	—	8
BaO	553.8	70.3	524.5	504.4	455.0	—	—	22
BaO$_2$	634.2	93.1	—	—	—	—	—	13
BeO	608.4	14.1	580.0	560.7	513.4	466.9	412.9	13
Bi$_2$O$_3$	570.7	151.6	496.6	—	—	—	—	17
CO	110.5	198.0	138.2	155.7	199.50	243.3	287.2	4
CO$_2$	393.77	213.9	394.4	394.8	395.2	395.7	396.1	4
CaO	634.3	39.8	603.7	584.1	534.7	481.9	405.7	4
CaO$_2$	659.4	—	—	—	—	—	—	4
CdO	259.4	54.8	229.7	212.1	204.1	106.6	—	4
Ce$_2$O$_3$	1 821.7	150.7	1 733.7	1 676.9	—	—	—	8
CeO$_2$	1 089.4	62.4	1 025.8	983.9	882.2	—	—	13
CoO	239.1	52.96	212.7	198.9	163.3	127.7	—	4
Co$_3$O$_4$	905.6	102.6	777.1	705.9	525.0	—	—	13
Cr$_2$O$_3$	1 130.4	81.2	1 051.3	991.0	861.6	731.0	601.2	17
CrO$_2$	582.8	51.0	—	—	—	—	—	8
CrO$_3$	579.9	71.9	—	—	—	—	—	10.5
Cs$_2$O	317.8	127.6	—	—	—	—	—	13
Cu$_2$O	167.5	93.8	144.9	129.8	95.5	59.5	—	4
CuO	155.3	42.7	127.3	108.9	66.6	—	—	4
Dy$_2$O$_3$	1 866.5	149.9	1 773.9	1 712.4	—	—	—	13
Er$_2$O$_3$	1 899.1	153.2	1 808.7	1 748.4	—	—	—	8
Fe$_{0.95}$O	264.6	58.82	241.2	228.6	197.2	165.8	—	13
Fe$_3$O$_4$	1 117.5	151.6	1 015.3	950.0	780.8	—	—	25
Fe$_2$O$_3$	821.9	87.5	—	695.0	556.8	—	—	21
Ga$_2$O	347.5	—	—	—	—	—	—	13
Ga$_2$O$_3$	1 083.5	84.78	992.3	—	—	—	—	8
Gd$_2$O$_3$	1 817.1	150.7	1 730.4	1 651.7	—	—	—	13
GeO	30.7(g)	224.0(g)	54.2(g)	69.2(g)	105.7(g)	—	—	17

(*continued*)

Table 8.8e   OXIDES—*continued*

Compound	$-\Delta H_{298}$ 25°C	S_{298} 25°C	$-\Delta G_{300}$ 27°C	$-\Delta G_{500}$ 227°C	$-\Delta G_{1000}$ 727°C	$-\Delta G_{1500}$ 1227°C	$-\Delta G_{2000}$ 1727°C	Accuracy ±kJ
GeO_2	580.2	39.7	525.3	488.7	397.1	—	—	8
H_2O	286.0(m)	70.13(m)	237.0(m)	—	—	—	—	0.8
H_2O	242.0(g)	188.8(g)	228.6(g)	219.0(g)	192.6(g)	164.1(g)	134.4(g)	1.3
H_2O_2	187.1(m)	109.5(m)	118.1(m)	—	—	—	—	2.1
H_2O_2	135.2(g)	226.9(g)	—	—	—	—	—	1.3
HfO_2	1 113.7	59.5	1 053.4	1 014.0	919.4	828.1	739.0	4
HgO (red)	90.9	70.3	58.32	36.59	—	—	—	4
Ho_2O_3	1 882.4	158.3	1 792.4	1 732.1	—	—	—	13
In_2O_3	927.4	—	834.0	771.6	—	—	—	13
IrO_2	241.5	56.5	186.1	149.2	57.0	—	—	17
IrO_3	−13.4(g)	290.5(g)	—	—	—	85.0	—	76
K_2O	363.3	—	320.7	295.2	229.4	163.3	—	3
KO_2	284.6	122.6	—	—	—	—	—	2.1
La_2O_3	1 794.1	128.1	1 706.2	1 648.4	1 509.0	—	—	8
Li_2O	596.6	37.93	560.2	534.2	467.7	—	—	4
Li_2O_2	635.6	—	—	—	—	—	—	8
MgO	601.6	26.97	571.1	550.1	498.6	440.5	329.9	6.3
MnO	385.2	59.9	363.0	348.8	312.3	275.9	234.5	13
Mn_3O_4	1 387.5	148.6	1 280.3	1 209.6	1 035.8	—	—	13
Mn_2O_3	960.5	110.5	882.2	821.9	703.8	—	—	13
MnO_2	520.4	53.2	462.2	424.5	334.9	—	—	13
Mn_2O_7	728.9	—	—	—	—	—	—	10.5
MoO_2	588.7	50.0	533.4	496.5	404.3	—	—	4
MoO_3	746.1	77.9	668.6	617.1	486.9	—	—	4
Na_2O	415.2	75.1	376.1	350.1	285.0	—	—	8
Na_2O_2	515.0	94.6	—	—	—	—	—	6.3
NaO_2	261.7	116.0	—	—	—	—	—	4
NbO	419.8	46.0	391.2	373.1	326.5	—	—	17
NbO_2	799.3	54.55	743.2	705.5	612.1	518.3	425.0	8
Nb_2O_5	1 900.8	137.3	1 766.0	1 676.4	—	—	—	13
Nd_2O_3	1 809.1	158.6	—	—	—	—	—	17
NiO	240.7	38.1	213.1	194.7	150.7	108.4	57.4	13
NpO_2	1 030.0	80.4	—	—	—	—	—	42
OsO_4	393.9	136.9	—	—	—	—	—	13
P_2O_5	1 493.0	229.0	1 381.6	—	—	—	—	25
PbO	219.4	66.3	188.8	168.3	117.2	—	—	13
Pb_3O_4	719.0	211.4	600.9	519.7	325.4	—	—	25
PbO_2	274.6	71.9	215.3	175.8	76.9	—	—	8
PdO	112.6	39.3	82.2	62.0	11.3	—	—	17
Pr_2O_3	1 828.8	158.6	—	—	—	—	—	6
$PrO_{1.72}$	937.8	—	—	—	—	—	—	6.3
PrO_2	974.9	80.0	—	—	—	—	—	17
PtO_2	−168.7(g)	256.0(g)	—	—	—	157.8	—	8
PuO_2	1 058.4	82.5	1 003.6	968.8	882.2	795	—	21
Rb_2O	330.3	—	—	—	—	—	—	13
Rb_2O_3	527.5	—	—	—	—	—	—	42
RbO_2	284.7	—	—	—	—	—	—	25
ReO_2	432.9	62.8	379.0	343.0	253.2	—	—	13
ReO_3	611.3	80.8	—	—	—	—	—	21
Re_2O_7	1 249.1	207.5	1 073.7	959.0	—	—	—	8
Rh_2O_3	383.0	92.1	—	—	—	—	—	17
RhO_2	195.9(g)	263.8(g)	—	—	—	157.8	—	4
RuO_2	304.4	60.7	252.3	217.6	130.9	44.2	—	8
RuO_3	78.3(g)	276.2(g)	—	—	—	—	—	—
RuO_4	180.9(g)	290.1(g)	—	—	—	—	—	—
SO	−5.0(g)	222.3(g)	—	73.2(g)	70.3(g)	79.1(g)	82.0(g)	8
SO_2	297.05(g)	248.07(g)	340.8(g)	326.6(g)	290.1(g)	254.1(g)	217.7(g)	2.1
SO_3	395.2(g)	256.2(g)	—	376.4(g)	295.6(g)	214.4(g)	—	6.3
Sb_2O_3	699.2	123.1	—	—	—	—	—	13

(*continued*)

Table 8.8e　OXIDES—*continued*

Compound	$-\Delta H_{298}$ 25°C	S_{298} 25°C	$-\Delta G_{300}$ 27°C	$-\Delta G_{500}$ 227°C	$-\Delta G_{1000}$ 727°C	$-\Delta G_{1500}$ 1 227°C	$-\Delta G_{2000}$ 1 727°C	Accuracy ±kJ
SbO_2	454.0	63.6	—	—	—	—	—	25
Sb_2O_5	1 008.0	125.2	—	—	—	—	—	63
Sc_2O_3	1 906.7	77.0	1 818.7	1 756.8	—	—	—	—
SeO_2	236.1	66.7	—	—	—	—	—	4
SiO	98.4(g)	211.64(g)	126.9(g)	144.5(g)	185.1(g)	223.7(g)	—	13
SiO_2	910.9(q)	41.5(q)	—	—	726.9	640.0	538.5	13
Sm_2O_3	1 833.0	151.1	1 744.2	1 685.2	—	—	—	13
SnO	286.4	56.5	—	—	—	—	—	4
SnO_2	580.7	52.3	519.4	478.3	—	—	—	4
SrO	592.3	55.5	562.4	542.3	492.3	—	—	13
SrO_2	633.7	59.0	—	—	—	—	—	17
Ta_2O_5	2 047.3	143.2	1 910.9	1 822.1	1 612.8	1 413.0	1 220.0	21
Tb_2O_3	1 828.8	—	—	—	—	—	—	8
$TbO_{1.71}$	934.9	—	—	—	—	—	—	4
$TbO_{1.8}$	947.9	—	—	—	—	—	—	4
TcO_2	433.3	58.6	—	—	—	—	—	13
Tc_2O_7	1 113.7	184.2	996.0	917.7	722.2	—	—	17
TeO_2	322.4	74.1	268.3	232.2	138.6	—	—	8
ThO_2	1 227.6	65.3	1 173.6	1 135.0	1 048.8	960.9	—	21
TiO	542.9	34.8	509.5	491.9	447.1	402.7	352.9	17
Ti_2O_3	1 521.6	77.3	1 427.1	375.8	245.7	1 116.5	986.4	17
Ti_3O_5	2 457.2	129.4	2 309.4	2 224.4	2 010.1	1 797.0	1 581.8	13
TiO_2	944.1	50.2	862.1	827.3	739.5	652.7	564.8	16
Tl_2O	167.4	134.3	—	—	—	—	—	4
Tl_2O_3	390.6	137.3	—	—	—	—	—	25
Tm_2O_3	1 889.9	—	—	—	—	—	—	8
UO_2	1 085.2	77.9	1 032.0	996.9	913.6	827.7	740.6	13
U_4O_9	4 513	336.2	—	—	—	—	—	42
U_3O_8	3 575.9	282.6	—	—	—	—	—	21
UO_3	1 230.7	98.8	—	—	—	—	—	21
VO	432.0	38.9	413.5	398.5	360.8	322.9	—	25
V_2O_3	1 219.4	98.4	1 147.0	1 099.7	982.0	864.4	—	33
VO_2	713.7	51.5	678.4	643.8	567.2	—	—	21
V_2O_5	1 551.3	131.0	1 454.1	1 373.7	1 172.8	—	—	50
WO_3	838.6	75.9						
WO_2	589.9	50.6	—	480.6	405.7	334.5	—	13
Y_2O_3	1 906.7	99.2	1 817.9	1 759.3	1 618.6	1 482.1	1 348.1	21
Yb_2O_3	1 815.8	133.1	—	—	—	—	—	8
ZnO	350.8	43.5	320.7	301.0	256.6	174.6	82.0	8
ZrO_2	1 101.3	50.7	1 042.2	1 003.7	909.9	818.2	729.5	17

Table 8.8f　SULPHIDES

Heats and free energies of formation in kJ and standard entropies.
In calculating the heats of formation, rhombic sulphur is taken as the standard state. For the free energies of formation the perfect diatomic gas S_2 is taken as the standard state at all temperatures.

Compound	$-\Delta H_{298}$ 25°C	S_{298} 25°C	$-\Delta G_{300}$ 27°C	$-\Delta G_{500}$ 227°C	$-\Delta G_{1000}$ 727°C	$-\Delta G_{1500}$ 1 227°C	$-\Delta G_{2000}$ 1 727°C	Accuracy ±kJ
Ag_2S	31.8(α)	144.4(α)	180.4(α)	70.8(β)	53.2(β)	32.2(m)	–	2.1
Al_2S_3	723.9	123.5	—	—	—	—	—	17
As_2S_3	167.4	169.6	—	—	—	—	—	23
B_2S_3	252.4	92.1	—	—	—	—	—	9
BaS	443.8	78.3	481.9	463.9	419.1	—	—	21
BeS	234.0	35.2	272	247	201	—	—	21
Bi_2S_3	201.8	200.5	—	—	—	—	—	—

(continued)

Table 8.8f SULPHIDES—*continued*

Compound	$-\Delta H_{298}$ 25°C	S_{298} 25°C	$-\Delta G_{300}$ 27°C	$-\Delta G_{500}$ 227°C	$-\Delta G_{1000}$ 727°C	$-\Delta G_{1500}$ 1227°C	$-\Delta G_{2000}$ 1727°C	Accuracy ±kJ
CS_2	−87.9(m)	151.6(m)	15.1(g)	16.3(g)	19.7(g)	23.4(g)	—	4
COS	138.5	231.5	207.2	208.9	212.4	218.5	223.1	13
CaS	476.4	56.5	—	491.9	444.2	393.1	319.9	8
CdS	149.4	69.1	185.5	—	—	—	—	4
CeS	456.7	78.3	496.5	479.8	437.9	396.0	354.2	21
Ce_3S_4	1 653.5	255.3	—	—	—	—	—	17
Ce_2S_3	1 188.8	180.4	—	—	—	—	—	13
$CoS_{0.89}$	94.6	52.3	125.2	110.5	73.3	—	—	8
Co_3S_4	359.2	184.6	—	427.9	—	—	—	13
CoS_2	153.2	69.1	—	—	—	—	—	17
CrS	155.7	64.0	—	—	—	—	—	13
Cs_2S	339.5	—	—	—	—	—	—	33
Cu_2S	79.6	120.9	128.5	118.1	100.9	—	—	8
CuS	52.3	66.6	—	—	—	—	—	4
FeS	100.5	60.3	140.2(α)	126.4(β)	97.3(β)	—	—	4
FeS_2	171.6	53.2	—	299.8	90.9	—	—	8
GaS	209.3	57.7	—	—	—	—	—	13
Ga_2S_3	514.0	139.8	—	—	—	—	—	13
GeS	76.2	66.2	46.5	36.7	—	—	—	33
H_2S	20.1(g)	205.6(g)	73.7(g)	65.10(g)	41.24(g)	16.3(g)	−8.0(g)	2.1
HgS	53.4	82.5	—	—	—	—	—	6.3
InS	133.9	69.1	—	—	—	—	—	13
In_5S_6	774.4	174.6	—	—	—	—	—	84
In_2S_3	355.8	163.7	—	—	—	—	—	17
Ir_2S_3	244.5	97.1	—	—	124.8	10.5	—	33
IrS_2	144.8	61.6	—	—	—	—	—	25
K_2S	428.7	—	—	—	—	—	—	14.7
La_2S_3	1 222.3	165.1	1 235.5	—	—	—	—	42
Li_2S	446.6	60.7	—	—	—	—	—	8
MgS	351.6	50.4	—	—	296.8	—	—	8
MnS	213.5	80.4	254.3	241.7	211.0	128.5	143.2(m)	6.3
MnS_2	223.9	99.9	—	—	—	—	—	10.5
Mo_2S_3	410.2	113.0	—	—	309.4	195.1	—	29
MoS_2	275.4	62.6	—	—	203.1	125.2	—	21
Na_2S	374.6	79.5	—	388.1	322.4	206.8	—	17
NaS	201.3	44.8	—	—	—	—	—	14.7
NaS_2	206.0	83.7	—	—	—	—	—	14.7
Ni_3S_2	216.0	134.0	—	252.0	—	—	—	13
NiS	94.2	53.0	—	—	—	—	—	6.3
OsS_2	100.5	—	181.7	144.9	66.2	−8.8	—	17
PbS	98.4	91.3	134.6	120.6	81.7	—	—	4
PtS	82.5	55.06	116.4	96.3	47.7	0.8	—	13
PtS_2	110.1	74.73	179.6	141.1	49.4	—	—	13
Rb_2S	348.3	133.1	—	—	—	—	—	25
ReS_2	139.0	60.7	211.4	177.1	97.1	21.4	—	25
RuS_2	206.0	—	242.8	207.7	125.2	46.9	—	21
S	−277.1(g)	167.9(g)	—	−146.5(g)	−115.1(g)	−83.7(g)	−53.2(g)	0.5
S_2	−129.8(g)	228.2(g)	0(g)	0(g)	0(g)	0(g)	0(g)	4
S_6	−103.0(g)	354.0(g)	179.6	121.8	−23.4	—	—	13
S_8	−99.6(g)	423.3(g)	258.7	174.2	−37.7	—	—	33
Sb_2S_3	205.1	182.1	285	230	—	—	—	29
SiS_2	213.5	80.4	278.4	—	—	—	—	25
SnS	108.4	77.0	143.6	—	—	—	—	6.3
SnS_2	153.6	87.5	—	—	—	—	—	17
SrS	452.6	69.1	488.6	468.5	419	—	—	42
Th_2S_3	1 082.7	—	—	—	—	—	—	10.5

(*continued*)

Table 8.8f SULPHIDES—*continued*

Compound	$-\Delta H_{298}$ 25°C	S_{298} 25°C	$-\Delta G_{300}$ 27°C	$-\Delta G_{500}$ 227°C	$-\Delta G_{1000}$ 727°C	$-\Delta G_{1500}$ 1 227°C	$-\Delta G_{2000}$ 1 727°C	Accuracy ±kJ
TiS	272.1	56.5	—	—	—	—	—	33
Tl$_2$S	95.0	159.0	—	—	—	—	—	4
WS$_2$	202.6	83.7	278.4	243.7	162.4	85.4	—	17
ZnS	205.3	57.8	235.4	219.1	174.3	90.4	−4.9	17

Table 8.8g HALIDES

Heats and free energies of formation in kJ and standard entropies.
For the free energies of formation the standard states for the halogens at all temperatures are the perfect diatomic gases at a pressure of one atmosphere.

Compound	$-\Delta H_{298}$ 25°C	S_{298} 25°C	$-\Delta G_{300}$ 27°C	$-\Delta G_{500}$ 227°C	$-\Delta G_{1000}$ 727°C	$-\Delta G_{1500}$ 1 227°C	$-\Delta G_{2000}$ 1 727°C	Accuracy ±kJ
AgF	206.0	83.7	185.5	173.8	161.2(m)	146.5(m)	—	8
AgF$_2$	359.6	—	304.8	276.3	213.5(m)	—	—	13
AgCl	126.9	96.3	109.3	98.8	80.4(m)	69.9(m)	—	0.8
AgBr	99.2	107.2	97.1	86.7	69.5(m)	65.7(m)	—	0.8
AgI	62.4	115.6	77.0	66.6	50.2	46.1	—	0.8
AlF	265.4(g)	215.2(g)	—	—	—	—	—	4
AlF$_3$	1 511.1(g)	66.6	1 418.5	1 377.9	—	—	—	0.8
AlCl	51.5(g)	227.8(g)	75.3(g)	94.6(g)	133.5(g)	164.9(g)	196.1(g)	2
AlCl$_3$	705.9	110.1	572.3(g)	566.5(g)	334.7(g)	500.7(g)	—	4
Al$_2$Cl$_6$	1 275.3(g)	475.8	—	1 238.0(g)	1 194.9(g)	—	—	2
AlBr$_3$	511.3	180.3	515.0	—	—	—	—	2
AlI$_3$	310.2	189.6	—	—	—	—	—	6
AsF$_3$	958.0(m)	180.5(m)	—	—	—	—	—	4
AsF$_3$	921.7(g)	289.3(g)	—	—	—	—	—	4
AsCl$_3$	335.8(m)	233.6(m)	294.8(m)	278.0(g)	254.1(g)	217.7(g)	—	25
AsBr$_3$	132.1(g)	363.9(g)	—	—	—	—	—	6
AsI$_3$	64.9	213.1	94.6	61.5	—	—	—	10.5
AuF$_3$	365.4	114.3	—	—	—	—	—	36
AuCl	34.8	92.9	17.6	5.4	—	—	—	4
AuCl$_3$	115.1	148.2	57.4	18.8	—	—	—	8
AuBr	18.4	119.0	16.3	4.6	—	—	—	4
AuBr$_3$	54.4	—	44.0	4.2	—	—	—	21
AuI	−0.8	—	12.6	0.0	—	—	—	4
BF$_3$	1 136.1(g)	254.6(g)	—	—	—	—	—	2.1
BCl$_3$	427.0(m)	206.0(m)	—	—	—	—	—	13
BCl$_3$	403.1(g)	290.1(g)	—	—	—	—	—	13
BBr$_3$	238.6(m)	228.9(m)	—	—	—	—	—	2
BBr$_3$	204.2(g)	324.5(g)	—	—	—	—	—	2
BaF$_2$	1 207.6	96.3	1 146.3	1 112.4	1 035.8	952.5	—	13
BaCl$_2$	860.8	123.7	—	—	—	—	—	3
BaBr$_2$	755.3	148.6	736.0	702.5	622.2	553.5	—	13
BaI$_2$	602.9	165.2	617.6	584.5	507.9(m)	447.2	—	21
BeF	175.0(g)	205.78(g)	236.6(g)	255.0(g)	299.8(g)	341.6(g)	378.5(g)	17
BeF$_2$	1 027.4	53.4	963.4	—	—	—	—	13
BeCl	−8.3(g)	217.7(g)	16.3(g)	35.6(g)	81.6(g)	125.2(g)	163.3(g)	—
BeCl$_2$	494.0	75.8	452.2	427.1	385.2(m)	—	—	8
BeBr$_2$	353.7	106.3	322	297	—	—	—	25
BeI$_2$	192.5	120.6	186.7	159.9	—	—	—	25
BiF$_3$	910.8	122.6	821	775	—	—	—	25
BiCl	−25.1(g)	255.1(g)	—	—	52.3(g)	—	—	17
BiCl$_3$	379.3	171.6	328.5	294.7	250.7	174.5	—	8
BiI$_3$	150.7	224.8	131.5	88	—	—	—	13

(continued)

Table 8.8g HALIDES—*continued*

Compound	$-\Delta H_{298}$ 25°C	S_{298} 25°C	$-\Delta G_{300}$ 27°C	$-\Delta G_{500}$ 227°C	$-\Delta G_{1000}$ 727°C	$-\Delta G_{1500}$ 1 227°C	$-\Delta G_{2000}$ 1 727°C	*Accuracy* ±kJ
CF_4	933.5(g)	262.5(g)	888(g)	858(g)	782(g)	707(g)	—	21
CCl_4	135.3(m)	214.8(m)	68.2	—	—	—	—	2.1
CCl_4	103(g)	309.4(g)	—	36.0	—	—	—	4
CBr_4	−50.2(g)	358.4(g)	—	—	—	—	—	17
CaF_2	1 220.2	68.87	1 167.1	1 132.1	1 050.7	970.3	—	9
$CaCl_2$	796.2	104.6	752.5	724	653	599(m)	—	5
$CaBr_2$	680.4	134.0	663.6	633.0	562.3	502	—	10.5
CaI_2	535.1	142	551.4	519.6	498.4	385.6(m)	—	21
CdF_2	697.1	83.6	648.1	613.4	538.0	—	—	5
$CdCl_2$	391.0	115.3	—	—	—	—	—	2.1
$CdBr_2$	314.8	139.0	301.4	272.1	202.2	—	—	8
CdI_2	204.2	158.4	—	—	—	—	—	4
$CeCl_3$	1 058.4	147.8	981.8	932.0	818.5	—	—	8
CeI_3	650.4	214.7	679.1	631.4	520.4	424.1(m)	—	21
CoF_2	692.9	82.1	646.4	618.4	547.6	—	—	13
CoF_3	811.2	95.4	744.3	706.3	602.3	—	—	33
$CoCl_2$	310.2	109.3	425.5	239.5	174.2	150.3	—	5
$CoBr_2$	221.0	134.0	203.9	177.1	122.7(m)	(87.1)(m)	—	13
CoI_2	87.9	153.2	118.9	85.0	30.6(m)	−5.0(m)	—	21
CrF_2	779.9	89.7	—	—	—	—	—	13
CrF_3	1 113.9	93.95	1 043	996	879	762(m)	—	21
CrF_4	1 199.5	—	1 118	1 065.5	—	—	—	25
$CrCl_2$	395.6	115.3	351.3	327.0	272.1	233.6(m)	252.9(g)	13
$CrCl_3$	556.7	123.01	491.9	445.0	327.4	—	—	21
CrI_2	158.3	—	179.6	151.1	88.3	46.5(m)	—	21
CrI_3	205.1	199.6	—	—	—	—	—	21
CsF	555.1	88.3	527.5	507.0	458.8	416.9	—	8
$CsCl$	433.3	100.1	404.4	384.3	336.6	298.9	—	5
$CsBr$	394.8	113.5	382.3	361.7	318.2	284.7	—	13
CsI	337.0	—	340.4	318.6	273.0	250.0	—	10.5
CuF_2	549.2	—	502.2	475.9	411.3	—	—	17
$CuCl$	137.3	87.1	117.2	105.9	93.8	81.6	—	8
$CuCl_2$	217.6	108.4	163.3	132.7	—	—	—	13
$CuBr$	104.3	96.3	101.3	89.6	70.8	55.7	—	4
$CuBr_2$	141.9	133.9	126.9	98.4	—	—	—	18.8
CuI	67.8	96.7	79.5	65.7	39.4	23.9	—	4
CuI_2	7.1	—	—	—	—	—	—	10.5
$ErCl_3$	959.2	146.9	—	—	—	—	—	4
FeF_2	711.6	87.1	667.6	640.8	574.3	—	—	25
FeF_3	1 042.3	98.4	975	929	824	728(m)	—	54
$FeCl_2$	341.2	120.2	304.4	279.7	222.7(m)	209.88	—	1.7
$FeCl_3$	400.7	142.3	—	—	—	—	—	4
$FeBr_2$	247.4	140.0	239.1	213.1	160.8	127.3(m)	—	4
$FeBr_3$	265.9	173.7	250.4	213.5	—	—	—	17
FeI_2	116.4	170.0	149.5	122.7	68.2(m)	32.7(m)	—	17
$GaCl_3$	523.4	135.2	466.8	431(m)	—	—	—	6.3
$GaBr_3$	386.9	180.0	375.1	339(m)	—	—	—	4
GaI_3	239.4	203.8	248.3	209(m)	—	—	—	13
$GdCl_3$	1 005.3	146.1	—	—	—	—	—	4
GeF_4	1 192.5(g)	302.9(g)	—	—	—	—	—	2
$GeCl_4$	504.8(g)	347.5(g)	463	—	—	—	—	—
$GeBr_4$	330.8	396.9	309.0	266.3(g)	—	—	—	10.5
GeI_4	37.7(g)	429.1(g)	—	—	—	—	—	13
HF	272.7(g)	173.8(g)	273.4(g)	274.7(g)	277.2(g)	278.8(g)	279.7(g)	8
HCl	92.4(g)	186.94(g)	95.0(g)	96.7(g)	100.5(g)	103.4(g)	107.2(g)	1.3
HBr	36.4(g)	198.9(g)	57.65(g)	59.58(g)	63.35(g)	66.65(g)	69.75(g)	1.3
HI	−26.4(g)	206.8(g)	7.70(g)	9.67(g)	13.57(g)	17.2(g)	21.52(g)	1.3
HfF_4	1 931.8	136.0	—	—	—	—	—	8

(*continued*)

Table 8.8g   HALIDES—*continued*

Compound	$-\Delta H_{298}$ 25°C	S_{298} 25°C	$-\Delta G_{300}$ 27°C	$-\Delta G_{500}$ 227°C	$-\Delta G_{1000}$ 727°C	$-\Delta G_{1500}$ 1 227°C	$-\Delta G_{2000}$ 1 727°C	Accuracy ±kJ
HfCl$_4$	991.9	190.9	898.9	—	—	—	—	17
HgF	246.2	—	219.0	—	—	—	—	21
HgF$_2$	397.7	—	—	—	—	—	—	42
HgCl	131.9	98.4	104.7	85.4	—	—	—	1.3
HgCl$_2$	230.3	144.4	183.4	153.2	—	—	—	4
HgBr	103.4	111.4	92.1	72.9	—	—	—	1.3
HgBr$_2$	169.6	162.9	152.0	121.0	—	—	—	1.3
HgI	60.3	121.4	66.2	48.6	—	—	—	0.8
HgI$_2$	105.5	170.8	119.3	87.5	—	—	—	1.7
InCl	186.3	95.0	180	163(m)	—	—	—	10.5
InCl$_2$	362.9	122.2	315.1	—	—	—	—	18
InCl$_3$	537.5	141.0	465	425.0	327(m)	—	—	13
InBr	173.8	113.9	—	—	—	—	—	8
InBr$_3$	411.1	178.7	352	306	—	—	—	13
InI	116.4	131.5	—	—	—	—	—	8
InI$_3$	249.1	—	285	247(m)	—	—	—	13
IrCl	67.0	—	50	42	8	—	—	13
IrCl$_2$	138.2	—	99.2	71	8	—	—	13
IrCl$_3$	254.5	114.9	—	—	—	—	—	13
KCl	436.9	82.5	409.1	389.0	338.7	267.1(m)	223.6(g)	2
KBr	394.0	96.7	379.7	360.9	313.6	280.1(m)	—	3
KF	568.9	66.6	532.6	510.4	455.9	416.2(m)	—	3
KI	327.8	104.3	332.0	312.8	264.6(m)	228.2(m)	—	1.3
LaCl$_3$	1 071.4	(144.4)	999.8	945.2	814.1	757.4(m)	—	8
LaI$_3$	657.3	214.7	680.8	632.2	519.2	423(m)	—	18.8
LiF	617.5	35.6	589.1	565.6	521.2	518.1(m)	—	8
LiCl	405.7	59.0	381.4	364.3	323.6(m)	288.5(m)	—	8
LiBr	349.2	—	340.8	324.1	283.9	255.8(m)	—	8
LiI	271.3	—	279.7	265.0	227.8	201.8(m)	—	8
MgF$_2$	1 113.7	57.4	1 068.1	1 034.6	949.6	—	—	8
MgCl$_2$	642.3	89.6	591.6	559.8	484.8(m)	421.6(m)	349.6(g)	6.3
MgBr$_2$	524.5	117.2	509.4	473.1	404.4(m)	360(m)	—	13
MgI$_2$	360.1	129.8	377.6	347.5	289.7(m)	242.0(m)	—	13
MnF$_2$	795.5	93.16	754	724	657	611(m)	—	25
MnF$_3$	996.5	—	—	—	—	—	—	29
MnCl$_2$	482.3	118.28	442.5	417.0	360.9(m)	324.9(m)	—	6.3
MnBr$_2$	385.1	138.1	373.0	336.2	273.4(m)	231.5(m)	—	8
MnI$_2$	242.8	—	262.9	233.6	172.9(m)	131.0(m)	—	13
MoF$_6$	1 559.6(g)	350.6(g)	1 473.4	—	—	—	—	8
MoCl$_5$	527.4	238.6	—	—	—	—	—	10.5
NaF	575.6	51.5	542.1	521.6	469.3	—	—	4
NaCl	412.8	72.9	384.8	365.1	315.7	253.7	—	2
NaBr	361.95	86.9	351.4	333.6	288.3	259.8(m)	—	4
NaI	288.0	98.4	295.2	276.3	232.8(m)	202.2(m)	—	7
NbF$_5$	1 814.7	160.4	1 702	—	—	—	—	50
NbCl$_2$	407.4	117	—	—	—	—	—	8
NbCl$_{2.67}$	538.4	(137.3)	—	—	—	—	—	8
NbCl$_{3.13}$	601.2	(151.6)	—	—	—	—	—	8
NbCl$_4$	695.0	(184.2)	—	—	—	—	—	—
NbCl$_5$	797.6	(226)	—	—	—	—	—	—
NbBr$_5$	556.4	(306)	—	—	—	—	—	8
NdCl$_3$	1 028.3	—	957.9	913	808	729(m)	—	8
NdI$_3$	628.9	—	652.3	603	490	394(m)	—	13
NiF$_2$	657.2	73.7	612.3	582	515	452	—	6.3
NiCl$_2$	305.6	97.76	259.2	232.4	163.7	109.7	—	2.1
NiBr$_2$	212.2	136.0	202.2	174.2	100.9	71.6	—	8
NiI$_2$	78.3	154.0	116.4	87.1	9.6	−15.5(m)	—	8
NpF$_3$	1 507.2	118.5	—	—	—	—	—	33

(continued)

Table 8.8g   HALIDES—*continued*

Compound	$-\Delta H_{298}$ 25°C	S_{298} 25°C	$-\Delta G_{300}$ 27°C	$-\Delta G_{500}$ 227°C	$-\Delta G_{1000}$ 727°C	$-\Delta G_{1500}$ 1 227°C	$-\Delta G_{2000}$ 1 727°C	Accuracy ±kJ
NpCl₃	904.3	160.3	—	—	—	—	—	21
NpCl₄	199.7	—	—	—	—	—	—	21
PF₃	920.9	272.9(g)	—	—	—	—	—	33
PCl₃	321(m)	218.1(m)	—	—	—	—	—	3.3
PCl₃	288(g)	312.3(g)	268.7(g)	254.4(g)	—	—	—	4
PCl₅	367.1(g)	364.6(g)	294.5(g)	—	—	—	—	13
PBr₃	132.3(g)	348.8(g)	—	—	—	—	—	13
PI₃	18.0(g)	374.6(g)	—	—	—	—	—	21
PbF₂	677.3	161.5	630.8	597.8	530.8	463.8	—	5
PbF₄	930.7	—	845.3	791	—	—	—	18.8
PbCl₂	359.2	136.5	312.8	284.5	220.6	196.8	—	2
PbBr₂	277.6	161.6	263.8	234.9	171.2(m)	—	—	4
PbI₂	175.4	176.7	195.1	166.2	111.4	—	—	8.3
PdCl₂	180.0	103.8	138	105	25	—	—	13
PdBr₂	103.4	—	93.8	64.9	—	—	—	21
PrCl₃	1 057.4	(144.4)	985.2	940.4	833.6	754.0(m)	—	2
PrI₃	654.7	235.3	679.7	631.2	520.7	424.2	—	18.8
PtCl₂	110.9	(129.8)	69.9	41.0	−29	—	—	13
PtCl₄	236.9	(205.1)	152	100.0	—	—	—	17
PtBr₂	80.4	(154.9)	—	—	—	—	—	13
PtBr₄	140.7	(251.2)	117	63	—	—	—	17
PtI₂	16.7	—	—	—	—	—	—	13
PtI₄	72.9	(281.3)	97.1	15.5	—	—	—	33
PuF₃	1 553.0	113.0	—	—	—	—	—	17
PuCl₃	949.4	164.1	—	—	—	—	—	4
PuBr₃	831.9	191.3	—	—	—	—	—	4
RbF	553.4	73.7	527.7	508.4	466.3(m)	427.4(m)	—	8
RbCl	430.8	91.7	403.2	383.9	335.8(m)	298.5(m)	—	8
RbBr	389.4	108.9	378.5	360.9	316.9(m)	279.3(m)	—	8
RbI	328.7	118.1	334.5	317.4	273.0(m)	235.3(m)	—	8
ReF₆	1 163.9(m)	—	1 068(m)	—	—	—	—	50
ReCl₃	263.8	—	200.1	157.4	—	—	—	13
ReBr₃	164.5	—	145.7	101.7	—	—	—	2
RhCl	83.7	83.7	66.2	56.5	21	—	—	13
RhCl₂	163.3	121.4	123.1	94.2	29	—	—	13
RhCl₃	275.4	123.9	—	—	—	—	—	25
RhBr₃	209.3	188.4	—	—	—	—	—	25
RuCl₃	253.3	127.7	182.1	136.1	26.4	—	—	13
RuCl₄	93.4(g)	374.7(g)	64.9(g)	45.6(g)	−5.0(g)	—	—	17
SbF₃	915.9	127.2	860.5	818.5	—	—	—	21
SbCl₃	382.7	187.1	—	297.7	—	—	—	2.1
SbCl₅	438.8(m)	—	357.1(m)	306.9(m)	—	—	—	13
SbBr₃	260.0	210.1	249.5	228.6(m)	—	—	—	38
SbI₃	100.5	215.5	133.6	111.0	—	—	—	33
ScF₃	1 658	90	—	—	—	—	—	50
ScCl₃	900.2	—	828.6	782.9	670	586(m)	—	50
ScBr₃	711.8	—	686.2	638.5	531.7	450.1(m)	—	50
SeF₆	1 117.4(g)	314.4(g)	1 014(g)	—	—	—	—	5
SeCl₄	188.4	194.6	—	—	—	—	—	6.3
SiF₄	1 607.7(g)	282.2(g)	—	—	—	—	—	25
SiCl₂	162.4(g)	282.0(g)	—	—	—	—	—	5
SiCl₄	695.8(m)	239.9(m)	598.3(m)	571.9(m)	—	—	—	8
SiBr₄	461.7(m)	279.6(m)	435(m)	389(m)	—	—	—	29
SiI₄	199.2	265.6	—	—	—	—	—	6
SmCl₂	818.9	—	—	—	—	—	—	4.6
SnCl₂	350.0	—	309.8	283.0	230.7(g)	—	—	6.3
SnCl₄	529.1(m)	258.7(m)	473.9(m)	436.3(g)	360.9(g)	—	—	6.3

(continued)

Table 8.8g   HALIDES—*continued*

Compound	$-\Delta H_{298}$ 25°C	S_{298} 25°C	$-\Delta G_{300}$ 27°C	$-\Delta G_{500}$ 227°C	$-\Delta G_{1000}$ 727°C	$-\Delta G_{1500}$ 1227°C	$-\Delta G_{2000}$ 1727°C	Accuracy ±kJ
$SnBr_2$	266.3	—	256.7	229.9	—	—	—	17
$SnBr_4$	348.2(g)	412.8(g)	322.6(g)	—	—	—	—	11
SnI_2	144.0	167.8	165.4	136.9	—	—	—	21
SrF_2	1 217.7	82.2	—	—	—	—	—	5
$SrCl_2$	829.0	114.9	780.8	755.7	686.6	628	—	5
$SrBr_2$	716.4	143.6	699.2	667.8	597.0	539.7	—	5
SrI_2	563.1	159.1	580.3	548.9	486.9	428.3	—	5
$TaCl_3$	553.5	(155)	—	—	—	—	—	—
$TaCl_4$	708.8	(193)	—	—	—	—	—	—
$TaCl_5$	860.4	(235)	—	—	—	—	—	—
$TaBr_5$	602.9	(306)	—	—	—	—	—	13
TeF_6	1 318.8(g)	337.9(g)	1 223	1 164	—	—	—	5
$TeCl_4$	324.0	200.9	—	—	—	—	—	10.5
$TeBr_4$	195.1	243.6	2 017.2	1 955.1	—	—	—	13
ThF_4	2 112.2	142.14	2 017.2	1 955.1	—	—	—	13
$ThCl_4$	1 187.1	190.5	1 095.7	1 032.0	879	—	—	4
$ThBr_4$	966.1	228.1	933	870.2	721.6	—	—	7
ThI_4	665	—	—	—	—	—	—	50
TiF_3	1 436.2	87.9	—	—	—	—	—	—
TiF_4	1 650.1	131.1	—	—	—	—	—	—
$TiCl_2$	515.7	87.4	—	—	—	—	—	—
$TiCl_3$	722.1	139.8	—	—	—	—	—	—
$TiCl_4$	804(m)	252(m)	—	—	—	—	—	—
$TiBr_2$	407.4	119.7	—	—	—	—	—	—
$TiBr_3$	550.2	171.7	—	—	—	—	—	—
$TiBr_4$	619.6	239.5	—	—	—	—	—	—
TiI_2	241.1	138.1	—	—	—	—	—	—
TiI_4	375.9	246.1	—	—	—	—	—	—
TlF	325.6	95.7	286.6	272.1	251.2	228.2	—	5
TlF_3	573.2	—	—	—	—	—	—	13
$TlCl$	205.2	113.0	185.1	170.4	150.7(m)	—	—	1.7
$TlCl_3$	315.2	152.4	—	—	—	—	—	13
$TlBr$	172.5	125.2	169.2	156.2	136.1(m)	—	—	2.9
TlI	124.3	127.7	137.3	123.5	102.6(m)	—	—	2.5
UF_4	1 884	151.9	1 793.2	1 736.7	1 595.6	1 464.1	—	25
UF_5	2 057.8	188.4	—	—	—	—	—	25
UF_6	2 189.7	227.8	—	1 999.2(g)	—	—	—	25
UCl_3	893.9	159.1	827.7	785.0	678.3	—	—	13
UCl_4	1 051.7	198.5	962.5	904.3	773.3(m)	—	—	8
UCl_5	1 094.8	(243)	—	—	—	—	—	13
UCl_6	1 133.4	285.9	—	—	—	—	—	21
UBr_3	721.4	(188)	695.4	651.0	541.4	—	—	17
UBr_4	826.9	—	794.2	734.8	606.2(m)	—	—	13
UI_3	478.1	(239)	—	—	—	—	—	8
UI_4	529.6	—	565.2	512.9	410(m)	—	—	17
VF_4	1 404.0	121.4	—	—	—	—	—	13
VF_5	1 481.0(m)	209.3(m)	—	—	—	—	—	25
VCl_3	561.0	—	—	—	—	—	—	13
VCl_4	570.2(m)	235.3(m)	498.2(m)	—	—	—	—	13
VBr_2	347.5	125.6	—	—	—	—	—	25
VBr_3	447.9	142.4	—	—	—	—	—	25
VBr_4	393.6(g)	334.9(g)	—	—	—	—	—	25
VI_2	263.8	(147)	—	—	—	—	—	25
VI_3	280.5	(203.1)	—	—	—	—	—	25
WF_6	1 748.3(m)	(251.2)	1 643.7	—	—	—	—	29
WCl_6	414	(217)	310	211.9	100(g)	8(g)	−88(g)	29
YCl_3	974.3	(136.9)	907.7	863.3	769.1	685.4	—	13
YI_3	618.2	(207)	640.4	588.9	490.0	383.8(m)	—	42

(*continued*)

Table 8.8g　HALIDES—*continued*

Compound	$-\Delta H_{298}$ 25°C	S_{298} 25°C	$-\Delta G_{300}$ 27°C	$-\Delta G_{500}$ 227°C	$-\Delta G_{1000}$ 727°C	$-\Delta G_{1500}$ 1 227°C	$-\Delta G_{2000}$ 1 727°C	Accuracy ±kJ
$YbCl_2$	800.0	130.6	—	—	—	—	—	13
ZnF_2	764.9	73.7	721.0	700.0	622.6	—	—	13
$ZnCl_2$	416.6	108.4	369.7	339.1	278.8(m)	338.7(g)	—	1.3
$ZnBr_2$	327.8	136.9	313.6	284.7	237.4	—	—	2.1
ZnI_2	209.3	161.2	231.1	201.4	153.7	—	—	1.7
ZrF_4	1 912.5	104.7	1 820.0	—	—	—	—	13
$ZrCl_4$	982.6	186.3	893.0	835.3	—	—	—	4
$ZrBr_4$	760.3	—	—	—	—	—	—	8
ZrI_4	485.3	—	—	—	—	—	—	8

Table 8.8h　SILICATES AND CARBONATES

Heats and free energies of formation from the constituent oxides in kJ and standard entropies.
The standard state of carbon dioxide is the perfect gas under 1 atm pressure; that of silica is quartz.

Compound	$-\Delta H_{298}$ 25°C	S_{298} 25°C	$-\Delta G_{300}$ 27°C	$-\Delta G_{500}$ 227°C	$-\Delta G_{1000}$ 727°C	$-\Delta G_{1500}$ 1 227°C	Accuracy ±kJ
Ag_2CO_3	81.2	167.5	30.6	−2.9	—	—	13
Al_2SiO_5 (kyanite)	8.4	83.7	—	—	—	—	5
(andalusite)	5.9	99.4	—	—	—	—	5
(sillimanite)	3.0	96.3	—	—	—	—	5
$Al_6Si_2O_{13}$ (mullite)	17.2	275.0	—	—	—	—	3
$BaCO_3$ (witherite)	269.4	112.2	214.9	183.1	97.3	—	17
$BaSiO_3$	159.1	—	112.2	112.2	111.8	111.8	13
Ba_2SiO_4	270.0	—	270.0	270.5	—	—	13
Be_2SiO_4	19.7	64.5	—	—	—	—	3
$CaCO_3$ (calcite)	179.2	92.9	131.0	98.8	18.4	—	8
$CaSiO_3$ (wollastonite)	90.0	82.1	89.2	89.2	88.8	88.3	2.1
Ca_2SiO_4	136.9	120.6	128.1	128.9	131.5	134	6.3
Ca_3SiO_5	113.0	168.7	117.2	118.9	122.7	—	8
$CdCO_3$	98.8	—	48.1	14.6	—	—	13
$CoCO_3$	78.9	88.7	—	—	—	—	21
$CuCO_3$	47.5	87.9	—	—	—	—	4
$FeCO_3$ (siderite)	82.7	92.9	35.6	−0.4	—	—	13
Fe_2SiO_4 (fayalite)	34.3	145.3	—	—	25.1	18.0(m)	13
K_2CO_3	393.7	155.6	342.9	308.6	224.8	—	10
K_2SiO_3	274.2	169.6	260.8	260.4	259.6	—	13
$K_2Si_4O_9$	307.4	265.8	—	—	—	—	13
$K_2Si_2O_5$	299.3	182.1	—	—	—	—	13
Li_2CO_3	226.7	90.2	—	—	37.7	—	15
Li_2SiO_3	139.8	80.4	—	—	—	—	13
$MgCO_3$	117.0	65.7	66.6	32.7	−52.3	—	17
Mg_2SiO_4 (forsterite)	63.2	95.2	63.2	63.2	63.2	63.2	6.3
$MgSiO_3$ (clinoenstatite)	36.4	67.8	36.0	34.8	32.7	30.6	4
$MnCO_3$	116.8	85.8	60.3	22.6	—	—	8
Mn_2SiO_4 (tephroite)	47.3	—	—	—	—	—	8
$MnSiO_3$ (rhodonite)	24.7	89.2	20.9	18.4	12.1	5.9	4
Na_2CO_3	322.0	136.1	280.9	253.7	185.9	—	17
Na_4SiO_4	360.0	195.9	316.9	319.5	325.3	—	25
Na_2SiO_3	232.4	113.9	232.4	232.8	233.2	233.6	33
$Na_2Si_2O_5$	230.6	165.0	233.5	235.2	—	—	13
$NiCO_3$	46.9	85.4	—	—	—	—	13
$PbCO_3$ (cerussite)	87.29	131.0	42.3	12.1	—	—	13
Pb_2SiO_4	16.7	186.7	—	23.4	30.1	39.7	9
$PbSiO_3$	17.6	109.7	—	18.0	19.2	26.4	4

(*continued*)

Table 8.8h   SILICATES AND CARBONATES—*continued*

Compound	$-\Delta H_{298}$ 25°C	S_{298} 25°C	$-\Delta G_{300}$ 27°C	$-\Delta G_{500}$ 227°C	$-\Delta G_{1000}$ 727°C	$-\Delta G_{1500}$ 1 227°C	*Aceuracy* ±kJ
Rb$_2$CO$_3$	404.9	—	—	—	—	—	29
SrCO$_3$	234.9	97.1	183.4	149.1	63.6	−22.2	17
SrSiO$_3$	130.6	—	—	—	—	—	4
Sr$_2$SiO$_4$	209.3	—	—	—	—	—	8
ZnCO$_3$	70.8	82.5	18.4	−16.7	—	—	4
Zn$_2$SiO$_4$	32.6	131.5	31.8	31.4	31.1	28.8	8
ZrSiO$_4$	24.0	84.6	—	—	—	—	13

Table 8.8i   COMPOUND (DOUBLE) OXIDES

Heats and free energies of formation from the constituent oxides in kJ and standard entropies.

Compound	$-\Delta H_{298}$ 25°C	S_{298} 25°C	$-\Delta G_{300}$ 27°C	$-\Delta G_{500}$ 227°C	$-\Delta G_{1000}$ 727°C	$-\Delta G_{1500}$ 1 227°C	*Accuracy* ±kJ
Al$_2$TiO$_5$	8.4	109.7	10.9	12.6	16.7	20.9	8
BaAl$_2$O$_4$	100.5	148.6	108.7	114.2	127.8	141.6	7
Ba$_3$Al$_2$O$_6$	188.4	—	—	—	—	—	9
BaH$_7$O$_3$	128.5	122.4	126.3	124.7	121.0	115.2	17
BaMoO$_3$	208.5	144.5	207.4	206.6	204.6	—	11
BaUO$_4$	205.1	168.7	204.9	204.8	204.6	204.3	18
BeAl$_2$O$_4$	13.0	66.3	13.5	13.8	14.5	15.3	13
CaV$_2$O$_6$	143.6	179.2	146.2	148.0	152.4	(156.8)	3
Ca$_2$V$_2$O$_7$	262.5	220.6	265.6	267.8	273.0	(240.3)	3
Ca$_3$V$_2$O$_8$	322.3	275.0	329.8	334.8	347.4	(314.7)	3
Ca$_3$Al$_2$O$_6$	8.4	205.5	18.9	25.9	43.6	61.1	5
CaAl$_2$O$_4$	15.0	114.1	22.0	26.6	38.2	49.8	3
CaAl$_4$O$_7$	12.6	177.9	23.4	30.6	48.6	66.6	3
CaTiO$_3$	80.8	93.8	81.9	82.6	84.4	86.2	5
Ca$_3$Ti$_2$O$_7$	209.3	234.8	213.7	216.7	224.1	231.5	50
Ca$_4$Ti$_3$O$_{10}$	297.2	328.6	362.7	306.4	315.7	324.9	50
CaZrO$_3$	30.6	93.8	31.6	32.3	33.9	35.5	10
CaHfO$_3$	31.4	100.5	31.8	32.0	32.6	33.2	20
CaMoO$_4$	166.6	122.6	168.1	169.1	171.6	(150.0)	5
CaWO$_4$	146.1	126.4	149.3	151.5	156.8	162.1	11
CaUO$_4$	135.6	143.2	137.0	137.9	140.2	142.5	20
CaFe$_2$O$_5$	40.2	188.8	48.3	53.8	67.4	81.0	5
Ca$_2$Fe$_2$O$_4$	20.9	145.2	36.5	46.8	72.8	98.8	11
CaSnO$_3$	72.8	83.7	70.3	68.6	64.4	60.2	3
Ca$_2$SnO$_4$	71.6	128.3	70.5	69.8	68.0	66.2	3
CaTiO$_3$	27.6	105.1	27.6	27.6	27.5	27.4	2
CdAl$_2$O$_4$	−15.1	125.2	−9.3	−5.5	4.1	13.7	5
CdWO$_4$	79.1	154.9	86.3	92.1	103.2	115.3	8
CeAlO$_3$	(13.6)	106.7	15.4	16.5	19.6	22.5	15
CoTiO$_3$	24.7	96.9	22.8	21.5	18.3	15.1	5
CoWO$_4$	62.0	126.4	61.3	60.7	59.5	58.3	5
CoAl$_2$O$_4$	33.5	99.6	32.3	31.3	29.1	26.9	5
CoCr$_2$O$_4$	59.9	126.8	57.7	56.2	52.6	48.9	3
CoFe$_2$O$_4$	27.6	142.7	28.3	28.8	29.9	30.1	3
CuAlO$_2$	11.9	67.0	10.4	9.4	6.8	4.3	3
CuAl$_2$O$_4$	−18.4	111.3	−13.1	−9.6	−0.8	8.0	3
CuGaO$_2$	7.1	83.3	5.4	4.3	1.5	—	3
CuGa$_2$O$_4$	−17.2	146.5	−11.5	−7.7	1.8	10.3	3
CuFeO$_2$	36.4	88.9	36.2	35.7	35.0	34.3	10
CuFe$_2$O$_4$	−10.0	146.4	(−5.0)	(−1.7)	6.6	14.9	10

(*continued*)

Table 8.8i COMPOUND (DOUBLE) OXIDES—*continued*

Compound	$-\Delta H_{298}$ 25°C	S_{298} 25°C	$-\Delta G_{300}$ 27°C	$-\Delta G_{500}$ 227°C	$-\Delta G_{1000}$ 727°C	$-\Delta G_{1500}$ 1 227°C	Accuracy ±kJ
$CuCr_2O_4$	−3.1	129.8	(−1.3)	(−0.2)	2.8	5.7	3
$FeTiO_3$	29.7	105.9	28.7	28.1	26.4	24.8	5
$FeAl_2O_4$	27.8	106.7	26.9	26.2	24.5	22.9	2
Fe_2ZnO_4	5.0	134.9	6.1	6.8	8.6	10.4	5
$FeMoO_4$	54.4	129.3	52.2	50.8	47.1	(20.0)	20
$FeWO_4$	75.3	131.9	74.4	73.9	72.4	71.0	13
Fe_2NiO_4	22.6	126.0	22.7	22.8	23.0	23.2	10
Fe_2CoO_4	27.6	142.7	28.3	28.7	29.9	31.0	5
$FeCr_2O_4$	52.1	146.9	54.1	55.6	58.0	61.5	3
FeV_2O_4	24.3	157.0	24.3	24.3	24.3	—	3
Li_2TiO_3	129.3	91.7	130.3	132.0	132.7	—	12
$LiAlO_2$	54.0	53.2	56.8	58.3	62.6	66.9	7
$LiFeO_2$	17.8	75.3	21.6	24.1	30.4	36.7	3
Mg_2TiO_4	—	103.8	—	—	—	—	—
$MgTiO_3$	23.0	74.5	22.2	21.6	20.2	18.8	3
$MgTi_2O_5$	—	127.3	—	—	—	—	—
$MgAl_2O_4$	35.6	80.6	36.4	36.9	38.2	39.5	3
$MgFe_2O_4$	15.5	123.9	18.3	20.8	25.5	30.2	3
$MgMn_2O_4$	11.3	—	—	—	—	—	—
$MgCr_2O_4$	41.9	105.9	41.2	40.7	39.6	38.5	5
$MgMoO_4$	54.0	118.9	58.2	61.0	68.1	(61.0)	5
$MgWO_4$	73.3	100.9	72.7	72.1	71.3	70.3	3
$Mg_2V_2O_7$	81.8	200.1	86.5	89.6	97.4	105.2	3
MgV_2O_6	49.2	160.7	50.3	51.0	52.8	—	3
$MnAl_2O_4$	37.7	103.8	35.5	34.2	30.6	27.1	5
$MnCr_2O_4$	69.1	134.0	67.0	65.6	62.0	58.5	5
$MnMoO_4$	61.1	136.0	60.6	60.3	59.5	(52.0)	12
Na_2TiO_3	213.5	121.8	212.4	211.7	209.9	(206.0)	20
$Na_2Ti_2O_5$	232.3	173.7	231.7	232.3	230.2	(224.0)	25
$Na_2Ti_3O_7$	237.3	234.0	239.7	241.2	246.1	(247.0)	25
$NaAlO_2$	87.5	70.7	89.8	91.3	95.2	99.0	5
Na_2CrO_4	334.5	185.9	346.1	353.9	373.3	392.6	14
$NaCrO_2$	101.5	83.3	103.0	103.1	106.6	109.4	2.5
$NaFeO_2$	44.0	88.3	46.1	47.5	51.0	54.5	10
Na_2MoO_4	306.0	159.5	308.0	309.3	312.6	(315.9)	50
$NaVO_3$	162.0	113.9	165.3	167.5	173.0	—	8
$Na_4V_2O_7$	534.6	318.6	545.9	553.5	572.4	—	18
Na_3VO_4	357.5	189.6	361.0	363.3	369.2	—	13
$NiAl_2O_4$	6.1	98.4	8.9	10.7	15.3	19.9	2
$NiTiO_3$	18.0	80.2	15.5	13.9	9.7	5.6	9
$NiWO_4$	45.2	118.0	46.5	47.3	49.4	51.6	9
$NiCr_2O_4$	13.6	124.3	15.1	16.1	18.6	21.1	5
$NiGa_2O_4$	4.4	130.2	6.6	8.1	11.8	15.7	2
$PbTiO_3$	34.6	112.0	33.2	32.4	29.9	27.5	20
$Pb_2V_2O_7$	146.9	—	—	—	—	—	7
$Pb_3V_2O_8$	170.8	—	—	—	—	—	9
$PbMoO_4$	41.9	166.2	48.5	52.9	63.9	—	15
$PbWO_4$	58.6	167.4	66.1	71.2	83.8	96.4	9
$SrTiO_3$	134.0	108.4	134.8	135.3	136.5	138.8	4
Sr_2TiO_4	159.1	159.1	158.4	157.9	156.7	155.5	9
$SrHfO_3$	78.3	113.0	77.7	77.3	76.3	75.3	20
$SrAl_2O_4$	16.6	108.8	17.3	17.7	18.8	19.9	25
$SrMoO_4$	212.2	128.9	210.9	210.0	207.8	—	9
$SrWO_4$	186.3	134.0	187.0	187.5	188.8	190.0	20
Zn_2TiO_4	3.3	148.2	3.6	8.8	13.8	—	3
$ZnTiO_3$	6.7	95.0	7.0	7.2	7.7	—	3
$ZnFe_2O_4$	5.0	134.9	6.1	6.8	8.6	10.4	5
$ZnWO_4$	41.9	144.4	49.3	54.3	66.7	79.1	3
$ZnAl_2O_4$	45.0	87.1	42.7	41.2	37.3	33.5	2
$ZnCr_2O_4$	62.8	116.4	60.3	58.6	54.3	50.1	2

Table 8.8j   PHOSPHIDES

Heats of formation in kJ and standard entropies.

Compound	$-\Delta H_{298}$ 25°C	S_{298} 25°C	Accuracy ±kJ	Compound	$-\Delta H_{298}$ 25°C	S_{298} 25°C	Accuracy ±kJ
AgP_2	44.8	87.9	9	InP	75.3	59.8	9
AgP_3	69.1	105.5	11	Mg_3P_2	464.6	77.4	80
AlP	164.5	47.3	3	MnP	96.3	52.3	17
Au_2P_3	97.5	150.7	15	MnP_3	174.1	96.7	31
Ca_3P_2	506.5	123.9	25				
				Ni_3P	200.2	106.3	11
Co_2P	157.5	77.4	15	Ni_5P_2	391.4	185.0	19
CoP	125.6	50.2	21	Ni_2P	164.5	77.4	18
CoP_3	204.7	98.4	21	Ni_6P_5	556.3	276.3	60
Cu_3P	129.0	119.3	18	NiP_2	129.3	73.3	15
CuP_2	90.1	81.6	11				
				NiP_3	157.8	98.4	17
Fe_3P	164.1	101.7	9	SiP	62.0	32.7	15
Fe_2P	160.3	72.4	9	ThP	361.3	70.3	40
Fe_1P	138.1	—	13	Th_3P_4	1 195.1	247.0	110
FeP_2	221.0	—	17	Zn_3P_2	159.0	150.7	18
GaP	122.2	52.3	9				
				ZnP_2	101.7	60.3	13
GeP	27.2	61.1	11				

The standard state is white phosphorus and for P white → P red, $\Delta H = -17.4\,\text{kJ g-atom}^{-1}$. With respect to red phosphorus, heats of formation are thus less negative per g-atom P by this amount.

Table 8.9a   PHOSPHIDES

Dissociation pressures in mmHg*.

Dissociating phase	Other condensed phase	Temp. °C	Phosphorus pressure mmHg p_{P_2}	p_{P_4}	Σp
$AgP_3(s)$	$AgP_2(s)$	411–501		$\log p = -\dfrac{7\,766}{T} + 12.74$	
$AgP_2(s)$	Ag(s)	420–506		$\log p = -\dfrac{7\,186}{T} + 11.60$	
$Au_2P_3(s)$	Au(s)	501–700		$\log p = -\dfrac{9\,124}{T} + 12.21$	
$CoP_3(s)$	CoP(s)	934	15	27	42
		994	54	113	167
		1 037	120	280	400
CoP(s)	$Co_2P(s)$	1 215	—	—	24
$CuP_2(s)$	$Cu_3P(s)$	632	—	—	18
		672	—	—	65
		712	—	—	200
		762	—	—	698
$FeP_2(s)$	FeP(s)	892	16.5	78.5	95
		922	34	166	200
		953	63	329	392
		973	92	516	608
FeP(s)	$Fe_2P(s)$	1 175	12.5	—	13
		1 215	25.5	—	27
GeP(s)	Ge(s)	500	—	—	64
		524	—	—	156
		559	—	—	394
$MnP_3(s)$	MnP(s)	580–680		$\log p = -\dfrac{11\,856}{T} + 14.948$	

(continued)

Table 8.9a   PHOSPHIDES—*continued*

Dissociating phase	Other condensed phase	Temp. °C	Phosphorus pressure mmHg		
			p_{P_2}	p_{P_4}	Σp
$NiP_3(s)$	$NiP_2(s)$	597	—	135	135
		644	—	460	460
		667	—	850	850
$NiP_2(s)$	$Ni_6P_5(s)$	740	4	109	113
		761	7	193	200
		780	11	323	334
		804	19	585	604
$Ni_6P_5(s)$	$Ni_2P(s)$	742	2	27	29
		761	3	43	46
		780	5	75	80
		806	11	172	183
		823	16	262	278
$OsP_2(s)$	$Os(s)$	1 190	—	—	8
$ReP_3(s)$	$ReP_2(s)$	863	11	57	68
		904	31	186	217
		935	63	430	493
		956	101	762	863
$ReP_2(s)$	$ReP(s)$	935	19	39	58
		986	52	120	172
		1 028	111	278	389
		1 059	187	494	681
$ReP(s)$	$Re(s)$	1 110	8.5	—	9
		1 141	14	—	15
		1 172	23	—	24.5
		1 214	43	—	46.5
$RhP_3(s)$	$RhP_2(s)$	968	19	22	41
		1 012	39	45	84
		1 038	59	71	130
		1 067	98	125	223
$RuP_2(s)$	$RuP(s)$	1 190	34	—	34
$SiP(s)$	$SiP_{0.2}(s)$	1 010	33	34	67
		1 055	76	86	162
		1 068	96	115	211
		1 101	169	220	389
$TaP_2(s)$	$TaP(s)$	787	7	111	118
		816	14	221	235
		836	23	397	420
		846	29	494	523
$VP_2(s)$	$VP(s)$	657	—	148	148
		670	—	195	195
		680	—	298	298
		689	—	402	402
$WP_{0.96}$	$WP_{0.65}$	750	1.65×10^{-5}	—	1.65×10^{-5}
		850	4.6×10^{-4}	—	4.6×10^{-4}
		950	7.2×10^{-4}	—	7.2×10^{-4}
$ZrP_2(s)$	$ZrP(s)$	729	2	26	28
		774	5	74	79
		814	12	165	177
		848	25	357	382

Ref. 5.
*To convert the pressures given in mmHg to pascals multiply by 133.322.

Table 8.9b   SULPHIDES

Dissociation pressures in mmHg[‡]
A = Dissociating phase. B = Other condensed phase.

Phases	Temp. °C	Vapour pressure mmHg p_{S_2}	Σp	Phases	Temp. °C	Vapour pressure mmHg p_{S_2}	Σp
A As$_2$S$_2$(g)	755–1 075	$\log K_p$		A PtS(s)	1 060	30	30
		$= \log \dfrac{p_{S_2}^4 \cdot p_{As_4} \cdot p_{As_2} \cdot p_{As^2}}{p_{As_2S_2^4}}$		B Pt(s)	1 110	75	75
					1 186	249	249
		$= -\dfrac{312\,900}{T} + 56.7$		A ReS$_2$(s)	1 110	13	13
				B Re(s)	1 189	55	55
A CoS$_{1.92}$(s)	700	—	38		1 225	96	96
	730	—	97	A Rh$_2$S$_5$(s)	715	—	44
	760	—	226		757	—	130
ACoS$_{1.5}$(s)	700	—	40	B Rh$_2$S$_3$(s)	790	—	300
	730	—	86		830	—	610
	760	—	200	A Rh$_2$S$_3$(s)	953	—	19
A CoS$_{1.3}$(s)	700	—	31		1 003	—	56
	730	—	79	B Rh$_3$S$_4$(s)	1 043	—	137
	760	—	162		1 083	—	300
A CuS(s)	460	19	90	A Rh$_3$S$_4$(s)	953	—	8
B Cu$_2$S(s)	474	34	210		1 003	—	24
A FeS$_{1.94}$(s)	629	—	66	B Rh$_9$S$_8$(s)	1 043	—	54
B FeS$_{1.12}$(s)	649	—	170		1 083	—	120
	659	—	258	A RuS$_2$(s)	1 123	3.0	3.0
	669	—	445		1 153	5.3	5.3
A Ir$_3$S$_8$(s)	880	—	400	B Ru(s)	1 184	9.7	9.7
B IrS$_2$(s)	—	—	—		1 208	15.3	15.3
A IrS$_2$(s)	880	—	52	A Th$_3$S$_7$(s)	651	—	43
B Ir$_2$S$_3$(s)	904	—	75		676	—	87
	944	—	153	B ThS$_2$(s)	713	—	163
A Ir$_2$S$_3$(s)	1 020	—	55		754	—	381
B Ir (s)	1 056	—	109	A TiS$_3$(s)	500	38.5	91[†]
	1 073	—	148		525	83	223[†]
A MnS$_2$(s)	408	—	424	B TiS$_2$(s)	538	122	349[†]
B MnS(s)	—	—	—		551	173	429
A NiS$_2$(s)	650	—	42	A US$_3$(s)	500	—	~2
	700	—	153		608	—	34
	730	—	325	B US$_2$(s)	650	—	155
	760	—	650		700	—	470
A OsS$_2$(s)	944	—	22	A VS$_4$(s)	390	—	48
	994	—	85	B V$_2$S$_3$(s)	412	—	131
B Os(s)	1 044	—	214		440	—	350
	1 094	—	490	A ZrS$_{2.6}$(s)	783	52	52
A PdS$_2$(s)	451	16	88*		814	98	98
B PdS(s)	476	35	210*	B ZrS$_2$(s)	847	187	187
	501	72	425*		874	311	311
A PtS$_2$(s)	616	—	38				
B PtS(s)	651	—	98				
	691	—	240				

*t°C	p_{S_6}	p_{S_8}	†t°C	p_{S_6}	p_{S_8}
451	52	20	500	43	9.5
476	125	50	525	112.5	27.5
501	254	99	538	181	46
			551	280	76

Ref. 6.
[‡]To convert the pressures given in mmHg to pascals multiply by 133.322.

8.6   Molar heat capacities and specific heats

Table 8.10   ELEMENTS

Molar heat capacities.

The molar heat capacity C_p may be given empirically by an equation of the form:

$$C_p = 4.186\,8(a + 10^{-3}bT + 10^5cT^{-2})\,\mathrm{J\,K^{-1}\,mol^{-1}}$$

where a, b and c are constants and T = temperature in K.
The values of the constants are given below.

Element	a	b	c	Remarks	Temp. range K
Ag(s)	5.09	2.04	0.36		298–m.p.
Ag(l)	7.3	—	—		m.p.–1 600
Al(s)	4.94	2.96	—		298–m.p.
Al(l)	7.0	—	—		m.p.–1 273
As(s)	5.54	1.32	—		273–1 090
Au(s)	5.66	1.24	—		298–m.p.
Au(l)	7.0	—	—		m.p.–1 575
B (amorph.)	3.835	2.39	−1.50		298–1 240
B (cryst.)	4.735	1.38	−2.20		298–1 100
Ba(β)	−1.36	19.2	—		673–m.p.
Ba(l)	10.5		—		m.p.–1 105
Be(s)	4.54	2.12	−0.82		298–1 173
Be(l)	6.08	0.515	—		1 560–2 200
Bi(s)	4.49	5.40	—		298–m.p.
Bi(l)	4.78	1.47	5.05		m.p.–820
Br$_2$(l)	17.2	—	—		300
Br$_2$(g)	8.93	0.11	−0.31		298–1 600
C (graph)	0.026	9.307	−0.354	$-4.155 \times 10^{-6}T^2$	298–1 100
	5.841	0.104			1 100–4 000
C (diam)	2.18	3.16			298–1 200
Ca(α)	6.064	−1.736	—	$+5.67 \times 10^{-6}T^2$	273–720
Ca(β)	−0.086	9.86	—		720–m.p.
Ca(l)	7.00	—	—		m.p.–1 250
Cd(s)	5.31	2.94	—		298–m.p.
Cd(l)	7.1	—	—		m.p.–1 100
Ce(α)	5.613	2.485	—	$+0.97 \times 10^{-6}T^2$	298–1 003
Ce(β)	9.05	—	—		1 003–m.p.
Ce(l)	9.35	—	—		m.p.–1 373
Cl$_2$(g)	8.82	0.06	−0.68		298–3 000
Co(α)	5.11	3.42	−0.21		298–650
Co(β)	3.30	5.86	—		718–1 400
Co(γ)	9.60	—	—		1 400–m.p.
Co(l)	9.65	—	—		m.p.–1 900
Cr(s)	5.84	2.36	−0.88		298–1 823
Cr(l)	9.4	—	—		m.p.
Cs(s)	7.42	—	—		298–m.p.
Cs(l)	7.62	—	—		m.p.–400
Cu(s)	5.41	1.40	—		298–m.p.
Cu(l)	7.5	—	—		m.p.–1 600
F$_2$(g)	8.29	0.44	−0.80		—
Fe(α, β, δ)	8.873	1.474	—	$-56.92T^{-1/2}$	298–m.p.
Fe(γ)	5.85	2.02	—		1 187–1 664
Fe(l)	10.0	—	—		m.p.–
Ga(s)	6.25	—	—		298–m.p.
Ga(l)	6.65	—	—		295–373
Ge(s)	5.16	1.40	—		298–m.p.
Ge(l)	6.60	—	—		m.p.–1 573
H$_2$(g)	6.52	0.78	0.12		298–3 000
Hf(s)	5.61	1.82	—		298–1 346

(continued)

Table 8.10   ELEMENTS—*continued*

Element	a	b	c	Remarks	Temp. range K
Hg(l)	6.61	—	—		298–b.p.
I_2(s)	9.59	11.90	—		298–387
I_2(l)	19.2	—	—		387–456
I_2(g)	8.89	—	—		456–1 500
In(s)	5.81	2.50	—		298–m.p.
In(l)	7.24	−0.33	—		m.p.–800
Ir(s)	5.56	1.42	—		298–1 800
K(s)	6.04	3.12	—		298–m.p.
K(l)	8.886	−4.57	—	$+2.94 \times 10^{-6} T^2$	m.p.–1 037
La(s)	6.17	1.60	—		298–800
Li(s)	3.33	8.21	—		273–m.p.
Li(l)	5.85	1.31	2.07	$-467 \times 10^{-6} T^2$	m.p.–580
Mg(s)	5.33	2.45	−0.103		298–m.p.
Mg(l)	7.80	—	—		m.p.–1 100
Mn(α)	5.70	3.38	−0.375		298–1 000
Mn(β)	8.33	0.66	—		1 108–1 317
Mn(γ)	6.03	3.56	−0.443		1 374–1 410
Mn(δ)	11.10	—	—		1 410–1 450
Mn(l)	11.0	—	—		m.p.–b.p.
Mo(s)	5.77	0.28	—	$+2.26 \times 10^{-6} T^2$	298–2 500
N_2(g)	6.67	1.0	—		298–2 000
Na(s)	19.71	−88.27	—	$+150 \times 10^{-6} T^2$	273–m.p.
Na(l)	8.965	−4.594	—	$+2.542 \times 10^{-6} T^2$	m.p.–451
Nb(s)	5.66	0.96	—		298–1 900
Nd(α)	3.503	6.434	+1.07		298–1 128
Nd(β)	10.65	—	—		1 128–m.p.
Ni(α)	4.06	+7.04	—		298–630
Ni(β)	6.00	+1.80	—		630–m.p.
Ni(l)	9.20	—	—		m.p.–2 200
O_2(g)	7.16	1.00	−0.40		298–3 000
Os(s)	5.69	0.88	—		298–1 900
P (yell.)	4.57	3.78	—		273–m.p.
P (red)	4.05	3.56	—		298–800
P(l)	6.29	—	—		m.p.–370
Pb(s)	5.63	2.33	—		298–m.p.
Pb(l)	7.75	−0.74	—		m.p.–1 300
Pd(s)	5.80	1.38	—		298–1 828
Pr(α)	6.21	0.3	—		298–1 068
Pr(β)	9.19	—	—	$+3.11 \times 10^{-6} T^2$	1 068–m.p.
Pt(s)	5.80	1.28	—		298–2 043
Pu(α)	5.91	5.8			298–395
Pu(β)	5.21	7.05			395–480
Pu(γ)	2.98	11.1			480–588
Pu(δ)	9.0	—	—		588–753
Pu(ε)	8.4	—	—		753–913
Pu(l)	10.0	—	—		913–2 000
Rb(s)	7.27	—	—		298–m.p.
Rb(l)	7.85	—	—		m.p.–373
Re(s)	5.80	0.95	—		298–2 300
Rh(s)	5.49	2.06	—		298–1 900
Ru(α)	5.28	1.1	—		298–2 000
S (rhomb.)	3.58	6.24	—		273–369
S (monocl.)	3.56	6.95	—		273–392
S(l)	5.4	5.5	—		m.p.–b.p.
S_2(g)	8.54	0.28	−0.79		298–2 000
Sb(s)	5.51	1.78	—		273–m.p.
Sb(l)	7.5	—	—		m.p.–1 000

(continued)

Table 8.10 ELEMENTS—*continued*

Element	a	b	c	Remarks	Temp. range K
Se(s)	4.53	5.5	—		273–490
Se(l)	7.00	—	—		490–600
Si(s)	5.72	0.59	−0.99		298–1 200
Si(l)	6.12	—	—		m.p.–1 873
Sm(α)	6.00	5.84	−0.61		298–1 190
Sm(β)	11.22	—	—		1 190–1 345
Sm(l)	12.57	—	—		m.p.–1 398
Sn(s)	5.16	4.34	—		298–m.p.
Sn(l)	8.29	−2.2	—		m.p.–800
Ta(s)	6.65	−0.52	−0.45	$+0.47 \times 10^{-6} T^2$	298–2 300
Te(s)	4.58	5.25	—		273–m.p.
Te(l)	9.0	—	—		m.p.–873
Th(s)	5.63	3.04	—		298–1 300
Ti(α)	5.28	2.4	—		298–1 155
Ti(β)	4.74	1.90	—		1 155–1 350
Tl(α)	3.74	6.04	+0.67		298–506
Tl(β)	5.00	5.00	—		506–m.p.
Tl(l)	7.2	—	—		m.p.–1 760
U(α)	2.61	8.95	1.17		298–941
U(β)	10.0	—	—		941–1 048
U(γ)	9.1	—	—		1 048–m.p.
V(s)	4.90	2.58	+0.2		298–1 900
W(s)	5.74	0.76	—		298–2 000
Y(α)	5.72	1.805	+0.08		298–1 758
Y(β)	8.37	—	—		1 758–m.p.
Y(l)	9.51	—	—		m.p.–1 950
Zn(s)	5.35	2.40	—		298–m.p.
Zn(l)	7.5	—	—		m.p.–b.p.
Zr(α)	5.25	2.78	−0.91		298–1 135
Zr(β)	5.55	1.11	—		1 135–m.p.

Table 8.11 ALLOY PHASES AND INTERMETALLIC COMPOUNDS

Molar heat capacities and specific heats.

Generally, Neumann and Kopp's rule applies better to intermetallic compounds than to inorganic salts. Thus, the molar heat capacity of an alloy may be calculated additively from the atomic heat capacities of the components. This relationship is useful as only a few specific heats of alloys have been measured.

In the following cases Neumann–Kopp's rule is obeyed to within ±3% in the temperature range 0–500°C:

Ag–Au; Ag_3Al; γ-Ag–Al; β-Ag–Mg; AlCu; Al_2Cu; $AlCu_3$; Al_3Mg_4; Bi–Cd; Co_2Sn; Cu_2Mg; $MgZn_2$; Mg_2Si; Pd–Sb

The heat capacities of heterogeneous alloys must always be calculated additively from those of the components.

In other cases the calculated values differ from the determined values by more than 3%. This result may be due in part to experimental errors.

C_p may be represented empirically by an equation of the form:

$$C_p = 4.1868(a + 10^{-3}bT + 10^5 cT^{-2}) \, \mathrm{J\,K^{-1}\,mol^{-1}}$$

where a, b and c are constants and T is the temperature in K. The values of the constants are given below.

Phase	a	b	c	Temp. range K
AuCd	7.66	10.36	—	298–900
AuCd (liq.)	14.4	—	—	900–1 100
AuCu		See ref. 3		320–900
$AuCu_3$		See ref. 3		298–1 200
$AuPb_2$ (liq.)	23.1	—	—	527–800

(*continued*)

Table 8.11 ALLOY PHASES AND INTERMETALLIC COMPOUNDS—*continued*

Phase	a	b	c	Temp. range K
$AuSb_2(\alpha)$	17.12	4.64	—	298–628
AuSn	11.13	3.8	—	298–691
AuSn (liq.)	14.6	—	—	691–900
$AuZn(\beta)$	11.87	2.85	—	723–m.p.
AuZn (liq.)	13.6	—	—	m.p.–1 200
Bi_3Tl_2	31.5	—	—	298–487
Bi_3Tl_2 (liq.)	35.8	—	—	487–700
Cd–Cu		See ref. 2		292–373
Cd_3Cu_2 (liq.)	41.5	—	—	835–1 000
Cd_3Mg		See ref. 2		298–543
CdMg		See ref. 2		298–543
$CdMg_3$		See ref. 2		298–543
Co_7W_6	68.86	29.7	—	293–1 145
$Cr_{0.48}Fe_{0.52}(\alpha)$	5.02	2.78	—	1 100–1 400
$Cr_{0.48}Fe_{0.52}$	8.06	—	—	870–1 100
Cu–Ni		79–94% Ni *see Figure 8.1*		
CuPd	12.02	1.96	−1.16	298–1 200
Cu_3Pd	20.98	8.80	−1.16	298–1 200
Cu_2Sb	16.38	6.60	—	298–600
$Cu_3Sb(\beta)$	21.79	9.00	—	273–573
Cu–Zn		See Figure 8.2		
Fe_7W_6	72.08	24.0	—	293–1 145
FeW_2	28.29	9.90	−1.60	298–1 250
$MgNi_2$	15.67	7.30	—	298–900
$Ni_3Sn(\delta)$	20.78	10.2	—	273–943
Ni_4W	25.7	11.14	—	293–1 100
$PtSb_2$	15.27	5.06	—	298–900
PtSn	11.26	2.20	—	298–1 400

Cr–Ni. Atomic heats of alloys with 1.8–11 at. % Cr obey Kopp's law above the Curie temperature.

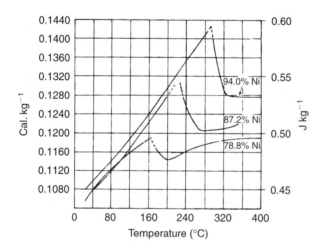

Figure 8.1 *Specific heat of* Cu–Ni *alloys*

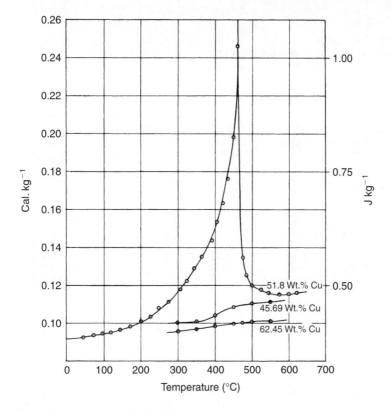

Figure 8.2   *Specific heat of* Cu–Zn *alloys*

Table 8.12a   BORIDES

Molar heat capacities.

C_p may be represented empirically by an equation of the form:

$$C_p = 4.186\,8(a + 10^{-3}bT + 10^5 cT^{-2})\,\mathrm{J\,K^{-1}\,mol^{-1}}$$

where a, b and c are constants and $T =$ temperature in K. The values of the constants are given below.

Boride	a	b	c	Temp. range K
CrB	10.12	3.83	−2.40	298–1 200
CrB$_2$	9.63	10.7	—	298–1 200
Mo$_2$B	18.42	1.3	—	298–800
MoB	9.77	3.07	−1.13	298–1 200
MoB$_2$	7.92	(13)	—	600–1 200
NbB$_2$	11.01	9.38	−1.78	298–1 200
NiB	10.26	3.5	−2.69	298–1 300
Ni$_3$B$_4$	37.28	11.74	−9.03	298–1 300
TaB	7.82	5.8	—	500–1 200
TaB$_2$	14.21	4.49	−3.60	298–3 370
TiB$_2$	13.48	6.48	−4.17	298–3 190
UB$_2$	25.58	−9.62	−8.48	298–2 300
W$_2$B	18.46	1.44	−3.14	298–1 200
WB	13.89	−0.78	−5.02	298–1 200
W$_2$B$_5$	39.27	2.08	−16.72	298–1 200
ZrB$_2$	15.79	4.22	−3.51	298–1 200

Table 8.12b   CARBIDES

Molar heat capacities.

C_p may be represented empirically by an equation of the form:

$$C_p = 4.186\,8(a + 10^{-3}bT + 10^5 cT^{-2})\,\mathrm{J\,K^{-1}\,mol^{-1}}$$

where a, b and c are constants and $T =$ temperature in K. The values of the constants are given below.

Carbide	a	b	c		Temp. range K
Al_4C_3	36.97	6.866	−10.02		298–1 800
B_4C	22.99	5.40	−10.72		298–1 100
Be_2C	27.37	3.60	−14.23		430–1 200
$CaC_2(\alpha)$	16.40	2.84	−2.07		298–720
$CaC_2(\beta)$	15.40	2.00	—		720–1 275
Cr_3C_2	30.03	5.58	−7.4		298–1 500
Cr_7C_3	57.00	14.38	−10.1		298–1 500
Cr_4C	29.35	7.40	−5.02		298–1 700
$Fe_3C(\alpha)$	19.64	20.0	—		298–463
$Fe_3C(\beta)$	25.62	3.0	—		463–1 500
$CH_4(g)$	5.65	11.44	−0.46		298–1 500
$Mn_3C(\alpha)$	25.26	5.60	−4.07		298–1 310
$Mn_3C(\beta)$	38.00	—	—		1 310–1 500
$NbC_{0.5}$	7.94	1.50	−1.025		298–1 703
$NbC_{0.75}$	8.95	2.25	−1.26		298–1 763
NbC	10.79	1.73	−2.15		298–1 790
SiC	12.14	0.47	−11.76	$1.96 \times 10^8 T^{-3}$	298–3 260
Ta_2C	15.88	3.33	−2.05		298–3 775
TaC	10.35	2.71	−2.10		298–4 270
$ThC_{1.94}$	15.17	2.89	−2.21		298–1 700
UC	13.40	1.02	−1.46		298–2 073
U_2C_3	29.9	3.06	−3.71		298–2 050
UC_2	16.5	2.04	−2.25		298–2 050
$VC_{0.88}$	8.69	3.18	−1.7		298–2 000
WC	10.37	2.06	−2.23	$0.24 \times 10^{-6} T^2$	298–2 500

Table 8.12c   NITRIDES

Molar heat capacities.

C_p may be represented empirically by an equation of the form:

$$C_p = 4.186\,8(a + 10^{-3}bT + 10^5 cT^{-2})\,\mathrm{J\,K^{-1}\,mol^{-1}}$$

where a, b and c are constants and $T =$ temperature in K. The values of the constants are given below.

Compound	a	b	c	Remarks	Temp. range K
AlN	8.22	4.05	−2.00		298–1 500
BN (hexag.)	16.29	0.675	—	$-854T^{-1/2}$	298–1 400
BN (cubic)	8.10	3.52	−5.51		298–1 200
Be_3N_2	12.89	24.75	−4.22		298–430
Ca_3N_2	33.17	3.70	−6.29		298–1 468
CeN	11.1	1.65	−1.73		298–2 000
Cr_2N	15.24	6.8	—		298–800
CrN	9.84	3.9	—		298–800
Cu_3N	21.7	—	—		298–373
Fe_4N	26.8	8.16	—		298–373
Fe_2N	14.9	6.1	—		298–373
NH_3	7.11	6.00	−0.37		298–1 800
Li_3N	11.73	23.00	—		298–800
$Mg_3N_2(\alpha)$	22.81	7.30	—		298–823
$Mg_3N_2(\beta)$	29.60	—	—		823–1 061

(continued)

Table 8.12c NITRIDES—*continued*

Compound	a	b	c	Remarks	Temp. range K
$Mg_3N_2(\gamma)$	29.54	—	—		1 061–1 300
Mn_4N	21.16	30.5	—		298–800
Mn_5N_2	30.55	38.40	—		298–800
Mn_3N_2	22.32	22.40	—		298–800
Mo_2N	11.19	13.8	—		298–800
NbN	8.69	5.40	—		298–600
Si_3N_4	16.83	23.6	—		298–900
Ta_2N	16.85	4.22	−1.69		298–3 000
TaN	13.21	0.65	−3.02		298–3 360
ThN	11.34	2.28	−1.14		298–2 000
Th_3N_4	27.78	31.8	—		298–800
TiN	11.91	0.94	−2.96		298–1 800
VN	10.94	2.10	−2.21		298–1 611
Zn_3N_2	19.00	22.5	—		298–700
ZrN	11.10	1.68	−1.72		298–1 700

Table 8.12d SILICIDES

Molar heat capacities.

C_p may be given empirically by an equation of the form:

$$C_p = 4.186(a + 10^{-3}bT + 10^3cT^{-2})\,\text{J}\,\text{K}^{-1}\,\text{mol}^{-1}$$

The values of the constants a, b and c are given below.

Compound	a	b	c	Temp. range K
CoSi	11.75	2.89	−1.80	298–1 733
CoSi (liq.)	20.88	—	—	m.p.–1 900
$CoSi_2$	16.935	4.46	−2.37	298–m.p.
$CoSi_2$ (liq.)	27.75	—	—	m.p.–1 900
Cr_5Si_3	47.46	11.78	−6.12	298–1 300
CrSi	12.43	2.09	−2.01	298–1 700
$CrSi_2$	15.68	5.38	−1.855	298–1 730
$CrSi_2$(liq.)	21.5	—	—	1 730–1 900
FeSi	10.72	4.30	—	298–900
Mg_2Si	17.52	3.58	−2.11	298–873
Mn_3Si	24.11	12.45	−3.52	298–950
Mn_5Si_3	48.13	12.94	−4.68	298–m.p.
MnSi	11.79	3.05	−1.53	298–m.p.
$Mn_{0.37}Si_{0.63}$	6.37	4.085	−1.16	298–1 425
Mo_3Si	20.52	5.42	+0.076	298–2 200
Mo_5Si_3	43.82	8.37	−2.87	298–2 200
$MoSi_2$	16.22	2.86	−1.57	298–2 200
Nb_5Si_3	45.21	7.36	−3.60	298–2 000
$NbSi_2$	15.10	3.67	−0.67	298–2 000
Ni_2Si	15.8	3.29	—	298–1 582
NiSi	11.65	1.47	−1.56	298–1 265
$Ni_{0.35}Si_{0.65}$	5.98	0.88	−0.86	298–1 200
Re_5Si_3	45.60	10.8	−3.36	298–1 500
Ta_5Si_3	42.95	9.35	−2.13	298–2 000
$TaSi_2$	17.51	1.84	−2.17	298–2 200
V_3Si	22.41	4.37	−1.66	298–1 400
VSi_2	17.08	2.78	−2.25	298–1 950
W_5Si_3	42.94	9.36	−2.13	298–2 200
WSi_2	16.21	2.64	−1.46	298–2 200

Table 8.12e OXIDES

Molar heat capacities.

C_p may be given empirically by an equation of the form:

$$C_p = 4.186\,8(a + 10^{-3}bT + 10^5cT^{-2})\,\mathrm{J\,K^{-1}\,mol^{-1}}$$

The values of the constants a, b and c are given below.

Compound	a	b	c	Remarks	Temp. range K
Ag_2O	14.18	9.75	−1.0		298–500
Al_2O_3	25.48	4.25	−6.82		298–m.p.
As_2O_3	14.30	42.0	—		298–m.p.
B_2O_3 (cryst.)	13.63	17.45	−3.36		298–723
B_2O_3 (glass)	2.28	42.10	—		298–723
B_2O_3 (liq.)	30.45	—	—		900–1 800
BaO	12.74	1.04	−1.984		298–1 270
BeO	9.94	2.44	−4.15	$-32 \times 10^{-6}T^2$	298–2 835
$Bi_2O_3(\alpha)$	24.74	8.0	—		298–800
$Bi_2O_3(\beta)$	35.0	—	—		978–1 097
CO	6.79	0.98	−0.11		298–2 500
CO_2	10.55	2.16	−2.04		298–2 500
CaO	11.86	1.08	−1.66		298–1 177
CdO	11.53	1.525	−1.17		298–1 500
CeO_2	15.49	4.23	−1.815		298–1 250
CoO	11.54	2.04	0.4		298–1 800
Co_3O_4	30.84	17.08	−5.72		298–1 000
Cr_2O_3	28.53	2.20	−3.74		350–1 800
Cu_2O	14.90	5.70	—		298–1 200
CuO	9.27	4.80	—		298–1 250
$Eu_2O_3(\alpha)$	29.60	6.48	−2.08		298–895
Eu_2O_3 (β)	31.06	4.16	—		895–1 802
Eu_2O_3 (cubic)	31.90	4.38	−3.04		298–1 350
$Fe_{0.947}O$	11.66	2.00	−0.67		298–m.p.
$Fe_{0.947}O$ (liq.)	16.30	—	—		m.p.–1 800
$Fe_3O_4(\alpha)$	21.88	48.2	—		298–900
$Fe_3O_4(\beta)$	48.00	—	—		900–1 800
$Fe_2O_3(\alpha)$	23.49	18.6	−3.55		298–950
$Fe_2O_3(\beta)$	36.0	—	—		950–1 050
$Fe_2O_3(\gamma)$	31.71	1.76	—		1 050–1 750
Gd_2O_3 (monocl.)	27.28	3.54	−2.54		298–1 802
Gd_2O_2 (cubic)	28.72	2.84	−3.88		298–1 550
H_2O (liq.)	18.03	—	—		298–373
$H_2O(g)$	7.17	2.56	0.08		298–2 500
$H_2O_2(g)$	12.50	2.84	−2.84		298–1 300
HgO	9.00	6.0	—	estim.	298–?
HfO_2	17.39	2.08	−3.48		298–1 800
IrO_2	17.33	1.93	−3.68		298–1 400
La_2O_3	28.86	3.08	−3.28		298–1 771
$Li_2O(s)$	14.94	6.08	−3.38		298–1 045
MgO	11.71	0.75	−2.80		298–3 098
MnO	11.11	1.94	−0.88		298–1 800
$Mn_3O_4(\alpha)$	34.64	10.82	−2.20		298–1 445
$Mn_3O_4(\beta)$	50.20	—	—		1 445–1 800
$Mn_2O_3(s)$	24.73	8.38	−3.23		298–1 350
MnO_2	16.60	2.44	−3.88		298–780
MoO_3	20.07	5.90	−3.68		298–1 068
$Na_2O(\alpha)$	13.26	16.78	0.99	$-7.3 \times 10^{-6}T^2$	298–1 023
$Na_2O(\beta)$	19.67	3.05	—		1 023–1 243
$Na_2O(\gamma)$	20.28	2.56	—		1 243–m.p.
Na_2O (liq.)	25.0	—	—		m.p.–2 200
NbO	10.04	2.35	−0.78		298–1 700
$NbO_2(\alpha)$	14.68	6.16	−2.42		298–1 040
$NbO_2(\beta)$	22.20	—	—		1 040–1 200
Nb_2O_5	38.76	3.54	−7.32		298–1 780

(continued)

Table 8.12e OXIDES—*continued*

Compound	a	b	c	Remarks	Temp. range K
$Nd_2O_3(\alpha)$	27.67	7.12	−2.84		298–1 395
$Nd_2O_3(\beta)$	37.20	—	—		1 395–1 795
$NiO(\alpha)$	−4.99	37.58	3.89		298–525
$NiO(\beta)$	13.88	—	—		525–565
$NiO(\gamma)$	11.18	2.02	—		565–1 800
P_2O_5	17.90	38.8	−3.73		298–700
$P_4O_{10}(g)$	73.6	—	—		631–1 400
PbO (red)	10.95	3.75	−1.0		298–1 900
PbO (yell.)	9.05	6.40	—		298–1 000
PbO_2	12.7	7.8	—	estim.	298–?
PdO	10.83	1.68	−0.80	$+0.09 \times 10^{-6}T^2$	298–1 200
Pr_6O_{11}	95.29	26.16	−9.31		298–1 172
Rh_2O	15.59	6.47	—		298–973
RhO	9.84	5.33	—		298–1 023
Rh_2O_3	20.74	13.8	—		298–973
SO(g)	7.70	0.84	−0.65		298–2 000
$SO_2(g)$	10.38	2.54	−1.42		298–1 800
$SO_3(g)$	13.70	6.42	−3.12		298–1 200
Sb_2O_3	19.10	17.1	—	estim.	298–930
SbO_2	11.30	8.1	—	estim.	298–1 198
Sb_2O_5	10.95	57.57	—		298–500
Sc_2O_3	23.17	5.64	—	estim.	298–2 500
α-quartz	10.49	0.24	−1.44		298–848
β-quartz	14.08	2.4	—		848–2 000
α-cristobalite	4.28	21.06	—		298–523
β-cristobalite	17.39	0.31	−9.90		523–2 000
α-tridymite	3.27	24.80	—		298–390
β-tridymite	13.64	2.64	—		390–2 000
SiO_2 (glass)	13.38	3.68	−3.45		298–2 000
$Sm_2O_3(\alpha)$	30.75	4.64	−4.30		298–1 195
$Sm_2O_3(\beta)$	36.90	—	—		1 195–1 798
Sm_2O_3 (cubic)	30.64	5.08	−3.96		298–1 150
SnO_2	17.66	2.40	−5.16		298–1 500
SrO	12.34	1.12	−1.806		298–1 270
Ta_2O_5	37.0	6.56	−5.92		298–1 800
TeO_2	15.58	3.48	−1.20		298–m.p.
TeO_2 (liq.)	26.95	0.52	—		m.p.–1 146
ThO_2	16.65	2.13	−2.24		298–2 500
$TiO(\alpha)$	10.57	3.6	−1.86		298–1 264
$TiO(\beta)$	11.85	3.0	—		1 264–1 800
$Ti_2O_3(\alpha)$	7.31	53.52	—		298–473
$Ti_2O_3(\beta)$	34.68	1.3	−10.2		473–1 800
$Ti_3O_5(\alpha)$	35.47	29.50	—		298–450
$Ti_3O_5(\beta)$	41.60	8.0	—		450–1 400
rutile	17.97	0.28	−4.35		298–1 800
anatase	17.83	0.50	−4.23		298–1 300
UO_2	19.20	1.62	−3.96		298–1 500
U_3O_8	67.5	8.83	−11.94		298–900
UO_3	22.1	2.64	−2.65		298–900
VO	11.32	3.22	−1.26		298–1 700
V_2O_3	29.35	4.76	−5.42		298–1 800
$VO_2(\alpha)$	14.96	—	—		298–345
$VO_2(\beta)$	17.85	1.70	−3.95		345–1 818
VO_2 (liq.)	25.5	—	—		1 818–1 900
V_2O_5	46.54	−3.90	−13.22		298–943
V_2O_5 (liq.)	45.60	—	—		943–1 500
WO_3	17.48	6.79	—	estim.	298–1 550
$Y_2O_3(\alpha)$	29.60	1.20	−4.78		298–1 330
$Y_2O_3(\beta)$	31.50	—	—		1 330–1 800
ZnO	11.71	1.22	−2.18		298–1 600
$ZrO_2(\alpha)$	16.64	1.80	−3.36		298–1 478
$ZrO_2(\beta)$	17.80	—	—		1 478–1 850

Table 8.12f   SULPHIDES, SELENIDES AND TELLURIDES

Molar heat capacities.

C_p may be given empirically by an equation of the form:

$$C_p = 4.186\,8(a + 10^{-3}bT + 10^5 cT^{-2})\,\mathrm{J\,K^{-1}\,mol^{-1}}$$

The values of the constants are given below.

Compound	a	b	c	Remarks	Temp. range K
$Ag_2S(\alpha)$	10.13	26.40	—		298–452
$Ag_2S(\beta)$	21.64	—	—		452–850
$Ag_2Se(\alpha)$	23.29	0.39	−4.0		298–406
$Ag_2Se(\beta)$	20.4	—	—		406–460
Bi_2S_3	26.25	9.8	—	estim.	—
Bi_2Te_3	36.0	13.05	−3.12		373–m.p.
$CS_2(l)$	18.4	—	—		298–b.p.
$CS_2(g)$	12.45	1.60	−1.80		298–1 800
CaS	10.80	1.85	—	estim.	298–2 000
CdS	10.65	3.30	—	estim.	273–1 273
$CdTe$	12.55	4.54	−1.76		298–m.p.
CoS	10.6	2.51	—	estim.	273–1 373
$Cu_2S(\alpha)$	19.50	—	—		298–376
$Cu_2S(\beta)$	23.25	—	—		376–623
$Cu_2S(\gamma)$	20.32	—	—		623–1 400
CuS	10.6	2.64	—	estim.	273–1 273
$Cu_2Se(\alpha)$	14.0	18.5	—		298–395
$Cu_2Se(\beta)$	20.1	—	—		383–488
$Cu_2Te(\zeta)$	20.9	—	—		841–950
$Cu_2Te(\epsilon)$	26.12	—	—		633–841
$Cu_2Te(\delta)$	32.0	—	—		590–633
$Cu_2Te(\gamma)$	27.0	—	—		531–590
$Cu_2Te(\beta)$	14.45	12.8	—		433–531
$Cu_2Te(\alpha)$	14.3	12.8	—		298–433
$FeS(\alpha)$	5.19	26.40	—		298–411
$FeS(\beta)$	17.40	—	—		411–598
$FeS(\gamma)$	12.20	2.38	—		598–1 468
$FeS(l)$	17.0	—	—		1 468–1 500
FeS_2	17.88	1.32	−3.05		298–1 000
$H_2S(g)$	7.81	2.96	−0.46		298–2 000
$H_2Se(g)$	7.59	3.50	−0.31		298–2 000
MnS	11.40	1.80	—		298–1 803
$MnS(liq.)$	16.00	—	—		1 803–2 000
$MoS_2(s)$	11.2	13.5	—	estim.	298–729
NiS	9.3	6.4	—		273–670
$NiTe(s)$	11.57	3.30	—		298–700
PbS	10.63	4.01	—		273–873
PtS	11.95	2.6	−2.12	estim.	298–1 100
PtS_2	14.07	7.08	−0.4	estim.	298–1 000
Sb_2S_3	24.2	13.2	—	estim.	273–821
$SnS(\alpha)$	8.53	7.48	0.9		298–875
$SnS(\beta)$	9.78	3.74	—		875–m.p.
$SnS(l)$	17.90	—	—		1 153–1 250
SnS_2	15.51	4.20	—		298–1 000
$SnTe$	16.88	3.67	—		273–603
$TiS_2(\alpha)$	8.08	27.34	—		298–420
$TiS_2(\beta)$	14.99	5.14	—		420–1 010
$ZnS(s)$	12.16	1.24	−1.36		298–1 200
$ZnTe$	11.11	2.61	—	—	298–m.p.

Table 8.12g   HALIDES

Molar heat capacities.

C_p may be given empirically by an equation of the form:

$$C_p = 4.186\ 8(a + 10^{-3}bT + 10^5 cT^{-2})\ \text{J K}^{-1}\ \text{mol}^{-1}$$

The values of the constants are given in the following table.

Compound	a	b	c	Remarks	Temp. range K
AgCl(s)	14.88	1.0	−2.70		298–728
AgCl(l)	16.0	—	—		728–900
AgBr(s)	7.93	15.40	—		298–703
AgBr(l)	14.9	—	—		m.p.–836
AgI(α)	5.82	24.10	—		298–423
AgI(β)	13.5	—	—		423–600
AlCl$_3$(s)	13.25	28.00	—		273–m.p.
AlCl$_3$(l)	31.2	—	—		m.p.–504
AlCl(g)	9.0	—	−0.68		298–2 000
AlBr$_3$(s)	18.74	18.66	—		273–m.p.
AlBr$_3$(l)	29.5	—	—		m.p.–407
AlI$_3$(s)	16.88	22.66	—		273–m.p.
AlI$_3$(l)	28.8	—	—		m.p.–480
AlF$_3$(α)	17.27	10.96	−2.30		298–727
AlF$_3$(β)	20.93	3.0	—		727–1 400
AlF(g)	8.9	—	−1.45		298–2 000
AsCl$_3$(l)	31.9	—	—	estim.	286–371
AsCl$_3$(g)	19.62	0.24	−1.42		298–1 000
AsF$_3$(l)	30.2	—	—		298
AsF$_3$(g)	18.18	1.66	−2.66		298–1 000
BaCl$_2$(α)	22.20	0.76	−4.0		273–1 198
BaCl$_2$(β)	26.61	—	—		1 198–m.p.
BaCl$_2$(l)	24.96	—	—		m.p.–1 339
BaBr$_2$(s)	15.96	6.22	—		487–1 126
CCl$_4$(g)	24.90	0.48	−4.74		298–1 000
CBr$_4$(α)	31.7	—	—		295–320
CBr$_4$(β)	33.0	—	—		320–m.p.
CBr$_4$(g)	25.03	0.60	−3.03		298–1 000
CF$_4$(g)	16.64	7.84	−4.00		298–1 200
CaCl$_2$(s)	17.18	3.04	−0.6		600–1 055
CaCl$_2$(l)	24.70	—	—		1 055–1 700
CaBr$_2$(s)	13.96	7.86	—		434–m.p.
CaBr$_2$(l)	27.38	—	—		m.p.–1 132
CaF2(α)	14.30	7.28	0.47		298–1 424
CaF2(β)	25.81	2.5	—		1 424–m.p.
CaF$_2$(l)	23.88	—	—		m.p.–1 800
CoCl$_2$(s)	14.41	14.60	—		298–1 000
CrCl$_2$(s)	15.23	5.96	—		298–m.p.
CrCl$_3$(s)	19.44	7.03	—		298–m.p.
CsBr(s)	11.6	2.59	—	estim.	273–909
CsCl(α)	12.78	1.23	−0.46		293–743
CsCl(β)	0.81	17.64	−0.89		743–m.p.
CsCl(l)	13.86	4.28	—		m.p.–1 170
CsF(s)	11.3	2.71	—	estim.	273–957
CsI(s)	7.06	10.36	1.93		298–m.p.
CsI(l)	−4.29	20.5	—		m.p.–1 170
CuCl(s)	5.87	19.2	—		298–703
CuCl(l)	15.8	—	—		703–1 200
CuCl(g)	8.92	—	−0.47		298–2 000
CuCl$_2$(s)	15.42	12.0	—		298–800
CuI(s)	12.1	2.86	—	estim.	273–675
CuI$_2$(s)	20.1	—	—	estim.	274–328
FeCl$_2$(s)	18.94	2.08	−1.17		600–950
FeCl$_2$(l)	24.42	—	—		950–1 100
FeCl$_3$(s)	29.56	—	−6.11		298–m.p.

(continued)

Table 8.12g   HALIDES—*continued*

Compound	a	b	c	Remarks	Temp. range K
HF(g)	6.43	0.82	0.26		298–2 000
HCl(g)	6.34	1.10	0.26		298–2 000
HBr(g)	6.25	1.40	0.26		298–1 600
HI(g)	6.29	1.42	0.22		298–2 000
HfF$_4$(s)	31.9	74.8	9.0		273–1 103
HfCl$_4$(s)	31.47	—	−2.38		298–485
HgCl(s)	11.05	3.70	—	estim.	273–798
HgCl$_2$(s)	15.3	10.3	—	estim.	273–553
HgI(s)	11.4	4.61	—	estim.	273–563
HgI$_2$(α)	18.50	—	—		298–403
HgI$_2$(β)	20.2	—	—		403–m.p.
HgI$_2$(l)	25.0	—	—		m.p.–603
KCl(s)	9.89	5.20	0.77		298–1 043
KCl(l)	16.0	—	—		1 043–1 200
KCl(g)	8.94	0.18	−0.24		298–2 000
KBr(s)	12.84	2.50	−2.84		600–1 000
KBr(g)	8.94	—	−0.17		298–2 000
KI(s)	21.10	−8.38	−10.38		600–1 000
KF(s)	11.02	3.12	—		298–1 130
LiCl(s)	11.0	3.39	—	estim.	273–887
LiBr(s)	11.5	3.02	—	estim.	273–825
LiF(s)	9.14	5.19	—		298–m.p.
LiF(l)	15.50	—	—		m.p.–1 170
LiI(s)	12.5	2.08	—	estim.	273–723
MgCl$_2$(s)	18.90	1.42	−2.06		600–987
MgCl$_2$(l)	22.10	—	—		987–1 500
MgF$_2$(s)	16.93	2.52	−2.20		298–m.p.
MgF$_2$(l)	22.57	—	—		1 538–1 800
MnCl$_2$(s)	18.04	3.16	−1.37		600–923
MnCl$_2$(l)	22.60	—	—		923–1 200
MnF$_2$(s)	14.79	5.70	−0.47		298–m.p.
NaCl(s)	10.98	3.90	—		298–1 073
NaCl(l)	16.0	—	—		m.p.–1 300
NaCl(g)	8.93	—	−0.41		298–2 000
NaBr(s)	11.87	2.10	—		298–550
NaBr(g)	8.93	—	−0.29		298–2 000
NaF(s)	10.40	3.88	−0.33		298–m.p.
NaI(s)	12.5	1.62	—	estim.	273–936
NH$_4$Cl(α)	11.80	32.0	—		298–457.7
NH$_4$Cl(β)	5.00	34.0	—		457.7–500
NiCl$_2$(s)	17.50	3.16	−1.19		298–1 303
NiCl$_2$(l)	24.00	—	—		1 303–1 336
PF$_3$(g)	17.18	1.92	−3.88		298–2 000
PCl$_3$(l)	28.7	—	—	estim.	284–371
PCl$_3$(g)	19.15	0.74	−1.91		298–1 000
PCl$_5$(g)	31.42	0.20	−4.27		298–2 000
PBr$_3$(g)	19.81	—	−1.43		298–1 000
PbCl$_2$(s)	16.1	4.0	—		298–771
PbCl$_2$(l)	28.2	—	—		m.p.–851
PbBr$_2$(s)	18.59	2.20	—		298–643
PbBr$_2$(l)	27.6	—	—		m.p.–860
PbI$_2$(s)	18.00	4.70	—		298–685
PbI$_2$(l)	32.3	—	—		m.p.–776
PbF$_2$(s)	16.5	4.12	—	estim.	273–1 091
RbF(s)	7.97	9.2	1.21		298–m.p.
RbF(l)	−11.30	0.88	350.7		m.p.–1 200
RbCl(s)	11.5	2.49	—	estim.	273–987
RbBr(s)	11.6	2.55	—	estim.	273–954
RbI(s)	11.6	2.63	—	estim.	273–913
SbCl$_3$(s)	10.3	51.1	—	estim.	273–346

(*continued*)

Table 8.12g HALIDES—*continued*

Compound	a	b	c	Remarks	Temp. range K
SbBr$_3$(s)	17.2	29.3	—	estim.	273–370
SiCl$_4$(g)	24.25	1.64	−2.75		298–1 000
SiBr$_4$(g)	25.19	0.64	−1.94		298–1 000
SiF$_4$(g)	21.86	3.15	−4.70		298–1 000
SnCl$_2$(s)	16.2	9.26	—	estim.	273–520
SnCl$_4$(l)	39.5	—	—	estim.	286–371
SnCl$_4$(g)	25.57	0.20	−1.87		298–1 000
SnBr$_4$(g)	25.80	—	−0.97		298–1 000
SnI$_4$(s)	19.4	36.0	—		298–b.p.
SnI$_4$(l)	40.1	—	—		b.p.–443
TeCl$_4$(s)	33.1	—	—		298–m.p.
TeCl$_4$(l)	53.2	—	—		m.p.–538
TeF$_6$(g)	35.33	1.62	−7.00		298–2 000
TiCl$_4$(l)	35.7	—	—	estim.	273–372
TiCl$_4$(g)	25.45	0.24	−2.36		298–2 000
TiBr$_4$(s)	28.0	—	—		298–m.p.
TiBr$_4$(l)	32.75	15.66	—		m.p.–423
TlCl(s)	12.00	2.00	—		298–700
TlCl(l)	14.2	—	—		m.p.–803
TlCl(g)	8.94	—	−0.25		298–2 000
TlBr(s)	11.07	4.95	—		298–733
TlBr(l)	25.25	−9.04	—		m.p.–800
UF$_4$(s)	25.7	7.00	−0.06		298–1 309
UF$_6$(s)	12.6	92.0	—		273–337
UF$_6$(g)	22.3	28.5	—		273–400
UCl$_3$(s)	20.8	7.75	1.05		298–900
UCl$_4$(s)	27.2	8.57	−0.79		298–800
UCl$_4$(l)	25.8	14.4	—		890–920
UBr$_4$(s)	31.4	4.92	−3.15		350–750
UI$_4$(s)	34.8	2.38	−4.72		380–720
UI$_4$(l)	39.6	—	—		820–870
VCl$_2$(s)	17.25	2.72	−0.71		298–1 200
VCl$_3$(s)	22.99	3.92	−1.68		298–900
ZnCl$_2$(s)	14.5	5.5	—		298–m.p.
ZnCl$_2$(l)	24.1	—	—		m.p.–1 000
ZnBr$_2$(s)	12.6	10.4	—		298–m.p.
ZnBr$_2$(l)	27.2	—	—		m.p.–1 000
ZrCl$_4$(s)	31.92	—	−2.91		298–604
ZrBr$_4$(s)	25.5	15.1	—	estim.	298–630

8.7 Vapour pressures

Table 8.13 ELEMENTS

Vapour pressures.

The vapour pressure p (mmHg) of an element may be represented by an equation of the type:

$$\log p = -\frac{A}{T} + B + C \log T + 10^{-3}DT$$

The values of the constants A, B, C and D in this equation are given below.

Element	A	B	C	D	Temp. range K
Ag	14 710	11.66	−0.755	—	298–1 234
	14 260	12.23	−1.055	—	1 234–2 400
Al	16 450	12.36	−1.023	—	1 200–2 800

(*continued*)

Table 8.13   ELEMENTS—*continued*

Element	A	B	C	D	Temp. range K
Am	13 700	13.97	−1.0	—	1 103–1 453
As₄	6 160	9.82	—	—	600–900
Au	19 820	10.81	−0.306	−0.16	298–1 336
	19 280	12.38	−1.01	—	1 336–3 240
B	29 900	13.88	−1.0	—	1 000–m.p.
Ba	9 730	7.83	—	—	750–983
	9 340	7.42	—	—	983–1 200
Be	10 734	9.067	—	−0.145	900–1 557
Bi	10 400	12.35	−1.26	—	m.p.–b.p.
Bi₂	10 730	18.1	−3.02	—	m.p.–b.p.
Ca	10 300	14.97	−1.76	—	713–m.p.
	9 600	12.55	−1.21	—	m.p.–b.p.
Ce	20 305	8.305	—	—	1 611–2 038
Cd	5 908	9.717	−0.232	−0.284	450–594
	5 819	12.287	−1.257	—	594–1 050
Co	22 210	10.817	—	−0.223	1 000–1 772
Cr	20 680	14.56	−1.31	—	298–m.p.
Cs	4 075	11.38	−1.45	—	280–1 000
Cu	17 870	10.63	−0.236	−0.16	298–1 356
	17 650	13.39	−1.273	—	1 356–2 870
Eu	8 980	8.16	—	—	696–900
Fe	21 080	16.89	−2.14	—	900–1 812
	19 710	13.27	−1.27	—	1 812–3 000
Ga	14 700	10.07	−0.5	—	m.p.–b.p.
Ge	20 150	13.28	−0.91	—	298–m.p.
	18 700	12.87	−1.16	—	m.p.–b.p.
Hf(α)	32 000	11.81	−0.5	—	298–2 023
Hf(β)	31 630	11.63	−0.5	—	2 023–m.p.
Hf	29 830	9.20	—	—	m.p.–b.p.
Hg	3 308	10.373	−0.8	—	298–630
I₂	3 578	17.72	−2.51	—	298–m.p.
	3 205	23.65	−5.18	—	m.p.–b.p.
In	12 580	9.79	−0.45	—	m.p.–b.p.
Ir	35 070	13.18	−0.7	—	298–m.p.
K	4 770	11.58	−1.370	—	350–1 050
La	22 120	10.39	−0.33	—	298–m.p.
	21 530	9.89	−0.33	—	m.p.–b.p.
Li	8 415	11.34	−1.0	—	m.p.–b.p.
Mg	7 780	11.41	−0.855	—	298–m.p.
	7 550	12.79	−1.41	—	m.p.–b.p.
Mn	14 850	17.88	−2.52	—	993–1 373
	13 900	17.27	−2.52	—	m.p.–b.p.
Mo	34 700	11.66	−0.236	−0.145	298–m.p.
Na	5 700	11.33	−1.718	—	400–1 200
Na₂	6 540	10.7	—	—	—
Nb	37 650	8.94	+0.715	−0.166	298–m.p.
Ni	22 500	13.60	−0.96	—	298–m.p.
	22 400	16.95	−2.01	—	m.p.–b.p.
P₄ (yell.)	3 530	19.09	−3.5	—	298–317
P₄	2 740	7.84	—	—	317–553
Pb	10 130	11.16	−0.985	—	600–2 030
Pd	19 800	11.82	−0.755	—	298–m.p.
	17 500	4.81	+1.0	—	m.p.–b.p.
Pr	17 190	8.10	—	—	1 425–1 692
Pt	29 200	13.24	−0.855	—	298–m.p.
	28 500	14.30	−1.26	—	m.p.–b.p.
Pu	17 590	7.90	—	—	1 392–1 793
Rb	4 560	12.00	−1.45	—	312–952
Re	40 800	14.20	−1.16	—	298–3 000
Rh	29 360	13.50	−0.88	—	298–m.p.

(*continued*)

Table 8.13   ELEMENTS—*continued*

Element	A	B	C	D	Temp. range K
Ru	33 550	10.76	—	—	2 000–2 500
S_2	6 975	16.22	−1.53	−1.0	m.p.–b.p.
S_x	4 830	23.88	−5.0	—	m.p.–b.p.
Sb_x	11 560	22.40	−3.52	—	298–m.p.
Sb_2	11 170	18.54	−3.02	—	m.p.–b.p.
$Sc(\beta)$	19 700	13.07	−1.0	—	1 607–m.p.
Se_x	4 990	8.09	—	—	493–958
Si	20 900	10.84	−0.565	—	m.p.–b.p.
Sm	11 170	13.76	−1.56	~	298–m.p.
Sn	15 500	8.23	—	—	505–b.p.
Sr	9 450	13.08	−1.31	—	813–m.p.
	9 000	12.63	−1.31	—	m.p.–b.p.
Ta	40 800	10.29	—	—	298–m.p.
Te_2	9 175	19.68	−2.71	—	298–m.p.
	7 830	22.29	−4.27	—	m.p.–
Th	30 200	12.95	−1.0	—	298–m.p.
$Ti(\beta)$	24 400	13.18	−0.91	—	1 155–m.p.
Ti	23 200	11.74	−0.66	—	m.p.–b.p.
Tl	9 300	11.10	−0.892	—	700–1 800
Tm	12 550	9.18	—	—	807–1 219
U	25 580	18.58	−2.62	—	298–1 405
	24 090	13.20	−1.26	—	1 405–4 200
V	26 900	10.12	+0.33	−0.265	298–m.p.
W	44 000	8.76	+0.50	—	298–m.p.
Y	22 230	11.835	−0.66	—	298–m.p.
	22 280	16.13	−1.97	—	m.p.–b.p.
Zn	6 883	9.418	−0.050 3	−0.33	473–692.5
	6 670	12.00	−1.126	—	692.5–1 000
Zr	31 820	11.78	−0.50	—	1 125–m.p.
	30 300	9.38	—	—	m.p.–b.p.

Table 8.14   HALIDES AND OXIDES

Vapour pressures.

$$\log p = \frac{A}{T} + B + C \log T + 10^{-3} DT \ \text{(mmHg)}$$

Substance	A	B	C	D	Temp. range K
AgCl	−11 830	12.39	−0.30	−1.02	298–m.p.
AgCl	−11 320	17.34	−2.55	—	m.p.–b.p.
AgBr	−12 400	19.33	−2.97	—	m.p.–b.p.
AgI	−10 250	20.09	−3.52	—	m.p.–b.p.
AlF_3	−16 700	23.27	−3.02	—	298–s.p.
Al_2Cl_6	−6 360	9.66	3.77	−6.12	298–s.p.
Al_2Br_6	−5 280	20.81	−1.75	−4.08	298–m.p.
Al_2Br_6	−5 280	46.70	−12.59	—	m.p.–b.p.
Al_2I_6	−7 150	17.76	0.12	−4.96	298–m.p.
Al_2I_6	−6 760	46.67	−11.89	—	m.p.–b.p.
AmF_3	−24 600	36.87	−7.05	—	1 100–1 300
AsF_3	−4 150	61.38	−18.26	—	265–292
AsF_5	−1 088	7.72	—	—	m.p.–b.p.
$AsCl_3$	−2 660	24.76	−5.83	—	m.p.–b.p.
As_4O_6*	−5 282	10.91	—	—	373–573

* claudetite

(continued)

Table 8.14   HALIDES AND OXIDES—*continued*

Substance	A	B	C	D	Temp. range K
As$_4$O$_6$[†]	$-5\,452$	11.468	—	—	488–573
As$_4$O$_6$	$-3\,130$	7.16	—	—	m.p.–b.p.
BCl$_3$	$-2\,115$	27.56	-7.04	—	m.p.–b.p.
BBr$_3$	$-2\,710$	28.36	-7.04	—	m.p.–b.p.
BI$_3$	$-3\,342$	24.31	-5.4	—	m.p.–b.p.
B$_2$O$_3$	$-16\,960$	6.64	—	—	1 300–1 650
BaF$_2$	$-20\,330$	28.04	-5.03	—	m.p.–b.p.
BaO	$-21\,900$	9.99	—	—	1 200–1 700
BeF$_2$	$-13\,000$	24.56	-3.79	—	m.p.–b.p.
Be$_2$Cl$_4$	$-8\,970$	37.0	-7.65	—	298–m.p.
BeCl$_2$	$-7\,870$	27.15	-5.03	—	298–m.p.
BeCl$_2$	$-7\,220$	26.28	-5.03	—	m.p.–b.p.
Be$_2$Br$_4$	$-8\,320$	35.9	-7.65	—	298–s.p.
BeBr$_2$	$-7\,650$	27.15	-5.03	—	298–m.p.
BeBr$_2$	$-6\,570$	25.63	-5.03	—	m.p.–b.p.
Be$_2$I$_4$	$-8\,520$	35.9	-7.65	—	298–m.p.
BeI$_2$	$-7\,000$	26.5	-5.03	—	298–m.p.
BeI$_2$	$-5\,800$	24.96	-5.03	—	m.p.–b.p.
BiCl$_3$	$-6\,200$	12.83	—	—	298–m.p.
BiCl$_3$	$-5\,980$	31.38	-7.04	—	m.p.–b.p.
BiBr$_3$	$-6\,190$	31.40	-7.04	—	m.p.–b.p.
CCl$_4$	$-2\,400$	23.60	-5.30	—	m.p.–b.p.
CBr$_4$	$-2\,650$	8.78	—	—	298–m.p.
CBr$_4$	$-2\,330$	7.89	—	—	m.p.–b.p.
CaF$_2(\alpha)$	$-23\,600$	27.41	-4.525	—	298–1 424
CaF$_2(\beta)$	$-23\,350$	27.23	-4.525	—	1 424–m.p.
CaF$_2$	$-21\,800$	26.31	-4.525	—	m.p.–b.p.
CaCl$_2$	$-13\,570$	9.22	—	—	1 110–1 281
CdF$_2$	$-16\,170$	27.50	-5.03	—	m.p.–b.p.
CdCl$_2$	$-9\,270$	17.46	-2.11	—	298–m.p.
CdCl$_2$	$-9\,183$	25.907	-5.04	—	m.p.–b.p.
CdBr$_2$	$-8\,250$	18.15	-2.5	—	298–m.p.
CdBr$_2$	$-7\,150$	16.85	-2.5	—	m.p.–b.p.
CdI$_2$	$-7\,530$	18.01	-2.5	—	298–m.p.
CdI$_2$	$-6\,720$	16.79	-2.5	—	m.p.–b.p.
CeCl$_3$	$-18\,750$	36.38	-7.05	—	298–m.p.
CeBr$_3$	$-18\,000$	36.49	-7.05	—	298–m.p.
CoCl$_2$	$-14\,150$	30.10	-5.03	—	298–m.p.
CoCl$_2$	$-11\,050$	27.06	-5.03	—	m.p.–b.p.
CrCl$_2$	$-14\,000$	15.14	-0.62	-0.58	298–m.p.
CrCl$_2$	$-13\,800$	27.70	-5.03	—	m.p.–b.p.
CrCl$_3$	$-13\,950$	17.49	-0.73	-0.77	298–s.p.
CrI$_2$	$-16\,080$	25.92	-3.53	—	298–m.p.
CrO$_2$Cl$_2$	$-3\,340$	34.94	-9.08	—	m.p.–b.p.
CsF	$-10\,930$	17.51	-2.12	—	298–m.p.
CsF	$-9\,950$	18.62	-2.84	—	m.p.–b.p.
CsCl	$-10\,800$	19.99	-3.02	—	700–m.p.
CsCl	$-9\,815$	20.38	-3.52	—	Temp.–b.p.
CsBr	$-10\,950$	20.02	-3.02	—	700–m.p.
CsBr	$-10\,080$	20.56	-3.52	—	m.p.–b.p.
CsI	$-10\,420$	19.70	-3.02	—	600–m.p.
CsI	$-9\,678$	20.35	-3.52	—	m.p.–b.p.
CuCl	$-10\,170$	8.04	—	—	1 000–1 900
Cu$_3$Cl$_3$	$-3\,750$	4.90	—	—	900–1 800
CuBr	$-7\,700$	7.69	—	—	1 000–1 480
Cu$_3$Br$_3$	$-4\,010$	4.88	—	—	1 000–2 000
Cu$_3$I$_3(\alpha)$	$-9\,463$	11.14	—	—	629
Cu$_3$I$_3(\beta)$	$-8\,351$	9.41	—	—	643–670
Cu$_3$I$_3(\gamma)$	$-7\,853$	8.68	—	—	684–770
FeCl$_2$	$-9\,890$	11.10	—	—	670–740

[†] arsenolite

(*continued*)

Table 8.14   HALIDES AND OXIDES—*continued*

Substance	A	B	C	D	Temp. range K
FeBr$_2$	−10 220	11.95	—	—	670–740
FeI$_2$	−12 180	29.59	−5.03	—	298–m.p.
FeI$_2$	−8 750	27.185	−5.535	—	m.p.–b.p.
Fe$_2$Cl$_6$	−9 540	45.53	−9.5	—	298–m.p.
Fe(CO)$_5$	−2 075	8.42	—	—	298–b.p.
GaCl$_2$	−4 886	29.14	−6.44	—	m.p.–b.p.
GaBr$_3$	−4 700	28.69	−6.44	—	m.p.–b.p.
GeCl$_4$	−2 940	34.27	−9.08	—	m.p.–b.p.
GeBr$_4$	−3 690	35.00	−9.05	—	m.p.–b.p.
GeI$_4$	−4 920	22.73	−4.02	—	298–m.p.
H$_2$O	−2 900	22.613	−4.65	—	m.p.–b.p.
H$_2$O$_2$	−3 560	29.68	−7.04	—	m.p.–
HfCl$_4$	−5 200	11.71	—	—	476–681
HfI$_4$(α)	−10 700	19.56	—	—	575–597
HfI$_4$(β)	−7 360	13.97	—	—	598–645
HfI$_4$(γ)	−6 173	12.13	—	—	648–678
HgCl$_2$	−4 580	16.39	−2.0	—	298–m.p.
HgBr$_2$	−4 500	11.47	0.05	−1.51	298–m.p.
HgBr$_2$	−4 370	24.18	−5.03	—	m.p.–b.p.
HgI$_2$	−5 690	30.27	−6.47	—	298–m.p.
HgI$_2$	−4 620	25.72	−5.33	—	m.p.–b.p.
InCl	−4 640	8.03	—	—	m.p.–b.p.
InCl$_3$	−8 270	13.62	—	—	500–s.p.
InBr	−6 470	16.31	−2.01	—	298–m.p.
InBr$_2$	−4 480	7.48	—	—	m.p.–b.p.
InBr$_3$	−5 670	11.67	—	—	500–s.p.
InI	−6 730	15.74	−1.97	—	298–m.p.
IrF$_6$	−1 657	7.952	—	—	m.p.–b.p.
KF	−12 930	17.30	−2.06	—	298–m.p.
KF	−11 570	16.90	−2.32	—	m.p.–b.p.
KCl	−12 230	20.34	−3.0	—	298–m.p.
KCl	−10 710	18.91	−3.0	—	m.p.–b.p.
KBr	−11 110	16.60	−2.0	—	298–m.p.
KBr	−10 180	18.67	−3.0	—	m.p.–b.p.
KI	−11 000	16.99	−2.0	—	298–m.p.
KI	−10 050	20.41	−3.52	—	m.p.–b.p.
LaCl$_3$	−19 040	36.20	−7.05	—	298–m.p.
LaBr$_3$	−18 780	36.83	−7.05	—	298–m.p.
LaI$_3$	−18 390	37.00	−7.05	—	298–m.p.
LiF	−14 560	23.56	−4.02	—	m.p.–b.p.
LiCl	−10 760	22.30	−4.02	—	m.p.–b.p.
LiBr	−10 170	20.55	−3.52	—	m.p.–b.p.
LiI	−11 110	21.70	−3.52	—	m.p.–b.p.
MgF$_2$	−19 700	27.80	−5.03	—	m.p.–b.p.
MgCl$_2$	−10 840	25.53	−5.03	—	m.p.–b.p.
MgBr$_2$	−10 930	26.07	−5.03	—	m.p.–b.p.
MgI$_2$	−8 090	25.18	−5.03	—	m.p.–b.p.
MnF$_2$	−17 400	22.06	−3.02	—	m.p.–b.p.
MnCl$_2$	−10 606	23.68	−4.33	—	m.p.–b.p.
MoF$_5$	−2 772	8.58	—	—	m.p.–b.p.
MoF$_6$	−1 500	7.77	—	—	m.p.–b.p.
MoOF$_4$	−2 854	9.21	—	—	313–m.p.
MoOF$_4$	−2 671	8.716	—	—	m.p.–b.p.
MoCl$_5$	−5 210	13.1	—	—	298–m.p.
MoO$_3$	−15 230	27.16	−4.02	—	298–m.p.
MoO$_3$	−12 480	24.60	−4.02	—	m.p.–b.p.
NaF	−14 960	17.53	−2.01	—	298–m.p.
NaF	−13 500	17.93	−2.52	—	m.p.–b.p.
NaCl	−12 440	14.31	−0.90	−0.46	298–m.p.
NaCl	−11 530	20.77	−3.48	—	m.p.–b.p.

(*continued*)

Table 8.14   HALIDES AND OXIDES—*continued*

Substance	A	B	C	D	Temp. range K
NaBr	−12 100	20.39	−3.0	—	298–m.p.
NaBr	−10 500	18.81	−3.0	—	m.p.–b.p.
NaI	−10 740	20.96	−3.52	—	m.p.–b.p.
NbF$_5$	−4 900	14.397	—	—	298–m.p.
NbF$_5$	−2 780	8.37	—	—	m.p.–b.p.
NbCl$_4$	−6 870	12.30	—	—	577–651
NbCl$_5$	−4 370	11.51	—	—	403–m.p.
NbCl$_5$	−2 870	8.37	—	—	m.p.–b.p.
NbBr$_5$	−4 085	9.33	—	—	m.p.–b.p.
NbOCl$_3$	−5 333	8.79	—	—	298–s.p.
NdCl$_3$	−18 220	36.27	−7.05	—	298–m.p.
NdBr$_3$	−17 650	36.51	−7.05	—	298–m.p.
NdI$_3$	−17 490	36.61	−7.05	—	298–m.p.
NiF$_2$	−14 650	20.28	−3.02	—	298–m.p.
NiCl$_2$	−13 300	21.88	−2.68	—	298–s.p.
NiBr$_2$	−13 110	16.68	−1.71	−0.35	298–s.p.
Ni(CO)$_4$	−1 530	7.73	—	—	298–b.p.
NpF$_6$	−2 892	18.48	−2.7	—	273–m.p.
	−1 913	14.61	−2.35	—	m.p.–350
OsF$_5$	−3 429	9.75	—	—	m.p.–b.p.
OsF$_6$	−1 858	8.726	—	—	273–m.p.
OsF$_6$	−1 473	7.47	—	—	m.p.–b.p.
OsO$_4$*	−2 955	9.64	—	—	273–329
OsO$_4$†	−2 580	10.70	—	—	273–315
OsO$_4$	−2 065	8.01	—	—	m.p.–b.p.
PCl$_3$	−2 370	22.74	−5.14	—	273–b.p.
PCl$_5$	−3 520	11.035	—	—	373–432
POCl$_3$	−1 830	7.72	—	—	273–b.p.
P$_4$O$_{10}$‡	−4 350	9.81	—	—	298–s.p.
PbF$_2$	−11 800	26.48	−5.03	—	m.p.–b.p.
PbCl$_2$	−9 890	15.36	−0.95	−0.91	298–m.p.
PbCl$_2$	−10 000	31.60	−6.65	—	m.p.–b.p.
PbBr$_2$	−9 320	18.44	−2.08	−0.34	298–m.p.
PbBr$_2$	−9 540	31.67	−6.76	—	m.p.–b.p.
PbI$_2$	−9 340	19.68	−2.35	−0.32	298–m.p.
PbI$_2$	−10 000	39.80	−9.21	—	m.p.–b.p.
PbO	−13 480	14.36	−0.92	−0.35	298–m.p.
PbO	−13 310	19.47	−2.77	—	m.p.–b.p.
PrCl$_3$	−18 490	36.31	−7.05	—	298–m.p.
PrBr$_3$	−17 800	36.53	−7.05	—	298–m.p.
PrI$_3$	−17 470	36.66	−7.05	—	298–m.p.
PuF$_3$	−24 950	36.91	−7.05	—	298–m.p.
PuF$_3$	−23 500	34.47	−6.45	—	m.p.–200
PuCl$_3$	−18 270	32.60	−5.34	—	298–m.p.
PuCl$_3$	−15 490	31.76	−6.45	—	m.p.–b.p.
PuBr$_3$	−17 460	31.32	−5.34	—	298–m.p.
PuBr$_3$	−15 030	32.34	−6.45	—	m.p.–b.p.
RbF	−11 230	18.26	−2.66	—	m.p.–b.p.
RbCl	−11 670	20.157	−3.0	—	298–m.p.
RbCl	−10 300	18.77	−3.0	—	m.p.–b.p.
RbBr	−11 510	20.155	−3.0	—	298–m.p.
RbBr	−10 220	18.805	−3.0	—	m.p.–b.p.
RbI	−10 280	20.64	−3.52	—	m.p.–b.p.
ReF$_5$	−3 037	9.024	—	—	m.p.–b.p.
ReF$_6$	−1 489	7.732	—	—	m.p.–b.p.
ReF$_7$	−2 206	13.045	−1.47	—	259–m.p.
ReF$_7$	−244	−21.585	+9.91	—	m.p.–b.p.
ReOF$_4$	−3 888	11.88	—	—	323–m.p.

* yellow   † white   ‡ hexagonal

(*continued*)

Table 8.14   HALIDES AND OXIDES—*continued*

Substance	A	B	C	D	Temp. range K
ReOF$_4$	−3 206	10.09	—	—	m.p.–b.p.
ReOF$_5$	−1 959	8.62	—	—	303–m.p.
ReOF$_5$	−1 679	7.727	—	—	m.p.–b.p.
Re$_2$O$_2$F	−3 437	10.36	—	—	m.p.–b.p.
Re$_2$O$_7$	−7 300	15.000	—	—	273–m.p.
Re$_2$O$_7$	−3 950	9.10	—	—	m.p.–b.p.
RuCl$_3$	−16 750	30.53	−4.63	—	298–1 000
SO$_3\alpha$	−2 680	11.44	—	—	273–m.p.
SO$_3\beta$	−2 860	11.97	—	—	273–m.p.
SO$_3\gamma$	−3 610	14.00	—	—	273–m.p.
SO$_3$	−2 230	9.90	—	—	m.p.–b.p.
SO$_2$Cl$_2$	−1 660	7.65	—	—	m.p.–b.p.
SbF$_5$	−2 364	8.567	—	—	282–416
SbCl$_3$	−3 460	2.81	3.88	−5.6	298–m.p.
SbCl$_3$	−3 770	29.48	−7.04	—	m.p.–b.p.
SbCl$_5$	−2 530	8.56	—	—	m.p.–350
SbBr$_3$	−2 860	7.97	—	—	435–561
SbI$_3$	−3 450	7.99	—	—	510–629
Sb$_4$O$_6$*	−10 360	12.195	—	—	742–839
Sb$_4$O$_6$†	−9 625	11.312	—	—	742–914
Sb$_4$O$_6$	−3 900	5.137	—	—	929–1 073
ScCl$_3$	−14 200	14.37	—	—	1 065–1 233
ScBr$_3$	−13 780	14.35	—	—	1 042–1 200
ScI$_3$	−13 340	14.17	—	—	1 010–1 180
SeF$_4$	−2 457	9.44	—	—	m.p.–b.p.
SeO$_2$	−6 170	21.40	−3.02	—	298–s.p.
SiCl$_4$	−1 572	7.64	—	—	273–333
SiI$_4$	−3 863	23.38	−5.0	—	m.p.–b.p.
SnCl$_4$	−1 925	7.865	—	—	298–b.p.
SnBr$_4$	−3 510	27.63	−6.5	—	303–b.p.
SnI$_4$	−3 990	10.08	—	—	298–m.p.
SnI$_4$	−2 975	7.666	—	—	m.p.–b.p.
SrF$_2$	−21 660	28.04	−5.03	—	m.p.–b.p.
TaCl$_5$	−6 275	34.305	−7.04	—	298–m.p.
TaCl$_5$	−2 975	8.68	—	—	m.p.–b.p.
TaBr$_5$	−7 320	34.85	−7.04	—	298–m.p.
TaBr$_5$	−3 260	8.14	—	—	m.p.–b.p.
TaI$_5$	−6 660	31.61	−7.04	—	298–m.p.
TaI$_5$	−3 955	7.72	—	—	m.p.–b.p.
Tc$_2$O$_7$	−7 205	18.28	—	—	298–m.p.
Tc$_2$O$_7$	−3 570	9.00	—	—	m.p.–b.p.
TeF$_4$	−3 174	9.093	—	—	298–m.p.
TeF$_4$	−1 787	5.640	—	—	m.p.–467
TeF$_6$	−1 460	9.13	—	—	194–241
TeCl$_2$	−3 350	8.51	—	—	m.p.–b.p.
TeO$_2$	−13 940	23.51	−3.52	—	298–m.p.
ThCl$_4$	−12 900	14.30	—	—	974–1 043
ThCl$_4$	−7 980	9.57	—	—	1 043–1 186
ThBr$_4$	−9 630	11.73	—	—	903–951
ThBr$_4$	−7 550	9.56	—	—	955–1 126
ThI$_3$	−6 890	9.09	—	—	856–1 107
ThO$_2$	−34 890	10.87	—	—	2 500–2 900
TiF$_4$	−5 332	19.51	−2.57	—	298–s.p.
TiCl$_2$	−15 230	19.36	−2.51	—	298–m.p.
TiCl$_2$	−13 110	17.93	−2.51	—	m.p.–b.p.
TiCl$_3$	−9 620	21.47	−3.27	—	298–m.p.
TiCl$_4$	−2 919	25.129	−5.788	—	298–b.p.
TiBr$_4$	−3 706	27.08	−6.24	—	m.p.–b.p.
TiI$_2$	−12 500	16.90	−1.51	—	298–1 000
TiI$_4$	−3 054	7.576	—	—	430–643

* cubic   † orthorhombic

(*continued*)

Table 8.14   HALIDES AND OXIDES—*continued*

Substance	A	B	C	D	Temp. range K
TlF	−7 710	17.66	−2.18	—	298–m.p.
TlCl	−7 370	16.49	−2.11	—	298–m.p.
TlCl	−6 650	16.92	−2.62	—	m.p.–b.p.
TlBr	−7 420	16.18	−2.0	—	298–m.p.
TlBr	−6 840	18.26	−3.02	—	m.p.–b.p.
TlI	−7 270	15.85	−2.01	—	298–m.p.
TlI	−6 890	18.20	−3.02	—	m.p.–b.p.
UF_4	−16 400	22.60	−3.02	—	298–m.p.
UF_4	−15 300	29.05	−5.03	—	m.p.–b.p.
UF_6	−2 858	16.36	−1.91	—	273–s.p.
UCl_4	−11 350	23.21	−3.02	—	298–m.p.
UCl_4	−9 950	28.96	−5.53	—	m.p.–b.p.
UCl_6	−4 000	10.20	—	—	298–450
UBr_3	−16 420	22.95	−3.02	—	298–m.p.
UBr_3	−15 000	27.54	−5.03	—	m.p.–(b.p.)
UBr_4	−10 800	23.15	−3.02	—	298–m.p.
UBr_4	−8 770	27.93	−5.53	—	m.p.–b.p.
UI_4	−12 330	26.62	−3.52	—	298–m.p.
UI_4	−9 310	28.57	−5.53	—	m.p.–b.p.
UO_2	−33 120	25.69	−4.03	—	1 500–2 800
VF_5	−2 423	10.43	—	—	m.p.–b.p.
VCl_4	−2 875	25.56	−6.07	—	298–b.p.
$VOCl_3$	−1 921	7.70	—	—	298–b.p.
V_2O_5*	−7 100	5.05	—	—	m.p.–1 500
WF_6	−1 380	7.635	—	—	m.p.–b.p.
$WCl_4(\alpha)$	−3 996	9.615	—	—	458–503
$WCl_4(\beta)$	−3 588	8.795	—	—	503–554
WCl_4	−3 253	8.195	—	—	555–598
WCl_5	−3 670	9.50	—	—	413–m.p.
WCl_5	−2 760	7.72	—	—	m.p.–b.p.
$WCl_6(\alpha)$	−4 580	10.73	—	—	425–t.p.
$WCl_6(\beta)$	−4 080	9.73	—	—	t.p.–m.p.
WCl_6	−3 050	7.87	—	—	m.p.–b.p.
WOF_4	−3 605	10.96	—	—	298–m.p.
WOF_4	−3 125	9.69	—	—	m.p.–b.p.
WO_3	−24 600	15.63	—	—	1 000–m.p.
ZnF_2	−13 650	26.90	−5.03	—	m.p.–b.p.
$ZnCl_2$	−8 500	16.61	−1.50	—	298–m.p.
$ZnCl_2$	−8 440	26.37	−5.03	—	m.p.–b.p.
$ZnBr_2$	−7 120	16.21	−2.01	—	298–m.p.
ZnI_2	−6 450	14.70	−1.76	—	298–m.p.
ZnS	−13 980	8.98	—	—	970–1 280
ZrF_4	−14 700	30.80	−5.03	—	298–s.p.
$ZrCl_4$	−5 400	11.765	—	—	480–689
$ZrBr_4$	−6 780	19.60	−1.76	−1.65	298–s.p.
ZrI_4	−7 680	20.87	−2.164	−1.344	298–s.p.

* Apparent vapour pressures V_2O_5 loses oxygen with increasing temperature.

REFERENCES—THERMOCHEMICAL DATA

1a. R. Hultgren, P. O. Desai, D. T. Hawkins, M. Gleiser, K. K. Kelley and D. D. Wagman, 'Selected Values of the Thermodynamic Properties of the Elements', American Society for Metals, Metals Park, Ohio 44073, 1973.
1b. R. Hultgren, P. O. Desai, D. T. Hawkins, M. Gleiser and K. K. Kelley, 'Selected Values of the Thermodynamic Properties of Binary Alloys', American Society for Metals, Metals Park, Ohio 44073, 1973.
2. O. Kubaschewski and C. B. Alcock, 'Metallurgical Thermochemistry', 5th edn, Pergamon, Oxford, 1979.
3. O. Kubaschewski and J. A. Catterall, 'Thermochemical Data of Alloys', Pergamon, Oxford, 1956.
4. A. Schneider and G. Heymer, NPL Symposium No. 9, 'Metallurgical Chemistry', HMSO, London, 1958.
5. E. F. Strotzer and M. Zumbusch, *Z. Anorg. Chem.,* 1941, **247**, 415 (concluding paper of a series).
6. M. Zumbusch and W. Biltz, *Z. Anorg. Chem.,* 1942, **249**, 1 (concluding paper of a series).

9 Physical properties of molten salts

In the following tables are given densities, electrical conductivities, surface tensions and coefficients of viscosity of pure molten salts, molten binary salt systems and other ionic melts. For comprehensive data and treatment, reference should be made to G. J. Janz, 'Molten Salts Handbook', 1967, Academic Press, New York/London.

Table 9.1 DENSITY OF PURE MOLTEN SALTS

The density of most pure molten salts varies almost linearly with temperature, and may be represented by the equation:

$$d_t = a - 10^{-3} \cdot bt$$

where d_t is the density in g cm^{-3}, a and b are constants, and t is the temperature in °C over appreciable ranges of temperature. Values of the constants and the appropriate temperature ranges are given below. Principal references are in **bold type**.

Substance	a	b	Range of observations* °C	References
AgBr	6.025	1.04	m.p. to 820	**91, 60, 33**, 10
AgCl	5.257	0.849	467 to 637	**140, 97, 91**, 33
AgClO$_3$	4.262 6	1.742	m.p. to 250	**18**
AgI	6.139	1.01	600 to 800	**33**
AgNO$_3$	4.167	1.00	m.p. to 410	**108, 100, 97, 18**, 112, 105, 93, 34, 10
AlBr$_3$	2.875	2.314	m.p. to 225	**89, 88, 38**
AlCl$_3$	1.805	2.5	194 to 250	**38**
AlI$_3$	3.70	2.5	m.p. to 250	**38**
Al$_2$O$_3$	5.259	1.127	2 102 to 2 352	**121**
AsBr$_3$	3.455	2.6	50 to 100	**96**, 34, 14
AsCl$_3$	2.205	2.18	m.p. to 130	**87**, 8, 1
AsF$_3$	$d_1 = 2.665\ 9 = 3.839 \cdot 10^{-3}t + 4.35 \cdot 10^{-8}t^2$		0 to 60	**8**
		$d_{25} = 3.01$		78
AsF$_5$	2.047	5.34	m.p. to -53	**68**
BBr$_3$		$d_0 = 2.650$		15
BCl$_3$		$d_{11} = 1.349$		52
BI$_3$		$d_{50} = 3.3$		12
B$_2$O$_3$	1.609	0.086 7	1 030 to 1 310	**101, 114**, 19
BaBr$_2$		$d_{850} = 4.00$		102
BaCl$_2$	3.829 2	0.681 3	966 to 1 081	**122**, 104, 103, 85, 19
BaF$_2$	5.502	0.999	1 327 to 1 727	**123**
Ba(NO$_2$)$_2$	3.448	0.70	—	**124**
BeCl$_2$	1.976	1.1	430 to 475	**47**
BiBr$_3$	5.248	2.6	270 to 330	**34**
BiCl$_3$	4.42	2.20	240 to 350	**120, 40, 34, 23**
Bi$_2$(MoO$_4$)$_3$		$d_{723} = 5.170$		125

(continued)

Table 9.1 DENSITY OF PURE MOLTEN SALTS—*continued*

Substance	a	b	Range of observations* °C	References
$CaCl_2$	2.410 8	0.422 5	787 to 950	**122**, 99, 86, 22, 19, 92
CaF_2	3.072	0.391	1 367 to 2 027	**123**, 111
$CdBr_2$	4.688	1.08	m.p. to 720	**91**
$CdCl_2$	3.858	0.825	m.p. to 800	**102**, 91
CdI_2	4.828	1.12	m.p. to 700	**102**
CeF_3	5.997	0.936	1 427 to 1 927	**123**
$CsBr$	3.911	1.22	m.p. to 860	**113**, 34
$CsCl$	3.478 5	1.065	667 to 907	**113**, 34, **140**
CsF	4.548 9	1.280 6	712 to 912	**122**, 34
CsI	3.918	1.18	645 to 855	**113**, 34
$CsNO_3$	3.302 3	1.160 0$_5$	415 to 491	**126**, 34
Cs_2SO_4	2.956	0.586	1 027 to 1 477	**34**
Cu_2Cl_2	4.010	0.79	m.p. to 585	**47**
Cu_2S	6.76	0.75	1 150 to 1 400	**119**
FeS	3.85	0	1 250 to 1 450	**119**
$GaBr_2$	3.753	1.69	160 to 175	**127**
$GaBr_3$	3.507	2.95	m.p. to 230	**70**
$GaCl_2$	2.652	1.36	166 to 177	**127**
$GaCl_3$	2.223	2.05	m.p. to 195	**116**, 70, 9
Ga_2I_4	4.380	1.688	181 to 265	**128**
Ga_2I_6	4.128	2.377	185 to 255	**128**, 70
$HfCl_4$		$d_{435} = 1.71$		**129**
$HgBr_2$	5.888 9	3.233 1	238 to 319	**130**, 26
$HgCl_2$	5.157 7	2.862 4	277 to 304	**130**, 26
Hg_2Cl_2	8.00	4.0	m.p. to 580	**47**
HgI_2	6.060 3	3.235 1	259 to 354	**130**, 26, 107
ICl	3.186	3.0	m.p. to 100	**8**, 5
$InBr_3$	3.674	1.5	450 to 530	**48**
$InCl$	4.055	1.4	269 to 365	**48**
$InCl_2$	3.43	1.6	268 to 437	**48**
$InCl_3$	3.37	2.1	597 to 666	**48**
InI_3	4.135	1.5	230 to 360	**48**
KBr	2.733	0.825 3	747 to 927	**113**, 102, **140**
KCl	1.976 7	0.583 1	777 to 947	**146**, 140, **102**, 110, **104**, 98
K_2CO_3	2.293 4	0.442	907 to 1 007	**141**
$K_2Cr_2O_7$	2.563 3	0.695	420 to 535	**142**
KF	2.468 5	0.651 5	881 to 1 037	**122**, 34
$KHSO_4$	2.232	0.767	207 to 230	**94**
KI	3.098 5	0.955 7	682 to 904	**110**, 102
K_2MoO_4	2.992 2	0.549 1	935 to 988	**134**, 34
KNO_2	2.005	0.700	440 to 500	**106**, 124
KNO_3	2.116	0.729	m.p. to 600	**106**, 102, **77**, 55, **34**, 21, **18**, 124, **126**
KOH	1.893	0.44	400 to 600	**45**, 37
KPO_3	2.455	0.43	990 to 1 200	**34**
K_2SO_4	2.472	0.545	1 100 to 1 300	**34**
K_2WO_4	3.844 7	0.727 3	944 to 1 053	**135**, 34
$LaBr_3$	5.008 9	0.096 0	796 to 912	**122**
$LaCl_3$	3.877 3	0.777 4	873 to 973	**122**, 47
LaF_3	5.607	0.682	1 477 to 2 177	**123**
$LiBr$	2.888	0.652	m.p. to 740	**113**, 16, 60
$LiCl$	1.766 0	0.432 8	627 to 777	**110**, 102, **98**, 34, **16**
$LiClO_4$	2.171 2	0.622 3	—	**136**
Li_2CO_3	2.10	0.373	737 to 847	**141**
LiF	2.224 3	0.490 2	876 to 1 047	**122**, 34
LiI	3.540	0.918	m.p. to 670	**113**

(*continued*)

Table 9.1   DENSITY OF PURE MOLTEN SALTS—*continued*

Substance	a	b	Range of observations* °C	References
Li_2MoO_4	3.200 8	0.415 2	781 to 963	**134**
$LiNO_3$	1.924	0.548	m.p. to 550	**34, 18**
Li_2SO_4	2.352 9	0.407	860 to 1 214	**143,** 34
Li_2WO_4	4.905 5	0.805 3	764 to 901	**135**
$MgCl_2$	1.894	0.302	727 to 967	**144,** 47, 83, 133
MgF_2	3.092	0.524	1 377 to 1 827	**123**
$MnCl_2$	2.639	0.44	650 to 850	**137**
MoF_6	2.637	4.91	m.p. to 34	**64**
MoO_3	4.443 6	1.498 3	821 to 918	**134,** 109
Na_3AlF_6	3.036	0.94	m.p. to 1 100	*See* Binary System Na_3AlF_6–NaF
NaBr	2.952	0.817	m.p. to 945	**113, 34, 16**
NaCl	1.991 1	0.543	803 to 1 030	**102, 147, 140, 146, 110,** 98
Na_2CO_3	2.357	0.449	867 to 1 007	**141**
NaF	2.502	0.560	997 to 1 057	**34,** 45
NaI	3.368	0.949	m.p. to 915	**113, 102, 34**
Na_2MoO_4	3.235	0.629	700 to 1 400	**35,** 109
$NaNO_2$	2.004	0.690	280 to 500	**106,** 102
$NaNO_3$	2.124 8	0.715	317 to 427	**106, 100, 102, 77, 34, 21,** 18
NaOH	1.937 4	0.478 4	320 to 450	**145**
$NaPO_3$	2.545	0.44	905 to 1 010	**34,** 19
Na_2SO_4	2.495	0.48	900 to 1 050	**34,** 16
Na_2WO_4	4.519 9	0.906 7	714 to 880	**135,** 35, 16
$NdBr_3$	4.762 6	0.777 9	695 to 860	**122**
NH_4NO_3	1.536	0.60	170 to 200	**118**
$Ni(CO)_4$	$d_1 = 1.356\,1 - 2.213 \cdot 10^{-3}t - 4.10^{-6}t^2$		0 to 36	**13**
$OsSO_4$	4.504	4.17	43 to 150	**62**
PBr_3	2.924	2.48	0 to 200	**8, 1**
PCl_3	1.612	1.86	−80 to 75	**87, 29, 8, 39,** 1
$PbBr_2$	5.036	1.45	377 to 497	**91,** 21
$PbCl_2$	5.702	1.5	502 to 710	**91,** 85, 61, 60, 21
$PbCl_4$		$d_0 = 3.18$		**50**
PbI_2		$d_{383} = 5.625$		**10**
$PbMoO_4$		$d_{1\,107} = 5.213$		**125**
RbBr	3.446 4	1.072	700 to 910	**113, 34**
RbCl	2.880	0.883	m.p. to 925	**113, 47, 34**
RbF	3.707	1.011	820 to 1 005	**34**
RbI	3.638	1.14	655 to 905	**113, 34**
$RbNO_3$	2.782	0.97	350 to 550	**34**
Rb_2SO_4	3.260	0.665	1 100 to 1 310	**34**
ReF_6	3.776	8.51	m.p. to 47.6	**79, 76**
Re_2O_7		$d_{331} = 4.30$		**67**
ReO_3Cl	3.94	4.0	16 to 37	**67**
$ReOF_4$	3.921	5.1	40 to 60	**79**
S_2Cl_2	1.710	1.57	0 to 136	**87, 8,** 2
$SOCl_2$	1.677	1.97	0 to 69	**8**
SO_2Cl_2	1.708	2.11	0 to 70	**8**
$SbBr_3$		$d = 3.691$ at m.p.		**81, 31, 27, 6,** 3
$SbCl_3$†	2.849	2.268	m.p. to 165	**84, 54, 41, 20,** 3
$SbCl_5$	2.392	2.04	m.p. to 80	**65, 54, 32,** 17
SbF_5		$d_{22.7} = 2.993$		**17**
$ScCl_3$		$d_{1\,000} = 1.63$		**47**
Se_2Cl_2		$d_{25} = 2.774$		**51**

(*continued*)

Table 9.1   DENSITY OF PURE MOLTEN SALTS—*continued*

Substance	a	b	Range of observations* °C	References
SeF_4		$d = 2.8$ at room temp.		**56**
SeF_6		$d = 2.3$ at m.p.		**69**
$SeOBr_2$		$d_{50} = 3.38$		**36**
$SeOCl_2$	2.478	2.08	m.p. to 80	**63, 57**, 43
$SeOF_2$		$d = 2.7$ at room temp.		**56**
$SiBr_4$	2.812	2.63	—	**8**
$SiCl_4$	1.523	2.08	−30 to 60	**87, 8,** 1
$SnCl_2$	3.6739	1.253	247 to 407	**47, 34**
$SnCl_4$	2.273	2.62	−19 to 113	**8,** 1
SnI_4	4.145	2.45	145 to 275	**53**
$SrCl_2$	3.2318	0.5781	893 to 1 037	**122,** 19
SrF_2	4.579	0.751	1 477 to 1 927	**123**
$TeCl_4$	2.965	1.64	230 to 430	**59**
TeF_6	2.442	6.02	m.p. to −10	**69**
$ThCl_4$		$d = 3.32$ at about 830		**47**
$TiBr_4$	3.043	2.25	40 to 120	**115**
$TiCl_4$	1.761	1.72	m.p. to 135	**73, 71, 8, 1,** 90, 58
TiI_4	3.755	2.19	166 to 270	**70**
$TlBr$	6.908_4	1.922_0	493 to 750	**131**
$TlCl$	6.402	1.80	m.p. to 640	**47**
$TlNO_3$	5.267_2	1.75	210 to 430	**34**
UCl_4	2.50	1.7	—	**138**
V_2O_5		$d_{1000} = 2.5$		**139**
$VOCl_3$	1.865	1.83	0 to 125	**25, 11, 8, 4**
WF_6	3.529	5.84	m.p. to 19	**64**
YCl_2	2.87	0.5	m.p. to 845	**47**
$ZnBr_2$	3.851_2	0.959	397 to 627	**144,** 72
$ZnCl_2$	2.693	0.515	m.p. to 630	**117, 47,** 61
$ZrCl_4$		$d_{448} = 1.54$		**129**

* Melting points will be found in Table 9.3.
† Between m.p. and 375°C density of $SbCl_3$ given by:

$$d_t = 2.622 - 2.268 \cdot 10^{-8}(t - 100) - 0.32 \cdot 10^{-8}(t - 100)^3.$$

Between 375°C and 505°C:

$$d_t = 2.622 - 2.268 \cdot 10^{-3}(t - 100) - 0.32 \cdot 10^{-8}(t - 100)^3 + 8.8 \cdot 10^{-12}(t - 100)^4 - 3.4 \cdot 10^{-14}(t - 100)^5.$$

REFERENCES TO TABLE 9.1

1. I. Pierre, *Ann. Chim. (Phys.)*, 1845, **15**, 325.
2. H. Kopp, *Ann. Chem.*, 1855, **95**, 307.
3. H. Kopp, *ibid.*, 1855, **95**, 350.
4. Roscoe, *Phil. Trans. R. Soc.*, 1868, **158**, 1.
5. Hannay, *J. Chem. Soc.*, 1873, **26**, 815.
6. R. W. E. MacIvor, *Chem. News*, 1874, **29**, 179.
7. Ditte, *Compt. Rend.*, 1877, **85**, 1069.
8. Nat. Res. Council, USA, 'International Critical Tables', Vol. 3, p. 23, 1928.
9. Lecoq de Boisbaudran, *Chem. News*, 1881, **44**, 166; *Compt. Rend.*, 1881, **93**, 294, 329, 815.
10. Rodwell, *Phil. Trans. R. Soc.*, 1882, **173**, 1125.
11. L'Hôte, *Compt. Rend.*, 1885, **101**, 1151.
12. H. Moissan, *ibid.*, 1891, **112**, 718.
13. Mond and Nasini, *Z. Phys. Chem.*, 1891, **8**, 150.
14. J. W. Retgers, *ibid.*, 1893, **11**, 328.
15. Ghira, *ibid.*, 1893, **12**, 765.
16. E. Brunner, *Z. Anorg. Chem.*, 1904, **38**, 350.
17. O. Ruff and W. Plato, *Ber. Dt. Chem. Ges.*, 1904, **37**, 679.
18. H. M. Goodwin and R. D. Mailey, *Phys. Rev.*, 1907, **25**, 469.
19. K. Arndt and A. Gessler, *Z. Elektrochem.*, 1908 **14**, 665.
20. Z. Klemensiewicz, *Bull. Int. Acad., Cracovie*, 1908, 487.

21. R. Lorenz, H. Frei and A. Jabs, *Z. Phys., Chem.*, 1908, **61**, 468.
22. K. Arndt and W. Löwenstein, *Z. Elektrochem.*, 1909, **15**, 789.
23. A. H. W. Aten, *Z. Phys. Chem.*, 1909, **66**, 641.
24. A. H. W. Aten, *ibid.*, 1910, **73**, 578.
25. Prandtl and Bleyer, *Z. Anorg. Chem.*, 1910, **66**, 152.
26. E. B. R. Prideaux, *J. Chem. Soc.*, 1910, **97**, 2032.
27. Izbekov and Plotnikov, *Z. Anorg. Chem.*, 1911, **71**, 328.
28. K. Arndt and Kunze, *Z. Elektrochem.*, 1912, **18**, 994.
29. Körber, *Ann. Physik.*, 1912, **37**, 1014.
30. Sackur, *Z. Phys. Chem.*, 1913, **83**, 297.
31. N. S. Kurnakov, Krotkov and Oksmann, *J. Russ. Phys. Chem. Soc.*, 1915, **47**, 558.
32. E. Moles, *Z. Phys. Chem.*, 1915, **90**, 74.
33. R. Lorenz and A. Höchberg, *Z. Anorg. Chem.*, 1916, **94**, 288.
34. F. M. Jaeger, *ibid.*, 1917, **101**, 16.
35. F. M. Jaeger and B. Kapma, *ibid.*, 1920, **113**, 27.
36. V. Lenher, *J. Am. Chem. Soc.*, 1922, **44**, 1668.
37. Meyer and Heck, *Z. Phys. Chem.*, 1922, **100**, 316.
38. W. Blitz and A. Voigt, *Z. Anorg. Chem.*, 1923, **126**, 39.
39. Timmermans, *Bull. Soc. Chim. Belg.*, 1923, **32**, 299.
40. A. Voigt and W. Blitz, *Z. Anorg. Chem.*, 1924, **133**, 277.
41. N. S. Kurnakov, *ibid.*, 1924, **135**, 86.
42. Pascal and Allendorff, *ibid.*, 1924, **135**, 327.
43. C. W. Muehlberger and V. Lenher, *J. Am. Chem. Soc.*, 1925, **47**, 1843.
44. Samsoen, *Compt. Rend.*, 1925, **181**, 354.
45. K. Arndt, *Z. Phys. Chem.*, 1926, **121**, 448.
46. H. V. A. Briscoe, P. L. Robinson and Stephenson, *J. Chem. Soc.*, 1926, 39.
47. W. Klemm, *Z. Anorg. Chem.*, 1926, **152**, 235.
48. W. Klemm, *ibid.*, 1926, **152**, 252.
49. W. Blitz and W. Klemm, *ibid.*, 1926, **152**, 267.
50. Friedrich, quoted by W. Blitz and W. Klemm, *ibid.*, 1926, **152**, 267.
51. V. Lenher and C. H. Kao, *J. Am. Chem. Soc.*, 1926, **48**, 1550.
52. H. V. A. Briscoe, P. L. Robinson and H. C. Smith, *J. Chem. Soc.*, 1927, 282.
53. Dortmann and Hildebrand, *J. Am. Chem. Soc.*, 1927, **49**, 737.
54. S. Sugden and A. Freiman, *J. Chem. Soc.* 1927, 1185.
55. Dantuma, *Z. Anorg. Chem.*, 1928, **175**, 33.
56. E. B. R. Prideaux and C. B. Cox, *J. Chem. Soc.*, 1928, 740, 1606.
57. W. J. R. Henley and S. Sugden, *ibid.*, 1929, 1064.
58. F. B. Garner and S. Sugden, *ibid.*, 1929, 1298.
59. J. H. Simons, *J. Am. Chem. Soc.*, 1930, **52**, 3491.
60. E. Salstrom, *ibid.*, 1930, **52**, 4647.
61. A. Wachter and J. Hildebrand, *ibid.*, 1930, **52**, 4656.
62. E. Ogawa, *Bull. Chem. Soc. Japan*, 1931, **6**, 315.
63. T. W. Parker and P. L. Robinson, *J. Chem. Soc.*, 1931, 1316.
64. O. Ruff and A. Ascher, *Z. Anorg. Chem.*, 1931, **196**, 417.
65. J. H. Simons and G. Jessop, *J. Am. Chem. Soc.*, 1931, **53**, 1265.
66. E. van Aubel, *Bull. Acad. Belg.*, 1932(5), **18**, 692.
67. H. V. A. Briscoe, P. L. Robinson and A. J. Rudge, *J. Chem. Soc.*, 1932, 2675.
68. O. Ruff, A. Brader, O. Bretschneider, W. Menzel and H. Plaut, *Z. Anorg. Chem.*, 1932, **206**, 59.
69. W. Klemm and P. Henkel, *ibid.*, 1932, **207**, 73.
70. W. Klemm and W. Tilk, *ibid.*, 1932, **207**, 161.
71. H. Ulich, E. Hertel and W. Nespital, *Z. Phys. Chem.*, 1932, **B17**, 369.
72. E. Salstrom, *J. Am. Chem. Soc.*, 1933, **55**, 1031.
73. T. Sugawa, *Sci. Rep. Tôhoku Univ.*, 1933, **22**, 959.
74. British Aluminium Co. Ltd., 1934, private commun.
75. S. Karpachev, A. Stromberg and O. Poltoratzkaya, *J. Phys. Chem. (USSR)*, 1934, **5**, 793.
76. W. Kwasnik, Diss, Breslau, T. H., *ibid.*, 1934, p. 13.
77. K. Laybourne and W. Madgin, *J. Chem. Soc.*, 1934, 1.
78. M. G. Malone and A. L. Ferguson, *J. Chem. Physics*, 1934, **2**, 99.
79. O. Ruff and W. Kwasnik, *Z. Anorg. Chem.*, 1934, **219**, 65.
80. P. Drossbach, 'Electrochemistry of Fused Salts', Berlin, 1938.
81. N. S. Kurnakov, N. K. Voskresenskaja and G. D. Gurovic, *Bull Acad. Sci., U.R.S.S., Ser Chim.*, 1938, 396.
82. Lundina, quoted by V. P. Mashovets, 'The Electro-Metallurgy of Aluminium', Russia, 1938.
83. W. Treadwell *et al.*, *Helv. Chim. Acta*, 1939, **22**, 445.
84. D. I. Zuravlev, *J. Phys. Chem. (USSR)*, 1939, **13**, 684.
85. V. P. Barzakovskii, *Bull. Acad. Sci., U.R.S.S.* (Cl. Sci. Chim.), 1940, 825.
86. V. P. Barzakovskii, *J. Appl. Chem. (USSR)*, 1940, **13**, 1117.
87. S. T. Bowden and A. R. Morgan, *Phil. Mag.*, 1940, **29**, 367.
88. E. Ya Gorenbein, *J. Gen. Chem. (USSR)*, 1947, **17**, 873.
89. E. Ya. Gorenbein, *ibid.*, 1948, **18**, 1427.

90. R. de Malleman and F. Suhner, *Compt. Rend.*, 1948, **227**, 546.
91. N. K. Boardman, F. H. Dorman and E. Heymann, *J. Phys. Chem.*, 1949, **53**, 375.
92. G. Fuseya and K. Ouchi, *J. Electrochem. Soc. Japan*, 1949, **17**, 254.
93. I. M. Bokhovkin, *J. Gen. Chem. (USSR)*, 1950, **20**, 397.
94. S. E. Rogers and A. R. Ubbelohde, *Trans. Faraday Soc.*, 1950, **46**, 1051.
95. A. Vayna, *Alluminio*, 1950, **19**, 541.
96. E. Ya. Gorenbein and E. E. Kriss, *J. Phys. Chem. (USSR)*, 1951, **25**, 791.
97. R. C. Spooner and F. E. W. Wetmore, *Canad. J. Chem.*, 1951, **29**, 777.
98. J. D. Edwards, C. S. Taylor, A. S. Russell and L. F. Maranville, *J. Electrochem. Soc.*, 1952, **99**, 527.
99. R. W. Huber, E. V. Potter and H. W. St. Clair, *Rep. Invest. US Bur. Mines No. 4858*, 1952.
100. J. Byrne, H. Fleming and F. E. W. Wetmore, *Canad. J. Chem.*, 1952, **30**, 922.
101. E. F. Riebling, *J. Am. Ceram. Soc.*, 1964, **47**, 478.
102. H. Bloom. I. W. Knaggs, J. J. Molloy and D. Welch, *Trans. Faraday Soc.*, 1953, **49**, 1458.
103. I. P. Vereshchetina and N. P. Luzhnaya, *Izvest. Sekt. Fiziko-Khim. Anal.*, 1954, **25**, 188.
104. J. S. Peake and M. R. Bothwell, *J. Am. Chem. Soc.*, 1954, **76**, 2653.
105. V. D. Polyakov, *Izvest. Sekt. Fiziko-Khim. Anal.*, 1955, **26**, 147.
106. V. D. Polyakov, and S. I. Berul, *ibid.*, 1955, **26**, 164.
107. V. D. Polyakov, *ibid.*, 1955, **26**, 191.
108. N. P. Popovskaya and P. I. Protsenko, *Zh. Fiz. Khim.*, 1955, **29**, 225.
109. K. B. Morris, M. I. Cook, C. Z. Sykes and M. B. Templeman, *J. Am. Chem. Soc.*, 1955, **77**, 851.
110. E. R. van Artsdalen and I. S. Yaffe, *J. Phys. Chem.*, 1955, **59**, 118.
111. T. Baak, *Acta. Chem. Scand.*, 1955, **9**, 1406.
112. N. P. Luzhnaya, N. N. Evseeva and I. P. Vereshchetina; *Zh. Neorg. Khim.*, 1956, **1**, 1490.
113. I. S. Yaffe and E. R. van Artsdalen, *J. Phys. Chem.*, 1956, **60**, 1125.
114. J. D. Mackenzie, *Trans. Faraday Soc.*, 1956, **52**, 1564.
115. J. M. Blocher, R. F. Rolsten and I. E. Campbell, *J. Electrochem. Soc.*, 1957, **104**, 553.
116. N. N. Greenwood and K. Wade, *J. Inorg. Nuclear Chem.*, 1957, **3**, 349.
117. F. R. Duke and R. A. Fleming, *J. Electrochem. Soc.*, 1957, **104**, 251.
118. S. Toshiaki and T. Ishibashi, *Sci. Papers Coll. Gen. Educ., Univ. Tokyo*, 1957, 7, 53.
119. M. Bourgon, G. Derge and C. M. Pound, *Trans. Amer. Inst. Min. Met. Eng.*, 1958, **212**, 338.
120. F. J. Keneshea and D. Cubicciotti, *J. Phys. Chem.*, 1958, **62**, 843.
121. A. D. Kirshenbaum and J. A. Cahill, *J. Inorg. Nucl. Chem.*, 1960, **14**, 283.
122. I. S. Yaffe and E. R. van Artsdalen, *Chem. Semi-Ann. Progr. Rep. No. 2159, Oak Ridge Natl Lab.*, 1956 p. 77.
123. A. D. Kirshenbaum, J. A. Cahill and C. S. Stokes, *J. Inorg. Nucl. Chem.*, 1960, **15**, 297.
124. P. I. Protsenko and A. Ya. Malakhova, *Zh. Neorg. Khim.*, 1961, **6**, 1662.
125. K. B. Morris, M. McNair and G. Koops, *J. Chem. Engng Data*, 1962, 7, 224.
126. N. V. Smith and E. R. van Artsdalen, *Chem. Semi-ann. Progr. Rep. No. 2159, Oak Ridge Natl Lab.*, 1956, p. 80.
127. N. N. Greenwood and I. J. Worrall, *J. Chem. Soc.*, 1958, 1680.
128. E. F. Riebling and C. E. Erickson, *J. Phys. Chem.*, 1963, **67**, 307.
129. L. A. Nisel'son, *Zh. Neorg. Khim.*, 1961, **6**, 1242.
130. G. J. Janz and J. D. E. McIntyre, *J. Electrochem. Soc.*, 1962, **109**, 842.
131. E. R. Buckle, P. E. Tsaoussoglou and A. R. Ubbelohde, *Trans. Faraday Soc.*, 1964, **60**, 684.
132. A. D. Kirshenbaum, J. A. Cahill, P. J. McGonigal and A. V. Grosse, *J. Inorg. Nucl. Chem.*, 1962, **24**, 1287.
133. J. N. Reding, *J. Chem. Engng Data*, 1965, **10**, 1.
134. K. B. Morris and P. L. Robinson, *J. Phys. Chem.*, 1964, **68**, 1194.
135. K. B. Morris and P. L. Robinson, *J. Chem. Engng Data*, 1964, **9**, 444.
136. J. Padova and J. Soriano, *ibid.*, 1964, **9**, 510.
137. I. G. Murgulescu and S. Zuca, *Acad. Rep. Pop. Romaine, Studii Cerc. Chim.*, 1959, 7, 325.
138. T. Kuroda and T. Suzuki, *J. Electrochem. Soc. Japan*, 1961, **29**, E215.
139. B. M. Lepinskikh, O. A. Esin and G. A. Teterin, *Zh. Neorg. Khim.*, 1960, **5**, 642.
140. H. Schinke and F. Saverwald, *Z. Anorg. Allgem. Chem.*, 1956, **287**, 313.
141. G. J. Janz and M. R. Lorenz, *J. Electrochem. Soc.*, 1961, **108**, 1052.
142. J. P. Frame, E. Rhodes and A. R. Ubbelohde, *Trans, Faraday Soc.*, 1959, **55**, 2039.
143. A. Kvist and A. Lunden, *Z. Naturforsch.*, 1965, **20a**, 235.
144. J. O'M. Bockris, A. Pilla and J. L. Barton, *Rev. Chim. Acad. Rep. Populaire. Roumaine*, 1962, 7, 59.
145. V. D. Polyakov, *Izv. Sektova. Fiz. Khim. Analiza Inst. Obshch. Neorgan, Khim. Acad. Nauk SSSR*, 1955, **26**, 173, 191.
146. E. Vogel, H. Schinke and F. Saverwald, *Z. Anorg. Allgem. Chem.*, 1956, **284**, 131.
147. J. O'M. Bockris, A. Pilla and J. L. Barton, *J. Phys. Chem.*, 1960, **64**, 507.

Table 9.2 DENSITIES OF MOLTEN BINARY SALT SYSTEMS AND OTHER MIXED IONIC MELTS

The density $(\mathrm{g\,cm^{-3}})$ at temperature $t(°C)$ and composition p(wt. %) of the first-named constituent is given as d_t, or the constants a and b in the equation $d_t = a - 10^{-3}\,bt$ or A, B and C in the equation $d_t = 10^6 At^{-2} - 10^3 Bt^{-1} + C$ are given together with the temperature range $r(°C)$. Principal references are in **bold type**.

AgBr–AgCl	p	0	27.8	46.9	71.6	100	
Ref. **25**	a	5.262	5.523	5.678	5.832	6.025	
	b	0.94	1.08	1.12	1.07	1.04	
	r	480–630	440–580	420–590	420–580	440–600	
AgBr–KBr	p	0	50.8	70.4	85.7	100	64.6
Ref. **64, 25, 7**	a	2.706	3.686	4.376	5.156	6.023	4.077
	b	0.80	0.98	1.03	1.12	1.05	1.03
	r	750–800	593–700	380–600	380–600	440–600	546–629
AgBr–LiBr	p	68.5					
Ref. **65**	a	4.504					
	b	0.877					
	r	517–555					
AgBr–NaBr	p	64.6					
Ref. **66**	a	4.311					
	b	0.9					
	r	607–619					
AgBr–RbBr	p	53.2					
Ref. **67**	a	4.470					
	b	1.23					
	r	514–624					
AgCl–AgNO₃	p	0	9.00	12.5	15.5	20.0	30.0	40.0
Ref. **33**	a	4.167	4.242	4.279	4.297	4.338	4.431	4.497
	b	1.00	1.00	1.00	1.00	1.00	1.00	0.90
	r				310–330			
AgCl–KCl	p	0	63.8	80.3	88.9	100	
Ref. **25, 7**	a	1.988	3.286	4.001	4.463	5.263	
	b	0.60	0.88	0.96	0.95	0.94	
	r	785–880	560–745	385–640	433–670	480–630	
AgCl–PbCl₂	p	0	11.0	19.4	24.7	31.0	41.0	66.9
Ref. **25**	a	5.702	5.660	5.634	5.582	5.543	5.520	5.387
	b	1.50	1.45	1.42	1.34	1.28	1.26	1.08
	r	516–710	520–700	470–680	445–670	444–680	380–700	478–660
AgI–AgNO₃	p	20	30	40	50	60	70
Ref. **54, 30**	a	4.53	4.97	5.12	5.31	5.44	5.53
	b	1.2	2.2	2.1	2.1	1.8	1.5
	r			about 150–300			
AgNO₃– Cd(NO₃)₂	p	32.4	37.0	46.8	57.2	68.4	74.2
Ref. **50**	a	1.721	1.762	1.843	1.930	2.003	2.032
	b	1.01	1.05	1.10	1.15	1.15	1.14
	r	210–290	160–290	160–290	190–290	190–290	210–290
AgNO₃–HgI₂	p	13.8	19.9	27.2	35.9	46.6	52.8	67.9
Ref. **49**	a	5.826	5.732	5.584	5.364	5.188	5.074	4.930
	b	1.58	1.46	1.46	1.46	1.40	1.34	1.28
	r	160–240	160–240	100–240	120–240	100–240	100–240	120–240
AgNO₃–KNO₃	p	15.7	29.6	41.9	52.8	62.7	79.7	93.8
Ref. **57, 46**	a	2.284	2.534	2.654	2.856	2.986	3.516	3.765
	b	0.76	0.97	0.84	0.89	0.76	1.08	0.85
	r	350–400	300–400	250–400	250–400	250–400	170–350	200–400
AgNO₃–NaNO₃	p	0	10	20	30	40	60	80
Ref. **37**	a	2.124	2.226	2.357	2.486	2.616	2.974	3.439
	b	0.70	0.71	0.77	0.79	0.79	0.70	0.90
	r			about 290–370			
AgNO₃–NH₄NO₃	p	0	5	10				
Ref. **58**	d_{170}	1.432	1.479	1.529				
	d_{180}	1.426	1.474	1.523				

(continued)

Table 9.2   DENSITIES OF MOLTEN BINARY SALT SYSTEMS AND OTHER MIXED IONIC MELTS—*continued*

		25	45	50	60	75	90	100
AgNO$_3$–TlNO$_3$	p	25	45	50	60	75	90	100
Ref. 26	d_{100}	—	4.671	4.630	4.554	—	—	—
	d_{150}	—	4.575	4.526	4.435	—	—	
	d_{200}	4.638	4.452	4.410	4.319	4.183	4.074	
	d_{225}	4.579	4.406	4.351	4.261	4.132	4.024	3.922
AlBr$_3$–HgBr$_2$	p	59.7	65.4	71.2	74.9	80.5	89.6	100
Ref. 21, 4	d_{110}	3.577	3.415	3.276	3.173	3.030	2.827	2.624
	d_{140}	3.504	3.359	3.202	3.097	2.961	2.755	2.555
AlBr$_3$–KBr	p	81.8	83.2	84.7	86.2	87.5	88.0	
Ref. 28, 4	d_{110}	2.818	2.815	2.810	2.804	2.798	2.790	
	d_{140}	2.775	2.771	2.764	2.755	2.749	2.741	
2AlBr$_3$–KCl	a	2.846						
Ref. 23	b	1.445						
	r	80–170						
AlBr$_3$–NaBr	p	83.8	84.7	85.6	86.8	87.9	88.9	
Ref. 28	d_{120}	2.827	2.820	2.817	2.809	2.798	2.786	
	d_{140}	2.797	2.792	2.787	(2.777)	2.767	2.756	
2AlBr$_3$.NaBr	a	3.005						
Ref. 23	b	1.5						
	r	110–170						
AlBr$_3$–NH$_4$Br	p	84.5	87.9	89.0	89.4	90.8	100	
Ref. 21	a	2.842	2.865	2.877	2.881	2.889	2.877	
	b	1.30	1.40	1.45	1.50	1.60	2.30	
	r				110–150			
2AlBr$_3$.NH$_4$Br	a	2.848						
Ref. 23	b	1.36						
	r	110–160						
AlBr$_3$–SbBr$_3$	p	0	34.1	42.5	67.4	71.9	91.3	100
Ref. 20, 4	d_{100}	3.697	3.402	3.318	3.034	2.971	2.737	2.644
	d_{140}	3.594	3.311	3.224	2.941	2.881	2.641	2.556
AlBr$_3$.SbBr$_3$	a	3.541						
Ref. 32, 23	b	2.3						
	r	80–170						
AlBr$_3$.SbBr$_3$–	p	0	16.3	59.9	67.9	77.3	86.8	100
AsBr$_3$	d_{85}	3.232	3.231	3.286	3.304	3.310	3.330	3.346
Ref. 32	d_{100}	3.193	3.191	3.247	3.263	3.274	3.295	3.313
AlBr$_3$–ZnBr$_2$	p	70.5	74.5	77.7	82.5	85.0	88.4	100
Ref. 20	d_{100}	3.014	(2.957)	2.914	2.849	2.811	2.768	2.644
	d_{150}	2.915	2.850	2.811	2.740	2.704	2.657	2.534
2AlBr$_3$.ZnBr$_2$	t	100	140	180				
Ref. 23	d	3.01	2.93	2.81				
AlCl$_3$–KCl**	p	50.4	53.7	62.5	64.2	71.2	76.1	84.2
Ref. 63, 45, 29, 19	a	1.787	1.785	1.755	1.937	1.859	1.819	1.820
	b	0.590	0.605	0.600	0.935	0.902	0.789	0.850
	r	600–800	600–800	600–800	260–350	240–280	190–270	175–225
AlCl$_3$–LiCl**	p	75.9	79.3	82.5	85.4	87.9	90.4	
Ref. 63, 19	a	1.735	1.737	1.757	1.759	1.768	1.741	
	b	0.77	0.68	0.80	0.84	0.90	0.80	
	r	180–330	175–225	175–225	175–225	175–225	175–225	
AlCl$_3$–NaBr	p	58.6	61.0	65.7	70.3	74.8	79.3	
Ref. 63	a	2.158	2.119	2.064	2.017	1.979	1.944	
	b	0.92	0.86	0.92	0.96	1.02	1.12	
	r		175–225				200–275	
AlCl$_3$–NaCl**	p	69.5	71.9	73.5	75.8	79.5	84.1	90.1
Ref. 63, 45, 16, 19	a	1.848	1.858	1.839	1.829	1.792	1.787	1.810
	b	0.812	0.910	0.849	0.844	0.715	0.840	1.04
	r	220–280	160–210	150–210	150–210	150–210	175–225	175–225
AlCl$_3$.NH$_4$Cl	t	284	293	311	315	324	354	
Ref. 19	d	1.475	1.470	1.445	1.440	1.425	1.420	

(*continued*)

Table 9.2 DENSITIES OF MOLTEN BINARY SALT SYSTEMS AND OTHER MIXED IONIC MELTS—*continued*

AlCl$_3$–RbCl Ref. 63							
p	71.9	76.7					
a	1.992	1.975					
b	0.96	1.00					
r	175–225						

AlF$_3$–Na$_3$AlF$_6$* Ref. 62, 38, 31, 24, 22, 15, 14, 13, 10, 9, 6							
p	0	5	10	15	20	25	30
$d_{1\,000}$	2.096	2.078	2.048	2.015	1.977	1.930	1.873
$d_{1\,100}$	2.002	1.987	1.965	1.935	1.894	1.839	1.775

Al$_2$O$_3$–Na$_3$AlF$_6$ Ref. 62, 38, 14, 31, 6							
p	0	2.5	5	7.5	10	12.5	15
$d_{1\,000}$	2.096	2.076	2.060	2.048	2.039	2.033	2.028
$d_{1\,100}$	2.002	1.985	1.974	1.966	1.960	1.957	1.954

B$_2$O$_3$–BaO Ref. 44							
p	20.7	30.0	39.8	49.2	59.7	67.9	
a	4.038	5.006	4.780	4.422	3.704	3.392	
b	—	1.32	1.40	1.34	0.96	0.96	
r	1120	1000–1200	1000–1100	850–1100	850–1000	850–1100	

B$_2$O$_3$–CaO Ref. 44							
p	55.3	59.8	62.8	67.9	70.2	73.3	
a	2.926	2.901	2.876	2.871	2.932	2.844	
b	0.44	0.45	0.46	0.51	0.59	0.56	
r	1160–1200	1110–1210	1140–1190	1100–1200	1060–1100	900–1200	

B$_2$O$_3$–K$_2$O Ref. 40							
p	51.5	55.7	65.6	75.3	84.9	94.8	98.5
a	2.690	2.534	2.412	2.185	2.057	1.855	1.737
b	0.905	0.687	0.500	0.310	0.295	0.255	0.217
r	900–1000	800–1000	700–1000	800–1000	700–1000	600–1000	500–1000

B$_2$O$_3$–Li$_2$O Ref. 40							
p	71.3	85.2	89.4	93.5	97.2	100.0	
a	2.340	2.400	2.281	2.103	1.901	1.662	
b	0.467	0.467	0.402	0.335	0.260	0.153	
r	800–1000	800–1000	700–1000	600–1000	600–1000	600–1200	

B$_2$O$_3$–Na$_2$O Ref. 40							
p	64.0	69.2	77.6	85.8	94.4	99.1	100.0
a	2.540	2.778	2.413	2.137	1.898	1.759	1.662
b	0.57	0.86	0.435	0.278	0.235	0.222	0.153
r	900–1000	700–800	700–1000	700–1000	600–1100	600–1000	600–1200

B$_2$O$_3$–SrO Ref. 44							
p	44.9	49.6	56.8	61.6	65.2	70.0	
a	3.472	3.513	3.402	3.230	3.124	2.948	
b	0.43	0.56	0.61	0.57	0.57	0.51	
r	1150–1200	1120–1220	960–1100	950–1100	950–1100	920–1120	

BaBr$_2$–KBr Ref. 41							
p	39.2	45.1	55.5	62.5	73.0	82.2	88.2
a	3.318	3.324	3.558	3.720	3.932	4.117	4.425
b	0.937	0.849	0.896	0.930	0.906	0.875	1.02
r	660–850	640–850	630–850	630–850	630–850	690–850	750–850

BaCl$_2$–CdCl$_2$ Ref. 25, 1							
p	0	18.99	39.00	57.35	100		
a	3.870	3.996	4.018	4.069	3.672		
b	0.84	0.93	0.93	0.96	0.52		
r	582–725	597–700	580–700	600–690	above 1 000		

BaCl$_2$–KCl Ref. 43							
p	20.4	29.6	47.4	58.2	70.1	82.8	91.4
a	2.199	2.237	2.554	2.744	3.035	3.268	3.456
b	0.64	0.59	0.63	0.70	0.77	0.65	0.68
r	790–900	790–890	790–890	800–890	790–880	820–910	880–940

BaCl$_2$–MgCl$_2$ Ref. 69, 35, 70							
p	9.95	25.0	50.2	75.1	90.0		
a	2.073	2.333	2.835	3.400	(3.642)		
b	0.36	0.47	0.64	0.75	0.71		
r		800–900					

BaCl$_2$–NaCl Ref. 71, 17							
p	44.0	54.4	62.7	69.7	75.3	80.1	84.4
d_{725}	—	2.204	2.350	2.491	2.620	—	—
d_{750}	—	2.192	2.334	2.471	2.590	—	—
d_{775}	1.992	2.180	2.318	2.452	2.560	2.650	2.760
d_{800}	1.986	2.169	2.302	2.432	2.531	2.616	2.730

BaCl$_2$–NH$_4$NO$_3$ Ref. 58		
p	0	1.5
d_{180}	1.426	1.436

(continued)

Table 9.2   DENSITIES OF MOLTEN BINARY SALT SYSTEMS AND OTHER MIXED IONIC MELTS—*continued*

BaCl₂–PbCl₂							
Ref. **25**, 1	p	0	10.71	15.57	24.83	100	
	a	5.702	5.430	5.356	5.156	3.662	
	b	1.50	1.35	1.36	1.27	0.52	
	r	516–710	565–700	575–690	660–710	above 1 000	

BaF₂–Na₃AlF₆					
Ref. **15**, 10	p	0	33.1	55.4	71.5
	a	3.069	3.429	3.978	3.804
	b	0.96	0.838	0.996	0.447
	r	1 020–1 130	920–1 175	960–1 155	910–1 120

BaF₂–NaF					
Ref. **15**	p	0	45.2	67.5	80.7
	a	2.590	3.291	3.739	4.195
	b	0.628	0.626	0.600	0.624
	r	1 010–1 120	955–1 160	970–1 165	955–1 180

Ba(NO₂)₂–			
Ba(NO₃)₂	p	5.7	13.4
Ref. **72**	d_{300}	3.240	3.243
	d_{320}	3.226	3.212
	d_{340}	3.212	3.216

Ba(NO₂)₂–KNO₂							
Ref. **72**	p	59.0	73.7	80.0	86.8	92.5	97.6
	d_{280}	2.274	2.492	—	—	—	3.154
	d_{320}	2.245	2.458	2.615	2.780	2.950	3.126
	d_{340}	2.132	2.440	2.577	2.750	2.921	3.099
	d_{360}	2.218	2.422	2.577	2.750	2.921	3.099

Ba(NO₃)₂–KNO₃								
Ref. **72**, 36, 73	p	9.54	25.8	33.1	39.7	45.9	51.5	57.0
	d_{340}	1.914	2.014	2.068	2.118	—	—	—
	d_{380}	1.884	1.984	2.036	2.088	2.143	2.201	—
	d_{420}	1.885	1.953	2.004	2.056	2.111	2.168	2.228
	d_{460}	—	—	1.974	2.026	2.080	2.136	2.196

Ba(NO₃)₂–				
NH₄NO₃	p	0	2.5	5.0
Ref. **58**	d_{170}	1.432	1.453	1.474
	d_{180}	1.426	1.447	1.466

BaO–SiO₂						
Ref. **74**	p	22.1	39.0	52.3	63.0	72.0
	a	2.580	3.016	3.401	3.748	4.044
	b	0.000	0.048	0.068	0.075	0.080
	r			1 600–1 950		

BeF₂–LiF		
Ref. **75**	p	51.5
	a	2.09
	b	0.27
	r	500–800

Bi–BiBr₃		
Ref. **68**	p	52.1
	d_{543}	6.69

Bi–BiCl₃								
Ref. **61**	p	0	7.5	14.7	19.7	23.8	97.9	100
	a	4.42	4.61	4.87	5.07	5.24	10.39	10.39
	b	2.20	2.06	2.08	2.13	2.15	1.29	1.29
	r	240–330	290–400	270–420	290–450	310–440	330–440	310–44

CaCl₂–KCl								
Ref. **8**	p	0	20	40	50	60	80	100
	d_{800}	1.495	1.573	1.671	1.725	1.780	1.896	2.057
	d_{900}	1.434	1.517	1.613	1.667	1.734	1.850	2.009

CaCl₂–MgCl₂						
Ref. **35**	p	22.6	27.7	42.6	59.4	76.6
	a	2.16	2.115	2.217	2.319	2.365
	b	0.47	0.40	0.44	0.48	0.45
	r	723–895	753–907	728–886	730–902	746–898

CaCl₂–NaCl**								
Ref. **71**, 27, 18, 8	p	17.5	32.3	44.9	65.5	74.0	88.5	94.5
	d_{625}	—	—	—	1.899	1.944	—	—
	d_{700}	—	—	1.767	1.855	1.912	2.005	—
	d_{775}	1.612	1.679	1.730	1.830	1.879	1.974	2.013
	d_{850}	1.575	1.646	1.693	1.798	1.846	1.944	1.985

CaF₂–CaO							
Ref. **53**	p	92.9	94.6	95.3	97.1	98.6	100
	$d_{1\,545}$	2.63	2.50	2.53	2.59	2.68	2.75

(*continued*

Table 9.2   DENSITIES OF MOLTEN BINARY SALT SYSTEMS AND OTHER MIXED IONIC MELTS—*continued*

CaF$_2$–Na$_3$AlF$_6$	p	0	10	20	30	40	50
Ref. **38, 31, 15**	$d_{1\,000}$	2.096	2.162	2.223	2.283	2.334	2.366
	$d_{1\,100}$	2.002	2.070	2.135	2.200	2.256	2.294
CaF$_2$–NaF	p	0	26.9	48.0	65.0		
Ref. **15**	a	2.602	2.692	2.804	2.929		
	b	0.64	0.58	0.57	0.54		
	r	1 010–1 120	930–1 185	870–1 165	1 050–1 170		
Ca(NO$_3$)$_2$–	p	0	5	10			
NH$_4$HO$_3$	d_{170}	1.432	1.460	—			
Ref. **58**	d_{180}	1.426	1.453	1.480			
CaO–SiO$_2$	p	28.6	38.4	48.4	58.5		
Ref. **74**	a	2.578	2.651	2.758	2.840		
	b	0.066	0.064	0.084	0.103		
	r			1 600–1 950			
CdBr$_2$CdCl$_2$	p	0	38.55	55.46	73.64	100	
Ref. **25**	a	3.870	4.138	4.255	4.390	4.687	
	b	0.84	0.90	0.91	0.93	1.08	
	r	582–725	580–680	590–710	606–705	580–720	
CdCl$_2$–CdI$_2$	p	0	14.2	33.2	59.9	100	
Ref. **41**	a	4.828	4.638	4.367	4.187	3.839	
	b	1.12	1.06	0.88	0.87	0.80	
	r	380–700	360–700	440–700	520–700	560–700	
CdCl$_2$KCl	p	0	44.77	62.10	78.10	92.36	100
Ref. **25, 7**	a	1.988	2.495	2.791	3.176	3.625	3.870
	b	0.60	0.72	0.82	0.95	0.96	0.84
	r	above 750	604–750	460–680	464–680	534–700	582–725
CdCl$_2$–LiCl	p	0	59.1	81.2	92.9	100	
Ref. **41**	a	1.731	2.572	3.245	3.593	3.839	
	b	0.382	0.577	0.845	0.825	0.800	
	r	600–750	560–750	510–750	520–750	560–750	
CdCl$_2$–NaCl	p	0	62.08	71.38	79.64	85.23	91.66	100
Ref. **25, 7**	a	2.053	2.896	3.090	3.315	3.543	3.678	3.870
	b	0.63	0.83	0.86	0.92	1.04	0.95	0.84
	r	above 800	580–690	500–690	570–680	540–680	580–700	582–725
CdCl$_2$–PbCl$_2$	p	0	27.62	52.14	75.66	85.40	100
Ref. **25**	a	3.870	4.305	4.726	5.222	5.402	5.702
	b	0.84	1.02	1.18	1.39	1.43	1.50
	r	582–725	540–680	515–700	480–680	545–680	510–710
CdI$_2$–KI	p	0	28.0	52.5	68.8	81.6	92.6	100
Ref. **41**	a	3.108	3.469	3.769	4.120	4.431	4.645	4.828
	b	0.96	1.09	1.12	1.22	1.29	1.17	1.12
	r	680–800	540–800	400–800	190–700	300–700	360–700	380–700
Cd(NO$_3$)$_2$–KNO3	p	50.0	55.7	65.7	74.1	81.2	
Ref. **50**	a	2.436	2.475	2.583	2.689	2.820	
	b	0.92	0.89	0.93	0.93	0.95	
	r	260–300	220–300	200–300	180–300	200–300	
Ce–CeCl$_3$	p	0.28	0.57	1.14	2.31	3.50	4.70
Ref. **76**	d_{850}	3.165 5	3.169 5	3.180 0	3.205 5	3.239 0	3.287 5
	d_{900}	3.127 5	3.132 7	3.144 0	3.170 0	3.204 1	3.249 0
	d_{950}	3.087 7	3.093 2	3.104 0	3.133 3	3.174 0	3.228 8
Cu$_2$Cl$_2$–KCl	p	0	11.49				
Ref. **5**	d_{800}	1.51	1.62				
Cu$_2$S–FeS	p	0	50	100			
Ref. **60**	a	3.85	5.10	6.76			
	b	—	0.62	0.75			
	r	1 250–1 450	1 100–1 500	1 150–1 400			
FeO–SiO$_2$	p	73.7	78.3	82.8	88.1	91.5	95.7	100
Ref. **55**	$d_{1\,300}$	3.67	3.81	4.00	4.16	4.32	4.61	4.90

(continued)

Table 9.2   DENSITIES OF MOLTEN BINARY SALT SYSTEMS AND OTHER MIXED IONIC MELTS—*continued*

KBr–KNO$_3$	p	0	17.2	37.0	54.1	70.2	86.9	100
Ref. **41**	a	2.116	2.186	2.285	2.415	2.476	2.635	2.725
	b	0.729	0.714	0.727	0.779	0.730	0.801	0.794
	r	330–600	330–600	340–600	440–600	530–800	650–800	740–800

KBr–NaCl	p	67.2	100
Ref. **41**	a	2.447	2.725
	b	0.723	0.794
	r	750–800	740–800

KBr–TlBr	p	5.7	15.2	37.0	41.4	57.6	60.5	74.6
Ref. **77**	a	6.493_5	6.908_4	—	5.732_8	—	—	3.978_7
	b	1.934_2	3.421_4	—	3.084_1	—	—	1.893_6
	A	—	—	1.781_6	—	3.218_3	6.425_3	—
	B	—	—	3.878_9	—	6.631_9	14.160_5	—
	C	—	—	4.192_8	—	6.087_0	10.347_8	—
	r	490–628	584–732	707–759	750–799	721–798	747–857	758–797

KBr–ZnSO$_4$	p	34.7	41.0	42.6	55.2	59.6	63.9	68.9
Ref. **42, 34**	a	3.090	3.003	2.985	3.105	2.939	2.940	2.965
	b	0.46	0.40	0.42	0.80	0.52	0.60	0.74
	r				500–550			

KCl–KI	p	0	2.8	7.8	13.3	26.8	41.2	64.4
Ref. **52**	a	3.098	3.030	2.914	2.873	2.684	2.498	2.274
	b	0.956	0.930	0.864	0.890	0.826	0.755	0.690
	r	680–900	680–900	710–900	640–910	620–900	680–920	710–900

KCl–KNO$_3$	p	6.4	15.3	23.5
Ref. **78**	a	2.101	2.090	2.069
	b	0.728	0.721	0.680
	r	349–540	429–580	485–633

KCl–LiCl**	p	0	28.2	42.6	55.8	72.2	87.6	100
Ref. **52, 12**	a	1.766	1.835	1.856	1.885	1.923	1.968	1.977
	b	0.433	0.489	0.507	0.528	0.561	0.509	0.583
	r	620–780	530–750	460–600	390–590	590–750	690–850	780–940

KCl–MgCl$_2$	p	10.7	20.4	31.2	41.5	49.7	59.0	80
Ref. **35, 69, 11**	a	1.896	1.993	1.990	1.924	1.946	1.944	1.946
	b	0.31	0.41	0.48	0.44	0.50	0.52	0.54
	r	706–886	695–890	756–901	734–894	707–880	707–881	704–876

KCl–MnCl$_2$	p	12.6	23.2	32.1	36.6	41.3	51.0	63.5
Ref. **79**	d_{500}	—	2.210	2.111	2.063	2.009	1.943	—
	d_{600}	2.337	2.157	2.051	2.007	1.955	1.881	1.795
	d_{700}	2.274	2.104	1.998	1.958	1.900	1.818	1.727
	d_{800}	2.212	2.051	1.941	1.898	1.845	1.755	1.668

KCl–NaBr	p	41.8	100
Ref. **41**	a	2.449	1.986
	b	0.728	0.582
	r	750–800	770–800

KCl–NaCl**	p	0	18.7	31.1	40.6	54.8	64.8	82.9
Ref. **52, 17, 8**	a	1.991	1.989	1.985	1.982	1.976	1.977	1.979
	b	0.543	0.554	0.560	0.557	0.568	0.575	0.581
	r	800–1 030	780–920	710–920	710–930	670–910	680–910	720–930

KCl–NaI	p	0	8.0	19.8	33.0	49.7	73.6	100
Ref. **41**	a	3.412	3.118	2.971	2.724	2.482	2.195	1.986
	b	1.00	0.824	0.900	0.815	0.727	0.625	0.582
	r	670–800	600–800	520–800	540–800	570–800	690–800	770–800

KCl–PbCl$_2$	p	0	5.51	13.21	22.94	100
Ref. **25, 7**	a	5.702	5.145	4.513	3.960	1.988
	b	1.50	1.42	1.28	1.13	0.60
	r	516–700	565–700	580–680	490–680	above 750

KCl–ZnCl$_2$	p	5.4	9.3	20.0	31.1	49.2	55.9	70.1
Ref. **59**	a	2.653	2.588	2.542	2.448	2.272	2.217	2.080
	b	0.55	0.50	0.62	0.66	0.60	0.60	0.57
	r	460–670	450–660	450–660	450–640	440–660	450–650	690–720

(continued)

Table 9.2   DENSITIES OF MOLTEN BINARY SALT SYSTEMS AND OTHER MIXED IONIC MELTS—*continued*

KCl–ZnSO₄							
Ref. **42, 34**	*p*	23.2	29.0	31.4	38.2	47.0	54.3	56.1
	a	3.024	2.884	2.821	2.681	2.619	2.511	2.491
	b	0.68	0.64	0.58	0.56	0.72	0.70	0.70
	r				475–550			

KF–LiF–NaF		
eutectic	*p*KF	59.0
Ref. **80**	*p*LiF	29.2
	a	2.47
	b	0.68
	r	600–800

KI–NaCl							
Ref. **41**	*p*	33.6	55.0	74.1	86.9	94.2	100
	a	2.253	2.484	2.727	2.893	3.034	3.108
	b	0.632	0.706	0.819	0.858	0.965	0.960
	r	720–800	680–800	560–800	580–800	640–800	680–800

K₂MoO₄–MoO₃							
Ref. **81**	*p*	17.5	27.5	43.0	52.2	60.0	79.4	91.0
	a	3.9974	3.9371	3.7813	3.6534	3.4282	3.1637	3.0360
	b	1.1337	1.0997	1.1189	1.0529	0.9951	0.7146	0.6340
	r	761–869	614–750	637–792	632–779	574–766	783–933	882–978

KNO₂–KNO₃					
Ref. **48**, 72	*p*	0	26.5	35.9	50.7	100
	a	2.116	2.068	2.062	2.052	2.005
	b	0.73	0.69	0.70	0.70	0.70
	r	340–500	340–500	340–500	380–500	440–500

KNO₂–NaNO₂							
Ref. **48**	*p*	0	17.8	29.1	45.1	55.2	74.2	100
	a	2.004	1.959	1.951	1.961	1.951	1.963	2.005
	b	0.69	0.59	0.58	0.59	0.58	0.61	0.70
	r	380–500	260–500	260–500	260–500	350–500	350–500	440–500

KNO₂–NaNO₃							
Ref. **48**	*p*	0	15	35	50	65	85	100
	a	2.121	2.117	2.073	2.057	2.028	1.993	2.005
	b	0.68	0.70	0.66	0.68	0.66	0.63	0.70
	r	350–500	300–500	200–500	200–500	240–500	350–500	440–500

KNO₃–LiNO₃							
Ref. **78**, 82	*p*	16.7	26.7	40.9	59.5	71.0	81.9	92.2
	a	1.954	1.974	1.997	2.033	2.055	2.083	2.102
	b	0.599	0.623	0.638	0.683	0.696	0.729	0.735
	r	257–403	264–350	219–445	328–476	306–454	294–425	318–452

KNO₃–NaNO₂							
Ref. **48**	*p*	0	20.5	44.0	59.4	73.1	89.3	100
	a	2.004	1.991	2.028	2.048	2.061	2.100	2.116
	b	0.69	0.61	0.66	0.67	0.67	0.72	0.73
	r	380–500	300–500	200–500	200–500	200–500	240–500	340–500

KNO₃–NaNO₃**						
Ref. **48, 47, 36,**	*p*	0	20	40	60	80	100
82, 8, 3	*a*	2.121	2.127	2.127	2.126	2.126	2.116
	b	0.68	0.71	0.72	0.73	0.72	0.73
	r	350–500	290–500	250–500	240–500	290–500	350–500

KNO₃–NH₄NO₃					
Ref. **58**	*p*	0	2.5	5.0	7.5	10.0
	d_{170}	1.432	1.441	1.451	1.462	1.473
	d_{190}	1.420	1.429	1.439	1.450	1.460

KNO₃–Pb(NO₃)₂							
Ref. **36**	*p*	40	50	60	70	80	90	100
	a	3.080	2.881	2.679	2.483	2.362	2.218	2.113
	b	1.00	1.00	0.908	0.750	0.787	0.709	0.730
	r	275–345	230–365	275–380	300–390	320–405	330–420	350–460

KNO₃–Sr(NO₃)₂						
Ref. **73, 36**	*p*	52.9	60	70	80	90	100
	a	2.403	2.360	2.285	2.226	2.167	2.113
	b	0.721	0.760	0.748	0.752	0.753	0.730
	r	421–452	365–445	300–420	305–465	325–475	350–460

K₂O–SiO₂						
Ref. **56, 39**	*p*	23.9	26.7	29.9	32.9	38.7	43.6
	a	2.353	2.320	2.377	2.380	2.464	2.504
	b	0.14	0.11	0.16	0.16	0.23	0.27
	r			1 000–1 400			

K₂SO₄–ZnSO₄							
Ref. **34**	*p*	26.59	32.66	35.15	36.77	41.41	49.92	58.86
	d_{500}	(2.841)	2.751	2.731	2.728	2.680	2.592	2.509
	d_{550}	2.812	2.727	2.708	2.692	2.641	2.556	2.485

(continued)

Table 9.2   DENSITIES OF MOLTEN BINARY SALT SYSTEMS AND OTHER MIXED IONIC MELTS—*continued*

K₂WO₄–WO₃							
K_2WO_4–WO_3 Ref. **83**	p	37.4	50.5	58.3	67.4	76.6	85.0	92.2
	a	5.889 8	5.486 5	5.332 6	4.955 3	4.532 2	4.244 4	4.077 1
	b	1.553 8	1.519 6	1.562 0	1.334 1	1.015 7	0.907 4	0.845 2
	r	897–999	772–943	655–783	682–851	803–910	860–987	904–1 029
$LiCl$–$LiNO_3$ Ref. **78**	p	6.4	13.3	20.8				
	a	1.911	1.903	1.899				
	b	0.537	0.538	0.524				
	r	278–446	340–497	378–497				
$LiClO_4$ in KNO_3–$NaNO_3$ eutectic (57.3% KNO_3) Ref. **84**	p_{LiClO_4}	5	10	15				
	a	2.120 1	2.127 5	2.139 4				
	b	0.711 0	0.723 2	0.744 1				
	r	—	230–400	—				
$LiClO_4$–$LiNO_3$ Ref. **73**	p	34.0	57.3	82.3				
	a	2.014	2.088	2.134				
	b	0.610	0.629	0.629				
	r	240–357	198–347	225–336				
Li_2MoO_4–MoO_3 Ref. **81**	p	19.5	33.9	51.5	60.5	71.3	85.2	91.0
	a	3.993 2	3.895 1	3.735 9	3.564 1	2.985 6	3.455 9	3.352 0
	b	0.950 2	0.845 1	0.772 3	0.644 8	0.031 7	0.593 0	0.531 7
	r	766–921	755–924	760–934	802–962	799–950	781–905	825–924
$LiNO_3$–NH_4NO_3 Ref. **58**	p	0	5					
	d_{180}	1.426	1.430					
Li_2O–SiO_2 Ref. **56, 39**	p	16.1	21.8	25.9	31.9	42.8	48.1	
	a	2.311	2.355	2.359	2.344	1.980	1.955	
	b	0.13	0.19	0.20	0.22	—	—	
	r		1 100–1 400		1 250–1 400		1 400	
Li_2WO_4–WO_3 Ref. **83**	p	54.6	59.7	69.0	77.5	85.2	92.3	
	a	5.971 8	6.077 6	5.769 6	5.527 5	5.281 9	5.016 0	
	b	1.162 8	1.341 1	1.190 3	1.112 4	0.991 3	0.851 1	
	r	809–939	764–968	755–978	736–930	714–923	733–959	
$MgCl_2$–$NaCl$ Ref. **35**	p	20.0	45.7	60.6	75.8			
	a	1.961	1.963	2.002	2.000			
	b	0.49	0.47	0.48	0.43			
	r	750–888	732–890	722–896	712–893			
MgF_2–Na_3AlF_6 Ref. **62**	p	3	6	9				
	d_{950}	—	—	2.14				
	d_{980}	2.12	2.13	—				
MgO–SiO_2 Ref. **74**	p	32.8	35.5	38.3	41.1	44.0		
	a	2.547	2.579	2.597	2.655	2.670		
	b	0.049	0.056	0.052	0.071	0.070		
	r			1 600–1 950				
MoO_3–Na_2MoO_4 Ref. **51**	p	21.4	26.4	51.2	62.2	71.2	83.7	86.2
	a	2.79	3.63	3.92	4.02	3.88	3.74	3.81
	b	—	1.0	1.2	1.2	1.0	0.8	0.9
	r	700–830	690–790	660–760	650–810	650–750	730–840	780–880
NH_4Cl–NH_4NO_3 Ref. **58**	p	0	2.5	5	7.5	10		
	d_{170}	1.432	1.426	1.419	1.412	1.403		
	d_{180}	1.420	1.414	1.407	1.403	—		
NH_4NO_3–$(NH_4)_2SO_4$ Ref. **58**	p	92.5	95.0	97.5	100			
	p_{180}	—	1.435	1.430	1.427			
	d_{200}	1.425	1.421	1.417	1.416			
NH_4NO_3–$Pb(NO_3)_2$ Ref. **58**	p	89.9	95.0	100				
	d_{170}	—	1.481	1.432				
	d_{180}	1.527	1.475	1.426				
NH_4NO_3–$Sr(NO_3)_2$ Ref. **58**	p	95	100					
	d_{170}	1.466	1.432					
	d_{190}	1.453	1.420					

(*continue*

Table 9.2   DENSITIES OF MOLTEN BINARY SALT SYSTEMS AND OTHER MIXED IONIC MELTS—*continued*

NaCl–NH$_4$NO$_3$ Ref. **58**

p	0	1.5
d_{180}	1.426	1.432

NaCl–NaNO$_3$ Ref. **71**

p	1.4	2.8	4.2	5.7	7.1	8.6	10.1
d_{350}	1.879	1.878	1.879	1.875	1.875	—	—
d_{400}	1.842	1.841	1.843	1.839	1.839	1.840	1.836
d_{450}	1.806	1.804	1.806	1.802	1.803	1.803	1.800

NaCl–PbCl$_2$ Ref. **17**

p	0	2.69	5.85	9.42	13.32	17.5
d_{500}	4.96	4.72	4.34	4.21	3.96	3.66
d_{600}	4.82	4.57	4.22	4.08	3.77	3.50

NaF–Na$_3$AlF$_6$† Ref. **38, 31, 24, 22, 15, 14, 13,** 10, 9, 6

p	0	10	20	40	60	80	100
$d_{1\,000}$	2.096	2.110	2.108	2.085	2.042	1.998	1.957
$d_{1\,100}$	2.002	2.017	2.019	2.002	1.970	1.933	1.895

NaF–UF$_4$–ZrF$_4$ Ref. **85**

P_{NaF}	19.0
P_{UF4}	11.4
a	3.93
b	0.93
r	600–800

NaF–ZrF$_4$ Ref. **86**

p	17.05	22.0	27.3	31.8	51.1
a	3.83	3.71	3.61	3.52	3.23
b	0.91	0.89	0.87	0.86	0.81
r	300–800				

NaNO$_2$NaNO$_3$ Ref. **48, 41**

p	0	21.3	44.8	65.4	82.1	100
a	2.121	2.094	2.066	2.043	2.028	2.004
b	0.68	0.68	0.68	0.68	0.70	0.69
r	310–500	250–500	220–500	270–500	270–500	280–500

NaNO$_3$–NH$_4$NO$_3$ Ref. **58**

p	0	2.5	5.0
d_{170}	1.432	1.443	1.452
d_{180}	1.426	1.436	1.446

NaNO$_3$–Pb(NO$_3$)$_2$ Ref. **36**

p	40	50	60	70	80	90	100
a	3.171	2.867	2.703	2.498	2.373	2.238	2.117
b	1.12	0.806	0.856	0.726	0.746	0.695	0.670
r	340–365	305–375	285–390	295–400	305–420	310–430	320–460

Na$_2$O–P$_2$O$_5$ Ref. **87**

$d_{p,t} = 2.372 + 0.204p/(100 - p) - 0.338 \times 10^{-3}t$	
r_p	30.4–48.5
r_t	liquidus–1 070

Na$_2$O–SiO$_2$ Ref. **56, 39**

p	20.0	30.8	33.6	36.9	50.0
a	2.312	2.380	2.436	2.456	2.516
b	0.10	0.14	0.18	0.20	0.26
r	1 100–1 400		900–1 400		1 050–1 400

Na$_2$WO$_4$–WO$_3$ Ref. **83**

p	45.9	50.9	55.9	65.1	74.2	83.0	91.6
a	5.963 6	5.711 3	5.882 5	4.415 9	5.232 5	4.997 6	3.934 4
b	1.424 4	1.228 9	1.557 0	−0.031 2	1.226 2	1.116 7	−0.051 0
r	782–926	775–898	758–899	737–882	688–880	654–815	685–880

Nd(NO$_3$)$_2$ in KNO$_3$–NaNO$_3$ eutectic (57.3% KNO$_3$) Ref. **84**

$P_{Nd(NO_3)_2}$	1.3	6.4	12.0	21.4	29.1	35.3
a	2.128 5	2.158 7	2.218 6	2.274 1	2.337 1	2.418 3
b	0.731 7	0.725 4	0.757 1	0.686 7	0.708 7	0.753 3
r	230–400					

Nd(NO$_3$)$_2$ in KNO$_3$–LiClO$_4$–NaNO$_3$ (45%–10%–45%) Ref. **84**

$P_{Nd(NO_3)_2}$	35.3
a	2.412 8
b	0.707 1
r	230–400

PbBr$_2$–PbCl$_2$ Ref. **25, 2**

p	0	24.46	57.19	87.88	100
a	5.702	5.840	6.025	5.264	6.338
b	1.50	1.52	1.55	1.71	1.65
r	516–570	492–620	465–640	410–600	505–600

(*continued*)

Table 9.2   DENSITIES OF MOLTEN BINARY SALT SYSTEMS AND OTHER MIXED IONIC MELTS—*continued*

$PbCl_2$–$ZnCl_2$ Ref. **89**	p d_{510} d_{553}	67.1 3.733 3.703						
PbO–B_2O_3 Ref. **88**	p $d_{1\,050}$	57.9 2.8	68.1 3.5	76.3 4.0	82.9 4.6	88.2 5.3	92.9 6.0	96.6 6.6
PbO–SiO_2 Ref. **88**	p $d_{1\,050}$	61.4 4.2	71.2 4.9	78.8 5.7	84.8 6.0	89.7 6.7	93.7 7.0	97.1 7.4
PbO–V_2O_5 Ref. **90**	p $d_{1\,000}$	15.3 3.1	39.6 4.0	52.8 5.0	65.4 5.0	79.0 6.2	85.1 6.8	92.1 7.8
PbO–SiO_2–V_2O_5 Ref. **90**	P_{PbO} P_{SiO_2} $d_{1\,000}$	28.5 10.3 3.8	38.6 10.4 3.8	58.4 10.4 4.0	60.0 10.1 4.0	68.5 10.6 4.0	80.0 10.0 5.2	
SrO–SiO_2 Ref. **74**	p a b r	48.1 3.129 0.058	53.4 3.247 0.059	58.5 3.384 0.074 1 600–1 950	63.3 3.493 0.079	67.8 3.612 0.072		
$TlCl$–$ZnSO_4$ Ref. **34**	p d_{450} d_{500}	55.23 (4.116) 4.076	57.99 4.146 4.094	59.86 4.196 4.155	61.04 4.223 4.177	63.55 4.277 4.227	69.24 4.421 4.376	74.82 (4.573) 4.512

*   *See also* NaF–Na_3AlF_6.
†   *See also* AlF_3–Na_3AlF_6.
**   *See also* 'Physical Properties Data Compilations Relevant to Energy Storage II Molten Salts: Data on Single and Multi-Component Salt Systems', Janz *et al.*, NSRDS–NBS 61.

REFERENCES TO TABLE 9.2

1. K. Arndt and A. Gessler, *Z. Electrochem.*, 1908, **14**, 665.
2. R. Lorenz, H. Frei, and A. Jabs, *Z. Phys. Chem.*, 1908, **61**, 468.
3. Smith and Menzies, *Proc. R. Soc.*, Edinburgh, 1910, **30**, 432.
4. Izbekov and Plotnikov, *J. Russ. Phys. Chem. Soc.*, 1911, **43**, 18.
5. Sackur, *Z. Phys. Chem.*, 1913, **83**, 297.
6. Pascal and Jouniaux, *Bull. Soc. Chim.*, France, 1914, **15**, 312; *Z. Elektrochem.*, 1916, **22**, 71.
7. F. M. Jaeger, *Z. Anorg. Chem.*, 1917, **101**, 175.
8. C. Sandonnini, *Gazz. Chim. Ital.*, 1920, **51**, 289.
9. J. D. Edwards, F. C. Frary and Z. Jeffries, 'The Aluminium Industry', New York, 1930, p. 308.
10. N. Kameyama and A. Naka, *J. Soc. Chem. Ind.*, Japan, 1931, **34**, 140.
11. S. V. Karpachev, A. G. Stromberg and O. Poltoratzkaya, *J. Phys. Chem.* (*USSR*), 1934, **5**, 793.
12. S. V. Karpachev, A. G. Stromberg and V. N. Podchainova, *J. Gen. Chem.* (*USSR*), 1935, **5**, 1517.
13. G. A. Abramov, *Legkie Metally*, 1936, **11**, 27.
14. Z. F. Lundina, *Trans. All Union Aluminium and Magnesium Inst.*, 1936, **13**, 5.
15. G. A. Abramov and P. A. Kozunov, *Trans. Leningrad Indust. Inst.*, 1939, *No. 1*, 60.
16. A. I. Kryagova, *J. Gen. Chem.* (*USSR*), 1939, **9**, 2061.
17. V. P. Barzakovskii, *Bull. Acad. Sci.*, *U.R.S.S.* (Class sci chim), 1940, 825.
18. V. P. Barzakovskii, *J. Appl. Chem.* (*USSR*), 1940, **13**, 1117.
19. Y. Yamaguti and S. Sisido, *J. Chem. Soc. Japan*, 1941, **62**, 304.
20. E. Ya. Gorenbein, *J. Gen. Chem.* (*USSR*), 1945, **15**, 729.
21. E. Ya. Gorenbein, *ibid.*, 1947, **17**, 873.
22. T. G. Pearson and J. Waddington, *Disc. Faraday Soc.*, 1947, No. 1, 307.
23. E. Ya. Gorenbein, *J. Gen. Chem.* (*USSR*), 1948, **18**, 1427.
24. V. P. Mashovets, 'The Electrometallurgy of Aluminium,' 1948.
25. N. K. Boardman, F. H. Dorman and E. Heymann, *J. Phys. Chem.*, 1949, **53**, 375.
26. I. M. Bokhovkin, *J. Gen. Chem.* (*USSR*), 1949, **19**, 805.
27. G. Fuseya and K. Ouchi, *J. Electrochem. Soc. Japan*, 1949, **17**, 254.
28. E. Ya Gorenbein and E. E. Kriss, *J. Gen. Chem.* (*USSR*), 1949, **19**, 1978.
29. H. Grothe, *Z. Elektrochem.*, 1949, **53**, 362.
30. I. M. Bokhovkin, *J. Gen. Chem.* (*USSR*), 1950, **20**, 397.
31. A. Vayna, *Alluminio*, 1950, **19**, 541.
32. E. Ya Gorenbein and E. E. Kriss, *J. Phys. Chem.* (*USSR*), 1951, **25**, 791.
33. R. C. Spooner and F. E. W. Wetmore, *Canad. J. Chem.*, 1951, **29**, 777.
34. I. P. Vereshchetina and N. P. Luzhnaya, *J. Appl. Chem.* (*USSR*), 1951, **24**, 148.
35. R. W. Huber, E. V. Potter and H. W. St. Clair, *U.S. Bur. Mines, Rep. Invest.* 4858, 1952.
36. K. Laybourne and W. M. Madgin, *J. Chem. Soc.*, **1934**, 1.

37. J. Byrne, H. Fleming and F. E. W. Wetmore, *Canad. J. Chem.*, 1952, **30**, 922.
38. J. D. Edwards, C. S. Taylor, L. A. Cosgrove and A. S. Russell, *Trans, Electrochem. Soc.*, 1953, **100**, 508.
39. L. Shartsis, S. Spinner and W. Capps, *J. Am. Ceram. Soc.*, 1952, **35**, 155.
40. L. Shartsis, W. Capps and S. Spinner, *J. Am. Ceram. Soc.*, 1953, **36**, 35.
41. H. Bloom, I. W. Knaggs, J. J. Molloy and D. Welch, *Trans. Faraday Soc.*, 1953, **49**, 1458.
42. N. P. Luzhnaya and I. P. Vereshchetina, *Izvest. Sekt. Fiziko-Khim. Anal.*, 1954, **24**, 192.
43. J. S. Peake and M. R. Bothwell, *J. Am. Chem. Soc.*, 1954, **76**, 2653.
44. L. Shartsis and H. F. Shermer, *J. Am. Ceram. Soc.*, 1954, **37**, 544.
45. R. Midorikawa, *J. Electrochem. Soc. Japan*, 1954, **23**, 310.
46. V. D. Polyakov, *Izvest. Sekt. Fiziko-Khim. Anal.*, 1955, **26**, 147.
47. A. G. Bergman, I. S. Rassonskaya and N. E. Schmidt, *ibid.*, 1955, **26**, 156.
48. V. D. Polyakov and S. I. Berul, *ibid.*, 1935, **26**, 164.
49. V. P. Polyakov, *ibid.*, 1955, **26**, 191.
50. N. P. Popovskaya and P. I. Protsenko, *Zh. Fiz. Khim.*, 1955, **29**, 225.
51. K. B. Morris, M. I. Cook, C. Z. Sykes and M. B. Templeman, *J. Am. Chem. Soc.*, 1955, **77**, 851.
52. E. R. van Artsdalen and I. S. Yaffe, *J. Phys. Chem.*, 1955, **59**, 118.
53. T. Baak, *Acta. Chem. Scand.*, 1955, **9**, 1406.
54. N. P. Luzhnaya, N. N. Evseeva and I. P. Vereshchetina, *Zh. Neorg. Khim.*, 1956, **1**, 1490.
55. S. I. Popel and O. A. Esin, *Zhur. Priklad. Khim.*, 1956, **29**, 651.
56. J. O'M. Bockris, J. W. Tomlinson and J. L. White, *Trans. Faraday Soc.*, 1956, **52**, 299.
57. H. Bloom and D. C. Rhodes, *J. Phys. Chem.*, 1956, **60**, 791.
58. S. Toshiaki and T. Ishibashi, *Sci. Papers Coll. Gen. Educ., Univ. Tokyo*, 1957, **7**, 53.
59. F. R. Duke and R. A. Fleming, *J. Electrochem. Soc.*, 1957, **104**, 251.
60. M. Bourgon, G. Derge and C. M. Pound, *Trans. Amer, Inst. Min. Met. Eng.*, 1958, **212**, 338.
61. F. J. Keneshea and D. Cubicciotti, *J. Phys. Chem.*, 1958, **62**, 843.
62. E. Vatslavik and A. I. Belyaev, *Zh. Neorg. Khim.*, 1958, **3**, 1044.
63. R. H. Moss, *Univ. Microfilms (Ann Arbor. Mich.)*, 1955, No. 12, 730.
64. E. J. Salstrom, *J. Am. Chem. Soc.*, 1931, **53**, 3385.
65. E. J. Salstrom and J. H. Hildebrand, *ibid.*, 1930, **52**, 4650.
66. E. J. Salstrom and J. H. Hildebrand, *ibid.*, 1931, **53**, 1794.
67. E. J. Salstrom and J. H. Hildebrand, *ibid.*, 1932, **54**, 4252.
68. L. E. Topol and F. Y. Lieu, *J. Phys. Chem.*, 1964, **68**, 851.
69. J. N. Reding, *J. Chem. Engng Data*, 1965, **10**, 1.
70. N. V. Bondarenko and K. L. Strelets, *Zh. Prikl. Khim.*, 1962, **35**, 1271.
71. I. P. Vereshchetina and N. P. Luzhnaya, *Izvest. Sekt. Fiziko-Khim. Anal.*, 1954, **25**, 188.
72. P. I. Protsenko and A. Ya. Malakhova, *Zh. Neorg. Khim.*, 1961, **6**, 1662.
73. G. F. Petersen, W. M. Ewing and G. P. Smith, *J. Chem. Engng Data*, 1961, **6**, 540.
74. J. W. Tomlinson, M. S. R. Heynes and J. O'M. Bockris, *Trans Faraday Soc.*, 1958, **54**, 1822.
75. B. C. Blanke, E. N. Bousquet, M. L. Curtis and E. L. Murphy, *Mound Lab., Miamisburg, Ohio. Memo, 1086* (1956).
76. G. W. Mellors and S. Senderoff, *J. Phys. Chem.*, 1960, **64**, 294.
77. E. R. Buckle, P. E. Tsaoussoglou and A. R. Ubbelohde, *Trans Faraday Soc.*, 1964, **60**, 684.
78. G. P. Smith and G. F. Petersen, *J. Chem. Engng Data*, 1961, **6**, 493.
79. I. G. Murgulescu and S. Zuca, *Studii, Cerc. Chim.*, 1959, **7**, 325.
80. M. Blander, W. R. Grimes, N. V. Smith and G. M. Watson, *J. Phys. Chem.*, 1959, **63**, 1164.
81. K. B. Morris and P. L. Robinson, *ibid.*, 1964, **68**, 1194.
82. P. C. Papaioannou and G. W. Harrington, *ibid.*, 1964, **68**, 2424.
83. K. B. Morris and P. L. Robinson, *J. Chem. Engng Data*, 1964, **9**, 444.
84. J. Padova and J. Soriano, *ibid.*, 1964, **9**, 510.
85. W. R. Grimes, N. V. Smith and G. M. Watson, *J. Phys. Chem.*, 1958, **62**, 862.
86. J. H. Shaffer, W. R. Grimes and G. M. Watson, *ibid.*, 1959, **63**, 1999.
87. C. F. Callis, J. R. Van Wazer and J. S. Metcalf, *J. Am. Chem. Soc.*, 1955, **77**, 1468.
88. J. O'M Bockris and G. W. Mellors, *J. Phys. Chem.*, 1956, **60**, 1321.
89. A. Wachter and J. H. Hildebrand, *J. Am. Chem. Soc.*, 1930, **52**, 4655.
90. B. M. Lepinskikh, O. A. Esin and G. A. Teterin, *Zh. Neorg. Khim.*, 1960, **5**, 642.

Table 9.3   DENSITY OF SOME SOLID INORGANIC COMPOUNDS AT ROOM TEMPERATURE

Compound	Density g cm^{-3}	Compound	Density g cm^{-3}	Compound	Density g cm^{-3}	Compound	Density g cm^{-3}
AgBr	6.47	CsI	4.51	LaCl$_3$	3.84	Rb$_2$SO$_4$	3.61
AgCl	5.56	CsNO$_3$	3.69	LiBr	3.46	ReF$_6$	4.25
AgClO$_3$	4.43	Cs$_2$SO$_4$	4.24			Re$_2$O$_7$	6.10
AgI(α)	5.683			LiCl	2.07	ReOF$_4$	4.03
AgNO$_2$	4.35	Cu$_2$Cl$_2$	4.14	Li$_2$CO$_3$	2.11	SbBr$_3$	4.15
		Cu$_2$O	6.0	LiF	2.64		
AlBr$_3$	2.64	GaBr$_3$	3.69	LiNO$_3$	2.38	SbCl$_3$	3.14
AlCl$_3$	2.44	GaCl$_3$	2.47	Li$_2$SO$_4$	2.22	SnCl$_2$	3.95
AlI$_3$	3.98	GaI$_3$	4.15			SnI$_4$	4.47
AsBr$_3$	3.54			MgCl$_2$	2.32	SrBr$_2$	4.22
BI$_3$	3.35	H$_3$BO$_3$	1.44	MgF$_2$	3.0	SrCl$_2$	3.05
		HgBr$_2$	6.11	MgO	3.58		
B$_2$O$_3$	1.84	Hg$_2$Br$_2$	7.31	MnO$_2$	4.9	SrF$_2$	4.24
BaBr$_2$	4.78	HgCl$_2$	5.44	Na$_3$AlF$_6$	2.90	SrI$_2$	4.55
BaCl$_2$	3.856	Hg$_2$Cl$_2$	7.15			SrO	4.7
BaCO$_3$	4.43			Na$_2$B$_4$O$_7$	2.37	TeCl$_4$	3.26
BaF$_2$	4.89	HgF$_2$	8.95	NaBr	3.20	ThCl$_4$	4.59
		HgI$_2$	6.36	NaCl	2.17		
BaI$_2$	5.15	Hg$_2$I$_2$	7.7	Na$_2$CO$_3$	2.53	TiBr$_4$	2.6
BaO	5.68	InBr$_3$	4.74	NaF	2.56	TiF$_4$	2.80
BaO$_2$	4.96	InCl	4.19			TiI$_4$	4.3
BaSO$_4$	4.50			NaI	3.67	TiO$_2$	4.26
BeBr$_2$	3.47	InCl$_2$	3.66	NaNO$_3$	2.26		rutile
		InCl$_3$	3.46	NaOH	2.13		3.84
BeCl$_2$	1.90	InI$_3$	4.69	Na$_4$P$_2$O$_7$	2.53		anatase
BeF$_2$	1.99	KBF$_4$	2.55	Na$_2$SO$_4$	2.70		
BeO	3.00	KBr	2.75			TlBr	7.56
BiBr$_3$	4.72			Na$_2$WO$_4$	4.18	TlCl	7.00
BiCl$_3$	4.75	KCl	1.98	NH$_4$Cl	1.53	TlI	7.1
		KCN	1.52	Ni(CO)$_4$†	1.32	TlNO$_3$	5.8
C$_2$Cl$_6$	2.09	K$_2$CO$_3$	2.43	NiO	6.67	WCl$_5$	3.88
CaBr$_2$	3.353	K$_2$Cr$_2$O$_7$	2.68	OsO$_4$	4.91	WCl$_6$	3.52
CaCl$_2$	2.15	KF	2.48				
CaCO$_3$	2.71*			PBr$_3$	2.85	WO$_3$	7.2
CaF$_2$	3.18	KHSO$_4$	2.31	PbBr$_2$	6.66	UCl$_4$	4.87
		KI	3.13	PbCl$_2$	5.85	YCl$_3$	2.8
CaO	3.25/3.38	KMnO$_4$	2.91	PbCl$_4$†	3.18	ZnBr$_2$	4.20
CaSO$_4$	2.96	KNO$_3$	2.11	PbI$_2$	6.16	ZnCl$_2$	2.91
CdBr$_2$	5.19	KOH	2.04				
CdCl$_2$	4.05			RbBr	3.35	ZnI$_2$	4.74
CsBr	4.43	K$_3$PO$_4$	2.56	RbCl	2.80		
		K$_2$SiF$_6$	3.08 (hex)	RbF	3.56		
CsCl	3.99		2.67 (cub)	RbI	3.55		
CsF	4.1	K$_2$SO$_4$	2.66	RbNO$_3$	3.11		

* Calcite.   † Liquid.

REFERENCES TO TABLE 9.3

1. 'Handbook of Chemistry and Physics', 58th edn, Cleveland, Ohio, 1977–78.
2. Gmelin, 'Handb. anorg. Chem.', 8th edn, Berlin.
3. J. W. Mellor, 'A Comprehensive Treatise on Inorganic Chemistry', London, 1922–1937.

Table 9.4   ELECTRICAL CONDUCTIVITY OF PURE MOLTEN SALTS

The conductivity in $\Omega^{-1}\,cm^{-1}$ is given at the melting point θ_m, and at 50° intervals beginning at θ, or as an equation in the temperature t, or at the temperatures given in brackets. All temperatures are in °C. Extrapolated conductivities are also bracketed. Principal references are in **bold type**.

Substance	θ_m °C	θ °C	θ_m	θ	$\theta+50$	$\theta+100$	$\theta+150$	$\theta+200$	$\theta+250$	References
					Conductivity $\Omega^{-1}\,cm^{-1}$					
AgBr	430	450	2.8561	2.903	3.016	3.119	3.214	3.3	3.378	**133, 96, 19, 18**, 22
AgCl	455	500	3.817	3.972	4.131	4.276	4.407	4.523	4.626	**133, 96, 76, 19, 18,**
AgClO$_3$	231	250	0.417	0.474						107, 22, **13**
AgI	557	600	2.3	2.35	2.40	2.45	2.5	2.55		**19**, 12
AgNO$_3$	210	250	0.67	0.84	1.05	1.24				**106, 100, 88, 87, 86, 85, 79, 76, 55**, 71, 22, 13, 4
AlBr$_3$	97	150	$(<10^{-8})$	(0.02×10^{-6})	0.10×10^{-6}	0.19×10^{-6}	(0.26×10^{-6})			27, 70, 68, 66, 62
AlCl$_3$	193	200	4.6×10^{-7}	5.5×10^{-7}	11.1×10^{-7}					37, 60, 53, 52, 40
AlI$_3$	191	200	(1.205×10^{-6})	1.852×10^{-6}	5.645×10^{-6}	7.1×10^{-6}				36, 37
AsBr$_3$	31	35	1.0×10^{-7}	2.6×10^{-7}						59, 7
AsCl$_3$	−16	—								39, **6**, 38
AsF$_3$	−13	0	1.56×10^{-5}	1.84×10^{-5}	2.92×10^{-5}					74
AsI$_3$	142					Poor				8
BCl$_3$	−107					Non-conductor				6
BF$_3$	−128.7	−120		$<5.0\times10^{-10}$						74
B$_2$O$_3$	570	600	—	1.2×10^{-6}	3.0×10^{-6}	5.2×10^{-6}	1.5×10^{-5}	2.3×10^{-5}	3.6×10^{-5}	**101, 99, 81**, 12, 10
		900		6.6×10^{-5}	9.7×10^{-5}	1.5×10^{-4}	2.1×10^{-4}	2.9×10^{-4}	4.0×10^{-4}	
		1200		5.3×10^{-4}						
BaBr$_2$	847	850	—	1.178	1.307	1.440	1.572	1.705		108
BaCl$_2$	960	975	—	2.085		2.176 (1000)	2.264 (1025)	2.354 (1050)		108, 109, 107, 105, 78, 64, 56, 9
						2.472 (1075)	2.532 (1100)			
BaI$_2$	711	750	—	0.784	0.910	1.024	1.136	1.248	1.361	108
Ba(NO$_2$)$_2$	—	280	—	0.182						110
BeCl$_2$	405	450	0.87×10^{-3}	2.986×10^{-3}	0.016	0.235 (300)	0.260 (310)			111
BeF$_2$	c. 800					Poor Non-conductor				34
BeI$_2$	510	250								5
BiBr$_3$	—	550		0.3214	0.3813	0.4237	0.4487	0.4562	0.4462	111
				0.320	0.298	0.260	0.224			
BiCl$_3$	230	250		0.4122	0.4910	0.546	0.5772	0.5845	0.5680	134, 32
				0.515	0.468					
BiI$_3$	550	400		0.285	0.300	0.310	0.309	0.305	0.295	111
	—	700		0.285	0.265	0.250	0.227	0.205		

Table 9.4   ELECTRICAL CONDUCTIVITY OF PURE MOLTEN SALTS—*continued*

Substance	θ_m °C	θ °C	θ_m	θ	$\theta+50$	$\theta+100$	$\theta+150$	$\theta+200$	$\theta+250$	References
					Conductivity Ω^{-1} cm^{-1}					
Bi$_2$(MoO$_4$)$_3$	—	708	—	0.274	—	0.333(747)	0.433(794) 0.550(847)	0.480(817)	—	112
CaBr$_2$	—	750	—	1.420	1.573	1.727	1.882	2.037	2.194	108
CaCl$_2$	—	800	—	2.116	2.342	2.515	2.778	3.006	—	108, 105, 104, 78, 64, 63, 57, 56, 46, 42, 36, 22, 14, 12, 9, 107, 109
CaF$_2$	1420	1500	—	3.9	4.1	—	—	—	—	90
CaI$_2$	—	800	—	1.172	1.292	1.390	1.486	1.566	—	108
CdBr$_2$	565	600	—	1.125	1.229	1.324	1.419	1.517	—	108, 65
CdCl$_2$	568	600	—	1.950	2.085	2.206	2.317	2.425	—	108, 91, 86, 82, 65, 31, 22
CdI$_2$	390	400	—	0.210	0.316	0.425	0.533	0.640	0.750	108, 82
CeCl$_3$	—	828	—	1.12	—	1.17(844) 1.21(858) 1.23(868) 1.29(886) 1.30(895) 1.38(931)				113
CeI$_3$	761	796	—	0.448	—	0.470(814) 0.499(836) 0.523(860)				114
CsBr	635	650	—	0.84	0.97	1.09	1.21	1.31	—	98
CsCl	645	650	—	1.12	1.27	1.42	1.57	1.71	1.83	98, 31, 115
CsF	—	—	—	$\kappa_t = -4.511 + 1.642 \times 10^{-2}t - 7.632 \times 10^{-6}t^2$ at 725–921						109
CsI	630	650	—	0.69	0.80	0.91	1.00	1.09	—	98
CsNO$_3$	417	—	—	$\kappa_t = -0.2452 + 1.8878_5 \times 10^{-3} t$ at 415–491						116, 21
CuCl	430	450	3.26	3.32	3.46	3.56	3.66	—	—	86, 31
Cu$_2$S	1130	1150	60	70	80	90	100	120	130	97, 16
		1450		140	160					
ErCl$_3$	—	801	—	0.468	0.496(818) 0.531(839)					117
FeS	1200	1250	1560	1540	1530	1520	1510	1500	1490	97
GaBr$_2$	—	—	—	$\log \kappa_t = 1.142 - 865/((t+273))$ at 167–189						118
GaBr$_3$	125	—	5.0×10^{-8}	—	—	—	—	—	—	41
GaCl$_2$	—	—	—	$\log \kappa_t = 1.180 - 784/((t+273))$ at 167–176						118, 2
GaCl$_3$	77	—	10^{-8} (approx.)	—	—	—	—	—	—	41

Salt							κ_t (expressions / values)			Refs
Ga$_2$I$_4$	—						$\log \kappa_t = 2.887 - 1401/(t + 273)$ at 150–211			119
Ga$_2$I$_6$	—						$\log \kappa_t = 1.292 - 1114/(t + 273)$ at 211–279 $\log \kappa_t = 0.460 - 653.4(t + 273)$ at 279–350 $\log \kappa_t = -0.068 - 1041/(t + 273)$ at 185–222 $\log \kappa_t = -1.807 - 624.4/(t + 273)$ at 222–284 $\log \kappa_t = -2.660 - 147.6/(t + 273)$ at 284–352 $\log \kappa_t = -4.211\,2 + 822.6/(t + 273)$ at 352–400			119, 41
GdBr$_3$	—	800	—	0.444	—		0.476(823) 0.500 (842)		—	117
GdCl$_3$	—	629	—	0.382	—		0.408(649) 0.465(675) 0.506(698)		—	117
HgBr$_2$	235	240	—	1.38×10^{-4}			2.08×10^{-4}(280) 2.85×10^{-4}(320)		—	108, 120, 75, 23
HgCl$_2$	277	280	—	3.20×10^{-5}			3.57×10^{-5}(290) 3.96×10^{-5}(300)		—	108, 120, 55, 23
Hg$_2$Cl$_2$	525	550	1.00	1.03			—	—	—	36
HgI$_2$	250	260	—	0.029 9	0.024 0		0.488(776) 0.526(800) 0.562(819)	0.018 8(350)	—	108, 120, 83, 55, 23
HoCl$_3$	—	747	—	0.431					—	117
InBr$_3$	436	450	0.168	0.167	0.162	0.156	—	—	—	37
InCl	225	250	0.88	1.04	1.33	1.62	(1.89)	—	—	37
InCl$_2$	235	250	0.23	0.26	0.36	0.46	0.56	0.64	0.72	37
InCl$_3$	498	500	(0.50)	(0.50)	(0.46)	0.41	0.37 (0.100)	0.32	(0.28)	37
InI$_3$	210	250	0.052	0.066	0.080	0.092	—	—	—	37
K$_3$AlF$_6$	1035	1050	2.33	2.38	1.180	1.243	—	—	—	102
KBF$_4$	742	550	—	1.075	—		—	—	—	121
KBr	742	—					$\kappa_t = -1.327_1 + 5.710_0 \times 10^{-3}t - 2.313_9 \times 10^{-6}t^2$ at 760–980			**122**, 98, 82, 31, 21, 70, 12, 4
KCl	770	800 / 1100	2.12	2.22 / 2.79	2.35	2.46	2.56	2.65	2.73	**92, 82, 78, 77, 63, 44,** 107, 68, 56, 54, 49, 37, 22
KClO$_3$	368	—	4.19	—			—	—	—	**1**
K$_2$CO$_3$	895	900	2.020	2.035	2.1783	2.3221	—	—	—	**135**
K$_2$Cr$_2$O$_7$	398	400	0.200	0.204	0.298	0.389	(0.474)	—	—	**83, 11**
KF	857	900	—	4.4728	4.6017	4.6975	4.7603 $\kappa_t = -3.493 + 1.480 \times 10^{-2}t - 6.608 \times 10^{-6}t^2$ at 869–1040	—	—	109, 121, 102, 47, 42, 29, 21

(continued)

Table 9.4   ELECTRICAL CONDUCTIVITY OF PURE MOLTEN SALTS —*continued*

Substance	θ_m °C	θ °C	Conductivity Ω^{-1} cm^{-1}							References
			θ_m	θ	$\theta+50$	$\theta+100$	$\theta+150$	$\theta+200$	$\theta+250$	
KHSO$_4$	212	301	0.049	0.141	—	—	—	—	—	72
KI	682	700	1.82	1.32	1.42	1.52	1.60 $\kappa_t = -21.0210 + 4.4823 \times 10^{-2}t - 22.506\,6 \times 10^{-6}t^2$ at 931–988	1.66	—	92, **82, 31,** 21, 12, 4
K$_2$MoO$_4$	—	—	—	—	—	—	—	—	—	123
KNH$_2$	330	340	—	0.389	—	—	—	—	—	20
KNO$_3$	337	350	0.62	0.66	0.82	0.96	1.10	1.24	1.38	**106, 93, 87, 86, 84,** **83, 82, 61, 55, 22, 21,** 116, 15, 11, 1
KOH	410		(1.78)		*See* ref.					35
K$_2$SO$_4$	1069	1100		1.84	1.94	0.981				9
K$_2$TaF$_7$	—	750	—	0.750	0.919	0.981	(1.052)	—	—	121
K$_2$TiF$_6$	—	850	—	1.359	1.462	1.559	(1.650)	—	—	121, 124
K$_2$WO$_4$	—				$\kappa_t = -8.2953 + 1.7633 \times 10^{-2}t - 8.0295 \times 10^{-6}t^2$ at 946–1024					125
LaCl$_3$	885	900	1.550	1.602	1.748	1.8531	1.9170	—	—	36, 109
Li$_3$AlF$_6$	—	800	—	3.45	3.65	3.80	3.95	—	—	102
LiBr	550	600	4.69	4.97	5.23	5.49	5.73	—	—	98, 65
LiCl	614	650	5.7306	5.9219	6.1714	6.4021	6.6139	6.807	6.9813	92, 82, 77, 49, 31, 107, 60, 42 126
LiClO$_3$	—	131.8	—	0.1151		0.1231(135.7) 0.12581(136.5) 0.1370(140.8) 0.1420(143.0)				
LiF	844	875	—	8.663		8.889(915) 9.058(958) 9.216(1008) 9.306(1037)				121, 109, 102, 42
LiI	465	500 / 800	—	3.800 / 4.830	4.067 / 4.947	4.250 / 5.045	4.417	4.570	4.703	127, 98
Li$_2$MoO$_4$	—	—	—	—	$\kappa_t = -43.2596 - 10.71 \times 10^{-2}t + 72.668 \times 10^{-6}t^2$ at 790–939	—	—	—	—	123
LiNO$_3$	254	300	0.8038	1.0545	1.3357	1.6259	1.9251	2.2333	—	138, 13, 21
Li$_2$WO$_4$	—	—	—	—	$\kappa_t = 7.5067 + 1.8733 \times 10^{-2}t - 8.8034 \times 10^{-6}t^2$ at 762–903	—	—	—	—	125
MgBr$_2$	—	750	—	0.766	0.857	0.950	1.045	1.138	—	108
MgCl$_2$	714	750	1.014	1.085	1.183	1.283	1.383	1.483	—	**108,** 91, 78, 63, 54, 48, 44, 31
MgI$_2$	—	650	—	0.408	0.496	0.585	0.675	0.765	0.856	108
MnCl$_2$	—	850	—	(1.436)	1.578	1.690	—	—	—	107, 128
MoCl$_5$	194	200	—	0.3×10^{-6}	6.3×10^{-6}	—	—	—	—	32

Salt	m.p. °C	t						References
MoO$_3$	795	—						**123, 89**
Na$_3$AlF$_6$	1000	1040	2.799	2.901	$\kappa_f = -5.7537 + 1.4563 \times 10^{-2}t - 8.1949 \times 10^{-6}t^2$ at 823–891	2.997	—	**102, 80**, 77, 73, **69**, 51, 50, 29
NaBr	745	800	2.87	3.06	3.22	3.37	3.49	**98, 12, 4**
NaCl	800	850	3.58	3.74	3.87	4.02	4.16	4.39 — **104, 92, 78, 77, 63, 107‡**
NaCN	562	700	—	1.15	1.27	1.39	—	42
Na$_2$CO$_3$	850	900	2.8347	3.029	3.222	3.222	—	**135**
NaF	992	1003	—	4.960	4.985(1018) 5.082(1047) 5.111(1059)	5.179(1086) 5.271(1099) 5.335(1138)	—	121, 102, 80, 50, 29
NaHSO$_4$	182	200	0.067	0.094	0.168	0.241	—	**72**
NaI	655	700	2.30	2.41	2.55	2.68	2.80	2.92 — 1.74 — **98, 82, 12, 4**
Na$_2$MoO$_4$	687	750	(1.15)	1.26	1.37	1.49	1.57	1.66 — 2.14 — 2.22 — **89, 21**
		1050		1.82	1.90	1.98	2.07	
		1350		2.30	2.37			
NaNH$_2$	208	210	—	0.593	1.62	1.89	2.16	**20**
NaNO$_2$	270	300	—	1.34	—	1.55	1.73	**91, 82**
NaNO$_3$	310	350	0.94	1.16	1.36	—	—	**95, 93, 86, 82, 79,** 55, **21**, 22, 11
NaOH	322	350	2.125	2.377	2.827	3.277	—	**35**
NaPO$_3$	625	650	0.36	0.425	0.55	0.675	0.80	0.925 — 1.05 — **12**
		950		1.175	1.30	1.42	1.54	
Na$_2$SO$_4$	885	900	(2.19)	2.23	2.37	2.50	2.64	2.77 — (2.91) — **9**
Na$_2$TaF$_7$	—	750	—	0.750	0.919	0.981	(1.052)	**121**
Na$_2$WO$_4$	696	—	—	0.22 × 10^{-6}	$\kappa_f = -1.4108 + 0.3451 \times 10^{-2}t - 0.2058 \times 10^{-6}t^2$ at 706–871			**125, 21**
NbCl$_5$	210	228	—	0.466	—	—	—	**26**
NdBr$_3$	—	713	—	—	0.498(731) 0.519(743) 0.541(759) 0.563(772) 0.603(797)	—	—	**117**
NdCl$_3$	761	800	(0.587)	0.692	0.821	0.945	(1.058)	0.416(818) 0.440(842) — **32**
NdI$_3$	787	799	0.079	0.396	—	—	—	**114**
NH$_4$HSO$_4$	145	150	0.31	0.086	0.154	—	—	**72**
NH$_4$NO$_3$	169.6	200	—	0.41	(0.53)	—	—	**4, 1**
PbBr$_2$	370	400	(0.57)	0.68	0.85	1.03	1.19	**96, 11, 22**

(continued)

Table 9.4   ELECTRICAL CONDUCTIVITY OF PURE MOLTEN SALTS—*continued*

Substance	θ_m °C	θ °C	Conductivity Ω^{-1} cm^{-1}							References
			θ_m	θ	$\theta+50$	$\theta+100$	$\theta+150$	$\theta+200$	$\theta+250$	
PbCl$_2$	498	500	5.8295	5.8398	6.088	6.3166	—	—	—	136, **137**, **96**, **107**, **11**
PbCl$_4$	−15	0	—	8.0×10^{-7}						32
PbI$_2$	—	450	—	0.4326	0.4628	0.4939	0.5306	0.3906(402.7)	—	127
PbMoO$_4$	—	1098	—	0.938		0.950(1107)	0.958(1116)	0.963(1118)		112
PrCl$_3$	—	—				$\kappa_t = -1.189 + 2.75\times10^{-3}\,t$ at 800–860				129, 32
PrI$_3$	738	763	—	0.399			0.426(786)	0.452(809)		114
RbBr	692	700	—	1.132	1.263	1.360	1.454	1.526	1.586	127, **98**
		1000	—	1.637	1.679	1.716	1.749	1.778	1.806	
RbCl	722	750	1.50	1.58	1.72	1.85	1.97			98, **31**
RbI	647	650	—	0.86	0.96	1.04	1.12	1.19	1.25	98
RbNO$_3$	316	350	0.431	0.509	0.618	0.72	0.8162	—		21
SbBr$_3$	97	100	228×10^{-6}	236×10^{-6}	355×10^{-6}	450×10^{-6}	—	—	—	23, 45
SbCl$_3$	73	—	0.5×10^{-6}	—	—					43, **33**, 25
SbF$_3$	290	300		0.065	—	—	—			94
SbF$_5$	6	50	0.4×10^{-8}	5.85×10^{-8}	—	41.0×10^{-8} (80)	—	—		74
SbI$_3$	170	200	3.91×10^{-8}	3.91×10^{-8}	4.83×10^{-8}	5.68×10^{-8}	6.44×10^{-8}	(7.11×10^{-8})		23
ScCl$_3$	960	980	0.56	0.63	0.66(1000)					37, 32, 28
SiO$_2$	—	—			ca.10^{-5} – ca.10^{-4} at 1800–2600					130
SnCl$_2$	247	250	0.89	0.90	1.11	1.38	1.67	(1.93)		86, **36**, 3
SrBr$_2$	—	650		0.728	0.980	1.142	1.295	1.446	1.600	108
SrCl$_2$	872	900	1.989	2.096	2.285	2.477	2.670	0.989	1.100	108, 9
SrI$_2$	—	550	—	0.516	0.637	0.758	0.875			108
		850	—	1.212	1.320	1.422	1.525	—	—	

Salt	m.p. (°C)	t (°C)							Ref.
TaCl$_5$	221	235	—	0.3×10^{-6}	—	—	0.145	—	26
TeCl$_2$	175	200	(0.005)	0.034	0.089	—	—	—	32
TeCl$_4$	224	250	0.100	0.131	0.186	—	—	—	32
ThCl$_4$	770	800	(0.54)	(0.61)	0.71	(0.86)	—	—	37
TiI$_4$	150	—	—	—	—	$1 \times 10^{-7} - 6 \times 10^{-7}$ (180–240)	—	—	41
TlBr	460	—	—	—	—	—	—	—	131, 18
TlCl	430	450	1.088	1.164	1.350	1.532	1.700	—	83, 17, 22
TlCl$_3$	c. 60	—	$<5.0 \times 10^{-5}$	—	—	—	—	—	37
TlI	440	450	(0.530)	0.551	0.651	0.747	0.840	—	17
TlNO$_3$	206	250	(0.35)	0.47	0.60	—	—	—	84, 55, 22
Tl$_2$S	448	—	—	—	—	See ref.	—	—	24
UCl$_4$	567	—	—	—	—	—	—	—	132, 32
WCl$_5$	248	250	0.61×10^{-6}	0.67×10^{-6}	1.84×10^{-6}	—	—	—	32, 25
WCl$_6$	275	300	1.86×10^{-6}	2.60×10^{-6}	4.05×10^{-6} (320)	—	6.94×10^{-6} (330)	—	32, 25
YCl$_3$	721	750	0.42	0.48	0.58	0.69	0.79	—	37
ZnBr$_2$	—	400	—	0.020	0.043	0.085	0.145	0.215 / 0.310	108
ZnCl$_2$	331	350 / 650	—	0.007 / 0.350	0.020 / 0.462	0.050	0.090	0.156 / 0.243	108, 103, 31, 1
ZnI$_2$	—	450	—	0.060	0.118	0.190	0.267	—	108

TlCl: $\kappa_t = -0.461_1 + 3.105_0 \times 10^{-3}t - 0.732_6 \times 10^{-6}t^2$ at 480–705

WCl$_5$: $\kappa_t = -0.318 + 1.22 \times 10^{-3}t$

‡ See also 64, 60, 58, 57, 56, 54, 53, 52, 48, 46, 42, 31, 22, 4.

REFERENCES TO TABLE 9.4

1. Foussereau, *J. Phys. Radium*, 1885 **4**, 189.
2. W. Hampe, *Chem. Zeit.*, 1888, **11**, 1109.
3. L. Graetz, *Wied Ann.*, 1890, **40**, 28.
4. M. Poincaré, *Ann. Chim.* (*Phys.*), 1890, **21**, 289.
5. P. Lebeau, *Compt. Rend.*, 1898, **126**, 1272; *Ann. Chim.* (*Phys.*), 1899, (7) **16**, **491**.
6. P. Walden, *Z. Anorg. Chem.*, 1900, **25**, 209, 227.
7. P. Walden, *ibid.*, 1902, **29**, 371.
8. J, H. Mathews, *J. Phys. Chem.*, 1905, **9**, **641**.
9. K. Arndt, *Z. Elektrochem.*, 1906, **12**, 337.
10. K. Arndt, *Ber. deut. Chem. Ges.*, 1907, **40**, 2938.
11. R. Lorenz and H. T. Kalmus, *Z. Physikal. Chem.*, 1907, **59**, 17.
12. K. Arndt and A. Gessler, *Z. Elektrochem.*, 1908, **14**, 662.
13. H. M. Goodwin and R. D. Mailey, *Phys. Rev.*, 1908, **26**, 28.
14. K. Arndt and W. Löwenstein, *Z. Elektrochem.*, 1909, **15**, 789.
15. A. H. W. Aten, *Z. Physikal. Chem.*, 1911, **78**, 1.
16. Gornemann and von Rauschenplat, *Metallurgie*, 1912, **9**, 473, 505.
17. C. Tubandt, 'Nernst Festschrift', Halle: 1912, p. 446.
18. C. Tubandt and E. Lorenz, *Z. Physikal. Chem.*, 1914, **87**, 513.
19. R. Lorenz and A. Höchberg, *Z. Anorg. Chem.*, 1916, **94**, 305.
20. Wöhler and Stang-Lund, *Z. Elektrochem.*, 1918, **24**, 261.
21. F. M. Jaeger and B. Kapma, *Z. Anorg. Chem.*, 1920, **113**, 27.
22. C. Sandonnini, *Gazz. Chim. Ital.*, 1920, **50**, 289.
23. G. von Hevesy, *Kgl. Danske Vid. Selsk. Medd.*, 1921, *III*, 13.
24. Pélabon, *Compt. Rend.*, 1921, **173**, 142.
25. W. Biltz, *Z. Phys. Chem.*, 1922, **100**, 52.
26. W. Biltz and A. Voigt, *Z. Anorg. Chem.*, 1922, **120**, 71.
27. W. Biltz and A. Voigt, *ibid.*, 1923, **126**, 39.
28. W. Biltz and W. Klemm, *ibid.*, 1923, **131**, 22.
29. K. Arndt and W. Kalass, *Z. Elektrochem.*, 1924, **30**, 12.
30. K. Arndt and G. Ploetz, *Z. Phys. Chem.*, 1924, **110**, 237.
31. W. Biltz and W. Klemm, *ibid.*, 1924, **110**, 318.
32. A. Voigt and W. Biltz, *Z. Anorg. Chem.*, 1924, **133**, 277.
33. Klemensiewicz, *Z. Physikal. Chem.*, 1924, **113**, 28.
34. B. Neumann and H. Richter, *Z. Elektrochem.*, 1925, **21**, 484.
35. K. Arndt and G. Ploetz, *Z. Phys. Chem.*, 1926, **121**, 439.
36. W. Klemm and W. Biltz, *Z. Anorg. Chem.*, 1926, **152**, 225.
37. W. Biltz and W. Klemm, *ibid.*, 1926, **152**, 267.
38. M. Ussanowitsch. *Z. Phys. Chem.*, 1929, *A***140**, 429.
39. V. S. Finkelstein, *J. Russ. Phys.-Chem. Soc.*, 1930, **62**, 161.
40. V. A. Plotnikov and P. T. Kalita, *ibid.*, 1930, **62**, 2195.
41. W. Klemm and W. Tilk, *Z. Anorg. Chem.*, 1932, **207**, 161.
42. E. Ryschkewitsch, *Z. Elektrochem*, 1933, **39**, 531.
43. M. Ussanowitsch and F. Terpugov, *Z. Phys. Chem.*, 1933, *A***165**, 39.
44. S. Karpachev. A. Stromberg and O. Poltoratzkaya, *J. Phys. Chem.* (*USSR*), 1934, **5**, 793.
45. M. Ussanowitsch and V. Serebrennikov, *J. Gen. Chem.* (*USSR*), 1934, **4**, 230.
46. V. P. Barzakovskii, *Sbornik Trudov Pervoi Vsesoyuznoi Konferentzii Nevodnuim Rastvoram Ukrain. Akad. Nauk. Inst. Khim.* (Proc. 1st All-Union Conf. Nonaqueous Solutions), 1935, 143.
47. K. P. Batashev and A. Zhurin, *Metallurg.* (*USSR*), 1935, **10**, 67.
48. K. P. Batashev, *ibid.*, 1935, **10**, 100.
49. S. Karpachev, A. Stromberg and V. N. Podchainova, *J. Gen. Chem.*, Moscow, 1935, **5**, 1517.
50. K. P. Batashev, *Legkie Metally*, 1936, **10**, 48.
51. J. W. Cuthbertson and J. Waddington, *Trans. Faraday Soc.*, 1936, **32**, 745.
52. Y. Yamaguti and S. Sisido, *J. Chem. Soc., Japan*, 1938, **59**, 1311.
53. A. I. Kryagova, *J. Gen. Chem.* (*USSR*), 1939, **9**, 2061.
54. A. A. Scherbakov and B. F. Markov, *J. Phys. Chem.* (*USSR*), 1939, **13**, 621.
55. A. G. Bergman and I. M. Chagin, *Bull. Acad. Sci. URSS*, Cl. Sci. Chim., 1940, 727.
56. V. P. Barzakovskii, *ibid.*, 1940, 825.
57. V. P. Barzakovskii, *J. Appl. Chem.* (*USSR*), 1940, **13**, 1117.
58. N. A. Belozerskii and B. A. Freidlina, *ibid.*, 1941, **14**, 466.
59. A. Bernshtein, *J. Gen. Chem.* (*USSR*), 1941, **11**, 901.
60. Y. Yamaguti and S. Sisido, *J. Chem. Soc. Japan*, 1941, **62**, 304.
61. E. R. Natsvilischvili and A. G. Bergman, *Bull. Acad. Sci., URSS*, Cl. Sci. Chim., 1943, 23.
62. E. Y. Gorenbein, *J. Gen. Chem.* (*USSR*), 1945, **15**, 720.
63. E. K. Lee and E. P. Pearson, *Trans. Electrochem. Soc.*, 1945, **88**, 171.
64. A. F. Alabyshev and N. Ya Kulakovskaya, *Trudy Leningrad. Tekhnol. Inst. im. Leningrad. Soveta*, 1946 No. 12, 152.
65. H. Bloom and E. Heymann, *Proc. Roy. Soc.*, 1946–7, *A***188**, 392.

66. E. Ya Gorenbein, *J. Gen. Chem. (USSR)*, 1947, **17**, 873.
67. N. M. Tarasova, *J. Phys. Chem. (USSR)*, 1947, **21**, 825.
68. E. Ya. Gorenbein, *J. Gen. Chem. (USSR)*, 1948, **18**, 1427.
69. M. Frejaques, *Bull. Soc. Franc. Elec.*, 1949, **9**, 684.
70. E. Ya. Gorenbein and E. E. Kriss, *J. Gen. Chem. (USSR)*, 1949, **19**, 1978.
71. I. M. Bokhovkin, *ibid.*, 1950, **20**, 397.
72. S. E. Rogers and A. R. Ubbelohde, *Trans. Faraday Soc.*, 1950, **46**, 1051.
73. A. Vayna, *Alluminio*, 1950, **19**, 215.
74. A. A. Woolf and N. N. Greenwood, *J. Chem. Soc.*, 1950, 2200.
75. G. Jander and K. Brodersen, *Z. Anorg. Chem.*, 1951, **264**, 57.
76. R. C. Spooner and F. E. W. Wetmore, *Canad. J. Chem.*, 1951, **29**, 777.
77. J. D. Edwards, C. S. Taylor, A. S. Russell and L. F. Maranville, *J. Electrochem. Soc.*, 1952, **99**, 527.
78. R. W. Huber, E. V. Potter and H. W. St. Clair, *Rep. Invest. US Bur. Mines*, No. 4858, 1952.
79. J. Byrne, H. Fleming and F. E. W. Wetmore, *Canad. J. Chem.*, 1952, **30**, 922.
80. J. D. Edwards, C. S. Taylor, L. A. Cosgrove and A. S. Russell, *Trans. Electrochem. Soc.*, 1953, **100**, 508.
81. L. Shartsis, W. Capps and S. Spinner, *J. Am. Ceram. Soc.*, 1953, **36**, 319.
82. H. Bloom, I. W. Knaggs, J. J. Molloy and D. Welch, *Trans. Faraday Soc.*, 1953, **49**, 1458.
83. I. N. Beylaev, *Izvest. Sekt. Fiziko-Khim. Anal.*, 1953, **23**, 176.
84. P. I. Protsenko and N. P. Popovskaya, *Zhur. Obshch. Khim.*, 1954, **24**, 2119.
85. P. I. Protsenko and N. P. Popovskaya, *Zh. Fiz. Khim.*, 1954, **28**, 299.
86. K. Sakai, *J. Chem. Soc. Japan, Pure Chem. Sect.*, 1954, **75**, 182.
87. V. D. Polyakov, *Izvest. Sekt. Fiziko-Khim. Anal.*, 1955, **26**, 147.
88. P. I. Protsenko, *ibid.*, 1955, **26**, 173.
89. K. B. Morris, M. I. Cook, C. Z. Sykes and M. B. Templeman, *J. Am. Chem. Soc.*, 1955, **77**, 851.
90. T. Baak, *Acta. Chem. Scand.*, 1955, **9**, 1406.
91. K. Sakai and S. Hayashi, *J. Chem. Soc. Japan, Pure Chem. Sect.*, 1955, **76**, 101.
92. E. R. van Artsdalen and I. S. Yaffe, *J. Phys. Chem.*, 1955, **59**, 118.
93. C. Kroger and P. Weisgerber, *Z. Phys. Chem. (Frankfurt)*, 1955, **5**, 192.
94. A. A. Woolf, *J. Chem. Soc.*, 1955, 279.
95. Yu K. Delimarskii, I. N. Sheiko and V. G. Fenchenko, *Zh. Fiz. Khim.*, 1955, **29**, 1499.
96. B. S. Harrap and E. Heymann, *Trans. Faraday Soc.*, 1955, **51**, 259.
97. G. M. Pound, G. Derge and G. Osuch, *Trans. Amer. Inst. Min. Met. Eng.*, 1955, **203**, 481.
98. I. S. Yaffe and E. R. van Artsdalen, *J. Phys. Chem.*, 1956, **60**, 1125.
99. J. D. Mackenzie, *Trans. Faraday Soc.*, 1956, **52**, 1564.
100. H. C. Cowen and H. J. Axon, *ibid.*, 1956, **52**, 242.
101. W. C. Phelps and R. E. Grace, *Trans. Amer. Inst. Min. Met. Eng.*, 1957, **209**, 1447.
102. E. W. Yim and M. Feinleib, *J. Electrochem. Soc.*, 1957, **104**, 626.
103. F. R. Duke and R. A. Fleming, *ibid.*, 1957, **104**, 251.
104. J. B. Story and J. T. Clarke, *Trans. Amer. Inst. Min. Met. Eng.*, 1957, **209**, 1449.
105. V. A. Kochinashvili and V. P. Barzakovskii, *Zhur. Priklad. Khim.*, 1957, **30**, 1755.
106. F. R. Duke and R. A. Fleming, *J. Electrochem. Soc.*, 1958, **105**, 412.
107. H. Winterhager and L. Werner, *ForschBer. Wirtn.-u. VerkMinist. NRhein.-Westf.*, 1956, **341**.
108. J. O'M. Bockris, E. Crook, H. Bloom and N. E. Richards, *Proc. R. Soc.*, 1960, *A***255**, 558.
109. I. S. Yaffe and E. R. van Artsdalen, *Chem. Semi-Ann. Progr. Rep.* No. 2159, *Oak Ridge Natl. Lab.*, 1956, p. 77.
110. P. I. Protsenko and O. N. Shokina, *Zh. Neorg. Khim.*, 1960, **5**, 437.
111. L. F. Grantham and S. J. Yosim, *J. Chem. Phys.*, 1963, **38**, 1671.
112. K. B. Morris, M. McNair and G. Koops, *J. Chem. Engng Data*, 1962, **7**, 224.
113. H. R. Bronstein, A. S. Dworkin and M. A. Bredig, *J. Phys. Chem.*, 1962, **66**, 44.
114. A. S. Dworkin, R. A. Sallach, N. F. Bronstein, M. A. Bredig and J. D. Corbett, *ibid.*, 1963, **67**, 1145.
115. B. F. Markov and V. D. Prusyazhnyi, *Ukr. khim. Zh.*, 1962, **28**, 268.
116. N. V. Smith and E. R. van Artsdalen, *Chem. Semi-Ann. Progr. Rep.* No. 2159, *Oak Ridge Natl. Lab.*, 1956, p. 80.
117. A. S. Dworkin, H. R. Bronstein and M. A. Bredig, *J. Phys. Chem.*, 1963, **67**, 2715.
118. N. N. Greenwood and I. J. Worrall, *J. Chem. Soc.*, 1958, 1680.
119. E. F. Riebling and C. E. Erickson, *J. Phys. Chem.*, 1963, **67**, 307.
120. G. J. Janz and J. D. E. McIntyre, *J. Electrochem. Soc.*, 1962, **109**, 842.
121. H. Winterhager and L. Werner, *ForschBer. Wirt.-u. VerkMinist. NRhein.-Westf.*, 1957, **438**.
122. E. R. Buckle and P. E. Tsaoussoglou, *J. Chem. Soc.*, 1964, 667.
123. K. B. Morris and P. L. Robinson, *J. Phys. Chem.*, 1964, **68**, 1194.
124. F. M. Kolomitskii and V. D. Ponomarev, *Izv. Akad, Nauk, kazakh, SSR, Ser. Metall. Obog. Ogneu.*, 1959, **1**, 21.
125. K. B. Morris and P. L. Robinson, *J. Chem. Engang Data*, 1964, **9**, 444.
126. A. N. Campbell, E. M. Kartzmark and D. F. Williams, *Can. J. Chem.*, 1962, **40**, 890.
127. W. Karl and A. Klemm, *Z. Naturf.*, 1964, **19***A*, 1619.
128. I. G. Murgulescu and S. Zuca, *Studii Cerc. Chim.*, 1959, 7, 325.
129. A. S. Dworkin, H. R. Bronstein and M. A. Bredig, *J. Phys. Chem.*, 1962, **66**, 1201.
130. M. Panish, *ibid.*, 1959, **63**, 1337.
131. E. R. Buckle and P. E. Tsaoussoglou, *Trans. Faraday Soc.*, 1964, **60**, 2144.
132. T. Kuroda and T. Suzuki, *J. Electrochem. Soc. Japan*, 1961, **29**, E215.

133. Y. Doucet and M. Bizouard, *Compt. Rend.*, 1960, **250**, 73.
134. L. F. Grantham, *J. Chem. Phys.*, 1965, **43**, 1415.
135. G. J. Janz and M. R. Lorenz, *J. Electrochem. Soc.*, 1961, **108**, 1052.
136. H. Bloom and E. Heymann, *Proc. Roy. Soc.* (*London*), 1947, *A*188, 392.
137. M. F. Lantratov and O. F. Moiseeva, *Zh. Fiz. Khim.*, 1960, **34**, 367.
138. L. A. King and F. R. Duke, *J. Electrochem. Soc.*, 1964, **111**, 712.

Table 9.5 ELECTRICAL CONDUCTIVITY OF MOLTEN BINARY SALT SYSTEMS AND OTHER MIXED IONIC MELTS

The electrical conductivity in $\Omega^{-1}\,cm^{-1}$ at temperature $t(^{\circ}C)$ and composition p (wt.%) of the first-named constituent is given as κ_t, or empirical equations for κ as a function of temperature within the stated range are given at each composition. Principal references are in **bold type**.

AgBr–AgCl	p	0	24.7	46.7	66.9	84.0	100
Ref. **59**, 3	κ_{450}	—	3.52	3.33	3.16	3.00	2.89
	κ_{500}	3.90	3.70	3.50	3.31	3.16	3.02
	κ_{550}	4.07	3.86	3.64	3.44	3.28	3.13
	κ_{600}	4.21	3.99	3.77	3.55	3.38	3.21

AgBr–KBr	p	52.2	61.9	78.0	90.1	100
Ref. **59**	κ_{400}	—	—	1.42	1.98	—
	κ_{500}	—	1.29	1.70	2.23	3.03
	κ_{600}	1.43	1.55	1.91	2.43	3.21

AgBr–KCl	p	38.6	62.6	79.0	91.0	100.0
Ref. **77**, 78	κ_{500}	—	—	1.678	2.210	3.020
	κ_{600}	—	1.680	1.942	2.432	3.215
	κ_{700}	1.872	1.933	2.172	2.602	3.374
	κ_{800}	2.127	2.150	2.340	2.742	3.505

AgCl–AgNO$_3$	p	0	9.00	12.5	15.5	20.0	30.0	40.0
Ref. **36**	κ_{220}	0.706	0.705	0.710	0.712	—	—	—
	κ_{270}	0.926	0.926	0.931	0.932	0.950	1.005	1.092
	κ_{320}	1.132	1.132	1.138	1.138	1.161	1.223	1.312

AgCl–AlBr$_3$	p				13.29			
Ref. **13**	t	100	110	125	130	140	155	170
	κ	1.07	1.21	1.61	1.68	1.90	2.67	3.22

AgCl–KBr	p	23.2	44.6	64.4	83.0	100.0
Ref. **77**, 78	κ_{500}	—	—	1.742	2.600	3.965
	κ_{600}	1.400	1.570	2.008	2.871	4.277
	κ_{700}	1.637	1.805	2.215	3.065	4.526
	κ_{800}	1.815	1.980	2.390	3.220	4.716

AgCl–KCl	p	50.2	63.0	72.5	87.9	100
Ref. **59**	κ_{400}	—	—	—	2.30	—
	κ_{500}	—	—	1.93	2.70	3.90
	κ_{600}	—	2.01	2.24	3.00	4.21
	κ_{700}	2.16	2.31	2.51	3.22	4.46

AgCl–PbCl$_2$	p	0	11.2	25.6	43.5	76.3	100
Ref. **59**	κ_{400}	—	—	1.27	1.67	2.62	—
	κ_{500}	1.43	1.59	1.82	2.20	3.11	3.90
	κ_{600}	1.92	2.07	2.27	2.61	3.46	4.21
	κ_{700}	2.33	2.51	2.68	2.96	3.73	4.46

AgCl–TlCl	p	0	20.6	37.0	58.0	77.2	100
Ref. **3**	κ_{500}	1.215	1.470	1.711	2.260	2.925	3.653

AgI–AgNO$_3$	p	0	10	20	30	40	50	60
Ref. **62**, 33	κ_{150}	—	—	—	0.38	0.44	0.54	0.65
	κ_{200}	—	—	0.58	0.60	0.66	0.75	0.86
	κ_{250}	0.80	0.78	0.78	0.80	0.85	0.93	1.03
	κ_{300}	1.00	0.98	0.98	1.00	1.05	1.12	1.21

AgNO$_3$–Cd(NO$_3$)$_2$	p	23.5	32.3	41.7	51.8	62.6	74.1	100
Ref. **46**	κ_{150}	—	—	0.07	0.15	—	—	—
	κ_{200}	—	0.11	0.19	0.28	0.38	0.54	—
	κ_{250}	0.15	0.23	0.34	0.44	0.55	0.65	0.83
	κ_{300}	0.27	0.38	0.49	0.61	0.73	0.86	—

(*continued*)

Table 9.5 ELECTRICAL CONDUCTIVITY OF MOLTEN BINARY SALT SYSTEMS AND OTHER MIXED IONIC MELTS—*continued*

AgNO$_3$–HgI$_2$ Ref. 53, 19

p	4.0	13.8	27.2	35.9	46.6	59.9	77.0
κ_{100}	—	—	—	0.08	0.07	0.08	0.08
κ_{150}	0.02	0.11	0.12	0.20	0.15	0.22	0.31
κ_{200}	0.09	0.24	0.22	0.30	0.27	0.36	0.51

AgNO$_3$–KNO$_3$ Ref. 71, 61, 51, 79

p	0	20	40	60	80	90	100
κ_{250}	—	—	0.47	0.54	0.66	0.74	0.84
κ_{300}	—	0.56	0.64	0.73	0.85	0.93	1.05
κ_{350}	0.66	0.73	0.81	0.91	1.04	1.13	1.24

AgNO$_3$–LiNO$_3$ Ref. 61

p	0	38.0	62.0	78.6	90.7	100
κ_{250}	0.78	0.79	0.80	0.81	0.82	0.83
κ_{300}	1.05	1.04	1.04	1.04	1.04	1.04

AgNO$_3$–NaNO$_3$ Ref. 38

p	0	10	20	30	40	60	100
κ_{290}	—	0.900	0.907	0.910	0.920	0.941	1.010
κ_{330}	1.076	1.088	1.094	1.104	1.103	1.118	1.173
κ_{370}	1.251	1.267	1.264	1.277	1.272	1.289	—

AgNO$_3$–RbNO$_3$ Ref. 52

p	22.3	33.0	43.4	63.3	72.9	82.2	91.2
κ_{200}	—	0.24	0.28	0.37	0.43	0.49	0.57
κ_{250}	0.33	0.38	0.42	0.54	0.61	0.69	0.77
κ_{300}	0.47	0.52	0.58	0.71	0.78	0.87	0.96

AgNO$_3$–TlNO$_3$ Ref. 29, 3

p	10	35	45	50	60	65	90
κ_{100}	—	—	0.137	0.123	0.145	—	—
κ_{200}	0.336	0.405	0.424	0.427	0.468	0.488	0.560
κ_{300}	0.607	0.712	0.735	0.753	0.808	0.824	0.943

AlBr$_3$–HgBr$_2$ Ref. 25, 3

p	59.7	60.5	63.9	68.0	79.0	89.6	92.2
$10^2 \cdot \kappa_{110}$	1.38	1.34	1.22	1.00	0.419	0.064	0.013
$10^2 \cdot \kappa_{1\,140}$	2.45	2.34	2.00	1.61	0.573	0.064	0.013

AlBr$_3$–KBr Ref. 30, 23, 2

p	81.8	83.2	84.7	86.2	87.5	88.0
κ_{110}	0.036 0	0.033 1	0.030 1	0.026 9	(0.024 2)	0.023 3
κ_{140}	0.060 1	0.054 6	0.049 3	0.043 9	(0.039 6)	0.037 9

2AlBr$_3$·KCl Ref. 28, 13

t	90	100	140	160	170
κ	0.023 2	0.029 5	0.058 7	0.074 2	0.082 9

4AlBr$_3$·LiCl Ref. 13

t	80	100	120	140	170	180
κ	0.011 0	0.013 9	0.017 0	0.020 4	0.024 2	0.025 1

AlBr$_3$–NaBr Ref. 30

p	83.8	84.7	85.6	86.8	87.9	88.9
κ_{120}	0.057 8	0.053 3	0.049 2	0.044 1	0.038 2	0.034 6
κ_{130}	0.067 4	0.061 3	0.056 5	0.050 6	0.044 4	0.040 2
κ_{140}	0.077 8	0.068 7	0.064 1	—	0.050 0	0.045 7

2AlBr$_3$·NaBr Ref. 28

t	110	130	150	170
κ	0.051 5	0.069 2	0.088 7	0.108 7

AlBr$_3$–NaCl Ref. 13

p	90.73	93.69	94.83	at $p = 80.1$, $\kappa_{170} = 0.036\,5$
κ_{110}	0.025 2	0.026 2	0.022 0	
κ_{130}	0.041 5	0.031 0	0.025 2	
κ_{150}	—	0.042 8	0.031 3	

AlBr$_3$–NH$_4$Br Ref. 25

p	84.50	86.90	88.98	89.42	90.81
κ_{110}	0.022	0.019	0.017	0.016	0.014
κ_{130}	0.033	0.027	—	0.022	0.020
κ_{150}	0.043	0.036	0.031	0.029	0.025

2AlBr$_3$·NH$_4$Br Ref. 28

t	110	160
κ	0.022 0	0.048 4

AlBr$_3$–SbBr$_3$ Ref. 23, 2

p	0	7.58	19.74	26.65	42.45	74.69	86.91
$10^3 \cdot \kappa_{100}$	0	12	18	16	11	4	1
$10^3 \cdot \kappa_{140}$	0	16	25	26	20	7	1

AlBr$_3$·SbBr$_3$ Ref. 28

t	100	150	170
$10^3 \cdot \kappa$	11.6	26.3	32.2

AlBr$_3$–SbBr$_3$–AsBr$_3$ Ref. 35

p	16.3	32.9	41.4	67.9	77.3	86.8	100
$10^2 \cdot \kappa_{85}$	0.089	0.121	0.252	0.766	0.855	0.892	0.803
$10^2 \cdot \kappa_{100}$	0.082	0.116	0.264	0.938	1.125	1.228	1.155

(*continued*)

Table 9.5   ELECTRICAL CONDUCTIVITY OF MOLTEN BINARY SALT SYSTEMS AND OTHER MIXED IONIC MELTS—*continued*

$AlBr_3$–$ZnBr_2$	p	73.43	78.03	82.57	87.03	91.42		
Ref. **23**	κ_{100}	5.5	5	3	1	0		
	κ_{130}	12.5	10.5	8	—	0		
	κ_{150}	20	17	11	2	0		
$2AlBr_3 \cdot ZnBr_2$	t	100	140	180				
Ref. **28**	$10^3 \cdot \kappa$	5.15	15.12	30.62				
$AlCl_3$–KCl	p	64.73	73.30	80.11	81.55	89.14		
Ref. **21**	κ_{200}	—	0.200	0.165	0.157	0.105		
	κ_{250}	0.346	0.280	0.228	0.214	0.143		
	κ_{300}	0.458	0.357	0.290	(0.270)	(0.180)		
$AlCl_3$–KCl**	p	50.4	53.7	62.5				
Ref. **31**	κ_{600}	0.790	0.811	0.858				
	κ_{700}	0.900	0.935	1.005				
	κ_{800}	1.010	1.060	1.155				
$AlCl_3 \cdot KCl$	t	250	300	500	600	700	800	
Ref. **31, 21**	κ	0.353	0.479	0.706	0.862	1.018	1.174	
$AlCl_3 \cdot LiCl$**	t	174	184	217	259	303.5	327	351
Ref. **21**	κ	0.354	0.380	0.468	0.553	0.647	0.687	0.731
$AlCl_3$–$NaCl$**	p	69.78	71.03	72.01	76.06	76.95	81.25	82.51
Ref. **15, 14, 5**	κ_{200}	0.436	0.262	0.280	0.160	0.170	0.090	0.070
	κ_{250}	0.532	0.378	0.380	—	0.300	0.178	—
$AlCl_3$–NH_4Cl	p	71.37	78.89	85.46	90.89			
Ref. **21**	κ_{200}	—	—	0.177	0.114			
	κ_{300}	0.479	0.387	0.302	0.194			
AlF_3–Na_3AlF_6*	p	0	2	4	6	8	10	20
Ref. **72, 65, 39,**	$\kappa_{1\,000}$	2.8	2.7	2.7	2.6	2.6	2.5	(2.3)
34, 12	$\kappa_{1\,050}$	2.9	2.9	2.8	2.8	2.7	2.6	(2.4)
Al_2O_3–Na_3AlF_6	p	0	3	6	9	12	15	20
Ref. **72, 65, 39, 34,**	$\kappa_{1\,000}$	2.8	2.6	2.5	2.3	2.2	2.1	(1.8)
80, 12, 4	$\kappa_{1\,100}$	3.0	2.9	2.7	2.6	2.4	2.3	(2.0)
Al_2O_3–Li_3AlF_6—	$p_{Al_2O_3}$	6.4	3.2	8.8	3.6	9.8	4.2	
Na_3AlF_6	$p_{Na_3AlF_6}$	46.8	58.2	64.0	67.6	72.2	76.6	
Ref. **80**	$\kappa_{1\,000}$	3.44	—	3.10	3.24	2.43	2.67	
	$10^3\alpha$	5.2	—	4.7	4.8	5.5	5.4	
	$\kappa_t = \kappa_{1\,000} + \alpha(t - 1\,000)$							
Al_2O_3–SiO_2	p	3	5.5	8	10	12		
Ref. **81**	a	−0.30	0.08	−0.40	−0.30	−1.00		
	$10^{-3}b$	5.9	5.7	4.1	4.8	3.0		
			$\log \kappa_t = a - b(t + 273)$ at $1\,600$–$1\,800$					
B_2O_3–BaO	p	30.0	35.0	44.5	49.2	59.7	67.9	
Ref. **50**	κ_{900}	—	—	0.007	0.005	0.002	0.002	
	$\kappa_{1\,000}$	0.10	0.07	0.036	0.024	0.013	0.008	
	$\kappa_{1\,100}$	0.19	0.15	0.089	0.062	0.037	0.026	
	$\kappa_{1\,200}$	0.28	0.23	—	—	—	0.056	
B_2O_3–CaO	p	55.3	59.8	62.8	67.9	73.3		
Ref. **50**	$\kappa_{1\,000}$	—	—	—	0.025	0.013		
	$\kappa_{1\,100}$	—	0.12	0.10	0.069	0.047		
	$\kappa_{1\,200}$	0.23	0.22	0.20	0.151	0.100		
B_2O_3–K_2O	p	61.6	70.6	79.6	89.0	94.8	98.5	
Ref. **75, 42**	κ_{600}	0.016	0.003	—	1.8×10^{-4}	8.3×10^{-5}	1.1×10^{-5}	
	κ_{800}	0.186	0.056	0.022	6.0×10^{-3}	2.2×10^{-3}	2.2×10^{-4}	
	$\kappa_{1\,000}$	0.513	0.309	0.112	3.0×10^{-3}	9.1×10^{-3}	1.1×10^{-3}	
B_2O_3–Li_2O	p	79.9	87.1	92.0	95.5	97.2	98.5	
Ref. **75, 42**	κ_{700}	0.08	—	—	0.003	0.001	—	
	κ_{800}	0.87	0.10	0.03	0.011	0.004	0.0012	
	κ_{900}	3.02	0.26	0.10	0.026	0.011	0.0028	
	$\kappa_{1\,000}$	5.25	0.40	0.17	0.051	0.019	0.0052	

(*continued*)

Table 9.5 ELECTRICAL CONDUCTIVITY OF MOLTEN BINARY SALT SYSTEMS AND OTHER MIXED IONIC MELTS—*continued*

B_2O_3–Na_2O Ref. 75, 42

p	64.0	68.2	77.6	85.8	94.4	99.1
κ_{600}	—	—	—	6.5×10^{-4}	1.5×10^{-4}	1.6×10^{-5}
κ_{800}	—	0.33	0.08	0.022	4.8×10^{-3}	3.5×10^{-4}
κ_{1000}	1.74	1.32	0.40	0.107	2.1×10^{-2}	1.7×10^{-3}

B_2O_3–PbO Ref. 69

p	88.2	92.7
κ_{650}	5×10^{-7}	4×10^{-7}
κ_{750}	2×10^{-5}	0.6×10^{-5}
κ_{850}	8.5×10^{-5}	2.0×10^{-5}

B_2O_3–SrO Ref. 50

p	44.9	49.6	56.8	61.6	65.2	70.0
κ_{1000}	—	—	0.019	0.012	0.010	0.005
κ_{1100}	0.10	0.08	0.062	0.047	0.036	0.021
κ_{1200}	0.21	0.17	—	—	—	—

$BaBr_2$–KBr Ref. 43

p	0	39.2	55.5	62.5	73.0	88.2	93.4
κ_{700}	—	1.17	1.03	0.85	0.89	—	—
κ_{750}	1.63	1.29	1.16	0.96	1.00	0.93	—
κ_{800}	1.76	1.41	1.27	1.06	1.13	1.06	0.98
κ_{850}	1.87	1.52	1.39	1.17	1.25	1.19	1.12

$BaCl_2$–$CaCl_2$ Ref. 68

p	17.3	32.0	44.6	55.6	65.3	73.8	88.3
κ_{800}	1.84	1.80	1.76	1.71	1.65	1.62	—
κ_{900}	2.23	2.15	2.08	2.05	1.98	1.93	1.82
κ_{1000}	2.57	2.51	2.44	2.38	2.32	2.27	2.14
κ_{1100}	2.86	2.81	2.74	2.66	2.60	2.56	2.44

$BaCl_2$–$MgCl_2$† Ref. 37

p	13.4	31.0	49.7	73.2	75.7
κ_{800}	1.405	1.620	1.70	1.61	1.57
κ_{900}	1.590	1.810	1.91	1.83	1.785
κ_{1000}	(1.775)	2.000	2.12	2.05	2.000

$BaCl_2$–$NaCl$ Ref. 49, 17

p	28.4	47.1	59.8	70.4	78.1	84.2	93.4
κ_{800}	3.00	2.55	2.26	2.00	1.54	1.20	—
κ_{900}	3.26	2.90	2.68	2.50	2.32	2.25	2.00
κ_{1000}	3.48	3.14	2.92	2.76	2.64	2.58	2.32

$BaCl_2$–$NaNO_3$ Ref. 40

p	1.17	1.98	6.14	9.12
κ_{350}	1.124	1.107	1.073	1.024
κ_{355}	1.143	1.128	1.093	1.052
κ_{360}	1.163	—	—	—

BaF_2–NaF Ref. 11

p			66.6		
t	900	950	1 000	1 050	1 100
κ	4.027	4.335	4.602	5.051	5.319

$Ba(NO_2)_2$–KNO_2 Ref. 82

p	47.8	59.0	68.3	76.3	83.5	89.6	95.0
κ_{300}	—	0.494	0.438	0.393	—	—	—
κ_{310}	—	0.528	0.472	0.428	0.381	—	0.311
κ_{320}	—	0.560	0.505	0.464	0.415	0.386	0.345
κ_{330}	—	0.594	0.542	0.498	0.450	0.418	0.377
κ_{340}	0.670	0.628	0.572	0.534	0.486	0.452	0.411
κ_{350}	0.702	0.660	0.610	0.570	0.520	0.484	0.445

$Ba(NO_2)_2$–KNO_3 Ref. 83

p	26.8	31.3	43.8	54.8	64.5	73.3	81.0
κ_{320}	—	0.50	—	0.42	—	0.39	—
κ_{340}	—	0.55	—	0.48	—	0.42	0.40
κ_{360}	0.65	0.60	0.58	0.55	0.50	0.49	—
κ_{380}	0.73	0.68	0.62	0.60	—	—	—

$Ba(NO_2)_2$–$NaNO_2$ Ref. 82

p	19.4	39.2	53.9	64.0	73.9	84.5	94.5
κ_{240}	—	—	0.598	0.488	0.372	0.258	—
κ_{260}	—	0.852	0.700	0.588	0.464	0.334	0.200
κ_{280}	1.120	0.957	0.802	0.688	0.554	0.410	0.260
κ_{300}	1.232	1.064	0.910	0.788	0.646	0.486	0.320
κ_{320}	1.343	1.170	1.016	0.888	0.738	0.564	0.380
κ_{340}	1.454	1.275	1.120	0.985	0.828	—	—

$Ba(NO_3)_2$–KNO_2 Ref. 83

p	50.1	61.0	70.2
κ_{260}	(0.38)	0.30	(0.28)
κ_{280}	0.41	0.36	(0.31)
κ_{300}	0.49	0.41	0.35
κ_{320}	0.55	0.46	0.40

(*continued*)

Table 9.5 ELECTRICAL CONDUCTIVITY OF MOLTEN BINARY SALT SYSTEMS AND OTHER MIXED IONIC MELTS—*continued*

BeCl$_2$–NaCl	p	39.1	45.2	49.2	58.3	63.6	70.8	83.2
Ref. **58**	κ_{300}	—	—	0.69	0.58	0.49	—	—
	κ_{400}	—	1.00	1.10	0.98	0.85	0.56	0.19
	κ_{500}	1.35	1.28	1.38	—	—	—	—
Bi–BiI$_3$	p	3.8	8.1	13.1	34.7	45.1	58.8	76.4
Ref. **84**	κ_{500}	0.5	0.8	2.1	200	600	1 400	3 100
Ca–CaCl$_2$	p	0.22	0.29	0.36	0.47	0.62	0.88	1.10
Ref. **85**	κ_{855}	2.50	2.52	2.55	2.58	2.61	2.68	2.73
CaCl$_2$–KCl	p	24.6	40.7	66.0	81.5			
Ref. **37, 3**	κ_{800}	1.770	1.500	1.415	1.560			
	κ_{900}	2.035	1.790	1.685	1.890			
CaCl$_2$–MgCl$_2$†	p	10.8	19.0	28.0	41.2	62.1	77.1	
Ref. **37**	κ_{700}	(1.18)	1.34	1.47	1.65	(1.79)	(1.87)	
	κ_{800}	1.37	1.53	1.68	1.90	2.07	2.20	
	κ_{900}	1.56	1.72	1.89	2.15	2.35	2.53	
CaCl$_2$–NaCl**	p	0	20	40	60	75	90	100
Ref. **67, 49, 17, 18,**	κ_{100}	—	2.78	2.31	1.92	1.71	—	—
8, 6, 3	κ_{800}	3.58	3.10	2.64	2.27	2.09	2.01	2.02
	κ_{900}	3.79	3.37	2.96	2.60	2.40	2.32	2.32
	$\kappa_{1\,000}$	4.00	3.58	3.22	2.92	2.71	2.67	2.68
CaF$_2$–CaO	p	90.7	92.6	94.5	96.4	98.2	100	
Ref. **55**	$\kappa_{1\,500}$	4.3	4.6	5.3	4.2	4.1	3.9	
	$\kappa_{1\,545}$	4.6	4.7	7.2	5.2	4.4	4.1	
CaF$_2$–Na$_3$AlF$_6$	p	0	5	10	20	30	40	
Ref. **65, 39, 26,**	$\kappa_{1\,000}$	2.8	2.8	2.7	2.6	2.5	2.4	
34, 12	$\kappa_{1\,100}$	3.0	3.0	3.0	2.9	2.8	2.7	
CaF$_2$–NaF	p	0	10					
Ref. **4, 11**	$\kappa_{1\,000}$	3.15	3.0					
	$\kappa_{1\,020}$	3.3	3.1					
Ca(NO$_3$)$_2$–KNO$_3$	p	0	16.9	24.0	34.2	44.8	54.9	65.5
Ref. **22**	κ_{300}	—	—	0.375	0.330	0.270	0.195	—
	κ_{350}	0.666	0.570	0.525	0.460	0.390	0.315	0.255
	κ_{400}	0.818	0.705	0.660	0.605	0.522	0.430	0.355
CaO–SiO$_2$	p	20	30	35	40	45	50	55
Ref. **81**	a	2.17	2.19	2.22	2.51	3.19	2.39	2.06
	$10^{-3}b$	5.7	5.7	5.2	5.7	6.7	5.0	4.1
	$\log\kappa_t = a - b/(t+273)$ at 1600–1800							
CdBr$_2$–CdCl$_2$	p	0	27.1	49.7	69.0	85.6	100	
Ref. **24**	κ_{600}	1.95	1.7	1.5	1.35	1.2	1.05	
	κ_{640}	2.1	1.85	1.65	1.5	1.35	1.25	
CdCl$_2$–CdI$_2$	p	0	14.2	33.2	59.9	100		
Ref. **43**	κ_{400}	0.22	0.35	—	—	—		
	κ_{500}	0.41	0.56	0.78	—	—		
	κ_{600}	0.66	0.79	1.03	1.32	1.96		
	κ_{700}	0.95	1.05	1.30	1.57	2.09		
CdCl$_2$–KCl	p	40	50	60	70	80	90	100
Ref. **47, 24, 3, 27**	κ_{500}	—	1.10	1.11	1.19	1.33	1.52	
	κ_{600}	—	1.24	1.27	1.36	1.51	1.71	1.96
	κ_{700}	1.55	1.48	1.48	1.60	1.77	1.97	2.18
	κ_{800}	1.81	1.73	1.66	1.78	1.94	2.09	2.25
CdCl$_2$–LiCl	p	0	59.1	81.2	92.9	100		
Ref. **43**	κ_{550}	—	—	2.8	2.3	—		
	κ_{650}	6.0	4.1	3.2	2.6	2.1		
	κ_{750}	6.4	4.5	3.5	3.0	2.3		
CdCl$_2$–NaCl	p	43.9	67.6	82.5	92.6	100		
Ref. **24**	κ_{550}	(2.25)	1.95	1.9	1.85	1.8		
	κ_{600}	(2.4)	2.15	2.1	2.0	1.9		
	κ_{700}	2.7	2.45	2.4	2.3	2.2		

(*continued*)

Table 9.5 ELECTRICAL CONDUCTIVITY OF MOLTEN BINARY SALT SYSTEMS AND OTHER MIXED IONIC MELTS—*continued*

CdCl$_2$–PbCl$_2$ Ref. 56, 27, 24

p	0	20	40	60	80	100
κ_{500}	1.44	1.68	—	—	—	—
κ_{600}	1.88	1.94	1.97	1.97	1.94	1.88
κ_{700}	2.28	2.35	2.36	2.31	2.26	2.20

CdCl$_2$–TlCl Ref.44, 3

p	0	10	25	40	60	80	100
κ_{500}	1.35	1.30	1.12	1.24	1.41	—	—
κ_{600}	1.70	1.66	1.56	1.52	1.66	1.81	1.97
κ_{700}	1.95	1.86	1.76	1.72	1.86	1.99	2.10

CdI$_2$–KI Ref. 43

p	0	19.7	35.5	64.3	75.3	89.8	100
κ_{300}	—	—	—	0.30	0.31	—	—
κ_{500}	—	—	—	0.66	0.64	0.57	0.41
κ_{700}	1.32	1.21	1.09	1.06	0.99	0.89	0.95
κ_{800}	1.52	1.45	1.29	1.26	—	—	—

Cd(NO$_3$)$_2$–CsNO$_3$ Ref. 45

p	39.4	44.6	54.8	64.5	73.8	82.9
κ_{200}	—	—	0.09	0.07	—	—
κ_{250}	—	0.17	0.16	0.15	0.12	—
κ_{300}	0.26	0.25	0.24	0.23	0.20	0.15

Cd(NO$_3$)$_2$–KNO$_3$ Ref. 46

p	36.9	50.0	60.9	70.0	77.8	84.4
κ_{200}	—	—	0.14	0.12	0.08	0.04
κ_{250}	—	—	0.26	0.23	0.18	0.12
κ_{300}	0.46	0.44	0.39	0.35	0.31	0.23

Cd(NO$_3$)$_2$–RbNO$_3$ Ref. 46

p	15.1	28.6	40.8	51.7	61.6	70.6	78.9
κ_{200}	—	—	0.12	0.10	0.10	0.08	—
κ_{250}	—	0.22	0.21	0.20	0.19	0.16	—
κ_{300}	0.31	0.34	0.32	0.31	0.29	0.26	0.20

Cd(NO$_3$)$_2$–TlNO$_3$ Ref. 45

p	9.0	18.1	27.5	37.1	47.0	57.1	67.4
κ_{150}	—	—	0.14	0.11	0.08	—	—
κ_{200}	—	0.29	0.25	0.22	0.17	—	—
κ_{250}	0.47	0.31	0.38	0.34	0.28	0.23	—
κ_{300}	0.60	0.54	0.50	0.46	0.40	0.34	0.26

Ce–CeCl$_3$ Ref. 95, 96

p	1.05	1.8	2.8	3.4	3.9	4.7	5.3
κ_{855}	1.56	2.02	2.59	3.26	3.81	4.45	5.35

Ce–CeI$_3$ Ref. 86

p	0.55	1.03	1.43	2.07	2.87	3.7	6.1
κ_{820}	1.00	1.92	2.74	4.75	8.88	15.5	c. 60

CsBr–NaCl Ref. 87

p	8.3	19.5	52.1	59.2
κ_{800}	2.40	1.69	1.40	1.21
κ_{850}	2.55	1.79	1.50	1.30

CsBr–ZnSO$_4$ Ref. 62

p	39.8	63.8	79.9
κ_{505}	0.03	0.09	0.20
κ_{550}	0.05	0.13	0.26

CsCl–NaBr Ref. 87

p	29.4	52.2	71.0	87.0
κ_{800}	2.05	1.61	1.50	1.45
κ_{850}	2.15	1.70	1.58	1.55

CsCl–TiCl$_3$ Ref. 88

p	52.1	62.3	72.0	81.5	90.8
κ_{800}	0.7	0.7	0.8	0.9	1.2

CuCl–KCl Ref. 47

p	25.8	42.0	47.6	57.5	76.0	85.9	100
κ_{450}	—	2.00	2.08	2.23	2.71	2.93	3.32
κ_{500}	—	2.07	2.15	2.32	2.82	3.06	3.43
κ_{550}	—	2.18	2.26	2.44	2.94	3.20	3.54
κ_{600}	2.22	2.32	2.40	2.59	3.08	3.34	3.66

CuO–V$_2$O$_5$ Ref. 74

p	10	30	40	50
κ_{800}	0.6	3.2	3.8	5.5
κ_{900}	1.4	5.0	6.0	10.0
$\kappa_{1\,000}$	2.2	—	—	—

Cu$_2$S–FeS Ref. 60

p	0	25	35	50	65	75	100
$\kappa_{1\,100}$	—	9.30	8.20	4.60	3.30	2.20	0.50
$\kappa_{1\,300}$	1.530	9.40	8.30	5.20	4.60	3.80	1.30
$\kappa_{1\,500}$	1.490	9.50	8.30	—	—	—	1.70

(continued)

Table 9.5   ELECTRICAL CONDUCTIVITY OF MOLTEN BINARY SALT SYSTEMS AND OTHER MIXED IONIC MELTS—*continued*

Dy–DyCl$_3$	p	0.0	0.9	2.7	4.9	7.8	12.0	16.7
Ref. **89**	κ_{700}	0.42	0.46	0.54	0.65	0.76	0.90	0.98
Er–ErCl$_3$	p	0.0	1.0	3.0	3.8			
Ref. **89**	κ_{820}	0.50	0.55	0.67	0.69			
FeO–SiO$_2$	p	55.9	63.3	68.6	72.4	75.4	78.7	84.9
Ref. **41**	$\kappa_{1\,300}$	0.5	0.9	1.4	2.1	3.2	4.2	9.0
	$\kappa_{1\,400}$	0.8	1.6	2.3	4.0	5.0	7.8	—
FeO–TiO$_2$	p	32.6	35.7	56.5	62.1	91.9	97.2	
Ref. **63**	$\kappa_{1\,300}$	48	50	40	38	145	—	
	$\kappa_{1\,400}$	48	47	34	33	155	225	
Fe$_2$O$_3$–V$_2$O$_5$	p	15.0	19.4	30.5				
Ref. **74**	κ_{900}	1.4	3.8	6.8				
	$\kappa_{1\,000}$	2.0	5.8	11.6				
	$\kappa_{1\,100}$	3.6	10.6	—				
Gd–GdBr$_3$	p	0.0	0.4	1.3	2.3	2.7		
Ref. **89**	κ_{820}	0.47	0.58	0.87	1.27	1.46		
Gd–GdCl$_3$	p	0.0	0.3	0.6	1.0			
Ref. **89**	κ_{650}	0.41	0.43	0.45	0.49			
HgBr$_2$–NH$_4$Br	p	61.19	78.63	84.66	89.57	93.64	100	
Ref. **32**	κ_{250}	0.4	0.5	0.6(2)	0.6	0.45	0	
	κ_{300}	0.55	0.7	0.8(2)	0.75	0.6	0	
HgCl$_2$–HgI$_2$	p	0	20	40	50	60	80	100
Ref. **19**	κ_{225}	—	0.029 5	0.016 0	0.009 4	0.005 3	—	—
	κ_{250}	—	0.028 1	0.016 0	0.009 7	0.005 6	0.001 1	—
	κ_{300}	0.028 0	0.025 2	0.016 0	0.010 4	0.006 3	0.002 0	0.000 8
HgCl$_2$–NH$_4$Cl	p	71.42	81.82	83.54	86.12	92.21	93.84	95.31
Ref. **32**	κ_{250}	0.7	0.9	0.7	0.8	0.55	0.78	0.4
	κ_{300}	0.95	1.15	1.1	1.15	0.75	1.0	0.6
HgCl$_2$–TlNO$_3$	p	0	10	20	40	50	60	75
Ref. **19**	κ_{225}	0.416	0.344	0.274	0.184	0.150	0.126	—
	κ_{250}	0.476	0.397	0.320	0.226	0.186	0.156	0.100
	κ_{275}	0.536	0.450	0.368	0.269	0.222	0.188	0.126
HgI$_2$–KI	p	73.3	80.4	86.5	91.6	96.1	100	
Ref. **44**	κ_{270}	—	0.48	0.64	0.64	0.43	0.03	
	κ_{290}	0.50	0.56	0.69	0.69	0.45	0.03	
	κ_{310}	0.56	0.64	0.75	0.75	—	—	
HgI$_2$–NH$_4$I	p	67.64	75.82	82.47	87.98	92.62	96.58	100
Ref. **32**	κ_{250}	0.3	0.4	0.5	0.6	0.55	0.37	0.03
	κ_{350}	0.6	0.5	0.73	0.77	0.7	0.4	0.03
HgI$_2$–TlNO$_3$	p	0	20	40	50	60	80	100
Ref. **19**	κ_{250}	0.476	0.302	0.174	0.154	0.136	0.142	—
	κ_{275}	0.536	0.348	0.210	0.182	0.161	0.156	0.030 5
	κ_{300}	0.596	0.395	0.248	0.214	0.187	0.171	0.028 0
Ho–HoCl$_3$	p	0.0	0.7	3.2	6.3	10.6		
Ref. **89**	κ_{800}	0.53	0.56	0.69	0.84	1.05		
K–KBr	p	1	2	3	4	6	8	
Ref. **73**	κ_{763}	5.6	11.7	22.4	40.8	104.7	229.2	
	κ_{870}	6.8	16.3	29.5	51.3	129.0	263.0	
K–KCl	p	1	2	4	6	8	10	
Ref. **73**	κ_{816}	3.8	6.2	14.1	31.6	64.6	117.5	
K–KF	p	1.35	2.72	4.10	4.80			
Ref. **90**	κ_{900}	4.5	6.7	12.0	18.0			
K–KI	p	0.5	1.0	1.5	2.0	—	—	—
Ref. **90, 91**	κ_{900}	4.5	8.5	13.5	19.0	—	—	—
	p	0.5	1.2	2.6	13.5	26.0	48.7	64.9
	κ_{700}	3.7	7.9	19.7	500	1 500	4 200	7 300
KBF$_4$–KF	p	30	40	50	60	70	80	90
Ref. **92**	κ_{600}	—	—	—	—	1.0	2.0	1.5
	κ_{700}	—	—	3.0	5.3	6.8	7.2	6.7
	κ_{800}	5.0	8.0	10.0	11.5	12.5	12.8	12.6

(continued)

Table 9.5 ELECTRICAL CONDUCTIVITY OF MOLTEN BINARY SALT SYSTEMS AND OTHER MIXED IONIC MELTS—*continued*

KBr–KNO$_3$	p	0	17.2	37.0	54.1	70.2	86.9	100
Ref. 43	κ_{400}	0.82	0.79	0.78	—	—	—	—
	κ_{600}	1.38	1.34	1.29	1.30	1.31	—	—
	κ_{700}	—	—	—	—	1.53	1.53	—
	κ_{800}	—	—	—	—	1.76	1.76	1.76
KBr–NaCl	p	33.7	57.6	85.9	89.0			
Ref. 93	κ_{800}	2.97	2.51	2.20	1.95			
	κ_{850}	3.05	2.60	2.29	2.05			
KBr–RbCl	p	19.7	39.5	59.5	81.3			
Ref. 94	κ_{770}	1.64	1.64	1.64	1.65			
	κ_{800}	1.70	1.70	1.70	1.72			
	κ_{850}	1.80	1.80	1.80	1.82			
KBr–TlBr	p	2.2	5.3	9.7	18.4	30.3	53.9	72.0
Ref. 97	a	0.489$_1$	0.441$_9$	2.008$_7$	2.390$_4$	0.475$_4$	0.808$_8$	0.426$_5$
	$10^3 b$	3.239$_7$	2.957$_2$	7.546$_5$	8.864$_5$	3.284$_7$	4.267$_7$	5.332$_0$
	$10^6 c$	0.835$_7$	0.499$_1$	3.845$_7$	4.904$_9$	0.845$_0$	1.478$_0$	0.865$_2$
	t range	537–701	473–640	650–762	619–717	626–791	687–796	718–783
				$\kappa_t = a + bt - ct^2$				
KBr–ZnSO$_4$	P	35.8	40.0	44.4	50.2	60.0	64.7	73.0
Ref. 48	κ_{475}	0.10	—	—	0.13	0.21	0.24	0.34
	κ_{500}	0.12	0.15	0.14	0.16	0.25	0.29	0.41
	κ_{550}	0.16	0.21	0.19	0.23	0.33	0.39	0.55
KCl–KI	p	2.8	7.8	13.3	26.8	41.2	64.4	
Ref. 57	κ_{700}	1.31	1.38	1.41	1.48	1.56	1.66	
	κ_{800}	1.52	1.56	1.61	1.70	1.80	1.96	
	κ_{900}	1.66	1.72	1.76	1.88	1.97	2.14	
KCl–LiCl**	p	0	28.2	42.6	55.8	72.2	87.6	100
Ref. 57, 10	κ_{500}	—	—	2.31	1.87	—	—	—
	κ_{600}	—	3.51	2.87	2.40	1.90	—	—
	κ_{700}	6.17	4.10	3.42	2.88	2.35	2.04	—
	κ_{800}	6.62	4.56	3.82	3.24	2.72	2.36	2.24
KCl–MgCl$_2$	p	10	20	30	50	60	70	80
Ref. 56, 37, 16,	κ_{600}	—	—	0.97	0.98	0.97	0.98	—
9, 7	κ_{700}	1.10	1.17	1.21	1.24	1.27	1.34	1.48
	κ_{800}	1.33	1.37	1.39	1.44	1.48	1.56	1.76
	κ_{900}	1.50	1.54	1.56	1.62	1.67	1.75	1.87
KCl–NaBr	p	22.0	32.0	51.4	74.0			
Ref. 93	κ_{800}	2.64	2.40	2.29	2.21			
	κ_{850}	2.76	2.53	2.39	2.32			
KCl–NaCl**	p	18.7	32.1	40.6	54.8	64.8	82.9	
Ref. 57, 16, 9, 6, 3	κ_{700}	—	—	2.52	2.35	2.21	—	
	κ_{800}	3.23	2.98	2.85	2.64	2.53	2.34	
	κ_{900}	3.50	3.24	3.11	2.88	2.76	2.56	
KCl–NaI	p	0	8.0	19.8	33.0	49.7	73.6	100
Ref. 43	κ_{600}	—	1.88	1.68	1.54	1.51	—	—
	κ_{700}	2.41	2.20	2.04	1.86	1.84	1.87	—
	κ_{800}	2.68	2.51	2.37	2.16	2.44	2.17	2.22
KCl–PbCl$_2$	p	2.8	4.5	6.1	14.8	20.7	37.9	51.2
Ref. 98, 47, 27, 24	κ_{450}	—	1.176	1.142	1.034	0.937	—	—
	κ_{500}	1.414	1.401	1.345	1.205	1.106	1.038	(1.110)
	κ_{550}	1.645	1.609	1.558	1.364	1.273	1.174	1.230
	κ_{600}	1.868	1.807	1.723	1.515	1.417	1.306	1.350
	κ_{650}	2.086	2.018	1.913	1.684	1.572	1.467	1.519
	κ_{700}	—	—	—	—	1.716	1.624	1.670
KCl–RbBr	p	9.9	22.7	39.8	63.9			
Ref. 94	κ_{770}	1.42	1.58	1.70	1.88			
	κ_{800}	1.50	1.65	1.77	1.98			
	κ_{850}	1.59	1.73	1.89	2.09			

(continued)

Table 9.5 ELECTRICAL CONDUCTIVITY OF MOLTEN BINARY SALT SYSTEMS AND OTHER MIXED IONIC MELTS—*continued*

KCl–SnCl$_2$	p	0.0	4.1	8.7	14.1	20.3	27.8	31.9
Ref. 99	κ_{300}	1.115	1.079	0.995	0.865	0.688	0.472	—
	κ_{350}	1.388	1.331	1.215	1.060	0.871	0.645	0.702
KCl–TiCl$_3$	p	32.6	42.0	53.0	65.9	81.4		
Ref. 88	κ_{800}	1.0	1.05	1.3	1.5	1.8		
KCl–ZnCl$_2$	p	0	10.5	17.8	26.3	35.9	52.6	58.7
Ref. 66	κ_{500}	0.09	0.40	0.64	0.70	0.76	0.78	0.79
	κ_{550}	0.16	0.51	0.77	0.82	0.89	0.92	0.95
	κ_{600}	0.26	0.63	0.90	0.95	1.01	1.05	1.09
	κ_{650}	0.37	0.76	1.02	1.06	1.14	1.19	1.24
KCl–ZnSO$_4$	p	19.0	24.0	29.3	33.7	39.8	50.0	56.8
Ref. 48	κ_{475}	—	0.09	0.13	0.16	0.21	0.30	0.36
	κ_{500}	0.09	0.12	0.17	0.21	0.25	0.35	0.41
	κ_{550}	0.14	0.16	0.24	0.27	0.32	0.44	0.51
K$_2$Cr$_2$O$_7$–KNO$_3$	p	0	24.4	42.1	55.5	74.4	87.1	100
Ref. 44	κ_{350}	0.56	0.52	0.42	0.35	0.22	0.17	—
	κ_{400}	0.72	0.66	0.64	0.48	0.34	0.26	0.20
	κ_{475}	0.95	0.88	0.74	0.67	0.52	0.41	0.35
KF–NaCl	p	33.20	49.85	66.54	100			
Ref. 6	κ_{750}	2.6	2.9	3.2	—			
	κ_{850}	3.4	3.6	3.7	4.1			
	κ_{950}	4.2	4.3	(4.4)	4.5			
KI–NaCl	p	33.6	55.0	74.1	86.9	94.2	100	
Ref. 43	κ_{600}	—	—	1.56	1.35	—	—	
	κ_{700}	—	2.24	1.87	1.64	1.47	1.32	
	κ_{800}	3.06	2.58	2.20	1.91	1.73	1.52	
KI–ZnSO$_4$	p	47.4	50.9	54.8	60.4	62.9	67.3	
Ref. 48	κ_{440}	—	—	0.11	0.13	0.15	0.21	
	κ_{460}	0.09	0.08	0.13	0.16	0.18	0.24	
	κ_{480}	0.12	0.10	0.15	0.18	0.20	0.28	
K$_2$MoO$_4$–MoO$_3$	p	17.5	27.5	43.0	52.2	60.0	79.4	91.0
Ref. 100	a	-3.6606	-1.3120	-1.9820	-1.2083	-2.7280	6.0644	4.2506
	$10^2 b$	0.9428	0.3365	0.5345	0.3333	0.8329	-1.2999	-0.8537
	$10^6 c$	-5.0101	-1.0589	-2.4426	-1.2061	-5.3073	8.1731	5.5075
	t range	760–869	614–804	654–783	642–794	583–705	882—934	877–995

$$\kappa_t = a + bt + ct^2$$

KNO$_2$–NaNO$_2$	p	12.0	23.6	32.7	45.1	55.3	74.2	83.3
Ref. 82	κ_{300}	1.262	1.176	1.108	1.052	0.998	—	—
	κ_{340}	1.472	1.380	1.308	1.244	1.185	1.040	—
	κ_{380}	—	1.586	1.510	1.434	1.366	1.204	1.130
	κ_{420}	—	—	—	1.625	1.548	1.365	1.287
	κ_{460}	—	—	—	—	—	1.525	1.442

KNO$_3$–LiNO$_3$ p 66.0
Ref. 101
$\kappa_t = -0.3788 + 3.40 \times 10^{-3}t$ at 160–240
$\kappa_t = -0.4037 + 3.47 \times 10^{-3}t$ at 240–400

KNO$_3$–NaNO$_3$**	p	0	20	40	50	60	80	100
Ref. 19, 101, 79, 1	κ_{300}	—	0.820	0.706	0.652	0.625	0.572	—
	κ_{400}	1.364	1.160	1.024	0.966	0.931	0.874	0.818
	κ_{500}	1.720	1.496	1.351	1.284	1.234	1.178	1.107
KNO$_3$–RbNO$_3$	p	7.1	18.6	31.4	40.7	56.0	73.3	86.0
Ref. 45	κ_{300}	—	0.40	0.42	—	—	—	—
	κ_{350}	0.51	0.52	0.55	0.57	0.59	0.63	0.64
	κ_{400}	0.63	0.65	0.68	0.70	0.73	0.77	0.79
K$_2$SO$_4$–ZnSO$_4$	p	31.6	41.8	52.0				
Ref. 62	κ_{450}	0.04	0.06	0.05				
	κ_{500}	0.06	0.08	0.08				
	κ_{550}	0.07	0.12	0.12				
K$_2$TaF$_7$–Ta$_2$O$_5$	p	78.0	83.4	89.0	94.5	100.0		
Ref. 102	κ_{800}	0.71	0.80	0.87	0.91	0.92		
	κ_{900}	0.85	0.94	1.00	1.04	1.05		

(*continued*)

Table 9.5 ELECTRICAL CONDUCTIVITY OF MOLTEN BINARY SALT SYSTEMS AND OTHER MIXED IONIC MELTS—*continued*

K_2TiF_6–TiO_2	p	81.9	87.5	92.3	94.5	96.6	98.3	100
Ref. **102**	κ_{870}	1.18	1.22	1.28	1.30	1.31	1.39	1.40
	κ_{920}	1.29	1.33	1.37	1.39	1.40	1.48	1.49
K_2WO_4–WO_3	p	37.4	50.5	58.3	67.4	76.6	85.0	92.2
Ref. **103**	a	2.8156	2.8809	2.3111	2.3562	5.4936	7.5436	38.80
	$10^2 b$	0.5601	0.6150	0.5492	0.5650	1.2501	1.6402	7.9646
	$10^6 c$	2.0875	2.5011	2.4324	2.3923	5.8822	7.5248	39.636
	t range	891–1010	759–912	754–770	674–819	773–955	842–1016	885–1029
				$\kappa_t = -a + bt - ct^2$				
La–$LaBr_3$	p	0.0	0.4	1.2	2.5	4.1	5.5	6.5
Ref. **89**	κ_{820}	0.70	0.94	1.55	3.15	7.35	15.8	25.9
La–$LaCl_3$	p	0.3	1.1	2.0	3.0	4.6	6.2	
Ref. **85**	κ_{910}	1.51	1.84	2.46	3.21	5.02	7.21	
La–LaI_3	p	0.5	1.4	2.4	3.5	4.4		
Ref. **86**	κ_{840}	0.87	2.39	5.54	12.2	20.9		
Li_3AlF_6–Na_3AlF_6	p	20	30	50	80			
Ref. **80**, **65**	κ_{1000}	3.05	3.38	3.80	4.33			
	$10^3 \alpha$	4.8	4.8	4.8	4.8			
		$\kappa_t = \kappa_{1000} + \alpha(t - 1000)$ at 762–936						
Li_2MoO_4–MoO_3	p	19.5	33.9	51.5	60.5	71.3	85.2	91.0
Ref. **100**	a	10.4708	7.4905	4.0311	4.6127	2.7958	21.3359	13.2305
	$10^2 b$	2.4981	1.8153	1.0289	1.1832	0.7539	5.3686	3.2259
	$10^6 c$	13.536	9.3589	4.4897	5.2117	2.887	30.734	16.322
	t range	778–917	800–907	768–917	799–945	845–956	764–872	809–908
				$\kappa_t = -a + bt - ct^2$				
$LiNO_3$–$RbNO_3$	p	10.5	16.7	23.8	31.9	52.3	65.2	80.8
Ref. **52**	κ_{200}	—	0.20	0.24	—	—	—	—
	κ_{250}	—	0.33	0.36	0.41	0.53	0.61	—
	κ_{300}	0.43	0.47	0.51	0.56	0.72	0.82	0.96
Li_2WO_4–WO_3	p	54.6	59.7	69.0	77.5	85.2	92.3	
Ref. **103**	a	6.0152	2.1491	7.4648	0.6063	5.2597	5.7237	
	$10^2 b$	1.3336	0.4558	1.7642	0.1086	1.2773	1.4048	
	$10^6 c$	5.5948	0.3923	7.6694	−1.9898	4.8237	5.7978	
	t range	805–952	769–924	757–938	735–890	716–897	734–907	
				$\kappa_t = -a + bt - ct^2$				
$MgCl_2$–$NaCl$	p	23.2	39.1	57.2	69.0	81.5	90.3	
Ref. **37**, **56**, **16**, **9**	κ_{700}	2.80	2.31	1.97	1.78	1.63	1.39	
	κ_{800}	2.96	2.49	2.16	1.98	1.82	1.57	
	κ_{900}	3.12	2.67	2.35	2.18	2.01	1.75	
MgF_2–Na_3AlF_6	p	5	10					
Ref. **72**	κ_{950}	—	2.2					
	κ_{975}	2.5	—					
MnO–SiO_2	p	40	50	55	60	65	70	80
Ref. **81**	a	2.32	2.26	1.85	2.25	2.94	2.21	2.00
	$10^{-3} b$	4.8	4.1	3.1	3.4	4.6	2.5	1.7
			$\log \kappa_t = a - b/(t + 273)$ at 1600–1800					
MoO_3–Na_2MoO_4	p	21.4	26.4	51.2	62.2	71.2	83.7	86.2
Ref. **54**	κ_{700}	1.06	1.07	0.86	0.80	0.75	0.63	—
	κ_{800}	1.45	1.41	1.17	1.11	1.08	0.75	0.91
	κ_{900}	—	—	—	—	—	—	1.16
Na–NaBr	p	0.25	0.5	0.75	1.0	1.5	2.0	
Ref. **73**	κ_{803}	4.0	6.3	7.4	8.2	—	—	
	κ_{894}	4.8	7.7	9.7	11.5	14.7	17.5	
Na–NaCl	p	0.25	0.5	0.75	1.0	1.5		
Ref. **73**	κ_{848}	4.3	4.9	5.3	5.7	6.4		
	κ_{893}	4.4	5.0	5.5	5.9	6.7		

(continued)

Table 9.5   ELECTRICAL CONDUCTIVITY OF MOLTEN BINARY SALT SYSTEMS AND OTHER MIXED IONIC MELTS—*continued*

Na–NaI Ref. 90						
p	0.4	0.6	1.0	1.3	1.6	2.0
κ_{700}	3.7	—	—	—	—	—
κ_{800}	4.5	3.5	6.5	—	—	—
κ_{900}	5.0	6.5	8.0	10.0	12.2	15.0

Na₃AlF₆–NaF‡ Ref. 65, 39, 34, 12							
p	0	20	40	60	80	90	100
$\kappa_{1\,000}$	5.4	4.9	4.4	3.9	3.3	3.1	2.8
$\kappa_{1\,050}$	5.7	5.1	4.6	4.0	3.5	3.2	2.9
$\kappa_{1\,100}$	5.9	5.4	4.8	4.2	3.6	3.3	3.1

NaBF₄–NaF Ref. 92						
p	40	50	60	70	80	90
κ_{500}	—	—	1.5	3.0	3.5	4.0
κ_{600}	—	3.0	5.0	6.0	6.5	7.0
κ_{700}	—	5.0	8.0	10.0	10.5	10.5
κ_{800}	6.5	12.0	14.0	15.0	15.5	15.0

NaBr–NaOH Ref. 70						
p	17.7	25.0	30.4	39.8	48.4	52.3
κ_{300}	1.18	1.14	1.03	0.95	0.83	0.79
κ_{400}	2.00	1.96	1.83	1.75	1.63	1.59

NaCl–Na₂CO₃ Ref. 6		
p	0	52.45	68.81
κ_{700}	—	2.2	2.3
κ_{850}	2.4	3.1	3.0
$\kappa_{1\,000}$	2.8	3.1	3.4

NaCl–NbCl₅ Ref. 20						
p	43.8	60.4	60.7	71.8	73.6	100
κ_{800}	1.32	2.20	2.19	—	—	3.34
κ_{850}	1.83	2.60	2.59	2.98	2.94	3.60

NaCl–PbCl₂ Ref. 17			
p	0	2.55	9.01	20.05
κ_{500}	1.445	1.51	1.58	1.645
κ_{600}	1.91	1.95	2.005	2.07

NaCl–TiCl₃ Ref. 88				
p	27.5	36.2	47.0	60.3	77.4
κ_{800}	1.75	1.75	1.9	2.3	2.7

NaCl–ZrCl₄ Ref. 20, 104						
p	11.5	34.9	59.7	75.1	88.1	88.7
κ_{350}	0.46	—	—	—	—	—
κ_{800}	—	1.45	1.74	2.41	2.88	2.68
κ_{825}	—	—	1.83	2.83	3.08	—
κ_{850}	—	—	1.97	3.01	3.34	3.36

NaF–SrF₂ Ref. 11					
p	—	—	33.3	—	—
t	900	950	1000	1050	1100
κ	4.441	4.798	4.961	5.407	5.642

NaI–NaOH Ref. 70			
p	21.5	35.0	44.8	52.8
κ_{300}	1.14	0.87	0.72	0.63
κ_{400}	1.90	1.53	1.32	1.23

NaNO₂–NaNO₃ Ref. 43						
p	0	21.3	44.8	65.4	82.1	100
κ_{300}	—	1.02	1.11	1.19	1.25	1.34
κ_{350}	1.16	1.27	1.34	1.44	1.51	1.62
κ_{400}	1.36	1.51	1.58	1.69	1.78	1.89
κ_{450}	1.55	1.73	1.79	1.92	2.02	2.10

Na₂O–TiO₂ Ref. 64							
p	28.3	41.7	44.6	51.7	60.0	65.5	71.6
$\kappa_{1\,000}$	—	0.32	0.38	0.47	0.56	0.70	0.86
$\kappa_{1\,100}$	0.32	0.45	0.56	0.80	0.90	0.94	1.15
$\kappa_{1\,200}$	0.74	1.03	1.16	1.55	1.76	2.10	2.54

Na₂WO₄–WO₃ Ref. 103							
p	45.9	50.9	55.7	65.1	74.2	83.0	91.6
a	4.4509	7.0013	4.5634	7.2754	3.1368	6.0743	2.7481
$10^2 b$	0.9625	1.5476	1.0326	1.7201	0.7399	1.5506	0.6718
$10^6 c$	4.0364	7.0320	4.4555	8.5600	2.4220	7.1943	2.4958
t range	780–923	750–888	767–905	741–881	689–850	653–803	701–869

$$\kappa_t = -a + bt - ct^2$$

Nd–NdBr₃ Ref. 89						
p	0.4	1.5	3.6	7.5	11.4	13.9	15.4
κ_{730}	0.53	0.61	0.79	0.94	1.01	0.96	0.92

(continued)

Table 9.5 ELECTRICAL CONDUCTIVITY OF MOLTEN BINARY SALT SYSTEMS AND OTHER MIXED IONIC MELTS—*continued*

Nd–NdCl$_3$	p	1.7	2.4	3.8	7.3	11.0	14.6	18.6
Ref. 85	κ_{855}	1.20	1.24	1.32	1.54	1.67	1.77	1.84
Nd–NdI$_3$	p	0.3	0.9	2.1	5.2	8.1	11.1	11.5
Ref. 86	κ_{820}	0.476	0.585	0.759	1.033	1.102	0.980	0.930
PbBr$_2$–PbCl$_2$	p	0	23.7	45.3	66.6	83.8	100	
Ref. 59, 3	κ_{450}	—	—	1.02	0.97	0.91	0.84	
	κ_{500}	1.43	1.34	1.25	1.17	1.10	1.03	
	κ_{550}	1.69	1.56	1.45	1.36	1.26	1.19	
PbO–SiO$_2$	p	61.4	71.2	78.8	84.8	89.7	93.7	97.1
Ref. 105	$\log \kappa_{900}$	−2.9	−2.0	−1.2	−0.7	−0.3	0.0	0.2
	$\log \kappa_{1\,150}$	−1.9	−1.2	−0.6	−0.2	0.0	0.2	0.6
Pr–PrBr$_3$	p	0.5	1.1	2.5	4.1	5.8	8.6	8.7
Ref. 89	κ_{740}	0.66	0.80	1.10	1.47	1.90	2.38	2.67
Pr–PrI$_3$	p	0.3	1.3	2.6	4.2	6.3	8.1	8.9
Ref. 86	κ_{780}	0.61	1.28	2.48	4.24	8.25	14.8	21.0
RbCl–TiCl$_3$	p	43.9	54.0	64.6	75.8	87.6		
Ref. 88	κ_{800}	0.95	0.95	1.0	1.2	1.4		
VCl$_2$ in KCl–NaCl	p	9.6	12.8	17.9	22.4	29.7	39.7	52.0
(equi-molar:	κ_{600}	0.22	0.48	2.34 (610)	1.30	1.18 (605)	1.46	0.050
56.1% KCl)	κ_{800}	3.68 (810)	3.77 (790)	3.94 (790)	3.94	3.85	3.68 (795)	2.93
Ref. 106								

*See also Na$_3$AlF$_6$–NaF.

**See also 'Physical Properties Data Compilations Relevant to Energy Storage II Molten Salts: Data on single and Multi-Component Salt Systems', Janz *et al.*, NSRDS–NBS 61, Part II.

† Melts may contain up to 1% MgO.

‡ See also AlF$_3$–Na$_3$AlF$_6$.

REFERENCES TO TABLE 9.5

1. H. M. Goodwin and R. D. Mailey, *Phys. Rev.*, 1908, **26**, 28.
2. V. A. Izbekov and V. A. Plotnikov, *J. Russ. Phys. -Chem. Soc.*, 1911, **43**, 18.
3. C. Sandonnini, *Gazz. Chim. Ital.*, 1920, **50**, 289.
4. K. Arndt and W. Kalass, *Z. Elektrochem.*, 1924, **30**, 12.
5. V. A. Plotnikov and P. T. Kalita, *J. Russ. Phys. -Chem. Soc.*, 1930, **62**, 2195.
6. E. Ryschkewitsch, *Z. Elektrochem.*, 1933, **39**, 531.
7. S. V. Karpachev, A. G. Stromberg and O. Poltoratzkaya, *J. Phys. Chem.* (*USSR*), 1934, **5**, 793.
8. V. P. Barzakovskii, Proc. 1st All-Union Conf. on Non-Aqueous Solutions, 1935, 153.
9. K. P. Batashev, *Metallurg.* (*USSR*), 1935, **10**, 100.
10. S. V. Karpachev, A. G. Stromberg and V. N. Podchainova, *J. Gen. Chem.* (*USSR*), 1935, **5**, 1517.
11. M. de Kay Thompson and A. L. Kaye, *Trans. Electrochem. Soc.*, 1935, **67**, 169.
12. K. P. Batashev, *Legkie Metal.*, 1936, **10**, 48, quoted by V. P. Mashovets, 'The Electrometallurgy of Aluminium', 1938.
13. Ya. P. Mezhennii, *Mém. Inst. Chem., Acad. Sci. Ukrain S.S.R.*, 1938, **4**, 413.
14. Y. Yamaguti and S. Sisido, *J. Chem. Soc. Japan*, 1938, **59**, 1311.
15. A. I. Kryagova, *J. Gen. Chem.* (*USSR*), 1939, **9**, 2061.
16. A. A. Sherbakov and B. F. Markov, *J. Phys. Chem.* (*USSR*), 1939, **13**, 621.
17. V. P. Barzakovskii. *Bull. Acad. Sci. U.R.S.S.*, Class Sci. Chim., 1940, 825.
18. V. P. Barzakovskii, *J. Appl. Chem.* (*USSR*), 1940, **13**, 1117.
19. A. G. Bergman and I. M. Chagin, *Bull. Acad. Sci. U.R.S.S.*, Class Sci. Chim., 1940, 727.
20. N. A. Belozerskii and B. A. Freidlina, *J. Appl. Chem.* (*USSR*), 1941, **14**, 466.
21. Y. Yamaguti and S. Sisido, *J. Chem. Soc. Japan*, 1941, **62**, 304.
22. E. R. Natsvilishvili and A. G. Bergman, *Bull Acad. Sci., U.R.S.S.*, Class Sci. Chim., 1943, 23.
23. E. Ya. Gorenbein, *J. Gen. Chem.* (*USSR*), 1945, **15**, 729.
24. H. Bloom and E. Heymann, *Proc. R. Soc.*, 1947, **188***A*, 392.
25. E. Ya. Gorenbein, *J. Gen. Chem.* (*USSR*), 1947, **17**, 873.
26. T. G. Pearson and J. Waddington, *Faraday Soc. Discussion*, 1947, No. 1, 307.
27. N. M. Tarasova, *J. Phys. Chem.* (*USSR*), 1947, **21**, 825.
28. E. Ya. Gorenbein, *J. Gen. Chem.* (*USSR*), 1948, **18**, 1427.
29. I. I. Bokhovkin, *ibid.*, 1949, **19**, 805.
30. E. Ya. Gorenbein and E. E. Kriss, *ibid.*, 1949, **19**, 1978.

31. H. Grothe, *Z. Elektrochem.*, 1949, **53**, 362.
32. I. N. Belyaev and K. E. Mironov, *Dokl. Akad. Nauk SSSR*, 1950, **73**, 1217.
33. I. I. Bokhovkin, *J. Gen. Chem. (USSR)*, 1950, **20**, 397.
34. A. Vayna, *Alluminio*, 1950, **19**, 215.
35. E. Ya. Gorenbein and E. E. Kriss, *J. Phys. Chem. (USSR)*, 1951, **25**, 791.
36. R. C. Spooner and F. E. W. Wetmore, *Canad. J. Chem.*, 1951, **29**, 777.
37. R. W. Huber, E. V. Potter and H. W. St. Clair, *Rep. Invest. US Bur. Mines, No. 4858*, 1952.
38. J. Byrne, H. Fleming and F. E. W. Wetmore, *Canad. J. Chem.*, 1952, **30**, 922.
39. J. D. Edwards, C. S. Taylor, L. A. Cosgrove and A. S. Russell, *Trans. Electrochem. Soc.*, 1953, **100**, 508.
40. A. Bogorodski, *J. Soc. Phys.-Chim. Russe*, 1905, **37**, 796.
41. K. Mori and Y. Matsushita, *Tetsu to Hagane*, 1952, **38**, 365.
42. L. Shartsis, W. Capps and S. Spinner, *J. Am. Ceram. Soc.*, 1953, **36**, 319.
43. H. Bloom, I. W. Knaggs, J. J. Molloy and D. Welch, *Trans. Faraday Soc.*, 1953, **49**, 1458.
44. I. N. Belyaev, *Invest. Sekt. Fiziko-Khim. Anal.*, 1953, **23**, 176.
45. P. I. Protsenko and N. P. Popovskaya, *Zh. Obshch. Khim.*, 1954, **24**, 2119.
46. P. I. Protsenko and N. P. Popovskaya, *Zh. Fiz. Khim.*, 1954, **28**, 299.
47. K. Sakai, *J. Chem. Soc. Japan, Pure Chem. Sect.*, 1954, **75**, 186.
48. N. P. Luzhnaya and I. P. Vereshchetina, *Izvest. Sekt. Fiziko-Khim. Anal.*, 1954, **24**, 192.
49. I. P. Vereshchetina and N. P. Luzhnava, *ibid.*, 1954, **25**, 188.
50. L. Shartsis and H. F. Shermer, *J. Am. Ceram. Soc.*, 1954, **37**, 544.
51. V. D. Polyakov, *Izvest. Sekt. Fiziko-Khim. Anal.*, 1955, **26**, 147.
52. P. I. Protsenko, *ibid.*, 1955, **26**, 173.
53. V. D. Polyakov, *ibid.*, 1955, **26**, 191.
54. K. B. Morris, M. I. Cook, C. Z. Sykes and M. B. Templeman, *J. Am. Chem. Soc.*, 1955, **77**, 851.
55. T. Baak, *Acta. Chem. Scand.*, 1955, **9**, 1406.
56. K. Sakai and S. Hayashi, *J. Chem. Soc. Japan, Pure Chem. Sect.*, 1955, **76**, 101.
57. E. R. van Artsdalen and I. S. Yaffe, *J. Phys. Chem.*, 1955, **59**, 118.
58. Yu. K. Delimarskii, I. N. Sheiko and V. G. Fenchenko, *Zh. Fiz. Khim.*, 1955, **29**, 1499.
59. B. S. Harrap and E. Heymann, *Trans. Faraday Soc.*, 1955, **51**, 259.
60. G. M. Pound, G. Derge and G. Osuch, *Trans. Amer. Inst. Min. Met. Eng.*, 1955, **203**, 481.
61. H. C. Cowen and H. J. Axon, *ibid.*, 1956, **52**, 242.
62. N. P. Luzhnaya, N. N. Evseeva and I. P. Vereshchetina, *Zh. Neorg. Khim.*, 1956, **1**, 1490.
63. K. Mori, *Tetsu to Hagane*, 1956, **42**, 1024.
64. B. M. Lepinskikh, O. A. Esin and S. V. Sharrin, *Zh. Priklad, Khin.*, 1956, **29**, 1813.
65. E. W. Yim and M. Feinleib, *J. Electrochem. Soc.*, 1957, **104**, 626.
66. F. R. Duke and R. A. Fleming, *ibid.*, 1957, **104**, 251.
67. J. B. Story and J. T. Clarke, *Trans. Amer. Inst. Min. Met. Eng.*, 1957, **209**, 1449.
68. V. A. Kochinashvili and V. P. Barzakovskii, *Zh. Priklad. Khim.*, 1957, **30**, 1755.
69. W. C. Phelps and R. E. Grace, *Trans. Amer. Inst. Min. Met. Eng.*, 1957, **209**, 1447.
70. S. Okado, S. Yashizawa, N. Watanabe and Y. Omota, *J. Chem. Soc. Japan, Ind. Chem. Sect.*, 1957, **60**, 670.
71. F. R. Duke and R. A. Fleming, *J. Electrochem. Soc.*, 1958, **105**, 412.
72. E. Vatslavik and A. I. Belyaev, *Zh. Neorg. Khim.*, 1958, **3**, 1044.
73. H. R. Bronstein and M. A. Bredig, *J. Am. Chem. Soc.*, 1958, **80**, 2077.
74. O. A. Esin and V. L. Zyazcv, *Izvest. Akad. Nauk, S.S.S.R.*, 1958, No. 6, 7.
75. K. A. Kostanyan, *Izvest. Akad. Nauk. Armyan. S.S.S.R.*, 1958, **11**, 65.
76. R. H. Moss, *Univ. Microfilms (Ann. Arbor, Mich.)*, 1955, No. 12, 730.
77. Y. Doucet and M. Bizouard, *Compt. Rend.*, 1960, **250**, 73.
78. B. F. Markov and V. D. Prusyazhnyi, *Ukr. Khim. Zh.*, 1962, **28**, 653.
79. Y. Doucet and M. Bizouard, *Compt. Rend.*, 1959, **248**, 1328.
80. V. P. Mashovets and V. I. Petrov, *Zh. Prikl. Khim.*, 1959, **32**, 1528.
81. J. O'M. Bockris, J. A. Kitchener, S. Ignatowitz and J. W. Tomlinson, *Discuss. Faraday Soc.*, 1948, **4**, 265.
82. P. I. Protsenko and O. N. Shokina, *Zh. Neorg. Khim.*, 1960, **5**, 437.
83. P. I. Protensko and A. Ya Malakhova, *ibid.*, 1960, **5**, 2307.
84. L. F. Grantham and S. J. Yosim, *J. Chem. Phys.*, 1963, **38**, 1671.
85. A. S. Dworkin, H. R. Bronstein and M. A. Bredig, *Discuss. Faraday Soc.*, 1961, **32**, 188.
86. A. S. Dworkin, R. A. Sallach, H. R. Bronstein, M. A. Bredig and J. D. Corbett, *J. Phys. Chem.*, 1963, **67**, 1145.
87. B. F. Markov and V. D. Prusyazhnyi, *Ukr. Khim. Zh.*, 1962, **28**, 268.
88. R. V. Chernov and Yu K. Delimarskii, *Zh. Neorg. Khim.*, 1961, **6**, 2749.
89. A. S. Dworkin, H. R. Bronstein and M. A. Bredig, *J. Phys. Chem.*, 1963, **67**, 2715.
90. H. R. Bronstein and M. A. Bredig, *ibid.*, 1961, **65**, 1220.
91. H. R. Bronstein, A. S. Dworkin and M. A. Bredig, *J. Chem. Phys.*, 1961, **34**, 1843.
92. V. G. Selivanov and V. V. Stender, *Zh. Neorg. Khim.*, 1959, **4**, 2058.
93. B. F. Markov and V. D. Prusyazhnyi, *Ukr. Khim. Zh.*, 1962, **28**, 130.
94. B. F. Markov and V. D. Prusyazhnyi, *ibid.*, 1962, **28**, 419.
95. H. R. Bronstein, A. S. Dworkin and M. A. Bredig, *J. Phys. Chem.*, 1962, **66**, 44.
96. G. W. Mellors and S. Senderoff, *ibid.*, 1960, **64**, 294.
97. E. R. Buckle and P. E. Tsaoussoglou, *Trans. Faraday Soc.*, 1964, **60**, 2144.
98. M. F. Lantratov and O. F. Moiseeva, *Zh. Fiz. Khim.*, 1960, **34**, 367.

99. V. V. Rafal'skii, *Ukr. Khim. Zh.*, 1960, **26**, 585.
100. K. B. Morris and P. L. Robinson, *J. Phys. Chem.*, 1964, **68**, 1194.
101. P. C. Papaioannou and G. W. Harrington, *ibid.*, 1964, **68**, 2424.
102. H. Winterhager and L. Werner, *ForschBer. Wirt.-u. Verk Minist. N Rhein.-Westf.*, 1957, **438.**
103. K. B. Morris and P. L. Robinson, *J. Chem. Engng Data*, 1964, **9**, 444.
104. L. J. Howell and H. H. Kellogg, *Trans. Am. Inst. Min. Engrs*, 1959, **215**, 143.
105. J. O'M. Bockris and G. W. Mellors, *J. Phys. Chem.*, 1956, **60**, 1321.
106. Yu. U. Samson, L. P. Ruzinov, N. S. Rezhemnukova and V. E. Baru, *Zh. Fiz. Khim.*, 1964, **38**, 481.

Table 9.6   SURFACE TENSION OF PURE MOLTEN SALTS

The surface tension (mNm^{-1}) at temperature t (°C) is given as γ_t, or the constants a, b and t_0 in the equatior $\gamma_t = a - b(t - t_0)$ are given for the temperature range r. Principal references are in **bold type**.

AgBr	γ_{500}	152	Ca(NO$_3$)$_2$	γ_{560}	101.5
Ref. **16**, 1	γ_{600}	148	Ref. **22**		
	γ_{700}	146	CdBr$_2$	γ_{600}	67
AgCl	γ_{500}	176	Ref. **16**	γ_{700}	65
Ref. **16**, 5	γ_{600}	171	CdCl$_2$	γ_{600}	83
	γ_{700}	166	Ref. **16**	γ_{700}	81
AgI	γ_{600}	115		γ_{800}	78
Ref. **16**			CsBr	γ_{650}	82.5
AgNO$_3$	a	152.1	Ref. **28**, 29	γ_{800}	73.35
Ref. **22**, 27, 19	b	0.082		$\gamma_{1\,000}$	61.147
	t_0	212	CsCl	γ_{700}	87.50
	r	244–400	Ref. **29**, 6, 12	γ_{900}	71.18
Al$_2$O$_3$	$\gamma_{2\,050}$	700		$\gamma_{1\,100}$	54.86
Ref. **28**			CsF	γ_{700}	106.1
B$_2$O$_3$	γ_{700}	67	Ref. **28**, 29	γ_{900}	92.26
Ref. **21**	$\gamma_{1\,000}$	83		$\gamma_{1\,100}$	78.46
	$\gamma_{1\,300}$	99	CsI	γ_{650}	73.47
BaBr$_2$	γ_{900}	150	Ref. **28**, 29	γ_{800}	64.77
Ref. **20**	γ_{950}	147		$\gamma_{1\,000}$	53.17
	$\gamma_{1\,000}$	143	CsNO$_3$	a	92.5
BaCl$_2$	γ_{966}	169.2	Ref. **22**, 6	b	0.069
Ref. **23**, 15, 12, 9, 2	γ_{979}	168.3		t_0	414
	γ_{984}	168.6		r	421–597
	γ_{996}	168.1	Cs$_2$SO$_4$	$\gamma_{1\,000}$	114.25
	$\gamma_{1\,000}$	167.8	Ref. **30**	$\gamma_{1\,250}$	98.84
	$\gamma_{1\,050}$	164		$\gamma_{1\,500}$	83.75
Ba(NO$_3$)$_2$	a	134.8	FeO	$\gamma_{1\,400}$	584.0
Ref. **22**	b	0.015	Ref. **28**		
	t_0	595	GaCl$_2$	a	56.6
	r	600–660	Ref. **24**	b	0.18
BiBr$_3$	γ_{250}	66		t_0	170
Ref. **6**	γ_{350}	56		r	166–170
	γ_{450}	45	GaCl$_3$	γ_{80}	27
BiCl$_3$	γ_{275}	67	Ref. **17**	γ_{110}	24
Ref. **6**	γ_{325}	59		γ_{140}	21
	γ_{375}	53	GeO$_2$	$\gamma_{1\,000}$	241
BrF$_3$	γ_{12}	37	Ref. **21**	$\gamma_{1\,200}$	253
Ref. **18**	γ_{27}	36		$\gamma_{1\,400}$	265
	γ_{45}	34	HgBr2	γ_{241}	64
BrF$_5$	γ_9	24	Ref. **8**	γ_{276}	60
Ref. **18**	γ_{33}	22	HgCl$_2$	γ_{293}	56
CaBr$_2$	γ_{780}	117.31	Ref. **8**		
Ref. **20**	γ_{800}	116.39	IF$_5$	γ_{18}	31
CaCl$_2$	γ_{800}	150.57	Ref. **18**	γ_{28}	28
Ref. **28**, 29	$\gamma_{1\,000}$	140.07	KBr	γ_{750}	90
CaI$_2$	γ_{800}	85	Ref. **16**, 11, 6, 27	γ_{800}	86
Ref. **20**	γ_{900}	83		γ_{850}	82
	$\gamma_{1\,000}$	81		γ_{900}	78

(continued)

Table 9.6   SURFACE TENSION OF PURE MOLTEN SALTS—*continued*

KCl	γ_{800}	98	NaCl	γ_{800}	116.42
Ref. **19, 16, 15, 12, 9, 6**, 27	γ_{900}	90	Ref. **6**, 27	γ_{900}	107.12
	$\gamma_{1\,000}$	82		$\gamma_{1\,000}$	97.82
	$\gamma_{1\,100}$	75		$\gamma_{1\,100}$	88.52
$K_2Cr_2O_7$	γ_{420}	140.28	NaF	$\gamma_{1\,000}$	185.21
Ref. **30**	γ_{470}	138.28	Ref. **6**, 31	$\gamma_{1\,200}$	168.81
	γ_{520}	136.28		$\gamma_{1\,400}$	152.41
KF	γ_{900}	138.12		$\gamma_{1\,500}$	144.21
Ref. **28, 29**, 30	$\gamma_{1\,100}$	123.12	NaI	a	147.4
	$\gamma_{1\,300}$	108.12	Ref. **27**	b	0.090
KI	a	138.7		t_0	0
Ref. **27**	b	0.087		r	(m.p. + 10)–
	t_0	0			(m.p. + 210)
	r	(m.p. + 10)–	Na_2MoO_4	γ_{700}	211.68
		(m.p. + 210)	Ref. **30**	γ_{900}	196.28
K_2MoO_4	γ_{900}	151.47		$\gamma_{1\,100}$	180.88
Ref. **30**	$\gamma_{1\,200}$	132.87		$\gamma_{1\,200}$	173.18
	$\gamma_{1\,500}$	114.27	$NaNO_2$	a	121.2
KNO_2	a	107.6	Ref. **22**, 27	b	0.041
Ref. **22**	b	0.080		t_0	277
	t_0	435		r	291–384
	r	445–501	$NaNO_3$	γ_{310}	116.21
KNO_3	γ_{350}	108.61	Ref. **27**	γ_{400}	112.70
Ref. **27**	γ_{450}	102.21		γ_{500}	108.80
	γ_{600}	92.61		γ_{600}	104.90
	γ_{800}	79.81	$NaPO_3$	γ_{800}	200.74
KPO_3	γ_{900}	193.54	Ref. **6**, 30	$\gamma_{1\,000}$	186.34
Ref. **30**	$\gamma_{1\,200}$	171.94		$\gamma_{1\,200}$	171.94
	$\gamma_{1\,500}$	150.34		$\gamma_{1\,500}$	150.34
K_2SO_4	$\gamma_{1\,100}$	143	Na_2SO_4	γ_{900}	193
Ref. **13**	$\gamma_{1\,300}$	130	Ref. **13**, 6	γ_{950}	190
	$\gamma_{1\,500}$	116		$\gamma_{1\,000}$	186
K_2WO_4	γ_{900}	162.41		$\gamma_{1\,050}$	183
Ref. **30**	$\gamma_{1\,200}$	134.51	Na_2WO_4	γ_{700}	201.46
	$\gamma_{1\,500}$	106.61	Ref. **6**, 30	$\gamma_{1\,000}$	182.0
La_2O_3	$\gamma_{2\,320}$	560		$\gamma_{1\,300}$	162.46
Ref. **7**				$\gamma_{1\,600}$	143.0
LiCl	γ_{600}	137.14	$PbCl_2$	γ_{500}	137.12
Ref. **28, 29**	γ_{800}	123.22	Ref. **27**	γ_{600}	126.12
	$\gamma_{1\,000}$	109.30		γ_{700}	115.12
LiF	γ_{850}	250.46	P_2O_5	γ_{100}	60
Ref. **28, 29**	$\gamma_{1\,050}$	228.60	Ref. **21**	γ_{200}	58
	$\gamma_{1\,250}$	206.74		γ_{400}	54
$LiNO_2$	a	115.4	RbBr	γ_{700}	99.77
Ref. **22**, 6	b	0.053	Ref. **28, 29**	γ_{900}	88.17
	t_0	255		$\gamma_{1\,100}$	76.57
	r	276–425	RbCl	γ_{750}	94.90
Li_2SO_4	γ_{900}	223	Ref. **28, 29**	γ_{900}	82.49
Ref. **13**, 6	$\gamma_{1\,000}$	216		$\gamma_{1\,000}$	74.22
	$\gamma_{1\,100}$	209		$\gamma_{1\,150}$	61.82
	$\gamma_{1\,200}$	202	RbF	γ_{800}	163.83
NH_4NO_3	a	101.9	Ref. 28, 29	γ_{900}	153.60
Ref. **22**	b	0.105		$\gamma_{1\,000}$	143.37
	t_0	170	RbI	γ_{700}	76.86
	r	170–220	Ref. **28, 29**	γ_{900}	63.18
Na_3AlF_6	$\gamma_{1\,000}$	134.06		$\gamma_{1\,000}$	56.34
Ref. **31**	$\gamma_{1\,010}$	132.78	$RbNO_3$	γ_{300}	109.53
	$\gamma_{1\,020}$	131.50	Ref. **32**	γ_{500}	95.53
NaBr	γ_{750}	106		γ_{700}	81.53
Ref. **16**, 6, 27	γ_{800}	103	Rb_2SO_4	$\gamma_{1\,100}$	130.8
	γ_{950}	92	Ref. **30**	$\gamma_{1\,300}$	120.4
	$\gamma_{1\,150}$	79		$\gamma_{1\,500}$	110

(continued)

Table 9.6 SURFACE TENSION OF PURE MOLTEN SALTS—*continued*

SiO_2	$\gamma_{1\,000}$	278	$Sr(NO_3)_2$	γ_{615}	128.4
Ref. **21**	$\gamma_{1\,500}$	295	Ref. **22**		
	$\gamma_{2\,000}$	313	$TINO_3$	a	94.8
$SnCl_2$	γ_{300}	96.74	Ref. **22**, 6	b	0.078
Ref. **28, 29**	γ_{400}	88.44		t_0	206
	γ_{500}	80.14		r	226–458
$SrBr_2$	γ_{700}	147	Tl_2S	a	215.6
Ref. **20**	γ_{800}	143	Ref. **25**	b	0.035 6
	γ_{900}	138		t_0	445
	$\gamma_{1\,000}$	134		r	500–700
$SrCl_2$	γ_{900}	168	V_2O_5	$\gamma_{1\,000}$	86
Ref. **20**	γ_{950}	165	Ref. **26**		
	$\gamma_{1\,000}$	162			
SrI_2	γ_{600}	112			
Ref. **20**	γ_{750}	106			
	γ_{850}	102			
	γ_{950}	98			

REFERENCES TO TABLE 9.6

1. A. Gradenwitz, *Ann. Physik.*, 1899, **67**, 467.
2. Z. Motylewski, *Z. Anorg. Chem.*, 1904, **38**, 410.
3. R. Lorenz and F. Kaufler, *Ber. Dt. Chem. Ges.*, 1908, **41**, 3727.
4. R. Lorenz and A. Liebmann, *Z. Phys. Chem.*, 1913, **83**, 459.
5. R. Lorenz, A. Liebmann and A. Hochberg, *Z. Anorg. Chem.*, 1916, **94**, 301.
6. F. M. Jaeger, *Z. Anorg. Chem.*, 1917, **101**, 1.
7. H. V. Wartenberg, G. Wehner and E. Suran, *Nachr. Ges. Wiss. Göttingen*, 1936, **2**, 65.
8. E. B. R. Prideaux and J. R. Jarrett, *J. Chem. Soc.*, 1938, 1203.
9. V. P. Barzakovskii, *Bull. Acad. Sci. URSS. Classe Sci. Chim.*, 1940, 825.
10. P. P. Kozakevich and A. F. Kononenko, *J. Phys. Chem. (USSR)*, 1940, **14**, 1118.
11. K. Semenchenko and L. P. Shikhobolova, *ibid.*, 1947, **21**, 613.
12. K. Semenchenko, *ibid.*, 1947, **21**, 707.
13. K. Semenchenko, *ibid.*, 1947, **21**, 1387.
14. A. Vajna, *Alluminio*, 1951, **20**, 29.
15. J. S. Peake and M. R. Bothwell, *J. Am. Chem. Soc.*, 1954, **76**, 2625.
16. N. K. Boardman, A. R. Palmer and E. Heymann, *Trans. Faraday Soc.*, 1955, **51**, 277.
17. N. N. Greenwood and K. Wade, *J. Inorg. Nuclear Chem.*, 1957, **3**, 349.
18. M. T. Rogers and E. E. Carver. *J. Phys. Chem.*, 1958, **62**, 952.
19. J. L. Dahl and F. R. Duke, *US Atomic Energy Comm.*, 1958, ISC-923.
20. R. B. Ellis, J. E. Smith and E. B. Baker, *J. Phys. Chem.*, 1958, **62**, 766.
21. W. D. Kingery, *J. Am. Ceram. Soc.*, 1959, **42**, 6.
22. C. C. Addison and J. M. Coldrey, *J. Chem. Soc.*, 1961, 468.
23. I. D. Sokolova and N. K. Voskresenskaya, *Zh. Prikl. Khim.*, 1962, **36**, 955.
24. N. N. Greenwood and I. J. Worrall, *J. Chem. Soc.*, 1958, 1680.
25. V. B. Lazarev and M. N. Abdusalyamova, *Izv. Akad. Nauk SSSR, Ser. Khim.*, 1964, 1104.
26. B. M. Lepinskikh, O. A. Esin and G. A. Teterin, *Zh. Neorg. Khim.*, 1960, **5**, 642.
27. H. Bloom, F. G. Davis and D. W. James, *Trans. Faraday Soc.*, 1960, **56**, 1179.
28. O. K. Sokolov, *Izv. Akad. Nauk SSSR Met. Gorn. Delv*, 1963, **4**, 59.
29. R. B. Ellis and W. S. Wilcox, Work performed under U. S. At. Energy Comm; T-10-7622, 1962, pp. 128–36.
30. 'International Critical Tables', McGraw-Hill, New York, 1933.
31. H. Bloom and B. W. Burrows, 'Proc 1st Australian Conf. Electrochem' (J. A. Friend and F. Gutman eds), p. 882. Pergamon Press, Oxford, 1964.
32. S. D. Gromakov and A. I. Kostromin, *Univ. in V. I U P Yanova-Lenina Khim.*, 1955, **115**, 93.

Table 9.7   SURFACE TENSION OF MOLTEN BINARY SALT SYSTEMS AND OTHER MIXED IONIC MELTS

The surface tension (γ in mN m^{-1}) at temperature t (°C) and composition p (wt.%) of the first-named constituent is given as γ_t (or σ_t), or the constants a, b and t_0 in the equation $\gamma_t = a - b(t - t_0)$ are given for the temperature range r. Principal references are in **bold type**.

AgBr–AgCl	p	0	24.7	46.7	66.4	84.0	100
Ref.16	σ_{500}	176	169	164	158	155	152
	σ_{600}	171	164	160	155	152	148
AgBr–AgI	p	0	16.7	34.8	54.5	76.2	100
Ref. 16	σ_{500}	—	122	127	133	141	152
	σ_{600}	115	118	123	130	138	148
AgBr–KBr	p	0	28.3	51.3	70.3	86.3	100
Ref. 16	σ_{700}	—	96	99	104	113	146
	σ_{750}	90	92	96	101	110	146
AgCl–KCl	p		32.6	56.3	72.6	88.6	100
Ref. 16	σ_{600}		112	117	123	136	171
	σ_{700}		105	110	117	129	166
AgCl–PbCl$_2$	p	0	11.4	25.6	43.7	67.4	100
Ref. 16	σ_{500}	137	139	144	150	160	176
	σ_{600}	127	129	134	142	152	171

AgNO$_3$–CsNO$_3$	p	22.5	46.5	72.4	88.8			
Ref. 19	a	126.1	130.3	140.1	148.4			
	b	0.075	0.072	0.073	0.072			
	t_0	0	0	0	0			
	r	liquidus–c.	400					
AgNO$_3$–KNO$_3$	p	20	40	60	80	95	100	
Ref. 18, 24. 19	σ_{250}	—	—	124	129	139	147	
	σ_{300}	—	118	121	126	136	144	
	σ_{350}	113	114	117	122	133	140	
AgNO$_3$–LiNO$_3$	p	45.0	71.2	88.1	100			
Ref. 19	a	137.8	145.7	153.5	163.7			
	b	0.064	0.068	0.067	0.066			
	t_0	0	0	0	0			
	r	liquidus–c.	400					
AgNO$_3$–NaNO$_3$	p	24.1	33.3	60.0	75.0	90.0	95.0	
Ref. 18, 24, 19	σ_{300}	—	—	126	130	136	137	
	σ_{350}	119	120	123	127	133	135	
	σ_{400}	115	116	—	—	—	—	
AgNO$_3$–RbNO$_3$	p	27.8	53.5	77.7	91.3			
Ref. 19	a	134.0	136.9	143.4	151.7			
	b	0.077	0.073	0.070	0.069			
	t_0	0	0	0	0			
	r	liquidus–c.	400					

AlF$_3$–Na$_3$AlF$_6$	p	2.5	5	10	15			
Ref. 11, 1	$\sigma_{1\,000}$	142	137	127	118.			
Al$_2$O$_3$–FeO	p	5.6	6.9					
Ref. 4	$\sigma_{1\,410}$	598	604					
Al$_2$O$_3$–Na$_3$AlF$_6$	p	2.5	5	10	15			
Ref. 11, 1	$\sigma_{1\,000}$	143	140	136	134			
B$_2$O$_3$–BaO	p	20.5	34.6	44.4	54.7	64.2	75.1	84.5
Ref. 14	σ_{900}	—	—	239	197	146	88	79
	$\sigma_{1\,100}$	315	257	229	187	138	92	87
	$\sigma_{1\,300}$	294	243	221	181	135	106	96
B$_2$O$_3$–CaO	p	51.2	62.8	73.3	83.1	93.7	98.1	
Ref. 14	σ_{900}	—	—	176	—	81	81	
	$\sigma_{1\,100}$	302	235	150	87	87	88	
	$\sigma_{1\,300}$	297	221	140	96	95	96	
B$_2$O$_3$–K$_2$O	p	51.5	55.7	65.6	75.3	84.9	94.8	98.5
Ref.13	σ_{700}	—	—	—	140	107	80	75
	σ_{900}	139	153	160	143	112	85	80
	$\sigma_{1\,100}$	—	—	149	142	118	93	88

(continued)

Table 9.7　SURFACE TENSION OF MOLTEN BINARY SALT SYSTEMS AND OTHER MIXED IONIC MELTS—*continued*

System	Prop							
B_2O_3–Li_2O	p	71.3	79.9	87.1	92.0	95.5	100	
Ref. 13	σ_{700}	—	221	—	120	84	74	
	σ_{900}	231	207	158	121	92	80	
	σ_{1100}	—	—	152	125	101	88	
B_2O_3–Na_2O	p	64.0	73.6	82.2	85.8	94.4	99.1	
Ref. 13	σ_{700}	—	196	146	—	83	77	
	σ_{900}	197	185	147	130	91	81	
	σ_{1100}	—	—	149	134	99	89	
B_2O_3–PbO	p	0	15.9	33.6	50.9	65.6	78.4	100
Ref. 10, 8	σ_{700}	—	162	160	—	—	—	—
	σ_{900}	132	163	144	92	78	79	80
	σ_{1000}	135	163	—	—	—	—	83
B_2O_3–SrO	p	35.4	49.6	61.6	70.0	89.0	96.0	
Ref. 14	σ_{900}	—	194	137	80	—		
	σ_{1100}	303	247	178	123	87	86	
	σ_{1300}	292	237	166	122	95	93	
B_2O_3–ZnO	p	20.5	29.4	35.4	41.4	50.0	90.0	100
Ref. 9	σ_{900}	—	—	235	184	80	78	80
	σ_{1100}	—	283	232	179	88	86	86
	σ_{1300}	355	287	238	185	—	95	94
$BaCl_2$–KCl	p	0	20.4	47.4	59.0	67.2	82.2	92.8
Ref. 15	σ_{850}	93	98	106	110	114	129	—
	σ_{900}	89	94	103	106	112	128	144
	σ_{950}	—	—	—	—	—	121	141
$BaCl_2$–Li_2SO_4	p	0	9.0	17.3	44.6	65.3	84.9	100
Ref. 6	σ_{1000}	220	198	192	167	163	161	175
	σ_{1050}	216	190	180	164	158	159	172
$BaCl_2$–$NaCl$	p	0	28.4	47.1	65.7	78.1	93.4	100
Ref. 3, 20	σ_{900}	107	109	117	126	133	160	—
	σ_{1000}	100	103	111	120	126	154	171
$CaCl_2$–$NaCl$*	p	0	17.4	44.8	65.5	88.3	100	
Ref. 3	σ_{800}	114	115	121	126	137	148	
	σ_{1000}	100	100	108	116	125	140	
CaF_2–Na_3AlF_6	p	2.5	5	10	15			
Ref. 11	σ_{1000}	149	150	152	155			
CaO–FeO	p	0	5.5	15.4	22.3			
Ref. 4	σ_{1410}	585	555	543	573			
$CdBr_2$–$CdCl_2$	p	0	20.9	41.3	61.7	80.9	100	
Ref. 16	σ_{600}	83	78	74	71	68	67	
	σ_{700}	81	76	72	68	66	65	
$CdCl_2$–KCl	p		38.1	62.1	78.7	90.7	100	
Ref. 16	σ_{600}		100	93	90	88	82	
	σ_{700}		94	87	84	84	81	
$CdCl_2$–$NaCl$	p	0	44.0	67.6	82.5	92.6	100	
Ref. 16	σ_{700}	—	108	98	90	85	81	
	σ_{800}	114	101	92	85	81	78	
$CdCl_2$–$PbCl_2$	p	0	14.1	31.8	49.7	72.4	100	
Ref. 16	σ_{600}	126	116	107	98	90	83	
	σ_{700}	116	108	101	93	87	81	
Cr_2O_3–FeO	p	0	3.1					
Ref. 4	σ_{1420}	585	588					
$CsCl$–Li_2SO_4	p	0	1.6	3.2	4.8	15.2	61.8	100
Ref. 7, 6	σ_{900}	224	205	193	189	163	102	72
	σ_{1000}	220	201	188	185	160	92	64
	σ_{1100}	211	194	181	177	154	—	—
$CsCl$–$PbCl_2$	p	12.4	27.4	38.6	49.0	64.7		
Ref. 18	σ_{500}	116	—	—	—	103		
	σ_{600}	—	97	—	93	94		
	σ_{625}	—	94	91	90	92		

(continued)

Table 9.7   SURFACE TENSION OF MOLTEN BINARY SALT SYSTEMS AND OTHER MIXED IONIC MELTS—*continued*

CsNO₃–KNO₃ Ref. 19						
p	39.1	66.0	85.3			
a	133.7	129.4	125.1			
b	0.079	0.077	0.074			
t_0	0	0	0			
r		liquidus–c.	400			

CsNO₃–LiNO₃ Ref. 19						
p	48.5	73.9	89.4			
a	124.3	124.3	122.9			
b	0.070	0.076	0.075			
t_0	0	0	0			
r		liquidus–c.	400			

CsNO₃–NaNO₃ Ref. 19						
p	43.1	67.6	87.3			
a	130.5	127.3	123.4			
b	0.068	0.074	0.072			
t_0	0	0	0			
r		liquidus–c.	400			

Cs₂O–SiO₂ Ref. 21					
p	47.2	54.0	69.0	78.9	
$\gamma_{1\,300}$	166.1	165.1	144.3	120.5	
$\gamma_{1\,400}$	163.8	162.5	—	—	

FeO–MnO Ref. 4					
p	90.5	94.0	100		
$\sigma_{1\,410}$	555	567	585		

FeO–SiO₂ Ref. 4					
p	68.1	76.9	84.1	95.8	100
$\sigma_{1\,410}$	409	468	503	563	585

FeO–TiO₂ Ref. 4					
p	81.9	85.1	100		
$\sigma_{1\,410}$	510	522	585		

KBr–KCl Ref. 16						
p	0	28.6	51.7	70.7	86.5	100
σ_{750}	—	97	95	93	92	90
σ_{800}	99	95	93	89	87	86

KBr–NaBr Ref. 16						
p	0	22.4	43.5	63.4	82.2	100
σ_{700}	—	103	98	96	95	—
σ_{800}	103	96	92	88	87	86

KBr–Na₂SO₄ Ref. 5					
p	0	4.2	21.8	45.6	100
σ_{900}	193	181	140	116	81

KCl–K₂SO₄ Ref. 7				
p	0	0.9	2.2	4.5
$\sigma_{1\,075}$	144	140	135	133

KCl–Li₂SO₄ Ref. 7, 6							
p	0	7.0	26.6	40.2	66.8	85.8	100
σ_{900}	224	182	145	128	109	97	91
$\sigma_{1\,000}$	220	178	142	123	101	90	85
$\sigma_{1\,100}$	211	173	135	116	92	81	75

KCl–MgCl₂ Ref. 17						
p	8.0	16.4	34.4	54.1	75.9	87.6
σ_{700}	76	79	83	85	93	97
σ_{800}	69	72	76	78	86	90
σ_{900}	62	65	69	71	79	83

KCl–NaCl* Ref. 18, 17, 3						
p	0	24.2	46.0	65.6	83.6	100
σ_{700}	—	—	114	111	—	—
σ_{750}	—	115	111	107	104	—
σ_{800}	114	111	107	104	101	96

KCl–NaI Ref. 24						
p	7.9	19.5	32.6	49.0	73.2	84.9
a	135.2	128.0	121.7	124.4	127.2	128.5
b	0.066	0.063	0.057	0.063	0.069	0.070
t_0	0	0	0	0	0	0
r		(m.p. + 10) – (m.p. + 210)				

KCl–Na₂SO₄ Ref. 5					
p	0	2.7	14.8	34.2	100
σ_{900}	193	183	153	129	92

KCl–PbCl₂ Ref. 18, 16					
p	0	5.5	15.2	28.6	51.7
σ_{500}	137	125	118	117	118
σ_{600}	117	114	109	107	107

KI–Na₂SO₄ Ref. 5						
p	0	5.8	11.5	28.0	53.9	100
σ_{900}	193	177	157	127	100	73

(continued)

Table 9.7 SURFACE TENSION OF MOLTEN BINARY SALT SYSTEMS AND OTHER MIXED IONIC MELTS—*continued*

System								
KNO₂–KNO₃	*p*	17.4	35.8		55.9	77.0		
Ref. 24	*a*	129.6	131.3		130.6	131.8		
	b	0.059	0.061		0.055	0.052		
	t₀	0	0		0	0		
	r		(m.p. + 10) − (m.p. + 210)					
KNO₃–LiNO₃	*p*	0	32.9		59.5	81.5	100	
Ref. 19	*a*	129.9	127.9		129.6	133.6	139.8	
	b	0.055	0.056		0.062	0.070	0.081	
	t₀	0	0		0	0	0	
	r		liquidus–c.		400			
KNO₃–NaNO₃*	*p*	0	40		60	100		
Ref. 18, 16, 19	σ₃₅₀	117	115		114	112		
	σ₄₀₀	114	112		111	109		
	σ₄₅₀	111	109		107	105		
KNO₃–RbNO₃	*p*	0	18.6		40.7	67.4		
Ref. 19	*a*	134.3	135.0		136.9	138.5		
	b	0.083	0.082		0.083	0.083		
	t₀	0	0		0	0		
	r		liquidus–c.		400			
K₂O–SiO₂	*p*	23.9	26.7		29.9	32.9	38.7	43.6
Ref. 12, 21	σ₁₀₀₀	—	227		227	227	225	—
	σ₁₂₀₀	222	223		220	219	218	215
	σ₁₄₀₀	219	219		216	214	210	206
K₂SO₄–RbCl	*p*	92.9	96.5		98.6	100		
Ref. 7	σ₁₀₇₅	134	139		141	143		
LiCl–PbCl₂	*p*	6.1	7.8		12.8	13.4	24.9	27.1
Ref. 18	σ₅₀₀	133	134		134	133	—	—
	σ₅₅₀	130	129		129	129	128	—
	σ₆₀₀	124	—		—	—	—	124
LiCl–RbCl	*p*	0	75.9	86.9	91.9	94.5	97.2	100
Ref. 7	σ₇₅₀	96	113	118	119	121	122	127
Li₂O–SiO₂	*p*	12.9	17.8	23.9	27.8	33.5	43.1	49.1
Ref. 12, 21	σ₁₁₀₀	—	315	328	338	—	—	—
	σ₁₃₀₀	311	317	328	334	352	369	381
	σ₁₄₀₀	316	317	328	332	349	364	374
LiNO₃–RbNO₃	*p*	13.5	31.8	58.3				
Ref. 19	*a*	130.0	129.6	125.0				
	b	0.075	0.074	0.059				
	t₀	0	0	0				
	r		liquidus–c.	400				
Li₂SO₄–NaCl*	*p*	0	38.8	65.5	81.6	94.5	98.7	100
Ref. 7, 6	σ₉₀₀	109	131	148	168	198	208	224
	σ₁₀₀₀	104	125	143	164	194	204	220
	σ₁₁₀₀	95	116	134	157	187	196	211
Li₂SO₄–RbCl	*p*	0	38.0	57.9	68.1	78.6	89.2	100
Ref. 7, 6	σ₉₀₀	83	105	126	140	155	179	224
	σ₁₀₀₀	74	97	120	134	149	174	220
	σ₁₁₀₀	66	92	114	—	145	168	211
MgCl₂–NaCl	*p*	17.9	33.0	56.7	74.7	88.7	94.7	
Ref. 17	σ₇₀₀	110	103	95	87	79	75	
	σ₈₀₀	104	97	89	81	73	69	
	σ₉₀₀	98	91	83	75	67	63	
Na₃AlF₆–NaF	*p*	85	90		95	97.5		
Ref. 11, 1	σ₁₀₀₀	161	159		155	152		
NaCl–NaNO₃	*p*	0	9.2					
Ref. 5	σ₄₇₀	110	110					
NaCl–Na₂SO₄	*p*	0	0.8		1.7	4.3		
Ref. 7	σ₉₀₀	192	189		184	179		

(*continued*)

Table 9.7   SURFACE TENSION OF MOLTEN BINARY SALT SYSTEMS AND OTHER MIXED IONIC MELTS—*continued*

NaCl–PbCl$_2$	p	0	4.9	11.7	15.1	18.2		
Ref. **18**, 3, 24	σ_{500}	—	133	131	—	—		
	σ_{550}	131	127	124	—	—		
	σ_{575}	128	124	122	122	122		
NaF–NaNO$_3$	p	0	9.8					
Ref. **5**	σ_{560}	108	108					
NaI–NaNO$_3$	p	0	10.1					
Ref. **5**	σ_{312}	119	119					
NaI–Na$_2$SO$_4$	p	0.5	1.1	2.1	3.2	5.3	10.5	
Ref. **2**	σ_{900}	168	167	160	158	154	141	
NaNO$_2$–NaNO$_3$	p	16.9	35.4	55.0	76.3			
Ref. **24**	a	127.9	127.1	128.9	126.8			
	b	0.039	0.035	0.038	0.030			
	t_0	0	0	0	0			
	r		(m.p. + 10)–(m.p. + 210)					
NaNO$_3$–RbNO$_3$	p	16.1	36.5	63.4				
Ref. **19**	a	132.5	133.7	135.4				
	b	0.076	0.073	0.068				
	t_0	0	0	0				
	r	liquidus–c.	400					
Na$_2$O–P$_2$O$_5$	$\gamma_t = 150.6 + 155p/(100 - p) - 0.037\,9t$							
Ref. **22**	r_p	30.4–39.5						
	r_t	liquidus–1050						
Na$_2$O–SiO$_2$	p	20.0	30.8	33.6	36.9	50.0		
Ref. **12, 2**, 21	$\sigma_{1\,000}$	277	284	286	288	—		
	$\sigma_{1\,200}$	276	280	281	283	295		
	$\sigma_{1\,400}$	273	274	274	276	284		
Na$_2$SO$_4$RbCl	p	0	1.7	3.4	4.3	8.6		
Ref. **7**	$\sigma_{1\,050}$	183	173	171	169	162		
PbCl$_2$–RbCl	p	46.7	62.8	69.0	83.1	90.3	98.2	
Ref. **7**	σ_{475}	—	112	112	116	124	—	
	σ_{525}	—	107	107	111	118	130	
	σ_{575}	104	104	104	107	113	124	
PbO–SiO$_2$	p	65.1	69.9	75.7	82.5	84.7	90.6	96.8
Ref. **8**, 23	σ_{900}	—	—	217	199	192	174	134
	$\sigma_{1\,100}$	234	232	221	204	196	179	145
	$\sigma_{1\,300}$	235	230	223	209	—	183	158
PbO–V$_2$O$_5$	p	15.3	39.6	52.8	65.4	79.0	85.1	92.1
Ref. **23**	$\gamma_{1\,000}$	92	135	168	174	205	202	192
PbO–SiO$_2$–V$_2$O$_5$	p_{PbO}	28.5	38.6	58.4	60.0	68.5	80.0	
Ref. **23**	p_{SiO_2}	10.3	10.4	10.4	10.1	10.6	10.0	
	$\gamma_{1\,000}$	128	142	150	190	210	202	
Rb$_2$O–SiO$_2$	p	39.1	43.9	49.5	59.5	67.3		
Ref. **21**	$\gamma_{1\,200}$	—	—	—	175.1	155.0		
	$\gamma_{1\,300}$	200.1	192.7	188.0	173.4	146.3		
	$\gamma_{1\,400}$	197.1	188.8	183.5	170.9	—		

* *See also:* 'Physical Properties Data Compilations Relevant to Energy Storage II Molten Salts: Data on Single and Multi-Component Salt Systems', Janz *et al.*, NSRDS–NBS 61, Part II.

REFERENCES TO TABLE 9.7

1. E. Elchardus, *Compt. Rend.*, 1938, **206**, 1460.
2. C. W. Parmelee and C. G. Harman, *Univ. Illinois Eng. Exptl. Sta. Bull.*, 1939, No. 311, 29.
3. V. P. Barzakovskii, *Bull Acad. Sci. U.R.S.S., Classe Sci. Chim.*, 1940, 825.
4. P. P. Kozakevich and A. F. Kononenko, *J. Phys. Chem.* (*USSR*), 1940, **14**, 1118.
5. K. Semenchenko and L. P. Shikhobolova, *ibid.*, 1947, **21**, 613.
6. K. Semenchenko, *ibid.*, 1947, **21**, 707.
7. K. Semenchenko, *ibid.*, 1947, **21**, 1387.

8. L. Shartsis, S. Spinner and A. W. Smock, *J. Am. Ceram. Soc.*, 1948, **31**, 23.
9. L. Shartsis and R. Canga, *J. Res. Nat. Bur. Stand.*, 1949, **43**, 221.
10. S. Carlen, *Trans. Chalmers Univ. Tech. Gottenburg*, 1949, No. 85.
11. A. Vajna, *Alluminio*, 1951, **20**, 29.
12. L. Shartsis and S. Spinner, *J. Res. Nat. Bur. Stand.*, 1951, **46**, 385.
13. L. Shartsis and W. Capps, *J. Am. Ceram. Soc.*, 1952, **35**, 169.
14. L. Shartsis and H. F. Shermer, *ibid.*, 1954, **37**, 544.
15. J. S. Peake and M. R. Bothwell, *J. Am. Chem. Soc.*, 1954, **76**, 2656.
16. N. K. Boardman, A. R. Palmer and E. Heymann, *Trans. Faraday Soc.*, 1955, **51**, 277.
17. O. G. Desyatnikov, *Zh. Priklad. Khim.*, 1956, **29**, 870.
18. J. H. Dahl and F. R. Duke, *J. Phys. Chem.*, 1958, **62**, 1498.
19. G. Bertozzi and G. Sternheim, *ibid.*, 1964, **68**, 2908.
20. I. D. Sokolova and N. K. Voskresenskaya, *Zh. Prkl. Khim.*, 1962, **36**, 955.
21. A. A. Appen and S. S. Kayalova, *Dokl. Akad. Nauk., SSSR, Ser. Fiz. Khim.*, 1962, **145**, 592.
22. C. F. Callis, J. R. Van Wazer and J. S. Metcalf, *J. Am. Chem. Soc.*, 1955, **77**, 1468.
23. B. M. Lepinskikh, O. A. Esin and G. A. Teterin, *Zh. Neorg. Khim.*, 1960, **5**, 642.
24. H. Bloom, F. G. Davis and D. W. James, *Trans. Faraday Soc.*, 1960, **56**, 1179.

Table 9.8   VISCOSITY OF PURE MOLTEN SALTS

The viscosity (centipoise) at temperature t (°C) is given as η_t, or the constants a and b in the equation $\log \eta_t = a + b/(t + 273)$ are given for the temperature range r. Principal references are in **bold type**.

AgBr Ref. **22**, **11**, **9**	η_{450}	3.3	**KI** Ref. 30	η_{750}	1.362
	η_{550}	2.4		η_{800}	1.205
	η_{650}	1.7		η_{900}	0.973
	η_{800}	1.2	**KNO$_2$** Ref. **33**	a	−0.87
AgCl Ref. **22**, **11**, **9**	η_{500}	2.05		b	960
	η_{600}	1.60		r	418–450
	η_{700}	1.25	**KNO$_3$** Ref. 27, 16, 31, **34**, 35	η_{350}	2.705
	η_{800}	1.05		η_{400}	2.090
AgI Ref. **11**, **9**	η_{600}	3.0		η_{450}	1.673
	η_{700}	2.3		η_{550}	1.163
	η_{800}	1.7	**KOH** Ref. **10**	η_{400}	2.3
AgNO$_3$ Ref. **25**, **21**, **17**, **11**, **9**	η_{230}	4.10		η_{500}	1.3
	η_{280}	3.05		η_{600}	0.8
	η_{330}	2.40	**LiBr** Ref. 7, **14**	η_{550}	1.815
	η_{350}	2.20		η_{700}	1.096
B$_2$O$_3$ Ref. **24**, **20**, **4**	η_{600}	158 000		η_{850}	0.757
	η_{800}	25 100	**LiI** Ref. **14**	η_{450}	2.50
	η_{1000}	6 300		η_{550}	1.70
	η_{1200}	2 000		η_{650}	1.30
BaCl$_2$ Ref. **2**, **6**	η_{1000}	4.506	**LiNO$_3$** Ref. **17**, **12**, **11**, **3**	η_{260}	5.5
BiCl$_3$ Ref. **5**	η_{260}	32		η_{300}	4.0
	η_{300}	23		η_{350}	2.9
	η_{340}	18	**Na$_3$AlF$_6$** Ref. **19**	η_{1007}	6.7
CaCl$_2$ Ref. **6**, **7**	η_{800}	3.021		η_{1017}	6.5
	η_{900}	1.870		η_{1050}	6.0
	η_{1000}	1.248	**NaBr** Ref. **14**, 16	η_{762}	1.345
CdBr$_2$ Ref. **18**	η_{600}	2.60		η_{766}	1.332
	η_{640}	2.35		η_{780}	1.288
	η_{680}	2.10	**NaCl** Ref. 16, **34**, 35	η_{800}	1.463
CdCl$_2$ Ref. **23**, **18**, **14**	η_{600}	2.35		η_{900}	1.009
	η_{700}	1.85		η_{1000}	0.737
	η_{800}	1.55	**NaI** Ref. 30	η_{650}	1.581
CuCl Ref. **14**	η_{500}	2.80		η_{750}	1.168
	η_{600}	1.95		η_{900}	0.818
	η_{700}	1.40	**NaNO$_2$** Ref. **33**	a	−1.07
HgBr$_2$ Ref. **8**, **13**	η_{255}	2.196		b	868
	η_{265}	2.008		r	282–310
	η_{275}	1.843	**NaNO$_3$** Ref. 31, **34**, 36, 37	η_{300}	3.156
HgCl$_2$ Ref. **32**	η_{281}	1.768		η_{400}	1.901
	η_{287}	1.738		η_{500}	1.305
	η_{292}	1.694	**NaOH** Ref. **10**	η_{350}	4.0
	η_{299}	1.600		η_{450}	2.2
	η_{306}	1.543		η_{550}	1.5
HgI$_2$ Ref. **32**	η_{268}	2.669	**NaPO$_3$** Ref. **28**, **4**	η_{650}	1 250
	η_{292}	2.244		η_{750}	440
	η_{314}	1.995		η_{850}	210
	η_{334}	1.715		η_{900}	150
	η_{358}	1.458	**PbBr$_2$** Ref. **22**, **17**, **1**	η_{360}	10.5
KBr Ref. 7, **14**, 16	η_{750}	1.150		η_{400}	7.3
	η_{800}	1.022		η_{450}	5.0
	η_{900}	0.831		η_{550}	3.0
KCl Ref. **14**, 16, 26	η_{800}	1.094	**PbCl$_2$** Ref. **22**, **1**	η_{500}	4.6
	η_{900}	0.841		η_{600}	2.8
	η_{1000}	0.673		η_{700}	1.9
K$_2$Cr$_2$O$_7$ Ref. 27, 29	η_{400}	13.79	**TlNO$_3$** Ref. **33**	a	−1.04
	η_{450}	9.665		b	565
	η_{500}	7.091		r	207–250

REFERENCES TO TABLE 9.8

1. R. Lorenz and H. T. Kalmus, *Z. Phys. Chem.*, 1907, **59**, 244.
2. V. T. Slavyanskii, *Dokl. Akad. Nauk. SSSR*, 1947, **58**, 1077.
3. H. M. Goodwin and R. D. Mailey, *Phys. Rev.*, 1908, **26**, 28.
4. K. Arndt, *Z. Chem. Apparat.*, 1908, **3**, 549.
5. A. H. W. Aten, *Z. Phys. Chem.*, 1909, **66**, 641.
6. G. J. Janz and R. D. Reeves, *Advan. Electrochem. Eng.*, 1967, **5.**
7. S. Karpachev and A. Stromberg, *Zh. Fiz. Khim.*, 1938, **11**, 852.
8. R. S. Dantuma, *Z. Anorg. Allgem. Chem.*, 1938, **175**, 1.
9. R. Lorenz and A. Hoechberg, *Z. Anorg. Chem.*, 1916, **94**, 317.
10. K. Arndt and G. Ploetz, *Z. Phys. Chem.*, 1926, **121**, 439.
11. E. van Aubel, *Bull. Sci. Acad. Roy. Belg.*, 1926, **12**, 374.
12. R. S. Dantuma, *Z. Anorg. Allg. Chem.*, 1928, **175**, 1.
13. G. Jander and K. Broderson, *Z. Anorg. Allgem. Chem.*, 1951, **264**, 57.
14. I. G. Murgulescu and S. Zuca, *Z. Physik. Chem. (Leipzig)*, 1961, **218**, 379.
15. A. G. Stromberg, *Zh. Fiz. Khim.*, 1939, **13**, 436.
16. C. E. Fawsitt, *Proc. Roy. Soc. (London)*, 1908, 93.
17. K. S. Evstropev, *Akad. Nauk. SSSR., Otdel. Tekh. Nauk. Inst. Mash. Sov.*, 1945, **3**, 61.
18. H. Bloom, B. S. Harrap and E. Heymann, *Proc. R. Soc.*, 1948, **A194**, 237.
19. A. Vajna, *Alluminio*, 1950, **19**, 133.
20. L. Shartsis, W. Capps and S. Spinner, *J. Am. Ceram. Soc.*, 1953, **36**, 319.
21. F. A. Pugsley and F. E. W. Wetmore, *Canad. J. Chem.*, 1954, **32**, 839.
22. B. S. Harrap and E. Heymann, *Trans. Faraday Soc.*, 1955, **51**, 259.
23. B. S. Harrap and E. Heymann, *ibid.*, 1955, **51**, 268.
24. J. D. Mackenzie, *ibid.*, 1956, **52**, 1564.
25. N. P. Luzhnaya, N. N. Evseeva and I. P. Vereshchetina, *Zh. Neorg. Khim.*, 1956, **1**, 1490.
26. S. Karpachev, *Zh. Obshch. Khim.*, 1935, **5**, 625.
27. R. Lorenz and T. Kalmus, *Z. Physik. Chem.*, 1907, **59**, 244.
28. G. G. Nozadze, *Soobshch. Akad. Nauk. Gruzin. S.S.S.R.*, 1957, **19**, 567.
29. J. P. Frame, E. Rhodes and A. R. Ubbelohde, *Trans. Faraday Soc.*, 1959, **55**, 2039.
30. I. G. Murgulescu and S. Zuca, *Rev. Roumaine. Chim.*, 1965, **10**, 123.
31. H. M. Goodwin and R. D. Mailey, *Phys. Rev.*, 1906, **23**, 22; *ibid.*, 1907, **25**, 469; *ibid.*, 1908, **26**, 28.
32. G. J. Janz and J. D. E. McIntyre, *J. Electrochem. Soc.*, 1962, **109**, 842.
33. J. P. Frame, E. Rhodes and A. R. Ubbelohde, *Trans. Faraday Soc.*, 1959, **55**, 2039.
34. R. S. Dantuma, *Z. Anorg. Allgem. Chem.*, 1938, **175**, 1.
35. K. Ogawa, *Nippon Kinzoku Gakkaishi*, 1950, **14B**, 49.
36. R. Lorenz and H. T. Kalmus, *Z. Physik. Chem.*, 1907, **59**, 17.
37. C. E. Fawsitt, *Proc. Roy. Soc. (London)*, 1908, **A80**, 290.

Table 9.9 VISCOSITY OF MOLTEN BINARY SALT SYSTEMS AND OTHER MIXED IONIC MELTS

The viscosity (centipoise) at temperature t (°C) and composition p (wt. %) of the first-named constituent is given as η_t, or the constants a and b in the equation $\log \eta_t = a + b/(t + 273)$ are given for the temperature range r. Principal references are in **bold type**.

AgBr–AgCl	p	0	32.5	46.7	66.9	84.0	100
Ref. 12	η_{440}	—	2.58	2.75	2.97	3.12	3.38
	η_{520}	1.98	2.10	2.18	2.35	2.46	2.69
	η_{600}	1.68	1.74	1.74	1.98	2.09	2.28

AgBr–KBr	p	57.1	66.0	78.0	86.0	100
Ref. 12	η_{400}	—	—	3.55	3.55	—
	η_{500}	—	2.55	2.40	2.46	2.83
	η_{600}	1.73	1.74	1.78	1.86	2.27

AgCl–KCl	p	54.3	61.1	80.4	88.9	100
Ref. 12	η_{500}	—	—	2.10	2.12	2.08
	η_{600}	1.63	1.61	1.56	1.62	1.66
	η_{700}	1.28	1.24	1.24	1.28	1.40

AgCl–PbCl$_2$	p	0	9.8	24.1	44.5	67.6	100
Ref. 12	η_{500}	4.56	4.13	3.47	2.95	2.47	2.08
	η_{600}	2.75	2.59	2.30	2.12	1.84	1.66
	η_{700}	1.87	1.99	1.84	1.76	1.65	1.40

AgI–AgNO$_3$	p	0	25.7	47.9	67.4
Ref. 17	η_{150}	—	9.6	13.6	19.6
	η_{200}	—	6.4	8.6	12.6
	η_{250}	4.2	5.2	6.5	8.7
	η_{300}	3.8	4.8	5.5	7.0

AgNO$_3$–HgI$_2$	p	9.8	13.8	19.9	27.2	35.9	46.6	56.2
Ref. 15, 21	η_{110}	—	2.3	—	—	19.5	13.8	9.6
	η_{150}	3.2	—	7.4	7.0	5.6	4.2	3.2
	η_{200}	1.5	2.3	2.4	2.2	1.9	1.8	1.4

AgNO$_3$–KNO$_3$	p	29.6	41.9	52.8	62.7	71.6	79.7	83.5
Ref. 14, 21	η_{150}	—	—	—	18.2	19.1	18.1	—
	η_{200}	—	—	—	8.9	9.6	9.2	8.9
	η_{250}	—	7.4	6.4	5.6	6.0	6.4	6.0
	η_{300}	5.8	5.1	4.8	4.1	4.3	4.6	4.5

AlF$_3$–Na$_3$AlF$_6$	p	2.5	5	10	15
Ref. 7	$\eta_{1\,000}$	6.5	6.3	5.7	4.8

Al$_2$O$_3$–Na$_3$AlF$_6$	p	2.5	5	10	15
Ref. 7	$\eta_{1\,000}$	6.9	6.9	7.1	10.9

B$_2$O$_3$–BaO	p	44.5	49.2	54.9	59.7	64.4	67.9
Ref. 10	η_{850}	—	30 900	35 500	35 500	33 900	25 100
	η_{900}	2 750	6 170	8 710	9 550	9 550	7 080
	η_{950}	980	1 900	3 160	3 710	3 240	2 950
	$\eta_{1\,000}$	—	—	—	1 260	1 260	1 180

B$_2$O$_3$–K$_2$O	p	61.6	70.6	84.9	89.0	94.8	98.5	100
Ref. 9	η_{600}	6.8×10^5	5.1×10^6	—	1.5×10^5	0.9×10^5	1.3×10^5	1.6×10^5
	η_{800}	780	4 680	4 360	3 900	5 130	13 800	21 400
	η_{900}	—	960	1 350	1 290	2 400	6 310	11 500
	$\eta_{1\,000}$	—	—	—	450	1 120	3 710	6 460

B$_2$O$_3$–Li$_2$O	p	85.2	87.1	92.0	95.5	97.2	100
Ref. 9	η_{600}	—	—	—	355 000	141 000	158 000
	η_{800}	3 630	3 020	5 370	3 630	7 240	21 400
	η_{900}	460	520	790	980	3 090	11 500
	$\eta_{1\,000}$	—	—	250	320	1 440	6 460

B$_2$O$_3$–Na$_2$O	p	68.2	69.2	77.6	85.8	94.4	99.1	100
Ref. 9	η_{600}	—	3.1×10^6	—	1.9×10^6	1.3×10^5	1.3×10^5	1.6×10^5
	η_{800}	930	1 290	6 920	5 890	3 800	12 000	21 400
	η_{900}	—	—	870	1 380	1 580	5 750	11 500
	$\eta_{1\,000}$	—	—	—	—	810	2 630	6 460

B$_2$O$_3$–NaPO$_3$	p	0	50	95	99.5
Ref. 18	η_{900}	150	450	4 730	11 000

(*continued*)

Table 9.9 VISCOSITY OF MOLTEN BINARY SALT SYSTEMS AND OTHER MIXED IONIC MELTS—*continued*

BaCl₂–NaCl Ref. 19, 5, 21							
p	45.5	55.6	63.8	70.5	75.3	80.1	84.4
η_{725}	—	3.2	3.42	4.05	4.6	—	—
η_{775}	2.40	2.58	2.82	3.22	3.48	3.70	3.85
η_{825}	2.00	2.23	2.50	2.80	3.00	3.15	3.28
η_{875}	1.75	2.02	2.36	2.55	2.72	2.84	3.05

BaO–SiO₂ Ref. 16					
p	31.7	46.1	53.1	63.2	71.7
$\eta_{1\,500}$	—	11 000	3 200	1 700	—
$\eta_{1\,700}$	19 500	2 140	850	540	190
$\eta_{1\,800}$	10 000	1 250	530	400	150

CaCl₂–NaCl* Ref. 5, 19						
p	0	17.4	44.8	65.5	88.3	100
η_{800}	1.59	1.65	2.59	3.49	4.36	4.92
η_{900}	1.00	1.16	2.11	2.95	3.69	4.22
$\eta_{1\,000}$	0.70	1.00	1.79	2.60	3.31	3.74

CaF₂–Na₃AlF₆ Ref. 7				
p	2.5	5	10	15
$\eta_{1\,000}$	6.9	7.0	7.3	8.0

CaO–SiO₂ Ref. 11, 3						
p	29.1	37.1	41.9	47.9	51.9	55.9
$\eta_{1\,500}$	—	1440	765	—	288	—
$\eta_{1\,600}$	—	730	405	218	157	113
$\eta_{1\,700}$	1360	392	235	133	96	74
$\eta_{1\,800}$	850	250	150	88	66	54

CdBr₂–CdCl₂ Ref. 6						
p	0	27.2	49.9	69.1	85.6	100
η_{600}	2.3	2.4	2.4	2.5	2.5	2.6
η_{660}	2.0	2.0	2.1	2.1	2.1	2.2

CdCl₂–KCl Ref. 13							
p	40.3	59.5	67.3	71.9	82.4	88.8	100
η_{500}	—	2.53	2.54	2.62	2.97	3.17	—
η_{600}	1.76	1.67	1.67	1.73	1.92	2.15	2.31
η_{700}	1.23	1.21	1.20	1.24	1.40	1.57	1.83

CdCl₂–NaCl Ref. 13						
p	53.9	75.8	79.0	82.5	90.2	100
η_{500}	—	2.68	2.71	2.84	2.99	—
η_{600}	—	1.78	1.80	1.88	2.02	2.31
η_{700}	1.37	1.30	1.27	1.39	1.50	1.83

CdCl₂–PbCl₂ Ref. 13						
p	0	18.0	26.8	37.9	65.9	100
η_{520}	4.22	3.75	3.50	3.36	—	—
η_{600}	2.75	2.60	2.47	2.38	2.26	2.31
η_{680}	2.00	2.02	1.88	1.85	1.85	1.90

KCl–MgCl₂ Ref. 2, 21						
p	37.6	48.5	58.9	59.9	69.1	78.8
η_{500}	3.26	—	2.78	2.69	—	—
η_{600}	2.19	1.55	1.51	1.70	1.44	—
η_{700}	—	1.22	—	—	0.99	1.25

KCl–NaCl* Ref. 5				
p	0	56.0	79.3	100
η_{800}	1.59	1.17	1.07	1.13
η_{900}	1.00	0.90	0.81	0.89

KCl–PbCl₂ Ref. 13							
p	0	5.2	8.8	18.3	21.8	34.7	43.3
η_{500}	4.56	4.02	3.66	3.34	3.37	—	—
η_{600}	2.75	2.42	2.27	2.06	2.10	2.15	—
η_{700}	1.87	1.65	1.56	1.42	1.47	1.48	1.50

K₂O–SiO₂ Ref. 16, 8							
p	3.9	9.5	15.9	23.9	29.9	38.7	43.6
$\eta_{1\,000}$	—	—	—	32 800	15 100	7 940	4 570
$\eta_{1\,200}$	—	—	—	2 240	1 350	661	355
$\eta_{1\,400}$	—	—	—	372	200	81	47
$\eta_{1\,600}$	468 000	107 000	24 000	—	—	—	—

Li₂O–SiO₂ Ref. 16, 8							
p	12.0	16.1	21.8	25.9	28.9	33.2	37.9
$\eta_{1\,100}$	—	50 100	12 000	4470	—	—	—
$\eta_{1\,300}$	42 700	8 320	2 190	1 000	510	220	72
$\eta_{1\,400}$	17 800	3 890	1 120	580	300	140	52
$\eta_{1\,500}$	8 320	2 040	710	400	200	100	41

MgO–SiO₂ Ref. 16					
p	34.8	35.5	36.2	40.1	41.4
$\eta_{1\,650}$	610	460	350	250	190
$\eta_{1\,750}$	360	270	200	150	120
$\eta_{1\,800}$	280	210	180	120	110

Na₃AlF₆–NaF Ref. 7				
p	85	90	95	97.5
$\eta_{1\,000}$	4.8	5.9	6.4	6.7

(continued)

Table 9.9   VISCOSITY OF MOLTEN BINARY SALT SYSTEMS AND OTHER MIXED IONIC MELTS—*continued*

NaCl–NaNO$_3$	p	1.4	2.8	4.2	5.7	7.1	8.6	10.1
Ref. **19**	η_{310}	3.20	3.27	3.68	—	—	—	—
	η_{350}	2.57	2.67	2.79	2.83	3.00	—	—
	η_{400}	2.06	2.21	2.32	2.18	2.11	2.20	2.39
	η_{450}	1.72	1.91	1.98	1.96	1.94	1.73	2.18
Na$_2$O–P$_2$O$_5$	p	30.4	32.0	36.0	39.6	43.3	46.5	48.4
Ref. **20**	a	−0.84	−0.63	−0.26	0.11	0.38	0.58	0.59
	$10^{-3}b$	3.57	3.19	2.45	1.87	1.45	1.10	1.00
	r	741–1013	707–1001	657–1017	822–1002	942–1064	1030–1077	995–1 070
Na$_2$O–SiO$_2$	p	15.2	20.0	30.8	36.9	40.9	45.2	50.0
Ref. **16, 8, 4, 1**	η_{900}	—	4.6×10^6	2.4×10^5	3.2×10^5	1.8×10^5	—	—
	$\eta_{1\,100}$	—	263 000	50 100	26 900	15 100	4070	890
	$\eta_{1\,300}$	118 000	33 100	7 940	4 270	2 340	690	160
	$\eta_{1\,400}$	41 700	16 600	4 900	2 570	1 350	390	100
PbBr$_2$–PbCl$_2$	p	0	28.4	63.1	80.2	100		
Ref. **12**	η_{450}	—	—	5.53	5.24	4.83		
	η_{500}	4.56	4.45	4.03	3.94	3.73		
	η_{550}	3.54	3.38	3.14	3.05	2.97		
SiO$_2$–SrO	p	36.4	42.0	46.0	57.8	62.9	69.7	
Ref. **16**	$\eta_{1\,650}$	47	360	620	1 380	4 220	—	
	$\eta_{1\,750}$	160	210	370	780	2 240	5 820	
	$\eta_{1\,800}$	130	180	300	580	1 660	4 300	

* *See also* 'Physical Properties Data Compilations Relevant to Energy Storage II Molten Salts: Data on Single and Multi-Component Salt Systems', Janz *et al.*, NSRDS–NBS, 61.

REFERENCES TO TABLE 9.9

1. C. L. Babcock, *J. Am. Ceram. Soc.*, 1934, **17**, 329.
2. S. Karpachev and A. Stromberg, *Z. Anorg. Allg. Chem.*, 1935, **222**, 78.
3. J. R. Rait and R. Hay, *J. R. Tech. Coll. (Glasgow)*, 1938, **4**, 252.
4. E. Preston, *J. Soc. Glass Tech.*, 1938, **22**, 45.
5. V. P. Barzakovskii, *Bull. Acad. Sci. URSS, Classe, Sci. Chim.*, 1940, 825.
6. H. Bloom, B. S. Harrap and E. Heymann, *Proc. R. Soc.*, 1948, **A194**, 237.
7. A. Vajna, *Alluminio*, 1950, **19**, 133.
8. L. Shartsis, S. Spinner and W. Capps, *J. Am. Ceram. Soc.*, 1952, **35**, 155.
9. L. Shartsis, W. Capps and S. Spinner, *ibid.*, 1953, **36**, 319.
10. L. Shartsis, S. Spinner, and H. F. Shermer, *ibid.*, 1954, **37**, 544.
11. J. O'M. Bockris and D. C. Lowe, *Proc. R. Soc.*, 1954, **A226**, 1167.
12. B. S. Harrap and E. Heymann, *Trans. Faraday Soc.*, 1955, **51**, 259.
13. B. S. Harrap and E. Heymann, *ibid.*, 1955, **51**, 268.
14. V. D. Polyakov, *Izvest. Sekt. Fiziko-Khim. Anal.*, 1955, **26**, 147.
15. V. D. Polyakov, *ibid.*, 1955, **26**,191.
16. J. O'M. Bockris, J. D. Mackenzie and J. A. Kitchener, *Trans. Faraday Soc.*, 1955, **51**, 1734.
17. N. P. Luzhnaya, N. N. Evseeva and I. P. Vereshchetina, *Zh. Neorg. Khim.*, 1956, **1**, 1490.
18. G. G. Nozadze, *Soobshch. Akad. Nauk. Gruzin. SSSR*, 1957, **19**, 567.
19. I. P. Vereshchetina and N. P. Luzhnaya, *Izvest. Sekt. Fiziko-Khim. Anal.*, 1954, **25**, 188.
20. C. F. Callis, J. R. Van Wazer and J. S. Metcalf, *J. Am. Chem. Soc.*, 1955, **77**, 1471.
21. G. J. Janz, 'Molten Salts Handbook', Academic Press, London, 1967.

10 Metallography

Metallography is the branch of science dealing with the study of the constitution and structure of metals and alloys, its control through processing, and its influence on properties and behaviour. The original implementation of this science was limited by the resolution of the reflected light microscope used to study specimens. This limitation has been overcome by the development of transmission and scanning electron microscopies (TEM and SEM). The analysis of X-rays generated by the interaction of electron beams with atoms at or near the surface, with wavelength or energy-dispersive spectroscopy (WDS, EDS) with SEM, electron microprobe analysis (EMPA) or TEM has added quantitative determination of local compositions, e.g., of intermediate phases, to the deductions based upon observations. Introduction of metrological and stereological methods, and the development of computer-aided image analysers, permits measurement of microstructural features. Crystallographic data can be obtained using classic X-ray diffraction methods using a diffractometer, or diffraction analysis can be performed with the TEM using selected area or convergent-beam electron diffraction (SAD and CBD) techniques, and more recently with the SEM with the orientation-imaging (EBSD) procedure. There is a wide variety of very sophisticated electron or ion devices that can be utilised to characterise surfaces and interfaces, but these devices are generally restricted in availability due to their high cost.

Conventional light-optical techniques are still the most widely used and are capable of providing the information needed to solve most problems. Examination by light optical microscopy (LOM) should always be performed before use of electron metallographic instruments. LOM image contrast mechanisms are different from electron microscopy (EM) imaging modes. Natural colour can not be seen with EM devices. Microstructures are easier to study at low magnification with the LOM than with the SEM. The LOM examination may indicate the need for SEM or TEM analysis and determine the locations for such work. Interpretation of LOM examination results is enhanced and reinforced by the use of electron metallographic techniques. The SEM has become ubiquitous in the metallographic laboratory.

Important Safety Note: The metallographic laboratory can involve a number of serious hazards, that can lead potentially to fatalities. For example, many of the reagents used in the preparation of metallographic etchants are highly corrosive (hydrofluoric acid severely attacks the bone) and/or extremely toxic. These reagents are frequently inflammable and in some cases are explosive (e.g., picric acid). It is especially important to understand that, often, the etchant can possess hazards that are greater and/or different from those of the reagents from which it was prepared (e.g., some perchloric acid based solutions pose a much greater explosion risk than others). Some etchants (e.g., those containing hydrofluoric acid) can be extremely hazardous even when dilute. All etchants have a limited shelf life and some require special storage precautions. In some cases (e.g., glyceregia), etchants cannot be stored safely and must be made up freshly at each use. Identifying a safe and legal method of disposal of used or discarded etchants requires consideration of all of their ingredients.

Given the risks involved, metallographic specimen preparation requires both suitable operating conditions/safety equipment and appropriate training for the metallographer. This chapter points out some of the hazards associated with specific etchants, but is not intended as a substitute for either a detailed laboratory safety manual or reading of the regulations in force in the metallographer's location. A discussion of etchant safety may be found in Petzow[1] and various other references given in the present section.

10.1   **Macroscopic examination**

For examination of large-scale features, known as macrostructure[2] which are visible with the unaided eye, discs are cut from cast or wrought products (in the case of wrought products, usually before extensive hot deformation is performed). The discs must be representative of the product and are taken usually from prescribed test locations. Mechanical sawing or abrasive sectioning is used to obtain the disc, which may be ground to various surface finish levels, depending upon the nature of the detail that must be observed. The discs are cleaned and hot acid etched to reveal the solidification structure, deformation structure, segregation, soundness, etc. Discs may also be subjected to contact printing methods, such as sulphur printing (see 2, ASTM E 1180 and ISO 4968).

The required disc is obtained by sawing, abrasive sectioning or machining with adequate cooling and lubrication, and is finished normally by grinding. Due to the size of these discs, workshop (machine shop) grinders are used. In the case of a small section, laboratory practices may be utilised. Grinding on abrasive cloth or paper, such as an 'endless belt grinder', is often used, but the platen beneath the belt must be kept flat. For most macroexamination work, a ground surface is adequate. In a few cases, chiefly dependant upon the nature of the examination and the desired etchant, a polished surface is required. Conventional metallographic procedures are adequate. The main problem is obtaining flatness over a section area that is rather large compared to standard metallographic specimens. But, with modern grinder/polisher devices, this can be accomplished readily.

Capturing images of macroetched components can be quite difficult due to the need for obtaining proper uniform illumination, especially if the surface is as-polished, as might be the case when documenting porosity, cracks or other voids in sections. Although film-based documentation is becoming less common, compared to digital imaging, film is still preferred at this time in failure analysis work involving litigation. Regardless of the technology used, illumination still must be uniform. With an as-polished surface, 'hot spots' and reflections are a major problem that can be overcome by a variety of 'tricks' utilising vertical illumination and passing light through translucent material.

Etching reagents for macroscopic work are given in many national and international standards, such as ASTM E 340, and textbooks.[2] The commonly used reagents are listed in Table 10.1. Directions for sulphur-printing are given in ISO 4968 and ASTM E 1180. This technique is used to show the distribution of sulphur in steel and is also described in Table 10.1.

Table 10.1   ETCHING REAGENTS FOR MACROSCOPIC EXAMINATION

See general safety note at the start of this Chapter.

Material		*Reagent**		*Remarks*
A. *Aluminium base*				
1.	Aluminium and its alloys	(a)	Concentrated Keller's Reagent	Can be diluted with up to 50 ml water
			Nitric acid (1.40) 100 ml	
			Hydrochloric acid (1.19) 50 ml	
			Hydrofluoric acid (40%)§ 1½ ml	
		(b)	Nitric acid (1.40) 30 ml	Widely applicable, but very vigorous
			Hydrochloric acid (1.19) 30 ml	
			2% conc. Hydrofluoric acid§ 30 ml	
		(c)	Tucker's Reagent	Use fresh
			Nitric acid (1.40) 15 ml	
			Hydrochloric acid (1.19) 45 ml	
			Hydrofluoric acid (40%)§ 15 ml	
			Water 25 ml	
		(d)	10% sodium hydroxide in water	Use at 60–70°C
2.	Unalloyed Aluminium and Al–Cn alloys	(e)	Flick's Reagent	Wash in warm water after etching and clear by dipping in concentrated nitric acid
			Hydrochloric acid 15 ml	
			Hydrofluoric acid§ 10 ml	
			Water 90 ml	
3.	Aluminium–silicon	(f)	Hume-Rothery's Reagent	For high-silicon alloys Fine polish undesirable. Immerse specimen 5–10s, remove, and brush
			Cupric chloride 15 g	
			Water 100 ml	

(continued)

Table 10.1   ETCHING REAGENTS FOR MACROSCOPIC EXAMINATION—*continued*

Material	Reagent*		Remarks
			away deposited copper or remove it with 50% nitric acid in water
4. Aluminium–copper	(g) Keller's Reagent 2½ % nitric acid (1.40) 1½ % hydrochloric acid (1.19) ½ % hydrofluoric acid (40%)§ Remaining water		More frequently used as micro-etch
5. Aluminium–magnesium	(h) 5% cupric chloride 3% nitric acid (1.40) Remaining water		Clear surface with strong nitric acid
6. Aluminium–copper–silicon	(g) Keller's Reagent (as above) (i) Nitric acid (1.40) Hydrochloric acid (1.19) Hydrofluoric acid (40%)§ Water	15 ml 10 ml 5 ml 70 ml	
7. Aluminium–copper–magnesium–nickel	(j) Zeerleder's Reagent Hydrochloric acid (1.19) Nitric acid (1.40) Hydrofluoric acid (40%)§ Water	20 ml 15 ml 5 ml 60 ml	
B. *Copper base* 1. Copper and copper alloys generally	(a) Alcoholic ferric chloride Ethanol Ferric chloride (anhydrous) Hydrochloric acid (1.19)	96 ml 59 g 2 ml	Avoid use of water for washing, or staining may result. Use alcohol or acetone instead. Grain contrast
	(b) Acid aqueous ferric chloride Ferric chloride Hydrochloric acid (1.40) Water	25 g 25 ml 100 ml	(a) and (b) require moderately high standard of surface finish
	(c) Concentrated nitric acid (1.40) Nitric acid (1.40) Water	50 ml 50 ml 10 ml	A rapid etch, suitable for roughly prepared surfaces. Addition of a trace of silver nitrate (5%) enhances contrast
	(d) 10% ferric chloride in water 5% chromium trioxide in saturated brine 20% acetic acid in water	10 ml 10 ml 20 ml	To reveal strains in brasses
	(e) A. 1% mercuric nitrate in distilled water B. 1% nitric acid (1.40) in water. Mix A and B in equal proportions		Time required to induce cracks is indication of residual stress
	(f) Chromium trioxide Ammonium chloride Nitric acid (1.40) Sulphuric acid (1.84) Distilled water	40 g 7.5 g 50 ml 8 ml 100 ml	Good for alloys with silicon and silicon bronzes
C. *Iron and steel*			
	(a) 50% hydrochloric acid in water		Use hot (70–80°C) for up to 1 h. Shows segregation porosity, cracks; useful for examination of welds for soundness
	(b) 20% sulphuric acid in water		Use hot (80°C) for 10–20 min. Scrub lightly to remove carbonaceous deposit. Purpose

(*continued*)

Table 10.1   ETCHING REAGENTS FOR MACROSCOPIC EXAMINATION—*continued*

Material	Reagent*		Remarks
Iron and steel—continued			
			as (a). Mixtures of (a) and (b) are also used similarly
(c)	25% nitric acid in water		Purposes as (a) and (b). May be used cold if more convenient
(d)	10% ammonium persulphate in water		Grain contrast etch. Apply with swab. Reveals grain growth and recrystallisation at welds
(e)	Stead's Reagent		For revealing phosphorus segregation and primary dendritic structure of cast steels. Dissolve the salts in the acid with addition of a minimum of water. Phosphorus segregate unattacked, also eutectic cells in cast iron
	Cupric chloride	10 g	
	Magnesium chloride	40 g	
	Hydrochloric acid (1.19)	20 ml	
	Alcohol to 1 litre		
(f)	Fry's Reagent		To reveal strain lines in mild steel. Heat to 150–250°C for 15–30 min before etching specimen. Etch for 1–3 min while rubbing with a soft cloth. Rinse with alcohol.
	Cupric chloride	90 g	
	Hydrochloric acid	120 ml	
	Water	100 ml	
(g)	Humphrey's Reagent		Reveals dendritic structure of cast steels. First treat surface with 8% copper ammonium chloride solution and then with (g) for ½–1½ h. Remove copper deposit (loosely adherent), dry and rub surface lightly with abrasive
	Copper ammonium chloride	120 g	
	Hydrochloric acid (1.19)	50 ml	
	Water	1 litre	
(h)	5–10% nitric acid in alcohol		Etch for up to ½ h. Reveals cracks and carbon segregation. More controlled than aqueous acids
(j)	Sulphur printing 3% sulphuric acid in water		Soak photographic printing paper in the acid and remove surplus acid with blotting paper. Lay paper face down on the clean steel surface and 'squeegee' into close contact. After 2 min remove paper, wash it and fix in 6% sodium thiosulphate in water. Brown colouration on the paper indicates local segregation of sulphides
(k)	Dithizone process for lead distribution		*See* p. **10**–49, *Lead in steels.* Analogous to sulphur printing
(l)	Marble's Reagent		Austenitic steels. High temperature steels. Fe–Cr–Ni casting alloys. Also shows depth of nitriding
	Hydrochloric acid (1.19)	50 ml	
	Saturated aqueous solution of cupric sulphate	25 ml	
(m)	Oberhoffer's Reagent		Good surface preparation needed. Steel castings. Darkens Fe-rich areas, reveals segregation and
	Hydrochloric acid (1.19)	42 ml	
	Ferric chloride	30 g	
	Stannous chloride	0.5 g	

(continued)

Table 10.1   ETCHING REAGENTS FOR MACROSCOPIC EXAMINATION—*continued*

Material		Reagent*		Remarks
Iron and steel—continued				
		Water	500 ml	primary cast structure
		Ethanol (acid added last)	500 ml	
		Rinse in 20% hydrochloric acid in ethanol		
	(n)	Klemm's Reagent		Phosphorus distribution
		Saturated aqueous solution of sodium thiosulphate	50 ml	in cast steel and cast iron. Grain contrast
		Sodium metabisulphite (can be increased for contrast)	1 g	
D. *Lead base*				
Lead and lead alloys generally	(a)	Russell's Reagent		Grain contrast etch; removes deformed layer
		A. 80 ml nitric acid (1.40) in 220 ml water		Mix equal parts of A and B immediately
		B. 45 g ammonium molybdate in 300 ml water		before use. Swab for 10–30 s. Rinse in water
	(b)	Ammonium molybdate	10 g	Bright etch revealing grain
		Citric acid	25 g	structure, defects, etc.
		Water	100 ml	
	(c)	Worner and Worner's Reagent		Chemical polish revealing
		Acetic acid, glacial '100 vol.'	75 ml	defects, etc. Specimen
		Hydrogen peroxide (30% conc.)	25 ml	must be dry and water content of solution as low as possible *N.B.*— Avoid all heating, as lead alloys recrystallise very readily
	(d)	Nitric acid (1.40)	20 ml	Immerse 5–10 min. Grain
		Water (distilled)	80 ml	contrast, laminations, welds. Up to 50% nitric acid can be used
	(e)	Glacial acetic acid	20 ml	Macrostructure of alloy
		Nitric acid (1.40)	20 ml	with Ca, Sb and Sn. Use
		Glycerol	80 ml	fresh only. Several minutes needed
	(f)	Glacial acetic acid	20 ml	2–10 s by swabbing.
		Nitric acid (1.40)	20 ml	Good for alloys with Bi,
		Hydrogen peroxide (30%)	20 ml	Te or Ni
		Water (distilled)	50 ml	
E. *Magnesium base*				
	(a)	Picric acid‡ (64%) saturated in ethanol (96%)	50 ml	Grain size. Flow lines in forging (wash precipitate
		Glacial acetic acid	20 ml	in hot water). Etch for
		Distilled water	20 ml	up to 3 min
	(b)	Ammonium persulphate	2 ml	Flow lines in forgings
		Distilled water	98 ml	
	(c)	Nitric acid (1.40)	20 ml	Internal defects in casts.
		Water	80 ml	Useful for Mg–Mn and Mg–Zr. Etch for up to 3 min
	(d)	Glacial acetic acid	10 ml	General defects; flow
		Water	90 ml	lines, segregation. Etch for up to 3 min
F. *Nickel base*				
	(a)	Nitric acid (1.40)	50 ml	Welds, Ni–Cr–Fe alloys
		Acetic acid	50 ml	
	(b)	Aqua regia		As (a)
		Nitric acid (1.19)	25 ml	
		Hydrochloric acid (1.19)	75 ml	

(continued)

Table 10.1   ETCHING REAGENTS FOR MACROSCOPIC EXAMINATION—*continued*

Material	Reagent*		Remarks
G. *Tin base*	(a) Sat. soln of ammonium polysulphide in water (wipe off surface film)		Grain structure; suitable most tin alloys (etching time 20–30 min)
	(b) FeCl Hydrochloric acid (1.18) Water	10 g 2 ml 100 ml	Sn–Sb alloys (up to 3 min)
H. *Zinc base* Zinc and zinc alloys	(a) Concentrated hydrochloric acid (1.19)		Good grain contrast
Zinc-rich alloys	(b) 5% hydrochloric acid in alcohol		HCl can be increased to 50% Wash under running water to remove reaction product
	(c) Sodium sulphate (3.5 g if hydrated) Chromium trioxide Water	1.5 g 20 g 100 ml	Better than above for Zn–Cu alloys
I. *Other metals* Many of these require etching in aggressive solutions comprising various mixtures of HCl, HNO_3 and HF	(a) Hydrochloric acid (1.18) Nitric acid (1.40) Hydrofluoric acid (40%)§	50 ml 20 ml 30 ml	Platinum metals group, especially Ru, Os, Rh
	(b) Hydrochloric acid (1.19) Nitric acid (1.40) Hydrofluoric acid (40%)§	30 ml 15 ml 30 ml	Cr, Mo, W, V, Nb, Ta
Nitric acid/HF etches: These do not appear to be very sensitive to composition. HF should be 5–10%. Heating to 60–80°C will accelerate etching, e.g., for Ti	(c) Nitric acid (1.40) Hydrofluoric acid (40%)§ Water	30–45 ml 10 ml 60–45 ml	Highly alloyed Ti, Hf, Zr; also Cr, W, Mo, V
Aqueous HCl, HNO_3 etches. The reactivity can be reduced by adding water	(d) Hydrochloric acid (1.19) Nitric acid (1.40)	66 ml 34 ml	Au, Pt, Pd. Used for Co alloy if added to 34 ml water
Acidified hydrogen peroxide etch	(e) Hydrofluoric acid§ Hydrogen peroxide (30%) Water	10 ml 45–60 ml 45–30 ml	Dilute Ti, Hf, and Zr alloys
Nitric acid in alcohol ('Nital')	(f) Nitric acid (1.40) Methanol	10 ml 90 ml	Silver. (*Note*: It is dangerous to *store* more than 5% HNO_3 in ethanol in a tightly stoppered bottled or >33% HNO_3 in methanol). In an open dish, 5% or 10% HNO_3 in ethanol is safe.
Hydrochloric acid etches	(g) Hydrochloric acid (1.19) Water Ammonium chloride Picric acid‡	10 ml 90 ml 2 or 4 g 2 g	Be and its alloys, especially for large grain sizes
	(h) Hydrochloric acid (1.19) Water 60–80°C for 30–60 min	50 ml 50 ml	Cobalt alloys

---

* Acids are concentrated, unless otherwise indicated, e.g., with specific gravity.

‡ Special handling is required for picric acid to avoid the risk of fire and/or explosion and local regulations must be consulted before use of this reagent.

§ Hydrofluoric acid produces irreversible bone damage and presents a range of other hazards. Even dilute hydrofluoric acid solutions should be handled with great care. Note that hydrofluoric acid attacks laboratory glassware.

10.2 Microscopic examination

Metallographic specimens are normally prepared for examination with the light microscope by cutting out the piece to be examined (preferably not more than about 30 mm diameter) using a laboratory abrasive cut-off machine or a precision saw. Sometimes, specimens are cut in the shop or in the field using more aggressive methods, such as power hacksaws, dry abrasive cutting, and even by flame cutting. These techniques introduce a great deal of damage into the structure adjacent to the cut. It is generally best to re-section the specimen with a laboratory abrasive cut-off machine, rather than to try to grind through the damaged layer. In many cases, the specimen is encapsulated in a polymeric compound, either a compression mounting resin or a castable resin. Larger specimens of uniform shape may be prepared without mounting, but the edge retention may not be adequate for examination above 100X. The specimens are then subjected to at least one grinding step and one or more diamond abrasive steps, followed by one or more polishing steps with other abrasives, such as alumina or colloidal silica. However, the steps must be designed to remove the damage from sectioning. Moreover, each abrasive does produce damage proportional to its particle size. So, each step must remove the damage from the previous step, so that at the final step, the damage depth is so thin that etching will remove it. At the same time, the preparation procedure must keep the specimen surface flat and other problems (e.g., pull out, drag, smear, embedding, etc.) are prevented. The surface must be more than just reflective in nature if the true microstructure is to be revealed.

For some purposes, e.g., the study of slip processes involving individual dislocations using transmission electron microscopical studies of fine structure, and microindentation hardness testing under light loads, electropolishing has been considered in the past to be almost indispensable (specimen preparation for nanoindentation techniques is beyond the scope of the present section). In the case of microindentation, this is certainly unnecessary today with proper mechanical preparation methods. Preparation of specimens for replica work requires a proper metallographic preparation procedure, but this can be done mechanically. Preparation of TEM thin foils normally employs jet-electropolishing (or similar alternative procedures) to perforate the specimen. But, this is the final step after initial mechanical grinding. Electropolishing does have certain advantages, but also disadvantages. It is best for single phase metals, particularly for relatively pure metals. In alloys containing two or more phases, the rate of electropolishing varies for each phase and flat surfaces may be difficult to obtain. Even for single phase metals, the surface is often wavy rather than flat which makes high magnification examination difficult or impossible. Electropolishing solutions are often quite aggressive chemically, and may be dangerous, even potentially explosive under certain operating conditions, or under careless operating procedures.

The most frequent problems are: failure to completely remove the distorted metal produced at the original cut surface; alteration of the structure by overheating the specimen; contamination of the succeeding steps by carrying abrasives over, due to poor cleaning practices, from a coarse stage of polishing, to a finer one; and, the development of false structures by staining through faulty drying after etching or bleed out from shrinkage gaps between the specimen and mounting material. Preparation of an unfamiliar material should be guided by checking the progress of the grinding and polishing steps during the preparation sequence to determine if the scratches from the previous step have been removed and to detect any problems, such as pull out, drag, embedding, and so forth. After the preparation has been completed, through to about a 1 μm abrasive size, the specimen can be etched (or examined with polarised light if it is an optically anisotropic metal) to examine the development of the structure. This will aid the metallographer in determining if the preparation sequence has been adequate. Hard metals and ceramics or sintered carbides can often be adequately examined after preparing down to only a 1 μm finish, although going to a finer abrasive size will yield improved results.

Many metals and alloys can be prepared with essentially the same procedure with more than adequate results. However, there are many metals and alloys, and non-metallic materials, that do require quite different preparation approaches. These will be described in the text where appropriate. Etching, however, is quite specific to the metal under examination and the feature of the structure to be investigated. There are no truly universal etchants, although a few have wide applicability. However, because of the specific corrosion aspects of different metals and alloys, it is unlikely that any universal etch would produce optimal results for more than a very limited range of compositions.

MOUNTING

Specimens that are of irregular shape, or are fragile or small in size are best encapsulated or 'mounted' in polymeric materials. Several specimens, if of similar materials, may be prepared in the same mount, with a saving of time, although etching may become a problem. In general, it is best to prepare as small a specimen as feasible, for example a 100 mm^2 square area is ideal. With the introduction of automated equipment, specimen sizes have gradually become larger due to the ease of preparation compared to manual preparation. To obtain flatness out to the extreme edge, e.g., for examination of platings, coatings and other surface treatments, mounting is required. The preparation process can be optimised for obtaining good edge retention ('edge preservation'). Compression mounting resins, particularly thermosetting resins, yield best results. A modern press that cools the specimen automatically to near ambient temperature after polymerisation has been completed, yields much better results than the former 'hot ejection' method because shrinkage gaps are reduced. If a shrinkage gap is formed between specimen and mounting compound, a 'free edge' exists and it will become bevelled by abrasives getting into the gap. This is one reason why a protective plating, such as electroless nickel, is effective in providing good edge retention. Modern automated polishing devices yield better edge retention than manual polishing. Rigid grinding discs used with coarse diamond abrasives and hard, woven, napless clothes for the subsequent polishing steps, yield much better flatness than the older clothes, such as canvas, felt or billiard cloth.

Modern mounting presses have the heating elements and cooling channels built in so that one no longer needs to remove the heating element from around the mould cylinder after the polymerisation cycle has been completed (thermosetting resin) or after the resin melts (thermoplastic resin) and place copper chill blocks around the cylinder, required for a thermoplastic resin, for cooling the polymer to below 70°C under pressure. These presses were introduced in the mid 1970s, chiefly to facilitate moulding of thermoplastic resins. It was not until the 1990s that metallographers realised that cooling a thermosetting mount, already polymerised, towards ambient temperature under pressure helped reduce shrinkage gap formation which greatly improved edge retention.

To mount a specimen, the specimen is placed on the lower ram which is lowered into the cylinder about 25 mm (1 inch). The desired powdered resin (see Table 10.2) is added to a depth of at least 25–37 mm (1–1.5 inches). The ram is lowered all the way to the bottom and the top ram is placed into the cylinder hole and pushed downward and locked in place. The press is then turned on. Modern presses recognise the mould size (25, 30, 40 or 50 mm or 1, 1.25, 1.5 or 2 inches in diameter) and are set to the required pressure and temperature for either thermosetting or thermoplastic resins. Further, many can be set to have a pre-heat cycle where the temperature is brought up to the set point without applied pressure. This is helpful for more delicate specimens that might be folded over of bent by the initial application of pressure. The cooling cycle is controlled by either setting a cooling time, or a desired final temperature. Upon reaching this time or temperature, the cooling will be stopped, a bell rings, and the press is turned off automatically.

It is essential to verify that the structure of the metal will not be materially affected by any heat and pressure applied in forming the mount. In cases where the temperature of moulding, typically

Table 10.2   PLASTICS USED FOR MOUNTING

Plastics	Type	Remarks
Phenolic	Thermosetting	Needs controlled heat and pressure. Sufficiently inert to most solvents. Normal grades good for general work but have high shrinkage; mineral-filled type preferable for edge sections. If curing insufficient, e.g., too low a temperature, the mount is soft and is attacked by acetone. Badly degraded by high temperature etchants. Least expensive thermosetting resin.
Epoxy resin	Thermosetting	Cures under heat and pressure yielding excellent mounts with superior edge retention and resistance to solvents and boiling reagents. Good polishing characteristics, low shrinkage.
Diallyl phthalate	Thermosetting	Needs controlled heat (130–140°C) and pressure. Low shrinkage, good polishing characteristics.[3] Expensive resin, provides only marginal improvement over phenolics.
Methyl methacrylate	Thermoplastic*	Needs controlled heat and pressure. Gives clear mount. Attacked by acetone. Rather soft. High shrinkage.
Acrylic resins	'Cold' setting	Mix resin and hardener; cures very quickly (<10 min.). Cheap but poor edge retention. High shrinkage.
Polyester	'Cold' setting	Several ingredients to be mixed for each batch. Not used commonly today. Inert to usual solvents.
Epoxy resins	'Cold' setting	Liquid resin and hardener are mixed; gives good mounts without heat or pressure. Inert to usual solvents, although the surface can become sticky in alcohol.

* Must be cooled under pressure to low temperature to solidify before ejection.

about 150°C, cannot be tolerated (e.g., low-melting point metals and alloys, alloys where aging will occur, etc.), the so called 'cold' mounting resins are used. It must be pointed out that obtaining a very low exotherm (from the heat of polymerisation) is not automatic when using these resins. Acrylic resins are used widely because they are very inexpensive and they cure very quickly, usually in 5–8 minutes. However, they can generate considerable heat during polymerisation, enough to burn the metallographer's hand. Polyester materials are not utilised much in metallography. Cast epoxy resins are quite popular due to several characteristics, although they are relatively expensive and do have a shelf life of about a year. The shelf life can be improved by storing resin and hardener in a refrigerator (but do not freeze the resin or hardener). It is best to buy only a quantity that is expected to be used in about six months. The epoxy resin and hardener are mixed, usually on a weight basis (more accurate than by volume), stirred gently for about a minute, then poured into a mould in which the specimen resides face down. Moulds can be made of plastic, phenolic ring forms, silicone rubber, glass or metal. The phenolic ring forms are usually allowed to adhere to the polymer, while the others are coated with a mould release agent and removed after the mount cures. Low viscosity epoxy can be drawn into cracks and voids in specimens under vacuum. This greatly facilitates specimen preparation as the voids are supported by the epoxy and liquids cannot enter the openings and bleed out later.

Some plastics used for encapsulation, and their characteristics, are listed in Table 10.2.

Thermoplastics, such as methyl methacrylate, and thermosetting resins, such as phenolic, are convenient for routine work because they are available as powders immediately ready for use, but they require a press, and normally only one size of cylindrical mould may be available. On some presses, it is rather simple to change moulds, while on others it may take more time and effort. Cold-setting resins may be formed simply in a container consisting of a short piece of tube standing on a glass plate, and are therefore suitable for occasional use and odd shapes and sizes.

To examine a surface edge on a cross section, use the best mounting compound (thermosetting epoxy containing a filler, e.g., Epomet® resin). The edge can be protected by plating (e.g., with copper or electroless nickel) or, if that is not possible, by applying an evaporated coating before mounting. Mount specimens with a press that incorporates water cooling after the polymerisation cycle to minimise or eliminate shrinkage gaps between specimen and mount. If a 'cold' mounting compound must be used, use epoxy, never an acrylic or a polyester. Use a slow curing epoxy, rather than a fast curing epoxy. Soft (about 750–800 HV) ceramic shot ('flat-edge filler' shot) can be added to the epoxy at the polishing face to reduce shrinkage and help support the edge. Hard ceramic shot, such as made from alumina, is incompatible in grinding and polishing characteristics with metals and should be avoided. Use automated grinding/polishing equipment with central force loading, use 'hard' napless cloths or pads or rigid grinding discs (usually with coarse diamond abrasive sizes), and use proper loads and times. Do not use napped cloths, especially with diamond abrasive, except in the final step with a very fine abrasive.

Conductive mounting resins have also been developed to simplify electrolytic polishing and etching and for SEM or EMPA work. Copper-filled resins have been used for some time, although they are rather expensive. For example, Probemet™ conductive resin consists of very fine thermosetting epoxy with a silica filler mixed about equally with very fine copper flake. Flake shaped particles are better for conductivity than spherical particles. This resin, besides exhibiting zero resistance between specimen and mount, also yields excellent edge retention. Carbon-filled phenolic resins provide a lower-cost alternative, but are not as conductive and do not provide as good edge retention. Conductive nickel flake can be added to cast resins, such as epoxy, for conductivity.

GRINDING

Silicon carbide abrasive bonded to waterproof paper is used commonly for grinding. Silicon carbide paper is preferred over emery paper because SiC is harder, has sharper particles and cuts at a faster rate. The simplest technique, used generally only by students, employs rolled strips about 50 mm wide where a portion is unrolled and laid flat on a hard, tilted surface and held mechanically along the edges. Water is run onto the top end of the paper to wet the surface. The specimen is rubbed up and down on the paper strip. Simple devices exist to hold four or five rolls of graded paper, for example, 120 (P120), 240 (P280), 320 (P400), 400 (P800) and 600 (P1200) grit sizes. Usually, grinding begins with the coarsest grit, and the specimen is rubbed until all traces from cutting are removed. Then, turn the specimen 45–90° and rub the specimen with the next finer grit until the first set of scratches are removed. Repeat at least once, because the depth of the deformed layer is several times the depth of the residual scratches. Then, progress to the next finer paper or cloth, turning the specimen through 45–90°, and again rub until the previous scratches are removed, then to the next finer paper similarly, until grade 600 silicon carbide paper is reached. Some soft metals are prone to embedding problems, that is, the finer SiC particle sizes will break off the paper and become embedded in the alloy surface. This is a particular problem with the low melting point alloys of lead, tin, bismuth and cadmium. To counter this tendency, wax can be rubbed onto the paper surface before grinding. Candle wax appears to be better for embedding prevention than bee's wax.

A more suitable grinding practice is to use round SiC paper discs attached to a platen, usually made from aluminium. Copper-based alloys have been used in the past, but they are more expensive and are unsuitable when attack polishing agents are used in the polishing stage. The discs can be applied to the platen using a pressure-sensitive adhesive (psa) backing.* This is an excellent practice as the disc will not move under the applied force during grinding. The other main alternative is to use plain-backed SiC paper discs. Water is placed on the platen surface. The disc is placed over the wet platen and a hold down ring is placed around the periphery of the platen. The motor-driven platen is turned on and set to 240–300 rpm, in most cases. When the specimen is forced against the SiC disc, the disc stalls for a few seconds until suction builds up and holds the disc against the platen. This method is satisfactory for manual polishing but has some disadvantages in automated polishing. With an automated specimen holder, the holder must be positioned over the platen so that its periphery does not strike the hold down ring. This restricts grinding out to the edge and prevents the user from setting up the head position so that it cannot sweep slightly over the edge of the SiC paper on the platen. This type of hold down system generally eliminates use of smaller diameter platen formats, e.g., an 200 mm (8 inch) diameter platen system cannot be effectively used as the holder must be made smaller than 100 mm (4 inch) diameter, so that grinding and polishing can be performed inside the hold down ring without crossing the centre of the platen (when that happens, differential grinding and polishing results and specimens are not flat). If the holder must be <100 mm (4 inch) in diameter, it cannot hold many specimens and the specimen diameters must be quite small. Thus, the metallographer must choose a larger diameter platen system where the cost of consumables is higher.

For very hard metals, sintered carbides and ceramics, diamond hones[4] and laps[5] have been used for grinding, although today different products are more common. The original metal-bonded diamond discs used a metal plate over which diamond of a specific size was spread and was held in place with electroless nickel. While these discs would provide effective cutting, they could not be used manually as the surface tension between specimen and disc was so great that the specimen could not be hand held. These discs were largely unsuitable for metals, except the very hardest, but they worked adequately for sintered carbides and ceramics. Lapping discs became popular for a period and they work well as long as the surface remains flat. Their use requires a careful adjustment of the specimen holder so that the entire lap surface is in contact with the specimens. In the majority of cases, diamond slurries or suspensions were used. Procedures using one or two 'lapping' stages were developed. It must be stated that these lapping platens were not actually lapping in the true sense of the word. Lapping implies that the abrasive is free to roll between the lap surface and the sample surface. But, with these laps, the abrasive would become embedded in the lap surface soon after being sprayed onto the platen. Hence, cutting resulted rather than true lapping (lapped surfaces appear to be 'hammered' by the abrasive, that is, the surface is made flat and specular more by deformation than by cutting, and this is not desired as the true microstructure will not be observed).

In recent years, the rigid grinding disc (RGD) has been introduced. These are basically quite similar to the former laps that were popular from about 1975–1990, except that they are thinner and some extra relief has been introduced to reduce surface tension. The RGD costs less to produce and, because it is thin, usually wears out before it becomes too non-planar to use. One disc style incorporates round spots of epoxy containing fine particles of Fe and Cu with a fair amount of the disc surface not covered by the spots. This reduces the surface tension (specimens can be easily held manually using this disc) while actually providing a higher removal rate than a similar disc with greater surface coverage. Such a disc is used for rough polishing with diamond abrasive and sizes from 45 to 3 μm have been used on specimens with a hardness of about 150 HV or higher. Although sheet metal specimens softer than this have been successfully prepared using a RGD in the practice, and their hardness is well below 150 HV, some deformation may persist when somewhat harder solution annealed stainless steel or nickel-based superalloys are prepared. While titanium alloys can be prepared with such a disc, commercial purity titanium cannot, unless extra steps are taken to remove the damage. Consequently, a second disc was developed using a different epoxy and the filler is fine particles of tin, much softer than Fe and Cu. This disc is less aggressive than the previous disc and can prepare softer metals, but is still useful for very hard metals, sintered carbides and ceramics. The RGD can be used to replace the rough grinding step with SiC, or after the initial grinding step with SiC abrasive, or for the first and second steps. Specimens prepared using RGDs in one or more steps are noted for exceptional flatness and edge retention. However,

---

* Magnetic platens are also available and allow rapid exchange of the grinding/polishing disc.

the scratch pattern produced using diamond of a specific size with a RGD is more pronounced than when the same diamond is used on a hard, woven cloth or a chemotextile pad. The latter are less aggressive in removal rate than the RGD. There are other alternatives to grinding with SiC papers. Alumina abrasive can be obtained on waterproof backings, such as paper or Mylar. Alumina is a better abrasive for Fe-based metals than SiC, but these papers are less readily available and more expensive. Alumina is slightly softer than SiC (about 2000 versus 2200 HV, respectively), but is more than hard enough to grind any steel. Alumina is also tougher and less prone to embed in soft metals, such as Pb, Sn, Bi or Cd.

MECHANICAL POLISHING

Mechanical polishing is often done in two stages, with coarse and fine abrasives (one or more of each may be used). The division between coarse and fine is somewhat arbitrary, but is usually at 3 μm. The coarse polishing stage is carried out at 150 rpm, or less, and uses a napless cloth such as selected silk, nylon, polyester or a synthetic chemotextile pad. Napped cloths should not be used, except in the final step, as they promote poor edge retention, relief at constituents and problems such as pull-out or drag and smearing. Napped cloths, such as canvas, felt, synthetic suede and others, were used with coarse and fine diamond abrasives for both rough and fine polishing, but in more recent years they have been shown to be poor for all but the final step. In coarse polishing, the cloth is charged with an abrasive, chiefly diamond (except for those metals where diamond is not very effective). The author of this section prefers to charge a cloth with diamond paste first. The amount added must be sufficient to obtain good cutting; many people add far too little diamond. Turn the platen on (with the cloth attached using a psa backing, not stretched, as it will be easily ripped using an automated specimen holder) to a speed of 120–150 rpm. Hold the diamond syringe against the centre of the cloth and press on the applicator to squeeze diamond paste out of the tube onto the centre of the cloth. Then, while pressing on the syringe, pull the tip towards the periphery of the cloth, while 'laying' a concentric track of diamond paste. Turn off the platen. Take the tip of the index finger (which must be clean) and spread the diamond over the cloth surface. Then, squirt on some of the lubricant, that is compatible with the diamond carrier paste, and start polishing. This gets the cutting action started quickly. If, on the other hand, a fresh cloth is charged with a liquid diamond slurry or suspension of the same particle size, the cutting rate will be much lower until the diamond particles become embedded in the cloth and start cutting. If the polishing step is more than a minute or two, add some slurry or suspension to the cloth periodically during the cycle to keep the cutting rate high. The cloth must be kept moist during polishing. If a slurry or suspension is not added during the cycle (these have the lubricant built in), then the lubricant must be added periodically. If the cloth gets too dry, smear, drag and pull-outs may result. If it gets too wet, the cutting rate can drop due to hydroplaning effects, but this does not damage the structure.

Fine polishing is carried out with abrasives smaller than 3 μm in diameter and includes diamond, alumina and amorphous colloidal silica, plus some proprietary blends. MgO, used formerly to final polish Al and Mg alloys is employed rarely today, as it is very difficult to use. MgO is available usually only down to a 1 μm size and getting really good quality MgO for polishing is difficult. Further, magnesium carbonates will form in the cloth after use, which ruins the cloth, unless the cloth is soaked in a dilute HCl solution, which is quite inconvenient. Iron oxides and chromium oxides have been used occasionally in metallography, but they have little overall value. Cerium oxide has been used to polish glass but has little value with metals. Polishing is conducted at a lower rotational speed (80–150 rpm) than for grinding using medium or low-nap cloths, such as synthetic suede bonded to a waterproof backing, or with napless polyurethane pads. The polishing agent should have a cutting action but it may produce a 'flowed' layer on the surface or both. The author of this section has made Laue patterns of coarse grained specimens and seen the sharpness of the diffraction spots decrease after final polishing with alumina slurries, indicating that some smearing is occurring. Because of this, it is common practice to etch the specimen after final polishing. If the final step is repeated, the re-etched structure will appear to be sharper and crisper in detail due to removal of the smeared metal. As an alternative, the specimen can be given a final polish using a vibratory polisher with the same cloth and abrasive as used for the final step. As little as 20 minutes is required with this device. Vibratory polishing is noted for producing excellent, deformation-free surfaces without sacrificing edge retention and relief control. However, if carried out too long, relief will be observed.

Some metals or coatings are readily stained or corroded in the presence of water, and for these a non-aqueous polishing mixture is preferred. Diamond abrasives suspended in oil are available and are effective, but cleaning is more difficult. Oil-based diamond is very effective as the oil is a fine lubricant. Final polishing can be performed with alumina powder suspended in alcohol, purified kerosene or mineral spirits. Again, cleaning is more difficult than with water-based products. A few proprietary abrasive suspensions are available that have a low water content and appear to work satisfactorily with magnesium alloys where water-based products are best avoided, at least in the final step. Galvanised steel and cadmium-plated steels cannot be final polished with aqueous suspensions. If they are, the coating is attacked heavily. Non-aqueous abrasives must be used in the final step, perhaps with some loss of fine scratch control. Some specimens are prone to staining around inclusions, such as sulphides. In this case, the use of distilled water, rather than tap water, helps to avoid staining. Also, water temperature may be important. Hot water is more reactive than cooler water.

Colloidal silica containing amorphous, spherical silica particles in a basic suspension (pH of 9.5 to 10, usually), has become a very popular final polishing abrasive and has replaced MgO as the final polish for aluminium and its alloys. Because the particles are nearly spherical, its action is more chemical than mechanical and measurements of its removal rate reveal very low values. However, it does often produce the best final polished surfaces. Its use is not without problems. If the cloth is allowed to dry out, the silica becomes crystalline and the cloth is ruined. So, after use, the cloth must be cleaned carefully. Also, specimen surfaces can be more difficult to clean as the colloid contains ions, such as sodium ions. A whitish smut may be seen on the surface when the specimen is not cleaned carefully. This will make etching a disaster. To counter these problems, when using an automated system, stop adding colloidal silica with about 20 s left in the polishing step. With 10 s left, direct the water jet onto the platen surface. This will wash off the cloth and clean the specimen simultaneously. Then, the specimens can be rinsed under running water, wet with ethanol, and blow dried under warm air without retaining the surface smut. A similar approach can be used for manual work.

Metallographers may notice that etch response can be different when using colloidal silica. This can be good or bad. Colour etchants used with Cu-based alloys generally show softer, more pastel-like colours after using colloidal silica and more gaudy, harsher colours after using standard alumina suspensions. Experiments with specimens such as Fe–Ni alloys reveals a change in etch nature with repeated use of colloidal silica. For example, an Fe—36% Ni specimen was prepared and etched with Marble's reagent after final polishing with colloidal silica. The specimen exhibited a grain-contrast etch appearance, that is, light and dark grain twin areas. With repeated final polishing steps and re-etching, the grain contrast appearance changed to a flat, grain boundary/twin boundary etch. This switch has been observed with other metals. Detrimental effects of colloidal silica can be obtained using austenitic stainless steels or duplex stainless steels. After final polishing, etchants such as glyceregia or Vilella's reagent may be used and the etch response is generally rather slow. However, sometimes after final polishing with colloidal silica, etching is extremely rapid. As soon as the cotton swab touches the surface, it darkens. This effect is called 'flashing' by metallographers. Examination of the surface reveals a heavily crazed scratch pattern and the scratches are far too deep to remove with the final polishing agent. Instead, the specimen must be completely reprepared. This appears to be due to surface passivation, probably by the ions in the colloidal silica adhering to the polished stainless steel surface. Flashing never occurs when electrolytic etching is performed. A properly electrolytically etched surface can then be etched with glyceregia or Kalling's No. 2 reagent, for example, and it may flash. Flashing appears to be most common in etchants that contain Cl^- ions and careful cleaning appears to reduce its occurrence. To prevent this problem, the author of this section uses a two-step final polishing procedure. A timer is set for 3 minutes. Polishing begins with colloidal silica. After about 90 s direct a water jet onto the cloth and flush off most of the colloidal silica; then add alumina. The present author has used a special 0.05 μm alumina suspension made by the sol-gel process (called Masterprep™ alumina). Unlike calcined aluminas, it is totally free of agglomerates. Polishing continues until about 10 s remaining. At this point, direct the water jet onto the cloth and wash both the cloth and the specimens before the platen stops. Then, etching can be conducted without flashing and the benefits of both polishing agents are obtained.

Over the years a generic preparation practice was developed that is commonly called the 'traditional method' today. It consists of the following steps and can be performed manually or with automated grinder/polisher units:

The 'Traditional Method'

1. Grind with waterproof SiC abrasive paper starting with 120 grit (P120) paper, water cooled, 240–300 rpm, 25 or 30 N (6 lb) pressure/specimen until the cutting damage is removed and all specimens in the holder are at the same plane.
2. Grind with 220 or 240 grit (P240 or P280) SiC paper, as in step 1, for 1 minute.
3. Grind with 320 grit (P400) SiC paper, as in step 2.
4. Grind with 400 grit (P800) SiC paper, as in step 2.
5. Grind with 600 grit (P1200) SiC paper, as in step 2.
6. Rough polish with 6 μm diamond paste on canvas, 150 rpm, 25 or 30 N (6 lb) per specimen, for 2 minutes. Use complementary motion with an automated device but contrarotation with manual work. Some people used 3 μm diamond rather than 6 μm, and others used nylon cloths.

7. Fine polish with 1 μm diamond paste on felt or billiard cloth, as in step 6. This step is not used by all metallographers.
8. Fine polish with 0.3 μm α-alumina aqueous slurry on a synthetic suede cloth at 150 rpm, 25–30 N (6 lbf) load per specimen, for 2 minutes. Head/platen directions as in step 6. This was often an optional step if 7 was used.
9. Fine polish with 0.05 μm γ-alumina aqueous slurry, as in step 8.

In this method, complementary means that the head and platen are both rotating in the counter clockwise direction while contra means that the head and platen rotate in opposite directions. Complementary rotation cannot be done manually. This method works reasonably well for many materials, but not all. It is slow because of the many steps. It is not adequate for edge retention and may lead to excessive relief. As a result, 'contemporary' preparation methods are largely replacing the traditional method, because they are more efficient and yield better control of flatness.

The following is a generic example of a contemporary preparation practice. It requires use of a good laboratory sectioning machine with the proper wheel to minimise cutting damage.

The 'Contemporary Preparation Method'

1. Grind with waterproof SiC paper as fine as possible, usually 180, 220, 240 or 320 grit (P180, P240, P280, or P400), at 240–300 rpm, 25–30 N (6 lbf) load per specimen with water cooling, until the cutting damage is removed and the specimens are at the same plane.
2. Rough polish with 9 μm polycrystalline diamond on a psa-backed selected silk cloth (e.g., an Ultra-Pol™ cloth) at 150 rpm, 25–30 N (6 lbf) per specimen, complementary or contrarotation, for 5 min.
3. Rough polish with 3 μm polycrystalline diamond on a psa-backed chemotextile pad (e.g., a Texmet® 1000 pad), as in step 2, but for 4 min.
4. Fine polish with 1 μm polycrystalline diamond on a psa-backed polyester woven cloth (e.g., a Trident™ cloth), as in step 2, but for 3 min. This is often an optional step but is used for more difficult specimens.
5. Fine polish with 0.05 μm colloidal silica or sol-gel alumina (Masterprep™ alumina) slurries, on a synthetic suede (rayon) cloth (e.g., a Microcloth® pad) or a polyurethane pad (e.g., a Chemomet® pad), as in step 2, but for 2–3 min. Contrarotation is preferred for this step (see note in the next paragraph about head speed).

Choice of the final polishing solution in step 5 is often a matter of personal preference, although there are some materials that are not prepared properly with colloidal silica (Mg alloys tend to be etched, precious metals are not affected, even if an attack polish agent is added; pearlitic cast irons often have small etch spots; and, austenitic stainless steels and Ni-base superalloys may 'flash' when etched, as discussed above). Contrarotation should not be used when the head speed is >100 rpm, as the abrasive will be thrown off the wheel and will hit the metallographer and walls. If the head speed is 60 rpm or less, the abrasive stays nicely on the work surface. Contra is slightly more aggressive than complementary rotation.

Variations can be made to the contemporary method in each step depending upon personal preferences and what is available for use. Several alternatives, mentioned above, exist for the first step using SiC. Rigid grinding discs can be substituted for either step one or two. For some materials, it is possible to use as little as three steps with excellent results. These modifications are discussed when dealing with the individual metals and alloys.

After polishing by any method, the specimen must be thoroughly washed and dried. It is best to etch immediately after polishing, particularly for those metals and alloys that form a tight oxide (or other film) on the surface with air exposure, for example, Al, Cr, Nb, Ni, stainless steels, Ti and some precious metals. After the specimen is washed under running water, it may be necessary to scrub the surface carefully with cotton soaked in a soapy solution. A mild dish washing detergent can often be used, or a proprietary detergent such as liquid alconox. Then, wash again with clean running water, rinse with ethanol to displace the water, and dry with a blast of hot air. If the etch contains HF, it is best to rinse the surface in a neutralising bath to remove any adherent HF that might damage the microscope optics (see ASTM E 407).

Attack polishing is a method of improving polishing action by the addition of a dilute etching agent to the abrasive suspension. For instance, ammonia may be used with advantage on the pad in polishing copper alloys. Hydrogen peroxide (30% conc.) is useful for refractory metals and precious metals to improve metal removal. Not only is the damaged layer at the surface removed more effectively, but also scratch control in these metals is enhanced over final polishing with the same abrasive without the added chemical attack. Table 10.3 gives reagents for use with various metals by this method. Several solutions have also been proposed for magnesium alloys.[6] Attack polishing is often performed using polyurethane pads or with synthetic suede cloths (the latter do not hold up as well, however).

Table 10.3   ATTACK POLISHING CONDITIONS FOR VARIOUS METALS AND ALLOYS

See general safety note at the start of this Chapter.

Material	Solution*		Time (min)	Remarks
Uranium	CrO₃ H₂O HNO₃ (1.40)	50 g 100 ml 100 ml	20–30	Medium contrast under polarised light, no pitting, good resistance to oxidation
Zirconium	HNO₃ (1.40) Glycerol	50 ml 150 ml	1–10	Good contrast under polarised light. Slight grain relief
Bismuth	HNO₃ (1.40) Glycerol	50 ml 150 ml	3–5	Good contrast under polarised light. Requires less pressure than usual
Chromium	Oxalic acid H₂O	15 g 150 ml	5–10	Bright polish revealing oxides, etc.
Molybdenum and tungsten	Pot. Ferricyanide Sodium hydroxide Water	3.5 g 1 g 300 ml		

* Acids are concentrated, unless otherwise indicated, e.g., with specific gravity.

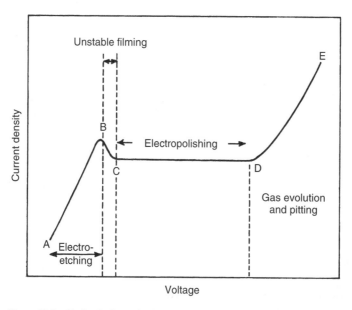

Figure 10.1   *Idealised relationship between current density and voltage in an electropolishing cell*

ELECTROLYTIC POLISHING

Extensive reviews have been given by Jacquet,[7] Tegart,[8] Petzow[1] and Vander Voort[2] which may be consulted for individual references (see also ASTM E 1558). A comparison with mechanical methods has been made by Samuels.[9]

The specimen is made the anode in a suitable solution, and conditions are adjusted so that the hills on the surface are dissolved much more rapidly than the valleys. When enough metal has been removed a smooth surface is obtained. The condition for polishing often corresponds to a nearly flat (i.e., constant current) region in the curve for cell current versus voltage. As the voltage is increased (see Figure 10.1), etching (AB) is replaced by film formation (BC). The voltage then increases and the current falls slightly as the film disappears and polishing conditions are established (CD). At higher voltages, gas evolution occurs with pitting. Near E gas evolution is rapid and polishing continues but the region just below D is preferred. By reducing the voltage to below B, the specimen can be etched in the same operation. For many specimens, electropolishing leads to a great saving in time, and it reliably produces surfaces free from strain provided sufficient metal is removed in the process. It tends to exaggerate porosity and is unsuitable for highly porous specimens. Inclusions are often removed, though not invariably, and their place taken by severe pits. Many two-phase and complex alloys, however, can be successfully polished.

Apparatus. To cover the widest range of applications, a d.c. supply of 4–5 A at voltages variable up to at least 60 V is required, but some solutions require only 2 V. Accurate voltage regulation is essential, and a rectifier set fed from a variac, a tapped battery or a potentiometer circuit across a constant d.c. source is recommended. Published recommendations for particular solutions sometimes state the voltage, and sometimes the current density, required. It is preferable to work on voltage, as the current density for a given electrode condition is much affected by temperature and other variables. If both are stated, but cannot be simultaneously obtained, the solution is probably wrong; if it is not, the current density should be disregarded. Two general cell arrangements are used: with electrodes in a beaker of still or gently stirred solution and with flowing or pumped electrolyte.[10–12] The first arrangement is easily set up and often suffices; the second is more powerful but requires more complicated apparatus (obtainable commercially, however). The characteristics are quite different: with flowing electrolyte a good polish may be obtained with more strongly conducting solutions, and hence with higher current densities, and it is therefore frequently possible to remove more metal in polishing and to start with a more roughly prepared surface. A small area of an article may be electropolished by the use of electrolyte flowing from a vertical jet above the article, the jet itself containing a projecting wire to act as cathode.[12] In suitable conditions, polishing of an area already ground with SiC paper may be completed in 3–10 s. Apparatus for this method is also available commercially.

Jacquet has described a device (the 'Ellapol') in which an electrolyte is applied to the surface by a small swab surrounding the cathode. The device can conveniently be used to polish a small area of a large component *in situ* (*see* e.g., refs. 13–15).

Solutions for electropolishing particular metals are listed in Table 10.4 [*see also* refs. 1, 2, 7, 8 and ASTM E 1558]. Table 10.4 is not a complete list, but should cover most requirements. Minor differences between solutions are often a consequence of the cell used. The most widely useful solutions are methanol—nitric acid mixtures, strong solutions of phosphoric acid and mixtures of perchloric acid with alcohol, acetic acid or acetic anhydride. Mixtures of perchloric acid with acetic anhydride, although frequently the best polishing agents, can be explosive and should be avoided. Electrolytes containing perchloric acid must be kept cold in use; plastics (especially cellulose) and bismuth must be kept away from them, as they may cause explosions, and they must not be stored in the laboratory as they are liable to explode without apparent reason. Note that the safe use of perchloric acid requires a dedicated fume hood, designed specifically for perchloric acid work, with an uninterrupted vertical stack and a stack wash down system. Quantities larger than a few hundred millilitres should not be used, in order to minimise the effects of an explosion, if one was to occur. The limits of the dangerous mixtures, according to Jacquet, the originator,[7,16] are indicated in Figure 10.2. Perchloric acid must always be added to the acetic anhydride-water mixture to avoid compositions in the detonation zone. Explosions outside the danger zone in this diagram have occurred. Consequently, the use of perchloric acid-acetic anhydride based electrolytes should be avoided (their use is forbidden by law in some locations). Safe operating practices in the metallography laboratory is a complex subject. In general, the metallography laboratory is a reasonably safe place, when well managed, but problems can occur, particularly in 'open' laboratories and at schools, where the personnel may not be as well versed in safe operating practices. It is impossible to list all possible safety issues in any text, as no one can envision all of the potential mis-uses that humans can create. Each laboratory must develop a comprehensive safety program based upon the materials that they use, their equipment, their particular circumstances and relevant regulatory requirements. The basic aspects of safe metallographic practices are given in references 2, 17, 18 and ASTM E 2014.

Because of the considerable number of solutions published in the literature, a selection has been made on the basis of (a) wide usage, (b) simplicity of composition, and (c) least danger. Temperatures should be in the range 15–35°C (or below). Cooling should be used to avoid temperatures above 35°C unless stated otherwise.

CHEMICAL POLISHING

Chemical polishing is adopted usually as a quick method of obtaining a passable result, rather than as a method of preparing a perfect surface. However, where it is difficult to prepare a deformation-free surface by other means, as with some very soft metals or where other difficulties are encountered, it may provide the best method of preliminary or final preparation. Chemical polishing of refractory

Table 10.4a   ELECTROLYTIC POLISHING SOLUTIONS FOR VARIOUS METALS AND ALLOYS

See general safety note at the start of this Chapter. Because of the considerable number of solutions published in the literature, a selection has been made on the basis of (a) wide usage, (b) simplicity of composition, and (c) least danger. Temperatures should be in the range 15–35°C (or below). Cooling should be used to avoid temperatures above 35°C unless stated otherwise.

	Composition of solution		Usage	Cell voltage	Time	Cathode
1	Ethanol	800 ml	Al alloys (not Al–Si)	30–80	15–60 s	Stainless steel
	Distilled Water	14 ml	Most steels	35–65	15–60 s	Stainless steel
	Perchloric Acid (1.61)§	60 ml	Lead alloys	10–35	15–60 s	Stainless steel
			Zinc alloys	20–60	15–60 s	Stainless steel
			Magnesium alloys	20–40	up to 2 min	Nickel
2	Ethanol	800 ml	Al alloys	35–80	15–60 s	Stainless steel
	Perchloric Acid (1.61)§	200 ml	Stainless steels	35–80	15–60 s	Stainless steel
			Lead alloys	15–35	15–60 s	Stainless steel
			Zinc alloys and many other metals	20–60	15–60 s	Stainless steel
3	Ethanol	940 ml	Stainless steel	30–45	15–60 s	Stainless steel
	Distilled Water	6 ml	Thorium	30–40	15–60 s	Stainless steel
	Perchloric Acid (1.61)§	54 ml				
4	Ethanol	700 ml	Al alloy			
	Water	120 ml	Steel, cast iron			
	2-Butoxyethanol	100 ml	Ni, Sn, Ag, Be	30–65	15–60 s	Stainless steel
	Perchloric acid (1.61)§	80 ml	Ti, Zr, U, Pb			
	Glycerol (100 ml) can replace butoxyethanol		Complex steels and nickel alloy general use			
5	Ethanol	760 ml	Al alloys including Al–Si alloys			
	Distilled water					
	Ether	30 ml	Fe–Si alloys	30–60	15–60 s	Stainless steel
	Perchloric acid (1.61)§	190 ml	Preferred solution for Al alloys			
		20 ml				
6	Methanol	590 ml	Germanium and silicon	25–35	30–60 s	Stainless steel
	Water (distilled)	6 ml	Titanium	~ 60	45 s	Stainless steel
	2-Butoxyethanol	350 ml	Vanadium	~ 30	3–5 s*	Stainless steel
	Perchloric acid§	54 ml	Zirconium	~ 70	15 s	Stainless steel
7	Glacial acetic acid	940 ml	Cr, Ti, U, Zr, Fe	20–60	up to 5 min	Stainless steel
	Perchloric acid (1.61)§	60 ml	Cast iron, all steels, V, Re and many other metal			
8	Glacial acetic acid	900 ml	Ti, Zr, U steels	10–60	up to 2 min	Stainless steel
	Perchloric acid (1.61)§	100 ml	Superalloys			
9	Glacial acetic acid	800 ml	U, Ti, Zr, Al steels	40–100	up to 15 min	Stainless steel
	Perchloric acid (1.61)§	200 ml	Superalloys			
10	Glacial acetic acid	700 ml	Nickel, Pb, especially Pb–Sb alloys	40–100	up to 5 min	Stainless steel
	Perchloric acid(1.61)§	300 ml				
11	Phosphoric acid (1.75)		Cobalt, Fe–Si alloys	1–2	up to 5 min	Stainless steel
12	Distilled water	300 ml	Cu, Cu alloys (not Cu–Sn)	1–1.6	10–40 min	Copper
	Phosphoric acid (1.75)	700 ml	Stainless steels rinse in 20% H_3PO_4			
13	Distilled water	600 ml	Brasses, Cu–Fe, Cu–Co	1–2	up to 15 min	Copper or stainless steel
	Phosphoric acid	400 ml	Co, Cd			

(continued)

Table 10.4a ELECTROLYTIC POLISHING SOLUTIONS FOR VARIOUS METALS—*continued*

Composition of solution		Usage	Cell voltage	Time	Cathode
14	Distilled water 200 ml Ethanol 400 ml Phosphoric acid (1.75) 400 ml	Al, Mg, Ag	25–30 at 40°C	4–6 min	Aluminium
15	Ethanol 300 ml Glycerol 300 ml Phosphoric acid (1.75) 300 ml	U (preferred solution)	20–30	4–6 min	Aluminium
16	Ethanol 500 ml Glycerol 250 ml Phosphoric acid (1.75) 250 ml	Mn, Mn–Cu	18	up to 10 min	Stainless steel
17	Ethanol 625 ml Phosphoric acid (1.75) 375 ml	Mg, Zn alloy	1.5–2.5	up to 30 min	Stainless steel
18	Ethanol 445 ml Ethylene glycol 275 ml Phosphoric acid (1.75) 275 ml	U alloys	18–20	up to 15 min	Stainless steel
19	Distilled water 750 ml Sulphuric acid (1.75) 250 ml	Stainless steel, iron, nickel Molybdenum	1.5–6.0 1.5–6.0	up to 10 min 1 min	Stainless steel Stainless steel
20	Methanol 875 ml Sulphuric acid (1.84) 125 ml	Molybdenum Keep below 27°C	6–18	1 min	Stainless steel
21	Distilled water 830 ml Chromium trioxide 170 g	Zn, Al bronze Brass	1.5–12	up to 1 min	Stainless steel
22	Distilled water 450 ml Phosphoric acid (1.75) 390 ml Sulphuric acid (1.84) 160 ml	Tin Tin bronzes (high tin) (rinse in 20% H_3PO_4)	2	up to 15 min	Copper
23	Distilled water 330 ml Phosphoric acid (1.75) 580 ml Sulphuric acid (1.84) 90 ml	Tin Tin bronzes (low tin <6%) (rinse in 20% H_3PO_4)	2	up to 15 min	Copper
24	Distilled water 170 ml Chromium trioxide 105 g Phosphoric acid (1.75) 460 ml Sulphuric acid (1.84) 390 ml	Stainless steel (use at 35–40°C)	2	up to 60 min	Stainless steel
25	Distilled water 240 ml Chromium trioxide 80 g Phosphoric acid (1.75) 650 ml Sulphuric acid (1.84) 130 ml	Stainless steel Alloy steels (use at 40–50°C)	2	up to 60 min	Stainless steel
26	Hydrofluoric acid (40%)‡ 100 ml Sulphuric acid (1.84) 900 ml	Tantalum Niobium (use at ~40°C)		5–15 min	Graphite
27	Glycerol 750 ml Glacial acetic acid 125 ml Nitric acid (1.40) 125 ml (*Warning*: This solution will decompose vigorously if kept, especially if cathode left in it. Safely discard solution as soon as finished with)	Bismuth	12	1–5 min	Stainless steel
28	Methanol 685 ml Hydrochloric acid (1.19) 225 ml Sulphuric acid (1.84) 90 ml Keep cool below 2°C Avoid water contamination	Molybdenum	19–35	30 s	Stainless steel

(continued)

Table 10.4a   ELECTROLYTIC POLISHING SOLUTIONS FOR VARIOUS METALS—*continued*

	Composition of solution		Usage	Cell voltage	Time	Cathode
29	Ethanol n-Butyl alcohol Hydrated aluminium trichloride Anhydrous Zn chloride	885 ml 100 ml 109 g 250 g	Ti and most other alloys	25–50	5 min	Stainless steel
30	The above diluted with 120 ml distilled water		Zinc	20–40	up to 3 min	Stainless steel
31	Glycerol Hydrofluoric acid (40%)‡ Nitric acid (1.40) As 27—will decompose on standing and must be discarded safely as soon as possible	870 ml 43 ml 87 ml	Zirconium	9–12	up to 10 min	Stainless steel
32	Potassium cyanide Potassium carbonate Gold chloride Distilled water to 1000 ml	80 g 40 g 50 g	Gold, silver	7–5	2–4 min	Graphite
33	Sodium cyanide Potassium ferrocyanide Distilled water to 1000 ml	100 g 100 g	Silver	2–5	up to 1 min	Graphite
34	Sodium hydroxide Distilled water to 1000 ml	100 g	Tungsten lead	6	10 min	Graphite
35	Methanol Nitric acid (1.40) (*Warning*: Do not keep longer than necessary. May become explosive. On no account substitute ethanol for methanol)	600 ml 330 ml	Ni, Cu, Zn, Ni–Cu Cu–Zn, Ni–Cr Stainless steel, In, Co Very versatile	40–70	10–60 s	Stainless steel

* With vanadium, give several 3–5 s bursts and avoid heating.

‡ Hydrofluoric acid produces irreversible bone damage and presents a range of other hazards. Even dilute hydrofluoric acid solutions should be handled with great care. Note that hydrofluoric acid attacks laboratory glassware.

§ A specially designed, dedicated fume hood, with a vertical stack and a wash down system is required for perchloric acid. Note that some solutions containing perchloric acid present grave explosion hazards and there are restrictions on the use of these solutions in some locations.

Table 10.4b   RECOMMENDED ELECTROPOLISHING SOLUTION FROM TABLE 10.4(a) FOR SPECIFIC METALS AND ALLOYS

Alloy	*Electrolyte* (No. in Table 10.4 (a))
Aluminium	1, 2, 4, 9, 14
Aluminium–silicon	5
Antimony	5
Beryllium	4
Bismuth	27
Cadmium	13
Cast iron	4, 7
Chromium	7
Cobalt	11, 13, 35
Copper and alloys	12, 13, 35
Copper–tin alloys	22, 23
Copper–zinc alloys	13, 21
Germanium	6
Gold	32
Hafnium	4

(*continued*)

Table 10.4b   RECOMMENDED ELECTROPOLISHING SOLUTION FROM
TABLE 10.4(a) FOR SPECIFIC METALS AND ALLOYS—*continued*

Alloy	*Electrolyte* (No. in Table 10.4 (a))
Indium	35
Iron-base alloys	4, 7, 8, 9, 19
Lead	1, 2, 4, 10, 34
Magnesium	1, 14, 17
Manganese	16
Molybdenum	19, 20, 28
Nickel and superalloys	4, 8, 9, 10, 19, 35
Niobium	26
Rhenium	7
Silver	4, 14, 32, 33
Stainless steels	1, 2, 3, 4, 7, 8, 9, 12, 19, 24, 25, 35
Steels: carbon and alloy	1, 4, 7, 8, 9, 19, 25
Tantalum	26
Thorium	3
Tin	4, 22, 23
Titanium	4, 6, 7, 8, 9, 29
Tungsten	34
Uranium	4, 7, 8, 9, 15, 17
Vanadium	6, 7
Zinc	1, 2, 17, 21, 30, 35
Zirconium	4, 6, 7, 8, 9, 31

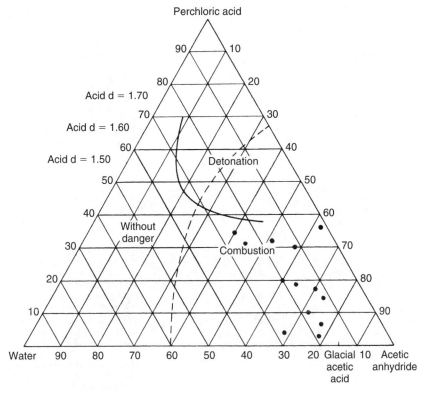

Figure 10.2   *Characteristics of perchloric acid/acetic anhydride/water solutions (after Jacquet)*[7,16].
• = *Typical electrolytes*

metals is often performed after mechanical polishing to improve polarised light response (e.g., for Zr, Hf), or to remove minor deformation (e.g., Nb, Ta, V).

In general, a ground specimen is immersed in the polishing agent, or swabbed with the solution, until a polish is obtained, and it is then etched or washed and dried, as appropriate. Reagents are listed in Table 10.5.

Table 10.5   REAGENTS FOR CHEMICAL POLISHING[1,2,8,19]

Metal	Reagent*		Time	Temp. (°C)	Remarks
Aluminium and alloys	Sulphuric acid (1.84) Orthophosphoric acid Nitric acid	25 ml 70 ml 5 ml	30 s–2 min	85	Very useful for studying alloys containing intermetallic compounds, e.g., Al–Cu, Al–Fe and Al–Si alloys
Beryllium	Sulphuric acid (1.84) Orthophosphoric acid (1.75) Chromic acid Water	1 ml 14 ml 20 g 100 ml	Several min	49–50	Rate of metal removal is approx. $1 \mu m\,min^{-1}$. Passive film formed may be removed by immersion for 15–30 s in 10% sulphuric acid
Cadmium	Nitric acid (1.4) Water	75 ml 25 ml	5–10 s	20	Cycles of dipping for a few seconds, followed immediately by washing in a rapid stream of water are used until a bright surface is obtained
Copper	Nitric acid Orthophosphoric acid Glacial acetic acid	33 ml 33 ml 33 ml	1–2 min	60–70	Finish is better when copper oxide is absent
Copper alloys	Nitric acid Hydrochloric acid Orthophosphoric acid Glacial acetic acid	30 ml 10 ml 10 ml 50 ml	1–2 min	70–80	Specimen should be agitated
Copper–zinc alloys	Nitric acid (1.40) Water	80 ml 20 ml	5 s	40	Use periods of 5 s immersion followed immediately by washing in a rapid stream of water. Slight variations in composition are needed for α–β and β–γ brasses to prevent differential attack. With β–γ alloys, a dull film forms and this can be removed by immersion in a saturated solution of chromic acid in fuming nitric acid for a few seconds followed by washing
Germanium	Hydrofluoric acid‡ Nitric acid Glacial acetic acid	15 ml 25 ml 15 ml	5–10 s	20	—
Hafnium	Nitric acid Water Hydrofluoric acid‡	45 ml 45 ml 8–10 ml	5–10 s	20	As for zirconium
Iron	Nitric acid Hydrofluoric acid (40%)‡ Water	3 ml 7 ml 30 ml	2–3 min	60–70	Dense brown viscous layer forms on surface; layer is soluble in solution. Low carbon steels can also be polished, but the cementite is attacked preferentially
Irons and steels	Distilled water Oxalic acid ($100\,g\,l^{-1}$) Hydrogen peroxide (30%)	80 ml 28 ml 4 ml	15 min	35	The solution must be prepared freshly before use. Careful washing is necessary before treatment. A microstructure is obtained similar to that produced by mechanical polishing, followed by etching with nital

(continued)

Table 10.5 REAGENTS FOR CHEMICAL POLISHING[1,2,8,19]—*continued*

Metal	Reagent*		Time	Temp. (°C)	Remarks
Lead	Hydrogen peroxide (30%)	80 ml	Periods of 5–10 s	20	Use Russell's reagent (Table 10.1) to check that any flowed layer has been removed before final polishing in this reagent
	Glacial acetic acid	80 ml			
Magnesium	Fuming nitric acid	75 ml	Periods of 3 s	20	The reaction reaches almost explosive violence after about a minute, but if allowed to continue it ceases after several minutes, leaving a polished surface ready for examination. Specimen should be washed immediately after removal from solution
	Water	25 ml			
Nickel	Nitric acid (1.40)	30 ml	½–1 min	85–95	This solution gives a very good polish
	Sulphuric acid (1.84)	10 ml			
	Orthophosphoric acid (1.70)	10 ml			
	Glacial acetic acid	50 ml			
Silicon	Nitric acid (1.40)	20 ml	5–10 s	20	1:1 mixture also used
	Hydrofluoric acid (40%)‡	5 ml			
Tantalum	Sulphuric acid (1.84)	50 ml	5–10 s	20	Solution is useful for preparing surfaces prior to anodising
	Nitric acid (1.40)	20 ml			
	Hydrofluoric acid (40%)‡	20 ml			
Titanium	Hydrofluoric acid (40%)‡	10 ml	30–60 s	—	Swab until satisfactory
	Hydrogen peroxide (30%)	60 ml			
	Water	30 ml			
	Hydrofluoric acid (40%)‡	10 ml			Few seconds to several minutes according to alloy
	Nitric acid (1.40)	10 ml			
	Lactic acid (90%)	30 ml			
Zinc	Fuming nitric acid	75 ml	5–10 s	20	As for cadmium
	Water	25 ml			
	Chromium trioxide	20 g	3–30 min	20	Solution must be replaced frequently
	Sodium sulphate	1.5 g			
	Nitric acid (1.40)	5 ml			
	Water to 100 ml				
Zirconium (also Hafnium)	Acid ammonium fluoride	10 g	½–1 min	30–40	Rate of dissolution varies markedly with temperature and is about 20–60 μm min⁻¹ in the given range
	Nitric acid (1.40)	40 ml			
	Fluosilicic acid	20 ml			
	Water	100 ml			
	Nitric acid (1.40)	40–45 ml	5–10 s repeated	—	Reaction is vigorous at air/ solution interface, and specimen is therefore held near surface of liquid. Hydrogen peroxide (30%) can be used in place of water.
	Water	40–45 ml			
	Hydrofluoric acid (40%)‡	10–15 ml			

* Acids are concentrated, unless otherwise indicated.
‡ Hydrofluoric acid produces irreversible bone damage and presents a range of other hazards. Even dilute hydrofluoric acid solutions should be handled with great care. Note that hydrofluoric acid attacks laboratory glassware.

ETCHING

Specimens should first be examined after polishing and before etching. This reveals features that have a significant difference in reflectivity from the main structure, or differences in colour and relief due to phases of large difference in hardness from the matrix. Nonmetallic inclusions, nitrides, graphite, cracks, pores, voids and various kinds of pits can be recognised clearly and should be recorded. These features can be quite difficult to see after etching.

In order to obtain the maximum resolution from the light optical microscope (LOM) with minimum reflections from stray light, the microscope must be set up using the 'Köhler' principle of illumination.[20] Most modern microscopes are constructed to achieve this principle and it is only necessary to adjust the two iris diaphragms. The first of these, usually called the field diaphragm, should project an image (of the aperture) sharply in focus on the specimen and should be adjusted so that the image just lies outside the field of view. The aperture diaphragm should be sharply in focus on the rear of the objective lens. It can be viewed by removing the eyepiece and should be adjusted (through an auxiliary lens on some microscopes) so that the image is centrally located on the rear of the objective and it illuminates 90% of the lens area. If after these adjustments the image is too bright, it should be dimmed by either reducing the light intensity or interposing a filter. Reduction of the aperture size reduces the resolution achieved by the lens, emphasises differences in level and can introduce artefacts. Bright field illumination is the most commonly used illumination mode in metallographic work.

To emphasise small differences in surface topography, or to take advantage of the optical anisotropy of certain metals with non-cubic crystal structures, several techniques exist that are usually available on good light optical microscopes. These include:

1. Dark field illumination[19,21]

By this technique a specimen is illuminated by an annulus of light which passes up the outside of the objective and is focused as a cone by a concave reflector. Thus, the normal beam of light is not used to form the image. Instead, the light scattered by angled surfaces is focused to make the image and thus the contrast is reversed. Cracks, inclusions and defects are seen as bright features on a black background. In some cases, the colour of particles may be quite different in dark field vs. bright field. Cuprous oxide in tough pitch copper is pale blue grey in bright field, as is the sulphide, but the oxide appears bright ruby red in dark field while the sulphide is invisible. Manganese sulphide, calcium sulphide and Mn,Ca sulphide all appear to be dove grey in colour in bright field. However, in dark field MnS is dark, although the interface with the matrix may be visible, but the Ca-containing sulphides are bright, with pure CaS the brightest. So, dark field can be quite useful for studying calcium-treated steels. Dark field is also quite useful for the examination of polymers and many minerals.

2. Interference microscopy[22–25]

Interference microscopes have been described by several authors but the most sensitive and useful techniques have been developed by Tolansky.[22] His multiple-beam interferometer can be used with conventional microscopes, at magnifications of up to about 250 times. Monochromatic light is essential and a parallel beam normal to the surface is used. An optical flat, silvered or aluminised to give about 95% reflectivity, is placed in contact with the specimen and is slightly tilted to produce a thin wedge between the two. The light is repeatedly reflected between the specimen and the optical flat, and interference takes place to produce very thin, sharp, dark fringes. The spacing can be varied by the tilt of the plate. Where a change in the surface of the specimen occurs, e.g., a step or depression, the interference fringe is displaced one way or the other. A total fringe displacement corresponds to a change in height of half the wavelength of light used; it is possible to measure surface displacements of about 5–250 nm. To obtain sharp fringes, the reflectivity of the metallic surface should be the same as the reference plate, i.e., approximately 95%, and it may be necessary to aluminise the specimen surface for the best results, or use a lower reflectivity reference mirror.

Normarski has designed a very sensitive microscope for detecting height variations on the surface of a specimen.[21,25] He used a conventional polarising microscope, into which a double quartz prism is inserted between the objective and the analyser. If a step is present, this produces two images slightly displaced to one another and interference between these produces light and dark fringes, the spacing of which can be varied by adjusting the prism. A modification of the technique produces interference contrast in images. This is often called DIC—differential interference contrast.[26] DIC can be a very effective tool for viewing microstructures that have slight height differences that may be invisible in bright field but show up clearly using this method.

3. Polarised light

Polarised light is an extremely powerful illumination method for studying inclusions and structures of unetched, electropolished or mechanically polished surfaces of metals and alloys with non-cubic crystal structures, for example, Be, Hf, α-Ti, U and Zr. In certain cases, metals with cubic crystal structures may be examined more effectively after etching using polarised light.

The equipment needed includes a strong source of illumination, a polariser (prism or polariod filter) that can be rotated to change the plane of polarisation, and an analyser of comparable material which can also be rotated. At least one of these, usually the polariser, must be adjustable and the other can be fixed. When the analyser is oriented so that its plane of polarisation is at 90° to that of the polariser, an isotropic specimen will appear black if the objectives and the microscope are well adjusted and no depolarisation occurs. The objectives should be strain free for the highest sensitivity. Then, if a properly prepared metal with a non-cubic crystal structure is examined, the structure will be revealed.

Bausch and Lomb developed the Foster prism[20] which used a special calcite crystal that gave excellent polarised light images. However, the original source for this material has been long exhausted. This system has a fixed crossed position and cannot be adjusted to move slightly off the crossed position. It is now available again on certain microscopes.

With isotropic metals and the polariser and analyser set in the 'crossed' position, no light reaches the eyepiece. However, with an optically anisotropic material, e.g., a hexagonal metal like beryllium, magnesium or zinc, the reflected beam becomes elliptically polarised and the intensity of the component normal to the plane of polarisation of the incident light depends on the orientation of the anisotropic structure. Thus, the intensity of light which passes through the analyser will be dependent on the orientation of the structure to the surface of the specimen and to the incident beam. The image will vary from dark to bright, according to orientation and the grain structure of an anisotropic material will be revealed. On rotating the specimen, the intensity of light passing through the analyser from any one grain will pass from minimum to maximum intensity every 45° to give four maxima per revolution at 90° to each other, with a minimum intensity at 45° to each maximum. If the analyser and polariser are not quite crossed and differ by a few degrees, only two maxima and minima occur. The difference in contrast between the maxima and minima is much greater under these conditions and improves the sensitivity of the method, especially for weakly anisotropic or pleochroic materials.

The method can be used for studying grain structures, twins and martensite. In uranium alloys, for instance, the isotropic gamma phase can be distinguished from weakly anisotropic retained beta phase and the strongly anisotropic variants of the alpha phase. Martensite can be produced in eutectoidal aluminium bronze (Cu—11.8% Al) when quenched from the beta field. This martensite is easily observed in as-polished specimens using polarised light. Martensite can be made in certain nonferrous shape-memory alloys and observed with polarised light after polishing. In other systems, such as aluminium, the grain structure can be revealed by anodising with Barker's reagent and is observed with polarised light. If a sensitive tint filter (sometimes called a full-wave plate, a first-order red plate, or a λ plate) is added to the light path, the grey-contrast image is converted into colour contrast. Some anodised alloys apparently produce an anisotropic oxide film which has orientations related to the underlying lattice orientations. But, anodising of aluminium with Barker's reagent[27] does not produce such a film, as often claimed, but disproved by Perryman and Lack.[28] If such a film is present, colour should be observed with bright field illumination due to interference effects, as with heat tinting or colour etching. However, colour is not observed. Instead, a roughened surface is produced that creates elliptical polarisation effects. The grains are coloured according to variations in crystal orientation but Barker's reagent does not reveal variations in composition within grains, even in as-cast specimens, as colour tint etches do.

The other useful application is in the identification of inclusions, although this is rarely used today due to the widespread availability of energy-dispersive or wavelength-dispersive spectroscopy systems with scanning electron microscopes and electron microprobes. Glassy inclusions such as silicates display the so called 'optical cross' in polarised light. Other inclusions are optically isotropic or anisotropic, e.g., MnO and MnS can be distinguished in steels. Both look similar in bright field but under polarised light, the oxide is bright and the sulphide is dark. Other inclusions are pleochroic and display characteristic colours under crossed polars. Examples are the grey-blue Cu_2O phase in tough-pitch copper which is ruby red under polarised light, and Cr_2O_3 which changes from blue grey to a beautiful, emerald green. References 21, 29–31 should be consulted for more detail.

The anisotropic metals include:

Antimony	Tin
Beryllium	Titanium
Cadmium	Uranium
Hafnium	Zinc
Magnesium	Zirconium

If prepared correctly, these metals will reveal their structure under crossed polarised light, although the quality of the images, and the strength of any observed colours, does vary.

The following metals can be made to respond to polarised light by anodising to produce an anisotropic film (see the note above about Barker's reagent), or by deep etching to produce an uneven surface with etch pits.

Aluminium	Molybdenum
Chromium	Nickel
Copper	Tungsten
Iron	Vanadium
Manganese	

Any fine lamellar structure, when etched, will reveal colour under polarised light because the etched fine lamellar structure produces elliptical polarisation. In a similar manner, solution annealed and aged beryllium copper produces beautiful colours under polarised light when etched due to the roughness created on the etched surface due to the overlapped coherency strains associated with the strengthening precipitates.

4. Colour[2,30,31]

Only two metals demonstrate natural colour: gold and copper. A few intermetallic phases exhibit natural colour, for example, $AuAl_2$, known as the 'purple plague' in the electronics industry. Others can be rendered in colour by:

(a) Optical methods of examination, such as polarised light, especially of etched structures with a sensitive tint plate inserted between polariser and analyser. This is available on most microscopes today and converts shades of grey into shades of colour. Unaffected regions that are white appear as magenta with a fixed sensitive tint filter. The colour pattern can be altered if the sensitive tint filter is adjustable (available on certain microscopes). Brighter colours are obtained in etched structures. Nomarski DIC also produces coloured surfaces where the colours correspond to height differences and can be adjusted by altering the Wollaston prism setting. These methods change shades of grey to colour contrast because of the use of the sensitive tint filter.

(b) By producing interference films by heat tinting or by chemical processes ('tint etching'). They can colour phases according to their reactivity or grains according to their crystal orientation. Most frequently, the reagents used deposit thin films of oxide, sulphide, chromate, phosphate or molybdate which cause colouration by interference of light. The colour depends on the thickness of films. Films that form on the less noble phase, usually the matrix, are produced by an anodic tint etch. Films that are preferentially deposited on the more noble phase, usually the second-phase particles, are produced by cathodic tint etches. A few tint etches, complex tint etches, colour both matrix and second phases. The anodic tint etches are most common. Heat tinting usually colours the matrix phase. The films grow according to the crystal orientation of the matrix phase, and growth is influenced by chemical segregation and residual deformation. A few tint etchants will colour MnS inclusions in steels white. Several etchants are known to preferentially attack or colour intermetallic compounds in Al and Mg alloys, but these are not tint etchants. They are usually called selective etchants. Some useful etchants are given in Tables 10.6 to 10.8. More details and examples are given in references 2, 30, 32–37.

Interference films can also be produced by vapour deposition of a compound to produce a thin dielectric film with a high refractive index. This increases the minor reflectivity differences between matrix and second-phase particles rendering them visible in colour contrast. Materials such as ZnTe, ZnSe, TiO_2, and ZnS have been commonly used for this purpose. The method was develop by Pepperhoff (1960) and is described in great detail by Bühler and Hougardy.[38] Gas contrast and reactive sputtering can also produce interference films.

5. Physical methods

(a) Cathodic vacuum etching (ion etching). The specimen is made a cathode in a high voltage gas discharge. High energy ions such as argon are accelerated at voltages of 1–10 kV and gas pressures of 10^{-2} Pa . This bombardment removes atoms at a rate dependent on orientation, presence of grain boundaries and intermetallic compounds.

(b) Thermal etching. On heating specimens, e.g., in vacuum or inert atmosphere, atoms are lost from weakly bound regions, e.g., grain boundaries. Surface tension forces lead to changes in surface topography at grain boundaries, leaving a structure characteristic of the high temperature. Thermal etching is used commonly with ceramics.

6. Chemical and electrochemical etching

Etching is usually an oxidation process. In general, elements with electrode potentials more negative than hydrogen will dissolve in many solutions, the rate depending on the local environment,

Table 10.6 COLOUR ETCHES FOR STEEL

See general safety note at the start of this Chapter.

Reagent			Procedure	Comments
1	Saturated sodium thiosulphate	50 ml	Immerse up to 2 min.	Klemm's I for carbon and alloys steels (also colours Cu alloys)
	Potassium metabisulphite	1 g		Colours ferrite, bainite and martensite, but not cementite
2	Saturated sodium thiosulphate	50 ml	Immersion up to 90 s	Klemm's II for austenitic Mn steels (also colours Cu alloys)
	Potassium metabisulphite	5 g		α martensite—brown, ε martensite—white Austenite—yellow to brown or shades of blue
3	0.6–10% HCl in water	100 ml	Immersion up to 60 s	Carbon, alloy and tool steels
	Ammonium bifluoride	0–1 g		Colours ferrite, bainite and martensite
	Potassium metabisulphite	0.5–1 g		Carbides unaffected
4	10–20% HCl in water	100 ml	Immerse up to 120 s	Stainless and maraging steels
	Ammonium bifluoride	0–2.4 g		Use lower HCl, no NH_4FHF for martensitic grades
	Potassium metabisulphite	0.5–1 g		Colours martensite and austenite but not carbides or nitrides

Table 10.7 COLOUR ETCHES FOR CAST IRON

See general safety note at the start of this Chapter.

Reagent			Procedure	Comments
1	HCl (35%)	2 ml	5–6 min immersion	Fe_3C—red/violet
	Selenic acid	0.5 ml	2–3 min if pre-etched in 2% Nital	Ferrite—white
	Ethanol	100 ml		Phosphides—blue/green
2*	Water	100 ml	Pre-etch in 2% nital	Phosphides—yellow/brown
	Sodium thiosulphate	24 g	Immerse until surface	Sulphides—white
	Citric acid	3 g	blue/violet	Ferrites and carbides—blue/violet
	Lead acetate	2.4 g		
3*	Water	100 ml	Pre-etch in 2% Nital	Ferrite—red/violet
	Sodium thiosulphate	24 g	(a) Immerse 20–40 s	Phosphides—orange/brown
	Citric acid	3 g	(b) Immerse 50–90 s	Carbides—blue/violet
	Cadmium chloride	2 g		Ferrite—yellow

* Mix in order tested. Age etchant for 24 h in a dark bottle before use. Large quantities can be made as a stock solution.

Table 10.8 COLOUR ETCHES FOR ALUMINIUM ALLOYS

See general safety note at the start of this Chapter.

Reagent			Procedure	Comments
1	Sodium molybdate	3 g	Immersion (time	AlCuFeMn script—blue
	Ammonium bifluoride	2 g	selected by trial	FeSiAl—brown/blue
	HCl (35%)	5 ml	and error)	Ni_3Al, FeNiAl—brown
	Water	100 ml		$CuAl_2$—pale blue
2	Chromic trioxide	20 g	Immerse 5–30 s	If alloy susceptible to intergranular
	Sulphuric acid (1.84)	20 ml		corrosion, grain boundaries
	Ammonium bifluoride	5 g		coloured black
	Water	100 ml		
3	Sulphuric acid (95–97%)	20 ml	Immerse 30 s at 70°C	AlCuFeMn—brown
	Water	80 ml		Mg_2Si—brown to black
				Outlines other phases
4	Sodium hydroxide	10 ml	Immerse 5 s at 70°C	$CuAl_2$, Mg_2Si—brown
	Water	90 ml		
5	Potassium permanganate	4 g	Immerse for up to 20 s	Weck's reagent
	NaOH	1 g		Colours alpha matrix
	Water	100 ml		

resulting in grain boundary attack or outlining of phases or other structures. For elements with electrode potentials more positive than hydrogen, or for elements which polarise, solutions containing oxidising agents are needed. Making the specimen the anode in a low voltage cell has the same effect. Indeed, most electropolishing solutions will cause etching if the voltage is reduced, usually by a factor of 10.

Detailed etching procedures for individual metals are given below but, as a general principle, iron alloys are usually etched in dilute oxidising agents, e.g., nitric acid or picric acid in ethanol (see safety note regarding picric acid later in this chapter). Stainless steels are usually etched in weak oxidising reagents in alkaline solution, e.g., alkaline ferricyanides. Virtually the same result is achieved electrolytically in a 1% potassium hydroxide solution.

In the case of copper alloys, most etchants require an ammoniacal atmosphere plus an oxidising reagent such as air (by swabbing), hydrogen peroxide or dichromates, permanganates, persulphates, etc. Electrolytic etching in an oxidising acid, e.g., 1% chromium trioxide often suffices, the control of etching being achieved by varying the voltage.

Nearly all pure metals are notoriously difficult to etch, e.g., aluminium. Reduction reactions can also be used and these cause staining or colouration of phases, especially intermetallic compounds.

A wide variety of etching reagents have been devised by empirical methods. Some of these are reproduced here but the perceptive reader will see that these can be modified easily to cope with new compositions. Small amounts of some components are needed to control pH or potential. Some solutions are not stable and their effectiveness will change with time. Others containing mixtures of oxidising agents and organic chemicals can become very dangerous with time and should only be retained for short times and discarded in an appropriate fashion, immediately after use.

Reference is given to simple electrolytic etches. They can be used with simple direct current power supplies. This gives more control over etching than provided by simple immersion or swabbing. For further control of the etch process it is recommended that these etchants be used with a potentiostat[39,40] to control the potential at a known and reproducible value. This will yield selective, reliable and consistent etching.[41–43]

Although most etching work is conducted to reveal the phases and constituents in metals and alloys, there are etchants that will reveal dislocations. Etching to reveal dislocations in minerals goes back to at least 1817 with a publication by Daniels. In 1927, Honess listed etchants used to reveal dislocations in minerals. However, it was not until the work of Shockley and Read in 1949 that dislocations were studied scientifically in metals. The first deliberate use of etch pitting reagents to reveal dislocations was published in 1952 by Horn and Gevers. This was the first direct proof that dislocations (the concept of dislocations was proposed in the 1930s to explain the apparent low strength of metals compared to their theoretical strength) could be revealed by etching. A wide range of etching techniques exist for revealing dislocations and many are listed in Table 10.9 [see also 2].

7. Washing and drying

On completion of etching (with most etches) the specimen should at once be flooded with water, washed free from water with ethanol (avoid using ether or methanol) and dried in a blast of hot air from a hand dryer, hair dryer or equivalent. With non-aqueous etches, it may be preferable to do the initial washing with alcohol or acetone rather than water.

Porous specimens are often easily stained by residual etching solution seeping out of the pores. This trouble can usually be avoided by prolonging the washing and drying stages and in severe cases by prolonged soaking of the specimen in alcohol or acetone before final drying, or by ultrasonic cleaning. When working with porous specimens, it is best to infiltrate the pores, if interconnected to the surface, with a low viscosity epoxy resin using a vacuum impregnation system. If the pores are not interconnected, place the polished surface in a vacuum impregnation chamber, face up, apply a thin layer of a low viscosity epoxy to the surface and drive the epoxy into the surface pores under vacuum, or by applying pressure (this requires a special pressurisation chamber). After it hardens, carefully remove the excess epoxy from the top surface and final polish the specimen. But, very little metal can be removed or the impregnated layer will be removed. Some metallographers have used molten wax to fill surface pores, then scrapped off the hardened wax on the surface and completed polishing. As the pores and voids are filled, bleed out does not occur. Gifkins demonstrated that EDTA (ethylene-diaminetetra-acetic acid) could be used to remove tarnish from the surface of polished lead specimens. He also found that cracked specimens of steel and brass could be etched stain-free by adding a few drops of EDTA to the etchant. An aqueous 2% EDTA solution can be used ultrasonically to remove stains. This solution can etch certain metals, so the time in the ultrasonic cleaner must be minimised [see 2 for more details].

Table 10.9 ETCHING REAGENTS FOR DISLOCATIONS

Taken largely from Lovell, Vogel and Wernick[44]. See general safety note at the start of this Chapter.

Metal or alloy	Reagent*		Remarks
Aluminium (99.99%) (see ref. 62)	Hydrochloric acid Nitric acid Hydrofluoric acid$^\otimes$	50% 4% 3%	–
	Hydrochloric acid Nitric acid Hydrofluoric acid$^\otimes$	50 ml 47 ml 3 ml	Lacombe and Beaujard's reagent[45,46]
	Hydrofluoric acid$^\otimes$ Hydrochloric acid Nitric acid Hydrogen peroxide (29% w/v)	37 ml 18 ml 9 ml 36 ml	Make A, 49 : 51 HF : H_2O_2 Make B, 65 : 35 HCl : HNO_3 Mix in ratio A:B = 5:2, care required Keep at 0–15°C in use (ref. 47)
Antimony	Hydrofluoric acid$^\otimes$ Nitric acid Acetic acid Bromine	3 pts 5 pts 3 pts 3 drops	Electrolytic etch on cleaved surface 2–3 s
	Hydrofluoric acid$^\otimes$ Superoxol	1 pt 1 pt	Electrolytic etch. 1 s
Bismuth	1% Iodine in methanol		Cleaved surface. 15 s
Brass (65% Cu–35% Zn)	0.2% Sodium thiosulphate		Electrolytic etch. $10\,A\,dm^{-2}$ 18–20°C. Remove film with hydrochloric acid
Brass (Alpha)[48]	Saturated aqueous ferric chloride Hydrochloric acid	50 ml 2 drops	—
Brass (Beta)[48]	Saturated aqueous ammonium molybdate Hydrochloric acid	30 ml 6–7 drops	Immerse the electrolytically polished surface for 30 min
Columbium (Niobium)	Sulphuric acid Hydrofluoric acid$^\otimes$ Water Superoxol$^{\ddagger\otimes}$	10 ml 10 ml 10 ml a few drops	Agitate specimen in solution
Copper (pure)	Saturated ferric chloride solution	4 pts	Rinse in ammonia solution
	Hydrochloric acid Acetic acid Bromine	4 pts 1 pt a few drops	See Ruff[49] for further solutions and references
Germanium (also 0.2 at. % boron, 6.0 at. % silicon, 0.2 at. % tin)	Hydrofluoric acid$^\otimes$ Nitric acid Acetic acid Bromine	3 pts 5 pts 3 pts 3 drops	3–5 s. Polish etch. 600 grit (P1200) carborundum ground surface
Germanium	Potassium ferricyanide Potassium hydroxide Water	8 g 12 g 100 ml	600 grit (P1200) carborundum ground surface. 2–5 min. Boiling solution
Iron (99.96%)	4% metanitrobenzosulphonic acid in ethanol		Long etch. Rinse in alcohol. Result questionable
Iron	(a) 1% nitric acid in ethanol (b) 0.5% picric acid in methanol†		1 min in (a) followed by rinse in methanol and 5 min in (b). Pits appear only in specimen cooled slowly from 750° to 800°C‡
	Fry's reagent Table 10.1, C(f)		10 s etch of chemically polished surface
Iron (99.99%)	Disa Electropol solution A–Z		Electrolytic etch. Observation by electron microscopy
Iron	2% Nital containing 2% of saturated picral		15 min
	Saturated picral		4 min. Anneal to decorate dislocations‡

(*continued*)

Table 10.9   ETCHING REAGENTS FOR DISLOCATIONS—*continued*

Metal or alloy	Reagent*		Remarks
Iron–silicon (3.25 Si)	Acetic acid	133 ml	Electrolytic etch. 3 A dm^{-2}. 17–19°C. Decorate with 0.004% carbon at 770°C or above in low-pressure acetylene atmosphere[‡]
	Chromium trioxide	25 g	
	Water	7 ml	
Nickel–manganese	Orthophosphoric acid	100 ml	Electrolytic etch. 2 min. 200 A dm^{-2} Copper cathode 40°C
	Ethanol	100 ml	
Silicon	Hydrofluoric acid$^{⊗}$	1 pt	15 min or longer. Chemically polished surface
	Nitric acid	3 pts	
	Acetic acid	12 pts	
	Hydrofluoric acid$^{⊗}$	4 pts	600 grit (P1200) carborundum ground surface
	Nitric acid	2 pts	
	3% aqueous mercuric nitrate	4 pts	
	Hydrofluoric acid$^{⊗}$	3 pts	600 grit (P1200) carborundum ground surface.
	Nitric acid	5 pts	Use de-ionised water
	Acetic acid	3 pts	Utensils and specimen must be dry
	3% aqueous mercuric nitrate	1.5–2 pts	2 min
	Hydrofluoric acid$^{⊗}$	160 ml	600 grit (P1200) carborundum ground surface
	Nitric acid	80 ml	
	Water	160 ml	
	Silver nitrate	8 g	
	Hydrofluoric acid$^{⊗}$		Use in ratio:
	Chromium trioxide	50 g	2:1 by vol. for large etch pits
	Water	per 100 g	1:1 for medium etch pits 2:3 for small etch pits Time 15 s[50]
Tellurium	Hydrofluoric acid$^{⊗}$	3 pts	1 min etch. Cleaved surface
	Nitric acid	5 pts	
	Acetic acid	6 pts	
Zinc	Chromium trioxide	160 g	Immerse with mild agitation for 1 min Chemically polish before etching. Dip in solution of 320 g chromium trioxide per litre after etching, to remove stain Decorate with 0.1 atomic % cadmium Anneal at 300–400°C.[§] Age 1 week at room temperature
	Hydrated sodium sulphate	50 g	
	Water	500 ml	
Zinc with 0.002% tin[51]	Saturated aqueous ammonium tungstate	35 ml	Etch by immersing for about 5 min Agitate to remove adherent layer
	Saturated aqueous ammonium molybdate	5 drops	Quench from 400°C and anneal 100–400°C[‡]
	Hydrochloric acid	5 drops	

* Acids are concentrated, unless otherwise indicated.
‡ Superoxol contains hydrogrn peroxide (30%) 1 pt, hydrofluoric acid (40%) 1 pt, water 4 pts.
§ Note that this heat treatment must alter the dislocation structure.
† Special handling is required for picric acid to avoid the risk of fire and/or explosion and local regulations must be consulted before use of this reagent.
$^{⊗}$Hydrofluoric acid produces irreversible bone damage and presents a range of other hazards. Even dilute hydrofluoric acid solutions should be handled with great care. Note that hydrofluoric acid attacks laboratory glassware.

TAPER SECTIONING

Taper sectioning[20,21] is used to give an apparent magnification of as much as 10 times by using a plane of section to cut the surface at a shallow angle, typically 6° (i.e., $\sin^{-1} 1/10$) so that the vertical features of the surface are given an apparent relative magnification of about 10 times. This magnification is only in the direction perpendicular to the coated surface when the taper angle is made as indicated in Figure 10.3.

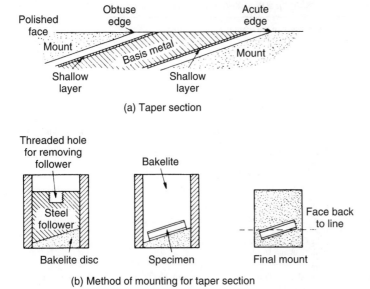

(a) Taper section

(b) Method of mounting for taper section

Figure 10.3   *Taper sectioning (from Vickers Projection Microscope Handbook)*

Taper sectioning requires plating of the surface to be examined, followed by mounting in plastic at a suitable angle. The plating may be done with nickel or copper from conventional plating baths. Assuming that the surface is ready to be examined, it must not be altered during the plating procedure; cleaning before plating must be confined to methods such as solvent washing and cathodic alkaline degreasing, which have no appreciable effect on the particular surface under examination. The mounting procedure is evident from Figure 10.3.

This method may also be used for examining thin intermediate layers, for example in electro-plating, or in the study of diffusion couples.

10.3   Metallographic methods for specific metals

10.3.1   Aluminium

PREPARATION

Aluminium and its alloys are soft and easily scratched or distorted during preparation. For cutting specimens, a laboratory abrasive cut-off saw should be used, with the correct wheel, and with light pressure and proper cooling, to avoid excessive heating and minimise deformation. Specimens are ground with waterproof SiC papers by the usual methods, but with slightly lower loads than in the contemporary method. Use 20 N (5 lbf) per specimen. Fine silicon carbide papers (600 grit (P1200) and finer) may cause embedding, particularly in the softer alloys. For high-purity aluminium, it may be advisable to coat the finer papers (if used) with paraffin wax. After the first grinding step (see the contemporary method described above) polishing is carried out in three or four steps, depending upon the alloy. Rigid grinding discs specially made for softer metals can be used for the second step, if substantial material must be removed. If heavy stock removal is not required, cloths can be used for step 2. Natural silk (with a psa backing or a magnetic platen set up) will yield the best surface finish, with excellent relief control and a reasonable removal rate. For steps 3 and 4 (if step 4 is chosen), use a woven polyester cloth rather than the chemotextile pad. For final polishing, either the napless polyurethane pad or a synthetic suede (rayon), medium nap cloth, can be used. The preferred final abrasive is colloidal silica. The sol-gel alumina suspension works nearly as well, and certainly far better than standard calcined aluminas. MgO is used rarely today as the final abrasive due to the many problems associated with its use, as discussed previously. The contemporary preparation

Table 10.10   MICROCONSTITUENTS WHICH MAY BE ENCOUNTERED IN ALUMINIUM ALLOYS

Microconstituent	Appearance in unetched polished sections
Al_3Mg_2	Faint, white. Difficult to distinguish from the matrix
Mg_2Si	Slate grey to blue. Readily tarnishes on exposure to air and may show iridescent colour effects. Often brown if poorly prepared. Forms Chinese script eutectic
$CaSi_2$	Grey. Easily tarnished
$CuAl_2$	Whitish, with pink tinge. A little in relief, usually rounded
$NiAl_3$	Light grey, with a purplish pink tinge
Co_2Al_9	Light grey
$FeAl_3$ (see note 1)	Lavender to purplish grey; parallel-sided blades with longitudinal markings
$MnAl_6$	Flat grey. The other constituents of binary aluminium–manganese alloys ($MnAl_4$, $MnAl_3$ and 'δ') are also grey and appear progressively darker. May form hollow parallelograms
$CrAl_7$	Whitish grey; polygonal. Rarely attacked by etches
Silicon	Slate grey. Hard, and in relief. Often primary with polygonal shape—use etch to outline
α(Al–Mn–Si) (see note 2)	Light grey, darker and more buff than $MnAl_6$
β(Al–Mn–Si) (see note 2)	Darker than α(Al–Mn–Si), with a more bluish grey tint. Usually occurs in long needles
Al_2CuMg	Like $CuAl_2$ but with bluish tinge
Al_6Mg_4Cu	Flat, faint and similar to matrix
(Al–Cu–Mn) (see note 3)	Grey
α(Al–Fe–Si) (see note 4)	Purplish grey. Often occurs in Chinese-script formation. Isomorphous with α(Al–Mn–Si)
β(Al–Fe–Si) (see note 4)	Light grey. Usually has a needle-like formation
(Al–Cu–Fe) (see note 5)	Grey α phase lighter than phase (*see* Note 5)
(Al–Fe–Mn) (see note 6)	Flat grey, like $MnAl_6$
(Al–Cu–Ni)	Purplish grey
(Al–Fe–Si–Mg) (see note 7)	Pearly grey
$FeNiAl_9$	Very similar to and difficult to distinguish from $NiAl_3$
(Al–Cu–Fe–Mn)	Light grey
$Ni_4Mn_{11}Al_{60}$	Purplish grey
$MgZn_2$	Faint white; no relief

The constituents are designated either by formulae denoting the compositions upon which they appear to be based, or by the elements, in parentheses, of which they are composed. The latter nomenclature is adopted where the composition is unknown, not fully established, or markedly variable.

(1) On very slow cooling under some conditions, $FeAl_3$ decomposes into Fe_2Al_7 and Fe_2Al_5. The former is micrographically indistinguishable from $FeAl_3$. The simpler formula is retained for consistency with most of the original literature.
(2) α(Al–Mn–Si) is present in all slowly solidified aluminium–manganese–silicon alloys containing more than 0.3% of manganese and 0.2% of silicon, while β(Al–Mn–Si), a different ternary compound, occurs above approximately 3% of manganese for alloys containing more than approximately 1.5% of silicon. α(Al–Mn–Si) has a variable composition in the region of 30% of manganese and 10–15% of silicon. The composition of β(Al–Mn–Si) is around 35% of manganese and 5–10% of silicon.
(3) (Al–Cu–Mn) is a ternary compound with a relatively large range of homogeneity based on the composition $Cu_2Mn_3Al_{20}$.
(4) α(Al–Fe–Si) may contain approximately 30% of iron and 8% of silicon, while β(Al–Fe–Si) may contain approximately 27% of iron and 15% of silicon. Both constituents may occur at low percentages of iron and silicon.
(5) The composition of this phase is uncertain. Two ternary phases exist. α(Al–Cu–Fe) resembles $FeAl_3$; β(Al–Cu–Fe) forms long needles.
(6) The phase denoted as (Al–Fe–Mn) is a solid solution of iron in $MnAl_6$.
(7) This constituent is only likely to be observed at high silicon contents.

method, with these minor changes, yields excellent surfaces with no relief, good edge retention, and freedom from artefacts, so that the true structure can be revealed. Many different second-phase particles, or intermetallic phases, can be encountered in aluminium and its alloys, see Table 10.10, and their polishing characteristics can vary somewhat.

Avoid using slurry or suspensions of fine diamond abrasives when polishing aluminium and its alloys as they are likely to become embedded in the surface. This is chiefly a problem with 1 μm diamond and finer sizes. When these fine sized diamond abrasives are added as a slurry, the particles roll between the specimen and the woven cloth or pad and can become embedded in the specimen as easily as in the cloth. This problem is not observed with coarser diamond added as slurries or suspensions. Fine diamond is best added as a paste where it is embedded in the cloth even before the wheel is turned on. See the previously described practice for charging a cloth with diamond paste.

Many aluminium alloys contain the reactive compound Mg_2Si. If this constituent is suspected, non-aqueous solvents, such as mineral spirits, purified kerosene or ethanol may be substituted for water during the final polishing step to avoid loss of the reactive particles by corrosion.

It should be noted that some aluminium alloys are liable to undergo precipitation reactions at the temperatures used to cure thermosetting mounting resins; this applies particularly to aluminium-magnesium alloys, in which grain boundary precipitates may be induced. A 'cold' setting resin should be used, but care must be taken with the set up to minimise the heat generated in polymerisation (see previous discussion under Mounting).

Electropolishing is often rapid and convenient (*see* Tables 10.4(a) and 10.4(b) but there may be nonuniform polishing or attack of the intermetallic phases.

ETCHING

Aluminium alloys now in use cover many complex alloying systems. A relatively large number of etching reagents have been developed; only those whose use has become more or less standard practice are given in Table 10.11. Many etches are designed to render the distinction between the many possible microconstituents easier, and the type of etching often depends on the magnification to be used. The identification of constituents, which is best accomplished by using cast specimens where possible, depends to a large extent on distinguishing between the colours of particles, so that the illumination should be as near as possible to daylight quality. It is recommended that a set of specially prepared standard specimens, containing various known metallographic constituents, be used for comparison. Because of these problems, and with the development of EDS and WDS techniques available on scanning electron microscopes and electron microprobes, the use of etchants to identify precipitates is infrequent today. However, as an aid to selecting constituents for image analysis measurements, these etchants have value as they can enhance the detectability of the desired particles.

Some etching reagents for aluminium require the use of a high temperature; in such cases the specimen should be preheated to this temperature by immersion in hot water before etching. For washing purposes, a liberal stream of running water is advisable.

Electrolytic etching of aluminium alloys

In addition to the reagents given for aluminium in Table 10.11, the following solutions have been found useful for a restricted range of aluminium-rich alloys:

1. The following solution has been used for grain orientation studies:
Orthophosphoric acid (density 1.65)	53 ml
Distilled water	26 ml
Diethylene glycol monoethyl ether	20 ml
Hydrofluoric acid (48%)	1 ml

   The specimen should be at room temperature and electrolysis is carried out at 40 V and less than $0.1\,A\,dm^{-2}$ An etching time of 1.5–2 min is sufficient for producing grain contrast in polarised light after electropolishing.

2. The solution below is also used for the same purpose and is more reliable for some alloys:
Ethanol	49 ml
Water	49 ml
Hydrofluoric acid	2 ml (quantity not critical)

   The specimen is anodised in this solution at 30 V for 2 min at room temperature. A glass dish must be used. Not suitable for high-copper alloys.

3. For aluminium alloys containing up to 7% of magnesium:
Nitric acid (density 1.42)	2 ml
40% hydrofluoric acid	0.1 ml
Water	98 ml

   Electrolysis is carried out at a current density of $0.3\,A\,dm^{-2}$ and a potential of 2 V. The specimen is placed 7.6 cm from a carbon cathode.

4. For cast duralumin:
Citric acid	100 g
Hydrochloric acid	3 ml
Ethanol	20 ml
Water	977 ml

   Electrolysis is carried out at $0.2\,A\,dm^{-2}$ and a potential of 12 V.

Table 10.11 ETCHING REAGENTS FOR ALUMINIUM AND ITS ALLOYS

See general safety note at the start of this Chapter.

No.	Reagent		Remarks
1	Hydrofluoric acid (40%)§ Hydrochloric acid (1.19) Nitric acid (1.40) Water (Keller's etch)‡	0.5 ml 1.5 ml 2.5 ml 95.5 ml	15 s immersion is recommended. Particles of all common micro-constituents are outlined. Colour indications:

Mg_2Si and $CaSi$: blue to brown
α(Al–Fe–Si) and (Al–Fe–Mn): darkened
β(Al–Cu–Fe): light brown
$MgZn_2$, $NiAl_3$, (Al–Cu–Fe–Mn),
Al_2CuMg and Al_6CuMg: brown to black
α(Al–Cu–Fe) and (Al–Cu–Mn): blackened
Al_3Mg_2: heavily outlined and pitted

The colours of other constituents are little altered. Not good for high Si alloys

2	Hydrofluoric acid (40%)§ Water	0.5 ml 99.5 ml	15 s swabbing is recommended. This reagent removes surface flowed layers, and reveals small particles of constituents, which are usually fairly heavily outlined. There is little grain contrast in the matrix. Colour indications:

Mg_2Si and $CaSi_2$: blue
$FeAl_3$ and $MnAl_6$: slightly darkened
$NiAl_3$: brown (irregular)
α(Al–Fe–Si): dull brown
(Al–Cr–Fe): light brown
Co_2Al_9: dark brown
(Al–Fe–Mn): brownish tinge
α(Al–Cu–Fe), (Al–Cu–Mg)
and (Al–Cu–Mn): blackened
α(Al–Mn–Si), β(Al–Mn–Si) and (Al–Cu–Fe–Mn) may appear light brown to black
β(Al–Fe–Si) is coloured red brown to black
The remaining possible constituents are little affected

3	Sulphuric acid (1.84) Water	20 ml 80 ml	30 s immersion at 70°C; the specimen is quenched in cold water. Colour indications:

Mg_2Si, Al_3Mg_2 and $FeAl_3$: violently attacked, blackened and may be dissolved out
$CaSi_2$: blue
α(Al–Mn–Si) and β(Al–Mn–Si): rough and attacked
$NiAl_3$ and (Al–Cu–Ni): slightly darkened
β(Al–Fe–Si): slightly darkened and pitted
α(Al–Fe–Si), (Al–Cu–Mg)
and (Al–Cu–Fe–Mn): outlined and blackened

Other constituents are not markedly affected

4	Nitric acid (1.40) Water	25 ml 75 ml	Specimens are immersed for 40 s at 70°C and quenched in cold water. Most constituents (not $MnAl_6$) are outlined. Colour indications: β(Al–Cu–Fe) is slightly darkened Al_3Mg_2 and AlMnSi: attacked and darkened slightly Mg_2Si, $CuAl_2$, (Al–Cu–Ni) and (Al–Cu–Mg) are coloured brown to black

5	Sodium hydroxide Water	1 g 99 ml	Specimens are etched by swabbing for 10 s. All usual constituents are heavily outlined, except for Al_3Mg_2 (which may be lightly outlined) and (Al–Cr–Fe) which is both unattacked and uncoloured. Colour indications:

$FeAl_3$ and $NiAl_3$: slightly darkened
(Al–Cu–Mg): light brown
α(Al–Fe–Si): dull brown*
α(Al–Mn–Si): rough and attacked; slightly darkened*
$MnAl_6$ and (Al–Fe–Mn): coloured brown to blue (uneven attack)
$MnAl_4$: tends to be darkened
The colours of other constituents are only slightly altered

(continued)

Table 10.11   ETCHING REAGENTS FOR ALUMINIUM AND ITS ALLOYS—*continued*

No.	Reagent		Remarks
6	Sodium hydroxide Water	10 g 90 ml	Specimens immersed for 5 s at 70°C, and quenched in cold water. Colour indications:

β(Al–Fe–Si): slightly darkened
$Mn_{11}Ni_4Al_{60}$: light brown
β(Al–Cu–Fe): light brown and pitted
$CuAl_2$: light to dark brown
$FeAl_3$: dark brown

($FeAl_3$ is more rapidly attacked in the presence of $CuAl_2$ than when alone)

$MnAl_6$, $NiAl_3$, (Al–Fe–Mn),
$CrAl_7$ and AlCrFe: blue to brown
α(Al–Fe–Si), α(Al–Cu–Fe),
$CaSi_2$ and (Al–Cu–Mn): blackened

No.	Reagent		Remarks
7	Sodium hydroxide Sodium carbonate (in water)	3%–5% 3%–5%	Useful for sensitive etching where reproducibility is essential. In general, the effects are similar to those of Reagent 5, but the tendency towards colour variations for a given constituent is diminished. Particularly useful for distinguishing $FeNiAl_9$ (dark blue) from $NiAl_3$ (brown). Potassium salts can be used.
8	Nitric acid Hydrofluoric acid[§] Glycerol	20 ml 20 ml 60 ml	A reliable reagent for grain boundary etching, especially if the alternate polish and etch technique is adopted. The colours of particles are somewhat accentuated
9	Nitric acid, 1% to 10% by vol. in alcohol		Recommended for aluminium–magnesium alloys. Al_3Mg_2 is coloured brown. 5–20% chromium trioxide can be used
10	Picric acid[†] Water	4 g 96 ml	Etching for 10 min darkens $CuAl_2$, leaving other constituents unaffected. Like Reagent 4
11	Orthophosphoric acid Water	9 ml 91 ml	The reagent is used cold. Recommended for aluminium–magnesium alloys in which it darkens any grain boundaries containing thin β-precipitates. Specimen is immersed for a long period (up to 30 min). Mg_2Si is coloured black, Al_3Mg_2 a light grey, and the ternary (Al–Mn–Fe) phase a dark grey
12	Nitric acid		10 s immersion colours Al_6CuMg_4 greenish brown and distinguishes it from Al_2CuMg, which is slightly outlined but not otherwise affected
13	Nitric acid (density 1.2) Water Ammonium molybdate, $(NH_4)_6Mo_7O_{21},4H_2O$	20 ml 20 ml 3 g	20 ml of reagent are mixed with 80 ml alcohol. Specimens are immersed and well washed with alcohol after etching. Brilliant and characteristic colours are developed on particles of intermetallic compounds. The effects depend on the duration of etching, and for differentiation purposes standardisation against known specimens is advised
14	Sodium hydroxide (various strengths, with 1 ml of zinc chloride per 100 ml of solution)		Generally useful for revealing the grain structure of commercial aluminium alloy sheet[27]
15	Hydrochloric acid (37%) Hydrofluoric acid (38%)[§] Water	15.3 ml 7.7 ml 77.0 ml	Recommended (30 s immersion at room temperature) for testing the diffusion of copper through claddings of aluminium, aluminium–manganese–silicon, or aluminium–manganese on aluminium–copper–magnesium sheet. Zinc contents up to 2% in the clad material do not influence the result[52]
16	Ammonium oxalate Ammonium hydroxide, 15% in water	1 g 100 ml	Develops grain boundaries in aluminium–magnesium–silicon alloys. Specimens are etched for 5 min at 80°C in a solution freshly prepared for each experiment

[*] These are isomorphous and the colour depends on the proportion of Mn and Fe.

[‡] Sodium fluoride can be used in place of HF in mixed acid etches.

[§] Hydrofluoric acid produces irreversible bone damage and presents a range of other hazards. Even dilute hydrofluoric acid solutions should be handled with great care. Note that hydrofluoric acid attacks laboratory glassware.

[†] Special handling is required for picric acid to avoid the risk of fire and/or explosion and local regulations must be consulted before use of this reagent.

5. For commercial aluminium:

Hydrofluoric acid (40%)	10 ml
Glycerol	55 ml
Water	35 ml

This reagent, used for 5 min at room temperature, with a current density of $1.5\,A\,dm^{-2}$ and a voltage of 7–8 V, is suitable for revealing the grain structure after electropolishing.[53]

6. For distinguishing between the phases present in aluminium-rich, aluminium–copper– magnesium alloys, electrolytic etching in either ammonium molybdate solution or 0.880 ammonia has been recommended. In both cases, Al_2CuMg is hardly affected, $CuAl_2$ is blackened, Al_6Mg_4Cu is coloured brown, while Mg_2Al_3 is thrown into relief without change of colour.[54]

Note: Hydrofluoric acid produces irreversible bone damage and presents a range of other hazards. Even dilute hydrofluoric acid solutions should be handled with great care. Note that hydrofluoric acid attacks laboratory glassware.

Anodising

Etching to reveal the grain structure of aluminium alloys is not always easy. In cases where standard etchants do not reveal the grain structure, anodising solutions are utilised. Of these, Barker's reagent[27] is the most widely used. This is an electrolytic etching method where an aluminium cathode is used; its area should be substantially greater than the polished surface to be etched. The specimen is the anode and is parallel to the cathode with about a 10 mm spacing. A dc current of 20–45 V is placed across the anode and cathode while they are immersed in the electrolyte, a 1.8% solution of fluoroboric acid in water. Voltage is not too critical. Good results have been obtained with 2 min of etching at 20 or 30 V dc. This roughens the specimen surface and no colour is observed in bright field. Contrary to many publications, Barker's reagent does not create a film at the polished face. High magnification examination of the anodised surface with the SEM reveals a fine roughening. When viewed with polarised light, the grain structure is revealed with grains in different shades from white to black. If a sensitive tint plate is insert in the light path, then the grains exhibit colours. Anodising does not reveal chemical segregation ('coring') within the grain structure, only grain orientation differences. There are other anodising solutions for aluminium, and for other metals [see 2 for a complete listing]. Some of these may produce an interference layer but Barker's does not.

GRAIN COLOUR ETCHING[55]

For many aluminium alloys containing copper, and especially for binary aluminium-copper alloys, it is found that Reagent No. 1 of Table 10.11 gives copper films on cubic faces which are subject to preferential attack and greater roughening of the surface. Subsequent etching with 1% caustic soda solution converts the copper into bronze-coloured cuprous oxide, and a brilliant and contrasting representation of the underlying surfaces is obtained. The technique is of use in orientation studies in so far as the films are dark and unbroken on {100} surfaces, but shrink on drying on other surfaces. In particular {111} faces have a bright yellow colour with a fine network on drying, which has no preferred orientation, while {110} faces develop lines (cracks in film) which are parallel to a cube edge.

ETCHING TO PRODUCE ETCH PITS[45]

The orientation of crystal grains may be determined by developed etch pits on specimens previously electropolished (see Table 10.4). The following reagents develop pits with facets parallel to the {100} cube planes:

	a	*b*	*c*
Fuming nitric acid	15 ml	15 ml	47 ml
Pure hydrochloric acid	45 ml	46 ml	50 ml
Hydrofluoric acid	15 ml	10 ml	3 ml
Distilled water	25 ml	29 ml	—

Note: Hydrofluoric acid produces irreversible bone damage and presents a range of other hazards. Even dilute hydrofluoric acid solutions should be handled with great care. Note that hydrofluoric acid attacks laboratory glassware.

Mixture *c* is recommended where only a few large well-formed etch figures are preferred. Etching with dry gaseous hydrochloric acid gives pits with facets parallel to {111} planes.

10.3.2   Antimony and bismuth

PREPARATION

Antimony, as usually obtained, is hard and brittle, and easily breaks up into fragments. Mounting in a press may cause fracturing. It is safer to mount specimens in a castable resin. Preparation can be conducted using either the traditional or contemporary methods. Pure antimony is not a commonly encountered metallographic subject but alloys containing Sb are. These are easier subjects than the pure metal. Sb is often present in alloys made up of the low-melting point metals, e.g., Pb, Bi or Cd. These alloys do not respond well to diamond abrasives and are better prepared with the traditional method using candle wax on the SiC papers, especially the fine papers. Embedded SiC particles can be easily removed when polishing with aqueous alumina slurries. Copper alloys may contain antimony as a minor element.

Bismuth is a soft metal, but also brittle, although not very difficult to prepare. The pure metal is encountered less than alloys containing bismuth. The preparation follows the same guidelines as for antimony and for low-melting point alloys containing Bi.

ETCHING

1. (a) Water   22 ml   Mix equal   Sb and Sb-Bi alloys
   HNO$_3$ (1.40)   8 ml   quantities of (a) and (b)
   (b) Water   30 ml   immediately before use
   Ammonium molybdate   4.5 g   ~1 min

2. Water   100 ml   ~1 min   Grain structure of Sb
   Citric acid   25 g   and Bi alloys
   Ammonium   10 g
   molybdate

10.3.3   Beryllium

PREPARATION

Due to its toxicity, preparation of Be and its alloys is restricted usually to laboratories designed to handle Be and suitably inspected and licensed. Be is not easy to prepare as it is easily deformed (mechanical twinning) by handling. Conventional methods must be modified. All operations which produce Be dust must be done in a glove box. Wet cutting and grinding are absolutely necessary to reduce damage to the specimens and to prevent dust formation. However, the sludge created must be properly disposed of. All metallic dusts, regardless of the chemical element, are dangerous if inhaled but Be dust is among the most dangerous dusts. The metal deforms easily by mechanical twinning, so light loads must be used during grinding and polishing. Some authors claim that water must be avoided, even when grinding Be, while others state that no difficulties were encountered using water as the coolant or lubricant. A four-step procedure can be used to prepare beryllium and its alloys. Start with 320 grit (P400) SiC paper, at 15–20 N (4 lbf) load, 240–300 rpm, with water cooling, until cutting damage has been removed. Next, use a natural silk cloth with a psa backing (or a magnetic platen setup) set up and 6 μm diamond paste, 120–150 rpm, same load, for 5 minutes. Follow with 3 μm diamond paste on a woven polyester cloth, used as in the second step, but for 4 minutes. Final polish with colloidal silica or with the sol-gel alumina suspension, using 10–15 N (3 lbf) per specimen, 80–120 rpm, for 2 minutes. An attack-polish agent is often added to the final polishing suspension to facilitate damage removal and enhance polarised light response. Aqueous 5% oxalic acid and 30% concentration hydrogen peroxide have been used as the attack polishing agents. Be is difficult to etch. However, due to its hcp crystal structure, excellent results are obtained with as-polished surfaces viewed with cross-polarised light.

Other methods of preparation have been discussed[56,57] including *electropolishing* (see Table 10.4).

ETCHING

1. Hydrofluoric acid (40%)   10 ml
   Ethanol   90 ml

   On immersion for 10–30 s microconstituents are in general outlined, and some colour differentiations are observed. The identifications of constituents commonly met with in commercial beryllium, both in the etched and unetched condition, are as shown in Table 10.12.

2. Water   95 ml   1–15 s   Be alloys
   Sulphuric acid (1.84)   5 ml

Table 10.12   MICROCONSTITUENTS IN COMMERCIAL BERYLLIUM

Constituent	Appearance unetched	Effect of etching
Carbide	Hard angular grey particles, staining in air to many colours, finally brown	No effect on colour
Silicon-rich phase	Light blue grey	Not stained, but outlined
Nitride	Needle-like particles, darker grey than carbide	Unaffected
$TiBe_{12}$	Angular particles, slightly pink	Unaffected
$CaBe_{13}$	Light yellow, usually angular	Unaffected
$FeBe_{12}$	Almost invisible unetched	Coloured reddish brown
$MnBe_{12}$	Similar to iron, but slightly pink	Outlined
Boride	Reddish particles in grain boundaries	Coloured blue or purple
$MoBe_{22}$	Almost invisible unetched	Coloured chocolate brown
UBe_{13}	Invisible unetched	Phase revealed as a yellow or green dendritic structure
Aluminium	Bright yellow, often appearing speckled. Soft and difficult to polish	Outlined
Magnesium	Very soft, and white in appearance	Removed
$ZrBe_{13}$	Barely distinguishable dendritic phase	Outlined and coloured light blue

3. Water	100 ml	2–16 min	Outlines precipitates first, then grain
Oxalic acid	10 g		boundaries. Better used electrolytically

Note: Hydrofluoric acid produces irreversible bone damage and presents a range of other hazards. Even dilute hydrofluoric acid solutions should be handled with great care. Note that hydrofluoric acid attacks laboratory glassware.

10.3.4   Cadmium

PREPARATION

The preparation of cadmium for metallographic examination should be carried out in the same manner as for zinc. Cd and its alloys are soft and easily deformed by mechanical twinning if cutting and grinding are too aggressive. When possible, use a precision saw to cut the specimens or a good quality laboratory abrasive cut-off saw. Pure Cd and high-Cd alloys do not respond well to diamond abrasive and it is better to use the 'traditional' method of preparation (see mechanical polishing), but substitute aqueous alumina slurries for diamond. Grinding with SiC can be continued to grit sizes finer than 600 (P1200) and then polish with 3 and 1 μm aqueous alumina slurries, before finishing polishing with colloidal silica (works very well with cadmium). Cd plated steels and other metals can be challenging to prepare as water must be avoided, at least in the final polishing step. Consequently, colloidal silica cannot be used. Final polishing is performed usually with alumina powders mixed in mineral spirits, purified kerosene or ethanol. These are less satisfactory but there is no attack of the coating, as with an aqueous suspension.

Electrolytic polishing is satisfactory[20,58] for pure Cd (*see* Table 10.4).

ETCHING

Reagent		Conditions	Remarks
1. Ethanol	2 ml	Few seconds	Most alloys
Nitric acid (1.40)	98 ml	to a minute	Thallium also
2. Water	100 ml	Up to a minute	Eutectics of Cd
Hydrochloric acid (1.19)	25 ml		
Ferric chloride	8 g		
3. Water	40 ml		
Hydrofluoric acid (40%)	10 ml	5–10 s	Most Cd alloys, also
Hydrogen peroxide (30%)	10 ml		thallium and indium

Note: Hydrofluoric acid produces irreversible bone damage and presents a range of other hazards. Even dilute hydrofluoric acid solutions should be handled with great care. Note that hydrofluoric acid attacks laboratory glassware.

10.3.5   Chromium

PREPARATION

Pure Cr is soft but brittle; however, plated Cr is hard and brittle and is characterised by an internal crack pattern. Chromium and its alloys are not too difficult to prepare but they are very difficult to etch. Mechanical polishing using the contemporary approach can be followed by a brief electrolytic polish or an attack polish can be used in the last step (*see* Tables 10.3 and 10.4). Hard or decorative chromium plate may be mounted in a thermosetting epoxy resin to enhance edge retention and prepared using the contemporary method. Water is not a problem.

ETCHING

Table 10.13   ETCHING REAGENTS FOR CHROMIUM

See general safety note at start of this Chapter.

No.	Reagent*		Remarks
1	Dilute hydrofluoric acid‡		After electropolishing with the reagent listed in Table 10.4, the specimen is agitated for a few seconds in dilute hydrofluoric acid‡
2	Hydrochloric acid (concentrated)		Shows striations in electrodeposits
3	Nitric acid	10 ml	Suitable for alloys. Also used electrolytically; specimen is made
	Hydrochloric acid	20 ml	anode at 4 V, 45 s. Proportions 1:3:2 also used. This decomposes
	Glycerol	30 ml	on standing, especially in contact with stainless steel. *Beware!*
4	Sulphuric acid (10%)		Used hot with swabbing

* Acids are concentrated unless otherwise indicated.
‡ Hydrofluoric acid produces irreversible bone damage and presents a range of other hazards. Even dilute hydrofluoric acid solutions should be handled with great care. Note that hydrofluoric acid attacks laboratory glassware.

10.3.6   Cobalt

PREPARATION

Cobalt and its alloys are more difficult to prepare than stainless steels or nickel-based superalloys and much more difficult to etch. Cobalt is a tough metal with an hcp crystal structure. However, it does not respond well under polarised light. While it is sensitive to deformation by mechanical twinning, this is not as big a problem as with Be or Mg. Preparation of cobalt and its alloys can be done by the contemporary method. Always cut specimens with a high quality laboratory abrasive cutter using a softly bonded wheel, such as used to section Ti. Co cuts slowly, like refractory metals, and can be damaged by aggressive cutting. Two steps of SiC may be needed, but start with as fine a grit as possible, 220 or 240 grit or 320 grit SiC (P240–P400). Three diamond steps are advised with 9, 3 and 1 μm diamond. Final polishing is best conducted using an attack polishing agent added to colloidal silica for up to 3 minutes. An etch-repolish-etch cycle may be needed to remove the last bit of disturbed metal. Some metallographers follow mechanical polishing with a brief chemical polish.

ETCHING

Cobalt may be etched with some of the reagents used for nickel and iron alloys, but some alloys are strongly resistant to attack. *See* Table 10.14.

10.3.7   Copper

PREPARATION

Pure copper is low in hardness, extremely ductile and highly malleable. There are a number of relatively pure Cu grades made, with minor differences in deoxidation, for electrical applications. Preparation of these grades can be rather challenging, particularly for scratch removal and obtaining a high degree of grain boundary delineation. Cu–Zn alloys (brasses) are also challenging to prepare

Table 10.14   ETCHING REAGENTS FOR COBALT AND ITS ALLOYS

See general safety note at the start of this Chapter.

No.	Reagent*		Remarks
1	Hydrochloric acid Nitric acid Acetic acid Water	60 ml 15 ml 15 ml 15 ml	The solution should be aged for 1 hour. Grain boundaries and the general structure of alloys are revealed. May also be used electrolytically. Of wide application, including hard metals
2	Nitric acid Hydrochloric acid Glycerol	10 ml 20 ml 30 ml	Reveals general structure. May be used electrolytically (*see* Table 10.13, Reagent 3. Same precautions)
3	Potassium ferricyanide Potassium hydroxide Water	10 g 10 g 100 ml	For hard cobalt–chromium alloys containing carbon. Used at approx. 70°C, 10–20 s. Can be replaced by electrolytic 3% KOH. Suitable for carbides
4	Chromium trioxide Sulphuric acid Water	2–10 g 10 ml 90 ml	Used electrolytically, the specimen being made the anode at 6 V

* Acids are concentrated unless otherwise indicated.

scratch-free and difficult to bring out all the grain boundaries. This is a common problem with all face-centred cubic metals and alloys, such as α-brass that exhibit annealing twins. Sectioning is a critical step as damage from sectioning can be extensive. The least damaging cutting methods are needed. Damage from band sawing can extend to nearly a mm in depth. Always cut Cu and its alloys with a high quality abrasive cut-off saw with the proper wheel, good coolant and low pressure. Use the contemporary preparation method described above, and in Ref. [33]. After careful sectioning, start grinding with the finest possible SiC grit, e.g., 240 (P280) or 320 (P400) grit, to remove the cutting damage. Do not grind with worn paper. With an automated machine using six specimens at a time, a sheet of SiC paper is worn out (i.e., its removal rate is reduced to a very low value) in about 60 s. For good scratch control, use of a psa-backed (or a magnetic platen setup) selected silk cloth (an Ultra-Pol™ cloth specifically) in step 2 is critical. Final polishing can be effectively performed using a medium nap synthetic suede cloth (e.g., a Microcloth® pad), without need for an attack polish. Attack polishing agents have been used, and can be helpful, but they are unnecessary if sectioning is done with minimal damage and the contemporary practice is used. To study inclusions, examine the specimens before etching. Tough-pitch copper (Cu—0.4 wt. % O) contains numerous Cu_2O particles but can also contain copper sulphides. Both types of particles look similar in bright field illumination. Under dark field illumination, the cuprous oxide particles are bright ruby red while the sulphides are invisible. There are several alloys, cast and wrought, with lead additions. These can be easily observed with bright field and are well retained using the contemporary method.

ETCHING

Cast copper alloys are easy to etch due to their chemical inhomogenity, although grain boundaries may be difficult to reveal. Dendrites are usually easily revealed, particularly using colour tint etches.[2,32–37] Most homogeneous copper alloys have much the same etching characteristics as copper, and the same etching reagents as are used for copper may be tried. Reagents 1 and 2, and several other variations, are used widely to reveal the structure of most coppers and copper alloys. Alloys containing more than one phase usually etch easily, but careful attention must be paid to the time of etching. Most etchants reveal one phase, but not the other. Etchants can be used to selectively colour or darken specific phases in metals and alloys,[59] and Cu alloys are no exception. For example, when alpha-beta alloys, such as naval brass are etched, there are a few etchants that will colour or darken the alpha phase only, many that will darken only the beta phase, and a few that will colour both alpha and beta. For example, equal parts of NH_4OH and H_2O_2 (3% conc.) will darken the alpha phase; Klemm's I (see Table 10.6) will colour beta phase; and, Beraha's selenic acid reagent for copper (similar to etch 1 in Table 10.7, but 300 ml ethanol) will colour both phases. There are a number of excellent colour 'tint' etchants for copper and its alloys[2,33] and these are very useful for Cu alloys.

LEADED COPPER ALLOYS

Difficulties can arise in the preparation and examination of leaded alloys, owing to the different hardnesses of the constituents, and also because specimens are particularly susceptible to smearing of the lead over the copper matrix. The contemporary method of preparation, as described above and in the section on Mechanical Polishing, works well on leaded brasses and bronzes. It may be helpful, if there is any flowed metal after polishing, to lightly etch the specimen, then repeat the final polishing step before examination. Colloidal silica is quite good for minimising smeared surface metal and works well with leaded bronzes. If it is necessary to etch, Reagents 31 and 32 of Table 10.15 may be used to etch the lead, or ferric chloride or chromic acid reagents to attack the copper-rich matrix.

The dithizone process described under *Iron and steel* (p. **10**–45–**10**–50) is also applicable to brasses. (*See also electrolytic etching*, below.)

Table 10.15   ETCHING REAGENTS FOR COPPER AND ITS ALLOYS

See general safety note at the start of this Chapter.

Principal use	No.	Reagent		Remarks
Copper, copper alloys in general. Brass, bronze and nickel–silver	1	Ammonium hydroxide Water Hydrogen peroxide (3% conc.)	50 ml 50 ml 20 ml	Used for copper, and many copper-rich alloys. Gives a grain boundary etch, and also tends to darken the α solid solution, leaving the β solid solution lighter. The hydrogen peroxide content may be varied. Less is required the lower the copper content (*see* Reagent 2)
	2	Ammonium hydroxide Water Hydrogen peroxide (3% conc.)	50 ml 50 ml 10 ml	Used for bronze, 70:30 and 60:40 brasses; this etch may with advantage be followed by a ferric chloride etch (Reagent 6) to darken the β areas in duplex alloys
	3	Ammonium hydroxide Water Ammonium persulphate, 2½% solution	10 ml 10 ml 10 ml	Recommended for polish attack on copper and copper alloys
	4	Ammonium persulphate Water	10 g 100 ml	Used cold or boiling for copper, brass, bronze, nickel–silver and aluminium–bronze. Tends to produce relief effects. Good for aluminium bronzes followed by Reagent 23 to darken β martensite (retained β is pink) and then Reagent 24 to etch $\gamma_2(\delta)$ brown
	5	Dilute ammonium hydroxide solution (10–50%)		May be used for polish or swab attack on brass and bronze. The oxidising action of atmospheric oxygen is necessary for the process
	6	Ferric chloride, various strengths and compositions. To 100 parts of water are added:		Used as a general reagent for copper, brass, bronze, nickel–silver, aluminium–bronze and other copper-rich alloys. It darkens the β constituent in brasses and gives grain contrast following ammoniacal or chromic acid etches. The most suitable composition should be found by trial and error in specific cases. This reagent generally emphasises scratches in imperfectly prepared specimens, and tends to roughen the surface. For sensitive work it is frequently a great advantage to replace the water in the reagent by a 50:50 water–alcohol mixture or by pure alcohol
		Hydrochloric acid (1.19)	Ferric chloride (g)	
		20	1	
		10	5	
		50	5	
		25	8	
		6	19	
		25	25	
		1	10*	
		10	3‡	
	7	Ethanol (commercial) Ferric chloride (anhydrous) Hydrochloric acid	96 ml 59 g 2 ml	Dilute 5:1 with alcohol. Wash with alcohol or acetone. More delicate and controllable than aqueous solutions

(*continued*)

Table 10.15 ETCHING REAGENTS FOR COPPER AND ITS ALLOYS—*continued*

Principal use	No.	Reagent		Remarks
Copper, copper alloys in general. Brass, bronze and nickel–silver (*continued*)	8	Chromic acid CrO_3 saturated in water		Used for copper, brass, bronze and nickel–silver. The etching time is 1 to 1½ min. Grain boundaries are attacked and β constituents are coloured pale yellow, while the primary solid solution is coloured dark yellow
	9	Chromic acid, 100–150 g l^{-1} of water, 1–2 drops of hydrochloric acid are added immediately before use to a 50 ml portion		Used for copper, brass, bronze, nickel–silver, and recommended for revealing the silicides of Ni, Co, Cr and Fe in certain alloys. Alternate polishing and etching is recommended when a grain contrast is obtained. The primary solid solution is not attacked. The β copper–zinc phase is coloured light yellow while the δ copper–tin phase is coloured brown to black. The reagent may be followed by a ferric chloride etch
	10	Chromic acid Nitric acid (1.40) Water	25 g 40 ml 35 ml	This reagent is useful for distinguishing the general constituents of many copper alloys. The γ and δ phases are particularly well shown up as shining blue crystals
	11	Sulphuric acid (1.84) Potassium dichromate saturated in water	5 ml 100 ml	Etching time ½ to 1 min. Suitable for copper, but attacks oxide inclusions strongly
	12	Nitric acid, various strengths		Used for any purpose requiring deep etching and the removal of a thick layer of surface material. The times of etching are short and difficult to control
	13	Sulphuric acid (1.84) Hydrogen peroxide (10 vol.)	5 ml 100 ml	Etching time is to 1–1½ min. Used for pure copper, but attacks oxide
	14	Silver nitrate, 100 g l^{-1} of water, following exposure to hydrogen sulphide		This technique has been used for 70:30 and 60:40 brasses, and produces satisfactory contrast
	15	Silver nitrate, 20 g l^{-1} of water		Immerse for 30 s, and wash off silver stain under water. Useful in certain cases for pure copper
	16	Ammonium hydroxide Ammonium oxalate saturated in water	10 ml 30 ml	Used for high-zinc brasses
	17	Ammonium hydroxide Potassium arsenate saturated in water	10 ml 30 ml	Used for high-zinc brasses
	18	Copper ammonium chloride. 100 g l^{-1} of water, plus ammonium hydroxide to slight alkalinity		Copper, brass and nickel–silver. Specially recommended for darkening in large β areas duplex brasses
	19	Ammonia Potassium permanganate (0.4%) in water	20 ml 30 ml	Used for pure copper. Etching time 2 to 3 min. Liable to produce staining
	20	Bromine water, saturated		Etching time 30 to 60 s. May be used, satisfactorily with pure copper if the coating which forms is removed by washing in strong ammonia

(*continued*)

Table 10.15   ETCHING REAGENTS FOR COPPER AND ITS ALLOYS—*continued*

Principal use	No.	Reagent		Remarks
Aluminium–bronze Copper alloys with beryllium, silicon, manganese and chromium	4	See above		As for Reagent 4
	6	See above		As for Reagent 6. Darkens β
	10	See above		As for Reagent 10. Also generally useful for aluminium-bronzes
	21	Chromic acid Nitric acid (1.40) Water	20 g 50 ml 30 ml	Recommended for aluminium–bronze after pretreatment with 10% hydrofluoric acid solution in order to remove surface oxide films[§]
	22	Chromic acid Nitric acid (1.40) Water	20 g 5 ml 75 ml	As for Reagent 21
	23	Nitric acid (1.40) Hydrogen peroxide (100 Vol.)	0.5 ml 99 ml	Darkens martensite
	24	Sod. hydroxide Water	10 g 100 ml	$\gamma_2(\delta)$ etched brown
	25	Ferric nitrate Ammonium nitrate Nitric acid Water	20 g 20 g 2 ml 500 ml	Good for complex aluminium bronzes
	26	Ammonium hydroxide Water Hydrogen peroxide, 3% solution	25 ml 25 ml 20 ml	Age 1 week in loosely stoppered bottle before use. To distinguish eutectoid $(\alpha + \gamma_2)$ from acicular β. γ_2 attacked
Copper alloys with beryllium, silicon, manganese and chromium	8	—		As for Reagent 8. Is also useful for manganese bronze
	27	Potassium dichromate Water Sodium chloride (saturated) Sulphuric acid (s.g. 1.84)	2 g 100 ml 4 ml 8 ml	Used for copper, and copper alloys with beryllium, manganese and silicon. Also suitable for nickel–silver, bronzes and chromium–copper alloys. This reagent should be followed by a ferric chloride etch to give added contrast
High-nickel alloys	28	Nitric acid (s.g. 1.42) Glacial acetic acid Water	50 ml 25 ml 25 ml	Recommended for high-nickel alloys which might prove resistant to attack by the more usual reagents, and for bright copper electro-deposits. 1:1 nitric acid : acetic acid is also used
Copper–nickel– aluminium	29	Acetic acid, 75% Nitric acid Acetone	30 ml 20 ml 30 ml	Recommended for copper-rich copper–nickel–aluminium alloys. NiAl shows as dove-grey rectangular needles; Ni_3Al is globular and darker grey. The γ_2 phase of the aluminium–copper system is a pale grey; (*see also* Reagent 24)
Copper-silicon alloys	30	Hydrogen peroxide (30 vol.) Water Potassium hydroxide (20%) Ammonium hydroxide (s.g. 0.90)	20 ml 25 ml 5 ml 50 ml	Used specifically to distinguish the κ phase in copper-silicon alloys, and in ternary and more complex alloys based on this system.
Leaded copper and bearing metals of this material	31	Nitric acid Glacial acetic acid Glycerol	20 ml 20 ml 80 ml	General reagent for leaded copper, bronzes, etc. *See* below for further notes on leaded coppers. Darkens lead
	32	Trichloroacetic acid Water Ammonium hydroxide to make 100 ml	20 g 20 ml	As for Reagent 31. Etches and outlines lead constituent, bringing it into prominence. Monochloroacetic acid may be substituted

(*continued*)

Table 10.15   ETCHING REAGENTS FOR COPPER AND ITS ALLOYS—*continued*

Principal use	No.	Reagent	Remarks
Macro-etching	33	Heat tinting	This technique occasionally gives useful results. Phosphor bronzes react the most favourably
	34	No-etching	Several constituents which may be present may be observed in unetched sections. Thus the selenide and telluride, appear blue grey, while the oxide appears blue. Metallic bismuth appears a very pale blue grey. Cuprous oxide, unless the particles are very small, appears ruby-red in polarised light or with dark-field illumination; selenide and telluride remain dark. Zinc oxide (in brass) appears transparent and anisotropic in polarised light, blue-grey with ordinary illumination

* Usually used with 1 part of chromic acid CrO_3.
‡ Used with one part of cupric chloride and 0.05 parts of stannous chloride.
§ Hydrofluoric acid produces irreversible bone damage and presents a range of other hazards. Even dilute hydrofluoric acid solutions should be handled with great care. Note that hydrofluoric acid attacks laboratory glassware.

ELECTROLYTIC ETCHING OF COPPER ALLOYS

Copper alloys are particularly suitable for electrolytic etching, and this technique frequently gives good results with alloys (e.g., high-nickel alloys) which are otherwise difficult to etch. Solutions which have been found effective are given in Table 10.16; potential differences and current densities must be adjusted to suit specific materials. Reagents listed for electrolytic polishing of copper (Table 10.4) may also be tried (at about one-tenth the voltage).

A very sensitive etching of homogeneous copper alloys may be obtained by electrolytically polishing in phosphoric acid solution ($1 \, kg \, l^{-1}$), and short-circuiting the electrodes when polishing is complete. The polarisation current set up gives anodic action at the crystal boundaries only.

10.3.8   Gold

PREPARATION

Gold and its alloys, as well as the other precious metals, are very soft, easily deformed (they are exceptionally malleable), and are quite difficult to prepare for metallographic examination. Control of deformation, smeared metal and scratch removal is a challenge. As one might expect, they are among the most difficult metals to etch due to their great corrosion resistance. Embedding of abrasive can be a problem with precious metals. Nevertheless, they can be prepared using a variation of the contemporary practice.

Sectioning must be done with a good metallographic laboratory cut-off saw, with a precision saw, or a low-speed saw, with the proper blade and low pressure to minimise the depth of damage. Gold and its alloys can be mounted in a thermosetting resin or in cast epoxy. Start by grinding with 220, 240 or 320 grit SiC (P240 to P400) waterproof paper, 10–15 N (3 lbf) load per specimen, 150–250 rpm with water cooling, until the cutting damage has been removed. Polishing is done only with diamond paste using chemotextile pads (known variously as Texmet® 1000, Pellon® or Pan W), with only a minor amount of water as the lubricant. The waxiness of the carrier in the paste will be the main lubricant. Loads in all steps are low, as in the SiC step, at 3 lbf per specimen. The rotational speed is in the range 150–250 rpm. Use three diamond steps in this way with 9, 3 and 1 μm diamond paste for 5, 3 and 2 minutes, respectively.

Final polishing is conducted using a polyurethane pad, such as a Chemomet® pad with a 0.05 μm, sol-gel alumina suspension (Masterprep™ alumina) at 10 N (2 lbf) per specimen, 100–150 rpm, for 2 minutes. Colloidal silica is useless with precious metals. Its amorphous silica particles are spherical in shape and produce little, if any, cutting action, while its pH of 9.5–10 is inadequate to yield a chemical attack-polish action. Ordinary calcined alumina abrasives do not yield as good

Table 10.16 ELECTROLYTIC ETCHING OF COPPER ALLOYS

See general safety note at the start of this Chapter.

Principal use	No.	Reagent		Remarks
General	*1*	Ferrous sulphate Sodium hydroxide Sulphuric acid (1.84) Water	30 g 4 g 100 ml 1900 ml	May be used to develop contrast after the use of ammoniacal hydrogen peroxide reagents. β in brasses is darkened. A current density of 0.1 A dm^{-2} at 8–10 V is suitable
Cupro–nickel and Nickel–silvers	*2*	Citric acid	100 g/l	Also useful for brasses
	3	Ammonium molybdate in excess of ammonia		Also useful for brasses
	4	Glacial acetic acid Nitric acid (1.40) Water	5 ml 10 ml 85 ml	This reagent tends to minimise the effect on the microstructure of the coring which usually occurs with these alloys
Brasses	*5*	Ammonium acetate 100 g l^{-1} of water		Current density 0.3 A dm^{-2}
	6	Ammonium sulphate 10 g l^{-1} of water		Current density 0.3 A dm^{-2}
	7	0.10 M ammonium acetate 0.50 M sodium thiosulphate 14 M ammonium hydroxide Distilled water	10 ml 30 ml 30 ml 30 ml	Carried out at 31 A dm^{-2}. The etching time is approximately inversely proportional to the copper content of the material
	8	Chromic acid CrO$_3$ 170 g l^{-1} of water		Useful for distinguishing the γ and ε phases of the copper-zinc system. At current densities above 23 A dm^{-2} the γ phase is attacked, but not the ε phase. At low current densities the order of the attack is reversed. The zinc-rich solid solution is attacked under both conditions.
Aluminium–bronze and copper–beryllium alloys	*9*	Chromic acid Water	10 g 900 ml	This reagent is satisfactory for all stages of heat treatment, and is useful for following the stages of precipitation in the age-hardening of copper-beryllium alloys. Distilled water must be used as tap water leads to staining. A potential of 6 V, with an aluminium cathode, is satisfactory
Leaded copper and brass	*10*	Dilute sulphuric acid (up to 10% by vol.)		Electrolytic etching at 6 V, with a carbon cathode, has been recommended for lead-bearing copper and brass, in order to avoid, misinterpretation of the structure due to surface flow

a surface finish as the sol-gel alumina. For high Au content specimens, such as 18-karat gold and higher ($\geq$75% Au), it is necessary to add an attack polish agent to the suspension for scratch removal. Several attack polishing agents can be used. CrO$_3$ in water is excellent and has been used in several concentrations from 5 to 20 g per 100 ml of water. This solution does contain undesirable Cr^{+6} ions which may pose a problem with its disposal (users must consult their local regulations on how to properly dispose the residue). Hydrogen peroxide (30% conc.) presents few environmental disposal problems and also works well. Generally, the attack polishing agent is added to the sol-gel alumina slurry in a 1:5 ratio by volume. If the slurry thickens too much, add some distilled water. When an attack polishing agent is used, the metallographer must avoid any physical contact with the abrasive and must properly clean up the equipment after use so that residues of it do not affect the next user of the equipment. It is advisable to wear proper personal protective equipment. When an attack polishing agent must be used, due to the higher purity of the alloy, polishing may be conducted for 3–6 minutes.

Table 10.17 ETCHING REAGENTS FOR GOLD AND ITS ALLOYS

See general safety note at the start of this Chapter.

No.	Reagent		Remarks
1	Potassium cyanide, 10% in water	10 ml	Used for gold and its alloys. A fresh solution, warmed
	Ammonium persulphate, 10% in water	10 ml	if necessary, must be used for each experiment. The etching time varies from ½ to 3 min. The attack may be speeded up by the addition of 2% of potassium iodide, but this is liable to give staining effects (see text below for details of its use)
2	Tincture of iodine, 50% solution in aqueous potassium iodide		Used for gold alloys. With silver–gold alloys a silver iodide film may form. This may be removed by immersion in potassium cyanide solution
3	Aqua regia (20 ml conc. nitric acid + 80 ml conc. hydrochloric acid)		The hot solution is used. If too much silver is present a silver chloride film may form; this may be removed by ammonium hydroxide or potassium cyanide solutions. Use fresh only; liable to decompose with evolution of chlorine and NO_2
4	Potassium sulphide solution		Solution used hot; particularly useful for gold-nickel alloys
5	Chromic acid (chromium trioxide)	3 g	Gold-rich alloys (up to 1 min)
	Hydrochloric acid (1.19)	100 ml	

ETCHING

Etching of gold and its alloys, and all precious metals, is challenging due to their great corrosion resistance. Etch 1 works well with gold, but its use is unusual. The two solutions must not be pre-mixed. One person holds the specimen with tongs and a tuff of cotton with a second pair of tongs. Another person takes two eyedroppers and fills each with one of the solutions. That person then drips equal amounts of the two etchants onto the specimen surface while the first person swabs the surface. Dripping and swabbing is continued until the correct degree of etching has been obtained. This must be done over something that will collect anything that drips from the specimen, so it requires careful pre-planning and execution. NaCN can be used in place of KCN and hydrogen peroxide (30% conc.) can be substituted for the ammonium persulphate solution with equivalent results. But, this etch becomes less effective for gold contents above 18 karat.

ETCHING

The five solutions given in Table 10.17 are of general suitability.

ELECTROLYTIC ETCHING OF GOLD

The microstructure of gold may be developed by anodic treatment in concentrated hydrochloric acid to which a little ferric chloride has been added. Dilute solutions of hydrochloric acid, potassium cyanide or potassium cyanide plus potassium iodide may also be used.

10.3.9 Indium

PREPARATION

The pure metal and many of its alloys are very soft and should be prepared by the methods recommended for lead or cadmium. It is frequently sufficient to cut suitable specimens with a sharp razor blade; however, the damage introduced must be removed. Pure In and its alloys can be more difficult to prepare than lead and its alloys, although some In-based intermetallic compounds are relatively easy to prepare. See the discussions on preparing Cd or Pb.

ETCHING

Etching reagents containing hydrofluoric acid, nitric acid, or mixtures of the two acids, are in general satisfactory, and the following are recommended for indium-rich alloys.[60,61] Vilella's reagent for steels, etch 2 (below), has been used successfully.

1. Potassium dichromate   1.3 g
   Sulphuric acid conc. (1.84)   4.5 ml
   Saturated sodium chloride solution   2.7 ml
   Hydrofluoric acid (40%)   17.7 ml
   Nitric acid conc. (1.40)   8.8 ml
   Water   66.3 ml   Use up to 1 min
2. Hydrochloric acid (1.19)   20 ml   For alloys containing bismuth it is
   Picric acid   4 g   necessary to increase the proportions
   Ethanol   400 ml   of alcohol and hydrochloric acid.

Note: Special handling is required for picric acid to avoid the risk of fire and/or explosion and local regulations must be consulted before use of this reagent.

Hydrofluoric acid produces irreversible bone damage and presents a range of other hazards. Even dilute hydrofluoric acid solutions should be handled with great care. Note that hydrofluoric acid attacks laboratory glassware.

10.3.10   Iron and steel

PREPARATION

Iron and low-carbon steels are more susceptible to surface deformation during preparation than higher alloyed steels and tool steels so sectioning and grinding must be performed carefully. The development of excessive heat at the surface must be avoided. Pure iron, and sheet steels, are soft and ductile. Removing all deformation and scratches is difficult because the structure is mainly ferrite. Higher hardness steels are easier to deal with and are more forgiving. However, some of the highly alloyed stainless steels can be rather challenging. It is not uncommon to see deformation-induced martensite in an austenitic stainless steel due to overly aggressive preparation. Preparing tool steels is usually easy, but there are situations where they can be quite difficult; for example, preparing as-quenched specimens or over-austenitised specimens. An as-quenched tool steel frequently exhibits hardnesses in the 60–70 HRC range. Sectioning must be done with the greatest care, low pressure and a very softly bonded wheel, to avoid introducing heat, or worse yet, burning. Some metallographers prefer to break the specimen in three-point bending to avoid the heat introduced during cutting an as-quenched tool steel. Over-austenitised tool steels generally contain excessive amounts of retained austenite because more carbon than desired was put into solution, which depresses the martensite start and finish temperatures. Excessive heat generated in section and grinding, and even polishing, can alter the retained austenite content. Austenitic stainless steels must be cut with low applied pressure as they can be work hardened dramatically. Hadfield manganese steels, also austenitic, have an even greater ability to be work hardened. Austenitic Fe–Ni magnetic alloys can be very difficult to prepare scratch free. Graphite retention is a problem with certain cast iron grades (grey iron, compacted graphite iron, ductile iron, malleable iron). Metallographers used to advocate grinding specimens with SiC paper up to 600 grit (P1200) with water as the coolant, and then repeat the 600 grit (P1200) step without water, claiming this enhanced graphite retention. That approach is unnecessary. A 4 or 5 step contemporary practice (see Mechanical Polishing), gives perfect surfaces with excellent graphite retention. White cast iron is easier to prepare but one must avoid any procedure that would create relief, that is, height differences between different phases. The use of the hard, woven napless cloths or synthetic napless pads in polishing ensure that relief is absent, when the correct load is used.

Galvanised steel (Zn coated) is a frequent metallographic subject. Final polishing steps must be performed without water, or the Zn coating will be heavily attacked. A similar technique may be employed for tin-plate, although Sn-coated layers are thinner and more difficult to study by light microscopy, and for Cd-coated steels (see section on cadmium).

The contemporary preparation practice works very well for irons and steels (see mechanical polishing). This practice can be altered depending upon needs and materials being prepared. Rigid grinding discs can be used for steps 1 or 2. For example, high hardness tool steels and alloy steels can be prepared perfectly in as little as 3 steps. Step 1 uses SiC paper for grinding to remove the cutting damage. Grits from 120–320 (P120–P400) can be used. Step 2 uses the Buehler Hercules H disc, or equivalent, with a 3 μm polycrystalline diamond suspension at 120–150 rpm, 25–30 N (6 lbf) load per specimen, for 5 minutes. Step 3 uses either a 0.05 μm sol-gel alumina suspension or a 0.05 μm colloidal silica at 120–150 rpm, with a similar load, for 5 minutes, on either a medium nap synthetic suede cloth (for e.g., a Microcloth® pad) or a polyurethane pad. If the specimens are mounted in the thermosetting epoxy resin, with a modern mounting press that cools the specimen under pressure after polymerisation so that there are no shrinkage gaps, then edge retention will be superb despite the 5 minute time, even if the medium napped cloth is used.

MICROGRAPHIC CONSTITUENTS OBSERVED

The metallography of iron and steel is complex, and the various constituents likely to be observed may be very briefly summarised as follows:

1. *Ingot iron and wrought iron.* These consist mainly of *ferrite* (α-iron; body centred cubic crystal structure), the etching characteristics of which may be affected by phosphorus, manganese or silicon in solid solution, or by the presence of slag inclusions (wrought iron is no longer made commercially). Carbon is usually present in low amounts and carbon present above the very low solid solubility in ferrite is in the form of *cementite* (iron carbide, Fe_3C). Sulphur may be present as *iron sulphide* if Mn is not present, but otherwise as *manganese sulphide.* Several forms of relatively pure iron are made, such as: *electrical iron, interstitial-free steel,* and sheets steels of various compositions. These are all essentially ferritic in content with minor amounts of non metallic inclusions and cementite. Dual-phase sheet steels have a more complicated structure, as they contain small grains consisting of martensite and retained austenite.

2. *Normalised and annealed carbon steels.*[*] The microconstituents present vary according to the carbon content. Below 0.8% of carbon, specimens consist of *ferrite* and an interlamellar eutectoidal mixture of *ferrite* and *cementite* which is known as *pearlite.* At 0.8% of carbon, specimens are entirely *pearlitic.* Above 0.8% carbon, the constituents are a network of massive *cementite* surrounding *pearlitic* areas. Normalised and annealed steels (<0.8% C) differ in the character and extent of the pearlitic areas, which are usually coarser in the latter case. Very rapid cooling gives very fine, unresolvable (by LOM) pearlite. Very slow cooling gives a coarse, easily resolved interlamellar spacing in the pearlite. Special annealing cycles can 'spheroidise' the cementite leading to the formation *of globular cementite,* producing a softer, more ductile condition than offered by fine pearlite. Prolonged heating of high carbon steels at 600–800°C decomposes cementite into *ferrite* and free *graphite,* which must be considered as a possible constituent.

3. *Hardened and tempered carbon and alloy steels.* At a high temperature (above that at which the transformation of α-iron into face centred cubic, γ-iron takes place), carbon steels consist of *austenite,* a solid solution of carbon in γ-iron. Some austenite can be retained in highly alloyed steels, such as tool steels, depending upon the austenitising temperature used in heat treatment. In most steels a variety of decomposition products may arise depending on the severity of quenching locally, the alloy content ('hardenability'), and the section size. The structures obtained are:

   (a) Martensite, which requires rapid quenching (the rate required is a function of the steel's hardenability and the section size). In high-carbon steels, particularly when too much carbon is put in solution, this structure appears as an intersecting system of parallel or lenticular 'needles' (often called, erroneously, 'acicular' martensite). In high-carbon steels, particularly when over-austenitised, some residual austenite may be retained in between the 'needles'. The three-dimensional shape, however, is not acicular but looks like two dinner plates facing each other, hence, the more correct name, 'plate' martensite. Electron microscopy is needed for a detailed study of martensite. Plate martensite exhibits a mid-rib twin. In low-carbon alloy steels and maraging steels, *lath martensite* is formed that has a parallel-banded structure containing high densities of dislocations. Medium-carbon steels have mixtures of lath and plate martensite. When the prior-austenite grain size is fine, as it should be, the structure of martensite is difficult to reveal with the light microscope.

   (b) Bainite is formed at lower rates of quenching. It is a non-lamellar aggregate of ferrite and cementite.

   Upper bainite comprises bundles of parallel laths of ferrite between which carbide precipitates. It can form by continuous cooling, or during isothermal decomposition of austenite at 400–500°C (it is usually easier to discern bainite types that are formed isothermally). It etches up more darkly, the lower the temperature at which it forms, but the structure is only seen in clear detail by electron microscopy.

   Lower bainite is more usually found in alloy steels with high Cr, Ni and/or Mo contents that were quenched too slowly to form fully martensitic structures or were isothermally transformed to deliberately form lower bainite. Like upper bainite, it is diffusion controlled (although in both cases, the actual γ to α transformation may be diffusionless with diffusion occurring behind the transformation front), but forms at lower temperatures, usually

---

[*] By 'normalising' is meant the reheating of a steel to a temperature at which it consists of a solid solution of carbon in γ-iron, followed by free cooling in air. By 'annealing' is meant the reheating of a steel to a similar temperature for an appreciable time, followed by slow cooling, usually in a furnace. See Chapter 29 for a more detailed discussion of steel heat treatment.

below 350°C. It is difficult to distinguish from a tempered martensite by optical examination and needs electron microscopy for full identification. It is acicular or plate-like with subsidiary plates or needles nucleated from existing plates (unlike upper bainite, which is nucleated from austenitic boundaries). Lower bainite comprises ferrite plates or needles with carbides precipitated internally on one orientation. (In tempered martensite, two or more orientations of carbides are found.) Lower bainite is also free from twins.

(c) *Pearlite.* This is a lamellar eutectoid of α-iron and cementite, the spacing of which is dependent on the temperature at which it forms. It nucleates as nodules which grow from prior-austenitic grain boundaries or in low-carbon steels from ferrite/austenite interfaces.

(d) In alloy steels and commercial plain carbon steels, manganese sulphides are present, not iron sulphide, which would melt at ordinary hot working temperatures and cause 'hot shortness'.

(e) In cast irons, similar structures are found, but the higher carbon content results in either primary cementite or flakes of graphite depending on alloy content and/or cooling rate. Graphite can also be present as a eutectic or as nodules. By careful preparation, the internal structure of graphite can be revealed by polarised light. Phosphorus introduces a characteristic eutectic of ferrite and Fe_3P or a ternary eutectic with Fe_3C called Steadite.

ETCHING OF IRON, STEEL AND CAST IRONS

Etching reagents are used for many purposes, such as:

1. To reveal the general structure of the steel for which a few reagents can be used for most steels.
2. To differentiate between various carbides or to differentiate carbides from nitrides (the latter remain unetched). For these, an enormous variety of quite complex reagents has been developed. It may be difficult to reproduce the effects claimed by those skilled in the use of these reagents, particularly the colours produced. Some of these reagents are used hot. In general, most reagents are based on an alkaline solution containing a mild, oxidising reagent and a chemical likely to cause selective staining of the minor phases. Phenolic mounts do not hold up well in boiling reagents, but the thermosetting epoxy resins are not degraded. In Table 10.18 a few tried and reliable etches are given which should be used before considering the more complex etches in the literature.
3. To reveal the prior-austenite grain boundaries in heat treated steels. The most successful etch is water saturated with picric acid (Note that special handling is required for picric acid to avoid the risk of fire and/or explosion and local regulations, must be consulted before use of this reagent) plus a wetting agent. Sodium dodecylbenzene sulphonate is very effective. A small

Table 10.18   ETCHING OF IRON STEEL AND CAST IRONS

See also refs. 1 and 19 for extended list of etchants. See general safety note at the start of this Chapter.

No.	Etchant		Conditions	Remarks
1	Nitric acid (1.40) Ethanol (Nital)	1.5–5 ml to 100 ml	5–30 s depending on steel	Ferrite grain boundaries in low carbon steels. Darkens pearlite and gives contrast with ferrite or cementite network. Etches martensite and its decomposition products in many steels. Better than picral for ferritic grain boundaries
2	Picric acid* Ethanol (Picral)	4 g 100 ml	5–30 s depending on steel	Similar to Reagent 1 but gives more uniform etch of pearlite. Better for detail of pearlite and bainite. Reveals undissolved carbides in martensite and gives better distinction of carbides in spheroidised steels. Differentiates pearlite and bainite. Does not etch ferrite grain boundaries or as-quenched martensite

(continued)

Table 10.18   ETCHING OF IRON STEEL AND CAST IRONS—*continued*

No.	Etchant		Conditions	Remarks
3	50/50 mixture of Reagents 1 and 2*		5–30 s depending on steel	Used for low alloy steels. Gives lower contrast but more even etching than Reagent 1
4	HCl (1.19)	1–5 ml	5–10 s	May attack prior-austenite boundaries; increased contrast between grains. (Attack of boundaries in martensite alloys increased if tempered for 30 min at 310°C). Good for tool steels and martensitic stainless steels
	Picric acid*	1–4 g		
	Ethanol (5 ml HCl and 1 g picric acid* usually referred to as Vilella's reagent)	to 100 ml		
5	Chromium trioxide	16 g	10–30 min in boiling solution	Reveals intergranular oxidation of medium carbon alloy (nickel) steels as a white layer
	Water	145 ml		
	Sodium hydroxide (add carefully)	80 g		
6	Sodium metabisulphite	1–20 g	5–60 s	Increases contrast in martensitic steels; distinguishes pearlite, bainite and martensite in high carbon alloy steels
	Water	to 100 ml		
7	Picric acid*	2 g	Either 10 s to 2 m. boiling soln. or electrolytically cold at 5–22 Am^{-2} for 2 min (stainless steel cathode)	Blackens cementite. No effect on M_7C_3, $M_{23}C_6$, M_6C or MC
	Sodium hydroxide	25 g		
	Water	to 100 ml		
8	Potassium ferricyanide	10 g	2–20 min at 20–50°C (or electrolytically as Reagent 7)	M_7C_3, $M_{23}C_6$ and iron phosphide stained. In cast irons, immerse 10 s at 80°C to darken iron phosphide, then 30 s, at 80°C to darken cementite unaffected by 10 s immersion. In stainless steels, carbides darken, sigma phase blue, ferrite yellow
	Potassium or sodium hydroxide	10 g		
	Distilled water to 100 ml (Murakami's reagent)			
9	Nitric acid (1.40)	10 ml	Immerse up to 30 s or use electrolytically with stainless steel cathode *Do not keep, discard safely when yellow; gives off chlorine and NO_x do not leave in contact with stainless steel*	Etches high chromium cast irons, stainless steel and high chromium steels. A mixture of 15 ml HCl, 10 ml glycerol and 5 ml HNO_3, known as glyceregia, is used to reveal the structure of stainless steels by swabbing.
	Hydrochloric acid (1.19)	20 ml		
	Glycerol (also known as Vilella's reagent but do not confuse with Reagent 4)	30 ml		
10	Sodium hydroxide	20–45 g to 100 ml	5–60 s Electrolytic with 1–3 V d.c. stainless steel or Pt cathode	In stainless steels, sigma and chi phases yellow to reddish brown (chi etches before sigma), ferrite blue grey or tar, carbides outlined after longer etch
	Water			
11	Potassium permanganate	4 g	1–10 min in boiling soln. Electrolytic for 5–30 s as Reagent 10	In tool steels, blackens MO_2C, M_6C and $Cr_{23}C_6$
	Sodium hydroxide	4 g to		
	Water	100 ml		
12	Ferric chloride	2 g	Immerse 1–5 min	In tool steels attacks ferrite and martensite, outlines carbides, leaves austenite unattacked
	Hydrochloric acid (1.19)	5 ml		
	Water	30 ml		

* Special handling is required for picric acid to avoid the risk of fire and/or explosion and local regulations must be consulted before use of this reagent.

teaspoon full is added to about 200 ml of the solution. The tridecyl version is also very effective as the wetting agent, but is harder to obtain. If the steel contains some Cr, it may be necessary to add a few drops of HCl per 100 ml of the etch. This etch works only when there is P in the steel, usually >0.005 wt. % is needed. The phosphorus segregates to the austenite grain boundaries during austenitisation. Good results are obtained with as-quenched steels and with steels tempered up to about 550°C. Tempering in the range of about 350–550°C will also drive P to the prior-austenite grain boundaries, with the maximum rate at about 454–482°C. This etch is less effective or ineffective for steels with less than about 0.2 wt. % C. Reference 2 describes the use of etchants for revealing prior-austenite grain boundaries in great detail and lists many etchants for this purpose.

4. To reveal a phase selectively or to reveal segregation or dendrites in castings. These are specialised tasks that can often be done effectively with colour etchants. In stainless steels, one might want to preferentially colour or darken delta ferrite so that it can be measured. Other such phases include sigma and chi. In quality studies, a longitudinal section might be prepared to evaluate the degree of chemical segregation. In carbon and alloy steels, segregation might promote a banded, or layered pattern of alternating ferrite and pearlite, and in the pearlitic band other constituents, such as bainite and martensite, might be observed. In heat treated steels, the structure may be fully martensitic, nevertheless, chemical variations can be detected. In austenitic stainless steels, chemical variations can be detected in fully austenitic structures. To achieve these goals, colour tint etches are used. Dendritic structures in cast steels and cast irons are well revealed using colour etchants.

5. To reveal austenite grain boundaries in austenitic stainless steels, Fe–Ni magnetic alloys or austenitic Hadfield manganese steels. This requires special etchants. When annealing twins are present, it would facilitate grain size measurement if the twins were not revealed while bringing up the grain boundaries. This is possible in basic austenitic stainless steels using 60% HNO_3 in water, 20°C, 1.4 V dc, Pt cathode, 60–120 s. Other less noble materials work for the cathode, but voltage control may be different. If the voltage increases, twins will be revealed. This etch also has the advantage of revealing virtually all of the grain boundaries, while 'standard' stainless steel etchants used by immersion or swabbing, do not reveal a high percentage of the grain boundaries, and also bring up twin boundaries (this makes grain size measurements more difficult).

LEAD IN STEELS

Lead particles in free-cutting steels can be revealed by three reagents:

1. 10% ammonium acetate in water,   Brown stains on lead particles after 30 s immersion.
2. 30 g potassium dichromate 225 ml water (hot to aid solution) 30 ml acetic acid added to cold solution   10–20 s etch reveals lead as yellow to gold particles under cross-polars with polarised light. (Steel not etched)
3. 1 g potassium cyanide 100 ml water, mixed with 0.25 g diphenyl diacarbazone in 10 ml chloroform   Etch first in picrate. Rinse and dry. Then swab with this solution. Lead coloured red, especially under polarised light. Note: this is a very dangerous etch and some of the ingredients may be illegal to use.

The general distribution of lead can be displayed by the following:

1. Etched ground surface in nitric acid, wash and dry.
2. Cover for 5 min with absorbent paper soaked in mixture of 50% glacial acetic acid and 50% of 10% chromium trioxide in water.
3. Strip paper and wash with 10% acetic acid leaving yellow lead chromatic colouration. Wash with water.
4. Develop paper in either the third etchant above or in 1% potassium cyanide in 100 ml water plus 10 ml of 0.1% solution of dithizone in chloroform. After washing and drying, the distribution of lead is indicated by red spots. (This is generally applicable to distribution of lead in metals, e.g., in brasses.)

Today, most metallographers examine leaded steels by means of SEM using a backscattered electron image, as the strong atomic number contrast produces excellent images.

INCLUSIONS AND PRECIPITATES IN STEEL

Many of these can be recognised by shape and colour without etching and can be identified positively by X-ray analysis with scanning electron microscopes. The following notes are a useful guide to a quick optical assessment.

Iron phosphide	Brilliant white, especially in cast irons. Distinguished from cementite (Fe_3C) by alkaline potassium ferricyanide (Murakami's reagent) which darkens phosphide before cementite at $20°C$ or by alkaline sodium picrate which only darkens cementite.
Iron nitride	Bluish grey. Not attacked by alkaline potassium ferricyanide. Coloured yellow in 4% picric acid in ethanol. See earlier comments regarding the hazards of using picric acid.
Iron sulphide	Brownish yellow (only present in steels made before the 20th C).
Manganese sulphide	Dove grey. Ferrous sulphide is attacked by 1% oxalic acid in water; manganese sulphide is attacked by 10% chromic acid in water. Both deformed by hot rolling.
Chromium oxide	Dark bluish grey; brilliant green under polarised light (crossed polars).
Silicates	Iron silicate dark grey; manganese silicate somewhat lighter, greenish tint. Often glassy as spheres (as-cast condition) which show 'optical cross' under polarised light.
Titanium nitride	Sections of golden yellow cubes. If large amounts, dendritic growth from corners of cube but this is rare.
Zirconium nitride	As TiN but more lemon yellow.
Alumina	Angular particles in groups not elongated by rolling, usually bluish grey.
Silica	Dark angular particles, often not elongated by rolling (only sand entrapped in a casting will contain silica).

FERROMAGNETIC ANALYSIS

To distinguish between magnetic and non-magnetic phases, e.g., ferrite from austenite in iron–chromium–nickel alloys, a thin film of a colloidal suspension of magnetic particles is applied to the surface of a specimen subject to a magnetic field. The particles concentrate on the magnetic constituent, usually in a characteristic banded or mosaic pattern. The following suspension has been recommended.[62]

A coarse flaky precipitate of magnetite is prepared by dissolving 2 g of $FeCl_2 \cdot 4H_2O$ and 5.4 g of $FeCl_3 \cdot 6H_2O$ in 300 ml of hot water, and adding, with constant stirring, 5 g of caustic soda in 50 ml of water. The precipitate is filtered, washed with water and with 0.1 N hydrochloric acid.

Table 10.19   ELECTROLYTE ETCHING OF STEELS

See general safety note at the start of this Chapter. Some etches in Table 10.18 can be used electrolytically. The following are used exclusively electrolytically

No.	Etchant		Conditions	Remarks
1	Lead acetate Water	10 g to 100 ml	2 V d.c. Stainless steel cathode 5–20 s	In Fe–Cr–Ni cast alloys Sigma phase: blue red Ferrite: Dark blue Austenite: Pale blue Carbides: Yellow
2	Ammonium hydroxide (0.88)		2–6 V d.c. Pt cathode 30–60 s	Stainless and high alloy steels. Etches carbides; sigma phase unattacked
3	Chromium trioxide Water	10 g to 100 ml	3–6 V d.c. Pt cathode 5–60 s	Stainless steels. Carbides outlined and attacked. Austenite, ferrite, phosphide attacked in that order
4	Oxalic acid Water	10 g to 100 ml	6 V d.c. Stainless steel cathode 5–60 s	Outlines carbides first then grains in stainless steels
5	Potassium or sodium hydroxide Water	1 g to 100 ml	1–6 V d.c. Stainless steel cathode 5–30 s	By control of voltage, can be made to etch various phases sequentially. High voltages outline carbides and grains. Low voltages selectively stain phases in sequence. Establish behaviour on a given composition (which applies to most electrolytic etches)

On transferring the precipitate to 1 l of 0.5% soap solution and boiling, a colloidal suspension is obtained, which should be filtered from the unsuspended residue. The suspension may now be applied to the surface of the specimen, which is placed in the field of an electromagnet. The magnetic properties of the particles deteriorate gradually owing to oxidation. The addition of small amounts of photographic reducing agents has been found an advantage in this connection.[62] Commercially prepared ferromagnetic solutions ('Ferrofluid') can be obtained.

10.3.11   Lead

PREPARATION

The preparation of lead and its alloys for micrographical examination is difficult, owing to the softness of the metal. Lead preparation is similar in many aspects to preparing the other soft, low-melting point metals and alloys, such as Bi, Sn, In and Cd. All require careful sectioning with a metallographic laboratory abrasive cut-off saw or a low-speed or precision saw. In past practices, cutting with a hack saw was recommended, but this produces substantial deformation and a rough surface. Old preparation methods never removed the deformed metal and created new deformation so that the etchants used were more akin to deep etches or chemical polishes that removed a great deal of matrix. Examination by light microscopy revealed a black matrix with white second phase particles, but the etch depth was so extensive that only low magnifications could be used. Those practices should be avoided. Lead and its alloys can be prepared properly. Because diamond abrasives are largely ineffective with lead and its alloys, it is better to use the traditional method, perhaps employing SiC grits finer than 600 (P1200), followed by aqueous alumina slurries on a medium napped synthetic suede cloth, such as a Microcloth® pad. After proper cutting, to minimise surface deformation and roughness, the specimen can be mounted (cast resins may be preferred if the press temperature can cause phase changes, recrystallisation, grain growth, or even melting; but use a moulding practice and resin that gives the lowest possible heat of polymerisation).

Start grinding with the finest possible SiC paper, e.g., 320 or 400 grit (P400 or P800) with low loads 20 N (4 lbf) per specimen, 150–250 rpm, for about 30 s per sheet of paper (when doing 6 specimens at the same time). The SiC paper loads up faster when grinding these soft metals and alloys and has a shorter effective life. For grits of 600 (P1200) and finer, embedding is a problem. There are several things that can be done. First, waterproof alumina paper, if available, embeds less as the alumina is tougher. Secondly, one can coat the surface with candle wax which helps reduce embedding. It has been scientifically shown that candle wax is much more effective than natural bee's wax. If embedding occurs, whether or not the metallographer coats the paper, it can be removed by polishing with aqueous alumina slurries on the medium napped synthetic suede cloth. Diamond polishing will not remove embedded SiC particles. Grinding with 400, 600 and 800 grit (P800, P1200 and P1500) SiC paper, with water cooling, low loads, 150–250 rpm, for 30 s each, is adequate.

For rough polishing, one or two steps of aqueous calcined deagglomerated alumina slurries, using 3 and 1 or only a 1 μm particle size, on a synthetic suede medium napped cloth, at 20–25 N (5 lbf) per specimen, 120–50 rpm, for 4–5 minutes is adequate. If the metallographer grinds only to a 600 grit (P1200) SiC paper, use both steps. Final polishing can be performed with the same type cloth using either the 0.05 μm sol-gel alumina slurry (Masterprep™ alumina), or colloidal silica, at 15–20 N (4 lbf) per specimen, 80–150 rpm, for 4 minutes. If the automated polishing head speed is <100 rpm, use contrarotation. Examine the specimen and then etch with a mild etchant, such as Pollack's reagent (100 ml water, 10 g citric acid and 10 g ammonium molybdate). Remove the etch by repeating the final polish, then re-etch. Even better results can be obtained by following this procedure with a brief vibratory polish using colloidal silica on a synthetic suede medium napped cloth. Perfect surfaces can be obtained and fine detail of the structure, revealed using Pollack's reagent, can be examined at 1000X.

ETCHING

Rapid tarnishing (oxidation) of polished and etched samples may be avoided by the use of EDTA in the etchant, or as a rinse after etching with a 2% solution (see: washing and drying under Etching). Table 10.20 lists many etchants that have been used for lead and its alloys. Bear in mind that nearly all were developed for use on specimens that were poorly prepared and the etchant was designed for heavy attack. This makes examination of the matrix phase almost impossible and permits examination of second phases only at low magnification. Pollack's reagent (100 ml water, 10 g citric acid, 10 g ammonium molybdate) is a leaner version of etch 15 and produces superb results. 2% nital is also acceptable.

Table 10.20   ETCHING REAGENTS FOR LEAD AND ITS ALLOYS

See general safety note at the start of this Chapter.

No.	Etchant*		Remarks
1	Nitric acid		Recommended for pure lead. Specimen is alternately etched and washed until the desired result is obtained. Macro etch mainly
2	Acetic acid 5% in alcohol		Slow-acting grain contrast etch for lead
3	Acetic acid Hydrogen peroxide (30 vol.)	30 ml 10 ml	This is recommended as a general-purpose reagent for lead, and its alloys with tin, antimony, calcium, sulphur, selenium and tellurium, as well as many other metals. The proportions may have to be adjusted slightly to suit individual cases. The reagent may be made somewhat less vigorous by dissolving a little lead in it before use. The usual etching time is 3–5 s. The action is exothermic and may give rise to recrystallisation and grain growth if continued for too long a time. The reagent should not be kept for more than 1 h. It is recommended that the surface of the specimen be cleaned after etching with nitric acid. (*See also* Table 10.5)
4	Acetic acid Nitric acid Water	30 ml 40 ml 160 ml	Used at 40–42°C for lead, and lead–tin alloys rich in lead
5	Acetic acid Nitric acid Glycerol	10 ml 10 ml 40 ml	This reagent is recommended when the alternate etching and polishing technique is employed. Used for lead, lead–calcium, lead–antimony and lead–cadmium alloys. Also used for ternary lead–cadmium–antimony. The lead-rich matrix is grey, the compound SbCd blue, the antimony-rich solid solution whitish yellow, and the cadmium-rich solid solution brownish yellow. This reagent may be used hot (up to 80°C). (Beware of recrystallisation!)
6	Hydrochloric acid Nitric acid Alcohol	10 ml 5 ml 85 ml	Used for eutectic lead–tin alloys
7	Hydrochloric acid Ferric chloride Water	30 ml 10 g to 150 ml	Used mostly for lead–antimony alloys
8	Hydrochloric acid (1.19)		General reagent for grain boundary etching. Usually good for alloys containing antimony
9	Hydrochloric acid, 1–5% in alcohol		Used for lead–tin alloys. For higher tin contents the 5% reagent is preferable
10	Nitric acid, 5% in alcohol		Produces grain contrast in lead-rich alloys
11	Nitric acid, 10% in water		Used for lead-rich alloys. May be improved by the addition of a little chromic acid
12	Sodium hydroxide saturated in water		Used for lead–tin alloys
13	Silver nitrate, 5–10% in water		Swab etch used for lead–tin and lead–antimony alloys. Useful for anti-friction alloys in general
14	Molybdic acid Ammonium hydroxide Water Filter solution and add to nitric acid	100 g 140 ml 240 ml 60 ml	This reagent, applied by the alternate swab and wash technique, gives a rapid etching which is very effective for removing thick layers of cold worked metal. It may be followed by Reagent 5 An alternative is Russell's reagent (*see* Table 10.1)
15	Citric acid Ammonium molybdate Water	25 g 10 g 100 ml	A useful grain boundary etch for lead and dilute lead alloys. Often used after chemical polish

* Acids are concentrated, unless otherwise indicated.

10.3.12 Magnesium

PREPARATION

Magnesium and its alloys are among the most difficult metals to prepare that are routinely encountered by metallographers. Magnesium and its alloys are low in hardness, although precipitates can be present that are substantially harder and relief control can be difficult. Mg is an hcp metal and deformation twinning occurs readily during handling, even in compression mounting. Sectioning must be done with a proper laboratory cut-off saw, or with a low speed or precision saw, using the correct wheel and ample coolant to minimise damage. There is considerable debate in the literature over avoiding or not avoiding water during preparation. Some say that water must be avoided in the last step, others say water must be avoided in all steps, still others say water need not be avoided. Pure Mg is attacked rather slowly by water, while Mg alloys exhibit a greater attack rate. Experiments by the present author have showed that careful, restricted use of water in cleaning after the final step (and all preceding steps), was not harmful. Some metallographers use a 3:1 mixture of glycerol to ethanol as the coolant in all steps, including grinding. Always use a coolant when grinding as all metal dusts, including Mg dust, is harmful if inhaled.

The contemporary preparation practice can be used, but with certain modifications. After obtaining a proper cut, mount the specimen. If precipitation reactions are possible, use a castable resin and a practice that minimises the heat of polymerisation (see Mounting). Grind the specimens with water cooled SiC paper starting with the finest grit size possible, 220–320 grit (P240–P400), 20 N (5 lbf) per specimen, 250 rpm, until the cutting damage is removed and the samples in the holder are on the same plane. Oil-based diamond slurries can be used in place of aqueous slurries, for example. Oil is a better lubricant, so the surface finish is better after each step, with less drag, but cleaning is more difficult than with water-based diamond suspensions. Use three steps of oil-based diamond slurries with 9, 3 and 1 μm particles sizes, 120–150 rpm, contrarotation (if the head speed is <100 rpm) if possible, for 6, 5 and 4 minutes, respectively, on a synthetic chemotextile pad (known variously as Texmet®, 1000, Pan W or Pellon®). Cleaning after each step was done by holding the specimen holder under running tap water (use lukewarm water) for about a second, just enough to remove much of the residue. Then, scrub the samples and the holder with ethanol using cotton, rinse with ethanol and blow dry under warm air.

The final step can be performed with several abrasives. Colloidal silica, although aqueous, works well with pure Mg, but it etches the alloys. The sol-gel alumina suspension, Masterprep™ alumina, is an aqueous suspension, but does not seem to attack constituents in Mg alloys. A proprietary suspension called Masterpolish®, which is nearly water free, has also been used and the results were excellent. All were utilising 0.05 μm particle size abrasives. They were used on a polyurethane synthetic pad with light load, 10–15 N (3 lbf) per specimen, 120–150 rpm (slower speeds are fine), for 90–180 s. The same washing procedure was used; brief rinse with lukewarm tap water, scrubbing with ethanol, rinse with fresh ethanol, and blow dry with hot air. Examine the specimen as-polished to see if the surface scratch control and relief control were adequate. Etch with one of the common reagents. If the structure is not crisp enough, repeat the last polishing step and re-etch. It is best to use surgical grade cotton as imperfections in cosmetic cotton puffs have been known to scratch polished Mg surfaces. Pure Mg, with its hcp crystal structure, will respond to polarised light. The response may be improved by using a polish-etch-polish cycle, or by using the vibratory polisher.

ETCHING

The general grain structure of alpha-Mg is revealed by examination under crossed polarised light. This will also reveal mechanical twins formed before or during preparation. A selection of etching reagents suitable for magnesium and its alloys is given in Table 10.21. Of these, 4 and 1 are the most generally useful reagents for cast alloys, while 16 is a useful macro-etchant and, followed by 4, is invaluable for showing up the grain structure in wrought alloys. The acetic-picral solutions are quite useful.

The appearance of constituents after etching. The micrographic appearances of the commonly occurring microconstituents in cast alloys are as given in Table 10.22.

ELECTROLYTIC ETCHING OF MAGNESIUM ALLOYS

The following has been recommended for forged alloys. The specimen is anodically treated in 10% aqueous sodium hydroxide containing 0.06 g l^{-1} of copper. A copper cathode is used and a current

Table 10.21   ETCHING REAGENTS FOR MAGNESIUM AND ITS ALLOYS

See general safety note at the start of this Chapter.

No.	Etchant*		Remarks
1	Nitric acid Diethylene glycol Distilled water	1 ml 75 ml 24 ml	This reagent is recommended for general use, particularly with cast, die-cast and aged alloys. Specimens are immersed for 10–15 s, and washed with hot distilled water. The appearance of common constituents following this treatment is outlined in Table 10.22. Mg-RE (rare earth) alloys also
2	Nitric acid Glacial acetic acid Water Diethylene glycol	1 ml 20 ml 19 ml 60 ml	Recommended for solution-heat-treated castings, and wrought alloys. Grain boundaries are revealed. The proportions are somewhat critical. Use 1–10 s
3	Citric Acid Water	5 g 95 ml	This reagent reveals grain boundaries, and should be applied by swabbing. Polarised light is an alternative
4	Nitric acid, 2% in alcohol		A generally useful reagent
5	Nitric acid, 8% in alcohol		Etching time 4–6 s. Recommended for cast, extruded and rolled magnesium–manganese alloys
6	Nitric acid, 4% in alcohol		Used for magnesium-rich alloys containing other phases, which are coloured light to dark brown
7	Nitric acid, 5% in water		Etching time 1–3 s. Recommended for cast and forged alloys containing approximately 9% of aluminium
8	Oxalic acid, 20 g l^{-1} in water		Etching time 6–10 s. Used also for extruded magnesium–manganese alloys
9	Acetic acid, 10% in water		Etching time 3–4 s. Used for magnesium–aluminium, alloys with 3% of aluminium
10	Tartaric acid, 20 g l^{-1} of water	Etching time 6 s	These reagents are recommended for magnesium alloys with 3 to 6% of aluminium
11	Orthophosphoric acid, 13% in glycerol	Etching time 12 s	
12	Tartaric acid, 100 g l^{-1} of water		Used for wrought alloys. Mg$_2$Si is roughened and pitted. 10 s to 2 min for Mg–Mn–Al–Zn
13	Citric acid and nitric acid in glycerol		Used for magnesium-zirconium alloys. The magnesium-rich matrix is darkened and other phases left white
14	Orthophosphoric acid‡ Picric acid‡ Ethanol	0.7 ml 4 g 100 ml	Recommended for solution-heat-treated-castings. The specimen is lightly swabbed, or immersed with agitation for 10–20 s. The magnesium-rich matrix is darkened, and other phases (except Mg$_2$Sn) are little affected. The maximum contrast between the matrix and Mg$_{17}$Al$_{12}$ is developed. The darkening of the matrix is due to the development of a film, which must not be harmed by careless drying
15	Picric acid saturated in 95% alcohol‡ Glacial acetic acid	10 ml 1 ml	A grain boundary etching reagent; especially for Dow metal (Al 3% Zn 1%, Mn 0.3%). Reveals cold work and twins
16	Picric acid, 5% in ethanol‡ Glacial acetic acid Distilled water	50 ml 20 ml 20 ml	Useful for magnesium–aluminium–zinc alloys. On etching for 15 s, an amorphous film is produced on the polished surface. When dry, the film cracks parallel to the trace of the basal plane in each grain. The reagent may be used to reveal changes of composition within grains, and other special purposes
17	Picric acid, 5% in ethanol‡ Glacial acetic acid Distilled water	50 ml 16 ml 20 ml	As for reagent 16, but suitable for a more restricted range of alloy composition[63]
18	Picric acid, 5% in ethanol‡ Glacial acetic acid Nitric acid (1.40)	100 ml 5 ml 3 ml	General reagent[64]
19	Picric acid, 5% in ethanol‡ Distilled water	10 ml 10 ml	Mg$_2$Si is coloured dark blue and manganese-bearing constituents are left unaffected[63]

(continued)

Table 10.21 ETCHING REAGENTS FOR MAGNESIUM AND ITS ALLOYS—*continued*

No.	Etchant*		Remarks
20	Hydrofluoric acid (40%)§ Distilled water	10 ml 90 ml	Useful for magnesium–aluminium–zinc alloys. $Mg_{17}Al_{12}$ is darkened, and $Mg_3Al_2Zn_3$ is left unetched. If the specimen is now immersed in dilute picric acid solution (1 vol. of 5% picric acid in alcohol and 9 vol. of water) the matrix turns yellow, and the ternary compound remains white[63]‡
21	Picric acid‡, 5% in ethanol Distilled water Glacial acetic acid	100 ml 10 ml 5 ml	Reveals grain-boundaries in both cast and wrought alloys. This reagent is useful for differentiating between grains of different orientations, and for revealing internally stressed regions[63]
22	Nitric acid conc.		Recommended for pure metal only. Specimen is immersed in the cold acid. After 1 min a copious evolution of NO_2 occurs, and then almost ceases. At the end of the violent stage, the specimen is removed, washed and dried. Surfaces of very high reflectivity result, and grain boundaries are revealed

* Acids are concentrated, unless otherwise indicated, e.g., with specific gravity.
‡ Special handling is required for picric acid to avoid the risk of fire and/or explosion and local regulations must be consulted before use of this reagent.
§ Hydrofluoric acid produces irreversible bone damage and presents a range of other hazards. Even dilute hydrofluoric acid solutions should be handled with great care. Note that hydrofluoric acid attacks laboratory glassware.

Table 10.22 THE MICROGRAPHIC APPEARANCE OF CONSTITUENTS OF MAGNESIUM ALLOYS

Microconstituent	Appearance in polished sections, etched with Reagent 1 (zirconium-free alloys)
$Mg_{17}Al_{12}$ (see note 1)	White, sharply outlined and brought into definite relief
$MgZn_2$ (see note 2)	Appearance very similar to that of $Mg_{17}Al_{12}$
$MgAl_2Zn_3$ (see note 3)	Appearance similar to those of $Mg_{17}Al_{12}$ and $MgZn_2$
Mg_2Si (see note 4)	Watery blue green; the phase usually has a characteristic, Chinese-script formation, but may appear in massive particles. Relief less than for manganese
Mg_2Sn (see note 4)	Tan to brown or dark blue, depending on duration of etching. Individual particles may differ in colour
Manganese (see note 5)	Grey particles, usually rounded and in relief. Little affected by etching
(MgMnAl) (see note 5)	Grey particles, angular in shape and in relief. Little affected by etching

Microconstituent	Appearance in polished sections, etched with Reagent 4 (zirconium-bearing alloys)
Primary Zr (undissolved in molten alloy)	Hard, coarse, pinkish grey rounded particles, readily visible before etching
Zinc-rich particles (see note 6)	Fine, dark particles, loosely clustered and comparatively inconspicuous before etching
Mg_9Ce	Compound or divorced eutectic in grain boundaries. Appearance hardly changed by few percent of zinc or silver
Mg_5Th	Compound or divorced eutectic in grain boundaries (bluish). Appearance hardly changed by few percent of zinc if Zn exceeds Th
Mg–Th–Zn	Brown acicular phase. Appears in Mg–Th–Zn–Zr alloys when Th $\geq$ Zn
$MgZn_2$	Compound or divorced eutectic in grain boundaries. Absent from alloys containing rare earths or Th

(1) This is the γ-phase of the magnesium–aluminium system; it is also frequently called Mg_4Al_3 or Mg_3Al_2.
(2) Although the phase MgZn may be observed in equilibrium conditions, $MgZn_2$ is frequently encountered in cast alloys.
(3) This ternary compound occurs in alloys based on the ternary system magnesium–aluminium–zinc and may be associated with $Mg_{17}Al_{12}$.
(4) Blue unetched.
(5) These constituents are best observed in the unetched condition.
(6) Alloys of zirconium with interfering elements such Al, Fe, Si, N and H, separating as a Zr-rich precipitate in the liquid alloy. Co-precipitation of various impurities makes the particles of indefinite composition.
Note: The microstructure of all zirconium—bearing cast alloys with satisfactory dissolved zirconium content is characterised by Zr-rich coring in the centre of most grains. In the wrought alloys zirconium is precipitated from the cored areas during preheating or working, resulting in longitudinal striations of fine precipitate which become visible on etching.
For descriptions of the metallography of magnesium alloys references 66 and 67 should be consulted.

density of $0.53\,A\,dm^{-2}$ is applied at 4 V. After etching, the specimen is successively washed with 5% sodium hydroxide, distilled water and alcohol, and is finally dried.

NON METALLIC INCLUSIONS IN MAGNESIUM-BASED ALLOYS[64,65]

The detection and identification of accidental flux and other inclusions in magnesium alloys involves the exposure of a prepared surface to controlled conditions of humidity, when corrosion occurs at the site of certain inclusions, others being comparatively unaffected. The corrosion product or the inclusion may then be examined by microchemical techniques.

MAGNESIUM ALLOYS

The surface to be examined should be carefully ground and polished as described above. Alcohol or other solvents capable of dissolving the flux must be avoided. As soon as possible the prepared specimens are placed in a humidity chamber, having been protected in transit by wrapping in paper. A suitable degree of humidity is provided by the air above a saturated solution of sodium thiosulphate. The presence of corrosive inclusions is indicated by the development of corrosion spots. At this stage the corroded area may be lightly ground away to expose the underlying structure for microexamination so that the micrographic features which are holding the flux become visible. With other specimens, or with the same specimens re-exposed to the humid conditions, identification of the inclusions may be proceeded with, as follows:

1. Detection of chloride
The corrosion product is scraped off, and dissolved on a microscope slide in 5% aqueous nitric acid. A 1% silver nitrate solution is then added, and a turbidity of silver chloride indicates the presence of the chloride ion. The solution of the corrosion product should preferably be heated before adding the silver nitrate to remove any sulphide ions, which also gives rise to turbidity. Alternatively, a 10% solution of chromium trioxide may be added directly to the corrosion spot, when chloride is indicated by an evolution of gas bubbles from the metal surface, and the development of a brown stain. This method is less specific than the silver nitrate method, and may give positive reactions in the presence of relatively large amounts of sulphates and nitrates.

2. Detection of calcium
Scrapings of corrosion product are dissolved in a small watch glass on a hot plate in 2 ml water and one drop of glacial acetic acid. A few drops of saturated ammonium oxalate solution are added to the hot solution. The presence of calcium is indicated by turbidity or precipitation. Spectroscopic identification of calcium in the solution is also possible.

3. Detection of boric acid in inclusions
Scrapings of corrosion product and metal are placed in a test tube with 1 ml of water. The inclusion dissolves, and complete solution of the sample is effected by adding a small portion of sulphuric acid (density 1.84) from 9 ml carefully measured and contained in a graduated cylinder. When solution is complete, the remainder of the acid is added and the mixture is well shaken; 0.5 ml of a 0.1% solution of quinalisarin in 93% (by weight) of sulphuric acid is now added, mixed in, and allowed to stand for 5 min. A blue colour indicates the presence of boric acid. The colour in the absence of boric acid varies from bluish violet to red according to the dilution of the acid, which must thus be carefully controlled as described.

4. Detection of nitride
A drop of Nessler's solution applied directly to the metal surface in the presence of nitride, gives an orange brown precipitate, which may take about 1 min to develop. This test should be made on freshly prepared surfaces on which no water has been used, since decomposition of nitride to oxide occurs in damp air.

5. Detection of sulphide
The corrosion product is added to a few drops of water slightly acidified with nitric acid. A drop of the solution placed on a silver surface gives rise to a dark stain if sulphide was present in the corrosion product. Sulphur printing may also be applied.

6. Detection of iron
The corrosion product is dissolved in hydrochloric acid. A drop of nitric acid is added with several drops of distilled water. In the presence of iron, the addition of a crystal of ammonium thiocyanate develops a blood-red colouration.

In all the above tests, a simultaneous *blank* test should be carried out.

Iron-printing, analogous to sulphur-printing, can be applied using cleaned photographic paper impregnated with a freshly prepared solution of potassium ferricyanide and potassium ferrocyanide acidified with nitric or hydrochloric acid.

10.3.13 Molybdenum

PREPARATION

Refractory metals and alloys are difficult to prepare as they exhibit low grinding and polishing rates and are easily deformed or smeared during preparation. Regardless of their actual indentation hardness, which is not high, cutting of refractory metals, such as Mo and its alloys, is slow and tedious. Avoid production cutting devices, such as power hack saws. Use a good laboratory abrasive cut-off saw with a very softly bonded wheel, designed for cutting refractory metals (such as Ti, Zr, Hf, etc.). Mo and its alloys can be tough or brittle, depending upon composition. They are susceptible to damage production during cutting and grinding. The contemporary method (see Mechanical Polishing) can be used, but requires modification. After sectioning, start with the finest possible grit size SiC paper, e.g., 320 grit (P400) SiC. Use moderate loads, 25–30 N (6 lbf) per specimen, 150–250 rpm, and grind until the cutting damage is removed and all surfaces are at the same plane. Next, use 9 μm diamond on a psa-backed selected silk cloth, such as an Ultra-Pol™ cloth, at the same load, 150–200 rpm, for 10 minutes. Charge the cloth with diamond paste and add the proper lubricant. During the 10 minute cycle, periodically add the same size diamond in slurry or suspension form. Keep the cloth properly lubricated during diamond polishing to minimise flowed metal. Next, use 3 μm diamond, in like manner, with a woven polyester pad, such as a Trident™ cloth, with the same load, 150–200 rpm, for 8 minutes. If the sample holder's head speed is <100 rpm, run these steps in contrarotation mode. A 1 μm diamond step, similar to the 3 μm step, can be used for the more difficult grades. Final polishing uses either the medium-nap, synthetic suede cloth, or a polyurethane pad, with colloidal silica at the same load, 120–150 rpm, for 5 minutes. For best results, an attack polish agent must be added to the colloidal silica. Many solutions have been recommended.[2] A 1:5 mixture of hydrogen peroxide (30% conc.) to colloidal silica is effective. A similar mix of 5–20% CrO_3 in water to colloidal silica also works well (see comments regarding this attack polish reagent under preparation of gold). Diluted Murakami's reagent has also been used.

ETCHING

Molybdenum is not easy to etch, but there are good etchants available. An excellent etchant for Mo consists of water, hydrogen peroxide (30% conc.) and sulphuric acid in a 7:2:1 ratio. If the specimen is immersed for 2 minutes, the grain structure is revealed in colour. If the etch is used by swabbing, a grain boundary etch results. Listed below are two additional useful etchants.

Etchant		*Conditions*	*Remarks*
1. (a) Potassium hydroxide	10 g	Mix equal amounts of	Grain boundary etch
Water	to 100 ml	(a) and (b) as needed	
(b) Potassium ferricyanide	10 g		
Water	to 100 ml		
2. Ammonia (0.88)	50 ml	Boil for up to 10 min	General etch
Hydrogen peroxide (3%)	50 ml		
Water	50 ml		

10.3.14 Nickel

PREPARATION

Nickel is a face-centred cubic metal and shares many of the preparation problems of other fcc metals. Damage is easily introduced during cutting and grinding. Pure Ni is more difficult to prepare than its alloys. Ni–Fe and Ni–Cu alloys can be rather challenging subjects while the Ni-based superalloys are simpler. The latter are prepared much like austenitic stainless steels and share common etchants. The contemporary method works well for Ni and its alloys (see Mechanical Polishing), although it must be modified somewhat. For pure Ni, Ni–Cu and Ni–Fe alloys, start with the best possible cut surface, i.e., least amount of induced deformation and smoothest surface. Always cut with a laboratory abrasive cut-off saw using the correct wheel, low pressure and ample coolant. Grind with 220, 240 or 320 grit (P240–P400) SiC paper, or alternative surfaces with similar abrasive sizes. Use 20–25 N (5 lbf) load per specimen, 200–300 rpm, until the cutting damage is removed and all the specimens are on the same plane.

Polish with three steps of diamond abrasive, 9, 3 and 1 μm diamond. Charge the cloth with diamond paste (see mechanical polishing) and add the recommended lubricant. Add the same diamond size as a slurry or suspension during the cycle. Use napless, woven cloths that yield the best possible surface finish, e.g., a psa-backed selected silk with the 9 μm diamond and a psa-backed (or a magnetic platen setup) polyester cloth with both 3 and 1 μm diamond sizes. Use 25–30 N (6 lbf) load per specimen, 100–150 rpm, and 5, 3 and 2 minutes for the 9, 3 and 1 μm steps, respectively. Final polishing is conducted with 0.05 μm colloidal silica or sol-gel alumina suspensions using the medium napped, synthetic suede cloth or a polyurethane cloth, both psa-backed (or a magnetic platen setup), at the same load, 80–150 rpm, for up to 2 minutes. Contrarotation is preferred if the head speed is <100 rpm. Examine the specimen, then etch it. If the surface quality is inadequate, remove the etch by repeating the last step, and re-etch. Alternatively, particularly with these alloys, remove the etch with the last step, then place the specimens on the vibratory polisher using a psa-backed synthetic suede cloth and colloidal silica for 20–30 minutes. This will facilitate removal of the last remnants of damage. The Ni-based superalloys are prepared in a similar manner, although the 1 μm step can usually be dropped, especially for solution annealed and aged specimens. If the 1 μm step is dropped, increase the 3 μm step time to 5 minutes.

ETCHING

Since nickel is generally resistant to corrosive media, etching involves the use of rather aggressive reagents, which tend to form etch pits and may attack inclusions. The difficulty of etching increases with the purity of the metal. Some etchants for pure Ni, such as No. 10, are rather dangerous to use. They must be used under a properly working chemical fume hood by personnel well-versed in such etchants' potential danger. In general, etching reagents (Table 10.23) either reveal the grain boundaries ('flat etch') or produce grain contrast (less common). Low-nickel copper–nickel alloys may be etched by the reagents recommended for the purpose under *Copper*. Heat-resisting nickel-base alloys are etched with the solutions used for stainless steels (Table 10.18).

ELECTROLYTIC ETCHING OF NICKEL AND ITS ALLOYS

This technique frequently gives better results than ordinary immersion or swab etching. Suitable solutions are summarised in Table 10.24.

Table 10.23   ETCHING REAGENTS FOR NICKEL AND ITS ALLOYS

See general safety note at the start of this Chapter.

No.	Etchant*		Remarks
1	Nitric acid (1.40) Hydrochloric acid (1.19) Glycerol	10 ml 20 ml 30 ml	Pure nickel, and nickel–chromium alloys. Grain boundaries etched
2	Nitric acid (1.40) Acetic acid Acetone	10 ml 10 ml 10 ml	Used for pure nickel, cupro–nickel, Monel metal and nickel–silver
3	Nitric acid (1.40) Glacial acetic acid Water	50 ml 25 ml 25 ml	Useful for nickel and most nickel-rich alloys
4	Nitric acid, 2% in alcohol, mixed in various proportions with hydrochloric acid, 2% in alcohol		Used for pure nickel

(continued)

Table 10.23   ETCHING REAGENTS FOR NICKEL AND ITS ALLOYS—*continued*

No.	Etchant*		Remarks
5	Hydrochloric acid Ferric chloride Water	30 ml 10 g 120 ml	Pure nickel
6	Hydrochloric acid (1.19) Ferric chloride Water	20 ml 10 g 30 ml	Pure nickel
7	Sulphuric acid (1.84) Hydrogen peroxide (10%)	1 ml 10 ml	Pure nickel
8	Sulphuric acid (1.84) Potassium dichromate (saturated in water)	10 ml 50 ml	Pure nickel
9	Ammonium persulphate, $100 \, \text{g} \, \text{l}^{-1}$ of water		Cast nickel
10	Ammonium persulphate, 10% in water Potassium cyanide, 10% in water	10 ml 10 ml	Pure nickel. Also recommended for cross-sections of nickel plated steels. The steel is not affected but may be etched by Reagent 2 for *Iron and steel*
11	Ammonium hydroxide		Used as a polish attack for nickel plate
12	Ammonium hydroxide Hydrogen peroxide (30%)	85 ml 15 ml	Used for nickel–silvers

* Acids are concentrated, unless otherwise indicated, e.g., with specific gravity.

Table 10.24   ELECTROLYTIC ETCHING OF NICKEL ALLOYS

See general safety note at the start of this Chapter.

No.	Etchant*		Conditions	Remarks
1	Sulphuric acid (1.84) Hydrogen peroxide (30%) Water	22 ml 12 ml to 100 ml	6 V d.c. Pt cathode 5–30 s	General etch
2	Sulphuric acid (1.84) Water	5 ml to 100 ml	6 V d.c. Pt cathode 5–15 s	All Ni-base alloys, especially Ni–Cu and Ni–Cr. Delineates carbides
3	Ammon. Persulphate Water	10 g to 100 ml	6 V d.c. Nickel cathode	Most alloys, especially Ni–Fe, Ni–Cr cast alloys
4	Chromium trioxide Water	6 g to 100 ml	As etchant 3	As etchant 3
5	Nitric acid (1.40) Glacial acetic acid Water	10 ml 5 ml 85 ml	1.5 V d.c. Pt cathode 20–60 s *(do not keep)*	Grain contrast for all nickel alloys
6	Oxalic acid Water	10 g 100 ml	6 V d.c. Stainless steel cathode 10–15 s	Superalloys, Ni–Au, Ni–Cr, Ni–Mo alloys
7	Phosphoric acid (1.71) Water (if necessary, sulphuric acid (1.84))	70 ml 30 ml 15 ml	2–10 V d.c. Ni cathode 5–60 s	Superalloys, Ni–Cr, Ni–Fe
8	Phosphoric acid (1.71) Sulphuric acid (1.84) Chromium trioxide	85 ml 5 ml 8 g	10 V d.c. Pt cathode 5–30 s	Ni-base superalloys. Gamma prime. Ti and Nb segregation

* Acids are concentrated, unless otherwise indicated, e.g., with specific gravity.

10.3.15   Niobium

PREPARATION

Niobium is soft and tough while its alloys are harder and simpler to prepare. Use the method described for molybdenum. Final polishing is performed using an attack-polishing agent, such as 30% conc. hydrogen peroxide. The following attack polishing solution can be added (exercising great care with respect to both operation and subsequent clean up in view of the hazards of hydrofluoric acid) to aqueous alumina slurries:

Hydrofluoric acid	2 ml
Nitric acid	5 ml
Lactic acid	30 ml

Note: Hydrofluoric acid produces irreversible bone damage and presents a range of other hazards. Even dilute hydrofluoric acid solutions should be handled with great care. Note that hydrofluoric acid attacks laboratory glassware.

Attack polishing using manual methods is quite tedious and increases the risk that the metallographer will have physical contact with the chemicals, even when using the correct personal protective equipment. An automated polisher should be used to reduce the risk of accidents.

Chemical polishing solutions are often used after mechanical polishing refractory metals and alloys to further improve the surface quality. A good solution consists of 30 ml water, 30 ml nitric acid, 30 ml HCl and 15 ml HF. Use at room temperature by swabbing or by immersion. Another solution contains 120 ml water, 6 g ferric chloride, 30 ml HCl and 16 ml HF. Immerse the specimen 2 minutes. This chemical polish is also recommended for V (immerse 1 minute) and for Ta (immerse for 3 minutes).

Table 10.25   ETCHING REAGENTS FOR NIOBIUM ALLOYS

See general safety note at the start of this Chapter.

No.	Etchant*		Conditions	Remarks
1	Hydrochloric acid (1.19) Sulphuric acid (1.84) Nitric acid (1.40) Water	15 ml 15 ml 8 ml 62 ml	Immerse 10–60 s	The only general-purpose etch free from HF
2	Hydrofluoric acid (40%)‡ Nitric acid (1.40) Lactic acid	10 ml 10 ml 30 ml	Immerse 15–20 s	General etch
3	Hydrofluoric acid (40%)‡ Nitric acid (1.40) Water	10 ml 20 ml 70 ml	Immerse up to 3 min *or* Use electrolytically 12–30 V d.c. Pt cathode up to 1 min	Grain boundary etch, especially at low voltages, electrolytically

* Acids are concentrated, unless otherwise indicated, e.g., with specific gravity.

‡ Hydrofluoric acid produces irreversible bone damage and presents a range of other hazards. Even dilute hydrofluoric acid solutions should be handled with great care. Note that hydrofluoric acid attacks laboratory glassware.

10.3.16   Platinum group metals

PREPARATION

The preparation of these metals is similar to that of gold (see section on Gold). The method described to prepare gold and high gold alloys works well for the other precious metals and will produce surfaces free of damage.

ETCHING

The reagents given in Table 10.25 will produce satisfactory microstructures if the preparation and polishing have been done carefully. These reagents are very strong and great care must be exercised in their use. They should only be used by people well trained in the use of dangerous chemicals.

Table 10.26   ETCHING REAGENTS FOR PLATINUM GROUP

See general safety note at the start of this Chapter.

No.	Etchant*		Conditions	Remarks
1	Aqua Regia Nitric acid (1.4) Hydrochloric acid (1.19)	 34 ml 66 ml	Up to 1 min. May need to be warmed	Pt, Pd, Rh alloys
2	Potassium ferricyanide Sodium hydroxide Water	3.5 g 1 g 150 ml	Several minutes	Most alloys, including Os alloys
3	Hydrochloric acid (1.19) Hydrogen peroxide (3%) Water	20 ml 1 ml 80 ml	Several minutes	Ru alloys
4	Potassium cyanide Water	5 g 100 ml	Electrolytic 1–5 V a.c. Pt cathode 1–2 min	Pt alloys
5	Hydrochloric acid (1.19) Sodium chloride Water	20 ml 25 g 65 ml	Electrolytic with Pt or graphite electrode. 25 s 10 V a.c. 1 min 1.5 V a.c. 1–2 min 20 V a.c. 1 min 6 V a.c. 1 min 5–20 V a.c.	 Rh alloys Pt-Rh alloys Ir alloys Pt alloys Ru-base alloys
6	Hydrochloric acid (1.19) Ethanol	10 ml 90 ml	Electrolytic 30 s, 10 V a.c. Graphite cathode	Os, Pd, Pt, Ir
7	Hydrochloric acid (1.19)		1–2 min electrolytic 5 V a.c. Pt or graphite electrode	Rh, Pt. Grain contrast

* Acids are concentrated, unless otherwise indicated, e.g., with specific gravity.

10.3.17   Silicon

PREPARATION

Silicon, which is very hard and brittle, is not often prepared for metallographic examination except in the area of electronic devices. Grinding Si with coarse grit SiC paper produces extensive damage to both the leading and trailing edges and should be avoided. Cut the specimen as close as possible to the desired plane-of-polish. Mounting in cast epoxy is preferred due to the brittleness of silicon. Si can be prepared in three steps, assuming that it has been sectioned using a low-speed saw or a precision saw. The first step uses a type A diamond lapping film, (e.g., an Ultra-Prep type A film) with 15 μm diamond abrasive and water cooling, with firm pressure, 100–150 rpm, until the cutting damage is removed, the surface is flat and has reached the desired location. Step 2 uses a synthetic chemotextile pad, such as Texmet® 1000, Pan W or Pellon®, with 3 μm diamond paste, 30 N (7 lbf) load, 240–300 rpm, for 2 minutes. A unique aspect of this step is the use of the 0.05 μm sol-gel alumina suspension as the lubricant with the diamond, rather than the traditional lubricant. This produces a dual polishing action that is very beneficial. The third, and final, step is with colloidal silica on a synthetic polyurethane pad, 20–25 N (5 lbf) load per specimen, 100–150 rpm for 90 s. Silicon responds well to colloidal silica. Specific details about preparing integrated circuits and other devices containing Si can be found in ref. [18].

ETCHING

Polished surfaces of silicon may be etched with 5% aqueous hydrofluoric acid, to which various amounts of concentrated nitric acid may be added. Microstructures developed in this way show angular grains, with twin markings, together with the particles of other constituents due to the presence of impurities (e.g., iron, aluminium and calcium).

Commercial silicon may contain inclusions of slag and unreduced quartz. Silicate inclusions may be detected by etching a polished section for 3 h in a stream of chlorine. The specimen is then immersed in concentrated hydrofluoric acid for 10 min in order to etch and attack the quartz

Table 10.27   ETCHING REAGENTS FOR SILVER AND ITS ALLOYS

See general safety note at the start of this Chapter.

No.	Etchant*		Remarks
1	Ammonium hydroxide and hydrogen peroxide—various proportions		The silver-rich matrix is in general unaffected. Other phases (e.g., the β silver–antimony phase) are often coloured blue to brown
2	Ammonium hydroxide Hydrogen peroxide (3%)	50 ml 10–30 ml	Recommended for silver, silver–nickel and silver–palladium alloys. Also useful for the examination of silver-soldered joints
3	Sulphuric acid (10% in water) to which a few crystals of chromic acid CrO_3 have been added (2 g)		This reagent reveals the grain structure of silver and silver-rich alloys
4	Solution containing 7.6 g l^{-1} of chromic acid (CrO_3) and 8 g l^{-1} of sulphuric acid		Useful general etching reagent. Used as a sensitive etching reagent for silver–copper alloys
5	Ferric chloride 20 g l^{-1} of water		Recommended for silver solders
6	Potassium cyanide, 10% in water Ammonium persulphate, 10% in water	10 ml 10 ml	Etch for pure silver and dilute alloys. Duration 1–2 min
7	Solution A: 50:50 nitric acid in water Potassium dichromate Solution B: Chromic acid CrO_3 Sodium sulphate Water	 100 ml 2 g 20 g 1.5 g 100 ml	Used for silver alloys in general. Solution A is diluted to 20 vol. and an equal amount of solution B added. The reagent is applied by gentle swabbing or with a camel hair brush. A loose film of silver chromate should form if the reagent is working correctly. If the film is adherent, more of solution A should be added. If no film forms, more of solution B is required
8	Chromic acid, CrO_3, 0.2% and sulphuric acid, 0.2% in water		Used (1 min immersion) for silver and silver-rich alloys
9	Potassium dichromate saturated in water Sodium chloride saturated in water Sulphuric acid	100 ml 2 ml 10 ml	Silver and silver-rich alloys. Silver solders
10	Chromic acid and hydrogen peroxide in water. Various proportions		General reagent. Composition adjustments must be made to suit specific cases
11	Sodium hydroxide (10%) Potassium ferricyanide (30%)	10 ml 10 ml	5–15 s. Ag alloys with W, Mo and WC (dilute with equal vol. of water if too fast)

* Acids are concentrated, unless otherwise indicated, e.g., with specific gravity.

inclusions (*see also* Table 10.9). This is a very dangerous etching procedure that requires properly designed equipment, correct ventilation, and well trained personnel.

10.3.18   Silver

PREPARATION

Silver is very soft and ductile, and is prepared in the same manner as gold and its alloys. The method is described in the section for gold. Embedding of abrasives can be a problem, as for gold.

Pure silver and dilute silver alloys are difficult to etch, but several solutions will give good results on duplex or more complex alloys. The strengths of the reagents, unless otherwise noted in Table 10.27, should be adjusted to the specific alloys to be examined.

ELECTROLYTIC ETCHING OF SILVER

In many cases excellent grain boundary and grain contrast etching is obtained by electrolysis (specimen as anode) in one of the solutions given in Table 10.28.

See also reagents listed for electrolytic polishing of silver, Table 10.4b. The silver cyanide–potassium carbonate reagent in Table 10.4b is particularly convenient, because, on reduction of the polishing potential from 1.5 to 0.5 V for 90 s, etching occurs.

Table 10.28   ELECTROLYTIC ETCHING OF SILVER ALLOYS

See general safety note at the start of this Chapter.

No.	Etchant*	Remarks
1	Citric acid $100\,g\,l^{-1}$ of water	15 s to 1 min. 6 V d.c. Ag cathode. Most alloys
2	Ammoniacal ammonium molybdate	Molybdic acid is dissolved in an excess of strong ammonia. The composition is not critical provided ammonia is in excess. The optimum potential and current density should be experimentally established for each case
3	Potassium cyanide $50\,g\,l^{-1}$ of water	This reagent is used particularly for silver when it is in contact with other metals, as in plated articles. Optimum conditions should again be established experimentally
4	Hydrofluoric acid + a little stannous chloride‡	Used for silver–tin alloys containing more than 73% of silver

* Acids are concentrated, unless otherwise indicated, e.g., with specific gravity.
‡ Hydrofluoric acid produces irreversible bone damage and presents a range of other hazards. Even dilute hydrofluoric acid solutions should be handled with great care. Note that hydrofluoric acid attacks laboratory glassware.

10.3.19   Tantalum

PREPARATION

The preparation of tantalum (pure and commercial) follows the same procedures as described for Mo and Nb (see these sections). Attack polishing agents are often used (see previous comments on safety issues). The following mixture has been used as the polishing medium:

Alumina, carefully levigated	35 g
Hydrofluoric acid (60%)	20 ml
Ammonium fluoride	20 g
Distilled water	1000 ml

Note: Hydrofluoric acid produces irreversible bone damage and presents a range of other hazards. Even dilute hydrofluoric acid solutions should be handled with great care. Note that hydrofluoric acid attacks laboratory glassware.

Chemical polishing is often used after mechanical polishing to further improve the surface quality. Kelly recommends the following mixture: 25 ml lactic acid, 15 ml nitric acid and 5 ml HF. Swab vigorously for 2 minutes.

ETCHING

The reagents given in Table 10.29 have been found useful.

10.3.20   Tin

PREPARATION

Tin is allotropic with a body-centred tetragonal crystal structure at room temperature. It will respond to polarised light when properly polished. Because of its relative softness, the same precautions should be observed as for lead and its alloys. The method described for lead, works well for tin.

For further details of methods for tin and tin alloys, refs.[1, 19, 69–71] may be consulted.

ETCHING

Tin is anisotropic and the grain structure can be revealed under crossed polarised light. The reagents of Table 10.30 have been recommended for tin-rich alloys. In general, etching times are not critical and the progress of the treatment should be judged visually.

Table 10.29   ETCHING REAGENTS FOR TANTALUM

See general safety note at the start of this Chapter.

No.	Etchant*		Remarks
1	Ammonium fluoride, 20% in water Hydrofluoric acid (43%), 60% in water‡	10 ml 10 ml	Used at 50–60°C (ineffective when cold). Immersion for 1 min etches up the structure of tantalum without colouring inclusions of tantalum sulphide, Ta_2S_5
2	Ammonium fluoride, 20% in water Sulphuric acid (1.84)	10 ml 20 ml	Used at 60°C for 1–2 min. Etching effects are similar to those for Reagent 1
3	Ammonium fluoride, 20% in water Nitric acid	10 ml 10 ml	Used at 60°C; the matrix is usually not adequately etched but Ta_2S_5 is blackened. This reagent may be used after Reagent 1 or 2 to identify Ta_2S_5
4	Ammonium fluoride, 20% in water		Used for 5–6 min at 80°C. The grain structure of the matrix is developed, and Ta_2S_5 is not affected
5	Ammonium fluoride, 20% in water Hydrogen peroxide	20 ml 10 ml	This reagent is used boiling, and colours Ta_2S_5 brown. The matrix is not affected
6	Sulphuric acid Nitric acid Hydrofluoric acid‡	25 ml 10 ml 10 ml	For general structure[68]

* Acids are concentrated, unless otherwise indicated, e.g., with specific gravity.
‡ Hydrofluoric acid produces irreversible bone damage and presents a range of other hazards. Even dilute hydrofluoric acid solutions should be handled with great care. Note that hydrofluoric acid attacks laboratory glassware.

ELECTROLYTIC ETCHING OF TIN ALLOYS

A 10% aqueous solution of hydrochloric acid or 20% sulphuric acid may be used at a very low current density; this is especially useful for tin-iron alloys. Satisfactory results are also obtained from:

Glacial acetic acid   130 ml
Perchloric acid   50 ml†

used at a current density of 3–$6\,A\,dm^{-2}$ (use with care; see Section 10.2, p. **10**–14, *Electrolytic Polishing*).

Note: A specially designed, dedicated fume hood, with a vertical stack and a wash down system is required for perchloric acid. Note that some solutions containing perchloric acid present grave explosion hazards and there are restrictions on the use of these solutions in some locations.

10.3.21   Titanium

PREPARATION

Pure titanium is soft and ductile. Because of its hcp crystal structure, it deforms by mechanical twinning, and this can be caused by rough handling. If improperly sectioned, the depth of damage can be extensive. Sectioning is very difficult, as cutting rates are low. While their indentation hardness values are low compared to tool steels and high speed steels, Ti and its alloys are much more difficult to cut and require a more softly bonded abrasive wheel than the hardest steels. Grinding and polishing rates are also lower than for steels, or most other metals. Some authors have stated that Ti and its alloys should not be mounted in phenolic resins as hydrogen can be absorbed from the resin to form TiH; or, if hydrides are present (as in a specimen in a H-rich service environment), the heat from compression moulding may dissolve some, or all, of the hydride. The author of the present section has never observed hydrides in specimens mounted with different thermosetting resins. However, in a field failure of commercial purity (CP) titanium that contained very large amounts of titanium hydrides, the mounting temperature was observed to dissolve some of the hydrides. A mount made using a low-viscosity, slow curing epoxy resin revealed the greatest amount of TiH. Mounts made with a fast curing epoxy (cures in about 45 min.), which generates a high heat of polymerisation, had a reduced amount of TiH, similar to the amount obtained using a thermosetting resin. So, in failure analysis work, if hydrides may be present, it is best to use a slow curing epoxy resin with a conductive mounting approach (see the section on *Mounting*), to minimise the exotherm produced during polymerisation and the chance of dissolving any TiH.

Table 10.30   ETCHING REAGENTS FOR TIN AND ITS ALLOYS

No.	Etchant*		Remarks
1	Nitric acid, 2% in alcohol		This reagent is a general one for tin-rich alloys, and particularly for tin–cadmium, tin–antimony and tin–iron alloys. The tin-rich matrix is darkened, and the intermetallic compounds usually little affected
2	Nitric acid, 5% in alcohol		Tin, tin–cadmium and tin–iron alloys
3	Nitric acid Acetic acid Glycerol	10 ml 30 ml 50 ml	Mainly used as a reagent for tin. Used at 38–40°C
4	Nitric acid Acetic acid Glycerol	10 ml 10 ml 80 ml	Used at 38–40°C for tin-lead alloys, especially tin-rich alloys Pb is blackened
5	Picric acid‡ and nitric acid in alcohol—proportions variable		Useful for tin-iron alloys in contact with steel
6	Hydrochloric acid, 1–5% in alcohol		Useful for tin–lead, tin–cadmium, tin–iron, tin–antimony–copper alloys
7	Hydrochloric acid Ferric chloride Water	2 ml 10 g 95 ml	Useful for tin-rich alloys in general and for Babbitt metal. It is often an advantage to add alcohol to this reagent. Any cadmium-rich solid solution present is stained black, while the β-phase of the tin–cadmium system turns brown
8	Hydrochloric acid Ferric chloride Water Alcohol	5 ml 2 g 30 ml 60 ml	Reveals general structure of tin and alloys without lead. Tin–iron and tin–copper compounds unattacked
9	Stannous chloride in acid solution		Tin-rich alloys in general. The composition may be varied to suit individual alloy systems
10	Potassium dichromate, in dilute acidified aqueous solution (composition variable)		Recommended for tin–cadmium alloys
11	Potassium ferricyanide in caustic soda (composition variable)		Used for tin–cadmium–antimony alloys. It distinguishes SbSn (with tin and cadmium in solid solution), which is not affected, from CdSb in tin–cadmium–antimony alloys
12	Ammonium persulphate Water	5–10 g 100 ml	Used particularly for tinplate. The tin is heavily darkened, leaving the basis metal unattacked. Most intermetallic compounds of tin are also unattacked. More dilute solution gives grain boundary etch with tin and alloys
13	Nitric acid and hydrofluoric acid in glycerol§		Various strengths are recommended for tinplate. The contrast between plating and steel may be improved by the use of Reagent 2 (Table 10.17)
14	Sodium sulphide 20% in water, with a few drops of hydrochloric acid		Useful for tin-rich tin–antimony–copper alloys. The phase SbSn is not affected, but the phase Cu_6Sn_5 is coloured brown[72]
15	Silver nitrate Water	5 g 100 ml	Darkens primary and eutectic lead in lead-rich lead–tin alloys[69]

* Acids are concentrated, unless otherwise indicated.
‡ Special handling is required for picric acid to avoid the risk of fire and/or explosion and local regulations must be consulted before use of this reagent.
§ Hydrofluoric acid produces irreversible bone damage and presents a range of other hazards. Even dilute hydrofluoric acid solutions should be handled with great care. Note that hydrofluoric acid attacks laboratory glassware.

To prepare Ti and its alloys, a three step contemporary procedure is used that has been shown to be capable of producing perfect images of CP Ti in crossed polarised light, and a vast range of alpha, alpha-beta and beta alloys. Of course, we must start with a properly sectioned piece, and this is extremely important with Ti. Use only a laboratory abrasive cut-off saw, or a precision saw, with the proper wheel (extremely critical), with low pressure and copious cooling. Mount the specimens as discussed above. Step 1 uses psa-backed 320 grit (P400) SiC paper, water cooled, with 25–30 N (6 lbf) load per specimen, 240–300 rpm, until the cutting damage is removed and the specimens are at the same plane. This may require more than one sheet of paper.

Step 2 employs a psa-backed (can use a magnetic platen set up) selected silk cloth, such as an Ultra-Pol™ cloth, with 9 μm diamond, 25–30 N (6 lbf) load per specimen, 120–150 rpm, for 10 minutes. The cloth is critical to the success of this method. Selected silk yields a good removal rate and a better surface finish than any other cloth while minimising relief, and maximising edge retention. If other cloths or rigid grinding discs are used, the results will be inadequate for CP Ti examination by polarised light (unless an additional step is added). If the polishing head rotates at <100 rpm, use contrarotation mode. Start by charging the cloth with paste, then add the appropriate lubricant. During the 10 minute cycle, periodically add the same size diamond in slurry or suspension format to keep the cutting rate high.

Step 3 uses 0.05 μm colloidal silica on a medium napped synthetic suede cloth, same load, 120–150 rpm, for 10 minutes. This method works much better if contrarotation can be used. This requires a polisher with a low head speed, ideally 30–60 rpm. If the head speed exceeds 100 rpm, the abrasive will be thrown off the polishing cloth onto the metallographer and the walls. This is especially bad, as the colloidal silica must be mixed in a 5:1 ratio with hydrogen peroxide (30% conc.). If the addition of hydrogen peroxide causes the colloidal silica to stiffen, add distilled water. With 20 s left in the third step, stop adding any abrasive. With 10 s left in the cycle, turn the water jet onto the polishing cloth surface and wash off both the abrasive from the cloth and from the specimen. Then, rinse the specimen and holder with water, scrub with cotton soaked in ethanol or in a soap solution, as desired, rinse in water, displace the water with ethanol, and blow dry under a stream of hot air. CP Ti specimens will yield excellent grain structure images under crossed polarised light. The image quality can be further enhance by a brief (20–30 minutes) vibratory polish using colloidal silica on the medium napped synthetic suede cloth.

ETCHING

The presence of surface oxide films on titanium and its alloys necessitates the use of strongly acidic etchants, usually containing HF. Those given in Table 10.31 are useful. HF is a very dangerous acid and skin contact must be avoided as it produces very severe burns involving irreversible damage to the bone, even in dilute solutions and has other potential fatal hazards. Reagents should be kept in polyethylene beakers, as glass is attacked by HF.

10.3.22 Tungsten

PREPARATION

Tungsten, compared to the other refractory metals, is one of the easier to prepare. The methods described for Mo and Nb are adequate for W. See these sections. The use of an attack-polish agent is helpful when preparing tungsten.[73]

ETCHING

The etching of tungsten is not difficult as Murakami's Reagent is quite satisfactory. Inclusions (e.g., thorium, uranium or calcium oxides) may be seen in the polished, unetched surface, while the general structure of filaments which have been heated *in vacuo* or in a reducing atmosphere may often be observed owing to evaporation effects. A selection of suitable etching reagents is given in Table 10.32.

ELECTROLYTIC ETCHING OF TUNGSTEN

This may be carried out in a mixture of 25 ml of normal aqueous sodium hydroxide and 20 ml of hydrogen peroxide. The current density and potential are somewhat critical, and should be carefully controlled after investigation to find the optimum conditions for specific cases. A sodium hydroxide solution of 0.025 normal strength has also been used, at a current density of 5 A dm.$^{-2}$

Table 10.31   ETCHING REAGENTS FOR TITANIUM AND ITS ALLOYS

See general safety note at the start of this Chapter.

No.	Etchant*		Conditions	Remarks
1	Hydrofluoric acid (40%)§ Nitric acid (1.40) Lactic acid	1–3 ml 10 ml 30 ml	5–30 s	Mainly unalloyed titanium; reveals hydrides
2	Hydrofluoric acid (40%)§ Nitric acid (1.40) Lactic acid	1 ml 30 ml 30 ml	5–30 s	As Etchant 1
3	Hydrofluoric acid (40%)§ Nitric acid (1.40) Water (Kroll's reagent)	1–3 ml 2–6 ml to 100 ml	3–10 s	Most useful general etch
4	Hydrofluoric acid (40%)§ Nitric acid (1.40) Lactic acid	10 ml 10 ml 30 ml	5–30 s	Chemical polish and grain boundary etch
5	Potassium hydroxide (40%) Hydrogen peroxide (30%) Water (can be varied to suit alloy)	10 ml 5 ml 20 ml	3–20 s	Useful for α/β alloys. α is attacked or stained. β unattacked
6	Hydrofluoric acid (40%)§ Nitric acid Glycerol	20 ml 20 ml 40 ml	5–15 s	General purpose, TiAlSn alloys
7	Hydrofluoric acid§ Nitric acid (1.40) Glycerol Water	1 ml 25 ml 45 ml 20 ml	3–20 s	TiAlSn alloys

* Acids are concentrated, unless otherwise indicated.
§ Hydrofluoric acid produces irreversible bone damage and presents a range of other hazards. Even dilute hydrofluoric acid solutions should be handled with great care. Note that hydrofluoric acid attacks laboratory glassware.

Table 10.32   ETCHING REAGENTS FOR TUNGSTEN

See general safety note at the start of this Chapter.

No.	Etchant*		Remarks
1	Sodium hydroxide, 10% in water Potassium ferricyanide, 10% in water (Murakami's reagent)	10 ml 10 ml	This reagent is used cold and, on immersion of the specimen for approximately 10 s, develops grain boundaries
2	Hydrogen peroxide, 3% in water		This reagent develops grain boundaries, but only after some 30–90 s in the boiling reagent
3	Potassium ferricyanide Caustic soda Water	305 g 44.5 g 1000 ml	Recommended for deep etching of single crystal bars and wires in order to produce etch-pits for the investigation of orientation
4	Hydrofluoric acid§ Nitric acid (1.40) Lactic acid	5 ml 10 ml 30 ml	Swab 10–20 s, rinse and dry. Follow with Murakami's reagent (Etchant 1)

* Acids are concentrated, unless otherwise indicated, e.g., with specific gravity.
§ Hydrofluoric acid produces irreversible bone damage and presents a range of other hazards. Even dilute hydrofluoric acid solutions should be handled with great care. Note that hydrofluoric acid attacks laboratory glassware.

10.3.23   Uranium

PREPARATION

Conventional methods are used but as the metal is toxic, pyrophoric when finely divided, and an α-particle emitter, it is essential to keep the metal wet during cutting and to carry out the operation

Table 10.33 ETCHING REAGENTS FOR URANIUM ALLOYS

See general safety note at the start of this Chapter.

No.	Etchant		Conditions	Remarks
1	Nitric acid (1.40) Glacial acetic acid Glycerol	30 ml 30 ml 30 ml	5–60 s	Most U alloys; distinguishes precipitated phases, e.g., U_2Ti U_2Mo, UZr_2, eutectoid nucleation at grain boundaries (like pearlite in steel)
2	Orthophosphoric acid Diethylene glycol Ethanol	30 ml 30 ml 40 ml	Electrolytic 20 V stainless steel cathode 5–30 s	Will electropolish at high voltages (30–40V), stains phases differentially at 10–20 V
3	Hydrofluoric acid (40%)* Nitric acid (1.40) Glycerol	10 ml 40 ml 40 ml	5–10 s	U–Mo, U–Nb, U–Zr alloys (gamma phase)
4	Hydrofluoric acid (40%)* Nitric acid (1.40) Lactic acid (or distilled water)	1 ml 30 ml 30 ml	5–30 s	U–Be, U–Nb, U–Zr, U–Mo. Also U–Al alloys. UAl_2 UAl_3, UAl_4 Stained differentially according to time

* Hydrofluoric acid produces irreversible bone damage and presents a range of other hazards. Even dilute hydrofluoric acid solutions should be handled with great care. Note that hydrofluoric acid attacks laboratory glassware.

behind a screen so that particles cannot be ingested. Preparation of any radioactive metal can only be done in specially designed facilities that have been licensed to work with such metals, and requires personnel properly trained to deal with the hazards involved.

Methods utilising attack-polish solutions are preferred and give structures which can be examined under polarised light with crossed polars, which is the most useful method of examination.[56]

Recommended attack-polish solutions are:

1. Dilute hydrofluoric acid and nitric acid (1 ml + 5 ml respectively in 100 ml water).
2. 50 g chromium trioxide, 10 ml nitric acid, 100 ml water is less aggressive but, like the other, needs careful handling to avoid contact with the skin. This gives the best results for polarised light examination.

Note: Hydrofluoric acid produces irreversible bone damage and presents a range of other hazards. Even dilute hydrofluoric acid solutions should be handled with great care. Note that hydrofluoric acid attacks laboratory glassware.

ETCHING

Polarised light is the most useful technique.

α-uranium—high grain contrast under crossed polars about 1° off the extinction position. The various transformation products, granular α, martensite, etc., can be distinguished.
β-uranium—weakly anisotropic when retained but clearly distinguishable from α and γ.
γ-uranium—optically isotropic and therefore distinguished from α and β. In uranium alloys with molybdenum, niobium and rhenium, γ_0 is tetragonal with $c/a = 0.5$, is optically active, and the grain structure is visible under crossed polars.

On ageing, these alloys develop γ_0 with $c/a < 0.5$ and this is readily detected by polarised light metallography.

Note that chemical etching may give quite different structures, e.g., in U–Ti alloys. Chemical etching is responsive to the manner of decomposition of γ uranium and will reveal the sequence of decomposition, e.g., distribution of U_2Ti, UZr_2, etc., whereas polarised light reveals the final and true grain structure which may be quite different from the etched structure.[74]

10.3.24 Zinc

PREPARATION

Though somewhat harder than lead and tin, zinc and its alloys must be carefully prepared, as improper cutting and grinding produces mechanical twinning in a comparatively thick surface layer. The depth

Table 10.34   ETCHING REAGENTS FOR ZINC AND ITS ALLOYS

See general safety note at the start of this Chapter. These reagents are of general applicability to zinc and many zinc-rich alloys. Etchant 3 is suitable for examination of microconstituents at high magnifications.

No.	Etchant		Remarks
1	Hydrochloric acid, 1% in alcohol		For Zn and lead alloys; immerse for up to several minutes.
2	Sodium hydroxide, 100 g l^{-1} of water		For pure Zn, Zn–Co and Zn–Cu alloys; swab or immerse for up to 15 sec.
3	Nitric acid, 1–2% in alcohol		Rinse in 20% chromic acid in water to avoid stains. Good for Zn–Fe layers in galvanised samples
4	Nitric acid	94 ml ⎱ Stock	A few drops of this stock solution are added to 100 ml
	Chromic acid	6 ml ⎰ solution	of water immediately before use. The resulting solution is generally useful, particularly for the recognition of small amounts of other microconstituents in zinc
5	Chromic acid, CrO$_3$	200 g	Palmeton Reagent for commercial zinc and zinc alloys.
	Sodium sulphate, Na$_2$SO$_4$	15 g	If a film of stain results, this may be removed by
	Water	1000 ml	immersion in a 20% solution of chromic acid in water
6	Chromic acid, CrO$_3$	50 g	As for Etchant 5. This composition is recommended
	Sodium sulphate, Na$_2$SO$_4$	4 g	for die-castings. Modified Palmerton Reagent
	Water	1000 ml	
7	Chromic acid, CrO$_3$	200 g	Recommended for zinc-rich alloys containing
	Sodium sulphate, Na$_2$SO$_4$	7.5 ml	copper. Subsequent immersion in 20% chromic
	Water	1000 ml	acid solution is again helpful
8	Chromic acid, CrO$_3$	200 g	Recommended for die-casting alloys containing
	Sodium sulphate, Na$_2$SO$_4$	7 g	aluminium, as the aluminium-rich microconstituent
	Sodium fluoride	2 g	is satisfactorily etched with this reagent. In the
	Water	1000 ml	presence of copper, staining may result; to prevent this, immerse after etching, without washing, in a solution of 50 g chromic acid and 4 g sodium sulphate in 1000 ml, water (*see* Etchant 5).
9	Solution prepared thus: mix 51 ml of concentrated potassium hydroxide solution with 50 ml of water and 20 ml of concentrated copper nitrate solution. Stir in 25 g of powdered potassium cyanide. Filter, and add 2.5 ml of concentrated citric acid solution before use		Etch by immersion for 10–20 s. The zinc-rich phase is coloured dark brown to black. The iron-phase in commercial alloys usually appears as white rods, and the lead-phase as round white spots

of damage can approach at least 2 mm on a saw cut or filed surface. If such a surface layer experiences preparation-induced heating, it is likely to cause recrystallisation of the deformed surface layer.

Zinc, as with Pb, Sn, Sb, Bi and Cd, is not abraded well with diamond abrasives. It is better to use the traditional method with several stages of SiC paper. Follow the approach described for lead. Embedding can be a problem with Zn, although it is less of a problem than with Pb. Polish using aqueous alumina slurries, as described for Pb.

ETCHING

The reagents listed in Table 10.34 may be used for zinc and its alloys.

ELECTROLYTIC ETCHING OF ZINC ALLOYS

Comparatively little work has been done on electrolytic etching of zinc alloys because the microstructures are relatively simple and adequately brought out by the above listed reagents. For alloys containing copper, however, a 20% aqueous chromic acid solution has been recommended, the specimen being the anode.

Ternary alloys containing aluminium and copper may be conveniently etched electrolytically in a solution made by adding 20 drops of hydrochloric acid to 50 ml alcohol.

DEVELOPMENT OF MICROSTRUCTURES WITHOUT ETCHING

Zinc and many zinc-rich alloys develop satisfactory microstructures if the polished surface is allowed to remain exposed to the air for 1–3 days. Similar results may be obtained by heating at 100°C for a shorter time, but care must be taken to ensure that the use of the high temperature does not lead to any structural modification.

10.3.25   Zirconium (and Hafnium)

PREPARATION

Pure zirconium, and hafnium (which is quite similar to Zr), are soft, ductile hcp metals that can easily be deformed by mechanical twinning.[75] As with Ti, they must be sectioned with a laboratory abrasive cut-off machine using a very softly bonded wheel to minimise damage. As with all refractory metals, the grinding and polishing rates are low and scratch elimination can be difficult using standard methods. The contemporary preparation method is suitable for the alloys, but for the pure metals, a procedure similar to used with lead, with attack polishing additions followed by chemical polishing, works better. After proper cutting, the specimen is mounted and ground with 320 grit (P400) SiC paper with water cooling, 20–25 N (5 lbf) load, 200–250 rpm, until the surface damage is removed and the specimens are all ground to a common plane.

Step 2 uses 9 μm diamond on a psa-backed selected silk cloth, same load, 150–200 rpm, for 5 minutes. Step 3 uses 3 μm diamond on a woven polyester cloth, with the same load, same rpm, for 3 minutes. Step 4 uses 0.05 μm colloidal silica, with an attack polish agent, on a medium napped, synthetic suede cloth, at 25–30 N (6 lbf) load, 120–150 rpm, for 7 minutes. In these three polishing steps, contrarotation should be used, if the head speed is <100 rpm. See comments about this under Ti. A number of attack-polish agents[2] have been used, and those described for Ti are fine with Zr and Hf. Cain has developed several chemical polishing solutions that are used after polishing to improve polarised light response. Kelly's chemical polish solution also works quite well: 25 ml lactic acid, 15 ml nitric acid and 5 ml HF (swab vigorously for 2 minutes). For high-purity Zr or Hf, use the method for lead (coating the SiC with wax is not necessary) and add a 5 or 3 μm aqueous alumina step. Use an attack polish agent with each alumina step and follow with a chemical polish using Ann Kelly's solution. Vibratory polishing with colloidal silica on a medium napped synthetic suede cloth for 20–30 minutes also improves the polarised light response of Zr, Hf and their alloys.

ETCHING

When properly prepared, these metals and alloys can be examined in crossed polarised light. Hence, etching is often unnecessary. Etching reagents for zirconium, hafnium and their alloys are summarised in Table 10.35.

Table 10.35   ETCHING REAGENTS FOR ZIRCONIUM AND HAFNIUM

See general safety note at the start of this Chapter.

No.	Etchant		Remarks
1	Nitric acid (1.40)	20 ml	The reaction rate may be increased, by heating the sample in a
	Hydrofluoric acid (1.19)*	20 ml	stream of hot water before immersion in the reagent. Conversely
	Glycerol	60 ml	the reaction rate is decreased by chilling the specimen
2	Glycerol	60 ml	On etching for 3–5 s, microconstituents are outlined and
	Hydrofluoric acid (1.19)*	20 ml	differentiated, and carbides unattacked. In the presence of
	Nitric acid	10 ml	moisture the reagent tends to stain the specimen
3	Glycerol	16 ml	Etching times of 1–2 s are used. The reagent is useful for alloys
	Hydrofluoric acid (1.19)*	2 ml	which are not satisfactorily etched by Etchant 2
	Nitric acid (1.40)	1 ml	
	Water	2–4 ml	
4	Hydrofluoric acid (1.19)*	20 ml	Short etching times are necessary (1–2 s). The reagent is similar
	Water	80 ml	to Etchant 3 but is more drastic
	Nitric acid (1.40)	1 ml	

* Hydrofluoric acid produces irreversible bone damage and presents a range of other hazards. Even dilute hydrofluoric acid solutions should be handled with great care. Note that hydrofluoric acid attacks laboratory glassware.

ELECTROLYTIC ETCHING

Solutions for electrolytic etching are given in Table 10.36.

10.3.26   Bearing metals (lead-tin-antimony), low-melting point solders, and type metals

PREPARATION

Bearing metals consist of hard particles of intermetallic compounds set in a matrix of soft lead-rich or tin-rich material. In general, bearing metals are easier to prepare than Pb, Sn, Bi, Cd and Zn. Many times, the metallographer encounters these as layers on metals, usually steels. These specimens must be mounted for edge retention and can be prepared with the contemporary method for steel. Printing metals and solders are easier to prepare than the pure metals, but the method described for lead should be used.

ETCHING

All the reagents listed for lead and tin alloys may be tried for revealing the structure of bearing metals, the choice depending on whether the alloy is lead-rich or tin-rich. Of the reagents used for lead, the most useful for the present purpose are: Nos. 4, 6, 8, 11 and 12; of the reagents for tin, Nos. 1, 6 and 7 are suitable.

Further reagents are summarised in Table 10.37.

ELECTROLYTIC ETCHING

Bearing metals may be polished electrolytically in a solution containing 60% perchloric acid and acetic anhydride in the ratio of 1 part to 4 parts by volume (see the safety comments under Electrolytic

Table 10.36   ELECTROLYTIC ETCHING OF ZIRCONIUM

See general safety note at the start of this Chapter.

No.	Etchant		Remarks
1	Ethanol	70 ml	Suitable for cast zirconium
	Perchloric acid (density 1.2)*	20 ml	
	2-Butoxyethanol	10 ml	
2	Acetic acid	1000 ml	Suitable for worked and annealed material
	Perchloric acid (density 1.59)*	50 ml	
3	Ethanol	30 ml	Etching time 10–20 s at 1 A dm^{-2}
	Hydrochloric acid conc.	10 ml	
4	Ethanol	450 ml	As for Etchant 3
	Distilled water	70 ml	
	Perchloric acid*	25 ml	

* A specially designed, dedicated fume hood, with a vertical stack and a wash down system is required for perchloric acid. Note that some solutions containing perchloric acid present grave explosion hazards and there are restrictions on the use of these solutions in some locations.

Table 10.37   ETCHING REAGENTS FOR BEARING METALS

See general safety note at the start of this Chapter.

No.	Etchant	Remarks
1	Iodine in potassium iodide solution	Type metals containing small amounts of zinc
2	Iodine, 10% in alcohol	Type metals containing large amounts of zinc
3	Silver nitrate (2–5%) in water	Useful for bearing metals in general

Polishing and the discussion of Figure 10.2 as this mixture can be explosive). If the voltage is reduced after polishing, sensitive etches of most bearing metal materials may be obtained.

10.3.27   Cemented carbides and other hard alloys

PREPARATION

Cemented carbides are hard, usually >1200 HV. This alters the preparation procedure. Sectioning is usually performed with a low-speed saw or a higher speed, precision saw, some of which are called linear precision saws because the sample is fed into the blade at a controlled rate rather than being gravity-fed. These saws use a non-consumable diamond blade and yield excellent cuts with little damage and superb surface smoothness. Because of the high hardness of sintered carbides, SiC grinding paper (and alumina grinding paper), is not useful. Instead, diamond must be used for all steps. A wide variety of products can be employed. After sectioning, the specimens are embedded in a polymer and the thermosetting epoxy containing a silica filler (such as Epomet® Resin) is ideal.

Step 1 uses either a metal-bonded diamond disc with a coarse diamond size (there are several types available, e.g., an Ultra-Prep disc) with a coarse diamond, e.g., 45 μm size, which is water cooled, as with SiC paper, with a load of 25–30 N (6 lbf) per specimen, 240–300 rpm, for 5 minutes. Otherwise, a rigid grinding disc can be used. In this case, the diamond abrasive is added in suspension or slurry form. A particle size of 30 to 45 μm is used. The other parameters are the same. If the head speed is <100 rpm, use contrarotation.

Step 2 can also use the same two working surface options, but with a finer diamond size. For example, a 9 μm metal-bonded diamond disc can be used, with the same settings as for step 1, except for 4 minutes. Alternatives include another rigid grinding disc, or a psa-backed selected silk or nylon cloth. These are used with 9 μm diamond and the same settings. Step 3 uses 3 μm diamond on either of several cloths—selected silk, nylon, polyester or a chemotextile pad (all with psa backings or using a magnetic platen set up). As elsewhere, charge the cloth with paste, press the paste into the cloth surface with a finger tip, add some lubricant and begin polishing. During the cycle, add diamond of the same size in slurry or suspension form. For step 3, use the same load, 120–150 rpm, for 3 minutes. Again, use contrarotation if the head speed is <100 rpm. At this point the surfaces are more than adequate quality for routine examination.

If the sintered carbide is coated, and particularly when multiple CVD layers are applied, the resolution of these coatings will be enhanced if additional steps are used. A 1 μm diamond step, in the same manner as the 3 μm step can be added. That can be followed by final polishing with colloidal silica on a synthetic suede cloth or with the polyurethane pad, same load, 120–150 rpm, for 2 minutes. Again, use contrarotation if possible. Another option is to use a brief vibratory polish with colloidal silica, 20–30 minutes is adequate. These polishing steps enable complex coatings to be resolved clearly and give maximum detail of the cobalt binder phase. These features must be studied at the limits of the light microscope for resolution and magnification, so SEM examination is also widely used, particularly backscattered images as the different carbides and the binder phase can be revealed nicely in the range of 2000–5000X, where the limited resolution of backscattered electron image remains sufficient.

ETCHING

Graphite and eta phase are observed best before etching. A number of etching reagents have been developed for these materials, and a selection is given in Table 10.38.

ELECTROLYTIC ETCHING OF TUNGSTEN CARBIDE–COBALT AND TUNGSTEN CARBIDE–TITANIUM CARBIDE–COBALT ALLOYS

Two solutions have been recommended:

1. For tungsten carbide in a matrix of cobalt:

   | Sodium hydroxide | 10% | in water |
   | Potassium ferricyanide | 10% | |

   On electrolysis at 2 V, the carbide particles are attacked, and the cobalt binder phase is almost unaffected.

Table 10.38 ETCHING REAGENTS FOR CEMENTED CARBIDES, ETC.

See general safety note at the start of this Chapter.

No.	Etchant		Remarks
1	Potassium hydroxide, 20% in water Potassium ferricyanide, 20% in water		The cold solution is used as a general reagent (Murakami's reagent). Equal volumes are mixed NaOH can be substituted for KOH
2	Potassium hydroxide Potassium ferricyanide Water	10 g 10 g 10 ml	The boiling solution is used as a general reagent (Mod. Murakami's reagent) NaOH can be substituted for KOH
3	Aqua regia (20 ml nitric acid + 80 ml hydrochloric acid)		General reagent
4	Hydrofluoric acid§ Nitric acid conc.	70 ml 30 ml	Rapidly attacks eutectic material in iron–tungsten carbide alloys and iron–molybdenum carbide alloys
5	Picric acid, 2% in alcohol*		Develops the eutectic structure in iron–tungsten carbide alloys and iron–molybdenum carbide alloys. The carbides are differentiated
6	Nitric acid, 3% in alcohol		The crystal boundaries in iron–tungsten carbide and iron–molybdenum carbide alloys are developed
7	Mixture of 5 and 6		Generally effective for this class of material
8	Phosphoric acid Hydrogen peroxide	10 ml 10 ml	Used after successive treatments in Etchants 4 and 1 for molybdenum carbide–titanium carbide–cobalt materials. The titanium carbide with tungsten in solution and the cobalt are darkened; the tungsten carbide with titanium in solution is relatively unaffected[76]
9	Dilute ammonium sulphide		Used after successive treatments in Etchants 4, 1 and 8 for similar materials as recommended for Etchant 8. The carbides are unaffected, but the cobalt-rich matrix is darkened and differentiated from titanium carbide[76]
10	Potassium permanganate Potassium hydroxide Water	10 g 5 g to 100 ml	WC grey, TiC pink, TaC gold, β phase rapidly attacked

* Special handling is required for picric acid to avoid the risk of fire and/or explosion and local regulations must be consulted before use of this reagent.

§ Hydrofluoric acid produces irreversible bone damage and presents a range of other hazards. Even dilute hydrofluoric acid solutions should be handled with great care. Note that hydrofluoric acid attacks laboratory glassware.

2. For tungsten and titanium carbides in a matrix of cobalt:

Nitric acid conc.	10 ml
Hydrofluoric acid	10 ml

Titanium carbide is attacked, while tungsten carbide is not attacked. The cobalt-rich matrix is attacked and dissolved.

ELECTROLYTIC ETCHING OF IRON–TUNGSTEN–CARBON AND IRON–MOLYBDENUM–CARBON ALLOYS

The use of a 5% solution of potassium ferrocyanide in 5% sodium hydroxide has been suggested The metallographic effects are very similar to those of Etchant 5 in Table 10.38.

10.3.28 Powdered and sintered metals

PREPARATION

The preparation of powders, pressed compacts, and sintered metals is a specialised process, and requires special methods.

To examine powders, mix a reasonable amount of powder into a small amount (10–15 ml, for example) of cast epoxy that has been mixed in the correct ratio. Pour this into a mould cup of the desired diameter. Then, fill up the mould cup with clear epoxy, add an identification label, and allow it to cure. The powder particles are only in the bottom several mm of the mount. Grinding, polishing and etching are then carried out as for the metal involved, as described above. Coarse powder particles have been mixed with thermosetting resins and placed into the mounting press, and cured. Generally, the use of cast epoxy is preferred as epoxy is the only resin that physically adheres to the specimens.

In compacted, sintered, and even in hot isostatically pressed specimens, voids will be present. Vacuum impregnation can be used to draw cast, low-viscosity epoxy into the voids. Voids can pick up grinding and polishing materials, and absorb etching reagents and other fluids, creating preparation problems in all stages of the preparation and examination process. Of course, some voids are not interconnected, and are not filled with epoxy. In this case, one can interrupt the preparation process after the 3 or 1 μm step, place the specimen back in the vacuum evaporation chamber, pour a small amount of mixed epoxy over the partially prepared surface (laying horizontal), and then draw the epoxy into the surface holes by alternating vacuum and atmospheric pressure steps. After the epoxy dries, carefully remove the excess epoxy on the surface by repeating the last step (it may help to switch to a polisher with individual specimen loading to do this efficiently). Then, perform the final polishing step. Some metallographers pour melted wax over the specimen surface. When the wax hardens, they take a plastic card, like a credit card, and scrap off the excess wax, then complete the preparation procedure.

10.4 Electron metallography and surface analysis techniques*

The application of electron microscopy to the study of the microstructures of metals and alloys is termed 'electron metallography'. There are basically two types of electron microscopy: transmission electron microscopy (TEM) involves the passage of an electron beam through a thin sample, and scanning electron microscopy (SEM), the rastering of an electron beam across the surface of a sample. TEM and SEM can be combined in a scanning transmission electron microscope (STEM) in which fine probes (<1 nm diameter) can be formed for the chemical analysis of small particles and boundaries. There is a wealth of references for both the principles and applications of electron microscopy[77–84] as well as a range of information available on different websites.[85]

10.4.1 Transmission electron microscopy

The sample to be examined in transmission using TEM has to have a thickness on the order of 100 nm (this is rather small, when compared with say the wavelength range of visible light ∼ 400–700 nm), although for the quantitative application of some techniques the specimen has to be less than about 30 nm in thickness. The reason for this is that when a high energy electron, travelling usually at more than 90% the speed of light, passes through a sample it is scattered not only elastically but also inelastically. Both behaviours depend on the types of atom present as well as on the form and regularity of the structure and it should be remembered that to a first order everything is viewed in projection. There is a range of approaches that can be used to prepare electron transparent samples[86,87] and their usage is dependent upon the problem at hand as well as the material to be examined.

Perhaps amongst the oldest TEM specimen preparation techniques are the replica and extraction methods. The replica method can be used for the examination of a fracture surface or surface topography of a metal and involves their reproduction in a thin film of a substance (usually carbon) that does not decompose in the electron beam. Samples can be made either in a two-stage plastic/carbon or in a single-stage direct carbon replica process. For two-stage replicas, a 20–250 μm thick sheet of triacetate cellulose is softened in acetone, laid on to the pre-polished metal surface and, as the acetone evaporates the plastic hardens and contracts into the surface features. The plastic is then carefully removed from the metal, and a 20–30 nm layer of carbon evaporated on to the replica face, after which the plastic is dissolved in acetone leaving a secondary carbon replica. The more direct single stage method is to evaporate the carbon layer on to the pre-prepared metal surface, the carbon layer then being removed by re-etching the surface to dissolve away the metal. A particularly useful variation of this technique is the carbon extraction replica method, which can be used for the examination of multiphase alloys, in which etching of the specimen is controlled so that preferential dissolution of the matrix takes place. In this way, it is relatively easy to extract the second phase particles present in most alloys. The technique is of further attraction because the original particle

---

*For a discussion of electron sources, see Chapter 18.

distribution is retained. Details of replica techniques suitable for a wide range of metals and alloys are given in references.[88–90]

The second type of TEM specimen preparation procedure involves the use of electropolishing methods to reduce the thickness of electrically conducting samples to electron transparency.[91–96] A common procedure requires the use of grinding or machining methods to make a disc that is 3 mm in diameter and <0.2 mm in thickness. Cutting can involve the use of abrasive,[86] wire[97,98] or acid string[99,100] saws, followed by metallographic grinding on wet abrasive papers.[1,101] The disc is then electropolished (anodically dissolved) through the application of an applied voltage until a small hole is formed near its centre. Classic electropolishing curves can be used to adjust the polishing conditions between those in which etching or pitting occurs,[102,103] the polishing being sensitively dependent on factors such as temperature and electrolyte concentration. The thin foil technique has been used for the microstructural characterisation of an enormous range of metals and alloys, and some suitable preparation conditions are given in Table 10.39. Further comprehensive details

Table 10.39 THIN FOIL SPECIMEN PREPARATION TECHNIQUES FOR INDUSTRIAL ALLOYS[82]

Material	Technique	Solution	Conditions
Ag–Zn	Jet	9% KCN, 91% H_2O	
Al	Window	1 part nitric acid, 3 parts methanol	$-70°C$, 40 V
Al–Cu	Chemical	94 parts phosphoric acid, 6 parts nitric acid	
Al–Au alloys	Window	20% perchloric acid, 80% methanol	16–18 V, 100 cm^{-2}, -55 to $-65°C$
Au	Window (3 mm discs)	100 ml CH_3COOH, 20 ml HCl, 3 ml H_2O	Stainless steel cathode, 32 V, 15°C
	Electrolytic	(a) 15% glycerol, 35% 50% HCl	30 V, $-30°C$, 1.5 A cm^{-2}—fast polish
		(b) 133 ml glacial CH_3COOH, 25 g CrO_3, 7 ml H_2O	Finish after (a) at 22 V, 0.8 A cm^{-2}, 0°C
Be, Be–Cu	Window	20% perchloric acid, 80% ethanol	$-30°C$, 40 V
Co–Fe	(a) Chemical	50% phosphoric acid, 50% hydrogen peroxide	
	(b) Jet	20% perchloric acid, 80% methanol	$-20°C$
Cu–Ni–Fe alloys	Chemical	20 ml acetic acid, 10 ml nitric acid, 4 ml HCl	
Cu–Ti	Jet	750 ml acetic acid, 300 ml phosphoric acid, 150 g chromic acid, 30 ml distilled H_2O	30–40 V, 45–55 A
Fe–Al–C	Chemical	HF, H_2O, H_2O_2 (1:3:16)	
Fe_3Si	Window	1% perchloric acid, 2.5% hydrofluoric acid in methanol	$-77°C$
Gd–Ce alloys	Window	1% perchloric acid, 99% methyl alcohol	$-77°C$
Hf	(a) Jet	45 parts nitric acid, 45 parts H_2O, 8 parts HF	
	(b) Jet	2 parts perchloric acid, 98 parts ethanol	40–45 V, 16–20 mA, $-77°C$
In	Window: from 0.1 mm thick	33% HNO_3, 67% methanol	$-40°C$
Mg		(a) 20% $HClO_4$, 80% ethanol	Fast polish, 0°C
		(b) 375 ml H_3PO_4, 625 ml ethanol	Finish at 15 V, 0°C, lowest current
	Chemical	6% nitric acid, 94% H_2O	3°C

(continued)

Table 10.39   THIN FOIL SPECIMEN PREPARATION TECHNIQUES FOR INDUSTRIAL ALLOYS[82]—*continued*

Material	Technique	Solution	Conditions
Mo	Jet	1 part sulphuric acid, 7 parts methanol	10 V
Nb	Chemical	60% HNO_3, 40% HF	0°C
Nb alloys	Window	100 ml lactic acid, 100 ml sulphuric acid, 20 ml HF	40°C, ~0.4 A cm^{-2}
Ni steels Stainless steels	Chemical	50 ml 60% H_2O_2, 50 ml H_2O, 7 ml HF	Place specimen in H_2O_2 solution, then add HF until reaction starts; produces 1000 Å foils which can then be finished by electropolishing by standard chromicacetic electropolishing (courtesy Republic Steel Corp.)
Ni alloys Cu–Ni–Fe	Jet	1 part HNO_3, 3 parts CH_3OH	6–10 V, 18–30 mA, −30°C
Ni	Jet	20% perchloric acid in ethanol	
Ni–Al alloys	Jet	20% sulphuric acid, 80% methanol	Dished at 150 V, final polish in same solution at 10–12 V
Ni and Ni–Fe	Jet	8 parts 50% sulphuric acid, 3 parts glycerine	Less than 10°C, 1.3 A cm^{-2}
Rene 95	Jet	250 ml methanol, 12 ml perchloric acid, 150 ml butylcellusolve	−35°C, 65 V
Th	Window	5% perchloric acid, 95% methanol	−77°C, 3–10 V
Pt	Electrolytic	H_2SO_4 (96%), HNO_3 (65%), H_3PO_4 (80%)	0.2–0.5 A cm^{-2}, 20–30°C
Re	Window	Ethyl alcohol, perchloric acid, butoxyethanol (6:3:1)	−40°C, 35 V
Ta	Chemical	3 parts HNO_3, 1 part HF	−10°C, immerse in Petri dish after lacquering
Ta	Jet	1 part sulphuric acid, 5 parts methanol	−5°C, current density of 0.5 A cm^{-2}
Ti	Jet	1 part HF, 9 parts sulphuric acid	
Ti–Nb	Chemical	4 parts nitric acid, 1 part HF	
TiC	Jet	6 parts nitric acid, 2 parts HF, 3 parts acetic acid	100 V
Ti and Ti alloys	Jet	30 ml $HClO_4$, 175 ml *n*-butanol, 300 ml CH_3OH	15 V, 0.1 A cm^{-2}, may need ion thinning at end to remove contamination layer
U	Jet	1 part perchloric acid, 10 parts methanol, 6 parts butylcellusolve	−20°C, 35–45 V
UO$_2$ (sintered)	Chemical	10 ml HOAc, 20 ml sat. CrO_3, 5 ml 40% HF, 7 ml HNO_3 ($d = 1.42$)	
	or Chemical	20 ml H_3PO_4 ($d = 1.75$), 10 ml HOAc, 2 ml HNO_3 ($d = 10.42$)	
V	Chemical	2 parts HF, 1 part nitric acid	Less than 10°C
V	Jet	20% sulphuric acid in methanol	
Zn–Al	Chemical	50–70% nitric acid in water	
Zn–Al	Window	90% methanol, 10% perchloric acid	
Zr–25% Ti	Chemical	50 ml nitric acid, 40 ml H_2O, 10 ml HF	
Zr alloys	Jet	5% perchloric acid, 90% ethanol	70 V, below −50°C

can be found in references.[77,86] The 'window technique'[104] is a low cost alternative to the use of commercial electropolishing machines and has enjoyed widespread application over the years. One of the attractions of electropolishing is that it is generally not strongly associated with the formation of artefacts, although defect structures can be altered through stress relaxation and some metastable alloys can undergo a phase transformation by shear. Titanium and zirconium alloys are also known to be susceptible to the formation of hydrides.[105]

Ion beam thinning[106–108] is a further method by which the thickness of a sample can be reduced to electron transparency. Ion milling involves the bombardment of the surface of a sample with high energy (4–6 keV) ions of an inert gas such as argon and has advantages over electropolishing methods in its use for non-conducting samples as well as those which are either difficult to electrothin or have complex geometries. Cross-sections of samples containing interfaces between materials such as metals and oxides can be prepared using ion beam thinning,[109] and there are now a number of techniques for the preparation of this class of specimen.[110,111] The ion milling process itself is a complex one and the thinning rate of a given material is dependent upon factors such as the accelerating voltage and ion current as well as the angle of incidence and distance from the source. Most commercial systems employ two ion guns into which the inert gas is introduced, subsequently ionised and then accelerated towards the surface of the sample. More specialised systems enable reactive ion milling to be performed[112] or allow the sample to be viewed during milling so as to thin specific target areas whilst the problems associated with differential milling have been combated by the development of machines with high-powered guns that can mill at very low angles of incidence (<5°). Even more specialised machines are now available which can mill at voltages as low as 200 V and are thus useful for the removal of the surface implant damage and amorphous layers created during thinning at higher potentials. Artefacts can be associated with localised compositional changes as well as phase transformations brought about by stress relief.[113,114]

Focused ion beam (FIB) thinning is a rather newer method of preparing samples for TEM examination and is a technique that has attracted considerable recent interest. The key advantages of FIB usage lie in the high spatial accuracy with which samples can be taken as well as the speed of the preparative procedure.[115] Samples are prepared in an ion microscope that generally uses a gallium liquid metal source to generate ions that are accelerated at voltages of between 10 and 50 keV.[116] Focusing is achieved through the use of two electrostatic lenses, and the beam current can be varied using apertures for which the spot size is some 7 nm at 1 pA and approximately 250 nm at 12 nA.[117,118] Gas assisted etching can be achieved by introducing a reactive gas such as iodine whilst metals can be locally deposited in the presence of Pt or W based organo-metallic precursor gases.[119] There are two general approaches that have been developed for TEM sample preparation. For the first, a small rectangular 'slab' of material that contains the site to be milled and which measures some 3 mm by 1 mm by 50 μm is prepared using standard metallographic procedures. The slab is fixed to a horseshoe shaped support grid and the site is located in the FIB microscope before its upper surface is protected by the deposition of a high atomic number 'strap'. Trenches are cut to the immediate sides of the target site at a high beam current, and the beam current subsequently decreased as the slice thickness is reduced to the point at which it is electron transparent.[120] The second approach uses a lift-out technique[121] to transfer an electron transparent section onto a support grid and does not involve loss of the integrity of the bulk sample through pre-FIB cutting, grinding and polishing. One of the disadvantages of FIB milling is associated with high rates of deposition and amorphisation of the surface of the TEM slice that may require removal using broad beam methods.[122]

Other methods that can be used for the preparation of TEM samples include crushing and grinding, ultramicrotomy and cleaving.[84] Ultramicrotomy is best suited to biological or man-made polymeric samples and requires considerable know-how whilst crushing methods can be used to disperse particles onto a support grid. Cleaving is relatively simple and can be used for layered structures that contain weak interfaces.

The fundamental strategy in the TEM characterisation of structures and compositions is based on using elastically scattered electrons to examine structure, and the various signals associated with inelastic scattering to characterise composition.[123] An understanding of the modern approaches that can now be applied centres on the realisation that such a divide is not necessarily essential. The conventional attitude is based on the idea that if a signal is to be used to provide a composition it has to come from a specific probed area, as it does in SEM.

The types of image that can be formed by selecting specific parts of the diffraction pattern are defined in Figure 10.4. A bright field image is obtained by placing an aperture around the central beam whilst a dark field image uses a beam scattered from only part of the illuminated area depending on its contribution to the selected beam. The 'structure image' uses a relatively large aperture in the back focal plane of the objective lens and, as long as two or more beams are present, interference of their amplitudes gives rise to fringes in the image which are related directly to the spacing of the associated planes in the object controlling the angles at which constructive interference takes

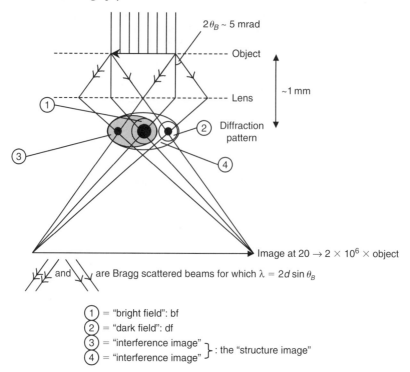

$2\theta_B \sim 5$ mrad

Object

Lens ~ 1 mm

Diffraction
pattern

Image at $20 \rightarrow 2 \times 10^6 \times$ object

and   are Bragg scattered beams for which $\lambda = 2d\sin\theta_B$

1 = "bright field": bf
2 = "dark field": df
3 = "interference image"   ⎫ : the "structure image"
4 = "interference image"   ⎭

Figure 10.4 *Types of image that can be formed using TEM*

place in the diffraction pattern. The bigger the aperture used the smaller the spacings seen in a structure image. It should be emphasised that there may be fringes where there are not planes (by the interference of beams symmetrically disposed on either side of the central beam). The intensities of such high resolution structure images are complicated, and depend on the operating conditions, the level of focus, the thickness of the specimen and the phase changes caused by the spherical aberration of the objective lens: it focuses electrons at an angle to its axis more powerfully than electrons passing along its axis.[124–126] High-resolution data can also be extracted from dark field images of relatively poor direct resolution. The displacement across an interface, for example, can be measured at much higher resolutions than are present in the image.[127] In a related way, defects such as dislocations are best viewed in weak beam dark field images.[128–130] The idea is to put the crystal in an orientation where as a whole it does not diffract strongly, and under these conditions the parts of the crystal that are rotated into a Bragg condition appear as white lines against a dark background. The approach can allow the measurement of segregation to such defects as well as the structural assessment of defect forms.[131] Structural data can also be obtained from diffraction patterns using convergent probes on the specimen.[132] Not only does this allow the 3-dimensional structure to be determined by contributions due to interference effects with off-plane reflections but also enables the magnitude of the Burgers vector of a dislocation to be assessed from characteristic disruptions to the scattering behaviour.[133] Convergent beam methods can similarly be used for the measurement of structure factors, Debye-Waller factors and charge density.[134]

Local compositions are analysed using a small probe to allow the inelastic scattering characteristics of the projected column to be assessed (see Figure 10.5). Energy dispersive X-ray analysis (EDX),[135–138] also known as energy dispersive X-ray spectroscopy (EDS), suffers from the signal coming from places other than that probed but is rather good for the comparative analysis of areas down to about 10 nm. Analysis of the energy losses of the electrons[139–141] that pass through the specimen (parallel electron energy loss spectroscopy, PEELS) provide rather better localisation (done to about 1 nm in a dedicated STEM) but can lead to degradation of the local area because of the high local variation in the electron flux. Specialised techniques can be used to obtain structural data

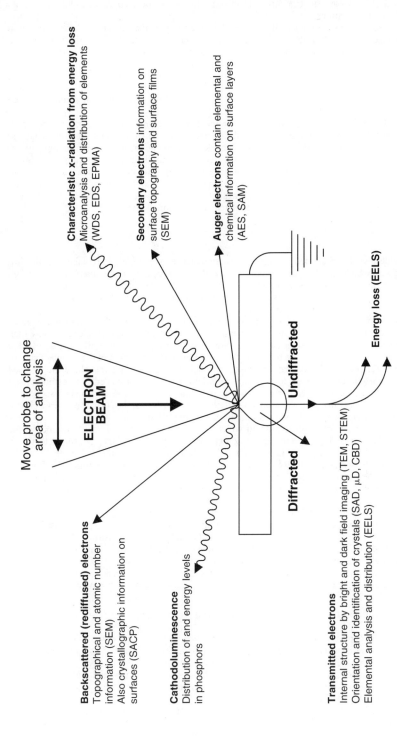

Characteristic x-radiation from energy loss Microanalysis and distribution of elements (WDS, EDS, EPMA)

Secondary electrons information on surface topography and surface films (SEM)

Auger electrons contain elemental and chemical information on surface layers (AES, SAM)

Move probe to change area of analysis

ELECTRON BEAM

Undiffracted

Energy loss (EELS)

Diffracted

Backscattered (rediffused) electrons Topographical and atomic number information (SEM) Also crystallographic information on surfaces (SACP)

Cathodoluminescence Distribution of and energy levels in phosphors

Transmitted electrons Internal structure by bright and dark field imaging (TEM, STEM) Orientation and identification of crystals (SAD, μD, CBD) Elemental analysis and distribution (EELS)

Figure 10.5   *Some information produced as a result of electron—specimen interactions. In part, this figure is after Watt[83] and this portion is used with the permission of Cambridge University Press*

from inelastic loss behaviour: fine structure in some losses can be related, for example, to band occupancies and separations.[142]

Compositional analysis is not usually performed by the analysis of bright and dark field structural images because inelastically scattered electrons make a complicated contribution to the elastic image. Various methods, however, have been developed for the determination of composition from images in which inelastically scattered electrons are present. One such approach relies on the analysis of pattern similarities rather than intensities in high-resolution structural interference images.[143] A second method[144,145] measures changes in the forward scattering potential at an interface and has been shown to be applicable for the quantitative description of the compositional abruptness of an interface at the atomic level.[146]

The situation has been changed by the development of spectrometers that enable images to be formed using either just the elastically scattered electrons or alternatively just the electrons scattered with any given loss. The principle is simple and relies on putting an image (or a diffraction pattern) at the entrance pupil of a sector magnet and then selecting the diffraction pattern at any chosen energy along the line in the exit pupil (defined by the energy dispersion of the magnet) to re-form the image.[147–149] The technology required for the reliable application of the technique has only relatively recently become commercially available and the advances mean that energy filtered structure images can be used to determine localised compositional changes.[150]

10.4.2 Scanning electron microscopy

SEM is a widely used method for the examination of the surfaces of materials. Part of the attraction lies in the ease with which samples can be prepared as well as the simplicity of the instrument operation and interpretation of the images obtained. Advances continue to be made with the introduction of digital storage systems and the improvement in image resolution brought about by the introduction of field emission guns and short-focus lenses.

SEM is analogous to the ion microscope described in 10.4.1. A beam of electrons is generated in either triode or field emission guns with energies varying between less than 1 keV to more than 30 keV. Two electromagnetic condenser lenses are used to produce a demagnified image of the electron source and a third lens then projects a focused spot on to the surface of the sample. Primary electrons are reflected and secondary electrons emitted from the surface and these are focused with an electrostatic electrode on to a biased scintillator. The light produced is transmitted to a photomultiplier and the signal generated is used to modulate the brightness of an oscilloscope spot which traverses a raster in exact synchronism with the electron beam at the surface of the specimen. Details of the principles of SEM can be found in references [151–154].

There are a number of different ways in which SEM can be performed. Most commonly, the emissive mode is used to examine surface topography using a secondary electron detector.[155] Atomic number contrast can be obtained through the detection of backscattered electrons[156] whilst more specialised approaches involve the analysis of magnetic[157] or voltage contrast.[158] Electron beam induced conductivity (EBIC) contrast[159] is limited to the examination of semiconductors, and cathodoluminescence[160] is another specialised mode of operation for the detection of defects and impurities. More typically, the SEM is used to obtain quantitative compositional data by either energy dispersive (EDX or EDS) or wavelength dispersive (WDX or WDS) spectroscopy.[161,162] Modern SEM usually enables fully automated computer measurement of the X-ray intensities to be made from which the relative emission factors are used to determine composition.

Improvements to image resolution have been brought about by the use of field emission guns and there are a number of commercial systems in which a resolution of 1.5 nm can be obtained at 30 keV using either cold emission or thermally assisted sources. Image resolution can also be improved with in-lens stages although at the expense of the size of the sample that can be examined.

10.4.3 Electron spectroscopy and surface analytical techniques

Electron spectroscopy is a technique for the investigation of the energy distribution of electrons ejected from a material when irradiated by a source of ionising irradiation.[163,164] X-ray photoelectron spectroscopy (XPS) refers to the technique when the ionising source is X-rays, and ultra-violet photoelectron spectroscopy (UPS) when ultra-violet light is used. Electron and high-energy X-ray beams can also cause the emission of auger electrons and the analysis of these electrons is known as Auger spectroscopy (AES).

XPS and AES can be used not only for the identification of an atomic species but also to provide information about its binding energy, charge and valency. Modern XPS enables quantification, depth profiling and imaging to be performed and is commonly applied in studies of interfaces as

Table 10.40   COMPARATIVE CAPABILITIES OF DIFFERENT SURFACE ANALYTICAL TECHNIQUES

Technique	Type of information	Elemental range	Effective probing depth	Lateral resolution	Sensitivity (at. %)
XPS	Elemental Chemical	Li–U	1–10 nm	10 μm–2 mm	0.01–1
AES	Elemental Chemical	Li–U	2–6 nm	10 nm	0.1–1
SAM	Elemental	Li–U	2 nm	5 nm	0.2
LEED	Crystallographic	All	3 nm	0.01 nm	–
SEM-EDS	Elemental	B–U	1–5 μm	1 μm	0.1–1
TEM-EDS	Elemental	Li–U	50 nm	2 nm	0.1–1
TEM-PEELS	Elemental	All	50 nm	1 nm	0.1–1
SIMS	Elemental	All	<5 nm	30 μm Profiling	1e14–1e17 at/cc
RBS	Elemental Depth Profiling	Be–U	2 nm–2 μm	5 μm–1 mm	10^{-6}

well as phenomena as diverse as adhesion[165] and corrosion.[166] AES is similarly a surface based technique but it is again possible to obtain quantitative data and as a function of depth. Scanning Auger Microscopy (SAM) involves rastering an electron beam over the surface of a sample to obtain elemental maps through the detection of Auger electrons for specific elements. Surface structural data can be obtained through the use of low energy electron diffraction (LEED) in which crystallographic and morphological information can be acquired. Rutherford backscattered spectroscopy (RBS) and secondary ion mass spectrometry (SIMS) are both popular techniques for the investigation of the surface composition of materials.[89] A comparison of the capabilities of a range of analytical techniques is given in Table 10.40.

10.5   Quantitative image analysis

Quantitative metallography has long been employed. Traditional, manual methods (such as mean linear intercept for grain size determination) are extremely labour intensive and do not permit a wide range of different measurements. In contrast, computerised image analysis is a rapid, flexible and (at least in recent years) sophisticated tool for quantitative metallography. At one time, dedicated image analysis hardware was employed. However modern image analysis software runs on standard personal computers (PCs).

Image analysis involves three stages:

Stage 1: image acquisition (and post-processing)
Stage 2: thresholding
Stage 3: measurement(s)

Image acquisition can involve image capture by the PC hosting the image analysis software or the use of images acquired elsewhere. High quality digital cameras are available at a (relatively) reasonable cost for light microscopes. Modern scanning electron microscopes employ digital imaging (and older instruments can be retrofitted for digital imaging). Digital cameras are readily available for transmission electron microscopes, although high resolution cameras, in particular, remain expensive.

In some cases, it may be necessary to process the acquired image (prior to thresholding). However, this should be done with great care to avoid distorting the input data (industrial computerised image processing is far behind the level of sophistication of similar operations performed by the human brain).

Thresholding separates the feature(s) of interest from the background, thus allowing measurements to be performed on the former and is a key step in the image analysis process. Thresholding instructs the software to ignore all parts of the image that lie outside a defined band, typically of grey levels (although other information, such as colour, can also be considered). For example, if the features of interest are mid-tones on a light and dark background, the lower threshold would be set to exclude the dark portions of the background and the upper threshold to exclude the light portions of the background. Unfortunately, real world images are rarely as simple as this example and grey shades often overlap between the features and the background. Fortunately, a number of quite sophisticated routines exist to process the thresholds, for example for edge detection purposes (as with image processing, processing of the thresholds should be performed with care, although thresholding mistakes are often quite easy to identify). In the case, for example, of grain size measurements, thresholding may require the accommodation of uneven etching response (after etching, not all of the boundaries may be etched to the point where these join up and further etching might serve merely to attack the entire sample).

Measurements* fall into two basic categories, namely: field measurements and object measurements. Field measurements, such as an area fraction of a given phase, are measurements of the overall characteristics of an entire region of interest and as such do not require separation of individual objects. In contrast, determination of, for example, a particle size distribution implies clearly that individual objects be identifiable separately. This can present challenges, as in the case, for example, of overlapping particles (although routines for separating overlapping objects during thresholding do exist). Some examples of object measurements (i.e., measurements involving individual objects, or groups of objects) include: length, diameter, area, perimeter, aspect ratio, circularity, angle, coordinate of object centre and extent of clumping.

Commercial image analysis software falls generally into one of two types. The first of these types is software designed to undertake a fixed palette of operations (typically, following ASTM defined methods), such as:

– determination of grain size;
– measurement of Knoop or Vickers hardness indentations;
– characterisation of the percentage of different graphite/carbide morphologies in cast irons;
– quantification of inclusion banding;
– measurement of dendrite arm spacing;
– characterisation of inclusion size and/or shape distributions.

Such software is ideal for rapid, routine measurements, as would be needed for quality control purposes and requires only a relatively low skill level on the part of the operator. The second type of software is highly configurable and supports a far wider range of measurements, but has a much steeper learning curve (of course the software's underlying image analysis engine can be the same in both cases). This second type of image analysis tool is ideal for research applications of metallography, where non-standard measurements may be required.

For further information, the extensive literature on image analysis can be consulted. However, it should be cautioned that a large portion of this literature is aimed at non-metallographic applications, such as machine vision. Two examples of recent texts on image analysis that focus on materials science applications are [167, 168].

10.6   Scanning acoustic microscopy

Scanning acoustic microscopy is a non-destructive analytical technique which uses the information from an ultrasonic probe which is scanned over the surface of the test sample to create a complete picture of the shapes, sizes and exact positions of defects, inclusions and other metallographic features which are displayed as full (false) colour screen or printed images.

The ultrasonic probe is used in pulse-echo mode as a **B**- or **C**-scan. The amplitude and phase of the reflected signal depend on the magnitude and the rate of change in acoustic impedance as it is reflected at an interface or other feature.

By time gating the returning acoustic wave train and detecting any change in phase, the variations can be encoded and analysed digitally to convert them into shades of colour or density to build up a two-dimensional image of the sample.

The scanning acoustic microscope comprises a scanning frame which carries an orthogonal *XYZ* probe transport system. Motion of the probe in the three axes is provided by three independent stepper motors. The system incorporates a pulser–receiver circuit which is driven by a computer with appropriate software.

The sample to be scanned is placed in a tank containing a liquid capable of transmitting acoustic waves from the probe into the sample. The liquid is usually water but it can be a suitable organic liquid. The motors then move the probe over the sample in a raster pattern. The computer can be programmed to follow curved as well as planar surfaces or interfaces.

---

* Statistics are clearly very important in quantitative metallography, but are not discussed here, in the interests of brevity. General information on statistical methods can be found in Chapter 2.

Scanning acoustic microscopy is suitable for imaging defects and inclusions in ceramic and composite as well as metallic materials. It is also useful for studying soldered and diffusion-bonded interfaces and for any internal artefact whether deliberately introduced or not.

10.7   Applications in failure analysis

Failure analysis is an important and significant area of metallurgy and finds application in both industry and research. The objectives of a metallurgical failure analysis typically are to determine the source or cause of the failure, and to recommend a solution or means of prevention of similar failures. A thorough failure analysis is a complex process and may incorporate many steps, including investigation and documentation of nominal service conditions, interviews with individuals that observed the failure, non-destructive evaluation of the component, stress or fracture mechanics analysis, etc. Several texts are available which detail the process and sequence of analysis,[169–175] a full description of which is beyond the scope of this book. This section briefly presents the use and relative merit of different microstructural techniques in what is considered the heart of metallurgical failure analysis—examination of the macro and microstructural features of the fractured surfaces, also known as fractography. Careful fractographic examination can conclusively identify the mode of failure, whether it was by overload, corrosion, or fatigue, and determine if material defects or unforeseen environmental factors led to the failure.

Nearly all of the techniques described in this chapter can be used in fractography, including stereo light microscopy, metallographic examination, scanning electron microscopy, transmission electron microscopy, X-ray spectroscopy and surface analysis techniques. Prior to performing analysis, however, it is important to insure that care is taken to preserve the fracture features in their original state. Techniques for preservation and cleaning of fracture surfaces are given in Brooks and Choudhury.[169] Analysis must begin with the most non-destructive techniques, and careful documentation of findings is critical.

Examination of failed components and their fracture surfaces by low-power light microscopes is a simple and powerful technique which can quickly and easily reveal the fracture initiation point and mode of fracture. The advantages of this technique are its non-destructive nature, the lack of expensive equipment, and the ability to see and document colours. Light microscopic examination of fracture surfaces is typically performed at magnifications of up to 50X. The main disadvantage is the very limited depth of field, especially at higher magnifications. Translating images seen through the lens into quality photographs can also be difficult.

Light microscopy is frequently used in examination of metallographic sections containing the fracture surface or fracture origin, enabling an observation of the interaction of the fracture path with microstructural features such as grain boundaries or inclusions. Metallographic sections can also determine whether the microstructure reflects the intended metallurgical condition of the component alloy (hardness testing is also used) and identify microstructural defects that may have led to the failure. Fine resolution of the fracture path can be obtained through sections containing secondary cracks, since these cracks frequently are not exposed to post-fracture mechanical and environmental damage. This technique, however, is inherently destructive in nature, and the resolution of features is limited to about 0.1 μm.

The scanning electron microscope (SEM) has developed into the workhorse of fractography, due to its wide range of magnification (10X to over 20,000X), good depth of field, simplicity of sample preparation, and integrated microchemical analysis capability (through X-ray spectroscopy). It has essentially replaced the combination of light microscopy and transmission electron microscopy which was popular before the commercialisation of the SEM. As with the light microscope, examination of the fracture surface is used to identify the mode and origin of fracture, with the advantage that both the macroscopic and microscopic features can be simply and quickly recorded. The SEM can also be used to examine metallographic cross-sections at higher magnifications than in the light microscope, and with the added feature of atomic number contrast in backscattered electron imaging mode. The microchemical analysis capability is used to identify inclusions, corrosion products, and chemical alterations near surfaces or in the base metal. The main limitations of the SEM are the need for some sample preparation (to fit within the vacuum chamber, and to provide a conduction path through the specimen to ground), and the relative expense of the instrument, especially compared to light microscopes (colour information provided by light microscopy can be useful in some cases).

The transmission electron microscope (TEM) was widely used for observation of fine-scale fracture features until about 1970. Carbon replicas of the surface with heavy-metal shadowing to produce contrast from topographical features are examined. This technique has higher ultimate resolution than the SEM (2 nm for a single-stage replica), but the images require interpretation and low magnification observation is not possible. Only a limited area of the fracture surface can be examined at once, and considerable sample preparation is required. Examination of replicas is, however, non-destructive (replicas can also be examined in the SEM). Despite these limitations, TEM examination is still used in special cases, for example, to observe very fine fatigue striations.

Surface analysis techniques such as auger electron spectroscopy (AES) and secondary ion mass spectroscopy (SIMS) are occasionally used in fractography. Their principal strength lies in identification of corrosion products and surface interaction layers. They have a broader elemental analysis capability (especially for light elements), better elemental resolution, and can be used to obtain profiles of surface composition through ion sputtering. An application example is the identification of fine-scale segregation of tramp elements to grain boundaries in the temper embrittlement of steels.

REFERENCES

1. G. Petzow, Metallographic Etching, 2nd Edition, ASM International, Materials Park, OH, 1999.
2. G. F. Vander Voort, Metallography: Principles and Practice, ASM International, Materials Park, 1999.
3. C. A. Godden, *Met. Progress*, 1961, **79**, 121.
4. L. P. Tarasov and C. O. Lundberg, *Met. Progress*, 1949, **55**, 183.
5. L. E. Samuels, *J. Aust. Inst. Metals*, 1960, **5**, 63.
6. V. J. Haddrell, *J. Inst. Metals*, 1963–1964, **92**, 121.
7. P. A. Jacquet, *Met. Reviews*, 1956, **1**(2), 157.
8. W. J. McG. Tegart, The Electrolytic and Chemical Polishing of Metals in Research and Industry, 2nd edition, Pergamon Press, Oxford, 1959.
9. L. E. Samuels, *Metallurgia*, 1962, **66**, 187.
10. E. Knuth-Winterfeldt, *Trans. Instruments and Measurements Conf., Stockholm*, 1949, p. 223. (See also P. Jacquet, *ONERA Publication No. 51*, 1952, in French.)
11. E. C. Sykes, V. J. Haddrell, H. R. Haines and B. W. Mott, *J. Inst. Metals*, 1954–55, **83**, 166.
12. E. Knuth-Winterfeldt, *Mét. et Corrosion*, 1948, **23**, 5.
13. P. A. Jacquet, *Compt. Rend.*, 1956, **243**, 1066.
14. P. A. Jacquet, *Proc. Am. Soc. Test Mater.*, 1957, **57**, 1290.
15. P. A. Jacquet, *Rech. Aéronautique*, 1962, **90**, 15.
16. M. Médard, P. Jacquet and R. Sartorius, *Rév. Met.*, 1949, **46**, 549.
17. R. C. Nester and G. F. Vander Voort, ASTM Standardisation News, 1992, **20**(5), 34.
18. G. F. Vander Voort et al. Buehler's Guide to Materials Preparation, Buehler Ltd. 2002.
19. ASM Handbook Vol. 9: Metallography and Microstructures, ASM International, Materials Park, OH, 1985.
20. C. P. Shillaber, Photomicrography in Theory and Practice, Wiley, London, 1944.
21. K. W. Andrews, Physical Metallurgy: Techniques and Applications, Allen and Unwin, 1973.
22. S. Tolansky, Multiple Beam Interferometry of Surfaces and Films, Clarendon Press, 1948.
23. F. W. Cuckow, *J. Iron Steel Inst.*, Jan. 1949, **161**, 1.
24. D. McLean, *Met. Treatment*, Feb. 1951, 51.
25. G. Nomarski and A. R. Weill, *Rév. Met.*, 1955, **52**, 121.
26. A. W. Rowcliffe, *J. Inst. Metals*, 1966, **94**, 263.
27. L. J. Barker, *Iron Age*, 1949, **163**, 74.
28. E. C. W. Perryman and J. M. Lack, *Nature*, 1951, **167**, 479.
29. G. K. T. Conn and F. J. Bradshaw (Eds), Polarised Light in Metallography, Butterworth, London, 1952.
30. F. Beraha and B. Shpigler, Colour Metallography, *Amer. Soc. Metals*, 1977.
31. B. Bousfield, *Metals and Materials*, 1991, **7**, 555.
32. G. F. Vander Voort, *Metal Progress*, 1985, **127**, 31.
33. G. F. Vander Voort, *Advanced Materials & Processes*, 2000, **158**, 36.
34. G. F. Vander Voort, *Advanced Materials & Processes*, 2001, **159**, 37.
35. E. Weck and E. Leistner, Metallographic Instructions for Colour Etching by Immersion, Part I: Klemm Colour Etching, Vol. 77, D. V. S. Verlag GmbH, Düsseldorf, 1982.
36. E. Weck and E. Leistner, Metallographic Instructions for Colour Etchants by Immersion, Part II: Beraha Colour Etchants and Their Different Variants, Vol. 77/II, D.V.S. Verlag GmbH, Düsseldorf, 1983.
37. E. Weck and E. Leistner, Metallographic Instructions for Colour Etching by Immersion, Part III: Non-Ferrous Metals, Cemented Carbides and Ferrous Metals, Nickel-Base and Cobalt-Base Alloys, Vol. 77/III, D. V. S. Verlag GmbH, Düsseldorf, 1986.
38. H. E. Bühler and H. P. Hougardy, Atlas of Interference Layer Metallography, DGM, Oberursel, 1980.
39. A. Hickling, *Trans. Farad. Soc.*, 1941, **38**, 27.
40. M. H. Roberts, *Brit. J. Appl. Phys.*, 1954, **5**, 351.
41. C. Edeleanu, *Metallurgia*, 1954, **50**, 113.
42. C. Edeleanu, *J. Iron Steel, Inst.*, 1957, **185**, 482 and 1958, **188**, 122.
43. V. Cihel and M. Pražák, *J. Iron Steel, Inst.*, 1959, **193**, 360.
44. L. C. Lovell, F. L. Vogel and J. H. Wernick, *Met. Progress*, May 1959, **75**, 96.
45. P. Lacombe and L. Beaujard, *Rev. Mét.*, 1947, **44**, 71 and *J. Inst. Metals*, 1948, **74**, 1.
46. R. W. Cahn, *J. Inst. Metals*, 1949–50, **76**, 121.

47. D. J. Barber, *Philos. Mag.*, 1962, **7**, 1925.
48. G. Bassi and J. P. Hugo, *J. Inst. Metals*, 1958–59, **87**, 155.
49. A. W. Ruff, *J. Appl. Phys.*, 1962, **33**, 3392.
50. E. Stirl and A. Alder, Z. Metallkd., 1961, **52**, 529.
51. G. Bassi and J. P. Hugo, *J. Inst. Metals*, 1958–59, **87**, 376.
52. H. J. Seemann and M. Dudek, *Metalloberfläche*, 1948, **2**, 84.
53. N. S. Bromelle and H. W. L. Phillips, *J. Inst. Metals*, 1949, **75**, 529.
54. H. Kostron, Z. Metallkd, 1948, **39**, 333.
55. M. C. Udy, G. K. Manning and L. W. Eastwood, *Trans. AIMME*, 1949, **185**, 779.
56. B. W. Mott and H. R. Haines, *J. Inst. Metals*, 1951–52, **80**, 629.
57. A. R. Kaufmann, P. Gordon and D. W. Lillie, *Met. Progress*, 1949, **56**(5), 664.
58. H. J. Merchant, *J. Iron Steel Inst.*, 1947, **155**, 179.
59. G. F. Vander Voort, Applied Metallography, Van Nostrand Reinhold Co., 1986, 1.
60. F. N. Rhines, W. M. Urquhart and H. R. Hoge, *Trans. Am. Soc. Metals*, 1947, **39**, 694.
61. S. C. Carapella and E. A. Peretti, *Met. Progress*, 1949, **56**, 666.
62. P. F. Weinrich, Australasian Engineer, Nov. 1948, 42.
63. P. F. George, *Trans. Am. Soc. Metals*, 1947, **38**, 686.
64. E. F. Emley, *J. Inst. Metals*, 1949, **75**, 431.
65. E. F. Emley, *J. Inst. Metals*, 1949, **75**, 481.
66. E. F. Emley, Principles of Magnesium Technology, Pergamon, Oxford, 1966.
67. E. F. Emley, *Brit. Weld. J*, 1957, **4**, 307.
68. R. Bakish, *J. Electrochem. Soc.*, 1958, **105**, 574.
69. B. L. Eyre, *Metallurgia*, Aug. 1958, **58**, 95.
70. L. T. Greenfield and J. E. Davies, The Preparation of Tin and Tin Alloys for Microscopical Examination, Tin Research Institute, Middlesex, UK, 1951.
71. Annonymous, Metallography of Tin and Tin Alloys, *I.T.R.I. Pub. No. 580, Intl. Tin Res. Inst., Perivale*, 1982.
72. J. V. Harding and W. I. Pell-Walpole, *J. Inst. Metals*, 1948, **75**, 115.
73. H. Woods, *Met. Progress*, 1947, **51**, 261.
74. G. B. Brook and R. I. Saunderson, 2nd Charlottesville Conference, AMMRC, 1981.
75. A. H. Roberson, Mrt. Progress, 1949, **56**, 667.
76. R. Kieffer, *Metallforschung*, 1947, **2**, 236.
77. J. W. Edington, Practical Electron Microscopy in Materials Science, Van Nostrand Reinhold, New York, 1976.
78. P. J. Goodhew and F. J. Humphreys, Electron Microscopy and Analysis, 2nd Edition, Taylor and Francis, New York, 1988.
79. P. W. Hawkes and E. Kasper, Principles of Electron Optics, 1–3, Academic Press, New York, 1994.
80. P. B. Hirsch, A. Howie, R. B. Nicholson, D. W. Pashley and M. J. Whelan, Electron Microscopy of Thin Crystals, 2nd Edition, Kreiger, New York, 1977.
81. M. H. Loretto, Electron Beam Analysis of Materials, 2nd Edition, Chapman and Hall, New York, 1994.
82. G. Thomas and M. J. Goringe, Transmission Electron Microscopy of Metals, Wiley, New York, 1979.
83. I. M. Watt, The Principles and Practice of Electron Microscopy, Cambridge University Press, New York, 1985.
84. D. B. Williams and C. B. Carter, Transmission Electron Microscopy, Plenum Press, New York, 1996.
85. http://www.amc.anl.gov.
86. P. J. Goodhew, Thin Foil Preparation for Electron Microscopy, Elsevier, New York, 1985.
87. K. C. Thompson-Russell and J. W. Edington, Electron Microscope Specimen Preparation Techniques in Materials Science, Macmillan, Philips Technical Library, Eindhoven, The Netherlands, 1977.
88. G. A. Savanick, *Rev. Sci. Instrum.*, 1967, **38**, 43.
89. J. H. M. Wilson and A. J. Rowe, Replica, Shadowing and Freeze-Etching Techniques, in Practical Methods in Electron Microscopy, Vol. 8 (Editor: A. M. Glauert), North-Holland, Amsterdam, 1980.
90. A. Pranckevicius, *J. Appl. Phys.*, 1966, **37**, 3331.
91. R. C. Glenn and J. C. Raley, *ASTM Spec. Tech. Pub.*, 1963, **339**, 60.
92. E. H. Lee and A. F. Rowcliffe, *Microstruct. Sci.*, 1979, **7**, 403.
93. W. M. Stobbs, *J. Phys. E: Sci. Instrum.*, 1969, **2**, 202.
94. R. W. Gardiner and P. G. Partridge, *J. Sci. Instrum.*, 1967, **44**, 63.
95. B. L. Eyre and M. J. Sole, *Philos. Mag.*, 1964, **9**, 545.
96. A. R. Cox and M. I. Mountford, *J. Inst. Metals*, 1967, **95**, 347.
97. A. N. Rushby and J. Woods, *J. Phys. E: Sci. Instrum.*, 1970, **3**, 726.
98. C. W. Fountain and R. L. Peck, *J. Phys. E: Sci. Instrum.*, 1970, **3**, 725.
99. T. R. McGuire and R. T. Weber, *Rev. Sci. Instrum.*, 1949, **20**, 962.
100. M. D. Hunt, J. A. Spittle and R. W. Smith, *J. Sci. Instrum.*, 1967, **44**, 230.
101. L. E. Samuels, Metallographic Polishing by Mechanical Methods, Pitman, Belfast, 1971.
102. Y. F. Le Coadic, Electrochemical Preparation of Thin Crystalline Foils, Proc. 7th Int. Cong. Electron Microscopy, Grenoble, 1970, **1**, 333.
103. B. J. Ginn and E. D. Brown, *Brit. Welding J.*, 1965, **12**, 90.
104. H. M. Tomlinson, *Phil. Mag.*, 1958, **3**, 867.
105. C. G. Shelton, PhD Thesis, University of Cambridge, 1985.
106. J. J. Trillat, Ionic Bombardment, Theory and Applications, Gordon and Breach, New York, 1964.
107. D. J. Barber, *J. Mat. Sci.*, 1970, **5**, 1.
108. J. Franks, *Adv. Electronics Electron Phys.*, 1978, **47**, 1.
109. S. B. Newcomb, C. B. Boothroyd and W. M. Stobbs, *J. Microsc.*, 1985, **140**, 195.

110. J. C. Bravman, R. M. Anderson and M. L. McDonald (Editors), Specimen Preparation for Transmission Electron Microscopy of Materials, MRS, Pittsburgh, PA, 1988.
111. R. M. Anderson, B. Tracy and J. C. Bravman (Eds), Specimen Preparation for Transmission Electron Microscopy of Materials-III, MRS, Pittsburgh, PA, 1992.
112. A. G. Cullis, N. G. Chew and R. M. Anderson, *Ultramicroscopy*, 1985, **17**, 203.
113. J. C. Barry, J. L. Hutchinson and R. L. Segall, *J. Mater. Sci.*, 1983, **18**, 1421.
114. R. A. Camps, J. E. Evetts, B. A. Glowacki, S. B. Newcomb and W. M. Stobbs, *Nature*, 1987, **329**, 229.
115. D. M. Schraub and R. S. Rai, *Prog. Cryst. Growth and Char.*, 1998, **36**, 99.
116. R. J. Young, *Vacuum*, 1993, **44**, 353.
117. J. Melngailis, *J. Vac. Sci. Technol.*, B5, 1987, 469.
118. L. Bischoff and J. Teichert, Materials Science Forum, 1997, 248–249, 445.
119. P. D. Prewett, *Vacuum*, 1993, **44**, 345.
120. L. A. Giannuzzi and F. A. Stevie, *Micron*, 1999, **30**, 197.
121. R. M. Langford and A. K. Petford-Long, *J. Vac. Sci. Technol.*, 2001, **A19**, 982–985 and 2186.
122. D. Sutton, S. M. Parle and S. B. Newcomb, pp 377–380 in *Inst. Phys. Conf. Ser.*, (Editors: M. Aindow and C. J. Kiely), 168, IOP, Bristol, 2001.
123. D. C. Joy, J. I. Goldstein and A. D. Romig, Principles of Analytical Electron Microscopy, Plenum Press, New York, 1986, pp. 377–380.
124. P. R. Buseck, J. M. Cowley and L. Eyring, High Resolution Electron Microscopy and Associated Techniques, 1988, Oxford University Press, New York.
125. J. C. H. Spence, Experimental High Resolution Electron Microscopy, 2nd Edition, 1988, Oxford University Press, New York.
126. S. Horiuchi, Fundamentals of High Resolution Transmission Electron Microscopy, 1994, North-Holland, Amsterdam.
127. G. J. Wood, W. M. Stobbs and D. J. Smith, *Philos. Mag.* A, 1984, **50**, 375.
128. D. J. H. Cockayne, *Z. Naturforsch.*, 1972, 27a, 452.
129. D. J. H. Cockayne, I. L. F. Ray and M. J. Whelan, *Phil. Mag.*, 1969, 20, 1265.
130. P. B. Hirsch, A. Howie and M. J. Whelan, *Philos. Tr. R. Doc.* S A, 1960, **252**, 499.
131. W. M. Stobbs and C. H. Sworn, *Philos. Mag.*, 1971, **24**, 1365.
132. J. W. Steeds, Convergent Beam Electron Diffraction of Alloy Phases, Hilger, Bristol, 1984.
133. J. A. Eades, *J. Electron Microsc. Tech.*, 1989, **13**, 1.
134. J. C. H. Spence and J. M. Zuo, Electron Microdiffraction, Plenum Press, New York, 1992.
135. J. I. Goldstein, Introduction to Analytical Electron Microscopy, Plenum Press, New York, 1979.
136. D. B. Williams, Practical Analytical Electron Microscopy in Materials Science, Philips Electron Optics Group, New Jersey, 1987.
137. D. B. Williams and J. I. Goldstein, Electron Probe Quantitation (Editors: K. F. J. Heinrich and D. E. Newbury), Plenum Press, New York, 1991.
138. N. J. Zaluzec, Introduction to Analytical Electron Microscopy (Editors: J. J. Hren, J. I. Goldstein and D. C. Joy), Plenum Press, New York, 1979.
139. R. F. Egerton, Electron Energy-Loss Spectroscopy in the Electron Microscope, TMS, Pennsylvania, 1996.
140. O. L. Krivanek (Ed.), Ultramicroscopy, 1995, **59**, 1.
141. R. Brydson, Electron Energy Loss Spectroscopy, BIOS Scientific Publishers, Oxford, 2001.
142. B. Rafferty and L. M. Brown, *Phys. Rev. B*, 1998, **58**, 10326.
143. A. Ourmazd, D. W. Taylor, J. Cunningham and C. W. Tu, *Phys. Rev. Lett.*, 1989, **62**, 933.
144. J. N. Ness, W. M. Stobbs and T. F. Page, *Philos. Mag. A*, 1986, **54**, 679.
145. F. M. Ross and W. M. Stobbs, *Philos. Mag. A*, 1991, **63**, 1 & 37.
146. W. C. Shih and W. M. Stobbs, *Ultramicroscopy*, 1991, **35**, 197.
147. J. Bentley, *Microsc. Microanal.*, 1998, **4**, 158.
148. F. Hofer and P. Warblicher, Transmission EELS in Materials Science (Editors: M. M. Disko, B. Fultz and C. C. Ahn), Wiley, New York, 2001.
149. L. Reimer, Energy Filtering Transmission Electron Microscopy, Springer, Heidelberg, 1995.
150. B. Freitag and W. Mader, *J. Microsc.*, 1999, **194**, 42.
151. J. I. Goldstein and H. Yakowitz, Practical Scanning Electron Microscopy, Plenum Press, New York, 1975.
152. L. Reimer, T. Tamir and A. L. Schawlow, Scanning Electron Microscopy: Physics of Image Formation and Microanalysis, Springer-Verlag, Germany, 1998.
153. D. E. Newbury, Advanced Scanning Electron Microscopy and X-Ray Microanalysis, Plenum Press, New York, 1986.
154. S. L. Flegler, J. W. Heckman and K. L. Klomparens, Scanning and Transmission Electron Microscopy, Oxford University Press, New York, 1997.
155. T. E. Everhart and R. F. M. Thornley, *J. Sci. Instrum.*, 1960, **37**, 246.
156. D. H. Krinsley, K. Pye, S. Boggs and K. N. Tovey, Backscattered Scanning Electron Microscopy and Image Analysis, Cambridge University Press, New York, 1998.
157. D. C. Joy and J. P. Jakubovics, *J. Appl. Phys. D*, 1969, **2**, 1367.
158. G. S. Plows and W. C. Nixon, *J. Phys. E: Sci. Instrum.*, 1968, **1**, 595.
159. D. B. Holt and D. C. Joy, SEM Microcharacterisation of Semiconductors, Vol. 12 of Techniques of Physics, Academic Press, London, 1989.
160. G. Pfefferkorn and R. Blaschke, Proc. 7th Annual SEM Symposium (Editor: O. Johari), IITRI, Illinois, 1974.
161. J. I. Goldstein, D. E. Newbury, P. Echlin and D. C. Joy, Scanning Electron Microscopy and X-ray Microanalysis, Kluwer Academic Publishers, The Netherlands, 1981.

162. G. Lawes, Scanning Electron Microscopy and X-ray Microanalysis, Wiley, New York, 1987.
163. L. C. Feldman and J. W. Mayer, Fundamentals of Surface and Thin Film Analysis, North-Holland, New York, 1986.
164. H. Bubert and H. Jenett, Surface and Thin Film Analysis, Wiley-VCH, Germany, 2002.
165. C.-A. Chang, Y.-K. Kim and A. G. Scrott, *J. Vac. Sci. Technol.*, 1990, **A8**, 3304.
166. A. P. Pijpers and R. J. Meier, *Chem. Soc. Rev.*, 1999, **28**, 233.
167. L. Wojnar, Image Analysis: Applications in Materials Engineering, CRC Press, 1999.
168. Practical Guide to Image Analysis, ASM International, Materials Park, OH, 2000.
169. C.R. Brooks and A. Choudhury, Metallurgical Failure Analysis, McGraw-Hill, Inc., New York, 1993.
170. Failure Analysis and Prevention: ASM Handbook Vol. 11, ASM International, Materials Park, OH, 1986.
171. Fractography: ASM Handbook Vol. 12, ASM International, Materials Park, OH, 1987.
172. D. Hull, Fractography: Observing, Measuring, and Interpreting Fracture Surface Topography, Cambridge University Press, Cambridge, UK, 1999.
173. A.K. Das, Metallurgy of Failure Analysis, McGraw-Hill, Inc., New York, 1997.
174. R.B. Ross, Investigating Mechanical Failures: the Metallurgist's Approach, Chapman and Hall, New York, 1995.
175. V.J. Colangelo and F.A. Heiser, Analysis of Metallurgical Failures, 2nd Edition, John Wiley and Sons, New York, 1987.

11 Equilibrium diagrams

11.1 Index of binary diagrams

For numbers in parenthesis see 11.3 Acknowledgements.

Ag–Al **11**–7 (8,9)
Ag–As **11**–8 (8,9)
Ag–Au **11**–8
Ag–Ba **11**–9 (10)
Ag–Be **11**–9 (10)
Ag–Bi **11**–10 (8,9)
Ag–Ca **11**–10 (8,9)
Ag–Cd **11**–11 (8,9)
Ag–Ce **11**–11 (10)
Ag–Co **11**–12 (1)
Ag–Cr **11**–12 (10)
Ag–Cu **11**–13 (9)
Ag–Dy **11**–13 (10)
Ag–Er **11**–14 (1)
Ag–Eu **11**–14 (1)
Ag–Ga **11**–15 (10)
Ag–Gd **11**–15 (8,9)
Ag–Ge **11**–16 (8,9)
Ag–Hg **11**–16 (8,9)
Ag–Ho **11**–16 (1)
Ag–In **11**–17 (8)
Ag–La **11**–17 (1)
Ag–Li **11**–18 (8,9)
Ag–Mg **11**–18 (10)
Ag–Mn **11**–19 (8,9)
Ag–Na **11**–20 (8,9)
Ag–Nd **11**–19 (1)
Ag–Ni **11**–20 (8,9)
Ag–O **11**–20
Ag–P **11**–20 (10)
Ag–Pb **11**–21 (9)
Ag–Pd **11**–21
Ag–Pr **11**–22 (1)
Ag–Pt **11**–22 (9)
Ag–Rh **11**–23
Ag–Ru **11**–23 (9)
Ag–S **11**–23 (10)
Ag–Sb **11**–23 (8)
Ag–Sc **11**–24 (8)
Ag–Se **11**–24 (9)
Ag–Si **11**–24
Ag–Sm **11**–25 (10)
Ag–Sn **11**–25 (8,9)
Ag–Sr **11**–26 (10)
Ag–Tb **11**–26 (8,9)
Ag–Te **11**–27 (10)
Ag–Th **11**–27 (8,9)
Ag–Ti **11**–27 (10)
Ag–Tl **11**–28 (8,9)

Ag–U **11**–28 (8,9)
Ag–Y **11**–28 (8,9)
Ag–Yb **11**–29 (8,9)
Ag–Zn **11**–29 (8,9)
Ag–Zr **11**–30 (8,9)

Al–As **11**–30 (8,9)
Al–Au **11**–31 (10)
Al–B **11**–31 (10)
Al–Ba **11**–32 (10)
Al–Be **11**–32 (8,9)
Al–Bi **11**–33 (8,9)
Al–Ca **11**–33 (9)
Al–Cd **11**–33 (9)
Al–Ce **11**–34 (8,9)
Al–Co **11**–34 (10)
Al–Cr **11**–35 (10)
Al–Cs **11**–35 (8)
Al–Cu **11**–36 (8,9)
Al–Dy **11**–36 (8,9)
Al–Er **11**–37 (8,9)
Al–Fe **11**–37 (10)
Al–Ga **11**–38 (9)
Al–Gd **11**–38 (8,9)
Al–Ge **11**–39 (9)
Al–Hf **11**–39 (10)
Al–Hg **11**–39 (9)
Al–Ho **11**–40 (3,9)
Al–In **11**–40 (8,9)
Al–K **11**–41
Al–La **11**–41 (8,9)
Al–Li **11**–42 (1)
Al–Mg **11**–42 (1,9)
Al–Mn **11**–43 (10)
Al–Mo **11**–43 (10)
Al–Na **11**–44 (9)
Al–Nb **11**–44 (8,9)
Al–Nd **11**–45 (8,9)
Al–Ni **11**–45 (10)
Al–Pb **11**–46 (8,9)
Al–Pd **11**–46 (10)
Al–Pr **11**–47 (8,9)
Al–Pt **11**–47 (10)
Al–Pu **11**–48 (8,9)
Al–Re **11**–49
Al–Ru **11**–49 (10)
Al–Sb **11**–50 (8,9)
Al–Sc **11**–50 (8,9)

Al–Se **11**–51 (9)
Al–Si **11**–51 (8,9)
Al–Sm **11**–51 (8,9)
Al–Sn **11**–52 (8,9)
Al–Sr **11**–52 (10)
Al–Ta **11**–53
Al–Te **11**–53 (10)
Al–Th **11**–54 (8,9)
Al–Ti **11**–54 (10)
Al–Tl **11**–55 (10)
Al–U **11**–55 (8,9)
Al–V **11**–56 (10)
Al–W **11**–56 (8,9)
Al–Y **11**–57 (1)
Al–Yb **11**–57 (8,9)
Al–Zn **11**–58 (8,9)
Al–Zr **11**–58 (10)

Am–Pu **11**–59 (10)

As–Au **11**–59 (8,9)
As–Bi **11**–59 (9)
As–Cd **11**–60 (10)
As–Co **11**–60 (10)
As–Cu **11**–61 (9)
As–Eu **11**–61 (8)
As–Fe **11**–62 (9)
As–Ga **11**–62 (8,9)
As–Ge **11**–63 (1,9)
As–In **11**–63 (8,9)
As–K **11**–64 (8,9)
As–Mn **11**–64 (10)
As–Ni **11**–65 (9)
As–P **11**–66 (10)
As–Pb **11**–66 (9)
As–Pd **11**–67 (10)
As–Pt **11**–67 (9)
As–S **11**–68 (10)
As–Sb **11**–68 (9)
As–Si **11**–69 (1,9)
As–Sn **11**–69 (10)
As–Te **11**–70 (8,9)
As–Tl **11**–70 (9)
As–Zn **11**–70 (9)

Au–B **11**–71 (8,9)
Au–Be **11**–71 (10)
Au–Bi **11**–72 (8,9)

Au–Ca 11–72
Au–Cd 11–73 (10)
Au–Ce 11–73 (8,9)
Au–Co 11–74 (8,9)
Au–Cr 11–74 (8,9)
Au–Cs 11–75 (1)
Au–Cu 11–75 (10)
Au–Dy 11–76 (8,9)
Au–Er 11–77 (10)
Au–Fe 11–77 (1)
Au–Ga 11–78 (8)
Au–Gd 11–78 (8)
Au–Ge 11–79 (8,9)
Au–Hg 11–79 (10)
Au–In 11–79 (10)
Au–K 11–80 (8,9)
Au–La 11–80 (5)
Au–Li 11–81 (10)
Au–Lu 11–81 (10)
Au–Mg 11–82 (9)
Au–Mn 11–82 (10)
Au–Na 11–83 (8,9)
Au–Nb 11–83 (9)
Au–Ni 11–84 (10)
Au–P 11–84 (8)
Au–Pb 11–85 (10)
Au–Pd 11–85
Au–Pr 11–85 (5)
Au–Pt 11–86 (8,9)
Au–Rb 11–86 (5)
Au–Rh 11–87 (10)
Au–Ru 11–87 (1)
Au–S 11–88 (8)
Au–Sb 11–88
Au–Sc 11–88 (8)
Au–Se 11–89 (8,9)
Au–Si 11–89 (8,9)
Au–Sm 11–90 (10)
Au–Sn 11–90 (10)
Au–Sr 11–91 (8,9)
Au–Tb 11–91 (8,9)
Au–Te 11–91 (9)
Au–Th 11–92 (10)
Au–Ti 11–92 (5)
Au–Tl 11–93 (9)
Au–Tm 11–93 (10)
Au–U 11–93 (10)
Au–V 11–94 (8,9)
Au–Yb 11–94, 95 (8,9)
Au–Zn 11–96 (1)
Au–Zr 11–96 (9)

B–C 11–97 (9)
B–Ce 11–97 (1)
B–Co 11–98 (1)
B–Cr 11–98 (10)
B–Cu 11–99 (1,9)
B–Dy 11–99 (2,9)
B–Er 11–100 (2)
B–Eu 11–100
B–Fe 11–101 (10)
B–Gd 11–101 (2,9)
B–Ge 11–101 (8,9)
B–Hf 11–102 (8,9)
B–Ho 11–102 (2)

B–La 11–103 (9)
B–Lu 11–103 (10)
B–Mn 11–104 (10)
B–Mo 11–104 (10)
B–Nb 11–105 (8)
B–Nd 11–105 (9)
B–Ni 11–106 (1)
B–Pd 11–106 (10)
B–Pt 11–107 (10)
B–Pr 11–107 (2,9)
B–Re 11–108 (8,9)
B–Rh 11–108 (2)
B–Ru 11–109 (2)
B–Sc 11–109 (1)
B–Si 11–110 (1)
B–Sm 11–110 (8,9)
B–Ta 11–111 (10)
B–Ti 11–111 (9)
B–U 11–112 (10)
B–V 11–112 (2,9)
B–W 11–113 (2,9)
B–Y 11–113 (10)
B–Zr 11–114 (10)

Ba–Ca 11–114 (10)
Ba–Cd 11–115 (8,9)
Ba–Cu 11–115 (8)
Ba–Eu 11–116 (1)
Ba–Ga 11–116 (10)
Ba–Ge 11–117 (10)
Ba–H 11–117 (9)
Ba–Hg 11–118 (8,9)
Ba–In 11–119 (8,9)
Ba–Li 11–119 (9)
Ba–Mg 11–120 (9)
Ba–Mn 11–120 (9)
Ba–Na 11–121 (8,9)
Ba–Pb 11–121 (8,9)
Ba–Pd 11–122 (9)
Ba–Pt 11–122 (10)
Ba–Se 11–123 (2,9)
Ba–Si 11–123 (9)
Ba–Sn 11–123
Ba–Sr 11–124 (10)
Ba–Te 11–124 (2,9)
Ba–Tl 11–125 (2)
Ba–Yb 11–125 (1)
Ba–Zn 11–126

Be–Bi 11–126 (8)
Be–Co 11–126 (8,9)
Be–Cr 11–127 (8,9)
Be–Cu 11–127 (10)
Be–Fe 11–128 (10)
Be–Ga 11–128 (8,9)
Be–Hf 11–129 (10)
Be–In 11–129 (9)
Be–Nb 11–130 (10)
Be–Ni 11–130 (10)
Be–Pd 11–131 (1)
Be–Pu 11–131 (9)
Be–Ru 11–132 (9)
Be–Si 11–132 (9)
Be–Sn 11–133 (9)
Be–Th 11–133 (9)
Be–Ti 11–134 (1)

Be–U 11–134 (8,9)
Be–W 11–135 (1)
Be–Y 11–135 (9)
Be–Zr 11–136 (10)

Bi–Ca 11–136 (8,9)
Bi–Cd 11–136 (9)
Bi–Ce 11–137 (10)
Bi–Co 11–137 (10)
Bi–Cr 11–138 (10)
Bi–Cs 11–138 (2,9)
Bi–Cu 11–138 (8,9)
Bi–Fe 11–139 (10)
Bi–Ga 11–139 (8,9)
Bi–Gd 11–139 (10)
Bi–Ge 11–140 (9)
Bi–Hg 11–140 (9)
Bi–In 11–141 (10)
Bi–K 11–141 (8,9)
Bi–La 11–142 (8,9)
Bi–Li 11–142 (9)
Bi–Mg 11–143 (9)
Bi–Mn 11–143 (10)
Bi–Na 11–144 (10)
Bi–Nd 11–144 (10)
Bi–Ni 11–144 (9)
Bi–Pb 11–145 (8,9)
Bi–Pd 11–145 (10)
Bi–Pr 11–146 (1,9)
Bi–Pt 11–146 (10)
Bi–Pu 11–147
Bi–Rb 11–147 (9)
Bi–Rh 11–148 (10)
Bi–S 11–148 (9)
Bi–Sb 11–149
Bi–Se 11–149 (10)
Bi–Si 11–150 (9)
Bi–Sn 11–150 (8,9)
Bi–Sr 11–151 (9)
Bi–Te 11–151 (10)
Bi–Th 11–152 (10)
Bi–Ti 11–152 (9)
Bi–Tl 11–153 (10)
Bi–U 11–153 (8)
Bi–Y 11–154 (8,9)
Bi–Zn 11–154 (10)
Bi–Zr 11–155 (9)

C–Co 11–155 (9)
C–Cr 11–156 (10)
C–Cu 11–156 (10)
C–Fe 11–157 (10)
C–Ge 11–157 (9)
C–Hf 11–158 (4,9)
C–Ir 11–158 (2,9)
C–La 11–159 (8,9)
C–Li 11–160 (9)
C–Mn 11–160 (10)
C–Mo 11–161 (10)
C–Nb 11–161 (9)
C–Ni 11–162 (10)
C–Pu 11–162 (9)
C–Re 11–163 (10)
C–Si 11–163 (1)
C–Ta 11–164 (10)
C–Th 11–164 (10)

C–Ti **11**–165 (1)
C–U **11**–165 (10)
C–V **11**–166 (10)
C–W **11**–166 (9)
C–Y **11**–167 (1)
C–Zr **11**–167 (9)

Ca–Cd **11**–168 (10)
Ca–Co **11**–168 (9)
Ca–Cu **11**–169 (8,9)
Ca–Eu **11**–169 (8,9)
Ca–Hg **11**–169
Ca–In **11**–170 (8,9)
Ca–La **11**–170 (1)
Ca–Li **11**–171 (1)
Ca–Mg **11**–171 (1)
Ca–Mn **11**–172 (10)
Ca–N **11**–172 (10)
Ca–Na **11**–173 (9)
Ca–Nd **11**–173 (1)
Ca–Ni **11**–174 (2,9)
Ca–Pb **11**–174 (8,9)
Ca–Pd **11**–175 (3,9)
Ca–Pt **11**–175 (3,9)
Ca–Sb **11**–176 (8,9)
Ca–Si **11**–176 (8,9)
Ca–Sn **11**–177 (9)
Ca–Sr **11**–177 (1)
Ca–Ti **11**–178 (10)
Ca–Tl **11**–178 (10)
Ca–Yb **11**–179 (8,9)
Ca–Zn **11**–179 (8,9)

Cd–Cu **11**–180 (8,9)
Cd–Eu **11**–180 (8,9)
Cd–Ga **11**–181 (9)
Cd–Gd **11**–181 (8,9)
Cd–Ge **11**–182 (3,9)
Cd–Hg **11**–182 (10)
Cd–In **11**–183 (9)
Cd–K **11**–183 (10)
Cd–La **11**–184 (8,9)
Cd–Li **11**–185 (9)
Cd–Mg **11**–185 (10)
Cd–Na **11**–186 (1,9)
Cd–Ni **11**–186 (1)
Cd–Np **11**–187 (8)
Cd–Pb **11**–187 (9)
Cd–Pd **11**–187 (10)
Cd–Pt **11**–188 (10)
Cd–Pu **11**–188 (9)
Cd–S **11**–189
Cd–Sb **11**–189
Cd–Se **11**–189 (9)
Cd–Sm **11**–190 (8,9)
Cd–Sn **11**–190 (8,9)
Cd–Sr **11**–191 (8,9)
Cd–Te **11**–191 (9)
Cd–Th **11**–192 (3,9)
Cd–Ti **11**–191 (8)
Cd–Tl **11**–192 (9)
Cd–U **11**–193 (8,9)
Cd–Y **11**–193 (8,9)
Cd–Yb **11**–194 (10)
Cd–Zn **11**–194 (9)
Cd–Zr **11**–195

Ce–Co **11**–195 (10)
Ce–Cr **11**–196 (8,9)
Ce–Cu **11**–196 (1)
Ce–Fe **11**–197 (8,9)
Ce–Ge **11**–197 (2,9)
Ce–In **11**–198 (10)
Ce–Ir **11**–198 (10)
Ce–La **11**–199 (1)
Ce–Mg **11**–199 (1)
Ce–Mn **11**–200 (10)
Ce–Nb **11**–200 (10)
Ce–Ni **11**–201 (10)
Ce–Pb **11**–201 (10)
Ce–Pd **11**–202 (10)
Ce–Pu **11**–202 (9)
Ce–Ru **11**–203 (10)
Ce–Si **11**–203 (10)
Ce–Sn **11**–204 (10)
Ce–Ti **11**–205 (8)
Ce–Th **11**–204 (10)
Ce–Tl **11**–205 (8)
Ce–U **11**–205 (9)
Ce–Y **11**–206 (1)
Ce–Zn **11**–206 (10)

Co–Cr **11**–207 (10)
Co–Cu **11**–207 (1)
Co–Er **11**–208 (10)
Co–Fe **11**–208 (10)
Co–Ga **11**–209 (10)
Co–Gd **11**–209 (10)
Co–Ge **11**–210 (10)
Co–Hf **11**–210 (4)
Co–Ho **11**–211 (8,9)
Co–In **11**–211 (8,9)
Co–Ir **11**–212
Co–La **11**–212 (10)
Co–Mg **11**–213 (8,9)
Co–Mn **11**–213 (4,9)
Co–Mo **11**–214 (8)
Co–Nb **11**–214 (10)
Co–Ni **11**–215 (8,9)
Co–Os **11**–215 (8,9)
Co–P **11**–215 (9)
Co–Pb **11**–216 (9)
Co–Pd **11**–216
Co–Pr **11**–217 (10)
Co–Pt **11**–217 (10)
Co–Pu **11**–218 (9)
Co–Re **11**–219 (8,9)
Co–Rh **11**–219 (8,9)
Co–Ru **11**–219 (8,9)
Co–S **11**–220 (10)
Co–Sb **11**–220 (10)
Co–Sc **11**–221 (8,9)
Co–Se **11**–221 (10)
Co–Si **11**–222 (10)
Co–Sm **11**–222 (2,9)
Co–Sn **11**–223 (10)
Co–Ta **11**–223 (10)
Co–Te **11**–224 (10)
Co–Ti **11**–224 (10)
Co–U **11**–225 (9)
Co–V **11**–225 (10)
Co–W **11**–226 (10)
Co–Y **11**–226 (10)

Co–Zn **11**–227 (10)
Co–Zr **11**–227 (10)

Cr–Cu **11**–228 (1)
Cr–Fe **11**–228 (9)
Cr–Ga **11**–229 (8,9)
Cr–Gd **11**–229 (9)
Cr–Ge **11**–230 (3,9)
Cr–Hf **11**–230 (10)
Cr–Ho **11**–231 (2)
Cr–Ir **11**–231 (8)
Cr–La **11**–232 (10)
Cr–Mn **11**–232 (10)
Cr–Mo **11**–233 (10)
Cr–Nb **11**–233 (10)
Cr–Nd **11**–234 (3,9)
Cr–Ni **11**–234 (10)
Cr–Os **11**–235 (10)
Cr–P **11**–235 (8,9)
Cr–Pb **11**–236 (10)
Cr–Pd **11**–236 (10)
Cr–Pr **11**–237 (10)
Cr–Pt **11**–237 (10)
Cr–Pu **11**–238 (10)
Cr–Re **11**–238 (9)
Cr–Rh **11**–239 (10)
Cr–Ru **11**–239 (10)
Cr–S **11**–240 (9)
Cr–Sc **11**–240 (8,9)
Cr–Si **11**–240 (10)
Cr–Sn **11**–241 (10)
Cr–Ta **11**–241 (10)
Cr–Tb **11**–242 (8,9)
Cr–Ti **11**–242 (10)
Cr–U **11**–243 (8,9)
Cr–V **11**–243 (10)
Cr–W **11**–244 (1)
Cr–Y **11**–244 (10)
Cr–Zn **11**–245 (9)
Cr–Zr **11**–245 (1,9)

Cs–Ga **11**–246 (10)
Cs–Hg **11**–246 (9)
Cs–In **11**–247 (3,9)
Cs–K **11**–247 (9)
Cs–Na **11**–248 (9)
Cs–Rb **11**–248
Cs–S **11**–248
Cs–Sb **11**–249
Cs–Te **11**–249 (10)
Cs–Tl **11**–250 (8)

Cu–Er **11**–250 (8)
Cu–Eu **11**–251 (3)
Cu–Fe **11**–251 (8,9)
Cu–Ga **11**–252 (8,9)
Cu–Gd **11**–253 (3,9)
Cu–Ge **11**–253 (8,9)
Cu–Hf **11**–254 (4,9)
Cu–Hg **11**–254 (9)
Cu–In **11**–255 (9)
Cu–Ir **11**–255 (8,9)
Cu–La **11**–256 (10)
Cu–Li **11**–256 (9)
Cu–Mg **11**–257 (10)
Cu–Mn **11**–257 (10)

Cu–Nd **11**–258 (3,9)
Cu–Ni **11**–259 (9)
Cu–O **11**–258 (10)
Cu–P **11**–259 (9)
Cu–Pb **11**–260 (1)
Cu–Pd **11**–260 (8)
Cu–Pr **11**–261 (9)
Cu–Pt **11**–261 (10)
Cu–Pu **11**–262 (9)
Cu–Rh **11**–262 (8,9)
Cu–S **11**–263 (10)
Cu–Sb **11**–263 (10)
Cu–Se **11**–264 (9)
Cu–Si **11**–264 (8,9)
Cu–Sm **11**–265 (10)
Cu–Sn **11**–265 (8,9)
Cu–Sr **11**–266 (8,9)
Cu–Te **11**–266 (10)
Cu–Th **11**–267 (8,9)
Cu–Ti **11**–267 (10)
Cu–Tl **11**–268 (9)
Cu–U **11**–268 (8,9)
Cu–V **11**–269 (10)
Cu–Y **11**–269 (10)
Cu–Yb **11**–270 (8,9)
Cu–Zn **11**–270 (8,9)
Cu–Zr **11**–271 (10)

Dy–Er **11**–271 (8,9)
Dy–Fe **11**–272 (9)
Dy–Ga **11**–272 (2,9)
Dy–Ge **11**–273 (8,9)
Dy–Ho **11**–273 (8,9)
Dy–In **11**–274 (8,9)
Dy–Mn **11**–274 (8,9)
Dy–Ni **11**–275 (8,9)
Dy–Sb **11**–275 (8,9)
Dy–Zr **11**–276 (9)

Er–Fe **11**–276 (10)
Er–Ho **11**–276 (8)
Er–In **11**–277 (3,9)
Er–Ni **11**–277 (8,9)
Er–Rh **11**–278 (8,9)
Er–Tb **11**–278 (8)
Er–Ti **11**–278 (9)
Er–V **11**–278 (10)
Er–Y **11**–279 (8)
Er–Zr **11**–279 (10)

Eu–In **11**–280 (8,9)

Fe–Ga **11**–280 (8,9)
Fe–Gd **11**–281 (8,9)
Fe–Ge **11**–281 (10)
Fe–Hf **11**–282 (10)
Fe–Ho **11**–282 (8,9)
Fe–In **11**–283 (8,9)
Fe–Ir **11**–283 (9)
Fe–La **11**–284 (9)
Fe–Mg **11**–284 (8,9)
Fe–Mn **11**–285 (8,9)
Fe–Mo **11**–286 (10)
Fe–N **11**–286 (8,9)
Fe–Nb **11**–287 (10)
Fe–Ni **11**–287 (9)

Fe–O **11**–288 (10)
Fe–Os **11**–288 (9)
Fe–P **11**–289 (9)
Fe–Pb **11**–290
Fe–Pd **11**–290 (8,9)
Fe–Pt **11**–291 (10)
Fe–Pu **11**–291 (8,9)
Fe–Re **11**–292 (8,9)
Fe–Ru **11**–292 (8,9)
Fe–S **11**–293 (10)
Fe–Sb **11**–293 (9)
Fe–Sc **11**–294 (10)
Fe–Se **11**–294 (10)
Fe–Si **11**–295 (10)
Fe–Sm **11**–295 (10)
Fe–Sn **11**–296 (10)
Fe–Ta **11**–296 (8,9)
Fe–Ti **11**–297 (9)
Fe–U **11**–297 (8,9)
Fe–V **11**–298 (10)
Fe–W **11**–298 (10)
Fe–Y **11**–299 (8,9)
Fe–Zn **11**–299 (10)
Fe–Zr **11**–299 (10)

Ga–Gd **11**–300 (10)
Ga–Ge **11**–300 (8,9)
Ga–Hg **11**–300 (8,9)
Ga–In **11**–301 (10)
Ga–K **11**–301 (10)
Ga–La **11**–302 (10)
Ga–Li **11**–302 (10)
Ga–Mg **11**–303 (10)
Ga–Mn **11**–303 (10)
Ga–Na **11**–304 (10)
Ga–Nb **11**–304 (10)
Ga–Nd **11**–305 (2,9)
Ga–Ni **11**–305 (10)
Ga–P **11**–306
Ga–Pb **11**–306 (9)
Ga–Pd **11**–307 (10)
Ga–Pr **11**–307 (10)
Ga–Pu **11**–308 (10)
Ga–Rb **11**–308 (10)
Ga–Sb **11**–309 (8,9)
Ga–Se **11**–309 (2,9)
Ga–Si **11**–310 (9)
Ga–Sm **11**–310 (2,9)
Ga–Sn **11**–311 (9)
Ga–Te **11**–311 (10)
Ga–Tl **11**–311 (9)
Ga–Tm **11**–312 (3,9)
Ga–U **11**–312 (10)
Ga–V **11**–313 (10)
Ga–Y **11**–313 (2,9)
Ga–Yb **11**–314 (10)
Ga–Zn **11**–314 (1,9)

Gd–Ge **11**–315 (8,9)
Gd–In **11**–315 (10)
Gd–La **11**–316 (10)
Gd–Mg **11**–316 (9)
Gd–Mn **11**–317 (9)
Gd–Ni **11**–318 (10)
Gd–Pb **11**–318 (8,9)
Gd–Pd **11**–319 (3,9)

Gd–Rh **11**–319 (10)
Gd–Ru **11**–320 (8,9)
Gd–Sc **11**–320 (8,9)
Gd–Ti **11**–321 (10)
Gd–Y **11**–321 (9)
Gd–Zr **11**–322 (9)

Ge–Hf **11**–322 (10)
Ge–In **11**–323 (9)
Ge–Mg **11**–323 (8,9)
Ge–K **11**–324 (8,9)
Ge–Mn **11**–324 (8,9)
Ge–Mo **11**–325 (8,9)
Ge–Ni **11**–325 (10)
Ge–Pb **11**–326 (9)
Ge–Pd **11**–326 (8,9)
Ge–Pt **11**–327 (8,9)
Ge–Rb **11**–327 (8,9)
Ge–Rh **11**–328 (8,9)
Ge–Ru **11**–328 (8,9)
Ge–S **11**–329 (9)
Ge–Sb **11**–329 (9)
Ge–Se **11**–330 (10)
Ge–Si **11**–330 (9)
Ge–Sn **11**–331 (8,9)
Ge–Sr **11**–331 (8,9)
Ge–Te **11**–332 (9)
Ge–Ti **11**–332 (10)
Ge–U **11**–333 (9)
Ge–Y **11**–334 (8,9)
Ge–Zn **11**–334 (8,9)
Ge–Zr **11**–334 (8,9)

H–Nb **11**–335 (10)
H–Sr **11**–335 (8,9)
H–Ti **11**–336 (10)
H–Zr **11**–336 (10)

Hf–Ir **11**–337 (8,9)
Hf–Mn **11**–337 (10)
Hf–Mo **11**–338 (8,9)
Hf–Nb **11**–338 (10)
Hf–Ni **11**–339 (10)
Hf–O **11**–339 (10)
Hf–Re **11**–340 (9)
Hf–Sn **11**–340 (9)
Hf–Ta **11**–341 (10)
Hf–Th **11**–341 (9)
Hf–Ti **11**–342 (10)
Hf–U **11**–342 (8,9)
Hf–W **11**–342 (8,9)
Hf–Zr **11**–343 (1)

Hg–In **11**–343 (10)
Hg–K **11**–344 (10)
Hg–Li **11**–344 (10)
Hg–Mg **11**–345 (9)
Hg–Mn **11**–345 (8,9)
Hg–Na **11**–346 (8,9)
Hg–Pb **11**–346 (10)
Hg–Rb **11**–347 (9)
Hg–Rh **11**–347 (8)
Hg–Sb **11**–347 (8,9)
Hg–Sn **11**–348 (10)
Hg–Te **11**–348 (10)
Hg–Tl **11**–349 (10)

Hg–U **11**–349 (10)
Hg–Zn **11**–350 (10)

Ho–Tb **11**–350 (8)

In–K **11**–351 (10)
In–La **11**–351 (3,9)
In–Li **11**–352 (10)
In–Mg **11**–352 (10)
In–Mn **11**–353 (10)
In–Na **11**–353 (10)
In–Nd **11**–354 (3,9)
In–Ni **11**–354 (10)
In–Pb **11**–355 (10)
In–Pd **11**–355 (10)
In–Pr **11**–356 (10)
In–Sb **11**–356 (9)
In–Se **11**–357 (10)
In–Si **11**–357 (8,9)
In–Sn **11**–358 (10)
In–Sr **11**–358 (8,9)
In–Te **11**–359 (10)
In–Ti **11**–359 (9)
In–Tl **11**–360 (9)
In–Y **11**–360 (3,9)
In–Yb **11**–361 (9)
In–Zn **11**–361 (8,9)
In–Zr **11**–362 (10)

Ir–Mn **11**–362 (10)
Ir–Mo **11**–363 (10)
Ir–Nb **11**–363 (9)
Ir–Os **11**–364 (9)
Ir–Pd **11**–364 (9)
Ir–Pt **11**–365 (10)
Ir–Re **11**–365 (9)
Ir–Th **11**–366 (10)
Ir–Ti **11**–366 (10)
Ir–W **11**–367 (10)

K–Li **11**–367 (10)
K–Mg **11**–368 (9)
K–Na **11**–368 (9)
K–Pb **11**–368 (9)
K–Rb **11**–369 (9)
K–S **11**–369 (10)
K–Sb **11**–369 (10)
K–Se **11**–370 (10)
K–Sn **11**–370 (9)
K–Tl **11**–371 (10)
K–Zn **11**–371 (9)

La–Mg **11**–371 (10)
La–Mn **11**–372 (10)
La–Nd **11**–372 (9)
La–Ni **11**–373 (10)
La–Pb **11**–373 (10)
La–Rh **11**–374 (8,9)
La–Sb **11**–374 (9)
La–Sn **11**–375 (10)
La–Tl **11**–375 (10)
La–V **11**–376
La–Y **11**–376 (1)

Li–Mg **11**–377 (8,9)
Li–Mn **11**–377 (9)

Li–Na **11**–378 (10)
Li–Pb **11**–378 (8,9)
Li–Pd **11**–379 (10)
Li–S **11**–379 (9)
Li–Si **11**–380 (10)
Li–Sn **11**–380 (10)
Li–Sr **11**–381 (10)
Li–Tl **11**–381 (9)
Li–Zn **11**–382 (9)

Mg–Mn **11**–382 (8,9)
Mg–Na **11**–383 (9)
Mg–Ni **11**–383 (9)
Mg–Pb **11**–383 (9)
Mg–Pr **11**–384 (10)
Mg–Pu **11**–384 (10)
Mg–Sb **11**–385 (9)
Mg–Sc **11**–385 (8,9)
Mg–Si **11**–386 (10)
Mg–Sn **11**–386 (8,9)
Mg–Sr **11**–387 (1)
Mg–Th **11**–387 (9)
Mg–Ti **11**–388 (10)
Mg–Tl **11**–388 (10)
Mg–U **11**–389 (8,9)
Mg–Y **11**–389 (8,9)
Mg–Zn **11**–390 (10)
Mg–Zr **11**–390 (10)

Mn–N **11**–391 (10)
Mn–Nb **11**–391 (8,9)
Mn–Nd **11**–392 (10)
Mn–Ni **11**–392 (8,9)
Mn–O **11**–393 (10)
Mn–P **11**–393 (6,9)
Mn–Pb **11**–394 (6,9)
Mn–Pd **11**–394 (10)
Mn–Pt **11**–395 (8,9)
Mn–Pu **11**–395 (10)
Mn–Rh **11**–395 (6,9)
Mn–Ru **11**–396 (8,9)
Mn–S **11**–397 (8)
Mn–Sb **11**–397 (10)
Mn–Si **11**–398 (10)
Mn–Sm **11**–398 (8,9)
Mn–Sn **11**–399 (10)
Mn–Tb **11**–399 (8,9)
Mn–Ti **11**–400 (10)
Mn–Tl **11**–400 (10)
Mn–U **11**–401 (10)
Mn–V **11**–401 (10)
Mn–Y **11**–402 (8,9)
Mn–Zn **11**–403 (10)
Mn–Zr **11**–403 (10)

Mo–Nb **11**–404 (10)
Mo–Ni **11**–404 (8,9)
Mo–Os **11**–405 (8,9)
Mo–Pd **11**–405 (10)
Mo–Pt **11**–406 (10)
Mo–Re **11**–406 (10)
Mo–Rh **11**–407 (9)
Mo–Ru **11**–407 (8,9)
Mo–S **11**–408 (4,9)
Mo–Si **11**–408 (10)
Mo–Ta **11**–409 (9)

Mo–Th **11**–409
Mo–Ti **11**–410 (1,9)
Mo–U **11**–410 (10)
Mo–V **11**–411 (9)
Mo–W **11**–411 (8,9)
Mo–Y **11**–411 (10)
Mo–Zr **11**–412 (10)

N–Ti **11**–412 (10)

Na–Pb **11**–413 (10)
Na–Rb **11**–413 (9)
Na–S **11**–414 (10)
Na–Sb **11**–414 (9)
Na–Se **11**–415 (10)
Na–Sn **11**–415 (8,9)
Na–Sr **11**–416 (1)
Na–Te **11**–416 (10)
Na–Th **11**–417 (8,9)
Na–Tl **11**–417 (10)
Na–Zn **11**–418 (1,9)

Nb–Ni **11**–419 (10)
Nb–O **11**–419 (10)
Nb–Os **11**–420 (2)
Nb–Re **11**–420 (8,9)
Nb–Rh **11**–421 (9)
Nb–Ru **11**–422 (4,9)
Nb–Sc **11**–422 (2,9)
Nb–Si **11**–423 (10)
Nb–Sn **11**–423 (10)
Nb–Ta **11**–424 (10)
Nb–Th **11**–424 (10)
Nb–Ti **11**–425 (10)
Nb–V **11**–425 (8,9)
Nb–W **11**–426 (7)
Nb–Y **11**–426 (8,9)
Nb–Zn **11**–427 (10)
Nb–Zr **11**–427 (10)

Nd–Ni **11**–428 (8)
Nd–Pr **11**–428 (9)
Nd–Pt **11**–429 (10)
Nd–Rh **11**–429 (10)
Nd–Sb **11**–430 (8,9)
Nd–Sc **11**–430 (2,9)
Nd–Sn **11**–431 (10)
Nd–Ti **11**–431 (8,9)
Nd–Zn **11**–432 (8,9)

Ni–O **11**–433 (1)
Ni–P **11**–433 (10)
Ni–Pb **11**–434 (10)
Ni–Pd **11**–434
Ni–Pr **11**–435 (10)
Ni–Pt **11**–435 (1)
Ni–Pu **11**–436 (9)
Ni–Re **11**–437 (1,9)
Ni–Rh **11**–437 (9)
Ni–Ru **11**–438 (10)
Ni–S **11**–438 (10)
Ni–Sb **11**–439 (10)
Ni–Si **11**–439 (10)
Ni–Sm **11**–440 (8,9)
Ni–Sn **11**–440 (10)
Ni–Ta **11**–441 (1)

Ni–Th **11**–441 (10)
Ni–Ti **11**–442 (1)
Ni–Tl **11**–442 (10)
Ni–U **11**–443 (9)
Ni–V –**11**–443 (8,9)
Ni–W **11**–444 (10)
Ni–Y **11**–444 (8,9)
Ni–Zn **11**–445 (10)
Ni–Zr **11**–445 (9)

Np–Pu **11**–446 (10)
Np–U **11**–446 (8,9)

O–Pu **11**–447 (10)
O–Ti **11**–447 (8)
O–V **11**–448 (10)

Os–Ru **11**–448 (9)
Os–Ti **11**–449 (10)
Os–U **11**–449 (9)
Os–W **11**–450 (8,9)

P–Pd **11**–450 (10)
P–Pt **11**–451 (8,9)
P–Si **11**–451 (10)
P–Sn **11**–452 (10)
P–Tl **11**–452 (9)
P–Zn **11**–453 (8,9)

Pb–Pd **11**–453 (10)
Pb–Pr **11**–454 (10)
Pb–Pt **11**–454 (10)
Pb–Pu **11**–455 (8,9)
Pb–S **11**–455 (9)
Pb–Sb **11**–455 (8,9)
Pb–Se **11**–456 (10)
Pb–Si **11**–456 (10)
Pb–Sn **11**–457 (8,9)
Pb–Sr **11**–457 (10)
Pb–Te **11**–458 (10)
Pb–Th **11**–458 (10)
Pb–Ti **11**–458 (8,9)
Pb–Tl **11**–459 (10)
Pb–U **11**–459 (1,9)
Pb–W **11**–460 (8)
Pb–Y **11**–460 (8,9)
Pb–Zn **11**–461 (8,9)

Pd–Rh **11**–461 (10)
Pd–S **11**–462 (10)
Pd–Sb **11**–462 (10)
Pd–Si **11**–463 (10)
Pd–Sn **11**–463 (10)
Pd–Ta **11**–464 (8,9)
Pd–Te **11**–464 (10)
Pd–Th **11**–465 (10)
Pd–Ti **11**–465 (10)
Pd–Tl **11**–466 (8,9)
Pd–U **11**–466 (8,9)
Pd–V **11**–467 (9)
Pd–W **11**–467 (9)
Pd–Y **11**–468 (10)
Pd–Zn **11**–468 (8,9)
Pd–Zr **11**–469 (10)

Pr–Sn **11**–469 (10)
Pr–Tl **11**–470 (10)

Pr–Zn **11**–470 (10)

Pt–Rh **11**–471 (10)
Pt–Ru **11**–471 (8,9)
Pt–Sb **11**–472 (10)
Pt–Si **11**–472 (10)
Pt–Sn **11**–473 (8,9)
Pt–Te **11**–473 (10)
Pt–Ti **11**–474 (1)
Pt–Tl **11**–474 (8,9)
Pt–V **11**–475 (8,9)
Pt–W **11**–475 (8,9)
Pt–Zr **11**–476 (10)

Pu–Si **11**–476 (9)
Pu–Th **11**–477 (10)
Pu–Ti **11**–477 (10)
Pu–U **11**–478 (10)
Pu–Zn **11**–478 (10)
Pu–Zr **11**–479 (10)

Rb–S **11**–479
Rb–Sb **11**–480 (8,9)

Re–Rh **11**–480 (9)
Re–Ru **11**–481 (8,9)
Re–Si **11**–481 (10)
Re–Tb **11**–481 (8,9)
Re–U **11**–482 (10)
Re–V **11**–482 (10)
Re–W **11**–483 (9)

Rh–Ta **11**–483 (9)
Rh–U **11**–484 (8,9)
Rh–V **11**–484 (10)
Rh–W **11**–485 (9)

Ru–Pt **11**–485 (9)
Ru–W **11**–486 (10)

S–Sb **11**–486 (9)
S–Se **11**–487 (9)
S–Sn **11**–487 (9)
S–Te **11**–487 (10)

Sb–Se **11**–488 (10)
Sb–Si **11**–488 (9)
Sb–Sn **11**–489 (10)
Sb–Te **11**–489 (9)
Sb–Tl **11**–490 (10)
Sb–U **11**–490 (8,9)
Sb–Zn **11**–491 (10)
Sb–Zr **11**–491 (10)

Sc–Ti **11**–492 (9)
Sc–Y **11**–492 (9)
Sc–Zr **11**–492 (9)

Se–Sn **11**–493 (9)
Se–Te **11**–493 (8,9)
Se–Th **11**–494 (10)
Se–Tl **11**–494 (10)

Si–Sn **11**–495 (9)
Si–Sr **11**–495 (10)
Si–Te **11**–496 (8,9)

Si–Th **11**–496 (9)
Si–Ti **11**–497 (10)
Si–Tl **11**–497 (10)
Si–U **11**–498 (10)
Si–V **11**–498 (1)
Si–W **11**–499 (10)
Si–Y **11**–499 (10)
Si–Zn **11**–500 (10)
Si–Zr **11**–500 (10)

Sn–Sr **11**–501 (10)
Sn–Te **11**–501 (10)
Sn–Ti **11**–502 (4,9)
Sn–Tl **11**–502 (8,9)
Sn–U **11**–503 (10)
Sn–V **11**–503 (1,9)
Sn–Yb **11**–504 (3,9)
Sn–Zn **11**–504 (9)
Sn–Zr **11**–505 (10)

Sr–Zn **11**–505 (3,9)

Ta–Th **11**–506 (10)
Ta–Ti **11**–506 (1,9)
Ta–U **11**–507 (10)
Ta–V **11**–507 (10)
Ta–W **11**–508 (8,9)
Ta–Y **11**–508 (10)
Ta–Zr **11**–509 (1,9)

Tc–V **11**–509 (8,9)

Te–Tl **11**–510 (10)
Te–Zn **11**–510 (10)

Th–Ti **11**–511 (10)
Th–U **11**–511 (10)
Th–V **11**–512 (9)
Th–Y **11**–512 (10)
Th–Zn **11**–513 (8,9)
Th–Zr **11**–513 (9)

Ti–U **11**–514 (10)
Ti–V **11**–514 (8,9)
Ti–W **11**–515 (8,9)
Ti–Y **11**–515 (10)
Ti–Zn **11**–516 (10)
Ti–Zr **11**–516 (9)

Tl–Zn **11**–517 (10)

U–V **11**–517 (9)
U–W **11**–518 (8,9)
U–Zn **11**–518 (10)
U–Zr **11**–519 (1,9)

V–W **11**–519 (9)
V–Y **11**–520 (8,9)
V–Zr **11**–520 (10)

W–Zr **11**–521 (8,9)

Y–Zn **11**–522 (8,9)
Y–Zr **11**–523 (9)

Zn–Zr **11**–523 (10)

11.2   Equilibrium diagrams*

References to individual diagrams are given at the end of this section. When no reference is given the reader should consult constitution of Binary Alloys by M. Hansen and K. Anderko, McGraw-Hill (1958) and the supplements by R. P. Elliot (1965) and F. A. Shunk (1969).

Diagrams updated (Reference 9) or replaced (Reference 10) are used with permission of ASM International, Materials Park OH.

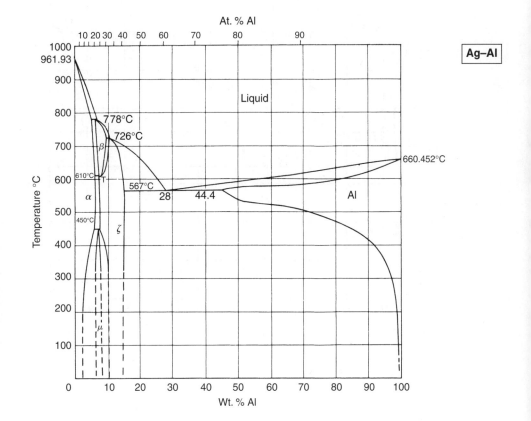

* *Note*: M.P. = Melting point.
        B.P. = Boiling point.

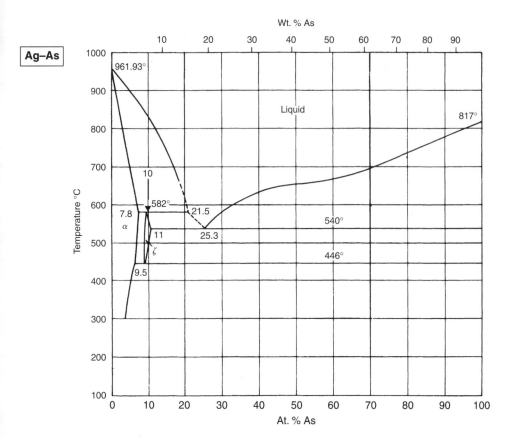

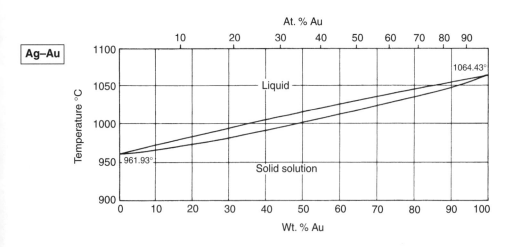

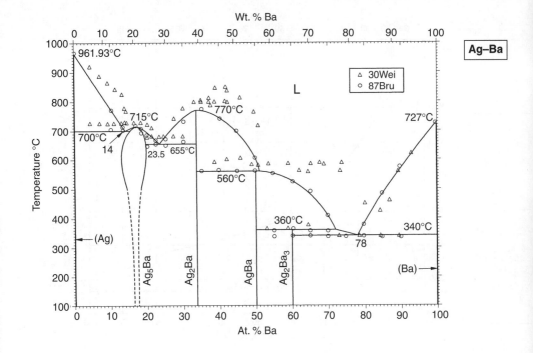

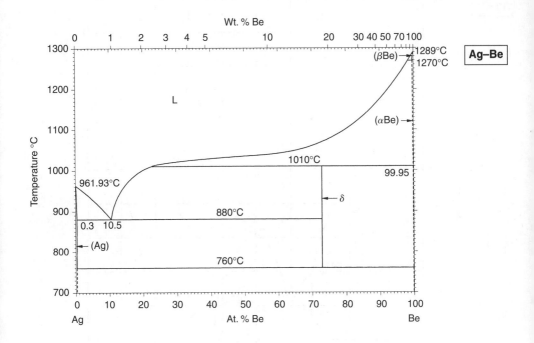

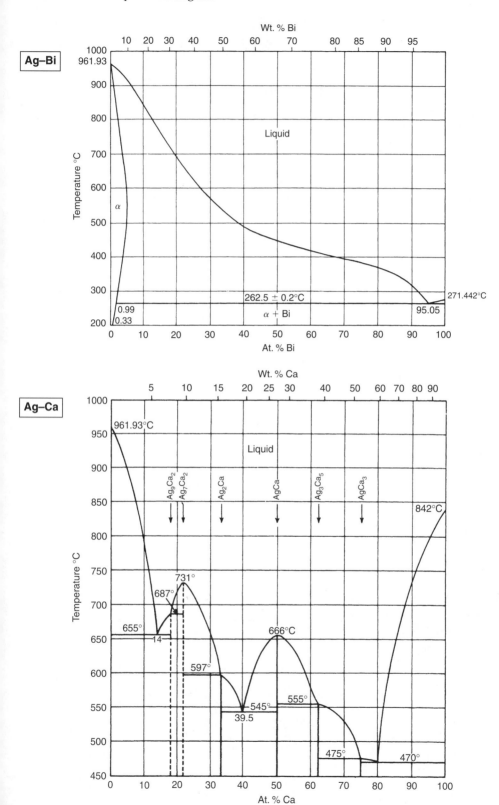

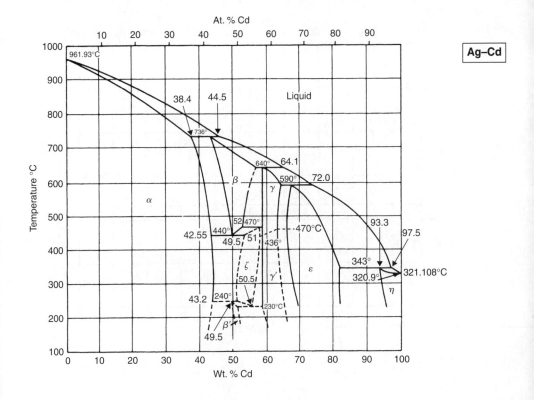

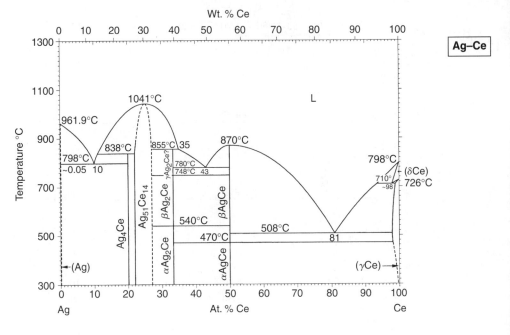

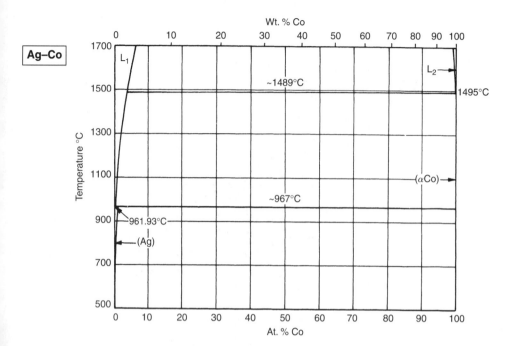

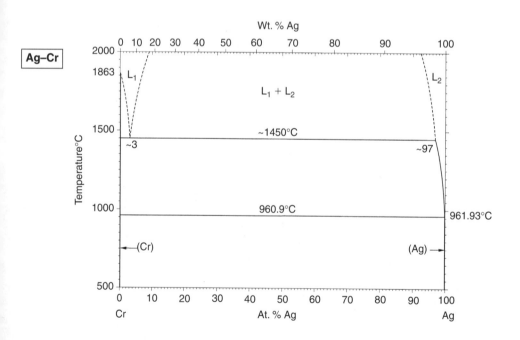

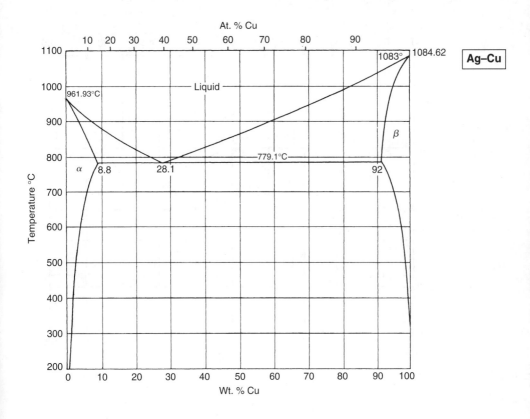

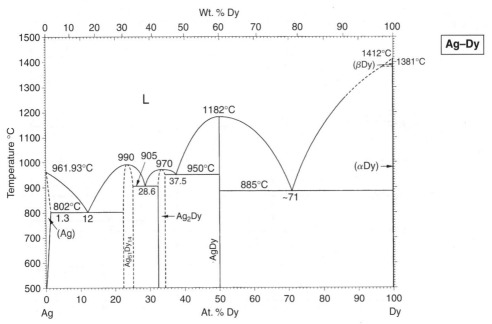

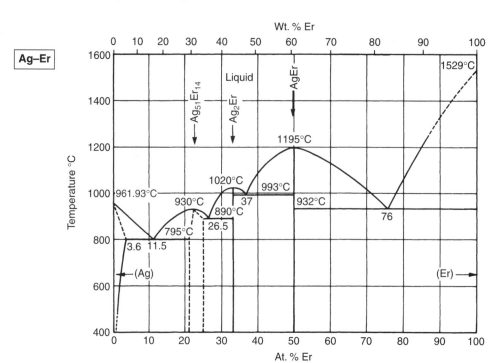

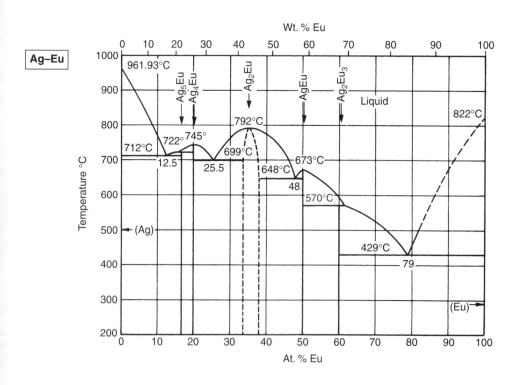

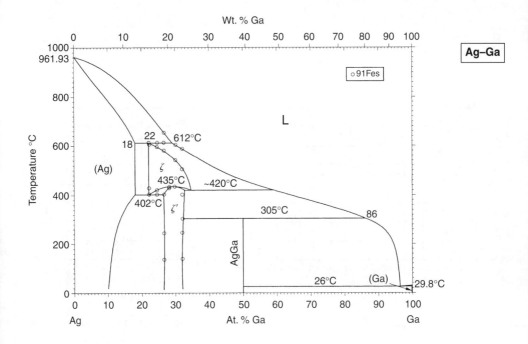

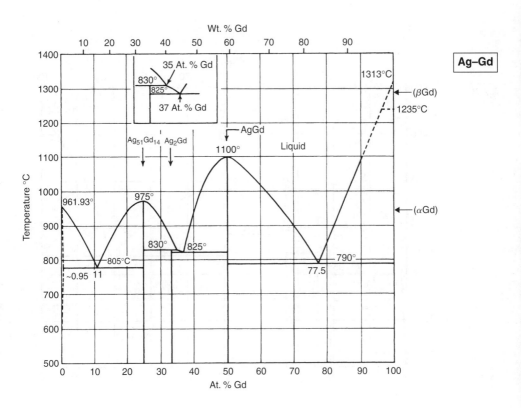

Ag–Ge

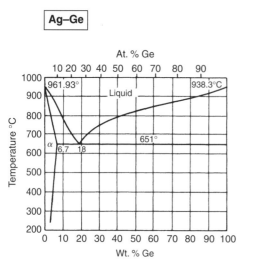

Ag–Hg

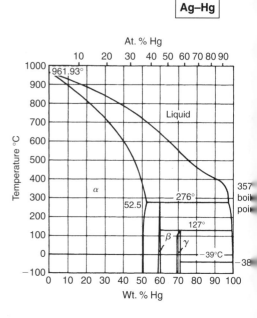

Ag–Ho

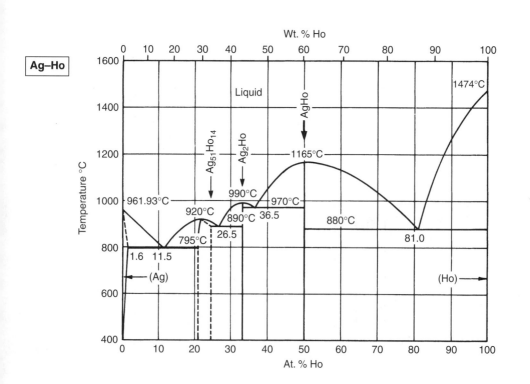

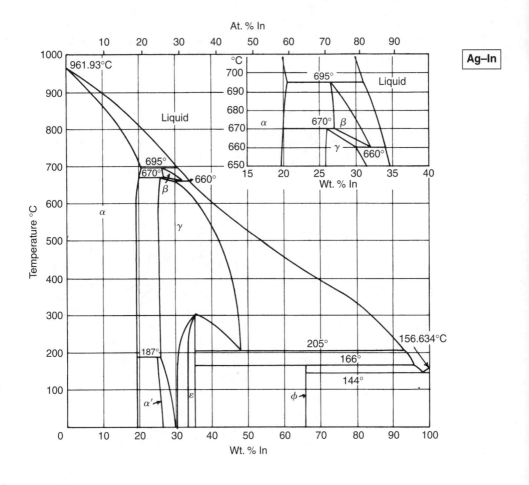

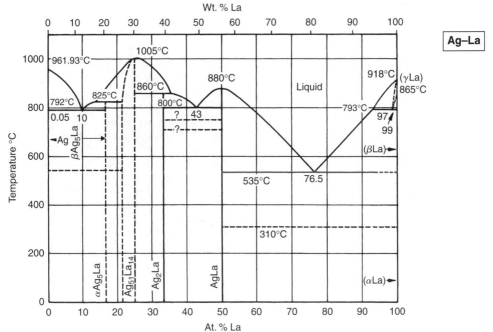

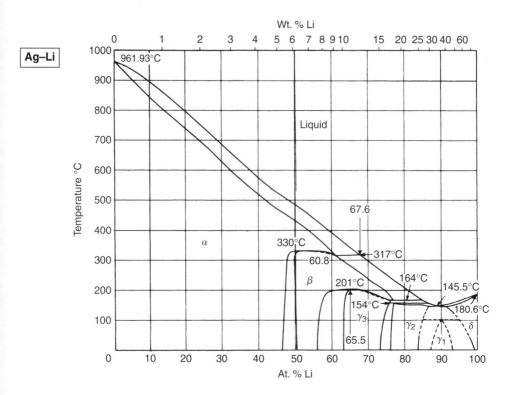

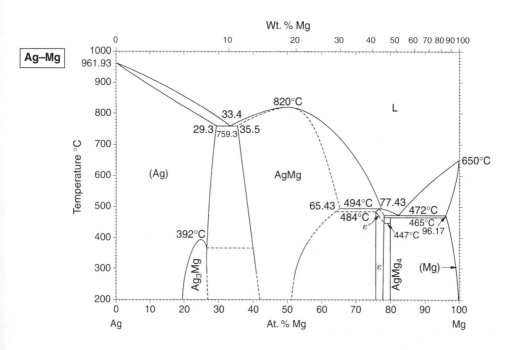

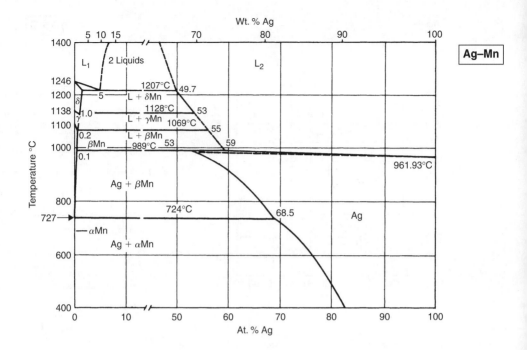

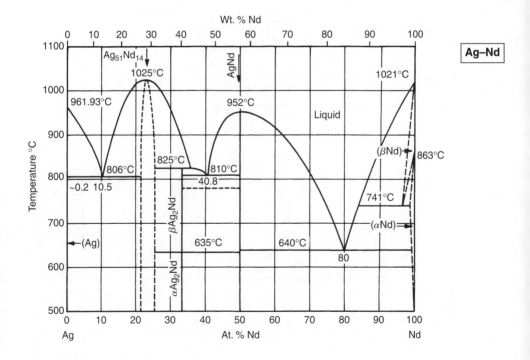

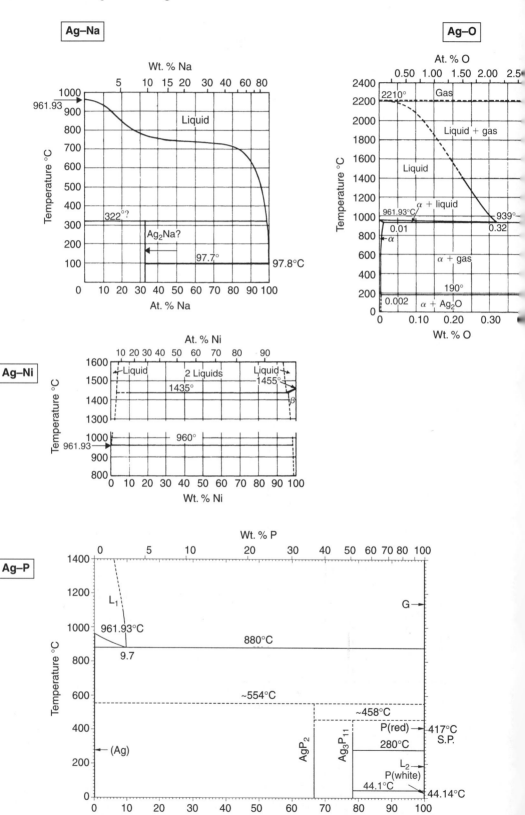

Ag–Na

Ag–O

Ag–Ni

Ag–P

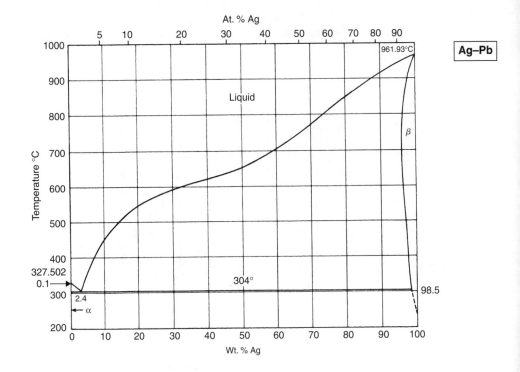

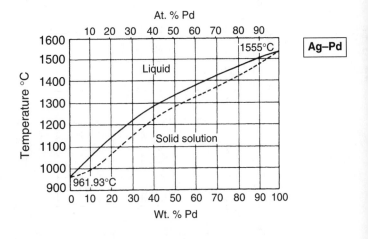

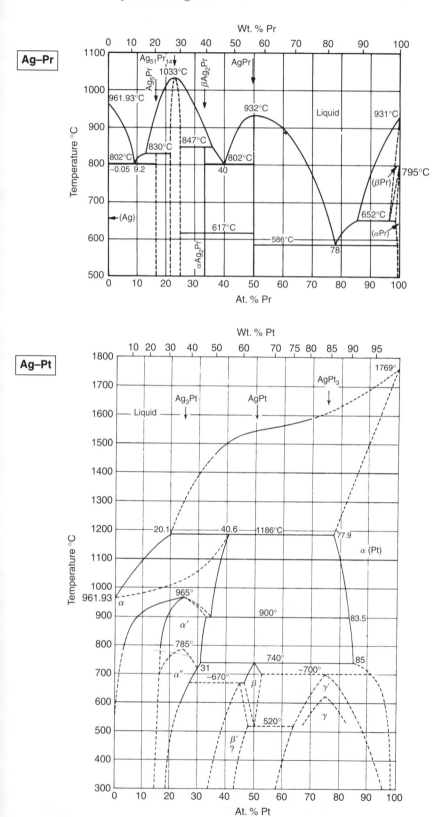

Ag–Rh

Ag–Ru

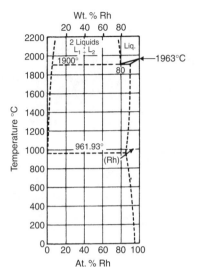

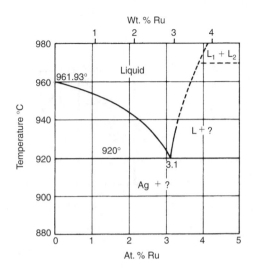

Ag–S

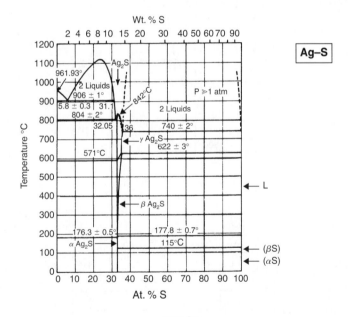

Ag–Sb

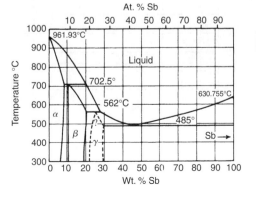

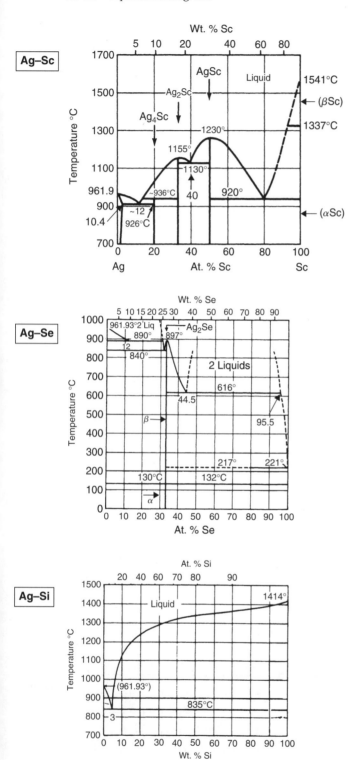

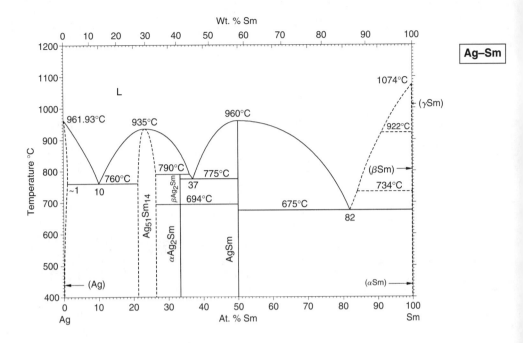

Ag–Sm

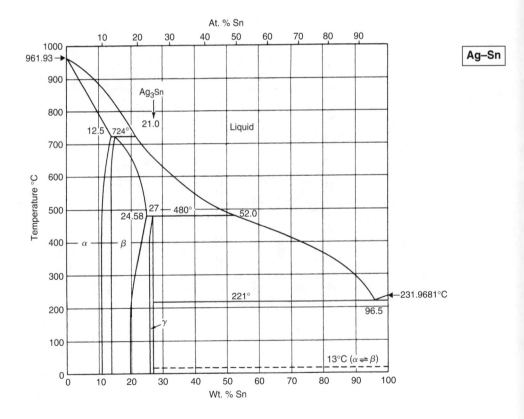

Ag–Sn

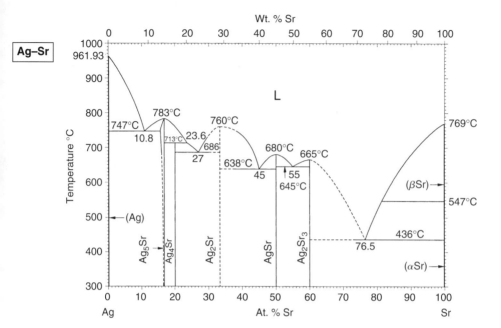

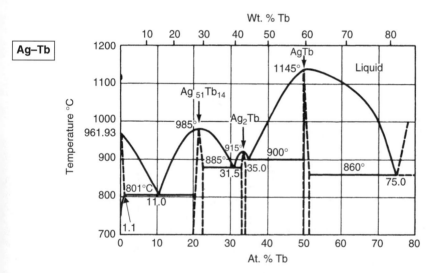

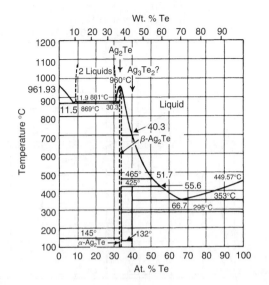

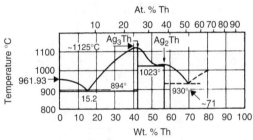

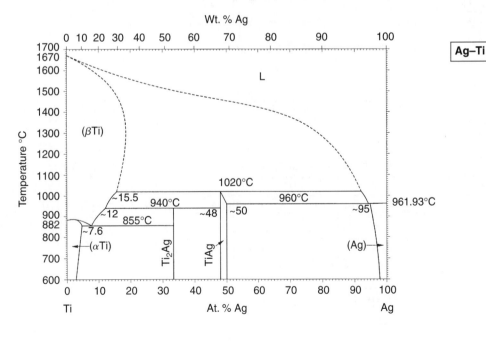

Ag–Tl

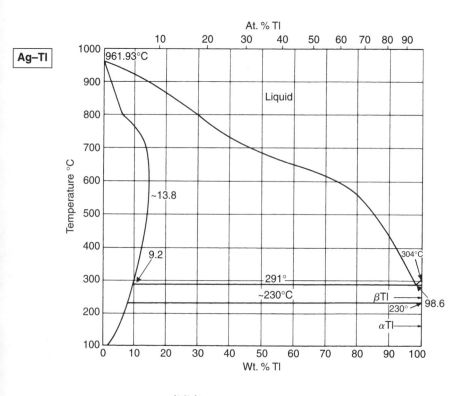

Ag–U

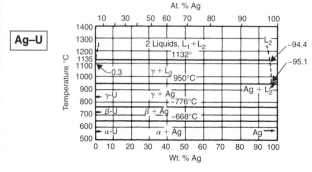

Ag–Y

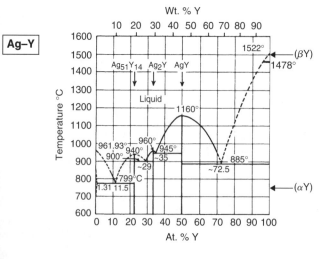

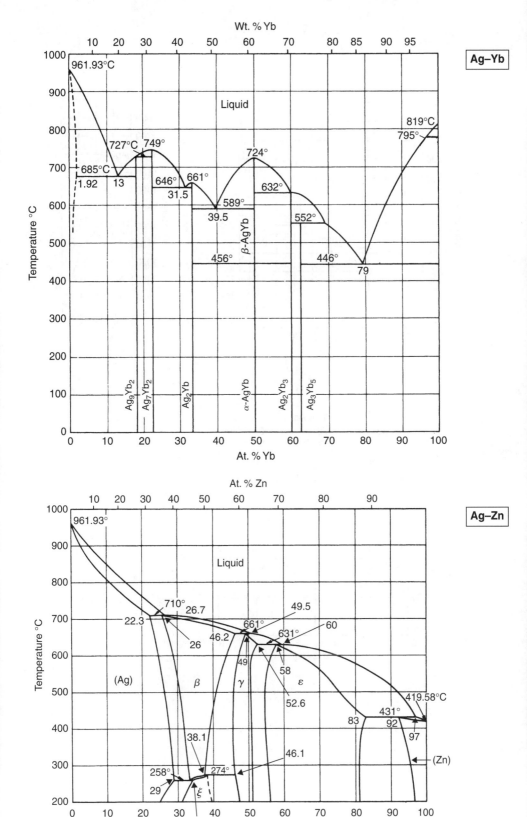

Ag–Zr

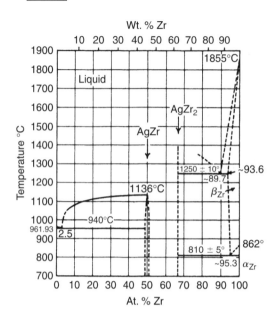

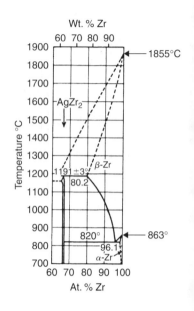

Al–As

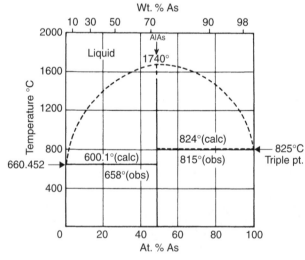

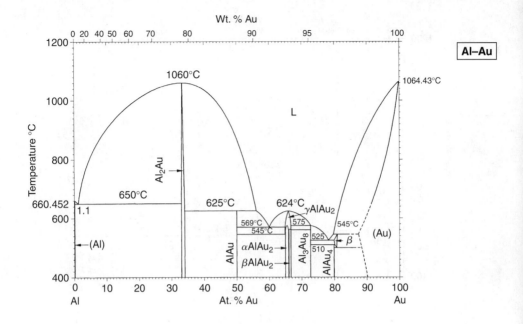

Al–Au

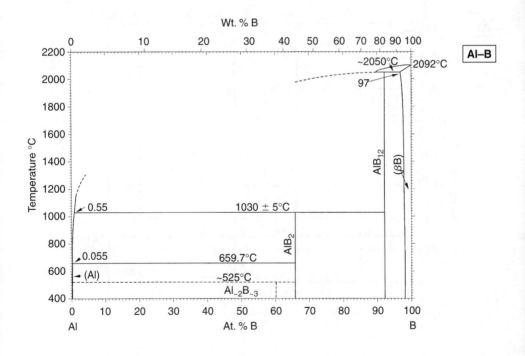

Al–B

Al–Ba

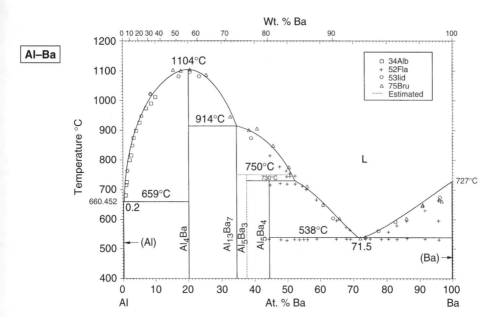

Al–Be

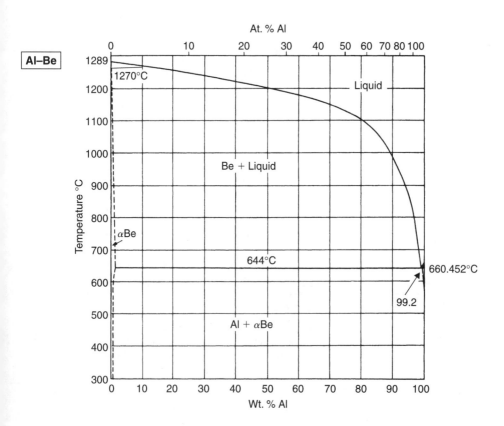

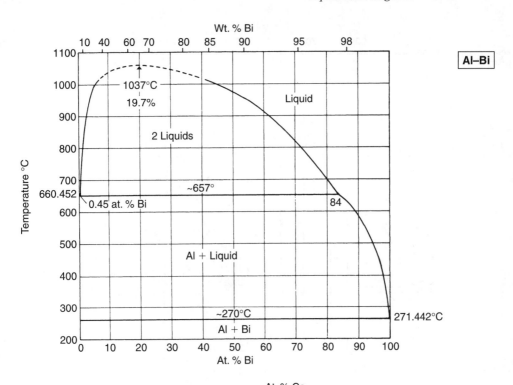

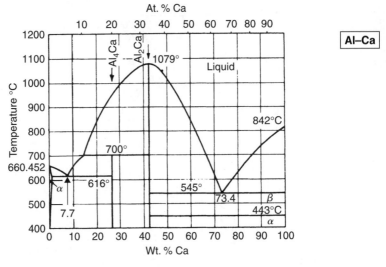

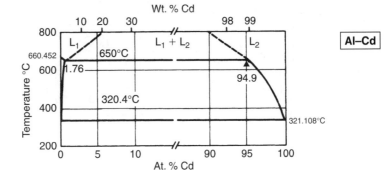

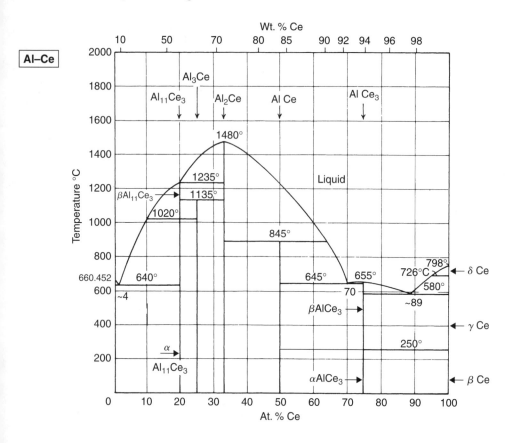

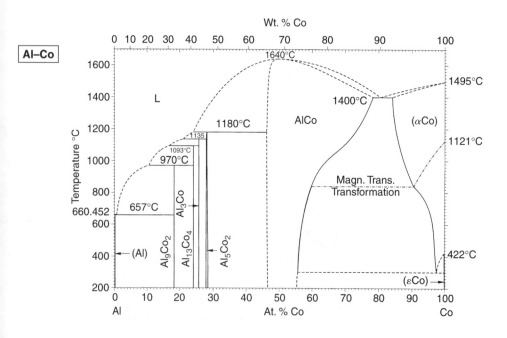

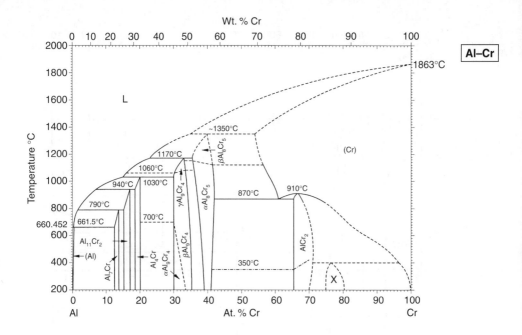

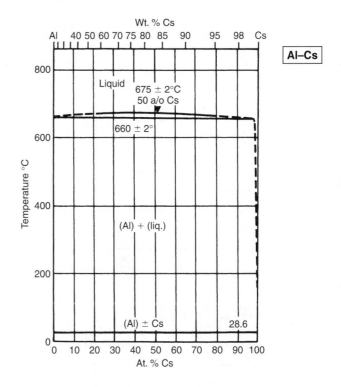

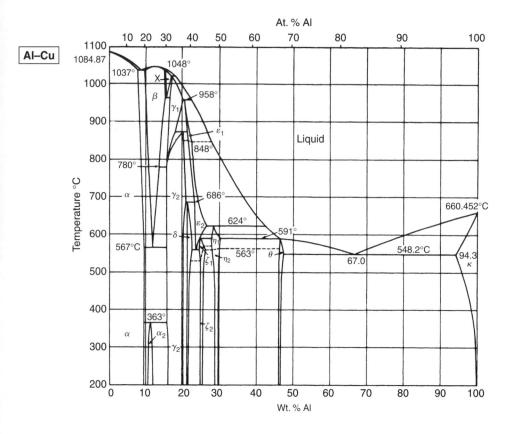

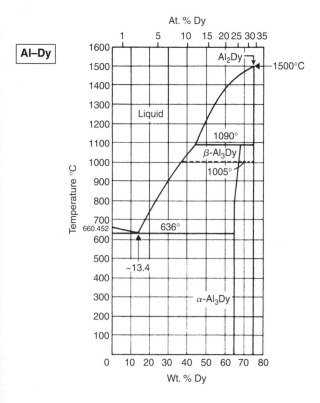

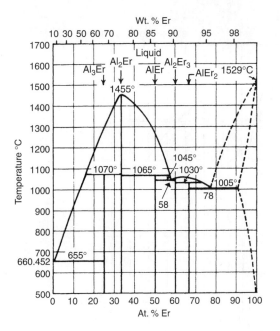

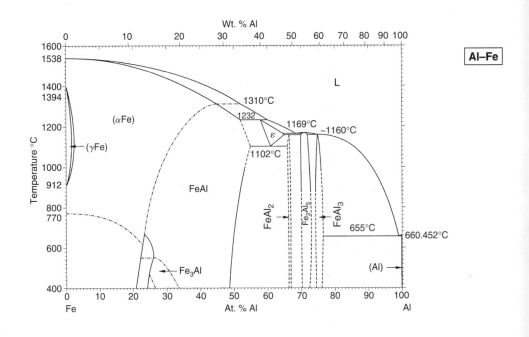

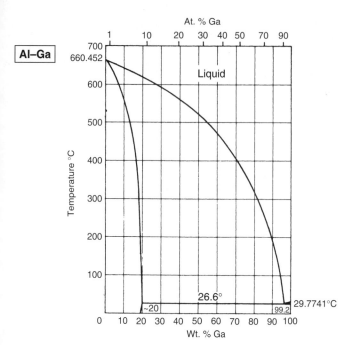

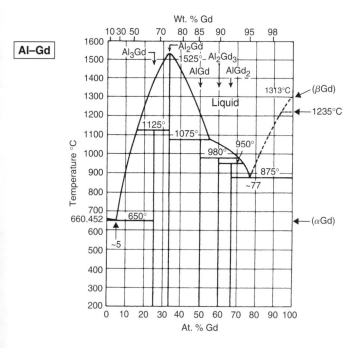

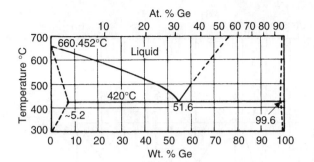

Al–Ge

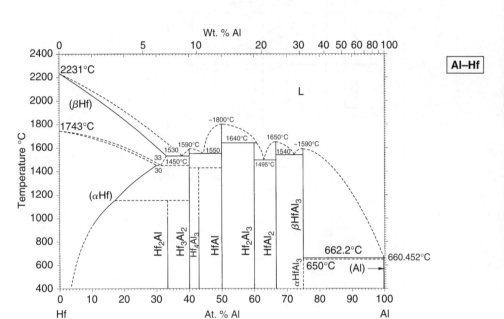

Al–Hf

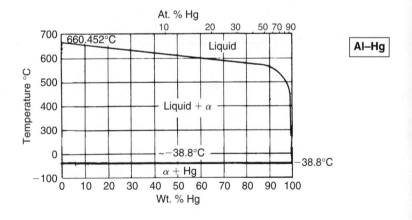

Al–Hg

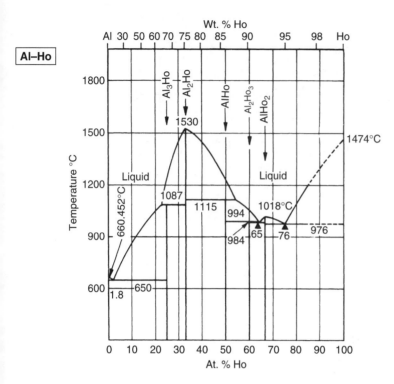

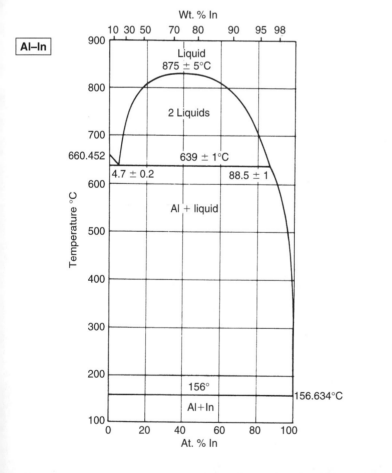

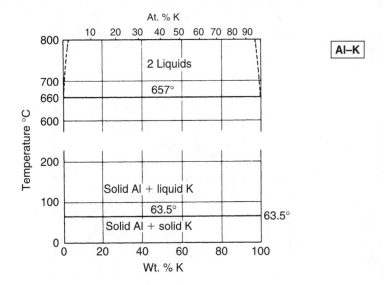

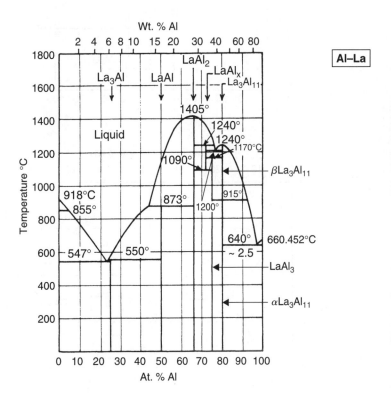

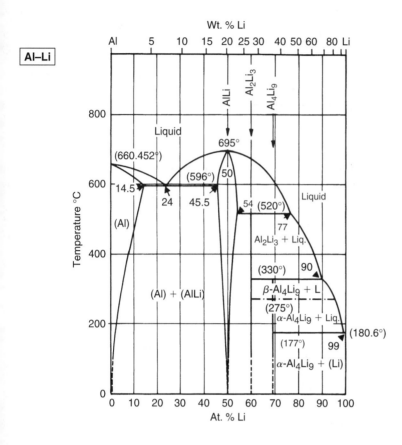

Al–Li

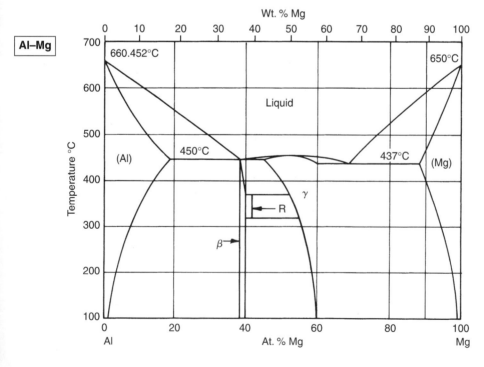

Al–Mg

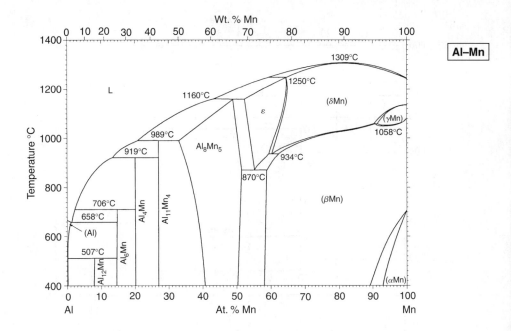

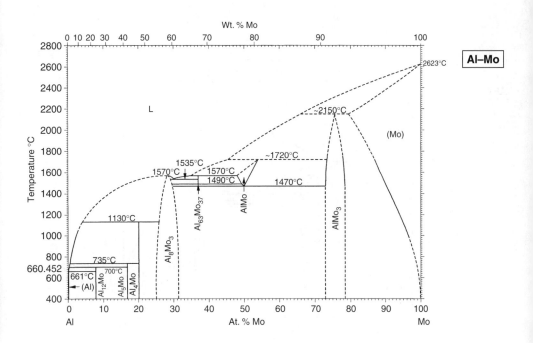

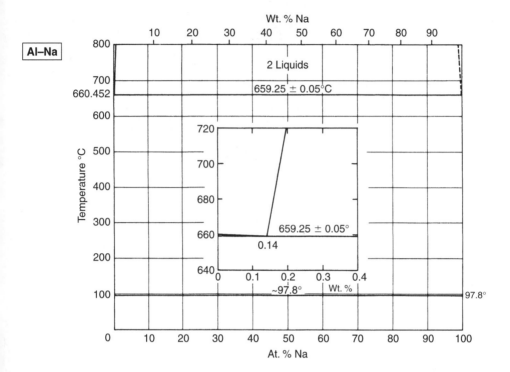

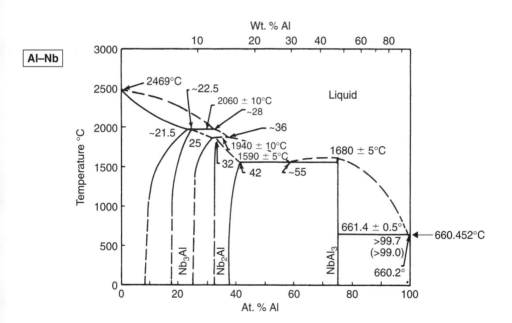

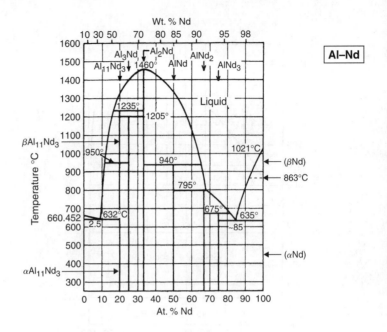

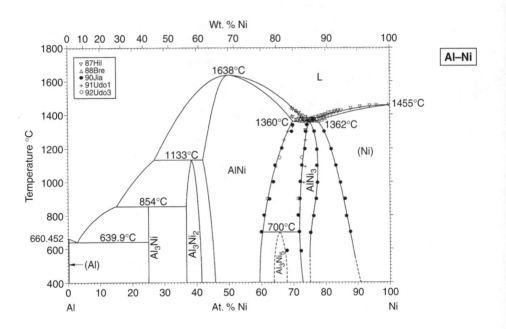

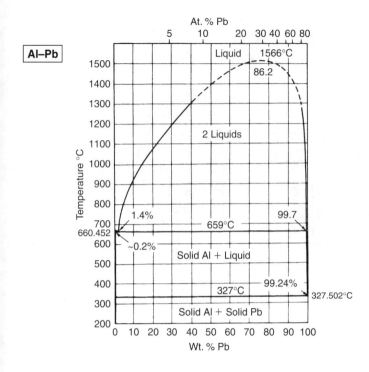

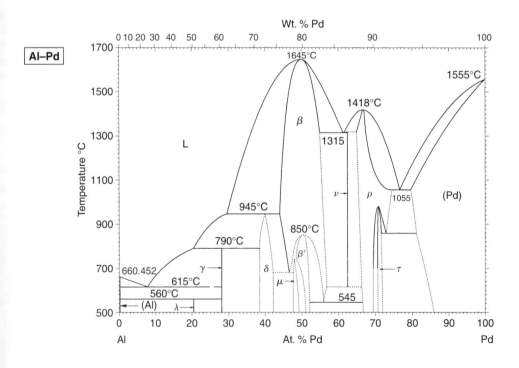

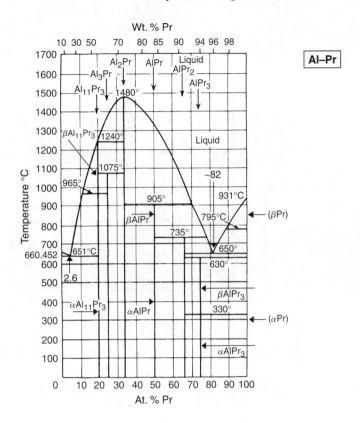

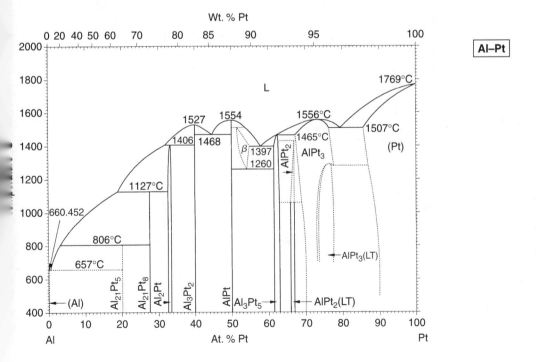

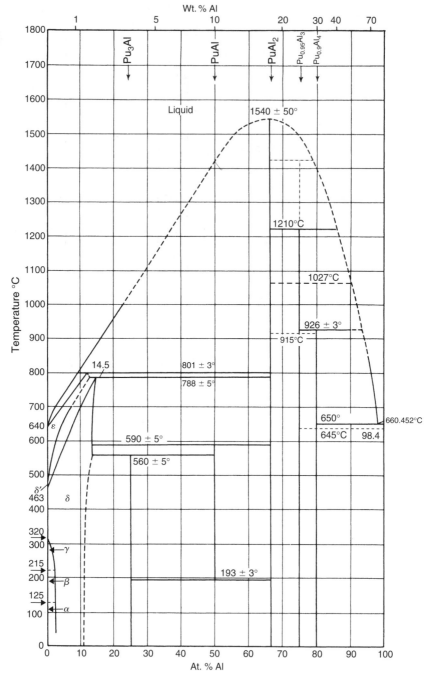

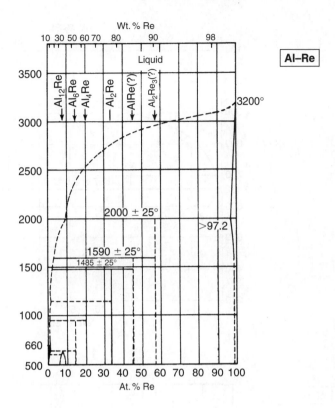

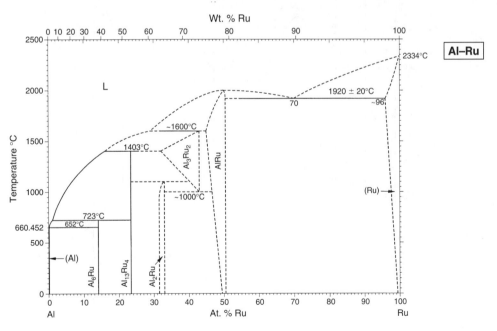

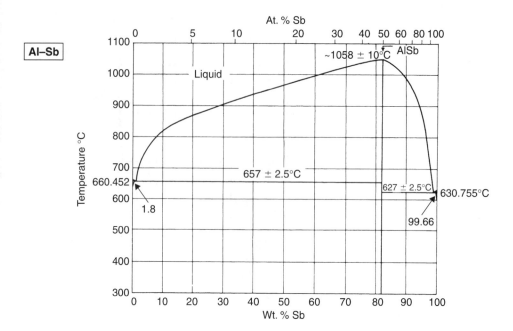

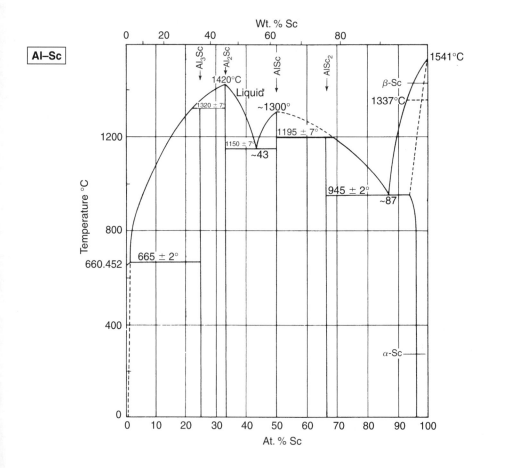

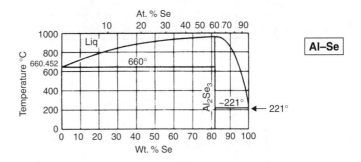

Al–Se

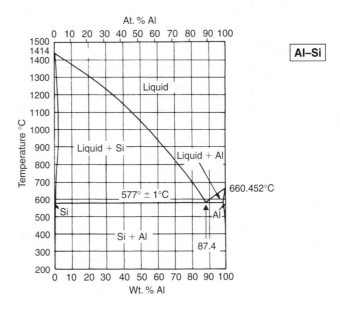

Al–Si

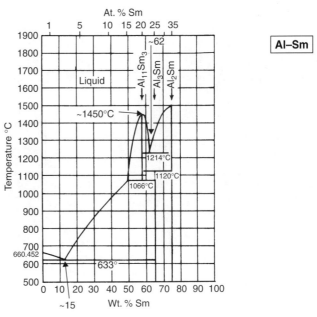

Al–Sm

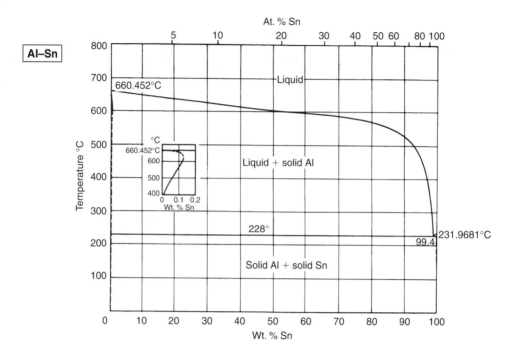

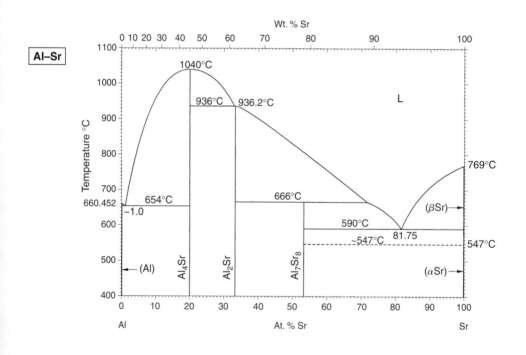

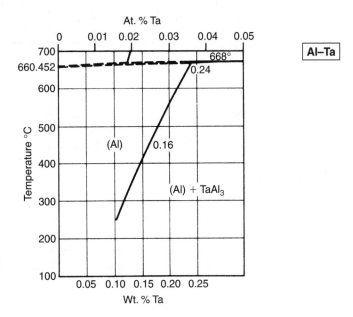

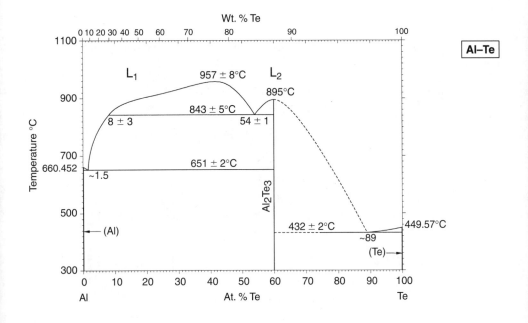

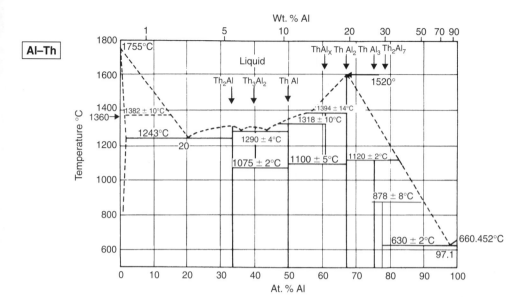

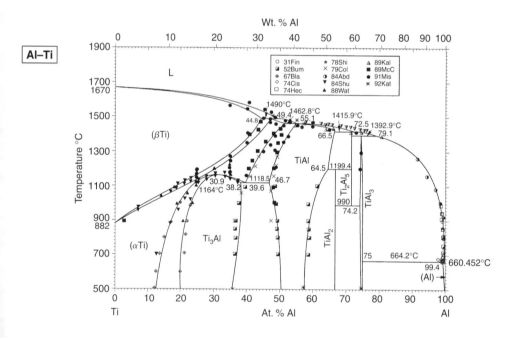

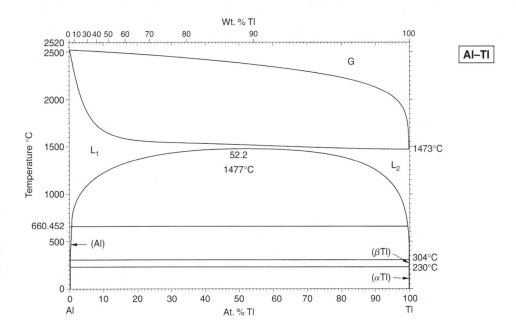

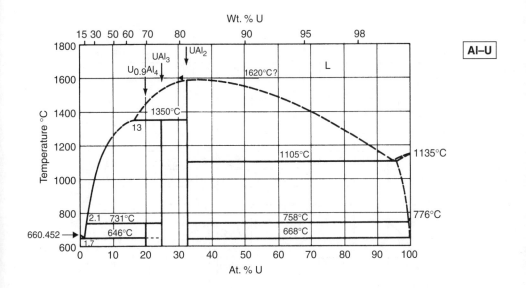

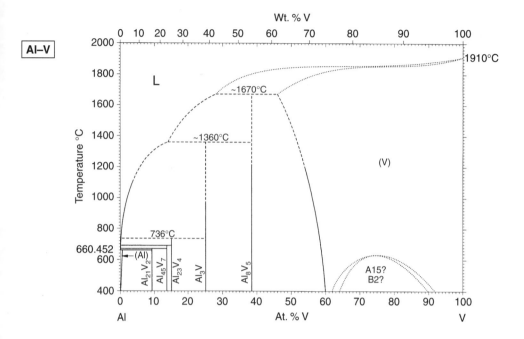

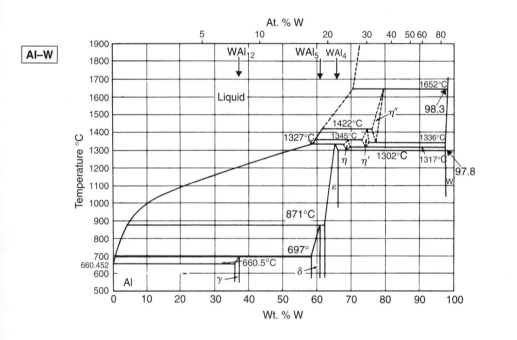

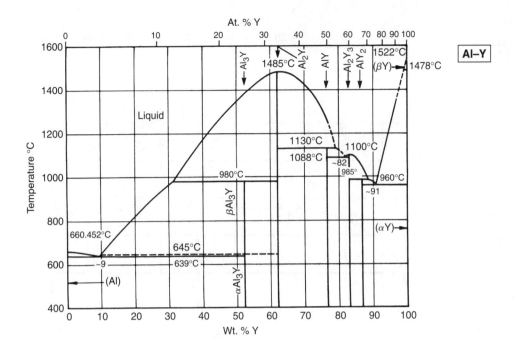

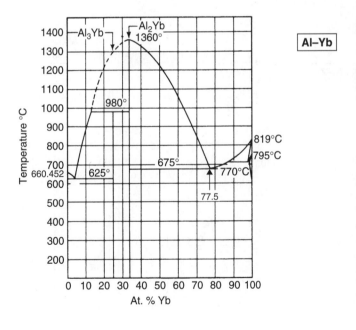

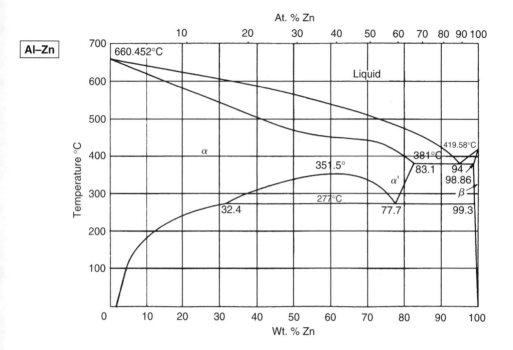

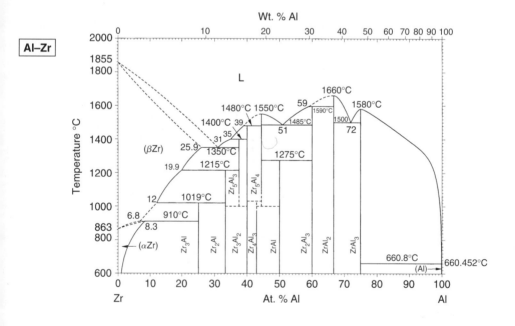

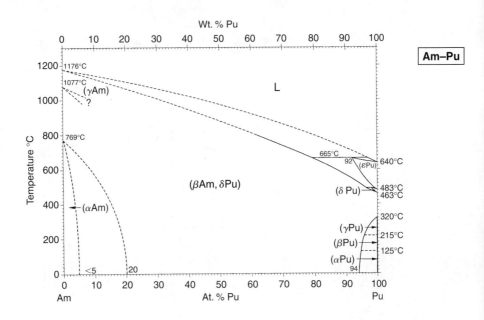

Am–Pu

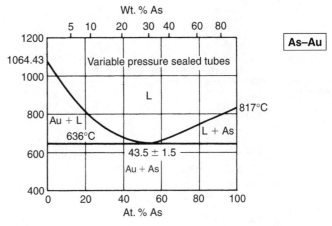

As–Au

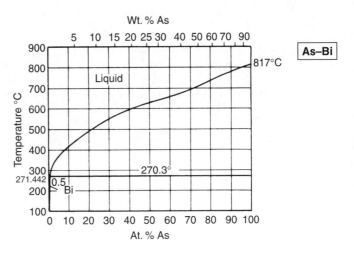

As–Bi

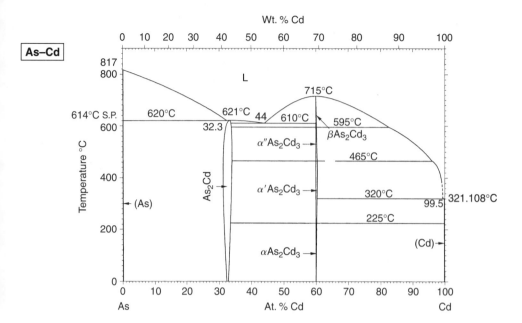

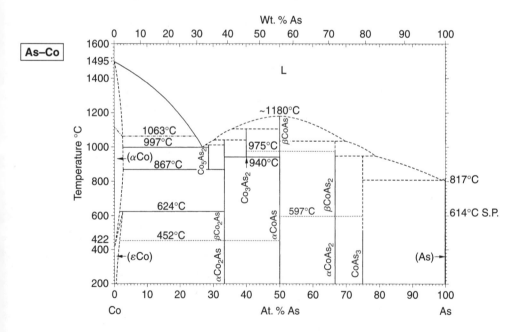

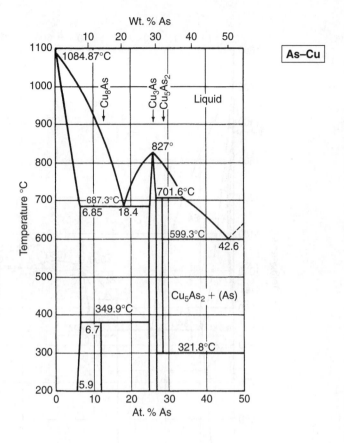

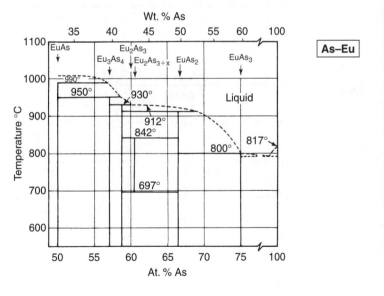

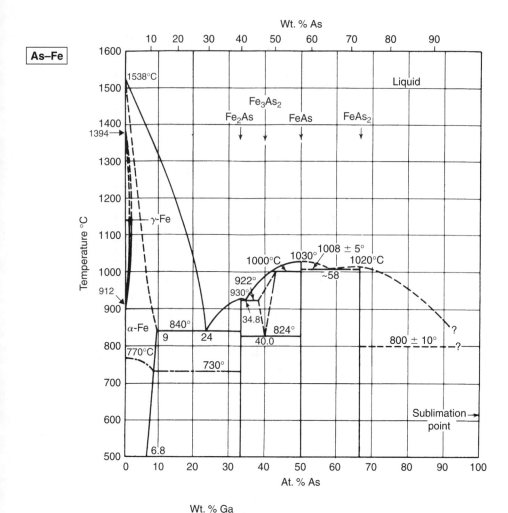

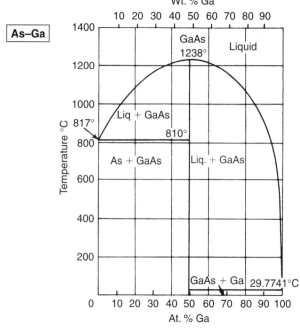

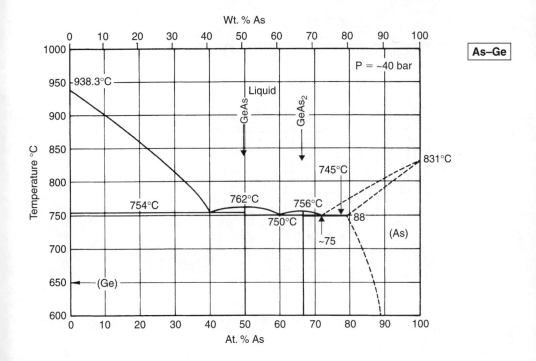

As–Ge

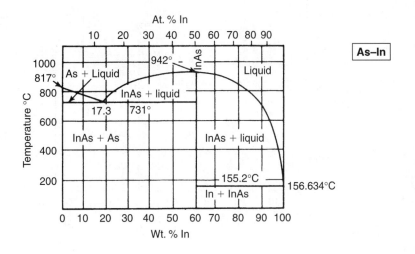

As–In

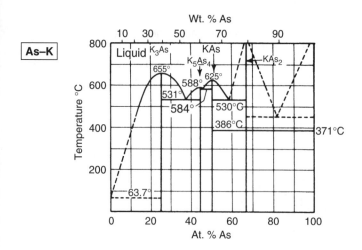

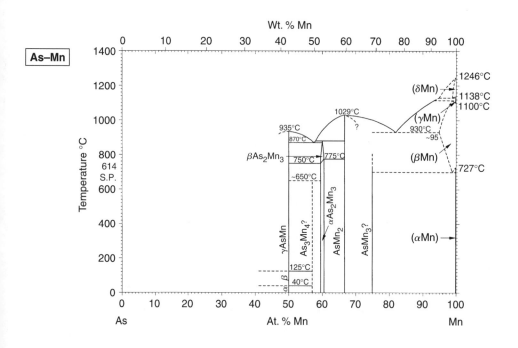

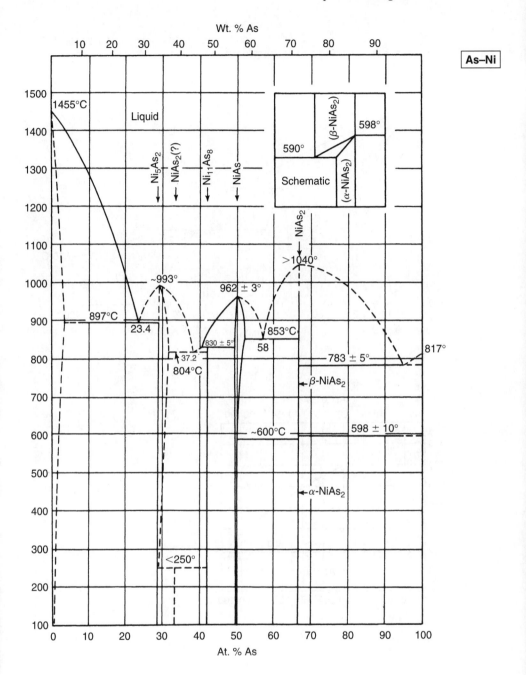

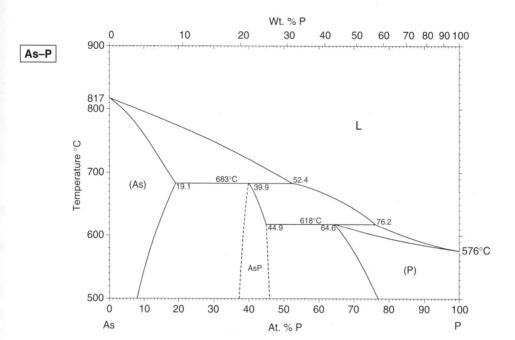

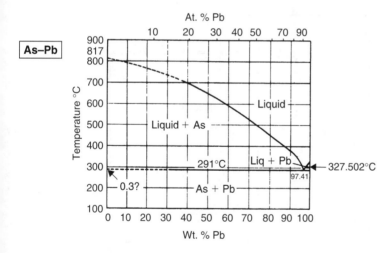

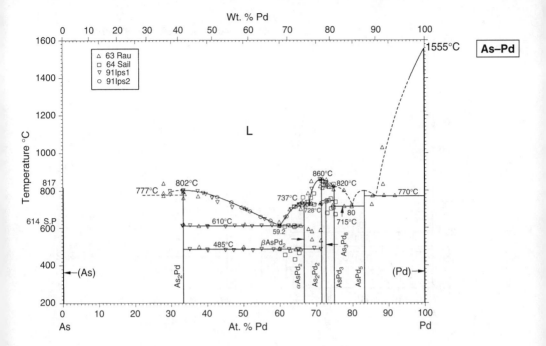

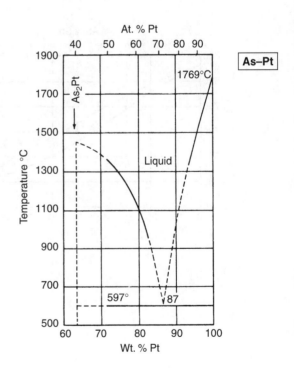

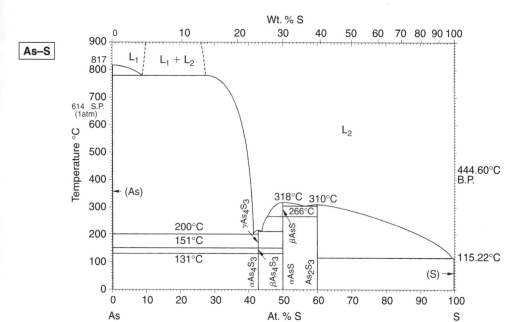

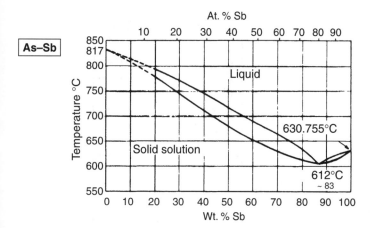

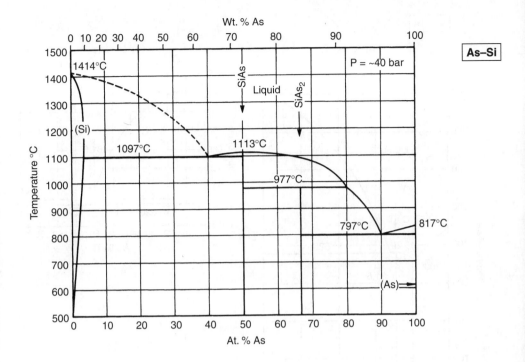

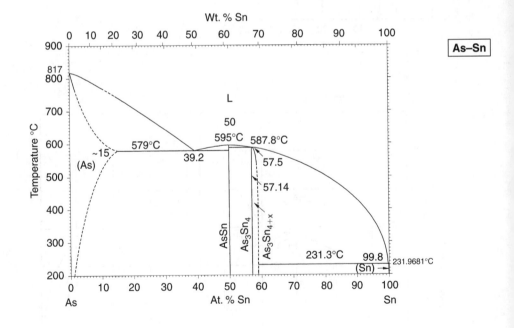

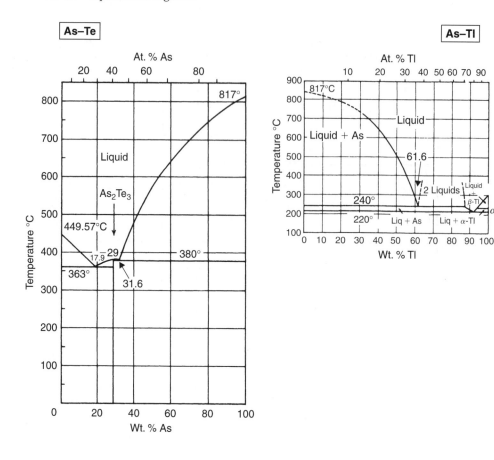

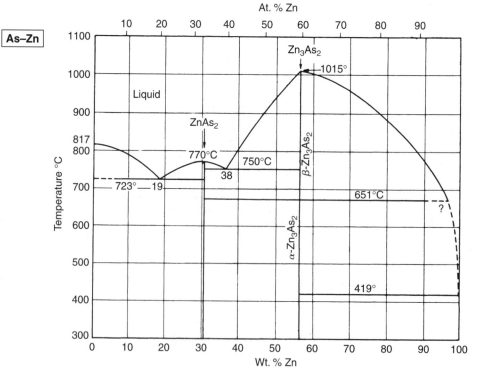

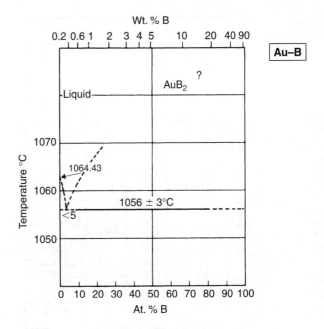

Au–B

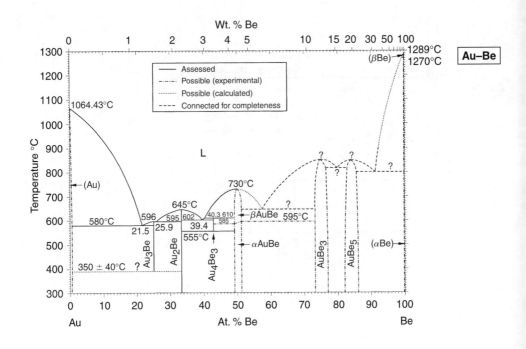

Au–Be

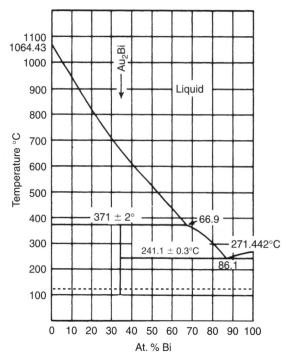

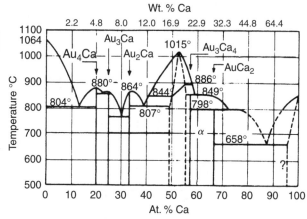

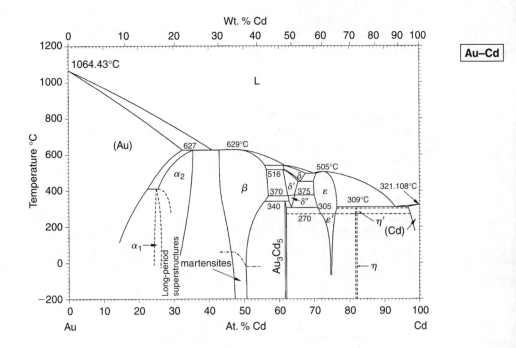

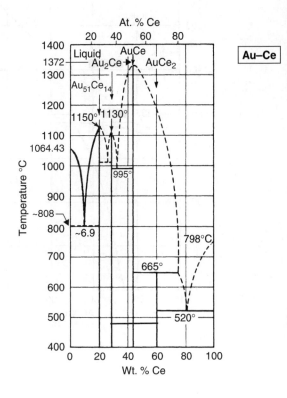

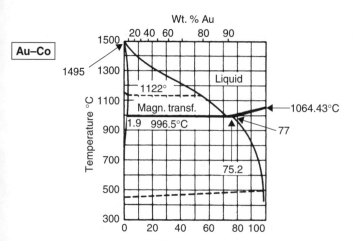

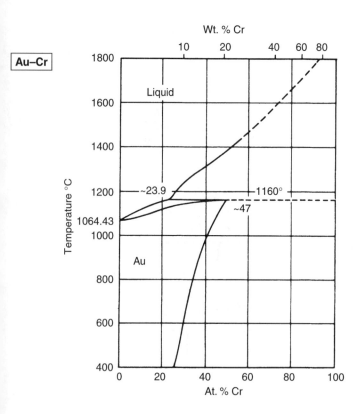

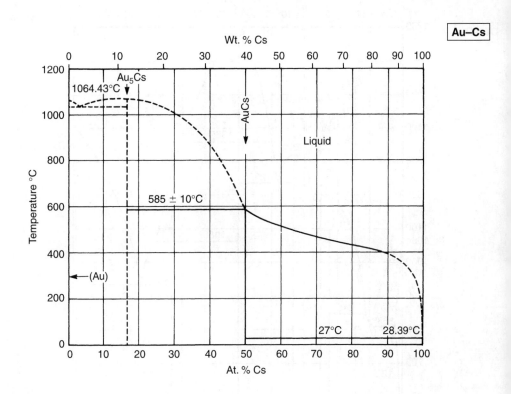

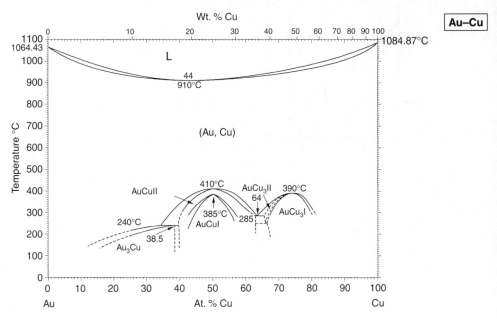

Au–Dy

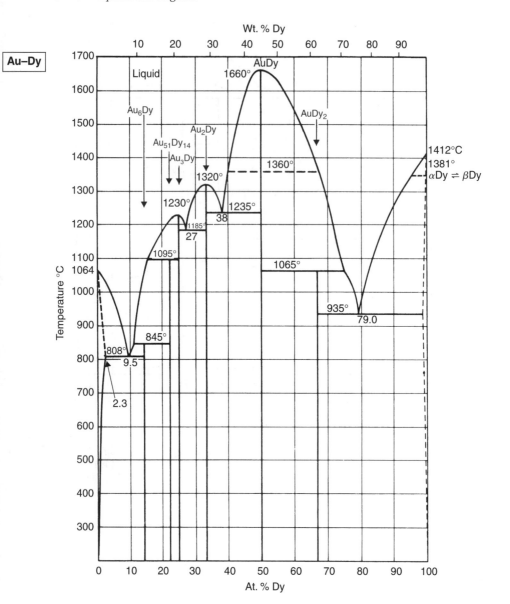

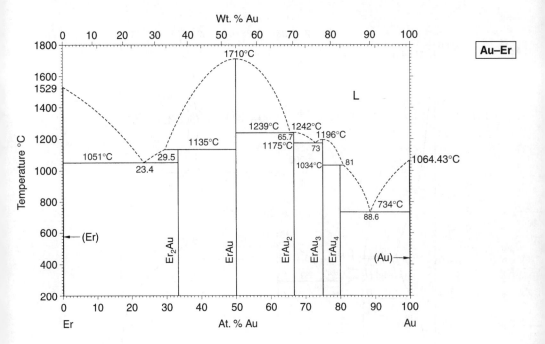

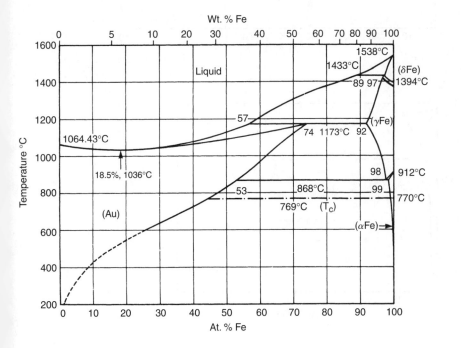

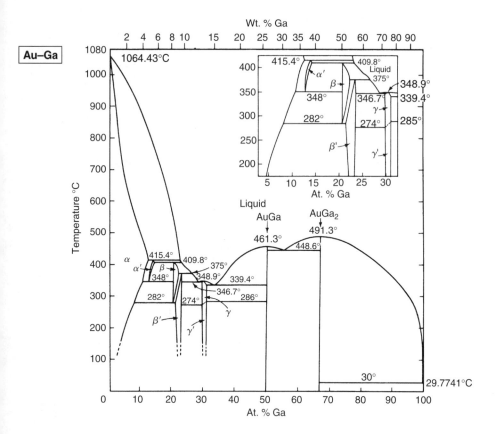

Au–Ga

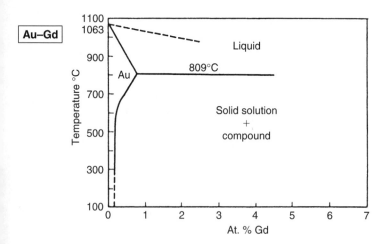

Au–Gd

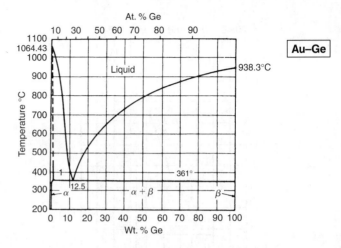

Au–Ge

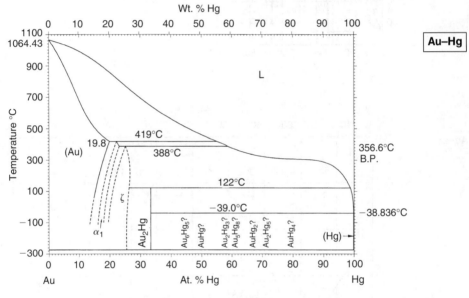

Au–Hg

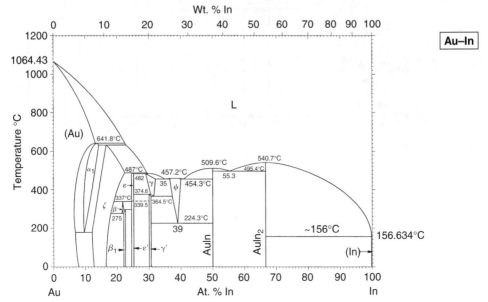

Au–In

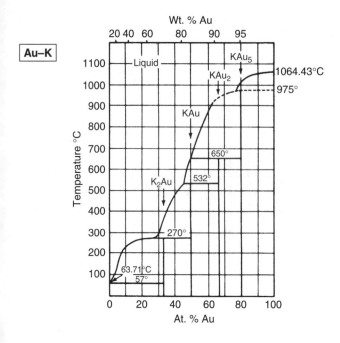

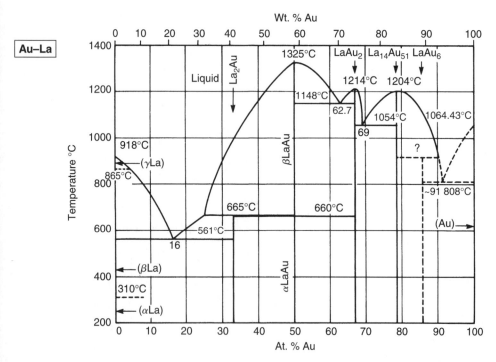

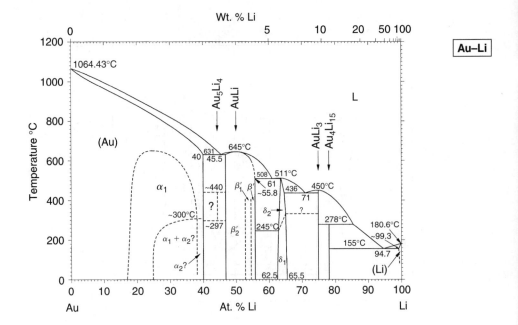

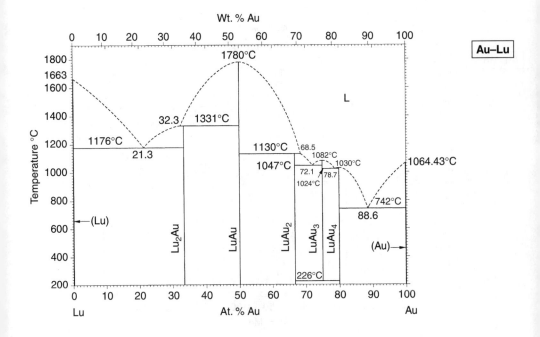

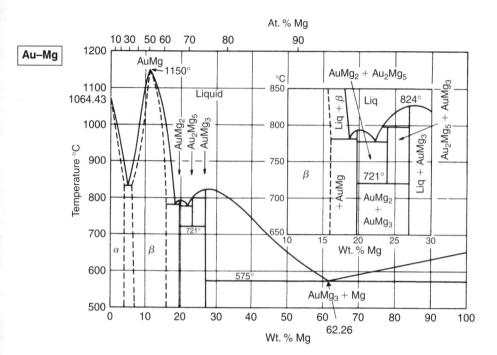

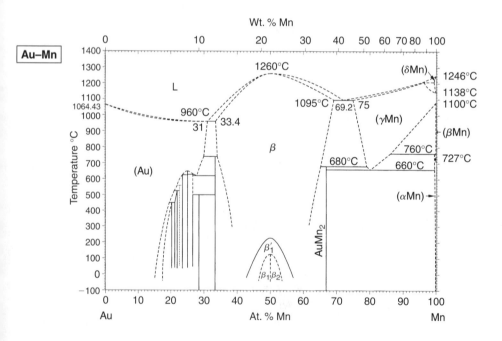

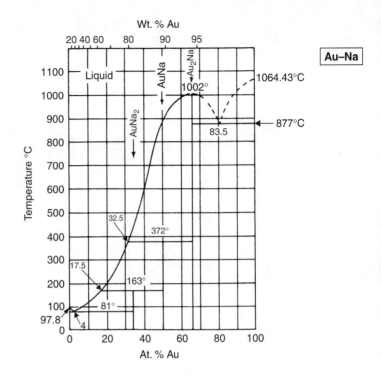

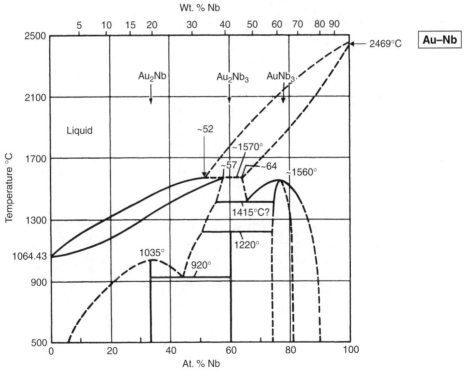

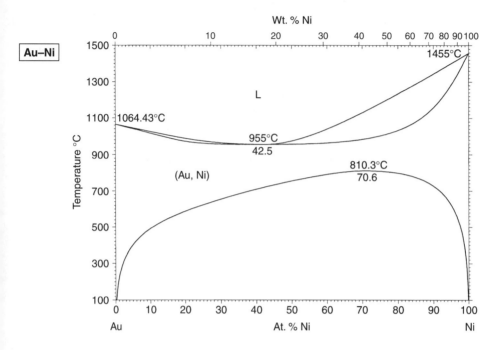

Au–Ni

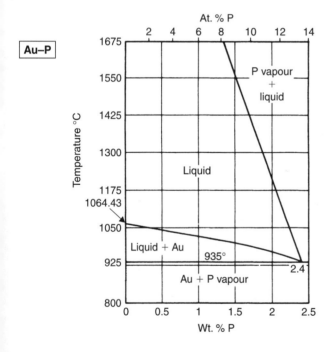

Au–P

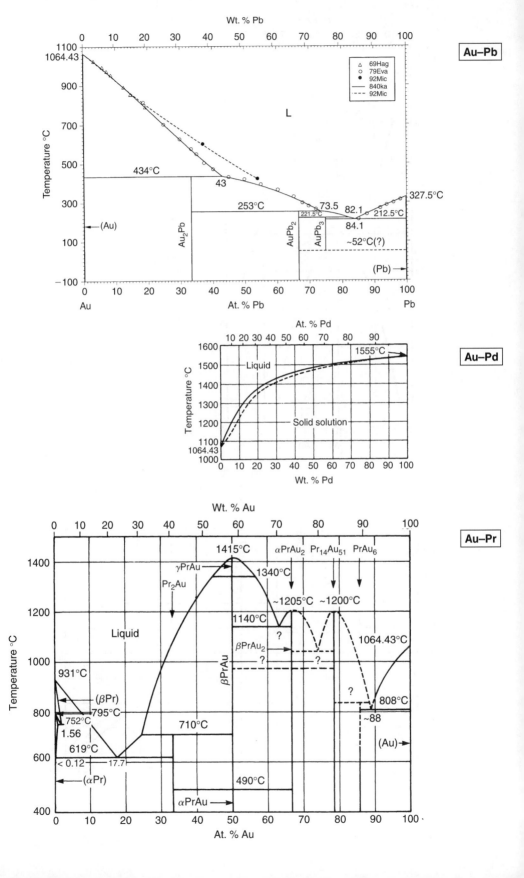

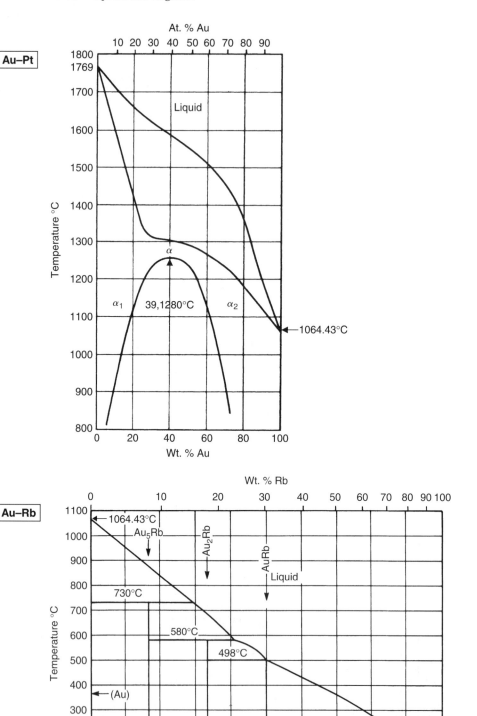

Au–Pt

At. % Au

Liquid

α

α_1 39,1280°C α_2

←1064.43°C

Wt. % Au

Au–Rb

Wt. % Rb

1064.43°C

Au$_5$Rb

Au$_2$Rb

AuRb

Liquid

730°C

580°C

498°C

(Au)

39.48°C

34°C

~98.6

(Rb)

At. % Rb

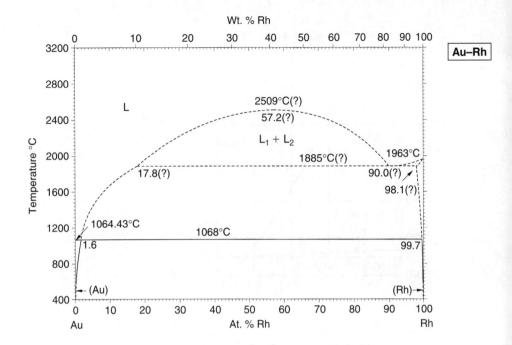

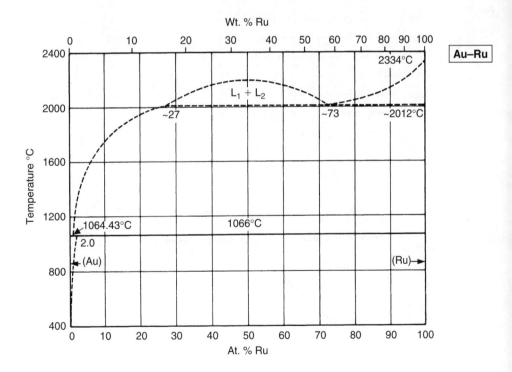

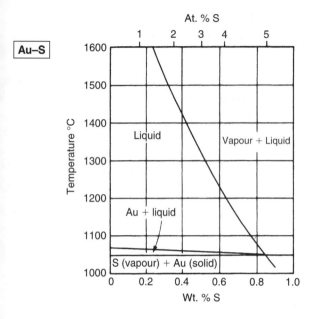

Au–S

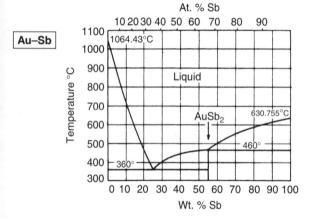

Au–Sb

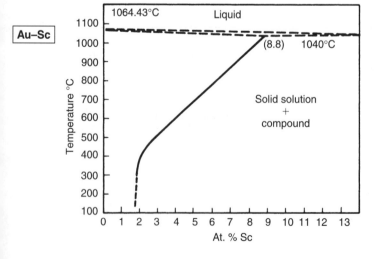

Au–Sc

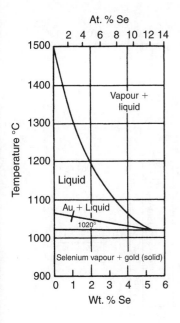

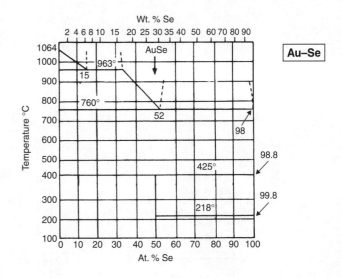

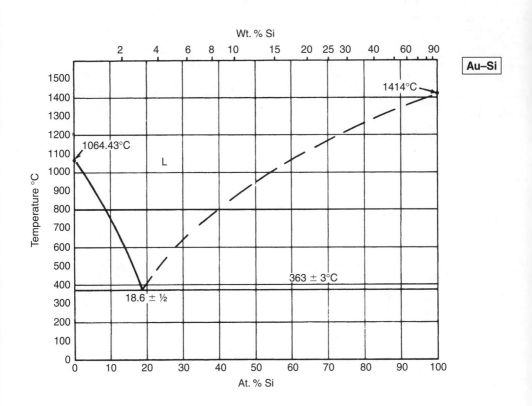

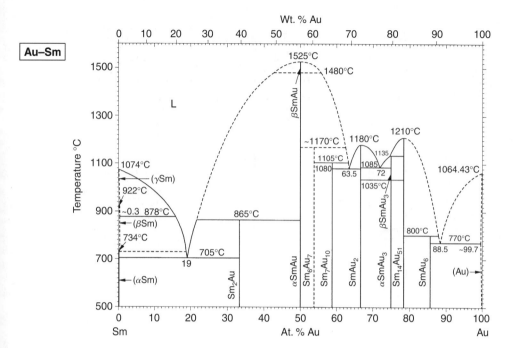

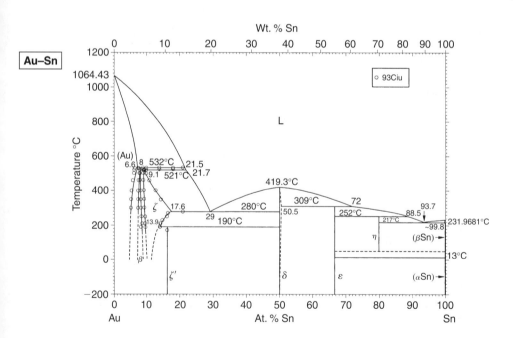

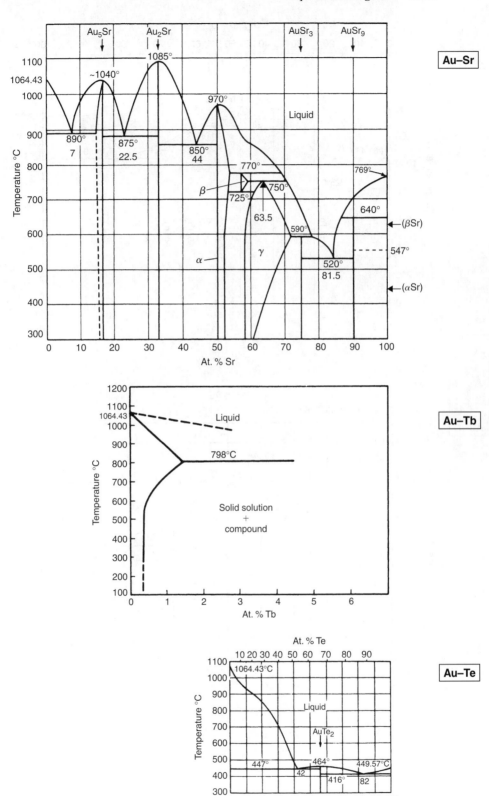

Au–Sr

Au–Tb

Au–Te

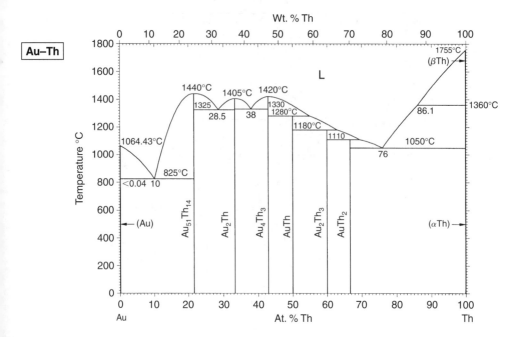

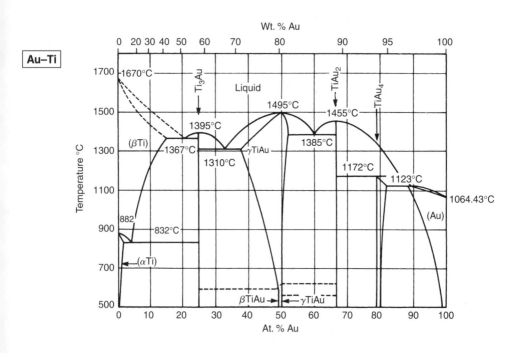

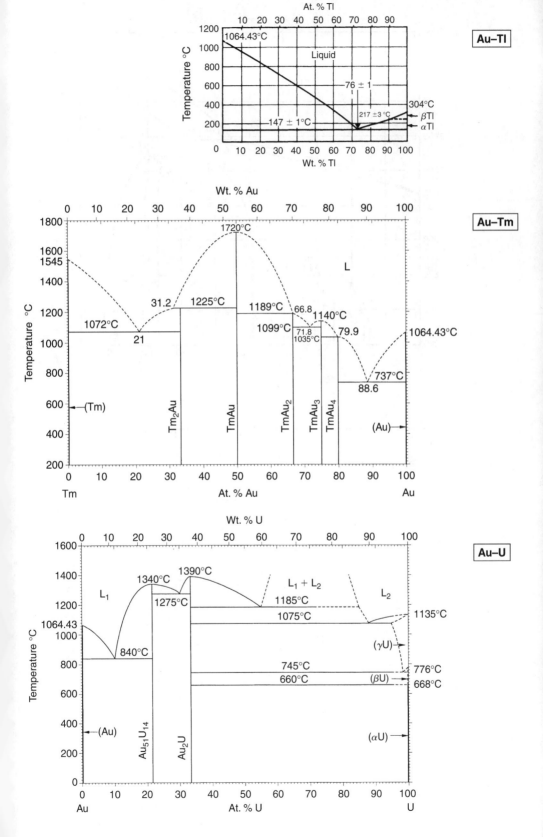

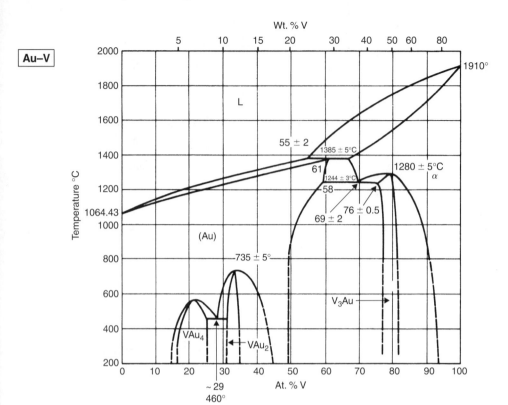

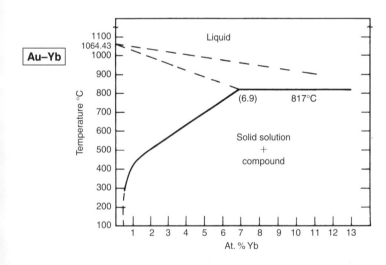

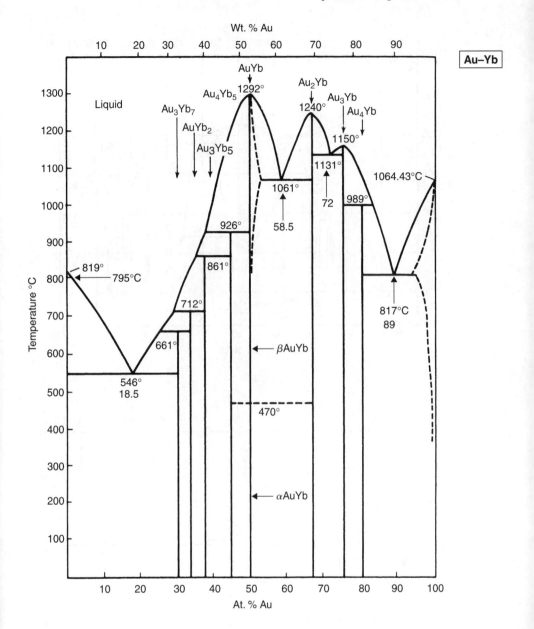

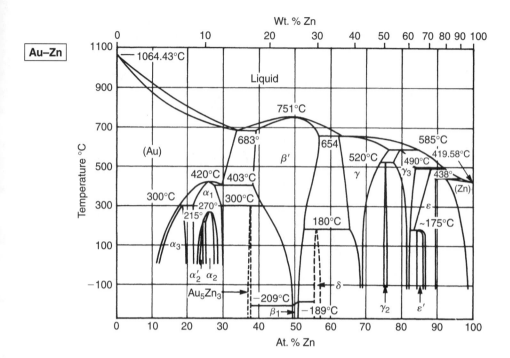

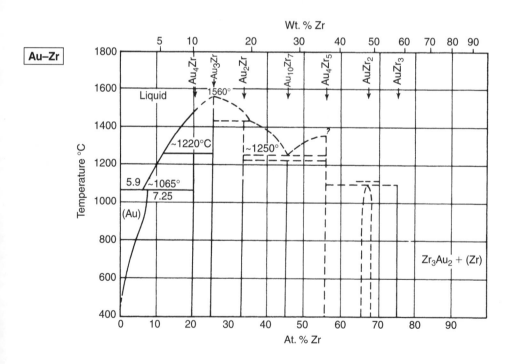

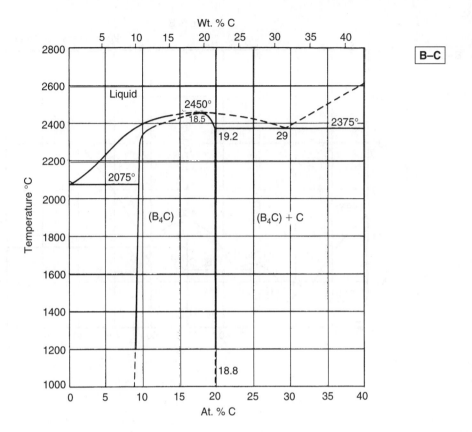

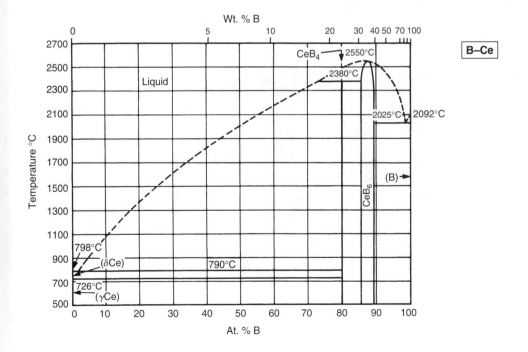

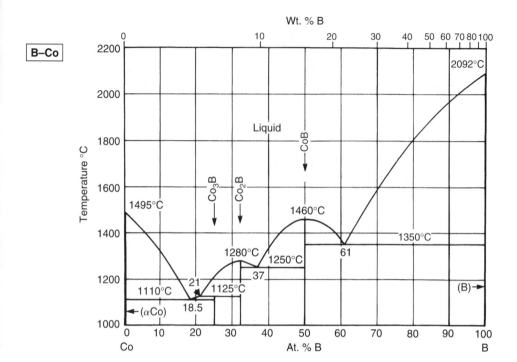

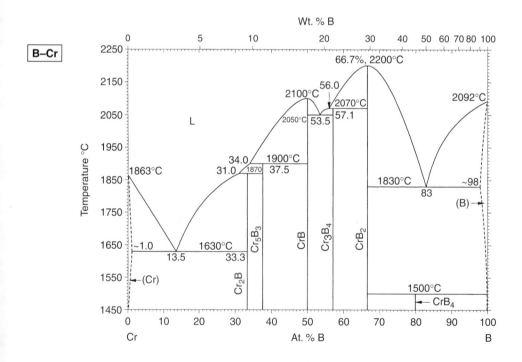

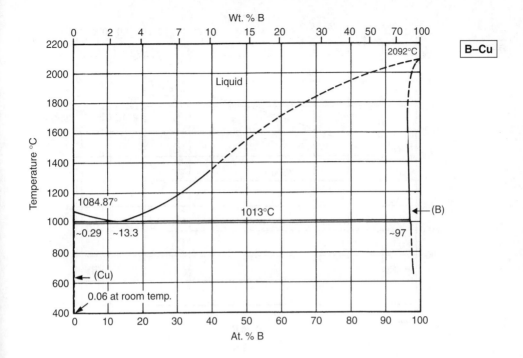

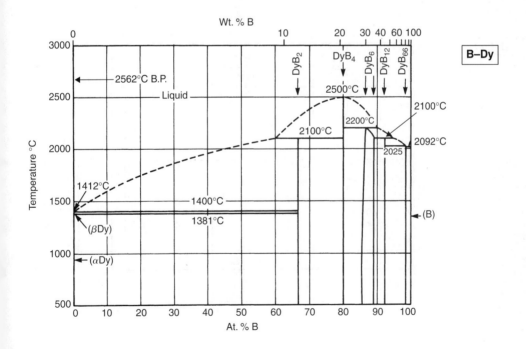

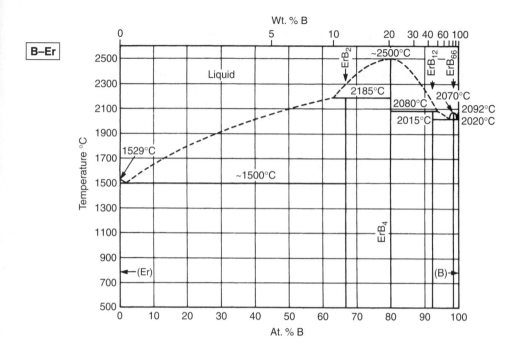

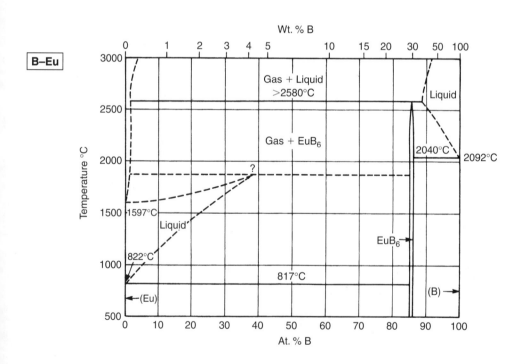

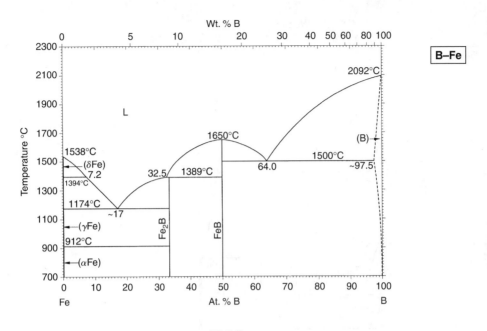

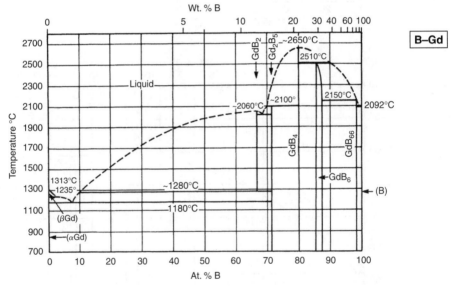

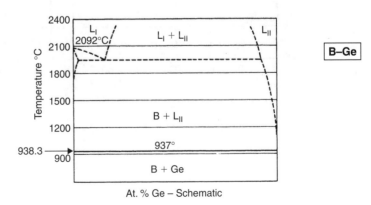

At. % Ge – Schematic

B–Hf

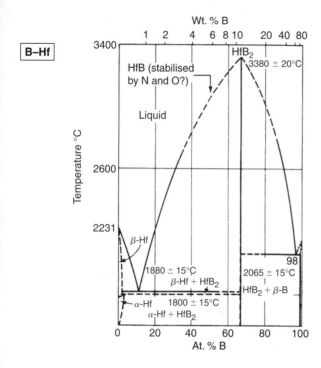

B–Ho

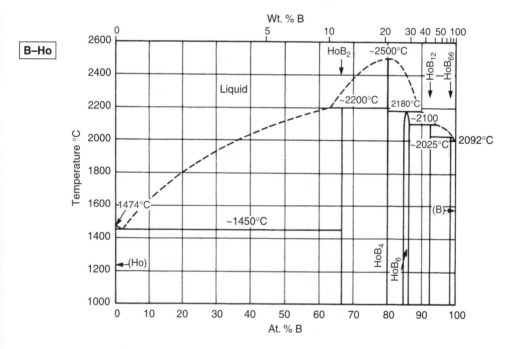

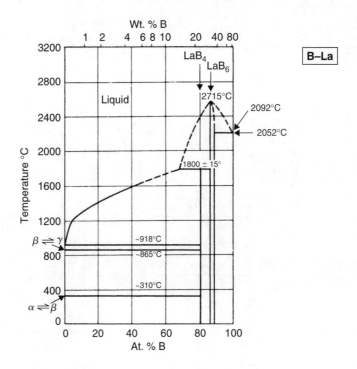

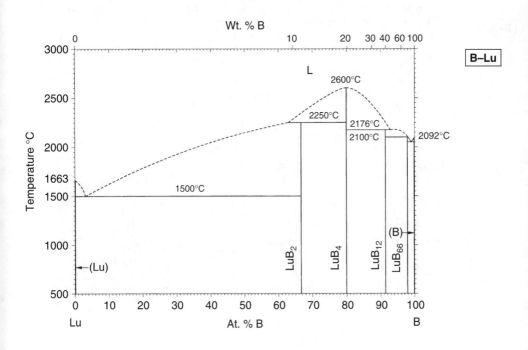

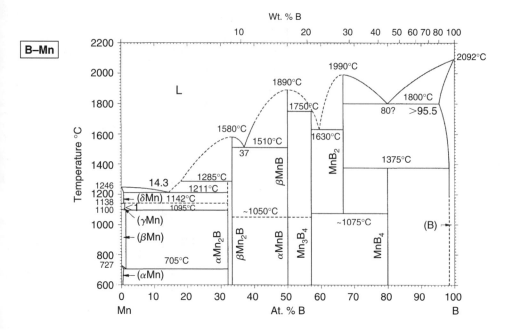

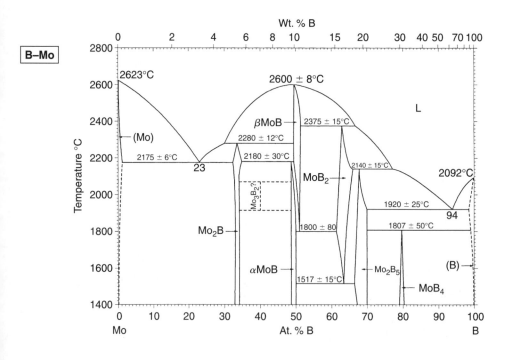

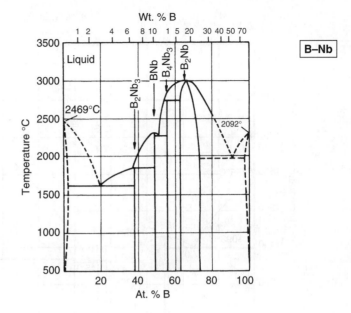

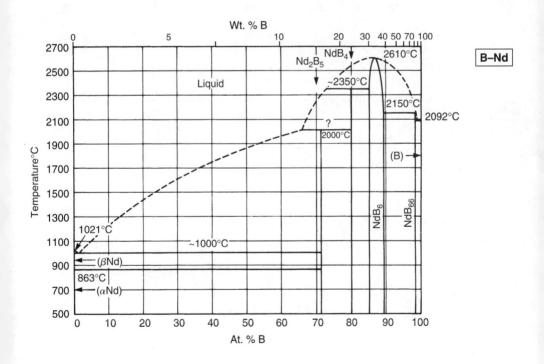

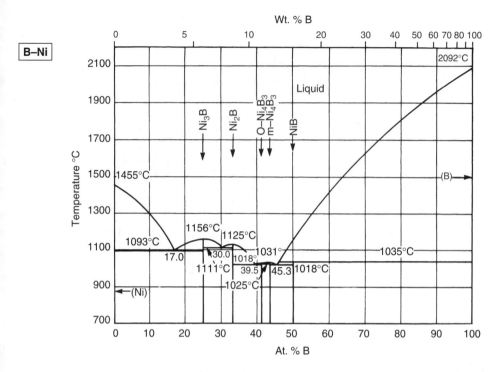

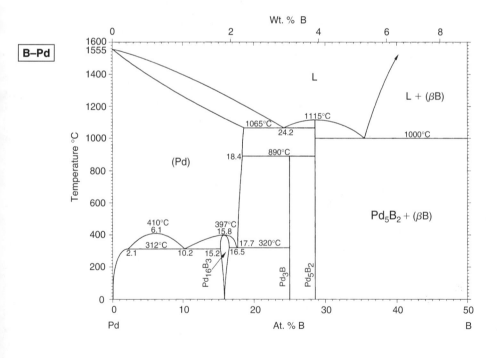

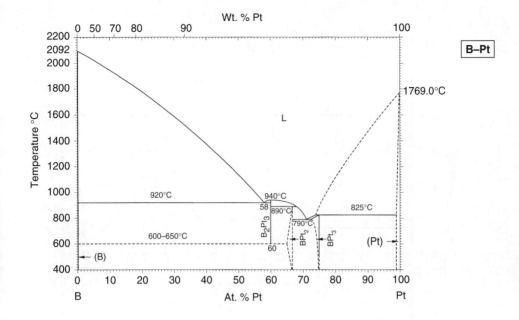

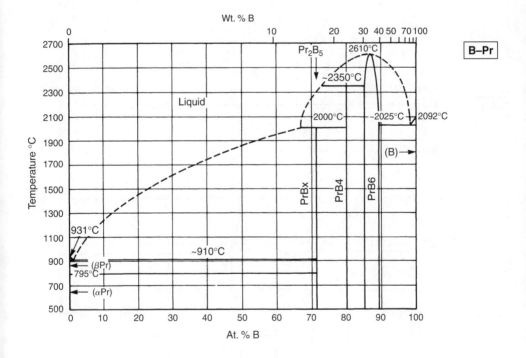

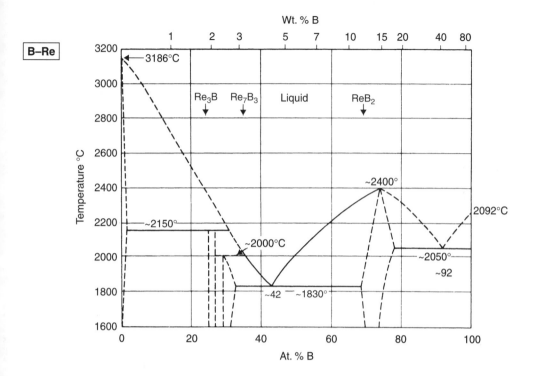

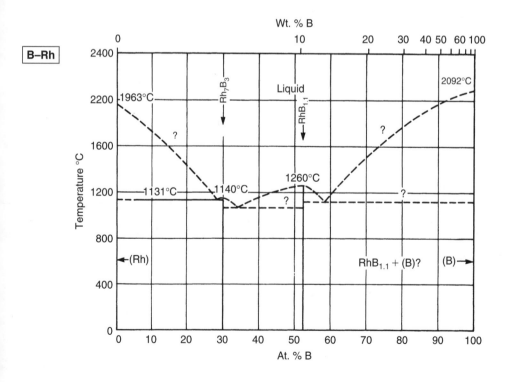

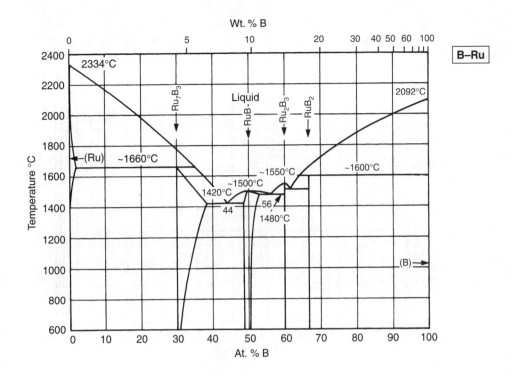

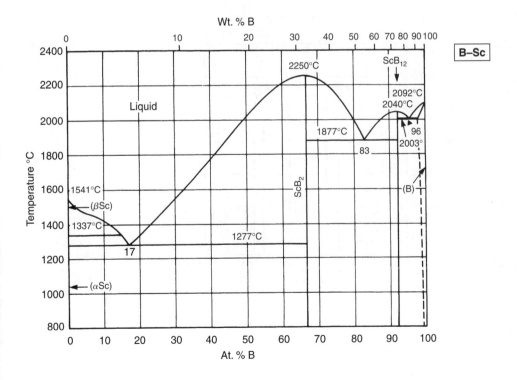

B–Si

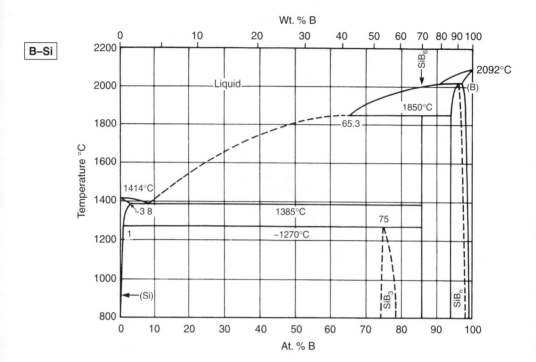

B–Sm

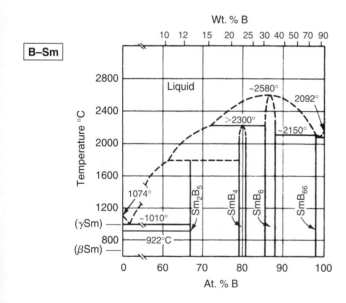

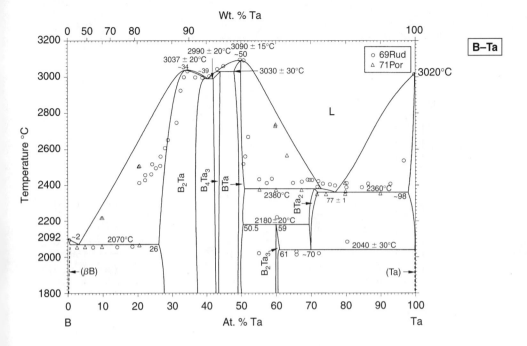

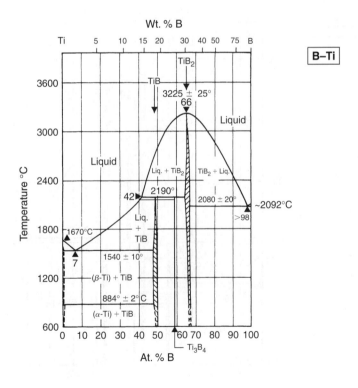

B–U

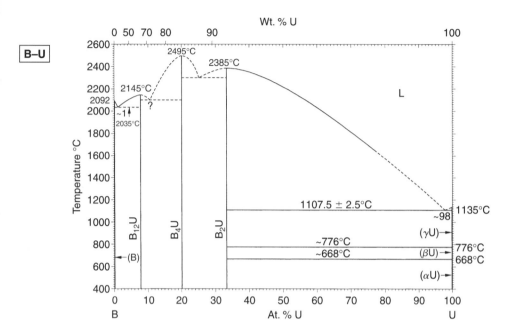

B–V

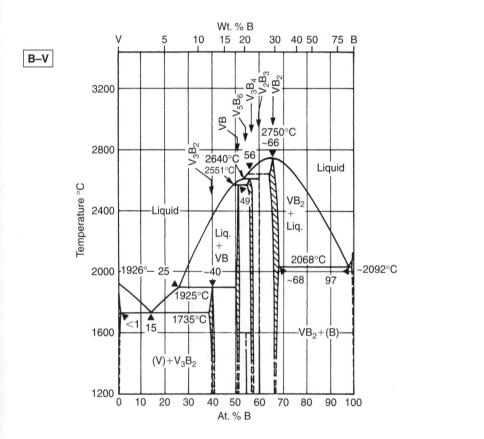

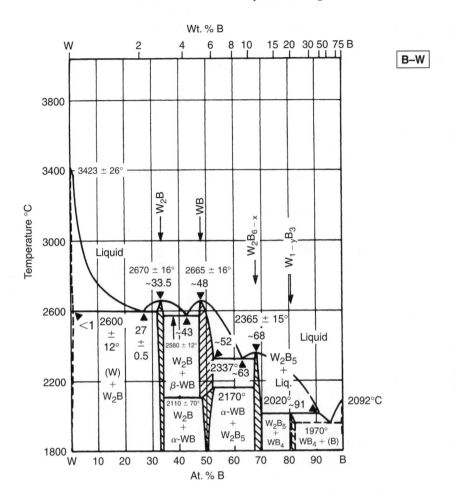

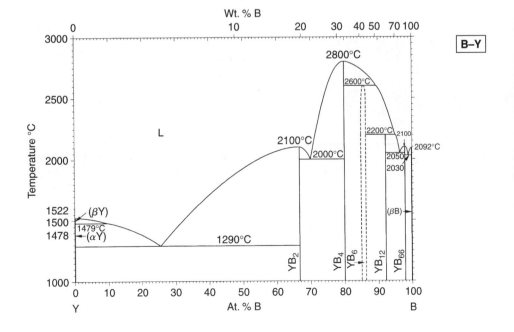

B–Zr

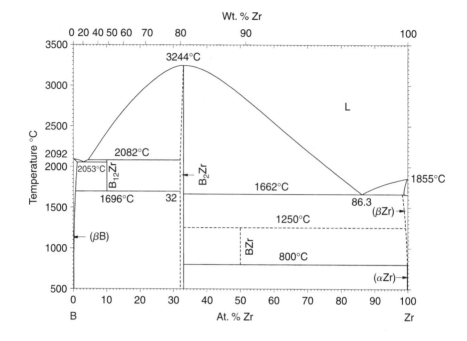

Ba–Ca

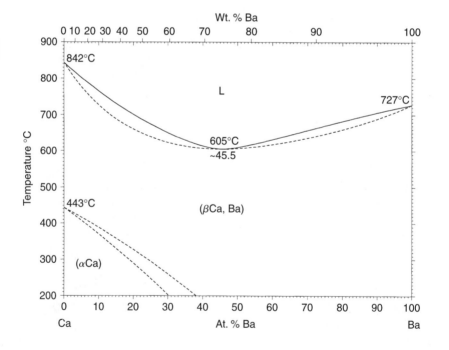

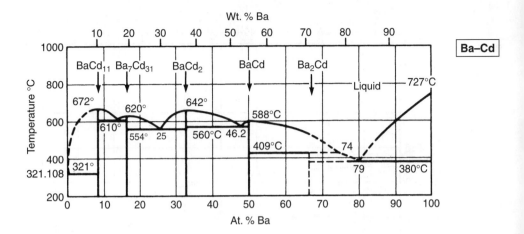

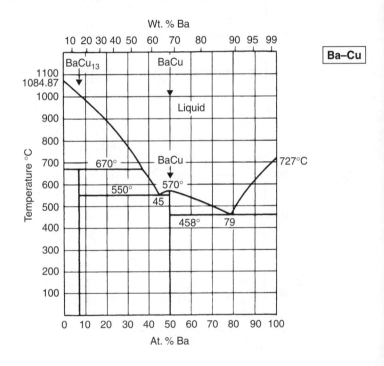

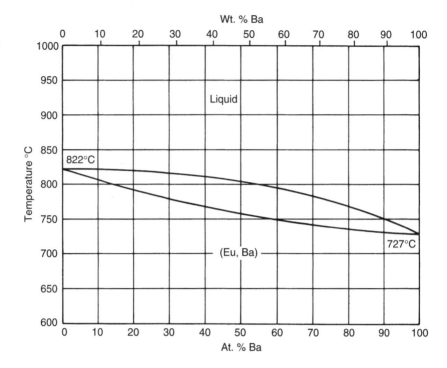

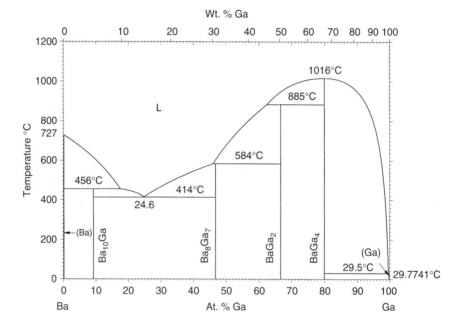

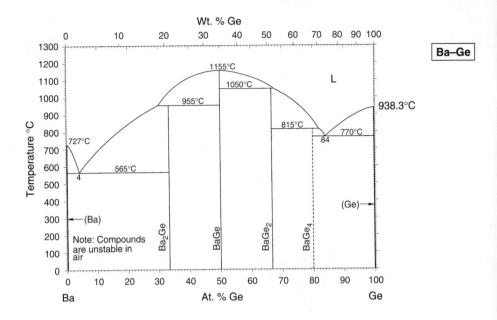

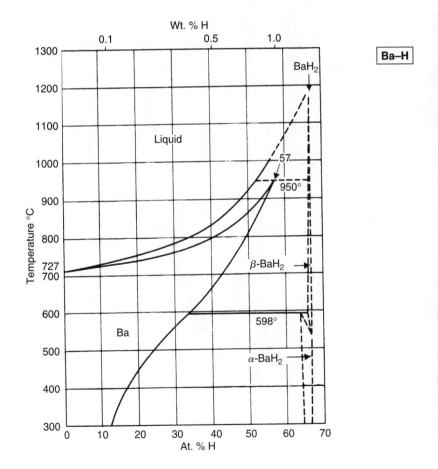

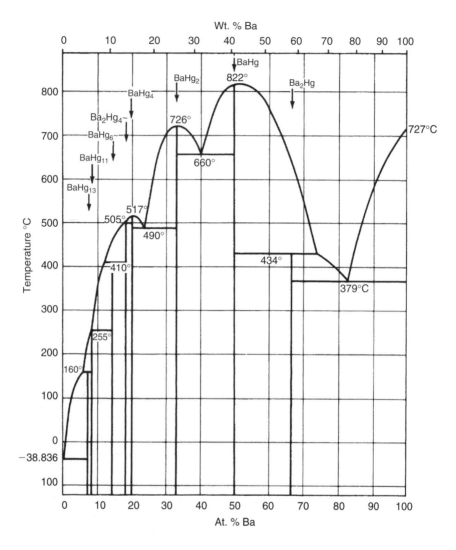

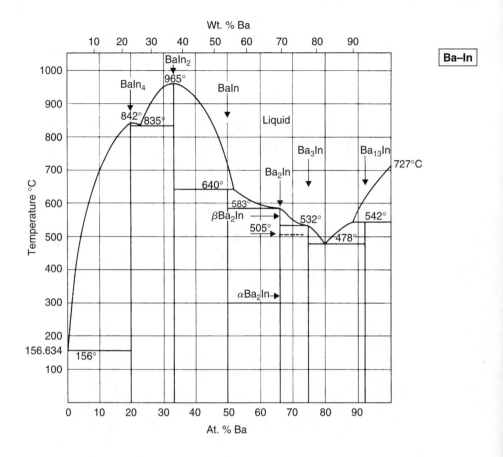

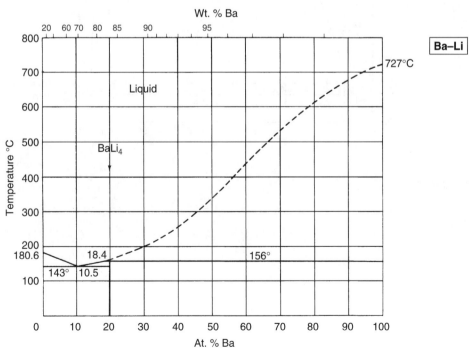

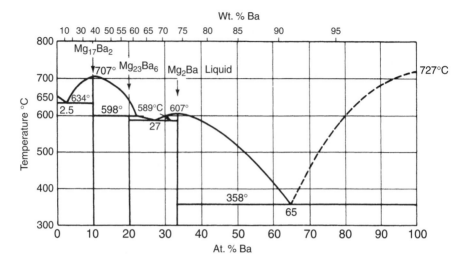

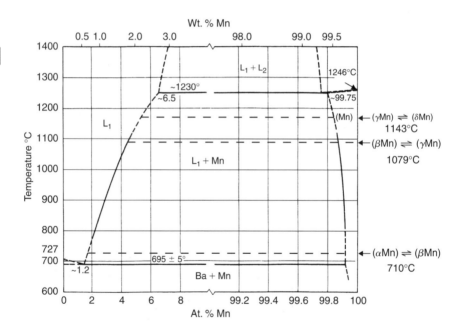

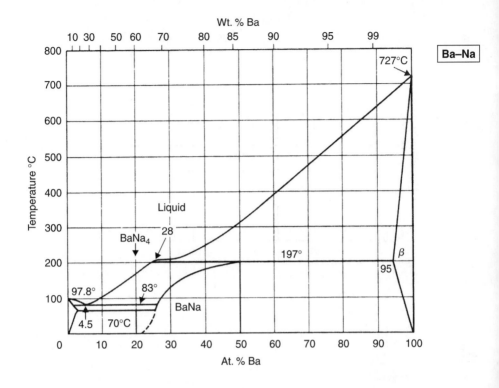

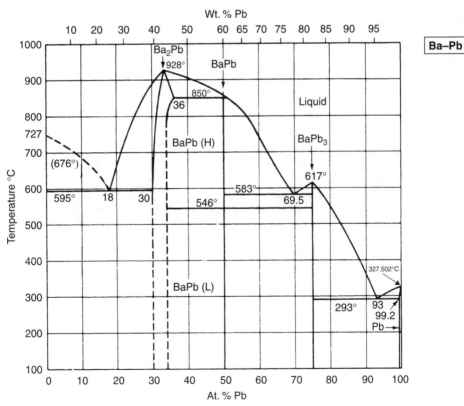

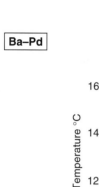

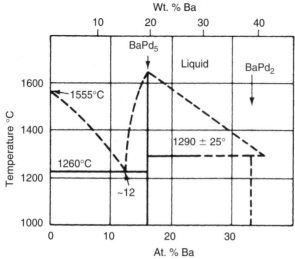

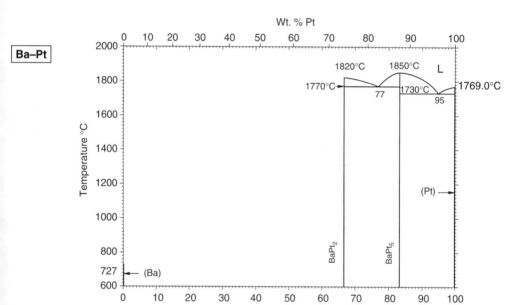

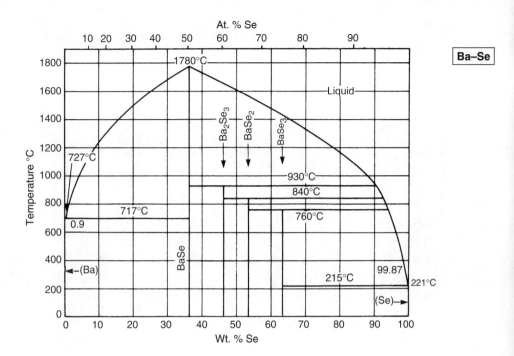

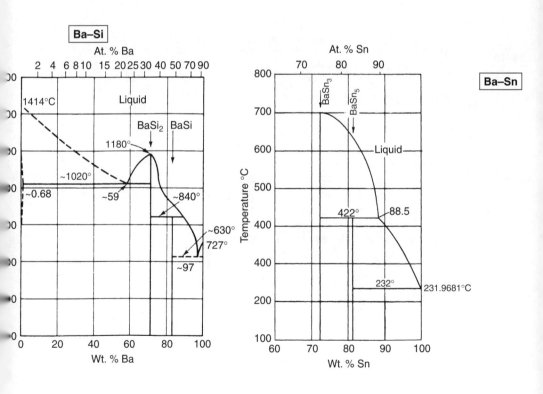

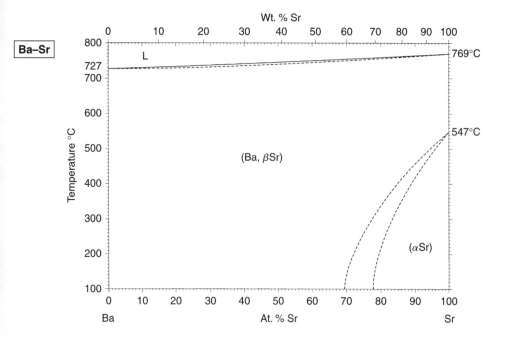

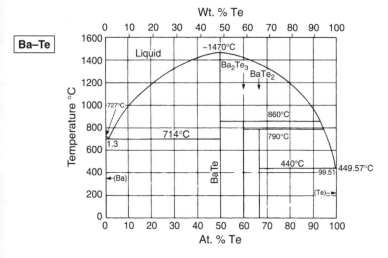

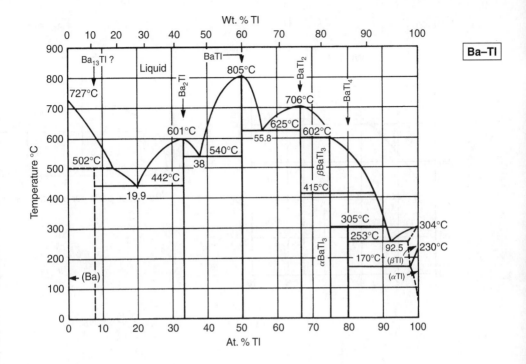

Ba–Tl

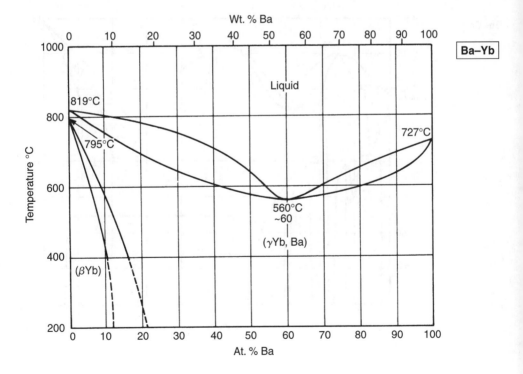

Ba–Yb

Ba–Zn

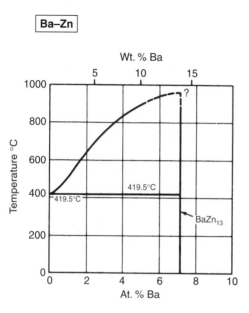

Be–Bi

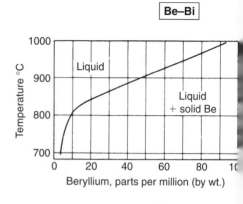

Be–Co

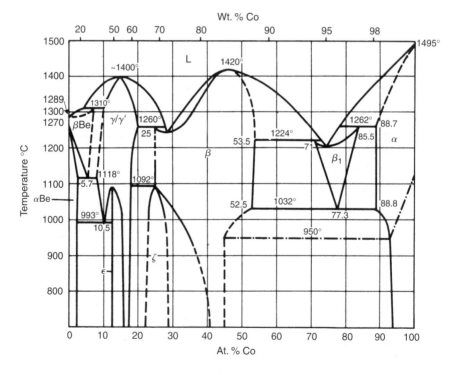

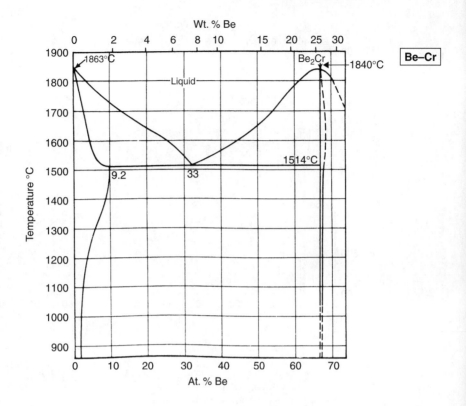

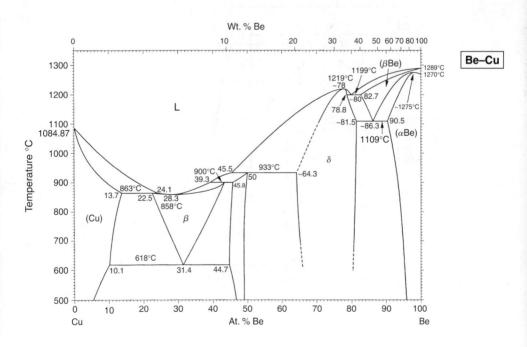

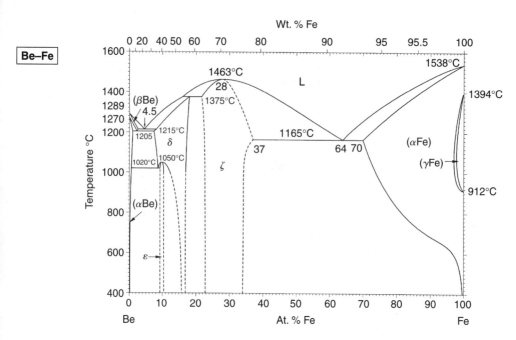

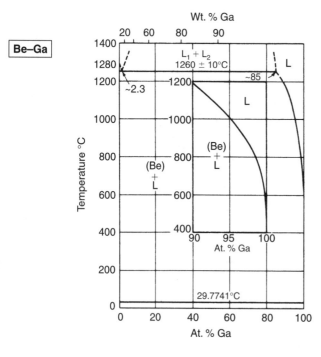

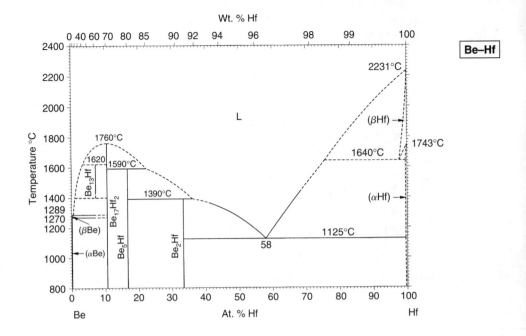

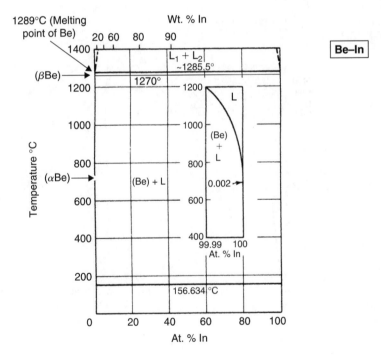

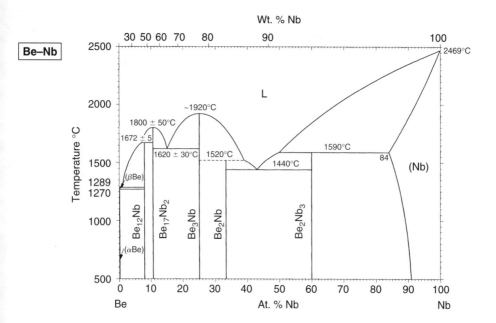

Be–Nb

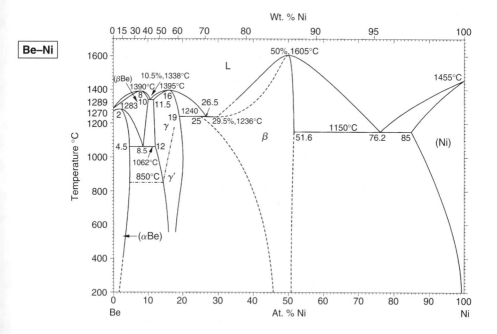

Be–Ni

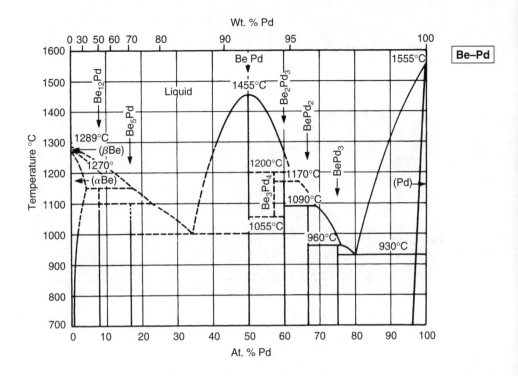

Be–Pd

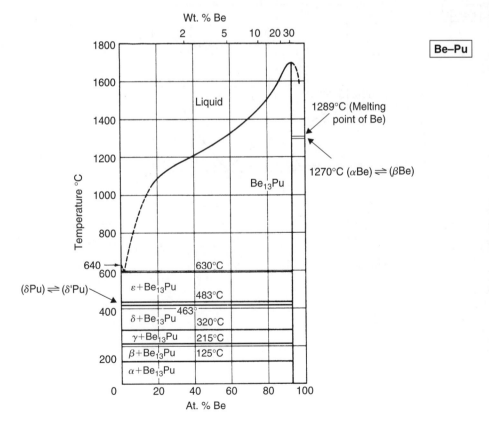

Be–Pu

Be–Ru

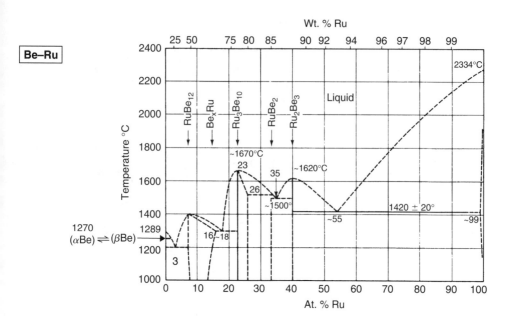

Be–Si

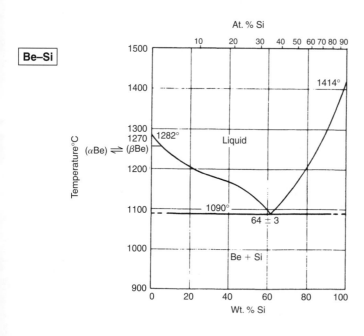

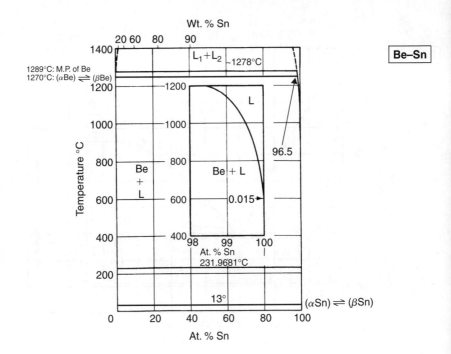

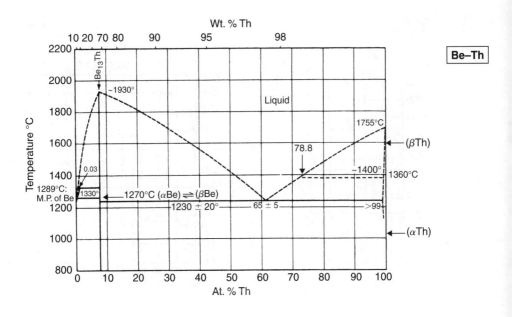

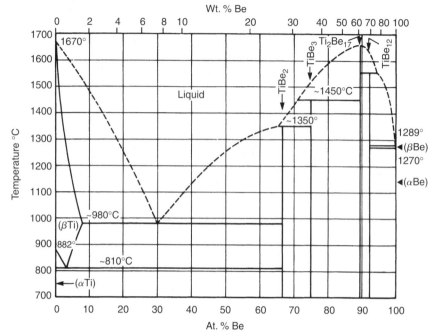

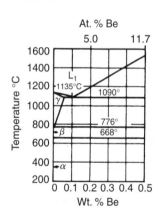

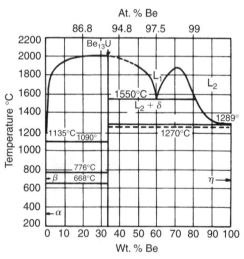

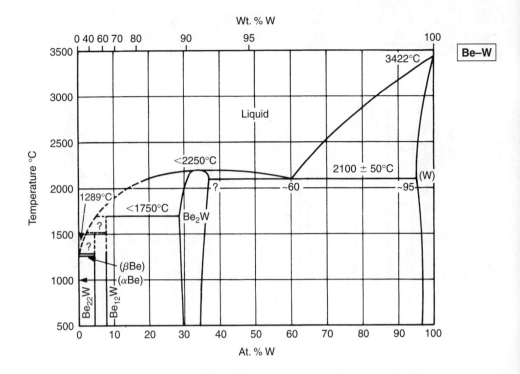

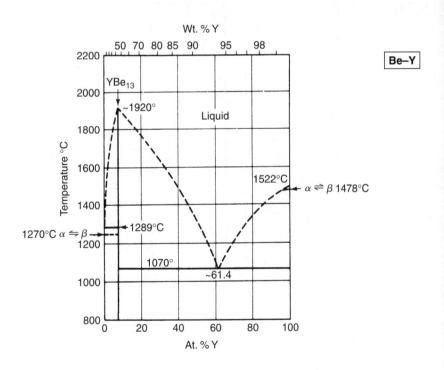

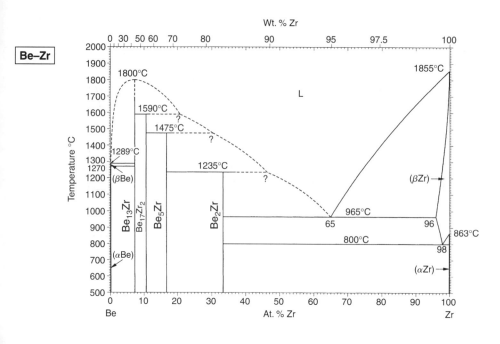

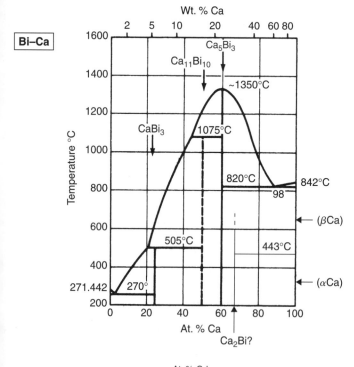

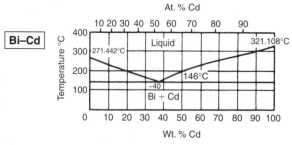

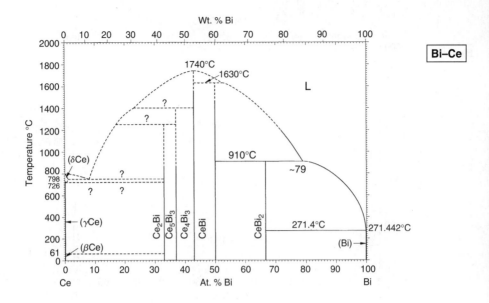

Bi–Ce

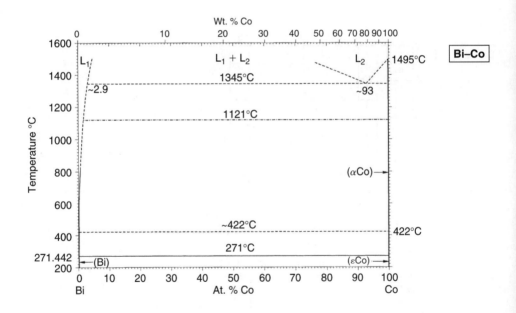

Bi–Co

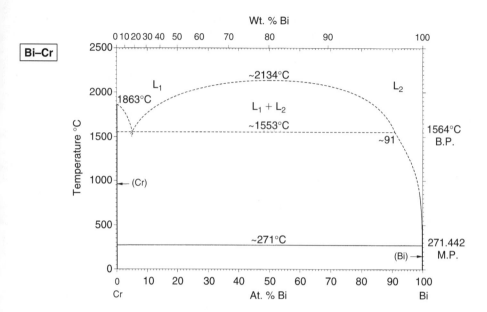

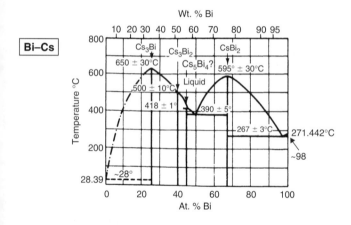

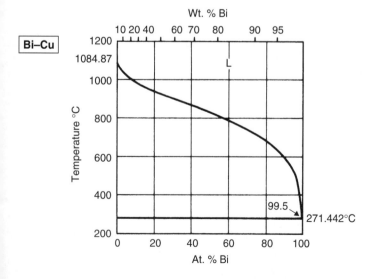

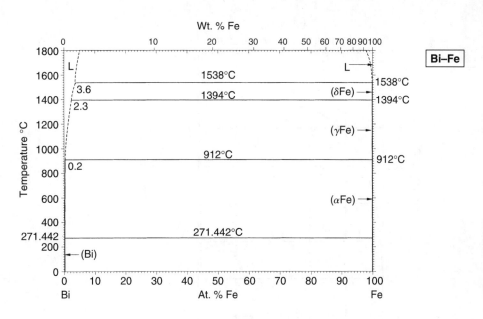

Bi–Fe

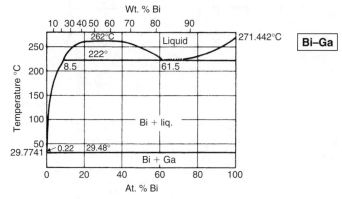

Bi–Ga

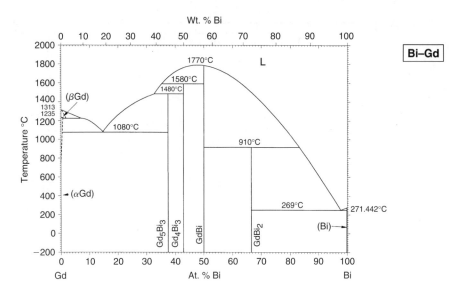

Bi–Gd

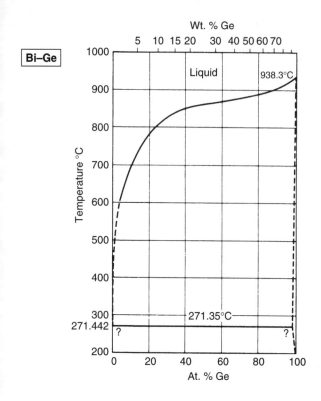

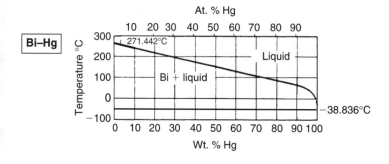

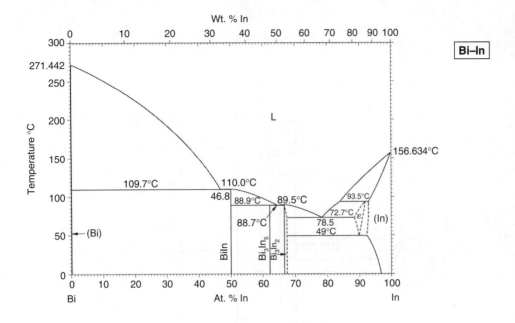

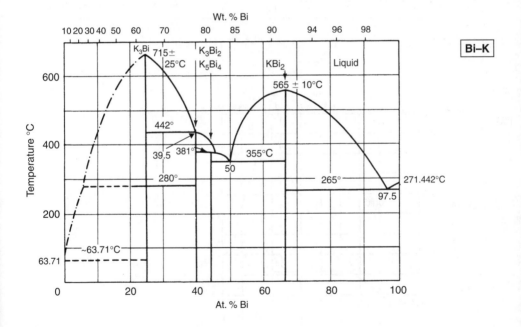

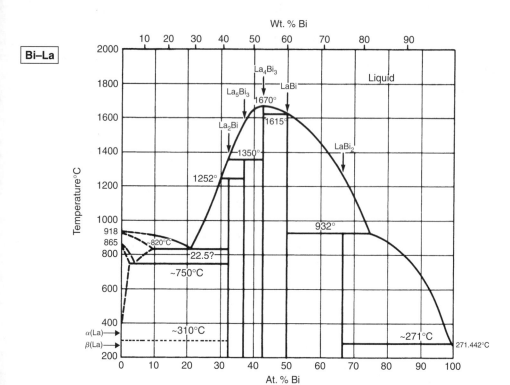

Bi–La

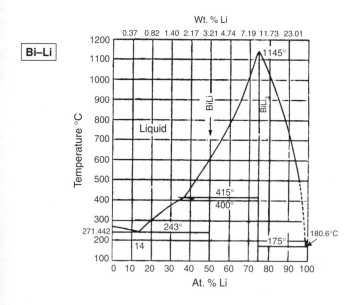

Bi–Li

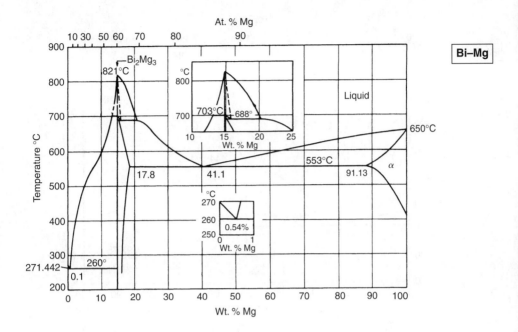

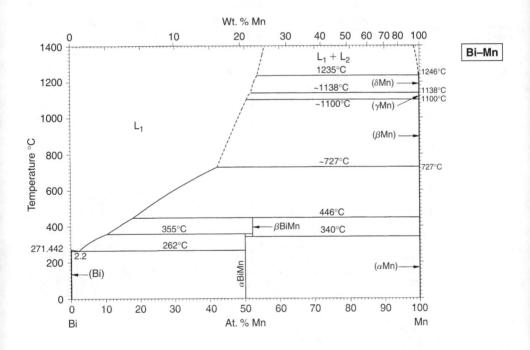

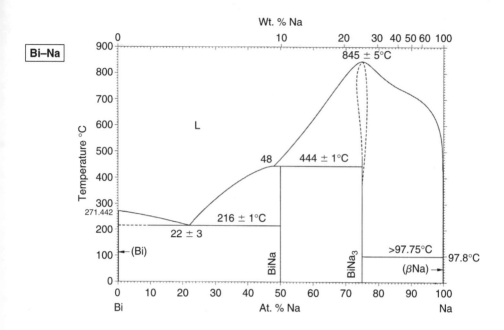

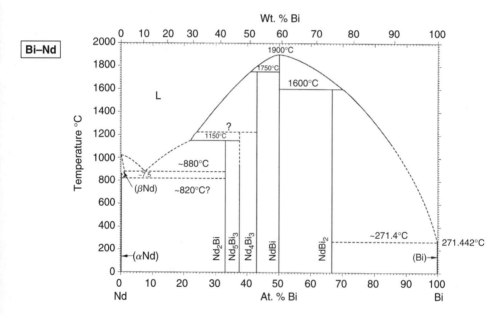

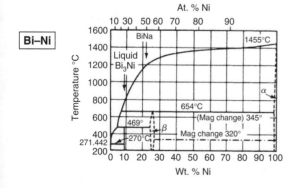

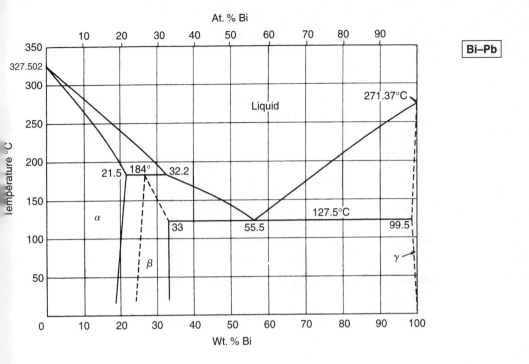

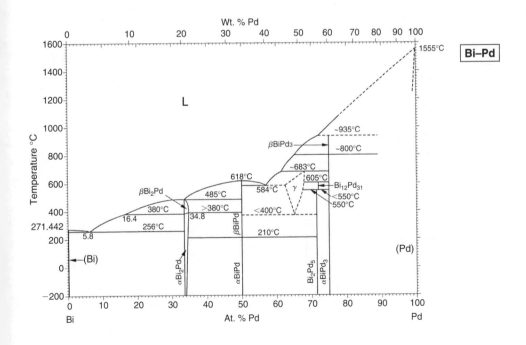

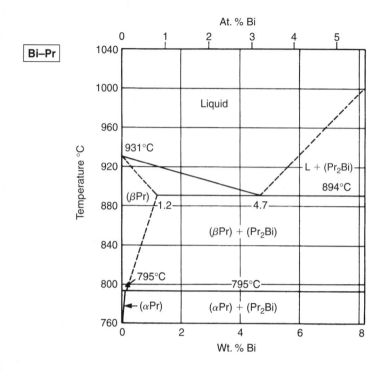

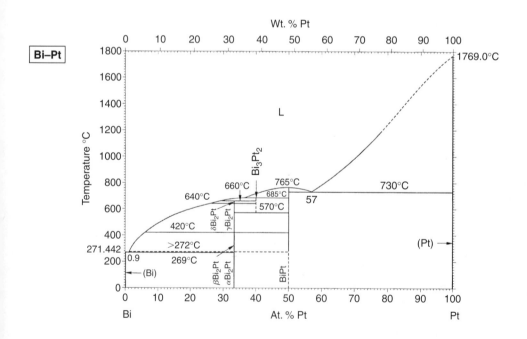

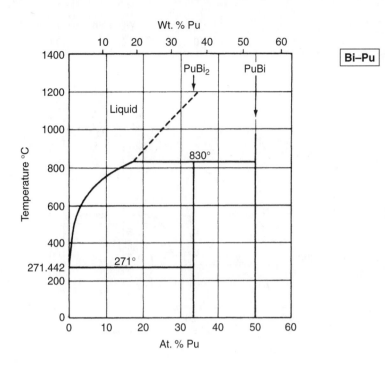

Bi–Pu

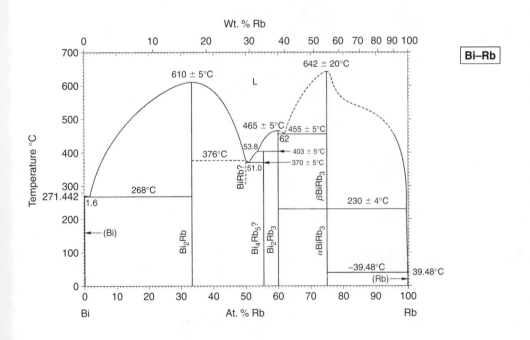

Bi–Rb

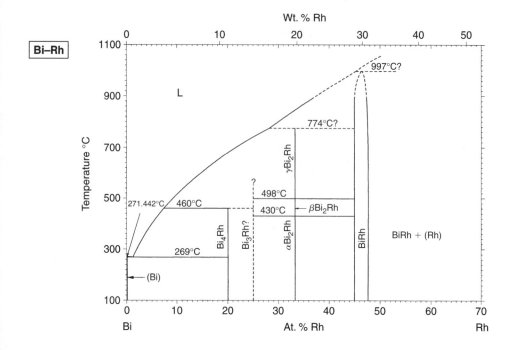

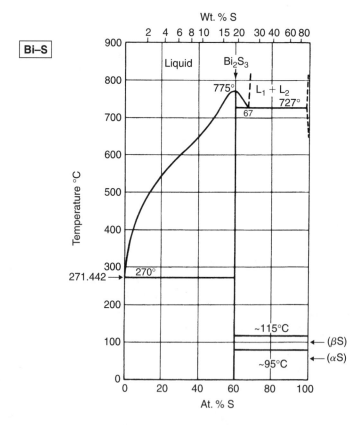

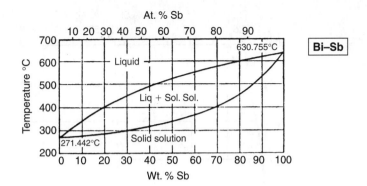

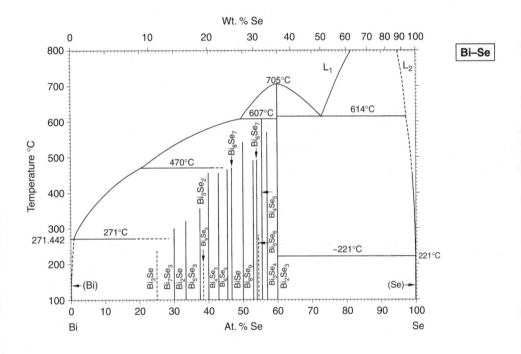

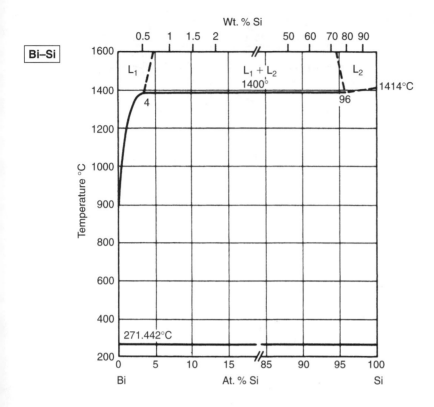

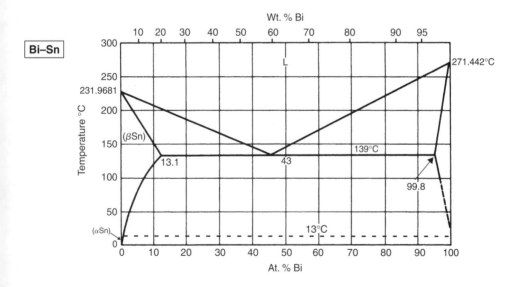

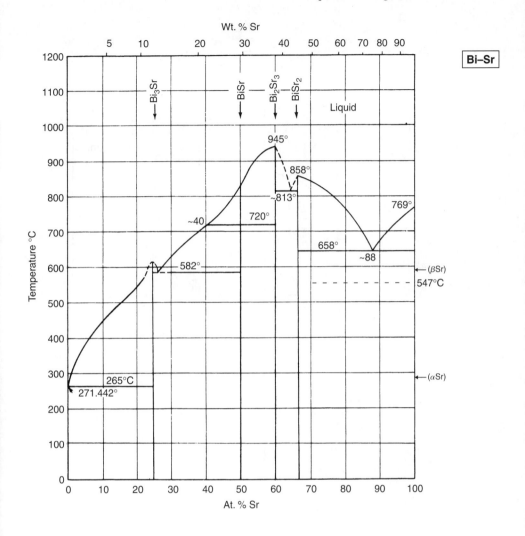

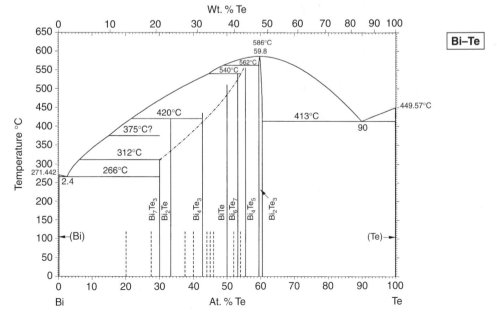

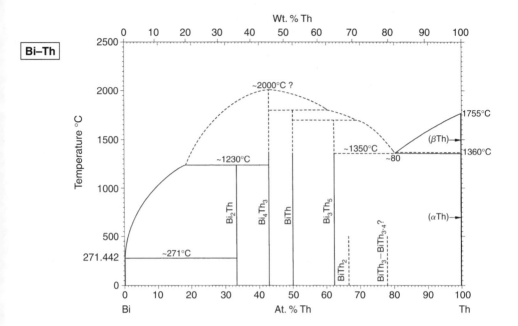

Bi–Th

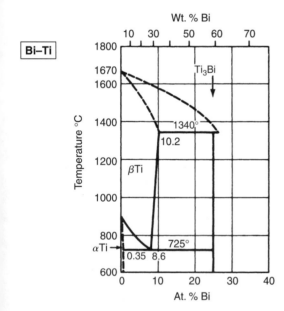

Bi–Ti

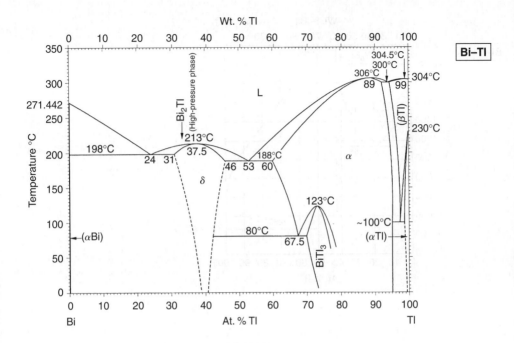

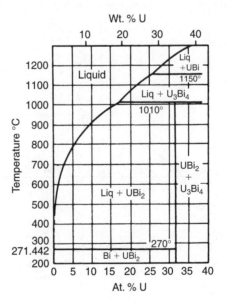

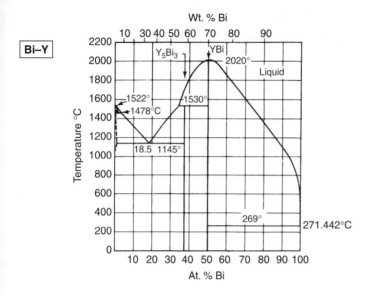

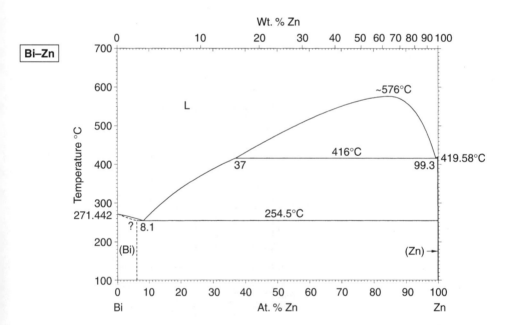

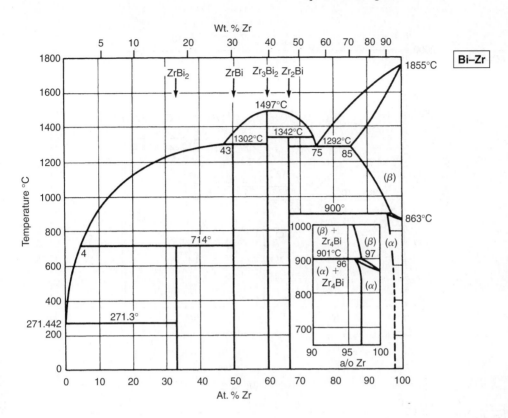

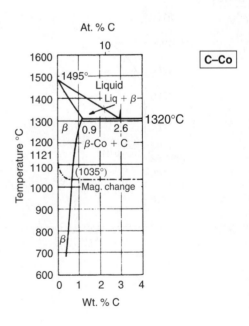

C–Cr

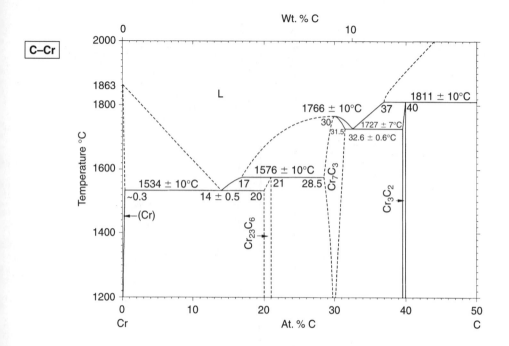

C–Cu

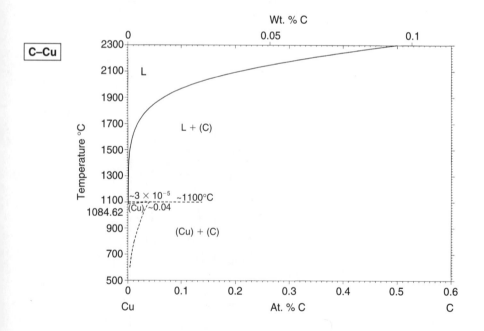

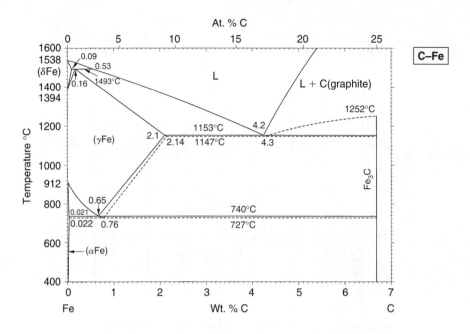

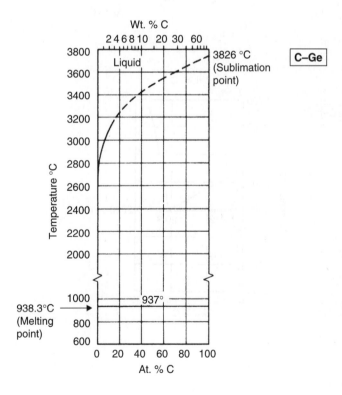

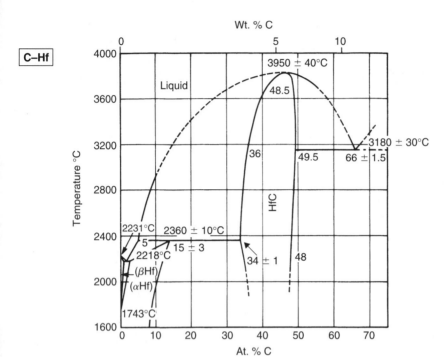

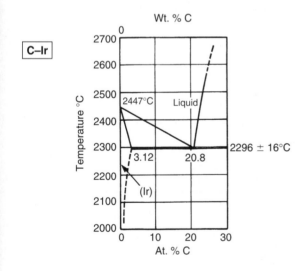

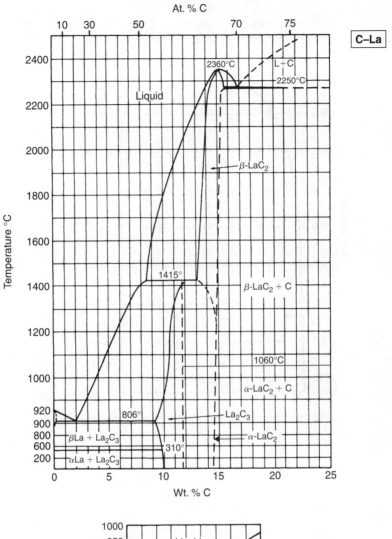

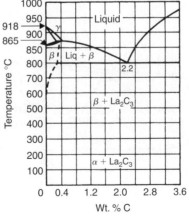

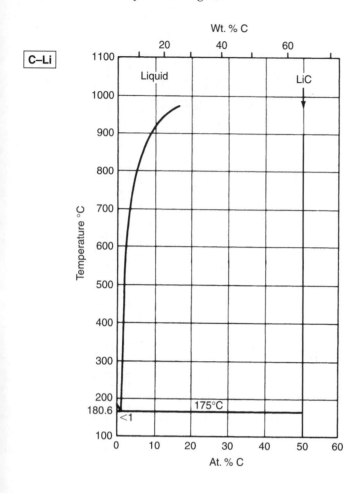

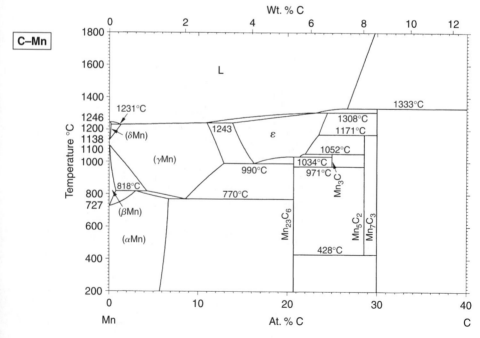

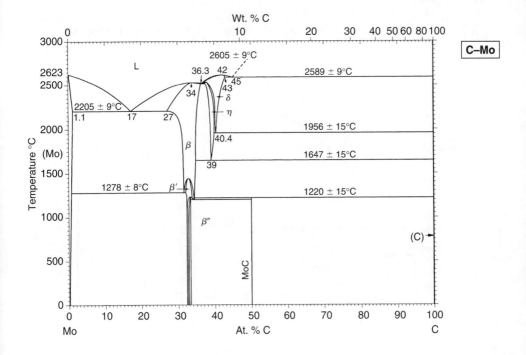

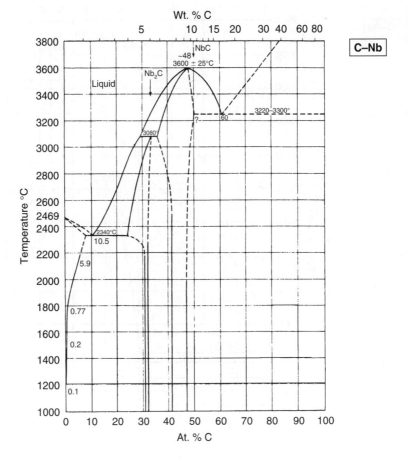

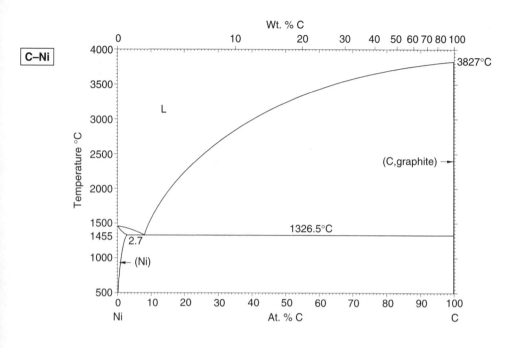

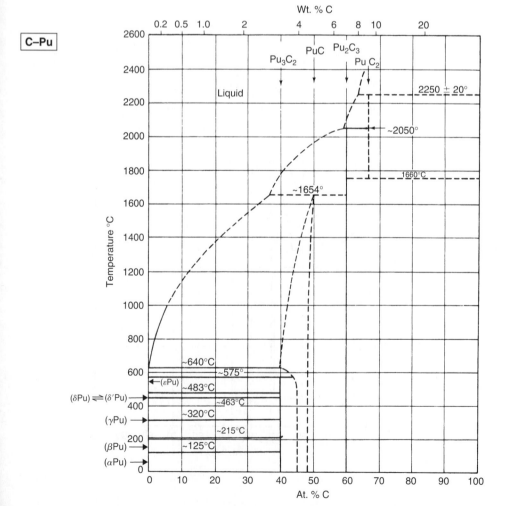

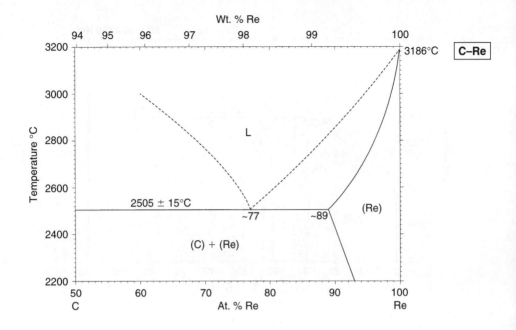

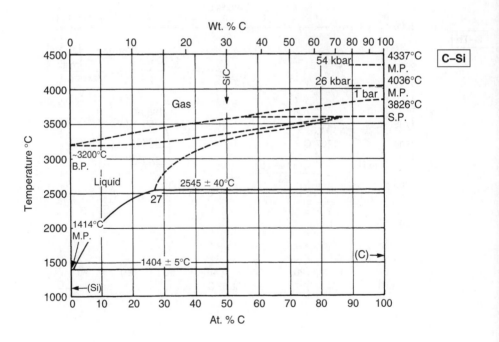

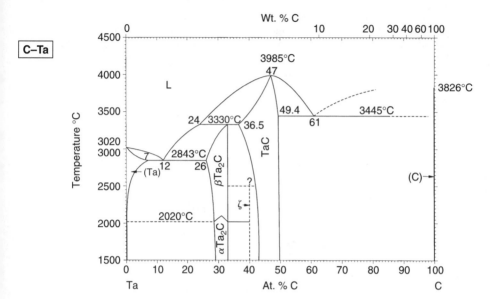

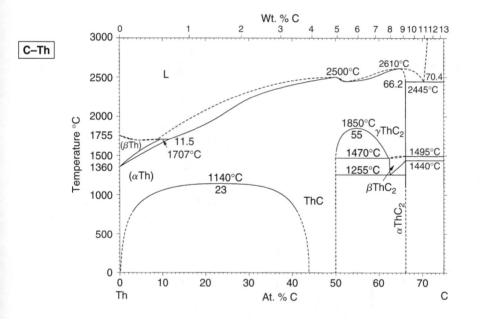

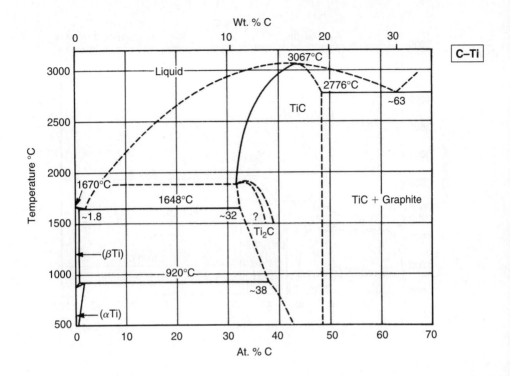

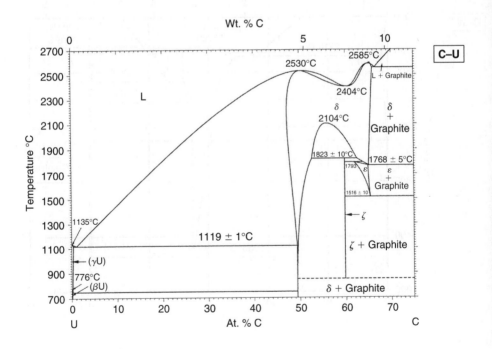

C–V

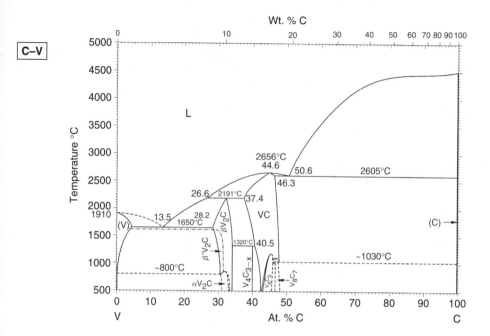

C–W

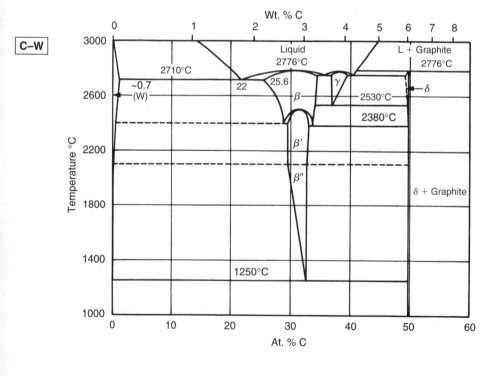

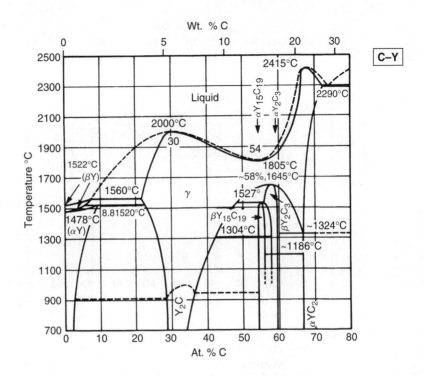

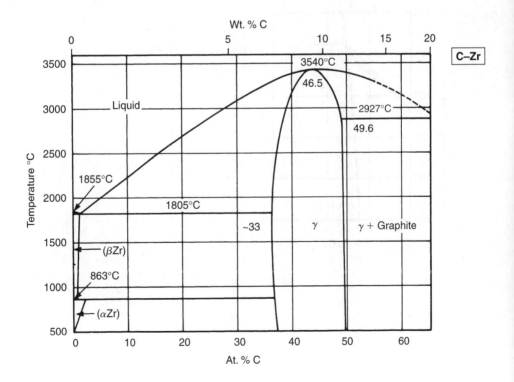

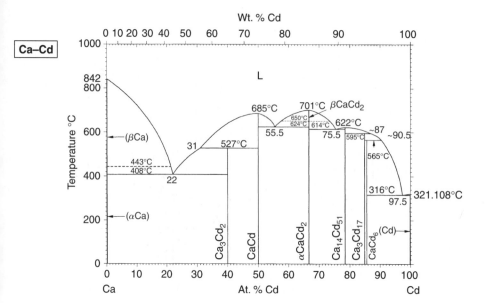

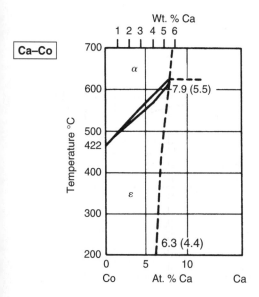

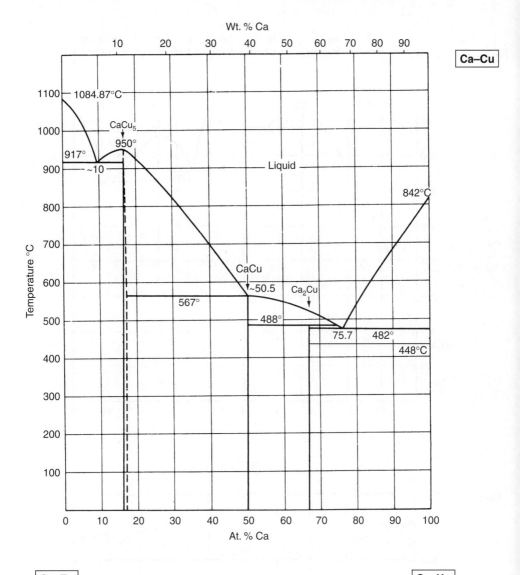

Ca–Cu

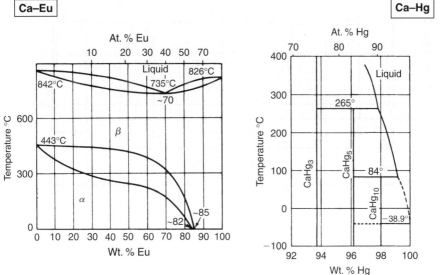

Ca–Eu

Ca–Hg

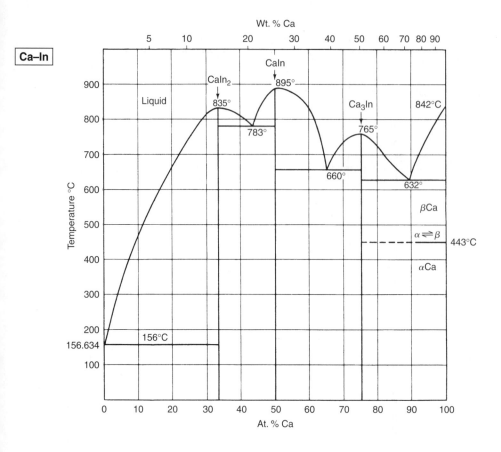

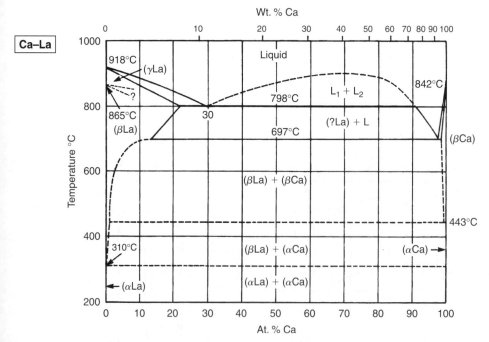

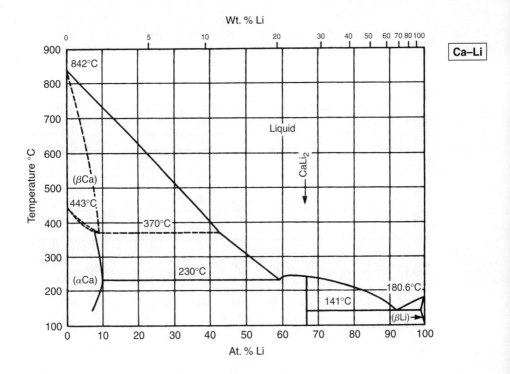

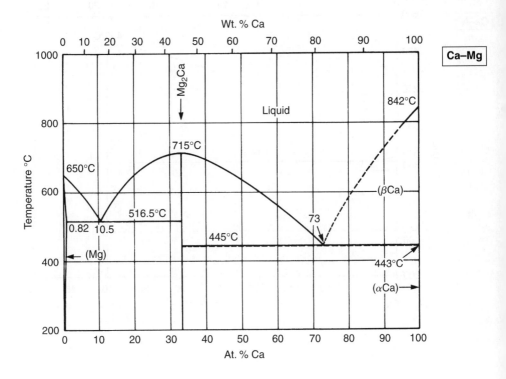

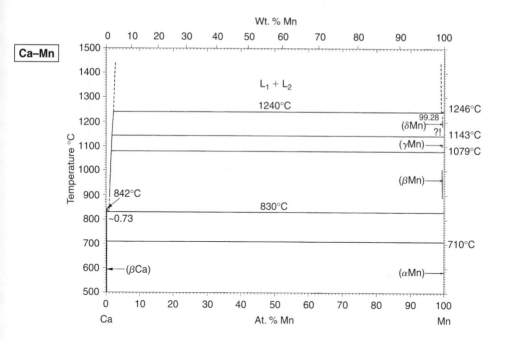

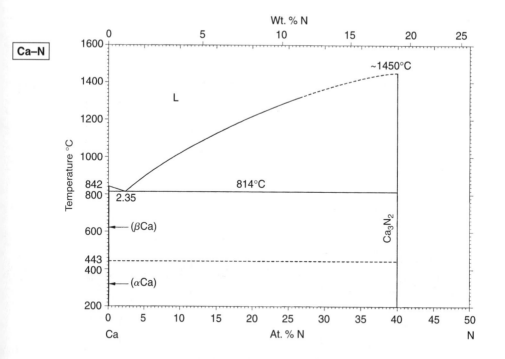

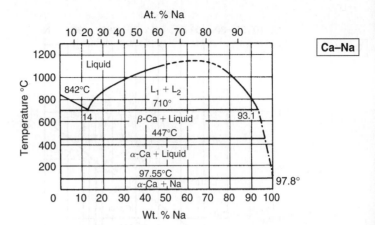

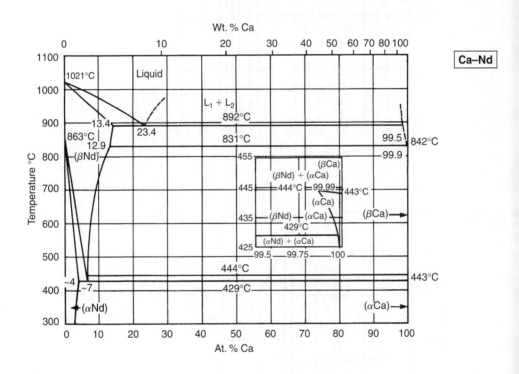

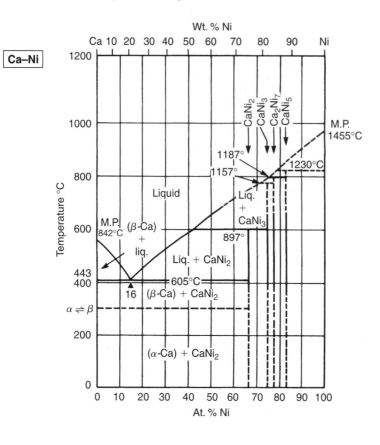

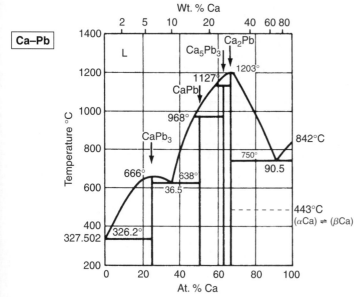

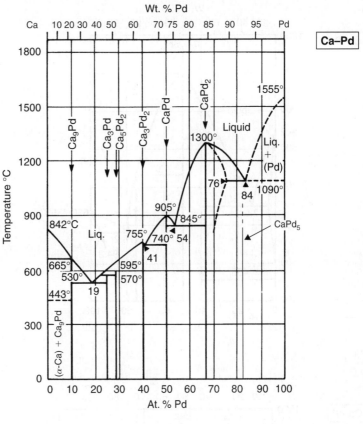

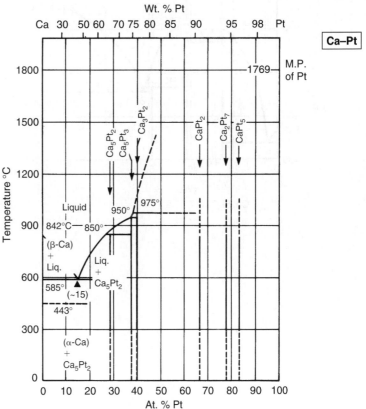

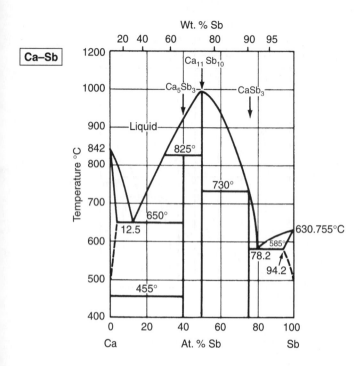

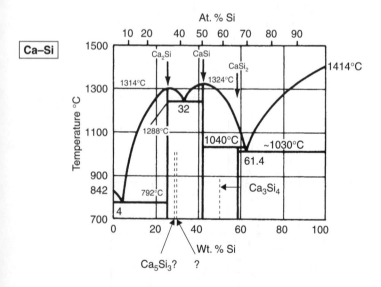

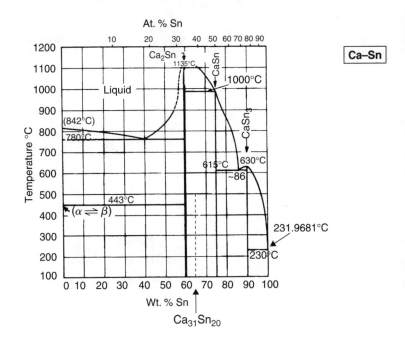

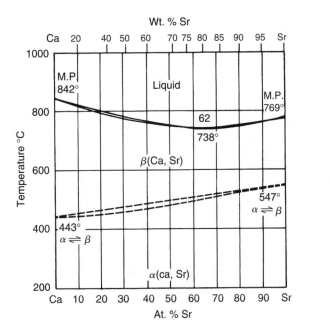

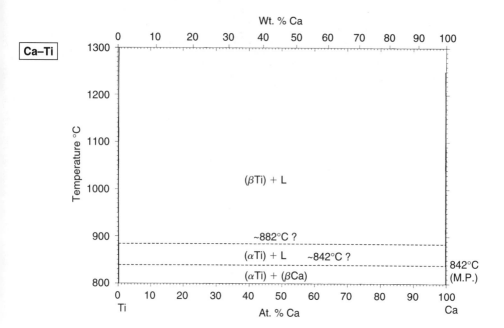

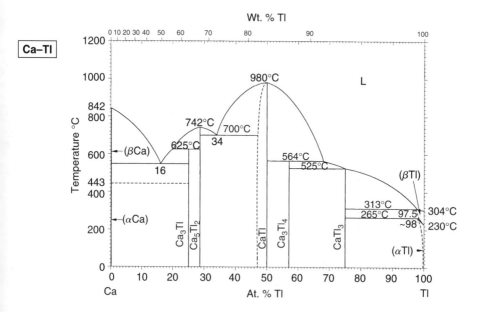

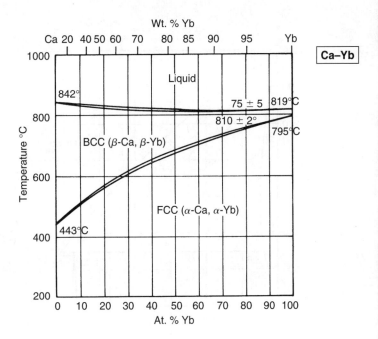

Ca–Yb

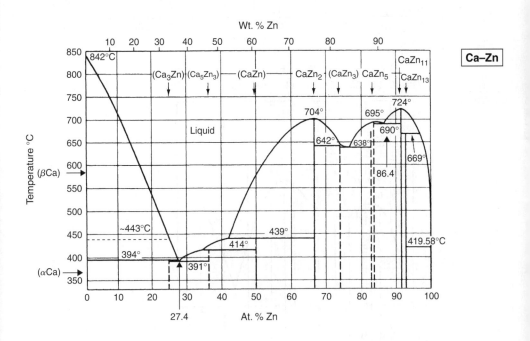

Ca–Zn

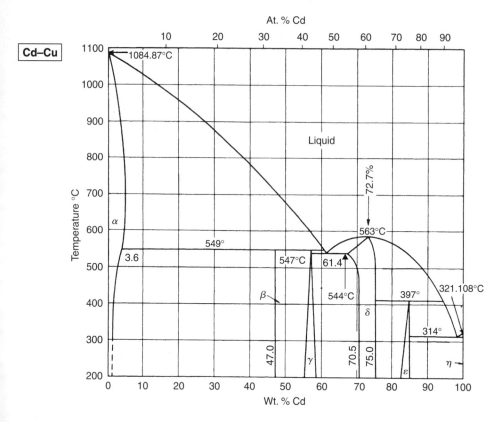

Cd–Cu

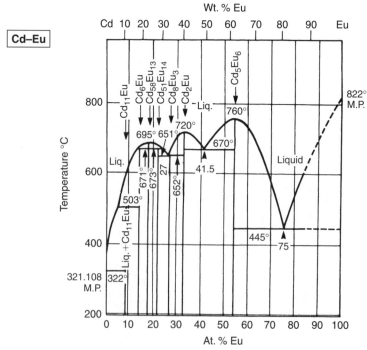

Cd–Eu

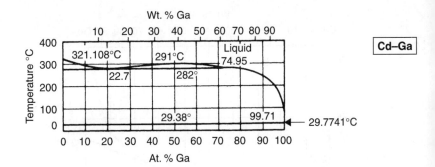

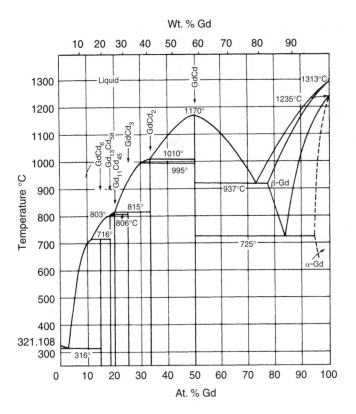

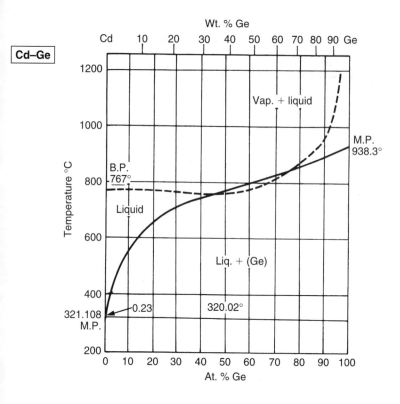

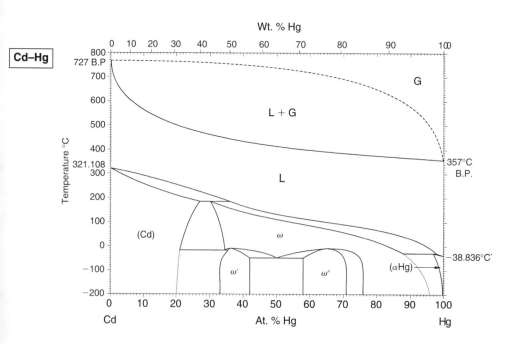

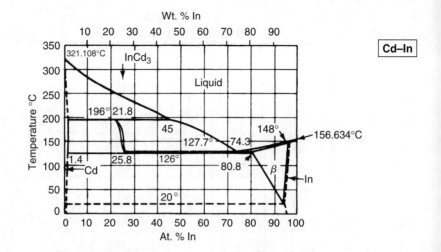

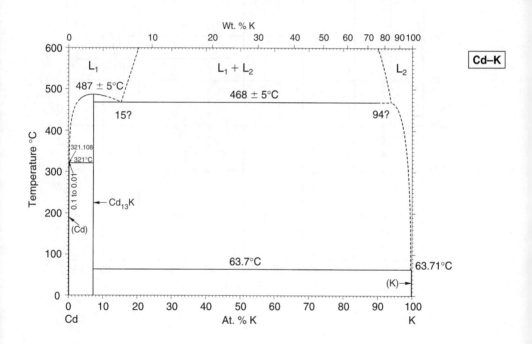

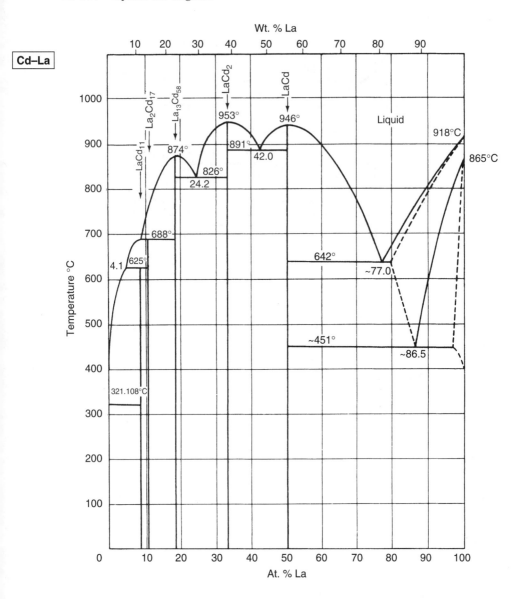

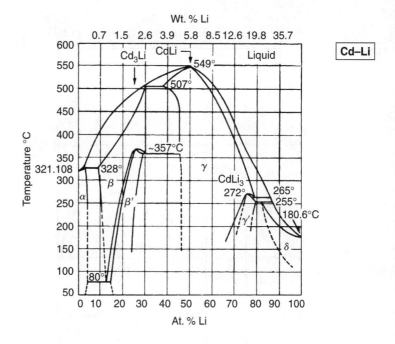

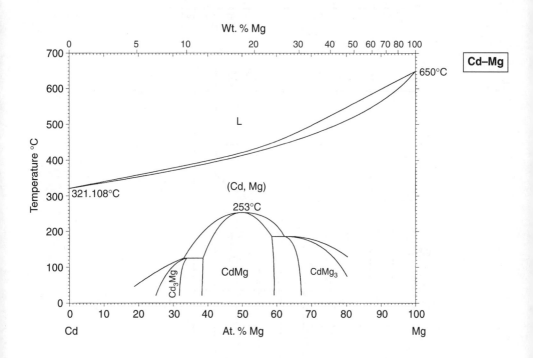

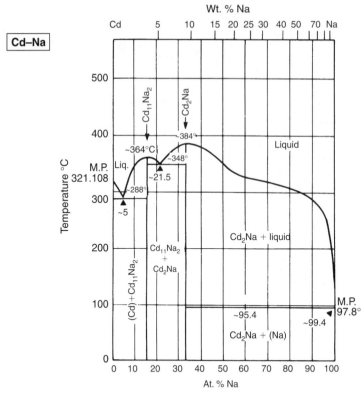

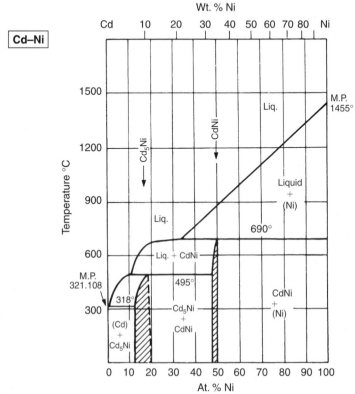

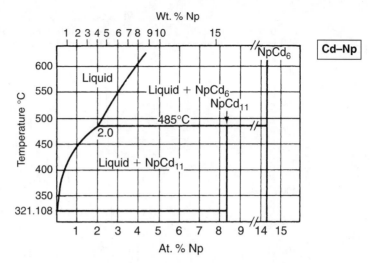

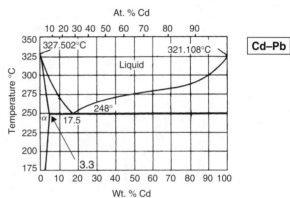

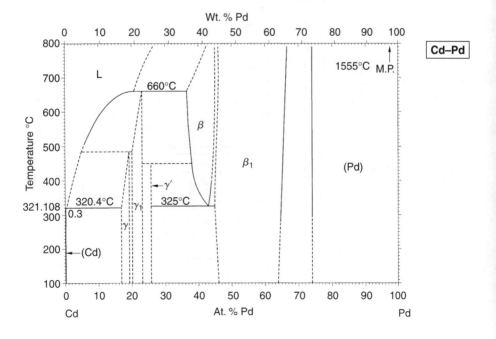

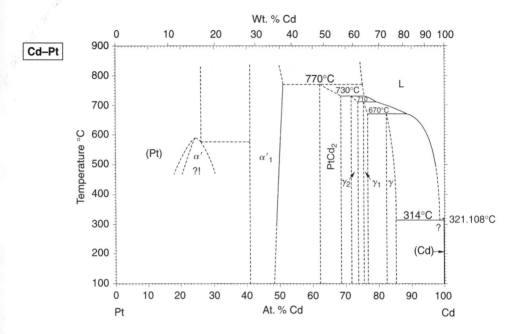

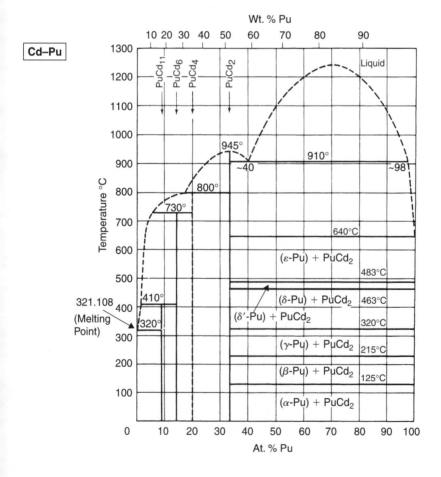

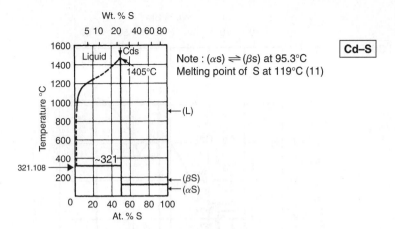

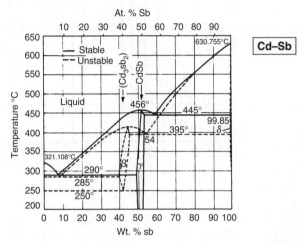

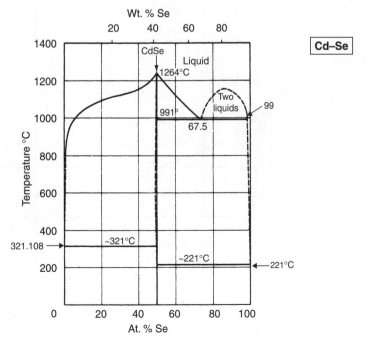

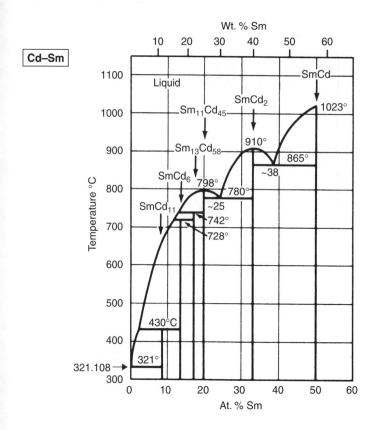

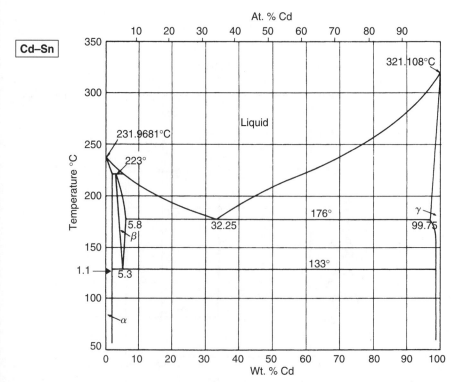

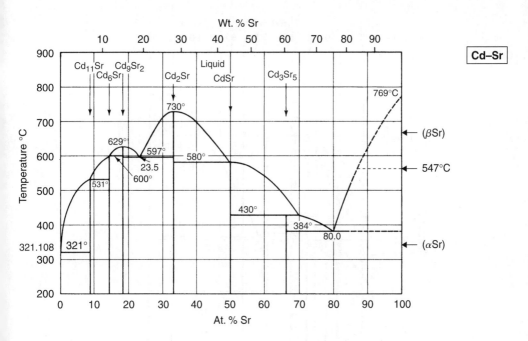

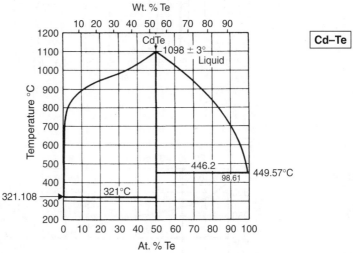

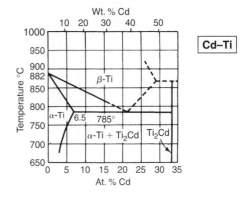

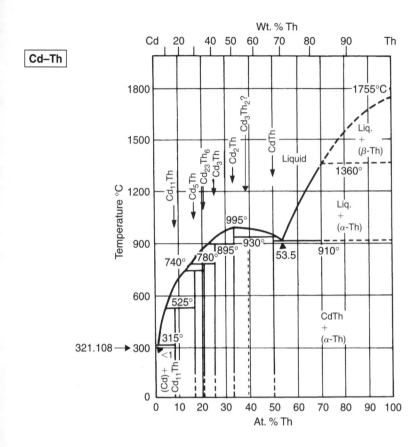

Cd–Th

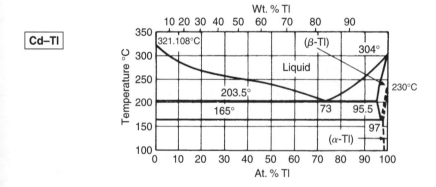

Cd–Tl

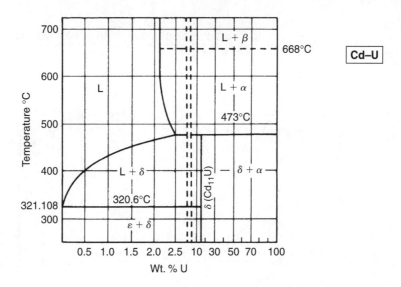

Cd–U

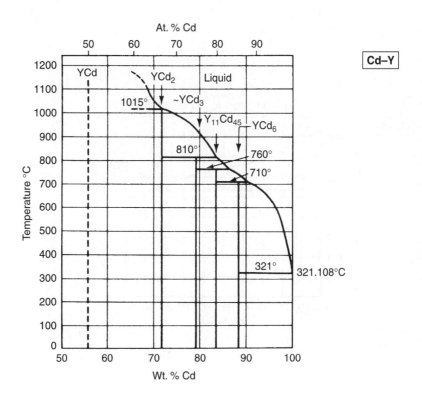

Cd–Y

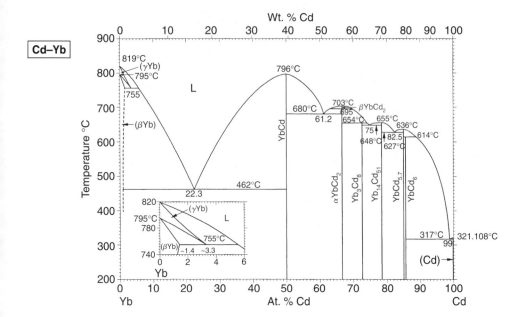

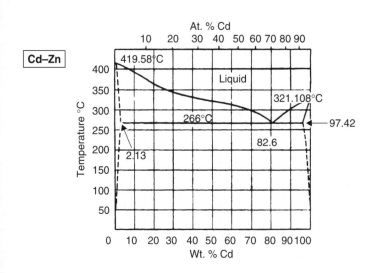

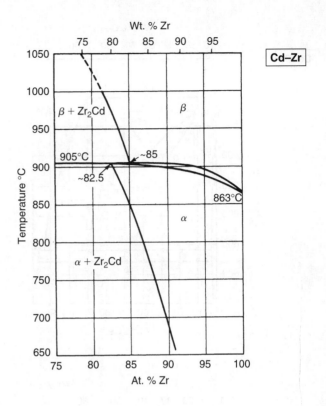

Cd–Zr

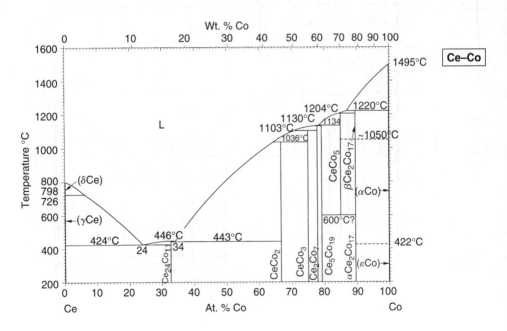

Ce–Co

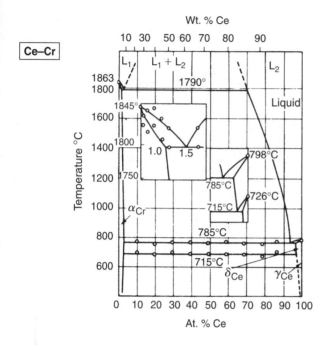

Ce–Cr

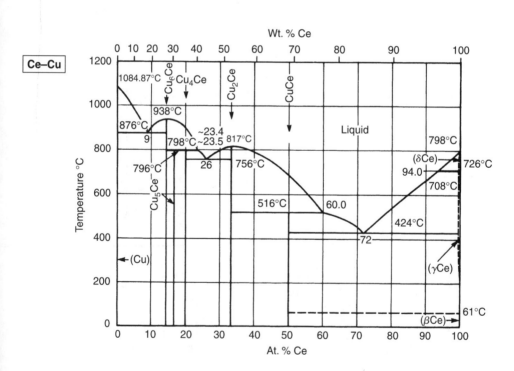

Ce–Cu

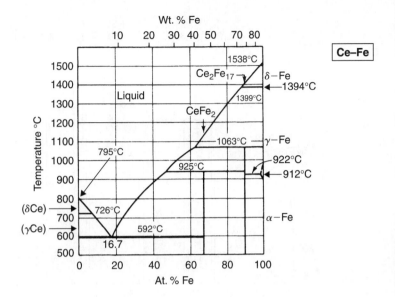

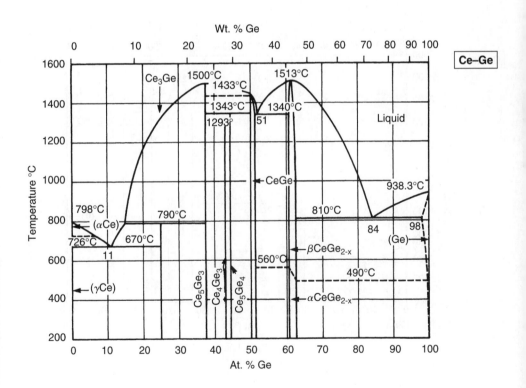

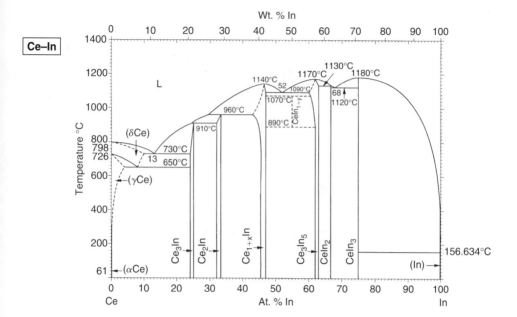

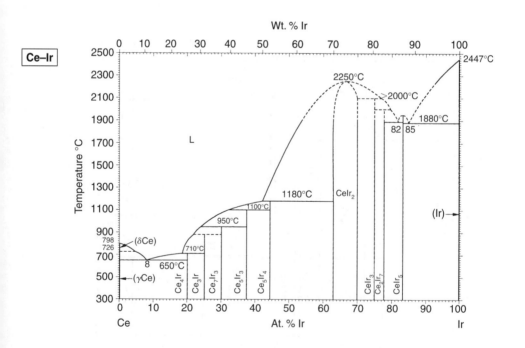

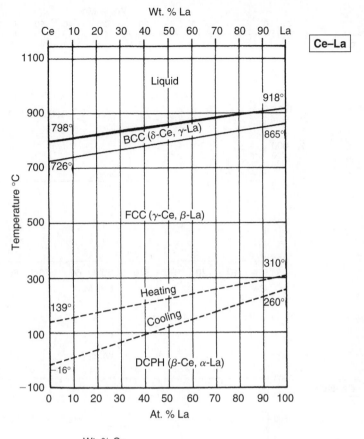

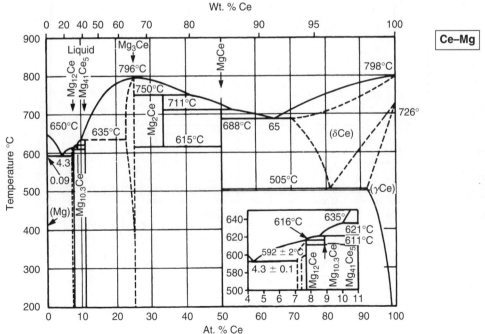

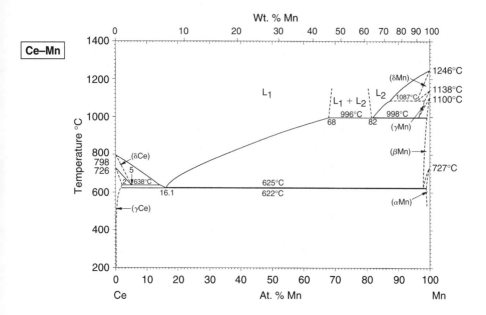

Ce–Mn

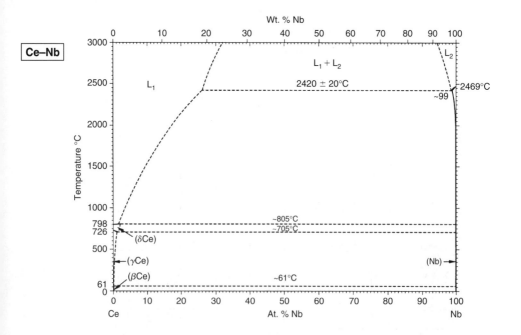

Ce–Nb

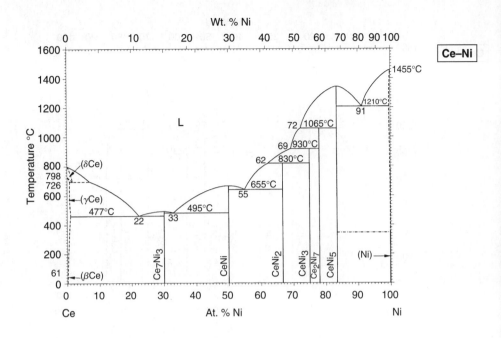

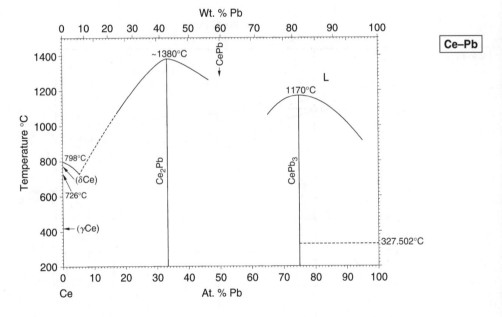

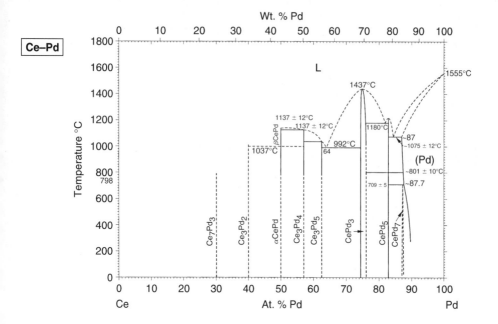

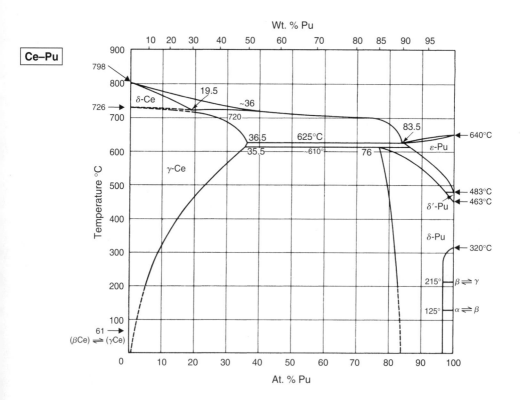

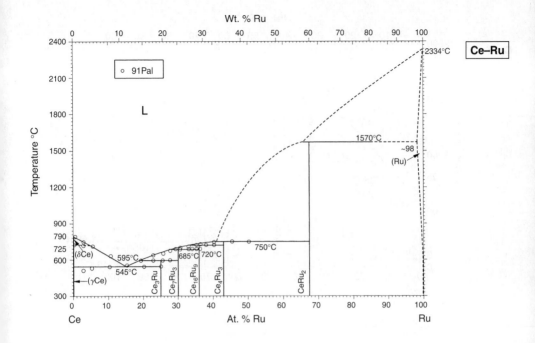

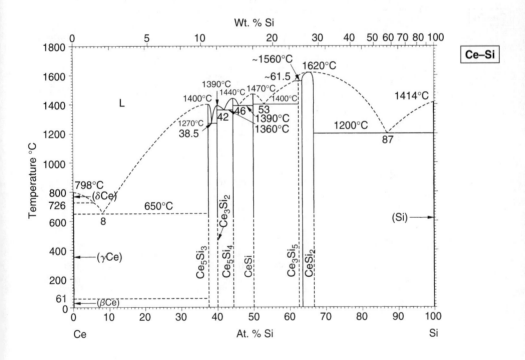

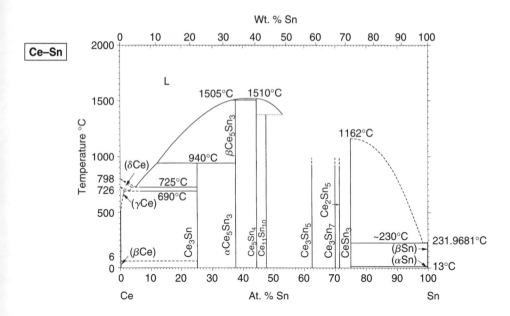

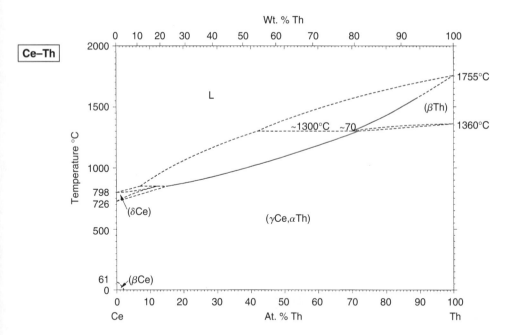

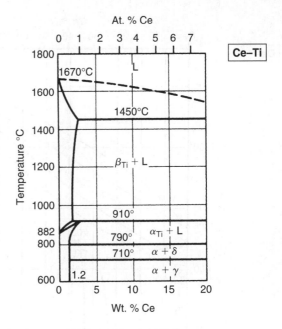

Ce–Ti

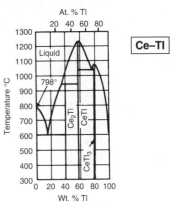

Ce–Tl

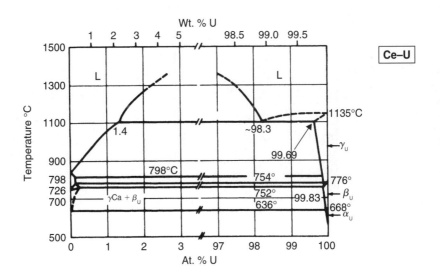

Ce–U

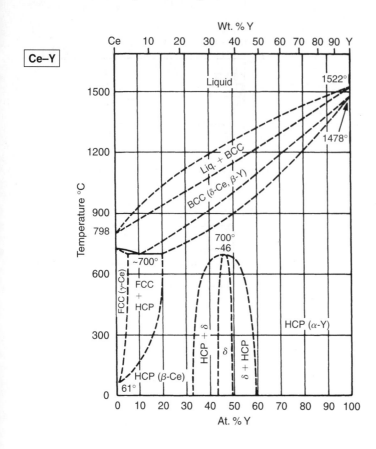

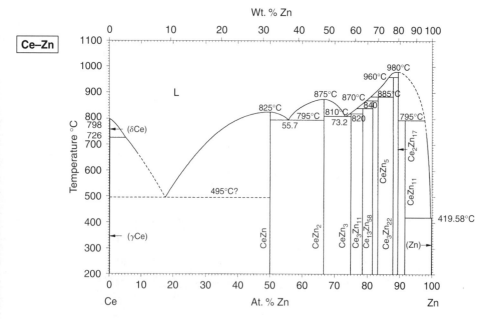

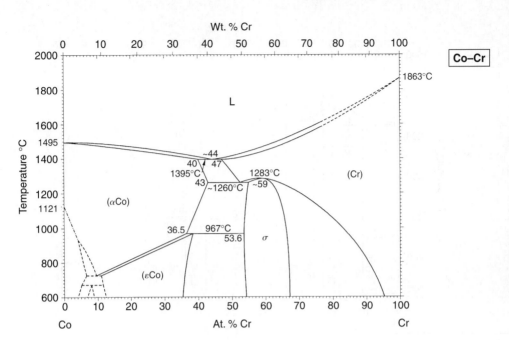

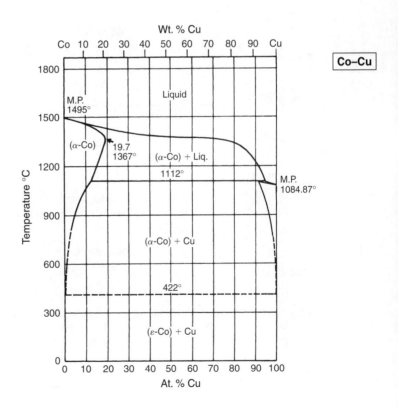

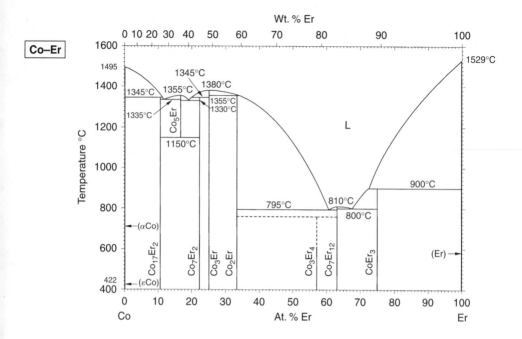

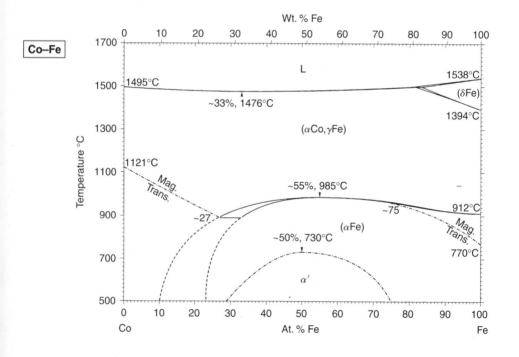

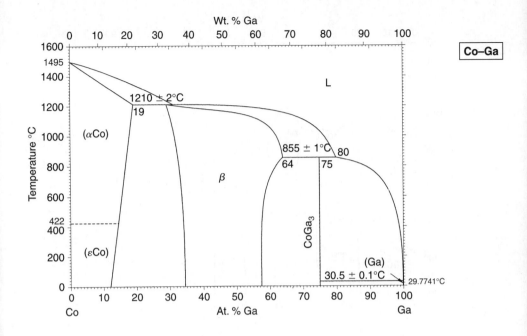

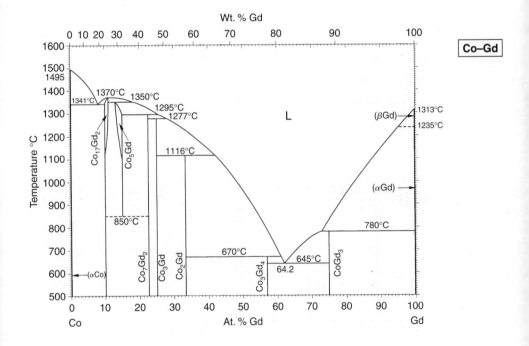

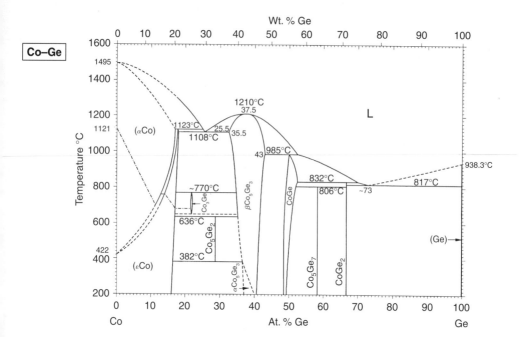

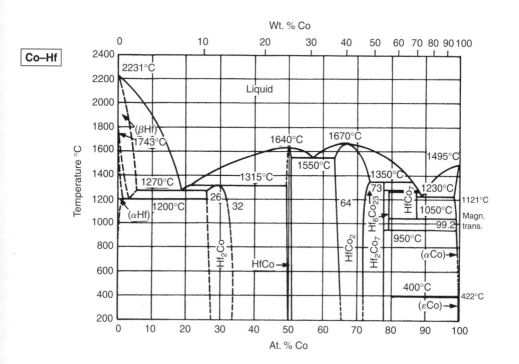

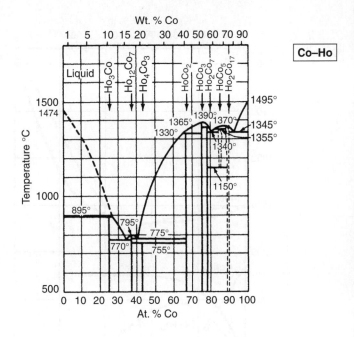

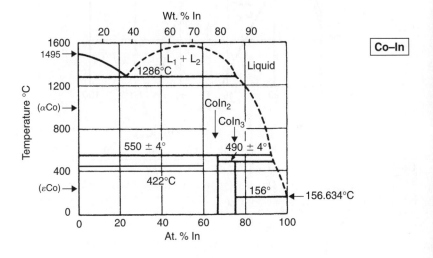

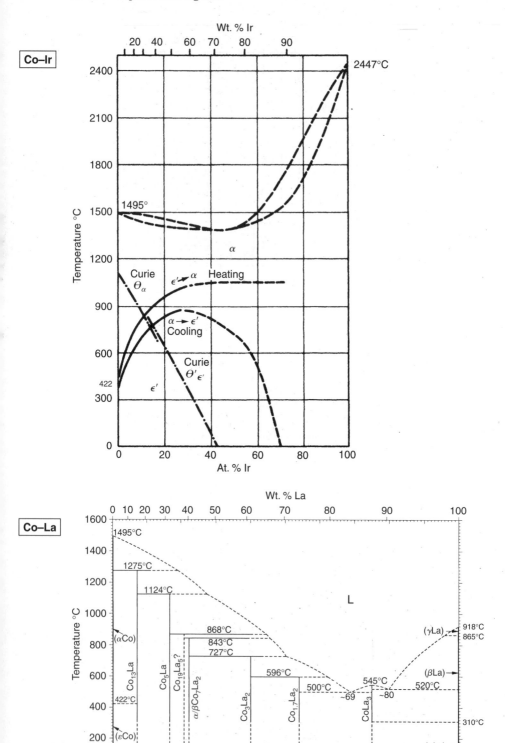

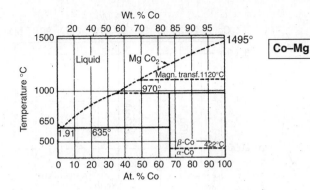

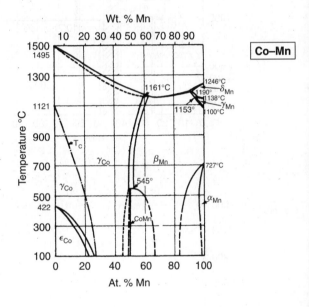

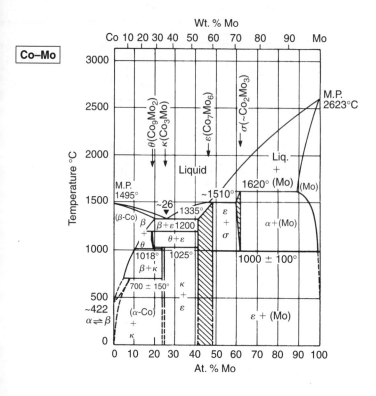

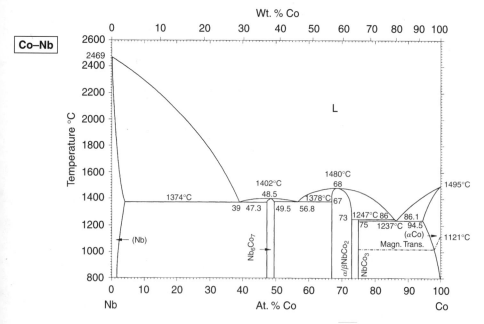

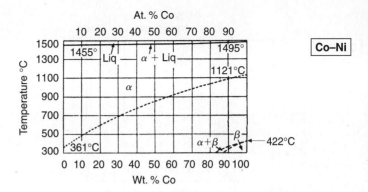

Co–Ni

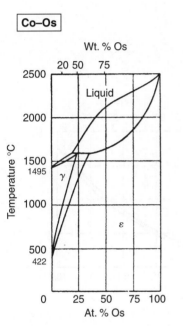

Co–Os

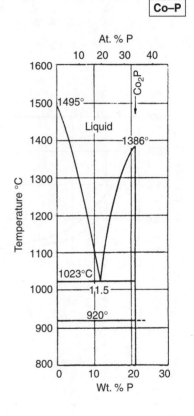

Co–P

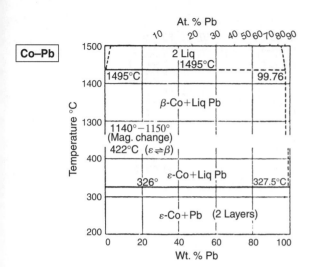

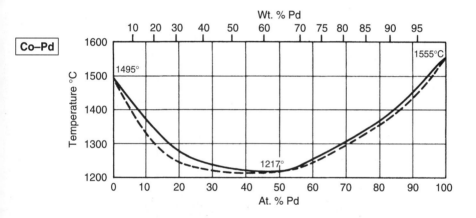

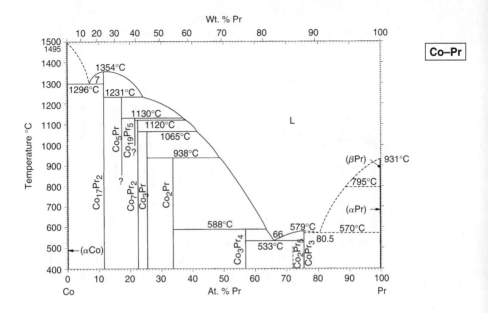

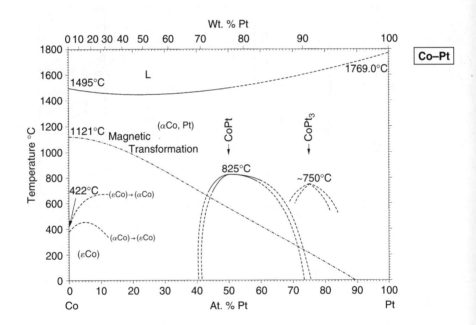

Co–Pu

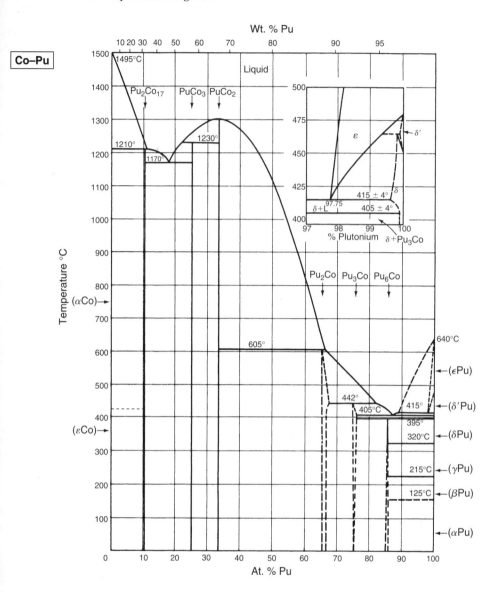

Co–Re

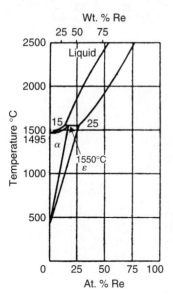

Co–Rh

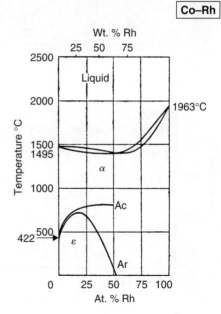

Co–Ru

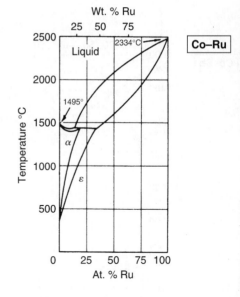

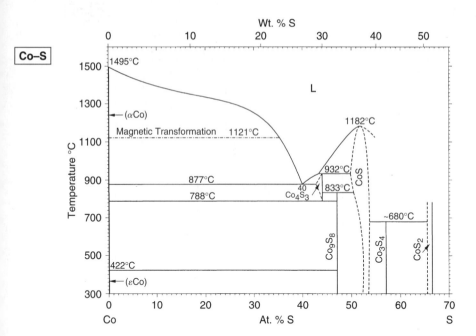

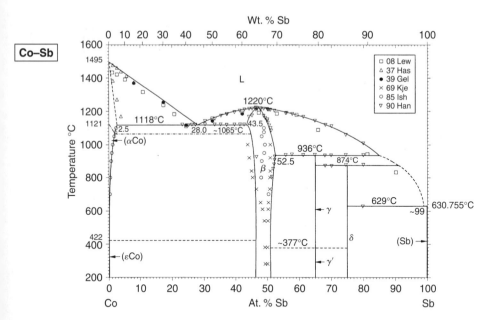

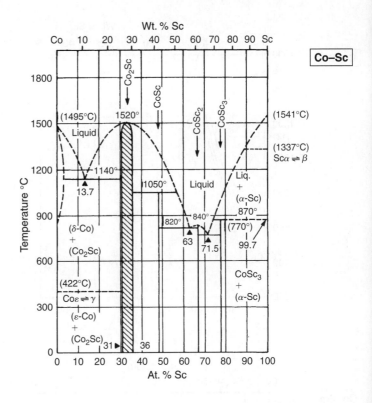

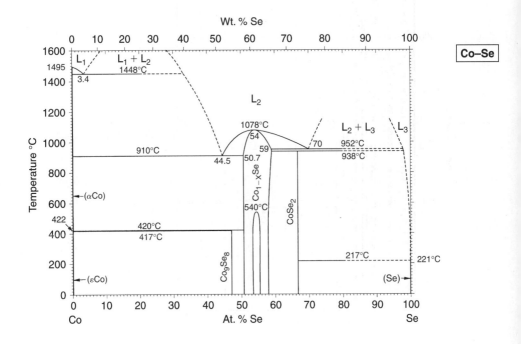

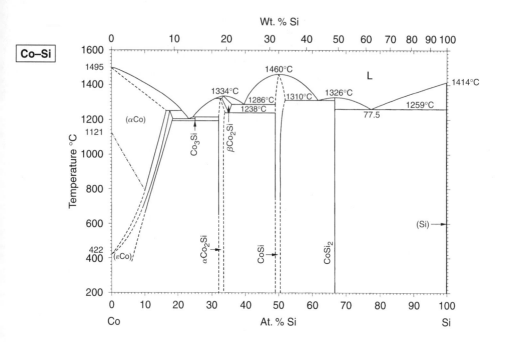

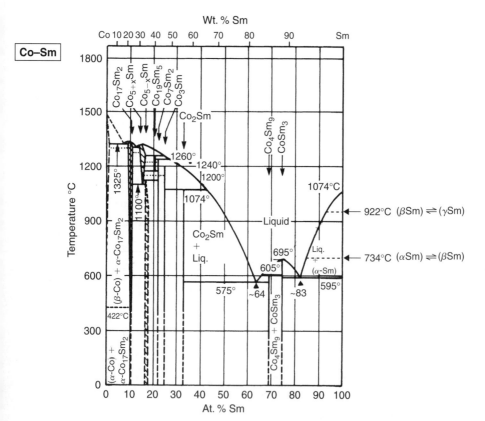

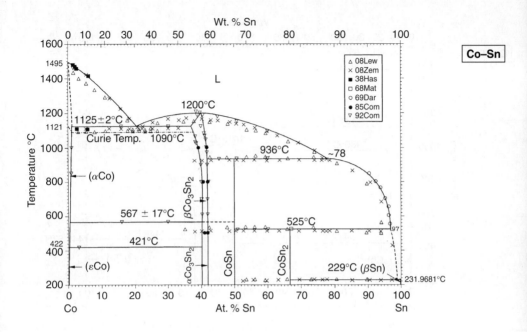

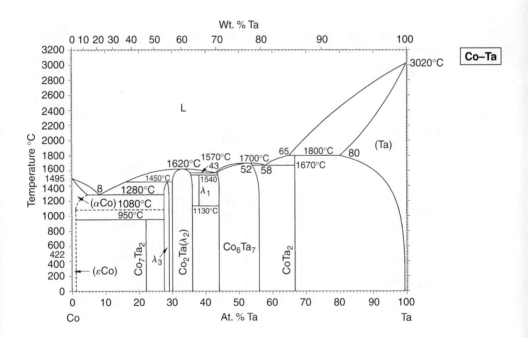

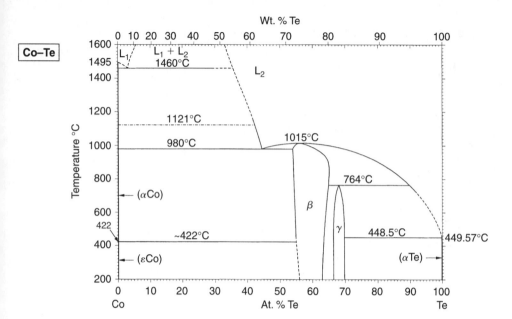

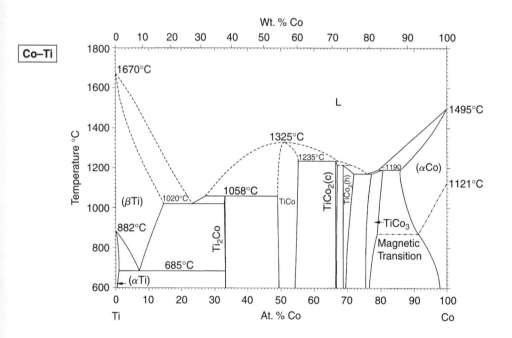

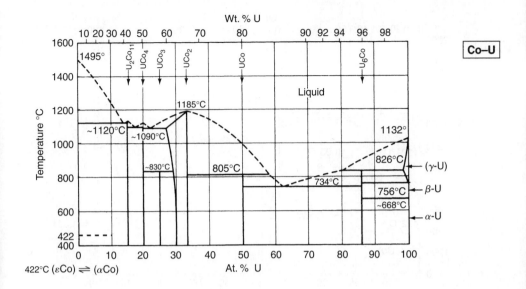

Co–U

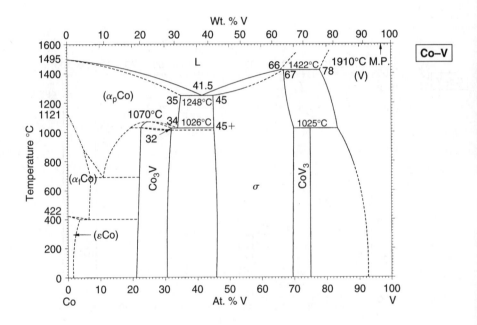

Co–V

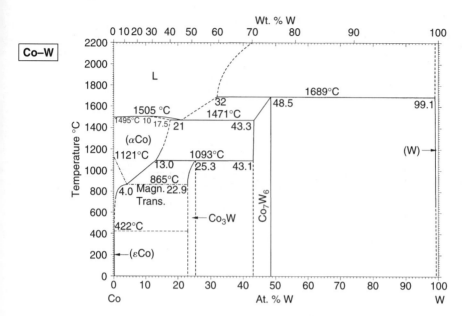

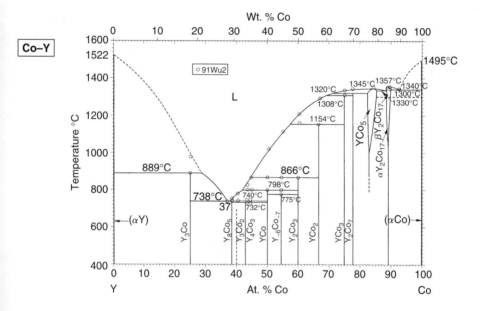

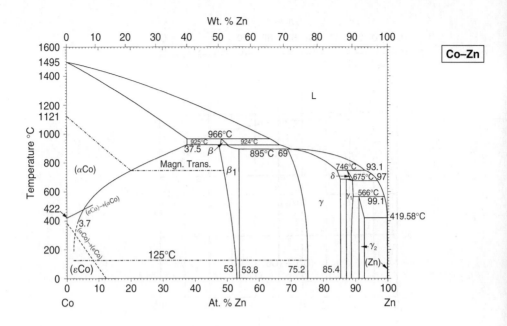

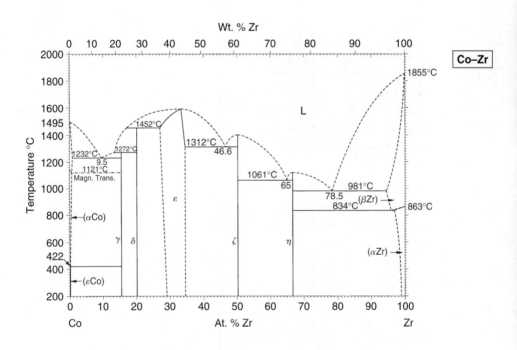

Cr–Cu

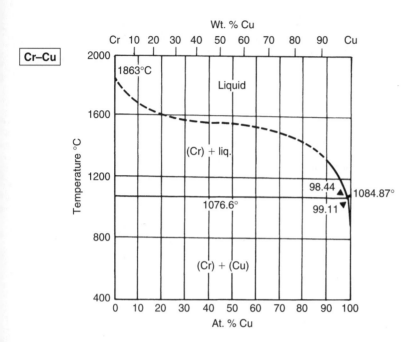

Cr–Fe

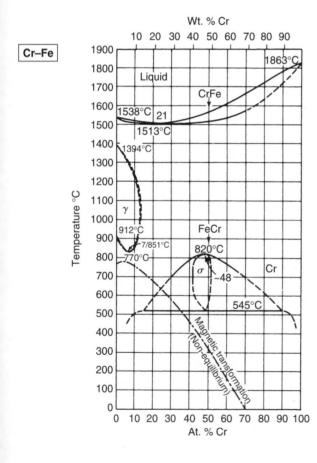

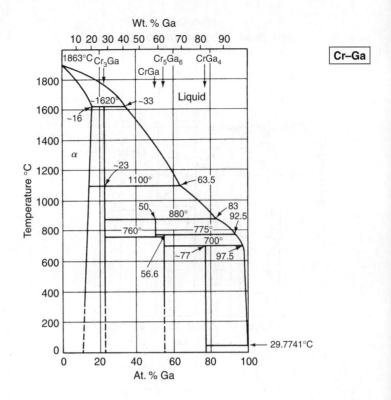

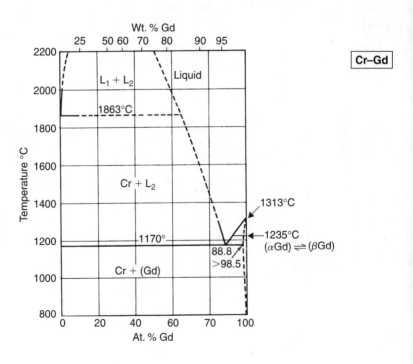

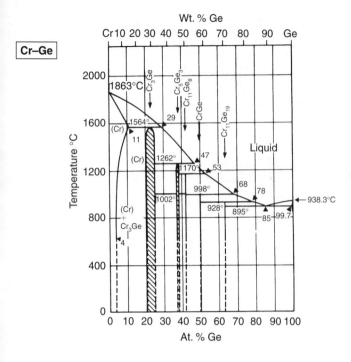

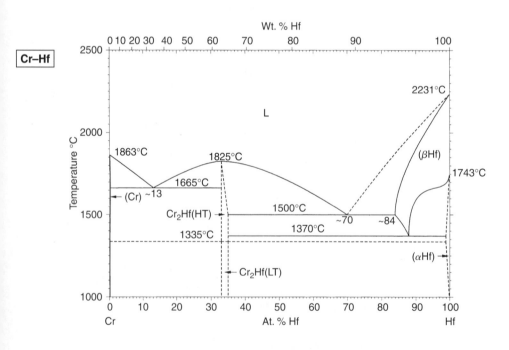

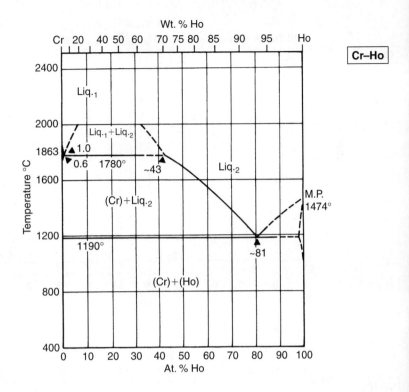

Cr–Ho

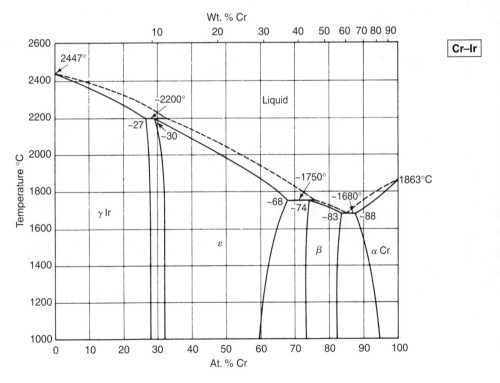

Cr–Ir

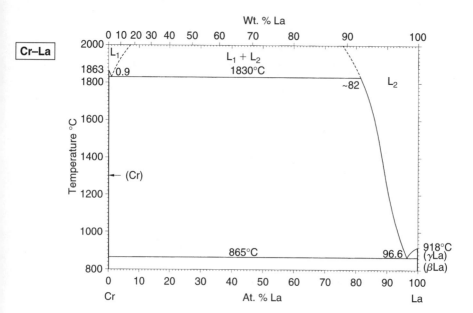

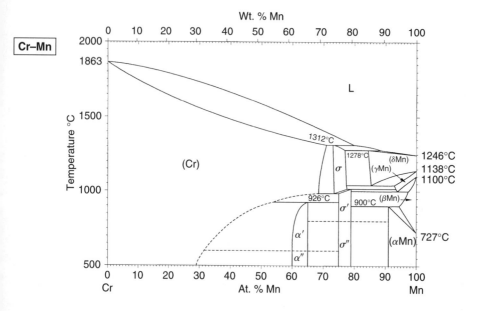

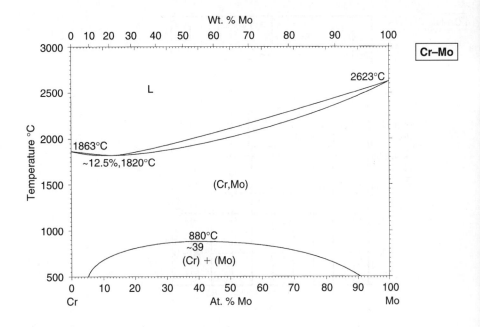

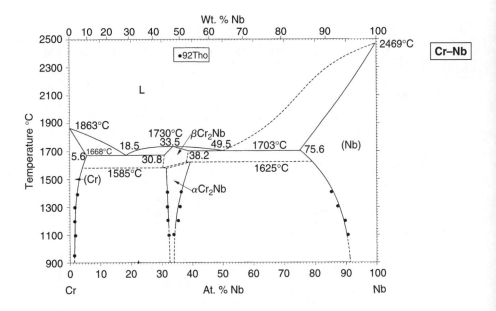

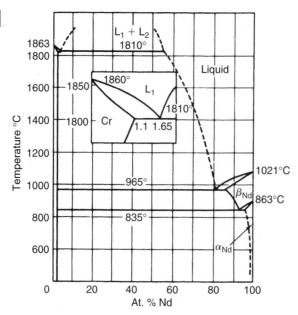

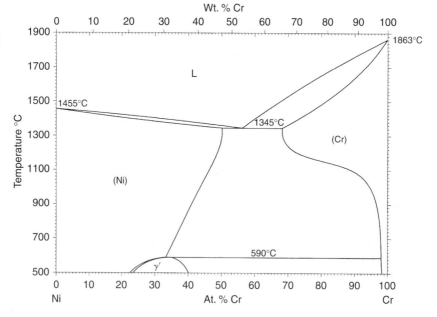

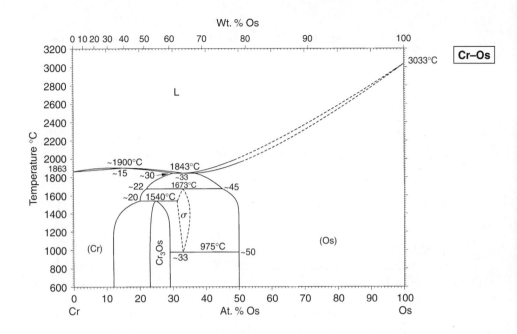

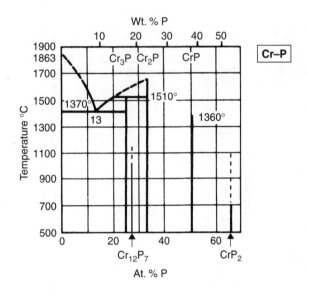

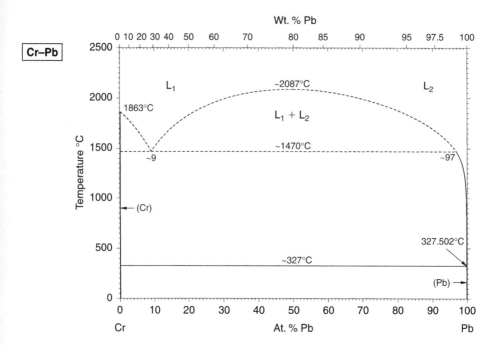

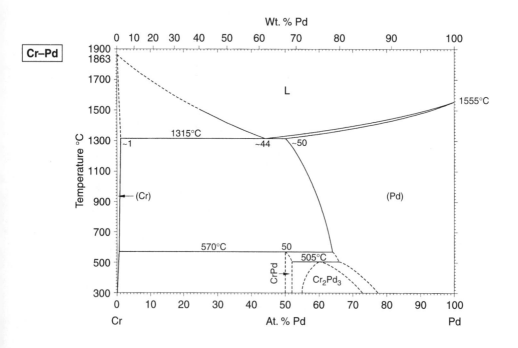

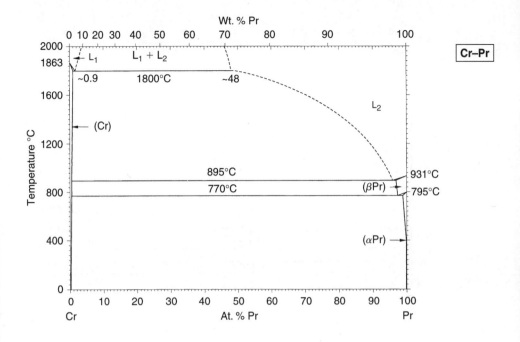

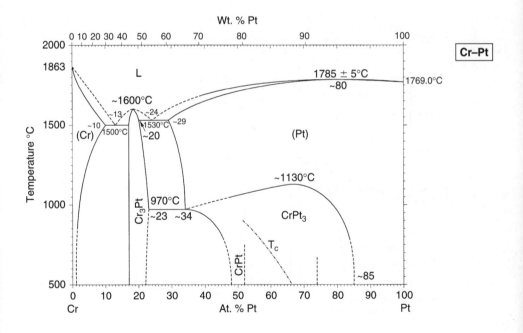

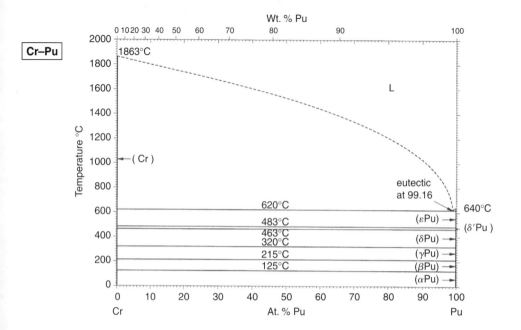

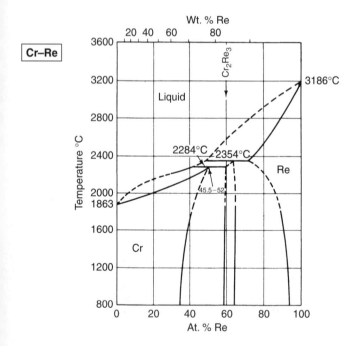

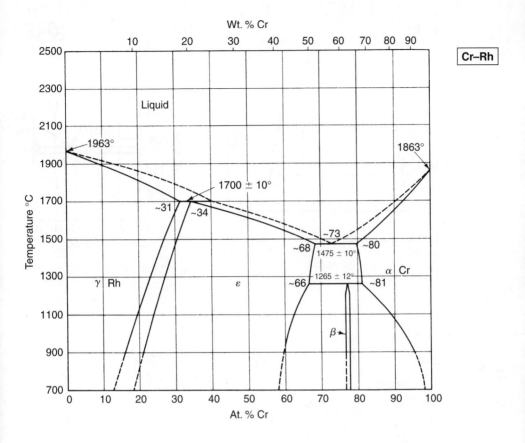

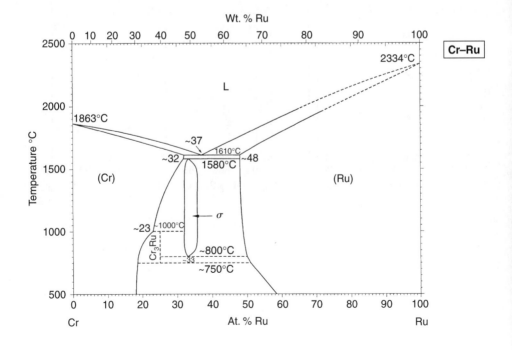

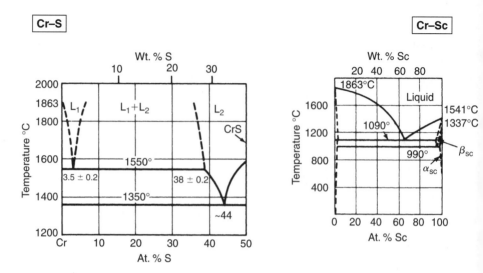

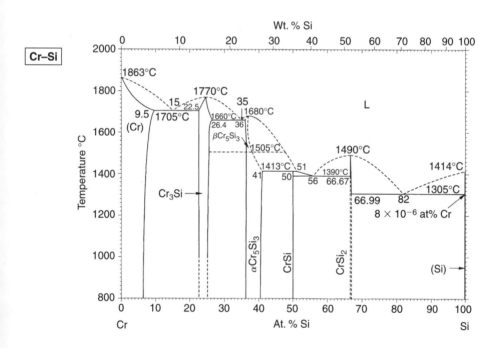

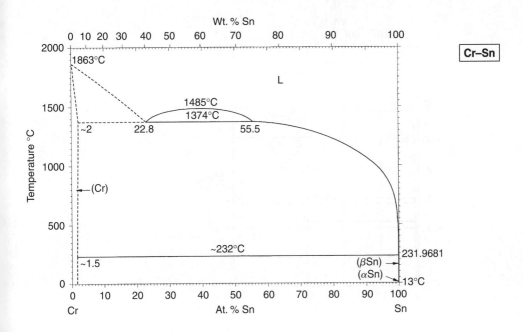

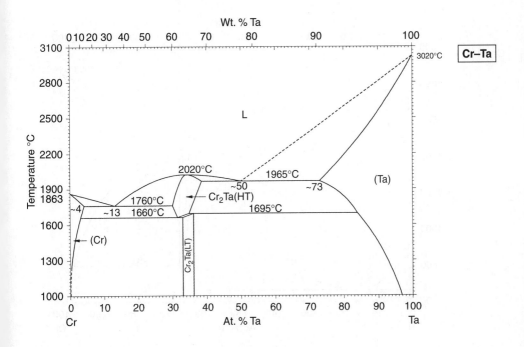

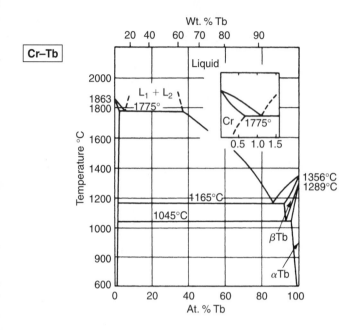

Cr–Tb

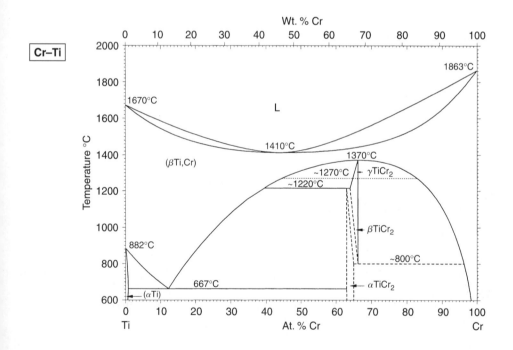

Cr–Ti

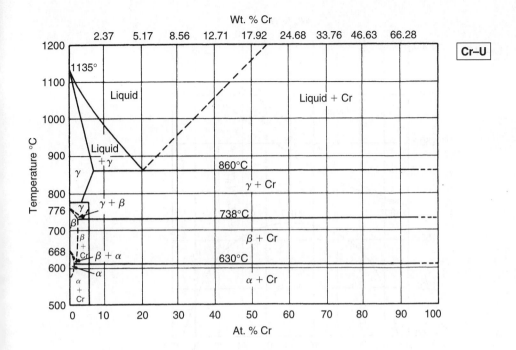

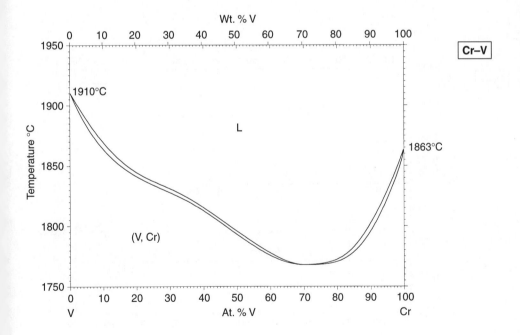

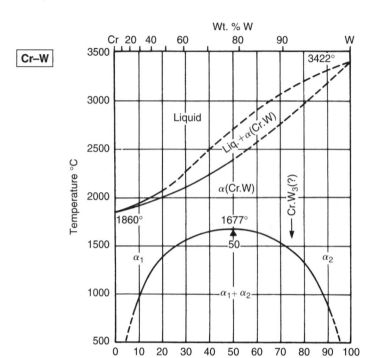

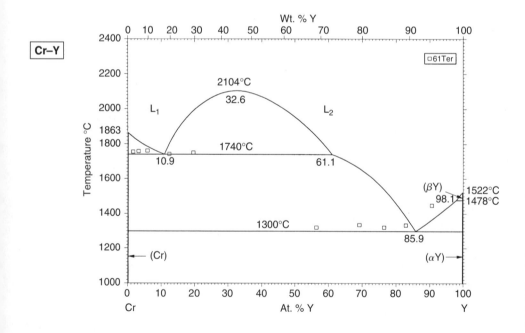

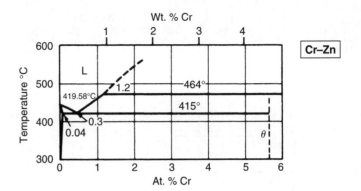

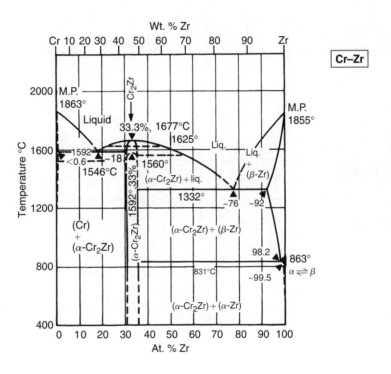

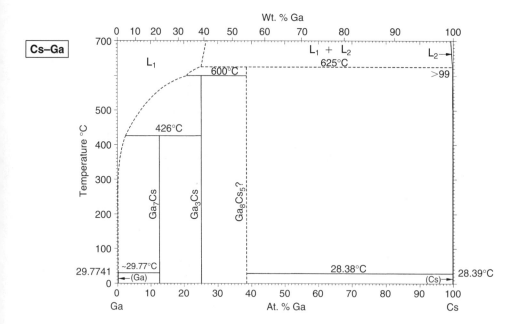

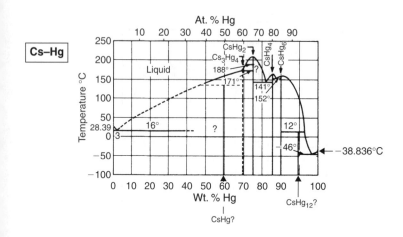

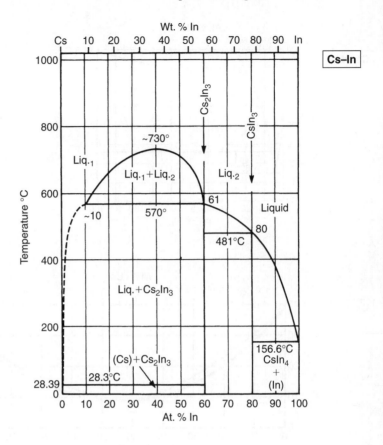

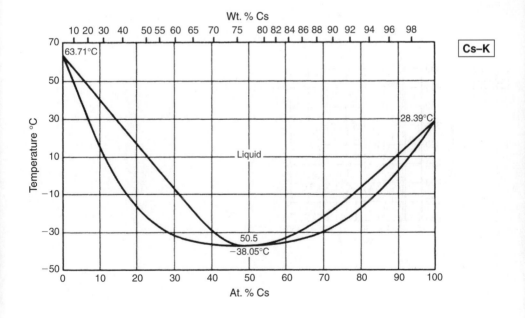

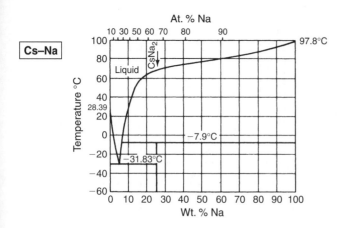

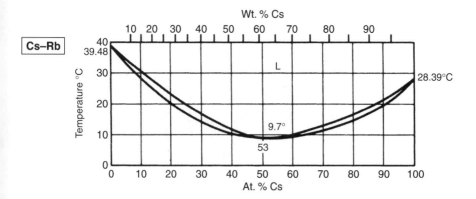

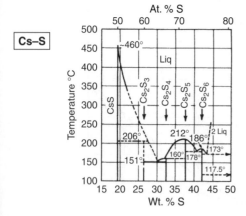

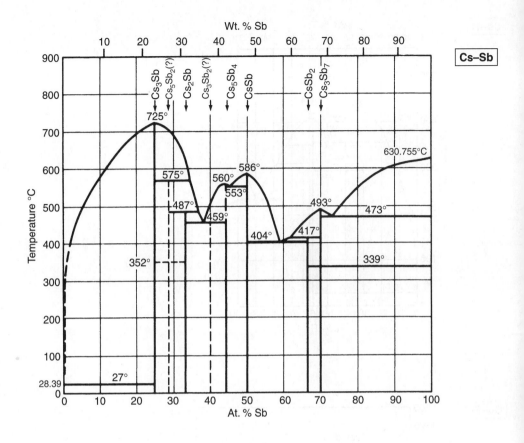

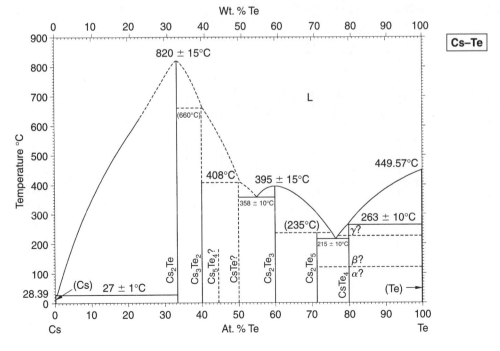

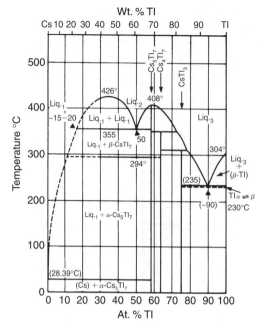

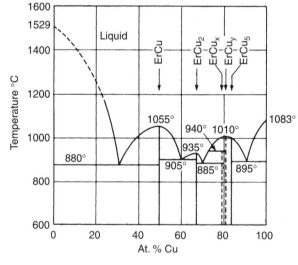

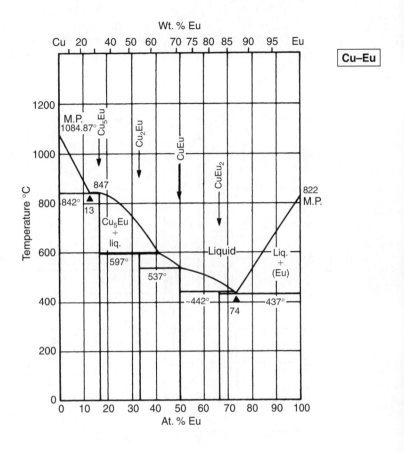

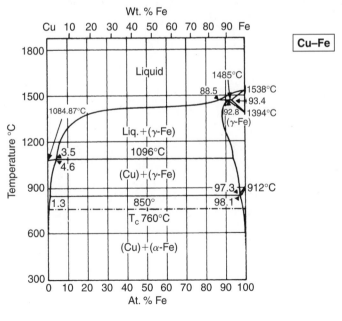

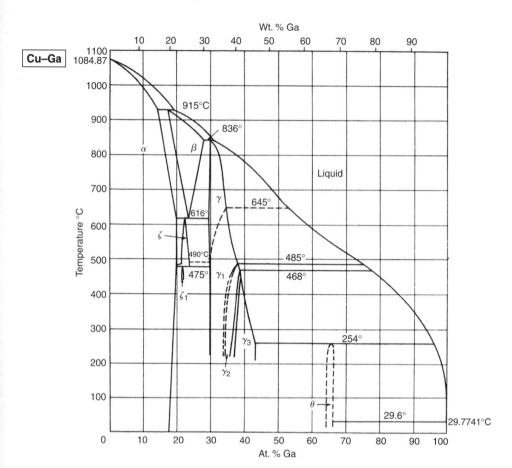

Cu–Ga

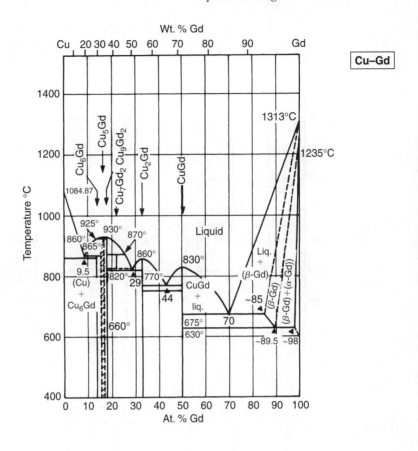

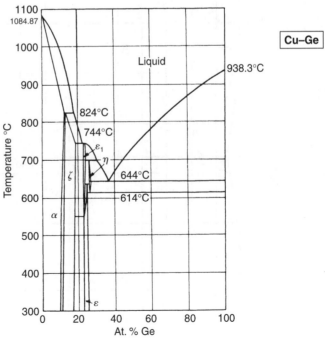

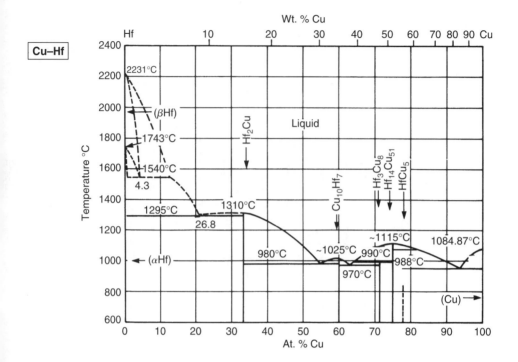

Cu–Hf

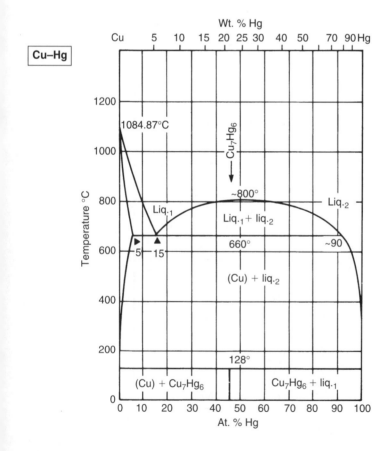

Cu–Hg

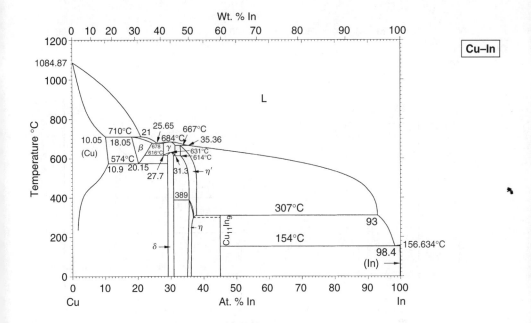

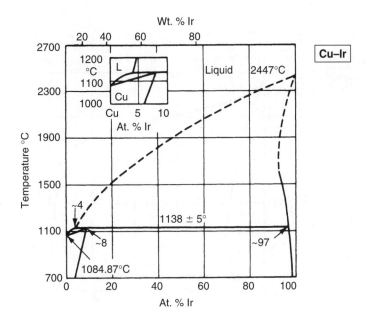

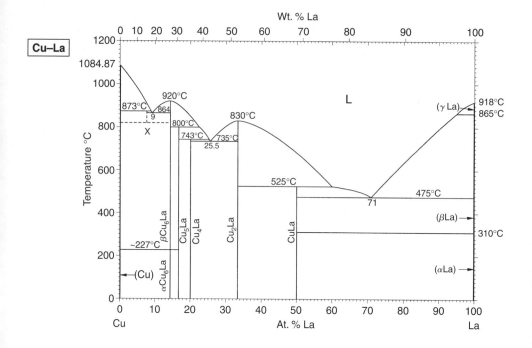

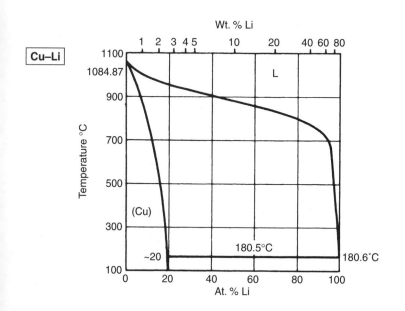

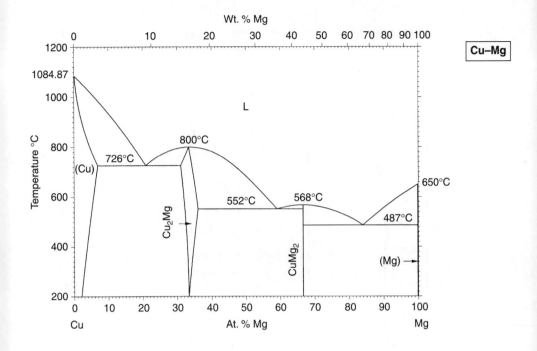

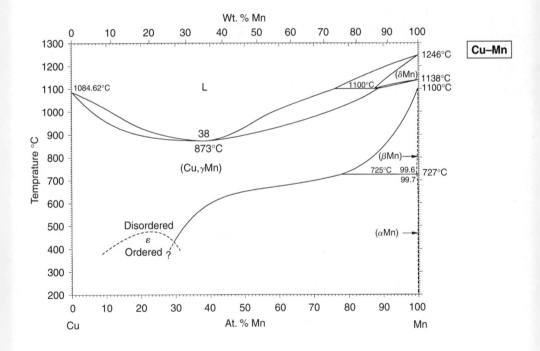

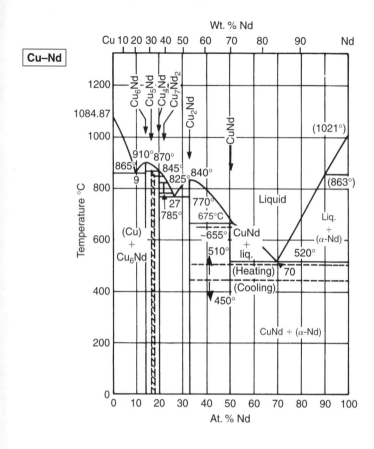

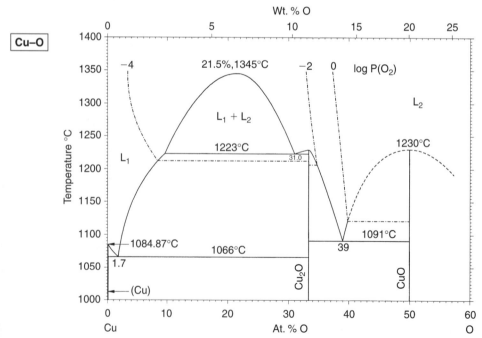

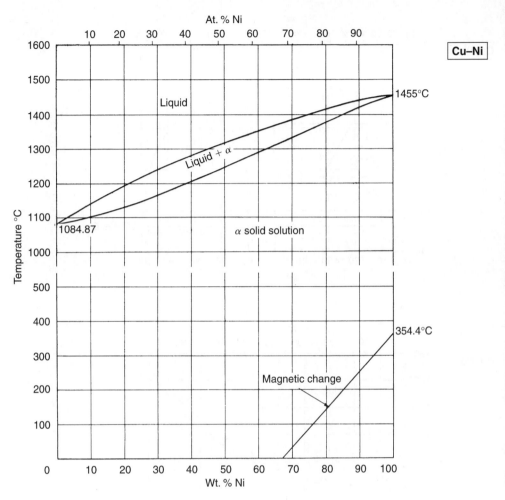

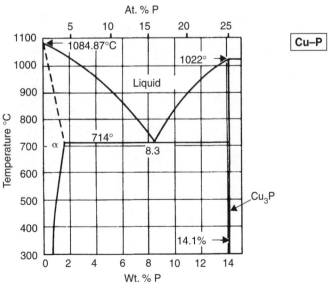

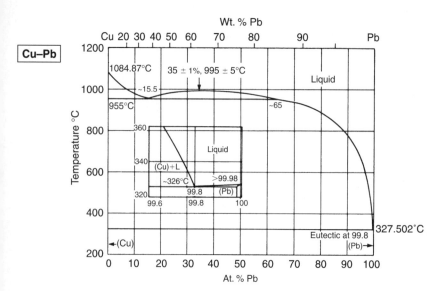

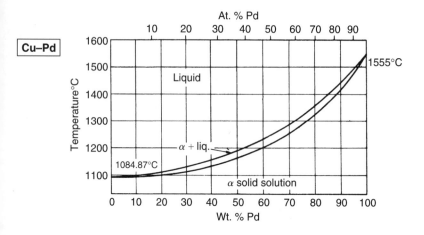

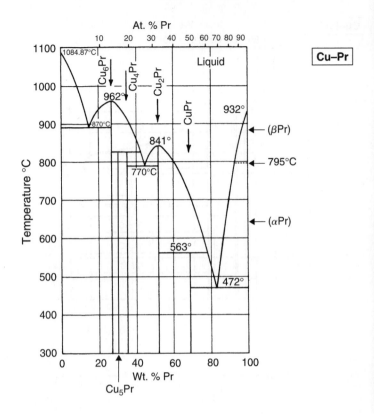

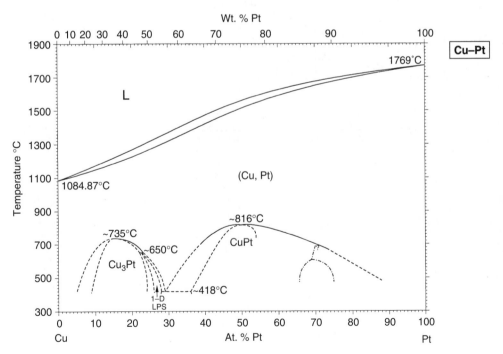

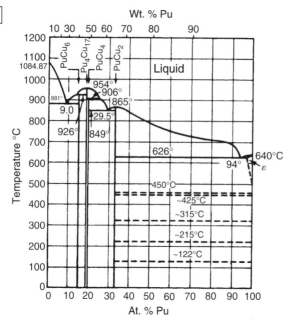

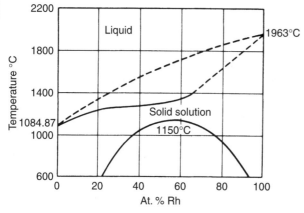

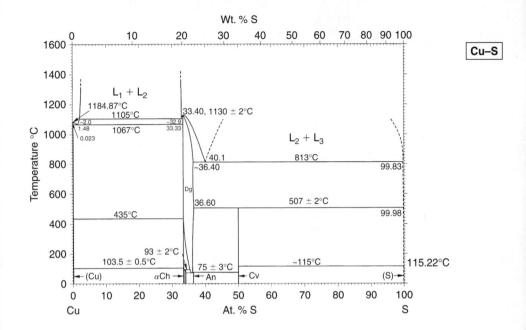

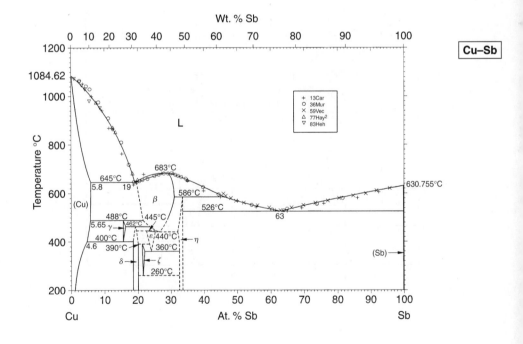

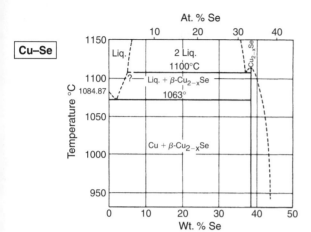

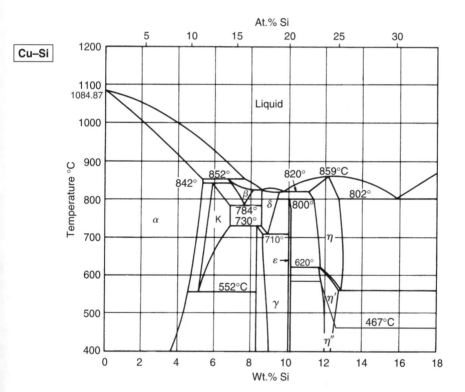

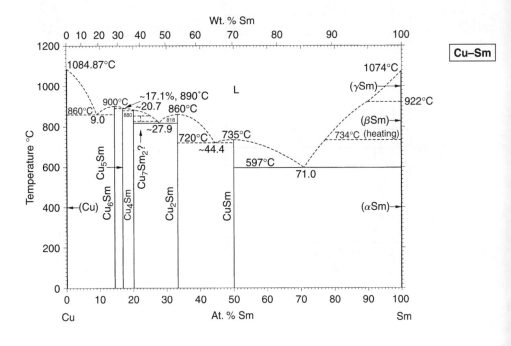

Cu–Sm

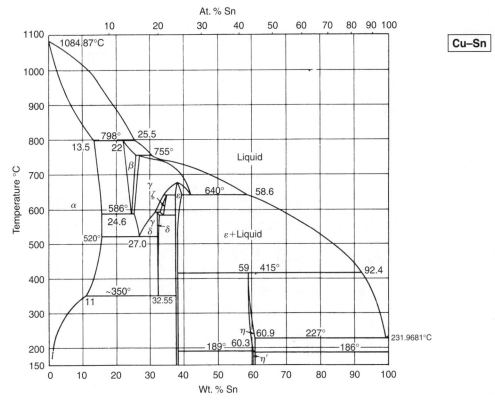

Cu–Sn

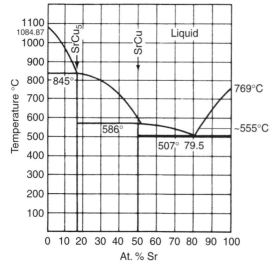

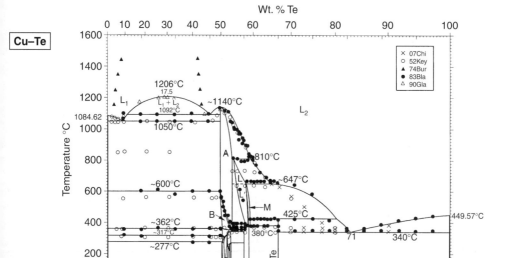

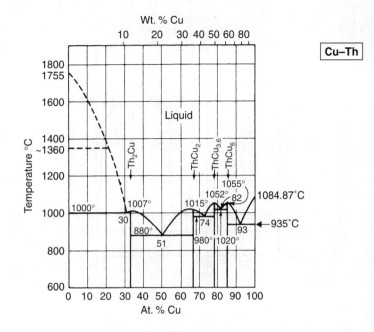

Cu–Th

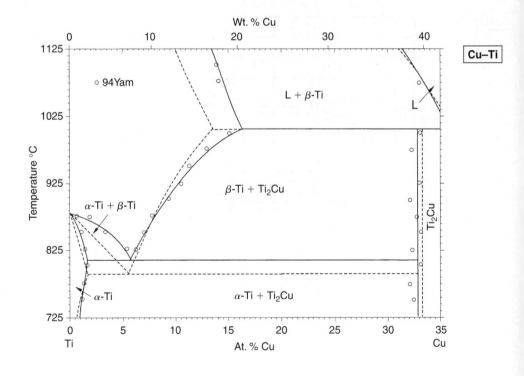

Cu–Ti

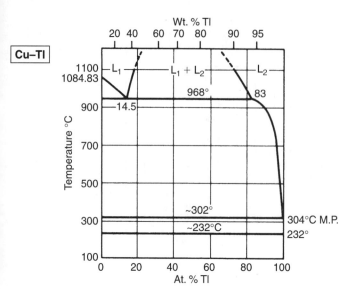

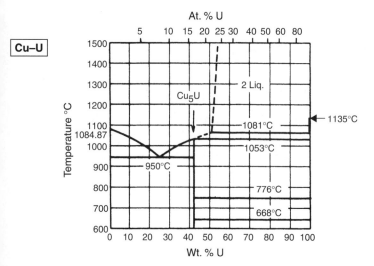

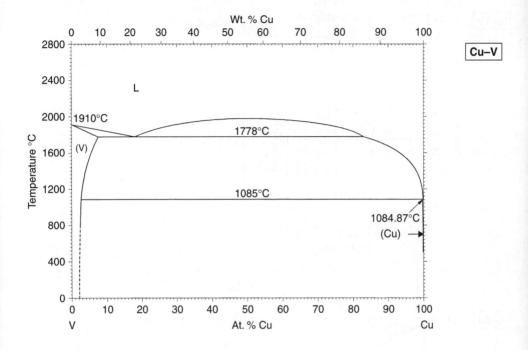

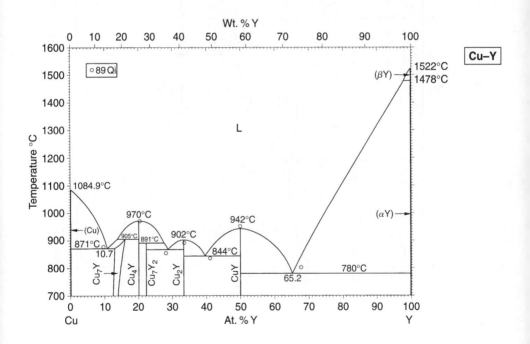

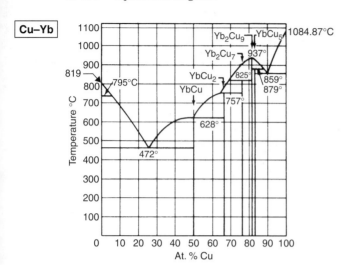

Cu–Yb

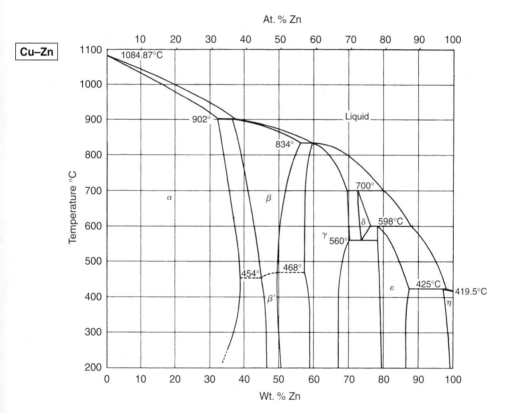

Cu–Zn

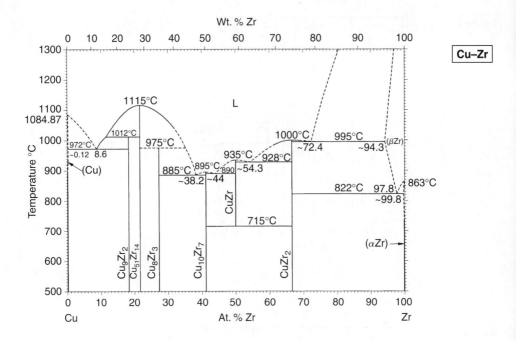

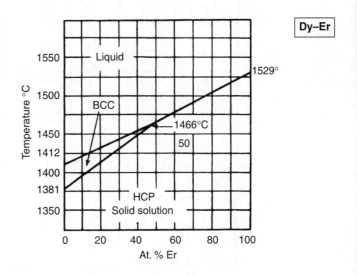

Dy–Fe

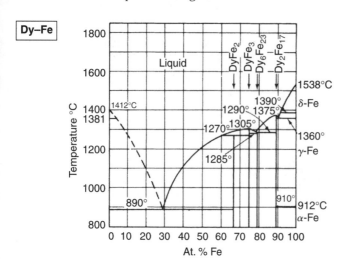

Dy–Ga

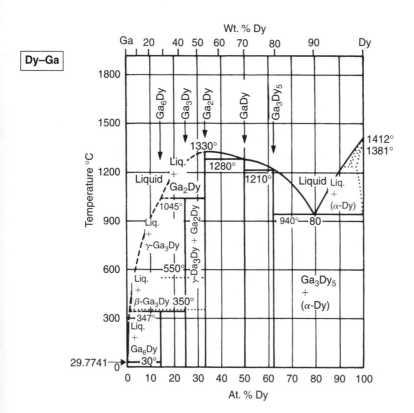

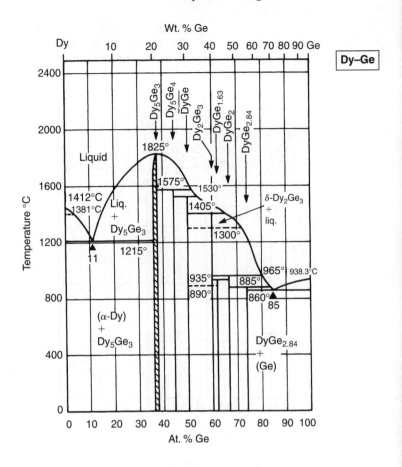

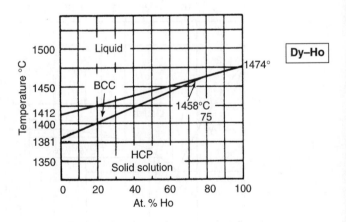

Dy–In

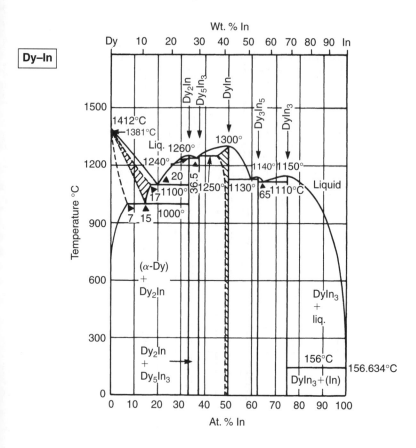

Dy–Mn

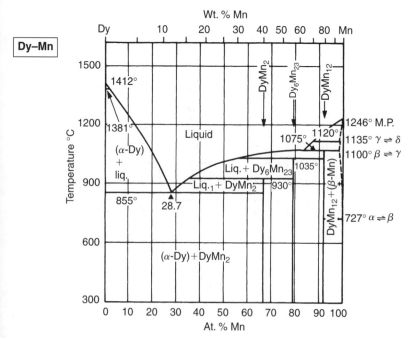

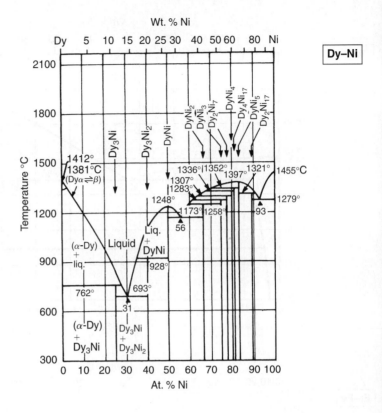

Dy–Ni

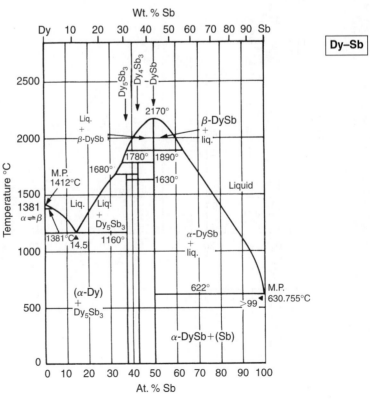

Dy–Sb

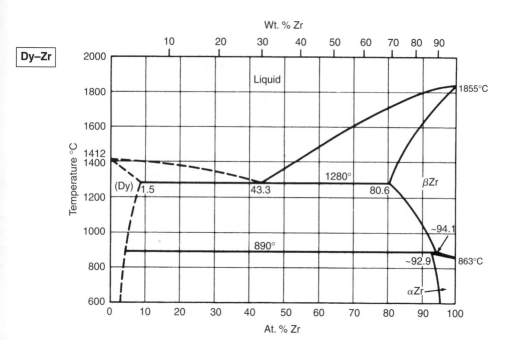

Dy–Zr

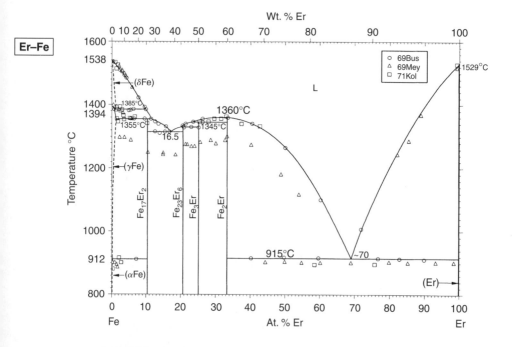

Er–Fe

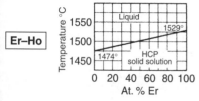

Er–Ho

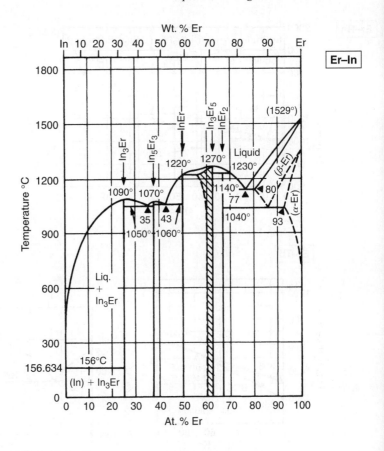

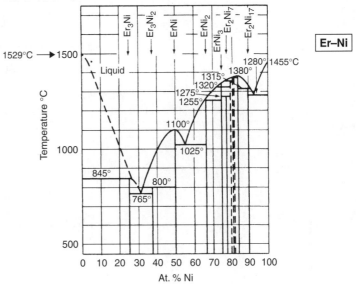

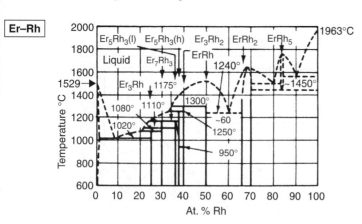

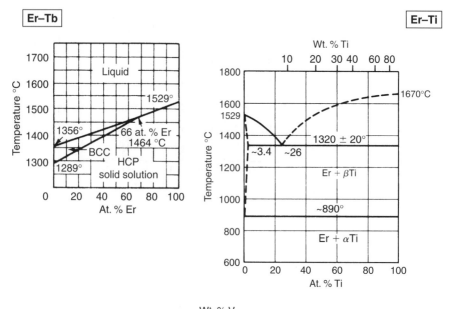

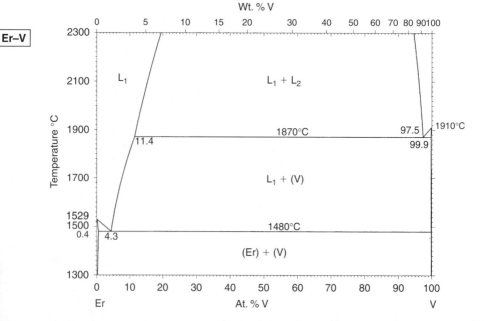

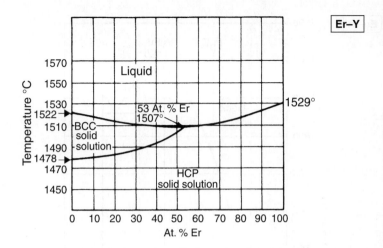

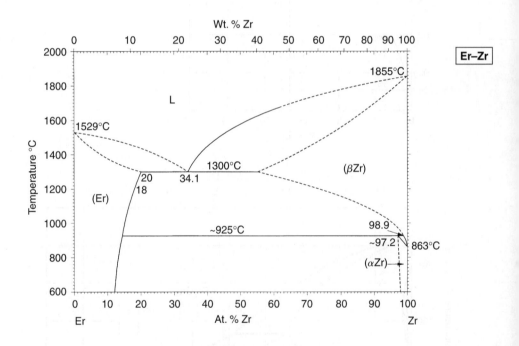

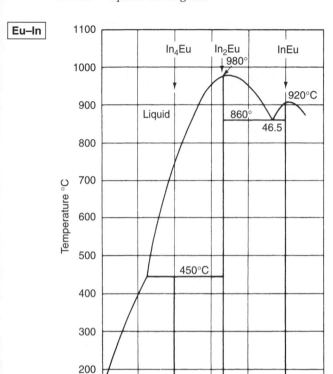

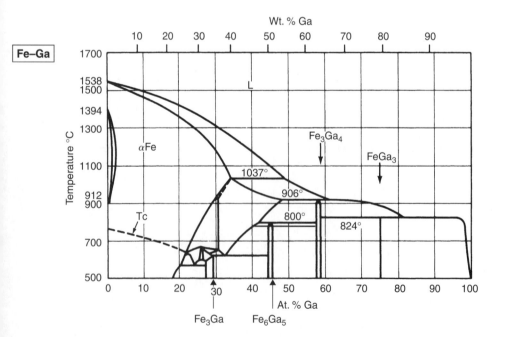

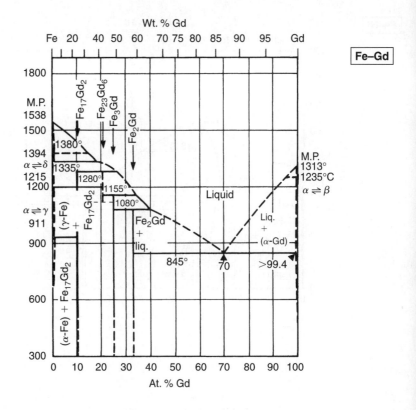

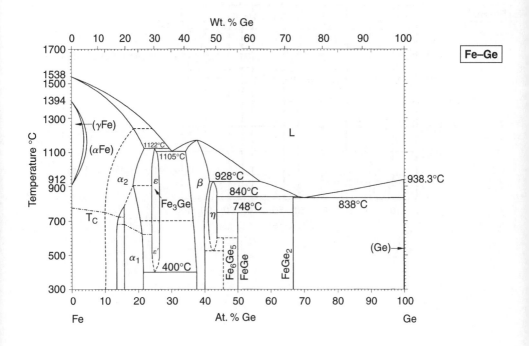

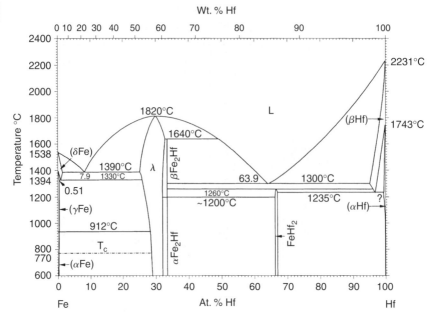

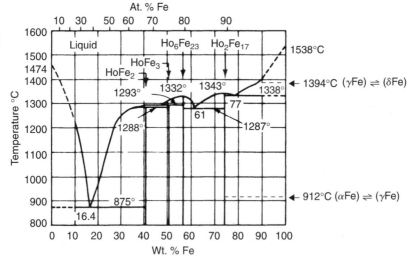

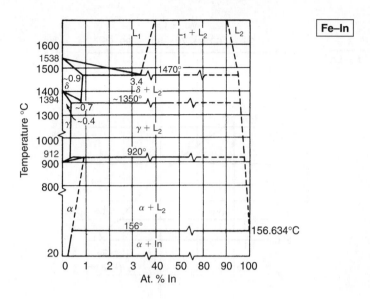

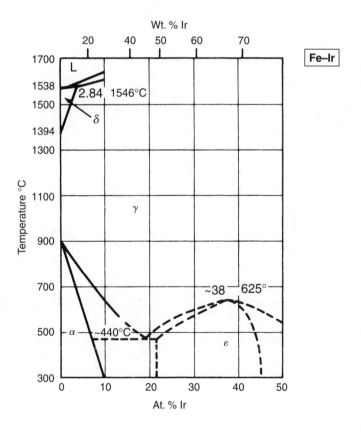

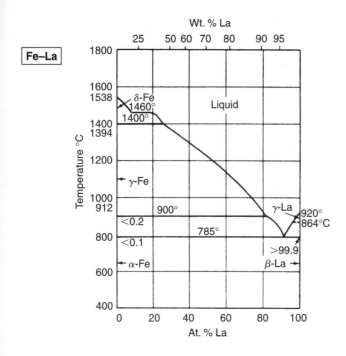

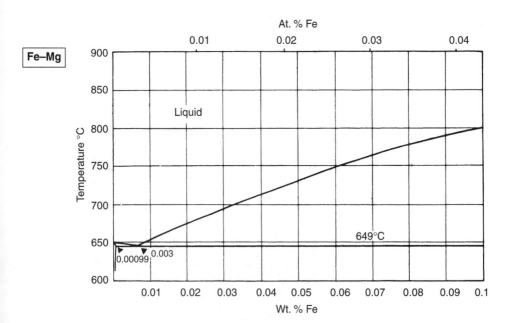

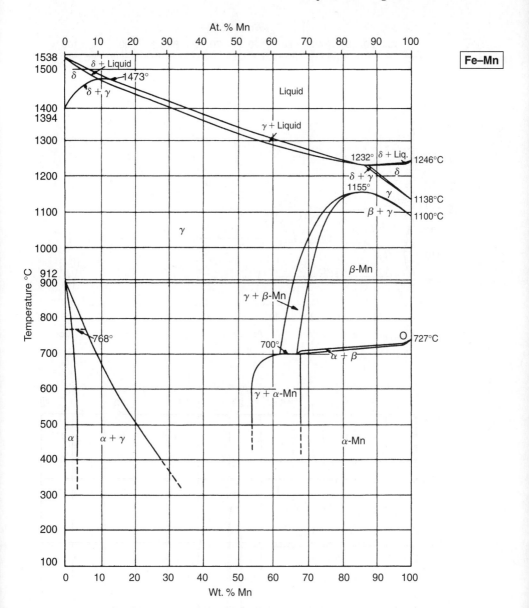

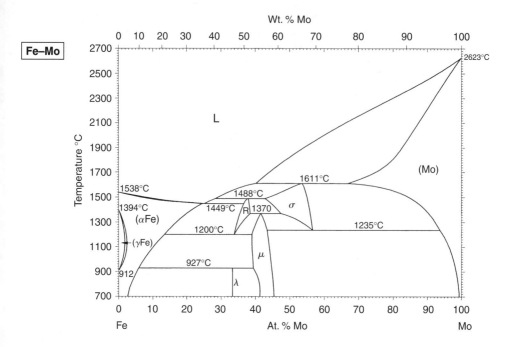

Fe–Mo

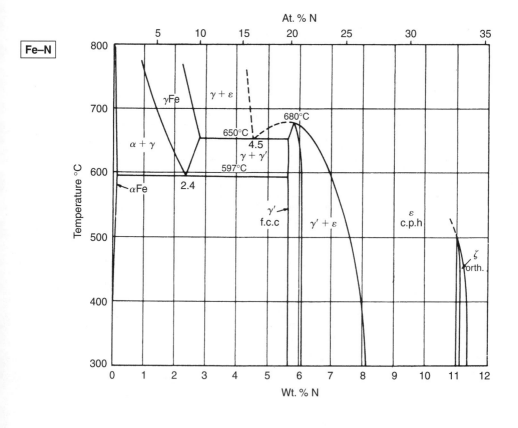

Fe–N

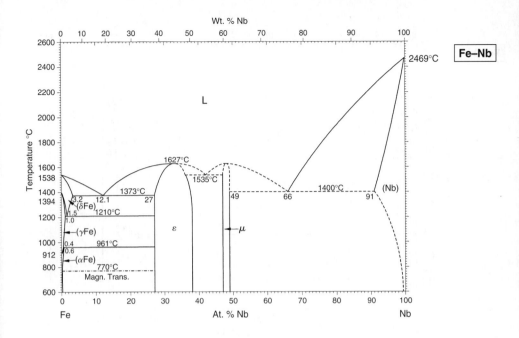

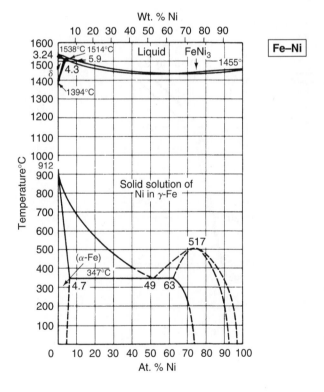

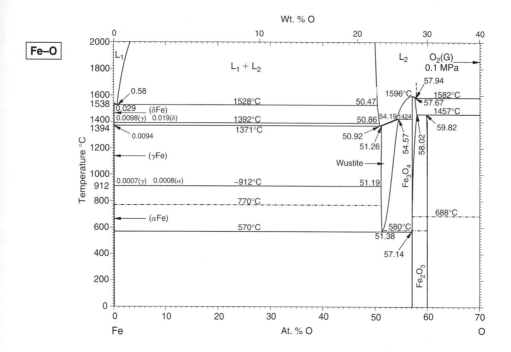

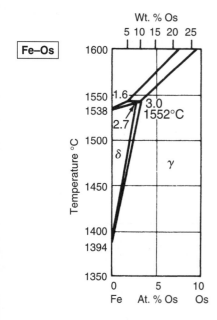

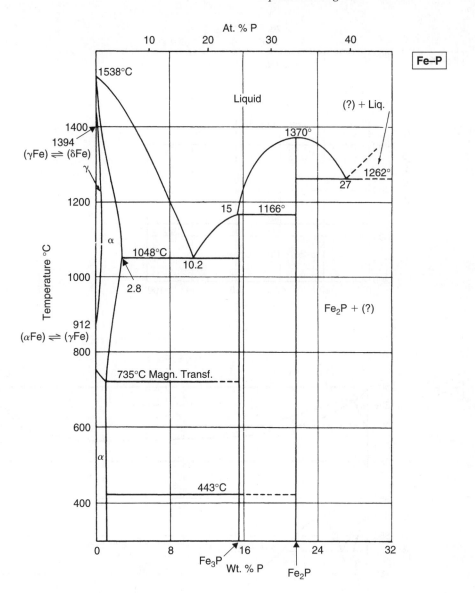

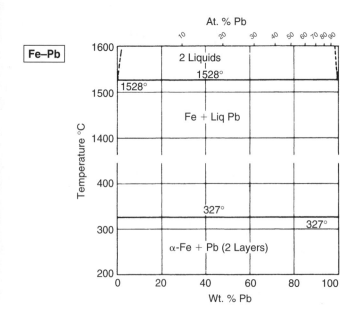

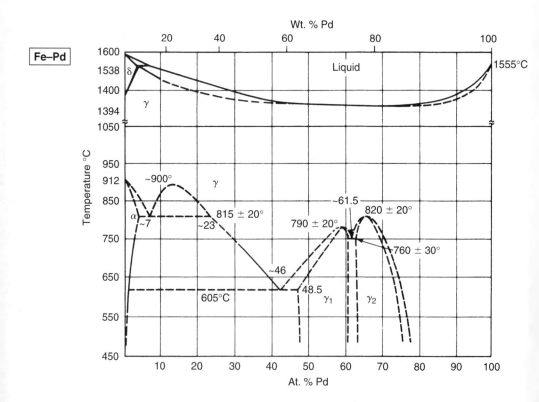

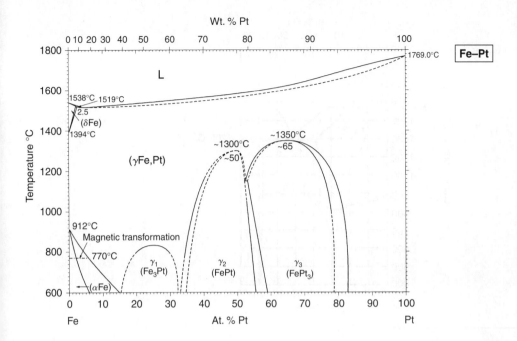

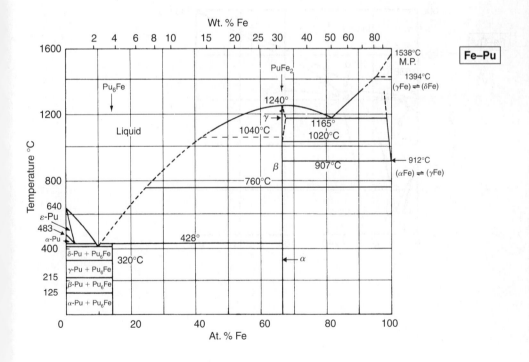

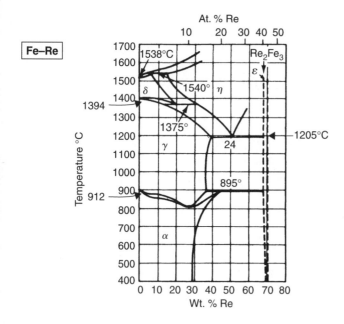

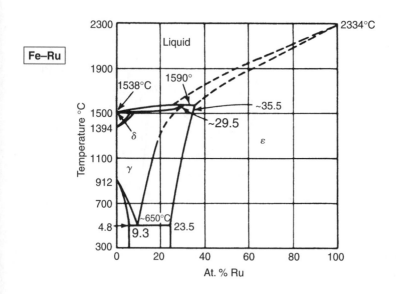

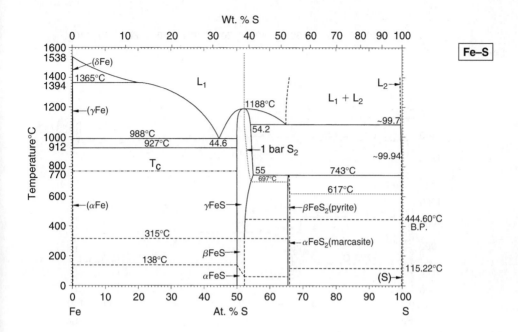

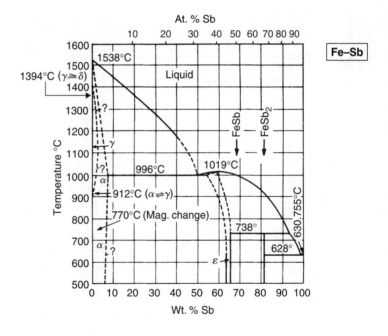

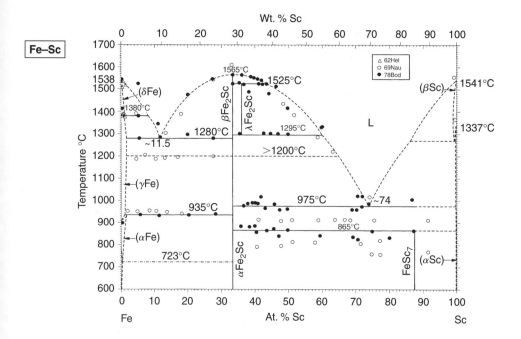

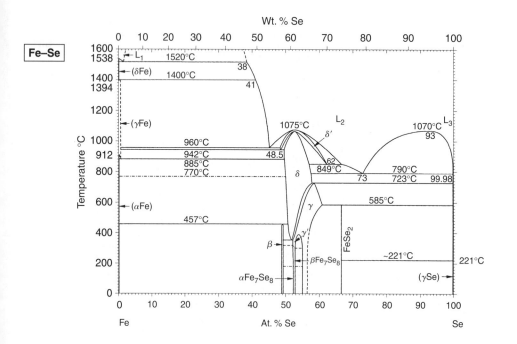

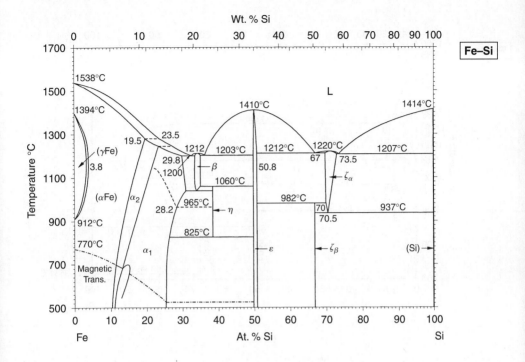

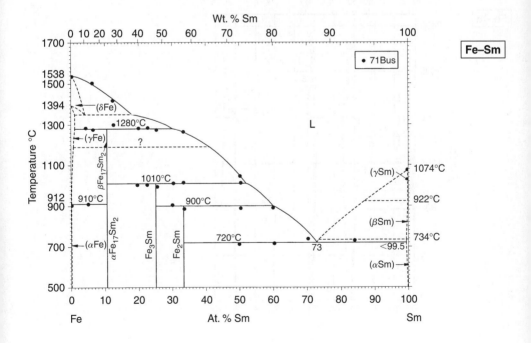

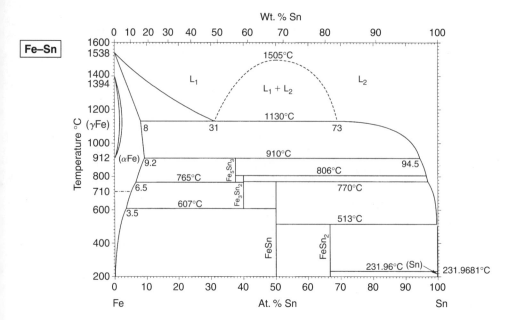

Fe–Sn

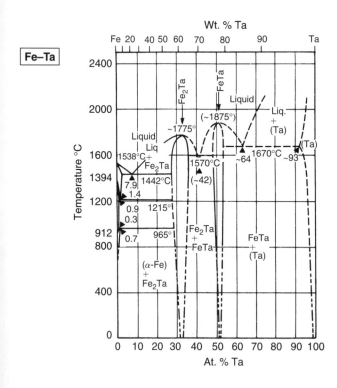

Fe–Ta

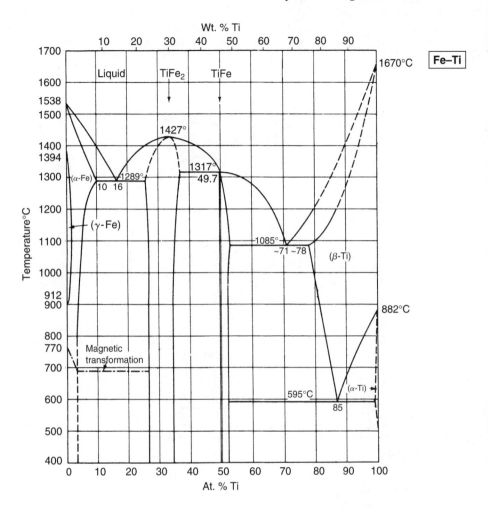

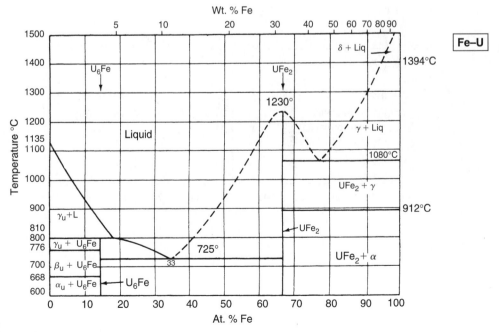

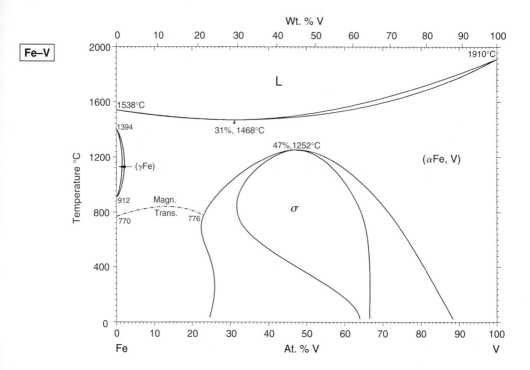

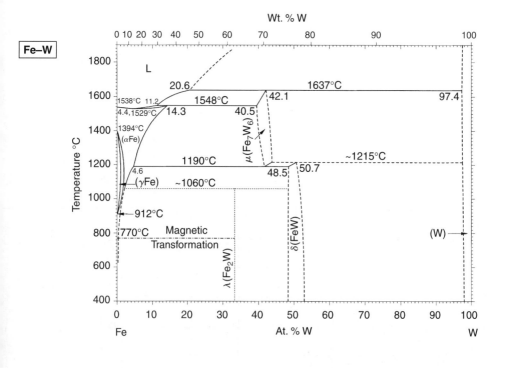

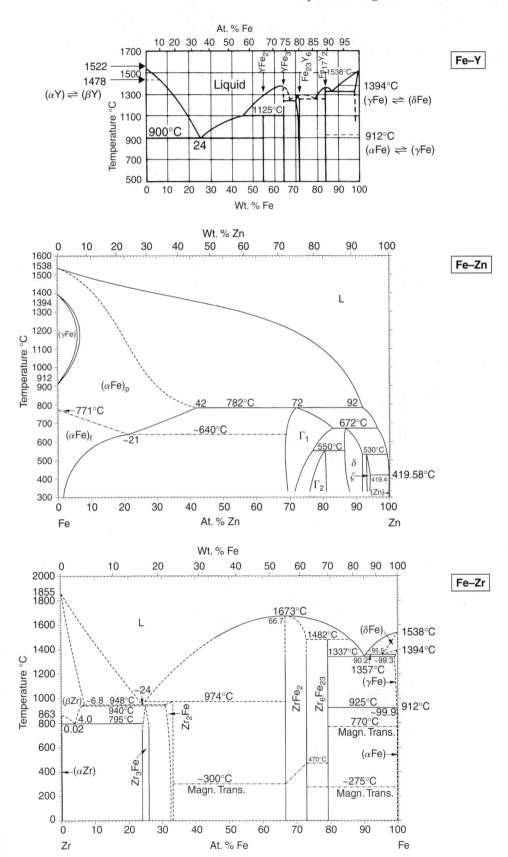

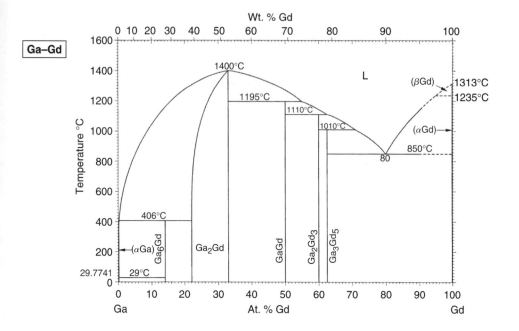

Ga–Gd

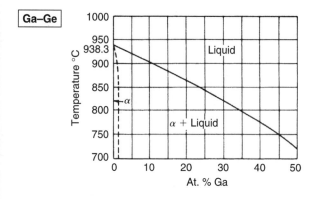

Ga–Ge

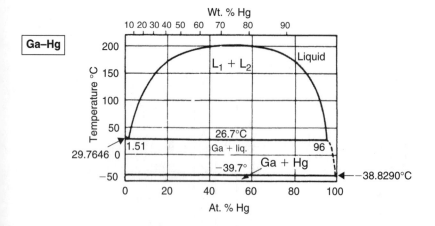

Ga–Hg

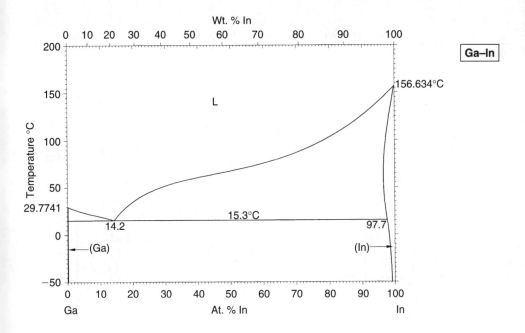

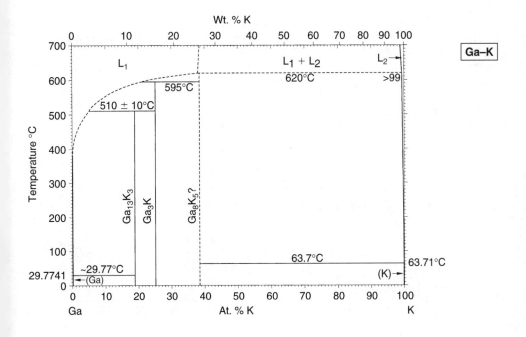

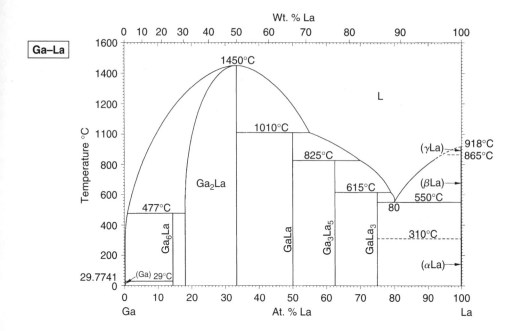

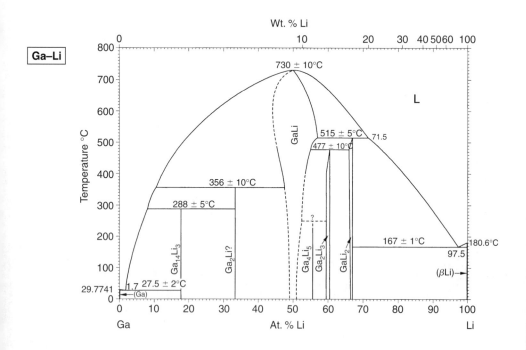

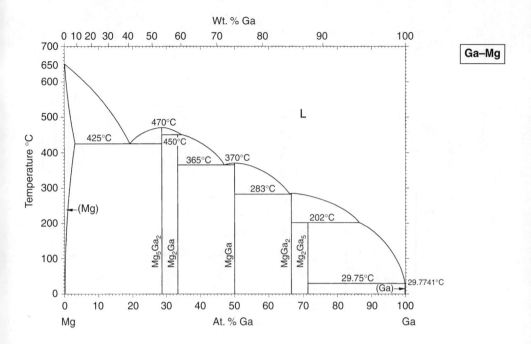

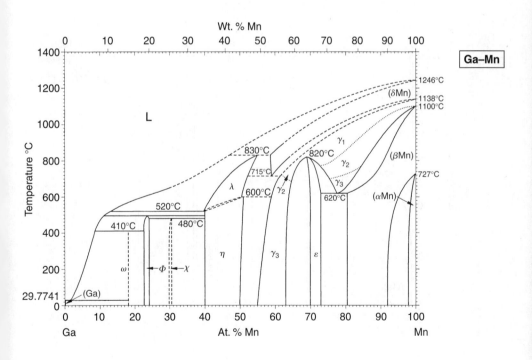

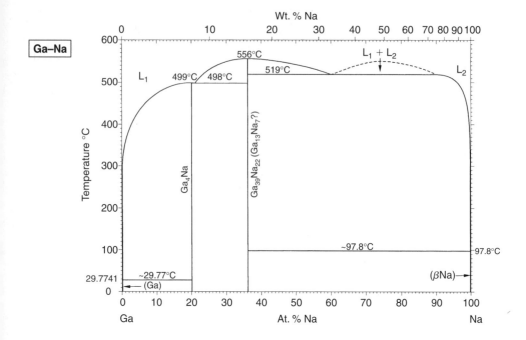

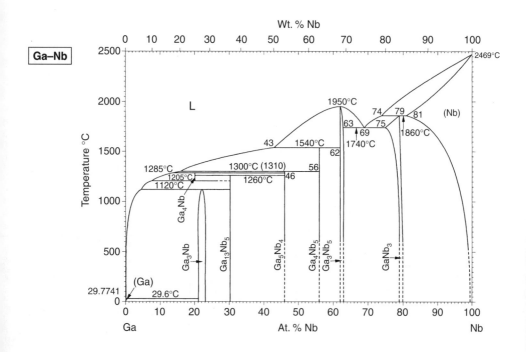

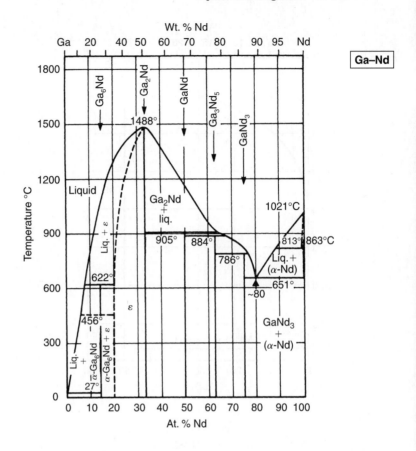

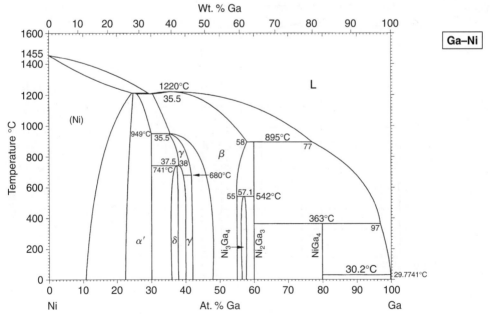

Ga–P

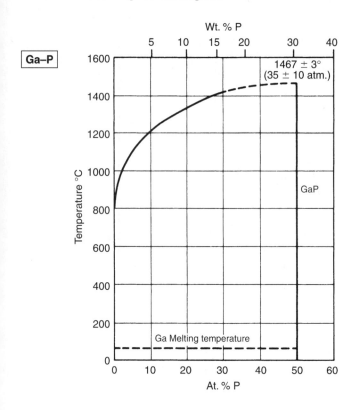

Ga–Pb

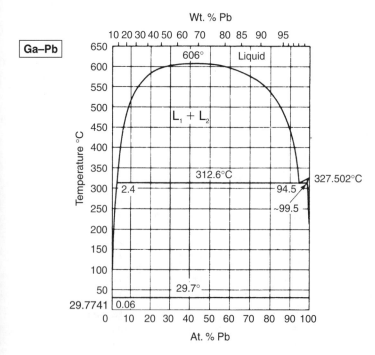

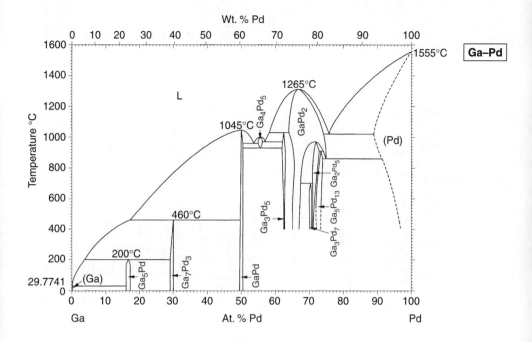

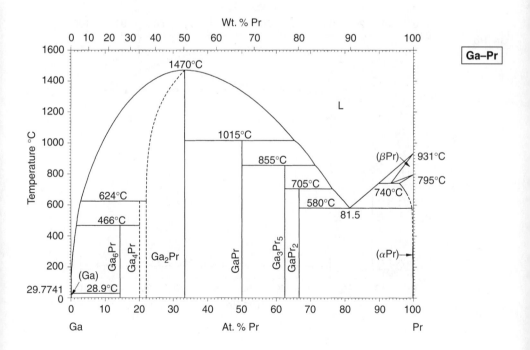

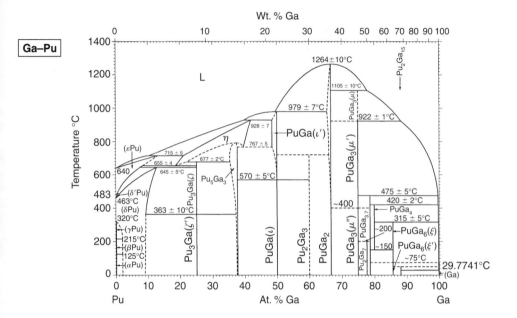

Ga–Pu

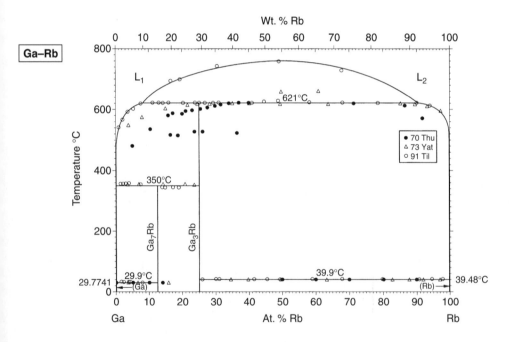

Ga–Rb

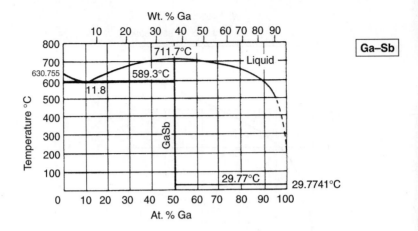

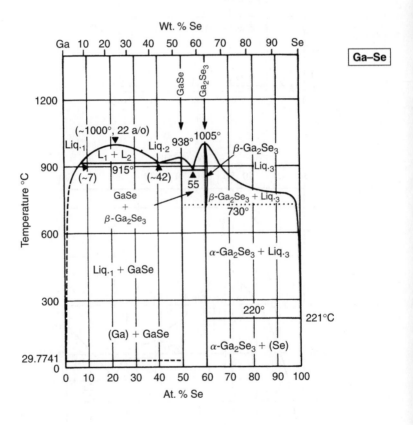

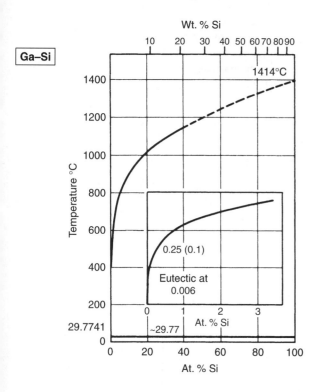

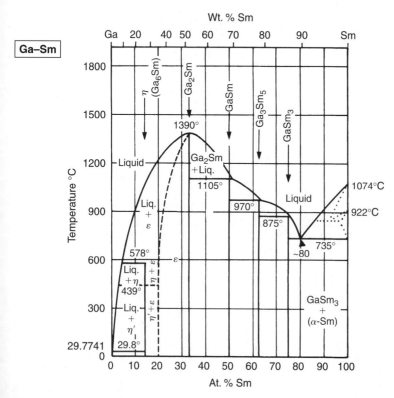

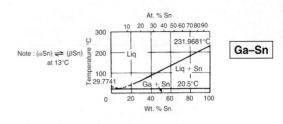

Ga–Sn

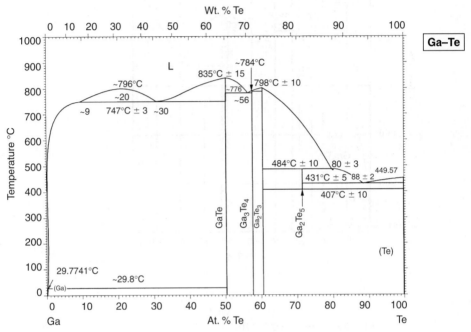

Ga–Te

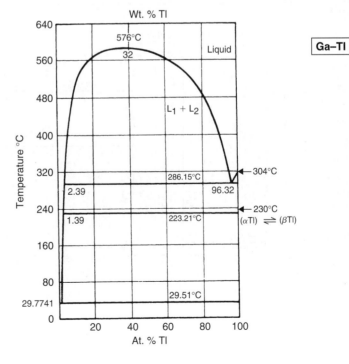

Ga–Tl

Ga–Tm

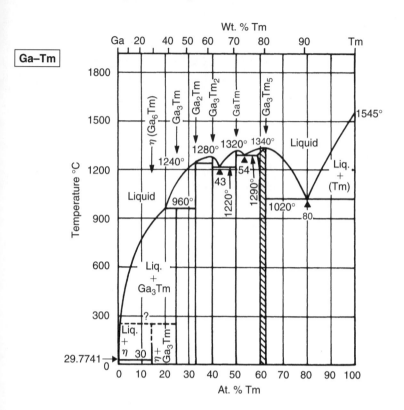

Ga–U

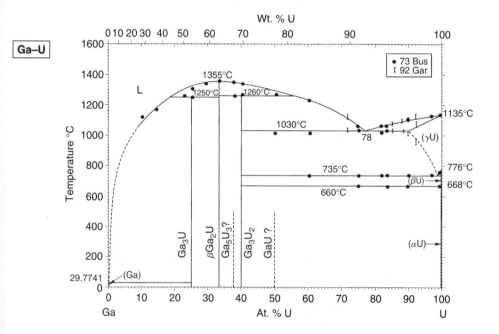

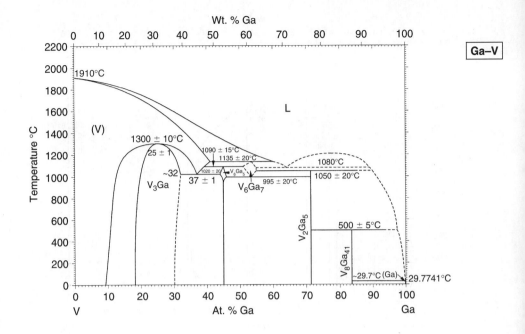

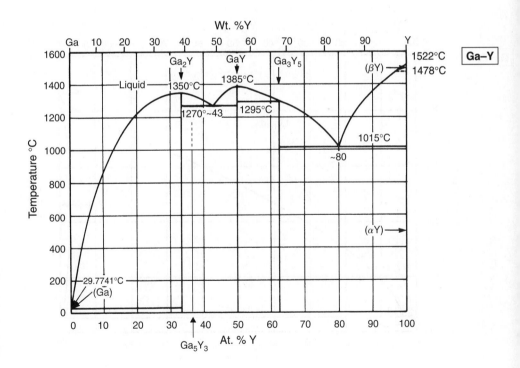

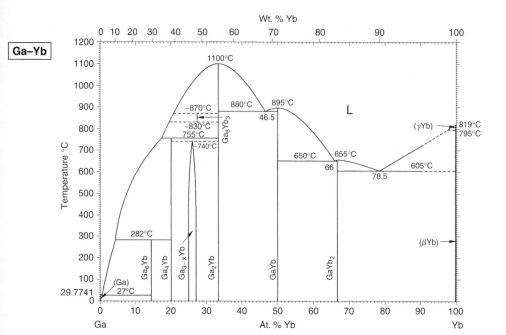

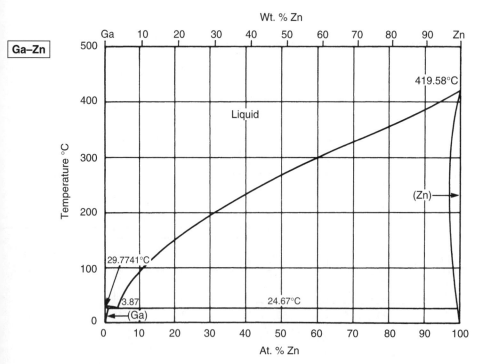

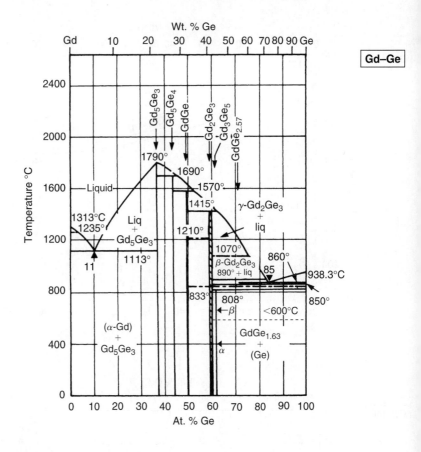

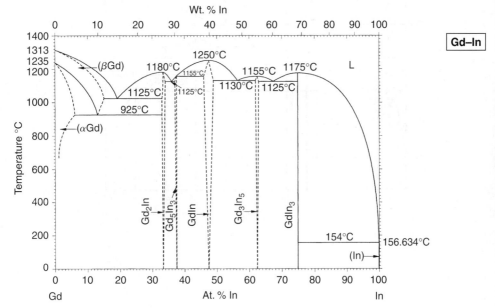

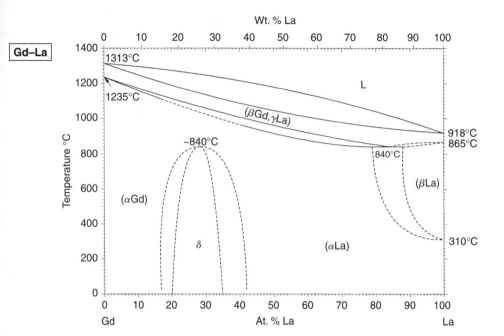

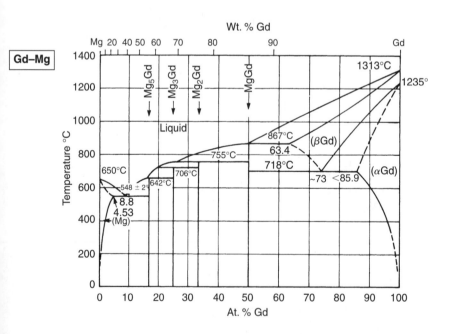

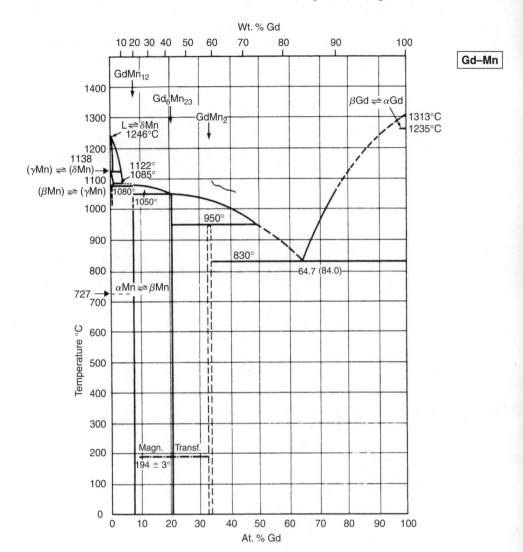

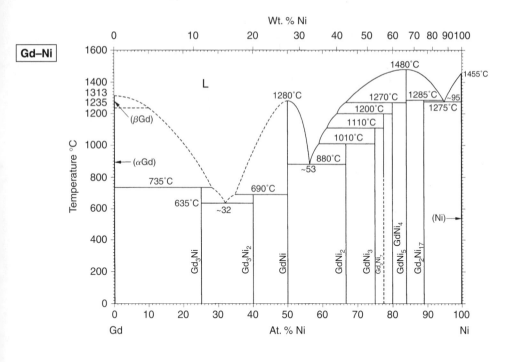

Gd–Ni

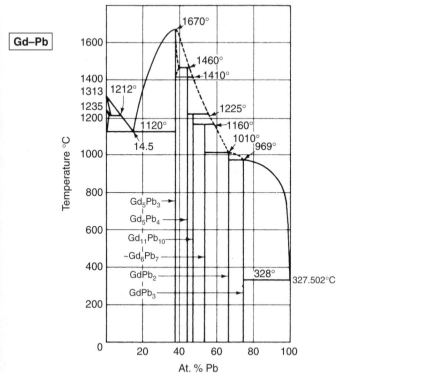

Gd–Pb

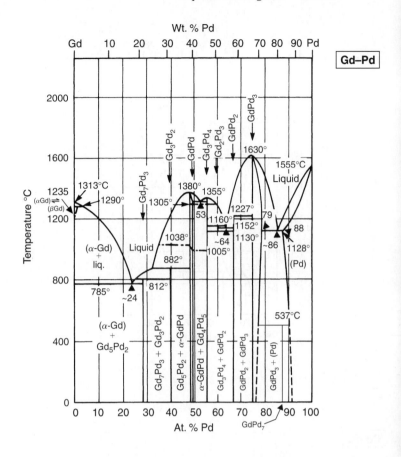

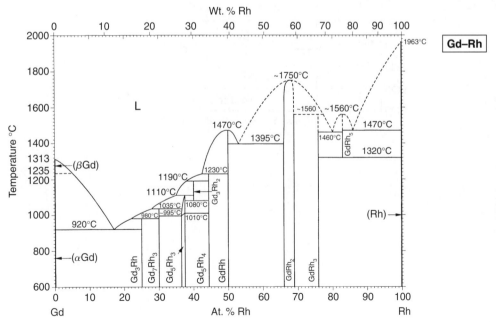

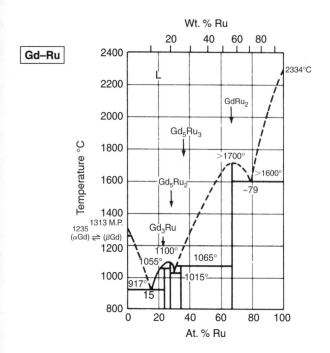

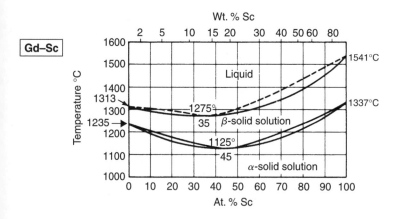

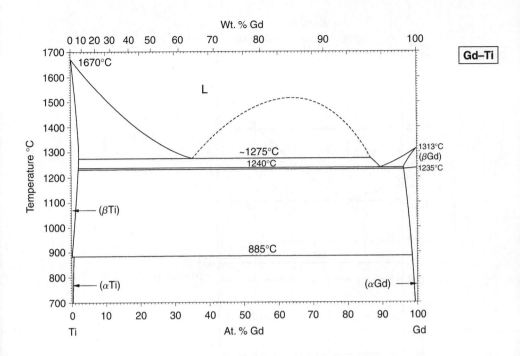

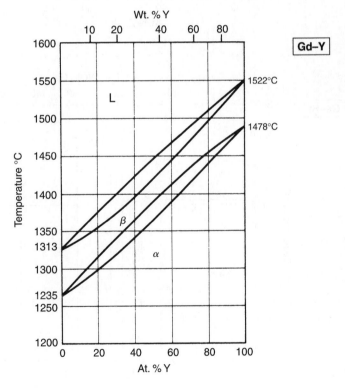

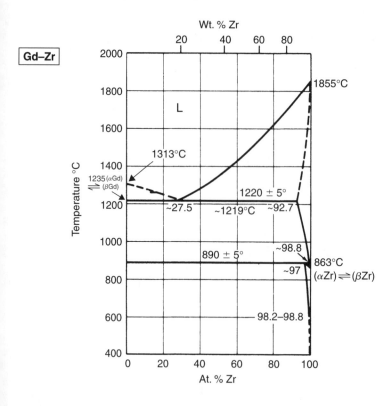

Gd–Zr

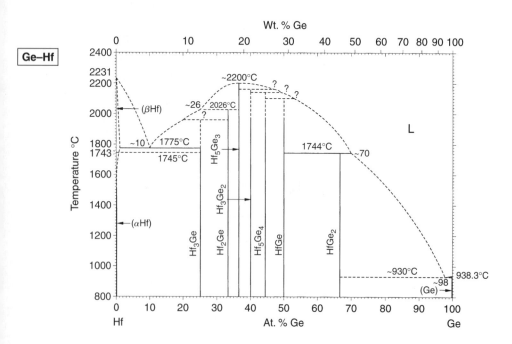

Ge–Hf

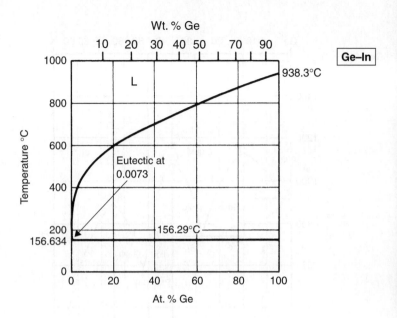

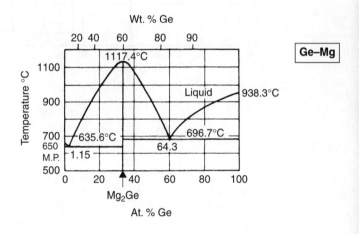

Ge–K

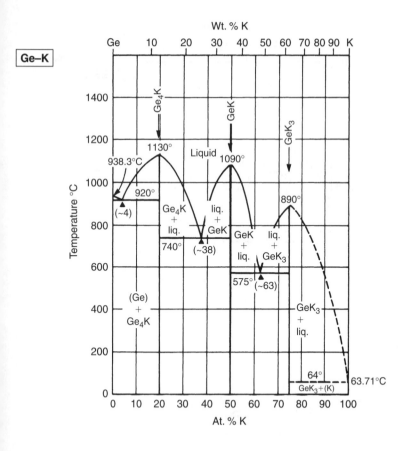

Ge–Mn

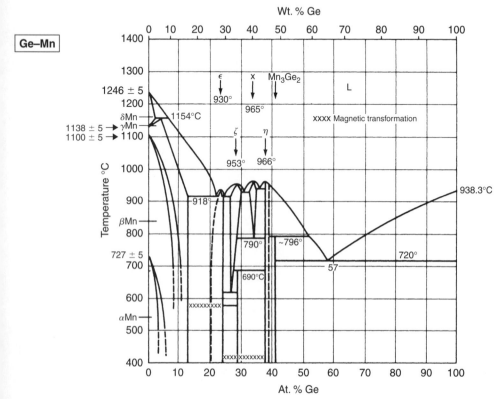

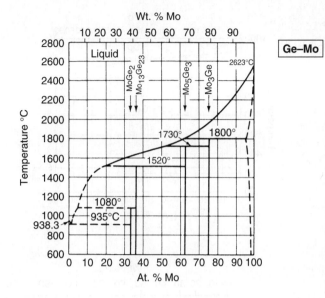

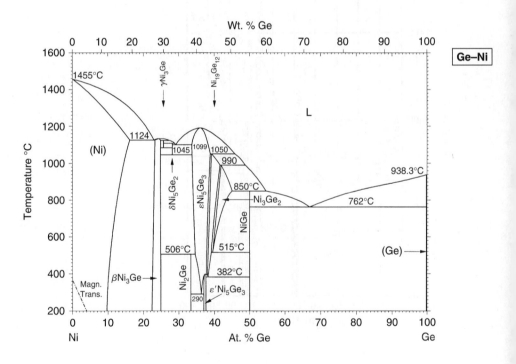

Ge–Pb

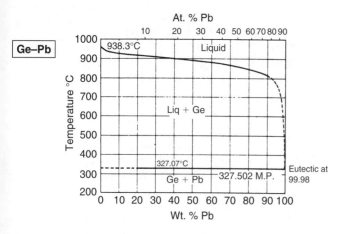

Ge–Pd

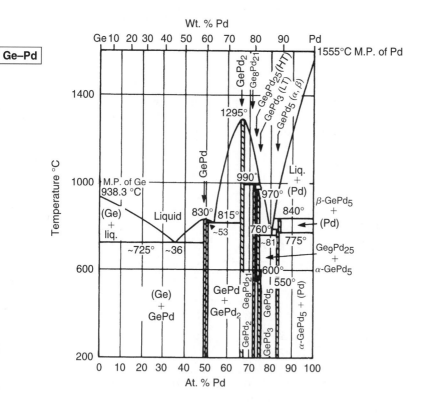

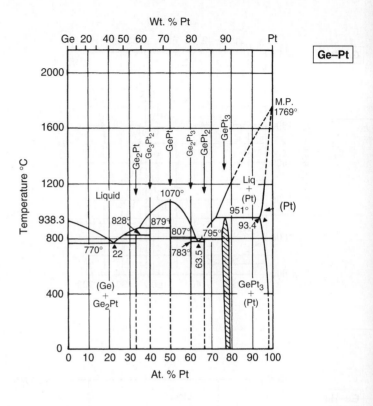

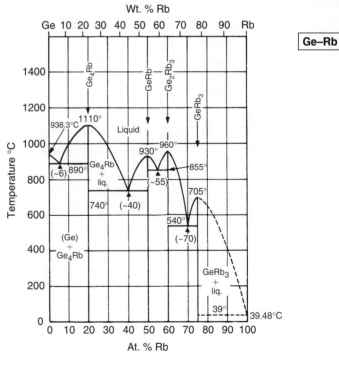

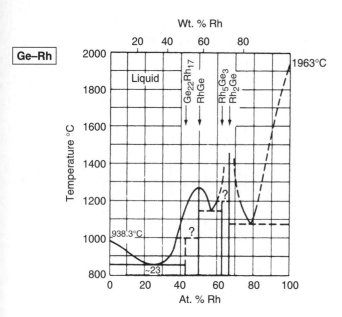

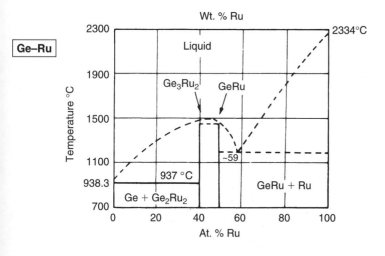

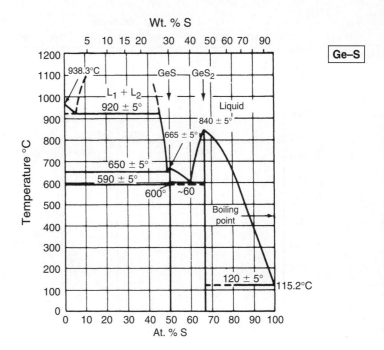

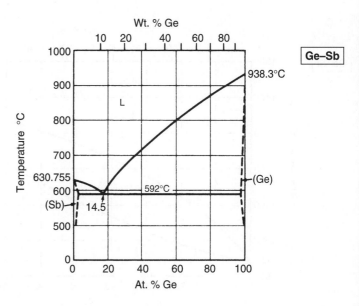

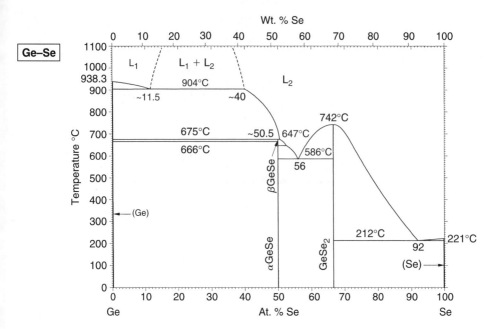

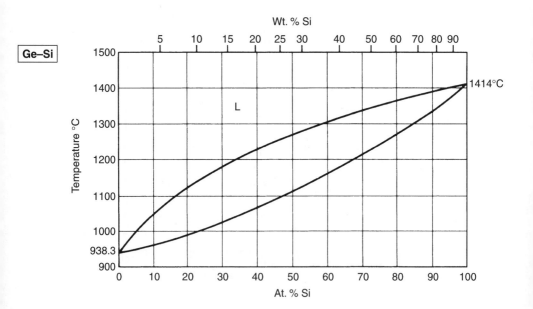

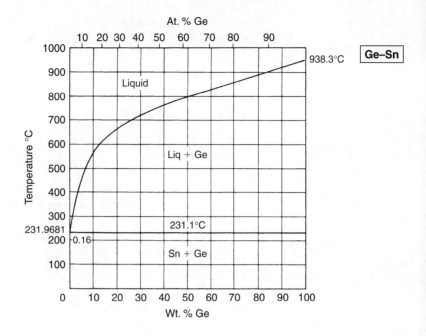

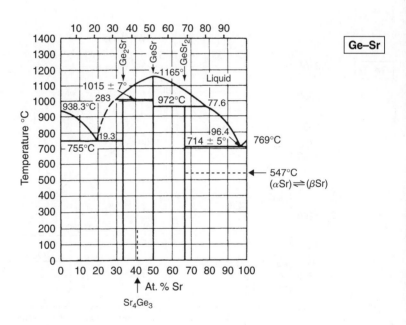

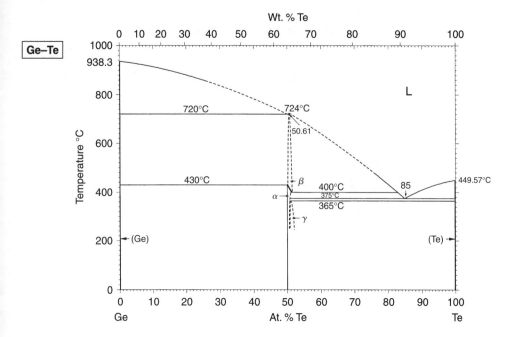

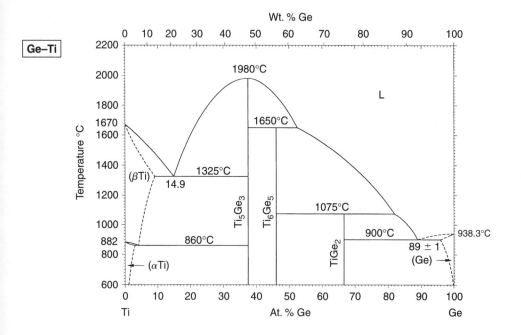

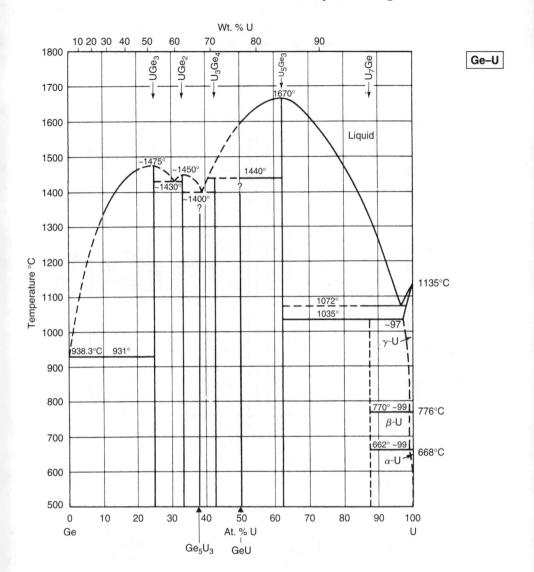

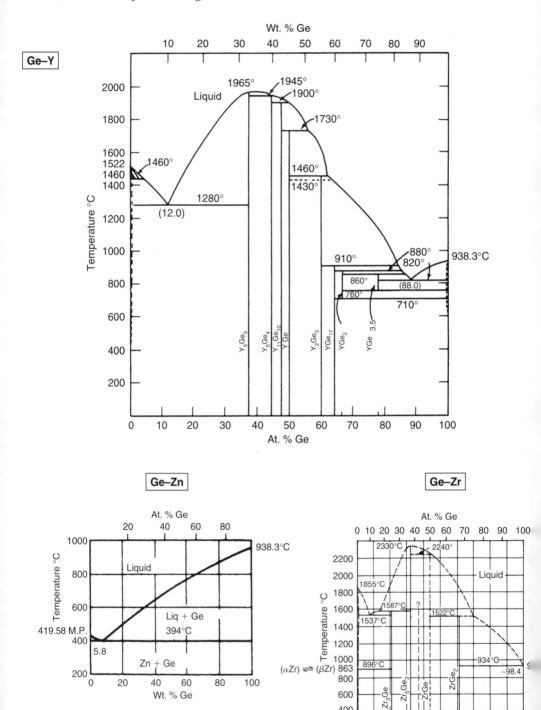

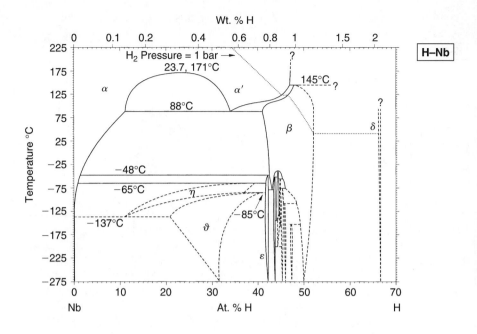

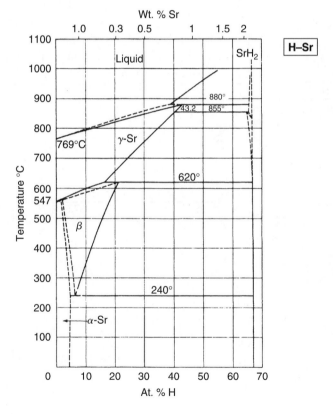

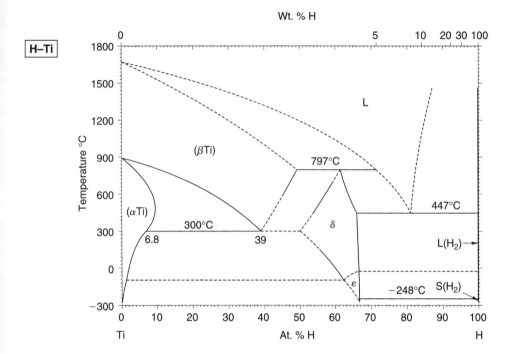

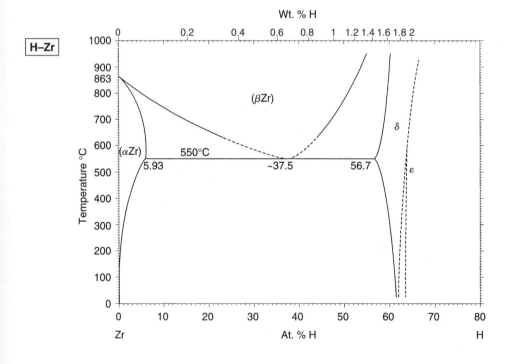

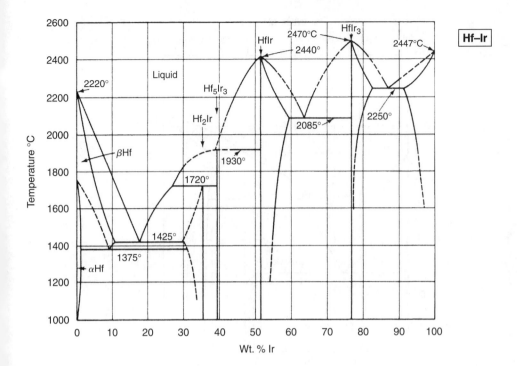

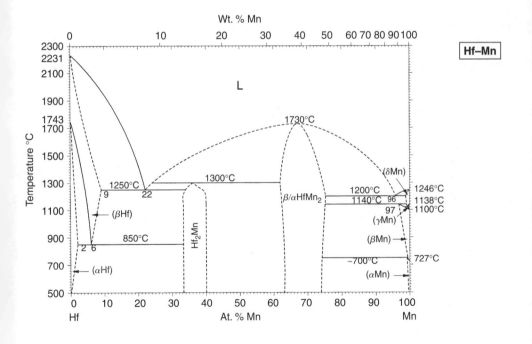

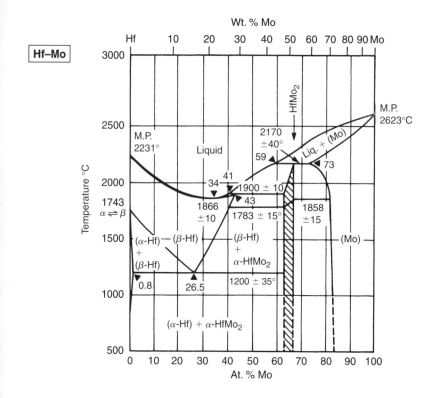

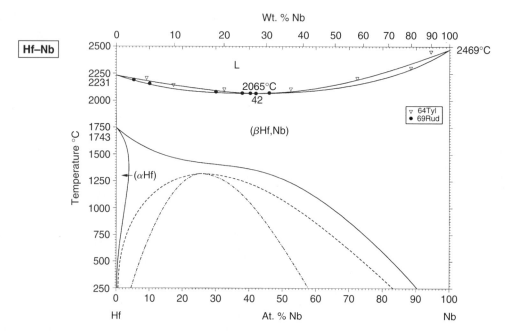

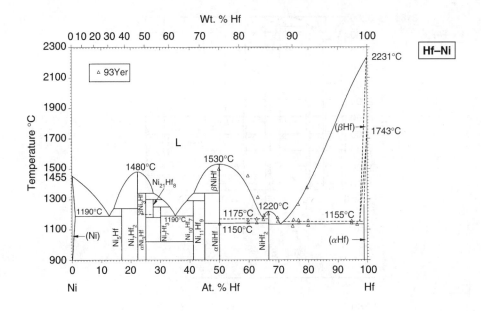

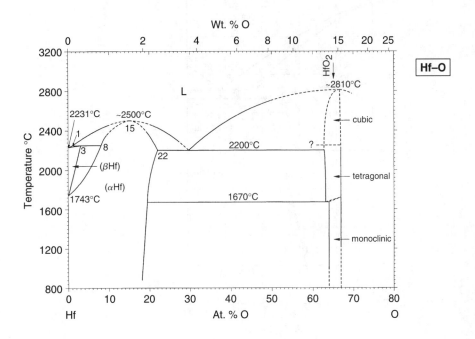

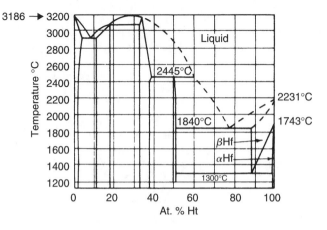

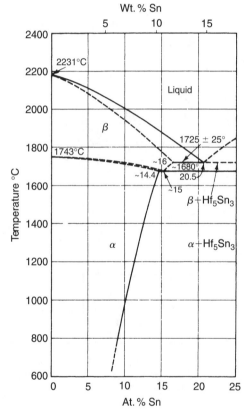

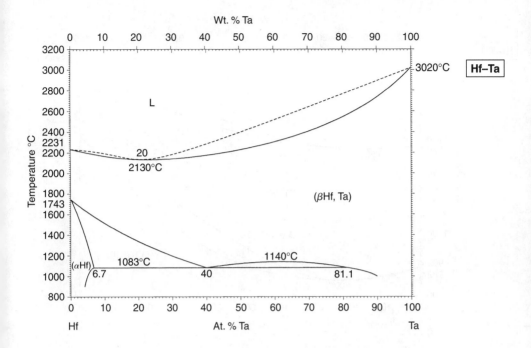

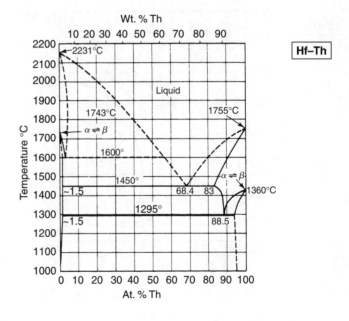

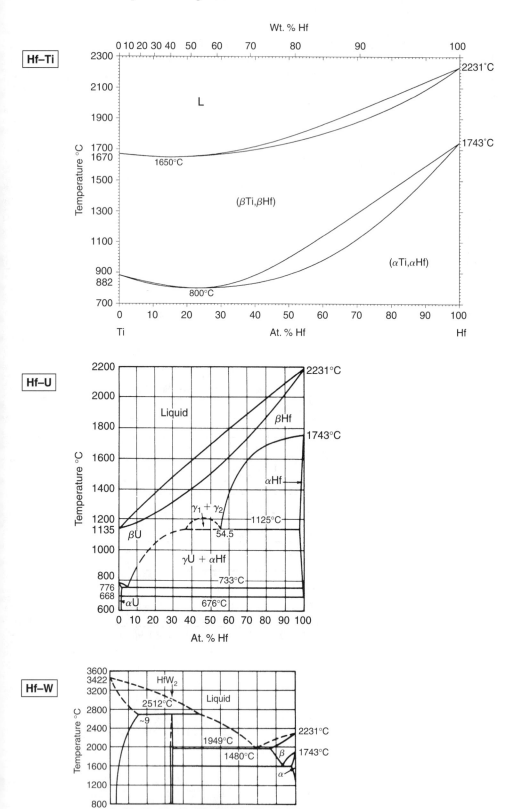

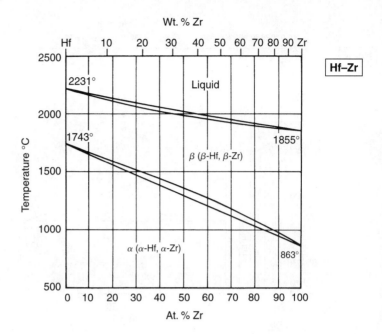

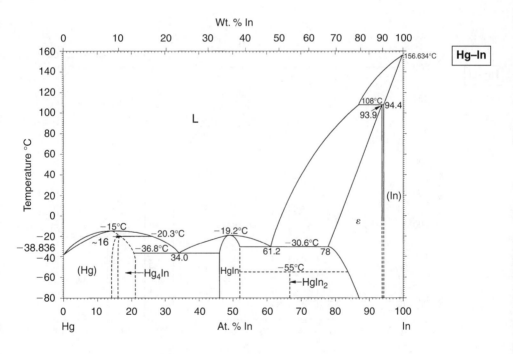

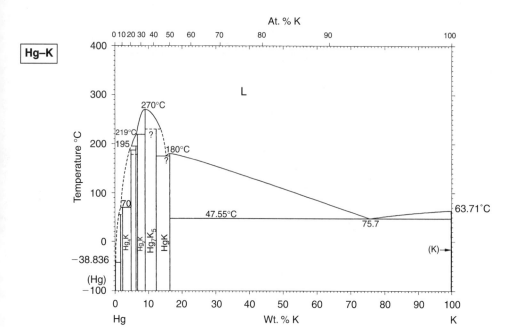

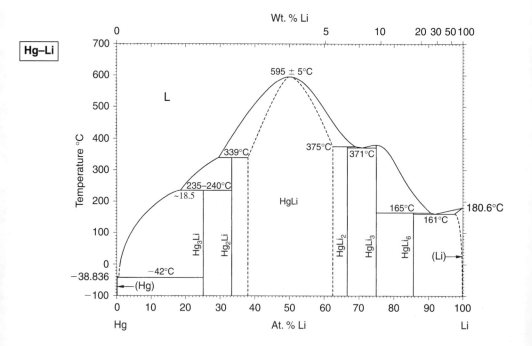

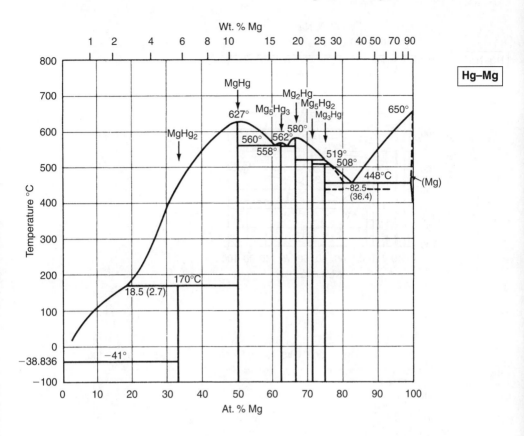

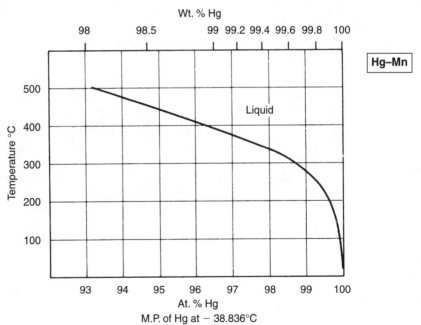

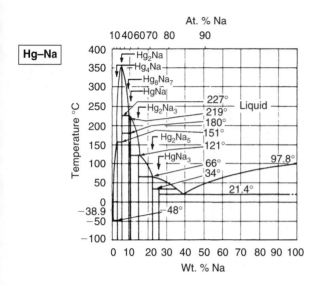

Hg–Na

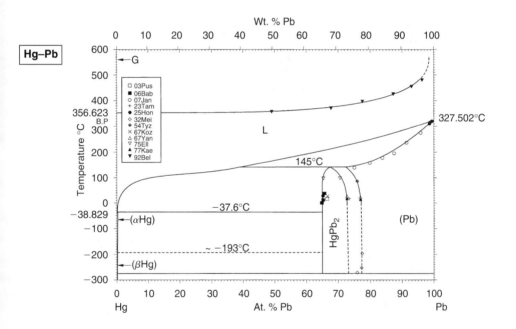

Hg–Pb

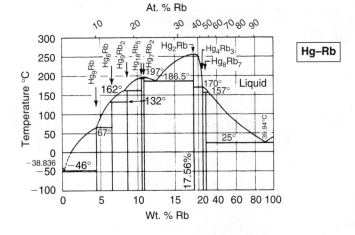

Hg–Rb

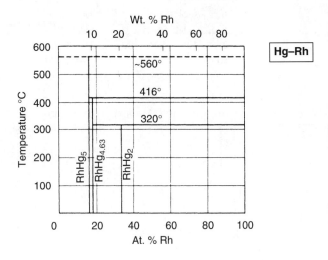

Hg–Rh

Hg–Sb

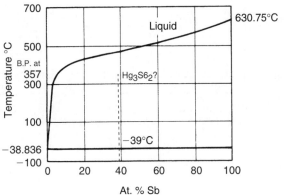

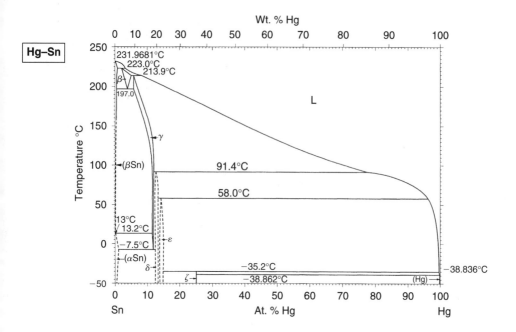

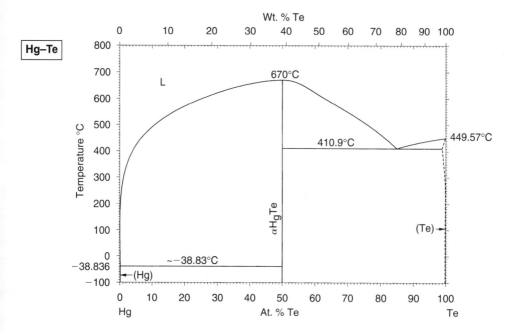

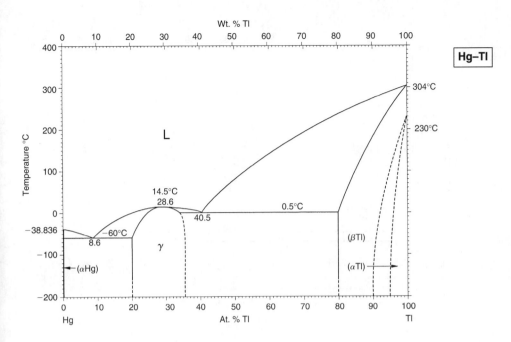

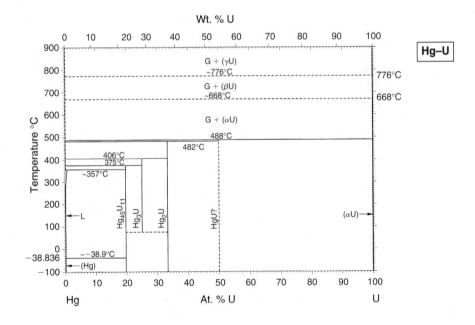

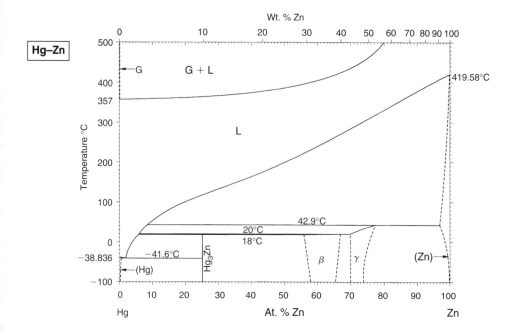

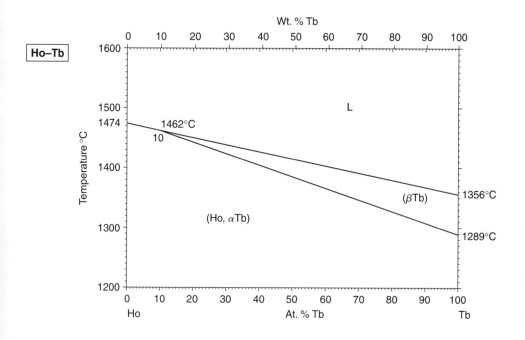

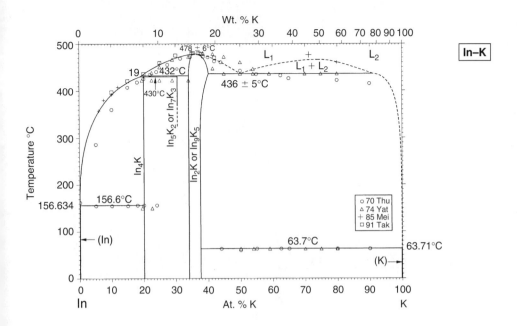

In–K

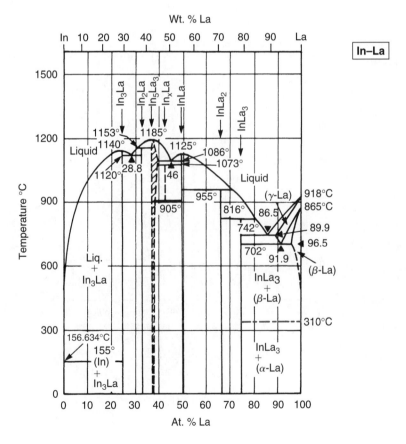

In–La

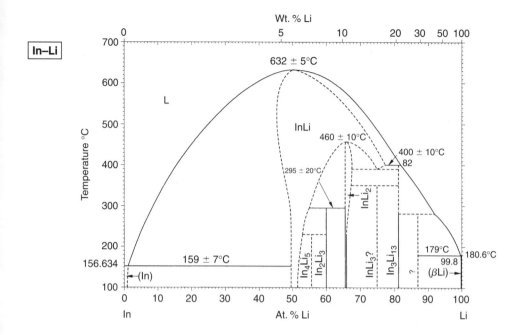

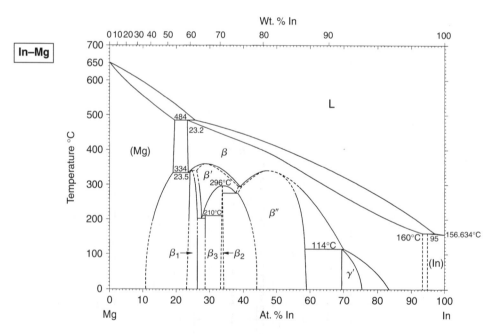

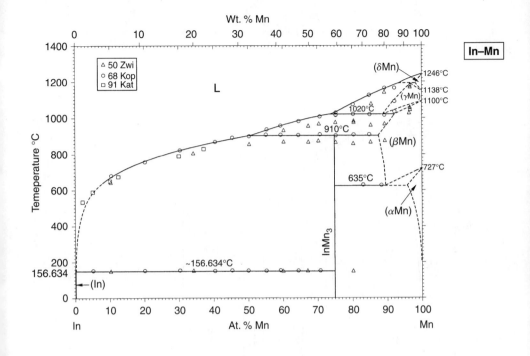

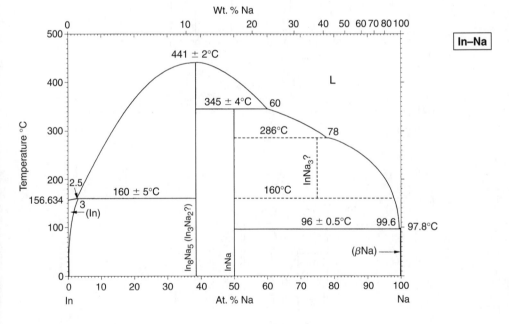

In–Nd

In–Ni

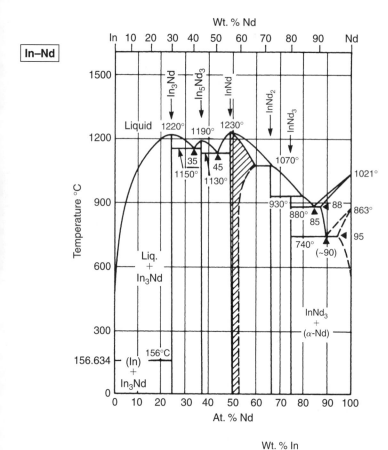

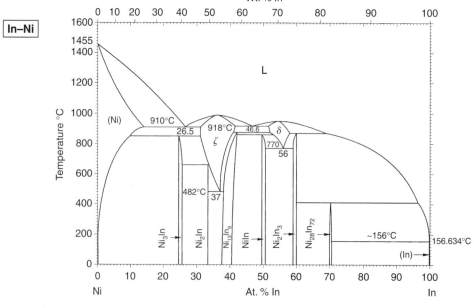

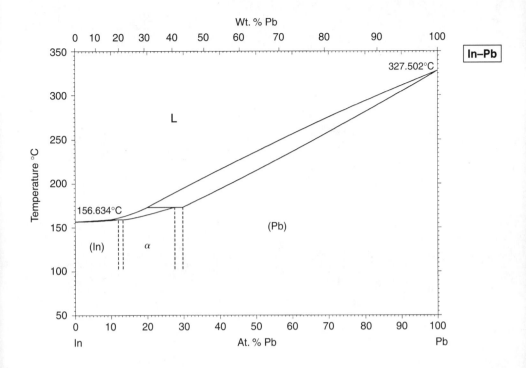

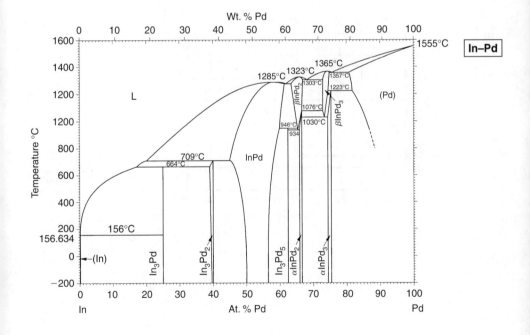

In–Pr

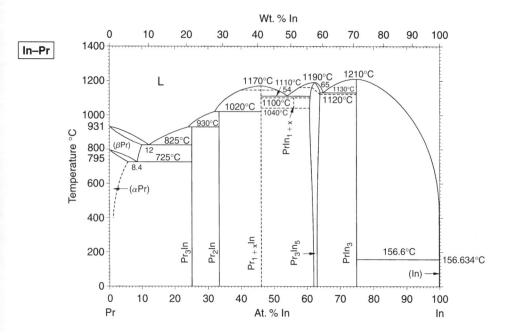

In–Sb

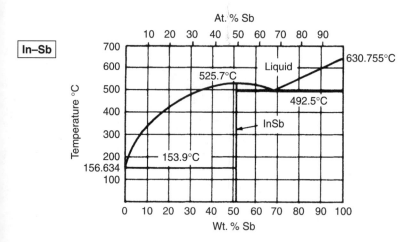

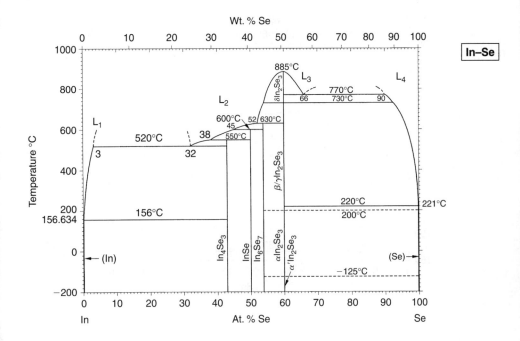

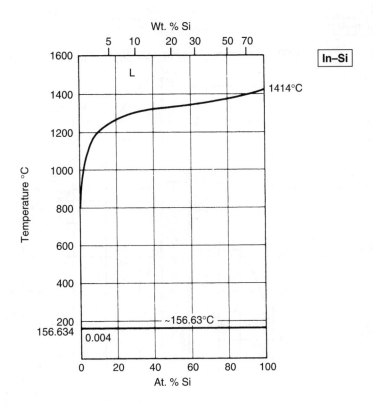

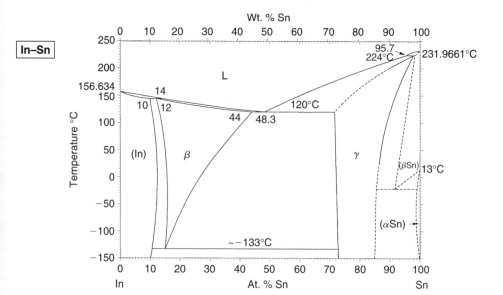

In–Sn

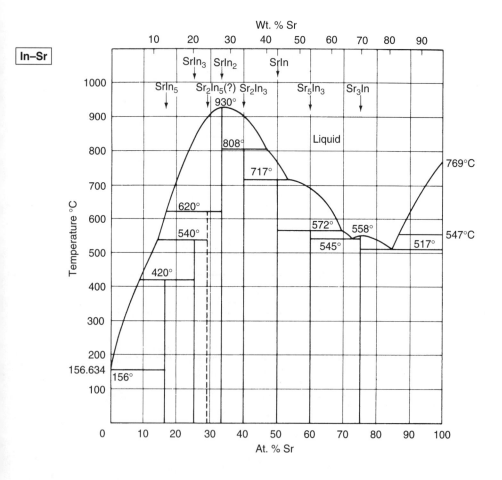

In–Sr

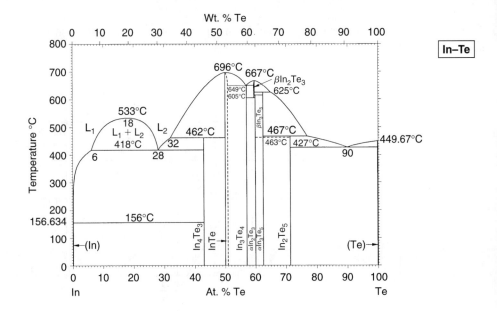

In–Te

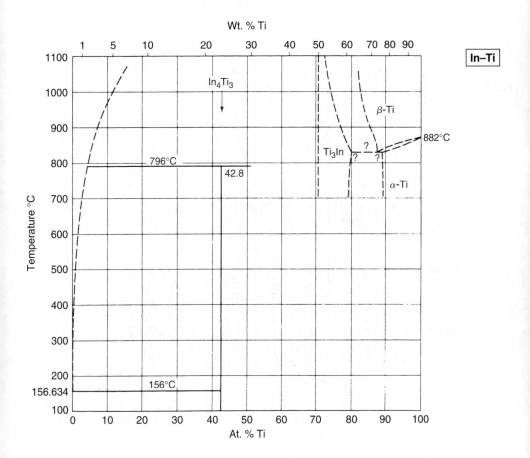

In–Ti

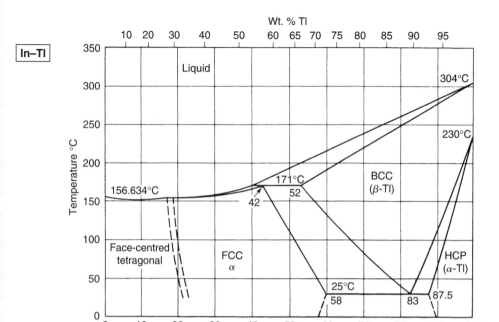

In–Tl

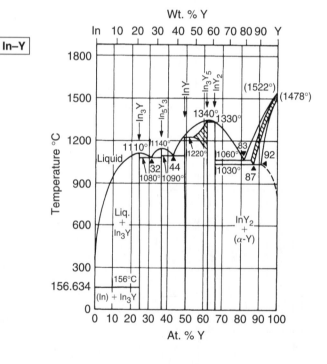

In–Y

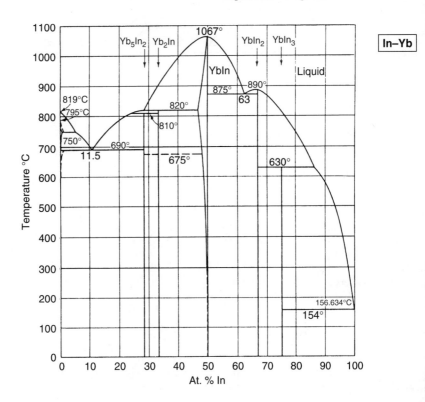

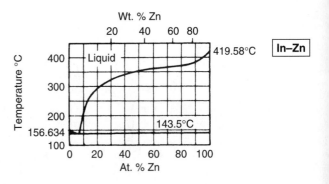

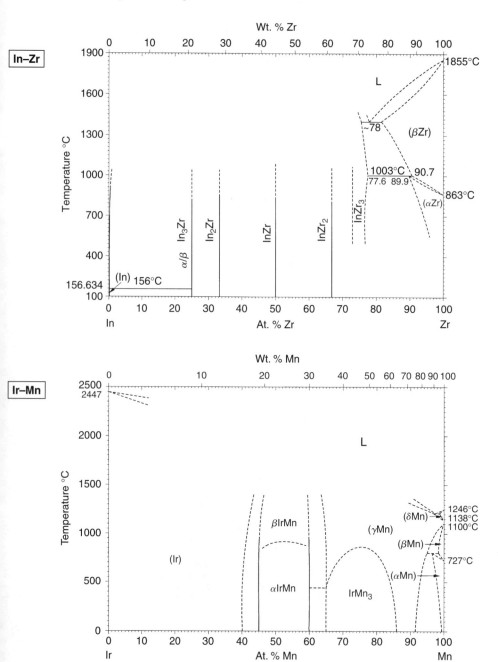

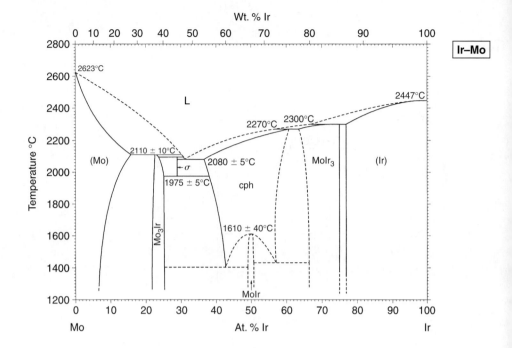

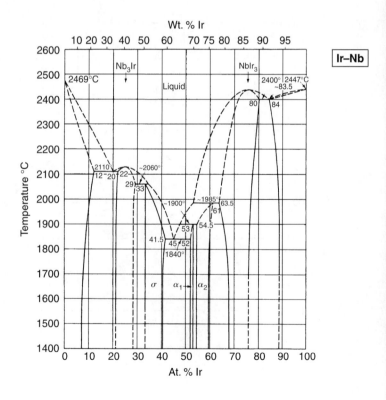

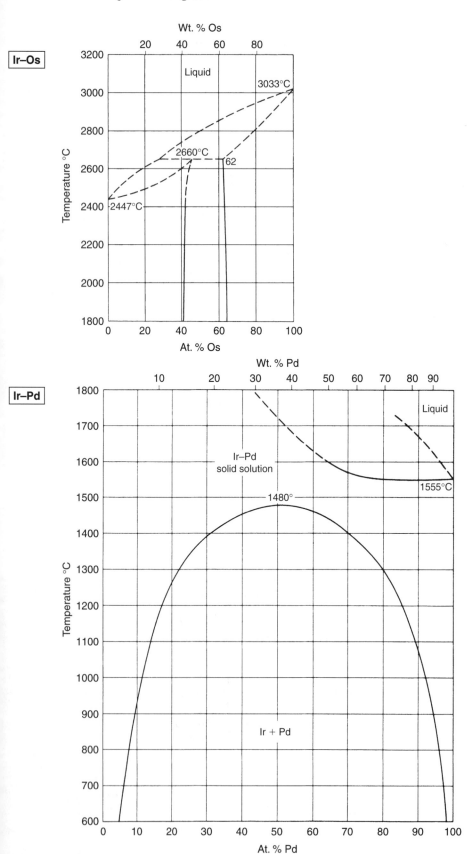

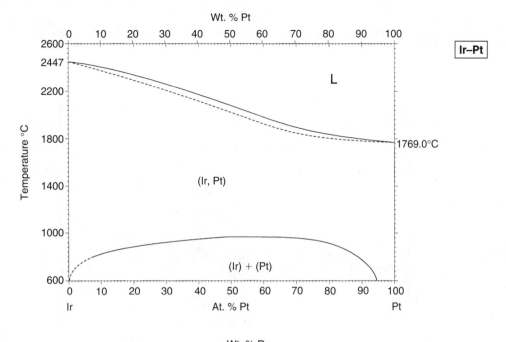

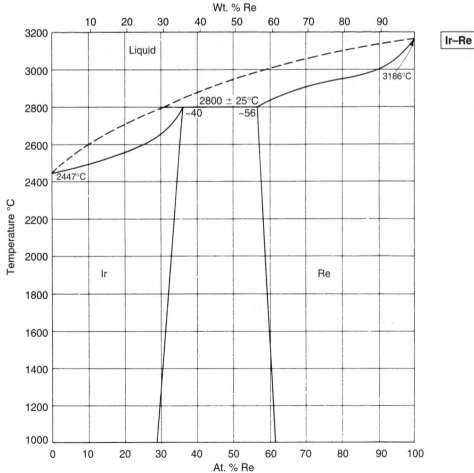

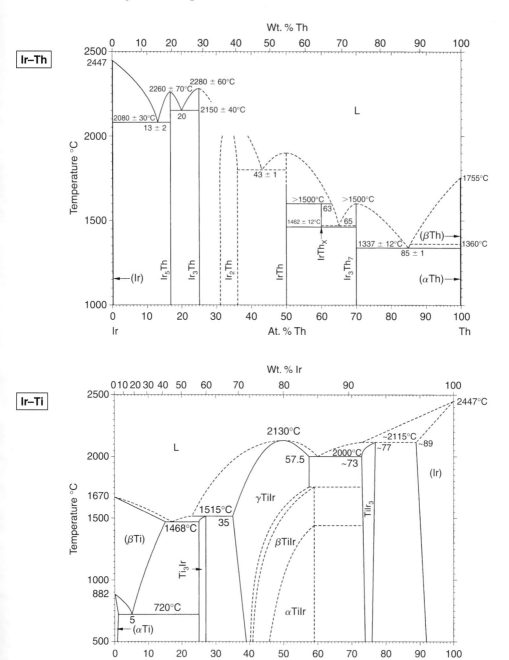

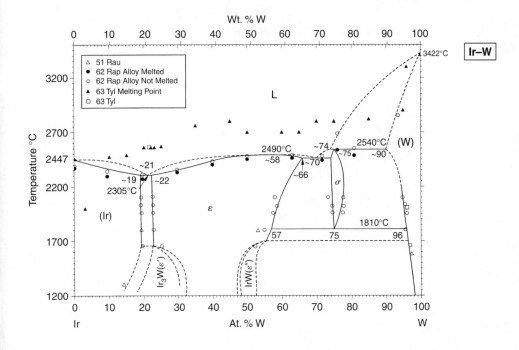

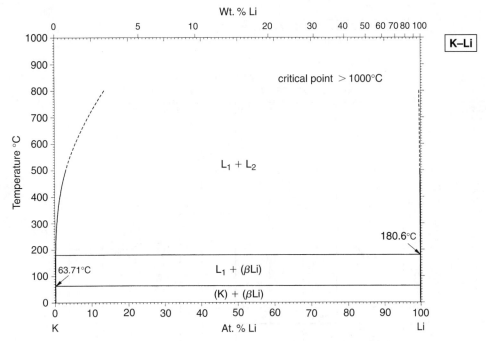

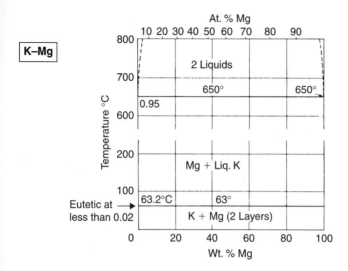

K–Mg

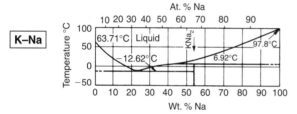

K–Na

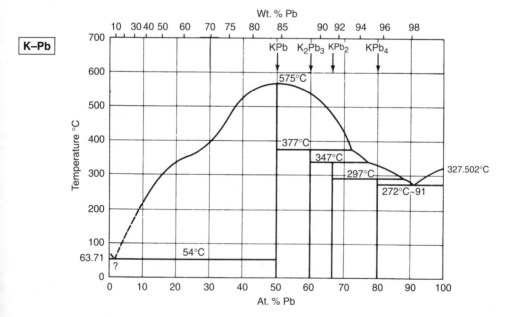

K–Pb

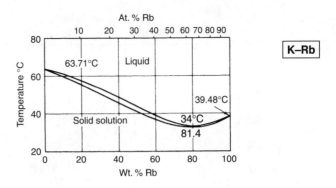

K–Rb

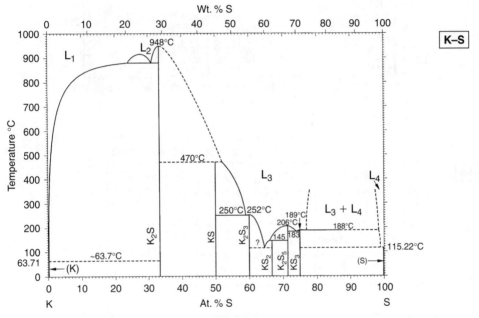

K–S

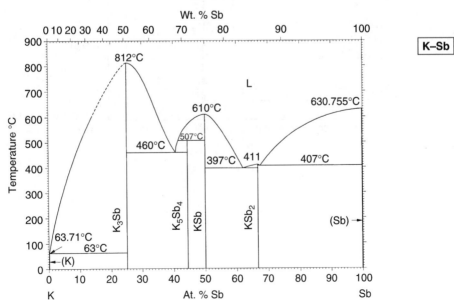

K–Sb

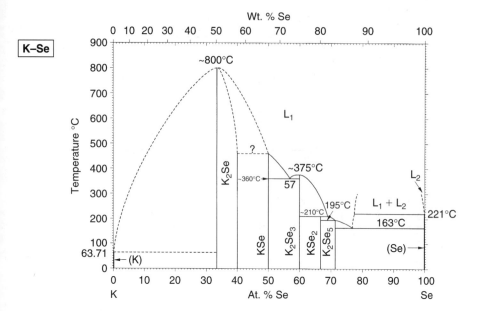

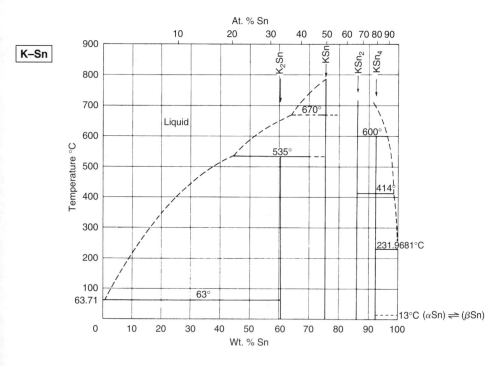

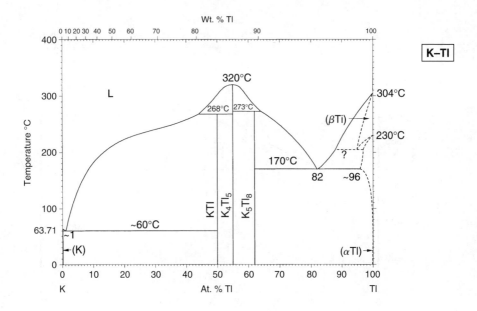

K–Tl

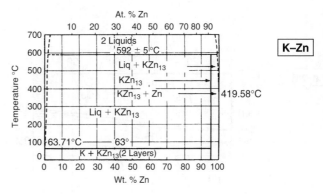

K–Zn

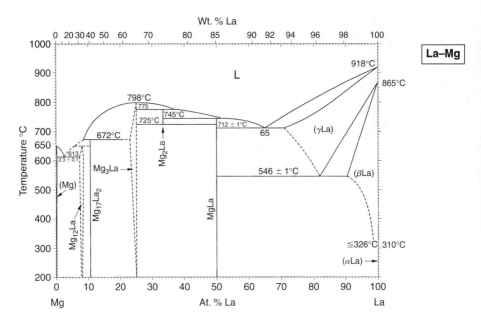

La–Mg

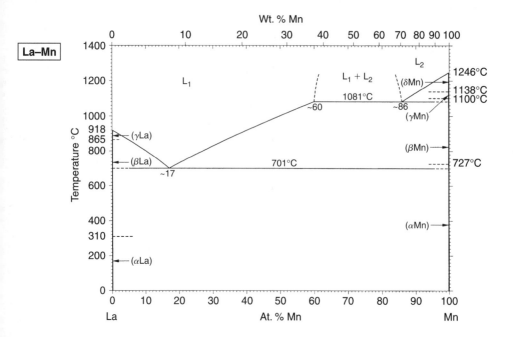

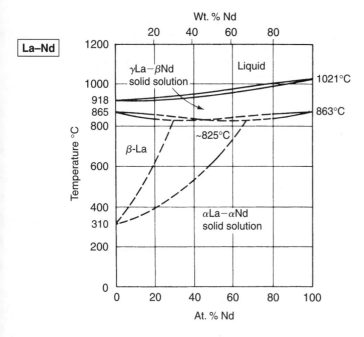

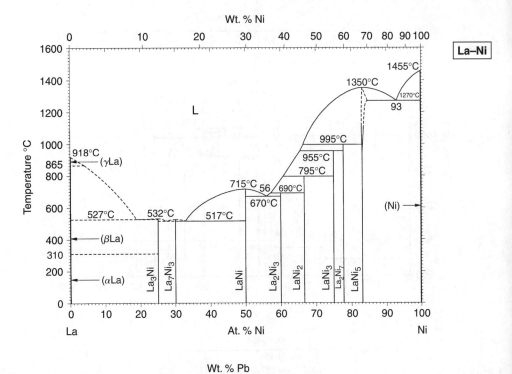

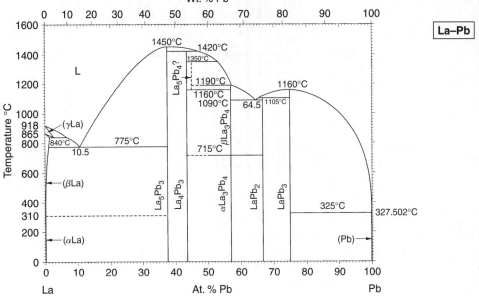

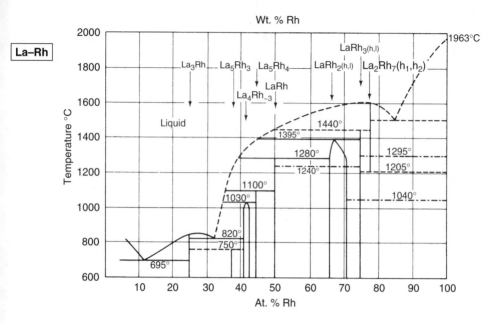

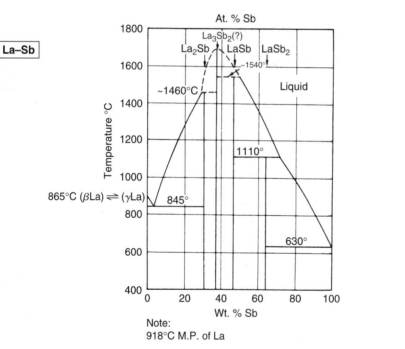

Note:
918°C M.P. of La

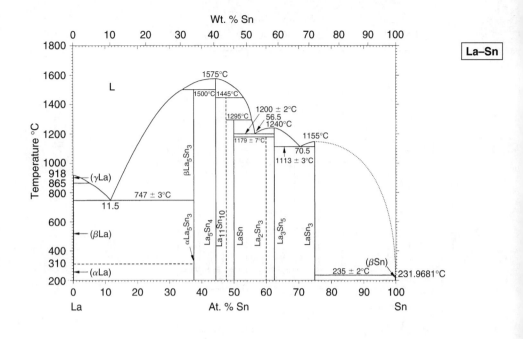

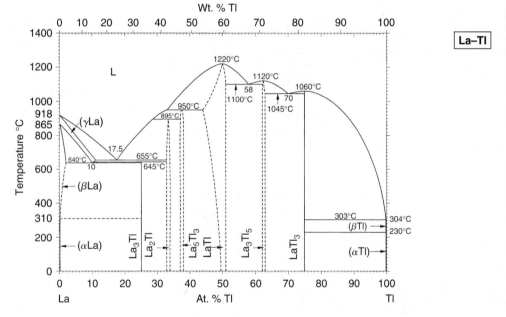

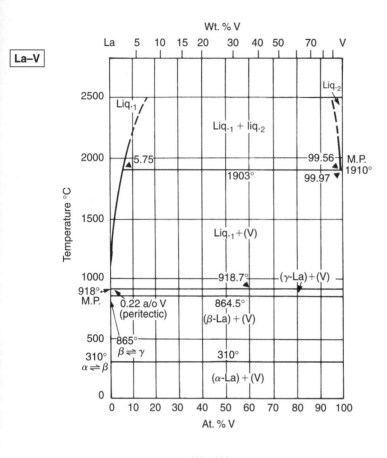

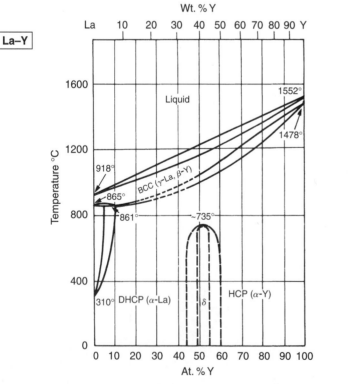

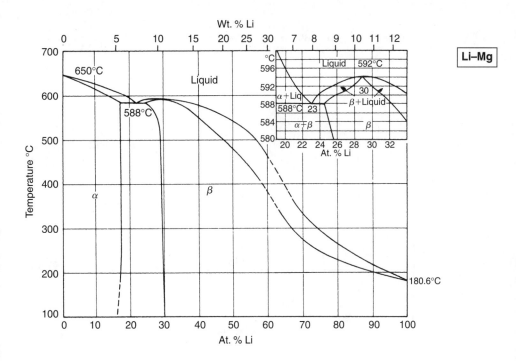

Li–Mg

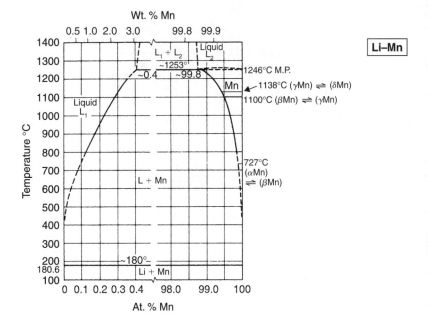

Li–Mn

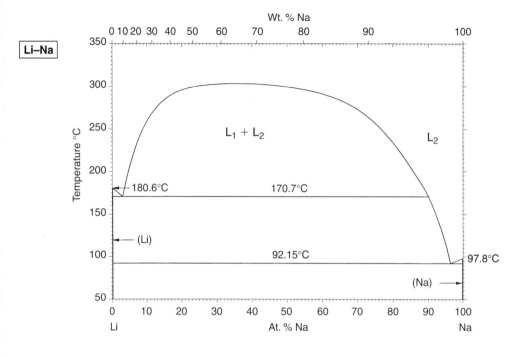

Li–Na

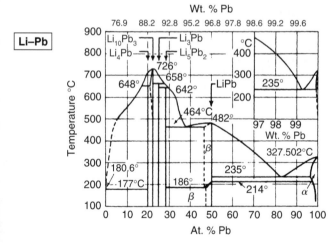

Li–Pb

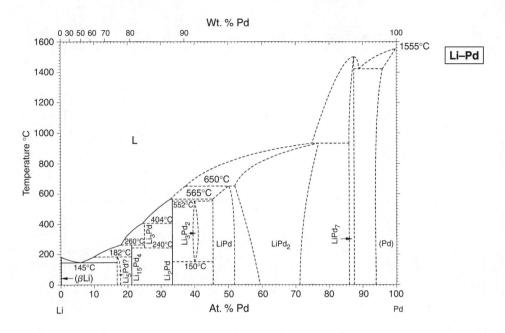

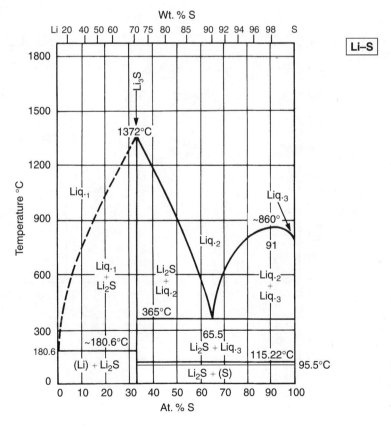

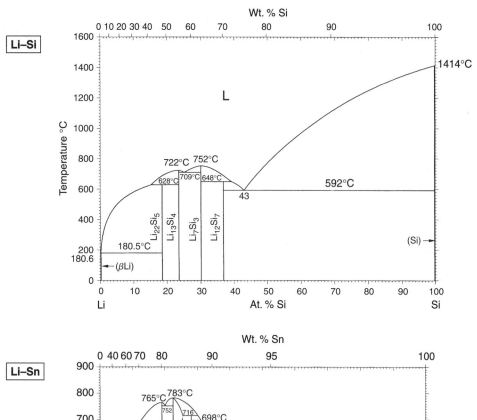

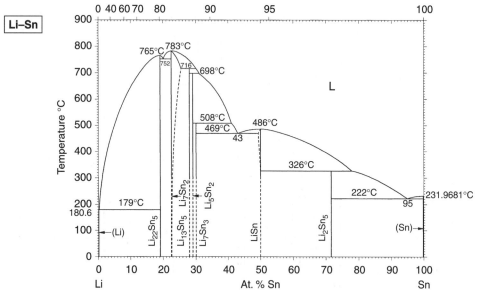

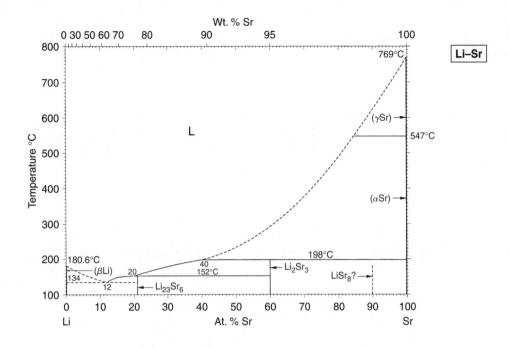

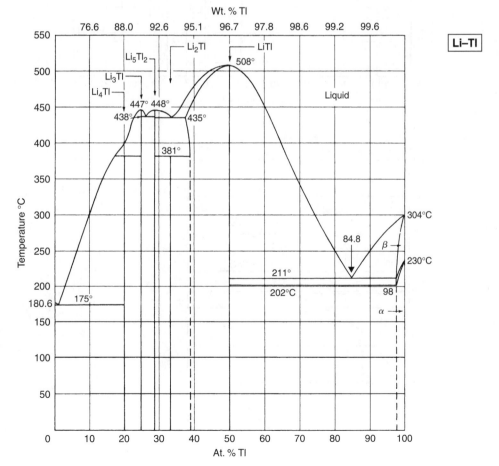

Li–Zn

Mg–Mn

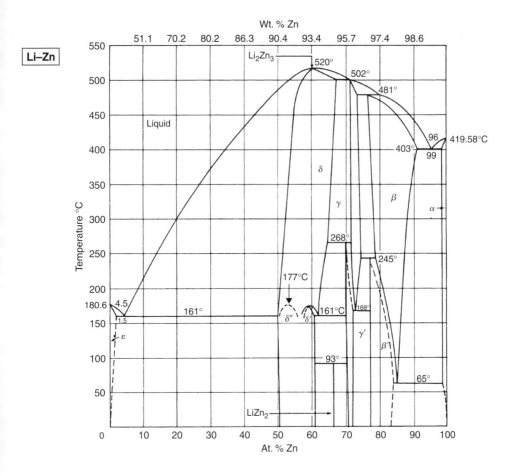

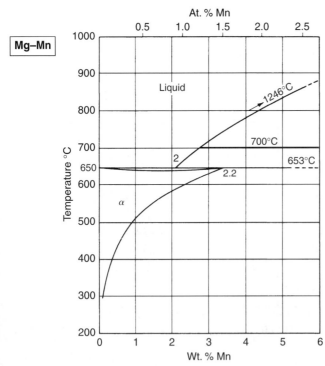

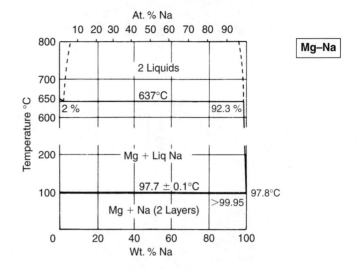

Mg–Na

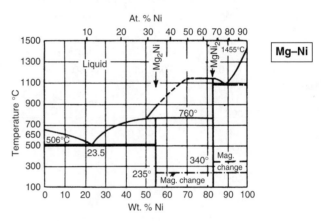

Mg–Ni

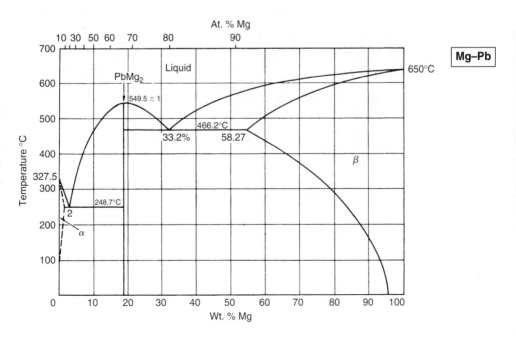

Mg–Pb

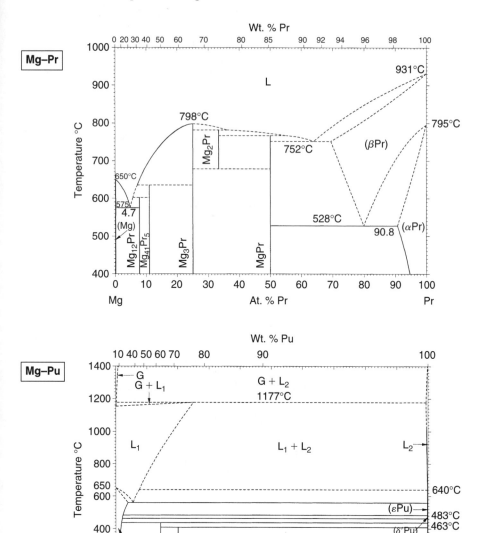

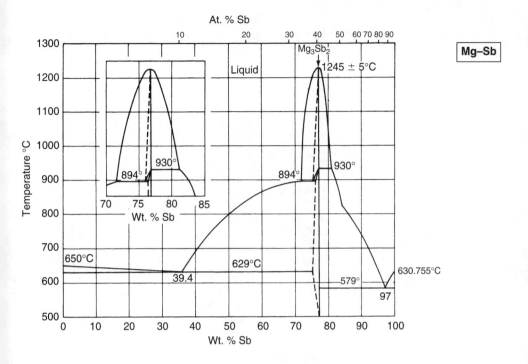

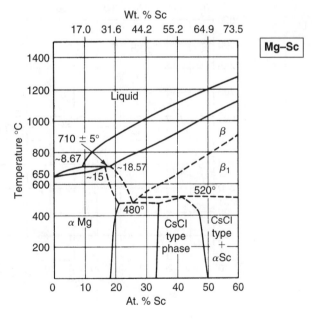

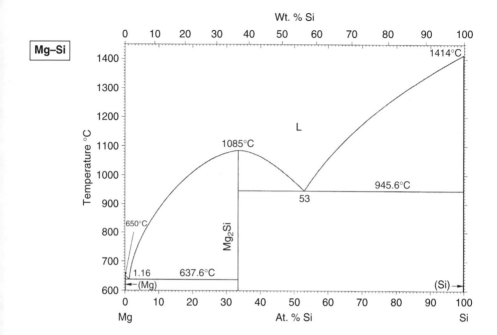

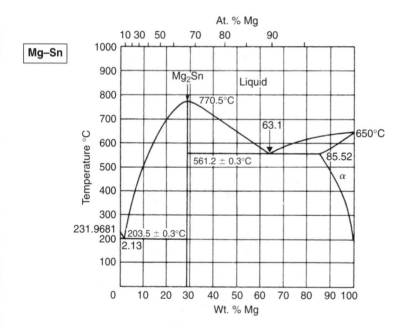

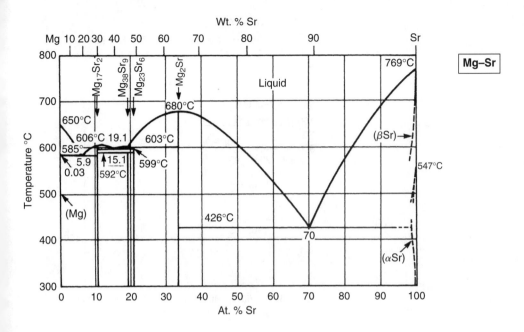

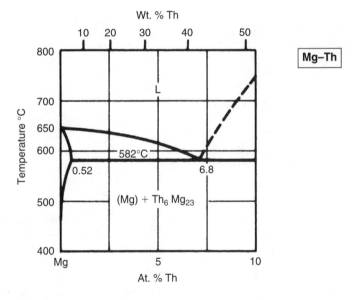

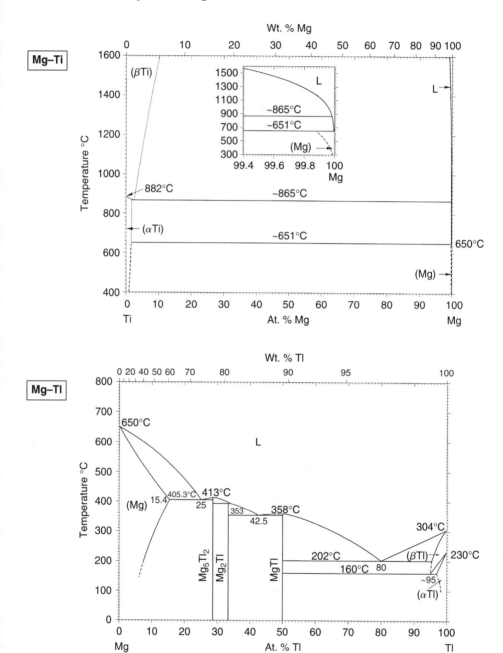

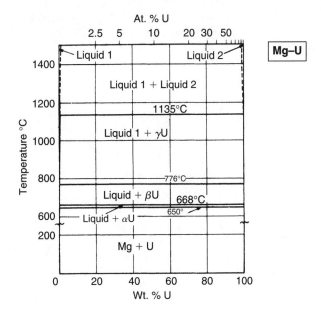

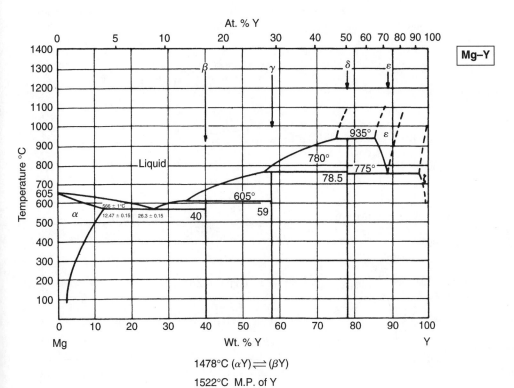

1478°C (αY) ⇌ (βY)

1522°C M.P. of Y

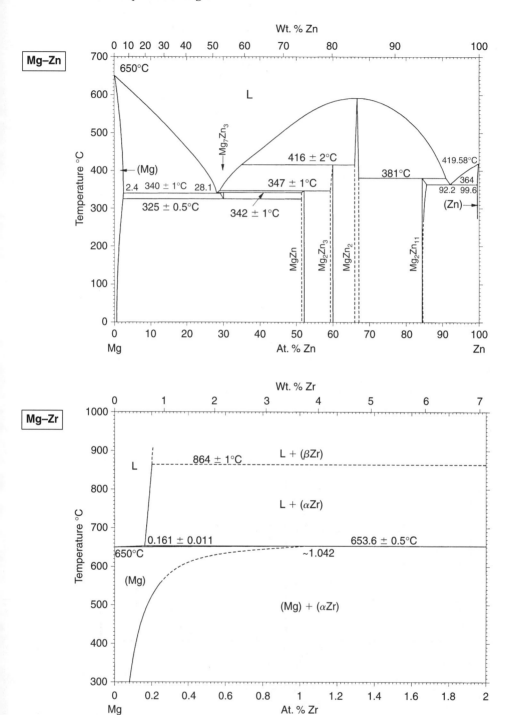

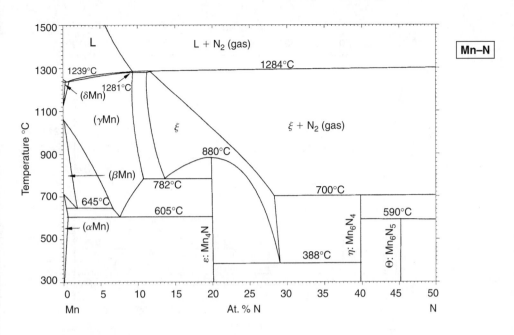

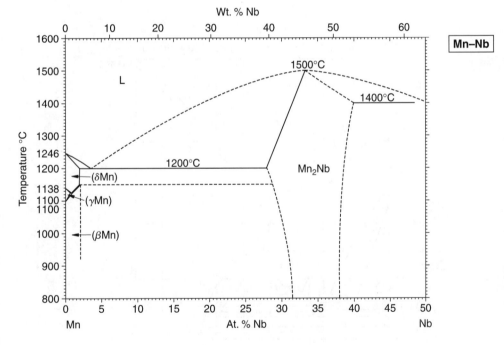

Mn–Nd

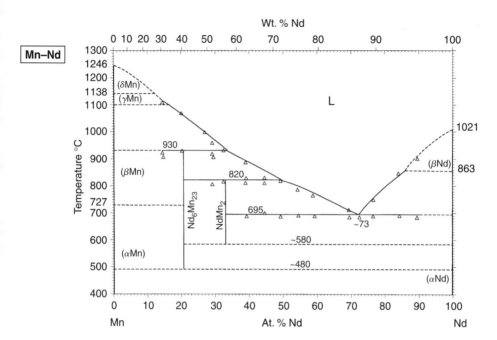

Mn–Ni

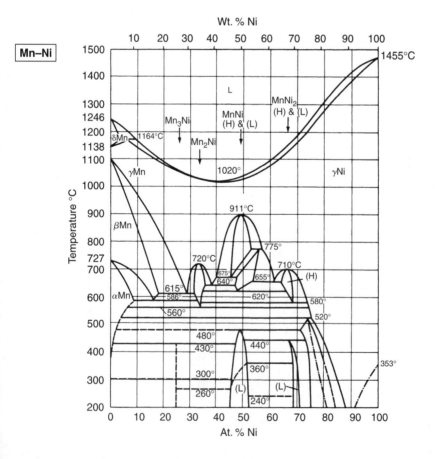

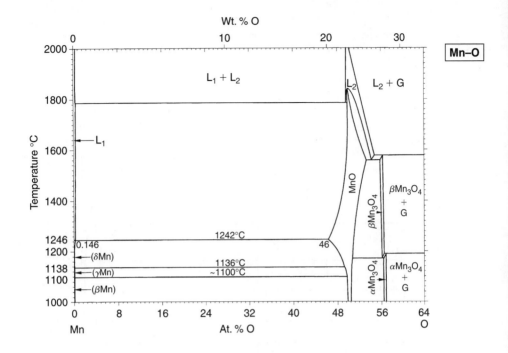

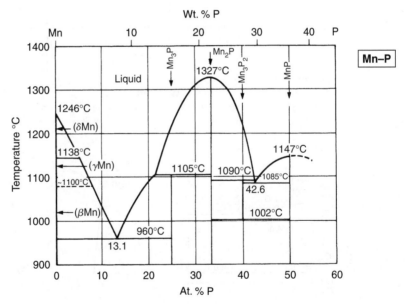

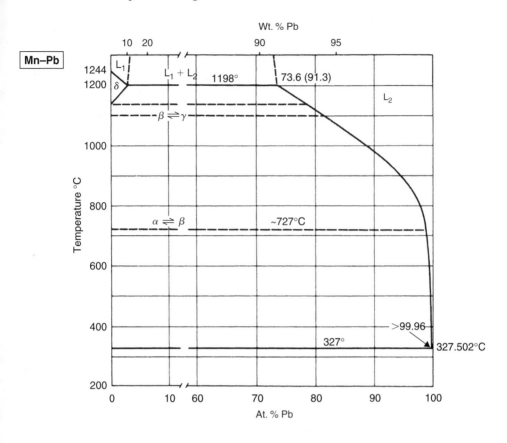

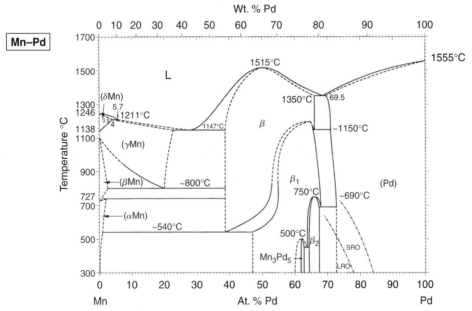

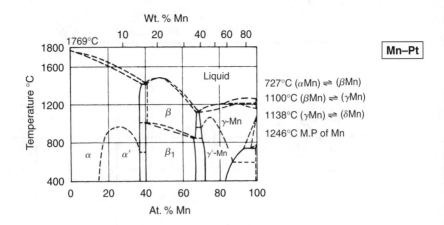

Mn–Pt

727°C (αMn) ⇌ (βMn)
1100°C (βMn) ⇌ (γMn)
1138°C (γMn) ⇌ (δMn)
1246°C M.P of Mn

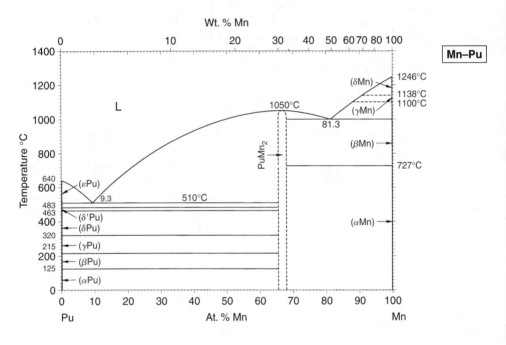

Mn–Pu

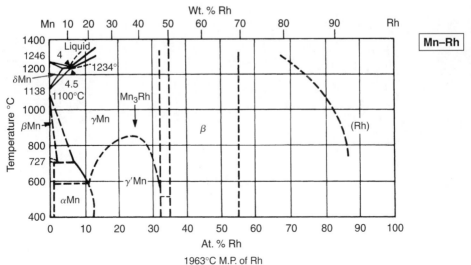

Mn–Rh

1963°C M.P. of Rh

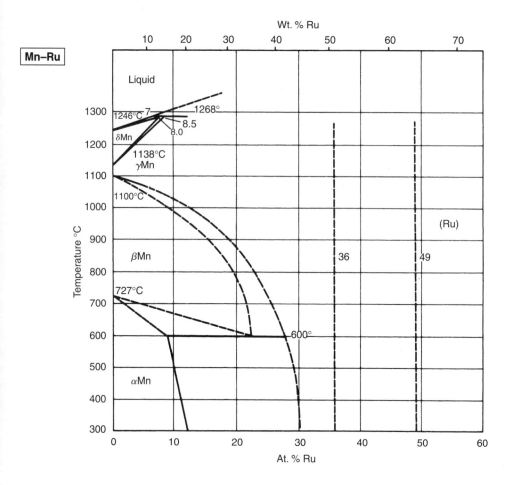

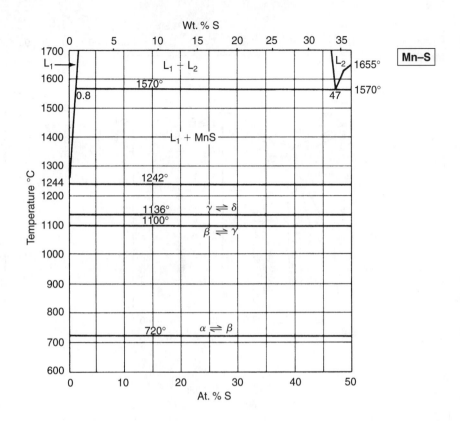

Mn–S

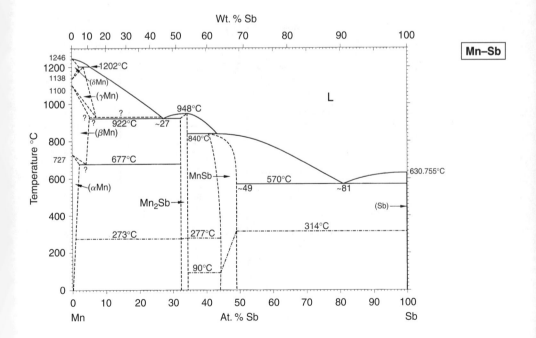

Mn–Sb

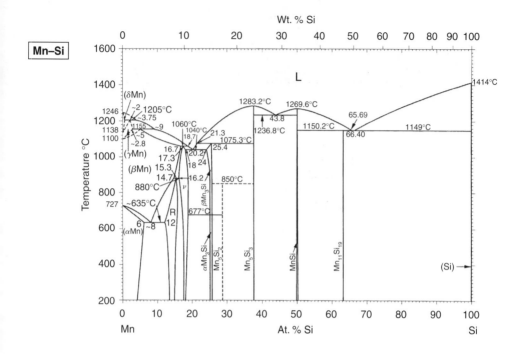

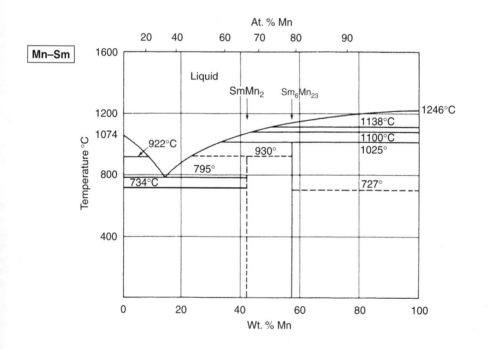

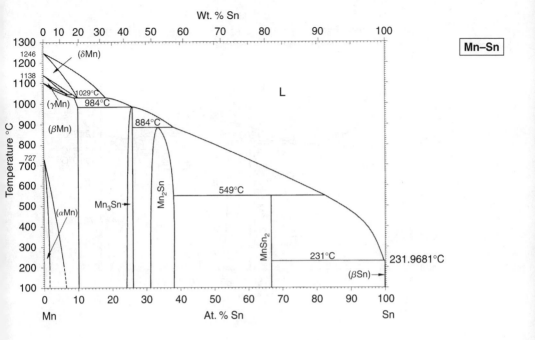

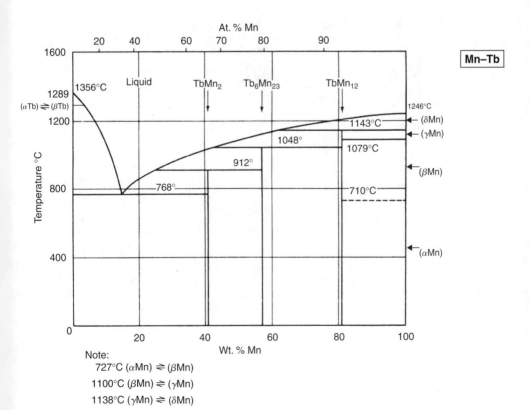

Note:
727°C (αMn) ⇌ (βMn)
1100°C (βMn) ⇌ (γMn)
1138°C (γMn) ⇌ (δMn)

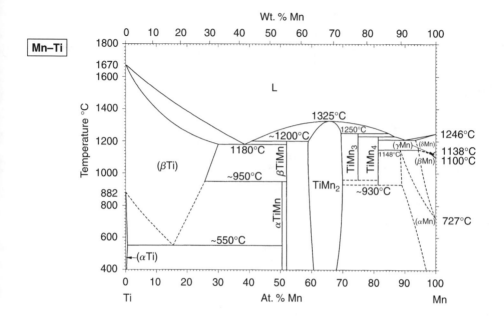

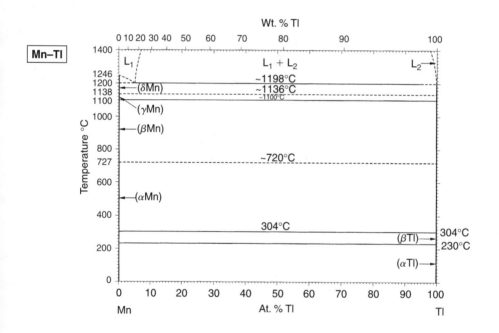

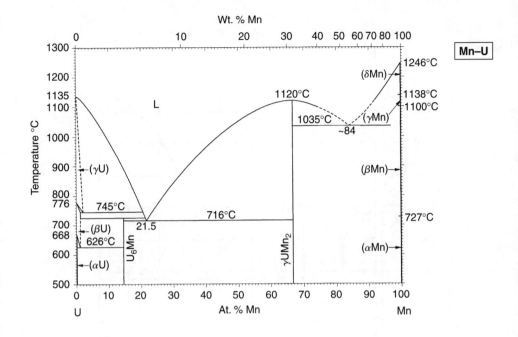

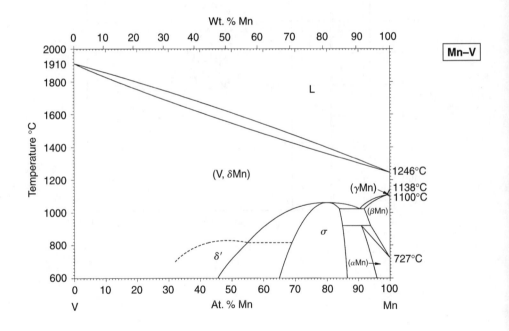

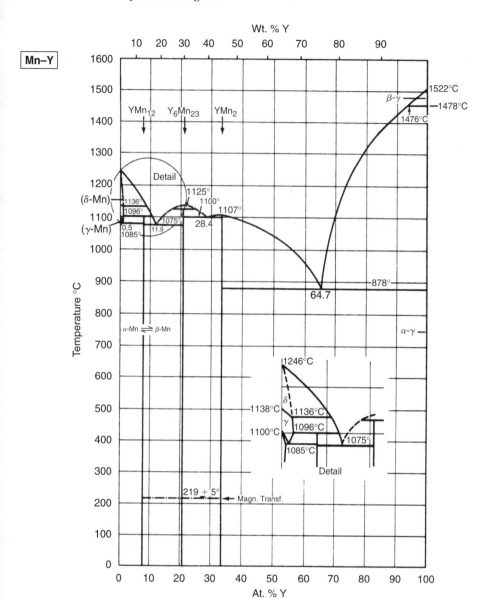

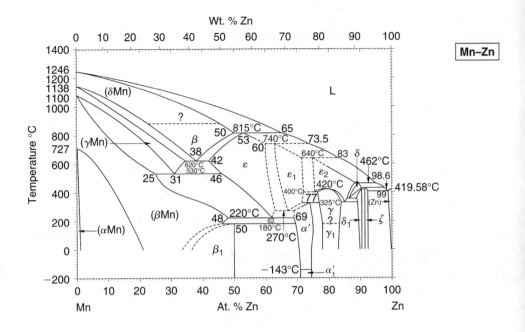

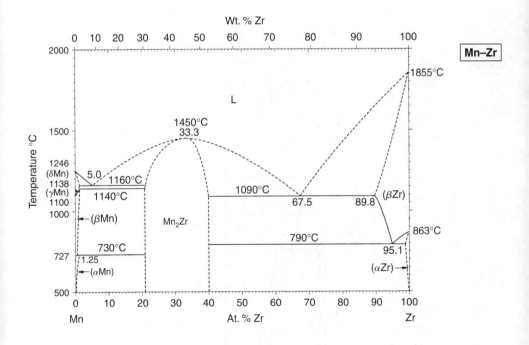

Mo–Nb

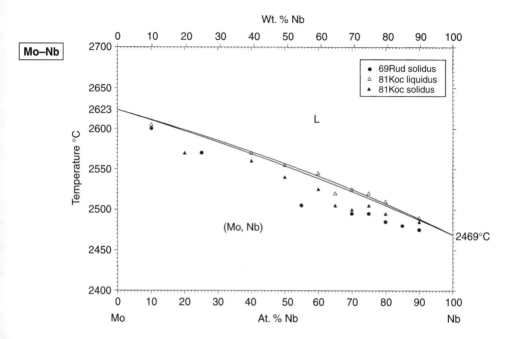

Mo–Ni

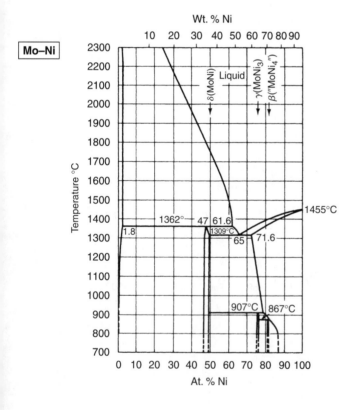

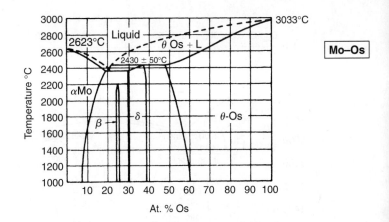

Mo–Os

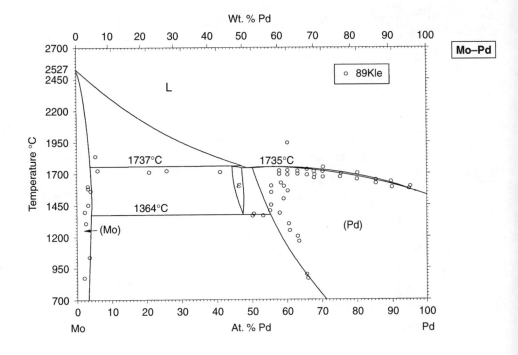

Mo–Pd

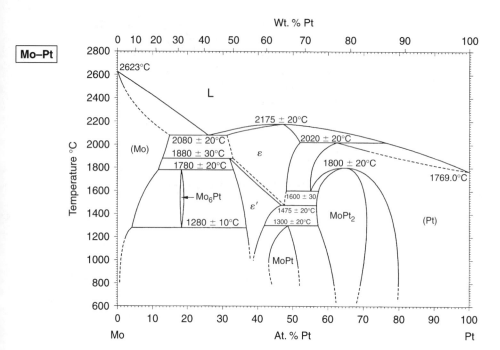

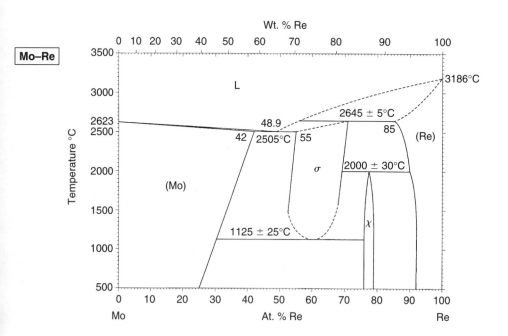

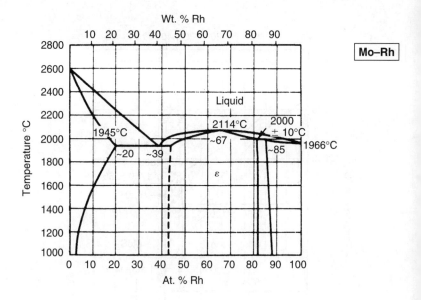

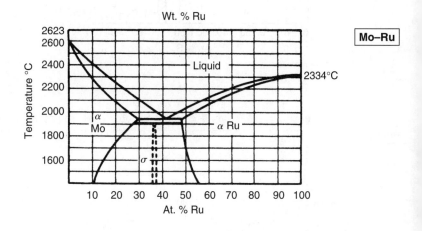

Mo–S

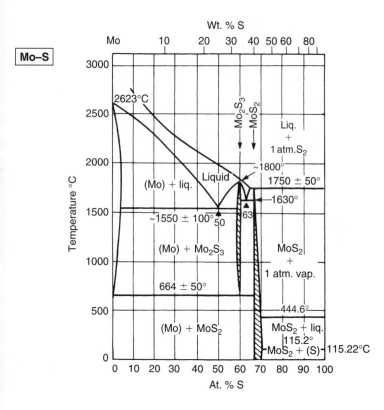

Mo–Si

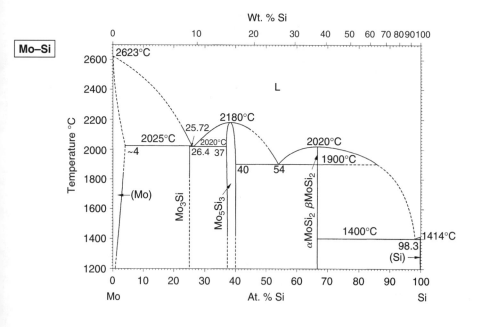

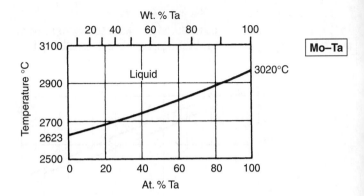

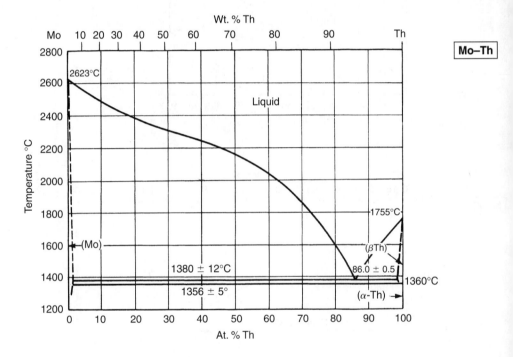

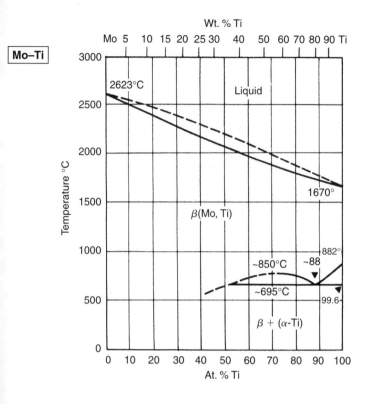

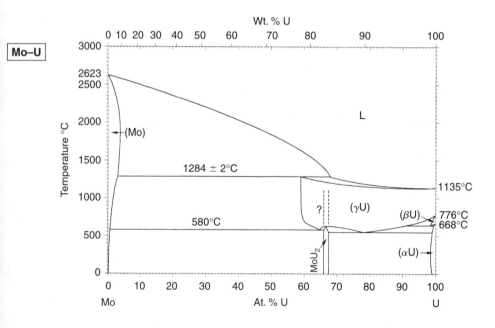

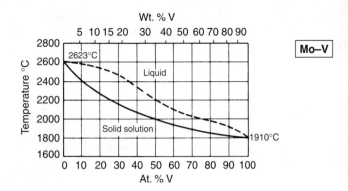

Mo–V

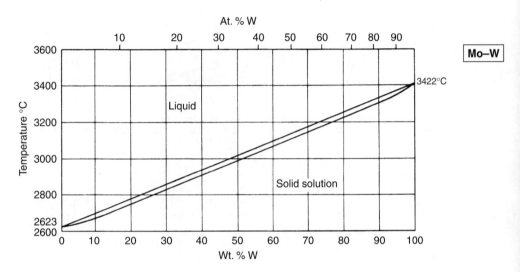

Mo–W

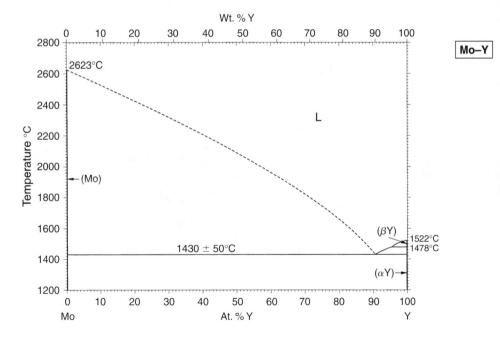

Mo–Y

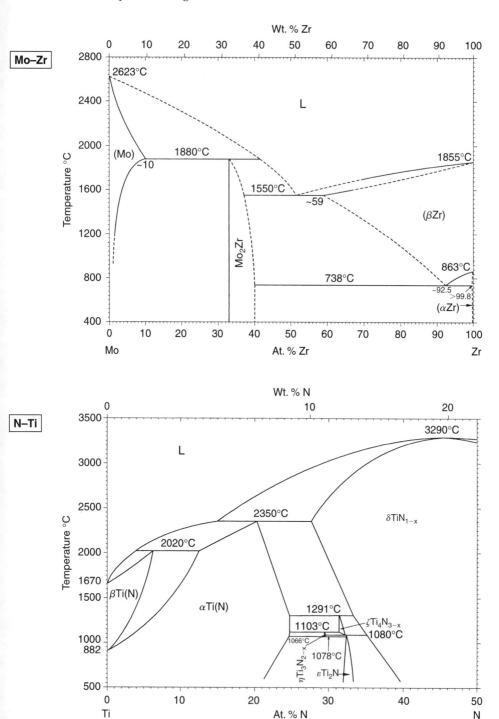

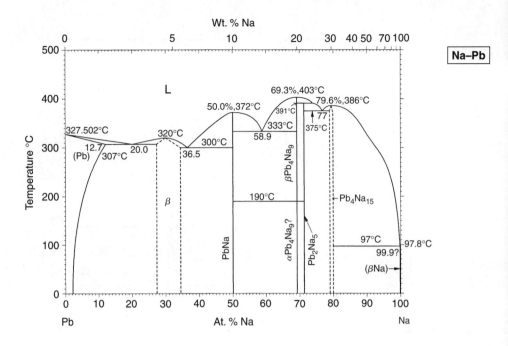

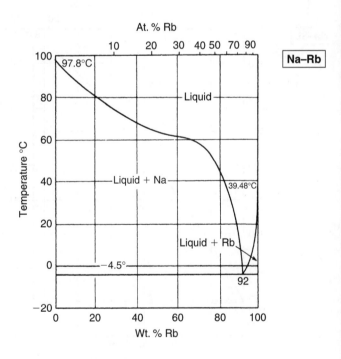

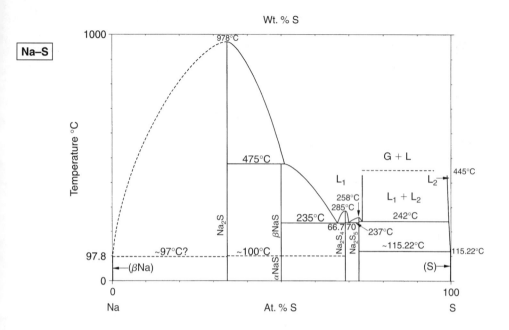

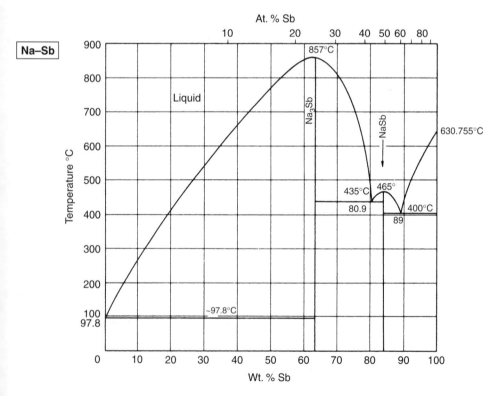

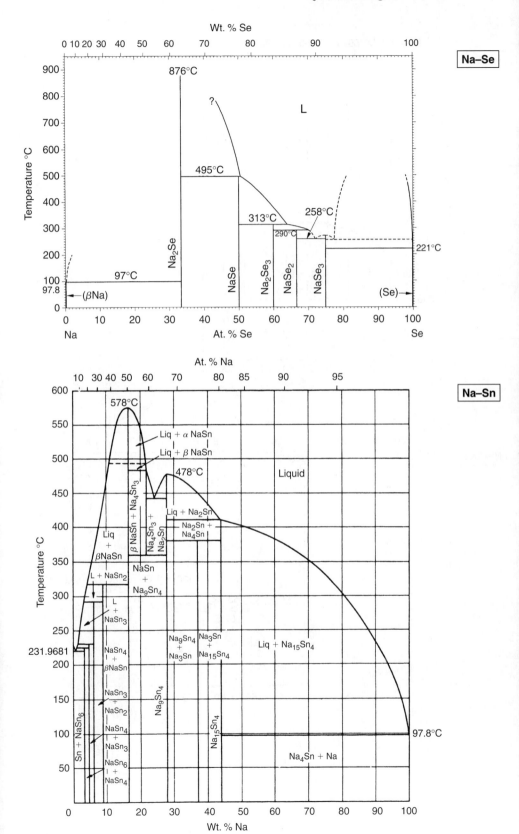

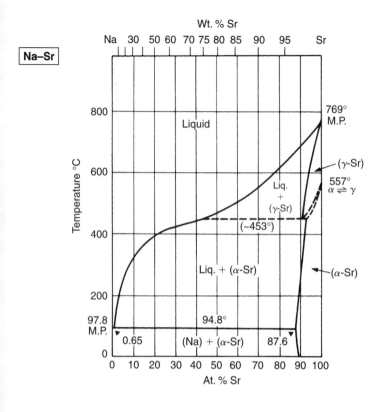

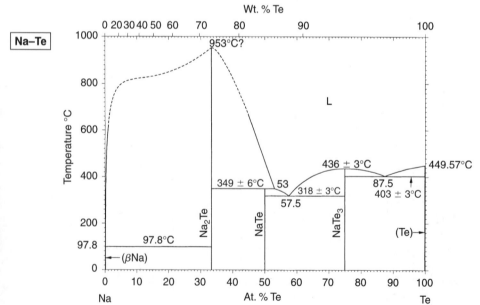

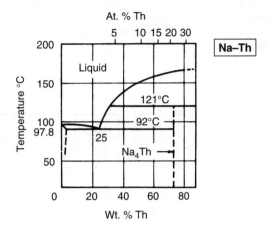

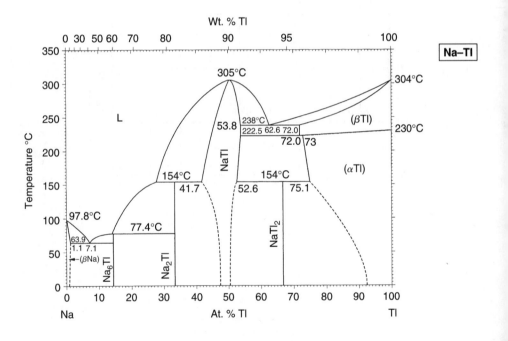

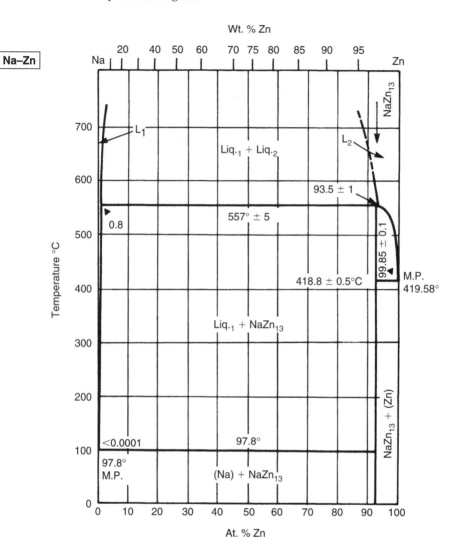

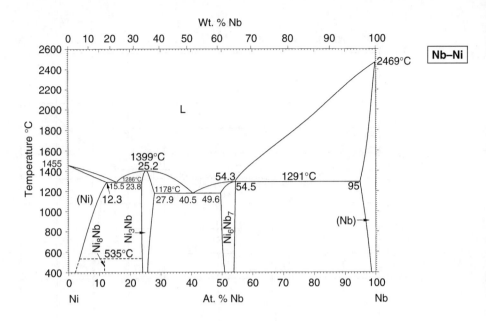

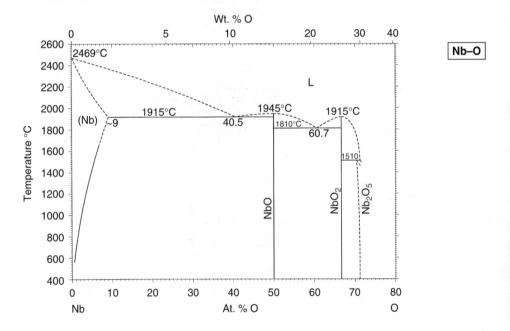

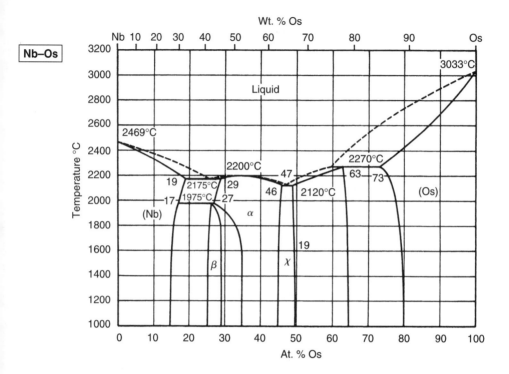

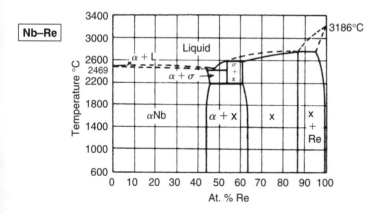

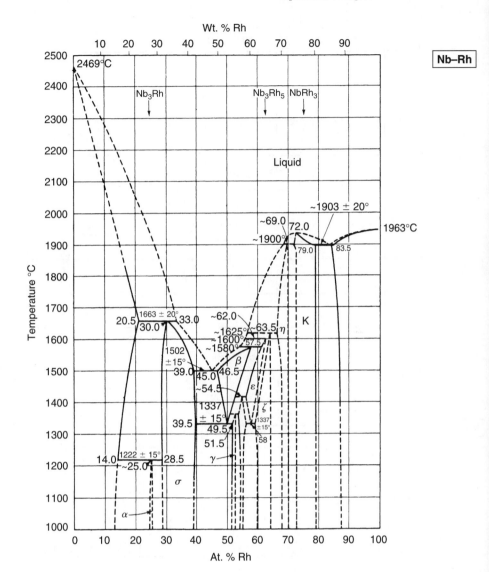

Wt. % Rh

Nb–Rh

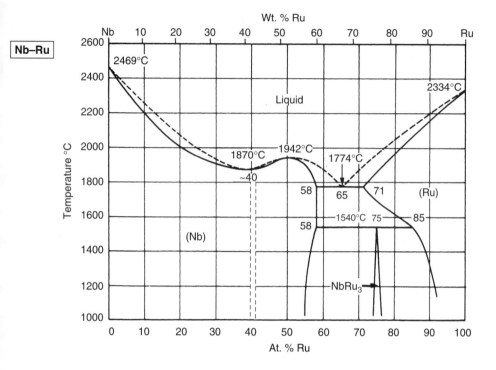

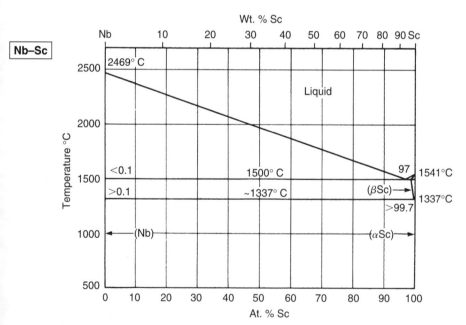

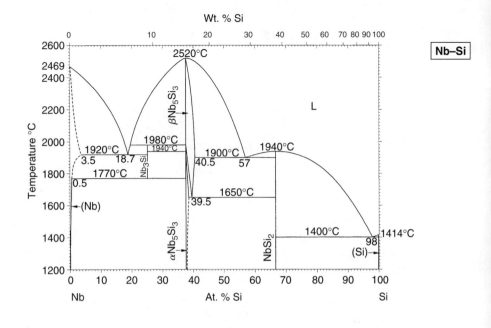

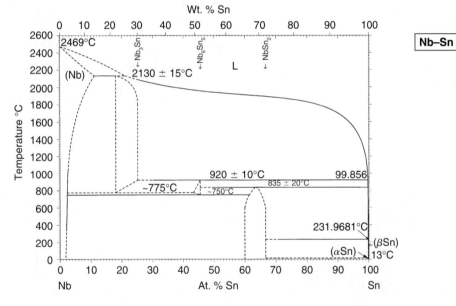

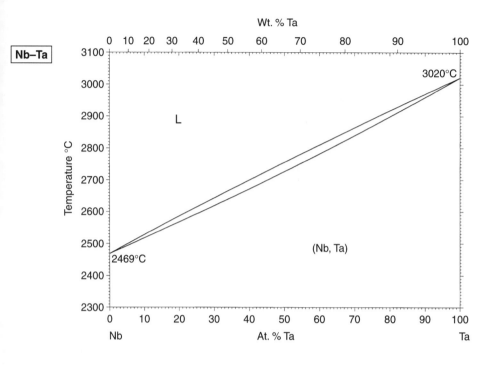

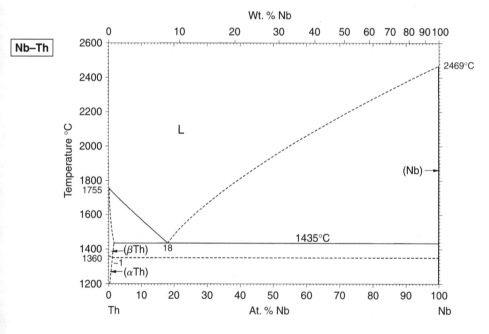

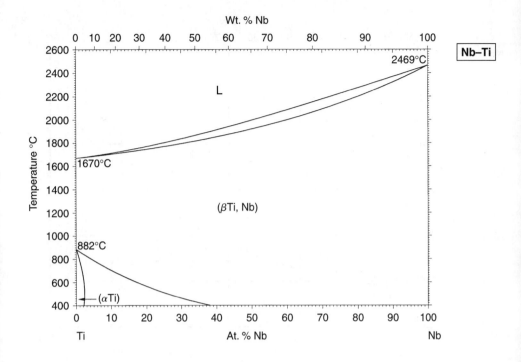

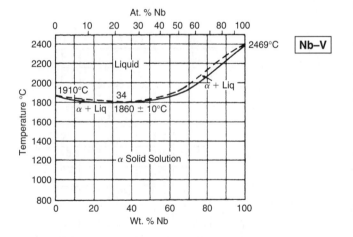

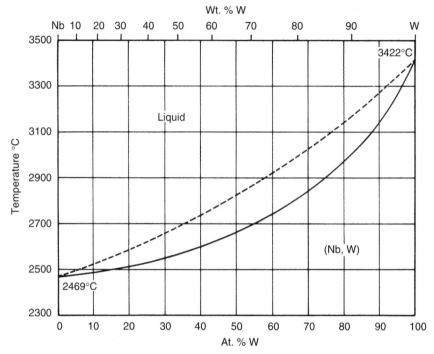

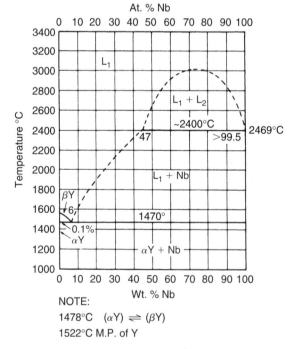

NOTE:

1478°C $(\alpha Y) \rightleftharpoons (\beta Y)$

1522°C M.P. of Y

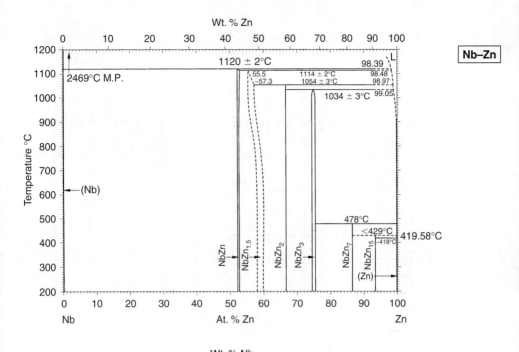

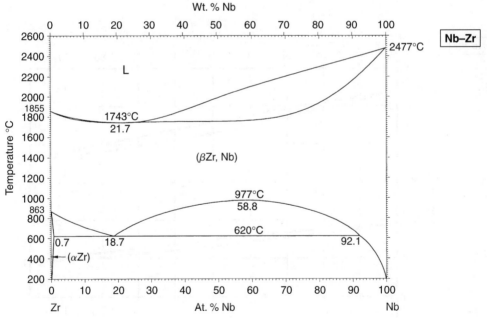

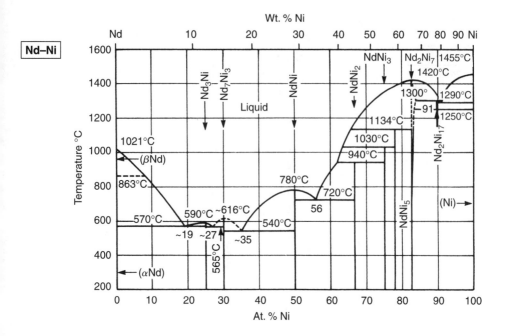

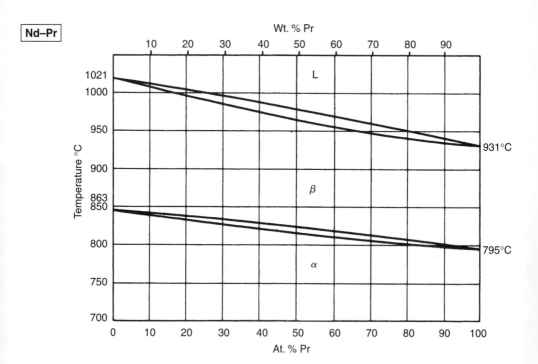

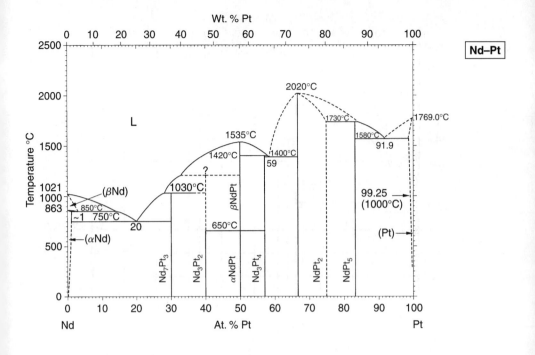

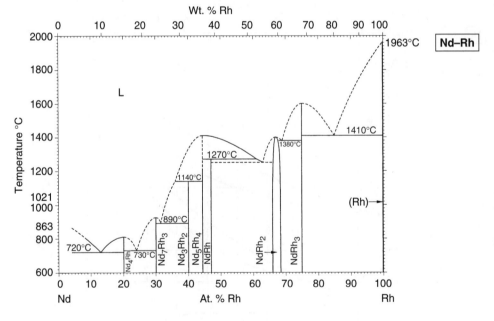

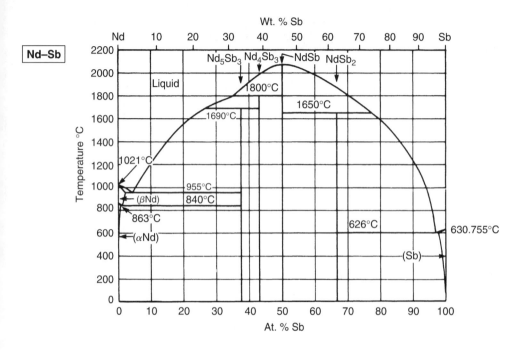

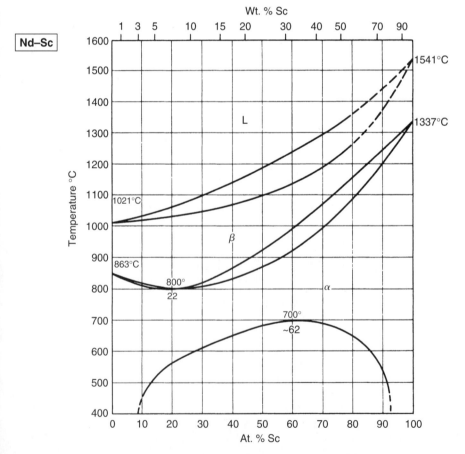

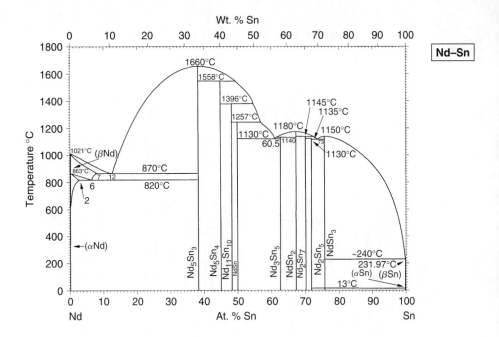

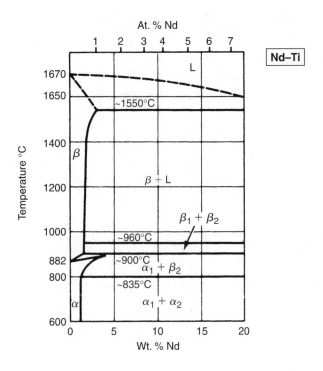

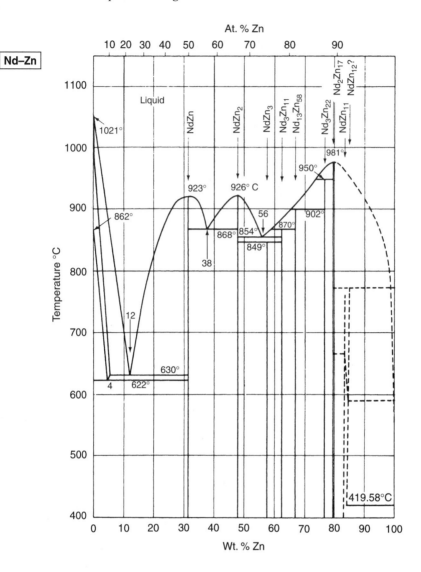

Nd–Zn

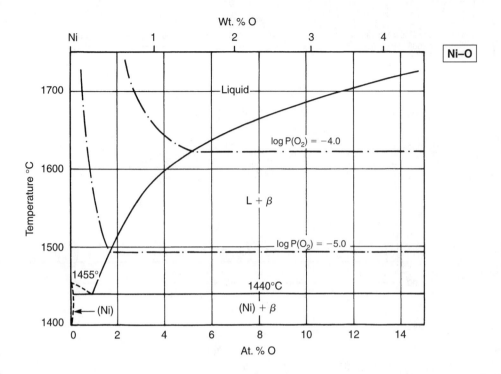

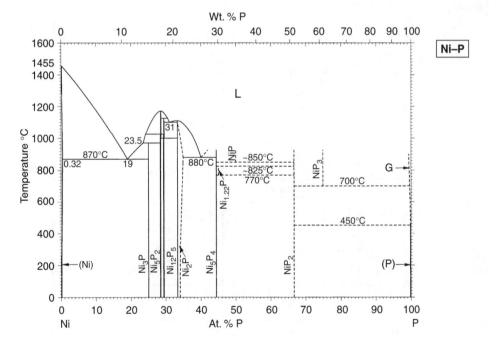

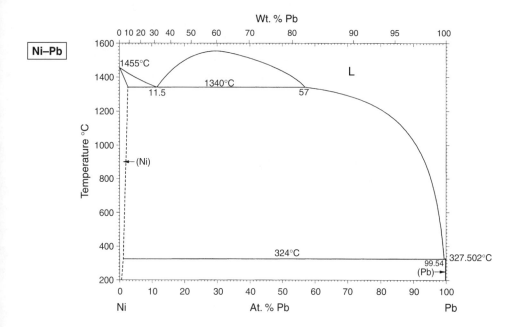

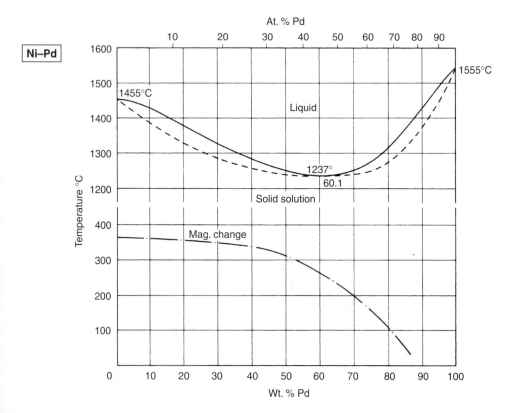

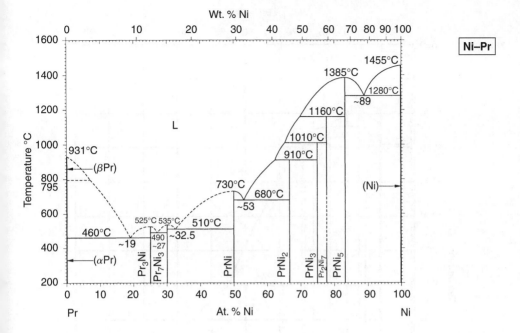

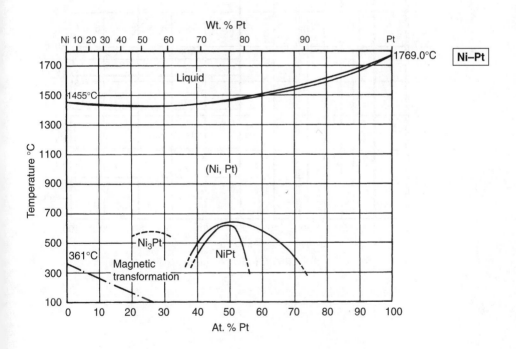

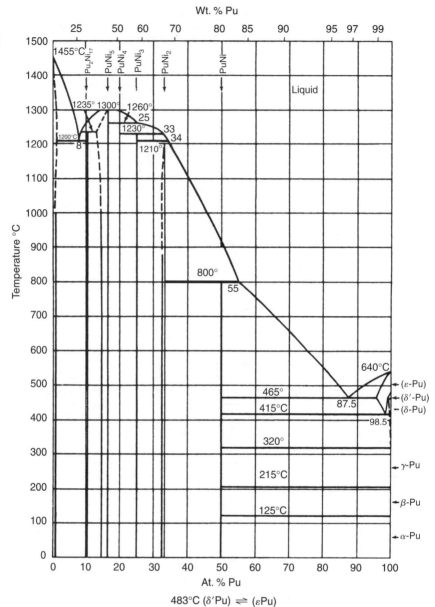

483°C (δ'Pu) ⇌ (εPu)

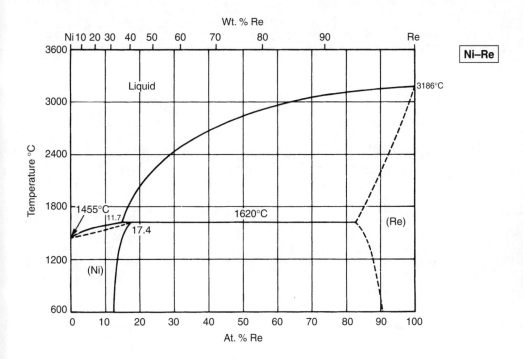

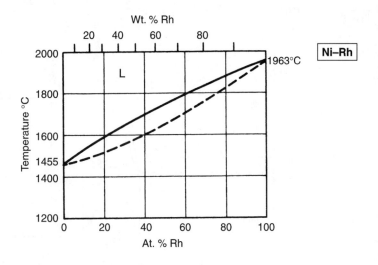

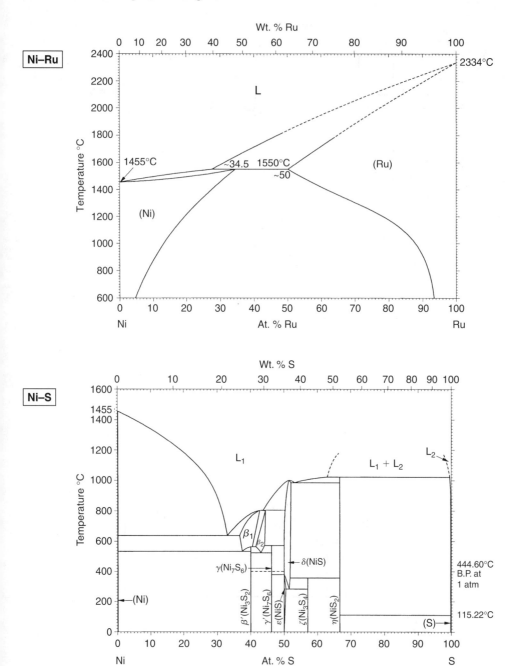

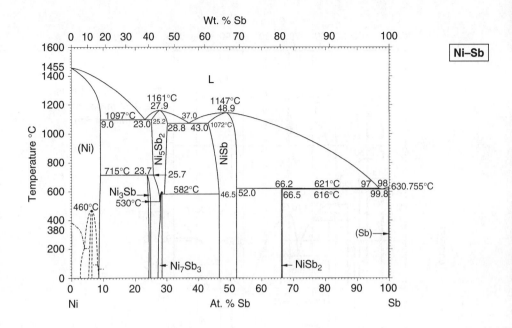

Ni–Sb

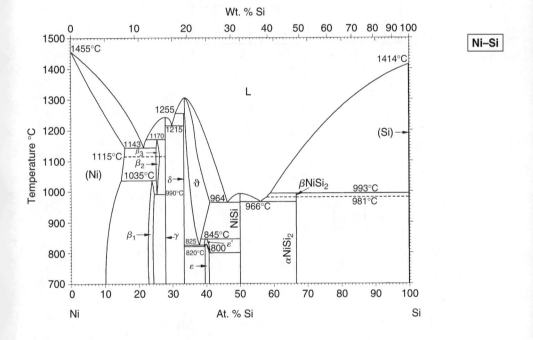

Ni–Si

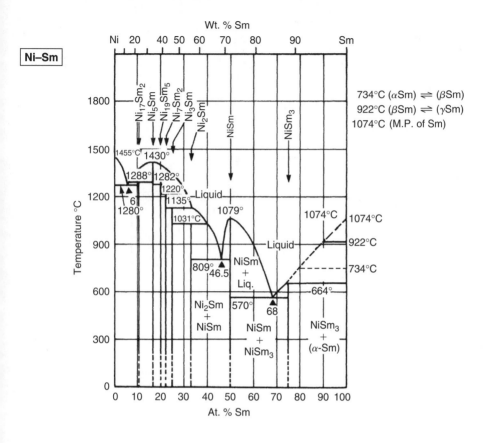

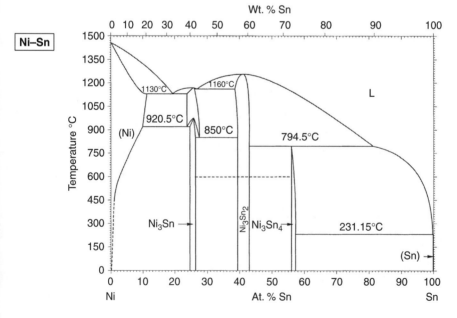

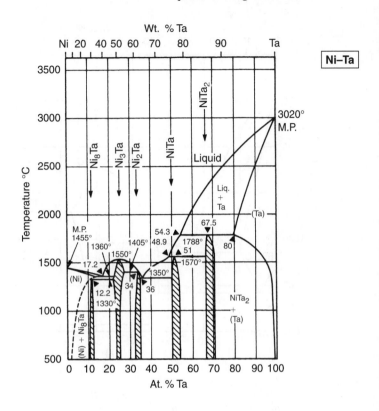

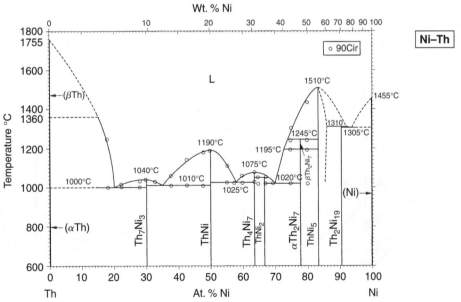

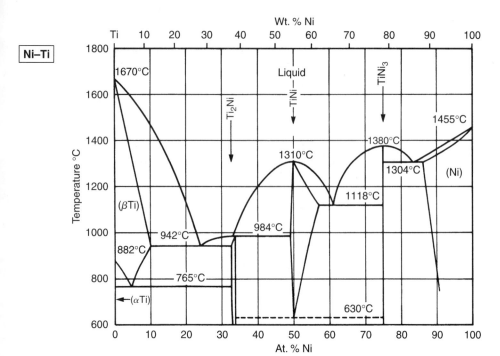

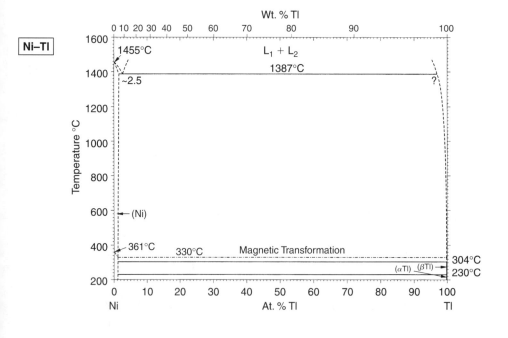

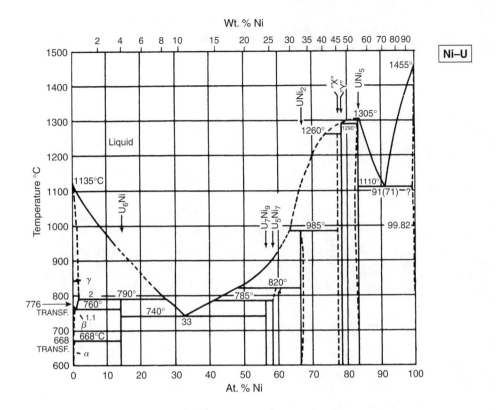

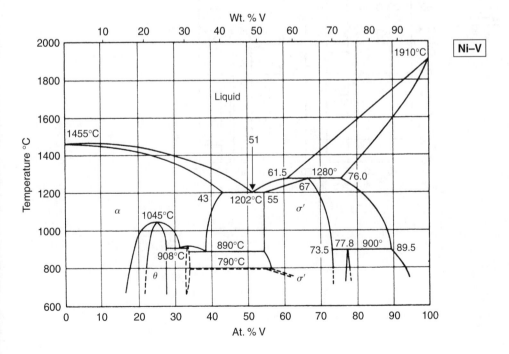

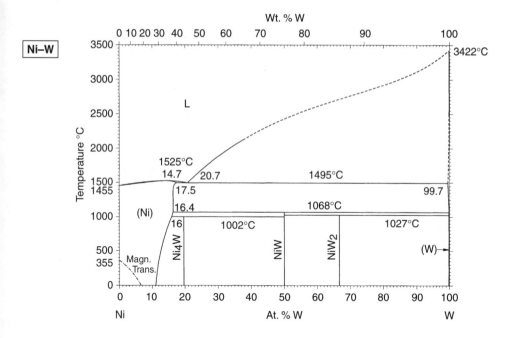

Ni–W

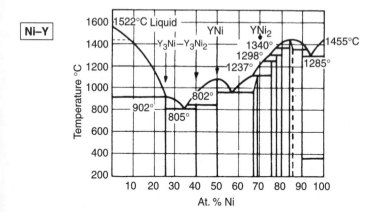

Ni–Y

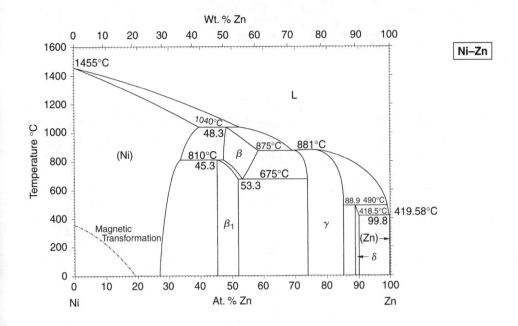

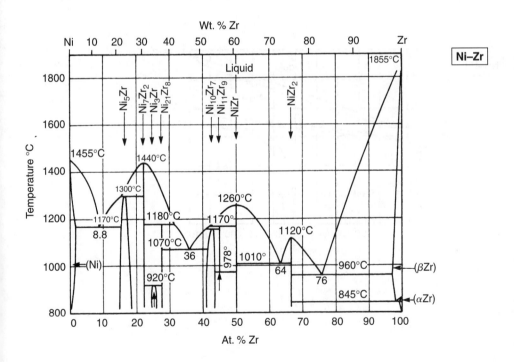

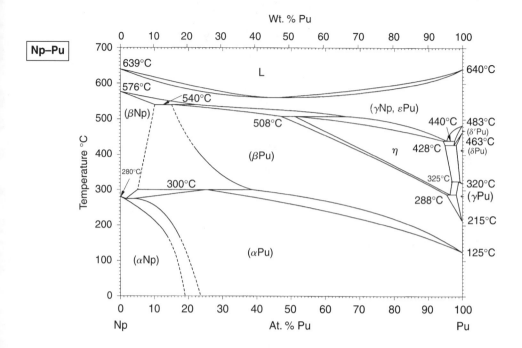

Np–Pu

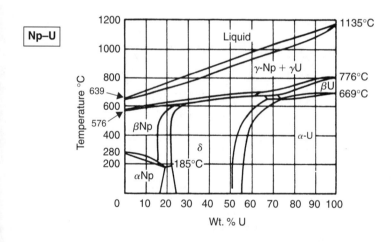

Np–U

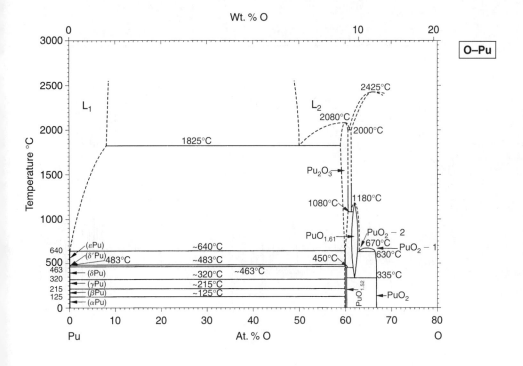

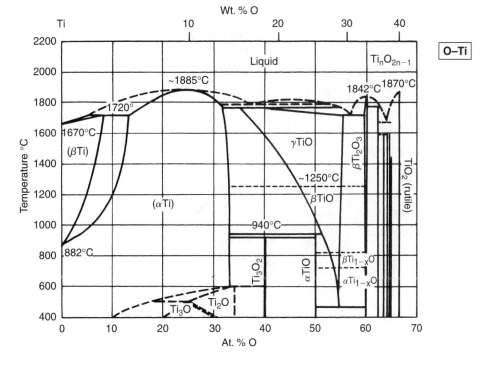

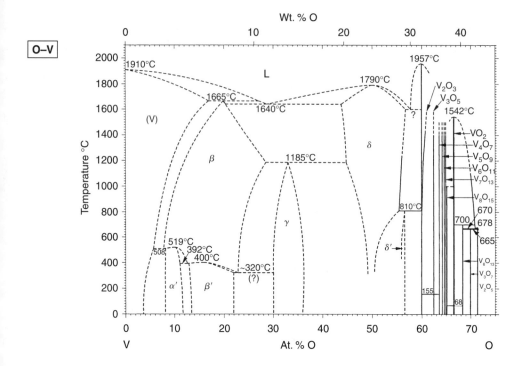

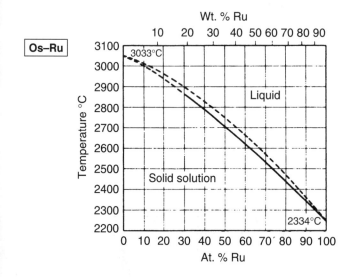

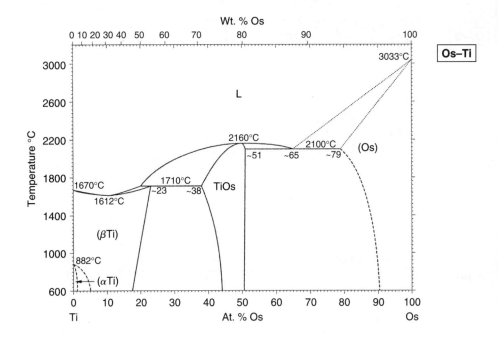

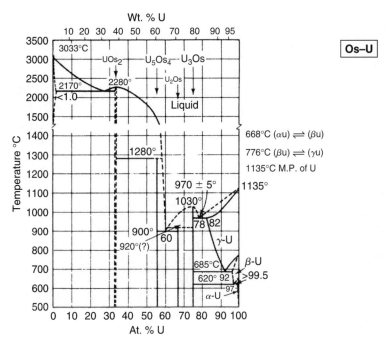

Os–W

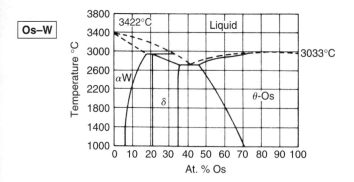

P–Pd

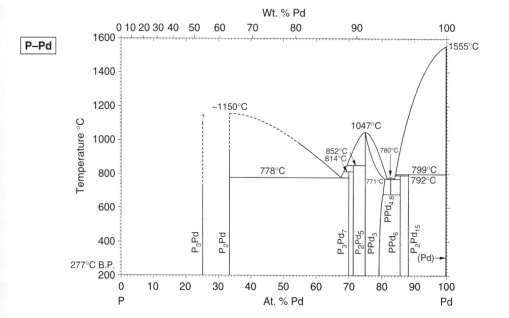

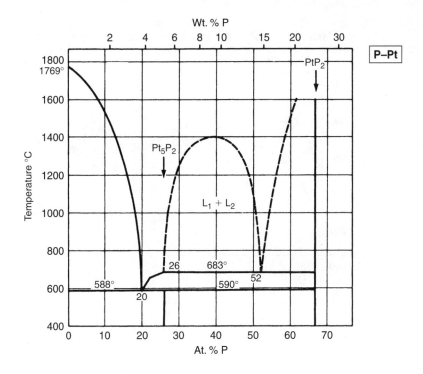

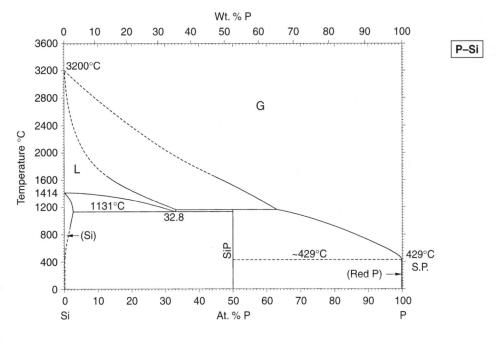

P–Sn

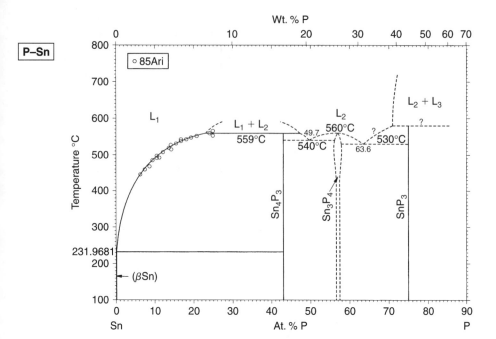

P–Tl

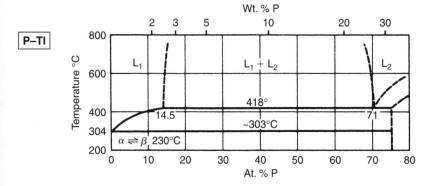

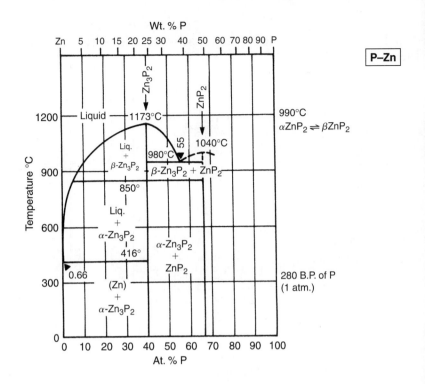

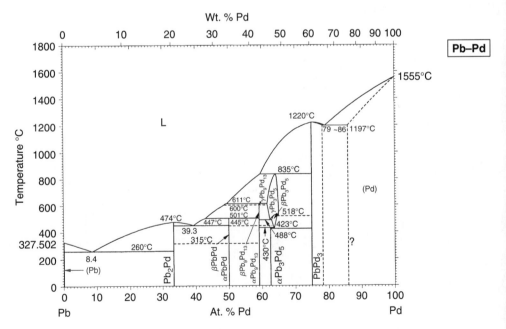

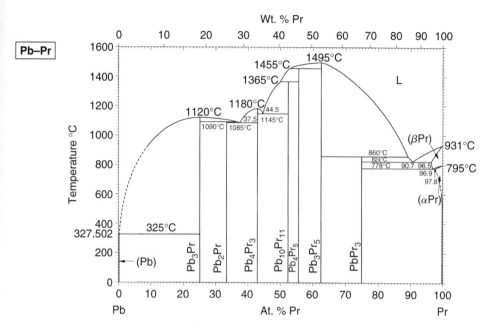

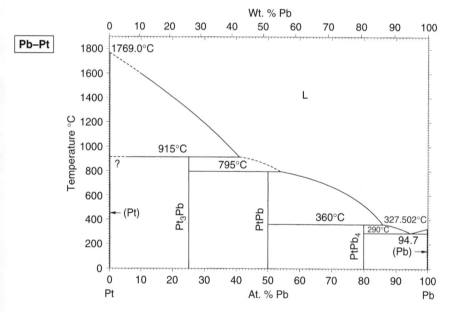

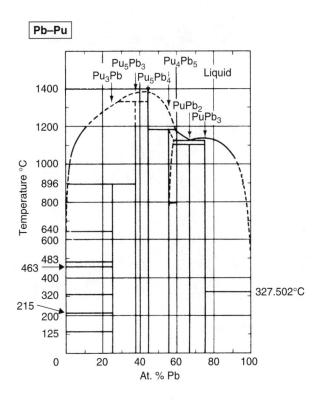

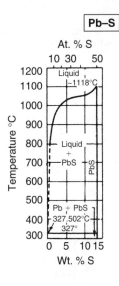

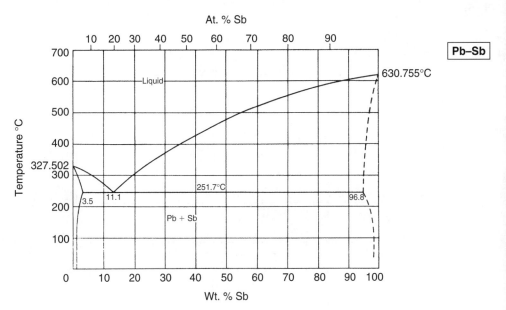

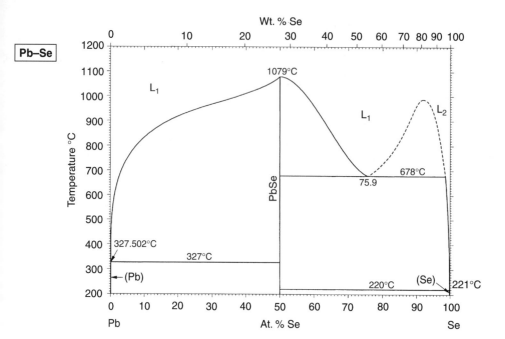

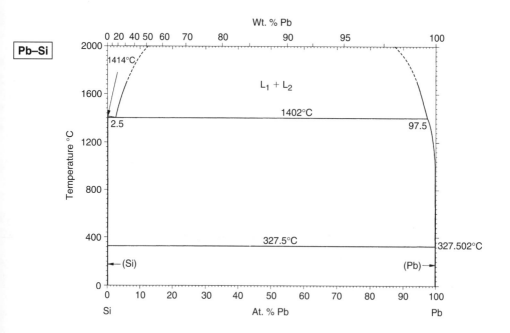

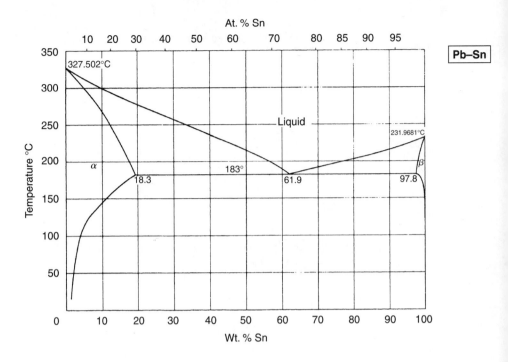

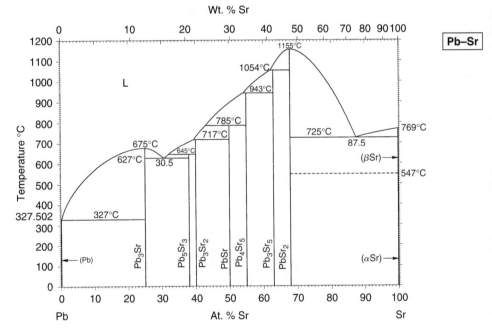

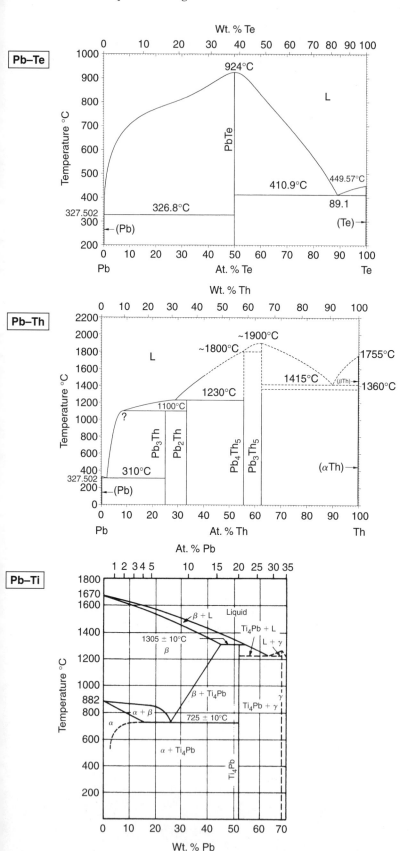

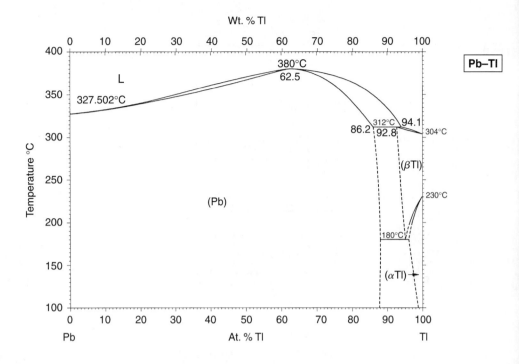

Pb–Tl

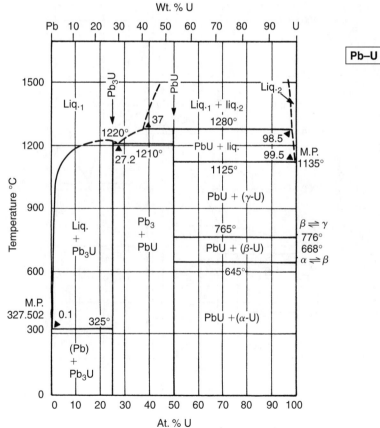

Pb–U

Pb–W

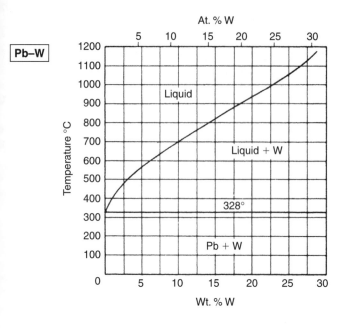

Pb–Y

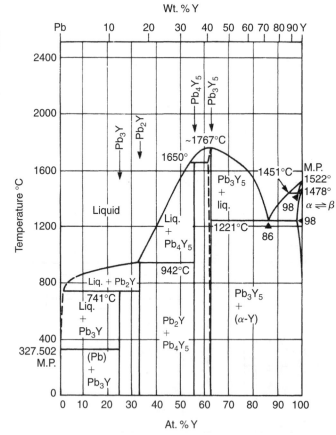

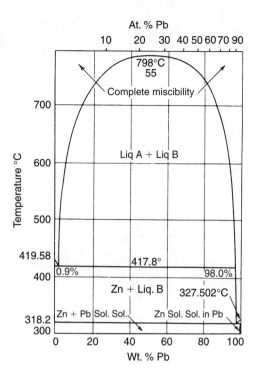

Pb–Zn

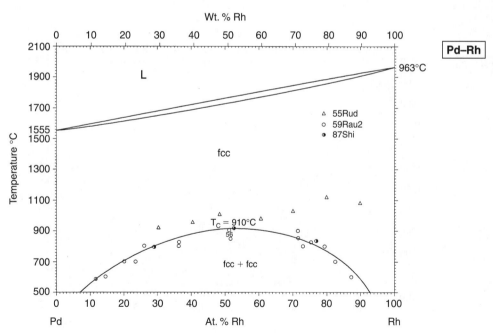

Pd–Rh

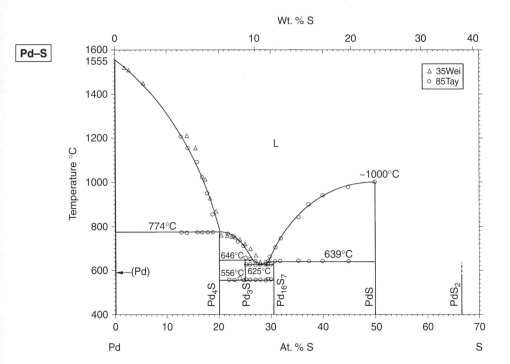

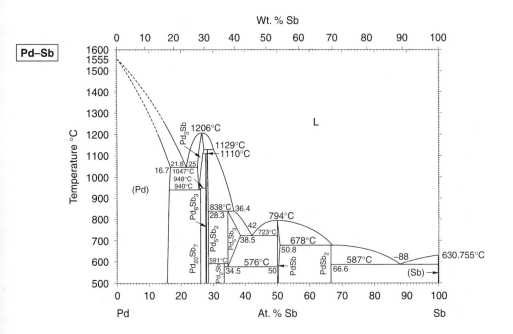

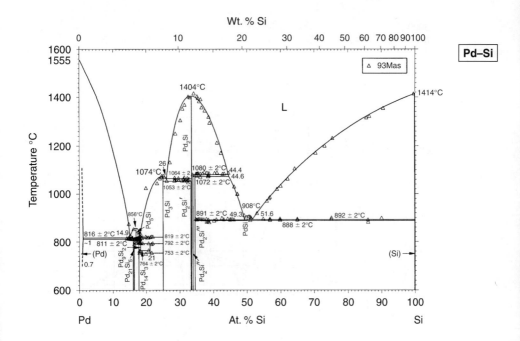

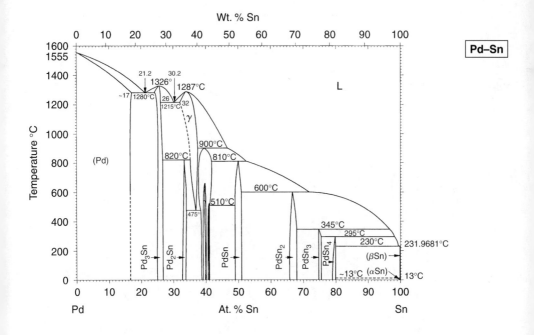

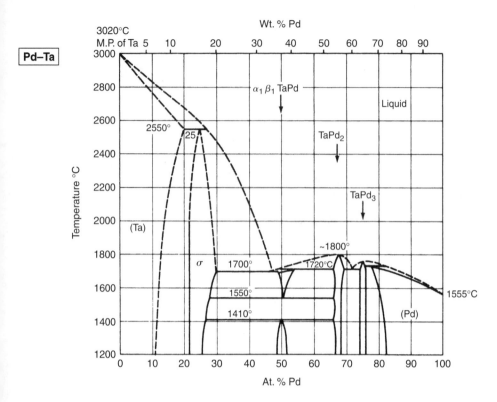

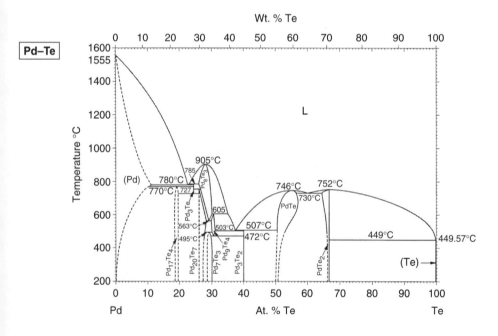

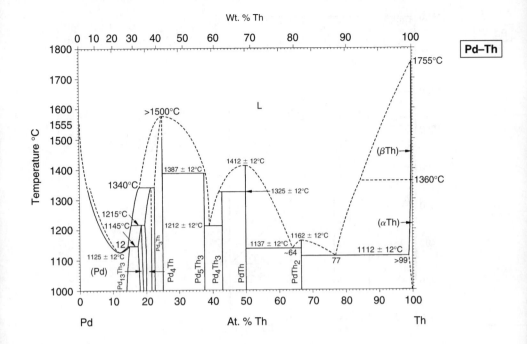

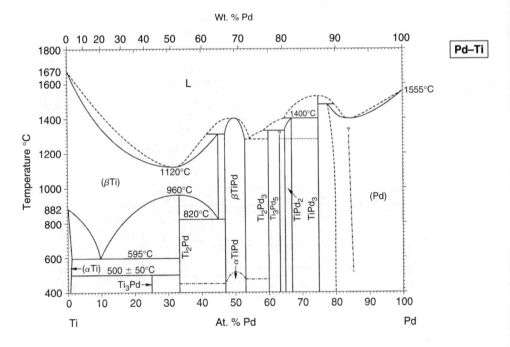

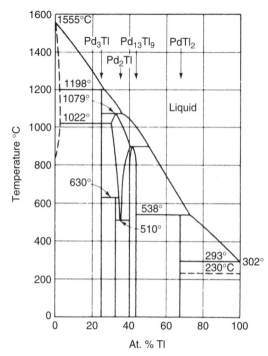

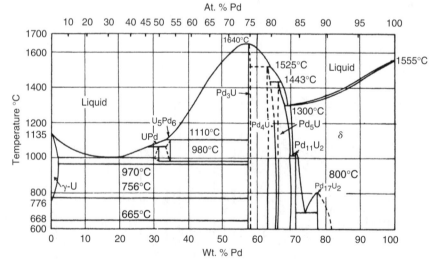

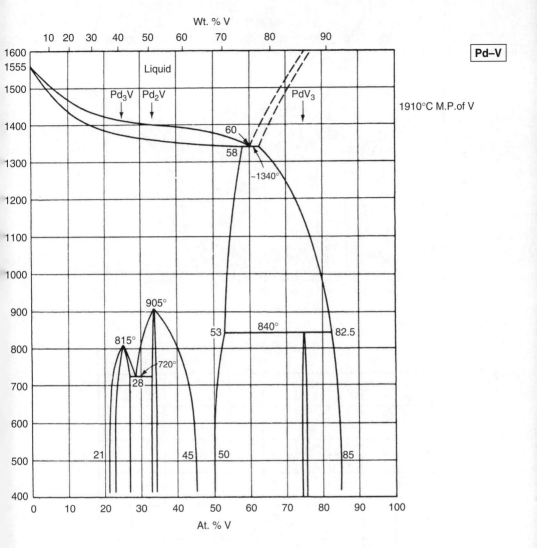

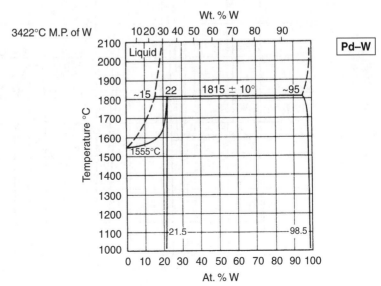

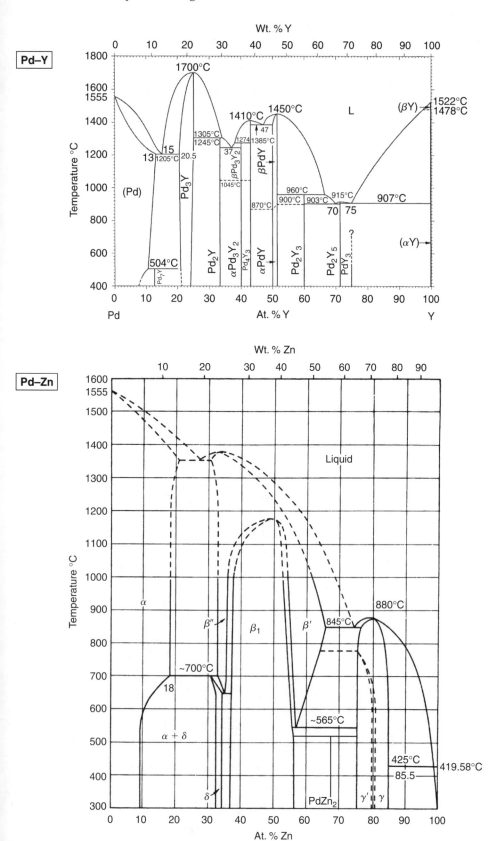

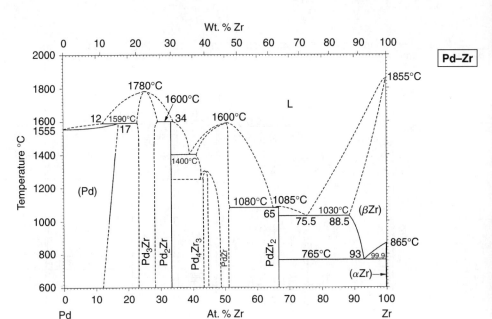

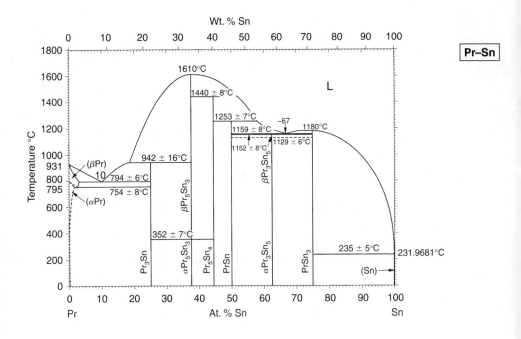

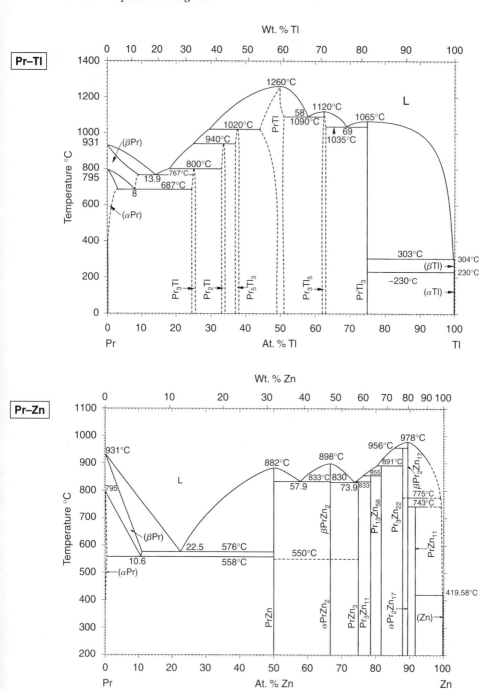

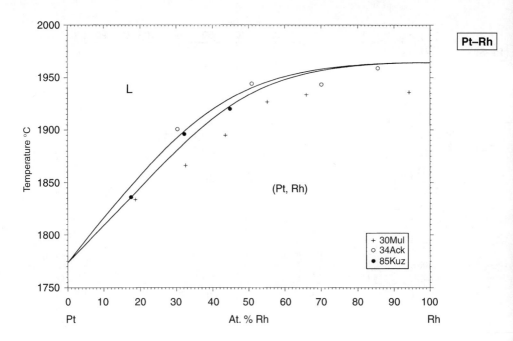

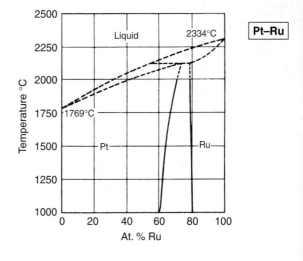

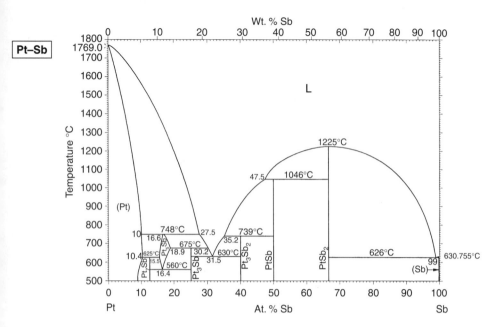

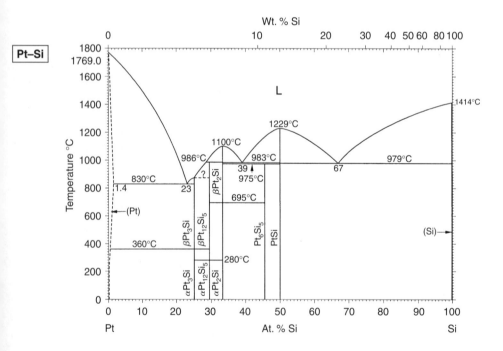

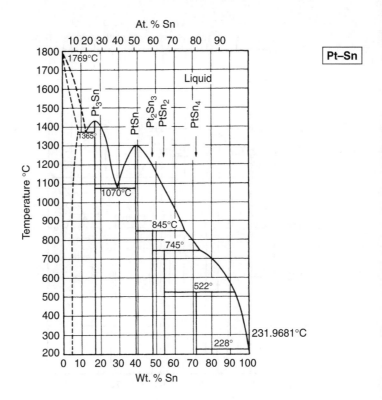

Pt–Sn

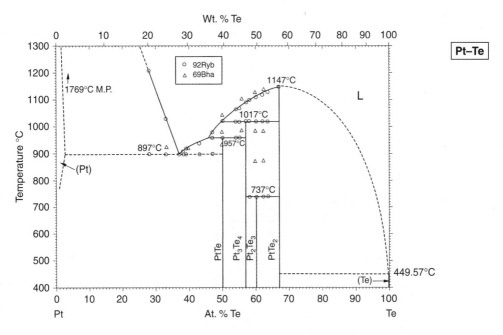

Pt–Te

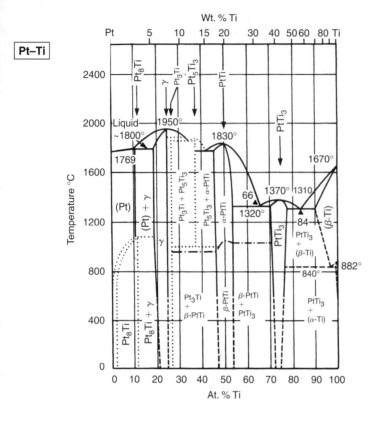

Pt–Ti

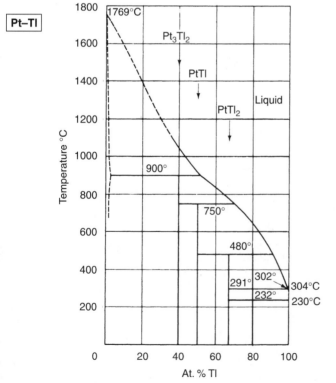

Pt–Tl

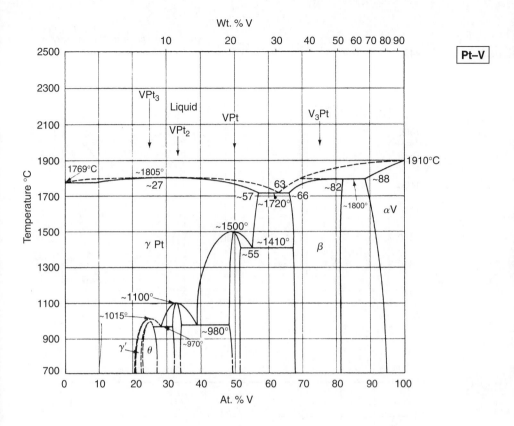

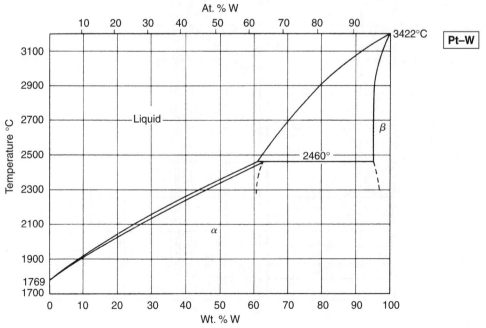

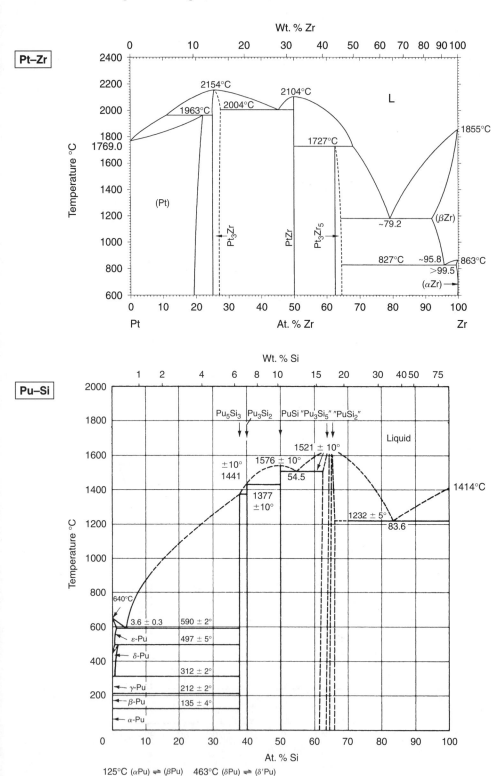

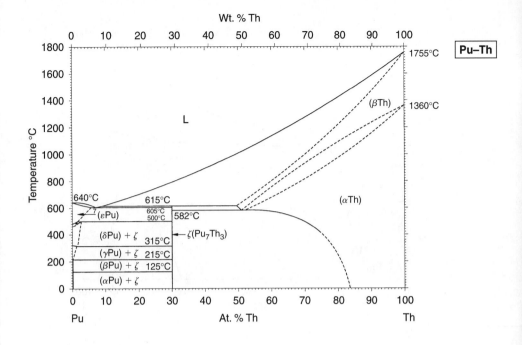

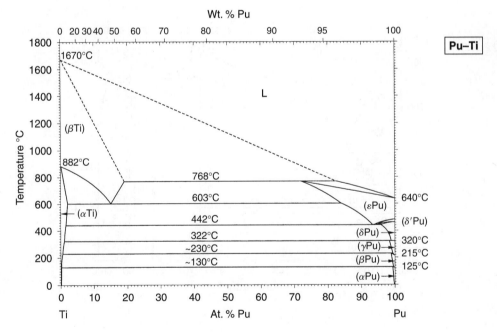

See Pu–Si phase diagram for Pu transformation temperatures.

Pu–U

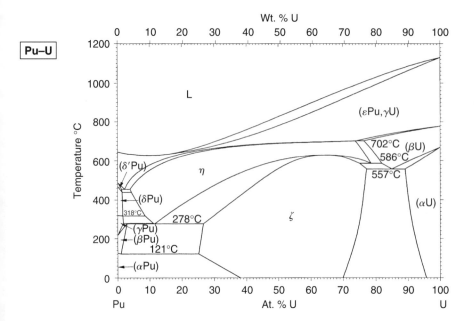

Pu–Zn

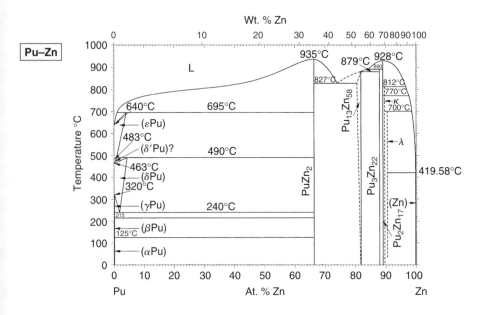

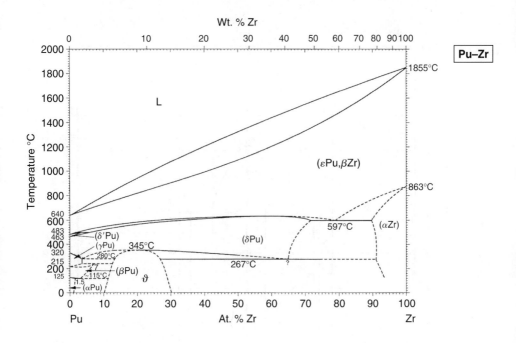

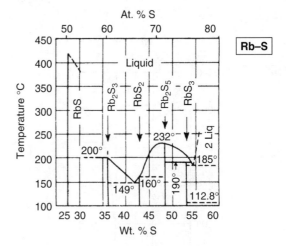

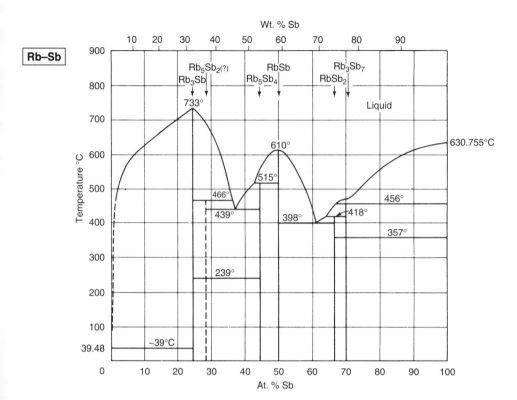

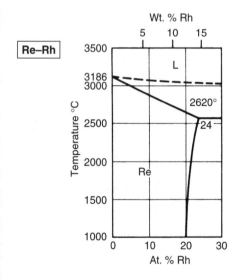

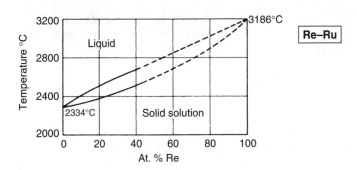

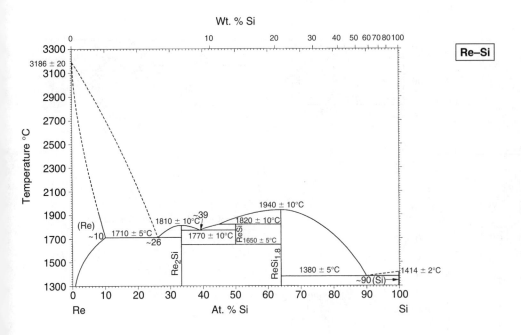

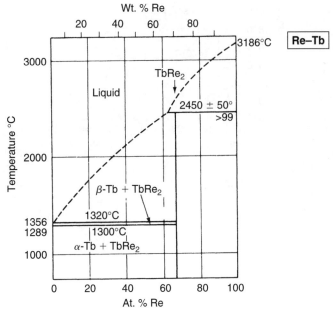

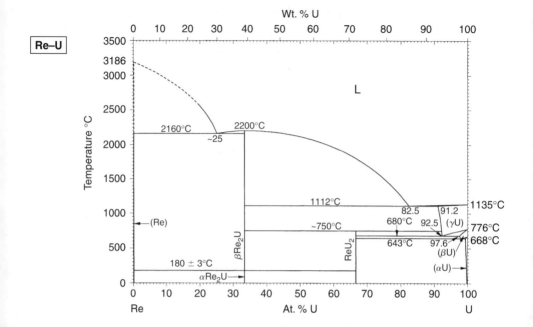

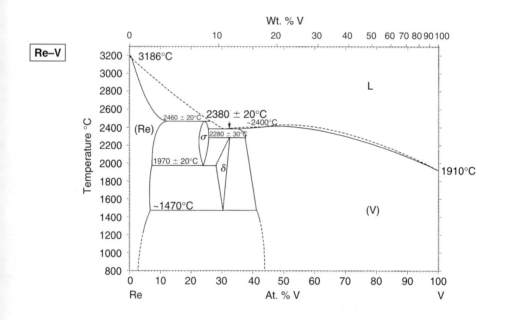

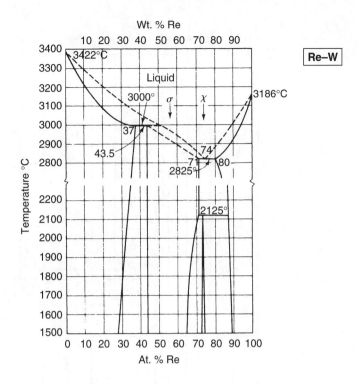

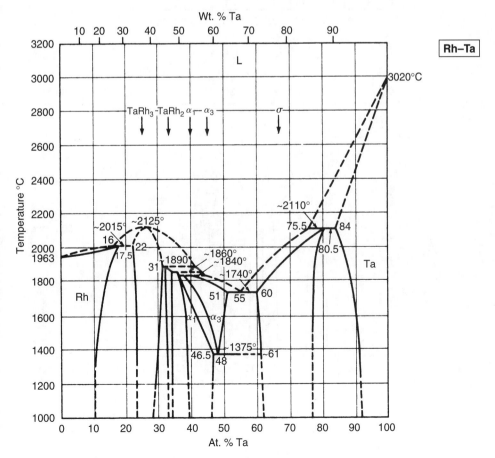

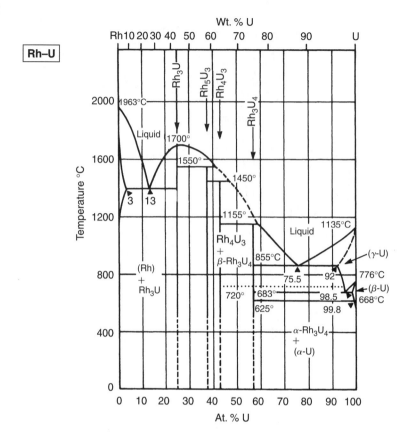

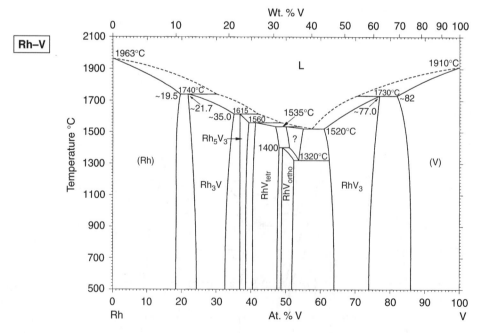

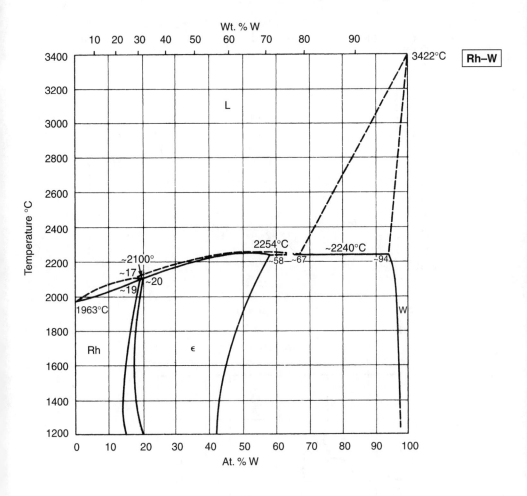

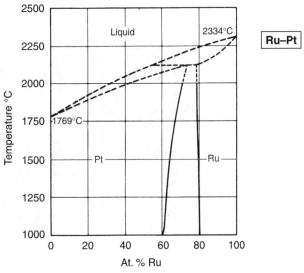

Ru–W

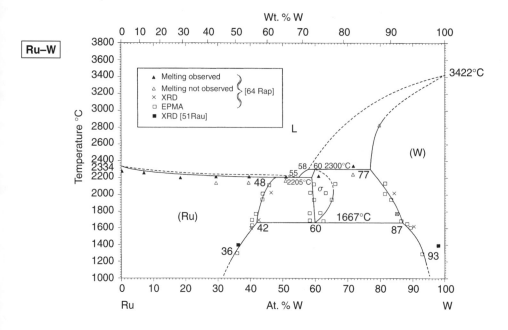

S–Sb

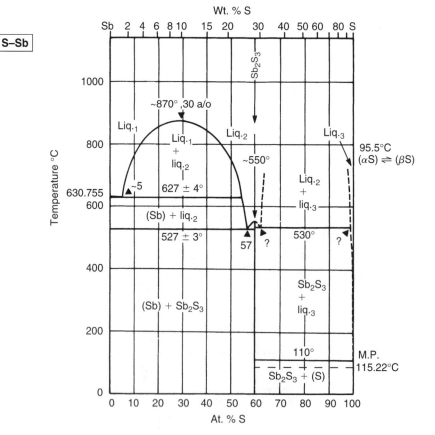

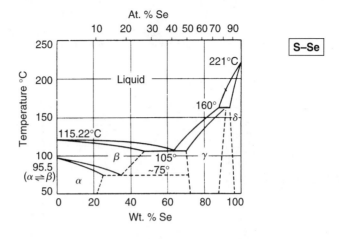

S–Se

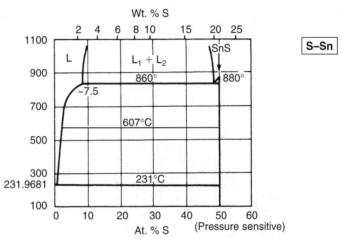

S–Sn

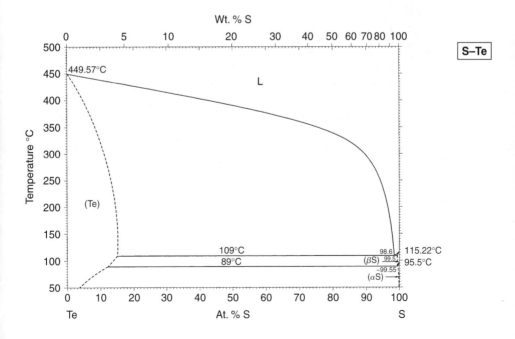

S–Te

Sb–Se

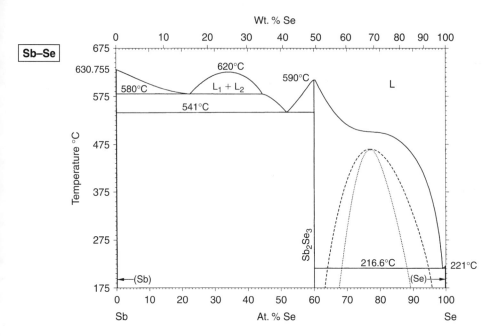

Sb–Si

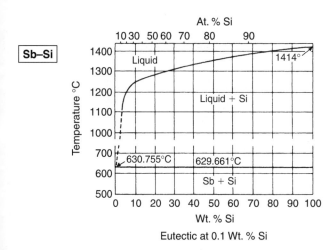

Eutectic at 0.1 Wt. % Si

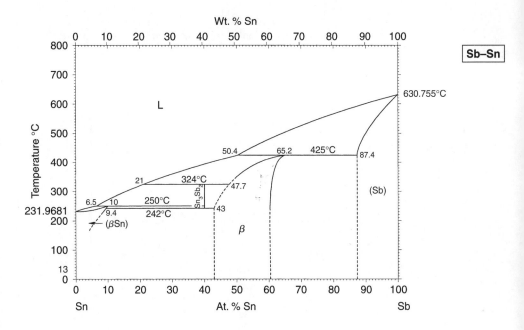

Sb–Sn

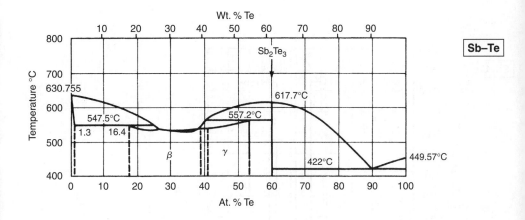

Sb–Te

Sb–Tl

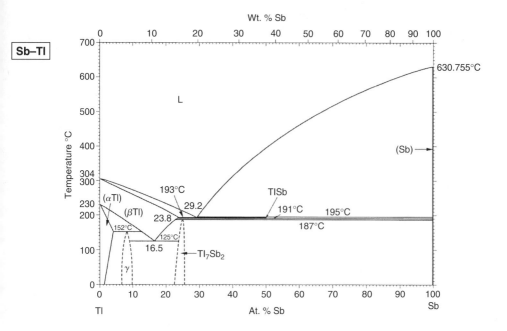

Sb–U

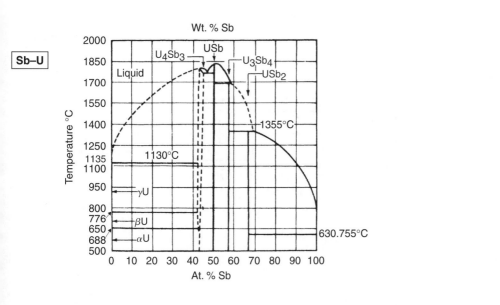

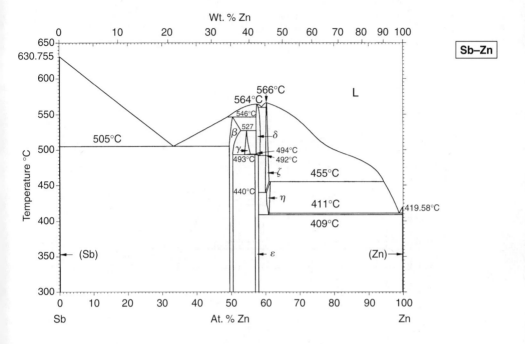

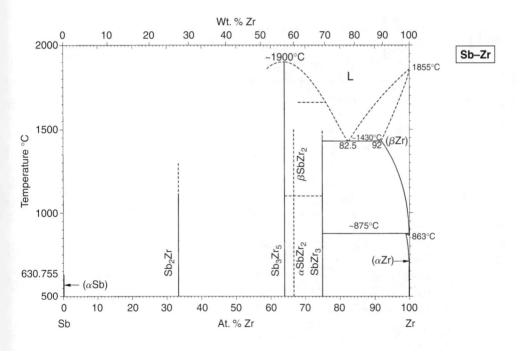

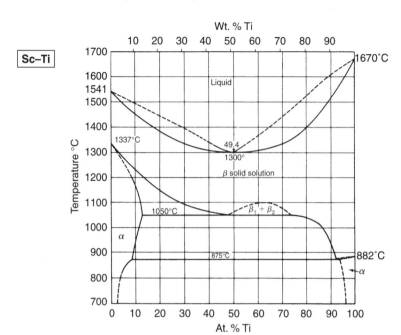

Sc–Ti

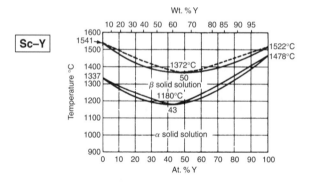

Sc–Y

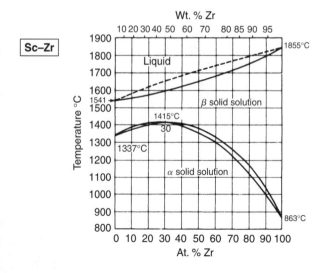

Sc–Zr

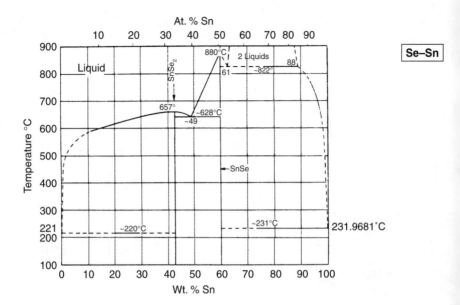

Se–Sn

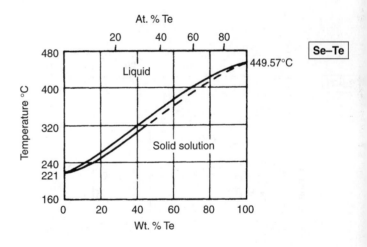

Se–Te

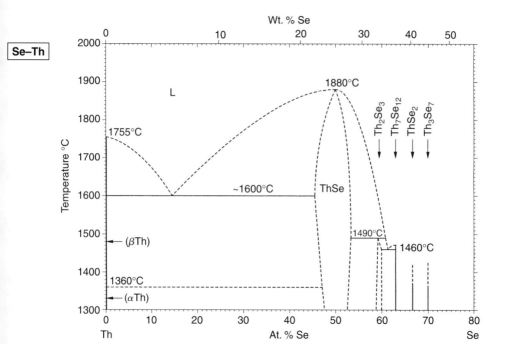

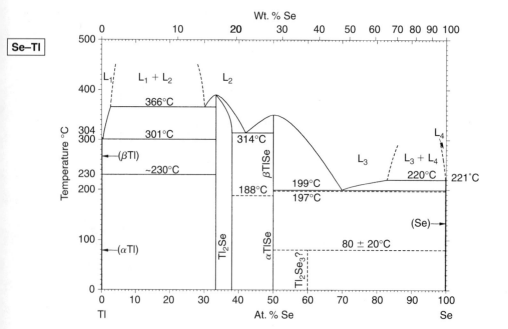

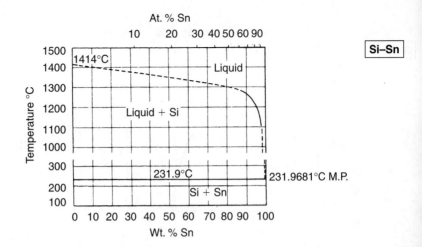

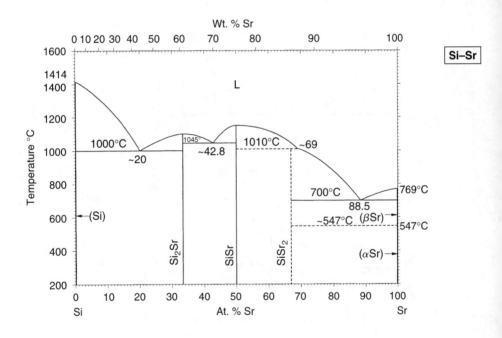

Si–Te

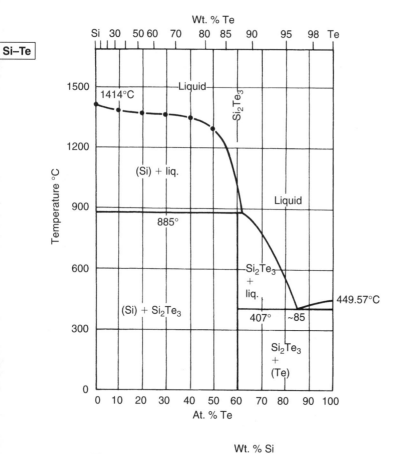

Si–Th

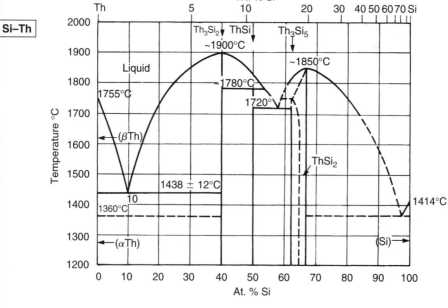

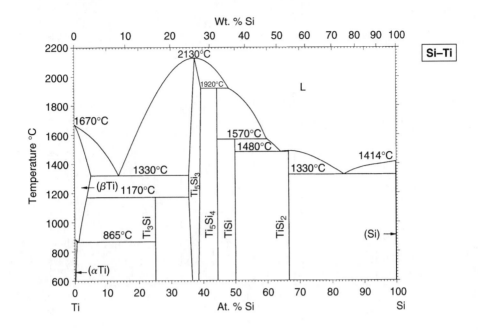

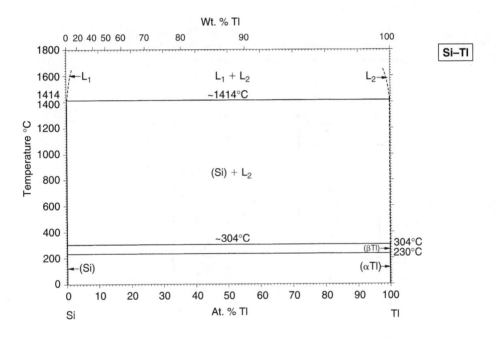

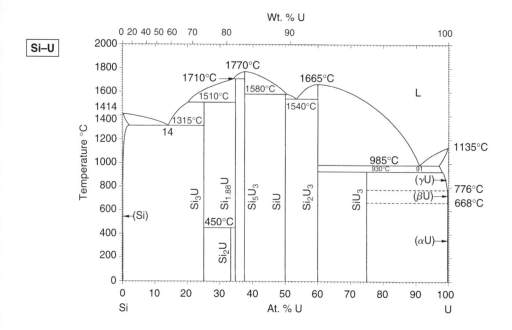

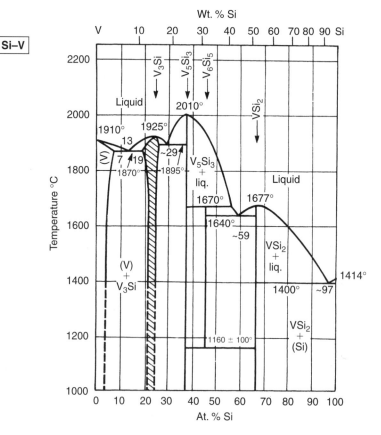

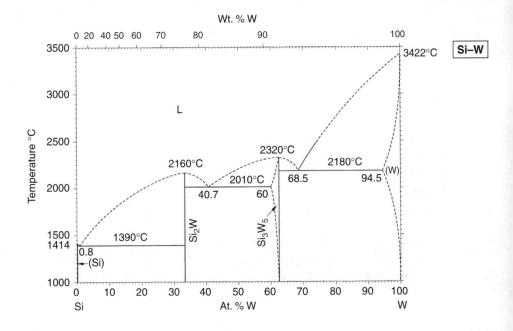

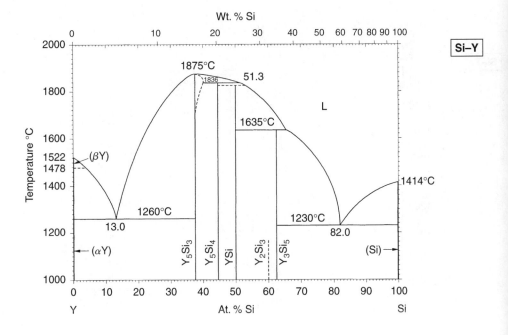

Si–Zn

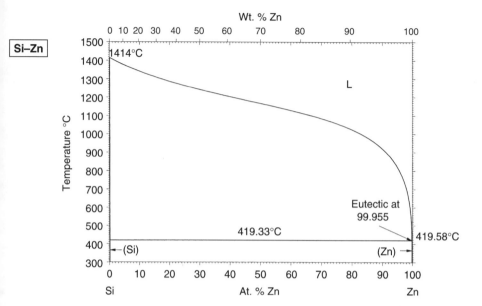

Si–Zr

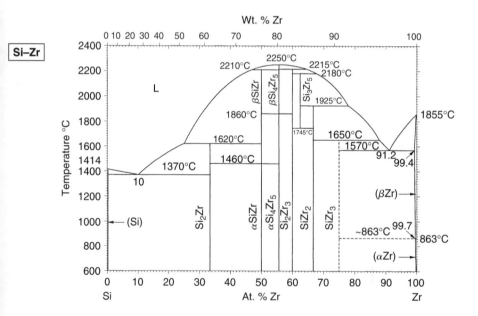

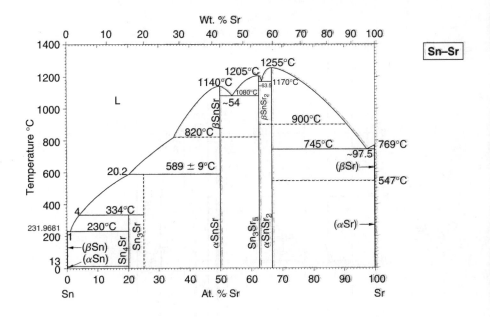

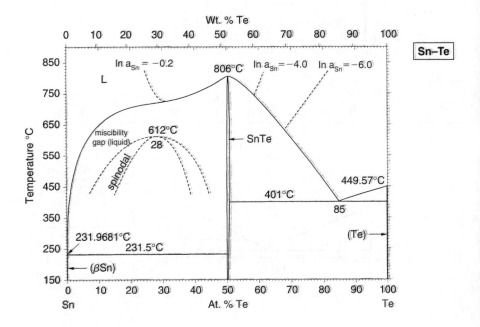

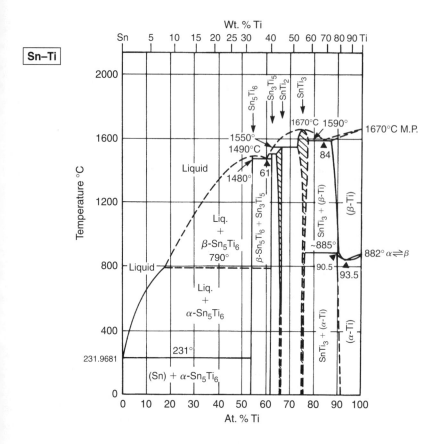

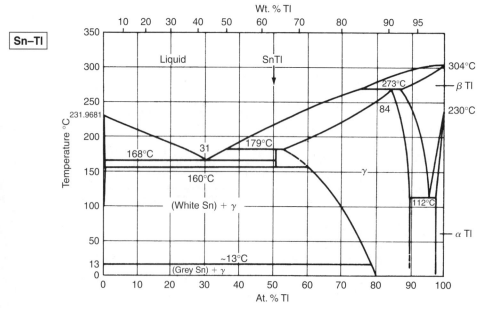

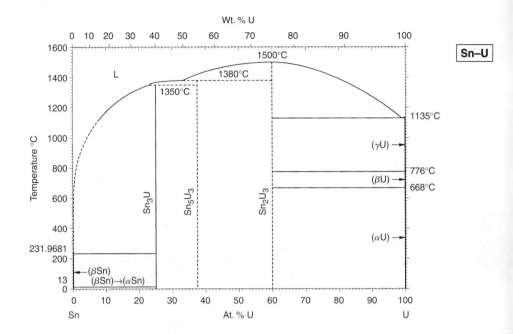

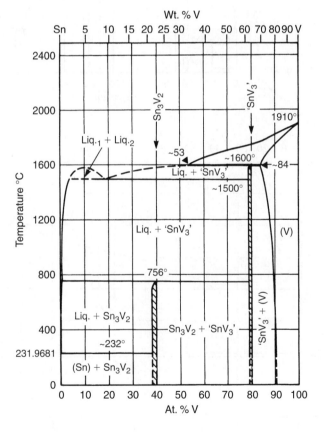

Sn–Yb

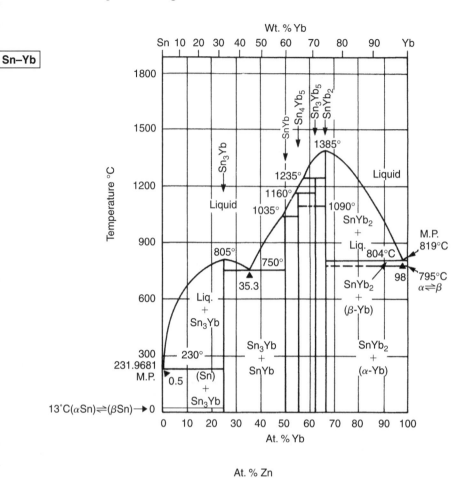

Sn–Zn

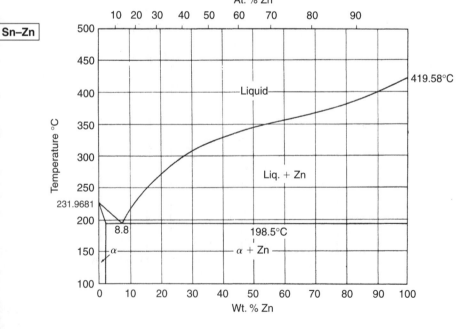

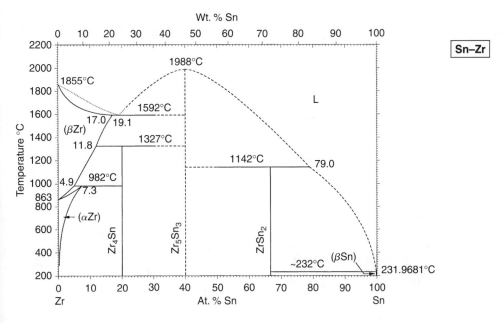

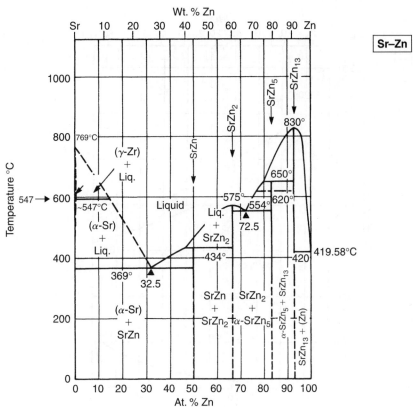

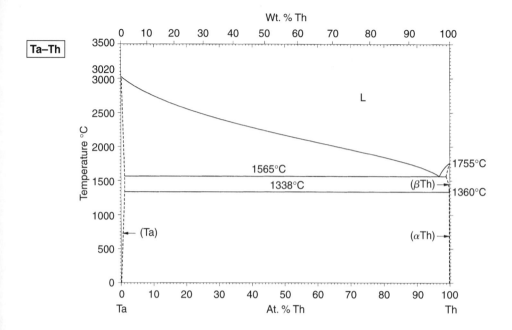

Ta–Th

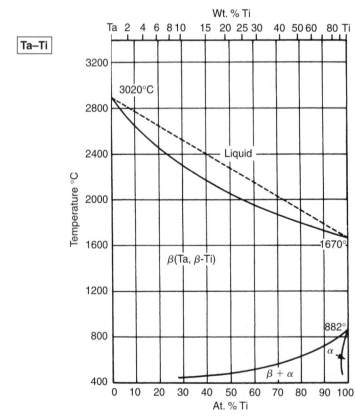

Ta–Ti

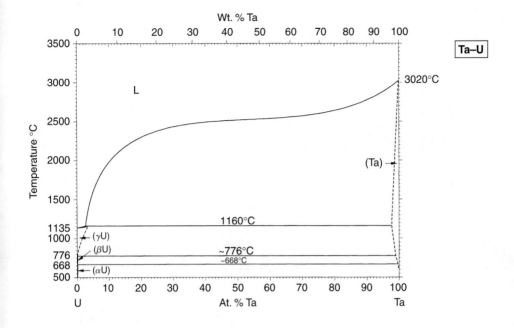

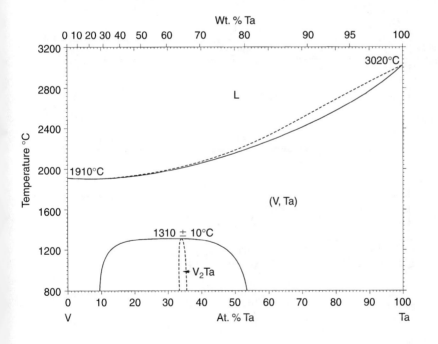

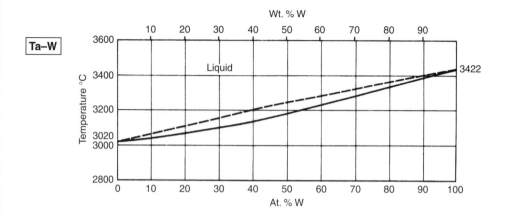

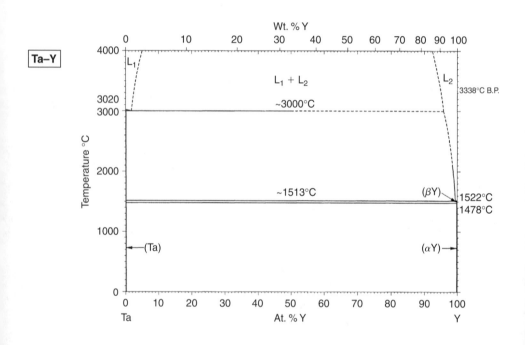

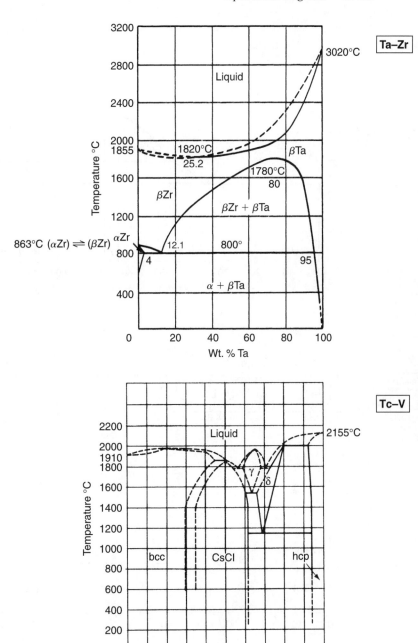

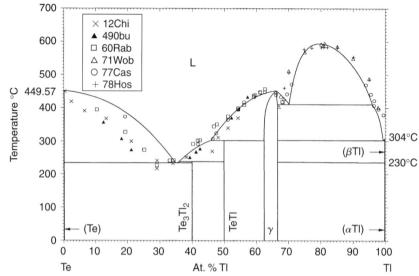

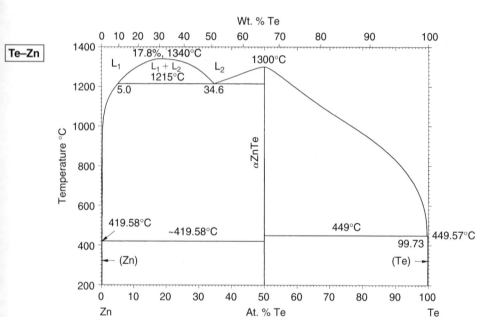

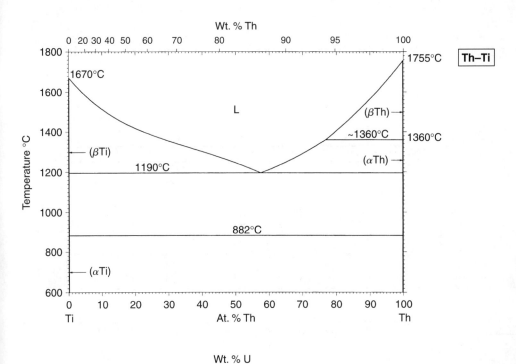

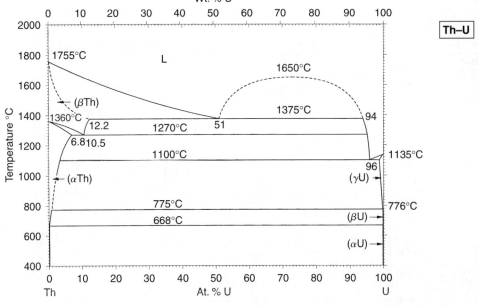

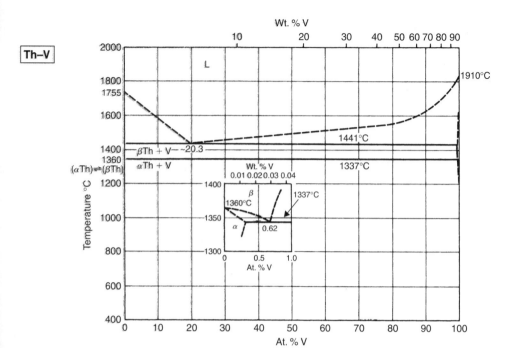

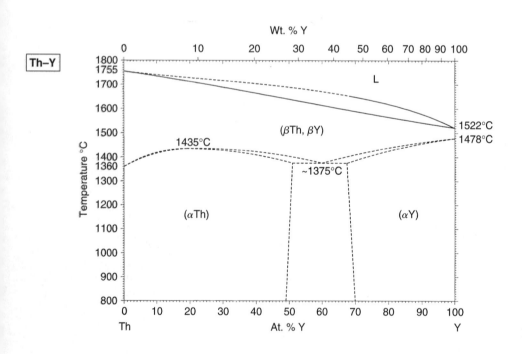

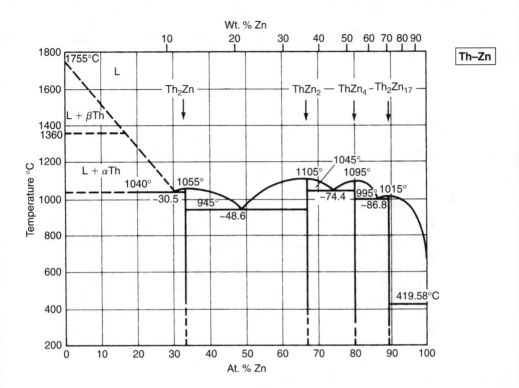

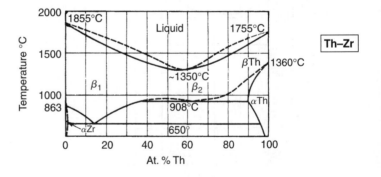

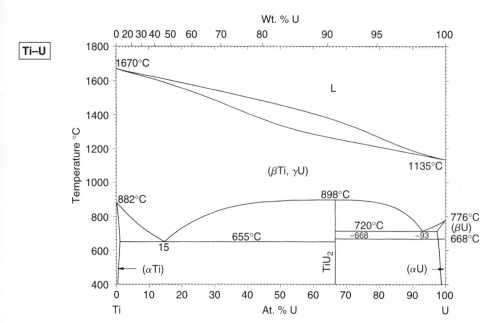

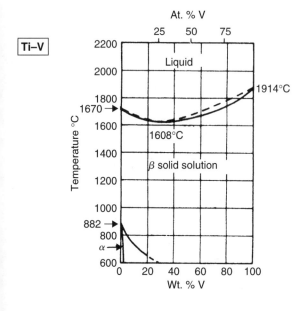

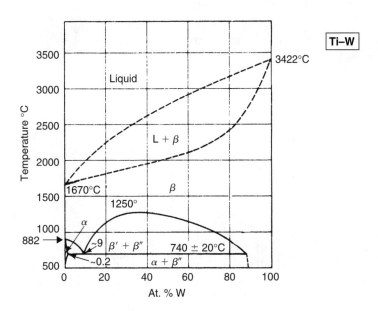

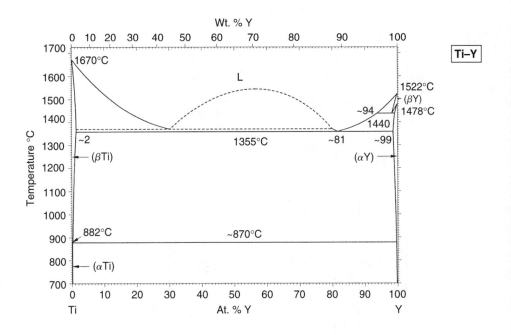

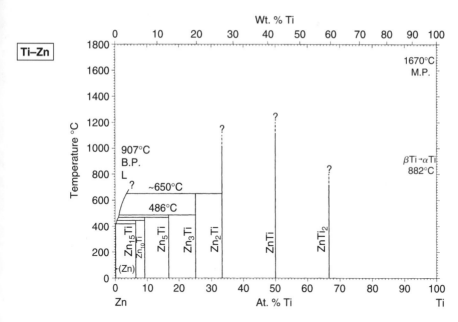

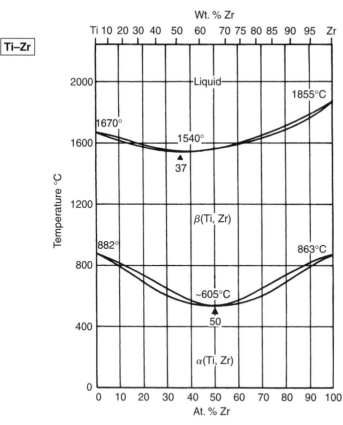

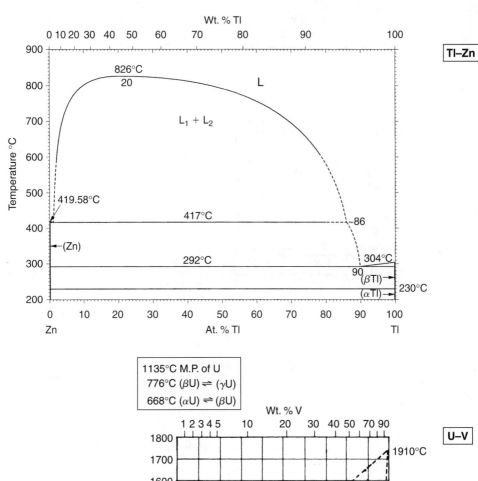

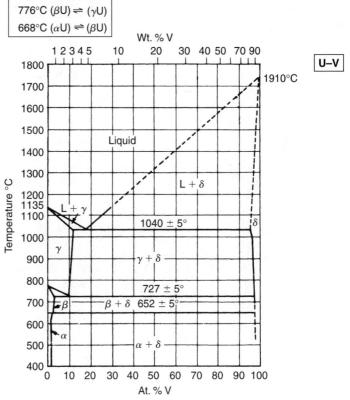

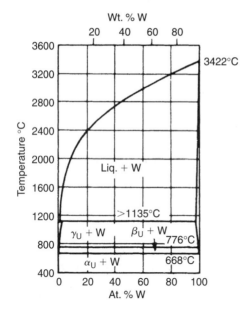

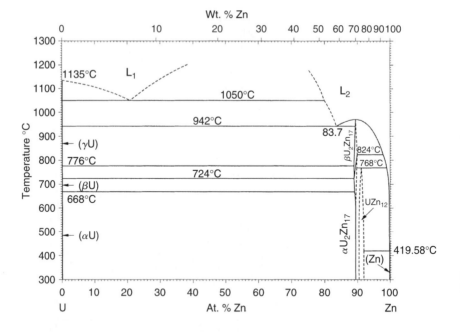

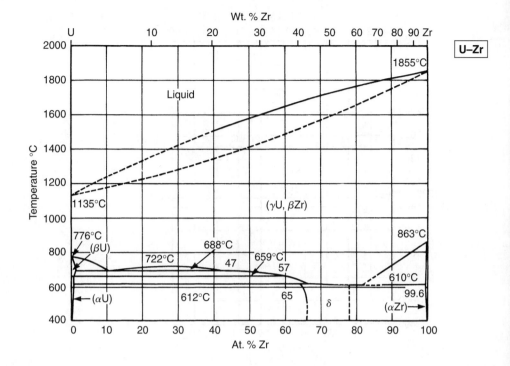

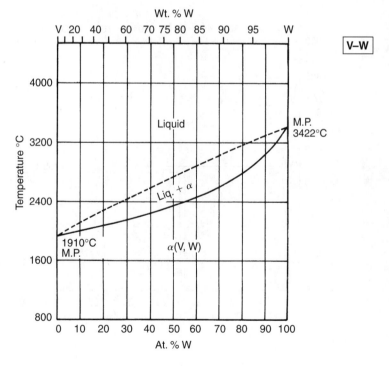

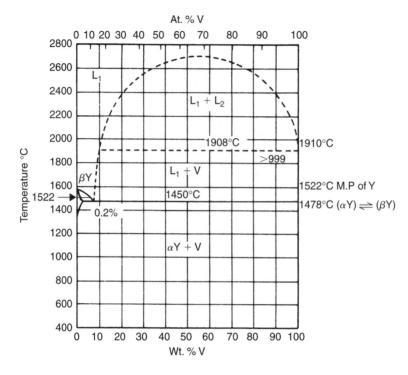

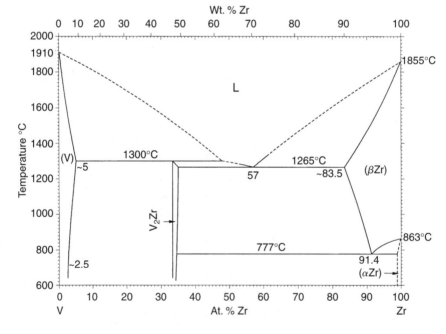

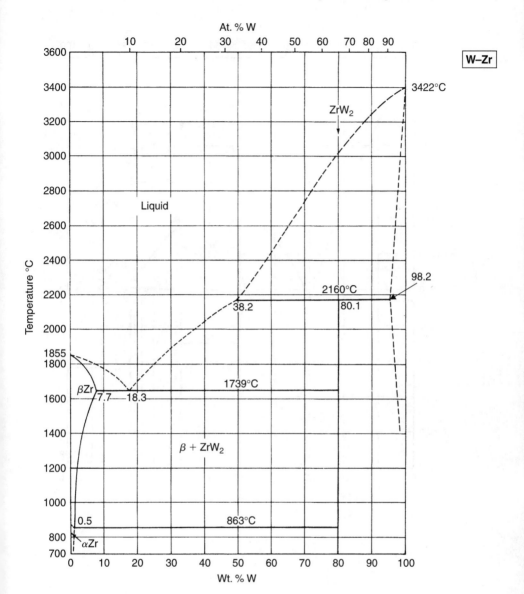

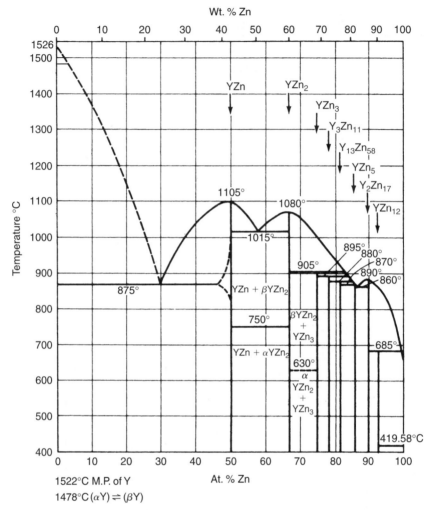

1522°C M.P. of Y

1478°C (αY) ⇌ (βY)

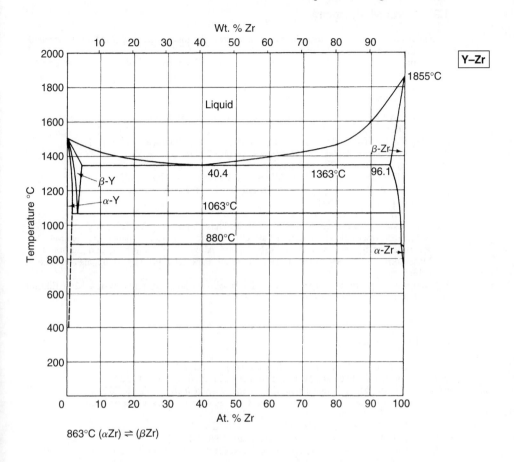

Y–Zr

863°C (αZr) ⇌ (βZr)

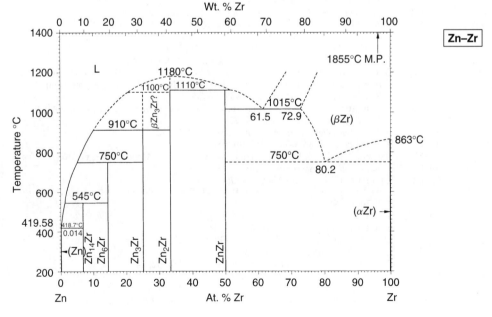

Zn–Zr

11.3 Acknowledgements

The diagrams of the binary systems have been taken from the following sources and thanks are made to the publishers and authors concerned.

The source references are shown in '11.1 Index of binary diagrams' as a number in parenthesis. For those marked (8) see list below. Those without a number are from material taken from 'Constitution of Binary Alloys', M. Hansen and K. Anderko, McGraw-Hill, New York, 1958 and from the Supplements by R. P. Elliott, 1965 and by F. A. Shunk, 1969.

(1) 'Binary Alloy Phase Diagrams' or 'Bulletin of Alloy Phase Diagrams', Massalski, ASM, Metals Park, Ohio, 44073, U.S.A.
(2) 'Handbook of Binary Diagrams', W. G. Moffatt, Genium Publications Corp., N.Y.
(3) *J. of less-common Metals.* For references see 11.3.2 below.
(4) *International Atomic Energy*, Vienna.
(5) 'Phase Diagrams of Binary Gold Alloys', K. A. Gschneider *et al.*, 1987.
(6) 'Manganese Phase Diagrams', The Manganese Centre, Paris, France.
(7) 'Metals Handbook', ASM.
(8) See below—in 11.3.1.
(9) Modified with data from ASM Binary Alloy Phase Diagrams, 2nd Edition T. B. Massalski *et al.*, ASM International, Materials Park OH, 1990 (plus updates to date of publication).
(10) Reprinted with permission from ASM Binary Alloy phase Diagrams, 2nd Edition T. B. Massalski *et al.*, ASM International, Materials Park OH, 1990 (plus updates to date of publication).
(11) Melting temperature data from CRC Handbook of Chemistry and Physics, 82nd Edition, D. R. Lide, CRC Press, Boca Raton, FL, 2001.

11.3.1 Binary systems

Ag–Al	H. W. L. Phillips, *Inst. Met. Ann. Eq. Diag. No.* 21.
Ag–As	Hansen; G. Eade and W. Hume-Rothery, *Z. Metallk.*, 1959, **50**, 123
Ag–Bi	Hansen; M. W. Nathans and M. Leider, *J. Phys. Chem.*, 1962, **66**, 2012
Ag–Ca	W. A. Alexander *et al.*, *Can. J. Chem.*, 1969, **47**, 611; A. N. Campbell and W. H. W. Wood, *Can. J. Chem.*, 1971, **49**, 1315
Ag–Cd	F. C. Kracek, 'Metals Handbook', Cleveland, Ohio, 1948
Ag–Dy	S. Delfino *et al.*, *J. less-common Metals*, 1976, **44**, 267
Ag–Ga	Hansen; W. Hume-Rothery and K. W. Andrews, *J. Inst. Metals*, 1942, **68**, 133; *Idem, Z. Metallk.*, 1959, **50**, 661
Ag–Gd	G. Kiessler *et al.*, *J. less-common Metals*, 1972, **26**, 293
Ag–Ge	Hansen; E. A. Owen and V. W. Rowlands, *J. Inst. Metals*, 1940, **66**, 371
Ag–Hg	H. M. Day and C. H. Mathewson, *Trans. Am. Min. Engrs*, 1938, **128**, 261
Ag–In	A. N. Campbell *et al.*, *Can. J. Chem.*, 1970, **48**, 1703
Ag–Li	W. G. Freeth and G. V. Raynor, *J. Inst. Metals*, 1953/4, **82**, 569
Ag–Mg	W. Hume-Rothery and K. W. Andrews, *J. Inst. Metals*, 1943, **69**, 488; G. F. Sager and B. J. Nelson, 'Metals Handbook', Cleveland, Ohio, 1948; J. L. Haughton and R. J. M. Payne, *J. Inst. Metals*, 1937, **60**, 351
Ag–Mn	R. Hultgren *et al.*, 'Selected Values of the Thermodynamic Props. of Binary Alloys', American Society for Metals, 1973
Ag–Na	Hansen; J. R. Weeks, *Trans. Quart. ASM*, 1969, **62**, 304
Ag–P	R. Vogel *et al.*, *Z. Metallk.*, 1959, **50**, 130
Ag–Sb	Hansen; P. W. Reynolds and W. Hume-Rothery, *J. Inst. Metals*, 1937, **60**, 365
Ag–Sc	I. Stapf *et al.*, *J. less-common Metals*, 1975, **39** (2), 219
Ag–Sn	Equilibrium Data for Sn Alloys, *Tin Res. Inst.*, 1949
Ag–Sr	F. Weibke, *Z. Anorg. Chem.*, 1930, **193**, 297; T. Heumann and N. Harmsen, *Z. Metallk.*, 1970, **61**, 906
Ag–Tb	S. Delfino *et al.*, *Z. Metallk.*, 1976, **67** (6), 392
Ag–Th	E. Raub, *Z. Metallk.*, 1949, **40**, 431
Ag–Ti	M. K. McQuillan, *J. Inst. Metals*, 1959/60, **88**, 235
Ag–Tl	E. Raub, *Z. Metallk.*, 1949, **40**, 432
Ag–U	R. W. Buzzard *et al.*, *J. Res. natn. Bur. Stand.*, 1954, **52**, 149
Ag–Yb	A. Palenzona, *J. less-common Metals*, 1970, **21**, 443
Ag–Zn	W. Hume-Rothery *et al.*, *Proc. Roy. Soc.*, 1941, **A 177**, 149
Ag–Zr	J. O. Betterton and D. S. Easton, *Trans. Am. Inst. Min. Engrs*, 1958, **12**, 470

Al–As   W. Köster and B. Thoma, *Z. Metallk.*, 1955, **46**, 291
Al–Be   H. W. L. Phillips, *Inst. Met. Ann. Eq. Diag. No.* 19
Al–Bi   R. Martin-Garin *et al.*, *Compte Rendu*, [C], 1966, **262**, 335
Al–Ce   K. H. J. Buschow and J. H. N. van Vucht, *Z. Metallk.*, 1966, **57**, 162
Al–Co   H. W. L. Phillips, *Inst. Met. Ann. Eq. Diag. No.* 20
Al–Cr   G. Falkenhagen and W. Hofmann, *Z. Metallk.*, 1950, **41**, 191
Al–Cs   V. D. Bushmanov and S. P. Yatsenko, *Russ. Met. (Metally)*, 1981, 5
Al–Cu   G. V. Raynor, *Inst. Met. Ann. Eq. Diag. No.* 4; R. P. Jewitt and D. J. Mack, *J. Inst. Met.*,
         1963/4, **92**, 59
Al–Dy   F. Casteels, *J. less-common Metals*, 1967, **12**, 210
Al–Er   K. H. J. Buschow and J. H. N. van Vucht, *Z. Metallk.*, 1965, **56**, 9
Al–Fe   H. W. L. Phillips, *Inst. Met. Ann. Eq. Diag.*, *No.* 13
Al–Gd   K. H. J. Buschow, *J. less-common Metals*, 1965, **9**, 453.
Al–Hf   E. M. Savitskii *et al.*, *Russ. Met.*, 1970, (1), 107
Al–In   B. Predel, *Z. Metallk.*, 1965, **56**, 791
Al–La   K. H. J. Buschow, *Philips Res. Reports*, 1965, **20**, 337.
Al–Mn   T. Godecke and W. Köster, *Z. Metallk.*, 1971, **62** (1), 727
Al–Nb   C. E. Lundin and A. S. Yamamoto, *Trans. Am. Inst. Min. Engrs*, 1966, **236**, 863
Al–Nd   K. H. J. Buschow, *J. less-common Metals*, 1965, **9**, 452
Al–Ni   W. O. Alexander and N. B. Vaughan, *J. Inst. Metals*, 1937, **61**, 250; H. Groeber and
         V. Hauk, *Z. Metallk.*, 1950, **41**, 283; H. W. L. Phillips, *Inst. Met. Ann. Eq.*
         Diag. No. 18
Al–Pb   E. Scheil and E. Jahn, *Z. Metallk.*, 1949, **40**, 319
Al–Pr   K. H. J. Buschow and J. H. N. van Vucht, *Z. Metallk.*, 1966, **57**, 162
Al–Pt   R. Huch and W. Klemm, *Z. anorg. allg. Chem.*, 1964, **329**, 123
Al–Pu   F. H. Ellinger *et al.*, *J. nucl. Mater.*, 1962, **5**, 165
Al–Ru   W. Obrowski, *Metall*, 1963, **17**, 108
Al–Sb   H. W. L. Phillips, *Inst. Met. Ann. Eq. Diag. No.* 15
Al–Sc   O. P. Naumkin *et al.*, *Metally Splavȳ Élektrotekh.*, 1965, (4), 176
Al–Si   H. W. L. Phillips, *Inst. Met. Ann. Eq. Diag. No.* 16
Al–Sm   F. Casteels, *J. less-common Metals*, 1967, **12**, 210
Al–Sn   R. C. Dorward, *Metal Trans. A*, 1976, **7A** (2), 308
Al–Sr   B. Closset *et al.*, *Metal Trans. A*, 1986, **17A**, 1250
Al–Th   J. R. Murray, *J. Inst. Metals*, 1958/9, **87**, 349
Al–U   P. R. Roy, *J. Nuc. Mat.*, 1964, **11** (1), 59
Al–V   R. Flükiger *et al.*, *J. less-common Metals*, 1975, **40** (1), 103
Al–W   W. D. Clark, *J. Inst. Metals*, 1940, **66**, 271
Al–Yb   A. Palenzona, *J. less-common Metals*, 1972, **29**, 289
Al–Zn   G. V. Raynor, *Inst. Met. Ann. Diag. No.* 1; G. R. Goldak and J. G. Parr, *J. Inst. Met.*,
         1963/4, **92**, 230
Al–Zr   W. L. Fink and L. A. Wiley, *Trans. Am. Inst. Min. Engrs*, 1939, **133**, 69;
         D. J. McPherson and M. Hansen, *Trans. Am. Soc. Metals*, 1954, **46**, 354

Am–Pu   F. H. Ellinger *et al.*, *J. nucl. Mater.*, 1966, **20**, 83

As–Au   B. Gather and R. Blacknik, *Z. Metallk.*, 1976, **67** (3), 168
As–Co   Hansen; W. Köster and W. Mulfinger, *Z. Metallk.*, 1938, **30**, 348
As–Eu   S. Ono *et al.*, *J. less-common Metals*, 1971, **25**, 287
As–Ga   W. Köster and B. Thoma, *Z. Metallk.*, 1955, **46**, 291
As–In   T. S. Liu and E. A. Peretti, *Trans. Am. Soc. Metals*, 1953, **45**, 677
As–K   F. W. Dorn *et al.*, *Z. anorg. allg. Chem.*, 1961, **309**, 204
As–Sn   E. A. Peretti and J. K. Paulsen, *J. less-common Metals*, 1969, **17**, 283
As–Te   J. R. Eifert and E. A. Peretti, *J. Mater. Sci.*, 1968, **3**, 293

Au–B   W. Obrowski, *Naturwiss*, 1961, **48**, 428; F. Wald and R. W. Stormont, *J. less-common*
         Metals, 1965, **9**, 423
Au–Bi   M. W. Nathans and M. Leider, *J. phys. Chem.*, 1962, **66**, 2012
Au–Cd   P. J. Durrant, *J. Inst. Metals*, 1929, **41**, 139; V. G. Rivlin *et al.*, *Acta met.*, 1962,
         11, 1143
Au–Ce   L. Rolla *et al.*, *Z. Metallk.*, 1943, **35**, 29
Au–Co   E. Raub and P. Walter, *Z. Metallk.*, 1950, **41**, 234
Au–Cr   E. Raub, *Z. Metallk.*, 1960, **51**, 290
Au–Cu   H. E. Bennett, *J. Inst. Metals*, 1962/3, **91**, 158; Hansen
Au–Dy   O. D. McMasters and K. A. Gschneidner, *J. less-common Metals*, 1973, **30**, 325

Au–Er	P. E. Rider *et al.*, *Trans. Am. Inst. Min. Engrs*, 1965, **233**, 1488
Au–Ga	C. J. Cooke and W. Hume-Rothery, *J. less-common Metals*, 1966, **10**, 42
Au–Gd	P. E. Rider *et al.*, *Trans. Am. Inst. Min. Engrs*, 1965, **233**, 1488
Au–Ge	R. I. Jaffee *et al.*, *Trans. Am. Inst. Min. Engrs*, 1945, **161**, 366
Au–Hg	C. Rolfe and W. Hume-Rothery, *J. less-common Metals*, 1967, **13**, 1
Au–K	G. Kienast and J. Verma, *Z. anorg. allg. Chem.*, 1961, **310**, 143
Au–Li	G. Kienast and J. Verma, *Z. anorg. allg. Chem.*, 1961, **310**, 143
Au–Lu	P. E. Rider *et al.*, *Trans. Am. Inst. Min. Engrs*, 1965, **233**, 1488
Au–Mn	E. Raub, *Z. Metallk.*, 1949, **40**, 359; E. Raub *et al.*, *Z. Metallk.*, 1953, **44**, 312
Au–Na	G. Kienast *et al.*, *Z. anorg. allg. Chem.*, 1961, **310**, 143
Au–P	R. Vogel *et al.*, *Z. Metallk.*, 1959, **50**, 130
Au–Pt	A. S. Darling *et al.*, *J. Inst. Metals*, 1952/3, **81**, 125
Au–S	R. Vogel and R. Gebhardt, *Z. Metallk.*, 1961, **52**, 318
Au–Sc	P. E. Rider *et al.*, *Trans. Am. Inst. Min. Engrs*, 1965, **233**, 1488
Au–Se	R. Vogel and R. Gebhardt, *Z. Metallk.*, 1961, **52**, 318; R. Rabenau *et al.*, *J. less-common Metals*, 1971, **24** (3), 291
Au–Si	R. P. Anantatmula *et al.*, *J. Electron. Mat.*, 1975, **4** (3), 445
Au–Sn	Equilibrium Data for Sn Alloys, *Tin Res. Inst.*, 1949
Au–Sr	M. Feller-Kniepmaier and T. Heumann, *Z. Metallk.*, 1960, **51**, 404
Au–Tb	P. E. Rider *et al.*, *Trans. Am. Inst. Min. Engrs*, 1965, **233**, 1488
Au–Tm	P. E. Rider *et al.*, *Trans. Am. Inst. Min. Engrs*, 1965, **233**, 1488
Au–U	R. W. Buzzard and J. J. Park, *J. Res. natn. Bur. Stand.*, 1954, **53**, 291
Au–V	R. Flükiger *et al.*, *J. less-common Metals*, 1975, **40**, 103
Au–Yb	A. Iandelli and A. Palenzona, *J. less-common Metals*, 1969, **18**, 221; P. E. Rider *et al.*, *Trans. Am. Inst. Min. Engrs*, 1965, **233**, 1488
B–Cr	K. I. Portnoi *et al.*, *Poroshk. Metall.*, 1969, No. 4, 51
B–Eu	E. M. Savitskii *et al.*, *Neorg. Materialy*, 1971, **7**, 617
B–Ge	L. R. Bidwell, *J. less-common Metals*, 1970, **20**, 19
B–Hf	K. I. Portnoi *et al.*, *Neorg. Materialy*, 1971, **7**, 1987
B–Mn	G. Pradelli and C. Gianoglio, *La Metallurgia Italiana*, 1974, **12**, 659
B–Mo	K. I. Portnoi *et al.*, *Metally. Splavy Élektrotekh.*, 1967, (4), 171
B–Nb	H. Novotny *et al.*, *Z. Metallk.*, 1959, **50**, 417
B–Re	K. I. Portnoi and V. M. Romashov, *Poroshk. Metall.*, 1968, (2) 41
B–Sm	G. I. Solovyev and K. E. Spear, *J. Am. Ceram. Soc.*, 1972, **55** (9), 475
B–Ta	K. I. Portnoi *et al.*, *Poroshk. Metall.*, 1971, No. 11, (107), 89; E. Rudy and I. Progulski, *Planseeber Pulvermetall*, 1967, **15**, 13; *see also* Elliott for alternative version
B–U	R. W. Mar, *J. Am. Ceram. Soc.*, 1975, **50** (3–4), 145
B–Zr	K. I. Portnoi *et al.*, *Poroshk. Metall.*, 1970, No. 7, (91), 68
Ba–Cd	R. T. Dirstine, *J. less-common Metals*, 1975, **39** (1), 181
Ba–Cu	G. Bruzzone, *J. less-common Metals*, 1971, **25**, 361
Ba–Ga	G. Bruzzone, *Boll. Sci. Fac. Industr. Bologna*, 1966, **24** (4), 113
Ba–Ge	V. G. Andrianov *et al.*, *Neorg. Materialy*, 1966, **2**, 2064
Ba–Hg	G. Bruzzone and F. Merlo, *J. less-common Metals*, 1975, **39**, 271.
Ba–In	G. Bruzzone, *J. less-common Metals*, 1966, **11**, 249
Ba–Na	F. A. Kanda *et al.*, *J. phys. Chem.*, 1965, **69**, 3867
Ba–Pd	E. M. Savitskii *et al.*, *Metally Splavy Élektrotekh.*, 1970, (6), 143
Ba–Pt	E. M. Savitskii *et al.*, *Metally Splavy Élektrotekh.*, 1971, (1), 157
Be–Bi	G. W. Horsley and J. T. Maskrey, *J. Inst. Metals*, 1957/8, **86**, 401
Be–Co	F. Aldinger and S. Jönsson, *Z. Metallk.*, 1977, **68** (5), 362
Be–Cr	A. R. Edwards and S. T. M. Johnston, *J. Inst. Met.*, 1955/6, **84**, 313
Be–Fe	R. J. Teitel and M. Cohen, *Trans. Am. Inst. Min. Engrs*, 1949, **185**, 285
Be–Nb	A. T. Grigoriev and I. I. Raevsky, *Russ. Met.*, 1968, (5), 134
Be–Ni	E. Jahn, *Z. Metallk.*, 1949, **40**, 399
Be–U	R. W. Buzzard, *J. Res. natn. Bur. Stand.*, 1953, **50**, 63
Bi–Ca	S. Hoesel, *Z. Phsik. Chem.* (*Leipzig*), 1962, **219**, 205
Bi–Co	R. Damm *et al.*, *Z. Metallk.*, 1962, **53**, 196
Bi–Cu	M. W. Nathans and M. Leider, *J. Phys. Chem.*, 1962, **66**, 2013
Bi–Cs	G. Gnutzmann and W. Klemm, *Z. anorg. allg. Chem.*, 1961, **309**, 181
Bi–Ga	B. Predel, *Z. Phys. Chem.*, 1960, **24**, 206

Bi–In	O. H. Henry and E. L. Baldwick, *Trans. Am. Inst. Min. Engrs*, 1947, **171**, 389
Bi–K	G. Gnutzmann and W. Klemm, *Z. anorg. allg. Chem.*, 1961, **309**, 181
Bi–La	K. Nomura *et al.*, *J. less-common Metals*, 1977, **52**, 259
Bi–Mn	A. U. Seybolt *et al.*, *Trans. Am. Inst. Min. Engrs*, 1956, **206**, 606 and 1406–8
Bi–Nd	G. F. Kobzenko *et al.*, *Neorg. Materialy*, 1971, **7**, 1438
Bi–Pb	G. O. Hiers, 'Metals Handbook', Cleveland, Ohio, 1948
Bi–Pt	N. N. Zhuravlev *et al.*, *Physics Metals Metallogr.*, 1962, **13**, 51
Bi–Rb	G. Gnutzmann and W. Klemm, *Z. anorg. allg. Chem.*, 1961, **309**, 181
Bi–Rh	R. G. Ross and W. Hume-Rothery, *J. less-common Metals*, 1962, **4**, 454
Bi–Sn	W. Oelsen and K. F. Golucke, *Arch. Eisenhuttenw.*, 1958, **29**, 689
Bi–U	P. Cotterill and H. J. Axon, *J. Inst. Metals*, 1958/9, **87**, 159
Bi–Y	F. A. Schmidt *et al.*, *J. less-common Metals*, 1969, **18**, 215
C–Cr	D. S. Bloom and N. J. Grant, *Trans. Am. Inst. Min. Engrs*, 1950, **188**, 141
C–Cu	M. B. Bever and C. F. Floe, *Trans. Am. Inst. Min. Engrs*, 1946, **166**, 128
C–La	F. H. Spedding *et al.*, *Trans. Am. Inst. Min. Engrs*, 1959, **215**, 192
C–Re	A. I. Evstyukhin *et al.*, *Met. i Metalloved. Chistikh Metallov, Moscow*, 1963, 149
C–Th	R. Benz and P. L. Stone, *High Temp. Sci.*, 1969, **1**, 114
C–V	E. K. Storms and R. J. McNeal, *J. phys. Chem.*, 1962, **66**, 1401
Ca–Cu	G. Bruzzone, *J. less-common Metals*, 1971, **25**, 361
Ca–Eu	V. F. Stroganova *et al.*, *Russ. Met.*, 1969, (6), 91
Ca–In	G. Bruzzone, *J. less-common Metals*, 1966, (11), 249
Ca–Mn	I. Obinata *et al.*, *Metallwiss. Technik*, 1963, (12), 1205
Ca–Pb	G. Bruzzone and F. Merlo, *J. less-common Metals*, 1976, **48**, 103
Ca–Sb	Z. U. Niyazora *et al.*, *Izv. Akad. SSSR Nedrg. Mater.*, 1976, **12** (7), 1293
Ca–Si	E. Schürmann, *et al.*, *Arch. Eisenhuttenw.*, 1975, **45** (6), 367
Ca–Ti	I. Obinata *et al.*, *Trans. Am. Soc. Metals*, 1960, **52**, 1072
Ca–Yb	S. D. Soderquist and F. X. Kayser, *J. less-common Metals*, 1968, **16**, 361
Ca–Zn	A. F. Messing *et al.*, *Trans. Am. Soc. Metals*, 1963, **56**, 345
Cd–Cu	Hansen; E. Raub, *Metallforschung*, 1947, **2**, 120
Cd–Eu	W. Köster and J. Meixner, *Z. Metallk.*, 1965, **56**, 695
Cd–Gd	G. Bruzzone *et al.*, *J. less-common Metals*, 1971, **25**, 295
Cd–La	G. Bruzzone and F. Merlo, *J. less-common Metals*, 1973, **30**, 303
Cd–Mg	W. Hume-Rothery and G. V. Raynor, *Proc. Roy. Soc.*, 1940, **A 174**, 471
Cd–Np	M. Krumpelt *et al.*, *J. less-common Metals*, 1969, **18**, 35
Cd–Sm	G. Bruzzone and M. L. Fornasini, *J. less-common Metals*, 1974, **37**, 289
Cd–Sn	D. Hansen and W. T. Pell-Walpole, *J. Inst. Metals*, 1936, **59**, 281
Cd–Sr	W. Köster and J. Meixner, *Z. Metallk.*, 1965, **56**, 695
Cd–Ti	W. M. Robertson, *Met. Trans*, 1972, **3**, 1443
Cd–U	A. E. Martin *et al.*, *Trans. Am. Inst. Min. Engrs*, 1961, **221**, 789
Cd–Y	E. Ryba *et al.*, *J. less-common Metals*, 1969, **18**, 49
Cd–Yb	A. Palenzona, *J. less-common Metals*, 1971, **25**, 367
Ce–Co	R. Vogel, *Z. Metallk.*, 1946, **37**, 98
Ce–Cr	V. M. Svechnikov *et al.*, *Dopov. Akad. Nauk. ukr. RSR*, 1969, (4), 354
Ce–Fe	J. O. Jepson and P. Duwez, *Trans. Am. Soc. Metals*, 1955, **47**, 543
Ce–In	R. Vogel and H. Klose, *Z. Metallk.*, 1954, **45**, 633
Ce–Mn	A. Landelli, *Atti Accad. Nazl. Lincei. Rend.*, 1952, **13**, 265; and B. J. Thamer, *J. less-common Metals*, 1964, **7**, 341 and 1965, **8**, 215
Ce–Pb	L. Rolla *et al.*, *Z. Metallk.*, 1943, **35**, 29
Ce–Pd	J. R. Thomson, *J. less-common Metals*, 1967, **13**, 307; and D. Rossi *et al.*, *J. less-common Metals*, 1975, **40**, 345
Ce–Ru	W. Obrowski, *Z. Metallk.*, 1962, **53**, 736
Ce–Sn	Equilibrium Data for Sn Alloys, *Tin Res. Inst.*, 1949
Ce–Th	R. T. Weiner *et al.*, *J. Inst. Metals*, 1957/8, **86**, 185
Ce–Ti	E. M. Savitskii and G. S. Burkhanov, *J. less-common Metals*, 1962, **4**, 301
Ce–Tl	L. Rolla *et al.*, *Z. Metallk.*, 1943, **35**, 29
Co–Cr	A. R. Elsea *et al.*, *Trans. Am. Inst. Min. Engrs*, 1949, **180**, 579
Co–Fe	W. C. Ellis and E. S. Greiner, *Trans. Am. Soc. Metals*, 1941, **29**, 415
Co–Ga	K. Schubert *et al.*, *Z. Metallk.*, 1959, **50**, 534
Co–Gd	V. F. Novy *et al.*, *Trans Am. Inst. Min. Engrs*, 1961, **221**, 588
Co–Ge	H. Pfisterer and K. Schubert, *Z. Metallk.*, 1949, **40**, 379

Co–Ho	K. H. J. Buschow and A. S. Van Der Goot, *J. less-common Metals*, 1969, **19**, 153
Co–In	J. D. Schöbel and H. Stadelmaier, *Z. Metallk.*, 1970, **61**, 342
Co–La	K. H. J. Buschow and W. A. J. J. Velge, *J. less-common Metals*, 1967, **13**, 11
Co–Mg	J. F. Smith and M. J. Smith, *Trans. Am. Soc. Metals*, 1964, **57**, 337
Co–Mo	T. F. J. Quinn and W. Hume-Rothery, *J. less-common Metals*, 1963, **5**, 314
Co–Nb	S. K. Bataleva, *Vest. Mosk. Univ. Khim.*, 1970, (4), 432
Co–Ni	C. E. Lacy, 'Metals Handbook', Cleveland, Ohio, 1948
Co–Os	W. Köster and E. Horn, *Z. Metallk.*, 1952, **43**, 444
Co–Pr	A. E. Ray, *Cobalt*, 1974, **13**
Co–Pt	W. Köster, *Z. Metallk.*, 1949, **40**, 431
Co–Re	W. Köster and E. Horn, *Z. Metallk.*, 1952, **43**, 444
Co–Rh	W. Köster and E. Horn, *Z. Metallk.*, 1952, **43**, 444
Co–Ru	W. Köster and E. Horn, *Z. Metallk.*, 1952, **43**, 444
Co–Sc	V. Markiv *et al.*, *Akad. Nauk. UKR Metall.*, 1978, **73**, 39
Co–Sn	Equilibrium Data for Sn Alloys, *Tin Res. Inst.*, 1949
Co–Ta	W. Köster and W. Mulfinger, *Z. Metallk.*, 1938, **30**, 348; M. Karchinsky and R. W. Fountain, *Trans. Am. Inst. Min. Engrs*, 1959, **215**, 1033
Co–V	W. Köster and H. Schmid, *Z. Metallk.*, 1955, **46**, 195
Co–W	S. Takeda, *Sci. Rep. Tôhoku Univ.*, 1936, Honda Anniv. Vol., p. 864
Co–Zn	J. Schramm, *Z. Metallk.*, 1941, **33**, 46
Co–Zr	S. K. Bataleva, *Vest. Mosk. Univ., Khim.*, 1970, **11** (5), 557
Cr–Ga	J. D. Bornand and P. Feschotte, *J. less-common Metals*, 1972, **29**, 81
Cr–Hf	E. Rudy and S. T. Windisch, *J. less-common Metals*, 1968, **15**, 13
Cr–Ir	R. M. Waterstrat and R. C. Manuszewski, *J. less-common Metals*, 1973, **32** (1), 79
Cr–Mn	S. J. Carlile *et al.*, *J. Inst. Metals*, 1949/50, **76**, 169; A. Hellawell and W. Hume-Rothery, *Phil. Trans. Roy. Soc.*, A **249**, 1957, 417
Cr–Mo	O. Kubaschewski and A. Schneider, *Z. Elektrochem.*, 1942, **48**, 671
Cr–P	R. Vogel and G. W. Kasten Arch. Eisenhuttenw., 1939, **12**, 387 and *J less common Metals*, 1962, **4**, 496
Cr–Pb	G. Hindricks, *JISI*, 1943, **148**, 428 and T. Alden *et al.*, *Trans. AIME*, 1958, **212**, 15
Cr–Pd	G. Grube and R. Knabe, *Z. Elektrochem.*, 1936, **42**, 793
Cr–Pr	V. G. Ivanchenko *et al.*, *Dopov. Akad. Nauk. ukr. RSR*, 1969. (A), (8), 748
Cr–Pt	R. M. Waterstrat, *Met. Trans.*, 1973, **4**, 1585
Cr–Rh	R. M. Waterstrat and R. C. Manuszewski, *J. less-common Metals*, 1973, **32**, 331
Cr–Sc	V. M. Svechnikov *et al.*, *Dopov. Akad. Nauk. ukr. RSR*, 1972, (A), (34), 266
Cr–Tb	V. G. Ivanchenko *et al.*, *Dopov. Akad. Nauk. ukr. RSR*, 1970, (A), (8), 758
Cr–Ti	F. B. Cuff *et al.*, *J. Metals*, 1952, 848
Cr–U	A. H. Daane and A. S. Wilson, *Trans. Am. Inst. Min. Engrs*, 1955, **203**, 1219
Cs–Ga	S. Yatsenko and K. Chuntonov, *Russ. Metal. (Metally)*, 1981, **6**, 1
Cs–Te	K. Chuntonov *et al.*, *Inorganic Materials*, 1982, **18** (7), 937
Cs–Tl	V. Bushmanov and S. Yatsenko, *Russ. Metal. (Metally)*, 1981, *No.* 5
Cu–Er	K. H. J. Buschow, *Phillips Res. Reports*, 1970, **25**, 227
Cu–Fe	O. Kubaschewski, 'Iron-Binary Phase Diagrams', Springer-Verlag, 1982
Cu–Ga	W. Hume-Rothery *et al.*, *Phil. Trans. Roy. Soc.*, 1934, A **233**, 1; J. O. Betterton and W. Hume-Rothery, *J. Inst. Metals*, 1951/2, **80**, 459
Cu–Ge	W. Hume-Rothery *et al.*, *J. Inst. Metals*, 1940, **66**, 221; J. Reynolds and W. Hume-Rothery, *J. Inst. Metals*, 1956/7, **85**, 120
Cu–In	R. O. Jones and E. A. Owen, *J. Inst. Metals*, 1953/4, **82**, 445
Cu–Ir	E. Raub and G. Röschel, *Z. Metallk*, 1969, **60**, 142
Cu–Mn	B. M. Loring, 'Metals Handbook', Cleveland, Ohio, 1948
Cu–O	F. N. Rhines and C. H. Mathewson, *Trans. Am. Inst. Min. Engrs*, 1934, **111**, 339
Cu–Pd	F. W. Jones and C. Sykes, *J. Inst. Metals*, 1939, **65**, 422
Cu–Rh	Ch. Raub *et al.*, *Metall*, 1972, **25**, 761
Cu–S	J. Nutting, *Inst. Met. Ann. Eq Diag. No.* 24
Cu–Sb	Hansen; J. C. Mertz and C. H. Mathewson, *Trans. Am. Inst. Min. Engrs*, 1937, **124**, 68
Cu–Si	C. S. Smith, 'Metals Handbook'. Cleveland, Ohio, 1948
Cu–Sm	K. Kuhn and A. Perry, *Met. Science*, 1975, **9**, 339
Cu–Sn	G. V. Raynor, *Inst. Met. Ann. Diag. No.* 2
Cu–Sr	G. Bruzzone, *J. less-common Metals*, 1971, **25**, 361–366
Cu–Th	R. J. Schiltz *et al.*, *J. less-common Metals*, 1971, **25**, 175

Cu–U   H. A. Wilhelm and O. N. Carlson, *Trans. Am. Soc. Metals*, 1950, **42**, 1311
Cu–Y   R. F. Domegala *et al.*, *Trans. Am. Soc. Metals*, 1961, **53**, 137
Cu–Yb   A. Iandelli and A. Palenzona, *J. less-common Metals*, 1971, **25**, 333
Cu–Zn   P. Chiotti *et al.*, *US Energy Com.* 13930, 1960, **78**
Cu–Zr   H. L. Burghoff, 'Metals Handbook', Cleveland, Ohio, 1948

Dy–Er   K. Mironov *et al.*, *Inorganic Mat. USSR*, 1980, **16** (1), 1332
Dy–Ge   V. Eremenko *et al.*, *Dop. Akad. Nauk, UKR Ser B*, 1977, *No.* 6, 516
Dy–Ho   F. H. Spedding *et al.*, *J. less-common Metals*, 1973, **31** (1), 1
Dy–In   D. Kuvandykov *et al.*, *Dokl. Akad. Nauk. Uzbek, SSR*, 1982, *No.* 2, 28
Dy–Mn   H. Kirchmayr and W. Lugshneider, *Z. Metallbunde*, 1967, **58** (3), 165
Dy–Ni   J. Zheng and C. Wang, *Acta. Phys. Sinica*, 1982, **31**, 668

Er–Fe   A. Meyer, *J. less-common Metals*, 1969, **18**, 41
Er–Ho   F. H. Spedding *et al.*, *J. less-common Metals*, 1973, **31**, 1
Er–Ni   K. H. J. Buschow, *J. less-common Metals*, 1968, **16**, 45
Er–Rh   R. H. Ghassem and A. Raman, *Met. Trans.*, 1973, **4**, 745
Er–Tb   F. H. Spedding *et al.*, *J. less-common Metals*, 1973, **31**, 1
Er–Y   F. H. Spedding *et al.*, *J. less-common Metals*, 1973, **31**, 1

Eu–In   W. Köster and J. Meixner, *Z. Metallk*, 1965, **56**, 695

Fe–Ga   W. Köster and T. Gödecke, *Z. Metallk.*, 1977, **68** (10), 661
Fe–Gd   V. F. Novy *et al.*, *Trans. Am. Inst. Min. Engrs*, 1961, **221**, 580
Fe–Ho   G. J. Roe and T. J. O'Keefe, *Met. Trans.*, 1970, **1**, 2565
Fe–In   C. Dasarathy, *Trans. Am. Inst. Min. Engrs*, 1969, **245**, 1838
Fe–Mg   A. S. Yue, *J. Inst. Metals*, 1962/3, **91**, 166
Fe–Mn   A. Hellawell, *Inst. Met. Ann. Eq. Diag. No.* 26
Fe–Mo   Hansen; A. K. Sinha *et al.*, *J. Iron Steel Inst.*, 1967, **205**, 191
Fe–N   K. H. Jack, *Acta Cryst.*, 1952, **5**, 404
Fe–Pd   W. S. Gibson and W. Hume-Rothery, *J. Iron Steel Inst.*, 1958, **189**, 243; E. Raub *et al.*,
        Z. Metallk, 1963, **54**, 549
Fe–Pt   A. Kussmann *et al.*, *Z. Metallk*, 1950, **41**, 470; R. A. Buckley and W. Hume-Rothery,
        J. Iron Steel Inst., 1959, **193**, 61
Fe–Pu   P. G. Marsden *et al.*, *J. Inst. Metals*, 1957/8, **86**, 166
Fe–Re   H. Eggers, *Mitt. K.-Wilhelm-Inst. Eisenforsch. Düsseld.*, 1938, **20**, 147
Fe–Ru   W. S. Gibson and W. Hume-Rothery, *J. Iron Steel Inst.*, 1958, **189**, 243; E. Raub and
        W. Plate, *Z. Metallk*, 1960, **51**, 477
Fe–S   Hansen; V. P. Buistrov *et al.*, *Tsvet. Metally*, 1971, (6), 21
Fe–Sc   O. P. Naumkin *et al.*, *Russ. Met.*, 1969, (3), 125
Fe–Se   W. Schuster *et al.*, *Monatschefte f. Chemie*, 1979, **110**, 1153
Fe–Si   W. Köster and T. Gödecke, *Z. Metallk*, 1968, **59**, 602
Fe–Sm   K. H. J. Buschow, *J. less-common Metals*, 1971, **25**, 131
Fe–Sn   Equilibrium Data for Sn Alloys, *Tin Res. Inst.*, 1949
Fe–Ta   A. K. Sinha and W. Hume-Rothery, *J. Iron Steel Inst.*, 1967, **205**, 671; Elliott
Fe–U   J. D. Grogan, *J. Inst. Met.*, 1950, **77**, 571
Fe–V   A. Hellawell, *Inst. Met. Ann. Eq. Diag. No.* 27
Fe–Y   R. F. Domegala *et al.*, *Trans. Am. Soc. Metals*, 1961, **53**, 137
Fe–Zn   S. Budurov *et al.*, *Z. Metallk*, 1972, **63**, 348 and G. F. Bastin *et al.*, *Z. Metallk.*, 1974,
        65 (10), 656

Ga–Ge   E. S. Greiner and P. Breidt, *Trans. Am. Inst. Min. Engrs*, 1955, **203**, 187
Ga–Hg   Bruno Predel, *Z. Phys. Chem.*, 1960, **24**, 206
Ga–In   J. P. Denny *et al.*, *J. Metals*, 1952, 39
Ga–Li   S. Yatsenko *et al.*, *Russ. Metal. (Metally)*, 1973, **1**, 131
Ga–Mg   H. Gröber and V. Hauk, *Z. Metallk.*, 1950, **41** (6), 191
Ga–Mn   E. Wachtel and K. J. Nier, *Z. Metallk.*, 1965, **56** (11), 779
Ga–Nb   R. E. Miller *et al.*, *Solid State Commun.*, 1971, **9** (20), 1769; *see also* L. L. Oden and
        R. E. Siemens, *J. less-common Metals*, 1968, **14**, 33 and V. V. Baron *et al.*,
        Russ. J. Inorg. Chem., 1964, **9** (9), 1172 for alternative versions
Ga–Ni   E. Hellner, *Z. Metallk.*, 1950, **41**, 480
Ga–Pd   K. Schubert *et al.*, *Z. Metallk.*, 1959, **50**, 534
Ga–Rb   S. Yatsenko *et al.*, *Russ. Metal. (Metally)*, 1973, **3**, 196
Ga–Sb   I. G. Greenfield and R. L. Smith, *Trans. Am. Inst. Min. Engrs*, 1955, **203**, 351

Ga–U   K. H. J. Buschow, *J. less-common Metals*, 1973, **31** (1), 165
Ga–V   J. H. N. van Vucht *et al.*, *Philips Res. Reports*, 1964, **19**, 407

Gd–Ge   V. N. Eremenko *et al.*, *Sov. Powder Met. & Ceramics*, 1980, **19**, 104
Gd–Ni   V. F. Novy *et al.*, *Trans. Am. Inst. Min. Engrs.*, 1961, **221**, 585
Gd–Pb   J. T. Demel and K. A. Gschneider, *J. Nucl. Mater.*, 1969, **29**, 11
Gd–Rh   O. Leobich and E. Raub, *J. less-common Metals*, 1976, **46** (1), 1
Gd–Ru   O. Leobich and E. Raub, *J. less-common Metals*, 1976, **46** (1), 7
Gd–Sc   B. J. Beauchy and A. H. Daane, *J. less-common Metals*, 1964, **6**, 322

Ge–K   M. E. Drits *et al.*, *Russ. Inorg. Mat.*, 1982, **18** (7), 969
Ge–Mg   'Gmelins Handbuch der anorganischen Chemie, Mg. A. 4', 8th edn, 1952
Ge–Mn   E. Watchel and E. T. Henig, *Z. Metallk.*, 1969, **60** (3), 243 and *J. less-common Metals*, 1970, **21**, 223
Ge–Mo   P. Stecher *et al.*, *Monatsch. Chem.*, 1963, **94**, 1154
Ge–Pd   K. Khalaff and K. Schubert, *Z. Metallkunde*, 1974, **65**, 379
Ge–Pt   Y. Oka and T. Suzuki, *Z. Metallkunde*, 1987, **78** (4), 295
Ge–Rb   M. E. Drits *et al.*, *Russ. J. Phys. Chem.*, 1977, **51** (5), 748
Ge–Rh   N. N. Zhuravlev and G. S. Zhdanov, *Kristallogra*, 1956, **1**, 205
Ge–Ru   E. Raub and W. Fritzsche, *Z. Metallk.*, 1962, **53**, 779
Ge–Sn   Equilibrium Data for Sn Alloys, *Tin Res. Inst.*, 1949
Ge–Sr   R. L. Sharkey, *J. less-common Metals*, 1970, **20**, 113
Ge–Ti   M. K. McQuillan, *J. Inst. Metals*, 1954/5, **83**, 485
Ge–Y   F. A. Schmidt *et al.*, *J. less-common Metals*, 1972, **26**, 53
Ge–Zn   E. Gebhardt, *Z. Metallk.*, 1942, **34**, 255
Ge–Zr   O. N. Carlson *et al.*, *Trans. Am. Soc. Metals*, 1956, **48**, 843

H–Sr   D. Petersen and R. P. Colburn, *J. Phys. Chem.*, 1966, **70**, 468
H–Ti   G. A. Lenning *et al.*, *Trans. Am. Inst. Min. Engrs*, 1954, **200**, 367
H–Zr   C. E. Ellis and A. D. McQuillan, *J. Inst. Metals*, 1956/7, **85**, 89

Hf–Ir   M. I. Copeland and D. Goodrich, *J. less-common Metals*, 1969, **18**, 347
Hf–Mn   A. K. Shufrin and G. P. Dmitrievna, *Dopov. Akad. Nauk. ukr. RSR*, 1969 (1), 67
Hf–Mo   A. Taylor *et al.*, *J. less-common Metals*, 1961, **3**, 265
Hf–Ni   M. E. Kirkpatrick and W. L. Larsen, *Trans. Am. Soc. Metals*, 1961, **54**, 580
Hf–O   E. Rudy and P. Stecher, *J. less-common Metals*, 1963, **5**, 75
Hf–Ta   L. L. Oden *et al.*, *U.S. Bur. Min. Rep. Invest.* 6521, 1964
Hf–U   D. T. Peterson and D. J. Beernstein, *Trans. Am. Soc. Metals*, 1959, **52**, 158
Hf–W   B. C. Giessen *et al.*, *Trans. Am. Inst. Min. Engrs*, 1962, **224**, 60; A. Braun and E. Rudy, *Z. Metallk.*, 1960, **51**, 362

Hg–Mn   G. Jangg and H. Palman, *Z. Metallk.*, 1963, **54**, 364
Hg–Rh   G. Jangg *et al.*, *Z. Metallk.*, 1967, **58**, 724
Hg–Sb   G. Jangg *et al.*, *Z. Metallk.*, 1962, **53**, 313
Hg–Sn   M. L. Gayler, *J. Inst. Metals*, 1937, **60**, 381
Hg–U   B. R. T. Frost, *J. Inst. Metals*, 1953/4, **82**, 456

Ho–Tb   F. H. Spedding *et al.*, *J. less-common Metals*, 1973, **31** (1), 1

In–K   S. P. Yatsenko *et al.*, *Struktura Faz, Faz Prev.: Diagm. Sos. Met. Sis.*
In–Li   W. A. Alexander *et al.*, *Can. J. Chem.*, 1976, **54** (7), 1052
In–Mg   G. V. Raynor, *Trans. Faraday Soc.*, 1948, **44**, 15
In–Mn   U. Zwicker, *Z. Metallk.*, 1950, **41**, 400
In–Na   G. J. Lamprecht and P. Crowther, *J. Inorg. Nucl. Chem.*, 1969, **31**, 925
In–Ni   E. Hellner, *Z. Metallk.*, 1950, **41**, 402
In–Pb   T. Heumann and B. Predel, *Z. Metallk.*, 1966, **57**, 50
In–Si   C. D. Thurmond and M. Kowalchik, *Bell System Tech. J.*, 1960, **39**, 169
In–Sn   J. C. Blade and E. C. Ellwood, *J. Inst. Metals*, 1956/7, **85**, 30
In–Sr   G. Bruzzone, *J. less-common Metals*, 1966, **11**, 249
In–Zn   S. Valentiner, *Z. Metallk.*, 1943, **35**, 250
In–Zr   J. O. Betterton and W. K. Noya, *Trans. Am. Inst. Min. Engrs*, 1958, **212**, 340
Ir–Mn   E. Raub and W. Mahler, *Z. Metallk.*, 1955, **46**, 282
Ir–Ti   V. N. Eremenko and T. D. Shtepa, *Russ. Met.*, 1970, (6), 127
La–Mg   R. Vogel and T. Heumann, *Z. Metallk.*, 1946, **37**, 1
La–Mn   L. Rolla and A. Landelli, *Ber. Deut. Chem. Gess.*, 1942, **75**, 2091

La–Ni R. Vogel, *Z. Metallk.*, 1946, **37**, 98
La–Rh P. P. Singh and A. Raman, *Trans. Am. Inst. Min. Engrs*, 1969, **245** (7), 1561
La–Sn Equilibrium Data for Sn Alloys, *Tin Res. Inst.*, 1949
La–Tl L. Rolla *et al.*, *Z. Metallk.*, 1943, **35**, 29

Li–Mg W. Hume-Rothery *et al.*, *J. Inst. Metals*, 1946, **72**, 538; W. E. Freeth and G. V. Raynor,
        J. Inst. Metals, 1953/4, **82**, 575
Li–Pb O. Loebich and Ch. J. Raub, *J. less-common Metals*, 1977, **55**, 67

Mg–Mn W. R. D. Jones, *Inst. Met. Ann. Eq. Diag. No.* 28
Mg–Pr L. Rolla *et al.*, *Z. Metallk.*, 1943, **35**, 29
Mg–Sc B. J. Beaudry and A. H. Daane, *J. less-common Metals*, 1969, **18**, 305
Mg–Sn Equilibrium Data for Sn Alloys, *Tin Res. Inst.*, 1949
Mg–U   P. Chiotti *et al.*, *Trans. Am. Inst. Min. Engrs*, 1956, **206**, 562
Mg–Y   E. D. Gibson and O. N. Carlson, *Trans. Am. Soc. Metals*, 1969, **52**, 1084; D. Mizer and
        J. A. Clark, *Trans. Am. Inst. Min. Engrs*, 1961, **221**, 207
Mg–Zn W. R. D. Jones, *Inst. Met. Ann. Eq. Diag. No.* 29
Mg–Zr J. H. Schaum and H. C. Burnett, *J. Res. Nat. Bur. Stds.*, 1952, **49**, 155

Mn–N   'Diagrammy Sov. Metall. Sis'., 1967, Moscow
Mn–Nb A. Hellawell, *J. less-common Metals*, 1959, **1**, 343 and V. M. Svechnikov, *Metallofiz.*,
        1976, **64**, 24
Mn–Nd H. Kirchmayr and W. Lugscheider, *Z. Metallk.*, 1970, **61** (1), 22
Mn–Ni K. E. Tsiuplakis and E. Kneller, *Z. Metallk.*, 1969, **60** (5), 433
Mn–O   G. Trömel *et al.*, *Erzmetall*, 1976, **29** (5), 234
Mn–Pd E. Raub and W. Mahler, *Z. Metallk.*, 1954, **45**, 430
Mn–Pt E. Raub and W. Mahler, *Z. Metallk.*, 1955, **46**, 282
Mn–Ru E. Raub and W. Mahler, *Z. Metallk.*, 1955, **46**, 282; and A. Hellawell, *J. less-common
        Metals*, 1959, **1**, 343
Mn–S   L. Staffansson, *Met. Trans. B*, 1976, **7B**, 131
Mn–Si 'Manganese Phase Diagrams'. The Manganese Centre, 1980
Mn–Sm H. Kirchmayr and W. Lugscheider, *Z. Metallk.*, 1970, **61** (1), 22
Mn–Sn 'Manganese Phase Diagrams', The Manganese Centre, 1980
Mn–Tb H. Kirchmayr and W. Lugscheider, *Z. Metallk.*, 1970, **61** (1), 22
Mn–U   H. A. Wilhelm and O. N. Carlson, *Trans. Am. Soc. Metals*, 1950, **42**, 1311
Mn–V   R. M. Waterstrat, *Trans. Am. Inst. Min. Engrs*, 1962, **224**, 240
Mn–Y   R. L. Myklebust and A. H. Daane, *Trans. Am. Inst. Min. Engrs*, 1962, **224**, 354
Mn–Zn O. Romer and E. Wachtel, *Z. Metallk.*, 1971, **62** (11), 820

Mo–Ni R. E. W. Casselton and W. Hume-Rothery, *J. less-common Metals*, 1964, **7**, 212
Mo–Os A. Taylor *et al.*, *J. less-common Metals*, 1962, **4**, 436
Mo–Re A. G. Knapton, *J. Inst. Metals*, 1958/9, **87**, 62
Mo–Ru E. Anderson and W. Hume-Rothery, *J. less-common Metals*, 1960, **2**, 443
Mo–U   P. C. L. Pfeil, *J. Inst. Metals*, 1950, **77**, 553; F. G. Streets and J. J. Stobo, *J. Inst. Metals*,
        1963/4, **92**, 171
Mo–W   W. P. Sykes, 'Metals Handbook', Cleveland, Ohio, 1948
Mo–Zr R. F. Domegala *et al.*, *Trans. Am. Inst. Min. Engrs*, 1953, **197**, 73

N–Ti   A. E. Palty *et al.*, *Trans. Am. Soc. Metals*, 1954, **46**, 312

Na–Sn Equilibrium Data for Sn Alloys, *Tin Res. Inst.*, 1949
Na–Th G. Grube and L. Botzenhardt, *Z. Elektrochem.*, 1942, **48**, 418

Nb–Re A. G. Knapton, *J. less-common Metals*, 1959, **1**, 480; B. C. Giessen *et al.*, *Trans. Am.
        Inst. Min. Engrs*, 1961, **221**, 1009
Nb–Si D. A. Deardorff *et al.*, *J. less-common Metals*, 1969, **18**, 11
Nb–Sn J. P. Charlesworth *et al.*, *J. Mater. Sci.*, 1970, **5**, 580
Nb–Th O. N. Charlson *et al.*, *Trans. Am. Inst. Min. Engrs*, 1956, **206**, 132
Nb–Ti M. Hansen *et al.*, *J. Metals*, 1951, 881
Nb–V   H. A. Wilhelm *et al.*, *Trans. Am. Inst. Min. Engrs*, 1954, **200**, 915
Nb–Y   C. E. Lundin and D. T. Klodt, *J. Inst. Metals*, 1961/2, **90**, 341

Nd–Ni Y. Y. Pan and P. Nash, *Acta Physica Sinica*, 1985, **34** (3), 384
Nd–Rh P. P. Singh and A. Raman, *Met. Trans.*, 1970, **1**, 236
Nd–Sb G. Kobzenko *et al.*, *Russ. Metall.*, 1972, **3**, 176
Nd–Ti E. M. Savitski and G. B. Burkanov, *J. less-common Metals*, 1962, **4**, 301
Nd–Zn J. T. Mason and P. Chiotti, *Met. Trans.*, 1972, **3**, 2851

Ni–Pr   R. Vogel, *Z. Metallk.*, 1946, **37**, 98
Ni–Rh   E. Raub and E. Röschel, *Z. Metallk.*, 1970, **61**, 113
Ni–Si   E. N. Skinner, 'Metals Handbook', Cleveland, Ohio, 1948
Ni–Sm   Y. Y. Pan and C. S. Cheng, *Acta Physica. Sinica*, 1983, **32** (1), 92
Ni–Sn   K. Schubert and E. Jahn, *Z. Metallk.*, 1949, **40** (8), 319
Ni–V   W. B. Pearson and W. Hume-Rothery, *J. Inst. Metals*, 1951/2, **79**, 643; L. R. Stevens and
         O. N. Charlson, *Met. Trans.*, 1970, **1**, 1267
Ni–W   F. H. Ellinger and W. P. Sykes, *Trans. Am. Soc. Metals*, 1940, **28**, 619
Ni–Y   B. J. Beaudry and A. H. Daane, *Trans. Am. Inst. Min. Engrs*, 1960, **218**, 854;
         R. F. Domegala *et al.*, *Trans. Am. Soc. Metals*, 1961, **53**, 137

Np–Pu   P. G. Mardon *et al.*, *J. less-common Metals*, 1961, **3**, 281
Np–U   P. G. Mardon *et al.*, *J. less-common Metals*, 1959, **1**, 467

O–Ti   T. H. Schofield and A. E. Bacon, *J. Inst. Metals*, 1955/6, **84**, 47
O–V   D. G. Alexander and O. N. Carlson, *Trans. Am. Soc. Metals*, 1971, **2**, 2805

Os–Ti   V. N. Eremenko *et al.*, *Russ. Met. (Metally)*, 1971, **1104**, 147
Os–W   A. Taylor *et al.*, *J. less-common Metals*, 1961, **3**, 333

P–Pd   L. O. Gullman, *J. less-common Metals*, 1966, **11**, 157
P–Pt   S. Rundquist, *Acta Chem. Scand.*, 1961, **15**, 451
P–Zn   M. Schneider and M. Krumnacher, *Neue Hutte*, 1973, **18**, 715

Pb–Pd   V. C. Marcotte, *Met. Trans. B*, 1977, **8B**, 185
Pb–Pr   L. Rolla *et al.*, *Z. Metallk.*, 1943, **35**, 29
Pb–Pu   D. H. Wood *et al.*, *J. Nucl. Mater*, 1969, **32** (2), 193
Pb–Sb   J. B. Clark and C. W. F. T. Pistorius, *J. less-common Metals*, 1975, **42**, 59
Pb–Sn   G. V. Raynor, *Inst. Met. Ann. Eq. Diag. No. 6*
Pb–Te   G. O. Hiers, 'Metals Handbook', Cleveland, Ohio, 1948
Pb–Ti   P. Farrar and H. Margolin, *Trans. Am. Inst. Min. Engrs*, 1955, **203**, 101
Pb–W   S. Inouye, *Mem. Coll. Sci. Kyoto Univ.*, 1920, **4**, 43
Pb–Y   O. N. Carlson *et al.*, *Trans. ASM*, 1967, **60** (2), 119
Pb–Zn   E. A. Anderson and J. L. Rodda, 'Metals Handbook', Cleveland, Ohio, 1939

Pd–Si   R. H. Willens *et al.*, *J. Metals*, 1964, **16**, 92
Pd–Ta   R. M. Waterstrat *et al.*, *Met. Trans. A*, 1978, **9A**, 643
Pd–Ti   V. N. Eremenko and T. D. Shtepa, *Poroshk. Metall.*, 1972, (3), 75
Pd–Tl   S. Bhan *et al.*, *J. less-common Metals*, 1968, **16**, 415
Pd–U   J. A Catterall *et al.*, *J. Inst. Metals*, 1956/7, **85**, 63; G. P. Pells, *J. Inst. Metals*, 1963/4,
         92, 416
Pd–Zn   W. Köster and U. Zwicker, *Heraus Festschrift*, 1951, 76
Pd–Zr   K. Anderko, *Z. Metallk.*, 1959, **60**, 681

Pr–Sn   Equilibrium Data for Sn Alloys, *Tin Res. Inst.*, 1949
Pr–Tl   L. Rolla *et al.*, *Z. Metallk.*, 1943, **35**, 29
Pr–Zn   J. T. Mason and P. Chiotti, *Met. Trans.*, 1970, **1**, 2119

Pt–Ru   J. M. Hutchinson, *Plantin Metals Rev.*, 1972, **16** (3), 88
Pt–Sn   K. Schubert and E. Jahn, *Z. Metallk.*, 1949, **40** (10), 399
Pt–Te   M. L. Gimpl *et al.*, *Trans. Am. Soc. Metals*, 1963, **56**, 209
Pt–Tl   S. Bhan *et al.*, *J. less-common Metals*, 1968, **16**, 415
Pt–V   R. M. Waterstrat, *Met. Trans.*, 1973, **4**, 455
Pt–W   R. I. Jaffee and H. P. Nielsen, *Trans. Am. Inst. Min. Engrs*, 1949, **180**, 603
Pt–Zr   E. G. Kendall *et al.*, *Trans. Am. Inst. Min. Engrs*, 1961, **221**, 445

Pu–Th   D. M. Poole *et al.*, *J. Inst. Met.*, 1957/8, **86**, 172
Pu–Ti   A. Languille, *Mém. Scient. Revue. Métall.*, 1971, **68** (6), 435
Pu–Zn   F. H. Ellinger *et al.*, 'Extractive and Physical Metallurgy of Plutonium and its Alloys',
         Interscience, N.Y., 1960, p. 169 et seq.
Pu–Zr   J. A. C. Marples, *J. less-common Metals*, 1960, **2**, 331

Rb–Sb   F. W. Dorn and W. Klemm, *Z. Anorg. Allg. Chem.*, 1961, **309**, 189

Re–Ru   E. Rudy *et al.*, *Z. Metallk.*, 1962, **53**, 90
Re–Tb   E. M. Savitskii and O. Kh. Khamidov, *Russ. Met.*, 1968, (6), 108

Rh–U   J. J. Park, *J. Res. Nat. Bur. Stds.*, 1968, **72** (1), 11
Rh–V   R. M. Waterstrat and R. C. Manuszewski, *J. less-common Metals*, 1977, **52**, 293

Sb–Sn   E. C. Ellwood, *Inst. Met. Ann. Eq. Diag.*, *No.* 23
Sb–U   B. J. Bauchy and A. H. Daane, *Trans. Am. Inst. Min. Engrs*, 1959, **215**, 199

Se–Te   E. Grison, *J. Chem. Phys.*, 1951, **19**, (9), 1109
Se–Th   R. W. M. D'Eye *et al.*, *J. Chem. Soc.*, 1952, 2555, 143; A. Brown and J. J. Norreys,
        J. Inst. Metals, 1960/1, **89**, 238

Si–Te   T. G. Davey and E. M. Baker, *J. Mat. Sci. Letters*, 1980 (15), 1601
Si–Ti   M. Hansen *et al.*, *Trans. Am. Soc. Metals*, 1952, **44**, 518
Si–U   A. Kaufmann *et al.*, *Trans. Am. Inst. Min. Engrs*, 1957, **209**, 23
Si–W   R. Kieffer *et al.*, *Z. Metallk.*, 1952, **43**, 284
Si–Zr   C. E. Lundin *et al.*, *Trans. Am. Soc. Metals*, 1953, **45**, 901

Sn–Tl   Oska Prefect., *Univ. Bull.*, 1966, **15**, 137
Sn–Zr   D. J. McPherson and M. Hansen, *Trans. Am. Soc. Metals*, 1953, **45**, 915

Ta–Th   O. D. MacMasters and W. D. Larsen, *J. less-common Metals*, 1961, **3**, 312
Ta–U   C. H. Schramm *et al.*, *Trans. Am. Inst. Min. Engrs*, 1950, **188**, 195
Ta–V   AFML–TR–65–2, Part V–Compendium of Phase Diagram Data, (June 1969), 113
Ta–W   AFML–TR–65–2, Part V–Compendium of Phase Diagram Data, (June 1969), 144
Ta–Y   C. E. Lundin and D. T. Klodt, *J. Inst. Metals*, 1961/2, **90**, 341

Tc–V   C. C. Koch and G. R. Love, *J. less-common Metals*, 1968, **15**, 43

Th–Ti   O. N. Carlson *et al.*, *Trans. Am. Inst. Min. Engrs*, 1956, **206**, 132
Th–U   J. R. Murray, *J. Inst. Metals*, 1958/9, **87**, 94
Th–Y   T. Eash and O. N. Carlson, *Trans. Am. Soc. Metals*, 1959, **52**, 1097, 301
Th–Zn   P. Chiotti and K. J. Gill, *Trans. Am. Inst. Min. Engrs*, 1961, **221**, 573

Ti–U   A. G. Knapton, *J. Inst. Metals*, 1954/5, **83**, 497
Ti–V   H. N. Aderstedt *et al.*, *J. Am. Chem. Soc.*, 1952, **44**, 990
Ti–W   E. Rudy and S. T. Windisch, *Trans. Am. Inst. Min. Engrs*, 1968, **242**, 953
Ti–Y   D. W. Bau, *Trans. Am. Soc. Metals*, 1961, **53**, 1
Ti–Zn   W. Heine and U. Zwicker, *Z. Metallk.*, 1962, **53**, 380

U–W   C. H. Schramm *et al.*, *Trans. Am. Inst. Min. Engrs*, 1950, **188**, 195
U–Zn   P. Chiotti *et al.*, *Trans. Am. Inst. Min. Engrs*, 1957, **209**, 51

V–Y   C. E. Lundin and D. T. Klodt, *J. Inst. Metals*, 1961/2, **90**, 341
V–Zr   J. T. Williams, *Trans. Am. Inst. Min. Engrs*, 1955, **203**, 345

W–Zr   R. F. Domegala *et al.*, *Trans. Am. Inst. Min. Engrs*, 1953, **197**, 73

Y–Zn   J. T. Mason and P. Chiotti, *Met. Trans. A*, 1976, **7A**, 289

Zn–Zr   P. Chiotti and G. R. Kilp, *Trans. Am. Inst. Min. Engrs*, 1959, **215**, 892

11.3.2   References to *J. less-common Metals* marked (3) in 11.1 Index of binary diagrams

Al–Ho	1966, **10**, 121	Cu–Gd	1983, **92**, 143	In–Nd	1983, **90**, 95
Ca–Pd	1982, **85**, 307	Cu–Nd	1983, **92**, 97	In–Y	1983, **90**, 95
Ca–Pt	1981, **78**, 49	Er–In	1983, **90**, 95	Mo–Pt	1968, **15** (2), 194
Cd–Ge	1985, **106**, 1	Ga–Tm	1979, **64**, 185	Ni–Th	1988, **142**, 311
Cr–Ge	1980, **72**, 7	Gd–In	1983, **90**, 95	Pd–Y	1973, **30**, 47
Cr–Nb	1961, **3** (1), 44	Gd–Pd	1973, **30**, 47	Sn–Yb	1976, **46**, 321
Cs–In	1982, **83**, 143	Ge–Ni	1980, **72**, 51	Sr–Zn	1983, **92**, 75
Cu–Eu	1985, **106**, 175	In–La	1974, **38**, 13		

11.4   Ternary systems and higher systems

No diagrams of these systems are given. The literature is very large. The following compendiums should be consulted for individual systems.

A. Prince, 'Multicomponent Alloy Constitution Bibliography, 1955–73'. London Metals Soc., 1978.

A. Prince, 'Multicomponent Alloy Constitution Bibliography, 1974–77', London Metals Soc., 1981.

V. Raghavan, 'Diagrams of Ternary Iron Alloys, Parts 1, 2, 3', Indian Inst. Metals, Part 1, 1987, Parts 2 and 3, 1988.

G. V. Raynor and V. G. Rivlin, 'Equilibria in Iron Ternary Alloys. Phase Diagrams of Ternary Iron Alloys Part 4', The Inst. of Metals, 1988.

Y. Austin Chang and Ker-Chang Hsieh, 'Diagrams of Ternary Copper–Oxygen Metal Systems', ASM International, 1989.

G. Petzow and G. Effenberg, Ternary Alloys, Vol. 1 Ag–Al–Au to Ag–Cu–P, VCH, 1988. Vol. 2 Ag–Cu–Pb to Ag–Zn–Zr, VCH, 1988. Vol. 3 Al–Ar–O to Al–Ca–Zn, VCH, 1990. Vol. 4 Al–Cd–Ce to Al–Cu–Ru, VCH, 1991.

A. Prince, G. V. Raynor and D. S. Evans, 'Phase Diagrams of Ternary Gold Alloys', Inst. of Metals, 1990.

K. P. Gupta, 'Ternary Nickel Alloys, Part 1', Indian Inst. of Metals, 1991.

12　Gas–metal systems

12.1　The solution of gases in metals

The gases which can be found in solution in measurable quantities in metals are the diatomic gases hydrogen, nitrogen and oxygen and also the noble gases in Group 8 of the Periodic Table.

12.1.1　Dilute solutions of diatomic gases

A diatomic gas dissociates on solution so that equilibrium between the solute and the gas phase is written:

$$X_2 \text{ (gas)} \rightleftharpoons 2X \text{ (dissolved in metal)}$$

In systems where no gas–metal compounds are formed and where also the solute does not contribute to the stability of the metallic phase, solutions are usually so dilute that Henry's law applies. If, also, the ideal gas laws are assumed for the gas phase and the standard states selected are (1) the infinitely dilute atomic fraction for the solute and (2) the pure element at a standard pressure $p^{\ominus}$ for the gas, the equilibrium constant is given by:

$$K = \frac{a_x^2}{a_{x_2}} = \frac{N_x^2 p^{\ominus}}{p} \tag{12.1}$$

where a_x and a_{x_2} are the activities of the solute and the diatomic gas, N_x is the atomic fraction of solute and p is the pressure of the gas. In a form rearranged to express the proportionality between the solute concentration, C, and the square root of the gas pressure, p, equation (12.1) is known as Sievert's relation:

$$C = S \left(\frac{p}{p^{\ominus}} \right)^{1/2} \tag{12.2}$$

where S is the solute concentration in equilibrium with the standard pressure, $p^{\ominus}$.

Values for solubilities in dilute solutions are usually presented as solute concentrations in equilibrium with the pure gas at one atmosphere pressure (101 325 Pa). The units of concentration in common use are cm^3 of gas measured at one atmosphere pressure and 273 K per 100 g of metal for hydrogen and mass % for nitrogen and oxygen.

Application of the Van't Hoff isochore gives the variation with temperature of the equilibrium constant at constant pressure and hence also of the corresponding equilibrium solute concentration:

$$\left(\frac{\mathrm{d}\ln K}{\mathrm{d}T} \right)_p = \frac{\Delta H^{\ominus}}{RT^2}$$

where $H^{\ominus}$ is the standard enthalpy of solution of the gas. Substituting for K from equation (12.1) gives:

$$\left(\frac{\mathrm{d}\ln N_x}{\mathrm{d}T} \right)_p = \frac{\Delta H^{\ominus}}{2RT^2}$$

For a small range of temperature it is permissible to disregard the temperature-dependence of $\Delta H^{\ominus}$ so that integration then yields:

$$\log\left(\frac{N_1}{N_2}\right) = \frac{-\Delta H^{\ominus}}{2.303 \times 2R}\left(\frac{1}{T_1} - \frac{1}{T_2}\right) \tag{12.3}$$

where N_1 and N_2 are the atomic fractions of solute in equilibrium at temperatures T_1 and T_2 with the same gas pressure. Standard enthalpies of solution $\Delta H^{\ominus}$ can be evaluated from experimental results fitted to equation (12.3), bearing in mind Kubaschewski's reminder[1] that thermochemical data derived from the temperature-dependence of equilibrium constants are uncertain.

12.1.2   Complex gas–metal systems

In systems where the solute contributes to the stability of the metallic phase or where near stoichiometric compounds separate, the information of interest extends to high solute concentrations where there is deviation from Henry's law and to equilibria between condensed phases. If complete, this information is most conveniently presented graphically as isotherms, i.e. expressing composition as a function of equilibrium pressure for selected temperatures. Regions where two condensed phases coexist in the system are manifest by pressure invariance between the compositions of the conjugate phases. For some systems, the available experimental results are limited to values for solute concentrations in equilibrium at selected temperatures with either a second phase or the gas at a selected pressure.

12.1.3   Solutions of hydrogen

Table 12.1 gives selected values of dissolved hydrogen concentrations in equilibrium with hydrogen at one atmosphere pressure for the metals silver, gold, aluminium, cobalt, chromium, copper, iron, magnesium, manganese, molybdenum, nickel, lead, platinum, silicon, tin, titanium, zinc and some of their alloys. These metals are endothermic occluders of hydrogen, so solubility of hydrogen increases with increasing temperature (except for α-manganese). The hydrogen solubility values in silver, aluminium, gold, magnesium and their alloys are based on combined regression analyses of the most reliable reported hydrogen solubility data.

The alkali and alkaline-earth metals form hydrides with non-metallic characteristics in which bonding is of predominantly ionic character with the hydrogen present as H^- anions. Table 12.2 gives hydrogen concentrations in equilibrium with the corresponding hydrides for barium, calcium, magnesium, sodium and strontium.

Certain of the metals in the transition, rare earth and actinide series have the remarkable ability to take hydrogen into interstitial solid solution until the atomic fraction of hydrogen approaches simple stoichiometric ratios but without loss of the metallic character of the phase so formed. Additional phases may also appear at higher hydrogen concentrations. The solute hydrogen increases the stability of the metallic phase, the solution process is exothermic and thus for a given hydrogen activity (i.e. for constant pressure of hydrogen in the gas phase) the stability decreases with temperature. Figures 12.1 to 12.11 present information for the systems formed by hydrogen with cerium, niobium, neodymium, palladium, praseodymium, plutonium, tantalum, thorium, titanium, vanadium and zirconium. Tables 12.3 and 12.4 give data for uranium and hafnium.

Table 12.1   HYDROGEN SOLUTIONS IN EQUILIBRIUM WITH GASEOUS HYDROGEN AT ATMOSPHERIC PRESSURE

Solvent metal		Hydrogen concentration, S at T°C cm³ of gas at atmospheric pressure and 0°C per 100 g of metal							References
Ag	T	400	500	600	700	800	900	961	16, 269
(solid)	S	0.001	0.002	0.006	0.014	0.029	0.051	0.069	
Al	T	350	400	500	600	660			17, 25, and
(solid)	S	0.001	0.003	0.009	0.026	0.042			270–274
(liquid)	T	660	700	800	900	1 000			17, 19, and
	S	0.70	0.92	1.66	2.72	4.12			275–282

(continued)

Table 12.1   HYDROGEN SOLUTIONS IN EQUILIBRIUM WITH GASEOUS HYDROGEN AT ATMOSPHERIC PRESSURE—*continued*

Solvent metal		Hydrogen concentration, S at T°C cm^3 of gas at atmospheric pressure and 0°C per 100 g of metal						References
Al–Cu	T		300	400	500	600		
(solid)	S	4% Cu	0.003	0.006	0.011	0.017		283
		10% Cu	0.003	0.008	0.018	0.033		283
		20% Cu	0.007	0.019	0.039	0.069		283
		33% Cu	0.007	0.025	0.064	0.13		283
		40% Cu	0.004	0.020	0.062	0.15		283
		50% Cu	0.013	0.038	0.088	0.17		283
Al–Cu	T		700	800	900	1 000		
(liquid)	S	2% Cu	0.74	1.42	2.44	3.84		19
		4% Cu	0.64	1.26	2.19	3.51		19
		8% Cu	0.50	1.01	1.79	2.91		19
		16% Cu	0.39	0.78	1.40	2.28		19
		32% Cu	0.35	0.67	1.14	1.79		19
Al–Fe	T		700[#]	800[#]	900[#]	1 000		
(liquid)	S	10% Fe	0.20	0.71	1.99	4.77		278
			[#](Liquid + Solid) region of alloy					
Al–Li	T		300	400	450	500	550	600	
(solid)	S	1% Li	0.87	1.04	0.62	0.70	0.77	0.84	279, 284
		2% Li	1.31	1.53	1.63	0.78	0.89	0.99	279, 284, 286
		3% Li	1.58	2.24	2.57		1.43	1.64	279, 284
Al–Li	T		700	800	900	1 000		
(liquid)	S	1% Li	2.48	3.98	5.88	8.19		279, 285
		2% Li	2.84	5.33	8.97	13.93		279, 285, 286
		3% Li	3.44	6.48	10.96	17.06		279, 285
Al–Mg	T		300	400	500	600		
(solid)	S	0.6% Mg			0.010			287
		2.2% Mg			0.016			287
		3.4% Mg			0.024			287
		5.3% Mg	0.002	0.008	0.022	0.048		287
		10.8% Mg			0.030			287
Al–Mg	T		700	800	900	1 000		
(liquid)	S	3% Mg	1.46	2.66	4.40	6.70		288
		6% Mg	2.37	4.30	7.04	10.66		288
		9.9% Mg	3.61					288
		20% Mg	6.82					289
Al–Si	T		300	400	500	600		
(solid)	S	1.5% Si			0.007			290
		2.4% Si			0.025			290
		5.7% Si	0.011	0.027	0.053	0.088		291
		6.5% Si	0.013	0.030	0.055	0.089		291
		9.4% Si	0.014	0.035	0.069	0.12		291
		11.4% Si	0.030	0.047	0.065	0.085		291
		14.3% Si	0.054	0.072	0.089	0.11		291
Al–Si	T		700	800	900	1 000		
(liquid)	S	2% Si	0.82	1.52	2.52	3.87		19
		4% Si	0.76	1.45	2.49	3.91		19
		7% Si	0.71	1.29	2.13	3.24		278
		8% Si	0.65	1.28	2.24	3.59		19
		16% Si	0.58	1.16	2.06	3.35		19
Al–Ti	T		700[#]	800[#]	900[#]	1 000[#]		
(liquid)	S	4% Ti	1.12	2.34	4.33	7.26		278
			[#](Liquid + Solid) region of alloy					
Al–Zn	T		425	500				
(solid)	S	3.4% Zn		0.05				292
		7% Zn	0.013	0.03				292

(*continued*)

Table 12.1   HYDROGEN SOLUTIONS IN EQUILIBRIUM WITH GASEOUS HYDROGEN AT ATMOSPHERIC
PRESSURE—*continued*

Solvent metal		*Hydrogen concentration, S at T°C* cm^3 *of gas at atmospheric pressure and 0°C per 100 g of metal*									*References*
Al–Zn	T		700	800	900	1 000					
(liquid)	S	1.8% Zn	0.72	1.45	2.61	4.27					277
		3.5% Zn	0.79	1.54	2.70	4.32					277
		4.4% Zn	0.78	1.35	2.13	3.13					277
		6% Zn	0.66	1.27	2.18	3.14					288
		8% Zn	0.68	1.36	2.41	3.89					288
Al–Zr	T		750								
(liquid)	S	0.41% Zr	4.42								293
		0.80% Zr	5.50								293
Au	T	500	600	700	800	900	1 000	1 063			
(solid)	S	0.002	0.004	0.006	0.008	0.011	0.014	0.016			269
Au	T	1 063	1 100	1 200	1 300	1 400	1 500				
(liquid)	S	0.07	0.11	0.36	0.98	2.38	5.24				278
Au–Cu	T		1 050	1 100	1 200	1 300	1 400				
(liquid)	S	4% Cu	0.25	0.33	0.55	0.86	1.27				278
		8% Cu	0.42	0.50	0.67	0.88	1.11				278
		10% Cu	0.21	0.23	0.28	0.33	0.38				278
		12% Cu	0.16	0.19	0.25	0.32	0.40				278
		13% Cu	0.65	0.69	0.77	0.85	0.93				278
		15% Cu	0.87	1.00	1.30	1.62	1.97				278
		20% Cu	1.49	1.60	1.84	2.06	2.29				278
		40% Cu	1.86	1.94	2.08	2.21	2.33				278
Au–Pd	T		1 100	1 200	1 300	1 400					
(liquid)	S	2% Pd	0.51	1.02	1.85	3.15					278
		4% Pd	0.77	1.35	2.20	3.39					278
		6% Pd	0.29	1.59	6.98	25.74					278
		8% Pd	1.16	2.64	5.42	10.19					278
		10% Pd	1.09	2.88	6.73	14.24					278
Co	T	600	700	800	900	1 000	1 100	1 200	1 300	1 400	1 492	
(solid)	S	0.9	1.22	1.85	2.51	3.30	4.31	5.40	6.7	7.75	8.65	20, 21
(liquid)	T	1 500	1 550	1 600	1 650							
	S	20.5	21.6	23.2	24.3							21, 22
Co–Fe		*See* Fe–Co										
Co–Ni		*See* Ni–Co										
Cr	T	400	500	600	700	800	900	1 000	1 100	1 200		24
(solid)	S	0.2	0.3	0.4	0.6	1.0	1.7	2.6	3.7	5.4		also 23
Cr–Fe		*See* Fe–Cr										
Cu	T	500	600	700	800	900	1 000	1 083				16, 25–27,
(solid)	S	0.10	0.21	0.37	0.58	0.85	1.17	1.46				269, 294–296
(liquid)	T	1 083	1 100	1 200	1 300	1 400						26–28
	S	5.60	5.88	7.67	9.66	11.85						
Cu–Ag	T		1 225	1 225	1 225	1 225	1 225					
(liquid)		% Ag	0	10	20	30	50					
	S		8.6	7.9	7.0	5.9	4.3					7
Cu–Al	T		700	800	900	1 000	1 050					
(solid)	S	1.43% Al	—	—	0.70	1.15	1.40					27
		3.3% Al	—	0.35	0.70	1.05	1.25					27
		5.77% Al	—	0.10	0.30	0.60	—					27
		6.84% Al	0.10	0.15	0.35	0.65	—					27
		8.1% Al	0.10	0.15	0.35	0.75	—					27

(*continued*)

Table 12.1 HYDROGEN SOLUTIONS IN EQUILIBRIUM WITH GASEOUS HYDROGEN AT ATMOSPHERIC PRESSURE—*continued*

Solvent metal		Hydrogen concentration, S at T°C — cm³ of gas at atmospheric pressure and 0°C per 100 g of metal										References
Cu–Al (liquid)	T		1 100	1 150	1 200	1 300	1 400					
	S	1.45% Al	5.1	5.8	6.6	—	—					27
		3.3% Al	4.4	5.1	5.7	—	—					27
		5.77% Al	3.3	3.9	4.5	—	—					27
		6.84% Al	3.0	3.5	4.1	—	—					27
		8.1% Al	2.7	3.2	3.5	4.5	5.9					27 also 7
Cu–Au (liquid)	T		1 225	1 225	1 225	1 225						
		% Au	0	10	30	50						
	S		8.6	7.9	5.6	3.4						7
Cu–Ni (liquid)	T		1 225	1 225	1 225							
		% Ni	0	10	20	See also Ni–Cu						
	S		8.6	13	17							7
Cu–Pt (liquid)	T		1 225	1 225	1 225							
		% Pt	0	10	20							
	S		8.6	9.6	9.7							7
Cu–Sn (liquid)	T		1 000	1 100	1 200	1 300						
	S	5.9% Sn	—	4.80	6.28	7.81						28
		11.5% Sn	3.09	4.11	5.35	6.85						28
		21.7% Sn	2.11	2.97	3.94	5.10						28
		40.2% Sn	0.53	0.94	1.50	—						28
		54.8% Sn	0.50	0.76	1.15	1.61						28 also 7
Fe–Zn (solid α)	T		500	600	700	800	875*					
	S	33% Zn	0.010	0.023	0.044	0.076	0.35					29
			(*α + β region)									
(solid α)	T		200	300	400	500	600	700	800	900		24
	S		—	0.1	0.2	0.6	1.2	1.7	2.4	3.0		also 30, 31, 32
(solid γ)	T		900	1 000	1 100	1 200	1 250	1 350	1 400			24
	S		4.7	5.4	6.4	7.4	8.4	—	9.3			also 21, 31, 33
(solid δ)	T		1 400	1 450								24
	S		6.1	6.4								also 21, 31, 33
(liquid)	T		1 535	1 550	1 600	1 700						59, 60, 61
	S		24.5	25.5	26.5	29.5						21 also 22, 31, 33 34, 35, 36
Fe–Co (liquid)	T		1 600	1 600	1 600	1 600	1 600					
		% Co	0	20	40	60	80					
	S		29.8	23.8	20.7	18.2	21.5					22
Fe–Cr (solid)	T		400	600	700	800	850	900	1 000	1 100	1 200	
	S	4% Cr	0.2	0.9	1.4	1.9	2.2	4.4	5.3	6.3	7.4	24
		10% Cr	0.2	0.7	1.2	2.1	5.0	6.0	7.1	8.3	9.3	24
		20% Cr	0.2	0.5	1.8	3.0	3.7	4.3	4.9	5.6	6.6	24
		44% Cr	0.2	0.5	1.2	2.3	3.5	4.3	5.4	6.2	6.9	24
		77% Cr	0.2	0.5	0.8	1.2	1.5	2.0	3.1	4.4	5.9	24 also 33
Fe–Cr (liquid)	T		1 600	1 600	1 600	1 600	1 600					
		% Cr	9.4	18.8	28.5	38.3	48.2					
	S		25	25	25	24	23					33
Fe–Nb (liquid)	T		1 560	1 685								
	S	*5.01% Nb	24.6	28.3								36
		8.97% Nb	29.2	31.7								36
		15.12% Nb	41.7	43.8								36
			*Estimated from results for equilibrium at 2 900 Pa									
Fe–Ni (solid)	T		300	400	500	600	700	800	900	1 000	1 100	1 200	
	S	3.3% Ni	—	0.3	0.7	1.3	2.0	3.9	5.0	6.2	7.3	8.4	24
		5.5% Ni	0.2	0.4	0.8	1.4	2.1	4.1	5.2	6.3	7.3	8.4	24

(*continued*)

Table 12.1   HYDROGEN SOLUTIONS IN EQUILIBRIUM WITH GASEOUS HYDROGEN AT ATMOSPHERIC PRESSURE—*continued*

Solvent metal		*Hydrogen concentration, S at T°C* cm³ *of gas at atmospheric pressure and 0°C per 100 g of metal*										*References*
		11.6% Ni	0.4	0.6	1.3	2.1	3.2	4.3	5.3	6.4	7.5	8.5	24
		21.0% Ni	0.5	0.7	1.4	2.2	3.3	4.3	5.4	6.5	7.6	8.8	24
		32.1% Ni	0.6	0.9	1.5	2.4	3.3	4.4	5.5	6.7	7.8	9.0	24
		53.2% Ni	0.8	1.2	1.8	2.7	3.7	4.7	5.8	6.9	8.1	9.3	24
		62.4% Ni	1.3	1.7	2.4	3.3	4.1	5.0	6.1	7.3	8.4	9.6	24
		72.4% Ni	1.6	1.8	2.6	3.4	4.3	5.2	6.3	7.4	8.7	9.9	24
		84.8% Ni	1.8	2.2	2.7	3.7	4.8	6.3	7.6	9.2	10.7	12.4	24 also 21
(liquid)	*T*		1 550	1 600	1 700								
	S	20% Ni	26.5	27.5	31								21
		40% Ni	28.5	30	32.5								21
		60% Ni	31.5	33	36								21
		80% Ni	36	38	41								21 also 22
Fe–Si	*T*		1 350	1 400	1 500	1 550	1 650						35
(liquid)	*S*	1.78% Si	—	—	25.5	27.5	31.6						35
		11.0% Si	11.5	12.5	14.4	15.4	17.4						35
		21.7% Si	7.0	7.6	8.9	9.5	10.7						35
		31.5% Si	—	6.0	6.5	6.8	7.3						35
		39.1% Si	—	7.0	8.5	9.1	10.3						35
		45.7% Si	9.3	9.9	11.1	11.8	13.1						35
		51.5% Si	12.7	13.1	14.0	14.4	15.3						35
		63.7% Si	20.9	21.4	22.4	22.9	23.8						35
Fe–Ti	*T*		1 560	1 685									
(liquid)	*S*	*0.18% Ti	24.3	26.7									36
		0.45% Ti	25.6	28.2									36
		0.70% Ti	28.2	30.6									36
		2.67% Ti	—	45.8									36
		3.14% Ti	50.0	52.9									36

* Estimated from results for equilibrium at 2 900 Pa

Solvent metal		*Hydrogen concentration, S at T°C* cm³ *of gas at atmospheric pressure and 0°C per 100 g of metal*										*References*
Fe–Cr–Ni	*T*		400	500	600	700	800	900	1 000	1 100	1 200		
(solid)	*S*	17.5% Cr + 9% Ni	0.2	0.8	1.8	2.4	3.4	4.9	6.4	7.7	8.8		24
		4.6% Cr + 10% Ni	0.6	1.0	1.7	2.8	3.6	4.4	5.5	6.6	8.1		24
Mg	*T*	300	400	500	600	651							29, 38–40
(solid)	*S*	5.90	11.14	17.82	25.61	29.89							and 297–3
Mg	*T*	651	700	800	900	1 000							38, 41, an
(liquid)	*S*	44.19	51.96	69.07	87.46	106.7							297–300
Mg–Ag	*T*		500	500	500	500							
(solid)		% Ag	1.0	3.0	5.0	10.0							
	S		18.4	18.3	17.8	17.6							289
Mg–Ag	*T*		700	700	700	700							
(liquid)		% Ag	1.0	3.0	5.0	10.0							
	S		51.0	49.5	48.0	46.6							289
Mg–Al	*T*		700	800	900	1 000							
(liquid)	*S*	5.3% Al	46.1	59.1	72.6	86.4							299
		8.3% Al	41.5	54.1	67.4	81.2							299
Mg–Mn	*T*		500	500	500								
(solid)		% Mn	0.15	1.6	2.2								
	S		19.2	17.3	16.5								289
Mg–Mn	*T*		700	700									
(liquid)		% Mn	0.15	1.6									
	S		52.8	51.8									289
Mg–Pb	*T*		500	500	500	500							
(solid)		% Pb	1.0	2.7	4.9	9.8							
	S		18.5	18.6	17.5	16.2							289

(continu

Table 12.1 HYDROGEN SOLUTIONS IN EQUILIBRIUM WITH GASEOUS HYDROGEN AT ATMOSPHERIC PRESSURE—*continued*

Solvent metal		*Hydrogen concentration, S at T°C* cm³ *of gas at atmospheric pressure and 0°C per 100 g of metal*									*References*
Ag–Pb (liquid)	T		700	700	700	700					
		% Pb	1.0	2.7	4.9	9.8					
	S		51.3	50.9	50.1	49.8					289
Ag–Sn (solid)	T		500	500	500	500					
		% Sn	0.8	3.0	4.8	9.0					
	S		19.0	18.5	18.5	16.1					289
Ag–Sn (liquid)	T		700	700	700	700					
		% Sn	0.8	3.0	4.8	9.0					
	S		52.7	50.6	49.8	47.7					289
Ag–Si (liquid)	T		700	800	900	1 000					
	S	1% Si	47.9	61.4	75.4	89.7					299
Ag–Sr (liquid)	T		700	800	900	1 000					
	S	0.5% Sr	51.9	67.6	84.2	101.3					299
		2% Sr	59.6	74.8	90.3	105.8					299
		3.4% Sr	68.7	83.6	98.3	112.7					299
Ag–Zn (liquid)	T		700	800	900	1 000					
	S	1.3%Zn	48.4	63.8	80.3	97.4					299
		3.2% Zn	49.1	62.3	75.8	89.5					299
		6.8% Zn	45.8	60.5	76.2	92.6					299
In (solid α)	T	25	100	200	300	400	500	600	Marked hysteresis during transition in range 600 < T < 800		43 also 42
	S	21.6	19.9	17.2	14.5	12.4	11.4	11.4			
(solid β and γ)	T	800	850	900	950	1 000	1 050	1 100	1 125	Hysteresis during transition in range 1 125 < T < 1 150	43 also 42
	S	28.6	29.3	30.1	31.4	32.8	34.2	40.0	42.2		
In (solid δ)	T	1 150	1 175	1 200	1 225	1 243					43 also 42
	S	41.1	41.7	42.8	44.4	46.6					
(liquid)	T	1 243	1 250	1 275	1 300						43 also 42
	S	46.6	50.0	58.3	60.2						
Io (solid)	T	500	600	700	800	900	1 000	1 100	1 200		23
	S	0.8	1.3	1.7	2.2	1.8	1.2	0.8	0.5		
i (solid)	T	300	400	500	600	700	800	900			21, 24 also 26, 32, 33
	S	2.0	2.5	3.3	4.3	5.6	7.0	8.5			
	T	1 000	1 100	1 200	1 300	1 400	1 453				21, 24 also 26, 32, 33
	S	10.0	11.5	13.0	14.0	16.5	18				
(liquid)	T	1 453	1 500	1 600	1 700						21 also 22, 26, 33, 34
	S	41	43	48	51						
i–Co (solid)	T		1 400	1 400	1 400	1 400					
		% Co	20	40	60	80					
	S		14	11.5	10.0	9.0					21
(liquid)	T		1 500	1 600	1 700						
	S	20% Co	37.5	40.5	43.5						21
		40% Co	31.5	34.5	36.5						21
		60% Co	28	30	33.5						21
		80% Co	23.6	26	29						21 also 22
–Cu (liquid)	T		1 500	1 600							
	S	20% Cu	41	44.5							21
		40% Cu	37	40							21
		60% Cu	31.5	34.5							21
		80% Cu	23.5	25.5							21
–Fe		*See* Fe–Ni									

(continued)

Table 12.1   HYDROGEN SOLUTIONS IN EQUILIBRIUM WITH GASEOUS HYDROGEN AT ATMOSPHERIC PRESSURE—*continued*

Solvent metal		Hydrogen concentration, S at T°C cm^3 of gas at atmospheric pressure and 0°C per 100 g of metal											References
Pb (liquid)	T	420	500	600	700	800	900	References 45 and 46 report					44
	S	—	0.11	0.25	0.45	0.80	1.25	no detected solution for T < 600					also 45, 46
Pb–Ca (liquid)		No solution detected in lead with 0.16 or 0.24% Ca at 420°C											45
Pb–Mg (liquid)	T	500	500	500	500	500	500	500	500	500	500	500	
	% Mg	1.0	2.1	3.1	5.5	6.5	7.6	7.8	8.1	8.5	9.0		17.5
	S	0.18	0.22	0.45	0.87	1.7	1.5	1.8	1.7	2.4	1.9	4.8	45
Pt (solid)	T	400	800	1 000	1 100	1 200	1 300						
	S	0.07	0.10	0.19	0.34	0.54	0.80						47
Si (solid)	T	1 200											
	S	0.001 6											48
Sn (liquid)	T	1 000	1 100	1 200	1 300								
	S	0.04	0.09	0.21	0.36	(results unreliable)							28
Zn (liquid)	T	516											
	S	<0.002											46

12.1.4   Solutions of nitrogen

Table 12.5 gives values of dissolved nitrogen concentration in equilibrium with nitrogen at one atmosphere pressure for the metals iron, cobalt, chromium, molybdenum, manganese, nickel, silicon and some of their alloys. The solutions are dilute and the solution process is endothermic, the solubility increasing with temperature.

Table 12.6 gives values for nitrogen concentrations in iron and chromium in equilibrium with nitrides, measured by methods including internal friction and calorimetry.

In the solid metals, the solute atoms are assumed to occupy interstitial sites, only a small proportion of the available sites being filled. If iron is cold-worked, the nitrogen solubility is enhanced by additional solute sites at lattice defects thereby introduced into the metal (*see* references 117 and 118).

Some transition metals dissolve nitrogen exothermically to form concentrated interstitial solid solutions of metallic character analogous to the corresponding hydrogen solutions. Figures 12.12 and 12.13 present information for the systems niobium–nitrogen and tantalum-nitrogen.

12.1.5   Solutions of oxygen

The free energies of formation of the lowest oxides of most metals are comparatively high, so that an oxide film is formed when these metals are exposed to oxygen, except for very low pressures or very high temperatures. The solubility data usually required is therefore the concentration of dissolved oxygen in equilibrium with the oxide phase. A few other metals (Os, Pt, Rh, Au, Hg, Pd, Ru and Ir) form less stable oxides so that no film is present when the metals are exposed to oxygen at atmospheric pressure at elevated temperature. Of these, only silver and palladium dissolve appreciable quantities of oxygen. Table 12.7 gives values for the dissolved concentrations of oxygen in equilibrium with the lowest oxide of the metal or with gaseous oxygen at atmospheric pressure as appropriate. These concentrations are often small and difficult to measure. The usual method of establishing equilibrium by allowing oxygen to diffuse inwards from a surface oxide phase or from the gas phase is liable to lead to erroneous results unless the metal is free from traces of impurities which form oxides more stable than its own oxide.

Some transition metals can dissolve large quantities of oxygen before a separate oxide phase appears. Figures 12.14 and 12.15 give isotherms for the systems niobium–oxygen and tantalum–oxygen.

Table 12.2 HYDROGEN SOLUTIONS IN EQUILIBRIUM WITH ALKALI AND ALKALINE EARTH METAL HYDRIDES

Solvent metal	Hydride		Atom fraction of hydrogen, N_H, and hydride dissociation pressure, P(Pa) at temperature, $T°C$								References
Ba (solid)	BaH$_2$	T	400	500	600	700					49
		N_H	0.18	0.24	0.35	0.50					
		P	—	—	—	—					
Ca (liquid)	CaH$_2$	T	780	800	830	860					50 also 51, 52
		N_H	0.26	0.29	0.31	0.32					
		P	1.87×10^3	3.63×10^3	6.27×10^3	1.08×10^4					
Mg (solid)	MgH$_2$	T	440	470	510	560					53 also 54 and Table 1
		N_H	0.020	0.031	0.034	0.093					
		P	3.90×10^6	6.55×10^6	1.19×10^7	2.36×10^7					
Na (liquid)	NaH	T	250	300	315	330	350	375	400	425	55, 56 also 58
		N_H	9.7×10^{-5}	5.1×10^{-4}	1.2×10^{-3}	2.4×10^{-3}	3.0×10^{-3}	5.8×10^{-3}	1.4×10^{-2}	2.8×10^{-2}	
		P	—							—	
Sr (solid $\alpha + \gamma$)	SrH$_2$	T	α Sr 212	γSr $\rightarrow$ 263	365	456	497	562	595		57
		N_H	0.048	0.062	0.096	0.12	0.14	0.19	0.19		
		P	—			—			—		
Sr (solid β)	SrH$_2$	T	668	740	802	810					57
		N_H	0.21	0.26	0.36	0.40					
		P	—			—					

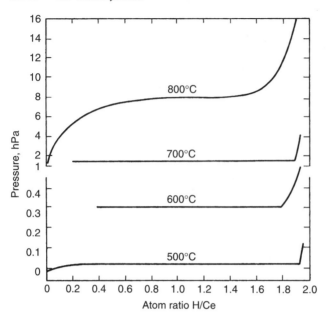

Figure 12.1   *The cerium–hydrogen system (Mulford and Holley.[64] See also Sieverts* et al.[65–67] *Ivanov and Stomakhin[68] and Edwards and Velekis[69])*

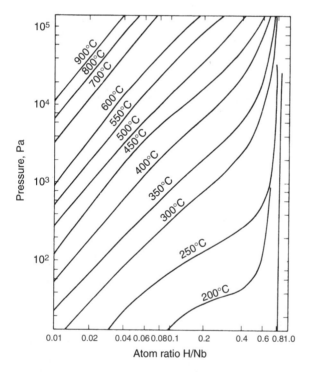

Figure 12.2   *The niobium–hydrogen system (Albrecht, Goode and Mallet.[73,74] See also Komjathy[75] and Walter and Chandler[72])*

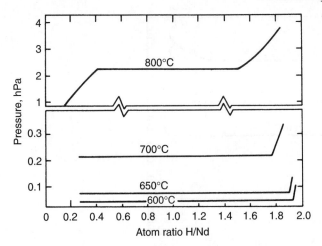

Figure 12.3 *The neodymium–hydrogen system (Mulford and Holley.[64] See also Sieverts and Roell[76])*

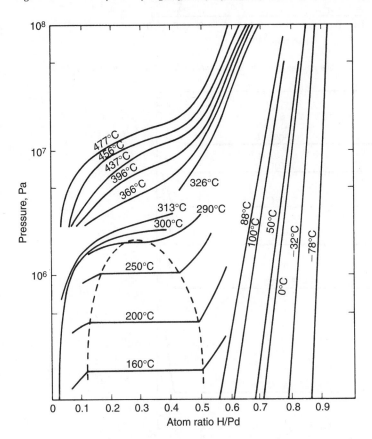

Figure 12.4 *The palladium–hydrogen system (Levine and Weal,[77] Gillespie et al.[78,79] and Perminov.[80] See also Everett and Nordon,[81] Flanagan,[82] Nakhutin and Sutyagina,[83] Mitacek and Aston,[84] Carson et al.,[85] Karpova and Tverdovsky,[86] Vert et al.[87] and Maeland and Flanagan[88])*

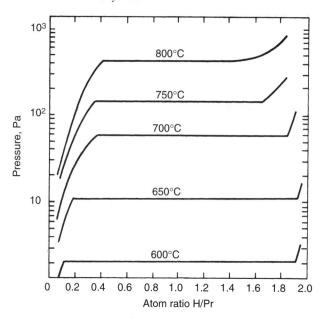

Figure 12.5   *The praseodymium–hydrogen system (Mulford and Holley.[64] See also Sieverts and Roell[76])*

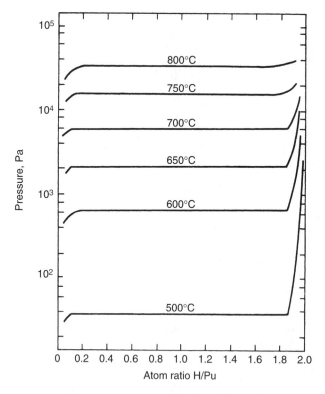

Figure 12.6   *The plutonium–hydrogen system (Mulford and Sturdy[89])*

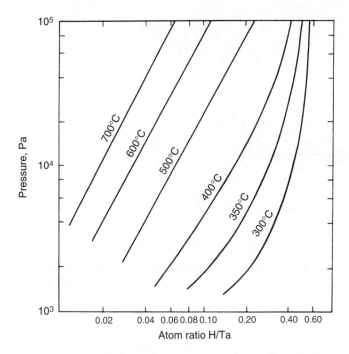

Figure 12.7 *The tantalum–hydrogen system (Mallet and Koehl.[92] See also Kofstad et al.[93] and Pedersen et al.[94])*

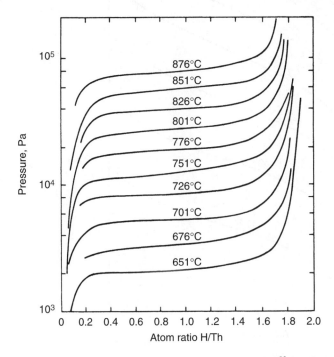

Figure 12.8 *The thorium–hydrogen system (Mallet and Campbell.[95] See also Peterson and Westlake[96])*

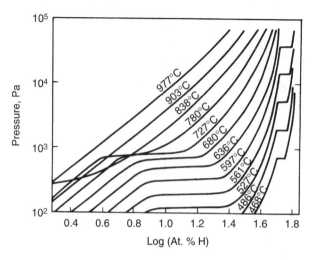

Figure 12.9   *The titanium–hydrogen systems (McQuillan.[97] See also Lenning et al.,[98] McQuillan,[99] Samsonov and Antonova[100] and Krylov[101])*

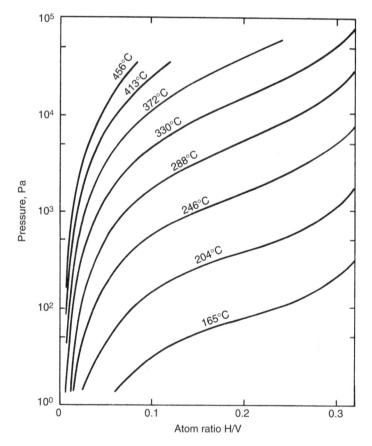

Figure 12.10   *The vanadium–hydrogen system (Kofstad and Wallace.[102] See also Brauer and Schnell[103] and Maeland[104])*

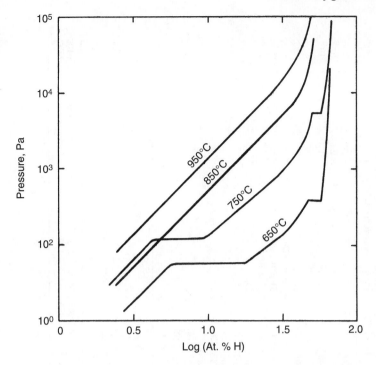

Figure 12.11 *The zirconium–hydrogen system (Private communication from McQuillan based on reference 105. See also Motz,*[106] *Schwartz and Mallet,*[107] *Gulbransen and Andrew,*[108] *Edwards et al.,*[109] *Mallet and Albrecht,*[110] *Espagno et al.,*[111] *Libowitz,*[112] *La Grange et al.,*[113] *Slattery,*[114] *Singh and Gordon Parr,*[115] *and Katz and Berger*[116])

12.1.6 Solutions of the noble gases

The solubilities in metals of the noble gases in Group 8 of the Periodic Table are so small that the quantities which dissolve by equilibrating metals with the pure gases are difficult to detect. For example, Kubaschewski's theoretical argument[1] predicts that at 600°C, only 3.5×10^{-10} atomic fraction of xenon will dissolve in liquid bismuth equilibrated with xenon gas at one atmosphere pressure. However, significant quantities of the noble gases can be inserted into metal lattices by very energetic processes such as nuclear fission or bombardment with accelerated ions. Examples of solutions produced in this manner are given in Table 12.8. Fuller information is given in a review by Blackburn.[14]

12.1.7 Theoretical and practical aspects of gas–metal equilibria

The equilibria between metals and gases are of a wide variety and the practical effects of absorbed gases in metals during industrial processes are diverse, usually deleterious and often difficult to assess. As a result, a vast amount of practical and theoretical effort has been applied in studying gas–metal interactions using numerous different approaches, as illustrated by the selection of reviews and papers of general or theoretical interest given in references 1–15.

Table 12.3   HYDROGEN SOLUTIONS IN URANIUM EQUILIBRATED WITH GASEOUS HYDROGEN AND WITH URANIUM HYDRIDE (Mallet and Trzeciak[62], also reference 63)

S, mass fraction of hydrogen in uranium equilibrated at $T°C$ with hydrogen gas at atmospheric pressure; Sp, mass fraction of hydrogen in uranium equilibrated at $T°C$ with uranium hydride. UH_3 P, UH_3 dissociation pressure (pascals) at $T°C$

U (solid α)

T	100	200	300	400	432	500	600	662
S	—	—	—	—	1.6×10^{-6}	1.8×10^{-6}	2.0×10^{-6}	2.2×10^{-6}
P	0.2	60	3×10^{3}	5×10^{4}	1.0×10^{5}	3.7×10^{5}	1.81×10^{6}	4.02×10^{6}
Sp	6×10^{-10}	2×10^{-8}	2×10^{-7}	1.1×10^{-6}	1.6×10^{-6}	3.5×10^{-6}	8.6×10^{-6}	1.35×10^{-5}

U (solid β)

T	662	700	725	750	769
S	7.8×10^{-6}	8.5×10^{-6}	9.0×10^{-6}	—	9.7×10^{-6}
P	4.02×10^{6}	6.25×10^{6}	8.24×10^{6}	1.05×10^{7}	1.31×10^{7}
Sp	4.9×10^{-5}	6.8×10^{-5}	8.1×10^{-5}	9.7×10^{-5}	1.11×10^{-4}

U (solid γ)

T	769	800	900	1000	1100	1129
S	1.47×10^{-5}	1.50×10^{-5}	1.56×10^{-5}	1.62×10^{-5}	1.67×10^{-5}	1.69×10^{-5}
P	1.31×10^{7}	2.16×10^{7}	4.07×10^{7}	8.04×10^{7}	1.49×10^{8}	1.77×10^{8}
Sp	1.68×10^{-4}	1.95×10^{-4}	3.12×10^{-4}	4.57×10^{-4}	6.43×10^{-4}	7.02×10^{-4}

U (liquid)

T	1129	1200	1300	1400
S	2.81×10^{-5}	2.93×10^{-5}	3.11×10^{-5}	3.27×10^{-5}
P	1.77×10^{8}	2.48×10^{8}	3.94×10^{8}	5.96×10^{8}
Sp	1.17×10^{-3}	1.45×10^{-3}	1.94×10^{-3}	2.52×10^{-3}

Table 12.4   HYDROGEN SOLUTIONS IN HAFNIUM EQUILIBRATED WITH GASEOUS HYDROGEN AT ATMOSPHERIC PRESSURE (Espagno, Azou and Bastien[70,71], also reference 69)

Temperature °C	100	300	500	700	900	950	1000	1050	1100
Atom ratio H/Hf	1.80	1.78	1.60	1.38	0.88	0.40	0.09	0.06	0.05

Table 12.5   NITROGEN SOLUTIONS IN EQUILIBRIUM WITH GASEOUS NITROGEN AT ATMOSPHERIC PRESSURE

Solvent metal		*Mass % of nitrogen at temperature, T°C*							*References*
Fe (solid α)	T	700	800	900					120
	Mass %	1.5×10^{-3}	2.3×10^{-3}	3.3×10^{-3}					also 119
Fe (solid γ)	T	900	1 000	1 100	1 200	1 300	1 400		
	Mass %	0.028	0.025	0.024	0.023	0.022	0.021		120, 121, 122, 145
Fe (solid δ)	T	1 400	1 450	1 500	1 535				
	Mass %	0.010 1	0.011 1	0.012 1	0.012 9				121
Fe (liquid)	T	1 600							
	Mass %	0.044	(average of 21 independent determinations)						22, 121, 124–147
Fe–Al (liquid)	T	1 600	1 600						
	% Al	0.25	0.5						140
	Mass %	0.044	0.044						
Fe–C (liquid)	T	1 600	1 600	1 600	1 600	1 600			
	% C	1	2	3	4	5			125, 128, 131, 132,
	Mass %	0.030	0.022	0.015	0.011	0.004	(averages of independent determinations)		139, 140, 144
Fe–Cr (solid γ)	T		1 000	1 100	1 200	1 300	1 400		
	Mass %	4.76% Cr	0.102	0.079	0.063	0.051	0.034*		146
		8.67% Cr	0.286	0.193	0.138	0.102	—		
		14.10% Cr	0.96	0.48	0.26	—	—		
			* δ phase						
Fe (liquid)	T	1 600	1 600	1 600	1 600	1 600	1 600	1 600	
	% Cr	5	10	15	20	40	60	70	127, 128, 133, 136,
	Mass %	0.07	0.12	0.18	0.29	1.00	2.3	3.5	140 also 143
	T	1 700	1 700	1 700	1 700	1 700	1 700	1 700	
	% Cr	10	20	40	50	70	80	90	
	Mass %	0.10	0.22	0.75	1.16	2.6	3.5	4.5	133
Fe–Co (liquid)	T	1 600	1 600	1 600	1 600	1 600	1 600		
	% Co	5	10	20	40	60	80		22, 137, 140, 144
	Mass %	0.040	0.035	0.028	0.023	0.013	0.010		
Fe–Cu (liquid)	T	1 600	1 600	1 600	1 600	1 600			
	% Cu	2	4	6	8	10			
	Mass %	0.044	0.042	0.042	0.040	0.039			137, 140, 144
Fe–Mn (solid γ)	T		1 050	1 200	1 300				
	Mass %	0.43% Mn	0.025	—	—				119
		0.53% Mn	0.024	—	—				
		1.46% Mn	0.025	—	0.020				
		12.98% Mn	0.066	0.046	—				
(liquid)	T	1 550	1 550	1 550	1 550	1 550	1 550		
	% Mn	5	10	20	40	60	80		
	Mass %	0.044	0.052	0.087	0.22	0.43	0.80		141, 142
	T	1 600	1 600	1 600	1 600	1 600			
	% Mn	1	2	5	10	20			
	Mass %	0.046	0.048	0.060	0.074	0.098			133, 140 also 143
Fe–Mo (liquid)	T	1 600	1 600	1 600	1 600	1 600			
	% Mo	2	4	6	8	10			135, 137, 140, 144
	Mass %	0.045	0.047	0.051	0.054	0.057			also 143
Fe–Nb (liquid)	T	1 600	1 600	1 600	1 600				
	% Nb	2	4	6	8				
	Mass %	0.060	0.085	0.11	0.16				140, also 143
Fe–Ni (solid)	T		918	999	1 217				
	Mass %	1.01% Ni	0.028 3	0.025 3	0.021 5				
		3.98% Ni	0.025 2	0.021 9	0.018 7				
		8.11% Ni	0.020 5	0.018 6	0.015 6				145 also 143
		15.46% Ni	0.013 2	0.012 5	0.011 1				
		26.8% Ni	0.006 1	0.006 2	0.006 1				
		40.7% Ni	0.001 9	0.002 3	0.002 7				

(*continued*)

Table 12.5   NITROGEN SOLUTIONS IN EQUILIBRIUM WITH GASEOUS NITROGEN AT ATMOSPHERIC PRESSURE—*continued*

Solvent metal		*Mass % of nitrogen at temperature, T°C*									References
Fe–Ni (liquid)	T	1 600	1 600	1 600	1 600	1 600	1 600				137
	% Ni	1	2	5	10	25	50				131, 132, 13
	Mass %	0.044	0.043	0.039	0.033	0.017	0.0067				135, 136, 14
											144 also 143
Fe–O (liquid)	Effect of up to 0.2% O is slight										140
Fe–S (liquid)	Effect of up to 0.3% S is slight										137, 140, 14
Fe–Si (solid α)	T		700	800	900	1 000	1 100				
	Mass % 2.83% Si	0.001 3	0.001 6	0.001 9	0.002 1	0.002 4					120
(solid γ)	T	1 000	1 050	1 100	1 200	1 300	1 350				
	Mass % 0.20% Si	—	0.024	—	0.022	—	0.020				
	0.58% Si	—	0.022	—	0.020	—	0.018				122, 147
	0.90% Si	0.024	—	0.022	0.021	0.020	—				
	1.26% Si	0.023	—	0.022	0.021	0.019	—				
(liquid)	T	1 600	1 600	1 600	1 600	1 600					
	% Si	2	4	6	8	10					126, 140, 14
	Mass %	0.035	0.026	0.020	0.013	0.010					also 143
Fe–Sn (liquid)	T	1 600	1 600	1 600	1 600	1 600					
	% Sn	2	4	6	8	10					
	Mass %	0.044	0.042	0.041	0.041	0.040					137, 140
Fe–Ta (liquid)	T	1 600	1 600	1 600	1 600	1 600					
	% Ta	2	4	6	8	10					
	Mass %	0.052	0.060	0.072	0.084	0.098					140
Fe–V (liquid)	T		1 600	1 700	1 800	1 900					
	Mass % 1% V		0.059	0.059	0.059	0.059					
	2% V		0.074	0.071	0.070	0.067					
	3% V		0.088	0.084	0.081	0.079					135, 140
	5% V		—	0.129	0.120	0.110					
	10% V		—	0.315	0.280	0.237					
Fe–W (liquid)	T	1 600	1 600	1 600	1 600	1 600	1 600	1 600			
	% W	2	4	6	8	10	12	14			
	Mass %	0.044	0.044	0.045	0.045	0.046	0.046	0.046			140
Fe–Cr–Ni (liquid)	T	1 600	1 600	1 600	1 600	1 600	1 600	1 600			
	% Cr	8.9	13.4	4.0	23.0	33.3	24.7	50.0			
	% Ni	6.5	13.4	29.5	6.0	33.1	50.6	25.0			136
	Mass %	0.083	0.255	0.041	0.316	0.476	0.762	1.35			
	For additional data *see* reference 136										
Fe–Mo–V	T	1 700	1 700	1 700	1 700	1 700	1 700	1 700	1 700	1 700	
	% Mo	1	3	5	1	3	5	1	3	5	
	% V	1	1	1	2	2	2	3	3	3	
	Mass %	0.054	0.058	0.063	0.069	0.071	0.077	0.089	0.093	0.099	135
	For additional data *see* reference 135										
Co (solid)	Not detectable for T 1 200										149, 150
(liquid)	T	1 600									
	Mass %	0.0047									151
Co alloys (liquid)	T	1 600	1 600	1 600	1 600	1 600					
	alloying element	0.8% Al	3.0% Cr	6.0% Cu	1.0% Fe	6.0% Mo					
	Mass %	0.004 4	0.006 4	0.005 3	0.004 6	0.005 3					151
	T	1 600	1 600	1 600	1 600	1 600	1 600	1 600			
	alloying element	5.5% Nb	3.5% Ni	1.8% Si	6.0% Ta	1.0% V	6.0% W				
	Mass %	0.009 8	0.003 9	0.002 8	0.007 2	0.005 9	0.005 3				151
Co–Fe	See Fe–Co										
Cr (liquid)	T	1 600	1 650	1 700	1 725	1 750					154 also
	Mass %	4.08	3.90	3.84	3.76	3.54					136, 153
Cr–Ni	See Ni–Cr										

(continue

ble 12.5   NITROGEN SOLUTIONS IN EQUILIBRIUM WITH GASEOUS NITROGEN AT ATMOSPHERIC PRESSURE—*continued*

lvent etal			*Mass % of nitrogen at temperature, T°C*							References
‑Si (quid)	*T*		1 600	1 650	1 700	1 750				
	Mass%	1.5% Si	3.83	3.68	3.54	3.08				
		7.5% Si	1.98	1.89	1.72	1.68				154
		10.0% Si	0.84	0.74	0.69	0.62				also 153
		20.0% Si	0.33	0.30	0.28	0.26				
·o (·lid)	*T*		950	1 000	1 050	1 100	1 150			
	Mass %		0.85	0.56	0.41	0.33	0.26			155
·n (quid)	*T*		1 245	1 300	1 400	1 500	1 600	1 700		
	Mass %		3.4	2.8	1.9	1.6	1.1	1.0		146, 156
n–Fe	*See* Fe–Mn									
	T		1 600							22, 136, 137, 144
(quid)	*Mass%*		<0.0025							
·–Cr (quid)	*T*		1 600	1 600	1 600	1 600	1 600	1 600	1 600	
	% Cr		10	20	30	40	50	60	70	136
	Mass%		0.016	0.068	0.17	0.44	1.0	1.7	2.6	
·–Fe	*See* Fe–Ni									
	T		1 420							
(·uid)	*Mass%*		~0.01							129

Some phase relationships discussed in references 171 and 172

·2.6   NITROGEN SOLUTIONS IN IRON AND CHROMIUM IN EQUILIBRIUM WITH NITRIDES

Phase in equilibrium	*Method of determination*		*Mass % of nitrogen at temperature, T°C*									*References*
·e Fe$_4$N	Calorimetry	*T*	200	240	300	330	400	450	575			
		Mass %	0.008	0.014	0.018	0.026	0.043	0.059	0.097			160
	Invariant pressure	*T*	450	500	550	590						
		Mass %	0.033	—	0.070	0.10						161
				0.06	—	—						117
	Internal friction	*T*	250	300	350	400	450	500	575	585*		
		Mass %	0.005	0.010	0.015	0.025	0.035	0.050	0.075	—		162
				0.008 4	—	0.025	—	0.055	—	0.095		163
			* eutectoid temperature									
			See also criticism in reference 164 and reply in reference 165									
	Fe$_8$N	Internal friction	*T*	20	100	150	200	250	300	400		
			Mass %	—	—	0.003 5	0.010	0.020	0.040	—		162
				1.4 × 10⁻⁵	5.2 × 10⁻⁴	—		0.008 8	—	0.055	0.20	163
	N$_2$ at 1 atm (for comparison)	Internal friction	*T*	500	585	700	800	900			
			Mass %	9.0 × 10⁻⁴	0.001 4	0.002 4	0.003 3	0.004 5				163
			See also Table 12.8									
Si Unidentified nitride	Internal friction	*T*	300	400	500	600	700	800	900	1 000		
		Mass %	0.001 0	0.001 5	0.002 4	0.004 0	0.006 1	0.010	0.014	0.019		166
	Cr$_2$N	*T*	1 100	1 200	1 300	1 400						
		Mass %	0.04	0.09	0.14	0.26						152
		P_{Cr_2N} Pa	2 × 10²	9 × 10²	2.5 × 10³	5.8 × 10³						

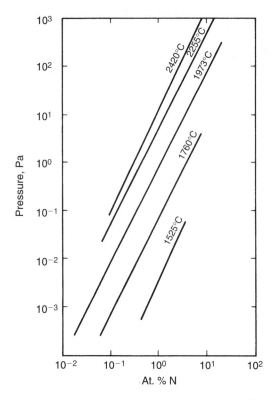

Figure 12.12   *The niobium–nitrogen system (Cost and Wert[168] See also Pemsler[167])*

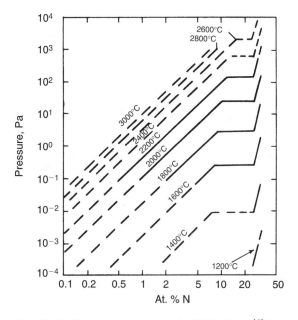

Figure 12.13   *The tantalum–nitrogen system (Gebhardt et al.[169] See also Gebhardt et al.[170] and Pemsler[167])*

Table 12.7   OXYGEN SOLUTIONS IN EQUILIBRIUM WITH OXIDES OR WITH GASEOUS OXYGEN AT ATMOSPHERIC PRESSURE

Solvent metal	Phase in equilibrium		Mass % of oxygen at temperature, T°C							References
Ag (solid)	O₂ (gas at atmospheric pressure)	T	300	400	500	600	700	800	900	174 also 173
		Mass %	3.0×10^{-5}	1.4×10^{-4}	4.4×10^{-4}	1.07×10^{-3}	2.16×10^{-3}	3.81×10^{-3}	6.14×10^{-3}	
Ag (liquid)	O₂ (gas at atmospheric pressure)	T	973	1 024	1 075	1 125				173
		Mass %	0.305	0.295	0.277	0.264				
Co (solid α)	CoO (solid)	T	600	700	810	875				175
		Mass %	0.6×10^{-2}	0.9×10^{-2}	1.6×10^{-2}	2.05×10^{-2}				
Co (solid β)	CoO (solid)	T	875	945	1 000	1 200				175
		Mass %	0.58×10^{-2}	0.7×10^{-2}	0.8×10^{-2}	1.3×10^{-2}				
Co (liquid)	CoO (solid)	T	1 550	1 600	1 650					178
		Mass %	0.13	0.16	0.23					
Cr (solid)	Cr₂O₃ (solid)	T	1 350							179
		Mass %	approximately 0.03%							
Cu (solid)	Cu₂O (solid)	T	600	700	800	900	950	1 000	1 050	181 also 180, 176, 177
		Mass %	1.6×10^{-3}	1.7×10^{-3}	2.1×10^{-3}	2.7×10^{-3}	3.4×10^{-3}	4.6×10^{-3}	7.7×10^{-3}	
Fe (solid)	FeO (solid)	T	<1 500							182, 189
		Mass %	<0.009							
Fe (liquid)	FeO (liquid)	T	1 550	1 600	1 650	1 700				183, 184, 185, 186, 187, 188, 267, 189, 268
		Mass %	0.18	0.23	0.28	0.34				
Hf	See references 190 and 191									
K (liquid)	K₂O (solid)	T	100	150	200	250	300			192
		Mass %	0.10	0.17	0.27	0.41	0.60			
Li (liquid)	Li₂O (solid)	T	250	400						193
		Mass %	0.0109	0.066						
Mo	See reference 194 for condensed-phase relationships									194
Na (liquid)	Na₂O (solid)	T	100	200	300	400	500	550		192
		Mass %	0.002	0.005	0.010	0.018	0.047	0.08		
Nb	See Figure 12.14 and references 167, 205, 206, 207 and 210									

(continued)

Table 12.7   OXYGEN SOLUTIONS IN EQUILIBRIUM WITH OXIDES OR WITH GASEOUS OXYGEN AT ATMOSPHERIC PRESSURE—*continued*

Solvent metal	Phase in equilibrium		Mass % of oxygen at temperature, T°C					References
Ni (solid)	NiO (solid)	T	600	800	1 000	1 200		195
		Mass %	0.020	0.019	0.014	0.012		
Ni (liquid)	NiO (liquid)	T	1 450	1 500	1 550	1 600	1 650	199
		Mass %	0.28	0.46	0.72	1.10	1.66	also 196, 197, 198
	See also references 178, 197 and 198 for activities of O in Ni–Fe and Ni–Co alloys							
Pb (liquid)	PbO (solid)	T	350	450				200
		Mass %	5.4×10^{-4}	8.6×10^{-4}				
Pd (solid)	O_2 (gas at atmospheric pressure)	T	1 200		550			201
		Mass %	<0.05		13.2×10^{-4}			
Pu	*See* reference 209 for plutonium–oxygen phase diagram							
Rh (solid)	O_2 (gas at atmosphere pressure)	Slight solubility						201
Si (solid)	SiO_2 (solid)	T	1 000	1 100	1 200	1 300	1 412	202, 203
		Mass %	2.8×10^{-4}	5.3×10^{-4}	9.1×10^{-4}	1.5×10^{-3}	2.2×10^{-3}	
Sn (liquid)	SnO_2 (solid)	T	536	600	700	751		204
		Mass %	0.000 18	0.000 55	0.002 8	0.004 9		
Ta	*See* Figure 12.15 and references 208 and 210–220							
Ti	*See* references 221–229							
U	*See* references 230–232							
V	*See* references 233–236							
Y	*See* reference 237							

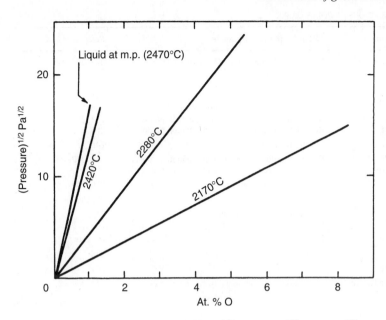

Figure 12.14   *The niobium–oxygen system (Pemsler.[167] Se also Elliot[210] and Fromm[208])*

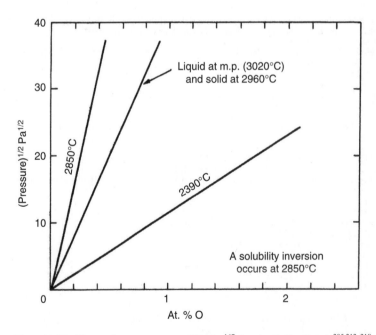

Figure 12.15   *The tantalum–oxygen system (Pemsler.[167] See also Gebhardt et al.,[205,212–218] Powers and Doyle,[219] Marcotte and Larsen,[206] Meussner and Carpenter[207] and Fromm[208])*

Table 12.8 SOLUTIONS OF NOBLE GASES

Solvent	Solutes	Method of introducing solute	References
Ag	Ar, Kr, Xe	Electric discharge on metal in gas at low pressure	243, 244, 246, 247, 249
	Ar	Injection of accelerated ions	245
	Kr	Equilibration with gas phase	248
Al	Ar	Injection of accelerated ions	245
	He	Injection of cyclotron-accelerated α-particles	250, 251, 252
Al–Li	He	Radioactive decay of neutron-irradiation products	253, 254
Al–U	Kr	Radioactive decay of neutron-irradiation products	255
Au	Ar	Injection of accelerated ions	245
Be	He	Radioactive decay of neutron-irradiation products	256, 257
	He	Injection of cyclotron-accelerated α-particles	258
Cu	He	Injection of cyclotron-accelerated α-particles	250, 259
Ge	He	Equilibration with gas phase	48, 260
Pb	Ar	Injection of accelerated ions	245
	K	Equilibration with gas phase	248
Si	He	Equilibration with gas phase	48, 260
Sn	Kr	Equilibration with gas phase	248
Ti	He		261
U	Ar, Kr, Xe	Electric discharge on metal in gas at low pressure	243, 244
	Kr	Injection of accelerated ions	245
	Kr, Xe	Nuclear fission of solvent	262, 263, 264, 265, 266
Zr	Ar, Kr, Xe	Electric discharge on metal in gas at low pressure	243, 244

REFERENCES

1. O. Kubaschewski, A. Cibula and D. C. Moore, 'Gases and Metals', Iliffe, London, 1970.
2. J. D. Fast, 'Interaction of Metals and Gases', Philips Technical Library, Eindhoven, 1965.
3. R. M. Barrer, *Discuss. Faraday Soc.*, 1948, No. 4, 68.
4. O. Kubaschewski, *Z. Electrochem.*, 1938, **44/2**, 152.
5. J. W. McBain, 'Sorption of Gases by Solids', London, 1932.
6. A. Nikuradse and R. Ulbricht, 'Das Zweistoffsystem Gas-Metal', Munich, 1950.
7. A. Sieverts, *Z. Metallk.*, 1929, **21**, 37.
8. E. Fromm and E. Gebhardt, 'Gases and Carbon in Metals'. Springer-Verlag, Berlin, 1976.
9. C. R. Cupp, *Prog. Metal Phys.*, 1954, **4**, 105.
10. D. P. Smith, 'Hydrogen in Metals', Chicago, 1948.
11. D. E. J. Talbot, 'Effects of Hydrogen in Aluminium, Magnesium, Copper and their Alloys', *International Met. Reviews*, 1975, **20**, 166.
12. P. Cotterill, *Prog. Mater. Sci.*, 1961, **9**, 205.
13. R. Fowler and C. J. Smithells, *Proc. R. Soc.*, 1937, **160**, 37.
14. R. Blackburn, 'Inert Gases in Metals', *Met. Reviews*, 1966, **11**, 159.
15. O. Kubaschewski and B. E. Hopkins, 'Oxidation of Metals and Alloys', 2nd Edition, Butterworth, London, 1962.
16. C. Thomas, *Trans. Met. Soc. AIME*, 1967, 239, 485.
17. C. E. Ransley and H. Neufeld, *J. Inst. Metals*, 1948, **74**, 599.
18. W. Eichenauer, K. Hattenbach and A. Pebler, *Z. Metallk.*, 1961, **52**, 682.
19. W. R. Opie and N. J. Grant, *Trans. AIMME*, 1950, **188**, 1237.
20. A. Sieverts and H. Hagen, *Z. Phys. Chem.*, 1934, **169**, 237.
21. H. Schenck and K. W. Lange, *Arch. Eisenhütt Wes.*, 1966, **37**, 739.
22. T. Busch and R. A. Dodd, *Trans. Met. Soc. AIMME*, 1960, **218**, 488.
23. E. Martin, *Arch. Eisenhütt Wes.*, 1929/30, **3**, 407.
24. L. Luckmeyer-Hasse and H. Schenk, *Arch. Eisenhütt Wes.*, 1932–33, **6**, 209.
25. W. Eichenauer and A. Pebbler, *Z. Metallk.*, 1957, **48**, 373.
26. A. Sieverts, *Z. Phys. Chem.*, 1911, **77**, 611.
27. P. Röntgen and F. Möller, *Metallwirt., Metallwiss., Metalltech.*, 1934, **13**, 81, 97.
28. M. B. Bever and C. F. Floe, *Trans. AIMME*, 1944, **156**, 149.
29. R. Eborall and A. J. Swain, *J. Inst. Metals*, 1952–53, **81**, 497.
30. W. Eichenauer, H. Kunzig and A. Pebler, *Z. Metallk.*, 1958, **49**, 220.
31. A. Sieverts, G. Zapf and H. Moritz, *Z. Phys. Chem.*, 1938, **183**, 19.

32. M. H. Armbruster, *J. Amer. Chem. Soc.*, 1943, **65**, 1043.
33. F. de Kazinczy and O. Lindberg, *Jernkont. Annlr*, 1960, **144**, 288.
34. H. Schenck and H. Wünsch, *Arch. Eisenhütt Wes.*, 1961, **32**, 779.
35. H. Liang, M. B. Bever and C. F. Floe, *Trans. AIMME*, 1946, **167**, 395.
36. M. M. Karnaukhov and A. N. Morazov, *Izv. Akad. Nauk*, (Otdelenie Tekh. Nauk), December 1948, 1845.
37. D. J. Carney, J. Chipman and N. J. Grant, *Trans. AIMME*, 1950, **188**, 404.
38. T. Watnabe, Y. C. Huang, and R. Komatsu, *J. Jap. Inst. Light Metals*, 1975, **46**, 76.
39. Z. D. Popovic and G. R. Piercy, *Metall. Trans A*, 1975, **6A**, 1915.
40. M. V. Sharov and V. V. Serebryakov, *Nauchnyye doklady vvyshe shkoly, Metallurgiya*, 1958, **2**, 37.
41. J. Koenman and A. G. Metcalf, *Trans. Amer. Soc. Metals*, 1959, **51**, 1072.
42. A. Sieverts and H. Moritz, *Z. Phys. Chem.*, 1938, **180/4**, 249.
43. E. V. Potter and H. C. Lukens, *Trans. AIMME*, 1947, **171**, 401.
44. W. R. Opie and N. J. Grant, *Trans. AIMME*, 1951, **191**, 244.
45. W. Mannchen and M. Bauman, *Z. Metallk.*, 1955, **9**, 686.
46. W. Hofmann and J. Maatsch, *Z. Metallk.*, 1956, **47**, 89.
47. A. Sieverts, *Ber. dt. Chem. Ges.*, 1912, **45**, 221.
48. A. van Wieringen and N. Warmoltz, *Physica*, 1956, **22**, 849.
49. D. T. Peterson and M. Indig, *J. Am. Chem. Soc.*, 1960, **82**, 5645.
50. W. D. Treadwell and J. Stecher, *Helv. Chim. Acta*, 1953, **36**, 1820.
51. W. C. Johnson, M. F. Stubbs, A. E. Sidwell and A. Pechukas, *J. Am. Chem. Soc.*, 1939, **61**, 318.
52. D. T. Peterson and V. G. Fattore, *J. Phys. Chem.*, 1961, **65**(11), 2062.
53. J. F. Stampfer, C. E. Holley and J. F. Shuttle, *J. Am. Chem. Soc.*, 1960, **82**, 3504.
54. J. A. Kenneley, J. W. Varwig and H. W. Myers, *J. Phys. Chem.*, 1960, **64**(5), 703.
55. C. C. Addison, R. J. Pulham and R. J. Roy, *J. Chem. Soc.*, 1965, 116.
56. D. D. Williams, J. A. Grand and R. R. Miller, *J. Phys. Chem.*, 1957, **61** 379.
57. D. T. Peterson and R. P. Colburn, *J. Phys. Chem.*, 1966, **70**, 468.
58. M. D. Banus, J. J. McSharry and E. A. Sullivan, *J. Am. Chem. Soc.*, 1955, **77**, 2007.
59. T. Bagshaw and A. Mitchell, *J. Iron Steel Inst.*, 1967, **205**, 769.
60. D. J. Carney, J. Chipman and N. J. Grant, *Trans. AIMME*, 1950, **207**, 597.
61. M. Weinstein and J. F. Elliot, *Trans. Met. Soc. AIMME*, 1963, **227**, 382.
62. M. W. Mallet and M. J. Trzeciak, *Trans. Am. Soc. Metals*, 1958, **50**, 981.
63. H. C. Mattrow, *J. Phys. Chem.*, 1955, **59**, 93.
64. R. N. R. Mulford and C. E. Holley, *J. Phys. Chem.*, 1955, **59**, 1222.
65. A. Sieverts and G. Muller-Goldegg, *Z. Anorg. Allg. Chem.*, 1923, **131**, 65.
66. A. Sieverts and E. Roell, *Z. Anorg. Allg. Chem.*, 1925, **146**, 149.
67. A. Sieverts and A. Gotta, *Z. Anorg. Allg. Chem.*, 1928, **172**, 1.
68. E. G. Ivanov, A. Ya. Stomakhin, G. M. Medveda and A. F. Filippov, *Chernaya Metallurgiya*, 1966, **5**, 69.
69. R. K. Edwards and E. Veleckis, *J. Phys. Chem.*, 1962, **66**, 1657.
70. L. Espagno, P. Azou and P. Bastien, *C. r. hebd. Séanc. Acad. Sci., Paris*, 1960, **250**, 4352.
71. L. Espagno, P. Azou and P. Bastien, *Mém. Scient. Rev. Metall.*, 1962, **59**, 182.
72. R. J. Walter and W. T. Chandler, *Trans. AIMME*, 1965, **233**, 762.
73. W. M. Albrecht, W. D. Goode and M. W. Mallet, *J. Electrochem. Soc.*, 1959, **106**, 981.
74. W. M. Albrecht, M. W. Mallet and W. D. Goode, *J. Electrochem. Soc.*, 1958, **105**, 219.
75. S. Komjathy, *J. Less-common Metals*, 1960, **2**, 466.
76. A. Sieverts and E. Roell, *Z. Anorg. Allg. Chem.*, 1926, **150**, 261.
77. P. L. Levine and K. E. Weal, *Trans. Faraday Soc.*, 1960, **56**, 357.
78. L. J. Gillespie and F. P. Hall, *J. Am. Chem. Soc.*, 1926, **48**, 1207.
79. L. J. Gillespie and L. S. Galstaun, *J. Am. Chem. Soc.*, 1936, **58**, 2565.
80. T. S. Perminov, A. A. Orlov and A. N. Frumkin, *Dokl. Akad. Nauk, SSSR*, 1952, **84**, 749.
81. D. H. Everett and P. Nordon, *Proc. R. Soc.*, 1960A, **259**, 341.
82. T. B. Flanagan, *J. Phys. Chem.*, 1961, **65**(2), 280.
83. I. E. Nakhutin and E. I. Sutyagina, *Fizica Metall.*, 1959, **7**, 459.
84. P. Nitacek and J. G. Aston, *J. Am. Chem. Soc.*, 1963, **85**(2), 137.
85. A. W. Carson, T. B. Flanagan and F. A. Lewis, *Trans. Faraday Soc.*, 1960, **56**, 1332 and 371.
86. R. A. Karpova and I. P. Tverdovsky, *Zh. Fiz. Khim.*, 1959, **33**, 1393.
87. Zh. L. Vert, I. P. Tverdovsky and I. A. Mosevich, *Zh. Fiz Khim.*, 1965, **39**, 1061.
88. A. Maeland and T. B. Flanagan, *Platin. Metals Rev.*, 1966, **10**, 20.
89. R. N. R. Mulford and G. Sturdy, *J. Am. Chem. Soc.*, 1955, **77**, 3449.
90. M. L. Lieberman and P. G. Wahlbeck, *J. Phys. Chem.*, 1965, **69**, 3973.
91. J. F. Stampfer, U.S. Atomic Energy Commission Rep., 1966 (LA-3473).
92. M. W. Mallet and B. G. Khoehl, *J. Electrochem. Soc.*, 1962, **109**, 611 and 968.
93. P. Kofstad. W. E. Wallace and L. J. Hyvönen, *J. Am. Chem. Soc.*, 1959, **81**, 5015.
94. B. Pedersen, T. Krogdahl and O. E. Stokkeland, *J. Chem. Phys.*, 1969, **42**, 72.
95. M. W. Mallet and I. E. Campbell, *J. Am. Chem. Soc.*, 1951, **73**, 4850.
96. D. T. Peterson and D. G. Westlake, *Trans. AIMME*, 1959, **215**, 445.
97. A. D. McQuillan, *Proc. R. Soc.*, 1951, **204**, 309.
98. G. A. Lenning, C. M. Craighead and R. I. Jaffee, *Trans. AIMME*, 1954, **200**, 367.
99. A. D. McQuillan, *J. Inst. Metals*, 1951, **79**, 73.
100. G. V. Samsonov, and M. M. Antonova, *Ukr. Khim. Zh.*, 1966, **32**, 555.

101. B. S. Krylov, *Izvest. Akad. Nauk, SSSR Metally*, 1966, **2**, 144.
102. P. Kofstad and W. E. Wallace, *J. Am. Chem. Soc.*, 1959, **81**, 5019.
103. G. Brauer and W. D. Schnell, *J. Less-common Metals*, 1964, **6**, 326.
104. A. J. Maeland, *J. Phys. Chem.*, 1964, **68**, 2197.
105. C. E. Ells and A. D. McQuillan, *J. Inst. Metals*, 1956–57, **85**, 89.
106. J. Motz, *Z. Metallk.*, 1962, **53**, 770.
107. C. M. Schwartz and M. W. Mallet, *Trans. Am. Soc. Metals*, 1954, **46**, 640.
108. E. A. Gulbransen and K. F. Andrew, *J. Electrochem. Soc.*, 1954, **101**, 474, and *J. Metals*, 1955, **7**, 136.
109. R. F. Edwards, P. Leversque and D. Cubicciotti, *J. Am. Chem. Soc.*, 1955, **77**, 1307.
110. M. W. Mallet and W. M. Albrecht, *J. Electrochem. Soc.*, 1957, **104**, 142.
111. L. Espagno, P. Azou and P. Bastien, *Mém. Scient. Rev. Metall.*, 1960, **57**, 254.
112. G. G. Libowitz, *J. Nucl. Mater.*, 1962, **5**, 228.
113. L. D. La Grange, L. J. Dijkstra, J. M. Dixon and U. Merten, *J. Phys. Chem.*, 1959, **63**, 2035.
114. G. F. Slattery, *J. Inst. Metals*, 1967, **95**, 43.
115. K. P. Singh and J. Gordon Parr, *Trans. Faraday Soc.*, 1963, **59**, 2248.
116. O. M. Katz and J. A. Berger, *Trans. Met. Soc. AIMME*, 1965, **233**, 1005 and 1014.
117. H. A. Wriedt and L. S. Darken, *Trans. Met. Soc. AIMME*, 1965, **233**, 111.
118. H. A. Wriedt and L. S. Darken, *ibid.*, 1965, **233**, 122.
119. A. Sieverts and G. Zapf, *Z. Phys. Chem.*, 1935, **174**, 359.
120. N. S. Corney and E. T. Turkdogan, *J. Iron Steel Inst.*, 1955, **180**, 344.
121. A. Sieverts, G. Zapf and H. Moritz, *Z. Phys. Chem.*, 1938, **183**, 19.
122. L. S. Darken, R. P. Smith and C. W. Filer, *Trans. AIMME*, 1951, **191**, 1174.
123. I. N. Milinskaya and I. A. Tomilin, *Dokl. Akad. Nauk, SSSR*, 1967, **174**, 135.
124. J. Chipman and D. Murphy, *Trans. AIMME*, 1935, **116**, 179.
125. L. Eklund, *Jernkont. Annlr.*, 1939, **123**, 545.
126. J. C. Vaughan and J. Chipman, *Trans. AIMME*, 1940, **140**, 224.
127. R. M. Brick and J. A. Creevy, *Metals Tech., AIMME*, Tech. Pub. No. 1165, April 10; 1940.
128. T. Kootz, *Arch. EisenhüttWes.*, 1941, **15**, 77.
129. C. R. Taylor and J. Chipman, *Trans. AIMME*, 1943, **154**, 228.
130. W. M. Karnaukoy and A. M. Marozov, *Bull. Acad. Sci. URSS Classe Sci. Tech.*, 1947, **735**, Brutcher Transl., no. 2029.
131. T. Saito, *Sci. Rep. Res. Insts., Tôhoku Univ.*, Ser. A., 1949, **1**, 411.
132. T. Saito, *ibid.*, 419.
133. H. Wentrup and O. Reif, *Arch. EisenhüttWes.*, 1949, **20**, 359.
134. Y. Kasamatu and S. Matoba, *Technology Rep., Tôhoku Univ.*, 1957, **22**, No. 1.
135. V. Kashyap and N. Parlee, *Trans. AIMME*, 1958, **212**, 86.
136. J. Humbert and J. F. Elliot, *Trans. Met. Soc. AIMME*, 1960, **218**, 1076.
137. H. Schenck, M. Frohberg and H. Graf, *Arch. EisenhüttWes.*, 1958, **29**, 673.
138. V. P. Fedotov and A. M. Samarin, *Dokl. Akad. Nauk, SSSR*, 1958, **122**, 597.
139. S. Maekawa and Y. Nakagawa, Tetsu-to-Hagané, *Abstr.*, 1959, **45**, 255.
140. R. D. Pehlke and J. F. Elliott, *Trans. Met. Soc. AIMME*, 1960, **218**, 1088.
141. S. Z. Beer, *Trans. Met. Soc. AIMME*, 1961, **221**, 2.
142. R. A. Dodd and N. A. Gokcen, *Trans. Met. Soc. AIMME*, 1961, **221**, 233.
143. P. H. Turnock and R. D. Pehlke, *Trans. Met. Soc. AIMME*, 1966, **236**, 1540.
144. H. Schenck, M. Frohberg and H. Graf, *Arch. EisenhüttWes.*, 1959, **30**, 533.
145. H. A. Wriedt and O. D. Gonzalez, *Trans. Met. Soc. AIMME*, 1961, **221**, 532.
146. E. T. Turkdogan and S. Ignatowicz, *J. Iron Steel Inst.*, 1958, **188**, 242.
147. E. T. Turkdogan and S. Ignatowicz, *J. Iron Steel Inst.*, 1957, **185**, 200.
148. E. T. Turkdogan, S. Ignatowicz and J. Pearson, *J. Iron Steel Inst.*, 1955, **181**, 227.
149. A. Sieverts and H. Hagen, *Z. Phys. Chem.*, 1934, **169**, 337.
150. R. Juza and W. Sachsze, *Z. Anorg. Allg. Chem.*, 1945, **253**, 95.
151. R. G. Blossey and R. D. Pehlke, *Trans. Met. Soc. AIMME*, 1966, **236**, 28.
152. A. U. Seybolt, and R. A. Oriani, *Trans. AIMME*, 1956, **206**, 556.
153. V. M. Berezhiani and B. M. Mirianashvili, *Trudy Inst. Metall., Tbilisi*, 1965, **14**, 163.
154. V. S. Mozgovoy and A. M. Samarin, *Dokl. Akad. Nauk, SSSR*, 1950, **74**, 729.
155. A. Sieverts and H. Brunig, *Arch. EisenhüttWes.*, 1933, **7**, 641.
156. N. A. Gokcen, *Trans. Met. Soc. AIMME*, 1961, **221**, 200.
157. F. Lihl, P. Ettmayer and A. Kutzelnigg, *Z. Metallk.*, 1962, **53**, 715.
158. V. P. Perepelkin, *Chernaya Metallurgia*, 1966, **3**, 88.
159. W. Kaiser and C. D. Thurmond, *J. Appl. Phys.*, 1959, **30**, 427.
160. G. Borelius, S. Berglund and O. Avsan, *Ark. Fys.*, 1950, **2**, 551.
161. V. G. Pararjpe, M. Cohen, M. B. Bever and C. F. Floe, *Trans. AIMME*, 1950, **188**, 261.
162. L. J. Dijkstra, *Trans. AIMME*, 1949, **185**, 252.
163. J. D. Fast and M. B. Verrijp, *J. Iron Steel Inst.*, 1955, **180**, 337.
164. H. U. Aström and G. Borelius, *Acta Met.*, 1954, **2**, 547.
165. J. D. Fast and M. B. Verrijp, *Acta Met.*, 1955, **3**, 203.
166. D. A. Leak, W. R. Thomas and G. M. Leak, *Acta Met.*, 1956, **3**, 501.
167. J. P. Pemsler, *J. Electrochem. Soc.*, 1961, **108**, 744.
168. J. R. Cost and C. A. Wert, *Acta Met.*, 1963, **11**, 231.

169. E. Gebhardt, H. D. Seghezzi and E. Fromm, *Z. Metallk.*, 1961, **52**, 464.
170. E. Gebhardt, H. D. Seghezzi and W. Dürrschnabel, *Z. Metallk.*, 1958, **49**, 577.
171. F. Anselin, *J. Nucl. Mater.*, 1963, **10**, 301.
172. R. Benz and M. G. Bowman, *J. Am. Chem. Soc.*, 1966, **88**, 264.
173. E. W. R. Steacie and F. M. G. Johnson, *Proc. R. Soc.*, 1926, **A112**, 542.
174. W. Eichenauer and G. Muller, *Z. Metallk.*, 1962, **53**, 321.
175. A. U. Seybolt and C. H. Mathewson, *Trans. AIMME*, 1935, **117**, 156.
176. N. G. Schmaal and E. Minzl, *Z. Phys. Chem.*, 1965, **314**, 142.
177. W. Hofmann and M. Klein, *Z. Metallk.*, 1966, **57**, 385.
178. V. V. Averin, A. Yu. Polyakov and A. M. Samarin, *Izvest. Akad. Nauk, SSSR (Tekhn.)*, 1957, **8**, 120.
179. D. Caplan and A. A. Burr, *Trans. AIMME.*, 1955, **203**, 1052.
180. F. N. Rhines and C. H. Mathewson, *Trans. AIMME*, 1934, **III**, 337.
181. A. Phillips and E. N. Skinner, *Trans. AIMME*, 1941, **143**, 301.
182. R. Sifferlen, *C. r. hebd. Séanc. Acad. Sci, Paris*, 1957, **244**, 1192.
183. L. S. Darken and R. W. Gurry, *J. Am. Chem. Soc.*, 1947, **68**, 798.
184. H. Schenke and E. Steinmetz, *Arch. EisenhüttWes.*, 1967, **38**, 813.
185. T. Fuwa and J. Chipman, *Trans. Met. Soc. AIMME*, 1960, **218**, 887.
186. T. P. Floridis and J. Chipman, *Trans. AIMME*, 1958, **212**, 549.
187. E. S. Tankins, N. A. Gokcen and G. R. Bolton, *Trans. Met. Soc. AIMME*, 1964, **230**, 820.
188. J. Skala, M. Kase and M. Mandl, *Hutn. Listy*, 1962, **17**, 841.
189. J. H. Swisher and E. T. Turkdogan, *Trans. Met. Soc. AIMME*, 1967, **239**, 427.
190. R. F. Dogmagala and R. Ruh, *Trans. Q. Am. Soc. Metals*, 1965, **58**, 164.
191. E. Rudy and P. Stecher, *J. less-common Metals*, 1963, **1**, 78.
192. D. D. Williams, J. A. Grand and R. R. Miller, *J. Phys. Chem.*, 1959, **63**, 68.
193. E. E. Hoffman, *Amer. Soc. Test. Mat. Symposium on Newer Materials*, 1959, **1960**, 195.
194. B. Phillips and L. L. Y. Chang, *Trans. Met. Soc. AIMME*, 1965, **233**, 1433.
195. A. U. Seybolt, Dissertation Yale University, 1936.
196. P. D. Merica and R. G. Waltenberg, *Trans. AIMME*, 1925, **71**, 715.
197. H. A. Wriedt and J. Chipman, *Trans. AIMME*, 1955, **203**, 477.
198. A. M. Samarin and V. P. Fedotov, *Izv. Akad Nauk, SSSR (Tekhn.)*, 1956, **6**, 119.
199. J. E. Bowers, *J. Inst. Metals*, 1961–62, **90**, 321.
200. K. W. Groshelm-Krisko, W. Hoffmann and H. Hanemann, *Z. Metallk.*, 1944, **36**, 91.
201. E. Raub and N. Plate, *Z. Metallk.*, 1957, **48**, 529.
202. H. J. Hrostowski and R. H. Kaiser, *Phys. Chem. Solids*, 1959, **9**, 214.
203. W. Kaiser and P. H. Keck, *J. Appl. Phys.*, 1957, **28**, 1427.
204. T. N. Belford and C. B. Alcock, *Trans. Faraday Soc.*, 1965, **61**, 443.
205. E. Gebhardt and R. Rothenbacher, *Z. Metallk.*, 1963, **54**, 623.
206. V. C. Marcotte and W. L. Larsen, *J. less-common Metals*, 1966, **4**, 229.
207. R. A. Meussner and C. D. Carpenter, *Corros. Sci.*, 1967, **2**, 115.
208. F. Fromm, *Z. Metallk.*, 1966, **57**, 540.
209. E. R. Gardner, T. L. Markins and R. S. Street, *J. Inorg. Nucl. Chem.*, 1965, **27**, 541.
210. R. P. Elliot, Amer. Soc. Metals Reprint, 1959, (143).
211. E. Gebhardt and H. D. Seghezzi, *Z. Metallk.*, 1959, **50**, 521.
212. E. Gebhardt and H. D. Seghezzi, *ibid.*, 1950, **50**, 248.
213. E. Gebhardt and H. D. Seghezzi, *ibid.*, 1955, **46**, 560.
214. E. Gebhardt and H. D. Seghezzi, *ibid.*, 1957, **48**, 430.
215. E. Gebhardt and H. D. Seghezzi, *ibid.*, 1957, **48**, 503.
216. E. Gebhardt and H. D. Seghezzi, *ibid.*, 1957, **48**, 559.
217. E. Gebhardt and Preisendanz, *Plansee Proc.*, 1955, 254.
218. E. Gebhardt and H. D. Seghezzi, *ibid.*, 1959, 280.
219. R. J. Powers and M. V. Doyle, *Trans. Met. Soc. AIMME*, 1959, **215**, 655.
220. M. Hoch and D. B. Bulrymowicz, *Trans. Met. Soc. AIMME*, 1964, **230**, 186.
221. I. Z. Kornilov and V. V. Glazova, *Izv. Akad. Nauk. SSSR Metally*, 1965, **1**, 189.
222. V. V. Glazova, *Dokl. Akad. Nauk, SSSR*, 1965, **164**, 567.
223. B. A. Bolachev, V. A. Livanov and A. A. Bukhanova, *Sov. J. Non-ferrous Metals*, 1966, **3**, 94.
224. P. Kofstad, P. B. Anderson and O. J. Krudtaa, *J. Less-common Metals*, 1961, **3**, 89.
225. F. Ehrlich, *Z. Anorg. Chem.*, 1941, **24**, 53.
226. M. K. McQuillan, *Corros. Anti-Corrosion*, 1962, **10**, 361.
227. T. Hurlen, *J. Inst. Metals*, 1960–61, **89**, 128.
228. M. T. Hepworth and W. B. Sample, *Trans. Met. Soc. AIMME*, 1962, **224**, 875.
229. M. T. Hepworth and R. Schuhmann, *Trans. Met. Soc. AIMME*, 1962, **224**, 928.
230. J. Besson, P. L. Blum and J. P. Morlevat, *C. r. hebd. Séanc. Acad. Sci., Paris*, 1965, **260**, 3390.
231. R. A. Smith, US Atomic Energy Commission Rep., 1966 (BMZ-17SS).
232. A. E. Martin and R. K. Edwards, *J. Phys. Chem.*, 1965, **69**, 1788.
233. N. P. Allen, O. Kubaschewski and O. V. Goldbeck, *J. Electrochem. Soc.*, 1951, **98**, 417.
234. W. Rostoker and A. S. Yamamoto, *Trans. Am. Soc. Metals*, 1955, **47**, 1002.
235. M. A. Gurevich and B. F. Ormont, *Zh. Neorg. Khim.*, 1957, **2**, 1566, 2581.
236. M. A. Gurevich and B. F. Ormont, *ibid.*, 1958, **3**, 403.
237. R. C. Tucker, E. D. Gibson and O. N. Carlson, *International Symposium on Compounds of Interest in Nuclear Reactor Technology*, Colorado, USA, 1964, and US Atomic Energy Commission Rep., 1964 (IS-812).

238. H. J. de Boer and J. D. Fast, *Recl. Trav. Chim. Pays-Bas Belg.*, 1936, **55**, 449.
239. O. Kubaschewski and W. A. Dench, *J. Inst. Metals*, 1955–56, **84**, 440.
240. B. Holmberg and T. Dagerhamn, *Acta Chem. Scand.*, 1961, **15**, 915.
241. E. Gebhardt, H. D. Seghezzi and W. Durrschnabel, *J. Nucl. Mater.*, 1961, **4**, 241, 255 and 269.
242. V. C. Marcotte, W. L. Larsen and D. E. Williams, *J. Less-common Metals*, 1964, **5**, 373.
243. G. Brebec, V. Levy and Y. Adda, *C. r. hebd. Séanc. Acad. Sci., Paris*, 1961, **252**, 722.
244. V. Levy *et al., C. r. hebd. Séanc. Acad. Sci., Paris*, 1961, **252**, 876.
245. C. W. Tucker and F. J. Norton, *J. Nucl. Mater.*, 1960, **2**, 329.
246. A. D. Le Claire and A. H. Rowe, *Revue Metall., Paris*, 1955, **52**, 94.
247. J. M. Tobin, *Acta Met.*, 1957, **5**, 398.
248. G. W. Johnson and R. Shuttleworth, *Phil. Mag.*, 1959, **4**, 957.
249. J. M. Tobin, *Acta Met.*, 1959, **7**, 701.
250. R. S. Barnes, *Phil. Mag.*, 1960, **5**, 635.
251. C. E. Ells and C. E. Evans, Atomic Energy of Canada Ltd., Rep. 1959 (CR Met–863).
252. C. E. Ells, *J. Nucl. Mater.*, 1962, **5**, 147.
253. G. T. Murray, *J. Appl. Phys.*, 1961, **32**, 1045.
254. D. W. Lillie, *Trans. Met. Soc. A.I.M.M.E.*, 1960, **218**, 270.
255. M. B. Reynolds, *Nucl. Sci. Engng.*, 1958, **3**, 428.
256. C. E. Ells and E. C. W. Perryman, *J. Nucl. Mater.*, 1959, **1**, 73.
257. V. Levy, *Bull. Inform. Sci. Tech.*, 1962, **62**, 56.
258. R. S. Barnes and G. B. Redding, *J. Nucl. Energy*, A, 1959, **10**, 32.
259. R. S. Barnes, G. B. Redding and A. H. Cottrell, *Phil. Mag.*, 1958, **3**, 97.
260. A. van Wieringen, Symposium: 'La Diffusion dans les Metaux', 1957, 107.
261. A. M. Rodin and V. V. Surenyants, *Fizika Metall.*, 1960, **10**, 216.
262. M. B. Reynolds, *Nucl. Sci. Engng.*, 1956, **1**, 374.
263. J. F. Walker, UK Atomic Energy Authority Publ. 1959 (IGR-TN/W-1046).
264. F. J. Norton, *J. Nucl. Mater.*, 1960, **2**, 350.
265. D. L. Gray, US Atomic Energy Commission Rep. 1960 (HW-62639).
266. N. R. Chellew and R. K. Steunenberg, *Nucl. Sci. Engng.*, 1962, **14**, 1.
267. H. Sawamura and S. Matoba, Sub. comm. for Phys. Chem. of Steelmaking, 19th Comm. 3rd Div. Jap. Soc. for promotion of Sci., July 4th, 1961.
268. Y. Sato, K. Suzuki, Y. Omori and K. Sanbongi, Tetsu-to-Hagané, *Abstr.*, 1968, **54**, 330.
269. R. B. McLellan, *J. Phys. Chem. Solids*, 1973, **34**, 1137.
270. M. Ichimura, M. Imabayashi, and M. Hayakawa, *Nippon Kinzoku Gakkaishi*, 1979, **43**, 876
271. E. Hashimoto and T. Kino, *J. Phys. F. Met. Phys.*, 1983, **13**, 1157.
272. W. Eichenauer, *Z. Metallkunde*, 1968, **59**, 613.
273. M. Ichimura, Y. Sasajima, and M. Imabayashi, *Materials Transactions, JIM*, 1991, **32**, 1109.
274. M. Ichimura, Y. Sasajima, and M. Imabayashi, *Materials Transactions, JIM*, 1992, **33**, 449.
275. W. Baukloh and F. Oesterlen, *Z. Metallkunde*, 1938, **30**, 386.
276. C. E. Ransley and D. E. J. Talbot, *Z. Metallkunde*, 1955, **46**, 328.
277. D. Stephenson, Ph.D. Thesis, 1978, Brunel University, Middlesex, England.
278. H. Shahani, Ph.D. Thesis, 1984, The Royal Institute of Technology, Stockholm, Sweden.
279. P. N. Anyalebechi, Ph.D. Thesis, 1985, Brunel University, Middlesex, England.
280. D. E. J. Talbot and P. N. Anyalebechi, Materials Science and Technology, 1988, **4**, 1.
281. J. Kocur, K. Tomasek, and L. Rabatin, *Hutnicke Listy*, 1989, **44**, 269.
282. H. Liu, M. Bouchard, L. Zhang, Light Metals 1995, edited by J. Evans, *The Minerals, Metals, and Materials Society*, 1996, 1285.
283. M. Ichimura, Y. Sasajima, and M. Imabayashi, *Jpn. Inst. of Light Metals*, 1989, **39**, 539.
284. P. N. Anyalebechi, D. E. J. Talbot and D. A. Granger: *Metall. Trans. B*, 1988, **19B**, 227.
285. P. N. Anyalebechi, D. E. J. Talbot and D. A. Granger: *Metall. Trans. B*, 1989, **20B**, 523.
286. M. Sargent, Ph. D. Thesis, 1989, Brunel University, Middlesex, England.
287. B. V. Levchuk and L. A. Andreev, *Met. Sci. Heat Treat.*, 1976, **18**, 591.
288. A. A. Grigoreva and V. A. Danelkin, *Tsevetnye Metally*, 1984, **1**, 90.
289. T. Watanabe, Y. Tachihara, Y. C. Huang, R. Komatsu: *Keikinzoku*, 1976, **26**, 167.
290. L. A. Andreev, B. V. Levchuk, A. A. Zhukhovitskii, & M. B. Zudin, *Izv. Vuz Tsve. Met.*, 1973, **5**, 127.
291. M. Ichimura and Y. Sasajima, *Jpn. Inst. of Light Metals*, 1993, **43**, 385.
292. B. V. Levchuk, L. A. Andreev, and V. A. Danilkin, *Technol. Legkih, Splavov., Nauchn. Tekhn. Byml. VILSa*, 1974, **5**, 22.
293. W. Dahl, W. Gruhl, W. G. Burchard, G. Ibe, and C. Dumitrescu: *Aluminum*, 1976, **52**, 616.
294. F. G. Jones and R. D. Pehlke, *Metall. Trans*, 1971, **2**, 2655.
295. W. Eichenauer *et al.*: *Z. Metallkunde*, 1965, **56**, 287.
296. M. Mokaram, Ph.D. Thesis, 1974, Brunel University, Middlesex, England. 1974.
297. Y. C. Huang, T. Watanabe, and R. Komatsu, Proc. 4th Intl. Conf. Vacuum Metallurgy, 1973, 173.
298. V. Shapavalov, N. P. Serdyuk, and O. P. Semik, *Dop. Akad. Nauk Ukr. RSR. Ser. A. Fiz.- Mat. Tekh. Nauki*, 1981, **6**, 99.
299. E. Overlid, P. Bakke, and T. A. Engh, *Light Metals 1997 Metaux Legers*, Canadian Inst. of Metallurgists, 1997, 141.
300. D. F. Chernega, Gotvjanskij and Prisjasznjokm Bestn. *Kiev. Politexn. In-Ta*, Kiev, 1959, **17**, 55.

13 Diffusion in metals

13.1 Introduction

In an isotropic medium the diffusion coefficient D^i of species i is defined through Fick's first law,

$$J^i = -D^i \operatorname{grad} c^i \tag{13.1}$$

J^i is the instantaneous net flux of species i, or diffusion current per unit area, and $\operatorname{grad} c^i$ is the gradient of the concentration c^i of i. If J and c are measured in terms of the same unit of quantity (e.g. J in $\mathrm{g\,cm^{-2}\,s^{-1}}$, c in $\mathrm{g\,cm^{-3}}$), D has the dimensions $(L^2 T^{-1})$. It has usually been expressed as $\mathrm{cm^2\,s^{-1}}$, although the units $\mathrm{m^2\,s^{-1}}$ are becoming more common. Generally, D depends on the concentration.

That matter is to be conserved at each point leads to Fick's second law,

$$\frac{\partial c^i}{\partial t} = \operatorname{div}(D^i \operatorname{grad} c^i) \tag{13.2}$$

giving the rate of the change of concentration with time to which diffusion gives rise.

The fluxes J^i are referred, at least for practical purposes, to axes fixed in the volume of the sample; but volume changes which take place as a result of diffusion lead to some ambiguity in the definition of such axes. Means have been proposed[3,6] for avoiding this by using axes scaled to the volume changes, but little use is made of these and it is more usual in accurate work to restrict the range of concentration employed so that volume changes are small or negligible.

When the concentration varies along only one direction, say the x axis, (13.1) and (13.2) become

$$J^i = -D^i \frac{\partial c^i}{\partial x} \tag{13.3}$$

$$\frac{\partial c^i}{\partial t} = \frac{\partial}{\partial x}\left(D^i \frac{\partial c^i}{\partial x}\right) \tag{13.4}$$

If, furthermore, D is independent of composition, and so also of position in the sample, (13.4) becomes

$$\frac{\partial c^i}{\partial t} = D^i \frac{\partial^2 c^i}{\partial x^2} \tag{13.5}$$

In anisotropic media diffusion rates vary with direction. In general, the diffusion flux is in the same direction as $\operatorname{grad} c$ *only* when $\operatorname{grad} c$ is along one of a set of orthogonal axes known as the 'principal axes of diffusion'. (These always coincide with axes of crystallographic symmetry so there is no difficulty in identifying them, except in cases of symmetry lower than orthorhombic.) For diffusion along principal axes equations like (13.3) may still be written.

$$\left.\begin{aligned} J^i_x &= -D^i_x (\partial c^i / \partial x) \\ J^i_y &= -D^i_y (\partial c^i / \partial y) \\ J^i_z &= -D^i_z (\partial c^i / \partial z) \end{aligned}\right\}$$

D_x, D_y and D_z are called 'principal coefficients of diffusion'.

In general grad c and J are not in the same direction. However, if l, m, n are the direction cosines of grad c then a diffusion coefficient for this direction may be defined as the ratio of the component of J along (l, m, n), divided by grad c. This is

$$D_{lmn} = l^2 D_x + m^2 D_y + n^2 D_z \tag{13.6}$$

Thus anisotropic diffusion can be completely described in terms of the three principal diffusion coefficients. In uniaxial crystals (tetragonal, trigonal, hexagonal) symmetry dictates, if the z axis is the unique axis, then $D_x = D_y$. Thus D is the same for all directions perpendicular to the unique axis and is often denoted $D_\perp$. D_z is then denoted as $D_\parallel$. Equation (13.6) may then be written

$$D_\theta = \sin^2 \theta \cdot D_\perp + \cos^2 \theta \cdot D_\parallel \tag{13.7}$$

where $\cos \theta \equiv n$.

Equations (13.4) and (13.5) still hold for anisotropic diffusion, with D given by (13.6) and (13.7).

Equation (13.1) provides a formal definition of a diffusion coefficient as the ratio of J^i to grad c^i. It also assumes that J^i is determined only by grad c^i. In the very large majority of diffusion measurements that have been made this holds true so that the above simple equations provide an adequate description of the diffusion process taking place. Such measurements are of three main types and these are discussed first and the nature of the diffusion coefficients they entail. They are:

1. Measurements which entail diffusion under a chemical concentration gradient (Chemical Diffusion Measurements—Table 13.4).

    (i) Diffusion of a single interstitial solute into a pure metal.
    (ii) Interdiffusion of two metals which form substitutional solid solutions (or interdiffusion between two alloys of the two metals).

2. Measurements which entail diffusion in essentially chemically homogeneous systems. These are possible through the use of radioactive or stable isotope tracers.

The diffusion of an interstitial solute in a pure metal [1(i)] is described by a single equation like (13.1) and the D has a simple and well-defined physical significance as describing diffusion of solute relative to the solvent lattice.

The same is true for the D for diffusion into a metal or alloy of any radioactive or stable tracer. The methods employed (*see* below) require such extremely small amounts and gradients of tracer that the system remains chemically homogeneous during diffusion. Any diffusion of other constituents is altogether negligible so that D refers simply to the diffusion of the tracer species relative to the solvent lattice.

For the interdiffusion of two metals or alloys [1(ii)] the situation is a little less simple. There would appear to be two diffusion coefficients required, one for each species, but *referred to volume fixed axes* these are equal because grad $c_1 = -$grad c_2 and J_1 must be equal and opposite to J_2. Again a single equation like (13.1) suffices to describe the diffusion process and the single D refers to the diffusion rate of either species relative to these axes. It is called the *chemical interdiffusion coefficient* and usually denoted $\tilde{D}$ (Table 13.4).

For many practical purposes $\tilde{D}$ is an adequate measure of the diffusion behaviour of a binary substitutional system. But of more fundamental physical interest are the rates of diffusion of the two species relative to local lattice planes. It is well established that generally these rates are not equal in magnitude. There is therefore a net total flux of atoms across any lattice plane, and if the density of lattice sites is to be conserved each plane in the diffusion zone must shift to compensate for this imbalance of the fluxes across it. At the same time lattice sites are created on one side of the sample and eliminated at the other, processes which are achieved by the creation and annihilation of vacancies. This shift of lattice planes, known as the Kirkendall effect, is observed experimentally as a movement of inert markers, usually fine insoluble wires, incorporated into the sample before diffusion. It is clear, then, that diffusion occurs on a lattice which locally is moving relative to the axes with respect to which $\tilde{D}$ was calculated. To provide a more complete description of binary substitutional diffusion it is therefore necessary to introduce diffusion coefficients D_A and D_B to describe diffusion of the two species relative to lattice planes. It is easy to show that these are related to $\tilde{D}$ by the equation

$$\tilde{D} = N_A D_B + N_B D_A \tag{13.8}$$

where N_A and N_B are the fractional concentrations of A and B. D_A and D_B, which are of more direct physical interest than $\tilde{D}$, are known as the *intrinsic* or *partial chemical diffusion coefficients*.

The velocity v of a marker is given by

$$v = (D_A - D_B)\partial N_A / \partial x, \tag{13.9}$$

where $\partial N_A / \partial x$ is the concentration gradient at the marker; so in principle D_A and D_B can be calculated separately when $\tilde{D}$ and v have been measured. In practice this is done usually only for markers placed at the original interface between the two interdiffusing metals or alloys: in this case a measurement of the *displacement* x_m of the marker after time t allows v to be obtained simply, for $v = x_m/2t$.

Equations (13.8) and (13.9) assume no net volume change and a compensation of the flux difference which is complete and which occurs by bulk motion along only the diffusion direction. These conditions are rarely met fully in practice, as is seen from the frequent occurrence of lateral changes in dimensions and of a porosity in the *side* of the diffusion zone suffering a net loss of atoms. This porosity, attributed to vacancies precipitating instead of being eliminated at sinks, suggests abnormal vacancy concentrations may be present in the diffusion zone. Because it is difficult to take into account the effect these abnormal conditions in the diffusion zone may have on the calculated values of $\tilde{D}$ and v, and hence on D_A and D_B, chemical interdiffusion experiments may provide results of limited accuracy and, for theoretical purposes, of limited significance: their effect is of course smaller the smaller the concentration gradients employed.

By contrast, radioactive tracer methods altogether avoid these difficulties and uncertainties associated with diffusion in a chemical gradient, and so are preferred in any investigation with a theoretical objective. They have the further advantage that the diffusion coefficients of each specie of an alloy can be determined separately and directly, rather than through any composite coefficient like $\tilde{D}$. These are referred to as *tracer diffusion coefficients* (Table 13.3) and will be denoted D_A^*, D_B^* etc. to distinguish them from the partial chemical diffusion coefficients D_A and D_B determined by chemical diffusion methods.[†]

Results on the diffusion coefficient D_A^* in very dilute alloys AB containing small concentrations C_B of B are frequently represented in terms of the solvent enhancement factors b_1, b_2, etc., in the equation

$$D_A^*(C_B) = D_A^*(C_B = 0)(1 + b_1 C_B + b_2 C_B^2 + \cdots) \tag{13.9a}$$

$D_A^*(C_B = 0)$ is of course just the self-diffusion coefficient of pure A.

A similar relation describes the diffusion of the solute B in A

$$D_B^*(C_B) = D_B^*(C_B = 0)(1 + B_1 C_B + B_2 C_B^2 + \cdots) \tag{13.9b}$$

B_1, B_2 are the solute enhancement factors and $D_B^*(C_B = 0)$ is the tracer impurity diffusion coefficient of B in A.

Except at vanishingly small concentrations of A, D_A and D_A^* differ fundamentally because the presence of the chemical concentration gradient under which D_A is measured imposes on the otherwise random motion of the atoms a bias, which makes atoms jump preferentially in one direction along the concentration gradient. Simple thermodynamic considerations lead to the relation

$$D_A = D_A^* \left(1 + \frac{\partial \ln \gamma_A}{\partial \ln N_A}\right) \tag{13.10}$$

between a partial chemical D_A and the corresponding tracer D_A^* measured at the same concentration. γ_A is the activity coefficient of A. In a binary system the bracket term is the same for both species (Gibbs–Duhem relation). Thus

$$\frac{D_A}{D_A^*} = \frac{D_B}{D_B^*} \tag{13.11}$$

(13.10) and (13.11) are approximate forms of more elaborate theoretical expressions, but are reasonably well obeyed experimentally. *See*, for example, references 4 and 7.

When D_A^* is measured at the extremely small concentrations of A that tracer methods permit (by diffusion of tracer into pure metal B it is called the *tracer impurity diffusion coefficient of A in B*

[†] D_A^* and D_B^* are sometimes referred to as the *self-diffusion coefficients of the alloy*. This is a perfectly acceptable alternative terminology. But there is a tendency nowadays to employ the term 'self' to the extent of describing tracer impurity diffusion coefficients as impurity self-diffusion coefficients. This latter term is ambiguous and misleading and its use is to be discouraged.

(Table 13.2). Such coefficients are of especial theoretical interest because of the particularly simple type of diffusion they describe (*see* footnote).

Finally, tracer methods are used as the commonest means of measuring *self-diffusion coefficients* in pure metals (Table 13.1, 13.5 and 13.6). By self-diffusion is meant the diffusion of a species in the pure lattice of its own kind.

For chemical diffusion in systems of more than two components, equation (13.1) and those following are inadequate. Experimentally it is found that when three or more components are present a concentration gradient of one species can lead to a diffusion flow of another, even if this is distributed homogeneously to start with. To cater for such cases Fick's first law is generalised by writing

$$J_i = \sum_{j=1}^{N} D_{ij} \frac{\partial c_j}{\partial x} \quad (j = 1, 2, \ldots, N) \tag{13.12}$$

But if there are *n* interstitial and $N-n$ substitutional components, and if the J_i are referred to volume-fixed axes then the relations

$$\sum_{j=n+1}^{N} J_i = 0 \quad \text{and} \quad \sum_{j=n+1}^{N} \partial c_j / \partial x = 0$$

allow (13.12) to be rewritten

$$J_i = \sum_{j=1}^{N-1} D_{ij} \frac{\partial c_j}{\partial x} \quad (j = 1, 2, \ldots, N - 1) \tag{13.13}$$

so that $(N - 1)^2$ coefficients suffice to describe the diffusion behaviour. The analogue of Fick's second law is

$$\frac{\partial c_i}{\partial t} = \sum_{j=1}^{N-1} \frac{\partial}{\partial x} \left(D_{ij} \frac{\partial c_j}{\partial x} \right) \tag{13.14}$$

These equations have been applied to a few ternary systems.[7,2]

It is possible to show from the principles of irreversible thermodynamics that not all the D_{ij} are independent and that a total of only $N(N - 1)/2$ coefficients are in fact sufficient to describe diffusion in an N-component system. No measurements in metals have employed this reduced scheme of coefficients, for to do so requires a knowledge of the thermodynamic properties of the system that is rarely available.

13.2   Methods of measuring D

13.2.1   Steady-state methods

These are based directly on Fick's first law. The usual procedure is to maintain concentrations of diffusant on the opposite sides of a sample, which is usually a thin sheet or a thin-walled tube, and to measure the resulting steady rate of flow J. This is generally practicable only when the diffusing element is a gas or can be supplied to and removed from the sample through a vapour phase. If the surface concentrations c_1 and c_2 in equilibrium with the ambient atmospheres are known, an average D over the concentration range is, for a sheet of thickness t for example, simply $\bar{D} = Jt/(c_1 - c_2)$ (Method Ib). Alternatively, if the steady concentration distribution across the sample is determined, $D(c)$ may be calculated from $D = -J(\partial c/\partial x)^{(-1)}$ (Method Ia). D may also be calculated from measurements of the time required to reach a steady state (Method Ic). These methods are used for measuring D only for interstitial solute diffusion: the Kirkendall effect complicates any attempt to apply it reliably to substitutional diffusion.

13.2.2 Non-steady-state methods

As a result of change in the concentration distribution in a sample D is deduced from a solution of Fick's second law [equations (13.2), (13.4), (13.5) or (13.14) appropriate to the conditions of the experiment]. There are three common types of experimental arrangements, two of which are usually employed in chemical diffusion coefficient measurements, the third in measurements of tracer diffusion coefficients.

(i) DIFFUSION COUPLE METHOD

Two metals, or two different homogeneous alloys of concentrations c_1 and c_2, are brought into intimate contact across a plane interface, say by welding. Diffusion is allowed to take place by annealing at a constant temperature for a time t. The distribution of concentration in the sample is then determined in some convenient manner, often by removal and subsequent analysis of a succession of thin layers cut parallel to the initial interface. It is usually arranged that the two halves of the couple be sufficiently thick that the diffusion zone does not extend to either end.

D generally varies with concentration, but no analytic solutions of (13.4) are available. So, a graphical method of analysis known as the Matano-Boltzmann method is employed. The concentration c is plotted against x and $D(c)$ determined graphically from

$$D(c) = (2t \cdot \partial c/\partial x)^{-1} \int_c^{c_1} x \, dc \quad \text{[Method IIa(i)]} \tag{13.15}$$

The origin of x is located by the condition

$$\int_{c_1}^{c_2} x \, dc = 0$$

and this may be shown to coincide, under ideal conditions, with the *initial* position of the interface between the two members of the couple. Thus it is $\tilde{D}$ which is measured in substitutional diffusion. Markers inserted at the interface locate its *final* position after diffusion. It has already been mentioned that measuring their displacement x_m from $x = 0$ allows the partial diffusion coefficients to be calculated.

If D varies little in the range c_1 to c_2, and this is often so if the range is sufficiently restricted, equation (13.5) may be used. The solution for this case is

$$\frac{c - c_2}{c_1 - c_2} = \frac{1}{2} \left\{ 1 - erf \left[\frac{x}{2\sqrt{(Dt)}} \right] \right\} \tag{13.16}$$

with $x = 0$ defined as before; D can then be calculated directly by a 'least squares' fit of the $c \sim x$ data to this or other appropriate equations [Method IIa(ii)].

Occasionally, the diffusion couple method is used to measure self-diffusion coefficients, one half of the couple being normal metal, the other enriched in one of its active or normal isotopes. It may also be used to measure diffusion coefficients in liquids (Shear-cell method).

With analytic solution like (13.16), D can be calculated by measuring c at one position only. This is sometimes done but it is not to be expected that values derived in this way will be as reliable as when derived from a complete $c \sim x$ curve (Method IIb).

The concentration range in a diffusion couple may span any number of phase regions in the equilibrium diagram of the system; the diffusion zone then consists of phase layers with concentration discontinuities across each boundary between two layers. In such cases equation (13.15) [Method IIa(i)] is still applicable. If D is assumed constant, analytic solutions are available and with these it is sometimes possible (Method IIc) to determine D from measurements only of the rates of movement of one or more phase boundaries and knowledge of the equilibrium concentrations at the boundary.[2,3,7]

(ii) IN-DIFFUSION AND OUT-DIFFUSION METHODS

Material is allowed to diffuse into, or out of, an initially homogeneous sample of concentration c_1 under the condition that the concentration at the surface is maintained at a constant and known value

c_0 by being exposed to a constant ambient atmosphere. c_1 is usually zero for in-diffusion experiments and so is c_0 for out-diffusion experiments.

D may be calculated from a measurement either of the total amount of material taken up by or lost from the sample (Method IIIb), or of the concentration distribution within the sample after diffusion (Method IIIa). The first method gives an average D over the range c_1 to c_0. For the second, equation (13.15) can be used again to give $D(c)$ or, if D is constant, it may be calculated from an appropriate analytic solution.

When the loss (or gain) of material from the sample entails the movement of a phase boundary, D can again be calculated from the rate of movement (Method IIIc). This method has been mostly used for interstitial solute diffusion, but also occasionally for substitutional diffusion measurements in systems with a sufficiently volatile component. A disadvantage is that conditions at the surface may not always be under adequate control so that c_0 is either ill-defined or not constant or both, with consequent uncertainty in D.

A common method of measuring liquid self-diffusion rates employs a type of out-diffusion method. A capillary tube, closed at one end and containing activated material, is immersed open-end uppermost in a large bath of inactive material. After the diffusion anneal the depleted activity content of the capillary is determined, and D calculated on the assumption that diffusion of the active species out of the tube is subject to zero concentration being maintained at the exit.

For determining the concentration distribution $c(x)$ in any of the above chemical diffusion methods a wide variety of techniques has been employed; these include the traditional methods of chemical and spectrographic analysis, X-ray and electron diffraction, X-ray absorption, microhardness measurements and so forth and the more recent, and often highly sensitive, methods of microprobe analysis, laser and spark source mass spectrometry nuclear reaction analysis and Rutherford back-scattering. Also may be mentioned are electrochemical methods where the diffusion sample is contrived as an electrolytic cell wherein diffusion fluxes may be measured as currents and/or surface concentrations as surface potentials. In this edition of the tables the method of analysis is not recorded for it is probably of less importance in assessing the reliability of a result than other features of the experimental procedure.

(iii) THIN LAYER METHODS

These are used now almost exclusively for the measurement of self and of tracer Ds. A very thin layer of radioactive diffusant, of total amount g per unit area, is deposited on a plane surface of the sample, usually by evaporation or electrodeposition. After diffusion for time t the concentration at a distance x from the surface is

$$c(x) = \frac{g}{(\pi Dt)^{1/2}} \exp\left(-\frac{x^2}{4Dt}\right) \tag{13.17}$$

provided the layer thickness is very much less than $(Dt)^{1/2}$. This condition is easy to satisfy because extremely small quantities suffice for studying the diffusion on account of the very high sensitivity of methods of detecting and measuring radioactive substances. For the same reason there is a negligible change in the chemical composition of the sample, so D is constant and equation (13.5), of which (13.17) is the solution for this case, is applicable.

After diffusion the activity of each of a series of slices cut from the sample may be determined and D calculated from the slope ($=1/4Dt$) of the linear plot of log activity in each slice against x^2 [Method IVa(i)]. Alternatively, such a plot may be constructed from intensity measurements made on an autoradiograph of a single section cut along or obliquely to the diffusion direction [Method IVa(ii)].

Another method is to calculate D from measurements made, after the removal of each slice, of the residual activity emanating from each newly exposed surface *of the sample*. [Residual activity method; Method IVb.]

Or, D may be determined by comparing the total activity from the surface $x = 0$ after diffusion with the original activity at $t = 0$ (surface decrease, Method IVc).

Methods IVb and IVc require an integration of equation (13.17). They are generally regarded as less reliable in principle than Method IVa because they necessitate a knowledge of the absorption characteristics of the radiation concerned. In addition Method IVc is particularly susceptible to errors arising from possible oxidation and from evaporation losses of the deposited material and so is rarely used nowadays.

A recent development has been the use of the electron-microprobe, and similarly sensitive methods (*see* above), to measure even impurity diffusion coefficients: instruments are now available with a sensitivity adequate to monitor diffusion from deposited layers of *inactive* diffusant thin enough to meet the requirements for use of equation (13.17) [Method IVa(iii)].

13.2.3 Indirect methods, not based on Fick's laws

In addition to macroscopic diffusion there are a number of other phenomena in solids which depend for their occurrence on the thermally activated motion of atoms. From suitable measurements made on some of these phenomena it is possible to determine a D. The more important of these are:[6,15]

1. Internal friction due to a stress-induced redistribution of atoms in interstitial solution in metals (Snoek effect and Gorsky effect, Method Va.)
2. A similar phenomenon occurring in substitutional solid solution and due, it is believed, to stress-induced changes in short range order (Zener effect, Method Va.)
3. Phenomena associated with nuclear magnetic resonance absorption, especially the 'diffusional narrowing' of resonance lines and a contribution, arising from atomic mobility, to the spin-lattice relaxation time T_1 (Method Vb).
4. Some magnetic relaxation phenomena in ferromagnetic substances (Method Vc).
5. The width of Mössbauer spectrum lines (Method Vd).
6. The intensity and shape of quasi-elastic neutron scattering spectra (Method Ve).

These methods are associated with atomic motion over only a few atomic distances, and so have the advantage of providing measurements of D at temperatures lower than are often practicable by conventional methods. However, some of them are of very limited application. For example, 3, 5, and 6 are obviously limited to diffusion of appropriate nuclei only. Since in every case measurements are made in homogeneous material the diffusion coefficients obtained are of the nature of tracer rather than chemical diffusion coefficients.

Most measurements of D are conducted at a series of temperatures so as to provide values of the constants A and Q occurring in the Arrhenius equation

$$D = A \exp\left(-Q/RT\right) \tag{13.18}$$

which usually describes very well the observed temperature dependence.* A is called the 'frequency factor' and Q the activation energy. Wherever possible, experimental measurements are reported in the tables in terms of A and Q alone. Occasionally, accurate measurements, particularly if over an extended temperature range, reveal slightly curved Arrhenius plots. These can usually be well represented by the sum of the Arrhenius terms

$$D = A_1 \exp\left(-Q_1/kT\right) + A_2 \exp\left(-Q_2/kT\right) \tag{13.19}$$

For such measurements A_2 and Q_2 are tabulated immediately below A_1 and Q_1. In Tables 13.3 and 13.4 A_2 and Q_2 are preceded by the sign $+$. (*see* e.g. CoGa in Table 13.3.)

Experiments may be made by any of the above methods either with single crystal or polycrystalline material. With polycrystals there is, in addition to diffusion through the grains (volume diffusion), diffusion at a more rapid rate locally through the disordered regions of grain boundaries. This can, however, be reduced to a negligible proportion of the whole by using large grain material and by working at relatively high temperatures because, since $Q_{gb} < Q_v$, grain boundary diffusion rates increase less rapidly with temperature than do volume diffusion rates. Obviously single crystals are to be preferred in accurate measurements of what is intended to be volume diffusion, but even in their case there may be, at too low temperatures, a contribution to D from diffusion along dislocations. Measured values of D will then tend to be above the values expected from an extrapolation of the high temperature data using (13.18) and when they do so to a noticeable extent are often discarded in estimating Q and A.

From measurements of the concentration distribution around a grain boundary—usually in a bicrystal into which material diffuses parallel to the boundary—a product $D'\delta$ may be deduced.[9,10] D' is the coefficient for diffusion *in* the boundary of width δ, an uncertain quantity generally assumed to be $\delta = 5.0 \times 10^{-8}$ cm. $D'\delta$ is found to depend on the orientation of the boundary and on the direction of diffusion within it. For an extensive range of data on $D'\delta$s, *see* reference 10.

---

* This is often true even of $\tilde{D}$, because Q_A and Q_B for the partial diffusion coefficients do not seem to differ very much.

13.3 Mechanisms of diffusion

Most theoretical discussions of diffusion are concerned with an understanding of A and Q rather than of D itself. On the basis of theoretical calculations of Q for various possible mechanisms of diffusion and comparison with observed values, it has been supposed for some time that in metals atoms diffuse substitutionally by thermally activated jumps into vacant lattice sites, i.e. by the 'vacancy mechanism'. This has been very convincingly confirmed, at least for f.c.c. metals, by thermal expansion and quenching experiments. At high temperatures, towards the melting point, there is believed to be a sizeable contribution to D from diffusion by bound divacancy pairs. While the same mechanism is usually thought to operate in most metal structures, there is considerable doubt at present whether this is in fact true for a number of so-called 'anomalous b.c.c. metals'— β-Ti, β-Zr, β-Hf, β-Pr, γ-U and δ-Ce—or at least whether the vacancy mechanism is the only one operating in their case. It is also believed that the noble metals and other low-valent solutes (Group II), plus the later transition elements, may dissolve interstitially, at least in part, and diffuse by an interstitial-type process in the alkali metals, in the high-valent Group III and IV elements and also in the *early* members of each of the transition groups, the lanthanide series and the actinide series of elements. This belief stems from the anomalously very large diffusion rates of these solutes in these solvents.[11,13,14]

REFERENCES

Textbooks

1. P. G. Shewmon, 'Diffusion in Solids', McGraw-Hill, New York, 1963.
2. W. Jst, 'Diffusion in Solids, Liquids and Gases', 2nd edn, Academic Press, New York, 1964.
3. Y. Adda and J. Philibert, 'Diffusion dans les Métaux', Presse Universitaire, Paris, 1966.
4. J. Manning, 'Diffusion Kinetics for Atoms in Crystals', Van-Nostrand Co. Inc., Princeton, New York, 1968.
5. C. P. Flynn, 'Point Defects and Diffusion', Clarendon Press, Oxford, 1972.
6. J. Crank, 'The Mathematics of Diffusion', 2nd edn, Clarendon Press, Oxford, 1975.
7. J. Philibert, 'Diffusion et Transport de Matière dans les Solides', Les Editions de Physique, Les Ulis Cedex, France, 1985.
8. R. J. Borg and G. J. Dienes, 'An Introduction to Solid State Diffusion', Academic Press, New York, 1988.
9. I. Kauer and W. Gust, 'Fundamentals of Grain and Interface Boundary Diffusion', Zeigler Press, Stuttgart, 1988.
10. I. Kaur, W. Gust and L. Kozma, 'Handbook of Grain Boundary and Interphase Boundary Diffusion Data', Ziegler Press, Stuttgart, Germany, 1989.

Reviews

11. Various papers in 'Diffusion in B.C.C. Metals' (eds J. A. Wheeler and F. R. Winslow), American Soc. Metals, 1965.
12. N. L. Peterson, *Solid St. Physics*, **22**, 409, Academic Press, 1968.
13. Various papers in 'Diffusion in Solids—Recent Developments', (eds A. S. Nowick and J. J. Burton), Academic Press, New York, 1975.
14. Various papers in *J. Nuclear Materials*, **69/70**, 1978.
15. Various Papers in DIMETA-82, Proc. Intl. Conf., Diffusion in Metals and Alloys, Tihany, 1982, Trans. Tech. Publs, Switzerland, 1983.
16. Various papers in 'Diffusion in Crystalline solids', (eds G. E. Murch and A. S. Nowick), Academic Press, New York, 1984.
17. 'Non-Traditional methods in diffusion' (eds G. E. Murch, H. K. Birnbaum and J. R. Cost), Metallurgical Soc. A.I.M.E., 1984.
18. Various papers in DIMETA-88, Proc. Intl. Conf., Diffusion in Metals and Alloys, Balotonfured, Hungary, 1988. Defect and Diffusion forum, 1990 **66/69**, pp. 1–1551, (eds F. J. Kedeves and D. L. Beke), Sci-Tech. Publs, Liechenstein and U.S.A.
19. 'Diffusion and Defect Forum'—previously 'Diffusion and Defect Data'—a data abstract and review journal published periodically by Sci-Tech. Publs, Liechtenstein and U.S.A.
20. 'Diffusion in Metals and Alloys', Landolt and Bornstein Critical Tables, Springer-Verlag, Berlin, 1991.

Summary of methods for measuring D

STEADY-STATE METHOD with

I.	(a)	Measurement of concentration distribution within the sample or,	Ia
	(b)	Average gradient calculated from c_1 and c_2 as deduced from equilibrium data or,	Ib
	(c)	Time-delay method (measurement of time to reach steady state)	Ic

NON-STEADY METHODS

II. *Diffusion couple methods*
    (a) With determination of $c \sim x$ curve and
        (i) Use of Matano-Boltzmann analysis to give $D(c)$ IIa(i)
        (ii) When it is evident (or assumed) that D is effectively constant,
             calculation of D from an analytic solution   IIa(ii)
        (iii) When it is evident that D is *not* constant and an analytic solution is used
             to calculate a D corresponding to each value of c—giving
             an approximate $D(c)$ IIa(iii)
    (b) D calculated from a single concentration measurement   IIb
    (c) D calculated from an analytic solution, assuming D constant, using
        measurements of rate of movement of phase boundaries and knowledge of
        equilibrium concentrations on the boundaries   IIc

III. *In-diffusion and out-diffusion methods* in-(i) out-(ii)
    (a) D calculated from $c \sim x$ curves   IIIa
    (b) D calculated from total gain or loss, or rate thereof   IIIb
    (c) D calculated from rate of phase boundary movement   IIIc

IV. *Thin layer methods*
    (a) With measurement of $c \sim x$ curve
        (i) By sectioning and counting—using radioactive diffusant   IVa(i)
        (ii) By autoradiography—using radioactive diffusant   IVa(ii)
        (iii) By electron-microprobe or similarly sensitive method—using
             non-radioactive diffusant   IVa(iii)
    (b) Residual activity method—using radioactive diffusant   IVb
    (c) Surface decrease method—using radioactive diffusant   IVc

V. *Indirect methods*
    (a) By internal friction   Va
    (b) By nuclear magnetic resonance   Vb
    (c) By ferromagnetic relaxation   Vc
    (d) From Mössbauer line spectra   Vd
    (e) From quasi-elastic neutron scattering   Ve

Notes on the tables

1. All measurements are reported whenever possible in terms of A and Q (*see* equation 13.18) or of $A_1 \cdot Q_1$, A_2 and Q_2 (equation 13.19). A in $cm^2 \, s^{-1}$: Q in $kJ \, mol^{-1}$: ($R = 8.31441 \, J \, mol^{-1} K^{-1}$ $eV = 96.4846 \, kJ \, mol^{-1}$).
2. The 'temperature range' is the range over which measurements were used to calculate A and Q. Extrapolation too far outside this range may not in some cases give reliable values for D.
3. All alloy concentrations are in atomic percentages unless otherwise stated. Purity of material is as quoted and is presumably in weight percentages, although this is not always stated explicitly in papers.
4. s.c. = single crystals; p.c. = polycrystals.
5. In Table 13.4 a single concentration denotes the concentration at which $D(c)$ was determined. Two concentrations separated by a hyphen denote the range of concentration over which measurements were made. Where this is followed by a single D value, or a single set of A and Q values, it is also the concentration range over which these values are averages.
6. Bold type in Table 13.4. This is used: (1) To indicate the species to which the D's, or A and Q values, refer in cases where there might be ambiguity–usually for interstitial solid solutions. Where there is no bold type the data refers to the interdiffusion coefficients of the first two substitutional species. (2) To indicate which component was used in the vapour phase in experiments employing methods I and III.
7. Where several measurements exist an attempt has been made to select what appear to be the most reliable one or two. Mostly these are later measurements and references to earlier work can usually be found by consulting the references quoted.

Table 13.1   SELF-DIFFUSION IN SOLID ELEMENTS

Element	A $cm^2 s^{-1}$	Q $kJ\,mol^{-1}$	Temp. range K	Method	Ref.
Group 1A					
Li	0.125	53.06	308–451	IVa(i), p.c., normal Li. ($\sim$8% Li^6)[a]	1 and 2
	$A_1 = 0.196$ $A_2 = 98$	$Q_1 = 53.7$ $Q_2 = 77.16$	227–451	Vb, p.c., Li^8 diffusion in Li[7(d)]	3
Na	$A_1 = 0.72$ $A_2 = 57 \times 10^{-4}$	$Q_1 = 48.15$ $Q_2 = 35.71$	195–371	IVa(i), p.c., Na^{22}	4
	0.12	41.5	349–371	Ve, s.c. and p.c.	102
K	0.16	39.19	221–335	IVa(i), p.c., K^{42}, 99.97%	5
Rb	0.23	39.4	250–313	Vb, p.c.	6
Group 1B					
Cu	$A_1 = 0.13$ $A_2 = 4.5$	$Q_1 = 197.8$ $Q_2 = 237.4$	573–1 334	IVa(i), s.c., Least squares fit to data of references[7]	8
Ag	$A_1 = 0.055$ $A_2 = 15.1$	$Q_1 = 171.1$ $Q_2 = 226.7$	550–1 228	IVa(i), s.c., Least squares fit to data of references[9]	8
Au	$A_1 = 0.025$ $A_2 = 0.83$	$Q_1 = 164.3$ $Q_2 = 212.3$	603–1 333	IVa(i), s.c., Least squares fit to data of references[10]	8
Group IIA					
Be	$\|c$ 0.62 $\perp c$ 0.52	165.0 157.4	836–1 343	IVb, s.c., Be[7]	13
Mg	$\|c$ 1.0 $\perp c$ 1.5	134.8 136.1	741–908	IVa(i), s.c., Mg^{28}, 99.9+%	14
	$\|c$ 1.78 $\perp c$ 1.75	139.0 138.2	773–903	IVa(i) and b, s.c., Mg^{28}, 99.99%	15
Ca	8.3	161.2	773–1 073	IVb, p.c., Ca^{45}, 99.95%	16
Group IIB					
Zn	$\|c$ 0.13 $\perp c$ 0.18	91.7 96.3	513–691	IVa(i), s.c., Zn$^{65/69}$, 99.999%	17
Cd	$\|c$ 0.118 $\perp c$ 0.183	77.92 82.02	420–600	IVa(i), s.c., Cd^{109}, 99.999%	18
Group IIIA and rare earths					
Y	$\|c$ 0.82 $\perp c$ 5.2	252.5 280.9	1 173–1 573	IVb, s.c., Y^{91}	19
β-La	1.5	188.8	933–1 113	IVa(i), p.c., La^{140}, 99.97%	20
γ-La	0.013	102.6	1 140–1 170	IV(i), p.c., La^{140}	88
	0.11	125.2	1 151–1 183	IVa(i), p.c., La^{140}, 99.85%	89
γ-Ce	0.55	153.2	801–965	IVa(i), p.c., Ce^{141}, 99.9%	21
δ-Ce	0.012	90.0	992–1 044		
β-Pr	0.087	123.1	1 323–1 473	IVa(i), Pr^{142}, p.c., 99.97%	22
Er	$\|c$ 3.71 $\perp c$ 4.51	301.66 302.58	1 479–1 684	IVa(i), s.c., Er^{169}, 99.91%	23
Eu	1.0	144.4	771–1 072	IVa(i), p.c., Eu^{152}, 99.75%	90
β-Gd	0.01	136.9	1 549–1 581	IVa(i), p.c., Gd^{159}, 99.57%	90
α-Yb	0.034	146.79	823–983	IVa(i), p.c., YB^{169}, 99.5%	91
γ-Yb	0.12	121.0	1 031–1 083		

(*continued*)

Table 13.1   SELF-DIFFUSION IN SOLID ELEMENTS—*continued*

Element	*A* cm² s⁻¹	*Q* kJ mol⁻¹	*Temp. range* K	*Method*	*Ref.*
Group IIIB					
Al	1.71	142.4	723–923	IVa(i), Al[26], p.c., 99.9%	24
	2.25	144.4	573–923	IVa(i), p.c.	25
	0.137	123.5	298–581[c]	Vb, p.c., 99.999%	27
	0.176	126.4	358–482	Void annealing rate, 99.9999%	28
Ga	$D = 5.3 \times 10^{-13}$	282.8 ⎫			
	$D = 5.3 \times 10^{-13}$	293.0 ⎪			
	$D = 7.8 \times 10^{-13}$	298.0 ⎬		IVa(i), s.c. and p.c., Ga[72], 99.9999%	30
	$D = 9.3 \times 10^{-13}$	300.5 ⎪			
	$D = 42 \times 10^{-13}$	302.7 ⎭			
In	∥c 2.7 ⊥c 3.7	78.3 ⎫ 78.3 ⎬	317–417	IVa(i), s.c., In[144], 99.995%	31
α-Tl	∥c 0.4 ⊥c 0.4	95.9 ⎫ 94.6 ⎬	423–498	IVa(i), s.c., Tl[204], 99.9+%	32
β-Tl	0.42	80.18	513–573	IVa(i), p.c., Tl[204], 99.999%	12
Group IVB					
C	0.4–14.1	682[e]	2458–2620	IIIb, C[14], natural graphite crystals	33
	∥a 0.91	657	2667–3175	IVb, C[14], columnar pyrocarbon	42
Si	1460	484.4	1318–1663	IVa(i), s.c., Si[31]	34
	154	448.4	1128–1448	IVa(iii) (IMS), s.c., Si[30]	44
Ge	24.8	303	822–1164	IVa(i), s.c., Ge[71], intrinsic Ge	36
	13.6	298	808–1177	IVa(i), s.c., intrinsic Ge	37
Sn	∥c 12.8 ⊥c 21.0	108.9 ⎫ 108.4 ⎬	425–500	IVa(i), s.c., Sn[113], 99.999%	38
	∥c 7.7 ⊥c 10.7	107.2 ⎫ 105.1 ⎬	433–501	IVa(i), s.c., Sn[113], 99.999%	39
Pb	0.995	107.4	473–596	IVa(i), s.c., Pb[210], 99.9999%	40 and 41
Group VB					
P	$3.6 \times 10^{9[c]}$	114.7	295–316 (α-P)	IVa(i), s.c., P[33], 99.999%	87
Sb	∥c 56 ⊥c 0.10	201.0 ⎫ 149.9 ⎬	773–903	IVa(i), s.c., Sb[124], 99.9999%	43
Bi		*See* footnote *f*			
Group VIB					
S	∥c — ⊥c 2×10^{17}	~290 ⎫ 215.2 ⎬	353–368[g]	IVa(i), s.c., S[35]	47
Se	∥c 0.2 ⊥c 100	116 ⎫ 135 ⎬	425–488	IVb, s.c., Se[75]	48
Te	∥c 130 ⊥c 39.1×10^{2}	168.52 ⎫ 195.48 ⎬	578–673	IVa(i), s.c., Te[127m], 99.9999%	49
	∥c 0.6 ⊥c 20	148 ⎫ 166 ⎬	496–640	IVa(i), s.c., Te[127m]	95
Group IVA					
α-Ti	6.6×10^{-5}	169.1	1013–1149	IVb, p.b., Ti[44], 99.97%	50
β-Ti	$A_1 = 3.58 \times 10^{-4}$ $A_2 = 1.09$	$Q_1 = 130.6$ ⎫ $Q_2 = 251.2$ ⎬	1171–1813	IVa(i), p.c., Ti[44], 99.9%	52
	$D = 3.5 \exp[(-39.4 \times 10^{3}/T)$ $+(1.548 \times 10^{7}/T^{2})]$		779–1128	IVa(i), p.c., Ti[44], 99.98%	51

(*continued*)

Table 13.1   SELF-DIFFUSION IN SOLID ELEMENTS—*continued*

Element	A cm^2 s^{-1}	Q kJ mol^{-1}	Temp. range K	Method	Ref.
α-Zr	*See* Figure 13.1		506–855	IVa(i), s.c., Zr^{95}, 99.99%	63
β-Zr	$A_1 = 8.5 \times 10^{-5}$ $A_2 = 1.34$	$Q_1 = 115.97$ $Q_2 = 272.98$	1 173–2 023	IVa(i), p.c., Zr^{95}, 99.94%	53 and 35
α-Hf	$\|c$ 0.86 $\perp c$ 0.28	370.1 348.3	1 493–1 883	IVa(i), s.c., Hf^{181}, Hf + 2.1% Zr	56
β-Hf	1.2×10^{-3} 1.1×10^{-3}	162.0 159.1	2 068–2 268 2 000–2 850	IVa(i), p.c., Hf^{181}, 99.99% IVa(i), p.c., —	96 97
Group VA					
V	$A_2 = 0.025$ 0.288 173	$Q_2 = 302.9$ 334.68 409.27	1 153–2 810 997–1 815 1 815–2 115	IVa(i), s.c./p.c., V^{48}, 99.99% IVa(i), s.c., V^{48}, 99.9+%	57 and 11 98
Nb	$A_1 = 8 \times 10^{-3}$ $A_2 = 3.7$	$Q_1 = 349.6$ $Q_2 = 438.4$	1 353–2 693	IVa(i), s.c., Nb^{95}, >99.92%	58
Ta	0.124 0.21	413.2 423.6	1 523–2 493 1 261–2 893	IVa(i), Ta^{182}, p.c. IVa(i), Ta^{182}, s.c.	59 99
Group VIA					
Cr	1 280	441.9	1 073–2 090	IVa(i), s.c., Cr^{51}, 99.986%	60 and 100
Mo	$A_1 = 0.126$ $A_2 = 139$	$Q_1 = 437.5$ $Q_2 = 549.3$	1 363–2 723	IVa(i), s.c., Mo^{99}	61
W	$A_1 = 0.04$ $A_2 = 46$	$Q_1 = 526.3$ $Q_2 = 666.1$	1 703–3 413	IVa(i), s.c., W^{187}, 99.999%	62
Group VIII					
α-Fe	*See* Figure 13.2$^{(k)}$ 118 2.76 121	281.52 250.58 281.6	Ferromag 1 048–1 087 (p) 1 052–1 148 (p) 1 067–1 169 (p)	IVa(i), s.c. and p.c. IVb, p.c., Fe^{59}, 99.78% IVa(i), p.c., Fe^{59}, 99.995% IVa(i), p.c., Fe^{59}, 99.98%	80 and 92 93 80 71
γ-Fe	0.49 4.085	284.12 311.1	1 443–1 634 1 223–1 473	IVa(i), p.c., Fe^{59}, 99.98% IVc, p.c., Fe^{55}, 99.98%	75 74
δ-Fe	2.01 6.8	240.7 258.3	1 701–1 765 1 407–11 788	IVa(i), p.c. and s.c., Fe$^{55·59}$ IVa(ii) p.c., Fe^{55}, 99.998%	72 77
Co	0.55 2.54$^{(b)}$	288.5 304	896–1 745 923–1 743	IVa(i), p.c., Co$^{57.60}$, 99.99% IVz(i), s.c., Co^{57}, 99.999%	76 78
Ir	0.36	438.3	2 298–2 937	IVa(i), s.c., Ir^{192}	103
Ni	$A_1 = 0.85$ $A_2 = 1350$	$Q_1 = 276.7$ $Q_2 = 400.3$	879–1 673	IVa(i) and b, s.c., Ni^{63}. Least squares fit to data of references^{79,94}	8
Pd	0.205	266.2	1 323–1 723	IVa(i), s.c., Pd^{103}, 99.999%	81
Pt	$A_1 = 0.034$ $A_2 = 88.6$	$Q_1 = 254.5$ $Q_2 = 390.6$	773–1 998	IVa(i) and c, p.c., Pt^{195m}. Least squares fit to data of references^{82,83,78R}	8
Transuranic elements					
α-Th	395 10^2–5 $\times 10^3$	299.8 347	963–1 183 1 373–1 673	IVc, p.c., Th^{208} Vd, s.c., 99.9%	54 55
β-Th	10^4–10^6	415	1 723–1 823		
α-U	2×10^{-3} $\|110$ $\|010$	167.5 $\|001$	853–923	IIa(ii), p.c., enriched U^{234}	64
$D \times 10^{14} =$ $\begin{cases} 36.7 & \leq 1.5 \\ 19.5 & \leq 1 \\ 6 & \leq 0.35 \\ 3.67 & \leq 0.35 \end{cases}$		42.9 19.5 7.5 3.84	925.9 (MC)$^{(P)}$ 898.5 (MC)$^{(P)}$ 860.4 (MC)$^{(P)}$ 587.4 (PC)$^{(P)}$	IVa(i), s.c., U$^{233/235}$	65

(continued)

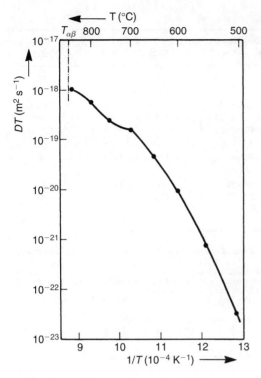

Figure 13.1   *Self diffusion coefficients of Zr*[63]

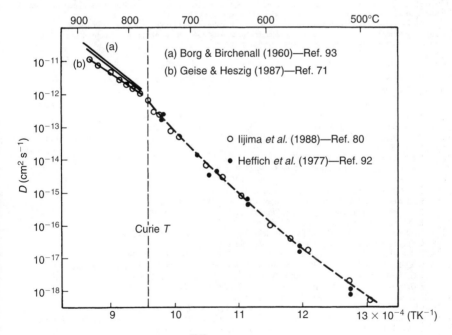

Figure 13.2   *Self diffusion coefficients of Fe*[80,92]

Table 13.1   SELF-DIFFUSION IN SOLID ELEMENTS—*continued*

Element	A cm^2 s^{-1}	Q kJ mol^{-1}	Temp. range K	Method	Ref.
β-U	1.35×10^{-2}	175.8	973–1 028	IIa(ii), p.c., enriched U^{234}	66
	2.8×10^{-3}	185.1	963–1 023	IVb, p.c., U^{235}	67
γ-U$^{(n)}$	1.19×10^{-3}	111.8	1 076–1 342	Various p.c.	68 and 69
	1.1×10^{-4}	150.7	1 123–1 323$^{(i)}$	IVb, p.c., 99.76%	70
	0.12	228.2	1 103–1 353$^{(j)}$		
ε-Pu	2.2×10^{-2}	77.5	773–895	IIa(ii), p.c., Pu^{240}	85
	4.5×10^{-3}	66.9	765–886	IVa(i), p.c., Pu^{233}	86
δ'-Pu	5.32×10^{32}	589	730–750		
δ-Pu	4.5×10^{-3}	99.6	623–713	IIa(ii), p.c., Pu^{238}	84
	0517	126.4	594–715	IVa(i), p.c., Pu^{238}	86
γ-Pu	0.38	118.4	484–564		
β-Pu	1.69×10^{-2}	108.0	408–454		

Notes:

(*a*) Reference 2 reports measurements of self-diffusion in Li of different isotopic compositions, from which are derived values of D for Li^6 and Li^7 diffusing in Li^6.

(*b*) Forced linear fit to slightly curved Arrhenius plot—equation 13–19.

(*c*) Although the higher Q values are confirmed by creep measurements, the indirect methods indicate lower values of Q at the lower temperatures at which they are applied. Collected results are discussed in references 29.

(*d*) Measures the 'macroscopic diffusion coefficient' D^{SD}. $D_{Tracer} = D^{SD} \times f$. f = the correlation factor ($\sim$0.5–0.7, depending on T). *See* reference 3 for details.

(*e*) Assumed in analysis of results that D in direction perpendicular to basal plane is negligible.

(*f*) *See* reference 45 for account of the highly anomalous self-diffusion behaviour of Bi and reasons for believing the often quoted results of Seith (reference 46) to be very suspect.

(*g*) Orthorhombic S crystals, $D_{||c}$ is 2 to 4 times less than $D_{\perp c}$.

(*h*) The results quoted from (68) are from a least squares fit to the three measurements of reference 69.

(*i*) Samples pre-annealed at diffusion temperature.

(*j*) Samples all pre-annealed at 1080°C.

(*k*) References 80 and 92 give expressions for D over the α range as a function of T and of the magnetisation.

(*l*) Measurements with P^{32} (reference 87) show larger D's due to radiation enhancement from the more energetic P^{32} radiation.

(*m*) These values are very close indeed to the values $A = 0.5$, $Q = 68.0$ previously chosen by Badia (76) as best representing the combined results of a number of other investigations over the range 1050/1400°C.

(*n*) $A = 1.27$, $Q = 67.2$ are given in reference 76 as the best fit to the combined results of references 79 and 80.

(*o*) Measurements with P^{32} (reference 42) show larger Ds due to radiation enhancement from the more energetic P^{32} radiation.

(*p*) MC refers to mosaic crystal and PC refers to perfect crystal.

REFERENCES TO TABLE 13.1

1. A. Ott, J. N. Mundy, L. Löwenberg and A. Loddin, *Z. Naturf.*, 1968, **23A**, 627.
2. A. Lodding, J. N. Mundy and A. Ott, *Phys. Status Solidi*, 1970, **38**, 559.
3. P. Heitjans, A. Körblein, H. Ackerman, D. Dubbers, F. Fujiwara and H. J. Stöckman, *J. Phys. F: Metal Phys.*, 1985, **15**, 41.
4. J. N. Mundy, *Phys. Rev. B*, 1971, **3**, 2431.
5. J. N. Mundy, T. E. Miller and R. J. Porte, *Phys. Rev. B*, 1971, 3, 2445.
6. D. F. Holcomb and R. E. Norberg, *Phys. Rev.*, 1955, **98**, 1074.
7. S. J. Rothman and N. L. Peterson, *Phys. Stat. Solidi*, 1969, **35**, 305.
   D. Bartdorff, G. Neumann and P. Reimers, *Phil. Mag.*, 1978, **38**, 157.
   K. Maier, C. Bassani and W. Schule, *Phys. Letters A*, 1973, **44**, 539.
   K. Maier, *Phys. Stat. Solidi (a)*, 1977, **44**, 567.
   N. Q. Lam, S. J. Rothman and L. J. Nowicki, *Phys. Stat. Solidi (a)*, 1974, **23**, K35.
8. G. Neumann and V. Tolle, *Phil. Mag. A*, 1986, **54**, 619.
9. S. J. Rothman, N. L. Peterson and J. T. Robinson, *Phys. Stat. Solidi*, 1970, **39**, 635.
   N. Q. Lam, S. J. Rothman, H. Mehrer and L. J. Nowicki, *Phys. Stat. Solidi (b)*, 1973, **57**, 225.
   J. G. E. M. Backus, H. Bakker and H. Mehrer, *Phys. Stat. Solidi (b)*, 1974, **64**, 151.
   J. Bihr, H. Mehrer and K. Maier, *Phys. Stat. Solidi (a)*, 1978, **50**, 171.
   G. Rein and H. Mehrer, *Phil. Mag. A*, 1982, **45**, 467.
10. Ch. Herzig, H. Eckseler, W. Bussmann and D. Cardis, *J. Nucl. Mat.*, 1978, **69/70**, 61.
    M. Werner and H. Mehrer, *DIMETA 82. Proc. Int. Conf.—Diffusion in Metals and Alloys, Tihany, 1982*, Trans. Tech. Publs, Switzerland (eds F. J. Kedves and D. L. Beke), 1983, p. 393.
11. H. M. Morrison, *Phil. Mag.*, 1975, **31**, 243.
12. R. Chiron and G. Faivre, *Phil. Mag. A*, 1985, **51**, 865.

13. J. M. Dupouy, J. Mathie and Y. Adda, *Mem. Scient. Revue Metall.*, 1966, **63**, 481.
14. P. G. Shewmon, *J. Metals, N. Y.*, 1956, **8**, 918.
15. J. Combronde and G. Brebec, *Acta Met.*, 1971, **19**, 1393.
16. L. V. Pavlinov, A. M. Gladyshev and V. N. Bykov, *Fiz. Metall.*, 1968, **26**, 823.
17. N. L. Peterson and S. J. Rothman, *Phys. Rev.*, 1967, **163**, 645.
18. C. Mao, *Phys. Rev.*, 1972, **5**, 4693.
19. D. S. Gorny and R. M. Altovski, *Fizica Metall.*, 1970, **30**, 85.
20. M. P. Dariel, G. Erez and G. M. J. Schmidt, *Phil. Mag.*, 1969, **19**, 1053.
21. M. P. Dariel, D. Dayan and A. Languille, *Phys. Rev.*, 1971, **B4**, 4348.
22. M. P. Dariel, G. Erez and G. M. J. Schmidt, *Phil. Mag.*, 1969, **19**, 1045.
23. F. H. Spedding and K. Shiba, *J. Chem. Phys.*, 1972, **57**, 612.
24. T. S. Lundy and J. F. Murdoch, *J. Appl. Phys.*, 1962, **53**, 1671.
25. M. Beyeler, *Thèse-Paris*, 1968; *J. Phys. (Fr.)*, 1968, **29**, 345.
26. S. L. Robinson and O. D. Sherby, *Phys. Status Solidi*, 1970, **al**, K199.
27. R. Messer, S. Dais and D. Wolf, *Proc. 18th Ampère Congress* (eds P. S. Allen, E. R. Andrew, C. A. Bates), Nottingham, 1974, **2**, 327.
28. T. E. Volin and R. W. Balluffi, *Phys. Status Solidi*, 1968, **25**, 163.
29. A. Seeger, D. Wolf and H. Mehrer, *Phys. Stat. Solidi B*, 1971, **48**, 481.
    S. Dais, R. Messer and A. Seeger, *Mat. Sci. Forum*, 1987, **15\18**, Pt. 1, 419.
30. A. C. Carter and C. G. Wilson *Br. J. Appl. Phys.*, 1968, **1**, 515.
31. J. E. Dickey, *Acta Met.*, 1959, **7**, 350.
32. G. A. Shirn, *Acta Met.*, 1955, **3**, 87.
33. M. A. Kanter, *Phys. Rev.*, 1957, **107**, 655.
34. H. J. Mayer, H. Mehrer and K. Maier, 'Rad. Effects in Semiconductors', 1976 (Inst. of Phys. Conference Series, 31), pp. 186–193.
35. G. V. Kidson, *Can. J. Phys.*, 1963, **41**, 1563.
36. G. Vogel, G. Hettich and H. Mehrer, *J. Phys. C: Sol. St. Phys.*, 1983, **16**, 6197.
37. M. Werner, H. Mehrer and H. D. Hochheimer, *Phys. Rev.*, 1985, **32**, 3930.
38. F. H. Huang and H. B. Huntington, *Phys. Rev. B*, 1974, **9**, 1479.
39. C. Coston and N. H. Nachtrieb, *J. Phys. Chem.*, 1964, **68**, 2219.
40. J. W. Miller, *Phys. Rev.*, 1969, **181**, 1095.
41. H. A. Resing and N. H. Nachtrieb, *J. Phys. Chem. Solids*, 1961, **21**, 40.
42. R. B. Evans, L. D. Love and E. H. Kobisk, *J. Appl. Phys.*, 1969, **40**, 3058.
43. H. Cordes and K. Kim, *J. Appl. Phys.*, 1966, **37**, 2181; *Z. Naturf.*, 1965, **20a**, 1197.
44. L. Kalinowski and R. Seguin, *Appl. Phys. Letters*, 1979, **35**, 211 and (Erratum) 1980, **36**, 171.
45. W. P. Ellis and N. H. Nachtrieb, *J. Appl. Phys.*, 1969, **40**, 472.
46. W. Seith, *Z. Electrochem.*, 1933, **39**, 538.
47. E. M. Hampton and J. N. Sherwood, *Phil. Mag.*, 1974, **29**, 762.
48. P. Brätter and H. Gobrecht, *Phys. Status Solidi*, 1970, **37**, 869.
49. R. N. Ghoshtagore, *Phys. Rev.*, 1967, **155**, 598.
50. F. Dyment in *Titanium 80. Proc. 46th Int. Conf. n Titanium, Kyoto, Japan 1980* (eds H. Kimura and O. Izumi) p. 519.
51. U. Köhler and Ch. Herzkg, *Phys. Stat. Solidi (b)*, 1987, **144**, 243.
52. J. F. Murdock, T. S. Lundy and E. E. Stansbury, *Acta Met.*, 1964, **12**, 1033.
53. J. I. Federer and T. S. Lundy, *Trans. Met. Soc. AIME*, 1963, **227**, 592.
54. F. Schmitz and M. Fock, *J. Nucl. Mater.*, 1967, **21**, 317.
55. C. J. Meechan, *2nd Nuclear and Eng. Sci. Conf.*, Phila. Pa., 1957, Paper No. 57, NESC.-7.
56. B. E. Davis and W. D. McMullen, *Acta Met.*, 1972, **20**, 593.
57. R. F. Peart, *J. Phys. Chem. Solids*, 1965, **26**, 1853.
58. R. E. Einziger, J. N. Mundy and H. A. Hoff, *Phys. Rev.*, 1978, **B17**, 440.
59. R. E. Pawel and T. S. Lundy, *J. Phys. Chem. Solids*, 1965, **26**, 937.
60. J. N. Mundy, C. W. Tse and W. D. McFall, *Phys. Rev.*, 1976, **B13**, 2349.
61. K. Maier, H. Mehrer and G. Réin, *Z. Metallk.* 1979, **70**, 271.
62. J. N. Mundy, S. J. Rothman, N. Q. Lam, H. A. Hoff and L. J. Nowicki, *Phys. Rev.*, 1978, **B18**, 6566.
63. J. Horvath, F. Dyment and H. Mehrer, *J. Nucl. Mat.*, 1984, **126**, 206.
64. Y. Adda and A. Kirianenko, *J. Nucl. Mater.*, 1962, **6**, 130.
65. S. J. Rothman, R. Bastar, J. J. Hines and D. Rokop, *Trans. Met. Soc. AIME*, 1966, **236**, 897.
66. Y. Adda, A. Kirianenko and C. Mairy, *J. Nucl. Mater.*, 1959, **1**, 300.
67. G. B. Federov, E. A. Smirnov and S. S. Moiseenko, *Met. Metalloved. Christ. Metal.*, 1968, No. 7, 124.
68. N. L. Peterson and S. J. Rothman, *Phys. Rev.*, 1964, **3A**, A842.
69. A. Bochvar, V. Kuznetsova and V. Sergeev, *Trans. 2nd Geneva Conf. on Peaceful Uses of At. Energy*, 1958, **VI**, 68.
    Y. Adda and A. Kirianenko, *J. Nucl. Mater.*, 1959, **1**, 120.
    S. J. Rothman, L. T. Lloyd and A. L. Harkness, *Trans. AIME*, 1960, **218**, 605.
70. G. B. Federov and E. A. Smirnov, *Met. Metalloved. Christ. Metal.*, 1967, No. 6, 181.
71. J. Geise and Ch. Herzig, *Z. Metallk.*, 1987, **78**, 291.
72. D. V. James and G. M. Leak, *Phil. Mag.*, 1966, **14**, 701.
73. R. J. Borg and D. Y. F. Lai, *Phil. Mag.*, 1968, **18**, 55.
74. I. G. Ivantsov and A. M. Blikein, *Phys. Met. Metallog.*, 1966, 22, (6), 68.

75. Th. Heumann and R. Imm, *J. Phys. Chem. Solids.*, 1968, 29, 1613.
76. W. Bussman, Ch. Herzig, W. Rempp, K. Maier and H. Mehrer, *Phys. Stat. Solidi (a)*, 1979, **56**, 87.
77. D. Graham and D. H. Tomlin, *Phil. Mag.*, 1963, **8**, 1581.
78. K. Hirano and Y. Iijima, *Defect and Diffusion Forum*, 1989, **66–69**, 1039.
79. H. Bakker, *Phys. Status Solidi*, 1968, **28**, 569.
80. Y. Iijima, K. Kimura and K. Hirano, *Acta Met.*, 1988, **36**, 2811.
81. N. L. Peterson, *Phys. Rev.*, 1964, **136**, 568.
82. G. V. Kidson and R. Ross, *Proc. 1st UNESCO Conf.*, 'Radioisotopes in *Sci. Res.*', Pergamon Press, 1958, **1**, 185.
83. F. Cattaneo, E. Germagnoli and F. Grasso, *Phil. Mag.*, 1962, **7**, 1373.
84. R. E. Tate and E. M. Cramer, *Trans. Met. Soc. AIME*, 1964, **230**, 639.
85. M. Dupuy and D. Calais, *Trans. Met. Soc. AIME*, 1968, **242**, 1679.
86. W. Z. Wade, D. W. Short, J. C. Walden and J. W. Magana, *Met. Trans. A*, 1978, **9A**, 965.
87. E. M. Hampton, P. McKay and J. N. Sherwood, *Phil. Mag.*, 1974, **30**, 853.
88. M. P. Dariel, *Phil. Mag.*, 1973, **28**, 915.
89. A. Languille, D. Calais and B. Coqblin, *J. Phys. Chem. Solids*, 1974, **35**, 1461.
90. M. Fromont and G. Marbach, *J. Phys. Chem. Solids*, 1977, **38**, 27.
91. M. Fromont, A. Languille and D. Calais, *J. Phys. Chem. Solids*, 1974, **35**, 1367.
92. G. Heffich, H. Mehrer and K. Maier, *Scripta Met.*, 1977, **11**, 795.
93. R. J. Borg and C. E. Birchenall, *Trans. Met. Soc. AIME*, 1960, **218**, 980.
94. K. Maier, H. Mehrer, E. Lessmann and W. Schüle, *Phys. Status Solidi*, 1976, **78**, 689.
95. M. Werner, H. Mehrer and H. Seithoff, *J. Phys. C: Sol. Stat. Phys.*, 1983, **16**, 6185.
96. F. R. Winslow and T. S. Lundy, *Trans. AIME*, 1965, **233**, 1790.
97. Ch. Herzig, L. Manke and W. Bussman, in 'Point Defects and Defect Interactions in Metals', Univ. of Tokyo Press (eds J. I. Takamura, M. Doyama and M. Kiritani) 1982, p. 554.
98. J. Pelleg, *Phil. Mag.*, 1974, **29**, 383.
99. D. Weiler, K. Maier and H. Mehrer, in *DIMETA 82. Proc. Int. Conf.–Diffusion in Metals and Alloys, Tihany 1982*, Trans. Tech. Publs, Switzerland (eds F. J. Kedves and D. L. Beke) p. 342.
100. J. Mundy, H. A. Hoff, J. Pelleg, S. J. Rothman, L. J. Nowicki and F. A. Schmidt, *Phys. Rev. B*, 1981, **24**, 658.
101. G. Rein, H. Mehrer and K. Maier, *Phys. Stat. Solidi (a)*, 1978, **45**, 253.
102. M. Ait-Salem, T. Springer, A. Heidmann and B. Alefeld, *Phil. Mag.*, 1979, **A39**, 797.
103. N. K. Arkhipova, S. M. Klotsman, I. P. Polikarpova, A. N. Timofeev and P. Shepatkovski, *Phys. Met. Metallog.*, 1986, **62**(6), 127.

Table 13.2 TRACER IMPURITY DIFFUSION COEFFICIENTS

Element	A cm^2 s^{-1}	Q kJ mol^{-1}	Temp. range K	Method	Ref.
In Ag					
Cu	1.23	193.0	990–1 218	IVa(i), s.c., Cu^{64}, 99.99%	13
	0.029	164.1	699–897	IVa(iii) (SIMS), s.c., 99.99%	258
Au	0.85	202.1	991–1 198	IVa(i), s.c., Au^{198}, 99.99%	18
Zn	0.54	174.6	916–1 197	IVa(i), s.c., Zn^{65}, 99.99%	14
	0.532	174.6	970–1 225	IVa(i), s.c., Zn^{65}, 99.999%	259
Cd	0.44	174.6	866–1 210	IVa(i), s.c., Cd^{115}, 99.99%	10
Hg	0.079	159.5	926–1 122	IVa(i), s.c., Hg^{203}, 99.99%	13
Al	0.13	159.5	873–1 223	IIb (X-ray), p.c.,—	211
Ga	0.42	162.9	873–1 213	IIb (X-ray), p.c.,—	260
In	0.41	170.1	886–1 209	IVa(i), s.c., In^{114}, 99.99%	10
	0.36	169.0	553–838	IVa(i), s.c., In^{114}, 99.999%	261
Tl	0.15	158.7	918–1 073	IVa(i), p.c., T1^{204},—	15
Ge	0.084	152.8	943–1 123	IVa(i)m p.c., Ge^{71}, —	15
Sn	0.25	165.0	865–1 121	IVa(i), s.c., Sn^{113}, 99.99%	10
Pb	0.22	159.5	973–1 098	IVa(i), p.c., Pb^{210}, —	19
As	0.42	149.6	915–1 213	IVa(iii) (EMPA), p.c., 99.98%	107
Sb	0.169	160.4	743–1 215	IVa(i), s.c., Sb^{124}, 99.99%	11
S	1.65	167.5	873–1 173	IVb, s.c., S^{35}, 99.999%	108
Se	0.285	157.4	759–1 109	IVa(i) (ion impl), s.c., Se^{75}, 99.999%	262
Te	0.21	154.7	650–1 169	IVa(i), s.c., Te^{121}, 99.999%	109
Ti	1.33	198	1 051–1 220	IIa(ii) (EMPA), p.c., 99.999%	263
V	2.72	209	1 012–1 218	IVb, p.c., V^{48}, 99.999%	263
Cr	3.29	210	1 023–1 215	IVb, p.c., Cr^{51}, 99.999%	263
	1.07	192.6	976–1 231	IVa(i), s.c., Cr^{51}, 99.9999%	264

(continued)

Table 13.2   TRACER IMPURITY DIFFUSION COEFFICIENTS—*continued*

Element	A cm^2 s^{-1}	Q kJ mol^{-1}	Temp. range K	Method	Ref.
Mn	4.29	196	883–1 212	IVb, p.c., Mn[54], 99.999%	263
Fe	2.6	205.2	1 073–1 205	IVa(i), s.c., Fe[59], 99.999%	106
Ru	180	275.5	1 066–1 219	IVa(i), s.c., Ru[103/106], 99.99%	12
Co	1.9	204.1	973–1 214	IVa(i), s.c., Co[60], 99.999%	106
Ni	21.9	229.3	1 022–1 223	IVa(i), s.c., Ni[63], 99.99%	16
	15	217.3	903–1 200	IVa(i), s.c., Ni[63], —	212
Pd	9.57	237.6	1 008–1 212	IVa(i), s.c., Pd[103], 99.999%	9
Pt	6.0	238.2	923–1 223	IIb (X-ray), p.c., —	210
	1.9	235.7	1 094–1 232	IVa(i), s.c., Pt[191/195], 99.9999%	265
In Al					
Cu	0.654	136.0	594–928	IVa(i) and b, s.c., Cu[67], 99.999%	362
Ag	0.118	116.5	644–928	IVa(i), s.c., Ag[110], 99.999%	26
	0.13	117.2	615–883	IVa(i), s.c., Ag[110], 99.999%	17
Au	0.131	116.4	642–928	IVa(i), s.c., Au[198], 99.999%	26
	0.077	113.0	696–882	IVa(i), s.c., Au[198], 99.999%	27
Zn	0.325	117.9	688–928	IVa(i), s.c., Zn[65], 99.999%	25 and 26
	0.245	119.6	614–920	IVa(i), p.c., Zn[65], 99.99%	131
Cd	1.04	124.3	714–907	IVa(i), s.c., Cd[115m], 99.999%	27
Hg	15.3	141.8	718–862	IVb, p.c., Hg[203], 99.999%	213
Ga	0.490	123.1	680–926	IVa(i), s.c., Ga[72], 99.999%	26
In	0.123	115.6	673–873	IVa(i) and b, p.c., In[114], 99.999%	28
	1.16	122.7	715–929	IVa(i), s.c., In[114], 99.999%	29
Tl	1.16	152.7	737–862	IVb(i), p.c., Tl[204], 99.999%	213
Si	2.48	137.0	753–893	IIa(i) [$D(c \rightarrow 0)$], p.c., 99.999%	215
Ge	0.481	121.3	674–926	IVa(i), Ge[71], s.c., 99.999%	26
Sn	0.245	119.3	673–873	IVa(i) and b, p.c., Sn[113], 99.999%	28
Pb	50	145.6	777–876	IVb, p.c., Pb[210], 99.999%	213
Sb	0.09	121.7	721–893	IVb, p.c., Sb[124], 99.995%	130
Li	0.35	126	803–923	Resistometric method, p.c., 99.993%	266
Na	6.7×10^{-4}	97.1	719–863	IVc, p.c., Na[24]	267
Cs	1.04×10^{-2}	99.2	453–573	IVb, p.c., Cs[137], 99.997%	268
Mg	0.062 3	115	598–923	IVa(i), s.c., Mg[28], 99.999%	215
	1.24	130.4	667–928	IVa(i), s.c., Mg[28], 99.999%	214
Zr	728	242	804–913	IVb, p.c., Zr[95], 99.999%	215
Cr	1.85×10^3	253.0[(a)]	859–923	IVa(i), s.c., Cr[51], 99.999%	26
Mo	14	250.0	898–928	IIa(ii) EMPA, p.c., 99.99%	269
Mn	104	211.4	730–933	IVa(i), s.c. and p.c., Mn[55/56], 99.999%	133
Fe	135	192.6	823–913	IVa(i), s.c., Fe[59], 99.999%	27
	53	183.4	793–930	IVa(i), p.c., Fe[59], 99.995%	134
Co	250	174.6	673–913	IVa(i) and b, p.c., Co[60], 99.995%	135
	506	175.7	724–930	IVa(i), s.c., Co[60], 99.999%	270 and 26
Ni	4.4	145.8	742–924	Resistometric method, p.c., 99.995%	256
U	0.1	117.2	798–898	IVa(ii), p.c., U[235], 99.995%	132

(*a*) Recalculated values

In Au					
Cu	0.105	170.2	973–1 179	IVa(iii) (EMPA), p.c., 99.99%	17
Ag	0.072	168.3	972–1 281	IVa(i), s.c., Ag[110], 99.99%	18
	0.086	169.3	1 004–1 323	IVa(i), s.c., Ag[101/105], 99.999%	271
Zn	0.082	158.1	969–1 287	IVa(i), s.c. and p.c., Zn[65], 99.999%	272
Hg	0.116	156.5	877–1 300	IIIa(i), p.c., Hg[203], 99.994%	20
Al	0.052	143.6	773–1 223	IIb (X-ray), p.c., —	216
In	0.075	153.7	973–1 273	IIa(ii) (EMPA), p.c., 99.999%	110
Ge	0.073	144.5	1 010–1 287	IVa(i), s.c. and p.c., Ge[68], 99.999%	272

(*continued*)

13–18   Diffusion in metals

Table 13.2   TRACER IMPURITY DIFFUSION COEFFICIENTS—continued

Element	A cm^2 s^{-1}	Q kJ mol^{-1}	Temp. range K	Method	Ref.
Sn	0.041 2	143.3	970–1 268	IIa(ii) (EMPA), p.c., 99.999%	110
Sb	0.011 4	129.4	1 003–1 278	IIa(ii) (EMPA), p.c., 99.999%	111
Te	0.063	141.1	909–1 145	IVa(i) (ion impl.), s.c., Te[121], 99.999%	262
Fe	0.082	174.2	1 027–1 221	IVb, p.c., Fe[59], 99.93%	22
	0.19	172.5	973–1 323	IIb (X-ray), p.c., —	273
Co	0.22	183.4	973–1 323	IIb (X-ray), p.c., —	273
	0.25	185.2	1 030–1 325	IVa(i), s.c., Co[57], 99.999%	274
Ni	0.30	192.6	1 153–1 210	IVa(i), p.c., Ni[63], 99.96%	275
	0.25	188.4	973–1 323	IIb (X-ray), p.c., —	276
Pd	0.076	195.1	973–1 273	IIb (X-ray), p.c., —	277
Pt	7.6	255.0	1 173–1 329	IVa(i), Pt[195], p.c., and s.c., 99.98%	21
	0.095	201.4	973–1 273	IIb (X-ray), p.c., —	277
In Be					
Te	∥c 0.38	198.6	733–1 273	Combined $T < 677$, IVa(i), Cu[64]	244
	⊥c 0.42	193.3	693–1 273	data, s.c. $T < 677$, IIa(ii)	217
Ag	6.2	193.0	923–1 183	IVb, s.c., Ag[110], 99.75%	82
	∥c 0.43	164.5		IVb, s.c., Ag[110], 99.75%	82
	⊥c 1.76	180.9	929–1 170		
Au	∥c $D = 1.5 \times 10^{-12}$				
	⊥c $D = 2.8 \times 10^{-12}$		938		
	∥c $D = 4.4 \times 10^{-11}$			IIIb(i), s.c., 99.95%	278
	⊥c $D = 6.5 \times 10^{-11}$		1 053		
C	3.2×10^{-5}	158.6	—	IVa(i), p.c., C[14]	422
Al	1.0	168.3	1 068–1 356	IVb, p.c., Al[26], 99.91%	218
V	29	243	1 173–1 423	IVb, p.c., V[48], 99.7%	219
Nb	2×10^4	359.6	1 318–1 513	IVb, p.c., Nb[95], 99.7%	219
Fe	0.53	216.9	973–1 349	IVb, p.c., Fe[59], 99.75%	82
	1.0	221.9	1 073–1 373	IIa(i) (EMPA), $D_{(c=0.05\ \text{wt% Fe})}$, p.c.	279
Co	27	287.2	1 253–1 493	Ib, p.c., Co[57], 99.8%	412
Ni	0.2	243	1 073–1 523	IVb, p.c., Ni[63], 99.7%	125
Ce	310	303.5	1 223–1 513	IVb, p.c., Ce[141], 99.7%	219
In Ca					
C	2.7×10^{-3}	97.5	773–1 073	IVb, p.c., C[14], 99.95%	328
Fe	3.2×10^{-5}	124.8	823–1 073	IVb, p.c., Fe[59], 99.95%	328
Ni	1.0×10^{-5}	121.0	823–1 073	IVc, p.c., Ni[63], 99.95%	328
U	1.1×10^{-5}	145.7	773–973	IVb, p.c., U[235], 99.95%	328
In Cd					
Ag	∥c 1.41	103.2	478–583	IVa(i), s.c., Ag[110], 99.999%	129
	⊥c 0.68	105.0			
Au	∥c 1.41	106.6	453–578	IVa(i), s.c., Au, 99.9999%	129
	⊥c 3.16	110.7			
Zn	∥c 0.13	75.5	428–588	IVa(i), s.c., Zn, 99.999%	129
	⊥c 0.084	75.4			
Hg ∥ and	⊥c 0.212	78.6	432–573	IVa(i), s.c., Hg[203], 99.999%	129
In	∥c 0.101	73.1	433–573	IVa(i), s.c., In[114m], 99.999%	129
	⊥c 0.090	70.9			
Pb	∥c 0.060	68.9	514–571	IVa(i), s.c., Pb[210], 99.999%	225
	⊥c 0.071	65.8			
In Ce					
Ag	2.5×10^{-2}	88.3	852–969 (γ)	IVa(i), p.c., Ag[110], 99.9%	209
	1.2×10^{-1}	92.9	996–1 049 (δ)		

(continued)

Table 13.2 TRACER IMPURITY DIFFUSION COEFFICIENTS—*continued*

Element	A cm^2 s^{-1}	Q kJ mol^{-1}	*Temp. range* K	*Method*	*Ref.*
Au	4.4×10^{-3}	62.4	823–973 (γ)	IVa(i), p.c., Au^{198}, 99.9%	209
	9.5×10^{-2}	85.8	999–1 048 (δ)		
La	3.8×10^{-2}	102.6	998–1 048 (δ)	IVa(i), p.c., La^{140}, 99.95%	221
Mn	*See* Figure 13.3$^{(a)}$		γ- and δ-range	IVa(i), p.c., Mn^{54}, 99.9%	220
Fe	1.7×10^{-2}	49.8	875–990 (γ)	IVa(i), p.c., Fe^{59}, 99.9%	220
	2.0×10^{-3}	32.2	1 005–1 046 (δ)		
Co	1.6×10^{-3}	35.6	<1 003 (γ)	IVa(i), p.c., —, —	280
	1.2×10^{-3}	33.5	1 003–1 073 (δ)		
Gd	1.2×10^{-2}	100.5	1 003–1 048 (δ)	IVa(i), p.c., —, —	280

(*a*) No numerical data reported.

In Co

Element	A cm^2 s^{-1}	Q kJ mol^{-1}	*Temp. range* K	*Method*	*Ref.*
Cu	~1.0	~275	1 158 and 1 273 (f) (Two temps. only)	IIIa(i) (EMPA) (0–5% Cu), p.c., 99.5%	281
Zn	0.12	266.7	1 081–$T_c^{(a)}$	IVb, s.c., Zn^{65}, —	223
	0.08	254.5	T_c–1 573		
C	8.72×10^{-2}	149.3	723–1 073 (p)	IVa(i)	363
	0.31	153.7	1 073–1 673 (f and p)	IVa(i)	400
S	1.3	226.1	1 423–1 523 (p)	IVb, p.c., S^{35}, 99.99%	282
V	$D = 3.41 \times 10^{-12}$		1 273 ⎫		
	$D = 6.56 \times 10^{-12}$		1 328 ⎪ (f) $^{(a)}$		
	$D = 2.46 \times 10^{-11}$		1 388 ⎪		
	$D = 9.34 \times 10^{-11}$		1 433 ⎬	IVb, p.c., V^{48}, 99.9985%	283
$^{(b)}$	$D = 1.65 \times 10^{-10}$		1 473 ⎪		
	$D = 2.39 \times 10^{-10}$		1 523 ⎪ (p) ⎭		
	$D = 3.29 \times 10^{-10}$		1 536		
Mn	3.15×10^{-2}	232.4	1 133–1 378 (f)	IVb, p.c., Mn^{54}, 99.952%	284
	1.10×10^{-2}	217.7	1 424–1 519 (b)		
Fe	0.11	253.3	1 223–1 643 (p)	IIa(i) (EMPA) [$D(c \to 0)$], p.c., 99.999%	34, 39 and 43
	0.34	259.6	1 081–T_c (f)	IVb, s.c., Fe^{59}, —	223
	0.16	248.7	T_c–1 573 (p)		
Ni	0.4	282.2	1 409–1 643 (p)	IVa(iii), and IIa(i) [($D(c \to 0)$] (EMPA), p.c., 99.999%	34
	0.34	269.2	1 045–1 321 (f)	IVb, p.c., Ni^{63}, 99.2%	33
	0.10	252.0	1 465–1 570 (p)		
Pt	0.65	279.3	1 354–1 481 (f and p)	IVb, p.c., Pt^{193m}, 99.99%	222

(*a*) f = ferromagnetic; p = paramagnetic; T_c = Curie temp. of Co (1 393 K)
(*b*) Each D is the mean of two values.

In Cr

Element	A cm^2 s^{-1}	Q kJ mol^{-1}	*Temp. range* K	*Method*	*Ref.*
C	8.74×10^{-3}	110.9	423–1 873	Va$_{(150/162°C)}$ and IIIa(i)$_{(1 150/1 600°C)}$	44
V	381	419	1 595–2 041	IVa(i), s.c., V^{48}	417
Mo	2.7×10^{-3}	242.8	1 373–1 693	IVb, p.c., Mo^{99}	46
Fe	0.47	332.0	1 518–1 686	IVb, p.c., Fe^{55}	61

In Cu

Element	A cm^2 s^{-1}	Q kJ mol^{-1}	*Temp. range* K	*Method*	*Ref.*
Ag	0.61	194.7	873–1 273	IVb, s.c. and p.c., Ag^{110}, 99.99%	228
Au	0.243	197.8	633–1 350	IVa(i), s.c., Au$^{196/198}$, 99.999%	286
Zn	0.34	190.9	878–1 322	IVa(i), s.c., Zn^{65}	1
	0.24	188.8	1 073–1 313	IVa(i), p.c., Zn^{65}, 99.99%	99

(*continued*)

Table 13.2 TRACER IMPURITY DIFFUSION COEFFICIENTS—*continued*

Element	A $cm^2 s^{-1}$	Q $kJ mol^{-1}$	Temp. range K	Method	Ref.
Cd	1.2	194.0	983–1 309	IVa(i), p.c., Cd^{115m}, 99.998%	7
	1.27	194.6	1 032–1 346	IVa(i), s.c., Cd^{109}, 99.99%	287
Hg	0.35	184.2	1 053–1 353	IVa(i), s.c., Hg^{203}	2
Ga	0.523	192.7	1 153–1 352	IVa(i), p.c., Ga^{67}, 99.99%	100
In	$A_1 = 0.29$ $A_2 = 3\,110$	$Q_2 = 179.6$ $Q_2 = 295.4$	602–1 354	IVa(i), s.c. and p.c. Least squares fit to data of 226, 287 and 298	5
Tl	0.71	181.3	1 058–1 269	IVa(i), s.c., Tl^{204}, 99.999%	36
Si	0.07	171.7	973–1 323	IIb (X-ray), p.c., —	289
Ge	0.315	185.5	1 110–1 326	IVa(i), p.c., Ge^{68}, 99.99%	100
	0.397	187.4	975–1 289	IVa(i), s.c., Ge^{63}, 99.998%	101
Sn	0.842	188.2	1 010–1 321	IVa(i), s.c., Sn^{113}, 99.99%	227
	0.67	184.4	1 018–1 355	IVa(i), p.c., Sn^{113}, 99.999%	290
Pb	0.862	182.4	1 008–1 225	IVa(i), s.c., Pb^{210}, 99.99%	231
P	3.05×10^{-3}	136.1	847–1 319	IVa(i), s.c., P^{32}, 99.999%	291
As	0.202	176.4	1 086–1 348	IVb, p.c., As^{73}, 99.99%	102
Sb	0.34	175.8	873–1 275	IVa(i), s.c., Sb^{124}, 99.99%	8
	0.48	179.5	1 049–1 349	IVa(i), p.c., Sb^{124}, 99.999%	290
Bi	0.766	178.1	1 074–1 348	IVa(i), s.c., Bi^{207}, 99.99%	231
S	23	206.6	1 073–1 273	IVb, s.c., S^{35}, 99.999%	105
Se	10	180.5	878–1 150	IVa(i) (ion impl.), s.c., Se^{75}, 99.999%	262
Te	0.97	180.5	822–1 214	IVa(i) (ion impl.), s.c., Te^{121}, 99.999%	262
Be	0.66	195.9	973–1 348	II(b) (X-ray), p.c.	289
Ti	0.693	196	973–1 283	IIa(i) [$D(c \rightarrow 0)$], p.c., 99.998%	292
V	2.48	215	995–1 342	IVb, p.c., V^{48}, 99.998%	103
Nb	2.04	251.5	1 080–1 179	IVb, p.c., Nb^{95}, 99.999%	104
Cr	0.337	195	999–1 338	IVb, p.c., Cr^{51}, 99.998%	103
	1.02	224.0	1 073–1 343	IVb, p.c., Cr^{51}, 99.995%	293
Mn	0.74	195.5	973–1 348	IIb (X-ray), p.c.	294
	1.42	204.3	773–976	IVa(i), s.c., Mn^{54}, 99.998%	295
Fe	1.4	216.9	1 103–1 347	IVa(i), s.c., Fe^{59} 99.998%	4
	1.01	213.3	989–1 329	IVa(i), s.c., Fe^{59}, 99.998%	6
Ru	8.5	257.5	1 221–1 335	IVa(i), s.c., Ru^{103}, 99.999%	229
Co	$A_1 = 0.74$ $A_2 = 736$	$Q_1 = 217.2$ $Q_2 = 312.8$	640–1 351	IVa(i), s.c., Co^{60}, 99.998% Combined data, 4 and 299	5
W	1.69	225.7	1 163–1 306	IVa(i), p.c., w	296
Rh	3.3	242.8	1 023–1 348	IIb (X-ray), p.c., —	297
Ir	10.6	276.4	1 183–1 303	IVa(i), s.c., Ir^{192}, 99.99%	230
Ni	$A_1 = 0.7$ $A_2 = 250$	$Q_1 = 225.0$ $Q_2 = 299.3$	613–1 349	Fit to data of 4, 99, 300 and 301	5
	0.76	224.8	600–1000		
	1.7	231.6	1172–1340	IV, p.c., Ni^{63}	425
	2.7	237.4	1016–1349		
	3.8	237.4	968–1334		
Pd	1.71	227.6	1 080–1 329	IVa(i), s.c., Pd^{103}, 99.999%	9
Pt	0.56	233	1 149–1 352	IVa(i), s.c., $Pt^{191/195}$, 99.999%	265
In Er					
Au	*See* Figure 13.4$^{(a)}$ $\|c$	15.3	1 270–1 485	IVa(i), s.c., Au^{198}, 99.91%	285
	$\perp c$	23.6	(Three Ts only)		
C	1.14×10^{-2}	117.2	953–1 473	IVb, p.c., C^{14}	401

(*a*) No values quoted for A.

(*continued*)

Table 13.2 TRACER IMPURITY DIFFUSION COEFFICIENTS—*continued*

Element	A cm^2 s^{-1}	Q kJ mol^{-1}	Temp. range K	Method	Ref.
In Fe[a]					
Cu	$D = 4.81 \times 10^{-14}$		963		
	$D = 9.55 \times 10^{-14}$		978		
	$D = 1.78 \times 10^{-13}$		993 $(\alpha$-f$)^{[a]}$	IVa(iii) (EMPA), s.c., 99.999%	155
	$D = 4.10 \times 10^{-13}$		1 008		
	$D = 8.26 \times 10^{-13}$		1 024		
	300	283.9	1 045–1 173 $(\alpha$-p$)$		
	0.19	272.6	1 198–1 323 (γ)		
	2.86	306.7	1 558–1 641 (γ)	IVa(i), s.c., Cu^{64}, 99.91%	302
	4.16	305.0	1 378–1 483 (γ)	IVb, p.c., Cu^{64}, 99.96%	303
Ag	1.95×10^3	288.9	1 021–1 161 (α)	IVa(i), p.c., Ag^{110}	158
	230	278	973–1 033 $(\alpha$-f$)$		
	38	259.2	1 053–1 173 $(\alpha$-p$)$	IVb, p.c., Ag, 99.97%	304
Au	$D = 9.58 \times 10^{-14}$		972.1		
	$D = 2.69 \times 10^{-13}$		997.8		
	$D = 5.49 \times 10^{-13}$		1 012.8 $(\alpha$-f$)$		
	$D = 1.58 \times 10^{-12}$		1 034.4	IVb, p.c., Au^{195}, 99.999%	40
	31.0	261.2	1 055–1 174 $(\alpha$-p$)$		
Zn	60	262.6	1 072–1 169 $(\alpha$-p$)$	Ia(iii) (EMPA), s.c.	418
C	Log $D = -0.906\,4 - 0.519\,9\chi$ $+1.61 \times 10^{-3}\chi^2$		$\chi = 10^4/T$ α-range	Best fit to 83 points from various sources	58
	0.234	147.81	γ-range	IIa(i) $[D(c \to 0)]\ D(\gamma)$ strongly dep. on T. *See* Table 13.4	305
N	Log $D = -1.948 - 0.433\,4\chi$ $+6.08 \times 10^{-4}\chi^2$		$\chi = 10^4/T$ α- and δ-range	Best fit to 52 points from various sources	58
	0.91	168.56	γ-range	Best fit to points from various sources	306
P	1.38×10^5	332	932–1 057 $(\alpha$-f$)$		
	2.87×10^2	271	1 078–1 153 $(\alpha$-p$)$	IVb, p.c., P^{32}	307
	6.3×10^{-2}	193.4	1 223–1 573 (γ)	Ib, p.c., P^{32}, 99.99%	308
As	4.3	219.8	1 223–1 653 (α)	IIa(ii) (EMPA) (α-stab. 0.5–5% As)	309
	0.58	246.6	1 323–1 573 (γ)	IIa(ii) (EMPA) 0–1.2% As	
Sb	80	269.9	773–873 $(\alpha$-f$)$	IIa(ii) ion impl. (NRA)	153
	440	270.0	1 040–1 173 $(\alpha$-p$)$	IVb, p.c. and s.c., Sb^{124}	159
Sn	5.4	232.4	973–1 033 $(\alpha$-f$)$	IVb, p.c., Sn^{113}	208
	2.4	221.9	1 073–1 183 $(\alpha$-p$)$		
	6.1×10^4	316.4	900–1 023 $(\alpha$-f$)$	IVb, s.c., Sn^{113}	310
	0.845	261.7	1 197–1 653 (γ)	IVa(i), p.c., Sn^{113}, >99.97%	311
S	34.6	231.5	973–1 173	IVb, p.c., S^{35}, >99.996%	312
	1.7	221.9	1 223–1 523 (γ)	Ib, p.c., S^{35}	150
Be	5.34	218.1	1 073–1 773 (α and β)	IVb, p.c., (α-stabilised-1 wt% Be) 99.9%	148
	0.1	241.2	1 373–123 (γ)	IVb, p.c., Be^7, 99.9%	149
Hf	3.6×10^3	407.4	1 371–1 626 (γ)	IVb, p.c., Hf^{181}, 99.98%	151
	9.0×10^4	473.1	1 438–1 593 (γ)	IVb, p.c., Hf^{185}, 99.95%	313
V	124	274	1 058–1 172 $(\alpha$-p$)$	IVa(i), p.c., V^{48}, 99.98%	314
	0.62	273.5	1 210–1 607 (γ)		
	0.75	264.2	1 393–1 653 (γ)	IVb, p.c., V^{48}, 99.98%	151
Nb	$D = 1.0 \times 10^{-12}$		993		
	$D = 5.4 \times 10^{-12}$		1 025 $(\alpha$-f$)$		
	50.2	252	1 059–1 162 $(\alpha$-p$)$	IVa(i), p.c., Nb, 99.98%	315
	0.83	266.5	1 210–1 604 (γ)		

(continued)

Table 13.2   TRACER IMPURITY DIFFUSION COEFFICIENTS—*continued*

Element	A cm² s⁻¹	Q kJ mol⁻¹	Temp. range K	Method	Ref.
Cr	8.52	250.8	1 070–1 150 (α)	IVb, p.c., Cr⁵¹, 99.98%	151
	10.80	291.8	1 233–1 669 (γ)		
	90	271	1 043–1 150 (α-p)	IVa(i), p.c., Cr⁵¹	316
Mn	1.49	233.6	973–1 033 (α-f)		
	0.35	219.8	1 073–1 173 (α-p)	IVb, p.c., Mn⁵⁴, 99.97%	152
	0.16	261.7	1 193–1 553 (γ)		
	0.76	224.6	(α and δ)	Combined data 317 [IIa(ii) (EMPA), p.c., 1719/1767 (δ)] and 152	317
Co	7.19	260.4	956–1 000 (α-f)	IVb, s.c., Co⁶⁰, 99.95%	
	6.38	257.1	1 081–1 157 (α-p)	IVb, s.c., and p.c., Co⁶⁰, 99.95%	154
	6.38	257.1	1 702–1 794 (δ)	IVa(i), p.c., Co⁶⁰, 99.95%	
	118	285.9	1 044–1 177 (α-p)	IVb, p.c., Co⁶⁰, 99.999%	40
	1.0	301.9	1 409–1 633 (γ)	IVa(iii), IIa(ii) (EMPA), p.c., 99.999%	34 and 42
	2.9×10^{-2}	247.4	1 233–1 493 (γ)	IVb, p.c., Co⁶⁰, 99.9%	318
	See Figure	13.5	785–1 036 (α-f) 1 058–1 164	IVa(i), s.c., Co⁶⁰, 99.997%	319 and 320
	See Figure	13.5	α-f and α-p	IVa(i), p.c., Co⁶⁰	316
Ni	1.4	245.8	873–953 (α-f)	IVb and c, s.c., Ni⁶³, 99.97%	
	1.3	234.5	1 083–1 173 (α-p)	IVb, s.c. and p.c., Ni⁶³, 99.97%	41
	0.77	280.5	1 203–1 323 (γ)	IVb, s.c., Ni⁶³, 99.97%	
	See Figure	13.6	972–1 032 (α-f)	IVb, p.c., Ni⁶³, 99.999%	90
	9.9	259.2	1 054–1 173 (α-p)		
	3.0	314	1 409–1 673 (γ)	IIa(ii) and IVa(iii) (EMPA), p.c., 99.999%	34 and 364
	See Figure	13.6	788–1 014 (α-f) 1 048–1 160 (α-p)	IVa(i), p.c., Ni⁶³	321
	9.7	267.5	1 748–1 767 (δ)	IIa(ii) (EMPA), p.c., 99.96%	361
Pd	0.41	280.9	1 373–1 573 (γ)	Ib, p.c., Pd¹⁰³, 99.97%	419
Pt	2.7	296	1 233–1 533 (γ)	IVb, p.c., Pt¹⁹³ᵐ	222
U	7×10^{-5}	133.2	1 223–1 348 (γ)	IVa(iii) and (fission fragment radiog), p.c.	160

(*a*) α-f = α-phase, ferromagnetic; α-p = α-phase, paramagnetic.

Element	A cm² s⁻¹	Q kJ mol⁻¹	Temp. range K	Method	Ref.
In Hf					
Al	170	357	1 023–1 173 (α)	IIa(ii) (ion impl. and NRA). 97% Hf + 3% Zr	323
C	74	312.3	1 393–2 033 (α)	IVb, p.c., C¹⁴, Hf + 1.5 wt% Zr	402
	0.8	211.4	2 073–2 373 (β)	IIa(ii), p.c., Hf + 3 wt% Zr	403
Co	5.3×10^{-3}	95.46	1 106–1 798 (α)	IVb, p.c., Co⁶⁰, 99.99%	235
Cr	0.14	213.9	1 183–2 173 (α and β)	IVb, p.c., Cr⁵¹, 99.99%	235
In Ho					
C	2.8×10^{-2}	125.6	873–1 430	IVb, p.c., ¹⁴C	161
In In					
Ag	∥c 0.11	48.1	298–423	IVa(i), s.c., Ag¹¹⁰ᵐ, 99.99%	94
	⊥c 0.52	53.6			
Au	9×10^{-3}	28.1	298–423	IVa(i), s.c.(*a*), Au¹⁹⁸, 99.99%	94
Tl	0.049	64.9	323–429	IVa(i), p.c., Tl²⁰⁴, 99.9%	53
Co	1.2×10^{-5}	25.1	383–423	IVa(i), s.c., Co⁶⁰, 99.997%	380

(*a*) Randomly oriented single crystals were used.

(*continued*)

Table 13.2 TRACER IMPURITY DIFFUSION COEFFICIENTS—*continued*

Element	A $cm^2 s^{-1}$	Q $kJ\,mol^{-1}$	Temp. range K	Method	Ref.
In K					
Au	1.29×10^{-3}	13.52	279–326	IVa(i), p.c., Au[198], 99.95%	123
Na	5.8×10^{-2}	31.19	273–335	IVa(i), p.c., Na[22], 99.95%	121
Rb	9.0×10^{-2}	36.76	272–333	IVa(i), p.c., Rb[86], 99.95%	124
In La					
Au	2.2×10^{-2}	75.8	873–1 073 (β)	IVa(i), p.c., Au[198], 99.97%	197
C	4.1×10^{-3}	83.7	723–1 128 (β)	IVb, p.c., C[14]	404
Ce	1.8×10^{-2}	104.7	1 139–1 170 (γ)	IVa(i), p.c., Ce[141], —	156
In Li					
Cu	4.7×10^{-2}	38.6	323–394	IVa(i), p.c., Cu[64], 99.98%	113
	0.3	41.87	363–420	IVa(i), p.c., Cu[64], 99.98%	324
Ag	0.37	53.72	340–434	IVa(i), p.c., Ag[110m], 99.98%	119
	0.54	53.72	323–423	IVa(i), p.c., Ag[110m], 99.98%	325
Au[a]	0.21	46.01	319–426	IVa(i), p.c., Au[195], 99.95%	118
Zn	0.57	54.34	330–446	IVa(i), p.c., Zn[65], 99.98%	117
Cd	0.62	62.80	355–449	IVa(i), p.c., Cd[115m], 99.98%	112
Hg	1.04	59.37	331–447	IVa(i), p.c., Hg[203], 99.98%	112
Ga	0.21	54.05	389–447	IVa(i), p.c., Ga[12], 99.98%	112
In	0.39	66.44	348–443	IVa(i), p.c., In[114m], 99.95%	115
Sn	0.62	66.32 [b]	380–447	IVa(i), p.c., —, 99.95%	114
Pb	1.6×10^{4} [b]	105.5	401–443	IVa(i), p.c., —, 99.95%	114
Sb	1.6×10^{12} [b]	173.8	413–449	IVa(i), p.c., —, 99.95%	114
Bi	5.3×10^{14} [b]	198.0	413–450	IVa(i), p.c., —, 99.95%	114
Na[a]	0.41	52.80	325–449	IVa(i), p.c., Na[22], 99.8%	116

(*a*) Refs. 326 and 327 report measurements in 95% Li[6] and in 92.5% Li[7].   (*b*) Recalculated values.

Element	A $cm^2 s^{-1}$	Q $kJ\,mol^{-1}$	Temp. range K	Method	Ref.
In Mg					
Ag	$\|\|c$ 3.62 $\perp c$ 17.9	133.1 148.2	752–913	IVa(i), s.c., Ag[110m], 99.99%	126
Zn	0.41	119.7	740–893	IVa(i), p.c., Zn[65], 99.875%	127
Cd	$\|\|c$ 1.29 $\perp c$ 0.46	140.7 132.7	733–898	IVa(i), s.c., Cd[109], 99.99%	126
In	$\|\|c$ 1.75 $\perp c$ 1.88	143.4 142.4	747–906	IVa(i), s.c., In[114m], 99.99%	126
C	2.1×10^{-7}	52.3	773–873	IVb, p.c., C[14]	405
Sn	$\|\|c$ 4.27 $D_{\perp c}/D_{\perp c} = 1$ $D_{\perp c}/D_{\perp c} = 1.13$	149.9	748–903 902.3 858.2	IVa(i), s.c., Sn[113], 99.99%	126
Sb	$\|\|c$ 2.57 $\perp c$ 3.27	137.3 138.2	781–896	IVa(i), s.c., Sb[124], 99.99%	126
Fe	4×10^{-6}	88.8	673–873	IVb, p.c., Fe[59], 99.95%	328
Ni	1.2×10^{-5}	95.9	673–873	IVb, p.c., Ni[63], 99.95%	328
U	1.6×10^{-5}	95.9	773–893	IVb, p.c., U[235], 99.95%	328
In Mo					
C	2.0×10^{-3}	115.85	493–543	Via C precipitation rate	389
	1.04×10^{-2}	139	1 533–2 283	IIIb(i), p.c.	388
P	0.19	337	2 273–2 493	IVb, s.c., P[32], 99.97%	167
S	32	422.9	2 493–2 743	IVa(iii), s.c., S[35], 99.97%	166
	3.4×10^{-2}	297.3	1 238–1 443	IVb, p.c., S[35]	331
Li	0.01	470.6	1 843–2 243	IIIa(i), s.c. natural Li	330
Y	1.8×10^{-4}	214.8	1 473–1 873	IVb, s.c., Y[91], 99.8/99.9%	175
V	2.9	473.1	1 803–1 998	IIa(i) (EMPA), p.c.	332

(*continued*)

Table 13.2 TRACER IMPURITY DIFFUSION COEFFICIENTS—*continued*

Element	A cm^2 s^{-1}	Q kJ mol^{-1}	Temp. range K	Method	Ref.
Nb	14	452.6	2 123–2 623	IVa(i), p.c., Nb^{95}, 99.98%	66
	2.9	569.4	1 998–2 453	IIa(i) (EMPA), p.c.	332
	1.7×10^{-2}	379.3	1 973–2 373	IVb, p.c., Nb^{95}	333
Ta	3.5×10^{-4}	347.5	1 193–1 423	IVb, p.c., Ta^{182}	334
	1.9	473.1	2 098–2 449	IIa(i), p.c.	332
Cr	1.88	342.5	1 273–1 423	—, s.c., Cr^{51}, 99.8%	164
	2.5×10^{-4}	226.1	1 273–1 773	IVb, p.c., Cr^{51}	334
W	1.7	460.5	1 973–2 533	IVa(i) and (ii), p.c., W^{185}	67 and 168
	4.5×10^{-4}	324.5	1 973–2 423	IVb, p.c., W^{185}	334
	140	569.4	2 093–2 453	IIa(i) (EMPA), p.c.	332
Re	9.7×10^{-2}	396.5	1 973–2 373	IVa(i), p.c., Re^{185}	60
Fe	0.15	346.2	1 273–1 623	IVb, p.c., Fe^{59}, 99.96%	232
	3.7×10^{-3}	291.8	1 200–1 478	IVb, p.c., Fe^{59}	331
Co	18	446.7	2 123–2 623	IVa(i), s.c. and p.c., Co^{60}, 99.98%	66
	6	324.5	1 273–1 773	IVb, p.c., Co^{60}	334
Ni	$D = 2.4$–3.2×10^{-12}		1 623	IVb, s.c., Ni^{63}	335
U	7.6×10^{-3}	319.9	1 773–2 273	IVb, p.c., U^{235}, 99.98%	165
	1.3×10^{-6}	316.5	2 073–2 373	IV, p.c., U^{235}	57
In Na					
Ag	1.5×10^{-2}	21.39	298–351	IVa(i), p.c., Ag^{110}, 99.95%	122
Au	3.34×10^{-4}	9.25	274–350	IVa(i), p.c., Au^{198}, 99.95%	120
Li	Graphical data only		297–353	IVa(i), p.c., Li^6, 99.95%	122
K	0.08	35.29	273–365	IVa(i), p.c., K^{42}, 99.95%	121
Rb	0.15	35.55	272–359	IVa(i), p.c., Rb^{86}, 99.95%	121
Cd	0.37	40.86	272–363	IVa(i), p.c., Cd^{115}, 99.95%	122
In	1.79	48.73	293–363	IVa(i), p.c., In^{114}, 99.95%	122
Sn	0.54	43.92	316–363	IVa(i), p.c., In^{113}, 99.95%	122
Tl	0.52	42.62	297–356	IVa(i), p.c., Tl^{204}, 99.95%	122
In Nb					
Cu	$D = 3.71 \times 10^{-10}$ $D = 1.02 \times 10^{-9}$		1 829 1 909	IVa(i), p.c., Cu^{64}, 99.9%	233
C	1×10^{-2}	141.92	403–2 613	Combined data from several sources	194
Al	450	430.1	1 700–2 000	IIIb(ii)	176
Sn	0.14	330.3	2 123–2 663	IVa(i), p.c., Sn^{113}, 99.85%	66 and 66a
P	5.1×10^{-2}	215.6	1 573–2 073	IVb, s.c., P^{32}, 99.9%	167
S	2.6×10^3	306.0	1 370×1 770	IVa(i), s.c. and p.c., S^{35}, 99.6%	177
Y	1.5×10^{-3}	232.8	1 473–1 873	IVb, s.c., Y^{91}, 99.9/99.8%	175
Ti	0.4	370.5	1 898–2 348	IVa(iii) and IIa(i) (EMPA), p.c., 99.9%	169
	9.9×10^{-2}	364.0	1 267–1 765	IVa(i), s.c., Ti^{44}, 99.98%	170
Zr	0.47	364	1 855–2 357	IVa(iii) and IIa(i) (EMPA), p.c., 9.9%	169
	0.85	379.4	1 923–2 523	IVa(i), s.c., Zr^{95}	336
V	0.47	377	1 898–2 348	IVa(iii) and IIa(i) (EMPA), p.c., 99.9%	169
	2.21	355.9	1 273–1 673	IVb, s.c., V^{48}, 99.98%	171
Ta	1.0	415.7	1 376–2 346	IVa(i), s.c., Ta^{182}, 99.76%	173

(*continued*)

Table 13.2   TRACER IMPURITY DIFFUSION COEFFICIENTS—*continued*

Element	A cm^2 s^{-1}	Q kJ mol^{-1}	Temp. range K	Method	Ref.
Cr	0.3	349.6	1 226–1 708	IVa(i), s.c., Cr^{51}, >99.98%	172
	0.13	337.5	1 220–1 766	IVa(i), p.c., Cr^{51}, 99.96%	337
Mo	92	511	1 998–2 455	IVa(iii) and IIa(i) (EMPA), p.c., 99.9%	169
	1.3×10^{-2}	350.4	1 973–2 298	IVb, p.c., Mo^{99}	333
W	5×10^{-4}	383.9	2 073–2 473	IVb, p.c., W^{185}, 99.9%	174
	7×10^4	653	2 175–2 473	IVa(iii) and IIa(i) (EMPA), p.c., 99.8%	169
Fe	1.5	325.3	1 663–2 373	IVa(ii), p.c., Fe^{55}, 99.74%	68
	0.14	294.3	1 663–2 168	IVa(i) and (iii), p.c., Fe^{59}, 99.9%	233
Ru	29.3	460.1	2 026–2 342	Ib, p.c., Ru^{103}, 99.9%	416
Co	4.18×10^{-2}	257.2	1 347–2 173	IVa(i), s.c., Co^{60}	234
	0.11	274.7	1 580–1 920	IVa(iii), p.c., Co^{60}, 99.9%	233
	0.74	295.2	1 834–2 325	IVa(ii), p.c., Co^{60}, 99.74%	68
Ni	9.3	336.6	1 261–1 519	IVb, p.c., Ni^{63}, 99.82%	338
	7.7×10^{-2}	264.2	1 433–2 168	IVa(ii) and b, p.c., Ni^{63}, 99.9%	233
U	8.9×10^{-6}	321.5	1 773–2 273	IVb, p.c., U^{235}, 99.55%	165
In Nd					
Mn	Graphical data only		α- and β-range	IVa(i), p.c., Mn^{54}, 99.9%	220
	See Figure	13.3			
Fe	4.6×10^{-3}	51.1	α-range $\rbrace$	IVa(i), p.c., Fe, 99.9%	220
	1.0×10^{-2}	56.9	β-range		
In Ni					
Cu	0.57	258.3	1 327–1 632	IVa(i), p.c., Cu^{64}, 99.955%	35
	0.61	255.0	1 080–1 613	IVa(i) (At. abs. analysis), p.c., 99.95%	340
	0.27	255.5	1 048–1 323	IVa(iii), s.c., Cu^{64}, 99.999%	45
Ag	8.25	282.2	1 123–1 323	IVb, s.c., Ag^{110}, 99.99%	341
	8.94	279.4	1 297–1 693	IVa(i), s.c., Ag$^{110/106}$, 99.98%	237
Au	2.0	272.1	1 173–1 373	IVa(ii), p.c., Au^{198}, 99.98%	37
Al	1.0	260	914–1 212	IVa(iii) (SIMS), s.c., 99.99%	342
In	$A_1 = 1.26$ $A_2 = 1.9 \times 10^4$	$Q_1 = 251.0$ $Q_2 = 397.8$ $\rbrace$	777–1 659	Least squares fit, data of 238 [IVa(i) s.c., In^{114}, 99.98%] and 343 [IVa(iii) (SIMS), s.c., 99.99%]	5
C	0.12	137.3	873–1 673	IIa(ii), p.c., C^{14}, 0.1 wt% C	50
Ge	2.1	264.0	939–1 675	IVa(i) (ion impl.), s.c., Ge^{68}, 99.99%	344
Sn	4.56	267.2	1 242–1 642	IVa(i), s.c., Sn^{113}, 99.98%	90
As	1.39	251.8	1 239–1 634	IVa(i), s.c., As^{73}, 99.98%	345
Sb	3.85	264.0	1 203–1 674	IVa(i), s.c., Sb^{125}, 99.98%	52
S	1.4	219.0	1 078–1 495	IVa(i), s.c., S^{35}, 99.98%	236
Te	2.6	254	1 135–1 553	IVa(i), s.c., Te^{123}, 99.99%	346
Be	1.9×10^{-2}	193.4	1 293–1 673	IVb, p.c., Be^7, 99.9%	147
Hf	1.8$^{(a)}$	251.0	1 023–1 423	IIa(ii) (EMPA), p.c., 99.99%	347
V	0.87	278.4	1 073–1 573	IVb, p.c., V^{48}, 99.99%	143
Cr	1.1	272.6	1 373–1 541	IVa(i), p.c., Cr^{51}, 99.95%	47
W	2.0	299.4	1 373–1 568	IVa(i), p.c., W^{185}, 99.95%	59
	2.87	308.1	1 346–1 668	IVa(i), s.c., W^{181}, 99.98%	49
Fe	1.0	269.4	1 478–1 669	IVa(i), s.c., Fe^{59}, 99.999%	203
	0.22	252.9	1 223–1 643	IVb and IIa(i) (EMPA), p.c., 99.999%	34 and 48
Co	0.59	269.6	1 123–1 643	IVb and IIa(i) (EMPA), p.c., 99.999%	34 and 33
	2.77	285.1	1 335–1 696	IVa(i), s.c., Co^{57}, 99.98%	49
Pt	2.5	286.8	1 354–1 481	IVb, p.c., Pt^{193}, 99.99%	222

(*continued*)

Table 13.2   TRACER IMPURITY DIFFUSION COEFFICIENTS—*continued*

Element	A cm^2 s^{-1}	Q kJ mol^{-1}	Temp. range K	Method	Ref.
Ce	0.66	254.6	973–1 370	IVa(i), p.c., Ce^{141}, 99.99%	144
Nd	0.44	250.5	973–1 373	IVa(i), p.c., Nd^{147}, 99.99%	144
U	1.0	236.1	1 248–1 348	IVa(ii), p.c., U^{235}, 99.998%	145
Pu	0.64$^{(b)}$	213.5	1 298–1 398	IIIa(i), p.c., Pu^{239} (autorad.), 99.997%	146

(*a*) Estimated values.
(*b*) An average value: 146 quotes $A = 0.14$–1.14 cm^2 s^{-1}.

In Pb

Element	A cm^2 s^{-1}	Q kJ mol^{-1}	Temp. range K	Method	Ref.
Cu	7.9×10^{-3}	33.6	498–598	IVa(i), p.c. and s.c., Cu^{64}, —	85
	8.6×10^{-3}	34.2	491–803	IVa(i), s.c., Cu^{54}, 99.9999%	51
Ag	4.6×10^{-2}	60.5	398–598	IVa(i), s.c., Ag^{110}, —	85
	4.8×10^{-2}	60.8	470–750	IVa(i), s.c., Ag^{110}, 99.999%	239 and 348
Au	4.1×10^{-3}	39.1	367–598	IVa(i), s.c., Au^{195}, 99.999%	31
	5.2×10^{-3}	38.6	334–563	IVa(i), s.c., Au^{195}, 99.9999%	349
	5.8×10^{-3}	40.2	441–693	IVa(i), s.c., Au^{195}, 99.9999%	239 and 356
Zn	1.65×10^{-2}	47.8	453–773	IVa(i), s.c., Zn^{65}, 99.9999%	242
Cd	0.409	88.9	423–593	IVa(i), s.c., Cd^{115}, 99.9999%	93
	0.92	92.8	523–823	IVa(i), s.c., Cd^{105}, 99.9999%	350
Hg	1.05	95.0	466–573	IVa(i), s.c., Hg^{203}, 99.9999%	141
	1.5	86.7	523–823	IVa(i), s.c., Hg^{203}, 99.9999%	350
In	33	112.2	437–493	IVa(iii) (EMPA), s.c., 99.999%	139
Tl	0.511	101.9	480–596	IVa(i), p.c., Tl^{204}, 99.99%	32
Sn	0.41	94.4	523–723	IVa(i), s.c., Sn^{113}, 99.9999%	243
	0.29	99.4	468–595	Resistivity method	351
Sb	0.29	92.9	461–588	IVa(i), s.c., Sb^{124}, 99.9999%	140
Bi	$D = 2.66 \times 10^{-10}$		563.4	IVa(i), p.c., Bi, 99.99%	32
	$D = 1.006 \times 10^{-9}$		595.7		
Co	9×10^{-3}	46.4	383–573	IVa(iii) (NRA), p.c., 99.9999%	241
Ni	9.4×10^{-3}	44.5	481–593	IVa(i), s.c., Ni^{63}, 99.9999%	240
	1.1×10^{-2}	45.4	432–523	IVa(i), p.c., Ni^{63}, 99.999%	352
Pd	3.4×10^{-3}	35.4	470–590	IVa(i), s.c., Pd^{109}, 99.9999%	239
Pt	1.1×10^{-2}	42.3	490–593	IVa(i), s.c., (conc. via M.Pt.) 99.9999%	353
Na	6.3	118.5	522–586	IVa(i), s.c., Na^{22}, 99.9999%	354

In Pd

Element	A cm^2 s^{-1}	Q kJ mol^{-1}	Temp. range K	Method	Ref.
Fe	0.18	260.0	1 373–1 523	IVa(i), p.c., Fe^{59}, 99.95%	357

In Pr

Element	A cm^2 s^{-1}	Q kJ mol^{-1}	Temp. range K	Method	Ref.
Cu	8.4×10^{-2}	75.8	926–1 059 (α)	IVa(i), p.c., Cu^{64}, 99.9%	199
	5.7×10^{-2}	74.5	1 086–1 187 (β)		
Ag	1.4×10^{-1}	106.3	886–1 040 (α)	IVa(i), p.c., Ag^{110}, 99.93%	200
	3.2×10^{-2}	90.0	1 085–1 195 (β)		
Au	4.3×10^{-2}	82.5	870–1 015 (α)	IVa(i), p.c., Au^{198}, 99.93%	200
	3.3×10^{-2}	84.2	1 075–1 185 (β)		
Au	$\|c$ $D = 4.4 \times 10^{-6}$		1 013 (α)	IVa(i), s.c., Au^{198}, 99.94%	355
	$\perp c$ $D = 3.7 \times 10^{-6}$				
	$\|c$ $D = 4.6 \times 10^{-6}$		1 053 (α)		
	$\perp c$ $D = 4.0 \times 10^{-6}$				
Zn	1.8×10^{-1}	103.8	876–1 040 (α)	IVa(i), p.c., Zn^{65}, 99.97%	202
	6.3×10^{-1}	113.0	1 095–1 194 (β)		
In	$D = 3.2 \times 10^{-9 (a)}$		1 039 (α)	IVa(i), p.c., In^{114}, 99.96%	201
	9.6×10^{-2}	121.0	1 075–1 200 (β)		

(*continued*)

Table 13.2 TRACER IMPURITY DIFFUSION COEFFICIENTS—*continued*

Element	A cm^2 s^{-1}	Q kJ mol^{-1}	Temp. range K	Method	Ref.
La	$D = 1.1 \times 10^{-9\,(a)}$ 1.8×10^{-2}	107.6	1 041 (α) 1 080–1 190 (β)	IVa(i), p.c., La^{140}, 99.96%	201
Hb	$D = 3.35 \times 10^{-9\,(a)}$ 9.5×10^{-3}	110.1	1 004 (α) 1 085–1 180 (β)	IVa(i), p.c., Ho^{166}, 99.96%	201
Mn	Graphical data only. *See* Figure	13.3	α- and β-range	IVa(i), p.c., Mn^{54}, 99.9%	220
Fe	2.1×10^{-3} 4.0×10^{-3}	39.4 43.5	885–1 060 (α) 1 075–1 180 (β)	IVa(i), p.c., Fe^{59}, 99.9%	220
Co	4.7×10^{-2} $D = 4.9 \times 10^{-5}$ $D = 5.0 \times 10^{-5}$	68.7	885–1 036 (α) 1 092 (β) 1 151 (β)	IVa(i), p.c., Co^{60}, 99.93%	200

(*a*) Values estimated from graphical data.

In Pt

Ag	0.13	258.1	1 473–1 873	IIa(ii) (EMPA), p.c., 99.99%	244
Au	0.13	252.0	850–1 265	IVa(i), s.c., Au^{199}, 99.99%	245
Al	1.3×10^{-3}	193.6	1 373–1 872	IIa(ii) (EMPA), p.c., 99.99%	244
Mn	7.3×10^{-7}	122	1 023–1 223	IVa(i), p.c., Mn^{54}	420
Fe	2.5×10^{-2}	243.4	1 273–1 673	IIa(ii) (EMPA), p.c., 99.99%	244
Co	19.6	310.7	1 023–1 323	IVc, p.c., Co^{60}, 99.99%	204

In Pu

Cu	1.0×10^{-3}	51.5	773–853 (ε)	IIa(ii), p.c.	247
Ag	$D = 1.08 \times 10^{-10}$ 4.9×10^{-15}	40.2	695 (δ) 772–884 (ε)	IVa(i), p.c., Ag^{110m}	247
Au	$D = 2.37 \times 10^{-10}$ 5.7×10^{-5}	43.1	713 (δ) 788–887 (ε)	IVa(i), p.c., Au^{198}	247
Co	1.2×10^{-2} 1.4×10^{-3}	53.2 41.4	617–699 (δ) 757–894 (ε)	IVa(i), p.c., Co^{60} IVa(i), p.c., Co^{60}	246 253

In Sc

C	4.5	205	1 273–1 573	IVb, p.c., C^{14}	423
Fe	1.5×10^{-19} $D = 4.1 \times 10^{-5}$ $D = 2.6 \times 10^{-5}$ $D = 3.4 \times 10^{-5}$ $D = 4.4 \times 10^{-5}$	54	1 241–1 528 (α) 1 643 (β) 1 702 (β) 1 755 (β) 1 790 (β)	IIa(ii) (SMLS), p.c., 99.96%	406

In Se

S	$\|c\ 1.10 \times 10^{-5}$ $\perp c\ 1.7 \times 10^{3}$	57.8 111.0	333–363	IVb, s.c., S^{35}	205
Tl	2×10^{-3}	69.5	333–423	IVc, p.c., Tl^{204}	206

In Sm

C	3.6	146.4	773–1 173 (α)	IVb, p.c., C^{14}	55

In Sn

Cu	$\|c\ D = 2 \times 10^{-6}$ $\perp c\ 2.4 \times 10^{-3}$	33.1	298 413–503	IVa(i), s.c., Cu^{64}	95
Ag	$\|c\ 7.1 \times 10^{-3}$ $\perp c\ 0.18$	51.5 77.0	408–498	IVa(i), s.c., Ag^{110}	87
Au	$\|c\ 5.8 \times 10^{-3}$ $\perp c\ 0.16$	46.1 74.1	408–498	IVa(i), s.c., Au^{198}	87
Zn	$\|c\ 1.1 \times 10^{-2}$ $\perp c\ 8.4$	50.2 89.2	408–496	IVa(i), s.c., Zn^{63}, 99.999%	198

(*continued*)

Table 13.2   TRACER IMPURITY DIFFUSION COEFFICIENTS—*continued*

Element	A cm^2 s^{-1}	Q kJ mol^{-1}	Temp. range K	Method	Ref.
Cd	$\parallel c$ 220 $\perp c$ 0.18	118.1 77.0	463–498	IVa(i), s.c., Cd^{109}, 99.9999%	198
Hg	$\parallel c$ 7.5 $\perp c$ 30	105.9 112.2	447–499	IVa(i), s.c., Hg^{203}, 99.9999%	136
In	$\parallel c$ 12.2 $\perp c$ 34.1	107.2 108.0	453–494	IVa(i), s.c., In^{114}, 99.998%	30
Tl	1.3×10^{-3}	61.5	410–489	IVa(i) and (ii), p.c., Tl^{204}, 99.999%	138
Sb	$\parallel c$ 7 $\perp c$ 73	121.8 123.1	466–499	IVa(i), s.c., Sb, 99.999%	198
Fe	4.8×10^{-4}	51.1	387–462	Vd, p.c., Fe^{57}, 99.9995%	359
Co	5.5	92.1	413–490	IVb, s.c. and p.c., Co^{60}, 99.999%	382
Ni	$\parallel c$ 1.92×10^{-2} $\perp c$ 1.87×10^{-2}	18.1 54.2	298–373 393–473	IVa(i), s.c., Ni^{63}, 99.999%	365

In Ta

Al	1.5	306.2	1 750–2 050	IIIb(ii)	176
C	6.7×10^{-3}	161.6	463–2 953	Combined data, several sources	194
S	0.01	293.1	1 970–2 110	IVb, S^{35}, p.c., 99.0%	162
Y	0.12	302.3	1 473–1 773	IVb, Y^{91}Y, s.c., 99.8/99.9%	175
Nb	$0.23^{(a)}$	413.2	1 194–2 757	IVa(i), Nb^{95}, s.c. and p.c., 99.7%	77
Mo	1.8×10^{-3}	339.1	2 923–2 493	IVb, Mo^{99}, p.c.	358
Fe	0.505	298.9	1 203–1 513	—, —, p.c., —	76
	5.9×10^{-2}	329.9	2 053–2 330	IVa(i), Fe^{59}, p.c.	360
Co	$D = 1.4 \times 10^{-9}$ $D = 8.0 \times 10^{-9}$	(b)	2 128 2 330	IVa(i), Co^{60}, p.c.	360
Ni	$D = 1.1 \times 10^{-9}$ $D = 1.14 \times 10^{-9}$	(b)	2 053 2 830	IVa(i), Ni^{60}, p.c.	360
U	7.6×10^{-5} 1.03×10^{-6}	353.4 117.2	1 873–2 243 2 186–2 530	IVb, U^{235}, p.c., — IVa(ii) (fission fragment,, radiography), nat. U, p.c., 99.9997%	57 329

(a) Average values—Arrhenius plot very slightly curved.
(b) Values estimated from graphical representation of results.

In Te

Hg	3.4×10^{-5}	78.3	543–713	IVa(i), p.c., Hg^{203}	81
Sb	$\parallel c$ 4.6×10^{-2} $\perp c$ 1.34	147 165	551–647	IVa(i), s.c., Sb^{122}	410
Se	2.6×10^{-2}	119.7	593–713	IVa(i), p.c., Se^{75}	81

In Th
α—Thorium

Fe	5×10^{-3}	80.8	1 238–1 558	IIa(i) (SSMS), p.c., 99.95%	248
Co	5×10^{-4}	55.3	1 238 × 1 558	IIa(ii) (SSMS), p.c., 99.95%	248
Ni	4×10^{-3}	77.9	1 238 × 11 558	IIa(ii) (SSMS), p.c., 99.95%	248
Pa	1.26×10^{2}	312.8	1 039–1 183	IVa (via α-emission spectra), Pa^{231}, p.c., 99.84%	207
U	2.21×10^{4}	332.0	963–1 150	IVa (via α-emission spectra), U^{233}, p.c., 99.84%	207

β—Thorium

C	2.2×10^{-2}	113.0	1 713–1 193	IIa(ii), p.c.	407
Zr	1.73×10^{4}	384	1 773–1 873	IIa(ii) (SMLS), p.c., 99.977%	390
Hf	$D = 1.09 \times 10^{-8}$ $D = 2.09 \times 10^{-7}$		1 693 1 963	IVa(i), p.c., Hf^{181}	391
V	1.9×10^{-2}	129.8	1 643–1 933	IIa(ii) (SSMS), p.c., 99.95%	249
Nb	0.5	201.8	1 643–1 933	IIa(ii) (SSMS), p.c., 99.95%	249

(continued)

Table 13.2   TRACER IMPURITY DIFFUSION COEFFICIENTS—*continued*

Element	A cm^2 s^{-1}	Q kJ mol^{-1}	Temp. range K	Method	Ref.
Ta	0.57	210.6	1 648–1 933	IIa(ii) (SSMS), p.c., 99.95%	249
Mo	15.1	216	1 698–1 873	IIa(ii) (SMLS), p.c., 99.977%	390
W	0.103	160	1 683–1 818	IIa(ii) (SMLS), p.c., 99.977%	390
Re	4.04×10^{-3}	84	1 663–1 943	IIa(ii) (SLMS), p.c., 99.977%	390
Fe	4×10^{-3}	71.6	1 613–1 898	IIa(ii) (SSMS), p.c., 99.95%	248
Co	4×10^{-3}	65.3	1 613–1 898	IIa(ii) (SSMS), p.c., 99.95%	248
Ni	4×10^{-4}	38.1	1 613–1 898	IIa(ii) (SSMS), p.c., 99.95%	248

In Ti
α—Titanium

Element	A cm^2 s^{-1}	Q kJ mol^{-1}	Temp. range K	Method	Ref.
Al	9.7×10^{-5}	115.1	973–1 123	'X-ray diff. method', p.c.	366
	7.4×10^{-7}	156.4	873–1 123	IIa(ii) (ion impl.) (N.R.A.), p.c., 99.9%	367
C	7.9×10^{-4}	127.7	873–1 073	IVb, p.c., C^{14}	83
Si	4.4×10^{-7}	105.2	923–1 073	IIIb(i) (ion impl.) (N.R.A.), p.c., 99.9%	368
P	$\|c\ 1.55 \times 10^{-1}$ $\perp c\ 4.70$	138.2 172.3 }	973–1 123	IVa(i), P^{32}, s.c. For Ts < 973, D is < as calc. from $D°$ and Q	369
Mn	$D = 1.42 \times 10^{-9}$ $D = 5.21 \times 10^{-10}$		1 124 1 072 }	IVa(i), Mn^{54}, p.c., 99.998%	370
	$\|c\ 6.0 \times 10^{-1}$ $\perp c\ 4.9 \times 10^{-2}$	189.2 160.5 }	878–1 135	IVa(i), Mn^{54}, s.c., 99.94%	371
Fe	$\|c\ 6.4 \times 10^{-2}$ $\perp c\ 4.7 \times 10^{-3}$	144.2 112.3 }	877–1 136	IVa(i), Fe^{59}, s.c., 99.96%	372
	1.2×10^{-4}	110.5	973–1 123	IVb, Fe^{59}, p.c., —	373
Co	$D = 6.65 \times 10^{-8}$ $D = 3.64 \times 10^{-8}$		1 129 1 072 }	IVa(i), Co^{60}, p.c., 99.998%	370
	$\|c\ 3.2 \times 10^{-2}$ $\perp c\ 1.9 \times 10^{-2}$	126.1 114.1 }	871–1 135	IVa(i), Co^{60}, s.c., 99.96%	374 and 375
Ni	$D = 2.1 \times 10^{-8}$ $D = 2.3 \times 10^{-8}$ $D = 7.6 \times 10^{-9}$ $D = 1.8 \times 10^{-9}$ $D = 6.7 \times 10^{-10}$		1 141 1 117 1 059 971 912 }	IVa(i), Ni^{63}, p.c., "high purity"	376
	$\|c\ 5.4 \times 10^{-2}$ $\perp c\ 5.6 \times 10^{-2}$	141.8 137.2 }	877–1 100	IVa(i), s.c., Ni^{63}, 99.96%	374
U	4.1×10^{-7}	114.5	1 020–1 124	—, —, —, —	377

β—Titanium

Element	A cm^2 s^{-1}	Q kJ mol^{-1}	Temp. range K	Method	Ref.
Cu	$A_1 = 11.3$ $A_2 = 2.1 \times 10^{-3}$	$Q_1 = 252$ $Q_2 = 122.3$ }	1 233–1 733	IVa(iii) (EMPA), p.c., 'iodide Ti'	88
Ag	3.0×10^{-3}	180.0	1 213–1 923	IVa(i), p.c., Ag^{110}, 99.95%	84
C	3.2×10^{-3}	79.1	1 223–1 923	IVb, p.c., C^{14}	366
Sn	$A_1 = 9.5$ $A_2 = 3.8 \times 10^{-4}$	$Q_1 = 289.7$ $Q_2 = 132.3$ }	1 226–1 868	IVa(i), Sn^{113}, p.c., 99.7%	65
P	$A_1 = 5.0$ $A_2 = 3.62 \times 10^{-4}$	$Q_1 = 236.6$ $Q_2 = 100.9$ }	1 218–1 873	IVa(i), P^{32}, p.c., 99.9%	65
Be	0.80	168.3	1 188–1 573	IVb, p.c., Be^7, 99.62%	187
Sc	4.0×10^{-3}	135.7	1 213–1 843	IVa(i), Sc^{46}, p.c., 99.95%	84 and 65
Zr	4.7×10^{-3}	148.2	1 193–1 773	IVb, Zr^{95}, p.c., 98.94%	70
	1.59×10^{-8}	316	1173–1773	IIa(i), p.c., Zr^{95}	424
	1.60×10^{-8}	322	1173–1773	IIa(i), p.c., Zr^{95}	424
	1.62×10^{-8}	329	1173–1773	IIa(i), p.c., Zr^{95}	424
	1.64×10^{-8}	334	1173–1773	IIa(i), p.c., Zr^{95}	424
V	$A_1 = 3.4$ $A_2 = 1.0 \times 10^{-3}$	$Q_1 = 257.5$ $Q_2 = 145.3$ }	1 173–1 813	IVa(i), V^{48}, p.c., 99.9%	63 and 65

(continued)

Table 13.2   TRACER IMPURITY DIFFUSION COEFFICIENTS—*continued*

Element	A cm^2 s^{-1}	Q kJ mol^{-1}	Temp. range K	Method	Ref.
Nb	$A_1 = 20$ $A_2 = 5.0 \times 10^{-3}$	$Q_1 = 305.6$ $Q_2 = 164.5$	1 273–1 923	IVa(ii), Nb^{95}, p.c., 99.7%	64
Ta	$A_1 = 13$ $A_2 = 3.0 \times 10^{-4}$	$Q_1 = 309.8$ $Q_2 = 140.3$	1 187–1 873	IVa(i), Ta^{182}, p.c., 'iodide Ti'	92
Cr	$A_1 = 4.9$ $A_2 = 5.0 \times 10^{-3}$	$Q_1 = 255.4$ $Q_2 = 147.8$	1 243–1 923	IVa(ii), Cr^{51}, p.c., 99.7–99.9%	64
Mo	$A_1 = 20$ $A_2 = 8.0 \times 10^{-3}$ 0.24 2.82×10^{-4}	$Q_1 = 305.6$ $Q_2 = 180.0$ 214.8 139.0	1 173–1 923 1 373–1 833 1 173–1 373	IVa(i), Mo^{99}, p.c., 99.7–99.9% IVb, Mo^{99}, p.c., 99.94%	64 70
W	3.6×10^{-3}	183.8	1 173–1 523	IVc, W^{185}, p.c., 98.94%	70
Mn	$A_1 = 4.3$ $A_2 = 6.1 \times 10^{-3}$	$Q_1 = 242.8$ $Q_2 = 141.1$	1 203–1 923	IVa(i), Mn^{54}, p.c., 99.7%	64
Fe	$A_1 = 2.7$ $A_2 = 7.8 \times 10^{-3}$ 5.6×10^{-3}	$Q_1 = 230.3$ $Q_2 = 132.3$ 131.0	1 193–1 923 1 273–1 473	IVa(ii), p.c., Fe^{55}, 99.9% IVb, Fe^{59}, p.c., —	64 373
Co	$A_1 = 2.0$ $A_2 = 1.1 \times 10^{-2}$	$Q_1 = 219.8$ $Q_2 = 128.1$	1 183–1 923	IVa(i), p.c., Co, 99.7%	64
Ni	$A_1 = 2.0$ $A_2 = 9.2 \times 10^{-3}$	$Q_1 = 219.8$ $Q_2 = 123.9$	1 203–1 923	IVa(ii), Ni^{63}, p.c., 99.7%	64
U	$A_1 = 2.0 \times 10^{-2}$ $A_2 = 1.6 \times 10^{-5}$ 5.1×10^{-4} 2.0×10^{-3}	$Q_1 = 192.6$ $Q_2 = 89.2$ 122.7 138.1	1 173–1 773 1 173–1 673 1 188–1 298	—, —, —, — IVb, U^{235}, p.c., 99.62% Ia(ii) (fission fragment, radiography), nat. U, p.c., 99.34%	377 195 160
Pu	1.4×10^{-6}	64.1	1 173–1 400	IIa (E.M.P.A. and α-radiography) p.c.3.16	196

In Tl

Element	A cm^2 s^{-1}	Q kJ mol^{-1}	Temp. range K	Method	Ref.
Ag	$\|c$ 2.7×10^{-2} $\perp c$ 3.8×10^{-2} 4.2×10^{-2}	46.9 49.4 49.8	388–493 (hcp) 513–573 (bcc)	IVa(i), s.c., Ag^{110}, 99.9999% IVa(i), p.c., Ag^{110}, 99.9999%	97
Ag	$\|c$ 2.0×10^{-5} $\perp c$ 5.3×10^{-4} 5.2×10^{-4}	11.7 21.8 25.1	363–493 (hcp) 513–573 (bcc)	IVa(i), s.c., Ag^{110}, 99.9999% IVa(i), p.c., Ag^{110}, 99.9999%	97

In U
α—Uranium

Element	A cm^2 s^{-1}	Q kJ mol^{-1}	Temp. range K	Method	Ref.
Fe	$D \simeq 3 \times 10^{-10}$		918	via rate of ppt. dissolution, p.c.	381

β—Uranium

Element	A cm^2 s^{-1}	Q kJ mol^{-1}	Temp. range K	Method	Ref.
Cr	$D = 3.6 \times 10^{-10}$ $^{(b)}D = 1.07 \times 10^{-9}$ $D = 1.77 \times 10^{-9}$		943 1 013 1 021.2	IVa(i), p.c., Cr^{51}, 99.98 IVa(i), p.c., Cr^{51}, —	193 80
Fe	$^{(a)}D = 8.71 \times 10^{-9}$ $^{(b)}D = 2.6 \times 10^{-8}$		974 1 033	IVa(i), p.c., Fe^{59}, 99.993%	80
Co	1.54×10^{-2}	114.9	964–1 036	IVa(i), p.c., Co^{60}, 99.98%	193

γ—Uranium

Element	A cm^2 s^{-1}	Q kJ mol^{-1}	Temp. range K	Method	Ref.
Cu	1.96×10^{-3}	100.7	1 059–1 312	IVa(i), p.c., Cu^{64}, 99.99%	78
Au	4.86×10^{-3}	127.3	1 057–1 280	IVa(i), p.c., Au^{195}, 99.99%	79
C	0.218	123	1 130–1 270	IIa(ii), p.c., C^{14}	408

(*continued*)

Table 13.2   TRACER IMPURITY DIFFUSION COEFFICIENTS—*continued*

Element	A cm^2 s^{-1}	Q kJ mol^{-1}	Temp. range K	Method	Ref.
Nb	4.87×10^{-2}	166.0	1 063–1 376	IVa(i), p.c., Nb[95], 99.99%	78
Cr	5.47×10^{-3}	102.4	1 070–1 311	IVa(i), p.c., Cr[51], 99.99%	78
Mn	1.81×10^{-4}	58.1	1 060–1 212	IVa(i), p.c., Mn[54], 99.99%	78
Fe	2.69×10^{-4}	50.3	1 059–1 263	IVa(i), p.c., Fe[59], 99.99%	78
Co	3.51×10^{-4}	52.6	1 056–1 263	IVa(i), p.c., Co[560], 99.99%	78
Ni	5.36×10^{-4}	65.57	1 059–1 313	IVa(i), p.c., Ni[63], 99.99%	78

(*a*) The mean of two values.
(*b*) Values reported at five other *T*s between 943 and 1 013, but results too scattered for Arrhenius.

In V

Element	A cm^2 s^{-1}	Q kJ mol^{-1}	Temp. range K	Method	Ref.
C	8.8×10^{-3}	116.364	333–2 098	Combined data—several sources	38
P	2.45×10^{-2}	208.5	1 473–1 723	IVb, p.c., P[32], 99.98%	163
S	3.1×10^{-2}	142.4	1 320–1 520	IVb, p.c., S[35], 99.8%	162
Ti	0.1	285.0	1 373–1 623 ⎱	IVa(i), p.c., Ti[44], 99.98%	265 and 383
	34.1	363.9	1 623–2 076 ⎰		
Zr	81	369.2	1 578–1 883	IVa(i), p.c., Zr[95], 99.95%	415
Ta	0.244	301.4	1 373–2 073	IVa(i), p.c., Ta[182]	251
Cr	9.54×10^{-3}	270.5	1 233–1 473	IVb, p.c., Cr[51], 99.8\5	62
Fe	0.373	297.3	1 233–1 618 ⎱	IVa(i), s.c., Fe[59]	56
	274	385.9	1 688–2 090 ⎰		
	2.48	318.7	1 473–1 823 ⎱	IVa(i) and (iii), s.c. and p.c., Fe[59]	384
	31.7	356.6	1 823–2 088 ⎰		
Co	1.12	295	1 298–2 126	IVa(i), s.c., Co[60]	250
Ni	0.18	266.24	1 175–1 948	IVb, s.c. and p.c., Ni[63]	385
U	1×10^{-4}	257.1	1 373–1 773	IVb, p.c., U[235]	57

In W

Element	A cm^2 s^{-1}	Q kJ mol^{-1}	Temp. range K	Method	Ref.
C	9.22×10^{-3}	169.1	2 073–3 073	IIa(ii), p.c., C[14]	378
	3.15×10^{-3}	172	373–673	Va, p.c.	89
P	26.8	510	2 153–2 453	IIIa(i), s.c., P[32], 99.99%	54
S	2.17×10^{-5}	292.2	2 153 × 2 453	IVb, s.c., S[35]	386
Y	6.7×10^{-3}	285.1	1 473–1 873	IVb, s.c., Y[91], 99.8/99.9%	175
Nb	3.01	576.1	1 578–2 640	IVa(i), s.c. and p.c., Nb[95]	192
Ta	3.05	585.7	1 578–2 648	IVa(i), s.c. and p.c., Ta[182]	192
Mo	0.05	506.6	2 273–2 673	IV, p.c., Mo[99]	191
	0.3	423	1 973–2 373	IVb, p.c., Mo[99]	334
	1.4	567.3	1 909–2 658	Ia(iii) (SIMS), s.c.	387
Re	275	681.6	2 939–3 501	IVa(i), s.c., Re[183/4], 99.99%	75
	4.0	597	2 110–2 900	IVa(i), s.c., Re[186]	190
Fe	0.014	276.3	1 213–1 513	—, p.c., Fe[59]	76
Os	0.64	538.4	2 105–2 928	IVa(i), s.c., Os[191]	190
Ir	0.32	506.2	2 007–2 960	IVa(i), s.c., Ir[192]	190
Ni	$D = 6 \times 10^{-11}$		1 913	IIa(i) (EMPA), p.c.	392
U	2×10^{-3}	433.3	1 973–2 473	IV, p.c., U[235]	57
	1.8×10^{-2}	389.4	2 245–3 000	IIIb(ii), p.c., 99.99%	186

In Y

Element	A cm^2 s^{-1}	Q kJ mol^{-1}	Temp. range K	Method	Ref.
Ag	5.4×10^{-3}	77	1 168–1 453 (α)	IIa(ii), p.c., 99 + %	339
C	1.7×10^4	272	1 273–1 733 (α)	IVb and IIa(ii), p.c. Combined data, recalculated	96 and 409
Zr	4.0×10^{-3}	159	1 273–1 573 (α)	IIIa(i), p.c.	411
Fe	1.8×10^{-2}	85	1 173–1 603 (α)	IIa(ii), p.c., 99 + %	339
Co	1.4×10^{-2}	83.3	1 290–1 620 (α)	IIa (LSMS), p.c., 99.6%	413
Ni	5.8×10^{-2}	96.5	1 290–1 580 (α)	IIa (LSMS), p.c., 99.6%	413

(*continued*)

Table 13.2   TRACER IMPURITY DIFFUSION COEFFICIENTS—*continued*

Element	A cm^2 s^{-1}	Q kJ mol^{-1}	Temp. range K	Method	Ref.
In Zn					
Cu	$\parallel c$ 2.22 $\perp c$ 2.00	123.6 125.3	611–688	IVa(i), s.c., Cu[64], 99.999%	86
Ag	$\parallel c$ 0.32 $\perp c$ 0.45	108.9 115.6	544–686	IVa(i), s.c., Ag[110], 99.999%	23
Au	$\parallel c$ 0.97 $\perp c$ 0.29	124.3 124.3	588–688	IVa(i), s.c., Au[198], 99.999%	24
Cd	$\parallel c$ 0.114 $\perp c$ 0.117	86.0 85.5	498–689	IVa(i), s.c., Cd[115], 99.999%	24
Hg	$\parallel c$ 0.056 $\perp c$ 0.073	82.5 84.5	533–686	IVa(i), s.c., Hg[203], 99.999%	98
Ga	$\parallel c$ 0.016 $\perp c$ 0.018	77.0 76.0	513–676	IVa(i), s.c., Ga[72], 99.999%	86
In	$\parallel c$ 0.062 $\perp c$ 0.14	80.0 82.1	444–689	IVa(i), s.c., In[114], 99.999%	23
C	1×10^{-5}	50.2	439–656	IVb, p.c., C[14]	96
Sn	$\parallel c$ 0.15 $\perp c$ 0.13	81.2 77.0	571–673	IVa(i), s.c., Sn[113], 99.999%	128
Ni	$\parallel c$ 8.1 $\perp c$ 0.43	136.6 121.5	564–664	IVa(ii), s.c., Ni[63], 99.999%	91
In Zr *α–Zirconium*					
Cu	$\parallel c$ 0.25 $\perp c$ 0.40	154.5 148.7	888–1 132	IVa(i), s.c., Cu[64], 99.95%	137
Ag	5.1×10^{-3}	187.1	1 037–1 120	IVa(i), p.c., Ag[110], 99.99%	254
	$\parallel c\ 6.7 \times 10^{-4}$ $\perp c\ 5.9 \times 10^{-2}$	173.8 212.4	1 063–1 118	IVa(i), s.c., Ag[110m] 99.999%	257
Au	$D = 1.3 \times 10^{-11}$		1 113	IVa(i), s.c. (unspec. orientation), Au[198], 99.93%	181
Zn	$D = 2.8 \times 10^{-11}$		1 099	IVa(i), s.c. (unspec. orientation), Zn[65], 99.93%	181
Al	$D = 3.4 \times 10^{-13}$		1 108	IVa(i), s.c. (unspec. orientation), Al[26], 99.93%	376
C	2.0×10^{-3}	151.6	873–1 123	IVb, p.c., C[14]	188
Sn	1.0×10^{-8}	92.1	'α-phase'	IVb, p.c., Sn[113], 99.0–99.9%	71
Sb	$D = 1.4 - 2.6 \times 10^{-13}$		1 120	IVa(i), s.c. (unspec. orientation), Sb[122], 99.93%	376
S	8.9	185.1	870–1 080	IVa(i), p.c., S[35], 99.94%	189
Rb	1.17×10^2	255.4	1 033–1 136	IIIb(ii), p.c. (poss. g.b. influence)	186
Be	0.33	133.6	933–1 120	IVa(i) and b, p.c., Be[7], 99.99%	255
Ti	$D = 9.4 \times 10^{-13}$		1 116	IVa(i), s.c. (unspec. orientation), Ti[44], 99.93%	376
V	1.12×10^{-8}	95.6	873–1 123	IVb, p.c., V[48], 99.84%	171
Nb	6.6×10^{-6}	131.9	1 013–1 130	IVb, p.c., Nb[95], 99.99%	183
Ta	1.0×10^2	293.1	973–1 073	IVb, p.c., Ta[182], 99.6%	73
Cr	$\parallel c$ 0.2 $\perp c$ 0.2	162.7 153.3	1 023–1 121	IVa(i), s.c., Cr[51], 99.99%	178
Mo	6.22×10^{-8}	103.7	873–1 123	IVb, p.c., Mb[99]	184
Mn	2.4×10^{-3}	126.4	893–1 083	IVb, p.c., Mn[54], 99.5 and 99.999%	179
Fe	$D = 3.7 \times 10^{-8}$ $D = 3.5 \times 10^{-7}$		973 1 071	IVa(i), s.c., Fe[59], 99.93%	180
	$D = 7.0 \times 10^{-7}$		1 113	IV(i), s.c. (unspec. orientation), Fe[59], 99.93%	376

(continued)

Table 13.2   TRACER IMPURITY DIFFUSION COEFFICIENTS—*continued*

Element	A cm^2 s^{-1}	Q kJ mol^{-1}	Temp. range K	Method	Ref.
Co	$\perp c$ 1.2×10^3	183.4	<873		
	$\perp c$ 37	145.8	>923	IVa(i), s.c., Co^{58}, 99.7 and 99.9%	394
	$\|c$ 4×10^4	191.2	860–990		
Ni	$D = 1.7 \times 10^{-7}$		971		
	$D = 4.0 \times 10^{-7}$		$1\,023$	IVa(i), s.c. (unspec. orientation),	180
	$D = 9.0 \times 10^{-7}$		$1\,074$	Ni^{63}, 99.93%	
	$D = 8.0 \times 10^{-7}$		$1\,103$		
	$\perp c$ $D = 1.6 \times 10^{-6}$		$1\,123$	IVa(i), s.c., Ni^{63}	395
	$\|c$ $D = 6.0 \times 10^{-6}$				
Ce	3.5×10^{-7}	106.2	923–$1\,123$	IVb, p.c., Ce^{141}	395

β—Zirconium

Element	A cm^2 s^{-1}	Q kJ mol^{-1}	Temp. range K	Method	Ref.
Ag	5.7×10^{-4}	136.9	$1\,224$–$1\,463$	IVa(i) and b, p.c., Ag^{110}, 99.99%	254
	$A_1 = 4.2 \times 10^{-4}$	$Q_1 = 132.3$	$1\,199$–$1\,988$	IVa(i), p.c., Ag^{110m}	396
	$A_2 = 190.5$	$Q_2 = 324.4$			
C	8.9×10^{-2}	133.1	$1\,143$–$1\,523$	IVb, C^{14}	188
	4.8×10^{-3}	111.8	$1\,173$–$1\,523$	IIa(ii), p.c.	157
Sn	5.0×10^{-3}	163.3	'β-phase'	IVb, p.c., Sn^{113}, 99.999%	71
P	0.33	139.4	$1\,223$–$1\,473$	IVb, p.c., P^{32}, 99.4%	163
S	27.6	162.4	$1\,428$–$1\,523$	IVa(i), p.c., S^{35}, 99.94%	189
Rb	8.8×10^{-4}	153.7	$1\,153$–$1\,303$	IIIb(ii), p.c.	186
Be	8.33×10^{-2}	130.2	$1\,188$–$1\,573$	IVb, p.c., Be^7, 99.7/99.99%	255 and 187
V	7.59×10^{-3}	191.8	$1\,143$–$1\,473$	IVb, p.c., V^{48}, 99.84%	171
	0.32	239.5	$1\,473$–$1\,673$		
	8.9×10^{-5}	116.5	$1\,166$–$1\,480$	IVa(i), p.c., V^{48}, 99.8%	421
Nb	$9 \times 10^{-6} \times$ $(T/1\,136)^{18.1}$	$105.1 +$ $146.6 \times$ $(T$–$1\,136)$	$1\,155$–$2\,031$	IVa(i), p.c., Nb^{95}, 99.94%	69
	7.8×10^{-4}	153.2	$1\,503$–$1\,908$	IVb, p.c., Nb^{95}	397
	1.23×10^{-4}	131.9	$1\,167$–$1\,433$	IVa(i), p.c., Nb^{95}, 99.77%	398
Ta	5.0×10^{-5}	113.0	$1\,173$–$1\,473$	IVb, p.c., Ta^{182}, 99.6%	73
Cr	4.17×10^{-3}	134.0	$1\,173$–$1\,473$	IVb, p.c., Cr^{51}, 99.7%	70
	7.0×10^{-3}	142.3	$1\,187$–$1\,513$	IVa(i), p.c., Cr^{51}, 99.92–99.999%	399
Mo	3.63×10^{-2}	185.9	$1\,173$–$1\,473$	IVb, p.c., Mo^{99}, 99.7%	70
	1.29	243.7	$1\,628$–$1\,833$		
	$A_1 = 1.99 \times 10^{-4}$	$Q_1 = 147.4$	$1\,173$–$1\,873$	IVb, p.c., Mo^{99}	185
	$A_2 = 2.63$	$Q_2 = 285.9$			
W	0.41	233.6	$1\,173$–$1\,523$	IVc, p.c., W^{185}, 99.7%	70
Mn	5.6×10^{-3}	138.2	$1\,225$–$1\,420$	IVb, p.c., Mn^{54}, 99.5 and 99.999%	179
Fe	9.1×10^{-3}	113.0	$1\,173$–$1\,673$	IVb, p.c., Fe^{59}, 99.7%	70
	7.4×10^{-3}	108	$1\,176$–$1\,886$	IVa(i), p.c., Fe^{59}	74
	$\|c$ $D = 2.22 \times 10^{-9}$		765		
	1.75×10^{-8}		834		
	4.70×10^{-8}		871		
	3.70×10^{-7}		934		
	5.5–7.1×10^{-7}		980.5		
	6.1×10^{-7}		983		
	1.68×10^{-6}		$1\,032$	IVa(i), s.c., Fe^{59}, 99.98%	414
	2.25×10^{-6}		$1\,093$		
	2.2–3.6×10^{-6}		$1\,131$		
	2.85×10^{-6}		$1\,133$		
	$\perp c$ $D = 1.0 \times 10^{-8}$		871		
	1.2×10^{-7}		980.5		
	3.4×10^{-7}		$1\,032$		
	9.9×10^{-7}		$1\,131$		

(*continued*)

Table 13.2   TRACER IMPURITY DIFFUSION COEFFICIENTS—*continued*

Element	A $cm^2\,s^{-1}$	Q $kJ\,mol^{-1}$	Temp. range K	Method	Ref.
Co	3.26×10^{-3}	91.36	1 193–1 878	IVa(i), p.c., Co^{60}, 99.99%	182
	3.3×10^{-3}	92	1 193–1 741	IVa(i), p.c., Co^{57}	74
Ce	$A_1 = 3.16 \times 10^{-2}$ $A_2 = 42.17$	$Q_1 = 173.3$ $Q_2 = 310.2$	1 153–1 873	IVb, p.c., Ce^{141}	185
U	8.15×10^{-5}	111.4	1 223–1 573	IVb, p.c., U^{235}, 99.61%	408
	$A_1 = 0.36$ $A_2 = 3.0 \times 10^{-6}$	$Q_1 = 242.8$ $Q_2 = 82.5$	1 223–1 773	IVa, p.c., U^{235}	72

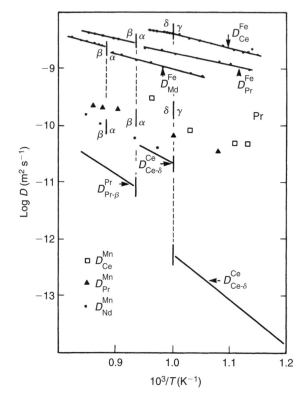

Figure 13.3   *Impurity diffusion of* Fe & Mn *in* CE, Pr & Nd

REFERENCES TO TABLE 13.2

1. J. Hino, C. Tomizuka and C. Wert, *Acta Met.*, 1957, **5**, 41.
2. C. Tomizuka. Quoted from D. Lazarus 'Solid State Physics' (1960) Vol. 10.
3. A. Ikushima, *J. Phys. Soc. Japan*, 1959, **14**, 1636.
4. C. A. Mackliet, *Phys. Rev.*, 1958, **109**, 1964.
5. M. Sakamoto, *J. Phys. Soc. Japan*, 1958, **13**, 845.
6. J. G. Mullen, *Phys. Rev.*, 1961, **121**, 1649.
7. G. Neumann and V. Tolle, *Phil. Mag. $*, 1988, **57**, 621.
8. K. Hoshino, Y. Iijima and K. Hirano, 'Point Defects and Defect Interactions in Metals', *Proceedings of the Yamada Conference V*, 1982, 562.
9. M. C. Inman and L. W. Barr, *Acta Met.*, 1960, **8**, 112.
10. N. L. Peterson, *Phys. Rev.*, 1963, **132**, 2471.

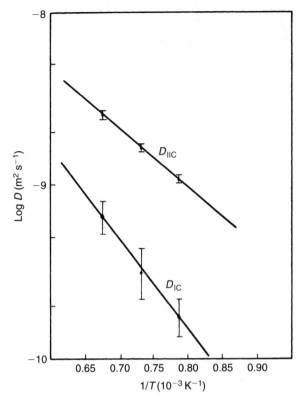

Figure 13.4   *Impurity diffusion of* Au *in* Er

11. C. T. Tomizuka and L. Slifkin, *Phys. Rev.*, 1954, **96**, 610.
12. E. Sonder, L. Slifkin and C. T. Tomizuka, *Phys. Rev.*, 1954, **93**, 970.
13. C. B. Pierce and D. Lazarus, *Phys. Rev.*, 1959, **114**, 686.
14. A. Sawatzky and F. E. Jaumot Jr., *J. Metals, N. Y.*, 1957, **9**, 1207.
15. A. Sawatzky and F. E. Jaumot Jr., *Phys. Rev.*, 1955, **100**, 1627.
16. R. E. Hoffmann, *Acta Met.*, 1958, **6**, 95.
17. T. Hirone, S. Miura and T. Suzuoka, *J. Phys. Soc. Japan*, 1961, **16**, 2456.
18. A. Vignes and J. P. Haeussler, *Men. Scient. Revue Metall.*, 1966, **63**, 1091.
19. W. C. Mallard, A. B. Gardner, R. F. Bass and L. M. Slifkin, *Phys. Rev.*, 1963, **129**, 617.
20. R. E. Hoffmann, D. Turnbull and E. W. Hart, *Acta Met.*, 1955, **3**, 417.
21. A. J. Mortlock and A. H. Rowe, *Phil. Mag.*, 1965, **11**, 1157.
22. A. J. Mortlock, A. H. Rowe and A. D. Le Claire, *Phil. Mag.*, 1960, **5**, 803.
23. D. N. Duhl, K. Hirano and M. Cohen, *Acta Met.*, 1963, **11**, 1.
24. J. H. Rosolowski, *Phys. Rev.*, 1961, **124**, 1828.
25. P. B. Ghate, *Phys. Rev.*, 1963, **131**, 174.
26. N. L. Peterson and S. J. Rothman, *Phys. Rev. B*, 1978, **17**, 4666.
27. N. L. Peterson and S. J. Rothman, *Phys. Rev.*, 1970, **B1**, 3264.
28. W. B. Alexander and L. M. Slifkin, *Phys. Rev.*, 1970, **B1**, 3274.
29. M. S. Anand and R. P. Agarwala, *Phys. Status Solidi*, 1970, **A1**, 41K.
30. G. M. Hood and R. S. Schultz, *Phys. Rev.*, 1971, **B4**, 2339.
31. A. Sawatsky, *J. Appl. Phys.*, 1958, **29**, 1303.
32. A. Ascoli, *J. Inst. Metals*, 1961, **89**, 218.
33. H. A. Resing and N. H. Nachtrieb, *J. Phys. Chem. Solids*, 1961, **21**, 40.
34. K. Hirano, R. P. Agarwala, B. L. Averbach and E. Cohen, *J. Appl. Phys.*, 1962, **33**, 3049.
35. M. Badia and A. Vignes, *Acta Met.*, 1969, **17**, 177.
36. K. Monma, H. Suto and H. Oikawa, *Nippon Kink. Gakk.*, 1964, **28**, 192.
37. S. Komura and N. Kunitomi, *J. Phys. Soc. Japan*, 1963, **18**, (Supp. II), 208.
38. A. D. Kurtz, B. L. Averbach and M. Cohen, *Acta Met.*, 1955, **3**, 442.
39. F. A. Schmidt and J. C. Warner, *J. Less Common Metals*, 1972, **26**, 325.

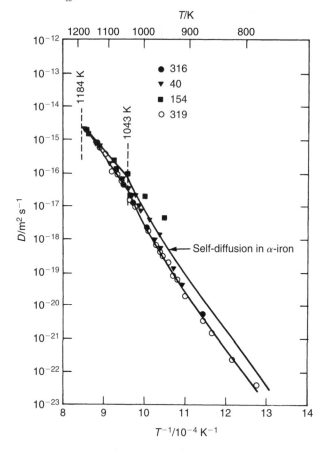

Figure 13.5 *Impurity diffusion of* Co *in* αFe *(replacing 13.2(b))*

40. H. W. Mead and C. E. Birchenall, *J. Metals*, 1955, **7**, 994.
41. R. J. Borg and D. Y. F. Lai, *Acta Met.*, 1963, **11**, 861.
42. K. Hirano, M. Cohen and B. L. Averbach, *Acta Met.*, 1961, **9**, 440.
43. T. Suzuoka, *Trans. Japan Inst. Metals*, 1961, **2**, 176.
44. M. Aucouturier and P. Lacombe, *Acta Met.*, 1965, **13**, 125.
45. S. V. Zemskiy and M. N. Spasskiy, *Fiz. Met. Metalloved.*, 1966, **21**, 129.
46. H. Helfmeier and M. Feller-Kniepmeier, *J. Appl. Phys.*, 1970, **41**, 3202.
47. P. L. Gruzin, S. V. Zemskii and I. B. Rodina, 'Metallurgy and Metallography of Pure Metals, No. 4, Moscow, 1963, 243, (AERE Trans. 1032).
48. K. Monma, H. Suto and H. Oikawa, *Nippon Kink Gakk.*, 1964, **28**, 188.
49. P. Guiraldenq. *Metaux. Corros. Inds.*, 1964, **39**, 347.
50. A. B. Vladimirov, V. N. Kaygorodov, S. M. Klotsman and I. Sh. Trakhtenberg, *Fiz. Met. Metalloved.*, 1978, **46**, 1232.
51. I. I. Kovenskiy, *Fiz. Met. Metalloved.*, 1963, **16**, 613.
52. D. L. Decker, *Phys. Rev. B*, 1975, **11**, 1770 and C. T. Candland, D. L. Decker and H. B. Vanfleet, *Phys. Rev. B*, 1972, **5**, 2085.
53. A. B. Vladimirov *et al.*, *Fiz. Met. Metalloved*, 1976, **41**, 429.
54. R. E. Eckert and H. G. Drickamer, *J. Chem. Phys.*, 1952, **20**, 13.
55. V. P. Iovkov, A. S. Panov and A. V. Ryabenko, *Izv. Akad Nauk SSSR, Met.*, 1978, **1**, 78. *Russian Metallurgy*, 1978, **78**.
56. K. Sato, *Trans. Japan Ins. Metals*, 1964, **5**, 91.
57. M. G. Coleman, C. A. Wert and R. F. Peart, *Phys. Rev.*, 1968, **175**, 788.
58. G. B. Federov *et al.*, *Sov. J. At. En.*, **31** (5), 1971, 1280.
59. J. R. G. da Silva and R. B. McLellan, *Mater. Sci. and Eng.*, 1976, **26**, 83.
60. K. Monma, H. Suto and H. Oikawa, *Nippon Kink Gakk.*, 1964, **28**, 197.

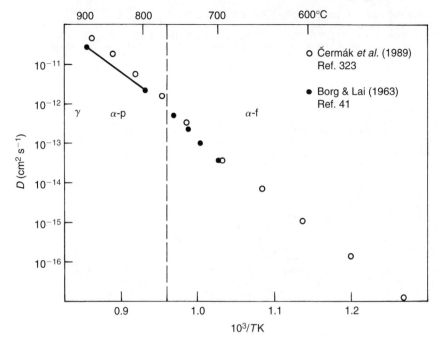

Figure 13.6 *Diffusion of* Ni *in* αFe *(replacing 13.2(a))*

61. M. B. Bronfin, 'Diffusion Processes, Structure and Properties of Metals' (Moscow 1964) Translation—Consultants Bureau, New York, 1965, 24.
62. R. A. Wolf and H. W. Paxton, *Trans. Met. Soc. AIME*, 1964, **230**, 1426.
63. R. A. Wolf and H. W. Paxton, *Trans. AIME*, 1964, **230**, 1426.
64. J. F. Murdock, T. S. Lundy and E. E. Stansbury, *Acta Met.*, 1964, **12**, 1033.
65. R. B. Gibbs, D. Graham and D. H. Tomlin, *Phil. Mag.*, 1963, **8**, 1269.
66. J. Askill and G. B. Gibbs, *Phys. Status Solidi*, 1965, **11**, 557. (Contains recalculated values of *A*s and *Q*s from the measurements on Cr, Mn, Fa, Co, Ni, Nb and Mo reported in ref. 64. However, since these seem, in some cases, to give a much inferior representation of the original data, none is quoted.)
67. J. Askill, Thesis Reading, 1964, and 'Diffusion in B.C.C. Metals' (American Society for Metals 1965) 247.
68. J. Askill, *Phys. Status Solidi*, 1965, **9**, K167.
69. S. Z., Bokshtein, M. B. Bronfin and S. T. Kishkin, 'Diffusion Processes Structure and Properties of Metals' (Moscow 1964). Translation—Consultants Bureau New York; 1965, 16.
70. R. F. Peart, D. Graham and D. H. Tomlin, *Acta Met.*, 1962, **10**, 519.
71. J. I. Federer and T. S. Lundy, *Trans. Met. Soc. AIME*, 1963, **227**, 592.
72. L. V. Pavlinov, *Phys. Met. Metallog.*, 1967, **24** (2), 70.
73. P. L. Gruzin, V. S. Emelyanov, G. G. Ryabora and G. B. Federov, *Geneva Conference Proceedings*, 1959, **19**, 187.
74. G. B. Federov, E. A. Smirnov, F. I. Zhomev, F. I. Gusev and S. A. Paraev, *Met. Metalloved. Christ. Metal.*, 1971, **9**, 30.
75. E. V. Borisov, Y. G. Godin, P. L. Gruzin, A. I. Eustyukhin and V. S. Emelyanov, *Met. i. Met. Izdatel Ak. Nauk. SSSR, Moscow*, 1959 and in Translation NP–TR–448 (1960) p. 196.
76. Ch. Herzig, J. Neuhaus, K. Vieregge and L. Manke, *Mater. Sci. Forum*, 1987, **15–18**, (Pt. 1), 481.
77. R. L. Andelin, J. D. Knight and M. Kahn, *Trans. Met. Soc. AIME*, 1965, **233**, 19.
78. Y. P. Vasil'ev, I. F. Kamardin, V. I. Skatskii, S. G. Chermomorchenko and G. N. Schuppe, *Trudy. Stred. Gos Univ. in u. i. Lenina*, 1955, **65**, 47.
79. R. E. Pawel and T. S. Lundy, *J. Phys. Chem. Solids*, 1965, **26**, 937.
80. N. L. Peterson and S. J. Rothman, *Phys. Rev.*, 1964, **136A**, 842.
81. S. J. Rothman, *J. Nucl. Mater.*, 1961, **3**, 77.
82. S. J. Rothman, N. L. Peterson and S. A. Moore, *J. Nucl. Mater.*, 1962, **7**, 212.
83. Sh. Merlanov and A. A. Kulier, *Soviet Phys. Solid St.*, 1962, **4**, 394.
84. M. C. Naik, J. M. Dupouy and Y. Adda, *Revue Metall.*, *Paris*, 1966, **63**, 488.
85. A. I. Nakanechnikov and L. V. Pavlinov, *Izv. Akad. Nauk. SSSR Metal*, 1972, **2**, 213.
86. J. Askill, *Phys. Status Solidi*, 1971, **B43**, 1K.
87. B. Dyson, T. R. Anthony and D. Turnbull, *J. Appl. Phys.*, 1966, **37** (6), 2370.

88. A. P. Batra and H. B. Huntington, *Phys. Rev.*, 1966, **145**, 542.
89. B. Dyson, *J. Appl. Phys.*, 1966, **37**, 2375.
90. O. Caloni, A. Ferrari and P. M. Strocchi, *Electrochem. Metall.*, 1969, **4**, 45.
91. V. Ya. Shchelkonogov, L. N. Aleksandrov, V. A. Piterimov and V. S. Mordyuk, *Phys. Met. Metallog.*, 1968, **25** (1), 68.
92. A. B. Vladimirov, V. N. Kaygorodov, S. M. Klotsman and I. Sh. Trakhtenberg, *Fiz. Met. Metalloved.*, 1979, **48**, 352.
93. A. J. Mortlock and P. M. Ewens, *Phys. Rev.*, 1967, **156**, 814.
94. J. Askill, *Phys. Status Solidi*, 1966, **16**, 63K.
95. J. W. Miller, *Phys. Rev.*, 1969, **181**, 1095.
96. T. R. Anthony and D. Turnbull, *Phys. Rev.*, 1966, **151** (2), 495.
97. B. F. Dyson, T. R. Anthony and D. Turnbull, *J. Appl. Phys.*, 1967, **38** (8), 3408.
98. R. M. Dubovstev, V. S. Dotov, T. I. Miroshnichenko and N. A. Nikolaev, *Fiz. Met. Metalloved.*, 1976, **42**, 1314.
99. T. R. Anthony, B. F. Dyson and D. Turnbull, *J. Appl. Phys.*, 1968, **39** (3), 1391.
100. A. P. Batra and H. B. Huntington, *Phys. Rev.*, 1967, **154**, 569.
101. K. J. Anusavic and R. T. de Hoff, *Met. Trans.*, 1972, **3**, 1279.
102. S. M. Klotsman, *et al.*, *Fiz. Met. Metalloved.*, 1971, **31**, 429.
103. F. D. Reinke and C. E. Dahlstrom, *Phil. Mag.*, 1970, **22**, 57.
104. S. M. Klotsman, *et al.*, *Fizica. Metall.*, 1970, **29**, 803.
105. K. Hoshino, Y. Iijima and K-I. Hirano, *Met. Trans. A*, 1977, **8A**, 469.
106. M. C. Savena and B. D. Sharma, *Trans. Indian Inst. Metals*, 1970, **23**, 16.
107. F. Moya, G. E. Moya and F. Cabanes-Brouty, *Phys. Status Solidi*, 1969, **35**, 893.
108. J. Bernardini and J. Cabane, *Acta Met.*, 1973, **21**, 1561.
109. T. Hehenkamp and R. Wübbenhorst, *Z. Metallk.*, 1975, **66**, 275.
110. N. Barbouth, J. Oudar and J. Cabane, *C. R. Hebd. Séanc. Acad. Sci.*, Paris, 1967, **C264**, 1029.
111. J. Geise, H. Mehrer, Ch. Herzig and G. Weyer, *Mater. Sci. Forum*, 1987, **15–18**, 443.
112. K. Dreyer, Chr. Herzig and Th. Heumann, 'Atomic Transport in Solids and Liquids' (Proc. Europhy. Conf.) (*Verlag Z. Naturforsch.* 1971), 237.
113. Ch. Herzig and Th. Heumann, *Z. Naturforsch*, 1972, **27a**, 613.
114. A. Ott, *Z. Naturforsch*, 1970, **25a**, 1477.
115. A. Ott, *J. Appl. Phys.*, 1969, **40**, 2395.
116. A. Ott, D. Lazarus and A. Lodding, *Phys. Rev.*, 1969, **188**, 1088.
117. A. Ott, *Z. Naturforsch.*, 1968, **23a**, 2126.
118. J. N. Mundy, A. Ott and L. Lowenberg, *Z. Naturforsch.*, 1967, **22a**, 2113.
119. J. N. Mundy, A. Ott, L. Lowenberg and A. Lodding, *Phys. Status Solidi*, 1968, **35**, 359.
120. A. Ott, *Z. Naturforsch.*, 1968, **23a**, 1683.
121. A. Ott and A. Norden-Ott, *Z. Naturforsch.*, 1968, **23a**, 473.
122. L. W. Barr, J. N. Mundy and F. A. Smith, *Phil. Mag.*, 1969, **20**, 389.
123. L. W. Barr, J. N. Mundy and F. A. Smith, *Phil. Mag.*, 1967, **16**, 1139.
124. L. W. Barr and F. A. Smith, *DIMETA 82, Proc. Intl. Conf. Diffusion in Metals and Alloys* (eds F. J. Kedves and D. I. Beke), Trans. Tech. Publs., Switzerland, 1983, p. 325.
125. F. A. Smith and L. W. Barr, *Phil. Mag.*, 1970, **21**, 633.
126. F. A. Smith and L. W. Barr, *Phil. Mag.*, 1969, **20**, 205.
127. V. M. Ananyn, *et al.*, *Sov. J. At. En.*, 1970, **29**, 941.
128. J. Combronde and G. Brebec, *Acta Met.*, 1972, **20**, 37.
129. K. Lal, Report CEA (Saclay) R3136, 1967.
130. J. S. Warford and H. B. Huntington, *Phys. Rev., B.*, 1970, **1**, 1867.
131. Mao, Chih-wen, *Phys. Rev. B*, 1972, **5**, 4693.
132. S. Badrinarayanan and H. B. Mathur, *Intl. J. Appl. Radiat. Isotopes*, 1968, **19**, 353.
133. D. L. Bdke, I. Gödény and F. J. Kedves, *Phil. Mag. A*, 1983, **47**, 281.
      I. Gödény, D. L. Beke and F. J. Kedves, *Phys. Stat. Solidi (a)*, 1972, **13**, K155.
      D. L. Beke, I. Gödény and F. J. Kedves, *Acta Met.*, 1977, **25**, 539.
134. J. J. Blechet, A. Van Craeynest and D. Calais, *J. Nucl. Mater.*, 1968, **27**, 112.
135. G. M. Hood and R. J. Schultz, *Phil. Mag.*, 1971, **23**, 1479.
136. G. M. Hood, *Phil. Mag.*, 1970, **21**, 305.
137. M. S. Anand and R. P. Agarwala, *Phil. Mag.*, 1972, **26**, 297.
138. W. K. Warburton, *Phys. Rev.*, 1972, **B6**, 2161.
139. G. M. Hood and R. J. Schultz, *Phys. Rev. B*, 1975, **B11**, 3780.
140. L. Bartha and T. Szalay, *Int. J. Appl. Radiat. Isotopes*, 1969, **20**, 825.
141. J. Kucera and K. Stransky, *Can. Met. Q.*, 1969, **8**, 91.
142. S. Nishikawa and K. Tsumuraya, *Phil. Mag.*, 1972, **26**, 941.
143. W. K. Warburton, *Phys. Rev.*, 1973, **B15**, 1330.
144. L. V. Pavlinov, A. M. Gladyshev and Y. N. Bykov, *Fiz. Met. Metalloved*, 1968, **26**, 946.
145. S. P. Murarka, M. S. Anand and R. P. Agarwala, *Acta Met.*, 1968, **16**, 69.
146. A. R. Paul and R. P. Agarwala, *Metal Trans.*, 1971, **2**, 2691.
147. J. P. Zanghi, A. Van Craeynest and D. Calais, *J. Nucl. Mater.*, 1971, **39**, 133.
148. J. J. Blechet, A. Van Craeynest and D. Calais, *J. Nucl. Mater.*, 1968, **28**, 177.
149. G. V. Grigorev and L. V. Pavlinov, *Fiz. Met. Metalloved.*, 1968, **25**, 836.
150. G. V. Grigorev and L. V. Pavlinov, *Fiz. Met. Metalloved.*, 1968, **26**, 946.

151. G. V. Grigorev and L. V. Pavlinov, *Fiz. Met. Metalloved*, 1968, **25**, 377.
152. A. Hoshino and T. Araki, *Trans. Nat. Res. Inst. Metals*, 1971, **13**, 99.
153. A. W. Bowen and G. M. Leak, *Met. Trans.*, 1970, **1**, 1695.
154. K. Nohara and K. Hirano, *Proc. Int. Conf. Sci. Tech. Iron and Steel*, 1970, **7**, 11.
155. S. M. Myers and H. J. Rack, *J. Appl. Phys.*, 1978, **49**, 3246.
156. D. W. James and G. M. Leak, *Phil Mag.*, 1966, **14**, 701.
157. G. Salje and M. Feller-Kniepmeier, *J. Appd. Phys.*, 1977, **48**, 1833.
158. M. Fromont, *J. Phys. (Paris) Lett.*, 1976, **37**, 117.
159. L. V. Pavlinov and B. N. Bykov, *Fiz. Met. Metalloved.*, 1966, **22**, 234.
160. Bondy and V. Levy, *C. R. Hebd, Séanc. Acad. Sci.*, *Paris*, 1971, **C272**, 81.
161. G. A. Bruggeman and J. A. Roberts Jr., *Met. Trans. A*, 1976, **6**, 755.
162. F. De Keroulas, J. Morey and Y. Quère, *J. Nucl. Mater.*, 1967, **22**, 276.
163. Ye. V. Deshkevish, R. M. Dubovtsev and V. S. Zotov, *Fiz. Met. Metalloved*, **55**, 1983, 186.
164. B. A. Vandyshev and A. S. Panov, *Izv. Akad. Nauk. SSSR Metally*, 1969, **1**, 244.
165. B. A. Vandyshev and A. S. Panov, *Izv. Akad. Nauk. SSSR Metally*, 1970, **2**, 231.
166. L. M. Mulyakaev, G. U. Scherbedinskii and G. N. Dubinin, *Metallov. Term. Obrab. Metall.*, 1971, **8**, 45.
167. L. U. Pavlinov, A. I. Makonechnikov and V. N. Bykov, *Soviet J. Atom. Ener.*, 1965, **19**, 1495.
168. B. A. Vandyshev and A. S. Panov, *Fiz. Metall.*, 1968, **25**, 321.
169. B. A. Vandyshev and A. S. Panov, *Fiz. Metall.*, 1968, **26**, 517.
170. J. Askill, *Phys. Status Solidi*, 1967, **23**, K21.
171. F. Roux and A. Vignes, *Rev. Phys. Appl. (Fr.)*, 1970, **5**, 393.
172. J. Pelleg, *Phil. Mag.*, 1970, **21**, 735.
173. R. P. Agarwala, S. P. Murarka and M. S. Anand, *Acta Met.*, 1968, **16**, 61.
174. J. Pelleg, *Phil. Mag.*, 1969, **19**, 25.
175. T. S. Lundy, *et al.*, *Trans. Met. Soc. AIME*, 1965, **233**, 1533.
176. G. B. Federov, F. J. Zhomov and E. A. Smirnov, *Metall. Metalloved. Chist. Metal*, 1969, **8**, 145.
177. D. S. Gornyi and R. M. Altovski, *Phys. Met. Metallog.*, 1971, **31** (4), 108.
178. G. I. Nikolaev and N. V. Bodrov, *Zh. Fiz. Khim.*, 1978, **52**, 143.
179. B. A. Vandyshev and A. S. Panov, *Izv. Akad. Nauk. SSSR.*, 1968, **1**, 206.
180. S. N. Balrt, N. Varela and R. Tendler, *J. Nucl. Mater.*, 1983, **119**, 59.
181. R. Tendler and C. F. Varotto, *J. Nucl. Mater.* 1973, **46**, 107.
182. G. M. Hood and R. J. Schultz, *Phil. Mag.*, 1972, **26**, 329.
183. G. M. Hood, 'Diffusion Processes', **1970**, Vol. 1, New York, Gordon and Breach, p. 361.
184. G. V. Kidson and G. J. Young, *Phil. Mag.*, 1969, **20**, 1057.
185. F. Dyment and C. M. Libanati, *J. Mater. Sci.*, 1968, **3**, 349.
186. R. P. Agarwala and A. R. Paul, *Proc. Nucl. Rad. Chem. Symp.*, 1967, **3**, 542.
187. A. R. Paul, *et al.*, *Int. Conf. Vac. and Insls. in Metals*, Julich, Sept. 1968, 1, 105.
188. E. Ch. Schwegler and F. A. White, *Intl. J. Mass Spectr. Ion Phys.*, 1968, 1, 191.
189. L. V. Pavlinov, G. V. Grigorev and G. O. Gromyko, *Izv. Akad. Nauk. SSSR Metal*, 1969, **3**, 207—*Russian Metall.*, 1969, **3**, 158.
190. R. P. Agarwala and A. R. Paul, *J. Nucl. Mat.*, 1975, **58**, 25.
191. B. A. Vandyshev, A. S. Panov and P. L. Gruzin, *Fiz. Metall.*, 1967, 23, 908.
192. N. K. Archipova, S. M. Klotsman, I. P. Polikarpova, G. N. Tartarinova, A. N. Timofeev and L. M. Veretennikov, *Phys. Rev. B*, 1984, **30**, 1788.
193. L. M. Larikov, V. M. Tyshkevich and L. F. Chorna, *Ukr. Fiz. Zh.*, 1967, **12**, 983.
194. R. E. Pawel and T. S. Lundy, *Acta Met.*, 1969, **17**, 979.
195. M. P. Dariel, M. Blumenfeld and G. Kimmel, *J. Appl. Phys.*, 1970, **41**, 1480.
196. F. A. Schmidt and O. N. Carlson, *J. Less Common Metals*, 1972, **26**, 247.
197. L. V. Pavlinov, *Phys. Met. Metallog.*, 1970, **30** (4), 129.
198. A. Languille, *Mém. Scient. Revue Metall.*, 1971, **68**, 435.
199. M. Dariel, G. Erez and G. M. J. Schmidt, *Phil. Mag.*, 1969, **19**, 1053.
200. F. H. Huang and H. B. Huntington, *Phys. Rev.*, 1974, **9B**, 1479.
201. M. Dariel, *J. Appl. Phys.*, 1971, **42**, 2251.
202. M. Dariel, G. Erez and G. M. J. Schmidt, *J. Appl. Phys.*, 1969, **40**, 2746.
203. M. Dariel, G. Erez and G. M. J. Schmidt, *Phil. Mag.*, 1969, **19**, 1045.
204. M. P. Dariel, *Phil. Mag.*, 1970, **22**, 563.
205. H. Backer, J. Backus and F. Waals, *Phys. Status Solidi (b)*, 1971, **45**, 633.
206. A. Kucera and T. Zemcik, *Can. Met. Q.*, 1968, **7**, 73.
207. P. Brätter and H. Gobrecht, *Phys. Status Solidi*, 1970, **41**, 631.
208. P. Brätter, H. Gobrecht and D. Wobig, *Phys. Status Solidi*, 1972, **11**, 589.
209. F. Schmitz and M. Fock, *J. Nucl. Mater.*, 1967, **21**, 353.
210. D. Treheux, *et al.*, *C. r. hebd. Séanc. Acad. Sci., Paris*, Series C. 1972, **274**, 1260.
211. M. P. Dariel, D. Dayan and D. Calais, *Phys. Status Solidi*, 1972, **A10**, 113.
212. R. L. Fogelson, Ya. Y. Ugai and I. A. Akimova, *Fiz. Met. Metalloved*, 1975, **39**, 447.
213. R. L. Fogelson, Ya. Y. Ugai and I. A. Akimova, *Izv. Vyssh. Uchebn Zaved., Tsvetn Metall.*, 1975, **2**, 142.
214. J. Ladet, J. Bernardini and F. Cabane-Brouty, *Scr. Met.*, 1976, **10**, 195.
215. F. Sawayanagi and R. R. Hasiguti, *J. Jap. Inst. Met.*, 1978, **42**, 1155.
216. S. J. Rothman, N. L. Peterson, L. J. Nowicki and L. C. Robinson, *Phys. Status Solidii B*, 1974, **63**, K29.
217. K. Hirano and S. Fujikawa, *J. Nucl. Mater.*, 1978, **69/70**, 564.

218. R. L. Fogelson and N. N. Trofimova, *Izv. Vyssh. Uchebn. Zaved. Tsvetz. Metall.*, 1978, **4**, 152.
219. S. M. Meyers, S. T. Picraux and T. S. Prevender, *Phys. Rev.*, 1974, **9**, 3953.
220. V. P. Gladkov, A. V. Svetlov, D. M. Skorov, V. I. Tenishev and A. N. Shabablin, *Atom. Ener.*, 1976, **40**, 257.
221. V. M. Anan'in, V. P. Gladkov, A. V. Svetlov and D. M. Skorov, *Sov. J. At. Energ.*, 1976, **40**, 304.
222. M. P. Dariel, *Acta Met.*, 1975, **23**, 473.
223. M. P. Dariel, *Phil. Mag.*, 1973, **28**, 915.
224. B. Million and J. Kucera, *Kovove Mater.*, 1973, **11**, 300.
225. A. Bristoti and A. R. Wazzan, *Rev. Bras. Fis.*, 1974, **4**, 1.
226. J. M. Dupouy, J. Mathie and Y. Adda., Proc. Int. Conf. Metallurgy of Be, Grenoble, p. 159, 1965.
227. D. C. Yeh, L. A. Acuna and H. B. Huntington, *Phys. Rev.*, 1981, **23**, 1771.
228. G. Krautheim, A. Neidhardt and V. Reinhold, *Krist. Techn.*, 1978, **13**, 1335.
229. V. A. Gorbachev *et al.*, *Fiz. Metal. Metalloved.*, 1973, **35**, 889.
230. G. Barreau, G. Brunnel, G. Ciceron and P. Lacombe, *C. R. Acad. Sci (Paris)*, 1970, **270**, 516.
231. J. Bernardini and J. Cabane, *Acta Met.*, 1973, **21**, 1561.
232. S. M. Klotsman *et al.*, *Fiz. Met. Metalloved.*, 1978, **45**, 1104.
233. V. A. Gorbachev *et al.*, *Fiz. Met. Metalloved.*, 1977, **44**, 214.
234. K. Nohara and K. Hirano, *Nippon Kinzoku Gakkaischi*, 1973, **37**, 731.
235. D. Ablitzer, *Phil. Mag.*, 1977, **35**, 1239.
236. J. Pelleg, *Phil. Mag.*, 1976, **33**, 165.
237. F. Dyment, *J. Nucl. Mat.*, 1976, **61**, 271.
238. A. B. Vladimirov *et al.*, *Fiz. Metal. Metalloved.*, 1975, **39**, 319.
239. A. B. Vladimirov *et al.*, *Fiz. Metal. Metalloved.*, 1978, **45**, 1015.
240. A. B. Vladimirov *et al.*, *Fiz. Metal. Metalloved.*, 1978, **45**, 1301.
241. D. L. Decker, C. T. Candland and H. B. Vanfleet, *Phys. Rev.*, 1975, **B11**, 4885.
242. C. T. Candland and H. B. Vanfleet, *Phys. Rev.*, 1973, **B7**, 575.
243. K. Kusunaki and S. Nishikawa, *Scripta Met.*, 1978, **12**, 615.
244. D. L. Decker *et al.*, *Phys. Rev.*, 1977, **B15**, 507.
245. D. L. Decker, J. D. Weiss and H. B. Vanfleet, *Phys. Rev.*, 1977, **B16**, 2392.
246. D. Bergner and K. Schwarz, *Neue Huette*, 1978, **23**, 210.
247. G. Rein, H. Mehrer and K. Maier, *Phys. Status Solidii*, 1978, **A45**, 253.
248. C. Charissoux, D. Calais and G. Gallet, *J. Phys. Chem. Sol.*, 1975, **36**, 981.
249. C. Charissoux and D. Calais, *J. Nucl. Mat.*, 1976, **61**, 317.
250. W. N. Weins and O. N. Carlson, *J. Less Common Met.*, 1979, **66**, 99.
251. F. A. Schmidt, R. J. Conzemius and O. N. Carlson, *J. Less Common Met.*, 1978, **56**, 53.
252. J. Pelleg, *Phil. Mag.*, 1975, **32**, 593.
253. J. Pelleg and M. Herman, *Phil. Mag.*, 1977, **35**, 349.
254. J. F. Murdock and C. J. McHargue, *Acta Met.*, 1968, **16**, 493.
255. C. Charissoux and D. Calais, *J. Nucl. Mat.*, 1975, **57**, 45.
256. R. Tendler and C. F. Varotto, *J. Nucl. Mat.*, 1974, **54**, 212.
257. R. Tendler, J. Abriata and C. F. Varotto, *J. Nucl. Mat.*, 1976, **59**, 215.
258. G. Erdélyi *et al.*, *Phil. Mag.*, 1978, **38**, 445.
259. G. Tobar and S. Balart, *DIMETA 88. Proc. Intl. Conf. Diffusion in Metals and Alloys* (eds F. J. Kedves and D. L. Beke) Defect and Diffusion Forum, 1989, **66–69**, 381.
260. P. Dorner, W. Gust, H. B. Hintz, A. Lodding, H. Odelius and B. Predel, *Acta Met.*, 1980, **28**, 291.
261. S. J. Rothman and N. L. Peterson, *Phys. Rev.*, 1967, **154**, 552.
262. R. L. Fogelson, Ya. A. Ugay and I. A. Akimova, *Izv. Vyssh. Uchebn. Zaved., Tsvertn. Metall.*, 1977, (1), 172.
263. H. Mehrer and D. Weiler, *Z. Metallk.*, 1984, **75**, 203.
264. G. Rummel and H. Mehrer, *DIMETA 88. Proc. Intl. Conf. Diffusion in Metals and Alloys* (eds F. J. Kedves and D. L. Beke) *Defect and Diffusion Forum*, 1989, **66–69**, 453.
265. F. Makuta, Y. Iijima and K. I. Hirano, *Trans. Jap. Inst. Metals*, 1979, **20**, 551.
266. G. Neumann, M. Pfundstein and P. Reimers, *Phys. Stat. Solidi (a)*, 1981, **64**, 225.
267. G. Neumann, M. Pfundstein and P. Reimers, *Phil. Mag. A*, 1982, **45**, 499.
268. Y. Minamino, T. Yamani and H. Araki, *Met. Trans. A*, 1987, **18**, 1536.
269. S. Sudar, J. Csikai and M. Buczko, *Z. Metallk.*, 1977, **68**, 740.
270. G. P. Tiwari and B. D. Sharma, *Indian J. Technol.*, 1973, **11**, 560.
271. N. V. Chi and D. Bergner, in *DIMETA 82. Proc. Intl. Conf., Diffusion in Metals and Alloys* (eds F. J. Kedves and D. L. Beke), Trans. Tech. Publs, Switzerland, 1983, p. 334.
272. G. M. Hood, R. J. Schultz and J. Armstrong, *Phil. Mag. A*, 1983, **47**, 775.
273. Ch. Herzig and D. Wolter, *Z. Metallk.*, 1974, **65**, 273.
274. D. Cardis, Thesis, Univ. of Münster, 1977.
275. R. L. Fogelson, N. N. Kazimirov and I. V. Soshnikova, *Fiz. Met. Metalloved*, 1977, **43**, 1105.
276. Ch. Herzig, H. Eckseler, W. Bussmann and D. Cardis, *J. Nuclear Mater.*, 1978, **69/70**, 61.
277. J. E. Reynolds, B. L. Averbach and M. Cohen, *Acta. Met.*, 1957, **5**, 29.
278. R. L. Fogelson, Ya. A. Ugay and I. A. Akimova, *Fiz. Met. Metalloved*, 1976, **41**, 653.
279. R. L. Fogelson, I. M. Vorinina and T. I. Somova, *Fiz. Met. Metalloved*, 1978, **46**, 190.
280. S. M. Meyers and R. A. Langley, *J. Appl. Phys.*, 1975, **46**, 1034.
281. G. Danz, R. Le Hazif, F. Maurice, D. Dutilloy and Y. Adda, *Comptes Rend. Acad. Sci. (Paris)*, 1962, **254**, 2328.
282. G. Marbach, C. Charissoux and C. Janot, 'La Diffusion dans les Milieux Condensés—Théorie et Applications', *Proc. Colloque de Métallurgie*, CEN Saclay, Vol. 1, 1976, p. 119. Report CEA-Conf. 36734.

283. M. Arita, M. Nakamura, K. S. Goto and Y. Ichinose, *Trans. Jap. Inst. Met.*, 1984, **25**, 703.
284. M. M. Pavlyuchenko and I. F. Kononyuk, *Dokl. Acad. Nauk. Belorusskoi SSR*, 1964, **8**, 157.
285. J. Kučera, B. Million and J. Růžičková, *Phys. Stat. Solidi (a)*, 1986, **96**, 177.
286. Y. Iijima and K. I. Hirano, *Phil. Mag.*, 1977, **35**, 229.
287. M. P. Dariel, L. Kornblit, B. J. Beaudry and ⊥. ®. × schneidner, *Phys. Rev. B*, 1979, **20**, 3949.
288. S. Fujikawa, M. Werner, H. Mehrer and A. Seeger, *Mater. Sci. Forum*, 1987, **15/18**, 431.
289. V. A. Gorbachev, S. M. Klotsman, Ya. A. Rabovskiy, V. K. Talinskiy and A. N. Timofeev, *Fiz. Met. Metalloved*, 1972, **34**, 879.
290. V. A. Gorbachev, S. M. Klotsman, Ya. A. Rabovskiy, V. K. Talinskiy and A. N. Timofeev, *Fiz. Met. Metalloved*, 1973, **35**, 143.
291. R. L. Fogelson, Ya. A. Ugay, A. V. Pokoev, I. A. Akimova and V. D. Kretinin, *Fiz. Met. Metalloved*, 1973, **35**, 1307.
292. G. Krautheim, A. Neidhardt, U. Reinhold and A. Zehe, *Phys. Letters A*, 1979, **72**, 181.
293. P. Spindler and K. Nachtrieb, *Phys. Stat. Solidi (a)*, 1976, **37**, 449.
294. Y. Iijima, K. Hoshino and K. I. Hirano, *Met. Trans. A*, 1977, **8**, 997.
295. G. Barreau, G. Brunel and G. Cizeron, *Comptes. Rend. Acad. Sci. (Paris)* C, 1971, **272**, 618.
296. R. L. Fogelson, Ya. A. Ugay and A. V. Pokoev, *Izv. Vyssh. Ucheb. Zaved. Chern. Met.*, (9), 1973, 136.
297. K. Maier, R. Kirchheim and G. Tölg, *Microchim. Acta Suppl.*, 1979, **8**, 125.
298. S. Badinarayan and H. B. Mathur, *Indian J. Pure Appl. Phys.*, 1972, **10**, 512.
299. R. L. Fogelson, Ya. A. Ugay and A. V. Pokoev, *Fiz. Met. Metalloved*, 1972, **34**, 1104.
300. W. Gust, C. Ostertag, B. Predel, U. Roll, A. Lodding and H. Odelius, *Phil. Mag. A*, 1983, **47**, 395.
301. R. Döhl, M. P. Macht and V. Naundorf, *Phys. Stat. Solidi (a)*, 1984, **86**, 603.
302. K. Monma, H. Suto and H. Oikawa, *J. Jap. Inst. Metals*, 1964, **28**, 192.
303. M. P. Macht, V. Naundorf and R. Döhl, *DIMETA 82, Proc. Intl. Conf., Diffusion in Metals and Alloys* (eds F. J. Kedves and D. L. Beke), Trans. Tech. Publs, Switzerland, 1983, 516.
304. S. J. Rothman, N. L. Peterson, C. M. Walter and L. J. Nowicki, *J. Appd. Phys.*, 1968, **39**, 5041.
305. K. Majima and H. Mitani, *Trans. Jap. Inst. Metals*, 1978, **19**, 663.
306. T. Egichi, Y. Iijima and K. Hirano, *Crystal Lattice Defects*, 1973, **4**, 265.
307. A. Agren, *Scripta Met.*, 1986, **20**, 1507.
308. P. Grieveson and E. T. Turkdogan, *Trans. Met. Soc. AIME*, 1964, **230**, 407.
309. T. Matsuyama, H. Hosokawa and H. Sato, *Trans. Jap. Inst. Metals*, 1983, **24**, 589.
310. P. L. Gruzin and V. V. Mural', *Phys. Met. Metallog.*, 1964, **17** (5), 154.
311. B. I. Božić and R. L. Lučić, *J. Mater. Sci.*, 1976, **11**, 887.
312. K. Hennesen, H. Keller and H. Viefhaus, *Scripta Met.*, 1984, **18**, 1319.
313. K. Kimura, Y. Iijima and K. Hirano, *Trans. Jap. Inst. Metals*, 1986, **27**, 1.
314. P. L. Gruzin, V. V. Mural' and A. P. Fokin, *Phys. Met. Metallog.*, 1972, **34** (6), 209.
315. B. Sparke, D. W. James and G. M. Leak, *J. Iron Steel Inst.*, 1965, **203**, 152.
316. J. Geise and C. Herzig, *Z. Metallk*, 1987, **78**, 291.
317. J. Geise and C. Herzig, *Z. Metallk*, 1985, **76**, 622.
318. K. Hirano and Y. Iijima, *DIMETA 88. Proc. Intl. Conf. Diffusion in Metals and Alloys* (eds F. J. Kedves and D. L. Beke), Defect and Diffusion Forum, 1989, **66–69**, 1039.
319. J. S. Kirkaldy, P. N. Smith and R. C. Sharma, *Metall. Trans. A*, 1973, **4**, 624.
320. G. Henry, G. Barreau and G. Cizeron, *Comptes Rendus Acad. Sci. (Paris)*, Series C, 1975, **280**, 1007.
321. H. Mehrer, D. Höpfel and G. Hettich, *DIMETA 82. Intl. Conf. Diffusion in Metals and Alloys*, 1982 (eds F. J. Kedves and B. L. Beke), Trans. Tech. Publs, Switzerland, 1983, p. 360.
322. J. Kucera, L. Kozak and H. Mehrer, *Phys. Stat. Solidi A*, 1984, **81**, 497.
323. J. Čermák, M. Lübbehusen and H. Mehrer, *Z. f. Metallk*, 1989, **80**, 213.
324. S. Fujikawa and K. Hirano, *DIMETA 88. Proc. Intl. Conf. Diffusion in Metals and Alloys* (eds F. J. Kedves and D. L. Beke), *Defect and Diffusion Forum*, 1989, **66–69**, 447.
325. J. Räisänen, A. Antilla and J. Keinonen, *Appd. Phys. A*, 1985, **36**, 175.
326. J. N. Mundy and W. D. McFall, *Phys. Rev. B*, 1963, **8**, 5477.
327. J. N. Mundy and W. D. McFall, *Phys. Rev. B*, 1973, 7, 4363.
328. A. Lodding and A. Ott, *Zeit. f. Naturforsch. (a)*, 1971, 81.
329. A. Ott, *J. Appl. Phys.*, 1971, **42**, 2999.
330. L. V. Pavlinov, A. M. Gladyshev and V. N. Bykov, *Fiz. Met. Metalloved*, 1968, **26**, 823.
331. C. S. Su, *Nucl. Inst. and Methods*, 1977, **145**, 361.
332. L. N. Larikov, V. I. Isaichev, E. A. Maksimenko and B. M. Belkov, *Dokl. Akad. Nauk. SSSR*, 1977, **237**, 315. *Sov. Physics Doklady*, 1977, **22**, 677.
333. B. Lesage and A. M. Huntz, *J. Less Common Metals*, 1974, **38**, 149.
334. R. Roux, Thesis, Nancy (France), 1972.
335. G. B. Federov, E. A. Smirnov, V. N. Gusev, F. I. Zhomov and V. L. Gorbenko, *Metallurgiya i Metalloved. Chystykh. Metallov*, 1973, No. 10, 62.
336. E. V. Borisov, P. L. Gruzin and S. V. Zemskii, *Zashch. Pokryt. Metal*, 1968, No. 2, *Protective Coatings on Metals*, 1970, **2**, 76 (Consultants Bureau).
337. N. A. Makhlin and L. I. Ivanov, *Izvest. Akad. Nauk. SSSR Metally*, 1971, No. 1, 222. *Russian Metallurgy*, 1971, No. 1, 152.
338. R. E. Einziger and J. N. Mundy, *Phys. Rev. B*, 1978, **17**, 449.
339. J. Pelleg, *J. Less Common Metals*, 1969, **17**, 319.
340. R. P. Agarwala and K. Hirano, *Trans. Jap. Inst. Metals*, 1972, **13**, 425.

341. J. E. Murphy, G. H. Adams and W. N. Cathay, *Met. Trans. A*, 1975, **6**, 343.
342. O. Taguchi, Y. Iijima and K. I. Hirano, *J. Jap. Inst. Metals*, 1984, **48**, 20.
343. D. Treheux, A. Heurtel and P. Guiraldenq, *Acta Met.*, 1976, **24**, 503.
344. W. Gust, H. B. Hintz, A. Lodding, H. Odelius and B. Predel, *Phys. Status Solidi (a)*, 1981, **64**, 187.
345. W. Gust, H. B. Hintz, A. Lodding, H. Odelius and B. Predel, *Phil. Mag. A*, 1981, **43**, 1205.
346. S. Mantl, S. J. Rothman, L. J. Nowicki and J. L. Lerner, *J. Phys. F*, 1983, **13**, 1441.
347. A. B. Vladimirov, S. M. Klotsman and I. Sh. Trakhtenberg, *Fiz. Met. Metalloved*, 1979, **43**, 1113.
348. P. Neuhaus, Thesis, Univ. of Münster, 1987.
349. D. Bergner, *Krist. Techn.*, 1972, **7**, 651.
350. H. R. Curtin, D. L. Decker and H. B. Vanfleet, *Phys. Rev.*, 1965, **139**, A1552.
351. D. L. Decker, J. G. Melville and H. B. Vanfleet, *Phys. Rev. B*, 1979, **20**, 3036.
352. H. B. Vanfleet, J. D. Jorgenson, J. D. Schmutz and D. L. Decker, *Phys. Rev. B*, 1977, **15**, 5545.
353. S. K. Sen and A. Ghori, *Phil. Mag. A*, 1989, **59**, 707.
354. C. E. Hu and H. B. Huntington, *Phys. Rev. B*, 1982, **26**, 2782.
355. H. B. Vanfleet, *Phys. Rev. B*, 1980, **21**, 4337.
356. C. W. Owens and D. Turnbull, *J. Appd. Phys.*, 1972, **43**, 3933.
357. M. P. Dariel, O. D. McMasters and K. A. Gschneidner, *Phys. Stat. Solidi (a)*, 1981, **63**, 329.
358. J. A. Weyland, D. L. Decker and H. B. Vanfleet, *Phys. Rev. B*, 1971, **4**, 4225.
359. J. Fillon and D. Calais, *J. Phys. Chem. Solids*, 1977, **38**, 81.
360. E. V. Borisov, P. L. Gruzin and S. V. Zemskii, *Zasch. Pokryt. Metal.*, 1968, No. 2, 104. *Protective Coatings on Metals*, 1970, **2**, 76 (Consultants Bureau).
361. M. Shimotomai, R. R. Hasiguti and S. Umeyama, *Phys. Rev. B*, 1978, **18**, 2097.
362. D. Ablitzer and M. Gantois, 'La Diffusion dans les Milieux Condensés. Théorie et Applications', CEN Saclay, 1976, Vol. 1, p. 299.
363. D. B. Moharil, I. Jin and G. R. Purdy, *Met. Trans.*, 1974, **5**, 59.
364. S. Fukikawa and K. Hirano, *DIMETA 88. Proc. Intl. Conf. Diffusion in Metals and Alloys* (eds F. J. Kedves and D. L. Beke) *Defect and Diffusion Forum*, 1989, **66–69**, 447.
365. J. Čeramác and H. Mehrer, To be published.
366. J. R. MacEwan, J. U. MacEwan and L. Yaffe, *Canad. J. Chem.*, 1959, **37**, 1629.
367. D. C. Yeh and H. B. Huntington, *Phys. Rev. Letters*, 1984, **53**, 1469.
368. A. V. Pokoev, V. M. Mironov and L. K. Kudryavtseva, *Izv. Vyssh. Uchebn. Zaved. Tsvetn. Metall., Soviet Non-Ferrous Met. Res.*, 1976, **4** (2), 81.
369. J. Räisänen, A. Anttila and J. Keinonen, *J. Appd. Phys.*, 1985, **57**, 613.
370. J. Räisänen and J. Keinonen, *Appd. Phys. Letters*, 1986, **49**, 773.
371. H. Nakajima, J. Nakazawa, Y. Minonishi and M. Koiwa, *Phil. Mag.*, 1986, **53**, 427.
372. E. Santos and F. Dyment, *Phil. Mag.*, 1975, **31**, 809.
373. Y. Nakamura, H. Nakijima, S. Ishioka and M. Koiwa, *Acta Met.*, 1988, **46**, 2787.
374. H. Nakajima, M. Koiwa and S. Ono, *Scripta Met.*, 1983, **18**, 1431.
375. L. G. Korneluk, L. M. Mirsky and B. S. Bokshtein, 'Titanium Science and Technology', Vol. II, 1973, p. 905.
376. H. Nakajima, M. Koiwa, Y. Minoshi and S. Ono, *Trans. Jap. Inst. Metals*, 1983, **24**, 655.
377. H. Nakajima, S. Ishioka and M. Koiwa, *Phil. Mag. A*, 1985, **53**, 743.
378. G. M. Hood and R. J. Schultz, *Acta Met.*, 1974, **22**, 459.
379. G. B. Federov and E. A. Smirnov, 'Diffuziya v. Reactornykh Materailakh' (Atomizdat Publs, Moscow, 1978). Trans: 'Diffusion in Reactor Materials' (Trans. Tech. Publs, Switzerland, 1984).
380. I. I. Kovenski, 'Diffusion in B.C.C. Metals', (ASM 1965), p. 283.
381. H. Nakajima and M. Koiwa, *Titanium Science and Technology. Proc. 5th International Conference on Ti*, 1985, p. 1759.
382. W. W. Albrecht, G. Frohberg and H. Wever, *Z. Metallk.*, 1974, **65**, 279.
383. M. Stelly and J. M. Servant, *J. Nucl. Mat.*, 1972, **43**, 269.
384. W. Chomka and J. Andruszkieicz, *Nuklecnik*, 1960, **5**, 611.
385. J. Pelleg, *Reviews High Temp. Mat.*, 1978, **IV**, 5.
386. D. Ablitzer, J. P. Haeusler, K. V. Sathyaraj and A. Vignes, *Phil. Mag. A*, 1981, **44**, 589.
387. J. Pelleg, *Phil. Mag. A*, 1986, **54**, L2i.
388. V. I. Iovkov, I. S. Panov and A. V. Ryabenko, *Phys. Met. Metallog.*, 1972, **34** (6), 204.
389. S. M. Klotsman, V. M. Koloskov, S. V. Osetrov, I. P. Polikarpova, G. N. Tatarinova and A. N. Timofeev, *Defect and Diffusion Forum*, 1989, **66–69**, 439.
390. G. Lorang and J. P. Langeron, *High Temp. High Press.*, 1978, **33**, 394.
391. K. Yoshioka and M. Kimura, *Acta Met.*, 1975, **23**, 1009.
392. F. A. Schmidt, M. S. Beck, D. K. Rehbein, R. J. Conzemius and O. N. Carlson, *J. Electrochem. Soc.*, 1984, **131**, 2169.
393. S. J. Rothman and N. L. Peterson, 'Diffusion in B.C.C. Metals' (ASM 1965), p. 183.
394. W. J. Muster, D. N. Yoon and W. J. Hippmann, *J. Less Common Met.*, 1979, **65**, 211.
395. F. A. Schmidt and O. N. Carlson, *Met. Trans. A*, 1976, **7**, 127.
396. G. V. Kidson, *Phil. Mag. A*, 1981, **44**, 401.
397. G. M. Hood and R. J. Schultz, *Materials Science Forum*, 1987, **15–18**, (Pt. 1), 475.
398. L. Manke and Ch. Herzig, *Acta Met.*, 1982, **30**, 2085.
399. G. B. Federov, E. A. Smirnov and S. M. Novikov, *Met. Metalloved. Christ. Metal.*, 1969, No. 8, 41.
400. G. P. Tiwari, M. C. Saxena and R. V. Patil, *Trans. Ind. Inst. Metals*, 1973, **26**, 55.
401. L. I. Nicolai and R. Tendler, *J. Nucl. Mater.*, 1979, **87**, 401.

402. Th. Hehenkamp, *Acta Met.*, 1966, **14**, 887.
403. Ye. V. Deshkevich, R. M. Dubovtsev and V. S. Zotov, *Fiz. Met. Metalloved*, 1985, **60**, 1206.
404. G. Ya. Meshcheryakov, R. A. Andriyevskiy and V. N. Zagryazkin, *Fiz. Metl. Metallov.*, 1968, **25**, 189.
405. O. N. Carlson, F. A. Schmidt and J. C. Sever, *Met. Trans.*, 1973, **4**, 2407.
406. R. M. Dubovtsev, V. S. Zotov and T. I. Miroshnichenko, *Fiz. Met. Metalloved*, 1982, **54**, 1128.
407. V. S. Zotov and A. P. Tsedilkin, *Sov. Phys. J.*, 1976, **14**, 1652.
408. S. C. Axtell, I. C. I. Okafor, R. J. Conzemius and O. N. Carlson, *J. Less Common Metals*, 1986, **115**, 269.
409. D. T. Peterson and T. Carnahan, *Trans. AIME*, 1969,**245**, 213.
410. L. V. Pavlinov, *Phys. Met. Metallog.*, 1970, **30**(2), 149.
411. O. N. Carlson, F. A. Schmidt and D. T. Peterson, *J. Less Common Metals*, 1966, **10**, 1.
412. M. Werner, H. Mehrer and H. Siethoff, *J. Phys. C: Sol. State Phys.*, 1983, **16**, 6185.
413. V. D. Rogazin, L. M. Gert and A. A. Babad-Zakhryapin, *Izvest. Akad. Nauk. SSSR Metally*, 1968, **3**, 228. *Russian Metallurgy*, 1968 (3), 159.
414. V. P. Gladkov, A. V. Svetlov, D. M. Skorov and A. N. Shabalin, *Fiz. Met. Metalloved*, 1979, **48**, 871.
415. I. C. I. Okafor and O. N. Carlson, *J. Less Common Metals*, 1982, **84**, 499.
416. H. Nakajima, G. M. Hood and R. J. Schultz, *Phil. Mag. B*, 1988, **58**, 319.
417. D. D. Pruthi and R. P. Agarwala, *Phil. Mag. A*, 1984, **49**, 263.
418. K. V. Sathyraj, D. Ablitzer and C. Demangeat, *Phil. Mag. A*, 1979, **40**, 541.
419. J. N. Mundy, C. W. Tse and W. D. McFall, *Phys. Rev. B*, 1976, **13**, 2349.
420. I. Richter and M. Feller-Kniepmeier, *Phys. Stat. Solidi (a)*, 1981, **68**, 289.
421. J. Fillon and D. Calais, *J. Phys. Chem. Solids*, 1977, **38**, 81.
422. D. Ansel, J. Barre, C. Mezière and J. Debuigne, *J. Less Common Metals*, 1979, **65**, 1.
423. D. D. Pruthi and R. P. Agarwala, *Phil. Mag. A*, 1982, **46**, 841.
424. V. P. Gladkov, V. S. Zotov, I. I. Papirov, D. M. Skorov and G. F. Tikhinski, 'Poluchenie i Issled. Svoitstv. Chistykh Metallov. (Kharkov)', *F.T.I. Akad. Nauk. Ukr. SSR*, 1970, **2**, 56.
425. V. S. Zotov, T. M. Miroshnichenko and A. M. Protasova, '*Diffusion Processes in Metals*' (Tul.skiy Politkh. Inst.), 1974, **2**, 73.
426. H. Araki, Y. Minamino, T. Yamane, T. Nakatsuka and Y. Miyamoto, '*Metall. Trans. A*', 1996, **27A**, 1807.
427. J. B. Adams, S. M. Foiles and W. G. Wolfer, *J. Mater. Res.*, 1989, **4**, 102.

Table 13.3 DIFFUSION IN HOMOGENEOUS ALLOYS
Tracer diffusion—Self-diffusion of alloys

Element 1 (purity) At.%	Element 2 (purity) At.%	A_1^* cm^2 s^{-1}	Q_1^* kJ mol^{-1}	A_2^* cm^2 s^{-1}	Q_2^* kJ mol^{-1}	Temp. range K	Ref.
Ag (—)	Al (—)		IVa(i)	p.c.			
	2.05	0.25	177.9	—			
	9.47	0.83	179.6	—	—}	973–1 123	1
	14.1	0.73	172.5	—	—		
(99.98) 0–9.4	(99.996) —	0.39	IVc 121.0	p.c. (indep. of conc.)		673–868	69
Ag (5N)	As (5N)	IVc (modified), p.c., Ag^{110}					
—	0–4.15	$b_1 = 57$	$b_2 = 3\,371$	—	—	909	
—	0–4.79	$= 40$	$= 2\,901$	—	—	954 }	26
—	0–3.49	$= 33$	$= 2\,257$	—	—	998	
—	0–3.19	$= 19$	$= 1\,554$	—	—	1 073	
Ag (99.99) 100	Au (99.99) 0	0.49	IVa(i) 186.2	s.c. 0.85	202.1		
	8	0.52	187.5	0.82	202.2		
	17	0.32	184.4	0.48	198.0		
	35	0.23	182.3	0.35	195.4		
	50	0.19	181.7	0.21	189.6 }	903–1 283	2
	66	0.11	174.7	0.17	186.4		
	83	0.09	171.2	0.12	180.2		
	94	0.072	168.6	0.09	176.2		
0	100	0.072	168.3	0.09	174.6		

(continued)

Table 13.3   DIFFUSION IN HOMOGENEOUS ALLOYS—*continued*

Element 1 (purity) At.%	Element 2 (purity) At.%	A_1^* cm² s⁻¹	Q_1^* kJ mol⁻¹	A_2^* cm² s⁻¹	Q_2^* kJ mol⁻¹	Temp. range K	Ref.
Ag (99.99)	Cd (99.999)		IVa(i)	s.c. and p.c.			
100	0	0.44	185.3	0.44	174.9		
	6.50	0.31	178.4	0.33	169.5		
	13.60	0.23	171.5	0.22	161.7	773–1 173	3
	27.5	—	—	0.25	150.5		
	28.0	0.16	156.0	—	—		
			IVa(i)	p.c.			
	31	—	~151	—	~147.4		
	34	—	~151[a]	—	~147.0	833–953	74
	37	—	~151	—	~146.1		
(—)	(—)	IVa(i), s.c., Ag^{110}					
—	0 ~ 3	$b_1 = 9.19$		—	—	1 133	
—		$= 13.69$		—	—	1 197	200
Ag (—)	Cu (—)		IVa(i)	p.c.			
	1.75	0.66	187.6	—	—		
	4.16	1.84	195.1	—	—	973–1 163	1
	6.56	0.51	182.1	—	—		
Ag (—)	Ge (—)		IVb	IVa(i) and		p.c.	
	1.50	0.55	184.2	D_{Ge} greater than for			
	3.00	1.59	189.7	infinite dilution by		973–1 123	1 and 7
	4.30	1.89	186.3	~15% At.% Ge			
	5.43	2.18	185.1				
Ag (—)	In (—)	IVa(i), s.c., Ag^{110m}					
—	0–0.94	$b_1 = 17.6$		—	—	1 054	151
(99.99)	(99.99)		IVa(i)	s.c. and p.c.			
100	0	0.44	185.3	0.41	170.8		
	4.40	0.36	178.7	—	—		
	4.70	—	—	0.45	168.7		
	12.40	—	—	0.57	160.7	773–1 173	3
	12.60	0.12	156.6	—	—		
	16.60	—	—	0.57	153.3		
	16.70	0.18	151.9	—	—		
Ag (99.95)	Mg (99.95)		IVb	p.c.			
	45.8	1.53	172.9	—	—		
	49.8	0.28	170.0	—	—	773–973	12
	52.0	0.134	159.1	—	—		
(99.98)	(99.9+)		IVa(i)	p.c.			
	41.10	0.095	139.0	—	—		
	43.60	0.15	147.8	—	—		
	48.48	0.37	165.4	—	—	773–973	13
	48.72	0.39	166.2	—	—		
	52.82	0.33	153.7	—	—		
	57.15	0.051	120.2	—	—	773–873	
	60.88	$D_{Ag}^* = 4.37 \times 10^{-9}$ at 773.5 K					
Ag (—)	Pb (—)		IVa(i)	p.c.			
	0.21	0.22	177.9	—	—		
	0.25	—	—	0.22	158.3		
	0.52	—	—	0.38	162.0		
	0.71	0.89	187.1	—	—	973–1 073	9 and 1
	1.30	0.70	182.1	—	—		
	1.32	—	—	0.46	161.2		
(—)	(99.999 9)		IVa(i)	s.c.			
0.09	—	D_{Pb}^*/D_{PurePb}^*	$= 1.115$				
0.18	—		$= 1.25$	$(b = 136.8)$	1 257.3		63

[a] Activation energies read from graph in Reference 74.

(*continued*)

Table 13.3 DIFFUSION IN HOMOGENEOUS ALLOYS—*continued*

Element 1 (purity) At.%	Element 2 (purity) At.%	A_1^* cm² s⁻¹	Q_1^* kJ mol⁻¹	A_2^* cm² s⁻¹	Q_2^* kJ mol⁻¹	Temp. range K	Ref.
Ag ('Spec. Pure')	Pd		IVa(i)	p.c.			
	0–21.8	$027e^{-8.2c}$	183.0	—(b)	—	988–1 215	4
	0–20.4	—(b)	—	$12.5e^{-7.5c}$	239.5	1 123–1 173	5
Ag (99.99)	Sb (99.99)		IVa(i)	s.c.			
	0.53	0.38	182.1	(c)	—		
	0.89	0.30	179.4	—	—	823–1 173	6
	1.42	0.275	175.8	—	—		
(5N)	(5N)	IVa(i), p.c., Ag¹¹⁰ᵐ					
—	0–6.5		*See* Figure 13.7			890–1 048	204
Ag (—)	Sn (—)			p.c.			
	0.18	0.13	174.6	—	—		
	0.48	0.13	171.2	—	—		24
	0.91	0.17	169.6	—	—		
	0.97	0.28	165.0	—	—	973–1 123	25
	2.8	0.1	161.6	—	—		
	4.56	0.23	166.2	—	—		24
	5.1	0.2	161.6	—	—		
	7.45	0.16	154.9	—	—		25
(—)	(—)		p.c.	IVb Ag¹¹⁰ᵐ, Sn¹¹³			
75	25	1.03×10^{-3}	93.8	4.01×10^{-4}	95.5	473–673	128
(—)	(—)		p.c. and s.c.	IVa Sn¹¹³, Ag¹¹⁰			
100	~0	1.0	191.3	0.17	160.8		
	1.7	—	—	0.125	156.8		
	0.108	0.13	172.9	—	—		
	0.8	0.12	168.3	—	—	893–1 073	129
	3.0	0.085	160.8	—	—		
	4.7	0.07	157.0	—	—		
	6.0	0.07	154.9	—	—		
( )	( )	IVa(i), s.c., Ag¹¹⁰ᵐ					
—	0–0.77	$b_1 = 20.2$		—	—	~1 052	151
(5N)	(5N)	IVb, p.c., Ag¹¹⁰ᵐ					

				$10^{10}D_{Ag}^*$			
		946 K	989 K	1 041 K	1 108 K	1 145 K	
—	0	0.237	0.666	2.47	8.51	16.6	
—	0.948	—	—	2.77	—	—	
—	0.963	—	—	—	—	17.6	
—	1.41	—	—	—	10.4	—	
—	1.44	0.373	0.990	—	—	—	
—	2.35	—	—	3.06	—	—	
—	2.54	—	—	—	—	25.6	
—	2.62	0.540	1.46	—	—	—	
—	3.0	—	—	—	16.1	—	
—	3.06	—	—	4.33	—	—	170
—	3.47	—	—	—	—	31.0	
—	4.32	—	—	—	19.6	—	
—	4.33	0.880	2.48	—	—	—	
—	4.48	—	—	—	—	42.4	
—	4.49	—	—	6.57	—	—	
—	5.71	—	—	—	30.6	—	
—	6.60	1.77	5.01	—	—	—	
—	7.04	—	—	—	41.1	—	
—	7.71	—	—	15.4	—	—	
—	8.67	3.36	7.32	—	—	—	

(b) c = conc. of Pd.
(c) In 0.7 At. % Sb alloys, D_{Sb}^* same as D in pure Ag.
  In At. % Sb alloys, D_{Sb}^* ~20% greater than D in pure Ag.

(*continued*)

Table 13.3 DIFFUSION IN HOMOGENEOUS ALLOYS—*continued*

Element 1 (purity) At.%	Element 2 (purity) At.%	A_1^* cm² s⁻¹	Q_1^* kJ mol⁻¹	A_2^* cm² s⁻¹	Q_2^* kJ mol⁻¹	Temp. range K	Ref.
Ag (—)	Tl (—)		IVa(i) and IVb p.c.				
100	0	0.724	190.5	0.15	158.7		7
	1.1	0.42	182.1	0.72	169.1	913–1 073	
	2.6	0.35	175.4	0.57	165.0		
	5.5	0.10	157.4	—	—		
Ag (99.999)	Zn (99.999)		IVa(i)	s.c.			
0.00		0.31	109.3	0.13	91.7		71 and 23
		$-\|c$ ... $\perp c$ 0.45	115.6	0.18	96.3		of Table 2
0.35		$-\|c$ —	—	—	—		
		$-\perp c$ 0.49	115.8	—	—		
0.57		$-\|c$ —	—	0.14	91.9		
		$\perp c$ —	—	0.22	96.8		
0.68		$-\|c$ 0.35	109.2	—	—	593–688	70
		$\perp c$ —	—	—	—		
0.89		$-\|c$ 0.42	110.1	—	—		
		$\perp c$ 0.69	117.3	—	—		
1.40		$-\|c$ —	—	0.17	92.4		
		$\perp c$ —	—	0.26	96.8		
(99.999)	(99.999)		IVa(i)	s.c.			
—	0	$D_1 = 1.43 \times 10^{-10}$		—	—	1 020	
		$= 17.2 \times 10^{-10}$		—	—	1 153	
—	1.10	$= 1.63 \times 10^{-10}$		—	—	1 020	
		$= 19.9 \times 10^{-10}$		—	—	1 153	
—	2.08	$= 1.80 \times 10^{-10}$		—	—	1 020	
		$= 21.8 \times 10^{-10}$		—	—	1 153	
—	3.10	$= 2.0 \ \times 10^{-10}$		—	—	1 020	
		$= 24.1 \times 10^{-10}$		—	—	1 153	
—	4.05	$= 2.28 \times 10^{-10}$		—	—	1 020	
		$= 26.2 \times 10^{-10}$		—	—	1 153	
(99.99)	(99.999)		IVa(i)	p.c.			
85	15		∼150.7	0.11	150.7		10
70	30	0.29	150.7	0.46	147.4	773–973	8
(Merck.)	(99.999)		IIb	p.c.			
52.4	47.6	4.55×10^{-3}	73.7	—	—	673–883	11
Al (5N)	Au (5N)		Vb, p.c., Al²⁷				
1.6	—	7×10^{-3}	135	—	—		
3.0	—	1.3×10^{-3}	123.5	—	—	550–1 150	112
5.0	—	3×10^{-4}	106	—	—		
Al (99.9)	Co (Carbonyl)		IVc	p.c.			
	10	—	—	2.65	282.6	1 313–1 493	14
42	—	—	—	1.84×10^2	355.9	1 273–1 473	
49	—	—	—	333×10^2	427.1	1 373–1 573	15
50.7	—	—	—	0.013×10^2	272.1	1 273–1 473	
	49/57		Individual D_{Co}^* values plotted at 1 250				16
Al (99.999)	Cu (—)		IVa(i)	s.c.			
—	1.0	—	—	$D = 4.02 \times 10^{-10}$		762	72
				$D = 7.92 \times 10^{-9}$		881	
( )	( )		IVb	p.c.			
0	—	—	—	0.43	203.1		
2.80	—	—	—	0.46	201.0		
5.50	—	—	—	0.30	196.8		
8.83	—	—	—	0.46	197.2	1 073–1 313	73
11.7	—	—	—	0.61	197.6		
14.5	—	—	—	4.2	213.9		

(*continued*)

Table 13.3 DIFFUSION IN HOMOGENEOUS ALLOYS—*continued*

Element 1 (purity) At.%	Element 2 (purity) At.%	A_1^* cm² s⁻¹	Q_1^* kJ mol⁻¹	A_2^* cm² s⁻¹	Q_2^* kJ mol⁻¹	Temp. range K	Ref.
(99.994)	()		Vb	p.c.			
—	0	0.10	127.7	—	—⎫	~603–733	127
	0.15	6×10^{-4}	100.9	—	—⎭		
Al ()	Cu ()		Vb, p.c., Al²⁷	—	—	750–1030	190
Al (99.99)	Cu (99.99)	IVa(i), p.c., Zn⁶⁵		Diffusion of Zn			
—	0	$D_{Zn}^*(c_i) = D_{Zn}^*(o)(1 + B_i^{Zn}c_i)$		0.16	116.9	714–894 ⎫	
—	0.69			0.13	115.2	714–894 ⎪	
—	1.23	$B_{Cu}^{Zn} = 13.3$ (Av. value)		0.28	120.0	714–872 ⎬ 41	
—	1.57			0.33	120.9	754–855 ⎪	
—	1.86			0.67	125.2	779–844 ⎭	

Element 1 At.%	Element 2 At.%	Zn (99.99)	IVa(i), p.c., Zn⁶⁵	Diffusion of Zn			
Al (99.99)	Cu (99.99)	(99.99)					
—	0	2.04		0.18	116.9	714–894 ⎫	
—	0.69	2.04	$B_{Zn}^{Zn} = 9.6$	0.20	116.9	714–854 ⎪	
—	1.23	2.04	(Av. values)	0.17	115.5	714–872 ⎬ 41	
—	1.57	2.04	$B_{Cu}^{Zn} = 11.7$	0.20	116.3	754–855 ⎪	
—	1.86	2.04		0.38	120.2	779–844 ⎭	

Element 1 (purity) At.%	Element 2 (purity) At.%	A_1^* cm² s⁻¹	Q_1^* kJ mol⁻¹	A_2^* cm² s⁻¹	Q_2^* kJ mol⁻¹	Temp. range K	Ref.
Al (—)	Fe (—)		IVa	p.c.			
		Diffusion of Co					
3.47	—	0.1	221.9	3.2	247.0	⎫	
7.95	—	1.9	234.5	4.5	251.2	⎪	
13.5	—	6.8	144.9	0.4	217.7	⎪	
20.6	—	22.0	251.2	32.0	263.8	⎪	
23.6	—	27.0	251.2	27.0	261.7	⎬ — 17	
35.5	—	210.0	280.5	—	—	⎪	
42.0	—	580.0	297.3	—	—	⎪	
47.3	—	6 300.0	330.8	—	—	⎪	
52.0	—	148.p	280.5	60.0	276.3	⎭	
Al (4N)	Fe (4N)		IVb, p.c., Fe⁵⁹				
6	—	—	—	0.42	197.7	1 088–1 478 (p) ⎫	
10	—	—	—	0.06	195.9	765–941 (f) ⎪	
10	—	—	—	0.02	183.5	1 156–1 450 (p) ⎬ 197	
18	—	—	—	0.01	197.7	816–953 (f) ⎪	
18	—	—	—	0.01	171.5	973–1 450 (p) ⎭	
(α phase)				p = paramagnetic,		f = ferromagnetic	
()	()		IVb	p.c.			
5.7	—	—	—	3.7	245.8	1 123–1 458	75
()	()		IVa(i)	p.c.			
'AlFe₃'		1.74×10^2	211.4	⎛Individual Re		823–903 ⎫	
'AlFe'		8.5×10^{-5}	105.1	⎜diffusion coefficients		823–903 ⎪	76
Al₂Fe		3.99×10^3	232.4	⎜are reported in		793–903 ⎬	
Al₃Fe		8.9×10^{-4}	117.2	⎝reference 77		823–873 ⎭	
Al	Li						
()	()		Vb, p.c., Li⁷				
50 (β)	50	—	—	2.63×10^{-6}	10.2	100–400	111
()	()		Vb, p.c., Li⁷				
51.7	48.3	—	—	2.54×10^{-5}	12.4	300–368	
50.6	49.4	—	—	1.56×10^{-6}	11.7	300–368	
50.0	50.0	—	—	8.34×10^{-6}	10.9	300–473	
48.1	51.9	—	—	2.34×10^{-6}	9.2	300–368	
46.9	53.1	—	—	2.58×10^{-7}	8.4	300–419	

(continued)

Table 13.3   DIFFUSION IN HOMOGENEOUS ALLOYS—*continued*

Element 1 (purity) At.%	Element 2 (purity) At.%	A_1^* cm^2 s^{-1}	Q_1^* kJ mol^{-1}	A_2^* cm^2 s^{-1}	Q_2^* kJ mol^{-1}	Temp. range K	Ref.
Al 45	Ll In 50 5	—	—	9.26×10^{-6}	11.1	300–353	150
Al 48 47	Li Ag 50 2 50 3	— —	— —	1.66×10^{-6} 3.63×10^{-7}	8.88 6.95	300–368 300–419	
Al (99.994)	Mg ( ) 0 2.2 5.5	0.10 0.93 0.21	Vb 127.7 128.1 116.0	p.c. — — —	— — — }	~603–733	127
Al (—)	Ni (—)	IVc		p.c.			
		Diffusion of Co^{60}					
—	47.3	47×10^{-2}	237.0	—	—		
—	48.5	9.3×10^{-2}	250.8	—	—		
—	49.4	4.4×10^{-3}	219.8	—	— }	1 323–1 623	18
—	50.7	57.7	337.5	—	—		
—	53.1	2.6	283.0	—	—		
—	55.1	7.2×10^{-3}	197.2	—	—		
( ) 24	( )	IVb, p.c., Ni^{63} —	—	132 $+1.1 \times 10^{-7}$	345 134.8		
25	(Ni$_3$Al)	—	—	146 $+1.1 \times 10^{-7}$ 105 $+2.0 \times 10^{-8}$ }	347 141.1 342 121.8	965–1 625	30
( )	( )	IVb		p.c.			
		Diffusion of In^{114m}					
—	48.3	—	—	1.2×10^{-4}	177.9		
—	48.6	1.29×10^{-3}	237.0	1.04×10^{-3}	209.8		
—	49.0	—	—	53×10^{-4}	200.5		
—	49.2	—	—	0.23	275.9		
—	50.0	3.98×10^{-3}	242.8	4.461	307.3	1 273–1 623	78 and 130
—	53.2	—	—	0.63	274.2		
—	54.5	—	—	0.15	250.4		
—	58.0	1.83×10^{-4}	170.0	0.035	216.5		
—	58.5	—	—	0.096	253.7		
—	58.7	—	—	0.725	250.4		
—	73.20	—	—	3.11	300.3)		
—	74.71	—	—	1.00	303.1 }		
—	76.20	—	—	4.41	306.3)		
Al ( ) 19.92 20.83 21.84	Ni Ti ( ) ( ) — 5.13 — 5.01 — 5.13	— — —	— — —	0.055 0.039 0.085	262.79 259.7 } 266.9	1 173–1 573	106
Al (99.99) — — — — —	Si (99.99) 0 0.05 0.10 0.15 0.20	IVa(i), p.c., Zn^{65} $B_{Cu}^{Zn} = 23$ (Av. value) (*See* Al Cu for definition)		Diffusion of Zn 0.26 0.23 0.30 0.32 0.29	119.1 118.2 119.5 } 120.2 119.8	673–837	41

(continued)

Table 13.3 DIFFUSION IN HOMOGENEOUS ALLOYS—*continued*

Element 1 (purity) At.%	Element 2 (purity) At.%	A_1^* cm^2 s^{-1}	Q_1^* kJ mol^{-1}	A_2^* cm^2 s^{-1}	Q_2^* kJ mol^{-1}	Temp. range K	Ref.
Al	Si	Zn					
(99.99)	(99.99)	(99.99)	IVa(i), p.c., Zn^{65}	Diffusion of Zn			
—	0	1.48		0.21	116.9		
—	0.05	1.48	$B_{Si}^{Zn} = 9$ (Av. value)	0.37	120.6		
—	0.10	1.48		0.40	121.1		
—	0.15	1.48		0.29	119.1		
—	0.20	1.48	$B_{Si}^{Zn} = 30$ (Av. value)	0.33	119.6		
—	0	2.13		0.31	119.4	673–837	41
—	0.05	2.13		0.21	116.7		
—	0.10	2.13		0.26	118.1		
—	0.15	2.13	$B_{Zn}^{Zn} = 7.5$ (Av. value)	0.43	121.2		
—	0.20	2.13		0.29	118.9		
Al	Zn						
(99.99)	(99.99)		IVa(i)	p.c.	Zn^{65}		
	0	—	—	0.27	120.4		
	1.16	—	—	0.25	119.0		
(At.%)	1.73	—	—	0.18	116.7		
	2.15	—	—	0.22	117.6	613–893	149
	2.80	—	—	0.22	117.1		
	3.29	—	—	0.24	117.5		
	3.76	—	—	0.23	116.9		
( )	( )		IVb, p.c., Zn^{65}				
—	7.06	—	—	0.170	112.46		
—	15.17	—	—	0.324	113.17		
—	24.24	—	—	0.209	108.27		
—	31.27	—	—	0.288	105.67		
—	41.51	—	—	0.229	103.54		
—	52.52	—	—	0.162	100.53	585–782	19
—	53.28	—	—	0.575	106.64		
—	55.04	—	—	0.692	108.10		
—	56.85	—	—	1.514	111.87		
—	57.28	—	—	0.575	106.81		
—	57.50	—	—	1.35	111.41		
(4N5)	( )		IVa(i), p.c., Zn^{65}				
—	20	—	—	0.25	109	621–783	
—	28.9	—	—	0.22	105	588–611	
—	40	—	—	0.17	101	632–714	198
—	50	—	—	0.23	100	558–611	
—	58.5	—	—	0.036	91	528–611	
(99.994)	( )		Vb	p.c.			
	0	0.10	127.7	—	—		
	8	0.087	122.7	—	—	~613–733	127
	11.5	0.035	117.2	—	—		
	18.8	0.025	113.5	—	—		
Al	Zn	Mg					
(99.99)	(99.99)	( )	IVa(i)	p.c.	Zn^{65}		
	0	1.11	0.34	121.5			
	0.71	1.12	0.25	119.2			
	1.49	1.13	0.23	118.0			
	2.11	1.14	0.31	119.5			
(At.%)	0	2.77	0.24	119.0		688–848	149
	0.71	2.80	0.27	119.0			
	1.48	2.83	0.21	116.8			
	2.11	2.86	0.17	115.1			
	0	4.43	0.28	118.6			
	0	6.63	0.50	121.1			

(continued)

Table 13.3   DIFFUSION IN HOMOGENEOUS ALLOYS—*continued*

Element 1 (purity) At.%	Element 2 (purity) At.%	A_1^* cm^2 s^{-1}	Q_1^* kJ mol^{-1}	A_2^* cm^2 s^{-1}	Q_2^* kJ mol^{-1}	Temp. range K	Ref.
Au (99.99)	Cd (99.95)		IVa(i)	s.c.			
50	50	0.17	116.8	0.23	117.2	573–863[a]	20
(99.999+)	(99.999+)		IVa(i)	s.c.			
	47.5	0.23	117.6	1.36	129.8	623–873⎫	
	49.0	0.61	125.6	1.50	130.6	713–823⎬	21
	50.5	0.12	109.7	0.22	113.5	713–823⎭	

(*a*) Between 863 K and the m.p. at 899 K there is marked upward curvature in the Arrhenius Plot.

Au ()	Cu ()		IVc	p.c.			
25	75	6.5×10^{-3}	159.9	—	—	823–1 173 (Disord.)	81
25	75	Diffusion of Co^{57} 4.2×10^{-2} 192.6				923–1 173	89

(5N)	(5N)		IVa(i), p.c., Cu^{64}, Au^{195}				
		$10^{10}D_{Au}^*$		$10^{10}D_{Cu}^*$			
1.25	–	1.7		1.83⎫			
2.5	–	1.8		2.03⎪			
5.0	–	2.08		2.48⎪			
7.48	–	2.40		3.02⎬		1 133	174
12.44	–	3.08		4.18⎪			
14.25	–	3.36		4.62⎪			
17.43	–	3.88		5.88⎭			

Au (99.96)	Ni (99.9)		IVa(ii)	p.c.	IVa(i)		
100	0	0.26	189.7	0.30	192.6		
	10	—	—	0.80	199.3		
	20	0.05	168.3	0.82	200.1	(Variable	
	35	—	—	1.10	205.6	For Ni,	32(Au)
	36	—	—	1.10	205.6	323–348	
	50	0.091	181.7	0.09	186.3	temp.	
	65	0.51	204.3	0.005	165.8	ranges	23(Ni)
	80	1.1	253.3	0.05	206.0	only)	
	90	—	—	0.04	213.9		
0		2.0	272.1	0.40	267.1	—	

Au ()	Pb (99.999 9)		IVa(i)	s.c.			
0.04		—	$D_{PB}^*/_{PurePB}^* = 2.2$	($b_1 = 5\,726$)		472.4⎫	
0.015		—	$= 0.55$			⎪	
0.04		—	$= 1.6$			⎪	
			$= 1.75^{(a)}$⎬	($b_1 = 4\,312$)		488.2⎬	63
0.06		—	$= 2.85$			⎪	
0.08		—	$= 3.8$			⎪	
			$= 4.05^{(a)}$⎭			⎭	

(*a*) Samples pre-annealed before diffusion.

Au ()	Zn ()		IVa(i)	s.c.			
—	49	0.19	133.6	0.84	144.9⎫		
—	50	0.33	138.6	1.93	148.2⎬	701–923 (Ordered β')	79
—	51	0.016	113.0	0.047	115.1⎭		

(continued)

Table 13.3   DIFFUSION IN HOMOGENEOUS ALLOYS—*continued*

Element 1 (purity) At.%	Element 2 (purity) At.%		A_1^* cm^{2}s^{-1}	Q_1^* kJ mol^{-1}	A_2^* cm^2 s^{-1}	Q_2^* kJ mol^{-1}	Temp. range K	Ref.
Be () — —	Ni () 1.68 7.9		— —	IVb — —	p.c. 0.41 0.23	247.0⎫ 188.4⎭	1 173–1 373	80
C () — —	Cr () 0.5 1.0	Fe ()	45 × 10^{-3} 20 × 10^{-3}	IVb 111.0 100.5	p.c. — —	—⎫ —⎭	1 673—1 873	85
C () —	Cr () —	Ta () 1.0	5 × 10^{-3}	IVb 120.6	p.c. —	—	1 573—1 773	85
Ce (—) 0 1.15 2.46 3.39 4.97 6.21	Fe (—)		— — — — — —	IVa(i) — — — — — —	p.c. 0.44 0.052 0.015 0.021 0.029 0.050	280.5⎫ 247.0⎪ 226.1⎪ 226.1⎬ 225.2⎪ 225.2⎭	1 273–1 573	28
Cu (—) c% c%	Fe (—) (γ)	Ni (—) 20 25	— —	IVb — —	p.c. 18.10$^{-0.92c}$ 71.10$^{-0.65c}$	314–25c 330.8–21c	1 073–1 573⎫ 1 323–1 603⎭	29

(*a*) These values also represent the best fit to the original (reference 26) and later measurements when plotted altogether. *See* reference 27.

C 0.104	Fe Si (99.97) (α) 5.5		V(a) D_c^* 'virtually' the same as in Si-free Fe	p.c.			26–70	67
Cd () 0–1.8	Cu (99.998)		—	—	$b_1 = 35$	$b_2 = 1\,400$	1 076	109
Cd (—) 75 75 25 25	Mg (—) 25 25 75 75		11.2 × 10^{-5} 0.074 4.10^{-5} 1.2 × 10^{-6}	IVb 51.9 (Ord.) 69.9 (Disord.) 68.2 (Ord.) 53.6 (Disord.)	s.c. — — — —	— — — —	327–363⎫ 368–473⎭ 393–428⎫ 428–554⎭	31 32
Cd ()	Pb (99.999 9)		(*See* eq. 13.9a)	IVa(i) Effect of Cd on D_{Pb}^* b_1	s.c. b_2	b_3		
0 to 5				45.197 30.028 19.138	438.1 407.7 1 070.2	— 44 286 —	472⎫ 521⎬ 574⎭	52
Cd (6N) 0–0.22	Pb (6N) —		IVa(i), p.c., Cd $B_1 = -1\,200$				470.7	180

With increasing Cd conc. D_{Cd}^* initially decreases and then beyond 0.022% monotonically increases.

(5N) 0 0.37 0.86 1.01 1.36 2.64 3.66	(5N) — — — — — — —	Diffusion of Ni $10^5D = 2.04$ $= 1.88$ $= 1.73$ $= 1.58$ $= 1.56$ $= 1.17$ $= 1.02$	IVa(i), p.c., Ni^{63}	—	—	506.7	186

(*continued*)

Table 13.3   DIFFUSION IN HOMOGENEOUS ALLOYS—*continued*

Element 1 (purity) At.%	Element 2 (purity) At.%		A_1^* cm² s⁻¹	Q_1^* kJ mol⁻¹	A_2^* cm² s⁻¹	Q_2^* kJ mol⁻¹	Temp. range K	Ref.
Co (—)	Cr (—)			IVb	p.c.			
	4		0.67	275.5	—	—⎫	1 373–1 623	33
	7		56.3	332.0	—	—⎭		
Co (—)	Cr (—)	Ni (—)		IVb	p.c.			
	9	26	6.3	301.9	—	—⎫	1 373–1 623	33
	18	26	0.4	268.8	—	—⎭		
Co (99.99)	Cu (99.99)	Si ( )			IVa(i), p.c., Co⁵⁸			
0–2.68	—	0.54		$B_1 = -9$			1 325	158
Co ( )	Fe ( )			IVa(i)	p.c.			
			⎧ 1.33	290.6	1.26	286.8	1 285–1 437 (γ) ⎫	
50	50		⎨ 2.0	251.2	0.25	230.3	1 068–1 218 (α) ⎬ 34	
			⎩ —	556.8	—	556.8	928–995 (CsCl) ⎭	
( )	( )			IVb	p.c.			
—	21		0.54	272.1	—	—	1 373–1 573	33
( )	(—)			IVa(iii)	p.c.			
—	0			Diffusion of Ni				
—	30.2			$D_{Ni} = 13.10^{-12}$				
—	68.5			$= 11.5$ ⎫				
—	88.3			$= 8$–8.8 ⎬			1 409	82
—	100			$= 5$ ⎭				
				$= 6$				
( ) 0	( )			IVb	p.c.	(F = Ferromagnetic)		
	(bcc)		1.83	234.0	—	—	1 073–1 172	
	(fcc)		0.77	265.0	—	—	1 223–1 633	
3	(bcc)		9.17	266.3	—	—	903–1 023 (F)	
6.8	(bcc)		0.469	187.1	—	—	903–1 073 (F)	
			5.72×10^{-5}	146.5	—	—	1 153–1 193	
	(fcc)		0.109	326.2	—	—	1 283–1 583	
28.6	(bcc)		1.25×10^{-3}	198.0	—	—	903–1 083 (F)	
	(fcc)		3.36×10^{-2}	266.3	—	—	1 333–1 583	
49.6	(bcc)		6.59×10^{-2}	247.0	—	—	1 023–1 123 (F) ⎫ 107	
	(fcc)		0.154	349.6	—	—	1 333–1 583	
67.2	(bcc)		6.04×10^{-3}	190.9	—	—	903–1 153 (F)	
	(fcc)		3.15×10^{-2}	265.0	—	—	1 333–1 583	
89.6	(fcc)		6.44×10^{-2}	251.2	—	—	1 073–1 283 (F)	
			1.61×10^{-2}	234.0	—	—	1 333–1 153	
100	(fcc)		0.50	273.8	—	—	1 045–1 321 (F) ⎭	
			0.17	260.4	—	—	1 465–1 570	
(99.8) 0	(99.9)		IVb	p.c.	Co⁶⁰			
			0.029	247.4	—	—⎫		
8			20.54	321.5	—	—⎪		
10			15.65	369.0	—	—⎬ 1 233–1 493	131	
15			1.98	289.7	—	—⎪		
20			0.31	261.7	—	—⎭		
Co ( )	Fe ( )		IVb, s.c., Fe⁵⁹					
—	6		—	—	0.58	273.3⎫	1 081–T_c	
—	6		—	—	0.15	261.7⎪	T_c–1 573	
—	10		—	—	0.68	279.3⎬	1 081–T_c	176
—	10		—	—	0.18	263.3⎭	T_c–1 573	
Co ( )	Fe ( )	Ti ( )		IVb	p.c.			
	15	4	0.008	214.4	—	—	1 373–1 473	33
Co ( )	Fe ( )	V ( )		IVa(i), p.c. Co⁶⁰				
60	15	25	10.5	313[(a)]	—	—	1 277–1 570	152

(a) Reports *D*s below T_c but these scattered and affected by g.b. diffusion.

(*continued*)

Table 13.3 DIFFUSION IN HOMOGENEOUS ALLOYS—*continued*

Element 1 (purity) At.%	Element 2 (purity) At.%	A_1^* cm² s⁻¹	Q_1^* kJ mol⁻¹	A_2^* cm² s⁻¹	Q_2^* kJ mol⁻¹	Temp. range K	Ref.
Co (5N)	Ga (5N)		IVa(i), p.c., Co⁶⁰, Ga⁷²				
48.6	—	2.0×10^3	311.6	2.7×10^3	319.4	$1\,248 \times 1\,353$	
52.4	—	2.7×10^3	310.7	1.3×10^2	309.7	$1\,256$–$1\,386$	
54.3	—	2.0×10^3	305.9	1.5×10^3	307.8	$1\,250$–$1\,340$	175
57.2	—	2.0×10^2	275.9	1.4×10^2	275.9	$1\,193$–$1\,345$	
( )	( )		IVa(i), p.c., Co⁶⁰, Ga⁶⁷				
45.2	—	1.91	247	766×10^3	326	651–$1\,100$	
		$^{(a)} +4.68 \times 10^3$	326	$+0$			
48	—	2.10×10^{-3}	209	554	308	700–$1\,150$	
		$+1.5 \times 10^3$	308	$+2.65 \times 10^9$	494		
50	—	0.379	245	220	301	639–$1\,150$	
		$+878$	301	$+8.11 \times 10^6$	417		132
56	—	1.02	235	475	303	625–$1\,150$	
		$+1.60 \times 10^3$	303	$+2.30 \times 10^7$	430		
60	—	0.303	216	80.8	281	600–$1\,150$	
		$+343$	281	$+1.99 \times 10^6$	395		
Co ( )	Mn ( )	IVb	p.c.	Mn⁵⁴			
100	~0	—	—	3.15×10^{-2}	232.4	$1\,133$–T_c	
		—	—	1.1×10^{-2}	217.7	T_c–$1\,519$	
—	5.22	—	—	1.38	268.4	$1\,141$–T_c	133
		—	—	0.501	256.7	T_c–$1\,473$	
—	10.24	—	—	1.36	263.3	$1\,176$–$1\,421$	
Co ( )	Nb ( ) Zr ( )			IVa(i), p.c., Co⁵⁷			
2	31 —	1.8	194	—	—	$1\,125$–$1\,645$	156
Co (99.5)	Ni (99.98)		IVb	p.c.		Ferromag.$^{(a)}$	
100	0	0.5	273.8	0.34	269.2	$1\,045$–$1\,321$	
	11	0.61	280.9	0.46	274.7	$1\,137$–$1\,321$	
	20	5.96	307.7	1.66	291.8	$1\,118$–$1\,321$	
	30	1.16	287.2	2.01	286.8	$1\,045$–$1\,172$	
	51	0.096	257.5	0.36	266.3	974–$1\,072$	
						Paramag.$^{(a)}$	35
100	0	0.17	260.4	0.10	252.0	$1\,465$–$1\,570$	
	11	0.21	266.3	0.17	262.5	$1\,417$–$1\,570$	
	20	2.42	297.3	0.41	275.1	$1\,363$–$1\,519$	
	30	0.78	280.5	0.67	271.3	$1\,323$–$1\,523$	
	51	0.12	252.0	0.21	253.7	363–$1\,463$	
	100	0.75	270.9	1.70	285.1	973–$1\,463$	
(99.4)	(99.8)		IVb	p.c.			
		$^{(a)}$ $^{(b)}$					
99.5	—	1.66	287.6	3.35	297.3	$1\,320$–$1\,584$ (O) / 493–$1\,693$ (Ni)	
93.8		7.4	305.2	5.40	302.3	$1\,483$–$1\,643$	
89.0		2.52	290.1	6.42	304.0		
80.6		0.99	275.9	1.89	285.5	$1\,433$–$1\,683$	
72.6		0.70	270.5	1.60	281.4		
49.4		0.18	250.4	0.25	61.1	$1\,373$–$1\,623$	
43.3		0.52	262.1	0.69	267.1		
21.1		0.33	255.8	0.45	262.1	$1\,433$–$1\,683$	
10.8		0.66	263.8	1.47	276.3		83
4.3		0.49	261.3	2.86	284.7	$1\,483$–$1\,643$	
0.03		1.11	271.7	1.39	275.9	$1\,423$–$1\,663$	
( )	( )		IVb	p.c.			
		Diffusion of Fe⁵⁹					
50	50	125	320.3			$1\,273$–$1\,523$	84
	(99.4%)	Diffusion of C¹⁴			IVb	p.c.	
5.25		0.4	154.9	—	—	873–$1\,173$	36

(a) According to reference 64 measurements in pure Co, in Co + 11% Ni and Co + 20% Ni alloys do not convincingly demonstrate difference in A^* and Q^* values for the paramagnetic and ferromagnetic regions.

(b) All alloys contain 0.1–0.2 Mg.

(continued)

Table 13.3   DIFFUSION IN HOMOGENEOUS ALLOYS—*continued*

Element 1 (purity) At.%	Element 2 (purity) At.%		A_1^* cm² s⁻¹	Q_1^* kJ mol⁻¹	A_2^* cm² s⁻¹	Q_2^* kJ mol⁻¹	Temp. range K	Ref.
Co ()	Mn ()	Ni ()		IVb	p.c.			
19.5	20.3	60.0	0.86	253.3	—	—	1 293–1 393 ⎫	
40.6	20.45	38.9	0.22	240.7	—	—	1 293–1 433 ⎬	14
59.7	20.6	19.5	0.05	228.2	—	—	1 333–1 433 ⎭	
Co (99.97)	Ti (99.97)			IVa(i)	p.c.			
1.6			—	—	1.26×10^{-3}	145.3	1 233–1 777 ⎫	
3.3			—	—	1.58×10^{-3}	140.3	1 166–1 617 ⎬	121
4.9			—	—	1.41×10^{-2}	160.4	1 076–1 573 ⎪	
7.4			—	—	2.50×10^{-2}	162.9	1 076–1 484 ⎭	
Co (99.9)	Ti (99.8)		IVa(i), p.c., Co⁶⁰					
—	21.5		4.3×10^{-2}	203	—	—⎫		
(Co₃Ti)	22.8		1.2×10^{-2}	188	—	—⎬	1 074–1 323	199
—	24.0		3.7×10^{-3}	174	—	—⎭		
Co ()	Zr ()			IVa(i)	p.c.			
			$D^*_{\text{Co in pure Zr}}$		$D^*_{\text{Zr}}/D^*_{\text{Pure Zr}}$			
0.395	—		0.900		0.948			
0.747	—		—		1.322 8		⎫	
1.22	—		0.861		1.668		⎬ 1 266	118
1.61	—		—		2.143 3		⎪	
1.995	—		0.820		2.223 9		⎭	
Cr (99.9)	Fe (99.9)			IVb	p.c.			
9.13			—	—	9.27	230.7	548–999 (Ferro.α) ⎫	
			—	—	0.42	219.4	1 050–1 098 (Para.α) ⎬	86
			—	—	0.12	237.4	1 173–1 313	87
15.22			—	—	1.25	226.5	868–950 (Ferro.α) ⎫	
			—	—	0.27	215.6	999–1 050 (Para.α) ⎬	86
19.75			—	—	0.65	217.3	848–919 (Ferro.α) ⎪	
			—	—	0.18	208.1	963–1 098 (Para.α) ⎭	
()	()			IVb	p.c.			
2			3.21	244.5	—	—⎫		
6			1.21	237.4	—	—⎪		
13			0.64	231.9	—	—⎬ 1 073–1 673	37
16			0.19	218.1	—	—⎪		
19			0.18	216.9	—	—⎭		
15	2.0		238.6	—	—	—	88	
()	()			IVb	p.c.	Fe⁵⁹		
0			—	—	4.4	253.3 ⎫		
0.87	(wt.%)		—	—	4.3	252.9 ⎪		
1.43			—	—	4.2	252.5 ⎪		
3.09			—	—	4.0	251.6 ⎬		
5.05			—	—	3.7	250.0 ⎬ Paramag. α phase	134
6.68			—	—	3.6	249.1 ⎪		
8.18			—	—	3.3	247.9 ⎪		
11.93			—	—	2.9	244.9 ⎭		
(99.997)	(99.997)		Diffusion of S		IVb, p.c. S³⁵			
18	—		0.166	184				
26 wt.%	—		3.06×10^{-2}	171	—	—	1 123–1 473	164
34	—		6.18×10^{-2}	177				
(99.9)	(99.9)				IVb, p.c., Cr⁵¹, Fe⁵⁹			
0	—		1.16	276.7	0.72	278.7 ⎫		
0.93	—		1.30	286.0	—	— ⎪		
1.53	—		—	—	0.73	283.1 ⎪		
3.31	—		0.35	273.8	—	— ⎬ 1 223–1 473	192
5.4	—		—	—	0.41	276.4 ⎪		
7.2	—		0.13	261.6	—	— ⎪		
8.8	—		—	—	0.069	257.1 ⎭		

(continued)

Table 13.3 DIFFUSION IN HOMOGENEOUS ALLOYS—*continued*

Element 1 (purity) At.%	Element 2 (purity) At.%		A_1^* cm² s⁻¹	Q_1^* kJ mol⁻¹	A_2^* cm² s⁻¹	Q_2^* kJ mol⁻¹	Temp. range K	Ref.
(≥98)	(≥98)							
26			0.156	202.6	—	—⎫		
51			40.0	293.1	—	—⎬	1 223–1 593	38
69			24.6	316.1	—	—⎭		
('Electrolytic')				IVa(i)	p.c.			
27			—	—	0.195	211.0⎫		
51			—	—	249	312.3⎬	1 313–1 673	39
84			0.376	268.8	146	342.9⎭		
()	()			IVb	p.c.			

			Diffusion of Ni⁶³					
202	—		0.356	274.6		fcc⎫		
5.73			0.282	278.1		fcc⎪		
9.47			0.12	265.4		fcc⎬		95
14.1			0.016	237.4		bcc⎪		
20.1			0.007	224.4		bcc⎭		

Cr	Fe	Ni						
()	()	()		IVa(i)	p.c.			
19.9	(a)	24.7	0.19	346.2	1.74	284.3⎫	1 113–1 563	
				(Ni) IVb	p.c.	⎬		102
			4.06	282.6	p.c.	—⎭	1 113–1 568	
()	()	()		IV Cr⁵¹, Fe⁵⁹, Ni⁶³		p.c.		
17		12	0.13	264.2	0.37	279.7⎫	873–1 573	135
					$A_{Ni}=8.8$	$Q_{Ni}=251.2$⎭		
()	()	()		IVa(i)	p.c.			
17		12	6.3 × 10⁻²	243.3	—	—	1 023–1 473	136
()	()	()		IVb	p.c.			

			Diffusion of Ni					
19	—	10	1.4	301.4	2.5 × 10⁻³	217.7⎫		
		30	1.4	301.4	1.0	278.4⎪		
	(wt.%)	45	2 × 10⁻⁴	401.9	1.2	255.4⎬	1 173–1 473	137
		55	2.8	297.3	4.5	255.4⎪		
		65	2.6	297.3	1.0	272.1⎪		
		75	7.2 × 10⁻²	253.3	4.10²	343.3⎭		

(99.99)	(99.44)	(99.95)		IVb, p.c., Cr⁵¹, Fe⁵⁸				
10.1	—	78.7	2.26	278.2	1.73	281.3		
10.1	—	64.6	1.35	273.4	1.10	276.5		
9.9	—	55.1	5.22	290.1	4.91	295.2		
10.1	—	35.0	8.49	299.1	3.75	295.1	1 286–1 536⎫	
20.0	—	63.5	2.18	282.3	2.22	286.7	(Fe) ⎪	
19.9 wt.%	—	55.1	4.72	295.8	7.62	307.8	⎬	162
19.6	—	43.0	4.69	293.2	1.19	284.1	1 298–1 548⎪	
20.0	—	34.9	1.95	282.8	4.75	298.7	(Cr) ⎭	
19.1	—	19.1	0.425	269.8	0.545	278.4		
30.4	—	65.3	1.945	279.3	1.23	280.1		
30.4	—	54.8	0.536	264.3	0.817	273.0		
30.4	—	34.8	5.82	294.5	2.42	291.0		

(5N)	(5N)	()	Diffusion of S		IVb, p.c., S³⁵			
15	10 wt.%	—	0.46	213.6	—	—	1 173–1 473	166

()	()	()		IVa(i), p.c., Cr⁵¹				
(19)	— (a)	10.4	5.59	296.5	—	—	1 216–1 311	171

()	()	()		IVa(i), p.c., Cr⁵¹, Fe⁵¹, Fe⁵⁹, Ni⁵⁷			Diffusion of Ni	
15	—	20	8.3	309	5.3	308	1.5	300	
15	—	45	4.0	293	2.1	288	1.8	293	177
22	—	45	4.1	295	1.5	286	1.1	291	
15	—	20(b)	7.1	303	5.1	303	4.8	310	

			1 233–1 673 K					
()	()	()		IVa(i), p.c., Fe⁵⁹				
17	—	ii(c)	—	—	1.18 × 10⁻²	228.5	1 178–1 483	78

(*continued*)

Table 13.3   DIFFUSION IN HOMOGENEOUS ALLOYS—*continued*

Element 1 (purity) At.%	Element 2 (purity) At.%	A_1^* cm² s⁻¹	Q_1^* kJ mol⁻¹	A_2^* cm² s⁻¹	Q_2^* kJ mol⁻¹	Temp. range K	Ref.
( )	( ) (99.8)	IVb p.c.		Diffusion of C^{14}			
0.74	0.52 —	—	—	0.1	142.4	873–1 173 ⎫	
4.65	0.36 —	—	—	0.5	154.9	773–1 173 ⎬	36
18	— 8			6.18	186.8	723–1 473 ⎭	

(a) Plus 0.55 Nb, 0.74 Mn, 0.69 Si, 0.03 Ti.
(b) +1.33% Mn, 0.65% Si, 0.46% Ti, 0.44% Cu. (321 St. Steel).
(c) +2.3% Mo.

Element 1 (purity) At.%	Element 2 (purity) At.%	A_1^* cm² s⁻¹	Q_1^* kJ mol⁻¹	A_2^* cm² s⁻¹	Q_2^* kJ mol⁻¹	Temp. range K	Ref.
Cr ('Electrolytic')	Ni		IVb	p.c.			
4.9		—	—	0.15	253.3 ⎫		
6.35		0.01	211.4	—	—		
9.93		—	—	0.039	238.6 ⎬	1 223–1 523	122
11.69		0.037	229.0	—	—		
19.7		0.01	242.8	0.006 3	220.2 ⎭		
( )	(99.95)		IVa(i)	p.c.			
0		1.1	272.6	1.9	284.7	1 373–1 543 (D_{Cr}^*) ⎫	
10		1.4	278.4	3.3	293.4	1 313–1 548 (D_{Ni}^*) ⎬	42
20		1.9	283.4	1.6	286.8	1 040–1 404 (Ni Self D) ⎭	
(99.99)	(99.95)	IVb, p.c., Cr^{51}, Fe^{59}		Diffusion of Fe			
—	95.8	—	—	2.48	285.9		
—	87.1	—	—	4.63	296.5		
—	78.5	7.40	297.8	5.33	299.9		
wt.%	73.0	—	—	1.475	287.0	1 286–1 536	162
—	68.3	—	—	1.16	285.1		
—	64.4	—	—	1.50	287.3		
—	55.3	—	—	1.84	286.6		
( )	( )	IVb, p.c., Ni^{63}, Cr^{51}					
20	—	0.61	264	0.15	259	1 123–1 473	173
(99.9)	(99.9)	IVb, p.c., Cr^{51}, Ni^{63}					
0	—	8.51	292.1	3.44	290.0 ⎫		
4.7	—	6.37	292.1	1.32	279.5		
14.3	—	5.66	293.6	2.31	289.3		
23.6	—	6.10	295.5	2.95	293.8 ⎬	1 223–1 473	192
29.4	—	2.91	288.5	1.02	285.4		
34.4	—	2.42	288.0	1.35	289.5		
38.4	—	2.16	284.0	1.43	290.4		
47.7	—	2.64	284.2	1.74	288.9 ⎭		
( )	( )	IVa(i)	p.c.	IVa(ii)			
35 (wt.%)		0.2	245.3	4.10^{-3}	205.2 ⎫	1 223–1 633	
80 (wt.%)		0.28	259.6	—	— ⎬	1 423–1 688	68
Cr	Ti						
	(99.4)	p.c.	IVa(ii)				
10.0		0.02	168.3	—	— ⎫	1 198–1 453	44
18.0		0.09	186.3	—	— ⎬		
Cr	Zr						
(99.99)	(99.9)	IV	p.c.		Cr^{51}		
1.5	(wt.%)	0.19	191.3	—	—	1 233–1 373 (β)	49
(99.95)	(nuclear)	IVa(i), p.c., Cr^{51}, Zr^{95}					
0	—	4.53×10^{-3}	137.84	6.80×10^{-4}	145	1 218–1 518 ⎫	
2.05	—	—	—	5.16×10^{-3}	165.2	1 200–1 497	
3.49	—	—	—	2.05×10^{-2}	169	1 196–1 518 ⎬	195
4.08	—	—	—	1.38×10^{-2}	169.2	1 218–1 516	
7.86	—	—	—	1.59×10^{-2}	173.4	1 227–1 516 ⎭	

(continued)

Table 13.3 DIFFUSION IN HOMOGENEOUS ALLOYS—*continued*

Element 1 (purity) At.%	Element 2 (purity) At.%	A_1^* cm² s⁻¹	Q_1^* kJ mol⁻¹	A_2^* cm² s⁻¹	Q_2^* kJ mol⁻¹	Temp. range K	Ref.
Cu (99.998 5)	Fe ()		IVa(i)	p.c. and s.c.			
	0.2		$D_{Cu}^*/D_{Cu\ Pure}=0.99$			1 296	
	1.38		$=0.98$			1 296	
	1.38		$=0.91$			1 293	
	1.44		$=0.90$	$(b_1=-5\pm1.5)$		1 351	120
	1.45		$=0.94$			1 293	
	1.82		$=0.91$			1 265	
	2.40		$=0.90$			1 293	
Cu (99.998)	In (5N)		IVa(i), p.c., In¹¹⁴ᵐ				
—	0	—	—		$10^{10}D_{In}^*=7.0$		
—	0.8	—	—		$=9.8$		
—	1.7	—	—		$=13$	1 089	179
—	3.1	—	—		$=18$		
—	5.1	—	—		$=34$		
(4NB)	(5N)		IVa(i), p.c., Cu⁶⁴				
—	0	0.6	210	—	—		
—	0.4	2.0	220	—	—		
—	0.8	0.4	200	—	—	1 005–1 145	
—	1.2	0.6	200	—	—		
—	1.7	0.2	190	—	—		
		$b_1=42$	$b_2=3\,500$			1 005	191
		$=43$	$=2\,300$			1 089	
		$=48$	$=790$			1 145	
Cu (99.99)	Ni (99.95)	p.c.	IVa(i)				
0		0.57	258.3	1.9	284.7	1 323–1 633	
13.0		1.5	263.8	35.0	313.6	1 323–1 633	
45.4		2.3	252.5	17.0	279.7	1 258–1 483	32
78.5		1.9	231.5	0.063	208.1	1 133–1 343	
100.0		0.33	201.8	1.7	231.5	**0 000–0 000**	
()	()		Resistance method.	p.c.			
—	0	—	—	1.95	236.5		
—	1.08	—	—	0.95	233.1	1 053–1 310	182
—	2.81	—	—	0.23	225.5		

Cu () At.%	Ni () At.%	Zn () At.%	A_1^*	Q_1^*	A_2^*	Q_2^*	Temp. range K	Ref.
			IVa(i)		p.c.			
100	—	—	—	—	0.24	188.8	1 073–1 313	
89.9	—	10.1	—	—	0.64	190.9	1 022–1 252	
79.5	—	20.5	—	—	0.35	176.7	1 021–1 213	
69.8	—	30.2	—	Diffusion of	0.32	164.5	973–1 175	
90.7	9.3	—	—	Zn⁶⁵	0.36	200.1	1 068–1 313	90
80.4	—	10.3	—	—	0.49	195	1 023–1 278	
70.2	9.3	20.5	—	—	1.41	196.4	1 623–1 248	
60.1	9.1	30.8	—	—	0.39	173.3	973–1 173	
81.8	18.2	—	—	—	0.89	214.8	1 068–1 278	
70.8	18.8	10.4	—	—	0.36	199.7	1 073–1 313	
60.6	18.6	20.8	—	—	0.73	187.1	1 021–1 213	
71.4	28.6	—	—	Diffusion of	1.37	226.5	1 143–1 353	90
61.2	28.2	10.6	—	Zn⁶⁵	1.44	220.2	1 128–1 313	
50.8	28.2	21.0	—	—	1.17	208.9	800–1 005	
40.7	27.9	31.4	—	—	1.13	198.5	1 033–1 249	

Element 1 (purity) At.%	Element 2 (purity) At.%	D_{Cu}^*		D_{Sb}^*		Temp. range K	Ref.
Cu ()	Sb ()		IVc	p.c.			
(δ)	19.4	4.10^{-10}		3.10^{-11}			
(χ)	24.4	7×10^{-9}		3.1×10^{-10}		663	61
(γ)	33.4	$\sim10^{-9}$		2.7×10^{-9}			

(*continued*)

Table 13.3   DIFFUSION IN HOMOGENEOUS ALLOYS—*continued*

Element 1 (purity) At.%	Element 2 (purity) At.%	A_1^* cm² s⁻¹	Q_1^* kJ mol⁻¹	A_2^* cm² s⁻¹	Q_2^* kJ mol⁻¹	Temp. range K	Ref.
( )	( )		IVb	p.c.			
	21	8.57×10^{-3}	43.8	—	—		
(β)	25	1.99×10^{-3}	30.4	—	—	793–903	91
	29	5.80×10^{-3}	24.3	—	—		
(4N8)	(5N)		IVa(i), p.c., Cu^{64}				
—	0	0.6	210	—	—		
—	0.3	0.4	200	—	—	1 005–1 145	
—	0.5	0.6	200	—	—		
—	0.8	0.7	200	—	—		191
		$b_1 = 79$	$b_2 = 27\,000$			1 005	
		$= 100$	$= 14\,000$			1 089	
		$= 130$	$= 12\,000$			1 145	
Cu (99.999)	Sn (99.999)		IVa(i)	p.c.			
(δ)	20.5	4.7	129.4	2.4×10^3	208.1	713–848	
(γ)	$\{^{18.0}_{9.8}$	1.4×10^{-2}	74.5	0.33	122.3	873–998	92
		3.6×10^{-3}	84.6	9.2×10^{-2}	113.5		
(99.998)	(5N)		IVa(i), p.c., Sn^{113}				
—	0	—	—	$10^{10} D_{Sn}^* = 8.1$			
—	0.8	—	—	$= 9.6$			
—	1.7	—	—	$= 12$		1 089	179
—	3.0	—	—	$= 20$			
—	4.9	—	—	$= 26$			
(4N8)	(6N)		IVa(i), p.c., Cu^{64}				
—	0	0.6	210	—	—		
—	0.4	0.4	200	—	—		
—	0.8	0.07	180	—	—	1 005–1 145	
—	1.1	0.06	180	—	—		191
—	1.7	0.03	170	—	—		
		$b_1 = 40$	$b_2 = 9\,600$			1 014	
		$= 48$	$= 3\,300$			1 089	
		$= 54$	$= 2\,600$			1 145	
(5N)	(5N)		IVa(i), p.c., Cu^{64}, Sn^{113}				
—	16.6	8.3×10^{-3}	83	0.22	118	811–998	196
—	20.2	1.8×10^{-2}	82	3.5×10^{-2}	107	897–1 003	
Cu ('Spec. P')	Zn ('Spec. P')	s.c.	IVa(i)				
(α)	31	0.34	175.4	0.73	170.4	850–1 178	45
	(99.99)	s.c.	IVa(i)				
		$\{$ 0.011	92.3	0.003 5	78.6	Disord. 770–1 090	
		180	155.3	78.10^3	185.2	Ord. 653–723	
(β)	45.65 to 48.1	$\{$ 80	150.8	163.0	152.0	Ord. 537–653	46
		$\overline{\text{Diffusion of Sb}}$					
		0.08	98.4			Disord. 771–867	
(—)	(—)			p.c. IVa(i)			
			98.1	0.022	92.3	Disord. 773–1 073	
(β)	46.7	$\{$ 0.020					47
		0.08	$129.8^{(b)}$	1.0	134.0	Ord. around 573	
(—)	(—)	p.c.	IVa(i)				
		$\overline{\text{Diffusion of Ag}}$		$\overline{\text{Diffusion of Co}}$			
(β)	47.2	0.014	91.7	0.47	112.6	Disord. 743–973	48

(*a*) Arrhenius plots in the ordered region are curved. The values of *A* and *Q* reported described straight line approximations to the data over the temperature ranges indicated.

(*b*) Ditto. The values of *A* and *Q* given here refer to the data at the lower end of the ordered temperature range investigated, viz: 300 K.

(*c*) Values of D^* for the ordered region are shown only in graphical form in reference 48.

(continued)

Table 13.3   DIFFUSION IN HOMOGENEOUS ALLOYS—*continued*

Element 1 (purity) At.%	Element 2 (purity) At.%	A_1^* cm² s⁻¹	Q_1^* kJ mol⁻¹	A_2^* cm² s⁻¹	Q_2^* kJ mol⁻¹	Temp. range K	Ref.
( )	( )		IVa(i)	p.c.			
		$10^{10} \times D_{Cu}^*$		$10^{10} \times D_{Zn}^*$			
—	0	2.71		9.66		1 167	
		6.70		22.7		1 220	
—	0.62	2.81		—		1 167	
—	1.09	7.42		—		1 220	
—	2.17	3.12		—		1 167	93
—	2.68	3.23		—		1 167	
—	2.99	8.33		—		1 220	
—	4.06	3.71		—		1 167	
—	4.13	9.23		—		1 220	
(Electrolytic)	(99.99)		IVb	s.c.			
0		—	—	$\perp c$ 1.62	108.9		(*d*)
0		—	—	$\|c$ 0.013	79.5		
0.2		—	—	$\perp c$ 3.2	108.9		
0.2		—	—	$\|c$ 0.021	79.3		
0.3		—	—	$\perp c$ 3.4	108.9	823–903	56
0.3		—	—	$\|c$ 0.025	79.5		
0.4		—	—	$\perp c$ 3.4	108.9		
0.4		—	—	$\|c$ 0.029	79.5		
0.5		—	—	$\perp c$ 3.5	108.9		
0.5		—	—	$\|c$ 0.035	83.7		

(*d*) Additions of 0.5 at % Al to this range of CuZn alloys have a negligible effect on Q_Z^{n*} but decrease $A_\perp$ by ~20% and A by ~40%.

Element 1 (purity) At.%	Element 2 (purity) At.%	A_1^* cm² s⁻¹	Q_1^* kJ mol⁻¹	A_2^* cm² s⁻¹	Q_2^* kJ mol⁻¹	Temp. range K	Ref.
Fe ( )	Ge ( )		IVa	p.c.			
	4.8	4.8	242.8	—	—	1 173–1 473	75
Fe (99.97) At.%	Mn (99.94)		IVb	p.c.			
	1.04	9×10^{-2}	265.4	5.5×10^{-2}	249.5		
	2.03	1.05×10^{-1}	262.9	2.0×10^{-2}	235.3		
	2.97	5.8×10^{-2}	255.8	9.6×10^{-3}	222.3	Fe	
	4.90	6.6×10^{-2}	255.0	1.7×10^{-2}	229.4	1 263–1 573	
	7.04	1.1×10^{-1}	262.1	7.2×10^{-2}	248.3	Mn	
	10.41	3.5×10^{-1}	275.5	2.9×10^{-1}	266.7	983–1 573	98
	18.15	6.4×10^{-1}	282.2	1.9×10^{-1}	261.7		
	25.5	8.5×10^{-1}	277.6	1.2×10^{-1}	251.6		
	33.98	6.0×10^{-1}	276.7	7.3×10^{-2}	242.0		
( )	( )		IVb	p.c.			
	wt.%		Diffusion of Ni⁶³				
	0.42	0.495	282.4			fcc range	95
	1.26	0.364	279.8				
	4.60	0.144	270.5				
	9.7	0.132	267.5				
Fe (4N)	Mn Si (4N) ( )	IVa(i), p.c., Nb⁹⁴/⁹⁵		Diffusion of Nb			
—	1.5 0			0.66	264		
wt.%	1.5 0.6			1.4	271.7	1 354–1 474	184
—	0 0.6			1.29	270.0		
Fe ( )	Mo ( )		IVb, p.c., Fe⁵⁹				
	0.54	15.5	263.3	—	—		
(wt.%)	1.06	23.6	257.1	—	—	953–1 173	141
	1.50	23.6	257.1	—	—	Paramag.	
	2.50	47.7	264.2	—	—		
Fe ( )	Nb ( )		IVa(i), p.c., Nb⁹⁵				
0–1.24		—	—	—	$b_1 = 61.24$	2 029	
0–0.90		—	—	—	$= 56.8$	2 172	201
0–0.53		—	—	—	$= 55.0$	2 261	

(*continued*)

Table 13.3 DIFFUSION IN HOMOGENEOUS ALLOYS—*continued*

Element 1 (purity) At.%	Element 2 (purity) At.%	A_1^* cm² s⁻¹	Q_1^* kJ mol⁻¹	A_2^* cm² s⁻¹	Q_2^* kJ mol⁻¹	Temp. range K	Ref.
Fe	Ni						
( )	( )		IVb	p.c.			
	0.2	—	—	1.09	290.3		
	0.55	—	—	1.09	290.3		
wt.%	2.29	—	—	0.593	281.8	sol. sol. range	95
	9.21	—	—	0.497	278.0		
	19.34	—	—	0.409	273.9		
(Electrol.)	(Electrol.)		IVc	p.c.			
	5.8	—	—	2.11	307.7	1 433–1 663	50
	14.88	—	—	5.0	316.5		
( )	( )		IVb	p.c.			
	20	18	314.0	—	—	1 073–1 573	
	25	71	330.8	—	—	1 323–1 603	29
( )	( )		IVa(i)	p.c.			
	50	—	—	66	322.8	1 286–1 508	96
	80	—	—	4.8	286.4		
( )	( )		IVa(i)	s.c.			
0 to 1.46		—	—	D_{Ni}^* nearly independent of composition		1 226 and 1 326	97
(99.44)	(99.95)	IVb, p.c., Fe⁵⁹, Cr⁵¹		Diffusion of Cr			
—	0	0.88	279.7				
—	15.5	1.96	286.5	1.96	280.5		
—	30.7	5.35	298.9				
—	46.6	4.73	295.4	2.14	281.2		
wt.%	61.7	16.6	306.0			1 298–1 548	162
—	71.1	14.1	304.1				
—	76.2	8.61	299.1	5.0	287.4		
—	80.6	11.48	302.9				
—	90.5	12.88	303.4				
—	100	5.83	293.3				
(99.99)	(99.9)	IVb, p.c., Ni⁶³					
—	0	—	—	5×10^{-3}	232.4		
—	5	—	—	1.5×10^{-2}	241.2		
wt.%	10	—	—	1.9×10^{-2}	243.2	1 193–1 513	172
—	15	—	—	0.277	270.5		
—	20	—	—	0.797	281.8		
(99.57)	(99.97)	IVa(i), p.c., Fe⁵⁹, Ni⁶³, W¹⁸⁵					
—	0	0.72	278.7	1.12	283.5		
—	14.9	2.13	286.3	1.88	289.4		
—	29.7	9.98	305.7	2.36	291.9		
—	45.3	8.75	301.8	8.04	303.4		
—	60.5	28.77	311.3	7.76	300.5		
—	70.0	11.99	302.7	13.90	305.3		193
—	75.3	20.28	309.9	13.31	307.3		
—	79.8	12.30	304.2	8.73	301.5		
—	90.0	17.99	307.1	7.67	299.6		
—	100	9.21	297.7	3.44	290.0	1 258–1 578 K	

Diffusion of W (for the (99.57)/(99.97) section):

Element 2 At.%	A^*	Q^*
0	—	—
14.9	1.10	286.4
29.7	17.70	324.1
45.3	13.90	320.5
60.5	23.82	324.3
70.0	14.39	320.2
75.3	10.69	317.0
79.8	9.75	317.5
90.0	2.06	299.5
100	—	—

Element 1 (purity) At.%	Element 2 (purity) At.%	A_1^* IVa(i) Fe⁵⁹	Q_1^*	A_2^* IVb Pd¹¹³ p.c.	Q_2^*	Temp. range K	Ref.
Fe (99.97)	Pd (99.95)						
0		0.18	260.0	0.04	249.5		
10		0.91	275.9	0.37	268.4		
20		0.91	269.6	0.79	271.3		
30	(At.%)	0.60	259.6	0.73	266.7		
40		0.69	258.7	0.79	266.3		
45		0.79	260.0	0.67	263.8	1 373–1 523	142
50		0.95	262.5	0.70	264.6		
60		0.95	264.2	1.05	270.9		
70		0.66	262.5	1.66	279.3		
80		0.93	271.7	1.84	284.7		
90		0.79	277.6	0.70	278.8		
100		0.41	280.9	0.41	280.9		

(continued)

Table 13.3   DIFFUSION IN HOMOGENEOUS ALLOYS—*continued*

Element 1 (purity) At.%	Element 2 (purity) At.%	A_1^* cm² s⁻¹	Q_1^* kJ mol⁻¹	A_2^* cm² s⁻¹	Q_2^* kJ mol⁻¹	Temp. range K	Ref.
Fe ( )	Pt ( )		IVb p.c., Pt¹⁹³ᵐ				
	0	—	—	2.7	296.0 ⎫		
	15	—	—	1.1	264.2 ⎪		
	20	—	—	0.34	265.0 ⎪		
	25	—	—	1.17	265.0 ⎪		
	30	—	—	0.28	264.2 ⎪		
	34	—	—	0.15	264.2 ⎬	1 053–1 693	138
	40	—	—	1.3	289.7 ⎪		
	45	—	—	1.13	284.3 ⎪		
	50	—	—	2.1	292.2 ⎪		
	55	—	—	0.85	286.8 ⎪		
	60	—	—	0.34	280.9 ⎭		
Fe ( )	Sb ( )		IVa	p.c.			
	2.5	0.51	216.9			1 169–1 370	75
Fe (99.999)	Si (99.99)		IVa and b	p.c.			
	0	1.39	236.6	—	— ⎫		
	4.7	1.63	232.4	—	— ⎪		
	6.3	0.50	229.9	—	— ⎪		
	8.2	0.50	213.1	—	— ⎬	1 073–1 573	100
	11.3	1.11	213.9	—	— ⎪		
	11.7	1.46	216.5	—	— ⎭		
( )	( )		IVb	p.c.			
	2.9	0.44	218.5	—		1 240–1 689	123
( )	( )		IVa(i) s.c., Fe⁵⁹				
	7.64	1.38	228.2	—	— ⎫	1 173–1 373	139
(At.%)	11.1	0.63	212.3	—	— ⎭		
( )	( )		IVa(i), p.c., Fe⁵⁹				
—	0	2.73	242	—	— ⎫		
—	1.48	1.03	276	—	— ⎪		
—	1.87	76.7	276	—	— ⎪		
—	6.55	5.2	242	—	— ⎬	1 073–1 373	188
—	8.64	4.93	236	—	— ⎪		
—	12.1	0.8	213	—	— ⎭		
(99.95)	( )		IVb, Fe⁵⁹, s.c.				
—	5.5	7.2	250	—	— ⎫		
—	6.4	3.15	238	—	— ⎪		
—	7.8	8.6	244	—	— ⎪		
—	11.6	0.63	212	—	— ⎬	1 013–1 373	203
—	15.3	2.1	219	—	— ⎪		
—	192	3.4	216	—	— ⎭		
(4N)	(4N)		IVb, p.c., Fe⁵⁹				
—	10	5.59	219.1	—	— ⎫	875–957 (α-f)	197
—		0.25	209.6	—	— ⎭	1 005–1 178 (α-p)	
Fe (99.97)	Sn (5N)		IVa(i), p.c., Fe⁵⁹				
—	0–26	$b_1 = 98$		—	— ⎫	1 168	
		= 63		—	— ⎬	1 093	159
		= 63		—	— ⎭	1 009	
Fe ( )	Ti ( )		IVa	p.c.			
	2	2.8	242.0	—	—	1 173–1 473	75
( )	( )		IVa(i)	p.c.			
	2	0.56	216.5	—	— ⎫		
	4	0.27	204.7	—	— ⎬	1 273–1 673	99
	6	0.40	209.9	—	— ⎭		
( )	( )		IVa(i)	p.c.			
	33.32	11.7	314.0	—	— ⎫	1 123–1 373	125
[a]	50.0	0.135	239.5	—	— ⎭		

(*continued*)

Table 13.3   DIFFUSION IN HOMOGENEOUS ALLOYS—*continued*

Element 1 (purity) At.%	Element 2 (purity) At.%	A_1^* cm² s⁻¹	Q_1^* kJ mol⁻¹	A_2^* cm² s⁻¹	Q_2^* kJ mol⁻¹	Temp. range K	Ref.
(Spec. P.)	(99.7)	IVa(ii)	p.c.	Diffusion of Nb⁹⁵			
5		9.2×10^{-2}	165.8	1.82×10^{-3}	196.1	1 123–1 533 ⎫	
10		2.14	203.5	2.9×10^{-2}	172.9	1 123–1 473 ⎬	53
15		52.5	243.3	9.9	234.5	1 123–1 373 ⎭	
Fe	V						
(—)	(—)	IVb		p.c.			
(γ)	0.53	1.46	288.5	—	⎫		
(γ)	1.09	0.53	277.2	—	⎬ —	1 373–1 573	54
(α)	2.11	0.10	257.5	—	⎭		
(—)	(—)	IVa(i)		p.c.		(Paramag.)	
	18.0	7.0	258.3	3.9	244.9	1 153–1 473	55
(99.98)	(—)	IVb		p.c.			
—	2.1	Diffusion of Hf¹⁸¹ 1.31	290.1	3.92	241.2	1 273–1 673	94
(99.999)	( )	IVa		p.c.			
—	1.8	1.4	237.0	—	—	1 173–1 773 ⎱	75
—	5.3	1.87	240.3	—	—	1 173–1 466 ⎰	
( )	( )	IVb		p.c.			
	2	—	—	3.92	241.2 ⎫		
	5	—	—	3.00	238.6 ⎪		
	9	—	—	2.28	236.1 ⎬	1 273–1 723	99
	14	—	—	2.12	236.6 ⎪		
	19	—	—	1.66	234.0 ⎭		
( )	( )	Diffusion of Cr⁵¹					
	1.7	2.0	238.6	—	—	—	88
( )	( )	IVb, p.c., V⁴⁸					
—	0	—	—	2.04	229.8	1 233–1 511 ⎫	
—	5.4	—	—	2.10	231	1 255–1 511 ⎪	
—	10.9	—	—	2.2	234	1 255–1 511 ⎪	
—	16.25	—	—	2.02	235	1 233–1 466 ⎪	
wt.%	20.0	—	—	10	255	1 233–1 466 ⎬	183
—	27.65	—	—	1.8	238	1 255–1 511 ⎪	
—	47	—	—	45	293	1 503–1 593 ⎪	
—	65	—	—	530	346	1 233–1 466 ⎪	
—	75	—	—	3.47×10^4	403	1 233–1 466 ⎪	
—	90	—	—	164	355	1 233–1 466 ⎭	
( )	( )	IVa, p.c., V⁴⁸					
0–1.11	—	—	—	$b_1 = 18.7$		1 573 ⎫	
0–0.96	—	—	—	$= 10.0$		1 757 ⎬	169
0–1.11	—	—	—	$= 12.6$–15.6		2 001 ⎭	
(Marz)	(Marz)	Diffusion of C		IIa(ii), p.c., C¹⁴			
—	2.45	54.7	170.4	—	—	1 173–1 598	185
Fe	Zr						
( )	( )	IVa(i), p.c., Fe⁵⁹, Zr⁹⁵					
0–3.5	—	—	—		$b_1 = 47$ ⎱	1 221	156
		$B_1 \sim 0$		—	⎰	1 078–1 578	
(99.9)	(nuclear)	IVa(i), p.c., Fe⁵⁹, Zr⁹⁵					
0	—	4.47×10^{-3}	117.1	6.8×10^{-4}	145 ⎫	1 218–1 518	
0.98	—	—	—	1.52×10^{-3}	149.04 ⎪	1 258–1 518	
1.35	—	—	—	2.08×10^{-3}	150.9 ⎪	1 218 × 1 518	195
1.64	—	—	—	1.62×10^{-3}	145.97 ⎬	1 196–1 520	
3.54	—	—	—	2.9×10^{-3}	140.8 ⎪	1 188–1 470	
6.37	—	—	—	5.26×10^{-2}	155.45 ⎭	1 276–1 513	
Ga	Pu						
( )	( )	IVa(i)		p.c.			
1.0 (wt)	—	⎧ —	—	76.4	152.0	613–781 (δ) ⎱	114
		⎩ —	—	6.98×10^{-4}	56.1	847–917 (ε) ⎰	

(continued)

Table 13.3 DIFFUSION IN HOMOGENEOUS ALLOYS—*continued*

Element 1 (purity) At.%	Element 2 (purity) At.%	A_1^* cm^2 s^{-1}	Q_1^* kJ mol^{-1}	A_2^* cm^2 s^{-1}	Q_2^* kJ mol^{-1}	Temp. range K	Ref.
Ga (6N) 25	V (99.9) 75 V$_3$ (Ga)	IVa(i), p.c., V[48] —	—	1.52×10^4	414.8	1 298–1 449	160
Ge (3N6) 0–8	Ni (3N6) —	IVa(iii), p.c., Co[59] —	—	$b_1 = 6$ $= 10$ $= 3$ $= 2$	$b_2 = 421$ $= 264$ $= 221$ $= 137$	1 333 1 380 1 418 1 471	51

Enhancement not of D_{Ni}^* but of the closely similar (*sic*) D_{Co}^*.

| In () 7 12 0–30 | Pb () — — — | IVa(i), p.c., Pb[210] — — $b_1 = 64.5$ | — — | 7.2×10^{-4} 5.1×10^{-4} $b_2 = 285.3$ | 69.1 64.0 | 468–558 516 | 187 |
| () 0 2 7 12 20 30 45 | () — — — — — — — | IVa(i), p.c., Ag[110m] — — — — — — — | — — — — — — — | Diffusion of Ag 4.6×10^{-2} 60.3 2.4×10^{-2} 56.9 3.5×10^{-2} 54.8 3.66×10^{-2} 52.4 1.33×10^{-2} 45.0 6.5×10^{-3} 39.3 4.4×10^{-3} 36.8 | | 403–523 | 145 |

In (6N) 51	Pd (4N) 49	IVa(i), pc, Pd[109], In[114m] 84.10^{-3} $+5 \times 10^2$	192 314.5	0.12	207	1 056–1 270	
50	50	5.0×10^{-3} $+1.06 \times 10^3$	181 318	2.30	243	996–1 327	
47	53	1.6×10^{-2} $+20.0$	191 293	0.60	222	996–1 425	161
44	56	0.14	215	0.20	205.5	1 039–1 473	

K (99.97)	Na (99.999 5)		IVa(i) —	p.c. $D_{Na}^* \times 10^9$ —			
0				1.34			
0.13				1.38			
0.14				1.44			
0.19				1.54			
0.33				1.58		273	
0.49				1.81			
0.56				1.79			
0.67				1.90			
1.03				2.12			
				$D_{Na}^* \times 10^8$			108
0				2.16			
0.13				2.28			
0.18				2.26			
0.38				2.39		373	
0.55				2.61			
0.73				2.86			
1.25				3.40			

Mn () 18	Pt () —	IVa(i), p.c., Mo[54] 2.1×10^{-10}	53	—	—	1 020–1 284	
25	— (MnPt$_3$)	3.0×10^{-2}	222	—	—	1 026–1 283	181
35	—	2.3×10^{-10}	57	—	—	1 026–1 284	

(*continued*)

Table 13.3 DIFFUSION IN HOMOGENEOUS ALLOYS—*continued*

Element 1 (purity) At.%	Element 2 (purity) At.%	A_1^* cm² s⁻¹	Q_1^* kJ mol⁻¹	A_2^* cm² s⁻¹	Q_2^* kJ mol⁻¹	Temp. range K	Ref.
Mn (99.97)	Ti (99.97)		IVa(i)	p.c.			
9.7		—	—	1.9×10^{-2}	171.2	1 133–1 723	
13.3		—	—	2.06×10^{-2}	172.1	1 073–1 623	121
17.9		—	—	2.60×10^{-2}	176.3	1 070–1 573	
20.6		—	—	5.47×10^{-1}	207.7	1 083–1 523	
Mn ( )	Zr ( )	IVa(i)	p.c.	Mn⁵⁴	Zr⁹⁵		
0		5.38×10^{-3}	140.7	0.31×10^{-4}	105.3		
0.5		2.92×10^{-3}	135.7	0.71^{-4}	112.3		
1.0		1.38×10^{-3}	125.5	1.2×10^{-4}	116.7	1 173–1 473	143
1.5		4.6×10^{-4}		2.17×10^{-4}	121.8		
2.0		8.0×10^{-5}	104.6	3.36×10^{-4}	125.3		
Mo ( )	Ni ( )		IVa	p.c			
8		1.31	229.9	2.55	236.1	1 272–1 673	
16		1.3	228.6	0.63	218.6	1 273–1 623	
18		0.45	218.6	0.34	218.6	1 223–1 623	126
20		0.25	210.6	0.19	204.3	1 223–1 573	
23		0.20	207.2	0.12	198.5	1 373–1 573	
(99.4)	(99.4)	IVb	p.c.	Diffusion of C¹⁴			
2.94		—	—	1.0	159.1	873–1 173	36
Mo (99.98)	Ti (99.62)	IVb	p.c.	Diffusion of U²³⁵			
5				0.26	206.4		
10				1.45	237.4		
15				2.88	249.5	1 173–1 673	116
20				2.69	270.5		
25				6.03	288.9		
30				33.0	314.0		
Mo (—)	U (—)		IIa(ii)	p.c.			
0		—	—	1.8×10^{-3}	115.1	1 073–1 313	57
10		—	—	2.5×10^{-3}	138.2		
Mo ( )	W ( )		IVa	p.c.			
0.1		142	468.9	0.007 5	297.3	2 073–2 673	
15		265	448.8	1.4	305.6		
20		146	427.1	1.7	312.3	1 673–2 673	
25		47	397.7	2.2	322.4		
35		28	385.2	6.9	355.9	1 773–2 773	
50		12	368.4	14	397.7	1 973–2 773	104
65		1.3	360.1	16	427.1	2 073–2 873	
75		0.2	353.8	20	485.7		
80		0.11	342.3	22	498.2	2 073–3 073	
85		0.08	334.9	25	570.8		
99.9		0.002 5	326.6	24	544.3	2 473–3 073	
( )	( )		IVa	p.c.			
44		0.17	448.0	—	—	2 173–2 673	105
55		0.12	431.2	—	—		
Nb ( )	Ni ( )		IVb s.c., Ni⁶³, Nb⁹⁵				
25	75	2.4×10^2	448.0	—	—	1 543–1 623	144
		—	—	0.18	304.8	1 363–1 643	
(99.9)	(99.99)		IVa(i)	p.c.			
1.2		—	—	0.12	254.6		
8		—	—	0.20	260.4	1 303–1 503	113
10		—	—	1.80	280.9		

(continued)

Table 13.3   DIFFUSION IN HOMOGENEOUS ALLOYS—*continued*

Element 1 (purity) At.%	Element 2 (purity) At.%	A_1^* cm² s⁻¹	Q_1^* kJ mol⁻¹	A_2^* cm² s⁻¹	Q_2^* kJ mol⁻¹	Temp. range K	Ref.
Nb	Ti						
(99.6)	(99.7)	p.c.	IVa(ii)	Diffusion of Fe⁵⁵			
5		1.2×10^{-4}	125.2	7.9×10^{-3}	138.6		
10		5.8×10^{-4}	151.1	115×10^{-3}	146.1	1 123–1 573	53
15		1.5×10^{-3}	164.5	7.9×10^{-3}	146.5		
31		9×10^{-3}	209.3	—	—	1 323–2 073	
54		8×10^{-2}	268.0	—(a)	—	1 473–2 073	58
66		0.1	301.4	—	—	1 773–2 273	
89		1.0	381.0	—	—	1 973–2 473	
Nb	Ti						
( )	(99.97)		IVa(i), p.c.	Ti⁴⁴	Nb⁹⁵		
	100	2.9×10^{-4}	129.9	4.54×10^{-4}	131.0		
(At.%)	94.6	1.79×10^{-3}	160.0	1.27×10^{-3}	149.1	1 223–1 784	147
	80.4	1.18×10^{-2}	198.2	3.15×10^{-3}	175.7		
	64.3	2.98×10^{-1}	258.5	2.51×10^{-1}	247.1		
( )	( )	IVb	p.c.	Diffusion of U²³⁵			
5				0.022	170.4		
10				0.035	182.1		
15				0.11	201.0		
20				0.26	215.6	1 173–1 673	116
25				0.25	226.1		
30				0.42	235.7		
Nb	U						
( )	( )		IVb	p.c.			
0		1.2×10^{-5}	119.3	1.1×10^{-4}	150.7		
5				3.2×10^{-6}	142.4	1 223–1 323	
10				3.5×10^{-6}	147.4		
20				10^{-5}	165.8		
35		3.1×10^{-2}	288.1	1.25×10^{-4}	222.7	1 323–1 493	
50		3.1×10^{-2}	304.4	2.5×10^{-4}	239.5	1 475–1 123	153
65		7.6×10^{-2}	324.5	4.0×10^{-3}	274.7	1 473–1 773	
80		1.1	383.1	6.3×10^{-4}	280.5	1 823–2 073	
90		5.2	421.2	2.5×10^{-3}	305.6	1 973–2 173	
100		0.91	421.2	6.5×10^{-6}	321.1	2 073–2 273	
( )	( )		IIa(ii)	p.c.			
10	—	—	—	1.66×10^{-4}	118.1	1 073–1 313	57
Nb	W						
( )	( )		IVa(i), p.c., Nb⁹⁵, W¹⁸⁵, Zr⁹⁵				
—	0	$D_{Nb}^* = 1.434 \times 10^{-8}$		$D_w^* = 4.58 \times 10^{-9}$			
—	2	$= 1.277 \times 10^{-8}$		$= 4.35 \times 10^{-9}$			
—	5	$= 1.065 \times 10^{-9}$		$= 3.30 \times 10^{-9}$			
—	10	$= 7.60 \times 10^{-9}$		$= 2.38 \times 10^{-9}$			
		$b_1 = -5.7$		$B_1 = -5.7$		2 680	153
—	0		Diffusion of Zr	$D = 4.42 \times 10^{-8}$			
—	2			$= 3.86 \times 10^{-8}$			
—	5	$B_1^{Zr} = -5.7$		$= 3.28 \times 10^{-8}$			
—	10			$= 2.42 \times 10^{-8}$			
Nb	Zr						
(99.99)	(99.9)		IV p.c.	Nb⁹⁵			
2.3	wt.%	1.63×10^{-3}	162.4	—	—	1 173–1 433	49
( )	( )		IVa(i), p.c., Zr⁹⁵	$D_{Zr}^* = A_2 \exp(-Q_2/kT)\exp(B/kT^2)$			
				A_2 Q_2 B			
5.5	—	—	—	0.44 315 1.34×10^5			
16.3	—	—	—	0.43 285 8.30×10^4			194
28.1	—	—	—	0.13 261 5.50×10^4			
( )	( )	Diffusion of Ni		IVa(i), p.c., Ni⁶³			
2.5 wt. %	—	$D = 8.9 \times 10^{-12}$				1 120.8	155
		$D = 1.1 \times 10^{-13}$				893.6	

(continued)

Table 13.3 DIFFUSION IN HOMOGENEOUS ALLOYS—*continued*

Element 1 (purity) At.%	Element 2 (purity) At.%		A_1^* cm² s⁻¹	Q_1^* kJ mol⁻¹	A_2^* cm² s⁻¹	Q_2^* kJ mol⁻¹	Temp. range K	Ref.
( ) 32	( ) —		Diffusion of Co 8.7×10^{-2} 154		IVa(i), p.c., Co⁵⁷		1 125–1 645	156
Nb ( )	Zr ( )	Mo ( )		IVb	p.c.			
5	95	—	4.6×10^{-2}	209.5	4.1×10^{-3}	169.6 ⎫		
10	90	—	4.4×10^{-1}	230.3	2.8×10^{-2}	196.8		
15	85	—	5.5×10^{-1}	242.8	8.6×10^{-2}	211.4		
2.5	—	2.5	7.6×10^{-2}	213.5	9.0×10^{-3}	180.0		
5	—	5	3.1×10^{-1}	234.5	5.4×10^{-2}	207.2 ⎬	1 473–1 773	119
7.5	—	7.5	1.1	251.2	2.8×10^{-1}	230.3		
—	95	5	—	—	6.4×10^{-3}	175.8		
—	90	10	—	—	6.6×10^{-2}	205.2		
—	85	15	—	—	2.10^{-1}	224.0 ⎭		

(a) Values read from graphically plotted results.

Element 1 (purity) At.%	Element 2 (purity) At.%		A_1^* cm² s⁻¹	Q_1^* kJ mol⁻¹	A_2^* cm² s⁻¹	Q_2^* kJ mol⁻¹	Temp. range K	Ref.
Nb ( )	Zr ( )	O ( )		IV	p.c.	O¹⁸		
	0.5		—	—	1.83×10^{-3}	110.1	873–1 373 ⎫	
wt.%	0.8		—	—	1.07×10^{-3}	108.4	873–1 373 ⎬	146
	1.0		—	—	4.11×10^{-4}	103.4	873–1 373 ⎭	
					(O Diffusion)			
Ni ( )	Pb (6N)		IVb, s.c., Ni⁶³					
sat. sol.			$B_1 = 77\,000$		Fitted to		408 ⎫	
soln			$= 50\,000$		$D_{Ni}^* = 2A(c = 0)/[1 + (1 - 4B_1c)^{1.2}]$		416 ⎬	202
			$= 15\,000$				450	
			$= 7\,000$				515 ⎭	
Ni ( )	Sb ( )		IVb	p.c.				
71.7			5.4×10^{-4}	61.5	—	— ⎫		
72.9 (β)			4.9×10^{-4}	62.0	—	— ⎬	873–1 272	101
73.7			6.3×10^{-4}	66.2	—	—		
75.0			6.9×10^{-4}	69.1	—	— ⎭		
( )	( )		IVa(i), s.c. and p.c. Ni⁶³, Sb¹²⁴					
50 (γ)	50 p.c.		0.019	161	1.84	278 ⎫	860–1 373	
53 (γ)	47 p.c.		0.051	159	0.58	258 ⎬	(Ni)	154
53 (γ)	47 s.c.		$\perp c$ 0.095	163	3.4	281	1 150–1 373	
			$\|c$ 0.082	165	7.3	284 ⎭	(Sb)	
Ni (3N6)	Si (3N6)		IVa(iii), p.c., Co⁵⁹					
—	0–8		$b_1 = 3$	$b_2 = 156$			1 410 ⎱	51
—	—		$= 2$	$= 174$			1 465 ⎰	

Enhancement not of D_{Ni}^* but of the closely similar (*sic*) D_{Co}^*.

Element 1 (purity) At.%	Element 2 (purity) At.%		A_1^* cm² s⁻¹	Q_1^* kJ mol⁻¹	A_2^* cm² s⁻¹	Q_2^* kJ mol⁻¹	Temp. range K	Ref.
Ni (99.98)	Sn (5N)		IVb, p.c., Ni⁶³, Sn¹¹³					
—	0–2.0		$b_1 = 55$	$b_2 = 770$		⎱	1 373	163
—	0–4.0		$D_{Sn}^* = 3.1 \times 10^{-10} \exp(0.191)N_{Sn}(\text{at.\%})$			⎰		
(3N6)	(3N6)		IVa(iii), p.c., Co⁵⁹					
—	0–90		$b_1 = 3$	$b_2 = 276$			1 419 ⎱	51
—	0–7.3		$= 3$	$= 225$			1 472 ⎰	

Enhancement not of D_{Ni}^* but of the closely similar (*sic*) D_{Co}^*.

(continued)

Table 13.3　DIFFUSION IN HOMOGENEOUS ALLOYS—*continued*

Element 1 (purity) At.%	Element 2 (purity) At.%	A_1^* cm²s⁻¹	Q_1^* kJ mol⁻¹	A_2^* cm²s⁻¹	Q_2^* kJ mol⁻¹	Temp. range K	Ref.
Ni ()	Ti ()		IVa(i)	p.c.			
	2.44	2.24	287.2	—	—		
	4.86	1.51	282.6	—	—		
	7.24	0.91	277.2	—	—	1 202–1 501	124
	9.63	6.6	296.0	—	—		
	12.68	3.0	319.0	—	—		
	25	6.9×10^{-3}	389.0	—	—		
Ni (99.95)	W		IVa(i)	p.c.			
	0	1.9	284.7	2.0	299.4	1 369–1 668 (D_{Ni})	
	1.7	30.0	320.3	2.2	306.1		66
	5.3	58.0	337.5	17.0	337.0	1 373–1 568	
	9.2	1.1	294.3	1.4	311.9	(D_w)	
Pb (5N)	Sn (5N)	Diffusion of Ag		IVa(i), p.c., Ag¹¹⁰ᵐ			
—	0	4.6×10^{-2}	61				
—	5	4.8×10^{-2}	60			423–573	168
—	8.4	2.0×10^{-2}	56				
—	12	2.2×10^{-2}	55				
Pb (99.99) 100	Tl (99.99)		IVa(i)	p.c. (except 62.6%)			
	0	1.372	109.1	0.511	101.9		
	5.21	1.108	107.8	0.364	100.0		
	10.27	0.880	106.6	0.361	99.8		
	20.2	0.647	104.9	0.353	99.6		
	34.6	0.367	102.7	0.193	96.8		
	50.3	0.231	102.3	0.091	94.3	479–596	59
	62.4	0.393	107.3	0.101	96.0		
	62.6 (s.c.)	0.287	105.9	0.126	97.1		
	74.5	0.691	112.3	0.194	99.9		
	76.2	0.862	113.6	0.330	102.5		
	81.8	2.575	118.2	0.957	106.2		
	87.1	17.0	124.4	1.20	106.9		
Pu () At.%	Zr ()		IVb, p.c., Pu²⁴⁰				
	40	0.04	124.1	—	—	913–1 113	148
	10	$D_1 = 1.05 \times 10^{-7}$		—	—	923	
Sc ()	Zr ()		IVa(i), p.c., Zr⁹⁵	$D_{Zr}^* = A_2 \exp(-Q_2/kT)\,\exp(B/kT^2)$			
				A_2　Q_2　B			
6.7	—	—	—	1.0　336　1.63×10^5		1 300–1 906	194
13.5	—	—	—	1.3　343　1.72×10^5		1 416 × 1 873	
0	(99.6)	—	IVb	p.c.			
		—	—	5.9×10^{-2}		923–1 100	
1.0		—	—	5.0	259.6		60
1.85		—	—	2.1×10^{-3}	314.0	740–827	
2.75		—	—	10	268.0		
()	(99.99)	Diffusion of Cr		IVa(i), p.c., Cr⁵¹			
0	—	7.4×10^{-3}	143				
0.45	—	4.5×10^{-3}	137.8			1 255–1 513	165
1.75	—	2.5×10^{-3}	132.3				

(*continued*)

Table 13.3 DIFFUSION IN HOMOGENEOUS ALLOYS—*continued*

Element 1 (purity) At.%	Element 2 (purity) At.%	A_1^* cm² s⁻¹	Q_1^* kJ mol⁻¹	A_2^* cm² s⁻¹	Q_2^* kJ mol⁻¹	Temp. range K	Ref.
Ti (see Fig. 13.8a)	V						
( ) 0–1.04	( )	IVa(i), p.c., V[48] $b_1 = 1.68$		—	—	1 960	169
(—) 0	(99.98)	IVa(i)	p.c.				
10	—						
20	—						
30	—	Curved Arrhenius plots		Curved Arrhenius plots			
40	—	*See* Figure 13.8a		*See* Figure 13.8b			
50	—	The lines of Figures 13.3a and 13.3b are drawn at				1 173–2 073	115
60	—	10 at.% intervals of composition as shown in column 1					
70	—						
80	—						
90	—						
( )	( )	IVa(i), p.c., V[48]		$D_v^* = A_2\exp(-Q_2/kT)\exp(B/kT^2)$			
				A_2 $\quad$ Q_2 $\quad$ B			
—	9.4	—	—	1.1 $\quad$ 307 $\quad$ 1.08×10^5			115 and
—	19	—	—	1.9 $\quad$ 310 $\quad$ 1.00×10^5			194
( )	( )	IVb	p.c.	Diffusion of U[235]			
—	5			0.034	172.5		
—	15			0.063	182.1		
—	15			0.047	182.1		
—	20			0.009 6	164.5		
—	25			0.025	177.9		
—	30			0.089	195.5	1 173–1 673	116
—	40			0.12	206.8		
—	50			0.65	235.7		
—	60			2.7	288.9		
—	70			5 900	365.1		
U	Zr						
(—)	(—)		IIa(ii)	p.c.			
	10	1.26×10^{-4}	92.1	—	—	1 073–1 313	57
( )	( )		IIa(ii)	p.c.	IVa(i)		
	95	$D_U^* =$ 1.5×10^{-9}	—	$D_{Zr}^* =$ 3.2×10^{-9}	—	1 273	62
		$D_U^* =$ 1.85×10^{-9}	—	$D_{Zr}^* =$ 4.2×10^{-9}	—	1 323	
( )	( )	IVa(i)		p.c.			
0	—	5.7×10^{-4}	127.7				
11	—	7.5×10^{-4}	141.9	0.12	205.6		
27	—	3.65×10^{-3}	160.8	2.8×10^{-2}	190.5		
39	—	7.10^{-3}	168.7	3.9×10^{-4}	146.5		
59	—	8.96	245.3	2.4×10^{-5}	118.5	1 173–1 338	117
78	—			3.8×10^{-6}	99.2		
85	—			7.5×10^{-7}	83.7		
100	—			1.6×10^{-7}	68.7		
V	Zr						
(99.95)	(nuclear)	IVa(i), p.c., V[48], Zr[95]					
—	0	118	389.3	81	369.2		
—	0.5	86	383.2	115	373.0		
—	1.0	72	379.4	153	376.0	1 578–1 888	
—	1.5	68	376.7	195	378.6		167
—	2.0	58	373.0	242	380.8		
		1 578 K $\quad$ 1 633 K $\quad$ 1 688 K		1 738 K	1 783 K $\quad$ 1 848 K $\quad$ 1 888 K		
		$b_1 = -28.7 \quad 26.5 \quad 24.4$		19.9	15.0 $\quad$ 16.9 $\quad$ 17.7		
(99.95)	(nuclear)	IVa(i), p.c., V[48], Zr[95]					
0	—	8.9×10^{-5}	116.55	3.1×10^{-5}	105.25		
0.5	—	5.5×10^{-5}	112.47	4.5×10^{-5}	109.02		
1.0	—	3.0×10^{-5}	106.88	5.9×10^{-5}	111.95	1 166–1 480	
1.5	—	1.5×10^{-5}	100.55	9.3×10^{-5}	116.83		
2.0	—	0.7×10^{-5}	94.15	14.0×10^{-5}	120.79		
		1 167 K $\quad$ 1 223 K $\quad$ 1 281 K		1 318 K	1 375 K $\quad$ 1 428 K $\quad$ 1 476 K		
		$b_1 = -4.14 \quad 0 \quad 2.8$		5.09	8.71 $\quad$ 10.41 $\quad$ 13.60		
( )	( )	IVa(i), p.c., V[48]					
—	0.5	0.64	312.6	—	—	1 428–2 078	189

(*continued*)

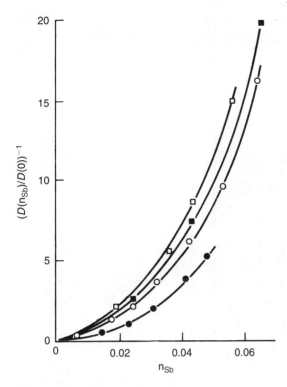

Figure 13.7

REFERENCES TO TABLE 13.3

1. R. E. Hoffman, D. Turnbull and E. W. Hart, *Acta Metall.*, 1955, **3**, 417.
2. W. C. Mallard, A. B. Gardner, R. F. Bass and L. M. Slifkin, *Phys. Rev.*, 1963, **129**, 617.
3. A. Schoen, *Ph. D. Thesis*, University of Illinois, 1958.
4. N. H. Nachtrieb, J. Petit and J. Wehrenberg, *J. Chem. Phys.*, 1957, **26**, 106.
5. R. L. Rowland and N. H. Nachtrieb, *J. Phys. Chem.*, 1963, **67**, 2817.
6. E. Sonder, *Phys. Rev.*, 1955, **100**, 1662.
7. R. E. Hoffman, *Acta Metall.*, 1958, **6**, 95.
8. D. Lazarus and C. T. Tomizuka, *Phys. Rev.*, 1956, **103**, 1155.
9. R. E. Hoffman and D. Turnbull, *J. Appl. Phys.*, 1952, **23**, 1409.
10. C. T. Tomizuka. Unpublished data.
11. T. Heumann and P. Lohman, *Z. Electrochem.*, 1955, **59**, 849.
12. W. C. Hagel and J. H. Westbrook, *Trans. Metall., Soc. AIME*, 1961, **221**, 951.
13. H. A. Domian and H. I. Aaronson, *ibid.*, 1964, **230**, 44.
14. S. D. Gertsricken and I. Y. Dekhtyar, *Proc. 1955 Geneva Conf.*, 1955, **15**, 99.
15. S. D. Gertsricken and I. Y. Dekhtyar, *Fizika Metall. Metallov.*, 1956, **3**, 242.
16. F. C. Nix and F. E. Jaumot, *Phys. Rev.*, 1951, **83**, 1275.
17. S. D. Gertsricken *et al.*, *Issled. Zharpr. Splav.*, 1958, **3**, 68.
18. A. E. Berkowitz, F. E. Jaumot and F. C. Nix, *Phys. Rev.*, 1954, **95**, 1185.
19. J. Čermák, K. Ciha and J. Kučera, *Phys. Stat. Solidi A*, 1980, **62**, 467.
20. H. B. Huntington, N. C. Miller and V. Nerses, *Acta Met.*, 1961, **9**, 749.
21. D. Gupta, D. Lazarus and D. S. Liebermann, *Phys. Rev.*, 1967, **153**, 863.
22. A. D. Kurtz, B. L. Averbach and M. Cohen, *Acta Metall.*, 1955, **3**, 442.
23. J. E. Reynolds, B. L. Averbach and M. Cohen, *ibid.*, 1957, **5**, 29.
24. M. Yanitskaya, A. A. Zhukhavitskii and S. Z. Bokstein, *Dokl. Akad. Nawk. SSSR*, 1957, **112**, 720.
25. S. D. Gertsricken and T. K. Yatsenko, *Vop. Fiz.*, 1957, **8**, 101.
26. C. Köstler, F. Faupel and T. Hehenkamp, *Acta Met.*, 1987, **35**, 2273.
27. R. P. Smith, *Trans. Metall. Soc., AIME*, 1962, **224**, 105.
28. H. W. Mead and C. E. Birchenall, *J. Metals.*, 1956, **8**, 1336.
29. P. L. Gruzin and E. V. Kuznetsov, *Dokl. Akad. Nauk. SSSR*, 1953, **93**, 808.

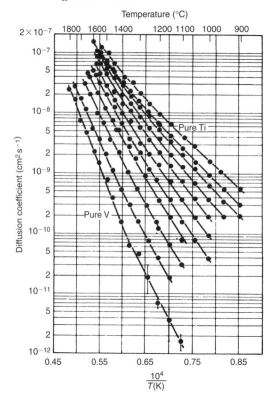

Figure 13.8a *Temperature dependence of diffusion of* Ti[44] *in titanium–vanadium alloys*[115] *at 10 at.% intervals*

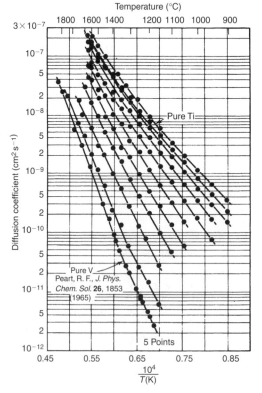

Figure 13.8b *Temperature dependence of diffusion of* V[48] *in titanium–vanadium alloys*[115] *at 10 at.% intervals*

30. K. Hoshino, S. J. Rothman and R. S. Averbach, *Acta Met.*, 1988, **36**, 1271.
31. B. Khomka, *Acta Physica Polonica*, 1963, **24**, 669.
32. B. Khomka, *Nukleonika*, 1963, **8**, 185.
33. P. L. Gruzin and B. M. Noskov, 'Problems of Metallography and Physics of Metals,' **4**, 509 (Moscow, 1955) and Aec. tr 2924, p. 355.
34. S. G. Fishman, D. Gupta and D. S. Liebermann, *Phys. Rev.*, 1970, **B2**, 1451.
35. K. Hirano, R. P. Agarwala, B. L. Averbach and M. Cohen, *J. Appl. Phys.*, 1962, **33**, 3049.
36. D. L. Gruzin, Yu. A. Polikarpov and G. B. Federov, *Fizika Metall. Metallov.*, 1957, **4**, 94.
37. A. W. Bowen and G. M. Leak, *Metal Trans.*, 1970, **1**, 2767.
38. H. W. Paxton and T. Kunitake, *Trans. Metall Soc. AIME*, 1960, **218**, 1003.
39. L. I. Ivanov and N. P. Ivanchev, *Izv Akad. Nauk. SSSR, Otdel, Tekhn. Nauk.*, 1958, **8**, 15.
40. A. Ya Shinyayev, Conference on Uses of Isotopes and Nuclear Radiations, *Met. Metallogr.*, 1958, p. 299, Moscow.
41. D. L. Beke, I. Gödény and F. J. Kedves, *Phil. Mag. A*, 1983, **47**, 281.
42. K. Monma, H. Suto and H. Oikawa, *Nippon Kink. Gakk.*, 1964, **28**, 188.
43. K. Monma, H. Suto and H. Oikawa, *ibid.*, 1964, **28**, 192.
44. A. J. Mortlock and D. H. Tomlin, *Phil. Mag.*, 1959, **4**, 628.
45. J. Hino, C. Tomizuka and C. Wert, *Acta Metall.*, 1957, **5**, 41.
46. A. B. Kuper, D. Lazarus, J. R. Manning and C. T. Tomizuka, *Phys. Rev.*, 1956, **104**, 1536.
47. P. Camagni, *Proc. 2nd Geneva Conf. Atomic Energy*, P/1365, Vol. 20, Geneva, 1958.
48. C. Bassani, P. Camagni and S. Pace, *Il Nuovo Cim.*, 1961, **19**, 393.
49. G. P. Tiwari, M. C. Saxena and R. V. Patil, *Trans. Ind. Inst. Met.*, 1973, **26**, 55.
50. J. R. MacEwan, J. U. MacEwan and L. Yaffe, *Can. J. Chem.*, 1959, **37**, 1629.
51. F. Faupel, C. Köstler, K. Bierbaum and T. Hehenkamp, *J. Phys. F*, 1988, **18**, 205.
52. J. W. Miller, *Phys. Rev.*, 1969, **181**, 1095.
53. R. F. Peart and D. H. Tomlin, *Acta Metall.*, 1962, **10**, 123.
54. M. S. Zelinski, B. M. Moskov, P. V. Pavlov and E. V. Shitov, *Physics Metals Metallogr.*, 1959, **8** (5), 79 and *Fiz. Metall. Metallov.*, 1959, **8**, 725.
55. J. Stanley and C. Wert, *J. Appl. Phys.*, 1961, **32**, 267.
56. I. A. Naskidashvili, *Soobshoheniya Akad. Nauk. Gauzin, SSSR*, 1955, **16**, 509.
57. Y. Adda and A. Kirianenko, *J. Nucl. Mater.*, 1962, **6**, 135.
58. G. B. Gibbs, D. Graham and D. H. Tomlin, *Phil. Mag.*, 1963, **8**, 1269.
59. H. A. Resing and N. H. Nachtrieb, *Physics Chem. Solids*, 1961, **21**, 40.
60. V. S. Lyashenko, V. N. Bykov and L. V. Pavlinov, *Physics Metals Metallogr.*, 1959, (3) **8**, 40.
61. T. Heumann and F. Heinemann, *Z. Electrochem.*, 1956, **60**, 1160.
62. Y. Adda, C. Mairy and J. M. Andreu, *Revue Métall.*, Paris, 1960, **57**, 549.
63. J. W. Miller, *Phys. Rev.*, 1970, **B2**, 1624.
64. R. J. Borg, *J. Appl. Phys.*, 1963, **34**, 1562.
65. S. D. Gertsricken and I. Ya. Dekhtyar, *Proc. 1955 Geneva Conf. Peaceful Uses of Atomic Energy*, 1955, **15**, 124.
66. K. Monma, H. Suto and H. Oikawa, *Nippon Kink. Gakk.*, 1964, **28**, 197.
67. D. A. Leak and G. M. Leak, *J. Iron St. Inst.*, 1958, **189**, 256.
68. J. Askill, *Phys. Status Solidi*, 1971, **a8**, 587.
69. Th. Heumann and H. Böhmer, *J. Phys. Chem. Solids*, 1968, **29**, 237.
70. C. J. Santoro, *Phys. Rev.*, 1969, **179**, 593.
71. S. J. Rothman and N. L. Peterson, *Phys. Rev.*, 1967, **154**, 552.
72. W. B. Alexander and L. M. Slifkin, *Phys. Rev.*, 1970, **B1**, 3274.
73. J. Kucera and B. Million, *Metal Trans.*, 1970, **1**, 2599.
74. A. B. Gardner, R. L. Sanders and L. M. Slifkin, *Phys. Status Solidi*, 1968, **30**, 93.
75. D. Y. F. Lai and R. J. Borg, USAEC Rept. UCRL 50314, 1967.
76. L. N. Larikov, *Avtom. Svarka*, 1971, **24**, 71.
77. L. N. Larikov, *et al.*, *Prot. Coat. Metals*, 1970, **3**, 91.
78. G. F. Hancock and B. R. McDonnell, *Phys. Status Solidi*, 1971, **4**, 143.
79. D. Gupta and D. S. Liebermann, *Phys. Rev. B*, 1971, **4**, 1070.
80. V. M. Ananin, *et al.*, *Soviet J. At. Ener.*, 1970, **29**, 220.
81. S. Benci, *et al.*, *J. Phys. Chem. Solids*, 1965, **26**, 687.
82. M. Badia and A. Vignes, *C. R. Hebd. Séanc. Acad. Sci., Paris*, 1967, **264C**, 858.
83. A Hässner and W. Lange, *Phys. Status Solidi*, 1965, **8**, 77.
84. P. Guiraldenq and P. Poyet, *C. R. Hebd. Séanc. Acad. Sci., Paris*, 1970, **C270**, 2116.
85. E. V. Borisov, D. L. Gruzin and S. V. Zemskii, *Prot. Coat. Metals*, 1968, **2**, 104.
86. S. P. Ray and B. D. Sharma, *Acta Met.*, 1968, **16**, 981.
87. S. P. Ray and B. D. Sharma, *Trans. Indian Inst. Met.*, 1970, **23**, 77.
88. L. V. Pavlinov, E. A. Isadzanov and V. P. Smirnov, *Fiz. Metall.*, 1968, **25**, 836.
89. S. Benci, G. Gasparrini and T. Rosso, *Phys. Letters.*, 1967, **24**, 418.
90. H. Oikawa, *et al.*, *ASM Trans. Quart.*, 1968, **61**, 354.
91. Th. Heumann, H. Meiners and H. Stüer, *Z. Naturforsch.*, 1970, **25a**, 1883.
92. R. Ebeling and H. Wever, *Z. Metall.*, 1968, **53**, 222.
93. N. L. Peterson and S. J. Rothman, *Phys. Rev.*, 1970, **B2**, 1540.
94. A. W. Bowen and G. M. Leak, *Metal Trans.*, 1970, **1**, 1695.

95. G. F. Hancock and G. M. Leak, *Met. Sci. J.*, 1967, **1**, 33.
96. A. Ya. Shinyaev. *Izv. Akad. Nauk. SSSR, Met.*, 1969, **4**, 182.
97. H. Bakker, J. Backus and F. Waals, *Phys. Status Solidi*, 1971, **B45**, 633.
98. K. Nohara and K. Hirano, *Proc. Int. Conf. Science Tech. Iron and Steel*, Tokyo, 1970, Sect. 6 1267; and *J. Jap. Inst. Met.*, 1973, **37**, 51.
99. A. W. Bowen and G. M. Leak, *Metal Trans.*, 1970, **1**, 2767.
100. D. Y. F. Lai and R. J. Borg, UCRL Rept. No 50516, 1968; *J. Appl. Phys.*, 1970, **41**, 5193.
101. Th. Heumann and H. Stüer, *Phys. Status Solidi*, 1966, **15**, 1966.
102. A. F. Smith and G. B. Gibbs, *Metal Sci. J.*, 1969, **3**, 93; 1968, **2**, 47.
103. R. P. Agarwala, *et al.*, *J. Nucl. Mater.*, 1970, **36**, 41.
104. I. N. Frantsevich, *et al.*, *J. Phys. Chem. Solids*, 1969, **30**, 947.
105. Larikov, *et al.*, *Ukr. Fiz. Zh.*, 1967, **12**, 983.
106. G. F. Hancock, *Phys. Status Solidi*, 1971, **A7**, 535.
107. K. Hirano and M. Cohen, *Trans. Jap. Inst. Met.*, 1972, **13**, 96.
108. J. N. Mundy and W. D. McFall, *Phys. Rev.*, 1972, **B5**, 2835.
109. K. Hoshino, Y. Iijima and K. Hirano, *Proc. Yamada Conference, Point Defects and Defect Interactions in Metals*, N. Holland, 1982, p. 562.
110. G. B. Federov, E. A. Smirnov and V. N. Gusev, *Atom. Ener.*, 1972, **32**, 11.
111. K. Kishio, J. R. Owers-Bradley, W. P. Halperin and J. O. Brittain, *J. Phys. Chem. Solids*, 1981, **42**, 1031.
112. B. Günther, O. Kanert and W. Tietz, *J. Phys. F.*, 1986, **16**, 1639.
113. A. Ya. Shinyaev, *Izv. Akad. Nauk. SSSR Metal*, 1968, **No. 1**, 203.
114. W. Z. Wade, *J. Nucl. Mater.*, 1971, **38**, 292.
115. J. F. Murdock and C. J. McHargue, *Acta Met.*, 1968, **16**, 493.
116. L. V. Pavlinov, *Fiz. Met. Metalloved.*, 1970, **30**, 379.
117. G. B. Federov, E. A. Smirnov and F. I. Zhomov, *Met. Metalloved. Chist. Metal.*, 1968, **No. 7**, 116.
118. G. V. Kidson and J. S. Kirkaldy, *Phil. Mag.*, 1969, **20**, 1057.
119. G. B. Federov, E. A. Smirnov and S. M. Novikov, *Met. Metalloved. Chist. Metal*, 1969, **No. 8**, 41.
120. J. L. Bocquet, *Acta Met.*, 1972, **20**, 1347.
121. E. Santos and F. Dyment, *Phil. Mag.*, 1975, **31**, 809.
122. G. B. Federov, E. A. Smirnov and F. I. Zhomov, *Met. Metalloved. Chist. Metal.*, 1963, **No. 4**, 110.
123. B. Mills, G. K. Walker and G. M. Leak, *Phil. Mag.*, 1965, **12**, 939.
124. A. Ya. Shinyaev, *Phys. Met. Metallogr.*, 1966, **21**, 76.
125. A. Ya. Shinyaev, *Izv. Akad. Nauk. SSSR Metal*, 1971, **No. 5**, 210.
126. I. N. Frantsevich, Kalinovich, I. I. Kovenski and M. D. Smolin, 'Atomic Transp. in Solids and Liquids' (Verlag Z. Naturforsch, Tubingen, 1971). p. 68.
127. Stoebe, *et al.*, *Acta Met.*, 1965, **13**, 701.
128. T. Okabe, R. F. Hochman and M. E. McLain, *J. Biomed. Mater. Res.*, 1974, **8**, 381.
129. P. Gas and J. Bernardini, *Scr. Met.*, 1978, **12**, 367.
130. A. Lutze-Birk and H. Jacobi, *Scr. Met.*, 1975, **9**, 761.
131. G. Henry, G. Barreau and G. Cizeron, *C. R. Hebd. Acad. Sci. Sevie C*, 1975, **280**, 1007.
132. N. A. Stolwijk, M. Van Gend and H. Bakker, *Phil. Mag.*, 1977, **42**, 783.
133. Y. Iijima and K. I. Hirano, *Phil. Mag.*, 1977, **35**, 229.
134. J. Kuceta *et al.*, *Acta Met.*, 1974, **22**, 135.
135. R. A. Perkins, R. A. Padgett and N. K. Tunali, *Met. Trans.*, 1973, **4**, 1665, 2535.
136. A. F. Smith, *Metal Sci.*, 1975, **9**, 375.
137. P. Guiraldenq and P. Poyet, *Mem. Sci. Rev. Met.*, 1973, **70**, 715.
138. J. Kucera and B. Million, *Phys. Stat. Sol. A*, 1975, **31**, 275.
139. H. V. M. Mirani *et al.*, *Phys. Stat. Sol. A*, 1975, **29**, 115.
140. A. T. Donaldson and R. D. Rawlings, *Acta Met.*, 1976, **24**, 285.
141. J. Ruzickova and B. Million, *Kovove Mater.*, 1977, **15**, 140.
142. J. Fillon and D. Calais, *J. Phys. Chem. Solids*, 1977, **38**, 81.
143. D. D. Pruthi, M. S. Anand and R. P. Agarwala, *Phil. Mag.*, 1979, **39**, 173.
144. Y. Muramatsu, *Trans. Natl. Res. Inst. Met.*, 1975, **17**, 21.
145. J. Shi, S. Mei and H. B. Huntington, *J. Appd. Phys.*, 1987, **62**, 451.
146. R. A. Perkins and R. A. Padgett, *Acta Met.*, 1977, **25**, 1221.
147. A. E. Pontau and D. Lazarus, *Phys. Rev. B*, 1979, **B19**, 4027.
148. J. P. Zanghi and D. Calais, *J. Nucl. Met.*, 1976, **60**, 145.
149. D. Beke, I. Gödeny, F. J. Kedves and G. Groma, *Acta Met.*, 1977, **25**, 539.
150. S. C. Chen, J. C. Tarczon, W. P. Halperin and J. O. Brittain, *J. Phys. Chem. Solids*, 1985, **46**, 895.
151. U. Köhler, P. Neuhaus and C. Herzig, *Z. f. Metallk.*, 1985, **76**, 170.
152. S. Mantl, S. J. Rothman, L. J. Nowicki and D. Braski, *Phil. Mag. A*, 1984, **50**, 591.
153. J. N. Mundy, S. T. Ockers and L. C. Smedskjaer, *Phys. Rev. B*, 1986, **33**, 847.
154. R. Hähnel, W. Miekeley and H. Wever, *Phys. Stat. Solidi A*, 1986, **97**, 181.
155. G. M. Hood and R. J. Schultz, *Mater. Sci. Forum*, 1987, **15/18**, 475.
156. Ch. Herzig, J. Neuhaus, K. Vieregge and L. Manke, *Mater. Sci. Forum*, 1987, **15/18**, 481.
157. D. D. Pruthi and R. P. Agarwala, *Phil. Mag. A*, 1982, **46**, 841.
158. D. L. Beke, I. Gödény, G. Erdelyi, F. J. Kedves and B. Albert, *DIMETA 82. Proc. Intl. Conf. Diffusion in Metals and Alloys* (eds F. J. Kedves and D. L. Beke), Trans. Tech. Publs, Switzerland, 1983, p. 374.
159. A. Kumagai, Y. Iijima and K. Hirano, *DIMETA 82 (see ref. 158)*, p. 389.
160. A. Van Winkel, M. P. H. Lemmens, A. E. Weeber and H. Bakker, *DIMETA 82 (see reference 158)*, p. 473.

161. H. Hahn, G. Frohberg and H. Wever, *Phys. Stat. Solidi A*, 1983, **79**, 559.
162. B. Million, J. Růžičková and J. Vřešťál, *Mater. Sci. Eng.*, 1985, **72**, 85.
163. Y. Iijima, K. Hoshino, M. Kikuchi and K. Hirano, *Trans. Jap. Inst. Met.*, 1984, **25**, 234.
164. C. Fillastre, N. Barbouth and J. Oudar, *Scripta Met.*, 1982, **16**, 537.
165. L. I. Nicolai, R. Migoni and R. H. de Tendler, *J. Nuclear Mat.*, 1983, **115**, 39.
166. H. Talah, N. Barbouth and P. Markus. *J. Nuclear Mat.*, 1987, **148**, 61.
167. D. D. Pruthi and R. P. Agarwala, *Phil. Mag. A*, 1984, **49**, 263.
168. C. K. Hu and H. B. Huntington, *Phys. Rev. B*, 1982, **26**, 2782.
169. D. Ablitzer, J. P. Haeussler and K. V. Sathyaraj, *Phil. Mag.*, 1983, **47**, 515.
170. Th. Hehenkamp and F. Faupel, *Acta Met.*, 1983, **31**, 691.
171. H. S. Daruvala and K. R. Bube, *J. Nuclear Mat.*, 1979, **87**, 211.
172. G. Henry and C. S. Cizeron, *Ann. Chim.*, 1978, **3**, 167.
173. D. Delauney, A. M. Huntz and P. Lacombe, *Scripta Met.*, 1979, **13**, 419.
174. Th. Heumann and T. Rottwinkel, *J. Nuclear Mat.*, 1978, **69/70**, 567.
175. A. Bose, G. Frohberg and H. Wever, *Phys. Stat. Solidi A*, 1979, **52**, 509.
176. A. Bristoti and A. R. Wazan, *Rev. Bras. Fís.*, 1974, **4**, 1.
177. S. J. Rothman, L. J. Nowicki and G. E. Murch, *J. Phys. F. Met. Phys.*, 1980, **10**, 383.
178. R. V. Patil and B. D. Sharma, *Met. Sci.*, 1982, **16**, 389.
179. Y. Iijima, K. Hoshino and K. Hirano, *Trans. Jap. Inst. Met.*, 1984, **25**, 226.
180. P. T. Carlson and R. A. Padgett, *Scripta Met.*, 1979, **13**, 355.
181. D. Ansel, J. Barre, C. Meziere and J. Debuigne, *J. Less Common Met.*, 1979, **65**, P1.
182. M. B. Dutt, S. K. Sen and A. K. Barua, *Phys. Stat. Solidi A*, 1976, **56**, 149.
183. K. Obrlík and J. Kučera, *Phys. Stat. Solidi A*, 1979, **53**, 589.
184. S. Kurokawa, J. E. Ruzzante, A. M. Hey and F. Dyment, *Met. Sci.*, 1983, **17**, 433.
185. S. N. Tewari and J. R. Cost, *J. Mater. Sci.*, 1982, **17**, 1639.
186. H. Nakajima, *Scripta Met.*, 1981, **15**, 577.
187. L. Cheriot and H. B. Huntington, *Acta Met.*, 1987, **35**, 1649.
188. D. Treheux, L. Vincent and P. Guiraldena, *Acta Met.*, 1981, **29**, 931.
189. J. Pelleg, *Phil. Mag.*, 1981, **43**, 273.
190. G. Günther and O. Kanert, *Solid State Comm.*, 1981, **38**, 643.
191. K. Hoshino, Y. Iijima and K. Hirano, *Acta Met.*, 1982, **30**, 265.
192. J. Růžičková and B. Million, *Mater. Sci. Eng.*, 1981, **50**, 59.
193. B. Million, J. Růžičková, J. Velíšek and J. Vřešťál, *Mater. Sci. Eng.*, 1981, **50**, 43.
194. C. Herzig and U. Köhler, *Materials Sci. Forum*, 1987, **15/18**, 301.
195. R. V. Patil, G. P. Tiwari and B. D. Sharma, *Phil. Mag. A*, 1981, **44**, 717.
196. N. Prinz and H. Wever, *Phys. Stat. Solidi A*, 1980, **61**, 505.
197. V. S. Raghunathan and B. D. Sharma, *Phil. Mag. A*, 1981, **43**, 427.
198. I. Gödény, D. L. Beke, F. J. Kedves and G. Groma, *Z. f. Metallk.*, 1981, **72**, 97.
199. H. Nakajima, Y. Nakamura, M. Koiwa, T. Tabasugi and O. Izumi, *Scripta Met.*, 1988, **22**, 507.
200. S. Bharati and A. P. B. Sinha, *Phys. Stat. Solidi (a)*, 1977, **44**, 391.
201. D. Ablitzer and A. Vignes, *J. Nucl. Mat.*, 1978, **69/70**, 97.
202. H. Amenzou-Badrour, G. Moya and J. Bernardini, *Acta Met.*, 1988, **36**, 767.
203. B. Million, *Czech. J. Phys.*, 1977, **B27**, 928.
204. T. Hehenkamp, W. Schmidt and V. Schlett, *Acta Met.*, 1980, **28**, 1715.

Table 13.4   CHEMICAL DIFFUSION COEFFICIENT MEASUREMENTS

Element 1 At.%	Element 2 At.%	A cm^2 s^{-1}	Q kJ mol^{-1}	D cm^2 s^{-1}	Temp. range K	Method	Ref.
A V. small	Ag ~ 100	0.12	140.7	—	873–1 073	IIIb(ii)	1
A V. small	Mg ~ 100	10^4	218	—	603–813	IIIb(ii)	104
Ag 0.5	Al	0.21	120.6				
1.0		0.30	123.5				
1.5		0.33	124.8	A few partial D_{Ag}'s			
2.0		0.55	128.95	and D_{Al}'s calculated.			
2.5		0.78	131.5	Very roughly	773–868	IIa(i)	2
3.0		1.50	136.1	$\tilde{D} = D_{Ag} \sim 2D_{Al}$			
3.5		3.0	141.1				
6.5		11.0	154.9				
8.5		16.0	159.1				
—	0–20	1.63	153	—	648–793		
			$\log \tilde{D} = -13.360 + 7.316c$		648	Electrochem.	292
(c = mol fraction)			$\log \tilde{D} = -10.601 - 1.084c + 11.392c^2$		723	method	
			$\log \tilde{D} = -10.322 + 1.799c$		793		

(continued)

Table 13.4   CHEMICAL DIFFUSION COEFFICIENT MEASUREMENTS—*continued*

Element 1 At.%	Element 2 At.%	A $\mathrm{cm^2\,s^{-1}}$	Q $\mathrm{kJ\,mol^{-1}}$	D $\mathrm{cm^2\,s^{-1}}$	Temp. range K	Method	Ref.
Ag 0–8.77	Au	0.0242	154.9		1 079–1 290	IIa(ii)	3
45.0				4.7×10^{-9}			
54.0				4.1×10^{-9}			
64.0				3.7×10^{-9}	1 213	IIIa(i)	4
75.0				2.8×10^{-9}			
85.0				1.9×10^{-9}			
50.8		0.14	174.6		1 036–1 238	IIa(i)	5
Ag	Cd						
—	0–25			*See* Figures 13.9 and 13.10	873–1 073		
—	8.6			$D_{cd}/D_{Ni} = 150$	1 073	IIa(i), p.c.	291
—	16.5			1.54			
—	3.4	—	—	$11.2 \times 10 \times 10^{-9}$	1 179.5		
—	10.4	—	—	4.3×10^{-9}	1 072.4		
—	10.7	—	—	$^{(a)}5.4 \times 10^{-9}$	1 087		
—	16.45	—	—	$^{(a)}6.71 \times 10^{-9}$	1 073.5		
—	17.1	—	—	9.36×10^{-9}	1 073.5		
				$10^9 D_{Ag}$ $\quad 10^9 D_{Cd}$		IIa(i), s.c.	51
—	3.4			1.85 $\quad$ 11.53	1 179.5		
—	10.8			1.67 $\quad$ 5.76	1 086.6		
—	16.5			1.32 $\quad$ $^{(a)}7.79$	1 073.5		
—	17.1			2.77 $\quad$ 10.72	1 073.5		

(*a*) Average values.

Element 1 At.%	Element 2 At.%	A $\mathrm{cm^2\,s^{-1}}$	Q $\mathrm{kJ\,mol^{-1}}$	D $\mathrm{cm^2\,s^{-1}}$	Temp. range K	Method	Ref.
Ag 0–3	Cu	0.012	149.1	—	990–1 140	IIa(ii)	6
	0–2	0.52	183.8	—	1 023–1 073	IIa(ii)	156
Ag	Ga						
	1.9–9.5	0.42	162.9	—	1 023–1 073	IIb	219
Ah	H						
	0–sol. limit	2.82×10^{-3} (*D* indep. of conc.)	31.4	—	611–873	IIb(ii)	7
Ag ~100	Kr	V. small	1.05	146.5	—	773–1 073	IIIb(ii)	8
Ag —	Mn 0–8.5	0.18	179.6	*D* indep. of conv.	849–1 206	IIa	151
Ag ~100	Ne V. small	2.5	249.1	—	1 073–1 213	IIb(ii)	105
Ag	O 0–sol. limit	3.66×10^{-3}	46.1	—	685–1 135	IIIb(ii)	9
Ag 0–0.12	Pb		7.4×10^{-2}	63.6	—	493–558	IIa(ii)	10
Ag 50	Pd 50	1.5×10^{-6}	103.0	—	873–1 173	II	152
Ag	S Sol. soln	2.34×10^{-3}	110.1	—	877–1 025		220
Ag ~100	Xe V. small	0.036	157.0	—	773–1 073	IIIb(ii)	13
Ag (β)	Zn 50	0.016 4	69.1	$D_{Zn} \sim 3$ to $4D_{Ag}$ Figure 13.11	673–883	IIc IIa(i)	11
	40–55						
(α)	26.8			8.7×10^{-9}	at 973	IIIa(i)	12
	27.6			2.45×10^{-9} $D_{Zn} \sim 1.5$–$2.2D_{Ag}$	at 923		
α-phase	0–18	*See* Figures 13.12 and 13.13		—	823–1 023	IIa(i), p.c.	312
(ε)	74–87	*See* Figure 13.19			583–673	IIa(i)	221

(*continued*)

Table 13.4   CHEMICAL DIFFUSION COEFFICIENT MEASUREMENTS—*continued*

Element 1 At.%	Element 2 At.%	A cm² s⁻¹	Q kJ mol⁻¹	D cm² s⁻¹	Temp. range K	Method	Ref.
Al 33.33 'α-phase' (AuAl)	Au 66.66	6.8 — —	105.5 ~138	— 2.6×10^{-13} 1.1×10^{-10}	393–553 573⎱723⎰	IIc, p.c. p.c., IIa(ii)	263 265
Al	Be 0.015 0.022 0.03	52 126 550	163.3⎱168.7⎰180.5	—	773–908	IIa(iii) and IIc	14
Al 0–10.2	Co —	10.5	295	—	1 273–1 473	IIa(i), p.c.	257
Al —	Cr 0–6	1.3×10^{6}	240	—	659–726	IIa(i), p.c.	276
Al	Cu 0–0.215 0–~2 (sol. sol. range)	0.29 0.18	130.29 126.02	— —	778–908 775–811	IIa(ii) IIc	15 154
0 2 4 6 8 10 12		0.131 0.231 0.287 0.364 0.588 1.033 1.293	185.22 187.74 187.57 187.19 (a) 189.37 194.43 191.46	—	985–1 270	IIa(i)	153
~25	(β) (γ₂) (δ) (ζ₂) (η₂) (θ)	0.19 0.85 2.1 1.6×10^{6} 2.2 0.56	115.1 136.1 138.2 230.7 148.6 127.7	— —	919–1 023 673–808	IIc IIc	17 155
11–13 wt.%	β-phase	0.65 $A_{Al} = 0.13$ $A_{Cu} 2.2$	176.7 $Q_{Al} = 162.9$ $Q_{Cu} = 181.7$	—⎱—⎰—	1 073–1 223	IIa(ii), p.c.	264

(a) $\log A = -0.864\,9 + 0.082\,9\,C_{Al}$⎱$Q = 44.27 + 0.14\,C_{Al}$⎰ least squares fit to the data of reference 153.

Al 9 17 25 33	Fe	2.7×10^{-1} 3.6×10^{-3} 7×10^{-2} 7.3×10^{-2}	188.0 142.4 167.1 129.8	— — — —	1 193–1 483 (Disordered) 1 373–1 483 (Disordered)	IIa	8
33		2.1×10^{4}	290.6	—	1 073–1 273 (Ordered)		
41 0–~52 0–15	(α) —	1.5×10^{5} 30.1 1.6	305.2 234.5 306	— — —	1 073–1 193 1 223–1 373 1 048–1 173	IIa(ii) Ion impl. c(x) via NRA	120 283
Al	H 0–sol. limit	0.11	110.95	—	633–873	IIIb(ii)	19
Al ~ 100	He V. small	3.0	152.8	—	—	IIIb(ii)	105
Al (α-phase AlLi)	Li 0–sol. limit	4.5 0.155	139.4 119.2	— $673–873$	690–870 Electrochem.	IIa(ii) 293 method	20
Al	Mg 0 0–10	1 4.4 12	129.8 140.3 143.6	—⎱—⎰—	523–713⎱623–693⎰	IIa	157
0–20 β phase γ phase		2.4×10^{-4} 9.9×10^{-1}	56.9 117.6	—⎱—⎰	598–698	IIc	213

(continued)

Table 13.4   CHEMICAL DIFFUSION COEFFICIENT MEASUREMENTS—*continued*

Element 1 At.%	Element 2 At.%	A cm² s⁻¹	Q kJ mol⁻¹	D cm² s⁻¹	Temp. range K	Method	Ref.
Al	Mg						
—	0	0.42	125				
—	1	0.49	124	@ 0.101 Mpa.			
—	2	0.61	127	Data also at	690–818	IIa(i), p.c.	273
—	3	0.32	122	2.2 and 3.3 Gpa			
—	4.06	0.45	122				
Al	Mn						
0.02–0.15	—	—		Figure 13.14	873–923	IIa(iii)	21
Al	Na						
0–0.002	1.1		134.0	—	823–923	IIIb(i)	22
Al	Nb						
33	67	2 × 10⁻³	230.3	—⎫	1 427–1 773	IIa	222
25	75	2.5	366.3	—⎬			
Al	Ni						
0–0.7		1.87	268.0	—	1 373–1 553	IIa(ii)	23

$10^{11}\tilde{D}$ @

		1 223	1 273	1 323	1 373	1 423 K			
0.7	—	—	2.2	—	11.0				
5.0	—	1.5	3.0	7.9	16.0	37.0			
10.0	(σ-phase) 2.4	3.8	11.5	24.0	55.0	$Q \sim 234$			
14.0	—	—	4.9	—	33.0	—			
25.0	(ε-phase) —	6.3	29.5	36.0	116.0	$Q \sim 266$	IIIa(i)	288
38.0	(δ-phase) —	62.0	—	270	—	$Q \sim 208$		

$\tilde{D}(\delta)$ varies rapidly with composition, with minimum close to the stoichiometric composition (50%).
Max $Q = 60$ @ 45% Al, decreasing to 44 @ 36%, 32 @ 53%.

Element 1 At.%	Element 2 At.%	A cm² s⁻¹	Q kJ mol⁻¹	D cm² s⁻¹	Temp. range K	Method	Ref.
0–10.2	—	1.3	257	—	1 273–1 473	IIa(i), p.c.	157
—	0	ln A = 3.1 × 10⁻²$\tilde{Q}$	284$^{(a)}$	—	1 273–1 573	IIa(i), p.c.	
—	16	−4.49	258	—	1 273–1 573	IIa(i), p.c.	313

(a) $\tilde{Q}$ decreases linearly with increase in concentration over this range.

Element 1 At.%	Element 2 At.%	A cm² s⁻¹	Q kJ mol⁻¹	D cm² s⁻¹	Temp. range K	Method	Ref.
Al	Pu						
3–9.1	(δ)	2.25 × 10⁻⁴	106.8	—	623–790	IIa(ii)	186
Al	Si						
	0–0.5	0.346	123.9	—	617–904	IIa(ii)	224
	0–0.5	2.02	136.1				
		A_Si = 3.95	Q_Si = 140.1		753–893	IIa(i)	24
		A = 5.07	Q_Al = 143.1				
	0–0.7	—	—	Figure 13.15	723–853	IIa(iii)	21 and 25
Al	Ti						
2.0	(β)	1.4 × 10⁻⁵	91.7	$\tilde{D}$ increases	1 256–1 523		
12.0		9.0 × 10⁻⁵	106.8	linearly with c			
10.0	(a)	1.6 × 10⁻⁵	99.2		1 107–1 173	IIa(i)	26
				$D_{Al} = 14.11 \times 10^{-9}$	1 523		
				$D_{Ti} = 4.61 \times 10^{-9}$			
Al	Zn						
—	0–4.49	0.27	118	—	383–433	X-ray analysis of g.b. depletion	272
—	0	0.406	124⎫		723–881	IIa(i), p.c.	277
—	3.5	0.280	121⎬				

Measurements also at 2 and 3 Gpa pressure.

	$10^{11} \times \tilde{D}$@330°C	360°C	400°C	440°C	485°C	540°C	
~ 0	1.84	3.98	12.7	49.2	149	610	
9.0	1.95	4.85	19.3	69.6	174	610	
18.1	3.64	6.12	20.0	74.8	2.2	—	27
37.6	—	1.10	—	51.1	—	—	

To very good approximation $D_{Zn} = \tilde{D}/c_{Al}(c_{Al} = $ fractional at. conc.)

(continued)

Table 13.4 CHEMICAL DIFFUSION COEFFICIENT MEASUREMENTS—*continued*

Element 1 At.%	Element 2 At.%	A cm^2 s^{-1}	Q kJ mol^{-1}	D cm^2 s^{-1}		Temp. range K	Method	Ref.
Al	Zr							
8		9.2×10^{-3}	169.1	—				
10		2.3×10^{-2}	179.2	—				
12		5.2×10^{-2}	189.2	—		1 373–1 573		
14		7.6×10^{-2}	192.2	—				
16		8×10^{-2}	192.2	—		1 433–1 573	IIa(i)	223
	Al$_3$Zr$_5$	9.2	283.0	—		1 373–1 573		
	Al$_2$Zr$_3$	3.4	272.1	—		1 273–1 573		
	Al$_3$Zr$_4$	1.6×10^5	382.3	—		1 273–1 573		
As	Fe							
0.6–4.6		4.3	219.8	—	(α)	1 223–1 653	IIa(i)	225
		0.58	246.6	—	(γ)			
Au	Cu							
	10–90	—	—		Figures 13.12 and 13.13	1 006–1 130	IIa(i)	160
'Au Cu'	50	$\begin{cases}2.36 \times 10^{-6}\\ 7.94 \times 10^{-8}\end{cases}$	57.1 / 44.9	(Disordered) (Ordered)		823–973 / 573–623	IIa(i)	167

		$10^{10}\tilde{D}$	$10^{10}D_{Au}$	$10^{10}D_{Cu}$				
1.25	—	1.84	1.8	4.2				
2.50	—	2.10	2.0	4.5				
5.0	—	2.70	2.4	6.4				
7.48	—	3.41	2.8	9.0		1 133	IIa(i)	267
12.44	—	5.40	3.1	16.9				
14.25	—	6.30	2.5	22.5				
17.43	—	8.20	3.0	25.2				
—	0.5	8.99×10^{-3}	133.1					
—	4.2	1.24×10^{-2}	135.9					
—	7.2	1.56×10^{-2}	138.0					
—	11.2	2.17×10^{-2}	140.9					
—	14.6	2.99×10^{-2}	143.7					
—	21.8	5.36×10^{-2}	148.9	—		659–827	Electro-chemical method	280
—	28.3	8.91×10^{-2}	153.5					
—	38.6	0.214	161.1					
—	55.3	0.828	173.2					
—	63.9	1.690	179.4					
—	79.0	6.092	190.7					
Au	Fe							
0–18.3		1.16×10^{-4}	102.2	—		1 023–1 273	IIa(ii)	28
Au	**H**							
		2.08×10^{-3}	20.8	—		523–1 073	Ib	29

Au	In		$\tilde{D}$ at 142°C	$\tilde{D}$ at 151°C ($\times 10^{-12}$)			
3			0.47	24.0			
33	(Au In$_2$)		7.0	29.0			
50	(Au In)		2.6	6.6	—	IIa(i)	30
69	(Au$_9$ In$_4$)		5.8	9.8	(Values of D_{In} listed,		
80	(Au$_4$ In)		0.49	0.68	calculated assuming		
91			0.24	0.28	$D_{Au} \ll D_{In}$)		

Au	Ni	A cm^2 s^{-1}	Q kJ mol^{-1}		Temp. range K	Method	Ref.
	2	4.3×10^{-2}	173.3				
	10	3.9×10^{-2}	173.8				
	20	9.5×10^{-2}	183.0				
	30	7.8×10^{-2}	183.8				
	40	2.2×10^{-2}	175.8				
	50	1.3×10^{-2}	177.5	—	1 123–1 248	IIa(i)	31
	55	5.9×10^{-2}	174.6				
	60	6.8×10^{-2}	204.3				
	65	1.4×10^3	305.6				
	70	2.0×10^7	402.8				
	75	6.2×10^8	439.6				
	98	1.8×10^4	356.7				

(*continued*)

Table 13.4   CHEMICAL DIFFUSION COEFFICIENT MEASUREMENTS—*continued*

Element 1 At.%	Element 2 At.%	A cm^2 s^{-1}	Q kJ mol^{-1}	D cm^2 s^{-1}	Temp. range K	Method	Ref.
Au 0–~0.09	Pb 0	0.35	58.6	—	386–573	IIa(ii)	32
Au 0–17.1	Pd	1.13×10^{-3}	156.6	—	1 000–1 243		33
~50		3.2×10^{-4}	152.8	—	873–1 323	IIa	152
Au 98	Pt	0.62	228.6				
96		1.0	234.5				
94		0.73	231.9				
92		0.67	231.9				
90		0.60	231.1				
88		0.58	231.1				
86		0.53	230.7	—	1 198–1 328	IIa(i)	34
84		0.52	231.1				
82		0.47	230.7				
80		0.43	230.3				
2–8		0.37	262.1				
95		$A_{Au} = 0.32$ $A_{Pt} = 0.09$	$Q_{Au} = 190.9$ $Q_{Pt} = 226.1$				
B 0–0.009 5	Fe (α)	10^6	259.6	—	973–1 108	IIIa(ii)	35
0–0.02	(γ)	0.002	87.9	—	1 123–1 573	IIIa(ii)	36

(Analysis of 'Fe' for both experiments (in wt.%) 0.003 8B, 0.43C, 1.64Mn, 0.02P, 0.019S, 0.37Si, 0.04Cr, 0.01Ni, 0.01Mn)

Element 1 At.%	Element 2 At.%	A cm^2 s^{-1}	Q kJ mol^{-1}	D cm^2 s^{-1}	Temp. range K	Method	Ref.
Ba	H	4×10^{-3}	19.0	—	473–893	IIa(ii)	250
Ba 0–sol. limit	U (γ)	0.112	170.8	—	1 123–1 313	IIc	85
Be 0–~15	Cu (α)	0.19	173.8	—	823–1 157		
~33	(β)	0.084	115.1	—	923–1 157		
~48	($\gamma(\beta')$)	0.054	129.8	—	823–1 157	IIa(i)	37
~75	(δ)	0.0012	138.2	—	823–1 157		
~33	(β)	$A_{Be} = 0.035$ $A_{Cu} = 0.045$	$Q_{Be} = 121.4$ $Q_{Cu} = 104.7$				
Be	Fe 0–0.2	1.0	226.1	—	1 073–1 373	IIa(i)	38

Very little variation of $\tilde{D}$ with c

Element 1 At.%	Element 2 At.%	A cm^2 s^{-1}	Q kJ mol^{-1}	D cm^2 s^{-1}	Temp. range K	Method	Ref.
Be	H	2.3×10^{-7}	18.42	—	473–1 273	IIIb	86
Be 0–sol. limit	Mg	8.06	157.0	—	773–873	IIIa(i)	212
Bi 0–2.0	Pb	0.018	77.0	—	493–558	IIa(i)	39
C	Co	8.72×10^{-3}	149.3	—	723–1 073	IIa(i), p.c.	158
~0.1–0		0.31	153.7	—	1 073–1 673	IIIa(i)	216

Element 1 At.%	Element 2 At.% (Co)	(Fe)	A cm^2 s^{-1}	Q kJ mol^{-1}	D cm^2 s^{-1}	Temp. range K	Method	Ref.
C	78.5		0.472	157.0	—	1 123–1 373	IIIb(ii)	41
0.48		89.4	0.442	157.0	—	1 123–1 385		
0–0.7 (wt.%)	0 5.8 10.6 20.2	(γ)	$0.04 + 0.08 c$ $0.04 + 0.08 c$ $0.03 + 0.1 c$ $0.03 + 0.06 c$	131.3 127.7 125.2 120.8		1 273–1 473	IIa(i)	146

$(c = \text{wt.\%C})$

Element 1 At.%	Element 2 At.%	A cm^2 s^{-1}	Q kJ mol^{-1}	D cm^2 s^{-1}	Temp. range K	Method	Ref.
C 0–0.1	Fe (α)	3.94×10^{-3}	80.2	—	313–623	Various	215
γ-range	$D = 4.53 \times 10^{-3}[1 + y_c(1 - y_c)8339.9/T]\exp[-(T^{-1} - 2.221 \times 10^{-4} \times 17767 - 26436y_c)]$				233–1 140	Combined	42
	$y_c = x_c/(1 - x_c),$		$x_c = $ mol fraction of C			data, several	40
α-range	$\text{Log } D = -0.9064 - 0.5199\chi + 1.61 \times 10^{-3}\chi^2$					sources	
	$\chi = 10^4/T$						

(*continued*)

Table 13.4   CHEMICAL DIFFUSION COEFFICIENT MEASUREMENTS—*continued*

Element 1 At.%	Element 2 At.%		A cm^2 s^{-1}	Q kJ mol^{-1}	D cm^2 s^{-1}	Temp. range K	Method	Ref.
C	Fe	Mn						
		0	0.07	131.3				
0.4	(γ)	1.0	0.08	132.3	—	1 273–1 473	IIa(i)	148
(wt.%)		12.1	0.19	141.9				
		19.2	0.41	151.1				
C	Fe	Ni						
		0	$0.04 + 0.08\,c$	131.3				
0–0.7	(γ)	3.9	$0.3 + 0.1\,c$	129.8	—	1 273–1 473	IIa(i)	147
(wt.%)		9.2	$0.03 + 0.1\,c$	127.9				
		17.3	$0.02 + 0.1\,c$	124.8				
				($c = $ wt.% C)				
		23	0.322	148.1				
		40	0.20	139.8				
0–0.1	(γ)	60.4	0.296	145.2	—	1 123–1 373	IIIa(i)	56
wt. %		70	0.372	148.5				
		80	0.680	156.0				
		100	0.366	149.3				
C	Fe	Si						
Sol. soln	(γ)	0–2.35	—	—	Figure 13.27	1 153–1 223	IIc	145
range								
C	Mo							
Sol. soln			0.034	171.7	—	2 053–2 243	IIIa(i)	198
range								
C	Nb							
0.3–1.0			1.8×10^{-2}	159.1	—	1 873–2 373	IIIb(ii)	128
C	Ni							
			0.12	137.3	—	873–1 673	IIa(ii), p.c.	43
0–0.1 wt%			0.366	149.3	—	1 123–1 373	IIIa(i)	56
C	Re							
2 or 3–0			0.1	221.9	—	1 503–2 003	IIb(ii)	205
C	T							
			6.7×10^{-3}	161.6	—	463–2 953	Various	249
C	Th							
					7.4×10^{-9} 1 273			
~7.2			—	~159.1	2.8×10^{-8} 1 372		IIIa(i)	44
					5.7×10^{-8} 1 473			
100–400	(β)		0.022	113.0	—	1 713–1 953	IIa(ii)	159
p.p.m.								
C	Ti							
0.14–sol.	(α)		5.06	182.1	$\tilde{D}$ indep. of conc.	1 009–1 108	IIc	45
limit								
	(β)		6×10^{-3}	94.6	—	1 613–1 873	IIa(ii)	252
C	V		8.8×10^{-3}	116.4	—	333–2 098	Various	247
C	Zr							
	(β)		0.0048	111.8	—	1 173–1 523	IIa	209
Cd	Cu							
0–0.5			3.5×10^{-3}	122.3	—	773–1 123	IIa(ii)	6
Cd	Pb							
0–1.0			1.85×10^{-3}	64.5	—	440–525	II	46
Cd	**Hg**							
	0–4.0		2.57	82.1	—	429–475	III(ii)	46
Ce	Mg							
Sol. soln			450	175.8	—	773–871	IIc	210 and 174
range in Mg								
Ce	Pu							
3.74–7.17	(δ)		1.31×10^{-2}	123.9	—	676–801	IIa(ii)	187

(*continued*)

Table 13.4 CHEMICAL DIFFUSION COEFFICIENT MEASUREMENTS—*continued*

Element 1 At.%	Element 2 At.%	A cm^2 s^{-1}	Q kJ mol^{-1}	D cm^2 s^{-1}	Temp. range K	Method	Ref.
Ce 0–sol. limit	U (γ)	3.92	278.0	—	1 073–1 273		
Co	Cr 0–15.2	0.084	253.7	—	1 273–1 573	IIa(ii)	47
	0–40	0.443	266.3	—	1 273–1 643	IIa(iii)	48
		(D reported f(c), but concentration dependence very slight)					
Co 0.1	Cu	0.6	213.5	—⎫			
2	(wt%)	5.7	242.8	—⎬	800–1 073	IIa(i)	256
—	0–5	~1.0	~275	—	1 158–1 273	IIIa(i), p.c.	274
Co 5–70 (b.c.c.)	Fe		See Figure 13.16a		993–1 228	IIa(i), p.c.	318
5–95 (f.c.c.)			See Figure 13.16b		1 273–1 573		
10		1.5×10^{-3}	219.0	—			
20		2.9×10^{-3}	215.2	—⎫			
30		4.4×10^{-3}	212.3	—			
40		5.8×10^{-3}	216.5	—			
50		7.0×10^{-3}	215.2	—⎬	1 273–1 673	IIa(i)	242
60		8.8×10^{-3}	217.3	—			
70		11.5×10^{-3}	218.1	—			
80		12.0×10^{-3}	218.1	—			
90		13.1×10^{-3}	219.0	—⎭			
Co	**H**						
	5 Atm.	8.3×10^{-3}	49.4	—	673–823 (α)⎫	Ic	195
		3.4×10^{-2}	57.8	—	473 × 673 (ε)⎭		
Co	Mn						
	5	7.79	296.2	(Ferro)	1 133–T_c		
	5	0.781	273.2	(Para)	T_c–1 423		
	10	3.07	284.2 ⎫				
	20	0.70	257.2 ⎪				
	30	0.721	248.2 ⎬		1 133–1 423		
	40	0.627	241.2 ⎭				
	33	$A_1 = 0.22$	$Q_1 = 263$				
		$A_2 = 0.98$	$Q_2 = 229$				
Co	Mo						
	0–10	0.231	262.9	—	1 273–1 573	IIa(ii)	47
	0–15	2.48	294.8	—	1 073–1 573	IIa(i)	226
Co 10–90	Ni	$\tilde{D} = 1.58 \exp(-0.95N) \exp(-278.3/RT)$			1 433–1 673	IIa(i), p.c.	242 and 161
5–95		$\tilde{D} = 0.22 \exp(-1.14N) \exp(-256.2/RT)$			1 263–1 423 ⎫		
		N = mol fraction of Co					
		A_{Co} $\quad$ Q_{Co} $\quad$ A_{Ni} $\quad$ Q_{Ni}					
15		0.388 $\quad$ 267 $\quad$ 0.0493 $\quad$ 232 ⎫				IIa(i), p.c.	161
20		0.410 $\quad$ 270 $\quad$ 0.136 $\quad$ 244 ⎪			1 263–1 423		
25		0.433 $\quad$ 273 $\quad$ 0.372 $\quad$ 256 ⎬					
30		0.457 $\quad$ 276 $\quad$ 1.02 $\quad$ 268 ⎭			⎭		
Co	**O**	67.8	241.2	—	1 323–1 573	Intern. oxid.	309
Co	Pd						
10							
20				13.7×10^{-11} ⎫			
30				55.0 ⎪			
40				120 ⎪			
50				145 ⎬	1 423	IIa(i)	144
60				120 ⎪			
70				70.7 ⎪			
80				28.2 ⎪			
90				13.9 ⎭			

(continued)

Table 13.4   CHEMICAL DIFFUSION COEFFICIENT MEASUREMENTS—*continued*

Element 1 At.%	Element 2 At.%	A cm² s⁻¹	Q kJ mol⁻¹	D cm² s⁻¹	Temp. range K	Method	Ref.
Co 0–100	Pt		Figure 13.17	—	1 398–1 573	IIa(i)	218
Co 4–8	Ti	15	280.5	—			
(Co₃Ti) 21		5.3×10^{-2}	167.5	—	1 173–1 413	II	227
(Co₂Ti) 30–32		0.28	217.7	—			
(Co Ti) 46–50		4.4×10^{-4}	173.3	—			
βTi 90–95		67	207.2	—	973–1 123		
Co 0–17	V	0.021	221.9	—	1 373–1 573	IIa(ii)	47
Co 0.5	W	0.008	238.2	—	1 373–1 573	IIa(ii)	47
Cr 10–20	Fe (α)	1.48	229.9	$\tilde{D}$ indep. of c (In the range 10–1%, $\tilde{D}$ increases by ~30%) $D_{Cr} \sim 1.5 D_{Fe}$	1 096–1 713	IIa(i)	49
37				80×10^{-10}			
42				56×10^{-10}			
52				27.8×10^{-10}			
62	(α)			11.7×10^{-10}	1 523	IIa(i)	50
72				7.1×10^{-10}			
81				4.9×10^{-10}			
91				4.6410^{-10}			
0–71	(γ)	0.0012	218.6	—	1 173–1 473	IIa(ii)	47
0	—	1690	299.3	—	1 044–1 124		
5	—	824	292.0	—	1 049–1 124		
10	—	221	279.2	—	1 039–1 124	IIa(i), Fe s.c.	258
15	—	256	280.4	—	984–1 124		
20	—	60	268.9	—	974–1 124		
25	—	15	258.7	—	974–1 124		
Cr	N Ti	1.6×10^{-2}	115.1	—	331–463	Vb	129
—	1 Atm. press. { 0	0.0096	119.3	—	1 273–1 673	Internal nitridation method	196
—	0.5	0.0090	118.1	—	1 272–1 673		
—	3.0	0.0054	111.4	—			
		0.0025	100.5				
Cr (γ)	Nb 2	34.0	409.9	—	1 373–1 897		
Cr₂Nb { 34		26.0	393.6	—	1 524–1 897		
38(a)		24.0	396.9(b)			IIa(i)	140
(α) 90		2.7	385.2	—	1 373–1 897		
95		0.31	360.1				

(a) The compound Cr₂Nb is reported to have a 5% solubility range.
(b) Arrhenius plots not always linear. $\tilde{D}$ and $\tilde{A}$ derived from measurements at the higher temperatures.

Cr 0–12	Ni	0.604	257.2	—	1 273–1 573	IIa(ii)	47
Cr 0–14	Ni wt.%	0.5(a)	260	—	1 373–1 523	IIa(i), p.c.	311

(a) Average values. $\tilde{D}$ increases linearly ~25% over the comp. range.

Cr 9	Ti (β)	—	—	3.6×10^{-9} $D_{Ti} = 2.8 \times 10^{-9}$ $D_{Cr} = 3.7 \times 10^{-9}$	1 258	IIa(i)	53
Cr 0–sol. limit	U (γ)	0.7	142.4	—	1 173–1 273	IIc	54

(*continued*)

Table 13.4   CHEMICAL DIFFUSION COEFFICIENT MEASUREMENTS—*continued*

Element 1 At.%	Element 2 At.%	A cm^2 s^{-1}	Q kJ mol^{-1}	D cm^2 s^{-1}	Temp. range K	Method	Ref.
Cu	Fe						
		A_{Cu}	Q_{Cu}	A_{Fe} $\quad$ Q_{Fe}			
ε Sol. soln		6.1	268.0	2.7 $\quad$ 265.9 ⎱	1 173–1 323		
γ Sol. soln		3.6	274.2	8.9 $\quad$ 314.0 ⎰		IIa(ii)	193
(γ-phase)		8.8×10^{-4}	204		1 323–1 583 ⎱	IIIa(i)	24
(δ-phase)		—	—	$5.7 \times 10^{-8(a)}$	1 173		
—	0–2.59	0.54	208.3	—	1 073–1 323	IIIa(ii), p.c.	303
wt.%	0–0.64	0.091	193.2	—	923–1 073		
Cu	Ga						
	0–3	0.58	193.8		973–1 323	IIb	219
	2.5	3×10^{-4}	134.0	—			
	4.9	1.8×10^{-3}	142.4	—			
	7.6	1.6×10^{-2}	154.9	—			
	10.3	1.8×10^{-1}	167.5	—	773–973	IIa(i)	228
	13.1	1.3×10^{-1}	157.0	—			
	15.9	8×10^{-2}	146.5	—			
Cu	**H**						
	(H) sol. limit	11.31×10^{-3}	38.88	—⎱			
	(D) sol. limit	7.30×10^{-3}	36.82	—	720–1 200	IIIb(iii)	162
	(T) sol. limit	6.12×10^{-3}	36.50	—⎰			
Cu	In						
—	0	0.73	189 ⎱				
—	1	0.93	188				
—	2	1.2	188				
—	3	1.8	189	—	949–1 119		
—	4	5.8	195				
—	5	9.8	198				
—	6	11	196				
—	7	18	198 ⎰				
		$10^{10}D_{Cu}$	$10^{10}D_{In}$				
—	0.9	0.71	1.5		1 005		
		4.4	8.85		1 089	IIa(i), p.c.	262
—	1.7	0.62	1.9		1 005		
		1.9	4.8		1 043		
		4.95	11		1 089		
—	2.9	0.24	1.25	—	970		
		0.62	2.9		1 005		
		1.65	7.4		1 043		
		5.4	16.5		1 089		
—	4.6	0.46	1.3		949		
		0.51	2.0		970		
		1.3	5.1		1 005		
Cu	Mn						
	0–28	0.58	177.5	D indep. of conc.	913–1 093	IIa	163
—	5	0.37	187 ⎱				
—	10	0.56	187				
—	15	0.51	183				
—	20	0.53	181				
—	25	0.66	181				
—	30	1.17	186				
—	35	1.33	189				
—	40	2.34	197	—	1 023–1 123	IIa(i), p.c.	243
—	45	10.7	216				
—	50	93.6	241				
—	55	41.0	239				
—	60	2.1	218				
—	65	0.14	198				
—	70	0.07	195				
—	75	0.16	204 ⎰				
Cu	Ni						
~0	~100	0.4	257.9				
	0–100			Figure 13.18 ⎱	1 038–1 339	IIa(i)	55
~100	~0	1.4	228.2				
—	0–7	$D = 6.09 \times 10^{-10} \exp(-5.94 N_{Ni})$ ⎱					
—	1.05–4.84	$D_{Ni} = 6.11 \times 10^{-10} \exp(-9.08 N_{Ni})$			1 273	IIa(i), p.c.	299
		$D_{Cu} = 2.3 \times 10^{-9} \exp(-6.13 N_{Ni})$ ⎰					

N_{Ni} = mol fraction of Ni.

(continued)

Table 13.4 CHEMICAL DIFFUSION COEFFICIENT MEASUREMENTS—*continued*

Element 1 At.%	Element 2 At.%	A cm^2 s^{-1}	Q kJ mol^{-1}	D cm^2 s^{-1}	Temp. range K	Method	Ref.
Cu	O						
		5.8×10^{-3}	57.4	—	873–1 273	Internal oxid.	197
Cu	Pd						
	50	0.48	224.0	—	1 073–1 323	IIa	152
	0–100	—	—	Figure 13.28	1 204–1 334	IIa(i)	160
Cu	Pt						
0–13.9		0.049	233.2	—	1 313–1 673	IIa(ii)	28
	Small	0.67	233.2	—	1 023–1 348	IIb	244
Cu	Sb						
—	0	0.82	185 ⎫				
—	1	0.32	174 ⎬		919–1 040		
—	2	0.094	160 ⎪				
—	3	0.030	147 ⎭			IIa(i), p.c.	281
		$10^{14}D_{Cu}$		$10^{14}D_{Sb}$			
—	1	1.8		2.9	1 005		
—	1	3.7		6.1	1 040		
—	1.7	0.5		2.2	970		
—	1.7	1.9		4.05	1 005		
—	1.7	3.7		7.8	1 040		
Cu	Si						
—	0	0.21	187 ⎫				
—	2	0.27	186 ⎪				
—	3	0.24	184 ⎪				
—	4	0.34	183 ⎪				
—	5	0.36	182 ⎬	—	998–1 173	IIa(i), p.c.	59
—	6	0.28	181 ⎪				
—	7	0.19	172 ⎪				
—	8	0.20	170 ⎪				
—	9.8	0.13	163 ⎭				
	α: sol. soln range	11.4	200.3	—	938–1 048	IIc	164
Cu (α)	Sn 0–7	$2 \times 10^{-2} \times 10^{0.133c_{Sn}}$ 156.1 ($c_{Sn} =$ at.%Sn)		$D_{Sn} > D_{Cu}$	1 000–1 100	II	16
— β-phase	14.5	9.11 $A_{Cu} = 0.258$ $A_{Sn} = 30.5$	134 $Q_{Cu} = 103$ $Q_{Sn} = 144$ ⎫⎬⎭	—	874–993	IIa(i), p.c.	275
(δ)	20.5	$\begin{Bmatrix} D_{Cu} \\ D_{Sn} \end{Bmatrix}$	Figure 13.20	$\begin{Bmatrix} 7.7 \times 10^{-10} \\ 3.6 \times 10^{-9} \\ 1.0 \times 10^{-8} \\ 1.1 \times 10^{-7} \\ 1.65 \times 10^{-7} \\ 1.3 \times 10^{-7} \end{Bmatrix}$	$\begin{matrix} 701 \\ 714 \\ 731 \\ 780 \\ 818 \\ 845 \end{matrix}$	IIc	60
(γ)	15–22	$\begin{Bmatrix} D_{Cu} \\ D_{Sn} \end{Bmatrix}$	Figure 13.22	Figure 13.21	979.5	IIa(i)	
(ε) (η)		1.43×10^{-4} 1.55×10^{-4}	70.8 64.9	—⎱ —⎰	433–493	II	229
Cu	Ti						
	0	0.693	196	— ⎫			
	0.5	0.934	199	— ⎪			
	1.0	1.41	204	— ⎬	973–1 283	IIa(i)	230
	1.5	1.92	208	— ⎭			
3 (α-phase)		1.32×10^{-5}	87.3 ⎫				
Cu$_7$Ti$_2$		7.15×10^{-6}	92.3 ⎪				
Cu$_3$Ti$_2$		1.05×10^{-4}	116 ⎪				
Cu$_4$Ti$_3$		2.25×10^{-3}	142 ⎬	—	1 023–1 123		278
CuTi		2.05×10^{-1}	166 ⎪				
CuTi$_2$		3.37×10^{-5}	112 ⎪				
β-phase		5.76×10^{-2}	182 ⎭				

(*continued*)

Table 13.4   CHEMICAL DIFFUSION COEFFICIENT MEASUREMENTS—*continued*

Element 1 At.%	Element 2 At.%	A cm² s⁻¹	Q kJ mol⁻¹	D cm² s⁻¹	Temp. range K	Method	Ref.
Cu	**Zn**						
	1	0.056	167.5				
	5	0.062	167.5		1 053–1 188		
	10	0.083	165.4	—			
(α)	16	0.095	159.1			IIa(i)	61
	20	0.09	152.9				
	25	0.031	136.1	—	997–1 188		
	28	0.016	124.3				
		D_{Cu} and D_{Zn}—Figures 13.23, 13.24 and 13.25					
—	0	0.412	196				
—	5	0.375	189		1 105–1 223	IIa(i), p.c.	266
—	9	0.285	182				
—	15	0.614	184[(a)]				
	10	0.13	170.8				
	15	0.21	170.8				
	20	0.36	170.8				
(α)	28	1.7	172.9	—	973–1 183	IIIa(i)	62
		2.1	172.5				
	28 $\begin{cases} D_{Zn} \\ D_{Cu} \end{cases}$	0.81	178.8				
(β)	46	—	—	$D_{Zn}/D_{Cu}=2.4$–3.6	873–1 073		
(β)	48	$\begin{cases} 0.006\,9 \\ 1\,440 \end{cases}$	78.7 150.7	(Disordered) (Ordered)	753–983 591–720	IIa	
(β)	44–48	0.018–0.013	83.3–76.2	—	773–1 073	IIa(i)	63
	59	2.45×10^{-2}	98.0				
	60	2.44×10^{-2}	95.9				
	61	1.71×10^{-2}	91.7				
	62	2.45×10^{-2}	91.7		648–923		
(γ)	63	1.14×10^{-2}	84.2	—			
	64	0.99×10^{-2}	80.8				
	65	0.62×10^{-2}	74.1			IIIa(i)	64
	65.5	0.19×10^{-2}	64.9	—	698–923		
	66.5	0.28×10^{-2}	64.1	—	798–923		
				D_{Zn}/D_{Cu}			
	65–66	—	—	9.4	648–748		
	67	—	—	11.4	798		
(γ)	68	—	—	8.6	848		
	68	—	—	5.7	923		
(ε)	79–86	*See* Figure 13.37		—	523–673	IIa(i)	251

(*a*) Results also for D's at pressures 2 and 3 Gpa.

Fe	**H**						
(α)		7.5×10^{-4}	10.13	—	230–1 100	Various	319
	At $T < \sim 200°$C $\tilde{D}$ is usually less than expected from extrapolation of higher T results and apparently depends on sample history. *See* reference 319						
(γ)		1.85×10^{-2}	50.02	—	1 184–1 667		65
(δ)		1.09×10^{-3}	12.54	—	1 667–1 811		66
Fe	**Mn**						
	5	5.95	314.0	—			
N	10	3.04	303.1	—			
	15	3.37	305.2	—			
	20	2.89	301.0	—			
	25	2.83	296.8	—			
(γ)	30	2.53	295.2	—	1 283–1 523		
	35	3.12	298.9	—			
	40	2.44	294.8	—			
	45	2.17	291.0	—			
	50	1.96^h	286.0	—		IIa(i)	166
	55	2.04	289.3	—			
	36.4	—	—	$\begin{aligned} D_{Fe} &= 3.48 \times 10^{-12} \\ D_{Mn} &= 6.22 \times 10^{-22} \end{aligned}$ 1 283			
	38.0	—	—	$\begin{aligned} D_{Fe} &= 5.29 \times 10^{-12} \\ D_{Mn} &= 1.55 \times 10^{-11} \end{aligned}$ 1 363			
				$\begin{aligned} D_{Fe} &= 9.07 \times 10^{-11} \\ D_{Mn} &= 2.42 \times 10^{-10} \end{aligned}$ 1 523			
	41.2	—	—	$\begin{aligned} D_{Fe} &= 3.36 \times 10^{-11} \\ D_{Mn} &= 8.04 \times 10^{-11} \end{aligned}$ 1 443			

(*continued*)

Table 13.4   CHEMICAL DIFFUSION COEFFICIENT MEASUREMENTS—*continued*

Element 1 At.%	Element 2 At.%	A cm^2 s^{-1}	Q kJ mol^{-1}	D cm^2 s^{-1}	Temp. range K	Method	Ref.
Fe	**Mn**						
	5	7.2×10^{-2}	250.8	—			
	10	1.75×10^{-2}	263.8	—			
	15	3×10^{-1}	269.6	—	1 123–1 573	IIa(i)	231
	20	1.63×10^{-1}	262.1	—			
	25	7.2×10^{-2}	248.7	—			
	30	1.20×10^{-1}	251.2	—			
Fe	Mn C (wt%)						
	4 0.02	0.57	277.2	—	1 323–1 723		
(γ)	14 0.02	0.54	273.8			IIa(i) and (ii)	67
	4 0.51	1.25	256.2	—	1 273–1 523		
	14 0.52	1.25	255.4				
		Empirically $\tilde{D}$ may be represented ($\pm$20%) over the ranges 0–20% Mn, 0–1.5 wt% C by $\tilde{D} = (0.486 + 0.011$ wt% Mn$)(1 + 2.53$ wt% C$) \exp(-66\,000/RT)$					
(γ)	0–60	—	—	Figure 13.26	1 473	IIa(i)	67
Fe (γ)	Mo 0–0.59	0.068	247.0	—	1 423–1 533	IIa(ii)	
		(Addition of 0.4 wt% C increases A to 0.091)					89
(α)	1.9–3.6	3.467	241.6	—	1 203–1 533	IIa(ii)	
	α sol. soln range	10	251.2	—	1 063–1 458	IIc	52
Fe (α)	**N**	4.88×10^{-3}	76.83	—	226–1 183	Combined data, several sources	68
(γ)	0–sol. in eq. with 0.95 atm. N$_2$	0.91	168.6	—	1 223–1 623	IIIa(i) and IIIb(ii)	70 and 71 69
Fe —	**Ni** 10.4–15.5	—	—	3.6×10^{-15}	1 073	IIa(ii), p.c.	270
		—	—	2.3×10^{-16}	1 030		
—	10	4.79×10^{-2}	247.8				
—	20	8.22×10^{-2}	250.3				
—	30	0.121	251.1				
—	40	0.109	245.3				
—	50	0.394	254.9	—	1 193–1 573	IIa(i), p.c.	141
—	60	1.92	270.0				
—	70	2.89	273.7				
—	80	3.03	274.6				
—	90	3.11	277.1				
	0–100	Figure 13.31					
	31–33	$A_{Fe} = 3.6$ $Q_{Fe} = 286.8$	$A_{Ni} = 1.6$ $Q_{Ni} = 303.5$	— —	1 403–1 629	IIa(i)	160
		Diff. of C					
wt.% — —	20	$\sim$0.1	$\sim$138				
	40–60	$\sim$0.01/$\sim$0.05	$\sim$117			IIa and IIIa(ii)	
	70	$\sim$0.05	$\sim$128	—	1223–1423		302
	80–90	$\sim$0.2/$\sim$0.4	$\sim$155				
	100	$\sim$0.15	$\sim$138				
Fe	**Ni**	C (wt.%)	A	Q	D K		
	4	0.03	0.44	283.4	1 370–1 723		
(γ)	16	0.03	0.51	281.8		IIa(i) and (ii)	72
	4	0.6	0.46	274.27	1 323–1 573		
	16	0.6	0.42	27.0			
		Empirically, $\tilde{D}$ may be represented ($\pm$20%) over the ranges 0–20 Ni, 0–1.5 wt% C by $\tilde{D} = (0.344 \pm 0.012$ wt% Ni$) \times (1 + 2.3$ wt% C$) \exp(-67\,500/RT)$					

(continued)

Table 13.4   CHEMICAL DIFFUSION COEFFICIENT MEASUREMENTS—*continued*

Element 1 At.%	Element 2 At.%	A cm^2 s^{-1}	Q kJ mol^{-1}	D cm^2 s^{-1}	Temp. range K	Method	Ref.
Fe *(a)*	**O** Sol. soln range	5.75	169.1	—	γ-range ⎱	Internal oxidation	168
	Sol. soln range	3.7×10^{-2}	98.0	—	α and δ ⎰ range		
(b)	Sol. soln range	0.4	167.1	—	973–1 123(α)	Internal	201

(a) Fe + 0.1%A1.   *(b)* Fe + 0.07%Si.

Element 1 At.%	Element 2 At.%	A cm^2 s^{-1}	Q kJ mol^{-1}	D cm^2 s^{-1}	Temp. range K	Method	Ref.
Fe (α and δ)	**P** Sol. soln range	~2.9	~23a3	—	1 125–1 148 and	IIa(i)	79
(γ)	Sol. soln range	28.3	292.2	—	1 683–1 731 1 523–1 623		
Fe 10	**Pd**	—	232.8				
20		—	228.2	—			
30		—	227.8	—			
40		—	227.8	—			
50		—	232.4	$D_{Fe} > D_{pd}$	1 373–1 523	IIa(i)	253
60		—	240.7	—			
70		—	242.4	—			
80		—	270.9	—			
90		—	273.8				
Fe (α)	**S** Sol. soln range	1.68	204.7	—	1 023–1 173	IIIa(i)	73
(α and δ)	Sol. soln range	1.35	202.6	—	1 023–1 173 1 673–1 723	IIa(i)	73 and 74
(γ)	Sol. soln range	2.42	223.6	—	1 473–1 625	IIa(i)	74
Fe 0	**Sb**	2.6×10^{-1}	264	—			
1		6.4×10^{-2}	249	—			
2		2.2×10^{-2}	237	*(a)* —	973–1 223	IIIa(i)	254
3		9.3×10^{-3}	228	—			
4		4.3×10^{-3}	220	—			

(a) D is a little lower ($\approx \times 2$) at temperatures below the magnetic transformation at 765°C.

Element 1 At.%	Element 2 At.%	A cm^2 s^{-1}	Q kJ mol^{-1}	D cm^2 s^{-1}	Temp. range K	Method	Ref.
Fe ()	**Si** 4.57.1	0.44	201.0	—	1 368 1 623	IIa(i)	75
(α)	0–$4.21^{(a)}$	$0.735 \times (1 + 0.124c_{Si})$	219	—	1 173–1 673	IIa(i)	169
	8.35	1.82	215.3	—			
	8.69	1.87	215.1	—			
	9.04	1.77	214.0	—			
(α)	9.38	1.62	212.8	—	1 173–1 373	IIa(i)	171
	9.73	1.52	211.8	—			
	10.07	1.55	211.6	—			
	10.41	1.66	212.1	—			
	~0	8	249.1	(Extrapolated)			
	4	17	249.1	—			
(α)	5	17	247.4		1 073–1 673	IIa(i)	170
	8	35	248.7				
(α Ferro.)	3–12	500	286.8		673–873 (Ferromagnetic)	—	172
(γ)	0–2		—	4×10^{-10}	1 479	IIa(ii)	75
			—	1.7×10^{-9}	1 566		

(a) +1.4% V to stabilise the α-phase.

(continued)

Table 13.4 CHEMICAL DIFFUSION COEFFICIENT MEASUREMENTS—*continued*

Element 1 At.%	Element 2 At.%	A cm²s⁻¹	Q kJ mol⁻¹	D cm²s⁻¹	Temp. range K	Method	Ref.
Fe	Sn						
—	0–1.6 wt.%	8×10^7	159.9	α-phase	1 033–1 235 ⎱	IIa(ii), p.c.	260
—	0–0.3 wt.%	2.32	185.5	γ-phase	1 235–1 473 ⎰		
α-and δ-phase		0.931	221		1 183–1 680 ⎱	IIIa(i), p.c.	297
γ-phase		5.0×10^{-5}	136		1 273–1 580 ⎰		
α-phase	1	$\ln A = 6.5 \times 10^{-2} \tilde{Q}$	228^(a)		1 073–1 373	IIIa(i), p.c.	81
	8	-13.19	221				

(*a*) $\tilde{Q}$ decreases linearly with increase in Sn composition over this range.

Element 1 At.%	Element 2 At.%	A cm²s⁻¹	Q kJ mol⁻¹	D cm²s⁻¹	Temp. range K	Method	Ref.
Fe	Ti						
(α)	~0.7–3.0	3.15	247.9	—	1 348–1 498	IIc ⎱	77
(γ)	0 ~ 0.7	0.15	251.2	—	1 348–1 498	IIa(i) ⎰	
	2	68	261.3	—	973–1 573 ⎱		
5	(β)	0.60	188.4	—	1 173–1 573 ⎰	IIa	76
10	(β)	0.77	193.0	—	973–1 573		
15	(β)	3.6	214.4	—	973–1 573 ⎰		
66.7–75	—	—	—	2.94×10^{-12}	1 323		
(TiFe₂)		—	—	7.38×10^{-12}	1 373		
				$D_{Fe} = 6.34 \times 10^{-12}$	1 323	IIa(ii), p.c.	300
				1.56×10^{-12}	1 373		
				$D_{Ti} = 1.24 \times 10^{-12}$	1 323		
				3.3×10^{-12}	1 373		
Fe	U						
0–sol. limit	(γ)	1.3	134.0	—	1 063–1 273	IIc	54
Fe	V						
	0.7	0.61	267.1				
	5.0	3.9	238.2				
	10.0	1.1	224.0				
(α) atm.	15.0	0.70	219.8	—	1 223–1 523		
	20.0	0.71	221.1				
	25.0	0.63	221.1				
	30.0	0.59	222.3				
	10.0(1, 2 and 40 kb pressure)			Figure 13.29			
(γ)	0.7(1, 20 and 40 kb)			Figure 13.30	1 223–1 573		
				5.8×10^{-12}	1 373		
				2.5×10^{-11}	1 435		
(γ)	2.0(40 kb)	—	—	2.9×10^{-11}	1 474	IIa(i)	78
				1.1×10^{-10}	1 548		
				2.6×10^{-10}	1 623		
				6.3×10^{-12}	1 373		
				2.0×10^{-11}	1 435		
				3.2×10^{-11}	1 474		
(γ)	3.0(40 kb)	—	—	1.6×10^{-10}	1 548		
				1.4×10^{-10}	1 565		
				2.8×10^{-10}	1 623		
Fe	W						
0–0.13		11.5	594.5		2 200	IIIb(ii)	82
	0–1.3	—	—	3.7×10^{-10}	1 553 ⎱	IIa	83
	0–1.2	—	—	2.4×10^{-9}	1 603 ⎰		
	0–3.4	—	—	1.0×10^{-9}	1 603		
Fe	Zn						
8.5		2.97×10^{-3}	64.5				
9.0		7.88×10^{-3}	72.4				
9.5		1.39×10^{-2}	80.4				
10.0		5.53×10^{-3}	79.5				
10.5	(δ₁)	2.82×10^{-3}	80.4				
11.0		1.02×10^{-3}	76.2		741–798	IIa(i)	232
11.5		5.81×10^{-4}	74.1				
12.0		6.98×10^{-4}	74.9				
12.5		1.26×10^{-3}	75.8				
13.0		8.21×10^{-3}	78.3				
21.5–22.5	(Γ₁)	2.04×10^{-4}	80.4				
31–32	(Γ)	1.05×10^{-3}	92.1				

(*continued*)

Table 13.4 CHEMICAL DIFFUSION COEFFICIENT MEASUREMENTS—*continued*

Element 1 At.%	Element 2 At.%	A cm^2 s^{-1}	Q kJ mol^{-1}	D cm^2 s^{-1}	Temp. range K	Method	Ref.
Fe 0–6	**Zn** (ζ)	2.28×10^{-2}	83.3		513–633	IIa	233
—	0	60	262.6				
—	2	38	256.2				
—	4	19	247.5				
—	6	11	239.9	—	1 068–1 169[a]	IIIa(i), s.c.	282
—	8	6.2	232.9				
—	10	3.9	226.4				
—	12	2.0	218.1				

(a) At $T < 1\,068$, D's decrease below extrapolation of high T Arrhenius line.

Element 1 At.%	Element 2 At.%	A cm^2 s^{-1}	Q kJ mol^{-1}	D cm^2 s^{-1}	Temp. range K	Method	Ref.
Ga	Pu						
3–7.9	(δ)	1.3	156.5	($\tilde{D}$ indep. of conc.)	623–790	IIa(ii)	184
0.48–2.6	(ε)	5.3×10^{-4}	55.3	—	833–913	IIa(i)	185
2.5–6.5	(δ)	0.098	138.2	—	673–807	IIa(i)	214
Ga	Ti						
Sol. soln range	(α)	4.4×10^{-4}	181.7	—	873–1 133		
25	(Ti$_3$Ga)	7.4×10^{-5}	183.4	—			
Ge	**H**						
	Range of c in eq. with H$_2$ gas over p, range 10–76 cm Hg	2.72×10^{-3}	364	—	1 073–1 183	Ic	84
Ge $\sim$100	**He** V. small	6.1×10^{-3}	67.0	—	1 068–1 145	Ic	106
Ge	Nb						
66.7	33.3	6.4×10^{-2}	161	(NbGe$_2$)			
40	60	2.27	282	(Nb$_3$Ge$_2$)	1 243–1 723	IIIa(i), p.c.	268
37.5	62.5	0.628	238	(Nb$_5$Ge$_3$)			
H	Mo	1×10^{-2}	58.6		1 173–1 773	Various	194
H 0.2–4.3	Nb	5×10^{-4}	10.24	—	273–573	Va	80
H	Ni	6.44×10^{-3}	40.2	—	297–1 393	Various	90
H D T	Sol. soln range	7.04×10^{-3} 5.27×10^{-3} 4.32×10^{-3}	39.5 38.7 38.1	— — —	673–1 273	IIIb(ii)	162
H	Pd	2.9×10^{-3}	22.2	—	$\sim$230–1 100	Combined data, several sources	319

Permeation of H through Pd very much affected by sample history,
— contamination, etc. *See* references 91–95

Element 1 At.%	Element 2 At.%	A cm^2 s^{-1}	Q kJ mol^{-1}	D cm^2 s^{-1}	Temp. range K	Method	Ref.
H 500 torr	Pt	6×10^{-3}	24.7	($D_H/D_D = 1.16$)	873–1 173	Ic	203
H 0–$\sim$10^{-8}	Si	9.4×10^{-3}	46.0	—	1 363–1 473	Ic	106
H 0.2–4.3	Ta	4.4×10^{-4} 2.0×10^{-6}	13.52[a] 3.86	— —	250–573 91–200	Va Combined data, several sources	80 319
H Sol. soln range	Th	2.92×10^{-3}	40.82	—	573–1 173	IIIa(i) and IIIb(ii)	96

(continued)

Table 13.4   CHEMICAL DIFFUSION COEFFICIENT MEASUREMENTS—*continued*

Element 1 At.%	Element 2 At.%	A cm^2 s^{-1}	Q kJ mol^{-1}	D cm^2 s^{-1}	Temp. range K	Method	Ref.
H	Ti						
	(α)	2.8×10^{-5}	23.85	—	293–500 ⎫		97
		3.3×10^{-2}	57.74	—	500–1 100 ⎭		
	(β)	2.46×10^{-3}	31.58	—	1 160–1 490		317 and 298
H	U						
(n.s.)	(α)	0.0195	46.47	—	663–903	IIIb(ii) ⎫	98
	(β)	—	—	$\begin{cases} 7.3 \times 10^{-5} \\ 1.0 \times 10^{-4} \end{cases}$	$\begin{matrix} 973 \\ 997 \end{matrix}$ ⎭	IIIb(ii)	
(n.s.)	(β)	3.3×10^{-4}	15.07	—	971–1 023	IIIb(ii) ⎫	99
	(γ)	1.5×10^{-3}	47.73	—	1 073–1 243	IIIb(ii) ⎭	
	(γ)	1.9×10^{-2}	48.5	—	1 020–1 250	IIIa(i)	316
H	V						
		3.1×10^{-4}	4.34	—	143–573	Combined data, several sources	319
H 10^{-8}– 600 torr	W	4.1×10^{-3}	37.68	—	1 103–2 493	IIIb(ii)	206

(*a*) True diffusion coefficients. Apparent *D*'s may be much smaller, with higher *Q* and *A*, due to surface effects. *See*, for example, references 87, 176, and 173.

Element 1 At.%	Element 2 At.%	A cm^2 s^{-1}	Q kJ mol^{-1}	D cm^2 s^{-1}	Temp. range K	Method	Ref.
H	Zn						
		$\perp c$ 8.5×10^{-2}	18.62	—	298–344	s.c.	315
H	Zr						
	(α)	4.0×10^{-2}	56.9	—	330–970		314 and 100
40–200 p.p.m.	(α)$^{(a)}$	7.0×10^{-3}	44.59	($D_{\|c} \approx 2D_{\perp c}$)	548–973	IIa(ii)	101
Sol. soln range	(α Zircalloy 2)	2.17×10^{-3}	35.09	—	533–833	IIIa(i)	102
0–41	(β)	5.32×10^{-3}	34.83	(D indep. of c)	1 033–1 283	III(i)	103
	(β)	7.37×10^{-3}	35.76	—	1 143–1 373	IIa(i)	100

(*a*) No significant difference in results for Zircalloy 2 and Zircalloy 4 from those for pure Zr. Values quoted are means of all measurements.

Element 1 At.%	Element 2 At.%	A cm^2 s^{-1}	Q kJ mol^{-1}	D cm^2 s^{-1}	Temp. range K	Method	Ref.
He 0–~6.10^{-9}	Mg	60.0	150.7	—	673–848	IIb(ii)	104
He 0–~4.10^{-10}	Si	0.11	121.4	—	1 443–1 480	Ic	106
He 0–0.13	Th (α)	10^{-3}–10^{-4}	159.1	—	1 173–1 723	IIIb(ii)	107
He 16	Ti (α)	1.1×10^{-9}	67.4	—	888–993	IIb(ii)	117
Hf$^{(a)}$ (α)	N	v. low	2.42×10^{-2}	242.3	—	823–1 173	IIa(ii), p.c.	305
(β)		v. low	8.0×10^{-3}	124.3	—	2 103–2 383	IIa(ii), p.c.	307

(*a*) Hf + 3 wt.% Zr.

Element 1 At.%	Element 2 At.%	A cm^2 s^{-1}	Q kJ mol^{-1}	D cm^2 s^{-1}	Temp. range K	Method	Ref.
Hf (α)	**O** Sol. soln range, in eq. with oxide	0.66	212.7	—	773–1 323	IIc and IIIb(i)	108
(β)	v. low	0.32	171.2	—	2 088–2 403	IIa(ii), p.c.	307

(*continued*)

Table 13.4   CHEMICAL DIFFUSION COEFFICIENT MEASUREMENTS—*continued*

Element 1 At.%	Element 2 At.%	A cm² s⁻¹	Q kJ mol⁻¹	D cm² s⁻¹	Temp. range K	Method	Ref.
Hf	Ti			*(see $10^9\tilde{D}$ table below)*		IIa(i), p.c.	301
Hf, 2–10	W	7.2×10^{-3}	197.6	—	1 823–2 173	IIa(i)	234
Hf, 0–100	Zr	*See Figure 13.35(a) (b)*		—	1 123–1 773	IIa(i)	235
Hg, 0–4	Pb	0.35	79.5	—	450–470	IIa(i)	46
In, 0–9	Ni	$0.027^{[a]}$	206.4	—	1 173–1 373	IIa(i), p.c.	296

For the Hf–Ti system the measured quantity is $10^9\tilde{D}$ at the temperatures shown (K):

Ti At.%	1 273	1 373	1 473	1 573	1 673	1 773	1 873	1 973	2 073	2 173	2 273
10								50	114	120	125
20							55.2	82	149		
30					22.0	33.8	53.6	100			
40				8.2	24.8	40.7	78.3	130			
50		2.6	5.4	8.6	23.2	49.6	82.4				
60	0.8	3.0	6.4	9.7	25.8	49.5	103.0				
70	1.1	3.3	7.1	11.0	29.3	54.8	106.0				
80	1.4	3.7	7.9	11.5	29.8	57.0	112.0				
90	1.6	4.1	8.5	12.7	30.6	55.0	105.0				

(*a*)   *A* estimated from a graph.

Element 1 At.%	Element 2 At.%	A cm² s⁻¹	Q kJ mol⁻¹	D cm² s⁻¹	Temp. range K	Method	Ref.
In, 0–1	Pb	—	—	2.3×10^{-10}	558	IIa(ii)	46
In, 0–3				3.5×10^{-11}	525		
				3.5×10^{-10}	593		
In, 20–60	Pd	*See Figure 13.36*		—	388–446	IIa(i)	245
Ir, 3	W (α)	3.9×10^{4}	$711.8^{[a]}$	—	1 573–2 383	IIa(i)	140
Ir, 24	(σ)	2.4×10^{-4}	254.6				
Ir, 50	(ε)	15.0	504.1				
Ir, 60	(ε)	15.0	504.1				
Ir, 90	(β)	1.1×10^{3}	608.8				

(*a*) Arrhenius plots non-linear. *Q* and *A* calculated from measurements at the highest temperatures.

Element 1 At.%	Element 2 At.%	A cm² s⁻¹	Q kJ mol⁻¹	D cm² s⁻¹	Temp. range K	Method	Ref.
La, Sol. soln range in La	Mg	0.022	102.2		813–871	IIc	210 and 174
La, Sol. soln range	U (γ)	117.0	233.2	—	1 123–1 363	IIc	109
Li, V. small	Si	2.5×10^{-3}	63.2	—	1 073 and 1 623	IIIb(ii)	110
Li $Li_{1+d}Sn$, $Li_{7+d}Sn_3$, $Li_{5+d}Sn_2$, $Li_{13+d}Sn_5$, $Li_{7+d}Sn_2$, $Li_{22+d}Sn_5$	Sn	Extensive graphical data for $\tilde{D}$ as a function of the departure *d* from stoichiometry for all six phases.			688	Electrochem. method	279
Li, n.s., but probably small	W	5.0	173.8	—	1 363–1 503	IIIb(ii)	111
Li, Sol. soln range	Zr	0.73	141.1	—	1 048–1 123	IIa(ii)	143

(*continued*)

Table 13.4 CHEMICAL DIFFUSION COEFFICIENT MEASUREMENTS—*continued*

Element 1 At.%	Element 2 At.%	A cm² s⁻¹	Q kJ mol⁻¹	D cm² s⁻¹	Temp. range K	Method	Ref.
Mg 0–<1	Ni	0.44	234.5	—	1 323–1 573	IIa(ii)	112
Mg 0.26	Pb	—	—	2.5–3.7×10^{-10}	523	IIa(iii)	
1.0		—	—	6.9×10^{-10}			
2.0		—	—	8.6×10^{-10}	523	IIa(i)	
3.0		—	—	1.1×10^{-9}			113
4.1		—	—	6.4–7.8×10^{-10}	523	IIa(iii)	
0.26		—	—	9.4×10^{-10}	543		
		—	—	1.2×10^{-10}	493	IIa(iii)	
1.9		—	—	1.3×10^{-10}	543		

Mg   Pu

At.%	$\tilde{D}$ at $T=420°C$ in units 10^{-11}	$475°$	$534°C$	Method	Ref.
0.01	6.1	25.0	130		
0.56	3.5	11.3	46.3	IIa(i)	118
1.12	3.1	9.3	49.7		
1.7	2.1	13.0	23.5		

Element 1 At.%	Element 2 At.%	A cm² s⁻¹	Q kJ mol⁻¹	D cm² s⁻¹	Temp. range K	Method	Ref.
Mg	U						
0.025		—	—	$\begin{cases} 1.2\times10^{-11} \\ 3.6\times10^{-11} \end{cases}$	400 500	IIa(i)	118
Mn 0–4	Ni	7.5	280.9	—	1 373–1 573	IIa(ii)	23
5	—	In $A =$	~271 [a]				
10	—	$2.25 \times 10^{-2}\tilde{Q}$	~262				
15	—	-4.19	~254	—	1 073–1 323		
20	—		~247				
25	—		~241				

At.%	$10^{11}\tilde{D}$	$10^{11}D_{Mn}$	$10^{11}D_{Ni}$		Method	Ref.
19.7	2.79	4.62	1.40	1 173	IIa(i), p.c.	285
	9.26	10.4	4.83	1 223		
	28.2	32.4	12.2	1 273		
	58.4	66.3	27.5	1 323		

(*a*) Values estimated from a graph.

Element 1 At.%	Element 2 At.%	A cm² s⁻¹	Q kJ mol⁻¹	D cm² s⁻¹	Temp. range K	Method	Ref.
Mn 8	Ti (β)	1.10^{-3}	147.4	—	1 103–1 463	IIa(i) and IIIa(ii)	26

(Very small dep. of $\tilde{D}$ on c in range 2–13%)

Element 1 At.%	Element 2 At.%	A cm² s⁻¹	Q kJ mol⁻¹	D cm² s⁻¹	Temp. range K	Method	Ref.
Mo	N						
	Sol. soln range	4.3×10^{-3}	108.9	—	1 573–2 273	IIIb(ii)	199
	Supersat. sol. soln	3×10^{-3}	115.6	—	1 773–2 273	IIIb(ii)	200
Mo ~0	Nb	1.10^{3}	552.7	$D_{Nb} \sim$			
50		1.10^{3}	573.6	3–$6\times$	2 073–2 438	IIa(i)	115
~100		1.10^{3}	573.6	D_{Mo}			
20		13.5	428.7	—			
40		3.8	413.2	—	1 673–2 648	IIa(i)	211
60		2.1	410.7	—			
80		0.052	345.4	—			
20–80		1.5	399.4	(Average representation of values above)			
Mo 0–0.93	Ni	3.0	288.5	—	1 423–1 673	IIa(ii)	112
0–9		0.853	269.6	—	1 273–1 573	IIa(ii)	47
Mo	O	3×10^{-2}	129.8	—	433	V[a]	58

(continued)

Table 13.4   CHEMICAL DIFFUSION COEFFICIENT MEASUREMENTS—*continued*

Element 1 At.%	Element 2 At.%	A cm^2 s^{-1}	Q kJ mol^{-1}	D cm^2 s^{-1}	Temp. range K	Method	Ref.
Mo	Pd						
	61	5.5×10^{-5}	188.4				
	66	4.0×10^{-5}	165.4	$\left(D_{pd} \approx \atop 10\text{–}20 \times D_{Mo}\right)$ [a]			
	71	5.0×10^{-5}	177.9				
	75	2.4×10^{-4}	200.5	—			
	80	1.6×10^{-3}	218.6	—	1 273–1 873	IIa(i)	175
	85	1.6×10^{-2}	253.3	—			
	90	9.0×10^{-1}	293.1	—			
	95	1.4×10^{-1}	282.6	—			

(*a*) At the original composition in pure Mo/pure Nb couples.

Element 1 At.%	Element 2 At.%	A cm^2 s^{-1}	Q kJ mol^{-1}	D cm^2 s^{-1}	Temp. range K	Method	Ref.
Mo	Ta						
	'Ta rich'	4.68×10^{-5}	251.2	—	2 175–2 573	IIa	190
	'Mo rich'	4.16×10^{-5}	234.5	—			
Mo	Ti						
0		$\sim 2 \times 10^{-2}$	196.8	—			
10		$\sim 2 \times 10^{-2}$	209.3	D_{Ti}/D_{Mo}			
20	(β)	$\sim 1 \times 10^{-2}$	217.7	~ 3 at 1 600°	1 483–1 873	IIa	190
				~ 13 at 820°			
30		$\sim 10^{-2}$	263.8	—			
40[a]		$\sim 10^{-2}$	255.4	—			
0–10	(β)	1.3×10^{-4}	138.6	—	1 173–1 573	IIa(ii)	114
Sol. soln range	(α)	3.5×10^{-8}	118.9	—	873–1 073		

(*a*) Results are reported in reference 115 for the whole composition range (0–100%), but for >40% Mo values vary greatly with type of couple used—incremental or pure metals.

Element 1 At.%	Element 2 At.%	A cm^2 s^{-1}	Q kJ mol^{-1}	D cm^2 s^{-1}		Temp. range K	Method	Ref.
Mo	U							
2		2.2	198.9					
4		9.58	191.8					
6		20.0	221.9					
8		16.0	230.3					
10		237.8	—					
12	(γ)	3.2	218.6	1.123–1.323				
16		0.096	191.3					
20		3.10^{-3}	165.0				IIa(i)	116
24		4.5×10^{-4}	161.2					
26		2.1×10^{-4}	142.4					
				D_*	D_{Mo}			
6.0		—	—	3.4×10^{-9}	5.2×10^{-10}	1 123		
				1.4×10^{-8}	2.1×10^{-9}	1 223		
8.0		—	—	1.6×10^{-8}	5.0×10^{-9}	1 273		
10.0		—	—	3.4×10^{-8}	1.3×10^{-8}	1 323		
Mo	W (wt%)							
	10	4.48	490.7					
	20	2.41	481.1	—				
	30	0.64	458.9	—				
	40	0.48	457.6	—				
	50	0.30	450.9	—		2 273–2 773	IIa(i)	119
	60	0.17	441.3	—				
	70	0.14	438.4	—				
	80	0.08	430.0	—				
	90	0.05	422.0	—				
Mo	Zr							
	0–10	1.6	449.2			1 923–2 108	IIa(ii)	
	0–10	—	—	1.3×10^{-11} to 3.7×10^{-11}		2 108	IIa(i)	115
(Mo$_2$Zr)	$33\frac{1}{3}$	1.10^{-3}	232.8	—		1 093–1 718	IIc	

(*continued*)

Table 13.4 CHEMICAL DIFFUSION COEFFICIENT MEASUREMENTS—*continued*

Element 1 At.%	Element 2 At.%	A $\mathrm{cm^2\,s^{-1}}$	Q $\mathrm{kJ\,mol^{-1}}$	D $\mathrm{cm^2\,s^{-1}}$	Temp. range K	Method	Ref.
N	Nb	6.3×10^{-2}	161.5	—	623–1 873	Combined data, several sources	247
N	Ni	3.0×10^{-6}	95.6	—	423–773	IIa(ii), ion implant	310
N	Re	1.4×10^{-1}	153.7		1 573–2 173	IIIb(ii)	57
N	Ta	5.21×10^{-3}	158.48	—	483–1 673	Combined data, several sources	121
N Sol. soln range	Th (α)	2.1×10^{-3}	94.2	—	1 118–1 763	IIb(i)	122
50–400 p.p.m.	(β)	3.2×10^{-3}	91.2	—	1 723–1 988	IIa(ii)	159
N v. low	Ti (α)	0.21	224.0	—	723–973	IIa(i), p.c.	304
Conc. range at diff. temp	(α)	0.012	176.9	—	1 173–1 843	IIIc(i)	123
Sol. soln range	(β)	0.035	141.5	—	1 173–1 1843	IIIc(i)	
Composition ranges at diff. temps	(α)	0.2	238.6	—	1 623–1 973	IIc	177
	(δ) (TiN)	20	376.8	—			
N	V	1.1×10^{-2}	145.1	—	440–630	Combined data, several sources	247
N 0–300 torr.	W	2.4×10^{-3}	118.9	—	1 673–2 473	IIIb(ii)	199
1–25 torr.		2.37×10^{-3}	150.3	—	1 273–2 073	IIIb(ii)	207
v. low		4.3	224	—	873–1 073	IIa(ii), p.c.	308
N v. low	Zr (α)	0.56	241.4	—	773–973	IIa(ii), p.c.	305
Comp. range at diff. temp.	(α)	0.3	238.6	—	1 623–2 023	IIc	177
Comp. range at. diff. temp.	(α)	0.15	226.5	—	923–1 123	IIIa(i)	178
Sol. soln range	(β)	0.015	128.5	—	1 193–1 913	IIIa(i)	124
Comp. range at diff. temp.	(δ)	0.06	251.2	—	1 623–1 973	IIc	177
N Sol. soln range	Zr Hf (β) 1.8–2.2	0.03	140.7	—	1 173–1 873	IIIa(i)	125
N Sol. soln range	Zr Sn (wt%) (β) 1.8	0.011	131.5	—	1 438–1 913		
	2.6	0.0014	129.4	—	1 373–1 803	IIIa(i)	126
	5.0	0.011	123.1	—	1 373–1 763		
Nb	O	5.86×10^{-3}	109.65	—	296–1 823	Combined data, several sources	121

Table 13.4   CHEMICAL DIFFUSION COEFFICIENT MEASUREMENTS—*continued*

Element 1 At.%	Element 2 At.%	A cm^2 s^{-1}	Q kJ mol^{-1}	D cm^2 s^{-1}	Temp. range K	Method	Ref.
Nb	Ta						
	10	1.1×10^{-2}	343.3	—			
	20	1.34×10^{-2}	351.7	—			
	25	1.0×10^{-2}	343.3	—			
	35	9.3×10^{-2}	347.5	—			
	40	1.0×10^{-2}	351.7	—			
	45	1.56×10^{-2}	360.1	—			
	55	1.25×10^{-2}	364.3	—	2 274–2 653	IIa(i)	246
	60	2.0×10^{-2}	373.6	—			
	65	3.16×10^{-2}	385.2	—			
	70	4.4×10^{-2}	393–6	—			
	75	5.6×10^{-2}	397–7	—			
	90	1.24×10^{-1}	414.5	—			
Nb	Ti$^{(a)}$						
	0	2.5×10^{-3}	293.1				
	20	2.5×10^{-3}	263.8				
	40	3.2×10^{-3}	238.6	$D_{Ti} \sim 2 \times D_{Nb}$	1 273–1 863	IIa(i)	115
	60	3.8×10^{-3}	209.3				
	80	3.8×10^{-3}	184.2				
	100	3.8×10^{-3}	167.5				

(*a*) Values taken from smoothed plots of A and Q against composition.

Nb	U(γ)$^{(a)}$						
	2	2.8×10^7	623.4		1 773–1 923		
	12	2.3×10^7	603.7	—			
	18	9.6×10^6	586.2				
	22	0.091	307.7		1 673–1 873		
	28	0.113	305.2	—			
	38	0.149	304.8	—	1 573–1 773		
	46	0.064	284.7	—	1 423–1 673		
	54	0.45	292.7	—	1 423–1 623		
	62	0.84	286.8		1 348–1 573	IIIa(i)	127
	68	1.94	291.8	—			
	74	0.82	252.9		1 223–1 448		
	78	1.16	252.5	—			
	82	1.19×10^{-4}	139.8	—	1 165–1 398		
	93	1.63×10^{-4}	126.4		966–1 298		
	97	2.31×10^{-4}	125.2	—			
	97	D_U— 124.8 3.82×10^{-3} D_{Nb}— 1 645 7.1×10^{-3}		—	966–1 298		
	4	—	—	$D_{Nb} \sim 30 \times D_U$			
	10–100	—	—	$D_U > D_{Nb}$			

(*a*) This is a representative selection from a larger table of values in reference 127.

Nb	V						
	0	1.6×10^{-2}	910.3				
	20	1.95×10^{-2}	343.3				
	40	2.3×10^{-2}	293.1	$D_V \sim 3$–$5 \times D_{Nb}$	1 678–2 023	IIa(i)	150
	60	2.8×10^{-2}	268.0				
	80	3.3×10^{-2}	263.8				
	100	3.8×10^{-2}	263.8				
Nb	W (wt %)						
	10	81.45	439.6	—			
	20	22.2	418.7	—			
	30	1.97	376.4	—			
	40	1.4×10^{-2}	280.1	—			
	50	7.4×10^{-3}	272.1	—	2 273–2 673	IIa(i)	119
	60	3×10^{-3}	252.0	—			
	70	1.8×10^{-3}	289.3	—			
	80	1.0×10^{-3}	236.1	—			
	90	6.0×10^{-4}	228.2	—			

(*continued*)

Table 13.4   CHEMICAL DIFFUSION COEFFICIENT MEASUREMENTS—*continued*

Element 1 At.%	*Element 2* At.%	*A* cm^2 s^{-1}	*Q* kJ mol^{-1}	*D* cm^2 s^{-1}	*Temp. range* K	*Method*	*Ref.*
Nb	Zr						
5		—	217.7 ⎫	—			
30		—	247.0 ⎬	—	1 173–1 873	IIa	179
95		—	332.9 ⎭	—			
0		~10^{-2}	~196.8 ⎫	—			
20		~4 × 10^{-2}	~209.3 ⎪	—			
40		~10^{-1}	~255.4 ⎬	—			
60	*(a)*	~3 × 10^{-1}	301.4 ⎪	—	1 718–1 963	IIa(i)	115
80		~2	~349.5 ⎪	—			
100		~10	~389.4 ⎭	—			
—	0–5	1.2	349.26	—	2 070–2 370	IIIa(ii), p.c.	269

(a) Values taken from smoothed plots of *A* and *Q* against composition.

Ni	O					Internal	
	(a)	7.9×10^4	309.4	(s.c.)	1 073–1 473 ⎫	oxidation	201
	(b)	9.5×10^4	311.5	(p.c.)	1 173–1 573 ⎭	method	
		12.1	241	—	623–1 273	IIIb(ii), p.c.	202

$^{(a)}$Ni + 0.58%Si. s.c.
$^{(b)}$Ni + 0.48%Si. p.c.

| Ni | Pb | | | | | | |
| 0–3 | | ~0.66 | 105.9 | — | 558–593 | II | 46 |

| Ni | Pb | | | | | | |

		859°C	$\tilde{D} \times 10^{11}$ 950°C	1 019°C	1 150°C		
10		0.68	1.30	1.79	18.6 ⎫		
20		1.03	2.01	3.37	30.0 ⎪		
30		1.52	3.54	7.39	70.4 ⎪		
40		1.72	5.37	11.7	98.2 ⎪		
50		1.97	6.61	16.1	1.58 ⎬	IIa(i)	144
60		1.40	5.42	15.6	131 ⎪		
70		0.84	3.06	11.3	104 ⎪		
80		0.56	1.90	5.58	58.7 ⎪		
90		0.50	1.25	2.30	35.0 ⎭		

Ni	Pd						
15.4		—	0.86	233.8	—		
23.1		—	1.55	234.2	—		
32.2		—	0.55	217.4	—		
42.2		—	0.38	209.3	—	900–1 200	II (EPMA)	320
52.1		—	0.42	209.9	—		
62.1		—	0.18	202.7	—		
73.5		—	0.12	204.0	—		
86.6		—	0.11	212.1	—		

Ni	Pt						
0–14.9		7.9 × 10^{-4}	180.5	—	1 316–1 674	IIa(ii)	28
	0–100		Figure 13.32		1 223–1 573	IIa(i)	218

Ni	Si						
0–< 1		1.5	258.3	—	1 393–1 573	IIa(ii)	112
wt. %	2.3	0.3$^{(a)}$	240	—	1 373–1 523	IIa(i), p.c.	311

Ni	Sn						
—	0	2.5	260 ⎫				
—	1	2.5	259 ⎪				
—	2	2.3	257 ⎪				
—	3	2.3	255 ⎬	—	1 223–1 473 ⎫		
—	4	1.9	251 ⎪				
—	5	2.5	253 ⎪		⎪		
—	6	2.4	251 ⎪		⎪		
—	7	2.3	250 ⎭		⎬	IIa(i), p.c.	261
		10$^{14}D_{Ni}$	10$^{14}D_{Sn}$		⎪		
—	0.8	1.9	3.4 ⎫		⎪		
—	2.2	2.2	3.7 ⎪				
—	2.2	1.8	4.2 ⎬	—	1373		
—	4.5	3.0	6.0 ⎭				

Ni	Ti						
0–0.9		0.86	257.1	—	1 373–1 573	IIa(ii)	23

(continued)

Table 13.4   CHEMICAL DIFFUSION COEFFICIENT MEASUREMENTS—*continued*

Element 1 At.%	Element 2 At.%	A cm² s⁻¹	Q kJ mol⁻¹	D cm² s⁻¹	Temp. range K	Method	Ref.
Ni	Ta						
Ta₂Ni		2.6×10^{-3}	230.7	—			
TaNi		2.1	306.9	—			
TaNi₂		0.1	250.4	—	1 423–1 573	II	236
TaNi₃		1.7×10^{-5}	133.6	—			
TaNi₈		0.9×10^{-2}	334.1	—			
Ni	U						
0–sol. limit	(γ)	2 500	192.6	—	1 123–1 273	IIc	54
Ni	V						
	0–16.5	0.287	247.9	—	1 373–1 573	IIa(ii)	47
Ni	W						
	0–1.5	11.1	321.5	—	1 423–1 563	IIa(ii)	23
	0–5	0.86	294.8	—	1 373–1 573	IIa(ii)	47
	1	2.24	303.0	—			
	2	2.16	303.5	—			
	3	2.11	303.8	—			
	4	2.07	304.2	—			
	5	2.04	304.5	—			
	6	2.01	304.8	—			
	7	1.98	305.2	—	1 273–1 589	IIa(i)	180
	8	1.95	305.6	—			
	9	1.94	805.9	—			
	10	1.92	306.3	—			
	11	1.90	306.7	—			
	12	1.89	367.2	—			
Ni	Zn						
5		$1.05 \times 10^3 \times \exp(-0.142C_{Ni})$	180.0	—	873–1 273	IIa(i)	181
95							
		A_{Ni} 8.3×10^{-2}	Q_{Ni} 182.1				
		A_{Zn} 0.176	Q_{Zn} 203.1				
δ		7.1×10^{-2}	85.0	—			
γ'		1.2×10^{-1}	90.9	—	483–873	II	237
γ''		3.0×10^{-2}	96.7[a]	—			
—	5		~265				
—	10	$\ln A =$	~252				
—	15	$2.4 \times 10^{-2}\tilde{Q}$	~242	—	1 073–1 323	IIIa(i)	286
—	20	-5.6	~236				
—	25		~219				
—	30		~208				

(*a*) Values estimated from a graph.

O	Pt						
Sol. soln range		9.3	326.6	—	1 708–1 777	Ic	204
O	Ta						
0–1.13		1.05×10^{-2}	110.43	—	298–1 673	Combined data, several sources	121
O	Th						
25–220 p.p.m.	(β)	1.3×10^{-3}	46.1	—	1 713–1 973	IIa(ii)	159
	(α)	1.3×10^2	209.2	—	1 273–1 473	IIa(ii)	248
O	Ti						
α-sol. soln		0.45	201	—	573–1 223	IIIa(i), p.c.	130
β-sol. soln		0.14	138.2	—	1 023–1 623	IIIa(i) and IIa(ii)	123 and 287

(*continued*)

Table 13.4 CHEMICAL DIFFUSION COEFFICIENT MEASUREMENTS—*continued*

Element 1 At.%	Element 2 At.%	A cm^2 s^{-1}	Q kJ mol^{-1}	D cm^2 s^{-1}	Temp. range K	Method	Ref.
O	**V**						
		2.46×10^{-2}	123.5	—	333–2 098	Various	247
O	**Zr**						
Sol. soln range	(α)	1.32	201.8	—	563–1 773 ⎤		131
	(α)	0.0661	184.2	—	563–923 ⎬ Various		
	(α)	16.5	229.0	—	923–1 773 ⎦		
Sol. soln range	(β)	0.977	171.7	—	1 322–1 473	III	183
	(β)	2.63×10^{-2}	118.1	—	1 273–1 573	IIa(ii) and IIc	306
Sol. soln range	(α) (Zircalloy)	0.196	171.7	—	1 273–1 773	IIIc(i)	132
Sol. soln range	(β) (Zircalloy)	0.045 3	118.1	—	1 273–1 773	IIIa(i)	132

(*a*) Reference 131 reviews all published data. Quoted A's and Q's are 'best mean values'. Data slightly better represented by two Arrhenius expressions, above and below 650°C.

Pb	**Sn**						
0–2		4.0	99.6	—	518–573	IIa(ii)	10
Sol. soln range			100.5	$\tilde{D}$ increases with conc. of Sn	443 and 454	IIa(i)	135
—	0	—	—	3×10^{-11} ⎤			
—	5	—	—	9.5×10^{-1} ⎬	523	IIa(i), p.c.	284
—	10	—	—	2×10^{-10} ⎦			
Pb	**Tl**						
0–2		0.025	81.2	—	493–558	IIa(ii)	29
0.53		1.03	103.0	Almost independent of conc.	533–588	IIa(i)	113
Pd	**Pt**						
0	—	$\sim4 \times 10^{-3}$	~54 ⎤				
10	—	$\sim4 \times 10^{-3}$	~54				
20	—	$\sim5 \times 10^{-3}$	~54.5				
30	—	$\sim5 \times 10^{-3}$	~55				
40	—	$\sim6 \times 10^{-3}$	~56				
50	— (*a*)	$\sim8 \times 10^{-3}$	~57 ⎬	—	1 335–1 676	IIa(i), p.c.	290
60	—	$\sim1.0 \times 10^{-2}$	~58				
70	—	$\sim1.3 \times 10^{-2}$	~59				
80	—	$\sim2.5 \times 10^{-2}$	~61				
90	—	$\sim5.0 \times 10^{-2}$	~64				
100	—	$\sim9.5 \times 10^{-2}$	~67 ⎦				

(*a*) Rough values estimated from graphs.

Pd	**Ti**						
	(β)	1.26×10^{-3}	131.9	—	973–1 273 ⎤		
	(γ)	1.6×10^{-8}	44.8	— ⎤			
	(δ)	3.6×10^{-4}	129.4	— ⎬	973–1 273	IIa(i)	238
	(ε)	1.6×10^{-6}	84.2	—			
	(η)	6.4×10^{-4}	745.3	— ⎦			
				$D_{Ti} = 3.54 \times 10^{-9}$	1 173		
				$D_{Pd} = 1.32 \times 10^{-9}$	1 173		
				D_{Ti}/D_{Pd}	1 073 ⎦		
Pt	**W**						
2	(β)	3.1×10^2	582.0 ⎤		1 573–2 016 ⎤		
50 ⎤	(γ)(*a*)	4.7×10^{-3}	350.0 ⎬	—			
55 ⎦		3.3×10^{-3}	343.7 ⎦				
65	(ε)(*a*)	4.4×10^{-2}	385.2	—	1 746–2 016 ⎬	IIa(i)	140
77 ⎤		1.8×10^{-2}	326.6 ⎤	—	1 573–2 016		
80 ⎬	(α)	1.2×10^{-2}	315.7 ⎦				
85 ⎦		1.3×10^{-2}	310.7		1 573–1 973 ⎦		

(*a*) The γ and ε are two new phases observed during the diffusion experiments and not previously reported.
(*b*) Arrhenius plots not always linear. Q and A derived from measurements at higher temperatures.

(*continued*)

Table 13.4　CHEMICAL DIFFUSION COEFFICIENT MEASUREMENTS—*continued*

Element 1 At.%	Element 2 At.%	A cm² s⁻¹	Q kJ mol⁻¹	D cm² s⁻¹	Temp. range K	Method	Ref.
Pu	**Ti**						
2	(β)	9.4×10^{-4}	123.9	—	1 173–1 373	IIa(ii)	217
15	(β)	2.3×10^{-3}	127.7				
Pu	**U**						
1.75		0.14×10^{-7}	56.1				
3.50		0.15×10^{-7}	57.4				
5.25		0.18×10^{-7}	59.0	Probably a			
7.0		0.28×10^{-7}	63.6	significant			
8.75	(α)	0.44×10^{-7}	68.2	contribution to	683–813	IIa(i)	134
10.50		0.88×10^{-7}	74.9	$\tilde{D}$ from g.b.			
12.25		1.18×10^{-7}	78.7	diffusion			
14.0		2.0×10^{-7}	83.7				
15.75		2.57×10^{-7}	86.2				
Pu	**Zr**						
20		7×10^{-1}	184.2	—	1 023–1 173		
30		1×10^{-2}	144.4	—	973–1 123		
40	($\varepsilon\beta$)	1.5×10^{-3}	119.3	—	IIa(i)		189
50		2.5×10^{-4}	98.4	—	973–1 143		
60		9×10^{-5}	77.5	—	973–1 143		
(δ)	4.2–11.4	5.89×10^{-6}	83.7	—	624–748	IIa	188
0.115		0.1	226.1	—			
	(α)			—	973–1 073	IIa(ii)	192
1.15		11.1	272.1				

		A_{Zr}	Q_{Zr}	A_{Pu}	Q_{Pu}		
20		8×10^{-1}	188.4	6×10^{-1}	184.2	Same T	
30	($\varepsilon\beta$)	7.5	205.2	4×10^{-1}	175.8	range as	189
40		2×10^{-1}	167.5	1×10^{-3}	113.0	for $\tilde{D}$	
50		1.5×10^{-4}	121.4	3×10^{-2}	117.2		

Element 1 At.%	Element 2 At.%	A cm² s⁻¹	Q kJ mol⁻¹	D cm² s⁻¹	Temp. range K	Method	Ref.
Rh	**W**						
3	(α)	1.3×10^{-6}	242.8				
60		1.5×10^{-6}	174.6				
70	(ε)	3.1×10^{-6}	181.7	—	1 573–2 073	IIa(i)	140
90	(β)	2.5×10^{-6}	174.2				
Ru	**W**		(a)				
5	(α)	5.5×10^{-3}	391.5	—	1 573–2 298		
39	(σ)	1.2×10^{-5}	255.4	—	2 058–2 298		
70		1.8×10^{-5}	207.2	—			
90	(β)	1.0×10^{-5}	239.5	—	1 573–2 298		

(a) Arrhenius plots not always linear. Q and A derived from measurements at highest temperatures.

Element 1 At.%	Element 2 At.%	A cm² s⁻¹	Q kJ mol⁻¹	D cm² s⁻¹	Temp. range K	Method	Ref.
S	**Ni**						
~0.01		2.3×10^{6}	376.8	—	1 273–1 473	IIIa(ii)	142
Si	**U**						
Sol. soln range	(γ)	20	188.4	—	1 123–1 323	IIc	54
Sn	**Ti**						
1.0	(β)	8.4×10^{-7}	64.1	Increases	1 273–1 523		
8.0		2.7×10^{-4}	124.8	linearly	1 363–1 523		
				with C		IIa(i)	26
2.0	(β)	—	—	$D_{Sn} = 9.18 \times 10^{-9}$ $D_{Ti} = 2.65 \times 10^{-9}$	1 523		
Sn	**Zr**						
Sol. soln range	(α)	3.10^{-4}	92.1	—	873–1 123	IIa(ii)	
0–3.9	(β)	6.9×10^{-4}	150.7	—	1 373–1 573	IIa(ii)	136
0–5		0.07	212	—	1 605–1 970	IIIa(ii), p.c.	271
Sr	**U**						
Sol. soln	(γ)	2.38×10^{-3}	196.8	—	1 073–1 273	IIc	109

(*continued*)

Table 13.4   CHEMICAL DIFFUSION COEFFICIENT MEASUREMENTS—*continued*

Element 1 At.%	Element 2 At.%	A cm² s⁻¹	Q kJ mol⁻¹	D cm² s⁻¹	Temp. range K	Method	Ref.
TA	W						
	'Ta rich'	1.78	498.2	—	2 373–2 773	IIa	190
	'W rich'	4.16×10^{-2}	418.7	—			
—	20–80	1.0	546.4				
30	70	$A_{Ta} = 1.8$ $A_W = 0.17$	$Q_{Ta} = 553.9$ $Q_W = 510.8$	—	1 573–2 373	IIa(ii), p.c.	259
Ti	U						
10.0		11.10^{-3}	153.2				
20.0		1.4×10^{-3}	138.2				
30.0		1.6×10^{-3}	145.7				
40.0		4.0×10^{-3}	160.8				
50.0	(γ)	9.5×10^{-3}	175.8				
60.0		2.6×10^{-3}	165.0	—	1 223–1 348		
70.0		2.6×10^{-3}	165.0				
80.0		2.2×10^{-3}	157.0			IIa(i)	137
90.0		1.1×10^{-3}	141.5				
95.0		0.46×10^{-3}	126.4				
				D_{Ti} D_U			
165.5		—		5.8×10^{-9} 2.2×10^{-8}	1 348		
				1.2×10^{-9} 4.7×10^{-8}	1 223		
18.0		—		2.9×10^{-9} 9.5×10^{-9}	1 273		
				4.1×10^{-9} 1.6×10^{-8}	1 323		
16.5–18		$Q_U = 161.2$	$Q_{Ti} = 167.5$				
Ti	V						
(α)	Sol. soln range	—	—	3.91×10^{-15}	873	IIa(ii)	114
				4.7×10^{-15}	973	IIa(ii)	
(β)	0–10	1.25×10^{-2}	173.3	—	1 173–1 573		
(β)	2.0	6.0×10^{-3}	165.8	Dep. on c in range 2–12% v. slight	1 173–1 523	IIa(i)	26
(β)	3.5	—	—	$D_{Ti} = 1.31 \times 10^{-9}$ $D_V = 14.9 \times 10^{-9}$	1 523		
10		8.3×10^{-4}	197.6	—			
20		1.5×10^{-3}	199.3	—			
30		4.4×10^{-3}	203.9	—			
40		1.3×10^{-2}	206.8	—			
50		2.4×10^{-2}	203.5	—	923–1 323	IIa(i)	240
60		1.1×10^{-2}	186.2	—			
70		8.1×10^{-4}	153.2	—			
80		4.1×10^{-4}	139.8	—			
90		1.6×10^{-4}	123.9	—			
—	20	1.6×10^{2}	191.3				
—	40	4.8×10^{-2}	212.7				
—	60	0.8	261.7	—	1 173–1 573		289
—	80	8.2	295.2				
Ti	Zr						
0–4	—	2.69	261	—	1 700–1 953	IIIa(ii), p.c.	271
(α)	0–10	1.7×10^{-12}	49.4	—	873–1 073	IIa(ii)	114
(β)	0–10	1.8×10^{-2}	167.9	—	1 173–1 573	IIa(ii)	114
10		1.4×10^{-2}	164.5	—	1 103–1 323		
25		3.3×10^{-3}	146.5	1 103–1 323			
		5×10^{-7}	65.7	—	923–1 103		
40		2.7×10^{-3}	142.8	—	1 103–1 323		
		1.2×10^{-6}	71.6	—	923–1 103		
50		2.4×10^{-3}	140.7	—	1 103–1 323		
		1.7×10^{-6}	74.1	—	923–1 103	IIa(i)	239
65		1.6×10^{-3}	136.5	—	1 103–1 323		
		2.0×10^{-6}	18.0	75.4	—		
80		1.5×10^{-3}	134.8	—	1 103–1 323		
		2.2×10^{-6}	75.4	—	923–1 103		
90		1.3×10^{-3}	131.9	—	1 103–1 323		
50.5				$D_{Zr} = 5.1 \times 10^{-10}$ $D_{Ti} = 3.2 \times 10^{-10}$	1 073		

(continued)

Table 13.4   CHEMICAL DIFFUSION COEFFICIENT MEASUREMENTS—*continued*

Element 1 At.%	Element 2 At.%	A $cm^2\,s^{-1}$	Q $kJ\,mol^{-1}$	D $cm^2\,s^{-1}$	Temp. range K	Method	Ref.
U 'low'	W	1.8×10^{-2}	389.4	—	2 243–3 003	IIIb	241
U (β)	Xe 0–10^{-6}	9×10^{-7}	96.3	—	973–1 323		
(γ)	0–10^{-6}	10^8	410.3	—	1 083–1 333	IIIb(ii)	149
U	Zr						
	10	9.5×10^{-4}	134.0				
	20	1.3×10^{-4}	119.7				
	30	0.35×10^{-4}	110.1				
	40	0.4×10^{-4}	114.7				
	50	0.8×10^{-4}	124.3				
(γ)	60	0.63×10^{-4}	124.3	—	1 223–1 348	IIa(i)	138
	70	0.55×10^{-4}	124.3				
	80	3.2×10^{-4}	143.6				
	90	78×10^{-4}	171.7				
	95	870×10^{-4}	196.8				
	10–95			D_U and D_{Zr} Figures 13.33 and 13.34	1 123–1 313	IIa(i)	139

REFERENCES TO TABLE 13.4

1. A. D. Le Claire and A. H. Rowe, *Rev. Métall.*, 1955, **52**, 94.
2. Th. Heumann and S. Dittrich, *Z. Electrochem*, 1957, **61**, 1138.
3. H. Ebert and G. Trommsdorf, *ibid.*, 1950, **54**, 294.
4. R. W. Baluffi and L. L. Seigle, *J. Appl. Phys.*, 1954, **25**, 607.
5. W. A. Johnson, *Trans. AIME*, 1942, **147**, 331.
6. O. Kubaschewski, *Trans. Faraday Soc.*, 1950, **46**, 713.
7. W. Eichenauer, H. Kunzi and A. Pebler, *Z, Metallk.*, 1958, **49**, 220.
8. J. M. Tobin, *Acta, Metall.*, 1957, **5**, 398.
9. W. Eichenauer and G. Müller, *Z. Metallk.*, 1962, **53**, 321; 1962, **53**, 700.
10. W. Seith and J. G. Laird, *ibid.*, 1932, **24**, 193.
11. T. Heumann and P. Lohmann, *Z. Electrochem.*, 1955, **59**, 849.
12. A. G. Guy, *Trans. Metall. Soc., AIME*, 1959, **215**, 279.
13. J. M. Tobin, *Acta Metall.*, 1959, **7**, 7101.
14. H. Buckle and J. Descamps, *Rev. Métall.*, 1951, **48**, 569.
15. J. B. Murphy, *Acta Metall.*, **9**, 563.
16. H. Oikawa and A. Hosoi, *Scr. Met.*, 1975, **9**, 823.
17. M. K. Asundia and D. R. F. West, *J. Inst. Met.*, 1964, **92**, 428.
18. K. Hirano and A. Hishunima, *J. Jap. Inst. Met.*, 1968, **32**, 516.
19. W. Eichenauer, K. Hattenbach and A. Pebler, *Z. Metallk.*, 1961, **52**, 682.
20. L. P. Costas, *USA Rep.*, TID-16676, 1962.
21. H. Bückle, *Z. Electrochem.*, 1943, **49**, 238.
22. C. E. Ransley and H. Neufeld, *J. Inst. Met.*, 1950, **78**, 25.
23. R. A. Swalin and A. Martin, *J. Met. Trans. AIME*, 1956, **206**, 567.
24. S. Fujikawa, K. Hirano and Y. Fukushima, *Met. Trans. A*, 1978, **9A**, 1811.
25. R. F. Mehl, F. N. Rhines and K. A. von den Steiner, *Met. Alloys*, 1941, **13**, 41.
26. D. Goold, *J. Inst. Met.*, 1960, **88**, 444.
27. J. E. Hilliard, B. L. Averbach and M. Cohen, *Acta Metall.*, 1959, **7**, 86.
28. O. Kubaschewski and H. Ebert, *Z. Electrochem*, 1944, **50**, 138.
29. V. A. Kurakin, A. A. Kurdyumov, V. N. Lyasnikov and M. I. Potapov, *Sov. Phys. Sol. St.*, 1979, **21**, (4), 616.
    . V. B. Denim, V. B. Vykhodets and P. V. Gel'd, *Phys. Met. Metallog.*, 1973, **35**, (4), 84.
30. G. W. Powell and J. D. Braun, *Trans. Metall, Soc. AIME*, 1964, **230**, 694.
31. J. E. Reynolds, B. L. Averbach and M. Cohen, *Acta Metall.*, 1957, **5**, 29.
32. W. Seith and K. Etzold, *Z. Electrochem.*, 1934, **40**, 829; 1935, **41**, 122.
33. W. Jost, *Z. Phys. Chem.*, 1933, **B21**, 158.
34. A. Bolk, *Acta Metall.*, 1961, **9**, 643.
35. P. E. Busby and C. Wells, *J. Met.*, 1954, **6**, 972.
36. P. E. Busby, M. E. Warga and C. Wells, *ibid.*, 1953, **5**, 1463.
37. R. Reinbach and F. Krietsh, *Z. Metallk.*, 1963, **54**, 173.
38. R. Le Hazif, G. Donze, J. M. Dupouy and Y. Adda, *Mem. Sci Rev. Met.*, 1964, **LXI**, 467.
39. W. Seith and F. G. Laird, *Z. Metallk.*, 1932, **24**, 193.
40. J. R. G. da Silva and R. B. McLellan, *Mater. Sci. Eng.*, 1976, **26**, 83.

41. R. P. Smith, *Trans. metall. Soc. AIME*, 1964, **230**, 476.
42. J. Agren, *Scripta Met.*, 1986, **20**, 1507.
43. I. I. Kovenskiy, *Fiz. Met. Metalloved.*, 1963, **16**, 613.
44. D. T. Peterson, *Trans. Am. Soc. Met.*, 1961, **53**, 765.
45. F. C. Wagner, E. J. Burcur and M. A. Steinberg, *ibid.*, 1956, **48**, 742.
46. W. Seith, E. Hofer and H. Etzold, *Z. Electrochem.*, 1934, **40**, 332.
47. A. Davin, V. Leroy, D. Coutsouradis and L. Habraken, *Rev. Metall.*, 1963, **60**, 275; *Cobalt*, June 1963, **19**.
48. J. W. Weeton, *Trans. Am. Soc. Met.*, 1952, **44**, 436.
49. T. Heumann and H. Bohmer, *Arch. Eisenhütt Wes.*, 1960, **31**, 749.
50. H. W. Paxton and E. J. Pasierb, *Trans. Metall. Soc. AIME*, 1960, **218**, 794.
51. D. B. Butrymowicz and J. R. Manning, *Met. Trans. A*, 1978, **9**, 947.
52. J. P. Pivot, A. Van Craeynest and D. Calais, *J. Nucl. Mater.*, 1969, **31**, 342.
53. R. F. Peart and D. H. Tomlin, *J. Phys. Chem. Solids*, 1962, **23**, 1169.
54. M. Mossé, V. Levy and Y. Adda, *C. R. Acad., Sci. Paris*, 1960, **250**, 3171.
55. G. Brunel, G. Cizeron and P. Lacombe, *C. R. Hebd. Séanc. Acad. Sci., Paris*, 1969, **269C**, 895.
56. R. P. Smith, *Trans. Met. Soc. AIME*, 1966, **236**, 1224.
57. H. Jehn, K. Hahloch and E. Fromm, *J. Less-common Met.*, 1972, **27**, 98.
58. V. I. Baranova *et al.*, *Fiz Khim. Obrab. Mater*, 1968, **2**, 61.
59. Y. Minamino, T. Yamane, T. Kimura and T. Takahashi, *J. Mater. Sci. Letters*, 1988, **7**, 365.
60. E. Starke and H. Wever, *Z. Metallk.*, 1964, **55**, 107.
61. G. T. Horne and R. F. Mehl. *Trans. Am. Inst. Min. Engrs.*, 1955, **203**, 88.
62. R. Resnick and R. W. Balluffi, *ibid.*, 1955, **203**, 1004.
63. U. S. Landergren, C. E. Birchenall and R. F. Mehl, *ibid.*, 1956, **206**, 73.
64. R. F. Mehl and C. F. Lutz, *Trans. Metall. Soc. AIME*, 1961, **221**, 561.
65. H. Bester and K. W. Lange, *Arch. Eisenhütt.*, 1976, **47**, 333.
66. H. Bester and K. W. Lange, *Arch. Eisenhütt.*, 1972, **43**, 207.
67. C. Wells and R. F. Mehl, *Trans. Am. Inst. Min. Engrs.*, 1941, **145**, 315.
68. A. E. Lord and D. N. Beshers, *Acta Met.*, 1966, **14**, 1659.
69. J. D. Fast and M. B. Verrijp, *J. Iron S. Inst.*, 1954, **176**, 24.
70. P. Grieveson and E. T. Turkdogan, *Trans. Metall. Soc. AIME*, 1964, **230**, 411.
71. L. S. Darken, R. P. Smith and E. W. Filer, *Trans. Am. Inst. Min. Engrs.*, 1951, **191**, 1174.
72. C. Wells and R. F. Mehl, *ibid.*, 1941, **145**, 129.
73. N. G. Ainslie and A. E. Seybolt, *J. Iron St. Inst.*, 1960, **194**, 341.
74. G. Seibel, *C. R. Acad. Sci. Paris*, 1962, **255**, 3182; *Mem. Sci. Rev. Métall.*, 1964, **61**, 413.
75. W. Baltz, H. W. Mead and C. E. Birchenall, *Trans. Metall. Soc. AIME*, 1952, **194**, 1070.
76. K. Hirano and Y. Ipposhi, *J. Jap. Inst. Met.*, 1968, **32**, 815.
77. S. H. Moll and R. E. Ogilvie, *Trans. Metall. Soc. AIME*, 1959, **215**, 613.
78. R. E. Hannemann, R. E. Ogilvie and H. C. Gates, *Trans. Metall. Soc. AIME*, 1965, **233**, 691.
79. G. Seibel, *C. R. Acad. Sci. Paris*, 1963, **256**, 4661; *Mem. Sci. Rév. Met.*, 1964, **61**, 413.
80. G. Schaumann, J. Völkl and G. Alefeld, *Phys. Stat. Sol.* 1970, **42**, 401.
81. T. Yamamoto, T. Takashima and K. Nishida, *J. Jap. Inst. Met.*, 1981, **45**, 985.
82. J. A. M. van Liempt, *Rec. Trav. Chim. Pays Bas*, 1945, **64**, 239.
83. G. Grube and K. Schneider, *Z. Anorg. Chem.*, 1927, **168**, 17.
84. R. C. Frank and J. E. Thomas, *J. Phys. Chem. Solids*, 1960, **16**, 144.
85. J. Tournier, Rep. CEA-R-2446, October 1964.
86. P. M. S. Jones and R. Gibson, *Rept AWRE*, 0-2/67, 1967.
87. W. M. Albrecht, W. D. Goode and M. W. Mallet, *J. Electrochem. Soc.*, 1959, **106**, 981.
88. M. L. Hill and E. W. Johnson, *Acta Metall.*, 1955, **3**, 566.
89. J. L. Ham, *Trans. Am Soc. Metals*, 1945, **35**, 331.
90. W. M. Robertson, *Z. F. Metallk.*, 1973, **64**, 436.
91. M. van Sway and C. E. Birchenall, *Trans. Metall. Soc. AIME*, 1960, **218**, 285.
92. W. D. Davis, *US Rep*. K.A.P. L. 1227, October, 1954.
93. O. M. Katz and E. A. Gulbransen, *Rev, Sci. Inst.*, 1960, **31**, 615.
94. W. D. Davis, *US Rep*. K.A.P. L. 1375, April 1955.
95. O. N. Salmon, D. Randall and E. A. Wilk, K.A.P. L., 1674, November 1956; K.A.P. L. 984, May 1954.
96. D. T. Peterson and D. G. Westlake, *J. Phys. Chem.*, 1960, **64**, 649.
97. O. P. Nazimov and L. N. Zhuravlev, *Izv. V.U.Z. Tsvetn. Met.*, No. 1, 1976, 160.
98. M. W. Mallet and M. J. Trzeciak, *Trans. Am. Soc. Met.*, 1958, **50**, 981.
99. H. W. Meyers, J. W. Varwig, J. L. Marshall, L. G. Weber and J. E. Kenelley, *U.S.A.E.C. Rep*. MCW-1439, December 1959.
100. M. Someno, *J. Jap. Inst. Met.*, 1960, **24**, 249.
101. J. J. Kearns, *J. Nucl. Mater.*, 1972, **43**, 330.
102. A. Sawatzky, *J. Nucl. Mater.*, 1960, **2**, 62.
103. V. L. Gelezunas, *J. Electrochem. Soc.*, 1963, **110**, 779.
104. H. R. Glyde, *Phil. Mag.*, 1965, **12**, 919.
105. H. R. Glyde, *J. Nucl. Mater.*, 1967, **23**, 75.
106. A. van Wieringen and N. Warmoltz, *Physica*, 1956, 22, 849.
107. A. Andrew, C. R. Davidson and L. E. Glasgow, *US Rep*. NAA-SR-1598, 1956.
108. J. P. Pemsler, *J. Electrochem. Soc.*, 1964, **111**, 1185.
109. Y. Adda, V. Levy, Z. Hadari and J. Tournier, *Rev. Métall.*, 1959, **57**, 278.
110. E. M. Pell. *Phys. Rev.*, 1960, **119**, 1014.
111. H. M. Love and G. M. McCracken, *Can. J. Phys.*, 1963, **41**, 83.
112. R. A. Swalin, A. Martin and R. Olsen, *Trans. Am Inst. Min. Engrs.*, 1957, **209**, 936.

113. W. Seith and J. Herrmann, *Z. Electrochem.*, 1940, **46**, 213.
114. R. P. Elliot, *US Rep.* AD. 290336, March 1962.
115. C. S. Hartley, J. E. Steedly and L. D. Parsons, *US Rep.* ML-TDR-64-316, December 1964, and 'Diffusion in B.C.C. Metals', *Am. Soc. Met*, 1965, p. 35.
116. Y. Adda and J. Philibert, *C. R. Acad. Sci. Paris*, 1958, **246**, 113; Rep C.E.A.-880, March 1958.
117. A. M. Rodin and V. V. Surenyants, *Phys. Met. Metallogr.*, 1960, **10**, (2), 58.
118. D. Calais, M. Beyeler, M. Mouchnino, A. van Craeynest and Y. Adda, *C. R. Hebd. Séanc. Acad. Sci. Paris*, 1963, **257**, 1285.
119. E. P. Nechiporenko *et al.*, *Fiz. Met. Metalloved.*, 1971, **32**, 89.
120. P. Gröbner, *Hutnické listy*, 1955, **10**, 200.
121. F. J. M. Baratto and R. E. Reed-Hill, *Mat. Sci. Eng.*, 1980, **43**, 97.
122. A. F. Gerds and M. W. Mallett, *J. Electrochem. Soc.*, 1954, **101**, 175.
123. I. F. Sokirianskii, D. V. Ignatov and A. Y. Shinyaev, *Fiz. Met. Metalloved.*, 1969, **28**, 287.
124. M. W. Mallett. J. Belle and B. B. Cleland, *J. Electrochem. Soc.*, 1954, **101**, 1.
125. M. W. Mallett, E. M. Baroody, H. R. Nelson and C. A. Papp, *J. Electrochem. Soc.*, 1953, **100**, 103.
126. M. W. Mallett, J. Belle and B. B. Cleland, *US Rep.* BM1-829, May 1953.
127. N. L. Peterson and R. E. Ogilvie, *Trans. Metall. Soc. AIME*, 1963, **227**, 1083.
128. G. Hoerz and K. Lindenmaier, *Z. Metallk.*, 1972, **63**, 240.
129. M. J. Klein, *J. Appd. Phys.*, 1967, **38**, 167.
130. D. David, G. Béranger and E. A. Garcia, *J. Electrochem. Soc.*, 1983, **130**, 3423.
131. I. G. Ritchie and A. Atrens, *J. Nucl. Mat.*, 1977, **67**, 254.
132. M. W. Mallett, M. W. Albrecht and P. R. Wilson, *ibid.*, 1959, **106**, 181.
133. G. Béranger, *C. R. Hebd. Séanc. Acad. Sci. Paris*, 1964, **259**, 4663.
134. M. Dupuy and D. Calais, *Mem. Sci Met.*, 1965, **LXII**, 721.
135. H. Cordus and M. Kukuk, *Z. Anorg. Allgem. Chemie*, 1960, **306**, 121.
136. R. Resnick and R. Balluffii, *US Rep.* S.E.P. 118, August, 1953.
137. Y. Adda and J. Philibert, *Acta Metall.*, 1960, **8**, 700.
138. Y. Adda, J. Philibert and Faraggi, *Rev. Métall.*, 1957, **54**, 597.
139. Y. Adda, C. Mairy and J. L. Andreu, *ibid.*, 1960, **57**, 550.
140. E. J. Rapperport, V. Merses and M. F. Smith, *US Rep.* ML-TDR-64-61, March 1964.
141. B. Million, J. Růžičková, J. Velíšek and J. Vřešťál, *Mater. Sci. Eng.*, 1981, **50**, 43.
142. I. Pfeiffer, *Z. Metallk.*, 1955, **46**, 516.
143. L. S. DeLuca, *US Rep.* KAPL-M-LSD-1, August, 1960.
144. I. B. Borovski, I. D. Marchukova and Yu. E. Ugaste, *Fiz. Met. Metalloved.*, 1966, **22**, 849.
145. M. A. Krishtal, *Dokl. Akad. Nauk. SSSR.* 1953, **92**, 951 and Nsf-tr 223.
146. M. Blanter, *Zhur. Tech. Phys. SSSR*, 1950, **20**, 1001.
147. M. Blanter, *ibid.*, 1950, **20**, 217.
148. M. Blanter, *ibid.*, 1951, **21**, 818.
149. M. B. Peraillon, V. Levy and Y. Adda, *Comm.* to 1964 Autumn Meetting, Société Française de Métallurgie.
150. R. C. Reiss, C. S. Hartley and J. E. Steedly, *J. Less-common Met.*, 1965, **9**, 309.
151. R. S. Barclay and P. Niessen, *Amer. Soc. Met. Qt.*, 1969, **62**, 721.
152. O. Neukman, *Galvanotechnick*, 1970, **61**, 626.
153. H. Oikawa, T. Obara and S. Karashima, *Met. Trans.*, 1970, **1**, 2969.
154. J. R. Cahoon, *Metal Trans.*, 1972, **3**, 1324.
155. Y. Funamizu and K. Watanabe, *Trans. Japan Inst. Met.*, 1971, **12**, 147.
156. J. R. Cahoon and W. V. Youdelis, *Trans. Met. Soc. AIME*, 1967, **239**, 127.
157. G. Moreau, J. A. Carnet and D. Calais, *J. Nuci. Mater.*, 1971, **38**, 197.
158. J. Čermak and H. Mehrer, In Press.
159. D. T. Peterson and T. Carnahan, *Trans. Met. Soc. AIME*, 1969, **245**, 213.
160. M. Badia, *Thesis*, Univ. of Nancy (France), 1969.
161. J. Kučera, K. Cíha and K. Stránský, *Czech. J. Phys.*, 1977, **B27**, 758 and 1049.
162. L. Katz, M. Guinan and R. J. Borg, *Phys. Rev.*, 1971, **B4**, 330.
163. O. Caloni and A. Ferrari, *Z. Metallk.*, 1967, **58**, 892.
164. H. I. Aeronson, H. A. Domain and A. D. Brailsford, *Trans AIME*, 1968, **242**, 738.
165. Yu. E. Ugaste and V. N. Pimenov, *Fiz. Met. Metalloved.*, 1971, **31**, 363.
166. A. Tsuji and K. Yamanaka, *J. Jap. Inst. Met.*, 1970, **34**, 486.
167. M. Khobaib and K. P. Gupta, *Scr. Med.*, 1970, **4**, 605.
168. J. H. Swisher and E. T. Turkdogan, *Trans. Met Soc. AIME*, 1967, **239**, 426.
169. R. J. Borg and D. Y. F. Lai, *J. Appl. Phys.*, 1970, **41**, 5193.
170. A. Vignes, *Trans. 2nd Nat. Conf. Electron Microprobe Analysis*, Boston, 1967, Paper No. 20.
171. H. V. M. Mirani and P. Maaskant. *Phys. Status Solidi*, 1972, **A14**, 521.
172. P. E. Brommer and H. A. 't Hooft, *Phys. Letters.*, 1967, **26A**, 52.
173. G. L. Holleck, *J. Phys. Chem.*, 1970, **24**, 1957.
174. K. Lal, *C.E.A.* (*France*), Rept. No. CEA-R. 3136, 1967.
175. W. Zaiss, S. Steeb and T. Krabichler, *Z. Metallk.*, 1972, **63**, 180.
176. T. O. Ogurtani, *Met. Trans.*, 1971, **2**, 3035.
177. V. S. Eremeev, Yu. M. Ivanov and A. S. Panov, *Izv. Akad. Nauk. SSSR Metal*, 1969, **4**, 262.
178. C. J. Rosa and W. W. Smeltzer, *Electrochem. Technol.*, 1966, **4**, 149.
179. G. N. Ronami *et al.*, *Vestn. Mos. Univ. Ser. 3*, 1970, **11**, 251.
180. J. M. Walsh and M. J. Donachie, *Met. Sci. J.*, 1969, **3**, 68.
181. M. Andreani, P. Azou and P. Bastien, *Mém. Scient. Revue Métall.*, 1969, **66**, 21.
182. C. J. Rosa, *Metal Trans.*, 1970, **1**, 2617.
183. J. Debuigne, *Métaux Corros. Inds*, 1967, **No. 501**, 186.

184. G. R. Edwards, R. E. Tate and E. A. Hakkila, *J. Nucl. Mater*, 1968, **25**, 304.
185. M. R. Harvey *et al.*, *J. Less-common Met.*, 1971, **23**, 446.
186. R. E. Tate, G. R. Edwards and E. A. Hakkila, *J. Nucl. Mater.*, 1969, **29**, 154.
187. M. R. Harvey, A. L. Rafalski and D. H. Riefenberg, *Trans. ASM*, 1968, **61**, 629.
188. M. R. Harvey, A. L. Rafalski and D. H. Riefenberg, *Trans. ASM*, 1969, **62**, 1014.
189. C. Remy, M. Dupuy and D. Calais, *J. Nucl. Mater.*, 1970, **34**, 46.
190. A. N. Ivanov, G. B. Krasilnikova and B. S. Mitin, *Fiz. Met. Metalloved.*, 1970, **12**, 291.
191. C. E. Shamblen and C. J. Rosa, *Met. Trans.*, 1971, **2**, 1925.
192. J. C. Lautier, A. van Craeynest and D. Calais, *J. Nucl. Mater.*, 1967, **23**, 111.
193. M. A. Krishtal *et al.*, *Fiz. Khim. Obrab. Mater.*, 1971, **No. 3**, 109.
194. R. Gibala and C. A. Wert, Rpt. COO-1676-3, 1967.
195. G. R. Caskey, R. G. Derrick and M. R. Louthan, *Scr. Met.*, 1974, **8**, 481.
196. J. L. Arnold and W. C. Hagel, *Met. Trans.*, 1972, **3**, 1471.
197. R. Kirchheim, *Acta Met.*, 1979, **27**, 869.
198. P. S. Rudman, *Trans. A.I.M.E.*, 1967, **239**, 1949.
199. H. Jehn and E. Fromm, *J. Less-common Met.*, 1970, **21**, 333.
200. J. H. Evans and B. L. Eyre, *Acta Met.*, **17**, 1109.
201. R. Barlow and P. J. Grundy, *J. Mater. Sci.*, 1969, **4**, 797.
202. S. P. Zholobov and M. D. Malev, *Zh. Tekh. Fiz.*, 1971, **41**, 677.
203. Y. Ebisuzaki, W. J. Kass and M. O'Keefe, *J. Chem. Phys.*, 1968, **49**, 3329.
204. L. R. Velho and R. W. Bartlett, *Met. Trans.*, 1972, **3**, 65.
205. R. Ducros and P. Le Groff, *C. R. Hebd. Séanc., Acad. Sci.*, Paris, 1968, **267C**, 704.
206. R. Frauenfelder, *J. Vac. Sci. & Technol.*, 1969, **6**, 388.
207. R. L. Wagner, *Metal Trans.*, 1970, **1**, 3365.
208. R. Frauenfelder, *J. Chem. Phys.*, 1968, **48**, 3966.
209. L. V. Pavlinov and V. H. Bykov, *Fiz. Met. Metalloved.*, 1965, **19**, 397.
210. K. Lal and V. Levy, *C. R. Hebd. Séanc. Acad. Sci.*, Paris, 1966, **262C**, 107.
211. B. S. Wyatt and B. B. Argent, *J. Less-common Met.*, 1966, **11**, 259.
212. V. F. Yerks, V. F. Zelensky and V. S. Krasnorutskiy, *Phys. Met. Metallogr., N. Y.*, 1966, **22**, 112.
213. F. Funamizu and K. Watanabe, *Trans. Japan Inst. Met.*, 1972, **13**, 278.
214. A. L. Rafalski, M. R. Harvey and D. H. Riefenberg, *Trans. Quarterly*, 1967, **60**, 721.
215. A. E. Lord and D. N. Beshers, *Acta Met.*, 1966, 14, 1659.
216. Th. Hehenkamp, *Acta Met.*, 1966, **14**, 887.
217. A. Languille, *Mem. Scient. Revue Métall.*, 1971, **68**, 435.
218. I. B. Borovskiy, I. D. Marchukova and Yu. E. Ugaste, *Fiz. Met. Metalloved.*, 1967, **24**, 436.
219. I. B. Borovskiy, I. D. Marchukova and Yu. E. Ugaste, *Izv. Vyssh. Uchebn. Zaved. Tsvetn. Metall.*, 1977, **1**, 172.
220. K. Fueki, K. Ota and K. Kishio, *Bull. Chem. Soc. Jpn.*, 1978, **51**, 3067.
221. T. Shimozaki and M. Onishi, *J. Jap. Inst. Met.*, 1978, **42**, 1083.
222. V. N. Agafonov *et al.*, *Vestn. Mosk. Univ. Khim.*, 1975, **16**, 121.
223. A. Gukelberger and S. Steeb, *Z. F. Metallk.*, 1978, **69**, 255.
224. D. Bergner and E. Cyrener, *Neue Huette*, 1973, **18**, 356.
225. Y. Iijima, O. Taguchi and K. I. Hirano, *Met. Trans. A.*, 1977, **8A**, 991.
226. C. P. Heijwegen and G. D. Rieck, *Acta Met.*, 1974, **22**, 1269.
227. P. J. M. Van der Straten *et al.*, *Z. Metallk.*, 1976, **67**, 152.
228. M. Wilhelm, *Z. Naturforsch. A.*, 1974, **29**, 733.
229. M. Onishi and H. Fujibuchi, *Trans. Jap. Inst. Met.*, 1975, **16**, 539.
230. Y. Iijima, K. Hoshino and K. Hirano, *Met. Trans. A.*, 1977, **8A**, 997.
231. K. Nohara and K. Hirano, *J. Jap. Inst. Met.*, 1973, **37**, 51.
232. Y. Wakamatsu, K. Samura and M. Onishi, *J. Jap. Inst. Met.*, 1977, **41**, 664.
233. Y. Wakamatsu, M. Onishi and H. Miura, *J. Jap. Inst. Met.*, 1975, **39**, 903.
234. B. A. Dainyak and V. I. Kostikov, *Izv. Vyssh. Uchebn. Zaved. Chern. Metall.*, 1976, **11**, 15.
235. E. A. Balakir *et al.*, *Izv. Vyssh. Uchebn. Zaved. Tsvetn. Metall.*, 1975, **4**, 162.
236. V. N. Pimenov, Y. E. Ugaste, K. A. Akkushkarova, *Izv. Akad. Nauk SSSR*, 1977, **1**, 184.
237. T. Shimozaki and M. Onishi, *J. Japan Inst. Met.*, 1978, **42**, 402.
238. P. Lamparter, T. Krabichler and S. Steeb, *Z. Metallk.*, 1973, **64**, 720.
239. A. Brunch and S. Steeb, *Z. Naturforsch. A.*, 1974, **29**, 1319.
240. A. Brunch and S. Steeb., *High Temp. High Press.*, 1974, **6**, 155.
241. E. C. Schwegler, *Intl. J. Mass Spec: Ion Physics*, 1968, **1**, 191.
242. T. Ustad and H. Sorum, *Phys. Stat. Sal.*, 1973, **A20**, 285.
243. J. L. Aubin, D. Ansel and J. Debuigne, *J. Less-common Met.*, 1985, **113**, 269.
244. R. L. Fogelson, Y. A. Ugai and A. V. Pokoev, *Fiz. Met. Metalloved*, 1972, **33**, 1102.
245. D. R. Campbell, K. N. Tu and R. E. Robinson, *Acta Met.*, 1976, **24**, 609.
246. A. Ya Shinyayev and N. I. Kopaleishvili, *Fiz. Met. Metalloved.*, 1974, **38**, 222.
247. J. Keinonen, J. Räisänen and A. Anttila, *Appd. Phys. A*, 1984, **34**, 49.
248. D. T. Peterson, *Trans. A.I.M.E.*, 1961, **221**, 924.
249. F. A. Schmidt and O. N. Carlson, *J. Less-common Met.*, 1972, **26**, 247.
250. D. T. Peterson and C. C. Hammerberg, *J. Less-common Met.*, 1968, **16**, 457.
251. Y. F. Funamizu and K. Watanabe, *Trans. Jap. Inst. Met.*, 1976, **17**, 59.
252. O. N. Carlson, F. A. Schmidt and R. R. Lichtenberg, *Met. Trans.*, 1975, **6A**, 725.
253. J. P. Gomez, C. Remy and D. Calais, *Mem Sci. Rev. Métall.*, 1973, **70**, 597.
254. K. Nishida, H. Murohashi and T. Yamamato, *Trans. Jap. Inst. Met.*, 1979, **20**, 269.
255. B. I. Bozic and R. J. Lucic, *J. Mat. Sci.*, 1976, **11**, 887.
256. F. J. Bruni and J. W. Christian, *Acta Met.*, 1973, **21**, 385.

257. A. Green and N. Swindells, *Mat. Sci. Techn.*, 1985, **1**, 101.
258. R. Braun and M. Feller-Kneipmeier, *Phys. Stat. Solidi A*, 1985, **90**, 553.
259. A. D. Romig and M. J. Cieslak, *J. Appd. Phys.*, 1985, **58**, 3425.
260. N. Sarafianos, *Mat. Sci. Eng.*, 1986, **80**, 87.
261. Y. Iijima, K. Hoshino, M. Kikuchi and K. Hirano, *Trans. Jap. Inst. Met.*, 1984, **25**, 234.
262. K. Hoshino, Y. Iijima and K. Hirano, *Phil. Mag. A*, 1981, **44**, 961.
263. R. A. Fouracra, *Thin Solid Films*, 1986, **135**, 189.
264. A. D. Romig, *J. Appd. Phys.*, 1983, **54**, 3172.
265. J. M. Vandenberg, F. J. A. Den Broeder and R. A. Hamm, *Thin Solid Films*, 1982, **93**, 277.
266. T. Takahashi, M. Kato, Y. Minamino, T. Yamane, T. Azukizawa, T. Okamoto, M. Shimada and M. Agawa, *Zeit F. Metallk.*, 1984, **75**, 440.
267. Th. Heumann and T. Rottwinkel, *J. Nucl. Mat.*, 1978, **69/70**, 567.
268. Y. Iijima, T. Igarishi and K. Hirano, *J. Mat. Sci.*, 1979, **14**, 474.
269. H. Jehn and E. Olzi, *High Temp. High Press.*, 1980, **12**, 85.
270. C. Narayan and J. I. Goldstein, *Met. Trans. A*, 1983, **14A**, 2437.
271. R. H. Zee, J. F. Watters and R. D. Davidson, *Phys. Rev. B*, 1986, **34**, 6895.
272. A. W. Nichols and I. P. Jones, *J. Phys. Chem. Solids*, 1983, **44**, 671.
273. Y. Minamino, T. Yamane, A. Shimomura, M. Shimada, M. Koizumi, N. Ogawa, J. Takahashi and H. Kimura, *J. Mater. Sci.*, 1983, **18**, 2679.
274. M. Arita, M. Nakamura, K. S. Goto and Y. Ichinose, *Trans. Jap. Inst. Met.*, 1984, **25**, 703.
275. M. Yakota, M. Nose and H. Mitani, *J. Jap. Inst. Metals*, 1980, **44**, 1007.
276. M. B. Chamberlain, *J. Vac. Sci. Techn.*, 1979, **16**, 339.
277. Y. Minamo, T. Yamane, M. Koizumi, M. Shimada and N. Ogawa, *Zeit. F. Metallk.*, 1982, **73**, 124.
278. W. Schatl, H.-J. Ullrich, K. Kleinstück, S. Dabritz, A. Herenz, D. Bergner and H. Luck, *Krist. Tech.*, 1978, **13**, 185.
279. C. J. Wen and R. H. Huggins, *J. Sol. State Chem.*, 1980, **35**, 376.
280. F. Lantelme and S. Belaidouni, *Electrochim. Acta*, 1981, **26**, 1225.
281. K. Hoshino, Y. Iijima and K. Hirano, *Trans. Jap. Inst. Met.*, 1981, **22**, 527.
282. I. Richter and M. Feller-Kniepmeier, *Phys. Stat. Solidi A*, 1981, **68**, 289.
283. J. Hirvonen and J. Räisänen, *J. Appd. Phys.*, 1982, **53**, 3314.
284. S. Mei, H. B. Huntington, C. K. Hu and M. J. McBride, *Scr. Met.*, 1987, **21**, 153.
285. M. Yokota, R. Harada and M. Mitani, *J. Jap. Inst. Metals*, 1979, **43**, 793 and 799.
286. M. Yamamoto, T. Takashima and K. Nishida, *J. Jap. Inst. Metals*, 1979, **43**, 1196.
287. D. V. Ignatov, M. S. Modeo, L. F. Sokirianskii and A. Y. Shinyaev, *Titanium Sci. and Tech. IV*, 1973, 2535.
288. S. Shankar and L. L. Seigle, *Met. Trans. A*, 1978, **9**, 1467.
289. E. I. Balakir, Yu. P. Zotov, E. B. Malysheva, V. I. Panchishnyi and V. P. Voevadin, *Izv. Vyssh. Uchebn. Zaved. Chern. Metall.*, 1977, (3), 5.
290. L. V. Yelokhina and V. I. Shalayev, *Phys. Met. Metallog.*, 1987, **63**, (5), 113.
291. T. Shimozaki, K. Ito and M. Onishi, *Trans. Jap. Inst. Metals*, 1987, **28**, 457.
292. Y. Chryssoulakis, F. Lantelme, A. Alexopoulou, S. Kalogeropoulou and M. Chemla, *Electrochim. Acta*, 1987, **32**, 699.
293. C. J. Wen, W. Weppner, B. A. Baukamp and R. A. Huggins, *Met. Trans. B*, 1980, **11**, 131.
294. M. Arita, M. Tanaka, K. S. Goto and M. Someno, *Met. Trans. A*, 1981, **12**, 497.
295. J. L. Arnold and W. C. Hagel, *Met. Trans.*, 1972, **3**, 1471.
296. S. J. Buderov, W. S. Boshinov and P. D. Kovatchev, *Krist. Tech.*, 1980, **15**, K19.
297. M. Arita, M. Ohyama, K. S. Goto and M. Someno, *Zeit. F. Metallk.*, 1981, **72**, 244.
298. W. M. Albrecht and M. W. Mallet, *Trans. AIME*, 1958, **212**, 204.
299. Th. Heumann and R. Damköhler, *Zeit. F. Metallk.*, 1978, **69**, 364.
300. M. Wein, L. Levin and S. Nadiv, *Phil. Mag. A*, 1978, **38**, 81.
301. L. Le Gall, D. Ansel and J. Debuigne, *Acta Met.*, 1987, **35**, 2297.
302. S. K. Bose and H. J. Grabke, *Zeit. F. Metallk.*, 1978, **69**, 8.
303. G. Salke and M. Feller-Kniepmeier, *J. Appld. Phys.*, 1978, **49**, 229.
304. A. Anttila, J. Räisänen and J. Keinonen, *Appd. Phys, Letters*, 1983, **42**, 498.
305. A. Anttila, J. Räisänen and J. Keinonen, *J. Less-common Met.*, 1984, **96**, 257.
306. R. A. Perkins, *J. Nucl. Mat.*, 1977, **68**, 254.
307. O. N. Carlson, F. A. Schmidt and J. C. Sever, *Met. Trans.*, 1973, **4**, 2407.
308. J. Keinonen, J. Räisänen and A. Anttila, *Appd. Phys. A*, 1984, **35**, 227.
309. P. J. Grundy and P. J. Nolan, *J. Mater. Sci.*, 1972, **7**, 1086.
310. R. Lappalainen and A. Anttila, **Appd. Phys. A, 1987, 42**, 263.
311. G. R. Johnston, *High Temp. High Press.*, 1982, **14**, 695.
312. T. Shimozaki, K. Itoh and M. Onishi, *Trans. Jap. Inst. Met.*, 1986, **27**, 160.
313. T. Yamamoto, T. Takashima and K. Nishida, *J. Jap. Inst. Met.*, 1980, **44**, 294.
314. F. M. Mazzolai and J. Ryll-Nardzewski, *J. Less Common Met.*, 1976, **49**, 323.
315. I. B. Kim and I. H. Moon, *J. Corrosion Sci. Soc. Korea*, 1972, **1**, 51.
316. G. L. Powell and J. B. Condon, *Anal. Chem.*, 1973, **45**, 2349.
317. M. Y. Lee, J. Y. Lee and S. S. Chung, *J. Korean Inst. Met.*, 1977, **15**, 265.
318. K. Hirano, Y. Iijima, K. Araki and H. Homma, *Trans. Iron Steel Inst. Jap.*, 1977, **17**, 194.
319. J. Völkl and G. Alefeld, *Topics in Applied Physics* (J. Springer, Berlin), 1978, **28**, *Hydrogen in Metals*, 321.
320. Van Dal, M.C.L.P. Pleumeekers, A. A. Kodentsov and F. J. J. Van Loo, *Acta Mater.*, 48, 2000, 385. 'Lab of Solid State and Materials Chemistry, Eindhoven University of Technology', 1999.

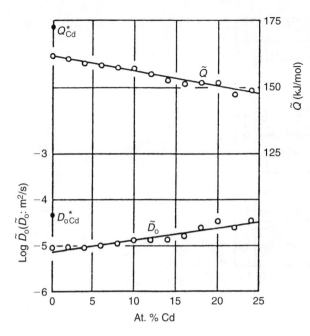

Figure 13.9   *The concentration dependences of $\tilde{Q}$ and $\tilde{D}$ in* Ag-Cd *alloys*[291]

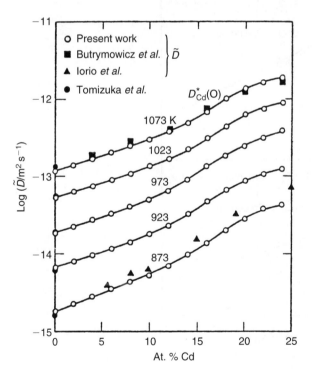

Figure 13.10   *Concentration dependence of the interdiffusion coefficients, $\tilde{D}$, at five temperatures between 873 and 1073 K determined by use of* Ag-25Cd *diffusion couples*[291]

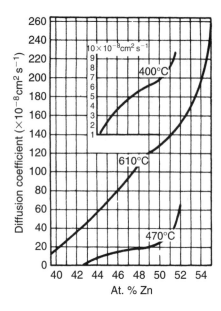

Figure 13.11   *Chemical diffusion coefficients in AgZn alloys*[11] *Table 13.4*

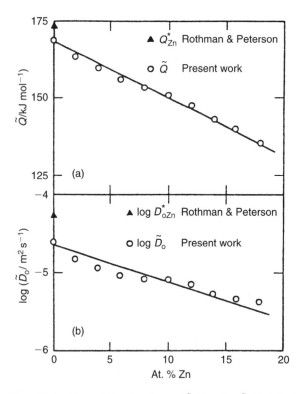

Figure 13.12   *Concentration dependence of $\tilde{Q}$ (a) and $\log \tilde{D}_o$ (b) in Ag-Zn α phase*[312]

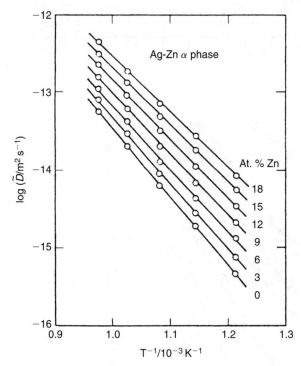

Figure 13.13   *Temperature dependence of the interdiffusion coefficients in Ag-Zn α phase*[312]

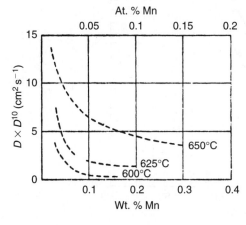

Figure 13.14   *Chemical diffusion in AlMn alloys*[21]

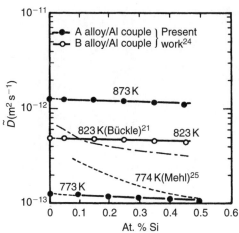

Figure 13.15   *Chemical diffusion in AlSi alloys*[21,25]

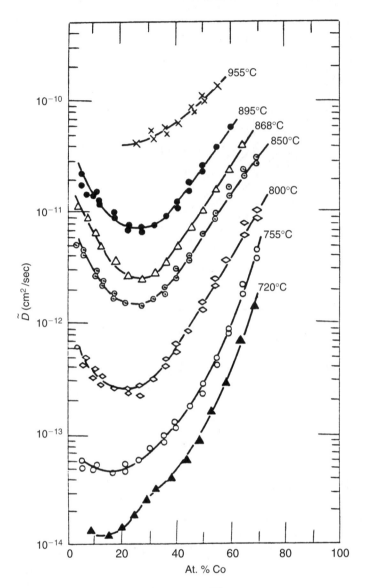

Figure 13.16a   *Interdiffusion in b.c.c.* FeCo *alloys*[318]

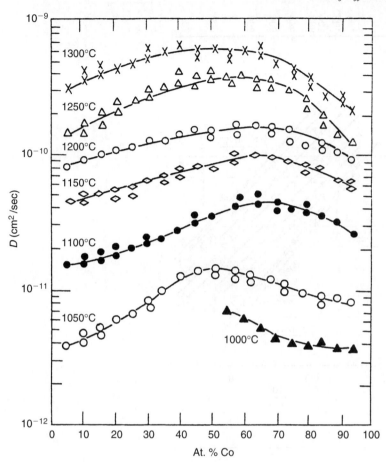

Figure 13.16b *Interdiffusion in f.c.c. FeCo alloys*[318]

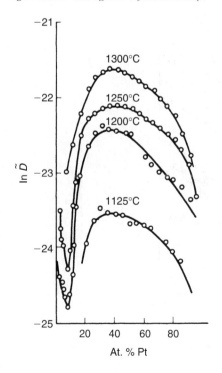

Figure 13.17 *Interdiffusion in Co-Pt*[218]

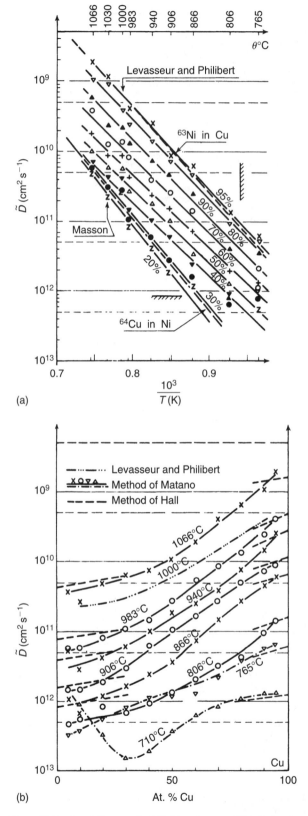

Figure 13.18   *Interdiffusion in Cu-Ni[55] (a) as function of 1/T; (b) As a function of composition*

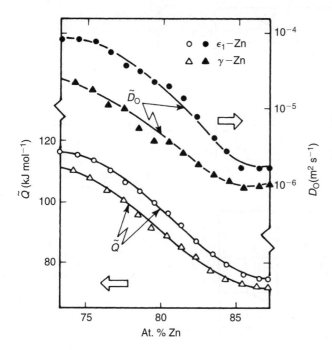

Figure 13.19   *Chemical diffusion* AgZn *alloys*[221]

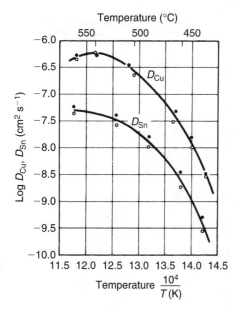

Figure 13.20   *Partial diffusion coefficients for* Cu *and* Sn *in* δ-*phase* CuSn *alloys*[60]

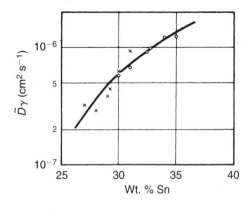

Figure 13.21 *Chemical diffusion coefficients for γ-phase of* Cu-Sn *systems at* 706.5°C[60]

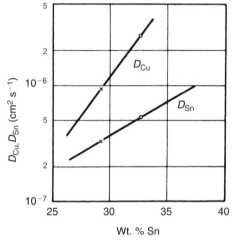

Figure 13.22 *Partial diffusion coefficients for* Cu *and* Sn *in* γ *CuSn alloys at* 706.5°C

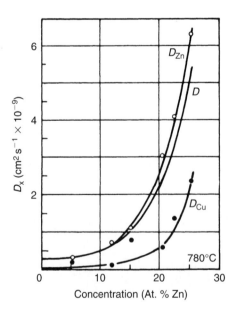

Figure 13.23 *Chemical and partial diffusion coefficients in* α-Cu-Zn *system at* 780°C[61]

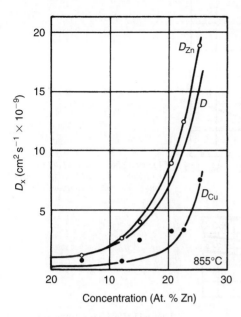

Figure 13.24 *Chemical and partial diffusion coefficients in α-Cu-Zn systems at 855°C[61]*

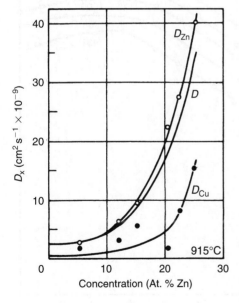

Figure 13.25 *Chemical and partial diffusion coefficients in α-Cu-Zn system at 915°C[61]*

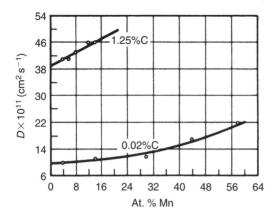

Figure 13.26   *Chemical diffusion coefficients in γ-Fe-Mn alloys with 0.02 and 1.25 wt%C. Temperature* 1200°C[67]

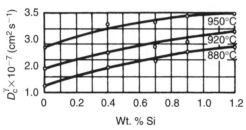

Figure 13.27   *Effect of* Si *content on chemical diffusion of* C *in* γFe[145]

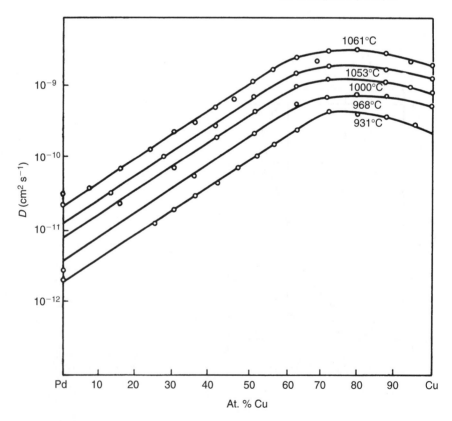

Figure 13.28   *Interdiffusion in the* Cu-Pd *system*[160]

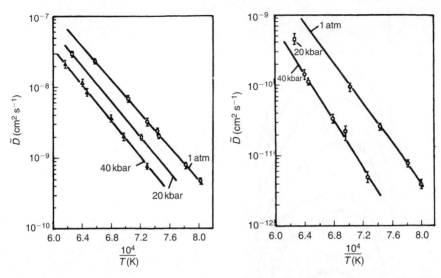

Figure 13.29   *Chemical diffusion in α-Fe V 10% alloy at* 1, 20 *and* 24 kbar *pressure*

Figure 13.30   *Chemical diffusion in γ* Fe V 0.7% *alloy at* 1, 20 *and* 40 kbar *pressure*[78]

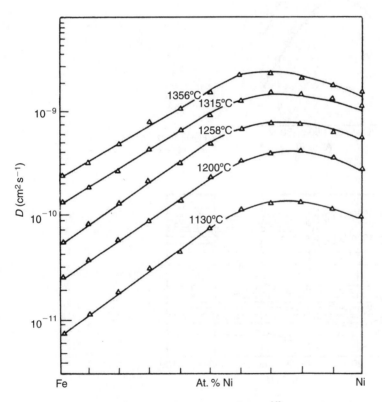

Figure 13.31   *Chemical diffusion coefficients in the* Fe-Ni *system*[160]

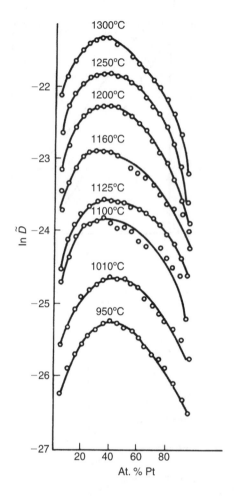

Figure 13.32   *Chemical diffusion coefficients in the* Ni-Pt *system*[218]

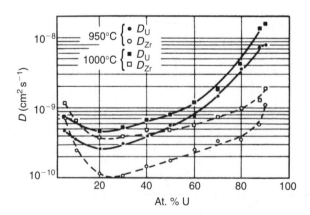

Figure 13.33   *Partial diffusion coefficients* D_U *and* D_{Zr} *in* γ U-Zr *alloys*[139]

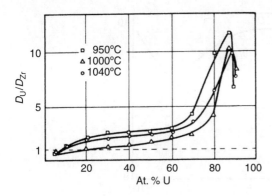

Figure 13.34 *Ratio of partial diffusion coefficients in γ U-Zr alloys*[139]

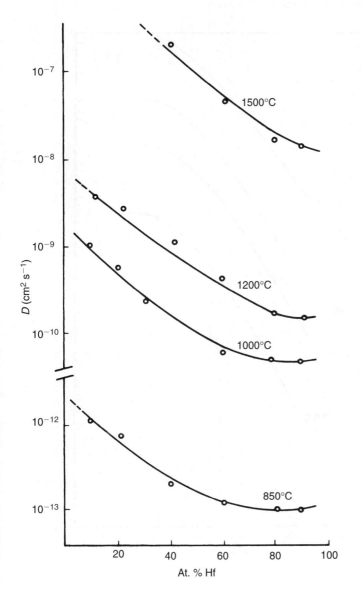

Figure 13.35a *Chemical diffusion in Hf-Zr alloys*

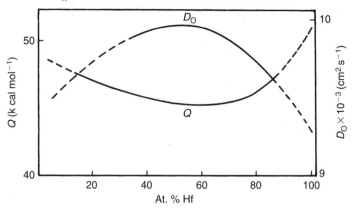

Figure 13.35b   *A and Q for chemical diffusion in* Hf-Zr *alloys*[235]

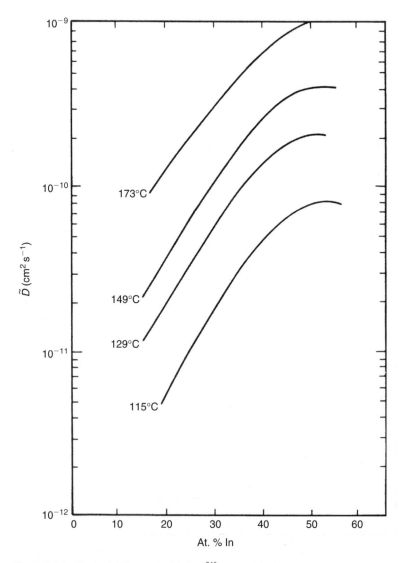

Figure 13.36   *Chemical diffusion in* In-Pd *alloys*[245]

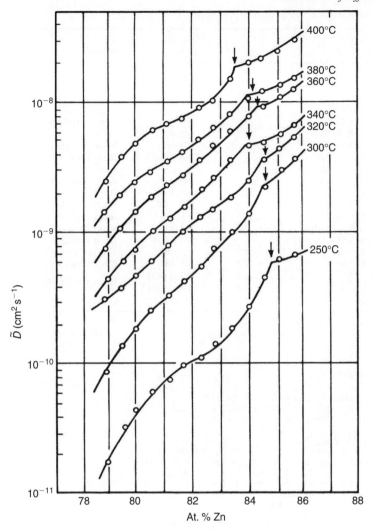

Figure 13.37 *Chemical diffusion in ε Cu-Zn alloys*[251]

Table 13.5 SELF-DIFFUSION IN LIQUID METALS

Element	A	Q	Temp. range °C	Method	Ref.
Li	1.44×10^{-3}	12.0	465–723	IIIb(ii), Capillary	1 and 2
Na	8.6×10^{-4}	9.29	375–563	IVa(i), Capillary	2
K	7.6×10^{-4}	8.46	355–563	IVa(i), Capillary	2
	$D = 5.344 \times 10^{-10} T^2 - 2.443 \times 10^{-5}$		354–868	IIa(ii), Capillary	22
Rb	6.6×10^{-4}	8.29	330–503	Electrotransport method	2 and 3
	$D = 3.824 \times 10^{-10} T^2 - 1.479 \times 10^{-5}$		337–856	IIa(ii), Capillary	22
Cs	4.8×10^{-4}	7.79	323–473	Calculated	2 and 4
Cu	1.46×10^{-3}	40.7	1 413–1 543	IIIb(ii), Capillary	5

(*continued*)

Table 13.5   SELF-DIFFUSION IN LIQUID METALS—*continued*

Element	A	Q	Temp. range °C	Method	Ref.
Ag	5.8×10^{-4}	32.1	1 243–1 573	IIIb(ii), Capillary	6
Zn	8.2×10^{-4}	21.3	723–893	IIIb(ii), Capillary	7
	1.2×10^{-3}	23.4	703–893	IIIb(ii), Capillary	8
Cd	3.62×10^{-4}	13.9	603–783	IIIb(ii), Capillary	9
	7.54×10^{-4}	18.5	603–773	IIIa(i), Capillary	23
Hg	$D = 4.48 \times 10^{-10} T^{1.854}$		273–567	IIIb(ii), Capillary	11
	$D = 4.34 \times 10^{-9} T^{3/2} - 4.81 \times 10^{-6}$		248–525	IIa(ii), shear cell	10
Ga	$D = 2.2 \times 10^{-10} T^2$		292–556	IIa(ii), Capillary	12
	$D = 6.01 \times 10^{-9} T^{3/2} - 1.60 \times 10^{-5}$		304–674	IIa(ii), Shear cell	13
In	2.89×10^{-4}	10.2	443–1 023	IIIb(ii), Capillary	14
Tl	3.17×10^{-4}	15.2	623–1 073	IIIb(ii), Capillary	15
Sn	3.24×10^{-4}	11.6	540–956	IIa(ii), Capillary	21
	$D = [1.8 + 0.012(T - T_M)]10^{-5}$		628–1 925	IIa(ii), Shear cell	16
	$D = 6.85 \times 10^{-11} T^2$		543–1 048	IIa(ii), Capillary	17
Pb	2.37×10^{-4}	24.7	623–783	IIIb(ii), Capillary	9
Bi	8.3×10^{-5}	10.5	540–980	IIIb(ii), Capillary	18
Fe	1×10^{-2}	65.7	(a) 1 610–1 680	IIb(ii), Capillary	19
	4.3×10^{-3}	51.1	(b) 1 510–1 630		
Te	1.36×10^{-3}	23.7	710–873	IVb(ii), Capillary	20

(*a*) Average values. *A* and *Q* both increase with temperature.
(*b*) Supercooled.

REFERENCES TO TABLE 13.5

1. A. Ott and A. Lodding, *Z. Naturforsch.*, 1965, **20a**, 1578.
2. S. J. Larsson, C. Roxbergh and A. Lodding, *Phys. Chem. Liquids*, 1972, **3**, 137.
3. A. Nordén and A. Lodding, *Z. Naturforsch.*, 1968, **22a**, 215.
4. A. Lodding, *Z. Naturforsch.*, 1972, **27a**, 873.
5. J. Henderson and L. Young, *Trans. Met. Soc. AIME*, 1961, **221**, 72.
6. V. G. Leak and R. A. Swalin, *Trans. Met. Soc. AIME*, 1964, **230**, 426.
7. N. H. Nachtrieb, E. Fraga and C. Wahl, *J. Phys. Chem.*, 1963, **67**, 2353.
8. W. Lange, W. Pippel and F. Bendel, *Z. Phys. Chem.*, 1959, **212**, 238.
9. M. Mirshamshi, A. Cosgarea and W. Upthegrove, *Trans. Metall. Soc., AIME*, 1966, **236**, 122.
10. E. F. Broome and H. A. Walls, *Trans. Met. Soc. AIME*, 1968, **242**, 2177.
11. R. E. Meyer, *J. Phys. Chem.*, 1961, **65**, 567.
12. S. Larsson *et al.*, *Z. Naturforsch.*, 1970, **25a**, 1472.
13. E. F. Broome and H. A. Walls, *Trans. Met. Soc. AIME*, 1969, **245**, 739.
14. A. Lodding, *Z. Naturforsch.*, 1956, **11a**, 200.
15. N. Petrescu and L. Ganovici, *Rev. Roum. de Chim.*, 1976, 21, 1293.
16. A. Bruson and M. Gerl, *Phys. Rev. B*, 1980, **21**, 5447 and *J. de Phys.*, 1980, **41**, 533.
17. U. Södervall, H. Odelius, A. Lodding, G. Frohberg, K.-H. Kratz and H. Wever, *Proc. 'SIMS V' Conference, Washington D.C., 1985.*
18. N. Petrescu and M. Petrescu, *Revue Roum. Chim.*, 1970, **15**, 189.
19. L. Yang, M. T. Simnad and G. Derge, *Trans. Met. Soc. AIME*, 1956, **206**, 1577.
20. L. Nicoloiu, L. Ganovici and I. Ganovici, *Rev. Roum. Chim.*, 1970, **15**, 1713.
21. G. Careri, A. Paoletti and M. Vincentini, *Nuovo Cimen.*, 1958, **10**, 1088.
22. M. Hseih and R. A. Swalin, *Acta Met.*, 1974, 22, 219.
23. I. Ganovici and L. Ganovici, *Rev. Roum. de Chim.*, 1970, **15**, 213.
24. M. Shimoji and T. Itami, *Atomic Transport in Liquid Metals. Diffusion and Defect Data*, 1986, **43**, 1.

14 General physical properties

14.1 The physical properties of pure metals

Many physical properties depend on the purity and physical state (annealed, hard drawn, cast, etc.) of the metal. The data in Tables 14.1 and 14.2 refer to metals in the highest state of purity available, and are sufficiently accurate for most purposes. The reader should, however, consult the references before accepting the values quoted as applying to a particular sample.

ble 14.1   THE PHYSICAL PROPERTIES OF PURE METALS AT NORMAL TEMPERATURES

tal	Melting point (°C) [i]	Boiling point (°C) [i]	Density (g/cm³) at 25°C [i]*	Thermal conductivity (W/m K) at 25°C [ii]	Mean specific heat (J/kg K) at 25°C [iii]	Resistivity (10⁻⁸ Ωm) at T (°C) [iv]		Temp. coeff. of resistivity (10⁻³ K⁻¹) 0–100°C	Coeff. of expansion at 25°C [i]
						T (°C)	Resistivity		
minium	660.323(*)	2 519	2.70	247	897	20	2.654 8	4.5	23.1
imony	630.63(*)	1 587	6.68	25.9	207	0	39.0	5.1	11.0
enic	{817}*	616*	5.727*	—	329	20	33.3	—	5.6
ium	727	1 897	3.62	18.4	204	0	6 000	—	20.6
yllium	1 287	2 471	1.85	210	1 825	20	4 (aa)	9.0	11.3
muth	271.40	1 564	9.79	8.2	122	0	106.8	4.6	13.4
dmium	321.069 (*)	767	8.69	97.5	232	0	6.83	4.3	30.8
sium	28.44	671	1.93	18.42	242	20	20	4.8	97
cium	842	1 484	1.54	196	647	0	3.91	4.57	22.3
ium	798	3 433	6.77	11.3	192	25	75	8.7	6.3
omium	1 907	2 671	7.15	67	449	0	12.9	2.14	4.9
alt	1 495 (*)	2 927	8.86	69.04	421	20	6.24	6.6	13.0
per	1 084.62 (*)	2 562	8.96	398	385	20	1.673	4.3	16.5
prosium	1 412	2 567	8.55	10.7	170	25	57	1.19	9.9
ium	1 529	2 868	9.07	14.5	168	25	107	2.01	12.2
olinium	1 313	3 273	7.90	10.5	236	25	140.5	0.9/1.76	9.4 (100°C)
lium	29.764 6 (*)	2 204	5.91	33.49	371		174 (bb)	—	18
manium	937	2 830	5.32	58.6	320		46 (cc)	—	5.75
d	1 064.18 (*)	2 856	19.3	317.9	129	20	2.35	4.0	14.2
nium	2 233	4 603	13.3	23	144	25	35.1	4.4	5.9
mium	1 474	2 700	8.80	16.2	165	25	87	1.71	11.2
um	156.598 5 (*)	2 072	7.31	83.7	233	20	8.37	5.2	32.1
um	2 446 (*)	4 428	22.5	147	131	20	5.3	4.5	6.4
d	1 538	2 861	7.87	80.4	449	20	9.71	6.5	11.8
thanum	918	3 464	6.15	13.4	195	0 to 26	57	2.18	12.1
d	327.462	1 749	11.3	33.6	129	20	20.648	4.2	28.9
ium	180.5	1 342	0.534	44.0	3 582	0	8.55	4.35	46
tium	1 663	3 402	9.84	16.4	154	25	79	—	9.9
nesium	650	1 090	1.74	155	1 023	20	4.45	4.25	24.8
ganese	1 246	2 061	7.3	7.79	479	20	185 (α)	—	21.7

(continued)

Table 14.1   THE PHYSICAL PROPERTIES OF PURE METALS AT NORMAL TEMPERATURES—*continued*

Metal	Melting point (°C) [i]	Boiling point (°C) [i]	Density (g/cm³) at 25°C [i]*	Thermal conductivity (W/m K) at 25°C [ii]	Mean specific heat (J/kg K) at 25°C [iii]	Resistivity (10⁻⁸ Ωm) at T (°C) [iv] T (°C)	Resistivity	Temp. coeff. of resistivity (10⁻³ K⁻¹) 0–100°C	Coeff. of expansion at 25°C [i]
Mercury	−38.83	356.73	13.533 6	8.21	140	50	98.4	1.0	60.4
Molybdenum	2 623	4 639	10.2	142	251	0	5.3	4.35	4.8
Neodymium	1 021	3 074	7.01	16.5	190	25	64	1.64	9.6
Nickel	1 455 (*)	2 913	8.90	82.9	444	20	6.84	6.8	13.4
Niobium	2 477	4 744	8.57	52.3	265	0	12.5	2.6	7.3
Osmium	3 033	5 012	22.59	—	130	20	9.5	4.1	5.1
Palladium	1 554.9	2 963	12.0	70	246	20	10.8	4.2	11.8
Platinum	1 768.4	3 825	21.5	71.1	133	20	10.6	3.92	8.8
Plutonium	640	3 228	19.7	6.5	142*	107	141.4	—	46.7
Polonium	254	962	9.20	—	—	—	—	—	23.5
Potassium	63.38	759	0.89	108.3	757	0	6.15	5.7	83.3
Praeseodymium	931	3 520	6.77	12.5	193	25	68	1.71	6.7
Radium	700	—	5	—	—	—	—	—	—
Rhenium	3 186	5 596	20.8	71.2	137	20	19.3	4.5	6.2
Rhodium	1 964	3 695	12.4	150	243	20	4.51	4.4	8.2
Rubidium	39.30	688	1.53	58.3	363	20	12.5	4.8	9
Ruthenium	2 334	4 150	12.1	—	238	0	7.6	4.1	6.4
Samarium	1 074	1 794	7.52	13.3	197	25	88	1.48	12.7
Scandium	1 541	2 836	2.99	15.8	568	22	61 (ave)	—	10.2
Selenium	220.5	685	4.79	2.48	321	0	12	—	37
Silicon	1 412	3 270	2.34	156	705	0	10	—	7.6
Silver	961.78 (*)	2 162	10.5	428	235	20	1.59	4.1	18.9
Sodium	97.72	883	0.97	131.4	1 228	0	4.2	5.5	71
Strontium	777	1 382	2.64	—	301	20	23	—	22.5
Tantalum	3 017	5 458	16.4	54.4	140	25	12.45	3.5	6.3
Terbium	1 356	3 230	8.23	11.1	182	20	11 600	—	10.3
Tellurium	450	988	6.24	5.98–6.02	202	23	4.36 × 10⁷	—	{ 1.7 ‖ c a 27.5 ⊥ c =
Thallium	304	1 473	11.8	47	129	0	18	5.2	29.9
Thorium	1 750	4 788	11.7	77	113	0	13	4.0	11.0
Thulium	1 545	1 950	9.32	16.9	160	25	79	1.95	13.3
Tin	231.928 (*)	2 602	7.26	62.8	228	0	11 (dd)	4.6	22.0
Titanium	1 668	3 287	4.51	11.4	523	20	42	3.8	8.6
Tungsten	3 422	5 555	19.3	160	132	27	5.65	4.8	4.5
Uranium	1 135	4 131	19.1	27.6	116	—	30 (ee)	3.4	13.9
Vanadium	1 910	3 407	6.0	31.0	489	20	24.8–26	3.9	8.4
Ytterbium	819	1 196	6.90	38.5	155	25	29	1.30	26.3
Yttrium	1 522	3 345	4.47	17.2	298	from 20–250	57 (ff)	2.71	10.6
Zinc	419.527 (*)	907	7.14	113	388	20	5.916	4.2	30.2
Zirconium	1 855	4 409	6.52	21.1	278	—	40	4.4	5.7

(*) Defined fixed point of ITS-90—see Ch. 16
{} Rare earths and rare metals
* Densities of higher allotropes not at 20°C

(aa) annealed, commercial purity
(bb) for a-axis; 8.1 for b-axis and 54.3 for c-axis
(cc) Ohm cm for intrinsic germanium at 300 K

(dd) for white tin
(ee) Crystallographic average
(ff) for polycrystalline material

[i] Reprinted with permission from Handbook of Chemistry and Physics 82nd Edition (12–219). Copyright CRC Press, Boca Raton, Florida
[ii] Ref. 47
[iii] Reprinted with permission from Handbook of Chemistry and Physics 82nd Edition (4–133). Copyright CRC Press, Boca Raton, Florida
[iv] Ref. 49

Table 14.2 THE PHYSICAL PROPERTIES OF PURE METALS AT ELEVATED TEMPERATURES†

Metal	Temperature $t°C$	Coefficient of expansion $20-t°C$ $10^{-6} K^{-1}$	Resistivity at $t°C$ $\mu\Omega$ cm	Thermal conductivity at $t°C$ $Wm^{-1} K^{-1}$	Specific heat at $t°C$ $J kg^{-1} K^{-1}$	References*
Aluminium	20	—	2.67	—	900	7, 8, 9, 50
	27	—	2.733	237	—	
	100	23.9	3.55	—	938	
	127	—	3.87	240	—	
	200	24.3	4.78	—	984	
	227	—	4.99	236	—	
	300	25.3	6.99	—	1 030	
	327	—	6.13	231	—	
	400	26.49	7.30	—	1 076	
	427	—	8.70	218	—	
Antimony	20	—	40.1	18.0	205	7, 10, 6
	100	8.4–11.0	59	16.7	214	
	500	9.7–11.6	154	19.7	239	
Beryllium	20	—	3.3	180	1 976	11, 50
	27	—	3.76	—	—	
	100	12	—	152	2 081	
	127	—	6.76	—	—	
	200	13	—	130.2	(2 215)	
	227	—	9.9	—	—	
	300	14.5	—	117.7	(2 353)	
	327	—	13.2	—	—	
	427	—	16.5	—	—	
	500	16	—	103.0	(2 621)	
	527	—	20.0	—	—	
	627	—	23.7	—	—	
	700	17	—	85.8	(2 889)	
Bismuth	20	—	117	8.0	121	7, 12
	100	13.4	156	7.5	130	
	250	—	260	7.5	147	
Cadmium	0	—	6.8	97.5	—	7, 13, 14, 50, 51
	20	—	—	84	230	
	27	—	—	96.8	—	
	100	31.8	—	87.9	239	
	127	—	—	94.7	—	
	227	—	—	92	—	
	300	(38)	—	104.7	260	
Chromium	20	—	13.2	—	444	7, 15, 16, 50
	27	—	12.7	93.7	—	
	100	6.6	18(152°C)	—	490	
	127	—	15.8	90.9	—	
	327	—	24.7	80.7	—	
	400	8.4	31(407°C)	—	582	
	527	—	34.6	71.3	—	
	700	9.4	47(652°C)	—	649	
	727	—	—	65.4	—	
Cobalt	20	—	5.86	—	434	42, 45
	100	12.3	9.30	—	453	
	200	13.1	13.88	—	478	
	300	13.6	19.78	—	502	
	400	14.0	26.56	—	527	
	600	—	40.2	—	575	
	800	—	58.6	—	716	
	1 000	—	77.4	—	800	
	1 200	—	91.9	—	883	

(continued)

Table 14.2 THE PHYSICAL PROPERTIES OF PURE METALS AT ELEVATED TEMPERATURES[†]—*continued*

Metal	Temperature $t°C$	Coefficient of expansion $20-t°C$ $10^{-6} K^{-1}$	Resistivity at $t°C$ $\mu\Omega$ cm	Thermal conductivity at $t°C$ $Wm^{-1} K^{-1}$	Specific heat at $t°C$ $J kg^{-1} K^{-1}$	References*
Copper	20	—	—	394	385	7, 17, 16, 18, 50
	27	—	1.725	401	—	
	100	17.1	—	394	389	
	127	—	2.402	393	—	
	200	17.2	—	389	402	
	227	—	3.090	386	—	
	427	—	5.262	366	—	
	500	18.3	—	341(538°C)	(427)	
	827	—	—	339	—	
	1 000	20.3	—	244(1037°C)	(473)	
Gold	20	—	2.2	293	126	7, 50
	27	—	2.271	317	—	
	100	14.2	2.8	293	130	
	127	—	3.107	311	—	
	500	15.2	6.8	—	142	
	527	—	6.81	284	—	
	900	16.7	11.8	—	151	
	927	—	—	255	—	
Hafnium	20	—	35.5	(22.2)	144	43, 44, 48, 50
	27	—	34.0	—	—	
	100	—	46.5	22.0	148	
	127	—	48.1	—	—	
	200	—	60.3	21.5	152	
	227	—	63.1	—	—	
	327	—	78.5	—	—	
	400	6.3	84.4	20.7	160	
	1 000	6.1	—	—	185	
	1 400	6.0	—	—	—	
	1 800	5.9	—	—	—	
Iridium	20	—	5.1	148(0°C)	130	19
	100	6.8	6.8	143	134	
	500	7.2	15.1	—	142	
	1 000	7.8	—	—	159	
Iron	20	—	10.1	73.3	444	7, 20, 50
	27	—	9.98	80.2	—	
	100	12.2	14.7	68.2	477	
	127	—	16.1	69.5	—	
	200	12.9	22.6	61.5	523	
	227	—	23.7	61.3	—	
	327	—	32.9	54.7	—	
	400	13.8	43.1	48.6	611	
	527	—	57.1	43.3	—	
	600	14.5	69.8	38.9	699	
	727	—	—	32.3	—	
	800	14.6	105.5	29.7	791	
Lead	20	—	20.6	34.8	130	7, 6, 11, 50
	27	—	21.3	35.3	—	
	100	29.1	27.0	33.5	134	
	127	—	29.6	34.0	—	
	200	30.0	36.0	31.4	134	
	227	—	38.3	32.8	—	
	300	31.3	50	29.7	138	
	327	—	—	31.4	—	

(*continued*)

Table 14.2 THE PHYSICAL PROPERTIES OF PURE METALS AT ELEVATED TEMPERATURES[†]—*continued*

Metal	Temperature $t°C$	Coefficient of expansion $20-t°C$ $10^{-6}\,K^{-1}$	Resistivity at $t°C$ $\mu\Omega\,cm$	Thermal conductivity at $t°C$ $Wm^{-1}\,K^{-1}$	Specific heat at $t°C$ $J\,kg^{-1}\,K^{-1}$	References*
Magnesium	20	—	4.2	167	1 022	7, 50
	27	—	4.51	156	—	
	100	26.1	5.6	167	1 063	
	127	—	6.19	153	—	
	200	27.0	7.2	163	1 110	
	227	—	7.86	151	—	
	327	—	9.52	149	—	
	400	28.9	12.1	130	1 197	
	427	—	11.2	—	—	
	527	—	12.8	146 (extp)	—	
Molybdenum	20	—	5.7	142	247	7, 21, 22, 23, 50, 51
	27	—	5.52	138	—	
	100	5.2	7.6	138	260	
	127	—	8.02	134	—	
	500	5.7	17.6	121	285	
	527	—	18.4	118	—	
	927	—	—	105	—	
	1 000	5.75	31	105	310	
	1 327	—	—	94.6	—	
	1 500	6.51	46	84	339 (mean)	
	1 727	—	—	88	—	
	2 500	—	77	—	—	
Nickel	20	—	6.9	88	435	24, 50
	27	—	7.20	90.7	—	
	100	13.3	10.3	82.9	477	
	127	—	11.8	80.2	—	
	200	13.9	15.8	73.3	528	
	227	—	17.7	72.2	—	
	300	14.4	22.5	63.6	578	
	327	—	25.5	65.6	—	
	400	14.8	30.6	59.5	519	
	427	—	32.1	—	—	
	500	15.2	34.2	62.0	535	
	527	—	35.5	67.6	—	
	727	—	—	71.8	—	
	900	16.3	45.5	—	595	
	927	—	—	76.2	—	
Niobium	0	—	15.2	53.3	—	
	20	—	14.6	—	268	42, 45, 50, 51
	27	—	—	53.7	—	
	200	7.19	25.0	—	271	
	227	—	—	56.7	—	
	400	7.39	36.6	—	284	
	527	—	—	61.3	—	
	600	7.56	48.1	—	292	
	727	—	—	64.4	—	
	800	7.72	59.7	—	301	
	927	—	—	67.5	—	
	1 000	7.88	71.3	—	310	
Palladium	20	—	10.8	75	243	19
	100	11.1	13.8	74	247	
	500	12.4	27.5	—	268	
	1 000	13.6	40	—	297	

(continued)

Table 14.2   THE PHYSICAL PROPERTIES OF PURE METALS AT ELEVATED TEMPERATURES[†]—*continued*

Metal	Temperature $t°C$	Coefficient of expansion $20-t°C$ $10^{-6}\,K^{-1}$	Resistivity at $t°C$ $\mu\Omega\,cm$	Thermal conductivity at $t°C$ $Wm^{-1}\,K^{-1}$	Specific heat at $t°C$ $J\,kg^{-1}\,K^{-1}$	References*
Platinum	20	—	10.58	72	134	7, 19, 23, 25, 50
	27	—	10.8	71.6	—	
	100	9.1	13.6	72	134	
	127	—	14.6	71.8	—	
	500	9.6	27.9	—	147	
	527	—	28.7	75.6	—	
	927	—	—	82.6	—	
	1 000	10.2	43.1	67	159	
	1 127	—	—	87.1	—	
	1 500	11.31	55.4	63	176	
	1 527	—	—	96.1	—	
Plutonium	20 $\alpha \to \alpha$	47	145.8	(8.4)	131	42, 45
	100 $\alpha \to \alpha$	203	141.6	—	138	
	200 $\alpha \to \beta$	173	107.8	—	145	
	300 $\alpha \to \gamma$	181	107.4	—	153	
	400 $\alpha \to \delta$	109	100.7	—	154	
	500 $\alpha \to \varepsilon$	101	110.6	—	144	
Rhenium	20	12.4 ∥-axis	18.7	48	134	26, 16, 27
	100	4.7 ⊥-axis	25		138	
	2 500	7.29(2 000°C)	132	—	209(2 527°C)	
Rhodium	20	—	4.7	149	243	19
	100	8.5	6.2	147	255	
	500	9.8	14.6	—	289	
	1 000	10.8	—	—	331	
Silver	20	—	16.3	419	234	7, 28
	27	—	1.629	429	—	
	100	19.6	2.1	419	222	
	127	—	2.241	425	—	
	500	20.6	4.7	(377)	(230)	
	527	—	4.91	396	—	
	900	22.4	7.6	—	(243)	
	927	—	—	361 (extp)	—	
Tantalum	20	—	13.5	57	138	7, 29, 30, 31, 32
	27	—	13.5	57.5	—	
	100	6.5	17.2	54	142	
	127	—	18.2	57.8	—	
	500	6.6	35	—	151	
	527	—	35.9	59.4	—	
	627	—	40.1	59.8	—	
	1 500	—	71	—	167	
	1 527	—	—	63.4	—	
	2 327	—	—	65.8	—	
	2 500	—	102	—	234(2 727°C)	
	2 727	—	—	66.5	—	
Thallium	20	—	16.6	46	134	33, 6
	100	30	—	45	138	
	200	—	—	45	142	
Tin	0	—	11.5	—	—	7, 34
	20	—	12.6	65	222	
	27	—	—	66.6	—	
	100	23.8	15.8	63	239	
	127	—	—	62.2	—	
	200	24.2	23.0	60	260	
	227	—	—	59.6	—	

(*continued*)

Table 14.2   THE PHYSICAL PROPERTIES OF PURE METALS AT ELEVATED TEMPERATURES[†]—*continued*

Metal	Temperature $t°C$	Coefficient of expansion $20-t°C$ $10^{-6}\,K^{-1}$	Resistivity at $t°C$ $\mu\Omega\,cm$	Thermal conductivity at $t°C$ $Wm^{-1}\,K^{-1}$	Specific heat at $t°C$ $Jkg^{-1}\,K^{-1}$	References*
Titanium	0	—	39	—	—	
	20	—	54	16	519	35, 36
	27	—	—	21.9	—	
	100	8.8	70	15	540	
	127	—	—	20.4	—	
	200	9.1	88	15	569	
	227	—	—	19.7	—	
	327	—	—	19.4	—	
	400	9.4	119	14	619	
	527	—	—	19.7	—	
	600	9.7	152	13	636	
	800	9.9	165	(13)	682	
	927	—	—	22.2	—	
Tungsten	20	—	5.4	167	134	37, 33, 38
	27	—	5.44	174	—	
	100	4.5	7.3	159	138	
	127	—	7.83	159	—	
	500	4.6	18	121	142	
	527	—	18.6	125	—	
	927	—	—	112	—	
	1 000	4.6	33	111	151	
	1 727	—	—	98	—	
	2 000	5.4	65	93	—	
	3 000	6.6	100	—	—	
Uranium	20 α	—	30	27	116	Expansion
	600 α	—	59	38	186	anisotropic
	700 β	—	55.5	40	176	42, 45
	800 γ	—	54	42.3	160	
Vanadium	20	—	24.8	—	492	42, 45
	27	—	20.2	—	—	
	100	8.3	31.5	31	505	
	127	—	28.0	—	—	
	500	9.6	—	36.8	570	
	527	—	53.1	—	—	
	627	—	58.7	—	—	
	700	—	—	35.2	603	
	900	10.4	—	—	636	
Zinc	20	—	5.96	113	389	39, 13, 40, 41, 6
	27	—	6.06	—	—	
	100	31	7.8	109	402	
	127	—	8.37	—	—	
	200	33	11.0	105	414	
	227	—	10.82	—	—	
	300	34	13.0	101	431	
	327	—	13.49	—	—	
	400	—	16.5	96	444	

* Items from the CRC Materials Science and Engineering Handbook, 3rd Edition, are reprinted with permission. Copyright CRC Press, Boca Raton, Florida.
[†] Data in this table are from multiple sources and may not be fully consistent.

REFERENCES TO TABLES 14.1 AND 14.2

1. W. Slough, *Private Communication*, Chemical Standards Division, NPL, 1972.
2. M. J. Swan, *Private Communication*, Electrical Science Division, NPL, 1972.
3. G. W. C. Kaye and T. H. Laby, 'Tables of Physical and Chemical Constants', Longmans, London, 1966.
4. 'Thermophysical Properties of Matter', TPRC Data Series, Volume 4.
5. R. J. Corrnecini and J. Gniewek, Nat. Bureau of Stds. Monograph; 1960.
6. 'Thermophysical Properties of Matter', TPRC Series Volume 1; 1970.
7. US Bur. Stds. Circular C447, 'Mechanical Properties of Metals and Alloys', Washington, 1952.
8. R. Hase, R. Heierberg and W. Walkenborst, *Aluminium*, 1940, **22**, 631.
9. T. G. Peason and H. W. L. Phillips, *Met. Rev.*, 1957, **2**, 305.
10. H. Tsutsumi, *Sci. Rep. Tôhoku Univ.* (1), 1918, **7**, 100.
11. R. W. Powell, *Phil. Mag.*, 1953, **44**, 657.
12. E. F. Northrup and V. A. Suydam, *J. Franklin Inst.*, 1913, **175**, 160.
13. Saldau, *Z. Metallogr.*, 1915, **7**, 5.
14. S. Grabe and E. J. Evans, *Phil. Mag.*, 1935, **19**, 773.
15. R. W. Powell and R. P. Tye, *J. Inst. Metals*, 1957, **85**, 185.
16. C. F. Lucks and H. W. Deem, ASTM Special Tech. Pubn., 1958, No. 227.
17. C. S. Smith and E. W. Palmer, *Trans. AIMME*, 1935, **117**, 225.
18. C. J. Meechan and R. R. Eggleston, *Acta Met.*, 1954, **2**, 680.
19. R. F. Vines, 'The Platinum Metals and their Alloys', New York, 1941.
20. BISRA, 'Physical Constants of Some Commercial Steels at Elevated Temperatures', London, 1953.
21. C. Zwikker, *Physica*, 1927, **7**, 73.
22. E. P. Mikol, US Atomic Energy Comm. Publ. ORNL–1131, 1952.
23. O. H. Kirkorian, UCRL–6132, TID–4500, Sept. 1960, USA.
24. Mond Nichel Co. Ltd., *Nickel Bull.*, 1951, **24**, 1.
25. K. S. Krishnan and S. C. Jain, *Br. J. Appl. Phys.*, 1954, **5**, 426.
26. C. Agte, H. Alterthum *et al.*, *Z. Anorg. Chem.*, 1931, **196**, 129.
27. R. E. Taylor and R. A. Finch, US Atomic Energy Com. Rep., 1961 NAA–SR–6034.
28. US Bur. Min. Circular C412, 'Silver, its Properties and Industrial Uses', Washington, 1936.
29. L. Malter and D. B. Langmuir, *Phys. Rev.*, 1939, **55**, 743.
30. M. Cox, *Phys. Rev.*, 1943, **64**, 241.
31. I. B. Fieldhouse *et al.*, WADC Tech. Rep. 55–495, 1956.
32. N. S. Rasor and J. D. McClelland, WADC Tech. Rep. 56–400, 1957.
33. A. E. van Arkel, 'Reine Metalle', Berlin, 1939.
34. Int. Tin R. and D. Council. Tech. Publ. B1, 1937.
35. E. S. Greiner and W. C. Ellis, Metals Tech., Sept., 1948.
36. L. Silverman, *J. Metals, N.Y.*, 1953, May, p. 631.
37. C. J. Smithells, 'Tungsten', London, 1952.
38. V. S. Gurneyuk and V. V. Lebeden, *Fizica Metall.*, 1961, **11**, 29.
39. F. L. Uffelmann, *Phil. Mag.* (7), 1930, **10**, 633.
40. Lees, *Phil. Trans. R. Soc.*, 1908, **A208**, 432.
41. Dewar and Fleming, *Phil. Mag.*, 1893, **36**, 271.
42. C. A. Hampel, 'Rare Metals Handbook', Chapman & Hall, London, 1961.
43. H. K. Adenstedt, *Trans. A.S.M.*, 1952, **44**, 949.
44. R. P. Cox *et al.*, *Ind. Eng. Chem.*, 1958, **50**, 141.
45. Thermochemical Data Section. *Met. Ref. Book.*
46. R. W. Powell, R. P. Tye and M. J. Woodman, *J. Less Common Metals*, 1967, **12**, 1.
47. CRC Handbook of Chemistry and Physics, 82nd edition, p. 12-219.
48. ASM Metals Handbook, p. 115.
49. CRC Handbook of Chemistry and Physics, 82nd edition, p. 4-133.
50. ASM Metals Reference Book, p. 143.
51. CRC Handbook of Chemistry and Physics, 82nd edition, p. 12-45, 12-221.
52. CRC Materials Science and Engineering Handbook, 3rd edition, p. 384–389.

14.2   The thermophysical properties of liquid metals*

Accurate and reliable information on the thermophysical properties of molten metals becomes increasingly significant as new casting techniques are developed and advancement is made in numerical modelling of these processes. To obtain precise thermophysical data for molten metals and alloys, it is necessary to know the special features of different techniques and to utilise the most suitable one for the specimen of interest.

14.2.1   Density and thermal expansion coefficient

Knowledge of the density of liquid metals is crucial in most theories related to the liquid state and for the simulation of the contraction that occurs during solidification. There are several methods for measuring the density of high-melting liquid metals: balanced columns, pycnometer, immersed-sinker, maximum-bubble, etc. However, the application of all these techniques is limited due to reaction between liquid metal sample and the apparatus. Therefore, an electromagnetic levitation technique can be considered as a good alternative for density measurements in molten metals. The density (ρ) of the specimen can, of course, be determined as

$$\rho = \frac{m}{V},$$

where m and V are the mass and the volume of the specimen, respectively. Assuming a spherical shape for the specimen, the volume can be determined from the radius of the droplet. Repeating the density measurements at different temperatures, the thermal expansion coefficient (β) can be simulated as

$$\beta = \frac{1}{V}\frac{\partial V}{\partial T}.$$

The determination of the thermal expansion coefficient of the materials requires the measurement of the linear size of the specimen as a function of the temperature.

It is experimentally found that the variation of the density of most liquid metals and alloys with temperature (T) is well represented by a linear equation

$$\rho = \rho_m + \frac{\partial \rho}{\partial T}(T - T_m),$$

where ρ_m is the density of the liquid metal or alloy at its melting point T_m. However, for certain metals (aluminium, gallium, antimony) this relationship is not linear. Table 14.3a presents density data available for liquid metals at their melting point and their temperature dependence ($\partial \rho / \partial t$) from [1, 2].

14.2.2   Surface tension

The surface tension is determined by the microscopic structure of the liquid near the surface. At a liquid–vapour interface the density changes severely from a high value in the liquid state to a very low value in the gas phase. Therefore, surface atoms experience an attraction toward the liquid phase, which is the cause of the surface tension. Due to its energetic and entropic origin, it is necessary to calculate the free energy of the system in order to determine the surface tension. Thus the surface tension is determined, as the additional free energy required to generate a unit surface area separating the liquid from its vapour phase. There are many techniques for surface tension measurements: sessile drop, pendant drop, maximum bubble pressure, maximum pressure in a drop, detachment or maximum pull, capillary-rise, drop weight, and oscillating drop methods. The sessile-drop technique has been widely used due to its many advantages. The sessile drop method utilises a molten drop resting on a horizontal ceramic substrate, and allows measurements over a wide range of temperatures. However, the surface tension data is affected by contaminants. To avoid the contamination effects the surface tension of a liquid droplet can be measured by exciting surface oscillations. The frequency of the oscillations is related to the surface tension.

---

* For the physical properties of molten salts, see Chapter 9.

Table 14.3a   THE THERMOPHYSICAL PROPERTIES OF LIQUID METALS

Density, surface tension and viscosity

		Density		Surface tension		Viscosity		
Metal	Temp. T_m K	ρ_m 10^3 kg m^{-3}	$-(\partial\rho/\partial T)$ 10^{-1} kg m^{-3} K^{-1}	γ_m mN m^{-1}	$-(d\gamma/dT)$ mN m^{-1}K^{-1}	η_{mp} mN s m^{-2}	η_0 mN s m^{-2}	E kJ mol^{-1}
Ag	1 233.7	9.33	9.1	966	0.19	3.88	0.453 2	22.2
Al	933	2.385	3.5	914	0.35	1.30	0.149 2	16.5
As	1 090	5.22	5.4	—	—	—	—	—
Au	1 336	17.36	15	1 169	0.25	5.0	1.132	15.9
B	2 350	2.08	—	1 060				
Ba	1 000	3.321	2.7	277	0.08	—	—	—
Be	1 556	1.690	1.2	1 390	0.29	—	—	—
Bi	544	10.05	11.8	378	0.07	1.8	0.445 8	6.45
Ca	1 138	1.365	2.2	361	0.10	1.22	0.065 1	27.2
Cd	594	8.01	12.2	570	0.26	2.28	0.300 1	10.9
Ce	1 077	6.685	2.3	740	0.33	2.88	—	—
Co	1 766	7.76	10.9	1 873	0.49	4.18	0.255 0	44.4
Cr	2 148	6.29	7.2	1 700	0.32	—	—	—
Cs	301.6	1.84	5.7	70	0.06	0.68	0.102 2	4.81
Cu	1 356	8.000	8.0	1 303	0.23	4.0	0.300 9	30.5
Fe	1 809	7.03	8.8	1 872	0.49	5.5	0.369 9	41.4
Fr	291	2.35	7.92	62	0.044	0.765	—	—
Ga	302.8	6.10	5.6	718	0.10	2.04	0.435 9	4.00
Gd	1 585	7.14	—	810	0.16	—	—	—
Ge	1 207	5.49	4.9	621	0.26	0.73	—	—
Hf	2 216	11.1	—	1 630	0.21	—	—	—
Hg	234.13	13.691	2.436	498	0.20	2.10	0.556 5	2.51
	273	13.595	—	—	—	—	—	—
	293	13.546	—	—	—	—	—	—
	373	13.352	—	—	—	—	—	—
In	429.6	7.03	6.8	556	0.09	1.89	0.302 0	6.65
Ir	2 716	20.0	—	2 250	0.31	—	—	—
K	336.5	0.827	2.4	115	0.08	0.51	0.134 0	5.02
La	1 203	5.955	2.4	720	0.32	2.45	—	—
Li	453.5	0.518	1.0	398	0.14	0.57	0.145 6	5.56
Mg	924	1.590	2.6	559	0.35	1.25	0.024 5	30.5
Mn	1 514	5.76	9.2	1 090	0.2	—	—	—
Mo	2 880	9.34	—	2 250	0.31	—	—	—
Na	369.5	0.927	2.35	191	0.10	0.68	0.152 5	5.24
Nb	2 741	7.83	—	1 900	0.24	—	—	—
Nd	1 297	6.688	5.3	689	0.09	—	—	—
Ni	1 727	7.905	11.9	1 778	0.38	4.90	0.166 3	50.2
Os	3 000	20.1	—	2 500	0.33	—	—	—
P	317	—	—	52	—	1.71	—	—
Pb	600	10.678	13.2	458	0.13	2.65	0.463 6	8.61
Pd	1 825	10.49	12.3	1 500	0.22	—	—	—
Pr	1 208	6.611	2.5	—	—	2.80	—	—
Pt	2 042	18.91	28.8	1 800	0.17	—	—	—
Pu	913	16.65	14.1	550	0.10	6.0	1.089	5.59
Rb	311.9	1.48	4.5	86	0.06	0.67	0.094 0	5.15
Re	3 431	18.8	—	2 700	0.34	—	—	—
Rh	2 239	10.8	—	2 000	0.30	—	—	—
Ru	2 700	10.9	—	2 250	0.31	—	—	—
S	392	1.819	8.00	61	0.07	12	—	—
Sb	903.5	6.483	8.2	367	0.05	1.22	0.081 2	22.0
Se	490	4.00	11.7	106	0.1	24.8	—	—
Si	1 683	2.53	3.5	865	0.13	0.94	—	—
Sn	505	6.98	6.1	560	0.09	1.85	0.538 2	—
Sr	1 043	2.37	2.6	303	0.10	—	—	—
Ta	3 250	15.0	—	2 150	0.25	—	—	—
Te	724	5.80	7.3	180	0.06	2.14	—	—
Th	1 964	10.5	—	978	0.14	—	—	—
Ti	1 958	4.13	2.3	1 650	0.26	5.2	—	—

(continued)

Table 14.3a   THE THERMOPHYSICAL PROPERTIES OF LIQUID METALS—*continued*

Metal	Temp. T_m K	Density ρ_m 10^3 kg m^{-3}	$-(\partial\rho/\partial T)$ 10^{-1} kg m^{-3} K^{-1}	Surface tension γ_m mN m^{-1}	$-(d\gamma/dT)$ mN m^{-1}K^{-1}	Viscosity η_{mp} mN s m^{-2}	η_0 mN s m^{-2}	E kJ mol^{-1}
Tl	575	11.35	13.0	464	0.08	2.64	0.298 3	10.5
U	1 406	17.27	10.3	1 550	0.14	6.5	0.484 8	30.4
V	2 185	5.36	3.2	1 950	0.31	—	—	—
W	3 650	17.6	—	2 500	0.29	—	—	—
Yb	1 097	—	—	—	—	1.07	—	—
Zn	692	6.575	9.8	782	0.17	3.85	0.413 1	12.7
Zr	2 123	5.8	—	1 480	0.20	8.0	—	—

The temperature dependence of surface tension is related to the surface entropy and the surface excesses. Therefore, the changes in the structure of the liquid specimen with temperature are reflected in the temperature coefficients of the surface tension. The variation of the surface tension with the temperature for most liquid metals can be expressed by a linear relationship

$$\gamma_T = \gamma_m - \frac{d\gamma}{dT}(T - T_m),$$

where γ_T and γ_m are the surface tensions at temperature T and at the melting point T_m, respectively. Experimental values of surface tension for liquid metals at their melting points, and their temperature coefficients of surface tension are presented in Table 14.3b. The data are taken from Allen.[3,2]

14.2.3   Viscosity

Viscosity is one of the most important transport properties of molten metals. It is related to the internal friction within the liquid and provides some information about the structure of the material. The most crucial hydrodynamic criteria such as the Reynolds number, the Rayleigh number, the Hartmann number, and the Marangoni number contain viscosity. The existing methods for measuring viscosities of liquids are restricted for liquid metals and alloys due to their low viscosities, high melting points, and chemical reactivity. The capillary, oscillating-vessel, rotational, and oscillating plate methods are most suitable techniques for determination of the viscosity of liquid metals and alloys. The viscosity η relates the shear stress τ to the shear rate $\dot{\gamma}$

$$\tau = \eta\dot{\gamma}.$$

By analogy to the Wiedemann–Franz–Lorenz law and the Stokes–Einstein relation, there is the following relationship between viscosity and surface tension

$$\frac{\gamma}{\eta} = \frac{15}{16}\sqrt{\frac{\kappa_B T}{m}},$$

where κ_B is the Boltzmann's constant and m is the atomic mass.

Since the viscosity measurements are not so responsive to convection as diffusion measurements, the diffusivity D may be estimated from the viscosity data using Stokes–Einstein theory

$$D = \frac{\kappa_B T}{6\pi R\eta}.$$

For most liquid metals and alloys the variation of viscosity with temperature may be determined as

$$\eta = \eta_0 \exp\left(\frac{E}{RT}\right),$$

where η_0 and E are constants, and are given in the Table 14.3a for liquid metals,[4,5] and R is the gas constant, $8.314\,4$ J K^{-1} mol^{-1}.

Table 14.3b   THE THERMOPHYSICAL PROPERTIES OF LIQUID METALS

Heat capacity, thermal conductivity and electrical resistivity

Metal	Temperature K	Heat capacity J g^{-1} K^{-1}	Thermal conductivity Wm^{-1} K^{-1}	Electrical resistivity μΩ m
Ag	1 233.7	0.283	174.8	0.172 5
	1 273	0.283	176.5	0.176 0
	1 373	0.283	180.8	0.184 5
	1 473	0.283	185.1	0.193 5
	1 573	0.283	189.3	0.202 3
	1 673	—	193.5	0.211 1
Al	933	1.08	94.05	0.242 5
	973	1.08	95.37	0.248 3
	1 073	1.08	98.71	0.263 0
	1 173	1.08	102.05	0.277 7
	1 273	—	105.35	0.292 4
As	1 090	—	—	2.10
Au	1 336	0.149	104.44	0.312 5
	1 373	0.149	105.44	0.318 0
	1 473	0.149	108.15	0.331 5
	1 573	0.149	110.84	0.348 1
	1 673	0.149	113.53	0.363 1
B	2 350	2.91	—	2.10
Ba	1 000	0.228	—	1.33
Be	1 556	3.48	—	0.45
Bi	544	0.146	17.1	1.290
	573	0.143	15.5	—
	673	0.147 5	15.5	—
	773	0.137 5	15.5	—
	873	0.133 6	15.5	—
	973	—	15.5	—
Ca	1 138	0.775	—	0.250
Cd	594	0.264	42	0.337
	673	0.264	47	0.343 0
	773	0.264	54	0.351 0
	873	0.264	61	0.360 7
Ce	1 077	0.25	—	1.268
	1 273	0.25	—	1.294
	1 473	0.25	—	1.310
Co	1 766	0.59	—	1.02
Cr	2 176	0.78	—	0.316
Cs	301.6	0.28	19.7	0.370
	373	0.265	20.2	0.450
	473	0.240	20.8	0.565
	673	0.21	20.2	0.810
	873	0.22	18.3	1.125
	1 073	0.25	16.1	1.570
	1 873	—	4.0	—
Cu	1 356	0.495	165.6	0.200
	1 373	0.495	166.1	0.202
	1 473	0.495	170.1	0.212
	1 673	0.495	176.3	0.233
	1 873	0.495	180.4	0.253
Fe	1 809	0.795	—	1.386
Fr	291	0.142	—	0.87
	973	0.134	—	—
Ga	302.8	0.398	25.5	0.26
	373	0.398	30.0	0.27
	473	0.398	35.0	0.28
	573	0.398	39.2	0.30
Gd	1 623	0.213	—	0.278
Ge	1 207	0.404	—	0.672
	1 273	0.404	—	0.727
Hf	2 500	—	—	2.18

(continued)

Table 14.3b THE THERMOPHYSICAL PROPERTIES OF LIQUID METALS—*continued*

Metal	Temperature K	Heat capacity $J\,g^{-1}\,K^{-1}$	Thermal conductivity $Wm^{-1}\,K^{-1}$	Electrical resistivity $\mu\Omega\,m$
Hg	234.13	0.142	6.78	0.905
	273	0.142	7.61	0.940
	293	0.139	8.03	0.957
	373	0.137	9.47	1.033
	773	0.137	12.67	1.600
	1 273	—	8.86	3.77
	1 733	—	~0.000 4	~1 000
Ho	1 773	0.203	—	1.93
In	429.6	0.259	42	0.323 0
	473	0.259	—	0.333 9
	673	0.259	—	0.436 1
	873	0.259	—	0.513 1
K	336.5	0.820	53.0	0.136 5
	373	0.810	51.7	0.154
	473	0.790	47.7	0.215
	773	0.761	37.8	0.444
	1 273	0.838	24.4	0.110
	1 773	—	15.5	
La	1 203	0.057 5	21.0	1.38
	1 273	0.057 5	—	1.43
	1 373	0.057 5	—	1.50
	1 473	0.057 5	—	1.56
Li	453.5	4.370	46.4	0.240
	473	4.357	47.2	—
	673	4.215	53.8	—
	873	4.165	57.5	—
	1 073	4.148	58.6	—
	1 273	4.147	58.4	—
	1 873	4.36	52.0	—
Mg	923	1.36	78	0.274
	973	1.36	81	0.277
	1 073	1.36	88	0.282
	1 273	1.36	100	—
Mn	397	0.838	—	0.40
Mo	2 880	0.57	—	0.605
Na	370	1.386	89.7	0.096 4
	373	1.385	89.6	0.099
	473	1.340	82.5	0.134
	673	1.278	71.6	0.224
	873	1.255	62.4	0.326
	1 073	1.270	53.7	0.469
	1 273	1.316	45.8	—
	1 473	1.405	38.8	—
Nb	2 741	—	—	1.05
Nd	1 297	0.232	—	1.26
Ni	1 727	0.620	—	0.850
P	317	—	—	2.70
Pb	600	0.152	15.4	0.948 5
	673	0.144	16.6	0.986 3
	773	0.137	18.2	1.034 4
	873	0.135	19.9	1.082 5
	1 073	—	—	1.169
	1 273	—	—	1.263
Po	527	—	—	3.98
Pr	1 208	0.238	—	1.38
Pt	2 043	0.178	—	0.73
Pu	913	—	—	1.33
Ra	1 233	0.136	—	1.71
Rb	311.8	0.398	33.4	0.228 3
	373	0.383	33.4	0.273 0
	473	0.364	31.6	0.366 5

(*continued*)

Table 14.3b   THE THERMOPHYSICAL PROPERTIES OF LIQUID METALS—*continued*

Metal	Temperature K	Heat capacity J g^{-1} K^{-1}	Thermal conductivity W m^{-1} K^{-1}	Electrical resistivity μΩ m
	773	0.348	26.1	0.689 0
	1 273	0.378	17.0	1.71
	1 773	—	8.0	5.32
Re	3 431	—	—	1.45
Ru	2 700	—	—	0.84
S	392	0.984	—	>10^{10}
Sb	903.5	0.258	21.8	1.135
	973	0.258	21.3	1.154
	1 073	0.258	20.9	1.181
	1 273	0.258	—	1.235
Sc	1 812	0.745	—	1.31
Se	490	0.445	0.3	~10^{6}
Si	1 683	1.04	—	0.75
	1 773	1.04	—	0.82
	1 873	1.04	—	0.86
Sm	1 345	0.223	—	1.90
Sn	505	0.250	30.0	0.472 0
	573	0.242	31.4	0.490 6
	673	0.241	33.4	0.517 1
	773	0.24	35.4	0.543 5
	1 273	0.26	—	0.670
Sr	1 043	0.354	—	0.58
Ta	3 269	—	—	1.18
Tb	1 638	—	—	2.44
Te	723	0.295	2.5	5.50
	773	0.295	3.0	4.80
	873	0.295	4.1	4.30
	1 073	0.295	6.2	3.9
	1 273	0.295	—	3.8
Ti	1 958	0.700	—	1.72
Tl	576	0.149	24.6	0.731
	673	0.149	—	0.759
	773	0.149	—	0.788
Tm	1 873	—	—	1.88
U	1 406	0.161	—	0.636
	1 473	0.161	—	0.653
	1 573	0.161	—	0.678
V	2 185	0.780	—	0.71
W	3 650	—	—	1.27
Y	1 803	0.377	—	1.04
Yb	1 097	—	—	1.64
Zn	692.5	0.481	49.5	0.374
	773	0.481	54.1	0.368
	873	0.481	59.9	0.363
	1 073	0.481	60.7	0.367
Zr	2 123	0.367	—	1.53

14.2.4   Heat capacity

Heat capacity (C) is one of the most essential thermodynamic properties of metals. A knowledge of the heat capacity and its temperature dependence allows prediction of the enthalpy and entropy of a material. It is determined as

$$C = \frac{\delta q}{dT},$$

where δq is an infinitesimal heat quantity added to (or withdrawn from) the matter and dT is the resulting infinitesimal temperature change. The specific heat is traditionally measured by adiabatic calorimetry techniques. The calorimeter is isolated from the environment, and requires long relaxation times. Heat capacities determined at constant volume or at constant pressure are of most general importance. Table 14.3b lists values of constant-pressure heat capacity for liquid elements at different temperatures.[6–8]

14.2.5   Electrical resistivity

Information on the electrical resistivity of molten metals and alloys is especially important in many metallurgical processes such as electroslug remelting, electromagnetic stirring in continuous casting, electrolysis, and induction melting in foundries. Due to the disordered arrangement of ions in the liquid state, molten metals and alloys exhibit higher ($\sim$1.5–2.3 times) electrical resistivity than those in the solid state. However, relatively few studies have been reported on the electrical resistivity of molten metals and alloys, particularly at elevated temperatures, since the measurements are extremely difficult. The methods of electrical resistivity measurements can be categorised into three groups: direct resistance measurements using contact probes, contactless inductive measurements, and non-contact containerless measurement techniques.

The technique of choice for solid materials is the direct resistance four-probe method, which is based upon application of Ohm's law. Although this direct method can be applied to low melting point, non-reactive liquid materials, reactions between the probes and the molten sample preclude using the four-probe method with high-melting point materials. Also, this technique is limited to materials with very narrow freezing ranges.

Inductive techniques for measuring electrical resistivity are contactless and thus prevent chemical reactions between molten samples and contacting probes as in the direct method. There are two different types of contactless methods. The rotating field method is usually based on the phenomenon that when a metal sample rotates in a magnetic field (or the magnetic field rotates around a stationary sample), circulating eddy currents are induced in the sample, which generate an opposing torque proportional to the electrical conductivity of the sample.[9,10] In liquid metals and alloys the applied magnetic field also causes significant rotation of the liquid in the crucible, which decreases the angular velocity between the field and the sample.

In both direct resistance and contactless inductive techniques chemically reactive components contained in the crucible may easily contaminate the molten sample, altering its electrical properties. The container may provide heterogeneous nucleants that generate early solid-phase nucleation. Electromagnetic levitation is another technique for containerless measurements of electrical conductivity in liquid metals.[11] It combines the containerless positioning method of electromagnetic levitation with the contactless technique of inductive electrical conductivity measurement.

The electrical resistivity ρ_e of liquid metals (except Cd, Zn, Hg, Te and Se) increases linearly with increasing temperature

$$\rho_e = aT + b,$$

where a and b are temperature coefficients of the resistivity, and are determined experimentally. The electrical resistivities of most liquid metallic elements at different temperatures are shown in Table 14.3b.[6–8]

14.2.6   Thermal conductivity

Accurate measurement of thermal conductivity of liquid metals and alloys is usually more difficult than the measurement of electrical conductivity and thermal diffusivity. The source of difficulty is mainly related to problems with making accurate heat flow measurements. Also there is possibility of some flows in the liquid sample. Thermal conductivity is directly related to the change in the atomic vibrational frequency. For a number of non-metallic substances it is found that

$$\frac{\lambda}{\sqrt{M}} = 2.4 \times 10^{-3},$$

where λ is the thermal conductivity, M is the molecular weight. Since free electrons are responsible for the electrical and thermal conductivities of conductors in both solid and liquid states, many researchers use the Wiedemann-Franz-Lorenz law to relate the thermal conductivity to the electrical resistivity:

$$\frac{\lambda \rho_e}{T} = \frac{\pi \kappa^2}{3e^2} \equiv L_0,$$

where κ is the Boltzmann constant, e is the electron charge. The constant

$$L_0 = \frac{\pi^2 \kappa^2}{3e^2} = 2.45 \times 10^{-8} \, W \, \Omega \, K^{-2},$$

is the Lorenz number. The validity of this relationship was confirmed experimentally with high accuracy by many researchers. The thermal conductivity values of various liquid metals at different temperatures are given in Table 14.3b.[6–8]

REFERENCES FOR SECTION 14.2

1. D. J. Steinberg, *Metall. Trans.*, 1974, **5**, 1341.
2. T. Iida and R. I. L. Guthrie, 'The Physical Properties of Liquid Metals', Clarendon Press, Oxford, UK, 1988.
3. B. C. Allen, 'Liquid Metals' (ed. S. Z. Beer), Marcel Dekker, New York, NY, 1972, 162–212.
4. R. T. Beyer and E. M. Ring, 'Liquid Metals' (ed. S. Z. Beer), Marcel Dekker, New York, NY, 1972, p. 450.
5. L. J. Wittenberg and D. Ofte, *Techniques of Metals Research*, 1970, **4**, 193.
6. A. V. Grosse, *Revue Hautes Temp. & Refrac.*, 1966, **3**, 115.
7. D. R. Stull and G. C. Sinke, 'Thermodynamic Properties of the Elements', *Amer. Chem. Soc.*, 1956.
8. R. Hultgren *et al.*, 'Selected Values of Thermodynamic Properties', Wiley, 1963.
9. S. I. Bakhtiyarov and R. A. Overfelt, *J. Materials Science*, 1999, **34**, 945–949.
10. S. I. Bakhtiyarov and R. A. Overfelt, *Acta Materialia*, 1999, **47**, 4311–4319.
11. S. I. Bakhtiyarov and R. A. Overfelt, Annals of New York Academy of Sciences, New York, NY, 2002.

14.3   The physical properties of aluminium and aluminium alloys

Table 14.4a   THE PHYSICAL PROPERTIES OF ALUMINIUM AND ALUMINIUM ALLOYS AT NORMAL TEMPERATURES

				Sand cast			
Material	*Nominal composition* %		*Density* g cm^{-3}	*Coefficient of expansion* 20–100°C 10^{-6} K^{-1}	*Thermal conductivity* 100°C Wm^{-1} K^{-1}	*Resistivity* μΩ m	*Modulus of elasticity* MPa × 10^3
Al	Al	99.5	2.70	24.0	218	3.0	69
	Al	99.0	2.70	24.0	209	3.1	—
Al–Cu	Cu	4.5	2.75	22.5	180	3.6	71
	Cu	8	2.83	22.5	138	4.7	—
	Cu	12	2.93	22.5	130	4.9	—
Al–Mg	Mg	3.75	2.66	22.0	134	5.1	—
	Mg	5	2.65	23.0	130	5.6	—
	Mg	10	2.57	25.0	88	8.6	71
Al–Si	Si	5	2.67	21.0	159	4.1	71
	Si	11.5	2.65	20.0	142	4.6	—
Al–Si–Cu	Si	10	2.74	20.0	100	6.6	71
	Cu	1.5					
	Si	4.5	2.76	21.0	134	4.9	71
	Cu	3					

(continued)

Table 14.4a   THE PHYSICAL PROPERTIES OF ALUMINIUM AND ALUMINIUM ALLOYS AT NORMAL
TEMPERATURES—*continued*

				Sand cast			
Material	*Nominal composition* %		*Density* g cm^{-3}	*Coefficient of expansion* 20–100°C 10^{-6} K^{-1}	*Thermal conductivity* 100°C Wm^{-1} K^{-1}	*Resistivity* μΩ m	*Modulus of elasticity* MPa × 10^3
Al–Si–Cu–Mg*	Si	17	2.73	18.0	134	8.6	88
	Cu	4.5					
	Mg	0.5					
Al–Cu–Mg–Ni	Cu	4	2.78	22.5	126	5.2	71
(Y alloy)	Mg	1.5					
	Ni	2					
Al–Cu–Fe–Mg	Cu	10	2.88	22.0	138	4.7	71
	Fe	1.25					
	Mg	0.25					
Al–Si–Cu–Mg–Ni	Si	12	2.71	19.0	121	5.3	71
(Lo-Ex)	Cu	1					
	Mg	1					
	Ni	2					
	Si	23	2.65	16.5	107	—	88
	Cu	1					
	Mg	1					
	Ni	1					

* Die cast.

Table 14.4b   THE PHYSICAL PROPERTIES OF ALUMINIUM AND ALUMINIUM ALLOYS AT NORMAL TEMPERATURES

					Wrought				
Specification	*Nominal composition* %		*Condition**	*Density* g cm^{-3}	*Coefficient of expansion* 20–100°C 10^{-6} K^{-1}	*Thermal conductivity* 100°C Wm^{-1} K^{-1}	*Resistivity* μΩ cm	*Temp. coeff. of resistance* 20–100°C	*Modulus of elasticity* MPa × 10^3
1199	Al	99.992	Sheet H111			239	2.68	0.004 2	69
			H18	2.70	23.5	234	2.70	0.004 2	69
			Extruded			239	2.68	0.004 2	69
1080A	Al	99.8	Sheet H111			234	2.74	0.004 2	69
			H18	2.70	23.5	230	2.76	0.004 2	69
			Extruded			230	2.79	0.004 1	69
1050A	Al	99.5	Sheet H111			230	2.80	0.004 1	69
			H18	2.71	23.5	230	2.82	0.004 1	69
			Extruded			226	2.85	0.004 1	69
1200	Al	99	Sheet H111			226	2.87	0.004 0	69
			H18	2.71	23.5	226	2.89	0.004 0	69
			Extruded			226	2.86	0.004 0	69
2014A	Cu	4.4	T4	2.8	22	142	5.3		74
	Mg	0.7	T6	2.8	22	159	4.5		
	Si	0.8							
	Mn	0.75							
2024	Cu	4.5	T3	2.77	23		5.7		73
	Mg	1.5	T6	2.77	23	151	5.7		73
	Mn	0.6							
2090	Cu	2.7	T8	2.59	23.6	88.2	9.59		76
	Li	2.3							
	Zr	0.12							
2091	Cu	2.1	T8	2.58	23.9	84	9.59		75
	Li	2.0							
	Mg	1.50							
	Zr	0.1							
3103	Mn	1.25	Sheet H111						
			H12						
			H14	2.74	23.0	180	3.9	0.003 0	69
			H16						
			H18						
			Extruded			151	4.8	0.002 4	—

(*continued*)

Table 14.4b   THE PHYSICAL PROPERTIES OF ALUMINIUM AND ALUMINIUM ALLOYS AT NORMAL TEMPERATURES—*continued*

				Wrought				
Specification	*Nominal composition* %	*Condition**	*Density* g cm^{-3}	*Coefficient of expansion* 20–100°C 10^{-6} K^{-1}	*Thermal conductivity* 100°C Wm^{-1} K^{-1}	*Resistivity* μΩ cm	*Temp. coeff. of resistance* 20–100°C	*Modulus of elasticity* MPa × 10³
5083	Mg 4.5 Mn 0.7 Cr 0.15	Sheet	H111 H12 H14	2.67	24.5	109	6.1	0.001 9	71
5251	Mg 2.0 Mn 0.3	Sheet Extruded	H111 H13 H16	2.69	24	155 147	4.7 4.9	0.002 5 0.002 3	70
5154A	Mg 3.5	Sheet Extruded	H111 H14	2.67	23.5	142 138 134	5.3 5.4 5.7	0.002 1 0.002 1 0.001 9	70 — —
5454	Mg 2.7 Mn 0.75 Cr 0.12	Sheet	H111 H22 H24	2.68	24	147	5.1		70
Al–Li	Li 2.0	Sheet	T6	2.56	—	—	—	—	77
Al–Mg–Li	Mg 3.0 Li 2.0	Sheet	T6	2.52	—	—	—	—	79
Al–Li–Mg	Li 3.0 Mg 2.0	Sheet	T6	2.46	—	—	—	—	84
6061	Mg 1.0 Si 0.6 Cu 0.2 Cr 0.25	Bar	H111 T4 T6	2.7 2.7 2.7	23.6 23.6 23.6	180 154 167			68.9 68.9 68.9
6063	Mg 0.5 Si 0.5	Extruded	T4 T6	2.70	23.0	193 201	3.5 3.3	0.003 3 0.003 5	71 —
6063A	Mg 0.5 Si 0.5	Bar	T4 T5 T6	2.7	24 24 24	197 209 201	3.5 3.2 3.3		69 69 69
6082	Mg 1.0 Si 1.0 Mn 0.7	Bar/Extruded	T4 T6	2.7 2.7	23 23	172 184	4.1 3.7	0.003 1 0.003 1	69 69
6082	Mg 1.0 Si 1.0	Sheet	T4 T6	2.69	23.0	188 193	3.6 3.4	0.003 3 0.003 5	69 —
6463	Mg 0.65 Si 0.4	Bar	T5 T6	2.71 2.71	23.4 23.4	209 201	3.1 3.3		69 69
Al–Cu–Mg–Si (Duralumin)	Cu 4.0 Mg 0.6 Si 0.4 Mn 0.6	Sheet	T6	2.80	22.5	147	5.0	0.002 3	73
	Cu 4.5 Mg 0.5 Si 0.75 Mn 0.75	Sheet	T4 T6	2.81	22.5	147 159	5.2 4.5	0.002 2 0.002 6	73 —
Al–Cu–Mg–Ni (Y alloy)	Cu 4.0 Mg 1.5 Ni 2.0	Forgings	T6	2.78	22.5	151	4.9	0.002 3	72
Al–Si–Cu–Mg (Lo–Ex)	Si 12.0 Cu 1.0 Mg 1.0 Ni 1.0	Forgings	T6	2.66	19.5	151	4.9	0.002 3	79
Al–Zn–Mg	Zn 10.0 Cu 1.0 Mn 0.7 Mg 0.4	Forgings		2.91	23.5	151	4.9	0.002 3	—
7075	Zn 5.7 Mg 2.6 Cu 1.6 Cr 0.25	Extrusion	T6	2.80	23.5	130	5.7	0.002 0	72
8090	Li 2.5 Cu 1.3 Mg 0.95 Zr 0.1	Plate		2.55	21.4	93.5	9.59		77

(*continued*)

Table 14.4b THE PHYSICAL PROPERTIES OF ALUMINIUM AND ALUMINIUM ALLOYS AT NORMAL TEMPERATURES—*continued*

				Wrought				
Specification	*Nominal composition* %		*Condition**	*Density* g cm^{-3}	*Coefficient of expansion* 20–100°C 10^{-6} K^{-1}	*Thermal conductivity* 100°C Wm^{-1} K^{-1}	*Resistivity* μΩ cm	*Temp. coeff. of resistance* 20–100°C	*Modulus of elasticity* MPa × 10^3
Al–Cu–Mg–Si (Duralumin)	Cu 4.0 Mg 0.6 Si 0.4 Mn 0.6		Sheet TF	2.80	22.5	147	5.0	0.002 3	73
	Cu 4.5 Mg 0.5 Si 0.75 Mn 0.75		Sheet TB TF	2.81	22.5	147 159	5.2 4.5	0.002 2 0.002 6	73 —
Al–Cu–Mg–Ni (Y alloy)	Cu 4.0 Mg 1.5 Ni 2.0		Forgings TF	2.78	22.5	151	4.9	0.002 3	72
Al–Si–Cu–Mg (Lo-Ex)	Si 12.0 Cu 1.0 Mg 1.0 Ni 1.0		Forgings TF	2.66	19.5	151	4.9	0.002 3	79
Al–Zn–Mg	Zn 10.0 Cu 1.0 Mn 0.7 Mg 0.4		Forgings	2.91	23.5	151	4.9	0.002 3	—
7075	Zn 5.7 Mg 2.6 Cu 1.6 Cr 0.25		Extrusion TF	2.80	23.5	130	5.7	0.002 0	72
8090	Li 2.5 Cu 1.3 Mg 0.95 Zr 0.1		Plate	2.55	21.4	93.5	9.59		77

*O = Annealed.
H111 = Annealed.
H12,22 = Quarter hard.

H14,24 = Half hard.
H16,26 = Three-quarters hard.
H18,28 = Hard.

T4 = Solution treated and naturally aged.
T6 = Solution treated and artificially aged.
See also pp. **22**–1 and **22**–2.

14.4 The physical properties of copper and copper alloys

Table 14.5 THE PHYSICAL PROPERTIES OF COPPER AND COPPER ALLOYS AT NORMAL TEMPERATURES

Material	*Composition* %		*Density* g cm^{-3}	*Melting point of liquidus* °C	*Coefficient of expansion* 25–300°C 10^{-6} K^{-1}	*Electrical conductivity* 20°C %IACS*	*Thermal conductivity* Wm^{-1} K^{-1}	*Refs.*
Oxygen-free high conductivity copper	Cu 99.99+		8.94	1 083	17.7	101.5	399	1
Tough pitch HC copper	O$_2$ 0.03		8.92	1 083	17.7	101.5	397	2,3,4
Phosphorus-deoxidised non-arsenical copper	P 0.005–0.012 P 0.013–0.050		8.94 8.94	1 083 1 083	17.7 17.7	85–96 70–90	341–395 298–372	5 5
Deoxidised arsenical copper	P 0.03 As 0.35		8.94	10.82	17.4	45	177	2
Silver bearing copper	O$_2$ 0.02 Ag 0.05		8.92	1 079	17.7	101	397	2,3

(*continued*)

Table 14.5 THE PHYSICAL PROPERTIES OF COPPER AND COPPER ALLOYS AT NORMAL TEMPERATURES—*continued*

Material	Composition %		Density g cm^{-3}	Melting point of liquidus °C	Coefficient of expansion 25–300°C 10^{-6} K^{-1}	Electrical conductivity 20°C %IACS*	Thermal conductivity Wm^{-1} K^{-1}	Refs.
Tellurium copper	Cu	99.5	8.94	1 082	17.7	98	382	2
	Te	0.5						
Chromium copper	Cu	99.4	8.89	1 081	17	45[1]	167	6
	Cr	0.6				82[2]	188	
Beryllium copper	Be	1.85	8.25	1 000	17	17[1]	84	7
	Co	0.25				23[2]	105	
	Be	0.5	8.75	1 060	17	23[1]	126	7
	Co	2.5				47[2]	210	
Cadmium copper	Cu	99.2	8.94	1 080	17	85	376	6
	Cd	0.8						
Sulphur copper	Cu	99.65	8.92	10.75	17	95	373	5
	S	0.35						
Cap copper	Cu	95	8.85	1 065	18.1	56	234	5
	Zn	5						
Gilding metals CuZn10	Cu	90	8.80	1 040	18.2	44	188	5
	Zn	10						
CuZn15	Cu	85	8.75	1 020	18.7	37	159	5
	Zn	15						
CuZn20	Cu	80	8.65	1 000	19.1	32	138	5
	Zn	20						
Brass CuZn30	Cu	70	8.55	965	19.9	28	121	5
	Zn	30						
CuZn33	Cu	67	8.50	940	20.2	27	121	5
	Zn	33						
CuZn37	Cu	63	8.45	920	20.5	26	125	5
	Zn	37						
CuZn40	Cu	60	8.40	900	20.8	28	126	5
	Zn	40						
Aluminium brass CuZn22Al2	Cu	76	8.35	1 010	18.5	23	101	5
	Zn	22						
	Al	2						
Naval brass CuZn36Sn	Cu	62	8.40	915	21.2	26	117	5
	Zn	37						
	Sn	1						
Free cutting brass CuZn39Pb3	Cu	58	850	900	20.9	26	109	3, 5
	Zn	39						
	Pb	3						
Hot stamping brass CuZn40Pb2	Cu	58	8.45	910	20.9	26	109	3, 5
	Zn	40						
	Pb	2						
High tensile brass	Cu	54–62	8.3–8.4	990 approx.	21 approx.	20–25	88–109	5
	Others 7 max.							
	Zinc—balance							
Nickel silver 10%	Cu	62	8.60	1 010	16.4	8.31	37	8, 9
	Ni	10						
	Zn	28						
12%	Cu	62	8.64	1 025	16.2	7.71	30	8, 9
	Ni	12						
	Zn	26						
15%	Cu	62	8.69	1 060	16.2	7.01	27	8, 9
	Ni	15						
	Zn	23						

(*continued*)

Table 14.5 THE PHYSICAL PROPERTIES OF COPPER AND COPPER ALLOYS AT NORMAL TEMPERATURES—*continued*

Material	Composition %		Density g cm^{-3}	Melting point of liquidus °C	Coefficient of expansion 25–300°C 10^{-6} K^{-1}	Electrical conductivity 20°C %IACS*	Thermal conductivity Wm^{-1} K^{-1}	Refs.
18%	Cu Ni Zn	62 18 20	8.72	1 100	16.0	6.3	28	8, 9
25%	Cu Ni Zn	62 25 13	8.82	1 160	17.0	5.1	21	8, 9
Phosphor bronze CuSn3P	Sn P	3.5 0.12	8.85	1 070	18.8	18.8	85	5, 10
CuSn5P	Sn P	5 0.09	8.85	1 060	18.0	16.8	75	5,10
CuSn7P	Sn P	7 0.12	8.80	1 050	18.5	14.0	67	5, 10
CuSn8P	Sn P	8 0.05	8.80	1 040	18.0	14.0	63	10
Copper-nickel CuNi5Fe	Ni Fe Mn	5.5 1.2 0.5	8.94	1 121	17.5	12.5	67	11
CuNi10FeMn	Ni Fe Mn	10.5 1.5 0.75	8.94	1 150	17.1	8.0	42	11
CuNi30FeMn	Ni Fe Mn	31.0 1.0 1.0	8.90	1 238	16.6	4.5	21	11
Silicon bronze	Si Mn	3 1	8.52	1 028	18.0	8.1	50	5
Aluminium bronze CuAl5	Cu Al	95 5	8.15	1 065	18.0	17.7	85	2
CuAl8Fe	Cu Al Fe	9 8 2	7.8	1 045	17.0	14.0	0	5
CuAl10Fe5Ni5	Al Fe Mn Ni	9.5 4.0 1.0 5.0	7.57	1 060	17.0	13	62	5

* The International Annealed Copper Standard is material of which the resistance of a wire 1 metre in length and weighing 1 gram is 0.153 28 ohm at 20°C. 100% IACS at 20°C = 58.00 MS m^{-1}.
(1) Solution heat treated.
(2) Fully heat treated (to maximum hardness).

REFERENCES TO TABLE 14.5

1. OFHC® Copper—Technical Information, Americal Metal Climax Inc., 1969.
2. R. A. Wilkins and E. S. Bunn, 'Copper and Copper Base Alloys', New York, 1943.
3. C. S. Smith, *Trans. AIMME*, 1930, **89**, 84.
4. C. S. Smith, *Trans. AIMME*, 1931, **93**, 176.
5. Copper Development Association, Copper and Copper Alloy Data Sheets, 1968.
6. Copper Development Association, High Conductivity Copper Alloys, 1968.
7. Copper Development Association, Beryllium Copper, 1962.
8. M. Cook, *J. Inst. Metals*, 1936, **58**, 151.
9. International Nickel Limited, Nickel Silver Engineering Properties, 1970.
10. M. Cook and W. G. Tallis, *J. Inst. Metals*, 1941, **67**, 49.
11. International Nickel Limited, Cupronickel Engineering Properties, 1970.

14.5   The physical properties of magnesium and magnesium alloys

Table 14.6   THE PHYSICAL PROPERTIES OF SOME MAGNESIUM AND MAGNESIUM ALLOYS AT NORMAL TEMPERATURE

Material	Nominal composition† %	Condition	Density at 20°C g cm⁻³	Melting point °C Sol.	Liq.	Coeff. of thermal expansion 20–200°C 10⁻⁶ K⁻¹	Thermal conductivity Wm⁻¹ K⁻¹	Electrical resistivity μΩ cm	Specific heat 20–200°C Jkg⁻¹ K⁻¹	Weldability by argon arc process‡	Relative damping capacity§
Pure Mag	Mg 99.97	T1	1.74	650		27.0	167	3.9	1050	A	
Mg–Mn	(MN70)Mn 0.75 approx.	T1	1.75	650	651	26.9	146	5	1050	A	
	(AM503)Mn 1.5	T1	1.76	650	651	26.9	142	5.0	1050	A	C
Mg–Al	AL80Al 0.75 approx. Be 0.005	T1	1.75	630	640	26.5	117	6	1050	A	
Mg–Al–Zn	(AZ31)Al 3 Zn 1	T1	1.78	575	630	26.0	(84)	10.0	1050	A	
	(A8)Al 8 Zn 0.5	AC	1.81	475*	600	27.2	84	13.4	1000	A	C
		AC T4	1.81			27.2	84	—	1000		
	(AZ91)Al 9.5 Zn 0.5	AC	1.83	470*	595	27.0	84	14.1	1000	A	C
		AC T4	1.83			27.0	84	—	1000		
		AC T6	1.83			27.0	84		1000		
	(AZM)Al 6 Zn 1	T1	1.80	510	610	27.3	79	14.3	14000	A	
	(AZ855)Al 8 Zn 0.5	T1	1.80	475*	600	27.2	79	14.3	1000	A	
Mg–Zn–Mn	(ZM21)Zn 2 Mn 1	T1	1.78			27.0			—	A	
Mg–Zn–Zr	(ZW1)Zn 1.3 Zr 0.6	T1	1.80	625	645	27.0	134	5.3	1000	A	A
	(ZW3)Zn 3 Zr 0.6	T1	1.80	600	635	27.0	125	5.5	960	C	
	(Z5Z)Zn 4.5 Zr 0.7	AC T6	1.81	560	640	27.3	113	6.6	960	C	
	(ZW6)Zn 5.5 Zr 0.6	T5	1.83	530	630	26.0	117	6.0	1050	C	

Alloy system	Alloy / composition (wt%)	Cond.	Temper									
Mg–Y–RE–Zr	(WE43) Y 4.0, RE(Δ) 3.4, Zr 0.6	AC	T6	1.84	550	640	26.7	51	14.8	966	A	
	(WE54) Y 5.1, RE(Δ) 3.0, Zr 0.6	AC	T6	1.85	550	640	24.6	52	17.3	960	A	
Mg–RE–Zn–Zr	(ZRE1) RE 2.7, ZN 2.2, Zr 0.7	AC	T5	1.80	545	640	26.8	100	7.3	1050	A	B
	(RZ5) Zn 4.0, RE 1.2, Zr 0.7	AC	T5	1.84	510	640	27.1	113	6.8	960	B	
	(ZE63) Zn 6, RE 2.5, Zr 0.7	AC	T6	1.87	515	630	27.0	109	5.6	960	A	
Mg–Th–Zn–Zr**	(ZTY) Th 0.8, Zn 0.5, Zr 0.6		T1	1.76	600	645	26.4	121	6.3	960	A	
	(ZT1) Th 3.0, Zn 2.2, Zr 0.7	AC	T5	1.83	550	647	26.7	105	7.2	960	A	(B)
	(TZ6) Zn 5.5, Th 1.8, Zr 0.7	AC	T5	1.87	500	630	27.6	113	6.6	960	B	
Mg–Ag–RE–Zr	(QE22) Ag 2.5, RE(D) 2.0, Zr 0.6	AC	T6	1.82	550	640	26.7	113	6.85	1000	A	
	(EQ21) RE(D) 2.2, Ag 1.5, Cu 0.07, Zr 0.7	AC	T6	1.81	540	640	26.6	113	6.85	1000	A	

(continued)

Table 14.6   THE PHYSICAL PROPERTIES OF SOME MAGNESIUM AND MAGNESIUM ALLOYS AT NORMAL TEMPERATURE—*continued*

Material	Nominal composition† %		Condition	Density at 20°C g cm⁻³	Melting point °C Sol.	Melting point °C Liq.	Coeff. of thermal expansion 20–200°C 10^{-6} K⁻¹	Thermal conductivity Wm⁻¹ K⁻¹	Electrical resistivity μΩ cm	Specific heat 20–200°C Jkg⁻¹ K⁻¹	Weldability by argon arc process‡	Relative damping capacity§
Mg–Zn–Cu–Mn	(ZC63)Zn Cu Mn	6.0 2.7 0.5	AC T6	1.87	465	600	26.0	122	5.4	962	B	
	(ZC71)Zn Cu Mn	6.5 1.3 0.8	T6	1.87	465	600	26.0	122	5.4	62	B	
MG–Ag–RE–** Th–Zr	(QH21)Ag RE(D) Th Zr	2.5 1.0 1.0 0.7	AC T6	1.82	540	640	26.7	113	6.85	1 005	A	—
Mg–Zr	(ZA)Zr	0.6	AC	1.75	650	651	27.0	(146)	(4.5)	1 050	A	A

AC Sand cast.
T4 Solution heat treated.
T5 Precipitation heat treated.
T6 Fully heat treated.
† Mg–Al type alloys normally contain 0.2–0.4% Mn to improve corrosion resistance.
** Thorium containing alloys are being replaced by alternative Mg alloys.

T1 Extruded, rolled or forged.
RE Cerium mischmetal containing approx. 50% Ce.
* Non-equilibrium solidus 420°C.
() Estimated value.
RE(D) Mischmetal enriched in neodynium.
RE(△) Neodynium + Heavy Rare Earths.

‡ Weldability rating:
A Fully weldable.
B Weldable.
C Not recommended where fusion welding is involved.

§ Damping capacity rating:
A Outstanding.
B Equivalent to cast iron.
C Inferior to cast iron but better than Al-base cast alloys.

14.6   The physical properties of nickel and nickel alloys

Table 14.7   THE PHYSICAL PROPERTIES OF WROUGHT NICKEL AND SOME HIGH NICKEL ALLOYS AT ROOM TEMPERATURE

Alloy*	Nominal composition %				Density g cm^{-3}	Coefficient of expansion 20–100°C 10^{-6} K^{-1}	Specific heat J kg^{-1} K^{-1}	Thermal conductivity W m^{-1} K^{-1}	Electrical resistivity μΩ cm
Nickel	99.4	Ni			8.89	13.3	456	74.9	9.5
Nickel 205	99.6	Ni			8.89	13.3	456	75.0	9.5
Monel† alloy 400	30 1.5 1.0	Cu Fe Mn			8.83	13.9	423	21.7	51.0
Monel 450	31.0 0.7 Rem.	Ni Fe Cu			8.91	15.5	380	29.4	41.2
Monel alloy K-500	29 2.8 0.5	Cu Al Ti			8.46	13.7	419	17.4	61.4
Cupro-nickel	55	Cu			8.88	14.9	421	19.5	52.0
Inconel† alloy 600	16 6	Cr Fe			8.42	13.3	460	14.8	103
Inconel 601	60.5 23.0 1.4 15.1	Ni Cr Al Fe			8.11	13.75	448	11.2	119
Inconel 617	55.7 21.5 12.5 9.0 1.2 0.1	Ni Cr Co Mo Al C			8.36	11.6	419	13.6	122
Inconel alloy 625	22 4 9	Cr Nb Mo	0.3 0.3	Ti Al	8.44	12.8	410	9.8	129
Inconel 718	52.5 19.0 18.8 5.2 3.1 0.9 0.5	Ni Cr Fe Nb Mo Ti Al			8.19	13	435	11.4	125
Inconel alloy X-750	15 7 2.5	Cr Fe Ti	0.6 0.8	Al Nb	825	12.6	425	12.0	122
Inconel MA 754	78.0 20.0 1.0 0.6	Ni Cr Fe Y2O3			8.3	12.2		14.26	107.5
INCO 330	35.5 44.8 18.5 1.2	Ni Fe Cr Si			8.08	14.9	460	12.4	101.7
INCO 020	35.0 38.4 20.0 3.5 2.5 0.6	Ni Fe Cr Cu Mo Nb			8.05	14.7	500	12.3	108

(*continued*)

Table 14.7   THE PHYSICAL PROPERTIES OF WROUGHT NICKEL AND SOME HIGH NICKEL ALLOYS AT ROOM TEMPERATURE—*continued*

Alloy*	Nominal composition %	Density g cm^{-3}	Coefficient of expansion 20–100°C 10^{-6} K^{-1}	Specific heat J kg^{-1} K^{-1}	Thermal conductivity W m^{-1} K^{-1}	Electrical resistivity μΩ cm
INCO G-3	49.0 Ni 22.5 Cr 19.5 Fe 7.0 Mo 2.0 Cu	8.3	12.2		14.26	107.5
INCO C-276	59.0 Ni 16.0 Mo 15.5 Cr 5.5 Fe 4.0 W	8.89	12.2	427	9.8	122.9
INCO-HX	48.3 Ni 22.0 Cr 18.5 Fe 9.0 Mo 1.5 Co 0.6 W 0.1 C	8.23	13.3	461	11.6	116
Incoloy† alloy 800 Incoloy alloy 800 H‡	45 Fe 0.4 Al 21 Cr 0.4 Ti	7.95	142.0	460	11.5	93
Incoloy alloy 825	32 Fe 2 Cu 21 Cr 1.0 Ti 3 Mo	8.14	14.0	441	11.1	113
Incoloy alloy DS	40 Fe 18 Cr 2 Si	7.91	14.2	450	12.0	108
Ni Span† alloy C-902	47 Fe 0.5 Al 5.5 Cr 2.5 Ti	8.10	7.6	502	12.1	101
Hastelloy B 2	28 Mo	9.22	10.3	373	11.1	137
Hastelloy C 4	16 Mo 16 Cr	8.64	10.8	406	10.1	125
Hastelloy alloy X	9 Mo 21 Cr 18 Fe	8.23	13.8	485	9.1	118
Nimonic† alloy 75	20 Cr 0.4 Ti	8.37	11.0	461	11.7	102
Nimonic alloy 80A	20 Cr 2.0 Ti 1.5 Al	8.19	12.7	460	11.2	117
Nimonic alloy 81	30 Cr 1.8 Ti 1.0 Al	8.06	11.1	461	10.9	127
Nimonic alloy 90	20 Cr 1.4 Al 17 Co 2.4 Ti	8.18	12.7	445	11.5	114
Nimonic alloy 105	15 Cr 5 Al 20 Co 1.2 Ti 5 Mo	8.01	12.2	419	10.9	131
Nimonic alloy 115	14 Cr 5 Al 13 Co 4 Ti 3 Mo	7.85	12.0	444	10.6	139

(*continued*)

Table 14.7 THE PHYSICAL PROPERTIES OF WROUGHT NICKEL AND SOME HIGH NICKEL ALLOYS AT ROOM TEMPERATURE—*continued*

Alloy*	Nominal composition %			Density g cm^{-3}	Coefficient of expansion 20–100°C 10^{-6} K^{-1}	Specific heat J kg^{-1} K^{-1}	Thermal conductivity W m^{-1} K^{-1}	Electrical resistivity μΩ cm
Nimonic alloy 263	20 20 6	Cr Co Mo	2 0.5	Ti Al	8.36	11.1	461	11.7	115
Nimonic alloy 901	13 35 6	Cr Fe Mo	3	Ti	8.16	13.5	419	—	—
Nimonic alloy PE16	16 32 3	Cr Fe Mo	1.0 1.0	Ti Al	8.02	11.3	544	11.7	110
Nimonic PK33	18.0 14.0 7.0 2.25 2.1	Cr Co Mo Ti Al			8.21	12.1	419	11.3	126
Astroloy	54.8 15.0 17.0 5.3 4.0 3.5	Ni Cr Co Mo Al Ti			7.91				
Rene 41	55.4 19.0 11.0 11.0 1.5 3.1	Ni Cr Co Mo Al Ti			8.25			9	130.8
Rene 95	61.5 14.0 8.0 3.5 3.5 3.5 2.5	Ni Cr Co Mo Nb Al Ti						8.7	
Udimet 500	53.7 18.0 18.5 4.0 2.9 2.9	Ni Cr Co Mo Al Ti			8.02			11.1	120.3
Udimet 700	55.5 15.0 17.0 5.0 4.0 3.5	Ni Cr Co Mo Al Ti			7.91			19.6	
Waspaloy	58.7 19.5 13.5 4.3 1.3 3.0	Ni Cr Co Mo Al Ti			8.19			10.7	124

* Where trade marks apply to the name of an alloy there may be materials of similar composition available from other producers who or may not use the same suffix along with their own trade names. The suffix alone e.g. Alloy 800 is sometimes used as a descriptive term for the type of alloy but trade marks can be used only by the registered user of the mark.

† Registered Trade Mark.

‡ A variant on alloy 800 having controlled carbon and heat treatment to give significantly improved creep-rupture strength.

14.7 The physical properties of titanium and titanium alloys

Table 14.8 PHYSICAL PROPERTIES OF TITANIUM AND TITANIUM ALLOYS AT NORMAL TEMPERATURES

Material IMI designation	Nominal composition %		Density g cm^{-3}	Coefficient of expansion 20–100°C 10^{-6} K^{-1}	Thermal conductivity 20–100°C Wm^{-1} K^{-1}	Resistivity 20°C μΩ cm	Temp. coefficient of resistivity 20–100°C μΩ cm K^{-1}	Specific heat 50°C J kg^{-1} K^{-1}	Magnetic suscept. 10^{-6} cgs units g^{-1}
CP Titanium	Commercially pure		4.51	7.6	16	48.2	0.002 2	528	+3.4
IMI 230	Cu	2.5	4.56	9.0	13	70	0.002 6	—	—
IMI 260/261	Pd	0.2	4.52	7.6	16	48.2	0.002 2	528	—
IMI 315	Al	2.0	4.51	6.7	8.4	101.5	0.000 3	460	+4.1
	Mn	2.0							
IMI 317	Al	5.0	4.46	7.9	6.3	163	0.000 6	470	+3.2
	Sn	2.5							
IMI 318	Al	6.0	4.42	8.0	5.8	168	0.000 4	610	+3.3
	V	4.0							
IMI 550	Al	4.0	4.60	8.8	7.9	159	0.000 4	—	—
	Mo	4.0							
	Sn	2.0							
	Si	0.5							
IMI 551	Al	4.0	4.62	8.4	5.7	170	0.000 3	400	+3.1
	Mo	4.0							
	Sn	4.0							
	Si	0.5							
IMI 679	Sn	11.0	4.84	8.0	7.1	163	0.000 4	—	—
	Zr	5.0							
	Al	2.25							
	Mo	1.0							
	Si	0.2							
IMI 680	Sn	11.0	4.86	8.9	7.5	165	0.000 3	—	—
	Mo	4.0							
	Al	2.25							
	Si	0.2							
IMI 685	Al	6.0	4.45	9.8	4.8	167	0.000 4	—	—
	Zr	5.0							
	Mo	0.5							
	Si	0.25							
IMI 829	Al	5.5	4.53	9.45	7.8	—	—	530	—
	Sn	3.5							
	Zr	3.0							
	Nb	1.0							
	Mo	0.3							
	Si	0.3							
IMI 834	Al	5.8	4.55	10.6	—	—	—	—	—
	Sn	4.0							
	Zr	3.5							
	Nb	0.7							
	Mo	0.5							
	Si	0.35							
	C	0.06							

14.8   The physical properties of zinc and zinc alloys

Table 14.9   PHYSICAL PROPERTIES OF ZINC AND ZINC ALLOYS

Material	Nominal composition	Density g cm^{-3}	Coefficient of expansion 10^{-6} K^{-1}	Thermal conductivity Wm^{-1} K^{-1}	Electrical conductivity % IACS 20°C	Condition	Melting point (liquids) °C
Zn Polycrystalline	99.993% Zn	7.13 (25°C)	39.7 (20–250°C)	113	28.27	Cast	419.46
ZnAlMg BS1004A	4% Al 0.04% Mg	6.7	27 (20–100°C)	113	27	Pressure die cast	387
ZnAlCuMg BS1004B	4% Al 1% Cu 0.04% Mg	6.7	27 (20–100°C)	109	26	Pressure die cast	388
ZnAlCuMg ILZRO 12 (ZA12)	11% Al 1% Cu 0.02% Mg	6.0	28 (20–100°C)	115	28.3	Chill cast	432
ZA27	27% Al 2.3% Cu 0.015% Mg	5.0	26 (20–100°C)	123	29.7	Chill cast	487

14.9   The physical properties of zirconium alloys

Table 14.10   PHYSICAL PROPERTIES OF ZIRCONIUM ALLOY

Alloy	Composition %	Density g cm^{-3}	Thermal cond. at 25°C Wm^{-1} K^{-1}	Coefficient of expansion 20–100°C 10^{-6}	Electrical resistivity $\mu\Omega$ cm
Zirconium 10	Commercially pure	6.50	21.1	5.04	—
Zirconium 30	Cu 0.55 Mo 0.55	6.55	25.3	5.93	—
Zircalloy II	Sn 1.5 Fe 0.12 Cr 0.10 Ni 0.05	6.55	12.3	5.67	—
Zr 702	Commercially pure with up to 4.5 Hf	6.51	22	5.89	39.7
Zr 704	Cr+Fe 0.2–0.4 Sn 1–2	6.57	—	—	—
Zr 705	Nb 2.5 O$_2$ 0.18	6.64	17.1	6.3	55
Zr 706	Nb 2.5 O$_2$ 0.16	6.64	17.1	6.3	55

See also Table 26.36 page **26**–52.

14.10   The physical properties of pure tin

Melting point	231.9°C
Boiling point	2 270°C
Vapour pressure at 727°C	7.4× 10^{-6} mmHg
1 127°C	4.4× 10^{-2} mmHg
1 527°C	5.6 mmHg
Volume change of freezing	2.7%
Expansion on melting	2.3%
Phase transformation $\alpha \rightleftharpoons \beta$	13.2°C

Density at 20°C	7.28 g/cm^3
Specific heat at 20°C	222 J kg^{-1} K^{-1}
Latent heat of fusion	59.6 kJ kg^{-1}
Latent heat of evaporation	2 497 J kg^{-1}
Linear expansion coefficient at 0–100°C	23.5 × 10^{-6} K^{-1}
Thermal conductivity at 0–100°C	66.8 Wm^{-1} K^{-1}
Electrical conductivity at 20°C	15.6 IACS
Electrical resistivity at 20°C	12.6 μΩ cm
Temperature coefficient of electrical resistivity at 0–100°C	0.004 6 K^{-1}
Thermal EMF against platinum cold junction at 0°C hot junction at 100°C	+0.42 mV
Superconductivity, critical temperature (T_c)	3.722 K
Viscosity	0.013 82 poise at 351°C
	0.011 48 poise at 493°C
Surface tension	548 mN m^{-1} at 260°C
	529 nM m^{-1} at 500°C
Gas solubility in liquid tin:	
Oxygen at 536°C	0.000 18%
Oxygen at 750°C	0.004 9%
Hydrogen at 1 000°C	0.04%
Hydrogen at 1 300°C	0.36%
Nitrogen	Very low

14.11   The physical properties of steels

Table 14.11   PHYSICAL PROPERTIES OF STEELS

Material and condition Composition %	Temperature °C	Specific gravity g cm^{-3}	*Thermal properties (see Notes)*		Thermal conductivity Wm^{-1} K^{-1}	Electrical resistivity μΩ cm
			Specific heat J kg^{-1} K^{-1}	Coefficient of thermal expansion 10^{-6} K^{-1}		
Carbon steels						
C 0.06	RT	7.87	—	—	65.3	12.0
Mn 0.4	100		48.2	12.62	60.3	17.8
	200		520	13.08	54.9	25.2
Annealed	400		595	13.83	45.2	44.8
	600		754	14.65	36.4	72.5
	800		875	14.72	28.5	107.3
	1 000		—	13.79	27.6	116.0
C 0.08	RT	7.86	—	—	59.5	13.2
Mn 0.31	100		482	12.19	57.8	19.0
	200		523	12.99	53.2	26.3
Annealed	400		595	13.91	45.6	45.8
	600		741	14.68	36.8	73.4
	800		960	14.79	28.5	108.1
	1 000		—	13.49	27.6	116.5
C 0.23 En 3	RT	7.86	—	—	51.9	15.9
Mn 0.6 060A22	100		486	12.18	51.1	21.9
	200		520	12.66	49.0	29.2
Annealed	400		599	13.47	42.7	48.7
	600		749	14.41	35.6	75.8
	800		950	12.64	26.0	109.4
	1 000		—	13.37	27.2	116.7

(*continued*)

Table 14.11 PHYSICAL PROPERTIES OF STEELS—*continued*

Material and condition Composition %	Temperature °C	Specific gravity g cm^{-3}	Thermal properties (see Notes)			
			Specific heat J kg^{-1} K^{-1}	Coefficient of thermal expansion 10^{-6} K^{-1}	Thermal conductivity Wm^{-1} K^{-1}	Electrical resistivity μΩ cm
C 0.42 ⎱ En 8	RT	7.85	—	—	51.9	16.0
Mn 0.64 ⎰ 060A42	100		486	11.21	50.7	22.1
	200		515	12.14	48.2	29.6
Annealed	400		586	13.58	41.9	49.3
	600		708	14.58	33.9	76.6
	800		624	11.84	24.7	111.1
	1 000		—	13.59	26.8	122.6
C 0.80 ⎱	RT	7.85	—	—	47.8	17.0
Mn 0.32 ⎰	100		490	11.11	48.2	23.2
	200		532	11.72	45.2	30.8
Annealed	400		607	13.15	38.1	50.5
	600		712	14.16	32.7	77.2
	800		616	13.83	24.3	112.9
	1 000		—	15.72	26.8	119.1
C 1.22 ⎱	RT	7.83	—	—	45.2	18.4
Mn 0.35 ⎰	100		486	10.6	44.8	25.2
	200		540	11.25	43.5	33.3
Annealed	400		599	12.88	38.5	54.0
	600		699	14.16	33.5	80.2
	800		649	14.33	23.9	115.2
	1 000		—	16.84	26.0	122.6
C 0.23 ⎱ En 14	RT	7.85	—	—	46.1	19.7
Mn 1.51 ⎰ 150M19	100		477	11.89	46.1	25.9
	200		511	12.68	44.8	33.3
Annealed	400		590	13.87	39.8	52.3
	600		741	14.72	34.3	78.6
	800		821	12.11	26.4	110.3
	1 000		—	13.67	27.2	117.4
C 0.13 ⎱	0	7.84	435			16.3
Mn 0.61 ⎰	100		494			22.6
Ni 0.12 ⎰	200		528			29.6
	400		599			48.2
Annealed	600		754			74.2
	800		833			110.0
	1 000		657			119.4
		Low alloy steels				
C 0.40 ⎱ 1% Ni	RT	7.85	*—	—	—	21.9
Mn 0.67 ⎰ En 12	100		486	11.90	49.4	26.4
Ni 0.80 ⎰	200		507	12.55	46.9	33.4
Hardened 850°C OQ	400		544	13.75	40.6	52.0
Tempered 600°C (1 h) OQ	600		586	14.45	34.8	77.5
C 0.37 ⎱ Mn–Mo	RT	7.85	*—	—	—	25.4
Mn 1.56 ⎰ En 16	100		456	12.45	48.2	30.6
Mo 0.26 ⎰ 605A37	200		477	13.20	45.6	39.1
Hardened 845°C OQ	400		532	14.15	39.4	60.0
Tempered 600°C (1 h)	600		599	14.80	33.9	88.5
C 0.37 ⎱ Mn–Mo	RT	7.85	*—	—	—	22.5
Mn 1.48 ⎰ En 17	100		482	12.45	45.6	27.2
Mo 0.43 ⎰ 608M38	200		494	13.00	44.0	34.3
Hardened 850°C OQ	400		519	13.90	39.4	52.5
Tempered 620°C (1 h) OQ	600		595	14.75	33.9	77.5

(*continued*)

Table 14.11   PHYSICAL PROPERTIES OF STEELS—*continued*

Material and condition Composition %	Temperature °C	Specific gravity g cm^{-3}	Specific heat J kg^{-1} K^{-1}	Coefficient of thermal expansion 10^{-6} K^{-1}	Thermal conductivity Wm^{-1} K^{-1}	Electrical resistivity μΩ cm
C 0.32 ⎤ 1% Cr	RT	7.84	—	—	48.6	20.0
Mn 0.69 ⎬ En 18B	100		494	12.16	46.5	25.9
Cr 1.09 ⎦ 530A32	200		523	12.83	44.4	33.0
Annealed	400		595	13.72	38.5	51.7
	600		741	14.46	31.8	77.8
	800		934	12.13	26.0	110.6
	1 000		—	13.66	28.1	117.7
C 0.39 ⎤ 1% Cr	RT	7.85	*—	—	—	22.8
Mn 0.79 ⎬ En 18D	100		452	12.35	44.8	28.1
Cr 1.03 ⎦ 530A40	200		473	13.05	43.5	35.2
Hardened 850°C OQ	400		519	14.40	37.7	53.0
Tempered 640°C (1 h) OQ	600		561	15.70	31.4	78.5
C 0.28/0.33 ⎤	0		—		42.7	—
Mn 0.4/0.6 ⎪	RT	7.85	—		—	22.3
Si 0.2/0.35 ⎬ 1% Cr–Mo	100		477		42.7	27.1
Cr 0.8/1.1 ⎪	200		515		—	34.2
Mo 0.15/0.25 ⎦	300		544		40.6	—
Hardened and tempered	400		595		—	52.9
	500		657		37.3	—
	600		737		—	78.6
	700		825		31.0	—
	800		883		—	110.3
	1 000		—		28.1	117.1
	1 200		—		30.1	122.2
C 0.41 ⎤ 1% Cr–Mo	RT	7.83	*—	—	—	22.2
Mn 0.67 ⎪ En 19	100		—	12.25	42.7	26.3
Cr 1.01 ⎬ 708A42	200		473	12.70	42.3	32.6
Mo 0.23 ⎦	400		519	13.70	37.7	47.5
Hardened 850°C OQ	600		561	14.45	33.1	64.6
Tempered 600°C (1 h) OQ						
C 0.4 ⎤	RT	7.85		—	—	—
Mn 0.4 ⎬ 1% Cr–Mo	100			12.3	41.9	
Cr 1.1 ⎬ En 20B	200			12.6	41.9	
Mo 0.7 ⎦	400			13.7	38.9	
Hardened and tempered	600			14.4	32.7	
	800			—	26.0	
C 0.4 ⎤	RT	7.83		—	—	—
Mn 0.6 ⎪ 3% Cr–Mo–V	100			12.5	37.7	
Cr 3.0 ⎬ En 40C	200			12.9	37.7	
Mo 0.8 ⎪ 897M39	400			13.5	34.8	
V 0.2 ⎦	600			14.0	31.0	
Hardened and tempered						
C 0.35 ⎤	RT	7.84	—	—	42.7	21.1
Mn 0.59 ⎪	100		477	12.67	42.7	27.1
Ni 0.20 ⎬ Low Ni–Cr–Mo	200		515	13.11	41.9	34.2
Cr 0.88 ⎪ En 19	400		595	13.82	38.9	52.9
Mo 0.20 ⎦	600		737	14.55	33.9	78.6
Annealed	800		883	11.92	26.4	110.3
	1 000		—	13.86	28.1	117.1

(continued)

Table 14.11   PHYSICAL PROPERTIES OF STEELS—*continued*

Material and condition Composition %		Temperature °C	Specific gravity g cm^{-3}	Thermal properties (see Notes)			
				Specific heat J kg^{-1} K^{-1}	Coefficient of thermal expansion 10^{-6} K^{-1}	Thermal conductivity Wm^{-1} K^{-1}	Electrical resistivity μΩ cm
C 0.23 Mn 0.45 Si 0.45 Cr 2.87 W 0.59 Mo 0.51 V 0.77	3% Cr–W–Mo–V	RT 100 200 400 600 800	7.83		— 11.9 12.4 13.1 13.6 14.1	38.5 33.6 33.1 30.6 29.3 28.9	35.5 39.0 46.2 63.0 85.4 —
Hardened and tempered							
C 0.32 Mn 0.55 Ni 3.47	3% Ni En 21	RT 100 200 400 600 800 1 000	7.85	— 482 523 590 749 604 —	— 11.20 11.80 12.90 13.87 11.10 13.29	36.4 37.7 38.9 36.8 32.7 25.1 27.6	25.9 32.0 39.0 56.7 81.4 112.2 118.0
Annealed							
C 0.33 Mn 0.50 Ni 3.4 Cr 0.8	3% Ni–Cr En 23	RT 100 200 400 600 800 1 000	7.85	— 494 523 599 775 557 —	— 11.36 12.29 13.18 13.72 10.69 13.11	34.3 36.0 36.8 36.4 31.8 26.0 27.6	25.6 31.7 38.7 56.7 81.7 111.5 117.8
Hardened and tempered							
C 0.41 Ni 1.43 Cr 1.07 Mo 0.26	1$\frac{1}{2}$% Ni–Cr–Mo En 24 817M40	RT 100 200 400 600	7.84		— — 12.40 13.60 14.30		24.8 29.8 36.7 55.2 79.7
Hardened 830°C OQ Tempered 630°C (1 h) OQ							
C 0.32 Ni 2.60 Cr 0.67 Mo 0.51	2$\frac{1}{2}$% Ni–Cr–Mo En 25 826M31	RT 100 200 400 600	7.85		— — 11.55 13.10 13.85		27.7 32.1 38.7 57.3 82.5
Hardened 830°C OQ Tempered 650°C OQ							
C 0.34 Mn 0.54 Ni 3.53 Cr 0.76 Mo 0.39	3% Ni–Cr–Mo En 27	RT 100 200 400 600 800 1 000	7.86	— 486 523 607 770 636 —	— 11.63 12.12 13.12 13.79 10.67 12.96	33.1 33.9 35.2 35.6 30.6 26.8 28.5	27.7 33.7 40.6 58.2 82.5 111.4 117.6
Hardened and tempered							
C 0.29 Ni 4.23 Cr 1.26	4$\frac{1}{4}$% Ni–Cr En 30A	RT 100 200	7.83		— 10.55 12.00	— 27.6 29.7	37.0 41.6 49.3
Hardened 820°C AC Tempered 250°C (1 h)							
C 0.18 Ni 1.76 Mo 0.20	2% Ni–Mo En 34 665A17	RT 100 200	7.85		— 12.50 13.10		24.9 29.6 37.1
Blank carburised 920°C Hardened 800°C OQ							

(*continued*)

Table 14.11   PHYSICAL PROPERTIES OF STEELS—*continued*

Material and condition Composition %	Temperature °C	Specific gravity g cm⁻³	Specific heat J kg⁻¹ K⁻¹	Coefficient of thermal expansion 10⁻⁶ K⁻¹	Thermal conductivity Wm⁻¹ K⁻¹	Electrical resistivity μΩ cm
C 0.15, Ni 4.25, Cr 1.18, Mo 0.20 — 4¼% Ni–Cr–Mo, En 39B, 835A15. Blank carburised 920°C. Hardened 800°C OQ	RT	7.85		—		36.3
	100			11.30		40.1
	200			12.55		46.7
C 0.39, Mn 1.35, Ni 0.65, Cr 0.48, Mo 0.17 — Low alloy steel, En 100, 945M38. Hardened 850°C OQ. Tempered 620°C (1 h) OQ	RT	7.86		—		24.7
	100			12.00		28.2
	200			12.75		34.0
	400			14.00		52.0
	600			14.75		74.7
C 0.39, Ni 1.39, Cr 1.02, Mo 0.14 — Low Ni–Cr–Mo, En 110, 816M40. Hardened 840°C OQ. Tempered 650°C (1 h) OQ	RT	7.84		—		24.8
	100			12.00		29.2
	200			12.65		35.6
	400			13.65		54.0
	600			14.30		78.0
C 0.17, Ni 0.86, Cr 0.71 — ¾% Ni–Cr, En 351, 635A14. Blank carburised 910°C. Hardened 820°C OQ	RT	7.85		—		29.1
	100			12.80		34.2
	200			13.10		41.1
C 0.17, Ni 1.25, Cr 1.02, Mo 0.15 — 1¼% Ni–Cr–Mo, En 353, 815A16. Blank carburised 910°C. Hardened 810°C OQ	RT	7.87		—		31.8
	100			11.30		36.6
	200			12.45		43.2
C 0.16, Ni 2.00, Cr 1.50, Mo 0.20 — 2% Ni–Cr–Mo, En 355, 822A17. Blank carburised 910°C. Hardened 810°C OQ	RT	7.84		—		34.5
	100			11.80		39.2
	200			12.30		45.7
C 0.48, Mn 0.90, Si 1.98, Cu 0.64 — 2% Si–Cu. Annealed	RT	7.73		—		41.9
	100		498	11.19	25.1	47.0
	200		523	12.21	28.5	52.9
	400		603	13.35	30.1	68.5
	600		749	14.09		91.1
	800		528	13.59		117.3
	1 000		—	14.54		122.3
C 0.05, Mn 0.3, Si 0.7, B 1.96, Al 0.03 — 2% B. Hot worked	0			—		24.9
	RT	7.72	461	—		—
	100			10.0		30.9
	200			11.0		38.7
	400			11.9		57.4
	600			11.8		81.9
	800			13.3		—

(continued)

Table 14.11   PHYSICAL PROPERTIES OF STEELS—*continued*

Material and condition Composition %	Temperature °C	Specific gravity g cm^{-3}	Thermal properties (see Notes) Specific heat J kg^{-1} K^{-1}	Coefficient of thermal expansion 10^{-6} K^{-1}	Thermal conductivity Wm^{-1} K^{-1}	Electrical resistivity μΩ cm
C 0.10, Mn 0.14, Si 0.43, B 4.2, Al 0.53 (4% B) As cast	0			—		39.9
	RT	7.40	523	—		—
	100			9.5		50.6
	200			—		61.5
	300			10.4		72.3
	400			—		83.3
	500			11.2		—
	600			—		106.5
	700			11.8		—
	800			—		129.4
	1 000			13.0		—
Typically C 0.10, Mo 0.5, B 0.004 (½% Mo–B 'Fortiweld') Normalised and stress-relieved 600°C	RT	7.86	*440	12.00	46.1	20.0
	100		465	12.55	45.2	24.5
	200		494	13.25	44.4	31.0
	400		557	14.30	41.5	48.5
	600		632	15.10	36.9	74.5
	700		674	15.40	35.2	88.0
C 0.10, Si 1.0 max, Cr 4.0/6.0 (5% Cr AISI 502) Annealed	30	7.7		—	36.0	
	100			11.0	—	
	200			11.6	35.2	
	400			12.6	—	
	600			13.3	—	
	800			—	26.8	
	1 200			—	26.8	

High alloy steels

Material and condition Composition %	Temperature °C	Specific gravity g cm^{-3}	Specific heat J kg^{-1} K^{-1}	Coefficient of thermal expansion 10^{-6} K^{-1}	Thermal conductivity Wm^{-1} K^{-1}	Electrical resistivity μΩ cm
C 0.45, Mn 0.5, Si 3.5, Cr 3.5 (3% Cr–3% Si) Hardened and tempered	RT	7.6		—	22.2	80
	100			13.0	—	
	300			13.0	—	
	500			13.0	—	
	700			14.0	—	
	900			—	31.4	
C 0.45, Mn 0.5, Cr 8.0, Si 3.4 (8% Cr–3% Si En 52 401S45) Hardened and tempered	RT	7.6		—	22.2	80.0
	100			13.0	—	—
	300			13.0	—	110.0
	500			13.0	—	
	700			14.0	—	
	900			—	31.4	
C 0.40, Mn 0.3, Cr 11.5 (11% Cr) Hardened and tempered	RT	7.75		—	23.5	60.0
	100			10.0	—	—
	300			11.0	—	—
	500			12.0	—	—
	700			12.0	—	—
	750			—	24.3	119.0
C 0.12, Cr 9.0, Mo 1.0 (9% Cr–Mo) Normalised and tempered	RT	7.78	*402	11.15	26.0	49.9
	100		427	11.30	26.4	55.5
	200		461	11.60	26.8	63.0
	400		528	12.10	27.6	79.5
	600		595	12.65	26.8	97.5
	700		624	12.85	26.8	106.5

(*continued*)

Table 14.11   PHYSICAL PROPERTIES OF STEELS—*continued*

Material and condition Composition %	Temperature °C	Specific gravity g cm^{-3}	Specific heat J kg^{-1} K^{-1}	Coefficient of thermal expansion 10^{-6} K^{-1}	Thermal conductivity W m^{-1} K^{-1}	Electrical resistivity μΩ cm
C 0.20 Mn 0.4 — 11% Cr–Mo–V–Nb Cr 11.0 Mo 0.5 V 0.7 Nb 0.15 Hardened and tempered	RT 100 200 400 600 800	7.75		9.3 10.9 11.5 12.1 12.2		
C 0.13 — 13% Cr Mn 0.25 — En 56B Cr 12.95 — 420S29 Ni 0.14 Annealed	RT 100 200 400 600 800 1 000	7.74	— 473 515 607 779 691 —	— 10.13 10.66 11.54 12.15 12.56 11.70	26.8 27.6 27.6 27.6 26.4 25.1 27.6	48.6 58.4 67.9 85.4 102.1 116.0 117.0
C 0.07 Mn 0.8 — 17% Cr Cr 17.0 Annealed	RT 100 200 300	7.7		— 10.0 11.0 12.0	21.8	62.0
C 0.06 Mn 0.8 — 21% Cr Cr 21.0 Annealed	RT 100 300 500 700 900	7.76	482	— 10.0 11.0 11.0 12.0 13.0	21.8	62.0
C 0.22 Cr 30.4 — 30% Cr–Ni Ni 0.26 Hardened and tempered	RT 100	7.90		— 10.0	12.6	80.0
C 1.22 — 13% Mn Mn 1.30 1 050°C Air-cooled	RT 100 200 400 600 800 1 000	7.87	— 519 565 607 704 649 673	— 18.01 19.37 21.71 19.86 21.86 23.13	13.0 14.6 16.3 19.3 21.8 23.5 25.5	66.5 75.7 84.7 100.4 110.0 120.4 127.5
C 0.28 Mn 0.89 — 28% Ni Ni 28.4 950°C, WQ	RT 100 200 400 600 800 1 000	8.16	— 502 519 540 586 586 599	— 13.73 15.28 17.02 17.82 18.28 18.83	12.6 14.7 16.3 18.9 22.2 25.1 27.6	82.9 89.1 94.7 103.9 111.2 116.5 120.6
C 0.10 — 12% Cr–4% Al Mn 0.60 — AISI 406 Cr 12.0 Al 4.5 Softened	RT 100 300 500 600 700 850	7.42	502	— 11.0 12.0 12.0 — 13.0 —	— 25.1 — 28.5 — — —	122 125 129 — 136 — 141

(*continued*)

Table 14.11 PHYSICAL PROPERTIES OF STEELS—*continued*

Material and condition Composition %	Temperature °C	Specific gravity g cm^{-3}	Specific heat J kg^{-1} K^{-1}	Coefficient of thermal expansion 10^{-6} K^{-1}	Thermal conductivity Wm^{-1} K^{-1}	Electrical resistivity μΩ cm
C 0.72, Mn 0.25, Ni 0.07, Cr 4.26, W 18.5 — 4% Cr–18% W, Annealed 830°C	RT	8.69	—	—	24.3	40.6
	100		410	11.23	26.0	47.2
	200		435	11.71	27.2	54.4
	400		502	12.20	28.5	71.8
	600		599	12.62	27.2	92.2
	800		716	12.97	26.0	115.2
	1 000		—	12.44	27.6	120.9
C 0.16, Mn 0.2, Ni 2.5, Cr 16.5 — 16% Cr–Ni, En 57, 431S29, Softened	RT	7.7	—	—	18.8	72.0
	100		482	10	—	—
	300			11	—	—
	500			12	24.3	103.0
C 0.08, Mn 0.3/0.5, Ni 8, Cr 18/20 — 18% Cr–8% Ni, En 58A, 302S25, 1 100°C WQ	RT	7.92	—	—	15.9	69.4
	100		511	14.82	16.3	77.6
	200		532	16.47	17.2	85.0
	400		569	17.61	20.1	97.6
	600		649	18.43	23.9	107.2
	800		641	19.03	26.8	114.1
	1 000		—	—	28.1	119.6
C 0.12, Mn 1.5, Ni 11.0, Cr 17.5, Nb 1.2 — 18% Cr–11% Ni (Nb stabilised), En 58G, 347S17, Softened	RT	7.9		—	15.9	72
	100			16.0	—	
	300			18.0	17.2	
	500			18.0	18.8	
	700			19.0	20.1	
C 0.22, Mn 0.6, Cr 20.0, Ni 8.5, Ti 1.2 — 20% Cr–8% Ni (Ti stabilised), Softened	RT	7.72		—		82
	100			15.0		
	300			15.0		
	500			16.0		
	700			17.0		
	900			18.0		
C 0.15, Mn 0.8, Ni 14, Cr 19, Nb 1.7 — 19% Cr–14% Ni (Nb stabilised), Softened	RT	7.92		—	—	
	100			17.0	15.1	
	200			17.2	16.8	
	400			17.6	20.1	
	600			18.6	24.3	
C 0.30, Mn 0.6, Si 1.5, Ni 8, Cr 20, W 4 — 18% Cr–8% Ni–W, En55, Softened	RT	7.8			13	85
	100			16.0		
	300			17.0		
	500			17.0		
	700			18.0		
	900			18.0		
	1 050			—	29	125
C 0.12, Mn 0.3, Ni 8.5, Cr 18.5, Ti 0.8, Al 1.4 — 18% Cr–8% Ni–Al (Ti stabilised), Normalised and tempered	RT	7.67		—	18.0	85
	100			15		
	300			15		
	500			15		
	700			16		
	900			17	26.0	125

(continued)

Table 14.11   PHYSICAL PROPERTIES OF STEELS—*continued*

Material and condition Composition %	Temperature °C	Specific gravity g cm^{-3}	Specific heat J kg^{-1} K^{-1}	Coefficient of thermal expansion 10^{-6} K^{-1}	Thermal conductivity Wm^{-1} K^{-1}	Electrical resistivity μΩ cm
				Thermal properties (see Notes)		
C 0.10 Mn 0.3 12% Cr–12% Ni Ni 12.5 En 58D Cr 12.5 Softened	RT 100	8.01	490	— 18	15.5 16.8	70 77
C 0.10 Mn 6.0 15/10/6/1 Cr 15.0 Cr–Ni–Mn–Mo Ni 10.0 Mo 1.0 Solution treated 1 100°C	RT 100 200 400 600 700	7.94	*477 494 511 536 557 565	14.80 15.70 16.75 18.25 18.95 19.30	12.6 13.8 15.4 18.8 21.8 23.0	74.1 80.0 86.7 99.4 108.4 114.4
C 0.27 Mn 1.25 11% Cr—36% Ni Ni 36 Cr 11 Softened	RT 100 300 500	8.08		— 14 15 16	12.1 — — 18.4	97 — — 117
C 0.1 Mn 1.3 Si 1.2 30% Cr–Ni Ni 1.8 Cr 29.0 Softened	RT 200 800 1 000 1 100	7.5		10 11 13 13	15.9 26.4	88 126
C 0.1 Mn 1.0 14% Cr–63% Ni Ni 63 Cr 14 Softened	RT 100 200 400 600 800 1 000	8.1		 12.0 12.5 13.5 14.5 15.5 16.5	12.6 28.9	105 110
C 0.30 Mn 3.0 17% Cr–17% Ni–Mo Ni 17.5 –Co–Nb Cr 16.5 Mo 3.0 Nb 2.5 Co 7.0 Softened	RT 100 300 500 700	8.0		— 15 16 16 17	12.6	93.8
C 0.4 Mn 0.9 13% Cr–13% Ni Si 1.4 –W–Nb Ni 13.0 Cr 13.0 W 2.3 Nb 0.9 Normalised	RT 100 200 400 600 800	8.03		— 16.8 17.3 18.3 18.9 19.3		
C 0.4 Mn 0.8 Si 1.0 13% Cr–13% Ni 13 Ni–W–Mo–Co–Nb Cr 13 W 2.5 Mo 2.0 Nb 3.0 Co 10.0 Solution treated	RT 100 200 400 600 800	8.13		— 15.6 15.8 16.9 17.3 18.0	— 13.4 17.2 18.8 22.2 25.5	

(continued)

Table 14.11 PHYSICAL PROPERTIES OF STEELS—*continued*

Material and condition Composition %		Temperature °C	Specific gravity g cm^{-3}	Thermal properties (see Notes)			
				Specific heat J kg^{-1} K^{-1}	Coefficient of thermal expansion 10^{-6} K^{-1}	Thermal conductivity Wm^{-1} K^{-1}	Electrical resistivity μΩ cm
C 0.27 Mn 0.77 Ni 10.5 20% Cr–10% Cr 19.1 Ni–46% Co Mo 2.2 Nb 1.4 V 3.0 Co 46.6 Solution treated and aged		RT 100 200 400 600 800	8.26	—	— 14.8 15.0 15.2 15.9 16.8	— 14.7 16.3 19.7 23.0 26.0	
Cast steels							
C 0.11 Mn 0.35	Plain carbon B.S. 1 617A A 950°C, N 950°C	100 200 300 400 500 600			12.2 12.6 13.2 13.6 13.9 14.2	48.6	19.5
C 0.4 Mn 0.5	Plain carbon BS 1 760 A 900°C, OQ 830°C T 650°C	100 200 300 400 500 600			11.8 12.4 12.8 13.3 13.7 14.2	42.3	23.5
C 0.17 Mn 0.74 Mo 0.50	Carbon Mo BS 1 398 A 920°C, SR 650°C	100 200 300 400 500 600			12.4 12.8 13.1 13.4 13.8 14.2		24.2
C 0.25 Mn 1.55	1½% Mn BŠ 1 456A A 950°C, WQ 910°C, T 660°C	100 200 300 400 500 600			13.2 13.3 13.7 14.1 14.7 15.2		
C 0.29 Cr 1.80 Ni 0.46 Mo 0.52	1½% Cr Ni Mo BŠ 1 458 OQ 900°C, T 660°C	100 200 300 400 500 600			12.5 12.7 13.0 13.4 13.9 14.4		27.6
C 0.34 Ni 2.82 Cr 0.74 Mo 0.42	2½ % Ni Cr Mo BŠ 1 459 OQ 850°C, T 640°C	100 200 300 400 500 600			12.0 12.3 12.6 13.0 13.5 13.9	39.4	27.3
C 0.24 Cr 3.23 Mo 0.51	3% Cr Mo BS 1 461 OQ 900°C, T 690°C	100 200 300 400 500 600			12.2 12.4 12.7 12.9 13.3 13.6		

(*continued*)

Table 14.11   PHYSICAL PROPERTIES OF STEELS—*continued*

Material and condition Composition %	Temperature °C	Specific gravity g cm^{-3}	Specific heat J kg^{-1} K^{-1}	Thermal properties (see Notes) Coefficient of thermal expansion 10^{-6} K^{-1}	Thermal conductivity Wm^{-1} K^{-1}	Electrical resistivity μΩ cm
C 0.1 5% Cr Mo Cr 4.06 BS 1 462 Mo 0.57 N 950°C, T 680°C	100 200 300 400 500 600			11.8 12.0 12.3 12.5 12.7 13.0		37.1
C 0.13 9% Cr Mo Cr 8.29 BS 1 463 Mo 1.1 OQ 900°C, T 690°C Si 1.1	100 200 300 400 500 600			11.9 11.6 11.7 11.7 11.8 11.9		
C 0.27 13% Cr Cr 12.1 BS 1 630 Ni 1.07 OQ 930°C, T 730°C Si 1.16	100 200 300 400 500 600			11.5 11.8 12.4 12.6 12.7 12.9	25.1	
C 0.47 Carbon Cr Cr 0.85 BS 1 956 A N 870°C, T 635°C	100 200 300 400 500 600			12.5 12.9 13.2 13.4 13.5 13.6		
C 0.1 $3\frac{1}{2}$% Ni Ni 3.35 BS 1 504–503 WQ 880°C, T 650°C	100 200 300 400 500 600			11.3 11.9 12.2 12.7 13.5 13.6		
C 0.19 $1\frac{1}{4}$ % Cr Mo Cr 1.13 BS 1 504–621 Mo 0.5 N 920°C, T 625°C	100 200 300 400 500 600			11.8 12.4 12.6 13.3 13.7 13.9	28.7	
Cast corrosion-resisting steels						
C 0.13 13% Cr Mn 0.80 BS 1 630 A Cr 12.5 Hardened and tempered	100 300 500 600	7.73	482	11.0 11.0 12.0	24.7 — — 27.6	56
C 0.25 13% Cr Mn 0.70 BS 1 630 C Cr 12.5 Hardened and tempered	100 300 500 600	7.75	482	11.0 11.0 12.0	24.3 — — 26.0	57
C 0.07 Si 0.70 Mn 0.80 18% Cr–8% Ni Ni 8.5 BS 1 631 A Cr 18.0 Normalised	100	7.93	502	17.0	16.3	72

(*continued*)

Table 14.11 PHYSICAL PROPERTIES OF STEELS—*continued*

Material and condition Composition %	Temperature °C	Specific gravity g cm⁻³	Thermal properties (see Notes)			Electrical resistivity μΩ cm
			Specific heat J kg⁻¹ K⁻¹	Coefficient of thermal expansion 10⁻⁶ K⁻¹	Thermal conductivity Wm⁻¹ K⁻¹	
C 0.08, Si 1.00, Mn 0.50, Ni 9.00, Cr 18.0, Nb 0.9 — 18% Cr, 8% Ni Nb BS 1 631 B Nb. Normalised	100 300 500 700	7.93	502	17.0 18.0 18.0 19.0	15.9 — — 20.1	
C 0.12, Si 1.50, Mn 0.80, Ni 9.00, Cr 19.00, Ti 0.6 — 18% Cr, 8% Ni Ti BS 1 631 B Ti. Normalised	100	7.78	444	17.0	15.5	70
C 0.06, Si 0.70, Mn 0.70, Ni 12.0, Cr 19.0, Mo 3.6 — 19% Cr, 12% Ni, $3\frac{1}{2}$% Mo BS 1 632 A. Water quenched	RT 100	7.96	502	— 16.0	16.3	
C 0.07, Si 1.00, Mn 1.00, Ni 10.5, Cr 18.0, Mo 2.75 — 18% Cr, 10% Ni $2\frac{1}{2}$% Mo BS 1 632 B. Normalised	RT 100 200 400 500 600 800	7.96	502	— 16.5 16.9 17.2 17.4 17.9 19.0	16.3	73
C 0.08, Si 1.00, Mn 0.50, Ni 10.5, Cr 18.0, Mo 2.75, Nb 0.90 — 18% Cr, 10% Ni, $2\frac{1}{2}$% Mo Nb BS 1 632 C Nb. Normalised	RT 100	7.96	502	— 16.0	16.3 17.6	73
C 0.10, Si 1.50, Mn 0.80, Ni 10.0, Cr 18.0, Mo 2.75, Ti 0.60 — 18% Cr, 10% Ni, $2\frac{1}{2}$% Mo Ti BS 1 632 C Ti. Normalised	RT 100	7.78	448	— 17.0	15.5	78
C 0.06, Si 0.70, Mn 0.60, Ni 8.5, Cr 18.0, Mo 2.5 — 18% Cr, 8% Ni, $2\frac{1}{2}$% Mo BS 1632 D. Normalised	RT 100	7.93	502	— 16.0	16.3	

(continued)

Table 14.11　PHYSICAL PROPERTIES OF STEELS—*continued*

Material and condition Composition %	Temperature °C	Specific gravity g cm^{-3}	Thermal properties (see Notes) Specific heat J kg^{-1} K^{-1}	Coefficient of thermal expansion 10^{-6} K^{-1}	Thermal conductivity Wm^{-1} K^{-1}	Electrical resistivity μΩ cm
Cast heat-resisting steels						
C 0.25	RT	7.75	482	—	24.3	57
Si 0.70　13% Cr	100			11.0	—	
Mn 0.70　BS 1 648 A	300			11.0	—	
Cr 12.5	500			12.0	—	
Hardened and tempered	600			—	26.0	
C 0.40	RT	7.63	482	—	20.9	70
Si 0.80　27% Cr	100			10.2		
Mn 0.90　BS 1 648 B	200			10.8		
Cr 29.0	400			11.0		
Tempered	600			11.5		
	800			12.4		
	1 000			13.3		
C 1.70	RT	7.63	482	—	20.9	70
Si 0.70　27% Cr	100			10.2		
Mn 0.70　BS 1 648 C	200			10.8		
Cr 27.0	400			11.0		
Tempered	600			11.5		
	800			12.4		
	1 000			13.3		
C 0.30	RT	7.74	502	—	—	80
Si 1.50　20% Cr 10% Ni	100			—	15.5	
Mn 1.50　BS 1 648 D	500			17.8	—	
Ni 10.0	800			18.5	26.8	
Cr 20.0	1 100			19.6	—	
C 0.35	RT	7.92	435	—	10.9	86
Si 1.50　21% Cr 8% Ni	100			13.6		
Mn 0.80　4% W	300			14.5		
Ni 7.0　BS 1 648 D	500			15.4		
Cr 21.0	700			16.5		
W 4.0	900			17.7	26.8	
	1 000			18.3	—	
C 0.20	RT	7.92	544	—	13.8	85
Si 1.20　25% Cr 12% Ni	100			16.5		
Mn 1.30　BS 1 648 E	200			16.6		
Ni 13.0	400			16.9		
Cr 25.0	600			17.6		
Normalised	800			18.2		
	1 000			18.7		
C 0.20	RT	7.90	502	—	12.6	87
Si 1.00　25% Cr 12% Ni	100			15.0		
Mn 0.80　3% W	300			16.0		
Ni 12.0　BS 1 648 E	500			16.0		
Cr 23.0	700			17.0		
W 3.0	900			19.0		
Normalised	1 000			—	29.3	
C 0.20	RT	7.90	544	—	15.9	90
Si 1.50　25% Cr 20% Ni	100			16.5		
Mn 1.00　BS 1 648 F	200			16.9		
Ni 20.0	400			17.5		
Cr 25.0	600			18.3		
Normalised	800			19.2		
	1 000			20.0		

(*continued*)

Table 14.11   PHYSICAL PROPERTIES OF STEELS—*continued*

Material and condition Composition %		Temperature °C	Specific gravity g cm^{-3}	Specific heat J kg^{-1} K^{-1}	Coefficient of thermal expansion 10^{-6} K^{-1}	Thermal conductivity Wm^{-1} K^{-1}	Electrical resistivity μΩ cm
C 0.35		RT	7.90	502	—	12.6	88
Si 0.90	25% Ni 15% Cr	100			15.0		
Mn 0.75	BS 1 648 G	300			16.0		
Ni 25.0		500			17.0		
Cr 15.0		700			17.0		
		900			18.0		
		1 000				29.3	
C 0.50		RT	7.93	460	—	—	100
Si 2.00		100			—	13.4	
Mn 1.50	35% Ni 15% Cr	500			16.0		
Ni 35.0		800			16.5		
Cr 15.0		1 100			17.6		
Cast							
C 0.50		RT	8.02	460	—	—	105
Si 2.0		100			—	13.4	
Mn 1.50	40% Ni 20% Cr	500			16.0	—	
Ni 40.0	BS 1 648 H	800			16.4	23.9	
Cr 20.0		1 100			17.4	—	
Cast							
C 0.50		RT	8.12	460	—	—	108
Si 2.00	60% Ni 15% Cr	100			—	13.4	
Mn 1.50	BS 1 648 K	500			14.2	—	
Ni 60.0		800			15.3	23.0	
Cr 15.0		1 100			16.5	—	
Cast							

Notes:
∗ The values are a mean from RT up to the temperature quoted.
1. Where *specific heats* are quoted at temperatures above RT the values have been determined over a range of 50°C up to the temperature quoted.
2. *Coefficients of expansion* are mean values from RT up to the temperature quoted.
3. *Electrical resistivity* values are uncorrected for dimensional changes of the specimen with temperature. Original dimensions as at RT.

REFERENCES TO TABLE 14.11

1. 'Metals Handbook', 4th edn.
2. J. Woolman and R. A. Mottram, 'Mechanical and Physical Properties of BS En Steels (BS 970, 1950), Pergamon Press.
3. Sundry technical information issued by industrial organizations e.g. British Steel Corporation, Mond Nickel Co. Ltd.

Table 14.12   SOME LOW TEMPERATURE THERMAL PROPERTIES OF A SELECTION OF STEELS

There is particular interest in the thermal properties (especially the thermal expansion) of steels used under conditions well below normal atmospheric temperature, and available information is set out below in respect of some such steels.

Material and condition Analyses %	Temperature °C	Coefficient of thermal expansion $10^{-6}\,K^{-1}$	Thermal conductivity $Wm^{-1}\,K^{-1}$
Typically	−200	−9.5	16.0
C 0.09 Ni 9 } 9 Ni	−150	−9.7	19.5
	−100	−9.9	23.0
Double normalised and tempered	−50	−10.2	26.5
	RT	10.5	29.5
	100	11.0	32.0
	200	11.7	34.0
	300	12.3	34.5
C 0.42 Ni 1.58 Cr 1.19 } $1\frac{1}{2}$ Ni–Cr–Mo Mo 0.24	−150	−10.4	
	−100	−11.2	
	−50	−11.8	
Hardened 840°C OQ Tempered 650°C (1 h)/AC	RT	12.1	
C 0.12 Mo 0.54 } $\frac{1}{2}$ Mo–B B 0.003	−150	−10.5	
	−100	−11.2	
	−50	−11.8	
Hardened 960°C OQ Tempered 700°C ($\frac{1}{2}$ h) AC	RT	12.2	
C 0.27 Cr 3.14 } 3 Cr–Mo Mo 0.49	−150	−9.8	
	−100	−10.3	
	−50	−10.8	
Hardened 900°C OQ Tempered 650°C (1 h) AC	RT	11.5	
C 0.09 Mn 6.23 Ni 9.88 Cr 14.88 } 15 Cr–10 Ni–6 Mn–Mo–V–B–Nb Mo 1.01 V 0.28 B 0.003 Nb 0.94	−150	−14.7	
	−100	−15.3	
	−50	−15.7	
1 150°C AC	RT	16.2	
C 0.13 Ni 4.16 } $4\frac{1}{2}$ Ni–Cr–Mo Cr 1.23 Mo 0.19	−150	−9.4	
	−100	−9.8	
	−50	−10.2	
Blank carburised 890°C AC 820°C ($\frac{1}{4}$ h), transferred to 580°C ($\frac{1}{2}$ h) OQ	RT	10.8	
C 0.41 Cr 3.14 } 3 Cr–1 Mo–V Mo 0.97 V 0.20	−150	−9.7	
	−100	−10.1	
	−50	−10.5	
Hardened 930°C OQ Tempered 700 ($\frac{1}{2}$ h) AC	RT	11.2	

(continued)

Table 14.12   SOME LOW TEMPERATURE THERMAL PROPERTIES OF A SELECTION OF STEELS—*continued*

Material and condition Analyses %	Temperature °C	Coefficient of thermal expansion $10^{-6}\,K^{-1}$	Thermal conductivity $Wm^{-1}\,K^{-1}$
C 0.12 ⎱	−150	−10.1	
Ni 3.10 ⎰ $3\frac{1}{2}$ Ni–Cr–Mo	−100	−10.6	
Cr 0.91	−50	−11.0	
Mo 0.16			
Blank carburised 910°C AC	RT		11.6
Hardened 840°C ($\frac{1}{4}$ h) OQ			
Tempered 760°C OQ			
C 0.99 ⎱ $1\,C$–$1\frac{1}{2}\,Cr$	−150	−9.6	
Cr 1.47 ⎰	−100	−10.6	
Hardened 850°C AC	−50	−11.6	
Tempered 650°C ($\frac{1}{2}$ h) AC	RT		12.3
C 0.17 ⎱	−150	−8.5	
Ni 1.74 ⎰ 2 Ni–Mo	−100	−9.5	
Cr 0.2 En 34	−50	−10.4	
Mo 0.22	RT		11.3
Blank carburised 910°C AC			
Hardened 870°C OQ			
Tempered 770°C OQ			
C 0.11 ⎱ 3 Ni	−150	−9.9	
Ni 3.04 ⎰	−100	−10.5	
Blank carburised 910°C AC	−50	−11.0	
Hardened 870°C OQ	RT		11.5
Tempered 770°C OQ			

∗ Thermal expansion values shown for temperatures other than RT are the mean values from RT to that temperature. For RT the instantaneous value is given.

REFERENCE TO TABLE 14.12

1. Sundry technical information issued by British Steel Corporation and Mond Nickel Co. Ltd.

15 Elastic properties, damping capacity and shape memory alloys

15.1 Elastic properties

The elastic properties of a metal reflect the response of the interatomic forces between the atoms concerned to an applied stress. Since the bonding forces vary with crystallographic orientation the elastic properties of metal single crystals may be highly anisotropic. However, polycrystalline metals and alloys with a randomly oriented grain structure behave isotropically. Table 15.1 lists elastic constants for polycrystalline metals and alloys in an isotropic condition. Any preferred orientation or texture resulting from rolling, drawing or extrusion, for example, will result in departures from the listed values to a degree that depends upon the elastic anisotropy of the individual crystals (which may be deduced from the single crystal elastic constants of Tables 15.2 to 15.6 that follow) and the nature and extent of the preferred orientation.

Since the elastic properties are determined by the aggregate response of the interatomic forces between all the atoms in the metal, the presence of small quantities of solute atoms in dilute alloys or their rearrangement by heat treatment will have relatively little effect on the magnitudes of their elastic constants. Consequently, the elastic constants of all the plain carbon and low alloy steels will be approximately the same unless some preferred orientation is present. Similarly with Cu-, Al- and Ni- base dilute alloys, etc. In the case of concentrated alloys there may be larger variations in elastic moduli, especially where there is a drastic change in the relative proportions of different phases in a multiphase alloy. In the case of ideal solid solutions the elastic moduli vary linearly with atom fraction. The elastic moduli of non-ideal solid solutions may show positive or negative deviations from linearity. Ordering produces an increase in elastic moduli.

Increase in temperature causes a gradual decrease in elastic moduli. The decrease is fairly linear over wider ranges of temperature but sharply increases in magnitude as the melting point is approached. Discontinuities are observed when structural transformations occur.

Ferromagnetic materials having a high degree of domain mobility may exhibit considerably higher elastic moduli below the Curie point in the presence of a high magnetic field. The lower elastic moduli in the absence of a magnetic field are due to magnetostrictive dimensional changes caused by stress-induced domain movement.

Table 15.1   ELASTIC CONSTANTS OF POLYCRYSTALLINE METALS AT ROOM TEMPERATURE

Metal	Young's modulus GPa	Rigidity modulus GPa	Bulk modulus GPa	Poisson's ratio	Ref.
Aluminium	70.6	26.2	75.2	0.345	1
Antimony	54.7	20.7	—	0.25–0.33	2, 3
	77.9	19.3	—	—	4
Barium	12.8	4.86	—	0.28	2
Beryllium	318	156	110	0.02	5
Bismuth	34.0	12.8	—	0.33	2
Brass 70Cu 30Zn	100.6	37.3	111.8	0.35	1
Cadmium	62.6	24.0	51.0	0.30	5
Caesium	1.7	0.65	—	0.295	2
Calcium	19.6	7.9	17.2	0.31	2, 6
Cast Iron—Grey, BS 1452:1977					
Grade 150	100	40	—	0.26	7, 8
Grade 180	109	44	—	0.26	7, 8
Grade 220	120	48	—	0.26	7, 8
Grade 260	128	51	—	0.26	7, 8
Grade 300	135	54	—	0.26	7, 8
Grade 350	140	56	—	0.26	7, 8
Grade 400	145	58	—	0.26	7, 8
—Blackheart malleable BS 310:1972					
Grades B340/12 to B290/6	169	67.6	—	0.26	7, 9
Pearlitic malleable BS 3333:1972					
Grades P4440/7 to P540/5	172	68.8	—	0.26	7, 9
Whiteheart malleable BS 309:1972					
Grades W340/3, W410/4	176	70.4	—	0.26	7, 9
Nodular BS 2789:1973					
Grades 370/17, 420/12	169	66	—	0.275	7, 10
Grades 500/7, 600/3	169–174	65.9	—	0.275	7, 10
Grades 700/2, 800/2 (pearlitic, normalised)	176	68.6	—	0.275	7, 10
pearlite 700/2, 800/2 (hardened, tempered)	172	67.1	—	0.275	7, 10
Cerium	33.5	13.5	—	0.248	11, 12
Chromium	279	115.3	160.2	0.21	1
Cobalt	211	82	181.5	0.32	13, 16
Constantan 45Ni 55Cu	162.4	61.2	156.4	0.327	1
Copper	129.8	48.3	137.8	0.343	1
Cupro-nickel 70Cu 30Ni	144	53.8	—	0.34	14
Duralumin	70.8	26.3	75.4	0.345	1
Gallium	9.81	6.67	—	0.47	2
Germanium	79.9	29.6	—	0.32	2
Gold	78.5	26.0	171	0.42	15, 16
Hafnium	141	56	109	0.26	17, 18
Incoloy 800 20Cr, 32Ni bal Fe	196	73	—	0.334	38
Indium	10.6	3.68	—	0.45	2
Invar 64Fe 36Ni	144	57.2	99.4	0.259	1
Iridium	528	209	371	0.26	16, 19, 20, 21
Iron (pure)	211.4	81.6	169.8	0.293	1
Lanthanum	37.9	14.9	—	0.28	2, 12
Lead	16.1	5.59	45.8	0.44	1
Lithium	4.91	4.24	—	0.36	23, 28
Magnesium	44.7	17.3	35.6	0.291	1
Manganese	191	79.5	—	0.24	2, 24

(continued)

Table 15.1   ELASTIC CONSTANTS OF POLYCRYSTALLINE METALS AT ROOM TEMPERATURE—*continued*

Metal	Young's modulus GPa	Rigidity modulus GPa	Bulk modulus GPa	Poisson's ratio	Ref.
Manganese–copper 70Mn 30Cu (high damping alloy)	93	22.4	—	—	25
Molybdenum	324.8	125.6	261.2	0.293	1
Monel 400 63–70Ni, 2Mn, 2.5Fe, bal Cu	185	66	—	0.32	26
Nickel	199.5	76.0	177.3	0.312	1
Nickel silver 55Cu, 18Ni, 27Zn	132.5	49.7	132	0.333	1
Nimonic 80A 20Cr, 2.3Ti, 1.8Al, bal Ni (fully heat-treated)	222	85	—	0.31	27
Niobium	104.9	37.5	170.3	0.397	1
Ni–span C902 (constant modulus alloy)	186	66	—	0.41	28
Osmium	559	223	373	0.25	16, 21, 29
Palladium	121	43.6	187	0.39	18, 19, 21, 29
Platinum	170	60.9	276	0.39	16, 19, 29, 30
Plutonium	87.5	34.5	—	0.18	31
Potassium (−190°C)	3.53	1.30 (room temp.)	—	0.35	15
Rhenium	466	181	334	0.26	2, 16, 32
Rhodium	379	147	276	0.26	16, 19, 29
Rubidium	2.35	0.91	—	0.30	2
Ruthenium	432	173	286	0.25	18, 21, 29
Selenium	58	—	—	0.447	15
Silicon	113	39.7	—	0.42	2, 33
Silver	82.7	30.3	103.6	0.367	1
Sodium	6.80	2.53	—	0.34	2, 23
Steel—Mild	208–209	81–82	160–169	0.27–0.3	34
0.75C	210	81.1	168.7	0.293	1
0.75C (hardened)	201.4	77.8	165	0.296	1
Tool 0.98C, 1.03 Mn, 0.65 Cr, 1.01 W	211.6	82.2	165.3	0.287	1
Tool 0.98C, 1.03 Mn, 0.65 Cr, 1.01 W (hardened)	203.2	78.5	165.2	0.295	1
Maraging Fe–18Ni 8Co 5Mo	186	72	—	0.30	35
Stainless austenitic (Fe–18Cr, 8–10 Ni)	190–201	74–86	—	0.25–0.29	36
Stainless, ferritic (Fe–13Cr)	200–206	78–79	—	0.27–0.3	36
Stainless, martensitic (Fe–13Cr, 0.1–0.3C)	200–215	80–83	—	0.27–0.3	1, 36
Stainless, martensitic (Fe–18Cr, 2Ni, 0.2C)	215.3	83.9	166	0.283	1
Strontium	15.7	6.03	12.0	0.28	2, 6
Tantalum	185.7	69.2	196.3	0.342	1
Tellurium	47.1	16.7	—	0.16–0.3	15
Thallium	7.90	2.71	28.5	0.45	2, 6
Thorium	78.3	30.8	54.0	0.26	2, 6
Tin	49.9	18.4	58.2	0.357	1
Titanium	120.2	45.6	108.4	0.361	1
Tungsten	411	160.6	311	0.28	1
Tungsten carbide	534.4	219	319	0.22	1
Uranium	175.8	73.1	97.9	0.20	37
Vanadium	127.6	46.7	158	0.365	1
Yttrium	66.3	25.5	—	0.265	12
Zinc	104.5	41.9	69.4	0.249	1
Zirconium	98	35	89.8	0.38	17, 18

REFERENCES TO TABLE 15.1

1. G. Bradfield, 'Use in Industry of Elasticity Measurements in Metals with the help of Mechanical Vibrations', National Physical laboratory. Notes on Applied Science No. 30, HMSO, 1964.
2. W. Köster, *Z. Electrochem. Phys. Chem.*, 1943, **49**, 233.
3. W. Köster, *Z. Metall.*, 1948, **39**, 2.
4. 'Metals Handbook', Amer. Soc. Metals, Vol. 1, 1961.
5. D. J. Silversmith and B. L. Averbach, *Phys. Rev.*, 1970, **B1**, 567.
6. S. F. Pugh, *Phil. Mag. Ser. 7*, 1974, **45**, 823.
7. H. T. Angus, 'Cast Irons, Physical and Engineering Properties', Butterworths, London, 1976.
8. 'Engineering Data on Grey Cast Irons', Brit. Cast Iron Res. Assoc., 1977.
9. 'Engineering Data on Malleable Cast Irons', Brit. Cast Iron Res. Assoc., 1974.
10. 'Engineering Data on Nodular Cast Irons', Brit. Cast Iron Res. Assoc., 1974.
11. M. Rosen, *Phys. Rev.*, 1969, **181**, 932.
12. J. F. Smith, C. D. Carlson and F. H. Spedding, *J. Metals*, 9; *Trans. AIME*, 1957, **209**, 1212.
13. 'Physical and Mechanical Properties of Cobalt', Cobalt Information Centre, Brussels, 1960.
14. 'Cupro-Nickel Alloys, Engineering Properties', Publ. 2969, Inco Europe, London, 1966.
15. Landolt-Börnstein, 'Zahlenwerte und Funktionen', Vol. 2, Part 1, Springer-Verlag, Berlin, 1971.
16. A. S. Darling, *Int. Met. Rev.*, 1973, 91.
17. Private communication, Imperial Metal Industries, Witton, Birmingham.
18. A. S. Darling, *Proc. Inst. Mech. Eng.*, 1965, Pt 3D, **180**, 104.
19. W. Köster, *Z. Metall.*, 1948, **39**, 1.
20. A. Roll and H. Motz, *Z. Metall.*, 1957, **48**, 272.
21. K. H. Schramm, *Z. Metall.*, 1962, **53**, 729.
22. P. W. Bridgman, *Proc. Amer. Acad. Arts Sci.*, 1922, **57**, 41.
23. O. Bender, *Ann. Phys.*, 1939, **34**, 359.
24. M. Rosen, *Phys. Rev.*, 1968, **165**, 357.
25. D. Birchon, *Engineering Mater and Design*, 1964, **7**, 606.
26. 'Wrought Nickel-Copper Alloys, Engineering Properties', Publ. 7011, Inco Europe, London, 1970.
27. 'Nimonic Alloy 80A', Publ. 3663, Henry Wiggin Ltd., Hereford, 1975.
28. 'Controlled Expansion and Constant Modulus Nickel–Iron Alloys', Publ. 6710, Inco Europe, London, 1967.
29. W. Köster, *Z. Metall.*, 1948, **39**, 111.
30. E. Grüneisen, *Ann. Phys.*, 1908, **25**, 825.
31. 'Plutonium Handbook', Ed. O. J. Wick, Gordon and Breach, New York, 1967, p. 39.
32. T. E. Tietz, B. A. Wilcox and J. W. Wilson, Standford Res. Instit. Calif., Report SU-2436, 1959.
33. R. L. Templin, *Metals and Alloys*, 1932, **3**, 136.
34. J. Woolman and R. A. Mottram, 'The Mechanical and Physical Properties of the British Standard En Steels', Vol. 1, Pergamon, Oxford, 1964.
35. '18% Nickel Maraging Steels', Publ. 4419, Inco Europe, London, 1976.
36. J. Woolman and R. A. Mottram, 'The Mechanical and Physical Properties of the British Standard En Steels', Vol. 3, Pergamon, Oxford, 1969.
37. 'Commercial Uranium', Brit. Nuclear Fuels Ltd., Warrington.
38. 'Incoloy 800', Publ. 3664, Henry Wiggin Ltd., Hereford, 1977.

15.1.1   Elastic compliances and elastic stiffnesses of single crystals

Single crystals are generally anisotropic and therefore require many more constants of proportionality than isotropic materials. The relations between stress and strain are defined by the generalised Hooke's law, which states that the strain components are linear functions of the stress components and vice versa.

That is,

$$\varepsilon_{xx} = S_{11}\sigma_{xx} + S_{12}\sigma_{yy} + S_{13}\sigma_{zz} + S_{14}\sigma_{yz} + S_{15}\sigma_{zx} + S_{16}\sigma_{xy}$$
$$\varepsilon_{yy} = S_{21}\sigma_{xx} + S_{22}\sigma_{yy} + S_{23}\sigma_{zz} + S_{24}\sigma_{yz} + S_{25}\sigma_{zx} + S_{26}\sigma_{xy}$$
$$\cdots\cdots\cdots\cdots\cdots\cdots\cdots\cdots\cdots\cdots\cdots\cdots\cdots\cdots\cdots\cdots\cdots\cdots\cdots\cdots\cdots$$
$$\varepsilon_{xy} = S_{61}\sigma_{xx} + S_{62}\sigma_{yy} + S_{63}\sigma_{zz} + S_{64}\sigma_{yz} + S_{65}\sigma_{zx} + S_{66}\sigma_{xy}$$

and correspondingly

$$\sigma_{xx} = C_{11}\varepsilon_{xx} + C_{12}\varepsilon_{yy} + C_{13}\varepsilon_{zz} + C_{14}\varepsilon_{yz} + C_{15}\varepsilon_{zx} + C_{16}\varepsilon_{xy}$$
$$\cdots\cdots\cdots\cdots\cdots\cdots\cdots\cdots\cdots\cdots\cdots\cdots\cdots\cdots\cdots\cdots\cdots\cdots\cdots\cdots\cdots$$
$$\sigma_{xy} = C_{61}\varepsilon_{xx} + C_{62}\varepsilon_{yy} + C_{63}\varepsilon_{zz} + C_{64}\varepsilon_{yz} + C_{65}\varepsilon_{zx} + C_{66}\varepsilon_{xy}$$

where

$\sigma_{xx}, \sigma_{yy}, \sigma_{zz}$ and $\sigma_{yz}, \sigma_{zx}, \sigma_{xy}$ represent normal and shear stresses, respectively;

$\varepsilon_{xx}, \varepsilon_{yy}, \varepsilon_{zz}$ and $\varepsilon_{yz}, \varepsilon_{zx}, \varepsilon_{xy}$ represent normal and shear strains, respectively.

The elastic constants S_{ij} and C_{ij} are called the elastic compliances and elastic stiffnesses, respectively. Many of the constants are equal, the number of independent constants decreasing with increasing crystal symmetry. For example, in the hexagonal system there are five independent constants, while in the cubic system there are only three independent elastic compliances S_{11}, S_{12}, S_{44} with corresponding elastic stiffnesses C_{11}, C_{12}, C_{44}.

The tensile and shear moduli will vary with orientation in a single crystal of a cubic metal according to

$$\frac{1}{E} = S_{11} - 2[(S_{11} - S_{12}) - \tfrac{1}{2}S_{44}]\,(l^2 m^2 + m^2 n^2 + l^2 n^2)$$

$$\frac{1}{G} = S_{44} - 2[(S_{11} - S_{12}) - \tfrac{1}{2}S_{44}]\,(l^2 m^2 + m^2 n^2 + l^2 n^2)$$

where l, m, n are the direction cosines of the specimen axis with respect to the crystallographic axes. For an isotropic crystal

$$S_{44} = 2(S_{11} - S_{12}) \text{ and } C_{44} = \tfrac{1}{2}(C_{11} - C_{12})$$

hence

$$E = \frac{1}{S_{11}} \quad \text{and} \quad G = \frac{1}{S_{44}}$$

Therefore, the degree of anisotropy is conveniently specified by

$$\frac{2(S_{11} - S_{12})}{S_{44}} \quad \text{or} \quad \frac{(C_{11} - C_{12})}{2C_{44}}$$

15.1.2   Principal elastic compliances and elastic stiffnesses at room temperature

The units are TPa^{-1} for S_{ij} (elastic compliances) and GPa for C_{ij} (elastic stiffnesses).

Table 15.2   CUBIC SYSTEMS (3 CONSTANTS)

Metal	S_{11}	S_{44}	S_{12}	C_{11}	C_{44}	C_{12}	Ref.
Ag	22.9	22.1	−9.8	123	45.3	92.0	1
Al	16.0	35.3	−5.8	108	28.3	62.0	1, 2
Au	23.4	23.8	−10.7	190	42.3	161	1
Ca	94.0	83.0	−31.0	16.0	12.0	8.0	3
Cr	3.08	9.98	−0.49	346	100	66.0	1
Cs (78 K)	1 676	676	−762	2.49	2.06	1.48	4
Cu	15.0	13.3	−6.3	169	75.3	122	1
Fe	7.67	8.57	−2.83	230	117	135	1
Ge	9.73	14.9	−2.64	129	67.1	48.0	1
Ir	2.24	3.72	−0.67	600	270	260	5
	2.28	3.90	−0.67	580	256	242	6
K	1 215	531	−558	3.71	1.88	3.15	7
	1 339	526	−620	3.69	1.90	3.18	8
Li	315	104	−144	13.4	9.60	11.3	9
Mo	2.71	9.0	−0.74	459	111	168	1
Na	549	233	−250	7.59	4.30	6.33	1
Nb	6.56	35.2	−2.29	245	28.4	132	1, 10, 11

(continued)

Table 15.2   CUBIC SYSTEMS (3 CONSTANTS)—*continued*

Metal	S_{11}	S_{44}	S_{12}	C_{11}	C_{44}	C_{12}	Ref.
Ni (zerofield)	7.67	8.23	−2.93	247	122	153	1, 12
Ni (saturation field)	7.45	8.08	−2.82	249	124	152	1
Pb	93.7	68.0	−43.0	48.8	14.8	41.4	1
Pd	13.7	14.0	−6.0	224	71.6	173	1
Pt	7.35	13.1	−3.08	347	76.5	251	13
Rb	1 330	625	−600	2.96	1.60	2.44	14
Si	7.74	12.6	−2.16	165	79.2	64	1
Sr	148	174	−60	14.7	5.74	9.9	15
Ta	6.89	12.1	−2.57	262	82.6	156	1, 16
Th	27.2	20.9	−10.7	75.3	47.8	48.9	17
	27.4	22.0	−10.9	77.0	45.5	50.9	18
Tl	101	91	−46	40.8	11.0	34.0	19
V	6.76	23.2	−2.32	230	43.2	120	1
W	2.49	6.35	−0.70	517	157	203	1

Table 15.3   HEXAGONAL SYSTEMS (5 CONSTANTS)

Metal		11	33	44	12	13	Ref.
Be	S	3.45	2.87	6.16	−0.28	−0.05	1
	C	292	349	163	24	6	1
Cd	S	12.2	33.8	51.1	−1.2	−8.9	1
	C	116	50.9	19.6	42	41	1
Co	S	5.11	3.69	14.1	−2.37	−0.94	1
	C	295	335	71.0	159	111	1
Dy	S	16.0	14.5	41.2	−4.6	−3.2	1
	C	74.0	78.6	24.3	25.5	21.8	1
Er	S	14.1	13.2	36.4	−4.2	−2.6	1
	C	84.1	84.7	27.4	29.4	22.6	1
Gd	S	18.3	16.1	48.3	−5.7	−3.8	20
	S	18.0	16.1	48.1	−5.7	−3.6	21
	C	66.7	71.9	20.7	25.0	21.3	20
	C	67.8	71.2	20.8	25.6	20.7	21
Hf	S	7.16	6.13	18.0	−2.48	−1.57	22
	C	181	197	55.7	77	66	22
Ho	S	15.3	14.0	38.6	−4.3	−2.9	1
	C	76.5	79.6	25.9	25.6	21.0	1
Mg	S	22.0	19.7	60.9	−7.8	−5.0	1
	C	59.3	61.5	16.4	25.7	21.4	1
Nd	S	23.7	18.5	66.5	−9.50	−3.90	23, 24
	C	54.8	60.9	15.0	24.6	16.6	23, 24
Pr	S	26.6	19.3	73.6	−11.3	−3.80	25, 26
	C	49.4	57.4	13.6	23.0	14.3	25, 26
Re	S	2.11	1.70	6.21	−0.80	−0.40	1
	C	616	683	161	273	206	1
Ru	S	2.09	1.82	5.53	−0.58	−0.41	20
	C	563	624	181	188	168	20
Sc	S	12.5	10.6	36.1	−4.30	−2.20	27
	C	99.3	107	27.7	39.7	29.4	27
Tb	S	17.4	15.6	46.0	−5.2	−3.60	21, 28
	C	69.2	74.4	21.8	25.0	21.8	21, 28
Tl	S	104	31.1	139	−83.0	−11.6	29, 30
	C	41.9	54.9	7.20	36.6	29.9	29, 30
Ti	S	9.69	6.86	21.5	−4.71	−1.82	1
	C	160	181	46.5	90.0	66.0	1
Y	S	15.4	14.4	41.1	−5.10	−2.70	31
	C	77.9	76.9	24.3	29.2	20.0	31
Zn	S	8.22	27.7	25.3	−0.60	−7.0	1, 32
	C	165	61.8	39.6	31.1	50.0	1
Zr	S	10.1	8.0	30.1	−4.0	−2.4	1
	C	144	166	33.4	74	67	1

Table 15.4 TRIGONAL SYSTEMS (6 CONSTANTS)

Metal		11	33	44	12	13	14	Ref.
As	S	30.6	140	45.0	20.5	−56.0	1.7	33
	C	130	58.7	22.5	30.3	64.3	−3.7	33
Bi	S	25.7	41.1	113	−7.8	−11.2	−21.4	1, 34
	C	62.3	37.0	11.5	23.1	23.4	7.3	1, 34
B	S	—	—	—	—	—	—	—
	C	467	473	198	241	—	15.1	35
Hg (83 K)	S	154	45.0	151	−119	−21	−100	36
	C	36.0	50.5	12.9	28.9	30.3	4.7	36
Sb	S	16.0	29.6	39.1	−6.1	−6.0	−12.4	1
	C	101	44.8	39.6	31.4	27.0	22.1	1
Se	S	131	41	112	−13	−40	56	1
	C	18.6	76.1	14.8	7.3	25.2	5.6	1
Te	S	53.4	24.3	52.1	−16.1	−13.6	26.7	1
	C	34.4	70.8	32.7	9.0	24.9	13.1	1

Table 15.5 TETRAGONAL SYSTEMS (6 CONSTANTS)

Metal		11	33	44	66	12	13	Ref.
In	S	149	199	154	83	−44	−96	1
	C	45.2	44.9	6.52	12.0	40	41.2	1
Sn	S	42.4	14.8	45.6	42.1	−32.4	−4.3	1
	C	73.2	90.6	21.9	23.8	59.8	39.1	1

Table 15.6 ORTHORHOMBIC SYSTEMS (9 CONSTANTS)

Metal		11	12	33	44	55	66	12	13	23	Ref.
Ga	S	12.2	14.0	8.49	28.6	23.9	24.8	−4.4	−1.7	−2.4	1
	C	100	90.2	135	35.0	41.8	40.3	37.0	33.0	31.0	1
U	S	4.91	6.73	4.79	8.04	13.6	13.4	−1.19	0.08	−2.61	37
	C	215	199	267	124	73.4	74.3	46.5	21.8	108	37

REFERENCES TO TABLES 15.2–15.6

1. Landolt-Börnstein, 'Numerical Data and Functional Relationships in Science and Technology', New Series, Group III, Vol. 2, Berlin, Springer-Verlag, 1979.
2. C. Gault, P. Boch, A. Dauger, *Phys. Stat. Solidi*, 1977, **a43**, 625.
3. M. Taut and H. Eschrig, *Phys. Stat. Solidi*, 1976, **b73**, 151.
4. F. J. Kollarits and T. Trivisonno, *J. Phys. Chem. Solids*, 1968, **29**, 2133.
5. H. G. Purwins, H. Hieber and J. Labusch, *Phys. Stat. Solidi*, 1965, **11**, k63.
6. R. E. Macfarlane, J. A. Rayne and C. K. Jones, *Phys. Letters*, 1966, **20**, 234.
7. P. A. Smith and C. S. Smith, *J. Phys. Chem. Solids*, 1965, **26**, 279.
8. G. Fritsch and H. Bube, *Phys. Stat. Solidi*, 1975, **a30**, 571.
9. H. C. Nash and C. S. Smith, *J. Phys. Chem. Solids*, 1959, **9**, 113.
10. E. Walker and M. Peter, *J. Appl. Phys.*, 1977, **48**, 2820.
11. D. M. Schlader and J. F. Smith, *J. Appl. Phys.*, 1977, **48**, 5062.
12. K. Salama and J. A. Alers, *Phys. Stat. Solidi*, 1977, **a41**, 241.
13. R. E. Macfarlane, J. A. Rayne and C. K. Jones, *Phys. Letters*, 1965, **18**, 91.
14. C. A. Roberts and R. Meister, *J. Phys. Chem. Solids*, 1966, **27**, 1401.
15. S. S. Mathur and P. N. Gupta, *Acustica*, 1974, **31**, 114.
16. W. L. Stewart *et al.*, *J. Appl. Phys.*, 1977, **48**, 75.
17. P. E. Armstrong, O. N. Carlson and J. F. Smith, *J. Appl. Phys.*, 1959, **30**, 36.
18. J. D. Greiner, D. T. Peterson and J. F. Smith, *J. Appl. Phys.*, 1977, **48**, 3357.
19. M. S. Shepard and J. F. Smith, *Acta Met.*, 1967, **15**, 357.
20. E. S. Fisher and D. Dever, *Trans. Met. Soc. AIME*, 1967, **239**, 48.
21. S. B. Palmer, E. W. Lee and M. N. Islam, *Proc. Roy. Soc.*, 1974, **A338**, 341.

22. E. S. Fisher and C. J. Renken, *Phys. Rev.*, 1964, **135A**, 482.
23. J. D. Greiner *et al.*, *J. Appl. Phys.*, 1976, **47**, 3427.
24. J. T. Lenkkeri and S. B. Palmer, *J. Phys.*, 1977, **F7**, 15.
25. J. D. Greiner *et al.*, *J. Appl. Phys.*, 1973, **44**, 3862.
26. S. B. Palmer and C. Isci, *Physica*, 1977, **86–88**, 45.
27. E. S. Fisher and D. Dever, *Proc. Rare Earth Res. Conf.*, Coronado Calif., 1968, Vol. 7, p. 237.
28. K. Salama, F. R. Brotzen and P. L. Donoho, *J. Appl. Phys.*, 1972, **43**, 3254.
29. R. Weil and A. W. Lawson, *Phys. Rev.*, 1966, **141**, 452.
30. R. W. Ferris, M. L. Shepherd and J. F. Smith, *J. Appl. Phys.*, 1963, **34**, 768.
31. J. F. Smith and J. A. Gjevre, *J. Appl. Phys.*, 1960, **31**, 645.
32. D. P. Singh, S. Singh and S. Chendra, *Ind. J. Phys.*, 1977, **A51**, 97.
33. N. G. Pace and G. A. Saunders, *J. Phys. Chem. Solids*, 1971, **32**, 1585.
34. A. M. Lichnowski and G. A. Saunders, *J. Phys.*, 1977, **C10**, 3243.
35. I. M. Silvestrova *et al.*, *Mater. Res. Bull.*, 1974, **9**, 1101.
36. H. B. Huntington, 'Solid State Physics', (ed. F. Seitz and D. Turnbull), New York, Academic Press, 1958, Vol. 7, p. 213.
37. E. S. Fisher and H. J. McSkimin, *J. Appl. Phys.*, 1958, **29**, 1473.

15.2   Damping capacity

The damping capacity of a metal measures its ability to dissipate elastic strain energy. The existence of this property implies that Hooke's law is not obeyed even at stresses well below the conventional elastic limit. In a perfectly elastic solid in vibration, stress and strain are always in phase and no energy is dissipated.

There are two important types of damping: anelastic and hysteretic.

In an anelastic solid there is a lag between the application of stress and the attainment of the resulting equilibrium strain; unless the stress changes exceedingly slowly. Processes with this characteristic give rise to an energy loss that reaches a peak at a critical frequency of vibration.

An hysteretic solid has a stress–strain curve on loading that does not coincide with that on unloading. The area between the two curves is proportional to the energy loss and does not vary with the frequency with which the load cycle is traversed but changes in a complex fashion with peak stress. Damping from this class of mechanism is often high and since it does not vary with frequency is of particular interest to the engineer since it can contribute to vibration and noise reduction and can limit the intensity of vibrational stress under resonant conditions and thus minimise fatigue failure.

Table 15.7 lists the specific damping capacity of a number of commercial alloys including some of very high damping that might be of interest to vibration engineers. In all cases the damping is predominantly of the hysteretic type.

Table 15.7   THE SPECIFIC DAMPING CAPACITY OF COMMERCIAL ALLOYS AT ROOM TEMPERATURE

The specific damping capacity which is normally measured on solid cylinders stressed in torsion is defined as the ratio of the vibrational strain energy dissipated during one cycle of vibration to the vibrational strain energy at the beginning of the cycle.

Alloy	Composition %	Specific damping capacity %	Surface shear stress MPa
Cast irons			
High carbon inoculated flake iron	2.5% C, 1.9% Si, 1.0% Mn, 20.7% Ni, 1.9% Cr, 0.13% P	19.3	34.5
Spun cast iron	3.54% C, 3.39% G.C., 1.9% Si, 0.4% Mn, 0.38% P	10.8	34.5
Non-inoculated flake iron	3.3% C, 2.2% Si, 0.5% Mn, 0.14% P, 0.03% S	8.5	34.5
Inoculated flake iron	3.3% C, 2.2% Si, 0.5% Mn, 0.14% P, 0.03% S	7.3	34.5
Austenitic flake graphite	2.5% C, 1.9% Si, 1% Mn, 20.7% Ni, 1.9% Cr, 0.03% P, 0.03% S	7.1	34.5
Alloyed flake graphite	3.14% C, 2% Si, 0.6% Mn, 0.7% Ni, 0.4% Mo, 0.14% P, 0.03% S	5.3	34.5
Nickel-copper austenitic flake	2.55% C, 1.9% Si, 1.25% Mn, 15.2% Ni, 7.3% Cu, 2% Cr, 0.03% P, 0.04% S	3.9	34.5
Undercooled flake graphite titanium/CO_2 treated	3.27% C, 2.2% Si, 0.6% Mn, 0.35% Ti, 0.14% P, 0.03% S	3.9	34.5

(continued)

Table 15.7   THE SPECIFIC DAMPING CAPACITY OF COMMERCIAL ALLOYS AT ROOM
TEMPERATURE—*continued*

Alloy	*Composition %*	*Specific damping capacity %*	*Surface shear stress* MPa
Annealed ferritic nodular	3.7% C, 1.8% Si, 0.4% Mn, 0.76% Ni, 0.06% Mg, 0.03% P, 0.01% S, <0.003% Ce	2.8	34.5
Pearlitic malleable	BS 3333/1961 Grade B.33/4	1.6	34.5
Blackheart malleable	BS 310/1958 Grade B.22/14	1.5	34.5
As cast pearlitic nodular	3.66% C, 1.8% Si, 0.4% Mn, 0.76% Ni, 0.06% Mg, 0.03% P, 0.01% S, <0.003% Ce	1.4	34.5
Steels			
BS 970 070M20 En3	0.17% C mild steel, normalised	1.5	34.5
BS 1407 (silver steel)	Spherodised	0.8	34.5
BS 1407 (silver steel)	Water quenched 800°C	0.5	34.5
BS 1407 (silver steel)	Water quenched 800°C aged 100°C $1\frac{1}{2}$ h	0.2	34.5
BS 970 653M31 En23T	3% Ni, 1% Cr, 0.3% C	0.8	34.5
BS 970 503M40 En12Q	1% Ni, 0.4% C	0.3	34.5
BS 970 709M40 En19U	1% Cr, 0.3% Mo, 0.4% C	0.15	34.5
BS 3S62	12% Cr, 0.2% C. Quenched tempered to 225 BHN	3.8	34.5
BS 970 321S20 En58B	18% Cr, 8% Ni, 0.6% Ti, 0.1% C. Solution treated 1 050°C, water quenched	1.8	34.5
NMC	0.62% C, 3.86% Cr, 8.6% Ni, 7.3% Mn. Solution treated 1 050°C, water quenched	0.7	34.5
BS 970 302S25 En58A	18% Cr, 8% Ni, 0.1% C. Solution treated 1 050°C, water quenched	0.3	34.5
Copper alloys			
Hidurel 6	As cast	1.35	34.5
Gunmetal	88% Cu, 10% Zn, 2% Sn	1.0	34.5
Brass (BS 265)	As extruded	0.4	34.5
Hidurel 5	As cast	0.4	34.5
Hidurel 7	As cast	0.25	34.5
High tensile brass	As cast	0.25	34.5
Novoston	—	0.25	34.5
Incramute	58% Cu, 40% Mn, 2% Al, aged 400°C	68	
Aluminium alloys			
Duralumin (HE 14)	—	0.25	34.5
RR57 (DTD 5004 WP)	—	0.20	34.5
RR58 (DTD 5014 WP)	—	0.10	34.5
Hiduminium 100 (SAP)	—	5.0	34.5
Magnesium alloys			
DTD 5005	Mg/Zn/Zr/Th	7.4	20.7
BS 1278	Mg/Zn/Mn	1.6	20.7
DTD 721A	Mg/Zn/Zr	0.65	20.7
Magnesium Elektron MSR Alloy	Mg/Ag/Zr	0.4	20.7
Manganese alloys			
Mn–Cu (quenched from 850°C aged 2 h at 425°C)	90% Mn–10% Cu	21	34.5
	85% Mn–15% Cu	27	34.5
	80% Mn–20% Cu	22	34.5
	70% Mn–30% Cu	42	34.5
	60% Mn–40% Cu	42	34.5
	50% Mn–50% Cu	33	34.5
Mn–Cu (sintered)	40wt % Cu–60wt % Mn	>40	35
Nickel alloys			
Ni–Ti (Nitinol)	55% Ni–45% Ti	26	69
T–D Nickel	2.5% Thoria	10.7	69
Mallory No-chat	—	9.4	69
Titanium alloys			
Ti–Al–V–Mo	Ti–5.5–6.75% Al–1–5% V–1–5% Mo (V + Mo >6%), all wt %	0.5	

Reference: D. Birchon, *Engineering Materials and Design*, Sept., Oct., 1964; also 307, 312.

15.2.1   Anelastic damping

Of interest to the physical metallurgist is the fact that a phase lag between stress and strain can give rise to a peak in energy dissipation or damping as a function of temperature or frequency. Several quite distinct atomic processes have been identified with damping peaks and measurements on these peaks in a wide variety of metals and alloys have been used to give diffusion data and to study precipitation, ordering phenomena and the properties of dislocations, point defects and grain boundaries. Table 15.8 identifies the damping peaks found in a number of pure metals and alloys with the relaxation process thought to be involved and also give an indication of the magnitude of the damping peak height. Detailed information on the specific mechanisms involved can be obtained from the reviews below and the references given for the respective damping peaks. The main types of peak that are observed are as follows. In cold worked pure metals movement of dislocation lines results in a number of low temperature peaks known as Bordoni peaks. Interaction of dislocations with point defects give rise to a further series of unstable peaks at higher temperatures; these are now called Hasiguti peaks. In alloys the stress-induced redistribution of solute atoms results in two types of peak, the Zener-type peak in substitutional solid solutions and the Snoek-type peak in interstitial solid solutions. The interaction of interstitial solute atoms with substitutional solute atoms give rise to a modified Snoek peak in ternary alloys. In cold worked alloys the interaction of interstitial solute atoms with dislocations results in the Köstler-type peaks at higher temperatures than the Snoek-type peaks. In pure metals and alloys the stress-induced migration of grain boundaries and/or polygonised (sub-grain) boundaries give rise to a further series of high temperature damping peaks.

In the ideal case, for a relaxation process having a single relaxation time (τ) the logarithmic decrement (δ) will be given by

$$\delta = \pi \Delta \frac{\omega \tau}{1 + \omega^2 \tau^2}$$

where ω is the angular frequency and Δ is the modulus defect. The relaxation time, being diffusion controlled, varies with temperature according to an Arrhenius equation of the form $\tau = \tau_0 \exp (H/RT)$ where τ_0 is a constant, H is the activation energy controlling the relaxation process and T is the absolute temperature. The condition for maximum damping (δ_p) is that $\omega \tau = 1$ and hence

$$\delta_p = \frac{\pi \Delta}{2} (= \pi Q^{-1} \mathrm{max})$$

The modulus defect which is a measure of the strength of the relaxation can be highly orientation dependent and therefore the values given in the table below must be interpreted with caution. It must also be noted that for many measured damping peaks a distribution of relaxation times is found to be present. This leads to a broader peak being observed than would be present if a single relaxation time were operative. The decrement will be given by

$$\delta = \pi \sum_i \Delta_i \frac{\omega \tau_i}{1 + \omega^2 \tau_i^2}$$

where each i refers to a component of the total peak that has the same form as that of a peak arising from a single relaxation time.

This theoretical aspect of the analysis of broad peaks in terms of a spectrum of τ's has been comprehensively dealt with by A. S. Nowick and B. S. Berry in *IBM Journal of Research and Development*, 1961, 5(4), 297–311, 312–20.

VIEWS

1. C. Zener, 'Elasticity and Anelasticity of Metals', Chicago: Chicago University Press, 1948.
2. K. M. Entwistle, 'Progress in Non-Destructive Testing' (edited by E. G. Stanford, J. H. Fearnon), vol. 2, p. 191, London: Heywood, 1960.
3. D. H. Niblett and J. Wilks, *Adv. Phys.*, 1960, **9**, 1.
4. K. M. Entwistle, *Metall. Rev.*, 1962, **7**, 175.
5. A. S. Nowick and B. S. Berry, 'Anelastic Relaxation in Crystalline Solids', Academic Press, 1972.

In Table 15.8, metals and alloys are listed in alphabetical order with the highest concentration constituent first. The values of the modulus defect are deduced using the equation

$$\Delta = \frac{2\delta_p}{\pi} = 2Q_{max}^{-1}$$

where δ_p is the peak decrement and Q_{max}^{-1} is the peak value of Q^{-1}. These relationships are valid only if the damping arises from a process having a single relaxation time. In most cases a distribution of relaxation times exists and the peaks are broader, but there is only rarely sufficient published data to permit this distribution to be deduced. As many authors do not make it clear which damping units they use, the quoted values of Modulus Defect in the table must not be interpreted too precisely. The aim is to record the existence of peaks and list the suggested mechanism giving rise to them, and the values of Δ serve to indicate the approximate strength of the relaxation. If detailed quantitative data are sought peaks should be analysed in particular cases using the method of Berry and Nowick.

Table 15.8 ANELASTIC DAMPING

Alloy	Composition and physical condition	Peak temp.	Frequency Hz	Modulus defect	Activation energy kJ mol^{-1}	Mechanism or type	Ref.
Ag	99.999% Ag	37 K	0.7	2.6×10^{-2}	0.84	Bordoni type	1, 2
	(CW* 16% at 4.2 K)	50 K	0.7	1.2×10^{-2}	0.84	Bordoni type	1, 2
Ag	99.999% Ag single crystal deformed at RT 43.7% [121] 5.4% [111]	~50 K	~600	13×10^{-4}	—	Bordoni type	3
	Single crystals deformed at (5.4% [111]) RT	~50 K	~600	6×10^{-3}	—	Bordoni type	3
Ag	99.99% Ag	173 K	10^3	—	21.3	Point defect/ dislocation interactions	4
	(CW* at RT)	200 K	10^3	—	35.6 etc.	Point defect/ dislocation interaction	4
Ag	99.998% Ag (84% area reduction, then annealed at 500°C for 1 h)	163°C	1.04	38×10^{-4}	92.0	Grain boundary	5
		356°C	0.98	39×10^{-4}	177	Grain boundary	5
Ag	99.999% Ag	150 K	1.5	1.3×10^{-2}	92	Grain boundary	6
Ag–Au	25–80 At. % Au 42 At. % Au	364–	1.0	$2 \times 10^{-3}–$	175.7–	Zener	7
		398°C		2×10^{-2}			
		460°C	0.36	0.104 max	165.3	Grain boundary	8
Ag–Cd	29 At. % Cd	~230°C	~1	—	152.3	Zener	9
	39 At. % Cd	—	~1	—	123.0	Zener	9
Ag–Cd	32 At. % Cd	220°C	0.6	4×10^{-2}	146.9	Zener	10
	0.9–32 At. % Cd	367–	1.5	$9.2 \times 10^{-2}–$	159–188	Grain boundary	6, 11
		452°C		1.38×10^{-1}			
Ag–In	9.6 At. % In–	580–	~1	$<3.5 \times 10^{-3}–$	152.8–	Zener	12
	17.9 At. % In	536 K		7×10^{-3}	130.5		
Ag–In	7.5 At. % In–	—	~1	$6.5 \times 10^{-4}–$	159	Zener	9
	15.6 At. % In	—		6×10^{-3}	133.1		
Ag–In	10.8 At. % In–	~500 K	$10^{-3}–$	$~20 \times 10^{-3}$	146.9–	Zener	13
	18.1 At. % In		10^{-1}		139.7		
Ag–In	16 At. % In	270°C	1.6	2.9×10^{-2}	—	Zener	14
Ag–In	1 At. % In–	355–	1.5	$9.5 \times 10^{-2}–$	172–188	Grain boundary	6, 11
	16 At. % In	450°C		1.27×10^{-1}			
Ag–Sb	6.3 At. % Sb	~510 K	$10^{-3}–10^{-1}$	$~5 \times 10^{-3}$	136	Zener	15
Ag–Sn	0.93 At. % Sn	200°C	1.5	1.3×10^{-3}	—	Zener	6, 11
Ag–Sn	8.1 At. % Sn	~550 K	$10^{-3}–10^{-1}$	$~5 \times 10^{-3}$	131.8	Zener	15
Ag–Sn	0.9 At. % Sn–	390–	1.5	$1.08 \times 10^{-1}–$	171–184	Grain boundary	6, 11
	8.0 At. % Sn	440°C		1.27×10^{-1}			

(continued)

Table 15.8   ANELASTIC DAMPING—*continued*

Alloy	Composition and physical condition	Peak temp.	Frequency Hz	Modulus defect	Activation energy kJ mol⁻¹	Mechanism or type	Ref.
Ag–Zn	15 At. % Zn–30 At. % Zn	280–232°C	0.68	1.2×10^{-2}–1.4×10^{-2}	142–136	Zener	16
Al	99.99% Al (CW* 3% at RT)	24 K	1.2×10^4	1.5×10^{-4}	2.3	Bordoni type	17, 1
Al	99.999% Al deformed at 20 K then at 80 K	70 K	3	—	11.6	Bordoni type	19
		100 K	3	—	18.3	Bordoni type	
		115 K	3	—	—	?	
		155 K	3	—	—	?	
Al	99.99% Al (CW* 6% at RT)	83 K	1.08×10^3	1.6×10^{-3}	16.3	Bordoni type	20
		119 K	1.06×10^3	2.4×10^{-3}	28.9	Bordoni type	
Al	99.999% Al 2 h at 470 K, then CW* by 0.5% at 77 K	110 K	4×10^{-2}	$\sim 10^{-2}$	24	Hasiguti type	21
		155 K	2×10^3	$\sim 2 \times 10^{-3}$	—	Hasiguti type	
Al	99.999% Al deformed at 85 K at 10^{-4} and cycled 10^2 times	130 K	~ 1	$\sim 30 \times 10^{-3}$	—	Bordoni type with contribution from impurity–dislocation interactions	22
Al	99.994% [111]	139 K	10^7–	—	4.1	Bordoni type	23
	[100]	153 K	5×10^7	—	6.0	Bordoni type	
	[110]	196 K		—	19.5	Bordoni type	
Al	99.999% Al (CW* 18% at RT)	213 K	10^3	—	38.5	Point defect/dislocation interaction	4
Al	99.6% Al reduced by 69%	270°C	5×10^{-1}	$\sim 1.2 \times 10^{-1}$	—	Relaxation of stresses by shear deformation and recrystallisation	24
Al	99.6% Al reduced by 99.3%	270°C	5×10^{-1}	$\sim 2 \times 10^{-1}$	—	Relaxation of stresses by shear deformation and recrystallisation	24
		400°C	5×10^{-1}	$\sim 10^{-1}$	—	Associated with grain boundaries	24
Al	99.999% Al deformed by 65%, annealed and deformed again	~ 300°C	~ 1	10^{-3}–10^{-2} depending on CW	48.2	Associated with dislocations	25
Al	99.96% Al area reduced by 75% annealed at 325°C for 2 h	340°C	2.32	7.85×10^{-2}	144.3	Grain boundary	5
Al	99.991% Al	275°C	0.69	1.4×10^{-1}	134	Grain boundary	26
Al	99.999% Al CW* 4% then irradiated by neutrons at 80 K, annealed 360 K	110 K	2.5	1.6×10^{-2}	—	Point defect/dislocation interaction	241
Al	99.9999wt% Al (bamboo boundaries)	~ 300°C	~ 1	1.44 eV		Grain boundary	251
Al	99.999wt% (3 h at 402°C)	396°C	8×10^{-3}	0.52	1.5 eV	Polygonisation	256
Al	99.999wt% (single crystal sheet)	400°C	1.3			Dislocation network	258
Al	99.9999wt% Al (polycrystalline wires)						272
	(grain size ≪ wire dia.)	210°C	~ 1			Grain boundary	
	(grain size ≫ wire dia.)	170°C	~ 1			Not known	
Al	99.999wt% (single crystal)	365°C	1		184 eV	Point defects/dislocation interaction	265
Al–Ag	20% Ag (quenched from 520°C, aged at 155°C)	140°C	0.25	1.2×10^{-2}	105–113	Stress induced change of local degree of precipitation	27

(*continue*

Table 15.8   ANELASTIC DAMPING—*continued*

Alloy	Composition and physical condition	Peak temp.	Frequency Hz	Modulus defect	Activation energy kJ mol^{-1}	Mechanism or type	Ref.
Al–Ag	2.5% Ag–30% Ag annealed	~140–~170°C	~1	~5 × 10^{-4} ~8 × 10^{-3}	155	Diffusion controlled relaxation of partial dislocations around precipitates	28
Al–Ag	15% Ag (quenched from 200°C)	160–210°C	0.45 0.45	~4 × 10^{-3} ~4 × 10^{-3}	— —	Associated with γ Associated with clustering	29
Al–Ag	Al + 30% Ag (1) quenched	410 K	—	8 × 10^{-3}	92	Zener in solid solution	247
Al–Ag	(2) aged at 520 K	420 K	—	14 × 10^{-3}	110	Zener in γ phase	247
	Al–2, 5, 10, 30wt% Ag (1 mm dia. wires of g.s. 0.75 mm)	400 K; 450 K; variable	~1			Zener; partial reversion of G.P. zones; gamma precipitation	301
Al–Cu	4% Cu (quenched from 520°C reverted at 200°C)	175°C	0.9	2 × 10^{-3}	128(117)	Zener	30, 31
Al–Cu	4% Cu (quenched, reverted, aged at 200°C for 144 h)	120°C	0.5	2.4 × 10^{-3}	92	Stress induced change of shape of precipitate	30
Al–Cu	Al–0.015wt% Cu (3 h at 402°C)	580°C	0.3	0.44		Polygonisation	256
Al–Cu– Mg–Si	Quenched from 500°C, aged at 50°C	20°C	2 × 10^3	6 × 10^{-4} max	56.5	Stress induced ordering of complex atom group	32
Al–Fe	Al–0.25% Fe (annealed at 600°C for 6 h)	~280°C ~310°C	20– 2 × 10^2	~5 × 10^{-4} ~2 × 10^{-3}	117 167	? ?	33 33
Al–Fe	Al–0.06% Fe	~280°C	~1	Depends on grain size	~140	Grain boundary	34
Al–Fe	Al–0.16% Fe–0.5% Fe	310–360°C 440–480°C	~1 ~1	Depends on grain size	~140 200	Grain boundary Related to relaxation of stresses and precipitation of Fe on grain boundaries	34 34
Al–Fe–Ce	Rapidly quenched A–8.6wt% Fe–3.8wt% Ce (extrusion compacted, rapidly solidified)	475 K	0.8	96 × 10^{-4}	150	Zener	263
Al–Mg– Si	Al–0.6% Mg– 0.6% Si (quenched from 480°C and aged at 230°C for $26\frac{1}{2}$ h)	215°C	2.25	5.5 × 10^{-2}	126	Stress induced diffusion of solutes (?)	35
Al–Mg	Al–0.03at% Mg (0.5% tensile strain at 208 K)	243 K; 333 K	~1		0.32 eV; 0.22 eV	Dragging of solute atoms and atom/vacancy pairs	299
Al–Mg	2% Mg	RT	2.5 × 10^4	1.2 × 10^{-5}	—	Thermoelastic	36
Al–Mg	5.45% Mg (quenched from 500°C, aged at 250°C)	165°C	1.5	7 × 10^{-4}	116.3	Zener	37
Al–Mg	7.5% Mg (annealed at 400°C then quenched)	~40°C	~2	—	66.9	Relaxation of solute clusters	38
		~80°C	~2	—	—	Zener	
		~120°C	~2	—	102.5	Zener	
		~227°C	~2	—	125.5	?	
Al–Mg	7.5% Mg (annealed at 400°C, cooled slowly)	~203°C	~2	—	125.5		
Al–Mg	0.93% Mg–12.1% Mg	~100– ~200°C	0.4–98	10^{-4}–2 × 10^{-3}	135.0	Zener	39
Al–Mn– Fe–Si	Al–1.07wt% Mn– 0.52wt% Fe–0.11wt% Si (cold worked and precipitated)	440 K	1			Ke, 3 times wider than single relaxation	305

(continued)

Table 15.8 ANELASTIC DAMPING—*continued*

Alloy	Composition and physical condition	Peak temp.	Frequency Hz	Modulus defect	Activation energy kJ mol^{-1}	Mechanism or type	Ref.
Al–Si	Insoluble Si particles in Al	420 K	~1		0.92 eV	Relaxation at Al–Si interface	298
Al–Si	(solution-treated and aged)	110–180°C	1			Migrates and falls to zero during ageing; vacancy/Si clusters	311 312
Al–Zn	3.7% Zn (quenched from 450°C to RT)	21°C	0.26	~2×10^{-3}	53	Stress induced ordering of zinc atom–vacancy complexes	40
A1050/SiC	Composite	300 K	0.07			Interface	252
Au	99.999% Au. (CW* 16% at 4.2 K)	43 K	0.5	1.4×10^{-1}	9.62	Bordoni type	1, 4
		65 K	0.5	1.6×10^{-1}	18.4	Bordoni type	1, 4
		77 K	0.5	1.6×10^{-1}	18.4	Bordoni type	1, 4
Au	99.999% Au (annealed at 1 170 K for 4 h, then CW* 3% at 77 K)	120 K	4×10^{-2}–	~10^{-4}	—	Hasiguti type	21
		180 K	2×10^{3}	~5×10^{-3}	35.7	Hasiguti type	21
		210 K		~10^{-4}	58 177 (two stages)	Hasiguti type	21
Au	99.999% Au (CW* 16% at 70 K)	130 K	4.0	4×10^{-3}	21.3	Hasiguti type	42
		190 K	4.0	2.2×10^{-2}	32.84	Hasiguti type	42
		210 K	4.0	6×10^{-3}	34.7	Hasiguti type	42
Au	99.999% Au (quenched) from 700°C)	160 K	~1	~2×10^{-3}	68(?)	Associated with dislocations	43
		230 K		~2×10^{-3}	—	Associated with dislocations	43
Au	99.999% Au (quenched from 800°C)	~290 K	~1	~4×10^{-3}	—	Associated with dislocations	43
Au	99.999% (quenched from 1 000°C)	210 K	~1	~4×10^{-4}	57.7	Stress induced reorientation of divacancies Hasiguti type	44
		220 K	~1	—	—		44
Au	99.999 9% Au (quenched from 1 000°C)	0°C	~10	7×10^{-4}– 8×10^{-3}	62.7	Stress induced reorientation of divacancies	45
Au	99.99% Au (CW*, annealed at 600°C)	330°C	0.7	Depends on grain size	141.4	Grain boundary	46
Au	99.999 8% Au (CW* 36% annealed, 650–870°C)	238°C	1.0	4.4×10^{-2}	144.3	Grain boundary	47
		404°C	1.0	3.2×10^{-2}	242.7	Associated with grain boundaries	47
Au	99.999 95% Au (annealed at 900°C)	~400°C	~1	~10^{-1}	435	Sliding at grain boundaries	48
Au–Ag–Zn	Au–42 At. % Ag– 15 At. % Zn	260°C	0.7	0.11	146	Zener	49
Au–Cu	10 At. % Cu– 90 At. % Cu	326– 392°C	1.0	3.8×10^{-3}– 0.57	114.6– 165.3	Zener	50
Au–Cu	10 At. % Cu– 90 At. % Cu	552– 753°C	1.0	0.14 0.76	201.7– 342.3	Adsorption of solute atoms on grain boundaries	48
Au–Cu	10 At. % Cu– 90 At. % Cu	175– 250°C	1.0	5×10^{-2}– 2×10^{-2}	—	Grain boundaries	48
Au–Cu	Au–25at% Cu (quenched)	490 K (quenched only)	~1			Point defects/order	267
	(quenched and annealed)	635 K (all)				Zener	
	(quenched and annealed)	760 K (all)				Grain boundary	
Au–Cu– Zn	42 At. % Cu– 15 At. % Zn	380°C	0.5	0.14	—	Order–disorder peak	49
Au–Cu– Zn	21 At. % Cu– 17 At. % Zn	300°C	0.5	0.50	—	Order–disorder peak	49
Au–Cu– Zn	63 At. % Cu– 17 At. % Zn	340°C	0.5	2.6×10^{-2}	—	Zener	49

(*continue*

Table 15.8 ANELASTIC DAMPING—*continued*

Alloy	Composition and physical condition	Peak temp.	Frequency Hz	Modulus defect	Activation energy kJ mol⁻¹	Mechanism or type	Ref.
Au–Fe	5% Fe	365°C–535°C	~1	2.4×10^{-3} 1.5×10^{-2}	159 177.8	Zener Precipitation of Fe	51 51
Au–Fe	7% Fe	380°C–564°C	~1	1.9×10^{-2}– 1.2×10^{-2}	159 177.8	Zener Precipitation of Fe	51 51
Au–Fe	10% Fe–27% Fe	390°C	~1	3.5×10^{-2}– 1.5×10^{-2}	151–151.9	Zener	51
Au–Ni	30 At. % Ni	397°C	1.0	—	182.0	Zener	52
Au–Ni	Au–30 At. % Ni (quenched)	~380°C	~0.5	$\sim 10^{-1}$	88.3	Zener, modified by quenched–in vacancies	53
Au–Ni	7.7 At. % Ni–90.8 At. % Ni	397°C 652°C	1.0	—	182.0–251.0	Zener	52
Au–Zn	15 At. % Zn	250°C	0.7	4.2×10^{-2}	218	Zener	49
Au–Zn	Stoichiometric AuZn annealed	260°C–290°C	0.2–1	$\sim 10^{-2}$	140	Concerned with short-range order (?)	54
Be		210 K 135°C	1.0 1.0	— —	51.9 101.7	Bordoni type Solute atom/defect interaction	55 55
Be	98.6% Be (annealed)	213 K	~1	$\sim 3 \times 10^{-4}$	50.2	Cold work induced line defect interaction (?)	56
		135°C	~1	$\sim 5 \times 10^{-4}$	100	Solute interaction with lattice (?)	56
Be–Fe–O	Be–0.4% Fe + interstitial impurities including oxygen	0.5°C	1.22	10^{-2}	63.6	Snoek type	57
Cd–Mg	Cd–29.3% Mg	20°C	0.75	0.26	80	Zener	58
Cd–Mg	5% Mg–30% Mg	20°C	1.0	0.29 max	79.5	Zener	59
Co	99.23% Co (0.69% N)	215 K (two peaks)	10^3	—	40.6	Bordoni type (?)	4
Co	99.23% Co (0.69% N) CW* at RT	263 K (two peaks)	10^3	—	51.0	Point defect/ dislocation interaction	4 4
		297 K	10^3	—	36.8	Twin boundaries (?)	
Co	99.999% Co (quenched)	410 K 600 K	10– 5×10^2	$\sim 2 \times 10^{-3}$	72.3	Movement of divacancies and dislocations	60
Co–C	Heated in C atmosphere at 1 050°C and quenched	280°C	$1–16 \times 10^5$	4×10^{-4}	159	Motion of C atom pairs in lattice	61
Co–Fe–Cr	Co–37.8% Fe–8.7% Cr. In magnetic field of 0.6×10^3 A/M	111°C	10^3	3×10^{-3}	—	Electron spin redistribution at Curie point	62
Co–Ni	Co–2% Ni	340°C 430°C	1.5– 1.8×10^4	$\sim 2.5 \times 10^{-2}$ $\sim 2.8 \times 10^{-2}$	— —	α–β phase transformation	63 63
Co–Ni	Co–23% Ni	150°C 310°C	1.5– 1.8×10^4	$\sim 2.9 \times 10^{-2}$ $\sim 3.1 \times 10^{-2}$	— —	α–β phase transformation	63 63
Co–Ni–Cr–W	Co–22wt% Ni–22wt% Cr–14wt% W (wrought alloy)	~370°C				Snoek effect (C–C pairs)	259
Cr	99.8% Cr	38°C	—	2×10^{-3}	—	Electron spin redistribution at Neel temp. (40°C)	64
Cr–N	Cr–0.004 5% N	160°C	1.0	1.5×10^{-3}	101.7	Snoek type	65
Cr–N	35 ppm N (quenched from 83°C)	~36°C	3	$\sim 4 \times 10^{-4}$	115.1	Magneto-mechanical damping	66
Cr–N	(Annealed in NH₃ at 1 150°C for 48 h)	155°C	1	up to 2.2×10^{-3}	85.8	Snoek type	67

(*continued*)

Table 15.8   ANELASTIC DAMPING—*continued*

Alloy	Composition and physical condition	Peak temp.	Frequency Hz	Modulus defect	Activation energy kJ mol^{-1}	Mechanism or type	Ref.
Cr–Re–N	Cr–35% Re (quenched from 1 000°C in NH$_3$ atmosphere)	130°C 190°C	1 1	$\sim$10^{-3} $\sim$10^{-3}	89.1 126.4	Snoek type Snoek type	68
Cu	99.999% Cu (CW* at 77 K	— —	$\sim$1 $\sim$1	— —	4.34 11.6	Niblett–Wilks type Bordoni type	69 69
Cu	99.999% Cu (single crystal) CW* 5% at 77 K	38 K 79 K	1.09 × 10^4 1.09 × 10^4	2 × 10^{-3} 4 × 10^{-3}	4.2 11.7	Bordoni type Bordoni type	19, 41, 70, 71, 19, 41, 70, 71
Cu	99.999% Cu (single crystal $\langle 100 \rangle$ orientation)	80°C	13	34 × 10^{-4}	—	Bordoni type	238
Cu	99.999% Cu (single crystal $\langle 110 \rangle$ orientation)	70°C	13	11.4 × 10^{-4}	—	Bordoni type	238
Cu	Electrolytic Cu (Fatigued, 4 × 10^5 cycles)	140 K	0.3	$\sim$4 × 10^{-4}	—	Dislocation-divacancies interaction	72
		225 K	0.3	$\sim$4 × 10^{-4}	—	Dislocation-vacancies interaction	72
		240 K	0.3	$\sim$2 × 10^{-4}	—	Dislocation-interstitials interaction	72
		165 K	0.3	$\sim$1 × 10^{-4}	—	Dislocation-interstitials interaction	72
Cu	99.999% Cu (deformed 2.5% quenched in liquid He)	30 K 70 K 190 K	6 × 10^2 6 × 10^2 6 × 10^2	$\sim$10^{-3} $\sim$3 × 10^{-3} $\sim$10^{-4}	— — —	Niblett–Wilks type Bordoni type Hasiguti type	73 73 73
Cu	99.999% Cu (CW* 5% at 77 K)	148 K	1.0	4 × 10^{-3}	31.0	Point defect, dislocation interaction	74
		170 K	1.0	1.2 × 10^{-2}	33.9	Point defect, dislocation interaction	74
		238 K	1.0	1 × 10^{-2}	41.4	Rotation of split interstitials (?)	74 75
Cu	99.999% Cu (annealed at 500°C for 4 h)	215°C	1.0	2.11 × 10^{-2}	157	Grain boundary	
Cu	99.99% Cu (area reduced 47%, annealed at 600°C for 2 h)	216°C	1.17	1.65 × 10^{-2}	132	Grain boundary	5
Cu	99.999 9% Cu	416°C	$\sim$1	$\sim$0.2	435	Sliding at grain boundaries	48, 76
		735°C	$\sim$1	$\sim$0.2	169.5	Sliding at grain boundaries	76
Cu	99.999% Cu	300°C	5.0	4 × 10^{-2}	156.9	Grain boundary	75
Cu–Ag	Cu–0.71% Ag	550°C	1	—	154.7	Grain boundary	237
Cu–Ag	Cu–0.1 At. % Ag CW* 5%	223 K 283 K	5 × 10^3 5 × 10^3	2.6 × 10^{-4} 2.5 × 10^{-4}	— —	Dislocation/ silver atom/ point defect interaction	250
Cu–Al	2% Al–10% Al (Deformed 3% at RT)	145 K	1.5 × 10^7	—	$\sim$24	Bordoni type	77
Cu–Al	Cu–15wt% Al (cold worked)	−60°C	$\sim$2	0.04			308

(continued

Table 15.8 ANELASTIC DAMPING—*continued*

Alloy	Composition and physical condition	Peak temp.	Frequency Hz	Modulus defect	Activation energy kJ mol^{-1}	Mechanism or type	Ref.
Cu–Al Cu–Al–Ni	Cu–16.8 At. % Al Cu–13wt% Al −7.9wt% Ni (solution-treated at 1 223 K and cooled at various rates)	360°C 323 K at all rates Others at high rates	0.66 ~520–680	7 × 10^{-3}	174.9	Zener Twin boundary relaxation of gamma-mart.	78 281
Cu–Au	Cu–25at% Au (annealed)	680 K 800 K	~1			Zener Grain boundary	266
Cu–Au	Cu–1.5at% Au (cold worked and annealed at 100°C)	−50°C	~1			Interaction between dislocations and Au clusters	310
Cu–Co	(Aged at 575°C for 3 min)	230°C	1	7 × 10^{-4} (depends on ageing)	184.9	Grain boundary	75
Cu–Ga	Cu–16 At. % Ga	330°C	1.0	2 × 10^{-2}	—	Zener	33
Cu–Fe	0.5% Fe–10% Fe	320°C– 350°C	~1	Depends on grain size	159	Grain boundary	79
		480°C– 580°C	~1	Depends on grain size	209	Connected with precipitation of Fe on Cu grains	79
Cu–Fe	Up to 1.5% Fe (Quenched from 820°C)	800°C– 850°C	~1	~10^{-1}	~125	Connected with ageing of alloy	80
Cu–Ni	25% Ni–75% Ni (Quenched from 720°C to 240°C)	34 K 150 K	~1 ~1	— —	79.9 111.3	Associated with precipitation Associated with precipitation	81 81
Cu–Ni	Cu–3% Ni (Annealed at 1 000°C)	~580°C	~1	2.1 × 10^{-3}	151	Zener	82
Cu–Ni	Cu–45% Ni (Reduced by 90% at RT)	~600°C ~800°C	~1 ~1	~2 × 10^{-2} ~5 × 10^{-2}	— 208	Associated with recrystallisation (?) Associated with dislocations	83 83
Cu–Ni	5.6 At. % Ni– 94.9 At. % Ni	590– 726°C	~1	0.129– 0.112	368–264	Grain boundary	84
Cu–Ni–	20 At. % Zn– 10 At. % Ni	381°C	1	5.3 × 10^{-3}	197	Zener	85
Cu–Ni–	Cu$_2$Ni$_{1.15}$ Zn$_{0.92}$ single crystal	76 K 71 K	6.9 6.9	3 × 10^{-2} 1.2 × 10^{-2}	— —	Zener Ordering	244 244
Cu–Pd	0.01% Pd–0.3% Pd	30–150 K	5×10^6	10^{-4}–10^{-3}	—	Overdamped resonance of dislocations	86
Cu–Pt	0.01% Pt–0.3% Pd	30–150 K	5×10^6	10^{-4}–10^{-3}	—	Overdamped resonance of dislocations	86
Cu–Si Cu–Si	Cu + 5.09 wt. % Si Cu − 5.09wt. % Si (precipitation treated)	200°C 200°C 366°C	3 3 3	7.5 × 10^{-3}	118 28 kcal mol^{-1} 1.74 eV	Precipitation of K^1 Relaxation at ppt. interfaces in stacking faults Grain boundary	248 306
Cu–Sn	3% Sn–9% Sn	490–500°C	1.8	0.14–0.152	151–205	Grain boundary	87
Cu–Zn	(α) Cu–31% Zn	R.T.	6 × 10^3	1.8 × 10^{-4}	—	Thermoelastic	88
Cu–Zn	10% Zn–30% Zn	290–350° C	0.7	$\left(\frac{At.\ conc\ Zn}{4}\right)$	159–178	Zener	12
Cu–Zn	17.6 At. % Zn– 29.4 At. % Zn	657–614 K	1.3	3.5 × 10^{-3}– 9.2 × 10^{-3}	182.0– 161.1	Zener	89
Cu–Zn	Cu–30% Zn	425°C	0.5	0.12	172(?)	Grain boundary	90

(continued)

Table 15.8   ANELASTIC DAMPING—*continued*

Alloy	Composition and physical condition	Peak temp.	Frequency Hz	Modulus defect	Activation energy kJ mol⁻¹	Mechanism or type	Ref.
Cu–Zn	(β) Cu–45 At. % Zn (Quenched from 400°C)	70°C	0.9	2.6×10^{-3}	69.5	Diffusion of Zn accelerated by vacancies	91
Cu–Zn	(β) Cu–45 At. % Zn	177°C	0.9	1.4×10^{-3}	130	Stress induced reorientation of Cu atom pairs	91 91
Cu–Zn	($\alpha - \beta$) Cu–43 At. % Zn	285°C	0.9	8×10^{-3}	159	Stress relaxation at β–α interfaces	91
Cu–Zn	($\beta - \gamma$) Cu–50 At. % Zn	190°C	0.9	2.2×10^{-3}	130	Stress relaxation at β–γ interfaces	91
Cu–Zn–Al	77% Cu–89% Cu 5% Zn–20% Zn 2% Al–8% Al	623–672 K	~1	2×10^{-3}– 8.5×10^{-3}	—	Zener	92
Cu–Zn–Al	Cu–29.5% Zn–2.4% Al	593 K	1	3.5×10^{-2}	150.6	Zener	93
Cu–Zn–Al Fe	Cu–17.0% Zn–9.0% Al (Re-electrolytic) (CW* 5% at RT, in magnetic field of 7.5×10^4 A/m)	615 K ~50 K	1 10^5	3.27×10^{-2} ~1.5×10^{-3}	163.2 —	Zener Associated with motion of kinks in dislocations	93 94
Fe	Armco (CW* at RT)	198 K	10^3	—	44.4	Point defect/dis-location interaction	4
Fe	Armco (CW* at RT)	230 K	10^3	—	54.8	Point defect/dis-location interaction	4
Fe	CW* 40% at RT	275°C	2.9	1.55×10^{-3}	174	Köster type	95
Fe	99.98% Fe ($<5 \times 10^{-5}$ CN)	526°C	1.03	Depends on grain size	192	Grain boundary	96
Fe	Fe pure (R ratio 1600) irradiated 2×10^{18} cm^{-2} at 20 K	110 K	1.4	4×10^{-4}	—	Magnetic relaxation of point defects	24·
		128 K	1.4	13×10^{-4}	—	Magnetic relaxation of point defects	24·
		155 K	1.4	55×10^{-4}	—	Relaxation of self-interstitials	24·
Fe–Al	Fe–40 At.% Al	180°C	0.6	3.5×10^{-3}	121	Movement of Al within tetrahedral lattice lattice	97
		320°C	0.6	6×10^{-3}	163	Movement of Al within tetrahedral lattice	
Fe–Al	Fe–40 At. % Al	440°C 550°C	0.6 0.6	1×10^{-3} 5×10^{-4}	184 —	Movement of Al within tetrahedral lattice Zener	97
Fe–Al	Fe–17 At. % Al	520°C	1.3	1×10^{-2}	234	Zener	98
Fe–Al–C	Fe–19.3 At. % Al–0.01% C (Quenched from 720°C, aged)	130°C	1.4	~1×10^{-3}	99.6	Stress induced dif-fusion of C in ordered Fe$_3$Al lattice	99
		168°C	1.4	~5×10^{-3}	—	(?)	
Fe–Al–C	Fe–0.7 At. % Al–0.01% C	41°C	1.2	~10^{-2}	—	Snoek type	99
Fe–Al–C Fe–B	Fe–9 At. % Al–0.01% C Fe–0.05% B	~100°C 79°C	1.2 14.0	~3×10^{-3} —	— 63	(?) Snoek type	99 1C
Fe–B	Fe–50 ppm B (C impurities)	260 K 50°C	~1	~10^{-4} ~3×10^{-3}	54 —	Snoek type due to B Snoek type due to C	1C
Fe–C	Fe–0.02% C	27°C	0.27	2×10^{-2}	84.1	Snoek type	1·
Fe–C	Fe–0.4% C (Martensite). (Quenched from 850°C, tempered at 300°C for 1 h)	~200°C	~1	~10^{-2}	84.5	Associated with dislocations	1C

(continu...)

Table 15.8   ANELASTIC DAMPING—*continued*

Alloy	Composition and physical condition	Peak temp.	Frequency Hz	Modulus defect	Activation energy kJ mol^{-1}	Mechanism or type	Ref.
Fe–C	Fe–0.02% C (Quenched from 700°C, CW* 52%, aged)	210°C	1	~5 × 10^{-3}	—	Dislocation movement between pinning points (precipitates)	104
Fe–C	Fe–0.01% C (CW* 25%)	235°C	2.20	8 × 10^{-3}	138	Köster peak	105
Fe–C	—	353°C	6.65 × 10^{6}	~3 × 10^{-3}	80.17	Snoek type	106
Fe–C	Fe–0.022% C	520°C	1.0	Depends on grain size	347	Grain boundary	107
Fe–Co–Cr	Fe–54% Co–10% Cr	380 K	6 × 10^{3}	~5 × 10^{-4}	—	Macro-eddy current peak	108
Fe–Cr	1.2% Cr–43% Cr	526°C–683°C	1.0	3 × 10^{-3}–10^{-2}	215.5–296.2	Grain boundary	109
Fe–Cr	Fe–16% Cr	661°C	~1	3.18 × 10^{-2}	233	Grain boundary	110
Fe–Cr	Fe–22.5% Cr	662°C	~1	0.55 × 10^{-2}	222	Grain boundary	110
Fe–Cr	Fe–22.5% Cr	~560°C	~1	1.1 × 10^{-2}	222	Zener	111
Fe–Cr	Fe–17wt% Cr–Deuterium (cold worked)	~250 K	300–2 000			Snoek	264
Fe–Cr–RE	Fe–13at% Cr–6at5 Al–2at% Mo–0.05at% Ce	450°C; 624°C; 690°C	0.67; 0.59; 0.55		—; 3.3 eV; 4.0 eV	Zener (Fe–Cr pairs); solute G.B. peak of Fe–Cr; ditto Fe–RE	294
	Fe–27at% Cr–7at% Al–2at% Mo–0.056at% Ce	523°C; 663°C; 713°C	0.61; 0.56; 0.51		—; 3.6 eV; 4.0 eV–;		
	Fe–23at% Cr–6at% Al–0.065at% Y	529°C 632°C; 690°C			3.3 eV; 3.9 eV		
Fe–Cr–C	Fe–(1.2% Cr–5.2% Cr)–C (Quenched from 750°C)	~320 K	~1	~10^{-2}	75.31–82.72	Snoek type	112
Fe–Cr–N	Fe–4.2% Cr–N	(Several peaks) 266 K–339 K	~1	Up to 2.3 × 10^{-3}	67.8–92.0	Motion of N atoms in various environments	113
Fe–Cr–N	Fe–16.6wt% Cr–130/200 ppm N (quenched from 1 000°C)	270/250°C	~0.9			Snoek	290
	Fe–16.6wt% Cr–480/580 ppm N	Peak broadened				Associated with martensitic phase	
Fe–Cr–Ni	304L stainless steel (cold worked)	335–340 K	~1 000		0.6 eV (other peaks at 0.4, 0.4 eV)	? but height depends on amount of cold work	289
Fe–Cr–Ni (–H)	304, 316 and 310S (+~10at% H) (electrolytically charged with H)	230 K 300 K	1.5 200–2 000		49.0 49.0	Stress-induced ordering of H pairs	303
Fe–Cr–Ni	Sus 310S, 316 and 304 (H charged only) (plus cold-work)	300 K 300 K+ 300 K (CW) and 230 K (H + CW)	~550			H-H Snoek Bordoni H-lattice defects	280
Fe–Cr–Ni	304L stainless steel (cold-worked)	325 and 360 K	~500			Gamma phase and alpha martensite resp.	286
Fe–H	Fe(<0.1 At. % H)	30 K	1.0	8 × 10^{-4}	—	Snoek type	114
Fe–H	Fe(<0.1 At. % H) (CW* 0.5%)	116 K	1.0	1.6 × 10^{-3}	—	Köster type	114
Fe–H	Fe–H (aged at 60°C for 5 h)	~150 K ~180 K	8 × 10^{4} 8 × 10^{4}	~2 × 10^{-4} ~6.5 × 10^{-4}	— —	Köster type Köster type	115

(continued)

Table 15.8 ANELASTIC DAMPING—*continued*

Alloy	Composition and physical condition	Peak temp.	Frequency Hz	Modulus defect	Activation energy kJ mol⁻¹	Mechanism or type	Ref.
Fe (–H)	Pure Fe electrolytically charged with H (deformed before charging)	130 K; 170 K; 330 K	1		0.22; 0.35; ?	Peaks at 130 and 170 K due to interaction between dislocations and aggregated H; thermally activated kink-pair on screw disloc.	296
αFe–D	(CW* 0.5%)	35 K	1.0	8×10^{-4}–	—	Snoek type	114
		120 K	1.0	10^{-3}	—	Köster type	114
Fe–Mn–C	Fe–18.5% Mn–0.7% C	250°C	2.0	1.1×10^{-2}	146	Diffusion of C in fcc Fe–Mn	116
Fe–Mn–C	Fe–15.5% Mn–0.35% C	285°C	1.01	1.6×10^{-3}	155	Stress induced ordering of C atoms	117
Fe–Mn–N	Fe–2% Mn–N (Quenched from 950°C)	(Several peaks) 267 K– 337 K	~1	Up to 1.5×10^{-2}	66.1–91.6	Motion of N atoms in various environments	113
Fe–Mn–N	Fe–1.6% Mn–N (Heated at 590°C in NH₃)	7°C	1	$\sim 4 \times 10^{-3}$	69.0	Snoek type in complex lattice	118
		23°C	1	$\sim 5 \times 10^{-4}$	77.4	Snoek type in Fe	
		34.5°C	1	$\sim 2 \times 10^{-3}$	81.6	N atoms jumping from Fe–Mn sites to Fe–Fe sites	
Fe–Mn–N	Fe–2% Mn–0.01% N	135°C	~1	$\sim 10^{-3}$	—		119
αFe–N	—	22°C	0.89	$\left(\frac{\text{wt. % N}}{0.63}\right)$	77.8	Snoek type	120
αFe–N	Fe–0.01% N (CW* 25%)	235°C	2.35	1.3×10^{-2}	138	Köster type	105
Fe–N	Fe–0.05% N (Quenched from 580°C, area reduced 21%)	250°C	1	$\sim 5 \times 10^{-3}$	—	Dislocation movement between pinning points (precipitates)	104
Fe–N	—	326°C	6.65×10^6	$\sim 2 \times 10^{-3}$	76.78	Snoek type	106
Fe–N	Fe–0.018% N	510°C	1.0	Depends on grain size	315.9	Grain boundary	107
Fe–Ni	Fe–31.5% Ni	415 K	6×10^3	$\sim 10^{-4}$	—	Macro-eddy current peak	108
Fe–Ni	Fe–54wt% Ni (hydrogen-charged)	260 K (sub-peak 225 K)	500		43.2	Snoek	275
	Fe–72.8wt% Ni (hydrogen-charged)	240 K (sub-peak 265 K)	500		42.3	Snoek	
Fe–Ni	Fe–5% Ni C impurities (CW* 40% at RT)	40°C	1.2	2×10^{-3}	—	Snoek type	95
		215°C	2.9	1.15×10^{-3}	135	Köster type	95
Fe–Ni	Fe–4% Ni (CW*, annealed)	500°C	1.0	0.24	213–259	Grain boundary	121
		750°C	1.0	0.50	247–289	(?)	121
Fe–Ni	Fe–36wt% Ni (Invar)–H	250 K	700				268
	Fe–36wt% Ni (Invar)–D	253 K	700		43	Snoek	
Fe–Ni–Cr	Fe–26.0% Ni–21.4 Cr (Annealed from 1 000°C)	629°C	~1	Depends on grain size	350	Grain boundary	122
		692°	~1		294	Grain boundary	122
		769°C	~1		296	Grain boundary	122
Ni–Cr steel	18CrNiWA steel (continuous cooling)	400–430°C	0.7			Bainitic transformation	260
	18CrNiWA steel (continuous cooling)	320°C					
	18CrNiWA steel (after decarburising)	600°C					
Fe–Ni–N	Fe–2.29% Ni–N	24.5°C	1.05	1.009×10^{-2}	77.8	Snoek type due to N	113
		40.5°C	1.05	8.8×10^{-4}	84.1	Snoek type due to C	113

(continued)

Table 15.8 ANELASTIC DAMPING—*continued*

Alloy	Composition and physical condition	Peak temp.	Frequency Hz	Modulus defect	Activation energy kJ mol^{-1}	Mechanism or type	Ref.
Fe–Ni–Cr–Mn	Fe–19wt% Cr–9wt% Ni (hydrogen-charged)	~300 K	700			Snoek	269
	Fe–17wt% Cr–4wt% Ni–6wt% Mn (hydrogen-charged)	~310 K	700			Snoek	
	Fe–15wt% Cr–12wt% Mn (hydrogen-charged)	~310 K	700			Snoek	
Fe–P	Fe–0.37% C–0.03% P	40°C	1.1	3.8×10^{-3}	146	Impurity peak due to P	123
Fe–Ni–P–B	Fe–40at% Ni–14at% P–6at% B (amorphous, cold-worked and H charged)	204 K (mag. after eff. 163 K)				Both peaks rise after ageing at 340 K	278
Fe–Si–N	Fe–2.83% Si–0.05% N	22.5°C	1.26	—	75.3	Snoek type	124
		37°C	1.26	—	50.2	N atom jump in vicinity of Fe–Si sites	124
		62°C	1.26	—	—	N atom jump in vicinity of nitride SiN	124
Fe–Ti–N	Fe–0.15% Ti–0.06% N	~120°C	~1	~10^{-3}	105	Interstitial N interacting with Ti	125
		~227°C	~1	~10^{-3}	130	Interstitial N interacting with Ti	125
Fe–V	Fe–18% V	625°C	0.73	1.6×10^{-2}	355.6	Zener	126
Fe–V–C	Fe–0.62% V (Carburised at 800°C for 4 h, quenched)	33°	0.95	3.1×10^{-3}	50	Snoek type due to C	127
		82°C	0.95	6.6×10^{-3}	46	C atom jumps associated with Fe–V sites	127
Fe–V–C	Fe–5.15% V–240 ppm C. (Quenched from 900°C)	77°C	451	9.8×10^{-5}	81.0	Interaction between V and C atoms in solution	128
		109°C	451	6.1×10^{-5}	83.9	Snoek type	
Fe–V–N	Fe–0.51 At. % V–0.58 At. % N	21.5°C	0.77	2×10^{-2}	79.5	Snoek type	129
		84°C	0.77	1.4×10^{-2}	—	N atom jump in vicinity of nitride VN	129
Fe–V–N	Fe–0.62% V–N. (In NH$_3$, 900°C for 14 h). (C impurities)	24°C	1	2×10^{-2}	75	Snoek type due to N	127
		35°C	1	3×10^{-3}	85.8	Snoek type due to C	127
		84°C	1	3×10^{-3}	75.3	N atom jumps associated with Fe–V sites	127
Ga–As	GaAs (Compound) [111] (with damaged surfaces)	229 K	3.7×10^4	4.0×10^{-5}	40.5	Associated with dislocations on surface	130
Ga–Sb	GaSb (Compound) (With damaged surfaces) [100]	225 K	9.46×10^4	2.0×10^{-5}	37.6	Associated with dislocations on surface	130
Gd	(Annealed at 900°C for 2 h)	190 K	~1	~10^{-3}	—	Micro-eddy current damping	131
		240 K	~1	~3×10^{-3}	—	Motion of 90° domain boundaries	131
Ge	Surface damaged by polishing [100]	150 K	~2×10^3	~2×10^{-5}	29	Associated with dislocations	132
	[111]	205 K	~2×10^{-3}	~2×10^{-5}	29	Associated with dislocations	132
Ge		209 K	3×10^3	1.3×10^{-7}	52.1	Associated with dislocations	133
Ge	Surface damaged [111]	228 K	5.49×10^4	4.1×10^{-5}	36.6	Associated with dislocations on surface	130

(*continued*)

Table 15.8 ANELASTIC DAMPING—*continued*

Alloy	Composition and physical condition	Peak temp.	Frequency Hz	Modulus defect	Activation energy kJ mol^{-1}	Mechanism or type	Re
Ge	Zone-refined. (Strained up to 12.2%)	35°C	$\sim 10^3$	$\sim 1.2 \times 10^{-4}$	60	Motion of geo-metrical kinks over jogs on dislocations	13
Ge	Zone-refined. (Strained up to 12.2%, annealed at 550°C)	180°C	$\sim 10^3$	$\sim 10^{-5}$	82	Motion of dis-locations in region of surface	1.
Ge	Single crystal	395°C	10^5	1.8×10^{-5}	77.4	Electronic relaxa-tion	1.
		770°C	10^5	4×10^{-7}	173.6	Impurity peak (?)	1.
Ge	(CW* at 800°C)	~ 450°C	~ 1	$\sim 10^{-4}$	~ 175	Bordoni type	1.
Ge–Sb	Ge, doped with Sb (CW* to produce screw dislocations)	~ 80°C	2.5×10^3	$\sim 3 \times 10^{-5}$	14.5	Kink motion	1.
		~ 180°C	2.5×10^3	$\sim 5 \times 10^{-5}$	24.1	Kink motion	1.
		~ 300°C	2.5×10^3	$\sim 1.5 \times 10^{-4}$	106	Kink motion	1.
Ge–Sb	Ge, doped with Sb (CW* to produce Edse dislocations)	~ 290°C	2.5×10^3	$\sim 3 \times 10^{-4}$	106	Kink/point defect interaction	1
Ge–Sb	Ge, doped with Sb annealed	~ 310°C	2.5×10^3	$\sim 10^{-4}$	106	Dislocation/point defect interaction	1
Ge–Sb	Ge–4×10^{17} atoms Sb/ millilitre	216 K	4.1×10^3	2.4×10^{-7}	55.0	Involving dislocations	1
αHf–O	—	490°C	0.95	—	—	O atom diffusion in vicinity of Zr impurity	1
Hg–Te	Mercury telluride	170–260 K	10^7–3×10^9	—	9.07	Bordoni type	1
In–Sb	InSb compound [111] with damaged surfaces	222 K	5.65×10^4	5.1×10^{-5}	32.8	Associated with dislocations on surface	1
In–Sb	InSb compound with surface damage [211]	231 K	3.54×10^4	3.8×10^{-5}	41.5	Associated with dislocations on surface	1
In–Tl	In–10 At.% Tl	270 K	2	$\sim 5 \times 10^{-4}$	67.4	Zener	1
		390 K	2	$\sim 2 \times 10^{-3}$	111.7	Reorientation of lattice with respect to grain boundaries	1
In–Tl	In–10 At.% Tl	273 K	~ 1	—	67	Zener	
		368 K	~ 1	0.5	29	Motion of twin boundaries	
Ir	99.96% Ir (Annealed at 1 720°C)	1 450°C	2.0	7.5×10^{-2}	339	Grain boundary	
K	(Nominal purity) annealed at RT cooled to 120 K	160 K	5×10^3	$\sim 10^{-3}$	~ 125	Stress induced change in size or shape of precipitates	
K–Rb	20 At. % Rb– 50 At. % Rb	193 K	3.7×10^3	2.25×10^{-4}	41.42–40.38	Zener	
		179.5 K	13×10^3	1.9×10^{-3}			
Li–Mg	Li–57 At.% Mg (single crystal)	127°C	1.38×10^3	1.7×10^{-2}	90.0	Zener	
Lu–H(–D)	Lu–<20at% H (annealed)	215–225 K				H–H Snoek	
	Lu <290at% H + D (deformed only)	250–260 K, 350 K				D-dislocation, dislocation/dislocation	
Mg	Mg–Small amounts of Al, Zn, Cd, Tl, In	Peaks from 0–300 K	1.5×10^{-2} — 0.75	Complex spectrum	—	Dislocation/ dislocation interaction	
Mg	High purity (99.9999wt% Mg) (deformed at 10 K)	40 K; 80 K	1			Bordoni, stable on annealing	
		105 K; 220 K				Hasiguti, removed by thermal cycling up to 300 K	

(continu

Table 15.8 ANELASTIC DAMPING—*continued*

Alloy	Composition and physical condition	Peak temp.	Frequency Hz	Modulus defect	Activation energy kJ mol^{-1}	Mechanism or type	Ref.
Mg	99.996% Mg	20 K	$\sim 10^4$	$\sim 4 \times 10^{-4}$	—	Associated with grain boundaries (?)	147
		40–240 K	$\sim 10^4$	$\sim 10^{-3}$	—	Dislocation movement in (0001) slip system	147
Mg	99.992% Mg (CW*)	20 K	4×10^4	3×10^{-3}	—	Bordoni type (?)	148 71
		100 K (very broad)	4×10^4	—	—	Bordoni type	71
Mg	—	37 K	1.5×10^7	—	0.9	Dislocation motion in basal plane	149
		106.5 K	1.5×10^7	—	8.7	Dislocation motion in non-basal plane	149
		155.5 K	1.5×10^7	—	40.5	?	149
Mg	99.99% Mg (CW* at RT)	180 K	10^3	—	28.9	Point defect/dislocation interaction	4
		225 K	10^3	—	42.3	Point defect/dislocation interaction	4
Mg	99.97% Mg	220°C	0.46	8×10^{-2}	—	Grain boundary	26
Mg	99.999 9% Mg	40 K	1–2	3×10^{-3}	—	Bordoni (B1)	242
	1–2% CW at 10 K	80 K	1–2	1×10^{-2}	—	Bordoni (B2)	242
Mg	High purity	15 K	~ 1			Pi sub 1	274
		40 K				Bordoni 1	
		80 K				Bordoni 2	
		105 K				Hasiguti 1	
		220 K				Hasiguti 2	
		420 K				P sub a	
Mg–Li	Mg–1.88% Li	51 K	1.5×10^7	—	1.74	Dislocation motion in basal plane	149
Mg–Li		99.5 K	1.5×10^7	—	9.6	Dislocation motion in non-basal plane	149
		154.5 K	1.5×10^7	—	126	?	149
Mg–N	Mg–0.004 8% N	46 K	1.5×10^7	—	1.0	Dislocation motion in basal plane	149
		116 K	1.5×10^7	—	22	Dislocation motion in non-basal plane	149
		157 K	1.5×10^7	—	35	?	149
Mn–Cu	Mn–12% Cu (Quenched from 925°C)	268 K	7×10^2	3.6×10^{-4}	—	Stress relaxation across twin boundaries	150
Mo	Zone refined (CW* 5% at RT)	80 K	5	3×10^{-3}	17.36	Bordoni type	151
		240 K	5	2.8×10^{-3}	44.4	Bordoni type	151
		285 K	5	2.8×10^{-3}	44.4	Bordoni type	151
Mo	High purity (deformation at high temp.)	130 K	25×10^3			Kink/pair formation and migration on disloc.	288
	(deformation at low temp.)	100 K; 140–200 K	25×10^3			Kink/pair on non-screw disloc.; kink migration on screw disloc.	
	(H anneal+deformation and low temp. anneal)	370 K	25×10^3			Interaction of H with screw dislocations	
Mo	99.9% Mo (CW* 30% at RT)	~ 110 K	$\sim 10^3$	$\sim 2 \times 10^{-3}$	—	Associated with dislocations	152
Mo	99.999 3% Mo (Deformed 50% at 450 K)	~ 140 K	$2–2 \times 10^7$	$\sim 10^{-2}$	—	Associated with dislocations	153
Mo	99.995% Mo polycrystalline (CW*)	145 K	1.9×10^3	$\sim 10^{-3}$	—	Dislocation type	154

(*continued*)

Table 15.8   ANELASTIC DAMPING—*continued*

Alloy	Composition and physical condition	Peak temp.	Frequency Hz	Modulus defect	Activation energy kJ mol⁻¹	Mechanism or type	Ref.
Mo	High purity Mo R/Ratio 7 500–8 000	~140 K	500	3×10^{-3}	—	Dislocation motion	245
Mo	Zone refined (CW* 5% at RT)	300 K	0.4	1.4×10^{-2}	—	Köster type	155
Mo	Impure	~100°C	~1	$\sim3.5 \times 10^{-3}$	109	Snoek type due to N	156
		~220°C	~1	$\sim7.2 \times 10^{-3}$	134	Snoek type due to O	156
		~310°C	~1	$\sim5.0 \times 10^{-3}$	163	Snoek type due to C	156
Mo	Zone refined (Annealed at 2 300°C for 20 min)	200–225°C	~1	$\sim2 \times 10^{-3}$	126–134	Dislocation type	157
Mo	Zone refined single crystal (annealed at 1 800°C for 40 min)	~800°C	~1	$\sim2 \times 10^{-2}$	272	Short range interaction between dislocations and C atom interstitials	158
Mo		820–870°C	1.0	—	—	Grain boundary	157
Mo	Commercial	1 000–1 150°C	~1	$\sim6 \times 10^{-2}$	372	Grain boundary	159
		1 300–1 600°C	~1	~0.13	54	Recrystallisation peak	159
Mo	99.98% Mo (Annealed at 1 340°C)	1 050°C	1.3	3.3×10^{-2}	389	Grain boundary	142
Mo	Zone refined, recrystallised, CW* RT, annealed at 1 900°C for 90 min	~1 200°C	5	$\sim10^{-2}$	—	Grain boundary	158
Mo–C	Mo–0.003 4%C	225°C	1.71	4.4×10^{-3}	126–134	Snoek type	157
Mo–N	Mo wire (nitrided at 2 400 K)	392 K, 498 K	~1			Dislocation interaction; Snoek	291
Nb	Zone refined (CW* 5% at RT)	130 K (α peaks)	8	6×10^{-4}	23.0	Bordoni type	151, 163, 164
		240 K (β peaks)	7.8	2.1×10^{-3}	45.2	Bordoni type	151, 163
Nb	99% Nb. (CW* 5%, annealed at 70°C for 2 h)	190 K	~1	$\sim10^{-4}$	—	Associated with point defect complexes	164
		200 K	~1	$\sim10^{-3}$	~95	Associated with point defect complexes	164
		220 K	~1	$\sim10^{-4}$	—	?	164
Nb	99.998% Nb (CW* 50% at RT)	220 K	$2–2 \times 10^7$	$\sim7.5 \times 10^{-3}$	—	Dislocation damping	153, 165, 166
Nb	Nb–0.005% N–0.005% O (Zone refined)	~580°C	8×10^3	$\sim7.8 \times 10^{-3}$	—	Snoek type due to N	167
		~380°C	8×10^3	$\sim3.5 \times 10^{-3}$	—	Snoek type due to O	167
		~470°C	8×10^3	$\sim1 \times 10^{-3}$	—		167
Nb	(CW*)	~630°C	8	—	—	Relaxation of ordered cluster of interstitials near dislocations	168
Nb–Al	Nb–0.29% Al (0.004% C, 0.003% N, 0.029% O) (Annealed)	152°C	1	$\sim9 \times 10^{-3}$	—	Snoek type due to N	169
		283°C	1	$\sim1 \times 10^{-3}$	—	Snoek type due to O	169
Nb–Al	Nb–0.29% Al (0.004% C, 0.003% N, >0.029% O) (CW*)	410°C	1	$\sim1 \times 10^{-3}$	—	Dislocation/solute interaction	169
Nb–C	Nb–0.014% C	259°C	0.57	4×10^{-3}	139.3	Snoek type	170 171
Nb	High purity single crystal + 180 at ppm H. CW at 320 K	270 K	1.3×10^3	3.0×10^{-2}	—	Snoek–Köster type	239

(*continued*)

Table 15.8   ANELASTIC DAMPING—*continued*

Alloy	Composition and physical condition	Peak temp.	Frequency Hz	Modulus defect	Activation energy kJ mol^{-1}	Mechanism or type	Ref.
Nb	Nb single crystal R. ratio 2 500	α_1 143 K	0.5	10×10^{-3}	29.7	Formation of kink-pairs in non-screw dislocation	249
		α_2 121 K	0.5	6×10^{-3}	19.8	Kink diffusion in screw dislocation	249
Nb–Mo	1.3% Mo–16% Mo (CW* 30% at RT)	180–220 K	$2–2 \times 10^7$	5×10^{-4}– 1×10^{-4}	24.1	Type of Bordoni mechanism	153, 166
Nb	99.97% Nb (CW*) (Superconducting type)	3.24 K	8×10^4	$\sim 2 \times 10^{-6}$	0.18	Motion of dislocations	160
	(Normal type)	2.08 K	8×10^4	$\sim 4 \times 10^{-6}$	0.15	Motion of dislocations	160
Nb	Zone refined (CW* 5% at RT)	11–19 K (Broad)	8.8	10^{-4}	—	Bordoni type (?)	161
Nb	99.9% Nb (CW*)	~ 30 K	2×10^4– 1×10^5	$\sim 2 \times 10^{-4}$	1.6	Motion of dislocations	162
		~ 180 K	2×10^4 1×10^5	$\sim 1 \times 10^{-4}$–	24.1	Interaction of dislocations and impurity atoms	162
Nb	High purity single crystal CW at 320 K	190 K	1.3×10^3	1.2×10^{-2}	—	Motion of dislocation	239
Nb–Mo–O	Nb–5.3 At. % Mo– 0.16 At. %O (Homogenised at 1 200°C for 15 min)	167°C	0.6	$\sim 10^{-2}$	—	Snoek type	172
Nb–N	Nb–0.018% N	274°C	0.55	1.1×10^{-2}	145.6	Snoek type	170, 171
Nb–N	Nb–0.066% N (CW* 10%)	500°C	0.31	2.4×10^{-2}	201	Köster type	173
Nb–N–T	Nb–0.50 at% N– 0.3at% T (Tritium)	57 K; 85 K	~ 1			N/H pairs; N/T pairs	297
Nb (–O)	'Pure' nobium (wire)	300–350 K	~ 1		0.6 eV	Extrinisic double-kink/O atoms	287
		350–360 K			0.8 eV	Intrinsic double-kink on screw disloc.	
		~ 500 K				Thermal unpinning of screw disloc. from O	
Nb–O	Nb–0.18 At. % O	152°C	0.6	$\sim 10^{-2}$	—	Snoek type	172
Nb–O	Nb–0.026% O	168°C	2.13	1.6×10^{-2}	114.2	Snoek type	174, 171
Nb–O	Nb–50 to 1 000 ppm O (worked polycrystals)	543–575 K	1		2.00–2.06 eV		270
		692–726 K	1		1.68–1.81 eV		
		618 K (low O)	1		1.49 eV		
		621 K (high O)	1		1.50 eV		
Nb–O–N	Nb–1.2 At. % O– 0.11 At. % N	420 K	~ 1	$\sim 5 \times 10^{-3}$	—	Segregation of O atoms to dislocations	175
	(CW* 34%)	500 K	~ 1	$\sim 1 \times 10^{-3}$	—	Segregation of N atoms to dislocations	175
Nb–O–T	Nb–0.56at% O– 0.3at% T (tritium)	50 K; 80 K	~ 1			O/H pairs; O/T pairs	297
Nb (–Ta)	Nb–0.03wt% Ta (70 ppm O, 30 ppm N) (neutron irradiated and annealed 1.25 mm wires)	470 K				O plus irradiation defects	283
		670 K				N plus irradiation defects	

(continued)

Table 15.8   ANELASTIC DAMPING—*continued*

Alloy	Composition and physical condition	Peak temp.	Frequency Hz	Modulus defect	Activation energy kJ mol⁻¹	Mechanism or type	Ref.
Nb–Ti	Nb–48% Ti	100°C	0.6	$\sim7 \times 10^{-3}$	100.0	Snoek due to O impurities	176
Nb–Ti	Nb–48% Ti (N atmosphere at 1 200°C for 1 h)	340°C	0.6	$\sim10^{-2}$	—	Snoek due to N	176
Nb–Ti	Nb–5at% Ti (plus D)	50 K	20×10^3				273
		100 K	20×10^3				
		170 K	20×10^3			Ti-D complexes reorientation	
Nb–V	Nb–0.3% V	~200 K	$2\text{-}2 \times 10^7$	2.5×10^{-3}	—	Dislocation damping	153
Nb–W–N	Nb–N (wires)	645 and 685 K	~1		143 and 153 respectively	Snoek	285
	Nb–2, 6 and 12 at% W–N	645, 685 and 745 K	~1		143, 153 and 167 respectively		
Nb–Zr	Nb–1.0% Zr	~200 K	$2\text{-}2 \times 10^7$	1×10^{-3}	—	Dislocation damping	153
Nb–Zr	Nb–1% Zr–O + traces of N	~500°C	5×10^4	—	110.9	Snoek due to O in Nb	177
		~500°C	5×10^4	—	111.3	Substitutional–interstitial process involving O	177
		~500°C	5×10^4	—	123.4	Substitutional–interstitial process involving O pairs	177
Nb–Zr	—	~500°C	5×10^4	—	146.4	Snoek due to N in Nb	177
		~500°C	5×10^4	—	147.3	Substitutional–interstitial process involving N	177
Ni	Zone refined (CW*)	138 K	3×10^{-4}	2×10^{-3}– 35×10^{-3}	—	Niblett and Wilks type	178
		248 K	3×10^4	1×10^{-2}–0.10	—	Bordoni type	178
Ni	Zone refined single crystal (CW*)	145–123 K	$\sim2 \times 10^4$	Up to 1.6×10^{-3}	—	Associated with dislocation reactions	179
		223–263 K	$\sim2 \times 10^4$	Up to 3.0×10	—	Bordoni type	179
Ni	99.99% Ni (CW* at RT)	155 K	10^3	—	29.7	Point defect/dislocation interaction	4
		350 K	10^3	—	51.0	Point defect/dislocation interaction	4
		397 K	10^3	—	69.4	Point defect/dislocation interaction	4
Ni	99.9% Ni	70°C	0.7	—	77.0	Stress induced re-orientation of interstitial Ni atom pairs	180, 1
Ni	99.999 9% Ni	150°C	1	~0.1	—	Magneto-mechanical damping	182
Ni	99.99% Ni. (Area reduced 90%, annealed at 905°C for 1 h)	432°C	1.41	3.4×10^{-2}	308	Grain boundary	5
Ni	99.98% Ni	440–460°C	0.5	0.10	—	Grain boundary	183
		630–720°C	0.5	0.12	—	Stress relaxation at polygonised boundaries	183
Ni–Al–C	Ni–2% Al–0.5% C (quenched)	280°C	0.5	—	—	Diffusion of C in Ni–Al	184

(continue

Table 15.8 ANELASTIC DAMPING—*continued*

Alloy	Composition and physical condition	Peak temp.	Frequency Hz	Modulus defect	Activation energy kJ mol^{-1}	Mechanism or type	Ref.
Ni–B	Ni (pure)	470°C; 670°C	~1			All grain boundary	300
	Ni–0.0035wt% B	550°C	~1				
Ni–C	(quenched)	230°C	0.5		—	Diffusion of C in Ni	184
Ni–C	(cold-worked)	~430°C	~1			Dislocation–C	276
Ni–Cr	0.5% Cr	530–800°C	2	Depends on grain size	—	Grain boundary	185
Ni–Cr	Ni–33.3at% Cr	390°C; 570°C	0.7–1.5			Long range order	282
Ni–Cr	Ni–20wt% Cr	~700°C	~1		370	Grain boundary	257
Ni–Cr–Ce	Ni–20wt% Cr–180 at ppm Ce	~700°C	~1		200	Grain boundary	257
Ni–Cr–C	Ni–20% Cr–1.87% C (quenched)	250°C	0.9	—	98.3	Diffusion of C in Ni–Cr	186
Ni–Cu	Ni–20% Cu	~140°C	1	~10^{-2}	—	Magneto-mechanical damping	182
		(Varies)	1	~10^{-2}	—	Magnetic ordering	182
Ni–Zr	0.1% Zr–0.5% Zr	~200°C	~1	2 × 10^{-3}–8 × 10^{-4}	—	Magneto-mechanical damping	187
		~450°C	~1	1 × 10^{-3}–3 × 10^{-4}	—	Grain boundary	187
		600–700°C	~1	<1 × 10^{-4}–2 × 10^{-4}	—	'Blocking' peak	187
Pd–Cu–Si	Amorphous Pd–6at% Cu–16.5at% Si (1 mm wire)	147°C			275 eV	Atomic diffusion	262
Pd–H	99.999% Pd (Annealed, electrolytically loaded with H) 40 At. % H–75 At. % H	70–80 K	2.7	~10^{-3}	12.34–16.19	Stress induced ordering of H pairs in β phase	188
Pd–H	PdH (β phase) (Strained)	120 K	~10^3	~2 × 10^{-3}	—	Snoek type due to H	189
		~150 K	~10^3	~5 × 10^{-3}	—	Pinning of dis-locations by inter-stitial impurities	189
Pd–D	99.999% Pd (Annealed, electrolytically loaded with D) 40 At. % D–73 At. % D	78–86 K	2.7	—	15.94–20.71	Stress induced ordering of D pairs in β phase	188
Pd–Si	80at% Pd–20at% Si with 1–2at% H (rapidly solidified amorphous)	180 K	100		0.31 eV	Short range diffusion of H	293
Pt	(Deformed 2.9%, annealed at 1 080 K for 1 h)	~70 K	5 × 10^3–6.5 × 10^4	5 × 10^{-6}	11.6	Associated with dislocations grouped into sub-grain boundaries	190
Pt	99.999% Pt (CW* 16% at 4.2 K)	125 K	0.8	8 × 10^{-3}	28.0	Bordoni type	1, 2
Pt	'Pure'	940–1 090 K	10^{-2}–1	—	275	Recrystallisation peak	191
Pu–Al	Pu–5 At. % Al	~65 K	10^7	—	—	Co-operative electron transition	192
Pu–Ce	Pu–6 At. % Ce	~65 K	10^7	—	—	Co-operative electron transition	192
Re	—	1 400 K	1.1	4 × 10^{-2}	586	Gain boundary	193
Se	(amorphous)	30 K	160 × 10^6			Attenuation of ultrasonic shear waves	302

(*continued*)

Table 15.8   ANELASTIC DAMPING—*continued*

Alloy	Composition and physical condition	Peak temp.	Frequency Hz	Modulus defect	Activation energy kJ mol^{-1}	Mechanism or type	Ref.
Si	Surface damaged [100] by polishing	160 K	$\sim 2 \times 10^{-5}$	$\sim 2 \times 10^{-5}$	29	Associated with dislocations	132
	[111]	200 K	$\sim 2 \times 10^{3}$	$\sim 4 \times 10^{-5}$	29	Associated with dislocations	132
Si	Pure n-type [111] (Quenched from 1 000°C to 77 K)	689 K	1.2×10^{3}	$\sim 2 \times 10^{-4}$	130	Migration of O-vacancy complex	194
Si	Single crystal	655°C	10^{5}	1.4×10^{-5}	134.7	Electronic relaxation	195, 196
Si	Si whiskers (annealed) (thermally cycled)	~ 900 K $\sim 1\,300$ K (max.) widened	~ 1 ~ 1			Relaxation in surface layers due to dislocation formation	277
Si–Cu	Si (Cu doped)	398 K	1.2×10^{3}	$\sim 3 \times 10^{-5}$	68	Migration of interstitial Cu	194
		626 K	1.2×10^{3}	$\sim 2 \times 10^{-4}$	96	Precipitation of Cu	194
Si–Li–B	Si–Li–0.01% B	210°C	1.74×10^{4}	6.5×10^{-5}	80.0	Reorientation of Li^{+} B^{-} pairs	197
Si–O	Single crystal	1 030°C	10^{5}	6×10^{-4}	246.0	Stress-induced diffusion of interstitial oxygen	195, 196
Sn	99.99% Sn	80°C	3×10^{2}	3×10^{-2}	79.5	Grain boundary	198
Ta	99.99% Ta (CW* 7%)	24.6 K	2.26×10^{4}	3.1×10^{-4}	3.7	Bordoni type	199
Ta	Zone refined (CW* 12% at RT)	124 K (α peaks)	0.8	9×10^{-4}	24.3	Bordoni type	151
		202 K (β peaks)	0.8	1.3×10^{-3}	41.4	Bordoni type	151
Ta	(Single crystal <111> orientation) CW* 3.1%	170 K	17	19×10^{-3}	—	Snoek–Köster type	239
Ta	(Single crystal)	150 K	0.7	—	4×10^{-3}	Dislocation kink formation	240
Ta	—	1 100°C	0.65	7×10^{-2}	418	Grain boundary	193
Ta	99.89% (Annealed at 1 700°C)	1 230°C	1.0	5.6×10^{-2}	406	Grain boundary	142
Ta–C	Ta–0.1% C	338°C	0.55	2.6×10^{-2}	161.1	Snoek type	200 171
Ta–H	99.9% Ta (CW* 1.7% H charged)	27 K	2×10^{4}– 10^{5}	$\sim 5 \times 10^{-5}$	1.6	Associated with dislocations	162
Ta–H	—	100 K	1.75×10^{2}	2.4×10^{-3}– 2.7×10^{-3}	—	Bordoni type (?)	201 162
		190 K	1.75×10^{2}	—	34.7	Stress induced ordering in the Ta$_2$H phase	201 162
Ta–H	8.5 At. % H–42.4 At. % H	1°C	1.75×10^{2}	6×10^{-3}– 1.6×10^{-2}	52.3	Stress induced ordering in Ta$_2$H phase	201
		36°C	1.75×10^{2}	3×10^{-3}– 1.9×10^{-2}	—	Long range ordering of H in Ta$_2$H phase	201
		54°C	1.75×10^{2}	2.6×10^{-3}– 8.5×10^{-3}	—	Short range ordering of H in Ta$_2$H phase	201
Ta–N	Ta–0.11% N	334°C	0.6	9×10^{-2}	156.9	Diffusion of interstitial N	202
		362°C	0.6	9×10^{-2}	167.4	Diffusion of interstitial N atom pairs	202
Ta–O	Ta–0.081% O	137°C	0.6	8×10^{-2}	104.6	Diffusion of interstitial O	203 204
		162°C	0.6	8×10^{-2}	104.6	Diffusion of interstitial O atom pairs	203 204

(continued)

Table 15.8 ANELASTIC DAMPING—*continued*

Alloy	Composition and physical condition	Peak temp.	Frequency Hz	Modulus defect	Activation energy kJ mol^{-1}	Mechanism or type	Ref.
Ta–O	Ta–0.01% O (CW* 30%)	340°C	0.5	4.4×10^{-3}	151	Köster type	205
Ta–Re–N	Ta–(1.3 At. % Re–3.8 At. % Re)–600 ppm N	~340°C ~380°C	0.8 0.8	$\sim 6 \times 10^{-3}$ $\sim 10^{-3}$	156.9 168.2	Snoek type due to N Snoek type due to N	206
Te	—	~400 K	1.44×10^4	$\sim 10^{-4}$	—	Recombination of election-hole pairs	207
Ti	99.7% Ti (CW* at RT)	220 K 305 K 336 K	10^3 10^3 10^3	— — —	33.9 42.3 51.9	Bordoni type (?) Point defect/dislocation interaction Point defect/dislocation interaction	4 4 4
Ti	High purity (cold worked)	133 K; 173 K 198 K; 223 K	1 1			Bordoni Hasiguti	309
Ti	99.6% Ti	775°C	1.0	0.38	201	Grain boundary	208
Ti–Al	0.04% Al–0.12% Al	675–725°C	1.0	2.5×10^{-2} –0.4	218–293	Grain boundary	208
Ti–Al–V	Ti–6wt% Al–4wt% V (alpha/beta quenched)	~−83 to −113°C ~−53 to 7°C	1 1			Hydride precipitation Dislocation/hydrogen	261
Ti–Al–V	Ti–6wt% Al–4wt% V (commercially heat-treated)	116 K 289 K	2×10^3 2×10^3		5.6 kcal mol^{-1} 11.1 kcal mol^{-1}	Snoek H peak in beta phase Snoek H peak in alpha phase	292
Ti–Au	Ti–0.05% Au	715°C	1.0	4×10^{-2}	259.4	Grain boundary	208
Ti–Cr	Ti–10 At. % Cr (Quenched from 1 000°C)	152 K	$\sim 10^5$	1.75×10^{-2}	29	Associated with vacancies in 2 coexisting electronic environments	209
Ti–H	Ti–0.15% H	273 K	1.2	1.6×10^{-2}	62.8	Diffusion of interstitial H	210
Ti	Ti–5wt% Mo–5wt% V–8wt% Cr–3wt% Al	130°C 45°C			0.29 eV 0.49 eV	H in beta phase Alpha phase	253
Ti–Mo–Zr–Sn	Ti–11.5wt% Mo–6wt% Zr–4.5wt% Sn (Beta III) (beta solid solution)	−170°C; 250°C	~1			Beta/omega shear trans; ditto diffusion	284
Ti–Nb	Ti–25 At. % Nb (Quenched from 1 000°C)	177 K	$\sim 10^5$	6.15×10^{-3}	29	Associated with vacancies in 2 coexisting electronic environments	209
Ti–Nb	0.04% Nb–0.12% Nb	625–675°C	1.0	2.7×10^{-2}–3.7×10^{-2}	213–264	Grain boundary	208
Ti–Ni	Ti–10 At. % Ni (Quenched from 1 000°C)	152 K	$\sim 10^5$	2.15×10^{-3}	29	Associated with vacancies in 2 coexisting electronic environments	209
Ti–Ni	TiNi Intermetallic compound. (Ni–49% Ti). Annealed at 800°C	203 K 223–313 K	~1 ~1	$\sim 10^{-2}$ $\sim 10^{-2}$	36.6 —	Dislocation motion (?) Fine structure of 203 K peak	211 211
Ti–Ni	—	350°C 600°C	~1 ~1	$\sim 10^{-3}$ $\sim 2 \times 10^{-2}$	— —	Impurity effect (?) Transition from TiNi (II) to TiNi (I)	211 211
Ti–O	Ti–2 At. % O	450°C	1.0	1.2×10^{-2}	200.8	Diffusion of interstitial O in presence of impurities	212, 213,

(*continued*)

Table 15.8   ANELASTIC DAMPING—*continued*

Alloy	Composition and physical condition	Peak temp.	Frequency Hz	Modulus defect	Activation energy kJ mol^{-1}	Mechanism or type	Ref.
Ti–O	Ti–0.6 At. % O	660–650°C	1.0	2.5×10^{-2}	188.3	Grain boundary	214
Ti–V	Ti–20 At. % V (Quenched from 1 000°C)	140 K	$\sim 10^5$	2.2×10^{-3}	29	Associated with vacancies in 2 co-existing electronic environments	209
Ti–V	Ti–50 At. % V (Quenched from 1 000°C)	161 K	$\sim 10^5$	2×10^{-3}	29	Associated with vacancies in 2 co-existing electronic environments	209
Ti–V	0.02% V–0.12% V	600–700°C	1.0	3×10^{-2}–0.42	230–335	Grain boundary	208
Ti–V–H	25at% V	84 K/123 K	2/1.5		16	Snoek type	254
Ti–Zr	0.02% Zr–0.12% Zr	650–700°C	1.0	3.1×10^{-2}–3.6×10^{-2}	251–502	Grain boundary	208
U	99.9% U (CW* at RT)	155 K	10^3	—	23.0	Bordoni type (?)	4
		202 K	10^3	—	42.3	Point defect/dislocation interaction	4
V–C	—	162°C	0.55	6.4×10^{-3}	114.2	Diffusion of interstitial C	215
V–H	99.99% V + 600 ppm H	170 K	500	13×10^{-4}	—	Point defect/dislocation interaction	243
V–H	—	18.5 K	2×10^2–10^5	$\sim 6 \times 10^{-5}$	1.1	Dislocation damping	216
		250 K	$\sim$	$\sim 10^{-5}$	—	Point defect/dislocation interaction	216
		285 K	$\sim$	$\sim 10^{-5}$	—	Point defect/dislocation interaction	216
V–H	99.99% + H in soln	203 K	2×10^2–10^5	$\sim 7 \times 10^{-4}$	50	Stress induced ordering in β phase	216
V–H	1.2 At. % H–14.5 At. % H	195 K	75	8×10^{-3}–7.2×10^{-2}	37.87	Diffusion of interstital H	217
V–N	—	272°C	1.0	2.2×10^{-2}	142.7	Diffusion of interstitial N	218
V–O	—	174°C	0.55	1.2×10^{-2}	122.6	Diffusion of interstitial O	215
W	99.999 8% W	—		—	469	Grain boundary	219
		—		—	46	Movement of vacancies	219
		150 K		—	33	Movement of divacancies	219
W	High purity (CW* 3% at 400°C)	165 K	1.5×10^4	10^{-3}	24.3	Bordoni type	151 200
W	99.99% W single crystal	~ 300°C	~ 1	$\sim 2 \times 10^{-4}$	146	Snoek type of uncertain origin	221
		~ 400°C	~ 1	$\sim 2 \times 10^{-4}$	188	Snoek type associated with C impurities	221
W	Commercial purity	1 250°C	0.94	0.28	481–523	Grain boundary	193 222
W	(CW* at RT)	1 535°C	~ 70	~ 0.2	—	Primary re-crystallisation peak	223
W	Zone refined	1 600–1 650 K	10^{-2}–0.25	~ 0.6	477	Recrystallisation peak	224
W	Commercial purity	1 900°C	0.35	0.56	619	Grain boundary	222 225
W	(CW* at RT, annealed at 3 000°C)	$\sim 2\,000$°C	~ 70	~ 0.1	—	Grain boundary	223

(*continue*

able 15.8   ANELASTIC DAMPING—*continued*

loy	Composition and physical condition	Peak temp.	Frequency Hz	Modulus defect	Activation energy kJ mol^{-1}	Mechanism or type	Ref.
–Re	W–20% Re	1 950°C	1.08	~5 × 10^{-2}	510	Grain boundary	223
n	99.999% Zn (Compressed 2.4% at RT)	~100 K	~2 × 10^7	—	5.8	Dislocation movements in basal plane	226
		~170 K	~2 × 10^7	—	15.4	Dislocation movements in prismatic plane	226
		~230 K	~2 × 10^7	—	19.2	Dislocation movements in pyramidal plane	226
1	99.999wt% Zn (worked single crystals)	140–220 K	1 × 10^3			Bordoni	271
1	99.999% Zn (Annealed at 100°C for 12 h)	383 K	~1	~0.3	95	Grain boundary	227
–Al	22wt% Al (2 h at 360°C, w.q.)	~25°C	1–3	9 × 10^{-2}			255
	22wt% Al–0.3wt% Cu (2 h at 360°C, w.q.)	~25°C	1–3	3 × 10^{-2}			
	As above + 0.021wt% Mg (2 h at 360°C, w.q.)	~25°C	1–3	1.44 × 10^{-2}			
	99.999% Zr (7 k bar pressure)	80 K	1.5 × 10^5	~5 × 10^{-4}	28	Bordoni type	228
		250 K	1.5 × 10^5	~2 × 10^{-3}	13	(?)	228
(–O)	High purity (deformed and annealed alpha Zr)	480°C; 530°C; 600°C	4			Dislocation/O interaction; longitudinal and transverse motion in dislocation core	295
	99.9% Zr (CW* at RT)	200–220 K	10^3	9 × 10^{-4}	33.9(?)	Bordoni type(?)	41, 229
		305 K	10^3	—	42.3	Point defect/dislocation interaction	4
	—	336 K	10^3	—	51.0	Point defect/dislocation interaction	4
	99.9% Zr	600°C	1.0	4.6 × 10^{-2}	218–243	Grain boundary	230, 231
		860°C	1.0	0.28	—	α–β transformation	230, 232
–Cu	Zr–up to 2.5% Cu (Annealed at 900°C for 1 h, CW* 10%)	~220 K	1.3 × 10^4	Up to 1.5 × 10^{-3}	33.8	Bordoni type	233
–Cu–	Zr–0.5% Cu–0.5% Mo (Deformed 10%)	~230 K	1.3 × 10^4	2 × 10^{-3}	33.8	Bordoni type	233
–H	Zr–0.89% H	228 K	1.0	4 × 10^{-3}	48.5	Diffusion of interstitial H	231
–H	Zr–1.28% H	5°C	1.0	2.4 × 10^{-2}	71.1	Diffusion of interstitial H atom pairs	231
–H	Zr–1.15% H	~50°C	~2 × 10^4	~1 × 10^{-3}	—	Associated with δ and γ phases	234
		~130°C	~2 × 10^4	~1 × 10^{-2}	—	Associated with δ and γ phases	234
–H	Zr–0.26% H	230°C	1.0	5 × 10^{-3}	—	Associated with ZrH precipitate	231
–Hf–O	Zr–(0.005% Hf–1% Hf)–O	530°C 540°C	~1	~7 × 10^{-3} ~10^{-2}	—	Grain boundary	235
		422°C	~1	2 × 10^{-4}– 4 × 10^{-4}	201	Diffusion of O in lattice	235
Mo	Zr–6 At. % Mo (Quenched from 1 000°C)	213 K	~10^5	1.36 × 10^{-2}	29	Associated with vacancies in 2 co-existing electronic environments	209
Nb	Zr–5% Nb	12 K	10^5	~2 × 10^{-4}	1.9	?	236

(*continued*)

Table 15.8   ANELASTIC DAMPING—*continued*

Alloy	Composition and physical condition	Peak temp.	Frequency Hz	Modulus defect	Activation energy kJ mol^{-1}	Mechanism or type	Ref.
Zr–Nb	5% Nb–25% Nb (CW*)	40 K		$\sim 6 \times 10^{-4}$	4.8	Jahn-Teller type	236
		160 K	10^5	Depends on CW	29	Stress induced re-orientation of atoms	236
Zr–Nb	12 At. % Nb– 75 At. % Nb (Quenched from 1 000°C)	163–207 K	$\sim 10^5$	1.5×10^{-4}– 5.9×10^{-3}	29	Associated with vacancies in 2 co-existing electronic environments	209
Zr–O	Zr–1.95% O	420°C	~ 1	Depends on O conc^n	201	Diffusion of O in lattice	235
Zr–V	Zr–10 At. % V (Quenched from 1 000°C)	193 K	$\sim 10^5$	1.5×10^{-4}	29	Associated with vacancies in 2 co-existing electronic environments	209

* Cold worked.

REFERENCES TO TABLES 15.7 and 15.8

1. S. Okuda, *J. Phys. Soc. Japan*, 1963, **18** (Suppl. 1), 187.
2. S. Okuda, *Appl. Phys. Letters*, 1963, **2**, 163.
3. B. M. Mecs and A. S. Nowick, *Phil. Mag.*, 1968, **17**, (147), 509.
4. R. R. Hasiguti, N. Igata and G. Kamoshita, *Acta Met.*, 1962, **10**, 442.
5. J. N. Cordea and J. W. Spretnak, *Trans. Met. Soc. AIME*, 1966, **236** (12), 1685.
6. L. Rotherham and S. Pearson, *J. Metals*, 1956, **8**, 881, 894.
7. T. J. Turner and G. P. Williams, Jr., *J. Phys. Soc., Japan*, 1963. **18** (Suppl. II), 218.
8. T. J. Turner and G. P. Williams, Jr., *Acta Met.*, 1960, **8**, 891.
9. B. Mills, *Phys. Status, Solidii*, 1971, **6a** (1), 55.
10. T. J. Turner and G. P. Williams, Jr., *Acta Met.*, 1962, **10**, 305.
11. S. Pearson, RAE Rep No Met., 1953, **71.**
12. B. N. Finkel'shteyn and K. M. Shtrakhman, *Phys. Met. Metallogr.*, 1964, **18** (4), 132.
13. G. P. Williams, Jr. and T. J. Turner, *Phys. Status Solidii*, 1968, **26** (2), 645.
14. B. G. Childs and A. D. Le Claire, *Acta Met.*, 1954, **2**, 718.
15. G. P. Williams, Jr. and T. J. Turner, *Phys. Status Solidii*, 1968, **26** (2), 645.
16. A. S. Nowick, *Phys. Rev.*, 1952, **88**, 925.
17. E. Lax and D. H. Filson, *Phys. Rev.*, 1959, **114**, 1273.
18. L. J. Bruner, *Phys. Rev.*, 1960, **118**, 399.
19. J. Völkl, W. Weinländer and J. Carsten, *Phys. Status Solidii*, 1965, **10** (2), 739.
20. W. J. Baxter and J. Wilks, *Acta Met.*, 1963, **11**, 978.
21. W. Benoit, B. Bays, P. A. Grandchamp, B. Vittoz, G. Fantozzi and P. Gobin, *J. Phys. Chem. Solids*, 1970, **31** (8), 1907.
22. J. L. Chevalier, P. Peguin, J. Perez and P. Gobin, *J. Phys.(D)*, 1972, **5** (4), 777.
23. M. Mongy, K. Salama and O. Beckman, *Solid State Commun.*, 1963, **1** (7), 234.
24. A. A. Galkin, O. I. Datsko, V. I. Zaytsev and G. A. Matinin, *Phys. Met. Metallogr.*, 1969, **28** (1), 207.
25. J. Perez and P. Gobin, *Phys. Status Solidi*, 1967, **24** (2), K167.
26. T. S.-Kê *Phys. Rev.*, 1947, **72**, 41.
27. A. C. Damask and A. S. Nowick, *J. Appl. Phys.*, 1955, **26**, 1165.
28. G. Schoeck and E. Bisogni, *Phys. Status Solidii*, 1969, **32** (1), 3.
29. R. E. Miner, T. L. Wilson and J. K. Jackson, *Trans. Met. Soc., AIME*, 1969, **245** (6), 1375.
30. B. S. Berry and A. S. Nowick, *NACA Tech. Note*, 1958, 4225.
31. I. N. Fitzpatrick, *Ph.D. Thesis*, University of Manchester, 1965.
32. K. M. Entwistle, *J. Inst. Met.*, 1953–1954, **82**, 249.
33. E. A. Attia, *Brit. J. Appl. Phys.*, 1967, **18** (9), 1343.
34. B. Ya Pines and A. A. Karmazin, *Phys. Met. Metallogr.*, 1970, **29** (1), 206.
35. K. J. Williams, *Acta Met.*, 1967, **15** (2), 393.
36. R. H. Randall and C. Zener, *Phys. Rev.*, 1940, **58**, 473.
37. W. G. Nilson, *Canad. J. Phys.*, 1961, **39**, 119.
38. B. N. Dey and M. A. Quader, *Canad. J. Phys.*, 1965, **43** (7), 1347.
39. J. Belson, D. Lemercier, P. Moser and P. Vigier, *Phys. Status Solidii*, 1970, **40**, 647.
40. H. Haefner and W. Schneider, *Phys. Status Solidii*, 1971, **4a**, K221.
41. S. Okuda, *J. Appl. Phys.*, 1963, **34**, 3107.

42. S. Okuda and R. R. Hasiguti, *Acta Met.*, 1963, **11**, 257.
43. C. H. Neuman, *J. Phys. Chem. Solids*, 1966, **27** (2), 427.
44. S. Okuda and R. R. Hasiguti, *J. Phys. Soc. Japan*, 1964, **19** (2), 242.
45. D. G. Franklin and H. K. Birnbaum, *Acta Met.*, 1971, **19** (9), 965.
46. W. Köster, L. Bangert and J. Hafner, *Z. Metall*, 1956, **47**, 224.
47. D. R. Mash and L. D. Hall, *Trans. AIMME*, 1953, **197**, 937.
48. M. E. De Morton and G. M. Leak, *Metal Sci. J.*, 1967, **1**, 166.
49. A. Pirson and C. Wert, *Acta Met.*, 1962, **10**, 299.
50. G. K. Mal'tseva, V. S. Postnikov and V. V. Usanov, *Phys. Met. Metallogr.*, 1963, **16** (2), 120.
51. B. A. Mynard and G. M. Leak, *Phys. Status Solidii*, 1970, **40** (i), 113.
52. C. Ang, J. Sivertson and C. Wert, *Acta Met.*, 1955, **3**, 558.
53. J. R. Cost, *Acta Met.*, 1965, **13** (12), 1263.
54. K. Mukherjee, *J. Appl. Phys.*, 1966, **37** (4), 1941.
55. C. Ang and K. T. Kamber, *J. Appl. Phys.*, 1963, **34**, 3405.
56. Choh-Yi Ang and K. T. Kamber, *J. Appl. Phys.*, 1963, **34** (11), 3405.
57. M. J. Elias and R. Rawlings, *J. Less-common Metals*, 1965, **9** (4), 305.
58. J. Lulay and C. Wert, *Acta Met.*, 1956, **4**, 627.
59. J. Enrietto and C. Wert, *Acta Met.*, 1958, **6**, 130.
60. R. Kamel and K. Z. Botros, *Phys. Status Solidii*, 1965, **12** (1), 399.
61. G. Mah and C. A. Wert, *Trans. Met. Soc. AIME*, 1968, **242** (7), 1211.
62. K. P. Belov, G. I. Katayev and R. Z. Levitin, *J. Appl. Phys.*, 1960, **31** (Suppl. 1), 1535.
63. V. N. Belko, B. M. Darinskiy, V. S. Postnikov and I. M. Sharshakov, *Phys. Met. Metallogr.*, 1969, **27** (1), 140.
64. M. E. Fine, E. S. Greiner and W. C. Ellis, *J. Metals, N. Y.*, 1951, **191**, 56.
65. M. E. De Morton, *J. Appl. Phys.*, 1962, **33**, 2768.
66. M. J. Klein, *J. Appl. Phys.*, 1967, **38** (2), 819.
67. M. J. Kelin and A. H. Claver, *Trans. Met. Soc., AIME*, 1965, **233** (11), 1771.
68. M. J. Klein, *Trans. Met. Soc., AIME*, 1965, **233** (1), 1943.
69. S. Okuda, *J. Appl. Phys.*, 1963, **34** (10), 3107.
70. D. H. Niblett and J. Wilks, *Phil. Mag.*, 1956, **1**, 415.
71. H. S. Sack, *Acta Met.*, 1962, **10**, 455.
72. P. Bajons and B. Weiss, *Scripta Metall.*, 1971, **5**, 511.
73. B. M. Mecs and A. S. Nowick, *Acta Met.*, 1965, **13** (7), 771.
74. M. Koiwa and R. R. Hasiguti, *Acta Met.*, 1963, **11**, 1215.
75. D. T. Peters, J. C. Bisseliches and J. W. Spretnak, *Trans. met. Soc. AIME*, 1964, **230** (3), 530.
76. M. E. De Morton and G. M. Leak, *Acta Met.*, 1966, **14** (9), 1140.
77. H. Kayano, K. Kamigaki and S. Koda, *J. Phys. Soc. Japan*, 1967, **23** (3), 649.
78. C. Y. Li and A. S. Nowick, *Phys. Rev.*, 1956, **103**, 294.
79. A. A. Karmazin and V. I. Startsev, *Phys. Met. Metallogr.*, 1970, **29** (6), 191.
80. V. S. Postnikov, S. A. Ammer, A. T. Kosilov and A. M. Belikov, *Phys. Met. Metallogr.*, 1966, **21** (5), 121.
81. B. N. Dey, *Scripta Metall.*, 1968, **2** (9), 501.
82. J. T. A. Roberts and P. Barrand, *Scripta Metall*, 1969, **3** (1), 29.
83. V. S. Postnikov, I. V. Zolotukhin and I. S. Pushkin, *Phys. Met. Metallogr.*, 1968, **26** (4), 147.
84. J. T. A. Roberts, *Metall. Trans.*, 1970, **1** (9), 2487.
85. M. G. Coleman and C. A. Wert, *Trans. Met. Soc. AIME*, 1966, **236** (4), 501.
86. A. Ikushima and T. Kaneda, *Trans. Japan Inst. Metals*, 1968, **9** (Suppl).
87. K. J. Marsh, *Acta Met.*, 1954, **2**, 530.
88. R. H. Randall, F. C. Rose and C. Zener, *Phys. Rev.*, 1939, **56**, 343.
89. K. M. Shtrakhman, *Phys. Met. Metallogr.*, 1967, **24** (3), 116.
90. T. S. Kê, *J. Appl. Phys.*, 1948, **19**, 285.
91. L. M. Clareborough, *Acta Met.*, 1957, **5**, 413.
92. K. M. Shtrakhman, YU. S. Logvinenko, V. F. Grishchenko and Yu. V. Piguzov, *Soviet Phys. solid St.*, 1971, **13** (5), 1238.
93. K. M. Shtrakhman, *Soviet Phys. Solid St.*, 1967, **9** (6), 1360.
94. K. Takita and K. Sakamoto, *Scripta Metall.*, 1970, **4** (5), 403.
95. A. I. Surin and M. S. Blanter, *Phys. Met. Metallogr.*, 1970, **29** (1), 199.
96. G. M. Leak, *Proc. Phys. Soc., Lond.*, 1961, **78**, 1520.
97. J. Delaplace, J. Hillairet and A. Silvent, *C.r. hebd. Séanc. Acad. Sci., Paris*, 1966 (c), **262** (4), 319.
98. D. B. Fishbach, *Acta Met.*, 1962, **10**, 319.
99. K. Tanaka, *J. Phys. Soc. Japan*, 1971, **30** (2), 404.
100. W. R. Thomas and G. M. Leak, *Nature, Lond.*, 1955, **176**, 29.
101. Y. Hayashi and T. Sugeno, *Acta Met.*, 1970, **18** (6), 693.
102. C. A. Wert, *Phys. Rev.*, 1950, **79**, 601.
103. R. Blackwell, *Nature, Lond.*, 1966, **211** (5050), 733.
104. P. Barrand and G. M. Leak, *Acta Met.*, 1964, **12** (10), 1147.
105. K. Kamber, D. Keefer and C. Wert, *Acta Met.*, 1961, **9**, 403.
106. A. E. Lord and D. N. Beshers, *Acta Met.*, 1966, **14** (12), 1659.
107. G. W. Miles and G. M. Leak, *Proc. Phys. Soc., Lond.*, 1961, **78**, 1529.
108. T. Maeda, *Japan J. Appl. Phys.*, 1971, **10** (10), 1299.
109. P. Barrand, *Acta Met.*, 1966, **14** (10), 1247.

110. P. Barrand, *Metal Sci. J.*, 1967, **1**, 127.
111. P. Barrand, *Metal Sci. J.*, 1967, **1**, 54.
112. C. R. Ward and G. M. Leak, *Metallurgical, ital.*, 1970, **62** (8), 302.
113. I. G. Ritchie and R. Rawlings, *Acta Met.*, 1967, **15** (3), 491.
114. W. R. Heller, *Acta Met.*, 1961, **9**, 600.
115. R. Gibala, *Acta Met.*, 1967, **15** (2), 428.
116. T. S. Kê and C. T. Tsien, *Phys. Met. Metallogr.*, 1957, **4** (2), 78.
117. V. Kandarpa and J. W. Spretnak, *Trans. Met. Soc. AIME*, 1969, **245** (7), 1439.
118. G. J. Couper and R. Kennedy, *J. Iron Steel Inst.*, 1967, **205** (6), 642.
119. E. T. Stephenson, *Metall. Trans.*, 1971, **2** (6), 1613.
120. J. D. Fast and M. B. Verrijip, *J. Iron Steel Inst.*, 1955, **180**, 337.
121. R. S. Lebyedev and V. S. Postnikov, *Phys. Met. Metallogr.*, 1959, **8** (2), 134.
122. D. Siddell and Z. C. Szkopiak, *Metall. Trans.*, 1972, **3** (7), 1907.
123. Yu. V. Grdina, Ye. E. Glikman and Yu. V. Piguzov, *Phys. Met. Metallogr.*, 1966, **21** (4), 90.
124. D. A. Leak, W. R. Thomas and G. M. Leak, *Acta Met.*, 1955, **3**, 501.
125. G. Szabó-Miszenti, *Acta Met.*, 1970, **18** (5), 477.
126. J. Stanley and C. Wert, *J. Appl. Phys.*, 1961, **32**, 267.
127. R. M. Jamieson and R. Kennedy, *J. Iron Steel Inst.*, 1966, **204** (2), 1208.
128. H. Sekine, T. Inoue and M. Ogasawara, *Japan. J. Appl. Phys.*, 1967, **6** (21), 272.
129. J. D. Fast and J. L. Meijering, *Philips Res. Rep.*, 1953, **8**, 1.
130. W. Hermann, *Solid State Commun.*, 1968, **6** (9), 641.
131. C. F. Burdett, *Phil Mag.*, 1968, **18** (154), 745.
132. B. M. Mecs and A. S. Nowick, *Appl. Phys. Letters*, 1966, **8** (4), 75.
133. A. Zuckerwar and W. Pechhold, *Z. Angew. Phys.*, 1968, **24** (3), 134.
134. K. Ohori and K. Sumino, *Phys. Status Solidii*, 1972(a), **9** (1), 151.
135. P. D. Southgate, *Proc. Phys. Soc. Lond.*, 1960, **76**, 385, 398.
136. L. N. Aleksandrov, Yu. N. Golobokov, V. N. Orlov and F. L. 'Edel' man, *Soviet Phys. Solid St.*, 1969, **10** (9), 2269.
137. F. Calzecchi, P. Gondi and S. Mantovani, *J. Appl. Phys.*, 1969, **40** (12), 4798.
138. E. Bisogni and C. Wert, US Air Force, Sci. Res. Rep., 1961, Contract AF49(638)672.
139. T. Alper and G. A. Saunders, *Phil. Mag.*, 1969, **20** (164), 225.
140. M. E. De Morton, *Phys. Status Solidii*, 1968, **126**, K73.
141. V. S. Postnikov, I. V. Zolotukhin, V. N. Burmistrov and I. M. Sharshakov, *Phys. Met. Metallogr.*, 1969, **28** (4), 210.
142. M. J. Murray, *J. Less-common Metals*, 1968, **15** (4), 425.
143. T. D. Gulden and J. C. Shyne, *J. Inst. Metals*, 1968, **96** (5), 139.
144. T. D. Gulden and J. C. Shyne, *J. Inst. Metals*, 1968, **96** (5), 143.
145. D. P. Seraphim and A. S. Nowick, *Acta Met.*, 1961, **9**, 85.
146. J. M. Roberts, *Trans. Japan Inst. Metals*, 1968, **9** (Suppl.), 69.
147. R. T. C. Tsui and H. S. Sack, *Acta Met.*, 1967, **15** (11), 1715.
148. H. L. Caswell, *J. Appl. Phys.*, 1958, **29**, 1210.
149. S. Koda, K. Kamigaki and H. Kayano, *J. Phys. Soc. Japan*, 1963, **18** (Suppl. 1), 195.
150. A. V. Siefert and F. T. Worrel, *J. Appl. Phys.*, 1951, **22**, 1257.
151. R. H. Chalmers and J. Schultz, *Acta Met.*, 1962, **10**, 466.
152. H. Mühlbach, *Phys. Status Solidii*, 1969, **36** (1), K33.
153. R. Gibala, M. K. Korenko, M. F. Amateau and T. E. Mitchell, *J. Phys. Chem. Solids*, 1970, **3** (8), 1889.
154. G. Rieu, J. De Fouquet and A. Nadeau, *C.r. hebd. Séanc. Acad. Sci., Paris*, 1970 (c), **270** (3), 287.
155. S. Z. Bokshtein, M. B. Bronfin, *et al.*, *Soviet Phys. solid St.*, 1964, **5** (11), 2253.
156. Yu. V. Piguzov, W. D. Werner and I. Ya. Rzhevskaya, *Phys. Met. Metallogr.*, 1967, **24** (3), 179.
157. R. H. Schnitzel, *Trans. Met. Soc. AIME*, 1964, **230** (3), 609.
158. M. J. Murray, *Phil. Mag.*, 1969, **20** (165), 561.
159. A. A. Belyakov, V. P. Yelyutin and Ye. I. Mozzhukhin, *Phys. Met. Metallogr.*, 1967, **23** (2), 115.
160. E. J. Kramer and C. L. Bauer, *Phys. Rev.*, 1967, **163** (2), 407.
161. J. Schultz, *Bull. Am. Phys. Soc.*, 1964, **9**, 214.
162. F. M. Mazzolai and M. Nuovo, *Solid State Commun.*, 1969, **7** (1), 103.
163. J. Filloux, H. Harper and R. H. Chalmers, *Bull. Am. Phys. Soc.*, 1964, **9**, 230.
164. M. W. Stanley and Z. C. Szkopiak, *J. Inst. Metals*, 1966, **94** (2), 79.
165. M. F. Amateau, R. Gibala and T. E. Mitchell, *Scripta Metall.*, 1968, **2** (2), 123.
166. M. F. Amateau, T. E. Mitchell and R. Gibala, *Phys. Status Solidii*, 1969, **36** (1), 407.
167. R. A. Hoffman and C. A. Wert, *J. Appl. Phys.*, 1966, **37** (1), 237.
168. F. Schlät, *Trans. Japan Inst. Metals*, 1968, **9** (Suppl.), 64.
169. E. Davenport and G. Mah, *Metall. Trans.*, 1970, **1** (5), 1452.
170. R. W. Powers and M. V. Doyle, *J. Metals, N.Y.*, 1957, **9**, 1285.
171. R. W. Powers and M. V. Doyle, *J. Appl. Phys.*, 1959, **30**, 514.
172. C. Vercaemer and A. Clauss, *C.r. hebd. Séanc. Acad. Sci., Paris* 1969 (c), **269** (15), 803.
173. D. H. Boone and C. Wert, *J. Phys. Soc. Japan*, 1963, **18** (Suppl. 1), 141.
174. C. Y. Ang, *Acta Met.*, 1953, **1**, 123.
175. D. J. Van Ooijen and A. S. Van Der Goot, *Acta Met.*, 1966, **14** (8), 1008.
176. G. Vidal and H. Bibring, *C.r. hebd. Séanc. Acad. Sci., Paris*, 1965, **260** (3), 857.

177. R. E. Miner, D. F. Gibbons and R. Gibala, *Acta Met.*, 1970, **18** (4), 419.
178. A. W. Sommers and D. N. Beshers, *J. Appl. Phys.*, 1966, **37** (13), 4603.
179. P. S. Venkatesan and D. N. Beshers, *J. Appl. Phys.*, 1970, **41** (1), 42.
180. A. Seeger, P. Schiller and H. Kronmüller, *Phil. Mag.*, 1960, **5**, 853.
181. P. Schiller, H. Kronmüller and A. Seeger, *Acta Met.*, 1962, **10**, 333.
182. J. T. A. Roberts and P. Barrand, *Acta Met.*, 1967, **15** (11), 1685.
183. O. I. Datsko and V. A. Pavlov, *Phys. Met. Metallogr.*, 1958, **6** (5), 122.
184. T. S. Kê, *Acta Phys. Sin., 1955*, **11** (5), 405.
185. V. N. Gridnev, A. I. Yefimov and N. P. Kushnareva, *Phys. Met. Metallogr.*, 1967, **23** (4), 142.
186. Y. S. Avraamov, L. N. Belyakov and B. G. Livshits, *Phys. Met. Metallogr.*, 1958, **6** (1), 104.
187. V. M. Azhazha, N. P. Bondarenko, M. P. Zeydlits and B. I. Shapoval, *Phys. Met. Metallogr.*, 1970, **29** (2), 101.
188. R. R. Arons, J. Bouman, M. Witzenbeek, P. T. A. Klaase, C. Tuyn, G. Leferink and G. De *Vries, Acta Met.*, 1967, **15** (1), 144.
189. R. R. Arons, C. Tuyn and G. De Vries, *Acta Met.*, 1967, **15** (10), 1673.
190. J. Coremberg and F. M. Mazzolai, *Solid State Commun.*, 1967, **6** (1), 1.
191. V. O. Shestopal, *Phys. Met. Metallogr.*, 1968, **26** (6), 176.
192. M. Rosen, G. Erez and S. Shtrikman, *J. Phys. Chem. Solids*, 1969, **30** (5), 1063.
193. R. Schnitzel, *J. Appl. Phys.*, 1959, **30**, 2011.
194. L. N. Aleksandrov, M. I. Zotov, R. Sh. Ibragimov and F. L. 'Edel' man, *Soviet Phys. Solid St.*, 1970. **11** (7), 1494.
195. P. D. Southgate, *Proc. Phys. Soc. Lond.*, 1957, **70** (B), 804.
196. P. D. Southgate, *Proc. Phys. Soc. Lond.*, 1960, **76**, 385, 398.
197. B. S. Berry, *J. Phys. Chem. Solids*, 1970, **13** (8), 1827.
198. L. Rotherham, A. D. N. Smith and G. B. Greenough, *J. Inst. Metals*, 1951, **79**, 439.
199. L. Verdini and L. A. Vienneau, *Canad. J. Phys.*, 1968, **46** (23), 2715.
200. R. W. Powers and M. V. Doyle, *J. Appl. Phys.*, 1957, **28**, 255.
201. P. Kofstad and R. A. Butera, *J. Appl. Phys.*, 1963, **34**, 1517.
202. R. W. Powers and M. V. Doyle, *Acta Met.*, 1956, **4**, 233.
203. R. W. Powers and M. V. Doyle, *Acta Met.*, 1955, **3**, 135.
204. R. W. Powers and M. V. Doyle, *Trans. AIMME*, 1959, **215**, 655.
205. G. Schoek and M. Mondino, *J. Phys. Soc. Japan*, 1963, **18** (Suppl. 1), 149.
206. A. A. Sagues and R. Gibala, *Scripta Metall.*, 1971, **5** (8), 689.
207. G. Arlt and W. Hermann, *Solid State Commun.*, 1969, 7 (1), 75.
208. J. Winter and S. Weinig, *Trans. AIMME*, 1959, **215**, 74.
209. J. E. Doherty and D. F. Gibbons, *Acta Met.*, 1971, **119** (4), 275.
210. W. Köster, L. Bangert and M. Evers, *Z. Metall.*, 1956, **47**, 564.
211. R. R. Hasiguti and K. Iwasaki, *J. Appl. Phys.*, 1968, **39** (5), 2182.
212. W. J. Bratina, *Acta Met.*, 1962, **10**, 332.
213. J. N. Pratt, W. J. Bratina and B. Chalmers, *Acta Met.*, 1954, **2**, 203.
214. D. Gupta and S. Weinig, *Acta Met.*, 1962, **10**, 292.
215. R. W. Powers and M. V. Doyle, *Acta Met.*, 1958, **6**, 643.
216. G. Cannelli and F. M. Mazzolai. *J. Phys. Chem. Solids*, 1970, **31** (8), 1913.
217. R. A. Butera and P. Kofstad, *J. Appl. Phys.*, 1963, **34**, 2172.
218. R. W. Powers, *Acta Met.*, 1954, **2**, 604.
219. L. N. Aleksandrov and V. S. Mordyuk, *Phys. Met. Metallogr.*, 1966, **21** (1), 101.
220. R. H. Chalmers and J. Schultz, *Phys. Rev. Letters*, 1961, **6**, 273.
221. R. H. Schnitzel, *Trans. Met. Soc. AIME*, 1965, **233** (1), 186.
222. L. H. Aleksandrov, *Phys. Met. Metallogr.*, 1962, **13** (4), 143.
223. I. Berlec, *Metall. Trans*, 1970, **1** (10), 2677.
224. V. O. Shestopal, *Phys. Met. Metallogr.*, 1968, **25** (6), 148.
225. V. P. Yelyutin and A. K. Natanson, *Phys. Met. Metallogr.*, 1963, **15** (5), 89.
226. H. Kayano, *J. Phys. Soc. Japan*, 1969, **26** (3), 733.
227. G. Roberts, P. Barrand and G. M. Leak, *Scripta Metall.*, 1969, **3** (6), 409.
228. J. E. Doherty and D. F. Gibbons, *J. Appl. Phys.*, 1971, **42** (11), 4502.
229. P. L. Gruzin and A. N. Semenikhin, *Phys. Met. Metallogr.*, 1963, **15** (5), 128.
230. W. J. Bratina and W. C. Winegard, *J. Metals, N.Y.*, 1956, **8**, 186.
231. K. Bungardt and H. Preisendanz, *Z. Metall.*, 1960, **51**, 280.
232. V. Y. Ivanov, B. I. Shapoval and V. M. Amonenko, *Phys. Met. Metallogr.*, 1961, **11** (1), 55.
233. P. Boch, J. Petit, C. Gasc and J. De Fouquet, *C.r. hebd. Séanc. Acad. Sci., Paris*, 1968 (c), **266** (9), 605.
234. H. L. Brown, P. E. Armstrong and C. P. Kempter, *J. Less-common Metals*, 1967, **13** (4), 373.
235. J. L. Gacougnolle, S. Sarrazin and J. De Fouquet, *C.r. hebd. Séanc. Acad. Sci., Paris* 1970 (c), **270** (2), 158.
236. C. W. Nelson, D. F. Gibbons and R. F. Hehemann, *J. Appl. Phys.*, 1966, **37** (13), 4677.
237. S. Karashima and K. Saito, *J. Jap. Inst. Metals*, 1973, **37**(3), 326.
238. H. Farman and D. H. Niblett, 'Proc. 3rd Euro. Conf. Int. Frict.', Manchester, 1980, Pergamon Press, p. 7.
239. H. Schulz, U. Rodrian and M. Maul, 'Proc. 3rd Euro. Conf. Int. Frict'., Manchester, 1980, Pergamon Press, p. 19.
240. H. E. Schaeffer, H. Schulz and H. P. Stark, 'Proc. 3rd Euro. Conf. Int. Frict'., Manchester, 1980, Pergamon Press, p. 25.

241. F. Baudraz and R. Gotthardt, 'Proc. 3rd. Euro. Conf. Int. Frict.', Manchester, 1980, Pergamon Press, p. 67.
242. S. M. Seyed Reihani, G. Fantozzi, C. Esnouf and G. Revel, *Scripta Met.*, 1979, **13(8)**, 1011.
243. H. Mizubayashi, S. Okuda and M. Daikubara, *Scripta Met.*, 1979 **13**(12), 1131.
244. A De Rooy, P. M. Bronsveld and J. Th M. De Hosson. 'Proc. 3rd. Euro. Conf. Int. Frict.', Manchester, 1980, Pergamon Press, p. 149.
245. J. N. Lomer and C. R. A. Sutton, 'Proc. 3rd Euro. Conf. Int. Frict.', Manchester, 1980, Pergamon Press, p. 199.
246. M. Weller and J. Diehl, 'Proc. 3rd Euro. Conf. Int. Frict.', Manchester, 1980, Pergamon Press, p. 223.
247. R. Schaller and W. Benoit, 'Proc. 3rd Euro. Conf. Int. Frict.', Manchester, 1980, Pergamon Press, p. 311.
248. M. Mondino and R. Gugelmeier, 'Proc. 3rd Euro. Conf. Int. Frict.', Manchester, 1980, Pergamon Press, p. 317.
249. R. Klam, H. Schulz and H. E. Schaeffer, *Acta Met.*, 1979, **278**, 205.
250. K. Iwasaki, K. Lücke and G. Sokolowski, *Acta. Met.*, 1980, **28**, 855.
251. B. L. Cheng and T. S. Ke, *Phys. Status Solidii*, 1988, **107**, 177.
252. H. Tezuka *et al.*, *J. Nucl. Mater.*, 1988, **155/7A**, 340.
253. S. J. Ding, W. B. Li and G. P. Yang, *Rare Met.* (China), **7**, 99.
254. K. Kato, O. Yoshinari and K. Tanaka, *Jpn Inst. Met.*, 1988, **29**, 251.
255. M. Tagami, T. Othani and T. Usami, *J. Jap. Inst. Light Met.*, 1988, **38** (2), 107.
256. S. C. Yan and T. S. Ke, *Phys. Status Solidii*, 1987, **104**, 715.
257. F. Cosandey *et al.*, *Scripta Metall.*, 1988, **22**, 395.
258. L. D. Zhang, J. Shi and T. S. Ke, *Phys. Status Solidii*, 1986, **98**, 151.
259. S. Wang, T. Dai and C. Shi, *Acta Metall. Sin.*, **22**, A441.
260. S. Chen, J., S. Zhang and Z. Xu, *Acta Met. Sin.*, **22**, 379.
261. T. Enjo and T. Kuroda, *Trans. JWRA*, 1986, **15**, 41.
262. S. Sinnema *et al.*, *Rapidly Quenched Metals*, 1985, **1**, 719.
263. K. E. Vidal, W. N. Weins and R. A. Winholz, *High Strength Powder Metallurgy Aluminium Alloys, II*, 1986, 255.
264. S. Asano and S. Tamaoka, *Scr. Metall*, 1986, **20**, 1151.
265. T. S. Ke and C. M. Su, *Phys. Status Solidii*, 1986, **94**, 191.
266. K. Iwasaki, *J. Phys. Soc. Jpn*, 1986, **55**, 546.
267. K. Iwasaki, *J. Phys. Soc. Jpn*, 1986, **55**, 845.
268. S. Asano and M. Kasaoka, *J. Jpn. Inst. Met.*, 1986, **50**, 391.
269. S. Asano and M. Usui, *J. Jpn. Inst. Met.*, 1985, **49**, 945.
270. G. Li, Z. Pan and J. Zhang, *Acta Metall. Sin.*, 1985, **21**, 21.
271. T. Yokoyama, *Scr. Metall.*, 1985, **19**, 747.
272. P. Cui, Q. Huang, T. S. Ke and S. C. Yan, *Phys. Status Solidii*, 1984, **86**, 593.
273. G. Canelli *et al.*, *J. Phys. F: Met. Phys.*, 1984, **14**, 2507.
274. C. Esnouf *et al.*, *Acta Metall.*, 1984, **32**, 2175.
275. S. Asano and H. Seki, *J. Jpn. Inst. Met.*, 1984, **48**, 694.
276. B. Purniah and R. Ranganathan, *Phil. Mag. A*, 1983, **47**, L23.
277. S. A. Antipov, A. I. Drozhzhin and A. M. Roshchupkin, *Fiz. Tverd. Tela*, 1983, **25**, 1392.
278. C. M. Mo and P. Moser, *Phys. Status Solidii* (a), 1983, **78**, 201.
279. J. N. Daou, P. Moser and P. Vajda, *J. Phys.* (Orsay), 1983, **44**, 543.
280. S. Asano and K. Oshima, *Trans. Jpn. Inst. Met.*, 1982, **23**, 530.
281. H. Mitani, N. Nakanishi and K. Suzuki, *J. Jpn. Inst. Met.*, 1980, **44**, 43.
282. V. F. Belostotskii, T. V. Golub and I. G. Polotskii, *Metalofizika*, 1982, **4**, 106.
283. K. Hakomori, N. Igata and K. Miyahara, *J. Jpn. Inst. Met.*, 1980, **44**, 474.
284. P. F. Gobin, J. Merlin and G. Vigier, *Ti and Ti Alloys*, 1982, **3**, 1691.
285. V. E. Bakhrushin, A. V. Novikov and Y. A. Pavlov, *Izv. V.U.Z. Chernaya Metall.*, 1982, **7**, 113.
286. H. B. Chen, N. Igata and K. Miyahara, *Scr. Metall.*, 1982, **16**, 1039.
287. N. Igata, M. Masamura and H. Murakami, *Mech. Props of B.C.C. Metals*, 1982, 75.
288. H. Muhlbach, *Phys. Status Solidii* (a), 1982, **69**, 615.
289. H. B. Chen, N. Igata, K. Miyahara and T. Uba, *7th Int. Conf. on Internal Friction and Ultrasonic Attenuation in Solids, 1981*, Inst. de Genie Atomique, Lausanne, Paper 1.B.11.
290. B. Dubois and M. Lebienvenu, *7th Int. Conf. on Internal Friction and Ultrasonic Attenuation in Solids, 1981*, Inst. de Genie Atomique, Lausanne, Paper 7.A.6.
291. G. Haneczok, J. Moron and T. Poloczek, *7th Int. Conf. on Internal Friction and Ultrasonic Attenuation in Solids, 1981*, Inst. de Genie Atomique, Lausanne, Paper 6.C.2.
292. J. Du, *7th Int. Conf. on Internal Friction and Ultrasonic Attenuation in Solids, 1981*, Inst. de Genie Atomique, Lausanne, Paper 6.B.4.
293. K. Agyeman, E. Armbruster, H. Guntherodt and H. U. Kunzi, *7th Int. Conf. on Internal Friction and Ultrasonic Attenuation in Solids, 1981*, Inst. de Genie Atomique, Lausanne, Paper 4.A.6.
294. C. Li, W. Li, Z. Lui and G. Yang, *7th Int. Conf. on Internal Friction and Ultrasonic Attenuation in Solids, 1981*, Inst. de Genie Atomique, Lausanne, Paper 3.C.4.
295. I. G. Ritchie and K. W. Sprungmann, *7th Int. Conf. on Internal Friction and Ultrasonic Attenuation in Solids, 1981*, Inst. de Genie Atomique, Lausanne, Paper 3.A.1.
296. K. Sakamoto and M. Shimada, *7th Int. Conf. on Internal Friction and Ultrasonic Attenuation in Solids, 1981*, Inst. de Genie Atomique, Lausanne, Paper 1.A.14.
297. R. Hanada, *Scr. Metall.*, 1981, **15**, 1121.

298. T. Mori, T. Mura and Mokabe, *Phil. Mag. A*, 1981, **44**, 1.
299. K. Qing-hu, G. Ting-sui, P. Zheng-liang and W. Zhong-guang, *Acta Phys. Sinica*, 1980, **25**, 1180.
300. S. Sato and H. Suto, *Trans. Jpn. Inst. Met.*, 1980, **21**, 83.
301. W. Bernoit and R. Schaller, *Mem. Sci. Rev. Metall.*, 1979, **76**, 521.
302. G. Bellessa and J. Y. Duquesne, *J. Phys. C: Solid State Phys.*, 1980, **13**, 215.
303. S. Asano, M. Shibata and Tsunoda, *Scr. Metall*, 1980, **14**, 377.
304. C. Esnouf, G. Fantozzi, G. Revel and S. M. Seyed-Reihani, *3rd European Conf. Internal Friction and Ultrasonic Attenuation in Solids, Manchester*, 1979, 1979.
305. C. Diallo and M. Mondini, *3rd European Conf. Internal Friction and Ultrasonic Attenuation in Solids, Manchester*, 1979, 1979.
306. R. Gugelmeier and M. Mondini, *3rd European Conf. Internal Friction and Ultrasonic Attenuation in Solids, Manchester*, 1979, 1979.
307. R. L. Crosby, J. L. Holman and L. A. Neumeier, U.S. Dept of Interior, Bureau of Mines Rep. Invest. No. 8383, 32pp.
308. M. Hirabayashi, M. Iseki and M. Koiwa, *6th Int. Conf. Internal Friction and Ultrasonic Attenuation in Solids, Tokyo*, 1977, 659.
309. A. Isore, L. Miyada, K. Tanaka and S. Watanabe, *6th Int. Conf. Internal Friction and Ultrasonic Attenuation in Solids, Tokyo*, 1977, 605.
310. K. Iwasaki, *J. Phys. Soc. Jpn*, 1978, **45**, 1583.
311. I. Brough, K. M. Entwistle and P. Fuller, *Acta Met.*, 1978, **26**, 1055.
312. N. Nagai, 1979 (1976), U.S. Patent No. 4, 134, 758.

15.3   Shape memory alloys*

15.3.1   Mechanical properties of shape memory alloys

Most shape memory alloys have compositions at which the crystallographic structure can change reversibly and reproducibly from a higher temperature phase with higher symmetry to a lower temperature phase with lower symmetry by a small change in temperature or by a change in mechanical stress at temperatures just above the transformation temperature at zero stress.

In most shape memory alloys (and in all the industrially useful alloys), the change of structure usually occurs over a narrow range of temperature by means of a self-accommodating martensitic transformation, during which a small amount of heat is evolved or absorbed depending on the direction of the temperature change. This usually gives rise to a thermal hysteresis of about 10 to 40°C over which the parent phase (usually referred to as 'austenite') and the martensite can co-exist.

If a stressed shape memory alloy is thermally cycled through its martensitic transformation temperature, the strain–temperature relationship will take the form of a closed hysteresis loop similar in shape to the *B–H* curve of ferromagnetic alloys. If a shape memory alloy is cooled to below its M_f temperature, it undergoes little change in shape or volume. If it is then deformed plastically to a new shape, it will recover its original undeformed shape on re-heating to a temperature above its A_s temperature. The amount of strain which can be recovered in this way is not unlimited but depends on the nature of the alloy. For example, the maximum recoverable strain is about 8% in Ti–Ni alloys and 10–12% in Cu–Zn alloys (although the latter cannot be achieved in industrially useful alloys).

On cooling through the transformation to the martensitic state, the temperatures at which the transformation starts and finishes at zero applied stress are denoted by M_s and M_f respectively. On re-heating, the temperatures at which the reverse transformation to the high temperature phase takes place are A_s and A_f respectively. These temperatures can be determined experimentally by thermal or dilatometric analysis or by changes in electrical resistivity. The M_s temperature is raised progressively by applied stress; the M_d temperature is the highest temperature at which the transformation can be induced by stress. Figure 15.1 illustrates these points.[11] If an alloy is deformed above the M_s but below the M_d temperature, a stress-induced martensitic strain can be obtained. This is completely recovered on unloading (*see* Figures 15.4 and 15.7).

Shape memory does not always appear to depend on the martensitic transformation. Small amounts of shape memory can be obtained in primary solid solutions of alloys of low stacking fault energy.[16] Examples are shown in Table 15.9, e.g. Cu–Al and Cu–Si primary solid solutions and some stainless steels. Although the latter undergo martensitic transformations to the alpha prime martensite, this is too brittle to deform. Shape memory is only found when the stainless steel is deformed at very low temperatures but above the M_s to form both delta and some alpha martensite. 20% deformation

---

* Additional references on this topic may be found in Chapter 38.

may be needed to obtain 1.0–1.5% shape memory strain and it has proved to be of little industrial relevance so far.

Though most work has concentrated on Ti–Ni and related alloys because of their industrial importance, the shape memory phenomenon has been demonstrated in a wide range of alloys, some of which are listed in Table 15.9.

M_s temperatures can be varied continuously by changing the composition. Examples include: (i) changing Ni content in Ti–Ni alloys; (ii) partially replacing Ni by Fe, Co or Pd in Ti–Ni alloys; and (iii) changing Al and Zn contents in Cu–Al–Zn alloys or by partially replacing either element by others such as Sn, Mn, etc. Such variations also change the character of the alloys. Table 15.9 shows typical compositions for which data are published but it is possible to derive additional alloys within the ternary and more complex systems.

Note that if shape memory alloys are cooled under stress, the M_s temperature is raised in direct proportion to stress below M_o (*see* Figure 15.7).

Note also that the M_s, etc temperatures are not exact in that for a given composition they can be changed by heat-treatment and by cold-working. Figure 15.8 illustrates an example of the extent to which the hysteresis can be widened in a Cu-based alloy.[37]

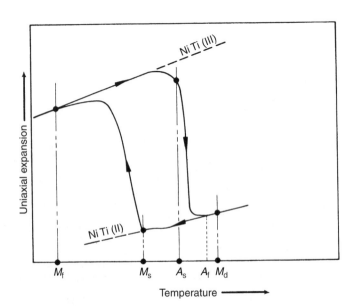

Figure 15.1 *Typical uniaxial dimensional change behaviour for drawn wire. Ti–55.0% Ni–0.07%C. After W. B. Cross et al.*[11]

Table 15.9   COMPOSITIONS AND TRANSFORMATION TEMPERATURES OF SHAPE MEMORY ALLOYS

Alloy composition wt%[(1)]	M_s *temperature at zero stress at* °C	A_s *temperature at zero stress* °C	*Maximum shape memory strain* %	*Reference*
Au–28at% Cu–46 At.% Zn	−15			6
Au–47.5% Cd	60	70		13
Au–12.9% Cu–25.5% Zn	−100		1.55	32
Au–15.2% Cu–28.0% Zn	−50		~1.0; brittle g.b. phase	32

(continued)

Table 15.9 COMPOSITIONS AND TRANSFORMATION TEMPERATURES OF SHAPE MEMORY ALLOYS—*continued*

Alloy composition wt%[1]	M_s temperature at zero stress at °C	A_s temperature at zero stress °C	Maximum shape memory strain %	Reference
Au–16.0% Cu–32.3% Zn	−118		ditto; Zn too high	32
Au–22.3% Cu–31.4% Zn	−64		2.15	32
Au–28.7% Cu–31.1% Zn	<−196		?; Zn too low	32
Cu–2.50% Al–31.75% Zn	−105		10.25	15
Cu–3.94% Al–25.60% Zn	54	48	>2.0	23
Cu–4.00% Al–26.10% Zn	24	23	14% (5–6% reversible)	20
Cu–6.00% Al–22.00% Zn	−50		4.8	15
Cu–7.50% Al–17.00% Zn	−10		4	15
Cu–11.75% Al–6.00% Zn	50		2.8	15
Cu–10.50% Al–7.25% Zn	140		3.9	15
Cu–11.25% Al–4.75% Zn	170		2	15
Cu–11.75% Al–2.50% Zn	250		1.9	15
Cu–4.90% Sn–31.25% Zn	−70		1	15
Cu–1.75% Si–34.50% Zn	−140		6.3	15
Cu–2.25% Si–31.25% Zn	−50		6	15
Cu–3.25% Si–27.50% Zn	75		2.95	15
Cu–12.00% Al–2.00% Mn	240		0.95	15
Cu–11.25% Al–4.25% Mn	160		0.45	15
Cu–10.75% Al–6.00% Mn	100		0.45	15
Cu–10.40% Al–7.00% Mn	83			21
Cu–10.60% Al–7.00% Mn	52			21
Cu–11.00% Al–7.00% Mn	5			21
Cu–11.10% Al–7.00% Mn	−10			21
Cu–12.50% Al–1.00% Fe	300		1.75	15
Cu–12.50% Al–8.00% Fe	250		1.1	15
Cu–13.25% Al–2.75% Ni	82		2.9	15
Cu–8.0% Al	N.A.	N.A.	1.05	16
Cu–4.0% Si	N.A.	N.A.	1.4	16
Cu–14.2% Al–4.3% Ni	−20	−15		8
Cu–40% Zn	−70	−120		8
Cu–34.7% Zn–3.0% Sn	−52	−50	4.5 (polycrystal) 8.5 (single crystal)	9
Fe–15% Cr–15% Ni–15% Co	<−196	−196 to 40	1.45	16
Fe–20% Cr–10% Ni–1% Al	<−196	−196 to 40	0.45	16
Fe–20% Mn–3.75% Ti	<−196	−196 to 40	1.4	16
Fe–20% Cr–15% Ni	<−196	−196 to 40	1.05	16
Fe–17% Cr–19% Ni	>−196	−196 to 40	0.9	16
Fe–30% Mn–6.5% Si	<20	20 to 400	2.1	22
Fe–24% Mn–3% Si	123	188	1.5–1.75	24
Fe–26% Mn–4% Si	106	182	1.5–1.75	24
Fe–27% Mn–3% Si	103	176	1.5–1.75	24
Fe–28% Mn–3% Si	102	164	1.5–1.75	24
Fe–28% Mn–4% Si	62	162	1.5–1.75	24
Fe–29% Mn–4% Si	88	164	1.5–1.75	24
Fe–32% Mn–4% Si	27	134	1.5–1.75	24
Fe–33% Mn–4% Si	22	131	1.5–1.75	24
Fe–34% Mn–4% Si	8	127	1.5–1.75	24
Fe–35% Mn–4% Si	−10	123	1.5–1.75	24
Ti–50.2at% Ni	35	50	1.4 (reversible)	17
Ti–50.2at% Ni	10	57		18
Ti–50.2at% Ni	~25	~55	5.5% (~4% reversible)	19
Ti–56.4% Ni	5			26
Ti–55.5% Ni	10			26
Ti–55.0% Ni	35			26

(continued)

Table 15.9   COMPOSITIONS AND TRANSFORMATION TEMPERATURES OF SHAPE MEMORY
ALLOYS—*continued*

Alloy composition wt%[1]	M_s temperature at zero stress at °C	A_s temperature at zero stress °C	Maximum shape memory strain %	Reference
Ti–53.7% Ni	75			26
Ti–54.8% Ni	62			27
Ti–52–56% Ni	~RT			1
Ti–53.5% Ni	98			2
Ti–54.0% Ni	140			2
Ti–54.5% Ni	170	175	8.0	2, 4
Ti–55.0% Ni	140			2
Ti–55.5% Ni	30			2
Ti–56.0% Ni	−25			2
Ti–56.5% Ni	−50			2
Ti–51at% Ni	28	62		7
TiNi[2]	166			3
TiCo	−238			3
TiNi$_x$Co$_{1-x}^{(3)}$	166 where $x = 1$ −238 where $x = 0$			3
TiFe	~ − 269			
Ti–35% Nb	−175	184	~2.5	10
Ti–51.4% Ni–3.57% Co				
0.3 in. rod	−51	−40	6 to 10	11
0.003 in. foil	−65	−45	6 to 10	11
0.01 in. wire	−73	−51	6 to 10	11
Ti–55.0% Ni–0.07% C				
0.2 in. rod	21	60	6 to 10	11
0.003 in. foil	27	43	6 to 10	11
0.01 in. wire	18	43	6 to 10	11
Ti–54.6% Ni–0.06% C				
0.625 in. rod	43	71	6 to 10	11
0.003 in. foil	38	66	6 to 10	11
0.01 in. wire	32	54	6 to 10	11
Ti–47.0% Ni–7% Cu	63			27
Ti–44.5% Ni–10% Cu	52			27
Ti–32.0% Ni–22% Cu	74			27
Ti–27.0% Ni–29% Cu	20			27
Ti–47.0at% Ni–3at% Fe	−180	−88		30
Ti–52.85% Ni–0.28% Fe	34	35	8 (0.094 rad in torsion)	14, 33, 34
Ti–47.0at% Ni–9at% Nb	−90	−56 up to 55		30
Ti–50at% Pd	533	573	~2.0	28, 29
Ti–50at% Pd	510	520		36
Ti–50at% Pd	563	580	2.39	35
Ti–40.0at% Pd–10.0at% Ni	403	419	2.66	35
Ti–30.0at% Pd–20.0at% Ni	241	230	4.38	35
Ti–20.0at% Pd–30.0at% Ni	95	90	1.84	35
Ti–10.0at% Pd–40.0at% Ni	−18	−26		35
Ti–50.0at% Ni	55	80		35
Ti–5.0at% Pt–45.0at% Ni	29	36		35
Ti–10.0at% Pt–40.0at% Ni	18	−27		35
Ti–20.0at% Pt–30.0at% Ni	300	263	2.17	35
Ti–30.0at% Pt–20.0at% Ni	619	626		35
Ti–50.0at% Pt	1 070	1 040		36
Ti–44at% Pd–6at% Fe	321	335	5.0	28, 29
Ti–42at% Pd–8at% Fe	293	250		28
Ti–40at% Pd–10at% Fe	173	178		28
Ti–38at% Pd–12at% Fe	96	99		28
Ti–36at% Pd–14at% Fe	25	25		28
Ti–34at% Pd–16at% Fe	−49	−45		28

(*continued*)

Table 15.9 COMPOSITIONS AND TRANSFORMATION TEMPERATURES OF SHAPE MEMORY ALLOYS—*continued*

Alloy composition wt%[1]	M_s temperature at zero stress at °C	A_s temperature at zero stress °C	Maximum shape memory strain %	Reference
Ti–42at% Pd–4at% Fe	228	216		28
Ti–40at% Pd–6at% Fe	119	124		28
Ti–38at% Pd–8at% Fe	18	3		28
ZrRu[4]	−233			3
ZrRh[4]	380			3
ZrPd[4]	727			3
U–4.0% Mo		200		31
U–4.5% Mo		80		31
U–5.0% Mo		50	3.2	31
U–6.0% Mo		−50		31
U–4.0% No		~250		31
U–5.0% Nb		~200		31
U–7.0% Nb		~100		31
U–14at% Nb	160	90	4.8	25
U–9at% Nb[5]	0	150		5
U–12at% Nb[5]	0	150		5
U–15at% Nb[5]	0	60		5
U–18at% Nb[5]	−196	0		5

[1] Unless otherwise stated.
[2] *See also* Figure 15.2.[3]
[3] *See* Figures 15.3.
[4] Non-linear interpolation between their compounds is possible. Relationship is of form $\log_e (M_s K) \propto x$ where x is in the range 0–1 in Zr–Ru_x–Rh_{1-x} etc.
[5] M_s and A_s not accurately determined but are within these temperature ranges.

Table 15.10 PHYSICAL AND MECHANICAL PROPERTIES OF A TITANIUM–55% NICKEL SHAPE MEMORY ALLOY

Property	Value
Density	6.45 g cm^{-3}
M.p.	1 310°C
Magnetic permeability	<1.002
Electrical resistivity	80 $\mu\Omega$ cm at 20°C
	132 $\mu\Omega$ cm at 900°C
Coefficient of thermal expansion (24–900° C)	10.4 × 10^{-6} °C^{-1}
Mechanical properties at 20°C (i.e. below M_s)	
0.2% yield stress	207 MPa
UTS	861 MPa
Elongation %	22
Reduction in area %	20
Impact (unnotched)	159 J at 20°C
	95 J at 80°C
Fatigue (rotating beam)	>25 × 10^6 cycles at 483 MPa

Note
If recovery is prevented mechanicaly, the TiNi will exert a stress and is capable of doing work when heated to above the A_s temperature. Samples of Ti–55.0% Ni–0.07%C were capable of exerting a stress of up to 758 MPa at 171°C (*see* Figure 15.5). The amount of mechanical work which this alloy was capable of doing was 17–20 J cm^{-3} on heating from 24°C to 171°C.

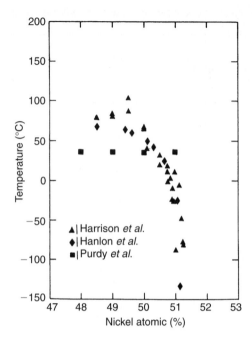

Figure 15.2   *The dependence of the transformation temperature M_s on composition of Ti–Ni alloys, after K. N. Melton*[39]

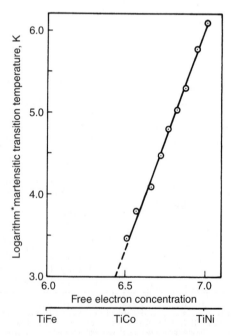

Figure 15.3   *See Table 15.9 after F. E. Wang and W. J. Buehler*[3] *Natural logarithm

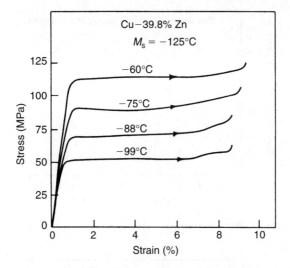

Figure 15.4 *Stress–strain curves for a Cu–Zn single crystal loaded in tension above M_S. As the M_S temperature is approached, the stress required to induce martensite is lowered, after C. M. Wayman and T. W. Duerig*[38]

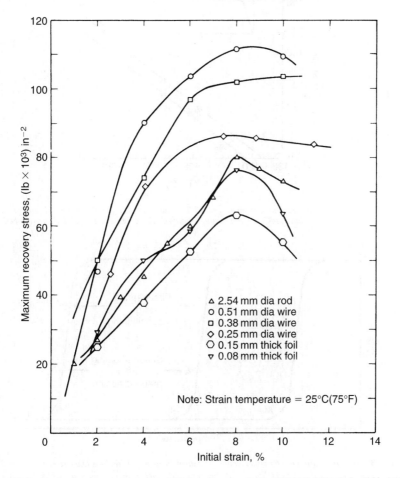

Figure 15.5 *Maximum recovery stress versus initial strain curves for* Ti–55.0% Ni–0.0% C. See Table 15.9, *after W. B. Cross et al.*[11]

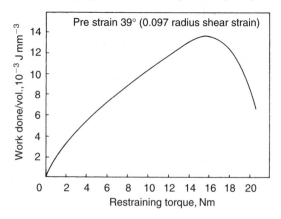

Figure 15.6   *The relationship between the capacity of* TiNi *alloy to do work in torsion and the restraining torque.*[14]

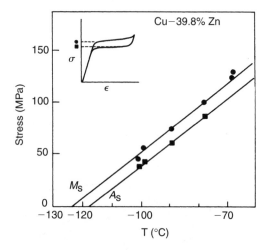

Figure 15.7   *Plotting the plateau stresses such as shown in Figure 15.4 as a function of temperature gives a linear plot which obeys the Clausius–Clapeyron relationship. The alloy's zero stress* A_s *and* M_s *are marked on the ordinate.*[38]

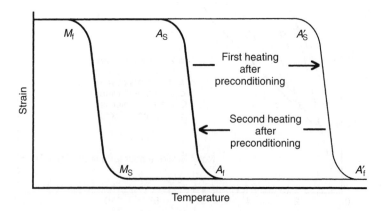

Figure 15.8   *The preconditioning process is a one-time displacement of* A_s *and* A_f. *Once recovery is complete, martensite can be reformed, after which* A_s *and* A_f *are restored to their original values, after Duerig et al.*[37]

REFERENCES FOR SECTION 15.3

For data on designs and applications using shape memory alloys, *see* 'Engineering Aspects of Shape Memory Alloys', edited by T. W. Duerig, K. N. Melton, D. Stöckel and C. M. Wayman, published by Butterworth–Heineman Ltd, London, 1990. For a general review of shape memory materials, their properties and applications, please refer to 'Shape Memory Materials', edited by K. Otsuka and C. M. Wayman, published by Cambridge University Press, 1998.

1. W. J. Buehler and R. C. Wiley, US Patent 3 174 851.
2. A. G. Rozner and W. J. Buehler, US Patent 3 351 463.
3. F. E. Wang and W. J. Buehler, US Patent 3 558 369.
4. G. B. Brook, unpublished data.
5. R. J. Jackson, J. F. Boland and J. L. Frankeng, US Patent 3 567 523.
6. N. Nakanishi, *et al.*, *Phys. Letters*, 1971, 37A, 61.
7. F. E. Wang, *et al.*, *J. Appl. Phys.*, 1968, **39**, 2166.
8. K. Otsuka, *et al.*, *Scripta Metall.*, 1972, **6**, 377.
9. J. D. Eisenwasser and L. C. Brown, *Met. Trans. AIME.*, 1972, **3**, 1359.
10. C. Baker, *Met. Sci. J.*, 1971, **5**, 92.
11. W. B. Cross, *et al.*, NASA Report CR—1433, Sept. 1969.
12. H. U. Schuerch, NASA Report CR—1232, Nov. 1968.
13. D. S. Lieberman, T. A. Read and M. S. Wechsler, *J. Appl. Phys.*, 1957, **28**, 532.
14. G. B. Brook, *et al.*, Fulmer Research Inst. Rep. No R662/4A, Feb., 1977.
15. G. B. Brook and R. F. Iles, British Patent Specification No. 1346047.
16. G. B. Brook, R. F. Iles and P. L. Brooks, in 'Shape Memory Effects in Alloys' (ed. J. Perkins), 1975, Plenum Press, New York, p. 477.
17. S. Edo, *J. Mat. Sci. Letters*, 1989, **24**, 3991.
18. H. Tamura, Y. Susuki and T. Todoroki, *Proc. Int. Conf. Martensitic Transform.*, 1986, p. 736 (Jap. Inst. Met.).
19. Y. Liu and P. G. McCormick, *Acta Met. Mat.*, 1990, **38**, 1321.
20. L. Contardo and G. Guenin, *Acta Met. Mat.*, 1990, **38**, 1267.
21. M. L. Blasquez, C. Lopez and C. Gomez, *Metallography*, 1989, **23**, 119.
22. J. S. Robinson and P. G. McCormick, *Scripta Met.*, 1989, **23**, 1975.
23. S. S. Leu and C. T. Hu, *Mat. Sci. and Eng.*, 1989, **A117**, 247.
24. M. Sade, K. Halter and E. Hornbogen, *J. Mat. Sci. Letters*, 1990, **9**, 112.
25. R. A. Vandermeer, J. C. Ogle and W. B. Snyder, *Scripta Met.*, 1978, **12**, 243.
26. V. N. Ermakov, V. I. Kolomytsev, V. A. Lobodyuk and L. G. Khandros, *Metall. Term. Obr. Metallov*, 1981, **5**, 57.
27. R. H. Bricknell, K. N. Melton and O. Mercier, *Met. Trans., AIME*, 1979, 10A, 693 (See also US Patent 4 144057, 1979 for additional compositions in Ti–Ni–Cu– (Co, Fe, Al, Cr) alloys).
28. K. Enami, T. Yoshida and S. Nenno, *Proc. Int. Conf. Martensitic Transform.*, 1986, p. 103 (Jap. Inst. Met.).
29. K. Enami, Y. Miyasaka and H. Takakura, *MRS Int. Conf. on Adv. Mat.*, 1989, **9**, 135.
30. J. A. Simpson, T. Duerig and K. M. Melton, European Patent, 1985, No. 0187452.
31. G. B. Brook and R. F. Iles, British Patent Specification, 1969, No. 1315653.
32. G. B. Brook and R. F. Iles, *Gold Bulletin*, 1975, **8**, 16.
33. D. Powley and G. B. Brook, *12th Aerospace Mechanisms Symp.*, NASA Conf. Publ. 2080, 1978, 119.
34. G. B. Brook, Inst. *Metallurgists Conf. on Phase Transf.*, 1979, Ser. 3, **2** (11), VI-1–3.
35. P. G. Lindquist and C. M. Wayman in 'Engineering Aspects of Shape Memory Alloys' (ed. T. W. Duerig *et al.*), 1990, Butterworth–Heinemann, London, p. 58.
36. H. C. Donkersloot and J. H. N. Van Vucht, *J. Less Common Metals*, 1970, **20**, 83.
37. T. W. Duerig, K. N. Melton and J. L. Proft in 'Engineering Aspects of Shape Memory Alloys' (ed. T. W. Duerig *et al.*), 1990, Butterworth–Heinemann, London, p. 130.
38. C. M. Wayman and T. W. Duerig in 'Engineering Aspects of Shape Memory Alloys' (ed. T. W. Duerig *et al.*), 1990, Butterworth–Heinemann, London, p. 3.
39. K. N. Melton in 'Engineering Aspects of Shape Memory Alloys' (ed. T. W. Duerig *et al.*), 1990, Butterworth–Heinemann, London, p. 21.

16 Temperature measurement and thermoelectric properties

16.1 Temperature measurement*

The unit of the fundamental physical quantity known as thermodynamic temperature, symbol T, is the Kelvin, symbol K, defined as the fraction 1/273.16 of the thermodynamic temperature of the triple point of water.[†]

Because of the way earlier temperature scales were defined, it remains common practice to express a temperature in terms of its difference from 273.15 K, the ice point. A thermodynamic temperature, T, expressed in this way is known as a Celsius temperature, symbol t, defined by:

$$t/°C = T/K - 273.15$$

The unit of Celsius temperature is the degree Celsius, symbol °C, which is by definition equal in magnitude to the Kelvin. A difference of temperature may be expressed in Kelvin or degrees Celsius.

The International Temperature Scale of 1990 (ITS-90) defines both International Kelvin Temperatures, symbol T_{90}, and International Celsius Temperatures, symbol t_{90}. The relation between T_{90} and t_{90} is the same as that between T and t, i.e.:

$$t_{90}/°C = T_{90}/K - 273.15$$

The unit of the physical quantity T_{90} is the Kelvin, symbol K, and the unit of the physical quantity t_{90} is the degree Celsius, symbol °C, as is the case for the thermodynamic temperature T and the Celsius temperature t.

The International Temperature Scale of 1990[1] was adopted by the International Committee of Weights and Measures at its meeting in 1989, in accordance with the request embodied in Resolution 7 of the 18th General Conference of Weights and Measures of 1987. This Scale supersedes the International Practical Temperature Scale of 1968 (amended edition of 1975) and the 1976 Provisional 0.5 K to 30 K Temperature Scale.

The ITS-90 extends upwards from 0.65 K to the highest temperature practicably measurable in terms of the Planck radiation law using monochromatic radiation. The ITS-90 comprises a number of ranges and subranges throughout each of which temperatures T_{90} are defined. Several of these ranges or subranges overlap, and where such overlapping occurs differing definitions of T_{90} exist: these differing definitions have equal status. For measurements of the very highest precision there may be detectable numerical differences between measurements made at the same temperature but in accordance with differing definitions. Similarly, even using one definition, at a temperature between defining fixed points two acceptable interpolating instruments (e.g. resistance thermometers) may give detectably differing numerical values of T_{90}. In virtually all cases these differences are of negligible practical importance and are at the minimum level consistent with a scale of no more than reasonable complexity: for further information on this point, see 'Supplementary Information for the ITS-90'.[2]

The ITS-90 has been constructed in such a way that, throughout its range, for any given temperature the numerical value of T_{90} is a close approximation to the numerical value of T according

---

* For details of emissivity, as needed for pyrometry—see Chapter 17.
[†] 'Comptes Rendus des Séances de la Treizième Conférence Générale des Poids et Mesures (1967–1968)', Resolutions 3 and 4, p. 104.

to best estimates at the time the scale was adopted. By comparison with direct measurements of thermodynamic temperatures, measurements of T_{90} are more easily made, are more precise and are highly reproducible.[3]

There are significant numerical differences between the values of T_{90} and the corresponding values of T_{68} measured on the International Practical Temperature Scale of 1968 (IPTS-68), see Figure 16.1 and Table 16.1. Similarly there were differences between the IPTS-68 and the International Practical Temperature Scale of 1948 (IPTS-68), and between the International Temperature Scale of 1948 (ITS-48) and the International Temperature Scale of 1927 (ITS-27).[3]

Between 0.65 K and 5.0 K T_{90} is defined in terms of the vapour-pressure temperature relations of ^{3}He and ^{4}He.

Between 3.0 K and the triple point of neon (24.556 1 K) T_{90} is defined by means of a helium gas thermometer calibrated at three experimentally realisable temperatures having assigned numerical values (defining fixed points) and using specified interpolation procedures.

Between the triple point of equilibrium hydrogen (13.803 3 K) and the freezing point of silver (961.78°C) T_{90} is defined by means of platinum resistance thermometers calibrated at specified sets of defining fixed points and using specified interpolation procedures.

Table 16.1   THE DIFFERENCES BETWEEN ITS-90 AND EPT-76, AND BETWEEN ITS-90 AND IPTS-68

$(T_{90} - T_{76})/\mathrm{mK}$

T_{90}/K	0	1	2	3	4	5	6	7	8	9
0						−0.1	−0.2	−0.3	−0.4	−0.5
10	−0.6	−0.7	−0.8	−1.0	−1.1	−1.3	−1.4	−1.6	−1.8	−2.0
20	−2.2	−2.5	−2.7	−3.0	−3.2	−3.5	−3.8	−4.1		

$(T_{90} - T_{68})/\mathrm{K}$

T_{90}/K	0	1	2	3	4	5	6	7	8	9
10					−0.006	−0.003	−0.004	−0.006	−0.008	−0.009
20	−0.009	−0.008	−0.007	−0.007	−0.006	−0.005	−0.004	−0.004	−0.005	−0.006
30	−0.006	−0.007	−0.008	−0.008	−0.008	−0.007	−0.007	−0.007	−0.006	−0.006
40	−0.006	−0.006	−0.006	−0.006	−0.006	−0.007	−0.007	−0.007	−0.006	−0.006
50	−0.006	−0.005	−0.005	−0.004	−0.003	−0.002	−0.001	0.000	0.001	0.002
60	0.003	0.003	0.004	−0.004	0.005	0.005	0.006	0.006	0.007	0.007
70	0.007	0.007	0.007	−0.007	0.007	0.008	0.008	0.008	0.008	0.008
80	0.008	0.008	0.008	−0.008	0.008	0.008	0.008	0.008	0.008	0.008
90	0.008	0.008	0.008	−0.008	0.008	0.008	0.008	0.009	0.009	0.009

T_{90}/K	0	10	20	30	40	50	60	70	80	90
100	0.009	0.011	0.013	0.014	0.014	0.014	0.014	0.013	0.012	0.012
200	0.011	0.010	0.009	0.008	0.007	0.005	0.003	0.001		

$(t_{90} - t_{68})/{}^{\circ}\mathrm{C}$

$t_{90}/{}^{\circ}\mathrm{C}$	0	−10	−20	−30	−40	−50	−60	−70	−80	−90
−100	0.013	0.013	0.014	0.014	0.014	0.013	0.012	0.010	0.008	0.008
0	0.000	0.002	0.004	0.006	0.008	0.009	0.010	0.011	0.012	0.012

$t_{90}/{}^{\circ}\mathrm{C}$	0	10	20	30	40	50	60	70	80	90
0	0.000	−0.002	−0.005	−0.007	−0.010	−0.013	−0.016	−0.018	−0.021	−0.024
100	−0.026	−0.028	−0.030	−0.032	−0.034	−0.036	−0.037	−0.038	−0.039	−0.039
200	−0.040	−0.040	−0.040	−0.040	−0.040	−0.040	−0.040	−0.039	−0.039	−0.039
300	−0.039	−0.039	−0.039	−0.040	−0.040	−0.041	−0.042	−0.043	−0.045	−0.046
400	−0.048	−0.051	−0.053	−0.056	−0.059	−0.062	−0.065	−0.068	−0.072	−0.075
500	−0.079	−0.083	−0.087	−0.090	−0.094	−0.098	−0.101	−0.105	−0.108	−0.112
600	−0.115	−0.118	−0.122	−0.125*	−0.08	−0.03	0.02	0.06	0.11	0.16
700	0.20	0.24	0.28	0.31	0.33	0.35	0.36	0.36	0.36	0.35
800	0.34	0.32	0.29	0.25	0.22	0.18	0.14	0.10	0.06	0.03
900	−0.01	−0.03	−0.06	−0.08	−0.10	−0.12	−0.14	−0.16	−0.17	−0.18
1 000	−0.19	−0.20	−0.21	−0.22	−0.23	−0.24	−0.25	−0.25	−0.26	−0.26

$t_{90}/{}^{\circ}\mathrm{C}$	0	100	200	300	400	500	600	700	800	900
1 000		−0.26	−0.30	−0.35	−0.39	−0.44	−0.49	−0.54	−0.60	−0.66
2 000	−0.72	−0.79	−0.85	−0.93	−1.00	−1.07	−1.15	−1.24	−1.32	−1.41
3 000	−1.50	−0.59	−1.69	−1.78	−1.89	−1.99	−2.10	−2.21	−2.32	−2.43

*A discontinuity in the first derivative of $(t_{90} - t_{68})$ occurs at a temperature of $t_{90} = 630.6°C$ when $(t_{90} - t_{68}) = -0.125°C$.

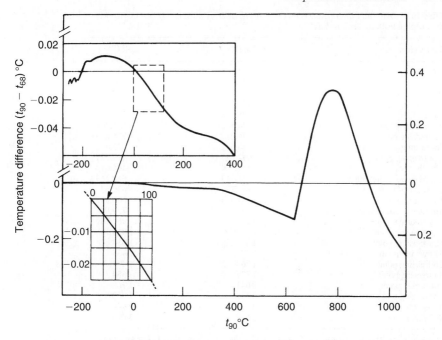

Figure 16.1   *The differences, $(t_{90} - t_{68})$, between ITS-90 and IPTS-68 in the range from –260°C to 1 064°C*

Above the freezing point of silver (961.78°C) T_{90} is defined in terms of a defining fixed point and the Planck radiation law.

The defining fixed points of the ITS-90 together with some selected secondary reference points are listed in Table 16.2. Full details of the practical realisation of this scale can be found in refs. 2 and 3.

Table 16.2   THE DEFINING FIXED POINTS OF ITS-90 AND SOME SELECTED SECONDARY REFERENCE POINTS

Equilibrium state	T_{90} (K)	t_{90} (°C)
Cd superconducting transition point	0.519	
Zn superconducting transition point	0.851	
Al superconducting transition point	1.179 6	
^{4}He lambda point	2.176 8	
In superconducting transition point	3.414 5	
^{4}He boiling point	4.222 1	
Pb superconducting transition point	7.199 6	
*Triple point of equilibrium hydrogen	13.803 3	
*Boiling point of equilibrium hydrogen	17.035 7	
at a pressure of 33 330.6 pascals		
(25/76 standard atmosphere)		
*Boiling point of equilibrium hydrogen	20.271 1	
*Ne triple point	24.556 1	
Ne boiling point	27.098	
*O_2 triple point	54.358 4	−218.791 6
N_2 triple point	63.150	−210.000
N_2 boiling point	77.352	−195.798
*Ar triple point	83.805 8	−189.344 2
O_2 condensation point	90.196	−182.954
Kr triple point	115.776	−157.374
CO_2 sublimation point	194.685	−78.465
*Hg triple point	234.315 6	−38.834 4
H_2O freezing point	273.15	0

(*continued*)

Table 16.2   THE DEFINING FIXED POINTS OF ITS-90 AND SOME SELECTED SECONDARY
REFERENCE POINTS—*continued*

Equilibrium state	T_{90} (K)	t_{90} (°C)
*H_2O triple point	273.16	0.01
*Ga melting point	302.914 6	29.764 6
H_2O boiling point	373.124	99.974
*In freezing point	429.748 5	156.598 5
*Sn freezing point	505.078	231.928
Bi freezing point	544.553	271.403
Cd freezing point	594.219	321.069
Pb freezing point	600.612	327.462
*Zn freezing point	692.677	419.527
S boiling point	717.764	444.614
Cu–Al eutectic melting point	821.313	548.163
Sb freezing point	903.78	630.63
*Al freezing point	933.473	660.323
Ag–Cu eutectic melting point	1 053.09	779.94
*Ag freezing point	1 234.93	961.78
*Au freezing point	1 337.33	1 064.18
*Cu freezing point	1 357.77	1 084.62
Ni freezing point	1 728	1 455
Co freezing point	1 768	1 495
Pd freezing point	1 827	1 554
Pt freezing point	2 041	1 768
Rh freezing point	2 235	1 962
Ir freezing point	2 719	2 446
W freezing point	3 693	3 420

* Defining point of ITS-90. For details of experimental techniques used in the realisation of these fixed
points see references 2 and 3. All except the triple points and the hydrogen boiling point at 33 330.6 Pa
are at a pressure of 101 325 Pa (1 standard atmosphere).

16.2   Thermocouple reference tables

The introduction of ITS-90 on 1st January 1990 has led to an international programme, coordinated
by the Comité Consultatif de Thermométrie, for the revision of the internationally agreed reference
tables for thermocouples. It will take a few years, however, for the new tables to be adopted by
national and international standards organizations. Meanwhile the following old tables (BS 4937
and IEC 584-1 1977) based upon IPTS-68 should be used together with the differences $t_{90} - t_{68}$
given in Table 16.1 and Figure 16.1.

Table 16.3   THERMAL ELECTROMOTIVE FORCE (millivolts) OF ELEMENTS RELATIVE TO PLATINUM*

A positive sign means that in a simple thermoelectric circuit the element is positive to the platinum at the
reference junction (0°C). The e.m.f. generated by any two elements, A and B, can also be found from this table.
It is the algebraic difference, $A_e - B_e$, between the values (A_e and B_e) for the e.m.f. generated by each relative to
platinum; a positive sign indicates that A is positive to B at the reference junction, and a negative sign that A is
negative to B.

Cold junction at 0°C

	Temperature of hot junction °C			
	−200	−100	+100	+200
Aluminium	+0.45	−0.06	+0.42	+1.06
Antimony	—	—	+4.89	+10.14
Bismuth	+12.39	+7.54	−7.34	−13.57
Cadmium	−0.04	−0.31	+0.91	+2.32
Caesium	+0.22	−0.13	—	—
Calcium	—	—	−0.51	−1.13
Carbon	—	—	+0.70	+1.54

(*continued*)

Table 16.3 THERMAL ELECTROMOTIVE FORCE (millivolts) OF ELEMENTS RELATIVE TO PLATINUM*—*continued*

Cold junction at 0°C

	Temperature of hot junction °C			
	−200	−100	+100	+200
Cerium	—	—	+ 1.14	+ 2.46
Cobalt	—	—	− 1.33	− 3.08
Copper	− 0.19	− 0.37	+ 0.76	+ 1.83
Germanium	−46.00	−26.62	+33.9	+72.4
Gold	− 0.21	− 0.39	+ 0.74	+ 1.77
Indium	—	—	+ 0.69	—
Iridium	− 0.25	− 0.35	+ 0.65	+ 1.49
Iron	− 3.10	− 1.94	+ 1.98	+ 3.69
Lead	+ 0.24	− 0.13	+ 0.44	+ 1.09
Lithium	− 1.12	− 1.00	+ 1.82	—
Magnesium	+ 0.31	− 0.09	+ 0.44	+ 1.10
Mercury	—	—	+ 0.06	+ 0.13
Molybdenum	—	—	+ 1.45	+ 3.19
Nickel	+ 2.28	+ 1.22	− 1.48	− 3.10
Palladium	+ 0.81	+ 0.48	− 0.57	− 1.23
Potassium	+ 1.61	+ 0.78	—	—
Rhodium	− 0.20	− 0.34	+ 0.70	+ 1.61
Rubidium	+ 1.09	+ 0.46	—	—
Silicon	+63.13	+37.17	−41.56	−80.58
Silver	− 0.21	− 0.39	+ 0.74	+ 1.77
Sodium	+ 1.00	+ 0.29	—	—
Tantalum	+ 0.21	− 0.10	+ 0.33	+ 0.93
Thallium	—	—	+ 0.58	+ 1.30
Thorium	—	—	− 0.13	− 0.26
Tin	+ 0.26	− 0.12	+ 0.42	+ 1.07
Tungsten	+ 0.43	− 0.15	+ 1.12	+ 2.62
Zinc	− 0.07	− 0.33	+ 0.76	+ 1.89

*The numerical values given in this table should be taken only as a guide since thermoelectric properties are very sensitive to impurities and state of anneal.

Table 16.4 THERMAL ELECTROMOTIVE FORCE (millivolts) OF SOME BINARY ALLOYS RELATIVE TO PLATINUM WITH JUNCTIONS AT 0° and 100°C*

Metal A: Metal B: %A	Lead Tin	Tin Copper	Zinc Copper	Gold Silver	Gold Palladium	Nickel Copper	Tin Bismuth	Antimony Cadmium	Antimony Bismuth
0	+0.44	+0.76	+0.76	+0.74	−0.57	+0.76	−7.34	+0.90	−7.34
10	+0.44	+0.53	+0.54	+0.55	−0.85	−2.63	+4.00	+1.52	−8.82
20	+0.44	+0.56	+0.53	+0.48	−1.25	−3.08	+3.52	+2.88	−7.31
30	+0.44	+0.65	+0.54	+0.47	−1.42	−3.54	+2.56	+6.4	−5.66
40	+0.45	+0.65	+0.51	+0.47	−1.69	−4.03	+2.10	+12.2	−4.05
50	+0.45	+0.69	+0.54	+0.48	−2.44	−3.64	+1.77	+23.1	−2.51
60	+0.44	+0.72	+0.47	+0.49	−2.97	−3.06	+1.14	+44.4	−1.06
70	+0.44	+0.62	+0.87	+0.49	−2.63	−2.54	+0.95	+21.5	+0.32
80	+0.43	+0.54	+0.66	+0.50	−0.46	−2.49	+0.78	+12.8	+1.79
90	+0.42	+0.48	+0.98	+0.59	−0.05	−1.93	+0.60	+8.1	+3.31
100	+0.42	+0.42	+0.76	+0.78	+0.78	−1.48	+0.42	+4.89	+4.89

*The numerical values given in these tables should be taken only as a guide since thermoelectric properties are very sensitive to impurities and state of anneal.

Table 16.5 ABSOLUTE THERMOELECTRIC POWER OF PLATINUM

Temperature (K)	300	400	500	600	700	800	900	1 000	1 100	1 200
Thermoelectric power (μVK^{-1})	−5.05	−7.66	−9.69	−11.33	−12.87	−14.38	−15.97	−17.58	−19.03	−20.56

Table 16.6   PLATINUM–10% RHODIUM/PLATINUM THERMOCOUPLE TABLES—TYPE S

Reference junction at 0°C

t_{68}/°C	0	10	20	30	40	50	60	70	80	90	t_{68}/°C
					e.m.f./μV						
0	0	−53	−103	−150	−194	−236					0
0	0	55	113	173	235	299	365	432	502	573	0
100	645	719	795	872	950	1029	1109	1190	1273	1356	100
200	1440	1525	1611	1698	1785	1873	1962	2051	2141	2232	200
300	2323	2414	2506	2599	2692	2786	2880	2974	3069	3164	300
400	3260	3356	3452	3549	3645	3743	3840	3938	4036	4135	400
500	4234	4333	4432	4532	4632	4732	4832	4933	5034	5136	500
600	5237	5339	5442	5544	5648	5751	5855	5960	6064	6169	600
700	6274	6380	6486	6592	6699	6805	6913	7020	7128	7236	700
800	7345	7454	7563	7672	7782	7892	8003	8114	8225	8336	800
900	8448	8560	8673	8786	8899	9012	9126	9240	9355	9470	900
1000	9585	9700	9816	9932	10048	10165	10282	10400	10517	10635	1000
1100	10754	10872	10991	11110	11229	11348	11467	11587	11707	11827	1100
1200	11947	12067	12188	12308	12429	12550	12671	12792	12913	13034	1200
1300	13155	13276	13397	13519	13640	13761	13883	14004	14125	14247	1300
1400	14368	14489	14610	14731	14852	14973	15094	15215	15336	15456	1400
1500	15576	15697	15817	15937	16057	16176	16296	16415	16534	16653	1500
1600	16771	16890	17008	17125	17243	17360	17477	17594	17711	17826	1600
1700	17942	18056	18170	18282	18394	18504	18612				1700

Table 16.7   PLATINUM–13% RHODIUM/PLATINUM THERMOCOUPLE TABLES—TYPE R

Reference junction at 0°C

t_{68}/°C	0	10	20	30	40	50	60	70	80	90	t_{68}/°C
					e.m.f./μV						
0	0	−51	−100	−145	−188	−226					0
0	0	54	111	171	232	296	363	431	501	573	0
100	647	723	800	879	959	1041	1124	1208	1294	1380	100
200	1468	1557	1647	1738	1830	1923	2017	2111	2207	2303	200
300	2400	2498	2596	2695	2795	2896	2997	3099	3201	3304	300
400	3407	3511	3616	3721	3826	3933	4039	4146	4254	4362	400
500	4471	4580	4689	4799	4910	5021	5132	5244	5356	5469	500
600	5582	5696	5810	5925	6040	6155	6272	6388	6505	6623	600
700	6741	6860	6979	7098	7218	7339	7460	7582	7703	7826	700
800	7949	8072	8196	8320	8445	8570	8696	8822	8949	9076	800
900	9203	9331	9460	9589	9718	9848	9978	10109	10240	10371	900
1000	10503	10636	10768	10902	11035	11170	11304	11439	11574	11710	1000
1100	11846	11983	12119	12257	12394	12532	12669	12808	12946	13085	1100
1200	13224	13363	13502	13642	13782	13922	14062	14202	14343	14483	1200
1300	14624	14765	14906	15047	15188	15329	15470	15611	15752	15893	1300
1400	16035	16176	16317	16458	16599	16741	16882	17022	17163	17304	1400
1500	17445	17585	17726	17866	18006	18146	18286	18425	18564	18703	1500
1600	18842	18981	19119	19257	19395	19533	19670	19807	19944	20080	1600
1700	20215	20350	20483	20616	20748	20878	21006				1700

Table 16.8   PLATINUM–30% RHODIUM/PLATINUM–6% RHODIUM THERMOCOUPLE TABLES—TYPE B

Reference junction at 0°C

t_{68}/°C	0	10	20	30	40	50	60	70	80	90	t_{68}/°C
					e.m.f./μV						
0	0	−2	−3	−2	−0	2	6	11	17	25	0
100	33	43	53	65	78	92	107	123	140	159	100
200	178	199	220	243	266	291	317	344	372	401	200
300	431	462	494	527	561	596	632	669	707	746	300
400	786	827	870	913	957	1002	1048	1095	1143	1192	400

(continued)

Table 16.8 PLATINUM–30% RHODIUM/PLATINUM–6% RHODIUM THERMOCOUPLE
TABLES—TYPE B—*continued*

Reference junction at 0°C

t_{68}/°C	0	10	20	30	40	50	60	70	80	90	t_{68}/°C
					e.m.f./μV						
500	1 241	1 292	1 344	1 397	1 450	1 505	1 560	1 617	1 674	1 732	500
600	1 791	1 851	1 912	1 974	2 036	2 100	2 164	2 230	2 296	2 363	600
700	2 430	2 499	2 569	2 639	2 710	2 782	2 855	2 928	3 003	3 078	700
800	3 154	3 231	3 308	3 387	3 466	3 546	3 626	3 708	3 790	3 873	800
900	3 957	4 041	4 126	4 212	4 298	4 386	4 474	4 562	4 652	4 742	900
1 000	4 833	4 924	5 016	5 109	5 202	5 297	5 391	5 487	5 583	4 680	1 000
1 100	5 777	5 875	5 973	6 073	6 172	6 273	6 374	6 475	6 577	6 680	1 100
1 200	6 783	6 887	6 991	7 096	7 202	7 308	7 414	7 521	7 628	7 736	1 200
1 300	7 845	7 953	8 063	8 172	8 283	8 393	8 504	8 616	8 727	8 839	1 300
1 400	8 952	9 065	9 178	9 291	9 405	9 519	9 634	9 748	9 863	9 979	1 400
1 500	10 094	10 210	10 325	10 441	10 558	10 674	10 790	10 907	11 024	11 141	1 500
1 600	11 257	11 374	11 491	11 608	11 725	11 842	11 959	12 076	12 193	12 310	1 600
1 700	12 426	12 543	12 659	12 776	12 892	13 008	13 124	13 239	13 354	13 470	1 700
1 800	13 585	13 699	13 814								

Table 16.9 NICKEL–CHROMIUM/COPPER–NICKEL THERMOCOUPLE—TYPE E
(Chrome–Constantan)

Reference junction at 0°C

t_{68}/°C	0	10	20	30	40	50	60	70	80	90	t_{68}/°C
					e.m.f./μV						
−200	−8 824	−9 063	−9 274	−9.455	−9 604	−9 719	−9 797	−9 835			−200
−100	−5 237	−5 680	−6 107	−6 516	−6 907	−7 279	−7 631	−7 963	−8 273	−8 561	−100
0	0	−581	−1 151	−1 709	−2 254	−2 787	−3 306	−3 811	−4 301	−4 777	0
0	0	591	1 192	1 801	2 419	3 047	3 683	4 329	4 983	5 646	0
100	6 317	6 996	7 683	8 377	9 078	9 787	10 501	11 222	11 949	12 681	100
200	13 419	14 161	14 909	15 661	16 417	17 178	17 942	18 710	19 481	20 256	200
300	21 033	21 814	22 597	23 383	24 171	24 961	25 754	26 549	27 345	28 143	300
400	28 943	29 744	30 546	31 350	32 155	32 960	33 767	34 574	35 382	36 190	400
500	36 999	37 808	38 617	39 426	40 236	41 045	41 853	42 662	43 470	44 278	500
600	45 085	45 891	46 697	47 502	48 306	49 109	49 911	50 713	51 513	52 312	600
700	53 110	53 907	54 703	55 498	56 291	57 083	57 873	58 663	59 451	60 237	700
800	61 022	61 806	62 588	63 368	64 147	64 924	65 700	66 473	67 245	68 015	800
900	68 783	69 549	70 313	71 075	71 835	72 593	73 350	74 104	74 857	75 608	900
1 000	76 358										

Table 16.10 IRON COPPER–NICKEL THERMOCOUPLE TABLES—TYPE J
(Iron–Constantan)

Reference junction at 0°C

t_{68}/°C	0	10	20	30	40	50	60	70	80	90	t_{68}/°C
					e.m.f./μV						
−200	−7 890	−8 096									−200
−100	−4 632	−5 036	−5 426	−5 801	−6 159	−6 499	−6 821	−7 122	−7 402	−7 659	−100
0	0	−501	−995	−1 481	−1 960	−2 431	−2 892	−3 344	−3 785	−4 215	0
0	0	507	1 019	1 536	2 058	2 585	3 115	3 649	4 186	4 725	0
100	5 268	5 812	6 359	6 907	7 457	8 008	8 560	9 113	9 667	10 222	100
200	10 777	11 332	11 887	12 442	12 998	13 353	14 108	14 663	15 217	15 771	200
300	16 325	16 879	17 432	17 984	18 537	19 089	19 640	20 192	20 743	21 295	300
400	21 846	22 397	22 949	23 501	24 054	24 607	25 161	25 716	26 272	26 829	400

(*continued*)

Table 16.10   IRON COPPER–NICKEL THERMOCOUPLE TABLES—TYPE J—*continued*

Reference junction at 0°C

$t_{68}/°C$	0	10	20	30	40	50	60	70	80	90	$t_{68}/°C$
					e.m.f./μV						
500	27 388	27 949	28 511	29 075	29 642	30 210	30 782	31 356	31 933	32 513	500
600	33 096	33 683	34 273	34 867	35 464	36 066	36 671	37 280	37 893	38 510	600
700	39 130	39 754	40 382	41 013	41 647	42 283	42 922	43 563	44 207	44 852	700
800	45 498	46 144	46 790	47 434	48 076	48 716	49 354	49 989	50 621	51 249	800
900	51 875	52 496	53 115	53 729	54 341	54 948	55 553	56 155	56 753	57 349	900
1 000	57 942	58 533	59 121	59 708	60 293	60 876	61 459	62 039	62 619	63 199	1 000
1 100	63 777	64 355	64 933	65 510	66 087	66 664	67 240	67 815	68 390	68 964	1 100
1 200	69 536										

Table 16.11   COPPER/COPPER–NICKEL THERMOCOUPLE TABLES—TYPE T
(Copper–Constantan)

Reference junction at 0°C

$t_{68}/°C$	0	10	20	30	40	50	60	70	80	90	$t_{68}/°C$
					e.m.f./μV						
−200	−5 603	−5 753	−5 889	−6 007	−6 105	−6 181	−6 232	−6 258			−200
−100	−3 378	−3 656	−3 923	−4 177	−4 419	−4 648	−4 865	−5 069	−5 261	−5 439	−100
0	0	−383	−757	−1 121	−1 475	−1 819	−2 152	−2 475	−2 788	−3 089	0
0	0	391	789	1 196	1 611	2 035	2 467	2 908	3 357	3 813	0
100	4 277	4 749	5 227	5 712	6 204	6 702	7 207	7 718	8 235	8 757	100
200	9 286	9 820	10 360	10 905	11 456	12 011	12 572	13 137	13 707	14 281	200
300	14 860	15 443	16 030	16 621	17 217	17 816	18 420	19 027	19 638	20 252	300
400	20 869										

Table 16.12   NICKEL–CHROMIUM/NICKEL–ALUMINIUM THERMOCOUPLE TABLES—TYPE K
(Chromel–Alumel)

Reference junction at 0°C

$t_{68}/°C$	0	10	20	30	40	50	60	70	80	90	$t_{68}/°C$
					e.m.f./μV						
−200	−5 891	−6 035	−6 158	−6 262	−6 344	−6 404	−6 441	−6 458			−200
−100	−3 553	−3 852	−4 138	−4 410	−4 669	−4 912	−5 141	−5 354	−5 550	−5 730	−100
0	0	−392	−777	−1 156	−1 527	−1 889	−2 243	2 586	−2 920	−3 242	0
0	0	397	798	1 203	1 611	2 022	2 436	2 850	3 266	3 681	0
100	4 095	4 508	4 919	5 327	5 733	6 137	6 539	6 939	7 338	7 737	100
200	8 137	8 537	8 938	9 341	9 745	10 151	10 560	10 969	11 381	11 793	200
300	12 207	12 623	13 039	13 456	13 874	14 292	14 712	15 132	15 552	15 974	300
400	16 395	16 818	17 241	17 664	18 088	18 513	18 938	19 363	19 788	20 214	400
500	20 640	21 066	21 493	21 919	22 346	22 772	23 198	23 624	24 050	24 476	500
600	24 902	25 327	25 751	26 176	26 599	27 022	27 445	27 867	28 288	28 709	600
700	29 128	29 547	29 965	30 383	30 799	31 214	31 629	32 042	32 455	32 866	700
800	32 277	33 686	34 095	34 502	34 909	35 314	35 718	36 121	36 524	36 925	800
900	37 325	37 724	38 122	38 519	38 915	39 310	39 703	40 096	40 488	40 879	900
1 000	41 269	41 657	42 045	42 432	42 817	43 202	43 585	43 968	44 349	44 729	1 000
1 100	45 108	45 486	45 863	46 238	46 612	46 985	47 356	47 726	48 095	48 462	1 100
1 200	48 828	49 192	49 555	49 916	50 276	50 633	50 990	51 344	51 697	52 049	1 200
1 300	52 398	52 747	53 093	53 439	53 782	54 125	54 466	54 807			1 300

Table 16.13 NICKEL–CHROMIUM–SILICON/NICKEL–SILICON (NICROSYL/NISIL) THERMOCOUPLE TABLES—TYPE N

Reference junction at 0°C

$t_{68}/°C$	0	10	20	30	40	50	60	70	80	90	$t_{68}/°C$
					e.m.f./μV						
−200	−3 990	−4 083	−4 162	−4 227	−4 277	−4 313	−4 336	−4 345			−200
−100	−2 407	−2 612	−2 808	−2 994	−3 170	−3 336	−3 491	−3 634	−3 766	−3 884	−100
0	0	−260	−518	−772	−1 023	−1 268	−1 509	−1 744	−1 972	−2 193	0
0	0	261	525	793	1 064	1 340	1 619	1 902	2 188	2 479	0
100	2 774	3 072	3 374	3 679	3 988	4 301	4 617	4 936	5 258	5 584	100
200	5 912	6 243	6 577	6 914	7 254	7 596	7 940	8 287	8 636	8 987	200
300	9 340	9 695	10 053	10 412	10 773	11 135	11 499	11 865	12 233	12 602	300
400	12 972	13 344	13 717	14 092	14 467	14 844	15 222	15 601	15 981	16 362	400
500	16 744	17 127	17 511	17 896	18 282	18 668	19 055	19 443	19 831	20 220	500
600	20 609	20 999	21 390	21 781	22 172	22 564	22 956	23 348	23 740	24 133	600
700	24 526	24 919	25 312	25 705	26 098	26 491	26 885	27 278	27 671	28 063	700
800	28 456	28 849	29 241	29 633	30 025	30 417	30 808	31 199	31 590	31 980	800
900	32 370	32 760	33 149	33 538	33 927	34 315	34 702	35 089	35 476	35 862	900
1 000	36 248	36 633	37 018	37 403	37 786	38 169	38 552	38 934	39 316	39 696	1 000
1 100	40 076	40 456	40 835	41 213	41 590	41 966	42 342	42 717	43 091	43 464	1 100
1 200	43 836	44 207	44 578	44 947	45 315	45 682	46 048	46 413	46 777	47 140	1 200
1 300	47 502										

Table 16.14 TUNGSTEN–5% RHENIUM/TUNGSTEN–26% RHENIUM—TYPE C[4]

Reference junction at 0°C

$t_{68}/°C$	0	10	20	30	40	50	60	70	80	90	$t_{68}/°C$
					e.m.f./μV						
0	0	135	272	412	554	698	845	993	1 144	1 296	0
100	1 451	1 607	1 765	1 925	2 087	2 250	2 415	2 581	2 749	2 918	100
200	3 089	3 261	3 434	3 609	3 785	3 962	4 140	4 319	4 500	4 681	200
300	4 863	5 047	5 231	5 416	5 601	5 788	5 975	6 163	6 352	6 541	300
400	6 731	6 921	7 112	7 304	7 496	7 688	7 881	8 074	8 267	8 461	400
500	8 655	8 849	9 044	9 239	9 434	9 629	9 824	10 019	10 215	10 410	500
600	10 606	10 801	10 997	11 192	11 388	11 583	11 778	11 974	12 169	12 364	600
700	12 558	12 753	12 947	13 142	13 336	13 529	13 723	13 916	14 109	14 302	700
800	14 494	14 686	14 877	15 069	15 260	15 450	15 640	15 830	16 020	16 208	800
900	16 397	16 585	16 773	16 960	17 147	17 333	17 519	17 704	17 889	18 073	900
1 000	18 257	18 440	18 623	18 805	18 987	19 168	19 349	19 529	19 709	19 888	1 000
1 100	20 066	20 244	20 421	20 598	20 774	20 950	21 125	21 299	21 473	21 647	1 100
1 200	21 819	21 991	22 163	22 334	22 504	22 674	22 843	23 012	23 180	23 347	1 200
1 300	23 514	23 680	23 846	24 010	24 175	24 339	24 502	24 664	24 826	24 988	1 300
1 400	25 148	25 308	25 468	25 627	25 785	25 943	26 100	26 256	26 412	26 568	1 400
1 500	26 722	26 876	27 030	27 183	27 335	27 486	27 637	27 788	27 938	28 087	1 500
1 600	28 236	28 384	28 531	28 678	28 824	28 969	29 114	29 259	29 402	29 546	1 600
1 700	29 688	29 830	29 971	30 112	30 252	30 391	30 530	30 668	30 805	30 942	1 700
1 800	31 078	31 214	31 349	31 483	31 617	31 749	31 882	32 013	32 144	32 274	1 800
1 900	32 404	32 533	32 661	32 788	32 915	33 041	33 166	33 291	33 415	33 538	1 900
2 000	33 660	33 782	33 902	34 022	34 142	34 260	34 378	34 494	34 610	34 725	2 000
2 100	34 839	34 953	35 065	35 177	35 288	35 397	35 506	35 614	35 721	35 827	2 100
2 200	35 932	36 036	36 138	36 240	36 341	36 441	36 539	36 637	36 733	36 828	2 200
2 300	36 922	37 015	37 107	37 197	37 286	37 374	37 460	37 545	37 629	37 711	2 300

REFERENCES

1. The International Temperature Scale of 1990, *Metrologia*, 1990, **27**, 3–10 and 167.
2. 'Supplementary Information for the ITS-90 and Techniques for Approximating the ITS-90', BIPM, Pavillion de Breteuil, F-92312 Sèvres Celex, France, 1990.
3. T. J. Quinn, 'Temperature', 2nd edn, Academic Press, London, 1990.
4. Values calculated using coefficients from: P. C. Fazio *et al.*, Annual Book of ASTM Standards, Volume 14.03, Temperature Measurement, ASTM, Philadelphia, PA, 1993.

16.3 Thermoelectric materials

16.3.1 Introduction

Thermoelectric devices convert thermal energy from a temperature gradient into electrical energy (Seebeck effect), or electrical energy into a temperature gradient (Peltier effect).[1,2] Thermoelectric materials are evaluated on the basis of their thermoelectric figure-of-merit, Z, which is defined by:

$$Z = \frac{\sigma S^2}{\kappa}$$

In this equation, σ is the electrical conductivity ($\Omega^{-1}\,\mathrm{m}^{-1}$), S is the Seebeck coefficient (VK^{-1}), and κ is the thermal conductivity ($\mathrm{Wm}^{-1}\,\mathrm{K}^{-1}$). Z has units of reciprocal temperature, and the product ZT is referred to as the dimensionless figure-of-merit. Good thermoelectric materials have large values of Z (or ZT). From a materials perspective, one seeks to maximise σ and S while minimising κ.

The Seebeck coefficient and the electrical conductivity are strong functions of the Fermi energy, which, in turn, depends on the free carrier concentration, the carrier effective mass and temperature. The Seebeck coefficient of most metals is typically too small for them to have any practical application beyond a simple thermocouple. On the other hand, many semiconductors have Seebeck coefficients on the order of hundreds of μV/K. Good thermoelectric materials must also have a reasonable electrical conductivity since charge transport is the basis by which heat is transferred, and resistive heating will lower the overall efficiency. In addition, the material must be a good thermal insulator to maintain the thermal gradient established by the motion of charge carriers. The relationships among the transport properties that influence the figure-of-merit are shown in Figure 16.2.

The dependence of Z on n is the basis of the general rule of thumb suggested by Ioffe that the best thermoelectric materials will be relatively highly doped semiconductors ($n \sim 10^{20}\,\mathrm{cm}^{-3}$).[3] The effect of doping is limited in practice by the fact that although the lattice part of the thermal conductivity is not very sensitive to the free carrier concentration, n, the electronic component clearly is. It is also well known that there is a correlation between the band gap and Z for many materials. The band gap rule states that $\mathrm{E_G} \sim 10\,\mathrm{k_B T_M}$, where $\mathrm{T_M}$ is the temperature where ZT is a maximum and $\mathrm{k_B}$ is Boltzmann's constant.[4] The choice of majority carrier is important in determining the sign of the Peltier and Seebeck coefficients.

In a recent review, Mahan offered several criteria for good thermoelectric materials.[5] In terms of their band structures, good thermoelectric materials will have multiple degenerate bands, high carrier mobilities, and large carrier effective masses. In addition, good thermoelectric materials must have a low thermal conductivity. There are three strategies to minimise the thermal conductivity: the use of heavy atoms, complex crystal structures and alloys. In general, heavy elements have lower thermal conductivities than lighter elements, heat is transferred less efficiently in crystals with a higher density of optical phonons, and alloying reduces the lattice thermal conductivity. Alloying also lowers the charge mobility, but the effect is greater for the lattice thermal conductivity.

Materials used today in thermoelectric devices can be grouped into three categories based on the device operation temperature: the Group V chalcogenides based on Bi_2Te_3-Sb_2Te_3 alloys for refrigeration (ca. 300 K); the Group IV chalcogenides based on PbTe for power generation (700 K); and Si-Ge solid solutions for power generation at high temperature (1 100–1 200 K). Examples of the best thermoelectric materials in these temperature ranges are given in Table 16.15.

Generally, thermoelectric devices suffer from poor efficiency. For example, current thermoelectric refrigerators are about one third as efficient as their conventional (Freon based) analogues. As a result, thermoelectric devices tend to be important in niche applications where the stability and long life afforded by an all solid-state device outweighs the efficiency issue.

16.3.2 Survey of materials

Bi_2Te_3 is composed of hexagonal, closest-packed Te layers with Bi atoms occupying two-thirds of the octahedral holes. The crystal consists of a sequence of atomic layers Te-Bi-Te-Bi-Te. Three such stacks sandwiched together along the c-axis of the crystal form the hexagonal unit cell, which repeats every nine Te layers. The band gap of Bi_2Te_3 is 0.15 eV, and increases upon doping. Both p- and n-type materials can be prepared by alloying with compounds such as Sb_2Te_3 or Bi_2Se_3, which have the same crystal structure as Bi_2Te_3. Bi_2Te_3 is the most important thermoelectric material for cooling applications at temperatures near 300 K. All devices that operate near room temperature are based on Bi_2Te_3 and its alloys. Alloys are used in order to lower the thermal conductivity and thereby increase ZT.

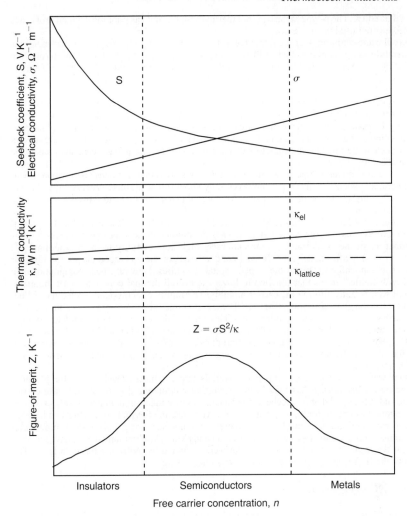

Figure 16.2   *Schematic depiction of the dependence of the Seebeck coefficient, electrical conductivity, thermal conductivity, and thermoelectric figure-of-merit on the free carrier concentration*

Table 16.15   THERMOELECTRIC MATERIALS FOR REFRIGERATION AND POWER GENERATION

Material	Type	$T(K)$	ZT
$Bi_{24}Sb_{68}Te_{142}Se_6$	p	300	0.96
$Bi_{0.5}Sb_{1.5}Te_{3.13}$	p	300	0.90
$Bi_{1.75}Sb_{0.25}Te_{3.13}$	n	300	0.96
PbTe	n	700	1.10
SiGe	p	1 100	0.505
SiGe	n	1 200	0.938

PbTe has the NaCl crystal structure and a band gap of 0.31 eV at room temperature. PbTe and its alloys are good thermoelectrics near 700 K and are used in power generation applications. PbTe frequently is alloyed with Sn to form $Pb_{1-x}Sn_xTe$, which is a good n-type thermoelectric with a ZT

of about 1.1 at 800 K. The p-type material is not thermally stable and has poor mechanical properties and therefore is not used in any devices.

TAGS (tellurium-antimony-germanium-silver) alloys make good p-type thermoelectric materials in a temperature range similar to that of PbTe (which is n-type), and they are often used together. In fact, TAGS were developed as a direct result of the search for a replacement for p-PbTe. TAGS is an alloy of GeTe with $AgSbTe_2$ (10–20%). The 20% alloy has a $ZT > 2$, which is among the highest ever reported. The measured value of Z was $3 \times 10^{-3} \, K^{-1}$ at $T = 700 \, K$.[6]

The transport properties of Si make it an excellent choice for high temperature power generation, although its thermal conductivity is too high. Typically, Si is alloyed with Ge to lower the thermal conductivity. Overall, Si-Ge solid solutions are the best materials for high temperature applications. The composition that gives a maximum ZT is approximately $Si_{0.7}Ge_{0.3}$. These materials are used in the thermonuclear thermoelectric generators that supply power for deep-space vehicles such as the Voyager spacecraft.

Bi-Sb is the only material other than Bi_2Te_3 that displays good thermoelectric performance near room temperature. The ZT is fairly low, however, only about 0.6 near 300 K. Interestingly, ZT can be increased by almost a factor of 2 by placing the material in a magnetic field. This is known as the thermomagnetic effect.

The skutterudites have the formula MX_3 where $M = Co$, Rh, Ir and $X = P$, As, Sb. The parent compound, $CoSb_3$, was first discovered in the town of Skutterude in Norway. One of the most interesting properties of the skutterudites is the extremely high hole mobility in the p-type materials (ca. $2\,000 \, cm^2/Vs$). Unfortunately, the thermal conductivities of these compounds are too high for thermoelectric applications. Due to their open crystal structures, however, these compounds may be doped with 'rattler' atoms that function to lower the overall thermal conductivity of the material without significantly altering the carrier mobility. The family of compounds called the 'filled' skutterudites has the general formula RM_4X_{12}. These compounds are cubic with a 34-atom unit cell. The rare earth atom (R) acts as the 'rattler' atom. These materials are promising candidates for high temperature applications. For example, Sales *et al.* report a value of ZT of about 1 for $LaFe_3CoSb_{12}$ at $T \sim 800 \, K$.[7] For comparison, the ordinary skutterudites have ZTs of about 0.3. The 3-fold increase in ZT in the filled skutterudites is attributed to the lowering of the thermal conductivity due to the lanthanum 'rattler' atom.

Quantum wells are interesting composite materials that are under development for potential thermoelectric applications. It has long been known that modulation doping can increase carrier mobilities, and it is hoped that the Seebeck coefficient can be enhanced in quantum well structures in an analogous fashion. In multiple quantum well structures, the thermal conductivity is lowered due to the presence of multiple interfaces that increase phonon scattering. This is an example of an 'electron transmitting/phonon blocking' structure. Recently Venkatasubramanian *et al.* have reported a ZT of 2.4 for Bi_2Te_3/Sb_2Te_3 multiple quantum wells prepared by molecular beam epitaxy.[8] The thickness of the Bi_2Te_3 layers in these materials was only 1.0 nm.

16.3.3   Preparation methods

Most thermoelectric materials are either compounds or solid solutions, and are easily prepared using standard solid-state synthesis procedures.[9] A variety of melt techniques, including both stoichiometric and non-stoichiometric melts, are commonly used, typical examples being the Bridgman method and liquid phase eptitaxy, respectively. Powder metallurgy, in which a compact powder is sintered at elevated temperature, is also often used. Other synthetic approaches, including mechanical alloying and PIES (pulverised and intermixed sintering method) have also been developed. For the fabrication of thin film structures, molecular beam epitaxy, liquid phase epitaxy as well as electrochemical deposition all can be used.

REFERENCES

1. T. J. Seebeck, Abh. K. Akad. Wiss. Berlin, 1821, 289.
2. J. C. A. Peltier, *Ann. Chem. Phys.*, 1834, **56**, 371.
3. A. F. Ioffe, Semiconductor Thermoelements and Thermoelectric Cooling, Infosearch, Ltd, London, 1957.
4. G. D. Mahan, *J. Appl. Phys.*, 1989, **65**, 1578.
5. G. D. Mahan, *Solid State Physics*, 1998, **51**, 81.
6. S. K. Plachkova, *Phys. Stat. Solid. A*, 1984, **83**, 349.
7. B. C. Sales, D. Mandrus and R. K. Williams, *Science*, 1996, **272**, 1325.
8. R. Venkatasubramanian, E. Siivola, T. Colpitts and B. O'Quinn, *Nature*, 2001, **413**, 597.
9. D. M. Rowe, Ed., CRC Handbook of Thermoelectrics, CRC Press, New York, 1995.

17 Radiative properties of metals

The ability of a surface to radiate energy is governed by the material of which the surface is composed and its physical condition. Any attempt, therefore, to place a numerical value on its radiating ability must be related to a definition of the surface condition. It is usual to choose smooth polished surfaces for this purpose and thus arrive at values which are comparable from one metal to another.

A perfect radiator (blackbody) provides a standard of comparison for defining the radiating ability of any other body or surface by determining the ratio of the emission of the surface to that of a blackbody when they are at the same temperature. An examination of the ratios thus obtained shows that the radiating ability of a metal surface varies with wavelength, temperature and angle of emission. The definition of the emittance, as this ratio is called, must therefore take into account these variations.

DEFINITIONS OF EMITTANCE

Spectral, directional emittance, ε_λ, of a surface is the ratio of the energy emitted over an infinitesimally small wavelength range at wavelength λ into a specified direction, per unit area of the surface, to the energy emitted by a unit area of a black surface at the same temperature. The emittance in a direction normal to the surface, called the *normal, spectral emittance*, $\varepsilon_{n\lambda}$, is most commonly employed.

The *spectral, hemispherical emittance*, ε_λ, is a directional average of the spectral, directional emittance, and gives the ratio of emitted radiative flux of the given surface to that of a blackbody at the same temperature. While normal emittances are the ones that are generally measured and tabulated, it is the hemispherical emittance that governs heat transfer rates between surfaces. For metals, the hemispherical emittance is always a little greater than its normal value, usually by about 10 to 20 percent.

Total emittance is a weighted spectral average of the spectral emittance, and is the ratio of radiative emission of a surface over the entire spectrum, compared that of a blackbody at the same temperature. Again, we distinguish between directional and hemispherical values; and again, the more important hemispherical values exceed the total, normal emittance by roughly 10 to 20 percent for metals.

The absorptance (α) and reflectance (ρ), for an opaque polished surface, are defined as the ratio of the rate of absorption or reflection of energy to the rate of incidence of radiative energy. Since the incident energy must be either reflected or absorbed, the sum of reflectance and absorptance must be unity. For spectral, directional values the absorptance is equal to the emittance and, with reasonable accuracy, this is often also true for hemispherical and/or total values.[1] Thus,

$$\varepsilon \cong \alpha = 1 - \rho$$

Hence the emittance may be derived from the reflectance, which is sometimes more convenient than a direct determination.

The following equation relates the normal reflectance of a well-polished surface to the *complex index of refraction*, given by $m = n - ik$, where n is the real part ('refractive index') and k is the complex part ('absorptive index' or 'extinction coefficient'):

$$\rho_{n\lambda} = \frac{n^2 + k^2 + 1 - 2n}{n^2 + k^2 + 1 + 2n}$$

where the absorptive index is related to the fraction of light transmitted perpendicularly through a layer of thickness d by $\exp(-4\pi dk/\lambda)$, λ being the wavelength in air.

Normal emittance follows as

$$\varepsilon_{n\lambda} = 1 - \rho_{n\lambda} = \frac{4n}{n^2 + k^2 + 1 + 2n}$$

For a homogeneous material the complex index of refraction can be related to material properties through Maxwell's electromagnetic wave theory,[1]

$$n^2 = \frac{1}{2}\left[\sqrt{\left(\frac{\varepsilon}{\varepsilon_0}\right)^2 + \left(\frac{\sigma\lambda}{2\pi c_0\varepsilon_0}\right)^2} + \frac{\varepsilon}{\varepsilon_0}\right]$$

$$k^2 = \frac{1}{2}\left[\sqrt{\left(\frac{\varepsilon}{\varepsilon_0}\right)^2 + \left(\frac{\sigma\lambda}{2\pi c_0\varepsilon_0}\right)^2} - \frac{\varepsilon}{\varepsilon_0}\right]$$

where σ is the electrical conductivity, ε and ε_0 are the electrical permittivity of the material and of vacuum, respectively, c_0 is the speed of light and λ the wavelength, both in vacuum or air. For metals with their large electrical conductivities, and for reasonably large wavelengths, $\sigma\lambda \gg 2\pi c_0\varepsilon$, leading to

$$n \approx k \approx \sqrt{\frac{\sigma\lambda}{2\pi c_0\varepsilon_0}} \gg 1$$

and

$$\varepsilon_{n\lambda} \cong \frac{0.365}{\sqrt{\sigma\lambda}} - \frac{0.0667}{\sigma\lambda}, \quad \sigma\lambda \text{ in } \Omega^{-1}, \tag{17.1}$$

which is commonly known as the Hagen-Rubens relation.[1]

The equations show that the spectral emittance of metals should increase as the wavelength decreases and this is in general agreement with experiment. However, these relations are valid only for sufficiently large wavelengths and, indeed, there generally is a lack of agreement between experimental and theoretical values in the visible and ultraviolet (see Fig. 17.1 for tungsten).

EFFECTS OF SURFACE TEMPERATURE

The Hagen-Rubens relation predicts that the spectral, normal emittance of a metal should be proportional to $1/\sqrt{\sigma}$. Since the electrical conductivity is approximately inversely proportional to temperature, the spectral emittance should, therefore, be proportional to the square root of absolute temperature for long enough wavelengths. This trend should also hold for the *spectral, hemispherical emittance*. Experiments have shown that this is indeed true for many metals. A typical example is given in Fig. 17.1, showing the spectral dependence of the hemispherical emittance for tungsten for a number of temperatures.[2,3] Note that the emittance for tungsten tends to increase with temperature only beyond a *crossover wavelength* of approximately 1.3 μm, while the temperature dependence is reversed for shorter wavelengths. Similar trends of a single crossover wavelength have been observed for many metals.

DIRECTIONAL DEPENDENCE OF EMITTANCE

The spectral, directional reflectance of a perfectly smooth surface is governed by the so-called *Fresnel's relations*.[1] For metals in the infrared this implies relatively small and constant values of emittance for most angles of emission, accompanied by a sharp increase in emittance at near-grazing angles, followed by a sharp drop back to zero at 90° off-normal. As an example, experimental results for platinum at 2 μm are compared with Fresnel predictions in Fig. 17.2.[1] For shorter wavelengths the directional behaviour follows the behaviour of nonmetals, i.e. relatively large and constant values of emittance for most angles of emission, accompanied by a gradual decrease beyond about 60° off-normal until zero emittance is reached at 90°.

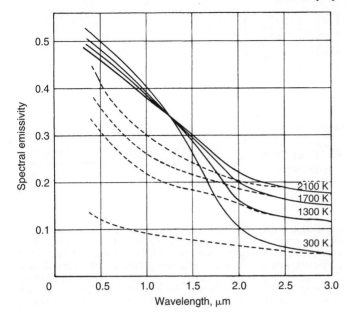

Figure 17.1 *Spectral emittance of tungsten as a function of wavelength for different temperatures. Dotted lines calculated from equation 17.1. References 2 and 3*

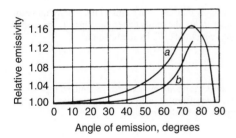

Figure 17.2 *Spectral, directional emittance of platinum at $\lambda = 2 \, \mu$m*

TOTAL EMITTANCES

The total, normal or hemispherical emittances are calculated by integrating spectral values over all wavelengths, with the blackbody emissive power as a weight function. Since the peak of the blackbody emissive power shifts toward shorter wavelengths with increasing temperature, it follows that hotter surfaces emit a higher fraction of energy at shorter wavelengths, where the spectral emittance is greater, resulting in an increase in total emittance.

The total, normal reflectance and emittance may be evaluated from the simple Hagen-Rubens relation. While this relation is not accurate across the entire spectrum, it does predict the emittance *trends* correctly in the infrared, and it does allow an explicit evaluation of total, normal emittance. Using Equation (17.1) in the weighted spectral averaging process and retaining the first two terms of the series expansion, leads to

$$\varepsilon_n = 0.578(T/\sigma)^{1/2} - 0.178(T/\sigma), \quad T \text{ in K}, \quad \sigma \text{ in } \Omega^{-1}\text{cm}^{-1} \tag{17.2}$$

The total, hemispherical emittance of a metal may be evaluated in a similar fashion, using Fresnel's relations for directional dependence, leading to slightly larger values, or[1]

$$\varepsilon(T) = 0.766(T/\sigma)^{1/2} - [0.309 - 0.089\,9\ln(T/\sigma)](T/\sigma), \quad T \text{ in K}, \quad \sigma \text{ in } \Omega^{-1}\text{cm}^{-1} \tag{17.3}$$

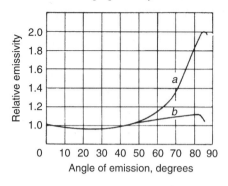

Figure 17.3   *Total, hemispherical emittance of various polished metals as function of temperature. Reference 4*

Because the Hagen-Rubens relation is only valid for large values of $(\sigma\lambda)$, this relation only holds for small values of (T/σ), i.e. the temperature of the surface must be such that only a small fraction of the blackbody emissive power comes from short wavelengths (where the Hagen-Rubens relation is not applicable). For pure metals, with its electrical conductivity inversely proportional to absolute temperature, the total emittance should be approximately linearly proportional to temperature. Comparison with experiment (Fig. 17.3)[4] shows that this nearly linear relationship holds for many metals up to surprisingly high temperatures; for example, for platinum $(T/\sigma)^{1/2} = 0.5$ corresponds to a temperature of 2700 K.

TEMPERATURE MEASUREMENT AND EMITTANCE

Radiation pyrometers, both spectral and total, are usually calibrated in terms of blackbody radiation and, thus, measure what is known as *radiance temperature*. The radiance temperature, T_r, measured by a total radiation pyrometer is related to the *true temperature*, T, by the formula

$$T = T_r/\varepsilon^{1/4}$$

where ε is the total emittance of the surface. True temperature *always* exceeds radiance temperature.
    For an optical pyrometer which measures irradiance from a narrow spectral interval only, the radiance temperature T_r is related to the true temperature by the equation

$$\frac{1}{T} = \frac{1}{T_r} - \frac{\lambda}{C_2} \ln \frac{1}{\varepsilon_\lambda}$$

where ε_λ is the spectral emittance and $C_2 = 1.438\,8$ cm K is the constant in Wien's approximation for the blackbody emissive power. For the same emittances the correction is considerably greater for total radiation than for spectral radiation. For metallic surfaces the difference in correction is even greater, since the spectral emittance in the visible region for a given temperature is always greater than the total emittance.

EMITTANCE VALUES

The values of emittance given in Tables 17.1 to 17.5 pertain, as far as is known, to emission from plane polished or plane unoxidised liquid metal surfaces. Emittance tends to be considerably higher if the surface is oxidised and/or rough. In any practical application, therefore, the values in the tables must be used with discretion and, where precise measurement is of importance, a determination of the emittance should be made for the prevalent surface conditions. As a rough guide the emittances for various oxidised surfaces are given in Tables 17.4 and 17.5.

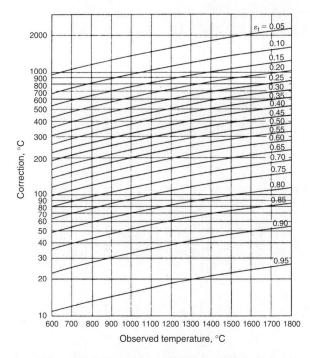

Figure 17.4 *Correction to radiation pyrometer readings for total emittance*

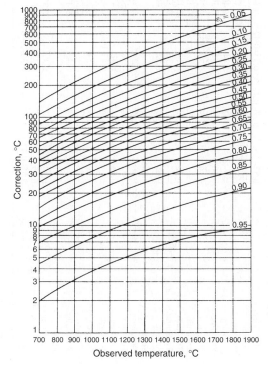

Figure 17.5 *Correction to optical pyrometer readings for spectral emissivity* $\lambda = 0.65\,\mu\text{m}$, $C_2 = 1.438\,\text{cm K}$

Table 17.1   SPECTRAL NORMAL EMITTANCE OF METALS FOR WAVELENGTH OF 0.65 μm

Metal	Temperature °C										*References*
	600	800	1 000	1 200	1 400	1 600	1 800	2 000	2 500	3 000	
Beryllium	—	—	0.37	0.37	—	—	—	—	—	—	5
Chromium	—	—	—	—	—	—	—	0.39[m]	—	—	6
Cobalt	—	—	0.33–0.38	0.34–0.37	0.39	0.37[m]	—	—	—	—	6, 7
Copper	—	0.11	0.10	0.10[m]	0.11[m]	0.12[m]	0.14[m]	—	—	—	8
Erbium	—	0.55	0.55	0.55	0.55	0.38[m]	—	—	—	—	6
Germanium	—	—	—	—	—	—	—	—	—	—	9
Gold	0.16–0.18	0.16–0.19	0.16–0.21	0.13[m]	—	—	—	—	—	—	5
Hafnium	—	—	—	—	0.45	—	—	—	—	—	10
Iridium	—	—	0.36	0.34	0.32	—	—	—	—	—	6, 11
Iron	—	0.37	0.36	0.35	0.35	0.37[m]	—	0.30	—	—	5
Manganese	—	—	—	0.59	0.59[m]	—	—	—	—	—	6
Molybdenum	0.36	0.37–0.43	0.36–0.42	0.35–0.42	0.34–0.41	0.34–0.41	0.33–0.40	0.32–0.39	0.31–0.37	—	5
Nickel	—	0.35	0.34	—	—	—	—	—	—	—	5
Niobium	—	—	0.37	0.37	0.37	0.37	0.37	0.37	0.40[m]	—	12
Osmium	—	—	0.52	0.44	0.40	0.38	0.38	0.38	—	—	13
Palladium	—	0.40	0.37	0.34	0.30	0.37[m]	—	—	—	—	14
Platinum	—	0.29–0.31	0.29–0.31	0.29–0.31	0.29–0.31	0.29–0.31	—	—	—	—	5
Rhenium	—	—	0.42	0.42	0.42	0.42	0.41	0.41	0.40	—	15
Rhodium	—	0.25	0.22	0.19	0.18	0.16	—	—	—	—	14
Ruthenium	—	—	—	0.35	0.32	0.31	0.31	0.31	—	—	13
Silicon	—	0.63	0.57	0.52	0.46	0.48[m]	—	—	—	—	9
Silver	0.47	0.055	0.055[m]	—	—	—	—	—	—	—	8
Tantalum	—	0.46	0.45	0.44	0.42	0.41	0.40	0.39	0.38	0.36	5
Thorium	—	—	0.38	0.38	0.38	—	—	—	—	—	12
Titanium	—	0.48	0.48	0.48	0.47	—	—	—	—	—	16, 17
Tungsten	—	—	0.46–0.48	0.43–0.48	0.42–0.47	0.42–0.47	0.41–0.47	0.40–0.47	0.38–0.46	0.36–0.45	5
Uranium	—	0.19–0.36	0.19–0.36	0.34[m]	0.34[m]	—	—	—	—	—	5
Yttrium	—	—	—	0.35	0.35	—	—	—	—	—	6
Zirconium	—	—	0.48	0.45	0.42	0.39	0.36	—	—	—	18
Alloys											
Cast iron	—	0.37	0.37	0.37	0.37	0.40[m]	—	—	—	—	—
Nichrome (in hydrogen)	—	0.35	0.35	0.35	0.35	—	—	—	—	—	5
Steel	—	0.35–0.40	0.32–0.40	0.30–0.40	—	0.37[m]	—	—	—	—	5

[m] Value for molten state.

Table 17.2   SPECTRAL EMITTANCE IN THE INFRA-RED OF METALS AT HIGH TEMPERATURES

Metal	Temperature °C	Wavelength μm												References
		1.0	1.2	1.4	1.5	1.6	1.8	2.0	2.5	3.0	3.5	4.0	4.5	
Cobalt	800	—	0.26	—	—	—	—	0.21	—	0.18	—	—	—	5
	1 000	—	0.26	—	—	—	—	0.21	—	0.19	—	—	—	5
	1 200	—	0.26	—	—	—	—	0.22	—	—	—	—	—	5
Copper	762	—	—	—	0.031	—	—	0.029	—	—	—	0.025	—	19
	901	0.049	—	—	0.079	—	—	0.065	0.052	0.043	0.038	0.032	—	20
	985	—	—	—	0.037	—	0.034	—	0.032	0.031	—	0.030	—	21
Iridium	827	0.229	0.203	0.185	—	0.167	0.152	0.140	—	—	—	—	—	22
	1 227	0.233	0.213	0.194	—	0.180	0.169	0.160	—	—	—	—	—	22
	1 727	0.243	0.228	0.210	—	0.199	0.188	0.180	—	—	—	—	—	22
	2 127	0.247	0.233	0.219	—	0.207	0.199	0.192	—	—	—	—	—	22
Iron	800	—	0.294	—	—	0.264	—	0.237	0.217	—	—	—	—	23
	1 000	—	0.294	—	—	0.267	—	0.245	0.227	—	—	—	—	23
	1 200	—	0.291	—	—	0.300	—	0.252	0.235	—	—	—	—	23
	1 245	0.340	0.316	0.298	0.290	0.282	0.268	0.260	0.248	0.240	0.235	0.225	0.218	23
Molybdenum	1 327	0.335	—	—	0.185	—	—	0.140	—	0.115	—	0.114	—	5
	1 727	0.300	—	—	0.195	—	—	0.170	—	0.155	—	0.145	—	5
	2 527	0.260	—	—	0.210	—	—	0.193	—	0.185	—	0.185	—	5
Nickel	800	—	0.295	0.267	—	0.250	0.230	0.215	—	—	—	—	—	23
	1 000	—	0.293	0.269	—	0.252	0.232	0.219	—	—	—	—	—	23
	1 200	—	0.290	0.271	—	0.253	0.235	0.223	—	—	—	—	—	23
	1 110	—	0.292	0.270	0.250	—	—	0.290	0.205	0.187	0.174	0.162	—	23
Niobium	827	0.345	—	—	0.23	—	—	0.19	—	—	—	—	—	5
	1 227	0.335	—	—	0.25	—	—	0.21	—	—	—	—	—	5
	1 727	0.320	—	—	0.26	—	—	0.23	—	—	—	—	—	5
	2 127	0.315	—	—	0.27	—	—	0.25	—	—	—	—	—	5

(continued)

Table 17.2   SPECTRAL EMITTANCE IN THE INFRA-RED OF METALS AT HIGH TEMPERATURES—*continued*

Metal	Temperature °C	Wavelength μm												References
		1.0	1.2	1.4	1.5	1.6	1.8	2.0	2.5	3.0	3.5	4.0	4.5	
Platinum	1 127	—	0.257	—	0.227	—	—	0.193	—	0.151	—	0.130	—	5
Rhenium	1 537	0.36	—	—	0.29	—	—	0.25	0.23	—	—	—	—	15
	2 118	0.36	—	—	0.30	—	—	0.27	0.24	—	—	—	—	15
	2 772	0.36	—	—	0.32	—	—	0.29	0.26	—	—	—	—	15
Tantalum	1 427	0.295	—	—	0.220	—	—	0.190	—	0.170	—	0.150	—	5
	1 927	0.310	—	—	0.245	—	—	0.215	—	0.192	—	0.180	—	5
	2 527	0.330	—	—	0.290	—	—	0.270	—	0.240	—	0.230	—	5
Titanium	750	0.490	—	0.510	0.500	—	—	0.455	—	0.525	0.575	0.600	—	24
Tungsten	1 327	0.385	—	—	0.28	—	—	0.21	—	0.13	—	0.095	—	5
	2 127	0.37	—	—	0.292	—	—	0.245	—	0.18	—	0.15	—	5
	2 527	0.36	—	—	0.30	—	—	0.26	—	—	—	—	—	5
Zirconium	1 127	0.46	—	—	0.422	—	—	0.386	0.360	0.348	—	0.325	—	25
	1 327	0.444	—	—	—	—	—	0.368	—	0.343	—	—	—	25
	1 727	0.442	—	—	0.375	—	—	0.357	0.351	0.342	0.330	—	—	25

Table 17.3 SPECTRAL NORMAL EMITTANCE OF METALS AT ROOM TEMPERATURE

Derived from reflectance data by formula $\varepsilon_\lambda = 1 - \rho_\lambda$

Metal	Wavelength μm							References
	10.0	9.0	5.0	3.0	1.0	0.6	0.5	
Aluminium	0.02–0.04	—	0.03–0.08	0.03–0.12	0.08–0.27	—	—	5
Antimony*	—	0.28	0.31	0.35	0.45	0.47	—	26
Bismuth	0.08	—	0.12	0.26	0.72	0.76	0.75	27
Cadmium	—	0.02	0.04	0.07	0.30	—	—	28
Chromium	—	0.08	0.19	0.30	0.43	0.44	0.45	26
Cobalt	—	0.04	0.15	0.23	0.32	—	—	28
Copper	0.021	—	0.024	0.026	0.030	0.080	0.36	5
Gold	0.015	0.015	0.015	0.015	0.020	0.080	0.45	5
Iridium	—	0.04	0.06	0.09	0.22	—	—	28
Iron	—	—	—	—	0.41	0.48	0.49	14
Lead	—	0.06	0.08	—	—	—	—	29
Magnesium*	—	0.07	0.14	0.20	0.26	0.27	0.28	26
Molybdenum	0.15	—	0.16	0.19	0.42	—	—	30
Nickel	—	0.04	0.06	0.12	0.27	—	—	31
Niobium	0.04	—	0.06	0.14	0.29	0.55	—	5
Palladium	—	0.03	0.10	0.12	0.28	0.37	0.42	18,14,32
Platinum	0.05	—	0.06	0.11	0.24	0.36	0.40	33
Rhodium	—	0.05	0.07	0.08	0.16	0.21	0.24	26
Silver	0.02	—	0.02	0.02	0.03	0.03	0.03	5
Tantalum	—	0.06	0.07	0.08	0.22	0.55	0.62	28
Tellurium	—	0.22	0.43	0.47	0.50	0.51	—	26
Tin	—	0.14	0.24	0.32	0.46	—	—	28
Titanium	0.05–0.12	—	0.10–0.18	0.25–0.33	0.37–0.49	—	—	5
Tungsten	0.03	—	0.05	0.07	0.40	0.44–0.49	—	5
Vanadium	0.06–0.09	—	0.07–0.11	0.10–0.17	0.36–0.50	0.42–0.57	0.43–0.59	5
Zinc	0.03	—	0.05	0.08	0.50–0.61	0.42–0.58	—	5

* Values for spectral, directional emittance at 15° only.

Table 17.4 SPECTRAL EMITTANCE OF OXIDISED METALS FOR WAVELENGTH OF 0.65 μm

Oxide formed on smooth surfaces. For oxides in the form of refractory materials, values of emittance widely different from those below may be given and will be dependent on the grain size.

Metal	$\varepsilon_{0.65}$	Metal	$\varepsilon_{0.65}$
Aluminium	0.30	Uranium	0.30
Beryllium	0.61	Vanadium	0.70
Chromium	0.60	Yttrium	0.60
Cobalt	0.77	Zirconium	0.80
Copper	0.70		
Iron	0.63		
Magnesium	0.20	*Alloys*	
Nickel	0.85	Cast iron	0.70
Niobium	0.71	Nichrome	0.90
Tantalum	0.42	Constantan	0.84
Thorium	0.57	Carbon steel	0.80
Titanium	0.50	Stainless steel	0.85

References 5 and 6.

Table 17.5   TOTAL NORMAL EMITTANCE OF METALS

Metal	Temperature °C										References
	20	100	500	1 000	1 200	1 400	1 600	2 000	2 500	3 000	
Aluminium	—	0.038	0.064	—	—	—	—	—	—	—	34
Beryllium	—	—	—	0.55	0.87	—	—	—	—	—	35
Bismuth	—	0.06	—	—	—	—	—	—	—	—	5
Chromium	—	0.08	0.11–0.14	—	—	—	—	—	—	—	5
Cobalt	—	0.15–0.24	0.34–0.46	—	—	—	—	—	—	—	5
Copper	—	—	0.02	—	0.12^{m}	—	—	—	—	—	5
Germanium	—	—	0.54	—	—	—	—	—	—	—	36
Gold	—	0.02	0.02	—	—	—	—	—	—	—	
Hafnium	—	0.07	0.14	—	—	0.31	0.32	—	—	—	37
Iron	0.05	—	—	0.24	0.30	—	—	—	—	—	5
Lead	—	0.63	—	—	—	—	—	—	—	—	38
Magnesium	—	0.12*	—	—	—	—	—	—	—	—	19
Mercury	—	0.12	—	—	—	—	—	—	—	—	
Molybdenum	0.065	0.08	0.13	0.19	0.22	0.24	0.27	—	—	—	5
Nickel	—	—	0.09–0.15	0.14–0.22	—	—	—	—	—	—	5
Niobium	—	—	—	0.12	0.14	0.16	0.18	0.21	—	—	39
Palladium	—	—	0.06	0.12	0.15	—	—	—	—	—	14
Platinum	—	—	0.086	0.14	0.16	—	—	—	—	—	5
Rhenium	—	—	—	0.22	0.25	0.27	0.29	—	—	—	40
Rhodium	—	—	0.035	0.07	0.08	0.09	—	—	—	—	14
Silver	0.03	0.02–0.03	0.02–0.03	0.11	0.13	0.15	0.18	—	—	—	5
Tantalum	—	0.04	0.06	—	—	—	—	0.23	0.28	—	5
Tin	—	0.07	—	—	—	—	—	—	—	—	41
Titanium	—	0.11	0.05	0.11	0.14	0.17	0.19	0.23	0.27	—	42
Tungsten	—	—	—	—	—	—	—	—	—	0.30	5
Uranium (α-phase)	—	—	0.33*	—	—	—	—	—	—	—	5
Uranium (γ-phase)	—	—	—	0.29–0.40*	—	—	—	—	—	—	5
Zinc	—	0.07	—	—	—	—	—	—	—	—	43
Zinc (galvanised iron)	—	0.21	—	—	—	—	—	—	—	—	41
Zirconium	—	—	—	0.22	0.25	0.27	—	—	—	—	44
Alloys											
Brass	—	0.059	—	—	—	—	0.29^{m}	—	—	—	41
Cast iron (cleaned)	—	0.21	—	—	—	—	—	—	—	—	5
Nichrome	—	—	0.95	0.98	—	—	—	—	—	—	5
Steel (polished)	—	0.13–0.21	0.18–0.26	0.55–0.80	—	—	—	—	—	—	5
Steel (cleaned)	—	0.21–0.38	0.25–0.42	0.50–0.77	—	—	—	—	—	—	5

* Value for total hemispherical emissivity.

Table 17.6 TOTAL NORMAL EMITTANCE OF OXIDISED METALS

The values depend on the degree of oxidation and the grain size.
Unless stated otherwise, the following are results obtained for metals oxidised in general above 600°C.

Metal	200°C	400°C	600°C	800°C	1 000°C	References
Alumium	0.11	0.15	0.19	—	—	38
Brass	0.61	0.60	0.59	—	—	38
Chromium	—	0.09	0.14–0.34	—	—	14
Copper (red heat for 30 min)	0.15	0.18	0.23	0.24	—	5
Copper (stably oxidised at 760°C)	—	0.40–0.50	0.60–0.66	—	—	5
Copper (extreme oxidation)	—	0.88	0.92	—	—	5
Cast iron	0.64	0.71	0.78	—	—	38
Cast iron (strongly oxidised)	0.95	—	—	—	—	—
Iron (red heat for 30 min)	0.45	0.52	0.57	—	—	5
Lead	0.63	—	—	—	—	—
Molybdenum (oxide volatile in vaccum above 540°C)	—	0.84	—	—	—	5
Monel	0.41	0.44	0.47	—	—	38
Nickel (stably oxidised at 900°C)	0.15–0.50	0.33–0.51	0.44–0.57	0.49–0.71	—	5
Nimonic (buffed, oxidised at 900°C)	—	0.46	—	—	—	45
Nimonic (buffed, oxidised at 1 200°C)	—	0.72	—	—	—	45
Niobium (oxidised and annealed)*	—	—	—	0.74	—	5
Palladium	0.03	0.05	0.076	—	0.124	14
Stainless steel (stably oxidised at high temperature)	—	0.80–0.87	0.84–0.91	0.89–0.95	—	5
Stainless steel (red heat in air for 30 min)	0.12–0.25	0.17–0.30	0.23–0.37	0.30–0.44	~	5
Stainless steel (buffed, stably oxidised at 600°C)	—	0.41	0.44	0.54	—	5
Stainless steel (polished, oxidised at high temperature)	—	—	0.65–0.70	—	0.73–0.83	5
Stainless steel (shot blasted stably oxidised at 600°C)	—	0.65	0.67	—	—	5
Tantalum (red heat for 30 min)	0.42	0.42	0.42	—	—	5
Zinc	—	0.11	—	—	—	38

*Value for total hemispherical emissivity.

REFERENCES

1. M. F. Modest, 'Radiative Heat Transfer', McGraw-Hill Publishing Company, New York, NY, 1993.
2. W. W. Coblentz, *Bull. US Bur. Stand.*, 1918, **14**, 312.
3. W. Weniger and A. H. Pfund, *Phys. Rev.*, 1919, **14**, 427.
4. W. J. Parker and G. L. Abbott, 'Symposium on Thermal Radiation of Solids', (ed. S. Katzoff), NASA SP-55, 1965, 11–28.
5. Y. S. Touloukian and D. P. DeWitt (editors), 'Thermophysical Properties of Matter', Vol. 7, 'Thermal Radiative Properties—Metallic Elements and Alloys', IFI/Plenum, New York, 1970.

6. G. K. Burgess and R. G. Wallenberg, *Bull. US Bur. Stand.*, 1915, **11**, 591.
7. H. B. Wahlin and H. W. Knop, Jr., *Phys. Rev.*, 1948, **74**, 687.
8. C. C. Bidwell, *Phys. Rev.*, 1914, **3**, 439.
9. F. G. Allen, *J. Appl. Phys.*, 1957, **28**, 1510.
10. M. L. Shaw, *J. Appl. Phys.*, 1966, **37**, 919.
11. O. K. Husmann, *J. Appl. Phys.*, 1966, **37**, 4662.
12. L. V. Whitney, *Phys. Rev.*, 1935, **48**, 458.
13. R. W. Douglass and E. F. Adkins, *Trans. Met. Soc. AIME*, 1961, **221**, 248.
14. H. T. Betz, O. H. Olsen, B. D. Schurin and J. C. Morris, WADC-TR-56-222 (Part 2), 1957, 1–184 (AD 202 493).
15. D. T. F. Marple, *J. Opt. Soc. Amer.*, 1956, **46**, 490.
16. F. J. Bradshaw, *Proc. Phys. Soc.*, 1950, **B63**, 573.
17. H. Seemuller and D. Stark, *Z. Phys.*, 1967, **198**, 201.
18. S. C. Furman and P. A. McManus, USAEC, GEAP-3338, 1960, 1–46.
19. W. G. D. Carpenter and J. H. Sewell, RAE Rpt: CHem-538, 1962, 1–6 (AD 295 648).
20. D. J. Price, *Proc. Phys. Soc.*, 1947, **59**, 118.
21. A. E. Anderson, Univ. of Calif., M.S. Thesis, 1962, 1–57.
22. T. B. Barnes, *J. Opt. Soc. Amer.*, 1966, **56**, 1546.
23. L. Ward, *Proc. Phys. Soc.*, 1956, **69**, 339.
24. J. G. Adams, Northrup Corp., Novair Div., 1962, 1–259 (AD 274 555).
25. J. A. Coffman, G. M. Kibler, T. F. Lyon and B. D. Acchione, WADD-TR-60-646 (Part 2), 1963, 1–183, (AD 297 946).
26. W. W. Coblentz, *Bull. US Bur. Stand.*, 1911, **7**, 197.
27. J. G. Adams, Northrup Space Labs., Hawthorne, Calif., NSL-62-198, 1962, 1–101.
28. W. W. Coblentz, Publ. Carneg. Instn. Wash. No. 65, 1906, p. 91; *Bull. US Bur. Stand.*, 1906, **2**, 457.
29. H. Schmidt and E. Furthmann, *Mitt. K. Wilhelm. Inst. Eisenforsch. Dusseld.*, 1928, **10**, 225.
30. R. V. Dunkle and J. T. Gier, Inst. of Eng. Res., Univ. of Calif., Berkeley, Progress Rpt., 1953, 1–73 (AD 16 830).
31. E. Hagen and H. Rubens. *Ann. Phys. Lpz.*, 1900 (4), **1**, 352.
32. E. O. Hulburt, *Astrophys. J.*, 1915, **42**, 203.
33. R. A. Seban, WADD-TR-60-370 (Part 2), 1962, 1–72 (AD 286 863).
34. E. Schmidt, V. A. W. Hauzeitschr and A. G. Erftwerk, *Aluminium*, 1930, **3**, 91.
35. S. Konopken and R. Klemm, NASA-SP-31, 1963, 505–513.
36. V. F. Brekhovskikh, *Inz.-fiz. Zh.*, 1964, **7** (5), 66.
37. D. L. Timrot, V. Yu. Voskresenskii and V. E. Peletskii, *High Temp.*, 1966, **4**, 808.
38. C. P. Randolf and M. J. Overholzer, *Phys. Rev.*, 1913, **2**, 144.
39. G. L. Abbott, WADD-TR-61-94 (Part 3), 1963, 1–30 (AD 435 825) (AD 436 887).
40. G. B. Gaines and C. T. Sims, *J. Appl. Phys.*, 1963, **34**, 2922.
41. T. T. Barnes, W. E. Forsythe and E. Q. Adams, *J. Opt. Soc. Amer.*, 1947, **37**, 804.
42. J. T. Bevans, J. T. Gier and R. V. Dunkle, *Trans. ASME*, 1958, **80**, 1405.
43. P. F. McDermott, *Rev. Sci. Instrum.*, 1937, **8**, 185.
44. D. L. Timrot and V. E. Peletskii, *High Temp.*, 1965, **3**, 199.
45. A. H. Sully, E. A. Brandes and R. B. Waterhouse, *Br. J. Appl. Phys.*, 1952, **3**, 97.

18 Electron emission*

Under normal conditions electrons are prevented from leaving a metal by a potential step at the surface. In the absence of an electric field, the height of this potential step is called the work function ϕ. Electrons can, however, escape if they are given enough energy. This energy can be supplied in a number of different ways, giving rise to the various types of electron emission.

18.1 Thermionic emission

When a metal is heated, some electrons with energies near the Fermi level are enabled to escape by acquiring extra thermal energy. An adjacent anode carrying a sufficiently positive potential will collect all the electrons emitted, and the saturated emission current will flow. A further increase of anode potential causes a positive field at the metal surface; this lowers the potential barrier slightly and increases the current. The 'zero field' saturated emission current per unit area of the cathode J, is related to the temperature according to the Richardson-Dushman equation.

$$J = AT^2 \exp\left(-e\phi/kT\right)$$

where A is a constant, e the magnitude of the electronic charge, k Boltzmann's constant and T the absolute temperature.

For a metal, the theoretical value of A is 1.2 MA m^{-2}. In practice ϕ usually has a temperature dependence and this results in a different value of A. The work function ϕ cannot be calculated reliably, but tends to increase with the density of the metal. The observed values of A and ϕ are shown in Table 18.1 for polycrystalline surfaces of a number of metals.

To calculate the emission from the values of A and ϕ in the table, the emission formula may be written as

$$\log_{10} J = \log_{10} A + 2\log_{10} T - 504\,0\,\phi T^{-1}$$

where J is in kA m^{-2}, T is in K.

The work function of a metal is lowered when a layer of a more electropositive material is adsorbed on its surface. This increases the thermionic emission, which has a maximum value when the adsorbed layer is approximately monatomic. Table 18.2 shows typical values of thermionic constants for various such surfaces. The emission may also be increased by a coating of a refractory metallic compound, usually about 100 μm thick. The thermionic emission follows the usual law but in the case of a semiconductor layer the quantities A and ϕ have a different significance.

Table 18.3 shows the emission constants for various carbides, borides and oxides, and the emission available at a particular operating temperature.

For a thermionic cathode to be technically useful it must have an adequate emission at a temperature where the rate of evaporation is not excessive. This limits the practical cathodes to a relatively small number. Table 18.4 gives the most important of these with their normal maximum operating temperatures and maximum operating emission densities for a generally acceptable life.

An 'L' cathode consists of a block of porous tungsten, the front emitting surface of which is activated with barium. A reaction between the tungsten and barium oxide at the rear surface produces free barium which diffuses through the porous tungsten.

---

* For applications in electron microscopy, see Section 10.4.

Table 18.1   THERMIONIC PROPERTIES OF THE ELEMENTS

Element	A kAm^{-2}K^{-2}	ϕ V	Element	A kAm^{-2}K^{-2}	ϕ V
Barium	600	2.11	Niobium	1 200	4.19
Beryllium	3 000	3.75	Osmium	1 100 000	5.93
Caesium	1 600	1.81	Palladium	600	4.9
Calcium	600	2.24	Platinum	320	5.32
Carbon	150	4.5	Rhenium	1 200	4.96
Chromium	1 200	3.90	Rhodium	330	4.8
Cobalt	410	4.41	Silicon	80	3.6
Copper	1 200	4.41	Tantalum	1 200	4.25
Hafnium	220	3.60	Thorium	700	3.38
Iridium	1 200	5.27	Titanium	—	3.9
Iron α	260	4.5	Tungsten	600	4.54
Iron γ	15	4.21	Uranium	60	3.27
Molybdenum	550	4.15	Zirconium	3 300	4.12
Nickel	300	4.61			

References: 1, 2, 3, 4 (General): 5 (Hf). 6 (Nb, Ta, Re, Os, Ir); 7 (Be, Cr, Cu).

Table 18.2   THERMIONIC PROPERTIES OF REFRACTORY METALS WITH ADSORBED ELECTROPOSITIVE LAYERS

Surface	A kAm^{-2}K^{-2}	ϕ V
Tungsten–barium	15	1.56
Tungsten–caesium	32	1.36
Tungsten–cerium	80	2.71
Tungsten–lanthanum	80	2.71
Tungsten–strontium	—	2.2
Tungsten–thorium	30	2.63
Tungsten–uranium	32	2.84
Tungsten–yttrium	70	2.70
Tungsten–zirconium	50	3.14
Molybdenum–thorium	15	2.59
Tantalum–thorium	15	2.52

Reference 8.

Table 18.3   THERMIONIC PROPERTIES OF REFRACTORY METAL COMPOUNDS

Compound	A kAm^{-2}K^{-2}	ϕ V	Emission kAm^{-2}
TaC	3	3.14	3 at 2 000 K
TiC	250	3.35	63 at 2 000 K
ZrC	3	2.18	40 at 2 000 K
ThC$_2$	5 500	3.5	40 at 2 000 K
SiC	640	3.5	4 at 2 000 K
UC	330	2.9	50 at 2 000 K
CaB$_6$	26	2.9	0.12 at 1 670 K
SrB$_6$	1.4	2.7	0.036 at 1 670 K
BaB$_6$	160	3.5	0.018 at 1 670 K
LaB$_6$	290	2.7	7 at 1 670 K
CeB$_6$	36	2.6	1.7 at 1 670 K
ThB$_6$	5	2.9	0.022 at 1 670 K
PrB$_6$	—	3.12	
NdB$_6$	—	4.6	
ThO$_2$	50	2.6	20 at 1 900 K
CeO$_2$	10	2.3	26 at 1 900 K
La$_2$O$_3$	9	2.5	8 at 1 900 K
Y$_2$O$_3$	10	2.4	13 at 1 900 K
BaO/SrO (oxide cathode)	1–10	1.0	See comments in text

References: 9 (Carbides), 10 (Borides), 11 (Oxides), 12 (UC).

Table 18.4 EMISSION AT THE NORMAL MAXIMUM OPERATING
TEMPERATURE OF PRACTICAL CATHODES

Cathode	Operating temperature K	Emission kA m^{-2}
Tungsten	2 500	3
Tantalum	2 400	8
Rhenium	2 400	0.5
BaO/SrO on nickel d.c.	1 100	10
'Oxide cathode' pulse	1 100	100
BaO/SrO Ni. Matrix type	1 150	20
'L' cathode	1 360	30
Impregnated tungsten	1 350	50
ThO$_2$ on W or Ir	1 900	10
LaB$_6$ on Re	1 450	0.5
LaB$_6$ bulk	1 900	50
Thoriated tungsten	1 900	10

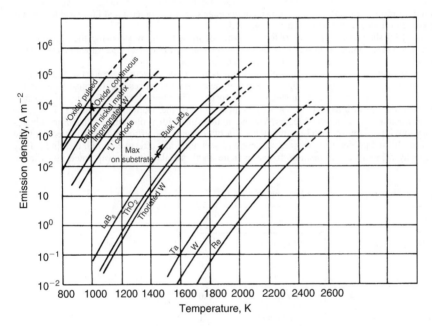

Figure 18.1 *Thermionic emission of practical cathodes as a function of temperature*

Impregnated tungsten is a porous tungsten block whose pores are filled with barium calcium aluminate by infiltration in the molten phase at 1 700°C. The emitting surface is partly the compound and partly barium activated tungsten.

The thoriated tungsten is a high density rod or wire containing about 1% thorium oxide. The surface is carburised to form a layer of W$_2$C. In operation a reaction between the oxide and the carbide produces free thorium, which activates the surface.

Figure 18.1 (reference 13) shows the saturated emissions available from some of these cathodes as a function of temperature. In practice it is usual to operate cathodes at rather less than the saturated emission. The full lines show the region where a useful life is obtainable. The cathode life is usually limited by the evaporation rate and lowering the temperature increases the life considerably. Typically, lowering the temperature of an 'L' or impregnated tungsten cathode by 80 K increases its life by an order of magnitude.

For an oxide cathode, the continuously drawn emission must be limited to about $10\,\text{kA m}^{-2}$, but if the current is drawn in microsecond pulses the pulse current may be increased to $100\,\text{kA m}^{-2}$.

The emissions shown on Figure 18.1 are obtainable only in an environment free from oxidising gases. In general the partial pressure of gases such as O_2, CO_2, and H_2O should not exceed approximately 10^{-5} Pa. The emission is unaffected by rare gases except when the cathode is bombarded excessively with positive ions; the presence of H_2 may counteract to some extent the effect of oxidising gases.

18.2   Photoelectric emission

When light of sufficiently high frequency is incident on the surface of a metal, electrons are emitted. In order that electrons shall be emitted with zero velocity from a metal at absolute zero of temperature, the energy of the light photons must equal the energy corresponding to the work function. Thus there is a threshold frequency v_0 at which $hv_0 = e\phi$ where h is Planck's constant. If the frequency of the light is v, where v is greater than v_0, the maximum energy of the emitted electrons is hv-hv_0. At temperatures above absolute zero the threshold is not sharp since light of a frequency less than v_0 can liberate a small number of electrons. An experiment measuring the energy of photoelectrons emitted for monochromatic light of known energy enables ϕ to be determined fairly accurately, the accuracy improving as the temperature is lowered.

Table 18.5 shows the values of ϕ for a number of metals determined using the photoelectric effect. The values obtained in this way should correspond to the thermionic values in Table 18.1; discrepancies are probably due to contamination of the surface.

For practical uses it is required to obtain the maximum photoelectric current for a given light flux. For extracting photocurrent, the efficiency of the conversion process (yield) must be separated from effects due to space charge in shielding the surface. The photoelectric efficiency of a surface may be defined in various ways, the most fundamental being the quantum efficiency Y. This is the ratio of the number of electrons released to the number of incident photons. For clean metals this is very low (10^{-4} approximately) so they are not often used. The efficient photoemitters as used in photocells and photomultipliers are semiconductors with a low effective photon threshold. They are usually formed by combination of one or more alkali metals with an evaporated thin film of

Table 18.5   PHOTOELECTRIC WORK FUNCTIONS

Surface	ϕ	Surface	ϕ
Aluminium	4.2	Molybdenum	4.2
Antimony	4.1	Nickel	4.9
Arsenic	5.1	Palladium	5.0
Barium	2.5	Platinum	5.3
Beryllium	3.4	Potassium	2.2
Bismuth	4.4	Rhenium	5.0
Boron	4.5	Rhodium	4.6
Cadmium	4.0	Rubidium	2.1
Caesium	1.9	Silicon	4.2
Calcium	2.9	Silver	4.7
Carbon	4.8	Sodium	2.2
Chromium	4.4	Strontium	2.7
Cobalt	4.0	Tantalum	4.1
Copper	4.5	Tellurium	4.8
Gallium	3.9	Thallium	3.8
Germanium	4.8	Thorium	3.5
Gold	4.8	Tin	4.3
Iron	4.4	Titanium	4.1
Iridium	4.6	Tungsten	4.5
Lead	4.0	Uranium	3.6
Lithium	2.4	Zinc	4.3
Manganese	3.8	Zirconium	3.8
Mercury	4.5		

References: 1, 2, 3 (General); 14 (Cu, Ag, Al); 15 (Re).

Table 18.6 PROPERTIES OF EFFICIENT PHOTOELECTRIC
EMITTING SURFACE

Surface	Photoelectric quantum efficiency Y	Photon threshold energy eV
Na_3Sb	0.02	3.1
K_3Sb	0.07	2.6
Rb_3Sb	0.10	2.2
Cs_3Sb	0.25	2.05
NaK_3Sb	0.30	2.0
$CsNaK_3Sb$	0.40	1.55

Reference 16.

antimony (apart from the type consisting of caesium on oxidised silver). Table 18.6 gives the value of maximum photoelectric yields for various such surfaces. This maximum is reached at photon energies 1–1.5 eV above the threshold value, which is also shown.

The corresponding wavelength of light λ in nm is related to the photon energy $e\phi$ in electron volts by the (non-relativistic) relationship

$$\lambda = 1.24 \times 10^3 \phi^{-1}$$

18.3 Secondary emission

When electrons (primaries) are incident upon a surface of a solid, electrons (secondaries) are produced which leave the surface in the direction from which the primaries arrive. The total flow of secondaries consists of:

1. Primaries elastically scattered.
2. Primaries reflected inelastically with an energy loss of some tens of volts.
3. True secondaries with an energy independent of the primary energy and a mean value of about 10 eV. Electrons with an energy up to about 50 eV are usually considered to be true secondaries.

The ratio of the total flow of secondaries to that of the primaries is called the secondary emission coefficient δ. As the primary electron energy is increased from zero, δ rises to reach a maximum value δ_{max} for a primary energy V_{max} in the range of 200–2 000 eV for metals, and it falls off more slowly at energies above V_{max}. The shape of the curve relating δ to V is approximately similar for most metals, and a (universal) curve normalised to δ_{max} and V_{max} is shown in Figure 18.2.

The values of δ_{max} and V_{max} are shown for most metals in Table 18.7. These values are for clean smooth polycrystalline surfaces. It is impossible to remove oxide films from many metals, such as aluminium or magnesium, by heating. These metals are usually deposited as clean layers either by evaporating in high vaccum or by sputtering. Alternatively the surface of the bulk metal may be cleaned by sputtering in an electric discharge in argon.

It should be noted that the total (integrated) secondary emission is reduced by roughening the surface, as some of the secondaries released in the valleys in the surface may be intercepted by the adjacent high spots. An example of this is carbon; the value of δ_{max} for polished graphite is approximately 1, while that for soot is only about 0.5.

The secondary emission of metal oxides is usually higher than that of metals. Surfaces with high values of δ are used in secondary electron multipliers and are prepared by oxidising metals containing small quantities, usually about 2%, of magnesium, beryllium or aluminium. Oxidised metal surfaces with caesium evaporated on to them also have high values of δ and are used in photomultipliers. Table 18.8 shows maximum (at $V/V_{max} = 1$) values of δ obtained from various such surfaces.

Table 18.9 shows the secondary emission from a number of insulating metal compounds either as evaporated films (e) or surface layers on the parent metal (s). These layers must be very thin to avoid accumulating a charge. The secondary electrons originate normally within 10 nm of the surface, so provided films are thicker than this a true value of δ will be obtained.

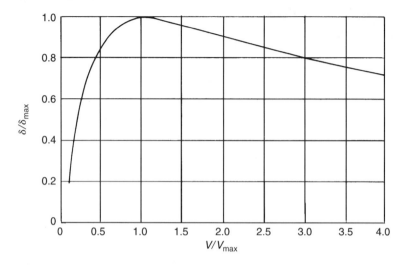

Figure 18.2   *Normalised curve of secondary emission as a function of primary voltage*

Table 18.7   MAXIMUM SECONDARY EMISSION COEFFICIENTS

Element	δ_{max}	V_{max}	Element	δ_{max}	V_{max}
Aluminium	0.97	300	Manganese	1.35	200
Antimony	1.30	600	Mercury	1.30	600
Barium	0.85	300	Molybdenum	1.20	350
Beryllium	0.5	200	Nickel	1.35	450
Bismuth	1.15	550	Niobium	1.20	350
Boron	1.2	150	Palladium	1.65	550
Cadmium	1.59	800	Platinum	1.60	720
Calcium	0.60	200	Potassium	0.53	175
Carbon (Graphite)	1.02	300	Rhenium	1.30	800
Caesium	0.72	400	Rubidium	0.90	300
Chromium	1.10	400	Ruthenium	1.40	570
Cobalt	1.35	500	Silicon	1.10	250
Copper	1.28	600	Silver	1.56	800
Dysprosium	0.99	900	Sodium	0.82	300
Erbium	1.05	1 100	Strontium	0.72	400
Gadolinium	1.04	600	Tantalum	1.25	600
Gallium	1.08	600	Terbium	1.02	900
Germanium	1.08	400	Thallium	1.40	800
Gold	1.79	1 000	Thulium	1.05	1 100
Holmium	1.02	900	Thorium	1.10	800
Indium	1.40	500	Tin	1.35	500
Iridium	1.55	700	Titanium	0.90	280
Iron	1.30	200	Tungsten	1.35	650
Lead	1.10	500	Ytterbium	1.04	800
Lithium	0.52	100	Zinc	1.40	800
Magnesium	0.97	275	Zirconium	1.10	350

References: 17; 18 (Pt, Ir, Ru); 19 (Rare earth metals).

18.4   Auger emission

When the energy distribution of secondaries is examined closely, small peaks can be seen superimposed on the basically smooth background. These peaks can be made much more visible by electronic differentiation, and have been shown to originate in Auger transitions, as follows (reference 22). A

Table 18.8 SECONDARY EMISSION FROM OXIDISED ALLOYS AND
PHOTOCELL SURFACES

	δ_{max}	V_{max}
Oxidised alloy		
Ag–Mg	10–16	600
Ag–Be	6	500
Ni–Be	5–10	600
Cu–Be	5	500
Cu–Mg	12.5	900
Cu–Mg–Al	10	700
Photocell surfaces		
Ag–O–Cs	6–10	500
Ni–O–Cs	5.7	500
Ag–O–Rb	5.5	800
Ag–Sb–Cs	8.0	500
Sb–Cs	12	450

Reference 20.

Table 18.9 SECONDARY EMISSION FROM INSULATING METAL
COMPOUNDS

Material	δ_{max}	V_{max}
Li F (e)	5.6	—
Na F (e)	5.7	—
NaCl (e)	6–6.8	600
KCl (e)	7.5–8.0	1 500
RbCl (e)	5.8	—
Cs Cl (e)	6.5	—
Na Br (e)	6.25	—
KI (e)	5.6	—
NAI (e)	5.5	—
Ca F_2 (e)	3.2	—
Ba F_2 (e)	4.5	—
Mg F_2 (e)	4.1	410
BeO (s)	3.4–8	200–400
MgO (s)	2.4–17.5	400–1 600
Al_2O_3 (s)	1.5–3.2	350–1 300
Cu_2O (s)	1.19–1.25	440
PbS (s)	1.2	500
MoS_2 (s)	1.10	—
WS_2 (s)	0.96–1.04	—
ZnS (s)	1.8	350
MoO_2 (s)	1.09–1.33	—
Ag_2O (s)	0.90–1.18	—
SiO_2 (e)	2.2	300
Cs_2O (s)	2.3–11	800

Reference 21.

primary electron removes an electron from an electron shell in an atom. Subsequently another electron in a higher energy shell transfers to the vacancy. The energy thus released is given to a third electron which is emitted. It may be seen that the energy of the third electron is independent of the energy of the primary and is characteristic of three energy levels in the excited surface atom. A large number of these Auger energies have been measured for many elements and they provide identification of the element concerned. Auger electrons have a very short mean free path and hence, in order to escape the surface, must originate within the top 10 nm of the surface. Fractions of a monolayer of an element can be detected using Auger electron spectroscopy. This relatively new technique of analysis appears to have a considerable number of possible uses.

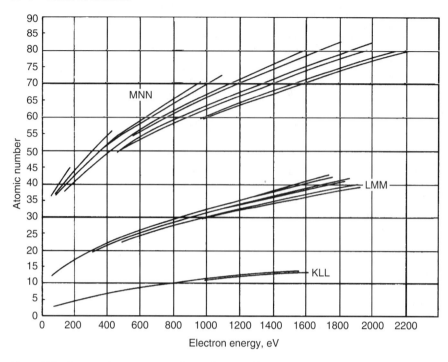

Figure 18.3 *Strongest Auger emission peaks as a function of the atomic number*

The energy of the primary bombarding electrons is usually greater than 2.5 keV, and the Auger peaks are detected from about 50 eV up to 2 kV. Figure 18.3 shows the Auger energies of the stronger peaks plotted against the atomic number Z. The letters are the electron levels involved in the Auger transition (reference 23).

18.5   Electron emission under positive ion bombardment

When an electrical discharge takes place between two electrodes in a low gas pressure the cathode is bombarded with positive ions and emits electrons. These electrons are essential for maintaining the discharge. The number of electrons released for each arriving ion is usually called γ, the second Townsend coefficient. The coefficient is generally approximately constant for positive ion energies from zero up to about 1 kV. The energy required to release the electron is supplied by neutralisation of the positive ion as follows. When the ion is very close to the metal surface the electrostatic force is sufficient to extract an electron which neutralises the positive ion. This releases a photon of energy $(I - \phi)e$, where I is the ionisation potential of the gas atom. This photon can then release a photoelectron from the metal provided $(I - \phi)e > \phi e$ or $I > 2\phi$.

The value of γ thus tends to increase with increasing I and decreasing ϕ. Values of γ for various inert gas ions and metals are shown in Table 18.10. At energies above a few keV the value of γ usually increases approximately linearly with energy, the extra electrons being released as a result of kinetic energy transfer.

18.6   Field emission

When a very high positive electric field is applied to the surface of a metal, the potential just outside the metal becomes more positive than the Fermi level in the metal. The work function barrier, instead of being a step becomes very thin, and electrons can 'tunnel' through the barrier and be emitted. This emission is usually called field emission (sometimes tunnel emission), and the emission density

Table 18.10 SECOND TOWNSEND COEFFICIENT γ ELECTRONS RELEASED PER POSITIVE ION ARRIVING

Metal	Ion	γ	Ion energy eV
Tungsten (outgassed)	Ne$^+$	0.25	0–1 000
Tungsten (outgassed)	He$^+$	0.24	0–1 000
Tungsten (outgassed)	A$^+$	0.10	0–1 000
Tungsten (outgassed)	Kr$^+$	0.05	0–1 000
Tungsten (outgassed)	Xe$^+$	0.02	0–1 000
Tantalum (outgassed)	A$^+$	0.02	100
Tantalum (gas covered)	He$^+$	0.2	500
Tantalum (gas covered)	He^{++}	0.7	500
Molybdenum (outgassed)	He^{++}	0.8	0–1 000
	He$^+$	0.22	0–1 000
	Ne^{++}	0.7	0–1 000
	Ne$^+$	0.2	0–1 000
	A^{++}	0.35	0–1 000
	A	0.08	0–1 000
	Kr	0.05	0–1 000
Nickel	He$^+$	0.7	800
	Ne$^+$	0.4	800
	A$^+$	0.1	800

Reference 24.

Table 18.11 FIELD EMISSION FROM TUNGSTEN

$\phi = 2.0$ eV			$\phi = 4.5$ eV			$\phi = 6.3$ eV		
Field 10^9 V m^{-1}	$\log_{10}J$ A m^{-2}	Current from 10^{-14} m^2 A	Field 10^9 V m^{-1}	$\log_{10}J$ A m^{-2}	Current from 10^{-14} m^2 A	Field 10^9 V m^{-1}	$\log_{10}J$ A m^{-2}	Current from 10^{-14} m^2 A
1.0	6.98	1×10^{-7}	2	0.67	4.7×10^{-14}	2	−8.0	10^{-22}
1.2	8.45	2.8×10^{-6}	3	5.57	3.4×10^{-9}	4	3.12	1.3×10^{-11}
1.4	9.49	3.1×10^{-5}	4	8.06	1.1×10^{-6}	6	7.25	1.8×10^{-7}
1.6	10.23	1.9×10^{-4}	5	9.59	3.4×10^{-5}	8	9.34	2.2×10^{-5}
1.8	10.89	7.8×10^{-4}	6	10.62	4.2×10^{-4}	10	10.66	4.6×10^{-4}
2.0	11.40	2.5×10^{-3}	7	11.36	2.3×10^{-3}	12	11.52	3.3×10^{-3}
2.2	11.82	6.6×10^{-3}	8	11.94	8.8×10^{-3}	14	12.16	1.5×10^{-2}
2.4	12.16	1.5×10^{-2}	9	12.39	2.4×10^{-2}	16	12.65	4.5×10^{-2}
2.6	12.45	2.8×10^{-2}	10	12.76	5.8×10^{-2}	18	13.04	1.1×10^{-1}
			12	13.32	2.1×10^{-2}	20	13.36	2.3×10^{-1}

Reference 25.

is related to the field by the Fowler-Nordheim law. Table 18.11 shows values of $\log_{10}$ (emission density) for various fields for clean tungsten ($\phi = 4.5$ V) and also for a barium contaminated surface ($\phi = 2.0$ V) and an oxygen contaminated surface ($\phi = 6.3$ V).

High fields which produce appreciable emissions usually result from field concentration at the tips of small projections, spikes or whiskers on metal surfaces. As these usually have submicron tip diameters the emission current in A for an area of 10^{-14} m^2 is also shown in Table 18.11.

The field at the tips of emitting projections is greater than the 'macroscopic' field at an electrode by a factor which varies between about 1 000 for a rough surface, down to less than 100 for a highly polished and voltage conditioned hard metal surface. When the field emission from a projection reaches an appreciable fraction of an ampere, the projection will melt and vaporise and this may initiate electrical breakdown between the electrodes in vacuum. Alternatively the field emission may heat the anode electrode and release gas or metal vapour which can also initiate a breakdown. From

Table 18.11 the necessary tip field is likely to be $2 \times 10^9 - 10^{10} \, \text{V m}^{-1}$, depending on work function while the macroscopic field will usually be of the order of 1% of this.

Single field emitting sources can be made, consisting of a point 0.1–$1.0 \, \mu\text{m}$ diameter etched on the end of a refractory metal wire. An anode electrode near the point and carrying a positive potential of a few kV is sufficient to cause field emission. The point usually emits over most of its approximately hemispherical tip. The intensity of the emission varies with direction, as different crystal planes on its surface have different work functions (reference 25). The close packed planes, e.g. the (110) plane of a body centred crystal such as tungsten, have the highest work function and lowest field emission. This effect is shown visually in a field electron microscope where the emission from the tip is viewed on a screen. This arrangement has been used extensively to study surface migration and adsorption phenomena. Measurements can however only be carried out in ultra high vacuum (pressure $<10^{-7}$ Pa) otherwise positive ions are formed in the gas and these quickly destroy the point by ion bombardment.

REFERENCES

1. C. Herring and M. H. Nichols, *Rev. mod. Phys.*, 1949, **21**, 232.
2. G. Herrman and S. Wagener, 'The Oxide Cathode', Vol. 2. Chapman & Hall; 1951.
3. H. H. Michaelson, *J. appl. Phys.*, 1950, **21**, 536.
4. A. Venema, 'Handbook of Vacuum Physics', Pergamon Press, 1966, **2**, 179–298.
5. D. L. Goldwater and W. E. Danforth, *Phys. Rev.*, 1956, **103**, 871.
6. R. C. Wilson, *J. appl. Phys.*, 1966, **31**, 3170.
7. R. C. Wilson, *J. appl. Phys.*, 1966, **31**, 2265.
8. A. L. Reimann, 'Thermionic Emission', 134, Chapman & Hall; 1934.
9. R. E. Haddad, D. C. Goldwater and F. H. Morgan, *J. appl. Phys.*, 1949, **20**, 886, 1130; ibid., 1951, **22**, 70.
10. J. M. Lafferty, *J. appl. Phys.*, 1951, **22**, 299.
11. D. A. Wright, *Proc. Inst. Elect. Engrs*, 1953, **100**, 125.
12. G. A. Haas and J. T. Jensen, *J. appl. Phys.*, 1960, **31**, 1231.
13. R. O. Jenkins, *Vaccum*, 1969, **19**, 353.
14. E. W. Mitchell and J. W. Mitchell, *Proc. R. Soc.*, 1951, **A210**, 70.
15. R. Levi and G. A. Espersen, *Phys. Rev.*, 1950, **78**, 231.
16. W. E. Spicer and F. Wooten, *Proc. IEEE*, 1963, **51**, 1119.
17. D. J. Gibbons, 'Handbook of Vacuum Physics', Pergamon Press; 1966, **2**, 319.
18. R. O. Jenkins (Unpublished).
19. H. Aspden, London University Ph.D. Thesis, 1968, 166.
20. D. J. Gibbons, 'Handbook of Vacuum Physics', Pergamon Press, 1966, **2**, 375.
21. D. J. Gibbons, 'Handbook of Vacuum Physics', Pergamon Press, 1966, **2**, 336.
22. J. J. Lander, *Phys. Rev.*, 1953, **91**, 1382.
23. P. W. Palmberg, 'Electron Spectroscopy', 838, North-Holland; 1972.
24. S. C. Brown, 'Basic Data of Plasma Physics', 222, Wiley; 1959.
25. R. H. Good and E. W. Müller, 'Encyclopedia of Physics', Springer, 1956, **21**, 188.

19 Electrical properties

19.1 Resistivity

The resistivities of a number of pure metals and alloys are given in Tables 19.1 to 19.4. Resistivity varies with the condition of the material and is sensitive to purity. In general, cold working increases and annealing decreases the resistivity. Common reactive gases such as oxygen, nitrogen and hydrogen may also affect the resistivity, either through selective chemical action with existing metallic impurities (which may even reduce the resistivity) or through solution in the host matrix itself.[1] Thermal cycling through phase transformation, quenching from high temperatures and irradiation, all introduce lattice defects which increase resistivity. These defects include vacancies, dislocations and interstitial atoms. Annealing will promote the movement and eventual removal of these defects. This recovery generally takes place in discrete stages at certain temperatures corresponding to the annealing out of each type of defect. Recovery is complete after treatment at the recrystallisation temperature.

Resistivity is often expressed approximately as the sum of the residual resistivity at absolute zero (arising from impurities and lattice defects) and a temperature-dependent intrinsic resistivity (arising from the effect of lattice vibrations upon conduction electrons). The form of temperature dependence is complex, and theories governing it in both solid and liquid metals have been recently discussed in, for example, references 2–15. Over a limited temperature interval, resistivity may be conveniently expressed as a linear relation of the form $\rho_1 = \rho_0(1 + \alpha T)$, where T is the interval between two temperatures T_1 and T_0, and α is the temperature coefficient of resistivity (TCR). In Table 19.1,

Table 19.1 RESISTIVITY AND TEMPERATURE COEFFICIENT OF PURE METALS

| Element | Room temperature | | | Melting point | | | Ref. |
| | Temp. K | Resistivity 10^{-8} Ωm | TCR 10^{-3} K^{-1} | Temp. K | Resistivity, 10^{-8} Ωm | | |
					Solid	Liquid	
Aluminium	293	2.61	4.2	933	—	24.2	7, 16
Americium	300	68.9	—	—	—	—	17
Antimony	293	37.6	5.1	913*	—	113.5	7, 18
Arsenic	293	31	—	—	—	—	19
Barium	300	34.3	5.0	1 002	276	306	20
Beryllium	300	3.76	8.0	—	—	—	20
Bismuth	293	115	4.6	573*	—	128	7, 18
Cadmium	293	6.6	4.3	603*	—	33.7	7, 21
Caesium	300	21.04	5.3†	301.6	21.2	36.9	22
Calcium	300	3.45	3.7	1 113	14.5	33.0	20
Cerium	293	75	0.9	1 068	—	125	23, 24
Chromium	300	12.9	5.9	2 148	—	80	25, 26
Cobalt	293	5.2	6.6	1 766	—	100	26, 27
Copper	293	1.58	4.3	1 356	—	20	26, 27
Dysprosium	293	97	—	—	—	—	23
Erbium	293	80	—	—	—	—	23
Europium	293	116	—	1 099	188	244	23, 24
Francium	300	34	7.2†	300.2	34	55	22

(continued)

19–1

Table 19.1   RESISTIVITY AND TEMPERATURE COEFFICIENT OF PURE METALS—*continued*

| | Room temperature | | | Melting point | | | |
| | Temp. K | Resistivity 10^{-8} Ωm | TCR 10^{-3} K^{-1} | Temp. K | Resistivity, 10^{-8} Ωm | | Ref. |
Element					Solid	Liquid	
Gadolinium	293	132	—	1 585	—	195	23, 24
Gallium	293	13.65	—	303	—	25.9	28, 29
Gold	293	2.01	4.0	—	—	—	27
Holmium	293	87	—	—	—	—	23
Indium	293	8.0	5.2	430	—	33.1	7, 30
Iridium	300	5.0	4.5	—	—	—	25
Iron	300	9.8	6.5	1 808	—	140	25, 26
Lanthanum	293	62	2.2	1 193	—	140	23, 24
Lead	293	19.3	4.2	673*	—	95	7, 31
Lithium	300	9.55	4.0	454	15.6	24.8	22
Lutetium	293	67	—	1 925	—	224	3, 23
Magnesium	300	4.51	3.7	—	—	—	20
Manganese	293	143.5	0.4	1 517	—	180	26, 32
Mercury	293	94.1	1.0‡	253	—	91	7, 31
Molybdenum	300	5.3	4.35	—	—	—	25
Neodymium	293	64.5	—	—	—	—	23
Nickel	293	6.2	6.8	1 725	—	85	26, 27
Niobium	293	13.27	2.6	—	—	—	33
Osmium	293	8.4	4.1	—	—	—	27
Palladium	300	10.5	4.2	1 825	48	83	25, 34
Platinum	300	10.4	3.9	2 042	—	90	25, 26
Potassium	300	7.47	6.0†	336.4	9.22	13.95	22
Praseodymium	293	66	—	—	—	—	23
Protactinium	298	19.3	—	—	—	—	17
Radium	300	88	6.5	—	—	—	20
Rhenium	293	16.9	4.5	—	—	—	27
Rhodium	293	4.37	4.4	—	—	—	27
Rubidium	300	13.32	6.3†	312.6	14.2	22.5	22
Ruthenium	293	6.7	4.1	—	—	—	27
Samarium	293	86	—	—	—	—	23
Silver	300	1.47	4.1	1 234	—	28.5	26, 27
Sodium	300	4.93	5.3†	371	6.86	9.43	22
Strontium	300	13.5	3.2	1 042	65.6	84.8	20
Tantalum	300	13.1	3.5	—	—	—	25
Terbium	293	107	—	—	—	—	23
Thallium	293	15.0	5.2	576	—	73.1	7, 31
Thorium	293	14.2	4.0	—	—	—	35
Thulium	293	95	—	—	—	—	23
Tin	293	10.1	4.6	—	—	—	36
Titanium	293	39	3.8	1 941	—	400	2, 27
Tungsten	300	5.3	4.8	—	—	—	25
Uranium	293	24.8	2.5	—	—	—	37
Vanadium	300	19.9	3.9	2 163	—	200	25, 26
Ytterbium	293	29	—	1 097	74	109	5, 23
Zinc	293	5.45	4.2	693	—	37.4	7, 38
Zirconium	293	38.9	4.4	—	—	—	27

* Liquid resistivity at temperature above melting point.
† TCR for the interval 250–300 K.
‡ TCR for the liquid phase.

TCR values are given for the temperature range 273–373 K, except for some lower-melting-point elements as indicated.

Resistivity values given in Table 19.1 are for bulk material. If the metal is deposited as a thin film, resistivity may deviate from the bulk value, being affected by parameters such as film thickness and grain size. Many recent investigations have been made on these effects (*see* references 39–41).

The resistivity and temperature coefficient of alloys is often dependent upon the method of preparation and heat treatment. The values given in Table 19.2 relate to particular samples and should not be assumed to apply accurately to other samples of similar composition. In systems which exhibit complete mutual solid solubility, there is often a resistivity maximum near 50/50 composition (e.g.

(*text continued on p* **19**–7)

Table 19.2 RESISTIVITY AND TEMPERATURE COEFFICIENT OF SOME ALLOYS

Alloy		*Nominal composition* wt. %†	*Temperature* K	*Resistivity* 10^{-8} Ωm	*TCR* 10^{-3} K^{-1}
Ag–Au	(normal silver)	Au 0.37	293	1.77	—
		Au 10	293	3.6	—
Ag–Cd–Zn–Cu		Cd 18 Zn 16.5 Cu 15	273	7.0	—
Ag–Cu	(standard silver)	Cu 7.5	293	1.9	—
		Cu 10–50	293	2.0–2.1	—
Ag–Cu–Zn		Cu 25 Zn 15	273	8.3	—
Ag–Mn*		Mn 6	293	14	0.2
		Mn 10	293	27	0.02
		Mn 12	293	33	−0.01
		Mn 16–20	293	42–46	−0.03
Ag–Pd		Pd 5	293	3.8	—
		Pd 10	293	5.8	—
		Pd 20	293	10.1	—
Al–Cu etc.		Cu 6	273	3.1	3.8
	(Duralumin)	Cu 4 Mn 0.6 Mg 0.6	293	5.0–5.3	2.3
		Cu 4.1 Mn 0.5 Mg 1.4 Fe 0.2	273	4.0–4.4	—
Al–Mg etc.		Mg 10	273	8.0	—
		Mg 4.75 Mn 0.63 Fe 0.2 Cr 0.13	273	5.66	—
Al–Mg–Si		Mg 0.5 Si 0.5	293	3.25	3.6
Al–Mn		Mn 1.25	293	3.4–4.4	—
Al–Si		Si 12	273	4.5	—
Al–Zn–Mg etc.		Zn 3.6 Mg 2.55 Mn 0.2 Cr 0.2	273	4.75	—
		Zn 5.6 Cu 1.6 Mg 2.5	273	3.7–5.0	—
Au–Ag		Ag 10	273	6.3	1.2
		Ag 33	273	10.8	0.65
Au–Co		Co 2.5	293	32.6	—
Au–Cr–Co		Cr 2.1 Co 0.25	293	39.8	+0.02
		Cr 2.1 Co 0.5	293	44.7	−0.06
		Cr 4.2 Co 0.4	293	57.8	−0.08
Au–Cr–Pd		Cr 2.1 Pd 3.5	293	36.9	0 ± 0.02
		Cr 2.1 Pd 9	293	38.7	0 ± 0.02
Au–Cr–Pt		Cr 2.1 Pt 2	293	34.7	0.07
		Cr 2.1 Pt 6	293	25.6	0.19
Au–Cu–Ag		Cu 78.3 Ag 14.3	273	3.6	1.8
		Cu 26.5 Ag 15.2	273	13.2	0.57
		Cu 15.5 Ag 18.1	273	14.6	0.53
Bi–Sn		Sn 2	273	24.4	—
		Sn 90.5	285	16	—
Bi–Sn–Pb	(Rose's metal)	Sn 23 Pb 28	273	64	2.0
	(Wood's metal)	Sn 12.5 Pb 25	273	52	2.0
Cr–Au[47]		Au 0.6 at %	300	19.2	—
Cu–Al		Al 3	273	8.3	1.0
		Al 10	273	12.6	3.2
Cu–Be		Be 2 (+Ni)	273	6.8–7.4	1.0–1.8
Cu–Mn*		Mn 0.98	273	4.83	—
		Mn 1.49	273	6.66	—
		Mn 4.2	293	17.9	0.25
		Mn 7.4	293	19.7	0.17

(*continued*)

Table 19.2   RESISTIVITY AND TEMPERATURE COEFFICIENT OF SOME ALLOYS—*continued*

Alloy		Nominal composition wt. %†	Temperature K	Resistivity 10^{-8} Ωm	TCR 10^{-3} K^{-1}
Cu–Mn–Al*		Mn 9 Al 3	293	38	0.010
		Mn 9 Al 5	293	42	0.012
		Mn 12 Al 3	293	48	−0.005
Cu–Mn–In*		Mn 12 In 1–3	293	42.5	0 ± 0.010
Cu–Mn–Fe*		Mn 23.2 Fe 6.2	273	77	0.01
		Mn 14.76 Fe 0.33	293	53.5	—
		Mn 7.1 Fe 1.9	273	20	0.1
Cu–Mn–Ni*		Mn 72 Ni 10	293	175	1.4
		Mn 24 Ni 3	273	48	−0.03
	(Manganin)	Mn 12 Ni 4	293	44	0.00
Cu–Ni		Ni 45 at %	293	49	0 ± 0.02
(*see* Figure 19.1)		Ni 30 at %	293	36.3	0.05
		Ni 20 at %	293	26.6	0.24
		Ni 10 at %	293	14.1	0.52
Cu–Ni–Fe[48]		Ni 45 Fe 2 at %	300	44	—
		Ni 51 Fe 1 at %	300	48	—
Cu–P		P 0.48	293	8.4	0.84
		P 0.93	293	15.2	0.50
Cu–Sn etc.		Sn 12	293	18	0.5
		Sn 5	273	9.5	—
		Sn 6 Zn 4	288	13.5	—
		Sn 5 Zn 5 Pb 5	273	10.5	—
		Sn 5 Ni 5 Zn 2	273	10.5–14	—
Cu–Zn etc.		Zn 10	293	3.8	—
		Zn 15	273	4.65	—
		Zn 30	273	6.65	1.6
		Zn 40	273	6.81	1.7
		Zn 39 Fe 1 Sn 1	273	7.03	—
	(German silver)	Zn 41 Ni 9 Pb 2	293	30.5	—
	(Admiralty brass)	Zn 27.6 Sn 1	273	6.93	—
Cu–Zn–Ni		Zn 32 Ni 8	273	72	—
		Zn 24 Ni 18	273	30.9	0.04
		Zn 26 Ni 30	273	47.6	0.04
		Zn 25 Ni 14	293	33	0.40
Fe–C etc.	(Mild steel)	C 1 Si 1.8	291	12.0	—
	(Cast iron)	C 3.4	293	66.0	—
Fe–Cr		Cr 20	293	62	—
		Cr 12	293	60	—
Fe–Cr–Ni etc.		Cr 18 Ni 8	293	73	—
		Cr 25 Ni 12	293	87	—
		Cr 25 Ni 20	293	88	—
		Cr 18 Ni 37	293	108	—
		Cr 17.7 Ni 7.4 Al 1.2 Mo 0.7 Si 0.4	273	71–102	—
		Cr 18.4 Ni 9.7 Mn 1.4 Si 0.6	273	70.4	—
		Cr 17.9 Ni 9.8 Mn 1.4 Si 0.6 Ti 0.4	273	73.9	—
		Cr 17.7 Ni 12.2 Mo 2.8 Mn 1.5 Si 0.7	273	76.5	—
		Cr 17.5 Ni 13.1 Mo 2.7 Mn 1.7 Si 0.4	273	71.8	—
		Cr 17.4 Ni 13.3 Mo 2.9 Mn 1.5 Si 0.5	273	77.6	—
		Cr 17 Ni 12.8 Mo 2.7 Mn 1.5 Si 0.4	273	75.0	—

(*continued*)

Table 19.2 RESISTIVITY AND TEMPERATURE COEFFICIENT OF SOME ALLOYS—*continued*

Alloy		Nominal composition wt. %†	Temperature K	Resistivity $10^{-8}\ \Omega m$	TCR $10^{-3}\ K^{-1}$
Fe–Ni		*See* Figures 19.2 and 19.3	—	—	—
Fe–Ni–Co		Ni 29 Co 17	293	49	3.7
Fe–Si		Si 25	293	45	
		Si 4	293	62	8.0
Fe–Ti etc.		Ti 2.5 C 0.15	293	16	—
Fe–V etc.		V 5 C 1.1	293	121	—
Fe–W etc.		W 20 C 0.2	293	24	—
		W 5 C 0.2	293	20	—
Mg–Al–Zn		Al 9 Zn 2	273	16	—
		Al 4.9 Zn 0.9	273	11.3	—
Na–K liquid[42,50]		K 20 at %	373	28.0	—
		K 40 at %	373	38.2	—
		K 60 at %	373	40.8	—
		K 80 at %	373	33.3	—
Ni–Al etc.		Al 50 at %[49]	300	9.3	—
		Al 2 Mn 2.5 Fe 0.5 Si 1	273	33.3	1.2
		Al 1.6 Si 1.2 Fe 0.1	273	28.1	2.4
Ni–Co–Cr etc.	‡(Udimet 700)	Co 19 Cr 15.2 Mo 5 Al 4.4 Ti 3.4 Fe 0.1	273	131.5	—
Ni–Cr[51]		Cr 5.5 at %	273	41	—
		Cr 11.5 at %	273	70	—
		Cr 27 at %	273	110	—
Ni–Cr etc.	‡(Evanohm)	Cr 20 Al 2.5 Cu 2.5	293	133	0.00
	‡(Karma)	Cr 20 Al 2.5 Fe 2.5	293	134	0.00
		Cr 15 Fe 7	273	98	—
		Cr 20 Fe 5–10	293	109	—
	‡(Hastelloy X)	Cr 22 Fe 20 Mo 9 C 0.15	273	113.8	—
	‡(Inconel X)	Cr 15.4 Fe 6.9 Ti 2.5 Al 0.9	273	124	—
		Cr 9.46 Fe 0.2 Si 0.4	273	70	0.4
Ni–Cu etc.	‡(Monel)	Cu 30 Fe 1.4 Mn 1	273	48	1.9
Ni–Mn		Mn 5	273	18	—
Ni–Mo–Fe		Mo 32 Fe 6	273	135	—
Pb–Sb		Sb 6	273	23	—
Pb–Te–Cu		Te 0.04 Cu 0.06	273	20	—
Pt–Ir		Ir 10	273	24.8	1.3
Pt–Rh		Rh 10	273	18.7	1.66
		Rh 13	273	19.0	1.56
Rh–Co[52]		Co 11 at %	4.2	3.07	—
		Co 20 at %	4.2	4.96	—
		Co 42 at %	4.2	18.98	—
Sn–Pb		Pb 10	288	13.5	—
		Pb 40	273	15	—
		Pb 66.7	288	16	—
Ti–Al–V etc.		Al 6.2 V 4 Fe 0.13	273	167.5	—
Ti–V–Cr–Al etc.		V 13.1 Cr 10.8 Al 3 Fe 0.17	273	149.2	—
Zn–Cu		Cu 1.05	273	6	—
Zn–Al–Cu etc.[53]		Al 4.1 Cu 3.1 Mg 0.05	295	7.2	—

* TCR applies over a restricted range at room temperature.
† Unless stated otherwise.
‡ Karma is a trade name of British Driver Harris Co. Ltd.
  Inconel and Monel are trade names of Henry Wiggin Co. Ltd.
  Udimet is a trade name of Special Metals Corp.
  Evanohm is a trade name of Wilbur B. Driver Co.
  Hastelloy is a trade name of Cabot Corporation.

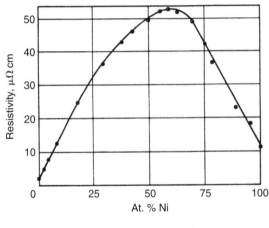

Figure 19.1   *Resistivity of copper-nickel alloys*

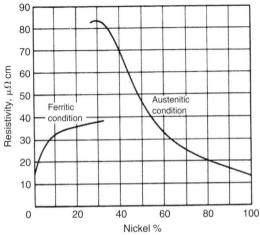

Figure 19.2   *Electrical resistivity of nickel–iron alloys*

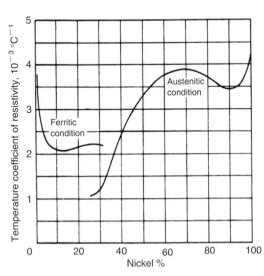

Figure 19.3   *Temperature coefficient of electrical resistivity of nickel–iron alloys*

Table 19.3 RESISTIVITY OF SOME SPECIFIC COPPER ALLOYS

Alloy No.	Composition or description	Resistivity at 293 K 10^{-8} Ωm
101, 102	OFHC	1.55–1.58
110	ETP	1.56–1.65
122	phos. deox.	2.0–2.1
220	10% Zn	3.8–3.9
510	5% Sn	10.5–11.5
706	10% Ni	16.5–19.0

Table 19.4 EFFECT ON THE RESISTIVITY OF EC ALUMINIUM OF SOME ADDITIONAL ELEMENTS

Added element 1 at. %	Increase in resistivity at 293 K 10^{-8} Ωm
Antimony	0.238
Cerium	0.049
Gadolinium	0.40
Hafnium	2.20
Lanthanum	0.147
Molybdenum	2.04
Niobium	0.44
Tantalum	1.32
Thorium	0.265
Tungsten	7.4
Yttrium	0.255
Zirconium	0.988
	Resistivity of EC aluminium 2.694×10^{-8} Ωm

Figure 19.1). The transition elements show complex resistivity and TCR behaviour upon alloying. Very low TCRs, which may be of importance industrially, are obtained in some quaternary Ni–Cr alloys after suitable heat treatment. TCR values given here are for the temperature interval 273–373 K.

Theoretical predictions for the resistivity of alloys have recently been discussed in references 42–46.

Dilute copper and aluminium alloys have important applications as electrical conductors. Table 19.3 lists specified resistivity values for certain copper alloys (*see* references 54–56). Table 19.4 shows the effect on the resistivity of EC aluminium (99.45 wt% Al) of the addition of 1 at.% of various solutes. When considering aluminium as an electrical conductor, it is important to achieve strengthening by the addition of suitable solutes without undue reduction in conductivity. The least detrimental additions form intermetallic compounds that remain out of solid solution (*see* references 57 and 57A).

19.2   Superconductivity

Below a transition temperature T_c some elements, compounds and alloys become superconducting. In this state they exhibit zero electrical resistance and some become perfectly diamagnetic. The application of a magnetic field whether applied externally or generated by a current passing along the superconductor in excess of a critical or limiting value restores the material to its normal resistive state.

Table 19.5 lists the elements which become superconducting when cooled to a sufficiently low temperature together with their transition temperatures and critical fields at 0 K (H_0).[58–60] The critical field H_c at temperature T is closely represented by the releationship

$$H_c = H_0 \left[1 - \left(\frac{T}{T_c} \right)^2 \right]$$

Table 19.5 TRANSITION TEMPERATURES AND CRITICAL FIELDS FOR SUPERCONDUCTING ELEMENTS

Element		T_c K	H_0 10^4 T	Element	T_c K	H_0 10^4 T
Aluminium		1.18	99	Osmium	0.61	82
Americium		1.06	—	Protactinium	<1.0	—
Beryllium		0.026	—	Rhenium	1.70	201
Cadmium		0.52	30	Ruthenium	0.49	66
Gallium		1.08	51	Tantalum	4.47	830
Hafnium		0.128	—	Technetium	7.8	—
Indium		3.41	283	Thallium	2.38	162
Iridium		0.11	—	Thorium	1.38	—
Lanthanum	α	4.88	—	Tin	3.72	306
	β	6.0	1 600	Titanium	0.4	100
Lead		7.20	803	Tungsten	0.02	—
Mercury	α	4.15	411	Vanadium	5.4	1 310
	β	3.95	340	Zinc	0.85	53
Molybdenum		0.92	—	Zirconium	0.61	47
Niobium		9.25	1 944			

Table 19.6 CRITICAL TEMPERATURES OF SOME SUPERCONDUCTING ALLOYS AND COMPOUNDS

Alloy or compound	T_c K	Ref.	Alloy or compound	T_c K	Ref.
$Al_{10}V$	1.6	61	Nb_3Rh	2.5	63
AuGa	1.2	62	$NbSi_2$	2.9	67
AuTl	1.92	62	Nb_3Sn	18.3	63
Cr_3Os	4.03	63	Ta_3Ge	8	63
Cr_3Ru	3.43	63	$TaGe_2$	2.7	67
$EuIr_2$	2.6	64	*Ta–42% Ir	6.6	68
$HfGe_2$	2.2	65	$TaSi_2$	4.4	67
$LaGe_2$	2.65	65	Ta_3Sn	6.4	63
La_3In	10.5	64	*Tc–3% Ti	10.2–10.9	70
Mo_3Ge	1.4	63	Ti_3Ir	4.6	63
Mo_3Ir	8.1	63	Ti_3Sb	5.8	63
Mo_3Os	11.68	63	U_6Co	2.5	64
Mo_3Pt	4.56	63	U_6Fe	3.9	64
Mo_3Si	1.3	63	U_6Mn	2.3	64
Mo_2Tc_3	13.5	63	V_3Al	9.6	63
Nb_3Al	18.9	63	V_3Au	3.2	63
Nb_3Au	11	63	V_3Ga	15.4	63
Nb_3Bi	2.25	63	V_3Ge	7	63
Nb_3Ga	20.3	63	V_3In	13.9	63
Nb_5Ga_3	1.35	66	V_3Ir	1.39	63
Nb_3Ge	23	63	V_3Os	5.15	63
$NbGe_2$	16	67	V_3Pb	3.7	63
Nb_3In	8	63	V_3Re_7	9.0	64
Nb_3Ir	1.76	63	V_3Si	17.1	63
*Nb–42% Ir	4–9	68	V_3Sn	4.3	63
*Nb–40% Ir–5% O	10.4–11.7	68	$YbGe_2$	3.8	65
NbN	14.5	64	$ZrGe_2$	8	65
Nb_3P	1.83	69	Zr_2Ir	7.3	71
Nb_3Pt	10	63	$ZrIr_2$	4.1	72

*at. %

The maximum axial current which a superconducting material of cylindrical form, radius r, will sustain while remaining superconducting is given by

$$I_c = 0.5 H_c r$$

Superconductivity is observed in indium and molybdenum only when these elements are very pure.[60] Traces of magnetic impurity, e.g. Fe > 0.02%, prevent its occurrence.

A large number of intermetallic compounds, non-stoichiometric alloys and solid solutions exhibit superconductivity. Among the compounds, the sodium chloride structure (e.g. NbN) and the A–15 or 'beta-tungsten' structure (e.g. Nb_3Sn) are two of the more common crystal types. There are currently nearly fifty A–15 compounds known to be superconducting. These alloys and compounds are potentially of considerable technological importance. Table 19.6 lists those with critical temperatures greater than 1 K.

Ternary additions can often lead to increase in critical temperature. Useful results have been discussed in references 68 and 73–77. Increased current-carrying capacity has also been seen to result from energetic particle irradiation.[78]

Superconducting compounds are usually brittle in bulk, and special techniques are required to fabricate them into useful forms. Powder metallurgical techniques may be used, while another typical method results in a composite of multifilamentary superconducting phase in a copper matrix. Superconductors may also be produced as thin films by chemical vapour deposition or sputtering (*see* references 79–83 and chapter 35).

Superconducting properties tend to be dependent upon the method of preparation and heat treatment; this observation is discussed in many of the references quoted.

For overall discussion of superconductivity, its applications, and the prospects of further developments of industrial importance, see references 64 and 84.

Much of the recent interest in superconductors has focused on high T_c ceramic materials. Ceramics are beyond the scope of the present work. However some reviews on high T_c materials may be found in references 85–89. Information on potential applications of these materials is contained in references 90–93.

REFERENCES

1. B. N. Aleksandrov., *Fiz. Metall.*, 1971, **31**, 1175.
2. J. S. Brown *et al.*, *J. Phys. F.*, 1978, **8** (8), 1703.
3. B. Delley and H. Beck, *J. Phys. F.*, 1979, **9** (3), 505 and 517.
4. H. N. Dunleavy and W. Jones, *J. Phys. F.*, 1978, **8** (7), 1477.
5. H. J. Güntherodt *et al.*, *J. Phys. F.*, 1976, **6** (8), 1513.
6. S. N. Khanna and A. Jain, *J. Phys. F.*, 1977, **7** (12), 2523.
7. F. R. Vukajlovic *et al.*, *Physica*, 1977, **92B + C** (1), 66.
8. Y. Waseda *et al.*, *J. Phys. F.*, 1978, **8** (1), 125.
9. S. N. Khanna and A. Jain, *J. Phys. Chem. Solids*, 1977, **38** (5), 447.
10. N. N. Sinha and P. L. Srivastava, *Phys. Status Solidi* (*b*), 1978, **90** (1), 369.
11. T. J. Bastow, *Phys. Lett.*, 1977, **60A** (5), 487.
12. F. J. Ohkawa, *J. Phys. Soc. Jpn.*, 1978, **44** (4), 1105.
13. M. Isshiki and K. Igaki, *Trans. Jpn Inst. Met.*, 1978, **19** (8), 431.
14. P. L. Rossiter, *J. Phys. F.*, 1979, **9** (5), 891.
15. J. Ziman, *Phil. Mag.*, 1961, **6**, 1013.
16. British Aluminium Company Ltd. Data Sheet.
17. R. O. A. Hall *et al.*, *J. Low Temp. Phys.*, 1977, **27** (1–2), 305.
18. G. K. White and S. B. Woods, *Phil. Mag.*, 1958, **3**, 342.
19. W. Meissner and B. Voigt, *Ann. Phys.*, 1930, **7**, 892.
20. T. C. Chi, *J. Phys. Chem. Ref. Data*, 1979, **8** (2), 439.
21. S. Gabe and E. G. Evans, *Phil. Mag.*, 1935, **19**, 773.
22. T. C. Chi, *J. Phys. Chem. Ref. Data*, 1979, **8** (2), 339.
23. M. V. Vedernikov *et al.*, *J. less-common Met.*, 1977, **52** (2), 221.
24. H. J. Güntherodt *et al.*, *Phys. Lett.*, 1974, **50A**, 313.
25. G. T. Meaden, 'Electrical Resistance of Metals', Plenum, New York, 1965.
26. K. Hirata *et al.*, *J. Phys. F.*, 1977, **7**, 419.
27. G. K. White and S. B. Woods, *Phil. Trans. R. Soc. London, Ser. A*, 1959, **251**, 273.
28. R. W. Powell, *Proc. Roy. Soc., London, Ser. A*, 1951, **209**, 525.
29. R. N. Lyon, 'Liquid Metals Handbook', 1952. Atomic Energy Commission, Washington DC.
30. R. W. Powell *et al.*, *Phil. Mag.*, 1962, **7**, 1183.
31. E. Grüneisen, *Ergebn. exakt. Naturw.*, 1945, **21**, 50.
32. G. T. Meaden and P. Pelloux-Gervais, *Cryogenics*, 1965, **5**, 227.
33. J. M. Abraham and B. Deviot, *J. less-common Met.*, 1972, **29**, 311.
34. B. C. Dupree *et al.*, *J. Phys. F.*, 1975, **5** (11), L200.
35. P. Haen and G. T. Meaden, *Cryogenics*, 1965, **5**, 194.
36. H. K. Onnes and W. Tuyn, *Proc. Acad. Sci. Amsterdam*, 1923, **25**, 443.
37. G. T. Meaden, *Proc. Roy. Soc., London, Ser. A*, 1963, **276**, 553.
38. W. Tuyn and H. K. Onnes, *Proc. Acad. Sci. Amsterdam*, 1933, **26**, 504.
39. F. Warkusz, *Thin Solid Films*, 1977, **41** (3), 261.
40. F. Warkusz, *J. Phys. D*, 1978, **11** (5), 689.

41. F. Warkusz, *Thin Solid Films*, 1978, **52** (2), 29.
42. L. N. Korochkina *et al.*, *Phys. Met. Metallogr.*, 1975, **40** (2), 1.
43. S. I. Masharov and N. M. Rybalko, *ibid.*, 5.
44. C. A. Rahim and R. D. Barnard, *J. Phys. F*, 1978, **8** (9), 1957.
45. L. V. Meisel and P. J. Cote, *J. Phys. F*, 1977, **7** (12), L321.
46. A. Fert and I. A. Campbell, *J. Phys. F*, 1976, **6** (5), 849.
47. A. Eroglu *et al.*, *Phys. Status Solidi (b)*, 1978, **87** (1), 287.
48. V. M. Beilin *et al.*, *Phys. Met. Metallogr.*, 1977, **43** (2), 68.
49. Y. Yoshitomi *et al.*, *Solid St. Commun.*, 1976, **20** (8), 741.
50. J. Hennephof *et al.*, *Physica*, 1971, **52**, 279.
51. Y. D. Yao *et al.*, *J. Low Temp. Phys.*, 1975, **21** (3–4), 369.
52. A. Tari, *J. Phys. F*, 1976, **6** (7), 1313.
53. K. Mori and Y. Saito, *Jpn J. Appl. Phys.*, 1976, **15** (10), 1997.
54. Copper Development Association, Standards Handbook.
55. K. Miska, *Mater. Eng.*, 1977, **85** (5), 28.
56. Y. T. Hsu and B. O'Reilly, *J. Met.*, 1977, **29** (12), 21.
57. M. Mujahid and N. N. Engel, *Scr. Metall.*, 1979, **13** (9), 887.
57A. A. Kutner *et al.*, *Aluminium*, 1976, **5**, 322.
58. B. W. Roberts, *J. Phys. Chem. Ref. Data*, 1976, **5**, 581 and NBS. Tech. Note, 983, 1978.
59. E. A. Lynton, 'Superconductivity', Methuen, London 1962.
60. A. C. Rose-Jones, 'Low Temperature Techniques', English Universities Press, London, 1964.
61. T. Claeson, *Commun. Phys.*, 1977, **2** (3), 53.
62. H. R. Khan and C. J. Raub, *Gold Bull.*, 1975, **8** (4), 114.
63. D. Dew-Hughes, *Cryogenics*, 1975, **15**, 435.
64. B. T. Matthias and P. R. Stein, Superconducting Materials, in 'Physics of Modern Materials', Vol. 2, International Atomic Energy Agency, Vienna, 1980.
65. A. K. Ghosh and D. H. Douglass, *Solid State Commun.*, 1977, **23** (4), 223.
66. E. E. Havinga *et al.*, *Phys. Lett.*, 1969, **29A**, 109.
67. C. M. Knoedler and D. H. Donglass, *J. Low Temp. Phys.*, 1979, **37** (1–2), 189.
68. W. L. Johnson and S. J. Poon, *J. less-common Met.*, 1975, **42** (3), 355.
69. J. O. Willis *et al.*, *Phys. Rev. B (Solid State)*, 1978, **17** (1), 184.
70. C. C. Koch, *J. less-common Met.*, 1976, **44**, 177.
71. D. P. Moiseev *et al.*, *Phys. Met. Metallogr.*, 1975, **39** (6), 33.
72. B. T. Matthias *et al.*, *J. Phys. Chem. Sol.*, 1961, **19**, 130.
73. N. Y. Alekseevskii *et al.*, *Phys. Met. Metallogr.*, 1977, **43** (1), 29.
74. M. Drys and N. Iliew, *J. less-common Met.*, 1976, **44**, 235.
75. C. H. Kopetskii and A. V. Pavlyuchenko, *Phys. Met. Metallogr.*, 1977, **43** (3), 38.
76. R. Somasundaram *et al.*, *J. Appl. Phys.*, 1976, **47** (10), 4656.
77. R. G. Sharma and N. E. Aleksivkii, *J. Phys. D*, 1975, **8** (15), 1783.
78. G. W. Cullen, Proc. Summer Study on Superconducting Devices and Accelerators, Brookhaven National Laboratory, 1968.
79. J. E. Kunzler *et al.*, *Phys. Rev. Lett.*, 1961, **6**, 89.
80. J. P. Harbison and J. Bevk, *J. Appl. Phys.*, 1977, **48** (12), 5180.
81. J. M. E. Harper *et al.*, *J. less-common Met.*, 1975, **43** (1/2), 5.
82. L. A. Pendrys and D. H. Douglass, *Solid State Commun.*, 1976, **18** (2), 177.
83. L. Schultz and R. Bormann, *J. Appl. Phys.*, 1979, **50** (1), 418.
84. T. Luhman and D. Dew-Hughes (eds). 'Metallurgy of Superconducting Materials', Treatise on Materials Science and Technology, Vol. 14, Academic Press, New York, 1979.
85. R. J. Cava, *J. Am. Ceram. Soc.*, 2000, **83** (1), 5–28.
86. D. P. Norton, *Annu. Rev. Mater. Sci.*, 1998, **28**, 299–343.
87. Z. Fisk and J. L. Sarrao, *Annu. Rev. Mater. Sci.*, 1997, **27**, 35–67.
88. D. Pines, *Physica C*, 1997, **282–287**, 273–278.
89. P. V. Reddy, *Physica C*, 2001, **364–365**, 232–234.
90. M. Campbell and D. A. Cardwell, *Cryogen.*, 1997, **37**, 567–575.
91. S. P. Hornfeldt, *Physica C*, 2000, **341–348**, 2531–2533.
92. J. Kellers and L. J. Masur, *Physica C*, 2002, in press.
93. P. Manuel, F. X. Camescasse, M. Coevoet, V. Leitloff, F. Lesur, E. Serres, P. Suau and J. A. Muraz, *Physica C*, 2002, in press.

20 Magnetic materials and their properties

20.1 Magnetic materials

All materials have magnetic properties. These characteristic properties may be divided into five groups: diamagnetic, paramagnetic, ferromagnetic, antiferromagnetic and ferrimagnetic. Only the ferromagnetic and ferrimagnetic materials have properties which are useful in practical applications.

Ferromagnetic properties are confined almost entirely to iron, nickel and cobalt in their alloys. The only exceptions are some alloys of manganese and some of the rare earth elements. Of these gadolinium has the highest Curie temperature, about 16°C (see the Appendix for definitions and conversion factors). Some of the magnetic properties of the ferromagnetic elements including the rare earths are given in Tables 20.1 and 20.2. Ferrimagnetism is the magnetism of the oxides of the ferromagnetic elements. These are variously called ferrites and garnets. The basic ferrite is magnetite (Fe_3O_4) which can be written as $FeO \cdot Fe_2O_3$. By substituting for the FeO with other divalent oxides a wide range of compounds with useful properties can be produced. The main advantage of these materials is that they have high electrical resistivity which minimises eddy currents when they are used at high frequencies.

Ferromagnetic and ferrimagnetic materials are characterised by moderate to high permeabilities (*see* Tables 20.4, 20.5 and 20.7–20.11). The permeability varies with the applied magnetic field, rising to a maximum at the knee of the *B–H* curve and reducing to a low value at very high fields. They also exhibit magnetic hysteresis whereby the intensity of magnetisation of the material varies according to whether the applied field is being increased in a positive sense or decreased in a negative sense. A typical hysteresis loop is shown in Figure 20.1 together with some of the more important derived magnetic properties. When the magnetisation is cycled continuously round a hysteresis

Table 20.1   FERROMAGNETIC ELEMENTS[9]

	$J_s(T)$		Curie temperature °C	σ_{4K}	$\sigma_{20°C}$
	20°C	4K			
Fe	2.153	2.193	770	221.7	217.6
Ni	0.617	0.656	358	58.6	55.1
Co	1.790	1.797	1 121	162.5	161.9
Gd	0	2.47	16	250	0

Table 20.2   RARE EARTH ELEMENTS[9]

	Gd	Tb	Dy	Ho	Er	Tm
$J_s(T)$ 4 K	2.47	3.43	3.75	3.81	3.41	2.70
σ_0 4 K	250	330	350	345	300	230
T_c K	289	222	85	20	20	25

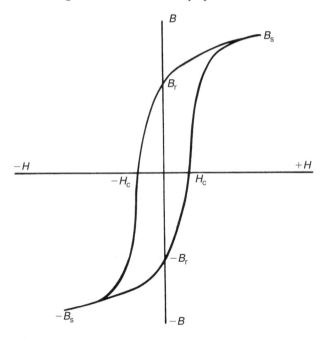

Figure 20.1 *Typical hysteresis loop*

loop, as for example when the applied field arises from an alternating current, there is an energy loss proportional to the area of the included loop. This is termed hysteresis loss, and is measured in joules per metre.[3] High hysteresis loss is associated with permanent magnetic properties exhibited by materials commonly termed 'hard' magnetic materials as these often have hard mechanical properties. Those materials with low hysteresis loss are termed 'soft' and are difficult to magnetise permanently. The ferromagnetic properties of these elements, alloys, and compounds disappear reversibly if heated above the Curie temperature, when they become paramagnetic.

Paramagnetic substances of elongated shape tend to align themselves with an applied magnetic field as do the ferromagnetics, but do not exhibit any permanent magnetic properties when the field is removed as hysteresis is absent. Their magnetic susceptibility is very small but positive, and consequently their relative permeability is only slightly greater than unity. The susceptibility is generally independent of field strength but may decrease with increasing temperature. Among the paramagnetic substances are included many iron salts and rare-earth elements salts, palladium and platinum, sodium, potassium, and oxygen, and the ferromagnetics above their Curie point.

The diamagnetic substances take on a very small magnetisation which is opposed to the sense of the applied field. Thus an elongated specimen will align itself transversely to a non-uniform field. The diamagnetics have negative susceptibilities and consequently permeabilities which are very slightly less than unity. Many of the metals and most non-metals are diamagnetic. It is usual to compare paramagnetic and diamagnetic substances by their susceptibilities rather than by permeability or saturation magnetisation as for the ferromagnetic materials. The mass susceptibilities of the elements at room temperature are given in Table 20.3.[5–9]

For bulk applications of magnetic material, iron, steel or cast iron are used. They have the advantage of cheapness and strength. These materials should only be used for d.c. applications as they have low electrical resistivity which would give rise to eddy currents if used in alternating fields. The properties of some steels and cast irons are given in Tables 20.4 and 20.5.[10–13]

For a list of current reviews, at the time of writing, on magnetic materials see references 14–19.

20.2 Permanent magnet materials[20,21]

The properties of a permanent magnet material are given by the demagnetisation curve, Figure 20.2, the second quadrant of the B–H hysteresis loop. This extends from the remanence B_r to the coercivity

Table 20.3   MASS SUSCEPTIBILITIES OF THE ELEMENTS[5-9]

Element	Atomic No.	Mass susceptibility 10^{-8} m^3 kg^{-1}	Element	Atomic No.	Mass susceptibility 10^{-8} m^3 kg^{-1}
H	1	−2.48	Ag	47	−0.23
He	2	−0.59	Cd	48	−0.23
			In	49	−0.14
Li	3	+0.63	Sn	50	−0.31
Be	4	−1.26	Sb	51	−1.09
B	5	−0.87	Te	52	−0.39
C	6	−0.62	I	53	−0.45
N	7	−1.00	Xe	54	−0.43
O	8	+133.5			
F	9		Cs	55	−0.28
Ne	10	−0.41	Ba	56	+1.13
			La	57	+1.02
Na	11	+0.64	Ce	58	+22.0
Mg	12	+0.69	Pr	59	+42.3
Al	13	+0.82	Nd	60	+48.0
Si	14	−0.16	Pm	61	—
P	15	−1.13	Sm	62	+11.1
S	16	−0.62	Eu	63	+27.6
Cl	17	−0.72?	Gd	64	*
Ar	18	−0.60	Tb	65	+1 360
			Dy	66	+545
K	19	+0.65	Ho	67	+549
Ca	20	+1.38	Er	68	+377
Sc	21	+8.80	Tm	69	+199
Ti	22	+4.21	Yb	70	+0.59
V	23	+6.28			
Cr	24	+4.45	Lu	71	+0.12
Mn	25	+12.2	Hf	72	+0.53
Fe	26	*	Ta	73	+1.07
Co	27	*	W	74	+0.39
Ni	28	*	Re	75	+0.46
			Os	76	+0.06
Cu	29	−0.10	Ir	77	+0.23
Zn	30	−0.20	Pt	78	+1.22
Ga	31	−0.30			
Ge	32	−0.15	Au	79	−0.18
As	33	−0.39	Hg	80	−0.21
Se	34	−0.40	Tl	81	−0.30
Br	35	−0.49	Pb	82	−0.15
Kr	36	−0.44	Bi	83	−1.70
			Po	84	—
Rb	37	+0.26	At	85	—
Sr	38	−0.25	Rn	86	—
Y	39	+6.66			
Zr	40	+1.66	Fr	87	—
Nb	41	+2.81	Ra	88	—
			Ac	89	—
Mo	42	+1.17	Th	90	+0.53
Tc	43	+3.42	Pa	91	+3.25
Ru	44	+0.54	U	92	+2.15
Rh	45	+1.32	Np	93	—
Pd	46	+6.57	Pu	94	+3.14
			Am	95	+5.15

H_{cB}. It can be shown that when a piece of permanent magnet material is put into a magnetic circuit, the magnetic field generated in a gap in the circuit is proportional to BHV, where B and H are the corresponding points at a point on the demagnetisation curve and V is the volume of permanent magnet. So to obtain a given field with the minimum volume of magnet material we require the product BH to be a maximum. The magnet is then designed so that its B,H value is as close as possible to the $(BH)_{max}$ value. It is also useful to use the BH_{max} value to compare the characteristics

Table 20.4 MAGNETIC PROPERTIES OF WROUGHT STEELS[12-13]

Grade*	En No	Material	Flux density T		Rel. permeability		Max. flux density T	Remanence T	Coercive force A m^{-1}
			H = 4 kA m^{-1}	H = 8 kA m^{-1}	Maximum at	H A m^{-1}	H = 20 kA m^{-1}		
220M07	1A	Free cutting carbon steel C 0.15 max., Si 0.40 max., Mn 0.90–1.30, S 0.20–0.35	1.50	1.76	1 200	800	1.95	0.87	240–400
040A04	2A/1	Low carbon steel C 0.08 max., Si 0.40 max., Mn 0.50 max., S and P 0.05 max.	1.67	1.77	2 400–5 000	400–160	1.96	0.5–1.1	40–120
070M20	3	Carbon steel C 0.25 max., Si 0.05–0.35, Mn 1.00 max., S and P 0.05 max.	1.57	1.74	1 100	800	1.96	1.00	240–400
080M30	5	'30' carbon C 0.25–0.35, Si 0.05–0.35, Mn 0.60–1.00, S and P 0.05 max.	1.52	1.73	1 000	800	1.88	—	800
070M55	9	'55' carbon C 0.50–0.60, Si 0.10–0.40, Mn 0.50–0.90, S and P 0.05 max.	1.45	1.67	400	2 000	1.81	1.30	1 200
Low alloy steels		C 0.20–0.60, Si 0.1–0.40, Mn 2.0 max., S and P 0.05 max., Ni 1.0 max., Cr 1.0 max.	1.40–1.65	1.70–1.90	400–650	—	1.75–1.90	0.88–1.55	600–1 600
3% Nickel steel		C 0.20–0.60, Si 0.10–0.40, Mn 0.8 max., S and P 0.05 max., Ni 2.5–3.0, Cr 1.0 max., Mo 0.5 max.	1.50–1.70	1.73–1.85	600	1 600	1.85–1.95	1.43	980
3% Chromium steel		C 0.20–0.60, Si 0.10–0.40, Mn 0.8 max., S and P 0.05 max., Cr 3.0, Ni 0.40 max.	1.65	1.75	680	1 600	1.75–1.85	1.50	880
Chromium rust resisting steels		C 0.35 max., Si 0.1 max., Mn 1.0 max., S and P 0.05 max., Ni 3.0 max., Cr 12–20	0.95–1.30	1.06–1.40	660	720	1.43–1.64	0.60–1.00	400–1 250

Information from 'The Mechanical and Physical Properties of British Standard En Steels' by J. Woolman and R. A. Mottram, volumes 1, 2 and 3, Pergamon, Oxford, 1964.[12]
*The numbered grades refer to BS 970:PART 1:1983 (see Ref. 13).

Table 20.5 SOME DATA ON THE MAGNETIC PROPERTIES OF CAST STEELS AND CAST IRONS

Material	Condition	Flux density T $H = 4\,\mathrm{kA\,m^{-1}}$	Flux density T $H = 8\,\mathrm{kA\,m^{-1}}$	Rel. permeability Maximum	at H A m^{-1}	Max. flux density T $H = 20\,\mathrm{kA\,m^{-1}}$	Remanence T $H = 20\,\mathrm{kA\,m^{-1}}$	Coercive force A m^{-1} $H = 20\,\mathrm{kA\,m^{-1}}$
Cast steel[11]								
0.10% carbon steel, Si, 0.33%; Mn, 0.67%	Annealed	1.62	1.74	2420	216	1.92	0.80	136
	Normalised	1.64	1.76	1950	248	1.96	0.85	168
	As-cast	1.61	1.74	2100	208	1.95	0.66	128
0.19% carbon steel, Si, 0.30%; Mn, 0.48%	Annealed	1.60	1.73	2100	224	1.88	0.87	156
	Normalised	1.57	1.72	1520	360	1.91	0.90	216
	As-cast	1.55	1.69	1720	232	1.89	0.73	168
0.34% carbon steel, Si, 0.44%, Mn, 0.55%	Annealed	1.53	1.66	1200	400	1.85	1.05	296
	Normalised	1.50	1.67	970	575	1.87	1.05	440
	As-cast	1.45	1.65	840	480	1.87	0.85	440
Manganese steel								
0.19% C, 0.48% Si, 1.14% Mn	A 925°C	1.51	1.66	1300	400	1.86	—	—
0.29% C, 0.29% Si, 1.40% Mn	A 950°C, N 880°C, T 600°C, a.c.	1.50	1.68	650	960	1.78	—	—
0.31% C, 0.52% Si, 1.37% Mn	A 950°C, OQ 860°C, T° 629°C, a.c.	1.55	1.65	750	960	1.70	—	—
Chromium-molybdenum steel								
0.31% C, 0.69% Mn	A 925°C, N 880°C, T 700°C, a.c.	1.45	1.59	—	—	1.76	—	—
1.16% Cr, 0.39% Mo	A 925°C, OQ 860°C, T 650°C, a.c.	1.54	1.66	—	—	1.81	—	—

For the cast-irons below, the column shown as "Max. flux density T" is Hysteresis loss J M^{-3} cycle^{-1}.

Material	Condition	Flux density T $H = 4\,\mathrm{kA\,m^{-1}}$	Flux density T $H = 8\,\mathrm{kA\,m^{-1}}$	Rel. permeability Maximum	at H A m^{-1}	Hysteresis loss J M^{-3} cycle^{-1}	Remanence T $H = 20\,\mathrm{kA\,m^{-1}}$	Coercive force A m^{-1} $H = 20\,\mathrm{kA\,m^{-1}}$
Cast-irons[10]								
Grey iron T.C. 3.12%, Si 2.2%, Mn 0.67%, P 0.13% (low phosphorus)	As-cast	0.70	0.86	315	640	2700	0.41	560
	Annealed	0.93	1.07	1560	200	700	0.44	200
Grey iron T.C. 3.3%, Si 2.04%, Mn 0.52%, P 1.03% (high phosphorus)	As-cast	0.79	0.97	281	1040	2730	0.43	720
	Annealed	0.85	1.00	760	320	1190	0.44	280
Whiteheart Malleable TC 1.01%, Si 0.66%, Mn 0.25%	Pearlitic center	1.32	1.42	730	560	1490	0.75	360
TC 0.46%, Si 0.66%, Mn 0.25%	Mainly ferritic	1.35	1.61	1455	280	840	0.74	200
Blackheart Malleable TC 1.26%, Si 0.83%, Mn 0.26%		1.40	—	2120	176	490	0.62	120
Spheroidal Graphite TC 2.9%, Si 2.61%, Mn 0.72%, Ni 2.18%	Pearlitic	0.93	1.12	290	1200	3040	0.54	920
	Annealed, ferritic	1.14	1.26	1150	320	700	0.46	200
Spheroidal Graphite TC 3.64%, Si 1.41%, Mn 0.3% (nickel free)	Annealed	1.23	1.42	2060	240	450	0.61	120

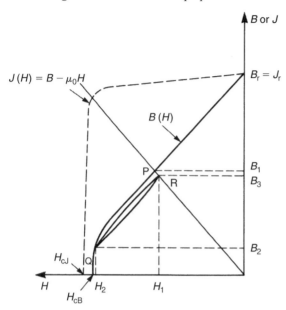

Figure 20.2   *Demagnetisation curve and recoil loop*

of materials. Generally the material with the highest $(BH)_{max}$ will be chosen but this has to be weighed against such considerations as cost, shape, manufacturing problems and stability.

When the magnet is fully magnetised in a completed magnetic circuit, the working point is the remanence B_r. When a gap is made in the circuit, the magnet will be partly demagnetised and its working point will fall to, say, the point P. If now a further demagnetising field is applied to the magnet corresponding to H_1-H_2 the flux density will be reduced to B_2. If this extra field H_1-H_2 is removed the field will recoil along the narrow loop QR to the point R. The corresponding flux density is B_3. This narrow loop is called the recoil loop and its slope is $\mu = \mu_0\mu_r$ where μ_r is the relative recoil permeability and μ_0 is the magnetic constant ($= 4\pi \times 10^{-7}$). This is an important consideration in dynamic applications such as motors, generators and lifting devices where the working point of the magnet changes as the magnetic circuit configuration changes. It is also important when considering premagnetising before assembly into a magnetic circuit. If μ_r is nearly equal to unity as it is for ferrites and for rare earth alloys then the magnets can be premagnetised before assembly into a magnetic circuit without much loss of available flux. However, if μ_r is considerably greater than unity, care must be taken to magnetise the magnet in the assembled magnetic circuit.

The dotted curve in Figure 20.2 is the J–H curve where $B = \mu_0 H + J$. For materials with B_r much greater than $\mu_0 H_{cB}$, e.g. Alnico, the two curves are almost identical, but for ferrites and rare earth alloys where B_r and $\mu_0 H_{cB}$ are close in value the curves become quite different. The J–H curve has only a very gradual slope from B_r; this indicates that the magnetisation in the material is nearly uniform and a new parameters H_{cJ} is introduced. This intrinsic coercivity H_{cJ} is a measure of the difficulty or ease of demagnetisation. As the value of H_{cJ} is increased the more the material will resist demagnetisation due to stray fields, etc. However, as a field of at least three times H_{cJ} must be applied to the magnet for magnetisation, difficulties may be encountered in attaining the full magnetisation if H_{cJ} is too high.

Magnets in the three main groups Alnico, ferrites and rare earth alloys are generally made anisotropic with the properties in one direction considerably better than in the other directions. This best direction is called the preferred direction of magnetisation. The curves and parameters supplied for such a material refer to the properties in this preferred direction.

To choose between these three materials for a particular application, Alnico must be used where magnetic stability is a requirement and this would apply in any type of instrument application. Ferrite will be used where cost is the main consideration and rare earth magnets will be used if the highest possible strength is required or if miniaturisation is needed.

20.2.1   Alnico alloys

A wide range of alloys with magnetically useful properties is based on the Al–Ni–Co–Fe system. They are characterised by high remanence, available energy and moderately high coercivity. They are very stable against vibration and have the widest useful temperature range (up to over 500°C) of any permanent magnet material. But they are mechanically hard, impossible to forge, and difficult to machine except by grinding and special methods such as spark erosion.

The preferred form of magnet is a relatively simple shape made by casting or sintering, precision ground on essential surfaces only. Soft-iron pole pieces may be clamped on or held by a suitable adhesive such as Araldite. Soldering, diecasting and even brazing (with precautions against overheating) may be used.

The commonest alloys contain 23–25% cobalt, 12–14% nickel, about 8% aluminium, a few percent copper and sometimes small additions of niobium and silicon, with the balance iron. They are cooled at a controlled rate in a magnetic field applied in the direction in which the magnets are to be magnetised. The properties are much improved in this direction at the expense of those in other directions. There is finally a fairly prolonged and sometimes rather complicated treatment at temperatures in the range 650–550°C.

The field cooling is much more effective in producing anisotropic properties if the magnets are cast with a columnar structure. Basically a columnar structure is produced by casting the alloy into a hot mould with a cold chill. The mould may be made of special materials which can be preheated to at least 1 250°C in a furnace, or alternatively contain chemicals which heat it by an exothermic reaction. Either method is inconvenient and expensive; there are also limitations on the lengths and shapes of magnets that can be made columnar. This Columax is therefore regarded as a rather expensive, quality material.

The coercivity of Alnico can be improved by a factor of two to three by increasing the cobalt content to 30–40% and adding 5–8% of titanium with possibly some niobium. For the best properties it is necessary to hold these alloys for several minutes in a magnetic field at a constant temperature, accurately controlled to ±10°C, instead of the usual field cooling process. This heat treatment as well as the high cobalt content makes these alloys expensive and they are only used where the higher coercivity is required. Columnar versions of these alloys can also be made but these are only used for very specialised applications.

The Alnico alloys can be produced as castings of up to 100 kg and down to a few grams in weight but it is found more economical to produce the smaller sizes of 50 g and less by the sintering process. In all cases it is advisable to contact the manufacturers before any design is considered.

20.2.2   Ferrite

The permanent magnet ferrites (also called ceramics) are mixed oxides of iron (ferric) oxide with a divalent heavy metal oxide, usually either barium or strontium. These ferrites have a hexagonal crystal structure, the very high anistropy of which gives rise to high values of coercivity, e.g. 150–300 kA m^{-1} (compared with about 110 kA m^{-1} for the best Alnico alloys). The general formula is MO · 5.9Fe$_2$O$_3$ where M is either barium or strontium, and the crystal structure is called magneto plumbite as it was originally found in the equivalent lead oxide compound. These ferrites are made by mixing together barium or strontium carbonate with iron oxide in the correct proportions. The mixture is fired in a mildly oxidising atmosphere and the resulting mixture is milled to a particle size of about 1 micron. This powder is then pressed in a die to the required shape (with a shrinkage allowance), and anisotropic magnets are produced by applying a magnetic field in the direction of pressing. After pressing, the compact is fired. This material is a ceramic and can only be cut by high speed slitting wheels. Ferrite magnets are produced in large quantities in a variety of sizes for different applications. Flat rings are made for loudspeakers ranging up to 300 mm in diameter with a thickness of up to 25 mm. Segments are made for motors with ruling diameters from 40 to 160 mm and rectangular blocks are made for separators with dimensions of up to 150 × 100 × 25 mm. These blocks can be built up into assemblies or cut down into smaller pieces for a variety of applications. In each of these cases the preferred direction of magnetisation is through the shortest dimension. Generally the magnetic length does not exceed 25 mm and if a longer length is required this is built up with magnets in series.

The great success of permanent magnet ferrites is due to the low price per unit of available magnetic energy, the high coercivities, the high restivities and the low density. The isotropic grades are the least expensive to manufacture and may be magnetised into complex pole configurations; they are used for a wide range of relatively small-scale applications. The largest application in terms of market volume is for magnets for loudspeakers; here the high-remanence anisotropic grades are used. The other high-volume application is for field magnets in small d.c. motors, particularly those used in the automotive industry, e.g. for windscreen wipers, blowers, etc. In this application the high-coercivity anisotropic ferrite is formed into an arc-shaped segment, magnetised radially; a pair of segments embrace the armature and provide the stator field. Other applications of anisotropic hard ferrites include magnetic chucks and magnetic filters and separators.

20.2.3 Rare earth cobalt alloys

These are the results of a fairly recent development and the first commercial material was $SmCo_5$. The alloy powder is produced either by reducing the rare earth powder together with cobalt powder or by preparing the alloy and powdering it down. The powder is compacted and pressed, almost invariably in a magnetic field, to produce an anisotropic compact. This is then sintered at $1\,155\pm15°C$ and followed by a tempering treatment at about $900°C$. It must be emphasised that during all the processing of these rare earth alloys a protective atmosphere is used to prevent the oxidation of the rare earth metals which are very reactive.

Another development in this series is the Sm_2Co_{17} composition which is really the generic name for a whole series of binary and multi-phase alloys with other transition elements replacing cobalt. These other elements include copper, chromium, manganese, zirconium and iron. The ratio of transition elements to rare earth may be less than the 8.5 given by the composition. An anisotropic compact is produced and sintered at $1\,150$ to $1\,250°C$, and this is followed by a solution treatment at $1\,100$ to $1\,250°C$. The compact is then step aged at temperatures from $900°C$ down to $400°C$ over a period of at least 20 h. This complicated heat treatment and difficulties with powder production have made the magnets rather difficult to produce with consistent properties. It is however the most stable of the rare earth alloys and has a higher remanence and $(BH)_{max}$ than the $SmCo_5$ alloy.

In both of these alloy systems the properties are due to a very strong magnetocrystalline anisotropy related to the hexagonal crystal structure. The range of sizes which can normally be produced is from about 1 mm cubes up to blocks of $50 \times 50 \times 25$ mm, the short dimension being the preferred direction. These magnets are very brittle and must be handled with care to prevent chipping or cracking. They can only be machined by grinding or slitting, and they cannot be drilled or turned. The magnets may be magnetised prior to assembly without flux loss because they combine a relative recoil permeability close to unity with a high coercivity. However, because of the handling problems with these brittle magnets it is often more convenient for the user to magnetise during assembly.

20.2.4 Neodymium iron boron

This alloy was announced in 1983 and it has generated a great deal of interest because at room temperature it has better properties than any previously known alloy. It is also a cobalt-free alloy and uses neodymium which is more abundant than samarium. These two factors make the raw material costs quite considerably less than those of the samarium cobalt alloys.

There are two main routes for the production of magnets from these alloys. The first is the conventional route similar to that used for $SmCo_5$. The constituents of the alloy are melted together, and the material is then powdered, pressed in a magnetic field and sintered. The basic alloy with this route is the atomic composition $Nd_{15}B_8Fe_{77}$. The second method is to use the melt spinning process where the molten alloy is ejected under pressure onto a water cooled wheel which is rotating at high speed. This cools the alloy very rapidly and an amorphous glass-like tape is produced. This tape is annealed at $650°C$, which recrystallises the alloy, and then it is mechanically broken into flake. The basic atomic composition of the alloy for this process is 13.5% Nd 4.75% B 81.75% Fe. There are advantages in each of these methods. To get the best properties the conventional sintering route must be used but the production of the powder is rather slow and expensive. Melt spinning allows the powder to be produced more easily and cheaply. This method is ideal for the production of flake for use in the manufacture of isotropic bonded material, this product being called Magnequench I. The production of anisotropic material from the flake is more difficult and expensive.

The mechanism which provides the high value of the coercivity is crystal anisotropy, in the tetragonal crystals of the alloy. The particle size required for optimum properties is about $3\,\mu m$. Powdering can be achieved by ball milling; however the hydrogen decrepitation process has been applied very successfully. This process, which involves the absorption of large volumes of hydrogen by the alloy followed by desorption, produces much less damage to the particles, resulting in better properties.

There are two disadvantages of these neodymium alloys when compared to the rare earth cobalt alloys. They suffer from corrosion and rusting, although this can be prevented by the use of a suitable coating. They also have a relatively low Curie temperature of 330°C resulting in instability with a rapid decrease in flux with increasing temperature. The stability can be improved by additions of dysprosium, niobium and/or vanadium, making the alloy useful up to 150°C. Similar comments to those for the rare earth cobalt alloys apply here regarding sizes, machining and magnetising but these neodymium alloys are much less brittle.

These alloys are being used in a wide range of rotating machine applications for both a.c. and d.c. generators and motors including stepper motors. They are also used for holding and lifting devices and for motors for compact discs. More specialised applications include body scanners using magnetic resonance imaging (MRI) and particle beam focusing devices.

20.2.5   Bonded materials

Each of the above materials can in the powdered form be mixed with a bond. These bonds can be rubbers, polymers or plastics, and they may be flexible or rigid. The flexible rubber bonded ferrites have found wide application in holding and display services and the best quality anisotropic material is used in small motors. Alnico and rare earth cobalt are also made in bonded forms and they have the advantages of a uniform level of properties and freedom from cracking. Their magnetic properties are not so good as the unbonded materials: at best the BH_{max} is about 50% lower.

For recent research, at the time of writing, in the field of bonded magnetic materials see references 22–24.

20.2.6   Other materials

There are a number of other permanent magnet materials with minor applications. These materials make up less than 1% of the total production. They mainly find applications because they can be mechanically formed into shape, which is not possible for Alnico, ferrites or rare earth alloys. It must be emphasised that these other materials are only available in a limited range of sizes and shapes. Steels which were the original permanent magnets are now almost obsolete because their properties are inferior to those of the materials mentioned above. They do, however, find applications in hysteresis motors where intermediate coercivities are required. Cunife (Cu Ni Fe) and Cr Fe Co are alloys which can be rolled and drawn into wire and the anisotropy is produced by this rolling or drawing. Cr Fe Co produced by rolling can have properties similar to those of anisotropic Alnico. Platinum cobalt (Pt Co), which has a very high coercivity, was much used in the early days of space research. For most applications it has now been replaced by rare earth alloys. However, it is still used for medical and other applications where the resistance to corrosion is required. Other materials which were used in the past are Lodex, which consisted of fine electrolysed particles of iron cobalt, Vicalloy (V Co Fe) and Mn Al C alloys. These materials are now obsolete because of cost, lack of demand and production difficulties.

Non-metallic magnetic materials are of great commercial importance. These are beyond the scope of the current work but are discussed in references 25–27.

20.2.7   Properties, names and applications

The range of properties for each class of material is given in Table 20.6. These ranges include deliberate variations in properties obtained by small changes in composition and heat treatment.

Applications of magnetic materials, especially in electronics, are changing rapidly. See references 28–31 for examples current at the time of writing. Trade names of these materials are not included here, as these are subject to change.

20.3   Magnetically soft materials

Materials or alloys which are classed as magnetically soft, and generally with a relatively high permeability, are used for a wide range of applications including transformer laminations, armature stampings and cores in electrical machinery. They also include materials of considerable importance in the electronic and communication industries. Many special materials of widely different chemical compositions have been developed to meet specialised requirements. These are not confined to metallic alloys and ceramic oxides have become increasingly important.

Table 20.6   CHARACTERISTICS OF PERMANENT-MAGNET MATERIALS

B_t	Remanence, T			
$(BH)_{max}$	Energy product, kJ m^{-3}			
H_{cB}	Coercivity, kA/m^{-1}			
μ_r	Relative recoil permeability			

Material	B_r	$(BH)_{max}$	H_{cB}	μ_r
Alnico				
Normal anisotropic	1.1–1.3	36–43	46–60	2.6–4.4
High coercivity	0.8–0.9	32–46	95–150	2.0–2.8
Columnar	1.35	60	60	1.8
Ferrites (ceramics)				
Barium isotropic[a]	0.22	8	130–155	1.2
anisotropic[a]	0.39	28.5	150	1.05
Strontium anisotropic[a]	0.36–0.43	24–34	240–300	1.05
Bonded ferrite				
Isotropic[a]	0.14	3.2	90	1.1
Anisotropic[a]	0.23–0.27	10–14	180	1.05
Rare earth				
SmCo$_5$ sintered[b]	0.9	160	640–700	1.05
SmCo$_5$ bonded[b]	0.5–0.6	56–64	400–460	1.1
Sm$_2$Co$_{17}$ sintered[c]	1.1	150–240	500–900	1.05
NdFeB[d]	1.0–1.2	200–275	700–835	1.05
Others				
Cunife	0.54	10	40	—
CrFeCo	1.3	40–46	46	3–4
PtCo	0.62	72	370	1.1

Intrinsic coercivity H_{cT}(kA m^{-1}) [a]160–3.10; [b]800–1 500; [c]600–1 300; [d]750–1 900.
The Curie temperature of Alnico is 800–8.50°C, of ferrite 450°C, of SmCo over 700°C and for NdFeB it is 300°C. The Alnicos have a resistivity of about 50×10^{-8} Ωm, for the ferrites it is about 10^4 Ωm and for the rare earths 90–140×10^{-3} Ωm.

Sometimes several conflicting attributes are required in magnetically soft materials for a given application. It is for this reason that a range of materials from iron through to silicon steels, high nickel–iron alloys and ferrites have emerged. Usually, the most important requirement is that the hysteresis loss due to cyclic magnetisation should be as low as possible. The eddy current loss, which is due to the induction of electric currents in the material by the changing magnetic flux, should also be small. Generally the magnetic permeability should be as high as possible, particularly the maximum permeability, and in certain special cases the initial permeability also. Sometimes the requirement is for a high permeability at low field strengths. In general, the saturation intensity should be as high as other conditions will permit. For magnetic temperature compensation there should be a linear relation, over a known range of temperature, between permeability and temperature.

Relevant data concerning a wide selection of materials are given in Tables 20.7–20.11 and are discussed in the corresponding sections.

20.3.1   Silicon–iron alloys

For the magnetic circuits of the larger pieces of electrical equipment normally operated at power frequencies of 50 Hz, such as power transformers, alternators and electric motors of all sizes, a magnetic material is essential which will have a low hysteresis, high resistivity and a high value of magnetic saturation. Its production in the form of thin sheet or strip must also be possible, and in view of the big tonnages essential for such purposes it must not be expensive. Silicon–iron alloys are the only ones which meet all these conflicting requirements.

Over the years, the quality of silicon steels has been continuously improved and more has been learned about the effects of silicon on the magnetic properties. These effects are briefly as follows:

1. Silicon precipitates the major portion of the oxygen dissolved in the steel, which reduces hysteresis loss and magnetic ageing.
2. Silicon increases the electrical resistivity and thus reduces the eddy current loss, which is proportional to the resistivity (*see* Figure 20.3).

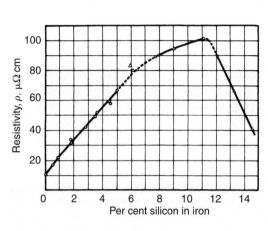

Figure 20.3 *Electrical resistivities of iron-rich iron–silicon alloys*

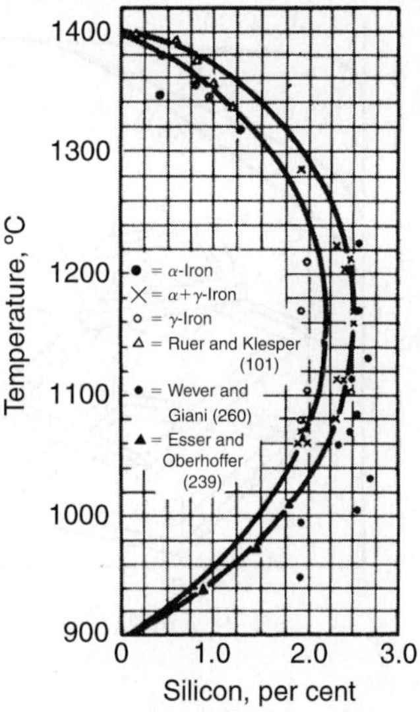

Figure 20.4 *The gamma phase region*

3. With over 2.5% silicon the alloys are outside the gamma loop on the equilibrium diagram, i.e. the alpha–gamma change point is suppressed (*see* Figure 20.4). Alloys of more than 2.5% silicon can therefore be annealed at high temperatures without recrystallisation occurring on cooling. As a result it is easier to produce large grain size in the finished sheet, with a corresponding reduction in hysteresis loss.
4. Silicon precipitates the carbon as graphite if it exceeds about 0.05%. Carbon present as graphite has no effect on hysteresis loss.

It is possible to orient the grains in silicon steels by cold rolling and heat treatments. Improvements in this grain-oriented sheet have been made as follows:

1. Limitation of the silicon content to about 3%. Above this value it is difficult to produce good quality sheet because of mechanical difficulties.
2. Reduction in the sheet thickness. The minimum thickness is 0.23 mm, which is limited by grain growth considerations.
3. Reductions of impurities to the minimum economic value. Impurities which increase hysteresis are those which occupy interstitial positions in the crystal lattice, e.g. carbon, oxygen, nitrogen and sulphur.
4. Improvements in the control of annealing equipment and protective atmospheres during final annealing.

The grain-oriented steel strip is used for transformers. The grain orientation is parallel to the rolling direction and in the transformer the flux is arranged so that it is parallel to this direction. This gives the advantage that the permeability is better and the hysteresis is lower with the direction of magnetisation parallel to the cube edge of the crystal lattice (*see* Figure 20.5).

During the annealing process a glass film is produced on the steel surface and this film gives a longitudinal tension in the sheet and provides insulation. This tension improves the magnetic properties of the steel, including a reduction in the magnetostriction which reduces the acoustic noise of the transformer. A phosphate coating is applied to complete the tensioning and insulation. The grain size of these sheets is comparatively large and the resulting domain boundaries are quite widely spaced. Artificial grain boundaries can be produced by laying down lines of ablated spots on the steel surface to give a stress and atomic disruption pattern which pins domain walls and leads to a smaller wall spacing. Various methods have been used to produce this ablating including spark and laser techniques.

In addition to the grain oriented material there is a need in rotating machinary applications for sheets which are comparatively isotropic and less expensive. The silicon content is generally lower as

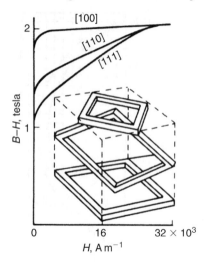

Figure 20.5   *Magnetisation curves for 3.85% silicon-iron single crystals at high field strengths in three directions*

the minimum core loss is less important than the high-saturation magnetisation needed for maximum torque.

Manufacturers offer a series of alloys whose properties are given in Table 20.7. These alloys conform to the British Standards and their international equivalents given in ref. 38. Methods of test are also given in these standards. In the past the Epstein square has been used by there is a rapidly growing trend towards the use of single-sheet testers which greatly reduce the labour involved.

20.3.2   Ferrites[34–36] and garnets[36]

Ferrites are iron oxide compounds usually containing at least one other metallic ion. The magnetic iron oxides Fe_3O_4 and Fe_3O_3, both of which are used in magnetic recording, are also included with the ferrites. The most important of the many ferrite materials now known are sintered ceramics of high resistivity, together with saturation magnetisations up to 0.5 T at room temperature. One structural class is the barium hexaferrites with high coercivities used as permanent magnets. The other two structural classes are the spinels and the garnets.

The spinel-structured ferrites are mixed oxides with the general composition $MO \cdot Fe_2O_3$ where M may be Fe, Ni, Mn, Zn, Co, Cu, Li or Al or mixtures of these. They are prepared from the constituent oxides which have been milled to a fine powder, preferably of the same particle size and intimately mixed. This mixture is calcined at about 1 000°C followed by crushing and milling and pressing the powder in a die or extrusion to the required shape. The resulting compact is a black brittle ceramic and any subsequent machining must be by grinding. They may be prepared with high permeability, and by virtue of the high resistivity, which renders eddy current levels negligible, can be used at frequencies up to 10 or 20 MHz as the solid core of inductors or transformers. The losses which do occur are a combination of hysteresis, eddy current, and residual losses which may be separately controlled by composition and processing conditions with due regard to the permeability required and frequency of operation.

The saturation flux density of ferrites is low, making them unsuitable for use in power or high flux transformers. Their use is therefore almost entirely in the electronic and telecommunications industry where they are now largely replacing laminated alloy and powder cores.

Typical properties of the commonly employed manganese–zinc and nickel–zinc ferrites are given in Table 20.8.

For higher frequencies of 100 mHz and above garnets are used. They have resistivities in excess of 10^8 Ωm compared with the ferrites, resitivities of about 10^3 Ωm at best. This means that eddy current losses which are dependent on resistivity are minimised. The basic Yttrium Iron Garnet (YIG) composition is $3Y_2O_3 \cdot 5Fe_2O_3$. This is modified to obtain improved properties such as very low loss and greater temperature stability by the addition of other elements, including aluminium

Table 20.7 TYPICAL PROPERTIES OF SILICON STEELS[32]

Grade identity	Thickness mm	Typical specific total loss at B = 1.5T; 50 Hz (or 1.7T; 50 Hz where stated) (W/kg)	Typical specific apparent power at B = 1.5T; 50 Hz (or 1.7T; 50 Hz where stated) (VA kg⁻¹)	Nominal silicon content %	Resistivity Ωm × 10⁸	Stacking factor %	$\hat{B}$ at H = 5 000 A m⁻¹ (or 1 000 A m⁻¹ where stated) T	Typical applications
Grain oriented: magnetic properties measured in direction of rolling only								
Unisil								
27 MOH	0.27	1.03 (1.7; 50)	1.4 (1.7; 50)	2.9	45	96	1.94 (1 000)	High-efficiency power transformers
30 M2H	0.30	1.12 (1.7; 50)	1.55 (1.7; 50)	2.9	45	96.5	1.92 (1 000)	
Unisil								
23M3	0.23	0.75	1.00	3.1	48	96	1.85 (1 000)	
27M4	0.27	0.84	1.16	3.1	48	96	1.85 (1 000)	
30M5	0.30	0.89	1.32	3.1	48	96.5	1.85 (1 000)	
35M6	0.35	1.00	1.39	3.1	48	97	1.84 (1 000)	
Non-oriented: magnetic properties measured on a sample comprising equal number of strips taken at 0° and at 90° to the direction of rolling								
Transil 300-35-A5	0.35	2.95	30	2.9	48	98	1.65	Large rotating machines
Losil 400-50-A5	0.50	3.60	19	2.4	44	98	1.69	Small transformers/chokes
Losil 800-65-A5	0.65	6.50	14	1.3	29	98	1.75	Motors and FHP motors
Non-oriented: grades supplied in the 'semi-processed' condition and which require a decarburising anneal after cutting/punching to attain full magnetic properties								
Newcor 800-65-D5	0.65	7.00	10	Nil	12	97	1.77	Motors and FHP motors
Newcor 1000-65-D5	0.65	8.80	11.5	Nil	12	97	1.74	
Tensile grades								
Tensiloy 250	1.60	—	—	Nil	—	—	1.60	Pole pieces large rotating machines

Table 20.8   TYPICAL PROPERTIES OF SOME MAGNETICALLY SOFT FERRITES*[34-36]

Parameter	Conditions		Ferrite grade					Units
	f kHz	$\hat{B}$ mT	(i)	(ii)	(iii)	(iv)	(v)	
Approximate composition:								
MnO/ZnO/Fe$_2$O$_3$			27/20/53	25/22/53	30/20/49	34/14/52	—	mol %
NiO/ZnO/Fe$_2$O$_3$			—	—	—	—	32/18/50	mol %
Initial permeability, μ_i	<10	<0.1	1 200–2 500	3 800–10 000	350–500	1 000–3 000	70–150	—
Temperature factor, $\Delta\mu_i/\mu^2\Delta\theta$ (5 to 55°C)	30 100	<0.1	0.5–1.5 0.8–2.0 1.5–3.0	5–10 10–50	20 30	—	0–8	°C^{-1} × 10^{-6}
Residual loss factor (tan δ_r)/μ_i	10^3 10 × 10^3	<0.1	—	—	75	—	20–40 60–100	10^{-6}
Hysteresis coefficient, η	10	1–3	0.5–1.0	0.1–1.0	0.23–0.27	0.4–0.5	2–10	mT^{-1} × 10^{-6}
Saturation flux density, B$_{\text{sat}}$ (at 1 kA m^{-1})			0.35–0.45	0.35–0.45		0.4–0.5	0.3–0.4	T
Curie point θ_c	<10	<0.1	130–250	90–220	130–160	180–280	250–400	°C
Resistivity	d.c.		1–5	0.05–0.5	1–5	1–5	>10^3	Ωm
Applications			Quality inductors	Wide-band or pulse transformers	Deflection yokes	High-power transformers	Quality inductors 2–12 MHz	

* At 25°C unless otherwise stated.

and gadolinium. The garnets are prepared by heating the mixture of oxides under pressure at over 1 300°C for up to 10 h. This material is used in microwave circuits in filters, isolators, circulators and mixtures. Some of their properties are given in Table 20.9.

Table 20.9 GARNET MATERIAL

	J_s T	T_c °C	ΔH (kA m^{-1})* Maximum
Y Al Fe	0.3–1.76	110–280	45–60
Y Gd Fe	0.67–1.22	280	110–250
Y Gd Al Fe	0.6–1.1	190–250	70–120

* This is the gyromagnetic resonant linewidth ΔH (−3 dB).
 The values of other parameters for these garnets are:
  dielectric constant ε' 13.8–15.2,
  $10^4 \tan \delta$ 2.5–3.

20.3.3 Nickel–iron alloys

The high permeability of nickel–iron alloys makes them very useful materials for the electronic communications and electrical power engineering industries. This high permeability at low fields (despite the low saturation magnetisation) gives scope for minimising core sizes in a wide range of applications.[32,33]
    There are three main groups of high-permeability nickel–iron alloys. The 80% nickel alloys variously called Mumetal and Permalloy have the highest permeability due to the almost zero crystalline anisotropy and magnetostriction at this composition. The commercial alloys have additions of chromium, copper and molybdenum to increase the resistivity and improve the magnetic properties. This material is used in specialised transformers, for magnetic screens and magnetic recording heads.
    The 50% nickel alloys have the highest saturation magnetisation of 1.6 T in this range of alloys. They have a higher permeability and better corrosion resistance than silicon–iron alloys but they are more expensive. Another advantage is they have a high incremental permeability over a wide range of polarising d.c. fields. A wide range of properties can be produced by various processing techniques. Severe cold reduction produces a cube texture and a square hysteresis loop in annealed strip. The properties can be tailored by annealing the material below the Curie temperature in a magnetic field. Applications of this material include chokes, relays and small motors.
    The 36% nickel alloys have higher resistivity and are less costly than the 50% alloy but the saturation magnetisation and permeability are lower. This alloy is of the Invar type and has a very low coefficient of expansion of 1×10^{-6}°C^{-1}. The main applications are relays, inductors and high-frequency transformers where the higher resistivity limits the eddy currents. The properties of the various nickel–iron alloys are given in Table 20.10.

Table 20.10 PROPERTIES OF NICKEL–IRON ALLOYS[32]

Alloy	36% Ni	45% Ni	50% Ni	54% Ni	65% Ni	77% Ni
Initial permeability $\times 10^{-3}$	3	6	0.5–11	65	—	60–200
Saturation flux density (T)	1.2	1.6	1.54–1.6	1.5	1.3	0.8
Coercivity force (Am^{-1})	10	10	3–10	2.5	2–10	0.5–1
Remanence (T)	0.5	1.0	1.1–1.5	0.7	1.25–0.2	0.5
Resistivity ($\mu\Omega$ m)	0.8	0.45	0.4	0.45	0.6	0.6
Density (kg m^{-3})	8 100	8 300	8 300	8 300	8 500	8 800
Curie temp. (°C)	280	530	530	550	520	350
Expansion coefficient (°C^{-1} $\times 10^6$)	1	8	10	11	12	13

20.3.4   Amorphous alloy material

The rapidly solidified alloy materials called metallic glasses are now an established group of soft magnetic materials. These materials are produced by rapid solidification of the alloy at cooling rates around a million degrees centigrade per second. The alloys solidify with a glass-like atomic structure which is a non-crystalline frozen liquid. The rapid solidification is achieved by causing the molten alloy to flow through an orifice on to a rapidly rotating water cooled drum. This can produce sheets of the alloy as little as 10 μm thick and up to a metre or more wide.

There are two main groups of amorphous alloys. The iron-rich magnetic alloys have the highest saturation magnetisation among the amorphous alloys and are based on inexpensive raw materials. The other group are the cobalt-based alloys which have very low or zero magnetostriction which leads to the highest permeability and the lowest core loss. All of these alloys have high electrical resistivity, higher than that of conventional crystalline electrical steels. Because of this, eddy current losses are minimised both at 50 Hz and at higher frequencies. They also have other advantages including flexibility without loss of hardness, high tensile strength and better corrosion resistance than similar crystalline alloys. Properties of typical amorphous alloys are given in Table 20.11.

Table 20.11   AMORPHOUS ALLOYS[2]

Amorphous material	J_s T	Curie temperature °C	Crystal- lisation °C	Electrical resistivity μΩ m	μ_{max}, d.c. annealed	Coercivity A m^{-1}	Saturation magneto- striction $\lambda_s \times 10^6$
Iron-based	1.28–1.80	310–415	430–550	1.23–1.37	35–600 × 10^3	1–1.6	19–35
Nickel–iron- based	0.88	350	410	1.38	8 × 10^5	0.4	12
Cobalt-based	0.55–0.70	200–365	520–550	1.36–1.42	6–10 × 10^5	0.15	<1

20.4   High-saturation and constant-permeability alloys

The addition of cobalt to iron increases the saturation intensity. A range of alloys with 24–50% cobalt has saturation up to 10% greater than that of iron. Alloys with around 25% cobalt are used for applications such as magnet pole tips where a combination of good ductility, ease of machining and high-saturation magnetisation are required. The permeability and loss characteristics are however inferior to the 50% cobalt alloy. The 49/49/2 Fe/Co/V alloy called Permendur, to which the vanadium has been added to improve the ductility and also substantially increase the resistivity, is used extensively for stators in lightweight generators.

Recent developments, which have improved its mechanical properties, have extended its use to rotor laminations. Other applications include small-volume transformers, relays, diaphragms, loudspeakers and magnet pole tips. The high cost of cobalt restricts the use of these alloys, but they are widely used in aircraft where weight is at a premium. The properties of the alloys are given in Table 20.12.

Table 20.12   IRON–COBALT ALLOYS[32]

Alloy	Saturation flux density T	Initial permeability	Maximum permeability	Coercive force A m^{-1}	Remanence T	Resistivity μΩ m	Curie temperature °C
24% Co	2.35	250	2 000	130	1.5	0.2	925
49/49/2 Co/Fe/V	235	800	7 000	950	1.6	0.4	975

Constant-permeability materials are required for applications where a constant inductance is essential for electrical filters or in loading coils for telephone lines. The original materials were in the ternary Fe/Ni/Co series but more recently they have been replaced by the 40–50% nickel Isoperms. They have an addition of 5–15% copper and by cold rolling and heat treatment a preferred orientation is produced. The permeability of 50–60 is constant over a range of fields up to 8 kA m^{-1}.

20.5   Magnetic powder core materials

The thickness of laminations of ferromagnetic alloy cores must be reduced as the operating frequency is increased above audio range if the increasing loss and reducing permeability caused by eddy currents are to be avoided. However, reduction of thickness of laminations causes increased cost. This may be avoided for many applications up to 200 kHz by the use of magnetic powder cores.[3]

The Permalloys are used in the form of compressed insulated powder cores, the effect of subdivision being to reduce eddy current losses to a minimum. The permeability is also reduced but is made extremely constant, becoming within a small percentage independent of magnetising force and unaffected by magnetic shock. The properties of such materials are in many ways similar to the constant permeability alloys described in the preceding section. *Sendust*, a brittle alloy of aluminium, silicon and iron, has been similarly used in Japan and Germany. Carbonyl iron powder is used for powder cores mainly at radio frequencies, specially low eddy current losses being obtained on account of the characteristic fine particle size. Some details of these materials are given in Table 20.13.

Table 20.13   MAGNETIC POWDER CORE MATERIALS

Material	*Composition*	*Effective initial relative permeability* μ_e	*Typical power loss* kw/m^3 (*at* 50 kHz, 100 mT, 25°C)	*Typical applications*
Molybdenum Permalloy	80% Ni–Fe +Mo	125 60	230 500	High Q fixed inductors 0.1 to 200 kHz
Radiometal (high flux)	50% Ni–Fe	125	600	Switch mode power supply chokes and transformers to 250 kHz
		69	650	Mains chokes and filters for suppression of radio frequency interference
Carbonyl iron powder (high frequency)	99.9% Fe	5–25	—	High Q fixed inductors 1–300 MHz. Wide band signal transformers up to 10 GHz. Chokes for RF suppression
Electrolytic iron powder (low frequency)	99% Fe	45–90	3 300	Switch mode power supply chokes to 50 kHz. Main chokes for RF suppression
			2 400	Dimmer chokes 50 Hz DC electromagnets
Sendust	9.5% Si, 5.5% Al, balance Fe	60	2 000	Chokes and inductors

20.6   Magnetic temperature-compensating materials

Alloys having a marked variation of permeability with temperature are used to compensate for errors in electrical instruments caused by changes in the ambient temperature.

The magnetic permeability of iron and iron alloys decreases with temperature (except for low values of the field), and this causes the readings of electrical meters which depend upon the fluxes maintained by a constant voltage or current or by a permanent magnet, to decrease with the ambient temperature. To compensate for such errors it is customary to shunt a certain proportion of the magnetic flux passing through the magnetic circuit of the meter by means of an alloy having an unusually high negative temperature coefficient between 0 and 100°C, so that, as the ambient temperature increases, the proportion of the shunted flux decreases, causing more flux to pass through the working air gap of the instrument than otherwise would be the case. By correctly proportioning the shunt component it is possible to render the readings almost completely independent of ambient temperature changes.

Compensation may also be provided for the changes of resistivity of eddy current devices. These alloys have a major application in the compensation of changes in the resistivity with temperature of the copper drag cup in speedometers.

The alloys used are in the nickel–iron series, about 30% nickel, sometimes with additions of chromium, manganese and/or silicon. Details of the characteristics of the alloy are given in Table 20.14.

Table 20.14   THERMOMAGNETIC TEMPERATURE COMPENSATING MATERIALS

	Magnetic flux density at $H = 8 \times 10^3 \, \mathrm{A \, m^{-1}}$ T								
	Temperature °C								
Material	−40	−20	0	20	40	60	80	100	120
Telecon 2799	0.53	0.43	0.33	0.22	0.13	0.05	—	—	—
Telecon 2800	—	—	0.76	0.67	0.59	0.50	0.40	0.29	0.19
Thermoflux 90/100 G	—	—	0.60	0.47	0.33	0.20	0.07	—	—
Thermoflux 65/100 G	—	—	0.43	0.31	0.17	—	—	—	—
Thermoflux 35/100 G	—	—	0.19	0.07	—	—	—	—	—
Carpenter 30	0.60	0.51	0.40	0.28	0.14	0.05	—	—	—
Carpenter 32	1.02	0.97	0.90	0.83	0.75	0.67	0.59	—	—

20.7   Non-magnetic steels and cast irons

For certain constructional purposes in electrical engineering where magnetic shunting effects have to be avoided there arises a requirement for almost non-magnetic materials possessing certain minimum mechanical properties.

The materials available comprise non-magnetic cast irons, austenitic steels, and certain other sensibly non-magnetic alloys; the choice of materials for any particular purpose is dependent upon the mechanical properties required; e.g. certain intricate stationary parts of electrical machinery would be suitably made from non-magnetic cast iron, while armature straps for high-speed rotors would be made in a high-tensile austenitic alloy steel in order to withstand the dynamic stresses encountered in practice. A high resistivity is sometimes beneficial as well as the possession of virtually non-magnetic properties.

There are two well-known non-magnetic cast irons, one called *Nomag*, a manganese-silicon-nickel-iron with a permeability of 1.04, and a resistivity of 1.40 microhm *m*, and the other *Ni-Resist*, containing lower proportions of manganese and silicon, but a higher proportion of nickel, also a little chromium and copper. These irons with flake graphite are brittle, and can only be used for parts of a simple nature, such as supports which are not subjected to severe mechanical stress. These non-magnetic irons are also made in the spheroidal graphite form e.g. *Nodumag* and *S. G. Ni-Resist*, having the characteristically improved mechanical properties of this type of cast iron.

As examples of austenitic steels. Hadfields 13% manganese steel and the chromium–nickel steels have found considerable use. The stainless steels of the chromium–nickel type are normally non-magnetic but if they are subjected to cold working appreciable magnetic properties may appear. This disadvantage can be overcome by using steels with higher alloy content than the well-known 18/8 type. Besides these materials other non-magnetic austenitic steels have been specially developed to give improved mechanical properties.

Although the originators of some of these steels and cast irons are no longer in business, it is possible to get steelmakers or foundries to produce them to order. Table 20.15 gives details of the non-magnetic cast irons and a number of non-magnetic cast steels.[37]

Table 20.15 NON-MAGNETIC STEELS AND CAST IRONS

Material	Nominal composition	Relative permeability	Resistivity μΩ m	Condition	Mechanical properties				
					Hardness HB	Proof stress MPa	Max stress MPa	Elongation L=4√A %	Reduction in area %
Wrought steels									
18/8 Stainless	8% Ni, 18% Cr and various minor constituents	1.06 or greater	0.72	Austenised	170	278	618	45	50
18/8 Stainless	8% Ni, 18% Cr and various minor constituents	1.4	—	20% cold reduction	280	—	927	25	—
18/8 Stainless	8% Ni, 18% Cr and various minor constituents	about 10.0	—	50% cold reduction	330	—	1 313	10	—
18/12 Stainless	12% Ni, 18% Cr etc.	1.03	0.80	Austenised	160	232	618	60	65
18/12 Stainless	12% Ni, 18% Cr etc.	1.1	—	50% cold reduction	—	—	1 313	—	—
18/12 Stainless	12% Ni, 18% Cr etc.	1.7	—	90% cold reduction	—	—	—	—	—
25/12 Stainless	12% Ni, 25% Cr etc.	1.01	0.90	Austenised	200	309	695	30	35
25/12 Stainless	12% Ni, 25% Cr etc.	1.03	—	90% cold reduction	—	—	—	—	—
Foxalloy 101	1% Si, 16–20% Mn, 12–16% Cr, 0.2–1% V	1.01 max	0.75	Austenitic	275	750	960	35	52

(continued)

Table 20.15   NON-MAGNETIC STEELS AND CAST IRONS—*continued*

Material	Nominal composition	Relative permeability	Resistivity $\mu\Omega$ m	Condition	Mechanical properties				
					Hardness HB	Proof stress MPa	Max stress MPa	Elongation L=4√A %	Reduction in area %
Cast steels									
Hadfields Mn	13% Mn, 1% Si	1.03	0.71	Water quenched 1 000°C	—	—	—	—	—
Clyde alloy non-magnetic	8% Ni, 8% Mn, 8% Cr	1.003	0.76	Water quenched 900°C	—	—	—	—	—
Clyde alloy non-magnetic	8% Ni, 4% Mn, 8% Cr	1.003	—	Water quenched 900°C	—	571	927	45	45
Firth Vickers NMC	10% Ni, 5% Mn, 4% Cr	1.008/1.03	—	—	—	618	927	45	45
Black alloy 525 (tool bit material)	48% Co, 35% Cr, 14% W carbide, 9% Nb/Ta	1.002/1.01	—	—	620	—	450	—	—
Cast irons									
Nomag	7% Mn, 11% Ni, 2.5% Si, 2.7% C	1.03/1.05	1.40	Stress relieved	120/150	—	~185	0	—
Nodumag	7% Mn, 11% Ni, 2.5% Si, 2.7% C	1.03	1.10	Stress relieved	120/150	309	432	15	—
Ni-Resist Type 2	Ni 20% Si 2.5%	1.03	1.70	Stress relieved	150	—	185	0	—
S. G. Ni-Resist Type D-2	Mn 1% Cr 2% C 2%	1.03	1.00	Stress relieved	140	232	415	15	—

Appendix

Units and definitions

Magnetic poles

The concept of magnetic poles is used to explain the behaviour of magnetised bodies acting on one another and in the earth's magnetic field. Like magnetic poles repel each other and unlike poles attract.

Magnetic dipole

A magnetic dipole is an elementary pair of unlike poles and can be represented by an infinitesimal current loop.

Magnetic moment m

For a magnetic dipole the magnetic moment is the product of the current and the loop area. For a magnetised body the magnetic moment is the sum of all the magnetic dipole moments and is equal to ml where l is the distance between the poles.

Magnetic field strength H

The magnetic field strength H at a point is an axial vector quantity associated with the magnetic flux density in a magnetic field. The unit of magnetic field is the ampere per metre $(\mathrm{A\,m^{-1}})$, which is the (axial) field inside an infinitely long solenoid of one turn per metre carrying a current of one ampere.

Magnetisation M

When a magnetic field is applied to a magnetic material the material becomes magnetised. The magnitude of this magnetisation M is equal to the total magnetic moment divided by the volume. The units of M are ampere per metre $(\mathrm{A\,m^{-1}})$.

Polarisation J

This is the name given to the product $\mu_0 M$, and it has the units of tesla (T).

Saturation magnetisation M$_s$ *and polarisation* J$_s$

These are the maximum possible values of the magnetisation and polarisation for a given material at a given temperature. The saturation polarisation J_s is sometimes called B_s.

Magnetisation per unit mass σ

The value of σ is usually the saturation value and is equal to M/ρ where ρ is the density. The units are joule/(tesla kg).

Magnetic flux density B (or magnetic induction)

The magnetic flux density B at a point is a vector quantity arising from the magnetic field and the magnetisation of the medium. It is given by:

$$B = \mu_0 H + \mu_0 M$$

or

$$B = \mu_0 H + J$$

where μ_0 is the magnetic constant. The units of B are tesla (T) or weber/(metre)^2 $(\mathrm{Wb\,m^{-2}})$.

Magnetic constant μ₀

This constant μ_0 is equal to $4\pi \times 10^{-7}$ Henries m^{-1}. For practical purposes this constant when multiplied by the field strength equals the flux density in vacuo. (It was in the past called the permeability of free space.)

Magnetic flux ϕ

The magnetic flux ϕ is the sum over an area A of the magnetic flux density and $\phi = BA$. The unit of flux is the weber (Wb).

Magnetic susceptibility κ

The magnetic susceptibility κ of a material is the ratio of the magnetisation M to the magnetic field H producing it. This quantity κ is the magnetic susceptibility per unit volume and is dimensionless.

Mass susceptibility χ

The magnetic susceptibility per unit mass or mass susceptibility χ is given by κ/ρ, where ρ is the density of the material. It has the units of $m^3\ kg^{-1}$.

Magnetic permeability μ

The permeability μ of a medium is the ratio of the magnetic flux B to the magnetic field H producing it. For practical purposes the value of μ is divided by the magnetic constant μ_0 to give the relative magnetic permeability μ_r which is dimensionless. For strongly magnetic materials the value of μ_r varies with the value of the magnetic field. The values of μ_i, the initial permeability, and μ_{max}, the maximum permeability, are widely quoted parameters for magnetic materials.

Curie point T_c

The Curie point, or temperature, (T_c) is the temperature below which a material is ferromagnetic or ferrimagnetic and above which it is paramagnetic.

Néel point N_c

The Neel point or temperature is the temperature below which a material is antiferromagnetic and above which it is paramagnetic.

Magnetic hysteresis

This is a phenomenon by which the magnetisation of a material depends not only on the magnetising field but also on the previous magnetic history. The magnetic flux B always lags behind magnetic field H in a material showing hysteresis. The complete B–H curve, or loop, produced by increasing H from zero to a maximum value, reducing to zero and then repeating with H in the negative sense is termed the hysteresis loop, and the dissipation of energy associated with it, the hysteresis loss. The hysteresis loss per unit volume is given by (area of the loop in $B \times H$) joules.

Remanence B_r

This is the value of the magnetic flux density remaining in a material in a closed magnetic circuit following magnetisation to saturation and subsequent removal of the magnetic field.

Coercivity H_c

This is the magnetic field strength when a material is brought from saturation to the point where the flux density B or magnetisation M is reduced to zero. In a graphical representation it is the value corresponding to the intersection of the B curve or the M with the H axis. Two cases must be distinguished. When the value of B is reduced to zero the coercivity is called H_{cB} and when the value of M (or J) is reduced to zero the value is called H_{cM} (or H_{cJ}). The quantity H_{cJ} is sometimes called H_{ci}.

Energy Product and $(BH)_{max}$

The energy product is the product of the flux density and the field strength at any point of a demagnetisation curve for a permanent magnet. The maximum value of the product is called $(BH)_{max}$. It is directly related to the stored energy per unit volume of the material. The units of $(BH)_{max}$ are joules/(metre)^3.

Magnetostriction

Elastic deformation which accompanies a change in the magnetisation of a material. This results in changes of length or volume.

Magnetic units and conversion factors

Authors and publishers of current literature on magnetism and magnetic materials have been very slow in adopting SI (systeme international) units in place of the centimetre gram second (c.g.s.) electromagnetic units previously used. Conversion of magnetic units between the two systems are therefore frequently required, as well as care in using equations which are different for the two systems. The relation between the relevant c.g.s. electromagnetic units and the SI (rationalised m.k.s.) units is given in Table 20.16.

Table 20.16   RELATION BETWEEN SI UNITS AND .C.G.S. UNITS OF MAGNETIC QUANTITIES

Quantity	*c.g.s. unit*	*SI unit*	*Conversion factor**
Magnetic field H	Oersted (Oe)	amperes per metre ($A\,m^{-1}$)	$\times 10^3/4\pi$
Magnetic flux ϕ	Maxwell (or line)	Weber (Wb)	$\times 10^{-8}$
Flux density B	Gauss (Gs)	Tesla (T)	$\times 10^{-4}$
Magnetomotive force	Oersted cm or Gilbert	amperes	$\times 10/4\pi$
Polarisation	I or J e.m.u.	J Tesla (T)	$\times 4\pi \cdot 10^{-4}$
Magnetisation	I or J e.m.u.	M amperes per metre	$\times 10^3$
Maximum energy product $(BH)_{max}$	M Gs Oe	$J\,m^{-3}$ or $T\,A\,m^{-1}$	$\times 10^5/4\pi$
Magnetic moment m	e.m.u.	Weber metre (Wb m)	$\times 4\pi 10^{-10}$
Magnetic constant μ_0	1	Henry m^{-1} ($H\,m^{-1}$)	$\times 4\pi 10^{-7}$
Resistivity ρ	ohm · cm	ohm · m	$\times 10^{-2}$

*The number by which the quantity in c.g.s. units must be multiplied to obtain the quantity in SI units.

REFERENCES

General reading

1. J. Crangle, 'Solid State Magnetism', Arnold, 1991.
2. D. Jiles, 'Introduction to Magnetism and Magnetic Materials', Chapman and Hall, 1991.
3. C. Heck, 'Magnetic Materials and their Applications', Butterworths, 1974.
4. R. S. Tebble and D. J. Craik, 'Magnetic Materials', Wiley, 1969.

Elements

5. L. F. Bates, 'Modern Magnetism', 4th Edn, Cambridge University Press, 1963.
6. 'C.R.C. Handbook of Chemistry and Physics', Ed. R. C. Weast, 69th Edn, C.R.C. Press, 1988.
7. G.W.C. Kaye and T. H. Laby, 'Tables of Physical and Chemical Constants', 15th Edn, Longman, 1986.
8. 'Landolt-Bornstein, Zahlenwerte und Functionen, Physik, Chemie', etc, II Band, 9 Teil, 'Magnetic Properties I', 1962.
9. 'American Institute of Physics Handbook', McGraw-Hill, 1972.

Steels and cast irons

10. M. H. Hillman, *BCIRA Journal of Research and Development*, **5**, 188, 1954.
11. W. J. Jackson, 'Effect of Heat Treatment on the Magnetic Properties of Carbon Steel Castings'. *J. Iron Steel Inst.*, **194**, 29, 1960.
12. J. Woolman and R. A. Mottram, 'Mechanical and Physical Properties of the B.S. En. Steels', Vol. 1, En 1–20, Vol. 2, En 21–29, Vol. 3, En 40–363, Pergamon, 1964–69.
13. 'Iron and Steels Specifications', 7th Edn, British Steel, 1989.

Reviews

14. H. H. Stadelmaier, 'Magnetic Properties of Materials', *Mater. Sci. and Eng. A*, **287**, 138–145, 2000.
15. S. T. Bramwell, 'Magnetism', *Annu. Rep. Prog. Chem., Section A: Inorganic Chemistry*, **97**, 461–474, 2001.

16. S. T. Bramwell, 'Magnetism', *Annu. Rep. Prog. Chem., Section A: Inorganic Chemistry*, **96**, 502–522, 2000.
17. S. T. Bramwell, 'Magnetism', *Annu. Rep. Prog. Chem., Section A: Inorganic Chemistry*, **95**, 467–480, 1999.
18. S. T. Bramwell, 'Magnetism', *Annu. Rep. Prog. Chem., Section A: Inorganic Chemistry*, **94**, 459–478, 1998.
19. S. T. Bramwell, 'Magnetism', *Annu. Rep. Prog. Chem., Section A: Inorganic Chemistry*, **93**, 457–487, 1997.

Permanent magnets

20. A. G. Clegg and M. McCaig, 'Permanent Magnets in Theory and Practice', 2nd Edn, Pentech and Wiley, 1987.
21. R. J. Parker, 'Advances in Permanent Magnetism'. Wiley, 1990.

Bonded magnets

22. M. Hamano, 'Overview and outlook of bonded magnets in Japan', *J. Alloys Compd.*, **222 (1–2)**, 8–12, 1995.
23. J. Schneider and R. Knehans-Schmidt, 'Bonded hybrid magnets', *J. Magnetis. Magnet. Mater.*, **157/158**, 27–28, 1996.
24. B. M. Ma, J. W. Herchenroeder, B. Smith, M. Suda, D. N. Brown and Z. Chen, 'Recent development in bonded NdFeB magnets', *J. Magnetis. Magnet. Mater.*, **239**, 418–423, 2002.

Non-metallic magnets

25. J. S. Miller and A. J. Epstein, 'Molecule-Based Magnets – An Overview', *MRS Bull.*, **25** (11), 21–28, 2000.
26. J. Veciana and H. Iwamura, 'Organic Magnets', *MRS Bull.*, **25** (11), 41–51, 2000.
27. K. Awaga, E. Coronado, and M. Drillon, 'Hybrid Organic/Inorganic Magnets', *MRS Bull.*, **25** (11), 52–59, 2000.

Applications

28. D. Weller and M. F. Doerner, 'Extremely High-Density Longitudinal Magnetic Recording Media', *Annu. Rev. Mater. Sci.*, **30**, 611–644, 2000.
29. K. A. Gschneider, Jr. and V. K. Pecharsky, 'Magnetocaloric Materials', *Annu. Rev. Mater. Sci.*, **30**, 387–429, 2000.
30. C. A. Ross, 'Patterned Magnetic Recording Media', *Annu. Rev. Mater. Sci.*, **31**, 203–235, 2001.
31. K. H. J. Buschow, 'Permanent-magnet materials and their applications', *Mater. Sci. Found.*, **5**, 1–82, 1998.

High-permeability materials (Si-Fe, Ni-Fe and Ferrites)

32. M. G. Say, G. R. Jones and M. A. Laughton, 'Electrical Engineers Reference Book', 15th Edn, Butterworths, 1993.
33. R. Boll, 'Soft Magnetic Materials: fundamentals, product, applications', Siemens, 1979.
34. E. C. Snelling, 'Soft Ferrites, Properties and Applications', 2nd Edn, Butterworths, 1989.
35. E. C. Snelling and A. D. Giles, 'Ferrites for Inductors and Transformers', Research Studies Press, 1986.
36. B. Lax and K. J. Button, 'Microwave Ferrites and Ferrimagnetics', McGraw Hill, 1962.

Non-magnetic steels and cast irons

37. C. B. Post and W. S. Eberly, 'Stability of austenite in stainless steel', *Trans. Am. Soc. Metals*, **39**, 868, 1947.

Standards

38. A series of International (IEC) and equivalent British Standards (BS) have been published for the Classification, Specification and Methods of Test for Magnetic Materials. The IEC series are numbered 404 and the equivalent BS series are 6404. In addition BS 4727, part I Group 07, 1991 is the standard for 'Magnetism terminology' and BS 5884, 1987. Methods for determination of relative magnetic permeability of feebly magnetic materials.

21 Mechanical testing

21.1 Hardness testing

Hardness tests are a quick and simple method to characterise a metal's resistance to deformation and can in some cases be related to an equivalent yield strength. A number of different hardness tests exist, which differ in the type of indenter used, the load applied, and the means used to quantify the extent of permanent deformation. Further information concerning the hardness of metals and hardness testing may be found in Refs [1–3]. Hardness values for some common metals and alloys are given in Chapter 22.

21.1.1 Brinell hardness

Relevant standards: ASTM E10-01, BS EN 10003-1:1996
An indenter comprising a hardened steel ball of diameter D mounted in a suitable holder is pushed into the material under test by a force F. The diameter of the indentation left in the surface of the material after removal of the load is measured in two directions at right angles. The area of the curved surface of the indentation is calculated from the mean diameter, d, the indentation being considered as a segment of a sphere of diameter D. The Brinell hardness is the quotient obtained by dividing the load, symbol f, expressed in kilograms-force, by the surface area of the indentation expressed in square millimetres.

Symbols:
> F = force (N)
> f = load symbol (kgf)
> D = diameter of ball in millimetres (mm)
> d = mean diameter of indentation in millimetres (mm)

$$\text{HBW} = \text{Brinell Hardness} = \frac{0.102 \times 2F}{\pi D[D - \sqrt{D^2 - d^2}]} = \frac{2f}{\pi D[D - \sqrt{D^2 - d^2}]}$$

$$h = \text{depth of indentation in millimetres} = \frac{f}{\pi D(\text{HBW})} \ (\text{mm})$$

The symbol HBW is supplemented by numbers indicating the diameter of the ball used and the load applied. Thus 226 HBW 10/3 000 indicates that a Brinell hardness of 226 was obtained by using a 10 mm diameter ball with a load symbol of 3 000 kgf. If the time of duration of load differs from the standard 10–15 sec, a further number is added to show the duration of the load in seconds.

RELATION OF LOAD TO BALL DIAMETER

The choice of force and ball diameter to be used in a Brinell test is determined by two factors:

1. the value of the ratio f/D^2, and
2. the size of the indentation which provides optimum accuracy.

The same value of f/D^2 will, in principle, give the same hardness value for different loads and the load used in practice depends on the nature and the hardness of the material under test. Five standard values of f/D^2, i.e. 30, 10, 5, 2.5 and 1, have been adopted. Recommended combinations of load and diameter for different materials are given in BS EN 10003-1:1996.

21.1.2   Rockwell hardness

Relevant standards: ASTM E18-00, BS EN ISO 6508-1:1999
An indenter of the standard type comprising a diamond cone or hardened steel ball mounted rigidly in a suitable holder, is forced into the test piece under a preliminary load F_0. When equilibrium has been reached, an indicating device which follows the movement of the indenter and so responds to changes in depth of penetration of the indenter is set to a datum position.

While the preliminary load is still applied, it is augmented by an additional load with resulting increase in penetration of the indenter. When equilibrium has again been reached, the additional load is removed but the preliminary load is maintained.

Removal of the additional load allows a partial recovery, so reducing the depth of penetration. The permanent increase in depth of penetration, e, resulting from application and removal of the additional load is used to deduce the Rockwell hardness number by means of the equation:

$$HR = E - e$$

where:

HR = Rockwell hardness number
E = a constant depending on the form of indenter:
    100 units when a diamond cone indenter is used
    130 units when a steel ball indenter is used

A variety of indenters and loads are used to give several scales of hardness, e.g. 60 HRC represents a Rockwell hardness of 60 on scale C. The various scales are defined in Table 21.1.

Table 21.1   ROCKWELL HARDNESS SCALES

Scale symbol		Total test force kN	kgf	Dial figures	Typical applications of scales
B	$\frac{1}{16}$ in (1.588 mm) ball	0.98	100	Red	Copper alloys, soft steels, aluminium alloys, malleable iron, etc.
C	Diamond	1.47	150	Black	Steel, hard cast irons, pearlitic malleable iron, titanium, deep case hardened steel and other materials harder than HRB100
A	Diamond	0.59	60	Black	Cemented carbides, thin steel, and shallow case-hardened steel
D	Diamond	0.98	100	Black	Thin steel and medium case hardened steel, and pearlitic malleable iron
E	$\frac{1}{8}$ in (3.175 mm) ball	0.98	100	Red	Cast iron, aluminium and magnesium alloys, bearing metals
F	$\frac{1}{16}$ in (1.588 mm) ball	0.59	60	Red	Annealed copper alloys, thin soft sheet metals
G	$\frac{1}{16}$ in (1.588 mm) ball	1.47	150	Red	Malleable irons, copper–nickel–zinc and cupro–nickel alloys. Upper limit HRG 92 to avoid possible flattening of ball
H	$\frac{1}{8}$ in (3.175 mm) ball	0.59	60	Red	Aluminium, zinc, lead
K	$\frac{1}{8}$ in (3.175 mm) ball	1.47	150	Red ⎫	
L	$\frac{1}{4}$ in (6.350 mm) ball	0.59	60	Red ⎪	
M	$\frac{1}{4}$ in (6.350 mm) ball	0.98	100	Red ⎪	
P	$\frac{1}{4}$ in (6.350 mm) ball	1.47	150	Red ⎬	Bearing metals and other very soft or thin materials. Use smallest ball and heaviest load that does not give anvil effect
R	$\frac{1}{2}$ in (12.70 mm) ball	0.59	60	Red ⎪	
S	$\frac{1}{2}$ in (12.70 mm) ball	0.98	100	Red ⎪	
V	$\frac{1}{2}$ in (12.70 mm) ball	1.47	150	Red ⎭	

Notes: (1) The preliminary load is 10 kgf for all scales.
    (2) The diamond indenter is conical with an included angle of $120 \pm 0.1°$ with a tip rounded to a radius of 0.20 mm.
    (3) ASTM E18-00 standard gives values in kgf. Newton values are conversions.

21.1.3 Rockwell superficial hardness

Relevant standards: ASTM E18-00, BS EN ISO 6508-1:1999
This test is similar in principle to the Rockwell test, but in order to keep the depth of impression small the minor load is restricted to 3 kg. It is used for thin specimens and for nitrided steels, etc. The following scales are used.

Table 21.2 ROCKWELL SUPERFICIAL HARDNESS SCALES

Scale	Penetrator	Preliminary Test Force		Total Test Force	
		kgf	N	kgf	kN
15-N	Diamond cone	3	29.4	15	0.14
30-N	Diamond cone	3	29.4	30	0.29
45-N	Diamond cone	3	29.4	45	0.44
15-T	$\frac{1}{16}$ in (1.588 mm) steel ball	3	29.4	15	0.14
30-T	$\frac{1}{16}$ in (1.588 mm) steel ball	3	29.4	30	0.29
45-T	$\frac{1}{16}$ in (1.588 mm) steel ball	3	29.4	45	0.44

21.1.4 Vickers hardness test

Relevant standards: BS EN ISO 6507-1:1998, ASTM E92-82 (1997)
A diamond indenter, in the form of a right pyramid with a square base and an angle of 136° between opposite faces, is forced into the material under a load F. The two diagonals, d_1 and d_2 of the indentation left in the surface of the material after removal of the load are measured and their arithmetic mean d calculated. The area of the sloping surface of the indentation is calculated, the indentation being considered as a right pyramid with a square base of diagonal d and vertex angle of 136°.

The Vickers hardness is the quotient obtained by dividing the load F, expressed in kilograms-force, by the sloping area of the indentation expressed in square millimetres.

Symbols:
F = load in kilograms force (kgf) where $1\,\mathrm{kgf} = 9.806\,65\,\mathrm{N}$
d = arithmetic mean of the two diagonals d_1 and d_2 in millimetres (mm)
HV = Vickers hardness
$= 2F \sin (136°/2)d^2$
$= 1.854F/d^2$

The loads employed vary from 1 to 100 kgf and are maintained from 10 to 15 seconds. Hardness numbers are expressed as 440 HV30 for example where a hardness of 440 is measured using a 30 kgf load.

21.1.5 Micro-indentation hardness testing

Relevant standards: ASTM E 384-99, BS EN ISO 6507-1:1997
Micro-indentation hardness testing is performed according to principles similar to the Vickers hardness test, but with considerably lower loads that produce microscopic indentations (typically 10–200 μm). Micro-indentation hardness testing, also referred to as 'microhardness' testing, is commonly used to measure hardness in very local areas, e.g. different microstructural phases or surface layers. As in the Vickers test, a diamond indenter with a specific geometry is impressed into the specimen surface with a known, calibrated force for a duration of 10–15 sec. Forces range from 1–1 000 gf. The indenter is removed from the surface and the size of the impression is measured using a microscope. Micro-indentation hardness testers are sold commercially.

Two indenter shapes are employed. The Vickers indenter is a square-based pyramid with face angles of 136°. The lengths of the two diagonals of the resulting square indent are averaged and the hardness, HV, is calculated according to:

$$HV = 1\,854.4\,P/d^2$$

where P is the applied force in gf and d is the mean diagonal length.

The Knoop indenter is a rhombic-based pyramid with included longitudinal edge angles of 172.5° and 130°; the long diagonal is approximately seven times the length of the short diagonal. The long diagonal is measured, and the hardness, HK, is calculated according to:

$$HK = 14\,229\,P/d^2$$

The narrow width of the Knoop indent makes it better suited for measuring steep hardness gradients since the indents can be more closely spaced.

Vickers hardness values are reported in the form: *value* HV_{force}, for example, $650\,HV_{100}$ corresponds to a Vickers hardness of 600 measured with a 100 gf force. The same form is used for Knoop hardness with HK instead of HV. In the European standard, force is reported in units of kgf.

When selecting the force to be used for testing, it is important to recognise that the measured hardness is not necessarily independent of test force, especially for forces less than 100 gf. Larger indents produce better accuracy due to reduced error in length measurement. The size of the indent, however, must be considerably smaller than the area of interest. In addition, the indenter must be perpendicular to the test piece, and the surface being measured must be free of deformation and damage resulting from preparation (e.g. grinding and polishing). Ref. [4] provides further information on microindentation hardness testing.

Recently, the techniques of microhardness indentation testing have been extended to smaller scales using lower loads and more precise instrumentation. This class of testing is referred to as nanohardness testing, nanoindentation testing, or instrumented indentation testing. These test techniques commonly employ a three-sided pyramidal indenter, rather than the four-sided Vickers indenter, and include the ability to measure and record the force on the indenter along with its vertical displacement. The latter feature of nanoindentation testing allows it to used to characterise a number of properties of surfaces and thin films, including elastic modulus, yield stress and strain-hardening behaviour, fracture toughness of brittle materials, and creep behaviour. Ref. [5] provides further information and references related to nanoindentation testing.

21.1.6   Hardness conversion tables

Table 21.3 only applies to steel of uniform chemical composition and uniform heat treatment, and is not recommended for non-ferrous metals or for case-hardened steels.

Table 21.3   APPROXIMATE CONVERSION OF HARDNESS VALUES

Non-austenitic steels

		Rockwell hardness No.			
Vickers hardness No. HV	*Brinell* 3 000 kg *load* 10 mm *ball* HBW	*C Scale* 150 kg *load* diamond cone HRC	*A scale* 60 kg *load* diamond cone HRA	*Superficial* 30–N 10 kg *load* diamond cone	*Scleroscope hardness No.*
100	95	—	43	—	—
120	115	—	46	—	—
140	135	—	50	—	21
160	155	—	53	—	24
180	175	—	56	—	27
200	195	—	58	—	30
220	215	—	60	—	31
240	235	20.3	60.7	41.7	34
260	255	24.0	62.4	45.0	37
280	275	27.1	63.8	47.8	40
300	295	29.8	65.2	50.2	42
320	311	32.2	66.4	52.3	45
340	328	34.4	67.6	54.4	47

(continued)

Table 21.3 APPROXIMATE CONVERSION OF HARDNESS VALUES—*continued*

Vickers hardness No. HV	Brinell 3 000 kg *load* 10 mm *ball* HBW	Rockwell hardness No.			Scleroscope hardness No.
		C Scale 150 kg *load* diamond cone HRC	A scale 60 kg *load* diamond cone HRA	Superficial 30–N 10 kg *load* diamond cone	
360	345	36.6	68.7	56.4	50
380	360	38.8	69.8	58.4	52
400	379	40.8	70.8	60.2	55
420	397	42.7	71.8	61.4	57
440	415	44.5	72.8	63.5	59
460	433	46.1	73.6	64.9	62
480	452	47.7	74.5	66.4	64
500	471	49.1	75.3	67.7	66
520	487	50.5	76.1	69.0	67
540	507	51.7	76.7	70.0	69
560	525	53.0	77.4	71.2	71
580	545	54.1	78.0	72.1	72
600	564	55.2	78.6	73.2	74
620	582	56.3	79.2	74.2	75
640	601	57.3	79.8	75.1	77
660	620	58.3	80.3	75.9	79
680	638	59.2	80.8	76.8	80
700	656	60.1	81.3	77.6	81
720	670	61.0	81.8	78.4	83
740	684	61.8	82.2	79.1	84
760	698	62.5	82.6	79.7	86
780	710	63.3	83.0	80.4	87
800	722	64.0	83.4	81.1	88
820	733	64.7	83.8	81.7	90
840	745	65.3	84.1	82.2	91
860	—	65.9	84.4	82.7	92
880	—	66.4	84.7	83.1	93
900	—	67.0	85.0	83.6	95
920	—	67.5	85.3	84.0	96
940	—	68.0	85.6	84.4	97

* Brinell values greater than 480 are determined with a carbide ball.

Table 21.4 APPROXIMATE CONVERSION OF HARDNESS VALUES

Aluminium and its alloys

HV 10 kg	Brinell* HBW	Rockwell superficial HR			HV 10 kg	Brinell* HBW	Rockwell superficial HR		
		15T	30T	45T			15T	30T	45T
20	19.4	2.3	—	—	80	74.6	75.3	44.1	—
24	23.1	18.1	—	—	84	78.3	76.5	46.9	—
28	26.8	29.5	—	—	88	82.0	77.6	49.4	—
32	30.5	38.2	—	—	92	85.7	78.6	51.8	—
36	34.2	45.0	—	—	96	89.4	79.5	54.0	—
40	37.8	50.3	—	—	100	93.0	80.3	55.9	—
44	41.5	54.8	—	—	104	96.7	81.1	57.7	—
48	45.2	58.6	5.2	—	108	100.4	81.8	59.4	—
52	48.9	61.8	12.6	—	112	104.1	82.5	60.9	—
56	52.6	64.5	19.1	—	116	107.8	83.1	62.4	—
60	56.2	66.9	24.5	—	120	111.4	83.7	63.8	39.0
64	59.9	69.0	29.4	—	124	115.1	84.3	65.0	41.2
68	63.6	70.8	33.7	—	128	118.8	84.8	66.2	43.3
72	67.3	72.5	37.6	—	132	122.5	85.3	67.4	45.2
76	71.0	74.0	41.0	—	136	126.2	85.7	68.5	47.1

(continued)

Table 21.4 APPROXIMATE CONVERSION OF HARDNESS VALUES—*continued*

HV 10 kg	Brinell* HBW	Rockwell superficial HR 15T	30T	45T	HV 10 kg	Brinell* HBW	Rockwell superficial HR 15T	30T	45T
140	129.8	86.2	69.4	48.8	172	159.3	88.3	75.8	59.6
144	133.5	86.6	70.4	50.4	176	163.0	89.2	76.4	60.7
148	137.2	86.9	71.3	51.9	180	166.6	89.4	77.0	61.7
152	140.9	87.3	72.1	53.3	184	170.3	89.7	77.6	62.7
156	144.6	87.6	72.9	54.7	188	174.0	89.9	78.1	63.6
160	148.2	88.0	73.7	56.0	192	177.7	90.1	78.7	64.5
164	151.9	88.3	74.4	57.3	196	181.4	90.3	79.2	65.4
168	155.6	88.6	75.1	58.4	200	185.0	90.6	79.7	66.2

* The ratio $\dfrac{\text{load}}{(\text{dia. of ball})^2}$ for the Brinell tests was 5 for Brinell hardnesses of 20–60 and 10 for hardnesses above 60.

Table 21.5 APPROXIMATE CONVERSION OF HARDNESS VALUES

Brass

HV	Rockwell hardness No. B scale, 100 kg load, $\frac{1}{16}$ in ball	F scale, 60 kg load, $\frac{1}{16}$ in ball	Rockwell superficial hardness No. 15-T scale, 15 kg load, $\frac{1}{16}$ in ball	30-T scale, 30 kg load, $\frac{1}{16}$ in ball	45-T scale, 45 kg load, $\frac{1}{16}$ in ball	Brinell hardness No. 500 kg load, 10 mm ball	HV	Rockwell hardness No. B scale, 100 kg load, $\frac{1}{16}$ in ball	F scale, 60 kg load, $\frac{1}{16}$ in ball	Rockwell superficial hardness No. 15-T scale, 15 kg load, $\frac{1}{16}$ in ball	30-T scale, 30 kg load, $\frac{1}{16}$ in ball	45-T scale, 45 kg load, $\frac{1}{16}$ in ball	Brinell hardness No. 500 kg load, 10 mm ball
45	—	40.0	—	—	—	42	120	67.0	95.5	—	61.0	41.0	106
47	—	45.0	—	—	—	44	124	69.0	96.5	—	62.5	43.0	110
49	—	49.0	54.5	—	—	46	128	71.0	97.5	—	63.5	45.0	113
52	—	53.5	57.0	—	—	48	132	73.0	98.5	84.5	65.0	46.5	116
56	—	58.8	60.0	15.0	—	52	136	74.5	99.5	85.0	66.0	48.0	120
60	10.0	63.0	62.5	20.5	—	55	140	76.0	100.5	85.5	67.0	50.0	122
64	15.5	66.8	65.0	22.5	—	59	144	77.8	101.5	86.0	68.0	51.5	126
68	21.5	70.0	67.0	30.0	—	62	148	79.0	102.5	—	69.0	53.0	129
72	27.5	73.2	69.0	34.0	—	64	152	80.5	103.0	—	—	54.0	133
76	32.5	76.0	70.5	38.0	4.5	68	156	82.0	104.0	87.0	70.5	55.5	136
80	37.5	78.6	72.0	41.0	10.0	72	160	83.5	—	—	71.5	56.5	139
84	42.0	81.2	73.5	44.0	14.5	76	164	85.0	105.5	—	72.0	58.0	142
88	46.0	83.5	75.0	47.0	19.0	79	168	86.0	106.0	88.0	73.0	59.0	146
92	49.5	85.4	76.5	49.0	23.0	82	172	87.5	106.5	—	73.5	60.0	149
96	53.0	87.2	77.5	51.5	26.5	85	176	88.5	107.0	—	—	61.0	152
100	56.0	89.0	78.5	53.5	29.5	88	180	90.0	107.5	—	75.0	62.0	156
104	58.0	90.5	79.5	55.0	32.0	92	184	91.0	—	—	75.5	63.0	159
108	61.0	92.0	—	57.0	34.5	95	188	92.0	—	89.5	—	64.0	162
112	63.0	93.0	81.0	58.5	37.0	99	192	93.0	—	—	77.0	65.0	166
116	65.0	94.5	82.0	60.0	39.0	103	196	93.5	110.0	90.0	77.5	66.0	169

Note: $\frac{1}{16}$ in $\equiv 1.588$ mm.

Table 21.6 APPROXIMATE CONVERSION OF HARDNESS VALUES

Hard metals

Vickers diamond hardness, 50 kg *load*	Rockwell A scale 60 kg load: diamond cone	Rockwell C scale 150 kg load: diamond cone
1 750	92.4	80.5
1 700	92.0	79.8
1 650	91.7	79.2
1 600	91.3	78.4
1 550	90.9	77.7
1 500	90.5	77.0
1 450	90.1	76.2
1 400	89.7	75.4
1 350	89.3	74.6
1 300	88.9	73.8
1 250	88.5	73.0
1 200	88.1	72.2
1 150	87.6	71.3
1 100	87.0	70.4
1 050	86.4	69.4
1 000	85.7	68.2
950	85.0	66.6
900	84.0	64.6
850	82.8	—

Table 21.7 APPROXIMATE CONVERSION OF HARDNESS VALUES FOR NICKEL AND HIGH NICKEL ALLOYS

Vickers hardness No. HV	Brinell hardness No. HBW	Rockwell hardness No.		
		A scale	*B scale*	*C scale*
Vickers indenter 1, 5, 10 *and* 30 kgf	10 mm *ball* 3 000 kgf	Diamond penetrator 60 kgf	$\frac{1}{16}$ in ball 100 kgf	Diamond penetrator 150 kgf
77	77	—	30	—
79	79	—	34	—
83	83	—	38	—
87	87	—	42	—
91	91	—	46	—
95	95	—	50	—
100	100	—	54	—
106	106	—	58	—
112	111	39.0	62	—
119	118	41.0	66	—
126	125	43.0	70	—
135	134	45.5	74	—
145	144	47.5	78	—
157	155	50.0	82	—
171	168	52.5	86	—
188	184	55.0	90	—
209	204	57.5	94	—
234	228	60.5	98	20.0
248	241	61.5	100	22.5
285	275	64.5	—	28.5
326	313	67.5	—	34.0
362	346	69.5	—	38.0
404	382	71.5	—	42.0
452	425	73.5	—	46.0
481	450	74.5	—	48.0
513	—	75.5	—	50.0

Note: $\frac{1}{16}$ in ≡ 1.588 mm.

21.2   Tensile testing

21.2.1   Standard test pieces

Relevant standards: ASTM E 8-00 and E 8M-00, BS EN 10002-1 through -5.

The tensile test is the most commonly used mechanical test for metals. A specimen is extended under a uniaxial tensile force until separation occurs. The force required to deform the specimen is recorded as a function of extension, and from these data a number of intrinsic properties may be derived, which include: modulus of elasticity, yield and ultimate tensile strengths, ductility, and strain hardening behaviour. Further information regarding the tensile behaviour of metals and tensile testing may be found in Refs [6–9]. Tensile properties of common metals and alloys are presented in Chapter 22.

A wide variety of specimen geometries may be employed. For certain product shapes (e.g. strips and bars), a full cross-section of the product may be tested. In most cases, however, a part of the product is machined into a specimen which has a central section of reduced area and grip zones at either end. Cross-sections of the reduced area are usually circular or rectangular. In the reduced section, a gauge length is defined, from which measurements are made to determine ductility (Fig. 21.1).

In the European test standard, two types of 'proportional' test pieces are defined, regardless of the cross-section shape. For these specimens the gauge length L_0 and cross-sectional area A_0 are related by:

$$L_0 = 5.65\sqrt{A_0} \text{ and}$$
$$L_0 = 11.3\sqrt{A_0}$$

Non-proportional test pieces are also defined for various shapes.

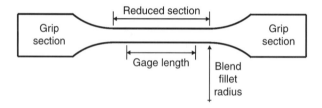

Figure 21.1   *Tensile test specimen*

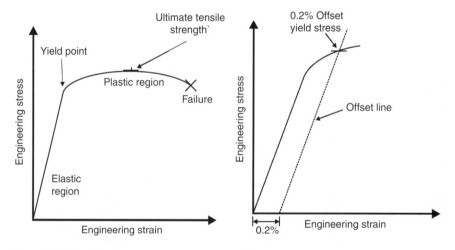

Figure 21.2   *Typical engineering stress–strain curve*   **Figure 21.3**   *Offset method for determining yield stress*

In the ASTM standard, specific dimensions are defined for the different cross-sectional shapes. The standard gage length and diameter for cylindrical specimens are 2.0 inch and 0.5 inch, respectively, while for rectangular cross-sections the gage length is 2.0 inch, the gage width 0.5 inch, and the thickness is variable. Sub-standard specimens proportionally based on the standard specimens are also defined. ASTM E 8M defines similarly sized specimens using metric units.

Several different grip end configurations are available. The specimen may be threaded, wedge-clamped, button-ended, or pin-loaded. The primary consideration for grips is that the specimen must not fail within the grip. The ability to maintain proper specimen alignment, so that the load is purely axial, is also important. Proper choice of grip arrangement is most critical for metals with limited ductility. If the specimen is machined from a product piece, the orientation of the long axis with respect to significant axes in the product (e.g. rolling direction) should be defined.

The force-extension data is gathered in a tensile test from transducers attached to the load frame and specimen. Strain-gauge based load cells built into the test frames are nearly exclusively used for force measurement. Extension is measured within the gauge length of the specimen, either mechanically (using clip-on strain gauges or linear variable differential transducers) or optically. Tensile behaviour is usually reported in terms of engineering stress and strain, which are calculated from the force and initial cross-sectional area (stress) and the extension and initial gauge length (strain). A typical stress–strain curve, shown in Fig. 21.2, shows an initial elastic region where stress and strain are linearly related, followed by non-linear plastic deformation at higher strains. Tensile behaviour may also be reported as true stress and strain, which are defined based on actual gauge length and area at a given applied force.

The strength of metal in a tensile test is quantified in terms of the yield strength and ultimate tensile strength (UTS). The yield strength reflects the stress needed to cause permanent plastic deformation; the exact value obtained from a given stress–strain curve depends on the definition used. The most common definition is the 0.2% offset strength, which is obtained from the intersection of the stress–strain curve and a line parallel to the initial elastic region but offset 0.2% in strain (Fig. 21.3). A second definition is the extension under load (EUL) which is defined as the stress needed to result in a given strain (typically 0.5%). The ultimate tensile strength is derived from the maximum load achieved in the tensile test, the highest point in the engineering stress–strain curve (see Fig. 21.2). For metals in which necking, a localised reduction in cross-section, occurs, the load on the specimen drops with continued extension beyond the UTS. The true stress in the neck does not diminish after the UTS is reached.

There are two measures of ductility assessed in a tensile test, elongation and reduction in area. Elongation is the increase in gauge length of the test piece after testing divided by the original gauge length. The increase may be manually measured by placing the two broken halves of the specimen back together and measuring the distance between gauge length marks placed on the specimen before testing, or by directly reading the extension at failure with an extensometer on the specimen. The two methods do not give equivalent results. Elongation values obtained are dependent on the initial gauge length of the specimen. This effect is particularly pronounced in materials which exhibit extensive necking. It is therefore important to report the gauge length along with tensile results.

Reduction in area is the difference between the original cross-sectional area and the smallest cross-sectional area after testing, expressed as a percentage of the original area. Reduction in area is not affected by gauge length differences, since measurements of area after testing are performed on any necked region.

The speed at which the test is conducted is an important parameter in tensile testing. Metal properties can exhibit a strong strain rate dependence, particularly at elevated temperatures. The strain rate is typically controlled through the rate of test machine cross-head displacement or the rate of stress increase, both of which can be roughly correlated with the strain rate. ASTM standard E 8 requires that the rate of stress increase be controlled to between 1.1 and 11 MPa/sec when determining the yield point. In any case, the speed of test (strain rate, stress rate, or cross-head displacement rate) should be reported with the tensile test data.

21.3   Impact testing of notched bars

Impact tests represent a simple way of measuring the toughness of a metal under rapid loading conditions and were the first fracture toughness tests devised. Despite the empirical nature of the data obtained, impact testing is still frequently used as a quality control measure and as a technique for determining the transition temperature from ductile to brittle behaviour in low-carbon steels. Further details on impact testing and the impact behaviour of metals may be found in Refs [6,10,11]. Data on impact energies for a number of common metals and alloys are presented in Chapter 22.

Table 21.8   DIMENSIONS OF IZOD IMPACT TEST PIECES—BRITISH STANDARD BS 131:PART 1:1961 (1996) AND AMERICAN STANDARD ASTM E23-01

	Square section		Circular section	
	Nominal dimension			
Item	mm	in *equivalent*	in	mm *equivalent*
Minimum overall length of test piece				
1 Notch	70[1]	2.75	2.8[1]	71
2 Notch	98	3.86	3.9	99
3 Notch	126[2]	4.96	5.0[3]	127
Width	10	0.394 ⎫		
Thickness	10	0.394 ⎭	0.45 dia.	11.48 dia.
Root radius of notch	0.25	0.010	0.010	0.25
Maximum depth below notch	8	0.315	0.32	8.1
Distance of plane of symmetry of notch from free end of test piece and from the adjacent notch	28	1.1	1.1	28
Angle of notch	45°		45°	

Notes: (1) American Standard overall length 75 mm (2.952 in).
　　　　(2) American Standard overall length 131 mm (5.157 in).
　　　　(3) American Standard overall length 137 mm (5.375 in).

21.3.1   Izod test

Relevant standards: ASTM E23-01, BS 131:Part 1:1961 (1996)
The test consists of measuring the energy absorbed in breaking a notched test piece by one blow from a striker carried by a pendulum. The test piece is gripped vertically with the root of the notch in the same plane as the upper face of the grips. The blow is struck on the same face as the notch and at a fixed height above it. Tests are usually performed at the ambient temperature of the test house.

TEST PIECES

The standard test pieces are either 10 mm square or 0.45 inches (11.4 mm) diameter cross-section. Complete dimensions are shown in Table 21.8. The notch profile is shown in Figure 21.2.

PRESENTATION OF RESULTS

When standard test pieces are used the following symbols are used in reporting the results of Izod tests:

　　I for Izod
　　S for square section
　　Rs for circular cross-section with straight notch
　　E.G. I120S: x ft lbf

An energy of x ft lbf was obtained from an Izod test with a striking energy of 120 ft lbf using a square section test piece.[†]

21.3.2   Charpy test

Relevant standards: ASTM E23-01, BS EN 10045-1:1990
The test consists of measuring the energy absorbed in breaking by one blow from a pendulum a test piece notched in the middle and supported at each end.

[†] BS 131:Part 2:1972 uses SI units with 1 J ≈ 4/3 ft lbf for striking energies (1 ft lbf = 1.355 82 J).

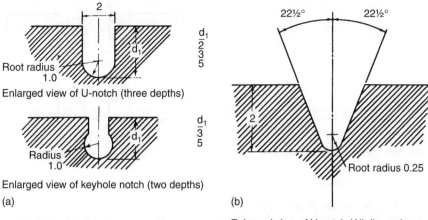

Figure 21.4   *Standard notch profiles for impact test pieces—see Tables 21.8 and 21.9*

Table 21.9   DIMENSIONS OF CHARPY IMPACT TEST PIECES—BRITISH STANDARDS BS EN 10045-1:1990; AMERICAN STANDARD ASTM E23-01

	Nominal dimension	
Item	mm	in *equivalent*
Length	55	2.165
Width standard test piece	10	0.394
subsidiary test piece[1]	7.5	0.295
subsidiary test piece	5.0	0.197
subsidiary test piece	2.5	0.098
Thickness	10	0.394
Root radius of V notch	0.25	0.010
Depth below V notch	8	0.315
Root radius of U notch	1	0.039
Depth below U notch	8 (2 mm notch)[2,3]	0.315
	7 (3 mm notch)[2,3]	0.276
	5 (5 mm notch)	0.197
Distance of notch from one end of test piece	27.5	1.083
Angle between plane of symmetry of notch and longitudinal axis of test piece	90°	
Angle of V notch	45°	

Notes: (1) Additional subsidiary test pieces are permitted by ASTM E23-01 as follows: 5 mm thick with 4 mm depth below the notch and 5, 10 or 20 mm width and 3 mm thick with 0.094 in (2.39 mm) depth below notch and 10 mm width.
(2) Not specified in ASTM E23-01.
(3) The 2, 3 and 5 mm notches are sometimes referred to as 'Mesnager', 'DVM' and 'Charpy', respectively.

TEST PIECES

The standard test piece has a 10 mm square cross-section with one of the three notch profiles shown in Figure 21.4. The test piece dimensions are shown in Table 21.9. Sub-standard test pieces are used where material thicknesses do not permit full-size specimens. It should be noted that the values obtained from subsidiary specimens cannot be compared with full-size specimens nor can the values obtained from different notches be compared. The current EN standard only includes the V- and U-notch profiles.

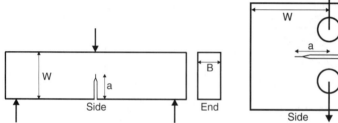

Figure 21.5   *Single-notch bend specimen*

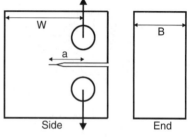

Figure 21.6   *Compact tension specimen*

PRESENTATION OF RESULTS

The report of the tests should include the following information:
Type of test: Charpy V or U notch; striking energy of the machine; size of test piece if sub-standard.
Nominal depth of notch and form ('U' or 'keyhole').
The energy absorbed (J) and the test temperature:

> e.g. C160V : x J at y °C

An energy of x J was recorded from a Charpy V notch specimen tested at y °C with a striking energy of 160 J:

> e.g. C320U2 : x J at y °C

In this case the specimen was a Charpy with a U notch 2 mm deep and the striking energy was 320 J.
Note: The above reporting form is used for the BS EN 10045-1 standard; there is no specified reporting form in ASTM E23-01.

21.4   Fracture toughness testing

Fracture toughness tests provide quantification of a material's resistance to crack extension using the principles of fracture mechanics in both linear-elastic and elastic–plastic forms. Tests are performed on specimens containing sharp, pre-existing defects formed by fatigue loading and are typically conducted at relatively slow loading rates (in contrast to impact tests, section 21.3). Information obtained in fracture toughness tests is used to determine the load-bearing capacity of materials and structures with defects. The following sections present information on common standardised fracture toughness tests for metals. Further information on fracture toughness testing is presented in Ref. [12]; the applicable standards must be conducted for complete test requirements.

21.4.1   Linear-elastic (K_{Ic})

Relevant standards: ASTM E 399-90, BS EN ISO 12737:1996
Linear elastic fracture toughness tests are used to measure resistance to crack extension for metals which are relatively strong and have relatively low toughness. For this case, the region of plastic deformation associated with the crack is small relative to the dimensions of the crack and the test specimen, and linear elastic fracture mechanics (LEFM) principles may be applied. In LEFM, the local stress field ahead of the tip of a sharp crack is characterised by the stress intensity parameter, K, which is dependent on the crack length, geometry, and applied load. The standard test measures K_{Ic}, the critical value of K for unstable crack propagation in an opening mode (mode I) with high constraint (plane-strain conditions). K_{Ic} is considered an intrinsic material property, and is believed to represent the lower limiting value of fracture toughness. For details on fracture mechanics and the applicability and use of K_{Ic}, consult Refs [13,14]. A limited amount of fracture toughness data is presented in Chapter 22.

Two specimen geometries are commonly used in fracture toughness testing, the single edge-notched bend and the compact tension specimens. These specimen names are abbreviated in ASTM standards as SE(B) and C(T), respectively; schematic diagrams of the two specimens are shown in Figs. 21.5 and 21.6. ASTM E 399 also defines three other specimen geometries with arc shapes

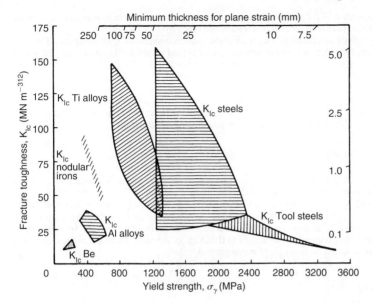

Figure 21.7 *Yield strength versus fracture toughness data for several alloy systems at 20°C. The upper scale of the diagram indicates the minimum thickness requirement for a given strength and toughness by projection of the line joining the origin and data point. (Source: Fulmer Materials Optimiser, Fulmer Laboratories Ltd, UK)*

Table 21.10 RECOMMENDED THICKNESS AND CRACK LENGTH FOR FRACTURE TOUGHNESS TEST PIECES

	Minimum recommended thickness and crack length	
σ_Y/E	mm	in
up to 0.005 0	100	4.0
0.005 0–0.005 7	75	3.0
0.005 7–0.006 2	63	2.5
0.006 2–0.006 5	50	2.0
0.006 5–0.007 1	38	1.5
0.007 1–0.008 0	25	1.0
0.008 0–0.009 5	13	0.5
0.009 5 or greater	6.5	0.25

that are less commonly used. All specimen types incorporate a sharp machined notch from which a fatigue crack may be grown. The SE(B) specimen requires more material for a given thickness than the C(T) specimen, but is simpler to machine.

The size of the specimen used is an important consideration, since certain size requirements must be met for the K_{Ic} test to be valid. The key dimensions of both specimen types are the thickness, B, and crack length, a (these are nominally equal). In order for a test to be valid, both must be larger than $2.5[(K_{Ic})/(\sigma_Y)]^2$, where σ_Y is the 0.2% yield stress for the material under the conditions of test, e.g. orientation, temperature, and loading rate. This requirement insures that the plastic zone ahead of the crack tip is sufficiently small with respect to these dimensions. Hence an initial measurement of σ_Y and an estimate of K_{Ic} must be made prior to selecting specimen size. Figure 21.7 can be used as a graphic guide to selection of thickness. In the absence of this information, the ratio of yield strength to elastic modulus may also be used to estimate specimen thickness (Table 21.10).

Prior to the actual fracture toughness test, a fatigue precrack must be grown from the starter notch. The standards define requirements for the length of fatigue crack which must be grown and the load levels used. The objective of fatigue precracking is to obtain a microscopically sharp crack with an insignificant region of plastic deformation ahead of the crack tip.

The test itself involves monotonically increasing the specimen load until fracture occurs. The load and displacement are recorded and used to generate a plot of load versus displacement. Displacement is measured at the crack mouth using a clip gauge. After the test, the load-displacement record is analyzed to determine P_Q, a load from which K_Q is computed. K_Q serves as a provisional value of K_{Ic} until the validity requirements have been demonstrated. A secant line through the origin is drawn with a slope 5% less than the original linear region of the load-displacement curve. The load P_Q is defined as the intersection of the secant line with the load-displacement curve, or any higher recorded value prior to the intersection. K_Q is computed using the equation:

$$K_Q = P\, f(a/W)/B\sqrt{W}$$

Expressions for $f(a/W)$ for the different specimen geometries are given in the standards. The crack length a is measured on the broken halves of the specimen at the end of the fatigue precrack.

A number of validity requirements must be met before the provisional K_Q may be accepted as K_{Ic}. In addition to the size requirements above (in which K_Q is substituted for K_{Ic}), the maximum load in the test must not exceed P_Q by more than 10%. This requirement limits the amount of stable crack extension. There are other requirements relating to the shape of the crack front and precracking load levels. The standards should be consulted for complete details.

21.4.2 K–R curve

Relevant standard: ASTM E 561-98
The K–R curve test provides a characterisation of fracture resistance for materials and specimen geometries in which stable, ductile crack extension occurs under elastic conditions. In the test, a plot of K, the stress intensity factor needed to drive the crack, as a function of crack extension, Δa, is developed, which is referred to as the R curve. An example R-curve is shown in Fig. 21.8. Once established, the R curve may be used to determine a value of K_c, the plane-stress fracture toughness, for various crack lengths and specimen geometries. The K–R test is frequently used to measure the fracture toughness of sheet materials, for which the thickness requirements (which ensure plane-strain conditions) of the K_{Ic} test (Section 21.4.1 above) may not be met. It is important to note that

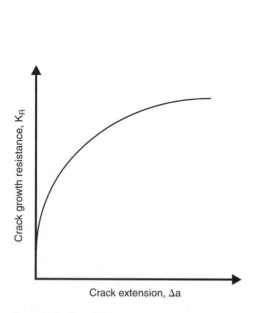

Figure 21.8 *Typical K–R curve*

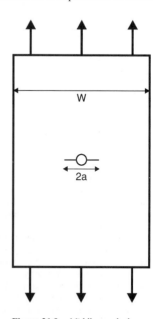

Figure 21.9 *Middle-cracked tension specimen*

results obtained in K–R curve testing are dependent on specimen thickness—therefore all results are given for a specified thickness.

The test involves loading a notched specimen with a fatigue precrack in tension to cause stable crack growth. The load is plotted as a function of crack mouth displacement. The effective crack length (accounting for plasticity ahead of the crack tip) is determined either by physical measurement or from the load-displacement record. The load and effective crack length are used to calculate K for a series of increasing effective crack lengths, which are then plotted as a function of crack extension (total effective crack length minus initial crack length). Specimens with the compact tension C(T) or middle cracked tension M(T) geometries may be used with this technique. The M(T) geometry is used for sheet materials—a schematic diagram is shown in Fig. 21.9.

An alternative technique for K–R curve determination allowed in the standard is the use of a crack-line-wedge-loaded compact specimen C(W). The geometry is similar to the C(T) specimen, except that the loading is applied via a wedge driven into the crack mouth, resulting in displacement-controlled loading. This loading technique is useful for lower-toughness materials. Full details on the use of this specimen may be found in the standard.

Data obtained in R-curve testing under this standard must be principally in the elastic loading regime. This requirement is expressed for the C(T) and C(W) specimens as:

$$b = (W - a) \geq (4/\pi)/(K_{max}\sigma Y)^2$$

where b is the uncracked ligament and K_{max} is the maximum level of K reached in the test. For the M(T) specimen the net section stress based on the uncracked ligament must be less than the yield stress.

21.4.3   Elastic-Plastic (J_{Ic}, CTOD)

Relevant standards: ASTM E 1820-01, BS 7448-1:1991, BS 7448-4:1997
For low strength, high-toughness materials (such as low-carbon steels) the requirements of LEFM (K_{Ic}) testing are not easily satisfied as extensive ductile crack extension with associated plasticity occurs prior to failure. To quantify the toughness behaviour in this regime, elastic-plastic fracture mechanics tests have been developed. Several separate tests which use similar specimen geometries and techniques exist. These tests have been recently consolidated by the ASTM (along with K_{Ic} testing) into a single test standard, E 1820-01. Two British Standards covering the same test procedures exist. The tests provide different measures of the conditions for initiation of stable crack extension (similar to K_{Ic}) and the resistance to further crack extension (R-curve behaviour). Details of the tests are found in the standards and Ref. [12].

The two parameters used in elastic-plastic fracture toughness testing are the J integral and the crack tip opening displacement (CTOD or δ). The J integral is a non-linear measure of the driving force available to extend a crack,[14] and is analogous to G, the strain energy release rate in the linear regime. The J integral is useful because it can be accurately calculated from experimental data or using numerical methods. The CTOD is the opening at the position of the original crack tip, a parameter that is believed to characterise the fracture process.[13] The CTOD is calculated from the measured displacement at the crack mouth (CMOD), which is simple to measure experimentally. Due to simplicity, the CTOD was the first parameter used to measure the fracture behaviour of materials beyond the LEFM regime. Its use, however, has to a large extent been superseded by the J integral, because of the assumptions which must be made in calculation of CTOD from CMOD.

Two specimen geometries are allowed in the standards—the C(T) and SE(B) specimens, both essentially identical to the geometries used for K_{Ic} testing. The notch of the C(T) specimen is widened for J-integral testing to allow for displacement measurements to be made along the load line. In addition, specimens used in J-integral testing may incorporate side grooves to keep the crack in its original plane. All specimens are tested after creation of a sharp precrack using fatigue loading, as is performed in K_{Ic} testing.

The basic test procedure for both J-integral and CTOD tests involves loading the precracked specimen while monitoring the CMOD and/or load-line displacement until unstable fracture occurs or a desired displacement level is reached. Depending on the behaviour of the material and the information desired, a number of different fracture parameters may be obtained. The exact test and analysis procedures are complex, and the applicable standards must be consulted for details.

The fracture toughness parameters measured in J-integral or CTOD testing are useful for comparing the behaviour of different materials or as a quality control measure, and may also serve as a basis for defect tolerance assessment of structural materials. However, particular caution must be used when applying data obtained with laboratory test specimens to actual service conditions.

21.5 Fatigue testing

Fatigue tests measure the resistance of a material to failure by application of cyclic stresses. A very broad range of stresses and number of cycles to failure are encountered, from less than 100 cycles for stresses well in excess of the yield strength, to millions of cycles for low, elastic stresses. Historical tests have used specimens with smooth surfaces or relatively blunt notches in which the number of cycles to initiate and propagate a crack resulting in failure were measured. With the advent of fracture mechanics and damage-tolerant design principles, tests to measure sub-critical crack propagation rates under cyclic loading were developed and are commonly used today.

This section presents basic information concerning fatigue tests which have been standardised, describing properties measured, specimen and machine requirements, basic test procedures, and important considerations in testing. The reader should always refer to the relevant test standard for complete details. For further information concerning the fatigue behaviour of metals, consults Refs [15–17]. Fatigue data for a number of common metals and alloys are presented in Chapter 22.

21.5.1 Load-controlled smooth specimen tests

Relevant standards: ASTM E 466-96, BS 3518-12:1962 (1997), BS 3518-3:1963 (1997)

Load-controlled fatigue tests are typically performed at stresses in the elastic range, resulting in relatively long fatigue lives (greater than 10^5 cycles to failure). The specimen surfaces are polished smooth or contain machined notches. Each fatigue test measures the number of cycles to failure at a single stress level; failure is nearly always defined as complete specimen separation. Several fatigue tests at different stress levels are required to define the fatigue behaviour in terms of a S-N curve, a plot of cycles to failure (x-axis) versus stress range (y-axis). The curve defines stress levels below which failure in a given number of cycles will not occur. The S-N curve may be used to derive a 'fatigue strength', or stress level below which failure does not occur in a very high number of cycles, usually 10^7 or 10^8.

Load-controlled fatigue tests are typically performed at cyclic frequencies greater than 10 Hz and sometimes as high as 1000 Hz to achieve failure in reasonable times. Loading may be either in simple tension or rotating bending. A number of different machine designs are commercially available, including servo-hydraulic and electromechanical. Rotating-bend tests are favoured for high frequencies, since these are easily achieved and the test machine is relatively simple and inexpensive, allowing a number of machines to be obtained and used simultaneously. Very high frequency tests ($\sim 10^4$ Hz) can be performed using ultrasonic excitation of the specimen.[18]

The applied load usually follows a sinusoidal waveform. The maximum and minimum loads in the cycle define the mean load applied, and the load ratio, R ($= P_{max}/P_{min}$). In the case of fully-reversed (equal tensile and compressive, R $= -1$) loads the mean load is zero. Most S-N curves are produced with fully-reversed loading; other types of diagrams may be constructed to account for the effect of non-zero mean loads.[17] Note that the waveform for rotating-bend tests is always sinusoidal and fully-reversed.

Dogbone-shaped specimens very similar in shape to the tensile specimens described in section 21.3 are defined in the standards. Cross-sections may be circular (necessary for rotating bend) or rectangular. The gauge sections of fatigue specimens need not be straight, but may incorporate a shallow radius as shown in Fig. 21.10. Because fatigue cracks are frequently surface-initiated, the

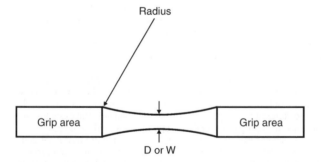

Figure 21.10 *Strain-controlled fatigue test specimen with continuous radius between ends*

finish of the specimen is very important. The standards define allowable machining removal rates and final polished roughness. The grip arrangement for fatigue specimens is also important, more so than in tensile tests. Threaded grip attachments are not recommended since the notches may initiate cracks.

Specimen alignment is critical in axial fatigue tests. The load must purely axial without the presence of significant bending stresses. Limits on the amount of bending stress are defined in the standards. Alignment is usually verified by loading a dummy specimen with several strain gages on the surface and comparing strain readings.

Load-controlled fatigue tests may be conducted at any desired temperature, provided that the gauge section of the specimen is uniformly heated or cooled. Several different furnace designs are available for use with fatigue testing machines, including clamshell resistance furnaces, induction furnaces, and radiative (lamp) furnaces. If the grips are heated or cooled along with the specimen they must be constructed from suitable materials. Temperature measurement of the specimen must be performed with care, so that surface defects are not introduced by the attachment of thermocouples.

The actual test procedure is simple, involving placing the specimen in the grips, applying a known cyclic force, and recording the elapsed cycles until failure occurs. A number of variables may influence the results obtained in load-controlled fatigue testing, and it is important to ensure that the conditions of test closely match those of the intended application. Variables to consider include: temperature, environment, humidity, specimen size, frequency, surface finish, and stress concentrators. Because of the statistical nature of fatigue phenomena, several specimens should be tested for each condition to assess the extent of scatter. This is especially important for lower-stress, longer-life conditions.

21.5.2 Strain-controlled smooth specimen tests

Relevant standards: ASTM E 606-92, BS 7270:1990
Tests in the so-called low cycle fatigue (LCF) regime usually involve significant plastic deformation of the specimen and are conducted with axial loading in strain control using servohydraulic testing machines. The use of strain as the test parameter is accomplished using extensometers attached to the specimen for feedback control of the test machine. Details on extensometry and testing machines used for LCF testing may be found in Refs [19–21]. The tests are otherwise similar to load-controlled tests, in that each specimen gives a single data point of cycles to failure at the applied strain range. Failure is not always defined as complete specimen separation, however, but may also be a specified drop in maximum load, observation of crack initiation, or a change in modulus ratio. Plots of cycles to failure as a function of applied strain range are constructed.

In addition to the number of cycles to failure, the cyclic stress–strain behaviour of the specimen may also be reported. This takes the form of hysteresis loops, plots of stress versus strain for a single fatigue cycle (Fig. 21.11). Changes in the hysteresis loop from the beginning to end of the test indicates whether material hardens or softens under cyclic loading. Instead of full hysteresis loops, the maximum and minimum loads may also be recorded. The load range at 50% of the cycles to failure is commonly reported.

LCF tests are typically performed at relatively low frequencies, less than 1 Hz, and at elevated temperatures. Strain waveforms are usually triangular rather than sinusoidal, so that loading is accomplished at a constant strain rate. Strain-rate effects are important at elevated temperatures due to the addition of creep as a deformation mechanism. Straining in LCF tests is nearly always fully-reversed, with equal tensile and compressive strains. The loads needed to impose these strains may not be equal. Smooth specimen LCF tests may be used to measure creep-fatigue behaviour, in which hold periods of constant strain are imposed in which creep occurs. Details of creep-fatigue testing may be found in Refs [20,21].

Specimens used in LCF testing must be designed in conjunction with the test machine gripping arrangements to ensure that the load profile in the transition from tension to compression is continuous. The grip ends must be securely clamped. As with load-controlled tests, surface finish is important, although it may be less so if significant specimen oxidation occurs at elevated temperatures during testing. Alignment of the specimen is again critical.

21.5.3 Fatigue crack growth testing

Relevant standards: ASTM E 647-00, BS 6835-1:1998
Fatigue crack growth testing measures rate of advance of a fatigue crack in terms of the applied driving force for growth using Linear Elastic Fracture Mechanics (LEFM) principles.[13–15] The range

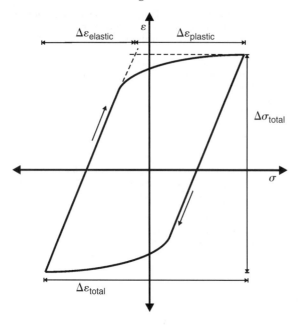

Figure 21.11   *Typical hysteresis loop obtained in strain-controlled low-cycle fatigue testing*

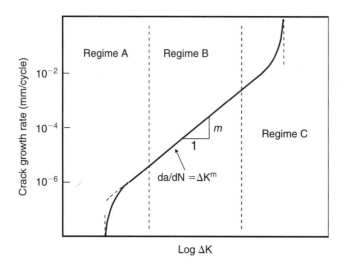

Figure 21.12   *General plot of crack growth rate versus stress intensity range showing three regimes of crack growth*

of stress intensity factor in the loading cycle (ΔK), is used as the driving force parameter. As described in Section 21.4, K is proportional to the applied load, the square root of the crack length, and a geometric factor. ΔK is simply the difference between the maximum and minimum K in each load cycle, and hence is proportional to the applied load range. The rate of crack growth, da/dN, where a is the crack length and N is the cycle number, is measured over a range of ΔK and a plot constructed. A generic plot of da/dN versus ΔK typical of metals is shown in Fig. 21.12. There are three characteristic regimes, the near-threshold region at low ΔK, the Paris regime where the logarithm of da/dN is linearly dependent on ΔK, and the stage III regime of high crack growth rates near failure. A threshold value of ΔK below which crack growth does not occur is commonly observed for metals. Further information on the fatigue crack growth behaviour of metals can be found in Ref. [15].

The fatigue test standards describe specimen, test machine, and reporting requirements for measurement of fatigue crack growth rates above 10^{-8} m/cycle. In the ASTM standard, procedures for measurement of fatigue threshold values are also given.

The specimens used in fatigue crack growth testing are identical to those used in other fracture mechanics tests (see section 21.4). In the ASTM standard three specimen types are defined: the compact tension, the single-edge bend, and the center-cracked tension. Similar specimen geometries are defined in the British standard. The specimens contain machined, sharp notches from which fatigue cracks are grown. Prior to the acquisition of valid crack growth rates, the specimens must be precracked to beyond the region of disturbed material at the notch root. Limits are set on the ΔK which may be used for precracking. Standard servo-hydraulic testing machines are used for crack growth testing, frequently equipped with computerised control and data acquisition systems to enable automated testing.

A critical aspect of fatigue crack growth testing is precise measurement of crack length. Three techniques are commonly used for crack length measurement: optical, compliance, and potential drop. In the optical technique the crack is measured at intervals using a travelling microscope. Strobe lighting may be used to allow easy observation of the crack while cycling. This is a historical technique not frequently used today due to the labour involved and lack of very fine resolution needed for threshold measurements. It is simple, however, and does not require computerised equipment.

The optical technique has been largely supplanted by compliance and potential drop techniques. In the compliance technique, the compliance of the specimen is computed from the crack-mouth opening displacement (measured with a clip-gauge mounted to the crack mouth) and the applied load. A calibrated curve of compliance as a function of crack length for the particular specimen geometry is used to derive crack length. In the potential drop method, a large DC current is passed through the specimen, and the potential generated across the crack mouth is measured. The potential drop increases with crack length, which can be quantified, again through a calibration curve. These techniques have the advantage of providing an accurate, essentially continuous measurement of crack length and are highly suited to automated control of the fatigue test. In fact, the complex nature of crack length derivations make automated, computerised measurement essentially a necessity.

Fatigue crack growth data in the higher growth rate regime (greater than 10^{-8} m/cycle) may be simply acquired by growing the crack at a constant load range expected to result in an initial growth rate near 10^{-8} m/cycle. ΔK increases as the crack grows, allowing a wide range of data to be obtained. Fatigue crack growth rates are calculated from the applied load range and crack length versus cycle number data using incremental polynomial or secant techniques.

To obtain near-threshold crack growth rate data (below 10^{-8} m/cycle), ΔK is decreased as the crack grows, a procedure referred to as 'load-shedding.' Proper load-shedding is complex because of three factors: ΔK depends on the crack length; the crack must be grown through the plastic zone of the previous ΔK level before the next reduction; and the threshold level must be reached before the crack has grown through the specimen. These considerations are accounted for in ASTM E 647-00; computerised, automated testing capability is again essential.

The frequencies used in fatigue crack growth testing are typically greater than 10 Hz to allow testing in reasonable timeframes. Fatigue crack growth rates may be a function of several material, environment, and test variables, depending on the metal and the regime of crack growth measured. A few examples include grain size, orientation, humidity, and load range. Requirements for analysing crack growth data and full reporting of fatigue results are given in the relevant standards. Fatigue crack growth testing may be performed at elevated temperatures or specific environments. Compliance and potential drop crack measurement techniques are especially amenable to testing in these situations.

21.6 Creep testing

Relevant standards: ASTM E 139-00, BS EN 10291:2000
Creep is defined as the time-dependent elongation of a metal which occurs at static stresses less than the nominal yield strength. For most metals creep only occurs at temperatures exceeding half of the absolute melting point. Creep testing measures the rate of specimen elongation as a function of applied stress, temperature, and test time. A number of tests must be conducted to fully characterise the behaviour of a given material. Details on the creep behaviour of metals may be found in Refs [22–24], and creep properties of common metals and alloys are presented in Chapter 22.

There are three types of creep tests: creep, creep-rupture, and stress-rupture tests. In a creep test, testing is only performed as long as necessary to obtain the minimum creep rate at the specified stress and temperature; the test is not performed to failure. Creep-rupture refers to a creep test conducted to failure. Creep rates are not measured in stress-rupture testing, only the time to failure under creep loading conditions. All three tests use similar specimens and load frames.

The specimens used for creep testing are essentially identical to those used in simple tensile tests (see Fig. 21.1), although creep specimens may have grooved attachment points on the shoulders of the specimen and longer gauge lengths to permit more accurate strain measurement. Creep testing may also be performed in compression; in this case the specimens may be either cylindrical or parallelepipeds with aspect ratios between 2 and 4.

Creep test frames are simple in design, with a static load applied with weights through a typically 20:1 lever arm. Clamshell-type resistance furnaces are used for heating. Due to the common use of high temperatures (frequently greater than 800°C), extensometers for creep testing are designed to translate the specimen extension out of the hot zone, where measurement is made using standard LVDT transducers. Pull rods, gripping devices, and extensometers must be constructed from suitable high-temperature materials. Due to the long test durations and the necessity to perform several tests to cover a range of stresses and temperatures, a creep testing laboratory usually includes a large number of frames, often with a central data collection system.

The test standards define specimen, load verification, and data reporting requirements for all types of creep tests. As with other axially-loaded mechanical tests, specimen alignment is critical and must be verified to demonstrate that bending stresses are negligible. An important additional parameter to consider in creep testing is the environment, which may have a strong influence at the high test temperatures commonly encountered. Specially designed retort systems are commercially available to enable testing in either inert or vacuum environments.

21.7 Non-destructive testing and evaluation

Non-destructive testing and evaluation (NDT, NDE) techniques permit examination of metal parts or samples without alteration or destruction. NDE encompasses all phases of the production, use, and eventual rejection of components and structures. The initial scope of NDT was the detection of damage in the form of flaws, cracks, inclusions, etc. Since the primary concern has been detection of the smallest possible flaw, sensitivity and probability of detection are important aspects of NDT technology. For materials which have many flaws, however, detecting an individual flaw is immaterial and the overall effect of the flaws on the residual strength of the structure is of greater importance. NDE has been more recently used for process control in manufacturing; a complex application since the inspection equipment must be compatible with process conditions, production equipment, and environment.

Another aspect of NDE, generally not considered, is the integration of NDE into the engineering design process.[25] Despite advances in NDE, it is still primarily used as an after-the-fact testing tool. There are numerous examples where neglecting NDE during the design stage has made its eventual use more costly. Components to be examined are either difficult to access or the available NDE tools are inadequate to perform an exhaustive evaluation. NDE must be an integral part of design to overcome these drawbacks. Various NDT & E techniques and their detection capabilities are listed in Table 21.11.

Table 21.11 BASIC NDT & E TECHNIQUES AND THEIR DETECTION CAPABILITIES

ND technique	Detection capability
Ultrasonic	Cracks, voids, inclusions, corrosion, anisotropy, thickness, elastic constants, grain size, porosity, crack propagation
Radiography	Cracks, voids, density variations, inclusions
Electromagnetic	Cracks, voids, alloy contents, inclusions, corrosion, anisotropy, electrical and thermal conductivity, thickness, heat treatment
Acoustic emission	Cracks, inclusions, plastic zone formation, damage propagation monitoring, leak detection
Thermal	Cracks, inclusions, voids, grain size, porosity, thermal conductivity, composition
Microwave	Thickness measurements, cracks, distributed inhomogeneity

21.7.1 Ultrasonic

Ultrasound is defined as acoustic waves with frequencies above the range of human hearing (>18 MHz). Wave velocity and attenuation are the two parameters measured in ultrasonic examination. Material discontinuities (e.g. voids, cracks, or inclusions) separate an impinging ultrasonic wave into three components: transmitted, reflected, and scattered. Each of these signals carries information about the material, such as stiffness, density, flaw location, and size[26]. An accurate measurement of the time of flight of the wave provides the exact location of the flaw. The reflected intensity is proportionate to the size of the flaw, therefore, measurement of the reflected signal amplitude and its spread effectively estimate the size of the flaw. Basic ultrasonic configurations are shown in Fig. 21.13.

The following relate the wave velocity of the sound to the stiffness E and Poisson's ratio v of the material:

$$E = \frac{\rho c_L^2(3c_L^2 - 4c_s^2)}{c_L^2 - c_s^2} \quad \text{and} \quad v = \frac{c_L^2 - 2c_s^2}{2(c_L^2 - c_s^2)}$$

where ρ is the density of the medium and c_L and c_S are the longitudinal and shear wave velocities in the medium, respectively. E and v can be calculated if the two wave velocities are measured accurately.[27,28] Since time, distance, and velocity are related, in a simple experiment at least one of the variables must be known. A variation of this technique, which simultaneously gives both velocity and thickness, is reported in Refs. [29,30]. Ultrasound can also be used to investigate anisotropic materials where the velocity measurements depend on the direction of the wave propagation.[31,32] Waves travelling in the plane of a plate, called Lamb waves or plate waves, are used to measure the elastic properties in the plane of the plate.[33] The scattered acoustic signal has been used to characterise microstructural features in titanium alloys.[34]

21.7.2 Radiography

Radiography is one of the most widely used NDE techniques.[35] The transmission of radiation (x-rays, gamma-rays, or neutrons) through a material is recorded on a photographic plate. Variation in the transmission through the material (due to inhomogeneity, defects, or thickness variations) results in contrast on the film. In determining defect size, magnification in the radiograph due to perspective must be accounted for, as shown in Fig. 21.14.

The radiograph image quality is defined by three factors: contrast, clarity, and graininess. *Contrast* results from varying transmission of the radiation through the object. Higher contrast images are easier to interpret. The edge *clarity* of the imaged feature depends on its size distance from the radiation source. The edges of the feature produce a shadow relative to the finite size of the source resulting in blurred boundaries, as shown in Fig. 21.14. In the last 15 years, advances in radiographic equipment and digital data processing have enabled the detection of features to reach new thresholds. The micro-focus x-ray source with a spot size of 5 microns has resulted in a significant improvement in resolution and clarity. The *graininess* of an image is a property of the imaging film. These fine grain-like spots can decrease the ability to detect small flaws. Digital recording and processing of data and digital enhancement can overcome these problems.[36]

An x-ray simulator (XRSIM) has been recently developed that simulates the entire process of the radiography, starting with a computer model of the object. This simulation aids the operator in deciding proper inspection angles and equipment calibration. The simulation can make NDE an

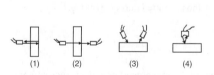

Figure 21.13 *Basic ultrasonic configurations: (1) reflection, (2) transmission, (3) surface or plate, (4) acoustic microscopy*

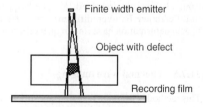

Figure 21.14 *Image magnification and shadow effect in radiography*

integral part of component design process, allowing NDE technicians to work with model developers to ascertain that components are easy to test during service.

21.7.3   Electrical and magnetic methods

The fundamental electrical and magnetic properties of materials are electrical conductivity, electrical permittivity, and magnetic permeability. These properties can be related to other characteristics such as microstructure, strength, hardness, and the presence of flaws and inclusions.[37,38] The primary electrical and magnetic NDE testing methods are: magnetic particle, magnetic flux, and eddy current testing.

MAGNETIC PARTICLE TESTING (MT)

This method is accomplished by using a yoke or a coil to induce a magnetic field, or by passing a current through a ferromagnetic material, and then dusting the surface with iron particles (either dry or suspended in liquid). Surface imperfections (defects such as cracks) distort the field. Concentration of iron particles due to field distortion reveals the defects.

MAGNETIC FLUX METHODS

Magnetic field distortion due to defects can be quantitatively assessed with devices such as Hall probes, magnetoresistive sensors, or coils.

EDDY CURRENT METHOD

The principal advantage of the eddy current method is that it does not require surface preparation, as there is no contact between the probe and the object under study.[39] Examples of feature which may be detected and measured using the eddy current method include surface and sub-surface cracks, the thickness of metallic plates, paint thickness on metal surfaces, and corrosion extent. Eddy currents are generated by an alternating magnetic field induced in the specimen with alternating currents in a coil separate from the specimen. The orientation of the coil is determined based on the assumed defect plane to obtain maximum sensitivity of detection. Imperfections or changes in the specimen's conductive properties will interrupt the flow of eddy currents, causing changes in the induced magnetic field that can be recorded.

21.7.4   Acoustic emission testing (AE)

The use of AE testing in the study of metals encompasses detection of crack propagation, yielding and plastic deformation, corrosion, and creep. AE is also used for continuous monitoring of structures like bridges, pipelines, dams, and cables, and for detection of in-service damage or leaks. When a metal is stressed, imperfections or cracking events emit short bursts of acoustic energy–high-frequency (100–300 kHz) transient elastic waves, which may be detected using piezoelectric sensors attached to the specimen. The acoustic emission technique is very sensitive and can easily detect defects that cannot be optically resolved.[40,41] The frequency and amplitude of the acoustic waves can be used to differentiate between different types of events and defects. For example, waves emitted during plastic deformation have low frequencies compared to those emitted during cracking.[42]

21.7.5   Thermal wave imaging

This method is useful in detecting subsurface variations of thermal properties, defects such as cracks or inclusions, and spatial variation of material properties.[43–45] It involves a burst of energy in the form of a flash lamp or a laser light that is impinged on the object under study. The resulting thermal waves are recorded using an infrared lens.

REFERENCES

1. D. Tabor, 'The Hardness of Metals', Clarendon Press, Oxford. 1951.
2. H. Chandler, Ed., 'Hardness Testing', 2nd ed., ASM International, Materials Park, OH. 1999.
3. M. O. Lai and K. B. Lim, *J. Mater. Sci.*, 1991, **16**, 2031.
4. G. F. Vander Voort, 'Metallography: Principles and Practice', ASM International, Materials Park, OH. 1999.
5. J. L. Hay and G. M. Parr, 'Instrumented Indentation Testing', in 'Mechanical Testing and Evaluation', ASM Handbook Vol. 8, ASM International, Materials Park, OH. 2000.
6. G. E. Dieter, 'Mechanical Metallurgy', 3rd ed., McGraw-Hill, New York, NY. 1986.
7. R. W. K. Honeycombe, 'The Plastic Deformation of Metals', 2nd ed., Edward Arnold, London. 1984.
8. P. Han, Ed., 'Tensile Testing', ASM International, Materials Park, OH. 1992.
9. J. M. Holt, 'Uniaxial Tensile Testing', in 'Mechanical Testing and Evaluation', ASM Handbook Vol. 8, ASM International, Materials Park, OH. 2000.
10. 'Impact Testing of Metals', STP 466, ASTM, Philadelphia, PA. 1970.
11. 'Impact Toughness Testing', in 'Mechanical Testing and Evaluation', ASM Handbook Vol. 8, ASM International, Materials Park, OH. 2000.
12. J. D. Landes, 'Fracture Toughness Testing', in 'Mechanical Testing and Evaluation', ASM Handbook Vol. 8, ASM International, Materials Park, OH. 2000.
13. D. Broek, 'Elementary Engineering Fracture Mechanics', Kluwer Academic Publishers, Dordrecht, The Netherlands. 1986.
14. T. L. Anderson, 'Fracture Mechanics', 2nd ed., CRC Press, Boca Raton, FL. 1995.
15. S. Suresh, 'Fatigue of Materials', Cambridge University Press, Cambridge, UK. 1991.
16. M. E. Fine and Y.-W. Chung, 'Fatigue Failure in Metals', in 'Fatigue and Fracture', ASM Handbook Vol. 19, ASM International, Materials Park, OH. 1996.
17. J. A. Bannantine, J. J. Comer and J. L. Handrock, 'Fundamentals of Metal Fatigue Analysis', Prentice-Hall, Englewood Cliffs, NJ. 1990.
18. 'Ultrasonic Fatigue Testing', in 'Mechanical Testing and Evaluation', ASM Handbook Vol. 8, ASM International, Materials Park, OH. 2000.
19. G. Sumner and V. B. Livesey (editors), 'Techniques for High Temperature Fatigue Testing', Elsevier Applied Science, Amsterdam, The Netherlands. 1985.
20. R. M. Wetzel and L. F. Coffin (editors), 'Manual of Low Cycle Fatigue Testing' STP 465, ASTM, Philadelphia, PA. 1969.
21. G. R. Halford and B. A. Lerch, 'Fatigue, Creep-Fatigue, and Thermomechanical Fatigue Life Testing, in 'Mechanical Testing and Evaluation', ASM Handbook Vol. 8, ASM International, Materials Park, OH. 2000.
22. O. D. Sherby and P. M. Burke, *Prog. Mater. Sci.*, **13**, 1967, 325.
23. J. C. Gibeling, 'Creep Deformation of Metals, Polymers, Ceramics, and Composites', in 'Mechanical Testing and Evaluation', ASM Handbook Vol. 8, ASM International, Materials Park, OH. 2000.
24. F. Garofalo, 'Fundamentals of Creep and Creep Rupture in Metals', The MacMillan Co., New York, NY, 1965.
25. L. W. Schmerr and D. O. Thompson, pp. 2325 to 2332 in 'Rev. of Prog. in QNDE', Vol. 12B (editors D.O. Thompson and D.E. Chimenti), Plenum Press, N.Y. 1993.
26. J. D. Achenbach, ed., 'The Evaluation of Materials and Structures by Quantitative Ultrasonics', Springer-Verlag, NY. 1993.
27. G. V. Blessing, A. Wolfenden, V. K. Kinra, Y. T. Chen, V. Dayal, M. R. Harmouche and P. Terranova, *J. Test. Eval.*, 1992, **20**, 321.
28. V. K. Kinra, M. S. Petraitis and S. K. Dutta, *Int. J. Solids Struc.*, 1980, **16**, 301.
29. V. Dayal, Experimental Mechanics, 1992, **32**, 197.
30. D. Fei, D. K. Hsu and M. Warchol, *J. of NDE*, 2001, **20**, 95.
31. V. K. Kinra and V. Dayal, Ch. 19 in 'Manual on Experimental Methods of Mechanical Testing of Composites' (ed. C.H. Jenkins), Society of Experimental Mechanics, Fairmount Press, GA.
32. D. E. Chimenti, *Appl. Mech. Rev.*, 1997, **50**, 247.
33. O. I. Lobkis, D. E. Chimenti and H. Zhang, *J. Acoust. Soc. Am.*, 2000, **107**, 1.
34. F. J. Margetan, R. B. Thompson and I. Yalda-Mooshabad, pp.1735 to 1742 in 'Rev. of Progress in QNDE', vol. 12B (eds. D. O. Thompson and D.E. Chimenti), Plenum Press, NY. 1993.
35. R. Halmshaw, 'Industrial Radiology: Theory and Practice', 2nd ed., New York, Chapman & Hall, 1995.
36. W. K. Pratt, 'Digital Image Processing', Wiley, New York, 1991.
37. J. Blitz, 'Electrical and Magnetic Methods of Non-Destructive Testing', New York, Chapman & Hall. 1997.
38. S. S. Udpa, *et al.*, editors, 'Electromagnetic Nondestructive Evaluation (IV)' IOS Press, Washington, D.C. 2000.
39. D. J. Hagemaier, 'Fundamentals of Eddy Current Testing' American Society for Nondestructive Testing, Columbus, OH. 1990.
40. M. N. Bassim, S. S. Lawrence and C. D. Liu, *Eng. Fracture Mech.*, 1994, **47**, 207.
41. J. G. Bakuckas Jr., W. H. Prosser and W. S. Johnson, *J. Comp. Mat.*, 1994, **28**, 305.
42. T. J. Holroyd, 'The Acoustic Emission and Ultrasonic Monitoring Handbook', Technical Standards Services Ltd., UK.
43. R. L. Thomas, L. D. Favro and P. K. Kuo, *Can. J. Phys.* 1986, **64**, 1234.
44. R. L. Thomas, L. D. Favro, P. K. Kuo, T.Ahmed, X. Han, L. Wang, X. Wang and S. M. Shephard, pp. 433–436 in 'Proc. 15th International Congress on Acoustics, Vol. 1' (ed. M. Newman).
45. L. D. Favro and X. Han, pp. 399–415 in 'Topics on Nondestructive Evaluation, Volume I, Sensing for Materials Characterisation, Processing, and Manufacturing' (editors G. Birnbaum and B.A. Auld), ASNT publication office, 1998.

22 Mechanical properties of metals and alloys

The following tables summarise the mechanical properties of important commercial metals and alloys.

The tables of tensile properties at ambient temperatures give the nominal composition of the alloys, followed by the typical specification numbers. Most specifications permit considerable latitude in both composition and properties, but the data given in these tables represent typical average values which would be expected from materials of the nominal composition, unless otherwise stated. For design purposes it is essential to consult the appropriate specifications to obtain minimum and maximum values and special conditions where these apply.

The data in the tables referring to properties at elevated and at sub-normal temperatures, and for creep, fatigue and impact strength have been obtained from a more limited number of tests and sometimes from a single study or test. In these cases the data refer to the particular specimens tested and may not reflect variations arising from sample-to-sample or testing conditions.

22.1 Mechanical properties of aluminium and aluminium alloys

The compositional specifications for wrought aluminium alloys involve a four-digit description of the alloy and is specified in the UK as BS EN 573, 1995. Registration of wrought alloys is administered by the Aluminum Association in Washington, DC. International agreement on temper designations has been achieved, and the standards agreed for the European Union, the Euro-Norms, are replacing the former British Standards. Thus BS EN 515. 1995 specifies in more detail the temper designations to be used for wrought alloys in the UK. At present, there is no Euro-Norm for cast alloys and the old temper designations are still used for cast alloys. While these designations are emphasised here, a cross-listing of some alloys with alternate designations can be found in Chapter 1 of this volume.

22.1.1 Alloy designation system for wrought aluminium

The first of the four digits in the designation indicates the alloy group according to the major alloying elements, as follows: 1XXX—aluminium of 99.0% minimum purity and higher; 2XXX—copper; 3XXX—manganese; 4XXX—silicon; 5XXX—magnesium; 6XXX—magnesium and silicon; 7XXX—zinc; 8XXX—other element, incl. lithium; 9XXX—unused; 1XXX Group—In this group the last two digits indicate the minimum aluminium percentage. Thus 1099 indicates aluminium with a minimum purity of 99.99%. The second digit indicates modifications in impurity or alloying element limits. 0 signifies unalloyed aluminium and integers 1 to 9 are allocated to specific additions; 2XXX—8XXX Groups—In these groups the last two digits are simply used to identify the different alloys in the groups and have no special significance. The second digit indicates alloy modifications, zero being allotted to the original alloy.

National variations of existing compositions are indicated by a letter after the numerical designation, allotted in alphabetical sequence, starting with A for the first national variation registered.

The specifications and properties for Cast Aluminium Alloys are tabulated in Chapter 26.

22.1.2   Temper designation system for aluminium alloys

The following tables use the internationally agreed temper designations for wrought alloys and the more frequently used ones are listed below.

A large number of variants in these tempers has been introduced by adding additional digits to the above designations. For example, the addition of the digit 5 after T1–9 signifies that a stress relieving treatment by stretching has been applied after solution heat-treatment. Some of the more common ones used in the following tables are given below.

T351   Solution heat-treated, stress-relieved by stretching a controlled amount (usually 1–3% permanent set) and then naturally aged. There is no further straightening after stretching. This applies to sheet, plate, rolled rod and bar and ring forging.

T3510   The same as T351 but applied to extruded rod, bar, shapes and tubes.

T3511   As T3510, except that minor straightening is allowed to meet tolerances.

T352   Solution heat-treated, stress-relieved by compressing (1–5% permanent set) and then naturally aged.

T651   Solution heat-treated, stress-relieved by stretching a controlled amount (usually 1–3% permanent set) and then artificially aged. There is no further straightening after stretching. This applies to sheet, plate, rolled rod and bar and ring forging.

T6510   The same as T651 but applied to extruded rod, bar, shapes and tubes.

T6511   As T6510, except that minor straightening is allowed to meet tolerances.

T73   Solution heat-treated and then artificially overaged to improve corrosion resistance.

T7651   Solution heat-treated, stress-relieved by stretching a controlled amount (Again about 1–3% permanent set) and then artificially over-aged in order to obtain a good resistance to exfoliation corrosion. There is no further straightening after stretching. This applies to sheet, plate, rolled rod and bar and to ring forging.

T76510   As T7651 but applied to extruded rod, bar, shapes and tubes.

T76511   As T7510, except that minor straightening is allowed to meet to tolerances.

Table 22.1

Designation	Meaning
F,M	As manufactured or fabricated
H111	Fully soft annealed condition

Strain-hardened alloys

H	Strain hardened non-heat-treatable material
H1x	Strain hardened only
H2x	Strain hardened only and partially annealed to achieve required temper
H3x	Strain hardened only and stabilised by low temperature heat treatment to achieve required temper
H12,H22,H32	Quarter hard, equivalent to about 20–25% cold reduction
H14,H24,H34	Half hard, equivalent to about 35% cold reduction
H16,H26,H36	Three-quarter hard, equivalent to 50–55% cold reduction
H18,H28,H38	Fully hard, equivalent to about 75% cold reduction

Heat-treatable alloys

T1	Cooled from an Elevated Temperature Shaping Process and aged naturally to a substantially stable condition
T2	Cooled from an Elevated Temperature Shaping Process, cold worked and aged naturally to a substantially stable condition
T3	Solution heat-treated, cold worked and aged naturally to a substantially stable condition
T4	Solution heat-treated and aged naturally to a substantially stable condition
T5	Cooled from an Elevated Temperature Shaping Process and then artificially aged
T6	Solution heat-treated and then artificially aged
T7	Solution heat-treated and then stabilised (over-aged)
T8	Solution heat-treated, cold worked and then artificially aged
T9	Solution heat-treated, artificially aged and then cold worked
T10	Cooled from an Elevated cTemperature Shaping Process, artificially aged and then cold worked

Table 22.2   ALUMINIUM AND ALUMINIUM ALLOYS—MECHANICAL PROPERTIES AT LOW TEMPERATURES

Specification	Nominal composition %	Form	Condition	0.2% Proof stress MPa	Tensile strength MPa	Elong.% on 50 mm ($\geq$2.6 mm) or $5.65\sqrt{S_0}$	Shear strength MPa	Brinell hardness ($P=5D^2$)	Fatigue strength (unnotched) 500 MHz MPa	Impact energy J	Fracture toughness (MPa m$^{1/2}$)	Remarks
						Wrought Alloys						
1199	Al 99.99	Sheet	H111	20	55	55	50	15	—	—	—	Highest quality reflectors
			H14	60	85	20	60	23	—	—	—	
			H18	85	110	12	70	28	—	—	—	
1080A	Al 99.8	Sheet	H111	25	70	50	60	19	—	—	—	Domestic trim, chemical plant
			H14	95	100	17	70	29	—	—	—	
			H18	125	135	11	70	29	—	—	—	
		Wire	H111	—	70	—	60	19	—	—	—	
			H14	90	105	—	70	30	—	—	—	
			H18	110–140	130–160	—	—	35–41	—	—	—	
1050A	Al 99.5	Sheet	H111	35	80	47	65	21	—	—	—	General purpose formable alloy
			H14	105	110	15	75	30	—	—	—	
			H18	130	145	10	85	40	—	—	—	
		Bars and sections as extruded	H15	50	75	38	65	22	—	—	—	
		Rivet stock	H111	125	140	—	—	—	—	—	—	
		Tubes	H111	—	75	—	65	21	—	—	—	
			H18<75 mm	120	125	—	75	—	—	—	—	
			H18>75 mm	110	115	—	70	—	—	—	—	
		Wire	H111	42	75	—	65	21	—	—	—	
			H14	100	115	—	75	30	—	—	—	
			H18	115–170	140–195	—	—	38–48	—	—	—	
1350	Al 99.5	Wire	H111	28	83	—	55	—	—	—	—	Electrical conductors
			H14	97	110	—	69	—	—	—	—	
			H18	165	186	—	103	—	48	—	—	
1200	Al 99.0	Sheet	H111	35	90	43	70	22	35	27	—	General purpose, slightly higher strength than 105A
			H13	95	105	20	75	31	40	—	—	
			H14	115	120	12	80	35	50	31	—	
			H16	125	135	11	90	38	60	—	—	
			H18	145	160	9	95	42	60	26	—	

(continued)

Table 22.2   ALUMINIUM AND ALUMINIUM ALLOYS—MECHANICAL PROPERTIES AT LOW TEMPERATURES—*continued*

Specification	Nominal composition %	Form	Condition	0.2% Proof stress MPa	Tensile strength MPa	Elong. % on 50 mm (≥ 2.6 mm) or $5.65\sqrt{S_0}$	Shear strength MPa	Brinell hardness $(P = 5D^2)$	Fatigue strength (unnotched) 500 MHz MPa	Impact energy J	Fracture toughness $(\text{MPa m}^{1/2})$	Remarks
						Wrought Alloys						
2011	Cu 5.5 Bi 0.5 Pb 0.5	Bars and sections as extruded		40	85	38	70	23	45*	27	—	Free machining alloy
		Tubes	H111	—	90	40	70	21	—	—	—	
			H > 75 mm	128	131	6	100	34	—	—	—	
			H < 75 mm	120	124	6	95	32	—	—	—	
2014	Cu 4.4 Mg 0.7 Si 0.8 Mn 0.75	Extruded bar	T3 25 mm	295	340	14	240	95	—	—	—	Heavy duty applications in transport and aerospace, e.g. large parts, wings
			T6 50–75 mm	260	370	16	240	100	—	—	—	
		Wire	T3 ≤ 10 mm	350	365	—	—	—	—	—	—	
		Plate	T451	290	425	22	260	108	140	—	—	
		Bar/tube	T651	415	485	10	290	139	125	—	—	
			T6510	440	490	8	—	—	—	—	—	
2014A	Cu 4.4 Mg 0.7 Si 0.8 Mn 0.75	Sheet	T4	270	450	20	260	115	130*	—	—	Aircraft applications (cladding when used 1070A)
			T6	430	480	10	295	135	130*	—	—	
		Clad sheet	T4	250	425	22	250	—	95*	—	—	
			T6	385	440	10	260	—	95*	—	—	
		Bars and sections	T4	315	465	17	—	115	140	22	—	
			T6	465	500	10	—	135	124	8	—	
		Tubes	T4	310	425	12	—	115	—	—	—	
			T6	415	480	9	—	135	—	—	—	
		Wire	T4	340	445	15	—	115	—	—	—	
			T6	425	465	—	—	135	—	—	—	
		River stock	T4	340	450	—	—	115	—	—	—	
		Bolt and screw stock	T6	425	460	—	—	135	—	—	—	
2024	Cu 4.5 Mg 1.5 Mn 0.6	Plate	T3	345	485	18	285	120	140	—	—	Structural applications, especially transport and aerospace
			T351	325	470	19	285	120	140	—	—	

Alloy	Composition (%)	Form	Temper									Applications
2024	Cu 4.5, Mg 1.5, Mn 0.6	Plate/sheet extrusions	H111	75	185	20	125	47	90	—	71	Aircraft structures
			T4	325	470	20	285	120	140	—	34	
			T6	395	475	10	—	—	—	—	—	
2117	Cu 2.5, Si 0.6, Mg 0.4	Sheet	T4	165	295	24	195	70	95	—	—	Vehicle body sheet
2090	Cu 2.7, Li 2.7, Zr 0.12	Plate	T81	517	550	8	—	—	—	—	—	High strength, low density aero-alloy Ref (Glazer and Morris)
		Plate (12.5 mm)	T81	535	565	11	—	—	—	—	—	
2091	Cu 2.1, Li 2.0, Mg 1.50, Zr 0.1	Plate (12 mm)	T8X51	310	420	14	—	—	—	—	—	Medium strength, low density aero-alloy in damage-tolerant temper
		Plate (40 mm)	T8X51	310	430	6	—	—	—	—	—	
		Extrusion (10 mm)	T851	505	580	7	—	—	—	—	35	
		Extrusion (30 mm)	T851	465	520	11	—	—	—	—	35	
		Plate (12 mm)	T851	460	525	10	—	—	—	—	43	Medium strength, low density aero-alloy
		Plate (38 mm)	T851	430	495	8	—	—	—	—	38	Medium strength, low density aero-alloy
		Sheet	T8	390	495	10	—	—	—	—	38	
2219	Cu 6, Mn 0.3, V 0.1	Plate/sheet/forgings	H111	75	170	18	—	—	—	—	—	Weldable, creep resistant, high-temperature aerospace applications
			T4	185	360	20	—	—	—	—	—	
			T6	290	415	10	—	—	105	—	—	
2004	Cu 6, Zr 0.4	Sheet	H111	150	230	15	—	—	100*	—	—	Superplastically deformable sheet
			T6	300	420	12	—	—	150*	—	—	
2031	Cu 2.3, Ni 1.0, Mg 0.9, Si 0.9, Fe 0.9	Forgings	T4	235	355	22	201	95	—	—	—	Aero-engines, missile fins
			T6	340	420	15	201	95	—	—	—	
2618A	Cu 2.0, Mg 1.5, Si 0.9, Fe 0.9, Ni 1.0	Forgings	H111	70	170	20	—	45	85*	—	—	Aircraft engines
			T6	330	430	8	295	130	170*	—	—	

(continued)

Table 22.2 ALUMINIUM AND ALUMINIUM ALLOYS—MECHANICAL PROPERTIES AT LOW TEMPERATURES—*continued*

Specification	Nominal composition %		Form	Condition	0.2% Proof stress MPa	Tensile strength MPa	Elong. % on 50 mm (≥ 2.6 mm) or $5.65\sqrt{S_0}$	Shear strength MPa	Brinell hardness ($P = 5D^2$)	Fatigue strength (unnotched) 500 MHz MPa	Impact energy J	Fracture toughness (MPa m$^{1/2}$)	Remarks
						Wrought Alloys							
3103	Mn	1.25	Sheet	H111	65	110	40	80	30	50	34	—	General purpose, holloware, building sheet
				H12	125	130	17	90	40	55	—	—	
				H14	140	155	11	95	44	60	29	—	
				H16	160	180	8	105	47	70	—	—	
				H18	185	200	7	110	51	70	20	—	
			Wire	H111	60	115	—	—	30	—	—	—	
				H14	135	155	—	—	45	—	—	—	
				H18	170–200	205–245	—	—	55–65	—	—	—	
3105	Mn	0.35	Sheet	H111	55	115	24	85	—	—	—	—	Building cladding sheet
	Mg	0.6		H14	150	170	5	105	—	—	—	—	
				H18	195	215	3	115	—	—	—	—	
3004	Mn	1.2	Sheet	H111	70	180	20	110	45	95	—	—	Steel metal work, storage tanks
	Mg	1.0		H14	200	240	9	125	63	105	—	—	
				H18	250	285	5	145	77	110	—	—	
3008	Mn	1.6	Sheet	H111	50	120	23	—	—	—	—	—	Thermally resistant alloy. Vitreous enamelling
	Fe	0.7		H18	270	280	4	—	—	—	—	—	
	Zr	0.3											
3003 clad with 4343	Mn	1.2	Sheet	H111	40	110	30	75	—	—	—	—	Flux brazing sheet
	Si	7.5		H12	125	130	10	85	—	—	—	—	
				H14	145	150	8	95	—	—	—	—	
				H16	170	175	5	105	—	—	—	—	
3003 clad with 4004	Mn	1.2	Sheet	Physical properties		as for 3003 clad with 4343								Vacuum brazing sheet
	Si	1.0											
	Mg	1.5											
4032	Si	12.0	Forgings	T6	240	325	5	—	115	110	—	—	Pistons
	Cu	1.0											
	Mg	1.0											
	Ni	1.0											

Alloy	Composition	Form	Condition									Applications
4043A	Si 5.0	Rolled wire	F	75	130	20	—	—	—	—	—	Welding filler wire
4047A	Si 12.0	Wire		189	225	8	—	—	—	—	—	Brazing rod
5657	Mg 0.8	Sheet	H111	40	110	25	75	28	—	—	—	High base purity, bright trim alloy
			H14	140	160	12	95	40	—	—	—	
			H18	165	195	7	105	50	—	—	—	
5005	Mg 0.8	Sheet	H111	40	125	25	75	28	—	—	—	Architectural trim, commercial vehicle trim
			H14	150	160	6	95	—	—	—	—	
			H18	195	200	4	110	—	—	—	—	
5251	Mg 2.25, Mn 0.25	Sheet	H111	95	185	22	125	45	110	50	—	Sheet metal work
			H14	230	245	7	145	70	125	29	—	
			H18	275	285	2	175	80	140	—	—	
5251	Mg 2.0, Mn 0.3	Bar	F	60	170	16	125	47	92	—	—	Marine and transport applications; good workability combined with good corrosion resistance and high fatigue resistance
		Sheet	H111	60	180	20	132	65	124	—	—	
			H22	130	220	8	139	74	—	—	—	
			H24	175	250	5	—	—	—	—	—	
			H28	215	270	4	—	—	—	—	—	
		Bars and sections as extruded (F)	H111	95	185	20	125	45	95*	49	—	
		Tubes	H111	100	200	20	—	—	—	—	—	
			H14	230	250	6	—	—	—	—	—	
			H18	255	270	5	—	—	—	—	—	
		Wire	H111	95	200	—	—	48	—	—	—	
			H18	260–290	280–310	—	—	75–85	—	—	—	
5154A	Mg 3.5, Mn 0.5	Sheet	H111	125	240	24	155	55	115	48	—	Welded structures, storage tanks, salt water service
			H22	245	295	10	175	80	125	—	—	
			H24	275	310	9	175	95	130	—	—	
		Bars and sections as extruded (F)	H111	125	230	25	145	55	140*	—	—	
			H14	125	225	20	—	55	—	—	—	
		Tubes		220	280	7	—	—	—	—	—	
		Wire	H111	125	240	—	—	55	—	—	—	
			H14	265	295	—	—	90	—	—	—	
			H18	310	355	—	—	100	—	—	—	
		Rivet stock	H111	125	250	—	—	—	—	—	—	
			H12	—	290	—	—	—	—	—	—	

(continued)

Table 22.2   ALUMINIUM AND ALUMINIUM ALLOYS—MECHANICAL PROPERTIES AT LOW TEMPERATURES—continued

Wrought Alloys

Specification	Nominal composition %	Form	Condition	0.2% Proof stress MPa	Tensile strength MPa	Elong.% on 50 mm (≥2.6 mm) or 5.65√S_0	Shear strength MPa	Brinell hardness ($P = 5D^2$)	Fatigue strength (unnotched) 500 MHz MPa	Impact energy J	Fracture toughness (MPa m$^{1/2}$)	Remarks
5454	Mg 2.7 Mn 0.75 Cr 0.12	Sheet	H111	105	250	22	159	65	115	—	—	Higher strength alloy for marine and transport, pressure vessels and welded structures
			H22	200	277	7	165	77	125	—	—	
			H24	225	297	5	179	85	130	—	—	
5083	Mg 4.5 Mn 0.7 Cr 0.15	Sheet	H111	170	310	21	170	72	—	—	—	Marine applications, cryogenics, welded pressure vessels.
			H24	290	370	9	210	110	—	—	—	
		Bars and sections as extruded (F)		180	315	19	180	77	—	—	—	
5083	Mg 4.5 Mn 4.5 Cr 0.15	Tube	H111	180	320	20	—	77	—	—	—	
			H14	300	375	7	—	—	—	—	—	
5556A	Mg 5	Wire	H14	250	330	12	—	—	—	—	—	Weld filler wire
5056A	Mg 5.0 Mn 0.5	Wire	H111	140	300	—	—	65	—	—	—	Rivets, bolts, screws
			H14	300	340	—	—	95	—	—	—	
			H18	340–400	400–450	—	—	110–120	—	—	—	
		Rivet stock	H111	140	300	—	—	65	—	—	—	
			H12	—	350	—	—	—	—	—	—	
		Bolt and screw stock	H14	300	340	—	—	—	—	—	—	
6060	Mg 0.5 Si 0.4	Bar	T4	90	150	20	—	—	—	—	—	Medium strength extrusion alloy for doors, windows, pipes, architectural use; weldable and corrosion-resistant
			T5	130	175	13	—	—	—	—	—	
			T6	190	220	13	—	—	—	—	—	

Alloy	Composition	Form	Temper									Application
6063	Mg 0.5 Si 0.5	Bars, sections and forgings	F	85	155	30	100	35	—	—	—	Architectural extrusions (fast extruding)
			T4	115	180	30	130	52	60	43	—	
			T6	210	245	20	160	75	70	31	—	
		Wire	H111	—	115	—	—	—	—	—	—	
			T4	115	180	—	—	50	—	—	—	
			T6	195	230	—	—	70	—	—	—	
6063A	Mg 1.0 Si 0.5	Bar	T6510	280	310	22	129	50	79	—	—	Transport, windows, furniture, doors and architectural uses, pipes (irrigation)
			T5	160	200	12	117	65	69	—	—	
			T6	210	240	12	152	78	85	—	—	
6061	Mg 1.0 Si 0.6 Cr 0.25 Cu 0.2	Bars and sections	T4	145	230	20	160	60	—	34	—	Intermediate strength extrusion alloy
			T6	280	310	13	200	90	—	27	—	
		Wire	T8 ≤ 6 mm	310–400	385–430	—	—	—	—	—	—	
			T8 (6–10 mm)	295–385	380–415	—	—	—	—	—	—	
		Bar	T6510	280	310	13	205	100	95	—	—	
		Bolt and screw stock	T8	290	340	—	—	—	—	—	—	
6082	Mg 1.0 Si 1.0 Mn 0.7	Bar/extrusion	T5	260	300	15	185	85	—	—	—	
			T6510	285	315	11	—	—	—	—	—	
		Plate	T451	150	240	19	—	68	—	—	—	
			T651	289	315	12	—	104	—	—	—	
	Mn 0.5	Bars, sections and forgings	T6	285	315	12	205	100	—	41	—	
			T4	160	240	25	180	65	—	34	—	
			T6	285	310	13	215	100	—	—	—	
		Tubes	T4	160	245	20	—	65	—	—	—	
			T6	285	325	10	—	95	—	—	—	
6463	Mg 0.55 Si 0.4	Bar	T4	130	180	16	150	55	70	—	—	Vehicle body sheet
			T6	215	240	12	150	79	70	—	—	
6009	Si 0.8 Mg 0.6 Mn 0.5 Cu 0.4	Sheet	T4	130	235	24	205	60	97	—	—	
			T6	325	345	12	150	—	115	—	—	

(continued)

Table 22.2 ALUMINIUM AND ALUMINIUM ALLOYS—MECHANICAL PROPERTIES AT LOW TEMPERATURES—continued

Specification	Nominal composition %	Form	Condition	0.2% Proof stress MPa	Tensile strength MPa	Elong.% on 50 mm (≥2.6 mm) or 5.65√S0	Shear strength MPa	Brinell hardness (P = 5D²)	Fatigue strength (unnotched) 500 MHz MPa	Impact energy J	Fracture toughness (MPa m^{1/2})	Remarks
					Wrought Alloys							
7020	Zn 4.5 Mg 1.2 Zr 0.15	Bars and sections	T4 T6	225 310	340 370	18 15	— —	100 126	— —	— —	— —	Transportable bridging
7075	Zn 5.6 Mg 2.5 Cu 1.6 Cr 0.25	Sheet/plate/ forgings/ extrusion	H111 T4 T73	105 505 435	230 570 505	17 11 13	150 330 —	60 150 —	— 160 —	— 7 —	— — —	Aircraft structures
7050	Zn 6.2 Mg 2.2 Cu 2.3 Zr 0.12	Thick section plate/ forgings	T736	455	515	11	—	—	220	—	—	Low quench sensitivity, high stress corrosion resistance. Aircraft structures
7475	Zn 5.7 Mg 2.2 Cu 1.5 Cr 0.2	Sheet/plate/ forgings	T61 T7351	525 —	460 —	12 —	— 270	— —	— 220	— —	— —	High base purity. High fracture toughness. Aircraft structures
7016	Zn 4.5 Mg 1.1 Cu 0.75	Extrusions	T6	315	360	12	—	—	—	—	—	Bright anodised vehicle bumpers
7021	Zn 5.5 Mg 1.5 Cu 0.25 Zr 0.12	Extrusion	H111 T6	115 395	235 435	16 13	— —	— —	— —	— —	— —	Bumper backing bars
8079	Fe 0.7	Foil	H111 H18	35 160	95 175	26 2	— —	— —	— —	— —	— —	Domestic foil

Alloy	Composition	Form	Temper	0.2% proof stress	Tensile strength	Elongation			Hardness		density	Condition
	Cu 1.3, Mg 0.95, Zr 0.1	Plate (38/65 mm)	T8151	387	476	6.5	—	—	—	—	42.5	Under-aged, damage-tolerant condition
			T8771	483	518	4.3	—	—	—	—	42	Peak-aged, medium strength condition
		Sheet	T81	360	420	11	—	—	—	—	42	Under-aged, damage-tolerant condition with a recrystallised grain structure
			T6	373	472	5.7	—	—	—	—	—	
			T8	436	503	5	—	—	—	—	75	Peak-aged, medium strength condition
		Extrusion	T81551	440	510	4	—	—	—	—	45	Damage-tolerant condition
		Extrusion (10 mm)	T82551	460	515	4.2	—	—	—	—	39	Peak-aged, medium strength condition
			T851	515	580	5	—	—	—	—	30	
		Extrusion (30 mm)	T851	460	520	9	—	—	—	—	40	
		Plate (12 mm)	T851	455	500	7	—	—	—	—	33	
8090	Li 2.4, Cu 1.2, Mg 0.50, Zr 0.14	Forging	T651	468	517	7	—	—	—	—	28.1	Peak-aged (32 h at 170°C). (Shrimpton)
			T651	400	453	7	—	—	—	—	36.7	Under-aged (20 h at 150°C). (Shrimpton)
8090	Li 2.5, Cu 1.2, Mg 0.66, Zr 0.12	Forging		420	499	7.8	—	—	—	—	16.98	Soln. trt, 530°C, WQ, aged 30 h at 170°C
8091	Li 2.4, Cu 1.9, Mg 0.85, Zr 0.1	—	—	—	—	—	—	—	—	—	—	
	Li 2.4, Cu 2.0, Mg 0.70, Zr 0.08	Plate (40 mm)		460	—	—	164	—	—	—	—	Peak-aged (6% stretch, 32 h at 170°C)
				408	—	—	159	—	—	—	—	Peak aged (no stretch, 100 h at 170°C)
				408	—	—	158	—	—	—	—	Duplex-aged (ditto, 24 h at RT, 48 h at 170°C)
	Li 2.3, Cu 1.7, Mg 0.64, Zr 0.13	Forging		436	503	8.2	—	—	—	—	20.72	soln. trt. 530°C, WQ, aged 20 h at 170°C

(continued)

Table 22.2   ALUMINIUM AND ALUMINIUM ALLOYS—MECHANICAL PROPERTIES AT LOW TEMPERATURES—*continued*

Specification	Nominal composition %	Form	Condition	0.2% Proof stress MPa	Tensile strength MPa	Elong.% on 50 mm (≥2.6 mm) or 5.65√S_0	Shear strength MPa	Brinell hardness ($P = 5D^2$)	Fatigue strength (unnotched) 500 MHz MPa	Impact energy J	Fracture toughness (MPa m$^{1/2}$)	Remarks
						Cast Alloys						
Al (LMO)	Al 99.0	Sand cast	F	30	80	30	55	25	30*	19	—	High conductivity, high ductility
		Chill cast	F	30	80	40	55	25	30*	19	—	
Al–Mg (LM5)	Mg 5.0 Mn 0.5	Sand cast	F	100	160	6	—	60	45	8	—	Very high corrosion resistance
		Chill cast	F	100	215	10	—	65	95	12	—	
(LM10)	Mg 10.0	Sand cast	T4	180	295	12	230	85	55	15	—	Strength + corrosion resistance
		Chill cast	T4	190	340	18	230	95	—	15	—	
Al–Si (LM18)	Si 5.0	Sand cast	F	60	125	5	90	40	55	1.5	—	Intricate castings
		Chill cast	F	70	155	6	120	50	85	2.5	—	
(LM6)(LM20)(LM6)	Si 11.5	Sand cast	F	65	170	8	110	55	45*	4	—	Very similar alloys, excellent casting characteristics and corrosion resistance. LM6 has slightly superior corrosion resistance
		Chill cast	F	75	215	10	130	60	60*	9.5	—	
Al–Si–Mg (2L99)	Si 7 Mg 0.4	Sand cast	T6	195	240	3	—	—	56	—	—	Good strength in fairly difficult castings. Cast vehicle wheels
		Chill cast	T6	210	290	6	—	90	90	—	—	
Al–Cu–Si (L154)	Cu 4.2 Si 1.2	Sand cast	T4	170	225	8	—	—	—	—	—	Aircraft castings
		Chill cast	T4	175	280	15	—	—	—	—	—	
(L155)		Sand cast	T6	215	295	5	—	85	—	—	—	
		Chill cast	T6	215	320	10	—	90	—	—	—	
Al–Cu–Si (LM24)	Cu 3.5 Si 8.0	Chill cast	F	110	200	3	—	85	—	—	—	Excellent die casting alloy
		Die cast	F	150	320	2	—	85	—	—	—	
(LM4)	Cu 3.0 Si 5.0	Sand cast	F	90	155	3	—	70	70*	0.7	—	General engineering, particularly sand and permanent mould castings
		Chill cast	F	100	170	3	—	80	75*	0.7	—	
		Sand cast	T6	230	260	1	—	105	—	—	—	
		Chill cast	T6	260	330	3	—	110	—	—	—	

Alloy	Desig.	Nominal composition %	Condition	0.2% Proof stress	Tensile strength	Elong. %		Hardness		Impact		Typical uses
	(LM22)	Cu 3.0 Si 5.0 Mn 0.5	Chill cast T4	115	260	9	—	75	—	4.5	—	Good combination of impact resistance and strength
	(LM2)	Cu 1.5 Si 10.0	Sand cast F	85	140	1	—	70	55	—	—	General purpose die casting alloy
			Chill cast F	95	185	2	—	80	60	—	—	
Al–Cu–Mg	(LM12)	Cu10.0 Mg 0.25	Chill cast F	155	185	1	—	85	—	0.9	—	Castings to withstand high hydraulic pressure
			Chill cast T6	285	310	—	—	130	60	—	—	
Al–Cu	(L119)	Cu 5.0 Ni 1.5	Sand cast T6	200	225	2	—	90	—	—	—	Sand castings for elevated temperature service
Al–Zn–Mg	(DTT) 5008B	Zn 5.3 Mg 0.6 Cr 0.5	Sand cast T4	—	220	5	—	—	—	—	—	Colour anodising alloy
Al–Cu–Si–Zn	(LM27)	Cu 2.0 Si 7.0	Sand cast F	85	155	2	—	75	—	—	—	Versatile general purpose alloy
			Chill cast F	100	180	3	—	80	—	—	—	
Al–Si–Cu–Mg	(LM30)	Si 17.0 Cu 4.5 Mg 0.6	Chill cast F	160	180	0.5	—	110	—	—	—	Die castings with high wear resistance, especially automobile cylinder blocks
			Die cast F	240	275	1	—	120	—	—	—	
			Die cast O	265	295	1	—	—	—	—	—	
	(LM16)	Si 5.0 Cu 1.0 Mg 0.5	Sand cast T4	130	210	3	200	80	70	1.5	—	Water-cooled cylinder heads and applications requiring leak-proof castings
			Chill cast T4	130	245	6	210	85	85	2.5	—	
			Sand cast T6	245	255	1	215	100	60	1	—	
			Chill cast T6	275	310	2	225	110	70	1.5	—	
Al–Si–Mg–Mn	(LM9)	Si 12.0 Mg 0.4 Mn 0.5	Sand cast T5	120	185	2	120	70	55*	1.5	—	Fluidity, corrosion resistance and high strength. Extensive use for low-pressure castings
			Chill cast T5	160	255	2.5	160	80	70*	2.5	—	
			Sand cast T6	235	255	1	200	100	70*	0.7	—	
			Chill cast T6	275	310	1	230	110	85*	1.5	—	

(continued)

Table 22.2　ALUMINIUM AND ALUMINIUM ALLOYS—MECHANICAL PROPERTIES AT LOW TEMPERATURES—*continued*

Specification	Nominal composition %	Form	Condition	0.2% Proof stress MPa	Tensile strength MPa	Elong.% on 50 mm (≥ 2.6 mm) or 5.65√S_0	Shear strength MPa	Brinell hardness ($P = 5D^2$)	Fatigue strength (unnotched) 500 MHz MPa	Impact energy J	Fracture toughness (MPa m$^{1/2}$)	Remarks
						Cast Alloys						
(LM25)	Si 7.0 Mg 0.3	Sand cast	F	90	140	2.5	—	60	—	—	—	The most widely used general purpose, high-strength casting alloy
		Chill cast	F	90	180	4	—	60	55	—	—	
		Sand cast	T5	135	165	1.5	—	75	—	—	—	
		Chill cast	T5	165	220	2.5	—	85	—	—	—	
		Sand cast	T7	95	170	3	—	65	75	—	—	
		Chill cast	T7	100	230	8	—	70	60	—	—	
		Sand cast	T6	225	255	1	—	105	95	—	—	
		Chill cast	T6	240	310	3	—	105	—	—	—	
Al–Cu–Mg–Ni (Y Alloy) (L35)	Cu 4.0 Mg 1.5 Ni 2.0	Sand cast	T6	220	235	1	—	115	80*	1.5	—	Highly stressed components operating at elevated temperatures
		Chill cast	T6	240	290	2	—	115	110*	4.5	—	
Al–Si–Cu–Mg–Zn (LM21)	Si 6.0 Cu 4.0 Mg 0.2 Zn 1.0	Sand cast	F	130	180	1	—	85	—	—	—	General engineering applications, particular crankcases
		Chill cast	F	130	200	2	—	90	—	—	—	
Al–Si–Cu–Mg–Ni	Si 23.0 Cu 1.0	Sand cast	T5	120	130	0.3	—	120	—	—	—	Pistons for high performance internal combustion engines
		Chill cast	T5	170	210	0.3	—	120	—	—	—	
(LM29)	Mg 1.0 Ni 1.0 Si 19.0 Cu 1.5	Sand cast	T6	120	130	0.3	—	120	—	—	—	High performance piston alloy
		Chill cast	T6	170	210	0.3	—	120	—	—	—	
		Chill cast	T5	170	190	0.5	—	120	—	—	—	
(LM28)	Mg 1.0 Ni 1.0 Si 11.0 Cu 1.0	Sand cast	T6	120	130	0.5	—	120	—	—	—	
		Chill cast	T6	170	200	0.5	—	120	—	—	—	
		Chill cast	T5	—	220	1	—	105	—	—	—	
(LM13)	Mg 1.0 Ni 1.0	Sand cast	T6	190	200	0.5	—	115	85*	1.4	—	Low expansion piston alloy
		Chill cast	T6	280	290	1	190	125	100*	—	—	

Alloy	Composition		Condition										Notes
Lo-Ex (LM26)	Si	9.0	Sand cast	T7	140	150	1	—	75	—	—	—	Piston alloy
	Cu	3.0	Chill cast	T7	200	210	1	—	75	—	—	—	
	Mg	1.0	Chill cast	T5	180	230	1	—	105	—	1.4	—	
	Ni	0.7											
Al–Cu–Si– Mg–Fe–Ni (3L52)	Cu	2.0	Sand cast	T6	260	285	1	—	120	80	—	—	Aircraft engine castings for elevated temperature service
	Si	1.5	Chill cast	T6	305	335	1	—	125	—	—	—	
	Mg	1.0											
	Fe	1.0											
	Ni	1.25											
Al–Cu–Si– Fe–Ni–Mg (3L51)	Cu	1.5	Sand cast	T5	135	170	2.5	—	70	—	—	—	Aircraft engine castings
	Si	2.0	Chill cast	T5	150	210	3.5	—	75	—	—	—	
	Fe	1.0											
	Ni	1.4											
	Mg	0.15											

*Fatigue Limit for 50 × 10⁶ cycles.

Table 22.3   ALUMINIUM AND ALUMINIUM ALLOYS—MECHANICAL PROPERTIES AT ELEVATED TEMPERATURE

Material (specification)	Nominal composition %	Condition	Temp. °C	Time at temp. h	0.2% Proof stress MPa	Tensile strength MPa	Elong. % on 50 mm or $5.65\sqrt{S_0}$
			Wrought Alloys				
Al	Al 99.95	Rolled rod H111	24	—	—	55	61
(1095)			93	—	—	45	63
			203	—	—	25	80
			316	—	—	12	105
			427	—	—	5	131
(1200)	Al 99	H111	24	10 000	35	90	45
			100	10 000	35	75	45
			148	10 000	30	60	55
			203	10 000	25	40	65
			260	10 000	14	30	75
			316	10 000	11	17	80
			371	10 000	6	14	85
		H14	24	10 000	115	125	20
			100	10 000	105	110	20
			148	10 000	85	90	22
			203	10 000	50	65	25
			260	10 000	17	30	75
			316	10 000	11	17	80
			371	10 000	6	14	85
		H18	24	10 000	150	165	15
			100	10 000	125	150	15
			148	10 000	95	125	20
			203	10 000	30	40	65
			260	10 000	14	30	75
			316	10 000	11	17	80
			371	10 000	6	14	85
Al–Mn	Mn 1.25	H111	24	10 000	40	110	40
(3103)			100	10 000	37	90	43
			148	10 000	34	75	47
			203	10 000	30	60	60
			260	10 000	25	40	65
			316	10 000	17	30	70
			371	10 000	14	20	70
		H14	24	10 000	145	150	16
			100	10 000	130	145	16
			148	10 000	110	125	16
			203	10 000	60	95	20
			260	10 000	30	50	60
			316	10 000	17	30	70
			371	10 000	14	20	70
		H18	24	10 000	185	200	10
		H18	24	10 000	185	200	10
			148	10 000	110	155	11
			203	10 000	60	95	18
			260	10 000	30	50	60
			316	10 000	17	30	70
			371	10 000	14	20	70
Al–Mg	Mg 1.4	H111	24	10 000	55	145	—
(5050)			100	10 000	55	145	—
			148	10 000	55	130	—
			203	10 000	50	95	—
			260	10 000	40	60	—
			316	10 000	30	40	—
			371	10 000	20	30	—

(*continued*)

Table 22.3   ALUMINIUM AND ALUMINIUM ALLOYS—MECHANICAL PROPERTIES AT ELEVATED
TEMPERATURE—*continued*

Material (specification)	Nominal composition %	Condition	Temp. °C	Time at temp. h	0.2% Proof stress MPa	Tensile strength MPa	Elong. % on 50 mm or $5.65\sqrt{S_0}$
			Wrought Alloys				
Al–Mg (*contd.*)		H14	24	10 000	165	195	—
			100	10 000	165	195	—
			148	10 000	150	165	—
			203	10 000	50	95	—
			260	10 000	40	60	—
			316	10 000	35	40	—
			371	10 000	20	30	—
		H18	24	10 000	200	220	—
			100	10 000	200	215	—
			148	10 000	175	180	—
			203	10 000	60	95	—
			260	10 000	40	60	—
			316	10 000	35	40	—
			371	10 000	20	30	—
Al–Mg–Cr (5052)	Mg 2.25 Cr 0.25	H111	24	10 000	90	195	30
			100	10 000	90	190	35
			148	10 000	90	165	50
			203	10 000	75	125	65
			260	10 000	50	80	80
			316	10 000	35	50	100
			371	10 000	20	35	130
		H14	24	10 000	215	260	14
			100	10 000	205	260	16
			148	10 000	185	215	25
			203	10 000	105	155	40
			260	10 000	50	80	80
			316	10 000	35	50	100
			317	10 000	20	35	130
		H18	24	10 000	255	290	8
			100	10 000	255	285	9
			148	10 000	200	235	20
			203	10 000	105	155	40
			260	10 000	50	80	80
			316	10 000	35	50	100
			371	10 000	20	35	130
(5154)	Mg 3.5 Cr 0.25	H111	24	10 000	125	240	25
			100	10 000	125	240	30
			148	10 000	125	195	40
			203	10 000	95	145	55
			260	10 000	60	110	70
			316	10 000	40	70	100
			371	10 000	30	40	130
		H14	24	10 000	225	290	12
			100	10 000	220	285	16
			148	10 000	195	235	25
			203	10 000	110	175	35
			260	10 000	60	110	70
			316	10 000	40	70	100
			371	10 000	30	40	130
		H18	24	10 000	270	330	8
			100	10 000	255	310	13
			148	10 000	220	270	20
			203	10 000	105	155	35
			260	10 000	60	110	70
			316	10 000	40	70	100
			371	10 000	30	40	130

(continued)

Table 22.3   ALUMINIUM AND ALUMINIUM ALLOYS—MECHANICAL PROPERTIES AT ELEVATED
TEMPERATURE—*continued*

Material (specification)	Nominal composition %	Condition	Temp. °C	Time at temp. h	0.2% Proof stress MPa	Tensile strength MPa	Elong. % on 50 mm or $5.65\sqrt{S_0}$
			Wrought Alloys				
Al–Mg–Mn (5056A)	Mg 5.0 Mn 0.3	As extruded F	20	1 000	145	300	25
			50	1 000	145	300	27
			100	1 000	145	300	32
			150	1 000	135	245	45
			200	1 000	111	215	56
			250	1 000	75	130	77
			300	1 000	50	95	100
			350	1 000	20	60	140
Al–Mg–Si (6063)	Mg 0.7 Si 0.4	T6	24	10 000	215	240	18
			100	10 000	195	215	15
			148	10 000	135	145	20
			203	10 000	45	60	40
			260	10 000	25	30	75
			316	10 000	17	20	80
			371	10 000	14	17	105
(6082)	Mg 0.6 Si 1.0 Cr 0.25	T6	24	10 000	230	330	17
			100	10 000	270	290	19
			148	10 000	175	185	22
			203	10 000	65	80	40
			206	10 000	35	45	50
			316	10 000	30	35	50
			371	10 000	25	30	50
(6061)	Mg 1.0 Si 0.6 Cu 0.25 Cr 0.25	T6	24	10 000	275	310	17
			100	10 000	260	290	18
			148	10 000	213	235	20
			203	10 000	105	130	28
			260	10 000	35	50	60
			316	10 000	17	30	85
			371	10 000	14	20	95
Al–Cu–Mn (2219)	Cu 6 Mn 0.25	Forgings T6	20	100	230	385	8
			100	100	—	365	—
			150	100	220	325	—
			200	100	185	280	—
			250	100	135	205	—
			300	100	110	145	—
			350	100	45	70	—
			400	100	20	30	—
Al–Cu–Pb–Bi (2011)	Cu 5.5 Pb 0.5 Bi 0.5	T4	24	10 000	295	375	15
			100	10 000	235	320	16
			148	10 000	130	195	25
			203	10 000	75	110	35
			260	10 000	30	45	45
			316	10 000	14	25	90
			371	10 000	11	17	125
Al–Cu–Mg–Mn (2017)	Cu 4.0 Mg 0.5 Mn 0.5	T4	24	10 000	275	430	22
			100	10 000	255	385	18
			148	10 000	205	274	16
			203	10 000	115	150	28
			260	10 000	65	80	45
			316	10 000	35	45	95
			371	10 000	25	30	100
(2024)	Cu 4.5 Mg 1.5 Mn 0.6	T4	24	10 000	340	470	19
			100	10 000	305	422	17
			148	10 000	245	295	17
			203	10 000	145	180	22
			260	10 000	65	95	45
			316	10 000	35	50	75
			371	10 000	25	35	100

(continued)

Table 22.3 ALUMINIUM AND ALUMINIUM ALLOYS—MECHANICAL PROPERTIES AT ELEVATED TEMPERATURE—*continued*

Material (specification)	Nominal composition %		Condition	Temp. °C	Time at temp. h	0.2% Proof stress MPa	Tensile strength MPa	Elong. % on 50 mm or $5.65\sqrt{S_0}$
				Wrought Alloys				
Al–Cu–Mg–Si–	Cu	4.4	T6	24	10 000	415	485	13
Mn	Mg	0.4		100	10 000	385	455	14
(2014)	Si	0.8		148	10 000	275	325	15
	Mn	0.8		203	10 000	80	125	35
				260	10 000	60	75	45
				316	10 000	35	45	64
				371	10 000	25	30	20
			Forgings T6	20	100	415*	480	10
				100	100	410	465	—
				150	100	400	430	—
				200	100	260	295	—
				250	100	85	110	—
				300	100	45	70	—
				350	100	35	50	—
Al–Cu–Mg–Ni	Cu	2.2	Forgings T6	20	100	325*	430	8
(2618)	Mg	1.5		150	100	340	440	—
	Ni	1.2		200	100	260	300	—
	Fe	1.0		250	100	170	210	—
				300	100	70	115	—
				350	100	30	50	—
				400	100	20	30	—
(2031)	Cu	2.2	Forgings T6	20	100	325*	430	13
	Mg	1.5		100	100	310	400	—
	Ni	1.2		200	100	255	310	—
	Fe	1.0		250	100	110	155	—
	Si	0.8		300	100	45	75	—
				350	100	30	40	—
Al–Si–Cu–Mg–	Si	12.2	Forgings T6	24	10 000	320	380	9
Ni	Cu	0.9		100	10 000	305	345	9
(4032)	Mg	1.1		148	10 000	225	255	9
	Ni	0.9		203	10 000	60	90	30
				260	10 000	35	55	50
				316	10 000	20	35	70
				371	10 000	14	25	90
Al–Zn–Mg–Cu	Zn	5.6	T6	24	10 000	505	570	11
(7075)	Cu	1.6		100	10 000	430	455	15
	Mg	2.5		148	10 000	145	175	30
	Cr	0.3		203	10 000	80	95	60
				260	10 000	60	75	65
				316	10 000	45	60	80
				371	10 000	30	45	65
				Cast alloys				
Al–Mg	Mg	5.0	Sand cast F	20	1 000	95	160	4
(LM 5)	Mn	0.5		100	1 000	100	160	3
				200	1 000	95	130	3
				300	1 000	55	95	4
				400	1 000	15	30	4
(LM 10)	Mg	10.0	Sand cast T4	20	1 000	180	340	16
				100	1 000	205	350	10
				150	1 000	154	270	0
				200	1 000	105	185	42
				300	1 000	40	90	85
				400	1 000	11	45	100

(*continued*)

Table 22.3   ALUMINIUM AND ALUMINIUM ALLOYS—MECHANICAL PROPERTIES AT ELEVATED
TEMPERATURE—*continued*

Material (specification)	Nominal composition %		Condition	Temp. °C	Time at temp. h	0.2% Proof stress MPa	Tensile strength MPa	Elong. % on 50 mm or $5.65\sqrt{S_0}$
				Cast Alloys				
Al–Si	Si	5.0	Pressure die	F 24	10 000	110	205	9
(LM 18)			cast	100	10 000	110	175	9
				148	10 000	103	135	10
				203	10 000	80	110	17
				260	10 000	40	55	23
(LM 6)	Si	12.0	Pressure die	F 24	10 000	145	270	2
			cast	100	10 000	145	225	2(1/2)
				148	10 000	125	185	3
				206	10 000	105	150	7
				260	10 000	40	75	13
Al–Si–Cu	Si	5.0	Sand cast	F 20	1 000	95*	155	2
(LM 4)	Cu	3.0		100	1 000	140	180	2
	Mn	0.5		200	1 000	110	135	2
				300	1 000	40	60	12
				400	1 000	20	30	27
Al–Si–Mg	Si	5.0	Chill cast	T6 20	1 000	270*	325	2
(LM 25)	Mg	0.5		100	1 000	255	290	2
				200	1 000	60	90	25
				300	1 000	25	40	65
				400	1 000	12	25	65
Al–Cu–Mg–Ni	Cu	4.0	Sand cast	T6 20	1 000	200*	275	1/2
(4L 35)	Mg	1.5		100	1 000	255	325	1/2
	Ni	2.0		200	1 000	150	135	1/2
				300	1 000	30	55	32
				400	1 000	15	40	60
Al–Si–Ni–Cu–Mg	Si	12.0	Chill cast	T6 20	1 000	275*	285	1/2
(LM 13)	Ni	2.5		100	1 000	280	320	1/2
	Cu	1.0		200	1 000	110	165	1/2
	Mg	1.0		300	1 000	30	60	15
				400	1 000	15	35	25
			Chill cast	T6 20	1 000	200*	275	1
			Special	100	1 000	195	250	1
				200	1 000	110	170	3
				300	1 000	35	60	15
				400	1 000	15	35	50

* 0.1% Proof stress.

Table 22.4   ALUMINIUM AND ALUMINIUM ALLOYS—MECHANICAL PROPERTIES AT LOW TEMPERATURES

Material (specification)	Nominal composition %		Condition	Temp. °C	0.2% Proof stress MPa	Tensile strength MPa	Elong. % on 50 mm or 50 mm	Reduction in area %	Fracture toughness MPa m$^{1/2}$	Reference
Al	Al	99.0	Rolled and	24	34	90	42.5	76.4	—	
(1200)			drawn rod	H111 −28	34	95	43.0	76.4	—	1
				−80	37	100	47.5	77.0	—	
				−196	43	170	56	74.4	—	
			H18 24	140	155	16	59.8	—		
				−28	144	155	152	59.4	—	1
				−80	147	165	18.0	65.3	—	
				−196	165	225	35.2	67.0	—	

(*continued*)

Table 22.4   ALUMINIUM AND ALUMINIUM ALLOYS—MECHANICAL PROPERTIES AT LOW TEMPERATURES—*continued*

Material (specification)	Nominal composition %	Condition		Temp. °C	0.2% Proof stress MPa	Tensile strength MPa	Elong.% on 50 mm or 50 mm	Reduction in area %	Fracture toughness MPa m$^{1/2}$	Reference
Al–Mn (3103)	Mn 1.25	Rolled and drawn rod	H111	24	40	110	43.0	80.6	—	
				− 28	40	115	44.0	80.6	—	1
				− 80	50	130	45.0	79.9	—	
				−196	60	220	48.8	71.2	—	
		Rolled and drawn rod	H18	24	180	195	15.0	63.5	—	
				28	185	205	15.0	64.4	—	1
				− 80	195	215	16.5	66.5	—	
				−196	220	290	32.0	62.3	—	
Al–Mg (5052)	Mg 2.5 Cr 0.25	Rolled and drawn rod	H111	24	97	199	33.2	72.0	—	
				− 28	99	201	35.8	74.2	—	
				− 80	97	210	40.8	76.4	—	
				−196	115	330	50.0	69.0		
			H18	24	235	275	16.6	59.1	—	
				− 28	230	280	18.3	63.2	—	1
				− 80	236	290	20.6	64.5	—	
				−196	275	400	30.9	57.4	—	
(5154)	Mg 3.5 Cr 0.25	Sheet	H111	26	115	240	28	66	—	
				− 28	115	240	32	72	—	
				− 80	115	250	35	73	—	
				−196	135	350	42	60	—	
			H18	26	275	330	9	—	—	
				− 80	280	340	14	—	—	2
				−196	325	455	30	—	—	
				−253	370	645	35	—	—	
(5056A)	Mg 5.0 Mn 0.2	Plate	H111	20	130	290	30.5	32.0	—	
				− 75	130	290	38.2	48.2	—	1
				−196	145	420	50.0	36.2	—	
Al–Mg–Si (6063)	Mg 0.7 Si 0.4	Extrusion	T4	26	90	175	32	78	—	
				− 28	105	190	33	75	—	
				− 80	115	200	36	75	—	
				−196	115	260	42	73	—	
		Extrusion	T6	26	215	240	16	36	—	
				− 28	220	250	16	36	—	
				− 80	225	260	17	38	—	
				−196	250	330	21	40	—	
Al–Mg–Si–Cr (6151)	Mg 0.7 Si 1.0 Cr 0.25	Forging	T6	24	300	320	15.2	38.8	—	
				− 28	310	352	12.0	34.0	—	1
				− 80	305	330	14.9	38.7	—	
				−196	330	385	18.3	34.7	—	
Al–Mg–Si–Cu–Cr (6061)	Mg 1.0 Si 0.6 Cu 0.25 Cr 0.25	Rolled and drawn rod	T6	24	270	315	21.8	56.4	—	
				− 28	280	330	21.5	52.5	—	1
				− 80	290	345	22.5	53.7	—	
				−196	315	425	26.5	46.5	—	
Al–Cu–Mg–Mn (2024)	Cu 4.5 Mg 1.5 Mn 0.6	Rolled and drawn rod	T4	24	300	480	23.3	31.8	—	
				− 28	305	500	24.4	33.1	—	1
				− 80	320	510	25.3	30.8	—	
				−196	400	615	26.7	26.3	—	
		Rolled and drawn rod	T8	24	400	500	14.5	25.8	–	
				− 28	405	502	12.7	21.5	–	1
				− 80	415	514	13.3	22.0	–	
				−196	460	605	14.0	19.7	–	

(*continued*)

Table 22.4   ALUMINIUM AND ALUMINIUM ALLOYS—MECHANICAL PROPERTIES AT LOW
TEMPERATURES—*continued*

Material (specification)	Nominal composition %	Condition		Temp. °C	0.2% Proof stress MPa	Tensile strength MPa	Elong.% on 50 mm or 50 mm	Reduction in area %	Fracture toughness MPa m$^{1/2}$	Reference
Al–Cu–	Cu 4.5	Rod	T4	26	290	430	20	28	–	
Si–Mg–	Si 0.8			− 28	290	440	22	28	–	
Mn	Mg 0.5			− 80	302	440	22	26	–	
(2014)	Mn 0.8			−196	380	545	20	20	–	
		Rod	T6	26	415	485	13	31	–	
				− 28	415	485	13	29	–	1
				− 80	420	495	14	28	–	
				−196	470	565	14	26	–	
		Forging	T6	26	410	465	12	24	–	
				− 80	460	510	14	24	–	2
				−196	530	610	11	22	–	
				−253	590	715	7	22	–	
(2090)	Cu 2.7	Plate		27	535	565	11	–	34	14
	Li 2.3	(12.5 mm)	T81	−196	600	715	13.5	–	57	
	Zr 0.12			−269	615	820	17.5	–	72	
(2091)	Cu 2.1	Plate		27	440	480	6	–	24	14
	Li 2.0	(38 mm)	T8	− 73	460	495	7	–	32	
	Mg 1.50			−196	495	565	10	–	32	
	Zr 0.1			−269	550	630	7	–	32	
Al–Zn–	Zn 5.6	Rolled and		24	485	560	15.0	29.1	–	
Mg–Cu	Mg 2.5	drawn rod	T6	− 28	490	570	15.3	26.2	–	1
(7075)	Cu 1.6			− 80	505	590	15.3	23.6	–	
				−196	570	670	16.0	20.1	–	

Table 22.5   ALUMINIUM ALLOYS—CREEP DATA

Material (specification)	Nominal composition %	Condition		Temp. °C	Stress MPa	Minimum creep rate % per 1 000 h	Total extension % in 1 000 h	Reference
Al	99.8	Sheet	H111	20	24.1	0.005	0.39	
(1080)				20	27.6	0.045	1.28	
				80	7.0	0.005	0.045	
				80	8.3	0.01	0.065	
				250	1.4	0.005	0.047	
				250	2.1	0.01	0.047	
				250	2.8	0.015	0.052	
				250	4.1	0.055	0.152	
Al–Mg	Mg	Sheet	H111	80	45	0.005	0.085	
(5052)								
(LM 5)	Mg 5.6	Cast		100	110	0.055	0.33	3
				100	115	0.17	0.57	
				100	125	0.21	1.19	
				200	30	0.08	0.21	
				200	45	0.20	0.39	
				200	60	0.62	0.92	
				300	3.90	0.045	0.10	
				300	7.7	0.12	0.25	
				300	15	0.35	0.60	

(*continued*)

Table 22.5 ALUMINIUM ALLOYS—CREEP DATA—*continued*

Material (specification)	Nominal composition %	Condition		Temp. °C	Stress MPa	Minimum creep rate % per 1 000 h	Total extension % in 1 000 h	Reference
(LM 10)	Mg 10	Cast		100	40	0.013	0.126	3
				100	55	0.022	0.107	
				100	75	0.046	0.174	
				150	7.5	0.126	0.413	
				150	15	0.147	0.647	
				200	7.5	0.107	0.341	
				200	15	0.273	0.658	
Al–Cu	Cu 4	Cast		205	17	0.04	—	5
				205	34	0.09	—	
				205	51	0.14	—	
				205	70	0.69	—	
				315	8.90	0.13	—	
				315	13.1	0.29	—	
	Cu 10	Cast		205	34	0.01	—	5
				205	68	0.11	—	
				315	8.90	0.12	—	
				315	13.1	0.43	—	
				315	17	0.99	—	
Al–Si (LM 13)	Si 13 Ni 1.7 Mg 1.3	Sandcast (modified)		100	45	0.016	0.190	3
				100	60	0.06	0.675	
				200	15	0.016	0.096	
				200	23	0.054	0.179	
				200	30	0.14	0.432	
				300	3.8	0.013	0.026	
				300	7.7	0.047	0.098	
				300	15	0.223	0.428	
Al–Mn (3 103)	Mn 1.25	Extruded rod		200	15	0.001	—	8
				200	31	0.022	—	
				200	34.8	0.040	—	
				200	38.6	0.060	—	
				200	42.5	0.13	—	
				200	46	0.15	—	
				200	54	0.73	—	
				300	7.5	0.007	—	
				300	15	0.39	—	
Al–Cu–Si (2 025)	Cu 4 Si 0.8	Extruded	T4	150	90	0.03	0.340	3
				150	125	0.045	0.395	
				150	155	0.325	0.722	
				200	30	0.035	0.107	
				200	45	0.1	0.204	
				200	60	0.040	0.700	
				250	15	0.02	0.156	
				250	23	0.07	0.176	
				250	30	2.36	—	
Al–Cu–Mg–Mn (2 024)	Cu 4.5 Mg 1.5 Mn 0.6	Clad sheet	T4	35	415	10.0	—	4
				100	344	1.0	—	
				100	385	10.0	—	
				150	276	1.0	—	
				150	327	10.0	—	
				190	140	1.0	—	
				190	200	10.0	—	

(continued)

Table 22.5   ALUMINIUM ALLOYS—CREEP DATA—*continued*

Material (specification)	Nominal composition %	Condition	Temp. °C	Stress MPa	Minimum creep rate % per 1 000 h	Total extension % in 1 000 h	Reference
Al–Cu–Mg–Mn (2 024) (*contd*)	Cu 4.5 Mg 1.5 Mn 0.6	Clad sheet	T6 35	424	1.0	—	4
			35	430	10.0	—	
			100	347	1.0	—	
			100	363	10.0	—	
			150	242	1.0	—	
			150	289	10.0	—	
			190	117	1.0	—	
			190	193	10.0	—	
Al–Cu–Mg–Ni (2 218)	Cu 4 Mg 1.5 Ni 2.2	Forged	T4 100	193	0.01	0.394	3
			100	232	0.02	0.440	
			100	270	0.04	0.835	
			200	77	0.028	0.173	
			200	108	0.16	0.345	
			300	7	0.037	0.078	
			300	15	0.5	0.640	
			400	1.5	0.05	0.110	
		Cast	T4 200	77	0.01	0.153	3
			200	116	0.08	0.287	
			300	7	0.018	0.072	
			300	15	0.08	0.151	
			400	1.50	0.06	0.132	
Al–Cu–Mg–Zn (7 075)	Zn 5.6 Cu 1.6 Mg 2.5	Clad sheet	T6 35	430	0.1	—	4
			35	480	1.0	—	
			35	495	10.0	—	
			100	295	0.1	—	
			100	355	1.0	—	
			100	370	10.0	—	
			150	70	0.1	—	
			150	170	1.0	—	
			150	245	10.0	—	
			190	45	0.1	—	
			190	75	1.0	—	
			190	125	10.0	—	
Al–Mg–Si–Mn (6 351)	Mg 0.7 Si 1.0 Mn 0.6	Extruded rod	100	193	0.007	—	8
			100	201	0.010	—	
			100	232	0.11	—	
			100	255	1.6	—	
			150	93	0.0087	—	
			150	108	0.023	—	
			150	154	0.22	—	
			200	31	0.011	—	
			200	46	0.040	—	
			200	62	0.13	—	
			200	77	0.28	—	

Table 22.6   ALUMINIUM ALLOYS—FATIGUE STRENGTH AT VARIOUS TEMPERATURES

Material (specification)	Nominal composition %	Condition	Temp. °C	Endurance (unnotched) MPa	MHz	Remarks	Reference
Al–Mg (5 056)	Mg 5.0	Extruded	−65	184	20	Rotating beam	
			−35	164			
			+20	133			

(*continued*)

Table 22.6 ALUMINIUM ALLOYS—FATIGUE STRENGTH AT VARIOUS TEMPERATURES—*continued*

Material (specification)	Nominal composition %	Condition		Temp. °C	Endurance (unnotched) MPa	MHz	Remarks	Reference
	Mg 7.0	Extruded rod		−65	182	20	Rotating	
				−35	178		beam	
				+20	173			
(LM 10)	Mg 10.0	Sand cast (oil quenched)		20	93	30	Rotating	9
				150	77		beam	
				200	40			
Al–Si (LM 6)	Si 12.0	Sand cast (modified)		20	51	50	Rotating	6
				100	43		beam, 24 h	
				200	35		at temp.	
				300	25			
Al–Cu (2219)	Cu 6.0	Forged	T6	20	117	120	Reverse	9
				150	65		bending	
				200	62		stresses	
				250	46			
				300	39			
				350	23			
Al–Si–Cu (LM 22)	Si 4.6 Cu 2.8	Sand cast		20	62	50	Rotating	6
				100	54		beam	
				200	60			
				300	42			
Al–Cu–Si–Mn (2014)	Cu 4.5 Si 0.8 Mn 0.8	Forgings	T6	148	65	100	Rotating	
				203	45		beam	
				260	25			
Al–Cu–Mn–Mg (2014)	Cu 4.0 Mn 0.5 Mg 0.5	Extruded rod	T4	25	103	500	Rotating	7
				148	93		beam,	
				203	65		100 days	
				260	31		at temp.	
Al–Cu–Mg–Si–Mn (2014)	Cu 4.4 Mg 0.7 Si 0.8 Mn 0.8	Forgings	T4	20	119	120	Reversed	9
				150	90		bending	
				200	62			
				250	54			
				300	39			
		Forgings	T6	20	130	120	Reversed	9
				150	79		bending	
				200	57			
				250	39			
				300	39			
Al–Cu–Mg–Ni (2218)	Cu 4.0 Mg 1.5 Ni 2.0	Forged		20	117	500	Rotating	
				148	103	100	beam after	
				203	65	100	prolonged	
				260	45	100	heating	
		Chill cast	T6	20	100	50	Rotating	6
				100	105		beam, 24 h	
				200	108		at temp.	
				300	80			
Al–Ni–Cu	Ni 2.5 Cu 2.2	Forged	T6	20	113	120	Reversed	9
				150	82		bending	
				200	70			
				250	59			
				300	39			
				350	39			

(continued)

Table 22.6   ALUMINIUM ALLOYS—FATIGUE STRENGTH AT VARIOUS TEMPERATURES—*continued*

Material (specification)	Nominal composition %		Condition		Temp. °C	Endurance (unnotched) MPa	MHz	Remarks	Reference
Al–Si–Cu–Mg–Ni (LM 13)	Si	12.0	Chill cast (Lo-Ex)		20	97	50	Rotating beam, 24 h at temp.	6
	Cu	1.0			100	107			
	Mg	1.0			200	97			
					300	54			
Al–Zn–Mg–Cu (7075)	Zn	5.6	Plate	T6	24	151	500	Reversed bending	—
	Mg	2.5			149	83			
	Cu	1.6			204	59			
	Cr	0.2			260	48			

REFERENCES

1. Bogardus, S. W. Steckley and F. M. Howell, N.A.C.A. Technical Note 2082, 1950.
2. 'A Review of Current Literature of Metals at Very Low Temperatures', Battelle Memorial Institute; 1961.
3. J. McKeown and R. D. S. Lushey, *Metallurgia*, 1951, **43**, 15.
4. A. E. Flanigan, L. F. Tedsen and J. E. Dorn, *Trans. Amer. Inst. Min. Met. Eng.*, 1947, **171**, 213.
5. R. R. Kennedy, *Proc. Am. Soc. Test. Mat.*, 1935, **33**, 218.
6. J. McKeown, D. E. Dineen and L. H. Back, *Metallurgica*, 1950, **41**, 393.
7. F. M. Howell and E. S. Howarth, *Proc. Am. Soc. Test Mat.*, 1937, **37**, 206.
8. N. P. Inglis and E. G. Larke, *J. Inst. Mech. Engrs.*, 1959.
9. P. H. Frith, 'Properties of Wrought and Cast Aluminium and Magnesium Alloys at Atmospheric and Elevated Temperatures', HMSO, 1956.
10. R. Grimes, T. Davis, H. J. Saxty and J. E. Fearon, '4th International Al-Li Conference, 10–12th June, 1987', *J. de Physique*, Sept. 1987, **48**, Colloque C3, p11.
11. G. Leroy *et al., ibid.*, C3, p33.
12. M. J. Birt and C. J. Beevers, *5th International AL-Li Conference, Williamsburg, Virginia, March 27–31, 1989* (ed. T.H. Sanders and E. A. Starke), Materials and Component Engineering Publications Ltd, UK, p983.
13. H. D. Peacock and J. W. Martin, *ibid.*, p1013.
14. J. Glazer and J. W. Morris, *ibid.*, p1471.
15. G. R. D. Shrimpton and H. C. Argus, *ibid.*, p1565.
16. A. F. Smith, *ibid.*, p1587.

FURTHER INFORMATION

1. *The Properties of Aluminium and its Alloys*, The Aluminium Federation, 9th Edition, Birmingham, UK, 2000.
2. ASM *Handbook*, Vol. 2, 10th Edition, ASM International, Materials Park, OH, 1990.

22.2   Mechanical properties of copper and copper alloys

A listing of related designations for copper alloys is given in Chapter 1.

Table 22.7   COPPER AND COPPER ALLOYS—TYPICAL MECHANICAL PROPERTIES AT ROOM TEMPERATURE

Condition of material is expressed in accordance with BS Nomenclature, viz:

O Material in the annealed condition

$\frac{1}{4}$H, $\frac{1}{2}$H, H, EH The various harder tempers produced by cold rolling / For certain of the materials in this schedule, these tempers may be produced by partial annealing

SH, ESH Spring hard tempers produced by cold rolling of thinner material

M Material in the 'as manufactured' condition. In this schedule confined to hot rolled or extruded material

W Material which has been solution heat treated and will respond effectively to precipitation treatment

W($\frac{1}{4}$H), W($\frac{1}{2}$H), W(H) Material which has been solution heat treated and subsequently cold worked to various harder tempers

WP Material which has been solution heat treated and precipitation treated

W($\frac{1}{4}$H)P, W($\frac{1}{2}$H)P, W(H)P Material which has been solution heat treated, cold worked and then precipitation treated

Material and composition	Condition	Limit of proportionality MPa	0.2% Proof stress MPa	UTS MPa	Elongation on 5d or 50 mm %	Shear strength MPa	Brinell hardness kg mm^{-2}	Vickers hardness 10 kgf	Modulus of elasticity GPa
Oxygen-free high conductivity copper Cu 99.95%+ Cu 99.99%+	Strip—O	15	48	216	48	162	42	51	117
	Strip—$\frac{1}{4}$H	198	176	263	32	170	82	90	
	Strip—$\frac{1}{2}$H	154	265	314	16	185	96	106	
Tough pitch copper O$_2$ = 0.03%	Strip—O	31	54	224	56	162	49	53	117
	Strip—$\frac{1}{2}$H	116	176	263	29	170	74	88	
	Strip—H	154	270	314	13	185	87	107	
	Wire and Rod—O	—	—	232	45	—	—	—	
	H (Over 5 mm dia.)	—	—	370	—	—	—	—	
	H (Under 5 mm dia.)	—	—	448	—	—	—	—	

(continued)

Table 22.7   COPPER AND COPPER ALLOYS—TYPICAL MECHANICAL PROPERTIES AT ROOM TEMPERATURE—*continued*

Material and composition	Condition	Limit of proportionality MPa	0.2% Proof stress MPa	UTS MPa	Elongation on 5d or 50 mm %	Shear strength MPa	Brinell hardness kg mm⁻²	Vickers hardness 10 kgf	Modulus of elasticity GPa
Phosphorus-deoxidised non-arsenical copper P 0.04	Plate M	31	54	239	58	162	54	58⎫	117
	Tube—½H	—	170	263	30	—	82	90⎭	
Deoxidised arsenical copper P 0.03% As 0.35%	Plate—M	31	54	239	56	162	50	54	117
Silver-bearing copper Ag 0.05%	Strip—O	31	54	224	56	162	55	59⎫	
	Strip—½H	—	180	286	25	178	75	80⎬	117
	Strip—H	—	263	316	10	185	90	100⎭	
Tellurium copper Te 0.5%	Rod—O	—	54	232	40	140	49	53⎫	117
	Rod—½H	—	232	278	15	185	80	85⎭	
Chromium copper Cr 0.6%	Rod—W	28	54	232	60	150	58	65⎫	
	Rod—WP	247	324	448	25	280	124	142⎬	—
	Rod—W(H)	309	479	541	14	—	140	160⎭	
Beryllium copper Be 1.85 Co 0.25	Strip—W	—	224	479	47	—	100	110⎫	
	Strip—WP	—	1066	1205	7	540	350	360⎬	159
	Strip—W(H)	—	1205	1313	2	—	363	380⎭	
Be 0.5 Co 2.5	Strip—W	—	178	324	27	220	67	75⎫	
	Strip—WP	—	618	757	11	360	205	210⎬	159
	Strip—WH	—	772	810	8	370	215	220⎭	
Cadmium copper Cd 0.8	Wire—H	—	600	649	—	350	—	—	124
Sulphur copper S 0.35	Rod—O	—	54	216	40	140	50	55⎫	117
	Rod—½H	—	231	263	12	185	80	85⎭	
Cap copper Cu 95 Zn 5	Strip—O	39	92	263	45	192	65	70⎫	
	Strip—½H	124	223	308	30	216	85	89⎬	124
	Strip—H	162	315	386	8	232	105	110⎭	
Gilding metals Cu Zn 10	Strip—O	39	100	262	60	216	65	70⎫	
	Strip—½H	124	247	317	24	224	85	90⎬	124
	Strip—H	173	324	386	8	247	100	110⎭	

Cu Zn 15	Strip—O	39	108	293	60	224	65	70	121
	Strip—½H	142	254	340	28	239	85	90	
	Strip—H	193	370	440	12	285	125	130	
Cu Zn 20	Strip—O	46	108	317	64	230	65	70	119
	Strip—⅓H	170	285	378	30	247	100	105	
	Strip—H	216	402	479	16	285	130	140	
Brasses Cu Zn 30	Strip—O	46	115	324	67	230	65	70	115
	Strip—⅓H	154	270	378	40	247	107	115	
	Strip—H	208	386	463	20	254	132	143	
Cu Zn 33	Strip—O	54	115	331	60	247	65	70	112
	Strip—½H	154	278	338	35	263	110	117	
	Strip—H	208	432	510	12	293	140	150	
Cu Zn 37	Strip—O	70	131	331	55	278	65	70	108
	Strip—½H	170	285	402	28	293	110	117	
	Strip—H	210	402	494	10	309	136	145	
Cu Zn 40	Rod—½H	—	309	424	20	300	120	125	102
	Plate—M	62	130	371	40	278	85	90	
Aluminium brass Cu Zn 22 Al 2	Tube—O	—	139	378	55	269	75	80	110
	Plate—M	—	154	393	50	285	95	100	
Naval brass Cu Zn 36 Sn 1	Rod—M	93	170	408	35	316	100	105	103
	Plate—M	77	124	386	40	285	90	95	
Free cutting brass Cu Zn 39 Pb 3	Rod—M	—	201	417	25	309	105	110	96
Hot stamping brass Cu Zn 40 Pb 2	Rod—M	—	224	402	30	316	95	100	96

(continued)

Table 22.7 COPPER AND COPPER ALLOYS—TYPICAL MECHANICAL PROPERTIES AT ROOM TEMPERATURE—*continued*

Material and composition	Condition	Limit of proportionality MPa	0.2% Proof stress MPa	UTS MPa	Elongation on 5d or 50 mm %	Shear strength MPa	Brinell hardness kg mm^{-2}	Vickers hardness 10 kgf	Modulus of elasticity GPa
High tensile brass Cu Zn 39 Fe Mn	Rod—M	93	210	440	30	309	95	150	103
High tensile brass Cu Zn 39 Al Fe Mn	Rod—M	139	290	520	30	340	—	160	103
Nickel silvers Cu Ni 10 Zu 27	Strip—O Strip—$\frac{1}{2}$H Strip—H	62 170 270	100 332 510	340 432 564	65 28 11	280 290 320	66 121 155	69 158 177	121
Cu Ni 12 Zn 24	Strip—O Strip—H	62 309	108 587	340 695	60 4	290 400	65 210	68 220	124
Cu Ni 15 Zn 21	Strip—O Strip—H	77 340	124 618	355 695	55 4	290 400	70 210	74 220	121
Cu Ni 18 Zn 20	Strip—O Strip—$\frac{1}{2}$H Strip—H	70 91 123	124 386 525	386 486 610	52 21 7	300 390 430	77 138 166	81 155 168	121
Cu Ni 25 Zn 18	Strip—O Strip—H	77 309	124 618	386 695	50 4	— —	75 201	80 210	121
Phosphor bronzes Cu Sn 3 P	Strip—O Strip—$\frac{1}{2}$H Strip—H	54 300 402	124 386 579	324 510 656	50 12 2	247 340 378	70 160 195	74 170 205	121
Cu Sn 5 P	Strip—O Strip—$\frac{1}{2}$H Strip—H	54 362 479	130 440 618	347 541 710	55 14 2	254 347 386	71 165 205	75 175 215	122
Cu Sn 7 P	Strip—O Strip—$\frac{1}{2}$H Strip—H	77 371 494	139 494 687	356 593 741	60 12 5	260 371 424	80 175 210	84 185 220	117
Cu Sn 8 P	Wire—O Wire—H	— —	— —	424 927	65 —	309 440	— —	— —	111
Copper-nickel Cu Ni 5 Fe	Tube—O Plate—M	— —	116 93	316 278	35 40	240 201	65 60	70 65	132

Material	Form								
Cu Ni 10 Fe Mn	Tube—O	—	139	331	38	247	70	74 ⎫	135
	Plate—M	—	108	324	40	240	65	70 ⎭	
Cu Ni 30 Fe Mn	Tube—O	—	170	417	42	309	90	95 ⎫	152
	Plate—M	—	161	355	38	278	90	95 ⎭	
Silicon bronze Cu 96 Si 3 Mn 1	Plate—M	—	77	410	55	—	—	—	103
	Rod—O	—		362	60	293	—	—	
Aluminium bronzes Cu Al 5	Plate—M	—	147	378	50	286	85	90	126
Cu Al 8 Fe 3	Plate—M	—	170	417	45	316	90	95 ⎫	123
	Rod—½H	—	479	602	12	378	160	170 ⎭	
Cu Al 10 Ni 5 Fe 4	Rod—H	—	417	757	18	571	200	210	131
Sn 7 P 0.1	Strip—O	77	139	356	60	260	80	84 ⎫	117
	Strip—½H	371	494	593	12	371	175	185 ⎬	
	Strip—H	494	687	741	5	424	210	220 ⎭	
Sn 8 P 0.5	Wire—O	—	—	424	65	309	—	—	111
	Wire—H	—	—	927	—	440	—	—	
Copper-nickels Ni 5.5 Fe 1.2 Mn 0.5	Tube—O	—	116	316	35	240	65	70 ⎫	132
	Plate—M	—	93	278	40	201	60	65 ⎭	
Ni 10.5 Fe 1.0 Mn 0.75	Tube—O	—	139	331	38	247	70	74 ⎫	135
	Plate—M	—	108	324	40	240	65	70 ⎭	
Ni 31.0 Fe 1.0 Mn 1.0	Tube—O	—	170	417	42	309	90	95 ⎫	152
	Plate—M	—	161	355	38	278	90	95 ⎭	
Silicon bronze Cu 96 Si 3 Mn 1	Plate—M	—	77	410	55	—	—	—	103
	Rod—O	—		362	60	293	—	—	
Aluminium bronzes Cu 95 Al 5	Plate—M	—	147	378	50	286	85	90	126
Cu 92 Al 8	Plate—M	—	170	417	45	316	90	95 ⎫	123
	Rod—½H	—	479	602	12	378	160	170 ⎭	
Cu 85.5 Al 9.5 Fe 3.0 Mn 1.0 Ni 1.0	Rod—H	—	417	757	18	571	200	210	131

Table 22.8   CAST COPPER—TYPICAL MECHANICAL PROPERTIES AT ROOM TEMPERATURE

Properties vary, dependent on composition, section size and foundry practice.

Material	Composition %		Condition	0.2% Proof stress N mm^{-2}	UTS N mm^{-2}	Elongation on 50 mm %	Brinell hardness
Brass for sand castings	Cu	64	Sand cast	77	232	25	45–60
	Zn	34					
	Pb	2					
Leaded gunmetal	Cu	85	Sand cast	108	216	15	65–75
	Pb	5	Continuous cast	124	254	20	75–95
	Sn	5					
	Zn	5					
High tensile brasses	Cu	56	Sand cast	201	494	20	100–150
	Al	1.5	Centrifugal cast	224	587	22	100–150
	Fe	1.5					
	Mn	1.5					
	Zn	39.5					
	Cu	56	Sand cast	386	710	12	150–230
	Al	5	Centrifugal cast	403	757	15	150–230
	Fe	2.5					
	Mn	2.5					
	Zn	34					
Gunmetal	Cu	88	Sand cast	139	286	18	50–70
	Sn	10	*Continuous cast	147	317	15	70–90
	Zn	2					
Leaded bronze	Cu	80	Sand cast	108	247	18	65–85
	Sn	10	*Continuous cast	170	293	20	80–90
	Pb	10					
Phosphor bronze	Cu	90	Sand cast	139	293	15	70–100
	Sn	10	*Centrifugal cast	185	370	16	100–150
Aluminium bronzes	Cu	87	Sand cast	201	541	25	90–140
	Al	10	*Centrifugal cast	216	571	28	120–160
	Fe	3					
	Cu	80	Sand cast	263	649	15	140–180
	Al	10	*Centrifugal cast	293	695	15	140–180
	Fe	5					
	Ni	5					
Aluminium silicon bronze	Cu	95.5	Sand cast	180	460	20	90–140
	Al	6					
	Si	2					

* Properties of continuously cast and centrifugally cast materials are generally similar.

Table 22.9   COPPER AND COPPER ALLOYS—TYPICAL TENSILE PROPERTIES AT ELEVATED TEMPERATURES

Limit of proportionality or proof stress is reported under a code, viz:
LP = Limit of proportionality
0.1 = 0.1% offset proof stress
0.2 = 0.2% offset proof stress
0.5 = 0.5% offset proof stress.

Material	Composition %	Condition (see Table 22.7)	Temperature °C	Limit of proportionality or proof stress		UTS MPa	Elongation on 50 mm %	Remarks	Reference
				MPa	Code (see above)				
Oxygen-free high conductivity copper	Cu 99.99+	Sheet—O	24	78	0.2	212	56.3	Average grain size 0.045 mm	2
			100	77	0.2	190	55.4		
			204	70	0.2	159	56.9		
Tough-pitch copper	O₂ 0.03	Sheet—O	24	68	0.5	214	57.8	Average grain size 0.043 mm	2
			100	68	0.5	187	57.4		
			204	67	0.5	157	56.9		
Phosphorus-deoxidised non-arsenical copper	P = 0.04	Sheet—Q	24	65	0.5	210	53.4	Average grain size 0.044 mm	2
			100	66	0.5	183	52.5		
			204	57	0.5	161	52.1		
Deoxidised arsenical copper	As 0.35 P 0.03	Plate—M	20	93	0.1	219	57	Hot rolled plate	2
			121	93	0.1	204	57		
			204	83	0.1	184	52		
Silver-bearing copper	Ag 0.05	Sheet—O	20	48	0.1	227	53	Average grain size 0.03 mm	2
			300	45	0.1	162	45		
			500	32	0.1	107	42		
Tellurium copper	Te 0.5	Rod—H	27	350	0.2	361	12.8		2
			260	265	0.2	266	4.7		
Chromium copper	Cr—0.6	Rod—O	20	—	—	259	35	Strain rate 0.1 in min⁻¹	4
			350	—	—	204	44		
			550	—	—	148	10		
		Rod—WP	20	—	—	374	21	Strain rate 0.1 in min⁻¹	4
			350	—	—	298	10		
			550	—	—	210	3		
Gilding metals Cu Zn 10	Cu 90 Zn 10	Rod—O	25	—	—	253	56	—	2
			375	—	—	137	9	—	
			635	—	—	59	17	—	
			875	—	—	19	16	—	
Cu Zn 15	Cu 85 Zn 15	Plate—O	20	79	0.1	297	50	—	2
			121	79	0.1	259	45	—	
			232	74	0.1	238	36	—	
Cu Zn 20	Cu 80 Zn 20	Rod—H	23	—	—	557	13	Cold worked 30%	2
			400	—	—	143	7		
			850	—	—	10	30		
Brasses Cu Zn 30	Cu 70 Zn 30	Strip—O	20	99	0.1	332	64	Grain size 0.03 mm	2
			200	93	0.1	293	54		
			300	88	0.1	238	32		
Cu Zn 37	Cu 63 Zn 37	Strip—O	20	125	0.1	347	60	Grain size 0.035 mm	2
			200	120	0.1	317	54		
			300	111	0.1	267	39		
Cu Zn 40	Cu 60 Zn 40	Plate—O	20	96	0.2	332	62	—	2
			121	96	0.2	312	62		
			204	105	0.2	297	62		

(continued)

Table 22.9   COPPER AND COPPER ALLOYS—TYPICAL TENSILE PROPERTIES AT ELEVATED TEMPERATURES—*continued*

Material	Composition %	Condition (see Table 22.7)	Temperature °C	Limit of proportionality or proof stress MPa	Code (see above)	UTS MPa	Elongation on 50 mm %	Remarks	Reference
Aluminium brass	Cu 76 Al 2 Zn 22	Tube—O	20 200 400	165 134 108	0.2 0.2 0.2	397 317 232	53.8 38.5 13.3	Elongation measured on 11.3√area	2
Naval brass	Cu 62 Sn 1 Zn 37	Plate—O	21 121 204	141 133 134	0.1 0.1 0.1	360 246 326	47 46 38	— —	2
Free cutting brass	Cu 58 Zn 39 Pb 3	Rod—M	21 482	266 —	LP —	477 76	33.5	—	2
Forging brass	Cu 59 Zn 39 Pb 2	Rod—O	20 200 400 610	— — — —	— — — —	364 319 163 34	50.2 42.2 25.4 22.5		2
High tensile brass	Cu 59.32 Zn 35.95 Fe 2.07 Al 1.21 Pb 0.71 Sn 0.67	Rod—O	27 232 427	207 193 —	0.5 0.5 —	479 314 208	13 28 47	Extruded and annealed 816°C for 20 min	5
Nickel silver 20% Ni	Cu 75 Ni 20 Zn 5	Rod—O	30 316 399	76 86 —	LP LP —	347 310 272	51.0 28.5 37.0	—	—
Phosphor bronze	Sn 5 P 0.1	Rod—O	17 260 500	— 100 65	— 0.2 0.2	337 278 141	84 34 6	—	2
	Sn 8 P 0.05	Tube—H	20 200 400	441 451 304	0.2 0.2 0.2	559 539 345	47 37 6	Temper hard and stress relieved	2
Silicon bronze	Si 3 Mn 1	Strip—O	20 200 300	104 94 92	0.1 0.1 0.1	371 309 276	66 54 52	Strain rate 2 in min⁻¹	—
Copper-nickels	Ni 5.5 Fe 1.2 Mn 0.5	Plate—O	20 177 316	158 134 133	0.1 0.1 0.1	301 255 230	40 38 34	—	2
	Ni 10.5 Fe 1.0 Mn 0.75	Tube—O	20 204 400	159 147 139	0.1 0.1 0.1	371 329 287	35 28 18	—	2
	Ni 31.0 Fe 1.0 Mn 1.0	Plate—M	20 232 371	114 96 86	0.1 0.1 0.1	369 304 283	50 46 63	Hot rolled	2
Aluminium bronze	Cu 89 Al 8 Fe 3	Rod—M	20 300 500	294 69 10	0.2 0.2 0.2	490 441 147	51 31 58	Extruded rod elongation measured on 11.3√area	
Cu Al 10 Fe 5 Ni 5	Cu 79.9 Al 10.0 Ni 4.8 Fe 4.8 Mn 0.4	Rod—M	20 200 400	556 510 263	0.2 0.2 0.2	848 817 310	17 18 35	Extruded rod elongation measured on 4.5√area	

Table 22.10   COPPER AND COPPER ALLOYS—TYPICAL TENSILE AND IMPACT PROPERTIES AT LOW TEMPERATURES

Limit of proportionality or proof stress is reported under a code, viz:

LP = Limit of proportionality
0.1 = 0.1% offset proof stress
0.2 = 0.2% offset proof stress
0.5 = 0.5% offset proof stress

Impact values are reported either as C = Charpy, V notch test or I = Izod test.

Material	Composition %		Condition	Temperature °C	Proof stress		UTS MPa	Elongation on 50 mm %	Impact value		Reference
					MPa	Code (see above)			Joules	Code (see above)	
Oxygen-free high conductivity copper	Cu	99.99+	Rod—O	Room	75	0.2	222	86.2	71.1	C	2
				− 78	80	0.2	270	84.5	77.1	C	
				−253	90	0.2	418	83.0	85.8	C	
Tough pitch copper	O₂	0.03	Rod—O	+ 20	59	0.1	216	48.0	58.3	I	2
				− 80	69	0.1	266	47.0	59.6	I	
				−180	79	0.1	351	57.6	67.7	I	
Beryllium copper	Cu	97.44	Rod—W	+ 20	171	0.1	525	36	55.5	I	
	Be	2.56		− 80	201	0.1	598	38	54.2	I	
				−180	344	0.1	769	41	54.2	I	
			Rod—WP	+ 20	865	0.1	1 287	2.6	2.7	I	6
				− 80	1 016	0.1	1 388	0.4	4.1	I	
				−180	1 069	0.1	1 480	3.0	4.1	I	
Gilding metals	Cu	90	Rod—O	+ 22	66	0.2	265	56 Measured	151.8	C	2
	Zn	10		−197	91	0.2	381	86 on 4.52	151.8	C	
				−269	147	0.2	470	91 √area	—		
Brasses	Cu	70	Rod—O	+ 20	194	0.1	352	49.4	88.8	I	2
	Zn	30		− 80	188	0.1	394	59.5	93.5	I	
				−180	204	0.1	507	74.6	106.4	I	
	Cu	60	Rod—O	+ 20	—	—	397	51.3	41.8	C	2
	Zn	40		− 78	—	—	421	53.0	42.0	C	
				−183	—	—	523	55.3	40.7	C	
			Rod—H (Cold worked 25% reduction)	+ 20	—	—	549	19.8	—		
				− 78	—	—	571	21.0	—		
				−183	—	—	669	24.4	—		
Free cutting brass	Cu	58	Rod—H	+ 20	—	—	559	27	—		2
	Zn	39		− 80	—	—	598	27	—		
	Pb	3		−195	—	—	735	26	—		

(continued)

Table 22.10 COPPER AND COPPER ALLOYS—TYPICAL TENSILE AND IMPACT PROPERTIES AT LOW TEMPERATURES —continued

Material	Composition %		Condition	Temperature °C	Proof stress		UTS MPa	Elongation on 50 mm %		Impact value		Reference
					MPa	Code (see above)				Joules	Code (see above)	
Forging brass	Cu	58	Rod—O	+ 20	—	—	364	50.2		21.5	C	2
	Zn	40		− 78	—	—	377	49.8		24.1	C	
	Pb	2		−183	—	—	475	50.6		22.6	C	
Phosphor bronzes	Cu	95	Sheet—O	+ 27	155	0.2	358	61		—	—	2
	Sn	9		− 40	173	0.2	393	73		—	—	
				− 73	180	0.2	427	76		—	—	
			Sheet—EH	+ 27	621	0.2	677	7		—	—	
				− 40	648	0.2	703	9.5		—	—	
				− 73	677	0.2	738	11		—	—	
	Cu	92	Rod—H	+ 24	772	0.2	807	20	(On 1 in gauge length)	—	—	2
	Sn	8		−196	964	0.2	986	30		—	—	
				−253	1 059	0.2	1 158	25		—	—	
Silicon bronze	Cu	96	Rod—H (Cold worked 42% reduction)	+ 25	—	—	511	39.8		—	—	6
	Si	3		− 80	—	—	571	31.7		—	—	
	Mn	1		−190	—	—	692	36.2		—	—	
Nickel silver	Cu	55.15	Sheet—O	+ 20	193	0.1	519	33		108.5	1	6
	Ni	30.5		−120	199	0.1	619	38		108.5	1	
	Zn	14.3		−190	196	0.1	718	41		118.0	1	
Copper-nickels	Ni	10.5	Rod—O	+ 22	147	0.5	342	37	On	154.6	C	2
	Fe	1.0		−197	208	0.5	569	50	4.52√area	155.9	C	
	Mn	0.5		−269	172	0.5	556	53		—	—	
	Ni	31.0	Rod—O	+ 22	129	0.5	398	47	On	155.9	C	2
	Fe	1.0		−197	218	0.5	619	52	4.52√area	154.6	C	
	Mn	1.0		−253	263	0.5	715	51		154.6	C	
Aluminium bronzes	Cu	92	Rod—O	+ 25	110	0.2	414	107	Gauge length not quoted	—	—	2
	Al	8		−196	134	0.2	558	77		—	—	
	Cu	79	Plate—M	+ 20	415	0.2	745	12	On	19.3	C	2

Table 22.11   COPPER AND COPPER ALLOYS—FATIGUE PROPERTIES AT ROOM TEMPERATURE

Note: Where the number of cycles is not given the value represents endurance limit.

Material	Composition %	Form and condition	Fatigue strength MPa	Number of cycles 10^6	Reference
Oxygen-free high conductivity copper	Cu 99.9+	Rod—cold worked 29.2%	117	300	2
Tough pitch copper	O_2—0.03	Rod—annealed Grain size—0.040 mm	62	300	2
		Wire—cold worked 37%	107	100	
Deoxidised non-arsenical copper	P = 0.04	Tube—annealed Grain size 0.050 mm	76	20	2
		Strip—cold rolled 60% reduction	128	100	
Silver-bearing copper	Ag—0.03	Strip-cold rolled 50% reduction	103	—	7
Chromium copper	Cr = 0.88 Si = 0.09 Fe = 0.07	Rod—cold drawn 90% reduction	178	300	7
		Rod—cold drawn 90% reduction—heat treated 3 h 400°C	193	300	
Beryllium coppers	Be = 1.85 Co = 0.25	Strip—solution heat treated	224	100	3
		Strip—rolled 'hard' after solution heat treatment	239	100	
		Strip—rolled 'hard' after solution heat treatment and then precipitation hardened	284	100	
	Be = 0.6 Co = 2.5	Strip—rolled '$\frac{1}{2}$ hard' after solution treatment	241	100	
Gilding metal	Cu = 90 Zn = 10	Strip—annealed Grain size 0.030 mm	69	100	2
		Strip—cold worked 21% reduction	110	100	
		Strip—cold worked 37% reduction	114	100	
		Strip—cold worked 60% reduction	124	100	
		Strip—cold worked 68% reduction	138	100	
	Cu = 80 Zn = 20	Strip—annealed Grain size 0.035 mm	97	100	2
		Strip—cold worked 60% reduction	152	100	
Brasses	Cu = 70 Zn = 30	Strip—annealed Grain size 0.025 mm	107	100	2
		Strip—cold rolled 21% reduction	124	100	
		Strip—cold rolled 60% reduction	152	100	
	Cu = 65 Zn = 35	Strip—annealed	104	100	2
		Strip—cold rolled 37.1% reduction	135	100	
		Strip—cold rolled 60.5% reduction	138	100	

(*continued*)

Table 22.11   COPPER AND COPPER ALLOYS—FATIGUE PROPERTIES AT ROOM TEMPERATURE—*continued*

Material	Composition %	Form and condition	Fatigue strength MPa	Number of cycles 10^6	Reference
Brasses (contd.)	Cu = 60	Rod—annealed	148	100	2
	Zn = 40	Rod—cold worked 25% and stress relieved at 275°C	210	100	
Aluminium brass	Cu = 76 Zn = 22 Al = 2	Rod—cold worked 20–25% and stress relieved	97	20	2
Naval brass	Cu = 62	Rod—hot rolled	128	100	2
	Zn = 38 Sn = 1	Rod—cold worked 27% reduction	183	100	
Phosphor bronze	Sn = 5 P = 0.1	Strip—annealed Grain size 0.035 mm	172	100	2
		Strip—cold rolled 69% reduction	221	100	
	Sn = 8 P = 0.1	Rod—annealed Grain size 0.020 mm	221	1 000	2
		Rod—cold worked 30.1% reduction	234	1 000	
	Sn = 10 P = 0.1	Rod—annealed Grain size—0.065/070 mm	172	1 000	2
		Rod—cold worked 30.1% reduction	159	1 000	
Silicon bronze	Cu = 96 Si = 3 Mn = 1	Rod—annealed Rod—hard	130 232	300 300	7
Copper-nickels	Ni = 5.5 Fe = 1.0 Mn = 0.5	Rod—annealed Rod—cold worked 25% reduction	131 173	100 100	2
	Ni = 10.5 Fe = 1.5 Mn = 0.75	Strip—cold rolled Hard temper	145	100	2
	Ni = 31.0 Fe = 1.0 Mn = 1.0	Tube—annealed Tube—cold worked and stress relieved	147 177	100 100	2
Aluminium bronze	Cu = 92 Al = 8	Rod—lightly worked	203	100	2
	Cu = Rem. Al = 9.7 Fe = 5.1 Ni = 5.3	Rolled rod	323	100	2
Nickel silver	Cu = 55 Ni = 18 Zn = 27	Strip—annealed Strip—cold rolled 70% reduction	114 173	100 100	—

Table 22.12   COPPER AND COPPER ALLOYS, IMPACT PROPERTIES

Material	Composition %		Form and condition	Temperature °C	Izid Joules	Charpy Joules	Reference
					\multicolumn{2}{c}{*Impact value*}		
Oxygen-free high conductivity copper	Cu	99.99+	Plate—hot rolled	0	—	62.4	1
				204	—	50.2	
				316	—	56.9	
				538	—	44.7	
				650	—	35.3	
Tough pitch copper	O_2	0.08	Plate—annealed	20	47.5	—	8
				200	37.9	—	
				300	35.3	—	
				500	31.2	—	
				600	28.5	—	
Phosphorus-deoxidised non-arsenical copper	P	0.06	Plate—annealed	20	61.0	—	8
				200	56.9	—	
				300	55.6	—	
				500	42.0	—	
				600	31.2	—	
Deoxidised arsenical copper	As P	0.36 0.07	Plate—annealed	20	62.4	—	8
				200	56.9	—	
				400	55.6	—	
				500	43.4	—	
				600	31.2	—	
Naval brass (American type)	Cu Sn Zn	60.25 0.75 Bal.	Annealed	20	—	82.4	6
				−50	—	79.9	
				−80	—	84.7	
				−115	—	80.6	
Phosphor bronze	Sn P	4 0.4	Rod–hard drawn	20	62.4	—	9
				−41	59.7	—	
Silicon bronze	Cu Si Mn	96 3 1	Rod—annealed	20	—	90.0	6
				−50	—	99.1	
				−80	—	93.8	
				−115	—	87.4	
Aluminium bronzes	Cu Al Fe	89.6 7.8 2.6	Rod—extruded annealed and roller straightened	24	—	98.9	
				−29	—	105.7	
				−59	—	103.0	
				−182	—	94.9	
	Cu Al Ni Mn	81.2 10.1 4.75 0.8	Rod—extruded annealed and roller straightened	24	—	20.3	
				−29	—	19.0	
				−59	—	17.6	
				−182	—	12.2	
Copper-nickels	Cu Ni	80 20	Annealed	20	104.4	—	6
				−80	107.1	—	
				−120	113.9	—	
				−180	115.2	—	
	Cu Ni	70 30	Annealed	20	—	89.9	6
				−30	—	80.5	
				−50	—	80.5	
				−80	—	79.6	
				−115	—	81.3	
Nickel silvers	Cu Ni Zn Mn	74.28 19.49 5.43 0.80	Annealed	20	—	91.5	6
				−30	—	76.5	
				−50			
				−80	—	75.7	
				−115	—	71.3	
	Cu Ni Zn	55.15 30.50 14.3	Annealed	20	—	108.4	—
				−40	—	118.0	
				−120	—	108.5	
				−180	—	118.0	

Table 22.13   WROUGHT COPPER AND COPPER ALLOYS, CREEP PROPERTIES

Notes: (1) All values relate to rod or wire products unless specified.
(2) Total extension = Initial extension + total creep.
       = Initial extension + intercept + (minimum creep rate × duration).

Materials and composition %	Condition	Test temperature °C	Applied stress MPa	Duration 1 000 h	Total extension %	Intercept %	Minimum creep rate in % per 1 000 h	Remarks
Oxygen-free copper Cu 99.99+	Annealed grain size 0.025 mm	149	14.5	6.4	0.053	0.024	0.001 7	Reference 2
			21.0	6.5	0.128	0.049	0.007 5	
			31.4	6.0	0.510	0.290	0.023	
			54.8	5.1	2.490	1.560	0.083	
		204	14.2	6.5	0.213	0.054	0.021	
			21.3	6.0	0.580	0.256	0.049	
			28.0	6.5	1.295	0.727	0.078	
			51.0	5.0	4.580	2.670	0.215	
	Hard drawn 84% reduction	149	54.6	6.0	0.102	0.002	0.008 2	
		204	14.5	3.2	0.157	0.090	0.014	
			28.0	6.0	0.422	0.185	0.034 5	
			48.3	6.5	3.80	2.50	0.19	
HC copper O₂ = 0.04		149	14.2	6.4	0.088	0.048	0.003 2	Reference 2
			20.7	6.5	0.257	0.133	0.013	
			41.4	6.5	1.875	1.120	0.057 5	
			55.9	6.5	3.475	1.795	0.088	
	Annealed grain size 0.025 mm	260	2.5	6.0	0.084	0.016	0.011	
			7.3	6.5	0.640	0.113	0.079 5	
			13.7	6.5	2.877	0.869	0.306	
	Drawn 84% reduction	149	52.0	6.4	0.118	0.041	0.004 9	
			68.9	6.5	0.167	0.042	0.010	
		204	7.3	6.5	0.064	0.045	0.001 1	
			28.0	6.5	1.080	0.409	0.097	
			49.0	6.5	5.418	2.47	0.44	
Deoxidised non-arsenical copper P = 0.008	Annealed grain size 0.032 mm	204	14.2	6.0	0.078	0.037	0.003 9	Reference 2
			21.0	7.08	0.355	0.164	0.018 5	
			35.1	7.08	1.378	0.660	0.051	
			55.4	6.0	3.334	1.120	0.120	
	Cold worked 84% reduction	204	24.4	7.7	0.126	−0.015 1	0.015 2	Reference 2
			34.7	7.08	0.119	−0.085	0.038	Accelerating creep rate
			62.4	7.08	0.534	−1.110	0.224	Accelerating creep rate
			103.5	0.58	0.169	0.034	0.055	
			103.5	4.2	2.813	−8.630	2.70	Accelerating creep rate
Deoxidised arsenical copper As = 0.35 P = 0.03	Annealed grain size 0.045 mm	149	20.7	6.5	0.055	0.013	<0.000 1	
			35.8	6.4	0.580	0.085	0.002 35	
			54.8	6.5	1.560	0.185	0.008 5	
			70.6	6.4	2.537	0.275	0.019	
		204	21.0	6.0	0.119	0.040	0.002 3	
			38.6	6.0	1.055	0.295	0.014	Reference 2
			49.0	6.0	1.648	0.365	0.022	
		260	14.2	6.86	0.107	0.049	0.005 5	
			24.4	6.86	0.678	0.335	0.024	
			43.1	6.0	2.584	0.700	0.152	
	Cold worked 84% reduction	149	85.6	6.4	0.089	0.017	0.000 24	
			138.8	6.5	0.153	0.033	0.000 5	
			207.6	6.5	0.282	0.089	0.001 6	
			275.8	9.45	0.452	0.118	0.007 8	
			325.4	0.5	0.750	0.160	0.42	
			325.4	0.75	0.910	—	—	
		260	10.7	6.0	0.029	0.001 5	0.001 45	
			14.2	6.0	0.075	−0.012	0.012	Accelerating creep rate
			24.4	7.3	0.605	−0.187	0.105	Accelerating creep rate
			34.4	6.5	0.933	−0.672	0.24	Accelerating creep rate

(continued)

Table 22.13 WROUGHT COPPER AND COPPER ALLOYS, CREEP PROPERTIES—*continued*

Materials and composition %	Condition	Test temperature °C	Applied stress MPa	Duration 1 000 h	Total extension %	Intercept %	Minimum creep rate in % per 1 000 h	Remarks
Silver-bearing copper Ag–0.086 O₂–0.02	Annealed strip grain size 0.030 mm	130	137.9	2.4	15.8	14.7	1.05	Reference 2
			96.5	2.6	7.0	6.7	0.35	
		175	137.9	2.4	27.4	23.2	1.85	
		225	41.2	3.0	0.9	0.6	0.02	
			96.5	2.5	10.6	8.0	1.1	
	Cold worked strip 25% reduction	130	55.1	4.75	0.08	0.075	0.002	Reference 2
			96.5	10.2	0.18	0.16	0.004	
			137.9	7.2	0.26	0.24	0.005	
		225	55.1	8.9	0.21	0.14	0.006 4	
			96.5	11.5	0.56	0.38	0.017	
	Cold worked strip 50% reduction	130	55.1	4.55	0.09	0.08	0.001 5	
			96.5	11.4	0.20	0.185	0.001 5	
			137.9	7.25	0.29	0.265	0.004	
		225	55.1	8.9	0.26	0.15	0.011	
			96.5	12.9	0.795	0.335	0.029	
			137.9	3.0	0.825	0.525	0.10	
Tellurium copper Te–0.5	Annealed grain size 0.025 mm	149	21.0	6.0	0.134 4	0.077	0.001 4	Reference 2
			35.8	6.0	0.251 5	0.121	0.007 8	
			59.0	6.0	1.553	0.737	0.022 3	
	Cold drawn 37%	149	47.6	6.0	0.089 5	0.034 4	0.000 85	
			68.9	6.0	0.149 4	0.068 1	0.001 9	
			103.5	6.0	0.223 4	0.088	0.005 2	
			137.5	6.0	0.390 5 5	0.155 5	0.011 5	
			201.3	6.0	1.133	0.479	0.080	
Cadmium copper Cd = 1.03	Cold worked 20% reduction	103	137.9	—	—	—	0.007	Reference 5
		205	55.1	—	—	—	0.011	
			137.9	—	—	—	0.075	
Chromium copper Cr = 0.73	Fully heat treated and drawn	343	68.9	—	—	—	0.008 9	Reference 5
			110.3	—	—	—	0.014	
			137.9	—	—	—	0.038	
Cap copper Cu 95 Zn 5	Cold drawn 51% reduction	200	98.1	3.305	0.123	—	—	Reference 2
			117.7	1.995	0.21	—	—	
			137.3	1.999	0.38	—	—	
			140.2	2.011	0.51	—	—	
Gilding metal Cu 85 Zn 15	Annealed grain size 0.060 mm	149	31.0	4.5	0.031	0.006	0.000 7	Reference 2
			48.0	5.1	0.087	0.024	0.002 9	
			67.6	4.5	1.07	0.44	0.026	
		260	13.9	5.0	0.091	0.026	0.008	
			24.2	5.3	0.241	0.037	0.030	
			34.3	5.0	0.507	0.083	0.073	
			41.3	5.14	0.958	0.180	0.138	
			47.9	1.6	1.100	0.510	0.32	
	Cold worked 84% reduction	149	66.9	4.4	0.096	0.029	0.000 7	
			136.5	5.1	0.197	0.054	0.000 26	
			273.7	4.5	0.445	0.108	0.011	
			371.0	5.1	0.860	0.250	0.033	
		260	4.2	5.54	0.096	0.038	0.010	
			6.6	5.3	0.270	0.080	0.034	
			13.4	2.95	0.678	0.104	0.19	
			20.5	3.65	2.715	−0.045	0.76	Accelerating creep rate
Brass Cu 70 Zn 30	Annealed grain size 0.022 mm	204	19.3	3.3	0.052	0.016	0.008 1	Reference 2
			39.2	1.6	0.104	0.025	0.036	
			59.2	3.3	0.430	0.035	0.11	
		260	7.7	4.8	0.115	0.014	0.021	
			13.3	3.72	0.357	0.061	0.074	
			18.8	4.08	0.908	0.090	0.195	
			25.3	3.72	1.557	0.147	0.37	

(*continued*)

Table 22.13   WROUGHT COPPER AND COPPER ALLOYS, CREEP PROPERTIES—*continued*

Materials and composition %	Condition	Test temperature °C	Applied stress MPa	Duration 1 000 h	Total extension %	Intercept %	Minimum creep rate in % per 1 000 h	Remarks
Brass Cu 70 Zn 30 (*contd.*)	Cold worked 84% reduction (fine grained)	149	68.9	5.2	0.156	0.058	0.002 5	Reference 2
			135.8	5.23	0.326	0.129	0.005 4	
			275.4	6.85	0.780	0.260	0.026	
			348.8	4.75	1.430	0.540	0.10	
		204	6.9	5.1	0.128	0.084	0.006 7	
			21.2	5.06	0.530	0.180	0.063 6	
			34.7	4.7	1.494	0.213	0.265	
		260	3.4	5.0	0.311	0.121	0.037	
			5.6	5.0	0.970	0.175	0.159	
			10.2	2.62	3.015	0.212	1.07	
Brass Cu 60 Zn 40	Annealed	149	34.4	6.43	0.053 5	0.009 5	0.001 1	Reference 2
			51.3	6.43	0.099	0.029	0.002 3	
			68.9	6.43	0.158	0.039	0.006	
			103.4	6.43	0.313	0.084	0.011 5	
			137.1	6.43	3.580	1.265	0.20	
		204	7.3	7.7	0.048	0.013 2	0.002 9	
			14.2	7.7	0.090	0.025	0.005 8	
			28.0	2.28	0.18	0.010	0.022	
			42.0	7.68	1.975	0.053	0.246	
Forging brass Cu 59.32 Zn 35.95 Pb 2.07	Extruded and annealed 816°C for 20 mm	149	68.9	—	—	—	0.052	Reference 5
		177	68.9	—	—	—	0.18	
		204	68.9	—	—	—	0.92	
		260	68.9	—	—	—	89.0	
		177	20.7	—	—	—	0.030	
		232	20.7	—	—	—	0.075	
		260	20.7	—	—	—	0.134	
		288	20.7	—	—	—	1.14	
Admiralty brass Cu 70 Zn 29 Sn 1	Annealed grain size 0.055 min	149	31.4	4.5	0.029	0.013	<0.001	Reference 2
			82.3	3.38	0.090	0.023	0.001 4	
			104.4	4.4	0.134	0.042	0.002 8	
		260	6.6	4.98	0.026	0.003	0.003 7	
			13.6	4.3	0.072	0.031	0.008	
			20.7	4.98	0.149	0.034	0.021	
			31.0	4.3	0.306	0.019	0.062	Accelerating creep rate
			42.5	6.3	1.528	−0.35	0.295	Accelerating creep rate
	Cold worked 60% reduction	149	86.5	5.2	0.164	0.066	0.001 6	Reference 2
			103.5	6.5	0.203	0.078	0.002 7	
			138.5	5.4	0.287	0.116	0.005 2	
			208.8	6.5	0.451	0.179	0.008 8	
			280.7	5.2	0.685	0.307	0.018	
			361.5	5.2	1.053	0.385	0.060	
		260	2.0	5.75	0.088	0.052	0.005 5	
			6.8	2.95	0.425	0.090	0.11	
			13.7	5.3	2.481	0.160	0.435	
			20.4	2.6	2.601	0.340	0.86	
Aluminium brass Cu 76 Zn 22 Al 2	Annealed grain size 0.030 mm	149	40.6	5.5	0.037 5	0.002	0.001 0	Reference 2
			68.3	6.5	0.099	0.024	0.003 8	
			137.1	6.5	0.450	0.146	0.029	
			151.7	9.45	0.979	0.367	0.051	
		260	3.8	6.86	0.097	0.043	0.007 1	
			7.3	6.86	0.298	0.082	0.030	
			10.7	6.5	0.524	0.128	0.059	
	Cold worked 37% reduction	149	132.0	6.4	0.208	0.074	0.002 8	
			207.6	6.5	0.349	0.128	0.006 3	
			275.1	11.1	0.590	0.233	0.010	
			344.7	6.5	1.169	0.498	0.054	
		260	7.3	6.0	0.159	0.081	0.012	
			14.4	6.0	0.577	0.197	0.061	
			21.0	6.0	1.243	0.117	0.184	
			34.7	3.48	2.322	−0.181	0.71	

(*continued*)

Table 22.13 WROUGHT COPPER AND COPPER ALLOYS, CREEP PROPERTIES—*continued*

Materials and composition %	Condition	Test temperature °C	Applied stress MPa	Duration 1 000 h	Total extension %	Intercept %	Minimum creep rate in % per 1 000 h	Remarks
Phosphor bronze Cu 95 Sn 5 P 0.2	Annealed grain size 0.050 mm	149	31.0	5.62	0.028	0.003	0.000 3	
			68.9	5.62	0.072	0.009	0.000 8	
			104.1	5.62	0.142	0.016	0.001 9	
			117.5	4.75	0.978	0.008	0.009 4	
		260	10.7	5.1	0.039	0.017	0.001 6	
			17.0	5.0	0.066	0.020	0.005	
			34.3	5.8	0.216	0.065	0.018	
			72.7	5.64	1.191	−0.100	0.213	Accelerating creep rate
	Cold worked 84% reduction	149	31.5	5.62	0.045	0.009	0.000 9	Reference 2
			103.0	5.62	0.170	0.050	0.002 6	
			136.5	5.65	0.235	0.075	0.003 7	
			205.4	5.62	0.367	0.112	0.008	
			344.4	4.75	0.726	0.250	0.027	
		260	2.2	5.1	0.113	0.025	0.016 4	
			3.7	5.64	0.285	0.090	0.033 6	
			6.80	8.15	0.797	0.269	0.063 4	
			20.8	5.75	2.511	−0.243	0.47	Accelerating creep rate
Silicon bronze Cu 96 Si 3 Mn 1	Annealed 450°C	204	86.5	—	—	—	0.065	Reference 11
			103.5	—	—	—	0.084	
		288	34.6	—	—	—	0.035	
			41.4	—	—	—	0.080	
			51.9	—	—	—	0.19	
			69.0	—	—	—	0.65	
Aluminium bronze Cu 95 Al 5	Annealed plate	200	107.8	3.0	0.2	—	0.000 07	
		300	33.4	2.0	0.2	—	0.000 1	
	Cold drawn (rod) Rockwell 92B	550	17.3	5.8	7.5	0.035	640	Rupture test results
			34.4	0.52	5.7	0.62	8 340	Reference 2
		600	6.8	85.3	23.5	0.090	79	
			17.3	2.9	8.5	0.316	1 250	
			34.4	0.14	8.0	0.0	23 500	
Aluminium bronze Cu 79 Al 10 Fe 5 Mn 1 Ni 5	Extruded	250	61.8	1.17	0.068	—	<0.004 2	Reference 2
			92.7	1.44	0.123	—	0.008 3	
			139.0	1.44	0.236	—	0.031 7	
			216.2	1.27	0.622	—	0.258 3	
			308.9	1.128	3.67	—	—	
		400	46.3	0.72	0.43	—	0.237 5	
			61.8	0.72	0.76	—	0.541 7	
			77.2	0.72	1.93	—	1.375	
Copper-nickel Ni 10.5 Fe 1.0 Mn 0.5	Annealed grain size 0.025 mm	149	103.5	6.0	0.870 55	0.1	<0.000 1	Reference 2
			137.9	6.0	2.131	0.242	0.000 16	
			172.4	6.0	4.705	0.163 7	0.000 22	
		260	63.5	6.0	1.090	0.143 6	0.000 61	
			90.7	6.0	0.516	0.253 8	0.001 7	
			126.6	6.0	1.803	0.175 6	0.003 8	
	Cold worked 21% reduction	149	138.2	6.0	0.139 1	0.018 8	<0.000 1	
			206.8	6.0	0.199	0.014 8	0.000 2	
			276.1	6.0	0.277	0.027 6	0.001 4	
			310.3	6.0	0.410	0.061	0.002 4	
			343.2	6.0	0.635	0.164	0.003 5	
		260	139.9	6.0	0.198 8	0.057 6	0.002 2	
			207.9	6.0	0.442 5	0.169	0.013 6	
			244.8	4.32	0.607	0.189	0.044	
			244.8	6.0	0.700	0.102	0.061 7	

(*continued*)

Table 22.13   WROUGHT COPPER AND COPPER ALLOYS, CREEP PROPERTIES—*continued*

Materials and composition %	Condition	Test temperature °C	Applied stress MPa	Duration 1 000 h	Total extension %	Intercept %	Minimum creep rate in % per 1 000 h	Remarks
Copper-nickel Ni 31 Fe 1 Mn 1	Cold worked and stress relieved	399	124.2	2.5	0.219	—	0.015	Reference 2
			172.4	2.5	0.319	—	0.032	
			206.8	1.5	0.359	—	0.055	
			241.4	1.0	0.490	—	0.17	
			275.8	1.0	0.818	—	0.40	
			310.3	1.0	3.25	—	1.0	
		454	48.3	2.5	0.142	—	0.019	
			96.5	1.5	0.339	—	0.072	
			172.4	1.0	0.993	—	0.61	
			206.8	1.5	9.80	—	2.2	
		510	13.7	1.5	0.096	—	0.032	
			41.4	1.0	0.292	—	0.18	
			124.2	1.5	11.2	—	3.4	
		566	10.3	0.5	0.185	—	0.3	
			68.9	0.5	7.20	—	7	
Nickel silver Cu 74.23 Ni 20.08 Zn 5.08 Mn 0.69	Annealed 650	316	34.4	—	0	—	—	Reference 11
			86.5	—	0.006	—	—	
			103.5	—	0.013	—	—	
			137.5	—	0.034	—	—	
			173.0	—	0.072	—	—	
		399	34.4	—	0	—	—	
			61.8	—	0.015	—	—	
			86.5	—	0.065	—	—	
			103.5	—	0.365	—	—	

Table 22.14   CAST COPPER ALLOYS—CREEP PROPERTIES

Material and composition %	Test temperature °C	Applied stress MPa	Minimum creep rate in % per 1 000 h	Reference
High tensile brass Cu 57.14 Al 0.44 Sn 0.10 Pb 0.49 Fe 1.72 Zn 40.11	149	69	0.118	
	177	69	0.69	11
	288	21	1.81	
Silicon brass Cu 81.6 Zn 13.95 Si 4.40 Fe 0.05	260	69	0.075	
	316	21	0.236	5
	371	21	0.73	
		69	35	
Aluminium bronze—sand cast Cu 90 Al 10	250	77	<0.146	
		131	<0.153	10
		185	<1.44	
		309	400	
Aluminium bronze—sand cast Cu 80 Al 10 Fe 5 Ni 5	250	77	<0.037 5	
		131	<0.121	10
		185	<0.321	
		309	<1.85	

(continued)

Table 22.14 CAST COPPER ALLOYS—CREEP PROPERTIES—*continued*

Material and composition %	Test temperature °C	Applied stress MPa	Minimum creep rate in % per 1 000 h	Reference
Leaded gun metal	232	55	0.007	
Cu 84.8		76	0.032	5
Pb 4.8		97	0.140	
Sn 5.1	260	41	0.007	
Zn 4.8		45	0.013	
		55	0.040	
Tin bronze	260	103	1.44	
Cu 90		55	0.042	
Sn 19				5
	316	69	2.0	
		55	0.89	
		28	0.225	
		7	0.030	
Admiralty gun metal	204	93	0.039	
Cu 88		120	0.75	
Sn 10				
Zn 2	260	42	0.019 5	
		62	0.023 5	
		76	0.06	12
	316	21	0.011	
		28	0.013	
		35	0.138	

Table 22.15 TENSILE AND CREEP PROPERTIES OF TOUGH PITCH COPPER-SILVER ALLOYS[13]

Silver %	Cold work %	Test temp. °C	0.2% Proof stress MPa	UTS MPa	Elong. on 50 mm %	Stress for 1% strain in 10^5 hours MPa	Stress for rupture in 10^5 hours MPa
0.002	0	RT	64	220	49	—	—
	0	100	63	193	50	42	107
	0	150	63	176	50	39	59
	30	RT	287	307	18	—	—
	30	100	262	270	10	—	120
	30	150	244	253	14	—	67
0.034	0	RT	73	225	53	—	—
	0	100	73	197	52	—	142
	0	150	62	182	52	—	112
	10	RT	199	241	35	—	—
	10	100	189	209	34	159	167
	10	150	178	192	36	134	137
	20	RT	291	297	13	—	—
	20	100	260	266	8	—	—
	20	150	244	248	7	—	—
	30	RT	296	303	11	—	—
	30	100	267	277	8	182	189
	30	150	253	260	8	139	147
0.065	0	RT	83	225	50	—	—
	0	100	81	199	52	70	151
	0	150	75	185	50	63	119
	10	RT	212	245	34	—	—
	10	100	204	215	30	168	179
	10	150	190	201	28	142	148
	20	RT	293	296	12	—	—
	20	100	275	278	7	—	—
	20	150	259	261	7	—	—
	30	RT	299	304	10	—	—
	30	100	277	282	7	203	215
	30	150	263	268	7	142	166

(continued)

Table 22.15   TENSILE AND CREEP PROPERTIES OF TOUGH PITCH COPPER-SILVER ALLOYS[13]—*continued*

Silver %	Cold work %	Test temp. °C	0.2% Proof stress MPa	UTS MPa	Elong. on 50 mm %	Stress for 1% strain in 10^5 hours MPa	Stress for rupture in 10^5 hours MPa
0.14	0	RT	67	230	51	—	—
	0	100	66	203	51	74	—
	0	150	69	189	53	65	144
	10	RT	205	249	36	—	—
	10	100	198	219	32	176	192
	10	150	189	203	30	153	166
	20	RT	288	296	12	—	—
	20	100	272	276	8	230	238
	20	150	261	263	7	200	204
	30	RT	328	337	7	—	—
	30	100	308	314	6	253	265
	30	150	290	306	5	198	217
0.32	0	RT	82	232	52	—	—
	0	100	74	206	52	83	—
	0	150	82	194	51	66	—
	10	RT	194	249	41	—	—
	10	100	184	221	36	—	—
	10	150	186	211	35	—	—
	20	RT	278	289	18	—	—
	20	100	267	272	10	—	—
	20	150	255	261	9	—	—
	30	RT	330	336	11	—	—
	30	100	314	317	7	282	280
	30	150	299	308	6	241	249

REFERENCES

1. OFHC Copper—Technical Information, American Metal Climax Inc.; 1969.
2. Copper Development Association, Copper and Copper Data Sheets.
3. Copper Development Association, High Conductivity Copper Alloys, 1968.
4. M. Cook and E. C. Larke, *J. Inst. Loco. Eng.*, 1938, **28**, 609.
5. Elevated Temperature Properties of Copper and Copper Base Alloys. ASTM Spl. Publication No. 181, 1956.
6. C. S. Smith, *Proc. Am. Soc. test. Mater.*, 1939, **39**, 642.
7. A. R. Anderson and C. S. Smith, *Proc. Am. Soc. test. Mater.*, 1941, **41**, 849.
8. M. Cook and E. C. Larke, *J. Inst. Metals*, 1942, **58**, 1.
9. H. W. Gillet, *Proc. Am. Soc., Project 13*, 1941.
10. E. Voce, *Metallurgia*, 1946, **35**, 3.
11. Compilation of Available High Temperature Creep Characteristics of Metals and Alloys. *Amer. Soc. Mech. Engineering*, 1938.
12. G. Chadwick, *J. Am. Soc. Naval Enging*, 1938, **50**, 52.
13. J. E. Bowers and R. D. S. Lushey, *Met. and Mat. Tech.*, 1978, **10**(7), 381.

22.3   Mechanical properties of lead and lead alloys

The mechanical properties of lead and its alloys, particularly the more dilute alloys, are extremely sensitive to variations in composition, grain size, metallurgical history and temperature and rate of testing. They are therefore rarely reproducible with any degree of accuracy, except on the same sample under identical test conditions and, even then, a delay of a few hours between tests may affect the results obtained.

The figures quoted in the following tables should therefore be considered only as *typical* and should not be used where accuracy is necessary. For these same reasons, the materials quoted in the tables are grouped according to their principal uses and the typical values quoted for various properties are merely intended as an indication of their suitability for those uses.

Lead alloys are also dealt with under 'Solders', Chapter 34.

Table 22.16   LEAD AND LEAD ALLOYS—TYPICAL MECHANICAL PROPERTIES

Common name	Nominal composition %	Hardness DPN	Tensile strength[1] MPa	Fatigue[2] strength MPa
A. Lead and lead alloys for chemical applications				
Type A lead	Pb > 99.99	4	16.8 (20°C)	±3.17 (20°C)
			12.1 (60°C)	±2.24 (60°C)
Copper lead	Cu 0.05–0.07	4.5	17.6 (20°C)	±4.96 (20°C)
			15.2 (60°C)	±4.69 (60°C)
Tellurium copper lead	Te 0.02–0.05 Cu 0.05–0.07	6	21.1 (20°C)	±7.24 (20°C)
			18.7 (60°C)	±6.55 (60°C)
Silver copper lead	Ag 0.003–0.005 Cu 0.003–0.005	5	16.4 (20°C)	±4.48 (20°C)
			13.4 (60°C)	±3.45 (60°C)
Antimonial lead	Sb 2.5–11.0			
	Sb 4.0	8–12	30.0 (20°C)	±10.62 (20°C)
			25.2 (60°C)	±10.04 (60°C)
	Sb 8.0	9–16	37.7 (20°C)	±14.82 (20°C)
			35.9 (60°C)	±12.06 (60°C)
Dispersion strengthened lead	PbO 1.5 dispersed phase	14	29.4	±13.4
	PbO 4.0 dispersed phase	—	35.5	±13.7
B. Lead alloys for cable sheathing				
	Sb 0.85	6–15 (depending on heat treatment)	31[3]	±8.3[4]
	Cd 0.075 Sn 0.2			±3.8[4]
	Sb 0.4 Sn 0.2	6	18.5[3]	±6.6[4]

Notes:
(1) Rate of testing: 0.4 mm mm^{-1} min^{-1}.
(2) 20×10^6 cycles.
(3) Testing rate: 0.1–0.25 mm mm^{-1} min^{-1}.
(4) 10^7 cycles.

Table 22.17   IMPORTANT LEAD ALLOYS WITH UNSPECIFIED MECHANICAL PROPERTIES[1]

Main application	Nominal composition %	Specific use
Soldering	Sn 64 or 60	Electrical and electronics
	Sn 40, 45, 50, 60	Can soldering, general engineering
	Sn 30, 35	Cable sheaths
	Sn 15, 20	Lamps, low service temperatures
	Sn 40 Sb 2.2	Heat exchangers, general dip soldering
	Sn 30, 32 Sb 1.6, 1.8	Plumbing, wiped joints
	Sn 28, 30 Sb 1.5, 1.6	General body soldering
	Sn 25 Sb 1.4	Radiator core dipping
	Sn 16, 18, 22 Sb 2.5, 1.0, 1.2	Body soldering, radiator dipping
	Sn 5.0 Sb 4.0	Hot dip coating, tubular radiator manufacture
	Sn 2.6, 10.2 Sb 5.1, 4.0 As 0.5	Body solders for shallow areas

(continued)

Table 22.17 IMPORTANT LEAD ALLOYS WITH UNSPECIFIED MECHANICAL PROPERTIES[(1)]—*continued*

Main application	Nominal composition %	Specific use
Printing		
	Sn 2–4	Electrobacking metal
	Sb 2–5	
	Sn 2–5	Slug casting metal
	Sb 10–13	
	Sn 5–10	Stereotype metal
	Sb 15	
	Sn 6–13	Monotype casting metal
	Sb 15–19	
	Sn 13–22	Cast type casting metal
	Sb 20–28	

Notes:
(1) In many uses of lead alloys, the strength, hardness, etc. of the alloy are not of prime importance. In general soldering, for example, the geometry of the joint and the materials being joined, are of greater importance than the strength of the bulk solder. In most applications of type metals, fluidity of the liquid alloy, contraction properties, and wear resistance of the solid alloy, are the most important characteristics in formulating an alloy. See also Chapter 34.

BIBLIOGRAPHY

1. L. I. Goff and G. Hewish, 'Creep Resistant Lead Sheet by the D.M. Process', *3rd Inter. Lead Conf.*, Venice, 1968, Publ'd. LDA, London.
2. L. I. Goff and R. D. Semmens, 'Lead Products by Continuous Casting', *2nd Inter. Lead Conf.*, Arnhem, 1965, Publ'd. LDA, London.
3. J. N. Greenwood, *Met. Rev.*, 1961, **6(23)**, 279–351.
4. W. Hofmann, 'Lead and Lead Alloys', English trans. of 2nd revised German edition, Springer-Verlag, Berlin. 'Lead Abstracts', 1962–80. LDA, London.
5. A. Lloyd and E. R. Newson, 'Dispersion Strengthened Lead Properties and Potentialities in Chemical Plant and some Early Trials.' *Second Inter. Lead Conf.*, Arnhem, 1965.
6. A. Lloyd and E. R. Newson, 'Dispersion Strengthened Lead—Developments and Applications in the Chemical Industry, *3rd Inter. Lead Conf.*, Venice, 1968, Publ'd, LDA, London.
7. *ASM Hand book*, vol. 2, 10th edition, ASM International, Materials Park, OH. 1990.

22.4 Mechanical properties of magnesium and magnesium alloys

Related designations for magnesium alloys are given in Chapter 1.

Table 22.18 MAGNESIUM AND MAGNESIUM ALLOYS (WROUGH)—TYPICAL MECHANICAL PROPERTIES AT ROOM TEMPERATURE

Material	Nominal* composition %	Form	Tension Proof stress 0.2% MPa	Tension UTS MPa	Tension Elong. %	Compression Proof stress 0.2% MPa	Hardness VPN 30 kg
Mg	Mg 99.9	Sheet, annealed	69	185	4	—	30–35
		Bar, extruded	100	232	6	—	35–45
Mg–Mn	Mn 1.5	Sheet	100	232	6	—	35–45
		Extruded bar (1 in. diam.)	162	263	7	124	45–55
		Extruded tube	154	247	6	—	45–55
Mg–Al–Zn	Al 3.0 Zn 1.0 Mn 0.3	Sheet, annealed	131	232	13	—	50–60
		half hard	170	263	10	100	55–70
		Extruded bar and sections	162	255	11	93	50–60
	Al 6.0 Zn 1.0	Forgings	183	293	8	147	60–70
		Extruded bar and sections	183	293	8	147	55–70
	Mn 0.3	Extruded tube	170	278	8	147	60–70
	Al 8.0 Zn 0.5 Mn 0.3	Forgings	208	293	8	185	65–75
Mg–Zn–Mn	Zn 2.0 Mn 1.0	Sheet, annealed	131	232	13		
		half breed	170	263	10		
		Extruded bar section	162	255	11		
Mg–Zn–Zr	Zn 1.0 Zr 0.6	Sheet	178	263	10	154	55–70
		Extruded bar and sections	208	293	13	177	60–75
		Extruded tube	193	278	7	—	60–75
	Zn 3.0 Zr 0.6	Sheet	185	270	8	154	60–70
		Forgings	224	309	8	193	60–80
		Extruded bar and sections (1 in. diam.)	239	309	18	213	65–75
	Zn 5.5 Zr 0.6	Bars and sections Heat treated	270	340	10	255	60–80
Mg–Zn–Cu–Mn	Zn 6.5 Cu 1.3 Mn 0.8	Bars and sections Heat treated	340	360	6	—	—
Mg–Th–Zn–Zr** (Creep resistant)	Th 0.8 Zn 0.5 Zr 0.6	Extruded bar and sections 5111	147	263	18	—	50–70
		Forgings 5111	147	232	13	—	50–70
Mg–Th–Mn** (Creep resistant)	Th 2.0 Mn 0.75	Sheet	165	247	9	179	—
	Th 3.0 Mn 1.2	Extruded bar and sections	227	287	8	185	—

* It is usual to add 0.2–0.4% Mn to alloys containing aluminium to improve corrosion resistance.
** Thorium-containing alloys are being replaced by alternative Mg alloys.

Table 22.19   MAGNESIUM AND MAGNESIUM ALLOYS (CAST) TYPICAL MECHANICAL PROPERTIES AT ROOM TEMPERATURE

Material	Nominal* composition %		Condition	Tension			Compression	
				Proof stress 0.2% MPa	UTS MPa	Elong. %	Proof stress 0.2% MPa	Brinell hardness[†] VPN 30 kg
Mg–Zr	Zr	0.6	AC	51	185	2.0	54	40–50
Mg–Al–Zn	Al	6.0	AC	97	199	5	97	50
	Zn	3.0	TB	97	275	10	97	55
			TF	131	275	5	131	73
	Al	8.0	AC	86	158	4	86	50–60
	Zn	0.4	TB	82	247	11	82	50–60
	Al	9.5	AC	93	154	2	93	55–65
	Zn	0.4	TB	90	232	6	90	55–65
			TF	127	239	2	124	75–85
			Die cast	111	216	3	108	60–70
	Al	9.0	AC	97	165	2	97	65
	Zn	2.0	TB	97	275	8	97	63
			TF	145	275	2	145	84
Mg–Zn–Zr	Zn	4.5	TE	161	263	6	162	65–75
	Zr	0.7						
Mg–Zn–RE–Zr	Zn	4.0	TE	150	216	5	139	55–75
	RE	1.2						
	Zr	0.7						
	Zn	6.0	TF§	190	295	7	190	70–80
	RE	2.5						
	Zr	0.7						
Mg–RE–Zn–Zr (Creep resistant to 250°C)	RE	2.7	TE	95	162	4.5	93	50–60
	Zn	2.2						
	Zr	0.7						
Mg–Th–Zn–Zr** (Creep resistant to 350°C)	Th	3.0	TE	93	216	7	93	50–60
	Zn	2.2						
	Z	r 0.7						
Mg–Zn–Th–Zr**	Zn	5.5	TE	167	270	8	162	65–75
	Th	1.8						
	Zr	0.7						
Mg–Th–Zr**	Th	3.0	TF	93	208	5	93	50–60
	Zr	0.7						
Mg–Ag–RE‡–Zr	Ag	2.5	TF	187	247	5	178	65–80
	RE	2.0‡		204	260	3	193	65–80
	Zr	0.6						
	Ag	2.5	TF	200	260	4	195	65–80
	RE	2.0‡						
	Zr	0.6						
Mg–RE(D)–Ag–Zr–Cu	RE(D)	2.2	TF	195	261	4	—	75–90
	Ag	1.5						
	Zr	0.6						
	Cu	0.07						
Mg–Ag–Th–RE‡–Zr**	Ag	2.5	TF	210	270	4	200	65–80
	RE	1.0‡						
	Th	1.0						
Mg–Y–RE(Δ)–Zr	Y	4.0	TF	185	265	7	—	75–90
	Zr	0.7						
	RE(Δ)	3.4						
	Zr	0.6						
	Y	5.1	TF	205	280	4	—	75–90
	RE(Δ)	3.0						
	Zr	0.6						
Mg–Zn–Cu–Mn	Zn	6.0	TF	158	242	4.5	—	55–65
	Cu	2.7						
	Mn	0.5						

* It is usual to add 0.2–0.4% Mn to alloys containing aluminium to improve corrosion resistance. RE = Cerium mischmetal containing approx. 50% cerium. RE(Δ) = Neodymium plus Heavy Rare Earth metals.
 RE(D) = Neodymium enriched mischmetal.
[†] Brinell tests with 500 kg on 10 mm ball for 30 s.
[‡] Fractionated rare earth metals: MSR-A contains 1.7%; MSR-B contains 2.5%.
§ Solution heat treated in an atmosphere of hydrogen.
AC = Sand cast. TE = Precipitation heat treated.
TB = Solution heat treated. TF = Fully heat treated.
** Thorium-containing alloys are being replaced by alternative Mg alloys.

Table 22.20   MAGNESIUM AND MAGNESIUM ALLOYS (EXCLUDING HIGH TEMPERATURE ALLOYS FOR WHICH SEE TABLE 22.21)—TYPICAL TENSILE PROPERTIES AT ELEVATED TEMPERATURES

				'Short-time' tension[†]			
Material	*Nominal composition* %*	*Form and condition*	*Test temp. °C*	*Young's modulus GPa*	*0.2% proof stress MPa*	*UTS MPa*	*Elong. %*
Mg	Mg 99.95	Forged	20	45	—	170	5
			100	—	—	128	8
			150	—	—	93	16
			200	—	—	54	43
Mg–Al–Zn	Al 8.0	Sand cast	20	45	86	158	4
	Zn 0.4		100	34	76	154	5
	(A8)		150	32	65	145	11
			200	25	62	100	20
			250	—	—	75	27
		Sand cast	20	45	82	247	11
		and	100	34	73	202	16
		solution	150	33	65	154	21
		treated	200	28	62	116	25
			250	—	—	85	21
	(AZ855)	Forged	20	45	221	309	8
			150	—	153	216	25
			200	—	102	154	28
	Al 9.5	Sand cast	20	45	93	154	2
	Zn 0.4		100	—	—	131	2
	(AZ91)		150	—	—	122	6
			200	—	—	108	25
			250	—	—	77	34
		Sand cast	20	45	90	232	6
		and	100	—	—	222	12
		solution	150	—	—	196	16
		treated	200	—	—	139	20
		Sand cast	20	45	127	239	2
		and	100	40	91	232	6
		fully heat	150	37	77	185	25
		treated	200	28	62	133	34
			250	19	46	103	30
Mg–Zn–Zr	Zn 4.5	Sand cast	20	45	161	263	6
	Zr 0.7	and heat	100	34	124	185	14
	(Z5Z)	treated	150	28	102	145	20
			200	22	79	113	23
			250	19	57	85	20
	Zn 3.0	Extruded	20	45	255	309	18
	Zr 0.6		100	40	162	182	33
	(ZW3)		200	22	46	127	56
			250	12	11	100	71
		Sheet	20	45	195	270	10
			100	40	120	165	33
			150	33	74	116	42
			200	—	—	76	51
			250	—	—	49	59
Mg–Zn–RE–Zr	Zn 4.0	Sand cast	20	45	150	216	4
	Re 1.2	Sand cast	20	41	134	195	6
	Zr 0.7	treated	150	40	120	167	19
	(RZ5)		200	38	99	131	29
			250	33	74	99	35

(continued)

Table 22.20   MAGNESIUM AND MAGNESIUM ALLOYS (EXCLUDING HIGH TEMPERATURE ALLOYS FOR WHICH SEE TABLE 22.21)—TYPICAL TENSILE PROPERTIES AT ELEVATED TEMPERATURES—*continued*

				'Short-time' tension[†]			
Material	*Nominal composition** %	*Form and condition*	*Test temp.* °C	*Young's modulus* GPa	*0.2% proof stress* MPa	*UTS* MPa	*Elong.* %
Mg–Zn–Th–Zr**	Zn 5.5	Sand cast	20	45	161	270	9
	Th 1.8	and heat	100	34	134	224	22
	Zr 0.7	treated	150	31	110	178	26
	(TZ6)		200	28	82	130	26
			250	26	52	91	25
Mg–Ag–	Ag 2.5	Sand cast	20	45	201	259	4
RE–Zr	RE(D)2.0	and fully	100	41	185	232	12
(D)	Zr 0.6	heat	150	40	171	210	16
	(QE22)	treated	200	38	154	185	20
			250	34	102	142	27
			300	31	68	88	59
Mg–RE(D)–	RE(D) 2.2	Sand cast	20	45	195	261	4
Ag–Zr–Cu	Ag 1.5	and fully	100	43	189	230	10
	Zr 0.6	heat	150	42	180	211	16
	Cu 0.07	treated	200	41	170	191	16
	(EQ21)		250	39	152	169	15
			300	35	117	132	10
Mg–Ag–RE(D)**	Ag 2.6	Sand cast	20	45	210	270	4
Th–Zr[‡]	RE(D) 1.0	and fully	100	41	199	242	17
	Th 1.0	heat	150	40	190	224	20
	Zr 0.6	treated	200	38	183	205	18
	(OH21)		250	37	167	185	19
			300	33	120	131	20
Mg–Y–RE(Δ)–Zr	Y 4.0	Sand cast	20	45	185	265	7
	RE(Δ) 3.4	and fully	150	42	175	250	6
	Zr 0.6	heat	200	39	170	245	11
	(WE43)	treated	250	37	160	220	18
			300	35	120	160	40
	Y 5.1	Sand cast	20	45	205	280	4
	RE(Δ) 3.0	and fully	100	43	197	260	4.5
	Zr 0.6	heat	150	42	195	255	5
	(WE54)	treated	200	41	183	241	6.5
			250	39	175	230	9
			300	36	117	184	14.5
Mg–Zn–Cu–Mn	Zn 6.0	Sand cast	20	45	158	242	4.5
	Cu 2.7	and fully	100	—	141	215	9
	Mn 0.5	heat	150	—	134	179	14
	(Zc63)	treated	200	—	118	142	11
	Zn 6.5	Extruded	20	45	325	350	6
	Cu 1.3	and fully	100	40	206	259	16
	Mn 0.8	heat	200	32	115	163	14
	(Zc71)	treated					

* It is usual to add 0.2–0.4% Mn to alloys containing aluminium to improve corrosion resistance.
[†] In accordance with BS1094: 1943; 1 h at temperature and strain rate 0.1–0.25 in^{-1} min^{-1}.
[‡] Tested according to BS4A4. RE = Cerium mischmetal containing approx. 50% Ce. RE(D) = Neodymium enriched mischmetal. RE(Δ) = Neodymium plus Heavy Rare Earth metals.
** Thorium-containing alloys are being replaced by alternative Mg alloys.

Table 22.21 HIGH TEMPERATURE MAGNESIUM ALLOYS—TENSILE PROPERTIES AT ELEVATED TEMPERATURE

Material	Nominal composition* %		Form and condition	Test temp. °C	*'Short-time' tension*[†]			
					Young's modulus GPa	0.2% proof stress MPa	UTS MPa	Elong. %
Mg–RE–Zn	RE	2.7	Sand cast	20	45	93	162	4.5
	Zn	2.2	and heat	100	40	79	150	11
	Zr	0.7	treated	150	38	76	139	19
	(ZRE1)			200	36	74	125	26
				250	33	65	107	35
				300	28	48	85	51
				350	21	26	56	90
Mg–Th–Zr**	Th	3.0	Sand cast	20	45	93	208	4
	Zr	0.7	and fully	100	40	88	188	10
	(HK31)		heat	150	38	86	174	13
	(MTZ)		treated	200	38	85	162	17
				250	36	83	150	20
				300	34	73	136	22
				350	29	56	103	23
Mg–Th–Zn–Zr**	Th	3.0	Sand cast	20	45	93	216	9
	Zn	2.2	and heat	100	36	88	159	23
	Zr	0.7	treated	150	34	79	131	27
	(ZT1)			200	33	65	108	33
				250	33	56	90	38
				300	31	49	76	41
				350	28	45	63	34
	Th	0.8	Sheet	20	45	181	266	10
	Zn	0.5		100	41	179	224	10
	Zr	0.6		150	41	176	201	11
	(ZTY)			200	40	165	171	15
				250	40	124	134	20
				300	34	73	96	27
				350	29	17	56	38
Mg–Ag–RE(D)–Zr	Ag	2.5	Sand cast					
	RE(D)	2.0	and fully					
	Zr	0.6	heat					
	(QE22)		treated					
Mg–RE(D)–Ag–Zr–Cu	RE(D)	2.2	Sand cast					
	Ag	1.5	and fully					
	Zr	0.6	heat					
	Cu	0.07	treated					
	(EQ21)							
Mg–Ag–RE(D)–Th–Zr**	Ag	2.5	Sand cast	High strength cast alloys with good elevated temperature properties—for which *see* Table 22.20				
	RE(D)	1.0	and fully					
	Th	1.0	heat					
	Zr	0.6	treated					
	(QH21)							
Mg–Y–RE(Δ)–Zr	Y	4.0	Sand cast					
	RE(Δ)	3.4	and fully					
	(WE43)		treated					
	Y	5.1	Sand cast					
	RE(Δ)	3.0	and fully					
	Zr	0.6	heat					
	(WE54)		treated					

* It is usual to add 0.2–0.4% Mn to alloys containing aluminium to improve corrosion resistance.
[†] In accordance with BS 1094: 1943; 1 h at temperature; strain rate 0.1–0.25 in^{-1} min^{-1}.
RE = Cerium mischmetal containing approx. 50% Ce. RE(D) = neodymium-enriched mischmetal.
RE(Δ) = Neodymium plus Heavy Rare Earths.
** Thorium containing alloys are being replaced by alternative Mg alloys.

Table 22.22   HIGH-TEMPERATURE MAGNESIUM ALLOYS—LONG-TERM CREEP RESISTANCE

Material	Nominal composition %		Form and Condition	Temp. °C	Time† h	Stress to produce specified creep strains %				
						0.05 MPA	0.1 MPa	0.2 MPa	0.5 MPa	1.0 MPa
Mg–RE–Zn–Zr	RE	2.7	Sand cast	200	100	52	66	71	—	—
	Zn	2.2	and heat		500	41	54	65	—	—
	Zr	0.7	treated		1 000	36	47	58	—	—
	(ZRE1)			250	100	23	28	32	36	—
					500	11	19	24	30	34
					1 000	—	14	20	26	30
				315	100	5.6	7.4	8	—	—
					500	—	5.2	6.5	—	—
					1 000	—	4.3	5.6	—	—
	Zn	4.0	Sand cast	100	100	—	97	111	117	—
	RE	1.2	and heat		500	—	—	106	117	—
	Zr	0.7	treated		1 000	—	—	103	116	—
	(RZ5)			150	100	77	86	97	101	107
					500	—	75	88	96	100
					1 000	—	70	83	91	97
				200	100	29	43	52	67	73
					500	22	28	37	52	64
					1 000	20	23	31	43	53
				250	100	6.2	12	19	32	39
					500	4.3	6.2	8.6	15	19
					1 000	3.9	5.4	6.9	12	15
Mg–Th–Zr**	Th	3.0	Sand cast	200	100	31*	45*	63*	97*	111*
	Zr	0.7	and fully		1 100	—	—	62*	100*	—
	(HK31)		heat	260	100	—	28*	43*	65*	—
	(MTZ)		treated		1 000	—	—	29*	45*	—
				315	100	9.3*	14*	19*	27*	32*
Mg–Th–Zn–Zr**	Th	0.8	Sheet	250	100	Stress of 46 MPa (3 tonf in⁻²)				
	Zn	0.5				produced 0.03% creep strain				
	Zr	0.6				Stress of 46 MPa (3 tonf in⁻²)				
	(ZTY)					produced 0.03% creep strain				
Mg–Th–Zn–Zr*	Th	3.0	Sand cast	250	100	42	50	56	63	66
	Zn	2.2	and heat		500	35	43	51	58	63
	Zr	0.7	treated		1 000	31	39	48	56	61
	(ZT1)			300	100	23	28	35	46	52
					500	19	21	25	36	41
					1 000	17	19	21	32	36
				325	100	14	19	24	29	36
					500	12	13	16	21	25
					1 000	10	12	13	15	20
				350	100	10	12	18	21	23
					500	—	9	10	12	14
					1 000	—	8	8	9	10
				375	100	—	8	11	12	13
					500	—	—	—	8	9
					1 000	—	—	—	—	8
	Zn	5.5	Sand cast	150	100	51	66	82	96	102
	Th	1.8	and heat		500	36	56	69	85	94
	Zr	0.7	treated		1 000	26	51	63	80	90
	(TZ6)			200	100	26	32	45	56	62
					500	15	22	26	40	49
					1 000	11	17	20	31	40

(continued)

Table 22.22 HIGH-TEMPERATURE MAGNESIUM ALLOYS—LONG-TERM CREEP RESISTANCE—*continued*

Material	Nominal composition %		Form and Condition	Temp. °C	Time[†] h	Stress to produce specified creep strains %				
						0.05 MPA	0.1 MPa	0.2 MPa	0.5 MPa	1.0 MPa
Mg–Ag– RE(D)–Zr	Ag	2.5	Sand cast and fully heat treated	200	100	55	74	88	—	—
	RE(D)	2.0			500	—	54	65	82	89
	Zr	0.6			1 000	—	46	56	73	79
	(QE22)			250	100	18	26	33	—	—
					500	—	15	22	28	31
					1 000	—	10	16	22	26
Mg–RE(D)–Ag– Zr–Cu	RE(D)	2.2	Sand cast and fully heat treated	200	100	—	78	95	116	—
	Ag	1.5			500	—	57	71	88	—
	Zr	0.6			1 000	—	48	62	76	—
	Cu	0.07								
	(EQ21)			250	100	—	29	36	42	—
					500	—	18	22	30	—
					1 000	—	14	19	24	—
Mg–Ag–RE(D)– Th–Zr**	Ag	2.5	Sand cast and fully heat treated	250	100	22	32	39	—	—
	RE(D)	1.0			500	—	20	26	32	36
	Th	1.0			1 000	—	—	21	26	30
	Zr	0.6								
	(QH21)									
Mg–Y–RE(Δ)–Zr	Y	4.0	Sand cast and fully heat treated	200	100	—	148	161	173	—
	RE(Δ)	3.4			500	—	—	115	148	—
	Zr	0.6			1 000	—	—	96	139	—
	(WE43)			250	100	—	44	61	—	—
					500	—	—	46	—	—
					1 000	—	—	39	—	—
	Y	5.1	Sand cast and fully heat treated	200	100	—	160	165	—	—
	RE(Δ)	3.0			500	—	120	140	—	—
	Zr	0.6			1 000	—	120	132	—	—
	(WE54)			250	100	—	47	61	81	—
					500	—	43	40	58	—
					1 000	—	16	32	48	—
Mg–Zn–Cu–Mn	Zn	6.0	Sand cast and fully heat treated	150	100	—	94	99	104	—
	Cu	2.7			500	—	82	92	98	—
	Mn	0.5			1 000	—	74	89	95	—
	(ZC63)			200	100	—	60	63	67	—
					500	—	51	55	61	—
					1 000	—	42	49	55	—

* Total strains.
[†] 4–6 h heating to test temperature followed by 16 h soaking at test temperature.
RE = Cerium mischmetal containing approx. 50% Ce.
RE(D) = Neodymium-enriched mischmetal.
RE(Δ) = Neodymium plus Heavy Rare Earth metals.
** Thorium-containing alloys are being replaced by alternative Mg alloys.

Table 22.23　HIGH-TEMPERATURE MAGNESIUM ALLOYS—SHORT-TERM CREEP RESISTANCE

| Material | Nominal composition % | Form and Condition | Temp. °C | Time† s | Stress to produce specified creep strains % | | | | | Stress to fracture MPa |
					0.05 MPA	0.1 MPa	0.2 MPa	0.5 MPa	1.0 MPa	
Mg–RE–Zn–Zr	Re 2.7 Zn 2.2 Zr 0.7 (ZRE1)	Sand cast and heat treated	200	30	—	—	98	118	130	136
				60	—	—	97	117	128	134
				600	—	—	96	116	125	129
			250	30	76	84	92	111	123	130
				60	74	83	91	110	120	129
				600	73	82	89	108	114	125
			315	30	52	59	73	80	85	90
				60	51	58	69	76	83	88
				600	42	49	56	62	68	73
	Zn 4.0 RE 1.2 Zr 0.7 (RZ5)	Sand cast and heat treated	200	30	100	107	116	127	—	136
				60	99	105	114	124	—	134
				600	86	99	103	114	—	125
			250	30	86	90	94	99	—	116
				60	83	88	91	96	—	113
				600	71	76	81	86	—	93
			315	30	62	69	76	79	83	86
				60	59	66	73	76	79	82
				600	48	53	59	64	67	69
Mg–Th–Zr*	Th 3.0 Zr 0.7 (HK31) (MTZ)	Sand cast and fully heat treated	250	30	96	103	119	138	—	145
				60	95	103	118	137	—	145
				600	94	102	117	137	—	144
			315	30	80	88	103	117	—	128
				60	78	86	102	116	—	127
				600	74	82	96	107	—	120
Mg–Th–Zn–Zr*	Th 0.8 Zn 0.5 Zr 0.6 (ZTY)	Sheet	250	30	—	—	110	159	163	165
				60	—	—	95	157	160	162
				600	—	—	—	145	149	151
			350	30	20	32	48	80	93	102
				60	18	28	40	67	82	98
				600	—	15	20	31	42	66
Mg–Th–Zn–Zr*	Th 3.0 Zn 2.2 Zr 0.7 (ZT1)	Sand cast and heat treated	200	30	—	—	—	100	118	125
				60	—	—	—	96	114	123
				600	—	—	—	85	103	114
			250	30	58	65	71	84	102	111
				60	57	64	69	81	99	107
				600	56	63	68	74	86	98
			315	30	55	60	64	73	76	82
				60	53	59	63	72	76	80
				600	50	59	61	71	74	77
	Zn 5.5 Th 1.8 Zr 0.7 (TZ6)	Sand cast and heat treated	200	30	96	113	120	128	137	144
				60	93	109	117	124	133	137
				600	63	90	102	110	114	119
			250	30	70	77	85	96	99	107
				60	65	74	80	90	94	99
				600	56	60	66	74	77	82
			315	30	54	59	64	70	74	76
				60	52	57	62	66	70	73
				600	44	49	53	56	58	59

† 1 h heating to test temperature followed by 1 h soaking at test temperature.
RE = cerium mischmetal containing approx. 50% Ce.
* Thorium-containing alloys are being replaced by alternative Mg alloys.

Table 22.24 MAGNESIUM AND MAGNESIUM ALLOYS—FATIGUE AND IMPACT STRENGTHS

Material	Nominal* composition %	Condition	‡ State	Temp. °C	Fatigue strength† at specified cycles						Impact strength§ for single blow fracture		
					10^5 MPa	5×10^5 MPa	10^6 MPa	5×10^6 MPa	10^7 MPa	5×10^7 MPa	Test temp. °C	Unnotched J	Notched J
Mg–Mn	Mn 1.5 (AM503)	Extruded	U	20	107	90	88	86	85	83	20	12–14	4–4.5
			N		76	90	54	51	50	48			
Mg–Al–Zn	Al 6.0 Zn 1.0 (AZM)	Extruded	U	20	161	139	133	125	124	120	20	34–43	7–9.5
			N		127	110	103	97	94	91			
	Al 8 Zn 0.4 (A8)	Sand cast	U	20	108	93	90	88	88	86	20	3–5	1.5–2
			Z		107	80	73	66	65	63			
		Sand cast and solution treated	U	20	124	102	97	91	90	90	20	18–27	4.5–7
			Z		108	86	82	74	73	69			
			U	150	93	69	66	59	57	57			
			U	200	71	52	48	38	36	31			
	Al 9.5 Zn 0.4 (AZ91)	Sand cast	U	20	114	91	89	88	86	85	–196	1.5	1–1.5
			Z		110	83	74	68	66	63	20	1.5–2.0	
		Sand cast and solution treated	U	20	124	93	93	93	93	—	20	7–9.5	3–4
			N		103	82	80	79	79	77			
		Sand cast and fully heat treated	U	20	117	90	80	79	77	76	20	3–4	1–1.5
			N		93	66	66	65	65	65			
Mg–Zn–Zr	Zn 3.0 Zr 0.6 (ZW3)	Extruded	U	20	151	137	134	128	127	124	20	23–31	9.5–12
			N		124	99	93	91	90	88			
	Zn 4.5 Zr 0.7 (ZSZ)	Sand cast and heat treated	U	20	111	86	85	82	80	77	20	7–12	3–4
			N		90	86	85	82	80	77	–196	0.8	
Mg–Zn–RE–Zr	Zn 4.0 RE 1.2 Zr 0.7 (RZ5)	Sand cast and heat treated	U	20	124	99	97	97	96	94	20	4–5.5	1–2
			N		108	93	91	88	86	83	–196	0.7	
			U	150	97	85	80	73	69	65			
			U	200	93	74	69	62	59	54			

(continued)

Table 22.24   MAGNESIUM AND MAGNESIUM ALLOYS—FATIGUE AND IMPACT STRENGTHS—*continued*

Material	Nominal* composition %	Condition	State ‡	Temp. °C	Fatigue strength† at specified cycles						Impact strength§ for single blow fracture		
					10^5 MPa	5×10^5 MPa	10^6 MPa	5×10^6 MPa	10^7 MPa	5×10^7 MPa	Test temp. °C	Unnotched J	Notched J
	Zn 2.2, RE 2.7, Zr 0.7 (ZRE1)	Sand cast and heat treated	U	20	100	82	80	79	77	74	20	6–7.5	1–2
			N		77	59	54	52	52	51			
			U	150	69	60	59	57	57	57			
			U	200	68	59	56	52	51	51			
			U	250	59	48	45	43	43	42	−196	0.5	
			U	300	49	39	37	37	36	34			
	Zn 6, RE 2.5, Zr 0.6 (ZE63)	Sand cast and fully heat treated**	U	20	144	131	127	121	119	117	20	12.9–17.6	2.3–2.7
			N		99	83	79	73	72	71			
Mg–Ag–RE(D)–Zr	Ag 2.5, RE(D) 2.5, Zr 0.6 (MSR–B)	Sand cast and fully heat treated	U	20	119	103	103	103	102	100			
			N		77	65	63	62	62	62			
			N	200	—	—	—	90	88	86			
			U	250	—	77	68	57	54	51			
Mg–Ag–RE(D)Th–Zr**	Ag 2.5, Re(D) 1.0, Th 1.0, Zr 0.6 (QH21)	Sand cast and fully heat treated	U	20	135	114	111	109	108	108			
			N		86	72	69	64	63	62			
			U	250	108	76	65	56	55	52			
Mg–Zn–Th–Zr**	Zn 5.5, Th 1.8, Zr 0.7 (TZ6)	Sand cast and heat treated	U	20	120	86	85	83	83	82	20	8–11	1.5–3
			N		100	86	80	77	76	76	−196	0.5	
Mg–Th–Zr**	Th 3.0, Zr 0.7 (MTZ)	Sand cast and fully heat treated	U	20	—	74	68	65	63	62			
			N		—	48	40	36	34	32			
			U	200	—	74	68	60	59	58			
			U	250	80	63	59	54	52	51			

Alloy system	Composition	Condition	Test	Temp. °C							Impact temp. °C	Impact
Mg–Th–Zn–Zr***	Th 0.7, Zn 0.5, Zr 0.6 (ZTY)	Extruded	U	20	100	86	83	79	76	74		1.5–3
			N	200	73	52	51	49	48	46		
			U	250	—	74	68	60	59	57		
	Th 3.0, Zn 2.2, Zr 0.7 (ZT1)	Sand cast and heat treated	U	20	80	63	59	54	52	51		
			N	200	97	82	79	74	71	68	20	7–8
			U	250	76	59	56	51	49	48		
			U	325	71	60	59	54	52	51		
					66	51	46	43	42	39		
					—	—	43	37	34	29		
Mg–RE(D)–As–Zr–Cu	RE(D) 2.2, Ag 1.5, Zr 0.6, Cu 0.07	Sand cast and fully heat treated	U	20	103	94	93	92	91	90	−196	0.8
Mg–Y–RE(Δ)–YZr	Y 4.0, RE(Δ) 3.4, Zr 0.6	Sand cast and fully heat treated	U	20	114	101	98	94	93	91		
			U	150	107	97	94	87	85	83		
			U	250	107	81	74	65	64	62		
	Y 5.9, RE(Δ) 3.0, Zr 0.6	Sand cast and fully heat treated	U	20	113	104	102	100	99	97		
			U	200	118	96	90	84	83	82		
			U	250	115	84	78	67	66	65		
Mg–Zn–Cu–Mn	Zn 6.0, Cu 2.7, Mn 0.5	Sand cast and fully heat treated	U	20	—	—	100	94	92	90		
			N		—	—	62	57	56	55		

*It is usual to add 0.2–0.4% Mn to alloys containing aluminium to improve corrosion resistance.

** Solution heat treated in an atmosphere of hydrogen.

† Wöhler rotating beam tests at 2 960 c.p.m.

‡ U = Unnotched.

N = Notched. Semi-circular notch of 0.12 cm (0.047 in) radius. Stress concentration factor 1.8.

§ Hounsfield balanced impact test, notched bar values are equivalent to Izod values.

RE(D) = Neodymium enriched mischmetal.

*** Thorium-containing alloys are being replaced by alternative Mg alloys.

RE(Δ) = Neodymium plus Heavy Rare Earths.

Table 22.25   HEAT TREATMENT OF MAGNESIUM ALLOY CASTINGS

Heat treatment conditions for magnesium sand castings can be varied depending on the particular components and specific properties required. The following are examples of the conditions typically used for each alloy.

Material		Nominal* composition %		Condition	Time h	Temperature °C
Mg–Al–Zn	(AZ80)	Al	8.0	TB	12–24	400–420
		Zn	0.4			
	(AZ91)	Al	9.5	TB	16–24	400–420
		Zn	0.4			
	(AZ91)			TF	16–24	400–420
						Air cool
					8–16	180–210
Mg–Zn–Zr	(Z5Z)	Zn	4.5	TE	10–20	170–200
		Zr	0.7			
Mg–Zn–RE–Zr	(RZ5)	Zn	4.0	TE	2–4	320–340
		RE	1.2			Air cool
		Zr	0.7		10–20	170–200
	(ZRE1)	RE	2.7	TE	10–20	170–200
		Zn	2.2			
		Zr	0.7			
Mg–Th–Zr†	(HK31)	Th	3.0	TF	2–4	560–570
		Zr	0.7			Air cool
					10–20	195–205
Mg–Zn–Th–Zr†	(TZ6)	Zn	5.5	TE	2–4	320–340
		Th	1.8			Air cool
		Zr	0.7		10–20	170–200
	(ZT1)	Th	3.0	TE	10–20	310–320
		Zn	2.2			
		Zr	0.7			
Mg–Ag–RE(D)Zr	(QE22)	Ag	2.5	TF	4–12	520–530
		RE(D)	2.0		Water/Oil Quench	
		Zr	0.6		8–16	195–205
Mg–RE(D)–Ag–Zr–Cu	(EQ21)	RE(D)	2.2	TF	4–12	515–525
		Ag	1.5		Water/Oil Quench	
		Zr	0.6		12–16	195–205
		Cu	0.07			
Mg–Ag–RE(D)–Th–Zr	(QH21)†	Ag	2.5	TF	4–12	520–530
		RE(D)	1.0		Water/Oil Quench	
		Th	1.0		12–20	195–205
		Zr	0.6			
Mg–Y–RE(Δ)–Zr	(WE43)	Y	4.0	TF	4–12	520–530
		RE(Δ)	3.4		Water/Oil Quench	
		Zr	0.6		12–20	245–255
	(WE54)	Y	5.1	TF	4.12	520–530
		RE(Δ)	3.0		Water/Oil Quench/Air Cool	
		Zr	0.6		12–20	245–255
Mg–Zn–Cu–Mn	(ZC63)	Zn	6.0	TF	4–12	435–445
		Cu	2.7		Water Quench	
		Mr	0.5		16–24	180–200

* It is usual to add 0.2–0.4% Mn to alloys containing aluminium to improve corrosion resistance.

    RE = Cerium mischmetal containing approximately 50% cerium.   TB = Solution heat treated.

    RE(D) = Neodymium-enriched mischmetal.   TE = Precipitation heat treated.

    RE(Δ) = Neodymium plus Heavy Rare Earth metals.   TF = Fully heat treated.

† Thorium-containing alloys are being replaced by alternative Mg alloys.

Note: Above 350°C, furnace atmospheres must be inhibited to prevent oxidation of magnesium alloys. This can be achieved either by:

    (i) adding $\frac{1}{2}$–1% SO_1 gas to the furnace atmosphere; or

    (ii) carrying out the heat treatment in an atmosphere of 100% dry CO_2.

Mechanical properties at subnormal temperatures

At temperatures down to −200°C tensile properties have approximately linear temperature coefficients: proof stress and UTS increase by 0.1–0.2% of the RT value per °C fall in temperature, and elongation falls at the same rate: modulus of elasticity rises approximately 19 MPa (2 800 lbf in^{-2}) per °C over the range 0° to −100°C. No brittle-ductile transitions have been found.

Tests at −70°C have suggested that the magnesium-zinc-zirconium alloys show the best retention of ductility and notched impact resistance at this temperature.

22.5 Mechanical properties of nickel and nickel alloys

Table 22.26 WROUGHT NICKEL AND HIGH NICKEL ALLOYS, STANDARD SPECIFICATIONS AND DESIGNATIONS

Note for 8th edition: In the case of the UK, France, and Germany, this table refers to designations and standards that may have been superceded by European (EN) and international (ISO) standards. Consult Chapter 1 for references to current standards.

Alloy	Nominal composition %		France AFNOR	Germany DIN	Werkstoff Nr.	UK BS and DTD	USA ASTM	ASME	AMS	AECMA	UNS
Nickel 200 201	Ni	99.6	—	17740; Ni 99.2 17750–4	2.4066/2.4060 2.4068/2.4061	3072–76:NA11 3072–76:NA12	B160–163 B725, B730	SB160–163	5533	—	NO2200
Nickel 205	—	99.6 min.	—	—	2.4061	—	F1–3, F9	5555	—	—	NO2205
Monel 400	Cu Fe Mn	30 1.5 1.0	NU 30	17743: Ni Cu 30 Fe 17750 17751 17752 17753 17754	2.4360 2.4361	3072–76:NA13	B127 B163–165 B564	SB127 SB163–165 SB564	4544 4574 4675 4730 4731 7233	—	NO4400
Monel 450	Ni Fe Cu	31.0 0.7 Rem.	—	—	—	—	B111, B122 B151, B171 B359, B395 B432, B466–7 B543, B552	SB111, SB117 SB359, SB395 SN466, SB4467 SB543	—	—	C71500
Monel K-500	Cu Al Ti	29 2.8 0.5	—	17743: Ni Cu 30 Al 17752 17754	2.4375	3072–76:NA 18	—	Boiler code sect. VIII	4676	—	NO5500
Inconel 600	Cr Fe	16 6	NC 15Fe	17742: Ni Cr 15 Fe 17750 17751 17752 17753 17754	2.4816	3072–76: NA 14	B163 B166–168 B516 B517 B564 B951	SB163 SB166–168 SB564	5540 5580 5665 5687 7232	—	NO6600
Inconel 601	Ni Cr Al Fe	60.5 23.0 1.4 15.1	—	17742 17750 17751 17752	2.4851	—	—	Boiler code VIII	5715 5870	—	NO6601

(continued)

Table 22.26 WROUGHT NICKEL AND HIGH NICKEL ALLOYS, STANDARD SPECIFICATIONS AND DESIGNATIONS—*continued*

Alloy	Nominal composition %		France AFNOR	Germany DIN	Werkstoff Nr.	UK BS and DTD	USA ASTM	ASME	AMS	AECMA	UNS
Inconel 617	Ni Cr Co Mo Al C	55.7 21.5 12.5 9.0 1.2 0.1	—	—	—	—	—	Boiler codes I and VIII	—	—	NO6617
Inconel 625	Cr Mo Nb	21.5 9.0 3.6	—	17744 17750 17751 17752 17754	2.4856	3072 3074 3076 (NA21)	B443 B444 B446 B565 B704 B705 B751	SB443 SB444 SB446 SB564	5666 5599 5837 5581	—	NO6625
Inconel 718	Ni Cr Fe Nb Mo Ti Al	52.5 19.0 18.8 5.2 3.1 0.9 0.5	NC 19 Fe Nb	—	2.4668	—	B637, B670	Boiler codes I and III	—	Pr En 2404 2405, 2407 2408, 2952 2961, 3219	NO7718
Inconel X-750	Cr Fe Ti Al Nb	15 7 2.5 0.6 0.8	NC 15Fe-T	—	2.4669	HR 505	B637	SA637	5542 5582/3 5598 5667/8/9 5670/1 568/9 5747, 7246	—	NO7750
Inconel MA 754	Ni Cr Fe Y_2O_3	78.0 20.0 1.0 0.6	—	—	—	—	—	—	—	—	NO7754
INCO 330	Ni Fe Cr Si	35.5 44.8 18.5 1.2	—	—	—	—	B511, B512 B535, B536 B546, B710	SB511, SB536 SB710 Boiler codes VIII, IX	5592 5716	—	NO8330

Alloy	Composition		German		3072	B	SB		Pr EN	N
INCO 020	Ni 35.0 Fe 38.4 Cr 20.0 Cu 3.5 Mo 2.5 Nb 0.6	—	—	—		B464-4 B468 B472-4 B751	SB462, SB464 SB468 Boiler codes III, VIII, IX	—	—	NO08020
INCO G-3	Ni 49.0 Cr 22.5 Fe 19.5 Mo 7.0 Cu 2.0	—	17744 17750–17752	2.4619	—	B581-2 B619, B622 B626, B751	SB581-2 SB619, SB622 SB626 Boiler codes VIII, IX	—	—	NO6985
INCO C-276	Ni 59.0 Mo 16.0 Cr 15.5 Fe 5.5 W 4.0	—	17744 17750–17752	2.4819	—	B574-5 B619, B622 B626, B751	SB574-5 SB619, SB622 SB626 Boiler codes I, III, VIII, IX	—	—	N10276
INCO-HX	Ni 48.3 Cr 22.0 Fe 18.5 Mo 9.0 Co 1.5 W 0.6 C 0.1	NC 22 Fe D	—	2.4665	HR 6, HR 204	B435, B572 B619, B622 B626, B751	SB435, SB572 SB619, SB622 SB626 Boiler codes III, VIII, IX	—	Pr EN 2182–2185	NO6002
Incoloy 800 Incoloy 800 HT	Fe 46 Cr 21 Ti 0.4 Al 0.4	X10Ni Cr AlTi 3220	1.4876		3072–76:NA 15	B163 B407–409 B564 B514–5 B751	SB163 SB407–409 SB564	5766 5871	—	NO8800 NO8811
Incoloy 825	Fe 30 Cr 21 Mo 3 Cu 2 Ti 2	WFe32C20DU	17744 17750–2 17754	2.4858	3072–76: NA 16	B163 B423–425 B704–5 B751	SB163 SB423–425	—	—	NO8825
Incoloy DS	Fe 40 Cr 18 Si 2	—	—	—	3072–76: NA 17	—	—	—	—	—

(continued)

Table 22.26 WROUGHT NICKEL AND HIGH NICKEL ALLOYS, STANDARD SPECIFICATIONS AND DESIGNATIONS—*continued*

Alloy	Nominal composition %	France AFNOR	Germany DIN	Werkstoff Nr.	UK BS and DTD	USA ASTM	ASME	AMS	AECMA	UNS
Ni-Span C-902	Ni 42 Fe Rem. Cr 5.5 Ti 2.5 Al 0.5	—	—	—	3127	—	—	5210 5221 5223 5225	—	NO9902
Hastelloy B-2	Mo 28	—	—	—	—	B333– B335– B619 B622	SB333 SB335 SB619 SB662	—	—	N10665
Hastelloy C-4	Mo 16 Cr 16	—	—	—	—	B574 B575 B619 B662	SB574 SB575 SB619 SB662	—	—	NO6455
Hastelloy X	Mo 9 Cr 21 Fe 18	NC 22FeD	LW2.4665	—	HR6 HR204	B435 B572 B619 B622 B626	—	5536, 5587 5754, 5588	Ni-P93–HT	NO6002
Nimonic 75	Cr 20 Ti 0.4	NC 20T	17742: Ni Cr 20 Ti LW2.4630 17750–2	2.4951 2.4630 0	Hr5 HR203 HR403 HR504	—	—	—	Pr EN 2293–4 2302 2306–8 2402 2411	NO6075
Nimonic 80A	Cr 20 Ti 2.0 Al 1.5	NC 20TA	Ni Cr 20 Ti Al 17742 17754	2.463 1 2.495 2	3076: NA 20 HR1 HR201 HR401 HR601	B637	—	—	Pr EN 2188–91 2396 2397	NO7080
Nimonic 90	Cr 20 Co 17 Ti 2.4 Al 1.4	NCK 20TA	—	2.463 2	HR2 HR202 HR402 HR501 HR502–3 BS3975–NA19	—	—	5829	Pr EN 2295–99 2401–2 2412 2669–70	NO7090

Alloy	Composition		Designation						
Nimonic 105	Cr Co Mo Al Ti	15 20 5 5 1.2	MK 20CDA	—	2.463 4	HR3	—	Pr EN 2179–81	—
Nimonic 115	Cr Co Mo Al Ti	14 13 3 5 4	NK 20ATD	—	2.463 6	HR4	—	Pr EN 2196 2197	—
Nimonic 263	Cr Co Mo Ti Al	20 20 6 2 0.5	NCK 20D	—	2.465 0	HR10 HR206 HR404	5872	Pr EN 2199–2203 2418	NO7263
Nimonic 901	Fe Cr Mo Ti	35 13 6 3	Z8 NC DT 42	—	2.466 2	HR53	5660 5661	Pr EN 2176–8	NO9901
Nimonic PE16	Fe Cr Mo Ti Al	32 16 3 1.0 1.0	NW 11 AC	—	—	HR55 HR207	—	—	—
Nimonic PK33	Cr Co Mo Ti Al	18 14 7 2.25 2.1	NC 19 KDu/v	—	—	DTD5057	—	—	—
Astroloy	Ni Cr Co Mo Al Ti	54.8 15.0 17.0 5.3 4.0 3.5	—	—	—	—	—	—	—

(continued)

Table 22.26   WROUGHT NICKEL AND HIGH NICKEL ALLOYS, STANDARD SPECIFICATIONS AND DESIGNATIONS—*continued*

Alloy	Nominal composition %		France AFNOR	Germany DIN	Werkstoff Nr.	UK BS and DTD	USA ASTM	ASME	AMS	AECMA	UNS
Rene 41	Ni	55.4	—	—	—	—	—	—	—	—	N07041
	Cr	19.0									
	Co	11.0									
	Mo	11.0									
	Al	1.5									
	Ti	3.1									
Rene 95	Ni	61.5	—	—		—	—	—	—	—	—
	Cr	14.0									
	Co	8.0									
	Mo	3.5									
	Nb	3.5									
	Al	3.5									
	Ti	2.5									
Rene 100	Ni	61.8	—	—		—	—	—	—	—	—
	Cr	9.5									
	Co	15.0									
	Mo	3.0									
	Al	5.5									
	Ti	4.2									
	V	1.0									
Udimet 500	Ni	53.7	—	—		—	—	—	—	—	N07500
	Cr	18.0									
	Co	18.5									
	Mo	4.0									
	Al	2.9									
	Ti	2.9									
Udimet 700	Ni	55.5	—	—		—	—	—	—	—	—
	Cr	15.0									
	Co	17.0									
	Mo	5.0									
	Al	4.0									
	Ti	3.5									
Waspaloy	Ni	58.7	—	—		—	—	—	—	—	N07001
	Cr	19.5									
	Co	13.5									
	Mo	4.3									
	Al	1.3									
	Ti	3.0									

Table 22.27 WROUGHT NICKEL AND HIGH NICKEL ALLOYS, MECHANICAL PROPERTIES[1] AT ROOM TEMPERATURES

Alloy[3]	Nominal composition %		Condition	0.2% Proof stress MPa	UTS MPa	Elonga- tion %	Brinell hardness	Izod impact J
Nickel–pure	Ni	99.9	Annealed	60	310	40	85	160
Nickel–comc.	Ni	99.0	Annealed	150	400	40	100	160
			Hot rolled	200	500	40	120	160
			Cold drawn	480	660	25	190	—
Nickel 205	Ni	99.6 min.	Annealed	90	345	45	80	—
Monel[1] 400	Cu	31.5	Annealed	240	550	40	125	140
	Fe	1.5	Hot rolled	490	620	35	150	140
	Mn	1.0	Cold drawn	570	700	25	190	110
Monel 450	Ni	31.0	Annealed	165	385	46	90	—
	Fe	0.7						
	Cu	Rem.						
Monel K-500	Cu	29	Annealed and aged	790	1 100	30	290	—
	Al	2.8						
	Ti	0.5	Hot rolled and aged	880	1 110	25	310	—
			Cold drawn and aged	850	1 110	20	310	—
Cupro-nickel[4]	Cu	55	Annealed	200	400	45	—	—
			Cold drawn	450	590	20	—	—
Inconel[1] 600	Cr	15.5	Annealed	310	655	45	150	160
	Fe	8	Cold drawn	700	880	20	200	110
Inconel 601	Ni	60.5	Annealed	210–340	550–790	70–40	110–150	—
	Cr	23.0						
	Al	1.4						
	Fe	15.1						
Inconel 617	Ni	55.7	Annealed	350	760	58	173	—
	Cr	21.5						
	Co	12.5						
	Mo	9.0						
	Al	1.2						
	C	0.1						
Inconel 625	Cr	22	Annealed	520	930	45	186	—
	Nb	4	Solution treated	352	810	50	157	—
	Mo	9						
Inconel 718	Ni	52.5	Precipitation hardened	1 180	1 350	17	382	—
	Cr	19.0						
	Fe	18.8						
	Nb	5.2						
	Mo	3.1						
	Ti	0.9						
	Al	0.5						
Inconel X-750	Cr	15	Fully heat treated	900	1 240	20	382	—
	Fe	7						
	Ti	2.5						
	Al	0.6						
	Nb	0.8						
Inconel MA 754	Ni	78.0						
	Cr	20.0						
	Fe	1.0						
	Y_2O_3	0.6						
INCO 330	Ni	35.5	Annealed	270	590	47	150	—
	Fe	44.8						
	Cr	18.5						
	Si	1.2						
INCO 020	Ni	35.0	Annealed	310	620	40	183	—
	Fe	38.4						
	Cr	20.0						
	Cu	3.5						
	Mo	2.5						
	Nb	0.6						

(continued)

Table 22.27   WROUGHT NICKEL AND HIGH NICKEL ALLOYS, MECHANICAL PROPERTIES[1] AT ROOM
TEMPERATURES—*continued*

Alloy[3]	Nominal composition %		Condition	0.2% Proof stress MPa	UTS MPa	Elonga- tion %	Brinell hardness	Izod impact J
INCO G-3	Ni	49.0	Annealed	320	690	50	146	—
	Cr	22.5						
	Fe	19.5						
	Mo	7.0						
	Cu	2.0						
INCO C-276	Ni	59.0	Annealed	415	790	50	184	—
	Mo	16.0						
	Cr	15.5						
	Fe	5.5						
	W	4.0						
INCO-HX	Ni	48.3	Annealed	340	790	45	184	—
	Cr	22.0						
	Fe	18.5						
	Mo	9.0						
	Co	1.5						
	W	0.6						
	C	0.1						
Incoloy[1] 800	Fe	46	Annealed	310	590	45	180	160
	Cr	21	Cold drawn	700	880	20	250	110
	Ti	0.4						
	Al	0.4						
Incoloy 825	Fe	32	Annealed	340	650	40	150	160
	Cr	21						
	Mo	3						
	Cu	2						
	Ti	1						
Incoloy DS	Fe	40	Annealed	220	600	61	190	140
	Cr	18	Cold drawn	850	1 000	10	300	110
	Si	2						
Ni-Span[1] C-902	Fe	47	Fully heat treated	770	1 200	25	300	—
	Cr	5.5						
	Ti	2.5						
	Al	0.5						
Hastelloy B-2	Mo	28	Solution treated	526	955	53	235	—
Hastelloy C-4	Mo	16	Solution treated	416	768	52	184	—
	Cr	16						
Hastelloy C-276	Mo	16	Solution treated	360	790	60	200	160
	Cr	15						
	W	4						
	Fe	5						
	C	0.02						
Hastelloy X	Mo	9	Solution treated	350	800	45	175	75
	Cr	21						
	Fe	18						
Brightvay C	Cr	20	Annealed	330	760	40	160	—
	Si	1						
Nimonic[1] 75	Cr	20	Annealed	240	750	40	170	110
	Ti	0.4						
Nimonic 80A	Cr	20	Fully heat treated	780	1 220	30	370	70
	Ti	2.0						
	Al	1.5						
Nimonic 90	Cr	20	Fully heat treated	750	1 175	30	380	70
	Co	17						
	Ti	2.4						
	Al	1.4						

(continued)

Table 22.27 WROUGHT NICKEL AND HIGH NICKEL ALLOYS, MECHANICAL PROPERTIES[1] AT ROOM TEMPERATURES—*continued*

Alloy[3]	Nominal composition %		Condition	0.2% Proof stress MPa	UTS MPa	Elonga-tion %	Brinell hardness	Izod impact J
Nimonic 105	Cr Co Mo Al Ti	15 20 5 5 1.2	Fully heat treated	780	1 140	22	380	16
Nimonic 115	Cr Co Mo Al Ti	14 13 3 5 4	Fully heat treated	865	1 240	27	400	—
Nimonic 263	Cr Co Mo Ti Al	20 20 6 2 0.5	Fully heat treated	580	970	39	320	—
Nimonic 901	Fe Cr Mo Ti	35 13 6 3	Fully heat treated	900	1 220	15	—	—
Nimonic PE16	Fe Cr Mo Ti Al	32 16 3 1.0 1.0	Fully heat treated	460	850	30	280	—
Nimonic PK 33	Cr Co Mo Ti Al	18 14 7 2.2 2.1	Fully heat treated	790	1 170	30	—	—
Astroloy	Ni Cr Co Mo Al Ti	54.8 15.0 17.0 5.3 4.0 3.5	Precipitation hardened	1 050	1 415	16	—	—
Rene 41	Ni Cr Co Mo Al Ti	55.4 19.0 11.0 11.0 1.5 3.1	Precipitation hardened	1 060	1 420	14	—	—
Rene 95	Ni Cr Co Mo Nb Al Ti	61.5 14.0 8.0 3.5 3.5 3.5 2.5	Precipitation hardened	1 310	1 620	15	—	—
Udimet 500	Ni Cr Co Mo Al Ti	53.7 18.0 18.5 4.0 2.9 2.9	Precipitation hardened	840	1 310	32	—	—

(*continued*)

Table 22.27   WROUGHT NICKEL AND HIGH NICKEL ALLOYS, MECHANICAL PROPERTIES[1] AT ROOM
TEMPERATURES—*continued*

Alloy[3]	Nominal composition %		Condition	0.2% Proof stress MPa	UTS MPa	Elonga-tion %	Brinell hardness	Izod impact J
Udimet 700	Ni	55.5	Precipitation hardened	965	1 410	17	—	—
	Cr	15.0						
	Co	17.0						
	Mo	5.0						
	Al	4.0						
	Ti	3.5						
Waspaloy	Ni	58.5	Precipitation hardened	795	1 275	25	—	—
	Cr	19.5						
	Co	13.5						
	Mo	4.3						
	Al	1.3						
	Ti	3.0						

(1) Registered Trade Mark.
(2) All values in the tables in this section are representative and not to be considered as minima or maxima not to be used for
   specification purposes. Conversion figures have all been rounded off and are not to be taken as precise equivalents.
(3) Where trade marks apply to the name of an alloy there may be materials of similar composition available from other producers
   who may or may not use the same suffix along with their own trade names. The suffix alone, e.g. Alloy 800 is sometimes used
   as a descriptive term for the type of alloy but trade marks can be used only by the registered user of the mark.
(4) Other copper–nickel alloys, cupro-nickels, will be found listed under copper base alloys.
(5) Fully heat-treated usually implies a solution treatment followed by some form of precipitation hardening cycle. The precise
   heat-treatment and the associated properties may be varied to suit particular applications or according to the form of the alloy
   as forging, bar, sheet or welded fabrication.
(6) There are many nickel chromium alloys with and without iron used for electrical resistance heating and for general
   high-temperature applications in furnaces and heat-treatment plant. This alloy is typical.

Table 22.28   WROUGHT NICKEL AND HIGH NICKEL ALLOYS, SHORT-TIME HIGH-TEMPERATURE TENSILE
PROPERTIES

Alloy	Nominal composition %		Condition	Test temperature °C	0.2% Proof stress MPa	UTS MPa	Elonga-tion %
Nickel	Ni	99.0	Hot rolled	20	170	490	50
				200	150	540	50
				400	140	540	50
				600	110	250	60
				800	—	170	60
Nickel 205	Ni	99.6 min.	Annealed	20	90	345	45
Monel* 400	Cu	30	Hot rolled	20	230	560	45
	Fe	1.5		200	200	540	50
	Mn	1.0		400	220	460	52
				600	120	260	30
				800	77	120	54
Monel 450	Ni	31.0	Annealed	20	165	385	46
	Fe	0.7		200	120	300	55
	Cu	Rem.		400	100	290	42
				600	75	150	40
				800	28	58	19
Monel* K-500	Cu	29	Fully heat treated	20	340	680	45
	Al	2.8		200	290	650	40
	Ti	0.5		400	260	600	30
				600	290	460	5
				800	—	185	30

(continued)

Table 22.28 WROUGHT NICKEL AND HIGH NICKEL ALLOYS, SHORT-TIME HIGH-TEMPERATURE TENSILE PROPERTIES—*continued*

Alloy	Nominal composition %		Condition	Test temperature °C	0.2% Proof stress MPa	UTS MPa	Elonga- tion %
Inconel* 600	Cr	16	Hot rolled	20	250	590	50
	Fe	6		400	185	560	50
				600	150	530	10
				800	95	250	20
				1 000	—	110	50
Inconel 601	Ni	60.5	Annealed	20	440	760	43
	Cr	23.0		200	410	750	38
	Al	1.4		400	380	740	38
	Fe	15.1		600	325	575	32
				800	170	205	92
				870	100	160	122
Inconel 617	Ni	55.7	Solution	20	330	720	63
	Cr	21.5	Annealed	200	260	640	70
	Co	12.5		400	230	600	72
	Mo	9.0		600	220	580	58
	Al	1.2		800	230	410	65
	C	0.1		1 000	80	160	92
Inconel 718	Ni	52.5	Precipitation	20	1 180	1 350	17
	Cr	19.0	hardened	200	1 100	1 280	16
	Fe	18.8		400	1 080	1 240	13
	Nb	5.2		600	960	1 160	16
	Mo	3.1		800	720	760	8
	Ti	0.9					
	Al	0.5					
Inconel X-750	Cr	15	Fully heat	20	620	1 110	24
	Fe	7	treated	400	590	1 100	28
	Ti	2.5		650	280	830	9
	Al	0.6		800	310	370	22
	Nb	0.8		800	100	170	90
Inconel MA 754	Ni	78.0	Annealed	20	560	940	20
	Cr	20.0		200	560	910	18
	Fe	1.0		400	540	860	16
	Y_2O_3	0.6		600	500	640	18
				800	230	280	30
				1 000	120	160	16
INCO 330	Ni	35.5	Annealed	20	280	580	48
	Fe	44.8		200	220	520	45
	Cr	18.5		400	200	500	44
	Si	1.2		600	165	430	49
				800	120	170	68
				980	70	80	74
INCO 020	Ni	35.0	Annealed	30	300	620	40
	Fe	38.4		200	280	575	40
	Cr	20.0		400	250	560	39
	Cu	3.5		600	185	505	41
	Mo	2.5		800	150	275	59
	Nb	0.6		870	140	220	72
INCO G-3	Ni	49.0	Solution	20	320	690	50
	Cr	22.5	Annealed	200	235	590	62
	Fe	19.5		400	190	555	68
	Mo	7.0		600	175	500	73
	Cu	2.0		800	170	355	66
				1 000	75	140	76

(*continued*)

Table 22.28   WROUGHT NICKEL AND HIGH NICKEL ALLOYS, SHORT-TIME HIGH-TEMPERATURE TENSILE
PROPERTIES—*continued*

Alloy	Nominal composition %		Condition	Test temperature °C	0.2% Proof stress MPa	UTS MPa	Elonga- tion %
INCO C-276	Ni	59.0	Annealed	20	415	790	50
	Mo	16.0		200	285	690	56
	Cr	15.5		400	240	660	64
	Fe	5.5		540	220	610	60
	W	4.0					
INCO-HX	Ni	48.3	Solution	20	340	790	45
	Cr	22.0	Annealed	200	300	745	40
	Fe	18.5		400	225	700	48
	Mo	9.0		600	220	600	46
	Co	1.5		800	180	370	48
	W	0.6		1 000	40	110	56
	C	0.1					
Incoloy* 800	Fe	45	Annealed	20	300	600	45
	Cr	21		600	210	440	40
	Ti	0.4					
	Al	0.4					
Incoloy DS	Fe	40	Annealed	20	300	700	45
	Cr	18		600	—	460	50
	Si	2		800	—	150	90
				1 000	—	75	100
Hastelloy* B-2	Mo	28	Solution treated	200	451	885	50
				320	426	864	49
				430	418	866	51
Hastelloy C-4	Mo	16	Solution treated	200	403	706	49
	Cr	16		320	371	675	52
				430	320	656	64
Hastelloy X	Mo	9	Solution treated	20	360	790	45
	Cr	21		600	280	620	40
	Fe	18		800	230	410	37
				1 000	110	160	45
Nimonic* 75	Cr	20	Annealed	20	420	800	35
	Ti	0.4		400	400	740	30
				600	310	590	30
				800	110	220	80
				1 000	50	90	58
Nimonic* 80A	Cr	20	Fully heat	20	740	1 240	24
	Ti	2.0	treated	400	680	1 150	26
	Al	1.5		600	620	1 080	20
				800	490	620	24
				1 000	70	120	120
Nimonic 90	Cr	20	Fully heat	20	750	1 175	30
	Co	17	treated	600	680	1 030	26
	Ti	2.4		800	530	900	18
	Al	1.4		1 000	48	76	130
Nimonic 105	Cr	15	Fully heat	20	780	1 140	22
	Co	20	treated	600	720	1 040	25
	Mo	5		800	680	810	25
	Al	5		1 000	150	175	42
	Ti	1.2					

(*continued*)

Table 22.28   WROUGHT NICKEL AND HIGH NICKEL ALLOYS, SHORT-TIME HIGH-TEMPERATURE TENSILE PROPERTIES—*continued*

Alloy	Nominal composition %		Condition	Test temperature °C	0.2% Proof stress MPa	UTS MPa	Elongation %
Nimonic 115	Cr	14	Fully heat treated	20	860	1 230	27
	Co	13		600	790	1 100	20
	Mo	3		800	760	1 020	19
	Al	5		1 000	200	420	26
	Ti	4					
					0.1% PS		
Nimonic 263	Cr	20	Fully heat treated	20	570	970	40
	Co	20		600	460	790	41
	Mo	6		800	390	560	20
	Ti	2		1 000	60	110	65
	Al	0.5					
Nimonic PE 16	Cr	16	Fully heat treated	20	490	880	37
	Fe	32		600	450	730	27
	Mo	3		800	290	390	53
	Ti	1.0		1 000	46	92	100
	Al	1.0					
Astroloy	Ni	54.8	Precipitation hardened	20	1 050	1 410	16
	Cr	15.0		540	965	1 240	16
	Co	17.0		650	965	1 310	18
	Mo	5.3		760	910	1 160	21
	Al	4.0		870	690	770	25
	Ti	3.5					
Rene 41	Ni	55.4	Precipitation hardened	21	1 060	1 420	14
	Cr	19.0		540	1 010	1 400	14
	Co	11.0		650	1 000	1 340	14
	Mo	11.0		760	940	1 100	11
	Al	1.5		870	550	620	19
	Ti	3.1					
Rene 95	Ni	61.5	Precipitation hardened	21	1 310	1 620	15
	Cr	14.0		540	1 250	1 540	12
	Co	8.0		650	1 220	1 460	14
	Mo	3.5		760	1 100	1 170	15
	Nb	3.5					
	Al	3.5					
	Ti	2.5					
Udimet 500	Ni	53.7	Precipitation hardened	21	840	1 310	32
	Cr	18.0		540	765	1 240	28
	Co	18.5		650	760	1 210	28
	Mo	4.0		760	730	1 040	39
	Al	2.9		870	495	640	20
	Ti	2.9					
Udimet 700	Ni	55.5	Precipitation hardened	21	965	1 410	17
	Cr	15.0		540	895	1 280	16
	Co	17.0		650	855	1 240	16
	Mo	5.0		760	825	1 030	20
	Al	4.0		870	635	690	27
	Ti	3.5					
Waspaloy	Ni	58.7	Precipitation hardened	21	795	1 280	25
	Cr	19.5		540	725	1 170	23
	Co	13.5		650	690	1 120	34
	Mo	4.3		760	675	795	28
	Al	1.3		870	515	525	35
	Ti	3.0					

* Registered trade mark.

Table 22.29   WROUGHT NICKEL AND HIGH NICKEL ALLOYS, CRYOGENIC PROPERTIES

Alloy	Nominal composition %		Condition	Test temp. °C	0.2% Proof stress MPa	UTS MPa	Elonga- tion %	Charpy impact J	Notched fatigue strength* MPa
Nickel	Ni	99.0	Annealed	20	160	500	48	230	—
				−75	170	560	58	230	120
				−200	230	710	54	230	140
Monel** 400	Cu	30	Annealed	20	220	540	52	260	—
	Fe	1.5		−180	340	800	50	240	—
	Mn	1.0							
Monel** K-500	Cu	29	Fully heat	20	670	1 060	22	74	—
	Al	2.8	treated	−80	740	1 140	24	68	300
	Ti	0.5		−200	830	1 260	30	—	330
				−255	940	1 380	28	—	330
Inconel** 600	Cr	16	Annealed	20	230	650	37	240	280
	Fe	6		−80	290	730	40	210	280
				−190	—	—	—	190	300
Inconel 718	Cr	19	Fully heat	20	1 400	1 500	—	—	—
	Nb	5	treated	−80	1 400	1 600	—	—	—
	Mo	3		−190	1 600	1 800	—	—	—
	Ti	0.9							
	Al	0.5							
Inconel X-750	Cr	15	Fully heat	20	700	1 200	24	—	—
	Fe	7	treated	−75	790	1 290	23	—	410
	Ti	2.5		−200	810	1 440	19	—	440
	Al	0.6		−255	900	1 450	16	—	460
	Nb	0.8							
Ni-Span** C-902	Fe	47	Fully heat	20	760	1 210	30	24	320
	Cr	5.5	treated	−130	860	1 410	28	23	310
	Ti	2.5		−255	1 000	1 690	25	23	390
	Al	0.5							

* 10^6 cycles Kt = 3.1.   ** Registered trade mark.

Table 22.30   WROUGHT NICKEL AND HIGH NICKEL ALLOYS, FATIGUE PROPERTIES

Alloy	Nominal composition %		Condition	Test temperature °C	Average endurance limit Rotating beam tests 10^8 cycles MPa
Nickel	Ni	99.0	Annealed	20	230
			Cold drawn	20	340
Monel** 400	Cu	30	Annealed	20	230
	Fe	1.5	Cold drawn*	20	300
	Mn	1.0			
Monel K-500	Cu	29	Annealed	20	260
	Al	2.8	Cold drawn and	20	320
	Ti	0.5	aged		
Inconel** 600	Cr	16	Annealed	20	270
	Fe	6		650	180
				850	75
Inconel 625	Cr	21.5	Hot rolled	29	460
	Mo	9.0		427	450
				538	425
	Nb	3.6		649	390
				760	310

(*continued*)

Table 22.30 WROUGHT NICKEL AND HIGH NICKEL ALLOYS, FATIGUE PROPERTIES—*continued*

Alloy	Nominal composition %		Condition	Test temperature °C	Average endurance limit Rotating beam tests 10^8 cycles MPa
Inconel 718	Cr	19	Fully heat treated	20	620
	Nb	5		650	500
	Mo	3			
	Ti	0.9			
	Al	0.5			
Inconel X-750	Cr	15	Fully heat treated	20	280
	Fe	7		700	340
	Ti	2.5			
	Al	0.6			
	Nb	0.8			
Incoloy** 800	Fe	45	Annealed	20	290
	Cr	21		500	260
	Ti	0.4		850	95

					$0 \pm$ stress. For lives of		
					MPa	h	cycles
Nimonic** 75	Cr	20	Annealed	20	260	65	10×10^6
	Ti	0.4		750	190	300	10×10^6
Nimonic 90	Cr	20	Fully heat treated	750	260	300	36×10^6
	Co	17			240	1 000	120×10^6
	Ti	2.4		870	140	300	36×10^6
	Al	1.4			110	1 000	120×10^6
Nimonic 105	Cr	15	Fully heat treated	20	350	50	3×10^7
	Co	20			250	500	3×10^8
	Mo	5		750	260	50	3×10^7
	Al	5			250	500	3×10^8
	Ti	1.2		870	240	50	3×10^7
					190	500	3×10^8
				980	170	50	3×10^7
					124	500	3×10^8

* Stress equalised 3 h at 260°C.   **Registered trade marks.

Table 22.31 WROUGHT NICKEL AND HIGH NICKEL ALLOYS, CREEP PROPERTIES

Alloy	Nominal composition %		Condition	Test temp. °C	Stress MPa *for creep extension of*								
					0.1%			0.2%			1%	Rupture	
					300 h	1 000 h	10 000 h	300 h	1 000 h	10 000 h	10 000 h	1 000 h	1 0000 h
Nickel	Ni	99.0	Cold drawn	400	—	—	77	—	—	—	220	—	—
Monel* 400	Cu	30	Hot rolled	400	—	—	140	—	—	—	215	—	—
	Fe	1.5											
	Mn	1.0											
Monel K-500	Cu	29	Cold drawn and aged	400	—	—	355	—	—	—	—	—	—
	Al	2.8											
	Ti	0.5											
Inconel* 600	Cr	16	Hot rolled	400	—	—	340	—	—	—	—	—	—
	Fe	6		600	—	—	38	—	—	—	77	—	—
Inconel 601	Ni	60.5	Annealed	650	195	—	—	—	—	—	—	—	—
	Cr	23.0		760	63	—	—	—	—	—	—	—	—
	Al	1.4		870	30	—	—	—	—	—	—	—	—
	Fe	15.1		980	14	—	—	—	—	—	—	—	—
				1 095	7	—	—	—	—	—	—	—	—

(*continued*)

Table 22.31 WROUGHT NICKEL AND HIGH NICKEL ALLOYS, CREEP PROPERTIES—*continued*

Alloy	Nominal composition %		Condition	Test temp. °C	Stress MPa for creep extension of								
					0.1%			0.2%			1%	Rupture	
					300 h	1 000 h	10 000 h	300 h	1 000 h	10 000 h	10 000 h	1 000 h	1 0000 h
Inconel 617	Ni	55.7	Solution Annealed	650	320	—	—	—	—	—	—	—	—
	Cr	21.5		760	150	—	—	—	—	—	—	—	—
	Co	12.5		870	58	—	—	—	—	—	—	—	—
	Mo	9.0		980	25	—	—	—	—	—	—	—	—
	Al	1.2		1 095	10	—	—	—	—	—	—	—	—
	C	0.1											
Inconel 718	Ni	52.5	Precipitation hardened	595	760	—	—	—	—	—	—	—	—
	Cr	19.0		650	590	—	—	—	—	—	—	—	—
	Fe	18.8		705	370	—	—	—	—	—	—	—	—
	Nb	5.2		760	170	—	—	—	—	—	—	—	—
	Mo	3.1											
	Ti	0.9											
	Al	0.5											
Incoloy* 800H	Fe	45	Solution treated	650	—	115	90	—	—	—	—	—	121
	Cr	21		870	—	24	20	—	—	—	—	—	24
	Tl	0.4		980	—	7	3.8	—	—	—	—	—	8.3
	Al	0.4											
Inconel X-750	Cr	15	Fully heat treated	595	—	—	—	—	—	—	—	630	—
	Fe	7		650	—	—	—	—	420	—	—	470	—
	Ti	2.5		730	—	—	—	—	—	—	—	260	—
	Al	0.6		815	—	—	—	—	110	—	—	110	—
	Nb	0.8		870	—	—	—	—	—	—	—	50	—
Inconel MA 754	Ni	78.0	Annealed	650	256	—	—	—	—	—	—	—	—
	Cr	20.0		760	199	—	—	—	—	—	—	—	—
	Fe	1.0		870	158	—	—	—	—	—	—	—	—
	Y₂O₃	0.6		980	129	—	—	—	—	—	—	—	—
				1 095	94	—	—	—	—	—	—	—	—
				1 150	78	—	—	—	—	—	—	—	—
INCO 330	Ni	35.5	Annealed	760	48	—	—	—	—	—	—	—	—
	Fe	44.8		870	21	—	—	—	—	—	—	—	—
	Cr	18.5		980	8.6	—	—	—	—	—	—	—	—
	Si	1.2		1 095	5.4	—	—	—	—	—	—	—	—
INCO-HX	Ni	48.3	Solution Annealed	760	110	—	—	—	—	—	—	—	—
	Cr	22.0		815	72	—	—	—	—	—	—	—	—
	Fe	18.5		870	45	—	—	—	—	—	—	—	—
	Mo	9.0		925	26	—	—	—	—	—	—	—	—
	Co	1.5		980	15	—	—	—	—	—	—	—	—
	W	0.6											
	C	0.1											
Nimonic* 75	Cr	20	Annealed	650	56	—	—	—	62	51	—	170	—
	Ti	0.4		750	25	—	—	—	29	25	—	50	—
				815	—	—	—	—	—	—	—	24	—
				870	—	—	—	—	—	—	—	15	—
				925	—	—	—	—	—	—	—	10	—
				980	—	—	—	—	—	—	—	8	—
Nimonic 80A	Cr	20	Fully heat treated	650	390	185	—	460	390	260	280	500	280
	Ti	2.0		705	—	—	—	—	—	—	—	350	—
	Al	1.5		750	170	120	51	200	150	66	90	220	85
				815	93	58	23	108	73	26	29	110	34
Nimonic 90	Cr	20	Fully heat treated	650	400	348	250	450	400	310	325	455	340
	Co	17		705	—	—	—	—	—	—	—	360	—
	Ti	2.4		750	185	137	65	220	175	85	110	240	135
	Al	1.4		815	85	57	28	108	77	31	38	150	54
				870	49	29	15	60	37	17	18	75	28

(*continued*

Table 22.31 WROUGHT NICKEL AND HIGH NICKEL ALLOYS, CREEP PROPERTIES—*continued*

Alloy	Nominal composition %	Condition	Test temp. °C	0.1% 300 h	1 000 h	10 000 h	0.2% 300 h	1 000 h	10 000 h	1% 10 000 h	Rupture 1 000 h	1 0000 h
Nimonic 105	Cr 15, Co 20, Mo 5, Al 5, Ti 1.2	Fully heat treated	750	285	230	—	325	280	193	—	363	208
			815	162	119	—	193	148	93	102	224	120
			870	86	57	—	108	83	34	45	134	57
			940	31	22	—	48	31	12	22	62	35
			980	—	—	—	—	—	—	—	32	—
Nimonic 115	Cr 14, Co 13, Mo 3, Al 5, Ti 4	Fully heat treated	750	355	310	—	385	340	—	—	450	250
			815	—	—	—	—	—	—	—	310	—
			850	154	116	51	222	137	53	74	—	108
			870	—	—	—	—	—	—	—	210	—
			950	51	34	14	63	43	15	17	—	52
			925	—	—	—	—	—	—	—	130	—
			980	—	—	—	—	—	—	—	80	—
			1 000	22	15	—	29	19	6	8	—	—
Astroloy	Ni 54.8, Cr 15.0, Co 17.0, Mo 5.3, Al 4.0, Ti 3.5	Precipitation hardened	650	770	—	—	—	—	—	—	—	—
			760	425	—	—	—	—	—	—	—	—
			870	170	—	—	—	—	—	—	—	—
			980	55	—	—	—	—	—	—	—	—
Rene 41	Ni 55.4, Cr 19.0, Co 11.0, Mo 11.0, Al 1.5, Ti 3.1	Precipitation hardened	650	705	—	—	—	—	—	—	—	—
			760	345	—	—	—	—	—	—	—	—
			870	115	—	—	—	—	—	—	—	—
Rene 95	Ni 61.5, Cr 14.0, Co 8.0, Mo 3.5, Nb 3.5, Al 3.5, Ti 2.5	Precipitation hardened	650	860	—	—	—	—	—	—	—	—
Udimet 500	Ni 53.7, Cr 18.0, Co 18.5, Mo 4.0, Al 2.9, Ti 2.9	Precipitation hardened	650	760	—	—	—	—	—	—	—	—
			760	325	—	—	—	—	—	—	—	—
			870	125	—	—	—	—	—	—	—	—
Udimet 700	Ni 55.5, Cr 15.0, Co 17.0, Mo 5.0, Al 4.0, Ti 3.5	Precipitation hardened	650	705	—	—	—	—	—	—	—	—
			760	425	—	—	—	—	—	—	—	—
			870	200	—	—	—	—	—	—	—	—
			980	55	—	—	—	—	—	—	—	—
Waspaloy	Ni 58.7, Cr 19.5, Co 13.5, Mo 4.3, Al 1.3, Ti 3.0	Precipitation hardened	650	615	—	—	—	—	—	—	—	—
			760	290	—	—	—	—	—	—	—	—
			870	110	—	—	—	—	—	—	—	—

[*] Registered trade mark.

BIBLIOGRAPHY

1. INCO Alloys International Ltd., Product Handbook, 1990.
2. G.W. Meetham (ed.), 'The Development of Gas Turbine Materials', Applied Science Publishers Ltd, Barking, Essex, 1981.
3. *ASM Handbook*, 10th edition, vols. 1 and 2, ASM International, Materials Park, OH, 1990.

Table 22.32   CAST NICKEL ALLOYS, TENSILE PROPERTIES

Alloy	Nominal composition wt. %, balance Ni		Test temperature °C	0.2% Yield strength MPa	Ultimate tensile strength MPa	Elongation %
Inconel 713C	Cr	12.5	21	740	850	8
	Mo	4.2	538	705	860	10
	Nb	2.0	1 038		255	19
	Al	6.1				
	Ti	0.8				
	C	0.12				
B-1900	Cr	8.0	21	825	970	8
	Co	10.0	538	870	1 005	7
	Mo	6.0	1 093	195	270	11
	Ta	4.3				
	Al	6.0				
	Ti	1.0				
Inconel 625	Cr	21.5	21	350	710	48
	Mo	8.5	538	235	510	50
	Nb	4.0	1 093			
	Al	0.2				
	Ti	0.2				
	Fe	2.5				
Inconel 718	Cr	18.5	21	915	1 090	11
	Mo	3.0	538			
	Nb	5.1	1 093			
	Al	0.5				
	Ti	0.9				
	Fe	18.5				
Inconel 100	Cr	10.0	21	850	1 018	9
	Co	15.0	538	885	1 090	9
	Mo	3.0	870	695	885	6
	Al	5.5	1 038	283	441	6
	Ti	4.7				
	V	1.0				
Inconel 738	Cr	16.0	21	950	1 095	6
	Co	8.5	732	841	1 000	3
	Mo	1.75	982	345	455	10
	W	2.6				
	Ta	1.75				
	Nb	0.9				
	Al	3.4				
	Ti	3.4				
MAR-M 200	Cr	9.0	21	840	930	7
	Co	10.0	538	880	945	5
	W	12.5	1 093	286	325	6
	Nb	1.8				
	Al	5.0				
	Ti	2.0				
MAR-M 246	Cr	9.0	21	860	965	5
	Co	10.0	538	860	1 000	5
	Mo	2.5	1 066	234	379	11
	W	10.0	1 093	138	207	5
	Ta	1.5				
	Al	5.5				
	Ti	1.5				
Nimocast 75	Cr	20.0	21	179	500	39
	Ti	0.5				
	C	0.12				
Nimocast 80	Cr	19.5	21	520	730	15
	Al	1.4				
	Ti	2.3				
	Fe	1.5				
	C	0.05				
Nimocast 90	Cr	19.5	21	520	700	14
	Co	18.0	538	420	595	15
	Al	1.4	1 000	25	53	
	Ti	2.4				
	Fe	1.5				
	C	0.06				

(continued)

Table 22.32 CAST NICKEL ALLOYS, TENSILE PROPERTIES—*continued*

Alloy	Nominal composition wt. %, balance Ni		Test temperature °C	0.2% Yield strength MPa	Ultimate tensile strength MPa	Elongation %
Rene 80	Cr	14.0	21	848	1 030	5
	Co	9.5	538	730	1 030	8
	Mo	4.0	1 093		180	13
	W	4.0				
	Al	3.0				
	Ti	5.0				
Udimet 500	Cr	18.5	21	815	930	13
	Co	16.5	538	725	895	13
	Mo	3.5				
	Al	3.0				
	Ti	3.0				
MAR-M 002 + Hf	Cr	9.0	21	825	1 035	7
	Co	10.0	538	860	1 035	5
	Ta	2.5	1 093	345	550	12
	Al	5.5				
	Ti	1.5				
	Hf	1.5				

Table 22.33 CAST NICKEL ALLOYS, CREEP PROPERTIES

Alloy	Nominal composition wt. %, balance Ni		Rupture strength, MPa 100 hr	Rupture strength, MPa 1 000 hr	Test temperature °C
Inconel 713C	Cr	12.5	370	305	815
	Mo	4.2	305	215	870
	Nb	2.0	130	70	980
	Al	6.1			
	Ti	0.8			
	C	0.12			
B-1900	Cr	8.0	510	380	815
	Co	10.0	385	250	870
	Mo	6.0	180	110	980
	Ta	4.3			
	Al	6.0			
	Ti	1.0			
Inconel 100	Cr	10.0	455	365	815
	Co	15.0	360	260	870
	Mo	3.0	160	90	980
	Al	5.5			
	Ti	4.7			
	V	1.0			
Inconel 738C	Cr	16.0	430	345	815
	Co	8.5	295	235	870
	Mo	1.75	140	90	980
	W	2.6			
	Ta	1.75			
	Nb	0.9			
	Al	3.4			
	Ti	3.4			
Inconel 738	Cr	16.0	462	338	815
	Co	8.5	324	228	870
	Mo	1.75	131	83	980
	W	2.6			
	Ta	1.75			
	Nb	0.9			
	Al	3.4			
	Ti	3.4			
MAR-M 200	Cr	9.0	495	415	815
	Co	10.0	385	295	870
	W	12.5	170	125	980
	Nb	1.8			
	Al	5.0			
	Ti	2.0			

(continued)

Table 22.33 CAST NICKEL ALLOYS, CREEP PROPERTIES—*continued*

Alloy	Nominal composition wt. %, balance Ni		Rupture strength, MPa 100 hr	1 000 hr	Test temperature °C
MAR-M 246	Cr	9.0	525	435	815
	Co	10.0	440	290	870
	Mo	2.5	195	125	980
	W	10.0			
	Ta	1.5			
	Al	5.5			
	Ti	1.5			
Nimocast 90	Cr	19.5	160	110	815
	Co	18.0	125	83	870
	Al	1.4			
	Ti	2.4			
	Fe	1.5			
	C	0.06			
Rene 80	Cr	14.0	620	496	760
	Co	9.5	350	240	870
	Mo	4.0	160	105	980
	W	4.0			
	Al	3.0			
	Ti	5.0			
Udimet 500	Cr	18.5	330	240	815
	Co	16.5	230	165	870
	Mo	3.5	90		980
	Al	3.0			
	Ti	3.0			

22.5.1 Directionally solidified and single crystal cast superalloys

Directionally solidified (DS) and single crystal (SC or SX) nickel-base superalloys were developed to increase the creep and thermal fatigue strength of first stage turbine blades in aero-engine gas turbines, and are now increasingly used in land-based gas turbines. Property improvements over those of polycrystalline superalloys result from the elimination of grain boundaries transverse to the principal stress axis and the alignment of the low-modulus <100> crystal direction with the stress axis. In the case of the SC alloys, further improvements were realised by optimising alloy chemistry in the absence of grain-boundary strengtheners such as C, B, and Zr.

Table 22.34 lists nominal compositions for common DS alloys; the so-called 'first-generation' alloys represent slight modifications from polycrystalline cast superalloy compositions (generally addition of Hf). The second-generation alloys were specifically designed for directional solidification and several incorporate Re to achieve creep strengths equivalent to first-generation single crystal alloys. Table 22.35 lists nominal compositions for common SC alloys. As mentioned above, the first-generation alloys eliminated the grain boundary strengtheners C, B, and Zr. Increasing levels of Re in the second and third generation alloys have led to even higher creep resistance, due to the potent solid-solution strengthening effect of Re at high temperatures.

Property data for the various alloys are not listed due to the proprietary and specialised nature of these alloys. Users should consult the developer or supplier for property data for their specific needs.

BIBLIOGRAPHY

1. K. Harris, G. L. Erickson and R. E. Schwer, pp. 994–1006 in 'ASM Handbook, vol. 1, 10th edition', ASM International, Materials Park, OH. 1990.
2. A. D. Cetel and D. N. Duhl, pp. 287–296 in 'Superalloys 1992' (eds. S.D. Antolovich, *et al.*), TMS, Warrendale, PA, 1992.
3. K. Harris, et al., pp. 297–306 in 'Superalloys 1992' (eds. S. D. Antolovich, *et al.*), TMS, Warrendale, PA. 1992.
4. E. W. Ross and K. S. O'Hara, pp. 257–258 in 'Superalloys 1992' (eds. S. D. Antolovich, *et al.*), TMS, Warrendale, PA, 1992.
5. E. Fleury and L. Remy, *Mater. Sci. Eng. A*, 1993, **A167**, 23.
6. G. L. Erickson, pp. 35–44 in 'Superalloys 1996' (eds. R.D. Kissinger, *et al.*), TMS, Warrendale, PA, 1996.
7. D. N. Duhl and M. L. Gell, United Technologies Corporation, G. B. Patent 21121812A–PWA 1483 alloy.
8. W. S. Walston, *et al.*, pp. 27–34 in 'Superalloys 1996' (eds. R. D. Kissinger, *et al.*), TMS, Warrendale, PA, 1996.

Nominal composition, wt. %

Alloy	Developer/Producer	Cr	Co	Mo	W	Re	Ta	Al	Ti	Nb	Hf	C	B	Zr	Ni	Reference
First-Generation																
MAR-M 002	Martin Metals	8	10	—	10	—	2.6	5.5	1.5	—	1.5	0.15	0.015	0.03	bal.	1
MAR-M 200 Hf	Martin Metals	8	9	—	12	—	—	5.0	1.9	1	2.0	0.13	0.015	0.03	bal.	1
MAR-M 247	Martin Metals	8	10	0.6	10	—	3	5.5	1.0	—	1.5	0.15	0.015	0.03	bal.	1
PWA 1422	Pratt and Whitney	10	10	—	12	—	—	5.0	2.0	1	1.5	0.14	0.015	0.1	bal.	2
Rene 80H	General Electric	14	9	4.0	4	—	—	3.0	4.7	—	0.8	0.16	0.015	0.01	bal.	1
Second-Generation																
CM 186 LC	Cannon-Muskegon	6	9	0.5	8	3	3	5.7	0.7	—	1.4	0.07	0.015	0.005	bal.	3
CM 247 LC	Cannon-Muskegon	8	9	0.5	10	—	3.2	5.6	0.7	—	1.4	0.07	0.015	0.01	bal.	1
PWA 1426	Pratt and Whitney	6.5	10	1.7	6.5	3	4	6.0	—	—	1.5	0.1	0.015	0.1	bal.	2
Rene 142	General Electric	6.8	12	1.5	4.9	2.8	6.4	6.2	—	—	1.5	0.12	0.015	0.02	bal.	4
Rene 150	General Electric	5	12	1.0	5	3	6	5.5	—	(2.2 V)	1.5	0.12	0.015	0.02	bal.	4

Table 22.35   SOME COMMON SINGLE CRYSTAL NICKEL-BASE SUPERALLOYS

Nominal composition, wt. %

Alloy	Developer/Producer	Cr	Co	Mo	W	Re	Al	Ti	V	Nb	Ta	Hf	C	B	Y	Ni	Reference
First-Generation																	
AM1	ONERA, France	8	6	2.0	6	—	5.2	1.2	—	—	8	—	—	—	—	bal.	5
CMSX-2	Cannon-Muskegon	8	5	0.6	8	—	5.6	1.0	—	—	6	—	—	—	—	bal.	6
CMSX-3	Cannon-Muskegon	8	5	0.6	8	—	5.6	1.0	—	—	6	0.1	—	—	—	bal.	6
CMSX-6	Cannon-Muskegon	10	5	3.0	—	—	4.8	4.7	—	—	2	0.1	—	—	—	bal.	6
CMSX-11B	Cannon-Muskegon	12.5	7	0.5	5	—	3.6	4.2	—	0.1	5	0.04	—	—	—	bal.	6
CMSX-11C	Cannon-Muskegon	15	3	0.4	4.5	—	3.4	4.2	—	0.1	5	0.04	—	—	—	bal.	6
PWA 1480	Pratt and Whitney	10	5	—	4	—	5.0	1.5	—	—	12	—	—	—	—	bal.	1
PWA 1483	Pratt and Whitney	13	9	2.0	4	—	3.6	4.0	—	—	4	—	—	—	—	bal.	7
Rene N4	General Electric	9	8	2.0	6	—	3.7	4.2	—	0.5	4	—	—	—	—	bal.	1
RR 2000	Rolls-Royce	10	15	3.0	—	—	5.5	4.0	1.0	—	—	—	—	—	—	bal.	1
SRR 99	Rolls-Royce	8	5	—	10	—	5.5	2.2	—	—	3	—	—	—	—	bal.	1
Second-Generation																	
CMSX-4	Cannon-Muskegon	6.5	9	0.6	6	3	5.6	1.0	—	—	6.5	0.1	—	—	—	bal.	6
PWA 1484	Pratt and Whitney	5	10	2.0	6	3	5.6	—	—	—	9	0.1	—	—	—	bal.	1
Rene N5	General Electric	7	7.5	1.5	5	3	6.2	—	—	—	6.5	0.15	0.05	0.004	0.01	bal.	8
SC 180	Garrett	5.3	10	1.7	5	3	5.2	1.0	—	—	8.5	0.1	—	—	—	bal.	6
Third-Generation																	
CMSX-10	Cannon-Muskegon	2	3	0.4	5	6	5.7	0.2	—	0.1	8	0.03	—	—	—	bal.	6
Rene N6	General Electric	4.2	12.5	1.4	6	5.4	5.7	—	—	—	7.2	0.15	0.05	0.004	0.01	bal.	8

22.6   Mechanical properties of titanium and titanium alloys

Table 22.36   TITANIUM AND TITANIUM ALLOYS. CORRESPONDING GRADES OR SPECIFICATIONS

Note for 8th edition: In the case of the UK, France, and Germany, this table refers to designations and standards that may have been superceded by European (EN) and international (ISO) standards. This table is retained for information only.

IMI designation	UK British Standards (Aerospace series) and Min. of Def. DTD series*	France AIR-9182, 9183, 9184	Germany BWB series[†]	AECMA recom- mendations	USA AMS series[‡]	USA ASTM series
IMI 115	BS TA 1, DTD 5013	T-35	3.7024	Ti-POI		ASTM grade 1
IMI 125	BS TA 2,3,4,5	T-40	3.7034	Ti-PO2	AMS 4902, 4941, 4942, 4951	ASTM grade 2
IMI 130	DTD 5023, 5273, 5283, 5293	T-50			AMS 4900	ASTM grade 3
IMI 155	BS TA 6				AMS 4901	ASTM
IMI 160	BS TA 7,8,9	T-60	3.7064	Ti-PO4	AMS 4921	grade 4
IMI 230	BS TA 21, 22, 23, 24, BS TA 52–55, 58	T-U2	3.7124	Ti-P11		
IMI 315	DTD 5043					
IMI 317	BS TA 14, 15, 16, 17	T-A5E		Ti-P65	AMS 4909, 4910, 4926, 4924, 4953, 4966	
IMI 318	BS TA 10, 11, 12, 13, 28, 56	T-A6V	3.7164	Ti-P63	AMS 4911, 4928, 4934, 4935, 4954, 4965, 4967	ASTM grade 5
IMI 318 ELI (extra low interstitial)	—	T-A6VELI			AMS 4907, 4930, 4931	ASTM grade 3, F. 136
IMI 325	—	T-A3V2.5	3.7194		AMS 4943 4944	ASTM grade 9
IMI 550	BS TA 45–51, 57	T-A4DE	3.7184	Ti-P68		
IMI 551	BS TA 39–42					
IMI 624	—	T-A6Zr4 DE	3.7144		AMS 4919, 4975, 4976	
IMI 646	—	—	—	—	AMS 4981	
IMI 662	—	—	3.7174	—	AMS 4918, 4971, 4978, 4979	
IMI 679	BS TA 18–20, 25–27				AMS 4974	
IMI 680	DTD 5213	T-E11DA				
IMI 685	BS TA 43, 44	T-A6ZD	3.7154	Ti-P67		
IMI 811	—	T-A8DV			AMS 4915, 4916	
IMI 834	—	T-A6E Zr4Nb				

* UK BS 3531 Part 1 (Metal Implants in Bone Surgery), and Draft British Standard for Lining of Vessels and Equipment for Chemical Processes, Part 9, also refer.
[†] Germany DIN 17850, 17860, 17862, 17863, 17864 (3.7025/35/55/65), and TUV 230-1-68 Group, I II, III and IV also refer.
[‡] USA MIL-T-9011, 9046, 9047, 14577, 46038, 46077, 05-10737 and ASTM B265-69, B338-65, B348-59T, B367-61T. B381-61T, B382-61T also refer.

Table 22.37   TYPICAL MECHANICAL PROPERTIES AT ROOM TEMPERATURE

Designation*	Grade	Condition	0.2% proof stress MPa	Tensile strength MPa	Elongation % on 50 mm	Elongation % on 5D	Red in area %	Specification bend radius 180° bend <1.83 mm	Specification bend radius <3.25 mm	Mod. of elasticity GPa	Mod. of rigidity GPa
Iodide	Pure, 60 HV										
IMI 115	Commercially pure	Annealed sheet	103	241	55		80	1t			
		Annealed rod	255	370	33	40	70		2t		
		Annealed wire	220	370	38						
				390							
IMI 125	Commercially pure	†Annealed sheet	340	460	30			1½t	2t		
		Annealed rod	305	460		28	57				
		Annealed tube	325	480	35						
IMI 130	Commercially pure	Annealed sheet	420	540	25		48	2t⁺	2½t	105	38
		Annealed rod	360	540		24					
		Annealed wire		550	24						
		Hard-drawn wire		700	11.5						
IMI 155	Commercially pure	Annealed sheet	540	640	24			2½t			
IMI 160	Commercially pure	Annealed rod	500	670	24		46		3t		
		Annealed wire		690		23					

* IMI Nomenclature.   † Up to 16.3 mm.

Table 22.38   TITANIUM ALLOYS TYPICAL MECHANICAL PROPERTIES AT ROOM TEMPERATURE

Designation*	Nominal composition %		Condition	0.2% proof stress MPa	Tensile strength MPa	Elongation % on 50 mm	Elongation % on 5D	Red. in area %	Specification bend radius 180°	Mod. of elasticity GPa	Mod. of rigidity GPa
IMI 230	Cu	6.0	Annealed sheet	520	620	24			2t (0.5–3 mm)	125	
			Aged sheet	670	770	20			2t (typical)		
			Annealed rod	500	630		27	45			
			Aged rod	580	740		22	41		125	
IMI 260	Pd	0.2	Similar to commercially Pure Titanium 115								
IMI 261	Pd	0.2	Similar to commercially Pure Titanium 125								
IMI 315	Al	2.0	Annealed rod	590	720		21	50		120	
	Mn	2.0									
IMI 317	Al	5.0	Annealed sheet	820	860	16			4t (<2 mm)	125	
	Sn	2.5							4½t (≤3 mm)		
			Annealed rod	930	1000		15	37		120	

(continued)

Table 22.38   TITANIUM ALLOYS TYPICAL MECHANICAL PROPERTIES AT ROOM TEMPERATURE—*continued*

Designation*	Nominal composition %		Condition	0.2% proof stress MPa	Tensile strength MPa	Elongation % on 50 mm	Elongation % on 5D	Red. in area %	Specification bend radius 180°	Mod. of elasticity GPa	Mod. of rigidity GPa
IMI 318	Al	6.0	Annealed sheet	1 110	1 160	10		40	5 t (≤3.25 mm)	106	46
	V	4.0	Annealed rod	990	1 050		15	40			
			Aged rod (fastener stock)	1 050	1 140		15				
			Hard-drawn wire		1 410	4					
IMI 550	Al	4.0	F.h.t. rod	1 070	1 200		14	42		116	
	Mo	4.0									
	Sn	2.0									
	Si	0.5									
IMI 551	Al	4.0	F.h.t.rod	1 140	1 300		12	40		113	43
	Mo	4.0									
	Sn	4.0									
	Si	0.5									
IMI 679	Sn	11.0	Quenched and aged rod	1 080	1 230		11	40		108	46
	Al	2.25	Air-cooled and	1 000	1 120		13	45			
	Mo	1.0	aged rod								
	Si	0.2									
IMI 680	Sn	11.0	Quenched and aged rod	1 200	1 350		12	37		115	
	Mo	4.0	Furnace-cooled and	1 080	1 160		14	47			
	Al	2.25	aged rod								
	Si	0.2									
IMI 685	Al	6.0	F.h.t. rod	920	1 020		11	22		124	47
	Zr	5.0									
	Mo	0.5									
	Si	0.25									
IMI 829	Al	5.5	F.h.t. rod	848	965		12	22		120	
	Sn	3.5									
	Zr	3.0									
	Nb	1.0									
	Mo	0.3									
	Si	0.3									
IMI 834	Al	5.8	F.h.t. rod	931	1 067		13	22		120	
	Sn	4.0									
	Zr	3.5									
	Nb	0.7									
	Mo	0.5									
	Si	0.35									
	C	0.06									

Table 22.39 COMMERCIALLY PURE TITANIUM SHEET, TYPICAL VARIATION OF PROPERTIES WITH TEMPERATURE

Designation*	Temperature °C	0.2% proof stress MPa	Tensile strength MPa	Elongation on 50 mm %	Mod. of elasticity GPa	Transformation temperature °C
IMI 115	−196	442	641	34		$\alpha/\alpha + \beta$
	−100	306	444	34		865
	20	207	337	40		
	100	168	296	43		
	200	99	218	38		
	300	53	167	47		
	400	42	131	52		
	450	36	120	49		
IMI 125	20	334	479	31		
	100	250	397	32		
	200	184	300	40		
	300	142	232	45		
	400	127	190	38		
	450	119	175	35		
IMI 130	−196	730	855	28		
	−100	590	737	28		
	20	394	547	28	108	
	100	315	462	29	99	$\alpha + \beta/\beta$
	200	205	331	37	91	915
	300	139	247	40	83	
	400	102	199	34	65	
	450	93	182	28		
	500				46	
IMI 155	20	460	625	25		
	100	372	537	26		
	200	219	386	32		
	300	151	281	36		
	400	110	221	33		
	450	96	202	26		

* IMI nomenclature.

Table 22.40   TITANIUM ALLOYS, TYPICAL VARIATION OF PROPERTIES WITH TEMPERATURE

Designation	Nominal composition %		Condition	Temperature °C	0.2% proof stress MPa	Tensile strength MPa	Elongation % on 50 mm	Elongation % on 5D	Red. in area %	Mod. of elasticity GPa	Transformation temperature °C
IMI 230	Cu	2.5	S.h.t. (trans.)	20	500	605	24				$\alpha/\alpha + \beta$
				100	410	540	29				790
				200	310	450	33				
				300	270	410	31				
				400	250	380	30				
				500	220	380	33				
			Aged sheet (trans.)	20	622	761	24				$\alpha + \beta/\beta$
				100	553	704	23				895 ± 10
				200	471	635	26				
				300	457	607	23				
				400	429	573	19				
				500	357	468	21				
			Aged	20	638	795		22	40	107	
				100	601	761		21	39	100	
				200	507	687		23	45	92	
				300	496	658		20	50	85	
				400	415	592		21	53	78	
				500	361	491		27	57	71	
IMI 260	Pd	0.2	Similar to IMI 115								
IMI 262	Pd	0.2	Similar to IMI 125								
IMI 315	Al	2.0	Annealed rod	20	618	757		18	41	110	$\alpha + \beta/\beta$
	Mn	2.0		100	510	649		21	46	107	915 ± 20
				200	386	525		22	48	97	
				300	293	432		19	50	86	
				400	278	417		18	56	76	
				500	201	340		22	72	62	
IMI 317	Al	5.0	Annealed rod	20	822	919		18	39	112	$\alpha/\alpha + \beta$
	Sn	2.5		100	692	798		19	40	109	950
				200	494	638		18	44	105	$\alpha + \beta/\beta$
				300	415	576		19	42	89	1025 ± 20
				400	374	522		18	41	84	
				500	346	485		21	57	81	

Alloy	Element	%	Condition	Temp (°C)	0.2% Proof stress (N/mm²)	Tensile strength (N/mm²)	Elong. (%)	R of A (%)	Modulus	Heat treatment
IMI 318	Al	6.0	Annealed rod	−196	1 560	1 675	6	29	106	α + β/β 1 000 ± 15
	V	4.0		−100	1 165	1 265	12	33	102	
				20	970	1 040	15	38	96	
				100	825	920	17	43	90	
				200	710	815	18	49	85	
				300	645	750	18	56	79	
				400	580	700	26	63		
				500	450	605	58	72		
				600	125	265	127	85		
				700	40	135		94		
			Heat-treated rod (fastener stock)	20	1 035	1 145	14			
				100	925	1 035	15			
				200	805	925	16			
				300	710	850	16			
				400	635	805	18			
				500	540	695	25			
IMI 550	Al	4.0	F.h.t. rod	20	1 081	1 220	15	49	116	α + β/β 980 ± 10
	Mo	4.0		100	965	1 130	15	49	112	
	Sn	2.0		200	805	960	16	60	106	
	Si	0.5		300	700	900	16	55	101	
				400	655	835	17	60	95	
				500	585	780	19	68	90	
				600	310	585	26	83	85	
IMI 551	Al	4.0	F.h.t. rod	20	1 250	1 390	10	27	113	α + β/β 1 050 ± 15
	Mo	4.0		100	1 125	1 300	11	29	108	
	Sn	4.0		200	925	1 145	14	38	103	
	Si	0.5		300	815	1 045	15	38	98	
				400	745	970	14	41	93	
				500	670	920	18	55	88	
				600	460	755	27	65	81	
IMI 679	Sn	11.0	Quenched and aged rod	20	1 050	1 230	10	37		α + β/β 950 ± 10
	Zr	5.0		100	940	1 145	11	43		
	Al	2.25		200	820	1 020	12	45		
	Mo	1.0		300	740	990	11	46		
	Si	0.2		400	710	940	11	46		
				450	680	910	11	46		
			Air-cooled and aged rod	20	1 020	1 095	14	41	108	
				100	895	995	16	47	103	
				200	770	900	16	49	99	
				300	695	865	14	49	94	
				400	665	850	14	48	90	
				500	600	795	15	48	85	

(continued)

Table 22.40   TITANIUM ALLOYS, TYPICAL VARIATION OF PROPERTIES WITH TEMPERATURE—*continued*

Designation	Nominal composition %	Condition	Temperature °C	0.2% proof stress MPa	Tensile strength MPa	Elongation % on 50 mm	Elongation % on 5D	Red. in area %	Mod. of elasticity GPa	Transformation temperature °C
IMI 680	Sn 11.0 Mo 4.0 Al 2.25 si 0.2	Quenched and aged rod	20	1 180	1 330		12	43	106	α + β/β 945 ± 15
			100	1 020	1 190		14	49	100	
			200	905	1 105		15	53	96	
			300	835	1 075		15	56	94	
			400	805	1 020		14	57	90	
			450	725	975		13	54	88	
IMI 685		Furnace-cooled and aged rod	−196	1 630	1 730		8½	36		α + β/β 1 020 ± 10
			−100	1 280	1 380		10	43		
			20	1 030	1 130		15	49		
	Al 6.0 Zr 5.0 Mn 0.5 Si 0.25	F.h.t. rod	−196	1 480	1 560		6	13		
			−100	1 140	1 270		10	18		
			20	890	1 030		12	22	124	
			100	800	935		13	22	120	
			200	720	850		15	24	114	
			300	650	800		16	27	108	
			400	595	750		18	31	102	
			500	535	695		19	37	95	
IMI 829	Al 5.5 Sm 3.5 Zr 3.0 Nb 1.0 Mo 0.3 Si 0.3	F.h.t. rod	20	895	1 028		$10\frac{1}{5}$	22	119	α + ββ 1 015 ± 15
			200	622	792		$14\frac{1}{2}$	28	110	
			500	501	665		$15\frac{1}{2}$	36	93	
			540	487	653		16	42	91	
			600	457	634		14	38	88	
IMI 834	Al 5.8 Sn 4.0 Zr 3.5 Nb 0.7 Mo 0.5 Si 0.35 C 0.06	F.h.t. rod	20	931	1 067		13	22	120	1 045
			100	840	962		13	23	116	
			200	746	885		14	27	112	
			300	700	832		14	32	106	
			400	662	790		14	36	102	
			500	609	764		15	42	96	
			600	505	656		16	50	92	

Table 22.41   COMMERCIALLY PURE TITANIUM—TYPICAL CREEP PROPERTIES

IMI designation	Temperature °C	Stress MPa *to produce* 0.1% *plastic strain in*		
		1 000 h	10 000 h	100 000 h
IMI 130	20	288	270	207
	50	243	221	165
	100	179	165	119
	150	140	133	96
	200	113	116	77
	250	96	101	66
	300	87	83	55
IMI 155	20	309	278	260
	50	252	232	213
	100	188	170	157
	150	145	131	122
	200	116	108	104
	250	102	97	94
	300	93	90	86

Table 22.42   TITANIUM ALLOYS—TYPICAL CREEP PROPERTIES

IMI designation	Nominal composition %		Condition	Temperature °C	Stress MPa *to produce* 0.1% *total plastic strain in*			
					100 h	300 h	500 h	1 000 h
IMI 230	Cu	2.5	Aged sheet	200	435	—	—	—
				300	375	—	—	—
				400	220	—	—	—
				450	109	—	—	—
			Annealed sheet	20	360	—	—	—
				100	279	—	—	—
				200	235	—	—	—
				300	202	—	—	—
				400	125	—	—	—
IMI 317	Al	5.0	Annealed rod	20	633	608	—	593
	Sn	2.5		100	474	463	—	458
				200	370	—	—	370
				300	359	—	—	359
				400	337	—	—	337
				500	162	119	—	88
IMI 318	Al	60	Annealed rod	20	832	818	—	788
	V	4.0		100	704	680	—	676
				200	638	636	—	635
				300	576	568	—	—
				400	287	144	—	102
				500	32	18	—	—
IMI 550	Al	4.0	Fully heat-	300	724	718	—	710
	Mo	4.0	treated bar	400	551	516	—	471
	Sn	2.0		450	254	174	—	101
	Si	0.5		500	82	51	—	31
IMI 551	Al	4.0	Fully heat-	400	621	575	540	501
	Mo	4.0	treated rod	450	307	217	—	—
	Sn	4.0						
IMI 679	Sn	11.0	Air-cooled and	20	896	880	—	880
	Zr	5.0	aged rod	150	703	695	—	672
	Al	2.25		300	664	664	—	649
	Mo	1.0		400	579	571	—	526
	Si	0.2		450	448	386	—	247
				500	131	93	—	62

(continued)

Table 22.42 TITANIUM ALLOYS—TYPICAL CREEP PROPERTIES—*continued*

IMI designation	Nominal composition %		Condition	Temperature °C	Stress MPa to produce 0.1% total plastic strain in			
					100 h	300 h	500 h	1 000 h
IMI 680	Sn	11.0	Quenched and	20	1 127	1 112	—	—
	Mo	4.0	aged rod	150	945	942	—	—
	Al	2.25		200	862	856	—	—
	Si	0.2		300	804	788	—	—
				400	555	540	—	—
				450	298	209	—	—
				500	88	51	—	—
			Furnace-cooled	300	570	—	—	—
			and aged rod	350	540	—	—	—
				400	490	—	—	—
IMI 685	Al	6.0	Heat-treated	200	599	—	592	589
	Zr	5.0	forgings	300	551	—	541	535
	Mo	0.5		400	497	—	480	462
	Si	0.25		450	461	—	431	426
				500	408	—	340	—
IMI 829	Al	5.5	Fully heat	450	478	—	—	—
	Sn	3.5	treated rod	500	420	—	—	—
	Zr	3.0		550	300	—	—	—
	Nb	1.0		600	130	—	—	—
	Mo	0.3						
	Si	0.3						
IMI 834	Al	5.8	Heat-treated	500	461	—	—	—
	Sn	4.0	forgings	550	339	—	—	—
	Zr	3.5		600	205	—	—	—
	Nb	0.7						
	Mo	0.5						
	Si	0.35						
	C	0.06						

Table 22.43 TITANIUM AND TITANIUM ALLOYS—TYPICAL FATIGUE PROPERTIES

IMI designation	Nominal composition %	Condition	Temperature °C	Tensile strength MPa	Details of test	Endurance limit for 10^7 cycles stated MPa
IMI 115	Commercial purity	Annealed rod	Room	354	Rotating bend Smooth $K_t = 1$	±193
				354	Notched $K_t = 3$	±123
IMI 125	Commercial purity	Annealed rod	Room	417	Rotating bend Smooth $K_t = 1$	±232
				417	Notched $K_t = 3$	±154
IMI 130	Commercial purity	Annealed rod	Room	550	Rotating bend Smooth $K_t = 1$	±270
				550	Notched $K_t = 2$	±170
				550	Notched $K_t = 3.3$	±170
				550	Direct stress (Zero mean) Smooth $K_t = 1$	±263
				550	Notched $K_t = 1, 5$	±247
				550	Notched $K_t = 2$	±170
				550	Notched $K_t = 3.3$	±116
				589	Smooth $K_t = 1$	±278
				589	Notched $K_t = 2$	±147
				589	Notched $K_t = 3$	±123
				589	Notched $K_t = 4$	±116

(continued)

Table 22.43 TITANIUM AND TITANIUM ALLOYS—TYPICAL FATIGUE PROPERTIES—*continued*

IMI designation	Nominal composition %		Condition	Temperature °C	Tensile strength MPa	Details of test	Endurance limit for 10^7 cycles stated MPa
						Direct stress (Zero mean)	
IMI 160	Commercial purity		Annealed rod	Room	674	Smooth K_t	±376
IMI 230	Cu	2.5	Annealed sheet	Room	564	Reversed bend	±390
			Aged sheet	room	772	Reversed bend	±490
						Direct stress (Zero minimum)	
			Aged sheet	room	761	Smooth $K_t = 1$	0→560
						Rotating bend	
			Annealed rod	room	598	Smooth $K_t = 1$	±370
				400		Smooth $K_t = 1$	±150
						Direct stress (Zero mean)	
			Annealed	Room	638	Smooth $K_t = 1$	±280
						Rotating bend	
			Aged rod	Room	700	Smooth $K_t = 1$	±450
				400	—	Smooth $K_t = 1$	±290
						Direct stress (Zero mean)	
			Aged rod	Room	792	Smooth $K_t = 1$	±470
						Notched $K_t = 3.3$	±200
IMI 260	Pd	0.2	Similar to IMI 115				
IMI 262	Pd	0.2	Similar to IMI 125				
						Rotating bend	Limits for this alloy 10^8 cycles
IMI 317	Al	5.0	Annealed rod	Room	—	Smooth $K_t = 1.0$	±371
	Sn	2.5				Notched $K_t = 2.0$	±263
						Notched $K_t = 3.3$	±239
						Direct stress (Zero mean)	
						Smooth $K_t = 1.0$	±433
						Notched $K_t = 1.5$	±278
						Notched $K_t = 2.0$	±201
						Notched $K_t = 3.3$	±154
						Rotating bend	
IMI 318	Al	6.0	Annealed rod	Room	960	Smooth $K_t = 1$	±470
	V	4.0			960	Notched $K_t = 2.7$	±230
						Direct stress (Zero minimum)	
					1 015	Smooth $K_t = 1$	0→750
					1 015	Notched $K_t = 1$	0→325
						Direct stress (Zero minimum)	
IMI 550	Al	4.0	Fully heat-treated rod	Room	1 180	Smooth $K_t = 1$	0→850
	Mo	4.0			1 180	Notched $K_t = 3$	0→350
	Sn	2.0					
	Si	0.5					
						Rotating bend	
						Rotating bend	
						Smooth $K_t = 1$	±587
						Notched $K_t = 2.4$	±394

(*continued*)

Table 22.43   TITANIUM AND TITANIUM ALLOYS—TYPICAL FATIGUE PROPERTIES—*continued*

IMI designation	Nominal composition %		Condition	Temperature °C	Tensile strength MPa	Details of test	Endurance limit for 10^7 cycles stated MPa
IMI 551	Al	4.0	Fully heat-treated rod	Room	—	Rotating bend Smooth $K_t = 1$	±750
	Mo	4.0			—	Notched $K_t = 3.2$	±430
	Sn	4.0					
	Si	0.5					
IMI 679	Sn	11.0	Air-cooled and aged rod	Room	—	Rotating bend Smooth $K_t = 1.0$	±641*
	Zr	5.0		200	—	Smooth $K_t = 1.0$	±510*
	Al	2.25		400	—	Smooth $K_t = 1.0$	±510*
	Mo	1.0		450	—	Smooth $K_t = 1.0$	±556
	Si	0.2		500	—	Smooth $K_t = 1.0$	±495
IMI 680	Sn	11.0	Quenched and aged rod	Room	1 272	Rotating bend Smooth $K_t = 1$	(Limits for 2×10^7 cycles) ±710
	Mo	4.0			1 272	Notched $K_t = 2$	±340
	Al	2.25			1 272	Notched $K_t = 3.3$	±293
	Si	0.2		Room	1 272	Direct stress (Zero mean) Smooth $K_t = 1$	(Limits for 2×10^7 cycles) ±695
						Notched $K_t = 2$	±371
						Notched $K_t = 3.3$	±232
				Room	—	Rotating bend Smooth $K_t = 1$	(Limits for 10^8 cycles) ±648
				200	—	Smooth $K_t = 1$	±495
				400	—	Smooth $K_t = 1$	±479
			Furnace-cooled rod	Room	1 100	Direct stress Zero Smooth $K_t = 1$	±680
IMI 685	Al	6.0	Fully heat-treated rod	20	—	Direct stress (Zero mean) Smooth $K_t = 1$	±440
	Zr	5.0		450	—	Smooth $K_t = 1$	±300
	Mo	0.5		520	—	Smooth $K_t = 1$	±260
	Si	0.25		450	—	Direct stress (Zero minimum) Smooth $K_t = 1$	0→475
				520	—	Smooth $K_t = 1$	0→425
			Fully heat-treated forging	Room	—	Direct stress (Zero minimum) Smooth $K_t = 1$	0→640
				Room	—	Notched $K_t = 3.5$	0→220
				475	—	Smooth $K_t = 1$	0→460
				475	—	Notched $K_t = 3.5$	0→210
IMI 829	Al	5.5	Fully heat-treated rod	Room	—	Direct stress (Zero minimum) Smooth $K_t = 1$	0→550
	Sn	3.5			—	Notched $K_t = 3$	0→260
	Zr	3.0					
	Nb	1.0					
	Mo	0.3					
	Si	0.3					
IMI 834	Al	5.8	Fully heat-treated rod	Room	—	Direct stress (Zero minimum) Smooth $K_t = 1$	0→577
	Sn	4.0			—	Notched $K_t = 2$	0→363
	Zr	3.5					
	Nb	0.7					
	Mo	0.5					
	Si	0.35					
	C	0.06					

* Limit for 10^8 cycles.

Table 22.44   IZOD IMPACT PROPERTIES OF TITANIUM AND TITANIUM ALLOYS

IMI designation	Nominal composition %	Condition	Izod value Joules(ft lbf)*							
			−196°C	−78°C	20°C	100°C	200°C	300°C	400°C	500°C
IMI 130[†]	Commercially pure	Annealed rod	—	62.4 (46)	61.0 (45)	62.4 (46)	72 (53)	82 $(60\frac{1}{2})$	84 (62)	82 $(62\frac{1}{2})$
IMI 317	Sn 5.0 Al 2.5	Annealed rod	17.6 (13)	20.3 (15)	27.1 (20)	35.2 (26)	52.8 (39)	63.7 (47)	70.5 (52)	71.8 (53)
IMI 318	Al 6.0 V 4.0	Annealed rod	13.5 (10)	14.9 (11)	20.3 (15)	25.7 (19)	40.6 (30)	65.0 (48)	83.5 (63)	92.0 (68)
IMI 550	Al 4.0 Mo 4.0 Sn 2.0 Si 0.5	Fully heat-treated rod	—	—	19.0 (14)	—	—	—	—	—

IMI designation	Nominal composition %	Condition	Charpy value Joules(ft lbf)							
			−196°C	−78°C	20°C	100°C	200°C	300°C	400°C	500°C
IMI 551	Al 4.0 Mo 4.0 Sn 4.0 Si 0.5	Fully heat-treated rod	13.5 (10)	19 (14)	20.3 (15)	21.7 (16)	24.4 (18)	26.5 $(19\frac{1}{2})$	28.5 (21)	31.2 (23)
IMI 679	Sn 11.0 Zr 5.0 Al 2.25 Mo 1.0 Si 0.2	Air-cooled and aged	10.8 (8)	13.5 (10)	14.9 (11)	16.3 (12)	19 $(14\frac{1}{2})$	25 $(18\frac{1}{2})$	30 (22)	33.9 (25)
IMI 680	Sn 11.0 Mo 4.0 Al 2.25 Si 0.2	Quenched and aged rod	8.1 (6)	8.8 $(6\frac{1}{2})$	10.8 (8)	12.2 (9)	14.9 (11)	17.6 (13)	20.3 (15)	25.7 (19)
IMI 685	Al 6.0 Zr 5.0 Mo 0.5 Si 0.25	Fully heat-treated rod	31.2 (23)	39.3 (29)	43.4 (32)	—	—	—	—	—

* BSS 131 (1) 0.45 in diameter straight notched test pieces. [†] Izod values of commercial purity titanium are appreciably affected by variation in hydrogen content within commercial limits (0.008% maximum) in Ti 130 rod.

22.7   Mechanical properties of zinc and zinc alloys

Table 22.45   MECHANICAL PROPERTIES OF ZINC ALLOYS AT ROOM TEMPERATURE

	Composition						Properties	
Zinc	Unalloyed zinc is generally used only in the wrought form. The data here are some typical values							
	Zn min	Pb	Cd max	Sn	Fe	Total Pb+ Cd + Sn + Fe + Cu (max)	Parallel to rolling direction	Across rolling direction
BS 3436 Zn 1	99.99	0.003	0.003	0.001	—	0.01	Zn 1	
Zn 2	99.95	0.03	0.02	0.001	0.01	0.05	Tensile strength	
Zn 3	99.5	0.35	0.15	0.001	0.03	0.5	MPa 120	150
Zn 4	98.5	1.35	0.15	0.02	0.04	1.5	Elongation % 60–80	40–60
							Hardness Vickers 30	

(continued)

Table 22.45　MECHANICAL PROPERTIES OF ZINC ALLOYS AT ROOM TEMPERATURE—*continued*

	Composition			Properties		
Zinc-copper-titanium	Cu	Ti	Balance			
	0.14	0.1–0.15	zinc of 99.995% purity	Tensile strength MPa	180	216
				Elongation %	35	20
				Hardness (Brinell)	40–45	

These alloys are used principally in pressure die castings and other castings. Properties are given in Chapter 26 Casting and foundry data.

22.8　Mechanical properties of zirconium and zirconium alloys

Table 22.46　MECHANICAL PROPERTIES OF ZIRCONIUM ALLOYS AT ROOM TEMPERATURE

Material	Nominal composition	Condition	0.1% proof stress MPa	UTS MPa	Elongation %	Macro-hardness HV	Reference
Zr (ex iodide)	>99.9% purity Impurities in ppm by wt O_2 65, N_2 15, H_2 12, Hf 35, Ni 20	Crystal bar, cold rolled and vacuum annealed 2 h at 750°C	100–130	170–210	40–45 on 1 in gauge	85–100	1
Zr (ex sponge)	>99.6% purity Impurities in ppm by wt O_2 1300, N_2 80, H_2 20, Hf 400, Ni 40	Sheet material, cold rolled and vacuum annealed 2 h at 750°C	250–310 in r.d.	350–390 in r.d.	23–31 on 1 in gauge	195–215	1
Zircaloy 2	Sn 1.2–1.7% Fe 0.07–0.2% Cr 0.05–0.15% Ni 0.03–0.08% Zr–remainder	Plate materials cold rolled and annealed 1 h at 750°C	340[†] in r.d. 490 in t.d.	450 in r.d. 520 in t.d.	29 in r.d. 23 in t.d. on 1 in gauge	205–220*	2 1
Zirconium 30*	Cu 0.46–0.66% Mo 0.50–0.60% Zr–remainder	Sheet and strip	220–320 in r.d.	470–550 in r.d.	20–31 2 in gauge	130–180	3
Zr/2½% Nb	Nb 2.55% Impurities in ppm by wt O_2 1050, N_2 20, H_2 10, Hf 70 N < 40	Plate, hot rolled at 750°C later annealed ½ h at 700°C	470 for forged product	590 for forged product 410–500 ppm O_2 600–2 500 ppm O_2	24 gauge unspecified[‡]	150–500 ppm O_2 230–2 500 ppm O_2	4 3
Zr 702	Commercially pure with up to Hf 4.5	Annealed sheet	207	379	16	5T	
Zr 704	Cr + Fe 0.2–0.4 Sn 1–2	Annealed sheet	241	413	14	5T	
Zr 705	Nb 2.5 O_2 0.18	Annealed sheet	379	552	16	3T	
Zr 706	Nb 2.5 O_2 0.16	Annealed sheet	345	510	20	2½T	

* Imperial Metal Industries Nomenclature.

Table 22.47 MECHANICAL PROPERTIES OF ZIRCONIUM ALLOYS AT ELEVATED TEMPERATURE

Material	Nominal composition	Test temperature °C	0.1% proof stress MPa	UTS MPa	Elongation %	Reference
Zr (ex sponge)	>99.6% purity Impurities in ppm by wt O_2 1 300, N_2 80, H_2 20, Hf 400, Ni 40	371	50[†]	110	57	3
Zircalloy 2	Sn 1.2–1.7% Fe 0.07–0.2% Cr 0.05–0.15% Ni 0.03–0.08% Zr–remainder	300	92–126	210–260	33–44 on 2 in gauge	3
Zirconium 30	Cu 0.46–0.66% Mo 0.50–0.60% Zr–remainder	300	160	250	34 on $5.65\sqrt{A}$ gauge	3
$Zr/2\frac{1}{2}\%$ Nb	Nb 2.55% Impurities in ppm by wt O_2 1 050, N_2 20, H_2 10, Hf 70, Ni < 40	300	210	340 280–500 ppm O_2^* 350–2 500 ppm O_2^*	33 gauge unspecified	3 4*
Zr 702	Commercially pure with up to Hf 4.5	100 200 300 400	296 262 207 145	413 351 275 227	32 37 47 54	
Zr 704	Cr + Fe 0.2–0.4	100 200 300 400	344 289 241 193	510 489 434 358	22 23 27 32	
Zr 705	Nb 2.5 O_2 0.18	100 200 300 400	455 379 317 269	565 489 427 372	26 31 33 33	

r.d. rolling direction.
t.d. transverse direction.
*Imperial Metal Industries nomenclature.
[†]0.2% proof stress.

REFERENCES TO TABLES 22.46 AND 22.47

1. B. J. Gill, Dept. of Metallurgy and Materials Tech., University of Surrey, Guildford, England, 1972.
2. W. Evans and G. W. Parry, *Electrochem. Tech.*, 1966, **4**, 225.
3. Imperial Metal Industries pubn 2 Ed/MK105/33/366.
4. J. Winton and R. A. Murgatroyd, *Electrochem. Tech.*, 1966, **4**, 358.

22.9 Tin and its alloys

The applications of tin are governed very largely by its low melting point, excellent fluidity when molten, relative softness, formability and readiness to form alloys with other metals.

The low melting point of tin means that at normal ambient temperature the metal is nearly 60% of its melting point on the absolute Temperature Scale. This results in rather low mechanical strength at room temperature, as would be expected from any metal tested at a temperature corresponding to such a high proportion of its Absolute melting point, i.e. 505 K. Because of this, pure tin recrystallises readily at room temperature. Consequently, unlike the majority of industrial metals, only slight work hardening occurs initially in tin, followed by work-softening with further deformation due to grain growth. A similar behaviour is found in tin-rich alloys. The mechanical properties recorded for tin-rich alloys are strongly dependent on impurity levels and the strain rate.

Within reasonable limits any tin-base alloy can be die cast successfully. The choice of alloy, therefore, rests on such matters as mechanical properties, wear resistance or cost rather than on any consideration of the casting behaviour.

PURE TIN—MECHANICAL PROPERTIES

Tensile strength (at 0.4 mm/mm min) at	20°C	14.5 MPa
	100°C	11.0 N mm^{-2}
	200°C	4.5 N mm^{-2}
Shear strength at 20°C		12.3 MPa
Hardness at 20°C		3.9 HB
100°C		2.3 HB
200°C		0.9 HB
Young's modulus at 20°C		49.9 GPa
Rigidity modulus at 20°C		18.4 GPa
Poisson's ratio		0.357
Creep strength at 15°C		
(approx. life at 2.3 N mm^{-2})		170 days
(approx. life at 1.4 N mm^{-2})		550 days
Fatigue strength for 10^8 reversals at 15°C		± 2.5 MPa
Elastic modulus (tension)		
As cast		41.6 GPa
Self annealed		44.3 GPa
Impact strength		
0°C		44.1 J
150°C		22.7 J

Note: Mechanical properties are very dependent on the rate of loading.

Table 22.48   TIN-RICH SOLDERS—MECHANICAL PROPERTIES (BULK)

	Nominal composition (wt %)					
Properties	63Sn37Pb	60Sn40Pb	50Sn50Pb	95Sn5Sb	96.5Sn3.5Ag	62Sn36Pb2Ag
Tensile strength (N mm^{-2}) (Tested at 0.05 mm/min)						
at 20°C	na	19	na	31	37	43
at 100°C	na	4	na	20	na	19
Elongation (%)						
at 20°C	na	135	na	25	31	7
at 100°C	na	>100	na	21	na	na
Hardness (HB)	17	16	14	15	15	na
Young's modulus (GPa)	na	29.99	na	49.99	na	22.96
Creep strength (MPa) for life of 1000 h						
at 20°C	na	29	na	21	22	na
at 100°C	na	4.5	na	9	na	27
Impact strength (J)	20	20	21	27	na	na

Note: Tensile strength, hardness etc., are very dependent on the rate of loading and temperature of testing.
na = not available.

Table 22.49 PROPERTIES OF SOME TIN-RICH ALLOYS

Mechanical properties

Nominal composition (wt. %)					UTS (MPa)	Elongation (%)	Compressive yield 0.125% set		Strength (MPa) 0.25% set		Hardness (HB)	Liquidus (°C)	Solidus (°C)	Density (g cm⁻³)
Sn	Sb	Cu	Pb	Bi			20°C	100°C	20°C	100°C				
Rem	4.5	4.5	—	—	64 Chillcast 62 Die cast	9 Chillcast 2 Die cast	30.3	17.9	88.2	47.6	17 Chillcast	371	223	7.34
Rem	7.5	3.5	—	—	77 Chillcast	18 Chillcast	42	21	103	60	24 Chillcast	354	241	7.39
Rem	8	8	—	—	69 Die cast	1 Die cast	46	21	117	6.8	27 Chillcast 30 Die cast	422	240	7.45
Rem	12	3	10	—	27 Chillcast	na	38.3	14.8	111.4	47.6	27 Chillcast	306	184	7.53
Rem	15	2	18	—	22.5 Chillcast	na	34	14	103	46	22.5 Chillcast	296	181	7.75
Rem	13	5	—	—	69 Die cast	1	na	na	na	na	29 Die cast	na	na	na
Rem	6	2	—	—	64.7 Chillcast 59 Annealed for 1 h air cooled 52 Cold rolled, 32% reduction	24 Chillcast 40 Annealed for 1 h air cooled 52 Cold rolled, 32% reduction	na	na	na	na	23.8 Chillcast 9.5 Annealed for 1 h air cooled 8 Cold rolled, 32% reduction	295	244	7.28
Rem	—	—	—	40	55	200 Slow loading	na	na	na	na	22	170	138	8.2

Note: Tensile strength, hardness, etc, are very dependent on the rate of loading and casting conditions.
Properties are typical values.
na = not available.

FURTHER READING

A. General

1. 'Tin and its Alloys', E. S. Hedges, Edward Arnold, London, 1960.
2. 'Tin and Its Alloys and Compounds', B. T. K. Barry and C. J. Thwaites, Ellis Horwood Ltd, Chichester, England, 1983.
3. 'Metals Handbook', 10th Edition, Vol. 2, 'Properties and Selection, Non Ferrous Alloys and Pure Metals', American Society for Metals, Ohio, 1990.
4. Goodfellow, 'Materials for Research and Development', Catalogue 8, 1987/88.

B. Bearing Alloys

5. 'Babbit Alloys for Plain Bearings', International Tin Research Institute (ITRI), Publication No. 149.
6. 'Developments in Plain Bearing Technology', *Tribologia e Lubrificazione*, Vol. 10, p94, 1975, ITRI Publication No. 513.
7. 'Materials for Plain Bearings', G. C. Pratt, *International Metallurgical Review*, no. 174, Vol. 18, p62, 1973.
8. 'Tribology Handbook', M. J. Neale, Butterworth, London, 1973.
9. 'Tin-Base Casting Alloys', C. J. Evans, *Engineering Materials and Design*, p15, Aug, 1975.
10. 'Tin in Plain Bearings', ITRI Publication no. 595.

C. Pewter

11. 'Working with Pewter', ITRI Publication no. 566.
12. 'Modern Pewter', S. Charron, Van Nostrand Rheinhold, New York, 1973.
13. 'The Techniques of Pewtersmithing', C. Hull and J. Murrel, B.T. Batsford Ltd., London, 1984.

D. Fusible Alloys

14. 'Fusible Alloys Containing Tin', ITRI Publication no. 175.
15. 'The Properties of Tin Alloys', ITRI Publication no. 155.
16. 'Fusible Alloys for Industry', C. J. Evans, *Tin and its Uses*, no. 133, 1982, p10.
17. 'Fusible Alloys', *Engineering*, 1981, **221**(11), ppl.

E. Solder Alloys

18. 'Soft Soldering Handbook', ITRI Publication no. 533.
19. 'Solder Alloy Data', ITRI Publication no. 656.
20. 'Soldering Handbook', B. M. Allen, Iliffe, London, 1969.
21. 'Solders and Soldering', H. H. Manko, McGraw-Hill, New York, 1964.
22. 'Soldering Manual' American Welding Society, New York, 2nd Edition 1979.
23. 'Das Löten für den/Praktiker', R. Strauss, Franzis-Verlag, München, 1978.

22.10   Steels

Note for 8th edition: the tables in this section refer to standards that have been recently superceded by European (EN) and international (ISO) standards. The older British standards have been retained for information. Chapter 1 list equivalent designations in other systems for some alloys. For full information the relevant specifications must be consulted.

Table 22.50   FORGED OR ROLLED STEELS—ROOM TEMPERATURE LONGITUDINAL MECHANICAL PROPERTIES

1. Carbon steels with up to 1.7% Mn content, including free cutting steels (Bs 970:Part 1:1983)

Material	British or other standard**	Composition %					Condition*	Limiting ruling section		Properties (minima unless otherwise stated)					
		C	Mn	Ni	Cr	Other elements	(Bs 970:Part 1:1983)	mm	in	UTS MPa	Yield stress MPa	Elong. (gl $5.65\sqrt{S_0}$) %	Impact J	Hardness HB	Remarks
0.10 C	BS 055M15 (En2) AISI 1010	0.2 max	0.8 max	—	—	0.05 max	AR	150	6	310	—	25	—	121	
0.20 C	BS 070M20 AISI 1020	0.16 0.24	0.50 0.90	— —	— —	P, S 0.050 max	N	152 254	6 10	430 400	215 200	21 20	— —	126/179 —	General constructional steel suitable for welding
							CD	13 76	$\frac{1}{2}$ 3	530 430	385 340	12 14	— —	— —	
							HT	19	$\frac{3}{4}$	540/690	355	20	41‡	—	
0.26	BS 070M26 AISI 1025	0.22 0.30	0.50 0.90	— —	— —	P, S 0.050 max	N	64 254	$2\frac{1}{2}$ 10	490 430	245 215	20 20	— —	— —	Medium strength engineering steel
							CD	13 76	$\frac{1}{2}$ 3	570 490	430 370	11 13	— —	— —	
							HT	13 29	$\frac{1}{2}$ $1\frac{1}{8}$	620/770 540/690	415 355	16 20	34‡ 41‡	— —	
0.30 C	BS 080M30 (En5) AISI 1030	0.26 0.34	0.60 1.00	— —	— —	P, S 0.050 max	N	150 250	6 10	490 460	245 230	20 19	— —	143/192 126/179	General engineering steels widely used in the bright drawn condition
							CD	13 76	$\frac{1}{2}$ 3	600 530	450 385	10 12	— —	— —	
							HT	19 64	$\frac{3}{4}$ $2\frac{1}{2}$	625/775 550/700	415 340	16 18	34‡ 34	179/229 152/207	
0.38 C	BS 080M36	0.32 0.40	0.60 1.00	— —	— —	P, S 0.050 max	N	64 254	$2\frac{1}{2}$ 10	540 490	280 245	16 18	27‡ —	— —	
							CD	13 76	$\frac{1}{2}$ 3	620 540	480 400	9 11	— —	— —	
							HT	13 29	$1\frac{1}{2}$ $1\frac{1}{8}$	789/850 625/775	465 400	16 16	34‡ 34‡	201/255 179/229	

‡ Grain controlled steel.
* AR—As rolled.   N—Normalised.   CD—Cold drawn.   HT—Hardened and tempered.
** The specification and standards cited may be obsolete, but have been retained because the designations are still in common use.

(continued)

Table 22.50   FORGED OR ROLLED STEELS—ROOM TEMPERATURE LONGITUDINAL MECHANICAL PROPERTIES—*continued*

Material	British or other standard**	Composition %					Condition*	Limiting ruling section		Properties (minima unless otherwise stated)					Remarks
		C	Mn	Ni	Cr	Other elements		mm	in	UTS MPa	Yield stress MPa	Elong. (gl $5.65\sqrt{S_0}$) %	Impact J	Hardness HB	
0.40 C Steel	BS 080M40 (En8) AISI 1040	0.36 0.44	0.60 1.00	— —	— —	P, S 0.050 max	N	150 250	6 10	550 510	280 245	16 17	20‡ —	152/207 146/197	Nuts and bolts, forgings and general engineering parts
							CD	13 76	½ 3	660 570	530 430	7 10	— —	— —	
							HT	19 64	¾ 2½	690/850 620/770	465 385	16 16	34‡ 34‡	201/255 179/229	
0.46 C Steel	BS 080M46	0.42 0.50	0.60 1.00	— —	— —	P, S 0.050 max	N	64 254	2½ 10	620 540	310 280	14 15	— —	— —	Nuts and bolts, forgings and general engineering parts
							CD	13 76	½ 3	690 620	555 480	7 9	— —	— —	
							HT	13 102	½ 4	770/930 620/770	525 370	14 16	— —	— —	
0.50 C Steel	BS 080M50 (En43A) AISI 1050	0.45 0.55	0.60 1.00	— —	— —	P, S 0.050 max	N	152 254	6 10	620 570	310 280	14 14	— —	— —	Gears and machined parts for flame or induction hardening
							CD	13 76	½ 3	730 650	585 510	8 10	— —	— —	
							HT	13 63	½ 2½	850/1 000 690/850	570 430	12 14	— —	— —	
0.55 C Steel	BS 070M55	0.50 0.60	0.50 0.90	— —	— —	P, S 0.050 max	N	64 254	2½ 10	700 600	355 310	12 13	— —	— —	General machine parts requiring higher wear resistance
							CD	13 76	½ 3	790 710	620 570	7 9	— —	— —	
							HT	19 102	¾ 4	850/1 000 690/850	570 415	12 14	— —	— —	
0.19 C 1.2 Mn Steel	BS 120M19 AISI 1518	0.15 0.23	1.00 1.40	— —	— —	P, S 0.050 max	N	102 254	4 10	490 460	295 260	20 19	34‡ —	— —	Armature shafts. High tensile bolts. Lifting gear chains. Automotive forgings
							CD	13 76	½ 3	600 530	450 385	11 12	— 34‡	— —	
							HT	19 102	¾ 4	690/850 540/690	510 355	16 18	34‡ 47‡	— —	

Type	BS (En)	C	Mn			P, S	Condition	Limiting section (mm)	(in)	Tensile (N/mm²)	Yield/Proof (N/mm²)	Elong. %	Izod	Hardness HB	Uses
0.19 C 1.5 Mn Steel	BS 150M19 (En14A)	0.15 / 0.23	1.30 / 1.70	—	—	0.050 max	N	150	6	550	325	18	41‡	152/207	Armature shafts. High tensile bolts. Lifting gear chains. Automotive forgings
							N	250	10	510	295	17		146/197	
							HT	29	1⅛	690/850	510	16	41‡	—	
							HT	152	6	540/690	340	18	54‡	—	
0.28 C 1.2 Mn Steel	BS 120M28	0.24 / 0.32	1.00 / 1.40	—	—	0.050 max	N	152	6	540	325	16	34‡	—	Not recommended for new designs
							N	254	10	530	310	17		—	
							CD	13	½	650	510	8		—	
							CD	76	3	570	430	10		—	
							HT	29	1⅛	690/850	510	16	34‡	—	
							HT	102	4	620/770	415	16	41‡	—	
0.28C 1.5 Mn Steel	BS 150M28 (En14B)	0.24 / 0.32	1.30 / 1.70	—	—	0.050 max	N	152	6	590	355	16	34‡	—	Not recommended for new designs
							N	254	10	560	325	16		—	
							HT	13	½	770/930	570	16	34‡	—	
							HT	152	6	620/770	400	16	47‡	—	
0.36C 1.2 Mn steel	BS 120M36 (En15B) AISI 1536	0.32 / 0.40	1.00 / 1.40	—	—	0.050 mm	N	150	6	590	355	15		174/223	General engineering steel
							N	250	10	570	340	16	34‡	163/217	
							CD	13	½	690	555	7		—	
							CD	76	3	620	480	9		—	
							HT	19	¾	770/930	570	14	34‡	—	
							HT	102	4	620/770	415	18	41‡	—	
0.36C 1.5 Mn Steel	BS 150M36 (En15)	0.32 / 0.40	1.30 / 1.70	—	—	0.050 max	N	150	6	620	385	14		179/229	
							N	250	10	600	355	15		170/223	
							HT	13	½	850/1 000	635	12	34‡	248/302	
							HT	150	6	620/770	400	18	47‡	179/229	
Low C Free cutting steel	BS 220M07 (En1A)	0.15 max	0.90 / 1.30			S 0.20 / 0.30	CD	13	½	460	380	7		—	Free cutting mild steels to be machined in high speed automatic lathes
							CD	76	3	370	240	10		—	
							HR	102	4	360	215	22		103 min	

‡ Grain controlled steel.
* N—Normalised. CD—Cold drawn. HT—Hardened and tempered.

(*continued*)

Table 22.50   FORGED OR ROLLED STEELS—ROOM TEMPERATURE LONGITUDINAL MECHANICAL PROPERTIES—*continued*

Material	British or other standard**	C	Mn	Ni	Cr	Other elements	Condition*	Limiting ruling section mm	in	UTS MPa	Yield stress MPa	Elong. (gl 5.65$\sqrt{S_0}$) %	Impact J	Hardness HB	Remarks
Low C Free cutting steel	BS 230M07 AISI 1213	0.15 max	0.90 1.30	— —	— —	S 0.25 0.35	CD HR	13 76 102	$\frac{1}{2}$ 3 4	460 370 360	350 225 215	7 10 22	— — —	— — 103 min	Free cutting mild steels to be machined in high speed automatic lathes
Low C Free cutting steels containing lead	DIN 9SMnPb28	0.14 max	0.90 1.30	— —	— —	S 0.24 0.32 Pb 0.15 0.30	CD	16 64 102	$\frac{5}{8}$–1 $2\frac{1}{2}$ 4	510/690 415/665 385/635	415 310 245	7 9 10	— — —	— — —	Free cutting mild steels to be machined in high speed automatic lathes
0.28 C Free cutting steel	BS 216M28	0.24 0.32	1.10 1.50	— —	— —	S 0.12– 0.20	CD HT	13 76 19 64	$\frac{1}{2}$ 3 $\frac{3}{4}$ $2\frac{1}{2}$	570 490 620/770 540/690	430 370 430 355	10 12 18 20	— — 34‡ 34‡	— — — —	Not recommended for new designs
0.36 C Free cutting steel	BS 212M36 (En8M)	0.32	1.00	—	—	S 0.12– 0.20	CD HT	13 76 13 100	$\frac{1}{2}$ 3 $\frac{1}{2}$ 4	620 540 700/850 550/700	480 400 495 340	7 9 16 20	— — 54‡ 34‡	— — 201/255 152/207	Free cutting medium strength engineering steels for machine parts and engine components, etc.
0.36 C Free cutting higher Mn	BS 216M36 (En15AM) AISI 1536	0.32 0.40	1.30 1.70	— —	— —	S 0.12– 0.20 P 0.060 Si 0.25 max	CW HT	16 76 102 29	$\frac{5}{8}$ 3 4 $1\frac{1}{8}$	650 570 550/700 700	510 420 340 480	7 9 20 16	— — 34‡ 34‡	— — 152/207 201/255	Free cutting medium strength engineering steels for machine parts and engine

Properties (minima unless otherwise stated)

Type	Designation	C	Mn		S, P, Si	Other	Cond.	Size (mm)		Tensile	Yield	Elong. %	Izod	Hardness	Application
0.36 C Free cutting	BS 212M36 (En8M)	0.40 0.48	1.00 1.40	— —	S 0.12–0.20, P 0.060, Si 0.25 max	— —	HT	100 13	4 $\frac{1}{2}$	550/700 770/850	340 495	20 16	28 28	152/207 201/255	Free cutting medium strength engineering steels

2. Low alloy high strength weldable steels (BS 4360:1986, 1501:Pt 2:1988, 1503:1980, ASTM Standards)

Type	Designation	C	Mn		S, P, Si	Other	Cond.	Size (mm)		Tensile	Yield	Elong. %	Izod	Hardness	Application
Mn steel	BS 4360–40B	0.20 max	1.50 max	—	P, S 0.05 max	—	AR	16	$\frac{5}{8}$	340/500	235	25	27	—	Structural steel
Mn steel	BS 4360–40C	0.18 max	1.50 max	—	P, S 0.04 max	—	AR	16	$\frac{5}{8}$	340/500	235	25	27	—	Structural steel
Mn steel	BS 4360–40D	0.16 max	1.50 max	—	P + S 0.04 max	—	N	16 40	$\frac{5}{8}$ $1\frac{1}{2}$	340/500 340/500	235 225	25 25	27 at −20°C 27 at −20°C	— —	Structural steel
Mn steel	BS 4360–43A	0.25 max	1.60 max	—	P + S 0.05 max	—	AR	16	$\frac{5}{8}$	430/580	275	22	—	—	Structural steel
Mn steel	BS 4360–43B	0.2 max	1.50 max	—	P + S 0.05 max	—	AR	16	$\frac{5}{8}$	430/580	275	22	27 at 20°C	—	Structural steel
Mn steel	BS 4360–43C	0.10 max	1.50 max	—	P + S 0.04 max	—	AR	16	$\frac{5}{8}$	430/580	275	22	27 at 0°C	—	Structural steel
Mn Nb (V) steel	BS 4360–43D†	0.16 max	1.50 max	—	Nb + V 0.10 max, P + S 0.04 max	—	N	16 40	$\frac{5}{8}$ $1\frac{1}{2}$	430/580 430/580	275 265	22 22	27 at −20°C 27	—	Structural steel
Mn steel	BS 4360–43EE	0.16 max	1.50 max	—	P 0.04, S 0.03	—	N	16 40 63	$\frac{5}{8}$ $1\frac{1}{2}$ $2\frac{1}{2}$	430/580 430/580 430/580	275 265 255	22 22 22	27 at −50°C — —	— —	Structural steel

† It is permissible for these steels to be supplied with no Nb or V. If no grain refining elements, or if grain refining elements other than Al, Nb or V are used, the manufacturer shall inform the purchaser at the time or order. If Nb and V are used the total shall be less than 0.10%.
‡ Grain controlled steel.
* CD—Cold drawn. HR—Hot rolled. HT—Hardened and tempered. CW—Cold worked. AR—As rolled. N—Normalised.
*** The specification and standards cited may be obsolete, but have been retained because the designations are still in common use.

(continued)

Table 22.50   FORGED OR ROLLED STEELS—ROOM TEMPERATURE LONGITUDINAL MECHANICAL PROPERTIES—*continued*

Material	British or other standard**	Composition % C	Mn	Ni	Cr	Other elements	Condition*	Limiting ruling section mm	in	Properties (minima unless otherwise stated) UTS MPa	Yield stress MPa	Elong. (g15.65√S0) %	Impact J	Hardness HB	Remarks
Mn Nb (V) Steel	BS 4360–50B†	0.2 max	1.50 max	—	P+S 0.05 max	Nb+V 0.10 max	AR	16	$\frac{5}{8}$	490/640	355	20	—	—	Structural steel. Plate, bars, sections, tubes
		0.20 max	1.50 max	—	—	"	N	100	4	490/640	325	20	—	—	
Mn Nb (V) Steel	BS 4360–50C†	0.20 max	1.50 max	—	—	"	AR	16	$\frac{5}{8}$	490/640	355	20	27 at 0°C	—	Structural steel. Plate bars, sections, tubes
		0.20 max	1.50 max	—	—	"	N	100	4	490/640	325	20		—	
Mn Nb (V) Steel	BS 4360–50D†	0.18 max	1.50 max	—	—	Nb+V 0.10 max P+S 0.04 max	N	16	$\frac{5}{8}$	490/640	355	20	27	—	Structural steel. Plate, bars, sections, tubes
							N	63	$2\frac{1}{2}$	490/640	340	20	at −30°C	—	
Mn Nb (V) Steel	BS 4360–50EE†	0.18 max	1.50 max	—	—	Nb+V 0.15 max	N	16	$\frac{5}{8}$	490/640	355	20	27 at −50°C	—	Structural steel
								40	$1\frac{1}{2}$	"	345	20	27	—	
								63	$2\frac{1}{2}$	"	340	20		—	
Mn Nb (V) Steel	BS 4360–55C†	0.22 max	1.60 max	—	P+S 0.04 max	"	AR	16	$\frac{5}{8}$	550/700	450	19	27 at 0°C	—	Structural steel. Plate, bars, sections, tubes
Mn Nb (V) Steel	BS 4360–55E†	0.22 max	1.60 max	—	"	"	N	16	$\frac{5}{8}$	550/700	450	19	27	—	Structural steel. Plate, bars, sections, tubes
							N	63	$2\frac{1}{2}$	550/700	400	19	at −50°C	—	

Type	Specification	C	Mn		Cr	Other elements	Condition*	Thickness (mm)	Thickness (in)	Tensile (N/mm²)	Yield (N/mm²)	Elong (%)	Impact (J)		Application
Cr–Cu Steel	Bs 4360 WR 50A	0.12 max	0.30/0.50 max	0.65 max	0.50 1.25	Cu 0.2 0.55 / P 0.070 0.150 / S 0.050	AR	12	$\frac{1}{2}$	480	345	21	27 at 0°C	—	Weather resisting plate
								38	$1\frac{1}{2}$	480	325	21	—	—	
Cr–CuV Steel	BS 4360 WR 50B	0.19 max	0.90 1.25	0.01 0.06	0.50 0.65	Cu 0.25 0.40 / V 0.02 0.10	AR	12	$\frac{1}{2}$	480	345	21	27 at 0°C	—	Weather resisting plate
								50	2	480	345	21	—	—	
Cr–Cu–V Steel	BS 4360 WR 50C	0.22 max	0.90 145	0.01 0.06	0.50 0.65	Cu 0.25 0.40 / V 0.02 0.10	AR or N	12	$\frac{1}{2}$	480	345	21	27 at −15°C	—	
								50	2	480	340	21	—	—	
Mn V Steel	ASTM A 242	0.22 max	1.25 max	—	—	V 0.02 min	AR	38	$1\frac{1}{2}$	420	310	21	—	—	Structural quality
Mn V Nb Steel	ASTM A 572	0.22 max	1.35 max	—	—	Nb 0.01 min and/or V 0.02 min N 0.015 max	AR	19	$\frac{1}{2}$	550	450	17	—	—	Structural quality
Mo B Steel	BS 1501–261 (Fortiweld)	0.10 0.17	0.40 0.80	0.30 max	0.25 —	Mo 0.40– 0.60 / B 0.001 0.005	N	89	$3\frac{1}{2}$	550/670	420	16 (transv)	20	—	Pressure vessel plate
1CrMo Steel	BS 1501–620	0.09 0.18	0.40 0.65	0.30 max	0.80 1.15	Mo 0.45– 0.60 / Al 0.02 / Cu 0.30	NT	150	6	430/550	255	16	—	—	Pressure vessel plate

† It is permissible for these steels to be supplied with no Nb or V. If no grain refining elements, or if grain refining elements other than Al, Nb or V are used, the manufacturer shall inform the purchaser at the time or order. If Nb and V are used, the total shall be less than 0.10%.

* AR—As rolled. N—Normalised. NT—Normalised and tempered.

‡ The specifications and standards cited are obsolete, but have been retained because the designations are still in common use.

(continued)

Table 22.50   FORGED OR ROLLED STEELS—ROOM TEMPERATURE LONGITUDINAL MECHANICAL PROPERTIES—*continued*

Material	British or other standard**	Composition %					Condition*	Limiting ruling section		Properties (minima unless otherwise stated)					Remarks
		C	Mn	Ni	Cr	Other elements		mm	in	UTS MPa	Yield stress MPa	Elong. (gl 5.65√S_0) %	Impact J	Hardness HB	
1CrMo Steel	BS 1501–622 –690	0.12 0.18	0.40 0.80	0.30 max	2.00 2.50	Mo 0.90 1.10, Al 0.02, Cu 0.30	NT	50	2	690/820	555	15	—	—	Pressure vessel plate
1CrMo Steel	BS 1503–620 –440	0.18	0.40 0.70	0.40 max	0.85 1.15	Mo 0.45 0.65, Al 0.020, Cu 0.30	NT, HT	—	—	440/590	275	19	—	—	Pressure vessel forgings
1¼CrMo Steel	BS 1501–621	0.09 0.17	0.40 0.65	0.30 max	1.00 1.50	Mo 0.45 0.60	NT	150 76	6 3	490/650 480/600	310 340	16 18	—	—	Pressure vessel steel
1¼CrMo Steel	BS 1503–621 –4460	0.18	0.40 0.70	0.40 max	1.10 1.40	Mo 0.45 0.65, Al 0.020, Cu 0.30	QT	—	—	460/610	275	18	41	—	Pressure vessel forgings
1¾CrMoV Steel	ASTM A 517 Grade E (SSS100)	0.12 0.20	0.40 0.70	— —	1.40 2.00	Cu 0.30 0.40, Mo 0.40 0.60, V 0.04, or Ti 0.10, +B	HT	64	2½	790/930	700	15	—	—	Pressure vessel plate
MnCrMoV Steel	BS 1503–271 –560	— 0.17	1.00 1.50	0.30 0.70	0.50 1.00	Mo 0.20 0.35, V 0.05 0.10, Al 0.020, Cu 0.30	NT	—	—	560/710	370	17	41	—	Pressure vessel forgings
MnCrMoV Steel	BS 1502–271	0.17 max	1.00 1.50	0.30 0.70	0.50 1.00	Mo 0.20 0.35, V 0.05 0.10, Al 0.02	NT or QT	160	6	560/710	370	19	—	—	Pressure vessel forgings

Type	Designation	C	Mn	Ni	Cr	Other	Condition						Impact	Remarks
MnNiMo Steel	ASTM A533 Grade B	0.25 max	1.15 1.50	0.40 0.70	—	Mo 0.45–0.60	HT	102	4	620/793	482	16	—	Pressure vessel plate
CrMoZr Steel	WG Proprietary steel (N-A-XTRA 70)	0.20 max	0.70– 1.10	—	0.60– 1.00	Si 0.50– 0.90, Mo 0.20– 0.60, Zr 0.06– 0.12	HT	25	1	790/930	700	16	—	—
NiCuMo Steel	UK Proprietary steel (Nicuage Type 1)	0.06 max	0.40 0.65	0.70 1.00	—	Cu 1.00– 1.30, Nb 0.02 min	AR	13	$\frac{1}{2}$	540/630	500/580	25	Charpy V 61–135 at −10°C 41–115 at −20°C	Structural steel resistant to atmospheric corrosion. Good low temperature toughness
							Aged at 500/570°C	13	$\frac{1}{2}$	630/740	590/680	20	41–115 at −10°C 34–81 at −20°C	
NiCuMo Steel	ASTM A588–87 Grade K	0.17 max	0.50 max	0.40 0.70	0.40 0.70	Mo 0.10 max, Cu 0.30 0.50, Nb 0.005 0.05	AR	102	4	480	34	21	—	—
NiCrMoVNb Steel	BS 1501–281	0.08 0.14	1.00 1.50	0.60 1.00	0.40 0.70	Mo 0.24 0.30, V 0.04– 0.12, Al 0.020 0.04, Cu 0.04 0.12	NT	150	6	590/690	430	16 (transv)	Charpy V 68 at −10°C 27 at −40°C	Pressure vessel plate
								25	1	640/760	500	16 (transv)		

* NT—Normalised and tempered.   HT—Hardened and tempered.   QT—Quenched and tempered.   AR—As rolled.

** The specification and standards cited may be obsolete, but have been retained because the designations are still in common use.

(*continued*)

Table 22.50   FORGED OR ROLLED STEELS—ROOM TEMPERATURE LONGITUDINAL MECHANICAL PROPERTIES—*continued*

Material	British or other standard**	Composition % C	Mn	Ni	Cr	Other elements	Condition*	Limiting ruling section mm	in	UTS MPa	Yield stress MPa	Elong. (gl 5.65√S₀) %	Impact J	Hardness HB	Remarks
NiCrMoVB Steel	ASTM A517F (T1)	0.10–0.20	0.60–1.00	0.70–1.00	0.40–0.65	Mo 0.40–0.60 V 0.03–0.08 Cu 0.15–0.50 –B	HT	63	2½	790/930	700	16	Charpy V 20 at –45°C	—	Pressure vessel plate
NiCrMo Steel	UK Proprietary steel	0.22 max	1.00–1.20	0.70–0.85	0.90–1.10	Mo 0.45–0.60	HT	—	—	660 (Typical values)	501 (Typical values)	20	Charpy V 120	—	Pressure vessel forgings
2¼CrMo Steel	BS 1501-622 –515	0.90–0.15	0.30–0.60	0.30 max	2.00–2.50	Mo 0.90–1.0 Al 0.020 Cu 0.30	NT	150	6	500/670	285	16	—	—	Pressure vessel plate
	ASTM A542	0.15 max	0.30–0.60	—	2.00–2.50	Mo 0.90–1.10	HT	102	4	710/835	590	16	—	—	Plate
	BS 1503-622 (Esshete CRM2)	0.08–0.15	0.40–0.70	0.40 max	2.00–2.50	Mo 0.90–1.10	NT or HT	—	—	540/700	370	19	—	—	Pressure vessel forgings
5CrMo Steel	BS 1503-625 –570	0.15 max	0.30–0.80	0.40 max	4.00–6.00	Mo 0.45–0.65 Al 0.020 Cu 0.30	NT or HT	—	—	520/670	365	18	27	—	Pressure vessel forgings
5CrMo Steel	BS 1503-625 –590	0.18	0.30–0.80	0.40 max	4.00–6.00	Mo 0.45–0.65	NT or HT	—	—	590/740	450	17	41	—	Pressure vessel forgings
3NiCrMo	ASTM A543 (HY 80)	0.18 max	0.40	2.25–3.25	1.00–1.50	Mo 0.45–0.60 V 0.03 max	HT	102	4	710/835	590	17	—	—	Plate

Properties (minima unless otherwise stated)

Steel	Specification**	C	Mn	Ni	Cr	Mo, V	Cond.*	Ruling section (mm)	(in)	Tensile (N/mm²)	Yield 0.2% (N/mm²)	Elong. (%)	Impact	Hardness HB	Form
	ASTM A543 (HY 100)	0.20 max	0.40 —	2.50 3.50	1.30 1.80	Mo 0.45–0.60 V 0.30 max	HT	102	4	790/930	700	16	—	—	Plate
3.25NiCrMo Steel	ASTM A508 Class 4a	0.23 max	0.20 0.40	2.75 3.90	1.50 2.00	Mo 0.40–0.60 V 0.03 max	HT	—	—	790	700	16	Charpy V 47 at −30°C	—	Pressure vessel forgings
3½NiCrMo Steel	BS 1501–503	0.15 max	0.80 max	3.25 3.75	0.30 max	Mo 0.10 max	NT or HT	38	1½	450	265	22	Charpy V 34 at −80°C 18 at −100°C	—	Pressure vessel plate
5NiCrMoV Steel	ASTM (HY 130)	0.12 max	0.60 0.90	4.75 5.25	0.40 0.70	Mo 0.30–0.65 V 0.05–	HT	102	4	990	900	15	Charpy V 88 at −20°C	—	Plate

3. Low alloy direct hardening steels including ultra high strength and steels suitable for nitriding
(BS 970:Part 1:1988, 4670:1971, DIN 17200:1969; 17211:1970, Stahl Eisen Werkstoffblatt 550–57)

Steel	Specification**	C	Mn	Ni	Cr	Mo	Cond.*	Ruling section (mm)	(in)	Tensile (N/mm²)	Yield 0.2% (N/mm²)	Elong. (%)	Impact	Hardness HB	Form
1Ni Steel	BS 503M40 (En12)	0.36 0.44	0.70 1.00	0.70 1.00	—	—	HT	22	7/8	770/930	585	15	40	—	Forgings
								254	10	620/770	430	17	27	—	
½Cr Steel	DIN 38Cr2	0.34 0.41	0.50 0.80	—	0.40 0.60	—	HT	16	5/8	790/930	540	14	—	—	
								102	4	590/740	340	17	—	—	
½Cr Steel	DIN 46Cr2	0.42 0.50	0.50 0.80	—	0.40 0.60	—	HT	16	5/8	880/1080	630	12	—	—	
								102	4	680/835	450	15	—	—	
¾Cr Steel	Bs 526M60 (En11)	0.55 0.65	0.50 0.80	—	0.50 0.80	—	HT	64	2½	1 000/1 160	740	8	—	—	
								102	4	850/1 000	620	11	—	—	
1Cr Steel	DIN 34Cr4	0.30 0.37	0.60 0.90	—	0.90 1.20	—	HT	16	5/8	880/1 080	680	12	—	—	
								102	4	680/835	465	15	—	—	
1Cr Steel	Bs 530M40 (En18)	0.36 0.44	0.60 0.90	—	0.90 1.20	—	HT	29	1⅛	850/1 000	680	13	50	248/302	Bright bar
								100	4	700/850	525	17	50	201/255	
1½MnMo Steel	BS 605M36 (En16)	0.32 0.40	1.30 1.70	—	—	Mo 0.22–0.32	HT	150	6	700/850	525	17	50	201/255	Bright bar
								19	¾	1 000/1 150	850	12	42	293/352	

* HT—Hardened and tempered.   NT—Normalised and tempered.
** The specification and standards cited may be obsolete, but have been retained because the designations are still in common use.

(continued)

Table 22.50   FORGED OR ROLLED STEELS—ROOM TEMPERATURE LONGITUDINAL MECHANICAL PROPERTIES—*continued*

Material	British or other standard**	Composition %					Condition*	Limiting ruling section		Properties (minima unless otherwise stated)					Remarks
		C	Mn	Ni	Cr	Other elements		mm	in	UTS MPa	Yield stress MPa	Elong. (gl 5.65 $\sqrt{S_0}$) %	Impact J	Hardness HB	
1½MnMo Steel	BS 606M36 (En16M)	0.32 0.40	1.30 1.70	— —	— —	Mo 0.22–0.32 S 0.15–0.25	HT	100 29	4 1⅛	700/850 850/1000	525 680	15 11	50 35	201/255 248/302	Free cutting steel
¾CrMo Steel	UK Proprietary	0.53 (Typical)	0.70	—	0.70	Mo 0.40	HT	508	20	805 (Typical values)	540 (Typical values)	14	—	—	Crankshaft forgings
1CrMo Steel	DIN 25CrMo4	0.22 0.29	0.50 0.80	— —	0.90 1.20	Mo 0.15–0.30	HT	40 150	1½ 6	790/930 635/790	585 415	14 16	— —	— —	
1CrMo Steel	DIN 34CrMo4	0.30 0.37	0.50 0.80	— —	0.90 1.20	Mo 0.15–0.30	HT	100 250	4 10	790/930 680/835	550 465	14 15	— —	— —	
1CrMo Steel	BS 708M40 (En19A)	0.36 0.44	0.70 1.00	— —	0.90 1.20	Mo 0.15–0.25	HT	150 29	6 1⅛	700/850 925/1075	525 755	17 12	50 42	201/255 269/331	
1CrMo Steel	BS 709M40 (En19)	0.36 0.44	0.70 1.00	0.40 max	0.90 1.20	Mo 0.25–0.35	HT	250 29	10 1⅛	700/850 1000/1150	495 850	15 12	28 42	201/255 293/352	
1CrMo Steel	Bs 711M40	0.36 0.44	0.60 1.00	0.40 max	0.90 1.50	Mo 0.25–0.40 max	HT	500 250	20 10	700/850 800/950	500 600	12/16 10/14	14/22 11/18	207/255 235/285	Forgings
3CrMo Steel	Bs 722M24 (En40B)	0.20 0.28	0.45 0.70	—	3.00 3.50	Mo 0.45–0.65	HT	150 254	6 10	925/1075 850/1000	755 650	12 13	42 35	269/331 245/302	Nitriding steel
3CrMo Steel	DIN 32CrMo12	0.28 0.35	0.40 0.70	0.30 max	2.80 3.30	Mo 0.30–0.50	HT	150 250	6 10	930/1130 880/1080	740 680	11 12	— —	— —	Nitriding steel
MnNiMo Steel	Bs 785M19	0.15 0.23	1.40 1.80	0.40 0.70	0.40 max	Mo 0.15 0.35	HT	250 500	10 20	600/750 600/750	440 440	14/18 14/18	20/34 24/41	174/223 174/223	Forgings
1½NiCrMo Steel	Bs 817M40 (En24)	0.36 0.44	0.45 0.70	1.30 1.70	1.00 1.40	Mo 0.20 0.35	HT	250 29	10 1⅛	850/1000 1075/1225	650 940	13 11	35 35	248/302 311/375	Bright bar or nitriding
1½NiCrMo Steel	Bs 818M40	0.36 0.44	0.45 0.85	1.30 1.80	1.00 1.50	Mo 0.20–0.40	HT	1000 250	39 10	800/950 950/1100	610 789	10/14 8/12	15/24 18/30	235/285 277/331	Forgings

Type	Specification	C	Mn	Ni	Cr	Other (%)	Condition	mm	in	Tensile (N/mm²)	Proof (N/mm²)	Elong. (%)	Izod (J)	Hardness	Remarks
1NiCrMo Steel	DIN 36CrNi Mo4	0.32 0.40	0.50 0.80	0.90 1.20	0.90 1.20	Mo 0.15 0.30	HT	100 250	4 10	880/1035 740/880	680 540	12 14	— —	— —	
2½NiCrMo Steel	Bs 826M40 (En26)	0.36 0.44	0.45 0.70	2.30 2.80	0.50 0.80	Mo 0.45 0.65	HT	100 250 100	4 10 4	1 550 925/1 075 1 550	123 740 1 235	7 12 7	11 28 33	444 269/331 444	Nitriding Nitriding Bright bar
3NiCrMo Steel	BS 830M31 (En27)	0.27 0.35	0.45 0.70	2.75 3.25	0.90 1.20	Mo 0.25 0.35	HT	254 64	10 2½	850/1 000 1 080/1 240	650 940	13 11	41 41	— —	
4NiCrMo Steel	BS 835M30 (En30B)	0.26 0.34	0.45 0.70	3.90 4.30	1.10 1.40	Mo 0.20 0.35	HT	150	6	1 550	1 235	7	16	444	
4NiCrMo Steel	BS 2S146	0.34 0.42	0.15 0.60	3.50 4.50	1.60 2.00	Mo 0.60	HT	100	4	1 760/1 960	1 420	8	15	530	Ultra high strength aircraft steel
3¼CrMo V Steel	BS 897M39 (En40C) DIN 39Cr Mo V139	0.35 0.43	0.45 0.70	— —	3.00 3.50	Mo 0.80 1.10 V 0.15 0.25	HT	63	2½	1 310	1 160	8	16	—	Nitriding steel Forgings
5CrMo V	Bs BH11 ASTM A579 Grade 41	0.32 0.42	0.40	—	4.75 5.25	Si 0.85 1.15 Mo 1.25 1.75 V 0.30 0.50	HT	—	—	—	—	229 max	—		Tool steel
1CrMoNiV Steel	ASTM A579 Grade 23 (D6AC)	0.45 (Typical)	0.75 (Typical)	0.55	1.05	Mo 1.00 V 0.10	HT	25	1	1 885 (Typical values)	1 680 (Typical values)	10	Charpy V 20	—	Ultra high strength aircraft steel
2NiCrMoV Steel	ASTM A579 Grade 32 (300M)	0.44 (Typical)	0.70 (Typical)	1.85	0.80	Si 1.60 Mo 0.35 V 0.06	HT	25 76	1 3	1 990 1 930 (Typical values)	1 685 1 620 (Typical values)	9 9	Charpy V 27 26	— —	Ultra high strength aircraft steel

* HT—Hardened and tempered.

** The specification and standards cited may be obsolete, but have been retained because the designations are still in common use.

(continued)

Table 22.50   FORGED OR ROLLED STEELS—ROOM TEMPERATURE LONGITUDINAL MECHANICAL PROPERTIES—*continued*

Material	British or other standard**	Composition %					Condition*	Limiting ruling section		Properties (minima unless otherwise stated)					Remarks
		C	Mn	Ni	Cr	Other elements		mm	in	UTS MPa	Yield stress MPa	Elong. (gl 5.65√S_0) %	Impact J	Hardness HB	
NiCrMoV Steel	UK Proprietary (NCMV)	0.44	0.45 (Typical)	1.80	1.50	Mo 0.90 V 0.25	HT	29	1 $\frac{1}{8}$	2070 (Typical values)	1700 (Typical values)	10	Charpy V 27	—	Ultra high strength aircraft steel
1 $\frac{1}{2}$ CrAlMo Steel	BS 905M39 (En41B)	0.35 0.43	0.40 0.65	— —	1.40 1.80	Mo 0.15 0.25 Al 0.90 1.30	HT	63 150	2 $\frac{1}{2}$ 6	850/1000 700/850	680 525	13 17	42 50	248/302 201/255	Nitriding steel§
1 $\frac{1}{2}$ CrAlMo Steel	DIN 41CrAlMo7	0.38 0.45	0.50 0.80	— —	1.50 1.80	Mo 0.25 0.40 Al 0.80 1.20	HT	102 152	4 6	930/1125 835/1035	740 635	12 14	— —	— —	Nitriding steel§
1 $\frac{1}{2}$ CrNiAlMo Steel	DIN 34CrAlNi Mo7	0.30 0.37	0.40 0.70	0.85 1.15	1.50 1.80	Mo 0.15 0.25 Al 0.80 1.20	HT	254	10	790/990	585	13	—	—	Nitriding steel§
1 $\frac{1}{2}$ MnNiCr Mo Steel	BS 945M38 (En100)	0.34 0.42	1.20 1.60	0.60 0.90	0.40 0.60	Mo 0.15 0.25	HT	29 250	1 $\frac{1}{8}$ 10	1000/1150 700/850	850 495	12 15	42 28	293/352 201/255	Nitriding or bright bar
3 $\frac{1}{4}$ NiCrMoV Steel	BS 976M33	0.28 0.38	0.20 0.60	2.90 3.60	0.90 1.70	Mo 0.45 0.65 V 0.08 0.15	HT	250 1000	10 39	1100/1250 850/1000	980 710	7/11 9/13	16/27 20/33	321/375 248/302	Forgings
4. Carburising steels (BS 970:Part 1:1983; DIN 17210:1969)															
$\frac{1}{2}$ Cr Steel	BS 523M15	0.12 0.18	0.30 0.60	— —	0.30 0.60	— —	OQ	19	$\frac{3}{4}$	620	—	13	28	—	Case hardening
$\frac{3}{4}$ Cr Steel	Bs 527M17	0.14 0.20	0.70 1.00	— —	0.60 0.90	— —	OQ	19	$\frac{3}{4}$	770	—	12	16	—	Case hardening
1 $\frac{1}{4}$ MnCr Steel	DIN 16MnCr5	0.14 0.19	1.00 1.30	— —	0.80 1.10	— —	OQ	11 30	$\frac{7}{16}$ 1 $\frac{3}{16}$	880/1175 790/1080	630 590	9 10	— —	— —	

Type of steel	Related specification**	C	Mn	Ni	Cr	Mo	Condition*	Ruling section (mm)	Ruling section (in)	Tensile strength (N/mm²)	Yield stress (N/mm²)	Elongation (%)	Izod (J)	Hardness	Remarks
1¼MnCr Steel	DIN 20MnCr5	0.17/0.22	1.10/1.40	—	1.00/1.30	—	OQ	11/30	7/16, 1 3/16	1 080/1 375, 990/1 280	740, 680	7, 8	—	—	
¾NiCr Steel	Bs 635M15 (En351)	0.12/0.18	0.60/0.90	0.70/1.10	0.40/0.80	—	OQ	19	3/4	770	—	12	27	—	
1NiCr Steel	BS 637M17 (En352)	0.14/0.20	0.60/0.90	0.85/1.25	0.60/1.00	—	OQ	19	3/4	930	—	10	16	—	
1½NiCr Steel	DIN 15CrNi6	0.12/0.17	0.40/0.60	1.40/1.70	1.40/1.70	—	OQ	11/30	7/16, 1 3/16	960/1 280, 880/1 175	680, 635	8, 9	—	—	
3¼NiCr Steel	BS 655M13 (En36A) AISI 3415	0.10/0.16	0.35/0.60	3.00/3.75	0.70/1.00	—	OQ	19	3/4	1 000	—	9	35	—	
4NiCr Steel	BS 659M15 (En39B)	0.12/0.18	0.25/0.50	3.90/4.30	1.00/1.40	Mo 0.15/0.30	OQ	19	3/4	1 310	—	8	28	—	Case hardening
½CrMo Steel	DIN 20MoCr4	0.17/0.22	0.60/0.90	—	0.30/0.50	Mo 0.40/0.50	OQ	11/30	7/16, 1 3/16	880/1 175, 790/1 080	635, 590	9, 10	—	—	Case hardening
½CrMo Steel	DIN 25MoCr4	0.23/0.29	0.60/0.90	—	0.40/0.60	Mo 0.40/0.50	OQ	11/30	7/16, 1 3/16	1 080/1 375, 990/1 280	740, 680	7, 8	—	—	Case hardening
1¼NiMo Steel	BS 655M17 (En34)	0.14/0.20	0.35/0.75	1.50/2.00	—	Mo 0.20/0.30	OQ	19	3/4	770	—	12	35	—	Case hardening
1¼NiMo Steel	Bs 655M20	0.17/0.23	0.35/0.75	1.50/2.00	—	Mo 0.20/0.30	OQ	19	3/4	850	—	11	22	—	Case hardening
1¾NiMo Steel	BS 665M23 (En35)	0.20/0.26	0.35/0.75	1.50	—	Mo 0.20/0.30	OQ	19	3/4	930	—	10	13	—	Case hardening
¾NiCrMo Steel	BS 805M17 (En361)	0.14/0.20	0.60/0.95	0.35/0.75	0.35/0.65	Mo 0.15/0.25	OQ	19	3/4	770	—	12	22	—	Case hardening
½NiCrMo Steel	BS 805M20 (En362)	0.17/0.23	0.60/0.95	0.35/0.75	0.35/0.65	Mo 0.15/0.25	OQ	19	3/4	850	—	11	16	—	Case hardening

§ Surface hardness after Nitriding 950HV.

* HT—Hardened and tempered. OQ—Oil quenched.

** The specification and standards cited may be obsolete, but have been retained because the designations are still in common use.

(continued)

Table 22.50 FORGED OR ROLLED STEELS—ROOM TEMPERATURE LONGITUDINAL MECHANICAL PROPERTIES—*continued*

Material	British or other standard**	Composition % C	Mn	Ni	Cr	Other elements	Condition*	Limiting ruling section mm	in	Properties (minima unless otherwise stated) UTS MPa	Yield stress MPa	Elong. (gl $5.65\sqrt{S_0}$) %	Impact J	Hardness HB	Remarks
½NiCrMo Steel	BS 805M22	0.19 0.25	0.60 0.95	0.35 0.75	0.35 0.65	Mo 0.15 0.25	OQ	19	$\frac{3}{4}$	930	—	10	11	—	Case hardening
1½NiCrMo Steel	BS 815M17 (En353)	0.14 0.20	0.60 0.90	1.20 1.70	0.80 1.20	Mo 0.10 0.20	OQ	19	$\frac{3}{4}$	1080	—	8	22	—	Case hardening
1½NiCrMo Steel	DIN 17CrNiMo6	0.14 0.19	0.40 0.60	1.40 1.70	1.50 1.80	Mo 0.25 0.35	OQ	11	$\frac{7}{16}$	1175/1420	835	7	—	—	Case hardening
							OQ	30	$1\frac{3}{16}$	1080/1330	790	8	—	—	Case hardening
1¾NiCrMo Steel	BS 820M17 (En354)	0.14 0.20	0.60 0.90	1.50 2.00	0.80 1.20	Mo 0.10 0.20	OQ	19	$\frac{3}{4}$	1160	—	8	22	—	Case hardening
2NiCrMo Steel	BS 822M17 (En355)	0.14 0.20	0.40 0.70	1.75 2.25	1.30 1.70	Mo 0.15 9.25	OQ	19	$\frac{3}{4}$	1310	—	8	22	—	Case hardening
3½NiCrMo Steel	BS 832M13 (En36C)	0.10 0.16	0.35 0.60	3.00 2.75	0.70 1.00	Mo 0.10 0.25	OQ	19	$\frac{3}{4}$	1080	—	8	28	—	Case hardening
4NiCrMo Steel	BS 835M15 (En39B)	0.12 0.18	0.25 0.50	3.90 4.30	1.00 1.40	Mo 0.15 0.30	OQ	19	$\frac{3}{4}$	1310	—	8	28	—	Case hardening

5. *High alloy steels*

Stainless, heat resisting and valve steels
BS 970: Parts 1 and 4; 1449:Part 2; 1983, and Aircraft steels DIN 17224:1968; 17440:1967, Stahl-Eisen Werkstoffblatt 390:61, 400:60, 470:60, 670:69, AICMA (Association International des Constructeurs de Material Aerospacial) Standards

Ferritic stainless and heat resisting steels

Material	British or other standard**	Composition % C	Mn	Ni	Cr	Other elements	Condition*	Limiting ruling section mm	in	UTS MPa	Yield stress MPa	Elong. (gl $5.65\sqrt{S_0}$) %	Impact J	Hardness HB	Remarks
12Cr Steel	BS 3361	0.12 max	1.00 max	1.00 max	11.5 13.5	—	HT	152	6	540/700	355	20	34	—	—
13Cr Steel	BS 403S17	0.08 max	1.00 max	0.50 max	12.0 14.0	—	S	51	2	420	200	20	—	121/183	B:F:P:Sh:St†
13CrSi Steel	DIN X10CrSi13	0.12 max	1.00 max	—	12.0 14.0	Si 1.9 2.4	S	—	—	540/680	340	15	—	—	Scaling resistant up to 950°C in air
13CrAl Steel	BS 405S17	0.08 max	1.0 max	0.50 max	12.0 14.0	Al 0.10 0.30	S	51	2	420	200	20	—	121/183	P:Sh:St†
13CrAl Steel	DIN X10CrAl13	0.12 max	1.00 max	—	12.0 14.0	Al 0.7 1.2	S	—	—	495/635	295	15	—	—	Scaling resistant up to 950°C in air

Type	Specification	C	Mn	Ni	Cr	Other elements	Condition	Section (mm)	Section (in)	Tensile (N/mm²)	Proof (N/mm²)	Elong. (%)	Izod	Hardness (HB)	Forms / Remarks
17Cr Steel	BS 430S17 (En60)	0.08 max	1.00 max	0.50 max	16.0 18.0	—	S	63	2½	430	280	20	—	170 max	B:F:P:Sh:St†
17CrTi Steel	DIN X8CrTi17	0.10 max	1.0 max	—	16.0 18.0	Ti 7× C min	S	116	5⁄8	450/585	265	20	—	—	—
17CrNb Steel	DIN X8CrNb17	0.10 max	1.00 max	—	16.0 18.0	Nb 12× C min	S	16	5⁄8	450/585	265	20	—	—	—
17CrMo Steel	BS 434S17	0.08 max	1.00 max	0.50 max	16.0 18.0	Mo 0.90 1.30	S	—	—	430	245	20	—	—	Sh:St†
17CrMo Steel	DIN X6CrMo17	0.07 max	1.00 max	—	16.0 17.5	Mo 0.90 1.20	S	16	5⁄8	450/635	295	25	—	—	—
17CrMoTi Steel	DIN X8CrMoTi17	0.10 max	1.00 max	—	16.0 18.0	Mo 1.50 2.00 7×Ti C min	S	—	—	495/635	295	20	—	—	Used in the chemical and textile industries
18CrSi Steel	DIN X10CrSi18	0.12 max	1.00 max	—	17.0 19.0	Si 1.90 2.40	S	—	—	540/680	340	15	—	—	Scaling resistant up to 1050°C in air
18CrAl Steel	DIN X10CrAl18	0.12 max	1.00 max	—	17.0 19.0	Al 0.70 1.20	S	—	—	495/635	295	12	—	—	Scaling resistant up to 1050°C in air
24CrAl Steel	DIN Z10CrAl24	0.12 max	1.00 max	—	23.0 25.0	Al 1.20 1.70	S	—	—	495/635	295	10	—	—	Scaling resistant up to 1200°C
29CrSi Steel	DIN X10CrSi29	0.12 max	1.00 max	—	28.0 31.0	Si 1.00 2.00	S	—	—	540/680	385	12	—	—	Scaling resistant up to 1150°C in air
16CrNiMo Nb Steel	UK (FV702)	0.03	0.60 (Typical)	2.5	16.0	Si 0.50 Mo1.00 Nb0.50	S	19	3⁄4	705 (Typical values)	500 (Typical values)	22	—	—	Resistant to stress corrosion

Martensitic stainless and heat resisting steels

Type	Specification	C	Mn	Ni	Cr	Other elements	Condition	Section (mm)	Section (in)	Tensile (N/mm²)	Proof (N/mm²)	Elong. (%)	Izod	Hardness (HB)	Forms / Remarks
13Cr Steel	BS 410S21 (En56A)	0.09 / 0.15 max	1.00 max	1.00 max	11.5 / 13.5 max	—	Ht	150	6	550/700	370	20	34	152/207	B:F:P:Sh:St†
						—		63	2½	700/850	525	15	34	201/255	

(continued)

* OQ—Oil quenched. HT—Hardened and tempered. S—Softened.
† B=Bars. F=Forgings. P=Plates. Sh=Sheets. St=Strips.
** The specification and standards cited may be obsolete, but have been retained because the designations are still in common use.

Table 22.50   FORGED OR ROLLED STEELS—ROOM TEMPERATURE LONGITUDINAL MECHANICAL PROPERTIES—*continued*

| Material | British or other standard** | Composition % | | | | | Condition* | Limiting ruling section | | Properties (minima unless otherwise stated) | | | | | |
		C	Mn	Ni	Cr	Other elements		mm	in	UTS MPa	Yield stress MPa	Elong. (gl 5.65√S_0) %	Impact J	Hardness HB	Remarks
13CrS Steel	BS 416S21 (En56AM)	0.09 0.15	1.50 max	1.00 max	11.5 13.5	(Mo 0.60 max) S 0.15 0.30	HT	15 63	6 2½	550/700 700/850	525	11	27	201/255	B:F† Free cutting steel
13CrSe Steel	BS 416S41 (En56AM)	0.09 0.15	1.50 max	1.00 max	11.5 13.5	Mo 0.60 max Se 0.15 0.30	HT	150	6	550/700	370	15	34	152/207	B:F† Free cutting steel
13Cr Steel	BS 420S29 (En56B)	0.14 0.20	1.00 max	1.00 max	11.5 13.5	— —	HT	150 29	6 1⅛	700/850 775/925	525 585	15 13	27 27	201/255 223/277	B:F†
13CrS Steel	BS 416S29 (En56BM)	0.14 0.20	1.50 max	1.00 max	11.5 13.5	(0.60 max) 0.15 0.35	HT	152 29	6	700/850 775/925	525 585	11 10	27 14	201/255 223/277	B:F† Free cutting steel
13Cr Steel	BS 3S62	0.18 0.25	1.00 max	1.00 max	12.0 14.0	— —	HT	64 150	2½ 6	700/850 700/850	525 525	15 15	34 27	201/255 201/255	—
13Cr Steel	BS S124	0.15 0.25	1.50 max	1.00 max	12.0 14.0	Mo 0.60 max Zr 0.60 max Mo 1.00 max +Zr S 0.15/0.40	HT	150	6	690/850	450	11	27	201/255	Free machining steel
13Cr Steel	BS 420S37 (En56C)	0.20 0.28	1.00 max	1.00 max	12.0 14.0	(Mo 0.60 max) S 0.15/0.35	HT	150	6	775/925	585	13	13.5	223/277	B:F†
13CrS Steel	BS 416S37 (En56CM)	0.20 0.28	1.50 max	1.00 max	12.0 14.0	(Mo 0.60 max) S 0.15/0.35	HT	150	6	775/925	585	10	13.5	223/277	B:F Free cutting steel
13Cr Steel	Bs 420S45	0.28 0.36	1.00 max	1.00 max	12.0 14.0	— —	HT	150	6	770/930	585	13	13.5	—	B:F:Sh:St†
17CrNi Steel	BS 431S29 (En57)	0.12 0.20	1.00 max	2.0 3.0	15.0 18.0	— —	HT	150	6	850/1 000	680	11	20	—	B:F†
17CrNiS Steel	BS 441S49	0.12 0.20	1.50 max	2.0 3.0	15.0 18.0	(Mo 060 max) S 0.15/0.30	HT	—	—	850/1 000	—	—	—	—	Free cutting steel round wire

Type	Designation													Application
17CrNiS Steel	BS 2S137	0.12 0.20	1.50 max	2.0 3.0	15.0 18.0	(Mo 0.60 max) S 0.15/0.30	HT	70	2¾	880/1080	690	11	255/321	Primarily intended for nuts: free machining
17CrNiS Steel	BS 441S49	0.12 0.20	1.50 max	2.00 3.00	15.0 18.0	(Mo 0.60 max) Se 0.15/0.30	HT	64	2½	850/1000	—	—	—	Free cutting round wire
12CrMo Steel	DIN X19CrMo12.1	0.15 0.23	0.30 0.80	0.80 max	11.0 12.5	Mo 0.80/1.20	HT	152	6	680/835	495	16	—	B:F† Creep resistant
12CrMoV Steel	DIN X20CrMoV12.1	0.17 0.23	0.30 0.80	0.30 0.80	11.0 12.5	Mo 0.80/1.20 V 0.25/0.35	HT	19	¾	680/835	495	16	—	P:Sh:St:T† Creep resistant Resistant to high pressure H_2
12CrMoV Steel	DIN X22CrMoV12.1	0.20 0.26	0.30 0.80	0.30 0.80	11.0 12.5	Mo 0.80/1.20 V 0.25/0.35	HT	152	6	790/925	590	14	—	B:F† Creep resistant
12CrMoVNb Steel	AICME Fe-PM36 (UK FV448 Jethete 160)	0.11 0.19	0.20 1.25	0.50 1.20	10.0 12.0	Mo 0.40/1.00 V 0.10/0.70 Nb 0.10/0.60 +N	HT	530 dia 89 thick disc	21 dia 3½ thick disc	975	835	16	—	Gas turbine components
12CrNiMoV Steel	AICME Fe-PM37 (UK FV66 (+Nb) Jethete M152/M154)	0.08 0.15	0.50 0.90	2.0 3.0	11.0 12.5	Mo 1.50/2.00 V 0.25/0.40 +N	HT	127 sq	5 sq	1065	850	18	77	Gas turbine components
										(Typical tangential values)				
12CrCoMoV Nb Steel	AICME Fe-PM38 (UK FV535)	0.05 0.12	0.20 1.35	0.20 1.20	9.8 11.5	Co 5.00/7.50 Mo 0.50/1.10 V 0.10/0.60 Nb 0.20/0.60 +B	HT	560 dia 102 thick	22 dia 4 thick	1065	925	17	—	Gas turbine components
										(Typical longitudinal values)				
										(Typical tangential values)				

* HT – Hardened and tempered.

† B = Bars. F = Forgings. P = Plates. Sh = Sheets. St = Strips. T = Tubes.

** The specification and standards cited may be obsolete, but have been retained because the designations are still in common use.

(continued)

Table 22.50   FORGED OR ROLLED STEELS—ROOM TEMPERATURE LONGITUDINAL MECHANICAL PROPERTIES—*continued*

Material	British or other standard**	Composition %					Condition*	Limiting ruling section		Properties (minima unless otherwise stated)					Remarks
		C	Mn	Ni	Cr	Other elements		mm	in	UTS MPa	Yield stress MPa	Elong. (g1 5.65$\sqrt{S_0}$) %	Impact J	Hardness HB	
Austenitic stainless and heat resisting steels															
12/12CrNi Steel	DIN X8CrNi 12 12 UK (FV DDq)	0.10 max	2.00 max	11.50 13.50	12.00 14.00	— —	S	—	—	495/635	200	50	—	—	Spoons and forks
14/10CrNi CuMoTi Steel	UK (FV467)	0.20	0.90	9.50 (Typical)	14.00	Mo 2.00 Cu 2.50 Ti 0.80	PH	—	—	680 (Typical values)	300 (Typical values)	52	135	—	B:F† Creep resistant
17/7CrNi Steel	BS 301S21	0.15 max	2.00 max	6.00 8.00	16.00 18.00	— —	S	—	—	540	215	40	—	220 max	Sh:St*. Also used as cold rolled strip or cold drawn wire for springs
18/8CrNi Steel	BS S205	0.15 max	0.50 2.00	7.50 9.00	17.0 19.0	— —	LT, HT	—	—	1 350/1 550 (Typical values)	—	—	—	—	Cold drawn and heat treated wire and springs
18/9CrNi Steel	BS 304S15 (En58E)	0.06 max	2.0 max	8.00 11.00	17.50 19.00	— —	HF or ST CD	19 44	$\frac{3}{4}$ $1\frac{3}{4}$	480 860 650	230 695 310	40 12 28	— — —	183 max — —	B:FSh:St†
18/9CrNiN Steel	UK (Hi-proof 304)	0.06 max	2.00 max	8.00 11.00	17.50 19.00	N 0.15 0.25	S	102	4	585	295	35	—	—	B:P†
18/19CrNiTi Steel	Bs 2S129	0.08 max	0.50 2.00	8.00 11.00	17.00 19.00	Mo 1.00 max Ti 5×C/0.8	S	150	6	540	210	35	69	183 max	
18/9CrNiNb Steel	BS 2S130	0.08 max	0.50 2.00	8.00 11.00	17.00 19.00	Mo 1.00 max Nb 10×C/1.1	S	150	6	540	210	35	69	133 max	
18/9CrNiTi Steel	BS 321S12	0.08 max	0.50 2.00	9.00 12.00	17.00 19.00	Ti 5× C/0.70	ST	19 5	$\frac{3}{4}$ 2	540 508	277 246	40 40	— —	149/217 137/201	

18/9CrNiS Steel	BS 303S21 (En58M)	0.12 max	1.00 2.00	8.0 11.0	17.0 19.0	S	0.15 0.30	ST	19 44	3/4 1 3/4	510 860	210 695	40 12	—	—	B:F† Free cutting
18/10CrNi Steel	BS S536	0.03	0.50 2.00	9.0 12.0	17.5 19.0	—		S	73	—	650	310	28	—	—	Sh:St†
18/10CrNi Steel	BS 304S12	0.03 max	0.50 2.00	9.0 12.0	17.5 19.0	—		ST	51	2	500/700	190	40	—	134/192	—
18/10CrNi N Steel	UK (Hi-proof 304L)	0.03 max	2.00 max	9.0 12.0	17.5 19.0	N	0.15 0.25	S	102	4	493	231	35	—	—	B:F†
18/10CrNi Steel	BS 304S15 (En58E)	0.06 max	2.00 max	8.0 11.0	17.5 19.0	—		S	160	—	585 480	295 230	40	—	183 max	Sh:St†
18/10CrNiTi Steel	BS S524, S526	0.08	0.50 2.00	9.0 11.0	17.0 19.0	Ti	5×C/0.70	CR S	3 3		800/1000 540	640 210	15 30	35	179 max	Sh:St†
18/10CrNiNb Steel	BS S525, S527	0.08	0.50 2.00	9.0 11.0	17.0 19.0	Nb	10×C/1.0	S	1.5 3.0		540 540	210 210	30 40	—	—	Sh:St†
18/11CrNi Steel	BS 305S19	0.10 max	2.00 max	11.0 13.0	17.0 19.0	—		S	—		460	185	40	—	185 max	Sh:St†
17/10CrNi Mo Steel	BS 315S16	0.07 max	2.00 max	9.0 11.0	16.5 18.5	Mo	1.25 1.75	S	—		510	205	40	—	205 max	B:F:Sh:St†
17/10CrNi MoNb Steel	UK (FV548)	0.08	1.00	11.5 (Typical)	16.5	Mo Nb	1.5 1.0	S	—		600 (Typical values)	225 (Typical values)	55	135	—	B:F† Creep resistant
17/11CrNi MoSe Steel	BS 326S36	0.08 max	0.50 2.00	10.0 13.0	16.5 18.5	Mo Se	2.25 3.00 0.15 0.30	S	—		510	210	40	—	—	B:F† Free cutting
17/12CrNi Mo Steel	BS 316S12	0.03 max	0.50 2.00	11.0 14.0	16.5 18.5	Mo	2.25 3.00	ST	51	2	508	246	40	—	137/201	B:F:P:Sh:St†
17/12CrNi MoN Steel	UK (Hi-proof 316L)	0.03 max	2.00 max	11.0 14.0	16.5 18.5	Mo N	2.25 3.00 0.15 0.25	S	102	4	620	315	35	—	—	B:F†

*S—Softened. PH—Precipitation hardened. LT—Low temperature. HT—Heat treatment. CR—Cold rolled.
† B = Bars. F = Forgings. P = Plates. Sh = Sheets. St = Strips. T = Tubes.
** The specification and standards cited may be obsolete, but have been retained because the designations are still in common use.

(continued)

Table 22.50   FORGED OR ROLLED STEELS—ROOM TEMPERATURE LONGITUDINAL MECHANICAL PROPERTIES—*continued*

Material	British or other standard**	Composition %					Condition*	Limiting ruling section		Properties (minima unless otherwise stated)					
		C	Mn	Ni	Cr	Other elements		mm	in	UTS MPa	Yield stress MPa	Elong. (g1 5.65$\sqrt{S_0}$) %	Impact J	Hardness HB	Remarks
17/12CrNi Mo Steel	BS 316S16	0.07 max	0.50 2.00	10.0 13.0	16.5 18.5	Mo 2.25 3.00	ST	52 19 44	2 3¼ 1¼	524 860 650	262 695 310	40 12 28	— — —	143/212 — —	B:F:P:Sh:St† Also used cold rolled strip or wire for springs
17/12CrNi MoN Steel	UK (Hi-proof 316)	0.07 max	2.00 max	10.00 13.0	16.5 18.5	Mo 2.25 3.00 N 0.15 0.25	S	102	4	620	315	35	—	—	B:F†
17/12CrNi MoTi Steel	BS 320S17	0.08 max	0.50 2.00	11.0 14.0	16.5 18.5	Mo 2.25 3.00 Ti 4× C/0.60	ST	52	2	524	270	40	—	143/212	B:F:P:Sh:St†
18/12 CrNiMo Nb Steel	BS (En58J-Nb) DIN X10CrNiMoNb 18 12	0.10 max	2.00 max	12.0 14.5	16.5 18.5	Mo 2.5 3.0 Nb 8×C min	S	64	2½	495/749	225	40	—	—	B:F:P:Sh:St†
16/13CrNi Nb Steel	DIN X8CrNiNb 16 13	0.04 0.10	1.5 max	12.0 14.0	15.0 17.0	Nb 10× C/1.2	S	152	6	510/680	200	35	—	—	Creep resistant
16/13CrNi MoVNb Steel	DIN X8CrNiMo VNb 16 13	0.04 0.10	1.5 max	12.5 14.5	15.5 17.5	Mo 1.1 1.5 V 0.60 0.85 Nb 10× C/1.2 +N	PH	152	6	540/740	255	30	—	—	Creep resistant. Resistant to high pressure hydrogen
16/16 CrNiMo Nb Steel	DIN X8CrNi MoNb 1616	0.04 0.10	1.50 max	15.5 17.5	15.5 17.5	Mo 1.6 2.0 Nb 10× C/1.2	S	152	6	525/680	215	35	—	—	Creep resistant

		C		Cr	Ni	Other elements	Condition			Tensile	Proof	Elong.		Hardness	Remarks
16/16CrNiW Nb Steel	DIN X6CrNiW Nb Steel	0.04 0.10	1.50 max	15.5 17.5	15.5 17.5	W 2.5 3.5 Nb 10× C/1.2	S	152	6	540/740	255	30	—	—	Creep resistant
16/16CrNi MoNbB Steel	DIN X8CrNiMo NbB 1616 Steel	0.04 0.10	1.50 max	15.5 17.5	15.5 17.5	Mo 1.6 2.0 Nb 10× C/1.2 B 0.05 0.10 +N	PH	152	6	540/740	280	30	—	—	Nuclear reactor components
18/15CrNi Mo Steel	BS 31S12	0.03 max	2.00 max	14.0 17.0	17.5 19.5	Mo 3.0 4.0	S	—	—	490	195	40	—	195 max	B:F:P:Sh:St†
18/13CrNi Mo Steel	BS 317S16	0.06 max	2.00 max	12.0 15.0	17.5 19.5	Mo 3.0 4.0	S	—	—	510	205	40	—	205 max	B:F:P:Sh:St†
20/18NiCr MoCuNb Steel	DIN X5CrNiMo CuNb 1818	0.07 max	2.00 max	19.0 21.0	16.5 18.5	Mo 2.0 2.5 Cu 1.8 2.2 Nb 8×C min	S	—	—	495/740	225	40	—	—	Resistant to sulphuric acid
20/12CrNiSi Steel	DIN X15CrNiSi 2012	0.20 max	2.00 max	11.0 13.0	19.0 21.0	Si 1.8 2.3	S	—	—	585/740	295	40	—	—	Resistant to scaling up to 1050°C in air
23/14CrNi Steel	BS 309S24, S522	0.15 max	2.00 max	13.0 16.0	22.0 25.0	—	S	—	—	510	205	40	—	205 max	P:Sh:St†
23/14CrNiTi Steel	BS S125, S528	0.15 max	0.50 2.00	13.0 16.0	22.0 25.0	Ti 4×C min	ST	152	6	540	215	28	68	—	B:F:Sh:St†
23/14CrNi Nb Steel	BS S126, S529	0.15 max	0.50 2.00	13.0 16.0	22.0 25.0	Nb 8×C min	ST	152	6	540	215	28	68	—	B:F:Sh:St†
24/17CrNiTi Steel	BS S127, S530	0.12 max	0.50 2.00	16.0 19.0	23.0 26.0	Ti 5× C/0.90	S	152	6	540	210	30	—	179 max	B:F:Sh:St*

* S—Softened. PH—Precipitation hardened.

† B = Bars. F = Forgings. P = Plates. Sh = Sheets. St = Strips. T = Tubes.

** The specification and standards cited may be obsolete, but have been retained because the designations are still in common use.

(continued)

Table 22.50 FORGED OR ROLLED STEELS—ROOM TEMPERATURE LONGITUDINAL MECHANICAL PROPERTIES—*continued*

Material	British or other standard**	Composition %					Condition*	Limiting ruling section		Properties (minima unless otherwise stated)					Remarks
		C	Mn	Ni	Cr	Other elements		mm	in	UTS MPa	Yield stress MPa	Elong. (gl 5.65$\sqrt{S_0}$) %	Impact J	Hardness HB	
24/17CrNi Nb Steel	BS S128, S531	0.12 max	0.50 2.00	13.0 16.0	23.0 26.0	Nb 10× C/1.40	S	152	6	540	210	30	—	179 max	B:F:Sh:St[†]
24/20CrNi Steel	BS 310S24	0.15 max	2.00 max	19.0 22.0	23.0 26.0	—	S	—	—	510	205	40	—	205 max	B:F:Sh:St[†] Resistant to scaling up to 1050°C in air
25/15NiCr TiMoVa1B Steel	AICMA FE-PA92HT (UK FV559 ASTM A286)	0.08 max	1.00 2.00	24.0 27.0	13.5 16.0	Si 0.4/1.0 Ti 1.9/2.3 Mo1.0/1.5 V 0.1/0.5 Al 0.35 max +B	PH	—	—	900	590	20	—	—	Gas turbine components. Also suitable for use at sub-zero temperatures
25/20CrNiSi Steel	DIN X15CrNiSi 25 20	0.20 max	2.00 max	19.0 21.0	24.0 26.0	Si 1.8 2.3	S	—	—	590/740	295	40	—	—	Scaling resistant up to 1 200°C in air
25/25 CrNiMoTi Steel	DIN X5CrNiMoTi 25 25	0.06 max	2.00 max	24.0 26.0	24.0 26.0	Mo2.0 2.5 Ti 5×C min	S	—	—	500/740	225	35	—	—	Used in the chemical industry
36/16NiCrSi Steel	DIN X12NiCrSi 36 16	0.15 max	2.00 max	34.0 37.0	15.0 17.0	Si 1.5 2.0	S	—	—	540/740	260	40	—	—	Resistant to scaling up to 1 100°C in air
41/12 NiCrMo TiCoAl B Steel	AICMA FE-PA99-HT	0.10 max	0.50 max	40.0 45.0	11.0 14.0	Mo5.0/6.5 Ti 2.6/3.1 Co 1.0 max Al 0.35 max +B	PH	—	—	1 130	820	10	—	—	Gas turbine components

Steel	Designation	C	Mn	Ni	Cr	Other elements	Condition	—	—	Tensile strength	Proof stress	Elongation %	Hardness	Application
20/20/20 CrNiCoMo WNbN Steel	AICMA FE-PA91-HT	0.08 0.16	1.00 2.00	19.0 21.0	20.0 22.5	Co 18.5/21.0, Mo 2.5/3.5, W 2.0/3.0, Nb 0.75/1.25, N 0.10/0.20	PH	—	—	680/960	340	30	—	Gas turbine components
12/11/6 CrNiMn Steel	DIN X15CrNiMn 12 10	0.05 0.20	5.5 6.5	9.0 11.0	10.5 12.5	— —	S CF	— —	— —	500/650 630/835	220 500	45 35	— —	—
15/10/6 CrNiMn-MoNbVb Steel	UK (Esshete 1250)	0.15 max	5.5 7.0	9.0 11.0	14.0 16.0	Mo 0.8/1.2, Nb 0.75/1.25, V 0.15/0.40, +B	S	—	—	495	180	30	—	Pressure vessel plate, super-heater tube, steam piping. Creep resistant
17/6/4 CrMnMiN Steel	AISI 201	0.15 max	5.5 7.5	3.5 5.5	16.0 18.0	N 0.25 max	S	—	—	790 (Typical values)	380 (Typical values)	40	—	—
17/8/5 CrMnNiM Steel	BS 284S16	0.07 max	7.00 10.00	4.0 6.5	16.5 18.5	N 0.15 0.25	S	—	—	630	300	40	200 max	P:Sh:St†
18/9/5 CrMnNiN Steel	AISI 202 DIN X8 CrMnNi 18 8	0.15 max	7.5 10.0	4.0 6.0	17.0 19.0	N 0.25 max	S	—	—	725 (Typical values)	380 (Typical values)	48	—	—
18/8/6 CrMnNiN Steel	DIN X7CrMnNiN 18 8	0.07	8.9	5.9 (Typical)	18.0	Mo 0.30, N 0.22	S	—	—	718 (Typical values)	370 (Typical values)	55	—	Used in the chemical industry
18/10/9 CrMnNi-MoN Steel	DIN X3CrMnNi-MoN 18 10	0.04	11.0	9.0 (Typical)	19.0	Mo 2.35, N 0.30	S	—	—	820	430	38	—	Used in the chemical industry
18/12/2 MnCrNi-Mo Steel	DIN X12MnCr 18 12	0.15 max	17.0 19.0	1.5 2.5	11.0 13.0	Mo 0.30, 0.80, P 0.08 max	S CF	— —	— —	630/790 790/990	295 500	45 25	— —	Used in the cutlery industry

*S—Softened.   PH—Precipitation hardened.   CF—Cold formed.
† B—Bars.   F—Forgings.   P—Plates.   Sh—Sheets.   T—Tubes.
** The specification and standards cited may be obsolete, but have been retained because the designations are still in common use.

(continued)

Table 22.50   FORGED OR ROLLED STEELS—ROOM TEMPERATURE LONGITUDINAL MECHANICAL PROPERTIES—*continued*

| Material | British or other standard** | Composition % | | | | | Condition* | Limiting ruling section mm   in | Properties (minima unless otherwise stated) | | | | | Remarks |
		C	Mn	Ni	Cr	Other elements			UTS MPa	Yield stress MPa	Elong. (gl 5.65 $\sqrt{S_0}$) %	Impact J	Hardness HB	
Austenitic-ferritic stainless and heat resisting steels														
25/5CrNi Steel	Swedish (UHB 45) AISI 327	0.10	Not given (Typical)	4.8	25.5	—	S	—	630 (Typical values)	460 (Typical values)	30	—	—	Resistant to scaling up to 1 075°C in air
25/4CrNiSi Steel	DIN X20CrNiSi 25 4	0.15 0.25	2.0 max	3.5 5.5	24.0 26.0	Si 0.8 1.3	S	—	590/740	390	26	—	—	Resistant to scaling up to 1 100°C in air
25/CrNiMo Steel	Swedish (UHB44) AISI 329	0.08	Not given (Typical)	5.3	25.0	Mo 1.50	S	—	660 (Typical values)	500 (Typical values)	30	—	—	Used in the chemical industry. Scaling resistant up to 1 075°C in air
Value steels (BS 970:Part 4:1970, Stahl Eisen Werkstoffblatt 490–52)														
9CrSi Steel	BS 401S45 (En52)	0.40 0.50	0.30 0.75	0.50 max	7.5 9.5	Si 3.00 3.75	HT	—	925 (Typical values)	680 (Typical values)	22	—	255	
14/14CrNiW Steel	BS 331S40 (En54)	0.35 0.50	0.50 1.00	12.0 15.0	12.0 15.0	Si 1.0/2.0 W 2.0/3.0	S	—	895 (Typical values)	510 (Typical values)	23	—	—	
14/14CrNi WMO Steel	BS 331S42, S111 (En54A)	0.37 0.47	0.50 1.00	13.0 15.0	13.0 15.0	Si 1.0/2.0 W 2.2/3.0 Mo 0.40/0.70	S	—	—	—	—	20	—	
18/9 CrNiWSi Steel	DIN X45CrNiW 189 (Similar to BS En55)	0.40 0.50	0.80 1.50	8.00 10.0	17.0 19.0	Si 2.0/3.0 W 0.8/11.2	AH	—	790/990	390	25	—	—	Resistant to lead oxide corrosion
20CrNiSi Steel	BS 443S65 (En59)	0.75 0.85	0.30 0.75	1.20 1.70	19.0 21.0	Si 1.75 2.25	HT	—	895/1 050 (Typical values)	710	7	—	269	

Steel	Specification**	C	Mn	Ni	Cr	Other elements	Heat treatment*	Tensile strength	Proof stress	Elong. %	Hardness HB	Remarks
21/11CrNiSi Steel	BS 381S34	0.15 / 0.25	1.50 max	10.5 / 12.5	20.0 / 22.0	Si 0.75/1.25 N 0.15/0.30	PH	—	—	—	197	Resistant to lead oxide corrosion
21/9/4 CrMnNiN Steel	BS 349S52	0.48 / 0.58	8.0 / 10.0	3.25 / 4.50	20.0 / 22.0	C+N 0.38/0.50 0.90 min	PH	1050 (Typical values)	620 (Typical values)	9	321/352 C+Ni	Resistant to lead oxide corrosion
21/9/4 CrMnNi-NS Steel	BS 349S54	0.48 / 0.58	8.0 / 10.0	3.25 / 4.50	20.0 / 22.0	N 0.38/0.50 S 0.035 0.080 C+N 0.90 min	PH	1050 (Typical values)	620 (Typical values)	8	321/352 C+Ni	Free cutting steel
21/9/4 CrMnNiNb Steel	BS 352S52	0.48 / 0.58	8.00 / 10.00	3.25 / 4.50	20.0 / 22.0	N 0.38/0.50 Nb 2.00/3.00 C+N 0.90 min	PH	—	—	—	285	
21/9/4 CrMnNiNb S Steel	BS 352S54	0.48 / 0.58	8.00 / 10.00	3.25 / 4.50	20.0 / 22.0	N 0.38/0.50 Nb 2.00/3.00 S 0.035 0.080 C+N 0.90 min	PH	—	—	—	—	Free cutting steel
23/5 CrNiMo Steel	DIN X45CrNi Mo 23 5	0.40 / 0.50	0.90 / 1.20	4.5 / 5.5	22.0 / 24.0	Si 1.0 1.3 Mo 2.5 3.0	PH	1235 (HRC40)	—	—	—	

High-strength stainless steels (ASTM A579-67, DIN 17 224:1968)
Semi-Austenitic steels (Typical compositions and mechanical properties)

Steel	Specification**	C	Mn	Ni	Cr	Other elements	Heat treatment*	Tensile strength	Proof stress	Elong. %	Hardness HB	Remarks
17/7CrNiAl Steel	AISI 631 ASTM A579 Grade 62 (17/7PH) DIN X7CrNiAl 177	0.07	0.8	7.0	17.0	Al 1.2	PH	1360	1250	10	—	Sheet and strip. Also used as cold-rolled strip or cold-drawn wire for springs at temperatures up to 350°C

* S—Softened. HT—Hardened and tempered. AH—Age hardened. PH—Precipitation hardened.
** The specification and standards cited may be obsolete, but have been retained because the designations are still in common use.

(*continued*)

Table 22.50   FORGED OR ROLLED STEELS—ROOM TEMPERATURE LONGITUDINAL MECHANICAL PROPERTIES—*continued*

Material	British or other standard**	Composition %					Condition*	Limiting ruling section		Properties (minima unless otherwise stated)					Remarks
		C	Mn	Ni	Cr	Other elements		mm	in	UTS MPa	Yield stress MPa	Elong. (g1.5.65√S0) %	Impact J	Hardness HB	
15/7CrNi-MoAl Steel	AISI 632 ASTM A579 Grade 63 (PH15/7Mo)	0.07	0.8	7.1	15.1	Mo 2.2 Al 1.2	PH	—	—	1515	1420	9	—	—	Pressure vessels, springs
14/5CrNiCu MoNb Steel	BS S145 (FV520B)	0.07	1.00	5.8	14.5	Mo 1.6 Cu 1.6 Nb 0.30	PH	—	—	1470	1030	10	20	—	Bars and forgings for aircraft parts
14/8CrNiMo Al Steel	USA (PH14/8Mo)	0.04	0.6	8.3	15.1	Mo 2.2 Al 1.2 +N	PH	—	—	1570	1450	8	—	—	Pressure vessels, aircraft parts
16/4CrNi-MoN Steel	AISI 633 (AM 350)	0.10	0.8	4.25	16.5	Mo 2.75 N 0.10	PH	—	—	1405	1175	11	Charpy V 20	—	Valves, piping, aircraft parts
15/4CrNi MoN Steel	AISI 634 ASTM A579 Grade 64 (AM 355)	0.13	0.8	4.25	15.5	Mo 2.75 N 0.10	PH	—	—	1480	1250	11	Charpy V 20	—	Aircraft parts, valves, turbine parts
15/5CrNi-CuMoTi Steel	BS S533 (FV520S)	0.05	1.3	5.5	16.0	Mo 1.8 Cu 2.00 Ti 0.10	PH	—	—	1370	980	11	—	—	Sheet and strip for aircraft parts
Martensitic steels															
17/4CrNiCu Nb Steel	AISI 630 ASTM A579 Grade 61 (17/4PH)	0.04	0.8	4.3	16.0	Cu 3.3 Nb 0.27	PH	—	—	1360	1265	12	—	—	Gears, springs, cutlery, aircraft parts, turbine components
15/5CrNi CuNb Steel	USA(15/5PH)	0.04	0.8	4.6	15.0	Cu 3.3 Nb 0.27	PH	—	—	1360	1265	12	—	—	Gears, cams, cutlery, aircraft parts
13/8CrNi-MoAl Steel	USA (PH13/8Mo)	0.03	0.10 max	8.2	12.8	Mo 2.2 Al 1.1 +N	PH	76	3	1465	1310	12	—	—	Aircraft parts, fasteners, shafts

Designation	Country (spec)	C				Other elements	Condition*						Impact	Applications
12/8CrNi-CuTiNb Steel	USA (Custom, 455)	0.03	0.25	8.5	11.7	Cu 2.2, Ti 1.2, Nb 0.3	PH	102	4	1420	1345	10	Charpy V 16	Fasteners, springs, pressure vessels, valve parts, forgings
14/13CrCoV Steel	USA(AFC77)	0.15	0.20	—	14.5	Co 13.0, Mo 5.0, V 0.40	PH	—	—	1630	1295	10	—	Die casting dies, glass moulds
14/15CrCo-NiMoTi Steel	USA(AM367)	0.02	Not given	3.5	14.0	Co 15.0, Mo 2.0, Ti 0.4	PH	—	—	1500	1465	11	—	Bars, forgings, sheet and strip
12/12CrCo-NiMoTiNb Steel	UK(D70)	0.02	Not given	4.0	12.0	Co 12.0, Mo 4.0, Ti 0.40, Nb 0.10, Al 0.10, +Zr, B	PH	—	—	1650	1600	9.5	16	Bars, forgings, sheet and strip
12/8/5CrNi-CoMoTi Steel	WG(Ultrafort 401)	0.01	Not given	8.1	12.6	Co 5.2, Mo 2.0, Ti 0.80, +Al, Zr, B	PH	—	—	1650	1570	11	—	Pressure vessels, springs, fasteners, aircraft parts, extrusion and stamping tools
Maraging nickel steels (ASTM A579-67) *Typical compositions*														
12/5NiCr-MoAlTi Steel	ASTM A579 Grade 75	0.02	Not given	12.0	5.0	Mo 3.0, Al 0.40, Ti 0.20	PH	—	—	1310 (Typical values)	1250 (Typical values)	15	Charpy V 74	Gears, fasteners, shafts, rocket and missile cases, aircraft parts, plastic mould dies, die holders, die casting die inserts, extrusion tools

* PH—Precipitation hardened.

** The specification and standards cited may be obsolete, but have been retained because the designations are still in common use.

(continued)

Table 22.50   FORGED OR ROLLED STEELS—ROOM TEMPERATURE LONGITUDINAL MECHANICAL PROPERTIES—*(continued)*

Material	British or other standard**	Composition %					Condition*	Limiting ruling section		Properties (minima unless otherwise stated)					Remarks
		C	Mn	Ni	Cr	Other elements		mm	in	UTS MPa	Yield stress MPa	Elong. (gl 5.65 $\sqrt{S_0}$) %	Impact J	Hardness HB	
18/8NiCo-MoTi Steel	ASTM A579 Grade 71	0.01	Not given	18.0	—	Co 8.5 Mo 3.0 Ti 0.20 Al 0.10 +Ca, Zr, B	PH	—	—	1 390/1 540	1 330/1 480	9/13	Charpy V 34/68	—	Gears, fasteners, shafts, rocket and missile cases, aircraft parts, plastic mould dies, die holders, die casting die inserts, extrusion tools
18/8NiCo-MoTi Steel	ASTM A579 Grade 72	0.01	Not given	18.0	—	Co 8.0 Mo 5.0 Ti 0.40 Al 0.10 +Ca, Zr, B	PH	—	—	1 680/1 910	1 160/1 820	8/10	Charpy V 20/41	—	Gears, fasteners, shafts, rocket and missile cases, aircraft parts, plastic mould dies, die holders, die casting die inserts, extrusion tools
18/8NiCo-MoTi Steel	ASTM A579 Grade 73	0.01	Not given	18.0	—	Co 9.0 Mo 5.0 Ti 0.60 Al 0.10 +Ca, Zr, B	PH	—	—	1 820/2 130	1 790/2 090	6/9	Charpy V 14/27	—	Gears, fasteners, shafts, rocket and missile cases, aircraft parts, plastic mould dies, die holders, die casting die inserts, extrusion tools
Other high alloy steels (BS 1501: Part 2:1988, 1503:1969, Stahl Eisen Werkstoffblatt 390-61, ASTM 579-67)															
12Mn Steel	DIN X120Mn12	1.1 1.3	11.5 13.5	—	—	P 0.1 max	S	—	—	790/1 080	340	40	—	—	Non-magnetic
18Mn Steel	DIN X35Mn18	0.30 0.40	17.0 19.0	—	—	P 0.1 max	S	—	—	680/930	250	30	—	—	Non-magnetic

Steel	Specification**	C	Mn	Ni	Cr	Other	Condition*			Tensile strength	0.2% proof stress	Elong. %	Impact (Charpy V)	Applications
18MnCr Steel	DIN X40MnCr18	0.30/0.50	17.0/19.0	—/—	3.0/3.5	P 0.1 max	S CF	— —	— —	740/930 990/1175	295 880	45 20	— —	Non-magnetic end bells for alternator rotors
23MnCr Steel	DIN X40MnCr23	0.30/0.50	21.0/24.0	—/—	3.0/3.5	P 0.1 max	S	—	—	680/880	310	45	—	Non-magnetic
9Ni Steel	BS 1 501-510	0.10 max	0.30/0.80	8.5/9.75	0.30 max	Mo 0.10 Al 0.015 0.055 Cu 0.30	QT	50	—	690	540	18	—	Pressure vessel plates
13Cr Steel	UK (Silver Fox 67)	0.60/0.70	0.50/1.0	—/—	12.0/13.5	—/—	A LR	— —	— —	770 1080	500 1050	18 3	— —	Razor and surgical blades, general cutting tools
9/4NiCoCr-MoV Steel	USA (HP9-4-20)	0.20	0.30 (Typical)	9.2	0.8	Co 4.5 Mo 1.0 V 0.10	HT	—	—	1420 (Typical values)	1235 (Typical values)	14	Charpy V 74	Rocket motor cases, seamless tubing, shafts, pressure vessels, piping
9/4NiCo CrMoV Steel	ASTM A579 Grade 81 (HP9-4-25)	0.25	0.25 (Typical)	8.0	0.45	Co 4.0 Mo 0.45 V 0.10	HT	6	152	1375 (Typical values)	1265 (Typical values)	13	Charpy V 47	Rocket motor cases, seamless tubing, shafts, pressure vessels, piping
9/4NiCoCr-MoV Steel	ASTM A579 Grade 82 (HP9-4-30)	0.39	0.25 (Typical)	7.5	1.0	Co 4.5 Mo 1.0 V 0.10	HT	5	126	1570 (Typical values)	1375 (Typical values)	11	Charpy V 34	Armour plate aircraft parts
9/4NiCo-MoV Steel	ASTM A579 Grade 83 (HP9-4-45)	0.45	0.25 (Typical)	7.5	0.30	Co 4.0 Mo 0.30 V 0.10	Salt bath quenched Hardened, subzero cooled and tempered	3 3	76 76	1850 1990 (Typical values)	1540 1710 (Typical values)	9 7	Charpy V 27 20	Aircraft parts, fasteners, connecting rods, valve-spring wires

* PH—Precipitation hardened.   S—Softened.   CF—Cold formed.   QT—Double normalised and tempered, or, hardened and tempered.   A—Annealed.
LR—Lightly rolled.   HT—Hardened and tempered.
** The specification and standards cited may be obsolete, but have been retained because the designations are still in common use.

(continued)

Table 22.50 FORGED OR ROLLED STEELS—ROOM TEMPERATURE LONGITUDINAL MECHANICAL PROPERTIES—*continued*

Material	British or other standard**	Composition %					Condition*	Limiting ruling section		UTS MPa	Yield stress MPa	Elong. (gl 5.65√S_0) %	Impact J	Hardness HB	Remarks
		C	Mn	Ni	Cr	Other elements		mm	in						
6. Spring Steels (BS 970:Part 1:1983, DIN 17221:1955, 17222:1955, 17225:1955)															
Carbon steels (cold rolled strip except for 060 A96)															
0.52C Steel	BS 080A52 (En43C) DIN Ck53	0.50 0.55	0.70 0.90	—	—	Si 0.10 0.35	HT	—	—	1175/1390	1035	7	—	—	Laminated springs
0.67C Steel	BS 080A67 (En43E) DIN Ck67	0.65 0.70	0.70 0.90	—	—	Si 0.10 0.35	S	—	—	—	—	—	—	229 max	Laminated springs
0.72C Steel	BS 070A72 (En42)	0.70 0.75	0.60 0.80	—	—	Si 0.10 0.35	HT	—	—	—	—	—	—	—	Laminated springs
0.78C Steel	DIN Mk75	0.75 0.82	0.60 0.80	—	—	Si 0.10 0.35	HT	—	—	1575/1775	1465	6	—	—	Laminated springs
0.96C Steel	DIN Mk101	0.93 1.00	0.50 0.70	—	—	Si 0.10 0.35	HT	—	—	1760/2315	1670	5	—	—	Coil springs up to 25 mm (1 in)
Manganese and manganese silicon steels															
2Mn Steel	DUN 50Mn7	0.45	1.80 (Typical)	—	—	—	HT	—	—	1175/1390	1030	7	—	—	Laminated springs
1¾ Si Steel	DIN 46Si7	0.45 0.50	0.50 0.80	—	—	Si 1.5 1.8	HT	—	—	1280/1465	1080	6	—	—	Elliptic or helical springs, laminated springs
2Si Steel	DN 51Si7	0.50 0.57	0.70 1.00	—	—	Si 1.70 2.10	HT	—	—	1280/1465	1080	6	—	—	Laminated springs, coil springs up to 25 mm (1 in) bar dia.
2Si Steel	DIN 55Si7	0.55 0.62	0.70 1.00	—	—	Si 1.70 2.10	HT	—	—	1280/1465	1080	6	—	—	Laminated springs, coil springs up to 25 mm (1 in) bar dia.

Steel	Specification	C	Mn	Cr	Other	Condition		Tensile strength		Elong. %		Applications
2Si Steel	DIN 66Si7	0.58 0.65	0.70 1.00	— —	Si 1.70 2.10	HT	—	1375/1540	1175	6	—	Laminated springs, coil springs over 25 mm (1 in) bar dia.
1SiMn Steel	DIN 60SiMn5	0.55 0.65	0.90 1.10	— —	Si 1.0 1.3	HT	—	1330/1540	1030	6	—	Laminated springs, ring springs
Low alloy steels (hot rolled)												
1¾MnV Steel	DIN 50MnV7	0.50 1.70 (Typical)		— —	0.10	HT	—	1280/1465	1130	7	—	Laminated springs, coil springs up to 25 mm (1 in) bar dia.
1¼SiCr Steel	DIN 67SiCr5	0.62 0.72	0.40 0.60	— —	Si 1.2 1.4	HT	—	1465/1690	1330	5	—	Coil springs and valve springs. Used up to 300°C
1CrV Steel	BS 735A50 (En47) DIN 50CrV4	0.46 0.54	0.60 0.90	0.80 1.10	V 0.15 min Si 0.10 0.35	HT	2.8	1685/1835	—	—	—	Round spring wire
1CrV Steel	DIN 58CrV4	0.55 0.62	0.80 1.10	0.90 1.20	V 0.07 0.12	HT	—	1465/1690	1330	6	—	Torsion bars over 40 mm (1 9/16 in) bar dia.
1CrMo Steel	DIN 50CrMo4	0.50 0.90 (Typical)		—	Mo 0.20	HT	—	1330/1540	1175	6	—	Laminated springs, coil springs

* HT—Hardened and tempered.   S—Softened.

** The specification and standards cited may be obsolete, but have been retained because the designations are still in common use.

(continued)

Table 22.50   FORGED OR ROLLED STEELS—ROOM TEMPERATURE LONGITUDINAL MECHANICAL PROPERTIES—*continued*

Material	British or other standard**	Composition %					Condition*	Limiting ruling section		Properties (minima unless otherwise stated)					Remarks
		C	Mn	Ni	Cr	Other elements		mm	in	UTS MPa	Yield stress MPa	Elong. (gl 5.65$\sqrt{S_0}$) %	Impact J	Hardness HB	
$\frac{1}{7}$ NiCrMo Steel	BS 8054A60	0.55 0.65	0.65 1.05	0.40 0.70	0.35 0.75	Mo 0.15 0.25 Si 0.15 0.40	HT	—	—	—	—	—	—	—	Hot formed springs
2SiCrMo Steel	BS925A60	0.55 0.65	0.70 1.00	— —	0.20 0.40	Si 1.70 2.10 Mo 0.20 0.30	HT	—	—	—	—	—	—	—	Hot formed springs
1CrMoV Steel	DIN 51CrMoV4	0.51 (Typical)	0.90	—	1.0	Mo 0.20	HT	—	—	1465/1690	1330	6	—	—	Torsion bars over 40 mm (1 $\frac{9}{16}$ in) bar dia.
1$\frac{1}{2}$ CrMoV Steel	DIN 45CrMoV67	0.40 0.50	0.60 0.80	— —	1.3 1.5	Mo 0.65 0.75 V 0.25 0.35	HT	—	—	1390/1690	Not given	Not given	—	—	Used up to 450°C
2$\frac{1}{4}$ CrWV Steel	DIN 30WCrV179	0.25 0.35	0.20 0.40	— —	2.2 2.5	W 4.0 4.5 V 0.50 0.70	HT	—	—	1390/1690	Not given	Not given	—	—	Used up to 500°C
High alloy steel															
8WCrMov Steel	DIN 65WMo348	0.63 0.68	0.30 appr	— —	3.5 4.0	W 8.0 9.0 Mo 0.80 0.90 V 0.60 0.80	HT	—	—	1390/1690	Not given	Not given	—	—	Used up to 550°C

* HT—Hardened and tempered.
** The specification and standards cited may be obsolete, but have been retained because the designations are still in common use.

Table 22.51 TYPICAL HOT TENSILE PROPERTIES OF FORCED OR ROLLED STEELS IN THE LONGITUDINAL DIRECTION

Material	British or other standards*	C	Mn	Ni	Cr	Other elements		Condition	Temperature °C	UTS MPa	Yield stress MPa	Elong. (gl 5.65 $\sqrt{S_0}$) %	RA %
1. Carbon steels													
Armco Iron	—	0.02	0.03	—	—	—		Normalised	RT	340	185	39	69
									200	448	185	23	54
									400	309	Not given	15	67
0.15C Steel	BS 040A12 (En2B)	0.13	0.50	—	—	—		Normalised	RT	417	247	33	Not given
									200	463	232	23	Not given
									400	386	193	33	Not given
									500	309	Not given	33	Not given
0.20C Steel	BS 070M20 (En3)	0.20	0.70	—	—	—		Normalised	RT	448	263	30	62
									200	463	224	24	54
									400	371	185	31	67
0.25C Steel	BS 070M26 (En4)	0.26	0.70	—	—	—		Normalised	RT	479	247	30	Not given
									200	510	232	22	Not given
									400	448	201	27	Not given
									500	340	185	28	Not given
0.35C Steel	BS 080M36	0.36	0.80	—	—	—		Normalised	RT	602	309	26	Not given
									200	633	309	17	Not given
									400	587	216	25	Not given
									500	432	Not given	28	Not given
0.40C Steel	BS 080M40 (En8)	0.40	0.80	—	—	—		Normalised	RT	602	340	29	53
									250	633	293	19	41
									450	479	Not given	31	59
2. Low alloy weldable steels													
$\frac{1}{2}$ Mo Steel	Similar to DIN 15Mo3	0.17	0.60	—	—	Mo 0.60		Normalised	RT	510	317	33	69
									400	510	232	30	73
									500	417	224	29	77
									600	293	185	34	77
$\frac{1}{2}$ MoV Steel	—	0.12	0.60	—	—	Mo 0.50 V 0.30		Normalised and tempered	RT	587	394	23	76
									200	556	386	22	73
									400	479	340	25	77
									600	278	232	32	81

*The specifications and standards cited in this table may be obsolete, but have been retained because the designations are still in common use.

(continued)

Table 22.51   TYPICAL HOT TENSILE PROPERTIES OF FORCED OR ROLLED STEELS IN THE LONGITUDINAL DIRECTION—*continued*

Material	British or other standards*	Typical composition %					Condition	Temperature °C	UTS MPa	Yield stress MPa	Elong. (gl 5.65 S_0) %	RA %
		C	Mn	Ni	Cr	Other elements						
1CrMo Steel	BS 1501–620	0.15	0.70	—	0.95	Mo 0.50	Normalised and tempered	RT	479	286	21	68
								400	595	263	23	57
								500	525	232	24	65
								600	340	216	27	82
2¼CrMo Steel	BS 1501–622	0.13	0.60	—	2.20	Mo 1.10	Normalised and tempered	RT	641	417	18	75
								535	479	402	28	79
								610	309	263	47	89
3. Low alloy direct hardening steels												
¾Mo V Steel	—	0.18	0.50	—	—	Mo 0.70 V 0.20	Hardened and tempered	RT	896	811	20	68
								400	741	680	21	62
								600	494	456	24	73
1CrMo Steel	—	0.40	0.40	—	1.10	Mo 0.70	Hardened and tempered	RT	1 004	903	19	38
								300	942	857	Not given	Not given
								400	842	741	20	44
								600	479	324	28	64
3CrMo Steel	BS 722M24 (En40B)	0.24	0.60	—	3.30	Mo 0.55	Hardened and tempered	RT	757	556	20	Not given
								400	664	510	21	Not given
								450	618	479	25	Not given
								500	556	463	27	Not given
3CrMoV Steel	—	0.28	0.40	—	3.30	Mo 0.50 V 0.20	Hardened and tempered	RT	788	649	21	63
								400	664	510	20	76
								600	510	386	24	76
3CrVMoW Steel	—	0.23	0.30	—	2.70	V 0.70 Mo 0.50 W 0.50	Hardened and tempered	RT	988	903	17	40
								200	896	826	18	38
								400	811	764	16	36
								600	571	510	16	33
								700	355	301	21	36

Steel	BS	C		Ni	Cr	Other	Condition	Temp				Izod
3½ Ni Steel	BS (En22)	0.34	0.60	3.50	—	—	Hardened and tempered	RT	788	571	21	Not given
								200	741	525	23	Not given
								400	726	510	22	Not given
								500	448	340	24	Not given
3½ NiCr Steel	—	0.30	0.50	3.50	0.60	—	Hardened and tempered	RT	896	757	20	Not given
								200	849	710	17	Not given
								400	849	726	16	Not given
								500	695	602	19	Not given
2½ NiCrMo Steel	BS 826M31 (En25)	0.31	0.55	2.70	0.60	Mo 0.55	Hardened and tempered	RT	988	880	21	Not given
								200	942	834	17	Not given
								400	911	834	17	Not given
								500	741	664	19	Not given
4NiCrMo Steel	—	0.30	0.30	4.10	2.00	Mo 0.30	Hardened and tempered	RT	927	664	20	61
								400	757	571	19	61
								600	355	170	40	88
3NiCrMoV Steel	—	0.29	0.40	3.10	0.70	Mo 0.50 V 0.20	Hardened and tempered	RT	788	649	20	60
								400	633	494	19	64
								600	386	232	27	84
4NiCrMoV Steel	—	0.36	0.60	3.90	2.20	Mo 0.50 V 0.20	Hardened and tempered	RT	927	695	17	56
								400	772	587	17	55
								550	525	293	28	77
4. High alloy steels												
8CrSi Valve steel	BS 401S45 (En52)	0.45	0.50	—	8.50	Si 3.50	Hardened and tempered	RT	942	718	23	53
								400	726	471	24	60
								600	270	178	52	93
13Cr Steel	BS 410S21 (En56A)	0.12	0.50	—	13.00	—	Hardened and tempered	RT	610	417	29	72
								200	502	363	28	76
								400	463	332	24	74
								600	309	178	43	90
13Cr Steel	BS 420S37 (En56C)	0.24	0.50	—	13.0	—	Hardened and tempered	RT	703	541	23	60
								200	595	456	23	62
								400	541	432	18	58
								600	293	293	38	80

* The specifications and standards cited in this table may be obsolete, but have been retained because the designations are still in common use.

(continued)

Table 22.51   TYPICAL HOT TENSILE PROPERTIES OF FORCED OR ROLLED STEELS IN THE LONGITUDINAL DIRECTION—*continued*

Material	British or other standards*	Typical composition %					Condition	Temperature °C	UTS MPa	Yield stress MPa	Elong. (gl 5.65 $\sqrt{S_0}$) %	RA %
		C	Mn	Ni	Cr	Other elements						
13Cr Steel	DIN X40Cr13	0.40	0.50	—	12.0	—	Hardened and tempered	RT	772	541	17	Not given
								400	687	Not given	14	49
								600	347	Not given	26	67
11CrVMoNb Steel	—	0.20	0.40	—	11.0	V 0.70 Mo 0.50 Nb 0.20	Hardened and tempered	RT	958	849	15	Not given
								400	795	741	10	Not given
								600	525	479	21	Not given
								650	448	378	23	Not given
11CrNiMoNbV Steel	—	0.20	0.80	1.20	10.5	Mo 0.70 Nb 0.60 V 0.20	Hardened and tempered	RT	1081	942	16	46
								400	803	741	16	48
								600	587	541	28	73
								700	363	324	30	81
20Cr Steel	BS 442S19	0.06	0.80	—	21.0	—	Softened	RT	541	409	30	60
								400	432	317	21	58
								600	216	208	45	79
14/8CrNiMoAl Steel	Similar to USA PH-13/8Mo	0.03	0.80	8.50	14.5	Mo 2.30 A 1.10	Precipitation hardened	RT	1683	1591	6	Not given
								31.5	1375	1236	4	Not given
								535	896	726	16	Not given
18/8CrNi Steel	BS 304S15 (En58E)	0.06	0.80	10.00	18.0	—	Softened	RT	571	216	52	70
								600	324	93	35	57
18/8CrNiTi Steel	BS 321S20 (En58B)	0.10	0.80	8.50	18.0	Ti 0.60	Softened	RT	656	255	46	68
								400	463	193	36	65
								600	378	162	31	67
18/11CrNiNb Steel	Similar to BS 347S17 (En58G)	0.08	1.50	11.0	17.5	Nb 1.20	Softened	RT	633	263	50	65
								400	432	178	33	64
								600	378	162	35	62
								750	239	Not given	42	73
17/13CrNiMo Steel	BS 316S16 (En58J)	0.07	1.40	12.50	16.50	Mo 2.80	Softened	RT	710	247	51	76
								600	463	124	43	62
								800	216	108	47	62

Material	Specification	C	Mn	Cr	Ni	Other	Condition	Temp (°C)				
14/14CrNiW Valve steel	BS 331S40 (En54)	0.42	0.70	14.00	14.00	W 2.50, Si 1.50	Softened	RT	973	556	22	24
								400	656	456	20	33
								600	525	301	24	55
								800	263	139	35	74
20/8CrNiW Valve steel	BS (En55)	0.30	0.80	7.50	20.00	W 4.00, Si 1.30	Softened	RT	834	463	33	36
								400	656	371	26	43
								600	510	263	30	41
								800	247	Not given	43	55
23/12CrNiW Valve steel	BS (En55)	0.20	0.40	11.50	23.00	W 3.00, Si 1.60	Softened	RT	718	402	28	34
								400	610	301	30	42
								600	456	247	29	36
								800	232	Not given	41	41
25/12CrNi Steel	—	0.10	1.50	12.00	25.00	—	Softened	RT	656	340	40	70
								400	556	293	37	58
								600	432	239	30	55
								800	170	139	Not given	Not given
24/20CrNi Steel	BS 310S24	0.14	0.80	21.00	24.00	Si 1.00	Softened	RT	649	317	43	63
								400	556	201	37	63
								600	463	193	28	42
								800	193	Not given	40	42
18/18/7 CrNiCoMo-CuTi Steel	—	0.21	0.80	17.50	17.00	Co 7.00, Mo 2.50, Cu 2.50, Ti 0.70	Precipitation hardened	RT	687	409	32	43
								400	579	355	28	41
								600	510	355	22	29
								800	332	293	20	29
20/20/20 CrNiCoMo-WNb Steel	AICMA FE-PA91-HT	0.12	1.50	20.00	21.00	Co 20.00, Mo 3.00, W 2.00, Nb 1.00	Precipitation hardened	RT	811	394	35	48
								400	734	386	30	47
								600	602	324	29	45
								800	340	208	29	42
35/15NiCr Steel	—	0.10	1.00	35.00	15.00	—	Softened	RT	571	293	41	70
								400	463	224	28	42
								600	355	170	22	27
								800	170	139	16	20

*The specifications and standards cited in this table may be obsolete, but have been retained because the designations are still in common use.

Table 22.52   TYPICAL FATIGUE STRENGTH OF STEELS ON SMOOTH SPECIMENS AT ROOM AND HIGHER TEMPERATURES

Material	British or other standards*	Typical composition % C	Mn	Ni	Cr	Other elements	Condition	Fatigue limit for 10^7 cycles of stress at temperatures indicated, MPa RT	100°C	200°C	300°C	400°C	500°C	600°C	650°C	Remarks
1. Forged or rolled steels tested in longitudinal direction																
Carbon steels with up to 1.5%Mn content																
Armco Iron	—	0.02	0.03	—	—	—	As rolled	±185	±170	±178	±232	±178	±116	—	—	—
0.20C Steel	BS070 M20 (En3)	0.20	0.70	—	—	—	Normalised	±193	±193	±193	±247	±232	±154	—	—	—
0.25C Steel	BS070 M26 (En4)	0.26	0.70	—	—	—	Normalised	±201	±193	±193	±247	±263	±185	—	—	—
0.30C Steel	BS080 M30 (En5)	0.30	0.80	—	—	—	Normalised	±232	—	—	—	—	—	—	—	—
0.40C Steel	BS080 M40 (En8)	0.40	0.80	—	—	—	Hardened and tempered	±278	—	—	—	—	—	—	—	—
0.55C Steel	BS070 M55 (En9)	0.55	0.65	—	—	—	Hardened and tempered	±293	—	—	—	—	—	—	—	—
CMn Steel	BS150M19 (En14A)	0.28	1.50	—	—	—	Normalised	±278	—	—	—	—	—	—	—	—
Low alloy weldable steel																
$\frac{1}{2}$Mo Steel	—	0.14	0.43	—	—	Mo 0.50	Normalised	±317	—	—	±402	±371	±278	—	—	—
Low alloy direct hardening steel																
2NiCrMo Steel	—	0.27	Not given	2.0	0.90	Mo 0.40	Hardened and tempered	±432	±432	±432	±440	±432	±247	—	—	—
$2\frac{3}{4}$CrMo VW Steel	—	0.23	0.30	—	2.70	Mo 0.50 V 0.70 W 0.50	Hardened and tempered	±440	—	—	—	—	±324	±247	—	4×10^7 cycles
$3\frac{1}{2}$Ni Steel	BS (En22)	0.40	0.80	3.50	—	—	Hardened and tempered	±525	±533	±556	±556	±517	±394	—	—	Rotating bending

Type	Designation	C	Mn	Ni	Cr	Other	Condition								
3CrMo Steel	BS722M24 (En40B)	0.24	0.60	—	3.30	Mo 0.55	Hardened and tempered	±293	—	—	—	—	—	—	—
3NiCr Steel	BS653M31 (En23)	0.31	0.60	3.0	0.90		Hardened and tempered	±432	—	—	—	—	—	—	—
3NiCrMoV Steel	Similar to BS976M33	0.29	0.40	3.1	0.70	Mo 0.50 V 0.20	Hardened and tempered	±486	—	—	—	—	—	—	—
4NiCrMo Steel	Similar to BS835M30 (En30B)	0.30	0.50	3.9	1.20	Mo 0.35	Hardened and tempered	±525	±432	—	—	—	—	—	—
4½NiCrMo Steel	—	0.20	0.57	4.65	1.40	Mo 0.58	Hardened and tempered	±571	±448	—	—	—	—	—	—
High alloy steels															
13Cr Steel	BS410S21 (En56A)	0.12	0.50	—	13.0	—	Hardened and tempered	±340	—	—	—	—	—	—	—
13Cr Steel	BS420S37 (En56C)	0.24	0.50	—	13.0	—	Hardened and tempered	±402	±394	±386	±355	±309	±201	—	—
17CrNi Steel	BS431S29 (En57)	0.16	0.50	2.50	16.50	—	Hardened and tempered	±371	—	—	—	—	—	—	—
18/10CrNi Steel	BS304S15 (En58E)	0.06	0.80	10.0	18.0	—	Softened	±263	—	—	—	—	—	—	—
18/9CrNi Steel	BS302S17	0.07	1.50	9.5	18.0	—	Softened	±278	—	—	—	±216	±216	±216	±208
18/10CrNi Steel	BS302S25 (En58A)	0.12	1.50	9.7	18.5	—	Softened	±293	—	—	—	±255	±263	±216	±216

(*continued*)

* The specifications and standards cited in this table may be obsolete, but have been retained because the designations are still in common use.

Table 22.52   TYPICAL FATIGUE STRENGTH OF STEELS ON SMOOTH SPECIMENS AT ROOM AND HIGHER TEMPERATURES—*continued*

| Material | British or other standards* | Typical composition % | | | | | Condition | Fatigue limit for 10^7 cycles of stress at temperatures indicated, MPa | | | | | | | | Remarks |
		C	Mn	Ni	Cr	Other elements		RT	100°C	200°C	300°C	400°C	500°C	600°C	650°C	
18/8CrNiTi Steel	BS321S20 (En58B)	0.10	0.80	8.5	18.0	Ti 0.60	Softened	±270	—	—	—	—	—	—	—	—
18/10CrNi-Nb Steel	Similar to BS347S17 (En58G)	0.08	1.50	11.0	17.5	Nb 1.20	Softened	±301	—	—	—	—	—	±208	±178	—
18/9CrNiMo Steel	BS315S16 (En58H)	0.07	1.00	9.5	18.0	Mo 1.25	Softened	±270	—	—	—	—	—	—	—	—
18/8CrNiMo Steel	—	0.07	0.50	8.0	18.0	Mo 2.70	Softened	±270	—	—	—	—	—	—	—	—
18/18CrNi-MoCuTi Steel	—	0.07	0.80	18.0	18.0	Mo 3.70 Cu 2.40 Ti 0.60	Precipitation hardened	±263	—	—	—	—	—	—	—	—
2. *Steel castings*																
Carbon steel castings	BS592 Grade B	0.32	0.77	—	—	—	Annealed	±229	—	—	—	—	—	—	—	—
							Normalised and tempered	±258								
1½Mn Steel Castings	BS1456 Grade B1	0.31	1.60	—	—	—	Normalised and tempered	±334	—	—	—	—	—	—	—	—
	BS1456 Grade B2	0.31	1.60	—	—	—	Hardened and tempered	±403	—	—	—	—	—	—	—	—
½NiCrMo Steel castings	BS1458 Grade A	0.36	0.89	0.38	0.59	Mo 0.32	Hardened and tempered	±372	—	—	—	—	—	—	—	—
1¾NiCrMo Steel castings	BS1458 Grade C	0.34	0.60	1.74	0.70	Mo 0.30	Hardened and tempered	±534	—	—	—	—	—	—	—	—

* The specifications and standards cited in this table may be obsolete, but have been retained because the designations are still in common use.

Table 22.53 TYPICAL HOT CREEP AND RUPTURE PROPERTIES OF FORGED OR ROLLED STEELS

Material	British or other standards†	Typical composition %					Condition	Temperature °C	Stress to produce 1.0% strain		Stress to cause rupture	
		C	Mn	Ni	Cr	Other elements			MPa in 10 000 h	MPa in 100 000 h	MPa in 10 000 h	MPa in 100 000 h
1. Carbon steels												
0.15C Steel	BS 040A12 DIN St35.8	0.13	0.50	—	—	—	Normalised	400 450 500	136 80 39	96 49 20	192 113 54	133 68 29
0.20C Steel	BS 070M20 DIN St45.8	0.20	0.70	—	—	—	Normalised	400 450 500	136 80 39	96 49 20	192 113 54	133 68 29
0.35C Steel	BS 080M36 DIN Ck35	0.36	0.80	—	—	—	Hardened and tempered	400 450 500	147 80 39	99 49 22	193 113 54	137 68 34
0.45C Steel	BS 080M46 DIN Ck45	0.46	0.80	—	—	—	Hardened and tempered	400 450 500	147 80 39	99 49 22	193 113 54	137 68 34
2. Low alloy weldable steels												
$\frac{1}{2}$Mo Steel	DIN 15Mo3	0.16	0.65	—	—	Mo 0.30	Normalised	450 500 530	216 133 85	167 74 37	304 176 105	246 93 48
1CrMo Steel	BS1501-620 DIN 13CrMo44	0.15	0.70	—	0.95	Mo 0.50	Normalised and tempered	450 500 560	247 162 63	193 100 31	371 232 93	286 139 40
$1\frac{1}{4}$CrMo Steel	BS1501-621	0.12	0.55	—	1.25	Mo 0.55	Normalised and tempered	450 500 560	309* 181* 69*	224* 119* 31*	432 258 100	314 168 40
$2\frac{1}{4}$CrMo Steel	BS1501-622 DIN 10CrMo910	0.13	0.60	—	2.20	Mo 1.10	Normalised and tempered	500 550 580	148 83 57	103 53 31	216 116 77	147 77 45
$1\frac{1}{4}$MnCrMoV Steel	BS1501-271	0.14	1.25	—	0.55	Mo 0.24 V 0.08	Normalised and tempered	450 500 550	263* 170* 74*	216* 100* 28*	371 235 102	309 139 39

* Estimated.
† The specifications and standards cited in this table may be obsolete, but have been retained because the designations are still in common use.

(continued)

Table 22.53   TYPICAL HOT CREEP AND RUPTURE PROPERTIES OF FORGED OR ROLLED STEELS—*continued*

Material	British or other standards†	Typical composition %					Condition	Temperature °C	Stress to produce 1.0% strain		Stress to cause rupture	
		C	Mn	Ni	Cr	Other elements			MPa in 10000 h	MPa in 100000 h	MPa in 10000 h	MPa in 100000 h
1NiCrMoV Steel	BS1501-281	0.12	1.10	0.85	0.55	Mo 0.24 V 0.08	Normalised and tempered	450 500 550	216* 139* 62*	178* 85* 23*	297 187 82	247 111 31
1½NiCrMoV Steel	BS1501-282	0.15	1.10	1.50	0.50	Mo 0.35 V 0.10	Normalised and tempered	450 500 550	216* 139* 62*	178* 85* 23*	297 187 182	247 111 31
3. Low alloy direct hardening steels												
1CrMo Steel	DIN 24CrMo5	0.24	0.65	—	1.00	Mo 0.25	Hardened and tempered	450 500 550	227 139 62	171 93 25	310 176 79	225 117 37
1½CrMoV Steel	DIN 24CrMoV 55	0.24	0.45	—	1.35	Mo 0.55 V 0.20	Hardened and tempered	450 500 550	303 193 93	239 128 56	405 256 139	321 184 74
1¼CrMoV Steel	DIN 21CrMoV 511	0.21	0.40	—	1.35	Mo 1.10 V 0.30	Hardened and tempered	450 500 550	340 230 120	276 165 65	423 303 156	349 212 93
2¼CrMoV Steel	DIN 30CrMoV92	0.30	0.55	—	2.50	Mo 0.20 V 0.15	Hardened and tempered	450 500 550	246 158 74	187 99 36	324 216 99	264 137 46
4. High alloy steels												
9CrMo Steel	—	0.12	0.44	—	9.50	Mo 0.95	Hardened and tempered	590 650	43 19	14 6	59* 26	19* 8*

Steel	DIN	C	Si	Ni	Cr	Other		Condition	Temp				
12CrMo Steel	DIN X19CrMo12 1	0.19	0.55	—	12.0	Mo	1.00	Hardened and tempered	500	241	187	300	235
									550	137	93	176	117
									600	59	34	83	46
12CrMoV Steel	DIN X22CrMoV12 1	0.22	0.55	—	12.00	Mo V	1.00 0.30	Hardened and tempered	500	289	221	338	275
									550	165	108	212	137
									600	79	45	103	59
18/9CrNi Steel	BS304S15 (En58E)	0.06	0.80	10.00	18.0	—		Softened	600	85	46	116*	62*
									650	57	29	77*	39*
									700	39	17	54*	23*
16/13CrNiNb Steel	DIN X8CrNiNb 16 13	0.07	1.00	13.0	16.00	Nb	10×C/1.2	Softened	600	113	79	158	108
									650	79	49	103	63
									700	49	26	63	34
									750	34	15	45	20
17/13CrNiMo Steel	BS316S16 (En58J)	0.07	1.40	12.50	16.50	Mo	2.80	Softened	600	139	77	185*	108*
									650	100	46	131*	62*
									700	63	29	85*	39*
									750	43	20	56*	26*
16/16 CrNiMoNb Steel	DIN X8CrNiMoNb 16 16	0.07	1.00	16.50	16.50	Mo Nb	1.80 10×C/1.2	Softened	600	158	108	225	151
									650	108	63	137	83
									700	63	34	83	45
									750	42	15	54	20
16/16 CrNiWNb Steel	DIN X6CrNiWNb 16 16	0.06	1.00	16.50	16.50	W Nb N	3.00 10×C/1.2 0.10	Softened	600	167	117	235	162
									650	113	68	147	90
									700	68	39	91	49
									750	39	17	54	23
25/13CrNi Steel	—	0.06	1.55	13.40	24.9	—		Softened	650	69	39	93*	54*
									700	48	17	62*	23*
24/20CrNi Steel	BS 310S24	0.14	0.80	21.00	24.0	Si	1.00	Softened	600	154	77	224*	103*
									650	103	56	139*	74*
									700	65	34	86	45*

* Estimated.
† The specifications and standards cited in this table may be obsolete, but have been retained because the designations are still in common use.

Table 22.54   MECHANICAL PROPERTIES OF FORGED OR ROLLED STEELS AT SUBZERO TEMPERATURES IN THE LONGITUDINAL DIRECTION

| Material | British or other standards† | Typical composition % | | | | | Conditions | Temperature °C | Typical properties | | | | | |
		C	Mn	Ni	Cr	Other elements			UTS MPa	Yield stress MPa	Elong. (gl 5.65 $\sqrt{S_0}$) %	RA %	Impact Test spec.	Impact J
1. Carbon steels														
Armco iron	—	0.02	0.03	—	—	—	Not given	RT	309	Not given	Not given	73	Izod	106
								−75	432	293	33	72	Izod	5.5
								−120	525	463	15	68		
0.1C Steel	BS 040A10 (En2A)	0.10	0.45	—	—	—	As rolled	RT	463	340	27	72	Charpy	80
								−30					Charpy	60
								−45					Charpy	8
								−160	649	556	24	62		
0.15C Steel	BS 040A12 (En2B)	0.13	0.50	—	—	—	Normalised	RT	463	355	28	67	Charpy	163*
								−65	710	571	30	58	Charpy	9.0*
								−160						
0.20C Steel	BS 070M20 (En3)	0.20	0.70	—	—	—	Normalised	RT	430	215	21	65	Charpy	99*
								−20					Charpy	13.5*
								−80						2.7*
								−160	726	587	Not given	56		
0.25C Steel	BS 070M26 (En4)	0.26	0.70	—	—	—	Normalised	RT	494	247	20	Not given	Izod	27
							Cold drawn	RT	571	432	11	Not given	Izod	3.4
								−40	757	Not given	11	Not given		
0.35C Steel	BS 080M36	0.36	0.80	—	—	—	Normalised	RT	541	278	27	62	Charpy	110*
								−55					Charpy	30*
								−80					Charpy	11*
								−100	741	571	28	57		
0.40C Steel	BS 080M40 (En8)	0.40	0.80	—	—	—	Hardened and tempered	RT	772	463	22	Not given		
								−185	1 112	Not given	9	Not given		
0.45C Steel	BS 080M46	0.46	0.80	—	—	—	Normalised	RT	618	309	15	Not given	Charpy	22*
								80	896	Not given	14	Not given	Charpy	11*
0.50C Steel		0.50	0.80	—	—	—	Hardened and tempered	RT	927	525	14	Not given	Charpy	68*
								−40					Charpy	36.5*
	BS 080M50 (En9) (En43A)						Normalised	RT	618	309	14	Not given		
								−30	803	Not given	12	Not given		

2. Low alloy weldable steels

Type	Designation	C	Mn	Ni	Cr	Other	Condition	Temp (°C)			Elong.		Impact test	Impact value
MnCrMoV Steel	BS1 501–271	0.14	1.40	—	0.70	Mo 0.28, V 0.10	Normalised and tempered	RT; 0; −60	649	463	18	Not given	Charpy V / Izod; Charpy V / Izod; Izod	41, 68; 27, 51.5; 16
1NiCrMoV Steel	BS 1 501–281	0.12	1.10	0.85	0.55	Mo 0.24, V 0.08, Nb 0.10	Normalised and tempered	RT; −10; −20; −40	649	417	18	Not given	Charpy V; Charpy V; Charpy V	68; 54; 27
1½NiCrMoV Steel	BS 1 501–282	0.15	1.10	1.50	0.50	Mo 0.35, V 0.10	Double normalised and tempered	RT; −10; −20; −40; −50	649	432	18	Not given	Charpy V; Charpy V; Charpy V; Charpy V	81; 68; 41; 27
3½Ni Steel	BS 1 501–503	0.15	0.55	3.50	—	—	Normalised and tempered	RT; −80; −100	448	263	20	Not given	Charpy V; Charpy V	34; 17.5
5Ni Steel	DIN 12Ni9	0.13	0.40	5.20	—	—	Hardened and tempered	RT; −150; −195	710; 1050; 1205	Not given; Not given; Not given	22; 22; 18	74; 57; 50	—; —; —	—; —; —

3. Low alloy direct hardening steels

Type	Designation	C	Mn	Ni	Cr	Other	Condition	Temp (°C)			Elong.		Impact test	Impact value
1CrV Steel	—	0.29	0.70	—	1.00	V 0.20	Hardened and tempered	RT; −75	896; 1035	Not given; Not given	13; 13	Not given; Not given	—; —	—; —
¾CrMo Steel	—	0.31	0.65	—	0.70	Mo 0.20	Hardened and tempered	RT; −40	880; 1035	Not given; Not given	17; 17	Not given; Not given	Izod; Izod	110; 109
1CrMo Steel	BS 709M40 (En19)	0.40	0.80	—	1.10	Mo 0.30	Hardened and tempered	RT; −70; −185	849; 1251	494; 1143	15; 19	70; 47	Charpy; Charpy	141*; 84*
1¼NiCr Steel	BS 640M40 (En111)	0.40	0.80	1.30	0.65	—	Hardened and tempered	RT; −185	1004; 1699	680; Not given	13; 4.5	60; 48	Izod; —	54; —

* Charpy specimen—Izod notch.   U = Unbroken.   V = Charpy specimen—V notch.   U = unbroken.
† The specifications and standards cited in this table may be obsolete, but have been retained because the designations are still in common use.

(continued)

Table 22.54 MECHANICAL PROPERTIES OF FORGED OR ROLLED STEELS AT SUBZERO TEMPERATURES IN THE LONGITUDINAL DIRECTION—*continued*

Material	British or other standards†	C	Mn	Ni	Cr	Other elements	Conditions	Temperature °C	UTS MPa	Yield stress MPa	Elong. (gl 5.65 $\sqrt{S_0}$) %	RA %	Test spec.	J
1NiCrMo Steel	—	0.47	0.80	1.10	1.00	Mo 0.20	Hardened and tempered	RT	1 081	Not given	12	Not given	—	—
								−76	1 220	Not given	11	Not given	—	—
2½NiCrMo Steel	BS 826M31 (En25)	0.31	0.55	2.70	0.60	Mo 0.55	Hardened and tempered	RT	1 004	741	13	65	Izod	34
								−60	1 127	988	12	63		—
								−180	1 390	1 266	12	63		—
2½NiCrMo Steel	—	0.34	0.50	2.30	1.90	Mo 0.40	Hardened and tempered	RT	1 158	1 019	15	65	—	—
								−70	1 251	1 127	15	63	—	—
								−185	1 560	1 390	17	62	—	—
3¼Ni Steel	BS (En21)	0.34	0.70	3.30	—	—	Hardened and tempered	RT	927	865	15	64	Charpy	109*
								−75	1 066	988	14	60	Charpy	38*
3NiCr Steel	BS 653M31 (En23)	0.31	0.60	0.90	0.90	—	Hardened and tempered	RT	1 004	757	12	60	Charpy	80
								−95					Charpy	61
								−185	1 683	1 668	5	49	—	
4½NiCr Steel	BS (En30A)	0.30	0.50	4.40	1.40	—	Hardened and tempered	RT	865	Not given	24	Not given	—	—
								−30	1 019	Not given	26	Not given	—	—
								−90	1 050	Not given	23	Not given	—	—
4NiCrMo Steel	BS 835M30 (En30B)	0.30	0.50	3.90	1.20	Mo 0.35	Hardened and tempered	RT	1 544	1 235	17	54	Charpy	34*
								−80	1 761	1 266	17	58		
								−100					Charpy	20.5*
								−196	2 039	1 544	17	50		—

4. High alloy steels

Material	British or other standards†	C	Mn	Ni	Cr	Other elements	Conditions	Temperature °C	UTS MPa	Yield stress MPa	Elong. (gl 5.65 $\sqrt{S_0}$) %	RA %	Test spec.	J
9Ni Steel	BS 1 501–509, 1 503–509 DIN X8Ni9 ASTM A353	0.10	0.55	9.25	—	—	Hardened and tempered	RT	695	525	18	Not given	Charpy	76
								−100					Charpy V	68
								−160					Charpy V	47
								−196	1 158	911	18	Not given	Charpy V	64
13Cr Steel	BS 420S45 (En56D)	0.32	0.40	—	13.5	—	Hardened and tempered	RT	927	587	13	37	Charpy V	34
								−185	1 792	1 452	3	4	—	—

Steel	Specification	C		Ni	Cr	Other		Condition	Test temp. (°C)	Tensile strength	0.2% Proof stress	Elong. (%)	Red. of area (%)	Impact test	Impact value
17Cr Steel	BS 430S15 (En60)	0.09	0.45	—	17.0	—	—	Softened	RT	432	278	25	71	Charpy	61*
									−20	—	—	—	—	Charpy	34*
									−60	—	—	—	—	Charpy	6.8*
									−185	1004	849	12	14	—	—
20CrCu Steel	—	0.25	0.50	—	20.0	Cu	1.0	Softened	RT	633	355	21	59	—	—
									−185	680	618	Not given	Not given	—	—
17CrNi Steel	BS 431S29 (En57)	0.14	Not given	2.50	17.5	—	—	Hardened and tempered	RT	1004	680	21	Not given	—	—
									−90	1050	Not given	18	Not given	—	—
18/10CrNi Steel	BS 302S35 (En58A) DIN X12Cr-Ni 18.9	0.12	0.50	9.7	18.5	—	—	Softened	RT	510	216	47	74	Izod	155U
									−60	942	340	52	74	Izod	146U
									−120	1236	494	43	63	Izod	146U
									−180	1544	510	40	55	Izod	146U
18/8Cr NiMo Steel	—	0.05	0.50	8.20	18.0	Mo	2.70	Softened	RT	710	Not given	58	Not given	—	—
									−30	896	Not given	51	Not given	—	—
									−55	1050	Not given	42	Not given	—	—
									−90	1174	Not given	35	Not given	—	—
14/14CrNi Steel	BS (En58D)	0.11	Not given	14.0	14.2	—	—	Softened	RT	602	232	44	76	Izod	159U
									−65	772	293	51	76	Izod	160U
									−120	958	417	49	72	Izod	159U
									−180	1328	819	45	60	Izod	157U
25/20CrNi Steel	BS 310S24	0.14	0.80	21.0	24.0	Si	1.00	Softened	RT	649	317	43	Not given	Charpy	203*
									−30	741	Not given	52	Not given	—	—
									−50	—	—	—	—	Charpy	190*
									−90	849	Not given	51	Not given	—	—
									−150	—	—	—	—	Charpy	122*
26Ni Steel	—	0.40	1.50	26.0	—	—	—	Softened	RT	695	371	33	52	—	—
									−185	1421	1097	10	10	—	—
36Ni Steel	—	0.16	0.90	36.0	—	—	—	Softened	RT	556	355	28	58	—	—
									−250	988	880	18	60	—	—

* Charpy specimen—Izod notch. U = Unbroken. V = Charpy specimen—V notch. U = unbroken.

† The specifications and standards cited in this table may be obsolete, but have been retained because the designations are still in common use.

Table 22.55 TOOL STEELS AND THEIR USES

Equivalent designations for some tool steels are presented in Chapter 1.

		Chemical composition limits %									
Material	*British or other standards***	C	Si	Mn	Cr	Mo	W	V	Co	*Other elements*	*Typical uses*
1. *Carbon tool steels*											
0.8C Steel	DIN C80W1	0.75 0.85	0.10 0.25	0.10 0.25	— —	— —	— —	— —	— —	— —	Cold heading dies, chisels, mandrels, punches, rivet snaps, vice jaws, drifts
0.9C Steel	BS BW1A*	0.85 0.95	0.30 max	0.35 max	0.15 max	0.10 max	— —	— —	— —	Ni 0.20 max	Gauges, chisels, punches, shear blades
1C Steel	BS BW1B*	0.95 1.10	0.30 max	0.35 max	0.15 max	0.10 max	— —	— —	— —	Ni 0.20 max	Large drills, reamers, cutters, cold heading tools, shear blades, wood working tools, lathe centres
1.2C Steel	BS BW1C*	1.10 1.30	0.30 max	0.35 max	0.15 max	0.10 max	— —	— —	— —	Ni 0.20 max	Twist drills, cutters, reamers, taps, files, screw gauges, wood-working tools
1CV Steel	BS BW2	0.95 1.10	0.30 max	0.35 max	0.15 max	0.10 max	— —	0.15 0.35	— —	Ni 0.20 max	Large drills, reamers, cutters, cold heading tools, shear blades, woodworking tools, lathe centres, when fine grained shallow case is required with increased toughness
2. *Hot work tool steels*											
$1\frac{3}{4}$NiCr-MoV Steel	DIN 55Ni-CrMoV6 Similar to BS No5 die steel	0.55	0.30	0.60	0.70	0.30 (Typical)	—	0.10	—	Ni 1.70	Solid dies for drop-hammer or press forging
3CrMoV Steel	Bs BH10	0.30 0.40	0.75 1.10	0.40 max	2.80 3.20	2.65 2.95	— —	0.30 0.50	— —	— —	Mandrels, hot extrusion and forging dies, punches, die inserts, gripper and header dies, hot shears aluminium die casting dies
3CrMoV Co Steel	BS BH10A	0.30 0.40	0.75 1.10	0.40 max	2.80 3.20	2.65 2.95	— —	0.30 1.10	2.80 3.20	— —	Close die forging tools
5CrMoV Steel	BS BH11	0.32 0.42	0.85 1.15	0.40 max	4.75 5.25	1.25 1.75	— —	0.30 0.50	— —	— —	Die casting dies and die inserts for light alloys, punches, piercing tools, mandrels, forging dies and die inserts, ejector pins, hot extrusion dies, sleeves, slides
5CrMoV Steel	BS BH13	0.32 0.42	0.85 1.15	0.40 max	4.75 5.25	1.25 1.75	— —	0.90 1.10	— —	— —	As for BH11, when higher performance is required

* BS 4659: 1971; omitted from BS 4659: 1989 but still referred to.

(*continued*)

** The specifications and standard cited in this table may be obsolete, but have been retained because the designations are still in common use.

Table 22.55   TOOL STEELS AND THEIR USES—*continued*

Material	British or other standards**	C	Si	Mn	Cr	Mo	W	V	Co	Other elements	Typical uses
						Chemical composition limits %					
5CrMo-WV Steel	BS BH12	0.30 0.40	0.85 1.15	0.40 max	4.75 5.25	1.25 1.75	1.25 1.75	0.50 max	— —	— —	Extrusion dies, gripper and header dies, forging die inserts, punches, mandrels, sleeves
4WCrV Steel	DIN X30-WCrV53	0.30	0.20	0.30	2.40	— (Typical)	4.30	0.60	—	—	Mandrels, hot extrusion and forging dies, die inserts, gripper and header dies, hot shears, aluminium die casting dies
4WCrCo-VMo Steel	BS BH19	0.35 0.45	0.40 max	0.40 max	4.00 4.50	0.45 max	4.00 4.50	2.00 2.40	4.00 4.50	— —	Extrusion dies and die inserts, forging die inserts, punches and mandrels when the highest performance is required
9WCr-MoV Steel	BS BH21	0.25 0.35	0.40 max	0.40 max	2.25 3.25	0.60 max	8.50 10.00	0.40 max	— —	— —	Mandrels, hot blanking dies, hot punches, extrusion dies and die casting dies for brass, piercer points, gripper dies, hot headers–when high performance is required
9WCrMo VNi Steel	BS BH21A	0.20 0.30	0.40 max	0.40 max	2.25 3.25	0.60 max	8.50 10.00	0.50 max	— —	Ni 2.0 2.5	Hot extrusion tools
18WCrMo VCo Steel	BS BH26*	0.50 0.60	0.40 max	0.40 max	3.75 4.50	0.60 max	17.50 18.50	1.00 1.50	0.60 max	— —	Closed die forging tools

3. *Plastic moulding steels*

Material	British or other standards**	C	Si	Mn	Cr	Mo	W	V	Co	Other elements	Typical uses
1½CrMo Steel	BS P20	0.28 0.40	0.4 0.6	0.65 0.95	1.5 1.8	0.35 0.55	—	—	—	0.40 max	Dies for plastic moulding
4NiCrMo Steel	BS P30	0.26 0.34	0.4 max	0.45 0.70	1.1 1.4	0.20 0.35	—	—	—	3.1 4.3	Dies for plastic moulding

4. *Shock resisting tool steels*

Material	British or other standards**	C	Si	Mn	Cr	Mo	W	V	Co	Other elements	Typical uses
2½WCrV Steel	BS BS1	0.48 0.55	0.70 1.00	0.30 0.70	1.20 1.70	— —	2.00 2.50	0.10 0.30	— —	— —	Bolt header dies, chipping and caulking chisels, concrete drills, forming dies, grippers, mandrels, punches, pneumatic tools, beading tools, scarfing tools, swaging dies, shear blades
1SiMoV Steel	BS BS2*	0.45 0.55	0.90 1.20	0.30 0.50	— —	0.30 0.60	— —	0.10 0.30	— —	— —	Hand and pneumatic chisels, forming tools, ejector pins, mandrels, shear blades, spindles, stamps, tool shanks

* BS 4659: 1971; omitted from BS 4659: 1989.
** The specifications and standard cited in this table may be obsolete, but have been retained because the designations are still in common use.

(*continued*)

Table 22.55　TOOL STEELS AND THEIR USES—*continued*

Material	British or other standards**	C	Si	Mn	Cr	Mo	W	V	Co	Other elements	Typical uses
					Chemical composition limits %						
2SiCrV Steel	BS BS5*	0.50 0.60	1.60 2.10	0.60 0.80	— —	0.30 0.60	— —	0.10 0.30	— —	— —	Hand and pneumatic chisels, forming tools, ejector pins, mandrels, shear blades, spindles, stamps, tool shanks, lathe collets, bending dies, punches, rotary shears
5. Hammer die steels											
1.5NiCr-Mo Steel	BS 224 BS 225	0.49 0.57	0.35 0.70	0.70 1.0	0.70 1.10	0.40 1.25	— —	— —	— —	Ni 1.25 1.80	Cold stamping, hammer dies, forming tools, punches, etc.
6. Cold work tool steels											
12CrV Steel	BS BD3	1.90 2.30	0.60 max	0.60 max	12.0 13.0	— —	— —	0.50 max	— —	Ni 0.40 max	Blanking dies, cold forming dies, thread rolling dies, shear blades, slitter knives, forming rolls, seaming rolls, burnishing tools, punches, gauges, crimping dies, swaging dies
12CrMoV Steel	Bs BD2	1.40 1.50	0.60 max	0.60 max	11.50 12.50	0.70 1.20	— —	0.25 1.00	— —	Ni 0.40 max	Blanking dies, cold forming dies, drawing dies, thread rolling dies, shear blades, slitter knives, forming rolls, burnishing tools, punches, gauge, knurling tools, lathe centres, broaches, cold extrusion dies, mandrels, swaging dies
13CrMoV Steel	BS BD2A	1.60 1.90	0.60 max	0.60 max	12.00 13.00	0.70 0.90	— —	0.25 1.00	— —	Ni 0.40 max	Coining dies
5CrMoV Steel	BS BA2	0.95 1.05	0.40 max	0.30 0.70	4.75 5.25	0.90 1.10	— —	0.15 0.40	— —	Ni 0.40 max	Thread rolling dies, extrusion dies, trimming, blanking and coining dies, mandrels, shear blades, spinning rolls, forming rolls, gauges, beading dies, burnishing tools, embossing dies, plastic moulds, stamping dies, bushes, punches, liners for brick moulds
2MnMo-Cr Steel	BS BA6	0.65 0.75	0.40 max	1.80 2.10	0.85 1.15	1.20 1.60	— —	— —	— —	Ni 0.40 max	Blanking, forming, coining and trimming dies, punches, shear blades, spindles, mandrels, plastic moulds

** The specifications and standard cited in this table may be obsolete, but have been retained because the designations are still in common use.

(*continued*)

Table 22.55 TOOL STEELS AND THEIR USES—*continued*

Material	British or other standards**	C	Si	Mn	Cr	Mo	W	V	Co	Other elements	Typical uses
$1\frac{1}{4}$MnCr-WV Steel	BS BO1	0.85 1.00	0.40 max	1.10 1.35	0.40 0.60	— —	0.40 0.60	0.25 max	— —	Ni 0.40 max	Blanking, drawing and trimming dies, plastic moulds, paper slitters, shear blades, taps, reamers, gauges, jigs, bending and forming dies, bushes, punches
$1\frac{3}{4}$MnV Steel	BS BO2	0.85 0.95	0.40 max	1.50 1.80	— —	— —	— —	0.25 max	— —	Ni 0.40 max	Blanking, stamping, trimming and forming dies and punches, threading dies, taps, reamers, gauges, plugs, jigs, broaches, circular cutters and saws, bushes
$1\frac{1}{2}$CrMn Steel	BS BL1	0.95 1.10	0.40 max	0.40 0.70	1.2 1.6	— —	— —	— —	— —	Ni 0.4 max	Ball bearings, taps, broaches, files, mandrels, cold rolls, gauges, drills
$1\frac{1}{2}$CrV Steel	BS BL3*	0.95 1.05	0.40 max	0.40 max	1.30 1.50	— —	— —	0.10 0.30	— —	— —	Mandrels, cold rolls, ball bearings, precision gauges, reamers, broaches, taps, drills, thread rolling dies, files
$1\frac{1}{2}$WCrV Steel	BS BF1*	1.15 1.35	0.40 max	0.40 max	0.25 0.50	— —	1.30 1.60	0.30 max	— —	— —	Taps, broaches, reamers, drills
3V Steel	DIN 145-V33	1.45	0.30	0.40	—	— (Typical)	—	3.30	—	—	Cold heading dies and punches

7. *High speed tool steels*

Material	British or other standards**	C	Si	Mn	Cr	Mo	W	V	Co	Other elements	Typical uses
3-3-2 WMoV Steel	DIN S3.3.2	0.95 1.03	0.40 max	0.40 max	3.80 4.50	2.50 2.80	2.70 3.00	2.20 2.50	— —	—	Hacksaw and circular saw blades for metal cutting
2.9.1 WMoV Steel	BS BM1	0.75 0.85	0.40 max	0.40 max	3.75 4.50	8.00 9.00	1.00 2.00	1.00 1.25	1.00 max	—	Drills, taps, reamers, milling cutters, hobs, punches, lathe and planer tools, form cutters, saws, chasers, broaches, routers, woodworking tools
2-9-2 WMoV Co Steel	BS BM34*	0.85 0.95	0.40 max	0.40 max	3.75 4.50	8.00 9.00	1.70 2.20	1.75 2.05	7.75 8.75	—	Cutting tools
6.5.2 WMoV Steel	BS BM2	0.82 0.92	0.40 max	0.40 max	3.75 4.50	4.75 5.50	6.00 6.75	1.75 2.05	1.20 max	—	Drills, taps, reamers, milling cutters, hobs, form cutters, saws, lathe and planer tools, chasers, broaches, boring tools, cold forming tools, e.g. punches for cold extrusion, cold heading die inserts

** The specifications and standard cited in this table may be obsolete, but have been retained because the designations are still in common use.

(*continued*)

Table 22.55   TOOL STEELS AND THEIR USES—*continued*

Material	British or other standards**	Chemical composition limits %								Other elements	Typical uses
		C	Si	Mn	Cr	Mo	W	V	Co		
6.5.2Co WMoVCo Steel	BS M35	0.85 0.95	0.40 max	0.40 max	3.75 4.50	4.75 5.25	6.00 6.75	1.75 2.15	4.6 5.2	— —	As for BM2, esp. cutting tools
6-5-4 WMoV Steel	BS BM4	1.25 1.40	0.40 max	0.40 max	3.75 4.50	4.25 5.00	5.75 6.50	3.75 4.25	1.00 max	— —	Heavy duty broaches, reamers, milling cutters, chasers, form cutters, lathe and planer tools, blanking dies and punches, swaging dies
6.5.4 WMoV Co Steel	BS BM15	1.45 1.60	0.40 max	0.40 max	4.50 5.00	2.75 3.25	6.25 7.00	4.75 5.25	4.5 5.5	— —	Cutting tools
6.5.2.5 WMo-VCo Steel	DIN S6-5-2-5	0.88 0.96	0.40 max	0.40 max	3.80 4.50	4.70 5.20	6.00 6.70	1.70 2.00	4.50 5.00	— —	Milling cutters, twist drills and taps, particularly suitable for intermittent cutting and drilling
2-9-1-8 WMo VCo Steel	BS BM42	1.00 1.10	0.40 max	0.40 max	3.50 4.25	9.00 10.00	1.00 2.00	1.00 1.30	7.50 8.50	— —	Heavy duty drills, reamers, form cutters, lathe tools, hobs, broaches, milling cutters, twist drills
18-0-1 WV Steel	BS BT1	0.70 0.80	0.40 max	0.40 max	3.75 4.50	0.70 max	17.5 18.5	1.00 1.25	1.00 max	— —	Drills, taps, reamers, hobs, lathe and planer tools, broaches, burnishing dies, chasers, form cutters, milling cutters
18-0-2 WV Steel	BS BT2*	0.75 0.85	0.40 max	0.40 max	3.75 4.50	0.70 max	17.50 18.50	1.75 2.05	0.60 max	— —	Cutting tools
18-1-1-5 WMo-VCo Steel	BS BT4	0.70 0.80	0.40 max	0.40 max	3.75 4.50	1.00 max	17.5 18.5	1.00 1.25	4.50 5.50	— —	As for BT1, when higher performance is required
19-1-2-10 WMoV Co Steel	BS BT5	0.75 0.80	0.40 max	0.40 max	3.75 4.50	1.00 max	18.50 19.50	1.75 2.05	9.00 10.00	— —	Cutting tools
20-1-2-12 WMo-VCo Steel	BS BT6	0.75 0.85	0.40 max	0.40 max	3.75 4.50	1.00 max	20.0 21.0	1.25 1.75	11.25 12.25	— —	Heavy duty lathes and planer tools, drills, cut-off tools, milling cutters, hobs
12-1-5-5 WMo-VCo Steel	BS BT15	1.40 1.60	0.40 max	0.40 max	4.25 5.00	1.00 max	12.0 13.0	4.75 5.25	4.50 5.50	— —	Heavy duty form cutters, milling cutters, broaches, blanking dies, lathe and planer tools
14-1 WV Steel	BS BT21	0.60 0.70	0.40 max	0.40 max	3.50 4.25	0.70 max	13.50 14.50	0.40 0.60	1.00 max	— —	Cutting tools
22.1 WV Steel	BS BT20*	0.75 0.85	0.40 max	0.40 max	4.25 5.00	1.00 max	21.00 22.50	1.40 1.60	0.60 max	— —	Cutting tools
9-3-3-9 WMo-VCo Steel	BS BT42	1.25 1.40	0.40 max	0.40 max	3.75 4.50	2.75 3.50	8.50 9.50	2.75 3.25	9.00 10.00	— —	Heavy duty milling cutters and form cutters when high performance is required

* BS 4659: 1971; omitted from BS 4659: 1989.
** The specifications and standard cited in this table may be obsolete, but have been retained because the designations are still in common use.

Table 22.56 TOOL STEEL HEAT TREATMENTS AND HARDNESS

Equivalent designations for some tool steels are presented in Chapter 1.

		Annealing	Hardening		Tempering	Hardness		
							Heat treated	
	British or other					*Annealed Max*	*Min*	
Material	*standards**	°C	°C	*Quenching medium*	°C	HB	HV	*Remarks*
1. Carbon tool steels								
0.8C Steel	DIN C80W1	680/710	780/810	Water or brine	150/350	207	790	The wear resistance increases but the toughness decreases with increasing carbon content
0.9C Steel	BS BW1A*	740/770	770/790	Water or brine	150/350	207	790	
1C Steel	BS BW1B*	740/770	770/790	Water or brine	150/350	207	790	
1.2C Steel	BS BW1C*	740/770	760/780	Water or brine	150/350	207	790	
1CV Steel	BS BW2	740/7	780/800	Water or brine	180/350	207	790	The V addition reduces the case depth, and raises the resistance to cracking
2. Hot work tool steel								
$1\frac{3}{4}$NiCr-MoV Steel	DIN 55NiCrMoV6 Similar to Bs No. 5 Die Steel	650/700	830/870	Oil	500/600	—	—	Tough and wear resistant
3CrMoV Steel	BS BH10	850/870	1 000/1 060$^\phi$	Oil or air	530/650	229	—	Wear resistance and fairly tough. Resistant to softening on tempering
3CrMoV Co Steel	BS BH10A	850/870	1 000/1 060$^\phi$	Air or oil	530/650	241	—	Wear resistant
5CrMoV steel	BS BH11	850/870	1 000/1 030$^\phi$	Air or oil	530/650	229	—	Wear resistant and tough
CrMoV Steel	BS BH13	850/870	1 020/1 030$^\phi$	Air or oil	530/650	229	—	More resistant to wear and softening on tempering than BH11
CrMo VV Steel	BS BH12	850/870	1 000/1 030$^\phi$	Air or oil	530/650	229	—	Wear resistant and tough
WCrV steel	X30WCrV53	750/800	1 060/1 100$^\phi$	Oil, air or salt bath of 500/550°C temperature	530/650	—	—	Less tough but more resistant to softening on tempering than BH11. Not suitable for water cooling in service
WCrCo-Mo teel	BS BH19	850/870	1 150/1 200$^\phi$	Air or oil	530/650	248	—	Highly resistant to wear and softening on tempering. Not suitable for water cooling in service
Wcr-1oV eel	BS BH21	870/890	1 100/1 180$^\phi$	Oil or air	560/675	235	—	Very resistant to softening on tempering. Not suitable for water cooling in service

*BS 4659: 1971; omitted from BS 4659: 1989.

*Pre-heat 800°C.

(continued)

Table 22.56   TOOL STEEL HEAT TREATMENTS AND HARDNESS—*continued*

Material	British or other standards*	Annealing °C	Hardening °C	Quenching medium	Tempering °C	Annealed Max HB	Heat treated Min HV	Remarks
9WcrMo V Steel	BS BH21A	870/890	1 100/1 170$^\phi$	Oil or air	560/675	235	—	Resistant to softening
18WCrMo V Steel	BS BH26*	870/890	1 180/1 260	Oil or air	550/570	—	—	Resistant to softening
3. Plastic moulding steels								
1½CrMo Steel	BS P20	—	850/880	Oil	180/650	—	—	Varies
4NiCrMo Steel	BS P30	640/660	810/830	Air or oil	180/650	—	—	Varies
4. Shock resisting tool steels								
2¼WCrV Steel	BS BS1	790/820	870/820	Oil	200/650	229	600	Tough and wear resistant
1SiMoV Steel	BS BS2*	790/820	850/900	Water or oil	175/425	—	—	Tough and wear resistant
2SiMoV Steel	BS BS5*	790/820	870/920	Oil	175/425	—	—	More resistant to wear, but less tough than BS2
5. Hammer die steels								
1½NiCrMo Steel	BS 224 225	850/870$^+$	820/840	Oil or water	520/660	—	A $^+$444/7 A 401/29 B 363/88 C 331/52 D 302/321 E 269/93	
6. Cold work tool steels								
12CrV Steel	BS BD3	850/870	950/1 000$^\phi$	Air or oil	150/220 450/550	255	763	Highly resistant to wear and abrasion. Resistant to corrosion and softening on tempering. Very little distortion aft air hardening. Low temper for hardne high temper for toughness
12CrMoV Steel	BS BD2	850/870	980/1 030$^\phi$	Air or oil	150/220 450/550	255	735	Similar to, but tougher than, BD
13CrMoV Steel	BS BD2A	850/870	980/1 030$^\phi$	Oil or air	150/220 450/550	255	763	Tough and wear resistant
5CrMoV Steel	BS BA2	850/870	950/980$^\phi$	Air	150/550	241	735	Tougher, but less corrosion and wea resistant than BD Very little distorti after hardening

* BS 4659: 1971; omitted from BS 4659: 1989.
$^\phi$ Pre-heat 800°C.

(*continue*

Table 22.56   TOOL STEEL HEAT TREATMENTS AND HARDNESS—*continued*

Material	British or other standards*	Annealing °C	Hardening °C	Quenching medium	Tempering °C	Hardness Annealed Max HB	Heat treated Min HV	Remarks
2MnMo-Cr Steel	BS BA6	730/750	830/850$^+$	Air	150/250	241	735	Less resistant to wear and softening on tempering than BD2. Very little distortion after hardening
$1\frac{1}{4}$MnCr-WV Steel	BS BO1	760/780	780/820	Oil	150/300	229	735	Hard and wear resistant. Little distortion after hardening
$1\frac{3}{4}$MnV Steel	BS BO2	760/780	760/780	Oil	150/300	229	735	Hard and wear resistant. Little distortion after hardening
$1\frac{1}{2}$CrV Steel	BS BL3*	790/810	820/840 790/810	Oil Water	175/320	—	760 min	Hard and wear resistant
$1\frac{1}{2}$WCrV Steel	BS BF1*	780/800	780/800	Oil or water	175/250	—	760 min	Hard and wear resistant
3V Steel	DIN 145V33	720/760	800/920	Water	100/250	—	—	Very hard and wear resistant. The case depth is controlled by the hardening temperature
$1\frac{1}{2}$CrMn Steel	BS BL1	780/820	800/850	Oil or water	150/300	—	—	
7. High speed steels								
3.3.2 WMoV Steel	DIN S3.3.2	800/840	1 180/1 220$^\phi$	Oil, air or salt bath at 500°C	520/550*	—	—	Steel for special applications, e.g. metal cutting saws
2-8-1 WMov Steel	BS BM1	850/870	1 190/1 210$^\phi$	Oil, air or salt bath at 500/560°C	530/550**	241	823	General purpose cutting tools
2-9-2 WMoV Steel	BS BM34*	870/900	1 215/1 235$^\phi$	Oil, air or salt bath at 500/560°C	530/550**	—	—	General purpose cutting tools
6.5.2 WMoV Steel	BS BM2	850/870	1 210/1 230$^\phi$	Oil, air or salt bath at 500/560°C	550/570**	248	836	General purpose cutting tools
6.5.2Co WMoVCo Steel	BS BM35	870/890	1 215/1 235$^\phi$	Oil, air or salt bath at 500/560°C	530/550†	269	869	
6.5.4 WMoV Steel	BS BM4	850/870	1 200/1 220	Oil, air or salt bath at 500/560°C	540/560**	255	849	Steel for special applications when very high resistance to wear is required

BS 4659: 1971, omitted from BS 4659: 1989.
* Double tempering.
Treble tempering.
Pre-heat 800°C.

(*continued*)

Table 22.56   TOOL STEEL HEAT TREATMENTS AND HARDNESS—*continued*

Material	British or other standards*	Annealing °C	Hardening °C	Quenching medium	Tempering °C	Annealed Max HB	Heat treated Min HV	Remarks
6.5.4 WMoV Co Steel	BS BM15	870/900	1 210/1 230$^\phi$	Oil, air or salt bath at 500/560°C	540/560†	227	869	High wear resistance
6.5.2.5 WMo-VCo Steel	DIN S6-5.2.5	800/840	1 210/1 240$^\phi$	Oil, air or salt bath at 500/560°C	540/570†	—	—	Steel of high hot hardness, particularly suitable for intermittent cutting and drilling
2-9-1-8 WMo-VCo Steel	BSμBM42	870/900	1 180/1 200$^\phi$	Oil, air or salt bath at 500/560°C	520/540†	269	897	Steel for special applications when very high hardness is required
18-0-1 WV Steel	BS BT1	870/890	1 270/1 290$^\phi$	Oil, air or salt bath at 500/560°C	550/570**	255	823	General purpose cutting tools
18-0-2 WV Steel	BS BT2*	870/890	1 270/1 290$^\phi$	Oil, air or salt bath at 500/560°C	550/570**	—	—	General purpose cutting tools
18-1-1-5 WMo-VCo Steel	BS BT4	880/900	1 280/1 300$^\phi$	Oil, air or salt bath at 500/560°C	550/570†	277	849	Steel of high hot hardness, is used at higher speed for heavier cuts than is BT1
19.1.2.10 MoWVCo Steel	BS BT5	880/900	1 290/1 310$^\phi$	Oil, air or salt bath at 500/560°C	550/570†	290	869	
20.1.2.12 WMo-VCo Steel	BS BT6	880/900	1 290/1 310	Oil, air or salt bath at 500/560°C	550/570†	302	869	Steel for special applications when the highest hot hardness is required
12-1-5-5 WMo-VCo Steel	BS BT15	870/890	1 230/1 250$^\phi$	Oil, air or salt bath at 500/560°C	550/570†	290	890	Steel for special applications when the highest wear resistance is required
14-1 WV Steel	BS BT21	850/870	1 270/1 290$^\phi$	Oil, air or salt bath at 500/560°C	550/570	255	798	Steel for less arduous service
9-3-3-9 WMo-VCo Steel	BS BT42	850/870	1 200/1 240$^\phi$	Oil, air or salt bath at 500/560°C	550/570†	277	912	Steel for special applications when both the highest wear resistance and highest hot hardness are required

* BS 4659: 1971; omitted from BS 4659: 1989.
† Treble tempering.
** Double tempering.
$^\phi$ Pre-heat 800°C.

2.11 Other metals of industrial importance

In Table 22.57 are summarised the mechanical properties of metals of industrial importance not included in previous tables. Unless otherwise indicated, the data refer, as far as possible, to the purest metals available.

Table 22.57 MECHANICAL PROPERTIES OF OTHER METALS OF INDUSTRIAL IMPORTANCE

Metal	Condition	Purity %	Heat treatment	Proof stress MPa	UTS MPa	Elong. %	Hardness HV	Impact strength J	Refs. [1]
Ag	Soft	99.9	Annealed 600°C	—	172	50	25	5	1
	Hard	99.9	Cold worked, 70% reduction	—	330	4	95	—	1
Ag–Cu (Sterling silver)	Soft	Ag > 92.5	Annealed 550°C	—	255	37	—	—	—
	Hard	Cu 7.5	Cold drawn 50%	—	430	5	—	—	—
Au	Soft	99.99	Annealed 450°C	—	130	40–50	20–30	—	2
	Hard	99.9	60% reduction	205	220	4	60	—	3, 4
Be	Soft	99.5	Hot pressed	240	310	2	(150)	1[2]	5
	Hard	99.5	Forged	345	550	10	(200)	3–5[2]	5
Ca	Soft	—	Cast	14	55	55	17HB	—	6
	Hard	—	Rolled	84.5	115	7	—	—	7
Co	Soft	99.9	Annealed	345–485	760	7–20	170	—	—
	Hard	99.9	25% reduction	—	1 135	0.5	320	—	—
Cr	V. pure soft	99.99	Iodide reduced; arc-cast swaged, annealed 600°C	—	415	44	230	—	8
	Soft	99.95[3]	Electrolytic; arc-cast, hot worked recrystallised above 900°C	—	103	0	130	—	8
	Hard	99.95[3]	Drawn ~500°C, 80–90% reduction	—	689	10	220	—	8
Hf	Soft	—	Iodide; hot rolled, 0–925°C; cold rolled, 5%; annealed 800°C for 1 h	240	445	43	(150–180)	—	9
	Hard	—	Iodide; extruded	365	745	20–27	—	—	7
Ir	Soft	—	Annealed	—	550–1 100	—	200–300	—	—
	Hard	—	Hot forged and swaged 800–1 500°C	—	1 200	13	650	—	10
Mo	V. pure, soft	99.99	Electron beam melted; hot rolled; recrystallised at 1 100°C for 1 h	345Y	435	5–25	—	—	8
	Soft	99.95[3]	Arc-cast; hot worked and recrystallised ~1 200°C	415–450Y	485–550	30–40	200	—	8
	Hard	99.95[3]	Rolled at about 1 000°C	550Y	620–690	10–20	250	—	8

(continued)

Table 22.57   MECHANICAL PROPERTIES OF OTHER METALS OF INDUSTRIAL IMPORTANCE—*continued*

Metal	Condition	Purity %	Heat treatment	Proof stress MPa	UTS MPa	Elong. %	Hardness HV	Impact strength J	Refs.[1]
Nb	V. pure, soft	99.97	Electron beam melted, swaged recrystallised ~1 100°C	170	240	50	65	122–230	8, 11, 12
	Soft	99.9[3]	Arc-melted; cold forged; recrystallised for 4 h at 1 200°C	240	330	50	115	16–115[1]	8
	Hard	99.9[3]	Cold rolled ~90%	550	585	5	160	—	8, 11
Nb–Hf Ti (WC-103 alloy)	Annealed	Nb 89.0 Hf 10.0 Ti 1.0	1 H at 1 315°C 20°C 1 500°C 2 500°C	262 124 69	372 276 83	20 16 50	— — —	— — —	— — —
Nb–Hf–W–Y (WC-129Y)	Annealed	Nb 79.9 Hf 10.0 W 10.0 Y 0.1	1 H at 1 315°C	414	552	20	—	—	—
Nb–Zr	Annealed	Nb 99.0 Zr 1.0	1 H at 1 315°C	103	241	20	—	207	19
Nb–W–Zr (Nb-752)	Annealed	Nb 87.5 W 10.0 Zr 2.5	1 H at 2 315°C	379	517	20	—	—	—
Np	Soft	99.1	Cast	—	—	—	355	—	13
Os	Soft	—	Wrought and annealed	—	—	—	350	—	6
	Hard	—	Cold worked 7%	—	—	—	1 000	—	4
Pd	Soft	99.9	Wrought and annealed	34.5	140–195	24–40	40	—	4
	Hard	99.9	Wrought and cold worked	205	325	15	100	—	4
Pt	Soft	99.9	Wrought and annealed	14–35	125–145	30–40	40	—	4
	Hard	99.9	Wrought and cold worked ~50%	185	195–205	3	100	—	4
Pu	Soft	99.9	Cast (αPu)	205–310	310–550	1	260–270	3	14
Re	Soft	99.9	Sintered, swaged and annealed	315	1 125	24	280	—	8
	Hard	99.9	Cold rolled sheet, 30% reduction	2 150	2 225	2	700	—	8
Rh	Soft	—	Annealed, 900°C	69–275	690–760	20–40	120	—	15
	Hard	—	Cold rolled	—	1 380–2 070	—	300	—	4, 7, 15
Ru	Soft	99.9	Wrought and annealed at 1 500°C	372	495	3	350	—	15
	Hard	99.9	Wrought and cold worked 10%	—	—	—	750	—	4
Sc	Soft	—	—	—	—	—	78HB	—	16
	Hard	—	Cold swaged 22%	—	—	—	136HB	—	16

(*continued*)

Table 22.57 MECHANICAL PROPERTIES OF OTHER METALS OF INDUSTRIAL IMPORTANCE—*continued*

Metal	Condition	Purity %	Heat treatment	Proof stress MPa	UTS MPa	Elong. %	Hardness HV	Impact strength J	Refs. [1]
Ta	V. pure, soft	99.98	Electron beam melted; cold rolled annealed 1 200°C	180	205	35–45	70–80	—	8
	Soft	99.95[3]	Recrystallised sheet	310–380	310–485	25–40	90	—	8, 11
	Hard	99.95[3]	Cold worked 95%	705	760	3	200	—	11
Th	Soft	~ 99.95	Iodide; wrought and annealed	48	115	36	38	41	7
	Hard	~ 99.95	Iodide, cold rolled	295	305	6	70	—	17
U	Soft	99.95	Coarse grained, cast	190	385	4	187	19	18
	Hard	99.95	Fine grained— rolled at 550°C —quenched from 722°C	250	580	9	250	15	18
U–Mo	As hot-rolled	Mo 2.0	Rolled 82% at 800°C	703	972	8	—	—	—
	As quenched	Mo 5.0	Quenched in water from 900°C	469	656	29	165	9.6	20
U–Mo– Re	As quenched	Mo 4.0 Re 1.0	Quenched in water from 900°C	239	409	3	169	3.7	—
U–Ti	Extruded	Ti 0.75	Extruded at 800°C	537	1 111	12	—	—	—
	Precipi-tation hardened		Quenched in water from 900°C and aged at 350°C	876	1 420	20	—	320	—
V	V. pure soft	99.95	Iodide reduced: arc-melted, swaged and recrystallised	103	190	39	55	—	8
	Soft	99.8[3]	Calcium reduced: arc-melted, cold worked recrystallised	170–450	260–585	28–34	80	10–136	8
	Hard	99.8[3]	Cold worked	515–690	530–730	1–3	150	—	Several
W	Soft	99.95	Sintered rod; swaged and annealed, 1 590°C	550	550–620	0	360	—	8, 11
	Hard	99.95	Sintered, cold worked	—	1 920	0	500	—	11
Y	Soft	—	—	57Y	130	(25)	30–60HB	24	16
	Hard	—	50% cold swaged	375Y	455	2	100–140HB	—	16

[1] Where data exist from more than one source, ranges of values are given or, alternatively, if the values are similar a probable average is given.

[2] Unnotched Charpy.

[3] Principal impurities are interstitial O, C, N.

[4] Properties very sensitive to impurity content.

Y = 'Yield stress'.

( ) Interpolated from another source.

HB = Brinell.

HV = Vickers PN.

REFERENCES TO TABLE 22.57

1. A. Butts and C. D. Coxe (Eds.), 'Silver', Van Nostrand, Princeton, 1967.
2. E. M. Wise, 'Gold', Van Nostrand, Princeton, 1964.
3. Engelhard Industries Inc., Newark, N.J.
4. Materials Selector, *Mater. Engng*, 1972, **76**, No. 4.
5. T. W. Farthing and J. R. Leech, *Proc. Inst. mech. Engrs*, 1956–66, **180**, Part 3D.
6. A. H. Everts and G. D. Bagley, *J. Electrochem. Soc.*, 1948, **93**, 265.
7. C. A. Hampel (Ed.), 'Rare Metals Handbook', 2nd Edn, Reinhold, New York, 1961.
8. T. E. Tietz and J. W. Wilson, 'Behaviour and Properties of Refractory Metals', Arnold, London, 1965.
9. USAEC, 'Reactor Handbook Materials—General Properties', McGraw-Hill, New York, 1955.
10. W. Betteridge, 5th Plansee Seminar, Reutte, 1964, paper 27.
11. R. Syne, 'Handbook of Properties of Nb, Mo, Ta and W', NATO, 1965.
12. 'Gmelins Handbuch der Anorganischen Chemie—Niob. Teil A', Verlag Chemie, Weinheim, 1969.
13. V. W. Eldred and G. C. Curtis, *Nature, Lond.*, 1957, **179**, 910.
14. E. Grison, W. B. H. Lord and R. D. Fowler (Eds.), 'Plutonium 1960', Cleaver-Hume, London, 1961.
15. International Nickel Ltd. Research Laboratories.
16. E. V. Kleber and B. Love, 'Technology of Scandium, Yttrium and Rare Earth Metals', Pergamon Press, Oxford, 1963.
17. A. B. Schwope, G. T. Muelenkamp and L. L. Marsh, BMI, 1952, 784.
18. J. H. Gittus, 'Uranium', Butterworths, London, 1963.
19. Teledyne Wah Chang Albany Data Brochure, TWCA-8204CB, 1985.
20. G. B. Brook. Unpublished information.

22. 12   Bearing alloys

Bearing alloys have to satisfy a number of criteria, depending on their application, of which the following are the more important.

WEAR RESISTANCE

The bearing must resist wear and more importantly must not cause wear of the mating surface. The stronger, harder bearing materials must therefore be operated against a hardened mating surface.

EMBEDDABILITY

When dirt or detritus from associated parts is present in the lubricant supply the bearing must minimise damage to the mating surface. Soft bearing materials can absorb more dirt than hard materials.

CONFORMABILITY

The misalignments inherent in mechanisms under load require an ability in the bearing to conform to the geometry of the mating surface. Softness and conformability in a bearing material also facilitate the creation of a hydrodynamic film of lubricant over the bearing surface under sparse lubrication conditions.

STRENGTH

Under static loads the bearing must have a yield strength greater than the applied load. Under dynamic fatigue or shock loads, a combination of high yield strength and ductility is required. The relative fatigue strengths of various bearing alloys can be seen from Figure 22.1 and mechanical properties from Table 22.58

CORROSION RESISTANCE

The bearing material must resist corrosion by the atmosphere or by the lubricant.

THERMAL EXPANSION

The expansion coefficient of the bearing material must be sufficiently close to that of the housing for there to be no significant loss of interference fit at bearing operating temperatures.

It will be apparent that a bearing material is a compromise between the conflicting requirements for softness and strength, and the optimum compromise depends on the application. High effective strength from a relatively weak bearing material may be achieved if the material is present as a thin lining on a steel backing, and the majority of bearings have this construction. The copper-based engine bearing linings and the stronger aluminium alloys are overlay plated with a thin layer of lead–indium. This soft, partly sacrifical layer provides surface embeddability and conformability, and is thin enough (0.01–0.03 mm) to have good load carrying capacity.

Table 22.58 MECHANICAL PROPERTIES OF BEARING ALLOYS

							Mechanical properties			
	Nominal composition and manufacturing process						*Proof stress*	*UTS*	*Elongation*	*Hardness*
Alloy	Al	Cu	Pb	Sb	Sb		MPa	MPa	%	HV5
Copper base	—	Rem	10	—	10	Sintered steel backed	249	303	5	120
alloys	—	Rem	22	—	4.5	Sintered steel backed	81	121	5	46
	—	Rem	25	—	1.5	Sintered steel backed	75	112	5	45
	—	Rem	30	—	—	Sintered steel backed	62	93	5	34
	—	Rem	—	—	10	P 0.5% max. Continuously cast	233	420	16	120
	—	Rem	9.5	—	5	Continuously cast	155	280	15	70
	—	Rem	10	—	10	Sand cast	101	233	10	75
	—	Rem	20	—	5	Continuously cast	124	233	10	70
	—	Rem	26	—	2	Rotary lined on steel	62	110	3	48
Aluminium base alloys	Rem	1	—	—	20	Continuously cast steel backed	42	120	25	37
	Rem	1	—	—	6	Ni 1.0% Continuously cast steel backed	50	140	28	45
	Rem	—	—	—	40	Continuously cast steel backed	40	67	26	27
	Rem	1	—	—	—	Si 10.6% continuously cast steel backed	87	194	17	56
	Rem	1	—	—	6	Ni 1.0% Cold worked 4% Continuously cast	140	147	22	50
	Rem	1.5	—	—	6	Ni 1.4% Mg 0.9% Si 0.5% Continuously cast	83	207	10	78
	Rem	1	0.9	—	—	Zn 5% Si 1.5%	—	216	15	64
Whitemetals tin base	—	3.3	—	7.5	Rem	Cast on steel strip	65	76	10	27
	—	3.3	—	7.5	Rem	Rotary lined on steel	39	70	15	31
	—	3	—	7.7	Rem	Cd 1.25% Rotary lined on steel	76	90	8	32
Whitemetals lead base	—	0.5	Rem	15	1	As 1% Cast on steel strip and annealed	25	70	6	17
	—	—	Rem	10	6	Cast on steel strip and annealed	30	42	33	16
	—	0.5	Rem	10	10	Rotary lined on steel	60	73	8	25

Rem = Remainder. Properties are typical values.

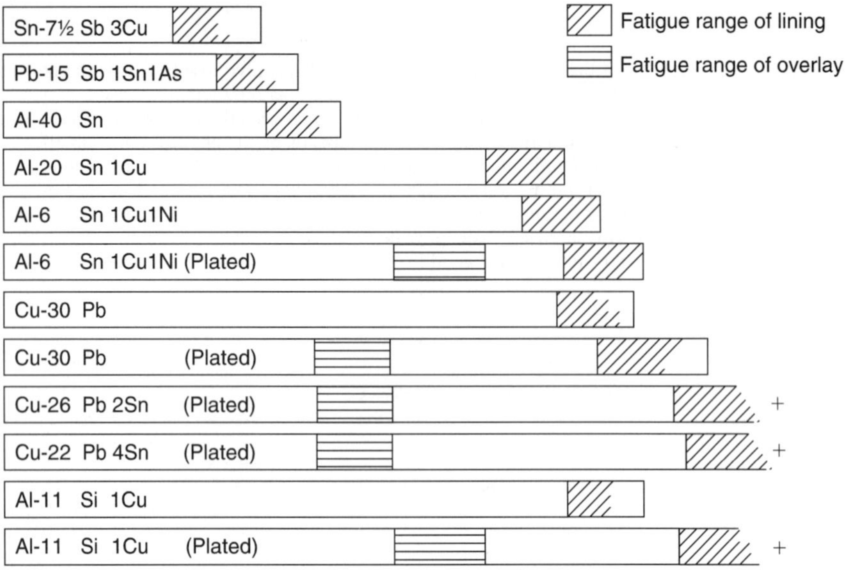

Sn-7½ Sb 3Cu	▨
Pb-15 Sb 1Sn1As	▨
Al-40 Sn	▨
Al-20 Sn 1Cu	▨
Al-6 Sn 1Cu1Ni	▨
Al-6 Sn 1Cu1Ni (Plated)	▤ ▨
Cu-30 Pb	▨
Cu-30 Pb (Plated)	▤ ▨
Cu-26 Pb 2Sn (Plated)	▤ ▨ +
Cu-22 Pb 4Sn (Plated)	▤ ▨ +
Al-11 Si 1Cu	▨
Al-11 Si 1Cu (Plated)	▤ ▨ +

▨ Fatigue range of lining
▤ Fatigue range of overlay

Figure 22.1   *Relative fatigue strength of bearing alloys. Back of bearing temperatures from* 70°C *to* 100°C

BIBLIOGRAPHY—BEARING ALLOYS

'Copper Alloy Bearing Materials', Copper Development Assoc., Tech. Note TN9, 1972.
R. Booser, 'Selecting Sleeve Bearing Materials', *Mater. Des. Engng.*, 1958, **48**, 119.
P. G. Forester, 'Bearing Materials', *Met. Rev.*, 1960, **5**, 507.
M. C. Shaw and F. Macks, 'Analysis and Lubrication of Bearings', New York, McGraw-Hill, 1949.
M. J. Neale, 'Tribology Handbook', Butterworths, London, 1973.
European Tribology Conf. London, *Inst. Mech. Engrs*, Sept. 1973.

23   Sintered materials

A wide range of useful metallic materials is made from powder by the process known as sintering. Such materials are commonly referred to as sintered materials, and individual engineering components made by the process are known as sintered parts, sintered components, or PM parts—PM or P/M being the acronym for powder metallurgy.

23.1   The PM process

This process consists of shaping the powder by means of pressure in a die or mould of prescribed shape; the compact thus formed must have sufficient green strength to allow it to be removed from the die and handled without fracture. This is an important property. After removal from the die, the compact is heated, normally in a protective atmosphere or vacuum, to a temperature usually below the melting point, such that the particles weld together and in some cases densify markedly to increase the strength. This is the step known as sintering. In some cases, especially when the sintered compact is to be subsequently rolled or extruded, isostatic compaction in a flexible mould is used. An exception to the compaction process is the production of filter elements from spherical powders, often of bronze, by what is known as loose powder sintering, in which powder is poured into a mould of the appropriate shape and sintered in the mould. Most recently shaping has embraced injection moulding and the use of polymers to lubricate the powder flow into a mould cavity. After moulding, the polymer is extracted and the powder is sintered to near full density.

Increasingly, sintered billets are being made for subsequent mechanical working such as hot extrusion or rolling. These processes yield wrought material to which the name sintered parts does not apply, but the bulk of PM products are items made individually to the final shape and size.

23.2   The products

The main classes of PM products are:

1. Engineering components, i.e. parts for machinery made by directly pressing or moulding in rigid dies of the required shape, and sintering with either no or only minor further shaping. This is the largest and most familiar class of PM product.
2. Refractory metals. These often have melting points that are inconveniently high and/or are difficult or impossible to work in the cast state because they are brittle. This group includes tungsten, molybdenum, tantalum, and related metals.
3. Intentionally porous materials for use as filters or as oil-retaining (self-lubricating) bearings.
4. Composites. These may consist of two or more metals that are insoluble in each other in the solid state, but the name is generally used to signify a metal in which is dispersed one or more non-metallic ingredients such as refractory oxide, carbide, or other compound that is insoluble in the metal. These are referred to as metal matrix composites. A very important member of this group, the hardmetals, is the subject of a later section.
5. High-duty alloys in wrought form having mechanical properties superior to those of the corresponding material made from ingot. In this category must be included certain composite materials having fine dispersions of oxide or the like designed to improve the strength, especially at elevated temperatures. In some other cases the principal reason for making wrought products from powder is that higher yields of usable material are possible. This applies to metals such as titanium alloys the scrap from which cannot readily be re-used.
6. Magnetic materials. This very important class of sintered material is dealt with in Chapter 20.

23.3   Manufacture and properties of powders

23.3.1   Powder manufacture

There is a large number of possible ways of making metal powders, the principal ones used on a commercial scale for powders for PM being:

— Solid state reduction of a compound of the metal, commonly the oxide. This often results in a loosely caked sponge which is converted into powder by milling.
— Thermal decomposition of a compound of the metal. This also may yield a sponge.
— Electrolysis. By suitable choice of the various parameters certain metals can be deposited in a spongey or powder form. Such powders are often dendritic. In the case of iron, a dense deposit is formed which is reduced to powder by comminution.
— Atomisation, in which molten metal is disintegrated into small drops which are caused or allowed to solidify out of contact with each other and with any solid surface. There are many ways in which the molten droplets may be produced, the most common being to allow a stream of molten metal to fall vertically from a tundish into a chamber where it is broken up by jets of high-pressure liquid or gas. Another process now gaining in importance is centrifugal atomisation, in which droplets are flung from a rapidly rotating pool of molten metal or from a disk on to which a thin stream of metal impinges. In another process the end of a rotating bar is progressively melted and if the heating is by laser beam or plasma arc the process can be carried out in near vacuum, thus substantially eliminating the risk of contamination. This is called REP (Rotating Electrode Process) and if a plasma arc is used PREP (Plasma REP). An important feature of atomisation is that it can produce homogeneous fully pre-alloyed powders.

Table 23.1 lists the powder production processes used for many of the metals used in PM.

— Rapid solidification. This means cooling molten metal at a very high rate such that on solidification an amorphous or other non-equilibrium structure results. In the case of certain aluminium alloys, for example, it is possible to hold in solution significantly higher percentages of alloying elements (see Table 23.2). If a rapidly solidified alloy is consolidated by a powder metallurgical process that does not involve heating to a temperature high enough to induced recrystallisation, wrought material with substantially improved mechanical properties may result. The processes

Table 23.1   PRODUCTION PROCESSES FOR METAL POWDERS

The most widely used method is put first.

Aluminium	Gas atomisation (usually air), Comminution
Beryllium	Comminution, electrolysis
Brass	Atomisation—water or gas
Bronze	Water atomisation for irregular powder, Air atomisation for spherical powder
Cobalt	Oxide reduction
Copper	Water atomisation, electrolysis (Both are in general use)
Iron: unalloyed	Oxide reduction, water atomisation, electrolysis, carbonyl decomposition
Iron: cast	Comminution
Molybdenum	Oxide reduction—by hydrogen
Nickel	Thermal decomposition of the carbonyl, reduction of a salt in solution
Nickel alloys inc. superalloys	Atomisation—inert gas, centrifugal
Platinum	Thermal decomposition of a salt
Silver	Precipitation from solution of a salt, usually by an organic reducing agent, electrolytic
Silver alloys	Atomisation
Steels	Atomisation—usually with water or nitrogen
Tin	Air atomisation
Titanium	Reduction of the chloride by magnesium, Thermal decomposition of the hydride
Tungsten	Reduction of the oxide—by hydrogen

Note: Powders of practically all metals can be produced by atomisation—the table lists the processes common in present day commercial practice.

Table 23.2   EXTENDED SOLUBILITY IN ALUMINIUM VIA RAPID SOLIDIFICATION

Additive	Maximum equilibrium solubility, at. %	Observed maximum by rapid quenching
Cu	2.5 (at 821 K)	18
Mn	0.7 (at 923 K)	9
Si	1.6 (at 850 K)	16

used mostly involve projecting molten metal to form a very thin film on a rotating water cooled cylinder to produce flake or ribbon which is crushed to powder.

— Mechanical alloying. Another way of achieving higher strength and especially retention of that strength at elevated temperatures is to include in the metal matrix a finely dispersed insoluble phase, commonly a stable oxide. Oxide Dispersion Strengthened (ODS) materials are increasingly of interest in aerospace technology, and they are of necessity made by PM. In general the performance improvement is greater the finer the particles of the dispersed phase, the mechanism being dislocation locking as in precipitation hardening. An approach near to ideal is mechanical alloying, in which a metal powder is mixed with a fine oxide powder and the mixture milled for several hours or even days usually in an attritor, during which process the oxide is beaten into the metal, the particles of which are repeatedly flattened, broken up, and reflattened while at the same time the oxide particles are broken down into progressively smaller fragments, typically less than 100 nm. Mechanically alloyed powders are generally consolidated and converted into wrought shapes.

23.3.2   Properties of metal powders and how they are measured

Chemical composition

Chemical composition is the main factor that determines the properties of the finished material. One feature that is specific to powders is the oxide layer on the surface of the particles. This has an important influence of the green strength of compacts but more especially on the sintering process which depends on the formation of true metallurgical bonds between adjacent particles. In the case of metals with easily reducible oxides the level of surface oxide is determined by heating a weighed sample in hydrogen under specified conditions, cooling in hydrogen, and weighing. The reduction in weight is commonly called the loss in hydrogen or hydrogen loss. The procedure gives a misleading figure if volatile material is present, such as organic additions (pressing lubricants), moisture, or volatile metals such as zinc. The process is not applicable to metals with very stable oxides that are not reducible in this way, nor to alloys containing a high percentage of metals such as aluminium, titanium, chromium, and stainless steels. Table 23.3 gives the specified reduction conditions for some common metals.

Table 23.3   CONDITIONS FOR DETERMINATION OF LOSS IN HYDROGEN

Metal powder	Reduction temperature °C	Reduction time min
Tin bronze	775 ± 15	30
Cobalt	1 050 ± 20	60
Copper	875 ± 15	30
Copper lead[1] and leaded bronze[1]	600 ± 10	10
Iron	1 150 ± 20	60
Alloyed steel	1 150 ± 20	60
Lead[1]	550 ± 10	30
Molybdenum	1 110 ± 20	60
Nickel	1 050 ± 20	60
Tin	550 ± 10	30
Tungsten	1 150 ± 20	60

[1] Results should be interpreted with caution.

Particle size and size distribution

The meaning of size is complicated by the enormous variety of shape of particles which makes quantitative assessment difficult. Volume, maximum linear dimension, average dimension, cross sectional area are different measurements which can be made. A considerable range of techniques is or has been used, including among others microscopical examination, elutriation, sedimentation, light obscuring, and light scattering, but for most PM powders the process normally used is laser light scattering. Historical size analysis was based on sieving. The screen size is designated by the number of strends of wire per unit length and has been superseded by specifying the aperture size in terms of the largest sphere that will pass through it. However the older designation is still used and Table 23.4 shows the relationship between the various standards. The less equiaxed the particles the less meaningful will be the result of a sieve test. For example in the extreme case of an acicular powder, the length rather than the diameter is likely to be the controlling dimension. Nevertheless, meaningful and useful results are obtained, especially when different batches of nominally similar powder are being compared. For very small subsieve size powders other methods have to be employed. This is the primary domain of laser light scattering. These data are supplemented by the BET method which is based on gas absorption. This measures in effect the total surface area.

Apparent density (AD)   This most important property is in a general sense the same as bulk density, i.e. it is the mass of a unit volume of powder, but the method and apparatus used to determine it need to be carefully specified if consistent results are to be obtained. The test normally specified—ISO 3923/1—is to allow the powder to emerge from a funnel through a specified orifice into a cylindrical cup of specified dimensions which overflows. The surplus is carefully removed by drawing a straight edge across the top of the cup, the powder in the cup then being weighed. The apparatus used is specified in ISO 3923/1 and illustrated in Figure 23.1. (The same apparatus is also used to determine the flow properties of powders.) AD is of importance in several ways but especially in that it determines the amount, i.e. mass, of powder in the die before compaction.

Tap density (TD)   This density is based on vibration of the powder to its highest packing density. The TD is an excellent measure of the maximum suspension density that can be expected in various polymer-assisted shaping technologies.

Flow factor   In the commercial production of compacts, powder is fed automatically into the die cavity and must, therefore, flow readily. The flowability is measured by timing the flow of 50 g of powder from a funnel of specified geometry and the result expressed in seconds. The procedure is laid down in ISO 4490, the apparatus being generally referred to as the Hall flowmeter after its designer (Figure 23.1). The properties of some widely used PM grade powders are given in Table 23.5.

23.4   Properties of powder compacts

Apart from dimensions which are, of course, of major importance in the manufacture of individual PM parts, the relevant properties of compacts are density and strength. Green density as it is called has a marked influence on green strength and can impact the final, i.e. sintered, density. It is determined by Archimedes principle. Green strength is the term used to indicate the strength of a compact and two standard methods are in general use for measuring it.

1. A three point bend test on a standard compact of rectangular cross section gives the Transverse Rupture Strength—ISO 3995.
2. A compression test on the diameter of a ring shaped compact gives the Radial Crushing Strength (K factor)—ISO 2739. This test applies especially to cylindrical bearings. The K factor is given by the formula: $(K = F(D - e)/Le^2$ N/mm^2, where $F =$ breaking load in Newtons, $L =$ the length in mm, $D =$ the external diameter in mm, $E =$ the wall thickness in mm.

Compressibility (compactibility)   For economic as well as technical reasons it is desirable to use as low a pressure as possible to achieve the required compact density, therefore the compressibility of the powder is of considerable importance. This factor is usually expressed either as the density obtained with a given compacting pressure (see Table 23.5) or is shown as a curve relating density to pressure. Compressibility as defined in this way is influenced by the AD of the powder and by the actual strength of the particles. Thus an alloy powder will have a lower compressibility than that of a pure metal. For this reason it is often preferred to use mixtures of elemental powders (for low alloys) rather than pre-alloyed powders of the composition ultimately required. However, today pre-alloyed powder is needed to satisfy most of the high performance applications.

Table 23.4   COMPARISON TABLES OF STANDARD SIEVES

mm	International	UK	USA	France	France	Germany	mm
3.35	4	5	6				3.35
				36	3.15	3.15	
2.80	2.8	6	7				2.80
				35	2.50	2.50	
2.40		7	8				2.40
2.00	2.0	8	10	34	2.00	2.00	2.00
1.68		10	12				1.68
				33	1.60	1.60	
1.40	1.4	12	14				1.40
				32	1.25	1.25	
1.20		14	16				1.20
1.00	1.0	16	18	31	1.00	1.00	1.00
μm 850		18	20				μm 850
				30	800	800	
710	710	22	25				710
				29	630	630	
600		25	30				600
500	500	30	35	28	500	500	500
420		36	35	27	400	400	420
355	355	44	45				355
				26	315	315	
300		52	50				300
250	250	60	60	25	250	250	250
212		72	70	24	200	200	212
180	180	85	80				180
				23	160	160	
150		100	100				150
125	125	120	120	22	125	125	125
106		150	140	21	100	100	106
90	90	170	170		90	90	90
				20	80	80	
75		200	200		71	71	75
63	63	240	230	19	63	63	63
					56	56	
53		300	270	18	50	50	53
45	45	350	325		45	45	45
				17	40	40	
37			400				37

Column standards:
- International — ISO 565-1972(E) Nominal size of aperture R20/3
- UK — BSS410 1969 mesh no.
- USA — American National Standard Z23, 1-1973 ASTM E-11-70 mesh no.
- France — AFNOR NF XII 1938 Common ref. no.
- France — AFNOR NF XII-504 1970 Nominal size of aperture ISO R20
- Germany — DIN 4188 Table 1 1969 Nominal size of aperture ISO R20

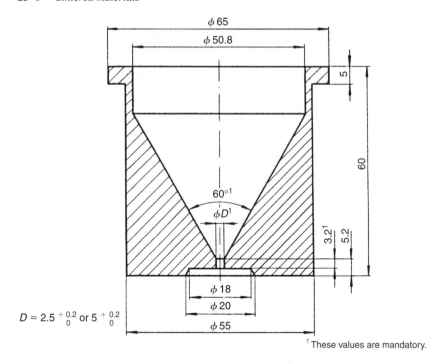

$D = 2.5 \, ^{+\,0.2}_{0} \text{ or } 5 \, ^{+\,0.2}_{0}$

[1] These values are mandatory.

Dimensions in millimetres

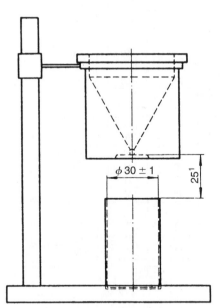

Figure 23.1   *The Hall flowmeter*

Table 23.5　PROPERTIES OF PM GRADE SPONGE IRON POWDER

	NC100.24 *Typical data*			*Testing method*

Apparent density/g/cm³

2.45	ISO 3923-1977

Flow/s/50 g

29	ISO 4490-1978

Sieve analysis/%

+208 µm	0	
+147 µm	5	ISO 4497-1983
−45 µm	19	

Compressibility/g/cm³

Comp. pressure	1	2	
4.2 t/cm²	6.55	6.63	
300 MPa	6.09	6.25	ISO 3927-1977
500 MPa	6.79	6.83	
700 MPa	7.12	7.0	

1. Lubricated die
2. 0.8 % Zn-st

Green strength/N/mm²

Comp. pressure	1	2	3
300 MPa		14	10
500 MPa	45	24	18
700 MPa		30	23

1. Lubricated die
2. 0.6% Aera wax
3. 0.8% Zn-st

Chemical analysis/%

Carbon	0.01	ISO 4491-1978
H₂-loss	0.20	ISO 7625-1983

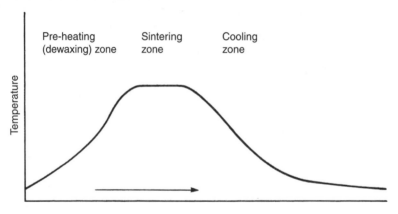

Figure 23.2

23.5   Sintering

Most PM components (other than hardmetals) are sintered in continuous furnaces normally having three major zones:

1. A pre-heating or de-waxing zone in which the lubricant is driven off.
2. The sintering zone proper, the temperature of which is called the sintering temperature.
3. A cooling zone normally surrounded by a water jacket. Figure 23.2 shows a typical temperature profile for the work load.

Compacts of pure metals and of fully pre-alloyed alloys shrink during sintering, the amount of shrinkage being dependent on the green density and the sintering regime, i.e. temperature and time. With mixtures of elemental powders or of alloys of different compositions, the situation is different. Diffusion effects can cause an expansion to occur initially and this feature can be used to provide mixes that neither shrink nor grow on sintering. The green density homogeneity is the major factor that determines the consistency of dimensions of the sintered component. It is in most cases the ability to produce components directly, i.e. without any subsequent machining, that provides the economic justification for making them from powder.

In the case of large compacts that are intended to be subsequently wrought into sheet, rod, or wire, these factors are of little or no importance.

23.6   Ferrous components

Iron-based materials account for the bulk of PM parts and they range in composition from nearly pure iron through low alloy steels to stainless steels and high-speed steels for cutting tools, etc. It should be noted, however, that as regards low-alloy steels the compositions used differ from those of ingot-based metal. The reason for this is that alloying elements that give rise to oxide films that are stable in normal sintering atmospheres and which therefore inhibit sintering are used in low concentrations. Such elements include manganese, chromium, and silicon. Favoured compositions for sintered steels contain, in addition to carbon, one or more of the elements copper, nickel and molybdenum, the oxides of which are readily reducible by hydrogen. In addition there are compositions containing up to 0.6% of phosphorus.

Sintered steels differ from wrought in strength in that the hardness, strength, and ductility depend not only on composition and state of heat-treatment, but also markedly on density (see Figure 23.3). Considerable effort has been put into ways of increasing the density of sintered parts in order to extend their use to more heavily stressed applications. Ways of doing this are:

1. Re-press and re-sinter.
2. Re-press at elevated temperature.
3. Hot forge.
4. Sinter density.

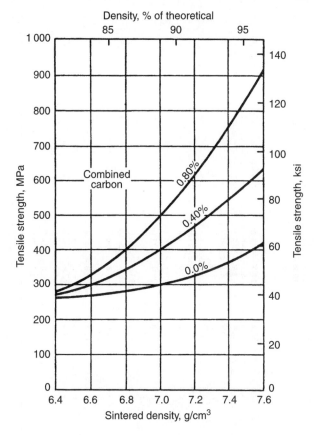

Figure 23.3 *Effect of density on strength of as-sintered 4% nickel steel*

In the third process, called sinter forging or powder forging, a sintered blank of appropriate geometry is hot forged in a closed die to the required shape and dimensions. In this way it is possible to achieve properties as good as those of wrought steels and in some respects better because of the complete homogeneity and isotropy of the material (see Table 23.6). Accuracy too may be superior because it is possible to maintain closer tolerances on the mass of the blank than is normal with blanks cut from bar. Flash should be completely absent. Another way of increasing the density is to fill the voids with copper or a dilute copper alloy which is melted on the surface and absorbed into the pores. This process, known as liquid infiltration, is liable to affect adversely the dimensional accuracy.

23.7   Copper-based components

There is a relatively limited market for structural components made of bronze, brass, or copper itself, copper usually being used for electrical applications. There is, however, a considerable market for oil-retaining bronze bearings, (porous) filter elements, art objects, and decorative hardware sintered from bronze. As regards bearings there is a conflict between the oil capacity which increases as the density decreases, and strength which increases as the density increases. The oil content normally lies between 15 and 30% by volume. Table 23.7 lists specified compositions and properties of several standard types of bearing including those made of iron or its alloys. Figure 23.4 shows the relationship between strength and sintering temperature for bronze bearings, and Figure 23.5 shows how the diameter varies with temperature.

Table 23.6 MECHANICAL PROPERTIES OF SINTER-FORGED STEEL FOR P/F-4600 MATERIALS (Mn 0.10–0.25, Ni 1.75–1.90, Mo 0.50–0.60)

Sintered at 1 120°C in dissociated ammonia unless otherwise noted.

Forging mode	Carbon, %	Oxygen, ppm	Ultimate tensile strength, MPa	0.2% offset yield strength, MPa	Elongation % in 25 mm	Reduction of area, %	Room temperature Charpy V-notch impact energy J	ft lb	Core hardness, HV30	Fatigue endurance limit, MPa	Ratio of fatigue endurance to tensile strength
Blank carburised											
Upset	0.24	230	1 565	1 425	13.6	43.2	16.3	12.0	487	565	0.36
Re-press	0.24	210	1 495	1 325	11.0	34.3	12.9	9.5	479	550	0.37
Upset [a]	0.22	90	1 455	1 275	14.8	46.4	22.2	16.4	473	550	0.38
Repress [a]	0.25	100	1 455	1 280	12.5	42.3	16.8	12.4	468	510	0.36
Upset [b]	0.28	600	1 585	1 380	7.8	23.9	10.8	8.0	513	590	0.37
Re-press [b]	0.24	620	1 580	1 305	6.8	16.9	6.8	5.0	464	455	0.29
Quenched and stress relieved											
Upset	0.38	270	1 985	1 505	11.5	33.5	11.5	8.5	554	—	—
Re-press	0.39	335	1 960	1 480	8.5	21.0	8.7	6.4	—	—	—
Upset	0.57	275	2 275	—	3.3	5.8	7.5	5.5	655	—	—
Re-press	0.55	305	1 945	—	0.9	2.9	8.1	6.0	—	—	—
Upset	0.79	290	940	—	0.0	0.0	1.4	1.0	712	—	—
Re-press	0.74	280	1 055	—	0.0	0.0	2.4	1.8	—	—	—
Upset	1.01	330	800	—	0.0	0.0	1.3	1.0	672	—	—
Repress	0.96	375	760	—	0.0	0.0	1.6	1.2	—	—	—
Quenched and tempered											
Upset [c]	0.38	230	1 490	1 340	10.0	40.0	28.4	21.0	473	—	—
Re-press [c]	—	—	1 525	1 340	8.5	32.3	—	—	—	—	—
Upset [d]	0.60	220	1 455	1 170	9.5	32.0	13.6	10.0	472	—	—
Re-press [d]	—	—	1 550	1 365	7.0	23.0	—	—	—	—	—
Upset [e]	0.82	235	1 545	1 380	8.0	16.0	8.8	6.5	496	—	—
Re-press [e]	—	—	1 560	1 340	6.0	12.0	—	—	—	—	—
Upset [f]	1.04	315	1 560	1 280	6.0	11.8	9.8	7.2	476	—	—
Repress [f]	—	—	1 480	1 225	6.0	11.8	—	—	—	—	—
Upset [g]	0.39	260	825	745	21.0	57.0	62.4	46.0	269	—	—
Upset [g]	0.58	280	860	760	20.0	50.0	44.0	32.5	270	—	—
Upset [h]	0.80	360	850	600	19.5	46.0	24.4	18.0	253	—	—
Upset [i]	1.01	320	855	635	17.0	38.0	13.3	9.8	268	—	—

[a] Sintered at 1 260°C in dissociated ammonia. [b] Sintered at 1 120°C in endothermic gas atmosphere. [c] Tempered at 370°C. [d] Tempered at 440°C. [e] Tempered at 455°C. [f] Tempered at 480°C. [g] Tempered at 680°C. [h] Tempered at 695°C. [i] Tempered at 715°C.

Table 23.7 OIL-REPLACING BEARINGS

Material grade	Type	Chemical composition					Bearing specifications				Related specifications				
		Cu %	Sn %	Graph. %	Fe %	Others % max.	Density g/cm^3	Typical oil content % v/v	K factor MPa min.	Maximum static load MN/m^2	ISO 5755/1	BS 2590	Sint	SAE	AFNOR
60H	Bronze	Bal.	10	1.5	—	2	5.8–6.2	27	120	52	P4021Z	A110	—	840	FU–E10–58
63H	Bronze	Bal.	10	1.5	—	2	6.1–6.5	24	155	69	P4022Z	A111	A51	841	FU–E10–62
68H	Bronze	Bal.	10	1.5	—	2	6.6–7.0	18	200	100	P4023Z	A112	B51	842	—
70Q	Bronze	Bal.	10	—	2	6.8–7.2	18	210	105	105	P4013Z	A113	850	—	—
581	Iron	—	—	—	Bal.	2	5.6–6.0	23	170	70	P1012Z	A200	A00	B50	FC10–56
58T	Iron/copper	2	—	—	Bal.	2	5.6–6.0	23	200	100	P2012Z	A300	A10	—	F10–U3–56
58J	Iron/bronze	45	5	—	Bal.	2	5.6–6.0	27	100	52	—	—	—	—	—

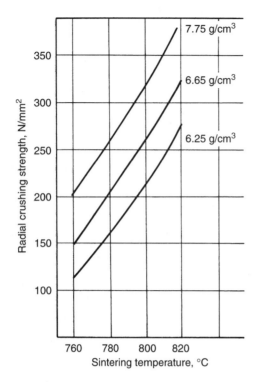

Figure 23.4   *Effect of sintering temperature on the strength of sintered bronze bearings*

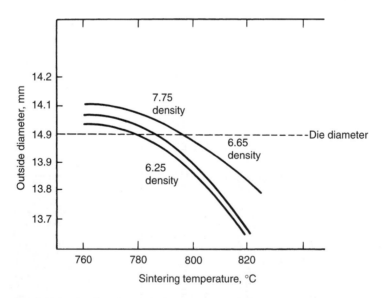

Figure 23.5   *The effect of sintering temperature on the diameter of bronze bearings*

23.8 Aluminium components

The sintering of pure aluminium is difficult by reason of the tenacious oxide film present on the surfaces of the particles which cannot be reduced by any feasible sintering atmosphere. However, if certain alloying elements are present the alloy may have a significant gap between the solidus and liquidus temperatures so that sintering may be carried out near the solidus and a liquid phase might be present; this breaks down the surface oxide films and allows interparticle welding to take place. This is a typical example of what is called liquid phase sintering. As with copper based materials, the quantity of sintered aluminium components is relatively small, but they find application where lightness, corrosion resistance, or electrical or thermal conductivity are controlling requirements. The properties of some sintered aluminium alloys are given in Table 23.8.

23.9 Determination of the mechanical properties of sintered components

The properties of interest to users are the same as those of cast or wrought metal, but the measurement of hardness especially requires some elaboration because of the porous nature of some sintered metals.

1. Hardness. A traditional indentor will make a larger impression and therefore record a lower hardness than that of dense metal of the same composition because the pore offer no resistance to deformation. A distinction is, therefore, made between apparent hardness as measured in a standard test, and true hardness, i.e. that of the same metal in the pore-free state. True hardness may be measured using a micro indentor and a carefully specified technique which minimises the risk of the result being falsified by the porosity. Apparent hardness is related to strength in the same way as with dense materials; true hardness is relevant to wear resistance.
2. Strength. Strength is determined by tensile test, but because sintered parts are normally used without machining, it is rarely desirable to cut test pieces from the parts themselves. It is standard practice, therefore, to produce separate test pieces compacted to the same density as a batch of parts and to sinter them along with the batch. Figure 23.6 shows the standard test piece.
3. Impact. The presence of porosity has a very deleterious effect on impact properties and it is not common practice to measure this property when porosity is over about 5%. For components sintered to full density, standard notched Charpy tests are employed, but for porous samples the convention is to use an unnotched Charpy bar to obtain higher apparent toughness.
4. Fatigue. In recent years there has been considerable concern over the fatigue properties of sintered ferrous alloys, since many of the applications involve components used in cyclic loading. Table 23.9 contains example fatigue endurance strengths as measured using rotating beam fatigue tests for several sintered compositions. These correspond to 50% survival at 10 million cycles. Residual pores act as cracks and greatly accelerate fatigue failure. Accordingly,

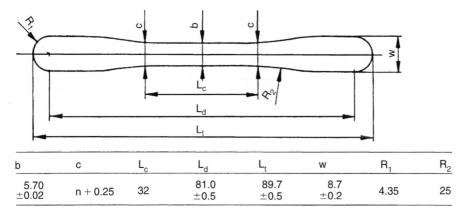

b	c	L_c	L_d	L_t	w	R_1	R_2
5.70 ±0.02	n + 0.25	32	81.0 ±0.5	89.7 ±0.5	8.7 ±0.2	4.35	25

Figure 23.6   *Standard tensile test piece (ISO 2740). Dimensions in mm*

Table 23.8   MECHANICAL PROPERTIES OF SINTERED ALUMINIUM ALLOYS

Family	Material grade	Composition, wt.% (Al = the remainder)						Density g/cm³	Relative density class	Hardness	Tensile strength MPa	Yield strength MPa	Elongation %	Impact strength[2] J	Fatigue limit MPa	Modulus GPa	Poisson's ratio	Symbol
		Mg	Si	Cu	Mn	Zn	Cr											
2000	*[1]	1	0.3	2	—	—	—	2.53	6	83 HRH	140	100	4	—	—	—	—	PA 506Y
	*T6[4]	1	0.3	2	—	—	—	2.53	6	73 HRE	230	190	2	—	—	—	—	PA 506A
	**[3]	—	—	4	—	—	—	2.64	7	—	160	75	10	—	—	57	0.25	PA 207Y
	**T6	—	—	4	—	—	—	2.64	7	—	225	145	7	—	—	57	0.25	PA 207A
	2014	0.5	0.8–1	4.4	<0.4	—	—	2.50	5	70 HRE	150	115	3	5	—	49	0.23	PA 615Y
	2014	0.5	0.8–1	4.4	<0.4	—	—	2.64	7	75 HRE	180	150	3	7	45	59	0.25	PA 617Y
	2014 T6	0.5	0.8–1	4.4	<0.4	—	—	2.50	5	85 HRE	250	235	1	3	—	49	0.23	PA 615A
	2014 T6	0.5	0.8–1	4.4	<0.4	—	—	2.64	7	90 HRE	300	280	2	6	52	59	0.25	PA 617A
6000	6061	1	0.6	0.25	—	—	0.1	2.42	6	62 HRE	100	65	4	7	—	47	0.23	PA 406Y
		1	0.6	0.25	—	—	0.1	2.55	8	65 HRH	125	80	6	—	35	56	0.25	PA 408Y
	6061 T6	1	0.6	0.25	—	—	0.1	2.42	6	72 HRH	140	130	0.5	2.5	45	47	0.23	PA 406A
		1	0.6	0.25	—	—	0.1	2.55	8	80 HRE	210	195	2	—	—	56	0.25	PA 408A
	***[3]	0.6	0.4	—	—	—	—	2.42	6	64 HRH	120	55	9	—	—	45	0.23	PA 306Y
		0.6	0.4	—	—	—	—	2.55	8	—	130	60	9	—	—	54	0.25	PA 308Y
	***T6	0.6	0.4	—	—	—	—	2.42	6	—	175	165	2	—	—	45	0.23	PA 306A
		0.6	0.4	—	—	—	—	2.55	8	—	185	170	3	—	—	54	0.25	PA 308A
7000	7075	2.5	—	1.6	—	5.6	0.2	2.51	6	90 HRH	205	150	3	—	—	49	0.23	PA 806Y
	7075 T6	2.5	—	1.6	—	5.6	0.2	2.51	6	80 HRE	310	275	2	—	—	49	0.23	PA 806A

Mechanical properties[1]

[1] Mechanical properties obtainable after sizing.
[2] IZOD impact test.
[3] Non-standard compositions.
[4] T6 = solution hardening and artificial ageing.

Table 23.9   FATIGUE STRENGTHS FOR SINTERED FERROUS ALLOYS

(50% Survival at 10^7 Fully Reversed Stress Cycles)

Composition, (*balance iron*) wt. %	Sintering conditions (*HT = heat treated*)	Density g/cm^3	Fatigue strength MPa
0.6C	1 120°C, 0.5 h	6.9	210
2Cu–0.5C	1 120°C, 0.5 h	7.1	125
2Cu–0.8C	1 120°C, 0.5 h	6.7	165
2Cu–0.8C	1 120°C, 0.5 h	7.0	235
2Cu–0.8C	1 330°C, 1 h	7.1	270
2Cu–2Ni–0.8C	1 120°C, 0.5 h, HT	7.0	240
3.5Mo	1 120°C, 0.5 h, HT	7.6	170
2Ni–0.8C	1 120°C, 0.5 h, HT	7.1	160
2Ni–0.8C	1 120°C, 0.5 h, HT	7.4	450
4Ni–1.5Cu–0.5Mo–0.6C	1 250°C, 0.5 h,	7.1	150
4Ni–1.5Cu–0.5Mo–0.6C	1 250°C, 2 h, HT	7.1	390
4Ni–1.5Cu–0.5Mo–0.8C	1 250°C, 2 h, HT	7.1	420
7Ni–2Cu–1Mo–0.6C	1 250°C, 0.5 h	7.0	390
12Cr (410L stainless)	1 230°C, 0.5 h	6.9	130
19Cr–11Ni (304L stainless)	1 290°C, 0.5 h	6.9	160
17Cr–13Ni–2Mo–0.3N (316LN stainless)	1 280°C, 0.5 h	6.6	120

attention on improved fatigue life demands densification improvements via techniques such as hot forging, high temperature sintering, and hot isostatic pressing.

The compositions and mechanical properties of the standard PM materials are shown in Tables 23.10–23.15.

23.10   Heat treatment and hardenability of sintered steels

Sintered steels can be hardened and tempered in the same way as wrought steels (see Figure 23.7), but for materials over approximately 5% porosity the use of salt baths is not recommended because of the difficulty of removing the salt from the pores. The porosity has a marked effect on hardenability as shown in Figure 23.8. This is a consequence of the lowering of the thermal conductivity.

23.11   Case hardening of sintered steels

Here also the porosity has a significant effect. In carburising, nitriding, and carbo-nitriding, the gas enters the pores so that for a given regime the depth of hardening is greater and the transition between case and core is much more gradual. As would be expected, these effects are more marked the greater the amount of porosity (*see* Figure 23.8). If the parts are first infiltrated (*see* Section 23.6) this effect largely disappears and a more nearly normal case is produced.

23.12   Steam treatment

A process peculiar to sintered ferrous components is to heat them in superheated steam which produces a layer of magnetite, Fe_3O_4, on the surface including those of the surface-connected pores. This significantly increases the surface hardness and wear resistance, and to some extent the density, as can be seen in Table 23.16.

23.13   Wrought PM materials

Substantial tonnages of wrought metals are made by PM. The products can be divided into three classes.

1. Metals that are difficult or impossible to produce in wrought form from ingot because of their inherent brittleness. Those of most commercial importance are tungsten, molybdenum, and the

(*text continued on p. **23**–23*)

Table 23.10   IRON, CARBON STEELS, AND COPPER STEELS

Materials	Grade	C combined %	Cu %	Fe %	Total other elements (max.) %	Density (min.) g/cm³	Ultimate tensile strength (min.) MPa	Apparent hardness (min.) HV 5	Relative density %	Yield strength (0.2% offset) MPa	Elongation[2] %	Apparent surface hardness after appropriate treatment[3] HV 5	Apparent hardness Rockwell
Iron	P1022-	<0.3	—	Balance	2	5.6	70	30	75	40	1		30 HRH
	P1023-					6.0	100	40	80	60	2		70 HRH
	P1024-					6.4	140	50	85	80	3		80 HRH
	P1025-					6.8	180	65	94	120	6	400	15 HRB
	P1026-					7.2	220	80	94	120	6	500	30 HRB
Carbon steel	P1033-	0.3 to 0.6	—	Balance	2	6.0	140	55	80	90	nm		20 HRB
	P1034-					6.4	190	75	85	120	1		45 HRB
	P103-					6.8	240	90	90	130	2	400	60 HRB
	P1042-	0.6 to 0.9	—	Balance	2	5.6	150	55	75	120	nm		35 HRB
	P1043-					6.0	200	80	80	160	nm		50 HRB
	P1044-					6.4	250	100	85	210	1		65 HRB
	P1045-					6.8	300	120	90	250	1	400	75 HRB
Copper steel	P2022-	<0.3	1 to 4	Balance	2	5.6	120	45	75	90	nm		70 HRH
	P2023-					6.0	160	55	80	120	1		80 HRH
	P2024-					6.4	200	65	85	140	2	300	15 HRB
	P2025-					6.8	240	75	90	170	3	450	25 HRB
	P2032-	<0.3	4 to 8	Balance	2	5.6	160	60	75	120	nm		80 HRH
	P2033-					6.0	200	75	80	140	nm		90 HRH
	P2034-					6.4	240	85	85	190	1		20 HRB
	P2035-					6.8	280	95	90	230	2	400	30 HRB
Copper carbon steel	P2043-	0.3 to 0.6	1 to 4	Balance	2	6.0	220	80	80	190	nm		45 HRB
	P2044-					6.4	280	100	85	230	nm	350	60 HRB
	P2045-					6.8	350	120	90	280	1	450	75 HRB
	P2053-	0.6 to 0.9	1 to 4	Balance	2	6.0	270	100	80	210	nm		60 HRB
	P2054-					6.4	340	120	85	270	nm	350	70 HRB
	P2055-					6.8	420	140	90	330	nm	450	80 HRB
	P2063-	0.3 to 0.6	4 to 8	Balance	2	6.0	250	90	80	210	nm		60 HRB
	P2064-					6.4	320	110	85	260	nm	350	70 HRB
	P2073-	0.6 to 0.9	4 to 8	Balance	2	6.0	300	110	80	240	nm		65 HRB
	P2074-					6.4	360	130	85	280	nm	350	75 HRB

[1] These materials may be supplied with additives to increase machinability, the properties given remain essentially unchanged. The values of all other properties, with the exception of density and copper content, will not then apply.
[2] nm = non-measurable.

Table 23.11 NICKEL STEELS AND NICKEL–COPPER STEELS (AS SINTERED)

		Mandatory values								Informative approximate values			
		Chemical composition					Mechanical and physical properties						
Materials	Grade	C combined %	Ni %	Cu %	Fe %	Total other elements (max.) %	Density (min.) g/cm³	Ultimate tensile strength (min.) MPa	Apparent hardness (min.) HV 5	Relative density %	Yield strength (0.2% offset) MPa	Elongation[2] %	Apparent hardness Rockwell
Nickel steels[1,2]	P3014-	<0.2	1 to 3	<0.8	Balance	2	6.4	200	50	85	140	6	35 HRB
	P3015-						6.8	250	60	90	170	8	40 HRB
	P3025-	<0.2	3 to 6	<0.8	Balance	2	6.8	300	80	90	200	6	60 HRB
Nickel–copper steels[2]	P3034-	<0.3	1 to 3	1 to 3	Balance	2	6.4	240	70	85	170	3	35 HRB
	P3035-						6.8	270	90	90	200	4	45 HRB
	P3044-	0.3 to 0.6	1 to 3	1 to 3	Balance	2	6.4	300	100	85	260	1	55 HRB
	P3045-						6.8	360	120	90	300	2	70 HRB
	P3054-	<0.3	3 to 6	1 to 3	Balance	2	6.4	250	70	85	190	3	40 HRB
	P3055-						6.8	290	90	90	220	4	55 HRB
	P3064-	0.3 to 0.6	3 to 6	1 to 3	Balance	2	6.4	320	100	85	280	1	60 HRB
	P3065-						6.8	380	130	90	320	2	75 HRB

[1] Weldable.
[2] Heat treatable.

Table 23.12 NICKEL–COPPER–MOLYBDENUM STEELS (AS SINTERED)

Materials	Grade	Mandatory values						Informative approximate values						
		Chemical composition						Mechanical and physical properties						
		C combined %	*Ni* %	*Cu* %	*Mo* %	*Fe* %	*Total other elements (max.)* %	*Density (min.)* g/cm³	*Ultimate tensile strength (min.)* MPa	*Apparent hardness (min.)* HV 5	*Relative density* %	*Yield strength (0.2% offset)* MPa	*Elongation²* %	*Apparent hardness Rockwell*
Nickel–copper–molybdenum steels[1]	P3074-	<0.3	1 to 3	1 to 3	0.3 to 0.7	Balance	2	6.4	240	80	85	170	3	45 HRB
	P3075-							6.8	270	100	90	200	4	60 HRB
	P3076-							7.0	290	110	90	220	5	70 HRB
	P3084-	0.3 to 0.6	1 to 3	1 to 3	0.3 to 0.7	Balance	2	6.4	330	120	85	300	2	70 HRB
	P3085-							6.8	440	150	90	360	3	80 HRB
	P3086-							7.0	480	160	90	390	4	90 HRB
	P3094-	0.6 to 0.9	1 to 3	1 to 3	0.3 to 0.7	Balance	2	6.4	350	140	85	330	nm	75 HRB
	P3095-							6.8	460	170	90	400	nm	85 HRB
	P3104-	0.3 to 0.6	3 to 6	1 to 3	0.3 to 0.7	Balance	2	6.4	410	150	85	350	nm	80 HRB
	P3105-							6.8	600	180	90	450	1	85 HRB
	P3106-							7.0	680	200	90	520	2	90 HRB

1 Heat treatable.
2 nm = non-measurable.

Table 23.13 MECHANICAL PROPERTIES OF SINTERED AND HEAT TREATED STEELS

Plain carbon steels

Carbon content %	Density g/cm³	Relative density class	Hardness	*Mechanical properties*						
				Ultimate tensile strength MPa	0.2% yield strength MPa	Elongation %	Impact strength* J	Fatigue limit MPa	En GPa	Poisson's ratio
0.25–0.6	6.55	3	30 HRA	400	380	0.5	—	150	110	0.21
0.25–0.6	6.9	5	50 HRA	560	520	0.5	—	210	135	0.23
0.6–1.0	6.1	2	100 HRB	400	380	<0.5	—	150	90	0.20
0.6–1.0	6.55	3	25 HRC	500	480	<0.5	—	190	110	0.21
0.6–1.0	6.9	5	30 HRC	650	630	<0.5	—	250	135	0.23

Copper steels

Composition, wt. %			Density g/cm³	Relative density class	Hardness HRC	*Mechanical properties*						
C	Cu	Ni				Ultimate tensile strength MPa	0.2% yield strength MPa	Elongation %	Impact strength* J	Fatigue limit MPa	En GPa	Poisson's ratio
0.25–0.6	1–3	—	6.5	3	30	580	550	<0.5	—	220	110	0.21
0.25–0.6	1–3	—	6.9	4	35	690	650	<0.5	—	260	130	0.23
0.6–1.0	1–3	—	6.1	2	25	380	360	<0.5	—	145	90	0.20
0.6–1.0	1–3	—	6.5	3	35	550	520	<0.5	—	210	110	0.21
0.6–1.0	1–3	—	6.9	4	45	690	650	<0.5	—	260	130	0.23

(*continued*)

Table 23.13   MECHANICAL PROPERTIES OF SINTERED AND HEAT TREATED STEELS—*continued*

Nickel steels

Composition, wt. %						Mechanical properties						
C	Cu	Ni	Density g/cm³	Relative density class	Hardness HRC	Ultimate tensile strength MPa	0.2% yield strength MPa	Elongation %	Impact strength* J	Fatigue limit MPa	En GPa	Poisson's ratio
0.25–0.6	—	2–4	6.6	3	32	670	550	0.5	7	270	115	0.22
0.25–0.6	—	2–4	7.1	5	42	910	550	0.5	15.5	360	145	0.24
0.25–0.6	—	2–4	7.3	6	45	1 070	890	1.5	28	400	160	0.26
0.6–1.0	—	2–4	6.6	3	35	700	650	0.5	8	280	115	0.22
0.6–1.0	—	2–4	7.1	5	45	930	880	0.5	—	370	145	0.24
0.6–1.0	—	2–4	7.3	6	47	1 100	1 050	0.5	26	410	160	0.26
0.6–1.0	—	4–6	6.6	3	35	650	550	0.5	5	250	115	0.22
0.6–1.0	—	4–6	7.1	5	42	750	650	0.5	9	350	145	0.24

Copper–nickel steels

Composition, wt. %							Mechanical properties						
C	Cu	Ni	Mo	Density g/cm³	Relative density class	Hardness HRC	Ultimate tensile strength MPa	0.2% yield strength MPa	Elongation %	Impact strength* J	Fatigue limit MPa	En GPa	Poisson's ratio
0.25–0.6	2–4	0.5–2	—	6.6	3	32	565	450	0.5	7.5	220	115	0.21
0.25–0.6	2–4	0.5–2	—	6.95	5	40	750	600	1	21	300	140	0.23
0.25–0.6	1–3	2–5	—	6.6	3	60 HRA	500	—	—	190	110	0.21	0.21
0.25–0.6	1–3	2–5	—	6.95	5	65 HRA	620	—	—	—	—	130	0.23

Copper–nickel–molybdenum steels

C	Cu	Ni	Mo	Density g/cm³	Relative density class	Hardness HRC	Ultimate tensile strength MPa	0.2% yield strength MPa	Elongation %	Impact strength* J	Fatigue limit MPa	En GPa	Poisson's ratio
0.6–1.0	1.5	1.75	0.5	6.75	4	32	650	550	0.5	5	250	117	0.22
0.6–1.0	1.5	1.75	0.5	7.1	5	42	900	750	1	9	350	145	0.24
0.25–0.6	1.5	4	0.5	6.75	4	24	715	560	0.5	—	285	117	0.22
0.25–0.6	1.5	4	0.5	7.1	5	38	980	770	1	—	390	145	0.24
0.6–1.0	1.5	4	0.5	6.75	4	30	790	660	0.5	—	320	117	0.22
0.6–1.0	1.5	4	0.5	7.1	5	40	1 080	890	1	—	420	145	0.24

* IZOD impact test.

Table 23.14 COMPOSITIONS AND PROPERTIES OF STAINLESS STEEL POWDERS AND PRODUCTS

Alloy type	Chemical compositions									Physical properties		Green properties			Sintered properties			
	C	Mn	Si	S	Cr	Ni	Mo	Fe	Cu	Apparent density g/cm³	Flow rate s/50 g	Compaction load MPa	Green density g/cm³	Green strength MPa	Sintered density g/cm³	Sintered strength MPa	Linear change %	Apparent hardness HV30 kg
303 L stainless	0.02	0.15	0.75	0.20	17.56	12.80	—	Balance	—	2.8	32	695	6.48	5.17	6.71	911	−1.07	156
304 L stainless	0.02	0.18	0.66	—	18.90	10.50	—	Balance	—	2.8	31	695	6.72	8.14	6.83	815	−0.52	165
316 L stainless	0.02	0.18	0.83	—	17.00	12.75	2.15	Balance	—	2.6	33	695	6.83	7.31	6.90	1.092	−0.49	140
347 L stainless	0.2	0.18	0.67	—	18.20	10.80	—	Balance	—	2.6	32	695	6.67	7.24	6.81	946	−0.52	180
410 L stainless	0.02	0.20	0.76	—	13.00	—	—	Balance	—	2.7	32	695	6.59	12.41	6.76	1.098	−0.78	163

Table 23.15   MECHANICAL PROPERTIES OF SINTERED COPPER ALLOYS

| Material | Composition, wt. % Cu = Remainder | | | | | Density g/cm³ | Relative density class | Mechanical properties | | | | | | |
	Sn	Zn	Ni	Pb	Others			Hardness	Tensile strength MPa	0.2% yield strength MPa	Elongation %	Impact strength* J	En GPa	Poisson's ratio
Bronze	9–11	—	—	—	<2 (Fe < 1)	5.8	1	20 HB	55	45	1	—	30	0.20
	9–11	—	—	—	2 (Fe < 1)	6.6	2	25 HB	95	75	1	—	40	0.22
	9–11	—	—	—	<2 (Fe < 1)	7.0	2	30 HB	120	100	2.5	—	50	0.24
	9–11	—	—	—	<2 (Fe < 1)	7.6	4	45 HB	200	150	10	—	70	0.26
	9–11	—	—	—	<2 (Fe < 1)	8.2	7	50 HB	300	220	20	—	87	0.30
Brass	—	8.3–12	—	—	—	7.4	4	57 HRH	140	100	8	—	65	0.25
	—	8.3–12	—	—	—	7.8	5	70 HRH	185	130	12	—	80	0.28
Brass	—	27.8–31.5	—	—	—	7.4	4	76 HRH	215	150	20	—	70	0.25
	—	27.8–31.5	—	—	—	7.8	6	85 HRH	285	180	26	—	85	0.28
Leaded brass	—	7–11	—	1–2	—	7.4	3	46 HRH	120	90	14	—	65	0.25
	—	7–11	—	1–2	—	7.8	5	60 HRH	175	120	20	—	78	0.28
Leaded brass	—	20	—	1–2	—	7.4	4	50 HB	165	90	13	12	63	0.25
	—	20	—	1–2	—	7.8	5	60 HB	195	105	19	19	76	0.28
	—	20	—	1–2	—	8.2	7	—	220	120	23	25	90	0.30
Leaded brass	—	26.5–30.5	—	1–2	—	7.4	4	65 HRH	195	100	22	—	65	0.25
	—	26.5–30.5	—	1–2	—	7.8	5	76 HRH	235	130	27	—	77	0.28
Nickel silver	—	18	18	—	—	7.8	5	70 HB	200	120	10	12	84	0.28
	—	18	18	—	—	8.2	7	80 HB	250	140	12	16	100	0.30
Leaded nickel silver	—	13.2–20	16.5–19.5	1–1.8	—	7.8	5	70 HB	190	120	10	11	80	0.28
	—	13.2–20	16.5–19.5	1–1.8	—	8.2	7	80 HB	240	130	12	14	95	0.30

[1] Compressive yield strength.
[2] Unnotched Charpy impact test.

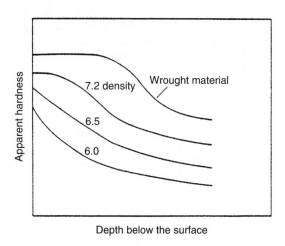

Figure 23.7 *Hardenability as a function of density and quench tests*

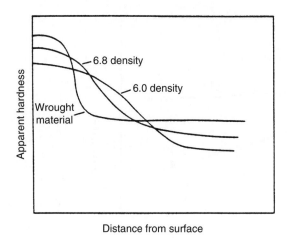

Figure 23.8 *Effect of porosity on the case hardening*

other refractory metals tantalum, niobium, and rhenium, as well as beryllium, chromium, and magnesium.

2. Metals containing a finely dispersed insoluble second phase. ODS materials and aluminium with a dispersion of silicon carbide have been referred to in Section 23.3 under 'Mechanical alloying'. In such cases the purpose is to produce materials with better properties than are available in standard wrought products of the same nominal composition but without the dispersed phase. The metals aluminium, copper, nickel alloys especially superalloys, molybdenum, and other refractory metals strengthened in this way are in commercial production. Additionally copper and silver containing dispersed oxide(s) are used for make and break electrical contacts combining the excellent conductivity of these metals with marked resistance to welding.

3. Metals and alloys that present no great production difficulty starting from ingot, but which can be produced more economically from powder, generally because of the higher yield of finished material and/or reduction in the number of annealing and pickling steps. Superior properties can sometimes be claimed and result from the greater homogeneity and freedom from pipe defects. Of importance commercially are austenitic stainless steels and superalloys.

Table 23.16 EFFECTS OF STEAM TREATING ON DENSITY AND HARDNESS OF FERROUS PM MATERIALS

Material	Density, g/cm³		Apparent hardness	
	Sintered	Steam treated	Sintered	Steam treated
Plain iron	5.8	6.2	7 HRF	75 HRB
	6.2	6.4	32 HRF	61 HRB
	6.5	6.6	45 HRF	51 HRB
Carbon steel	5.8	6.1	44 HRB	100 HRB
	6.2	6.4	58 HRB	98 HRB
	6.5	6.6	60 HRB	97 HRB
6–8% Ni steel	5.7	6.0	14 HRB	73 HRB
	6.35	6.5	49 HRB	78 HRB
	6.6	6.6	58 HRB	77 HRB
6–8% Ni, 2% Cu steel	5.7	6.0	52 HRB	97 HRB
	6.3	6.4	72 HRB	94 HRB
	6.6	6.6	79 HRB	93 HRB

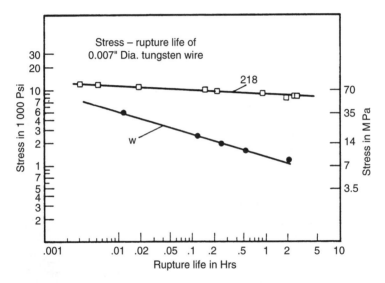

Figure 23.9 *Stress dependence of rupture life in hours of 0.007″ (178 μm) tungsten wire at 3 000 K. W is pure tungsten wire, 218 is doped tungsten wire*

23.13.1 Refractory metals

Tungsten wire was one of the first PM products to be made on a large scale, the use being filaments for incandescent electric light bulbs, for which purpose it is still made. In order to induce in the wire a particular elongated microstructure which increases remarkably the life of the filament, a small percentage of thorium oxide was included. It is to tungsten doped in this way that the properties shown in Figure 23.9 refer.

Molybdenum is used extensively in high-temperature furnaces as heating elements, radiation shields, and structural components. The high-temperature strength can be increased by ODS. ODS platinum/rhodium used for handling molten glass has much improved creep strength as is shown in Figure 23.10.

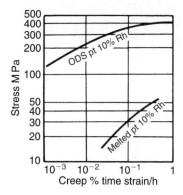

Figure 23.10   *Creep strength of ODS Pr/10% Rh*

23.13.2   Superalloys

This family of alloys is used for load-bearing applications at elevated temperatures, and is especially relevant to turbine discs and blades for aircraft engines. For manufacture by PM, spherical powders produced either by gas (usually argon) atomisation or centrifugal atomisation are used. The attractions of the PM route are that it is possible to achieve a fine homogeneous structure free from pipe-related defects and segregation, and also a better yield of usable material. Table 23.17 lists some relevant properties of one well known alloy. Most all of the current production of PM superalloys relies on hot isostatic pressing to obtain full density. Often the HIP consolidation is followed by hot forging.

Superalloys owe their good strength properties to precipitation hardening by a titanium–aluminium compound, but as the temperature rises, this compound begins to go into solution and loses its effect. This puts a ceiling on the working temperature. The mechanical alloying process to produce a fine dispersion of a ceramic material—yttria is favoured—that is virtually insoluble in the metal matrix, significantly increasing the temperature at which useful strength is retained. A list of ODS superalloy compositions is given in Table 23.18, and Figure 23.11 shows the improved performance of one of these alloys compared with that of a similar composition made conventionally.

23.13.3   Copper

One of the main users of copper is the electrical industry where conductivity is the primary consideration. Increasingly, power plant is required to operate at temperatures well above ambient where pure copper recrystallises and becomes very soft. Increasing the strength and recrystallisation temperature by alloying drastically reduces the conductivity. The inclusion in pure copper of a small percentage of finely dispersed aluminium oxide provides significantly better elevated temperature strength with only a small reduction in conductivity. An atomised powder of a dilute aluminium copper alloy is internally oxidised to give a very fine Al_2O_3 particles, the powder being then processed by compaction, sintering, and working. The improved strength at room temperatures enables smaller sections to be used in, for example, miniaturised systems. Figure 23.12 gives some results for a range of alumina contents.

23.13.4   Lead

This is another soft metal whose strength can be increased by a dispersed oxide phase. In this case lead oxide is used. The chief application is to chemical plant especially for handling sulphuric acid.

Table 23.17   MECHANICAL PROPERTIES OF NIMONIC ALLOY AP1 DISCS HIPPED AND FORGED (Inco Alloys International)

Disc size	Solution treatment	T_{test} °C	Yield strength (0.2% offset) MPa	Ultimate tensile strength MPa	Elongation %	Reduction of area %	Notched tensile strength MPa	Stress rupture 760 MPa at 705°C — Plane — Life h	Plane Elongation %	Notch Life h	Low fatigue at 600°C — Stress MPa	Life cycles
380 mm dia.	4 h/1 110°C/AC	650	906	1 316	30.4	30.5	1 627	51	11.5	699	1 080 +1 100 +1 120	125 200 21 510 1 390
380 mm dia.	4 h/1 160°C/AC	650	902	1 348	25.9	27.0	1 599	63	19.5	424	1 080	400 007
280 mm dia.	4 h/1 080°C/OQ	650	961	1 325	26.8	28.7	1 701	109	7.6	814	1 080	54 063
280 mm dia.	4 h/1 160°C/OQ	650	977	1 370	25.9	23.9	1 678	107	8.9	252	1 080 +1 100 +1 120 +1 140	123 200 20 000 20 000 4 430
200 mm dia.	4 h/1 080°C/OQ	R.T.	991	1 369	18.5	20.0						
200 mm dia.	4 h/1 160°C/OQ	R.T.	1 045	1 451	22.8	35.4						

Aged 24 h 650°C air cool and 8 h air cool; R.T.: Room temperature, AC: air cool, OQ: oil quench.

Composition of Nimonic alloy AP1 (wt. %)

C	Al	Co	Cr	Mo	Ti	Zr	B	Ni
0.02	4.0	17.0	14.8	5.0	3.5	0.04	0.02	bal

Table 23.18 NOMINAL COMPOSITIONS (WT. %) OF MECHANICALLY ALLOYED OXIDE DISPERSION STRENGTHENED SUPERALLOYS

	Ni	Fe	Cr	Al	Ti	C	Y_2O_3	Mo	W	Ta	B	Zr
INCOLOY* alloy MA 956		Bal	20	4.5	0.5	0.05	0.5					
INCOLOY alloy MA 957		Bal	14		1.0	0.05	0.25	0.3				
INCONEL* alloy MA 954	Bal	1.0	20	0.3	0.5	0.05	0.6					
INCONEL alloy MA 758	Bal	1.0	30	0.3	0.5	0.05	0.6					
INCONEL alloy MA 6000	Bal		15	4.5	2.5	0.05	1.1	2.0	4.0	2.0	0.01	0.15
INCONEL alloy MA 760	Bal		20	6.0		0.05	0.95	2.0	3.5		0.01	0.15

* INCOLOY and INCONEL are trademarks of the Inco family of companies.

Iron–chromium alloys: INCOLOY alloy MA 956 sheet, plate, bar, spinnings, rings and forgings have applications in the hot sections of gas turbines and diesel engines where the resistance of the alloy to creep, oxidation and sulphidation allow higher metal temperatures and longer component life. The alloy is being used to replace molybdenum in high-temperature vacuum furnaces for fixtures and heat-treatment trays. INCOLOY alloy MA 957 is intended for nuclear power applications, especially fuel cladding in liquid metal cooled reactors. Compared with 316 type stainless steel it has higher strength at 700°C and considerable resistance to irradiation damage.

Nickel–chromium alloys: INCONEL alloy MA 754 is used for brazed nozzle guide vane and band assemblies in advanced military aero engines. The principal advantages of the alloy for these applications are thermal fatigue resistance, long term creep strength and a high melting point. INCONEL alloy MA 758 is highly resistant to attack by molten glass and is used in spinnerettes for the production of fibre glass. The parts are formed by hot spinning plate.

Gamma prime ODS alloys: The immediate applications for INCONEL alloy MA 6000 are for first- and second-stage turbine vanes and blades machined from solid bar. Forced airfoil components have also been developed. The characteristics of INCONEL alloy MA 6000 allow blade cooling to be reduced or eliminated as the metal temperature can be increased by 100 K or more in engines where the stresses are medium or low. INCONEL alloy MA 760 is an industrial gas turbine derivative of INCONEL alloy MA 6000 having greater resistance to corrosion and oxidation. Initial applications are for machined vanes and blades but forged components are under development.

23.13.5 Aluminium

Section 23.3.1 referred to the mechanical alloying of Al with graphite to produce powder containing a carbide dispersion. This powder compacted into billets and extruded gives a product with much improved strength at elevated temperatures (see Figure 23.13).

23.13.6 Ferrous alloys

Powder metallurgy has found success at full density with ferrous alloys because of the improved microstructural homogeneity, especially important with high alloy systems. Table 23.19 provides mechanical property data for a few example full density ferrous alloys fabricated by hot isostatic pressing. Most of the applications for these materials are in demanding situations, such as military hardware, oil well drilling, and high performance automobiles. However, a few products have found their way into microelectronics processing, printers, and other applications demanding high performance.

23.13.7 Aluminium matrix composites

Mixed powders are used to form particulate or whisker reinforced composites. A common technique is to mix the two constituents and to hot consolidate the mixture, since sintering proves difficult with the

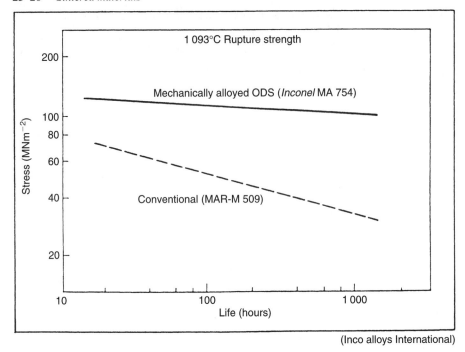

(Inco alloys International)

Figure 23.11 *Properties of mechanically alloyed Inconel (Inconel is a trademark of Inco Alloys International) compared with a conventional superalloy*

Composition of Inconel alloy MA 754:

Fe	Al	Cr	C	Y_2O_3	Ti	Ni
1.0	0.3	20	0.05	0.6	0.5	bal.

reinforcing phase. Up to approximately 50 vol.% of the inert phase can be fabricated by a combination of mixing and hot pressing. Two types of hot pressing are common based on uniaxial or isostatic pressures in a heated chamber. Table 23.20 gives the properties of Al–SiC composites fabricated by vacuum hot pressing using an aluminium alloy for the matrix. The tabulation shows how thermal and mechanical properties change with composition. The harder, stiffer carbide phase increases the elastic modulus, while lowering the thermal expansion coefficient and thermal conductivity. Note ductility falls rapidly with the addition of the silicon carbide.

23.14 Spray forming

This process is not powder metallurgy in the strict sense of the term in so far as the metal is at no stage in the form of powder. However, for reasons that will be apparent the PM world has adopted it. The process involves gas atomisation of a liquid metal, but instead of allowing the droplets to solidify as powder, the spray is caused to impinge on a solid surface where the droplets are collected as a semi-solid layer which solidifies as a dense metal. This layer may be built up to any desired thickness, and by suitable choice of design of the original target, the angle of the spray, and other parameters, near-net shapes can be produced. For example if the target is a cylinder rotating horizontally and capable of being moved in a controlled manner in the axial direction, a tube of dense deposited metal is formed. The deposit has all the advantages of dense metal produced from powder, i.e. complete absence of macro-segregation and pipe-related defects. A further merit of the process is that by injecting fine refractory powder particles, i.e. by entraining them in the atomising gas stream, ODS material can be deposited.

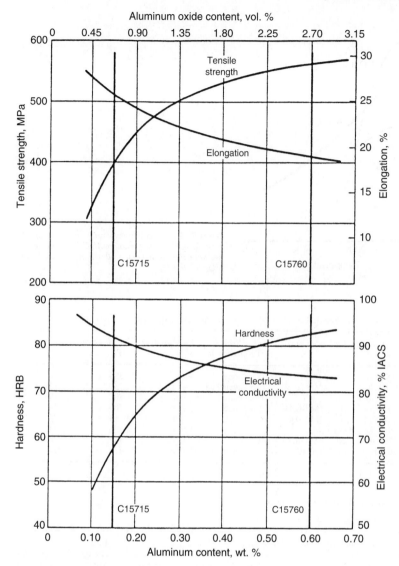

Figure 23.12   *Properties of dispersion-strengthened copper (C15715 and C15760 are compositions made by SCM Metal Products)*

23.15   Injection moulding

Commonly referred to as PIM (powder injection moulding), this process consists of mixing small ≤20 μm metal powder with a thermosetting polymer material to form a plastic mass that can be injected under pressure into a mould to form the equivalent of a compact. The compact is then carefully treated by solvents and/or heat to remove the polymer and then at a higher temperature to sinter the powder. Shrinkage of the order of 15% linear occurs but this can be predicted accurately and parts with very close dimensional tolerances and of quite complex shape can be made. It is widely used process because the advantages in eliminating expensive machining operations make it viable for a number of applications.

Unlike traditional powder metallurgy, the metal powder injection moulding route is followed by high temperature sintering. As a result, the sintered alloys are essentially full density, giving superior

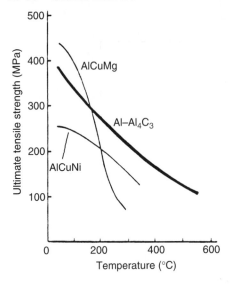

Figure 23.13   *Strength/temperature relationship of aluminium with a dispersion of Al$_4$C$_3$ produced by mechanical alloying compared with ingot-based high strength alloys*

Table 23.19   TENSILE PROPERTIES OF FULL DENSITY HOT ISOSTATICALLY PRESSED STEELS

Alloy	Composition (balance Fe) wt. %	Heat-treatment	Yield strength MPa	Tensile strength MPa	Elongation %
1080	0.8C	none	—	385	18
1080	0.8C	quenched	—	730	12
4600	1.8Ni–0.5Mo–0.5C	none	590	800	19
A286	30Ni–14Cr–2Ti–1.3Mo–0.4V	none	380	700	38
A286	30Ni–14Cr–2Ti–1.3Mo–0.4V	aged	740	1 115	25
Maraging	19Ni–8Co–5Mo–0.4Ti	aged	1 515	1 600	14
Tool steel	12Cr–1Mo–1Ni–0.6Mn–0.3V–0.24C	—	655	845	20
Tool steel	5Cr–1.4Mo–1V–1Si–0.4C	tempered	1 310	1 430	21

Table 23.20   PROPERTIES OF Al-SiC COMPOSITES

(6061 alloy matrix, mixed powders densified by vacuum hot pressing)

Composition vol. % SiC	Thermal expansion coefficient $10^{-6}/°C$	Thermal conductivity W/(°C m)	Elastic modulus GPa	Density g/cm^3	Yield strength MPa	Elongation %
0	22.5	166	69	2.71	430	20
15	18.7	138	97	2.77	435	6
20	17.4	130	103	2.80	450	5
25	16.2	121	114	2.83	475	4
30	13.1	143	121	2.85	510	0
40	10.4	135	138	2.91	379	<1

---

mechanical properties. Table 23.21 tabulates several example alloys processed by PIM, showing the superior densities and properties. Thus, the combination of greater shape complexity, full density, and net-shape manufacturing make the metal powder injection moulding route the fastest growing field in powder metallurgy.

Table 23.21 MECHANICAL PROPERTIES OF INJECTION MOULDED AND SINTERED STEELS

(HT = heat treated)

Composition, wt. %	Density g/cm³	Yield strength MPa	Ultimate tensile strength MPa	Elongation %	Hardness
Fe–1Cr–0.4C	7.4	1 240	1 380	2	40 HRC
Fe–2Cr–1Ni–1Mn–0.4C	7.6	480	620	6	20 HRC
Fe–2Ni–1Mo–0.4C	7.7	1 400	2 000	3	30 HRC
Fe–36Ni	7.8	240	425	40	65 HRB
Fe–8Ni	7.6	255	440	24	75 HRB
Fe–50Ni	8.0	170	420	20	50 HRB
Fe–2Ni–0.5C	7.4	1 230	1 230	1	45 HRC
Fe–2Ni–0.9C	7.6	450	650	9	90 HRB
Fe–0.6P	7.8	260	280	2	80 HRB
Fe–3Si	7.8	345	520	25	85 HRB
Fe–29Ni–17Co	8.3	350	520	42	60 HRB
Fe–16Cr–4Ni–4Cu	7.7	965	1 140	12	35 HRC
Fe–18Cr–8Ni	7.7	240	480	35	85 HRB
Fe–6W–5Mo–4Cr–2V–1C	8.1	—	2 000	0	66 HRC

23.16   Hardmetals and related hard metals

This family of PM materials consists of fine, hard, and usually brittle carbide particles bonded with a relatively tough binder phase which is normally metallic. The hard particles are generally between 1 and 5 μm, but even finer grades with particles below 1 μm are now being made. The original hard phase was tungsten monocarbide (WC) and the preferred binder phase was cobalt. The name *cemented carbide* was, and to a large extent still is, used to describe these materials. Carbides other than that of tungsten were later added, e.g. those of tantalum, niobium, and titanium, and binder metals other than cobalt have been used, but the WC/Co-based compositions still have the lion's share of the market. More recently, products have been developed in which the hard phase is not carbide, but nitride, boride, carbo-nitride, oxide, or combinations of these.

Uses

Hardmetals were originally developed as a substitute for diamond as wire drawing dies for tungsten, and they are still used for that purpose, but the largest single use today is as cutting tools and oil drilling or mining tools. It is for this application that the non-carbide alloys have been developed, but no cost effective substitute has been found for the WC/Co hardmetal where straightforward wear resistance is the primary requirement, and this includes cutting tools for non-ferrous metals and non-metallic materials. Other such cases are wire drawing dies, dies for the compaction of metal powders for the manufacture of PM parts, rolls for metal rolling mills, and other large abrasion-resistant parts.

Manufacture

The process for the production of hardmetal is a classic example of liquid phase sintering: a mixture of WC and cobalt powders is pressed and sintered at a temperature above the eutectic temperature. Good results require special procedures in the preparation of the powder mix. The ingredients are wet milled together to coat each carbide particle with cobalt, and to facilitate this, the cobalt powder must be extremely small. It is well known that very small powders do not flow readily, if at all, and to overcome this problem the WC/Co mixture is 'granulated', by which is meant the production of agglomerates. A favoured method of granulation is the spray drying of a slurry of the powder with a liquid containing also a pressing lubricant. Injection moulding is now being used to replace die compaction, because of the greater shape complexity and higher final tolerances. During sintering, the compact shrinks by as much as 50 vol% to become nearly 100% dense. The sintering temperature used in practice varies with the composition, being lowest (1400°C) when the cobalt content is high, and rising to 1600°C or higher with compositions having high proportions of the carbides of Ti, Ta, and/or Nb and low cobalt contents. In cases where it is not possible to get close to the required shape by direct pressing and sintering, the compact may be pre-sintered at a lower temperature so as to remove the lubricant and provide sufficient strength for handling. In this state, the object can be machined using tools of bonded diamond or other hard material.

Although it is usual to refer to the binder phase as being cobalt there is some mutual solubility between it and the carbide, and great care is needed to ensure that the carbon balance is maintained such that neither the brittle W/Co (eta) phase nor free graphite is formed. The toughness is affected also by the size of the carbide particles; the finer they are the harder but less shock-resistant is the final product. However, grades with carbide particle size well below 1 μm are reported to combine high hardness with toughness.

Hot isostatic pressing

Although the porosity of conventionally produced hardmetal is normally low, porosity can be completely eliminated by hot isostatic pressing (HIP). Toughness is, thereby considerably increased and the possibility of the rejection of large and expensive components at a late stage of grinding or polishing no longer presents a problem. HIP is now routinely applied to a large number of hardmetal parts including indexable cutting tool tips that are a major product. Recently it has been found possible to combine HIP with the sintering stage. The parts are sintered in vacuum to a density such that the porosity is sealed, and then high pressure gas, usually argon, is introduced into the furnace. The pressure required for full densification is lower than that needed for the HIP of already sintered parts, and the new process, referred to as Sinter-HIP or Pressure-Assisted Sintering is rapidly replacing the original two-stage process of vacuum sintering followed by HIP in a separate furnace, at least for cutting tool tips.

Compositions

Straightforward WC/Co hardmetal appears to be the most cost-effective material for many applications where wear resistance is the primary requirement, including the machining of non-ferrous metals. Additions of e.g. TaC improve the already good wear resistance by acting as grain growth inhibitors, and are especially valuable in applications involving high temperatures. When the use is the machining of ferrous materials at high speeds, the situation is different. In addition to abrasive wear, reaction between the carbide particles and the steel results in what is known as crater wear. The substitution of more stable carbides such as those of Ta, Nb, Ti, and Hf for some or all of the WC considerably improves the cratering resistance. Wear resistance is, as would be expected, markedly influenced by the amount of binder phase as shown in Figure 23.14, but reduction in the cobalt content also makes the alloy more brittle, so a compromise between toughness and wear resistance is necessary.

Table 23.22 gives compositions and properties of one manufacturer's range. Table 23.23 lists the ISI classification of carbides according to use.

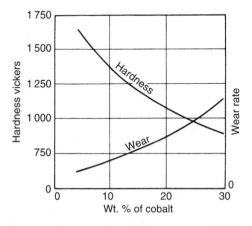

Figure 23.14   *Relationship between cobalt content, hardness and wear of WC/Co hard metal*

ISO codes	WC	TiC	Ta(Nb)C	Co	Ni/Cr	Average grain size μm	Transverse rupture strength MPa	Density g/cm³	Hardness HV30	Compressive strength GPa	Young's modulus of elasticity GPa	Poisson's ratio	Coercive force kA/m	Electrical resistivity μΩ cm	Thermal conductivity (20°C) W/mK	Coefficient of thermal expansion (20–400°C) 10⁻⁶/K
P05	60.5	22	11	6.5	—	2–3	1400	10.0	1705	5.3	490	0.22	11.94	91	25	7
P10	69	16	8	7	—	2–3	1650	11.3	1675	5.3	530	0.22	11.94	58	30	6
P20–25	71.5	9	11	8.5	—	2–3	1800	12.4	1575	5.2	540	0.22	11.94	37	40	6
P20–P30	69.5	6.5	14	10	—	2–3	1900	12.7	1555	5.1	550	0.22	13.13	30	50	6
P30–P40	76.5	4	7.5	12	—	2–3	2200	13.1	1390	5.0	560	0.22	11.14	25	60	5.5
P40	77	4	8	11	—	3	2400	13.3	1420	4.9	540	0.22	11.94	24	60	6
M10	83.5	5	5.5	6	—	1–2	1800	13.5	1685	5.7	620	0.22	14.92	24	60	5.5
M10–M20 / K10–K20	84	7.5	1	7.5	—	1	1800	12.9	1730	5.7	600	0.22	19.89	26	50	5.5
M15	86	2.5	5.5	6	—	3	1950	14.0	1575	5.3	590	0.22	12.73	25	60	5.5
M20	77	4	10	9	—	3	2050	13.3	1460	5.1	550	0.22	10.5	28	60	5.5
K01–K05	96.7	—	0.3	3	—	1	1600	15.3	1850	6.1	660	0.22	24	20	100	5
K05	87.5	1.5	6	5	—	1	1800	14.5	1830	6.1	650	0.22	19.5	23	100	5
K10	94.2	—	0.3	5.5	—	1–2	1900	14.8	1750	6.0	650	0.22	18.70	20	100	5
K10–K20	93.7	—	0.3	6	—	2	1950	14.9	1675	5.8	640	0.22	17.11	20	100	5
K20	93.2	0.2	0.6	6	—	1	2000	14.7	1615	5.7	640	0.22	15.12	20	100	5
K20–K30	90.2	—	0.3	9.5	—	2	2100	14.7	1545	5.5	630	0.22	14.72	18	90	5
K30	90.2	0.2	0.3	9	—	2	2200	14.6	1470	5.4	620	0.22	12.33	18	90	5
K40	87.2	0.2	0.6	12	—	2	2550	14.3	1340	5.0	600	0.22	10.74	17	90	5.5
K50	84.2	0.2	0.6	15	—	3	2800	14.0	1225	4.5	570	0.23	9.55	17	80	6
	80	—	—	20	—	3	3000	13.5	1030	3.8	520	0.24	7.00	17	70	6.5
	76	—	—	24	—	3	2900	13.2	920	3.5	500	0.24	5.81	16	60	6.5
	72	—	—	28	—	3	2800	12.8	855	3.2	480	0.24	5.25	16	60	6.5
	94	—	—	6	—	2–3	2050	14.9	1500	5.5	650	0.22	11.95	16	100	5
	91	—	—	9	—	3–4	2500	14.6	1215	4.2	630	0.22	7.3	19	90	5.5
	88	—	—	12	—	3–4	2550	14.2	1115	4.0	580	0.22	6.5	19	80	6
	85	—	—	15	—	3–4	2600	14.0	990	3.8	560	0.23	6.0	17	80	5
	94	—	—	—	6	1–2	2000	14.8	1500	5.5	630	0.19	—	—	70	4.8
	91	—	—	—	9	2	2300	14.4	1400	5.0	600	0.21	—	—	61	—
K05–K20 / P05–P25	86	2.5	5.5	6	—	3	—	—	Hardness of coating >2000 V	—	—	—	—	—	—	—
K05–K20 / P05–P20	86	2.5	5.5	6	—	3	—	—	—	—	—	—	—	—	—	—
K05–K20 / P15–P30	83.5	5	5.5	6	—	1–2	—	—	—	—	—	—	—	—	—	—
K10–K20 / P15–P30	77	4	10	9	—	3	—	—	—	—	—	—	—	—	—	—
K10–K20	77	4	10	9	—	3	—	—	—	—	—	—	—	—	—	—
P25–P40	76.5	4	7.5	12	—	2–3	—	—	—	—	—	—	—	—	—	—
P25–P40	76.5	4	7.5	12	—	2–3	—	—	—	—	—	—	—	—	—	—
P05–P25	86	2.5	5.5	6	—	3	—	—	—	—	—	—	—	—	—	—
P35–P45	91	—	—	—	—	3–4	—	—	—	—	—	—	—	—	—	—

* Coating/thickness 8 μm. ** Coating/thickness TiC, Ti(C, N), TiN 10–12 μm. *** Coating/thickness TiC, Ti(C, N), TiN 5 μm.

Table 23.23 ISO CLASSIFICATION OF CARBIDES ACCORDING TO USE

Main groups of chip removal			Groups of application			Direction of increase in characteristic of carbide	
Symbol	Broad categories of material to be machined	Distinguishing colours	Designation	Material to be machined	Use and working conditions	of cut	
			P 01	Steel, steel castings	Finish turning and boring, high cutting speeds, small chip section, accuracy of dimensions and fine finish, vibration-free operation		
			P 10	Steel, steel castings	Turning, copying, threading and milling, high cutting speeds, small or medium chip sections		
		BLUE	P 20	Steel, steel castings / Malleable cast iron with long chips	Turning, copying, milling, medium cutting speeds and chip sections, planing with small chip sections		
P	Ferrous metals with long chips		P 30	Steel, steel castings / Malleable cast iron with long chips	Turning, milling, planing, medium or low cutting speeds, medium or large chip sections, and machining in unfavourable conditions*		
			P 40	Steel / Steel castings with sand inclusion and cavities	Turning, planing, slotting, low cutting speeds, large chip sections, with the possibility of large cutting angles for machining in unfavourable conditions* and work on automatic machines		
			P 50	Steel / Steel castings of medium or low tensile strength, with sand inclusion and cavities	For operations demanding very tough carbides: turning, planning slotting, low cutting speeds, large chip sections, with the possibility of large cutting angles for machining in unfavourable conditions* and work on automatic machines		

Direction of increase in characteristic of carbide:
— Toughness →
← Wear resistance —
— Increasing feed →
← Increasing speed —

Group		Colour	Designation	Raw material or component to be machined	Type of machining operation
M	Ferrous metals with long or short chips and non-ferrous metals	YELLOW	M 10	Steel, steel castings, manganese steel. Grey cast iron, alloy cast iron	Turning, medium or high cutting speeds. Small or medium chip sections
			M 20	Steel, steel castings, austenitic or manganese steel, grey cast iron	Turning, milling. Medium cutting speeds and chip sections
			M 30	Steel, steel castings, austenitic steel, grey cast iron, high temperature resistant alloys	Turning, milling, planing. Medium cutting speeds, medium or large chip sections
			M 40	Mild free cutting steel, low tensile steel. Non-ferrous metals and light alloys	Turning, parting off, particularly on automatic machines
K	Ferrous metals with short chips, non-ferrous metals and non-metallic materials	RED	K 01	Very hard grey cast iron, chilled castings of over 85 Shore, high silicon aluminium alloys, hardened steel, highly abrasive plastics, hard cardboard, ceramics	Turning, finish turning, boring, milling, scraping
			K 10	Grey cast iron over 220 Brinell, malleable cast iron with short chips, hardened steel, silicon aluminium alloys, copper alloys, plastics, glass, hard rubber, hard cardboard, porcelain, stone.	Turning, milling, drilling, boring, broaching, scraping
			K 20	Grey cast iron up to 220 Brinell, non-ferrous metals: copper, brass, aluminium	Turning, milling, planing, boring, broaching, demanding very tough carbide
			K 30	Low hardness grey cast iron, low tensile steel, compressed wood	Turning, milling, planing, slotting, for machining in unfavourable conditions* and with the possibility of large cutting angles
			K 40	Soft wood or hard wood. Non-ferrous metals	Turning, milling, planing, slotting, for machining in unfavourable conditions* and with the possibility of large cutting angles

Arrows for group M: ←— Toughness —→ ; ←— Wear resistance ; —→ Increasing feed —→ ; ←— Increasing speed

Arrows for group K: ←— Toughness —→ ; ←— Wear resistance ; —→ Increasing feed —→ ; ←— Increasing speed

* Raw material or components in shapes which are awkward to machine: casting or forging skins, variable hardness etc., variable depth of cut, interrupted cut, work subject to vibrations.

Reproduced from ISO recommendation 513 by permission of the British Standards Instituion, 2 Park Street, London, W1A 2BS.

Alternative binders

Because of the high price of cobalt and the perceived instability of the countries that produce the bulk of it, continuing efforts have been made to find alternatives. Ni, Fe, and Ni/Mo have been used successfully in certain applications, and a recent entry into the field is Ni/Cr which now appears in some commercial grades. One grade is based on TiC with MoC and a Ni/Mo binder. Good results have been reported also with intermetallic and superalloy binders, which combine conspicuous high-temperature strength with toughness. Useful results have been reported also with a nickel binder containing ruthenium, but this member of the platinum group is costly.

Coatings

For tools for the machining of steels the most dramatic improvement has been the development of surface coatings. An ideal tool material from the cutting point of view would be a pure, very hard, and stable compound such as one of the more stable carbides, nitrides, or oxides, were it not for the fact that they are far too brittle. However, if they are applied as a very thin layer to a sufficiently strong and tough substrate, an approximation to the best of both worlds is possible. Such coatings can be applied successfully by chemical vapour deposition (CVD) or physical vapour deposition (PVD). Titanium nitride, TiN, is widely used and gives the tool a striking gold colour. An essential requirement, of course, is that the coating adheres firmly to the substrate and this is facilitated by using multiple coatings only a few micrometres thick. Another factor in the equation is the possibility of reaction between the coating and the substrate and between the coating and the steel workpiece. In addition to nitrides, carbides, carbo-nitrides, borides, and oxides—especially alumina—are used, and as many as ten layers are applied, so the permutations are manifold. Recent additions to the range of coatings are cubic boron nitride (CBN) and diamond, but the latter is not suitable for the high-speed cutting of steel because it reacts with the steel at the high temperatures that are reached. The importance of matching the tool to the application should be emphasised.

Tool life

In cutting applications much is made of the improved tool life of modern throw-away inserts achieved by careful design of substrate and coating, but a more important factor from the economic point of view is the improved cutting speed that is made possible. The cost of the tool insert is a small part only of the total machining cost, and if the cutting rate can be doubled and the tool life halved in consequence, the overall efficiency may well be much greater than that of prolonging the life of the tool.

23.17   Novel and emerging PM materials

Besides the traditional PM materials, largely dominated by WC-Co, iron and steel, and refractory metals, several useful and exotic materials are emerging with unique properties only available via sintering techniques. Many of these are targeted at applications such as sporting equipment, computer or electronic components, sensors and instruments, and military components. Sintering is performed at high temperatures for most of these applications to improve the density and properties. High sintering temperatures induce more rapid atomic motion, resulting in properties that rival wrought materials. Further, new compositions are possible via sintering that cannot be fabricated via alternative techniques such as casting. A good example of the latter is seen with the tungsten heavy alloys. The properties of several liquid phase sintered heavy alloys are given in Table 23.24. These are formed

Table 23.24   PROPERTIES OF LIQUID PHASE SINTERED TUNGSTEN HEAVY ALLOYS

Composition, wt. %	Density g/cm^3	Hardness HRA	Yield strength MPa	Tensile strength MPa	Elongation %
97W–2Ni–1Fe	18.6	65	610	900	19
93W–5Ni–2Fe	17.7	64	590	930	30
90W–7Ni–3Fe	17.1	63	530	920	30
86W–4Mo–7Ni–3Fe	16.6	64	625	980	24
82W–8Mo–8Ni–2Fe	16.2	66	690	980	24
74W–16Mo–8Ni–2Fe	15.3	69	850	1 150	10

Table 23.25   PROPERTIES OF P/M THERMAL MANAGEMENT MATERIALS

(compositions in wt. %)

Property	W–10Cu	W–20Cu	Mo–30Cu	Mo–50Cu	Al–40Si	Al–65SiC
Thermal expansion, ppm/°C	6.0	7.0	7.5	8.5	10.3	9.8
Thermal conductivity, W/(°C m)	209	247	183	234	125	165
Density, g/cm^3	17.0	15.1	9.7	9.5	2.5	3.0
Strength, MPa	500	570	—	—	350	500
Elastic modulus, GPa	340	290	220	215	340	205

Table 23.26   EXAMPLE MAGNETIC PROPERTIES OF P/M MATERIALS

(compositions in wt.%)

Property	Fe	Fe	Fe–0.8P	Fe–2Si	Fe–50Ni	Fe$_{14}$Nd$_2$B	SmCo$_5$
Density, g/cm^3	7.0	7.4	7.7	7.3	7.7	7.4	8.3
Maximum permeability	2 100–4 000	2 400–4 400	4 000–14 000	6 100	5 000–40 000	—	—
Induction*,	1.1 (r)	1.2 (r)	1.8 (s)	1.4 (s)	1.4 (s)	1.3 (r)	1.2 (r)
Tesla	1.5 (s)	1.6 (s)			0.4 (r)	1.4 (s)	
Coercive force, A/m	168	145	32	62	16	10^6–10^8	10^6

* either the remnant (r) or saturation (s) is given as appropriate.

by mixing tungsten, molybdenum, nickel, iron, or copper powders. The mixed powders are shaped by die compaction, cold isostatic pressing, or injection moulding, and then sintered at temperatures near 1500°C to give over 99% of theoretical density. The composite sintered microstructure consists of hard tungsten grains in a matrix of solidified liquid that provides toughness. These alloys find uses in projectiles, heat sinks, weights, and computer disk drives.

A related application are for liquid phase sintered composites relates to the need for high thermal conductivity and low thermal expansion in microelectronic packaging applications. These are for heat dissipation around semiconductors where computing performance is degraded by heat. For these applications, the nickel-iron matrix phase in the heavy alloys is replaced by copper to improve thermal conductivity. If weight is not an issue, then the W–Cu composites are the best combinations of properties, but as weight becomes more of a concern then Mo–Cu and Al-based compositions are more popular. However, as illustrated in Table 23.25, these latter alloys suffer with respect to thermal conductivity when compared to the W–Cu alloys.

Finally, another emerging area for sintered metals is in magnetic applications. Several alloys are available, ranging from pure iron for soft magnetic uses to iron-neodymium-boron rapidly solidified materials for hard magnetic applications. Table 23.26 gives a tabulation of the magnetic properties available from some of the most popular alloys.

ACKNOWLEDGEMENTS

Associazione Industriali Metallurgici Meccanici Affini (AMMA)	Tables 23.8, 23.13 and 23.15
American Society for Metals (ASM)	Figures 23.3 and 23.12
British Powder Metals Federation (BPMF) and Powder Metal Industries Federation (MPIF)	Tables 23.10, 23.11 and 23.12
BSA Metal Powders	Table 23.14
gkm Bound Brook	Table 23.7
Hoeganaes Corporation	Table 23.6
Hganäs AB	Table 23.5
Inco Alloys International	Tables 23.17 and 23.18
Makin Metal Powders	Table 23.4
Metal Powder Industries Federation	Tables 23.2 and 23.16, and Figure 23.9
Metallnormencentralen	Table 23.3, and Figure 23.1
Metallwerk Plansee	Table 23.22 (properties of hardmetals)

24 Lubricants

24.1 Introduction

Lubricants minimise friction and wear in rubbing contacts by reducing metal–metal contact, removing wear debris, and carrying away frictional heat. They may also prevent rusting and with liquid lubricants remove heat. Lubricants may be solid, such as graphite, molybdenum disulphide, polytetrafluoroethylene and talc; or gaseous, commonly air; but the principal lubricants are liquids such as mineral oil, or the semi-solid greases formed from liquids by the use of thickening agents.

24.1.1 Main regimes of lubrication

HYDRODYNAMIC LUBRICATION (HL)

A viscous fluid film between two solid surfaces moving in very close proximity to one another becomes pressurised and holds those surfaces apart, often against considerable loads. During the process, the surface may be deformed elastically (EHL – elastohydrodynamic lubrication) or plastically.

HYDROSTATIC LUBRICATION

In this case, two solid surfaces are separated by a thick fluid film supplied from an external pressure source, e.g. by an oil pump system.

BOUNDARY LUBRICATION

When the contact of asperities on sliding couples increases as the load increases, the sliding speed decreases or the fluid viscosity decreases, the friction significantly increases and the load is mainly supported by the asperity contact. Such lubrication condition is known as boundary lubrication.

MIXED LUBRICATION

This is the situation when several lubrication modes, such as HL and boundary lubrication, coexist.

24.2 Lubrication condition, friction and wear

In hydrodynamic lubrication friction is purely viscous and is directly dependent on the area of the film, the rate of shear and the viscosity of the lubricant. Coefficients of friction are as low as 0.001–0.003 and wear is negligible. When the speed of sliding and the oil viscosity are insufficient for the viscous film to carry the load, contact occurs between asperities on the opposing surfaces. Friction then rises and wear ensues. Where viscous effects are absent or negligible, lubrication is independent of the nominal area of contact and is said to be of boundary type. Where boundary and viscous effects occur together, the conditions are said to be those of 'mixed' lubrication.

When sliding speeds are very low a 'stick-slip' or jerky motion arises due in part to the elastic response of the drive and in part to the coefficient of static friction exceeding the coefficient of dynamic friction. This undesirable effect can be suppressed by the use of special lubricants containing fatty acids, acid phosphates, or similar materials able to react with the metal surfaces to produce soft, easily-sheared layers of soap which considerably reduce the coefficient of static friction.

For moderate sliding speed and load, boundary lubrication can exist with only the oxide film being worn away and replaced at a tolerable rate of wear. Under these conditions lubricants which produce low friction do not necessarily produce low wear. For most machine bearings, boundary lubrication is inadequate, since the oxide film wears away very rapidly and direct metal to metal contact occurs. The surface asperities weld together momentarily and are broken apart again to produce wear debris largely composed of metallic particles. This is called adhesive wear, or scuffing.

Boundary lubrication is the most important mode in the chipless-forming type of metal working operation since the local pressures have to be high enough to exceed the yield strength of the metal. In such operations the surface area of the workpiece is being enlarged and new, easily-weldable areas are being created. For severe operations, therefore, solid lubricants are used which can resist high pressures and are capable of extension to cover the new areas. Examples of such lubricants are waxes, graphite, molybdenum disulphide, talc, whiting and pastes of dry soap and fats. For more severe operations such as tube and bar drawing thick layers of lubricants may be built up before the operation by, for example, phosphating, baking on a coating lime, treating with special inorganic salts, or by coating with soft metals such as lead.

24.3   Characteristics of lubricating oils

24.3.1   Viscosity

Viscosity is probably the most important property of a lubricating oil or grease. For most fluids in laminar flow there is a linear relationship between the shear stress and the rate of shear. The constant of proportionality is the viscosity or, more specifically, the dynamic viscosity to distinguish it from the kinematic viscosity which is given by the ratio of viscosity to density.

Dynamic viscosity may be defined as the shear stress necessary to move a flat surface at unit speed over a parallel surface unit distance apart when the intervening space is filled with the fluid. In SI units, dynamic viscosity is expressed in $N\,s\,m^{-2}$, but the centipoises or $nN\,s\,m^{-2}$ is at present commonly used. Direct measurements are relatively few. Usually dynamic viscosities are derived from measurement of density and kinematic viscosity, which is easily measured in calibrated glass capillary tubes using gravity flow. The units of kinematic viscosity are m^2s^{-1} but centistokes ($1 \times 10^{-6}\,m^2\,s^{-1}$) are most commonly used. Obsolescent units such as Redwood and Saybolt seconds and Engler degrees derived originally from short efflux tube instruments are still to be found, but nowadays they are derived from measurement of kinematic viscosity with which they have almost linear relationships.

With increase of temperature, the viscosity of gases rises while that of liquids falls. A moderately viscous mineral oil falls from $1\,N\,s\,m^{-2}$ at 5°C to $0.01\,N\,s\,m^{-2}$ at 100°C. This property is commonly expressed as the Kinematic Viscosity Index (KVI) measured over the range 37.8–98.9°C (100–210°F) and expressed in values usually ranging between 0 and 100. However, a more rational Dynamic Viscosity Index, which may be measured over any convenient range, has been proposed. Oils with a KVI of 35 and below are known as Low Viscosity Index (LVI) oils, those between 35 and 80 are Medium Viscosity Index (MVI) oils, those between 80 and 110 are High Viscosity Index (HVI) oils and those over about 110 are Very High Viscosity Index (VHVI) oils.

With minerals oils and oils of similar molecular weight, viscosity is independent of the rate of shear, except possibly under very high pressures. Shear rates in elastohydrodynamic lubrication (or EHL) are said to be 'Newtonian'. Oils of very high molecular weight, such as silicones, exhibit a reduction in viscosity at quite moderate rates of shear and are said to be 'non-Newtonian'. While such loss in viscosity may be only temporary, there may also be permanent loss due to mechanical breakdown of very high molecular weight polymer molecules.

The lubricant viscosity usable in practice depends on the type of machine involved. Low viscosity oils are used for high speeds and high viscosities for low speeds. Very approximately $3\,N\,s\,m^{-2}$ is the maximum viscosity at which machines can be started up, while 0.002 is the minimum for maintaining hydrodynamic lubrication under running conditions.

24.3.2   Boundary lubrication properties

This property of a lubricant is essentially its ability to produce on one or both of the rubbing surfaces a layer of adequate thickness of coherent and adherent, low shear strength material which will minimise metal-to-metal contact and reduce friction. Lubricants exemplifying this property under very mild, slow speed conditions are the fatty acids which produce soap layers on metals with active oxide surfaces, such as copper.

Under such mild conditions the property is often known as 'lubricity'. Under more severe conditions of load, speed and temperature, the property is known as 'Extreme Pressure' or 'EP'. As an example, free sulphur is added to cutting oils to prevent excessive tool wear with tough steels.

24.3.3   Chemical stability

Lubricants are usually required to have an effective life of some hundreds or thousands of hours during which its necessary properties are sensibly unaltered. They may, however, change in whole or part by thermal decomposition, oxidation or by hydrolysis. Thermal decomposition followed by polymerisation results in the formation of materials of very high molecular weight, especially insoluble coke-like substances, as well as low molecular weight materials including gases. The viscosity and flash point of the lubricant therefore generally drops. Similar materials are also formed during oxidation, but in addition highly oxygenated species including lacquers and organic acids are formed and the viscosity generally increases. This viscosity increase, the amount of insoluble material and the increase in organic acidity, are conveniently used as expressions of the degree of oxidation. Lubricants based on esters and lubricants containing esters or salts as components, e.g. as additives, are subject to hydrolysis and here again acidity, perhaps with a corrosion test for a sensitive metal such as lead, are used as criteria of stability.

Water-containing lubricants require particularly clean working conditions as contamination may lead to bacterial attack and thus to unpleasant odour, corrosion, and reduced effectiveness. Systems should be prepared and regularly cleaned by flushing with 5% solutions of caustic soda, detergent solutions or both. Biostats or biocides may also be helpful.

24.3.4   Physical properties

The thermal capacity of an oil is particularly important as in many cases the flow of oil is used to remove heat. Thermal capacity varies from around $2\,000\,\mathrm{J\,kg^{-1}}\,\mathrm{K}$ for mineral oils to around $1\,500$ for silicones and triaryl phosphate esters, which compare with $4\,200\,\mathrm{Jkg^{-1}}\,\mathrm{K}$ for water.

In general, lubricants must not evaporate rapidly at the highest temperatures of usage. At well above their maximum usable temperatures, mineral oils and similar flammable oils reach their flash points at which, in particular apparatus, the evaporation is sufficient to reach the lower explosive limit with air. High volatility is, however, occasionally desirable. In some metalworking operations, working is followed by an annealing operation and any lubricant remaining must evaporate away cleanly.

At low temperatures paraffinic mineral oils reach a lower limit of usage at the pour point. At about $-7°\mathrm{C}$ a separate wax phase comes out of solution from the rest of the oil and at about $-10°\mathrm{C}$ sufficient needle-like crystals form to block the flow.

24.4   Mineral oils

The most important type of oil, the mineral oils, are made from petroleum. They are abundant, are comparatively low in cost and are available in a wide range of viscosities. The fatty oils are inferior in all respects except boundary friction properties which, where required, can easily be provided by the incorporation of a low concentration of fatty material. The various synthetic oils are of growing importance and find effective use in extreme applications where some particular property or properties justifies their high price but, in general, their scale of use is very limited.

There are, broadly speaking, two types of mineral oil: paraffinic, having comparatively long alkyl side chains in the molecule and consequently high KVI, together with a 'wax' pour point; and naphthenic, having comparatively short side chains, low to medium VI and a 'viscosity' pour point. With both types a wide range of viscosity grades can be made. Low viscosity mineral oils are frequently known as 'spindle oils' from their early use in textile machinery while the more viscous oils are known as 'cylinder oils' from their use in steam engines. Grades between these two extremes are called 'light machine' and 'heavy machine' oils. These basic grades are produced by refining and intermediate grades are produced by blending.

The wide range of viscosities available is indicated in Figure 24.1 for paraffinic oils of 95 KVI to BS 4231, in which each grade is designated by its mid-point viscosity at $37.8°\mathrm{C}$ ($100°\mathrm{F}$) expressed in

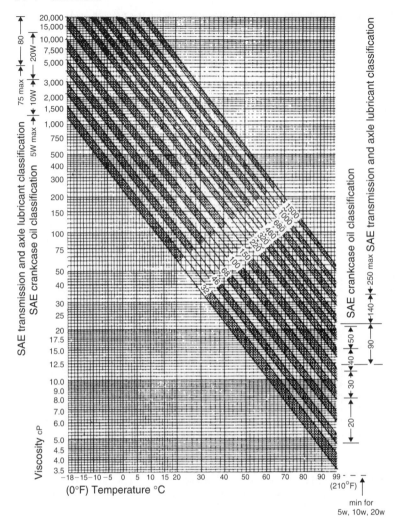

Figure 24.1 *BS 4231 ranges for 95 KVI oils with SAE limits*
NB: Such oils normally have a pour point of approximately −10°F, but the lines have been extrapolated to permit comparison with the SAE grades

centistokes (mm² s⁻¹). Figure 24.1 also indicates in its margins the important classification system of the Society of Automotive Engineers (SAE), the 'W' or winter grades of which are classified in dynamic units, the remainder in Saybolt seconds.

Oxidation stability may be significantly increased by the use of antioxidants as shown in Figure 24.2, which indicates the life obtainable for a moderate degree of deterioration of a well refined paraffinic oil.

Oxidised oils tend to be corrosive to active metals such as cadmium, zinc and, especially, lead but there is no difficulty in selecting suitable materials of construction since resistant paints and elastomers have been developed over many years.

Typical physical properties for mineral oils are given in Table 24.1 and Table 24.2. Additives are often added to the lubricating oil to improve their performance (more information on oil additives has been given in section 24.9).

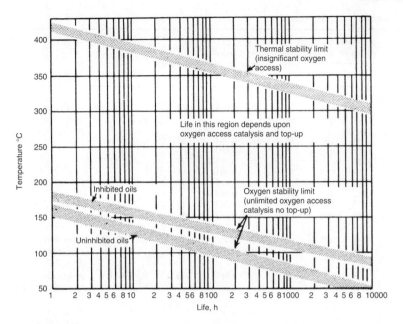

Figure 24.2 *Approximate life of well-refined mineral oils*

Table 24.1 TYPICAL PHYSICAL PROPERTIES OF HIGHLY REFINED MINERAL OILS

	Naphthenic oils			Paraffinic oils		
	Spindle	*Light machine*	*Heavy machine*	*Light machine*	*Heavy machine*	*Bright stock*
Density at 25°C	0.862	0.880	0.897	0.862	0.875	0.891
Viscosity (mN s m^{-2}) at						
30°C	18.6	45.0	171	42.0	153	810
60°C	6.3	12.0	31	13.5	34	135
100°C	2.4	3.9	7.5	4.3	9.1	27
Dynamic viscosity index	92	68	38	109	96	96
Pour point, °C	−43	−40	−29	−9	−9	−9
Pressure-viscosity coefficient (m^2 N^{-1} × 10^8) at						
30°C	2.0	2.8	2.8	2.2	2.4	3.4
60°C	1.6	2.0	2.3	1.9	2.1	2.8
100°C	1.3	1.6	1.8	1.4	1.6	2.2
Isentropic secant bulk modulus at 35 MN m^{-2} at						
30°C	—	—	—	198	206	—
60°C	—	—	—	172	177	—
100°C	—	—	—	141	149	—
Thermal capacity (J kg^{-1}°C) at						
30°C	1 880	1 860	1 850	1 960	1 910	1 880
60°C	1 990	1 960	1 910	2 020	2 010	1 990
100°C	2 120	2 100	2 080	2 170	2 150	2 120
Thermal conductivity Wm m^{-2} at						
30°C	0.132	0.130	0.128	0.133	0.131	0.128
60°C	0.131	0.128	0.126	0.131	0.129	0.126
100°C	0.127	0.125	0.123	0.127	0.126	0.123
Temperature (°C) for vapour pressure of 0.001 mm Hg	35	60	95	95	110	125
Flash point, open, °C	163	175	210	227	257	300

Table 24.2 COMMON PETROLEUM OILS

Type	Viscosity, mm²/s(= cST) 40°C	Viscosity, mm²/s(= cST) 100°C	Flush point, °C	Pour point, °C	Specific gravity at 15°C	Viscosity index	Additives*	Application
Automobile (SAE)								
10W	28	4.9	204	−28	0.878	106	R, O, D, VI, P, W, F, M	Automobiles, trucks and marine engines
20W	48	7.0	218	−24	0.884	103		
30	93	10.8	228	−20	0.890	100		
40	134	13.7	238	−16	0.895	97		
50	204	17.8	250	−10	0.901	94		
5W-30	65	10.4	210	−36	0.875	147		
10W-30	62	10.3	208	−36	0.880	155		
20W-40	138	15.3	246	−21	0.897	114		Railroad diesels
15W-40	108	15.0	218	−27	0.885	145		Diesels
Gear (SAE)								
80W-9Q	144	14.0	192	−22	0.900	93	EP, O, R, P, F	Automotive & industrial gear units
85W-140	416	27.5	210	−14	0.907	91		
Automatic transmission	38	7.0	188	−40	0.867	140	R, O, W, F, VI, P, M	Automotive hydraulic systems
Turbine								
Light	31	5.4	206	−10	0.863	107	R, O	Steam turbines, electric motors, industrial systems
Medium	64	8.7	220	−6	0.876	105		
Heavy	79	9.9	230	−6	0.879	103		
Hydraulic fluids								
Light	30	5.3	206	−24	0.868	99	R, O, W	Machine tools, hydraulic systems
Medium	43	6.5	210	−23	0.871	98		
Heavy	64	8.4	216	−22	0.875	97		
Extra low temp	14	5.1	96	−62	0.859	370	R, O, W, VI, P	Aircraft hydraulic systems
Aviation								
Grade 65	98	11.2	218	−23	0.876	100	D, P, F	Reciprocating aircraft engines
Grade 80	139	14.7	232	−23	0.887	105		
Grade 100	216	19.6	244	−18	0.898	100		
Grade 120	304	23.2	244	−18	0.893	95		

* R – rust inhibitor; O – oxidation inhibitor; D – detergent-dispersant; VI – viscosity index improver; P – pour-point depressant; W – antiwear; EP – extreme pressure; F – antifoam; M – friction modifier.
Source: Kennedy *et al.* (1998)

24.5 Emulsions

For many applications, particularly in metalworking and for the provision of fire-resistant lubricants, emulsions of water and lubricating oils are used. Because their thermal capacities are close to that of water, emulsions of 1–10% oil in water are used where heat dissipation is more important than lubrication, as for instance in high speed metal cutting, in grinding and in rolling. Emulsions are also used for fire-resistant hydraulic fluids where a low cost fluid is required. The water should be clean and free from acids with hardness preferably between 15 and 50 p.p.m. $CaCO_3$ equivalent. Very soft water may cause foaming while very hard water may reduce stability and corrosion protection. Oil/water emulsions, however, cannot be used as a direct replacement for oil in conventional hydraulic systems.

Water in oil emulsions with about 40% deionised water are used as cylinder lubricants for reciprocating compressors handling oil-soluble gases and as fire-resistant lubricants to replace oil, however, the presence of water accelerates the fatigue failure of heavily loaded rolling bearings.

Emulsions are non-Newtonian. At high rates of shear their viscosities are close to that of the base oil, which is usually a spindle oil. They are not broken down by shear. Their useful life is largely limited by emulsion stability. To limit loss of water and loss of emulsion stability the maximum continuous operating temperature is about 65°C.

24.6   Water-based lubricants

Aqueous solutions, either true or colloidal, of various substances are widely used as coolants and lubricants. The water-glycol type of fire-resistant lubricants are solutions of 50–65% polyglycols, sometimes including ethylene glycol. Two or three grades between 0.03 and 0.07 N s m^{-2} at 38°C (100°F) are usually available. Their VIs are very high and pour points very low, e.g. −40°C. They are completely shear stable but operating temperatures are usually limited to 65°C in order to avoid excessive loss of water.

Solutions of corrosion inhibitors and load carrying additives are used for high speed cutting and grinding operations.

24.7   Synthetic oils

Many synthetic lubricants have been developed which have found increased use, particularly for those applications requiring their unique properties, such as for extreme temperature and fire resistance. However, high cost limits their use. In addition, environmental problems in handling and disposing of some synthetic lubricants are also an issue.

24.7.1   Diesters

These were developed principally for aviation gas turbines because of their very high VI, and low pour point gives them a wide range of usage while the susceptibility to antioxidants allows them to work continuously at 120–150°C with bearing temperatures up to 250°C.

The load carrying capacity of the diesters in spur gear rigs is about twice as high as that of equiviscous mineral oil; under boundary conditions they appear to be equal to mineral oil while in heavily loaded ball and roller bearings the diesters appear to provide somewhat superior protection against surface fatigue.

24.7.2   Neopentyl polyol esters

This group, based on neopentyl alcohols and mixed aliphatic acids, has significantly better thermal stability and the possibility of rather higher viscosity (0.002–0.008 N s m^{-2}) than the diesters. These esters are suitable for continuous operation at 200°C with hot spots of 275°C but at these high operating temperatures it is necessary to avoid the use of cadmium, magnesium and silver because of corrosion and to use silicone and fluorinated types of elastomers.

The load-carrying properties of the neopentyl esters are better than those of the diesters, because of their higher viscosity. They apparently do not promote surface fatigue pitting in heavily loaded ball and roller bearings.

24.7.3   Triaryl phosphate esters

These esters find their greatest use as fire-resistant hydraulic fluids in diecasting machines, and like machines, as well as in the governor systems of large steam turbines. Viscosities range from 0.0036 to 0.008 N s m^{-2} at 100°C. Their oxidation stability is good, but thermal degradation is catalysed by steel. This limits their maximum operating temperature to about 120°C. The fire-resistance is on a par with that of the water-in-oil emulsions, and is generally adequate for industrial purposes.

The hydrolytic stability of these esters is rather poor and appears to be catalysed by acid impurities or similar substances developed by initial hydrolysis. If the acidity can be kept low by filtration through fullers earth the degradation can be very greatly retarded. Hydrolysis may result in some corrosion of aluminium and steel but the phosphate esters are not corrosive to cadmium, zinc or other common metals.

Triaryl phosphate esters tend to damage conventional rubber elastomers and therefore butyl, silicone or fluorinated types are preferred. Ordinary paints are also affected and those based on epoxy resins should be used.

These esters are good lubricants under both boundary and hydrodynamic (including elasto-hydrodynamic) conditions but they promote fatigue pitting of rolling element bearings and, for a life equivalent to that when using mineral oils, a 20% reduction in load may be required.

Toxicity is largely related to the amount of ortho-tolyl isomer in the oil which is accordingly kept to a low value. Particular attention should be paid to personal cleanliness and good ventilation wherever people come into contact with these oils.

24.7.4　Fluorocarbons

Usually these are polymers of trifluorovinyl chloride, the terminal groups being fluorine. The range of oils with pour points below 20°C is only 0.002–0.004 N s m^{-2}. Densities and volatilities are unusually high. They are exceptionally stable to strong oxidising agents such as fuming nitric acid, hydrogen peroxide, etc. Thermally, they are completely stable below 300°C and the degradation at high temperatures is depolymerisation so that carbonaceous deposits are not formed.

Fluorocarbons are non-corrosive to metals. Load carrying capacity under boundary conditions is rather better than that of equiviscous mineral oil, but they may have a lower protection against fatigue failure of heavily loaded rolling bearings.

24.7.5　Polyglycols

These oils are also known as polyalkylene glycols, polyoxyalkylenes, glycols and polyethers. The water soluble types are mainly polyethylene oxides and have high pour points and very high viscosity indices, while the mainly polypropylene oxides are water insoluble with low pour points and somewhat lower viscosity indices. A wide viscosity range is covered: from 0.008 to 19.5 N s m^{-2} at 38°C.

Polyglycols are very responsive to oxidation inhibitors and when inhibited are much more stable to oxidation than mineral oils. At about 250°C polyglycols exhibit rapid thermal decomposition, but as the products of decomposition are volatile they do not form deposits. Polyglycols are not corrosive to the usual metals, but since even the water-insoluble grades are slightly hygroscopic rust inhibited grades are preferred wherever moisture may enter the oil.

The water-soluble types have important uses as components of water-glycol type fire-resistant lubricants and automotive brake fluids and are very good lubricants under hydrodynamic and elasto-hydrodynamic conditions. Under boundary conditions they are not very good but may be provided with suitable properties by the addition of small amounts of long-chain fatty acids.

Typical physical properties of these synthetic lubricating oils are given in Table 24.3.

In summary, synthetic oils are used in place of mineral oils for the liquid phase where their special advantages outweigh their greater cost. Diesters are particularly useful where low volatility and good performance at low temperature are needed, e.g. in aircraft bearings. Polyglycols are used where good oxidation stability and good lubrication between steel and bronze are required, also for special cases where the liquid phase is required to evaporate at high temperature without passing through deposit-forming decomposition stages. Silicones are used where good stability is required at high temperatures without the conditions of load and speed being at all severe. Fluorocarbons are, however, preferred in spite of their very high cost where maximum resistance to oxidation is required, e.g. from contact with liquid oxygen or ozone.

24.8　Greases

24.8.1　Composition

The standard definition of a lubricating grease is 'A solid to semi-fluid product of dispersion of a thickening agent in a liquid lubricant. Other ingredients imparting special properties may be included.' The most common types of thickener are calcium and lithium metal soaps. Bentonite is used for high temperatures, above about 140°C. Esterified silica, vat dyestuffs and urea compounds are used for the most specialised applications.

Table 24.3 TYPICAL DATA ON FINISHED SYNTHETIC BASED OILS

Base oil	Organic acid esters							Silicones		
	di-(2-ethyl-hexyl) sebacate	di(iso-octyl) azelate	Mixed C₃–C₆ penta-erythritol ester	Mixed C₄–C₁₀ dipenta-erythritol ester	Triaryl phosphate ester	Fluorocarbon	Polyglycol	Poly-dimethyl (1000 cS at 25° C)	Medium phenyl	Chloro-phenyl
Density at 25°C	0.911	0.911	1.00	1.02	1.13	1.95	1.02	0.97	1.07	1.01
Viscosity, N s m⁻², at										
30°C	0.016	0.0165	0.032	0.087	0.087	0.280	0.220	0.140	0.170	0.046
60°C	0.0065	0.00677	0.012	0.025	0.0195	0.0385	0.070	0.083	0.060	0.027
100°C	0.00287	0.00301	0.005	0.083	0.0053	0.0103	0.0225	0.045	0.0225	0.0156
DVI	145	141	144	132	0	−27	164	200	175	197
Pour point, °C	−60	<−65	−60	−50	−18	−15	−25	−55	−50	<−73
Pressure–viscosity coefficient* (×10⁻⁸), m² N⁻¹, at										
30°C	1.40	—	1.8†	2.4†	3.3	4.4	1.76	1.81	2.0	2.0
60°C	1.28	1.38	1.2†	1.6†	2.0	3.7	1.43	1.81	1.9	
100°C	1.05	1.18			1.2	2.83	1.22	1.94	1.9	2.0
Thermal capacity, 1 kg°C, at										
30°C					1510	1340	1870			1490
60°C	1960	1960			1610		1970	1550		1550
100°C	2100	2100			1700		2100			1670
Thermal conductivity at										
30°C	0.154	0.151			0.127		0.150	0.162		0.150
60°C	0.149	0.148			0.127		0.148	0.159	0.146	0.145
100°C	0.142	0.144			0.126		0.146	0.155		0.140
Temperature for vapour pressure of 0.001 mmHg, °C	117	93			125	35		177	145	
Flash point, open, °C	230	243	260	300	—	none	277	316		>273
Approximate cost relative to mineral oil	5	5	10	15	5	300	5	30	60	70

* Average value over pressure range 0–5000 lbf in⁻² (0–34.5 M Pa).
† Estimated

The fatty acids of the metal soaps also influence the properties of the grease. Mixed acids from tallow, stearic acid and hydroxy stearic acid are probably the most widely used. Complex soaps formed by the co-crystallisation of two compounds permit operation at high temperatures.

A grease is usually 80–90% liquid lubricant, commonly low and medium viscosity mineral oil but high viscosity residual oils are used for high temperatures and low speeds. For special purposes, synthetic oils are used.

Additives are commonly used in greases for particular purposes as follows:

Solid lubricants	e.g. graphite, molybdenum disulphide for heavily loaded low speed applications where lubrication will be mainly of the boundary type.
Antioxidants	To prevent rapid oxidation during storage and in use.
Metal passivators	To reduce catalytic oxidation of the grease by cuprous-metals, e.g. in the cages of rolling element bearings. To prevent rusting, particularly of rolling element bearings, during storage and use.
Extreme pressure	To prevent scuffing and wear under boundary lubrication conditions, particularly those arising temporarily from shock loads.

24.8.2   Properties

The essential property of a grease is that it possesses a yield stress up to which it only deforms elastically and above which it flow plastically. When flow commences the ratio shear stress/rate of shear decreases smoothly until at shear rates in the region of 10^6 s^{-1} it closely approaches that ratio for the liquid phase of the grease, i.e. its viscosity. Above the yield stress greases are non-Newtonian liquids, and at any point the ratio shear stress/rate of shear is called its 'apparent viscosity', which is, in effect, the viscosity a Newtonian fluid would have if it exhibited the same shear stress at the same shear rate.

The significance of these properties, in relation to plain bearings is that under stationary conditions grease tends to remain in place in clearance spaces and at the ends of bearings. Thus lubricant is available immediately the machine starts up again, and grease clinging to the ends of bearings acts as a seal to exclude dirt.

Semi-fluid greases of negligible yield stress reduce leakage from gearboxes by virtue of their very high apparent viscosity at low rates of shear. They also permit feeding through long narrow bore piping, particularly at sub-zero temperatures.

The yield stress of a grease is not easily measured and for production quality control and other ordinary purposes the worked Penetration (IP 50/69, ASTM D217-C8) i.e. the depth in mm of the penetration in the grease of a special metal cone under its own weight, is used. The National Lubricating Grease Institute has classified greases according to their consistency after a specified amount of mechanical working as shown in Table 24.4.

No. 2 grade is popular since it combines satisfactory yield properties with easy pumpability, but where there are extreme vibration and shock loads a No. 3 grade is preferred. Grades more fluid than No. 0 or stiffer than No. 3 are not normally used for roller bearings. Softer grades (down to grade 000) are used for better feeding characteristics in some roller bearings and gear mechanisms.

Important properties of the thickener structure are temperature stability, resistance to water and mechanical stability. Table 24.5 lists the various types of thickeners and indicates the extent to which they have these properties.

Table 24.4   NLGI GREASE CLASSIFICATION

NLGI grade	Worked penetration range (0.1 mm)	Description	Approximate yield value, Pa
000	445–475	Very fluid	0
00	400–430	Fluid	90
0	355–385	Semi-fluid	130
1	310–340	Very soft	180
2	265–295	Soft	300
3	220-250	Semi-firm	560
4	175–205	Firm	1 300
5	130–160	Very firm	—
6	85–115	Hard	—

Source: ASTM D 217

Table 24.5 COMPARISON OF GREASE THICKNERS

Thickener type	Temperature stability	Water resistance	Mechanical stability
Calcium soap	Low	Excellent	Excellent
Sodium soap	Good	Poor	Fair
Lithium soap	Very good	Very good	Very good
Modified clay	Excellent	Good	Good

Table 24.6 TYPES OF ADDITIVES

Main types	Function and sub-types
Acid neutralisers	Neutralise contaminating strong acids formed for example by combustion of high sulphur fuels or, less often, by decomposition of active EP additives
Anti-foam	Reduce surface foam. Silicon polymers are often used to enhance separation of air bubbles from oil
Antioxidants	Reduce oxidation. Various types are: oxidation inhibitors, retarders; anti-catalyst metal deactivators, metal passivators
Antiwear agents	Reduce wear and prevent scuffing of rubbing surfaces under steady load operating conditions by the formation of low shear strength films on metallic surfaces to diminish friction and wear, usually including sulphur, phosphorous, zinc compounds or inorganic substances
Corrosion inhibitors	Type (1) Reduces corrosion of lead. Type (2) Reduces corrosion of cuprous metals
Detergents	Reduce or prevent deposits formed at high temperatures, e.g. in internal combustion engines
Dispersants	Prevents deposition of sludge by dispersing a finely divided suspension of insoluble material formed at low temperatures
Emulsifiers	Form emulsions either water in oil or oil in water according to type
Extreme pressure	Prevent scuffing of rubbing surfaces under shock load operating conditions mainly by formation of inorganic surface films
Lubricity	Reduce friction under boundary lubrication condition, increase load carrying capability especially where limited by frictional temperature rise, by formation of organic film. Examples are fatty acids and their esters
Pour point depressant	Reduces the temperature at which a mineral oil is immobilised by wax
Tackiness	Reduces loss of oil by gravity, e.g. from vertical sliding surfaces, or centrifugal forces
Viscosity index improvers	Reduce the decrease in viscosity due to increase of temperature Long-chain polymers are usually used in multi-grade oils

24.9 Oil additives

Plain mineral oils are used in many units and systems for the lubrication of bearings, gears and other mechanisms where their oxidation stability, operating temperature range, ability to prevent wear, etc. are adequate. The addition of fatty oils improves boundary lubrication properties at the expense of oxidation stability and demulsibility, but over the last 30 years oil-soluble chemical compounds called 'additives' have been developed which improve or confer a wide range of properties. The functions required of these 'additives' gives them their common names as indicated in Table 24.6.

24.9.1 Machinery lubricants

As shown in Tables 24.7 and 24.8 below, additives and oils are combined in various ways to provide the performance required. It must be emphasised, however, that indiscriminate mixing can produce undesirable interactions. Indeed some additives may be included in a blend simply to overcome problems caused by other additives.

Table 24.7   TYPES OF OIL REQUIRED FOR VARIOUS TYPES OF MACHINERY

Type of machinery	Usual base oil type	Usual additives	Special requirements
Food processing	Medicinal white oil	None	Safety in case of ingestion
Plain roll-neck bearings of rolling mills	HVI	None	Best demulsibility
Oil hydraulic	HVI down to −20°C MVIN below	Antioxidant Anti-rust Anti-wear Pour point depressant VI Improver Anti-foam	Minimum viscosity change with temperature Minimum wear of steel/ steel
Steam and gas turbines	HVI or MVIN distillates	Antioxidant Anti-rust	Ready separation from water, good oxidation stability
Steam engine cylinders	Unrefined or refined residual or high viscosity distillates	None or fatty oil	Maintenance of oil film on hot surfaces, resistance to washing away by wet steam
Air compressor cylinders	HVI or MVIN distillates	Antioxidant, anti-rust	Low deposit formation tendency
Gears (steel/steel)	HVI or MVIN	Anti-wear EP antioxidant Anti-foam Pour point depressant	Protection against wear and scuffing
Gears (steel/bronze)	HVI	Oiliness, tackiness	Maintains smooth sliding at very low speeds. Keeps film on vertical surfaces
Hermetically sealed refrigerators	MVIN	None	Good thermal stability, miscibility with refrigerant, low floc point
Diesel engines	HVI or MVIN	Detergent Dispersant Antioxidant Acid-neutraliser Anti-foam Anti-wear Corrosion inhibitor	Vary with type of engine thus affecting additive combination

Table 24.8   CHIP-FORMING METAL WORKING LUBRICANTS

Type of lubricant	Base lubricant	Additive	Remarks
Soluble oil (oil-in-water emulsion)	LVI oil	Emulsifiers Rust inhibitors	With 20–50 parts water For grinding and light cutting operation where cooling and absence of fuming important
Aqueous cutting solution	Water	Rust inhibitors	As for soluble oil
Inactive EP cutting oil	HVI	Mild EP but no free sulphur	For cutting yellow metal and other non-ferrous alloys where good lubrication without staining required
Active EP cutting oil	HVI	Mild EP and free sulphur	For heavy cuts on tough steel

24.9.2   Cutting oils

Factors entering into the selection of cutting oils are: the material of the workpiece; the speed and nature of the operation; whether cooling is more important than lubrication; and the compatibility of the cutting oil with the machine tool.

Table 24.8 gives a very general system of lubricant selection.

Table 24.9   TYPES OF LUBRICANTS FOR DRAWING, STAMPING AND PRESSING

Metal	Lubricant in order of severity of operation
Steel	Mineral oils of medium to heavy viscosities Fatty oil/mineral oil blends Soap solutions Soap/fat pastes Baked-on-lime coatings Soft metals, e.g. lead
Brass and copper	Dilute soap solutions Light mineral oil Soap/fat pastes with solid lubricants Dried on soap
Magnesium	Colloidal graphite in low volatile mineral oils Graphite in volatile solvents
Aluminium	Oil-in-water emulsions Mineral oils, viscosity increasing with severity Mineral oil with 10–15% fatty oil

Table 24.10   ROLLING OILS

Metal	Lubricant in order of severity of operation
Steel	Oil-in-water emulsions Mineral oil/fatty oil blends and with lubricity and EP additives Palm oil
Brass and copper	Oil-in-water emulsion Mineral oil Mineral oil with lubricity additive
Aluminium	Mineral oils of viscosity from 40 mN s m^{-2} at 20°C to 50 mN s m^{-2} at 40°C with lubricity additives

24.9.3   Lubricants for chipless-forming

Lubricants for chipless-forming probably present a greater range of diversity than any other branch of lubrication. Table 24.9 gives the types of lubricants used in drawing, stamping and pressing.

24.9.4   Rolling oils

Lubrication is often required in metal rolling but excessive lubricity causes 'lack of bite' or roll slippage. Lubrication affects surface finish, ease of application, removal, and uniformity of adherence to the surface. Table 24.10 gives selected lubricants used in rolling.

24.10   Solid lubricants

Many mechanical components are required to work under extreme conditions, such as high or ultra-low temperatures, high loads, high vacuum, nuclear radiation, oxidative environment, etc. Fluid lubricants may not be used under such conditions. Solid lubricants therefore play an important role in lubricant engineering.

Solid lubricants provide thin films of solid between two rubbing surfaces. In order to reduce friction and wear, the solid lubricants are required to have low shear strength. Table 24.11 lists several common solid lubricants.

Table 24.11   COMMON SOLID LUBRICANTS

| Material | Usage temperatures (°C) | | | | Average friction coefficient (f) | | Remarks |
| | Minimum | | Maximum | | | | |
	In air	In vacuum or N_2	In air	In vacuum or N_2	In air	In vacuum or N_2	
MoS_2	−240	−240	370	820	0.10–0.25	0.05–0.10	Low f with high-load carrying capability; may promote metal corrosion
Graphite	−240	unstable	540	unstable	0.10–0.30	0.02–0.45	Low f with high-load carrying capability in air; corrosion resistant, stable up to 450°C, low thermal expansion; electrically and thermally conductive
PTFE	−70	−70	290	290	0.02–0.15	0.02–0.15	Lowest f with moderate load carrying capability that decreases at elevated temperatures
Calcium fluoride-barium fluoride eutectic, Ca_2–BaF_2	430	430	820	820	0.10–0.25 above 540°C 0.25–0.40 below 540°C	0.10–0.25 above 540°C 0.25–0.40 below 540°C	Can be used at higher temperatures than other solid lubricants, high f below 540°C

Source: Kennedy *et al.* (1998)

Table 24.12   SOFT METALS USED AS SOLID LUBRICANTS

Metal	Mohs' hardness	Melting point °C	Examples of application
Gallium		30	Used in the form of coatings for bolts, dies, bearings, sliding and rolling contact in vacuum and at high temperatures. The coatings may be made using electroplating, hot dipping, and ion sputtering techniques.
Indium	1	155	
Thallium	1.2	304	
Lead	1.5	328	
Tin	1.8	232	
Gold	2.5	1 063	
Silver	2.5–3	961	

Source: Peterson *et al.* (1960)

Soft metals are also used as solid lubricants due to their low shear strength, high thermal conductivity, good adherence to metal substrates, ability to imbed abrasive debris and to act as sacrificial components in critical or expensive machinery. Table 24.12 and Table 24.13 respectively lists some soft metals and alloys often used for lubrication purpose.

The solid lubricants can be applied in different ways, such as:

a) Add solid lubricants in the form of powder to sintered work pieces and polymeric composite components;
b) Paint or spray mixtures of solid lubricants and adhesives onto the surface of work pieces;
c) Coat work pieces with solid lubricants using electroplating (e.g. soft metals and self-lubricated composites), vapour deposition, PVD, and CVD techniques;
d) Use solid lubricants as oil additives.

Table 24.13 SOFT METAL ALLOY BEARING MATERIALS

Materials	Composition	Hardness Vickers DPH	Thermal conductivity Watt/mK	Melting point °C
Al 1 100	Si & Fe 1, Cu .1, Mn .05, Zn .1, Al	30	221	649
Al 750	99	45	183	649
LEAD BABBITT 13	Sn 6.5, Cu 1, Ni 1, Al 91.5	20	24	240
LEAD BABBITT 15	Pb 83, Sb 10, Sn 6, As .25, Cu .5	20	—	281
LEAD BABBITT 7	Pb 82.5, Sb 15, Sn 1, As 1, Cu .6	22	24	240
LEAD BABBITT 8	Pb 75, Sb 15, Sn 10, As .5, Cu .5	11	24	240
TIN BABBITT 1	Pb 80, Sb 15, Sn 5, As .5, Cu .5	17	54	223
TIN BABBITT 2	Sn 91, Sb 4.5, Cu 4.5	24	52	240
TIN BABBITT 3	Sn 89, Sb 7.5, Cu 3.5	26	—	240
ZINC 12	Sn 84, Sb 8, Cu 8	93	116	404
ZINC 27	Zn 88.2, Al 11, Cu .75, Mg .02	107	125	427
	Zn 70.8, Al 27, Cu 2.2, Mg .015	—	—	—

Source: William A. Glaeser (1992)

BIBLIOGRAPHY

1. F. P. Bowden and D. Tabor, 'The Friction and Lubrication of Solids', Parts I and II, Oxford, 1954, 1964.
2. E. L. H. Bastian, 'Metalworking Lubricants', McGraw-Hill, New York, 1951.
3. 'Lubrication and Wear: Fundamentals and Application to Design', *Proc. Inst. mech. Engrs*, 1967–68, 182, Part 3A.
4. 'A Glossary of Petroleum Terms', Inst. Petroleum, London, 1961.
5. M.M. Khonsari and E.R. Booser, 'Applied Tribology – Bearing Design and Lubrication', John Wiley & Sons, New York, 2001.
6. W. A. Glaeser, 'Materials for Tribology', Elsevier, Amsterdam, 1992.
7. E.E. Kennedy, E.R. Booser, and D.F. Wilcock, 'Tribology, Lubrication, and Bearing Design', The CRC Handbook of Mechanical Engineering, CRC Press, Boca Raton, FL, 1998.
8. M.B. Peterson, S.F. Murray, and J.J. Florek, DTLE Transactions, 1960, 3, 225.

25 Friction and wear

The friction and wear characteristics of materials are not intrinsic properties but, rather, depend on a large number of variables including the physical, chemical and mechanical properties of the material and surfaces and the environment.

25.1 Friction

25.1.1 Friction of unlubricated surfaces

DEFINITION

Friction is the resistance to motion when two bodies in contact slide on one another. The frictional force F is the force required to initiate or maintain motion. If W is the normal reaction of one body on the other, the coefficient of friction μ is defined as $\mu = F/W$.

STATIC AND KINETIC FRICTION

If the force to initiate motion of one of the bodies is F_s and the force to maintain its motion at a given speed is F_k, there is a corresponding coefficient of static friction $\mu_s = F_s/W$ and a coefficient of kinetic friction $\mu_k = F_k/W$. In some cases these coefficients are approximately equal; in most cases $\mu_s > \mu_k$ and there is a tendency for intermittent or 'stick-slip' motion to occur.

BASIC LAWS OF FRICTION

The two classic laws of friction, which are valid over a wide range of experimental conditions, state that:

1. The frictional force F between solid bodies is proportional to the normal force between the surfaces, i.e. μ is independent of W.
2. The frictional force F is independent of the apparent area of contact.

Friction generally comes from two main sources: adhesion and deformation including ploughing and asperity deformation.[1] The coefficient of friction may be expressed as

$$\mu = \mu_a + \mu_d$$

where μ_a and μ_d are components corresponding to the adhesion and deformation effects, respectively. Higher adhesion between the two surfaces in contact would result in larger friction. Rough surfaces could favour asperity interlocking, thus resulting in increased friction. However, if surfaces are sufficiently smooth, the adhesive force could be large due to lack of air between the surfaces in contact, leading to consequently large friction.

25.1.2 Friction of unlubricated materials

When clean metal surfaces are placed in contact they do not touch over the whole of their apparent area of contact. The load is supported by surface irregularities (asperities) which deform plastically as the load is applied. The area of real contact is approximately proportional to the load and inversely proportional to the hardness of the surfaces.[1]

For very clean surfaces strong adhesion occurs at regions of real contact, a part of which may be due to cold-welding, and these junctions must be sheared if sliding is to take place. Thus, it is almost impossible to slide such surfaces in a vacuum and complete seizure often occurs as shown in Table 25.1. However, if the surfaces are contaminated the adhesion is much weaker because the formation of strong junctions is inhibited. For example, hydrogen or nitrogen atmospheres have little effect on in-vacuo coefficients of friction, but the smallest trace of oxygen or water vapour produces a profound reduction in friction (Table 25.1). A further reduction in the coefficient of friction often occurs at high sliding speeds, particularly at speeds sufficient to produce local hot-spots and surface melting,[4] e.g. ice at $0.1\,\mathrm{m\,s^{-1}}$ for which μ may be less than 0.1.

Friction values for metal couples in air are influenced by a number of factors. Principal ones are the tendency for formation of oxide films, the degree of deformation in sliding, the ability of oxide films to survive sliding contact and the tendency for transfer of material from one surface to the other. Table 25.2 shows the relative hardnesses of some common metals and their oxides and the load (for a spherical slider on polished surfaces) at which appreciable metallic contact occurs. For instance, the oxide on copper is not easily penetrated, whereas the very hard aluminium oxide on the soft aluminium substrate is readily shattered during sliding. Thick oxide films, such as produced by anodising aluminium, may be more protective because sliding deformation can be restricted entirely to the oxide. Similarly, with very hard metal substrates, such as chromium, the surface deformation may be so small that the oxide is never ruptured.

Coefficients of friction of a number of metals and alloys are shown in Table 25.3. It should be indicated that pick-up may occur on the surfaces in contact such that the sliding couple becomes the metal on itself. Harder materials generally have lower friction coefficients due to less mechanical interlocking, greater support of the surface oxide, and the effect of second phase such as carbides and graphite in reducing adhesion of junctions.

Very hard solids often have low coefficients when sliding on themselves or other materials because of the reduced asperity interlocking and limited surface deformation that occurs during sliding (Table 25.4).

Very low coefficients of friction may be obtained by plating hard metal substrates with thin soft metal films (Table 25.5). The substrate supports the load while sliding occurs within the soft film which has low shear strength. Typical film thicknesses are 1 to $10\,\mu\mathrm{m}$. The friction of materials

Table 25.1   STATIC FRICTION OF METALS (SPECTROSCOPICALLY PURE) IN VACUUM (OUTGASSED) AND IN AIR (UNLUBRICATED)

	Metals											
Conditions	Ag	Al	Co	Cr	Cu	Fe	In	Mg	Mo	Ni	Pb	Pt
μ_s metal on itself in vacuo	S	S	0.6	1.5	S	1.5	S	0.8	1.1	2.4	S	4
μ_s metal on itself in air	1.4	1.3	0.3	0.4	1.3	1.0	2	0.5	0.9	0.7	1.5	1.3

S: signifies gross seizure ($\mu = 10$).

Table 25.2   BREAKDOWN OF OXIDE FILMS PRODUCED DURING SLIDING

	Vickers hardness $(\mathrm{kg\,mm^{-1}})$		*Load* (g) *at which appreciable metallic contact occurs*
Metal	*Metal*	*Oxide*	
Gold	20	—	0
Silver	26	—	0.003
Tin	5	1 650	0.02
Aluminium	15	1 800	0.2
Zinc	35	200	0.5
Copper	40	130	1
Iron	120	150	10
Chromium plate	800	—	Never

Reference: 2.

Table 25.3   FRICTION COEFFICIENT DATA FOR METALS SLIDING ON METALS[3–10]

Fixed specimen	Moving specimen	μ_s	μ_k
Ag	Ag	0.50	—
	Au	0.50	—
	Cu	0.48	—
	Fe	0.49	—
Al	Al	0.57	—
	Ti	0.54	—
Al alloy 6061-T6	Al, alloy 6061-T6	0.42	0.34
	Cu	0.28	0.23
Au	Au	0.49	—
Brass 60Cu–40Zn	Tool steel	—	0.24
Cd	Cd	0.79	—
	Fe	0.52	—
Co	Co	0.56	—
	Cr	0.41	—
Cr	Co	0.41	—
	Cr	0.46	—
Cu	Co	0.44	—
	Cr	0.46	—
	Cu	0.55	—
	Fe	0.50	—
	Ni	0.49	—
	Zn	0.56	—
	4619 steel	—	0.82
Fe	Co	0.41	—
	Cr	0.48	—
	Fe	0.51	—
	Mg	0.51	—
	Mo	0.46	—
	Ti	0.49	—
	W	0.47	—
	Zn	0.55	—
In	In	1.46	—
Mg	Mg	0.69	—
Mo	Fe	0.46	—
	Mo	0.44	—
Nb	Nb	0.46	—
Ni	Cr	0.59	—
	Ni	0.50	—
	Pt	0.64	—
Pb	Ag	0.73	—
	Au	0.61	—
	Co	0.55	—
	Cr	0.53	—
	Fe	0.54	—
	Pb	0.90	—
	Steel	—	0.80
Pt	Ni	0.64	—
	Pt	0.55	—
Sn	Fe	0.55	—
	Sn	0.74	—
Steel	Cu	—	0.80
	Pb	—	1.40
1020 Steel	4619 steel	—	0.54
1032 Steel	Al alloy 6061-T6	0.47	0.38
	Cu	0.32	0.25
	1032 steel	0.31	0.23
	Ti-6Al-4V	0.36	0.32

(*continued*)

Table 25.3   FRICTION COEFFICIENT DATA FOR METALS SLIDING ON METALS[3–10]—*continued*

Fixed specimen	Moving specimen	μ_s	μ_k
52100 steel	Ni$_3$Al alloy IC-396M	—	1.08
	Ni$_3$Al alloy IC-50	—	0.70
	1015 steel, annealed	—	0.74
	DP-80 dual phase steel	—	0.55
	O$_2$ tool steel	—	0.49
Mild steel	Mild steel	—	0.62
M50 tool steel	Ni$_3$Al alloy IC-50	—	0.68
Stainless steel	Tool steel	—	0.53
304 stainless steel	Tool steel	0.23	0.21
Stellite	Tool steel	—	0.60
Ti	Al	0.54	—
	17-4PH stainless steel	0.48	0.48
	Ti	0.47	0.40
	Ti–6Al–4V	0.43	0.36
Ti–6Al–4V	Al alloy 6061-T6	0.41	0.38
	Cu–Al (bronze)	0.36	0.27
	Nitronic 60	0.38	0.31
	17-4PH stainless steel	0.36	0.31
	440C stainless steel	0.44	0.37
	Stellite 12	0.35	0.29
	Stellite 6	0.45	0.36
	Ta	0.53	0.53
	Ti–6Al–4V	0.36	0.30
W	Cu	0.41	—
	Fe	0.47	—
	W	0.51	—
Zn	Cu	0.56	—
	Fe	0.55	—
	Zn	0.75	—
Zr	Zr	0.63	—

Table 25.4   FRICTION OF VERY HARD SOLIDS

Bonded tungsten carbide (cobalt binder) slider

Material	μ_s
Tungsten carbide	0.2
Aluminium oxide	0.25
Copper	0.4
Cadmium	0.8–1.0
Iron	0.4–0.8
Cobalt	0.3

Table 25.5   FRICTION OF THIN METALLIC FILMS[11]

(Sliding on a 6 mm diameter steel sphere)

	Coefficient of static friction μ_s			
Load g	Indium film on steel	Indium film on silver	Lead film on copper	Copper film on steel
4 000	0.08	0.1	0.18	0.3
8 000	0.04	0.07	0.12	0.2

Table 25.6   FRICTION OF MATERIALS SLIDING ON THEMSELVES AT LOW AND HIGH TEMPERATURES[12,13]

| | *Coefficient of friction μ* | | | | | |
| | *Low temperatures in gaseous medium* | | | *High temperatures in air* | | |
Material	4K	77K	295K	315°C	650°C	980°C
Aluminium	1.52	1.45	1.49	—	—	—
Austenitic stainless steel	0.26	0.35	0.99	—	—	—
Carbon–graphite	—	—	—	0.18	—	—
Copper	0.81	0.78	0.70	—	—	—
Iron	0.97	0.84	0.75	—	—	—
Nickel	1.06	1.10	1.12	—	—	—
Silicon nitride	—	—	—	0.60	0.28	0.48
Tool steel (15 Mo 15 Co)	—	—	—	—	0.30	0.26
Zinc	0.43	0.39	0.52	—	—	—

changes with temperature largely due to changes in adhesion, mechanical behaviour, phase transformations, and the formation of oxide film. If there are not considerable changes in these factors with temperature, change in the friction of materials may not be significant (Table 25.6).

It needs to be pointed out that large fluctuations in μ may exist, due to the fact that friction is influenced by many factors such as roughness, temperature and surface deformation, which may significantly vary during sliding. However, μ values given in the tables reflect general friction behaviour of the sliding couples and can therefore be used as references.

Compared to that of metals, the friction of ceramic materials is relatively low. Ceramic materials generally have very limited slip systems and low adhesion due to the nature of their ionic or covalent bonds. These factors make ceramic materials never have very high coefficients of friction (Table 25.7). The friction coefficient of ceramic materials may be similar to the values of metallic materials with oxide films sliding in air. It should be indicated that the surfaces of many ceramics are susceptible to tribochemical reaction. Atmospheric composition and temperature may lead to variations in the friction coefficient of ceramics (Table 25.8).

The friction of many materials is little affected by high or low temperatures (see Table 25.7). Exceptions are when the plastic flow pressure changes significantly or when oxide films become very much thicker.

The friction behaviour of polymers differs from that of metals in three respects. First, the coefficient of friction tends to decrease with increasing load; it also tends to decrease if the geometric contact area is decreased. Second, if the surfaces are left in contact under load the area of true contact may increase with time because of creep and the starting friction may be correspondingly larger. Thirdly, the friction may show changes with speed which reflect the visco-elastic properties of the polymer but the most marked changes occur as a result of frictional heating. Even at speeds of only a few m s^{-1} the friction of unlubricated polymers can rise to very high values. On the other hand at extremely high speeds the friction may fall again because of the formation of a molten lubricating film. Table 25.9 lists friction coefficients of some polymeric sliding on various materials.

The main effect of speed of sliding is an increase in local temperature caused by frictional heating at the regions of real contact. Local hot-spots may produce phase changes or alloy formation at or near the sliding interface, they may greatly change the rate of surface oxidation and even produce local melting. At speeds of a few m/s these effects are not as marked as those at very high speeds (see Table 25.10) but may still be significant. At very high sliding speeds the friction generally falls off because of the formation of a very thin molten surface layer, which acts as a lubricant film.[2] Although this is, broadly speaking, the main trend, other factors may considerably change the behaviour. For example, with steel sliding on diamond the friction first diminishes and then increases, because at higher speeds the steel is transferred to the diamond so that the sliding resembles that of steel on steel. In some cases the metals may fragment at very high speeds, particularly if they are of limited ductility. Again, if appreciable melting occurs the friction may rise at high speeds because of the viscous resistance of the liquid interface: this occurs with bismuth.

Table 25.7 FRICTION COEFFICIENT DATA FOR CERAMICS SLIDING
ON VARIOUS MATERIALS[4,14–17]

Material		Friction coefficient	
Fixed specimen	*Moving specimen*	*Static*	*Kinetic*
Ag	Alumina	—	0.37
	Zirconia	—	0.39
Al	Alumina	—	0.75
	Zirconia	—	0.63
Alumina	Alumina	—	0.50
	WRA[a]	—	0.53
	WRZTA[b]	—	0.50
	ZTA[c]	—	0.56
Boron carbide	Boron carbide	—	0.53
Cr	Alumina	—	0.50
	Zirconia	—	0.61
Cu	Alumina	—	0.43
	Zirconia	—	0.40
Fe	Alumina	—	0.45
	Zirconia	—	0.35
Glass, tempered	Al alloy 6106-T6	0.17	0.14
	1032 Steel	0.13	0.12
	Teflon[d]	0.10	0.10
Silicon carbide	Silicon carbide	—	0.52
	Silicon nitride	—	0.53
Silicon nitride	Silicon carbide	—	0.54
	Silicon nitride	—	0.17
Steel, M50 tool	Boron carbide	—	0.29
	Silicon carbide	—	0.29
	Silicon nitride	—	0.15
	Tungsten carbide	—	0.19
Ti	Alumina	—	0.42
	Zirconia	—	0.27
Tungsten carbide	Tungsten carbide	—	0.34

[a] WRA, silicon carbide whisker-reinforced alumina.
[b] WRZTA, silicon carbide whisker-reinforced, zirconia-toughened alumina.
[c] ZTA, zirconia-toughened alumina.
[d] Teflon, polytetrafluoroethylene.

Table 25.8 HARD SOLIDS SLIDING ON THEMSELVES[4,18]

	Coefficient of friction μ_s		
		Outgassed and measured in vacuo	
Material	*In air at 20°C*	*20–1 000°C*	*Comments*
Aluminium oxide	0.2	—	
Boron carbide	0.2	0.9	Rises rapidly above 1 800°C
Silicon carbide	0.2	0.6	
Silicon nitride	0.2	—	
Titanium carbide	0.15	1.0	Rises rapidly above 1 200°C
Titanium monoxide	0.2	0.6	
Titanium sesquioxide	0.3	0.7	
Tungsten carbide	0.15	0.6	Rises rapidly above 1 000°C

Table 25.9   FRICTION COEFFICIENTS OF SELECTED POLYMERS SLIDING ON VARIOUS MATERIALS[4,7,18–22]

Material (see below)		Friction coefficient	
Fixed specimen	*Moving specimen*	*Static*	*Kinetic*
Polymers sliding on polymers			
Acetal	Acetal	0.06	0.07
Nylon 6/6	Nylon 6/6	0.06	0.07
PMMA	PMMA	0.80	—
Polyester PBT	Polyester PBT	0.17	0.24
Polystyrene	Polystyrene	0.50	—
Polysthyrene	Polysthyrene	0.20	—
Teflon	Teflon	0.08	0.07
Dissimilar pairs with the polymer as the moving specimen			
Nylon 6 (cast)	Steel, mild	—	0.35
Nylon 6 (extruded)	Steel, mild	—	0.37
Nylon 6/6	Polycarbonate	0.25	0.04
Nylon 6/6 (+PTFE)	Steel, mild	—	0.35
PA66	Steel, 52100	—	0.57
PA66 (+15% PTFE)	Steel, 52100	—	0.13
PA66 (PTFE/glass)	Steel, 52100	—	0.31
PEEK	Steel, 52100	—	0.49
PEEK (+15% PTFE)	Steel, 52100	—	0.18
PEEK (PTFE/glass)	Steel, 52100	—	0.20
PEI	Steel, 52100	—	0.43
PEI (+15% PTFE)	Steel, 52100	—	0.21
PEI (PTFE/glass)	Steel, 52100	—	0.21
PETP	Steel, 52100	—	0.68
PETP (+15% PTFE)	Steel, 52100	—	0.14
PETP (PTFE/glass)	Steel, 52100	—	0.18
Polyurethane	Steel, mild	—	0.51
POM	Steel, 52100	—	0.45
POM (+15% PTFE)	Steel, 52100	—	0.21
POM (PTFE/glass)	Steel, 52100	—	0.23
PPS	Steel, 52100	—	0.70
PPS (+15% PTFE)	Steel, 52100	—	0.30
PPS (PTFE/glass)	Steel, 52100	—	0.39
Teflon	Al, alloy 6061-T6	0.24	0.19
Teflon	Cr plate	0.09	0.08
Teflon	Cu	0.13	0.11
Teflon	Ni (0.001P)	0.15	0.12
Teflon	Steel, 1 032	0.27	0.27
Teflon	Ti–6Al–4V	0.17	0.14
Teflon	TiN (Magnagold)	0.15	0.12
UHMWPE	Steel, mild	—	0.14
Dissimilar pairs with the polymer as the moving specimen			
Steel, carbon	ABS resin	0.40	0.27
Steel, mild	ABS	0.30	0.35
Steel, mild	ABS + 15% PTFE	0.13	0.16
Steel, mild	Acetal	0.14	0.21
Steel, 52100	Acetal	—	0.31
Steel, 52100	HDPE	—	0.25
Steel, carbon	HDPE	0.36	0.23
Steel, carbon	LDPE	0.48	0.28
Steel, 52100	Lexan 101	—	0.60
Steel, mild	Nylon (amorphous)	0.23	0.32
Steel, carbon	Nylon 6	0.54	0.37
Steel, mild	Nylon 6	0.22	0.26
Steel, carbon	Nylon 6/6	0.53	0.38
Steel, mild	Nylon 6/6	0.20	0.28
Steel, carbon	Nylon 6/10	0.53	0.38
Steel, mild	Nylon 6/10	0.23	0.31
Steel, mild	Nylon 6/2	0.24	0.31
Steel, mild	PEEK 6/2	0.20	0.25

(*continued*)

Table 25.9   FRICTION COEFFICIENTS OF SELECTED POLYMERS SLIDING ON VARIOUS MATERIALS[4,7,18–22]—*continued*

Material (*see below*)		Friction coefficient	
Fixed specimen	*Moving specimen*	*Static*	*Kinetic*
Steel, carbon	Phenol formaldehyde	0.51	0.44
Steel, 52100	PMMA	—	0.68
Steel, carbon	PMMA	0.64	0.50
Steel, mild	Polycarbonate	0.31	0.38
Steel, mild	Polyester PBT	0.19	0.25
Steel, mild	Polyethylene	0.09	0.13
Steel, carbon	Polyimide	0.46	0.34
Steel, carbon	Polypropylene	0.30	0.17
Steel, carbon	Polypropylene	0.36	0.26
Steel, mild	Polypropylene	0.08	0.11
Steel, carbon	Polystyrene	0.43	0.37
Steel, mild	Polystyrene	0.28	0.32
Steel, mild	Polysulphone	0.29	0.37
Steel, carbon	PVC	0.53	0.38
Steel, carbon	PTFE	0.37	0.09
Al, alloy 6061-T6	Teflon	0.19	0.18
Cr plate	Teflon	0.21	0.19
Glass, tempered	Teflon	0.10	0.10
Ni (0.002P)	Teflon	0.22	0.19
Steel, 1032	Teflon	0.18	0.16
Ti–6Al–4V	Teflon	0.23	0.21
TiN (Magnagold)	Teflon	0.16	0.11

ABS: acrylonitrile butadiene styrene
HDPE: high-density polyethylene
LPDE: low-density polyethylene
Lexan: trademark of the General Electric Co. (polycarbonate)
PA: polyamide
PBT: polybutylene terephthalate
PEI: polyetherimide
PETP: polyethylene terephthalate
PMMA: polymethylmethacrylate
POM: polyoxymethylene
PPS: polyphenylene sulphide
PTFE: polytetrafluoroethylene
PVC: polyvinyl chloride
UHMPE: ultra high molecular weight polyethylene
Magnagold: product of General Magnaplate, Inc.
Teflon: trademark of E.I. Du Pont de Nemours & Co. Inc. (PTFE)

Table 25.10   KINETIC FRICTION OF UNLUBRICATED METALS AT VERY HIGH SLIDING SPEEDS (UP TO 600 m s^{-1}) SLIDING ON A SPHERE OF BALL-BEARING STEEL

Surface	*Duration of expt.* s	*Coefficient of friction* μ_k			
		9 m s^{-1}	45 m s^{-1}	225 m s^{-1}	450 m s^{-1}
Bismuth	1–10	0.25	0.1	0.05	—
Lead	1–10	0.8	0.6	0.2	0.12
Cadmium	1–10	0.3	0.25	0.15	0.1
Copper	1–10	>1.5	1.5	0.7	0.25
Molybdenum	1–10	1	0.8	0.3	0.2
Tungsten	1–10	0.5	0.4	0.2	0.2
Diamond	1–10	0.06	0.05	0.1	≈0.1

Reference: 2.

25.1.3   Friction of lubricated surfaces

DEFINITIONS

When moving surfaces are separated by a relatively thick film of lubricant the resistance to motion is entirely due to the viscosity of the interposed layer. The friction is extremely low ($\mu = 0.001 \sim .0001$) and there is no wear of the solid surfaces. These are the conditions of hydrodynamic lubrication under which bearings operate in the ideal case. If the pressures are sufficiently high or the sliding speeds sufficiently low the hydrodynamic film becomes so thin that it may be less than the height of the surface irregularities. The asperities then rub on one another and are separated by films only one or two molecular layers thick. The friction under these conditions ($\mu \approx 0.05$ to 0.15) is much higher than for ideal hydrodynamic lubrication and some wear of the surfaces occurs. This type of lubricated sliding is called 'boundary' lubrication.[24] Under boundary conditions as for unlubricated surfaces the frictional resistance is proportional to the load and independent of the size of the surfaces.

In certain circumstances a further type of lubrication, known as elastohydrodynamic lubrication, may obtain. It arises in the following way.[25–28] Under conditions of severe loading the moving surfaces may undergo appreciable elastic deformation: this not only changes the geometry of the surfaces, it also implies that very high pressures are exerted on the oil film. The main effect of this is to produce a prodigious increase in the viscosity of the oil. For example at contact pressures of 30, 60, 100 kg mm^{-2} (such as may occur between gear teeth of hardened steel) the viscosity of a simple mineral oil is increased by 200, 40 000 and 1 000 000 fold respectively. Thus the harder the surfaces are pressed together the more difficult it is to extrude the lubricant. Consequently effective lubrication may obtain under conditions where it would normally be expected to break down.

In general, elastohydrodynamic lubrication becomes effective when the oil film thickness is of the order of 10^{-1}–1 µm. This is very much thicker than the boundary film (1–10 nm) but it is very small in engineering terms. Consequently for practical exploitation of elastohydrodynamic lubrication the surfaces must be very smooth and carefully aligned.

25.1.4   Boundary lubrication

Boundary lubricants function by interposing between the sliding surfaces a thin film which can reduce metallic interaction and which is, in itself, easily sheared. The latter criterion restricts boundary lubricants almost exclusively to long chain organic compounds, e.g. paraffins, alcohols, esters, fatty acids and waxes. Radioactive tracer experiments show that while a good boundary lubricant may reduce the friction by a factor of about 20 (from $\mu \approx 1$ to $\mu \approx 0.05$) it may reduce the metallic transfer by a factor of 20 000 or more. Under these conditions the metallic junctions contribute very little to the frictional resistance: the friction is due almost entirely to the force required to shear the lubricant film itself. For this reason two good boundary lubricants may give indistinguishable coefficients of friction, but one may easily give 20 times as much metallic transfer (i.e. wear) as the other. Thus with good boundary lubricants the friction may be an inadequate indication of the effectiveness of the lubricant.

Most boundary lubricants are used as additives, dissolved as a few per cent in a mineral oil: they provide lubrication by adsorbing from solution on to the surfaces. As the temperature is raised the film may dissolve into the superincumbent fluid and lubrication may become ineffective at temperatures appreciably below the melting point of the film itself. The breakdown temperature depends on solubility and concentration, as well as on speed, load, and surface roughness.[29]

With more protracted heating, oxidation of the lubricant occurs and the behaviour is now determined by the properties of the oxidation products themselves. In the early stages these may be beneficial but later they lead to polymerisation, gumming and the formation of other deleterious products. The effect of lubrication on friction of some metallic materials is demonstrated in Tables 25.11–25.14.

25.1.5   Extreme pressure (EP) lubricants

Even the best boundary lubricants (e.g. long-chain acids or soaps) cease to provide any lubrication above about 200°C. Since localised hot-spots of very much higher temperature are often reached in running mechanisms it is necessary to use surface films that have a high melting point and which, as far as possible, possess a low shear strength. One obvious method is to coat the metal with a thin film of a softer metal. These films are effective up to their melting point but are gradually worn away with repeated sliding. Other materials which are very effective are listed in Table 25.15.

Table 25.11   LUBRICATION OF STEEL* SURFACES BY PARAFFINS, ALCOHOLS AND FATTY ACIDS. STATIC FRICTION

Lubricant	Length of chain	μ_s	Lubricant	Length of chain	μ_s
(a) Paraffins			(c) Fatty acids		
Decane	C_{10}	0.23	Valeric	C_5	0.17
Cetane	C_{16}	0.16	Capric	C_{10}	0.11
Triacontane	C_{30}	0.11	Lauric	C_{12}	0.11
(b) Alcohols			Palmitic	C_{16}	0.1
Octyl	C_8	0.23	Stearic	C_{18}	0.1
Decyl	C_{10}	0.16			
Cetyl	C_{16}	0.1			

*C 0.13, Ni 3.42.

Table 25.12   STATIC FRICTION OF VARIOUS METALS (SPECTROSCOPICALLY PURE) LUBRICATED WITH 1% SOLUTION OF LAURIC ACID (M.P. 44°C) IN PARAFFIN OIL AT ROOM TEMPERATURE

Metal	Coefficient of friction μ_s	
	Unlubricated	Lubricated
Aluminium	1.3	0.3
Cadmium	0.5	0.05
Chromium	0.4	0.34
Copper	1.4	0.10
Iron	1.0	0.15
Magnesium	0.5	0.10
Nickel	0.7	0.3
Platinum	1.3	0.25
Silver	1.4	0.55

Reference: 2.

Table 25.13   LUBRICATION OF STEEL SURFACES BY VARIOUS LUBRICANTS. STATIC FRICTION

Lubricant	Static friction μ_s		Lubricant	Static friction μ_s	
	20°C	100°C		20°C	100°C
None	0.58	—	Mineral oils		
Vegetable oils			Light machine	0.16	0.19
Castor	0.095	0.105	Thick gear	0.125	0.15
Rape	0.105	0.105	Solvent refined	0.15	0.2
Olive	0.105	0.105	Heavy motor	0.195	0.205
Coconut	0.08	0.08	BP paraffin	0.18	0.22
			Extreme pressure	$0.09 \sim 0.1$	$0.09 \sim 0.1$
Animal oils			Graphited oil	0.13	0.15
Sperm	0.10	0.10	Oleic acid	0.08	0.08
Pale whale	0.095	0.095	Trichlorethylene	0.33	—
Neatsfoot	0.095	0.095	Alcohol	0.43	—
Lard	0.085	0.085	Benzene	0.48	—
			Glycerine	0.2	0.25

Another approach is to form a protective film *in situ* by chemical attack, a small quantity of a suitable reactive compound being added to the lubricating oil. The most common materials are additives containing sulphur or chlorine or both. Phosphates are also used. The additive must not be too reactive, otherwise excessive corrosion will occur. The results in Table 25.16 are based on

Table 25.14 LUBRICATION OF METALS ON STEEL. STATIC FRICTION

Bearing surface	Rape oil μ_s	Castor oil μ_s	Mineral oil μ_s	Long chain fatty acids μ_s
Hard steel (axle steel)	0.14	0.12	0.16	0.09
Cast iron	0.10	0.13	0.21	—
Gun metal	0.15	0.16	0.21	—
Bronze	0.12	0.12	0.16	—
Pure lead	—	—	0.5	0.22
Lead-base white metal (Sb 15, Cu 0.5, Sn 6, Pb 78.5)	—	—	0.1	0.08
Pure tin	—	—	0.6	0.21
Tin-base white metal (Sb 6.5, Cu 4.2, Ni 0.1, Sn 89.2)	—	—	0.11	0.07
Sintered bronze	—	—	0.13	—
Brass (Cu 70, Zn 30)	—	0.11	0.19	0.13

Table 25.15 FRICTION OF METALS LUBRICATED WITH CERTAIN PROTECTIVE FILMS

Protective film	Coefficient of friction μ_s	Temperature up to which lubrication is effective
PTFE (Teflon)	0.05	~320°C
Graphite	0.07–0.13	~600°C
Molybdenum disulphide	0.07–0.1	~800°C

Table 25.16 EFFECT OF SULPHIDE AND CHLORIDE FILMS ON FRICTION OF METALS

Metal	Coefficient of friction (μ_s)				
		Sulphide films		Chloride films	
	Clean	Dry	Covered with lubricating oil	Dry	Covered with lubricating oil
Cadmium on cadmium	0.5	—	—	0.3	0.15
Copper on copper	1.4	0.3	0.2	0.3	0.25
Silver on silver	1.4	0.4	0.2	—	—
Steel on steel (0.13 C, 3.42 Ni)	0.8	0.2	0.05	0.15	0.05

Reference: 11.

Table 25.17 KINETIC FRICTION, INITIAL SEIZURE LOADS AND WELD LOADS OF BALL BEARING STEEL SURFACES LUBRICATED WITH TYPICAL EP ADDITIVES.[16] FOUR BALL MACHINE. FRICTION MEASUREMENTS AT 10 kg LOAD

Lubricant		Coefficient of friction μ_s	Initial seizure load kg	Weld load kg
Base oil	Additive			
Mineral oil	None	0.09	~45	~120
Mineral oil	Zinc di-secbutyl thio-phosphate (10% wt)	0.09	80	230
Mineral oil	Sulphurised sperm oil (5% wt)	0.095	80	250
	Sulphur	—	65	340
	Chlorinated additive (1% wt)	0.085	85	310
Mineral oil	Tributyl phosphate (1% wt)	—	80	150
Paraffin oil	Tributyl phosphate (1% wt)	—	40	125
Mineral oil	Tricresyl phosphate (1% wt)	—	75	140
Paraffin oil	Tricresyl phosphate (1% wt)	—	40	110

laboratory experiments in which metal surfaces were exposed to H_2S or HCl vapour and the frictional properties of the surface examined. The results show that the films formed by H_2S give a higher friction than those formed by HCl: however in the latter case the films decompose in the presence of water to liberate HCl and for this reason chlorine additives are less commonly used than sulphur additives.

The detailed behaviours of commercial additives depend not only on the reactivity of the metal and the chemical nature of the additive but also on the type of carrier fluid used (e.g. aromatic, naphthenic, paraffinic). Further, the chemical reactions which occur are far more complicated than originally supposed.

The differences in friction are not very marked showing that the friction is a very poor criterion of the effectiveness of an EP lubricant. Marked differences in seizure-preventing properties are often accompanied by almost indistinguishable coefficients of friction. The last four lines of the table also show that EP effectiveness depends to some extent on the nature of the base oil.

25.2   Wear

DEFINITIONS

Wear is the progressive loss of substance from the operating surface of a body occurring as a result of relative motion at the surface. Wear is usually detrimental, but in mild form may be beneficial, e.g. during the running-in of engineering surfaces. The major types of wear are abrasive wear, adhesive wear, erosive wear, fretting and corrosive wear. Abrasive wear is wear by displacement of material caused by hard protuberances or particles. Adhesive wear is, strictly, wear by transference of material from one surface to another due to the process of solid-phase welding. Erosive wear is loss of material from a solid surface due to relative motion in contact with a fluid which contains solid particles impingement by a flow of sand, or collapsing vapour bubbles. Fretting is a wear phenomenon occurring between two surfaces having oscillatory motion of small amplitude and is used, frequently, to include fretting corrosion, in which a chemical reaction predominates. Corrosive wear results from synergistic attack of wear and corrosion. The wear rate could be much higher than that caused by either wear or corrosion only.

25.2.1   Abrasive wear

Abrasive wear is caused by asperities of surfaces in contact that move in opposite directions (two-body abrasion) or by particles existing between two surfaces moving in opposite directions (3-body abrasion). Abrasion rates vary considerably for abrasives of different hardness, size, and shape. Wear rates increase approximately linearly with increase in applied load per unit area, up to loads at which extensive failure of the abrasive occurs. Figure 25.1 shows the major effect of relative hardness of the worn surface and abrasive on volume wear rate. Relative wear rates in practice may vary over a wide range, as shown in Tables 25.18–25.22. The wear resistance generally increases as the material hardness increases, except when materials are hardened by prior plastic deformation that decreases the strain at fracture. Archard's wear equation is often used to estimate the wear rate of engineering materials,[1] which states that the volume loss (V) of a material due to wear is inversely proportional to its hardness (H) while proportional to the normal load (L) and the distance slid (S), that is:

$$V = K \frac{L \cdot S}{H}$$

where K is a dimensionless constant known as the wear coefficient.

It should be indicated that for highly elastic materials the hardness only plays a partial role in resisting wear. The wear resistance of the materials also benefits from their elastic behaviour. Examples include pseudoelastic TiNi alloy and tool steels,[30] whose wear behaviour deviates from Archard's wear equation.

Since the hardness plays a predominant role in resisting wear for most engineering materials, surface modification through diffusion processes, surface alloying, and coating is often used to harden surfaces for enhanced wear resistance. Table 25.23 lists various processes of surface treatment.

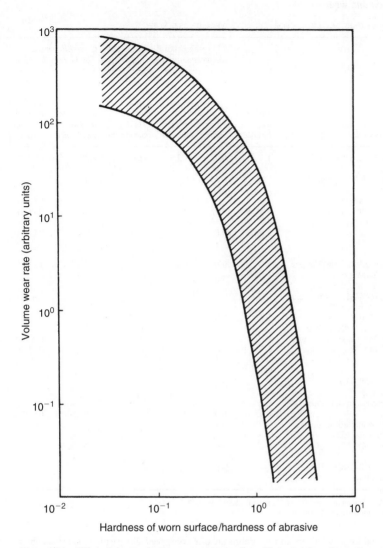

Figure 25.1 *Effect of abrasive hardness on wear rate of metallic materials and ceramics worn on* 80–400 μm *commercial bonded abrasives under an applied stress of* 1 MN m^{-2} *(Reproduced from the Fulmer Materials Optimiser by permission of Fulmer Research Institute Ltd.). Refs* [31–33]

25.2.2 Adhesive wear

Adhesive wear involves the contact and interaction of asperities on two surfaces with strong adhesive force. Adhesive wear is promoted by two major factors:

1. The tendency for different materials to form solid solutions or intermetallic compounds with one another. Thus, material combinations of different crystal structure and chemical properties tend to have lower adhesion. Low adhesion may result in low wear rate and friction, provided that other properties and surface geometries are the same. Figure 25.2 illustrates the tendency of metal couples to adhere together.
2. The cleanliness of the surface. Cleaner surfaces are more likely to bond together. Surfaces having a thick and adherent oxide film have low wear. Stainless steels and nickel alloys that do not form thick oxides have poor adhesive wear resistance.

Table 25.18a   COMMONLY USED MATERIALS FOR ABRASIVE WEAR RESISTANCE[32–40]

		Wear rates relative to 0.4% C low alloy steel quenched and tempered to about 500 Vickers hardness					
Type of material	*Typical commercially available materials*	*Sliding wear by coke*	*Wear by blast furnace sinter**		*Wear of ball mill media/ grinding quartz ores*	*Wear by flint stone sand loam agricul- tural soil*	*Wear in laboratory jaw crusher, siliceous ores*
			sliding	*impact*			
Cast irons	Low alloy 2.5–2.8% C, ~800 Vickers	1.2	0.8	1.2	~1.0	0.3	—
	Heat treated nodular graphite, 700 Vickers	1.0	0.15	0.5	—	1.5	—
	15/3 Cr/Mo Martensitic	—	—	—	0.8	—	0.04–0.3
	High Cr, 25–30%, Martensitic	0.55–0.8	0.15	0.3	0.8	0.3	0.08–0.6
	Ni-hard type (3% C, 4% Ni 2% Cr)	0.6	0.07	0.3	~1.0	0.4	0.08
Steel cast and rolled Steels	0.4% C, low alloy, ~500 Vickers	1.0	1.0	1.0	1.0	1.0	1.0
	0.8% C, ~800 Vickers	—	—	—	~1.0	0.5	—
	0.3% C, 0.6% Mn, 1.5% Cr, 0.75% Ni, 0.4% Mo, ~450 Vickers	0.3	—	—	—	—	—
	0.2% C, 1.2% Mn, 1.3% Cr, 0.25% Mo, 350 Vickers	1.2	0.85	—	—	—	—
	2% C, 12% Cr, ~700 Vickers	—	—	—	0.9	0.5	—
	1% C, 6% Mn, Cr/Mo, austenitic	—	—	—	~1.0	—	0.3–1.3
	1% C, 12–14% Mn, austenitic	0.7–1.1	—	1.1	1.2	0.9	0.35–1.4
Hard facings (*See also* I.D-MSC) and Table 25.20	3–5% C, 20–30% Cr, Co/Mo/V/W/B Mn/Nl ferrous alloys, manual are deposited	0.45–0.8	0.09	0.6	—	0.25–0.4	0.7
	Tungsten carbide/ferrous matrix, arc or gas tubular rods	—	0.2	—	—	~0.3	—
	3.5% C, 33% Cr, 13% W, Co alloy	—	—	—	—	0.2–0.5	—
	1% C, 1% Fe, 26% Cr, 4% Si, 3.5% B, Ni alloy	—	—	—	—	~0.3	—

* The Sinter was produced from foreign ore with ASTM $\frac{1}{4}$-strength index of about 47.
  Reproduced by courtesy of Fulmer Research Institute Ltd.

In metal-to-metal wear, two forms of wear debris are often observed; at very high loads the debris is mainly oxide, but at intermediate loads it is metallic. The transition from oxidative to metallic wear is accompanied by a rapid increase in wear rate. The transition load varies for different materials, microstructures, sliding speed and environment. Thus, wear rates of materials vary by several orders of magnitude (Table 25.24).

25.2.3   Erosive wear

Erosive wear due to the impact of a stream of solid particles is largely dependent on the size, hardness, velocity and angle of impact of the particles. Wear rate generally increases with increasing particle size, angularity, hardness and impact velocity. For strong and tough materials the maximum wear rate occurs at an impact angle of about 30°, but for hard and brittle materials it occurs at an impact angle of about 90° and for tough and elastic materials such as rubber, the maximum wear rate occurs at an impact angle close to 0° (Figure 25.3). The performance of selected materials in erosion are given in Tables 25.25 and 25.26.

Cavitation erosion is wear resulting from localised high impact stresses when bubbles within liquid collapse at or close to a solid surface. During cavitation erosion, corrosion could be involved. The cavitation erosion resistance of a range of materials is given in Table 25.27.

Ease and convenience of replacement	*Typical fields of application*	*Remarks*
Usually convenient with good design to facilitate replacement.	Cast irons are very suitable materials to resist medium to high stress abrasive wear due to their good wear resistance and reasonable cost. At very severe levels of impact abrasion, however, inadequate toughness can be a problem and only materials of the work-hardening type should be used. Also cheaper materials may be preferred due to the excessively high wear rates involved.	These materials have the merit that a combination of strength i.e. toughness and hardness, may be readily obtained by varying the alloying method of manufacture, and treatment; thus giving suitable combinations of these properties to suit a particular application and wear situation. Various techniques of surface hardening can also be
Usually convenient with good design to facilitate replacement.	Due to the very large quantity production involved, steels tend to be comparatively cheap. Thus steels with low wear rates become a competitive materials choice. Their main application lies in hardened steels to resist medium stress abrasion as very low wear rates can be obtained. Austenitic manganese steels can be used in more severe situations due to their work-hardening capability.	employed to improve resistance to abrasive types of wear. Other products are sintered metals and metal coatings, e.g. Cr plate and sprayed coatings.
Replacement can be difficult if applied *in situ*. These materials are often chosen because hard weld may be built up and worn away several times to its total depth under severe wear situations.	For medium and high stress abrasion hard-facings give low wear rates generally, and so are used in many situations to resist abrasive wear, e.g. excavator teeth and other earth moving applications.	

25.2.4 Fretting wear

Fretting wear occurs when two contacting surfaces are subject to very small oscillatory slip (of no more than 150 µm). Damage occurs when oxide films are disrupted locally, and may proceed by continuous formation and removal of the oxide, by the abrasive action of the oxide or by localised formation and failure of metal-to-metal adhesive bonds. The rate of fretting wear is normally very low—about 0.1 mg per 10^6 cycles, per MN m^{-2} normal load, per µm amplitude of slip for mild steel. However, localised cyclic stresses may enhance fatigue crack initiation, causing up to 80% reduction in fatigue strength.

Fretting damage is reduced by eliminating slip (by increasing the contact pressure or separating the surfaces) by lubrication (to separate surfaces and wash away debris) and by surface treatments such as electrodeposits of soft metals or chemical conversion coatings of phosphate and sulphidised coatings on steels and anodised coatings on aluminium alloys.

25.2.5 Corrosive wear

Corrosive wear is encountered in various industries. The synergistic attack of wear and corrosion may result in a significantly high wear rate. In such a wear mode, highly wear resistant materials, such as metal-matrix composites, may perform poorly. The microstructural inhomogeneity accelerates

Table 25.18b COMMONLY USED MATERIALS FOR ABRASIVE WEAR RESISTANCE

Type of material	Typical commercially available materials	Wear rates relative to 0.4% C low alloy steel quenched and tempered to about 500 Vickers hardness				Ease and convenience of replacement	Typical fields of application	Remarks
		Sliding wear by coke	Sliding wear by blast furnace sinter	Impact wear by blast furnace sinter	Wear on commerical bonded 384 μm flint abrasive			
Ceramics	Fusion cast 50% Al_2O_3, 32% ZrO_2, 16% SiO_2	0.1—0.2	~0.2	0.6	—	Convenient if ceramic is bolted in place. Less convenient if ceramic is fixed by adhesive or cement as long curing times may lead to unacceptably long down-times.	Possible to achieve very high hardness but brittleness tends to be a problem. Most suitable to resist low stress abrasion by low density materials and powders.	—
	Sintered 95–99% Al_2O_3	—	~0.2	~0.7	0.04–0.3			
	Reaction Bonded SiC	0.9	6.9	—	~0.02			
	Cast basalt	0.07	0.9	—	~3			
	Tungsten carbide/6% Co				0.007			
Glass	Plate glass	4.5	—	—	22	Used in sheet form where transparency is required.	Glass is brittle and so it is only used at the lowest levels of abrasive wear.	—
Concretes	Aluminous cement-based concrete with proprietary aggregates	3.5	15	—	—	Long curing times can lead to unacceptably long down-times. Can be messy and difficult under dirty conditions.	Useful to resist wear of irregularly shaped components and when abrasion is of low to medium stress.	Also useful in large flat areas, especially when curing time is no real problem, e.g. aircraft hanger flooring, etc.
	As above with 2% by volume 25 × 0.4 mm diam. wire fibres			2				
	Concrete tile—6 mm wear resistant surface		6.7					
Elastomers	Wear resistant rubbers, 55°–70° shore hardness	7.8				Bonded and bolted. Sticking with adhesive can be difficult under dirty conditions.	Very useful to resist impact abrasion—most wear resistant at 90° impact angles. Softer types of rubber are used for low stress impact abrasion. Resilient rubber for more severe impact.	Bonding of rubber to component is a very large problem in high stress abrasive wear. Good anti-sticking properties and low density.
	65° shore hardness rubber with saw tooth surface profile	15.1						
Plastics	Polyurethane	18.5	2.7			Usually used in sheet form. Difficult to bond plastic to component. Solid moulded components are superior but are limited to small sizes.	Low coefficient of friction, good anti-sticking properties. Best for low stress abrasion by fine particles. Resin bonded aggregates are trowellable and so are useful to resist wear or irregular shaped components.	Composite plastics are only as tough as their bonding matrix and therefore find more applications where low stress abrasion by powders or small particles takes place.
	High density polyethylene	15.5–31						
	Epoxy resin based PTFE	40						
	Calcined bauxite filled epoxy resin	11						

Reproduced by courtesy of Fulmer Research Institute Ltd.

Table 25.19 COMPARISONS OF RELATIVE WEAR RATES OF FERROUS MATERIALS[32,36,41]

Wear rates relative to 0.4% C, 1½% Ni/Cr/Mo steel, quenched and tempered to 500 kg mm⁻² (Vickers)

| Material | Vickers hardness kg mm⁻² | Practical wear environments | | | | | | | Laboratory wear environments | | | | | |
| | | *Agricultural soils* | | | *Quartz/feldspar Mo ores* | | | | *Commercial bonded abrasive discs* | | | | *Rubber wheel* | |
		Pumice	Stone free sand	Iron-stone loam/sand	Ball mill	Slusher scraper	Screen rods	Mine car wheels	84 μm Corundum 1 MN m⁻²	84 μm Flint 1 MN m⁻²	384 μm Flint 1 MN m⁻²	84 μm Glass 1 MN m⁻²	Dry quartz sand low stress	Wet quartz sand high stress
0.4% C, 1½% Ni/Cr/Mo steel	500	1.0	1.0	1.0	1.0	1.0	1.0	1.0	1.0	1.0	1.0	1.0	1.0	1.0
0.4% C steel	500	—	—	0.95	—	—	—	—	—	0.93	—	0.76	—	—
0.8% C steel	800	0.06	0.43	0.57	—	—	—	—	0.65	0.49	0.56	<0.01	—	—
0.95% C steel	550	—	—	—	—	—	0.77	—	—	0.85	—	0.34	0.81	0.95
2% C, 12% Cr steel	700	0.40	0.83	0.52	0.91	0.83	—	—	0.57	0.09	0.56	0.07	—	—
1% C, 12% Mn steel	210	—	—	0.92	1.2	—	—	0.59–0.83	0.72	0.63	0.79	—	—	—
18/8 Cr/Ni stainless	150	—	—	1.9	1.7	—	—	—	0.92	0.91	—	1.7	—	—
3% C chilled iron	600	0.06	0.10	0.43	0.83	—	—	2.0	0.65	0.23	0.63	<0.01	0.29	1.1
3% C, 30% Cr white iron	700	—	—	0.44	—	—	—	—	0.47	<0.01	0.44	—	0.07	0.51
15% Cr, 3% Mo white iron	900	—	—	—	0.77	—	—	—	—	0.26	0.67	—	—	—
Ni-hard type iron	700	—	—	0.58	1.0	—	—	—	0.66	0.17	0.67	—	0.17	0.47

Reproduced by courtesy of Fulmer Research Institute Ltd.

Table 25.20   RELATIVE WEAR RATES OF HARDFACING[39,41]

| | | Wear rates relative to 0.4% C, $1\frac{1}{2}$% Ni/Cr/Mo steel at 500 kg mm^2 *Vickers hardness* | | | |
| | | Commercial bonded abrasives | | Rubber wheel test | |
Material	Flint clay soil	34 μm flint 1 MN m^{-2}	384 μm flint 1 MN m^{-2}	Dry and low stress	Wet sand high stress
Tubular Fe/70% tungsten carbide, arc weld	0.29	0.04	0.35	—	—
Ni alloy/40% tungsten carbide, fusion spray	0.18	0.04	0.26	—	—
3.5%, 33% Cr austenitic iron, arc weld	0.24	0.08	0.59	~0.05	~0.95
High C/Cr martensitic iron arc weld	0.44	0.10	0.42	~0.05	~0.87
0.8% C, 3% Ni, 5% Cr, 12% Mn austenitic steel, arc weld	1.01	0.78	0.89	—	—
0.9% C, 4.5% Cr, 7.5% Mo, 1.6% V, 2% W, 1.5% Si, 1.3% Mn martensitic steel, arc weld	0.59	0.79	0.74	~0.29	~0.70
0.95% C, 26% Cr, 4% Si, 3.5% B Ni alloy, fusion spray	0.32	0.12	0.85	—	—
3.5% C, 33% Cr, 13% W Co alloy, gas weld	0.48	0.07	0.78	~0.08	~0.63
97.6% Al$_2$O$_3$, 2.5% TiO$_2$ plasma spray	—	1.64	4.33	—	—
82% Cr, 18% B paste, fused to substrate by gas weld	—	<0.01	0.03	—	—

Reproduced by courtesy of Fulmer Research Institute Ltd.

Table 25.21   RELATIVE WEAR RATES OF CERAMICS[33,42]

| | | | Wear rates relative to a sintered 95% alumina (*bracketed figures relative to 0.4% C, $1\frac{1}{2}$% Ni/Co/Mo steel at 500 kg mm^{-2} Vickers hardness*) | | | | | |
| | | | Commercial bonded flint abrasives | | | | | |
Material	Vickers hardness GN m^{-2}	Fracture toughness K_c MN m$^{-3/2}$	84 μm flint 1 MN m^{-2}	384 μm flint 1 MN m^{-2}	84 μm corundum 1 MN m^{-2}	84 μm SiC 1 MN m^{-2}	Diamond sawing
95% sintered Al$_2$O$_3$	12–13	4–6	1.0 (0.024)	1.0 (0.099)	1.0 (0.39)	1.0 (0.81)	1.0
Hot pressed Si$_3$N$_4$	~16	5–9	0.32	0.11	—	0.56	1.1
Reaction bonded Si$_3$N$_4$	~7	4–5	3.8	8.8	—	2.3	—
Reaction bonded SiC	16–20	7–9	0.15	0.11	—	0.41	—
Hot pressed B$_4$C	~30	6–9.5	0.21	0.06	—	0.04	0.65
Hot pressed Al$_2$O$_3$	14–20	5–7.5	0.3	0.15	—	0.27	—
97.5% sintered Al$_2$O$_3$	~15	~8	1.0	0.3	—	0.58	—
99.7% sintered Al$_2$O$_3$	~12	~4	1.8	1.4	—	1.2	—
Cast basalt	~5	~2	26	26	—	3	—
Soda-lime glass	4–5	2–4	84	168	100	4.5	—
Sintered TiO$_2$	~7	~2.5	16	12	—	4.5	—
ZrO$_2$	15	~2.5	—	—	—	—	1.3
Spinel	16	~1.7	—	—	—	—	2.9
MgO	7	~2.2	—	—	—	—	5.6

Reproduced by courtesy of Fulmer Research Institute Ltd.

Wear rates relative to 4–5% Co composites

Composition TiC+			Carbide grain size μm	Vickers hardness GN m⁻²	Fracture toughness K_c MN m⁻³ᐟ²	Laboratory wear tests						
						Rock drilling		Loose abrasive water slurry		Commercial bonded abrasives		
WC%	TaC%	Binder %				Percussive, granite	Rotary sandstone	0.2–0.5 mm Al_2O_3	0.1–0.5 mm SiC/Al_2O_3	84 μm flint 1 MN m⁻²	384 μm flint 1 MN m⁻²	84 μm SiC 1 MN m
96		4 Co	1–2	~18	~7	(1.0)	1.0	1.0				
96		4 Co	2–3	~16	~10		2.5					
95.5	0.5	4 Co		~17					1.0			
93	2	5 Co	1–2	~14	~7					1.0	1.0	1.0
94		6 Co	1–2	~16	~10	~1.7	1.7	2.0	~1.7	1.6	1.4	1.6
94		6 Co	2–3	~15	~13		5.0	2.5		2.0	1.1	1.5
93	1	6 Co	2–3	~13	~7				1.9			
92	2	6 Co		~16.5					5.3			
72	22	6 Co		~17					3.3–4.2			
93		7 Co		~14.5					0.5			
90	3	7 Co		~17					3.5			
81.5	12	7.5 Co		~16.5								
92		8 Co	2–3	~15	~12	5.0		5.0				
91		9 Co	1–2	~15	~9	4.0	3.3	10.0				
91		9 Co	2–3	~13	~13		6.7			2.6	1.9	1.3
84.5	6.5	9 Co		~15					6.3			
71	20	10 Co		~16					6.3			
50	40	11 Co		~17		10.0			7.3			
89		13 Co		~12					6.0–7.0			
87		13 Co		~11					8.7			
86	1	15 Co		~13					7.9			
85		15 Co	1–2	~11	18		10.0		~11.0			
85		16 Co	2–3	~11	19		25.0		13.0			
69	15	20 Co		~13				10.0	18.0			
80		10 Mo/10 Ni		~10.5					9.0			
75		25 Co		~9					20.0			
70		30 Co		~8.5					25.0			
0.4% C 1½ Ni/Cr/Mo steel				5					850–1500	530	150	6.3
2% C, 12–14% Cr tool steel				~7					140	48	84	~4

Reproduced by Courtesy of Fulmer Research Institute Ltd.

Table 25.23   VARIOUS PROCESSES FOR SURFACE MODIFICATION

Process	Substrate	Thickness or case depth (μm)	Hardness, HV
Carburised (case hardened)	Low-carbon steel	1 000–3 000	700–900
Nitrided or nitrocarburised	Low-carbon steel	5–10	400–600
	Tool steel	50–200	800–1 000
	Stainless steel	20–50	1 000–1 200
Boronised	Mild steel	10–20	500–700
	Low-alloy steel	20–30	800–1 000
	Stainless steel (316)	30–40	1 000–1 200
Chromised	Stainless steel (316)	20–50	300–400
Aluminised	Stainless steel (316)	20–50	400–500
Phosphated	Low-carbon	4–7	~200
Chromated	Various	1–2	—
	Oxidised steel	3–5	250–350
Ion implanted	Steel	0.1–1	—
PVD TiN	Various	1–5	2 000–3 000
PVD CrN	Various	2–20	1 800–2 500
Diamond-like carbon	Various	1–2	1 500–2 000
CVD chromium nitride	Stainless steel (316)	10–15	1 100–1 300
CVD chromium carbide	High-carbon steel	10–15	1 500–2 000
CVD alumina	Steel	5–10	1 500–2 000
Chromium plate	Various	5–250	800–1 000
Nickel plate	Various	10–1 000	250–650
Copper plate	Various	10–250	70–90
Cadmium plate	Various	5–10	~50
Zinc plate	Various	5–10	~50
Electroless nickel	Various	5–50	500–1 000
Electroless nickel/ceramic	Various	5–50	<1 300
Hot dip galvanising	Steel	20–250	70–250
Electrogalvanised steel strip	Low-carbon steel	5–10	~70
Hot dip aluminised steel strip	Low-carbon steel	5–10	~70
Thermally sprayed chromium oxide	Various	20–100	1 200–1 600
Thermally sprayed alumina	Various	20–100	1 500–1 800
Thermally sprayed tungsten carbide/cobalt	Various	20–100	1 100–1 600
Slurry/sinter formed ceramics	Steel	20–100	1 000–1 200
Hardfacing overlays: WC/Ni alloy WC/Co	Steel	1 000–50 000	600–1 800

the material dissolution, particularly at interphase boundaries, thus resulting in break-away of the reinforcing hard particles from the matrix and consequent high wear rate. Under corrosive wear attack, corrosion-resistant materials such as stainless steel may be as ineffective as carbon steel. Since the protective passive film is continuously damaged by wear during such a process, passive materials may lose their resistance to corrosion attack and thus perform poorly during corrosive wear. However, if the mechanical properties of the passive film and its bond to the substrate are improved, corrosive wear of the materials can be decreased.[51,52]. Table 25.28 lists wear rates of selected materials in a H_2SO_4 slurry.

Although the synergism of wear and corrosion has not yet been completely understood, general guidelines are available for selecting suitable materials against wear in corrosive environments:

a) Severe wear and minor corrosion:
   In this case, one may choose hard materials, such as composites and precipitation-hardened alloys that have high resistance to mechanical wear.

b) Severe corrosion and minor wear:
   Since corrosion is the major factor responsible for the material loss, corrosion-resistant materials such as stainless steel may therefore be an adequate choice. Although they may not perform

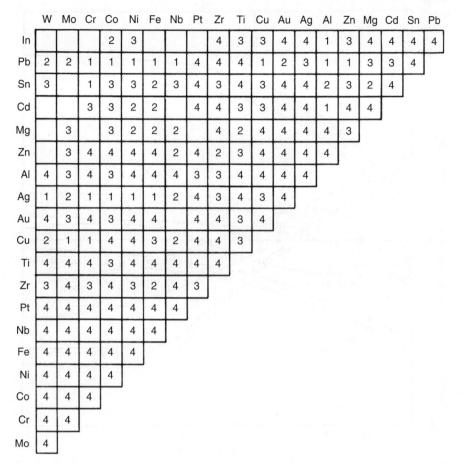

Figure 25.2 *Tendency of metal couples to adhere together. 1 represents the greatest resistance and thus the best combination for wear. 4 represents the least resistance and thus the worst combination for wear. 2 and 3 represent intermediate resistance.*

Table 25.24 WEAR RATES OF SOME COMMON ENGINEERING MATERIALS IN UNLUBRICATED SLIDING AT $1.8\,\mathrm{m\,s^{-1}}$ AND A LOAD OF 400 g

Material	Wear rate $\mathrm{mm^3\,mm^{-1}}$
Mild steel on itself	1.57×10^{-11}
60/40 leaded brass on hardened tool steel	2.4×10^{-12}
PTFE on hardened tool steel	2.0×10^{-13}
Stellite on hardened tool steel	3.2×10^{-14}
Ferritic stainless steel on hardened tool steel	2.7×10^{-4}
Polyethylene on hardened tool steel	3.0×10^{-15}
Tungsten carbide composite on itself	2.0×10^{-16}

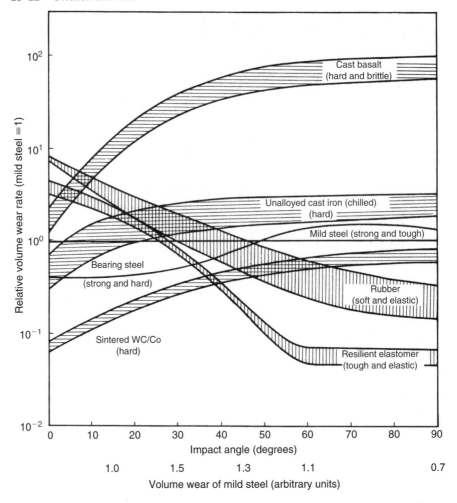

Figure 25.3 *Effect of impact angle on erosion wear of materials impacted with dry 0.2–1.5 mn quartz.*[47] *(Reproduced from The Fulmer Materials Optimiser by permission of Fulmer Research Institute Ltd.).*

as well as they do under pure corrosion attack, rapid re-passivation capability or high intrinsic resistance to electrochemical attack is beneficial to their resistance to surface failure under such a condition.

c) Severe wear and severe corrosion:

Under such a wear condition, one may select materials that have high strength with less microstructural inhomogeneity. Solution-strengthened materials having high electrochemical potential (i.e. intrinsically resistant to corrosion) could be suitable candidates.

d) Minor wear and minor corrosion:

In this case, there are no particular requirements for selecting materials.

Table 25.25 EROSION VALUES FOR SELECTED METALLIC AND CERAMIC MATERIALS EVALUATED AT ROOM TEMPERATURE AND AT ELEVATED TEMPERATURE.[48]

Test parameters: 90° impingement; 27 mm Al_2O_3 particles; 5 g/min particle flow; 170 m/s particle velocity; 3 min test duration; N_2 atmosphere

Material	Composition	Relative erosion factor (Ref)*	
		20°C	700°C
Metals			
Ti–6Al–4V	—	1.26	0.54
Haynes 93	17Cr–16Mo–6.3Co–3C–bal Fe	1.25	1.00
25Cr Iron	25Cr–2Ni–2Mn–0.5Si–3.5C–bal Fe	1.19	1.16
Stellite 6K	30Cr–4.5W–1.5Mo–1.7C–bal Co	1.08	1.06
Stellite 3	31Cr–12.5W–2.4C–bal Co	1.04	1.61
Stellite 6B	30Cr–4.5W–1.5Mo–12C–bal Co	1.00	1.00
Type 304 stainless steel	17Cr–9Ni–2Mn–1Si–bal Fe	1.00	0.73
Type 316 stainless steel	17Cr–12Ni–2Mn–1Si–2.5Mo–bal Fe	0.99	0.56
Haynes 188	22Cr–14.5W–22Ni–0.15C–bal Co	0.97	0.83
Haynes 25	22Cr–15W–10Ni–1.5Mn–0.15C–bal Co	0.96	0.85
Type 430 stainless steel	17Cr–1Mn–1Si–0.1C-bal Fe	0.93	0.62
HK-40	26Cr–20Ni–0.4C–bal Fe	0.93	0.78
Inconel 600	76Ni–15.5Cr–8Fe	0.92	0.61
Inconel 800H	19Cr–35Ni–1.5Mn–1.3Si–bal Fe	0.91	0.79
RA 330	19Cr–35Ni–1.5Mn–1.3Si–bal Fe	0.91	0.79
Beta III Ti	11.5Mo–6Zr–4.5Sn–bal Ti	0.90	0.57
Inconel 800	32.5Ni–46Fe–21Cr	0.83	0.57
RA 333	25Cr–1.5Mn–1.3Si–3Co–3Mo–3W–18Fe–bal Ni	0.80	0.80
Inconel 671	50Ni–48Cr–0.4Ti	0.77	0.62
Mild steel	0.15C–bal Fe	0.76	—
Molybdenum	—	0.52	—
Tungsten	—	0.48	0.17
Ceramics			
ZRBSC-M	50Ni–48Cr–0.4T	6.36	>5
Chromite	—	2.44	3.43
Refrax 20C	SiC–Si_3N_4 bond	0.91	1.15
HD 435	Recrystallised SiC	0.80	0.32
Carbofrax D	SiC-ceramic bond	0.49	1.38
HD 430	Recrystallised SiC	0.4	0.38
Si_3N_4	—	0.40	0.12
Norbide	B_4C	0.38	0.21
BT-9	2MgO–25TiB_2–3.5WC–bal Al_2O_3	0.37	0.36
BT-12	1.5MgO–49TiB_2–3.5WC–bal Al_2O_3	0.35	0.16
BT-11	1.7MgO–38TiB_2–3.5WC–bal Al_2O_3	0.33	0.26
ZRBSC-D	ZrB_2–SiC	0.32	0.07
BT-24	2MgO–30TiB_2–3.5WC–bal Al_2O_3	0.32	0.20
BT-10	2MgO–30TiB_2–3.5WC–bal Al_2O_3	0.30	0.25
Noroc 33	Si_3N_4–SiC	0.20	0.42
TiC–Al_2O_3	—	0.19	0.30
SiC	—	0.12	0.02
CBN	—	0	0
Diamond	—	0	0

* Volume loss material/volume loss Stellite 6B.

Table 25.26   RELATIVE RESISTANCE OF VARIOUS MATERIALS TO EROSIVE WEAR[49,50]

	Relative erosive wear resistance	
Materials	*In sandy water*	*Pneumatic conveying of minerals*
Ceramics	2–150	—
Aluminas	—	6.5–10
Weld overlays (hard facings)	1.1–20	—
Cast irons	0.5–6	1.2–10
Quenched and tempered steels	—	1.9–5
Rolled and forged steels	1–6	—
Mild steel	1	1
Cast steels	1–2.5	—
Titanium	1	—
Wrought copper alloys	0.3–1	—
Cast copper alloys	0.4–0.9	—
Aluminium alloys	0.1–0.5	—
Plastics	0.04–0.3	—
Elastomers	0.04–0.09	0.12–0.45
Hard Concrete	—	0.11–0.16

Table 25.27   RESISTANCE OF VARIOUS MATERIALS TO CAVITATION EROSION

Range of materials	*Typically used materials*		*High intensity relative cavitation erosion resistance (Aluminium bronze = 1)*	*Remarks*
	Material type	*Specification*		
Cast bronzes	Aluminium bronze	BS1400 AB2	1.3	Aluminium bronze also
	Aluminium bronze	BS1400 AB1	1.0 (standard)	has good corrosion resistance
	Novoston bronze	BS1400 CMA1	0.5	in sea water, but is
	Gunmetal	BS1400 G1	0.22	more difficult to cast than
	High tensile brass	BS1400 HTB1	0.18	Novoston bronze
	Phosphor bronze	BS1400 PB3	0.15	
	Nickel gunmetal	—	0.15	
	Leaded gunmetal	BS1400 LG3	0.14	
Wrought bronzes	Manganese bronze	—	0.2	—
	Everdur A	BS2872 CZ101	0.16	
	Aluminium brass	BS2872 CZ110	0.15	
	Free turning brass	BS2872 CZ121	0.14	
	Naval brass	BS2872 CZ112	0.12	
	Hot stamping brass	BS2872 CZ122	0.11	
	Tellurium copper	BS2872 CZ109	0.03	
Nickel alloys	Monel K-500 (aged)	BS3076 NA18	0.7	—
	S Monel	BS3071 NA3	0.5	
	Monel K-500 (cold drawn)	BS3076 NA18	0.3	
	Nimonic 75	DTD 703A	0.22	
	Nickel (hard)	BS3076 NA11	0.21	
	Nickel (annealed)	BS3076 NA11	0.2	
Wrought aluminium alloys	Al 4% Cu solution treated and aged	BS1476 HE15	0.08	
	Al 5% Mg As rolled	BS1476 NE6M	0.04	Low density
	Al 3.5% Mg As rolled	BS1476 NE5M	0.03	
Titanium alloys	Ti–2.25 Al–4 Mo– 0.255C–11Sn	DTD 5213	1.0	Excellent corrosion resistance
	Ti CP with medium interstitial content	DTD 5273	0.7	Low density. High cost
Cast irons and cast steels	Martensitic stainless	Hardened	1.8	Stainless steels are reasonably
	Austenitic stainless	Annealed	0.8	corrosion resistant.
	Ferritic stainless	Annealed	0.77	Martensitic stainless
	SG cast iron	Tempered	0.7	steels are less corrosion resista

(*continue*

Table 25.27 RESISTANCE OF VARIOUS MATERIALS TO CAVITATION EROSION—*continued*

Range of materials	Material type	Specification	High intensity relative cavitation erosion resistance (Aluminium bronze = 1)	Remarks
	Martensitic stainless	Tempered	0.5	than austenitic
	Alloy cast iron (D2 Ni-resist)		0.2	stainless steels
Wrought steels	HT low alloy	Hardened En24	2.5	
	Martensitic stainless	Tempered En57	0.67	
	Martensitic stainless	Tempered En56	0.62	
	Austenitic stainless	Annealed En58J	0.5	
	Mild steel	En1A	0.2	
Stellites (cobalt alloys)	Stellite 12 (cast)—29% Cr 9% W 1% C		19.3	Most resistant to high cavitation erosion
	Stellite 4 (cast)—31% Cr 14% W 1% C		15.0	
	Stellite 6 (cast)—26% Cr 5% W 1% C		5.3	
	Stellite 6 (wrought)		3.4	
	Stellite 7 (cast)—26% Cr 6%W 0.4%C		2.5	
Plastics	Nylons, acetals, high impact polyethylenes		1.7–3.0	Values of cavitation erosion resistance fall at sharply elevated temperatures. Generally, plastics with high impact strengths have good cavitation erosion resistance. Very suitable for mass manufacture of small components, e.g. boat propellers. Corrosion resistant.
	Polyethylenes Polypropylenes		0.9–1.6	
	PVC PTFE		0.15–3.5	
	Chlorinated polyether		0.04–0.1	
	Perspex		0.01–0.035	
Rubbers	Bonded rubber coatings		—	Rubber requires careful preparation and application. Rubber gives satisfactory resistance to mild cavitation erosion, and where relative fluid velocities are low. Failure usually occurs at bond to components at higher fluid velocities.
	Neoprene			

Reproduced by courtesy of Fulmer Research Institute Ltd.

Table 25.28 WEAR RATES OF SEVERAL MATERIALS IN 2 wt pct SILICA SAND SLURRY IN 1 N H_2SO_4 WITH SPEED = 16 m/s[53]

Material	Wear rate (mm³/h)	Relative wear, %	Hardness, RA
Iron	5.08	230	27.5
Fe–5Al	17.62	345	43.5
Fe–10Al	9.68	438	55.5
Fe–15Al	14.3	647	68.5
Fe–20Al	12.3	556	65.0
Fe–25Al	7.19	325	68.0
316 Stainless steel	2.38	108	41.1
A514 steel	2.21	100	62.0
20 Cr WCI	29.1	1 320	81.0
UHMWPE	0.84	38.1	67.02

20 Cr WCI: 20 Cr white cast iron
UHMWPE: ultrahigh molecular weight polyethylene

REFERENCES

1. I. M. Hutchings, 'Tribology—Friction and Wear of Engineering Materials', Edward Arnold, London, 1992.
2. F. P. Bowden and D. Tabor, 'Friction and Lubrication of Solids', Pt. II, Clarendon Press, Oxford, 1964.
3. I. E. Rabinowicz, *ASLE Trans.*, 1971, **14**, 198.
4. 'Friction Data Guide', General Magnaplate Corporation, 1988.
5. J. F. Archard, in 'ASME Wear Control Handbook' (eds M. B. Peterson and W. O. Winer), American Society of Mechanical Engineers, 1980.
6. A. W. Ruff, L. K. Ives and W. A. Glaeser, p. 235 in 'Fundamentals of Friction and Wear of Materials', ASM International, Materials Park, OH, 1981.
7. F. P. Bowden and D. Tabor, 'The Friction and Lubrication of Solids', Oxford Press, Oxford, UK, 1986.
8. P. J. Blau and C. E. DeVore, *Tribol. Int.*, 1990, **23**, 226.
9. P. J. Blau, *J. Tribology*, 1985, **107**, 483.
10. K. G. Budinski, p. 289 in 'Proc. of Wear of Materials', American Society of Mechanical Engineers, 1991.
11. F. P. Bowden and D. Tabor, 'Friction and Lubrication of Solids', Pt. I, revised reprint, Clarendon Press, Oxford, 1954.
12. E. F. Finkin, J. Calabrese and M. B. Peterson, *J. Amer. Soc. Lubr. Engrs.*, 1973, **29**, 197.
13. B. V. Elkonin, *Wear*, 1980, **61**, 169.
14. K. Demizu, R. Wadabayashim, and H. Ishigaki, *Tribol. Trans.*, 1990, **33**, 505.
15. P. J. Blau, Oak Ridge National Laboratory, personal communication.
16. C. S. Yust, p. 27 in 'Tribology of Composite Materials (eds. P. K. Rohatgi, P. J. Blau, and C. S. Yust), ASM International, Materials Park, OH, 1990.
17. B. Bhushan and B. K. Gupta, table in 'Handbook of Tribology', McGraw-Hill, New York, 1991.
18. 'Lubricomp Internally-Lubricated Reinforced Themoplastics and Fluoropolymer Composites', Bulletin 254–688, ICI Advanced Materials.
19. J. M. Thorp, *Tribol. Int.*, 1982, **15**, 69.
20. J. W. M. Mens and A. W. J. de Gee, *Wear*, 1991, **149**, 255.
21. R. P. Stein, *Metall. Eng. Quart.*, 1967, **7**, 9.
22. N. P. Sub, p. 226 in 'Tribophysics', Prentice-Hall, 1986.
23. M. B. Peterson and S. F. Murray, *Metals Eng. Quart.*, 1967, May 22.
24. W. B. Hardy, 'Collected Scientific Papers', 1936, Cambridge University Press, Cambridge, UK, 1936.
25. A. W. Crook, *Phil. Trans.*, 1958, **A250**, 387. *ibid.*, 1961, **A254**, 237.
26. J. F. Archard and M. T. Kirk, *Proc. R. Soc.*, 1961, **A261**, 532.
27. D. Dowson and G. R. Higginson, 'Elastohydrodynamic Lubrication', Pergamon Press, Oxford, UK, 1966.
28. A. Cameron, 'Basic Lubrication Theory', Longman, Harlow, 1971.
29. W. Hirst and J. V. Stafford, *Proc. Inst. Mech. Engrs*, 1972, **186**, 179.
30. R. Liu and D. Y. Li, *Wear*, 2001, **251**, 956.
31. G. K. Nathan and W. J. D. Jones, *Proc. I. Mech. E.*, 1966/67, **181**, 215.
32. R. C. D. Richardson, *Wear*, 1968, **11**, 245.
33. M. A. Moore and F. S. King, *Wear*, 1980, **60**, 123.
34. H. Hocke, *Iron & Steel Int.*, 1977, 361.
35. F. Borik and L. W. G. Scholz, *J. Materials*, 1971, **6**, 590.
36. T. E. Norman and E. R. Hall, ASTM STP No. 446, 1969.
37. W. Fairhurst and K. Rohrig, *Foundry Trade J.*, 1974.
38. R. C. D. Richardson, *J. Agric. Engng. Res.*, 1967, **12**, 22.
39. M. A. Moore, *J. Agric. Engng. Res.*, 1975, **20**, 167.
40. G. G. Brown and J. D. Watson, East Asian Iron & Steel Conf., 1977.
41. H. S. Avery, SAE Off-Highway Vehicle Meeting, Milwaukee, 1975.
42. A. G. Evans and T. R. Wilshaw, *Acta Met.*, 1976, **24**, 939.
43. E. Cuboni, *Metallurgia Italiana*, 1969, **12**, 593.
44. J. Larsen-Badse, *Powder Met.*, 1973, **16**, 1.
45. H. Feld, P. Walter and Z. Werkstoffe, 1976, **7**, 300.
46. K. Stevens, 'Surface Engineering to Combat Wear and Corrosion: A Design Guide', The Institute of Materials, London, UK, 1997.
47. K. Wellinger and H. Uetz, *Jernkont. Ann.*, 1963, **147**, 845.
48. P. 65 in 'Surface Engineering for Corrosion and Wear Resistance' (ed. J. R. Davis), ASM International, Materials Park, Ohio, 2001.
49. W. A. Stauffer, *Schinzer Archiv. fur Angewandte Wissenschaft und Technik*, 1958, **24**, 3.
50. E. Olsen, Bulk-Storage Movement Control, 1976, 48.
51. T. Zhang and D.Y. Li, p. 535 in 'Hydraulic Failure Analysis—Fluids, Components, and System Effects' (ed. G. E. Totten, D. K. Wills and D. G. Feldmann), ASTM STP 1339, 2001.
52. X. Wang and D. Y. Li, *Mater. Sci. & Eng. A*, 2001, **315**, 158.
53. B. W. Madsen, p. 49 in 'Wear-Corrosion Interaction in Liquid Media' (eds A. A. Sagues and E. I. Meletis), The Minerals, Metals & Materials Society, Warrendale, PA, 1991.

26 Casting alloys and foundry data[†]

26.1 Casting techniques*

Table 26.1 TECHNIQUES USING AN EXPENDABLE MOULD AND PATTERN

	Investment casting	*Full mould*
Equivalent terminology	Lost wax casting Precision casting	Cavity-less casting Lost foam
Pattern material	Metal ⎫ Rubber ⎬ die Plastic ⎭ Wax, thermoplastics, frozen mercury	Foamed polystyrene or polyurethane Foamed poly
Mould material	Silica base ceramic Slurry + stucco	Silica sand Full mould—
Binder	Ethyl silicate Magnesium or aluminium phosphate	Usually Na_2SiO_3 or resin (Furane) Lost foam—unbonded
Size range (kg)	Up to 4.5 except frozen Hg for which up to 45	Up to 5 000 in iron
Scope	Complex shapes and fine detail	One-off castings, e.g. jigs and press tools, complex shapes
Alloys cast	High temp Steel Cu base Al base alloys	iron and aluminium castings
Surface finish (CLA)	25 25–125 50–85 50–125	60–250
Minimum section for length 25 mm 150 mm	 1.5 1.0 1.5 12 1.5 1.5 1.5 12	
Dimensional tolerance[†] ± mm 25 mm 150 mm	 0.13 0.13 0.13 0.6 0.38 0.25 0.18 2.5	 0.20 0.61
Labour cost per kg Equipment cost per kg Mould material cost	High High High	Low Low Polystyrene pattern approximate cost one third that of wood. Low for unbonded. For other *see* Table 26.2.

* Report No. 187 *PERA*, March 1969.
R.A. Flinn, 'Fundamentals of Metal Casting', Addison–Wesley, New York, 1963.
[†] The following additional tolerances should be added if dimension chosen crosses parting line:
Sand ±0.38   Permanent ±0.38   Centrifugal ±0.25
Shell ±0.25   Plaster   ±0.25   Shaw   ±0.10 to 0.15 for a 25 mm casting.

[†] For additional alloy data, *see* Chapters 1 and 22. Fuels and furnace design are discussed respectively in Chapters 28 and 40.

Table 26.2   TECHNIQUES USING AN EXPENDABLE MOULD

	Ceramic mould casting	Plaster mould casting	Shell mould casting	Sand casting	Vacuum moulding
Equivalent terminology	Shaw process Ceramic shell process	'Antioch' process Gypsum moulding	'C' process Cronig process	—	'V' Process
Pattern material	Wood, plastic, metal	Wood, plastic, metal	Metal	Wood, plastic metal	Wood, plastic metal
Mould material	Molochite, sillimanite, mullite	Calcium sulphate	Silica or zircon sand	Silica, zircon, chromite, olivine, sand	Silica, zircon, chromite, olivine, sand
Binder	Ethyl silicate plus gelling agent, e.g. 50/50 NH_4OH/H_2O	Water (autoclave in saturated steam atmosphere at 0.1 $N\,mm^{-2}$ (15 psi) for 6–8 h-stand at amb. temp 14 h)	Phenol-formaldehyde resin (Novolac) + hexamethy-lenetetramine + heat (from pattern at *c.* 260°C)	Many variants, e.g. clay, CO_2-silicate, cement, resin self set, silicate self set, etc.	None
Size range (kg)	Usually up to 180	All sizes	Up to 136	Green sand up to 500, dry sand 500 and above, others to 500 and above	All sizes
Scope	Complex shapes and fine detail	Limited by pattern withdrawal. Intricate shapes, e.g. wave guides	Better surface finish and accuracy than conventional sand castings	Limited only by pattern work	Excellent surface Finish—No draft
Alloys cast	Most types: Mg base, Al base, Cu base, steels	Mainly aluminium alloys and some Mg and Cu base	Carbon and alloy steels. Manganese steel. Some stainless and HR steels	Grey, SG and malleable irons	Al, Cu and Mg base alloys	All common foundry alloys	Steel Cast-irons Cu base Al base	All common foundry alloys
Surface finish (CLA)	120 to 180, with fine zircon facing 80 or better	30–40	100–300 100–250 100	300–1 000	125–200
Minimum section for length 25 mm	2.5 2.5 3.0	Al 1.4 Cu 1.1	4.6 3.2 Al 1.6 Cu 2.3	6.3 3.2 2.3 3.2	—
150 mm	3.2–4.0 3.2 3.2–4.0	1.4 1.4	6.4 3.5 3.1 3.1	6.3 5.1 3.6 3.6	
Dimensional tolerance ±mm 25 mm	0.18 0.18 0.18	0.13 0.13	0.25 0.25 0.13 0.13	1.52 0.76 0.38 0.38	0.25
150 mm	0.30 0.25 0.25	0.18 0.25	0.76 0.76 0.33 0.33	2.54 1.27 0.38 0.38	0.51
Labour cost per kg	High	Medium	Low	Low	Medium
Equipment cost per kg	Moderately high	Moderately high	Medium to high	Low	Low
Mould material cost	Moderately high	Medium	Medium	Low	Low

26.3 TECHNIQUES USING A NON-EXPENDABLE MOULD (DIE)

	Permanent mould casting (*Gravity-die casting*)	Low-pressure die casting	High-Pressure die casting (*including 'Acurad'*)	Centrifugal casting	Continuous casting	Semi-solid casting
...valent ...nology	Gravity-die casting chill casting	—	Pressure-die casting Cold-chamber (Al-base) Hot-chamber (Zn base)	'Spun-cast'	'Semi'-continuous (in non-ferrous)	Rheocasting Thixocasting Squeeze casting
...n material	—	—	—	Conventional wood	—	—
...d material	Heat-resisting cast iron (expendable cores sometimes used)	Heat-resisting cast-iron (expendable cores sometimes used)	Heat resistant steel die	Metal die Conventional sand mould	Water-cooled metal die	Heat-resistant steel die
...r	—	—	—	—	—	
...ange (kg)	Usually 0.45–25 in aluminium but can be up to 250	Up to 11.5 in Al-alloy	Up to 2.25 in brass, 20 in Mg. 34 in Zn, 45 in Al	Pipes up to 1.2 m dia. × 11 m long. Rings up to 15.2 m dia.	Casting speeds vary between 0.04 and 0.11 m s^{-1}	Up to about 10 in Al-alloy
	Casting design must allow for removal from die	Complex shapes and pressure tight castings	Restricted internal shapes. Die withdrawal	Mainly grey iron pipes and cylinder liners and SG iron pipes and other symmetrical annular shapes	Semi-finished simple shapes (strip, rod, bar) in non-ferrous and grey iron and steel	Restricted internal shapes. Die withdrawal
...cast	All common non-ferrous foundry alloys + some grey iron	Almost exclusive to Al-alloys but steel, cast in USA using graphite mould	Mainly Zn and Al base but some brass and Mg base, Sn and Pb	Highly alloyed steels, grey and SG irons. Most Cu base	Alloy and plain carbon steels, Al and Cu base	All common non-ferrous foundry alloys
	Zn base Cu base Mg base Cast-iron Al base	Aluminium	Zn base Al base Mg base Cu base			
...e finish	100–250	40–100	40–100	Varies widely according to die used and metal cast. Typical range 100–500	100–200	40–100
...um section ...gth	2.5 2.5 6.4 / 4.1 5.0 10.2	1.3–6.4	1.3 0.8 / 2.0 1.5	Varies widely according to die and metal cast. Typical range 1.8–7.6	Down to 7.6	—
...sional ...ce* ± mm	0.18 0.38 0.76 / 0.25 0.53 0.76	0.10 / 0.36	0.10 0.10 / 0.60 0.51	Dependent on type of die. Usually in region of ±0.64	Usually within 0.13–25 mm	0.10 / 0.60
...cost	Low	Low	Low	Moderately low	Moderate	Low
...ent cost	High	High	High	High	High	High

...lowing additional tolerances should be added if dimension chosen crosses parting line:
...38 Permanent ±0.38 Centrifugal ±0.25
...25 Plaster ±0.25 Shaw ±0.10 to 0.15   for a 25 mm casting.

Table 26.4   MOULDING AND CORE-MAKING MATERIALS.[†] CLASSIFICATION SPECIFICATION AND PROPERTIES

(a) Typical chemical analyses of moulding and core-making sands

Origin	SiO_2	Al_2O_3	Fe_2O_3	CaO	Na_2O + K_2O	Loss on ignition	AFS GFN*	AFS clay content	Remarks
Naturally bonded sands									
Dullatur 40	90.20	5.74	0.91	0.07	1.45	1.56	40	16.4	For steel and heavy iron
Levenseat No. 9	82.63	9.28	1.65	0.31	0.22	4.45	55	20.3	For steel
Dursand High Bond	80.98	9.64	2.55	0.22	1.99	3.32	73	15.7	For steel
Weatherill	80.76	8.90	2.06	0.10	3.36	3.18	76	14.0	For steel
Bramcote	87.50	4.41	2.51	1.00	1.49	2.55	75	8.4	For iron and non-ferrous
Pickering	89.70	4.44	1.36	0.08	1.70	2.00	100	21.1	(Yellow) light iron non-ferrous
Swynnerton	84.10	7.69	1.91	0.11	3.27	1.80	115	11.4	For iron and non-ferrous
Mansfield Red	82.50	4.96	1.27	2.39 $\begin{bmatrix} MgO \\ 1.56 \end{bmatrix}$	2.30	2.15	137	13.6	Non-ferrous and some iron
Erith loam	83.80	6.92	2.91	0.52 $\begin{bmatrix} MgO \\ 0.83 \end{bmatrix}$	2.63	1.95	158	13.2	Iron and non-ferrous
Washed Silica Sands									
Biddulph + 36 mesh	97.65	0.38	0.65	0.10	0.03	0.37	24	—	Sub-angular/angular
Arnold No. 19 (dried)	98.76	0.34	trace	0.11	0.30	0.00	29	—	Rounded/sub-angular
Garside dried No. 21	98.01	0.43	0.90	0.17	0.37	0.00	36	—	Sub-angular
	97.91	1.13	0.50	0.11	0.72	0.21	44	—	Sub-angular
Erith silica	98.97	0.03	0.65	0.05	0.04	0.06	56	—	Sub-angular
Kings Lynn 60 (4F)	98.50	0.78	0.19	0.03	0.35	0.30	66	—	Sub-angular
Redhill 65 (F)	99.50	0.14	0.06	0.00	0.08	0.14	66	—	Sub-angular
New Windsor Rose	97.00	1.50	0.54	0.22	0.93	0.45	68	—	Sub-angular
Chelford Fine (95)	94.70	2.73	0.20	0.10	1.76	0.40	98	—	Sub-angular
Kings Lynn SS (100)	98.40	0.82	0.15	0.03	0.37	0.25	107	—	Sub-angular
Redhill H (110)	99.00	0.27	0.12	<0.01	0.04	0.35	105	—	Sub-angular/angular
Ryarsh	97.50	0.87	0.40	0.03	0.31	0.56	129	—	Sub-angular/angular
Non-siliceous sands									
		ZrO_2		TiO_2					
Zircoruf	33.19	64.90	0.14	0.64	—	0.16	70	—	Zirconium silicate
AMA zircon	32.00	63.40	2.40	0.08	—	0.23	110	—	Zirconium silicate
		Al_2O_3		CaO	MgO				
Olivine No. 2	41.35	0.60	6.25	1.51	48.75	0.33	59	—	Forsterite/Fayalite
Olivine No. $3\frac{1}{2}$	41.76	0.80	6.15	0.84	49.40	0.43	96	—	Forsterite/Fayalite
				Cr_2O_3					
FW chromite	3.10	15.60	26.50	42.00	10.40	under nitrogen 0.23		—	Chromite and other spinels
Fine grade chromite	3.01	12.20	21.50	49.50	12.30	under nitrogen 1.35	111	—	Chromite and other spinels

(*continued*)

Table 26.4 MOULDING AND CORE-MAKING MATERIALS.[†] CLASSIFICATION SPECIFICATION AND PROPERTIES—*continued*

(b) Typical mechanical analyses of moulding and core-making sands.

| Origin | Wt.% Retained on British Standard Sieve Mesh No.[‡] (*typical*) | | | | | | | | | Specific surface cm^2 g^{-1} |
	16	22	30	44	60	100	150	200	Thro' 200	
Naturally bonded sands										
Dullatur 40	—	31.8	23.2	16.7	10.3	8.4	3.6	2.0	4.0	—
Levenseat No. 9	0.9	5.3	12.0	15.1	13.9	23.4	5.1	1.4	2.4	122
Dursand High Bond	22.9	2.8	5.4	8.8	13.8	25.1	9.8	8.8	6.0	190
Weatherill	12.4	4.8	5.2	8.5	15.1	29.8	12.9	5.3	5.0	173
Bramcote	—	0.0	0.5	3.0	16.0	45.0	22.0	7.5	6.0	—
Pickering	—	1.9	0.8	0.9	1.5	17.5	57.1	14.2	6.1	—
Swynnerton	—	0.5	0.5	2.2	6.7	26.0	25.0	15.3	23.8	—
Mansfield Red	0.4	0.4	0.4	0.8	1.6	4.4	30.4	22.9	25.1	—
Erith loam	1.0	0.6	0.6	0.4	0.4	1.0	15.5	28.8	38.5	—
Washed silica sands										
Biddulph + 36 mesh	6.7	28.4	40.1	17.3	4.5	1.4	0.6	0.3	0.2	63
Arnold No. 19 (dried)	0.0	0.6	38.5	47.5	10.9	1.6	0.2	0.0	0.0	66
Garside dried No. 21	—	0.6	14.4	49.5	27.8	6.7	0.7	0.1	0.1	75
Chelford W.S. (50)	0.1	1.2	6.4	23.6	44.0	23.8	0.7	0.0	0.0	93
Erith silica	0.0	0.0	0.1	3.6	43.2	50.4	2.2	0.3	0.2	128
Kings Lynn 60 (4F)	0.0	0.0	1.0	7.5	24.4	40.6	13.1	2.3	0.6	162
Redhill 65 (F)	0.3	1.4	4.1	8.5	18.7	42.3	20.2	1.7	0.8	160
New Windsor Rose	0.0	0.1	0.5	3.4	20.9	58.8	14.5	1.3	0.4	133
Chelford Fine (95)	0.0	0.0	0.1	0.3	1.7	49.1	33.2	10.4	4.7	206
Kings Lynn SS (100)	0.0	0.0	0.1	0.2	0.6	16.4	63.6	16.6	2.2	220
Redhill H (110)	—	0.0	0.1	0.4	2.2	22.0	41.6	15.3	17.5	260
Ryarsh	0.0	0.0	0.0	0.2	0.2	1.3	59.0	29.1	7.1	320
Non-siliceous sands										
Zircoruf	—	—	—	0.2	16.0	70.7	12.8	0.1	0.2	73
AMA, zircon	—	—	—	—	0.0	6.8	72.6	19.5	1.0	113
Olivine No. 2	5.6	8.8	13.1	18.7	19.9	19.4	6.0	2.4	5.7	161
Olivine No. 3 $\frac{1}{2}$	0.2	2.1	5.9	15.1	18.7	24.2	12.4	6.1	14.6	274
FW, chromite	0.3	1.0	4.5	13.7	22.8	32.7	15.1	6.4	3.8	106
Fine grade chromite	—	0.1	0.7	6.0	14.0	28.0	23.5	15.5	12.3	187

* For comparison purposes only.

[†] 'Data sheets on Moulding Materials', SCRATA, 3rd Ed. 1972.

[‡] For aperture sizes *see* BS410, 1969—*see* Introductory Tables, Table 2.9.

[§] Included to give some indication of binder requirement, particularly liquid types.

Table 26.5 TYPICAL PROPERTIES OF SOME SODIUM SILICATES USED IN THE CARBON-DIOXIDE, SELF-SETTING SILICATE AND FLUID-SAND PROCESSES*

| Grade | Weight ratio $SiO_2 : Na_2O$ | Molecular ratio $SiO_2 : Na_2O$ | Typical analyses % by weight | | | Specific gravity at 20°C/20 | | | Approximate viscosity at 20°C cP |
			Na_2O	SiO_2	H_2O	SG	Degrees Twaddell	Degrees Baumé	
C112	2.0	2.05	15.2	30.4	54.4	1.56	112	51.8	850
C125	2.0	2.05	16.6	33.2	50.2	1.625	125	55.8	4 500
E100	2.21	2.28	13.2	29.2	57.6	1.50	100	48.1	220
H100	2.4	2.50	12.7	30.8	56.5	1.50	100	48.1	310
H112	2.4	2.5	13.7	33.3	53.0	1.56	112	51.8	2 500
M75	2.9	3.0	9.2	26.8	64.0	1.38	75	39.4	100

*K. E. L. Nicholas, 'The CO_2-Silicate Process in Foundries', BCIRA, 1972.

Table 26.6   TYPES OF MATERIAL USED IN GREEN AND DRY SAND MOULDING*

Class	Material	Approximate AFS/GFN
A	Naturally bonded sands of medium grain size (Dullatur 40 and other 'rotten rocks')	40
B	Naturally bonded sands of fine grain size (Bramcote)	80
C	Naturally bonded sands of very fine grain size (Mansfield Red)	Over 130
D	Silica sand of medium grain size (Arnold No. 19, Garside No. 21)	30–40
E	Silica sand of fine grain size (Redhill 65, Windsor Rose)	60–70
F	Silica sand of very fine grain size (Ryarsh, Redhill H)	> 100
G	Old sand renovated for further use. Grain fineness depends on original sand	
H	Added clay bond—bentonite	
I	Fuller's earth or other intermediate bonds with proprietory trade names	
J	Fire clay	
K	Organic binders—molasses, sulphite lye solution	
L	Dextrine and other cereal binders	
M	Special additions—coal dust (increasing fines and decreasing quantities for thinner castings), blacklead (graphite), blacking	
N	Sulphur	
P	Boric acid	
Q	Fluoride salts, e.g. ammonium bifluoride NH_4HF_2, ammonium silicofluoride $(NH_4)_2SiF_6$, ammonium borofluoride NH_4BF_4, ammonium fluoride NH_4F	
R	Crushed materials such as 'grog' firebrick or ganister	

* W. H. Salmon and E. N. Simons, 'Foundry Practice', Pitman, London, 1996.

Table 26.7   TYPICAL MIXTURES USED IN GREEN AND DRY SAND MOULDING*

The figures refer to weight percentages and the letters to the materials shown in Table 26.6

Nature of material	Sand	Clay	Additions	Added moisture %	Remarks
Cast-iron					
Green sand:					
Thin castings	60G 37C		3M	6	
Medium castings	76G 20B		4M	6	NB: Coal dust not used on
Thick castings	66G 28B		6M	7	castings for vitreous
Synthetic green sand:	91G				enamelling
Thin castings	54G 36E	3H	6M	$3\frac{1}{2}$	
Thick castings		4I	6M	$3\frac{1}{2}$	
Non-ferrous					
Copper alloys	70C 30F			7	
Phosphor bronze–dry sand	65G 25C		10M	7	
Aluminium alloys	90G 10C			$6\frac{1}{2}$	+5% P for high Mg alloys
Magnesium alloys	88E	4H	6N, 1P	4	
Steel[†]					
Synthetic green sand	73E 25D	$1\frac{1}{2}$H	$\frac{1}{2}$L	$3\frac{1}{2}$	
Synthetic dry sand	85G 14D	$\frac{1}{2}$H	$\frac{1}{2}$L	5	
					⎧ NB: For 'Compo' stove to
Naturally bonded dry sand	75A 25D			7	⎪ dull red-heat. Highly per-
Compo-dried mould	84R	15J	1M	8	⎨ meable refractory mixture
Loam	40B 40R	10J	10M	15–20	⎩ for very large castings

* W. H. Salmon and E. N. Simons, 'Foundry Practice', Pitman, London, 1966.
† Refractory sand essential for steel casting.

Table 26.8   TYPICAL STANDARD AFS TESTS ON CLAY-BONDED MOULDING SAND MIXTURES*

Casting	Moisture %	AFS green permeability number	Compressive strength 10^4 Pa	
			Green	Dry
Cast iron				
Stove plate	9.0	10	4.1	13.8
Radiators	7.0	35	3.4	34.5
Cylinder blocks	6.5	80	4.8	31.0
Average castings	7.0	35	4.1	41.4
Synthetic–thin	3.5	70	5.5	41.4
Non-ferrous copper base				
Small	6.0	25	4.8	28.0
Large	5.5	60	6.9	41.4
Aluminium				
Small castings	7.0	20	3.4	13.8
Large castings	7.0	35	3.4	20.7
Magnesium	4.0	80–150	5.5	69.0
Steel				
Green sand	3.5	180	5.5	48.3
Dry sand	7.5	120	6.2	103
Compo	7.5	40	8.3	172

* W. H. Salmon and E. N. Simons, 'Foundary Practice', Pitman, London, 1960.
Note: The tensile strength of sands is less than half the compressive strength.

Table 26.9   TYPICAL MIXTURES FOR THE CO_2-PROCESS AND TYPICAL PROPERTIES OF STANDARD AFS COMPACTS*

Sand	Silicate			Gassing time s	Compression strength (kPa)	
	Weight ratio $SiO_2 : Na_2O$	°TW	% Added		As gassed	After 24 h storage
Chelford 50	2:1	112	4	60	1 050	2 680
Redhill 65	2:1	112	4	60	860	2 140
Erith (56)	2:1	112	4	60	550	2 300
King's Lynn (95)	2:1	112	4	60	1 700	—
					Gassing time s	After 24 h storage
Windsor	2.0:1	112	4	40	1 340	30	5 600
Rose	2.4:1	112	4	40	2 040	24	1 790
(68)	2.9:1	100	6	40	1 620	18	1 035

*K. E. L. Nicholas, 'The CO_2-Silicate Process in Foundries', BCIRA, 1963.
CO_2 flow-rate in all cases 2.5/1 min^{-1}.

Table 26.10   SUMMARY OF CORE BINDING PROCESSES

Method	Binder	Equipment
Natural sand	Clay	Wood, metal, or resin core boxes
Synthetic sand core	Clay, usually bentonite	Metal, wood, or resin core boxes
Loam	Clay	Generally strickled on spindle or barrel
Cereals	Starch and corn products Dextrose	Wood, metal, or resin core boxes
Oil	Wide variety including linseed, cotton-seed and fish oils	Wood, metal, or resin core boxes
Cellulose	Water-soluble cellulose ethers and melamine resins	Wood, metal, or resin core boxes
CO_2	Carbon dioxide and sodium silicate plus breakdown agent, e.g. dextrose monohydrate or molasses	Wood, metal, or resin core boxes
Self-setting silicate	Sodium silicate plus one of: (a) Dicalcium silicate (b) Ferro-silicon (c) Organic esters (acetins or glycol diacetates)	Wood, metal, or resin core boxes. Special box paints preferred for (a) and (b)
Air set (no bake)	Oils, usually heat-treated Tung linseed and alkyd resin modified. Catalysts—sodium perborate, 'metal driers', e.g. cobalt naphthenate, and isocyanates	Wood, metal, or resin core boxes
Cold box	Resin (Novolak) isocyanate MDI plus trimethylamine/ air vapour	Gas dispensers Fume extraction Wood, metal, or resin core boxes
Air set (no bake)	(a) Urea-formaldehyde/furfuryl alcohol plus phosphoric acid (b) Phenol–formaldehyde/furfuryl alcohol or (c) Furfuryl alcohol polymer both plus paratoluene sulphonic acid (b) and (c) are nitrogen free	Wood, metal, or resin core boxes
Fascold	Resin (polyurethane) + catalyst (isocyanate) Chemical polymerisation at room temperature	Need special core-blowing machine (sand + resin to mix with sand + catalyst) Wood ⎫ Metal ⎬ core boxes Resin ⎭
Hot box	Phenolic furfuryl or other resin and catalyst (NH_4^+ salts or p.t.s.a.)	Metal core boxes on machine with heating and cooling cycle
Shell	Phenol formaldehyde resin Novalac type plus hexamine catalyst originally. Now usually 'precoated' or 'resin-coated' sands used	Metal core boxes on machine with heating and cooling cycle
Shaw	Ethyl silicate	Accurate and highly finished metal core boxes
Plaster	Gypsum	Accurate and highly finished metal core boxes
SO_2 process	UFFA or Furane polymer resin with peroxide mix	Gas dispensers. Fume extraction Wood, metal, or resin core boxes

(continued)

Process	Applications and limitations
Cores usually baked to develop the clay bond	Suitable for medium and large simple cores for jobbing foundries
Strength developed by baking clay bond	General use for jobbing foundry for all sizes of core. High green strength, good breakdown
Hand-made cores shaped by strickling on straw rope base, clay bond developed by baking	Suitable only for jobbing foundries for simple round and cylindrical cores
Bond developed by baking	Suitable for small and medium cores in jobbing repetition foundries
Oil oxidised by baking to develop bond	For general applications in all foundries but mostly on small and medium work. Poor green strength, and accuracy therefore difficult to maintain
Cores stoved to develop bond	Suitable for jobbing or repetition work. Good breakdown, finish and low gas content
No baking. Core gassed with CO_2 in core box	Suitable for jobbing or repetition work. Fast—accurate—stoving cores, low gas evolution. Main difficulties on breakdown, improved using breakdown agent, but storage life reduced
Silicate cures by either dehydration or for chemical gelling in 0.5–4 h	Suitable for jobbing foundry. Medium to large cores. Good finish. No fume. Breakdown little better than CO_2
Air drying process of 0.5–4 h duration plus 1–2 h baking	Suitable for jobbing foundry and large cores. Good finish, easy breakdown, good green strength, high accuracy
Almost instantaneous hardening in box. Gas passage, e.g. 3 s Air purge, e.g. 5 s	Sand air must be dry. Good dimensional accuracy. Knockout excellent. Efficient ventilation required. Cores of intermediate size in batches too small to warrant use of shell process
Resin cures after addition of acid in 10–4 min. No baking	Suitable for jobbing foundry and large cores. Good finish, easy breakdown. Good green strength. High accuracy
Both resin and catalyst liquid. Cured in 30–45 s Usable 30–60 min after strip	Machine operations are automatic Mass production of small cores—similar sizes to hot-box (Usually for iron-foundry)
Resin cures after addition of catalyst in 10–40 s baking at 180–300°C	Suitable only for large quantity repetition work and generally small cores.
Resin cured by heating cycle of 30–120 s at 200–400°C	Suitable for large quantity repetition work. Good finish, accuracy and breakdown. High permeability due to shell construction
Conversion of ethyl silicate to silica bond by hydrolysis	For accurate high-temperature metal application where surface finish and accuracy are of prime importance
Hydration of gypsum subsequently dried	For accurate low-temperature applications where surface finish and accuracy are of prime importance
Almost instantaneous hardening by SO_2 gas	Good dimensional accuracy. Knockout excellent. Not sensitive to water

(*continued*)

Table 26.10   SUMMARY OF CORE BINDING PROCESSES—*continued*

Method	Binder	Equipment
FRC (free radical cure)	Acrylic/epoxy resin + organic peroxide/sulphur dioxide gas	Gas dispensers. Fume extraction. Wood, metal, or resin core boxes
Alphaset	Alkalic phenolic resin self setting	Wood, metal, or resin core boxes
Betaset	Alkalic phenolic resin hardened by methyl formate gas	Gas dispenser. Fume extraction (although methyl formate itself is odourless and non-toxic) Wood, metal or resin core boxes

26.2   Patterns—crucibles—fluxing

Table 26.11   CONTRACTION ALLOWANCES OF COMMON ALLOYS (SAND CAST)*

Alloy	Pattern size m (in)		Type of construction	Contraction allowance Traditional	%
Grey cast iron	Up to 0.61	(2.4)	Open	1 in 96	1.04
(*see* Note 2)	0.64–1.22	(25–48)	Open	1 in 120	0.83
	Over 1.22	(Over 48)	Open	1 in 144	0.69
	Up to 0.61	(24)	Cored	1 in 96	1.04
	0.64–1.22	(25–48)	Cored	1 in 120	0.83
	Over 1.22	(Over 48)	Cored	1 in 144	0.69
Cast steel	Up to 0.61	(24)	Open	1 in 48	2.08
	0.64–1.83	(25–72)	Open	1 in 64	1.56
	Over 1.83	(Over 72)	Open	1 in 77	1.30
	Up to 0.46	(18)	Cored	1 in 48	2.08
	0.48–1.22	(19–48)	Cored	1 in 64	1.56
	1.24–1.68	(49–66)	Cored	1 in 77	1.30
	Over 1.68	(Over 66)	Cored	1 in 96	1.04
Manganese steel	—	—	—	1 in 38 decreasing to	2.63
				1 in 64 for long castings	1.56
Malleable cast iron	—		Section thickness		
(*see* Note 3)			mm (in)		
			1.6 ($\frac{1}{16}$)	1 in 70	1.43
			3.2 ($\frac{1}{8}$)	1 in 77	1.30
			4.8 ($\frac{3}{16}$)	1 in 80	1.25
			6.4 ($\frac{1}{4}$)	1 in 85	1.18
			9.5 ($\frac{3}{8}$)	1 in 96	1.04
			12.7 ($\frac{1}{2}$)	1 in 110	0.91
			15.9 ($\frac{5}{8}$)	1 in 128	0.78
			19.1 ($\frac{3}{4}$)	1 in 152	0.66
			22.2 ($\frac{7}{8}$)	1 in 256	0.39
			25.4 (1)	1 in 384	0.26
Pearlitic nodular iron, as cast	—	—	—	1 in 120 to 1 in 180	0.83 to 0.56
Pearlitic nodular iron, heat-treated to ferritic	—	—	—	1 in 120	0.83

(*continued*)

Process	Applications and limitations
Almost instantaneous hardening by SO_2 gas	Good dimensional accuracy. Knockout excellent. Not sensitive to water + long bench life of mixed sand; to resistant to distortion on casting
The resin is cured with a liquid ester	Environmentally pleasant. Resistant to finning and veining defects. Good surface finish for steel castings
The resin is cured with a gaseous ester: methyl formate	Environmentally pleasant. Resistant to finning and veining defects. Good surface finish for steel castings

Table 26.11   CONTRACTION ALLOWANCES OF COMMON ALLOYS (SAND CAST)*—*continued*

Alloy	Pattern size m (in)		Type of construction	Contraction allowance Traditional	%
Nodular iron (thin section) containing carbide as cast and annealed	—		—	1 in 120 contraction to 1 in 240 expansion	0.83 0.42
Aluminium alloys	Up to 1.22	(48)	Open	1 in 77	1.30
	1.24–1.83	(49–72)	Open	1 in 85	1.18
	Over 1.83	(Over 72)	Open	1 in 96	1.04
	Up to 0.61	(24)	Cored	1 in 77	1.30
	0.64–1.22	(25–48)	Cored	1 in 85 to 1 in 96	1.20 to 1.04
	Over 1.22	(Over 48)	Cored	1 in 96 to 1 in 192	1.04 to 0.52
Magnesium alloys (*see* Note 4)	Up to 1.22	(48)	Open	1 in 70	1.43
	Over 1.22	(Over 48)	Open	1 in 77	1.30
	Up to 0.61	(24)	Cored	1 in 77	1.30
	Over 0.61	(Over 24)	Cored	1 in 77 to 1 in 96	1.30 to 1.04
Brass	—		—	1 in 64	1.56
Bronze	—		—	1 in 96 to 1 in 48	1.04 to 2.08
Nickel alloys	—		—	1 in 48	2.08
Everdur (silicon bronze)	—		—	1 in 64	1.56
PMG (silicon bronze)	—		—	1 in 96 to 1 in 64	1.04 to 1.56
High tensile brass	—		—	1 in 120 to 1 in 64	0.83 to 1.56
Aluminium bronze	—		—	1 in 43	2.32
Zinc alloys	—		—	1 in 85	1.18

* From 'Cast Metals Handbook', American Foundrymen's Society.

Notes:
(1) Contraction varies with the casting design, type of metal, pouring temperature, and mould or core resistance. It may be necessary to use several different contraction allowances for the various dimensions of a single pattern.
(2) Standard pattern maker's allowance for common grey iron is 1 in 96. For higher strength alloys and white cast irons, the contraction allowance averages 1 in 77.
(3) Contractions shown for malleable irons are net values, e.g. white iron castings which shrink 1 in 48 when cast, expand 1 in 96 during the anneal, giving a net contraction of 1 in 96.
(4) Contraction varies with the alloy. Average values are:

A8, AZ91, C	1 in 85.
Z5Z, RZ5, TZ6, MSR–A, MSR–B	1 in 77.
ZREI, ZT1	1 in 64.

Table 26.12   WEIGHTS OF CASTINGS FROM PATTERN WEIGHTS

Multiply pattern weight by the factor shown for the type of wood/metal and the alloys. For wooden patterns, allowance must be made for any metal reinforcement in the pattern.

Pattern material				*Metal cast*				
	Cast steel	Cast iron	Brass	Gun metal or bronze	Zinc	Copper	Aluminium	Magnesium
Mahogany	9.5	8.5	9.5	10.0	8.2	10.1	3.1	2.05
White pine	16.3	14.7	16.5	17.3	14.3	17.5	5.3	3.5
Yellow pine	14.4	13.1	14.7	15.4	12.7	15.6	4.7	3.1
Oak	10.4	9.4	10.5	10.8	9.1	11.2	3.4	2.24
Aluminium alloys	2.87	2.56	2.87	2.98	2.45	3.08	0.93	0.61
Brass	0.98	0.84	0.95	0.99	0.81	1.04	0.31	0.21
Iron	1.09	0.97	1.09	1.13	0.93	1.17	0.35	0.23
White metal	1.08	0.96	1.08	1.11	0.92	1.15	0.34	0.23

Note:
Allowance must also be made for cores and coreprints. For round cores and prints multiply the square of the diameter by the length of the core and prints in inches, and the product by 0.014. This will give the weight of the white pine core in pounds to be deducted from the weight of the pattern.

26.2.1   Pattern materials

(1) Wood: Yellow pine (for limited production) and mahogany (for production patterns) represent the most usual practice.
(2) Metal:
    (a) Cast iron.
    (b) Brass.
    (c) Aluminium alloy
    (d) White metal (only to a limited extent because of softness and easily worn away by moulding sand. Used, for example, for lining stripping plates).

Notes:

(1) For small patterns, an alloy of 40% Sn, 18% Sb, 40% Pb, 2% Bi, is used, shrinkage being negligible. Similarly, use is made of a modified 'Wood's metal' (50% Bi, 26.7% Pb, 10% Cd, 13.3% Sn) for making master patterns. It is completely molten at about 70°C and can readily be poured at the temperature of boiling water into wood or plaster moulds.
(2) Contraction allowance for 'master' patterns for casting metal patterns is contraction allowance for casting material, plus contraction allowance for metal pattern material.
(3) Plaster of Paris and proprietory cements: mainly for odd-side impressions.
(4) Plastics: epoxy resins, often reinforced with fibreglass.

Table 26.13   RECOMMENDED STANDARD COLOURS FOR PATTERNS

Part of pattern		*Colour*
As-cast surfaces which are to be left unmachined		Red or orange
Surfaces which are to be machined		Yellow
Core prints for unmachined openings and end prints	Periphery	Black
	Ends	Black
Core prints for machined openings	'A' periphery	Yellow stripes on black
	'B' ends	Black
Pattern joint (split patterns)	'A' cored section	Black
	'B' metal section	Clear varnish
Touch core	Cored shape	Black
	Legend	'Touch'
Seats of and for loose pieces and loose core prints		Green
Stop offs		Diagonal black stripes with clear varnish
Chilled surfaces	Outlined in	Black
	Legend	'Chill'

Table 26.14 GUIDE TO MACHINING ALLOWANCE ADDITIONS TO PATTERN CONTRACTION ALLOWANCES*

Casting alloy	Pattern size m (in)	Bore mm (in)	Finish mm (in)
Cast iron	Up to 0.3 (12)	3.12 (0.125)	2.38 (0.09)
	0.3–0.6 (13–24)	4.76 (0.187 5)	3.18 (0.125)
	0.6–1.1 (25–42)	6.35 (0.25)	4.76 (0.187 5)
	1.1–1.5 (43–60)	7.94 (0.312 5)	6.35 (0.25)
	1.5–2.0 (61–80)	9.53 (0.375)	7.94 (0.312 5)
	2.0–3.0 (80–120)	11.1 (0.437 5)	9.53 (0.375)
Cast steel	Up to 0.3 (12)	4.76 (0.187 5)	3.18 (0.125)
	0.3–0.6 (13–24)	6.35 (0.25)	4.76 (0.187 5)
	0.6–1.1 (25–42)	7.94 (0.312 5)	7.94 (0.312 5)
	1.1–1.5 (43–60)	9.53 (0.375)	9.53 (0.375)
	1.5–2.0 (61–80)	12.7 (0.5)	11.1 (0.437 5)
	2.0–3.0 (80–120)	15.9 (0.625)	12.7 (0.5)
Malleable iron	Up to 0.15 (6)	1.58 (0.062 5)	1.58 (0.062 5)
	0.15–0.23 (6–9)	2.38 (0.093 75)	1.58 (0.062 5)
	0.23–0.30 (9–12)	2.38 (0.093 75)	2.38 (0.093 75)
	0.3–0.6 (12–24)	3.97 (0.156 25)	3.18 (0.125)
	0.6–0.9 (24–35)	4.76 (0.187 5)	4.76 (0.187 5)
Brass, bronze and	UP to 0.3 (12)	2.38 (0.093 75)	1.58 (0.062 5)
aluminium alloy	0.3–0.6 (13–24)	4.76 (0.187 5)	3.18 (0.125)
castings	0.6–0.9 (25–36)	4.76 (0.187 5)	3.97 (0.156 25)

* 'Pattern Makers Manual', American Foundrymen's Society. *Note:* Above pattern sizes quoted, need special instructions.

Table 26.15 MATERIALS FOR PARTING POWDERS AND LIQUIDS

Powder	Moisture proofer
Pulverised limestone*	
Precipitated calcium carbonate*	Paraffin wax
Fireclay grog	
Sillimanite	Calcium stearate
Phosphate rock	
Bone ash	Aluminium stearate
Walnut shells	
Ptfe (from aerosol spray)	
Lycopodium powder (very expensive, for art casting only)	
Aluminium powder (as in aluminium paint)	
Liquid—paraffin plus small addition Colza oil	

* React unsatisfactory with chemical binders such as acid-catalyzed resin binders.

MOULD DRESSING, POWDERS, PAINTS

Table 26.16 SURFACE POWDERS FOR GREEN SAND MOULDS

Non-ferrous	Flour, soapstone, talc (French chalk)
Cast iron	Plumbago, black lead, bituminous coal dust

Material dusted on to mould surface, rubbed in by hand or sleeked with a smooth tool.

Table 26.17 MATERIALS FOR MOULD DRESSINGS

Filler material	Suspending agent	Binder	Carrier
Silica flour	(a) For water base	Core oil	Water
Zircon flour	Western bentonite	Dextrine	Iso-propyl alcohol
Graphite	China clay	Fire clay	(<5% H_2O)
Olivine	Carboxy-methyl-cellulose	Linseed oil	Carbon tetrachloride
Mica	Sodium alginates	Phenolic resin	Toluene
Molochite	(b) For spirit base	Vinsol resin	Petroleum-ether
Magnesite	Sodium bentonite	Polyvinyl acetate	Methylated spirits
Chamotte	Bentone	Molasses	
Tale			
Plumbago			
Carbon black			

Table 26.18 MIXES FOR WATER-BASED MOULD DRESSINGS*

Alloy type	Mixture components (wt. % & component)				Added water wt. %
Copper base	20A	6.6B	6.6C		66.6
High lead or phosphor bronze	11.5D	23B	8.5C		57
Cores for heavy section bronze		34E			66
Aluminium	22A	11F	11C		56
Cast-iron					
Blackwash	22G	4H or C			74
Blackwash	21G	4L	4H		71
Heavy section	20J	6.6L	6.6B		66.6
Heavy section	25K	6L	3H	3M	63
Steel					
Light section	30K	1.5L	4.5M		64
Facing mix	45J	1.5D	4H		49.5
Manganese steel	42.5N	5L	2.5H		50
Zircon to 1.8Sp.G (65°CBé)	60P	4L	1Q		35

* W. H. Salmon and E. N. Simons, 'Foundry Practice', Pitman, London, 1966.

A.	Talc.	E.	Sodium silicate.	J.	Chamotte (200 mesh).	N.	Magnesite.
B.	Plumbago.	F.	Whitening.	K.	Silica flour.	P.	Zircon (200 mesh).
C.	Molasses.	G.	Blackening.	L.	Bentonite.	Q.	Core cream.
D.	China clay.	H.	Dextrine.	M.	Core oil.		

Table 26.19 TYPICAL 'SPIRIT–BASE' OR 'FLASH-OFF' DRESSINGS

Component	Polar solvent carrier		Non-polar solvent carrier	
	Description	p.b.w.*	Description	p.b.w.*
Refractory filler	Zircon flour (200 mesh)	69.0	Zircon flour (300 mesh)	29.0
Suspension agent	Bentone 18C + pure	2.8	Bentone 34 + Ind.	1.0
	grade toluene	1.5	meths (64 o.p.)	0.4
Resin binder	Novolac type	1.3	Cumarone/Indene type	2.0
Carrier	Iso-propyl alcohol	25.0	White spirit (SBP6)	31.6

* p.b.w.—parts by weight.

Table 26.20 MOULD COATINGS FOR SPECIAL APPLICATIONS

Cast metal	Coating constituent(s)	Purpose	For further information consult
Aluminium base alloys	Acetylene soot	Increased fluidity,	—
	Hexachlorethane-containing wash	Increased fluidity + refined grain	—
Copper base alloys	325 mesh molochite Powdered vinsol resin 99% Iso-propyl alcohol	Prevention of 'steam' reaction at metal–mould interface resulting in globular sub-surface porosity	BNFMTC, AFS
Phosphor-bronzes and gunmetals	Finely divided Al–MG alloy suspended in a solvent (e.g. toluene)	Prevention of pinhole porosity	BNFMTC, AFS
Grey iron	Finely divided bismuth	Suppresses micro-shrinkage porosity in 'hot spots'	BCIRA, AFS
Grey iron	Finely divided tellurium	Local chilling for additional hardness	BCIRA, AFS
Malleable irons	Zinc/ground coke or cobalt, cobalt/iron oxide	Prevents hot-checking on white-iron castings and modifies dendrite structure	BCIRA, AFS

BNFMTC—British Non-Ferrous Metals Technology Centre.
BCIRA—British Cast Iron Research Association.
AFS—American Foundry Society.

26.2.2 Crucibles and melting vessels

Refractory crucibles were originally produced from clay alone, but since the beginning of the century most melting crucibles have incorporated up to 30% graphite, bonded with clay, to improve thermal shock resistance, erosion from fluxes and conductivity. Additions of silicon carbide are also made where higher duty is required. The life of crucibles has been improved over recent years by development of glazes which protect the crucible against oxidation or 'perishing'.

Carbon-bonded silicon carbide crucibles made from silicon carbide and graphite bonded with tar, which on firing produces a carbon bond, have superior thermal shock resistance and improved performance against flux attack and oxidation under certain conditions. 80% of crucibles manufactured in recent years are produced from carbon-bonded silicon carbide. They are generally more expensive than clay-graphite crucibles.

TYPES OF CRUCIBLES

Three types of crucibles are in general use for non-ferrous work. These are:

1. *Crucibles used in pit or 'pull-out' furnaces*
   A and C shape which have to be removed from the furnace for metal to be poured.
   These vary in size from a few kilograms to greater than 1 000 kg brass capacity, though the latter sizes are not generally used on account of the difficulty of handling them. In oil- and gas-fired furnaces the crucible rests on a small stand or stool—of similar material to the crucible, and a ring or cylinder to accommodate and protect protruding solid metal is sometimes placed on the top of the crucible. The chief value of the pit furnace lies in its flexibility and low capital cost and it is ideal for small or medium-sized foundries where relatively small quantities of a variety of metals or alloys are required.
2. *Crucibles used in tilting furnaces*
   For convenience in pouring or where the size of crucible is such it cannot be readily handled, it is mounted permanently in the furnace, which is tilted to discharge its molten contents.
   Crucibles varying from 70 kg to greater than 760 kg brass capacity are in general use, though larger crucibles and furnaces are now available with capacities of up to 1 520 kg brass (590 kg aluminium).
3. *Maintaining crucibles*
   Crucibles or basins for holding molten metal for periodic casting or for process work are extensively used for purposes such as die casting.
   Those used for die casting are usually of basin shape and are wide in proportion to their height. They are made in sizes up to 600 kg aluminium capacity for aluminium die casting, or 135 to 180 kg brass capacity for aluminium bronze or brass die casting.
   Silicon carbide basins are particularly suited to maintaining molten aluminium alloys for die casting, as neither aluminium nor aluminium oxide attack the crucible material.

26.2.3 Iron and steel crucibles—fluxing

Steel crucibles are normally used for the melting of magnesium alloys. Low carbon steels are used and nickel and cobalt should be absent from the steel because of the adverse effect of even small traces of these elements on the corrosion resistance of magnesium alloys.

The crucibles may be of a pressed construction or made up of welded boiler plate.

For very large melts, cast-steel ($\sim$0.4% carbon) crucibles are used.

Although cast-iron crucibles were often used in holding furnaces for aluminium alloy die casting, they appear to have fallen out of favour and are now rarely found in use.

Where they are still used a refractory wash of similar nature and composition to the following:

Finely ground whitening	9 wt.%
Water	90 wt.%
Waterglass (hot conc. soln)	1 wt.%

applied daily will minimise iron contamination.

Silicon carbide or plumbago basins are used in preference to cast-iron pots nowadays, for aluminium melting.

Table 26.21 COMMON FLUXING AND INOCULATION PRACTICES FOR VARIOUS ALLOYS

Alloy group	Cover fluxes	Degassing and/or cleansing fluxes and treatments
Most aluminium alloys (except Al–Mg, *see* below)	(a) NaCl + NaF (or cryolite) + KCl + small amounts of $SO_4^{2-} + NO_3^-$ (b) $Na_2SiF_6 + NaCl^-$ $\downarrow$ also used for metal recovery from drosses (c) Reverberatory furnaces (typical) $\quad CaCl_2 \quad 50\%$ $\quad NaCl \quad 30\%$ $\quad KCl \quad 20\%$ (d) Rotary furances (typical) $\quad NaCl \quad 45\%$ $\quad KCl \quad 45\%$ } m. pt 607°C $\quad Cryolite \quad 10\%$	(a) Inert gases N_2, He, Ar [Response time slowest] [Use N_2 at <700°C or – AlN formation Use N_2 at ≤670°C with Al–Mg alloys or → Mg_3N_2 formation] (b) Inert active gas mixtures e.g. $90N_2/10Cl_2$ [Response time reasonably fast] (c) Active gases Cl_2, F_2 or compounds containing them, e.g. hexachlorethane C_2Cl_6 and Freon 12 CCl_2F_2 [Response time fast] (d) Metal chlorides $\quad ZnCl_2$ $\quad BCl_3$ $\quad CuCl_2$ } Careful selection to prevent contamination. All hygroscopic $\quad MnCl_2$ $\quad TiCl_2$
Aluminium–magnesium alloys (3–10% Mg)	Similar to magnesium alloys. Mixtures of chlorides and fluorides, e.g. $MgCl_2 + KCl + MgF_2$ or CaF_2 *NB:* No sodium compounds of any description if present → embrittlement, hot-shortness	Groups (b) and (c) above (20% Mg loss at 690°C 90% Mg loss at 710°C, 50% Mg loss with C_2Cl_6 regardless of temperature) Also used—K_2ZrF_6, Ti sponge
Magnesium alloys (a) Mg $\quad$ Mg–Al $\quad$ Mg–Mn	$CaCl_2 \quad 40\%$ $NaCl \quad 30\%$ $KCl \quad 20\%$ 'Melrasal Z' $MgCl_2 \quad 10\%$	Degas with (a) C_2Cl_6 $\quad$ (b) Cl_2 $\quad$ (c) N_2 'Flux bubbled' [$MgCl_2$ rich flux on crucible bottom 650–680°C] $MgCl_2 \quad 35\%$ $CaCl_2 \quad 15\%$ $NcCl \quad 10\%$ Inspissating flux $KCl \quad 10\%$ 'Melrasal E' $CaF_2 \quad 20\%$ $MgO \quad 10\%$ $CaCl_2 \quad 19\%$ $NaCl \quad 11\%$ For holding furnaces $KCl \quad 7.5\%$ $BaCl_2$ addition to increase $BaCl_2 \quad 37.5\%$ density $MgF_2 \quad 25\%$
	'Melrasal UE' all-purpose flux (melting, refining, covering) made up of 1 part 'Z' flux to 6 parts 'E' flux	
(b) Mg–Zr	$BaCl_2 \quad 39\%$ $MgCl_2 \quad 30\%$ $KCl \quad 10\%$ $BaF_2 \quad 21\%$ $BaCl_2 \quad 40\%$ $MgCl_2 \quad 35\%$ $KCl \quad 10\%$ for alloys containing thorium $MgF_2 \quad 15\%$	$BaCl_2 \quad 28\%$ $CaCl_2 \quad 26\%$ $KCl \quad 10\%$ for alloys containing rare earths $NaCl \quad 16\%$ $MgF_2 \quad 20\%$	Zr acts as its own degasser $Zr + 2[H]Mg \rightarrow ZrH_2 \downarrow$
Copper-base alloys HC Copper	Charcoal	Degas N_2 only if necessary Deoxidation: 15% phosphor-copper (equiv. to 0.005% P) followed by CaB_2, Li (0.003–0.01%) or 10% Be–Cu (0.005–0.02% Be)

Special purpose fluxes and inoculants

(1) Grain refinement
  (a) Combined additions of K_2TiF_6
          +
          KBF_4
          +
    (sometimes) C_2Cl_6 for better efficiency
    KBF_4 used alone if alloy contains Ti near maximum specified
  (b) Using a 'hardener', e.g.: Al–5% Ti–1% B

(2) 'Modification' of Al–Si alloys

  (a) Eutectic or near-eutectic (11–14% Si)
    Salt mixtures containing NaF,
    e.g. 88% NaCl–KCl (1:1) + 12% NaF (fuses at 607°C)
    Vacuum-melted sodium sealed in air-tight Al containers
    ⎧ Sr modification lasts for several hours
    or Sr ⎨ Sr modified alloys can be remelted without loss
    ⎩ or fine structure

  (b) Hypereutectic (17–25% Si)
    Red phosphorus introduced as particles (<60 μm) in a mixture of KCl and K_2TiF_6. Effect strengthened
    if subsequent Cl_2 treatment

Grain refinement with salt mixtures as in 1(a) above

Grain refinement:

Alloy type	Treatment	Degree of grain refinement achieved
Mg–Al(–Zn–Mn)	Carbon inoculation (C_2Cl_6)	Marked
	Superheat to ≥850°C	Marked
Mg–Al–Mn(–Zn)	$FeCl_3$	Marked
Mg–Zn(–RE–Mn)	$FeCl_3$, Zn–T% Fe alloy, or NH_3	Very marked
Mg–Mn	$Ca + N_2$	Mild
	Zr	Increases with falling Mn content
Mg–Zr	Use master alloy (proprietory) 'ZIRMAX' Max gr. ref. at max solubility of Zr (0.6%)	
	Zr will not grain refine in presence of: Al, Si, Sn, Ni, Fe, Co, Mn, Sb	Zr will grain refine in presnce of: Zn, Cd, Ce, Ag, Th, Tl, Cu, Bi, Pb, Ca

Protection:
(1) 'Dusting' fluxes
    S, NH_4BF_4,
    NH_4HF_2, NH_4F,
    H_3BO_4
    e.g. S + 5 to 30% H_3BO_4
(2) SO_2 atmosphere
    'Purge' mould prior
    to cast

(*continued*)

Table 26.21   COMMON FLUXING AND INOCULATION PRACTICES FOR VARIOUS ALLOYS—*continued*

Alloy group	Cover fluxes		Degassing and/or cleansing fluxes and treatments	
Gunmetals	Charcoal	C($\geq$25 mm thick)	Degas with dry N_2	
	Borocalcite	$4CaO \cdot 5B_2O_3 \cdot 9H_2O$	(e.g. 45 kg melts at 5–10 l min^{-1} for about 5 min)	
	Boric oxide	B_2O_3 (crucible attack)	or use 'oxidation-reduction' technique	
	Borax	$Na_2B_4O_7 10H_2O$		
	Soda ash	Na_2CO_3	*Oxidising agents*	*Deoxidants*
		(desulphuriser)		
	Common salt	NaCl	Cupric oxide CuO	15% P–Cu
	Calcium fluoride	CaF_2	Cuprous oxide Cu_2O^*	Lithium
	Silica sand	SiO_2	Manganese dioxide ⎫ $MnO_2^†$	10% Li–Cu
	Glass (green bottle)	—	Manganese ore ⎭	98% Zn–2% Na
	Barium carbonate	$BaCO_3$	Barium peroxide BaO_2	
		(desulphuriser)		
	Calcium oxide (lime)	CaO	⎡* Reduced by H_2 to Cu, does not evolve oxygen⎤	
			⎣† Also removes As, Fe, Si and S ⎦	
Tin bronzes	Typical: Fused borax 30, dry silica sand 50, cupric oxide 20:2.75 kg charge			
	Electric furnace melting: 3 parts lime +1 part fluorspar			
Brasses and high tensile brasses	Charcoal, borax, greenbottle glass		None usually practised	
	↓ ↓		(Zn, Al, Mn are all deoxidants)	
	Virgin All scrap melts			
	melts			
Aluminium bronzes	Manganese chloride (MnCl$_2$) ⎫ Both reduce		Degas with dry N_2 or CO_2	
	Cryolite (AlF$_3$–3NaF) ⎭ crucible life			
	(Do not use compounds containing boron)			
Silicon bronzes	Bottle glass + small % of fluoride salts		Degas with dry N_2 or CO_2	
	(Pb free)			
	Can be thickened before skim with silica		Deoxidation not required	
	sand. *NB:* Not charcoal			
Nickel silvers	Broken glass, 80:20 borax–boracic acid,		Deoxidation 0.1% Mn + 0.06% max. Mg	
	Borocalcite, all Pb free. *NB:* not charcoal.		(Mg also desulphurises)	
	Thicken for removal with powdered CaO			
Copper-nickels and Monels	As for Ni-Silvers + MnO$_2$ (1% charge wt),		Deoxidation Si 0.5–2.0% dependent on % Ni	
	NiO or Cu$_2$O		Mg 0.03–0.06% dependent on % Ni	
	NB: Not charcoal		e.g. for 30% Ni alloy 1.2% Mn followed by	
			0.6% Si followed by	
			0.03% Mg	
			Can degas with N_2 or dry air	
Cast-irons			Acid cupola: Limestone 28–32 kg tonne^{-1}	
			providing	
			(a) Standard cupola practice	
			(b) Coke ash content–10%	
			(c) $CaCO_3$ is 96% of limestone	
			'Energised' flux for increased efficiency	
			21 kg tonne^{-1} (14 kg limestone + 7 kg CaF$_2$)	
			Basic cupola: 80 to 90 kg tonne^{-1} of flux made up as	
			Limestone 30–40% coke wt	
			Dolomite 10% coke wt	
			*Fluorspar 25% of limestone	

* Can replace fluorspar by 50 : 50 fluorspar/soda ash.

Special purpose fluxes and inoculants

Grain refinement not usually beneficial with these alloys—encourages 'layer porosity'—although has been achieved with 0.03% Zr + 0.02% B, Ti, Co

Grain refinement of manganese bronzes by iron

Fe in alloy acts as a grain-refiner

—

Some grain refinement with Mn

—

Anhydrous soda ash or calcium carbide for desulphurisation of acid cupola metal
Calcium silicide: 'Meehanite iron'

Ferro–silicon (+1–2% Al and up to 1% Ca), zirconium etc., { for grain size control (eutectic cells) and reduction of 'chill depth'

75% FeSi (Al and Ca free) and 1–4% Sr
↓
elimination of 'pinholing' in green sand moulds

Nodularisation (spheroidisation): sulphur ≤0.02%
Magnesium for hypo- and hypereutectic irons introduced as: Mg impregnated coke
Mg–Ni(Si) alloys
Mg–FeSi alloys
Mg–Si alloys

Cerium for hypereutectic irons introduced as mischmetal

26.3 Aluminium casting alloys

Table 26.22 ALUMINIUM–SILICON ALLOYS

Specification BS 1490:1988	LM6M(Ge)	LM20M(Ge)	LM9M(SP)	LM9TE(SP)	LM9TE(SP)	LM13TF(SP)	LM13TF(SP)	LM13TF7(SP)
Related British Specifications	BS L33							
Composition (%) (single figure indicates maximum)								
Copper	0.1	0.4	0.1			0.7–1.5		
Magnesium	0.1	0.2–0.6	0.2			0.8–1.5		
Silicon	10.0–13.0	10.0–13.0	10.0–13.0			10.0–12.0		
Iron	0.6	1.0	0.6			1.0		
Manganese	0.5	0.5	0.3–0.7			0.5		
Nickel	0.1	0.1	0.1			1.5		
Zinc	0.1	0.2	0.1			0.5		
Lead	0.1	0.1	0.1			0.1		
Tin	0.05	0.1	0.05			0.1		
Titanium	0.2	0.2	0.2			0.2		
Other	—	—	—			—		
Properties of material								
Suitability for:								
Sand casting	E	E*	G			G		
Chill casting (gravity die)	E	E	E			G		
Die casting (press die)	G	G	G*			F*		
Strength at elevated temperature	P	P	G			E		
Corrosion resistance	E	G	E			G		
Pressure tightness	E	E	G			F		
Fluidity	E	E	G			G		
Resistance to hot shortness	E	E	E			E		
Machinability	F	F	F			F		
Melting range, °C	565–575	565–575	550–575			525–560		
Casting temperature range, °C	710–740	680–740	690–740			680–760		
Specific gravity	2.65	2.68	2.68			2.70		

Heat treatment							
Solution temperature, °C	—	—	—	520–535	—	515–525	515–525
Solution time, h	—	—	—	2–8	—	8 (minimum)	8 (minimum)
Quench	—	—	—	Cold water	—	Water, 70–80 °C	Water, 70–80°C
Precipitation temperature, °C	—	—	150–170	150–170	160–180	160–180	For pistons: 200–250
Precipitation time, h	—	—	16 (minimum)	16 (minimum)	4–16	4–16	4–6**
Stabilisation temperature, °C	—	—	—	—	—	—	—
Stabilisation time, h	—	—	—	—	—	—	—
Special properties	Suitable for thin and intricate castings, readily welded	Pressure casting alloy		Suitable for low-pressure casting. High strength and hardness	Low coefficient of expansion. Good bearing properties. Piston alloy		
Mechanical properties—sand cast—SI units (Imperial units in brackets)							
Tensile stress min., MPa (tonf in^{-2})	160(10.4)	—	170(11.0)	240(15.5)	—	170(11.0)	140(9.1)
Elongation min. %	5	—	1.5	0–1	—	0.5	1
Expected 0.2% proof stress,	60–70	—	110–130	220–250	—	160–190	130(8.4)
MPa (tonf in^{-2})	(3.9–4.5)	—	(7.1–8.4)	(14.2–16.2)	—	(10.4–12.3) HB 100–150	HB 65–85
Mechanical properties—chill cast—SI units (Imperial units in brackets)							
Tensile stress min., MPa (tonf in^{-2})	190(12.3)	190(12.3)	230(14.9)	295(19.1)	210(13.6)	280(18.1)	200(12.9)
Elongation, min. %	7	5	2	0–1	1	1	1
Expected 0.2% proof stress,	70–80	75–85	150–170	270–280	—	270–300	190(12.3)
MPa (tonf in^{-2})	(4.5–5.2)	(4.9–5.5)	(9.7–11.0)	(17.5–18.1)	HB90–120	(17.5–19.4) HB 100–159	HB 65–85

* Not normally used in this form.
† If Ti alone is used for grain refinement then Ti $\not>$ 0.05%.
‡ Fully heat-treated.
§ Refine with phosphorus—subject to examination under microscope.
** Or such time to give required BHN.

Notes:
Association of Light Alloy Refiners and Smelters Grading;
E—Excellent, F—Fair, G—Good, P—Poor, U—Unsuitable,
(Ge—General purpose alloy; SP—Special purpose alloy as per BS 1490:1988).

Table 26.22   ALUMINIUM–SILICON ALLOYS—*continued*

Specification BS 1490:1988 / Related British Specifications	LM18M(SP)	LM25M(Ge)	LM25TE(Ge)	LM25TB7(Ge)	LM25TF(Ge)	LM29TE(SP)	LM29TF(SP)
Composition % (Single figure indicates maximum)							
Copper	0.1		0.20				0.8–1.3
Magnesium	0.1		0.20–0.60				0.8–1.3
Silicon	4.5–6.0		6.5–7.0				22–25
Iron	0.6		0.5				0.7
Manganese	0.5		0.3				0.6
Nickel	0.1		0.1				0.8–1.3
Zinc	0.1		0.1				0.2
Lead	0.1		0.1				0.1
Tin	0.05		0.05				0.1
Titanium	0.2		0.2†				0.2
Other	—		—				Cr 0.6; Co 0.5, P§
Properties of material							
Suitability for:							
Sand casting	G		G				
Chill casting (gravity die)	G		E				
Die casting (press die)	G*		G*				
Strength at elevated temperature	P		G‡				
Corrosion resistance	E		E				
Pressure tightness	E		G				
Fluidity	G		G				
Resistance to hot shortness	E		G				
Machinability	F		F				
Melting range, C	565–625		550–615				520–770
Casting temperature range, °C	700–740		680–740				At least 830
Specific gravity	2.69		2.68				2.65

Heat treatment						
Solution temperature, °C	—	—	525–545	525–545	—	495–505
Solution time, h	—	—	4–12	4–12	—	4
Quench	—	—	Water, 70–80°C	Water, 70–80°C	—	Air blast
Precipitation temperature, °C	155–175	155–175	—	—	185	185
Precipitation time, h	8–12	8–12	—	—	To produce HB requirement	8
Stabilisation temperature, °C	—	—	250	—	185	—
Stabilisation time, h	—	—	2–4	—	8–12	—
Special properties	Readily welded	General purpose high-strength casting alloy			More suited to chill (grav. die) casting; Piston alloy	
Mechanical properties—sand cast—SI units (Imperial units in brackets)						
Tensile stress min., MPa (tonf in⁻²)	130(8.4)	150(9.7)	160(10.4)	230(14.9)	120(7.8)	120(7.8)
Elongation min.%	3	1	2.5	0–2	0.3	0.3
Expected 0.2% proof stress, MPa (tonf in⁻²)	55–60(3.6–3.9)	80–100(5.2–6.5)	120–150(7.8–9.7)	200–250(12.9–16.2)	120(7.8)	120(7.8)
					HB 100–140	HB 100–140
Mechanical properties—chill cast—SI units (Imperial units in brackets)						
Tensile stress min., MPa (tonf in⁻²)	140(9.1)	190(12.3)	230(14.9)	280(18.1)	190(12.3)	190(12.3)
Elongation, min. %	4	3	5	2	0.3	0.3
Expected 0.2% proof stress, MPa (tonf in⁻²)	60–70(3.9–4.5)	80–100(5.2–6.5)	130–200(8.4–12.9)	220–260(14.2–16.8)	170(11.0)	170–190 (110.–12.3)
MPa (tonf in⁻²)					HB 100–140	HB 100–140

* Not normally used in this form.
† If Ti alone is used for grain refinement then Ti ≯ 0.05%.
‡ Fully heat-treated.
§ Refine with phosphorus—subject to examination under microscope.
** Or for such time to give required BHN.

Note:
E—Excellent, F—Fair, G—Good, P—Poor, U—Unsuitable.
(Ge—General purpose alloy; SP—Special purpose alloy as per BS: 1490:1988).

Table 26.23   ALUMINIUM–SILICON–COPPER ALLOYS

Specification BS 1490:1988 *Related British Specifications*	LM2M(Ge)	LM4M(Ge)	LM4MTF (Ge)	LM16TB (SP)	LM16TF 3L78 (SP)	LM21M(SP)
Composition % (Single figures indicate maximum)						
Copper	0.7–2.5	2.0–4.0		1.0–1.5		3.0–5.0
Magnesium	0.30	0.15		0.4–0.6		0.1–0.3
Silicon	9.0–11.5	4.0–6.0		4.5–5.5		5.0–7.0
Iron	1.0	0.8		0.6		1.0
Manganese	0.5	0.2–0.6		0.5		0.2–0.6
Nickel	0.5	0.3		0.25		0.3
Zinc	2.0	0.5		0.1		2.0
Lead	0.3	0.1		0.1		0.2
Tin	0.2	0.1		0.05		0.1
Titanium	0.2	0.2		0.2*		0.2
Properties of material Suitability for:						
Sand casting	G†	G		G		G
Chill casting (gravity die)	G†	G		G		G
Die casting (press die)	E	G		F†		G†
Strength at elevated temp.	G‡	G		G		G
Corrosion resistance	G	G		G		G
Pressure tightness	G	G		G		G
Fluidity	G	G		G		G
Resistance to hot shortness	E	G		G		G
Machinability	F	G		G		G
Melting range, °C	525–570	525–625		550–620		520–615
Casting temperature range, °C	—	700–760		690–760		680–760
Specific gravity	2.74	2.73		2.70		2.81
Heat treatment						
Solution temperature, °C	—	—	505–520	520–530	520–530	—
Solution time, h	—	—	6–16	12 (min)	12 (min)	—
Quench	—	—	Water at 70–80°C	Water at 70–80°C	Water at 70–80°C	—
Precipitation temperature, °C	—	—	150–170	—	160–170	—
Precipitation time, h	—	—	6–18	—	8–10	—
Special properties	Alloy for pressure die castings	General engineering alloy Can tolerate relatively high static loading in TF condition		Pressure tight. High strength alloy in TF condition		Equally suited to all casting processes
*Mechanical properties—sand cast—*SI units (Imperial units in brackets)						
Tensile stress min. MPa (tonf in $^{-2}$)	—	140(9.1)	230(14.9)	170(11.0)	230(14.9)	150(9.7)
Elongation min. %	—	2	—	2	—	1
Expected 0.2% proof stress, MPa (tonf in $^{-2}$)	—	70–110 (4.5–7.1)	200–250 (12.9–16.2)	120–140 (7.8–9.1)	220–280 (14.2–18.1)	80–140 (5.2–9.1)
*Mechanical properties—chill cast—*SI units (Imperial units in brackets)						
Tensile strength min. MPa (tonf in $^{-2}$)	150(9.7)	160(10.4)	280(18.1)	230(14.9)	280(18.1)	170(11.0)
Elongation min.%	1	2	1	3	—	1
Expected 0.2% proof stress, MPa (tonf in $^{-2}$)	90–130 (5.8–8.4)	80–110 (5.2–7.1)	200–300 (12.9–19.4)	140–150 (9.1–9.7)	250–300 (16.2–19.4)	80–140 (5.2–9.1)

(*continued*)

Table 26.23   ALUMINIUM–SILICON–COPPER ALLOYS—*continued*

Specification BS 1490:1988 *Related British Specifications*	LM22TB (SP)	LM24M (Ge)	LM26TE (SP)	LM27M (Ge)	LM30M (SP)	LM30TS (SP)
Composition % (Single figures indicate maximum)						
Copper	2.8–3.8	3.0–4.0	2.0–4.0	1.5–2.5	4.0–5.0	
Magnesium	0.05	0.1	0.5–1.5	0.3	0.4–0.7	
Silicon	4.0–6.0	7.5–9.5	8.5–10.5	6.0–8.0	16–18	
Iron	0.6	1.3	1.2	0.8	1.1	
Manganese	0.2–0.6	0.5	0.5	0.2–0.6	0.3	
Nickel	0.15	0.5	1.0	0.3	0.1	
Zinc	0.15	3.0	1.0	1.0	0.2	
Lead	0.1	0.3	0.2	0.2	0.1	
Tin	0.05	0.2	0.1	0.1	0.1	
Titanium	0.2	0.2	0.2	0.2	0.2	
Properties of material Suitability for:						
Sand casting	G†	F†	G	G	U	
Chill casting (gravity die)	G	F†	G	E	F	
Die casting (press die)	G†	E	F†	G†	G	
Strength at elevated temp.	G	G‡	E	G	G	
Corrosion resistance	G	G	G	G	G	
Pressure tightness	G	G	F	G	F	
Fluidity	G	G	G	G	G	
Resistance to hot shortness	G	G	F	G	F	
Machinability	G	F	F	G	P	
Melting range, °C	525–625	520–580	520–580	525–605	505–650	
Casting temperature range,°C	700–740	—	670–740	680–740	Well above 650°C	
Specific gravity	2.77	2.79	2.76	2.75	2.73	
Heat treatment						
Solution temperature, °C	515–530	—	—	—	—	
Solution time, h	6–9	—	—	—	—	
Quench	Water at 70–80°C	—	—	—	—	*Stress relief*
Precipitation temperature, °C	—	—	200–210	—	—	175–225
Precipitation time, h	—	—	7–9	—	—	8(minimum)
Special properties	Chill casting alloy (grav. die)	Alloy for pressure die castings	Piston alloy, retains strength and hardness at elevated temps.	Excellent castability	Alloy for pressure die casting automobile engine cylinder blocks	

Mechanical properties—sand cast—SI units (Imperial units in brackets)

Tensile stress min., MPa (tonf in^{-2})	—	—	—	140(9.1)	—	—
Elongation min. %	—	—	—	1	—	—
Expected 0.2% proof stress, MPa (tonf in^{-2})	—	—	—	80–90 (5.2–5.8)	—	—

Mechanical properties—chill cast—SI units (Imperial units in brackets)
HB = 90–120

Tensile strength min. MPa (tonf in^{-2})	245(15.9)	180(11.7)	210(13.6)	160(10.4)	150(9.7)	160(10.4)
Elongation min. %	8	1.5	1	2	0.5	0.5
Expected 0.2% proof stress, MPa (tonf in^{-2})	110–120 (7.1–7.8)	100–120 (6.7–7.7)	160–190 (10.4–12.3)	90–110 (5.8–7.1)	150–200 (9.7–12.9)	160–200 (10.4–12.9)

* 0.05% min. if Ti alone used for grain refinement.
† Not normally used in this form.
‡ The use of die castings is usually restricted to only moderately elevated temperatures.
Notes:
E—Excellent, F—Fair, G—Good, P—Poor, U—Unsuitable.
(Numerical purpose alloy; SP—Special purpose alloy as per BS 1490:1970).

Table 26.24   ALUMINIUM–COPPER ALLOYS

Specification BS 1490:1988	LM12M(SP)	LM12TF(SP)*	[LM14-WP]†	[LM11-W]	[LM11-WP]		
Aerospace BSL series	—	—	4L35	2L91	2L92	361B	741A
DTD series	—	—	—	—	—	—	—
Composition % (Single figures indicate maximum)							
Copper	9.0–11.0		3.5–4.5	4.0–5.0		4.0–5.0	3.5–4.5
Magnesium	0.2–0.4		1.2–1.7	0.10		0.10	1.2–2.5
Silicon	2.5		0.6‡	0.25		0.25	0.5
Iron	1.0		0.6‡	0.25		0.25	0.5
Manganese	0.6		0.6	0.10		0.10	0.1
Nickel	0.5		1.8–2.3	0.10		0.10	0.1
Zinc	0.8		0.1	0.10		0.05	0.1
Lead	0.1		0.05	0.05		0.05	0.1
Tin	0.1		0.05	0.05		0.05	0.05
Titanium	0.2		0.25	0.25			
Other	—		—	—		Ti + Nb 0.05–0.30	Co 0.5–1.0 Nb 0.05–0.3
Properties of material							
Suitability for:							
Sand casting	F		F	F		F	F
Chill casting (gravity die)	G		G	P		P	G
Die casting (press die)	U		U	U		—	—
Strength at elevated temperature	G		E	F		F	F
Corrosion resistance	P		F	F		P	F
Pressure tightness	G		E	P		P	G
Fluidity	F		G	F		F	G
Resistance to hot shortness	G		G	P		P	G
Machinability	E		G	G		G	G
Melting range, °C	525–625		530–640	545–640		540–640	530–640
Casting temperature range, °C	700–760		700–750	680–700		675–750	710–725
Specific gravity	2.94		2.82	2.80		2.80	2.80

Heat treatment						
Solution temp., °C	515–520	500–520	525–545	525–545	525–545	495–505
Solution time, h	6	6	12–16	12–16	16 (minimum)	10 (minimum)††
Quench	Water at 70–80 °C	Boiling water	Water at 70–80 °C	Water at 70–80 °C	Water or oil	Oil at 80–90 °C
Precipitation temperature, °C	175–180	95–103§	120–140	120–170	160–70	195–205
Precipitation time, h	2 (minimum)	2**	1–2	12–14	8–16	4–5
Special properties	Piston alloy, now superseded by LM13 and LM26. Excellent machinability	Excellent props. at elevated temperatures Grav. die alloy	Good shock resistance		High strength alloy	
Mechanical properties–sand cast—SI units (Imperial units in brackets)						
Tensile stress min., MPa (tonf in⁻²)	—	220 (14.2)	220 (14.2)	280 (18.1)	324 (21.0)	263 (17.0)
Elongation %	—	—	7	4	—	—
Expected 0.2% proof stress, min., MPa (tonf in⁻²)	—	210–240 (13.6–15.5)	165–200 (10.7–12.9)	200–240 (12.9–15.5)	310 (20.1)	250 (16.2)
Mechanical properties–chill cast—SI units (Imperial units in brackets)						
Tensile stress min., MPa (tonf in⁻²)	170 278 (18.0)	280 (18.1)	265 (17.1)	310 (20.1)	402 (26.0)	340 (22.0)
Elongation %	—	—	13	9	4	—
Expected 0.2% proof stress, min., MPa (tonf in⁻²)	140–170 139–170 (9.0–11.0) HB 100–150	230–260 (14.9–16.8) HB 100–130	165–200 (10.7–12.9)	200–240 (12.5–15.5)	360 (23.3)	260 (16.8)

* Not included in BS 1490:1988.
† [ ] signifies obsolete specification:
‡ Si + Fe 1.0 max.
§ Or 5 days ageing at room temp.
** Can substitute stabilising treatment at 200–250 °C if used for pistons.
†† Allow to cool to 480 °C before quench.

Notes:
E—Excellent. F—Fair. G—Good. P—Poor. U—Unsuitable.
(Ge—General purpose alloy; SP—Special purpose alloy as per BS 1490:1988).

Table 26.25   MISCELLANEOUS ALUMINIUM ALLOYS

Specification BS 1400:1988	LM5M(SP)	—	LM10TB(SP)	—
Aerospace				
BSL series	—	—	4L53	L99
DTD series	—	5018A	—	—

Composition % (Single figures indicate maximum)				
Copper	0.1	0.2	0.1	0.1
Magnesium	3.0–6.0	7.4–7.9	9.5–11.0	0.20–0.45
Silicon	0.3	0.25	0.25	6.5–7.5
Iron	0.6	0.35	0.35	0.20
Manganese	0.3–0.7	0.1–0.3	0.10	0.10
Nickel	0.1	0.1	0.10	0.10
Zinc	0.1	0.9–1.4	0.10	0.10
Lead	0.05	0.05	0.05	0.05
Tin	0.1	0.05	0.05	0.05
Titanium	0.2	0.25	0.2[†]	0.20
Other	—	—	—	—

Properties of material				
Suitability for:				
Sand casting	F	F	F	G
Chill casting (gravity die)	F	F	F	E
Die casting (press die)	F[‡]	—	F[‡]	F[‡]
Strength at elevated temp.	F	F	F	—
Corrosion resistance	E	E	E	E
Pressure tightness	P	P	P	G
Fluidity	F	F	F	G
Resistance to hot shortness	F	G	G	G
Machinability	G	G	G	F
Melting range, °C	580–642	—	450–620	550–615
Casting temperature range, °C	680–740	680–720	680–720	680–740
Specific gravity	2.65	2.64	2.57	2.67

Heat treatment				
Solution temperature, °C	—	425–435[§]	425–435	535–545
Solution time, h	—	8	8	12
Quench	—	Oil at 160°C** or boiling water	Oil at no more[††] than 160°C	Water at 65°C min
Precipitation temp., °C	—	—	—	150–160
Precipitation time, h	—	—	—	4

Special properties	Good corrosion resistance in marine atmospheres	—	Good shock resistance and high corrosion resistance ***	Excellent castability with good mech. props.

Mechanical properties—sand cast—SI units (Imperial units in brackets)				
Tensile stress min, MPa (tonf in^{-2})	140 (9.1)	278	280 (18.0)	230 (14.9)
Elongation %	3	3	8	2
Expected 0.2% proof stress, min MPa (tonf in^{-2})	90–110 (5.8–7.1)	170 (11.0)	170–190 (11.0–12.3)	185 (12.0)

Mechanical properties—chill cast—SI units (Imperial units in brackets)				
Tensile stress min, MPa (tonf in^{-2})	170 (11.0)	309 (20.0)	310 (20.1)	280 (18.1)
Elongation %	5	10	12	5
Expected 0.2% proof stress, min MPa (tonf in^{-2})	90–120 (5.8–7.8)	170 (11.0)	170–200 (11.0–12.9)	200 (12.9)

*[ ] obsolete.

[†] 0.05% min, if Ti alone used for grain refinement.

[‡] Not normally used in this form.

[§] Or 8 h 435–445°C then raise to 490–500°C for further 8 h and quench as in table.

** Do not retain castings in oil for more than 1 h.

*** Not generally recommended since occasional brittleness can develop over long periods.

Table 26.25 MISCELLANEOUS ALUMINIUM ALLOYS—*continued*

Specification BS 1400:1988	LM28TE(SP)	LM28TF(SP)	[LM23P]*	[LM15WP]*	—
Aerospace					
BSL series	—	—	3L51	3L52	—
DTD series	—	—	—	—	5008B

Composition % (Single figures indicate maximum)

Copper		1.3–1.8	0.8–2.0	1.3–3.0	0.1
Magnesium		0.8–1.5	0.05–0.2	0.5–1.7	0.5–0.75
Silicon		17–20	1.5–2.8	0.6–2.0	0.25
Iron		0.7	0.8–1.4	0.8–1.4	0.5
Manganese		0.6	0.1	0.1	0.1
Nickel		0.8–1.5	0.8–1.7	0.5–2.0	0.1
Zinc		0.2	0.1	0.1	4.8–5.7
Lead		0.1	0.05	0.05	0.1
Tin		0.1	0.05	0.05	0.05
Titanium		0.2	0.25	0.25	0.15–0.25
Other		Cr 0.6 Co 0.5	—	—	Cr 0.4–0.6

Properties of material
Suitability for:

Sand casting		P	G	F	F
Chill casting (gravity die)		F	G	G	P
Die casting (press die)		—	G‡	U	U‡
Strength at elevated temp.		F	G	E	F
Corrosion resistance		G	G	G	E
Pressure tightness		F	G	F	F
Fluidity		F	F	F	F
Resistance to hot shortness		G	G	G	P
Machinability		P	G	G	G
Melting range, °C		520–675	545–635	600–645	572–615
Casting temperature range, °C		≮735	680–750	685–755	730–770
Specific gravity		2.68	2.77	2.75	2.81

Heat treatment

Solution temperature, °C	—	495–505	—	520–540	—
Solution time, h	—	4	—	4	—
Quench	—	Air blast	—	Water at 80–100 °C Oil or air blast	—
Precipitation temp., °C	185	185	150–175	150–180 (195–205)	175–185‡‡ (at least 24 h after cast)
Precipitation time, h	To produce required HB	8	8–24	8–24 (2–5)	10 (at least 24 h after cast)
Special properties	Piston alloy		Aircraft engine castings	High mechanical props. at elevated temps.	Good strength without heat treatment. *See*‡‡

Mechanical properties—sand cast—SI units (Imperial units in brackets)
Tensile stress min,

MPa (tonf in^{-2})	—	120 (7.8)	160 (10.4)	280 (18.1)	216 (14.0)
Elongation %	—	—	2	—	4
Expected 0.2% proof stress, min MPa (tonf in^{-2})	—	— HB 100–140	125 (8.1)	245 (15.9)	150 (9.7)

Mechanical properties—chill cast—SI units (Imperial units in brackets)
Tensile stress min,

MPa (tonf in^{-2})	170 (11.0)	190 (12.3)	200 (13.0)	325 (21.0)	232 (15.0)
Elongation %	—	—	3	—	5
Expected 0.2% proof stress, min MPa (tonf in^{-2})	— HB 90–130	160–190 (10.4–12.3) HB 100–140	140 (19.1)	295 (19.1)	180 (11.7)

* Can be furnace cooled to 385–395°C before quench. Do not retain in oil for more than 1 h. Further quench in water or air.
Alternative—room temp. age-harden for 3 weeks.
Note:
E—Excellent, F—Fair. G—Good. P—Poor. U—Unsuitable.
(Ge—General purpose alloy; SP—Special purpose alloy as per BS 1490:1988).

Table 26.26a   HIGH STRENGTH CAST Al ALLOYS BASED ON Al-4.5 Cu

European Designation	KO1				A-U5GT					
Aluminum Assoc. (USA)	201.0	201.2	A201.0	A201.2	204.0	204.2	206.0	206.2	A206.0	A206.2
Cu	4.0–5.2	4.0–5.2	4.0–5.0	4.0–5.0	4.2–5.0	4.2–4.9	4.2–5.0	4.2–5.0	4.2–5.0	4.2–5.0
Mg	0.15–0.55	0.20–0.55	0.15–0.35	0.20–0.35	0.15–0.35	0.20–0.35	0.15–0.35	0.20–0.35	0.15–0.35	0.20–0.35
Si	0.10	0.10	0.05	0.05	0.20	0.15	0.10	0.10	0.05	0.05
Fe	0.15	0.10	0.10	0.07	0.35	0.10–0.20	0.15	0.10	0.10	0.07
Mn	0.20–0.50	0.20–0.50	0.20–0.40	0.20–0.40	0.10	0.05	0.20–0.50	0.20–0.50	0.20–0.50	0.20–0.50
Ni	—	—	—	—	0.05	0.03	0.05	0.03	0.05	0.03
Zn	—	—	—	—	0.10	0.05	0.10	0.05	0.10	0.05
Sn	—	—	—	—	0.05	0.05	0.05	0.05	0.05	0.05
Ti	0.15–0.35	0.15–0.35	0.15–0.35	0.15–0.35	0.15–0.30	0.15–0.25	0.15–0.03	0.15–0.25	0.15–0.30	0.15–0.25
Ag	0.40–1.0	0.40–1.0	0.40–0.10	0.40–0.10	—	—	—	—	—	—
Others each	0.05	0.05	0.03	0.03	0.05	0.05	0.05	0.05	0.05	0.05
Others total	0.10	0.10	0.10	0.10	0.15	0.15	0.15	0.15	0.15	0.15

Data for tables 26.26a, 26.27 and 26.28 supplied by the Aluminum Association, Washington DC, USA and is used with permission.

Table 26.26b   MINIMUM REQUIREMENTS FOR SEPARATELY CAST TEST BARS

Alloy	Sand/Die	Treatment*	UTS MPa	0.2PS MPa	El % in 50 mm or 4×diam	Typical HB 500 kgf 10 mm
201.0	Sand	T6	414	345	5.0	110–140
	Sand	T7	414	345	3.0	130
204.0	Sand	T4	311	194	6.0	95
	Die	T4	331	220	8.0	

* For temper designations see Table 29.13.

Table 26.26c   TYPICAL PROPERTIES OF SEPARATELY CAST TEST BARS

Alloy	Sand/Die	Treatment*			UTS MPa	0.2PS MPa	El % in 50 mm or 4×diam	Typical HB 500 kgf 10 mm
201.0	Sand	T4			365	215	20	95
		T43			414	255	17	
	Sand	T6			448–485	380–435	7–8	135
	Sand	T7	Room	Temp.	460–469	415	4.5–5.5	130
		T7	150C*	0.5–100 h	380	360	6–8.5	
				1 000 h	360	345	8	
				10 000 h	315	275	7	
			205C	0.5 h	325	310	9	
				100 h	285	270	10	
				1 000 h	250	230	9	
				10 000 h	185	150	14	
			260C	0.5 h	195	185	14	
				100 h	150	140	17	
				1 000 h	125	110	18	
			315C	0.5 h	140	130	12	
				100 h	85	75	30	
				1 000 h	70	60	39	
				10 000 h	60	55	43	
A206.0	Sand	T4						118 HV
		T7						137 HV
			20C		436	347	11.7	
			120C		384	316	14.0	
			175C		333	302	17.7	

* Elevated temperature properties.

Table 26.27a   CHEMICAL COMPOSITION LIMITS FOR COMMONLY USED SAND AND PERMANENT MOULD CASTING ALLOYS ⓐ ⓑ

Aluminum Assoc. (USA) Alloy	Product ⓒ	Silicon	Iron	Copper	Manganese	Magnesium	Chromium	Nickel	Zinc	Titanium	Others Each	Others Total ⓚ
201.0	S	0.10	0.15	4.0–5.2	0.20–0.50	0.15–0.55	—	—	—	0.15–0.35	0.05 ⓗ	0.10
204.0	S&P	0.20	0.35	4.2–5.0	0.10	0.15–0.35	—	0.05	0.10	0.15–0.30	0.05 ⓘ	0.15
208.0	S&P	2.5–3.5	1.2	3.5–4.5	0.50	0.10	—	0.35	1.0	0.25	—	0.50
222.0	S&P	2.0	1.5	9.2–10.7	0.50	0.15–0.35	—	0.50	0.8	0.25	—	0.35
242.0	S&P	0.7	1.0	3.5–4.5	0.35	1.2–1.8	0.25	1.7–2.3	0.35	0.25	0.05	0.15
295.0	S	0.7–1.5	1.0	4.0–5.0	0.35	0.03	—	—	0.35	0.25	0.05	0.15
296.0	P	2.0–3.0	1.2	4.0–5.0	0.35	0.05	—	0.35	0.50	0.25	—	0.35
308.0	P	5.0–6.0	1.0	4.0–5.0	0.50	0.10	—	—	1.0	0.25	—	0.50
319.0	S&P	5.5–6.5	1.0	3.0–4.0	0.50	0.10	—	0.35	1.0	0.25	—	0.50
328.0	S	7.5–8.5	1.0	1.0–2.0	0.20–0.6	0.20–0.6	0.35	0.25	1.5	0.25	—	0.50
332.0	P	8.5–10.5	1.2	2.0–4.0	0.50	0.50–1.5	—	0.50	1.0	0.25	—	0.50
333.0	P	8.0–10.0	1.0	3.0–4.0	0.50	0.05–0.50	—	0.50	1.0	0.25	—	0.50
336.0	P	11.0–13.0	1.2	0.50–1.5	0.35	0.7–1.3	—	2.0–3.0	0.35	0.25	0.05	—
354.0	S&P	8.6–9.4	0.20	1.6–2.0	0.10	0.40–0.6	—	—	0.10	0.20	0.05	0.15
355.0	S&P	4.5–5.5	0.6 ⓓ	1.0–1.5	0.50 ⓓ	0.40–0.6	0.25	—	0.35	0.25	0.05	0.15
C355.0	S&P	4.5–5.5	0.20	1.0–1.5	0.10	0.40–0.6	—	—	0.10	0.20	0.05	0.15
356.0	S&P	6.5–7.5	0.6 ⓓ	0.25	0.35 ⓓ	0.20–0.45	—	—	0.35	0.25	0.05	0.15
A356.0	S&P	6.5–7.5	0.20	0.20	0.10	0.25–0.45	—	—	0.10	0.20	0.05	0.15
357.0	S&P	6.5–7.5	0.15	0.05	0.03	0.45–0.6	—	—	0.05	0.20	0.05	0.15
A357.0	S&P	6.5–7.5	0.20	0.20	0.10	0.40–0.7	—	—	0.10	0.04–0.20	0.05 ⓔ	0.15
359.0	S&P	8.5–9.5	0.20	0.20	0.10	0.50–0.7	—	—	0.10	0.20	0.05	0.15
443.0	S&P	4.5–6.0	0.8	0.6	0.50	0.05	0.25	—	0.50	0.25	—	0.35
B443.0	S&P	4.5–6.0	0.8	0.15	0.35	0.05	—	—	0.35	0.25	0.05	0.15
A444.0	P	6.5–7.5	0.20	0.10	0.10	0.05	—	—	0.10	0.20	0.05	0.15
512.0	S	1.4–2.2	0.6	0.35	0.8	3.5–4.5	0.25	—	0.35	0.25	0.05	0.15
513.0	P	0.30	0.40	0.10	0.30	3.5–4.5	—	—	1.4–2.2	0.20	0.05	0.15
514.0	S	0.35	0.50	0.15	0.35	3.5–4.5	—	—	0.15	0.25	0.05	0.15
520.0	S	0.25	0.30	0.25	0.15	9.5–10.6	—	—	0.15	0.25	0.05	0.15
535.0	S&P	0.15	0.15	0.05	0.10–0.25	6.2–7.5	—	—	—	0.10–0.25	0.05 ⓕ	0.15
705.0	S&P	0.20	0.8	0.20	0.40–0.6	1.4–1.8	0.20–0.40	—	2.7–3.3	0.25	0.05	0.15
707.0	S&P	0.20	0.8	0.20	0.40–0.6	1.8–2.4	0.20–0.40	—	4.0–4.5	0.25	0.05	0.15
710.0	S	0.15	0.50	0.35–0.65	0.05	0.6–0.8	—	—	6.0–7.0	0.25	0.05	0.15

Alloy	Type											
711.0	P	0.30	0.7–1.4	0.35–0.65	0.05	0.25–0.45	0.40–0.6	—	6.0–7.0	0.20	0.05	0.15
712.0	S	0.30	0.50	0.25	0.10	0.50–0.65	0.35	—	5.0–6.5	0.15–0.25	0.05	0.20
713.0	S&P	0.25	1.1	0.40–1.0	0.6	0.20–0.50	0.06–0.20	0.15	7.0–8.0	0.25	0.10	0.25
771.0	S	0.15	0.15	0.10	0.10	0.8–1.0	0.06–0.20	—	6.5–7.5	0.10–0.20	0.05	0.15
850.0	S&P	0.7	0.7	0.7–1.3	0.10	0.10	—	0.7–1.3	—	0.20	⑧	0.30
851.0	S&P	2.0–3.0	0.7	0.7–1.3	0.10	0.10	—	0.30–0.7	—	0.20	⑧	0.30
852.0	S&P	0.40	0.7	1.7–2.3	0.10	0.6–0.9	—	0.9–1.5	—	0.20	⑧	0.30

ⓐ The alloys listed are those which have been included in US Federal Specifications QQ-A-596d, ALUMINUM ALLOYS PERMANENT AND SEMI-PERMANENT MOULD CASTINGS, QQ-A-601E, ALUMINUM ALLOY SAND CASTINGS, and Military Specification MIL-A-21180c, ALUMINUM ALLOY CASTINGS, HIGH STRENGTH. Other alloys are registered with The Aluminum Association and are available. Information on these should be requested from individual foundries or ingot suppliers.

ⓑ Except for 'Aluminium' and 'Others,' analysis normally is made for elements for which specific limits are shown. For purposes of determining conformance to these limits, an observed value or calculated value obtained from analysis is rounded off to the nearest unit in the last right hand place of figures used in expressing the specified limit, in accordance with the following:
When the figure next beyond the last figure or place to be retained is less than 5, the figure in the last place retained should be kept unchanged.
When the figure next beyond the last figure or place to be retained is greater than 5, the figure in the last place retained should be increased by 1.
When the figure next beyond the last figure or place to be retained is 5 and
(1) there are no figures or only zeros, beyond this 5, if the figure in the last place to be retained is odd, it should be increased by 1; if even, it should be kept unchanged;
(2) if the 5 next beyond the figure in the last place to be retained is followed by any figures other than zero, the figure in the last place retained should be increased by 1; whether odd or even.

ⓒ S = Sand Cast; P = Permanent Mould Cast
ⓓ If iron exceeds 0.45 percent, manganese content shall not be less than one-half the iron content.
ⓔ Also contains 0.04–0.07 percent beryllium.
ⓕ Also contains 0.003–0.007 percent beryllium, boron 0.005 percent maximum.
ⓖ Also contains 5.5–7.0 percent tin.
ⓗ Also contains 0.40–1.0 percent silver.
ⓘ Also contains 0.05 max percent tin.
ⓚ The sum of those 'Others' metallic elements 0.010 percent or more each, expressed to the second decimal before determining the sum.

Table 26.27b   MECHANICAL PROPERTY LIMITS FOR COMMONLY USED ALUMINIUM SAND CASTING ALLOYS[1]

Aluminum Assoc. (USA) Alloy	Temper.[2]	Minimum properties				% Elongation in 2 inches or 4 times diameter	Typical Brinell hardness [4] 500—kgf load 10—mm ball
		Tensile strength					
		Ultimate		Yield (0.2% Offset)			
		ksi	(MPa)	ksi	(MPa)		
201.0	T7	60.0	(415)	50.0	(345)	3.0	110–140
204.0	T4	45.0	(310)	28.0	(195)	6.0	—
208.0	F	19.0	(130)	12.0	(85)	1.5	40–70
222.0	0	23.0	(160)	—	—	—	65–95
222.0	T61	30.0	(205)	—	—	—	100–130
242.0	0	23.0	(160)	—	—	—	55–85
242.0	T571	29.0	(200)	—	—	—	70–100
242.0	T61	32.0	(220)	20.0	(140)	—	90–120
242.0	T77	24.0	(165)	13.0	(90)	1.0	60–90
295.0	T4	29.0	(200)	13.0	(90)	6.0	45–75
295.0	T6	32.0	(220)	20.0	(140)	3.0	60–90
295.0	T62	36.0	(250)	28.0	(195)	—	80–110
295.0	T7	29.0	(200)	16.0	(110)	3.0	55–85
319.0	F	23.0	(160)	13.0	(90)	1.5	55–85
319.0	T5	25.0	(170)	—	—	—	65–95
319.0	T6	31.0	(215)	20.0	(140)	1.5	65–95
328.0	F	25.0	(170)	14.0	(95)	1.0	45–75
328.0	T6	34.0	(235)	21.0	(145)	1.0	65–95
355.0	T51	25.0	(170)	18.0	(125)	—	50–80
355.0	T6	32.0	(220)	20.0	(140)	2.0	70–105
355.0	T7	35.0	(240)	—	—	—	70–100
355.0	T71	30.0	(205)	22.0	(150)	—	60–95
C355.0	T6	36.0	(250)	25.0	(170)	2.5	75–105
356.0	F	19.0	(130)	—	—	2.0	40–70
356.0	T51	23.0	(160)	16.0	(110)	—	45–75
356.0	T6	30.0	(205)	20.0	(140)	3.0	55–90
356.0	T7	31.0	(215)	29.0	(200)	—	60–90
356.0	T71	25.0	(170)	18.0	(125)	3.0	45–75
A356.0	T6	34.0	(235)	24.0	(165)	3.5	70–105
357.0	—	—	—	—	—	—	—
A357.0	—	—	—	—	—	—	—
359.0	—	—	—	—	—	—	—
443.0	F	17.0	(115)	7.0	(50)	3.0	25–55
B433.0	F	17.0	(115)	6.0	(40)	3.0	25–55
512.0	F	17.0	(115)	10.0	(70)	—	35–65
514.0	F	22.0	(150)	9.0	(60)	6.0	35–65
520.0	T4[5]	42.0	(290)	22.0	(150)	12.0	60–90
535.0	F or T5	35.0	(240)	18.0	(125)	9.0	60–90
705.0	F or T5	30.0	(205)	17.0	(115)	5.0	50–80
707.0	T5	33.0	(230)	22.0	(150)	2.0	70–100
707.0	T7	37.0	(255)	30.0	(205)	1.0	65–95
710.0	F or T5	32.0	(220)	20.0	(140)	2.0	60–90
712.0	F or T5	34.0	(235)	25.0	(170)	4.0	60–90
713.0	F or T5	32.0	(220)	22.0	(150)	3.0	60–90
771.0	T5	42.0	(290)	38.0	(260)	1.5	85–115
771.0	T51	32.0	(220)	27.0	(185)	3.0	70–100
771.0	T52	36.0	(250)	30.0	(205)	1.5	70–100
771.0	T53	36.0	(250)	27.0	(185)	1.5	—
771.0	T6	42.0	(290)	35.0	(240)	5.0	75–105
771.0	T71	48.0	(330)	45.0	(310)	2.0	105–135
850.0	T5	16.0	(110)	—	—	5.0	30–60
851.0	T5	17.0	(115)	—	—	3.0	30–60
852.0	T5	24.0	(165)	18.0	(125)	—	45–75

[1] Values represent properties obtained from separately cast test bars and are derived from ASTM B-26, Standard Specification for Aluminum-Alloy Sand Castings; Federal Specification QQ-A-601e, Aluminum Alloy Sand Castings; and Military Specification MIL-A-21180c, Aluminum Alloy Castings, High Strength. Unless otherwise specified, the tensile strength, yield strength and elongation values of specimens cut from castings shall be not less than 75 percent of the tensile and yield strength values and not less than 25 percent of the elongation values given above. The customer should keep in mind that (1) some foundries may offer additional tempers for the above alloys, and (2) foundries are constantly improving casting techniques and, as a result, some may offer minimum properties in excess of the above. If quality level 4 castings are specified as described in Table 1 of AA-CS-M5-85, no tensile tests shall be specified nor tensile requirements be met on specimens cut from castings.

[2] F indicates 'as cast' condition; refer to AA-CS-M11 for recommended times and temperatures of heat treatment for other tempers to achieve properties specified.

[3] Footnote no longer in use.

[4] Hardness values are given for information only; not required for acceptance.

[5] The T4 temper of Alloy 520.0 is unstable; significant room temperature ageing occurs within life expectancy of most castings. Elongation may decrease by as much as 80 percent.

Table 26.27c MECHANICAL PROPERTY LIMITS FOR COMMONLY USED ALUMINIUM PERMANENT MOULD CASTING ALLOYS[①]

Aluminum Assoc. (USA) Alloy	Temper.[②]	Minimum properties				%Elongation in 2 inches or 4 times diameter	Typical Brinell hardness [③] 500—kgf load 10–mm ball
		Tensile strength					
		Ultimate		Yield (0.2% Offset)			
		ksi	(MPa)	ksi	(MPa)		
204.0	T4	48.0	(330)	29.0	(200)	8.0	—
208.0	T4	33.0	(230)	15.0	(105)	4.5	60–90
208.0	T6	35.0	(240)	22.0	(150)	2.0	75–105
208.0	T7	33.0	(230)	16.0	(110)	3.0	65–95
222.0	T551	30.0	(205)	—	—	—	100–130
222.0	T65	40.0	(275)	—	—	—	125–155
242.0	T571	34.0	(230)	—	—	—	90–120
242.0	T61	40.0	(275)	—	—	—	95–125
296.0	T6	35.0	(240)	—	—	2.0	75–105
308.0	F	24.0	(165)	—	—	2.0	55–85
319.0	F	28.0	(195)	14.0	(95)	1.5	70–100
319.0	T6	34.0	(235)	—	—	2.0	75–105
332.0	T5	31.0	(215)	—	—	—	90–120
333.0	F	28.0	(195)	—	—	—	65–100
333.0	T5	30.0	(205)	—	—	—	70–105
333.0	T6	35.0	(240)	—	—	—	85–115
333.0	T7	31.0	(215)	—	—	—	75–105
336.0	T551	31.0	(215)	—	—	—	90–120
336.0	T65	40.0	(275)	—	—	—	110–140
354.0	T61	48.0	(330)	37.0	(255)	3.0	—
354.0	T62	52.0	(360)	42.0	(290)	2.0	—
355.0	T51	27.0	(185)	—	—	—	60–90
355.0	T6	37.0	(255)	—	—	1.5	75–105
355.0	T62	42.0	(290)	—	—	—	90–120
355.0	T7	36.0	(250)	—	—	—	70–100
355.0	T71	34.0	(235)	27.0	(185)	—	65–95
C355.0	T61	40.0	(275)	30.0	(205)	3.0	75–105
356.0	F	21.0	(145)	—	—	3.0	40–70
356.0	T51	25.0	(170)	—	—	—	55–85
356.0	T6	33.0	(230)	22.0	(150)	3.0	65–95
356.0	T7	25.0	(170)	—	—	3.0	60–90
356.0	T71	25.0	(170)	—	—	3.0	60–90
A356.0	T61	37.0	(255)	26.0	(180)	5.0	70–100
357.0	T6	45.0	(310)	—	—	3.0	75–105
A357.0	T61	45.0	(310)	36.0	(250)	3.0	85–115
359.0	T61	45.0	(310)	34.0	(235)	4.0	75–105
359.0	T62	47.0	(325)	38.0	(260)	3.0	85–115
443.0	F	21.0	(145)	7.0	(50)	2.0	30–60
B443.0	F	21.0	(145)	6.0	(40)	2.5	30–60
A444.0	T4	20.0	(140)	—	—	20.0	—
513.0	F	22.0	(150)	12.0	(85)	2.5	45–75
535.0	F	35.0	(240)	18.0	(125)	8.0	60–90
705.0	T5	37.0	(255)	17.0	(120)	10.0	55–85
707.0	T7	45.0	(310)	35.0	(240)	3.0	80–110
711.0	T1	28.0	(195)	18.0	(125)	7.0	55–85
713.0	T5	32.0	(220)	22.0	(150)	4.0	60–90
850.0	T5	18.0	(125)	—	—	8.0	30–60
851.0	T5	17.0	(115)	—	—	3.0	30–60
851.0	T6	18.0	(125)	—	—	8.0	—
852.0	T5	27.0	(185)	—	—	3.0	55–85

[①] Values represent properties obtained from separately cast test bars and are derived from ASTM B-108, Standard Specification for Aluminum-Alloy Permanent Mould Castings; Federal Specification QQ-A-596d, Aluminum Alloy Permanent and Semi-Permanent Mould Castings; and Military Specification MIL-A-21180c, Aluminum Alloy Castings, High Strength. Unless otherwise specified, the average tensile strength, average yield strength and average elongation values of specimens cut from castings shall be not less than 75 percent of the tensile strength and yield values and not less than 25 percent of the elongation values given above. The customer should keep in mind that (1) some foundries may offer additional tempers for the above alloys, and (2) foundries are constantly improving casting techniques and, as a result, some may offer minimum properties in excess of the above.

[②] F indicates 'as cast' condition; refer to AA-CS-M11 for recommended times and temperatures of heat treatment for other tempers to achieve properties specified.

[③] Hardness values are given for information only; not required for acceptance.

Table 26.27d   TYPICAL PHYSICAL PROPERTIES OF COMMONLY USED SAND AND PERMANENT MOULD CASTING ALLOYS

These typical properties are not guaranteed, and should not be used for design purposes but only as a basis for general comparison of alloys and tempers with respect to any given characteristic.

Aluminum Assoc. (USA) Alloy	Temper	Specific gravity ⓐ	Density ⓐ lb. per cu. in.	Density (kg/m³)	Approximate Melting range °F	Approximate melting range °C	Electrical conductivity % IACS	Thermal conductivity at 25°C, CGS ⓑ	Coeff. of thermal expansion, per °F × 10⁻⁶		Coeff. of thermal expansion, per °C × 10⁻⁶	
									68–212°F	68–572°F	20–100°C	20–300°C
201.0	T6	2.80	0.101	2 800	1 060–1 200	571–649	27–32	0.29	19.3	24.7	34.74	44.46
	T7	2.80	0.101	2 800	1 060–1 200	571–649	32–34	0.29	—	—	—	—
204.0	T4	—	—	—	985–1 200	529–649						
208.0	F	2.79	0.101	2 790	970–1 160	521–627	31	0.30	12.4	13.4	22.32	24.12
222.0	T61 ⓒ	2.95	0.107	2 950	965–1 155	518–624	33	0.31	12.3	13.1	22.14	23.58
242.0	T571 ⓒ	2.81	0.102	2 810	990–1 175	532–635	34	0.32	12.6	13.6	22.68	24.48
	T77	2.81	0.102	2 810	990–1 175	532–635	38	0.36	12.6	13.6	22.68	24.48
295.0	T6	2.81	0.102	2 810	970–1 190	521–643	35	0.33	12.7	13.8	22.86	24.84
296.0	T6 ⓒ	2.80	0.101	2 800	970–1 170	521–632	33	0.31	12.2	13.3	21.96	23.94
308.0	F	2.79	0.101	2 790	970–1 135	521–613	37	0.35	11.9	12.9	21.42	23.22
319.0	F	2.79	0.101	2 790	960–1 120	516–604	27	0.26	11.9	12.7	21.42	22.86
328.0	F	2.70	0.098	2 700	1 025–1 105	552–596	30	0.29	11.9	12.9	21.42	23.22
332.0	T5 ⓒ	2.76	0.100	2 760	970–1 080	521–582	26	0.25	11.5	12.4	20.7	22.32
333.0	F ⓒ	2.77	0.100	2 770	960–1 085	516–585	26	0.25	11.4	12.4	20.52	22.32
	T5 ⓒ	2.77	0.100	2 770	960–1 085	516–585	29	0.28	11.4	12.4	20.52	22.32
	T6 ⓒ	2.77	0.100	2 770	960–1 085	516–585	29	0.28	11.4	12.4	20.52	22.32
	T7 ⓒ	2.77	0.100	2 770	960–1 085	516–585	35	0.33	11.4	12.4	20.52	22.32
336.0	T551 ⓒ	2.72	0.098	2 720	1 000–1 050	538–566	29	0.28	11.0	12.0	19.8	21.6
354.0	T61	2.71	0.098	2 710	1 000–1 105	538–596	32	0.30	11.6	12.7	20.88	22.86
355.0	T51	2.71	0.098	2 710	1 015–1 150	546–621	43	0.40	12.4	13.7	22.32	24.66
	T6	2.71	0.098	2 710	1 015–1 150	546–621	36	0.34	12.4	13.7	22.32	24.66
	T6 ⓒ	2.71	0.098	2 710	1 015–1 150	546–621	39	0.36	12.4	13.7	22.32	24.66
	T61	2.71	0.098	2 710	1 015–1 150	546–621	37	0.35	12.4	13.7	22.32	24.66
	T62 ⓒ	2.71	0.098	2 710	1 015–1 150	546–621	38	0.39	12.4	13.7	22.32	24.66
	T71	2.71	0.098	2 710	1 015–1 150	546–621	39	0.36	12.4	13.7	22.32	24.66
C355.0	T61	2.71	0.098	2 710	1 015–1 150	546–621	39	0.36	12.4	13.7	22.32	24.66
356.0	T51	2.68	0.097	2 680	1 035–1 135	557–613	43	0.40	11.9	12.9	21.42	23.22
	T6	2.68	0.097	2 680	1 035–1 135	557–613	39	0.36	11.9	12.9	21.42	23.22
	T6 ⓒ	2.68	0.097	2 680	1 035–1 135	557–613	41	0.38	11.9	12.9	21.42	23.22
	T7	2.68	0.097	2 680	1 035–1 135	557–613	40	0.37	11.9	12.9	21.42	23.22
	T7 ⓒ	2.68	0.097	2 680	1 035–1 135	557–613	43	0.40	11.9	12.9	21.42	23.22

Alloy	Temper											
A356.0	T61	2.67	0.097	2 670	1 035–1 135	557–613	39	0.36	11.9	12.9	21.42	23.22
357.0	F	2.67	0.097	2 670	1 035–1 135	557–613	39	0.36	11.9	12.9	21.42	23.22
A357.0	T61	2.67	0.097	2 670	1 035–1 135	557–613	39	0.36	11.9	12.9	21.42	23.22
359.0	T6	2.67	0.097	2 670	1 045–1 115	563–602	35	0.33	11.6	12.7	20.88	22.86
443.0	F	2.69	0.097	2 690	1 065–1 170	574–632	37	0.35	12.3	13.4	22.14	24.12
B443.0	F	2.69	0.097	2 690	1 065–1 170	574–632	37	0.35	12.3	13.4	22.14	24.12
A444.0	F	2.68	0.097	2 680	1 070–1 170	577–632	41	0.38	12.1	13.2	21.78	23.76
512.0	F	2.65	0.096	2 650	1 090–1 170	588–632	38	0.35	12.7	13.8	22.86	24.84
513.0	F ©	2.68	0.097	2 680	1 075–1 180	579–638	34	0.32	13.4	14.5	24.12	26.1
514.0	F	2.65	0.096	2 650	1 110–1 185	599–641	35	0.33	13.4	14.5	24.12	26.1
520.0	T4	2.57	0.093	2 570	840–1 120	449–604	21	0.21	13.7	14.8	24.66	26.64
535.0	F	2.62	0.095	2 620	1 020–1 165	549–629	23	0.23	13.1	14.8	23.58	26.64
705.0	F	2.76	0.100	2 760	1 105–1 180	596–638	25	0.25	13.1	14.3	23.58	25.74
707.0	F	2.77	0.100	2 770	1 085–1 165	585–629	25	0.25	13.2	14.4	23.76	25.92
710.0	F	2.81	0.102	2 810	1 105–1 195	596–646	35	0.33	13.4	14.6	24.12	26.28
711.0	F ©	2.84	0.103	2 840	1 120–1 190	604–643	40	0.37	13.1	14.2	23.58	25.56
712.0	F	2.81	0.101	2 810	1 135–1 200	613–649	35	0.33	13.7	14.8	24.66	26.64 ⓓ
713.0	F	2.81	0.100	2 810	1 100–1 180	593–638	30	0.29	13.4 ⓓ	14.6 ⓓ	24.12 ⓓ	26.28 ⓓ
771.0	F	2.81	0.102	2 810	1 120–1 190	604–643	37	0.33	13.7	14.8 ⓓ	24.66	26.64 ⓓ
850.0	T5 ©	2.88	0.104	2 880	435–1 200	224–649	47	0.43	13.0	ⓔ	23.4	ⓔ
851.0	T5 ©	2.83	0.103	2 830	440–1 165	227–629	43	0.40	12.6	ⓔ	22.68	ⓔ
852.0	T5 ©	2.88	0.104	2 880	400–1 175	204–635	45	0.41	12.9	ⓔ	23.22	ⓔ

ⓐ Assuming solid (void-free) metal. Since some porosity cannot be avoided in commercial castings, the actual values will be slightly less than those given.

ⓑ Cgs units equals calories per second per square centimeter per centimeter of thickness per degree centigrade.

© Chill cast samples; all other samples cast in green sand mould.

ⓓ Estimated value.

ⓔ Exceeds operating temperature.

Reference:
Aluminum, Volume I. Properties, Physical Metallurgy and Phase Diagrams, American Society for Metals, Metals Park, Ohio (1967). Data for alloy 771.0 supplied by the U.S. Reduction Company, East Chicago, Indiana.

Table 26.28 COMPOSITIONS AND PROPERTIES OF ALUMINIUM DIE CASTING ALLOYS

Designations (AA number)	A360.0	A380.0	383.0	B390.0	A413.0
Composition					
Silicon	9.0–10.0	7.5–9.5	9.5–11.5	16.0–18.0	11.0–13.0
Iron	1.3	1.3	1.3	1.3	1.3
Copper	0.6	3.0–4.0	2.0–3.0	4.0–5.0	1.0
Magnesium	0.4–0.6	0.10	0.10	0.45–0.65	0.10
Manganese	0.35	0.50	0.50	0.50	0.35
Nickel	0.50	0.5	0.30	0.10	0.50
Zinc	0.50	3.0	3.0	1.5	0.50
Tin	0.15	0.35	0.15	—	0.15
Titanium	—	—	—	0.10	—
Others (total)	0.25	0.50	0.50	0.20	0.25
Aluminium	Balance	Balance	Balance	Balance	Balance
Mechanical properties					
Ultimate tensile strength [ksi {MPa}]	46 {320}	47 {320}	45 {310}	46 {320}	42 {290}
Yield strength [ksi {MPa}]*	24 {170}	23 {160}	22 {150}	36 {250}	19 {130}
Elongation (% in 2 inches)	3.5	3.5	3.5	<1	3.5
Hardness (BHN)**	75	80	75	120	80
Shear strength [ksi {MPa}]	26 {180}	27 {190}	—	—	25 {170}
Impact strength (Notched charpy) [ft-lb {J}]	—	—	3 {4}	—	—
Fatigue strength [ksi {MPa}]	18 {120}	20 {140}	21 {145}	20 {140}	19 {130}
Young's modulus [ksi {GPa}]	10 300 {71}	10 300 {71}	10.3 {71}	11.8 {81.3}	—
Physical properties					
Density [lb/cu. in. {g/cc}]	0.095 {2.63}	0.098 {2.71}	0.099 {2.74}	0.098 {2.73}	0.096 {2.66}
Melting range	1 035–1 105°F 557–596°C	1 000–1 100°F 540–595°C	960–1 080°F 516–582°C	950–1 200°F 510–650°C	1 065–1 080°F 574–582°C
Specific heat [BTU/lb F {J/kg C}]	0.230 {963}	0.230 {963}	0.230 {963}	0.230 {963}	0.230 {963}
Coeff. of thermal expansion [μ in./in. F {μm/m K}]	11.6 {2.10}	12.1 {21.8}	11.7 {21.1}	10.0 {18.0}	11.9 {21.6}
Thermal conductivity [BTU/ft hr F {W/m K}]	65.3 {113}	55.6 {96.2}	55.6 {96.2}	77.4 {134}	70.1 {121}
Electrical conductivity (% IACS)	29	23	23	27	31
Poisson's ratio	0.33	0.33	0.33	—	—

Typical values based on 'as-cast' characteristics for separately die cast specimens, not specimens cut from production die castings.
* 0.2% offset
** 500 kg load, 10 mm ball
*** Rotary bend 5×10⁸ cycles
Adapted with permission from NADCA Product Specification Standards for Die Castings. © The North American Die Casting Association.

26.4 Copper base casting alloys

Table 26.29 GUNMETALS

Alloy	*Gunmetal* 88/10/2	*Nickel gunmetal* (*as cast*)	*Nickel gummetal** (*fully heat treated*)
Specification BS 1400:1969	G1	G3	G3WP
alloy grouping[†]	C	C	C

Composition % (Single figures indicate maximum)

Tin	9.5–10.5		6.5–7.5
Zinc	1.75–2.75		1.5–3.0
Lead	1.5		0.10–0.50
Nickel	1.0		5.25–5.75
Silicon	0.02		0.01
Bismuth	0.03		0.02
Aluminium	0.01		0.01
Iron + arsenic + antimony	0.20		0.20
Manganese	—		0.20
Total impurities	0.50		0.50

Properties of material
Suitability for:

Sand casting	2§		2
Chill casting	3		3
Die casting (gravity)	3		3
Centrifugal	2		2
Continuous	1		2

Pressure tightness for sand castings:

Thin sections	2**		1
Thick sections	2		1
Machining	2§§		2
Resistance to corrosion	All these alloys withstand atmospheric natural water and sea water corrosion to a degree depending largely upon the tin content (88/10/2 best)		
Fluidity	Addition of phosphorus in amounts of order of 0.05% for deoxidation gives high fluidity in all cases		
Hot shortness	All alloys are very hot short		
Bearing properties	All alloys suitable for bearing applications—choice dependent on conditions		
Density (g cm^{-3})	8.8		8.8

Casting temperature range, °C

Under 13 mm section	1 200 ⎫		
13–39 mm section	1 170 ⎬		1 080–1 200
Over 40 mm section	1 130 ⎭		

Mechanical properties—SI units (Imperial units in brackets) in order sand cast, chill cast, continuously cast, centrifugally cast[††] (BS 1400)

Tensile stress min,

MPa	270, 230, 300, 250	280, —, 340, —	430, —, 430, —
(tonf in^{-2})	(17.5, 14.9, 19.4, 16.2)	(18.1, —, 22.0, —)	(27.8, —, 27.8, —)

0.2% proof stress min

MPa	130, 130, 140, 130	140, —, 170, —	280, —, 280, —
(tonf in^{-2})	(8.4, 8.4, 9.1, 8.4)	(9.1, —, 11.0, —)	(18.1, —, 18.1, —)
Elongation 5.65$\sqrt{S_0}$%	13, 3, 9, 5	16, —, 18, —	3, —, 3, —
			HB 160 min

* Heat treatment 2 h at 790 ± 10°C air cool plus 6 h at 320 ± 10°C air cool.

[†] Group A alloys in common use; group B special purpose alloys; group C alloys in limited production.

[‡] Tin + $\frac{1}{2}$ Nickel content: 7.0–8.0%.

§ Grading 1 = Excellent.
          2 = Satisfactory.
          3 = Possible with special techniques.
          4 = Unsuitable.
          5 = Not applicable.

** Grading 1 = Suitable
          2 = Less suitable.
          3 = Unsuitable.

[‡] Values apply to samples cut from centrifugal castings made in metallic moulds. Min. props. of centrifugal castings made in sand moulds same as for other sand castings.

Table 26.29   GUNMETALS—*continued*

Alloy	Leaded gunmetal 83/3/9/5	Leaded gunmetal 85/5/5/5	Leaded gunmetal 87/7/3/3	Leaded semi-red brass 76/3/6/15
Specification BS 1400:1969	LG1	LG2	LG4	—
alloy grouping[†]	B	A	A	A

Composition % (Single figures indicate maximum)

Tin	2.0–3.5	4.0–6.0	6.0–8.0[‡]	2.5–3.5
Zinc	7.0–9.5	4.0–6.0	1.5–3.0	13.0–17.0
Lead	4.0–6.0	4.0–6.0	2.5–3.5	5.25–6.75
Nickel	2.0	2.0	2.0[‡]	1.0
Silicon	0.02	0.02	0.01	0.01
Bismuth	0.10	0.05	0.05	—
Aluminum	0.01	0.01	0.01	—
Iron + arsenic + antimony	0.75	0.50	0.40	0.55
Manganese	—	—	—	—
Total impurities	1.0	0.80	0.70	—

Properties of material

Suitability for:

Sand casting	1	1	1	1
Chill casting	2	2	2	2
Die casting (gravity)	3	3	3	3
Centrifugal	2	1	1	2
Continuous	2	1	1	2

Pressure tightness for sand castings:

Thin sections	1	1	2	2
Thick sections	2	2	1	2
Machining	1	1	1	2

Resistance to corrosion	All these alloys withstand atmospheric natural water and sea water corrosion to a degree depending largely upon the tin content (88/10/2 best)			
Fluidity	Addition of phosphorus in amounts of order of 0.05% for deoxidation gives high fluidity in all cases			
Hot shortness	All alloys are very hot short			
Bearing properties	All alloys suitable for bearing applications—choice dependent on conditions			
Density (g cm^{-3})	8.8	8.8	8.8	8.77

Casting temperature range °C

Under 13 mm section	1 180	1 200	1 200	1 150–1 260
13–39 mm section	1 140	1 150	1 160	1 110–1 220
Over 40 mm section	1 100	1 120	1 120	1 066–1 177

Mechanical properties—SI units (Imperial units in brackets) in order sand cast, chill cast, continuously cast, centrifugally cast[††] (BS 1400)

Tensile stress min MPa	180, 180, —, —	200, 200, 270, 220	250, 250, 300, 250	Sand cast 172
(tonf in^{-2})	(11.7, 11.7, —, —)	(13.0, 13.0, 17.5, 13.0)	(16.2, 16.2, 19.4, 16.2)	(Sand cast 11.6)
0.2% proof stress min, MPa	80, —, —, —	100, 110, 100, 110	130, 130, 130, 130	0.5% proof stress sand cast 83
(tonf in^{-2})	(5.2, —, —, —)	(6.5, 7.1, 6.5, 7.1)	(8.4, 8.4, 8.4, 8.4)	(Sand cast 5.4)
Elongation 5.65$\sqrt{S_0}$%	11.2, —, —	13, 6, 13, 8	16, 5, 13, 6	Sand cast 15 (50.8 mm GL)

[§§] Grading 1 = Excellent
    2 = Good
    3 = Satisfactory with special techniques.
(Comparison between copper alloys rather than with other metals).

Alloy	Phosphor bronze	Phosphor bronze for gear blanks	Phosphor bronze	Phosphor bronze	Leaded phosphor bronze
Specification BS 1400:1985	PB1	PB2	CT1	PB4	LPB1
Alloy grouping†	B	B	B	A	A
Composition % (Single figures indicate maximum unless otherwise stated)					
Tin	10.0 min	11.0–13.0	9.0–11.0	9.5 min	6.5–8.5
Phosphorus	0.50 min	0.15 min	0.15 max	0.40 min	0.30 min
Lead	0.25	0.50	0.25	0.75	2.0–5.0
Zinc	0.05	0.30	0.05	0.50	2.0
Nickel	0.10	0.50	0.25	0.50	1.0
Iron	0.10	0.15	—	—	—
Silicon	0.02	0.02	—	—	—
Aluminium	0.01	0.01	—	—	—
Total impurities	0.60	0.20	0.80	0.50	0.50
Copper	REM	REM	REM	REM	REM
Properties of material					
Suitability for:					
Sand casting	2‡	2	2	2	2
Chill casting	1	1	1	1	1
Die casting (gravity)	3	3	3	3	3
Centrifugal	1	1	1	1	2
Continuous	1	1	1	1	1
Pressure tight sand castings:					
Thin sections	3§	3	2	3	2
Thick sections	3	3	3	3	2
Machining	2‡‡	2	2	2	1
Corrosion resistance	All alloys are resistant to natural and sea water				
Fluidity	Good	Good	Good	Good	Good
Hot shortness	All alloys are very hot short				
Density (g cm⁻³)	8.8	8.8	8.8	8.8	8.8
Casting temperature range, °C					
Under 13 mm section	1 120	1 170	1 100	1 120	1 130
13–39 mm section	1 100	1 120	1 070	1 100	1 050
Over 40 mm section	1 040	1 070	1 040	1 060	1 030
Mechanical properties—SI units (Imperial units in brackets) in order sand cast, chill cast, continuously cast, centrifugally cast** BS 1400					
Tensile stress (min) MPa	210, 310, 360, 330	220, 270, 310, 280	230, 270, 310, 280	190, 270, 330, 280	190, 220, 270, 230
(tonf in⁻²)	(13.6, 20.1, 23.3, 21.4)	(14.2, 17.5, 20.1, 18.1)	(14.9, 17.5, 20.1, 18.1)	(12.3, 17.5, 21.4, 18.1)	(12.3, 14.2, 17.5, 14.9)
0.2% proof stress (min) MPa	130, 170, 170, 170	130, 170, 170, 170	130, 140, 160, 140	100, 140, 160, 140	80, 130, 130, 130
(tonf in⁻²)	(8.4, 11.0, 11.0, 11.0)	(8.4, 11.0, 11.0, 11.0)	(8.4, 9.1, 10.4, 9.1)	(6.5, 9.1, 10.4, 9.1)	(5.2, 8.4, 8.4, 8.4)
Elongation on 5.65$\sqrt{S_0}$%	3, 2, 6, 4	5, 3, 5, 3	6, 5, 9, 6	3, 2, 7, 4	3, 2, 5, 4

(continued)

Table 26.30   BEARING BRONZES—*continued*

Specification BS 1400:1985 *alloy grouping*†	LB1 C	LB2 A	LB4 A	LB5 B
Composition % (Single figures indicate maximum unless otherwise stated)				
Tin	8.0–10.0	9.0–11.0	4.0–6.0	4.0–6.0
Phosphorus	0.10	0.10	0.10	0.10
Lead	13.0–17.0	8.5–11.0	8.0–10.0	18.0–23.0
Zinc	1.0	1.0	2.0	1.0
Nickel	2.0	2.0	2.0	2.0
Iron	Sb 0.50	Sb 0.50 Fe 0.15	Sb 0.50	Sb 0.50
Silicon	0.02	0.02	0.02	0.01
Aluminium	—	0.01	—	—
Total impurities	0.30	0.50	0.50	0.30
Copper	REM	REM	REM	REM
Properties of material				
Suitability for:				
Sand casting	3	2	2	3
Chill casting	2	1	1	2
Die casting (gravity)	4	4	4	4
Centrifugal	2	2	2	3
Continuous	1	1	1	3
Pressure tight sand castings:				
Thin sections	2	2	2	2
Thick sections	3	2	2	3
Machining	1	1	1	1
Corrosion resistance	All alloys are resistant to natural and sea water			
Fluidity	Good	Good	Good	Good
Hot shortness	All alloys are very hot short			
Density (g cm^{-3})	9.1	9.0	9.0	9.2
Casting temperature range, °C				
Under 13 mm section	1 110	1 130	1 110	1 090
13–39 mm section	1 030	1 080	1 070	1 030
Over 40 mm section	1 010	1 030	1 040	1 010
Mechanical properties—SI units (Imperial units in brackets) in order sand cast, chill cast, continually cast, centrifugally cast** BS 1400				
Tensile stress (min)				
MPa	170, 200, 230, 220	190, 220, 280, 230	160, 200, 230, 220	160, 170, 190, 190
(tonf in^{-2})	(11.0, 13.0, 14.9, 14.2)	(12.3, 14.2, 18.1, 14.9)	(10.4, 13.0, 14.9, 14.2)	(10.4, 11.1, 12.3, 12.3)
0.2% proof stress (min)				
MPa	80, 130, 130, 130	80, 140, 160, 140	60, 80, 130, 80	60, 80, 100, 80
(tonf in^{-2})	(5.2, 8.4, 8.4, 8.4)	(5.2, 9.1, 10.4, 9.1)	(3.9, 5.2, 8.4, 5.2)	(3.9, 5.2, 6.5, 5.2)
Elongation on $5.65\sqrt{S_0}$ %	4, 3, 9, 4	5, 3, 6, 5	7, 5, 9, 6	5, 5, 8, 7

* This bronze is now primarily intended for shaped castings rather than bearings.
† *See footnote* ‡‡ *to* Table 26.29.   ** *See footnote* ‡ *to* Table 26.29.

Table 26.31 BRASSES

Alloy	Brass sand cast	Brass sand cast	Naval brass sand cast
Specification BS 1400:1985	SCB1	SCB3	SCB4
Alloy grouping	A	A	C

Composition % (Single figures indicate maximum unless otherwise stated)

Copper	70.0–80.0	63.0–70.0	60.0–63.0
Lead	2.0–5.0	1.0–3.0	0.5
Tin	1.0–3.0	1.5	1.0–1.5
Iron	0.75	0.75	—
Nickel	1.0	1.0	—
Aluminium	0.01	0.1	0.01
Others	—	—	—
Total impurities	1.0	1.0	0.75
Zinc	REM	REM	REM

Properties of material
Suitability for:

Sand casting	1‡	1	1
Chill casting	5	5	5
Die casting (gravity)	5	5	5
Centrifugal	3	3	3
Continuous	2	2	5

Pressure tight castings (sand):

Thin sections	1§	1¶	1
Thick sections	1	1¶	1
Machining	1§§	1	2
Corrosion resistance	Generally excellent in natural waters, DCB1, DCB3,PCB1, SCB1, SCB3 undergo dezincification in sea water		
Hot shortness	Alpha brasses tend to be hot short. Alpha/beta brasses relatively good		
Density (g cm^{-3})	8.5	8.4	8.3

Casting temperature range, °C

Under 13 mm section	1 150	1 100	1 100
13–39 mm section	1 100	1 050	1 050
Over 40 mm section	1 070	1 020	1 020

Mechanical properties—SI units (Imperial units in brackets)—sand cast props.** for alloys SCB1, 3, 4, 6—chill cast‡‡ for DCB1 and 3 and PCB1

Tensile stress (typical)

MPa	170–200	190–220	250–310
(tonf in^{-2})	(11.0–13.0)	(12.3–14.2)	(16.2–20.1)

0.2% proof stress (typical)

MPa	80–110	70–110	70–110
(tonf in^{-2})	(5.2–7.1)	(4.5–7.1)	(4.5–7.1)
Elongation on $5.65\sqrt{S_0}$% (typical)	18–40	11–30	18–40

* 0.1% Pb if required.
† Nickel be counted as copper.
‡§ *See* footnotes § and ** to Table 26.29.
¶ For pressure tight castings Al ≤ 0.02%

** On separately cast test bars.
†† Values based on 15–40 mm thick sections from castings.
‡‡ Values based on 15–40 mm sections for DCB1, DCB3 and PCB1.
§§ *See* footnote §§ to Table 26.29.

Alloy	Brass brazable castings	Brass die castings	Brass for diecasting	Brass for pressure diecasting
Specification BS 1400:1969	SCB6	DCB1*	DCB3†	PCB1
alloy grouping	A	A	A	A

Composition % (Single figures indicate maximum unless otherwise stated)

Copper	83.0–86.0	59.0–63.0	58.0–63.0	57.0–60.0
Lead	0.5	0.25	0.5–2.5	0.5–2.5
Tin	—	—	1.0	0.5
Iron		—	0.8	0.3
Nickel		—	1.0	—
Aluminium		0.5	0.2–0.8	0.5
Others	As 0.05–0.20	—	Mn 0.5 Si 0.05	—
Total impurities	1.0 (incl. Pb)	0.75	2.0 (excl. Ni + Pb + Al)	0.5
Zinc	REM	REM	REM	REM

(*continued*)

Table 26.31 BRASSES—*continued*

Alloy	*Brass brazable castings*	*Brass die castings*	*Brass for diecasting*	*Brass for pressure diecasting*
Specification BS 1400:1969	SCB6	DCB1*	DCB3†	PCB1
alloy grouping	A	A	A	A

Properties of material
Suitability for:

Sand casting	1	5	5	5
Chill casting	5	1	1	1
Die casting (gravity)	5	1	1	1
Centrifugal	3	2	2	2
Continuous	2	5	3	5

Pressure tight castings (sand):

Thin sections	1	—	—	—
Thick sections	1	—	—	—
Machining	3	2	1	2
Corrosion resistance	Generally excellent in natural waters. DCB1, DCB3, PCB1, SCB1, SCB3 undergo dezincification in sea water.			
Hot shortness	Alpha brasses tend to be hot short. Alpha/beta brasses relatively good			
Density (g cm^{-3})	8.6	8.3	8.3	8.3

Casting temperature range, °C

Under 13 mm section	1 150			Injection temp. 950
13–39 mm section	1 100	$\left\{1\,050\right\}$	$\left\{1\,050\right\}$	
Over 40 mm section	1 070			

Mechanical properties—SI units (Imperial units in brackets)—sand cast props., ** for alloys SCB1, 3, 4, 6—chill cast‡‡ for DCB1 and 3 and PCB1

Tensile stress (typical)

MPa	170–190	280–370	300–340	280–370
(tonf in^{-2})	(11.0–12.3)	(18.1–24.0)	(19.4–22.0)	(18.1–24.0)

0.2% proof stress (typical)

MPa	80–110	90–120	90–120	90–120
(tonf in^{-2})	(5.2–7.1)	(5.8–7.8)	(5.8–7.8)	(5.8–7.8)
Elongation on 5.65 $\sqrt{S_0}$% (typical)	18–40	23–50	13–40	25–40

Table 26.32 HIGH TENSILE BRASSES

Zinc equivalents

Guillet method		*American method*	
Element	*Coefficient*	*Element*	*Coefficient**
Silicon	10	Silicon	+10
Aluminium	6	Aluminium	+5.0
Tin	2	Tin	+1.0
Lead	1	Lead	—
Iron	0.9	Iron	−0.1
Manganese	0.5	Manganese	−0.5
Nickel	−1.2	Nickel	−2.3
Magnesium	2	Magnesium	+1.0

$$\text{Zinc equivalent} = \frac{[\%Zn + \Sigma(\%M \times \text{coeff})]100}{[\%Zn + \Sigma(\%M \times \text{coeff}) + \%Cu]}$$

$$\text{Zinc equivalent} = \left[100 - \frac{100 \times \%Cu}{100 + \Sigma(\%M \times \text{coeff})}\right]$$

Where
M = alloying element
% α-phase = 10 (46.6-zinc equivalent)

In practice, these factors operate with good accuracy, provided that the element considered is not present in amounts greater than 2%.

* These values are Guillet Coefficients mimus 1.

These values are employed as a guide to whether a complex brass will have either alpha, beta, or mixed alpha/beta structure. For instance equivalent zinc values of 46.6% and above will produce a beta structure. To obtain 15% alpha the zinc equivalent must be below 45%. As further examples, 1%Si will be equivalent to 10%Zn, and 1%Ni is equivalent to taking 1.2%Cu out (i.e. putting 1.2%Cu in).

Table 26.32   HIGH TENSILE BRASSES—*continued*

Alloy	*Alpha Beta* (460 MPa[†]) *High tensile brass*	*All Beta* (740 MPa[‡]) *High tensile brass*
Specification BS 1490:1985	HTB1[§]	HTB3
Alloy grouping	B	B

Composition % (Single figures indicate maximum)		
Copper	55.0 min	55.0 min
Manganese	3.0	4.0
Aluminium	0.5–2.5	3.0–6.0
Iron	0.7–2.0[¶]	1.5–3.25
Tin	1.0	0.20
Nickel	1.0	1.0
Lead	0.50	0.20
Silicon	0.10	0.10
Total impurities	0.20	0.20
Zinc	REM	REM

Properties of material Suitability for:		
Sand casting	2[††]	2
Chill casting	5	5
Gravity die casting	2	4
Centrifugal	2	2
Continuous	5	5

Pressure tight sand castings:		
Thin sections	1[‡‡]	1
Thick sections	1	1
Machining	3**	3
Corrosion resistance	α/β structure—not susceptible to stress corrosion cracking	All β alloy—susceptible to stress corrosion cracking
Density (g cm^{-3})	8.3	7.9
Casting temperature range, °C		
Under 13 mm section	1 060	1 060
13–39 mm section	1 020	1 020
Over 40 mm section	980	980

Mechanical properties—SI units (Imperial units in brackets) in order sand cast, chill cast, continuously cast, centrifugally cast[§§]

Tensile stress, min		
MPa	470, 500, —, 500	740, —, —, 740
(tonf in^{-2})	(30.4, 32.4, —, 32.4)	(47.9, —, —, 47.9)
0.2% proof stress, min		
MPa	170, 210, —, 210	400, —, —, 400
(tonf in^{-2})	(11.0, 13.6, —, 13.6)	(25.9, —, —, 25.9)
Elongation on 5.65 $\sqrt{S_0}$% min	18, 18, —, 20	11, —, —, 13

[†] Originally known as 30 ton HTB.
[‡] Originally known as 48 ton HTB.
[§] Micro-structure requirement; 15% α-phase min.
[¶] For grain refinement (grain dia. 0.5 mm or less on 29 mm sand cast test bars).
  Optimum iron for grain refinement approx.−1.1% in HTB 1.
[††] For grading *see* footnote § to Table 26.29.
[‡‡] For grading *see* footnote** to Table 26.29.
[§§] Values apply to samples cut from castings made in metallic moulds.
** *See* footnote [§§] to Table 26.29.

Table 26.33   MISCELLANEOUS COPPER-BASE ALLOYS

Alloy	Aluminium bronze	Aluminium bronze	Copper manganese aluminium	Copper manganese aluminium
	AB1	AB2	CMA1	CMA2
Specification BS 1400:1985 *Alloy grouping**	B	B	B	B
Composition % (Single figures indicate maximum)				
Aluminium	8.5–10.5	8.8–10.0	7.5–8.5	8.5–9.0
Iron	1.5–3.5	4.0–5.5	2.0–4.0	2.0–4.0
Nickel	1.0	4.0–5.5	1.5–4.5	1.5–4.5
Manganese	1.0	1.5	11.0–15.0	11.0–15.0
Zinc	0.50	0.50	0.50	0.50
Silicon	0.25	0.10	0.15	0.15
Tin	0.10	0.10	1.0	1.0
Lead	0.05†	0.05†	0.05	0.05
Phosphorus			0.05	0.05
Magnesium	0.05	0.05		
Chromium			—	—
Copper	REM	REM	REM	REM
Total impurities	0.30	0.30	0.30	0.30
Properties of material				
Suitability for:				
Sand casting	2‡	2	2	2
Chill casting	5	5	5	5
Gravity die casting	1	2	2	2
Centrifugal	2	2	2	2
Continuous	3	3	3	3
Pressure tight castings (sand):				
Thin sections	1§	1	1	1
Thick sections	1	1	1	1
Machining	3¶	3	3	3
Corrosion resistance	Very good except under reducing conditions**	Very good except under reducing conditions**	Very good	Very good
Fluidity	Fair	Fair	Good	Good
Strength at elevated temperatures	Very good	Excellent	Very good	Very good
Electrical conductivity % IACS at 15°C	13	8	3	3
Density (g cm^{-3})	7.6	7.6	7.5	7.5
Casting temperature range, °C				
Under 13 mm section	1 250	1 240	1 150	1 150
13–39 mm section	1 200	1 170	1 100	1 100
Over 40 mm section	1 150	1 120	1 050	1 050

Mechanical properties—SI units (Imperial units in brackets) in order sand cast, chill cast, centrifugally cast

	Nickel silver	High-conductivity copper	Copper chromium	High-strength cupro-nickel
Typical tensile stress MPa	500–590, 540–620, 560–650	640–700, 650–740, 670–730	650–730, 670–740, —	740–820, —, —
(tonf in $^{-2}$)	(32.4–38.2, 35.0–40.1, 36.3–42.1)	(41.4–45.3, 42.1–47.9, 43.4–47.3)	(42.1–47.3, 43.4–47.9, —)	(47.9–53.1, —, —)
Typical 0.2% proof stress MPa	170–200, 200–270, 200–270	250–300, 250–310, 250–310	280–340, 310–370, —	380–470, —, —
(tonf in $^{-2}$)	(11.0–13.0, 13.0–17.5, 13.0–17.5)	(16.2–19.4, 16.2–20.1, 16.2–20.1)	(18.1–22.0, 20.1–24.0, —)	(24.6–30.4, —, —)
Typical elongation 5.65√S 0%	18–40, 18–40, 20–30	13–20, 13–20, 13–20	18–35, 27–40, —	9–20, —, —
Typical hardness range HB	90–140, 130–160, 120–160	140–180, 160–190, 140–180	160–210, —, —	220–260, —, —

Alloy	Nickel silver	High-conductivity copper	Copper chromium	High-strength cupro-nickel
Specification BS 1400:1985	—	HCC1**	CC1-WP††	—
Alloy grouping*	—	B	B	B
Composition % (Single figures indicate maximum)				
Aluminium	—	—	—	—
Iron	1.0	—	—	1.0–1.4
Nickel	18–22	—	—	28.0–32.0
Manganese	1.0	—	—	1.0–1.4
Zinc	6.0–10.0	—	—	—
Silicon	0.15	—	—	0.35–0.50§§
Tin	2.5–4.0	—	—	—
Lead	4.0–6.0	—	—	0.01
Phosphorus	0.05	—	—	0.02
Magnesium	—	—	—	—
Chromium	—	—	0.6–1.2	0.3
Copper	REM	B.S. 1035–1037 grade	REM	REM
Total impurities	1.0 (incl. Fe)	—	—	¶

(continued)

Table 26.33 MISCELLANEOUS COPPER-BASE ALLOYS—*continued*

Alloy	Nickel silver	High-conductivity copper	Copper chromium	High-strength cupro-nickel
Specification BS 1400:1985	—	HCC1**	CC1-WP††	—
*Alloy grouping**	—	B	B	B
Properties of material				
Suitability for:				
Sand casting	2	2	2	2
Chill casting	—	5	5	2
Gravity die casting	—	3	3	—
Centrifugal	—	2	2	—
Continuous	—	3	3	—
Pressure tight castings (sand)				
Thin sections	—	1	2	1
Thick sections	—	2	2	2
Machining	2	3	3	2
Corrosion resistance	Natural and sea-water applications	Very good	Very good	Excellent
Fluidity	Fair	Good	Fair	Fair
Strength at elevated temperatures	—	Oxidise more readily than the other alloys specified in BS 1400		Good up to 300°C
Electrical conductivity % IAS at 15°C	5	90	80	5
Density (g cm^{-3})	8.7	8.9	8.85	—
Casting temperature range, °C				
Under 13 mm section	1 300	1 200	1 230	1 450
13–39 mm section	1 280	1 170	1 190	1 425
Over 40 mm section	1 250	1 130	1 150	1 400
Mechanical properties—SI units (Imperial units in brackets) in order sand cast, chill cast, centrifugally cast	Sand cast			Sand cast
Typical tensile stress				
MPa	309	160–190, —, —	270–340, —, —	540–618
(tonf in^{-2})	(20.0)	(10.4–12.3, —, —)	(17.5–22.0, —, —)	(35–40)
Typical 0.2% proof stress				
MPa	(0.5%) 185	—	170–250, —, —	(0.5%) 355–386
(tonf in^{-2})	(12.0)	—	(11.0–16.2, —, —)	(23–25)
Typical elongation 5.65 $\sqrt{S_0}$ %	(4$\sqrt{S_0}$) 15.0	23–40, —, —	(4$\sqrt{S_0}$) 18–30, —, —	(4$\sqrt{S_0}$) 20–25
Typical hardness range HB	80–110	—	Mandatory 100 min	160 (typical)

* *See* footnote † to Table 26.29.
† When castings to be welded lead not to exceed 0.01%
‡ *See* footnote § to Table 26.29.
§ *See* footnote ** to Table 26.29.
¶ *See* footnote §§ to Table 26.29.
** Maximum resistivity 0.019 μΩm^{-1}.
†† Maximum resistivity 0.022 μΩm^{-1}.
§§ Can be reduced to 0.2% to improve weldability under severe constraint.

Table 26.34 CHEMICAL COMPOSITION AND TYPICAL PROPERTIES OF SAND CAST COPPER-BASE ALLOYS

UNS number	Alloy name	Nominal chemical composition (%)								Mechanical properties						Physical properties			
		Cu	Al	Fe	Pb	Ni	Sn	Zn	Others	Tensile strength (ksi/MPa)	Yield strength # (ksi/MPa)	Elong. (%)	Brinell hardness [500]	Fatigue strength* (ksi/MPa)	Izod impact (ft.lb/J)	E (Msi/GPa)	Specific gravity	Thermal cond** % of Cu	Elect. Cond.***
C81100	Copper	99.7	—	—	—	—	—	—	—	25/172	9/62	40	44	9/62	0/0	17/117	8.94	200/346.1	92/0.538
C83600	Leaded red brass	85	—	—	5	—	5	5	—	37/255	17/117	30	60	11/76	10/14	13.5/93.1	8.83	41.6/72	15/0.087
C84400	Leaded semi-red brass	81	—	—	7	—	3	9	—	34/234	15/103	26	55	—	8/11	13/89.6	8.69	41.8/72.4	16/0.095
C85200	Leaded yellow brass	72	—	—	3	—	1.4	23.5	—	38/262	13/90	35	45	—	0/0	11/75.8	8.5	48.5/83.9	18/0.104
C86300	Manganese bronze	63	6.2	3	—	—	—	25	3.7 Mn	119/821	62/427	18	115	25/172	15/20	14.2/97.9	7.83	20.5/35.5	8/0.046
C87500	Silicon bronze & brass	82	—	—	—	—	—	14	4 Si	67/462	30/207	21	—	22/152	0/0	15.4/106	8.28	16/27.7	6/0.039
C90500	Tin bronze	87.5	—	—	—	—	10	2	—	45/310	22/152	25	75	13/90	10/13	15/103.4	8.72	43.2/74.8	11/0.064
C92200	Leaded tin bronze	88	—	—	1.5	—	6	4.5	—	40/276	20/138	30	65	11/76	0/0	14/96.5	8.64	40.2/69.6	14/0.083
C93700	High-leaded tin bronze	80	—	—	9.5	—	10	—	—	35/241	18/124	20	60	13/90	5/7	11/75.8	8.86	27.1/46.9	10/0.059
C95400	Aluminium bronze	83.2	10.8	4	—	—	—	—	1 Nb	85/586	35/241	18	—	28/93	16/22	15.5/107	7.45	33.9/58.7	13/0.075
C96400	Cupro-nickel 30%	68.2	—	0.9	—	30	—	—	—	68/469	37/255	28	—	18/124	0/0	21/145	8.94	16.4/28.5	5/0.028
C97600	Nickel silver	65	—	—	4	20.3	4	6	—	45/310	24/165	20	80	16/107	0/0	19/131	8.9	13/22.6	5/0.029

0.5% ext. under load.
* Endurance limit (100 million cycles).
** (Btu.ft/hr.ft². F) at 68°F/(W/m.K) at 20°C.
*** (% IACS at 68°F)/(MegaSiemens/cm at 20°C)
Current at the time of publication. For the most up-to-date data, contact the Copper Development Association or visit their website at http://www.copper.org
Copyright Copper Development Association, used with permission.

Table 26.35   TYPICAL COMPOSITIONS OF NICKEL CASTING ALLOYS

Alloy name	Chemical composition—wt% nominal or range—(not specification) bal. nickel													Spec.
	C	Si	Mn	Cu	Cr	Fe	Mo	Co	Ti	Al	W	Nb	Others	
Nickel	0.1/03	1/2	1/1.5	0.3	—	1	—	—	—	—	—	—	Mg 0.08/0.2, S 0.05 max Pb 0.005 max	ASTM 494 M3F2
Monel† Low Si	0.1/0.3	0.5/1.5	0.5/1.5	28/32	—	3	—	—	—	—	—	—	Mg 0.08/0.12, S 0.05, Pb 0.005 max	494 M3F2
Monel Med Si	0.15	2.5/3	0.5/1.5	28/32	—	3	—	—	—	—	—	—	Mg 0.008/0.12, S 0.05 max Pb 0.005 max	BS 3071 NA2
Monel High Si	0.15	3.5/4.5	0.5/1.5	28/32	—	3	—	—	—	—	—	—	Mg 0.08/0.12, S 0.05 max, Pb 0.005 max	BS 3071 NA3
Nichrome†	0.8	1.5/2.0	0.5/1.5	—	15/19	9/16	—	—	—	—	—	—	—	AMS 5396*
Hastelloy†	0.12	1	1	—	1	4/6	26/30	2.5	—	—	—	—	V 0.2/0.6, S 0.05 max, Pb 0.005 max	
Hastelloy† C	0.12	1	1	—	15.5/17.5	4.5/7	16/18	1.5	—	—	3.75/5.23	—	S 0.05 max, Pb 0.005 max	AMS 5388C/9A
Hastelloy† D	0.12	8.5/10	0.5/1.25	2/4	1	2	—	—	—	—	—	—	—	
Hastelloy† X	0.05/0.15	1	1	—	20/23	17/20	8/10	0.5/2.5	—	—	0.2/1	—	—	AMS 5390
Nimocast† 80	0.07	0.4	0.4	0.2	19.5	2	—	—	2.5	1.5	—	—	Ca 0.02 max, Pb 0.005 max	
Nimocast† 90	0.07	0.4	0.4	0.2	19.5	2	—	17	2.5	1.5	—	—	Ca 0.02 max, Pb 0.005 max	
Nimocast† 242	0.35	0.4	0.5	0.2	22	0.75	10.5	10	0.3	0.2	2.5	—	—	
Nimocast† PE10	0.05	0.25	0.3	0.2	20	3	6	—	—	6	—	7	Pb 0.005 max	AFNOR-NC 20NH
Nimocast† PD16	0.13	0.5	0.5	0.5	6	0.5	2	—	—	5.5	11	1.5	—	
Nimocast† 263	0.06	0.4	0.6	0.2	20	0.7	6	20	2	0.5	—	—	—	
IN-100	0.1/0.15	0.2	0.2	—	9.5/10.5	0.5	2.75/3.5	15	5	5.5	2	—	V 1.0	AFNOR-NK 15 CAT
IN-162	0.18	0.2	0.2	—	9/11	0.5	3.5/4.5	12	0.6/1.2	6.2/6.7	—	3	—	
IN-591	0.1	0.3	—	—	3	—	—	11/13	—	5.75	19	—	Ta 3, Zr 0.37, B 0.03	
IN-643	0.4/0.55	1	—	—	24/26	5	—	—	—	0.1	9	2	—	
IN-657	0.1	—	—	—	48/52	1	0.3/0.8	—	0.01/0.25	—	—	1.5	—	
IN-738LC	0.11	—	—	—	16	—	1.7	8.5	3.4	3.4	2.6	0.9	Ta 1.7, Zr 0.05, B 0.01	
IN-792	0.12	—	—	—	12.4	—	1.9	9	4.5	3.1	3.8	—	Ta 3.9, Zr 0.1, B 0.02	
IN-939	0.15	—	—	—	22.5	—	—	19	3.7	1.9	2	1	Ta 1.4, Zr 0.1, B 0.01	
IN-940	0.04	—	—	—	50	—	2.5	—	2	1	—	—	Zr 0.3, B 0.003	

Alloy	C			Cr		Mo	Co	Ti	Al	W	Nb	Others	
IN-6201	0.03	—	—	20	—	0.5	20	3.6	2.5	2.3	1	Ta 1.5, Zr 0.05, B 0.8	—
IN-6212	0.15	—	—	12	—	3	6	4.7	4.5	2	—	Zr 0.03, B 0.02	—
IN-6203	0.15	—	—	22	—		19	3.5	2.3	2	0.8	Ta 1.1, Hf 0.75, Zr 0.1, B 0.01	—
C242	0.34	—	—	20.5	—	10.5	10	0.3	0.2	—	—		
C263	0.06	—	—	20	—	5.8	20	2.2	0.5	—	—	Zr 0.04, B 0.008	
713C	0.12	—	—	12.5	—	4.2	—	0.8	6.1	—	2.2	Zr 0.1, B 0.012	
713LC	0.05	—	—	12	—	4.5	—	0.6	5.9	—	2	Zr 0.1, B 0.010	
B1990	0.1	—	—	8	—	6	10	1	6	—	—	Ta 4, Zr 0.1, B 0.015	
MAR-M002†	0.15	—	—	9	—	—	10	1.5	5.75	10	1	Zr 0.06, B 0.013, Hf 1.6	
NAR-M200†	0.15	—	—	9	—	—	10	2	5	12	1	Zr 0.05, B 0.015	
MAR-M200 + Hf†	0.14	—	—	9	—	—	10	2	5	12	1	Hf 1.8, Zr 0.08, B 0.02	
MAR-M246†	0.15	—	—	9	—	2.5	10	1.5	5.5	10	—	Ta 1.5, Zr 0.05, B 0.015	
MAR-M247†	0.15	—	—	8.3	—	0.7	10	1	5.5	10	—	Ta 3, Nf 1.5, Zr 0.05, B 0.015	
MAR-M421†	0.15	—	—	15.8	—	2	9.5	1.8	4.3	3.8	2	Zr 0.5, B 0.015	
MAR-M432†	0.15	—	—	15.5	—	—	20	4.3	2.8	3	2	Ta 2, Zr 0.05, B 0.015	
M-21	0.13	—	—	5.7	—	2	—	—	6	11	1.5	Zr 0.12, B 0.02	
M-22	0.13	—	—	5.7	—	2	—	—	6.3	11	—	Ta 3, Zr 0.6	
Rene 77†	0.07	—	—	14.6	—	4.2	15	4.3	4.3	—	—	Zr 0.04 B 0.016	
Rene 80†	0.17	—	—	14	—	4	9.5	5	3	4	—	Zr 0.03, B 0.015	
UDIMET 500†	0.07	—	—	18	—	4.2	19	3	3	—	—	Zr 0.05, B 0.007	
UDIMET 710†	0.07	—	—	18	—	3	15	5	2.5	1.5	—	Zr 0.05, B 0.020	
CMSX 2/3†	—	—	—	8	—	0.6	4.6	1	5.6	8	—	Ta 6, Hf 0.1	
CMSX 4†	—	—	—	6	—	0.6	10	1	5.6	6	—	Ta 6, Hf 0.1, Re 3	
CMSX 6†	—	—	—	9.8	—	3	5	4.7	4.8	—	—	Ta 2, Hf 0.1	
SRR 99	—	—	—	8.5	—	—	5	2.2	5.5	9.5	—	Ta 2.8	
A54	—	—	—	10	—	3	5	1.5	5	4	—	Ta 12	
SC 16	—	—	—	16	—	3	—	3.5	3.5	—	—	Ta 3.5	
INCONEL 625†*	0.2	—	—	21.6	—	8.7	—	0.2	0.2	—	3.9	—	
INCONEL 718†*	0.05	0.75	—	19	—	3	—	0.8	0.6	—	5.2	B 0.006	
INCOLOY 825†*	0.05	1.0	—	20/23	5	8/10	—	—	—	—	3 12/45	—	

† Registered trade mark.
* Strictly wrought but now commonly cast.

26.5   Nickel-base casting alloys

Table 26.36a   PHYSICAL PROPERTIES AND USE OF NICKEL CASTING ALLOYS

Alloy name	Casting characteristics	Melting range °C	Density kg dm⁻³	Uses
Nickel	C 0.2/0.3 for better castability	1 360–1 430	8.4	Corrosion resistant to alkales
Ferry Metal (60% Cu)	Castability improves as Si increases	1 315–1 350	8.6	Good corrosion resistance
Monel Med Si	Tends to be hot short	1 290–1 320	8.6	Good corrosion and galling resistance
Monel High S	Tends to be hot short	1 260–1 290	8.5	Corrosion resistant with max hardness and galling resistance
Nichrome†	Susceptible to H absorption	1 375–1 425	8.2	Heat and oxidation resistance. Furnace windings
Hastealloy B†	Less liable to H absorption than Nichrome	1 320–1 350	9.2	Good corrosion resistance especially to HCl
Hastealloy C†	Castability—moderate	1 270–1 310	8.9	Good corrosion resistance over wide range
Hastealloy D†	Similar to Monels	1 110–1 120	7.8	Resistance to hot conc. H₂SO₄
Hastealloy X†	Moderate	1 380–1 400	8.2	High temp. service, gas turbines
Nimocast 80†	—	1 310–1 380	8.17	High temp. service, gas turbines, furnace equipment
Nimocast 90†	Moderate	1 310–1 320	8.18	High temp. service, gas turbines, furnace equipment
Nimocast 242†	—	1 370–1 400	8.4	High temp. thermal starch resistance
Nimocast PE10†	Good fluidity	1 235–1 340	8.84	Gas turbines, diesel engines
Nimocast 263†	Vacuum cast	1 268–1 385	7.75	Gas turbine blades
IN-100	Vacuum cast	1 265–1 335	7.75	Gas turbine blades
IN-162	Vacuum cast	1 275–1 330	8.1	Gas turbine blades
IN-591			9.13	Nozzle guide vanes
IN-643	Moderate	1 450–1 500	8.7	Reformer tubes
IN-657	Moderate	1 304–1 314	7.96	Reformer tubes and general high temp. service
IN-738LC	Vacuum cast	1 230–1 315	8.11	Industrial and marine texture and nozzle guide vances
IN-792	Vacuum cast		8.25	Industrial and marine texture and nozzle guide vanes
IN-939	Vacuum cast	1 235–1 315	8.16	Turbine blades and nozzle guide vanes—gen. service
IN-6201	Vacuum cast	1 185–1 290	—	Turbine blades and nozzle guide vanes—gen. service
IN-6203	Vacuum cast	—	—	Turbine blades and nozzle guide vanes—gen. service
IN-6203	Vacuum cast	—	—	Turbine blades and nozzle guide vanes—gen. service

Alloy	Form	Temperature range	Density	Application
IN-6312	Vacuum cast	1 320–1 410	—	Turbine blades and nozzle guide vanes—gen. service
C242	Vacuum cast	1 225–1 340	8.4	Turbine blades and nozzle guide vanes—gen. service
C263	Vacuum cast	1 300–1 355	8.36	Turbine blades and nozzle guide vanes—gen. service
713C	Vacuum cast	1 260–1 290	7.91	Turbine blades and nozzle guide vanes—gen. service; Turbo charges
713LC	Vacuum cast	1 290–1 320	8.0	Turbine blades and nozzle guide vanes—gen. service; Turbo charges
B-1900	Vacuum cast	1 275–1 300	8.22	Servo gas turbine rotor blades
MAR-M002†	Vacuum cast	—	8.53	Aero gas turbine rotor and nozzle guide vanes
MAR-M002 + Hf	Vacuum cast	—	8.53	Aero gas turbine rotor blades
MAR-M246†	Vacuum cast	1 315–1 345	8.44	Aero gas turbine rotor blades
MAR-M247†	Vacuum cast	—	8.53	Aero gas turbine rotor blades
MAR-M421†	Vacuum cast	—	8.03	Industrial and marine gas turbine blades + nozzle guide vanes
MAR-M432†	Vacuum cast	—	8.16	Industrial and marine gas turbine blades + nozzle guide vanes
M-21	Vacuum cast	1 320–1 375	8.53	Aero gas turbine nozzle guide vanes
M-22	Vacuum cast	1 280–1 375	8.63	Aero gas turbine nozzle guide vanes
Rene 77†	Vacuum cast	—	7.92	Industrial and marine turbine blades and nozzle guide vanes
Rene 80†	Vacuum cast	—	8.16	Industrial and marine turbine blades and nozzle guide vanes
UDIMET 500†	Vacuum cast	1 300–1 395	8.02	Industrial and marine turbine blades and nozzle guide vanes
UDIMET 710†	Vacuum cast	—	8.08	Industrial and marine gas turbine rotor blades
CMSX 2/3†	Vacuum cast	—	8.56	Aero gas turbine rotor blades
CMSX 4†	Vacuum cast	—	8.70	Aero gas turbine rotor blades
CMSX 6†	Vacuum cast	—	7.98	Aero gas turbine rotor blades
SRR 99	Vacuum cast	—	8.3	Aero gas turbine rotor blades
454	Vacuum cast	—	8.7	Aero gas turbine rotor blades
SC 16	Vacuum cast	—	8.2	Industrial and marine gas turbine blades and nozzle guide vanes
INCONEL 625†*	Vacuum cast	—	8.44	General high temperature service
INCONEL 718†*	Vacuum cast	1 205–1 345	8.22	General high temperature service
INCONEL 825†*	Similar to stainless steel	1 370–1 400	8.1	For aggressive mineral acids, sour systems and stress corrosion cracking environments

† Registered trade mark.
* Strictly wrought but now commonly cast.

Table 26.36b   MECHANICAL PROPERTIES OF NICKEL CASTING ALLOYS

Alloy name	Heat treatment	Tensile stress MPa (tonf in⁻²)	0.1% proof stress MPa (tonf in⁻²)	Elongation %	Creep-rupture MPa 100 h life 870°C	980°C	1 000 h life 870°C	980°C
Nickel	None	390 (25)	125 (8)	20	—	—	—	—
Ferry Metal	None	495 (32)	155 (10)	25	—	—	—	—
Monel Med Si	Can be precipitation-treated to increase hardness	590 (38)	230 (15)	12	—	—	—	—
Monel High Si	None	695 (45)	—	Nil	—	—	—	—
Nichrome†	None	465 (30)	185 (12)	10	—	—	—	—
Hastelloy B†	Sand castings—air cool from 1 120°C	540 (35)	390 (25)	8	—	—	—	—
Hastelloy C†	Sand castings—air cool from 1 200°C	540 (35)	310 (20)	8	—	—	—	—
Hastelloy D†	Furnace cool from 1 050°C—handle with care	770 (50)	—	1	—	—	—	—
Hastelloy X†	Sand castings—air cool from 1 200°C	480 (31)	320 (21)	12	—	—	—	—
Nimocast 80†	4 h/1 080°C + 16 h/700°C	770 (50)	525 (34)	14	—	—	—	—
Nimocast 90†	4 h/1 080°C + 16 h/700°C	730 (47)	525 (34)	12	125	—	83	—
Nimocast 242†	None	480 (31)	260 (17)	7	—	—	—	—
Nimocast PE10†	4 h/1 100°C + 16 h/750°C	710 (46)	560 (36)	10	144	—	107	—
Nimocast PD 16†	None	750 (49)	680 (44)	3	—	—	—	—
Nimocast 263	None	1 000 (65)	650 (42)	—	—	—	—	—
IN-100	None	900 (58)	850 (55)	7	380	170	255	105
IN-162	None	930 (60)	770 (50)	6	—	—	—	—
IN-591	None	—	—	—	—	230	—	153
IN-643	None	600 (39)	280 (18)	12	—	(65)*	—	(50)*
IN-607	None	750 (49)	450 (29)	25	—	—	—	15
IN-738LC	2 h/1 120°C + 24 h/845°C	—	—	—	315	130	42	83
IN-792	2 h/1 120°C + 24 h/845°C	—	—	—	365	165	215	105
IN-939	4 h/1 160°C + 6 h/1 000°C + 24 h/900°C + 16 h/700°C	—	—	—	285*	102	200	64
IN-940	—	—	—	—	(85)*	—	(55)*	—
IN-6201	4 h/1 150°C + 16 h/900°C	—	—	—	(350)*	165	(240)*	95

Alloy	Heat treatment							
IN-6212	2 h/1 190°C + 24 h/845°C				380	160	250	100
IN-6203	8 h/1 160°C + 16 h/700°C				315	—	(225)*	—
C242	None				90	45	59	—
C263	None				—	—	—	—
713C	None				290	145	195	90
713LC	None				295	140	205	90
B1900	None				385	170	255	105
MAR-M002†	—				415	185	285	125
MAR-M002 + Hf†	19 h/1 230°C + 32 h/870°C				450	200	—	140
MAR-M246†	50 h/845°C				440	195	290	125
MAR-M247†	16 h/870°C				450	185	290	125
MAR-M421	2 h/1 150°C + 4 h/1 065°C + 16 h/780°C				310	125	215	83
MAR-M432	4 h/1 090°C + 16 h/760°C				295	140	215	87
M-21	None				380	188	(255)*	117
M-22	None				395	200	382	130
Rene 77†	4 h/1 165°C + 4 h/1 080°C + 24 h/925°C + 16 h/760°C				310	130	215	62
Rene 80†	2 h/1 120°C + 4 h/1 095°C + 4 h/1 050°C + 16 h/845°C				350	165	240	105
UDIMET 500†	4 h/1 150°C + 4 h/1 080°C				230	90	165	—
UDIMET 710†	2 h/1 150°C + 4 h/1 065°C + 7 h/760°C				305	150	215	76
CMSX 2/3†	3 h/1 302°C + 5 h/980°C + 20 h/870°C				480	240	350	170
CMSX 4†	—				580	275	435	190
CMSX 6†	3 h/1 238°C + 3 h/1 271°C + 3 h/1 277°C + 20 h/870°C				—	—	—	—
SRR 99	—				480	240	350	170
454	4 h/1 288°C + 4 h/1 080°C + 32 h/870°C				455	200	315	125
SC 16	3 h/1 250°C + 4 h/1 100°C + 24 h/850°C				(376)*	—	(276)*	—
INCONEL 625†*	None	0.2% 290/400 (19/26)	545 (35)	20	97	34	76	28
INCONEL 718†*	1 h/1 095°C + 1 h/955°C + 8 h/720°C CF + 8 h/620°C	0.2% 170/200 (11/13)	296/320 (19/21)	20/25	—	—	—	—
INCOLOY 825†*	None				—	—	—	—

† Registered trade mark.
* Wrought but not commonly cast.
( )* Extrapolated.

26.6 Magnesium alloys

Table 26.37   ZIRCONIUM-FREE MAGNESIUM ALLOYS

Grain refined (0.05–0.2 mm chill cast) when superheated to 850–900°C or suitably treated with carbon (as hexachlorethane)

Elektron designation ASTM designation	A8 AZ81		A8 (High purity) AZ81	
	MAG1M*(GP)†	MAG1TB*(GP)	MAG2M(SP)†	MAG2TB(SP)
Specifications BS 2970:1989				
BSS L series	—	3L.112	—	—
Equivalent DTD	—	—	684A	690A
Composition % (Single figures indicate maximum)				
Aluminium	7.5–9.0		7.5–9.0	
Zinc	0.3–1.0		0.3–1.0	
Manganese	0.15–0.4		0.15–0.7	
Copper	0.15		0.005	
Silicon	0.3		0.01	
Iron	0.05		0.00	
Nickel	0.01		0.001	
Cu + Si + Fe + Ni	0.40		—	
Material properties				
Founding	Good		Good	
Characteristics	Sand and permanent‡ mould		Special melting technique required	
Tendency to hot tearing	Little		Little	
Tendency to micro-porosity	Appreciable		Appreciable	
Castability§	A		A	
Weldability (Ar-Arc process)	Good		Good	
Relative damping capacity¶	C		C	
Strength at elevated temperature**	C		C	
Corrosion resistance	Moderate		Moderate	
Density, g cm^{-3}	1.81		1.81	
Liquids, °C	600		600	
Solidus, °C	475		475	
Non-equilibrium solidus, °C	420		420	
Castings temperature range, °C	680–800		680–800	
Heat treatment				
Solution				
Time, h	—	12 (min)	—	12 (min)
Temperature, °C	—	435 (max)‡‡	—	435 (max)‡‡
Cooling	—	Air, oil or water	—	Air, oil or water

	AZ91 AZ91			C alloy	
Elektron designation *ASTM designation*	MAG3M(GP)	MAG3TB(GP)	MAG3TF(GP)	MAG7M(GP)	MAG7TF(GP)
Specifications BS 2970:1989	—	—	—	—	—
BSS L series	—	3L.124	3L.125	—	—
Equivalent DTD	—	—	—	—	—
Composition % (Single figures indicate maximum)					
Aluminium		9.0–10.5			7.5–9.5
Zinc		0.3–1.0			0.3–1.5
Manganese		0.15–0.4			0.15–0.8
Copper		0.15			0.35
Silicon		0.3			0.40
Iron		0.05			0.05
Nickel		0.01			0.02
Cu + Si + Fe + Ni		0.40			0.75
Material properties					
Founding		Good			Good
Characteristics		Sand, permanent mould and die (pressure)			Sand, permanent mould and die (pressure)
Tendency to hot tearing		Little			Little
Tendency to micro-porosity		Less than MAG1			Less than MAG1
Castability§		A			A
Weldability (Ar-Arc process)		Good, but some difficulty with die castings			Good, but some difficulty with die castings
Relative damping capacity¶		C			C
Strength at elevated temperature**		C			C
Corrosion resistance		Moderate			Moderate
Density, g cm^{-3}		1.83			1.82
Liquidus, °C		600			595
Solidus, °C		470			475
Non-equilibrium solidus, °C		420			420
Castings temperature range, °C		680–800			680–800
Heat treatment *Solution:*					
Time, h	—		16 (min)	—	16 (min)
Temperature, °C	—		435 (max)‡‡	—	435 (max)‡‡
Cooling			Air, oil or water		Air, oil or water

(continued)

Table 26.37   ZIRCONIUM-FREE MAGNESIUM ALLOYS—*continued*

	AZ91E	AZ91(HP)	AZ91D	ZC63
Elektron designation	AZ91E	AZ91(HP)	AZ91D	ZC63
ASTM designation				ZC63
Specifications BS 2970:1989		MAG11(GP)		—
BSS L series				—
Equivalent DTD				
Composition % (Single figures indicate maximum)				
Aluminium		8.5–9.5		
Zinc		0.45–0.9		5.5–6.5
Manganese		0.15–0.40		0.25–0.75
Copper		0.015		2.4–3.0
Silicon		0.020		0.20
Iron		0.005		0.05
Nickel		0.0010		0.05
Cu + Si + Fe + Ni				0.01
Material properties				
Founding	Sand and permanent mould		High pressure die	Sand, permanent and high pressure die
Characteristics	Good		Good	Good
Tendency to hot tearing	Little		Little	Little
Tendency to micro-porosity	Less than MAG1		Little	Little
Castability§	A		A	B
Weldability (Ar-Arc process)	Good		Difficult	Good
Relative damping capacity¶	C		C	C
Strength at elevated temperature**	C		C	B
Corrosion resistance	Excellent		Excellent	Moderate
Density, g cm^{-3}	1.83		1.83	1.84
Liquidus, °C	595		595	635
Solidus, °C	470		470	465
Non equilibrium solidus, °C	420		420	—
Casting temperature range, °C	680–800		620–800	700–810
Heat treatment				
Time, h	16 (min)		Not suitable	8
Temperature, °C	435 (max)			440 (max)
Cooling	Air, oil or water			oil or water

Elektron designation ASTM designation	A8 AZ81		A8 (High purity) AZ81	
	MAG1M*(GP)†	MAG1TB*(GP)	MAG2M(SP)†	MAG2TB(SP)
Specifications BS 2970:1989 BSS L series Equivalent DTD	— —	3L.112 —	— 684A	— 690A
Heat treatment—continued Precipitation: Time, h Temperature, °C	— —	— —	— —	— —
Stress relief: Time, h Temperature, °C	2–4 250–330	— —	2–4 250–330	— —
Mechanical properties—sand cast—(SI units first, Imperial units following in brackets) Tensile strength (min), MPa (tonf in^{-2}) 0.2% proof stress (min), MPa (tonf in^{-2}) Elongation % min ($5.65\sqrt{S_0}$)	140 (9.1) 85 (5.5) 2	200 (13.0) 80 (5.2) 6	140 (9.1) 85 (5.5) 2	200 (13.0) 80 (5.2) 6
Mechanical properties—chill cast—(SI units first, Imperial units following in brackets) Tensile strength (min), MPa (tonf in^{-2}) 0.2% proof stress (min), MPa (tonf in^{-2}) Elongation % min ($5.65\sqrt{S_0}$)	185 (12.0) 85 (5.5) 4	230 (14.9) 80 (5.2) 10	185 (12.0) 85 (5.5) 4	230 (14.9) 80 (5.2) 10
Applications	Automobile road wheels	Good ductility and shock resistance	High-purity alloy—offers good corrosion resistance	

(*continued*)

Table 26.37 ZIRCONIUM-FREE MAGNESIUM ALLOYS—*continued*

Elektron designation ASTM designation	AZ91			C alloy		
	MAG3M(GP)	MAG3TB(GP)	MAG3TF(GP)	MAG7M(GP)	MAG7TB(GP)	MAG7TF(GP)
Specifications BS 2970:1989 BSS L series	—	3L.124	3L.125	—	—	—
Equivalent DTD	—	—	—	—	—	—
Heat treatment—continued Precipitation:						
Time, h	—	—	8 (min)	—	—	8 (min)
Temperature, °C	—	—	210 (max)	—	—	210 (max)
Stress relief:						
Time, h	2–4	—	—	2–4	—	—
Temperature, °C	250–330	—	—	250–330	—	—
Mechanical properties—sand cast—(SI units first, Imperial units following in brackets)[†]						
Tensile strength (min), MPa (tonf in^{-2})[††]	125 (8.1)	200 (13.0)	200 (13.0)	125 (8.1)	185 (12.0)	185 (12.0)
0.2% proof stress (min), MPa (tonf in^{-2})	95 (6.2)	85 (5.5)	130 (8.4)	85 (5.5)	80 (5.2)	110 (7.1)
Elongation % min (5.65$\sqrt{S_0}$)	—	4	—	—	4	—
Mechanical properties—chill cast—(SI units first, Imperial units following in brackets)[†]						
Tensile strength (min), MPa (tonf in^{-2})[††]	170 (11.0)	215 (13.9)	215 (31.9)	170 (11.0)	215 (13.9)	215 (13.9)
0.2% proof stress (min), MPa (tonf in^{-2})	100 (6.5)	85 (5.5)	130 (8.4)	85 (5.5)	80 (5.2)	110 (7.1)
Elongation % min (5.65$\sqrt{S_0}$)	2	5	2	2	5	2
Applications		For pressure tight applications Increased proof stress after full heat treatment		Principal alloy for commercial usage		

Elektron designation ASTM designation	AZ91E	AZ91(HP)	AZ91D	ZC63 ZC63
Specifications BS 2970:1989 BSS L series Equivalent DTD		MAG11(GP) — —		— — —
Heat treatment—continued Precipitation: Time, h Temperature, °C Stress relief: Time, h Temperature, °C	8 (min) 210 (max) — —		Not suitable — —	16 (min) 200 (max) — —
Mechanical properties—sand cast—(SI units first, Imperial units following in brackets) Tensile strength (min), MPa (tonf in^{-2}) 0.2% proof stress (min), MPa (tonf in^{-2}) Elongation % min (5.65 $\sqrt{S_0}$)	 200 130 2		 Typical high pressure die-cast properties — —	 210 125 3
Mechanical properties—chill cast—(SI units first, Imperial units following in brackets) Tensile strength (min), MPa (tonf in^{-2}) 0.2% proof stress (min), MPa (tonf in^{-2}) Elongation % min (5.65 $\sqrt{S_0}$)	 215 130 2		 200 150 1	 210 125 3
Applications		High purity alloy—offers excellent corrosion resistance. Max. temp. 120°C		Better foundability than AZ91 with superior elevated temperature properties

* M—as cast.
 TS—Stress relieved only.
 TE—Precipitation treated only.
 TB—Solution treated only.
 TF—Solution and precipitation treated.
† GP General purpose alloy.
 SP Special purpose alloy.
‡ Permanent mould = gravity die casting.
§ Ability to fill mould easily, A, B, C, indicate decreasing castability.

¶ Damping capacity ratings.
 A = Outstanding; better than grey cast iron.
 B = Equivalent to cast-iron.
 C = Inferior to cast-iron but better Al-base cast alloys.
** A = Particularly recommended.
 B = Suitable but not especially recommended.
 C = Not recommended where strength at elev. temps is likely to be an important consideration.
†† 1 MPa = 1 H mm^{-2} = 0.064 75 tonf in^{-2}.
‡‡ SO$_2$ or CO$_2$ atmosphere.

Table 26.38   MAGNESIUM-ZIRCONIUM ALLOYS
Inherently fine grained (0.015–0.035 mm chill cast)

	Z5Z	RZ5	ZRE1
Elektron designation			
ASTM designation	ZK51	ZE41	EZ33
Specifications BS 2970:1989			
BSS L series	MAG4TE*(GP)†	MAG5TE(SP)	MAG6TE(SP)
Equivalent DTD	2L 127	2L 128	2L 126
Composition % (Single figures indicate maximum)			
Zinc	3.5–5.5	3.5–5.5	0.8–3.0
Silver	—	—	—
Rare earth metals	—	0.75–1.75	2.5–4.0
Thorium	—	—	—
Zirconium	0.4–10	0.4–1.0	0.4–1.0
Copper	0.03	0.03	0.03
Nickel	0.005	0.005	0.005
Iron	—	—	—
Silicon	—	—	—
Manganese	—	—	—
Material properties			
Founding characteristics	Good in sand and permanent moulds§	Good in sand and permanent moulds	Excellent in sand and permanent moulds
Tendency to hot tearing	Marked	Some	Little
Tendency to micro-porosity	Very appreciable	Virtually none	None
Castability¶	B	A	A
Weldability (Ar-Arc Process)	Not recommended	Moderate	Very good
Relative damping capacity**	B/C	B/C	B
Strength at elevated temperature††	C	B	A
Resistance to creep at elevated temperature	Poor	Moderate	Good up to 250°C
Corrosion resistance	Moderate	Moderate	Moderate
Density, g cm⁻³ (20°C)	1.81	1.84	1.80
Liquidus, °C	640	640	640
Solidus, °C	560	510	545
Casting temperature range, °C	720–810	720–810	720–810

Heat treatment			
Solution:			
Time, h	16	—	—
Temperature, °C	180	—	—
Cooling	Air cool	—	—
Precipitation:			
Time, h	2	2 followed by 16	8
Temperature, °C	330	330 180	200
	to precede precipitation treatment	Air cool after each	Air cool
Post-weld stress relief:			
Time, h		Precipitation	10
Temperature, °C		treatment affords s/relief	250 max
			Air cool
*Mechanical properties—sand cast—*SI units (Imperial units in brackets)			
Tensile strength (min), MPa (tonf in^{-2})	230 (14.9)	200 (13.0)	140 (9.1)
0.2% proof stress (min), MPa (tonf in^{-2})	145 (9.4)	135 (8.7)	95 (6.2)
Elongation % min (5.65$\sqrt{S_0}$)	5	3	3
*Mechanical properties—chill cast—*SI units (Imperial units in brackets)			
Tensile strength (min), MPa (tonf in^{-2})	245 (15.9)	215 (13.9)	155 (10.0)
0.2% proof stress (min), MPa (tonf in^{-2})	145 (9.4)	135 (8.7)	110 (7.1)
Elongation % min (5.65$\sqrt{S_0}$)	7	4	3
Applications	High strength plus good ductility. Not suitable for spidery complex shapes	For high-strength pressure-tight applications	High degree of pressure tightness at room and elevated temperatures

Table 26.39   MAGNESIUM–ZIRCONIUM ALLOYS
Inherently fine grained

	ZT1‡‡‡	TZ6‡‡‡	ZE63
Elektron designation	ZT1‡‡‡	TZ6‡‡‡	ZE63
ASTM designation	HZ32	ZH62	ZE63
Specifications BS 2970:1989	MAG8TE(SP)	MAG9TE(SP)	—
BSS L series			
Equivalent DTD	5005A	5015A	5045
Composition % (Single figures indicate maximum)			
Zinc	1.7–2.5	5.0–6.0	5.5–6.0
Silver	—	—	—
Rare earth metals	0.10	0.20	2.0–3.0
Thorium	2.5–4.0	1.5–2.3	—
Zirconium	0.4–1.0	0.4–1.0	0.4–1.0
Copper	0.03	0.03	0.03
Nickel	0.005	0.005	0.005
Iron	0.01	0.01	0.01
Silicon	0.01	0.01	0.01
Manganese	0.15	0.15	0.15
Material properties			
Founding characteristics	As per MAG7 but more sluggish	Similar to MAG5	Good
Tendency to hot tearing	Little	Very little	Negligible
Tendency to micro-porosity	None	Low	Virtually none
Castability¶	C	B	A
Weldability (Ar-Arc process)	Very good	Fair	Very good***
Relative damping capacity**	B	C	B/C
Strength at elevated temperature††	A	B	C
Resistance to creep at elevated temperature	Good up to 350°C	Fair	Poor
Corrosion resistance	Moderate	Moderate	Moderate
Density, g cm^{-3} (20°C)	1.85	1.87	1.87
Liquidus, °C	645	630	625
Solidus, °C	550	520	516
Casting temperature range, °C	720–810	720–810	720–810
Heat treatment			
Solution:			
Time, h	—	—	30 for 12 mm sctn. 70 for 25 mm sctn.
Temperature, °C	—	—	480††
Cooling	—	—	Air blast or water spray

	MSR-A	MSR-B	MSR / QE22	MTZ‡‡‡ / HK31	EQ21	WE54	WE43
Precipitation:							
Time, h		16			2 followed by 16	48 or 72	
Temperature, °C		315			330 by 180	138 127	
		Air cool			Air cool after each	Air cool	
Post weld stress relief:							
Time, h		2			Pptn. treatment affords stress relief	—	—
Temperature, °C		350				—	—
		Air cool					
Mechanical properties—sand cast—SI units (Imperial units in brackets)							
Tensile strength min, MPa (tonf in⁻²)		185 (12.0)		255 (16.5)	255 (16.5)	275 (17.8)	
0.2% proof stress min, MPa (tonf in⁻²)		85 (5.5)		155 (10.0)	155 (10.0)	170 (11.0)	
Elongation, % ($5.65\sqrt{S_0}$) min		5		5	5	5	
Mechanical properties—chill cast—SI units (Imperial units in brackets)							
Tensile strength min, MPa (tonf in⁻²)		185 (12.0)			255 (16.5)	Sand	
0.2% proof stress min, MPa (tonf in⁻²)		85 (5.5)			155 (10.0)	Cast	
Elongation, % ($5.65\sqrt{S_0}$) min		5			5	Alloy	
Applications		Creep resistant alloy			For heavy duty structural usage	High strength with good ductility and excellent fatigue resistance. Structural parts aircraft, etc.	

Elektron designation	MSR-A	MSR-B	MSR	MTZ‡‡‡	EQ21	WE54	WE43
ASTM	—	—	QE22	HK31	EQ21	WE54	WE43
Specifications BS 2970:1972							
BSS L series	—	MAG12TF(SP)	—	—	MAG13TF(SP)	MAG14TF(SP)	—
Equivalent DTD	5025A	5035A	5055	—	—	—	—
Composition % (Single figures indicate maximum)							
Zinc	0.2	0.2	0.2	0.3	0.2	0.2	0.2
Silver	2.0–3.0	2.0–3.0	2.0–3.0	—	1.3–1.7	—	—
Rare earth metals	1.2–2.0‡	2.0–3.0‡	1.8–2.5‡	0.1	1.5–3.0‡	2.0–4.0¶¶¶	2.4–4.4¶¶¶
Thorium	—	—	—	2.5–4.0	—	—	—
Zirconium	0.4–1.0	0.4–1.0	0.4–1.0	0.4–1.0	0.4–1.0	0.4–1.0	0.4–1.0
Copper	0.03	0.03	0.03	0.03	0.05–0.10	0.03	0.03
Nickel	0.005	0.005	0.005	0.005	0.005	0.005	0.005
Iron	0.01	0.01	0.01	0.01	0.01	0.01	0.01
Silicon	0.01	0.01	0.01	0.01	0.01	0.01	0.01
Manganese	0.15	0.15	0.15	0.15	0.15	0.15	0.15
Yttrium	—	—	—	—	—	4.75–5.5	3.7–4.3

(*continued*)

Table 26.39   MAGNESIUM–ZIRCONIUM ALLOYS INHERENTLY FINE GRAINED—*continued*

	MSR-A	MSR-B	MSR QE22	MTZ‡‡‡ MK31	EQ21 EQ21	WE54 WE54	WE43 WE43
Elektron designation	MSR-A	MSR-B	MSR	MTZ‡‡‡	EQ21	WE54	WE43
ASTM designation	—	—	QE22	MK31	EQ21	WE54	WE43
Specifications BS 2970: 1989	—	MAG12TF (SP)	—	—	MAG13TF(SP)	MAG14TF(SP)	—
BSS L series							
Equivalent DTD	5025A	5035A	5055	—	—	—	—
Material properties							
Founding characteristics	Good	Good	Good	Less easy to found than MSR types	Good	Good	Good
Tendency to hot tearing	Little	Little	Little	Very little	Little	Very little	Very little
Tendency to micro-porosity	Slight	Slight	Slight	Negligible	Slight	Slight	Slight
Castability¶	B	B	B	C	B	B	B
Weldability (Ar-Arc process)	Very good	Very good	Very good	Very good	Very good	Very good	Very good
Relative damping capacity**	B/C	B/C	B/C	B/C	B/C	B/C	B/C
Strength at elevated temperature††	A	A	A	A	A	A	A
Resistance to creep at elevated temperature	Good up to 200°C	Good up to 200°C	Good up to 200°C	Good up to 350°C for short time applications	Good up to 200°C	Very good up to 250°C	Very good up to 250°C
Corrosion resistance	Moderate	Moderate	Moderate	Moderate	Moderate	Excellent	Excellent
Density, g cm⁻³ (20°C)	1.81	1.82	1.81	1.84	1.81	1.85	1.85
Liquidus, °C	640	640	640	645	640	640	640
Solidus, °C	550	550	550	590	545	550	550
Casting temperature range, °C	720–810	720–810	720–810	720–810	720–810	720–810	720–810
Heat treatment							
Solution:							
Time, h	8	8	8	2	8	8	8
Temperature, °C	525‡‡	525‡‡	525‡‡	565‡‡¶¶	520‡‡	525‡‡	525‡‡
Cooling	Water or oil	Water or oil	Water or oil	Air cool	Water or oil	Air cool	Water or oil
Precipitation:							
Time, h	16	16	16	16	16	16	16
Temperature, °C	200	200	200	200	200	250	250
	Air cool	Air cool	Air cool	Air cool	Air cool	Air cool	Air cool

	1 / 510 followed by above quench and age	1 / 510 followed by above quench and age	Repeat above cycle	1 / 505 followed by above quench and age	1 / 510 followed by above aircool and age	1 / 510 followed by above quench and age
Post-weld stress relief: Time, h / Temperature, °C	1 / 510 followed by above quench and age	1 / 510 followed by above quench and age	Repeat above cycle	1 / 505 followed by above quench and age	1 / 510 followed by above aircool and age	1 / 510 followed by above quench and age
Mechanical properties—sand cast—SI units (Imperial units in brackets)						
Tensile strength min, MPa (tonf in^{-2})	240 (15.5)	240 (15.5)	200 (13.0)	240	250	250
0.2% proof stress min, MPa (tonf in^{-2})	170 (11.0)	185 (12.0)	93 (6.0)	170	175	165
Elongation, % (5.65$\sqrt{S_0}$) min	4	2	5	2	2	2
Mechanical properties—chill cast—SI units (Imperial units in brackets)						
Tensile strength min, MPa (tonf in^{-2})	240 (15.5)	240 (15.5)	Usually sand	240	250	250
0.2% proof stress min, MPa (tonf in^{-2})	170 (11.0)	185 (12.1)		170 cast	175	165
Elongation, % (5.65$\sqrt{S_0}$) min	4	2	2		2	2
Applications	High strength in thick and thin section castings. Good elevated temperature (up to 250°C) short time tensile and fatigue props.	Similar to MSRA-B.	Superior short time tensile and creep resistance at temperatures around 300°C.	Similar to MSR alloys but less	Excellent strength up to 300°C for short time applications. Excellent corrosion resistance.	Excellent strength up to 250°C for long time applications. Excellent corrosion resistance.

* *See* footnote to Table 26.37.
† *See* footnote to Table 26.37.
‡ Neodymium-rich rare earths (others Ce-rich).
¶ *See* footnote to Table 26.37.
** *See* footnote to Table 26.37.
§ *See* footnote to Table 26.37.

†† *See* footnote to Table 26.37.
‡‡ SO_2 or CO_2 atmosphere.
¶¶ Castings to be loaded into furnace at operating temperature.
*** But only before hydriding treatment.
††† In hydrogen at atmospheric pressure.
‡‡‡ Thorium containing alloys are being replaced by alternative magnesium based alloys.
¶¶¶ Neodymium and heavy rare earths.

26.7   Zinc base casting alloys

Table 26.40   ZINC BASE ALLOYS—COMPOSITIONS

UNS Designation / Common / Designation	BS1004A	BS1004B	Z35636 ZA8	Z35631 ZA12	Z35841 ZA27	Z35541 No. 2. alloy, AC43A
Composition % (Single figure indicates maximum unless otherwise stated)						
Aluminium	3.8–4.3	3.8–4.3	8.0–8.8	10.5–11.5	25–28	3.5–4.3
Copper	0.10	0.75–1.25	0.8–1.3	0.5–1.2	2.0–2.5	2.5–3.2
Magnesium	0.03–0.06	0.03–0.06	0.015–0.030	0.015–0.030	0.01–0.02	0.03–0.06
Iron	0.10	0.10	0.10	0.075	0.10	0.075
Nickel	0.020	0.020	—	—		
Lead	0.005	0.005	0.005	0.005	0.005	—
Cadmium	0.005	0.005	0.004	0.004	0.004	Pb + Cd 0.009
Tin	0.002	0.002	0.003	0.003	0.003	0.002
Thallium	0.001	0.001	—	—	—	—
Indium	0.000 5	0.000 5	—	—	—	—
Titanium	—	—	—	—	—	—
Chromium	—	—	—	—	—	—
Zinc	Remainder	Remainder	Remainder	Remainder	Remainder	Remainder
Dimensional changes after casting (mm m^{-1})						
After 5 weeks	−0.32	−0.69				
After 6 months	−0.56	−1.03				
After 5 years	−0.73	−1.36				
After 8 years	−0.79	−1.41				
Dimensional changes after stabilising heat treatment (mm m^{-1}) 16 h at 100°C ± 5°C—air cool						
After 5 weeks	−0.20	−0.22				
After 3 months	−0.30	−0.26				
After 2 years	−0.30	−0.37				
			Shrinkage 0.03% after 100 days.	Shrinkage 0.005% after 30 days ambient, and 0.03% after 1 000 days. At 95°C, shrinkage followed by growth, i.e. zero change after 1 000 days.	Shrinkage 0.005% after 30 days ambient, and 0.015% after 1 000 days. At 95°C rapid shrinkage then growth to +0.1% after 1 000 days.	
Ageing	At RT all the alloys show little decrease in strength. Exposure to elevated temperature (100°C) may result in decrease of up to 30%					
Applications and uses	Extensive use where a large number of strong dimensionally accurate metal components required, e.g. components of cars, domestic appliances, business machines, record players, hydraulic and pneumatic valves, toy models.		Used for pressure diecastings requiring creep resistance. Gravity castings requiring excellent fluidity or very high quality electroplating.	General purpose sand and gravity casting alloy and a higher strength diecasting alloy. Good wear and bearing properties.	Strongest of the alloys. Used in the sand, gravity or pressure diecast form for applications where its strength is required. Best wear and bearing properties (comparable to SAE660 bronze).	Pressure die, sand and gravity die cast. For applications where hardness is an advantage. Sheet metal forming dies, moulds for plastics‡‡, zip fastener sliders.

Table 26.40a ZINC BASE CASTING ALLOYS—PROPERTIES

UNS Designation / Common Designation	BS1004A	BS1004B	Z35636 ZA8		Z35631 ZA12			Z35841 ZA27			Z35541 No. 2. alloy, AC43A	
Condition	Pressure die cast	Pressure die cast	Gravity cast	Pressure die cast	Sand cast	Gravity cast	Pressure die cast	Sand cast	Gravity cast	Pressure die cast	Sand cast	Pressure die cast
Density g/cm³	6.7	6.7	6.3	6.3	6.0	6.0	6.0	5.0	5.0	5.0	6.6	6.6
Melting range °C	382–387	379–388	375–404	375–404	377–432	377–432	377–432	375–484	375–484	375–484	390–379	390–379
Casting temp. °C	400–425	400–425	420–500	410–430	450–550	450–550	450–500	500–600	500–600	500–550	420–440	400–425
Coefficient of thermal expansion ($\mu m/m^{-1}/k^{-1}$)	27.4	27.4	23.2	23.2	24.1	24.1	24.1	26.0	26.0	26.0	27.8	27.8
Specific heat J/Kg/K	418	418	435	435	448	448	448	534	534	534	418	418
Thermal conductivity $W\,m^{-1}\,k^{-1}$	113	110	115	115	116	116	116	126	126	126	—	105
Electrical conductivity % IACS	26	26	27.7	27.7	28.3	28.3	28.3	29.7	29.7	29.7	25	25
Resistivity $\mu\Omega$ cm	6.4	6.5	6.2	6.2	6.1	6.1	6.1	5.8	5.8	5.8	6.8	—
Tensile strength MPa	283	324	240	375	300	330	400	440	320	425	252	350
Yield strength (0.2% offset) MPa	—	—	210	290	210	260	320	370	260	370	177	—
Young's modulus (GPa)	83	92	86	86	82	82	82	78	78	78	83	85
Elongation % in 2 in. (50 mm)	15	9	1–2	6–10	1–2	1.5–2.5	4–7	3–6	8–11	2.0–3.5	3	7
Hardness Brinell 500-10-30	83	92	90	100	95	90	100	115	95	120	100	100
Impact strength (un-notched RT)	58	57	—	40	25	—	30	44	60	12	7	47
Castability	Very good	Very good	Very good	Very good	Good	Good	Good (cold)	Good	Good	Good (cold) with proper feeding and chilling	Good	Very good

* Mechanical properties determined at thicknesses typical of the process and the alloy; they are intended for general guidance and comparison. (Source: International Lead Zinc Research Organization.) UNS = Unified Numbering System.

26.8   Steel castings

26.8.1   Casting characteristics

Casting is usually carried out at 1500–1700°C, or even higher, according to the type of steel and section of the casting. Highly refractory sands with high resistance to the molten metal stream are essential. Liquid shrinkage is high, so that generous risers are needed and the metal is hot short. Manganese steels, in particular, attack all furnace and ladle refractories, and only olivine sand has adequate resistance to constitute a satisfactory moulding material, tolerably resistant to burn-on.

26.8.2   Heat treatment

HOMOGENISING

High-temperature treatment intended to reduce interdendritic segregation.

ANNEALING

Heating above the critical range (Ac_3) for about 1 h per 25 mm of maximum cross-section and furnace cooling to nearly room temperature. Removes brittleness associated with coarse as-cast grain size or Widmanstätten structure, gives most complete relief of internal stresses, gives minimum yield stress and tensile strength, good ductility but poor impact values. Best magnetic properties in low carbon steels.

Table 26.41   STEEL CASTINGS FOR GENERAL ENGINEERING

Grade BS 3100	Description of steels	Composition (maxima unless stated)							
		C	Si	Mn	P & S	Cr	Mo	Ni	Others
A1		0.25	0.60	0.90[1]	0.060	0.25[2]	0.15[2]	0.40[2]	Cu 0.30[2]
A2	C General	0.35	0.60	1.00	0.060	—	—	—	—
A3		0.45	0.60	1.00	0.060	—	—	—	—
A4		0.18/0.25	0.60	1.20/1.60	0.050	—	—	—	—
A5	1½ Mn general	0.25/0.33	0.60	1.20/1.60	0.050	—	—	—	—
A6		0.25/0.33	0.60	1.20/1.60	0.050	—	—	—	—
AL1	C Low temp. use	0.20	0.60	1.10[5]	0.040	—	—	—	—
AM1	C-High mag.	0.15	0.60	0.50	0.050	0.25[2]	0.15[2]	0.40[2]	Cu 0.30[2]
AM2	permeability	0.25	0.60	0.50	0.050	0.25[2]	0.15[2]	0.40[2]	Cu 0.30[2]
AW1	C-Case hardening	0.10/0.18	0.60	0.60/1.00	0.050	0.25[2]	0.15[2]	0.40[2]	Cu 0.30[2]
AW2	Wear resistance	0.40/0.50	0.60	1.00	0.050	0.25[2]	0.15[2]	0.40[2]	Cu 0.30[2]
AW3	surface hardening	0.50/0.60	0.60	1.00	0.050	0.25[2]	0.15[2]	0.40[2]	Cu 0.30[2]
B1	C-Mo Elevated temperature	0.20	0.20/ 0.60	0.50/1.00	0.050	0.25[2]	0.45/0.65	0.40[2]	Cu 0.30[2]
B2		0.20	0.60	0.50/0.80	0.050	1.00/1.50	0.45/0.65	0.40[8]	Cu 0.30[8]
B3	Cr-Mo for	0.18	0.60	0.40/0.70	0.050	2.00/2.75	0.90/1.20	0.40[8]	Cu 0.30[8]
B4	elevated temps.	0.25	0.75	0.30/0.70	0.040	2.50/3.50	0.35/0.60	0.40[8]	Cu 0.30[8]
B5		0.20	0.75	0.40/0.70	0.040	4.00/6.00	0.45/0.65	0.40[8]	Cu 0.30[8]
B6		0.20	1.00	0.30/0.70	0.040	8.00/10.0	0.90/1.20	0.40[8]	Cu 0.30[8]
B7	Cr–Mo–V for elevated temps.	0.10/0.15	0.45	0.40/0.70	0.030	0.30/0.50	0.40/0.60	0.30[8]	Cu 0.30[8] Sn 0.058 V 0.22/ 0.30
BL1		0.20	0.60	1.00	0.040	—	0.45/0.65	—	—
BL2	Ferritic for low temperature	0.12	0.60	0.80	0.030	—	—	3.00/4.00	—

SUB-CRITICAL ANNEALING

Heating to a temperature below the Ac_1, and cooling slowly. Used for stress relief after welding, etc., and for softening high alloy steel castings.

NORMALISING

Similar to annealing except that cooling is in still air. Gives a smaller ferrite/pearlite grain size, higher yield stress, tensile strength and impact, but slightly lower elongation values than annealing. Medium and higher carbon steels which harden appreciably are usually tempered subsequently. Normalising may be preceded by annealing, or a prior normalising treatment at a higher temperature to break down the cast structure.

QUENCHING (OR HARDENING) AND TEMPERING

Heating above the critical range, but not so high as for annealing or normalising, and rapidly cooling in water, oil or air blast followed by reheating to a temperature below the critical range until the desired properties are achieved followed by air cooling, furnace cooling (if stress relief is also required) or cooling in water (to minimise temper embrittlement). Gives highest physical properties; high tempering temperatures usually gives lower strength but higher elongation and impact values. Normally preceded by annealing.

SURFACE HARDENING

Heating the surface layers by flame or induction to a temperature above the critical range and quenching by water jets following the heat source or by immersion.

Supply condition	Tensile strength R_m MPa	Lower yield stress or $R_{p0.2}$ MPa	El % on 5.65 $\sqrt{S_0}$	Bend test Bend angle	Bend radius	Impact Charpy V J	Hardness HB	Limiting section thickness mm
HT	430	230	22	120^4	1.5t	25^4	—	—
HT	490	260	18	90^4	1.5t	20^4	—	—
HT	540	295	14	—	—	18^3	—	—
HT	540/690	320	16	—	—	30	152/207	—
HT	620/770	370	13	—	—	25	179/229	100
HT	690/850	495	13	—	—	25	201/255	63
$—^6$	430	230	22	—	—	20 ($-40°C$)	—	—
A or N	340/430	185	22	120	1.5t	—	—	—
A or N	400/490	215	22	120	1.5t	—	—	—
C, A or N	460^7	—	12	—	—	25	—	—
Z, N or N & T	620	325	12	—	—	—	—	—
A, N or N & T	690	370	8	—	—	—	—	—
N & T	460	260	18	120^4	1.5t	20^4	—	—
N & T	480	280	17	120^4	1.5t	30^4	140/212	—
N & T	540	325	17	120^4	3t	25^4	156/235	—
N & T or H & T	620	370	13	120^4	3t	25^4	170/255	—
N & T or H & T	620	420	13	90^4	3t	25^4	179/255	—
N & T or H & T	620	420	13	—	—	—	179/255	—
N & T	510	295	17	120	3t	—	—	—
$—^6$	460	260	18	—	—	20 ($-50°C$)	—	—
$—^6$	460	280	20	—	—	20 ($-60°C$)	—	—

Table 26.41 STEEL CASTINGS FOR GENERAL ENGINEERING—*continued*

Grade BS 3100	Description of steels	C	Si	Mn	P & S	Cr	Mo	Ni	Others
BT1	Alloy-higher	—	—	—	0.050	—	—	—	—
BT2	tensile strengths	—	—	—	0.040	—	—	—	—
BT3		—	—	—	0.030	—	—	—	—
BW1	Alloy-case hardening	0.12/0.18	0.60	0.30/0.60	0.040	0.60/1.10	0.15/0.25	3.00/3.75	Cu 0.30[8]
BW2	Cr abrasion resistant	0.45/0.55	0.75	0.50/1.00	0.060	0.80/1.20	—	—	—
BW3		0.45/0.55	0.75	0.50/1.00	0.060	0.80/1.20	—	—	—
BW4	Cr-abrasion	0.55/0.65	0.75	0.50/1.00	0.060	0.80/1.50	0.20/0.40	—	—
BW10	Austenitic-Mn	1.00/1.25[9]	1.00	11.0 min	P 0.070 S 0.060	—	—	—	—
410C21[16]	13 Cr	0.15	1.00	1.00	0.040	11.5/13.5	—	1.00	—
420C29[16]		0.20	1.00	1.00	0.040	11.5/13.5	—	1.00	—
425C11	13 Cr 4Ni	0.10	1.00	1.00	0.040	11.5/13.5	0.60	3.40/4.20	—
302C25[16]		0.12	1.50	2.00	0.040	17.0/21.0	—	8.00 min	—
304C12[16]		0.03	1.50	2.00	0.040	17.0/21.0	—	8.00 min	—
304C15[16]	Austenitic Cr–Ni	0.08	1.50	2.00	0.040	17.0/21.0	—	8.00 min	—
347C17[13,16]		0.08	1.50	2.00	0.040	17.0/21.0	—	8.50 min	Nb8 × C/1.0[12]
315C16[16]		0.08	1.50	2.00	0.040	17.0/21.0	1.00/1.75	8.00 min	—
316C12[16]		0.03	1.50	2.00	0.040	17.0/21.0	2.00/3.00	10.0 min	—
316C16[16]	Austenitic Cr–Ni–No	0.08	1.50	2.00	0.040	17.0/21.0	2.00/3.00	10.0 min	—
316C71[16]		0.08	1.50	2.00	0.040	17.0/21.0	2.00/3.00	8.00 min	—
317C16[16]		0.08	1.50	2.00	0.040	17.0/21.0	3.00/4.00	10.0 min	—
318C17[16]		0.08	1.50	2.00	0.040	17.0/21.0	2.00/3.00	10.0 min	Nb8 × C/1.0[12]
452C11	Cr—for high temp. use	1.00	2.00	1.00	0.060	25.0/30.0	1.50	4.00	—
452C12		1.00/2.00	2.00	1.00	0.060	25.0/30.0	1.50	4.00	—
420C24		0.25	2.00	1.00	0.060	12.0/16.0	—	—	—
302C35		0.20/0.40	2.00	2.00	0.060	17.0/22.0	1.50	6.00/10.0	—
309C30		0.50	2.50	2.00	0.060	22.0/27.0	1.50	10.0/14.0	—
309C40	High alloy for	0.50	2.00	2.00	0.060	25.0/30.0	1.50	8.00/12.0	—
310C45	high temp. use	0.50	3.00	2.00	0.060	22.0/27.0	1.50	17.0/22.0	—
311C11		0.50	3.00	2.00	0.060	17.0/23.0	1.50	23.0/28.0	—
330C12		0.75	3.00	2.00	0.060	13.0/20.0	1.50	30.0/40.0	—
331C60		0.75	3.00	2.00	0.060	15.0/25.0	1.50	36.0/46.0	—
334C11		0.75	3.00	2.00	0.060	10.0/20.0	1.50	55.0/65.0	—
309C32	High alloy for	0.20/0.45	1.50	2.50	0.040	24.0/28.0	1.50	11.0/14.0	N 0.2
309C35	high temp. use	0.20/0.50	1.50	2.00	0.040	24.0/28.0	1.50	11.0/14.0	—
310C40		0.30/0.50	1.50	2.00	0.040	24.0/27.0	1.50	19.0/22.0	—
330C11		3.35/0.55	1.50	2.00	0.040	13.0/17.0	1.50	33.0/37.0	—
331C40		0.35/0.55	1.50	2.00	0.040	17.0/21.0	1.50	37.0/41.0	—

[1] For each 0.01% carbon below the max an increase of 0.04% Mn is permitted up to a max of 1.10%.
[2] Residual elements: total shall not exceed 0.80%.
[3] Only mandatory if specified by the purchaser.
[4] Either a bend test or an impact test may be specified.
[5] Mn/C ratio shall be greater than 3:1.
[6] The heat treatment to be applied shall be at the discretion of the manufacturer.
[7] Properties in the blank carburised and heat treated condition.
[8] Residual elements.
[9] The maximum carbon content may be increased to 1.35% by agreement.
[10] Not applicable to free machining steels.
[11] For free machining grades the minimum elongation shall be 12%.
[12] Gy agreement the Nb may be replaced by Ti in the range 5 × C/0.70%.
[13] If required with specified low temperature impact properties the carbon content shall be 0.06% max and the niobium 0.90% max.

Supply condition	Tensile strength R_m MPa	Lower yield stress or $R_{p0.2}$ MPa	EI % on 5.65 $\sqrt{S_0}$	Bend test Bend angle	Bend test Bend radius	Impact Charpy V J	Hardness HB	Limiting section thickness mm
H & T	690/850	495	11	—	—	35	201/255	—
H & T	850/1 000	585	8	—	—	25	248/302	—
H & T	1 000/1 160	695	6	—	—	20	293/341	—
A	1 000[7]	—	7	—	—	20	—	—
A, N & T or	—	—	—	—	—	—	—	—
H & T	—	—	—	—	—	—	201/255	—
A, N & T or	—	—	—	—	—	—	—	—
H & T	—	—	—	—	—	—	293 min	—
A, N & T or	—	—	—	—	—	—	—	—
H & T	—	—	—	—	—	—	341 min	—
WQ	—	—	—	—	—	—	—	—
HT	540	370	15	120[10]	2t	—	152/207	—
HT	690	465	11	—	—	—	201/255	—
HT	770	620	12	—	—	30	235/321	—
ST	480	240[15]	26[11]	—	—	—	—	—
ST	430	215[15]	26[11]	—	—	41[3,10] (−196°C)	—	—
ST	480	240[15]	26[11]	—	—	41[3,10] (−196°C)	—	—
ST	480	240[15]	22[11]	—	—	20[3,10] (−196°C)	—	—
ST	480	240[15]	26[11]	—	—	34[3,10] (−196°C)	—	—
ST	430	215[15]	26[11]	—	—	41[3,10] (−196°C)	—	—
ST	480	240[15]	26[11]	—	—	34[3,10] (−196°C)	—	—
ST	510	260[15]	26[11]	—	—	34[3,10] (−196°C)	—	—
ST	480	240[15]	22[11]	—	—	—	—	—
ST	480[14]	240[15]	18[11]	—	—	—	—	—
—[6]	—	—	—	—	—	—	—	—
—[6]	—	—	—	—	—	—	—	—
—[6]	—	—	—	—	—	—	—	—
—[6]	—	—	—	—	—	—	—	—
—[6]	—	—	—	—	—	—	—	—
—[6]	—	—	—	—	—	—	—	—
—[6]	—	—	—	—	—	—	—	—
—[6]	—	—	—	—	—	—	—	—
—[6]	—	—	—	—	—	—	—	—
—[6]	—	—	—	—	—	—	—	—
—	560	—	3	—	—	—	—	—
—	510	—	7	—	—	—	—	—
—	450	—	7	—	—	—	—	—
—	450	—	3	—	—	—	—	—
—	450	—	3	—	—	—	—	—

[14] The minimum value for a free version shall be 460 N/mm².

[15] 1% PS value.

[16] If a free machining grade is specified the sulphur content may be as high as 0.5% and/or other suitable elements may be present. Free machining grades shall be denoted by the letter F after the grade designation.

[17] Properties minima unless stated.

[18] HT—Heat treated, A—annealed, N—normalised. C—As cast, T—tempered, H & T—hardened and tempered, WA—water quenched, ST—solution treated.

Table 26.42a   STEEL CASTINGS FOR PRESSURE PURPOSES

Designation BS	Description of steels	Composition (*maxima unless stated*)							
		C	Si	Mn	P & S	Cr	Mo	Ni	Others
1504-161-430	Carbon steels	0.25	0.60	0.90[1]	0.050	0.25[3]	0.15[3]	0.40[3]	Cu 0.30[3]
1504-161-480		0.30	0.60	0.90[1]	0.050	0.25[3]	0.15[3]	0.40[3]	Cu 0.30[3]
1504-161-540	C-Mo	0.35	0.60	1.10[2]	0.050	0.25[3]	0.15[3]	0.40[3]	Cu 0.30[3]
1504-245		0.20	0.20/0.60	0.50/1.00	0.040	0.25[4]	0.45/0.65	0.40[4]	Cu 0.30[4]
1504-503LL60	3½ Ni	0.12	0.60	0.80	0.030	—	—	3.0/4.0	—
1504-621	1¼ Cr Mo	0.20	0.60	0.50/0.80	0.050	1.0/1.5	0.45/0.65	0.40[6]	Cu 0.30[6]
1504-622	2½ Cr Mo	0.18	0.60	0.40/0.70	0.050	2.0/2.75	0.90/1.2	0.40[6]	Cu 0.30[6]
1504-623	3 Cr Mo	0.25	0.75	0.30/0.70	0.040	2.5/3.5	0.35/0.60	0.40[6]	Cu 0.30[6]
1504-625	5 Cr Mo	0.20	0.75	0.40/0.70	0.040	4.0/6.0	0.45/0.65	0.40[6]	Cu 0.30[6]
1504-629	9 Cr Mo	0.20	1.00	0.30/0.70	0.040	8.0/10.0	0.90/1.20	0.40[6]	Cu 0.30[6]
1504-660	Cr Mo V	0.10/0.15	0.45	0.40/0.70	0.030	0.30/0.50	0.40/0.60	0.30[6]	Cu 0.30[6] Sn 0.05[6]
1504-420C29	13 Cr	0.20	1.0	1.0	0.040	11.5/13.5	—	1.0	Cu 0.30[6]
1504-425C11	13 Cr 4 Ni	0.10	1.0	1.0	0.040	11.5/13.5	0.60	3.4/4.2	—
1504-304C15	Austenitic Cr Ni	0.08	1.5	2.0	0.040	17.0/21.0	—	8.0 min	—
1504-304C12	Austenitic Cr Ni	0.03	1.5	2.0	0.040	17.0/21.0	—	8.0 min	—
1504-347C17	low C stabilised	0.08[11]	1.5	2.0	0.040	17.0/21.0	—	8.5 min	Nb8 × C/1.0[11,12]
1504-315C16	Austenitic Cr Ni 1½ Mo	0.08	1.5	2.0	0.040	17.0/21.0	1.0/1.75	8.0 min	—
1504-316C12		0.03	1.5	2.0	0.040	17.0/21.0	2.0/3.0	10.0 min	—
1504-316C16	Austenitic Cr Ni	0.08	1.5	2.0	0.040	17.0/21.0	2.0/3.0	10.0 min	—
1504-316C71	2½ Mo	0.08	1.5	2.0	0.040	17.0/21.0	2.0/3.0	8.0 min	—
1504-318C17		0.08	1.5	2.0	0.040	17.0/21.0	2.0/3.0	10 min	Nb8 × C/1.0[12]
1504-317C12	Austenitic Cr Ni	0.03	1.5	2.0	0.040	17.0/21.0	3.0/4.0	10.0 min	—
1504-317C16	3½ Mo	0.08	1.5	2.0	0.040	17.0/21.0	3.0/4.0	10.0 min	—
1504-364C11	Austenitic Cr Ni	0.07	2.5	2.0	0.030	20.0/24.0	3.0/6.0	20.0/26.0	Cu 2.0 Nb 0.50
1504-332C11	Mo Cu high alloy	0.07	1.5	2.0	0.040	19.0/22.0	2.0/3.0	26.5/30.5	Cu 3.0/4.0
1504-310C40	Austenitic Cr Ni High alloy	0.30/0.50	1.5	2.0	0.040	24.0/27.0	1.50[6]	19.0/22.0	—
1504-330C11	Austenitic Ni Cr High alloy	0.35/0.55	1.5	2.0	0.040	13.0/17.0	1.50[6]	33.0/37.0	—

Treatment	Tensile strength R_m MPa	Lower yield stress or $R_{p0.2}$ MPa	El A% on 5.65 $\sqrt{S_0}$	Bend test Bend angle	Bend radius	Impact Charpy V J	Hardness HB
A or N	430	230	22	120	1.5t[5]	25 @ RT[5] 20 @ −40°C[8]	—
A or N	480	245	20	90	1.5t[5]	20 @ RT[5]	—
A or N	540	280	13	90	1.5t[5]	20 @ RT[5]	—
N & T or Q & T	460	260	18	120	1.5t[5]	20 @ RT[5] 20 @ −50°C[8]	—
N & T or H & T	460	280	20	—	—	20 @ −60°C	134/187
N & T	480	280	17	120	1.5t[5]	30 @ RT[5]	140/212
N & T	540	325	17	120	3t[5]	25 @ RT[5]	156/235
N & T or H & T	620	370	13[7]	120	3t[5]	25 @ RT[5]	179/255
H & T	620	420	13[7]	90	3t[5]	25 @ RT[5]	179/255
H & T	620	420	13[7]	—	—	—	179/255
N & T	510	295	17	120	3t[5]	—	—
V 0.22/0.30 Al[9]							
H & T	620	450	13[7]	120	3t[5]	25 @ RT[5]	179/235
H & T	770	620	12[7]	—	—	30 @ RT[5]	235/321
ST	480	240[10]	26	—	—	41 @ −196°C[8]	—
ST	430	215[10]	26	—	—	41 @ −196°C[8]	—
ST	480	240[10]	22	—	—	20 @ −196°C[8]	—
ST	480	240[10]	26	—	—	34 @ −196°C[8]	—
ST	430	215[10]	26	—	—	41 @ −196°C[8]	—
ST	480	240[10]	26	—	—	34 @ −196°C[8]	—
ST	510	260[10]	26	—	—	34 @ −196°C[8]	—
ST	480	240[10]	18	—	—	—	—
ST	430	215[10]	22	—	—	—	—
ST	480	240[10]	22	—	—	—	—
ST	430	200[10]	20	—	—	—	—
ST	430	200	20	—	—	—	—
—	450	—	7	—	—	—	—
—	450	—	3	—	—	—	—

Table 26.42b   STEEL CASTINGS FOR PRESSURE PURPOSES—ELEMENTAL TEMPERATURE PROPERTIES—E DESIGNATION

Designation BS	Description of steels	Minimum lower yield stress R_{el} or 0.2% proof stress, $R_{p0.2}$, MPa at a temperature in °C of														
		20[13]	50[13]	100	150	200	250	300	350	400	450	500	550	600	650	700
1504-161-430E	Carbon steels	208	199	187	178	168	157	148	142	139	—	—	—	—	—	—
1504-161-480E		241	232	218	205	193	181	172	164	161	—	—	—	—	—	—
1504-245[14]	C–Mo	262	256	248	241	232	210	181	171	165	165	161	151	128	—	—
1504-622	$2\frac{1}{4}$ Cr Mo	335		323	312	305	296	290	280	273	258	240	211	180	—	—
1504-625	5 Cr Mo	480	465	443	430	423	416	411	398	382	353	314	—	—	—	—
1504-316C16	Austenitic Cr Ni $1\frac{1}{2}$ Mo	—	—	140	—	116	—	96	—	93	—	85	—	77	—	76
1504-316C71	Austenitic Cr Ni $2\frac{1}{4}$ Mo	—	—	162	—	137	—	116	—	113	—	102	—	93	—	81

Notes for Table 26.42a and b

[1] For each 0.01% carbon below the maximum an increase of 0.04% manganese will be permitted up to a maximum of 1.10%.
[2] For each 0.01% carbin below the maximum an increase of 0.05% manganese will be permitted up to a maximum of 1.60%.
[3] Ni + Cr + Mo + Cu = 0.80% max.
[4] Ni + Cr + Cu = 0.80% max.
[5] Either a bend test or an impact test may be specified but not both.
[6] Residual elements.
[7] This value may be increased to 15% min, subject to agreement at the time of enquiry and order.
[8] When specifically ordered.
[9] When used for deoxidation shall not exceed 0.25 kg/t.
[10] 1% Proof stress.
[11] C content 0.06 max and Nb 0.9 max when low temperature properties are specified.
[12] Ti 5XC/0.70 may be substituted for Nb by agreement.
[13] Values at 20°C and 50°C are included for design purposes only and are not subjected to verification. All values except those at 50°C that are obtained by interpolation are based on tests according to BS 3688 and thus values at 20°C differ from the corresponding room temperature values given in this specification.
[14] Only apply to sections <32 mm.
[15] Properties minima unless stated.
[16] A—Annealed, N—normalised, T—tempered, Q & T—quenched and tempered, H & T—hardened and tempered, ST—solution treated.

Table 26.43   WELDABLE CHROMIUM–NICKEL CENTRIFUGALLY CAST TUBES

BS 4534: 1969	Grade	Description	Composition (maxima unless stated)									Treatment	Tensile strength R_e MPa	El % on $5.65\sqrt{S_0}$	Maximum[3] wall thickness mm
			C	Si	Mn	P	S	Cr	Mo	Ni	Others				
	1	18/11 Cr Ni 0.07C	0.04/0.09	1.00	2.00	0.030	0.030	17.0/19.0	0.50	11.0/13.5	—	C	450	26	20
	1F[1]	18/8 Cr Ni 0.07C	0.04/0.09	1.50	2.00	0.040	0.040	17.0/20.0	—	8.0 min	—	C	480	26	20
	2	18/21/1 Cr Ni Nb 0.07C	0.04/0.09	1.00	2.00	0.030	0.030	17.0/19.0	0.50	11.5/14.0	Nb 8 × C/1.0	C	450	22	20
	2F[1]	18/8/1 Cr Ni Nb 0.07C	0.04/0.09	1.50	2.00	0.040	0.040	17.0/20.0	—	8.0 min	Nb 8 × C/1.00	C	480	26	20
	3	17/3/2.5 Cr Ni Mo 0.07C	0.04/0.09	1.00	2.00	0.030	0.030	16.0/18.0	2.00/2.75	12.5/15.0	—	C	450	26	20
	3F[1]	18/8/3 Cr Ni Mo 0.07C	0.04/0.09	1.50	2.00	0.040	0.040	17.0/20.0	2.00/3.00	8.00 min	—	C	480	22	20
	4	20/10 Cr Ni 0.3C	0.25/0.35	0.50/1.50	2.00	0.040	0.040	18.0/21.0	0.50	9.00/11.5	—	C	480	22	20
	5	20/12 Cr Ni 0.3C	0.25/0.40	0.50/1.50	2.00	0.030	0.030	23.0/27.0	0.50	11.5/14.0	—	Aged	510	9	20
	6	25/30 Cr Ni 0.4C	0.35/0.45	0.50/1.50	2.00	0.040	0.040	23.0/27.0	0.50	18.0/22.0	—	C	480	9	20
	7	25/20 Cr Ni 0.15C	0.15	0.50/1.50	2.00	0.040	0.040	23.0/27.0	0.50	19.0/23.0	—	C	450	22	20
	8	18/35 Cr Ni 0.45C	0.40/0.50[2]	0.50/1.50	2.00	0.040	0.040	17.0/21.0	0.50	33.0/37.0	—	C	450	7	20
	9	18/35 Cr Ni 0.5C	0.35/0.55	1.50/2.00	2.00	0.040	0.040	17.0/21.0	0.50	33.0/37.0	—	C	450	7	20
	10	18/35 Cr Ni 0.15C	0.15	0.50/1.50	2.00	0.040	0.040	17.0/21.0	0.50	33.0/37.0	—	C	420	22	20
	11	20/32 Cr Ni 0.1C	0.10	0.30/1.00	1.50	0.030	0.030	19.0/22.0	0.50	30.0/34.0	—	C	400	26	20

[1] For use where ferrite content is to be within a specified range to facilitate welding.
[2] By agreement this grade can be supplied to a C range of 0.35/0.45%.
[3] Above 20 mm wall thickness properties shall be agreed between manufacturer and purchaser.
[4] Properties minima unless stated.
[5] C—As cast.

Table 26.44   AEROSPACE SERIES—STEEL CASTINGS—1973/1974

Designation BS	Description of steels	Compositions (maxima unless stated) %							
		C	Si	Mn	P & S	Cr	Mo	Ni	Others
HC1	C 500 MPa	0.35	0.6	1.0	0.035	0.25[1]	0.15[1]	0.4[1]	Cu 0.3[1]
HC2	C Mn 550 MPa	0.25	0.6	1.7	0.035	0.25[1]	0.15[1]	0.4[1]	Cu 0.3[1]
HC3	1 Cr Mo Low alloy 700 MPa	0.3	0.6	0.8	0.35	1.2	0.4	—	Cu 0.3
HC4	3 Cr Mo 620/770 MPa	0.25	0.75	0.6	0.035	3.5	0.6	—	Cu 0.3
HC5	3 Ni Case hard. 700 MPa	0.18	0.6	0.6	0.035	0.25	0.15	3.5	Cu 0.3
HC6	3 Cr Mo Nitrid. 850/1000 MPa	0.3	0.75	0.6	0.020	3.5	0.7	0.4	Cu 0.3 V 0.02 Sn 0.03
HC7	3 Cr Mo 800/1 080 MPa	0.32	0.75	0.8	0.025	3.5	0.7	0.4	Cu 0.4 V 0.02 Sn 0.03
HC8	3 Cr Mo 1 150/1 300 MPa	0.32	0.75	0.8	0.025	3.5	0.7	0.4	Cu 0.4 V 0.02 Sn 0.03
HC9	Ni Cr Mo 880/1 080 MPa	0.34	0.6	0.8	0.025	1.3	0.7	3.0	Cu 0.4
HC10	Ni Cr Mo 1 150/1 300 MPa	0.34	0.6	0.8	0.025	1.3	0.7	3.0	Cu 0.4
HC101[2]	Pt. hard. Cr Ni Cu Mo 950 MPa	0.07	2.0	1.0	0.025	15.5	2.5	6.0	Cu 3.5 Nb 0.5
HC102[2]	Pt. hard. Cr Ni Cu Mo 1250 MPa	0.07	2.0	1.0	0.025	15.5	2.5	6.0	Cu 3.5 Nb 0.5
HC103	23 Cr Ni W corr. resist.	0.25	2.0	1.0	0.04	25.0	—	13.0	W 3.5
HC104	19 Cr 10 Ni, Vb corr. resist.	0.08	2.0	2.0	0.04	21.0	—	12.5	Nb 1.0
HC105	18 Cr 11 Ni 2.5 Mo Nb corr. resist.	0.08	1.5	2.0	0.04	20.0	3.0	12.5	Nb 1.0
HC106	Pt. hard. Cr Ni Cu corr. resist.	0.08	1.0	1.0	0.04	17.5	—	5.0	Cu 3.0 Ta + Nb 0.4
HC401	18 Ni Maraging Precision cast 1 600/1 850 MPa	0.03	0.1	0.1	0.01	0.25	4.9	17.5	Al 0.15 Co 11.0 Ti 0.6

Notes:
[1] Total residual elements shall not exceed 0.8%
[2] A restricted range of oncomposition may be agreed if welding or brazing is intended. Castings in composition range are subject to Patent applications.
[3] Hardness values are for maraged castings.
* Angle of bend 120°. Radius of bend 1.5t, t = thickness of test piece.
** Angle of bend 120°. Radius of bend 2t, t = thickness of test piece.

Table 26.45a   INVESTMENT CASTINGS—CARBON AND LOW ALLOY STEELS

Grade BS 3146 Part 1	Description of steels	Composition (maxima unless stated)							
		C	Si	Mn	P & S	Cr	Mo	Ni	Others
CLA1-A		0.15/0.25	0.20/0.60	0.40/1.00	0.035	0.30[1]	0.10[1]	0.40[1]	Cu 0.30[1]
-B	C steels	0.25/0.35	0.20/0.60	0.40/1.00	0.035	0.30[1]	0.10[1]	0.40[1]	Cu 0.30[1]
-C		0.35/0.45	0.20/0.60	0.40/1.00	0.035	0.30[1]	0.10[1]	0.40[1]	CI 0.30[1]
CLA2	1$\frac{1}{5}$ Mn	0.18/0.25	0.20/0.50	1.20/1.70	0.035	0.30[1]	0.10[1]	0.40[1]	Cu 0.30[1]
CLA3	Alloy steels 700–850 MPa	—	—	—	0.035	—	—	—	—
CLA4	Alloy steels 850–1 000 MPa	—	—	—	0.035	—	—	—	—
CLA5-A	1 000 MPa min	—	—	—	P 0.025)	—	—	—	—
-B		—	—	—	S 0.020)	—	—	—	—
CLA7	3 Cr Mo	0.15/0.25	0.30/0.80	0.30/0.60	0.035	2.50/3.50	0.35/0.60	0.40[4]	Cu 0.30[4]
CLA8	C Steels for surface hardening	0.37/0.45	0.20/0.60	0.50/0.80	0.035	0.30[1]	0.10[1]	0.40[1]	Cu 0.30[1]

Tensile strength R_{m} MPa	Lower yield $R_{\mathrm{p0.2}}$ MPa	El A%	Z %	KI ft/lb	Hardness			
					HB	HV	HRB	HRC
500	215	15	—	25	—	—	—	—
550	310	13	—	30	—	—	—	—
700	480	12	—	30	201	200	—	—
620	480	14	—	25	179	175	88	—
700	350	13	—	30	197	200	93	—
850	600	8	—	15	248	255	—	22
880	700	8	30	30	262	265	—	27
1 150	940	5	14	10	331	335	—	35
880	700	8	30	30	262	625	—	27
1 150	940	5	14	10	331	335	—	35
950	800	12	30	15	—	—	—	—
1 250	950	8	15	8	375	400	—	40
—	—	—	—	—	—	—	—	—
460	200	20	*	—	—	—	—	—
500	210	20	**	—	—	—	—	—
1 250	1 030	8	20	—	375	395	—	40
1 600	1 500	6	25	10	461[3]	495[3]	—	49[3]

Treatment	Tensile strength R_{m} MPa	$R_{\mathrm{p0.2}}$ MPa	El % on $5.65 \sqrt{S_0}$	Impact Izod ft. lbf	Hardness HB
N, N & T or H & T	430	195[3]	15	—	121/174[2]
N, N & T or H & T	500	215[3]	13	—	143/183[2]
N, N & T or H & T	540	245[3]	11	—	163/207[2]
N, N & T or H & T	550/700	310[3]	13	30	152/201[2]
H & T	700/850	495	11	25	201/255[2]
H & T	850/1 000	585	11	15	248/302[2]
H & T	1 000	880	9	30	269/321[2]
H & T	1 160	1 000	5	10	341/388[2]
H & T	620/770	480	14	25	179/223[2]
N or N & T	540	245[3]	15	—	—

(*continued*)

Table 26.45a   INVESTMENT CASTINGS—CARBON AND LOW ALLOY STEELS—*continued*

Grade BS 3146 Part 1	Description of steels	Composition (*maxima unless stated*)							
		C	Si	Mn	P & S	Cr	Mo	Ni	Others
CLA9	C Steel for case hardening	0.10/0.18	0.20/0.60	0.60/1.00	0.035	0.10[1]	0.40[1]	Cu 0.30[1]	
CLA10	3 Ni for case hardening	0.10/0.18	0.20/0.60	0.30/0.60	0.035	0.30[4]	0.10[4]	2.75/3.50	Cu 0.30[4]
CLA11	3 Cr Mo for nitriding	0.20/0.30	0.30/0.80	0.30/0.60	0.035	2.90/3.50	0.40/0.70	0.40[4]	Cu 0.30[4] Sn 0.03[4] V 0.02[4]
CLA12-A & B	Cr for abrasion resisting	0.45/0.55	0.30/0.80	0.50/1.00	0.035	0.80/1.20	0.10[4]	0.40[4]	Cu 0.30[4]
-C	Ni Mo for case	0.55/0.65	0.30/0.80	0.50/1.00	0.035	0.80/1.50	0.20/0.40	0.40[4]	Cu 0.30[4]
CLA13	hardening	0.12/0.20	0.20/0.60	0.30/0.70	0.035	0.30[4]	0.20/0.30	1.50/2.00	Cu 0.30[4]

Notes:
[1] Residual elements—total not to exceed 0.80%
[2] When specifically requested
[3] For information only
[4] Residuals.

Table 26.45b   INVESTMENT CASTINGS CORROSION AND HEAT RESISTING STEELS—*continued*

Grade BS 3146 Part 2	Description of steels	Composition (*maxima unless stated*)							
		C	Si	Mn	P & S	Cr	Mo	Ni	Others
ANC1-A		0.15	0.20/1.20	0.20/1.00	0.035[1]	11.5/13.5	—	1.00	—
-B	13 Cr	0.12/0.20	0.20/1.20	0.20/1.00	0.035[1]	11.5/13.5	—	1.00	—
-C	Martensitic	0.20/0.30	0.20/1.20	0.20/1.00	0.035[1]	11.5/13.5	—	1.00	—
ANC2	18 Cr 2 Ni Martensitic	0.12/0.25	0.20/1.00	0.20/1.00	0.035	15.5/20.0	—	1.50/3.00	—
ANC3-A	18 Cr 10 Ni	0.12	0.20/2.00	0.20/2.00	0.035[1]	17.0/20.0	—	8.00/12.0	—
	Austenitic	0.12	0.20/2.00	0.20/2.00	0.035[1]	17.0/20.0	—	8.50/12.0	Nb 8 × C/1.10
ANC4-A	18 Cr 11 Ni	0.08	0.20/1.50	0.20/2.00	0.035[1]	18.0/20.0	3.00/4.00	11.0/14.0	—
-B	3 Mo	0.08	0.20/1.50	0.20/2.00	0.035[1]	17.0/20.0	2.00/3.00	10.0 min	—
-C	Austenitic	0.12	0.20/1.50	0.20/2.00	0.035[1]	17.0/20.0	2.00/3.00	10.0 min	Nb 8 × C/1.10
ANC5-A		0.50	0.20/3.00	0.20/2.00	—	22.0/27.0	—	17.0/22.0	—
-B	Ni Cr Steels	0.50	0.20/3.00	0.20/2.00	—	15.0/25.0	—	36.0/46.0	—
-C		0.75	0.20/3.00	0.20/2.00	—	10.0/20.0	—	55.0/65.0	—
ANC6-A		0.15/0.30	0.75/2.00	0.20/1.00	0.035	20.0/25.0	—	55.0/65.0	—
-B	Cr–Ni Steels	0.15/0.30	0.75/2.00	0.20/1.00	0.035	20.0/25.0	—	10.0/15.0	W 2.50/3.50
-C		0.05/0.15	0.75/2.00	0.20/1.00	0.035	20.0/25.0	—	10.0/18.0	W 2.50/3.50
ANC20	14 Cr 5 Ni	0.07	0.20/2.00	0.20/1.00	0.025	12.5/15.5	0.50/2.50	3.00/6.00	Cu 1.00/3.50
A & B	2 Cu 1 Mo								Nb 0.50
ANC21	26 Cr 5 Ni	0.05	0.75	0.75	0.050	25.0/27.0	1.75/2.25	4.75/6.00	Cu 2.75/3.25
	3 Cu 2 Mo								N 0.10
ANC22 A, B & C	16 Cr 4 Ni 3 Cu	0.06	1.00	0.70	P 0.035 S 0.030	15.5/16.7	—	3.60/4.60	Cu 2.80/3.50 Nb + Ta 0.15/0.40 N 0.05

ANC 8-19 are specialised high alloy nickel and cobalt base castings—BS 3147 Part3:1976 are also very high alloy castings including Ni and Co base casting—for details *see* British Investment Casters Technical Association.

[1] Where specified S may be as high as 0.3% in free machining grades.

Treatment	Tensile strength R_m MPa	$R_{p0.2}$ MPa	El % on 5.65 $\sqrt{S_0}$	Impact Izod ft. lbf	Hardness HB
C or N	495[5]	215[3,5]	15[5]	20[5]	—
C or N	700[5]	350[3,5]	14[5]	30[5]	—
H & T	850/1 000	600	8	15	248/302[2]
N & T or H & T	700[6]	—	8[6]	—	207[6,7] min
N & T or H & T	—	—	—	—	341 min
C or N	700[5]	350[3,5]	14[5]	30[5]	—

[5] Properties quoted are for the final heat treated condition
[6] Grade A only
[7] Grade B hardness 293 HB min
[8] Properties minima unless stated
N—Normalised, T—tempered, H & T—hardened & tempered, C—As cast

Treatment	Tensile strength R_m MPa	$R_{p0.2}$ MPa	El % on 5.65 $\sqrt{S_0}$	Impact Izod ft. lbf	Hardness HB
H & T	540	340[3]	15	—	152/207
H & T	620	415[3]	13	—	183/229
H & T	695	435[3]	11	—	201/255
H & T	850/1 000	630[3]	8	—	248/320
ST	460	200	20	—	—
ST	460	200	20	—	—
ST	500	210[3]	12	—	—
ST	500	210[3]	12	—	—
ST	500	210[3]	12	—	—
C	—	—	—	—	—
C	—	—	—	—	—
C	—	—	—	—	—
C	460[2]	—	17[2]	—	—
C	460[2]	—	17[2]	—	—
C	460[2]	—	17[2]	—	—
PH (A)	950/1 200	800	12	15	—
PH(B)	1 250/1 500	950	8	8	—
C	700	500	18	10	—
PH (A)	1 230	1 030	8	—	361[2]
PH (B)	1 030	895	8	—	313[2]
PH (C)	900	830	8	—	294[2]

[2] When specifically requested
[3] For information only
[4] Properties minima unless stated
H & T—Hardened and tempered, ST—solution treated, C—as cast, PH—precipitation hardened.

Table 26.46 SUMMARY OF ASTM SPECIFICATION REQUIREMENTS FOR VARIOUS CARBON AND ALLOY STEEL CASTINGS

Cast alloy designation	Composition, wt. % [i]											Tensile strength [ii]		Yield point [ii]		Elongation in 2 in. (50 mm)	Reduction of area (%)
	C	Mn	Si	S	P	Ni	Cr	Mo	Al	V	Other	(MPa)	(ksi)	(MPa)	(ksi)		
ASTM A 27—Carbon steel castings for general application																	
60–30	0.30 [a]	0.60 [a]	0.80	0.06	0.05	—	—	—	—	—	—	415	60	205	30	24	35
65–35	0.30 [a]	0.70 [a]	0.80	0.06	0.05	—	—	—	—	—	—	450	65	240	35	24	35
70–36	0.35 [a]	0.70 [a]	0.80	0.06	0.05	—	—	—	—	—	—	485	70	250	36	22	30
70–40	0.25 [a]	1.20 [a]	0.80	0.06	0.05	—	—	—	—	—	—	485	70	275	40	22	30
ASTM A 148—High strength steel castings for structural application																	
80–40	**	**	**	0.06	0.05							550	80	275	40	18	30
80–50	**	**	**	0.06	0.05							550	80	345	50	22	35
90–60	**	**	**	0.06	0.05							620	90	415	60	20	40
105–85	**	**	**	0.06	0.05							725	105	585	85	17	35
115–95	**	**	**	0.06	0.05							795	115	655	95	14	30
130–115	**	**	**	0.06	0.05							895	130	795	115	11	25
135–125	**	**	**	0.06	0.05							930	135	860	125	9	22
150–135	**	**	**	0.06	0.05							1 035	150	930	135	7	18
160–145	**	**	**	0.06	0.05							1 105	160	1 000	145	6	12
165–150	**	**	**	0.020	0.020							1 140	165	1 035	150	5	20
210–180	**	**	**	0.020	0.020							1 450	210	1 240	180	4	15
260–210	**	**	**	0.020	0.020							1 795	260	1 450	210	3	6
ASTM A 216—Carbon steel castings suitable for fusion welding for high temperature service																	
WCA	0.25 [a]	0.70 [a]	0.60	0.045	0.04						[b]	415–585	60–85	205	30	24	35
WCB	0.30 [a]	1.00 [a]	0.60	0.045	0.04						[b]	485–655	70–95	250	36	22	35
WCC	0.25 [a]	1.20 [a]	0.60	0.045	0.04						[b]	485–655	70–95	275	40	22	35
ASTM A 217—Martensitic stainless and alloy steel castings for pressure-containing parts for high temperature service																	
WC1	0.25	0.50–0.80	0.60	0.045	0.04			0.45–0.65			[c] [d]	450–620	65–90	240	35	24	35
WC4	0.05–0.20	0.50–0.80	0.60	0.045	0.04	0.70–1.10	0.50–0.80	0.45–0.65			[d] [e]	485–655	70–95	275	40	20	35
WC5	0.05–0.20	0.40–0.70	0.60	0.045	0.04	0.60–1.00	0.50–0.90	0.90–1.20			[d] [e]	485–655	70–95	275	40	20	35
WC6	0.05–0.20	0.50–0.80	0.60	0.045	0.04		1.00–1.50	0.45–0.65			[d] [e]	485–655	70–95	275	40	20	35
WC9	0.05–0.18	0.40–0.70	0.60	0.045	0.04		2.00–2.75	0.90–1.20			[c] [d]	485–655	70–95	275	40	20	35
WC11	0.15–0.21	0.50–0.80	0.30–0.60	0.015	0.020		1.00–1.50	0.45–0.65	0.01		[c]	550–725	80–105	345	50	18	45
C5	0.20	0.40–0.70	0.75	0.045	0.04		4.00–6.50	0.45–0.65			[c] [d]	620–795	90–115	415	60	18	35
C12	0.20	0.35–0.65	1.00	0.045	0.04		8.00–10.00	0.90–1.20			[c] [d]	620–795	90–115	415	60	18	35
CA15	0.15	1.00	1.50	0.040	0.040	1.00	11.5–14.0	0.50			[c] [d]	620–795	90–115	450	65	18	30

ASTM A 389—Alloy steel castings for pressure containing parts for high temperature service

Grade	C	Mn	Si	P	S	Ni	Cr	Mo	V	Notes	T.S. (MPa)	T.S. (ksi)	Y.S. (MPa)	Y.S. (ksi)	Elong (%)	R.A. (%)
C23	0.20	0.30–0.80	0.60	0.045	0.04	—	1.00–1.50	0.45–0.65	0.15–0.25	[c]	483	70	276	40	18.0	35.0
C24	0.20	0.30–0.80	0.60	0.045	0.04	—	0.80–1.25	0.90–1.20	0.15–0.25	[c]	552	80	345	50	15.0	35.0

ASTM A 487—Steel castings for pressure service

Grade	C	Mn	Si	P	S	Ni	Cr	Mo	V	Notes	T.S. (MPa)	T.S. (ksi)	Y.S. (MPa)	Y.S. (ksi)	Elong (%)	R.A. (%)
1A	0.30	1.00	0.80	0.045	0.04	—	—	—	0.04–0.12	[c] [e]	585–760	85–110	380	55	22	40
1B	0.30	1.00	0.80	0.045	0.04	—	—	—	0.04–0.12	[c] [e]	620–795	90–115	450	65	22	45
2A	0.30	1.00–1.40	0.80	0.045	0.04	—	0.10–0.30	—	—	[c]	585–760	85–110	365	53	22	35
2B	0.30	1.00–1.40	0.80	0.045	0.04	—	0.10–0.30	—	—	[c]	620–795	90–115	450	65	22	40
6A	0.05–0.38	1.30–1.70	0.80	0.045	0.04	0.40–0.80	0.30–0.40	—	—	[e] [f]	795	115	550	80	18	30
6B	0.05–0.38	1.30–1.70	0.80	0.045	0.04	0.40–0.80	0.30–0.40	—	—	[e] [f]	825	120	655	95	12	25
8A	0.05–0.20	0.50–0.90	0.80	0.045	0.04	—	2.00–2.75	0.90–1.10	—	[e] [f]	585–760	85–110	380	55	20	35
8B	0.05–0.20	0.50–0.90	0.80	0.045	0.04	—	2.00–2.75	0.90–1.10	—	[e] [f]	725	105	585	85	17	30
10A	0.30	0.60–1.00	0.80	0.045	0.04	1.40–2.00	0.55–0.90	0.20–0.40	—	[e] [f]	690	100	485	70	18	35
10B	0.30	0.60–1.00	0.80	0.045	0.04	1.40–2.00	0.55–0.90	0.20–0.40	—	[e] [f]	860	125	690	100	15	35
12A	0.05–0.20	0.40–0.70	0.60	0.045	0.04	0.60–1.00	0.50–0.90	0.90–1.20	—	[e] [g]	485–655	70–95	275	40	20	35
12B	0.05–0.20	0.40–0.70	0.60	0.045	0.04	0.60–1.00	0.50–0.90	0.90–1.20	—	[e] [g]	725–895	105–130	585	85	17	35
13A	0.30	0.80–1.10	0.60	0.045	0.04	1.40–1.75	—	0.20–0.30	—	[e] [h]	620–795	90–115	415	60	18	35
13B	0.30	0.80–1.10	0.60	0.045	0.04	1.40–1.75	—	0.20–0.30	—	[e] [h]	725–895	105–130	585	85	17	35
14A	0.55	0.80–1.10	0.60	0.045	0.04	1.40–1.75	—	0.20–0.30	—	[e] [h]	825–1000	120–145	655	95	14	30

[i] Maximum percentage unless a range is given.
[ii] Minimum percentage unless a range is given.
[a] For each reduction of 0.01% carbon below the maximum specified, an increase of 0.04% manganese above the maximum specified will be permitted to the maximum given in each applicable specifications.
[b] Specified residual elements include 0.30% Copper, 0.50% Nickel, 0.50% Chromium, 0.20% Molybdenum, 0.03% Vanadium. All values are maximum with the sum of total not exceeding 1.00%.
[c] Maximum total amount of residual elements is 1.00%.
[d] Specified residual elements include 0.50% Copper and 0.10% Tungsten, maximum.
[e] Specified residual elements include 0.50% Copper, 0.10% Tungsten, 0.03% Vanadium, maximum.
[f] Maximum total amount of residual elements is 0.60%.
[g] Maximum total amount of residual elements is 0.50%.
[h] Maximum total amount of residual elements is 0.75%.
** Content selected by manufacturer to obtain specific mechanical properties.
Adapted with permission from the Annual Book of ASTM Standards. Copyright ASTM International.

26.9   Cast irons

Data Source BCIRA.

26.9.1   Classification of cast irons

Cast irons may be divided into two main groups, comprising the general purpose grades which are used for the majority of engineering applications and the special purpose or alloy cast irons which are used where the operating conditions involve extremes of heat, corrosion or abrasion.

26.9.2   General purpose cast irons

These materials may be further classified into four groups, depending on the graphite form.

British Standards specifications exist for each of these materials. New metric specifications incorporating SI units are available for malleable, grey and nodular irons. The grade number denotes the tensile strength and the % elongation, e.g. BS 2789 Grade 400/18 has a minimum tensile strength of 400 MPa and a minimum elongation of 18%. In the case of grey irons, which all fail with an elongation less than 1%, the grade number denotes tensile strength only in test specimens of standard dimensions.

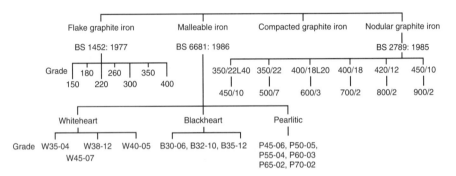

Figure 26.1   *Classification of general purpose cast irons*

GREY IRONS

The grade number denotes the minimum tensile strength requirement on a test piece machined from a 30 mm (1.2 in) diameter as-cast bar and this, rather than analysis, is the basis for specification of cast irons for engineering purposes.

For unalloyed irons the main constituents influencing tensile strength are carbon, silicon and phosphorus. Their combined effect may be expressed as a carbon equivalent value (CEV), where

$$CEV = \text{total carbon \%} + \frac{\text{phosphorus \% + silicon \%}}{3} \qquad (26.1)$$

Typical analysis and CEVs are given in Table 26.53.

Grey irons are section sensitive, their structure and mechanical properties depending on cooling rate in the mould, which in turn depends to a first approximation on section thickness. This effect is shown in Figure 26.2, with tensile and test bar diameter requirements in Table 26.54.

Excessive rates of cooling lead to the formation of free carbides, or chilling. This effect is more severe with the higher strength irons, and occurs first at free edges of castings, where the cooling rate is a maximum, as shown in Table 26.55.

Inoculation of high duty grey irons, from Grade 260 to 400, is carried out using a technique in which a small amount of inoculant, usually containing silicon, is added to the metal immediately before pouring so as to increase the silicon content by 0.3–0.5%. Some commonly used inoculants include ferrosilicon, calcium silicide, SMZ (silicon–manganese–zirconium).

Table 26.47 SUMMARY OF ASTM SPECIFICATION REQUIREMENTS FOR HEAT AND CORROSION RESISTANT STEEL CASTINGS

Cast alloy designation	Composition, wt. % *									Tensile strength		Yield point		Elongation in 2 in. (50 mm)	Reduction of A (%)
	C	Mn	Si	P	S	Cr	Ni	Mo	Other	(MPa)	(ksi)	(MPa)	(ksi)		
Heat Resistant Castings															
ASTM A 297															
HF	0.20–0.40	2.00	2.00	0.04	0.04	18.0–23.0	8.0–12.0	0.50	—	485	70	240	35	25	—
HI	0.20–0.50	2.00	2.00	0.04	0.04	26.0–30.0	14.0–18.0	0.50	—	485	70	240	35	10	—
HK	0.20–0.60	2.00	2.00	0.04	0.04	24.0–28.0	18.0–22.0	0.50	—	450	65	240	35	10	—
HT	0.35–0.75	2.00	2.50	0.04	0.04	15.0–19.0	33.0–37.0	0.50	—	450	65	—	—	4	—
HU	0.35–0.75	2.00	2.50	0.04	0.04	17.0–21.0	37.0–41.0	0.50	—	450	65	—	—	4	—
HW	0.35–0.75	2.00	2.50	0.04	0.04	10.0–14.0	58.0–62.0	0.50	—	415	60	—	—	—	—
HX	0.35–0.75	2.00	2.50	0.04	0.04	15.0–19.0	64.0–68.0	0.50	—	415	60	—	—	8	—
HD	0.50	1.50	2.00	0.04	0.04	26.0–30.0	4.0–7.0	0.50	—	515	75	240	35	—	—
HL	0.20–0.60	2.00	2.00	0.04	0.04	28.0–32.0	18.0–22.0	0.50	—	450	65	240	35	10	—
HP	0.35–0.75	2.00	2.50	0.04	0.04	24–28	33–37	0.50	—	430	62.5	235	34	4.5	—
Corrosion Resistant Castings															
ASTM A 351															
CF3	0.03	1.50	2.00	0.04	0.04	17.0–21.0	8.0–12.0	0.50	—	485	70	205	30	35.0	—
CF3M	0.03	1.50	1.50	0.04	0.04	17.0–21.0	9.0–13.0	2.0–3.0	—	485	70	205	30	30.0	—
CF8	0.08	1.50	2.00	0.04	0.04	18.0–21.0	8.0–11.0	0.50	—	485	70	205	30	35.0	—
CF8C	0.08	1.50	2.00	0.04	0.04	18.0–21.0	9.0–12.0	0.50	—	485	70	205	30	30.0	—
CF8M	0.08	1.50	1.50	0.04	0.04	18.0–21.0	9.0–12.0	2.0–3.0	—	485	70	205	30	30.0	—
CH20	0.04–0.20	1.50	2.00	0.04	0.04	22.0–26.0	12.0–15.0	0.50	—	485	70	205	30	30.0	—
CK20	0.04–0.20	1.50	1.75	0.04	0.04	23.0–27.0	19.0–22.0	0.50	—	450	65	195	28	30.0	—
CN7M	0.07	1.50	1.50	0.04	0.04	19.0–22.0	27.5–30.5	2.0–3.0	3.0–4.0 Cu	425	62	170	25	35.0	—
CD4MCu	0.04	1.00	1.00	0.04	0.04	24.5–26.5	4.75–6.00	1.75–2.25	2.75–3.25 Cu	690	100	485	70	16.0	—
CG8M	0.08	1.50	1.50	0.04	0.04	18.0–21.0	9.0–13.0	3.0–4.0	—	515	75	240	35	25.0	—
ASTM A 352															
CA6NM	0.06	1.00	1.00	0.04	0.03	11.5–14.0	3.5–4.5	0.4–1.0	—	760–930	110.0–135.0	550	80	15.0	35.0
ASTM A 743															
CF-20	0.20	1.50	2.00	0.04	0.04	18.0–21.0	8.0–11.0	—	—	485	70	205	30	30	—
CF-16F	0.16	1.50	2.00	0.04	0.04	18.0–21.0	9.0–12.0	—	—	485	70	205	30	25	—
CE-30	0.30	1.50	2.00	0.04	0.04	26.0–30.0	8.0–11.0	—	—	550	80	275	40	10	—
CA-15	0.15	1.00	1.50	0.04	0.04	11.5–14.0	1.00	0.50	—	620	90	450	65	18	30
CB-30	0.30	1.00	1.50	0.04	0.04	18.0–21.0	2.00	—	—	450	65	205	30	—	—
CC-50	0.50	1.00	1.50	0.04	0.04	26.0–30.0	4.00	—	—	380	55	—	—	—	—
CA-40	0.20–0.40	1.00	1.50	0.04	0.04	11.5–14.0	1.0	0.5	—	690	100	485	70	15	25

* Maximum percentage unless a range is given. Balance of composition is iron.
Adapted with permission from the Annual Book of ASTM Standards. Copyright ASTM International.

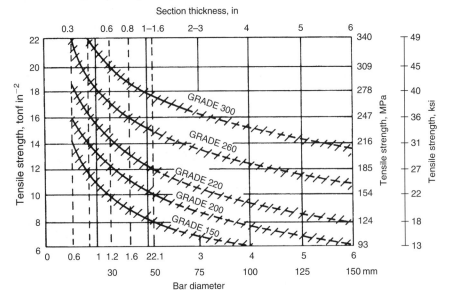

Figure 26.2 *Variation of tensile strength with section thickness*

Table 26.48 TYPICAL ANALYSES OF GREY IRONS

BS grade	150	180	220	260	300	350	400
Equiv. U.S. Class**	21	26	32	38	43	51	58
Total carbon %	3.1–3.4	3.2–3.5	3.2–3.4	3.0–3.2	2.9–3.1	3.1 max.	2.9 max.
Silicon % (final)	2.5–2.8	2.2–2.5	2.0–2.5	1.6–1.9	1.5–1.8	1.4–1.6	1.4–1.6
Manganese %	0.5–0.7	0.5–0.8	0.6–0.8	0.6–0.8	0.5–0.7	0.6–0.75	0.6–0.75
Sulphur % (max.)	0.15	0.15	0.15	0.15	0.12	0.12	0.12
Phosphorus %	0.9–1.2	0.6–0.9	0.1–0.5	0.3 max.	0.2 max.	0.15 max.	0.15 max.
Chromium %	—	—	—	—	0.4–0.6	—	—
Nickel %*	—	—	—	—	0.8–1.5	1.5	2.0
Molybdenum %	—	—	—	—	—	0.3–0.5	0.5–0.6
CEV	4.5–4.55	4.3–4.35	4.1–4.2	3.8–3.9	3.5–3.6	—	—

* Copper may be used partially to replace nickel.
** Tensile strength in ksi.

Table 26.49 SPECIFIED PROPERTIES OF GREY IRONS

Cross-sectional thickness of casting represented mm (in)	Diam, of as-cast bar mm (in)	Gauge diam. mm (in)	*Grade* *Minimum tensile strength* 10 MPa (ksi)	12 MPa (ksi)	14 MPa (ksi)	17 MPa (ksi)	20 MPa (ksi)	23 MPa (ksi)	26 MPa (ksi)
Up to 9.5 (up to $\frac{3}{8}$)	15.2 (0.6)	10.1 (0.399)	170 (24.7)	201 (29.1)	247 (35.8)	294 (42.6)	340 (49.3)	386 (56.0)	432 (62.6)
9.5–19.0 ($\frac{3}{8}$–$\frac{3}{4}$)	22.2 (0.875)	14.3 (0.564)	162 (23.5)	193 (28.0)	232 (33.6)	278 (40.3)	324 (47.0)	371 (53.8)	417 (60.5)
19.0–29.6 ($\frac{3}{4}$–1$\frac{1}{8}$)	30.5 (1.2)	20.3 (0.798)	154 (22.3)	185 (26.8)	216 (31.3)	262 (38.0)	309 (44.8)	355 (51.5)	402 (58.3)
29.6–41.3 (1$\frac{1}{8}$–1$\frac{5}{8}$)	40.6 (1.6)	28.6 (1.128)	147 (21.3)	178 (25.8)	209 (30.3)	247 (35.8)	294 (42.6)	340 (49.3)	386 (56.0)
over 41.3 (over 1$\frac{5}{8}$)	53.3 (2.1)	37.9 (1.493)	139 (20.2)	170 (24.7)	201 (29.1)	232 (33.6)	278 (40.3)	324 (47.0)	371 (53.8)

Table 26.50   CHILLING TENDENCY IN DIFFERENT GRADES OF IRON

BS grade	150	180	220	260	300
Approximate minimum chill free edge					
(in)	0.1–0.2	0.2–0.3	0.25–0.4	0.4–0.65	0.6–1.1
(mm)	2.5–5.0	5.0–7.5	6.3–10.0	10.0–16	15–30

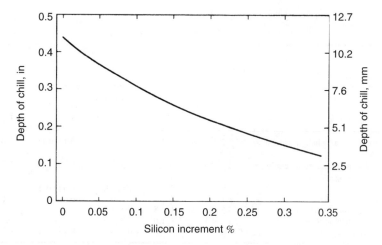

Figure 26.3   *Effect of inoculation on chill depth*

Graphite is also a powerful inoculant. Several other proprietary inoculants, most of which are based on one or more of these substances, are also effective. The efficiency of a graphite inoculant depends on high purity, but that of other types depends on the presence of minor constituents such as aluminium, calcium, barium, strontium and cerium.

The tensile strength of hypo-eutectic grey irons (i.e. CEV less than 4.3) can be significantly increased by inoculation, particularly with silicon-containing inoculants. When the CEV is in the range 3.9–4.1 maximum strength is obtained when the inoculant adds 0.2–0.3% silicon, corresponding to an increase of approximately 15 MPa (1 tonf in^{-2}) compared with uninoculated material. With CEV below 3.9 the addition of inoculant sufficient to give a 0.5% increase in silicon content can add up to 60 MPa (4 tonf in^{-2}) to the tensile strength, the effect being progressive with increasing additions up to at least this level.

The chilling tendency of grey irons is also reduced by inoculation, enabling significantly thinner chill free sections than those in Table 26.50 to be cast. The effect is progressive up to at least 0.4% of added silicon.

The chill reducing effect forms the basis for most control tests for inoculation. A sand cast wedge or a small block cast in sand with one face against a metal plate may be made from the metal before and after inoculation. The test pieces are then fractured and the change in depth of chill measured as an assessment of the success of the treatment. Typical results are illustrated in Figure 26.3.

In castings which are susceptible to shrinkage defects inoculation will accentuate these problems, and its use should be restricted only to those castings where chilling or strength considerations make it necessary.

Charge materials for production of grey irons include pig irons, cast iron scrap, steel scrap and ferro alloys.

The main purposes served by pig irons is the provision of carbon, silicon, manganese and, where required, phosphorus and alloying elements. They also limit the sulphur content of cupola melted iron.

Blast furnace pig irons are available in a variety of grades which differ mainly in phosphorus content. Each grade is generally available within different ranges of silicon and manganese contents.

The carbon content will generally be at the lower end of the above ranges when the silicon content is at the upper end of the corresponding range. The silicon content can be specified in increments of 0.25 or 0.50% within the ranges given.

Table 26.51   ANALYSES OF PIG IRONS

	Typical composition				
	TC %	Si %	Mn %	S % *max*	P %
Hematite	3.7–4.5	0.5–3.5	0.5–1.2	0.05	0.05 max
Low phosphorus	3.8–4.2	1.0–4.5	0.7–1.1	0.05	0.065 max
	3.8–4.2	1.0–4.5	1.1–1.5	0.05	0.065 max
	3.8–4.2	1.0–4.5	1.5–2.0	0.05	0.065 max
Medium phosphorus	3.5–4.0	2.0–3.5	0.8–1.0	0.05	0.5–0.7
High phosphorus	3.3–3.8	2.0–4.5	0.6–0.8	0.05	0.7–1.2
	3.3–3.8	2.0–4.5	0.8–1.2	0.05	0.7–1.2

Table 26.52   COMPOSITION OF SCRAP

	Approximate composition %				
Type of scrap	C	Si	Mn	S	P
Light section scrap, usually less than $\frac{1}{4}$ in thick	3.2–3.5	2.2–2.8	0.5–0.7	0.10–0.15	1.0–1.5
Textile and machine scrap, generally up to $1\frac{1}{2}$ in average section	3.0–3.3	1.8–2.2	0.5–0.8	0.10–0.15	0.5–1.0
Railway chairs	2.8–3.5	1.5–2.5	Up to 0.5	Up to 0.25	1.0–1.5
Automobile engine scrap	3.1–3.3	2.0–2.2	0.5–0.8	0.08–0.15	<0.2
Ingot mould scrap (heavy section)	3.5–3.8	1.4–1.8	0.5–1.0	0.08	<0.1
Blackheart malleable scrap	2.2–3.1	1.3–1.6	0.3–0.6	0.09–0.25	0.06–0.08
Whiteheart malleable scrap	0.2–2.3	0.5–0.8	0.2–0.3	0.15–0.25	0.06–0.08

Refined alloy and special pig irons are produced to the customer's specific requirements in a wide range of compositions. They may contain alloying elements if required, within specified ranges of alloy content.

Silvery pig iron (10–12% silicon, 12–14% silicon) provides silicon in a more dilute form than ferrosilicon. It is used to minimise variation in silicon content of cupola melted iron when a substantial quantity of silicon is added, and to reduce the risk of aluminium contamination of the metal which may arise if large additions of ferrosilicon are made.

Special irons, generally of high carbon content, but with a very low content of residual elements are also available, their main use being for the production of nodular graphite iron.

Return scrap is the best source of scrap for re-melting providing this is of known and consistent composition and this should be fully utilised.

Purchased scrap is available in several fairly readily identifiable types as given in Table 26.51.

In comparison with pig iron and cast iron scrap, steel scrap is low in carbon and silicon contents, and the phosphorus and sulphur contents are generally below 0.05%. It is used to lower carbon and silicon contents, particularly in the production of higher strength irons.

Ferro-alloys are used to remedy deficiencies of certain elements, particularly silicon and manganese in the charges.

FERROSILICON

The two most common grades, in lump form, contain 75–80% and 45–50% of silicon. Ferromanganese in lump form contains 75–80% manganese.

BRIQUETTES

For relatively small additions of silicon, manganese and chromium, briquettes may be used for convenience and consistency. They contain a fixed amount of the alloy and avoid the necessity of weighing the addition.

Table 26.53 LOSSES OF ALLOYING ELEMENTS

	Melted in cupola	*Added to ladle*
Copper	0 of amount charged	0
Nickel	0 of amount charged	0
Molybdenum	0–5% of amount charged	5% of amount added
Manganese	20–30% of amount charged	5–10% of amount added
Chromium	10% of amount charged	15% of amount added
Silicon	10–15% of amount charged	10% of amount added

Table 26.54 SUMMARY OF STRUCTURAL EFFECTS OF ALLOYING ELEMENTS ON CAST IRON

Element	*% used in pearlitic irons*	*'Chill'*	*Effect on carbides (at high temps)*	*Effect on graphite structure*	*Effect on combined carbon in pearlite*	*Effect on matrix*
Chromium chill inducing	0.15–0.5	Increases*	Strongly stabilises	Mildly refines	Increases	Refines pearlite
Vanadium	0.15–0.3	Increases	Strongly stabilises	Refines	Increases	and hardens
Boron	—	Strongly increases	—	—	—	—
Manganese mildly chill inducing	0.3–1.25	Mildly increases	Stabilises	Mildly refines	Increases	Refines pearlite and hardens
Molybdenum	0.3–1.0	Mildly increases	About neutral	Strongly refines	Mildly increases	Refines pearlite
Tin	0.1–0.2	About neutral	About neutral	About neutral	Increases	and strengthens
Copper mildly chill restraining	0.5–2.0	Mildly restrains	About neutral	About neutral	Mildly decreases	Hardens
Carbon chill restraining	—	Strongly restrains	Decreases stability	Coarsens	Strongly decreases	Produces ferrite and softens
Silicon	—	Strongly restrains	Decreases stability	Coarsens	Strongly decreases	Produces ferrite and softens
Aluminium	—	Strongly restrains	Decreases stability	Coarsens	Strongly decreases	Produces ferrite and softens
Nickel	0.1–3.0	Restrains[†]	Mildly decreases stability	Mildly refines	Mildly decreases and stabilises at eutectoid	Refines pearlite and hardens
Titanium	0.05–0.10	Restrains[†]	Decreases stability	Strongly refines[‡]	Decreases	Produces ferrite and softens
Zirconium	0.10–0.30	Restrains[†]	—	About neutral	—	Produces ferrite and softens

* Chill inducing effect of 1 part of chromium about balances chill restraining effect of $1\frac{1}{2}$ parts silicon or $2\frac{1}{2}$ parts nickel.

[†] Chill restraining effect of nickel about half that of silicon.

[‡] Strong refining action of titanium takes place when small amounts are added, particularly when oxygen is also present.

Table 26.55

Element	*Graphitising value* (Si = 1)	*Element*	*Graphitising value* (Si = 1)
Aluminium	0.5	Manganese	−0.25
Copper	0.35	Molybdenum	−0.35
Nickel	0.3–0.4	Chromium	−1.2

Melting range varies widely with composition, the melting temperature falling with increase of carbon content so that low carbon irons must be cast at considerably higher temperatures than high carbon irons. With grey irons containing high phosphorus, melting of the steadite (phosphide eutectic) begins at about 960°C, giving a relatively long melting range.

FLUIDITY OF CAST IRON

The fluidity of cast iron depends primarily upon composition and pouring temperature. The fluidity, as measured by a special test spiral casting, may be improved by increasing the carbon, silicon or phosphorus contents, or by increasing the pouring temperature. Of these variables, carbon content is the most effective from the point of view of composition, but an increase of 15–20°C in the pouring temperature improves the fluidity as much as an increase of 0.10% carbon, 0.30% silicon or 0.20% phosphorus.

Table 26.56   DENSITIES AND LIQUIDUS TEMPERATURES OF SOME TYPICAL GREY CAST IRONS (BASED ON 3.8 cm ($1\frac{1}{2}$ in) SECTION)

	Iron number						
	1	2	3	4	5	6	7
Density, g cm^{-3} (room temperature)	7.02	7.09	7.26	7.03	7.08	7.27	7.14
Density, g cm^{-3} (liquidus)	6.90	6.94	6.92	6.89	6.89	6.92	6.89
Liquidus temperature, °C	1 150	1 150	1 250	1 150	1 155	1 250	1 195
Liquid contraction per 100°C (%)	1.1	1.1	1.1	1.1	1.1	1.1	1.1
Composition, %							
Total carbon	3.69	3.67	3.10	3.39	3.27	3.08	2.90
Graphitic carbon	3.53	3.26	2.31	3.20	2.88	2.18	2.68
Silicon	2.87	2.10	1.69	2.86	2.87	1.68	2.88
Phosphorus	0.68	0.46	0.35	0.67	0.59	0.35	0.66
Manganese	0.59	0.54	0.48	0.58	0.52	0.44	0.44
Sulphur	0.03	0.05	0.04	0.03	0.03	0.04	0.03

Table 26.57   HARDNESS AND DENSITY OF MICRO-CONSTITUENTS OF CAST IRON

Constituent	Density g cm^{-3}	Brinell hardness No.	Remarks
Ferrite	7.86	70–75	Iron (electrolytic)
Silico-ferrite	—	88	Containing 0.82% silicon
Silico-ferrite	—	124	Containing 2.28% silicon
Silico-ferrite	—	150	Containing 3.4% silicon
Pearlite	7.846	240	—
Pearlite (silico-ferrite and cementite)	—	200–450	Depending on interlamellar distance
Graphite	2.55	—	—
Ledeburite (massive cementite and sat. austenite)	—	680–840	—
Steadite	7.32	—	—
Manganese sulphide	4.00	—	—
Iron sulphide	5.02	—	—
Divorced pearlite	—	130–150	—
Phosphide eutectic	—	—	400–600 HV
Iron carbide Fe$_3$C or (FeCr)$_3$C	7.66	550	800–1 200 HV
Iron carbide (CrFe)$_7$C$_3$	—	—	1 300–1 800 HV

Table 26.58   SPECIFIC GRAVITY OF TYPICAL IRONS OF BRITISH STANDARDS GRADES (BS 1452)

Grade	Up to 180	220	260	300	350	400
Specific gravity	6.8–7.15	7.2–7.3	7.25–7.35	7.3–7.4	7.3–7.4	7.4–7.6

Heat treatment of grey cast irons may be carried out to eliminate residual stresses, improve machinability or to increase wear resistance.

Stress relief heat treatment is carried out by slow heating at 50–100°C per hour to 600°C ± 10°C, holding at 600°C for one hour plus an additional half hour per cm (0.4 in) of maximum section thickness, and cooling at 50–100°C per hour to below 200°C followed by air cooling.

Annealing heat treatment is applied when improved machinability is required. There are two distinct sets of conditions in which annealing may be carried out.

1. To break down free carbide or chill formed as a result of failure to match carbon equivalent value to minimum free edge thickness. Annealing must then be carried out at 900°C for 1–5 h to ensure complete breakdown of carbide, followed by air cooling to ensure a pearlitic matrix. This treatment is carried out as an emergency measure to salvage otherwise unmachinable castings.
2. To provide castings which can be machined at a very high rate. Castings which are fully pearlitic can be annealed to give a mainly ferritic matrix with a hardness of 140–180 HB by holding at 780–820°C for 1–2 h followed by slow cooling.

Hardening by quenching and tempering is carried out where high surface hardness is required, with corresponding improvement in wear resistance. Hardening temperatures in the range 850–880°C are used, followed by oil quenching and tempering at 300°C to reduce internal stresses. Tensile strength is not increased to the same extent as hardness (Figure 26.4) and because of the risk of cracking during quenching, this process is usually restricted to small castings of simple shape.

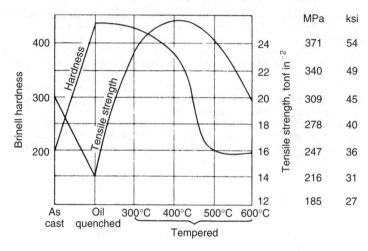

Figure 26.4   *Effect of heat treatment on the strength and hardness of alloy cast iron*

Surface hardening, by flame or induction heating, is widely used to improve the wear resistance of critical surfaces on large castings such as slideways on machine tools. For good response to this treatment the as-cast structure must be fully pearlitic, and the phosphorus content must be below 0.2% to avoid pitting. Hardness in the range 450–500 HV are obtained with depths of 1–3 mm (0.040–0.125 in).

MALLEABLE IRONS

Malleable irons are brittle as cast, their structure consisting of iron carbide in a pearlitic matrix. By suitable heat treatment the carbides are broken down resulting in a structure that consists of graphite aggregates (temper carbon) in a matrix which may be ferritic or pearlitic, depending on composition and heat treatment conditions.

The British Standards (BS 309, 310 and 3333) specify neither composition (apart from maximum phosphorus content of 0.12%) nor heat treatment.

In practice, whiteheart malleable irons have an initial total carbon content of about 3.5% which is reduced to the range 0.25–2.0% by heat treating the castings at 900°C in an oxidising environment,

Table 26.59   MECHANICAL PROPERTIES OF WHITEHEART MALLEABLE IRON

Grade BS 6681:1986	Diameter of test bar d. mm[†]	Tensile strength min. R_m MPa	0.2 Proof stress min. $R_{p0.2}$ MPa	Elongation min. A %	Hardness typical max HB
W35-04	9	340	—	5	230
	12	350	—	4	
	15	360	—	3	
W38-12*	9	320	170	15	200
	12	380	200	12	
	15	400	210	8	
W40-05	9	360	200	8	220
	12	400	220	5	
	15	420	230	4	
W45-07	9	400	230	10	220
	12	450	260	7	
	15	480	280	4	

* Although all grades of whiteheart malleable are weldable with correct procedure, W38-12 should be selected where strength and avoidance of post weld heat treatment is required.
† Test bar diameter should be representative of the important sectional thickness of the casting. Ref. British Cast Iron Research Association.

and slowly cooling. The required conditions may be produced by packing the castings in an oxidiser (e.g. hematite ore) or by use of a controlled atmosphere furnace. This results in a carbon gradient within the castings, the outer layer being normally ferritic and graphite-free while the core structure consists of temper carbon aggregates in a pearlitic matrix. Small castings of thin section may have a fully decarburised structure throughout, and this is sometimes referred to as a weldable grade of malleable iron.

Blackheart and pearlitic malleable irons have a lower initial total carbon content in the range 2–3%. Heat treatment in a neutral atmosphere at 850–875°C followed by slow cooling results in a uniform structure of temper carbon in a ferritic matrix.

Malleable irons Pearlitic have a structure consisting of temper carbon in a pearlitic matrix. This is produced either by fapid cooling after annealing or by the addition of 0.5% or more manganese. Pearlitic malleable irons have a good response to surface hardening by flame or induction heating, and hardness values of HV 500 can be consistently achieved in production.

NODULAR IRONS

In nodular (spheroidal graphite) irons free graphite is present as spheres or nodules in the as-cast condition. Graphite in this form has a much smaller weakening effect on the matrix than the dispersed graphite flakes in grey ions. Nodular irons therefore have considerably higher strength, ductility, and impact values than grey irons.

Cerium and magnesium additions both produce nodular structures, but the latter has been found to be more adaptable and economical. Both elements are desulphurisers and nodule formation is not possible until the sulphur content has been lowered to about 0.02%. Very small amounts of trace elements, such as 0.003% bismuth, 0.004% antimony, 0.009% lead and 0.12% titanium prevent nodule formation. The effect of these elements is additive, but it can be neutralised by the addition of sufficient cerium to give a residual content of 0.005–0.01%.

Magnesium may be added directly to the ladle as nickel–magnesium, nickel–silicon–magnesium or iron–silicon–magnesium alloy. Higher magnesium recovery is obtained using a plunging technique in which lower density, higher magnesium content additions such as magnesium impregnated coke are held below the liquid metal surface by means of a plunging head. Maximum recovery results from the addition of pure magnesium to the molten iron in a closed pressure-tight converter vessel. Because of equipment costs the use of this latter method is normally restricted to large-scale production.

In all cases the amount of magnesium to be added is given by:

$$Mg = \frac{\frac{3}{4}(\text{initial sulphur content}) + \text{residual magnesium content (usually } 0.03\text{–}0.05\%)}{\text{expected magnesium recovery}}$$

Nodular irons are inoculated with 0.4–0.8% silicon after nodulising to refine the structure and minimise chilling.

The carbon content of nodular irons is usually kept above 3.5% in the interests of good castability. Silicon, manganese and phosphorus should be below 2.3%, 0.4% and 0.06% respectively to give maximum ductility and impact value in the ferritic condition.

Nodular irons are slightly more prone to shrinkage defects than grey irons.

Adequate feed metal should be provided and moulds of high rigidity are to be preferred. Running systems should be designed to minimise turbulence, so as to prevent the entrapment of dross which tends to be formed as a result of the magnesium content.

Although nodular irons are much less section sensitive than grey irons, depending on the trace amounts of carbide stablising elements present, their matrix structures may range from fully pearlitic to completely ferritic, and chilling may occur in sections thinner than 5 mm (0.2 in).

By close control of analysis and inoculation practice nodular irons can be produced in the ascast condition over a wide range of section thicknesses with any required matrix structure from fully ferritic to fully pearlitic.

Alternatively, the matrix structure of nodular iron castings can be modified by appropriate heat treatments, since the presence of free carbon in the form of graphite enables diffusion of carbon to or from the graphite particles to take place. This is not possible with steels, which contain no free graphite. The effect of variation in matrix structure on mechanical properties is much more pronounced with nodular iron than with flake graphite cast iron, and by heat treatment of an iron of fixed composition foundries can produce castings conforming to the complete range of the grades of BS 2789:1985.

Practical heat treatments include:

Annealing Heat to 850–900°C where the matrix becomes completely austenitic and slow furnace cool at 20–35°C per hour to below 700°C. Alternatively, cool more rapidly to 700–720°C and hold for 4–12 h, followed by air cooling. Ferritic iron produced in this way confirm to BS 2789: 1985 grades 350/22 to 420/12.

Normalising This is carried out by air cooling from 850 to 900°C, and produces a mainly pearlitic matric conforming to grades 700/2 and 800/2 in castings of light and medium section. The use of alloying elements is often necessary to produce a pearlitic matrix in heavier section castings.

Hardened and tempered structures These are produced by oil quenching from 850 to 900°C and tempering at 550–600°C. Material conforming to BS 2789 Grade 800/2 and 900/2 is sometimes invariably produced by this method.

Austempering is carried out by heating castings to 850–950°C, followed by quenching to an isothermal treatment temperature within the range 230–400°C and holding this temperature typically for 1–2 h. A variety of bainitic structures can be obtained, resulting in combinations of strength, ductility and toughness which cannot be achieved in ductile irons by other means. These austempered ductile irons (ADI) are used for many different engineering applications such as gears, crankshafts, vehicle suspension components and parts of earthmoving equipment. Provisional specifications for various countries to cover ADIs are summarised in Table 26.63. The various grades can be typically produced from the same ductile iron by adjustment of the austempering time and temperature.

ADI exhibits section size sensitivity and test bars are not representative of heavy sections. At −40°C high toughness grades may show a drop of 15% in yield strength as well as a drop in tensile strength compared with room temperature values. No changes have been found in elevated temperature properties up to 300°C. Also of concern is the low temperature toughness and the ductile–brittle transformation temperature although notched specimen results indicate gradual toughness transition as temperature decreases.

Table 26.60 MECHANICAL PROPERTIES OF BLACKHEART MALLEABLE IRON

Grade BS 6681:1986	Diameter of test bar d. mm	Tensile strength min. R_m MPa	0.2% Proof stress $R_{p0.2}$ min. MPa	Elongation min. %	Hardness typical max HB
B30-06	12	300	—	6	150
	15	300	—	6	
B32-10	12	320	190	10	150
	15	320	190	10	
B35-12	12	350	200	12	150
	15	350	200	12	

Ref British Cast Iron Research Association.

Table 26.61   MECHANICAL PROPERTIES OF PEARLITIC MALLEABLE IRON

Grade BS 6681:1986	Diameter of test bar d. mm	Tensile strength min. R_m MPa	0.2% Proof stress $R_{p0.2}$ min. MPa	Elongation min. %	Hardness typical HB
P45-06	12	450	270	6	150/200
	15	450	270	6	
P50-05	12	500	300	5	160/220
	15	500	300	5	
P50-04	12	550	340	4	180/230
	15	550	340	4	
P60-03	12	600	390	3	200/250
	15	600	390	3	
P65-02	12	650	430	2	210/260
	15	650	430	2	
P70-02*	12	700	530	2	240/290
	15	700	530	2	

* If air quenched and subsequently tempered min 0.2% proof stress 530 MPa min.
Ref. British Cast Iron Research Association.

Mixed matrix structures Structures intermediate between the annealed and normalised grades have a range of mechanical properties depending on the ratio of ferrite to pearlite. The corresponding grades of BS 2789 are 400/10, 500/7 and 600/3. In practice these structures are produced by austenitising at 850–900°C followed by either controlled rapid cooling at lapproximately 100°C per hour through the critical temperature range of 720–800°C or by rapid air cooling from an appropriate intermediate temperature, e.g. 730°C, within the critical range.

26.9.3   Compacted graphite irons

Although the existence of compacted graphite containing irons has been known for many years, it is only relatively recently that they have been commercially exploited. Such irons are characterised by the graphite being present in the form of relatively short, thick flakes with rounded extremities and undulating surfaces.

In general compacted graphite irons have mechanical and physical properties intermediate between those of conventional flake graphite irons (grey irons) and nodular graphite irons. The outstanding characteristics are good thermal conductivity combined with useful ductility and higher tensile and fatigue strengths than for grey irons.

At present no British Standard exists for these materials.

The presence of certain element combinations in irons which would otherwise solidify with a conventional flake graphite form may promote the formation of the compacted type of graphite. At present the production of compacted graphite irons is largely based on additions of magnesium and/or cerium to irons of low sulphur content. The magnesium based treatments are similar to those employed for nodular graphite iron production while the cerium methods are usually simple ladle addition processes.

In the magnesium process, as the magnesium content is increased, the graphite structure changes from conventional flake, through compacted flake to fully nodular. The range of magnesium contents to facilitate compacted flake structures is very narrow and it is usually necessary to employ additions of titanium and cerium to extend the range and provide a practical ladle based process. Magnesium and titanium are usually present in the ranges 0.015–0.035 per cent and 0.08–0.15 per cent respectively with a trace of cerium.

Fully compacted graphite irons have combinations of properties intermediate between those of conventional flake and nodular irons, as shown in Table 26.64. They behave elastically over a range of stresses although their limit of proportionality is lower than nodular irons. The tensile properties of compacted graphite irons are less sensitive to variations in carbon equivalent than conventional flake irons (Fig. 26.5) but they are section sensitive (Fig. 26.6).

Compacted graphite irons have intermediate thermal conductivity values between flake and nodular iron and comparative figures are given in Table 26.65.

Table 26.62   MECHANICAL PROPERTIES OF NODULAR IRONS

Grade BS 2789: 1985	Tensile strength R_m min		Proof stress $R_{p0.2}$ min		Elongation A min	Hardness	Impact Charpy V-notch J*			Structure †
	MPa	ksi	MPa	ksi	%	HB	20°C	−20°C	−40°C	
350/22L40	350	51	220	32	22	≤ 160	—	—	9(12)	F
350/22	350	51	220	32	22	≤ 160	14(17)	—	—	F
400/18/L20	400	58	250	36	18	≤ 179	—	9(12)	—	F
400/18	400	58	250	36	18	≤ 179	11(14)	—	—	F
420/12	420	61	270	39	12	≤ 212	—	—	—	F
450/10	450	65	320	46	10	160/221	—	—	—	F/P
500/7	500	72	320	46	7	170/241	—	—	—	F/P
600/3	600	87	370	54	3	192/269	—	—	—	P/F
700/2	700	102	420	61	2	229/302	—	—	—	P
800/2	800	116	480	70	2	248/352	—	—	—	For TS
900/2	900	131	600	87	2	302/359	—	—	—	TM

Notes:
Verification of hardness and 0.2% proof stress is optimal.
* Individual value. Value in brackets is mean of 3 tests.
† F = ferrite, F/P = ferrite/pearlite, P/F = pearlite/ferrite, P = pearlite, TS = tempered structure, TH = tempered martensite.

Table 26.63   MECHANICAL PROPERTIES OF AUSTEMPERED DUCTILE IRON (ADI) (PROPOSED SPECIFICATIONS)

Grade ASTM A897M-90	Tensile strength R_m min		Yield strength min		Elongation A min	Hardness	Impact unnotched
	MPa	ksi	MPa	ksi	%	HB	Charpy J
850/550/10	850	123	550	80	10	269/321	100
1050/725/7	1 050	152	700	101	7	302/363	80
1200/850/4	1 200	174	850	123	4	341/444	60
1400/1 100/1	1 400	203	1 100	160	1	388/477	35
1600/1 300/-	1 600	232	1 300	188	*	444/555	*
VDG(FRG)							
GGG-80B	800	116	500	72	6	250/310	—
GGG-100B	1 000	145	700	101	5	280/340	—
GGG-120B	1 200	174	950	138	2	330/390	—
GGG-140B	1 400	203	1 200	174	1	440/442	—
GGG-150B	1 500	217	—	—	—	421/475	—
BCIRA (GB)							
950/6	950	138	670	97	6	300/310	—
1050/3	1 050	152	780	113	3	345/335	—
1200/1	1 200	174	940	136	1	390/400	—
KYMI-KYME/ENE (FIN)							
K-9007	900	130	730	106	6	280/310	—
K-1005	1 000	145	800	116	3	300/350	—
K-12003	1 200	174	1 000	145	1	380/430	—

* Elongation and impact not specified. Grades 1 400/1 100/1 and 1 600/1 300/- mainly used for gear and wear resistant applications.

Table 26.64   A COMPARISON OF THE MECHANICAL PROPERTIES OF FLAKE, COMPACTED AND NODULAR GRAPHITE CAST IRONS IN 30 mm BAR SECTIONS

	Flake graphite iron Grade 260	Compacted graphite iron* ($CEV = 4.3$)		Nodular graphite iron Grade 500/7
		Ferritic	Pearlitic	
Tensile strength, MN/m²	260	365	440	500
0.1% Proof stress, MN/m²	73	260	305	323
0.2% Proof stress, MN/m²	—	290	330	339
0.5% Proof stress, MN/m²	—	325	365	356
Elongation %	0.57	4.5	1.5	7 min–15
Modulus of elasticity, GN/m²	128	162	165	169
Hardness, HB 10/3000	185–226	140–155	225–245	~172

* Irons produced by magnesium–titanium–cerium treatment.

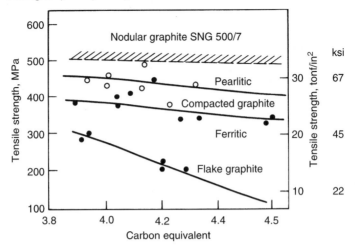

Figure 26.5   *Effect of carbon equivalent on the tensile strengths of flake, compacted and nodular graphite irons cast into 30 mm-diameter bars*

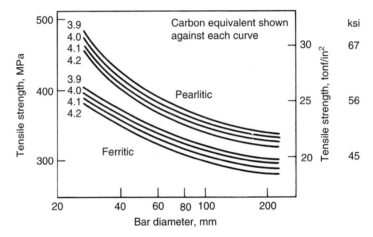

Figure 26.6   *Variation of tensile strength with cast-section size for ferritic and pearlitic compacted graphite irons of various carbon equivalents. Each curve is the centre of a band of variation in strength of about ±30 MPa (± 2 tonf/in²)*

Table 26.65   A COMPARISON OF THE THERMAL CONDUCTIVITIES OF FLAKE, COMPACTED AND NODULAR GRAPHITE CAST IRONS

	Thermal conductivity W/m K				
	100°C	200°C	300°C	400°C	500°C
Flake graphite iron (Grade 17)	48.8	47.8	46.8	45.8	44.8
Compacted graphite iron (CEV = 4.3)	41.0	43.5	41.0	38.5	36.0
Nodular graphite iron (Grade 500/7)	35.5	35.35	35.2	35.05	34.9

Table 26.66 AUSTENITIC CAST IRONS

Grade BS 3468:1986*	Composition %								Tensile strength R_m min		Proof strength $R_{p0.2}$ min		Elongation A^{**}
	C max	Si	Mn	Ni	Cr	Nb	P max	Other	MPa	ksi	MPa	ksi	%
General engineering													
F1 (FG)	3.0	1.5/2.8	0.5/1.5	13.5/17.5	1.0/2.5	—	0.2	Cu 5.5/7.5	170	25	—	—	—
F2 (FG)	3.0	1.5/2.8	0.5/1.5	18.0/22.0	1.5/2.5	—	0.2	Cu 0.5 max	170	25	—	—	—
S2 (SG)	3.0	1.5/2.8	0.5/1.5	18.0/22.0	1.5/2.5	—	0.08	Cu 0.5 max	370	54	210	30	7
S2W1 (SG)	3.0	1.5/2.2	0.5/1.5	18.0/22.0	1.5/2.5	0.12/0.2	0.05	Cu 0.5 max	370	54	210	30	7
S5S (SG)	2.2	4.8/5.4	1.0 max	34.0/36.0	1.5/2.5	—	0.08 max	—	370	54	210	30	7
Special purposes													
F3 (FG)	2.5	1.5/2.8	0.5/1.5	28/32	2.5/3.5	—	0.2	Cu 0.5 max	190	28	—	—	—
S2B (SG)	3.0	1.5/2.8	0.5/1.5	18/22	2.5/3.5	—	0.08	Cu 0.5 max	370	54	210	30	7 (4)
S2C (SG)	3.0	1.5/2.8	1.5/2.5	21/24	0.5 max	—	0.08	Cu 0.5 max	370	54	170	25	20 (20)
S2M1 (SG)	3.0	1.5/2.8	4.0/4.5	21/24	0.5 max	—	0.08	Cu 0.5 max	420	61	200	29	25 (15)
S3 (SG)	2.5	1.5/2.8	0.5/1.5	28/32	2.5/3.5	—	0.08	Cu 0.5 max	370	54	210	30	7
S6 (SG)	3.0	1.5/2.8	6.0/7.0	12/14	0.2 max	—	0.08	Cu 0.5 max	390	57	200	29	15

* (FG) = Flake graphite, (SG) = speroidal graphite.
** Value in bracket is Charpy V notch strength at 20°C in J.

Table 26.67   ALLOY CAST IRONS

Grade BS 4844: 1986	Composition %							Hardness			
	C	Si	Mn	Ni	Cr	P max	Other	HB		HV	
								(Section thickness mm)			
Low alloy grades											
1A	2.4/3.4	0.5/1.5	0.2/0.8	—	2.0 max	0.15	—	400(<50)	350(>50)	428(<50)	368(>50)
1B	2.4/3.4	0.5/1.5	0.2/0.8	—	2.0 max	0.5	—	400(<50)	350(>50)	428(<50)	369(>50)
1C	2.4/3.0	0.5/1.5	0.2/0.8	—	2.0 max	0.15	—	250(<50)	225(>50)	225(<50)	205(>50)
Nickel chromium grades											
2A	2.7/3.2	0.3/0.8	0.2/0.8	3.0/5.5	1.5/3.5	0.15	Mo 0.5	500(<125)	450(>125)	542(<125)	485(>125)
2B	3.2/3.6	0.3/0.8	0.2/0.8	3.0/5.5	1.5/3.5	0.15	Mo 0.5	550(<125)	500(>125)	599(<125)	542(>125)
2C	2.4/2.8	1.5/2.2	0.2/0.8	4.0/6.0	8.0/10.0	0.10	Mo 0.5	500(<125)	450(>125)	542(<125)	485(>125)
2D	2.8/3.2	1.5/2.2	0.2/0.8	4.0/6.0	8.0/10.0	0.10	Mo 0.5	550(<125)	500(>125)	599(<125)	542(>125)
2E	3.2/3.6	1.5/2.2	0.2/0.8	4.0/6.0	8.0/10.0	0.10	Mo 0.5	600(<125)	550(>125)	655(<125)	599(>125)
High chromium grades											
3A	1.8/3.0	1.0 max	0.5/1.5	2.0 max	14/17	0.10	Mo/Cu 2 max	600 min		655 min	
3B	3.0/3.6	1.0 max	0.5/1.5	2.0 max	14/17	0.10	Mo/Cu 2 max	650 min		712 min	
3C	1.8/3.0	1.0 max	0.5/1.5	2.0 max	17/22	0.10	Mo/Cu 2 max	600 min		655 min	
3D	2.0/2.8	1.0 max	0.5/1.5	2.0 max	22/28	0.10	Mo/Cu 2 max	600 min		655 min	
3E	2.8/3.5	1.0 max	0.5/1.5	2.0 max	22/28	0.10	Mo/Cu 2 max	600 min		655 min	
3F	2.0/2.7	1.0 max	0.5/1.5	2.0 max	11/13	0.10	Mo/Cu 2 max	600 min		655 min	
3G	2.7/3.4	1.0 max	0.5/1.5	2.0 max	11/13	0.10	Mo/Cu 2 max	650 min		712 min	

Table 26.68 CORROSION RESISTING HIGH SILICON CAST IRONS

Grade *BS* 1591: 1975	*Composition %*						
	C *max*	Si	Mn *max*	P *max*	S *max*	Cr	*Application*
Si10	1.2	10.0/12.0	0.5	0.25	0.1	—	Strength greater than S14
Si14	1.0	14.25/15.25	0.5	0.25	0.1	—	General corrosion
Si Cr 144	1.4	14.25/15.25	0.5	0.25	0.1	4.0/5.0	Cathodic protection anodes
Si16	0.8	16.0/18.0	0.5	0.25	0.1	—	Corrosion resistance at expense of strength

Heat treatment: Castings to be stripped from moulds while hot and as soon as possible after solidification, the hot castings to be charged to a furnace preheated to approximately 600°C and kept at this during charging. Then heat to not less than 750°C and not more than 850°C. Soak for 2 hours for castings of simple form and thickners less than 18 mm and for 8 hours for heavy castings. Cool slowly after soak to 300°C before unloading.

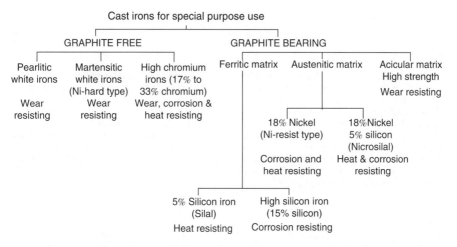

Figure 26.7 *Classification of special purpose cast irons*

The combination of relatively high strength and good thermal conductivity has resulted in compacted graphite irons primarily finding application where containment of stress at high temperature or under thermal cycling conditions are important. Applications for ingot moulds, cylinder heads, brakedrums and discs and manifold castings are typical.

British Standards are summarised for Austenitic irons BS 3468: 1986 Table 26.66, for alloy cast irons BS 4844: 1986 in Table 26.67 and for corrosion resisting high silicon irons BS 1591: 1975 in Table 26.68.

26.9.4 Applications of special purpose cast irons

HIGH TEMPERATURE

Irons in the general engineering group are suitable for applications up to at least 350°C where long-term dimensional stability is required. There are also many instances of the use of grey irons in the temperature range up to 700°C provided that appreciable growth and scaling can be tolerated.

For extended life at temperatures up to 850°C one of the alloyed irons, such as Ni-resist, Silal or Nicrosilal, may be used. The austenitic materials Ni-resist and Nicrosilal have good thermal shock resistance, but Silal is limited to applications where severe temperature gradients are absent. Above 850°C the most suitable material is 30% chromium iron, which has good oxidation resistance and a useful level of creep strength up to 1 050°C.

Table 26.69   TYPICAL ANALYSES OF OTHER SPECIAL PURPOSE CAST IRONS

Type or trade name	C %	Si %	Mn %	S %	P %	Ni %	Cr %	Mo %	Characteristics
Silal	2.2	5.5	0.6	0.1	0.1	—	—	—	Heat resistant
Pearlitic white cast iron	2.9	1.2	0.8	0.1	0.1	—	—	—	Wear resistant
Ni-hard*	3.0	0.8	0.8	0.1	0.3	4.5	2.1	—	Wear resistant
High chromium cast irons	1.0	0.4	0.4	0.03	0.05	—	30.0	—	Heat and corrosion resistant
	2.7	0.8	0.8	0.03	0.05	—	27.0	—	Wear and corrosion resistant
	3.0	0.8	0.8	0.03	0.05	—	15.0	3.0	Wear and corrosion resistant
Nomag†	3.0	1.5	7.0	0.03	0.05	11	—	—	Non-magnetic
Low expansion cast iron	2.2	1.5	0.8	0.03	0.05	35	2.0	—	Minimum thermal expansion

* Trade name–International Nickel Ltd.
† Trade name–Ferranti Ltd.

CORROSION

Ni-resist, Nicrosilal, high silicon iron and high chromium iron have good corrosion resistance in appropriate media. For example, Ni-resist and Nicrosilal have good resistance to sea water, strong alkalis, inorganic salts and weak acids. These materials are machinable without difficulty. High silicon irons have excellent resistance to sulphuric and nitric acids at all temperatures and concentrations, but have the disadvantage that they are brittle and can only be machined by grinding. High chromium (30%) irons are the only cast irons which can be regarded as stainless so far as atmospheric exposure is concerned. They develop a passive film under oxidising conditions and their outstanding characteristic is the ability to withstand attack by nitric acid at temperatures up to boiling point and concentrations up to 70%. This grade of iron is machinable without difficulty provided the carbon content is restricted to a maximum of 1.5%.

ABRASION

A graphite-free structure is essential for good resistance to abrasion. The irons which meet this requirement, as shown in Figure 26.7 are the unalloyed pearlitic white irons, martensitic white irons of the Ni-hard type and high carbon high chromium irons.

It is difficult to relate wear characteristics determined in a laboratory to practical service conditions, and there is considerable overlapping between the fields of application for these materials. In the absence of experience of similar conditions, evaluation of material must sometimes be made by trial. In general, martensitic white irons have better abrasion resistance than pearlitic white irons. High chromium, high carbon irons are particularly useful where abrasion is combined with impact loading and where abrasive and corrosive conditions exist together. *See* Chapter 25. Some proprietary special purpose cast irons are listed in Table 26.69.

26.10   Acknowledgements

Valuable assistance in the compilation of data on casting alloys and foundry data from the following organizations.

Steel: SCRATA—Steel Casting Research and Technical Association, ASTM—American Society for Testing and Materials (now called ASTM International)
Cast Iron: BCIRA—British Cast Iron Research Association
Nickel: INCO—INCO Engineering Products Ltd
Investment Casting: BICTA—British Investment Casters' Technical Association
Magnesium: MEL—Magnesium Electron Ltd
Zinc: ZDA—Zinc Development Association
Copper: CDA—Copper Development Association, AFS—American Foundry Society
Aluminium: The Aluminum Association
Die Castings: NADCA—North American Die Casting Association

27 Engineering ceramics and refractory materials*

27.1 Physical and mechanical properties of engineering ceramics

Intensive development of ceramic materials has increased the availability of well-characterised engineering ceramics capable of use over a range of temperatures and atmospheres. Table 27.1 summarises the properties of the most commonly used engineering ceramics while Table 27.2 gives more information about a range of ceramics. However, the data in the tables should be regarded as representative since actual values are a strong function of exact composition and microstructure. Furthermore, since these materials are often used at high temperatures the properties at use rather than room temperature (RT) are important.

---

* See Chapter 40 for applications in furnace design.

Table 27.1   SUMMARY OF PROPERTIES OF IMPORTANT ENGINEERING CERAMICS (see also Table 27.2)

Type	Melting or decomposition temperature, °C	Density g cm^{-3}	Thermal conductivity W/mK (measured temperature or range, °C)	Thermal expansion ($\times 10^{-6}$ K^{-1}) (measured temperature or range, °C)	Young's modulus GPa	Strength MPa (technique, measured temperature or range, °C)	K_{Ic} MPam$^{1/2}$
Silicon Carbide							
Hot Pressed	2 200	3.2	50 (600)	4.55	440–450	640–650 (bend, 20–1 400)	5.7
Sintered	—	3.1	55 (600)	4.5–4.9 (20–1 500)	395–410	430–450 (bend, 20–1 400)	3–5
Reaction Bonded	—	3	—	4.2–4.3 (20–1 500)	280–390	350–540 (bend, 20)	4.4–5
Silicon Nitride							
Hot Pressed	1 800d	3.2–3.9	15–50 (20)	2–3 (20–1 000)	280–320	400–1 000 (bend, 20)	3.4–8.2
Sintered	1 800d	3.13	—	3.5	245	420	4.8
Reaction Bonded	1 800d	2.2–3.2	3–30 (20)	2–3 (20–1 000)	100–220	190–400 (bend, 20)	1.5–3.6
SIALONS	1 800d	3.22–3.25	10–22 (20)	3–3.2 (20–1 200)	288	800–945 (bend, 20)	4–8
Alumina							
>99.9% Al$_2$O$_3$ Solid State Sintered	2 050	3.96–3.98	38.9 (20)	6.5–8.9 (200–1 200)	400–410	550–600 (bend, 20), >2 600 (compression, 20)	3.8–4.5
86 to 94.5% Al$_2$O$_3$ Liquid Phase Sintered	~1 200 (glass)	3.4–3.7	15–20 (20)	7–7.6 (20–800)	250–300	250–330 (bend, 20) 1 800–2 000 (comp, 20)	—
Zirconia							
Partially Stabilised Zirconia, PSZ (MgO)	2 600	5.9	1–2	6.8 (20–1 000)	170–210	440–720 (bend, 20)	6–20
PSZ (Y$_2$O$_3$ stabilised)	2 600	6.05	1–2	10.2	180–220	650–1 000 (bend, 20)	6–8
PSZ (CaO)	2 600	—	—	—	200–220	400–650 (bend, 20)	6–12
Tetragonal Zirconia Polycrystals, TZP (CeO$_2$) TZP (Y$_2$O$_3$)	2 600				140–200	500–800 (bend, 20)	6–30
Zirconia Toughened Alumina, TZP (30% ZrO$_2$)	2 050	4.54			140–200	800–1 300 (bend, 20)	6–15
					330	770	6
Zirconia Toughened Mullite (30% ZrO$_2$)	1 850	—			150	220	2.7

Table 27.2   PROPERTIES OF PURE CERAMIC MATERIALS

Material	Melting point °C	Bulk density g cm^{-3}	Thermal conductivity Wm^{-1} K^{-1} at temp. °C	Thermal expansion coefficient 10^{-6} to temp. °C		Strength MPa (technique, measured temp. °C)	Remarks
Chromium diboride, CrB$_2$	2 100	5.2	32 at 20	7.5	20	—	Stable in presence of carbon
Hafnium diboride, HfB$_2$	3 250	11.2	6.3 at 20	5.3 5.5	500 1 000	—	No reaction with basic slags for 6 minutes at 1 520°C. Stable in presence of carbon, no reaction with steel for 6 min at 1 620°C
Tantalum monoboride, TaB	2 340	14.0	—	—		—	Oxidation severe at 1 100–1 400°C. Unstable in presence of carbon
Tantalum diboride, TaB$_2$	3 200	12.4	10.9 at 20 13.9 at 200	—		—	Oxidised in air at 800°C. Stable in presence of carbon
Titanium diboride, TiB$_2$	2 980	4.5	26 at 20 26 at 200	7.6	1 000	—	Very stable, even in presence of carbon. Oxidation in air severe at 1 100–1 400°C. Max. working temperature 2 000°C reducing, 800°C oxidising
Zirconium diboride, ZrB$_2$	3 060	6.1	23 at 20 23–26 at 200	5.5–6.6 7.0	1 000 1 500	—	No reaction with basic slags at 1 520°C for 6 min. No reaction with carbon steel at 1 620°C for 2 h. Stable in presence of carbon. Oxidation in air severe at 1 100–1 400°C. Stable under inert or reducing conditions to over 2 000°C
Boron carbide, B$_4$C	2 350	2.51	29 at 20 84 at 425	4.8 5.5 6.5 7.1	500 1 000 2 000 2 500	2 900 (comp, 20), 155 (comp, 980), 300 (MOR*, 20)	Attacked by iron, Thermal shock resistance poor. Resistant to air up to 1 000°C but max. working temp. 2 000°C reducing, 600°C oxidising
Hafnium monocarbide, HfC	3 890	12.2	—	6.3 6.25	500 1 000	—	Oxidation in air severe at 1 100–1 400°C. Stable to 2 000°C in He
Silicon carbide, SiC	2 700	3.17	42 at 20 21 at 1 000	4.6 5.5 5.9	500 1 500 2 500	550 (comp, 20)	Quite resistant to oxidation by air up to 1 650°C. Thermal shock resistance very good. Reacts with Fe and MgO in basic slags to give silicides. Reacts with iron. High resistance to acid or neutral slags and coal ash: Used in gasifiers, zinc retorts. Also used as abrasive. Used in electrical heating elements. Max. working temp. 2 320°C reducing, 1 650°C oxidising (see also Table 27.1)
Tantalum monocarbide, TaC	3 880	14.7	22 at 20	6.3 6.7 8.4	500 1 000 2 500	—	Oxidation in air severe at 1 100–1 400°C. Useful in He to 3 760°C

* MOR, Modulus of Rupture

(continued)

Table 27.2   PROPERTIES OF PURE CERAMIC MATERIALS—*continued*

Material	Melting point °C	Bulk density g cm^{-3}	Thermal conductivity Wm^{-1} K^{-1} at temp. °C	Thermal expansion coefficient 10^{-6} to temp. °C		Strength MPa (technique, measured temp. °C)	Remarks
Titanium monocarbide, TiC	3 140	4.25	32 at 20 5.5 at 1 000	7.7 9.7	1 000 2 500	750 to 1 300 (comp, 20)	Oxidation in air becomes severe at 1 200°C. Max useful temp. 3 000°C in He
Tungsten carbide, WC	2 777	15.7	84 at 20	4.9 5.8 to 6.1	1 000 2 000	—	Oxidation in air severe 500–800°C useful to 2 000°C in He. Extremely hard (9+Mohs. Vickers pyramid 2 400) used in drill tips. Max. working temp. 2 000°C reducing, 550°C oxidising
Zirconium monocarbide, ZrC	3 540	6.7	21 at 20	6.1 6.6 7.6	500 1 000 2 000	1 640 (comp, 20)	Oxidation in air becomes severe at 1 100–1 400°C. Max. useful temp. 2 350°C in He
Graphite, C	3 650 (sub-limes)	1.50 to 2.25	63 to 210 at 20, parallel to grain. 42 to 130 at 20, perpendicular to grain. 47 at 1 300 34 at 2 500	Parallel: 1 to 4 20 Perpendicular: 2.5 20 to 4.5 4.0 1 000 to 9.8 5.5 1500 to 11 compression		Parallel 3.5 to 7.6 at 20°C Perpendicular 3.5 to 70 at 20°C compression 36 at 2 500°C compression	Thermal shock resistance very good. Not wetted by iron. Resistant to acidic and basic slags. Oxidised in air above 300°C. Resistant to non-oxidising gases. May be 'welded' using molybdenum disilicide. Excellent conductor of electricity
Aluminium nitride, AlN	2 230	3.26	150–180 at 20°C	4.8 5.5	500 1 000	300 (MOR, 20)	Oxidised by O_2 above 1 000°C. Unstable in water vapour. Stable in N_2–H_2 mixtures at 1 200–1 600°C
Boron nitride, BN (hexagonal)	2 330 (sub-limes)	2.1	15 at 20 27 at 1 000 (Perpendicular values are approx. half of these)	Parallel to pressing direction 2.0 1 000 Perpendicular 13.3 1 000		Parallel to pressing direction 310 (comp, 20) Compressive perpendicular 235 (comp, 20)	Thermal shock resistance very good when dry. No reaction with iron at 1 600°C for 30 min. Oxidation in air severe at 1 100–1 400°C. Stable to 1 000°C in O_2. Resists attack by molten metals and glasses. Low coeff. of friction. Fabricated by hot pressing. Machines easily. Max. working temp. 2 200°C reducing, 1 000°C oxidising
Silicon nitride, Si_3N_4	1 900 (sub-limes)	3.2	2.3 to 13 at 20 9.4 at 1 200	α phase 2.1 500 3.7 1 500 β phase 1.5 500 3.1 1 500		—	Thermal shock resistance good. Reacts with iron. Useful to 1 850°C in reducing or inert conditions. Stable in air to 1 200°C. Resistant to molten glasses, molten Al, Pb, Zn, Sn and to HCl, H_2SO_4 and HNO_3. Reacts with molten Cu. Slowly attacked by boiling water. May be partially nitrided, machined then fully nitrided. Used in aluminium handling thermo-couple sheaths, cutting tools, wear resistant parts. *See* Table 27.1 for different types

(*continued*

Table 27.2   PROPERTIES OF PURE CERAMIC MATERIALS—*continued*

Material	Melting point °C	Bulk density g cm^{-3}	Thermal conductivity Wm^{-1} K^{-1} at temp. °C	Thermal expansion coefficient 10^{-6} to temp. °C	Strength MPa (technique, measured temp. °C)	Remarks
SIALONS, α (glass bonded)	1 800d	3.22	10.9 (20)	3.2 (20–1200)	833 (bend, 20)	Sialons $\equiv$ Si$_3$N$_4$ with Al and O replacing some Si and N
β (glass bonded)	1 800d	32.5	22 (20)	3.04 (20–1200)	945 (bend, 20)	respectively. α-sialons incorporate cations e.g. Ca,Mg into crystal structure. Commercial sialons use 2nd phases e.g. glass or yttrium aluminium garnet to provide bonding. Good thermal shock resistance and better oxidation resistance than pure nitrides.
Titanium mononitride, TiN	2 900	5.3	29 at 20 8.5 at 1 000	—	—	No reaction with basic slags. Slightly wetted by carbon steel at 1 620°C. Poor oxidation resistance to O$_2$ at 600°C and to CO$_2$ at 1 200°C
Zirconium mononitride, ZrN	2 950	7.1	27 at 20 6.7 at 1 000	6.13 450 7.03 680	—	Slight reaction with cast iron at 1 450°C for 2 h. Oxidation in air severe at 1 100–1 400°C. Slow hydrolysis in water
Aluminium oxide, Al$_2$O$_3$	2 050	3.97	39 at 20 9.2 at 600 5.9 at 1 400 7.1 at 1 800	7.6 500 8.5 1 000 8.9 to 1 400 9.1	2 940 (comp, 20) 48 (comp, 1 600) 203 (shear, 20) 23 (shear, 1 500)	Thermal shock resistance fair. α stable form, range of metastable cubic transition aluminas (e.g. χ, δ, γ, θ, etc.) form e.g. on calcination of Gibbsite converting to α at high temperature. Good resistance to basic and acidic slags. Max. working temp. 1 900°C (see Table 27.1 for zirconia-toughened alumina)
Beryllium oxide, BeO	2 530	3.00	202 at 100 29 at 1 000 15 at 1 700	7.6 500 8.6 to 1 000 9.0 10.3 1 500 11.1 2 000	786 (comp, 20) 48 (comp, 1 600)	Becomes volatile at 2 100°C; very poisonous. Not reduced by carbon or hydrogen. Attacked by acids, fluxed by alumina
Calcium oxide, CaO	2 572	3.32	15.5 at 100 8 at 1 000	11.8 500 13.1 1 000 15.3 1 500	—	Thermal shock resistance fair. Poor resistance to attack by slags containing FeO and SiO$_2$. Subject to hydration
Cerium dioxide, CeO$_2$	2 600	7.3	—	8.2 500 8.9 1 000	—	Useful in air to 2 400°C. Not useful in reducing conditions. Subject to hydration
Dichromium trioxide, Cr$_2$O$_3$	2 435	5.21	—	8.4 500 8.6 1 000 8.8 1 500	—	Slag resistance good under oxidising conditions. Less resistant to basic slags Cr^{6+} known carcinogen may form in use.
Hafnium dioxide, HfO$_2$	2 810	9.68	—	Monoclinic 5.5 500 5.8 1 000 6.4 1 700 Tetragonal 1.3 1 700 3.0 2 000	— —	Useful in air to 2 400°C. Stable in H$_2$ to 1 925°C

(*continued*)

Table 27.2   PROPERTIES OF PURE CERAMIC MATERIALS—*continued*

Material	Melting point °C	Bulk density g cm^{-3}	Thermal conductivity Wm^{-1} K^{-1} at temp. °C	Thermal expansion coefficient 10^{-6} to temp. °C	Strength MPa (technique, measured temp. °C)	Remarks
Magnesium oxide, MgO	2 800	3.58	46 at 20 8.4 at 800 6.3 at 1 400 9.2 at 1 800	12.8 500 13.6 1 000 15.1 1 500 15.9 1 800	83 (shear, 20) 39 (shear, 1 300)	Thermal shock resistance poor; can be improved by small amounts of spinel. Resistance to both acidic and basic slags excellent. Limits of usefulness: to 1 600°C in vacuum; 1 700–1 980°C in reducing atmosphere; 2 400°C in air. Melts at 2 680°C in oxygen-free helium. Subject to hydration
Silicon dioxide SiO$_2$	1 710	2.32	1.5 at 20 2.5 at 1 600	α quartz 22.2 575 β quartz 27.8 575 14.6 1 000 Vitreous 0.55 1 000		Thermal shock resistance of vitrified silica is excellent. Polymorphic forms of silica are quartz, tridymite and cristobalite. The last two of these are metastable under ordinary conditions. Transition temps: quartz-tridymite 870°C. Tridymite-cristobalite 1 470°C. Vitreous silica devitrifies at 1 100°C
Thorium dioxide ThO$_2$	3 205	9.7	10 to 15 at 20 2.9 at 1 000 2.5 at 1 400	8.6 500 9.1 1 000 to 9.4 10.4 1 400	1 480 (comp, 20) 10 (comp, 1 500) 8.3 (shear, 1 300)	Thermal shock resistance poor. High resistance to basic slags. Reduced by carbon at high temp. Becomes volatile in He at 2 300°C
Zirconia, ZrO$_2$	2 690	5.75	2.0 at 25 2.3 at 800 2.7 at 1 400	Monoclinic 6.5 500 7.7 1 050 Tetragonal 7.9 600 8.3 1 400	2070 (comp, 20) 102 (comp, 20) 19 (comp, 1 500)	Zirconia phase changes monoclinic → cubic at 1 050°C. Thermal shock resistance influenced by volume change accompanying phase change. Stable in oxidising atmosphere. Fairly stable in reducing atmosphere. Excellent resistance to basic and acidic slags. See Table 27.1 for stabilised zirconias
Mullite, 3Al$_2$O$_3$·2SiO$_2$	1 830	3.16–3.22	7.1 at 25 4.0 at 800 3.8 at 1 400	5.1 to 5.8 1 000°C	150–500 (comp, 20) 16.6 (shear, 1 100)	Thermal shock resistance good. Max. working temp. 1 650°C (see also Table 27.1)
Magnesia spinel, MgO. Al$_2$O$_3$	2 135	3.51	18 at 25 8 at 600 5.5 at 1 200	8.4 1 000 to 8.6 9.4 1 400	1 370 (comp, 550) 59 (comp, 1 600) 65 (shear, 20) 37 (shear, 1 300)	Resistant to slags containing iron oxide. More resistant than alumina to action of reducing slags. Stable to 1 400°C in H$_2$. Thermal shock resistance fair. Max. working temp. 1 950°C

(continued)

Table 27.2 PROPERTIES OF PURE CERAMIC MATERIALS—*continued*

Material	Melting point °C	Bulk density g cm^{-3}	Thermal conductivity Wm^{-1} K^{-1} at temp. °C	Thermal expansion coefficient 10^{-6} to temp. °C		Strength MPa (technique, measured temp. °C)	Remarks
Zircon, ZrO$_2$·SiO$_2$	2 550	4.56	6.1 at 100 4.2 at 800 4.0 at 1 400	3.8 4.6 5.3	500 1 000 1 500	60 (shear, 20) 16 (shear, 1 300)	Dissociates above 1 700°C. Due to action of slags containing FeO, zircon would dissociate in steel-making environments. Thermal shock resistance good
Molybdenum disilicide, MoSi$_2$	2 030	5.95 to 6.24	31.5 at 20 to 200 17 at 1 100	7.8 8.5 9.0	500 1 000 1 500	2 280 (comp, 20)	Carbon reduces melting point to 1 870°C. Corrosion in air becomes severe at 1 700°C. No attack by O$_2$ to 1 100°C. Used in electrical heating elements
Tungsten disilicide, WSi$_2$	2 165 to 2 180	9.25		7.8 8.3	500 1 000		Corrosion in air severe above 1 950°C

See also: R. Morrell 'Handbook of Properties of Technical and Engineering Ceramics Part 2: Data Reviews' (Her Majesty's Stationary Office, London, UK, 1985, 1987), W. E. Lee and W. M. Rainforth 'Ceramic Microstructures, Property Control by Processing' (Chapman and Hall, London, UK, 1994).

27.2 Prepared but unshaped refractory materials

These materials are commonly used for installation and/or repair of refractory linings. By definition, unshaped or monolithic refractory materials are prepared mixtures for use either as delivered or after the addition of an appropriate liquid. This definition covers the refractory cements, mouldable and castable materials, ramming and gunning mixes. One major difference between materials in this group and brick or blocks (shaped refractories) is the considerably reduced number of joints in a structure, i.e. these materials tend towards a monolithic construction. Monolithics now account for over 50% of refractories sold worldwide. Both unshaped monolithic and shaped brick refractories have a microstructure consisting of large (often mm scale) isolated grain or aggregate phases held together by a much finer and more reactive bond or matrix system. Typical grain materials listed in Table 27.3 include bauxite, graphite and magnesite while bond systems may be clay based or use a range of fine powders such as fumed SiO$_2$, hydratable alumina or aluminous cements (section 27.3). In addition many steelmaking refractories are bonded with very fine carbon derived by pyrolysis of phenolic resin or pitch binders (Table 27.5). Table 27.4 lists properties of typical fired bricks while Table 27.5 gives properties of blocks which are installed unfired and then fired in situ. Properties of low thermal mass (LTM) refractories are given in Table 27.6.

27.3 Aluminous cements

Hydraulic aluminous cement is manufactured by fusing or sintering a mixture of bauxite and limestone. In general the silica content is kept as low as the raw materials permit and it is preferred that the iron should be in the ferric rather than ferrous condition. Commercial aluminous cements vary somewhat in composition but usually lie in the following ranges: SiO$_2$ 4–7%, Al$_2$O$_3$ 36–42%, Fe$_2$O$_3$ 8–12%, FeO 4–8%, TiO$_2$ 2–3% and CaO 36–42%. These cements depend on the presence of calcium aluminates for their properties and by adding various forms of alumina, calcined bauxite, chrome and magnesia, it is possible to produce hydraulic cements with excellent refractory properties.

   Calcium aluminate cements have unique properties in that they can be moulded, air-hardened and used directly as the bond system in refractories. They are relatively slow-setting (1–2 h) but rapid-hardening (24 h). In general, the higher the temperature of firing and within limits the higher the percentage of alumina they contain, the more refractory they become. They are notable for their relatively small shrinkage on heating to 400°C and small expansion on heating to 1 350°C (0.5–1.5%). Their resistance to heat is very good and the best products can be used up to 1 600°C

Table 27.3   RAW MATERIALS

Raw material	Al$_2$O$_3$	CaO	K$_2$O	Na$_2$O	MgO	Cr$_2$O$_3$	Fe$_2$O$_3$	SiO$_2$	TiO$_2$	ZrO$_2$	LOI*	C	S
Andalusite	58.6		5.6		—	—	—	33.8	—	—	—	—	—
Ball clay (Devon)	20 to 30	0.04	1.2 to 2.7	0.2 to 0.4	0.1 to 0.5	—	0.8 to 1.4	53 to 71	0.9 to 1.6	—	6 to 10	0.1 to 0.7	0.1 to 1.5
Bauxite (Calcined)	84.9	—	0.9	0.9	—	—	3.3	8.0	1.2	—	—	—	—
China clay	38.1	0.1	1.5	0.2	0.2	—	0.7 (FeO)	47.4	—	—	11.9	—	—
Chrome ore (Turkey)	15.0	0.7	—	—	14.0	56.8	12.1	1.5	—	—	—	—	—
Diatomite (Skye)	5.2	1.7	0.1	0.6	1.3	—	3.1	72.1	0.4	—	14.6	—	0.4
Dolomite (Salop)	1.6	31.7	—	—	19.3	—	0.5	1.2	—	—	45.6	—	—
Fireclay	34.9	0.4	1.6		0.9	—	1.9	46.8	1.5	—	11.2	—	—
Flint clay (English)	—	0.3	Trace	0.1	0.2	—	Trace	98.2	—	—	1.4	—	—
Foundry sand (Natural)	8.0	0.1	0.8		0.1	—	0.2	88.4	0.2	—	2.2	—	—
Ganister (Sheffield)	0.9	0.1	—	—	—	—	0.7	96.8	—	—	—	—	—
Graphite (Canada)	10.5	16.9	36	0.27	9.4	—	9.4	48.9	0.15	—	—	95.2	—
				ash composition (total 3.81%)									
Kyanite (Kenya)	59.7	0.3	—	—	0.1	—	0.6	37.6	1.1	—	0.6	—	—
Magnesite (Austrian)	0.8	1.8	—	—	88.4	—	6.8	1.4	—	—	—	—	—
Magnesite (Sea water)	0.5	2.2	—	—	93.7	—	1.4	2.1	—	—	—	—	—
Olivine (Norwegian)	0.7	—	—	—	49.0	0.4	6.5	42.0	—	—	1.2	—	—
Quartzite (Welsh)	0.6	0.1	—	—	—	—	0.4 (FeO)	97.8	—	—	—	—	—
Serpentine (Shetland)	2.5	—	—	—	37.6	4.4	6.9	33.2	—	—	15.2	—	—
Silica sand (Pure)	0.5	—	0.1	0.1	—	—	0.1	99.3	—	—	—	—	—
Sillimanite (Australia)	58.8	0.8	0.4	—	0.1	—	2.4	34.5	1.5	—	1.4	—	—
Zircon sand (Australia)	1.2	0.1	—	—	—	0.4	0.4	30.0	2.1	65.4	—	—	—

* Loss on ignition.

without softening. Calcium aluminate cements lose strength when heated to 400°C, but then remain unchanged up to 1 000°C and gradually recover again at higher temperatures.

27.4   Castable materials

These are mixtures of graded refractory aggregate and either a hydraulic cement or a chemical bonding agent. The material is usually supplied dry, and at the appropriate moisture content it may be cast. The mix is controlled to be either vibratable where vibration is used to remove bubbles increasing density or self flowing where no additional vibration is needed during installation. Modern castable compositions are designed to have reduced levels of calcium aluminate cement (termed low cement, ultra low or cement free castables) since the lime degrades high temperature properties. Properties of typical materials in this group are shown in Table 27.7. Castable materials are used increasingly in furnace linings, production of special shapes, covers of soaking pits, burner blocks, floors of aluminium holding vessels, cyclones and incinerators.

Table 27.4 PROPERTIES OF TYPICAL DENSE FIRED REFRACTORY BRICKS

Material	Chemical analysis (% by weight)										Bulk density g cm^{-3}	Apparent porosity %	0.19 MPa RUL* % deformation at °C	PLC %†	Free thermal expansion %	Thermal conductivity Wm^{-1}K^{-1}			Max. service temp. °C	Uses
	Al$_2$O$_3$	CaO	Cr$_2$O$_3$	Fe$_2$O$_3$	K$_2$O	MgO	Na$_2$O	SiO$_2$	TiO$_2$	ZrO$_2$						500	900	1300°C		
High Alumina	87	0.3	—	1.5 to 2.0	—	0.2	—	6.6 to 9.5	2.4 to 3.0	—	2.82 to 2.97	17 to 21	10 1730	0 to −2 2 h 1600°C	0.95 to 1400°C	1.3	1.6	1.9	1800	Arc furnace roof torpedo ladle
Fired dolomite	2	55	—	2	—	37	—	3	—	—	2.7	18	—	0 to 0.8 2 h 1700°C	1.4 to 1000°C	—			—	Rotary kilns
Firebrick	38	0.5	<0.1	2.9	0.6	0.55	0.5	56	1.4	—	2.1	18 to 25	5 1500	−0.7 2 h 1410°C	0.5 to 1000°C	—			—	Ladles, rotary kilns
Chrome–magnesite ‡	13 to 16	0.9 to 1.3	17 to 23	9 to 12	—	46 to 52	—	2.5 to 3.5	—	—	2.9 to 3.1	18 to 24	—	0 to +3 5 h 1700°C	1.5 to 1400°C	1.8 to 2.2			—	Rotary kilns, electric arc furnaces
Firebrick	41.7	0.5	—	2.7	0.6	0.5	0.1	52.3	1.7	—	1.9 to 2.0	21 to 26	—	0 to −3 2 h 1600°C	0.5 to 1000°C		1.3	1.5	—	Glass tank furnaces, blast furnace stack
Magnesite–chrome	10 to 12	0.8 to 1.1	14 to 18	8 to 10	—	60 to 65	—	1.8 to 2.3	—	—	3 to 3.2	16 to 20	—	−0.5 to +3.0 5 h 1800°C		2.2 to 2.7			—	Vacuum degassers
Magnesia–spinal	3 to 7	0.8 to 2	—	0.3 to 0.8	—	90 to 94	—	0.3 to 1	—	—	2.9 to 3.05	16 to 20	—		1–1.2 to 1000°C	2.5–3 (1 000)			—	Glass tank checkers Rotary cement kilns
Magnesia–dolomite	0.5	44	—	0.3	—	54.5	—	0.8	—	—	2.85	16 to 20	—		1.2 to 1000°C	42 (1 000)			—	AOD, LF Rotary cement kilns

(continued)

Table 27.4   PROPERTIES OF TYPICAL DENSE FIRED REFRACTORY BRICKS—*continued*

Material	Chemical analysis (% by weight)										Bulk density g cm⁻³	Apparent porosity %	0.19 MPa RUL* % deformation at °C		PLC%†	Free thermal expansion %	Thermal conductivity Wm⁻¹K⁻¹ 500 900 1300°C	Max. service temp. °C	Uses
	Al_2O_3	CaO	Cr_2O_3	Fe_2O_3	K_2O	MgO	Na_2O	SiO_2	TiO_2	ZrO_2									
Magnesite	0.2 to 0.4	1.8 to 2.3	—	0.15 to 0.3	—	95.5	—	0.7 to 0.9	—	—	2.9 to 3.0	15 to 19	—	—	−0.2 to −1.5 5 h 1800°C	2.1 to 1400°C	3.7 to 4.4	—	Electric arc furnaces, LD backing lining
Mullite	74.2	—	—	0.7	0.7	0.1	0.3	23.2	0.2	—	2.6 to 2.7	13 to 17	2	1700	−0.2 2 h 1700°C	0.63 to 1400°C	1.5 1.7 2.0	1700	Glass tank furnaces
Silica	0.6	1.7	—	0.5	0.2	0.2	0.2	96.5	0.1	—	1.7 to 1.8	21 to 25	10	1680	0 4 h 1600°C	1.3 to 1200°C	1.3 1.7	1700	Coke ovens, hot blast stoves
Zircon	1.2	—	—	0.3	—	—	—	31.9	0.3	64.2	3.7 to 3.9	14 to 18	0	1700	0 2 h 1600°C	0.7 to 1400°C	2.6 2.4 2.3	1700	Glass tank furnaces
Carbon‡	—	—	—	—	—	—	—	—	—	—	1.56 to 1.64	—	0	1700	−0.5 2 h 1500°C	0.55 to 1000°C	3.6	—	Blast furnace hearth and bosh
Pitch impregnated fired magnesite	0.15 to 0.25	1.9 to 2.3	—	0.15 to 0.30	—	96.0 to 97.0	—	0.7 to 0.9	—	—	2.85 to 3.0	14 to 18	—	—	−0.2 to −1.5 5 h 1800°C	2.1 to 1400°C	3.7 to 4.4	—	LD, Q-BOP

* Refractoriness under load.
† Permanent linear change.
‡ Typical range into which most products fall.
See also: G. Routschka (Editor) 'Pocket Manual Refractory Materials (Vulkan–Verlag, Essen, Germany 1997), 'Refractories Handbook' (Technical Association of Refractories, Tokyo, Japan 1998).

Table 27.5 PROPERTIES OF TYPICAL UNFIRED REFRACTORY BRICKS

Material	Chemical analysis (% by weight)								Bulk density g cm⁻³	PLC %*	Thermal conductivity Wm⁻¹K⁻¹	Residual carbon %	Loss on ignition %	Compressive strength MPa				Uses
	Al_2O_3	CaO	Fe_2O_3	K_2O	MgO	Na_2O	SiO_2	TiO_2						20°C	120°C	180°C	300°C	
Pitch bonded magnesite	0.5 to 0.7	1.5 to 3.0	0.5 to 1.0	—	92.5 to 96.5	—	0.8 to 1.3	—	2.9 to 3.1	—	3.9 to 4.6 at 900°C	1.8 to 2.5	3 to 5	—				LD
Chemically bonded high alumina	86 to 88	0.3	1.5 to 2.0	0.2	0.2	0.2	6.6 to 9.5	2.4 to 3.0	2.9 to 3.1	+1.5 to +3.5 2 h at 1 600°C	—	—	—	—				Aluminium melting vessels, electric arc furnaces
Pitch bonded dolomite (tempered)	0.6 to 0.9	53.0 to 57.0	1.8 to 2.4	—	39.0 to 41.0	—	1.0 to 1.6	—	2.80 to 2.95	—	2.4 to 2.9 at 900°C	1.8 to 2.2	3.8 to 4.8	21 to 42	2 to 6	1 to 6	1 to 4	BOS converters
Pitch bonded magnesia doloma (tempered)	0.4 to 0.7	21.0 to 26.0	1.0 to 1.5	—	70.0 to 75.0	—	0.8 to 1.2	—	2.87 to 3.02	—	3.5 to 4.1 at 900°C	2.5 to 3.2	4.2 to 5.2	20 to 40	2 to 6	2 to 6	1 to 4	BOS converters
Pitch bonded magnesia (tempered)	0.1 to 0.3	1.8 to 2.8	0.2 to 0.5	—	95.0 to 97.0	—	0.7 to 1.0	—	3.00 to 3.15	—	3.8 to 4.4 at 900°C	3.0 to 3.6	4.5 to 5.5	20 to 40	5 to 20	3 to 14	2 to 6	BOS converters
Resin bonded MgO–C	0.1 to 0.8	0.6 to ~3	0.2 to 2	—	92 to 99	—	0.2 to ~1.5	—	2.70 to 3.15	—	5 to 25 at 1 000°C	7 to 25	—	30 to 85				BOS EAF Ladles

* Permanent linear change.

Table 27.6 PROPERTIES OF TYPICAL LOW THERMAL MASS (LTM) REFRACTORY BRICKS

Material	Chemical analysis (% by weight)									Bulk density g cm⁻³	Apparent porosity %	MOR* at room temp. MPa	Thermal expansion %	Max. service temp. °C	Thermal conductivity Wm⁻¹ K⁻¹			PLC % at °C	Uses
	Al_2O_3	CaO	B_2O_3	Fe_2O_3	K_2O	MgO	Na_2O	SiO_2	TiO_2						400°C	600°C	1100°C		
Insulating firebrick	38 to 40	1.0 to 14.0	—	0.4 to 2.0	0.1 to 0.3	—	0.1 to 0.3	45 to 55	1.0 to 1.5	0.6 to 0.9	70 to 75	0.8 to 1.0	0.45 to 0.6 to 1 000°C	1 300 to 1 400	0.3 to 0.1	0.35 to 0.15	0.4 to 0.25	0 to −1.1 1 300	General insulation
Insulating firebrick	78 to 94	—	—	0.1 to 0.2	0 to 0.1	0 to 0.2	0 to 0.2	—	0 to 0.5	1.2 to 1.4	60 to 65	1 to 3	0.6 to 0.7 to 1 000°C	1 700	0.3 to 0.5	0.4 to 0.6	0.4 to 0.6	−0.4 to −0.7 1 700	High temperature insulation
Semi-insulating firebrick	30 to 40	0.2 to 1.6	—	1 to 2	0 to 1	0 to 1	0 to 1	51 to 61	1 to 2	0.6 to 1.0		1 to 3	—	1 320	0.25 to 0.35	0.35 to 0.45	0.45 to 0.55	0 to −0.5 1 300	Rotary kilns
Diatomite insulating brick	—	—	—	—	—	—	—	—	—	0.5 to 0.6	72 to 78	—	—	900	0.15 to 0.17	0.18 to 0.19			Low temperature insulation
Fibre blanket	47 to 65	0 to 0.2	0 to 1.2	0 to 0.3	0.1 to 0.4	0 to 0.1	0.1 to 0.4	38 to 50	0 to 0.1	—		Fibre blanket	—	1 300	0.05 to 0.09	0.09 to 0.22	0.11 to 0.45	−2.5 to −4.0 1 260°C	Low temperature insulation, vacuum degasser gaskets, jointing material, etc.

N.B. Fibre board is similar to fibre blanket but contains a rigidiser. *Modulus of rupture.

Table 27.7 TYPICAL PROPERTIES OF UNSHAPED REFRACTORIES

Material	Chemical analysis (% by weight)*								Bulk density g cm⁻³		PLC %** at °C	Max service temp.°C	Thermal conductivity Wm⁻¹k⁻¹			Refractoriness	MOR†† MPa
	Al_2O_3	CaO	Fe_2O_3	MgO	K_2O	Na_2O	SiO_2	TiO_2	Unfired	Fired			400°C	600°C	1100°C		
Lightweight insulating castable	30 to 49	9.5 to 12	5 to 8	0 to 1	0 to 1	0 to 1	30 to 50	1 to 2	1.4 to 1.6	1.2 to 1.4	−0.2 to −1.0 1200°C	1300	0.2 to 0.4	0.3 to 0.45		1350°C– 1450°C	—
Low cement castable (LCC)	90	1.4	0.1	—	—	—	8	—	—	3.1	−0.2 1600°C	1700	—			1700°C	30 at 20°C
Ultra Low cement castable (ULCC)	99	0.5	0.1	—	—	—	0.2	—	—	3.1	−0.3 1600°C	1800	—			1800°C	20 at 20°C
Ramming mix [Phosphate Bonded]	83 to 85	0.2	1.6	0.2	0.03	—	9 to 10.9	2.7	1.81	—	−1 1700°C	1800	—			1800°C+	3.0 at 20°C 14.2 at 1000°C 3.0 at 1400°C
High alumina mouldable	42 to 70	0 to 10	1 to 5	0 to 0.5	0.2 to 1	0 to 1	25 to 50	0 to 2	—	1.0 to 2.0	−0.5 to +1.5 1600°C	1600		0.35 to 0.6	0.45 to 0.7	—	—

* The chemical constituents of the binders are not included: these may include phosphates and organic materials.
** Permanent linear change.
†† Modulus of rupture.
See also: G. Routschka (Editor) 'Pocket Manual Refractory Materials' Vulkan–Verlag, Essen, Germany 1997), 'Refractories Handbook' (Technical Association of Refractories, Tokyo, Japan 1998).

27.5   Mouldable materials

These are mixtures of graded refractory aggregates and plasticisers, usually clay, supplied mixed with water in a workable condition. Chemical bonding agents may also be incorporated. The workability of the material is such that it may be placed by hand malleting. These materials usually have good thermal shock resistance, but a low compressive strength which does, however, improve after the production of a ceramic bond. The thermal conductivity is usually lower than that of the equivalent shaped and fired material.

Properties of a typical material from this group are shown in Table 27.7.

27.6   Ramming material

This is a mixture of graded refractory aggregate with or without the addition of a plasticiser and with or without water usually supplied at a consistency which requires a mechanical method of application. The material is placed in position by means of hand or pneumatic rammers. It is most important that all the material should be compacted to the same extent (i.e. an even density of packing) and that laminations should be avoided.

Properties of a phosphate-bonded ramming mix are given in Table 27.7.

27.7   Gunning material

Many of the materials which fall into the above groups may be suitably prepared for application by gunning techniques. The suitably prepared material is introduced into a high pressure compressed air line in a specially designed gun and the material is blasted at the desired area to be lined. In dry gunning water is mixed with the powder batch at the gun nozzle whereas in wet gunning or shotcreting the powder and water are pre-mixed. The benefit of the latter process is reduced dust hazard and material rebound.

27.8   Design of refractory linings

A refractory lining usually comprises a safety or backup layer behind the working lining in contact with the furnace contents. The whole body is encased in a metal shell. The temperature gradients in the system and the properties of all materials must be considered.

Essential to the designer of refractory structures is a knowledge of the mechanical properties of materials to be used (e.g. thermal expansion, thermal conductivity, Young's modulus and ultimate strength). Standard test methods do not necessarily give the most useful data. The cold crushing strength of a fired brick is generally much greater than its crushing strength at higher temperatures; rectangular blocks heated from one end do not expand as expected from the free thermal expansion; nor is the Young's modulus of a stressed block the same as that of an unstressed block.

Table 27.8 indicates the amount of data needed when installing a refractory. Since the values may change substantially with temperature it is important to have data over the range of temperature. In particular, the designer of any refractory lining should be aware of the free thermal expansion and hot strength of all materials used. This is most important in any system operating above 1 100°C. In order to establish shell temperatures, interface temperatures etc. thermal conductivity values should also be known. Finite element methods are frequently used to determine temperature/property profiles.

The first stage of design is to carry out a thermal analysis under the worst thermal conditions expected with the aim of ensuring that no material exceeds a safe operating temperature. There are several commercial computer programs that perform these calculations. Most materials should not be allowed to come within 100°C of their 'maximum service temperature' but for fibre materials the margin should be increased to 200°C. Once service temperatures are determined, stability of the lining should be considered. Linings in tilted vessels such as steel ladles or electric arc furnaces must not be allowed to come loose (bricks will fall out or metal penetration of joints will occur). On the other hand, expansion of refractories should not be such as to cause excess stresses leading to lining failure by cracking, shell fracture, anchor failure or other problems. It is not easy to devise a simple rule dealing with such conditions and if this may be a concern (especially for vessels more than 2 m wide) expert engineering advice should be sought. If high stresses are expected, expansion allowances should be used where possible. Pieces of cardboard, fibre blanket, wood etc. may be placed between bricks during installation to give expansion allowance. For monolithic (castable) linings, gaps are left between panels. These methods are more difficult when the lining will hold molten metal, because metal may penetrate the expansion allowances. For vessels holding molten metal, expansion allowances are often introduced by installing a layer of soft ramming material behind the working lining. During operation, the working lining expands outwards to compress the ramming layer. Where soft ramming layers are used, it is important to install the correct thickness of a suitable material, and this is another matter on which expert advice should be sought. When considering high-temperature properties the hot modulus of rupture (flexural or bend strength) of refractories is often given although refractoriness-under-load test data are sometimes available but should be interpreted carefully. An index of standards applicable to refractories is given in Table 27.9.

Table 27.8 THERMAL EXPANSION AND YOUNG'S MODULUS FOR FIRED MAGNESITE*

	Temperature °C												
	100	200	300	400	500	600	700	800	900	1 000	1 100	1 200	1 300
Free thermal expansion (%)	0.08	0.24	0.38	0.55	0.70	0.88	1.05	1.21	1.38	1.54	1.71	1.87	—
Thermal expansion measured under thermal gradient heating (%)	0.14	0.29	0.45	0.61	0.76	0.92	1.09	1.25	1.37	1.54	—	—	—
Young's modulus (sonic method) (GPa)	84	83	81	79	78	76	74	73	71	70	69	68	67
Young's modulus measured during restraint of thermal expansion (GPa)	73	28	16	11	9.3	7.8	7.0	6.2	5.0	3.4	1.7	0.81	0.38

* Data published by permission of Ceram Research, Stoke on Trent, UK.

Table 27.9 INDEX OF REFRACTORY STANDARDS

Subject	*BS*	*DIN*	*ASTM*	*ISO*
Abrasion resistance	1902	—	C704–88	—
Acid resistance	—	51102	—	8590–88
Alkali attack	—	51103	C767–86	—
Basic refractories	3056 Part 1	—	—	5417–86
	3056	—	—	5019–84
	1902	—	—	—
Bricks—application	—	1082 Bbl	—	—
Bricks—dimensions	3056 Part 1	—	C134–34	—
	3056	—	C861–77	—
	3056	—	C861–77	—
	—	—	C909–81	—
	—	—	C134–84	—
Bricks—End arches	—	1082	—	1145
—dimensions	—	—	—	—
Bricks—rectangular—dimensions	—	—	—	R475
Bricks—side arches—dimensions	—	1082	—	R1145
Bricks—skewbacks—dimensions	—	—	—	—
Carbon monoxide attack	1902	—	C288–87	—
Castable refractories	1902	—	C401–84	—
	—	—	C179–85	—
	—	—	C862–87	—
Carbon-containing refractories	—	—	C831	10060–93
	—	—	—	10081–91
Cement	4550	—	—	—
	—	—	C198–83	—
	—	—	C199–84	—

(*continued*)

Table 27.9   INDEX OF REFRACTORY STANDARDS—*continued*

Subject	BS	DIN	ASTM	ISO
Chemical analysis—alimina refractories	4140	51077	C573–81	P803–80
	1902	—	—	—
Chemical analysis—Aluminosilicate refractories	1902	51070	C573–81	—
	—	E51083	C575–81	—
Chemical analysis—carbon-containing refractories	—	—	C571–81	—
Chemical analysis—chrome refractories	1902	51074	C572–81	—
Chemical analysis—dolomite	1902	—	C574–84	10058–92
Chemical analysis—magnesia refractories	1902	51073	C574–84	10058–92
Chemical analysis—raw materials	—	—	C572–81	—
Chemical analysis—sample preparation	—	51062	—	—
Chemical analysis—silicon carbide	—	51075	—	—
refractories	—	51076	—	—
Chemical analysis—	—	—	C576–81	—
Chemical analysis—	—	—	C705–84	—
Chemically-bonded basic bricks	—	51050	—	—
Chimneys and flues	4207	1057	—	—
Chrome brick	—	—	C455–84	—
Chrome–Magnesite brick	—	—	C455–84	—
Classification	1902	—	—	1109–75/10080–
				90/10081–91
Coke ovens	999	1089	—	—
	4966	—	—	—
Cold crushing strength	1902	51050	C133–84	8895
	—	51067	C93–84	—
Concrete	1881	1048	C860–83	2736, 4109, 27
	—	—	C865–87	—
	—	—	C862–87	—
	—	—	C903–88	—
Corrosion resistance	—	V51069	C622–84	—
	—	—	C621–84	—
	—	—	C768–85	—
	—	—	C874–85	—
	—	—	C767–86	—
	—	—	C575–81	—
Creep Density	—	51053	C832–84	3187–89
	—	51050	C914–79	—
	1902	51057	C357–85	5018–83
				5016–86
				5017–88
	—	51065	C134–84	—
	—	—	C830–83	—
	—	—	C20–87	—
	—	—	C493–86	—
Drying shrinkage	—	—	C179–85	—
Fireclay products	3056 Part 3	1089	—	5407
	—	51060	C673–84	—
	3056	—	C27–84	—
	—	—	C605–87	—
Glass melting furnaces	4966	—	—	—
	3056 Part 2	—	—	—
Glossary	3446	—	C108–46	836[†]–68
	—	—	C71–85	2246
Grain size	—	51033	B430	—
Heat transmission	—	—	C108–46	—
High alumina	3056 Part 3	—	C673–84	5417
	3056	—	C27–84	—
	—	—	C1054–85	—
Hydration	—	—	C492–82	—
	—	—	C544–85	—
	—	—	C620–87	—
	—	—	C456–87	—
Insulating firebrick	—	—	—	2245–72
	—	—	C155–84	—
	—	—	C134–84	—
	—	—	C210–85	—
	—	—	C134–84	—
	—	—	C182–88	—

(*continued*)

Table 27.9 INDEX OF REFRACTORY STANDARDS—*continued*

Subject	BS	DIN	ASTM	ISO
Liquid absorption	—	—	C830–83	—
	—	—	C20–87	—
Magnesite brick	—	—	C455–84	—
Magnesite–chrome brick	—	—	C455–84	—
Modulus of rupture	1902	51048	C133–84	5014–86
				5013[†]–85
	—	—	C583–80	—
	—	—	C607–88	—
	—	—	C93–84	—
Moisture content	—	—	C92–88	—
Monolithic linings	4207	—	—	—
Mullite	—	—	C467–84	—
Non-destructive methods of test	1881	—	—	—
Nozzles, fireclay	—	—	C605–87	—
Pallets	—	—	—	—
Permanent linear change	1902	51066	C113–87	2477–87
	—	—	—	2478
	—	—	C436–83	—
	—	—	C605–87	—
	—	—	C210–85	—
	—	—	C179–85	—
Permeability	1902	51058	C577–87	8841–91
	—	51050	—	—
Porosity	1902	—	C20–87	5017–88
	—	—	C830–83	—
	—	—	C493–86	—
Pouring pit	—	—	C435–84	—
Pyrometric cone	—	51063	—	1146–88
Pyrometric cone equivalent	1902	51063	C24–84	528–83
Ramming mixes	—	—	C673–84	—
Refractoriness-under-load	1902	51053	—	1893[†]–89
	—	51064	—	—
Rotary cement kilns	3056 Part 3	—	—	5417–86
Sampling	1902	51061	—	—
	616	—	—	5022–79
				8656–88
Sedimentation	—	51033	—	—
Sieve analysis	1902	51033	C92–88	—
Silica	4966	1089	C911–87	—
	3056 Part 2	—	C416–84	—
	—	—	C575–81	—
	—	—	C439–61	—
Silicon carbide refractories	—	—	C863–83	—
Slag resistance	—	—	C768–85	—
	—	—	C874–85	—
Specific gravity	1902	—	C830–83	—
	—	—	C20–87	—
	—	—	C604–86	—
	—	—	C135–86	—
Strength testing	1902	—	C67–87	—
	—	—	C16–81	—
Tar-bonded refractories	—	—	C831–88	—
Tar-impregnated refractories	—	—	C831–88	—
Test methods	1902	1089	—	—
	—	V51046	—	—
	—	51048	—	—
Thermocouple reference tables	4937	—	—	—
Thermal conductivity	1902	V51046	C201–86	8894–90
	—	—	C202–86	—
	—	—	C767–86	—
	—	—	C182–88	—
	—	—	C113–89	—

(continued)

Table 27.9   INDEX OF REFRACTORY STANDARDS—*continued*

Subject	BS	DIN	ASTM	ISO
Thermal expansion	1902	51045	C832–84	—
Thermal shock resistance	1902	51068	C38–79	—
	—	E51068	C107–76	—
	—	—	C122–76	—
	—	—	C439–61	—
Unshaped refractory products	1902	51061	C673–84	1927–84
	—	—	C179–85	—
	—	—	C491–85	—
	—	—	C417–88	—
	—	—	C181–82	—
	—	—	C860–83	—
	—	—	C865–87	—
	—	—	C862–87	—
	—	—	C903–88	—
Warpage	1902	—	—	—
Young's modulus	—	—	C885–87	—
Zircon	—	—	C545–84	—

[†] Approx. equivalent

28 Fuels*

28.1 Coal

There are a number of different sets of National Standards in use and the major ones are the British Standards, ASTM (US), EN (Europe), and the Russian and Chinese standards. There are many common features amongst these standards many of which emanate from the British Standards that are listed below.

28.1.1 Analysis and testing of coal

SAMPLING FOR ANALYSIS

In order to be representative, the gross sample is compiled by collecting a number of increments spaced evenly throughout the mass of the consignment of coal and, for the specific method, BS 1017: 1989 should be consulted. The mass of an increment is determined by the maximum size of the coal shown in Table 28.1.

The minimum number of increments required may be obtained from Table 28.2 and is determined by the class of coal and the purpose of the sample. Separate general analysis and total moisture samples may be desirable, e.g. when the coal is very wet, otherwise a common sample can be taken from which both total moisture and general analysis samples are prepared.

Reference standards for the precision of measurements on the samples are given in Tables 28.3 and 28.4. The levels of precision represent the 95% probability limits of the deviation of any single value from the true value.

Sampling from a stopped belt is the ideal method for sampling commercial coal and it should be used whenever practicable as the standard against which other methods are checked.

PROXIMATE ANALYSIS CONSISTS OF THE FOLLOWING DETERMINATIONS (BS 1016: 1991):

Total moisture which accounts for the 'free' or adventitious moisture, together with the 'inherent' or original moisture always associated with the coal. Moisture is determined by heating the air-dried coal, ground to pass a 0.2 mm sieve, to 105–110°C in a vacuum oven or in a stream of nitrogen. The loss in weight is the moisture on the air-dried sample (BS 1016: Section 104.1; 1991).

Table 28.1 MINIMUM MASS OF INCREMENT AND SIZE OF ENTRY INTO SAMPLING IMPLEMENT

Nominal upper size of coal, mm	<10	10–25	25–50	50–75	75–100	100–125	125–150	>150
Minimum size of entry into sampling implement, mm	30	75	150	200	250	320	375	>375
Minimum mass of increment, kg	0.5	1.5	3.0	4.5	6.0	8.0	14.0	>14

---

* See Chapter 40 for applications in furnaces.

Table 28.2 INCREMENTS REQUIRED TO FORM A GROSS SAMPLE

Minimum number of increments to be collected from a consignment weighing up to 1 000 tonnes

Situation	Common sample			General analysis sample		Total moisture sample		Size analysis sample
	Sized coals— dry-cleaned or washed	*Washed smalls (<50 mm),*	*Blended part-treated, untreated, run-of-mine and 'unknown coals'*	*Sized coals— dry-cleaned or washed and unwahsed dry coals*	*Blended part-treated, untreated, run-of-mine and 'unknown coals'*	*Sized coals— dry-cleaned or washed and unwahsed dry coals*	*Washed smalls (<50 mm), blended, untreated, run-of-mine and 'unknown coals'*	*All coals*
Streams	20	35	35	20	35	20	35	40
Wagons and lorries Barges								
Sea-going ships (from conveyor during off-loading)	25	35	50	25	50	20	35	40
Sea-going ships (from the hold) Stockpiles	35	35	65	35	65	20	35	40

Table 28.3   REFERENCE STANDARDS OF PRECISION FOR MOISTURE AND ASH

True value	Total moisture	Ash (dry basis)
Below 10%	1% absolute	1% absolute
10% to 20%	0.1% of true value	0.1% of true value
Above 20%	2% absolute	2% absolute

Table 28.4   REFERENCE STANDARDS OF PRECISION FOR SIZE ANALYSIS: PERCENTAGE BETWEEN TWO SIEVES

Per cent in fraction	<5	5–10	10–20	20–30	30–50
Precision % absolute	0.8	1.8	2.7	3.2	3.5

Volatile matter is the % loss in weight corrected for moisture when 1 g of the less than 0.2 mm coal is heated in the absence of air to 900°C (BS 1016, Section 104.3; 1991).

Ash is determined by placing 1 g of the 0.2 mm coal in a muffle furnace at room temperature, raising the temperature to 500°C in 30 min, to 815°C in a further 60–90 min and maintaining this temperature until the residue, which is the ash, is constant in weight (BS 1016: Section 104.4; 1991).

% Fixed carbon is defined as:

$$100 - (\% \text{ moisture} + \% \text{ ash} + \% \text{ volatile matter})$$

ULTIMATE ANALYSIS

Requires in addition to ash and moisture determinations (described above) figures for carbon, hydrogen, nitrogen and sulphur.

CALORIFIC VALUE

For practical purposes may be expressed as the number of heat units liberated by the complete combustion of unit weight of coal in a bomb calorimeter. Corrections are made for the formation of nitric and sulphuric acid originally present as nitrogen and sulphur in the coal (BS 1016: Part 105; 1992).

CALCULATION OF CALORIFIC VALUE FROM ULTIMATE ANALYSIS

The calorific value of a fuel may be checked from the ultimate analysis. Usually the calculated value agrees to within 1–2% of the determined value.

In the following formula, which gives the *gross* calorific value (CV), the symbols used give the percentages of: C, carbon; H, hydrogen; O, oxygen; N, nitrogen; S, sulphur.
Grummell Davies:

$$CV = (0.015\,22\,H + 0.937)[C/3 + H - (O - S)/8]\,\text{MJ kg}^{-1}$$

Gross and net calorific values   All coals contain hydrogen and water, and in the determination of calorific value the water vapour resulting from the combustion of the hydrogen and the vaporisation of the original water is condensed to the liquid state. In boiler practice it is not possible to cool the flue gases to a temperature below the dew point and thus the latent heat of condensation of the steam is not recovered.

The *gross* calorific value as determined by the bomb is, therefore, corrected for boiler efficiency work by deducting 2.454 MJ kg^{-1} (1 055 Btu lb^{-1}) of water obtained on combustion. The corrected figure is termed the *net* calorific value.

BS SWELLING NUMBER

Provides a means of assessing the tendency of a coal to swell when it is carbonised or used in a combustion appliance. If 1 g of less than 0.2 mm coal is heated in a squat-shaped silica crucible of standard dimensions by a Meker burner (with rich gas) or a Teclu burner with coal gas or a specially designed electric furnace, a coke button of a definite size and shape is produced. By reference to standard profiles a BS swelling number may be assigned to the coal (BS 1016: Section 107.1; 1991).

ROGA TEST[1]

A mixture of 1 g of less than 0.2 mm coal and 5 g of a standard anthracite is carbonised in a crucible. The resulting coke button is tested in a Roga drum for its resistance to abrasion. From the results obtained the coal coking index (Roga index) is calculated.

International standard ISO 335-1994(E).

GRAY KING COKE TYPE

The caking properties of a coal or blend of coals is assessed by carbonising in a laboratory assay under standard conditions. The coke residue is classified by comparison with a series of described standard coke types (BS 1016: Section 107.2; 1991).

AUDIBERT-ARNU DILATOMETER TEST[1]

The test assesses the coking properties of coal or coal blends. A pencil of powdered coal is inserted in a narrow tube and topped by a steel rod which slides in the bore of the tube. The whole is heated at a constant rate. The displacement of the piston is recorded as a function of the temperature. The maximum dilatation and contraction are recorded as a percentage of the original length of the pencil, and the temperatures of the points of softening, maximum dilatation and maximum contraction are noted.

ASH FUSION POINT

Ash fusion point of a coal ash is considered to be a rough guide to its clinkering propensities. The fusion point in a reducing atmosphere is lower by about 40°C than that in an oxidising atmosphere.

Group 1   Fusion temperature 1 425–1 710°C, Clinkering troubles absent.
Group 2   Fusion temperature 1 200–1 425°C. Clinkering manageable.
Group 3   Fusion temperature 1 040–1 200°C. Clinkering troubles excessive, unless adequate precautions are taken.

The determination is made on a trilateral pyramid, a cube or a right cylinder prepared in a mould from finely ground ash (BS 1016: Part 113; 1995). The test-piece is heated in an oxidising or a reducing atmosphere defined as follows:

A reducing atmosphere   An atmosphere consisting by volume of 50% hydrogen and 50% carbon dioxide with a tolerance of ±5%.

or

An oxidising atmosphere   An atmosphere consisting of either carbon dioxide or air.

The temperatures determined are:
(a) *Deformation temperature.* The temperature at which the first sign of rounding of the tip of the test specimen occurs.
(b) *Hemisphere temperature.* The temperature at which the height of the specimen is equal to half the base, its shape being approximately hemispherical.
(c) *Flow temperature.* The temperature at which the height of the specimen is equal to one third of that at the hemisphere temperature.

CALCULATION OF THE MINERAL MATTER CONTENT OF COAL

All classification systems are based on coal free from mineral matter and moisture. The analytical data for the coal 'as received' thus require correction in the sense of the following equations. The symbols have the following meanings: W_a, moisture; A, ash; V, volatile matter; C, carbon; H, hydrogen; S, sulphur; N, nitrogen; M_b and M_k, estimates of the mineral matter content; all expressed as percentages on the air-dried basis. Q is the calorific value MJ kg^{-1} (Btu lb^{-1}), on air-dried basis and W_t is the total moisture % on the 'as received' or 'as fired' basis.

(a) To convert 'air dried' to 'as received':

Multiply A, V, C, H, S, N, Q by $\left[\dfrac{100 - W_t}{100 - W_a}\right]$

(b) To convert 'air dried' to 'dry ash-free':

Multiply V, C, H, S, N, Q by $\left[\dfrac{100}{100 - (W_a + A)}\right]$

(c) To convert 'air dried' to mineral matter-free basis:

Multiply $C, H; N, Q$ by $\left[\dfrac{100}{100 - (W_a + M_k)}\right]$

The volatile matter on a dry mineral matter-free basis is given by[2]

$$\frac{100(V - c)}{[100 - (W_a + M_k)]}$$

According to the analysis available the correction c is given by

$$c = 0.13A + 0.2S_{pyr} + 0.7CO_2 + 0.7Cl - 0.20$$
$$\text{or } c = 0.13A + 0.2S_{total} + 0.7CO_2 + 0.7Cl - 0.32$$
$$\text{or } c = 0.13A + 0.2S_{total} + 0.7CO_2 - 0.12$$

The mineral matter (M_k) is most accurately expressed by a modification of the King, Maries and Crossley[3] formula:

$$M_k = 1.13A + 0.5S_{pyr} + 0.8CO_2 - 2.8S_{ash} + 2.8S_{sulph} + 0.5Cl$$

in which A is the determined ash; CO_2, carbon dioxide; S_{pyr}, pyritic sulphur; S_{ash}, sulphur in ash; S_{sulph}, sulphate sulphur in coal; Cl, chlorine.

Where full analytical data are not available the mineral matter (M_b) may be approximately assessed for some British and US coals.[4]

$$M_b = 1.1A + 0.53S_{total} + 0.74CO_2 - 0.32 \simeq 1.15A$$

28.1.2 Classification

Three major classification systems have been devised based on the proximate analysis of coals.[1–5]

These are the Fuel Research Board/National Coal Board (NCB) classification,[2] the American ASTM classification,[5] and the International Classification[1] of Hard Coals by Type devised by the Economic Commission for Europe (ECE). The Fuel Research Board/NCB classification is described below and outlines are given of the others. The technological characteristics of coals are better defined in terms of their petrographical constituents and a coal classification system on this basis is in prospect.

FUEL RESEARCH BOARD/NCB CLASSIFICATION[2]

Coals are assigned code numbers according to their volatile matter content on a dry, mineral matter-free basis and their caking propensities are assessed by the Gray–King low temperature assay (BS 1016: Part 12; 1959). Clean coal must be used for determining the Gray–King coke type, and if the coal has initially a higher ash content than 10% it is floated at such a specific gravity as will give the maximum yield of coal with not more than 10% of ash.

Using the criterion of volatile matter alone, a first division into the following groups is obtained:

	Volatile matter	*Code no.*
Anthracites	Under 9.1%	100
Low-volatile steam coals	9.1–19.5%	200
Medium-volatile coals	19.6–32.0%	300
High-volatile coals	Over 32%	*See* Table 28.5

In the first three groups, i.e in coals of volatile matter up to 32%, there is a close relationship between volatile matter content and caking properties. Consequently, the effect of subdividing into progressive ranges of volatile matter content is also to produce classes with progressive ranges of caking power. The corresponding Gray–King coke types are given in Table 28.5 as an indication of caking properties.

Table 28.5   COAL CLASSIFICATION SYSTEM USED BY NATIONAL COAL BOARD (REVISION OF 1964)

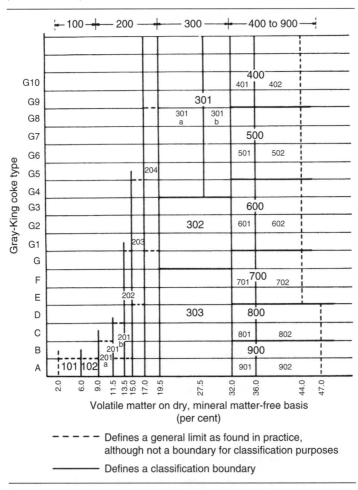

Volatile matter on dry, mineral matter-free basis
(per cent)

– – – – – Defines a general limit as found in practice, although not a boundary for classification purposes

————— Defines a classification boundary

Notes:
(1) Coals that have been affected by igneous intrusions ('heat-altered' coals) occur mainly in classes 100, 200 and 300, and when recognised should be distinguished by adding the suffix H to the coal rank code, e.g. 102H, 201bH.
(2) Coals that have been oxidised by weathering may occur in any class, and when recognised should be distinguished by adding the suffix W to the coal rank code, e.g. 801 W.

In the fourth group, i.e. in coals with more than 32% of volatile matter—there is a wide range of caking properties at any given volatile matter content, and subdivision has been made on the basis of the Gray–King coke type. Six ranges of caking properties, listed in Table 28.5, are recognised for these high-volatile coals.

Each of the 400–900 classes can be further subdivided according to volatile-matter content: a 1 in the third figure of the code number indicates that the volatile matter lies between 32.1 and 36.0, and a 2 that it is over 36%. Also, a subdivision is made of the 301 class into 301a and 301b with volatile ranges of 19.6–27.5% and 27.6–32% respectively.

A full list of code numbers, with ranges of volatile contents and caking properties, is given in Table 28.5.

ASTM CLASSIFICATION OF COALS BY RANK[5]

Coals are classified according to their fixed carbon and calorific value expressed in Btu lb^{-1} on a mineral matter-free basis. The higher rank coals are classified according to fixed carbon on the dry basis; the lower rank coals are classified according to calorific value on the moist basis. Agglomerating character is used to differentiate between certain adjacent groups. There are four classes: I anthracite, II bituminous, III sub-bituminous, IV lignite, each containing a number of named groups. The position of a coal in the scale of rank can be expressed in a condensed form, e.g. (62–146) in which the parentheses signify that the contained numbers are on a mineral matter-free basis. The first number represents the fixed carbon on the dry basis reported to the nearest whole per cent, the second the calorific value expressed as hundreds of Btu lb^{-1} to the nearest hundred.

ECE INTERNATIONAL CLASSIFICATION OF HARD COALS[1]

Coals are first placed in classes according to their volatile matter on the dry ash free basis. Then coals with a volatile matter greater than 33% are placed in classes according to their gross calorific value, Table 28.6.

Each class is further subdivided into groups according to their caking properties expressed either by their crucible swelling number or their Roga index, Table 28.7.

Each group is then subdivided into subgroups according to their coking properties assessed by their maximum dilatation in the Audibert–Arnu dilatometer test or by their Gray–King coke type, Table 28.8.

Table 28.6 DIVISION OF COALS INTO CLASSES (ECE)

Class number	Volatile matter % d.a.f.	Class number	Gross calorific value moist ash free kcal kg^{-1} (MJ kg^{-1})
1A	3–6.5	6	>7 750 (32.5)
1B	>6.5–10	7	>7 200–7 750 (30.1–32.5)
2	>10–14	8	>6 100–7 200 (25.5–30.1)
3	>14–20	9	>5 700–6 100 (23.9–25.5)
4	>20–28		
5	>28–33		
6–9	>33		

Table 28.7 DIVISION OF COAL CLASSES INTO GROUPS (ECE)

Group number	Crucible swelling number	Roga index
1	$0–\frac{1}{2}$	0–5
2	1–2	5–20
3	$2\frac{1}{2}–4$	20–45
4	>4	>45

Table 28.8 DIVISION OF COAL GROUPS INTO SUBGROUPS (ECE)

Subgroup number	Maximum dilatation	Gray–King coke type
0	Non-softening	A
1	Contraction only	B–D
2	0 and less	E–G
3	>0–50	G1–G4
4	>50–140	G5–G8
5	>140	>G8

A three-digit code number is used to describe the classified coal. The first digit indicates the class, the second digit the group, and the third digit the subgroup.

28.1.3 Physical properties of coal

Methods for predicting the specific heat, enthalpy, and entropy of coal, char, tar and ash as a function of temperature and material composition are presented in reference 7.

The mean thermal conductivity, k, of coking coals between $0°C$ and $t°C$ is given approximately by

$$k = 130 + 0.67t + 0.000\,67t^2 \text{ mW m}^{-1}\text{ K}^{-1}$$

Table 28.9 PHYSICAL PROPERTIES OF COAL

	Density kg m^{-3}	Bulk density* kg m^{-3} (lb ft^{-3})	Specific heat† kJ kg^{-1} K^{-1}	Coeff. linear thermal exp.‡ 10^6 K^{-1} 30°C	90°C	220°C	330°C
Fusain	—	—	0.88–0.92	—	—	—	—
Bituminous coal	1 250–1 450	600–670 (38–42)	1.00–1.09	33	—	45	60
Anthracite	1 400–1 700	700–790 (44–49)	0.92–0.96	—	—	—	—
Anthracite parallel to bedding plane	—	—	—	—	15	16.5	18
Anthracite perp. to bedding plane	—	—	—	—	27	29	29
Coal ash	—	—	0.67–0.71	—	—	—	—

* The approximate bulk densities refer to dry graded coals loosely packed in large containers. The bulk density is influenced by: (1) size and grading; (2) size of the container; (3) % 'free' moisture in excess of the inherent moisture; (4) shape of particles; (5) method of packing.
† Reference 6. The specific heat of coal increases with increase in volatile content and decrease in the carbon/hydrogen ratio.
‡ Data supplied by British Coal Utilisation Research Association.

28.2 Metallurgical cokes

28.2.1 Analysis and testing of coke

SAMPLING FOR ANALYSIS AND SHATTER TEST

Analysis In order to obtain a representative sample cokes are divided into four classes as follows:

Class 1: Large or graded gas coke from which breeze has been removed.
Class 2: Large or graded oven cokes from which breeze has been removed.
Class 3: Gas or oven cokes from which breeze has not been removed.
Class 4: Breeze.

In order to obtain a specified accuracy for any particular determination the number of increments required depends on the type of coke, its moisture, the degree of accuracy chosen and, in many

Table 28.10 TYPICAL ANALYSES OF SOLID FUELS

The table attempts to give analyses for fuels falling into each of the classes given but it should be understood that they can only be considered as a guide and that wide variations will be encountered among fuels belonging to each class

	Anthracite	Semi-anthracite	Semi-bituminous coals	Bituminous coals							Lignite	Peat	Wood	Charcoal	Coke	Semi-coke
Code numbers																
NCB	100a	201	204	301a	401	502	601	702	802	902	—	—	—	—	—	—
ECE	120A	221	344	445	545	844	843	832	821	921	—	—	—	—	—	—
Proximate analysis (air-dried basis)																
Moisture	1.0	1.0	1.0	0.9	0.9	1.9	2.0	5.8	8.6	13.8	15.0	20.0	15.0	2.0	2.5	2.5
Volatile matter less moisture	5.0	11.2	17.9	25.9	30.8	34.4	32.7	33.6	34.0	34.7	40.0	50.0	70.0	8.0	1.5	8.5
Fixed carbon	91.0	83.8	77.1	71.3	64.0	56.6	58.5	55.3	52.5	46.9	40.0	25.0	14.5	89.0	88.0	80.0
Ash	3.0	4.0	4.0	1.9	4.3	7.1	6.8	5.3	4.9	4.6	5.0	0.5	0.5	1.0	8.0	9.0
Volatile matter (dry, ash-free)	5.2	11.8	18.8	26.6	32.3	37.8	35.9	37.8	39.4	42.5	50.0	66.7	82.8	8.2	1.7	9.6
Ultimate analysis (air-dried coal)																
Carbon	89.4	86.9	86.0	86.6	83.5	76.7	77.0	74.4	70.0	64.6	55.2	43.1	42.4	90.4	85.1	82.3
Hydrogen	2.9	3.8	4.3	4.8	5.1	4.9	4.8	4.8	4.6	4.4	3.9	4.6	5.1	2.4	0.8	2.7
Nitrogen	1.1	1.2	1.3	1.6	1.5	1.6	1.5	1.5	1.2	1.3	0.7	0.6	0.3	0.8	0.9	0.9
Sulphur	0.9	1.0	1.0	0.8	1.1	2.6	1.5	1.1	0.9	0.6	0.6	1.3	0.3	0.7	0.7	1.1
Oxygen and errors	1.7	2.1	2.4	3.4	3.6	5.2	6.4	7.1	9.8	10.7	19.6	25.4	36.4	2.7	1.6	1.5
*Ultimate analysis (dry, mineral water-free basis)**																
Carbon	93.5	92.0	86.0	89.4	88.8	85.9	85.6	84.3	81.7	79.9	69.0	57.5	50.2	93.2	95.1	93.0
Hydrogen	3.0	4.0	4.5	5.0	5.3	5.4	5.3	5.3	5.4	5.4	4.9	6.1	6.0	2.5	0.9	3.1
Nitrogen	1.2	1.3	1.4	1.7	1.6	1.7	1.7	1.7	1.4	1.6	0.9	0.8	0.4	0.8	1.0	1.0
Sulphur†	0.9	1.1	1.1	0.8	0.8	1.2	1.2	0.8	1.1	0.8	0.7	1.8	0.4	0.7	1.2	1.2
Oxygen and errors	1.4	1.6	2.0	3.1	3.5	5.8	6.2	7.9	10.4	12.3	24.5	33.8	43.0	2.8	1.8	1.7
Caking and swelling tests																
BS swelling number	1	2	7	8	9	8	7	3½	1½	1	—	—	—	—	—	—
Gray–King coke type	A	B	G6	G9	G10	G6	G4	E	C	B	—	—	—	—	—	—
Calorific value MJ kg⁻¹ (Btu lb⁻¹)																
Air-dried coal	34.24 (14 720)	34.52 (14 840)	34.61 (14 880)	35.00 (15 050)	34.35 (14 770)	31.91 (13 720)	31.82 (13 680)	30.42 (13 080)	28.28 (12 160)	26.05 (11 200)	21.03 (9 040)	16.68 (7 170)	15.75 (6 770)	33.70 (14 500)	30.17 (12 970)	30.38 (13 060)
Dry, mineral matter-free coal	35.80 (15 400)	36.52 (15 700)	36.63 (15 750)	36.66 (15 760)	36.52 (15 700)	35.59 (15 300)	35.54 (15 280)	34.42 (14 800)	33.00 (14 190)	32.19 (13 840)	26.30 (11 300)	22.24 (9 560)	18.60 (8 010)	34.77 (14 950)	33.70 (14 490)	34.33 (14 760)

Note: Calorific value $MJ\,kg^{-1}$ ($Btu\,lb^{-1}$).

* Dry, ash-free basis for lignite, peat, wood and manufactured fuels.
† Organic sulphur for coals.

Table 28.11 NUMBER OF INCREMENTS FOR ACCURACY OF ±1%

	Moisture		
Class	*3% or less*	*3–5%*	*Over 5%*
1	32	32	48
2	48	72	108
3	100	150	225
4	16	16	16

Table 28.12 WEIGHT OF INCREMENT AND SAMPLING IMPLEMENTS

Maximum size of coke	*Sampling implement*
38 mm ($1\frac{1}{2}$ in) (i.e. not more that 5% over $1\frac{1}{2}$ in)	1.14 kg ($2\frac{1}{2}$ lb) scoop
76 mm (3 in) (i.e. not more than 5% over 3 in)	2.3 kg (5 lb) scoop
101 mm (4 in) (i.e. not more than 5% over 4 in)	4.5 kg (10 lb) scoop
Over 101 mm (4 in)	6.8 kg (15 lb) scoop

cases, the conditions under which the sampling is to be carried out. The number of increments is independent of the total weight of coke sampled.

The number of increments of the various classes of coke for ash and moisture determinations required to give an accuracy of ±1% at the 95% probability level is given in Table 28.11 and the weight of the increments in Table 28.12 according to BS 1017: 1989.

Shatter test According to BS 1016: 1994 the gross sample of 25 kg for the shatter test should be collected specifically for the test according to BS 1017: 1989 and should contain all the sizes over 51 mm (2 in) in approximately the same proportion as are found in the original size analysis.

GENERAL

The qualities of coke which have the most influence on metallurgical practice are purity, hardness and combustibility. The purity of any particular sample is determined by chemical tests, but the physical properties of hardness and combustibility may only be assessed by empirical tests.[8]

CHEMICAL ANALYSIS

Chemical analysis normally includes determinations of water, ash, volatile matter and sulphur. Phosphorus is important in the manufacture of acid pig iron.

The carbon content of a coke is an index of its thermal value and is roughly assessed by subtracting the 'impurities', as determined by chemical analysis, from 100.

OTHER TESTS

The size of coke is specified by the size of the square meshed screen through which it passes or on which it rests, the results being expressed as cumulative percentages on screens of decreasing sizes.

Bulk density of coke This is an indication of the weight of the lump material that will fill a known (large) volume. The cubical container used has a capacity of 0.1 m³ (2 ft³), 465 mm (15 in) side internally (BS 1016: 1991).

Apparent density is the ratio of the weight of a given volume of coke to the weight of an equal volume of water (BS 1016: 1991).

True density is the ratio of the weight of a given volume of dry coke passing a 0.2 mm test sieve to the weight of an equal volume of water at the same (atmospheric) temperature (BS 1016: 1991).

Porosity may be either 'apparent' or 'total'.

$$\% \text{ Apparent porosity} = \left[\frac{\text{Volume of open pores}}{\text{Volume of coke}} \right] \times 100$$

$$= \left[\frac{W_3 - W_1}{W_3 - W_2} \right] \times 100$$

where W_1 = weight of dried coke, W_2 = weight of coke saturated with water weighed in a tank of cold water, W_3 = weight of coke saturated with water.

It can also be shown that:

$$\% \text{ Total porosity} = \left[\frac{\text{Real specific gravity} - \text{Apparent specific gravity}}{\text{Real specific gravity}} \right] \times 100$$

The real specific gravity in the above expression is determined by the specific gravity bottle method on material passing a 0.2 mm sieve, care being taken, however, to boil the coke with water in order to remove air and to saturate it (BS 1016: 1991).

The apparent specific gravity may be obtained from weighings required by the apparent porosity:

$$\text{Apparent specific gravity} = \frac{W_1}{W_3 - W_2}$$

The micum indices of a coke should measure its liability to attrition in the blast furnace. It is determined as follows: 25 kg of coke over 60 mm in size and with less than 5% moisture is placed in a special drum and rotated for 100 revolutions. A size analysis of the coke is then made and the percentages of coke remaining on a 40 mm sieve (M_{40}) and passing through a 10 mm sieve (M_{10}) are normally reported (BS 1016: 1973).

The 'shatter index' is a measure of the liability of a coke to form breeze during loading, unloading and charging operations.

To determine this index 25 kg of greater than 51 mm (2 in) coke is dropped four times from a special box which is placed 1.83 m (6 ft) above a cast iron or steel plate. The shattered coke is then screened and the average of three tests of the percentages retained on 51 mm (2 in), 38 mm ($1\frac{1}{2}$ in), 25 mm (1 in) and 13 mm ($\frac{1}{2}$ in) square aperture screens are reported as respective shatter indices (BS 1016: Part 13; 1969).

The reactivity of coke determines its behaviour towards air or oxygen or the rate with which it reduces carbon dioxide to monoxide.

The 'critical air blast' (CAB), determines the reactivity of coke to air by finding, by trial and error, the minimum rate of blast which will maintain combustion in an ignited bed of 1.2–0.6 mm coke contained in a glass or quartz tube of specified dimensions (BS 1016: 1973).

The thermal value of the volatile matter remaining in the coke (volatile therms) is a measure of its ignitability and, indirectly, of its reactivity to air (BS 1016: 1973).

28.2.2　Properties of metallurgical coke

BLAST FURNACE COKE

A specification for blast furnace coke proposed by the BSC/BISRA iron-making panel of the Iron and Steel Institute (Publication P127: 1969) is shown in Table 28.13.

A size range of 20–80 mm as charged and with higher ash is more usual. Consistency of the properties of the coke supplied is of the utmost importance.

FOUNDRY COKE

The range of properties specified[9] for foundry coke from various plants in the UK is given in Table 28.15. These supersede the recommendations of TS 47 1959.[10]

Table 28.13   SPECIFICATION FOR BLAST-FURNACE COKE

Moisture content
This shall not exceed 3%; a mean of 2% is desired
Variation -2 to $+3$ on single samples
                    -0.3 to $+0.5$ on weekly average

Size
The overall size range shall be 19–64 mm ($\frac{3}{4}$–$2\frac{1}{2}$ in); this
implies pre-crushing of the coke

Shatter index
The 38 mm ($1\frac{1}{2}$ in) shatter index shall not be less than 90
Variation ± 2 on single samples
                    ± 0.3 on weekly average

Micum index
The M_{40} index shall not be less than 75
Variation ± 3 on single samples
                    ± 0.45 on weekly average

The M_{10} index shall not exceed 7
Variation ± 1 on single samples
                    ± 0.45 on weekly average

Ash content
The ash content should not exceed 3%
Variation ± 1.7 on single samples
                    ± 0.27 on weekly average

Sulphur content
The sulphur content should not exceed 0.6%
Variation ± 0.17 on single samples
                    ± 0.03 on weekly average

Table 28.14   PROPERTIES OF COKES

Real density, $kg\,m^{-3}$	1 700–2 000
Apparent density, $kg\,m^{-3}$	700–1 100
Total porosity, %	36–55
Apparent porosity, %	35–47
Calorific value, $MJ\,kg^{-1}$	33.14
Ash, % dry	8–11
Volatile matter	0.6–1.1
Sulphur	0.57–1.4
Phosphorus	0.01–0.14
Critical air blast, $1\,min^{-1}$	1.56–2.4

Table 28.15   SPECIFICATION FOR FOUNDRY COKE

Moisture, % maximum	3.0–5.5
Ash, % maximum	9
Volatile matter, % maximum	0.7–1.0
Sulphur, % maximum	0.85–1.0
Shatter index, 50 mm (2 in) minimum	90
Mean size minimum, mm	102–107
(in)	(4–4.2)
Undersize, not more than 4% less than 50 (2 in)	

    There are indications[11] that a narrow range of size of coke as charged is more important to the
operation of the cupola furnace than the mean size in the size range 40–110 mm and that there is
little to be gained by using large coke. Consistency in the properties of the coke supplied is most
important.

Table 28.16   CHARACTERISTICS OF FORMED COKE PROCESSES

Process	Forming	Feed	Binder
BBF	Hot briquetting	Any coal	30% caking coal
Consol–BNR	Hot pelletising	High-volatile coal	Caking coal
Iniex	Briquetting	Low-volatile, non-caking	Pitch
FMC	Briquetting	High-volatile coal	Pitch
Sapoznikov	Hot briquetting	Slightly caking	Caking coal
Guiprokoks	Hot briquetting	Low–medium volatile, weakly caking	High-volatile, weakly caking coal
DKS	Briquetting	Non-caking	Pitch and caking coal

Table 28.17   PROPERTIES OF FORMED COKE BRIQUETTES

	Process and coal source				
Property	BBF,* Germany	BBF,* UK	FMC, USA	DKS, Japan	GI, USSR
Analysis, dry%					
Fixed carbon	81.5	81.2	89.9	80.3	na
Ash	5.5	12.1	5.5	12.6	na
Volatile matter	9.1	6.0	3.9	6.5	1.5
Sulphur	0.9	1.0	0.7	0.5	na
Bulk density, $kg\,m^{-3}$	578	622	554	779	na
Strength					
M + 40	84	86	—	—	85–90
M + 30	—	—	95	—	—
M + 20	—	—	—	94	—
M − 20	9.3	10.9	5.1	5.6	7–8

* Non-calcined;   na—not available.

FORMED COKE

Processes developed to produce formed coke briquettes[12] from weakly coking coals are shown in Table 28.16. Properties of formed cokes used in blast furnace trials are given in Table 28.17.

BULK DENSITY OF COKE

A graded coke 20–40 mm with normal ash and moisture has a bulk density of 420–480 kg m^{-3}, and run of oven coke 460–510 kg m^{-3}.

SPECIFIC HEAT OF COKE[13]

The relationship between the specific heat of coke and the ash content is a linear one. Values for the mean specific heat between $21°$ and $t°C$ are given in Table 21.18, where A denotes % ash present in the coke.

28.3   Gaseous fuels, liquid fuels and energy requirements

28.3.1   Liquid fuels

Liquid fuels are easy to handle, store and control. The two main groups are derived from (1) petroleum and to a far lesser extent, (2) coal carbonisation. Distillate fuels contain practically no ash, and residual fuels contain very little ash in comparison with solid fuels. Sulphur in residual fuel oil depends mainly on the source of the crude oil from which it was obtained. High sulphur contents are usually undesirable metallurgically. Slagging troubles occur at over 700°C owing to the presence of

Table 28.18   MEAN SPECIFIC HEAT OF COKE

Temperature °C	Mean specific heat kJ kg⁻¹ K⁻¹
400	1.11–0.001 8 A
500	1.26–0.002 8 A
600	1.36–0.003 4 A
700	1.45–0.004 4 A
800	1.50–0.004 5 A
900	1.56–0.005 4 A
1 000	1.60–0.005 7 A
1 100	1.63–0.005 7 A
1 200	1.66–0.005 7 A
1 300	1.69–0.005 7 A

Table 28.19   THERMAL EXPANSION OF COKE

	Coefficient of linear thermal expansion 10^{-6} K⁻¹			
	Bituminous coal. Strongly coking carbonisation temp. °C		Bituminous coal. Weakly coaking carbonisation temp. °C	
Temperature of measurement °C	600	1 000	600	1 000
100	9.0	3.4	6.8	1.8
200	10.0	4.1	7.8	1.8
300	10.5	4.5	8.0	2.3
400	—	4.6	8.0	—

Na_2O, S and V_2O_5 in the fuel. The higher flame emissivity of very heavy fuel oils and coal tar fuels is an advantage in high temperature processes.[14]

British Standards specifications[15,16] for liquid fuels are intended as a guide but more details should be specified for metallurgical use of fuel. Tables 28.20 and 28.21 refer to typical properties of petroleum and tar fuels respectively.

(A) CALORIFIC VALUE

In the absence of a bomb calorimeter determination, approximate values may be calculated for petroleum oils, or tars. However, the correlations are separate for each group.

$$\text{Gross calorific value} = 51.91 - 8.79d^2 - \{0.519\,1 - 0.087\,9d^2(\%H_2O + \%ash + \%S)\}$$
$$+ 0.094\,2(\%S)\,\text{MJ kg}^{-1}$$
$$\text{or} \quad 22.320 - 378\,0d^2 - \{223 - 37.8d^2(\%H_2O + \%ash + \%S)\}$$
$$+ 40.5(\%S)\,\text{Btu lb}^{-1}$$
$$\simeq 59.91 - 8.79d^2\,\text{MJ kg}^{-1}$$
$$\text{or} \quad 22\,320 - 3\,780d^2\,\text{Btu lb}^{-1} \text{ (reference 17)}$$

$$\text{Net calorific value} = 46.5 + 3.14d - 8.84d^2\,\text{MJ kg}^{-1}$$
$$\text{or} \quad 20\,000 + 1\,350d - 3\,800d^2\,\text{Btu lb}^{-1} \text{ (reference 17)}$$

where d is the specific gravity (relative density) at 15.6°C (60°F) for petroleum oils.

The net calorific values are about 2.8 MJ kg⁻¹ (1 200 Btu lb⁻¹) less than the gross values for distillate fuels down to about 2.3 MJ kg⁻¹ (1 000 Btu lb⁻¹), less for heavy fuel oils.

Table 28.20   PETROLEUM LIQUID FUELS

	Kinematic viscosity cSt (10^{-2} $m^2 s^{-1}$)	I.B.P. °C	F.B.P. °C	specific gravity 15.6°/ 15.6°C 60°/ 60°F	C %	H %	O+N %	S %	Ash %	Calorific value MJ kg^{-1} (Btu lb^{-1}) gross
	at 37.8°C (100°F)									
Primary flash distillate	—	37	72	0.649	83.98	16.0	—	0.02	—	47.9 (20 600)
Primary flash distillate	—	32	163	0.704	84.87	15.0	—	0.03	—	47.1 (20 250)
Kerosine	1–2	160	185	0.78	85.9	14.0	—	0.08	—	46.5 (20 000)
Gas oil	3.3	190	—	0.844	85.3	13.2	—	0.30	0.001	45.4 (19 600)
Gas oil	3.3	190	—	0.833	85.0	13.05	—	0.95	0.001	45.6 (19 500)
	at 82.2°C (180°F)									
Light fuel oil	10	—	H$_2$O%	0.935	85.55	11.50	0.70	2.55	0.02	43.3 (18 600)
Light fuel oil	12.5	—	—	0.99	86.94	11.4	0.5	1.1	0.06	42.9 (18 450)
Medium fuel oil	30	—	0.05	0.967	84.12	11.50	0.8	3.5	0.03	42.6 (18 300)
Medium fuel oil	34	—	0.05	0.968	85.85	11.83	0.8	1.4	0.03	42.98 (18 480)
Heavy fuel oil	50	—	0.1	0.950	85.2	11.70	1.0	1.9	0.1	43.0 (18 500)
Heavy fuel oil	70	—	0.1	0.980	84.5	10.7	0.8	3.8	0.1	42.6 (18 300)
Heavy fuel oil	72	—	0.25	0.939	87.6	11.07	0.9	0.36	0.02	43.7 (18 780)

Table 28.21   COAL TAR FUELS

Type	CTF 50	CTF 100	CTF 200	CTF 250	CTF 300	CTF 400	Hard pitch	Crude tar
Specific gravity	1.010	1.025	1.145	1.175	1.205	1.245	1.22	1.165
Calorific value, gross								
MJ kg^{-1}	39.66	39.43	38.77	38.59	38.49	37.63	37.22	37.68
(Btu lb^{-1})	(17 050)	(16 950)	(16 670)	(16 590)	(16 550)	(16 180)	(16 000)	(16 200)
Calorific value, net								
MJ kg^{-1}	38.03	37.91	37.45	37.29	37.24	36.49	36.15	36.49
(Btu lb^{-1})	(16 350)	(16 300)	(16 100)	(16 030)	(16 010)	(15 690)	(15 540)	(15 690)
Viscosity								
cSt (10^{-2} $m^2 s^{-1}$)	13	24	300	35 000	25 000	—	—	~300
at temp °C	37.8	37.8	37.8	30	55			37.8
Analysis, %								
C	87.65	88.80	89.36	89.57	89.88	90.42	90.66	90.5
H	7.38	6.90	5.90	5.90	5.73	5.23	4.90	5.4
O (difference)	3.35	2.79	2.49	2.33	2.23	2.01	1.70	1.7
N	0.92	0.84	1.11	1.16	1.22	1.38	1.42	1.13
S	0.66	0.63	0.89	0.84	0.69	0.65	0.86	0.6
Ash	0.04	0.04	0.10	0.20	0.25	0.31	0.46	0.5

For tar fuels:

$$\text{Gross calorific value} = 0.337(\%C) + 1.44(\% H - \tfrac{1}{8}\% O) + 0.093(\%S)\,\text{MJ kg}^{-1}$$

$$\text{or} \quad 145(\%C) + 620(\%H - \tfrac{1}{8}\% O) + 40(\%S)\,\text{Btu lb}^{-1} \text{ (reference 18)}$$

$$\text{Net calorific value} = 0.75(\text{gross CV} + 1.09)\,\text{MJ kg}^{-1}$$

$$\text{or} \quad 0.75(\text{gross CV} + 4\,700)\,\text{Btu lb}^{-1} \text{ (reference 18)}$$

(B) SPECIFIC GRAVITY CORRECTION COEFFICIENTS PER 1°C

Table 28.22   CORRECTION COEFFICIENTS FOR PETROLEUM PRODUCTS[19]

Specific gravity 15.6/15.6°C (60/60°F)	*Correction coeff. per* 1°C	*Specific gravity* 15.6/15.6°C (60/60°F)	*Correction coeff. per* 1°C
0.605 0–0.613 3	0.001	0.742 5–0.753 7	0.000 79
0.613 4–0.621 9	0.000 99	0.753 8–0.764 9	0.000 77
0.622 0–0.632 0	0.000 97	0.765 0–0.776 0	0.000 76
0.632 1–0.641 9	0.000 95	0.776 1–0.786 9	0.000 74
0.642 0–0.653 0	0.000 94	0.787 0–0.798 8	0.000 72
0.653 1–0.664 9	0.000 92	0.798 9–0.812 4	0.000 70
0.665 0–0.677 5	0.000 90	0.812 5–0.828 3	0.000 68
0.677 6–0.689 9	0.000 88	0.828 4–0.859 9	0.000 67
0.690 0–0.702 5	0.000 86	0.860 0–0.925 0	0.000 65
0.702 6–0.716 6	0.000 85	0.925 1–1.024 9	0.000 63
0.716 7–0.730 0	0.000 83	1.025 0–1.074 9	0.000 61
0.730 1–0.742 4	0.000 81	1.075 0–1.124 9	0.000 59

Table 28.23   CORRECTION COEFFICIENTS FOR COAL TAR FUELS[18]

Type	*Specific gravity correction coeff. per* 1°C
CTF 50 + 100	0.000 76
CTF 200	0.000 63
CTF 250	0.000 58
CTF 300	0.000 50
CTF 400	0.000 49

(C) SPECIFIC HEAT

For petroleum oils.[17]

$$\text{Specific heat} = \frac{1.69 + 0.003\,4t}{\sqrt{d}}\,\text{kJ kg}^{-1}\,\text{K}^{-1}$$

$$+2\% \text{ for naphthenic base crudes}$$
$$-2\% \text{ for paraffin base crudes}$$

where

$$d = \text{Specific gravity } 15.6/15.6°C\ (60/60°F)$$
$$t = \text{Temperature, }°C$$

For coal tar fuels.[18]

$$\text{Specific heat} = 1.46\text{–}1.68\,\text{kJ kg}^{-1}\,\text{K}^{-1}$$
$$(0.35\text{–}0.40\,\text{Btu lb}^{-1}\,°F^{-1})$$

(D) THERMAL CONDUCTIVITY

For petroleum oils:[17]

$$K = \frac{0.1184 - 0.0000195T}{d} \text{ W m}^{-1} \text{ K}^{-1}$$

$$\text{or} \quad \frac{0.821 - 0.000244t}{d} \text{ Btu (ft}^2 \text{ h)}^{-1} (^\circ\text{F in}^{-1})^{-1}$$

where

$T = $ temperature, $^\circ$C

$t = $ temperature, $^\circ$F

For tar fuels,[18]

$$K = 0.138\text{–}0.147 \text{ W m}^{-1} \text{ K}^{-1}$$
$$\text{or } 0.96\text{–}1.02 \text{ Btu (ft}^2 \text{ h)}^{-1} (^\circ\text{F in}^{-1})^{-1}$$

(E) VISCOSITY[19]

British fuel oil kinematic viscosities are quoted in centistokes (cSt) at a specified temperature (BS 4708: 1971), and, formerly, by seconds Redwood I at 100°F (Redwood II is approximately 1/10 seconds Redwood I). Coal tar fuels are numbered according to the temperature in °F at which their viscosity is 100 seconds Redwood I, i.e. about 24 cSt at 37.8°C. Suitable handling temperatures are quoted for each class of fuel.

Table 28.24   VISCOSITY OF LIQUID FUEL

	Kinematic viscosity				Atomising
Class	*Temperature* °C (°F)	*cSt* (10^{-2} m^2 s^{-1})	*Storage* °C	*Pumping* °C	*temperature* °C
Gas oil	37.7 (100)	4	—	—	Ambient
Light fuel oil	82.2 (180)	12.5	10	10	55
Medium fuel oil	82.2 (180)	30	25	30	90
Heavy fuel oil	82.2 (180)	70	35	45	125
CTF 50	37.8 (100)	13	—	—	15
CTF 100	37.8 (100)	24	35	35	38–50
CTF 200	37.8 (100)	300	35	40	80–95
CTF 250	30	35 000	55	75	115–130
CTF 300	55	25 000	85	100	140–155
CTF 400	—	—	135	150	190–205

(F) FLASH POINT

Special precautions are required by statute[20] for liquids with flash points below 32°C (90°F). Typical values for fuels are:

	Flash point °C (°F)
Petrol	−40 (−40)
Kerosine	+43 (+110)
All coal tar fuels	over 65 (150)
Gas oil	77 (170)
Light fuel oil	82 (180)
Medium fuel oil	93 (200)
Heavy fuel oil	115 (240)

Table 28.25   PROPERTIES OF CONSTITUENTS OF GASEOUS FUELS

Gas	Formula	Specific gravity (air = 1)	Gas density (calc.) kg m⁻³ NTP dry	lb ft⁻³ dry	STP sat.	Calorific value MJ m⁻³ (st) dry	Calorific value Btu ft⁻³ STP sat. gross	STP sat. net
Oxygen	O_2	1.1044	1.428	0.08457	0.08393	—	—	—
Atmospheric nitrogen	N_2	0.9723	1.257	0.07446	0.07400	—	—	—
Air	—	1.0000	1.293	0.07657	0.07607	—	—	—
Carbon dioxide	CO_2	1.5185	1.963	0.11628	0.11509	—	—	—
Carbon monoxide	CO	0.9663	1.250	0.07400	0.07354	12.04	318	318
Hydrogen	H_2	0.06958	0.090	0.00533	0.00607	12.12	320	270
Methane	CH_4	0.5533	0.715	0.04237	0.04247	37.68	995	895
Ethane	C_2H_6	1.0371	1.341	0.07941	0.07887	65.52	1730	1580
Propane	C_3H_8	1.5210	1.966	0.11646	0.11526	93.87	2479	2282
Butane	C_4H_{10}	1.9358	2.503	0.14823	0.14648	117.23	3095	2848
Acetylene	C_2H_2	0.8980	1.161	0.06876	0.06840	55.00	1452	1402
Ethylene	C_2H_4	0.9675	1.251	0.07409	0.07363	59.08	1560	1460
Propylene	C_3H_6	1.4512	1.876	0.11113	0.11003	87.11	2300	2150
Butylene	C_4H_8	1.9350	2.501	0.14817	0.14643	115.13	3040	2840
Benzene	C_6H_6	2.6938	3.483	0.20626	0.20351	141.64	3740	3590
Water	H_2O	0.6218	0.804	—	0.04761	—	—	—
Hydrogen sulphide	H_2S	1.1762	1.521	0.09007	0.08933	23.86	630	580
Sulphur dioxide	SO_2	2.2115	2.860	0.16933	0.16721	—	—	—

Table 28.26   BLAST FURNACE GAS ANALYSES (MODERN FURNACES WITH OIL INJECTION)

	% Analysis by volume			Calorific value STP sat.	
CO	CO_2	H_2	N_2	MJ m⁻³	Btu ft⁻³
20–23	20–22	3–5	52–55	2.9–3.2	79–86

28.3.2   Gaseous fuels

Table 28.25 shows the properties of most fuel gas constituents.[21]

(A) BLAST FURNACE GAS

This is the byproduct of iron or ferromanganese production, obtained from the top of the furnace after cooling and suitable dust removal. Injection processes for steam, oil, coal, gas and oxygen tend to vary the typical analyses. Below 3.7 MJ m⁻³ (100 Btu ft⁻³) calorific value the gas should be enriched[22] or pre-heated before combustion.

The specific gravity relative to air varies from 1 to 1.07.

(B) LIQUEFIED PETROLEUM GAS, LPG

This type of industrial gas is a byproduct of the petroleum industry. These gases are transported in liquid form and vaporised before combustion. They may be used directly, mixed with other gases or distributed as a mixture with air.[23] The liquids have large coefficients of thermal expansion.

(C) NATURAL GAS

Natural gas, supplied is from many gas fields in the world. It is a rich hydrocarbon gas, predominantly methane, but they contain higher hydrocarbon and nitrogen. Large users have been supplied on an interruptible basis often necessitating the use of dual-fuel burners.[27] Its limits of inflammability are narrow (*see* Table 28.30) and flame speed low when compared with hydrogen containing fuels.

Table 28.27   TYPICAL RICH GASEOUS FUELS

	Natural gas	LPG[24,25] Propane	LPG[24,25] Butane	Refinery gas[26] Low	Refinery gas[26] Medium	Refinery gas[26] High
		% Analysis by volume				
H_2	—	—	—	56	22	0.5
CH_4	86–90	—	—	12	22	22
C_2H_6	2.9–5.3	2	0.5	13	22	25
C_3H_8	0.5–1.3	87	10	13	18	36
C_4H_{10}	0.2–0.3	6	85	2	3	3
C_2H_4	—	2	0.5	1	3	2
C_3H_6	—	2	3	1	7	9
C_4H_8	—	1	1	1	2	2
N_2	1.2–6.8	—	—	1	1	0.5
Specific gravity (air = 1.0)	0.585–0.631	1.528	1.872	0.531	0.876	1.180
Calorific value,						
Gross						
$\quad$MJ m^{-3} (st)	37.9–39.2	92.7	111.7	36.5	55.1	72.3
$\quad$Btu ft^{-3}	1 016–1 051	2 486	2 995	977	1 478	1 938
$\quad$MJ kg^{-1}	—	49.4	48.6	55.8	51.3	49.9
Net						
$\quad$MJ m^{-3}	—	85.4	102.9	32.5	50.4	66.4
$\quad$Btu ft^{-3}	—	2 289	2 758	872	1 352	1 780
$\quad$MJ kg^{-1}	—	45.5	44.8	49.8	46.9	45.8
Liquid specific gravity	—	0.51	0.58	—	—	—
Latent heat of vaporisation kJ kg^{-1}	—	426	391	—	—	—
Vapour pressure at 37.8°C bar (a)	—	16.1	5.86	—	—	—
Liquid coefficient of cubical expansion per °C		0.0016	0.0011			

(D) PRODUCER GAS

Producer gas[28] is formed by partial oxidation of a solid fuel bed. The process is modified by the introduction of steam (blue water gas) and by thermal cracking of a hydrocarbon fuel such as natural gas, propane, butane, petroleum distillate, gas oil or heavy fuel oil (carburetted water gas).

Normally oxidation of the fuel bed is carried out with air, but total gasification processes also use oxygen. Tar in hot raw producer gas may increase the calorific value to around 74 MJ m^{-3} (200 Btu ft^{-3}).

(E) REFINERY GAS

Refinery gas may be a byproduct of distillation, cracking or reforming of gas in the petroleum industry. Typical gases are shown in Table 28.27.

(F) TOWN GAS

Town gas is used in only a few location in the world and is made almost exclusively by coal carbonisation. It can be made of a complex mixture of gases,[29] including coke oven gas, natural gas, liquefied petroleum gas, water gas, refinery gas, reformed hydrocarbon fuels or hydrocarbon fuels diluted with air. To make gas safer the $CO + H_2O = CO_2 + H_2$ shift reaction may be used to reduce the quantity of CO present.[30] CO_2 may be removed to increase calorific value and to reduce the specific gravity.

Gases are grouped according to the Wobbe index.

$$WI = \frac{\text{Gross calorific value MJ m}^{-3} \text{ (st) dry}}{\sqrt{[\text{Specific gravity (air} = 1.0)]}} \quad \text{or}$$

$$\frac{\text{Gross calorific value Btu ft}^{-3} \text{STP (ISC) sat.}}{\sqrt{[\text{Specific gravity (air} = 1.0)]}}$$

	Wobbe index	
Gas group	Btu ft^{-3} STP *sat.*	MJ m^{-3} (st) *dry*
G3	800 ± 40	30.4 ± 1.52
G4	730 ± 30	27.7 ± 1.14
G5	670 ± 30	25.4 ± 1.14
G6	615 ± 25	23.3 ± 0.95
G7	560 ± 30	21.3 ± 1.14
Natural gas	1 335 ± 5%	50.68 ± 5%

Table 28.28   ANALYSES OF PRODUCER GAS

	Gas analysis (% volume)					*Specific gravity (air = 1.00)*	kJ m^{-3} *sat. at* 15.6°C (Btu/SCE *sat.*)	
Representative of type	CO	CO$_2$	H$_2$	CH$_4$	N$_2$		*Gross*	*Net*
Producer gas Coke, no steam	31.3	2.1	0.5	—	66.1	0.978	3 770 (101.2)	3 763 (101.0)
Coke, steam	29.3	5.6	12.4	—	52.7	0.889	4 952 (132.9)	4 720 (126.7)
Anthracite	24.0	7.5	16.5	1.2	50.8	0.858	5 253 (141.0)	4 900 (131.5)
Bituminous coal	27.0	4.5	14.0	3.0	51.5	0.857	5 980 (160.5)	5 707 (150.5)
Blue water gas Coke, steam	43.5	3.5	47.3	0.7	5.1	0.560	11 050 (296.7)	10 150 (272.4)
Fixed bed slagging gasifier	65.5	4.0	29.0	0.4	1.0	0.727	11 380 (305.5)	10 830 (290.6)

Table 28.29 shows the typical properties of some manufactured gaseous fuels.

For a given piece of equipment the gases within a group as defined by the Wobbe index are interchangeable[31] without burner and nozzle adjustment. However, when changing gaseous fuels, blow-off and flame speed and flash-back must be carefully considered.

(G) ANALYSIS OF FUEL GASES[32]

The following constituents may also be found in fuel gases.

Dust is found in insufficiently cleaned blast furnace and producer gas.

Hydrogen sulphide should be below 5 ppm in natural gas.

7–20 g m^{-3} may be present in unpurified coal gas.

45 g m^{-3} may be present in refinery gas.

Organic sulphur Up to 1 g m^{-3} (40 gr per 100 ft^3) may be present in town gas. This figure will be lowered by oil washing, benzole stripping, etc. Below 0.02% by weight sulphur is usually present in LPG. Natural gas contains little organic sulphur, 0.3–9.0 ppm v/v and a limit of about 30 mg m^{-3} will be set.

Condensable vapours are chiefly benzene and toluene in carbonisation gas, the quantity depending chiefly on temperature of carbonisation, volatile matter in coal, and degree of benzole removal.

Tar may be found in producer gas, made from coal.

Water vapour   The majority of fuel gases contain water vapour, hence the usual determinations used to assume that the gases were saturated with water vapour at 30 in Hg total pressure (1013.75 mbar) and at 60°F (15.6°C). Natural gas properties are quoted on a dry basis (st).

(H) LIMITS OF INFLAMMABILITY

The limits of inflammability[33] of gaseous fuels and vapours are shown approximately in Table 28.30.

These limits are affected by direction of propagation, temperature, pressure and inert diluents.

Table 28.29 PROPERTIES OF SOME MANUFACTURED GASEOUS FUELS

	Horizontal retort	inter-mittent vertical chamber	Coke oven	Con-tinuous vertical retort	Low temp.	Car-buretted water gas	Butane air	CWG, CO_2,CO conv. butane enriched
				% Analysis by volume				
CO	7.20	13.04	7.72	14.06	3.68	31.29	—	4.0
CO_2	1.72	3.01	2.39	3.49	4.89	4.59	—	21.0
H_2	56.07	50.24	48.02	55.99	16.58	36.48	—	49.7
CH_4	26.52	19.11	25.94	17.81	33.52	8.04	—	7.0
N_2	1.88	7.98	9.15	4.35	25.80	8.41	57.67	7.0
O_2	0.14	1.14	0.68	0.18	2.80	0.72	15.33	—
$C_nH_m^{st}$	4.0 ($C_{2.5}H_5$)	3.10 ($C_{2.5}H_5$)	4.34 ($C_{2.5}H_5$)	2.45 ($C_{2.5}H_5$)	3.04 (C_4H_8)	8.75 ($C_{2.5}H_5$)	—	8.6 ($C_{2.5}H_5$)
C_nH_{2n+2}	2.47 (C_2H_6)	2.38 (C_2H_6)	1.76 (C_2H_6)	1.67 ($C_{2.5}H_7$)	9.69 ($C_{2.5}H_7$)	1.72 ($C_2.H_6$)	27.0 (C_4H_{10})	2.7 (C_4H_{10})
Calorific value, Gross								
$\quad$ MJ m^{-3}	21.8	18.4	20.5	18.0	25.9	18.4	31.1	18.2
$\quad$ Btu ft^{-3}	586	494	550	483	695	495	836	488
Net								
$\quad$ MJ m^{-3}	19.5	16.5	18.4	6.1	23.5	16.9	28.6	16.4
$\quad$ Btu ft^{-3}	523	442	493	432	630	455	769	439
Specific gravity	0.375	0.465	0.455	0.422	0.771	0.655	1.252	0.655
Wobbe index								
$\quad$ MJ m^{-3}	35.6	27	30.4	27.7	29.5	27.3	27.8	22.5
$\quad$ Btu ft^{-3}	956	724	816	743	792	734	747	603

* C_nH_m refers to the average analysis of the mixture of remaining hydrocarbons.

Table 28.30 LIMITS OF INFLAMMABILITY

	% By volume in air	
	Lower limit	*Upper limit*
Acetylene	2.5	80
Blast furnace gas	35	74
Butane	1.9	8.5
Carbon monoxide	12.5	74
Carburetted water gas	5.5	36
Coal gas	5.3	32
Coke oven gas	4.4	34
Ethane	3.0	12.5
Hydrogen	4.0	75
Methane	5.3	15
Petrol	1.4	7.6
Producer gas	17	70
Propane	2.2	9.5
Water gas	7.0	72
Commercial butane	1.9	8.5
Commercial propane	2.4	9.5

(I) OXYGEN AND PRE-HEAT

Regeneration to about 1 100°C and recuperation to about 800°C as a mean of pre-heating combustion air are valuable methods of waste heat recovery in large furnaces. The recuperative burner can be employed on smaller furnaces.[34] Lean fuels such as blast furnace and producer gas can be pre-heated to about 500°C without cracking, to raise the adiabatic flame temperature of the fuel.

Oxygen is available from 200 to 1 700 tonnes per day units (700 m^3 t^{-1} STP). In bulk, oxygen must be handled in oil- and grease-free equipment. Pipelines are required to be of stainless or special carbon steel, the oxygen velocity should not exceed 8 m s^{-1}. Non-ferrous valves, flow controllers

Table 28.31   FLAME TEMPERATURES OF SOME GASES

	Grass calorific value of fuel		Flame temperature °C	
	MJ m^{-3}(st) dry	Btu per SCF sat.	In air maximum	In oxygen including dissociation
Acetylene	55.1	1 452	2 325	3 200
Butane	117.5	3 095	1 895	—
Butylene	115.4	3 040	1 930	—
Carbon monoxide	12.07	318	1 950	—
Ethane	65.7	1 730	1 895	—
Ethylene	59.2	1 560	1 975	—
Hydrogen	12.15	320	2 045	2 200
Methane	37.8	995	1 880	—
Propane	94.1	2 479	1 925	2 500
Propylene	87.3	2 300	1 935	—
Coal gas	21.1	560	2 045	2 100
Coal gas	18.0	475	2 045	—
Blue water gas	11.2	295	2 080	—
Producer gas	6.3	165	1 800	—
Producer gas	4.8	128	1 690	—
Blast furnace gas	3.5	92	1 460	—

and water-cooled oxygen-free copper lance tips are desirable for controlling the flow of oxygen in refining processes.

Tonnage oxygen for steelmaking is normally supplied with a purity of 99.5% O_2 for the following purposes:

(a) Oxidation refining processes.
(b) Raising flame temperature.
(c) To obtain extra output from a given furnace system by burning more fuel per unit combustion space.[35,36,37]
(d) Reducing metallic losses, e.g. electric steelmaking.
(e) Reducing nitrogen pick-up by metal from combustion air.

Tonnage oxygen for copper smelting, carried out outside the UK, is normally supplied with a purity of 95% O_2.

Use of oxygen lowers the volume of waste gases and hence the heat loss from a furnace as compared with atmospheric air for combustion. Oxygen enrichment of combustion air has a similar effect to increasing its pre-heat temperature.

The adiabatic flame temperature is a guide to the maximum furnace temperature obtainable from a fuel. Its value is lowered by the following conditions:

(a) Dilution with inert gases such as N_2, CO_2.
(b) Dissociation of H_2O and CO_2, particularly over 2000°C.
(c) Heat loss by radiation.

28.3.3   Energy use data for various metallurgical processes (data shown is for operations in the USA)

All data is from Stubbles.[38]

Table 28.32   ENERGY USAGE IN BLAST FURNACE HOT METAL PRODUCTION

Good practice, 1997–8.

	Energy requirement	
Input	(MBtu/*net ton of hot metal*)	(GJ/*net tonne of hot metal*)
Sintering	0.53	0.62
Pelletising	2.47	2.87
Coke plant	1.66	1.93
Coke	8.45	9.83
Tuyère injectants	3.75	4.36
Blast furnace	3.22	3.75
Subtotal	20.08	23.36
Top gas credit	−4.60	−5.35
Net total	15.48	18.01

Table 28.33   ENERGY USAGE IN BASIC OXYGEN STEELMAKING

Good Practice, 1997–8.

Input	Energy requirement	
	(MBtu/*net ton of cast product*)	(GJ/*net tonne of cast product*)
Hot metal	12.90	15.00
Oxygen	0.32	0.37
Lime	0.25	0.29
Electricity	0.26	0.30
Caster	0.15	0.17
All other	0.34	0.40
Total	14.22	16.54
Total excluding hot metal	1.32	1.54

Table 28.34   ENERGY USAGE IN ELECTRIC ARC FURNACE (EAF) STEEL PRODUCTION

Good Practice, 1998.

Input	Energy requirement	
	(MBtu/*ton of steel tapped*)	(GJ/*tonne of steel tapped*)
Electricity	5.25	6.11
Carbon	0.63	0.73
Oxygen	0.21	0.24
Natural gas	0.30	0.35
Lime	0.31	0.36
Total	6.70	7.79

Table 28.35   ENERGY USAGE IN INTEGRATED MILL OPERATIONS

Good Practice, 1998.

Process step	Energy requirement for process step		Average utilisation of product from process step (mass of product utilised/mass of steel shipped)	Contribution to total energy requirement	
	(MBtu/*ton of product of process step*)	(GJ/*tonne of product of process step*)		(MBtu/*ton of steel shipped*)	(GJ/*tonne of steel shipped*)
Ironmaking	15.48	18.01	0.96	14.86	17.29
Steelmaking/casting	1.32	1.54	1.09	1.44	1.67
Hot rolling	2.20	2.56	0.15	0.33	0.38
Cold rolling	4.20	4.89	0.25	1.05	1.22
Plate/structural	3.00	3.49	0.15	0.45	0.52
Hot dip galvanised	5.50	6.40	0.24	1.32	1.54
Electrogalvanised	7.00	8.14	0.06	0.42	0.49
Tinplate	7.80	9.07	0.07	0.55	0.64
Other	3.00	3.49	0.08	0.24	0.28
Total	49.50	57.58	—	20.66	24.03

REFERENCES

1. 'International Classification of Hard Coals by Type', Economic Commission for Europe, E/ECE/247, Geneva, 1956.
2. 'The Coal Classification System Used by the National Coal Board (Revision of 1964)'. The National Coal Board, London, 1964.
3. D. W. van Krevelen, 'Coal 3rd Edn', Elsevier, Amsterdam, 1993.
4. A. Williams, M. Pourkashanian J. M. Jones and N. Skorupska, 'Combustion and Gasification of Coal'. Taylor & Francis, New York, 2000
5. 'Classification of Coals by Rank', Amer. Soc. for Testing Materials ANSI/ASTM D388–77, 1977.
6. S. Coles, *J. Soc. Chem. Ind.*, 1923, **42**, 4351.
7. W. Eisermann *et al.*, 'Estimating Thermodynamic Properties of Coal, Char, Tar and Ash'. *Fuel Process. Tech.*, 1980, **3**, 39.
8. R. A. Mott and R. V. Wheeler, 'The Quality of Coke', London, 1939.
9. H. J. Leyshon, 'The Requirements of Coke for Iron Foundries'. Coke Oven Managers Year Book, 1975, p. 216.
10. Report of TS 47., *Brit. Foundryman*, 1959, **52**, 136.
11. J. Gibson, 'Recent Research and Development Work on Foundry Coke', *Brit. Foundryman*, 1973, **66**, 203.
12. E. W. Voice and J. M. Ridgion, 'Changes in Ironmaking Technology in relation to the Availability of Coking Coals'. *Iron and Steel Making*, 1974, **1**, 2.
13. E. Terres and A. Schaller, *Gas Wass.*, 1922, **65**, 761.
14. A. Williams 'Combustion of Liquid Fuel Sprays', Butterworths, London, 1990.
15. Oil Fuels, BS 2869: 1970.
16. Coal Tar Liquid Fuels, BS 1469: 1962.
17. C. S. Cragoe, US Bureau of Standards Misc. Publication No 97, 1929.
18. W. H. Huxtable, 'Coal Tar Fuels' 2nd Edn, Assoc. of Tar Distillers, London, 1961.
19. 'I.P. Petroleum Measurement Tables', *Am. Soc. Test. Mat.*, Philadelphia, PA.
20. 'The Highly Flammable Liquids and Liquefied Petroleum Gases Regulations.' 1972, Statutory Instrument No. 917, HMSO.
21. SI Units and conversion factors for use in the British Gas Industry, Gas Council and SBGI, 1971.
22. H. B. Lloyd, C. G. Miles and F. H. Dawes, 'The Use of Naphtha for Blast Furnace Gas Enrichment. *J. Iron Steel Inst.*, 1966. **204**, 203.
23. I. Carter. 'LPG Air Installations', *J. Inst. Fuel.* 1968, **41**, 366.
24. Specifications for Commercial Propane and Commercial Butane, BS 4250: 1997.
25. A. F. Williams and W. L. Lom, 'Liquefied Petroleum Gases: Guide to Props. Applications and Usage'. Ellis Horwood, London, 1974.
26. J. Burns, 'The Romford Reforming Plant', *Trans. Instn. Gas Engrs.*, 1959, **108**, 1260.
27. 'Interim Code of Practice for Large Gas and Dual Fuel Burners, The Gas Council Report No 764/70, 1971.
28. W. R. Bulcraig, 'Components of Raw Producer Gas', *J. Inst. Fuel*, 1961, **34**, 280.
29. A. L. Roberts, J. H. Towler and B. H. Holland, 'The Hydrocarbon Content of Fuel Gases'. *Trans. Instn. Gas Engrs.*, 1957, **106**, 378.
30. W. B. S. Newling and J. D. F. Marsh, 'The Partial Removal of Oxides of Carbon from Fuel Gases', *Trans. Instn. Gas Engrs.*, 1963, **3**, 143.
31. J. A. Prigg and D. E. Rooke, 'Utilisation Problems and Their Relation to New Methods of Gas Production', *Trans. Instn. Gas Engrs.*, 1963, **3**, 85.
32. Analysis of Fuel Gases, BS 3156: 1959.
33. H. F. Coward and G. W. Jones, 'Limits of Flammability of Gases and Vapours', Bur. of Mines Bulletin No 503, 1952.
34. E. F. Winter, 'The Optimum Industrial Utilisation of Gaseous Fuels', *J. Inst. Fuel*, 1978, **51**, 46.
35. T. C. Churcher, 'The Use of Oxygen in Combustion Processes', *J. Inst. Fuel*, 1960, **33**, 73.
36. H. G. Lunn and G. Waterhouse, 'Fuel Oil Injection into Blast Furnaces', *J. Inst. Fuel*, 1976, **49**, 70.
37. S. L. Rowland, 'Recent Experience in the Use of Oxygen in Cupola Operation', *Brit. Foundryman*, 1974, **67**, 187.
38. J. Stubbles, 'Energy Use in the U.S. Steel Industry: An Historical Perspective and Future Opportunities', U.S. Department of Energy, Office of Industrial Technologies, Washington, DC, 2000.

29   Heat treatment

29.1   General introduction and cross references

This chapter is composed of two sections; the first is concerned with the heat treatment of steels and the second with age hardenable aluminium alloys. All compositions are given as wt. % unless specified otherwise. Related information may be found at the following locations:

- Information on alloy specifications and designations—Chapter 1.
- Crystallographic data on some of the phases discussed here, information on relevant metallo-graphic techniques and phase diagrams—Chapters 6, 10 and 11 respectively.
- Diffusion data—Chapter 13.
- Data relevant to temperature measurement by thermocouple and pyrometer techniques—Chapters 17 and 18, respectively.
- Mechanical property data—Chapter 22.
- Furnace design and vacuum systems—Chapter 40.

29.2   Heat treatment of steel

29.2.1   Introduction

Heat treating is defined by the IFHTSE (International Federation for Heat Treating and Surface Engineering) as: 'a process in which the entire object, or a portion thereof, is intentionally submitted to thermal cycles and, if required, to chemical and additional physical actions, in order to achieve desired (change in the) structures and properties'.[1] Krauss has added the additional caveat that 'heat treatment for the sole purpose of hot-working is excluded from the meaning of this definition'.[2] The thermal cycles referred to in this definition are the various heat treatment steps which include: stress relieving, austenitising, normalising, annealing, quenching, and tempering. Steel is heat treated to: control the microstructure, increase the strength and toughness, release residual stresses and prevent cracking, control hardness (and softening), improve machinability and to improve mechanical properties including: yield and tensile strength, corrosion resistance and creep performance. Each step of the heat treatment process is performed for a particular purpose. Taken together, these heat treatment steps are like 'links in a chain'.[3] The acceptability of the final properties are limited by the weakest link.

In this section, an overview of the metallurgy involved in heat treatment will be provided which includes a discussion of common microstructures and phase diagram interpretation. The use of ITD (isothermal transformation diagrams), also known as TTT (time-temperature-transformation) curves, and CCD (continuous cooling diagrams), also known as CCT (cooling time-transformation) curves, to predict microstructure formation during heat treatment will be discussed. This will be followed by an overview of the different heat treatment steps including: austenitising, stress relief, normalising, annealing, quenching and tempering. Quantitative equations for estimating appropriate heat treatment temperatures and times will be given, where possible.

29.2.2   Transformations in steels

Properties such as hardness, strength, ductility and toughness are dependent on microstructure and grain size. The first step in the heat treating process is to heat the steel to its austenitising temperature.

The steel is then cooled rapidly to avoid the formation of ferrite and maximise the formation of martensite, which is a relatively hard transformation product, to achieve the desired as-quenched hardness.

The most common transformation products that may be formed from austenite in quench-hardenable steels are in order of formation with decreasing cooling rate: martensite, bainite, pearlite, and pearlite/proeutectoid ferrite. Each of these microstructures provides a unique combination of properties. The standard description of austenite and its transformation products used in ASTM E7 [4] and other sources will be used here (the transformation products are listed in order of their formation with increasing cooling velocity, temperatures and compositions are from Chapter 11):[5]

1. *Austenite*—A solid solution of carbon or other elements in face centred cubic gamma iron (γ-Fe). It is the desired solid solution microstructure produced prior to hardening. Gamma iron is the solid nonmagnetic phase of pure iron which is stable at temperatures between 912–1 394°C.[4]

2. *Ferrite*—This is a designation commonly applied to body centred cubic alpha iron (α-Fe) containing alloying elements in solid solution. Alpha iron is the solid phase of pure iron which is stable at temperatures below 912°C. It is ferromagnetic below 768°C. Increasing carbon content decreases markedly the high temperature limit of this phase at equilibrium. Fully ferritic steels are only obtained when the carbon content is very low. However, ferrite may frequently be found to nucleate on austenite grain boundaries, leaving a layer of ferrite at the prior location of these boundaries.

3a. *Pearlite*—A metastable microstructure formed, when local austenite areas undergo the eutectoid reaction, in alloys of iron and carbon containing greater than 0.022 percent but less than 6.67 percent carbon. The structure is an aggregate consisting of alternate lamellae of ferrite and cementite, formed on slow cooling, during the eutectoid reaction. In an alloy of given composition, pearlite may be formed isothermally at temperatures below the eutectoid temperature by quenching austenite to a desired temperature (generally above 550°C) and holding for a period of time necessary for transformation to occur. The interlamellar spacing is directly proportional to the transformation temperature, that is, the higher the temperature, the greater the spacing.[4]

3b. *Cementite*—A very hard and brittle compound of iron and carbon corresponding to the empirical formula of Fe_3C. It is commonly known as iron carbide and possesses a primitive orthorhombic lattice. In 'plain carbon steels', some of the iron atoms in the cementite lattice are replaced by manganese (manganese is added to plain carbon and other steels to 'tie-up' deleterious sulphur in the compound MnS). In 'alloy steels' the iron in cementite may be partially substituted by other elements such as chromium and tungsten. Cementite will often appear in hypo-eutectoid steels as distinct lamellae (as a constituent of pearlite) or as spheroids or globules of varying size (depending on the heat-treatment employed). Cementite is in metastable equilibrium and has a tendency to decompose into iron and graphite, although the reaction rate is very slow and is not normally observed in steels.[4] The highest cementite contents are observed in white cast irons.[5]

3c. *Ledeburite*—An intimate mixture of austenite and cementite in metastable equilibrium, formed on relatively rapid cooling during the eutectic reaction in alloys of iron and carbon containing greater than 2.14 percent but less than 6.67 percent carbon. Further slow cooling causes decomposition into ferrite and cementite (in the form of pearlite) as a result of the eutectoid reaction.[4] (An eutectic reaction, or equilibrium, is defined as a reversible univariant transformation in which a liquid that is stable only at a superior temperature, decomposes into two or more conjugate solid phases, for example: $L = \beta + \gamma$. In the case of a eutectoid reaction, the high-temperature phase is a solid, for example: $\alpha = \beta + \gamma$).

4. *Bainite*—A metastable microstructure resulting from the transformation of austenite at temperatures between those which produce pearlite and martensite. The formation of bainite can occur on continuous (slow) cooling if the transformation rate of austenite to pearlite is much slower than that of austenite to bainite. Ordinarily, bainite may be formed isothermally by quenching austenite to the desired temperature (below the range at which pearlite forms, but above that at which martensitic transformation is possible) and holding for a specific period of time necessary for transformation to occur. If the transformation temperature is just below that at which the finest pearlite is formed, typically 350°C (660°F), the bainite (upper bainite) has a feathery appearance. If the temperature is just above that at which martensite is produced, the bainite (lower bainite) is acicular, slightly resembling tempered martensite. At higher resolutions using a transmission electron microscope, upper bainite is observed to consist of plates of cementite in a matrix of ferrite. These discontinuous plates tend to have parallel orientation in the direction of the longer dimension of the bainite areas. Lower bainite consists of ferrite needles containing carbide platelets in parallel arrays cross-striating each needle axis at an

Table 29.1   FEATURES OF THE IRON–CARBON SYSTEM

Reprinted from reference 6, p. 4—courtesy of Marcel Dekker, Inc.

Phase or mixture of phases	Name
Solid solution of carbon in α-iron	Ferrite
Solid solution of carbon in γ-iron	Austenite
Iron carbide (Fe_3C)	Cementite
Eutectic mixture of carbon solid solution in γ-iron with iron carbide	Ledeburite
Eutectoid mixture of carbon solid solution in α-iron with iron carbide	Pearlite

angle of about 60°. Intermediate bainite resembles upper bainite; however, the carbides are smaller and more randomly oriented.[4]

5. *Martensite*—A metastable phase resulting from the diffusionless athermal decomposition of austenite below a certain temperature known as the M_s temperature (martensite start temperature). This is a mechanical shear transformation in which the parent and product phases have a specific crystallographic relationship. Martensitic transformation occurs during quenching from the austenitic condition, when the cooling rate of a steel is sufficiently high, such that the pearlite and bainite transformations are suppressed. Since the transformation is diffusionless, the composition of the martensite is identical with that of the austenite from which it is formed. Therefore, martensite is a supersaturated solid solution of carbon in distorted alpha iron (ferrite) having a body centred tetragonal lattice. At low austenite carbon contents, martensite forms by dislocation motion and the result is laths with a high dislocation density, dovetailed into packets (only the latter are visible in the light microscope). At high austenite carbon contents, martensite forms by twinning and the result is midrib twinned plates. Low carbon martensite packets may appear needle-like or vermiform in cross-section.[4] The extent of transformation depends on the martensitic temperature range (M_s–M_f) of the alloy concerned, since there is a distinct temperature where martensitic transformation begins (M_s) and ends (M_f).[5]

Table 29.1 provides a summary of the features of the iron–carbon system.[6]

IRON–CARBON (Fe–C) PHASE DIAGRAM

The fundamental elements of heat treatment design are derived from the Fe–C equilibrium phase diagram (as modified by the alloying additions for the steel of interest), since the science of heat treatment is dependent on the formation of the desired phases and microstructures from the transformation of austenite. For this discussion, the iron–carbon (Fe–C) phase diagram shown in Chapter 11 will be considered. Although commonly used, the term 'iron–carbon phase diagram' is strictly incorrect when applied to a diagram that contains cementite on the extreme right of the diagram. When cementite is involved, this diagram should be more properly designated as the iron–cementite (Fe–Cm) metastable equilibrium diagram. This distinction arises from the fact that cementite is not truly a stable equilibrium phase and degrades to iron and graphite over very long periods of time, typically much longer times than those encountered in commercial practice. The dashed lines in the Fe–C phase diagram show the metastable equilibrium between Fe_3C and different phases of iron. Solid lines show the equilibrium between iron and graphite, but graphitisation rarely occurs in steel. Thus, all references in the present section to the Fe–C diagram refer to the metastable Fe–Cm system. Thus, in the present section, the extent of the different phase fields as a function of temperature and composition are those governed by the metastable equilibrium between Fe and Fe_3C and not the stable equilibrium between Fe and graphite.

Note, both martensite and cementite are 'metastable' phases in the sense that they both have only the appearance of equilibrium after suitable cooling to room temperature and are capable of transformation into the equilibrium phases with a suitable thermal exposure.[5] However, the thermal exposure needed to temper martensite is of a much shorter duration than that required to transform cementite to iron plus graphite.

When referring to the Fe–C system, as with any phase diagram, it is important to properly characterise constituents, phases and microstructures. A constituent is defined as '. . . a phase or combination of phases that occurs in a characteristic configuration in an alloy microstructure'.[4] A phase is defined as 'a physically homogenous and distinct portion of a material system'.[4] A microstructure is defined as 'the structure of a material revealed by a microscope at magnifications greater than

Table 29.2   RECOMMENDED MAXIMUM HOT WORKING TEMPERATURES FOR VARIOUS STEELS

Property of the Timken Company, used with permission.

SAE No.	Temperature °C	Temperature °F	SAE No.	Temperature °C	Temperature °F
1 008	1 232	2 250	4 820	1 232	2 250
1 010	1 232	2 250			
1 015	1 232	2 250	5 060	1 177	2 150
1 040	1 204	2 200			
			5 120	1 232	2 250
1 118	1 232	2 250	5 140	1 204	2 200
1 141	1 204	2 200	5 160	1 177	2 150
1 350	1 204	2 200	51 100	1 121	2 050
			52 100	1 121	2 050
2 317	1 232	2 250			
2 340	1 204	2 200	6 120	1 232	2 250
			6 135	1 232	2 250
2 512	1 232	2 250	6 150	1 204	2 200
3 115	1 232	2 250	8 617	1 232	2 250
3 135	1 204	2 200	8 620	1 232	2 250
3 140	1 204	2 200	8 630	1 204	2 200
			8 640	1 204	2 200
3 240	1 204	2 200	8 650	1 204	2 200
3 310	1 232	2 250	8 720	1 232	2 250
3 316	1 232	2 250	8 735	1 204	2 200
3 335	1 232	2 250	8 740	1 204	2 200
			9 310	1 232	2 250
4 017	1 260	2 300			
4 032	1 204	2 200	302	1 204	2 200
4 047	1 204	2 200	303	1 204	2 200
4 063	1 177	2 150	304	1 204	2 200
			309	1 177	2 150
4 130	1 204	2 200	310	1 121	2 050
4 132	1 204	2 200	316	1 177	2 150
4 135	1 204	2 200	317	1 177	2 150
4 140	1 204	2 200	321	1 177	2 150
4 142	1 204	2 200	347	1 177	2 150
4 320	1 204	2 200			
4 337	1 204	2 200	410	1 204	2 200
4 340	1 204	2 200	416	1 204	2 200
			420	1 204	2 200
4 422	1 232	2 250	430	1 149	2 100
4 427	1 232	2 250	440A	1 149	2 100
4 520	1 232	2 250	440C	1 121	2 050
			443	1 149	2 100
4 615	1 260	2 300	446	1 038	1 900
4 620	1 260	2 300			
4 640	1 204	2 200	C-Mo	1 260	2 300
			DM	1 260	2 300
4 718	1 232	2 250	DM-2	1 260	2 300

25x'.[5] The phases shown in the Fe–C diagram include: molten alloy, austenite, ferrite, cementite and graphite. Pearlite and bainite are constituents and important microstructures but they are not phases!

The Fe–C diagram shows the effect of adding carbon to iron up to 7 percent by weight. Steels are iron alloys containing up approximately 2 percent of carbon by weight; most often the total carbon content is less than 1 percent by weight. If the total carbon content is >2 percent by weight, the alloy is classified as a cast iron. The presence of carbon stabilises austenite and expands the temperature range within which this phase is stable. Carbon goes into interstitial solid solution in both ferrite and austenite, but is much more soluble in austenite (maximum solubility 2.14 wt. % at 1 148°C) than in ferrite (maximum solubility 0.022 wt. % at 727°C).[7]

Carbon solubility in ferrite and austenite is temperature dependent, as shown in the Fe–C phase diagram. When the insertion of carbon atoms into interstices of the relevant Fe structure exceeds the

Table 29.3   CORRELATION OF HOT STEEL TEMPERATURE WITH COLOUR

Temperature		
°F	°C	*Hot steel colour*
752	400	Red: visible in the dark
885	474	Red: visible in twilight
975	525	Red: visible in daylight
1 077	581	Red: visible in sunlight
1 292	700	Dull red
1 472	800	Turning to cherry red
1 652	900	Cherry red
1 832	1 000	Bright cherry red
2 012	1 100	Orange red
2 192	1 200	Orange yellow
2 372	1 300	White
2 552	1 400	Brilliant white
2 732	1 500	Dazzling white
2 912	1 600	Bluish white

carbon solubility, a new primitive orthorhombic crystal structure of cementite or iron carbide (Fe_3C) which is capable of greater carbon solubilisation will be created.[7]

The austenite phase field shown in the phase diagram is the basis for selecting hot working and heat treating temperature limits for carbon steels. Annealing, normalisation and austenitisation processes are conducted in this region to facilitate the dissolution of carbon in iron. For example, the Fe–C phase diagram shows that if the steel is cooled slowly, the structure will change from austenite to ferrite and cementite (producing a ferrite + pearlite microstructure). With faster cooling martensite is formed. Austenite is only stable at elevated temperatures, but can be retained as an unstable phase in high-carbon martensite.

The temperature range designated as the 'critical temperature range' or 'transformation range' is defined as 'those ranges of temperature within which austenite forms during heating and transforms during cooling. . .'.[5] The transformation ranges on heating and cooling may overlap, but never coincide exactly and are dependent on the alloy composition and the rate of temperature change. Table 29.2 provides the recommended maximum hot working temperatures for various steels.[8] Table 29.3 provides a table showing a correlation of surface colours and approximate temperatures.

The transformation temperature indicates the limiting temperature of a transformation range. For irons and steels, the following standard terms are applied:[5]

Ac_{cm}—The temperature at which the transformation from cementite to austenite is completed during heating (in a hypereutectoid steel).

Ac_1—The temperature at which austenite begins to form during heating.

Ac_3—The temperature at which the transformation of ferrite to austenite is completed during heating (in a hypoeutectoid steel).

Ac_4—The temperature at which austenite transforms to delta ferrite during heating (in a hypoeutectoid steel).

Ar_{cm}—The temperature at which the precipitation of cementite starts during cooling (in a hypereutectoid steel).

Ar_1—The temperature at which transformation of austenite to ferrite or to ferrite and cementite is completed during cooling.

Ar_3—The temperature at which austenite begins to transform to ferrite during cooling (in a hypoeutectoid steel).

Ar_4—The temperature at which delta ferrite transforms to austenite during cooling (in a hypoeutectoid steel).

Ar'—The temperature at which transformation from austenite to pearlite begins during cooling.

Ar''—The temperature at which transformation from austenite to martensite begins during cooling.

A summary Ac_1, Ac_3, Ar_1 and Ar_3 values for different steels is provided in Table 29.4.[8]

STRUCTURAL CLASSIFICATION OF STEELS

The Fe–C phase diagram depicted in Chapter 11 illustrates an important characteristic in the steel composition range which is called the 'eutectoid' which refers to the composition of a solid phase

Table 29.4   APPROXIMATE CRITICAL TEMPERATURES AND M_s/M_f POINTS FOR CARBON AND ALLOY STEELS
Property of the Timken Company, used with permission.

SAE No.	Heating				Cooling				Quench temp.		M_s		M_f	
	Ac_1		Ac_3		Ar_3		Ar_1[3]							
	°C	°F	°C	°F	°C	°F	°C	°F	°C	°F	°C	°F	°C	°F
1015	743	1370	852	1565	841	1545	688	1270	—	—	—	—	—	—
1020	732	1350	846	1555	824	1515	688	1270	—	—	—	—	—	—
1030	732	1350	807	1485	796	1465	688	1270	—	—	—	—	—	—
1035	732	1350	802	1475	782	1440	688	1270	—	—	—	—	—	—
1040	732	1350	793	1460	771	1420	688	1270	—	—	—	—	—	—
1045	732	1350	782	1440	763	1405	688	1270	—	—	—	—	—	—
1050	727	1340	771	1420	754	1390	688	1270	—	—	—	—	—	—
1065	—	—	—	—	—	—	—	—	816	1500	274	525	149	300
1090	—	—	—	—	—	—	—	—	885	1625	216	420	79	175
1330	718	1325	799	1470	727	1340	627	1160	—	—	—	—	—	—
1335	713	1315	793	1460	727	1340	629	1165	843	1550	338	640	232	450
1340	727	1340	771	1420	710	1310	627	1160	—	—	—	—	—	—
1345	718	1325	771	1420	704	1300	627	1160	—	—	—	—	—	—
2317	696	1285	779	1435	685	1265	574	1065	—	—	—	—	—	—
2330	693	1280	738	1360	652	1205	488/566	910/1050	—	—	—	—	—	—
2340	696	1285	732	1350	641	1185	571	1060	788	1450	304	580	204	400
2345	685	1265	724	1335	607	1125	560	1040	—	—	—	—	—	—
2512	699	1290	760	1400	621	1150	571	1060	—	—	—	—	—	—
2515	682	1260	760	1400	627	1160	588	1090	—	—	—	—	—	—
3115	735	1355	816	1500	804	1480	671	1240	—	—	—	—	—	—
3120	732	1350	804	1480	785	1445	666	1230	—	—	—	—	—	—
3130	729	1345	793	1460	738	1360	660	1220	—	—	—	—	—	—
3140	735	1355	766	1410	691	1275	663	1225	843	1550	332	630	227	440
3141	735	1355	766	1410	704	1300	657	1215	—	—	—	—	—	—
3150	735	1355	749	1380	691	1275	657	1215	—	—	—	—	—	—
3310	724	1335	782	1440	668	1235	627	1160	—	—	—	—	—	—
3316	724	1335	785	1445	668	1235	627	1160	—	—	—	—	—	—
4027	738	1360	816	1500	760	1400	666	1230	—	—	—	—	—	—
4032	727	1340	816	1500	732	1350	677	1250	—	—	—	—	—	—
4042	727	1340	793	1460	727	1340	654	1210	816	1500	321	610	—	—
4053	710	1310	760	1400	716	1320	649	1200	—	—	—	—	—	—
4063	738	1360	754	1390	660	1220	643	1190	816	1500	229	445	—	—
4068	741	1365	757	1395	657	1215	646	1195	—	—	—	—	—	—
4118	752	1385	816	1500	766	1410	691	1275	—	—	—	—	—	—
4130	749	1380	802	1475	732	1350	677	1250	871	1600	377	710	288	550
4140	749	1380	793	1460	743	1370	693	1280	816	1500	338	640	—	—
4147	—	—	—	—	—	—	—	—	816	1500	310	590	—	—
4150	754	1390	788	1450	699	1290	674	1245	—	—	—	—	—	—
4160	—	—	—	—	—	—	—	—	857	1575	260	500	—	—
4320	735	1355	807	1485	721	1330	449/632	840/1170	—	—	—	—	—	—
4340	732	1350	774	1425	660	1220	385/654	725/1210	843	1550	288	550	166	330
4342	—	—	—	—	—	—	—	—	843	1550	277	530	—	—
4615	727	1340	807	1485	760	1400	649	1200	—	—	—	—	—	—
4620	704	1300	810	1490	724	1335	660	1220	—	—	—	—	—	—
4640	718	1325	760	1400	660	1220	468/610	875/1130	843	1550	338	640	254	490
4695[1]	—	—	—	—	—	—	—	—	843	1550	124	255	—	—
4718	696	1285	821	1510	766	1410	649	1200	—	—	—	—	—	—
4815	696	1285	788	1450	710	1310	460/599	860/1110	—	—	—	—	—	—
4820	699	1290	782	1440	682	1260	441/599	825/1110	—	—	—	—	—	—

(continued)

Table 29.4 APPROXIMATE CRITICAL TEMPERATURES AND M_s/M_f POINTS FOR CARBON AND ALLOY STEELS—*continued*

SAE No.	Heating Ac₁ °C	Ac₁ °F	Ac₃ °C	Ac₃ °F	Cooling Ar₃ °C	Ar₃ °F	Ar₁[3] °C	Ar₁[3] °F	Quench temp. °C	°F	Ms °C	°F	Mf °C	°F
5 045	738	1 360	777	1 430	707	1 305	679	1 255	—	—	—	—	—	—
5 060	743	1 370	766	1 410	707	1 305	696	1 285	—	—	—	—	—	—
5 120	749	1 380	829	1 525	793	1 460	707	1 305	—	—	—	—	—	—
5 140	738	1 360	788	1 450	729	1 345	666	1 230	843	1 550	332	630	238	460
51 100	752	1 385	768	1 415	716	1 320	704	1 300	—	—	—	—	—	—
52 100	727	1 340	824	1 515	716	1 320	688	1 270	849	1 560	174	345	—	—
52 100	—	—	—	—	—	—	—	—	899	1 650	152	305	—	—
52 100	—	—	—	—	—	—	—	—	949	1 740	127	260	—	—
6 117	760	1 400	849	1 560	777	1 430	688	1 270	899	1 650	152	305	—	—
6 120	766	1 410	832	1 530	782	1 440	704	1 300	949	1 740	127	260	—	—
6 140	—	—	—	—	—	—	—	—	843	1 550	327	620	238	460
6 150	749	1 380	788	1 450	746	1 375	691	1 275	—	—	—	—	—	—
8 615	738	1 360	843	1 550	791	1 455	685	1 265	—	—	—	—	—	—
8 620	732	1 350	829	1 525	760	1 400	649	1 200	—	—	—	—	—	—
8 630	732	1 350	804	1 480	727	1 340	654	1 210	871	1 600	366	690	282	540
8 640	732	1 350	779	1 435	691	1 275	632	1 170	—	—	—	—	—	—
8 650	718	1 325	754	1 390	671	1 240	646	1 195	—	—	—	—	—	—
8 695[2]	—	—	—	—	—	—	—	—	816	1 500	135	275	—	—
8 720	749	1 380	827	1 520	760	1 400	649	1 200	—	—	—	—	—	—
8 740	732	1 350	788	1 450	704	1 300	638	1 180	—	—	—	—	—	—
8 750	732	1 350	766	1 410	685	1 265	643	1 190	—	—	—	—	—	—
9 310	713	1 315	810	1 490	707	1 305	443/582	830/1 080	—	—	—	—	—	—
9 317	704	1 300	791	1 455	699	1 290	427	800	—	—	—	—	—	—
9 395[2]	—	—	—	—	—	—	—	—	927	1 700	77	170	—	—
9 442	732	1 350	779	1 435	693	1 280	643	1 190	857	1 575	327	620	210	410

[1] Represents the case of 4 600 grades of carburising steels.

[2] Represents the case of 8 600 and 9 300 grades of carburising steels respectively.

[3] When two temperatures are given for Ar_1, the higher temperature represents the pearlite reaction and the lower temperature the bainite reaction.

which, upon cooling, undergoes a univariant transformation into two, or more, other solid phases.[4] For a carbon steel, the eutectoid point occurs at 0.77 wt. % carbon. This is the basis of steel classification into: hypoeutectoid, eutectoid and hypereutectoid steels.

Hypoeutectoid steels are those steels with less than ~0.80 wt. % carbon (strictly 0.77 wt. % C, but a less demanding definition is used in commercial practice). Hypoeutectoid steels can, upon initial cooling from the austenite single phase field, exist as two different phases, proeutectoid ferrite and austenite, each with different carbon contents. Upon further cooling, the remaining austenite undergoes the eutectoid reaction to ferrite plus cementite (i.e. pearlite) and so the microstructure of these steels typically exhibits proeutectoid ferrite grains and pearlite islands.

Eutectoid steels contain 0.76 wt. % carbon (in practice, steels with 0.75–0.85 wt. % carbon are often classified as eutectoid steels). These steels form as a solid solution at any temperature in the austenitic range and all carbon is dissolved in the austenite. At the critical (eutectoid) temperature of the iron-cementite system (1 340°F, 727°C), there is a transformation from austenite to lamellar pearlite. However, if the steel is cooled slowly to a temperature just below Ar_1 and held for a suitable duration, spheroidal cementite particles in ferrite are obtained, instead of lamellar pearlite. This microstructure is called spheroidite and exhibits improved machinability and formability.

Hypereutectoid steels contain ~0.8–2.0 wt. % carbon. Upon cooling to Ar_{cm}, proeutectoid cementite separates from austenite. Below 1 340°F, 727°C, the remaining austenite transforms to pearlite. Hence, at room temperature, the microstructure consists of proeutectoid cementite and pearlite.

ISOTHERMAL AND CONTINUOUS COOLING TRANSFORMATION DIAGRAMS FOR STEELS

Microstructures that are formed upon cooling and the proportions of each are dependent on the austenitisation conditions (which influence the austenite grain size and the solutioning of alloying elements in austenite), the time and temperature cooling history of the particular alloy and composition of the alloy. The transformation products formed are illustrated typically with the use of transformation diagrams, which show the temperature-time dependence of the microstructure formation process, for the alloy being studied.

Two of the most commonly used transformation diagrams are firstly TTT (time-temperature-transformation), which is also referred to as an ITD (isothermal transformation diagram) and secondly CCT (continuous cooling transformation), which is also referred to as a CCD (continuous cooling diagram). When selected properly, either of these types of diagrams can be used to predict the microstructure and hardness of a given steel after heat treatment or they may be used to design a heat treatment process when the desired microstructure and hardness are known.

TTT diagrams (ITD)

TTT diagrams are generated by heating small samples of steel to the desired austenitising temperature and then cooling rapidly to a temperature intermediate between the austenitising and the M_s temperature. After holding at this temperature for a fixed period of time, the transformation products are then determined. At any given holding temperature, holding times up to those at which the transformation is complete need to be investigated. This is done repeatedly for a series of temperatures until a TTT diagram is constructed such as that shown in Figure 29.1. TTT diagrams can only be read along the isotherms.

The fraction transformed to ferrite, pearlite and bainite in isothermal processes can be calculated from:[9]

$$M = 1 - \exp(-bt^n)$$

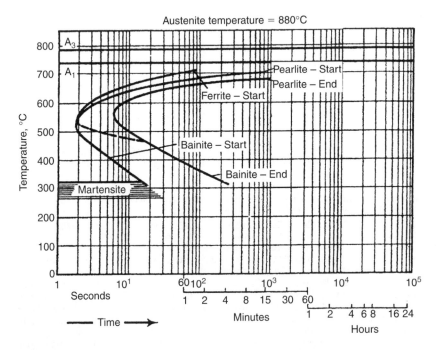

Figure 29.1 *Isothermal transformation diagram (ITD), also known as a time-temperature-transformation (TTT) diagram, of an unalloyed steel containing 0.5% carbon (Source: Verlag Stahleisen m.b.H. Düsseldorf).*

where M is the fraction of the phase transformed, t is the time in seconds, $b = 2 \times 10^{-9}$ and $n = 3$. By convention, the beginning of the transformation is defined as 1% of the parent phase transformed and the end is defined as 99% of this phase transformed.

Only martensite formation occurs without diffusion. The Hougardy equation may be used to predict the amount of martensite formed in structural steels:[9,10]

$$M = 1 - 0.929 \exp\left[-0.976 \times 10^{-2}(M_s - T)^{1.07}\right]$$

where $M =$ the amount of martensite, M_s is the martensite start temperature and T is the temperature of interest below the M_s temperature.

The accuracy of TTT diagrams with respect to the isothermal positions on the diagram is typically accepted to be $\pm 10°C$ ($\pm 20°F$) or $\pm 10\%$ with respect to time.

Examples of heat treatment processes where it is only appropriate to use a TTT diagram are: isothermal annealing, austempering and martempering. These processes are illustrated schematically in Figure 29.2.[9]

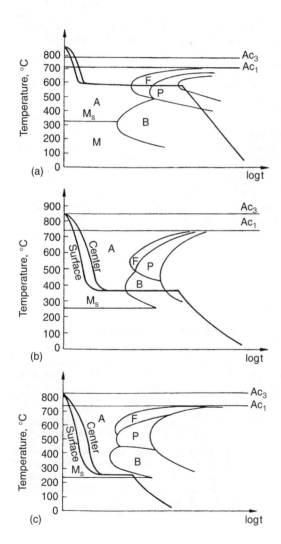

Figure 29.2 *Isothermal processes for which only ITDs (TTT diagrams) may be used (reprinted from reference 9, p. 546—courtesy of Marcel Dekker, Inc.): (a) isothermal annealing; (b) austempering; (c) martempering.*

CCT diagrams (CCD)

Steel may also be continuously cooled at different specified rates using a dilatometer allowing the proportion of transformation products formed after cooling to various temperatures intermediate between the austenitising temperature and the M_s temperature to be determined. These data can then be used to construct a CCT diagram. CCT curves correlate the temperatures for each phase transformation, the amount of transformation product obtained for a given cooling rate with time and the cooling rate necessary to obtain martensite. These correlations are obtained from CCT diagrams by using the different cooling rate curves.

The 'critical cooling rate' is the time required to avoid formation of pearlite for the particular steel being quenched. As a general rule, a quenchant must produce a cooling rate equivalent to, or faster than, that rate indicated by the 'nose' of the pearlite transformation curve, to maximise the formation of martensite.

If the temperature-time cooling curves for the quenchant and the CCT curves of the steel are plotted on the same scale, then they may be superimposed to select the steel grade which will provide the desired microstructure and hardness for a given cooling condition.[9] This assumption is limited to bars up to 100 mm in diameter quenched in oil and bars up to 150 mm quenched in water.

CCT diagrams may be constructed in various forms such as those shown in Figure 29.3. Figure 29.3a is a CCT diagram for an unalloyed carbon steel (AISI 1045) which provides curves for the beginning and ending of the different phase transformations.[11] Figure 29.3b, was generated for a DIN 50CrV4 (AISI 6145) steel.[9] In this figure, the fraction of the transformation product formed by a cooling curve is shown on the diagram and the resulting hardness is shown on the isotherm at the bottom of the diagram.

An alternative form of a CCT diagram is shown by Figure 29.3c.[9] This curve was not generated using a dilatometer but instead cooling curves were determined experimentally at different distances from the end of a Jominy test bar. The corresponding Jominy curve is shown along with a diagram for a particular quenchant and agitation condition which permits the prediction of cross-sectional hardness for a round bar.[9,12]

Another form of CCT diagram, originally developed by Atkins,[13] is illustrated in Figure 29.3d. This CCT diagram was generated by determining the cooling curves of round bars of the alloy represented in different quenchant media and then determining the corresponding transformation temperatures, microstructures and hardnesses.[9] The data represented by these curves refer only to the centre of the bar being quenched. A scale of cooling rates is provided at the bottom of the diagram. These diagrams are read along vertical lines with respect to different cooling rates. This diagram is especially useful to quickly identify the relative hardenability of different steels.

There are a number of heat treatment processes where only the use of a CCT diagram is appropriate. These include: continuous slow cooling processes such as normalising by cooling in air, direct quenching to obtain a fully martensitic structure and continuous cooling processes resulting in mixed microstructures as illustrated in Figure 29.4.[9]

The Rose-Strassburg cooling law can be used to predict cooling times and temperatures for steel samples whose cross-sectional area are not excessively large. Therefore their cooling is modelled by the following relationship:[14]

$$T = T_0 \exp[-\alpha t]$$

where T_0 = austenitising temperature, α = heat transfer coefficient and t = time.

A number of points should be noted:

- The CCT diagram is only valid for the steel composition for which it was determined.
- It is NOT correct to assume that the area of intersection of a cooling curve with the transformation product is equivalent to the amount of that product that is formed.
- Scheil has shown that transformation begins later in time for a continuous cooling process than for an isothermal process.[15] This is consistent with TTT and CCT curve comparison.
- Since increasing the austenitisation temperature will shift the curves to longer transformation times, it is necessary to use CCT diagrams generated at the desired austenitising temperature.

Caution: It is becoming increasingly common to see cooling curves (temperature-time profiles) for different cooling media (quenchants) such as oil, water, air and others, superimposed on either TTT or CCT diagrams. However, superimposition of such data, especially on a TTT diagram, is not a rigorously correct practice. Various errors are introduced into such analyses due to the inherently different kinetics of cooling used to obtain TTT or CCT diagrams versus the quenchants being represented. In particular, a continuous cooling curve can be superimposed on a CCT, but not on a TTT diagram.

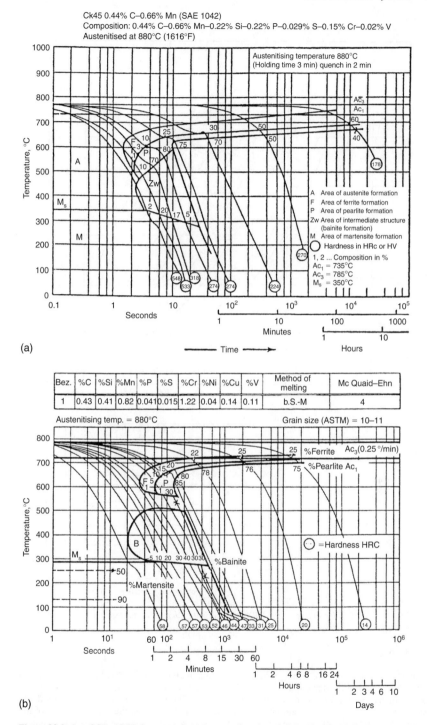

Ck45 0.44% C–0.66% Mn (SAE 1042)
Composition: 0.44% C–0.66% Mn–0.22% Si–0.22% P–0.029% S–0.15% Cr–0.02% V
Austenitised at 880°C (1616°F)

(a)

Bez.	%C	%Si	%Mn	%P	%S	%Cr	%Ni	%Cu	%V	Method of melting	Mc Quaid–Ehn
1	0.43	0.41	0.82	0.041	0.015	1.22	0.04	0.14	0.11	b.S.-M	4

Austenitising temp. = 880°C   Grain size (ASTM) = 10–11

(b)

Figure 29.3a,b *CCDs (CCT diagrams) for: (a) an unalloyed steel; DIN Ck45 (AISI 1045, reprinted from reference 11, p. 164—courtesy of Marcel Dekker, Inc.); (b) DIN 50CrV4 (AISI 6145, reprinted from reference 9, p. 539—courtesy of Marcel Dekker, Inc.)*

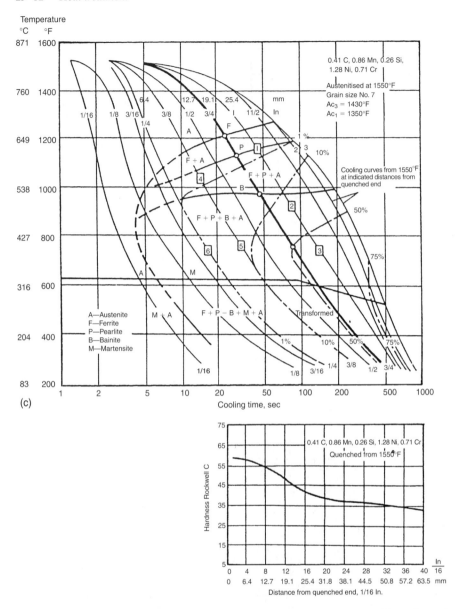

Figure 29.3c   *CCDs (CCT diagrams) for: AISI 3140, together with a Jominy curve (reprinted from reference 9, p. 548—courtesy of Marcel Dekker, Inc.)*

29.2.3   Hardenability

Hardenability has been defined as the ability of a ferrous material to develop hardness to a given depth after being austenitised and quenched. This general definition comprises two sub-definitions, the first of which is the ability to achieve a certain hardness.[16] The ability to achieve a certain hardness level is associated with the highest attainable hardness which depends on the carbon content of the steel and more specifically on the amount of carbon dissolved in the austenite after austenitising.

This is illustrated by considering the problem of hardening of high-strength, high-carbon steels. The higher the concentration of dissolved carbon in the austenitic phase, the greater the increase in

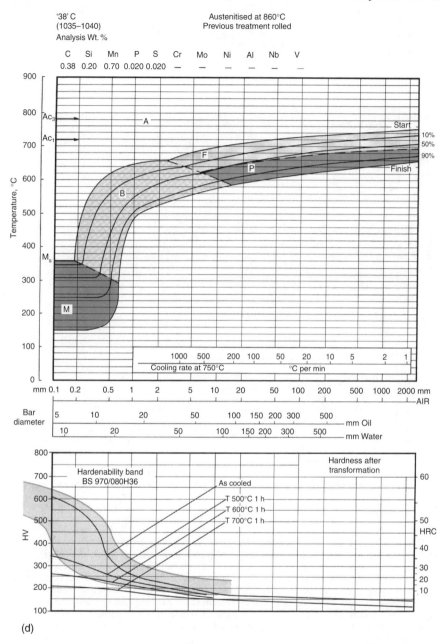

(d)

Figure 29.3d *CCDs (CCT diagrams) for: Comparison with the correlation of Jominy distance and cooling rate for round bars of various diameters quenched in into oil and water (© ASM International, Materials Park, OH, used with permission)*

mechanical strength after rapid cooling and transformation of the austenite into the metastable phase, martensite. Martensitic steels typically exhibit increasing hardness and strength with increasing carbon content, but they also exhibit relatively low ductility. However, with increasing carbon concentration, martensitic transformation from austenite becomes more difficult, resulting in a greater tendency for retained austenite and correspondingly lower strength.

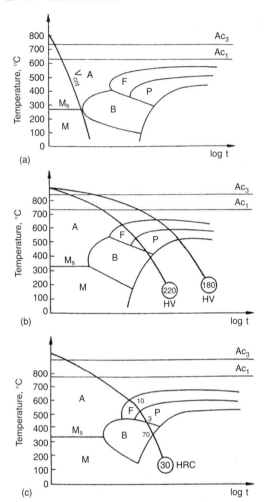

Figure 29.4   *Heat treatment processes where only CCDs (CCT diagrams) may be used (reprinted from reference 9, p. 544 courtesy of Marcel Dekker, Inc.): (a) direct quenching to obtain a fully martensitic microstructure; (b) slow cooling to obtain a ferrite-pearlite microstructure; (c) continuous cooling for a mixed microstructure*

The second sub-definition of hardenability refers to the hardness distribution within a cross-section from the surface to the core under specified quenching conditions. In this case, hardenability depends on the quantity of carbon which is dissolved interstitially in austenite and the amount of alloying elements dissolved substitutionally in the austenite during austenitisation. Therefore, as Figure 29.5 shows, carbon concentrations in excess of 0.6% do not yield correspondingly greater strength. Also, increasing the carbon content influences the M_f temperature relative to M_s during rapid cooling as shown in Figure 29.6.[17] In this figure, it is evident that for steels with carbon contents above 0.6%, the transformation of austenite to martensite will be incomplete if the cooling process is stopped at 0°C or higher.

The depth of hardening depends on the following factors:

- Size and shape of the cross-section.
- Hardenability of the material.
- Quenching conditions.

The cross-sectional shape exhibits a significant influence on heat extraction during quenching and, therefore, on the hardening depth. Heat extraction is dependent on the surface area exposed to the quenchant. Figure 29.7 can be used to convert square and rectangular cross-sections to equivalent circular cross-sections.[18]

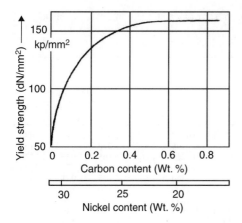

Figure 29.5 *Influence of the carbon content in steel on the yield strength ($\sigma_{0.6}$) after quench hardening. The yield strength values were obtained from compression tests; the additional variation of nickel content causes negligible solid-solution-hardening and was selected to obtain a constant M_s temperature for the start of martensitic transformation (used with the permission of the Association of Iron and Steel Engineers (AISE))*

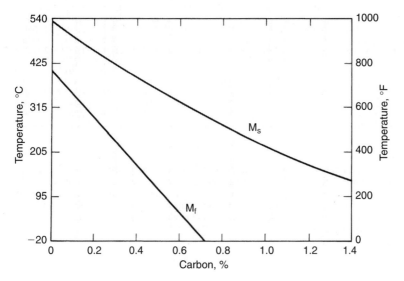

Figure 29.6 *Influence of the carbon content in plain-carbon steels on the temperature of the start of martensite formation (M_s) and the end of martensite formation (M_f)*

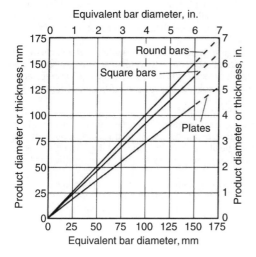

Figure 29.7 *Correlation between rectangular cross-sections and their equivalent round bar and plate sections*

Table 29.5 HARDENABILITY FACTORS FOR CARBON CONTENT, GRAIN SIZE AND SELECTED ALLOYING ELEMENTS IN STEEL

© ASM International, Materials Park, OH, used with permission.

| Carbon content (%) | Grain size no. (G) | | | | | | Alloying element | | | | |
| | 6 | | 7 | | 8 | | | | | | |
	mm	inch	mm	inch	mm	inch	Mn	Si	Ni	Cr	Mo
0.05	2.0676	0.0814	1.9050	0.0750	1.7704	0.0697	1.167	1.035	1.018	1.1080	1.15
0.10	2.9286	0.1153	2.7051	0.1065	2.5273	0.0995	1.333	1.070	1.036	1.2160	1.30
0.15	3.5890	0.1413	3.3401	0.1315	3.0785	0.1212	1.500	1.105	1.055	1.3240	1.45
0.20	4.1224	0.1623	3.8329	0.1509	3.5560	0.1400	1.667	1.140	1.073	1.4320	1.60
0.25	4.6228	0.1820	4.2621	0.1678	3.9624	0.1560	1.833	1.175	1.091	1.54	1.75
0.30	5.0571	0.1991	4.6965	0.1849	4.3180	0.1700	2.000	1.210	1.109	1.6480	1.90
0.35	5.4712	0.2154	5.0800	0.2000	4.6787	0.1842	2.167	1.245	1.128	1.7560	2.05
0.40	5.8420	0.2300	5.4102	0.2130	5.0190	0.1976	2.333	1.280	1.146	1.8640	2.20
0.45	6.1976	0.2440	5.7379	0.2259	5.3086	0.2090	2.500	1.315	1.164	1.9720	2.35
0.50	6.5532	0.2580	6.0452	0.2380	5.5880	0.2200	2.667	1.350	1.182	2.0800	2.50
0.55	6.934	0.273	6.375	0.251	5.867	0.231	2.833	1.385	1.201	2.1880	2.65
0.60	7.214	0.284	6.655	0.262	6.121	0.241	3.000	1.420	1.219	2.2960	2.80
0.65	7.493	0.295	6.934	0.273	6.375	0.251	3.167	1.455	1.237	2.4040	2.95
0.70	7.772	0.306	7.188	0.283	6.604	0.260	3.333	1.490	1.255	2.5120	3.10
0.75	8.026	0.316	7.442	0.293	6.858	0.270	3.500	1.525	1.273	2.62	3.25
0.80	8.280	0.326	7.696	0.303	7.061	0.278	3.667	1.560	1.291	2.7280	3.40
0.85	8.534	0.336	7.925	0.312	7.290	0.287	3.833	1.595	1.309	2.8360	3.55
0.90	8.788	0.346	8.153	0.321	7.518	0.296	4.000	1.630	1.321	2.9440	3.70
0.95	—	—	—	—	—	—	4.167	1.665	1.345	3.0520	—
1.00	—	—	—	—	—	—	4.333	1.700	1.364	3.1600	—

The effect of steel composition on hardenability may be calculated in terms of the 'ideal critical diameter', D_I, which is defined as the largest bar diameter that can be quenched to produce 50% martensite at the centre, after quenching in an 'ideal' quench, i.e. under 'infinite' quenching severity. The ideal quench is one that reduces surface temperature of an austenitised steel to the bath temperature instantaneously. Under these conditions, the cooling rate at the centre of the bar depends only on the thermal diffusivity of the steel.

The ideal critical diameter may be calculated from:

$$D_I = D_{I\,Base} \text{ (carbon concentration and grain size)} \times f_{Mn} \times f_{Si} \times f_{Cr} \times f_{Mo} \times f_V \times f_{Cu} \times f_{Ni} \times f_x$$

where f_x is a multiplicative factor for the particular substitutionally dissolved alloying element. The base value $D_{I\,Base}$ and one set of alloying factors are provided in Table 29.5.[19,20] (Note: This is not an exhaustive list of alloying factors but these are commonly encountered and they permit calculations to illustrate the effect of steel chemistry variation on hardenability.) D_I values for a range of steels with differing hardenability is provided in Table 29.6.[19,21]

Grain size refers to the dimensions of grains or crystals in a polycrystalline metal exclusive of twinned regions and subgrains when present. Grain size is usually estimated or measured on the cross-section of an aggregate of grains. Common units are: (1) average diameter, (2) average area, (3) number of grains per linear unit, (4) number of grains per unit area and (5) number of grains per unit volume.

Grain size may be determined according to ASTM Test Method E 112.[22] The procedures in Test Method E 112 describe the measurement of average grain size and include the comparison procedure, the planimetric (or Jeffries) procedure, and the intercept procedures. Standard comparison charts are provided. These test methods apply chiefly to single phase grain structures but they can be applied to determine the average size of a particular type of grain structure in a multiphase or multiconstituent specimen.

In addition, the test methods provided in ASTM E 112 are used to determine the average grain size of specimens with a unimodal distribution of grain areas, diameters, or intercept lengths. These distributions are approximately log normal. These test methods do not cover methods to characterise

Table 29.6 IDEAL CRITICAL DIAMETER (D$_I$) VALUES FOR VARIOUS STEELS

© ASM International, Materials Park, OH, used with permission.

Steel	D$_I$ mm	inch	Steel	D$_I$ mm	inch	Steel	D$_I$ mm	inch
1 045	22.9–33.0	0.9–1.3	4 135 H	63.5–83.8	2.5–3.3	8 625 H	40.6–61.0	1.6–2.4
1 090	30.5–40.6	1.2–1.6	4 140 H	78.7–119.4	3.1–4.7	8 627 H	43.2–68.6	1.7–2.7
1 320 H	35.6–63.5	1.4–2.5	4 317 H	43.2–61.0	1.7–2.4	8 630 H	53.3–71.1	2.1–2.8
1 330 H	48.3–68.6	1.9–2.7	4 320 H	45.7–66.0	1.8–2.6	8 632 H	55.9–73.7	2.2–2.9
1 335 H	50.8–71.1	2.0–2.8	4 340 H	116.8–152.4	4.6–6.0	8 635 H	61.0–86.4	2.4–3.4
1 340 H	58.4–81.3	2.3–3.2	4 620 H	35.6–55.9	1.4–2.2	8 637 H	66.0–91.4	2.6–3.6
2 330 H	58.4–81.3	2.3–3.2	4 620 H	38.1–55.9	1.5–2.2	8 640 H	68.6–94.0	2.7–3.7
2 345	63.5–81.3	2.5–3.2	4 621 H	48.3–66.0	1.9–2.6	8 641 H	68.6–94.0	2.7–3.7
2 512 H	38.1–63.5	1.5–2.5	4 640 H	66.0–86.4	2.6–3.4	8 642 H	71.1–99.1	2.8–3.9
2 515 H	45.7–73.7	1.8–2.9	4 812 H	43.2–68.6	1.7–2.7	8 645 H	78.7–104.1	3.1–4.1
2 517 H	50.8–76.2	2.0–3.0	4 815 H	45.7–71.1	1.8–2.8	8 647 H	76.2–104.1	3.0–4.1
3 120 H	38.1–58.4	1.5–2.3	4 817 H	55.9–73.7	2.2–2.9	8 650 H	83.8–114.3	3.3–4.5
3 130 H	50.8–71.1	2.0–2.8	4 820 H	55.9–81.3	2.2–3.2	8 720 H	45.7–61.0	1.8–2.4
3 135 H	55.9–78.7	2.2–3.1	5 120 H	30.5–48.3	1.2–1.9	8 735 H	68.6–91.4	2.7–3.6
3 140 H	66.0–86.4	2.6–3.4	5 130 H	53.3–73.7	2.1–2.9	8 740 H	68.6–94.0	2.7–3.7
3 340	203.2–254.0	8.0–10.0	5 132 H	55.9–73.7	2.2–2.9	8 742 H	76.2–101.6	3.0–4.0
4 032 H	40.6–55.9	1.6–2.2	5 135 H	55.9–73.7	2.2–2.9	8 745 H	81.3–109.2	3.2–4.3
4 037 H	43.2–61.0	1.7–2.4	5 140 H	55.9–78.7	2.2–3.1	8 747 H	88.9–116.8	3.5–4.6
4 042 H	43.2–61.0	1.7–2.4	5 145 H	58.4–88.9	2.3–3.5	8 750 H	96.5–124.5	3.8–4.9
4 047 H	45.7–68.6	1.8–2.7	5 150 H	63.5–94.0	2.5–3.7	9 260 H	50.8–83.8	2.0–3.3
4 047 H	43.2–61.0	1.7–2.4	5 152 H	83.8–119.4	3.3–4.7	9 261 H	66.0–94.0	2.6–3.7
4 053 H	53.3–73.7	2.1–2.9	5 160 H	71.1–101.6	2.8–4.0	9 262 H	71.1–106.7	2.8–4.2
4 063 H	55.9–88.9	2.2–3.5	6 150 H	71.1–99.1	2.8–3.9	9 437 H	61.0–94.0	2.4–3.7
4 068 H	58.4–91.4	2.3–3.6	8 617 H	33.0–58.4	1.3–2.3	9 440 H	61.0–96.5	2.4–3.8
4 130 H	45.7–66.0	1.8–2.6	8 620 H	40.6–58.4	1.6–2.3	9 442 H	71.1–106.7	2.8–4.2
3 132 H	45.7–63.5	1.8–2.5	8 622 H	40.6–58.4	1.6–2.3	9 445 H	71.1–111.8	2.8–4.4

the nature of these distributions. Characterisation of grain size in specimens with duplex grain size distributions is described in ASTM Test Methods E 1181. Measurement of individual, very coarse grains in a fine-grained matrix is described in Test Methods E 930. These test methods deal only with determination of planar grain size, i.e. characterisation of the two-dimensional grain sections revealed by the sectioning plane. Determination of spatial grain size, i.e. measurement of the size of the three-dimensional grains in the specimen volume, is beyond the scope of these test methods.

The test methods described in ASTM E 112 are techniques performed manually using either a standard series of graded chart images for the comparison method or simple templates for the manual counting methods. Utilisation of semi-automatic digitising tablets or automatic image analysers to measure grain size is described in ASTM Test Methods E 1382.[23]

The ASTM grain size number (G), referred to in Table 29.5, is a grain size designation bearing a relationship to the average intercept distance (L) at 100 diameters magnification, according to the equation:[4]

$$G = 10.00 - 2 \log_2 L$$

The smaller the ASTM grain size, the larger the diameter of the grains.

The effect of quenching conditions on the depth of hardening are not only dependent on the quenchant being used and its physical and chemical properties but also on process parameters such as bath temperature and agitation.

29.2.4 Hardenability measurement

There are numerous methods to estimate steel hardenability. However, two of the most common are: Jominy curve determination and Grossmann hardenability which will be discussed here.

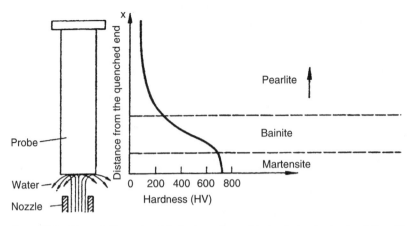

Figure 29.8   *Schematic illustration of the Jominy end-quench test and microstructural variation with increasing distance from the quenched end. The spacing of the pearlite lamellae increases with distance from the quenched end, i.e. along the direction of the arrow (reprinted from reference 16, p. 109—courtesy of Marcel Dekker, Inc.)*

JOMINY BAR END-QUENCH TEST

The most familiar and commonly used procedure for measuring steel hardenability is the Jominy bar end-quench test, as illustrated in Figure 29.8. This test has been standardised and is described in ASTM A 255, SAE J406, DIN 50191 and ISO 642. For this test, a 100 mm (4 inch) long by 25 mm (1 inch) diameter round bar is austenitised to the appropriate temperature, dropped into a fixture, and one end rapidly quenched with 24°C (75°F) water from a 13 mm (0.5 inch) orifice under specified conditions.[16] The austenitising temperature and time is selected according to the specific steel being studied. However, most steels are heated to temperatures of 870–900°C (1 600–1 650°F). The cooling velocity decreases with increasing distance from the quenched end.

After quenching, parallel flats are ground on opposite sides of the bar and hardness measurements made at 1/16 in. (1.6 mm) intervals along the bar as illustrated in Figure 29.9.[16] The hardness as a function of distance from the quenched end is measured and plotted and, together with measurement of the relative areas of the martensite, bainite and pearlite that is formed, it is possible to compare the hardenability of different steels using Jominy curves. As the slope of the Jominy curve increases, the ability to harden the steel (hardenability) decreases. Conversely, decreasing slopes (or increasing flatness) of the Jominy curve indicates increasing hardenability (ease of hardening).

The Jominy end-quench is used to define the hardenability of carbon steels with different alloying elements, such as chromium (Cr), manganese (Mn), or molybdenum (Mo) and having different critical cooling velocities. Jominy curves for different alloy steels are provided in Figure 29.10. These curves illustrate that the unalloyed, 0.4% carbon steel exhibits a relatively small distance for martensite (high hardness) formation. The 1% Cr and 0.2% Mn steel, however, can be hardened up to a distance of 40 mm. Figure 29.10 illustrates that steel hardenability is dependent on the steel chemistry. Unalloyed steels exhibit poor hardenability. Jominy curves provide an excellent indicator of relative steel hardenability.

The Jominy test provides valid data for steels having an ideal diameter from about 25 to 150 mm (1 to 6 in.). This test can be used for D_I values less than 25 mm (1 in.) but Vickers or microhardness tests must be used to obtain readings that are closer to the quenched end of the bar and closer together than generally possible using the standard Rockwell 'C' hardness test method.[16]

The austenitising time and temperature, extent of special carbide solution in the austenite and degree of oxidation or surface decarburisation during austenitising, care and consistency of surface flat preparation and bar positioning prior to making hardness measurements are important factors that influence test results. Therefore, all tests should be conducted in compliance with the standard being followed.[16]

Using the composition of the steel, it is possible to calculate the Jominy end-quench curve for a wide range of steels with excellent correlation to experimental results. In many cases, calculation is preferred over experimental determination.

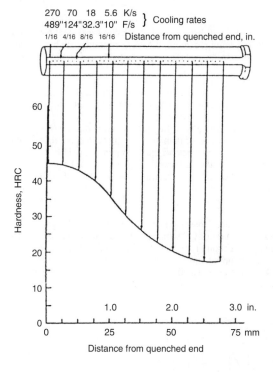

Figure 29.9 *Measuring hardness on the Jominy test specimen and plotting hardenability curves (reprinted from reference 16, p. 109—courtesy of Marcel Dekker, Inc.)*

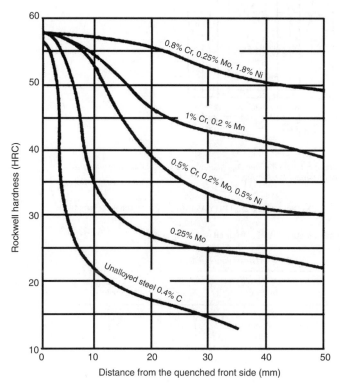

Figure 29.10 *Jominy curve comparison of the hardenability of one unalloyed and a number of other different alloy steels*

GROSSMANN HARDENABILITY

The Grossmann method of measuring hardenability utilises a number of cylindrical steel bars with different diameters each hardened in a given quenching medium.[16] After sectioning each bar at mid-length and examining it metallographically, the bar that has 50% martensite at its centre is selected, and the diameter of this bar is designated as the *critical diameter* D_{crit}. Other bars with diameters smaller than D_{crit} will have more martensite and correspondingly higher hardness values and bars with diameters larger than D_{crit} will attain 50% martensite only up to a certain depth as shown in Figure 29.11.[16] The D_{crit} value is valid only for the quenching medium and conditions used to determine this value.

To determine the hardenability of a steel independently of the quenching medium, Grossmann introduced the term *ideal critical diameter*, D_I, which is the diameter of a given steel bar that would produce 50% martensite at the centre when quenched in a bath of quenching intensity $H = \infty$. Here $H = \infty$ indicates a hypothetical quenching intensity that reduces the temperature of heated steel to the bath temperature in zero time. Alternatively, excellent correlations with reported H-values are potentially achievable using cooling rates obtained by cooling curve analysis with 13, 25, 38 and 50 mm (0.5, 1.0, 1.5 and 2.0 inch) Type 304 stainless steel probes.[24] (Ideal diameters for various steels are provided in Table 29.6.[19])

To identify a quenching medium and its condition, Grossmann introduced the Quenching Intensity (Severity) factor 'H'. Table 29.7 provides a summary of Grossmann H-Factors for different quench media and different quenching conditions.[16,25] Although this data has been published in numerous reference texts for many years, it is of relatively limited quantitative value. One of the most obvious reasons is that quenchant agitation is not adequately defined with respect to mass flow rate, directionality and turbulence and is often unknown, yet it exhibits enormous effects on quench severity during quenching.

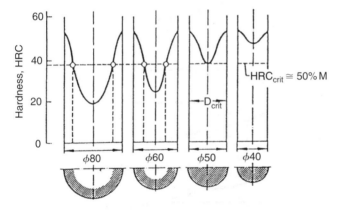

Figure 29.11   *Determination of critical diameter D_{crit} according to Grossmann (reprinted from reference 16, p. 97—courtesy of Marcel Dekker, Inc.)*

Table 29.7   EFFECT OF AGITATION ON QUENCH SEVERITY AS INDICATED BY GROSSMANN QUENCH SEVERITY FACTORS

H-Factors; © ASM International, Materials Park, OH, used with permission.

	Grossmann H-Factor			
Agitation	*Air*	*Oil and salt*	*Water*	*Caustic soda or brine*
None	0.02	0.25–0.3	0.9–1.0	2
Mild	—	0.30–0.35	1.0–1.1	2–2.2
Moderate	—	0.35–0.4	1.2–1.3	—
Good	—	0.4–0.5	1.4–1.5	—
Strong	—	0.5–0.8	1.6–2.0	—
Violent	—	0.8–1.1	4	5

The Grossmann value 'H' is based on the Biot (Bi) number which interrelates the interfacial heat transfer coefficient (α), thermal conductivity (λ) and the radius (R) of the round bar being hardened:

$$Bi = \alpha/\lambda R = HD$$
$$H = \alpha/(2\lambda)$$

Since the Biot number is dimensionless, this expression means that the Grossmann value, H, is inversely proportional to the bar diameter. This method of numerically analyzing the quenching process presumes that heat transfer is a steady state, linear (Newtonian) cooling process. However, this is seldom the case and almost never the case in vaporisable quenchants such as oil, water and aqueous polymers. Therefore, a significant error exists in the basic assumption of the method.

Another difficulty is the determination of the H-value for a cross-section other than one measured experimentally. In fact, H-values depend on the cross-sectional size. Values of H do not account for specific quenching characteristics such as composition, oil viscosity, or temperature of the quenching bath. Tables of H-values do not specify the agitation rate of the quenchant, either uniformly or precisely (*see* Table 29.7). Therefore, although H-values are used commonly, more current and improved procedures ought to be employed when possible. For example, cooling curve analyses and the various methods of cooling curve interpretation that have been reported are all significant improvements over the use of Grossmann Hardenability factors.[24,25]

29.2.5 Austenitisation

As indicated in the discussion thus far, the austenitisation process refers to the formation of austenite by heating the steel above the critical temperature for austenite formation. It is important to note that the term austenitisation means to completely transform the steel to austenite.[2] However, there are a number of critically important variables in the austenitisation process, two of which are heat rate and holding (soaking) time.

The heating rate is critical. There is a specific heating rate that cannot be exceeded without causing warpage or cracking since steels typically posses insufficient plasticity to accommodate increased thermal stresses in the temperature range of 250–600°C. Therefore, this is a particularly critical temperature range especially when the component has both thick and thin cross-sections. Therefore the heating rate is dependent on:[9]

- Size and shape of the component.
- Initial microstructure.
- Steel composition.

For this reason, steel is often heated to the final austenitising temperature in steps of 93–204°C (200–400°F) per hour.

Shapes with corners, or sharp edges, are also susceptible to cracking and this is known as the 'corner-effect'. If heating rates are excessive or non-uniform, the resulting thermal stresses may be sufficient to cause cracking.

The propensity for steel to crack is dependent on composition. For example, increasing carbon content increases the potential for cracking. The effect of composition on the potential for cracking can be modelled by calculation of the carbon equivalent (C_{eq}) using the following equation:[9]

$$C_{eq} = C + \frac{Mn}{5} + \frac{Cr}{4} + \frac{Mo}{3} + \frac{Ni}{10} + \frac{V}{5} + \frac{Si - 0.5}{5} + \frac{Ti}{5} + \frac{W}{10} + \frac{Al}{10}$$

where the elements shown represent wt. % concentrations in the steel. The limits of this equation are: $C \leq 0.9\%$, $Mn \leq 1.1\%$, $Cr \leq 1.8\%$, $Mo \leq 0.5\%$, $Ni \leq 5.0\%$, $V \leq 0.5\%$, $Si \leq 1.8\%$, $Ti \leq 0.5\%$, $W \leq 2.0\%$ and $Al \leq 2.0\%$. Crack sensitivity increases with the C_{eq} value. The following general rules were reported by Liscic:[9]

$C_{eq} \leq 0.4$	Steel not sensitive to cracking; may be heated quickly.
$C_{eq} = 0.4–0.7$	Moderate sensitivity to cracking.
$C_{eq} \geq 0.7$	Steel is very sensitive to cracking and should be preheated to a temperature close to Ac_1 and held until the temperature was uniform throughout to minimise thermal stresses when austenitising.

In addition to these effects, steel with high hardness and a non-uniform microstructure should be heated more slowly, due to its crack sensitivity, than a steel with low hardness and a uniform microstructure.

Table 29.8   SAE AMS 2759/1C RECOMMENDED ANNEALING, NORMALISING, AUSTENITISING
TEMPERATURES AND QUENCHANTS FOR VARIOUS STEELS

Material designation[1]	Annealing temperature (°C)	Normalising temperature (°C)	Austenitising temperature (°C)	Quenching medium[3]
1 025	885	899	871	w, p
1 035	871	899	843	o, w, p
1 045	857	899	829	o, w, p
1 095[4]	816	843	802	o, p
1 137	788	899	843	o, w, p
3 140	816	899	816	o, p
4 037	843	899	843	o, w, p
4 130	843	899	857	o, w, p
4 135	843	899	857	o, p
4 140	843	899	843	o, p
4 150	829	871	829	o, p
4 330V	857	899	871	o, p
4 335V	843	899	871	o, p
4 340	843	899	816	o, p
4 640	843	899	829	o, p
6 150	843	899	871	o, p
8 630	843	899	857	o, w, p
8 735	843	899	843	o, p
8 740	843	899	843	o, p

[1] SAE AMS 2759/1C should be consulted for detailed description of the overall heat treatment requirements
for these alloys.
[2] The cooling rate is not to exceed 111°C/h to below 538°C except for 4 330 V, 4 335 V and 4 340 to
below 427°C, and 4 640 to below 399°C.
[3] o = oil, w = water and p = an aqueous polymer quenchant.
[4] 1095 parts should be spheroidise annealed before hardening.

The austenitisation temperature for a given steel is typically specified, such as those values shown in Table 29.8. These values were selected to provide optimum hardness and grain size. As the austenitisation temperature increases, the grain size increases. This is important because the grain size affects heat treatment and subsequent performance under various working conditions. For example, increasing grain size increases the brittle to ductile transition temperature and increases the propensity for brittle fracture. Fine grained steels have greater fatigue strength than coarse grained steels. However, coarse grained steels have better machinability than fine grained steels. The Hall-Petch equation predicts the effect of grain size (d) on yield stress (σ_y):[26]

$$\sigma_y = \sigma_0 + K_y d^{-0.5}$$

where σ_0 and K_y are material-specific constants. Increasing the austenitising temperature also:[9]

- Increases the hardenability due to increased carbide solubilisation and increased grain size.
- Decreases the M_s temperature.
- Increases the (incubation) time for isothermal transformation to pearlite or bainite to begin.
- Increases the amount of retained austenite.

For unalloyed steels, the optimum austenitisation temperature is 30–50°C above the Ac_3 temperature for hypoeutectoid steels and 30–50°C above the Ac_1 for hypereutectoid steels. The alloying elements in alloy steels may shift the A_1 temperature either higher or lower and therefore appropriate references such as national standards must be consulted.

A 'rule of thumb' is often used to estimate the appropriate soaking time during austenitisation:

$$t = 60 + D$$

where t is the soaking time in minutes, and D is the maximum diameter of the component in millimetres. However, such rules of thumb are imprecise. Soaking times are dependent on: geometrical factors related to the furnace and the load, type of load, type of steel, thermal properties of the load, load and furnace emissivities, initial furnace and load temperatures, characteristic fan curves, and composition of the furnace atmosphere.

A. Monolayer, horizontally oriented, ordered loads
Packed   Spaced

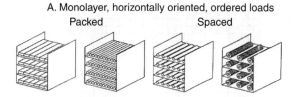

B. Monolayer, horizontally oriented, random loads

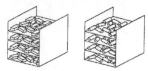

C. Multilayer ordered and random loads
Packed   Spaced   Bulk

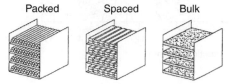

D. Vertically oriented loads

Figure 29.12 *Aronov load characterisation diagram for soaking time calculation*

Aronov developed a method for predicting furnace soaking times for batch loads based on 'load characterisation'.[27,28] Load characterisation diagrams are provided in Figure 29.12 and these models are based on the generalised characterisation equation for soaking time (t_s):

$$t_s = T_{Sb}K$$

where T_{Sb} is the baseline soak temperature condition taken from Figure 29.13 and K is a correction factor, the value of which depends upon the type of steel (K = 1 for low alloy steel and 0.85 for high alloy steel).[27,28]

29.2.6 Annealing

A primary purpose of annealing is to soften steel to enhance its workability and machinability. However, annealing may also be performed for:[26]

- Relief of internal stresses arising from prior processing including casting, forging, rolling, machining and welding.
- Improvement or restoration of ductility and toughness.
- Improvement of machinability.
- Grain refinement.
- Improvement of the uniformity of the dispersion of alloying elements.
- Achieving a specific microstructure.
- Reduction of gaseous content within the steel.

Annealing may be an intermediate step in an overall process or it may be the final process in the heat treatment of a component. Table 29.8 provides a summary of annealing temperatures recommended for some common steel alloys.

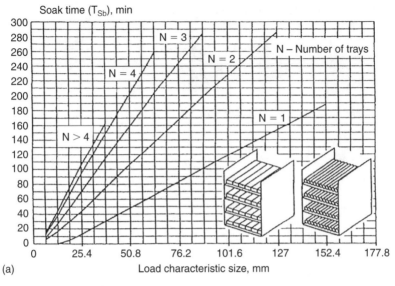

(a)

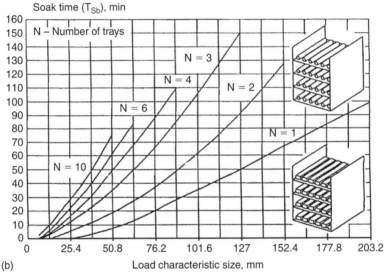

(b)

Figure 29.13a,b *Aronov soaking times for: (a) packed load; (b) spaced load*

Annealing processes may be classified as: full annealing, process (subcritical) annealing, isothermal annealing, recrystallisation annealing, spheroidising, and normalising. Partial (intercritical) annealing is a subclass of full annealing. Figure 29.14 provides an illustrative summary of these annealing processes.

FULL ANNEALING

Full annealing of steel involves heating the steel 30–50°C above the upper critical temperature (Ac$_3$) for hypoeutectoid steels, followed by furnace cooling through the critical temperature range, at a specified cooling rate, which is selected based on the final microstructure and hardness required.[29] Full annealing results in a relatively coarse microstructure. This process is typically used for steels

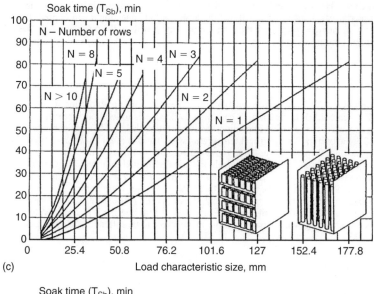

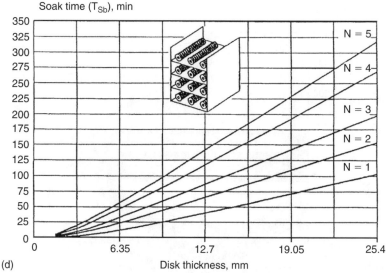

Figure 29.13c,d   *Aronov soaking times for:* (c) *vertical load;* (d) *disks*

with carbon contents of 0.30–0.60%, for example, to improve machinability. Unless otherwise noted, the term 'annealing' often refers to full annealing.[1]

PARTIAL (INTERCRITICAL) ANNEALING

Partial annealing, as illustrated in Figure 29.14, is conducted by heating the steel to a point within the critical temperature range (Ac_1–Ac_3) followed by slow furnace cooling. Partial annealing, also known as 'intercritical annealing', may be performed on hypereutectoid steels to obtain a microstructure of fine pearlite and cementite instead of coarse pearlite and a network of cementite at the grain boundaries, as observed in the case of full annealing.[26] For hypereutectoid steels, this results in grain refinement which usually occurs at 10–30°C above Ac_1. Partial annealing is performed to improve machinability. However, steels with a Widmanstätten or coarse ferrite/pearlite structure are unsuitable for this process. Krauss has noted that the term 'partial annealing' is an imprecise term and to be meaningful, the type of material, the time-temperature profile of the process and the degree of cold working must be specified.[2]

PROCESS (SUBCRITICAL) ANNEALING

Process annealing (Figure 29.14) is performed to improve the cold-working properties of low-carbon steels (up to 0.25% carbon) or to soften high-carbon and alloy steels to facilitate shearing, turning or straightening processes.[30,31] Process annealing involves heating the steel to a temperature below (typically 10–20°C below) the lower critical temperature (Ac$_1$) and is often known as 'subcritical' annealing. After heating, the steel is cooled to room temperature in still air. The process annealing temperatures for plain carbon and low alloy steels is typically limited to about 700°C to prevent partial reaustenitisation. In some cases this is limited to about 680°C for steel compositions, such as high-nickel containing steels, where the nickel further reduces the Ac$_1$ temperature.[31]

This process can be used to temper martensitic and bainitic microstructures to produce a softened microstructure containing spheroidal carbides in ferrite.[31] Fine pearlite is also relatively easily softened by process annealing, while coarse pearlite is too stable to be softened by this process.

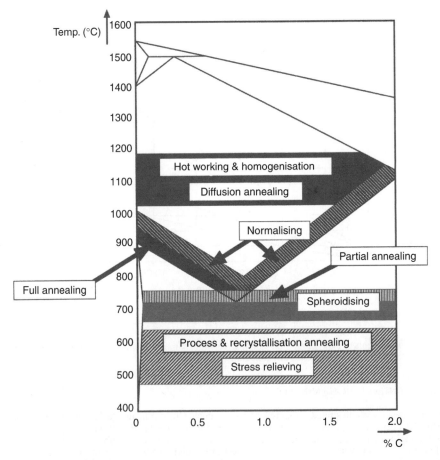

Figure 29.14 *Annealing-related processes in plain carbon steels*

RECRYSTALLISATION ANNEALING

Prior to cold working, steel microstructures that are either spheroidised or ferritic are highly ductile. However, when steels are cold worked, they become work hardened and ductility is reduced. Work hardening resulting from cold working can be removed by a recrystallisation process which produces strain-free grains, resulting in a ductile, spheroidised microstructure.[2] Nearly all steels which are heavily cold worked undergo recrystallisation annealing, which reduces hardness and increases ductility.[26]

Recrystallisation annealing is performed by heating the steel for 30 minutes to one hour, at a temperature above the recrystallisation temperature shown in Figure 29.14.[32] When heating is complete, the steel is cooled. As opposed to other annealing processes where the processing temperature is fixed, the recrystallisation annealing temperature is dependent upon composition, prior deformation, grain size and holding time.[26] Liscic reported a correlation between recrystallisation temperature (T_R) and the melting temperature (T_m) of the steel in Celsius:[9]

$$T_R = 0.4\,T_m$$

ISOTHERMAL ANNEALING

Isothermal annealing (Figure 29.15) is conducted by heating the steel within the austenite single-phase region (i.e. above Ac_3 for a hypoeutectoid steel, or above Ac_1 for a eutectoid steel) for a time sufficient to complete the solutionising process, yielding a completely austenitic microstructure. The steel is then cooled rapidly at a specified rate within the pearlite transformation range indicated by the TTT diagram for the steel (less than Ac_1, typically between 600–700°C) until complete transformation into ferrite plus pearlite (lamellar pearlite) occurs, at which time the steel is cooled rapidly to room temperature.[1,26] It should be cautioned that it is not rigorously correct to use TTT diagrams which are developed for austenitisation temperatures which do not match those for isothermal annealing.[32]

Isothermal annealing is used to achieve a more homogeneous microstructure within the steel and is faster and less expensive than full annealing. It is typically performed on hypoeutectoid steels and it is usually not performed on hypereutectoid steels.[26,30] When isothermal annealing is used in continuous production lines for small parts or for parts with thin cross-sections, it is called 'cycle annealing'.

SPHEROIDISING (SOFT ANNEALING)

Spheroidising involves the prolonged heating of steel starting from a temperature either just above, or just below the lower critical temperature (Ac_1) as illustrated in Figure 29.14.

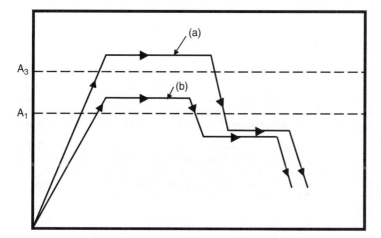

Figure 29.15 *Schematic of a heat treatment cycle for isothermal annealing of: (a) hypoeutectoid steel; (b) eutectoid steel. Note: the A_3 temperature shown in the figure refers to that of the hypoeutectoid steel*

In the simplest case, the steel is heated to just below Ac_1 and held for a protracted period. Since the steel remains in the ferrite plus cementite two-phase field, no phase transformations occur. However, the pearlite gradually spheroidises (driven by the resulting reduction in α–Fe_3C interfacial area and hence total interfacial energy).

In commercial practice, eutectoid steels are heated initially to 20–30°C above Ac_1 and hypereutectoid steels are heated initially to 30–50°C above Ac_1. The Ac_1 temperature can be determined from Table 29.4, obtained from the appropriate TTT diagram or calculated from:[9]

$$Ac_1(°C) = 739 - 22\,(\%C) + 2\,(\%Si) - 7\,(\%Mn) + 14\,(\%Cr) + 13\,(\%Mo)$$
$$+ 13\,(\%Ni) + 20\,(\%V)$$

(Hypereutectoid alloy steels may need to be heated at a higher temperature than suggested by this equation.)

Medium carbon steels may be spheroidised by heating just above or just below the Ac_1 temperature. Heating is followed by furnace cooling to a temperature just below Ar_1. Instead of pearlite, the resulting microstructure consists of ferrite plus fine spheroidal and/or globular cementite (with the cementite morphology depending in part upon the carbon content of the steel). The required cooling rate is given by:[9]

1. For plain carbon and low-alloy steels up to 650°C (1200°F)—cooling rate = 20–25 K/h (furnace cooling)
2. For medium alloy steels up to 630°C (1 166°F)—cooling rate = 15–20 K/h (furnace cooling)
3. For high alloy steels up to 600°C (1 112°F)—cooling rate = 10–15 K/h (furnace cooling)

For alloy steels, the spheroidising temperature (T) may be calculated from:[9]

$$T(°C) = 705 + 20\,(\%Si - \%Mn + \%Cr - \%Mo - \%Ni + \%W) + 100\,(\%V)$$

High carbon and alloy steels are spheroidised to enhance their machinability and ductility.[1] This process is desirable for cold-formed low-carbon and medium-carbon steels and for high-carbon steels that are premachined, prior to final machining.

DIFFUSION (HOMOGENISING) ANNEALING

Diffusion annealing (homogenising) is performed on steel ingots and castings to minimise chemical segregation. Chemical segregation defects occur as dendrites, columnar grains and chemical occlusions. The presence of these defects produces increased brittleness and reduced ductility and toughness. The homogenisation process, illustrated in Figure 29.14, is conducted by heating the steel rapidly to 1 100–1 200°C and holding for 8 to 16 hours. The steel is then furnace cooled to 800–850°C and subsequently cooled to room temperature in still air.[30] The defects are eliminated by solute diffusion.

NORMALISING

For normalising, hypoeutectoid steels are heated to a somewhat higher temperature (40–50°C above the Ac_3) than that used for full annealing. Hypereutectoid steels are heated above the Ac_m temperature as illustrated in Figure 29.14. The holding time depends on the size of the part. The minimum time is 15 minutes at temperature, with longer times being employed for larger parts, to ensure that the part is completely austenitised. Table 29.8 provides normalisation temperatures for various commonly encountered steels.

On completion of the required holding time, specified by the size of the part, the part is cooled in still air.[1,30] For a hypoeutectoid steel, the result will be a fine ferrite plus pearlite microstructure. Plain carbon and low-alloy steels should always be normalised.[32] Some alloy steels produce martensitic microstructures even with air cooling. Therefore, with such alloys, slower cooling rates are required to provide a uniform microstructure of ferrite plus pearlite.

Normalising is conducted:[30]

- To provide the desired microstructure and hence mechanical properties. For a given steel composition, normalised structures will be harder and stronger with lower ductility than if fully annealed.
- To improve hardening response by grain refinement and improved homogenisation.

- To improve machining characteristics, particularly for 0.15–0.40% carbon steels.
- To eliminate carbide networks in hypereutectoid steels.

There are various equations that may be used to calculate the hardness of normalised steel. One method is by using the Bofors equation.[32] The first step in this calculation is to determine the sum of the 'carbon potentials' – C_p:

$$C_p = C[1 + 0.5(C - 0.20)] + Si \times 0.15 + Mn[0.125 + 0.25(C - 0.20)]$$
$$+ P[1.25 - 0.5(C - 0.20)] + Cr \times 0.2 + Ni \times 0.1$$

where C is the carbon concentration in %. The ultimate tensile strength (in MPa) after normalisation is given by:

$9.81(27 + 56C_p)$ for hot-rolled steel

$9.81(27 + 50C_p)$ for forged steel

$9.81(27 + 48C_p)$ for cast steel

For steels which may be used in sub-zero conditions, a double normalising treatment may be specified.[32] In these cases, the steel is first heated to 50–100°C above the usual normalising temperature. This will produce greater dissolution of the alloying elements. The second normalisation step is conducted near the lower limit of the normalisation temperature range for the purpose of producing a finer grain structure.

STRESS RELIEVING

Stress relieving is used typically to remove residual stresses which have accumulated from prior manufacturing processes. Stress relief is performed by heating to a temperature below Ac_1 (for ferritic steels) and holding at that temperature for the required time, to achieve the desired reduction in residual stresses. The steel is then cooled sufficiently slowly to avoid the formation of excessive thermal stresses. No phase transformations occur during stress relief processing. Nayar recommends heating to:[30]

- 550–650°C for unalloyed and low-alloy steels;
- 600–700°C for hot-work and high-speed tool steels.

These temperatures are above the recrystallisation temperatures of these types of steels. Little or no stress relief occurs at temperatures <260°C and approximately 90% of the stress is relieved at 540°C. The maximum temperature for stress relief is limited to 30°C below the tempering temperature.[17]

The results of the stress relieving process are dependent on the temperature and time which are correlated through Holloman's parameter (P):[9]

$$P = T(C_{HJ} + \log t)$$

where T is the temperature (K), t is the time (h) and C_{HJ} is the Holloman–Jaffe constant which is calculated from:

$$C_{HJ} = 21.53 - (5.8 \times \%C)$$

P is a measure of the 'thermal effect' of the process and that processes with the same Holloman's parameter exhibit the same effect.

Another similar commonly used expression employed in evaluating the stress relief of spring steels is the Larson–Miller equation:[9]

$$P = T(\log t + 20)/1000$$

Stress relieving results in a significant reduction of yield strength in addition to a decrease in the residual stresses to some 'safe' value. Typically heating and cooling during stress relieving is performed in the furnace, particularly with distortion and crack-sensitive materials. Below 300°C, faster cooling rates can be used.

29.2.7   Quenching

QUENCHANT SELECTION AND SEVERITY

Quench severity, as expressed by the Grossman H-value (or number), is the ability of a quenching medium to extract heat from a hot steel workpiece. A typical range of Grossmann H-values (numbers) for commonly used quench media is provided in Table 29.7. Although Table 29.7 is useful to obtain a relative measure of the quench severity offered by different quench media, it is difficult to apply in practice because the actual flow rates for 'moderate', 'good', 'strong' and 'violent' agitation are unknown.

Alternatively, the measurement of actual cooling rates or heat fluxes provided by a specific quenching medium does allow a quantitative metric of the quench severity provided. Some illustrative values are provided in Table 29.9.[9,33] Typically, the greater the quench severity, the greater the propensity of a given quenching medium to cause increased distortion or cracking. This usually is the result of increased thermal stresses not transformational stresses. Specific recommendations for quench media selection for use with various steel alloys is provided by standards, such as AMS 2779. Some additional general comments regarding quenchant selection include:[34,35]

- Most machined parts made from alloy steels are oil quenched to minimize distortion.
- Most small parts or finish-ground larger parts are free-quenched.
- Larger gears, typically those over 8 inches are fixture (die)—quenched to control distortion.
- Smaller gears and parts such as bushings are typically plug-quenched on a splined plug typically constructed from carburised 8620 steel.
- Although a reduction of quench severity leads to reduced distortion, this may also be accompanied by undesirable microstructures such as the formation of upper bainite within carburised parts.
- Quench speed may be reduced by quenching in hot 149–204°C (300–400°F) oil. When hot oil quenching is used for carburised steels, lower bainite, which exhibits properties somewhat similar to those of martensite, is formed.
- Excellent distortion control is typically obtained with austempering, quenching into a medium just above the M_s temperature and then holding until the material transforms completely to bainite. The formation of retained austenite is a significant problem with austempering processes. Retained austenite is most pronounced where Mn and Ni are major components. The best steels for austempering are plain carbon, Cr and Mo alloy steels.[34]
- Aqueous polymer quenchants may often be used to replace quench oils but quench severity is still of primary importance.
- Gas or air quenching will provide the least distortion and may be used if the steel has sufficient hardenability to provide the desired properties.
- Low hardenability steels are quenched in brine or vigorously agitated oil. However, even with a severe quench, undesired microstructures such as ferrite, pearlite or bainite can form.

Table 29.9   COMPARISON OF TYPICAL HEAT TRANSFER RATES

Quench medium	Heat transfer rate $(Wm^{-2}\,K^{-1})$	Reference
Furnace	15	9
Still air	30	9
Compressed air	70	9
Nitrogen (100 kPa)	100–150	33
Salt bath or fluidised bed	350–500	33
Nitrogen (1 MPa)	400–500	33
Air-water mixture	520	9
Helium (1 MPa)	550–600	33
Helium (2 MPa)	900–1 000	33
Still oil	1 000–1 500	33
Liquid lead	1 200	9
Hydrogen (2 MPa)	1 250–1 350	33
Circulated oil	1 800–2 200	33
Hydrogen (4 MPa)	2 100–2 300	33
Circulated water	3 000–3 500	33

It is well known that the cracking propensity increases with carbon content. Therefore, the carbon content of the steel is one of the determining factors for quenchant selection. Table 29.10 summarises the mean carbon content limits for water, brine or caustic quenching of some steels.[33,35]

COMPONENT SUPPORT AND LOADING

Many parts, such as ring gears, may sag and creep under their own weight when heat treated, which is an important cause of distortion. Proper support when heating is required to minimise out-of-flatness and ovality problems which may result in long grinding times, excessive stock removal, high scrap losses and loss of case depth.[36] To achieve adequate distortion control, custom supports or press quenching may be required. Pinion shafts are also susceptible to bending along their length if they are improperly loaded into the furnace. When this occurs, the parts must be straightened, which will add to production cost.

SURFACE CONDITION

Quench cracking may be due to various steel related problems that are only observable after the quench but the root cause of which is not the quenching process itself. Examples include: prior steel structure, stress raisers from prior machining, laps and seams, alloy inclusion defects, grinding cracks, chemical segregation and alloying element depletion.[37] In this section, three surface condition related problems that may contribute to poor distortion control and cracking will be discussed: 'tight' scale formation, decarburisation and the formation of surface seams or 'non-metallic stringers'.

Tight scale problems are encountered with forgings hardened from direct-fired gas furnaces with high-pressure burners.[35,38] The effect of tight scale on the quenching properties of two steels, 1095 carbon steel and 18-8 stainless steel, is illustrated in Figure 29.16.[25] These cooling curves were obtained by still quenching into fast oil. A scale thickness of less than 80 μm (0.003 in.) increased the rate of cooling of 1095 steel, as compared with the rate obtained on a specimen without scale. However, a thick scale of 130 μm (0.005 in.) retarded the cooling rate. A very light scale, 13 μm (0.0005 in.) thick also increased the cooling rate of the 18-8 steel over that obtained with the specimen without scale.

In practice, the formation of tight scale will vary in depth over the surface of the part resulting in thermal gradients due to differences in cooling rates. This problem may yield soft spots and uncontrolled distortion and is particularly a concern with nickel-containing steels. Surface oxide formation can be minimised by the use of an appropriate protective atmosphere.

Another surface related condition is decarburisation which may lead to increased distortion or cracking.[39] At a given depth within the decarburised layer, the part does not harden as completely as it would at the same depth below the surface, if there was no decarburisation. This leads to non-uniform hardness which may contribute to increased distortion and cracking because the decarburised surface transforms at a higher temperature than the core (the M_s temperature decreases with carbon content).[35] This will lead to high residual tensile stresses at the decarburised surface or a condition of unbalanced stresses and distortion. Since the surface is decarburised, it will exhibit lower hardenability than the core. This will cause the upper transformation products to form early, nucleating additional undesirable products in the core. The decarburised side will be softer than the side that did not undergo decarburising, the greater amount of martensite in the latter leading to distortion. The solution to this problem is to restore carbon into the furnace atmosphere or machine off the decarburised layer.

Table 29.10   SUGGESTED CARBON CONTENT LIMITS FOR WATER, BRINE AND CAUSTIC QUENCHING

Hardening method	Shapes	Max. % Carbon
Furnace hardening	General usage	0.30
	Simple shapes	0.35
	Very simple shapes, e.g. bars	0.40
Induction hardening	Simple shapes	0.50
	Complex shapes	0.33

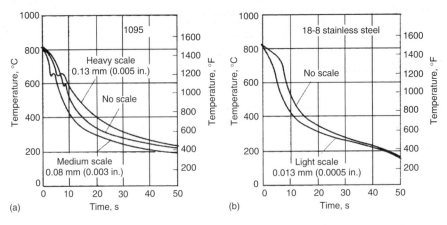

(a)   (b)

Figure 29.16   *Cooling curves illustrating the effect of scale on cylindrical steel bars (13 mm dia. × 64 mm, 0.5 in. dia. × 2.5 in.) when quenched into an accelerated oil without agitation (copyright ASM International, Materials Park, OH, used with permission): (a) AISI 1095 steel–oil temperature = 50°C (125°F); (b) 18-8 stainless steel–oil temperature = 25°C (75°F)*

Table 29.11   MINIMUM RECOMMENDED MATERIAL REMOVAL FROM HOT-ROLLED STEEL PRODUCTS TO PREVENT SURFACE SEAM AND NON-METALLIC STRINGER PROBLEMS DURING HEAT TREATMENT

	Minimum material removal per side[1]	
Condition	*Non-resulphurised*	*Resulphurised*
Turned on centres	3% of diameter	3.8% of diameter
Centreless turned or ground	2.6%	3.4%

[1] Based on bars purchased to special straightness, i.e. 3.30 mm in 1.52 m (0.13 inch in 5 feet) maximum.

Surface seams or non-metallic inclusions, which may occur in hot-rolled or cold-finished material, are defects that prevent the hot steel from welding to itself during the forging process. These defects act as stress raisers. To prevent this problem with hot rolled bars, a layer of material should be removed before heat treatment. Recommendations made earlier by Kern are provided in Table 29.11.[38]

HEATING AND ATMOSPHERE CONTROL

An important source of steel distortion and cracking during heat treatment is non-uniform heating without the appropriate protective atmosphere. For example, if steel is heated in a direct-gas-fired furnace with high moisture content, the load being heated may absorb hydrogen leading to hydrogen embrittlement and subsequent cracking which would not normally occur with a dry atmosphere.[5,40]

Localised overheating is a problem for inductively heated parts.[34,41] Subsequent quenching of the part leads to quench cracks at sharp corners and areas with sudden changes in cross-sectional area (stress raisers). Cracking is due to increases in residual stresses, at the stress raisers, during the quenching process. The solution to the problem is to increase the heating rate by increasing the power density of the inductor. The temperature difference across the heated zone is decreased by continuous heating or scanning of several pistons together on a single bar.[41]

For heat treating problems related to furnace design and operation, it is usually suggested that:[40]

a. The vestibules of atmosphere-hardening furnaces should be loaded and unloaded in the presence of a purge gas. Load transfer for belt and shaker hearth furnaces should only occur with thorough purging to minimise atmosphere contamination.
b. Hardening furnaces typically contain excessive loads prior to quenching. If the steel at quenching temperature is greater than 20% of the distance from discharge to charge door, this is too large a distance. Either the production rate can be increased or some of the burners can be turned off.

Table 29.12 DIMENSIONAL VARIATION IN HARDENED HIGH-CARBON STEEL WITH TIME AT AMBIENT TEMPERATURE

Steel type	Tempering temperature (°C)	Hardness (HRC)	Change in length (% × 10³) after time (days)			
			7	30	90	365
1.1% C	None	66	−9.0	−18.0	−27.0	−40.0
tool steel	120	65	−0.2	−0.6	−1.1	−1.9
790°C	205	63	0.0	−0.2	−0.3	−0.7
quench	260	62	0.0	−0.2	−0.3	−0.3
1% C/Cr	None	64	−1.0	−4.2	−8.2	−11.0
840 °C	120	65	0.3	0.5	0.7	0.6
quench	205	62	0.0	−0.1	−0.1	−0.1
	260	60	0.0	−0.1	−0.1	−0.1

RETAINED AUSTENITE

Dimensional changes, which are due to the specific volume of the transformation products formed as a result of quenching, may occur either slowly or rapidly. One of the most important, with respect to residual stress variation, distortion and cracking is the formation and transformation of retained austenite. For example, the data in Table 29.12[33] illustrates the slow conversion of retained austenite to martensite, which was still occurring days after the original quenching process, for the two steels shown.[42,43] This is a problem since dimensional control and stability is one of the primary goals of heat treatment. Therefore, microstructural determination is an essential component of any distortion control process.

QUENCHANT UNIFORMITY

Quench non-uniformity is perhaps the greatest contributor to quench cracking. Quench non-uniformity can arise from non-uniform flow fields around the part surface during the quench or non-uniform wetting of the surface.[33] Both lead to non-uniform heat transfer during quenching. Non-uniform quenching creates large thermal gradients between the core and the surface of the part. These two contributing factors, agitation and surface wetting will be discussed here.

Poor agitation design is a major source of quench non-uniformity. The purpose of the agitation system is not only to take hot fluid away from the surface to the heat exchanger, but also to provide uniform heat removal over the entire cooling surface of all of the parts throughout the load being quenched. Even though agitation is a critically important contributor to the performance of industrial quenching practice, relatively little is known about the quality and quantity of fluid flow encountered by the parts being quenched. Recently, agitation in various commercial quenching tanks has been studied by computational fluid dynamics (CFD) and in no case was optimal and uniform flow present, without subsequent modification of the tank.[44] Thus, identifying the sources of non-uniform fluid flow during quenching continues to be an important tool for optimising distortion control and minimising quench cracking.

The second source of non-uniform thermal gradients during quenching is related to interfacial wetting kinematics which is of particular interest for vaporisable liquid quenchants including, water, oil and aqueous polymer solutions. Most liquid vaporisable quenchants exhibit boiling temperatures between 100 and 300°C at atmospheric pressure. When parts are quenched in these fluids, surface wetting is usually time dependent which influences the uniformity of the cooling process and the achievable hardness and potential for the formation of soft spots. This is a problem with oil contaminated aqueous polymer quenchants, sludge contaminated oils and foaming.

29.2.8 Tempering

When steel is hardened, the as-quenched martensite is not only very hard but also brittle. Tempering, also known as 'drawing' (not to be confused with the metal forming process of the same name), is the thermal treatment of hardened (in general quenched, but as already noted some alloy steels will harden even if normalised) steels to obtain the desired mechanical properties which include: improved

Table 29.13 METALLURGICAL REACTIONS OCCURRING AT VARIOUS TEMPERATURE RANGES AND RELATED PHYSICAL CHANGES OF STEEL DURING TEMPERING

Stage	Temperature range	Metallurgical reaction	Expansion/ Contraction
1	0–200°C 32–392°F	Precipitation of ε-carbide Loss of tetragonality	Contraction
2	200–300°C 392–572°F	Decomposition of retained austenite	Expansion
3	230–350°C 446–662°F	ε-carbides decompose to cementite	Contraction
4	350–700°C 662–1 292°F	Precipitation of alloy carbides. Grain coarsening	Expansion

toughness and ductility, lower hardness and improved dimensional stability. During tempering, as-quenched martensite is transformed into tempered martensite which is composed of highly dispersed spheroids of cementite (or other carbides) dispersed in a soft matrix of ferrite, resulting in reduced hardness and increased toughness. The objective is to allow hardness to decrease to the desired level and then to stop the carbide decomposition by cooling. The extent of the tempering effect is determined by the temperature and time of the process.

The tempering process involves heating hardened steel to some temperature below the eutectoid temperature for the purposes of decreasing hardness and increasing toughness. (Tempering is performed as soon as possible after the steel has cooled to between 50–75°C and room temperature, to reduce the potential of cracking. If a tool steel cannot be tempered immediately after quenching, it is recommended that it be held at 50–100°C in an oven until it can be tempered.[32]) In general, the tempering process is divided into four stages, which are summarised in Table 29.13. The changes occurring during tempering include:[45]

1. transformation of the martensite,
2. transformation of retained austenite,
3. transformation and coarsening of the decomposition products of martensite.

The tempering process may be conducted at any temperature up to the lower critical temperature (Ac_1). Figure 29.17 illustrates the effect of carbon content of the martensite and tempering temperature on the hardness of carbon steels.[46] The specific tempering conditions that are selected are dependent on the desired strength and toughness. Nayar has recommended the following tempering conditions:[30]

- Heat the parts to 150–200°C to reduce internal stresses and increase toughness without a significant loss in hardness. This is often done with surface-hardened parts.
- To obtain the highest attainable yield strength and sufficient toughness, temper at 350–500°C.
- For optimum strength and toughness, heat to 500–700°C. (Note: when tempering at high temperatures, 675–705°C, precautions must be taken not to exceed the Ac_1 temperature, above which undesirable austenite may be formed, which upon cooling would transform to pearlite. This is a particular concern for nickel-containing steels, since nickel depresses the Ac_1 temperature.[47])

When steel is tempered in air, the heated oxide film on the surface of the steel exhibits a colour, known as a 'tempering colour' which is characteristic of the surface temperature. Table 29.14 provides a summary of characteristic surface temperatures for tempering and their colours.[32]

In addition to the four steps shown, there is another step referred to as 'refrigeration' or 'sub-zero' treatment. Sub-zero treatment is performed on steels to transform retained austenite to as-quenched martensite. Conversion of retained austenite in this way results in improved hardness, wear resistance and dimensional stability. Sub-zero treatment is performed using dry-ice or liquid nitrogen and involves cooling the steel to a temperature less than the M_f temperature of the steels which is typically between −30°C and 70°C (tool steels will elongate between about 0.5–2 μm per mm of the original length during heat treatment[3]). An immediate tempering step is required to remove residual stresses imparted to the steel by this process. Sub-zero treatments are not effective on steels that have been held at room temperature for several hours and therefore such treatments are typically performed immediately after hardening.[26] Although the general rule is to allow one hour per 25 mm (~1 inch) of the thickest cross-section, tool steels should be held at temperature for a minimum of two hours for each temper.[3]

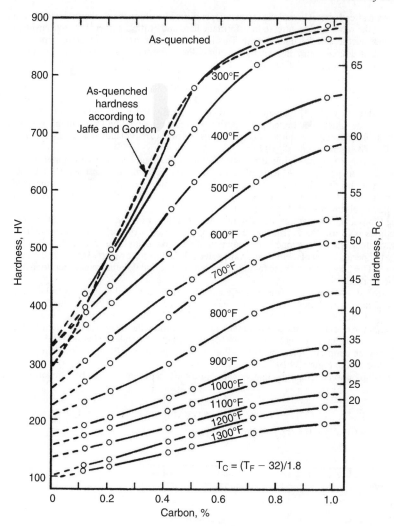

Figure 29.17 *Correlation of carbon content of martensite and hardness of different Fe–C alloys at different tempering temperatures. (T$_C$ and T$_F$ are the temperature in Centigrade and Fahrenheit respectively; reproduced with permission of Metallurgical and Materials Transactions)*

Table 29.14   TEMPER COLOURS OF STEEL

Abstracted from a much more detailed table in Thelning.[32]

Temperature (°C)	Range of temper colours
220–270	Straw yellow to brown
285–310	Purple to light blue
325–400	Various greys

Martensitic stainless steels and alloy steels that contain ≥0.4% carbon and which exhibit M$_s$ temperatures of about 300°C are particularly susceptible to cracking, especially if they are through-hardened.[32] In such cases, cooling may be interrupted at about 80°C followed by an immediate temper at about 170°C to stop the formation of martensite. However, significant amounts of untempered martensite remain after the steel is cooled to room temperature. Therefore, the steel must be tempered a second time, at the same temperature, to transform the hard and brittle, as-quenched martensite to a softer and more ductile, tempered martensite. This is one form of a 'double tempering' process.

Another form of double tempering occurs for high-alloy chromium steels and high-speed tool steels where significant amounts of retained austenite are transformed to martensite after tempering at about 500°C. Such steels should be retempered to obtain a tougher martensitic structure. Generally, this second tempering step is performed at about 10–30°C below the original tempering temperature.[32]

Steels that exhibit a high M_s temperature, $\geq 400°C$, typically those that contain <0.3% carbon, form martensite which may be tempered during the remaining cooling (quenching) process. This is called 'self-tempering' or 'auto-tempering' and such steels are typically not crack-sensitive, particularly if the M_f temperature is $\geq 100°C$.[32]

Typically, tempering times are a minimum of approximately one hour. Thelning has reported a 'rule of thumb' of 1–2 hours per 25 mm (1 inch) of section thickness, after the load has reached a preset temperature.[32] After heating, the steel is cooled to room temperature in still air. The recommended tempering conditions, in addition to the recommended heat treatment cycles, for a wide range of carbon and alloy steels are provided in SAE AMS 2759.

Tempering times and temperatures may also be calculated by various methods. One of the more common methods is to use the Larson–Miller equation discussed previously. The Larson–Miller equation, although originally developed for prediction of creep data, has been used successfully for predicting the tempering effect of medium/high alloy steels.[31] Bofors have reported that a Holloman–Jaffe constant of $C = 20$ was appropriate for all steels[3] but Grange and Baughman have reported that $C = 18$ should be used.[48]

Figure 29.18[49] shows that the Holloman–Jaffe constant varies with carbon content and desired hardness.[48] The incremental contribution to hardness of each alloying element in a steel may be determined from Table 29.15.[49] The Vickers hardness (HV) is calculated by multiplying the concentration of each of the alloying elements (within the range shown) by the factor for that element at a given constant (C) and then all of these values are added together to provide the hardness (HV). The relationship between tempering time and the Holloman–Jaffe parameter (P) at different tempering temperatures is shown in Figure 29.19.[49]

The relationship between tempering temperature, time and steel chemistry has been reported by Spies:[9,50]

$$HB = 2.84H_h + 75\,(\%\,C) - 0.78\,(\%Si) + 14.24\,(\%Mn) + 14.77\,(\%Cr) + 128.22\,(\%Mo)$$
$$- 54.0\,(\%V) - 0.55T_t + 435.66$$

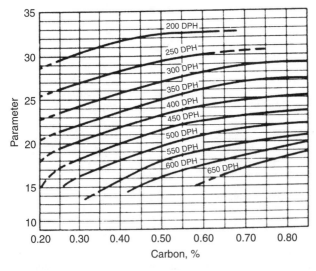

Figure 29.18 *Variation of the Holloman Parameter (P) at different hardness levels (© ASM International, Materials Park, OH, used with permission)*

Table 29.15   FACTORS FOR PREDICTING THE VICKERS HARDNESS OF TEMPERED MARTENSITE

© ASM International, Materials Park, OH, used with permission.

Element	Range (%)	*Factors at indicated parameter* (C) *value* [2]					
		20	22	24	26	28	30
Mn	0.85–2.1	35	25	30	30	30	25
Si	0.3–2.2	65	60	30	30	30	30
Ni	≤4	5	3	6	8	8	6
Cr	≤1.2	50	55	55	55	55	55
Mo	≤0.35	40	90	160	220	240	210
		(20)[1]	(45)[1]	(80)[1]	(110)[1]	(120)[1]	(105)[1]
V [3]	≤0.2	0	30	85	150	210	150

[1] If 0.5–1.2% Cr is present use this factor.
[2] Note: the boron factor is 0.
[3] May not apply if vanadium is the only carbide former present.

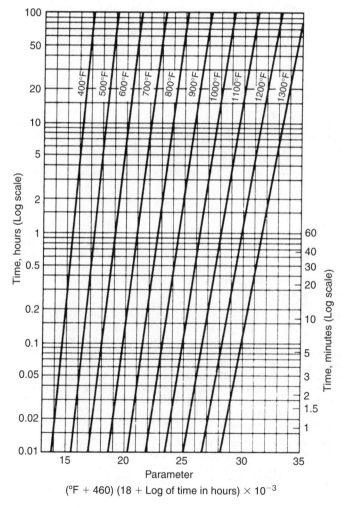

Figure 29.19   *Time-temperature versus Holloman Parameter chart for C = 18 (© ASM International, Materials Park, OH, used with permission)*

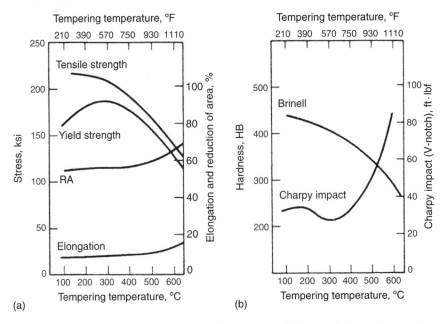

Figure 29.20   *Loss in room-temperature toughness due to temper embrittlement of oil-quenched wrought Ni–Cr–Mo (AISI 4130) steel; section size = 13 mm (0.5 in.)*

where HB is the Brinell hardness after hardening and tempering, H_h represents the Rockwell C (HRc) hardness after hardening and T_t is the tempering temperature in °C. This equation was developed for the following conditions: H_h = 20–65 HRc, C = 0.20–0.54%, Si = 0.17–1.40%, Mn = 0.50–1.90%, Cr = 0.03–1.20% and T_t = 500–650°C (932–1 202°F).

The German DIN 17021 standard provides a relationship between the as-quenched hardness (H_h) and the as-tempered hardness (H_t) :[9]

$$H_h(\text{HRc}) = (T_t/167 - 1.2)H_t - 17$$

where T_t = tempering temperature in °C.

Temper embrittlement may occur for some steels if they are tempered below 595°C, which is observable by a reduction in notch toughness as illustrated in Figure 29.20 for AISI 4130 (Ni–Cr–Mo), steel which was tempered at 260–370°C.[17] However, temper embrittlement may be avoided by tempering at higher temperatures, with subsequent quenching, to minimise the time the steel will spend in the intermediate temperature range.[17,47]

Tempering may be performed in convection furnaces, salt baths or even by immersion in molten metal. Induction tempering and flame heating are also used but will not be discussed here. Table 29.16 provides a comparative summary of the different heating media.[51] Of the different tempering systems shown, convection furnaces are the most common and it is important that they be equipped with fans and/or blowers to provide uniform heat transfer, when heating the load. Typically, convection tempering furnaces are designed for use in the 150–750°C range.

Salt baths may also be used for various heating processes over the temperature range 150–1 320°C[19] and they provide relatively rapid heat transfer, compared with convection furnaces, although the actual use temperature is dependent on the composition of the salt bath. A comparison of heating rates between salt baths and muffle furnaces is provided in Table 29.17.[32]

Sinha has classified salt baths into three groups:[19]

- Low-temperature salt baths may be used from 150–620°C. These baths are of two types: a binary mixture of equal parts of potassium nitrate and sodium nitrite, which may be used for heating to 150–500°C and binary mixtures of potassium nitrate and sodium nitrate which may be used for heating in the range of 260–620°C. In addition to tempering, these baths may also be used for cooling. It is essential however, that the baths are not contaminated with cyanides, organic compounds or water!

Table 29.16 TEMPERING TEMPERATURE RANGES ACHIEVABLE WITH DIFFERENT TEMPERING EQUIPMENT
© ASM International, Materials Park, OH, used with permission.

Equipment type	Temperature range		Use conditions
	°C	°F	
Convection furnace	50–750	120–1 380	For large volumes of similar parts; for variable loads. Temperature control more difficult.
Salt bath	160–750	320–1 380	Rapid uniform heating. Low to medium volume. Should not be used for complex, hard to clean parts.
Fluidised beds	100–750	212–1 380	Broad range of heat transfer rates are possible by varying the choice of fluidising gas, gas velocity, bed temperature and the bed particle size. More energy efficient than convection furnaces and they provide a safe and ecologically friendly alternative to salts and lead, with similar heat transfer rates.
Oil bath	≤250	≤480	Good if long exposure times are desired; special ventilation and fire control are required.
Molten metal bath	>390	>735	Very rapid heating; special fixturing required; molten metals may be toxic (Pb baths).

Table 29.17 HEAT-UP TIMES FOR 100 mm DIAMETER BARS TO 950–1 000°C

Abstracted from a more detailed table in Thelning.[32]

Salt bath (min)	Muffle furnace (min)
8	60

- Medium-temperature neutral baths which are suitable for use over the range of 650°C–1 000°C. These baths are binary or ternary mixtures of the following salts: potassium chloride, sodium chloride, barium chloride or calcium chloride. Two examples of typical binary compositions and working temperatures include: NaCl (45%)/KCl (55%) which is suitable for use at 675–900°C and NaCl (20%)/$BaCl_2$ (80%) which is suitable for use at 675–1 060°C. $BaCl_2$, if used at 100%, has a relatively narrow use range of 1 025–1 325°C.[32] An advantage of these baths is that when they are freshly prepared, the steel surface will be clean without surface carburisation or decarburisation.
- High-temperature salt baths are used in the range of 1 000–1 300°C and they typically contain mixtures of barium chloride, sodium tetraborate (borax), sodium fluoride and silicates. These baths may decarburise steels, as oxides build-up after use.

Molten metal, most typically lead, has been used in the past, but due to its toxicity, its use is now restricted to various heating operations where its outstanding heat transfer properties are essential. Lead baths are used from 327°C, which is the melting point of lead, up to 900°C.[19]

Fluidised beds are formed by passing a gas through solid particles such as aluminium oxide and silica sand, which causes the particles to behave like a bubbling liquid. The particles are generally inert and do not react with metal parts but act to facilitate heat transfer between the fluidising gas and the part being processed.

A broad range of heat transfer rates are possible over operating temperatures which may range from 100–1 050°C (212–1 920°F) with fluidised bed furnaces for tempering operations typically ranging from 100–750°C (212–1 380°F) (*see* Table 29.16). Fluidised bed furnaces are not only more energy-efficient than convection furnaces but also exhibit heat transfer efficiency similar to salt baths and lead pots, without the health and environmental safety hazards commonly associated with these systems.

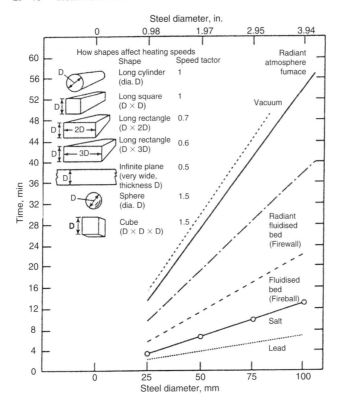

Figure 29.21   *Recommended heating times for various heating media and shapes*

Figure 29.21 provides a comparison of the relative heating rates that can be achieved with these different heating media.[27]

29.2.9   Austempering

Austempering requires that the cooling process be fast enough to avoid the formation of pearlite, as illustrated by the cooling curve for an austempering process, which is superimposed on to an ITD (TTT diagram) of a steel in Figure 29.22A.[9] The steel is cooled below the nose of the pearlite transformation curve for the steel being quenched and just above the M_s temperature, in a molten salt bath. The steel is held at this temperature until the transformation from austenite to bainite is complete. Upper or lower bainite may be formed depending on the molten salt temperature. Austempering eliminates the volumetric expansion due to martensite, which helps to eliminate cracking and provides greater toughness and ductility at a particular hardness.

In practice, austempering is performed by heating the steel to 790–870°C and then quenching into a molten salt bath at a temperature just above the M_s temperature (260–400°C). The steel is kept at the molten salt bath until the transformation is complete, at which time it is cooled to room temperature in air.[30]

The suitability of steels for austempering depends on:[52]

- The location of the nose of the pearlite start (P_s) curve on the ITD and the ability to cool the steel sufficiently fast to avoid pearlite formation.
- The time required for complete transformation from austenite to bainite, at the austempering temperature.
- The M_s temperature of the steel.

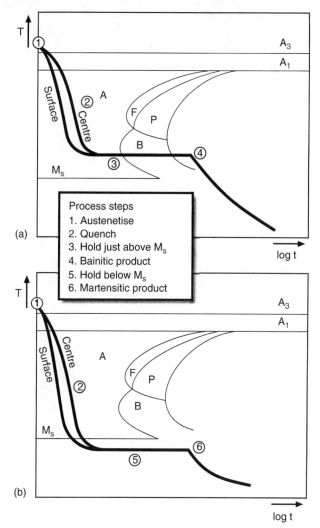

Figure 29.22 *Illustration of: (a) austempering; (b) martempering (redrawn, with modifications, after reference 9, p. 456—courtesy of Marcel Dekker, Inc.)*

29.2.10 Martempering

Martempering (also known as marquenching) is a cooling process used to minimise distortion and cracking and to reduce the amount of residual stress formation of a part upon cooling. Typically, martempering involves austenitising steel at a temperature above Ac_1 (815–870°C) and then quenching into hot oil (marquenching oil) or molten salt at a temperature just above the M_s temperature (usually at a temperature ranging from 260–400°C) followed by holding at this temperature until the entire steel workpiece is at the same temperature, at which time the workpiece is cooled to room temperature, at a rate sufficient to achieve a fully martensitic structure. Note that if the part is held at the martempering temperature too long, undesirable formation of bainite may occur.[2] Usually the quenching step is performed in air which minimises thermal gradients and related residual stresses throughout the part while cooling. This process is illustrated in Figure 29.22B.[9] After quenching, the part is tempered per specification.

A variation of this process is to cool the part to a temperature intermediate between the M_s and M_f temperature, until the temperature at the surface and core have completely equilibrated, at which

Table 29.18   TEMPERATURE RANGE OF MARTENSITE FORMATION OF SEVERAL CARBON AND LOW-ALLOY STEELS

Steel grade AISI	M_s temperature		50% Martensite formed		M_f—99% Martensite formed	
	°C	°F	°C	°F	°C	°F
1030	343	650	293	560	232	450
1065	277	530	219	425	149	300
1090	219	425	157	315	82	180
1335	343	650	293	560	232	450
2340	304	580	293	560	207	405
3140	335	635	288	550	221	430
4130	380	715	343	650	288	550
4140	343	650	299	570	227	440
4340	288	550	249	480	188	370
4640	343	650	299	570	249	480
5140	343	650	299	570	232	450
6140	327	620	293	560	232	450
8630	366	690	332	630	277	530
9440	330	625	282	540	207	405

Table 29.19   TYPICAL USE TEMPERATURES FOR MARTEMPERING OILS

Viscosity at 40°C, SUS	Minimum flash point		Use temperatures			
			Open air		Protective atmosphere	
	°C	°F	°C	°F	°C	°F
250–550	220	430	95–150	200–300	95–175	200–350
700–1 500	250	480	120–175	250–350	120–205	250–400
2 000–2 800	290	550	150–205	300–400	150–230	300–450

time the steel is removed from the bath and air cooled or placed directly in the tempering furnace. Table 29.18 provides a summary of M_s and M_f temperatures for a variety of common steels.[17]

Either martempering (hot quenching) oils or molten salt baths may be used for martempering. Molten salts were reviewed briefly above. Martempering oils will be discussed briefly here. Martempering oils are used at temperatures up to 205°C, sometimes as high as 235°C. Molten salts are used for martempering operations performed at 204–400°C.[52] The oils are usually formulated with solvent refined petroleum oils with a very high paraffinic fraction, to maximise oxidative and thermal stability, which is further enhanced by the addition of antioxidants. Accelerated and non-accelerated martempering oils are available. Typical temperature ranges for commercially available martempering oils are summarised in Table 29.19.[53] Because martempering oils are used at relatively high temperatures, a protective, non-oxidising atmosphere is often employed, which allows the use temperatures much closer to the flash point, than is generally recommended for open air conditions.

More hardenable alloy steels are martempered more often than less hardenable plain carbon steels. Steels that are most often martempered to full hardness include: AISI 1090, 4130, 4140, 4150, 4340, 4640, 5140, 6150, 8630, 8640, 8740, 8745. Carburised steels from 3312, 4620, 5120, 8620 and 9310 may also be martempered. Plain carbon steels 1008 through 1040 are insufficiently hardenable to be martempered. Thin sections of some steels of borderline hardenability such as AISI 1541 can be martempered.[52]

29.2.11   Carburising

THERMODYNAMICS OF THE CARBURISING PROCESSES

Carburising systems involve reversible chemical reactions of the form:

$$xA + yB \iff qC + rD$$

for which the equilibrium constant (K_{eq}) is:

$$K_{eq} = (p_C^q p_D^r)/(p_A^x p_B^y)$$

where p_A, p_B, p_C, p_D are the partial pressures of reactants A and B and products C and D and x, y, c and d represent the stoichiometry of the reaction. In the event that product C is dissolved in the steel, the activity of C in the steel (a_C) would be substituted for p_C.

These reactions attempt to approach equilibrium involving the atmospheric constituents and the substrate material (steel). At equilibrium, the change in Gibbs free energy (ΔG) is zero by definition. Hence:

$$\Delta G = \Delta G^\circ + RT \ln K_{eq} = 0$$

Thus, at equilibrium, the equilibrium constant, K_{eq}, is given by:

$$\Delta G^\circ = -RT \ln K_{eq}$$

where:

ΔG°—standard free energy of the reaction [kJ mol^{-1}],
R—Gas constant (8.314 J K^{-1} mol^{-1}),
T—Absolute temperature [K].

A carbon-rich atmosphere that is capable of carburising steel objects, when held at an appropriate process temperature, is known as a controlled carburising atmosphere. Decarburising/carburising is governed by the following reversible reaction:

$$C(\text{in } \gamma) + CO_2(g) \Longleftrightarrow 2CO(g)$$

which has an equilibrium constant of $K_B = p_{CO}^2/(p_{CO_2} a_C)$

When using methane as the process enrichment gas, the reaction involved is:

$$CH_4(g) \Longleftrightarrow C(\text{in } \gamma) + 2H_2(g)$$

for which the equilibrium constant is $K_M = (p_{H_2}^2 a_C)/p_{CH_4}$
As well as the following:

$$C(\text{in } \gamma) + H_2O(g) \Longleftrightarrow CO(g) + H_2(g)$$

with an equilibrium constant $K_W = p_{CO} p_{H_2}/(p_{H_2O} a_C)$, where K_B, K_M, K_W are the equilibrium constants of the reaction concerned, a_C is the carbon activity, p_{CO}, p_{CO_2}, p_{H_2}, p_{H_2O}, p_{CH_4} are the partial pressures of the species shown.

Thus, by selection of an appropriate atmosphere, carbon may be introduced into the austenite as a solid-solution. Alternatively, free carbon may be present in the furnace atmosphere, in the form of soot (graphite). If the atmospheric carbon concentration is high enough, graphitic soot will precipitate out of the atmosphere and begin to concentrate on the work piece, the walls of the furnace and mechanical parts within the furnace, such as the rails, hearth, rollers, burner tubes and elements.

The equilibrium constants for the reactions shown above are:

$$\log K_B = -8\,750/T + 9.022$$
$$\log K_M = -4\,768/T - 5.767$$
$$\log K_W = -6\,908/T + 7.457$$

KINETICS OF THE CARBURISATION PROCESS

Carbon deposition at the steel surface

Consider now the rate at which carburisation occurs. At the steel/gas phase interface the carburisation reaction will be dependent on the difference between the carbon potential (carbon activity) in the atmosphere and at the steel surface. Carbon will diffuse from the atmosphere to the steel surface when the activity of the carbon in the atmosphere is higher than the activity of carbon in the austenite surface, which is dependent on furnace temperatures and the initial carbon concentration in the steel. Thus, the number of carbon atoms, N, that penetrate a surface area S, in time dt, is dependent on the difference between the carbon activity in the gaseous medium α_{cg} and the carbon activity on the steel surface α_{cp}:

$$\frac{N}{S\,dt} = -\beta(\alpha_{cg} - \alpha_{cp})$$

If the carbon concentration is substituted for activity then:

$$\frac{M}{F\,dt} = -\beta(C_A - C_S)$$

where:

  M—mass of carbon deposited (g),
  F—surface area (m^2) penetrated by carbon of mass M,
  dt—time of carbon penetration (s),
  C_A—carbon potential of the atmosphere ($g\,m^{-3}$),
  C_S—surface carbon content ($g\,m^{-3}$),
  β—surface reaction rate constant ($m\,s^{-1}$).

The carbon transfer coefficient is defined in European norm EN10052: 1993[54] as: 'the transfer of a carbon mass from the carburising medium to the steel surface, through individual carbon molecules to the steel surface by the difference of potentials between carbon potential of medium, and the steel carbon potential at the surface at any given moment. The mass transfer stream of carbon from atmosphere to steel is quantifiable in the following formula':

$$\text{Carbon flux (in } g\,s^{-1}) = \beta_C\,(C_P - C_S)\,F$$

where:

  β_C—carbon transfer coefficient ($m\,s^{-1}$),
  C_P—carbon potential of atmosphere ($g\,m^{-3}$)
  C_S—carbon content of steel surface ($g\,m^{-3}$)
  F—surface area (m^2)

which is identical to the previous equation, apart from the change in nomenclature.

Assuming that steady-state conditions are established, the rate of carbon transport to the steel surface can be described by means of Fick's first law of diffusion, which states, using nomenclature appropriate to the diffusion of carbon, that:

$$J_C = -D_C\frac{\partial C}{\partial x}$$

In this equation, J_C is the carbon flux, i.e. the amount of carbon passing through 1 m^2 of reference plane per second ($g\,m^{-2}\,s^{-1}$). C represents the carbon concentration ($g\,m^{-3}$) and x is distance (m), so that the one dimensional carbon concentration gradient is ($\partial C/\partial x$). Note: any convenient measure of the amount of carbon can be used in place of grams in this equation, but a consistent unit must be employed for the amount of carbon in the C and J terms. D_C represents the carbon diffusion coefficient (diffusivity) in the medium in which the carbon is diffusing (and has units of $m^2\,s^{-1}$). The value of D_C will depend strongly on the carburisation temperature.

Re-expressing the carbon flux in Fick's first law in units of $g\,s^{-1}$ yields:

$$\text{Carbon flux (in } g\,s^{-1}) = D_C\frac{dC}{dx}F$$

Given continuity of carbon mass flow from atmosphere to steel and then into the steel surface, it is then possible to equate the two expressions for the carbon flux, which yields the following mass transfer formula:

$$\frac{D_C}{\beta_C} = \frac{C_P - C_S}{dC/dx}$$

The value of the carbon transfer coefficient β_C (and hence the capability to transport carbon to the steel surface) depends on the source and hence the carbon potential of the medium used for carburising. From the literature, the smallest carbon transfer coefficient β_C is $5.4 \times 10^{-10} \, m\,s^{-1}$, for the mixture CO–CO_2. A somewhat larger value of $1.79 \times 10^{-9} \, m\,s^{-1}$ has been reported for a mixture of CH_4–H_2–H_2O. Mixtures of CO–CO_2–CH_4–H_2–H_2O exhibit higher β_C values: $3.1 \times 10^{-7} \, m\,s^{-1}$ for a mixture of 50% CO–CO_2 and 50% CH_4–H_2–H_2O. An industrial endothermic atmosphere composed of fuel gases exhibits a β_C value of $1.2 \times 10^{-7} \, m\,s^{-1}$.

Diffusion of carbon into the bulk

In the last section, Fick's first law was used to describe the steady-state diffusion of carbon to the surface of the steel, such that a fixed carbon concentration C_S is established at the surface of the steel. Within the steel, however, non-steady state carbon diffusion occurs as the case develops and so it is necessary to employ Fick's second law of diffusion, i.e.:

$$\frac{\partial C}{\partial t} = D_C^\gamma \frac{\partial^2 C}{\partial x^2}$$

In this equation, t is time (in seconds) and D_C^γ is the diffusion coefficient of carbon in austenite at the carburisation temperature. Other terms are as defined previously (although, in the present case they refer to the carbon concentration within the steel). This equation does not permit a general analytical solution. However, the following result can be obtained for the carbon concentration as a function of distance and time, $C(x,t)$, during carburisation of a steel with an initial carbon concentration C_I prior to carburisation:

$$C(x,t) = C_I + (C_S - C_I) \left[1 - \mathrm{erf}\left(\frac{x}{2\sqrt{D_C^\gamma t}} \right) \right]$$

In this case, $x = 0$ is defined as the surface of the steel in contact with the carburising atmosphere, which has a fixed carbon concentration C_S. This equation may be solved with appropriate error function (erf) tables, or a spreadsheet that supports the error function. Alternatively, if all that is required is an order of magnitude estimate of the case depth, then the following approximation can be employed:

$$\text{Case depth} \sim \sqrt{D_C^\gamma t}$$

In both of these equations, the assumption is that the rate controlling step governing the formation of the case is the solid-state diffusion of carbon in austenite. The value of D_C^γ for use in either of these equations can, in turn, be calculated using the Arrhenius equation:

$$D_C^\gamma = (D_0)_C^\gamma \exp\left[-\frac{Q_C^\gamma}{RT} \right]$$

Notice that both the pre-exponential term $(D_0)_C^\gamma$, sometimes called the frequency factor (with units of $m^2 \, s^{-1}$) and the activation energy for diffusion, Q_C^γ (with units of $kJ \, mol^{-1}$), are material specific to the diffusing solute (carbon) and matrix (austenite) involved. In this equation, R is the gas constant ($8.314 \, J \, K^{-1} \, mol^{-1}$) and T is the absolute temperature (in K) at which carburisation is performed. Table 29.20 contains tabulated values for D_C^γ as a function of temperature. See Chapter 13 for diffusion data.

CLASSIFICATION OF CARBURISING PROCESSES

The most common processes that are encountered industrially are:

a. *Pack carburising*—The pack carburising process is typically conducted by surrounding the steel in a pit furnace or steel box furnace with granules of charcoal or charcoal plus coke. An 'activator' for the charcoal, such as barium borate ($BaBO_3$) is added to facilitate the release of CO_2, which then reacts with excess CO_2 to form CO which in turn reacts with the low-carbon steel surface to form carbon, which diffuses into the steel. Pack carburising is conducted typically at 920–940°C for 2–36 hours.

Table 29.20   EFFECT OF TEMPERATURE ON THE DIFFUSION
COEFFICIENT OF CARBON IN AUSTENITE (D_C^γ)

Temperature		
°C	°F	D_C^γ (m^2 s^{-1})
760	1 400	7.8×10^{-13}
788	1 450	1.2×10^{-12}
816	1 500	1.9×10^{-12}
843	1 550	2.8×10^{-12}
871	1 600	4.2×10^{-12}
899	1 650	6.0×10^{-12}
927	1 700	8.6×10^{-12}
954	1 750	1.2×10^{-11}
982	1 800	1.6×10^{-11}
1 010	1 850	2.2×10^{-11}
1 038	1 900	3.0×10^{-11}
1 066	1 950	4.0×10^{-11}
1 093	2 000	5.2×10^{-11}
1 121	2 050	6.8×10^{-11}

b. *Liquid carburising*—Liquid carburising is conducted typically in internally or externally heated molten salt pots containing a cyanide salt such as sodium cyanide (NaCN). There are generally two types of liquid carburising processes. One type is a *low temperature* process (840–900°C) which is conducted when low case depths of 75 to 760 µm (0.003–0.03 inches) are required. The second liquid carburising process is conducted at a *high temperature* (900–950°C) when case depths of 760 to 3 050 µm (0.03–0.12 inches) are desired. In either case, process times may be 1–4 hours.

c. *Gas carburising*—Currently, the most common carburising process is gas carburising which may potentially be performed with any carbonaceous gas such as: methane, ethane, propane, or natural gas. Carburising times of 4–10 hours are typical. The carburising temperature is greater than the upper critical temperature (in the austenite transformation region, >954°C). Case depths are typically less than 1.3 mm (0.05 inches). The conventional gas carburising process allows measurement of the gaseous carbon activity within the furnace process chamber, using the shim test method, dew point test method, CO/CO_2 test method and oxygen probe.

d. *Vacuum carburising*—Vacuum carburising is a clean method used to introduce carbon into the surface of the steel and also prevents grain boundary oxidation. Vacuum, or low-pressure, carburising is carried out in a vacuum furnace at pressures below normal atmospheric pressure. The principle of carburising is exactly the same as that of the gas carburising process, the main difference being the use of subatmospheric pressure.

e. *Ion carburising*—Carbonaceous gases such as methane have been used for vacuum carburising because of their widespread availability and current use in gas carburising. Although methane is reactive and controllable when used with endothermic atmospheres, methane alone is extremely stable, even at elevated temperatures. Therefore, when methane is used in vacuum carburising, relatively high pressures of 33 to 53 kPa (250–400 torr) are required. Furthermore, carburising processes conducted at these relatively high pressures produce significant amounts of carbon sooting. These deficiencies however, may be overcome by using an ion process. Methane ionisation is produced with a high-voltage (approximately 1 000 volts) at a relatively low pressure of 1.3 kPa (10 torr). When methane ionises by this 'ion carburising' process, a reactive gas blanket is formed in close proximity to the workpiece, without concurrent soot formation. Other hydrocarbons can be used similarly.

f. *Fluidised bed carburising*—Steel carburising processes may also be conducted in fluidised bed furnaces. Various atmospheres may be used, including conventional endo gas/hydrocarbon mixtures or nitrogen/methanol/hydrocarbon mixtures. Depending on the carburising atmosphere, fluidised bed temperatures of 850–975°C for 30 min to 3 hours may be used. Case depths of up to 700 µm are typical.

Table 29.21 provides a comparative summary of the relative advantages and disadvantages of these processes.

Table 29.21 COMPARISON OF CARBURISING METHODS

Description of method	Disadvantages	Advantages
A. Carburising with solids (pack carburising)		
1. The basic component of the carburising medium is ground wood charcoal (around 3–5 mm granules), which is mixed with carbonates of barium, sodium, calcium, lithium or potassium.	1. A long heat up time is necessary to reach the process temperature and to achieve temperature uniformity throughout the box.	1. Low capital equipment cost. cost.
2. The temperature used for carburisation is about 900°C.	2. Decarburisation of the steel surface will occur if the components are allowed to air cool without protection, or removed from the process box.	2. Simple procedure.
3. It is necessary to place the components into a steel box with a spacing of 25 mm between the components. The box design is very simple, although this does require a lid which can be sealed with clay to contain the liberated gas.	3. It is difficult (but possible) to harden directly from the carburising box. It is usual to allow the box to cool down and then reheat to the required austenitising temperature.	3. Low operating costs.
4. Starting the reactions during heating can be accomplished by: • Burning of a small quantity of the wood charcoal by the introduction of oxygen. • Reaction of an activator with wood charcoal.	4. The method is not reliable in terms of repeatability and cannot be controlled accurately. This is a slow production method.	
5. After the carburising, the carburising boxes can be cooled down within the furnace. Alternatively, the boxes can be removed from the furnace whilst hot and air cooled.	5. Grinding is necessary after the procedure, due to a slight potential for surface porosity.	
B. Liquid carburising		
1. A mixture of molten salts is the carburising medium. Usual mixtures are carbonates or chlorides and cyanides of alkaline metals, sometimes with an addition of SiC. A typical mixture would be: 75% Na_2CO_3, 15% NaCl, 10% SiC.	1. There is a large amount of sludge that collects in the bottom of the salt bath. It is mandatory to clean out the sludge on a frequent basis.	1. Possibility of direct hardening.
2. The temperature used for carburisation is usually between 900°C and 950°C.	2. High operating costs, due to post cleaning and effluent disposal. The process is labour intensive and involves long pre-wiring times for the components.	2. Uniform carburising of clean surfaces.
3. The steel components are placed directly into the molten salts after pre-heating. A cocoon of salt will immediately adhere to the steel surface, thus offering some thermal protection.	3. Carburising stop off is difficult.	3. Elimination of steel boxes that are used in pack carburising procedures.
4. In the presence of iron, the cyanide salt decomposes at the high temperature of the bath and cyanate (CN) is liberated, which further decomposes to provide carbon to diffuse into the steel.	4. Components need a long pre-heat time.	4. It has been possible to accelerate the process considerably by applying an electrolytic method.
	5. The probability of distortion remains.	

(*continued*)

Table 29.21   COMPARISON OF CARBURISING METHODS—*continued*

Description of method	Disadvantages	Advantages
C. Gas carburising		
1. Atmospheres for carburising are produced in special generators which produce a process gas from natural gas blended with air. The generators are known as endothermic generators. The natural gas will contain a large proportion of methane, plus lower concentrations of other hydrocarbon gases. The air that is mixed with the natural gas will contain moisture which will assist in controlling the carbon potential of the endothermic gas.	1. Limited speed of diffusive satiating, resulting from limitations of the furnace construction as well as the carbon potential of previous carburising atmospheres.	1. Easy to change the carbon potential of the furnace atmosphere, simply by adjusting the enrichment gas in relation to either the moisture present in the atmosphere or by the presence of free oxygen.
	2. Carburising atmospheres will contain oxygen in the form of moisture, which will cause intergranular oxidation and also create the potential for grain boundary corrosion. This will cause a deterioration of the fatigue strength of the carburised case.	2. Small waste of energy and economy of time.
2. The temperature used for carburisation is usually between 870–950°C.		3. Parts are relatively clean except from oil quenching if the furnace has an oil quench system, such as is seen on an integral quench furnace.
3. The furnace heating system will assist the gaseous atmosphere to ensure good temperature uniformity within the furnace process chamber. In addition to this, the carbon potential is usually very uniform throughout the process chamber.	3. Considerable emission to atmosphere of harmful substances (oxides of carbon and heated quench oil effluent resulting from quenching). It is advisable to install fume extraction systems to ensure adequate shop ventilation.	4. Possibility of hardening directly after carburising. However great care must be taken when considering this.
	4. Oil quenching may lead to distortion and possibly the risk of cracking.	
D. Vacuum carburising		
1. The process is conducted at sub-atmospheric (partial vacuum) pressures. Process gases of methane, propane, or acetylene are introduced into the process chamber.	1. High capital cost of equipment.	1. Fast process times at conventional carburising temperatures.
	2. High operating costs.	2. Advantage can be taken of high temperature carburising, up to 1 075°C. Since carburising process temperatures approximately 150°C higher than for conventional gas carburising temperatures can be used, this will result in a shorter gas phase transportation time, resulting in much faster carburising times. The superficial concentration of carbon as a result of an unequal process of break-up of hydrocarbons is, as a rule, very high. Both of these factors will accelerate considerably the diffusive saturation.
2. Atomic carbon is generated as a result of the break-up of the process gases. The process of vacuum the carburising involves following stages: • Saturation of the atmosphere with carbon at the process working pressure. • Diffusive transportation of excess carbon into the steel surface in high vacuum.	3. Difficult process control in terms of determining and controlling the atmospheric carbon potential.	
3. Single chamber vacuum furnaces can be used in any configuration. Also, front or rear cooling chambers can be fitted, that can facilitate controlled cooling of the processed batch after austenitising. High pressure gas quenching can be accomplished when using blended gaseous mixtures of either nitrogen and helium or nitrogen and hydrogen. Blended gas quenching can (depending on the gas blend and delivery pressure), equal the quench speed of oil.		3. The carburised layer shows the best mechanical properties. 4. Clean finished work surfaces that do not require post cleaning. 5. Environmentally friendly with no toxic gas emissions. Also, lower volumes of effluent gas than for conventional processes.

(continued)

Table 29.21　COMPARISON OF CARBURISING METHODS—*continued*

Description of method	Disadvantages	Advantages
D. Vacuum carburising—continued		
		6. Mechanical handling equipment can easily be installed into the system for part transportation within and outside of the furnace. Effective, energy-saving process.
		7. Hydrocarbons are the exclusive carrier of carbon. Therefore there is no risk of grain boundary oxidation.
		8. Cooling of the work load can be accomplished, if necessary, under nitrogen. This will eliminate the need for post washing.
		9. The quench gas direction can be manipulated to suit the part geometry, thus reducing the risk of distortion. In addition, the gas flow rate can be adjusted when using a two speed gas circulation drive motor. Thus, the risk of distortion is less than when quenching into oil.
E. Ion carburising		
1. The process depends on placing the steel in a vacuum chamber in a low pressure hydrocarbon atmosphere, with the simultaneous application of a high voltage, on heating.	1. High capital investment.	1. It is possible to control the thicknesses and structure of diffusive layers.
2. The furnace wall is the anode whilst the work load sits on the furnace hearth at the cathode's potential. A voltage is applied in the region of 450 to 800 volts (dependent on the chamber pressure and the workload surface area), which will cause a glow discharge in the process chamber. With the work load at cathode's potential, the process gas is immediately ionised and carburisation will begin to take place.	2. Analysing the carbon potential within the process chamber is difficult. Thus, determination of the carbon potential must usually be accomplished prior to each batch, using data acquired from previously carburised loads.	2. It has been possible to carburise and successfully treat (on a repeatable, continuous basis) stainless steels, heat-proof and acid-resistant steels.
3. The steel surface does not act as a catalyst as occurs with the more conventional carburising procedures. However, the process gases used are still methane, propane or acetylene, along with nitrogen and hydrogen.		3. Case uniformity is excellent, irrespective of the part geometry.

(*continued*)

Table 29.21 COMPARISON OF CARBURISING METHODS—*continued*

Description of method	Disadvantages	Advantages
F. Carburising in fluidised beds		
1. A fluidised bed is created usually as a result of the activation by a gas passing through a bed of particles such as sand or aluminium oxide. The particles of the bed are kept in suspension by hot process gas passing upwards and through the particles.	1. No major problems.	1. Simple, but efficient furnace design.
		2. Low operating cost.
		3. Ease of operation.
		4. Not labour intensive.
2. Heat transfer is rapid in a fluidised bed.		5. The parts are not wet as in a salt bath. There is no slag to handle, or desludging of the bed to perform.
3. The fluidised bed furnace can be heated directly or indirectly. Heating can be electrical or by gas, with the enrichment gases being added with the heating gas.		6. Good heat transfer up to the process temperature.
		7. It has been possible to apply direct hardening of components by reducing the carburising temperature to the appropriate austenitising temperature.
4. The method of operation is exactly the same as is with the salt bath method of heat treatment, with the parts simply being immersed into the fluidised particles.		

STEEL GRADES USED FOR CARBURISING

Construction alloy steels that can carburised successfully are many and varied. When selecting a steel for carburising, it is important to consider machinability, resistance to overheating, susceptibility to deformation during thermal processing, hardenability relative to the cross-sectional size and geometrical features, in addition to mechanical strength, not only in the case but also in the core.

Typically, steels selected for carburising contain <0.25% carbon. The alloy composition is selected to provide case and core hardenability. Plain carbon steels may be carburised but the carburising response is limited owing to the lack of alloying elements. This is illustrated by the selected listing of carburising steels provided in Table 29.22.[55]

Case depths of 75 to 6 350 µm (0.003–0.250 inches) with surface hardnesses of $R_C = 58$–62 are usually specified. An 'effective case depth' is typically defined, which may be determined in various ways. One method is to measure the case depth on a metallographic sample of the part, or of a test bar, by determining the microhardness at various depths from the surface. The required desired case depth is governed by the end application, such as those shown in Table 29.23.[55]

Table 29.24 provides a summary of the common carburising grades of steel used internationally.

HEAT TREATMENT AFTER CARBURISING

After carburising, austenitisation is necessary. If the workpiece is quenched directly from the carburising temperature, retained austenite will form, along with a coarse grain size (depending on the carburising time at temperature). Post-carburisation heat treatment is necessary to provide high surface hardness, that will resist abrasion and wear. Depending on the requirements of the carburised part, the post heat treatment temperatures are selected to provide not only a high surface hardness but also the appropriate core strength. Some options for heat treatment after carburising are shown in Figure 29.23.

The hardening methods summarised in Figure 29.23 are the most commonly used. However, the following procedures are also used, although less commonly:

- Hardening after pearlitic transformation will provide greater energy efficiency and will assure grain size reduction.

Table 29.22 SELECTED LIST OF STEEL COMMONLY USED FOR CARBURISING AND THEIR FEATURES

Steel grade	Features and benefits
4620	Relatively low cost, chromium/nickel/molybdenum steel. Used only where nominal hardenability and core response are required.
8620	Most commonly specified steel for carburising. Excellent carburising response, with good hardenability for most section sizes.
4320	Higher hardenability for improved core response in thicker cross-sections.
4820	Increased nickel content for improved core toughness, slower response results in longer processing times.
9310	Maximum nickel content for maximum core toughness, slower response results in longer processing times.

Table 29.23 REQUIRED CASE DEPTHS OF SELECTED APPLICATIONS FOR CARBURISED PARTS

Application	Case depth	
	μm	inches
High wear resistance, low to moderate loading. Small and delicate machine parts subject to wear.	≤508	≤0.020
High wear resistance, moderate to heavy loading. Light industrial gearing.	508–1 016	0.020–0.040
High wear resistance, heavy loading, crushing loads or high magnitude alternating bending stresses. Heavy duty industrial gearing.	1 016–1 524	0.040–0.060
Bearing surfaces, mill gearing and rollers.	1 524–6 350	0.060–0.250

- Hardening can be performed after initially normalising the structure of the outer layer and core; this method is similar to double hardening, except that the steel is cooled slowly after carburising. The core possesses a normalised structure with less strength but more ductility.

Depending on the alloy selected and the component shape, the carburised steel may be quenched in water, oil or a water polymer solution. Specific recommendations are provided in SAE Standard AMS 2759.

Given the various limitations of both oils and water, there is continuing research into alternative quench media in the heat treating industry. For example, there has been a long search for quenchants, which would exhibit faster cooling rates than those produced by many quench oils, to avoid pearlitic microstructure formation in many steels. This has led to the development and use of aqueous polymer quenchants, as an alternative to many quench oils.

Water quenching of gas carburised parts will provide cooling rates in excess of the critical velocity for martensitic transformation. However, water quenching is likely to cause hardening cracks. A new quenching procedure has been developed using water as the medium and it is known as intensive quenching.

Quenching in oil, on the other hand, reduces the risk of cracking, but often it is not possible to achieve the desired martensitic microstructure throughout the entire carburised case. Only water soluble polymer quenchant solutions provide cooling rates ranging between those attainable with water and those achievable with oil. The Grossmann H-factor obtained for polymer solutions may vary from $H = 0.2$–1.2 (water typically exhibits an H-factor between 0.9 and 2, whereas the H-factors for oil may vary from 0.25 to 0.8). Therefore, the use of an aqueous polymer quenchant will provide a relatively 'mild' quench severity, sufficient to provide the desired martensitic transformation for both case and core.[56–61]

After hardening the carburised case by quenching, it is necessary to temper the steel by selecting a low tempering temperature in the region of 180–275°C to reduce the residual stress caused by the phase transformation from austenite to martensite. A reduction in the quantity of retained austenite may be achieved by cryogenic treatment before tempering, especially after hardening variants '2' and '4' in Figure 29.23 in the case of alloy steels.

Table 29.24 SUMMARY OF COMMON INTERNATIONAL CARBURISING GRADES OF STEEL

Alloy group	Grade	Equivalent grade	ASTM norms	Average composition (wt. %)					Temperature (°C)		Minimum mechanical properties				Applications
				C	Mn	Cr	Na	Mo	Hardening	Tempering	UTS (MPa)	Yield stress (MPa)	Elongation (%)	Impact energy (J)	
Chromium	15 H	~5 117	ASTM A 322-91	0.15	0.7	0.9	—	—	880/800	180	690	490	12	70	Small dimension parts, subjected to moderate stresses e.g. camshafts, reels, bolts, sleeves, spindles.
	20 H	~5 120	ASTM A 322-91	0.20	0.7	0.9	—	—							
Chromium–manganese	16 HG	~5 120H	ASTM A 534-94	0.16	1.2	1.0	—	—	860	180	830	590	12	—	Relatively small dimension parts such as cogs, camshafts, screws, bolts.
	20 HG			0.20	1.3	1.2	—	—	880	180	1 080	740	7	—	Medium dimension cogs, shafts and other parts which are subjected to a high and variable stresses.
	18 HGT			0.18	1.0	1.2	—	—	870/820	200	980	830	9	80	High stress parts requiring a high durability core: cogs, shafts etc.
Chromium–manganese–molybdenum	15 HGM			0.15	1.0	1.0	—	0.2	840	180	930	780	15	80	Medium dimension parts like cogs, shafts, and other parts which bear significant and variable stresses.
	15 HGMA			0.15	1.0	1.0	—	0.2	840	180	930	780	15	80	Larger dimension parts like cogs, shafts and other parts which see a high and variable stresses.
	18 HGM			0.18	1.1	1.1	—	0.2	860	190	1 080	880	10	90	

Group	Grade	Standard												Application
Chromium–nickel–manganese	15 HGN		0.15	0.9	1.0	1.5	—	850	175	880	640	10	59	High stress, small dimension cogs, gears which work at full capacity, shafts.
	17 HGN		0.18	1.2	1.0	0.8	—	860	160	1030	830	11	70	Significantly loaded very small cogs, shafts, screws, bolts.
Chromium–nickel	15 HN		0.15	0.6	1.6	1.6	—	860	190	980	830	12	80	High stress, small dimension cogs, gears which work at full capacity, shafts.
	15 HNA		0.15	0.6	1.6	1.6	—	860	190	980	830	12	80	High stress cogs and rolls, gears that work at full capacity and those which bear variable loads.
	18 H2N2 / 12 HN3A	~E 3310	0.18 / 0.12	0.6 / 0.5	2.0 / 0.8	2.0 / 3.0	— / —	860 / 860/790	190 / 180	1230 / 930	890 / 690	7 / 11	— / 90	To make equipments which are especially loaded like air equipments and parts of internal combustion engine
	12 H2N4A / 20 H2N4A	ASTM A 837-91	0.12 / 0.20	0.5 / 0.5	1.5 / 1.5	3.5 / 3.5	— / —	860/790 / 860/780	180 / 180	1130 / 1270	930 / 1080	10 / 9	90 / 80	
Chromium–nickel–molybdenum	17 HNM		0.17	0.6	1.7	1.6	0.3	860	170	1180	830	7	—	High load cogs, camshafts that work at full capacity, shafts, and other parts subjected to a high stress.
	20 HNM	8620 H / ASTM A 534-94	0.20	0.8	0.5	0.6	0.2	880/820		—*	—*	—*	—*	Medium load cogs, shafts and other parts which take a high stress.
	22 HNM	8622 H / ASTM A 304-95	0.22	0.8	0.5	0.6	0.2	880/820		—*	—*	—*	—*	
Chromium–nickel–tungsten	18 H2N4WA		0.18	0.4	1.5	4.2	—	950/850	180	1130	830	12	100	Equipment which is specially loaded, like air equipment and parts of internal combustion engines.

* Based on purchase specification data.

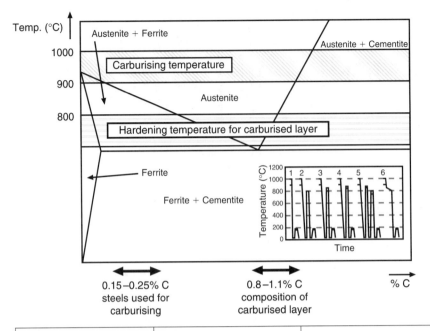

Treatment	Carburized surface	Core
1. Pot quenching (direct hardening).	Dissolved carbides, retained austenite.	Increased hardness.
2. Single hardening from a temperature optimised for the carburised surface.	Excess of undissolved carbides.	Soft and workable.
3. Single hardening from an intermediate temperature.	Slightly coarsened grains, limited extent of carbide dissolution.	Greater strength and more ductility than 1.
4. Single hardening from a temperature optimised for the core.	Slightly larger grains, well dissolved carbides, occurrence of retained austenite.	Peak hardened, with a better compromise between strength and ductility than 2.
5. Regenerative quenching (double hardening).	Limited extent of carbide dissolution, reduced occurrence of retained austenite.	Soft and workable, high level of ductility and resistance to high strain rate deformation (impulse strength).
6. Pot quenching with isothermal holding.	Treatment with single stage diffusion of carbon, excess of carbides decays, reduced retained austenite content, minimum of deformation.	Increased hardness.

Figure 29.23   *Heat treatment after carburising (not drawn to exact scale)*

Steels that are hardened after vacuum or ion carburising are often quenched in a high-pressure gas. The high-pressure gas may also be blended (helium/nitrogen) to achieve the desired cooling rates required for the alloy being quenched and the geometry of the part. The heat transfer coefficient of a gas quenching medium depends on the gas type, delivery pressure, velocity, turbulence and directionality. In addition to producing clean parts, a major advantage of gas quenching is that is does not harm the environment.

STRUCTURE–PROPERTY RELATIONSHIPS

Despite the progress and range of applications of carburising technology, the influence of case structure on properties continues to be controversial. One of the most significant issues is the influence, on properties, of retained austenite and carbides in the structure of case-hardened layers and of the carbon concentration in the surface zone of those layers. For example, an improvement in bending resistance has been observed for 10–30% retained austenite. However, others have reported an unfavourable influence without specifying the percentage values.[62–68] Reference 62 states that fatigue strength increases as the amount of retained austenite increases and recommends that the minimum retained austenite content should be about 25%. Other work has reported a detrimental influence without stating the bending resistance limit.[62–68]

The differences of opinion on the influence of retained austenite are due to a lack of understanding of the detailed conditions and mechanisms involved in microstructure formation and consequently of the structure–property relationships of case-hardened layers. Variable process methodologies that have been utilised also contribute to the different results obtained. The problem is related to the ways in which retained austenite forms in the surface zone of carburised layers and the need to account for the influence, on properties, of changes in other layer parameters of case-hardened steels. These additional parameters include: the carbon content in hardened structures, chemical composition of the matrix, size of former austenite grains, martensite morphology, internal stress state, etc.

There have been reports that carbide precipitation does not unfavourably influence properties.[62,64–67] However, others have reported detrimental effects. These reports should be regarded with great caution, since lower properties than expected may also be due to an overabundance of carbon or tensile stresses in the surface zone of the layer, related to this overabundance. Also, the manner in which the carburised part is loaded can have an effect on the observed properties. The presence of carbides in the structure of carburised cases, with appropriate hardness, allows the production of machine parts and tools which will have high abrasive wear and bending resistance. The carbide form and fraction in the case structure will depend on how the carburising process is conducted, especially on the carbon potentials of the atmosphere, and the temperature and time of the process (*see* Table 29.25).

One of the factors that exhibits a great influence on the properties of machined parts and tools is grain size. Austenite grain formation in the diffusion layer and in the core influences the morphology of martensite being formed.[62,66,69,70] Austenite grain formation also affects the plastic properties of the carburised steel. Grain size variation during carburisation influences the resistance to bending fatigue[71] or the value of the fracture toughness K_{IC}.[72] The small effect of austenite grain size in the case, or in the core, can be examined by controlling the formation of carbides, the chemical composition of the steel or through process control.

According to previous work[73,74] austenitising from a two-phase austenite–cementite region reduces the occurrence of plate martensite microcracking. Austenitising in the temperature range where a two phase structure, austenite plus carbide, is formed, leads to cracking of the martensite. Austenitising at this temperature is favoured for small austenite grain material. The favourable influence of austenitisation temperatures less than A_{cm} has been reported.[75,76] The influence of austenitisation temperature on the properties exhibited by the carburised case has also been addressed[63,77–79] (*see* Table 29.25). In addition, the effect of the mechanical loading mode on the relationship between structure and properties of the diffusion layer has been considered[66] (*see* Table 29.25).

A low M_s temperature results in a reduction of the temperature range for which self-tempering of martensite may occur, which favourably affects many properties. In other words, a low M_s temperature leads to significant inhibition of self-tempering processes occurring during hardening, and, therefore, to a high carbon content in solid solution (in martensite) which will increase the strength and hardness. A minimum M_s value corresponds to the highest compressive stresses.

The presence of hard carbides can have a significant effect on the hardness of the overall coating. Carbide microhardness in carburised carbon steel has been measured at 1 000 HV, and in 2% Ni–Cr steel $\sim$880 HV.[80] Therefore, carburised and hardened surfaces containing 25 to 50% of dispersed

Table 29.25   INFLUENCE OF DIFFERENT STRUCTURAL PARAMETERS ON THE CHARACTERISTIC MECHANICAL PROPERTIES OF CARBURISED LAYERS

Structural parameter	Mechanical property			
	Hardness H	*Fatigue strength σ_D*	*Contact fatigue strength*	*Intensity of cracking*	*Abrasive wear resistance*
Thickness of layer (t_c)	Hardness can be maximised throughout the entire thickness of the layer. It is possible to produce cases with a hardness in the range 700–900 HV.	Data has been reported for samples subject to bending (specifically bent-rotating loading). The optimum coating thickness according to Dawes and Cooksey is 0.08 of that of the part.[87] Tauscher gives a value in the range 0.014–0.21.[88] Weigand and Tolash specify values of 0.07–0.075.[89] This large dispersion of results is a consequence of varying levels of internal stress, strength of the core and internal oxygenation in these different samples.[90] Overall, it seems that the value of the optimum thickness of the layer relative to the thickness of part (i.e. that which correlates with the maximum σ_D) is in the range 0.06–0.07, when accompanied by a core strength of 1.080 GPa.[90]	In general, the most effective thickness is double the depth subjected to the maximum stress (however, other factors such as possible geometrical defects, overloads etc. must be considered).	Cracking generally decreases with increasing thickness. However, cracking depends on numerous parameters, in addition to the thickness of the layer carburised. These additional parameters include: the species present in the steel and the details of preliminary processing, as well as the nature of the applied load.[91,92] Processing parameters influencing the intensity of cracking are the following:[91,92] nickel content in the steel, hardening (especially in warm oil at 80–160°C) and tempering after carburising.	The optimum thickness of the layer is dependent on the nature of the abrasive employed.

Strength of core	An unequivocal relationship between the strength of the core and hardness is not observed.	Optimum for a UTS in the range of 1.080–1.240 GPa.[89]	Optimum for a UTS in the range of 850–1 080 MPa.[89]	Decreases, as the strength of the core increases.	An unequivocal relationship between the strength of the core and abrasive wear resistance is not observed.
Residual austenite γ_R	On increasing γ_R, hardness diminishes.	γ_R seems to be tolerated up to about 25%, but not above this value. The influence of residual austenite in carburised layers on fatigue strength is controversial. Some authors affirm improvement of the resistance to bending fatigue with a limited content of residual austenite of between 10 and 30%.[93] Other authors remark on an ufavourable influence of γ_R on σ_D. A γ_R of 80 to 85% induces a reduction in σ_D of about 25%[94] to 30%.[95] In accord with the above results, the acceptable content of residual austenite is in the range of 15 to 25%[96] or 20 to 30%.[97]	γ_R is tolerated at levels up to 50% Various studies show how residual austenite can have a beneficial influence on contact fatigue resistance, for rolling or rolling with a slide.[98–101]	The residual austenite content has only a small influence on the intensity of cracking.[95]	Abrasive wear resistance falls if γ_R increases.
Carbides	The hardness is slightly sensitive to which carbide phase forms, but depends more on the quantity of carbides.	It has been observed that σ_D can fall when carbides are present.[101] When meshes of interlinked carbides are produced, these lower the fatigue strength.[102] In contrast, the presence of globular carbides (either dispersed, or even when these form a network), has only a small influence on σ_D.[103]	Globular carbides are acceptable and have only a small influence on contact fatigue strength, even when these are superficial.[99]	An unequivocal relationship between the intensity of cracking and the presence of carbides is not observed.	Carbides are an acceptable part of the structure, but not when these are present in the form of a network.

(continued)

Table 29.25   INFLUENCE OF DIFFERENT STRUCTURAL PARAMETERS ON THE CHARACTERISTIC MECHANICAL PROPERTIES OF CARBURISED LAYERS—*continued*

Structural parameter	Mechanical property			
	Hardness H	*Fatigue strength* σ_D	*Contact fatigue strength*	*Intensity of cracking*	*Abrasive wear resistance*
Internal oxygenation	The hardness drops, as the thickness of the region with internal oxygenation grows.	σ_D drop, as the thickness of the region with internal oxygenation grows. Changes in σ_D are proportional to the thickness of the region of internal oxidation. Thus it may be possible to tolerate thickness for this region of up to 6 to 10 μm. [104] However, for a thickness of 13 μm, the reduction in the fatigue limit can reach 20–25% and 45% for a depth of 30 μm. [105]	The thickness of the region of internal oxygenation has only a small influence on contact fatigue resistance.	An unequivocal relationship between the intensity of cracking and internal oxygenation is not observed.	Internal oxygeneration has an unfavourable influence on abrasive wear resistance.

carbides should exhibit higher hardness, 65 to 67 HR_C (830 to 900 HV),[81] than carburised surfaces without carbides.

The presence of carbides in the case also has an indirect influence over the occurrence of internal stresses. The nature and extent of carbide precipitation's influence on the matrix depends on the quenching method. This influence can occur either as a result of the effect that carbide formation has on the chemical composition of the carbides' surroundings (a micro-scale effect), or the influence that the carbides themselves have on the overall stress state in the coating (a macro-scale effect). With slow cooling the micro-scale influence of the carbides on the matrix dominates, while for fast cooling the macro-scale influence is the important consideration. If the matrix of a case containing carbides transforms into martensite, the resulting macrostresses will probably be compressive. The magnitude of these stresses depends on the amount of retained austenite and on the martensite type (plate or lath) (*see* Table 29.25). Slow formation of large carbides will leave the eutectoidal carbon content in the matrix, which for high-alloy steels is a relatively low content. If the formation of increased amounts of plate martensite in the matrix is favoured during quenching, this results in higher compressive macrostresses.

It has also been shown[82] that surfaces containing large amounts of carbides result in the formation of lower compressive stresses (-60 MPa) than a surface without carbides (-500 MPa). Other work has shown how various carbides influence surface internal stresses (*see* Table 29.25).[83] In cases where the formation of carbides has led to lower compressive stresses, the extent to which this is desirable depends on the operating conditions under which the hardened carburised case is used.

Contact fatigue life for four different carburised surfaces is presented in reference [84], based on the results of sliding and rolling tests on carburised and tempered 2% Cr–Mn steel (*see* Table 29.25). Tests showed that coarse carbides can also (like coarse-grain martensite) influence contact fatigue (*see* Table 29.25).

The presence of sufficient amounts of hard carbides might be expected to guarantee that surfaces will have both abrasive and adhesive wear resistance. Indeed, it is generally found that increasing the surface carbide content will increase abrasive wear resistance, in cases where spheroidised carbides appear in a non-martensitic matrix. In contrast, when the structure of the matrix consists of martensite and retained austenite in various amounts, the amount of carbides present does not significantly influence wear resistance.[85] A microstructure that produces the best contact fatigue resistance also increases abrasive wear resistance.[86] It should be further noted that contact fatigue and abrasive wear resistance are also influenced by the quality and condition of lubrication employed in the system.

Carbide network continuity, carbide coarsening and excessive penetration depth reduce fatigue resistance. A high manganese fraction in a low-alloy steel may be the cause of carbide networks in carburised cases. The form and amount of carbides are influenced not only by the diffusion process temperature and time, but also temperature and duration of subsequent treatments (e.g. annealing, or hardening of the system—*see* Table 29.25).

Table 29.25 provides factors relating structural and working properties of the carburised case.[62]

29.2.12   Carbonitriding

THE PHYSICAL–CHEMICAL BASIS OF THE CARBONITRIDING PROCESSES

Process overview

Carbonitriding is dependent on the simultaneous diffusion of carbon and nitrogen into the surface layers of the steel. The process is usually conducted in the temperature range of 850–880°C. The process gas for the treatment is based upon the use of either a nitrogen/methanol blended gas or an endothermic atmosphere, with the addition of a hydrocarbon enrichment gas as a source of carbon, plus ammonia to supply the nitrogen.

The carbonitriding process combines simultaneously nitriding and carburising, and while these two processes are distinguishable, they are not completely independent.

Deposition of carbon and nitrogen and diffusion into the steel

The diffusion of carbon from the process atmosphere into the steel surface is controlled by Fick's laws of diffusion (as discussed earlier in this chapter, with respect to carburising). However, the simultaneous diffusion of two interstitial solutes (i.e. carbon and nitrogen) in the austenite does raise the possibility of interactions between these, which might lead to a change in the diffusivity of each solute. Furthermore, both carbon and nitrogen are derived from the process atmosphere and undergo adsorption onto the steel surface (followed by diffusion into the surface) and so the deposition reactions might act in concert. The above raises a question, 'what ability (if any) does nitrogen have to facilitate the carbon adsorption process and does the nitrogen assist carbon absorption into the steel surface?'

At the carbonitriding process temperature, the primary equilibrium thermal disassociation processes producing carbon and nitrogen are:

$$C + CO_2 \Longleftrightarrow 2CO \qquad \text{Reaction 1}$$
$$CH_4 \Longleftrightarrow C + 2H_2 \qquad \text{Reaction 1}$$
$$C + H_2O \Longleftrightarrow CO + H_2 \qquad \text{Reaction 2}$$
$$2NH_3 \Longleftrightarrow N_2 + 3H_2 \qquad \text{Reaction 3}$$
$$NH_3 \Longleftrightarrow N + 3/2H_2 \qquad \text{Reaction 4}$$
$$1/2 N_2 \Longleftrightarrow N \qquad \text{Reaction 5}$$
$$HCN \Longleftrightarrow C + N + 1/2H_2 \qquad \text{Reaction 6}$$
$$CO + 2NH_3 \Longleftrightarrow CH_4 + H_2O + N_2 \qquad \text{Reaction 7}$$
$$CO + NH_3 \Longleftrightarrow HCN + H_2O \qquad \text{Reaction 8}$$
$$CO_2 + H_2 \Longleftrightarrow CO + H_2O \qquad \text{Reaction 9}$$
$$CH_4 + H_2O \Longleftrightarrow CO + 3H_2 \qquad \text{Reaction 10}$$
$$CH_4 + CO_2 \Longleftrightarrow 2CO + 2H_2 \qquad \text{Reaction 11}$$

Reactions 1 to 6 play a direct and active part in the production of both carbon and nitrogen. In creating atomic carbon for the diffusion reactions, reactions 1–3 have the greatest significance. Atomic nitrogen is created in reactions 4–6 and is available to the steel for diffusion, together with the carbon. A key point is that, at the process temperature of the carbonitriding procedure, molecular nitrogen will decompose to create active, atomic nitrogen (N) in accordance with reaction 6. In accelerating the penetration of both carbon and nitrogen, part of the process will create very small quantities of prussic acid (hydrocyanic acid) on the steel surface. The remaining direct or indirect reactions are inconsequential.

Atomic hydrogen is emitted and is directed to the surface of the steel and will intensify both the reduction and dissociation of surface oxides, as well as converting the oxygen emitted as a result of the carburising reaction. When using 'real-world' furnace atmospheres, oxygen will always be present, as a result of moisture from the endothermic gas, or from unavoidable furnace leakages.

The source of nitrogen for diffusion into the steel is derived from the introduction of ammonia, together with the hydrocarbon enrichment gas. At the process temperature, ammonia dissociation will occur rapidly, which begins in reaction 4 and ends in the final decomposition described in reaction 5. Atomic nitrogen is produced almost immediately as result of the decomposition of the ammonia, followed by the decomposition of molecular into atomic nitrogen. Note that the volume of ammonia should be between 4% to 8% of the total gas flow into the process furnace.

The size of the N_2 molecule is too large to permit this to diffuse into the austenite matrix of the steel and so nitrogen will not dissolve into the steel as molecular nitrogen (the solubility is limited to 0.025%). It is only the active or atomic nitrogen, resulting from the dissociated ammonia, that will penetrate the steel surface and diffuse into the austenite. The limit of solubility of atomic nitrogen in iron, at a temperature of 600°C, is approximately six percent.

The quantity of dissociated ammonia, which is available for diffusion, is proportional to the un-dissociated ammonia present in the furnace atmosphere. It is not possible to quantify the un-dissociated ammonia on the basis of thermodynamic dependence, resulting from the conditions of equilibrium between ammonia and other products present in the furnace atmosphere. This is because such an equilibrium will only occur in the range of temperatures between 400 to 600°C. Between this range of temperatures, dissociation is almost total and the quantity of un-dissociated ammonia does not exceed 0.1%.

The factors influencing the degree of dissociation of ammonia, which occurs during the carbonitriding process are:

- Process temperature.
- Atmosphere changes per hour within the furnace.
- Circulation and distribution of the process gas within the furnace.
- Furnace size (process chamber volume).
- Furnace heating elements versus a gas heating system.

The surface properties of the carbonitrided layers are essential in determining the concentration of nitrogen in the superficial layer of the steel. This concentration will depend directly on that established during the early reaction stages of the ammonia (at the gas–metal interface) and indirectly on other influencing factors such as the speed of the reaction. Therefore for the control of the process, the superficial concentration of nitrogen is usually a function of the concentration of ammonia in the atmosphere, for different temperatures.

At even higher temperatures of 850 to 930°C, the amount of ammonia will be increased, as a result of the temperature. This addition of ammonia will, to a small degree, increase the concentration of nitrogen in the surface layer of the steel. If this is allowed to occur, then the potential for the formation of nitride networks is extremely high. Conversely, if the process temperature is reduced, the opposite will occur. Thus, even small additions of ammonia will cause a dramatic increase in the surface layer nitrogen concentration. Great care should be taken with the addition of ammonia, as a source of atomic nitrogen, as this can cause nitride networks to occur within the formed case.

Steels that include nitride forming elements (for example: aluminium, boron or silicon) will react readily with nitrogen. The alloying elements that will react favourably with nitrogen to form stable nitrides.

It should be noted that chromium, when combined with manganese, will display higher concentrations of nitrogen in the surface layers in comparison with steels that do not include these elements. Furnace atmospheres with higher carbon potentials can saturate the steel surface excessively. Consideration must be given not only to the gaseous reactions, but also to suppression, by the nitrogen activity at the steel surface, of the potential for retained austenite formation. Excess carbon will also encourage carbide formation in the surface of the steel, and nitrogen will assist in diminishing the retention of retained austenite, by allowing a lower austenitising temperature to be selected. An added benefit of the nitrogen reaction (particularly with the nitride forming alloying elements), is that there is an increase in the resulting surface hardness, as well as a reduction of the potential for distortion.

ROLE OF NITROGEN IN THE CARBONITRIDING PROCESS

The direct influence of carbon on the diffusivity of nitrogen in solid solution in the steel is comparatively insignificant. However, there are many other very essential influences of carbon on the process, which allow nitrogen to enter the steel. In the process of carbonitriding, a role of nitrogen is to facilitate the solutioning of carbon in iron. Nitrogen will also contribute to a significant increase in the speed of the carbon diffusion process.

Changes of carbon potential in the furnace atmosphere

Ammonia and the products of ammonia dissociation will change the equilibrium conditions of the main carburising reaction, thus continually changing the carbon potential of the furnace process atmosphere. The products of the dissociation of ammonia (nitrogen and hydrogen) are in accord with reaction (4) and this influences the partial pressures of the carburising components of the atmosphere, compared with non-carburising components and has the effect of raising the carbon potential of the furnace atmosphere. In contrast, ammonia will also react with oxides of carbon and will produce water vapour, which will lower the carbon potential of the furnace atmosphere. Thus, the completion of the ammonia dissociation reaction (and hence its ability to change the atmospheric carbon potential) is very important.

Ammonia will influence the dependence between the carbon potential of the endothermic atmosphere (and enrichment gas) and the temperature of the dew point. Seemingly insignificant additions of ammonia to a carburising atmosphere will cause a decrease in the carbon potential of the furnace atmosphere, which also means a reduction of the atmosphere's dew point.

Increasing the diffusion coefficient of carbon in austenite

The diffusion of nitrogen into steel raises the diffusion coefficient of carbon in austenite. This offers the possibility of utilising low process temperatures. A carbonitride layer developed, at a given temperature, can have a thickness that is similar to that of a carburised layer produced at a temperature that is approximately 50°C higher. Creation of the carbonitrided layer will begin below 850°C and the case will begin to form faster, than for carburising at a temperature of 900°C.

The activity of carbon in austenite

Nitrogen will increase the activity of carbon in austenite, as a result of the superficial concentration of carbon in the primary layer from the furnace atmosphere. This provides an additional reason why, in comparison with carburising, the process of carbonitriding can utilise a lower process temperature.

Changes in the equilibrium conditions in the Fe–Fe$_3$C *system*

Changes in the equilibrium conditions in the iron–cementite system are caused by the presence of nitrogen. Nitrogen acts as a γ stabiliser and reduces both the A$_1$ and A$_3$ temperatures. This assists the diffusion of carbon into the steel surface (the carbon solubility in austenite is much higher than that in ferrite) and gives the possibility of applying lower process temperatures for carbonitriding, as well as utilising lower temperatures for hardening.

Response to quenching

The introduction of nitrogen into the surface of the steel improves the ability to quench the steel without damage. Addition of nitrogen delays diffusional decomposition of the austenite and hence allows the use of a reduced cooling rate.

By the utilisation of lower austenitising temperatures, oil can be employed successfully as the quench medium, thus considerably reducing the potential for distortion. Because of nitrogen diffusion into the steel surface, the dimensional stability of the treated part is greatly improved.

Quenching conditions will determine the surface hardness, the type and properties of induced internal stresses and mechanical properties, such as torsional and tensile strength. Because of potential defects related to water quenching, it is generally recommended that carbonitrided steels be quenched in oil. The use of specially blended quench oils, that would facilitate martensitic transformation of the surface, whilst allowing a pearlitic core structure, within a specific range of oil cooling speeds is recommended. However, it is typically necessary to provide exhaust ventilation for gaseous waste products and there is often a concern for the environmental impact related to the use of quench oils.

It has been found that aqueous solutions of polyalkylene glycol (PAG) based quench media will reduce the potential environmental impact and by varying the concentration of the polymer quenchant in water, a wide range of cooling speeds is possible. The general impact of quenchants and cooling rates on the properties of carbonitrided steels will be discussed subsequently.

Reduction of the martensite start temperature

The presence of nitrogen in the carbonitrided layer will reduce the martensite start temperature (M$_s$), as well as significantly raising the martensite finish (M$_f$) temperature. Thus, the M$_s$ temperature of the carbonitrided steel will be lower than if the same steel was carburised and will also cause a reduction in the compressive stress through the case.

Reduction of retained austenite

Lowering the martensite start temperature will reduce the amount of residual austenite in the carbonitrided layer. Residual austenite will normally be found when the surface carbon potential is in excess of 0.7% to 0.9%, with concentrations of nitrogen in solution. If the concentration of both carbon and nitrogen is controlled in the case, then the retention of any untransformed austenite will be minimal, thus producing a relatively dimensionally stable case.

Control of carbon and nitrogen within the diffusion layer

The production of available carbon for the process is derived principally from the enrichment gas, which is added to the endothermic gas, or the nitrogen/methanol plus enrichment gas. The objective is to ensure a surface carbon potential (0.7% to 0.9% carbon) around the eutectoid composition. The carbon potential can be controlled by accepted methods, such as: dew point, shim analysis, oxygen probe or CO/CO$_2$ analysis. The nitrogen is normally derived from the ammonia enrichment gas added to the furnace. This is extremely difficult to control, as there is no commercial system that can be used to control accurately the decomposition of ammonia to release nascent nitrogen. Therefore one must rely on the relationship of the volumetric flow of the process gas (carrier gas and enrichment gas) to the ammonia. Generally, the ammonia flow should be between 4% to 8% (by volume) of the total gas flow into the furnace.

THE CARBONITRIDING PROCEDURE

The carbonitriding process very closely resembles the process of gas carburising. The procedure involves the following steps:

- Conditioning of the furnace.
- Loading the work into the furnace.

- The carbonitriding process.
- Quenching the carbonitrided work.

The carbonitriding process times are much shorter than soaking times experienced with the carburising process.

During the heating cycle of the batch, the introduction of the carbonitriding atmosphere into the furnace takes place. It is not necessary to introduce the process gas into the furnace during the heat up cycle. This is because during the heat up of the batch and particularly in the temperature range of 550°C to 700°C, ammonia can cause the formation of nitrides at the steel surface. These nitrides may not be stable at the carbonitriding process temperature and may undergo transformation.

The selection of the carbonitriding process temperature requires careful consideration, with due regard to the atmospheric composition employed, as well as the heating method of the furnace. This is to ensure a well conducted process that is able to produce the required case depth, as well as the appropriate carbon and nitrogen concentrations within the case. When selecting either low carbonitriding process temperatures or long cycle times to produce deep cases, the probability of high surface nitrogen concentrations is extremely high.

It is not usual to select deep carbonitrided cases. Generally the case for carbonitriding has a maximum depth of 500 μm (0.020 in.). Deep case formation will give rise to a high probability of the retention of un-transformed austenite (retained austenite). If ammonia is introduced into the furnace, for example during the last one-third of the process cycle, then the risk of high surface nitrogen concentrations is reduced almost to the point of elimination. On completion of holding of the workload in the furnace atmosphere, having allowed sufficient time at the process temperature to form the desired case, the load is quenched from the appropriate austenitising temperature. Care must be taken when selecting the austenitising temperature, due to the potential risk of retained austenite and of course the potential for the occurrence of distortion.

HEAT TREATMENT AFTER CARBONITRIDING

Once the process of carbonitriding is completed, this is usually followed by a cooling procedure, down to the hardening temperature. A light tempering process at 180–200°C usually follows this. This is intended to temper the martensitic case and reduce the potential for case/surface cracking.

The cool-down procedure is selected to accomplish the formation of the structures that are needed to produce the properties required by the engineering design for the core and surface of the steel. The steel can be:

- Cooled within the furnace down to the case austenitising temperature in a controlled manner, followed by quenching.
- Cooled down to, perhaps, 500°C, removed from the furnace and cooled externally, followed by reheating to the austenitising temperature and then by a quenching procedure which will harden the carbonitrided case.

Examples of heat treatments after carbonitriding may be found in Figure 29.24.

THE INFLUENCE OF STRUCTURAL FACTORS ON THE PROPERTIES OF CARBONITRIDED LAYERS

Table 29.26 provides information on the relationship between structural factors and the properties of carbonitrided materials.[62]

Table 29.26 The influence of structural factors on the properties of carbonitrided layer.[77,106–108]

STEELS USED IN CARBONITRIDING PROCESSES

Some steels used internationally for carbonitriding may be found in Table 29.27.

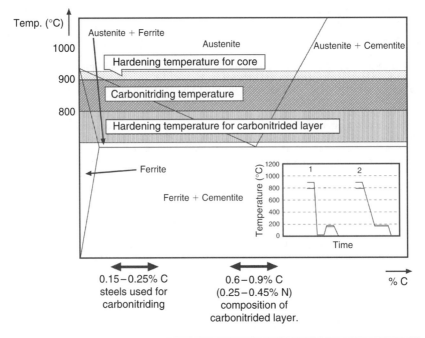

Treatment	Carbonitrided surface	Core
1. Pot quenching (direct quenching).	Dissolved carbides, retained austenite.	Increased hardness.
2. Pot quenching with isothermal holding.	Treatment with single diffusion of carbon and nitrogen, excess of carbides or carbonitrides decays, reduced content of retained austenite, minimum of deformation.	Increased hardness.

Figure 29.24   *Heat treatment after carbonitriding (not drawn to exact scale)*

Table 29.26 THE INFLUENCE OF STRUCTURAL FACTORS ON THE PROPERTIES OF CARBONITRIED LAYER[77,106-108]

Structural factor	Hardness H	Fatigue strength σ_D	Contact fatigue strength	Resistance to cracking	Wear resistance
			Properties		
Thickness of case t_c	The hardness of case generally remains within the range 700–900 HV.	Optimum fatigue strength when: $t_c/t_p \sim 0.7$ (t_p = thickness of plastick zone).	Contact fatigue strength is maximised when: $t_c \sim 2t_z$ (t_z = thickness of maximum stress zone).	Resistance to cracking is reduced as the thickness of the layer is increased.	Thickness limits wear resistance.
UTS of core	—	Optimum fatigue strength when the UTS is 1.3 to 1.5 GPa, with a carbon content in the core of 0.2 to 0.25%.	Optimum contact fatigue strength when the UTS is 850 to 1 080 MPa.	Resistance to cracking drops if the strength of the core is increased.	—
Retained austenite γ_R	If γ_R increases, the hardness diminishes.	Fatigue strength is enhanced by 25 to 60% γ_R.	Contact fatigue strength is enhanced by γ_R up to 70%.	—	—
Choice of carbonitriding process	Little effect on hardness.	—	Small influence on contact fatigue strength.	—	Increased wear resistance
Internal oxygenation	Hardness is reduced.	—	Small of harmful influence on contact fatigue strength.	—	Harmful influence on wear resistance.
Pearlite or bainite	Hardness is reduced.	Fatigue strength falls if the thickness of the layer of pearlite or bainite is increased.	—	—	Harmful influence on wear resistance.
Porosity	Hardness is reduced.	Harmful.	—	—	—

Table 29.27   THE CHEMICAL COMPOSITION OF STEELS USED INTERNATIONALLY IN THE
CARBONITRIDING PROCESSES

| | *Average composition (%)* | | | | | |
Grade	C	Mn	Ni	Cr	Mo	*Country*
20MoCr$_4$	0.20	0.70	—	0.40	0.45	
25MoCr$_4$	0.25	0.70	—	0.40	0.45	
20CrMo$_2$	0.20	0.70	—	0.60	0.35	Germany
4028	0.24	0.82	—		0.25	
20NiMoCr$_6$	0.20	0.70	1.60	0.50	0.45	
23CrMoB$_3$	0.23	0.80	—	0.70	0.35 + B	
8620	0.20	0.80	0.55	0.50	0.20	
4024/28	0.23/0.28	0.80	—	—	0.25	Great Britain
Mod8822	0.22	0.85	0.55	0.50	0.55/0.65	
18Co$_4$	0.18	0.75	—	1.05	0.22	France
8620	0.20	0.80	0.55	0.50	0.20	
8620	0.20	0.80	0.55	0.50	0.20	Italy
18NiCrMo$_5$	0.18	0.70	1.35	0.80	0.20	

29.3   Heat treatment of aluminium alloys

In this section, a brief review of aluminium alloy physical metallurgy, as affected by heat treating, is provided. Defects arising from heat treating operations are also discussed. The importance of selecting the proper quenchant, and quenching parameters are explained.

29.3.1   Introduction to aluminium heat treating

Aluminium in many forms has been used in aircraft since their very beginning. This is because aluminium alloys can be heat-treated to relatively high strengths, while maintaining low weight. These alloys are easy to bend and machine, and the cost of the material is relatively low. Because of these advantages, aluminium is the most common material used in aerospace today. It is used in the manufacture of advanced commercial aircraft such as the Boeing 777, Airbus 380, and military aircraft such as the Boeing UCAV or F/A-18 E/F.

Typically, the aluminium alloys used in aerospace structures are the heat treatable grades, such as 2XXX, 6XXX and 7XXX. 7XXX alloy grades such as 7075, 7040 and 7050 predominate in aerospace structures. Some of the Al–Li alloys such as 2090 and 8090 have also found application. These alloys are used commonly because they are easily available and readily formed and heat-treated to yield high strength, corrosion resistant components.

However, the use of aluminium is not limited to the aerospace industry. It is used extensively in, for example, sporting equipment and in the automotive industry.

The aluminium alloys most commonly used are 7XXX wrought products. The thickness of parts used ranges from 0.6 mm to 250 mm. Aluminium alloy 7075 and 7050 extrusions are used extensively for stringers and other structural requirements. Some extrusions of 2024 and 2014 are also used. Sheet, both clad and non-clad, of 2024, 7475 and 7075 is used for wing and fuselage skins. Sheet is also formed and built-up to produce bulkheads and other structural requirements. For heavy loading, large forgings of 7050 and 7040 are used commonly, particularly in military aircraft.

Recently, there has been much interest in the use of heat treated 7050 and 7040 plate to minimise the cost of forgings. In this case, the plate (heat treated at the aluminium mill) is supplied to the airframe manufacturer. There the plate is heavily machined to fabricate large ribs and bulkheads. The advantage of this is that it avoids heat-treating small parts and may serve to improve distortion and residual stress control. Unfortunately, there is a significant property variation through the thickness of thick plate. This variation has been studied in detail and the microstructure development in thick sections has been researched extensively.[108a]

Primarily, because of affordability, castings have also been used. However, the use of castings has been limited because of design factors and their limited ductility. Typically casting alloys such as A356 and A357 are used for casting. Applications of castings are simple, non-flight critical components, such as door handles and avionics cabinets.

Heat-treating of aluminium requires stringent controls. To achieve repeatable results, and to provide a quality product, the airframe manufacturers, suppliers and heat-treaters have developed a series of specifications. The most widely used specification is AMS-2770 'Heat Treatment of Aluminum Parts'.[109] This specification details solution treating times, temperatures, quenchants and ageing practices. It also defines the required documentation for heat treat lot traceability. Furthermore, this specification establishes the quality assurance provisions needed to ensure that a quality product is provided to the airframe manufacturer.

29.3.2   A brief description of aluminium physical metallurgy

There are four aluminium alloy types that dominate the heat treatable alloys used in industry today. These are: Al–Cu, Al–Cu–Mg, Al–Mg–Si and Al–Zn–Mg–Cu. A brief explanation of the precipitation sequence, for each alloy type follows.

29.3.2.1   Al–Cu *alloys*

The aluminium–copper system has been reviewed in detail.[110] The equilibrium phase diagram[111] contains an eutectic, at 548°C and approximately 32% copper, involving the face centred cubic Al solid solution phase (α) in equilibrium with $CuAl_2$ (θ). The extent of solid solubility at the aluminium rich end is approximately 5.7% copper. Commercial alloys of this type are 2219, 2011 and 2025.

The precipitation sequence was originally established by Guinier[112,113] and Preston.[114,115] Hornbogen further examined the precipitation in Al–Cu and confirmed the results of Guinier and Preston.[116] The precipitation sequence, on ageing after rapid quenching, has been accepted as being Guinier–Preston zones (GPZ) in the form of plates parallel $\{001\}_{Al}$, transforming to the coherent precipitate θ'', followed by semi-coherent θ'' plates parallel $\{001\}_{Al}$. The final equilibrium precipitate is $\theta(Cu_2Al)$. Silcock *et al.*,[117] examined this progression of precipitates and showed multiple stages in precipitation, evidenced by changes in hardness and Laue reflections.

29.3.2.2   Al–Cu–Mg *alloys*

Aluminium–copper–magnesium alloys were the first precipitation hardenable alloys discovered.[118] The first precipitation hardenable alloy was a precursor to alloy 2017 (4% Cu, 0.6% Mg and 0.7% Mn). A very popular alloy in this group is 2024.

The addition of magnesium greatly accelerates precipitation reactions. In general, the precipitation sequence, starting from a supersaturated solid-solution (SSSS), is:

$$SSSS \longrightarrow GP\ zones \longrightarrow S'(Al_2CuMg) \longrightarrow S\,(Al_2CuMg)$$

The GP zones in this system are generally considered to be collections of Cu and Mg atoms collected as disks on the $\{110\}_{Al}$ planes. S$'$ is incoherent and can be observed directly in the transmission electron microscope (TEM). S$'$ precipitates heterogeneously on dislocations. These precipitates appear as laths on the $\{210\}_{Al}$, oriented in the <001> direction.[119] Since S$'$ precipitates on dislocations, cold working after quenching increases the number density of S$'$ and produces a fine distribution of precipitates in the matrix.

29.3.2.3   Al–Mg–Si *alloys*

The aluminium–magnesium–silicon alloy system forms the basis for the 6XXX series aluminium alloys. In this heat treatable alloy system, magnesium is generally in the range of 0.6–1.2% Mg and silicon is in the range of 0.4–1.3% Si. The sequence of precipitation is the formation of GP zones, followed by metastable β' (Mg_2Si), followed by the equilibrium β (Mg_2Si). The GP zones in this case, are needles oriented in the <001> direction, with β' and β showing similar orientations.

29.3.2.4   Al–Zn–Mg–Cu *alloys*

This aluminium–zinc–magnesium–copper series of alloys is probably the most popular and readily used. These alloys are used extensively in the aerospace industry because they have excellent corrosion resistance, fracture toughness and strength, compared with the other age hardenable aluminium alloys. In 7XXX Al–Zn–Mg–Cu alloys, several phases have been identified that occur in Al–Zn–Mg–Cu alloys as a function of precipitation sequence. Four precipitation sequences have been identified. These are shown schematically below:

$$\alpha_{ssss} \Rightarrow S$$
$$\alpha_{ssss} \Rightarrow T' \Rightarrow T$$
$$\alpha_{ssss} \Rightarrow VRC \Rightarrow GPZ \Rightarrow \eta' \Rightarrow \eta$$
$$\alpha_{ssss} \Rightarrow \eta$$

In the first precipitation sequence, the S phase, Al_2CuMg, is precipitated directly from the supersaturated α solid solution (α_{ssss}). The S phase is reported to be orthorhombic,[120] with a space group of Cmcm, and 16 atoms per unit cell. The lattice parameters are:[121] a = 401 pm, b = 925 pm, and c = 715 pm. This phase has been identified[122] as a coarse intermetallic that is insoluble in typical Al–Zn–Mg–Cu alloys at 465°C and as a fine lath precipitate in Al–4.5%Zn–2.7%Cu–2.2%Mg–0.2%Zr alloys. No orientation relationships to the matrix have been identified in the literature.

In the second precipitation sequence, an intermediate phase T' occurs in the decomposition of the supersaturated solid solution. Bernole and Graf first identified this phase.[123] Auld and McCousland[124] suggested that the structure was hexagonal with the reported lattice parameters a = 1.39 nm and c = 2.75 nm. It was further suggested that the orientation of the hexagonal cell to the aluminium matrix is:

$$(0001)_{T'}//(111)_{Al} \quad (10\bar{1}0)_{T'}//(11\bar{2})_{Al}$$

Further on, in the second precipitation sequence, the equilibrium T phase forms. This phase was reported[125] to be cubic, space group Im3, with 162 atoms in the unit cell. It was indicated that the lattice parameter varies from 1.41 to 1.47 nm, with this variation being due to compositional variations. Bergman *et al.*[126] have proposed that the chemical formula $Mg_{32}(Al,Zn)_{49}$ is appropriate. The T phase was found to be incoherent with the aluminium matrix. This phase has been rarely reported in substantial quantities, even though commercial heat treatments up to 150°C lie in the $Al + MgZn_2 + Mg_{32}(Al,Zn)_{49}$ phase field. In general, the T phase only precipitates above 200°C.[128]

In the third sequence of precipitation, the supersaturated solid solution decomposes to form vacancy-rich clusters, Guinier–Preston zones, η' and then η. Guinier–Preston zones have been inferred in Al–Zn–Mg alloys, based on small increases in electrical conductivity and an increase in hardness during the initial stages of ageing.[134]

The η' phase is an intermediate step towards the precipitation of the equilibrium phase η ($MgZn_2$). Direct evidence of η' is rare, and difficult to obtain. It has recently been accepted that the η' phase is hexagonal, however, the reported lattice parameters vary widely.

29.3.3   Defects associated with heat treatment

During the production of a part, defects can occur. These can come from operations before heat treatment, such as midline porosity or inclusions that are formed during casting the ingot. Further defects can form during homogenisation of the ingot, such as segregation, the formation of hard intermetallics and other second phase particles.

Defects associated with the heat treatment of aluminium can occur during solution treatment, quenching or ageing. Solution treatment defects include oxidation, incipient melting and underheating. Defects that occur during quenching are typically distortion of the part or inadequate properties caused by a slow quench, resulting in precipitation during quenching and inadequate supersaturation. Defects that can occur during ageing include growth or shrinkage of the part. Problems can also arise from an inadequate response, by the material, to the ageing treatment. Heat-treatment-related defects are discussed in more detail below, in the sections dealing with the relevant process step.

29.3.4   Solution treatment

Aluminium alloys are classified as either heat treatable or not heat treatable, depending on whether the alloy responds to precipitation hardening. In the heat treatable alloy systems, such as 7XXX,

Table 29.28 SOLUTION TREATMENT TEMPERATURE RANGE AND
EUTECTIC MELTING TEMPERATURE FOR 2XXX ALLOYS

Alloy	Solution treatment temperature range (°C)	Eutectic melting temperature (°C)
2014	496–507	510
2017	496–507	513
2024	488–507	502

6XXX and 2XXX, the alloying elements show greater solubility at elevated temperatures than at room temperature. This is illustrated for the Al–Cu phase diagram (see Chapter 11).[127]

The Al–Cu phase diagram shows that holding a 4.5% Cu alloy at 515 to 550° will cause all the copper to go into solution completely. The temperature used to achieve complete solutioning is known typically as the 'solution heat treating temperature' or 'solution treatment temperature'. If the alloy is cooled slowly, then the equilibrium structure of α (saturated solid solution of Cu in Al) + $CuAl_2$ will form. The $CuAl_2$ that forms is large, coarse and incoherent. However, if the alloy is cooled rapidly, there will be inadequate time for the $CuAl_2$ to precipitate. Hence, in the rapidly cooled alloy, all the solute is held in a supersaturated condition. Controlled precipitation of the solute, as finely dispersed particles, at room temperature (natural ageing) or at elevated temperatures (artificial ageing) is used to develop the optimised mechanical and corrosion properties of these alloys.

Solution treatment involves heating the aluminium alloy to a temperature slightly below the eutectic melting temperature. Solution treatment develops the maximum amount of solute into solid solution. This requires heating the material close to the eutectic temperature and holding at temperature long enough to allow close to complete solid solutioning. After solution treatment, the material is quenched to maintain the solute in supersaturated solid solution.

Because the solution treatment temperature is so close to the eutectic melting temperature, temperature control is critical. This is especially true for 2XXX series alloys. In this alloy group, the eutectic melting temperature is only a few degrees centigrade above the maximum recommended solution treatment temperature (Table 29.28).

29.3.4.1 *Oxidation*

If parts are exposed to temperature for too long a time, what is commonly referred to in the industry as 'high-temperature oxidation' could become a problem. Note: the term high-temperature oxidation is really a misnomer, in the context of the heat-treatment of aluminium alloys. Instead, in the present case, the culprit is actually moisture present in the air, during solution treatment. This moisture is a source of hydrogen, which diffuses into the base metal. Voids form at inclusions or other discontinuities. The hydrogen gas gathers and forms a surface blister on the part. In general, 7XXX alloys are the most susceptible (particularly 7050), followed by the 2XXX alloys. Extrusions are the most prone to blistering, followed by forgings.

Eliminating the moisture will minimise the problem of surface blistering. This is accomplished by the sequencing of doors over quench tanks and thoroughly drying and cleaning furnace loads prior to solution treatment. It is also important to make sure that the load racks used for solution treatment are dry. However, it is not always possible to eliminate high humidity in the air, in order to prevent surface blistering. Often the ambient relative humidity is high, so that other measures may have to be taken.

Ammonium fluoroborate is typically used to prevent blistering on 7XXX extrusions and forgings. An amount equivalent to 5 g per m^3 of workload space is usually employed to prevent surface blistering. This is applied as a powder, in a shallow pan, hanging from the furnace load rack. This material is very corrosive and requires operators to wear the appropriate personal protective safety equipment. Because the material is corrosive at temperature, it is recommended that the inside panels in the furnace be manufactured from stainless steel. The use of stainless steel panels will reduce corrosion and hence maintenance.

Anodising parts prior to solution treatment is an alternative to ammonium fluoroborate. This is generally practical for larger extrusions and forgings, where the cost of anodising is small compared with the cost of the part. However, for small parts, the additional cost does not generally justify the possible benefit of anodising prior to solution treatment.

29.3.4.2   *Eutectic melting*

Non-equilibrium conditions can occur because of localised solute concentrations. Because of the increased concentration of solute, the eutectic temperature could be decreased, causing localised melting. This is often called *incipient melting*. When this occurs, significant decreases in properties result. Properties most affected include toughness and tensile properties (both strength and ductility).

Local melting can also occur if the material is heated too quickly. This is particularly true of 2XXX alloys. In this alloy system, there are local concentrations of Al_2Cu. At slow heating rates, the Al_2Cu dissolves slowly into the matrix. At high heating rates, there is inadequate time for the Al_2Cu to dissolve. Local concentrations cause the local eutectic temperature to drop, resulting in localised melting. If inadequate time is allowed for this metastable liquid to dissolve into the matrix, then in general, there is no decrement in properties.

Based on the equilibrium solidus temperature, 7XXX series aluminium alloys should be safe from eutectic melting. However, these alloys can exhibit eutectic melting under certain circumstances. In the 7XXX series alloys, there are two soluble phases, $MgZn_2$ and Al_2CuMg. Al_2CuMg is very slow to dissolve during solution treatment. Local concentrations of Al_2CuMg can cause non-equilibrium melting, between the temperatures of 485 and 490°C (905 to 910°F) and this is a problem, since the work may be brought to this temperature range, or a higher temperature.[128] Homogenisation practice of the ingot is the primary source of S phase in these alloys. Because of the hazards of eutectic melting, it is imperative that the 7XXX alloys be homogenised.

29.3.4.3   *Under-heating*

Under-heating during solution treatment can cause problems, by not allowing enough solute to go into solid solution. This means that less solute is available during subsequent precipitation hardening reactions. As an illustration of this, Figure 29.25 shows the effect of solution-treating temperature on the yield strength and ultimate tensile strength. As the temperature is increased for both alloys shown in the figure, the tensile strength is also increased. For 2024-T4, it can be seen that there is a change in slope and a rapid rise in final properties as the solutionising temperature is increased past about 488°C.

29.3.4.4   *Furnaces for solution treatment*

Furnaces used for the solution treatment of aluminium alloys are typically of two types. Either 'drop-bottom' furnaces or salt bath furnaces are used. Both are batch type operations. There are several continuous types of aluminium furnaces, but these are typically limited to smaller parts.

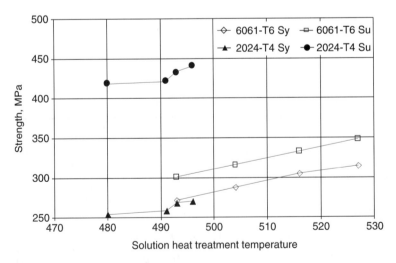

Figure 29.25   *Tensile strength as a function of solution-treating temperature for 6061 and 2024 (after Hatch)*[128]

Solution treating and quenching of these alloys is typically accomplished in large high-temperature ovens. In some applications, the furnace is supported above the quench tank. The quench tank moves under the furnace on rails. Sometimes there is more than one quench tank, with each tank containing a different quenchant.

Drop bottom furnaces are typically arranged with a moveable quench tank underneath. Often, there is more than one quench tank, one for water and the other for polymer quenchants. These quench tanks move back and forth as desired, and the selected quench tank is positioned under the furnace. The door on the bottom of the furnace opens and the workload is immersed into the quenchant. Sometimes, because the quench tank is so large, it is impractical to move the tank. Therefore, the furnace is moved in a similar fashion to a gantry furnace.

29.3.5   Quenching

In this section, the effects of quenching on the structure and properties of aluminium alloys are discussed. Further detail on quenching media may be found in the Appendix (Section 29.3.7).

The key consideration during quenching is to prevent precipitation, without damaging the part. The problem of undesired precipitation during quenching can be understood in terms of the kinetics of nucleation and growth, during diffusional phase transformations.[128] As with other diffusional phase transformations, the kinetics of precipitation occurring during quenching are dependent on the degree of solute supersaturation, versus the rate of solute diffusion, both of which are functions of temperature. The degree of supersaturation increases with undercooling and hence with decreasing temperature. Hence the driving force for nucleation increases as the temperature decreases. Conversely, solute diffusion coefficients, and hence the rate of diffusion-controlled growth of precipitates, increase with increasing temperature. When either the solute supersaturation or diffusivity is low, the overall rate of precipitation is low. At intermediate temperatures, the amount of supersaturation is relatively high, as is the rate of diffusion. Therefore the overall rate of precipitation is greatest at intermediate temperatures. This is shown schematically in Figure 29.26. The amount of time spent in this critical temperature range is governed by the quench rate.

Precipitation occurring during quenching reduces the amount of subsequent hardening possible. This occurs because solute is precipitated from solution during quenching, is unavailable for any further precipitation reactions. Precipitation during quenching results in lower as-aged tensile strength, yield strength, ductility and fracture toughness.

Quantifying quenching and the cooling effect of different quenchants, has been studied extensively.[129–132] The first systematic attempt to correlate properties to the quench rate in Al–Zn–Mg–Cu alloys was performed by Fink and Wiley[133] for thin, 1.6 mm (0.064″) sheet. A time-temperature-tensile property curve was created and was probably the first instance of a TTT-type diagram for aluminium alloys. It was determined that the critical temperature range for 75S is 400°C to 290°C. This is similar to the critical temperature range found for Al–Zn–Mg–Cu alloys.[134]

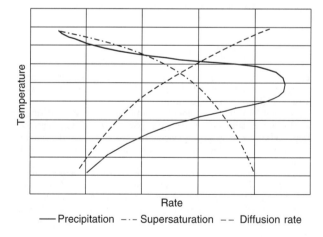

Figure 29.26   *Schematic showing the interrelationship between the amount of solute supersaturation and diffusion rate on the amount of heterogeneous precipitation occurring during quenching*

At quench rates exceeding $450°C\,s^{-1}$, it was determined that maximum strength and corrosion resistance were obtained. At intermediate quench rates, from 450 down to $100°C\,s^{-1}$, the strength obtained was lowered (using the same ageing treatment), but the corrosion resistance was unaffected. Between 100 and $20°C\,s^{-1}$, the strength decreased rapidly, and the corrosion resistance was at a minimum. At quench rates below $20°C\,s^{-1}$, the strength decreased rapidly, but the corrosion resistance improved. However, for a given quenching medium, the cooling rate through the critical temperature range was invariant no matter what the solution treatment temperature.

One method that quantifies the quench path and the material kinetic properties is called the 'quench factor' and was originally described by Evancho and Staley.[135] This method is based on the integration of the area between the time-temperature-property curve and the quench path. Wierszykkowski[136] provided an alternative explanation of the underlying principles of the quench factor. However, his discussion is more generally applied to the thermal path prior to isothermal transformation. The procedures for developing the quench factor have been well documented.[137] This procedure could be used to predict tensile properties,[143] hardness[144] and conductivity.[138] It was found that the quench factor could not be used to predict elongation because of its strong dependence on grain size[138]. This method tends to overestimate the loss of toughness.[142] The quench factor method also can be used to determine the critical quench rate for property degradation.[145]

29.3.5.1 *Quench rate effects*

The objective of quenching is to preserve the solid solution formed during solution-treating. This is accomplished by cooling rapidly to room temperature. This rapid cooling maintains the vacancy concentration necessary to enable the low temperature diffusion required for GP zone formation. Solute that precipitates, at either the grain boundaries, or dispersoids, is lost and cannot be utilised for further strengthening. Hence, the best properties (with respect to both strength and corrosion resistance) are usually achieved by employing the highest possible quench rate.

As an illustration of quench rate effects in a 7050 alloy, a Jominy end quench bar was fabricated from extruded 7050, and quenched using the Jominy apparatus.[108a] Quench rates ranging from in excess of $1\,000°C\,s^{-1}$ to $1–2°C\,s^{-1}$ were obtained. This bar was sectioned and examined at specific locations along the quenched bar, using a TEM. Distinct differences in the microstructure were observed. At fast quench rates, the GP zones are small, and exhibit strain fields around the zones. There is little evidence of precipitation at the grain boundaries. As the quench rate is decreased, the strain fields are still present, however they are not as intense. There is also evidence of precipitation on the grain boundaries, although these precipitates are small. Some precipitation also occurs within the grains. Again, these precipitates are small. As the quench rate is further decreased, larger and coarser precipitates are evident. These precipitates occur at the grain boundaries and in the interior of the grains. The lack of a strain field around the precipitates indicates that the precipitates are incoherent with the matrix. The solute contained in the coarse precipitates at the grain boundaries and in the interior of the grain are unrecoverable, and will not participate in any strengthening of the alloy. Further, the presence of the precipitates at the grain boundaries can serve as localised galvanic cells, providing a preferred path for intergranular and stress-corrosion-cracking.

29.3.5.2 *Warpage and distortion*

Of all the possible 'defects' occurring during the heat treatment of aluminium, distortion during quenching is the most common. It is probably responsible for most of the non-value-added work (straightening) and costs associated with aluminium heat-treating.

Distortion during quenching is caused by differential cooling, and differential thermal strains developed during quenching. These thermal strains could be developed centre-to-surface, or surface-to-surface. This differential cooling can be caused by large quench rates, so that the centre is cooled much slower than the surface (non-Newtonian cooling) or by non-uniform heat transfer across the surface of the part.

Aluminium alloys are more prone to quenching distortion than steels. This is because solution treatment temperatures are so close to the liquidus temperature in the former case. Aluminium alloys exhibit less strength and greater plasticity than steels at the solution treatment temperature (or austenitising temperature for steels). Furthermore, much higher quench rates are necessary in aluminium alloys, than in steels, to prevent premature precipitation from occurring during quenching and to maintain supersaturation of the solute.

In steels, there is a phase transformation from austenite to martensite, on quenching. This causes around a 3% volume change, during quenching. There is no analogous phase transformation in aluminium alloys that can cause cracking or distortion. However, the coefficient of linear expansion of aluminium is approximately twice that of steel ($2.38 \times 10^{-5}\,K^{-1}$ compared with $1.12 \times 10^{-5}\,K^{-1}$). This causes much greater changes in length or volume as a function of temperature, and increases the probability that distortion will occur.

Proper racking of the parts is critical. The parts should be fully supported, with the loads spread out over a large area, since the creep strength of aluminium is poor. Parts should be wired loosely to prevent the parts from hitting each other during solution treatment. If wired too tightly, the wire could cut the parts. The use of pure aluminium wire minimises this problem.

Because of the poor strength of the solution-treated aluminium parts, distortion of the parts can occur as they enter the quenchant. As a general rule, parts should enter the quenchant aerodynamically to avoid distortion of the part, before it even enters the quenchant. The parts should enter the quenchant smoothly—they should not 'slap' the quenchant.

Racking a part so that it enters the quenchant smoothly also offers the benefit that it is more likely to have uniform heat transfer across the part. Distortion is more likely to occur because of horizontal changes in heat transfer than by vertical differences in heat transfer.

29.3.5.3   *Grain boundary precipitation*

Grain boundary precipitation and the formation of a precipitate free zone (PFZ), are the result of solute coming out of solution during quenching. With a perfect quench, all solute would be held in supersaturated solid solution and no precipitation would occur during quenching. However, a perfect quench is rare with real parts of the types used typically in contemporary practice. Since there is little likelihood of a perfect quench, some precipitation almost invariably occurs. The degree of precipitation depends on the alloy and the quench rate.

As already discussed, the rate of precipitation during quenching is based on two competing factors: supersaturation and diffusion. As temperature is decreased during quenching, the amount of super-saturation increases, providing increased driving force for precipitation. In contrast, at the beginning of quenching, the temperature is high, increasing the rate of diffusion. The Avrami precipitation kinetics for continuous cooling can be described by:[135]

$$\zeta = 1 - \exp(k\tau)^n \tag{1}$$

where ζ is the fraction transformed, k is a constant and τ is defined as:

$$\tau = \int \frac{dt}{C_t} \tag{2}$$

where τ is the quench factor, t is the time (s) and C_t is the critical time (s). The collection of the C_t points for continuous cooling, also known as the C-curve, is analogous to the time-temperature-transformation curve for isothermal transformation. These equations can be used to predict the volume fraction that has precipitated, and how that precipitation will affect the properties.

To avoid excessive precipitation during quenching, three requirements must be met. First, the transfer time from the solution treatment furnace into the quench tank must be minimised. Second, the properties of the quenchant must be selected to enable a quench that is fast enough to ensure that proper supersaturation is achieved, so that the desired properties can be obtained during subsequent ageing. Lastly, the quench tank must have adequate thermal inertia so that the quenchant does not heat excessively, causing an interrupted quench. In addition, the quenching system must extract heat uniformly to minimise property variations.

The quench delay time, or the transfer time from the furnace to the quench tank, is defined differently for air furnaces and salt baths. For air furnaces, the quench delay time is defined as the time interval, from when the furnace door first begins to open, until the last corner of the workload is immersed into the quench tank. For salt baths, the quench delay time is defined as the time interval from when the first corner of the workbasket is exposed, to the time at which the last corner of the workload is immersed into the quench tank. In general, this time interval is independent of the alloy, but depends on the solution treatment temperature, the velocity of movement and the emmisivity of the workload. Table 29.29 shows typical allowable quench delay times for various thicknesses.

The specification for the maximum allowable quench delay time is based on the assumed amount of cooling of the workload, before it enters the quenchant. In general, the maximum allowable quench delay times can be exceeded if it can be demonstrated that the part temperatures do not fall below approximately 413°C before immersion. An exception to this is for AA2219, where the part temperatures can not fall below 482°C before immersion.

Table 29.29 TYPICAL MAXIMUM QUENCH DELAY TIMES

Minimum thickness		Maximum time
mm	inch	(s)
Up to 0.41	Up to 0.016	5
Over 0.41–0.79	Over 0.016–0.031	7
Over 0.79–2.29	Over 0.031–0.090	10
Over 2.29	Over 0.090	15

It is difficult to measure directly and control the temperature drop during transfer of the workload from the solution treating furnace to the quench tank. However, the quench delay time is easily controlled using only a stopwatch. This is augmented with the results from routine tensile testing and intergranular corrosion testing.

29.3.6 Ageing (natural and artificial)

29.3.6.1 *Natural ageing*

Some heat treatable alloys, especially 2XXX alloys, harden appreciably at room temperature to produce the useful tempers T3 and T4. These alloys, when naturally aged to the T3 or T4 tempers, exhibit high ratios of ultimate tensile strength/yield strength and also have excellent fatigue and fracture toughness properties.

Natural ageing, and the resulting increase in properties, occurs by the rapid formation of GP (Guinier–Preston) zones from the supersaturated solid solution, by means of solute diffusion involving quenched-in vacancies. Strength increases rapidly, with properties becoming stable after approximately 4–5 days. The T3 and T4 tempers are based on natural ageing for 4 days. For 2XXX alloys, improvements in properties after 4–5 days are relatively minor, and become stable after one week.

The Al–Zn–Mg–Cu and Al–Mg–Cu alloys (7XXX and 6XXX), harden by the same mechanism of GP zone formation. However, the properties resulting from natural ageing are less stable. These alloys still exhibit significant changes in properties, even after many years.

The natural ageing characteristics vary from alloy to alloy. The most notable differences are the initial incubation time for changes in properties to be observed, and the subsequent rate of change in properties. Ageing effects are suppressed with lower than ambient temperatures. In many alloys, such as 7XXX alloys, natural ageing can be nearly completely suppressed by holding at −40°C.

Because of the very ductile and formable nature of as-quenched alloys, retarding natural ageing, increases scheduling flexibility for forming and straightening operations. It also allows for uniformity of properties during the forming process. This contributes to a quality part. However, refrigeration at the temperatures employed normally, does not completely suppress natural ageing. Some precipitation still occurs. Table 29.30 shows typical temperature and time limits for refrigeration.

29.3.6.2 *Artificial ageing*

After quenching, most aerospace aluminium alloys are aged artificially. This is a complex process and requires an understanding of vacancies, and the interaction of precipitation and metastable phases. In general, the sequence of precipitation occurs by clustering of vacancies, formation of GP zones, nucleation of a coherent precipitate, precipitation of an incoherent precipitate and finally the coarsening of the precipitates.

Precipitation hardening is the mechanism where the hardness, yield strength and ultimate strength dramatically increase with time at a constant temperature (the ageing temperature) after rapidly cooling from a much higher temperature (the solution treatment temperature). This rapid cooling or quenching results in a supersaturated solid solution and provides the driving force for precipitation on ageing. The age hardening phenomenon was first discovered by Wilm,[118] who found that the hardness of aluminium alloys with small quantities of copper, magnesium, silicon, and iron, increased with time, after quenching from a temperature just below the melting temperature.

Table 29.30 TYPICAL TIME AND TEMPERATURE LIMITS FOR REFRIGERATED PARTS STORED IN THE AS-QUENCHED (AQ) CONDITION

| Alloy | Maximum delay time after quenching (min) | Maximum storage time for retention of the AQ condition (days) | | |
		−12°C (10°F) Max.	−18°C (0°F) Max.	−23°C (−10°F) Max.
2014 2024 2219	15	1	30	90
6061 7075	30	7	30	90

Precipitation hardening (ageing) involves heating the alloyed aluminium to a temperature in the 95 to 230°C (200–450°F) range. In this range, the supersaturated solid solution, created by quenching from the solution treatment temperature, begins to decompose. Initially there is a clustering of the solute atoms. Once sufficient solute atoms have diffused to these initial clusters, coherent precipitates form. Because the clusters of solute atoms have a lattice parameter mismatch with the aluminium matrix, a strain field surrounds the solute clusters. As more solute diffuses to the clusters, eventually the matrix can no longer accommodate the matrix mismatch. At this point, a semi-coherent precipitate forms.

Finally, after the semi-coherent precipitate grows to a large enough size, the matrix can no longer support the crystallographic mismatch, and the equilibrium phase forms as incoherent precipitates.

Heating the quenched material in the range of 95–205°C accelerates precipitation in heat treatable alloys. This acceleration is completely not due to changes in reaction rate. As was shown above, structural changes occur that are dependent on time and temperature. In general, the increase in yield strength that occurs during artificial ageing increases faster than the ultimate tensile strength. This means that the alloys lose ductility and toughness. T6 properties are higher than T4 properties, but ductility is reduced. Overageing decreases the tensile strength, and increases the resistance to stress-corrosion-cracking. It also enhances the resistance to fatigue crack growth and imparts dimensional stability to the part.

Precipitation hardening curves have been developed for all the most common alloys. Figure 29.27 shows ageing curves for 2024 and 6061. Both alloys show evidence of reversion of GP zones, as indicated by initial reductions in hardness. This reduction in hardness is caused by the re-solutioning of small GP zones that are below the critical size required for stability. Similar ageing curves have been developed for 7075 and casting alloys.

The ageing curves for different alloys vary; however, generally the higher the ageing temperature, the shorter the time required to attain maximum properties. When high ageing temperatures are used, properties are reached very rapidly. For this reason, ageing temperatures are usually lower to assure that the entire load is brought to the required ageing temperature without risk of reduced properties caused by over-ageing of the fastest heated components.

29.3.7 Appendix: Quenchants

In section 29.3.5 the quenching process step was discussed. In this section, the nature of interactions between the hot part and the quenching medium are discussed and used to explain the choice of quenchants for industrial heat treatment.

For any quenchant, there are typically three phases that occur while the part cools: the vapour phase; nucleate boiling and finally convection. Each of these stages has very specific characteristics and heat transfer mechanisms.

In the vapour phase, a stable gas film of superheated quenchant surrounds the part. The stability of this vapour film depends on several factors, including surface roughness, the boiling temperature of the quenchant and viscosity of the quenchant. Heat transfer is very slow through this film, as heat transfer occurs primarily by radiation and conduction. Because of the relatively low temperatures involved in aluminium alloy solution treatment, radiation heat transfer through the film is negligible. Conduction is also negligible because of the poor conductive heat transfer characteristics of gases.

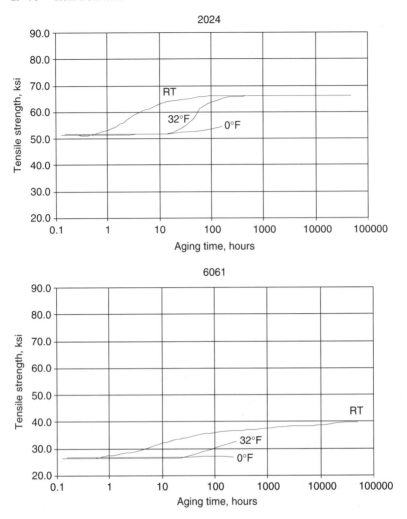

Figure 29.27 *Ageing curves for 2024 and 6061*

As the part cools, the stability of the vapour film also decreases, until the collapse of the vapour film occurs. At this point, nucleate boiling occurs. This transition between stable film boiling, and nucleate boiling is called the 'Ledenfrost temperature'.

Nucleate boiling is the fastest regime of cooling during quenching. This is where the vapour stage starts to collapse and all liquid in contact with the component surface erupts into bubbles as boiling occurs. The high heat extraction rates are due to the carrying away of heat from the hot surface and the transferring of it further into the liquid quenchant, which allows cooled liquid to replace it at the surface. In many quenchants, additives have been included to enhance the maximum cooling rates obtained with a given fluid. The boiling stage stops when the temperature of the component's surface reaches a temperature below the boiling point of the liquid. For distortion prone components, high boiling temperature oils or liquid salts could be used, if cooling in these media is fast enough to harden the material, but both of these quenchants see relatively little use in induction hardening.

The final stage of quenching is the convection stage. This occurs when the component has reached a temperature below that of the quenchant's boiling point. Heat is removed by convection and is controlled by the quenchant's specific heat and thermal conductivity, and the temperature differential between the component's temperature and that of the quenchant. The convection stage is usually the slowest of the three quenching stages. Typically, it is this stage where most distortion occurs.

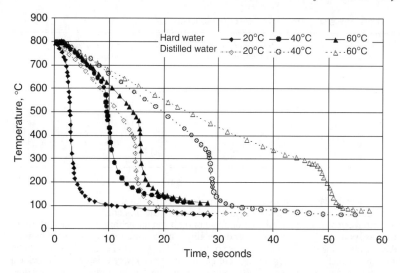

Figure 29.28 *Cooling curve of distilled water showing extended stable vapour phase*

Obtaining the desired properties with low distortion is usually a balancing act. Often, optimal properties are obtained at the expense of high residual stresses or excessive distortion. Conversely, low distortion or residual stresses are usually obtained with a sacrifice in properties. Therefore, the optimum quench rate is one where properties are just met. As a general rule, this usually provides the minimum distortion.

There are several types of quenchants used for heat treating aluminium. The most common are cold water, hot water and polymer quenchants.

29.3.7.1 Cold water quenching

Water is the most common of all quenchant materials. It is easy and inexpensive to obtain, and is readily disposed of, unless severely contaminated. In general, as the temperature of the water is raised, the stability of the vapour phase increases and the onset of nucleate boiling in a stagnant fluid is suppressed. Hence, both the maximum and overall rates of cooling are decreased.

Cold water quenching is the most severe of commonly used quenchants. In an early study using cooling curves,[146] it was shown that quenching into still water caused rapid heat transfer. This study found that heat transfer at the surface of the part was very turbulent at the metal/water interface. The same investigation also showed that there was a marked difference between hard water and distilled water. Distilled water produced an extensive vapour blanket that extended to very low temperatures (Figure 29.28).

29.3.7.2 Hot water quenching

The cooling rate produced by water quenching is independent of material properties, such as thermal conductivity and specific heat. Instead, the cooling rate is primarily dependent on water temperature and agitation.[147] Water temperature is the largest primary variable controlling the cooling rate. As already discussed, with increasing water temperature, the cooling rate decreases. The maximum cooling rate also decreases, as the water temperature is increased. In addition, the temperature of maximum cooling decreases with increasing water quenchant temperature. The length of time and stability of the vapour barrier increases, with increasing water temperature. This is shown in Table 29.31.

Often, the application of cold water quenchants causes excessive distortion, residual stresses or cracking. Elevated temperatures up to approximately 80°C (180°F) are sometimes used. However, the extended vapour phase that results tends to cause the properties of the workpiece to be compromised, while doing nothing to reduce distortion. In this application, the uniformity of agitation is critical.

Table 29.31   EFFECT OF WATER TEMPERATURE ON COOLING RATES[148]

Water temperature (°C)	Maximum cooling rate (°C s^{-1})	Temperature corresponding with maximum cooling rate (°C)	Cooling rate (°C s^{-1}) at workpiece temperature		
			704°C	343°C	232°C
40	153	535	60	97	51
50	137	542	32	94	51
60	115	482	20	87	46
70	99	448	17	84	47
80	79	369	15	77	47
90	48	270	12	26	42

Agitation must be able to reach all portions of the part to be effective. Otherwise, vapour can collect and serve as insulation. This will result in soft spots with inadequate properties.

Quenching into water at <50–60°C often produces non-uniform quenching. This non-uniformity manifests itself as spotty hardness, distortion and cracking. This non-uniformity is caused by relatively unstable vapour blanket formation. Because of this difficulty, it was necessary to identify an alternative to water quenching. Polyalkylene glycol quenchants (PAG) were developed to provide a quench rate in between that of water and oil. By control of agitation, temperature and concentration, quench rates similar to water and thick oil can be achieved.

29.3.7.3   Polymer quenching

Polyalkylene glycol polymer quenchants are used in the aerospace industry to control and minimise the distortion occurring during the quenching of aluminium. Typically these quenchants are governed by AMS 3025, and are either Type I or Type II. Type I quenchants are single polyalkylene glycol polymers, while Type II quenchants are multiple molecular weight polyalkylene glycol polymers. Each offers different benefits. Because of the higher molecular weight of the Type II PAG quenchants, lower concentrations can be used. However, Type II polymers have a lower cloud point temperature, which can cause higher drag-out, if parts are removed from the quenchant before they reach the quenchant temperature, typically 25–40°C (80–100°F). Typical concentrations used for Type I polymer quenchants are shown in Table 29.32. A more extensive listing of recommended concentrations is provided in AMS 2770 (Wrought products and forgings) and AMS 2771 (castings).

Polyalkylene glycols exhibit inverse solubility in water. They are completely soluble in water at room temperature, but insoluble at elevated temperatures. The inverse solubility temperature can range from 60°C to 90°C depending on the molecular weight of the polymer, and the structure of the polymer. This phenomenon of inverse solubility modifies the conventional three-stage quenching mechanism[149] and provides great flexibility to control the cooling rate.

When a component is first immersed, the solution in the immediate vicinity of the metal surface is heated to above the inverse solubility temperature. The polymer becomes insoluble and a uniform polymer-rich film encapsulates the surface of the part. The stability and longevity of this polymer film is dependent on the temperature, concentration and amount of agitation present. The stable polymer-rich film eventually collapses uniformly, and cool quenchant comes into contact with the hot metal surface. Nucleate boiling results, with high heat extraction rates.

As the period of active boiling subsides, cooling occurs by conduction and convection into the quenchant. When the surface temperatures fall below the inversion temperatures, the polymer dissolves, forming a homogeneous solution again.

The cooling rate of these polymers can be readily varied to suit the specific application by changing the concentration, quenchant temperature, and the amount of agitation. Typically, for most applications, the agitation is usually fixed, while the concentration is changed.

The concentration of the polymer influences the thickness of the polymer film that is deposited on the surface of the part during quenching. As the concentration increases, the maximum rate of cooling, and the cooling rate in the convection phase decrease.

Agitation has an important effect on the quenching characteristics of the polymer quenchant. It ensures uniform temperature distribution within the quench tank and it also affects the quench rate. As the severity of agitation increases, the lifetime of the polymer-rich phase decreases and eventually disappears, and the maximum rate of cooling increases. Agitation has comparatively little effect on the cooling rate during the convection stage.

Table 29.32 TYPICAL CONCENTRATION LIMITS FOR QUENCHING IN PAG

Alloy	Form	Maximum thickness		Concentration (vol. %)
		inches	mm	
2024	Sheet	0.040	1.02	34 Max.
		0.063	1.60	28 Max.
		0.080	2.03	16 Max.
2219	Sheet	0.073	1.85	22 Max.
6061	Sheet, Plate	0.040	1.02	34 Max.
7049		0.190	4.83	28 Max.
7050				
7075		0.250	6.35	22 Max.
6061	Forgings	1.0	25	20–22
7075		2.0	50	13–15
		2.5	64	10–12
7049	Forgings	3.0	76	10–12
7149				
7050	Forgings	3.0	76	20–22
6061	Extrusions	0.250	6.35	28 Max.
7049				
7050	Extrusions	0.375	9.52	22 Max.
7075				

REFERENCES

1. A. K. Sinha, 'Chapter 10—Basic Heat Treatment', *Ferrous Physical Metallurgy*, 1989, Butterworths, Boston, MA, p. 403–440.
2. G. Krauss, 'Glossary of Selected Terms', *Steels: Heat Treatment and Processing Principles*, 1990, ASM International, Materials Park, OH, p. 453–468.
3. E. Taney, 'Heat Treatment of Tool Steels', Tooling & Production, 2000, May, p. 102–104.
4. ASTM E7-01 'Standard Terminology Relating to Metallography', ASTM International, West Conshohocken, PA.
5. J. R. Davis, *ASM Materials Engineering Dictionary*, 1992, ASM International, Materials Park, OH.
6. A. V. Sverdlin and A. R. Ness, 'Chapter 1—Fundamental Concepts in Steel Heat Treatment', *Steel Heat Treatment Handbook*, Eds G. E. Totten and M. A. H. Howes, 1997, Marcel Dekker, Inc., New York, NY, p. 1–44.
7. G. Krauss, 'Phases and Structures', *Steels: Heat Treatment and Processing Principles*, 1990, ASM International, Materials Park, OH, p. 1–16.
8. *Practical Data for Metallurgists*, 1996, The Timken Company, Canton, OH.
9. B. Liscic, 'Chapter 8—Steel Heat Treatment', *Steel Heat Treatment Handbook*, Eds G. E. Totten and M. A. H. Howes, 1997, Marcel Dekker Inc., New York, NY, p. 527–662.
10. H. P. Hougardy, 'Die Darstellung des Umwandlungsverhaltens von Stählen in den ZTU-Schaubildern', *Häterei-Tech, Mitt.*, 1978, Vol. 33, No. 2, p. 63–70.
11. H. M. Tensi, A. Stich and G. E. Totten, 'Chapter 4—Quenching and Quenching Technology', *Steel Heat Treatment Handbook*, Eds G. E. Totten and M. A. H. Howes, 1997, Marcel Dekker Inc., New York, NY, p. 157–249.
12. K.-E. Thelning, 'Chapter 4—Hardenability', *Steel and Its Heat Treatment—Second Edition*, 1984, Butterworths, London, p. 144–206.
13. M. Atkins, *Atlas of Continuous Transformation Diagrams for Engineering Steels*, ASM International, Materials Park, OH, 1980.
14. A. Rose and W. Strassburg, 'Anwendung des Zeit-Temperatur-Umwandlungs- Schaubildes für kontinuierliche Abkülung auf Fragen der Wärmbehandlung, *Archiv. Eissenhüttenwes.*, 53, Vol. 24, No. 11/12, p. 505–514.
15. E. Scheil, *Arch. Eisenhüttenwes.*, 1934/35, Vol. 8, p. 565–567.
16. B. Liscic, 'Chapter 3—Hardenability', *Steel Heat Treatment Handbook*, Eds G. E. Totten and M. A. H. Howes, 1997, Marcel Dekker Inc., New York, NY, p. 93–156.
17. G. E. Totten, C. E. Bates and N. A. Clinton, 'Chapter 1—Introduction to the Heat Treating of Steel', *Handbook of Quenchants and Quenching Technology*, 1993, ASM International, Materials Park, OH, p. 1–33.
18. G. E. Totten, C. E. Bates and N. A. Clinton, 'Chapter 2—Measuring Hardenability and Quench Severity', *Handbook of Quenchants and Quenching Technology*, ASM International, Materials Park, OH, 1993, p. 35–68.

19. A. K. Sinha, 'Chapter 11—Hardening and Hardenability', *Ferrous Physical Metallurgy*, 1989, Butterworths, Boston, MA, p. 441–522.
20. *Hardenability Slide Rule*, United States Steel Corp., 1970.
21. G. Krauss, *Principles of Heat Treatment of Steel*, ASM, 1980.
22. ASTM Standard Test Method E 112-96E1 'Standard Test Methods for Determining Average Grain Size', ASTM International, West Conshohocken, PA.
23. ASTM Standard Test Method E 1382-97, 'Standard Test Methods for Determining Average Grain Size Using Semiautomatic and Automatic Image Analysis', ASTM International, West Conshohocken, PA.
24. G. E. Totten, C. E. Bates and N. A. Clinton, 'Chapter 3—Cooling Curve Analysis', *Handbook of Quenchants and Quenching Technology*, ASM International, Materials Park, OH, 1993, p. 69–128.
25. C. E. Bates, G. E. Totten and R. L. Brennan, 'Quenching of Steel', *ASM Handbook Vol. 4—Heat Treating*, 1991, ASM International, Materials Park, OH, p. 67–120.
26. T. V. Rajan, C. P. Sharma and A. Sharma, 'Chapter 5—Heat Treatment Processes for Steels', *Heat Treatment: Principles and Techniques*—Revised Edition, 1994, Prentice Hall of India, Pvt. Ltd., New Delhi, India, p. 97–123.
27. G. E. Totten, G. R. Garsombke, D. Pye and R. W. Reynoldson, 'Chapter 6 – Heat Treating Equipment', *Steel Heat Treatment Handbook*, Eds G. E. Totten and M. A. H. Howes, 1997, Marcel Dekker, Inc., New York, NY, p. 293–481.
28. M. A. Aronov, J. F. Wallace and M. A. Ordillas, 'System for Prediction of Heat-Up and Soak Times for Bulk Heat Treatment Processes', *Heat Treating: Equipment and Processes*, Eds G. E. Totten and R. A. Wallis, Proc. of 1994 Conference, Shaumberg, IL, 1994, ASM International, Materials Park, OH, p. 55–61.
29. K. Kasten, 'A Primer of terminology for Heat Treat Customers', *Heat Treating*, 1980, February, p. 32–39.
30. A. Nayar, 'Chapter 3.4—Heat Treatment of Steel', *The Metal Databook*, 1997, McGraw-Hill, New York, NY.
31. D. J. Naylor and W. T. Cook, 'Heat Treated Engineering Steels', Materials Science and Technology, 1992, Vol. 7, p. 435–488.
32. K.-E. Thelning, 'Chapter 5—Heat Treatment—General', *Steel and Its Heat Treatment—Second Edition*, 1984, Butterworths, London, p. 207–318.
33. M. Narazaki, G. E. Totten and G. M. Webster, 'Hardening by Reheating and Quenching', *Handbook of Residual Stress and Deformation of Steel*, Eds G. E. Totten, M. A. H. Howes and T. Inoue, 2002, ASM International, Materials Park, OH, p. 248–295.
34. F. Legat, 'Why Does Steel Crack During Quenching', *Kovine, Zlitine Technologie*, 1998, Vol. 32, Nos. 3 and 4, p. 273–276.
35. R. Kern, 'Distortion and Cracking II: Distortion from Quenching', *Heat Treating*, 1985, March, p. 41–45.
36. P. C. Clarke, 'Close Tolerance Heat Treatment of Gears', *Heat Treatment of Metals*, 1998, Vol. 25, No. 3, p. 61–64.
37. R. R. Blackwood, L. M. Jarvis, D. G. Hoffman and G. E. Totten, 'Conditions Leading to Quench Cracking Other Than Severity of Quench', *Heat Treating Including the Liu Dai Symposium*, Proc. 18th Conference, Eds R. A. Wallis and H. W. Walton, 1998, ASM International, Materials Park, OH, p. 575–585.
38. R. Kern, 'Distortion and Cracking III: How to Control Cracking', *Heat Treating*, 1985, April, p. 38–42.
39. H.-H. Shao, 'Analysis of the Causes of Cracking of a 12% Cr Steel Cold Die During Heat Treatment', *Jinshu Rechuli*, 1995, No. 11, p. 43.
40. R. F. Kern, 'Thinking Through to Successful Heat Treatment', *Met. Eng. Q.*, 1971, Vol. 11, No. 1, p. 1–4.
41. X. Sheng and S. He, 'Analysis of Quenching Cracks in Machine Tool Pistons under Supersonic Frequency Induction Hardening', *Heat. Treat. Met. (China)*, 1991, No. 4, p. 51–52.
42. W. T. Cook, 'Review of Selected Steel-Related Factors Controlling Distortion in Heat Treatable Steels', *Heat Treat. Met.*, 1999, Vol. 26, No. 2, p. 27–36.
43. Y. Toshioka, 'Heat Treatment Deformation of Steel Products', *Mater. Sci. Tech.*, 1985, Oct. Vol. 1, p. 883–892.
44. H. M. Tensi, G. E. Totten and G. M. Webster, 'Proposal to Monitor Agitation of Production Quench Tanks', *Heat Treating Including the 1997 Induction Heat Treating Symposium—Proc. 17th Conf.*, Eds D. L. Milam, D. A. Poteet, G. D. Pfaffmann, V. Rudnev, A. Muehlbauer and W. B. Albert, 1997, ASM International, Materials Park, OH, p. 443–441.
45. G. Krauss, 'Tempering of Steel', *Steels: Heat Treatment and Processing Principles*, 1990, ASM International, Materials Park, OH, p. 206–261.
46. R. A. Grange, C. R. Hribal and L. F. Porter, Hardness of Tempered Martensite in Carbon and Low-Alloy Steels, *Met. Trans A*, 1977, Vol 8A, p. 1775–1785.
47. M. A. Grossmann and E. C. Bain, 'Chapter 5 Tempering After Quench Hardening', *Principles of Heat Treatment*, 1964, American Society for Metals, Metals Park, OH, p. 129–175.
48. A. K. Sinha, 'Chapter 12 Tempering', *Ferrous Physical Metallurgy*, 1989, Butterworths, Boston, MA, p. 523–608.
49. R. A. Grange and R. W. Baughman, *ASM Trans.*, 1956, Vol. 48, p. 165.
50. H. J. Spies, G. Münch and A. Prewetz, 'Möglichkeiten der Optimierung der Auswahl vergütbarer Baustähle durch Berechnung der Härt-und-vergütbarkeit', *Neue Hütte*, 1977, Vol. 8, No. 22, p. 443–445.
51. 'Heat Treating of Steel', *Metals Handbook—Desk Edition, Second Edition*, Ed. J. R. Davis, 1998, ASM International, Materials Park, OH, p. 970–982.
52. H. E. Boyer and P. R. Cary, 'Chapter 4—Molten Quenching Methods: Martempering and Austempering', *Quenching and Control of Distortion*, 1988, ASM International, Materials Park, OH, p. 71–88.

53. G. E. Totten, C. E. Bates and N. A. Clinton, 'Chapter 4—Quenching Oils', *Handbook of Quenchants and Quenching Technology*, 1993, ASM International, Materials Park, OH, p. 129–160.
54. PN-EN 10052:1999 (EN10052: 1993), 'Dictionary of term of heat treatment of iron alloy IDT EN 10052:1993', Polish Committee for Standardization, Warsaw, Poland.
55. Source: Treat All Metals Inc. Website: http://www.treatallmetals.com/gas.htm, November 4, 2002.
56. G. E. Totten, M. E. Dakins and L. M. Jarvis, 'How H-factors can be used to characterise polymers', *Journal of Heat Treating*, December 1989, p. 28–29.
57. J. Grum, S. Bozic and R. Lavric, 'Influence of Mass of Steel and a Quenching Agent on Mechanical Properties of Steels,' *Journal of Heat Treating*, 1998.
58. N. I. Kobasko, G. E. Totten, G. M. Webster and C. E. Bates, 'Compression of Cooling Capacity of Aqueous Poly (Alkylene Glycol) Quenchants with Water and Oil,' *Heat Treating*, 1998.
59. M. E. Dakins, C. E. Bates and G. E. Totten, 'Estimating Quench Severity With Cooling Curves', *Journal of Heat Treating*, April 1992, p. 24–26.
60. G. E. Totten, M. E. Dakins and R. W. Heins, 'Cooling Curve Analysis of Synthetic Quenchants—A Historical Perspective', *Journal of Heat Treating*, 1988, Vol. 6, No. 2, p. 87–95.
61. G. E. Totten, Y. Sun, G. M. Webster, L. M. Jarvis and C. E. Bates, 'Quenchants Selection', *Journal of Heat Treating*, 1998, p. 183–191.
62. J. P. Peyre and C. Tournier, Choix des traitements thermiques superficiels; Center technique des Industries Mécaniques, Paris, 1985.
63. M. Przyłęcka, Materiałowo—technologiczne aspekty trwałości łożysk tocznych., Politechnika Poznańska, Seria Rozprawy Nr202, Poznań,1988, 300 str.
64. M. Przyłęcka, 'The modeling of structure and properties of carburised low-chromium hypereutectoid steels', *Journal of Materials Engineering and Performance*, April 1996, Vol. 5(2), p. 165–191, ASM International, Materials Park, Ohio.
65. H.-Y. Lee, *Journal of the Korean Institute of Metals and Materials* (South Korea), Feb. 1996, Vol. 34, No. 2, p. 150–157.
66. G. Parrish, *Carburising: Microstructure and Properties*, ASM International, 1999, Metals Park, OH, p. 99–170.
67. T. Burakowski and T. Wierzchoń, *Surface Engineering of Metals*, CRC Series in Materials Science and Technology, CRC Press, New York, 1999.
68. W. J. Moon, C. Y. Kano and J. M. Suno, *Journal of the Korean Institute of Metals and Materials*, Mar 1997, Vol. 35, No. 3, p. 297–304.
69. N. Murai, T. Takayama *et al.*, *Journal of the Iron and Steel Institute of Japan*, Mar 1997, Vol. 83, No. 3, p. 215–220.
70. A. K. Tikhonov, YUM. Palagin, *Metal Science and Heat Treatment*, May 1995, Vol. 36, No. 11–12, p. 655–657.
71. M. A. Balter and J. S. Dukarevicz, Vliyaniye kachyestva matyeriala na nadyeznost I dolgoviechnost zubchatih kolyes, Metallovedenie i Termicheskaya Obrabotka Metalloy, 1985, **7**, p. 50–53.
72. A. P. Gulalev and L. N. Serebriusikov, Vliyanie raznozyernistosti na myehanichyeskiye svoystva stali 18H2N4MA, Metallovedenie i Termicheskaya Obrabotka Metalloy, 1977, **4**, p. 2–5.
73. M. G. Menditatta, J. Sasser and G. Krauss, 'Effect of dissolved carbon on microcracking in martensite of an Fe–1.39% C alloy', *Metal. Transactions.*, 1969, **62**, p. 351–353.
74. R. A. Grange, *Metallurgical and Materials Transactions*, 1969, **62**, p. 1024–1027.
75. F. A. Apple and G. Krauss, 'Microcraking and fatigue in carburised steel', *Metallurgical and Materials Transactions*, 1973, **4**, p. 1195–1200.
76. G. Krauss, 'The microstructure and fatigue of carburised steel', *Metallurgical and Materials Transactions*, 1978, **9A**, p. 1527–1535.
77. M. Przyłęcka and W.Gęstwa, 'The Modeling of Residual Stresses after Direct Hardening of Carburised and Carbonitrided Low-Chromium Hypereutectoid Steels (ŁH15)', 'MAT-TEC 97—Analysis of residual stresses from materials to bio-materials', IITT International, 1997, p. 117–124.
78. J. Lesage, D. Chicot, M. Przyłęcka, M. Kulka and W. Gęstwa, Role du chrome sur la cementation Hyper-Austenitique d'un acier a roulement, Traitement Thermique, 1994, **276**, p. 42–46.
79. M. Przyłęcka and W. Gestwa, 'The modelling of structure and properties of carburising or carbonitriding layers, as well as hardening in different quenchning mediums', *20th ASM Heat Treating Society Conference Proceedings, 9–12 October 2000, St.Louis, MO*, ASM International, 2000, p. 624–634.
80. B. Vinokur, The Composition of the Solid Solution, Structure and Contact Fatigue of Case Hardened Layer, *Metallurgical and Materials Transactions A*, May 1993, Vol. 24, p. 1163–1168.
81. R. F. Kern, 'Supercarburising', *Journal of Heat Treatment*, Oct 1986, p. 36–38.
82. E. L. Gyulikhandanov and V. G. Khoroshailov, 'Carburising of Heat Resistant Steels in a Controlled Endothermic Atmosphere', *Met. Sci. Heat Treat. (USSR)*, Aug 1971, Vol. 13 (No. 8), p. 650–654.
83. J. Wang, L. Qia, J. Zhau, 'Formation and Properties of Carburised Case with Spheroidal Carbides,' *International Congress on 5th Heat Treatment of Materials Proceedings, Budapest, Hungary, 20–24 October 1986*, International Federation for the Heat Treatment of Materials (Scientific Society of Mechanical Engineers) 1986, p. 1212–1219.
84. A. L. Geller and L. G. Lozhushnik, 'Contact Fatigue Limit of Carburised 25Kh2GNTA Steel', *Met. Sci. Heat Treat. (USSR)*, No. 6, Plenum Publishing Corp., June 1968, p. 474.
85. V. S. Sagaradze, 'Effect of Heat Treatment on the Properties of High Carbon Alloyed Steels', *Met. Sci. Heat Treat. (USSR)*, Dec 1964, No. 12, p. 720–724.

86. M. Przyłęcka, 'The effect carbon on utility properties of cemented bearing steel', *International Congress on 5th Heat Treatment of Materials Proceedings, Budapest, Hungary, 20–24 October 1986*, International Federation for the Heat Treatment of Materials (Scientific Society of Mechanical Engineers) 1986, p. 1268–1275.

87. C. Dawes and R. S. Cooksat, 'Surface treatment of engineering components', *Heat Treatment of Metals*, Special report 1995, 77, 92. The Iron and Steel Institute, 1966.

88. H. Tauscher, 'Relationship between carburised case depth stock thickness and fatigue strength in carburised stell', *Symposium on fatigue Damage in Machine Parts, Prague 1960* (translation BISI 11, 340).

89. H. Weigand and G. Tolasch, Fatigue behaviour of case hardened samples. Translation BISI 6329 from *Härterei Technische Mitteilungen*, déc. 1962, Vol. 22, No. 4, 330–338.

90. G. Parrish, The influence of microstructure on the properties of case carburised components. Part 6. Core properties and case depth. *Heat Treatment of Metals*, 1977, **2**, p. 45–54.

91. M. Lacoude, 'Propriétés d'emploi des aciers de cémentation pour pignonerie', *Aciers Spéciaux*, No. 12, 1970, p. 21–29.

92. B. Champin, L. Seraphin and R. Tricot, 'Effets comparés des traitements de cémentation et de carbonitruration sur les propriétés d'emploi des aciers pour engrenages. Mémoires Scientifiques', *Revue de Métallurgie*, fév. 1977, p. 77–90.

93. Wyszykowski, H. Preignitz, E. Gozdzik and A. Ratliewcs, 'Influence de l'austénite résiduelle sur quelques propriétés de l'acier cémenté', *Revue de Métallurgie*, Vol. 68, No. juin 1971, p. 411–422.

94. C. Razim, 'Influence de l'austénite résiduelle sur la résistance mécanique d'éprouvettes cémentées soumises à des efforts alternés', *Revue de Métallurgie*, fév. 69, p. 147–157.

95. W. Beumelburg, 'Comportement d'éprouvettes cémentées présentant divers états superficiels, des teneurs en carbone variables en surface, lors d'essais de flexion rotative, flexion statique et resilience', *Exposé présenté à la réunion ATTT 26/3/75*.

96. G. Parrish, 'The influence of microstructure on the properties of case hardened components Part 4. Retained Austenite', *Heat Treatment of Metals*, 1976, **4**, p. 101–109.

97. B. Champin, 'Commentaires sur la conférence de M. BEUMELBURG', *Réunion ATTT 26/3/75*.

98. J. P. Sheehan and M. A. Howes, 'The effect of case carbon content and heat treatment on the pitting fatigue of 8620 steel SAE 720 268', *Automotive Engineering Congress Detroit Michigan*, January 10–14, 1972.

99. C. Razim, 'Einfluss des Randgefüge einsatzgehärteter Zahnräder auf die Neigung zur Grübchenbildung', *Härterei Technische Mitteilungen*, 1974, Heft 4, p. 317–325.

100. M. A. Turbalter and M. L. Turovskh, 'Résistance of case hardened steel to contact fatigue', *Metal Science and Heat Treatment*, March 1966, No. 3, p. 177–180.

101. A. Diament, R. El Haik, R. Lafont and R. Wyss, 'Tenue en fatigue superficielle des couches carbonitrurées et cémentées en relation avec la répartition des contraintes résiduelles et les modification du réseau cristallin apparaissant en cours de fatigue', *Traitement Thermicque*, 1974, No. 87, p. 87–97.

102. K. Z. Shepelyakovskii, V. D. Kal'ner and V. F. Mikonov, 'Technology of heat treating steel with induction heating', *Metal Science and Heat Treatment*, November 1970, No. 11, p. 902–908.

103. G. H. Robinson, *The effect of surface condition on the fatigue esistance of hardened steel. Fatigue durability of carburised steel* 11.46. American Society for Metals, 1957.

104. W. Beumelburg, 'The effect of surface oxidation on the rotating bending strength and static bending strength of case hardened specimens', *Hart. Tech. Mitt.*, Oct. 1970, Vol. 25, No. 3, 191–194.

105. G. Parrish, 'The influence of microstructure on the properties of case carburised components', *Heat Treatment of metals*, 1976, **2**, p. 49–53.

106. G. W. P. Rengstorff, M. B. Bever and C. F. Floe, *Transaction of the Society for Metals*, 1951, 43.

107. W. Luty, The parts of ŁH15(52100) steel carbonitrided in atmosphere with liquid of organic compounds, The work of bearing industry no. 1/23/, Poland, 1972.

108. E. V. Vasilewa, C. H. Sawiceve and J. V. Krjukowa, Porysenie iznosojkosti stali SCH15 ionnoj implantacjej. Metalowedienie i termiceskaja obrabotka metallov, Nr 1, 1987, p. 59–62.

108a. D. S. MacKenzie, 'Quench Rate and Aging Effects in Aluminum Al–Zn–Mg–Cu Alloys', PhD Dissertation, University of Missouri-Rolla, Dec. 2000.

109. AMS 2770 'Heat Treatment of Aluminum Alloy Parts', Society of Automotive Engineers, Warrendale, PA.

110. L. Mondolfo, *Aluminum Alloys, Structure and Properties*, (Butterworths, 1976) London.

111. Metals Handbook, 8th Ed., Vol. 8, ASM (Metals Park: 1973) p.259.

112. A. Guinier, 'Arrangement for obtaining intense diffraction diagrams of crystalline powders with monochromatic radiation', *Compt. Rend.*, 1937, **204**, p. 1115–1116.

113. A. Guinier, 'Structure of age-hardened aluminium–copper alloys', *Nature*, 1938, **142**, 569–570.

114. G. D. Preston, 'The diffraction of x-rays by age-hardening aluminium–copper alloys', *Proc. Roy. Soc. A*, 1938, **167**(931), p. 526–538.

115. G. D. Preston, 'Structure of age-hardened aluminium copper alloys', *Nature*, 1938, **142**, p. 570.

116. E. Hornbogen, 'Investigation by means of the electron microscope of precipitation in aluminium–copper solid solutions. II. The formation of θ'', θ and θ' phases in aluminium–copper solid solutions', *Aluminium*, 1967, **43**(2), p. 115–121.

117. J. M. Silcock, T. J. Heal and H. K. Hardy, 'Structural ageing characteristics of binary aluminium–copper alloys', *J. Inst. Metals*, 1954, **82**, p. 239–248.

118. A. Wilm, 'Physical–metallurgical investigation of aluminium–magnesium alloys', *Metallurgie*, 1911, **8**, p. 225–227.

119. R. N. Wilson and P. G. Partridge, 'The nucleation and growth of s' precipitates in an aluminium–2.5 percent copper–1.2 percent magnesium alloy', *Acta Metall.*, 1965, **13**(12), p. 1321–1327.

120. G. C. Weatherly, *Ph.D. Dissertation*, Cambridge, 1966.
121. C. J. Peel, *RAE Technical Report 75062*, 1975.
122. M. V. Hyatt, *Proc. Int. Conf. Aluminum Alloys*, Torino, Italy, October 1976.
123. M. Bernole and R. Graf, 'The influence of zinc on the decomposition of the supersaturated aluminium–magnesium solid solution', *Mem. Sci. Rev. Metall.*, 1972, **69**(2), p. 123–142.
124. J. Auld and S. McK. Cousland, 'The metastable τ' phase in Al–Zn–Mg and Al–Ag–Mg alloys', *Met. Sci.*, 1976, **10**(12), p. 445–448.
125. H. Löeffler, I. Kovács and J. Lendvai, 'Decomposition processes in Al–Zn–Mg', *J. Mater. Sci.*, 1983, **18**(8), p. 2215–2240.
126. G. Bergman, J. L. T. Waugh and L. Pauling, 'Crystal structure of the intermetallic compound $Mg_{32}(Al,Zn)_{49}$ and related phases', *Nature*, 1952, **169**, p. 1057–1058.
127. Metals Handbook, 8th Ed. Metallography and Phase Diagrams, ASM.
128. J. E. Hatch (Editor), *Aluminum: Properties and Physical Metallurgy*, ASM (Metals Park: 1984) 154.
129. M. A. Grossman, *Metal Progress*, 1938, **4**, 373.
130. H. Scott, '*Quenching Mediums*', Metals Handbook, ASM 1948, 615.
131. F. Wever, *Archiv fur das Eisenhuttenwesen*, 1936, **5**, 367.
132. M. Dakins, Central Scientific Laboratory, Union Carbide, Report CSL-226A.
133. W. L. Fink and L. A. Wiley, 'Quenching of 75S-aluminium alloy', *Trans. AIME*, 1948, **175**, p. 414–427.
134. H. Suzuki, M. Kanno, H. Saitoh and K. Itoi, 'Effects of zirconium addition on quench sensitivity of Al–Zn–Mg–Cu alloys', *J. Jpn. Inst. Light Met.*, 1983, **33**(1), p. 29–37.
135. J. W. Evancho and J. T. Staley, 'Kinetics of precipitation in aluminum alloys during continuous cooling', *Metall. Trans.*, 1974, **5**, p. 43–47.
136. I. A. Wierszykkowski, 'The effect of the thermal path to reaction isothermal temperature on transformation kinetics', *Metall. Trans. A*, 1991, **22A**, p. 993–999.
137. C. E. Bates and G. E. Totten, 'Procedure for quenching media selection to maximise tensile properties and minimise distortion in aluminum alloy parts', *Heat Treatment of Metals*, 1988, **15**(4), p. 89–97.
138. L. Swartzenruber, W. Beottinger, I. Ives *et al.*, *National Bureau of Standards Report NBSIR* 80-2069, 1980.
139. C. E. Bates, T. Landing and G. Seitanakis, 'Quench factor analysis: a powerful tool comes of age', *Heat Treat.*, 1985, **17**(12), p. 13–17.
140. C. E. Bates, '*Recommended Practice for Cooling Rate Measurement and Quench Factor Calculation*', *ARP 4051 Aerospace Materials Engineering Committee (SAE)*, 1987, 1.
141. J. T. Staley, R. D. Doherty and A. P. Jaworski, 'Improved model to predict properties of aluminum alloy products after continuous cooling', *Metall. Trans. A.*, 1993, **24A**(11), p. 2417–2427.
142. J. T. Staley, 'Quench factor analysis of aluminium alloys', *Mater. Sci. Technol.*, 1987, **3**(11), p. 923–935.
143. D. D. Hall and I. Mudawar, 'Predicting the impact of quenching on mechanical properties of complex-shaped aluminum alloy parts', *J. Heat Transfer*, 1995, **117**(2), p. 479–488.
144. J. S. Kim, R. C. Hoff and D. R. Gaskell, *Materials Processing in the Computer Age*, Ed. V. R. Vasvev 1991, 203.
145. C. E. Bates, 'Quench Factor-Strength Relationships in 7075-T73 Aluminum' *Southern Research Institute* 1 (1987).
146. K. G. Speith and H. Lange, 'The quenching power of liquid hardening agents', *Mitt. aus dem Kaiser-Wilhelm-Institut fuer Eisenforschung zu Düsseldorf*, 1935, **17**, p. 175–184.
147. A. Rose, *Arch. Eisenhullennes*, 1940, **13**, 345.
148. C. E. Bates, G. E. Totten, R. J. Brenner, ASM Handbook, V4 Heat Treating, ASM 1991, p. 51.
149. G. E. Totten, Handbook of Quenching and Quenching Technology (ASM: Metals Park OH) 1987.

30 Metal cutting and forming

30.1 Introduction and cross-references

The noted British mathematician R. Hill once described the state of metal cutting and forming operations with the statement that 'it is notorious that the extant theory of the mechanics of machining do not agree well with experiment'.[1] As this has largely continued to be the case, this chapter makes no attempt to review those theories. Rather, a short summary of the principal processes is provided along with a detailed list of reference recommendations. Professionals with an interest in the theory of metal cutting are referred to several of the many texts in this area.[2-5] Metal forming experts are referred to the excellent text by Altan *et al.*,[6] although numerous texts exist in this area e.g. [7-10]. For a comprehensive manufacturing approach to the processes with many accompanying illustrations, the reader is referred to the text by DeGarmo *et al.*,[11] or the text by Stephenson *et al.*[12] These texts provide detailed, comprehensive illustrations of many of the processes discussed below with comprehensive supporting details.

The processes discussed in this chapter represent a very brief survey of metal cutting operations, grinding, deburring, metal forming, machinability, lubricants/coolants, and emerging non-traditional machining techniques. Although occupational safety guidelines that apply to these operations are incorporated as they are discussed, a short, concise reference section on the occupational safety standards that may apply is provided for the professional engineer to refer to as he/she reviews and selects the appropriate processes. Related material on lubricants, superplasticity and diffusion bonding (used commonly in conjunction with superplastic forming) may be found respectively in Chapters 24, 36 and 33.

30.2 Metal cutting operations

30.2.1 Turning

In turning, a cutting tool is fed into the rotating workpiece to generate a cylindrical or conical surface that is concentric with the axis of rotation. Traditionally, this has been done using a lathe, which is one of the oldest and most versatile of the conventional machine tools. A very large number of lathes have been developed over the years and they are well documented in the literature.[13,14]

30.2.1.1 *Hard turning*

A special case of turning is hard turning, in which heat-treated metals are machined using very hard tools. This produces very fine finishes and tolerances and avoids the growing United States National Institute for Occupational Safety and Health (NIOSH)/industry concerns with metal working fluids.[15-18]

30.2.2 Boring

The boring operation is equivalent to turning, but is performed exclusively on internal surfaces. It is used for roughing (sometimes referred to as 'hogging'), semi-finishing, or finishing of castings or drilled holes. Finish boring is particularly noted for its ability to achieve dimensional and surface finish tolerances. Specialised versions of the boring machine include vertical and horizontal as well as jig borers.[11]

30.2.3   Drilling

Drilling is the standard process for producing holes and as such, it is the most common metal cutting process. The drill has a complex geometry, and the process is complex, often being the key bottleneck process in high volume manufacturing. Critical holes are often drilled in two or more passes to promote chip removal and dimensional accuracy, each tool penetrating to only a fraction of the final depth. Often, boring or reaming operations will follow drilling. There are a great many types of highly specialised drilling tool designs.

30.2.3.1   *Conventional drilling*

Most people are familiar with conventional drilling methods. Process machines include upright machines (drill presses), radial machines, and specialised single cycle automatics amongst other variations of the basic machines.[19,20]

30.2.3.2   *Deep hole drilling*

A deep hole is one with a depth-to-diameter ratio of more than 5:1. Special machines are often required to drill these deep holes with adequate accuracy and to ensure proper chip removal. Various lubricants and coolants are frequently used in these processes and are a special concern in the emerging area of metal working fluid hazards defined by NIOSH.[17] Three commonly used deep hole drilling methods in use are solid drilling, trepanning, and counter-boring.[19–21]

30.2.3.3   *Microdrilling*

Microdrilling is the machining of small diameter holes (typically less than 0.5 mm) with a depth-to-diameter ratio greater than 10.[22] The small diameter of the holes precludes the use of coolants due to the lack of coolant channels. This requires frequent tool withdrawal in a process known as 'pecking', which increases the cycle time of the process.

30.2.4   Reaming

Reaming is a specialised process used to enlarge a hole and/or improve its roundness and surface finish. Although reaming is superficially similar to boring, there are notable differences in the design of the tool, making it a separate classification.[11,20] No special machines are required. The same machine that was used in drilling the hole can be used to ream the hole after a tool exchange.

30.2.5   Milling

During milling operations, material is removed from the workpiece by a rotating cutter with multiple teeth. In general, the workpiece is fed into the rotating (but stationary) cutter. The material (metal) removal rate is quite high compared to other machining operations and the surface finish can be very good. The cutter's teeth will be intermittently engaging the workpiece, so the cutting is interrupted, frequently causing chatter and vibration problems for the process.

30.2.5.1   *Face milling*

Face milling generates a surface normal to the axis of rotation. It is generally used for wide flat surfaces. The peripheral portions of the teeth do most of the metal cutting.[11]

30.2.5.2 *End milling*

This is a specific type of face milling operation which is used for milling pockets, facing, profiling and slotting operations.[11]

30.2.5.3 *Peripheral (or plain) milling*

Peripheral milling generates a surface parallel to the axis of rotation. Both flat and formed surfaces can be produced by this method, with the cross section of the resulting surface corresponding to the axial contour of the cutter. The process is sometime referred to as 'slab' milling.[11]

30.2.5.4 *Up-milling and down-milling*

The terms up-milling and down-milling can be applied to all milling procedures. Up milling is the traditional way to mill, and is often called 'conventional milling'. The cutter rotates against the direction of feed rate of the workpiece. In climb, or down milling, the cutter rotation is in the same direction as the feed rate. Although the cutting tool is identical, the method of chip formation is entirely different. In addition, the cutting forces in climb are different from down milling, since the climb milling process undergoes extensive interrupted cutting when entering the work piece, creating large impact loading, cyclic heating and cyclic forces not seen in other cutting operations.[11]

30.2.5.5 *Planing and shaping*

Although commonly used in woodworking, planing and shaping are not as commonly used as a metal cutting process. They are single point cutting processes and have largely been displaced by high speed milling operations. In a planing operation, the workpiece reciprocates while the tool is fed across the workpiece to provide the relative motion. In shaping, the tool reciprocates across the workpiece. Generally, only small or medium size parts are made by shaping. Broaching tools offer a more economical choice over planing and shaping machines.

30.2.5.6 *Broaching*

Broaching resembles sawing since a tool is moved (drawn) across the workpiece.[23,24] However, broaching uses a series of tool edges set at increasing heights. As the tool is drawn over the work surface, the next (longer) tooth of the broach removes successively deeper cuts from the workpiece. A wide variety of finishes (rough to very fine) are thus possible in a single operation, along with a large variety of shapes. Rapid exchange of broach tools allows the broach to be one of the most economical and versatile tools in industry.

30.2.5.7 *Tapping and threading*

Many mechanical assemblies attempt to take advantage of either internal or external screw threads. Also threads may be produced by grinding operations, the most common means are tapping and threading operations.[11,25]

Tapping

When a pre-existing hole exists, tapping is generally used to form the internal threads. A hardened, specially formed threading tool is fed into a hole drilled in a previous operation by one of the drilling processes described earlier.

Thread turning

Used since the earliest lathes,[26] thread turning, usually with a single point tool, remains one of the most widely used methods for producing internal and external threads. A skilled technician may use this technique as a quick way to produce threaded surfaces on conventional or computer numerical control (CNC) machines.

Thread milling

Thread milling, typically done in a machine commonly referred to as a 'screw machine', uses milling cutters to rapidly produce both internal and external threaded surfaces in large quantities. This is now an old technology with many manufacturers of these relatively inexpensive machines.[27,28]

Die threading

This simple process, which may be done manually, forms the threads by using a hard die on softer materials.

Thread rolling

This is an external cold forming process[6] for producing threads. The die tool extrudes the metal from the part surface to form the threads.

30.3 Abrasive machining processes

Abrasive machining is unquestionably the oldest of the basic machining processes,[29–32] with many texts written on the subject. Basically, chips are formed by very small cutting edges that are integral parts of abrasive machining processes, usually man-made. The formed chips may be extremely small, mandating the use of eye protection at all times.

30.3.1 Surface grinding

In this process, the workpiece is typically moved past the grinding surface by either reciprocating or rotating it perpendicular to the grinding surface while it is fed laterally into the grinding surface.[33] This method competes directly with planing and milling operations that require much greater cutting forces.

30.3.2 Cylindrical grinding

These machines often resemble highly specialised machine lathes, since they are equipped with many of the same functional parts (headstock, tailstock, etc). They remove material from external cylindrical surfaces by rotating both the workpiece and the grinding surface in opposite directions. It is typically used for parts requiring a fine surface finish.[32]

30.3.3 Centreless grinding

This process is similar to cylindrical grinding, but the workpiece is allowed to 'float'[32] between the rollers of the grinding surfaces. The workpiece is allowed to find its own centre. Centreless grinding is used when it is not possible to place centering holes in the part or when extreme precision is required in the cylindricity of the part.

30.3.4 Plunge grinding

This is a specialised form of cylindrical grinding in which the wheel is moved continuously into the workpiece rather than traversing. It may be used to part material using modern composite grinding wheels in lieu of more traditional band-saw technology.

30.3.5 Creep feed grinding

In this specialised process,[31] a surface or cylindrical grinding wheel removes a full depth of cut in a single pass on the workpiece at a very slow feed rate. It is commonly used with brittle materials such as ceramics and for materials for which the extent of plastic deformation during machining is critical, such as nickel-base superalloys.

30.3.6   Honing

Honing[34] uses fine abrasive stones to remove very small amounts of metal. It is often the final machining process, removing common errors left by boring or machining, or to remove the tool marks left by other grinding techniques. The honing may be external, or internal, manual or automated. It is limited only by the imagination of the individual designing the honing process.

30.3.7   Microsising

This specialised honing technique uses modern super-abrasives[31] to improve the accuracy of internal cylindrical surfaces. A pre-set, barrel-shaped tool or a highly specialised expandable tool with abrasive stones around its periphery is used. The tool, the workpiece, or both, may rotate. It provides excellent control of the bore.

30.3.8   Belt grinding

As the ability of materials engineers to attach grinding abrasives to flexible belts has increased, the use of belt grinding in metal forming has expanded. It may now be use in place of many of the other grinding processes. It is generally viewed as more efficient than other techniques of grinding.[35]

30.3.9   Disc grinding

This is a face grinding method in which all or most of the flat face of a grinding wheel contacts the workpiece. The wide surface area, which this geometry allows, generates surfaces with low flatness errors, provided that the backing plate is itself quite flat and the grinding surface is regular in thickness.

30.4   Deburring

Burrs are undesired projections of material beyond the edge(s) of the workpiece, which are usually due to plastic deformation during machining and or metal forming. Many deburring methods are widely used. These include barrel tumbling, centrifugal barrel tumbling, vibratory deburring, water jet and abrasive jet deburring, sanding, liquid honing and many others.[36–45] There are many conferences each year to address this important manufacturing topic.

30.5   Metal forming operations

30.5.1   Introduction

In metal forming, an initially simple part, usually a metal billet or sheet blank, is plastically deformed between tools (or dies) to obtain the desired final configuration. Thus, a simple initial geometry is transformed into a more complex one, whereby the tools 'store' the desired geometry and impart pressure on the deforming material through the tool/material interface. Metal forming processes usually produce little or no scrap and generate the final part geometry in a very short time, usually in one or a few strokes of a press or hammer. They are generally considered to offer the greatest savings in both energy and material. Additionally, for a given weight, parts produced by metal forming exhibit better mechanical and metallurgical properties and reliability than do those manufactured by casting or machining.[6] Generally, metal forming is viewed as having two major subdivisions, the massive forming operations and the sheet metal forming processes. Table 30.1 provides a summary listing of these operations.

30.5.2   Massive forming operations

In 'massive' forming operations, the input material is in billet, rod or slab form, and a large increase in the surface to volume ratio occurs in the formed part. Sheet metal operations do not undergo this

Table 30.1 SUMMARY LISTING OF METAL FORMING OPERATIONS

For much more detailed explanations, with excellent diagrams, refer to the metal forming text by Altan *et al.*,[6] or the manufacturing text by Degarmo *et al.*,[11] or many of the other cited references.

Typical classifications of the forming processes			
Massive forming		**Sheet metal forming**	
Extrusion	*Rolling*	*Bending*	*Deep recessing*
Nonlubricated hot extrusion	Sheet rolling	Brake bending	Spinning
Lubricated direct hot extrusion	Shape rolling	Roll bending	Roller flanging
Hydrostatic extrusion	Tube rolling		Deep drawing
	Ring rolling	*Contouring sheets*	Rubber pad forming
Forging	Rotary tube piercing	Stretch forming	Marform process
Coining	Gear rolling	Androforming	Rubber diaphragm
Closed die with flash	Roll forging	Age forming	hydroforming
Closed die w/o flash	Cross rolling	Creep forming	Superplastic forming
Forward extrusion forging	Surface rolling	Die-quench forming	
Backward extrusion forging	Shear forming	Vacuum forming	*Shallow recessing*
Hobbing	Tube reducing	Bulging	Dimpling
Nosing		Superplastic forming	Electromagnetic forming
Isothermal forging	*Drawing*		Explosive forming
Electro upsetting	Drawing		Joggling
Open-die forging	Drawing with rolls		
Orbital forging	Ironing		
P/M forging	Tube sinking		
Radial forging			
Upsetting			
Swaging			

large, massive change in surface to volume ratio. Processes which fall under the massive forming processes have the following features:

- The workpiece undergoes large plastic deformation, resulting in a marked change in shape and/or cross section.
- The portion of workpiece that undergoes permanent deformation is much larger in general than the portion undergoing elastic deformation. Elastic recovery after deformation is essentially negligible.

Examples of massive forming processes are extrusion, forging, rolling and drawing.

30.5.2.1 *Extrusion*

In the extrusion process, metal is compressed and forced to flow through a suitably shaped die to form a product with reduced but constant cross section. It may be performed either hot or cold, although hot extrusion is commonly employed to reduce the forces required, eliminate cold-working effects and reduce directional properties. Basically, the extrusion process is like squeezing toothpaste out of a tube. Extrusion types include nonlubricated hot extrusion, lubricated direct hot extrusion and hydrostatic extrusion.[46,47]

30.5.2.2 *Forging*

Forging is the term applied to a family of processes where the deformation is induced by local compressive forces. It is generally regarded as the oldest metal working process (e.g. the village blacksmith). The equipment can be manual or power hammers, presses, or special forging machines. While the deformation can be done in the hot, cold, warm or isothermal mode, the term *forging* usually implies hot forging done above the recrystallisation temperature. A metal may be *drawn out* to increase its length while decreasing its cross section, *upset* to decrease the length and increase the cross section, or *squeezed* in closed impression dies to produce multidirectional flow. Common forging processes include closed die forging with flash, closed die forging without flash, coining, electro-upsetting, forward extrusion forging, backward extrusion forging, hobbing, isothermal forging, nosing, open die forging, orbital forging, powder metal (P/M) forging, radial forging, upsetting and swaging.[6,47–49]

30.5.2.3   *Rolling*

Rolling is often the first process that is used to convert material into a finished product. Fundamentally, in the basic rolling process, the material being rolled is passed between two rollers rolling in opposite directions. It may be done hot or cold, although most materials undergo both processes during their life cycle. During hot-rolling, thick starting stock can be rolled into blooms, billets, or slabs. The initial shape may be obtained directly from a continuous casting process as input to the rolling process. A bloom has a square or rectangular cross section, with a thickness greater than six inches ($\sim$152 mm) and a width no greater than twice that thickness. A billet is usually smaller than a bloom with a square or circular cross section. A slab is a rectangular solid where the width is greater than twice the thickness. Slabs are typically further rolled to produce plate, sheet and strip. It is at this point that cold rolling becomes practical with the reduced dimensions of the plates, sheets and strips. Types of cold and hot rolling include sheet rolling, shape rolling, tube rolling, ring rolling, rotary tube piercing, gear rolling, roll forging, cross rolling, surface rolling, shear forming (or flow turning) and tube reducing.[8,10,11,50–52]

30.5.2.4   *Drawing*

Drawing is a plastic deformation process in which a flat sheet or plate is formed into a recessed, three-dimensional part with a depth more than several times the thickness of the metal. As a punch descends into a mating die (or the die moves upward over a mating punch), the metal assumes the desired configuration. Hot drawing is used to form relatively thick walled parts of simple geometries, usually cylindrical, with considerable thinning of the material as it passes through the dies. In contrast, cold working uses relatively thin metal, changes the thickness very little or not at all, and produces a wide variety of shapes. Generally, drawing is classified as drawing, drawing with rolls, ironing and tube sinking.[8,47,51]

30.5.3   Sheet-metal forming operations

In sheet forming operations, a relatively thin sheet blank is plastically deformed into a three dimensional object without any significant changes in sheet thickness and surface characteristics. Characteristics of sheet-metal forming include:

- Workpiece is a sheet or a part fabricated from a sheet.
- Deformation usually causes significant changes in shape, but not cross section, of the sheet.
- The magnitudes of plastic and recoverable elastic deformations are comparable. Springback, during elastic recovery, may be very significant as a control issue in the manufacturing process.

Examples of processes, which fall under the category of sheet forming processes, include bending and straight flanging, surface contouring of a sheet, linear contouring, deep recessing and flanging, and shallow recessing. The large number of techniques within these categories includes: bulging, joggling, dimpling, marforming, vacuum forming, and many other processes.[6,10,47,53–58]

30.5.4   Superplasticity

Some alloys, under proper conditions, can undergo extraordinary tensile deformation (up to 300%) without fracturing. This has led to a specialised form of sheet metal forming referred to as 'superplastic forming' which is normally combined with some type of diffusion bonding process to form complex aircraft frames and shapes for example. A range of materials can exhibit superplasticity including some aluminium, titanium and nickel alloys, certain steels and even a few ceramics.[59–67]

Table 30.1 provides a summary listing of the many metal-forming processes. For the most complete description, refer to the references cited in the paragraphs above.

30.6   Machinability and formability of materials

Machinability and formability have nearly as many meanings as there are processes and materials.[47,68–86] Malleability, workability, and formability all refer to a material's suitability for plastic deformation processing. Each material's formability must be evaluated for specific combinations of material, process and process parameters. Machinability generally refers to the ease with which a material can be machined to an acceptable surface finish.

30.6.1   Definitions

The three principal definitions of machinability are entirely different from each other.

30.6.1.1   *Machinability as a material property*

Machinability can be defined as the ease with which the material is machined in terms of specific energy, specific horsepower, or shear stress. In general, the larger the shear stress or specific power values, the more difficult the material is to machine and form, requiring greater forces and lower speeds. In this definition, the material is the key.[11]

30.6.1.2   *Machinability as a factor in tool life*

An important economic measure of a materials machinability is defined by the relative cutting speed for a given tool life of the tool cutting some material, compared to a standard material cut with the same tool material. Thus tool-life curves may be used to express the machinability ratings. For example, the steel industry has established the B1112 steel as the comparative standard, using an expected tool life of 60 minutes at a cutting speed of 100 sfpm ($\sim$0.15 m^2 s^{-1}). Thus, if another steel X has a rating of 70% of B1112, you adjust the cutting speed to 70% to achieve a 1-hour tool life. Each metal association (e.g. the Copper Metal Working Association) defines these comparative ratings for the various alloys for which they have authority as the standard setter.

30.6.1.3   *Machinability in terms of the cutting speed*

Machinability may be defined in terms of the maximum speed at which a tool can provide satisfactory performance for a specified time under specified conditions in specified machines (e.g. ASTM Standard E-618-81: 'Evaluating machining performance of ferrous metals using an automatic screw/bar machine').

30.6.1.4   *Other factors, criteria, tests and indices*

Other factors that are sometimes used to characterise machinability include tool wear rates, chip form, achievable surface finish, achievable tolerance, functional or surface integrity, cutting temperature, and mechanical properties (e.g. hardness) of the material being machined.

30.6.2   Formability

The Metals Handbook[19] defines formability as the relative ease with which a metal can be shaped through plastic deformation, and uses it as a synonym for workability of the material. This simplistic definition opens the door to the entire library of plastic deformation theories, none of which are perfect, when applied to forming operations (e.g. Chow[87]). Many represent the best compromise, based on needs of the end user who is forming the material.[77–87] Over the years, many formal tests, specified by ASTM, ISO, Euronorm *et al.*, have been developed which a manufacturer may be required to provide documentation for in the engineering specifications (e.g. [88, 89]). The most common tests usually develop some type of flow curve (tensile, compression etc), although impact tests, fracture tests, fatigue tests, hardness and corrosion testing are also frequently required. One of the most widely specified tests is the uniaxial compression test, or upset test. In the test, a right cylindrical specimen is compressed between flat, parallel dies. From this, flow stress values are derived for use in diagramming flow curves.[83]

30.7 Coolants and lubricants (see also Chapter 24)

A significant fraction of metal cutting and forming operations are performed wet, using what the United States National Institute of Occupational Safety and Health classifies as Metal Working Fluids (MWFs). Closely allied with the definitions of machinability and formability, their use permits the use of increased speeds and feed rates, while reducing wear and friction. In machining, the fluid is often critical in the removal of the formed chips. The application of coolants and lubricants represent a significant portion of the part manufacturing cost, and an increasing cost of the health insurance costs in the manufacturing sector. Cutting fluids may be applied in a liquid, gaseous or mixed liquid-gaseous form.[90–93] There are significant occupational risks associated with each type, requiring an informed trade-off in benefit versus risk.

30.7.1 Liquid metal working fluids

Cutting liquid fluids are classified into four types: oil based, soluble oils, semi-synthetics, and synthetics. The last three are generally referred to as water-based coolants. Water-based coolants provide the best cooling action, while oil based fluids provide the best lubrication effects. Many of the hydrocarbon based metal working fluids have been proven to be a Type B carcinogen, whereas there is a growing awareness of the biological hazards associated with the water-based metal working fluids. Most types have been shown to be a cause of dermatitis.

30.7.1.1 *Oil based fluids*

Oil based fluids include straight mineral oils as well as fatty oils and compound oils. They reduce friction in the cutting zone and are generally most effective in grinding, where they are best able to improve the surface finish.[94] NIOSH has recommended airborne limits for their usage,[17] based on dermatitis and thoracic cancer concerns.

30.7.1.2 *Soluble or emulsifiable oils*

These are special types of mineral oils diluted in water at a typical concentration of 20:1, with emulsifying agents added to form stable dispersions. They are generally regarded as superior to oil based metal working fluids at higher cutting speeds because of the added cooling effect of water. However, the additives, particularly those with chlorinated paraffin additives and carboxylates, increase the risks identified by NIOSH.[17,95]

30.7.1.3 *Semi-synthetics*

These are surface-active water based cutting fluids that consist of fine colloidal dispersions of organic and inorganic materials (oil and chemicals) in water. In addition to the biological concerns of NIOSH, these cutting fluids frequently leave a residue on the machined surface and foam excessively.[96] Disposal of the waste material is carefully regulated because of the use of chemical additives.

30.7.1.4 *Synthetic cutting fluids*

These are water-based fluids consisting of chemical lubricating agents, wetting additives, disinfectants and stabilisers. They are oil free, but contain alkanolamines and organic esters, for which NIOSH has identified exposure levels.[96] They are excellent coolants because of the large percentage of water, which encourages biologic growth, requiring additives to limit such growth. Their disadvantages include reduced lubricity due to the absence of petroleum products, a tendency to leave hard crystalline deposits, high alkalinity, a tendency to foam, and disposal problems.[12]

30.7.2   Gaseous fluids and gaseous-liquid mixtures

Types of gaseous fluids include air, Freon, helium, and nitrogen, with air being the most common. High-pressure air can be used to eject chips if proper machine guarding techniques are used. Noise generation is a significant concern with high velocity and/or pressure air systems, frequently exceeding the permissible noise exposure levels.[12]

30.8   Non-traditional machining techniques

Most of the technology discussed in this chapter existed prior to the end of World War II, although it continues to dominate the manufacturing world at the start of the 21st century. These techniques remain the most cost effective way to produce most manufactured goods. However, with the drive towards electronics and miniaturisation, new technologies have emerged which are 'non-traditional' in the sense that they have not been 'on-the-scene' as long as the other processes. The six processes to alter material in non-traditional machining are electricity, water, new abrasives, chemicals, ionised gas, and light.

30.8.1   Electrical Discharge Machining (EDM)

Wire EDM uses a travelling wire electrode that physically passes through the workpiece. The wire is monitored precisely by a CNC system. Like any other machining tool, wire EDM removes metal; but wire EDM removes metal with electricity by means of spark erosion. Rapid DC electrical pulses are generated between the wire electrode and the workpiece, using a dielectric shield between the two of deionised pure water. When sufficient voltage is applied, the fluid ionises and a controlled spark erodes a small section of the workpiece, causing it to melt and vaporise. High air turnover rates in the workspace are required to prevent accumulation of metal fumes, with periodic monitoring by a certified industrial hygienist.[97,98] Extremely complicated shapes can be cut economically, automatically, and precisely, even in materials as hard as carbide. A specialised type of EDM referred to as ram EDM machining (or sometimes as sinker EDM, die sinker, vertical EDM and plunge EDM) may be employed to produce blind cavities in materials.[99]

30.8.2   Fast Hole EDM drilling

Although it is possible to drill holes with conventional EDM machines such as described in the preceding paragraph, highly specialised EDM machines have been developed that are capable of drilling holes at speeds up to two inches per minute ($\sim$850 μm s^{-1}). Basically, a servo system maintains a gap between the tip of the electrode and the workpiece. If the electrode touches the workpiece, a short occurs and the servo resets the gap. This 'Fast Hole' EDM drilling technique is used for putting holes in turbine blades, fuel injectors, cutting tool coolant channels, hardened punch ejector holes, plastic mould vent holes, wire EDM starter holes, and other complex operations that require very exact hole placement and sizing.[99]

30.8.3   Waterjet and abrasive waterjet machining

Manufacturers are increasingly turning to waterjet and abrasive waterjet machining in their quest for productivity and quality. Extremely powerful abrasive waterjets are capable of cutting through 3-inch ($\sim$76 mm) tool steel at a rate of 2 inches per minute ($\sim$850 μm s^{-1}), 10-inch ($\sim$254 mm) reinforced concrete at more than 1 inch per minute, and 1-inch ($\sim$25 mm) plate glass at 2 feet per minute ($\sim$10 mm s^{-1}). Pure water, or water mixed with an abrasive is supplied at high pressure through a sapphire, ruby or diamond orifice cutting head which periodically requires replacement. Some ceramics have been developed which are capable of acting as cutting heads.[99–101] When manufacturers order metal stock, there is a high probability that it was resized for delivery using waterjet technology.

30.8.4   Plasma cutting

Plasma cutting systems use a highly conductive gas to cut electrically conductive materials. The interior of a plasma-cutting torch normally consists of a consumable nozzle, an electrode, and either gas or water to provide cooling. Electric arcs are generated inside the nozzle, which ionises the coolant and produces an electrically conductive plasma arc. A pressurised gas flowing through the nozzle creates a pilot arc. As the heated plasma arc strikes the workpiece, the arc vaporises the metal. The flowing gas forces out the molten metal and pierces the workpiece. After piercing, the motion machine begins to advance the torch in the pre-programmed directions for the cut.[99]

30.8.5 Photochemical machining

Photochemical machining (PCM), which is also known as photoetching, photolithography, and photochemical milling, is a process that etches out parts by means of chemicals. Protected areas are masked, and chemicals are then used to dissolve the unmasked areas. Parts are typically designed in a computer-aided design (CAD) system, and then projected with a precision camera system onto the work surface. The metal, which has been coated with a photosensitive material, is then chemically processed using a corrosive solution. Where the photo resistant material has been exposed, the metal is etched away. It can produce elaborate, stress free parts, eliminates hard tooling and provides rapid turn-around (just in time ordering). It always leaves a bevelled edge though, and is therefore not suitable for processes that require straight, cornered lines.[102]

30.8.6 Electrochemical machining

Electrochemical machining (ECM) is one of the least used non-traditional methods despite a high, stress free metal removal rate because it requires a high level of craftsmanship in allowing for the flow effects of the electrolyte. The process is used for various automotive and aircraft components, gun barrel rifling, steam turbine casings, and for deburring. With electrochemical machining, a pressurised conductive salt solution, the electrolyte, flows around the workpiece as an electrical–chemical reaction de-plates material from the workpiece. It is the opposite of the classic plating process. When plating, metal is applied with the use of electrical current. ECM removes metal with electric current. Since it is done chemically, material hardness and toughness have no effect.[103–106]

30.8.7 Ultrasonic machining

Ultrasonic machining is a highly specialised process used to machine detailed and complex parts, especially in graphite materials. The process can also be used for polishing and machining glass, carbide and ceramics. A metal forming tool, the 'sonotrode' is fabricated to the desired pattern. Then the sonotrode is mounted in a recessed opening in the ultrasonic machine and the target is lowered from above into the recessed sonotrode, with an abrasive slurry mixture being injected between them. The sonotrode vibrates at speeds in excess of 20 000 cycles per second (20 kHz), causing the abrasive particles to erode the target and produce a mirror image of the form tool. Very delicate details are possible for objects (such as coinage dies) that may be sized to these configurations.[107,108]

30.8.8 Lasers

Lasers are everywhere in modern society, from the checkout stand to the manufacturing floor. With rapid advances in laser technology (and falling prices), laser cutting is becoming the first choice for many manufacturers. Typical cutting speeds range from 1 to 1 000 inches per minute ($\sim$400 μm s^{-1} to $\sim$400 mm s^{-1}). They have high accuracy, speed and minimum workpiece distortion. They can economically produce round holes, square cut-outs, radii, tapers, and undercuts to any imaginable shape, without expensive tooling costs as with turret or power presses. Generally, a desired shape is programmed on a CAD system, post-processed with a computer-aided manufacturing (CAM) system, which converts it into a CNC language, and downloaded into the cutting laser machine. The laser physically vaporises the material which is carried away by a combination purge and exhaust system. Although tooling and lubricants/coolants are eliminated, control of potentially dangerous fumes is an ever-present concern with these systems. In addition to the fumes from the material being cut, which may vary widely, lasers are a high voltage threat and pose a threat to human eyes. With increasing strength and efficiency, lasers will continue to increase their use in manufacturing, requiring more operator training to avoid exposure.[109–113]

30.8.8.1 *Lasers in rapid prototyping and freeforming*

Lasers are being used more and more frequently to develop rapid prototypes, which can then be analyzed quickly for potential design/failure problems. In stereolithography, laser equipment translates

CAD designs into solid objects through a combination of laser, photochemistry and software technologies. A computer-driven laser solidifies a vat of liquid resin, layer by layer, into a three-dimensional epoxy prototype part. In another method, a selective laser sintering station [SLS] uses a plastic powder to create prototype parts. The powder is melted, layer-by-layer, by a computer-directed heat laser. Additional powder is deposited on top of each solidified layer and again sintered. SLS allows for great diversity in material selection, including nylon, glass filled nylon, SOMOS [a rubber like compound] and Trueform [used in investment casting]. One company has invented a technique, laminated object manufacturing, which builds parts by laminating and laser trimming the material delivered in sheet form. Each layer arrives, is glued in place, then trimmed by the laser. The workpiece moves down away from the laser and work surface just before delivery of the next layer, permitting a rapid build-up of the desired part from an initial CAD drawing. These techniques are frequently referred to as formless manufacture, since the parts do not use traditional machining or mould fabrication techniques.[110,114–126]

30.9 Occupational safety issues

In the area of metalworking, which includes both metal cutting and forming, three major occupational safety concerns exist. These include machine guarding, workers exposure to hazardous materials, and excessive exposure to noise level. Safety requirements are frequently a hidden cost in operations.

30.9.1 Machine guarding

Rotating machinery operating at high voltage may represent a significant crushing, nipping or shock hazard to personnel. Most countries have established occupational safety standards detailing minimum protection requirements, which include machine guards, eye protection and operating procedures.[127–131]

30.9.2 Hazardous materials

Workers are generally entitled by law to know the physical dangers of the materials around which they work.[132–142] Professionals should consider these life cycle costs in their selection of manufacturing processes.

30.9.3 Noise exposure

A general rule of thumb amongst occupational safety professionals is that 'if you must raise your voice to be heard, you are probably in a high noise area' (>85 db in the United States). There are strict medical protocols and legally required programs, depending upon the measured level and local government which should be reviewed for their applicability.[143,144]

30.9.4 Ergonomics

There are many complex ergonomic issues, including tool hand vibration, whole body vibration, lifting of loads, tool positioning and others, which the engineer selecting a major manufacturing tool or system must consider.[145–159] Relatively simple and inexpensive ideas, incorporated during the tool selection process, may provide a large, long-term return on investment.

The many occupational safety issues mentioned in this chapter and above would fill the professional's library very quickly. One of the less expensive, smallest (back pocket), yet most comprehensive references is a spiral bound book produced by the American Conference of Governmental Industrial Hygienists.[160] The title is somewhat misleading, since it addresses nearly every occupational issue of interest to the modern managing engineer.

REFERENCES

1. R. Hill, *Journal of the Mechanics and Physics of Solids*, 1954, **3**, 47.
2. E. M. Trent and P. K. Wright, 'Metal Cutting', Butterworth-Heinemann, Boston, 2000.
3. M. C. Shaw, 'Metal Cutting Principles', Clarendon Press, New York, 1984.

4. P. L. B. Oxley, p. 31, in 'On the Art of Cutting Metals-75 Years Later'. 1982, ASME.
5. E. J. A. Armarego and R. H. Brown, 'The Machining of Metals', Prentice-Hall, Englewood Cliffs, NJ, 1969.
6. T. Altan, S.-I. Oh, and H. L. Gegel, 'Metal Forming, Fundamentals and Applications', American Society for Metals, Metals Park, OH, 1983.
7. B. Avitzur, 'Metal Forming; Processes and Analysis', McGraw-Hill, New York, 1968.
8. R. H. Wagoner and J. L. Chenot, 'Fundamentals of Metal Forming', Wiley, New York, 1997.
9. R. H. Wagoner and J. L. Chenot, 'Metal Forming Analysis', Cambridge University Press, New York, NY, 2001.
10. Schuler Gmbh., 'Metal Forming Handbook', Springer-Verlag, Berlin, New York, 1998.
11. E. P. Degarmo, J T. Black, R. A. Kosher, B. E. Klamecki, 'Materials and Processes in Manufacturing', John Wiley and Sons, Hoboken, NJ, 2003.
12. D. Stephenson and J. S. Agapiou, 'Metal Cutting Theory and Practice', Marcel Dekker, Inc, New York, NY, 1996.
13. S. S. Heineman and Genevro, 'Machine Tools—Processes and Applications', Canfield Press, San Francisco, CA, 1979.
14. 'Tool and Manufacturing Engineers Handbook', SME, Dearborn, MI, 1983.
15. W. Konig, *CIRP Annals*, 1984, **32**, 417.
16. G. T. Smith, 'Advanced Machining—the Handbook of Cutting Technology', Springer Verlag, New York, 1989.
17. B. V. Subcommittee, 'American National Standard Technical Report: Mist Control Considerations for the Design, Installation and Use of Machine Tools Using Metalworking Fluids', American National Standards Institute, New York, NY, 1997.
18. M.W.F.S.A. Committee, 'Standard Recommendation with Pel of 0.5 mg/m^2, NIOSH, Hunt Valley, MD, 1999.
19. 'Metals Handbook', 9th Edition, Vol. 16, ASM International, Metal Park, Ohio, 1989.
20. S. O. M. Engineers. 1983, SME: Dearborn, MI.
21. H. J. Swinehart, 'Gundrilling, Trepanning, and Deep Hole Machining', Springer Verlag, New York, NY, 1967.
22. A. Feifer, Tooling and Production, 1989, p. 58.
23. W. W. Burden and Broaching Tool Institute New York, 'Broaches and Broaching', Broaching tool institute, New York, NY, 1944.
24. E. W. Kokmeyer and Society of Manufacturing Engineers, 'Better Broaching Operations', Society of Manufacturing Engineers: Publications/Marketing Division, Dearborn, Mich., 1984.
25. F. H. Colvin and F. A. Stanley, 'Drilling and Surfacing Practice; Drilling, Reaming, Tapping, Planing, Shaping, Slotting, Milling, and Broaching', McGraw-Hill Book Co., New York, 1948.
26. J. S. Murphy, 'Screw Thread Production', Machinery Pub. Co., Brighton Eng., 1953.
27. R. A. Warren, 'Warren's Book on Automatic Screw Machine Operations', Chicago, 1945.
28. D. T. Hamilton, 'Automatic Screw Machine Practice', The Industrial press, New York, 1912.
29. M. C. Shaw, 'Principles of Abrasive Processing', Clarendon Press, New York, 1996.
30. S. F. Krar, 'Grinding Technology', Delmar Publishers, Albany, 1995.
31. S. C. Salmon, 'Modern Grinding Process Technology', McGraw-Hill, New York, 1992.
32. J. A. Borkowski and A. Szyma Nski, 'Uses of Abrasives and Abrasive Tools', Ellis Horwood, New York, Warsaw, 1992.
33. R. I. King and R. S. Hahn, 'Handbook of Modern Grinding Technology', Chapman and Hall, New York, 1986.
34. A. J. Cox, 'Modern Honing Practice', Machinery Publishing, Brighton, 1969.
35. W. Konig *et al.*, *CIRP Annals*, 1986, **35**, 487.
36. Society of Manufacturing Engineers, 'Controlling Deburring Costs through Standards, Research, and Automation: May 25–27, 1976, Cleveland, Ohio', Society of Manufacturing Engineers, Dearborn, Mich., 1976.
37. Society of Manufacturing Engineers, 'Deburring: An Assessment of Capabilities Conference, March 2–4, 1976, Culver City, California', Society of Manufacturing Engineers, Dearborn, Mich., 1976.
38. Society of Manufacturing Engineers, 'Deburring: An International Evaluation Seminar, April 9–10, 1975, Detroit, Michigan', Society of Manufacturing Engineers, Dearborn, Mich., 1975.
39. Society of Manufacturing Engineers, 'Deburring and Surface Conditioning '89: February 13–16, 1989, San Diego, California', SME, Dearborn, Mich., 1989.
40. G. J. Wiens *et al.*, 'Manufacturing Science and Engineering, 1997: Presented at the 1997 ASME International Mechanical Engineering Congress and Exposition, November 16–21, 1997, Dallas, Texas', American Society of Mechanical Engineers, New York, 1997.
41. American Society of Mechanical Engineers. Winter Meeting (1989: San Francisco Calif.) *et al.*, 'Mechanics of Deburring and Surface Finishing Processes: Presented at the Winter Annual Meeting of the American Society of Mechanical Engineers, San Francisco, California, December 10–15, 1989', American Society of Mechanical Engineers, New York, NY, 1989.
42. L. K. Gillespie, 'Deburring and Edge Finishing Handbook', Society of Manufacturing Engineers, New York, 1999.
43. L. K. Gillespie and Society of Manufacturing Engineers, 'Deburring Technology for Improved Manufacturing', Society of Manufacturing Engineers, Dearborn, Mich., 1981.
44. L. K. Gillespie and R. E. King, 'Robotic Deburring Handbook', Society of Manufacturing Engineers, Dearborn, Mich., 1987.
45. A. F. Scheider, 'Mechanical Deburring and Surface Finishing Technology', M. Dekker, New York, 1990.
46. W. Johnson and H. Kudo, 'The Mechanics of Metal Extrusion', Manchester University Press, Manchester, 1962.
47. E. P. Degarmo, J. T. Black and R. A. Kohser, 'Materials and Processes in Manufacturing', Prentice Hall, Upper Saddle River, NJ, 1997.

48. A. M. Sabroff, F. W. Boulger and H. J. Henning, 'Forging Materials and Practices', Reinhold Book Corp., New York, 1968.

49. J. A. Cross, 'Metal Forging and Wrought Iron Work: A Manual for Schools', Mills & Boon, London, 1967.

50. K. Lange, 'Handbook of Metal Forming', Society of Manufacturing Engineers, Dearborn, Mich., 1995.

51. W. F. Hosford and R. M. Caddell, 'Metal Forming: Mechanics and Metallurgy', Prentice Hall, Englewood Cliffs, NJ, 1993.

52. K. Lange, 'Handbook of Metal Forming', McGraw-Hill, New York, 1985.

53. M. Y. Demeri, Minerals Metals and Materials Society. Shaping and Forming Committee, and Minerals Metals and Materials Society. Meeting, 'Sheet Metal Forming Technology: Proceedings of a Symposium Held at the 1999 TMS Annual Meeting in San Diego, California, February 28–March 4, 1999', TMS, Warrendale, Pennsylvania, 1999.

54. International Deep Drawing Research Group. Congress (15th: 1988: Dearborn Mich.), 'Controlling Sheet Metal Forming Processes', ASM International: North American Deep Drawing Research Group, Metals Park, Ohio, 1988.

55. D. P. Koistinen and N.-M. Wang, 'Mechanics of Sheet Metal Forming: Material Behavior and Deformation Analysis: Proceedings', Plenum Press, New York, 1978.

56. Z. Marciniak and J. L. Duncan, 'Mechanics of Sheet Metal Forming', Edward Arnold, London, 1992.

57. R. Pearce, 'Sheet Metal Forming', Hilger, Bristol; Philadelphia, 1991.

58. N.-M. Wang, S. C. Tang and Metallurgical Society of AIME. Detroit Section, 'Computer Modeling of Sheet Metal Forming Process: Theory, Verification, and Application: Proceedings of a Symposium Sponsored by the Metallurgical Society and the TMS Detroit Section, Held at the 12th Automotive Materials Symposium, Ann Arbor, Michigan, April 29–30, 1985', Metallurgical Society, Warrendale, Pa., 1985.

59. N. Chandra *et al.*, 'Advances in Superplasticity and Superplastic Forming: Proceedings of a Symposium Sponsored by the Mdmd Shaping and Forming Committee, Held at the TMS/ASM Materials Week, 2–5 November, 1992', Minerals Metals & Materials Society, Warrendale, Pa., 1993.

60. A. K. Ghosh *et al.*, 'Superplasticity and Superplastic Forming, 1998: Proceedings of a Conference on Superplasticity and Superplastic Forming Sponsored by the TMS Shaping and Forming Committee and Held as Part of the TMS Annual Meeting in San Antonio, Texas, February 16–19, 1998', The Minerals Metals & Materials Society, Warrendale, Penn., 1998.

61. S. P. Agrawal and American Society for Metals. Los Angeles Chapter, 'Superplastic Forming: Proceedings of a Symposium', ASM, Metals Park, Ohio, 1985.

62. C. H. Hamilton *et al.*, 'Superplasticity and Superplastic Forming: Proceedings of an International Conference on Superplasticity and Superplastic Forming', TMS, Warrendale, Pa., 1988.

63. H. C. Heikkenen, T. R. Mcnelley and Metallurgical Society (U.S.). Shaping and Forming Committee., 'Superplasticity in Aerospace: Proceedings of a Symposium Sponsored by the Shaping and Forming Committee and Held at the Annual Meeting of the Metallurgical Society in Phoenix, Arizona, January 25–28, 1988', The Society, Warrendale, Pa., 1988.

64. N. E. Paton *et al.*, 'Superplastic Forming of Structural Alloys: Proceedings of a Symposium', Metallurgical Society of AIME, Warrendale, Pa., 1982.

65. Society of Manufacturing Engineers., 'Superplastic Forming and Bonding of Metallic Alloys: June 16–17, 1992, Dayton, Ohio. Practical Results from Integrating Total Quality Management (Tqm): Lessons Learned to Improve Competitiveness: October 27–28, 1992, Rosemont, Illinois', Society of Manufacturing Engineers, Dearborn, Mich. (1 SME Dr., Dearborn 48121), 1992.

66. H. Tan, P. Gao and J. Lian, *Mater. Manuf. Process*, 2001, **16**(3): p. 331.

67. R. Whittingham, *Mater. Sci. Forum*, 2001, **357-3**, p. 29.

68. P. L. B. Oxley, 'The Mechanics of Machining: An Analytical Approach to Assessing Machinability', Halsted Press, New York, 1989.

69. R. W. Thompson and American Society for Metals, 'The Machinability of Engineering Materials: Proceedings of an International Conference on Influence of Metallurgy on the Machinability of Engineering Materials, Rosemont, Illinois, 13–15 September 1982', American Society for Metals, Metals Park, Ohio, 1983.

70. B. Mills and A. H. Redford, 'Machinability of Engineering Materials', Applied Science Publishers, New York, 1983.

71. Machinability Data Center, 'Machining Data Handbook', MDC, Cincinnati, Ohio, 1980.

72. American Society for Metals. Machinability Activity, *et al.*, 'Machinability Testing and Utilisation of Machining Data: Proceedings from an International Conference', American Society for Metals, Metals Park, Ohio, 1979.

73. V. A. Tipnis and American Society for Metals. Machinability Activity, 'Influence of Metallurgy on Machinability: Proceedings from an International Symposium', The Society, Metals Park, Ohio, 1975.

74. N. R. Parsons, 'N/C Machinability Data Systems', Society of Manufacturing Engineers, Dearborn, Mich., 1971.

75. N. E. Woldman and R. C. Gibbons, 'Machinability and Machining of Metals', McGraw-Hill, New York, 1951.

76. J. E. Thompson, 'Designing for Machinability', Bunhill publications Ltd., London, 1942.

77. G. E. Dieter and American Society for Metals. Metal Workability Group, 'Workability Testing Techniques', American Society for Metals, Metals Park, Ohio, 1984.

78. International Deep Drawing Research Group. and Centre National De Recherches Métallurgiques., 'The Study and Classification of Sheet Metal Forming Operations. 4th Biennial Sic Colloquium of the International Deep Drawing Research Group, I.D.D.R.G., Liège 1966. Sheet Metal Forming and Formability. June 7th 1966. Sheet Metal Testing. June 8th 1966 Organisation: Centre National de Recherches Métallurgiques', Centre national de recherches métallurgiques, Liège, 1966.

79. S. S. Hecker *et al.*, 'Formability, Analysis, Modeling, and Experimentation: Proceedings of a Symposium Held in Chicago, Illinois, October 24 and 25, 1977', Metallurgical Society of AIME, New York, 1978.

80. B.-J. Jungnickel and W. J. Bartz, 'Solid State Forming of Polymers', Mechanical Engineering Publications, London, 1992.
81. K. Pöhlandt, 'Materials Testing for the Metal Forming Industry', Springer-Verlag, Berlin; New York, 1989.
82. A. K. Sachdev *et al.*, 'Formability and Metallurgical Structure: Proceedings of a Symposium Co-Sponsored by the Mechanical Metallurgy and Shaping and Forming Committees of TMS-AIME and Held in Orlando, Florida October 5–9, 1986, at the Fall Meeting of the Metallurgical Society', The Society, Warrendale, PA, 1987.
83. S. L. Semiatin and J. J. Jonas, 'Formability and Workability of Metals: Plastic Instability and Flow Localisation', American Society for Metals, Metals Park, Ohio, 1984.
84. E. Shapiro, F. N. Mandigo and International Copper Research Association, 'Forming Limit Analysis for Enhanced Fabrication', International Copper Research Association, New York, NY, 1981.
85. R. H. Wagoner, K. S. Chan, and S. P. Keeler, 'Forming Limit Diagrams: Concepts, Methods, and Applications: A Reference Book on the Available Experimental and Analytical Methods for Determination of Forming Limit Diagrams', TMS, Warrendale, Pa., 1989.
86. D. Banabic, 'Formability of Metallic Materials: Plastic Anisotropy, Formability Testing, Forming Limits', Springer, Berlin, New York, 2000.
87. C. L. Chow, 'Analysis of Sheet Metal Formability', Ford Motor Company, Dearborn, Michigan, 1998.
88. ASTM Committee on Terminology, 'Compilation of ASTM Standard Definitions', ASTM, Philadelphia, PA, 1994.
89. ASTM Standardization News, ASTM, Philadelphia, PA.
90. Society of Manufacturing Engineers, 'Cutting Fluids & Lubricants: Cutting Fluids, Lubricants, Cutting Fluid Systems, Metalforming Compounds, Treatment, and Application Equipment', Society of Manufacturing Engineers Publications Development Department, Dearborn, Mich., 1985.
91. British Petroleum Company, 'Machine Tools, Metals and Cutting Fluids', B.P. Trading Ltd., London, 1973.
92. W. J. Olds, 'Lubricants, Cutting Fluids, and Coolants', Cahners Books, Boston, 1973.
93. P. J. C. Gough and Machine Tool Industry Research Association, 'Swarf and Machine Tools: A Guide to the Methods Used in the Handling and Treatment of Swarf and Cutting Fluids', Hutchinson, London, 1970.
94. P. V. Kotvis, *Journal of the American Society of Lubrication Engineers*, 1986, **42**(6), 363.
95. E. O. Bennett and D. L. Bennett, 'Cutting Fluids and Odors', in *Clinic on Metalworking Coolants*. 1987. Dearborn, Michigan.
96. United States Department of Health and Human Services, 'What you need to know about occupational exposure to metal working fluids', National Institute for Occupational Safety and Health, 1998, Atlanta, GA.
97. C. Sommer and S. Sommer, 'Wire Edm Manual', Technical Advance Pub., Houston, TX, 1992.
98. Point Control Company, 'SMARTCAM Advanced Wire EDM', 1994, Point Control Company, Eugene, Oregon.
99. C. Sommer, 'Non-Traditional Machining Handbook', Advance Pub., Houston, TX, 1999.
100. R. K. Miller, 'Waterjet Cutting: Technology and Industrial Applications', Fairmont Press, Lilburn, GA, 1991.
101. R. K. Miller and T. C. Walker, 'Survey on Waterjet Cutting Technology and Markets', Future Technology Surveys, Madison, GA (123 West Washington St., Madison, GA 30650), 1989.
102. D. M. Allen, 'The Principles and Practice of Photochemical Machining and Photoetching', A. Hilger, Bristol, England, Boston, 1986.
103. American Society of Tool and Manufacturing Engineers., 'Electrochemical Machining Seminar; Technical Papers', American Society of Tool and Manufacturing Engineers, Dearborn, Mich., 1967.
104. A. E. De Barr and D. A. Oliver, 'Electrochemical Machining', American Elsevier Pub. Co., New York, 1968.
105. C. L. Faust, Electrochemical Society. Electrodeposition Division, and Electrochemical Society. Electrothermics and Metallurgy Division, 'Fundamentals of Electrochemical Machining', Electrodeposition Division and Electrothermics and Metallurgy Division Electrochemical Society, Princeton, NJ, 1971.
106. J. A. Mcgeough, 'Principles of Electrochemical Machining', Halsted Press Division, New York, 1974.
107. L. D. Rozenberg, 'Ultrasonic Cutting', Consultants Bureau, New York, 1964.
108. A. I. Markov, 'Ultrasonic Machining of Intractable Materials', Iliffe, London, 1966.
109. N. N. Rykalin, A. A. Uglov and A. Kokora, 'Laser Machining and Welding', Pergamon Press, Oxford; New York, 1978.
110. L. Lü, J. Y. H. Fuh and Y. S. Wong, 'Laser-Induced Materials and Processes for Rapid Prototyping', Kluwer Academic Publishers, Boston, 2001.
111. S. Jahanmir *et al.*, 'Machining of Advanced Materials: Proceedings of the International Conference on Machining of Advanced Materials, July 20–22, 1993, Gaithersburg, Maryland', U.S. Dept. of Commerce National Institute of Standards and Technology, for sale by the Supt. of Docs. U.S. G.P.O., Gaithersburg, MD, Washington, D.C., 1993.
112. M. L. Gaillard and A. Quenzer, 'High Power Lasers and Laser Machining Technology: Proceedings: 25–28 April 1989, Paris, France', SPIE—the International Society for Optical Engineering, Bellingham, Wash., 1989.
113. G. Chryssolouris, 'Laser Machining: Theory and Practice', Springer-Verlag, New York, 1991.
114. F. Abe and K. Osakada, *Int. J. Jpn. S. Prec. Eng.*, 1996, **30**(3), 278.
115. F. Abe *et al.*, *Int. J. Jpn. S. Prec. Eng.*, 1998, **32**(3), 221.
116. K. A. Bartels *et al.*, *J. Microsc-Oxford*, 1993, **169**, 383.
117. J. Beuth and N. Klingbeil, *Jom-J. Min. Met. Mat. S.*, 2001, **53**(9), 36.
118. P. Calvert and R. Crockett, *Chem. Mater.*, 1997, **9**(3), 650.
119. J. G. Conley and H. L. Marcus, *J. Manuf. Sci. E-T ASME*, 1997, **119**(4B), 811.
120. K. Daneshvar, M. Raissi and S. M. Bobbio, *Journal of Applied Physics*, 2000, **88**(5), 2205.
121. P. Dickens, p. 141, in 'Proceedings of Rapid Prototyping and Manufacturing Conference'. 1998, SME: Dearborn, MI.

122. R. Irving, *Int. J. Powder Metall.*, 2000, **36**(4), 69.
123. D. Kochan, 'Solid Freeform Manufacturing', Elsevier, Amsterdam, 1993.
124. A. E. Lange and M. Bhavnani, *Sampe. J.*, 1994, **30**(5), 46.
125. G. K. Lewis and E. Schlienger, *Mater. Design*, 2000, **21**(4), 417.
126. Y. Z. Zhang *et al.*, *Rare Metal Mat. Eng.*, 2000, **29**(6), 361.
127. T. Hanson and J. J. Keller and Associates, 'Workplace Safety in Action', J. J. Keller & Associates, Neenah, Wis., 1997.
128. J. K. Blundell, 'Machine Guarding Accidents: Trial Lawyer's Guide', Hanrow Press, Columbia, Md. (P.O. Box 1083, Columbia 21044), 1983.
129. New Zealand. Dept. of Labour, 'The Ergonomics of Machine Guarding: A Guide to the Principles Which Should Be Followed in the Construction of Machine Guards, Based on Human Measurements', Wellington, 1997.
130. V. L. Roberts and Institute for Product Safety, 'Machine Guarding: A Historical Perspective', Institute for Product Safety, Durham, N.C., 1980.
131. F. R. Spellman and N. E. Whiting, 'Machine Guarding Handbook: A Practical Guide to Osha Compliance and Injury Prevention', Government Institutes, Rockville, Md., 1999.
132. United States. Occupational Safety and Health Administration, 'Osha's Right to Know Hazardous Chemicals: A Consolidated List', Hazardous Materials Information Center, Middletown, Conn., 1986.
133. Safety-Kleen (Firm), 'Introduction to Regulatory Compliance: A Guide to Osha, Epa, and Dot Requirements', Safety-Kleen Corp., Elgin, IL (1000 N. Randall Rd., Elgin 60123), 1994.
134. M. F. Fingas, 'The Handbook of Hazardous Materials Spills Technology', McGraw-Hill, New York, 2001.
135. C. Hawley, 'Hazardous Materials Air Monitoring and Detection Devices', Delmar Thomson Learning, Albany, 2001.
136. F. R. Spellman, J. Drinan, and N. E. Whiting, 'Transportation of Hazardous Materials: A Practical Guide to Compliance', Government Institutes, Rockville, MD, 2001.
137. J. B. Sullivan and G. R. Krieger, 'Clinical Environmental Health and Toxic Exposures', Lippincott Williams & Wilkins, Philadelphia, 2001.
138. L. Zhang *et al.*, *Journal of Physics-Condensed Matter*, 2001, **13**(26), 5947.
139. G. F. Batalha and M. Stipkovic, *Journal of Materials Processing Technology*, 2001, **113**(1–3), 732.
140. United States. General Accounting Office, 'Hazardous Materials Training: Dot and Private Sector Initiatives Generally Complement Each Other: Report to Congressional Requesters', The Office, Washington, D.C. (P.O. Box 37050, Washington, D.C. 20013), 2000.
141. J. M. Liu and S. S. Chou, *Materials Science and Technology*, 2000, **16**(9), 1037.
142. G. Woodside, 'Hazardous Materials and Hazardous Waste Management', John Wiley & Sons, New York, 1999.
143. R. L. Stepkin, R. E. Mosely and American Society of Safety Engineers, 'Noise Control: A Guide for Workers and Employers', American Society of Safety Engineers, Park Ridge, Ill., 1984.
144. National Institute for Occupational Safety and Health, 'Occupational Noise Exposure', U.S. Dept. of Health and Human Services Public Health Service Centers for Disease Control and Prevention National Institute for Occupational Safety and Health, Cincinnati, Ohio, 1998.
145. D. Meister and T. P. Enderwick, 'Human Factors in System Design, Development, and Testing', Lawrence Erlbaum Associates, Mahwah, NJ, 2002.
146. M. Hanson, 'Contemporary Ergonomics 2001', Taylor & Francis, London, 2001.
147. F. Violante, T. J. Armstrong, and Å. Kilbom, 'Occupational Ergonomics: Work Related Musculoskeletal Disorders of the Upper Limb and Back', Taylor & Francis, London, 2000.
148. United States. Safety and Health Administration, 'Ergonomics, the Study of Work', U.S. Dept. of Labor Occupational Safety and Health Administration, [Washington, D.C.?], 2000.
149. D. Macleod, 'The Rules of Work: A Practical Engineering Guide to Ergonomics', Taylor & Francis, New York, 2000.
150. G. Cranz, 'The Chair: Rethinking Culture, Body, and Design', W. W. Norton, New York, 2000.
151. N. Stanton and M. S. Young, 'A Guide to Methodology in Ergonomics: Designing for Human Use', Taylor & Francis, London; New York, 1999.
152. D. Meister, 'The History of Human Factors and Ergonomics', Lawrence Erlbaum Associates Publishers, Mahwah, NJ, 1999.
153. W. Karwowski and W. S. Marras, 'The Occupational Ergonomics Handbook', CRC Press, Boca Raton, Fla., 1999.
154. S. Kumar, 'Biomechanics in Ergonomics', Taylor & Francis, London; Philadelphia, PA, 1999.
155. P. A. Hancock, 'Human Performance and Ergonomics', Academic, San Diego, Calif., 1999.
156. International Labor Office, 'Your Health and Safety at Work', ILO, 1999, Geneva, Switzerland.
157. D. K. Claiborne, N. J. Powell, and K. Reynolds-Lynch, 'Ergonomics and Cumulative Trauma Disorders: A Handbook for Occupational Therapists', Singular Pub. Group, San Diego, 1999.
158. J. Anshel, 'Visual Ergonomics in the Workplace', Taylor & Francis, Bristol, PA, 1998.
159. Wiley Interscience (Online Service), 'Human Factors and Ergonomics in Manufacturing', Wiley, New York, NY, 1997.
160. Acgih, '2001 TLVs and BEIs', American Conference of Governmental Industrial Hygienists, Cincinnati, OH, 2001.

31 Corrosion

31.1 Introduction

It is generally accepted that ten unique types of corrosion exist. Unfortunately, identification of a specific type of corrosion mechanism often occurs only in forensic analysis after a component fails. Design considerations must account for corrosion through the use of corrosion resistant materials, or by increasing the bulk of a material with a known corrosion characteristic to account for loss during its service life. Actuarial science and economics govern this choice in the design phase, but corrosion failures often arise in service operations not anticipated in the design phase. There are numerous remedies in existence to counteract specific corrosion mechanisms and increase the service life of specific components. The goal of this chapter is to summarise each of the individual corrosion mechanisms and describe tests to identify a material's susceptibility to a specific corrosion mechanism.

31.2 Types of corrosion

Corrosion mechanisms can be divided into two categories: (1) general or uniform corrosion and (2) non-uniform or localised forms of corrosion. Though uniform corrosion accounts for the majority of metal consumed (raw tonnage), localised corrosion is much more insidious because it is difficult to detect and its rates are often larger by several orders of magnitude. Due to these accelerated rates, breach of the material and subsequent failure are rapid and most often only detected by post-mortem failure analysis.

31.3 Uniform corrosion

Uniform corrosion is recognised universally as the most prevalent of all forms of corrosion, but the least damaging in terms of number of failures. Like all forms of corrosion, uniform corrosion is the result of electrochemical reactions occurring at the surface of a metal. Uniform corrosion prevails when there are no preferential sites on the surface that are more favoured for attack than others: thus all sites are equivalent and uniform dissolution takes place. Atmospheric corrosion is probably the most visible form of uniform corrosion.

There are two aspects of uniform corrosion that need to be considered more carefully, *corrosion susceptibility* and *corrosion rate*. Corrosion susceptibility is the tendency of a particular material to corrode and is a function of thermodynamics, specifically the potential energy that is required to oxidise and reduce species present in a corrosion cell. The most quoted example used to demonstrate corrosion is rust formation on iron metal. Rust, the orange-red coloured powder found commonly at the surface of iron, is the hydrated complex $Fe_2O_3 \cdot H_2O$. It is formed by the anodic oxidation of iron, $Fe\,(s) = Fe^{+2}\,(aq) + 2e^-$, and the cathodic reduction of oxygen, $O_2(g) + 4H^+\,(aq) + 4e^- = 2H_2O$ (l). The standard potential versus a normal hydrogen electrode (NHE) for the oxidation of Fe is $-0.44\,V$. The potential for the reduction of O_2 is $1.229\,V$ vs. NHE. The corrosion cell thus possesses a potential of $1.669\,V$ vs. NHE. Since $V > 0$, the overall process will proceed. The susceptibility of a material determines if it will corrode, while the second parameter, corrosion rate, determines how much a material will corrode in a given time interval.

The *corrosion rate* is the single most important measure for a corrosion engineer, especially for uniformly corroding systems. In these systems, the corrosion rate is measured primarily by monitoring mass loss or by using electrochemical techniques. Relevant ASTM test standards include:

- G 1–90 Standard Practice for Preparing, Cleaning, and Evaluating Corrosion Test Specimens
- G 31–72 Standard Practice for Laboratory Immersion Corrosion Testing of Metals
- G 50–76 Standard Practice for Conducting Atmospheric Corrosion Tests on Metals

The mass loss corrosion rate is calculated as follows:

$$\text{Corrosion Rate} = (K \times W)/(A \times T \times \rho)$$

where K is a constant, 8.76×10^4 (in mm/year), W is the mass lost in grams, A is the specimen area in cm^2, T is the time of exposure in hours, and ρ is the density of the material in grams per cm^3. Mass loss measurements have the advantage of high accuracy, as long as the specimen and exposure medium are truly representative of the service conditions. A disadvantage is that the time required to determine a significant mass loss is usually quite long. Electrochemical methods overcome the problem of excessive test periods.

Electrochemical techniques are preferred to mass loss determinations because they (1) are accomplished in shorter periods, (2) measure corrosion rates in real-time, and (3) provide more information about the corrosion rate—time relationship. There are several electrochemical tests that can be performed to provide information about corrosion processes. In addition to the test specimen and electrolyte solution, all electrochemical tests require a potentiostat/galvanostat, a test flask, a calibrated reference electrode, and suitable counter electrodes. These techniques range from potentiostatic and potentiodynamic measurements to more complex impedance measurements. There are disadvantages in using electrochemical techniques, however, electrochemical techniques thermodynamically destabilise a test environment in order to measure a response. This destabilisation, or perturbation from the equilibrium condition, may drive a system to extremes that are not truly representative of the service environment. Therefore, it is important to reasonably modify standard test procedures to more accurately reflect actual service conditions. ASTM standard test procedures for electrochemical techniques include:

- G 5–94 Standard Reference Test Method for Making Potentiostatic and Potentiodynamic Anodic Polarisation Measurements
- G 59–97 Standard Test Method for Conducting Potentiodynamic Polarisation Resistance Measurements
- G 102–89 Standard Practice for Calculation of Corrosion Rates and Related Information from Electrochemical Measurements

31.3.1 Galvanic corrosion

A galvanic cell is formed when two dissimilar metals are electrically connected while both are immersed in an electrolyte solution. The less noble of the two materials will corrode preferentially with respect to the other. In order to predict galvanic activity, it is useful to consider where each metal lies, relative to the other, in a galvanic series. A galvanic series provides rest potentials of various metals in a specific environment. Figure 31.1 displays a generic galvanic series of metals and alloys in order from the most active to the least active (most noble). While it is tempting to believe that greater relative separation among metals within the galvanic series will yield more aggressive corrosion, in reality this must be more carefully addressed by considering the coupling within mixed potential theory.

Mixed potential theory is best demonstrated by an example Evans Diagram of two metals. Figure 31.2 shows a galvanic couple between an active metal, M, and a more noble metal, N, in a dilute corrosion solution. According to mixed potential theory, the coupled potential, E_{couple}, is determined at the point where the sum anodic current (total oxidation) is equivalent to the sum cathodic current (total reduction). In the case where N is truly inert, e.g. platinum, only the active side of the couple need be considered. At E_{couple} the anodic dissolution rate for M, the anode in the couple, increases from $I_{corr,M}$ to $I_{couple,M}$, and that for N, the cathode decreases from $I_{corr,N}$ to $I_{couple,N}$.

Electrochemical analysis is frequently chosen as a method to measure galvanic corrosion processes. One of the simplest methods employs a zero-resistance ammeter (ZRA) connected in series with an external wire attached to each of the dissimilar metals. The ZRA is a sensitive instrument that allows direct measurement of current flowing between the two materials. Additionally, galvanic series determinations can be used in materials selection to weed out materials with a propensity for galvanic corrosion. These are implemented by individually placing the materials of interest into a simulated service environment and monitoring their electrochemical potential over time. The material with the more electronegative potential will become the anode and experience accelerated degradation when

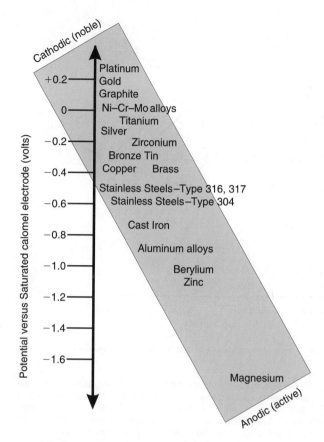

Figure 31.1 *Galvanic series of select pure metals and alloys in seawater*

placed in a galvanic couple. Galvanic corrosion can also be examined by overlaying polarisation curves for each metal. ASTM test standards for assessing galvanic corrosion include the following:

- G 71–81 Standard Guide for Conducting and Evaluating Galvanic Corrosion Tests in Electrolytes
- G 149–97 Standard Practice for Conducting the Washer Test for Atmospheric Galvanic Corrosion

31.3.2   Erosion, cavitation, and fretting corrosion

The common link between erosion, cavitation, and fretting attack is that each is potentially damaging to the surface films that protect materials. Erosion originates from either solid particles or liquid droplets impinging on the surface of a material. The amount of damage depends upon the impingement angle, the velocity of the particle or droplet, the size of the particle or droplet, and the type of material being affected. Solid particles impacting ductile materials at low impingement angles produce greater damage than more normal angles. The opposite is true for brittle materials. In ductile materials, solid particles impinge on the surface and create a crater. In brittle materials, the energy of the impacting particle is focused and creates a microcrack. Loss of material occurs when these cracks coalesce and material chips away. Erosion-corrosion tests are quite simplistic and generally involve directing a jet of gas-entrained particles or liquid at the surface of a material and monitoring the weight loss over some time interval.

Cavitation is a phenomenon closely connected with fluid dynamics and the formation and collapse of gas bubbles near the surfaces of materials. One of the most common examples is the cavitation

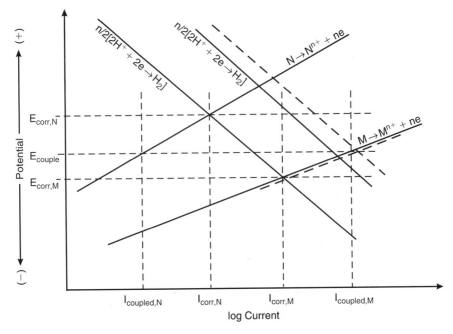

Figure 31.2   *Evans diagram of generic mixed potential galvanic couple between active material* M *and less active metal* N

of ship propellers. The collapse of the bubble near the propeller tip is sufficiently rapid and violent that it physically shocks the material surface and results in pitting. Cavitation results from highly localised pressure discontinuities that lead to a massive instability in the bubble interface. In cavitation corrosion tests, a specimen coupon is vibrated while immersed in a liquid to induce cavities at its surface, then the weight loss is monitored as a function of both velocity and exposure time.

Fretting corrosion is a tribological effect of two surfaces in contact and in relative motion to one another. There are a number of factors that control fretting: the number of cycles, the load normal to each surface, the amplitude of translational motion, the frequency of motion, and the temperature of the system. Fretting corrosion tests involve bringing two surfaces in contact under a known load and contact geometry, moving one surface relative to the other under an oscillation with small amplitude. The weight loss is measured as a function of time. ASTM test standards that apply to cavitation and erosion include the following:

- G 32–03 Standard Test Method for Cavitation Erosion Using Vibratory Apparatus
- G 134–95 Standard Test Method for Erosion of Solid Materials by a Cavitation Liquid Jet

31.4   Localised forms of corrosion

Localised corrosion refers to any type of material dissolution that is heterogeneous, i.e. there are preferential sites with accelerated degradation relative to surrounding areas. Localised forms of corrosion are commonly much more aggressive than uniform corrosion, because the dissolution flux is concentrated in a much smaller area. Many distinct individual mechanisms of localised corrosion have been identified. In contrast with uniform corrosion, localised corrosion is more probabilistic, more aggressive, and less predictable.

31.4.1   Crevice corrosion

The term crevice corrosion refers to enhanced dissolution in discrete areas of a material that are physically occluded, or shadowed. The mechanism that drives this enhanced dissolution is inhibition of electrolyte diffusion by a lack of linear transport paths, usually resulting from the presence of a physical barrier such as a crevice. This situation is often associated with deposits on a surface and at fabricated joints. The crevice corrosion mechanism is broken down into four stages: (1) deoxygenation, (2) increasing localised concentration of acids and salts, (3) depassivation, and (4) propagation of the crevice.

Crevice corrosion is a direct result of impeded mass transport at an occluded site on a material surface. Convection between the solution trapped within the occluded area and the bulk medium is slow, so the dissolved oxygen in the occluded region is rapidly consumed. A local galvanic cell evolves from the differential between the less oxygenated crevice region (anode) and the surrounding bulk oxygenated surface (cathode). This differential aeration cell may also serve to attract aggressive anions such as Cl^-.

A relatively simple test for crevice corrosion involves the use of a crevice former, such as trifluoroethylene (TFE) blocks, held tightly against the surface of the material of interest. The test specimen is immersed in a ferric chloride ($FeCl_3$) solution for 72 hours. After the test period, the area beneath the TFE blocks is examined for indications of crevice corrosion. Another method relies upon the determination of critical temperatures to rank alloys. The most common method for this is electrochemical critical temperature (ECT) testing. In ECT tests for crevice corrosion specimens are typically rectangular coupons with a hole machined in their center. A threaded support runs through the hole and crevice formers—Delrin round nuts with machined flat steps—are fastened tightly to each face of the specimen. With the proper combination of heat and applied potential, localised corrosion will initiate and result in a rapid increase of measured current density at the specimen. The temperature at which this takes place is referred to as the electrochemical critical crevice corrosion temperature (ECCT).

Table 31.1 provides a listing of degrees of crevice corrosion resistance of some common materials. ASTM test standard relating to crevice corrosion include the following:

- G 48–00 Standard Test Methods for Pitting and Crevice Corrosion Resistance of Stainless Steels and Related Alloys by Use of Ferric Chloride Solution
- G 78–01 Standard Guide for Crevice Corrosion Testing of Iron-Base and Nickel-Base Stainless Alloys in Seawater and Other Chloride-Containing Aqueous Environments.

31.4.2 Dealloying corrosion

Dealloying is the selective dissolution of one or more components of a solid solution alloy. Dealloying is known to occur in a wide range of conditions and is characterised by the formation of a

Table 31.1   RESISTANCE TO CREVICE CORROSION

Metal	Very resistant →	Moderate corrosion →			Severe corrosion
Mild steel					
Low alloy steel		Sea water			
Cupro nickel Cu 10Ni 1.5Fe	Sea water	Neutral solutions			
Cu and Cu/Zn	√				
Titanium and Ti alloys	√ (ambient temp)	95°C			Halide* and sulphate solns
Stainless steel 10% H_2SO_4 RT	321	316	304	302	13% Cr
10% H_2SO_4+NaCl	316	321	302	304	16% Cr
Fe 18Cr 13Ni 3Mo 2Si NNG	√				
Fe 18Cr 14Ni 2Mo NiTi	√				
Fe 18Cr 24Ni 3Mo 2Cu	√				
Fe 20Cr 25Ni 5Mo 1.5Cu	√				
Fe 18Cr 10Ni 2.5Mo 2.5Si	√				
Fe 25Cr 2Mo(duplex)	√				
FE 25Cr 6.5Ni 3Mo(duplex) 0.3 W	√				
Nickel alloys					
Hastalloy 'C'	<60°C				
Inconel 625	<60°C				
Cobalt alloys					
Vitallium	<60°C				

* Ti 0.2% Pt 2% Mo(or 2% Ni) relatively more resistant.

Table 31.2   VARIOUS DEALLOYING PROCESSES

Alloy	Environment	Element removed
Brasses	Waters, acid chlorides	Zn
Ni–Cu alloys	HF and other acids	Cu in some acids, Ni in others
Au–Ag amalgam	Human saliva	Ag
High-Nickel Alloys	Molten salts	Cr, Fe, Mo, W
High-Carbon steels	Oxidising conditions, H_2 at high temperatures	C
Fe–Cr Alloys	High temperature oxidising conditions	Cr
Ni–Mo Alloys	High temperature oxygen	Mo

porous microstructure resembling a sponge. It is common to observe a high degree of interconnec-tivity between these pores with islands of the more noble metals remaining intact. Often alloys that undergo dealloying are also susceptible to stress corrosion cracking (SCC). The porous microstruc-ture that forms during dealloying often localises at pre-existing crack tips, which when subjected to a stress, promotes ductile fracture within the corroded region and may propagate into the unaffected surrounding material. There are a number of mechanistic models for dealloying corrosion. They are the bulk diffusion model, the vacancy diffusion model, the surface diffusion model, the dissolution and reprecipitation model, the oxide formation model, and the percolation model. On the basis of measured corrosion rates, the surface diffusion model seems the most plausible.

The most common example of dealloying corrosion is the dezincification of brass. Selective dissolution of zinc in brass (Cu–Zn alloys) occurs in extended exposures to chloride containing solutions. The $\alpha + \beta$ two-phase alloys are most susceptible with significant attack occurring in the zinc concentrated β phase. Dezincification results in a porous, highly weakened layer of copper and a copper oxide corrosion product. At room temperature dezincification occurs in slightly acidic water or acid chlorides and is more uniform throughout the β phase. The effect at higher temperatures is more locally pronounced with very high penetration rates. Dezincification can be prevented by controlling the purity of process streams, by limiting the amount of zinc, or by substituting it with more corrosion resistant components. Table 31.2 lists some environments leading to dealloying corrosion.

31.4.3   Environmental cracking—stress corrosion cracking and corrosion fatigue cracking

Environmental cracking is a general term that is utilised to describe localised corrosion mechanisms that couple the simultaneous actions of mechanical stress (typically static and tensile) and a chem-ically reactive environment. Isolated, each of these actions is of little consequence to a material, but in their combined form they often lead to unpredictable catastrophic failures. Three distinct subset mechanisms of environmental cracking are stress corrosion cracking (SCC), corrosion fatigue cracking (CFC), and hydrogen induced cracking (HIC). HIC will be highlighted within its own section (31.4.4)

SCC is the brittle failure of a material in a corrosive environment at relatively low tensile stress. Four conditions need to be present to activate SCC. These include a corrosive environment, a suscep-tible material, the electrochemical state of a system relative to its surroundings, and a tensile stress. Pure metals are more resistant, but they are not immune. SCC can occur in either transgranular or intergranular fashion, with the crack following a path normal to the tensile stress. In transgranular failure, the cracks usually propagate through grains with a particular crystallographic misorienta-tion, whereas in intergranular corrosion cracks simply follow grain boundaries with no regard to misorientation.

Alloy chemistry has a considerable effect on SCC, particularly with respect to passivity and metallurgical condition. From an electrochemical perspective, the chemical composition and microstructural condition are very important in determining its stability. Heterogeneities associ-ated with microstructure and second phases with varying susceptibility to cracking can produce very different electrochemical properties. In addition to the metallurgical parameters related to electrochemical character, both grain size and yield strength can strongly affect SCC behaviour.

The physical, chemical, and mechanical characteristics of the service environment all affect SCC. Physical characteristics such as temperature and nuclear irradiation can affect the susceptibility and magnitude of SCC. A given alloy system is usually sensitive to SCC in very specific chemical environments—these range from supercritical sulphuric acid (in the case of the hydrogen-producing sulphur-iodide thermochemical cycle), via natural marine seawater, to steam. Mechanical stress is the major variable that separates SCC from other localised forms of corrosion. A tensile stress is required for SCC to occur.

It is generally accepted that SCC occurs via three stages: initiation, growth, and failure. The time duration of each of these stages is dependent on the fracture toughness of the material, the geometry and surface condition of the specimen, and the loading condition. ASTM standards related to characterisation and measurement of SCC include the following:

- B 858–01 Standard Test Method for Ammonia Vapour Test for Determining Susceptibility to Stress Corrosion Cracking in Copper Alloys
- G 30–97 Standard Practice for Making and Using U-Bend Stress-Corrosion Test Specimens
- G 38–01 Standard Practice for Making and Using C-Ring Stress-Corrosion Test Specimens
- G 123–00 Standard Test Method for Evaluating Stress-Corrosion Cracking of Stainless Alloys with Different Nickel Content in Boiling Acidified Sodium Chloride Solution
- G 168–00 Standard Practice for Making and Using Precracked Double Beam Stress Corrosion Specimens

Corrosion fatigue cracking (CFC) is a cumulative effect arising from load cycling in an aggressive environment. It originates from the interaction of irreversible cyclic plastic deformation with localised electrochemical reactions. CFC damage typically accumulates with increasing number of loading cycles (N) and in four stages: (1) cyclic plastic deformation; (2) initiation of microcracks; (3) crack growth and coalescence; and (4) macrocrack propagation. Hydrogen typically plays an important role in CFC. Atomic hydrogen adsorbs to crack surfaces as a result of the electrochemical reduction of hydrogen or water, and mass transport carries the hydrogen to the crack tip where it can act to locally embrittle the metal or acidify the environment, leading to enhanced crack propagation.

Passive film rupture is another CFC mechanism. In passive film rupture, a fissure in the protective film is followed by anodic dissolution at the surface initiation site of crack tip. Localised plastic straining arising from mechanical cycling aids passive film rupture. CFC growth in this mechanism depends upon anodic charge passed per load cycle. Other CFC mechanisms have been proposed based upon interactions between dislocations and localised electrochemical environments present at initiation sites and crack tip surfaces.

31.4.4   Hydrogen damage

Hydrogen atoms readily diffuse in metal lattices and can pose a serious threat to mechanical strength and ductility. Hydrogen can be made available to the metal surface from a variety of sources, including the electrochemical reduction of hydrogen or water:

$$2\,H^+ + 2e \longrightarrow H_2$$

or

$$2\,H_2O + 2e \longrightarrow H_2 + 2\,OH^-$$

These reactions are often present during corrosion, but may also occur during cleaning procedures and cathodic protection. Hydrogen may also enter the lattice during heat treatment, welding and various post-fabrication processes. Another ready source of hydrogen is steam, as either a by-product of welding, or in a service environment such as piping.

There are many types of mechanical damage and failures associated with the presence of hydrogen in metals. At room temperature, hydrogen in steel may lead to a decrease in ductility (embrittlement), blistering, and cracking that occurs at lower loads than measured in the absence of hydrogen. At higher temperatures, hydrogen is known to decarburise steel by forming methane bubbles that trap and coalesce, exerting significant internal pressure within the lattice.

Testing for hydrogen damage can take on many forms. Methods such as tritium radiography can be employed to determine hydrogen concentrations in a lattice, and mechanical testing can report changes in ductility or greater propensity to cracking. ASTM test standards related to hydrogen embrittlement include the following:

- F 519–97 Standard Test Method for Mechanical Hydrogen Embrittlement Evaluation of Plating Processes and Service Environments
- F 1459–93 Standard Test Method for Determination of the Susceptibility of Metallic Materials to Gaseous Hydrogen Embrittlement

- F 1624–00 Standard Test Method for Measurement of Hydrogen Embrittlement Threshold in Steel by the Incremental Step Loading Technique
- F 2078–01 Standard Terminology Relating to Hydrogen Embrittlement Testing
- G 142–98 Standard Test Method for Determination of Susceptibility of Metals to Embrittlement in Hydrogen Containing Environments at High Pressure, High Temperature, or Both

31.4.5   Intergranular corrosion

Intergranular corrosion (IGC) is strongly associated with the properties and microstructure of a metal. A well-known example of IGC is the sensitisation of austenitic 18Cr—8Ni stainless steels. In the temperature range of 538–927°C, insoluble chromium carbides, $Cr_{23}C_6$, precipitate at the grain boundaries. This precipitate is a product of the reaction between chromium and the carbon diffusing along the grain boundaries. Below 538°C, the carbon remains relatively immobile, and above 927°C the chromium carbides are soluble. Formation of chromium-rich precipitates quickly depletes chromium from a region immediately adjacent to the grain boundary. This reduces the alloy's ability to generate chromium oxide protective films and leads to an increased susceptibility for localised attack in the grain boundary region.

Welding is a common source of localised heating that can sensitise a material. IGC is often observed within the heat affected zone after welding a material susceptible to heat sensitisation. Figure 31.3 illustrates IGC coincident to a weld bead. Note that the sensitisation band occurs on each side of the weld, parallel to its axis of propagation. The displacement and width of the IGC regions is dependent upon the thermal conductivity of the material and the duration that it resides in the critical sensitisation temperature range.

Knifeline attack is a highly localised form of IGC that has been observed in titanium and niobium doped austenitic stainless steels, Types 321 and 327, respectively. The region of attack is typically only a couple of grain diameters from the weld bead. Knifeline attack results from dissolution of titanium or niobium carbides at high temperatures, >1230°C, followed by rapid cooling.

IGC usually requires a strongly oxidising condition. Limiting the use of sensitised materials to weakly corrosive environments will help insure their stability. Likewise it is possible to control oxidation by injecting reducing gases such as H_2, or by operating at lower acidity. Metallurgical techniques are more common for preventing IGC. Typically, those employed are: a) decreasing the level of carbon loading; b) solution annealing; and c) stabilisation using titanium or niobium.

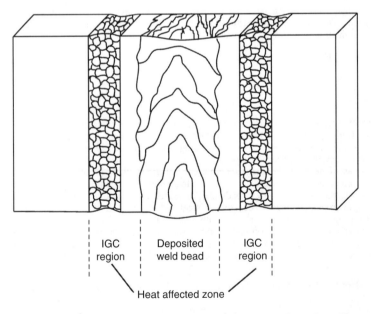

Figure 31.3   *Illustration of IGC region and heat affected zone coincident with a weld*

IGC testing is accomplished by determining the susceptibility of a material to heat sensitisation, then exposing the material to a strongly oxidising environment. Most involve an exposure to sulphuric acid, nitric acid, oxalic acid, or a solution of sulphuric acid and copper sulphate. ASTM standards relating to IGC include the following:

- A 262–02 Standard Practices for Detecting Susceptibility to Intergranular Attack in Austenitic Stainless Steels
- A 763–93 Standard Practices for Detecting Susceptibility to Intergranular Attack in Ferritic Stainless Steels
- G 28–02 Standard Test Methods of Detecting Susceptibility to Intergranular Corrosion in Wrought, Nickel-Rich, Chromium-Bearing Alloys
- G 108–94 Standard Test Methods of Detecting Susceptibility to Intergranular Corrosion in Wrought, Nickel-Rich, Chromium-Bearing Alloys
- G 110–92 Standard Practice for Evaluating Intergranular Corrosion Resistance of Heat Treatable Aluminium Alloys by Immersion in Sodium Chloride + Hydrogen Peroxide Solution

31.4.6   Pitting corrosion

Pitting corrosion is one of the most studied areas of corrosion science. One motivation for studying pitting is its lack of predictability. In general pitting, requires that a material initially be passive in its environment. For pitting to initiate, there must be process that leads to the breakdown of this passivity. Often, this process involves either a structural or chemical heterogeneity in the passive film at the surface of the material or within the solution medium. Following the initial breakdown in passivity, the pit grows, then it potentially repassivates.

There are a variety of valid mechanistic theories that consider the initial breakdown of passivity. In the penetration mechanism, aggressive anions transfer through the passive layer to the metal-oxide interface. The film breaking mechanism relies on ruptures in the passive film. This rupture is caused by sudden changes in the electrode potential that cause excessive stresses in the surface film. The adsorption mechanism starts with the formation of surface complexes that lead to enhance transport between the metal-oxide interface and the electrolyte.

During pit growth, accumulation of corrosion products enhances the precipitation of a salt film. When the salt film forms, the pit growth rate slows down and the pit morphology becomes hemispherical. The general requirement for pit stability is the ability to transport aggressive ions that accumulate about the pit. If the transport of aggressive anions out of pits is efficient then there is a strong probability that the corrosion pit will repassivate. Accumulation of aggressive anions prevents formation of a passive layer at the active pit surface.

Coupon testing can be utilised to quantify pitting corrosion. Typically this involves measured, cleaned, and weighed coupons immersed into the medium of interest. After some determined exposure time, the coupons are removed, examined for surface corrosion, then cleaned and weighed to determine any loss. The major disadvantage in using this method is that the exposure times required to measure an appreciable change are rather long.

Electrochemical techniques offer greater utility because one is able to utilise a solution representative of the service environment and that the testing period is relatively short. The most common electrochemical test is cyclic polarisation. The specimen is anodically polarised from its corrosion potential, E_{corr}, to a potential where the onset of localised corrosion is indicated by a large increase in the measured applied current. The onset point is referred to as the breakdown potential, E_{BD}. The polarisation is then reversed (cathodic direction) to determine the point at which the current returns to background. This point is referred to as the repassivation potential, E_{RP}, the potential where pits repassivate. An analogue to cyclic polarisation is critical pitting temperature determination. Instead of polarising the specimen, the temperature of the system is increased to the point where an increase in current is measured.

Electrochemical noise (ECN) measurements are gaining popularity as a non-destructive, real-time monitoring technique. In electrochemical noise measurements, variations in potential and current transients are recorded for two identical specimen electrodes in a system of interest. Large amplitude peaks in the measured signals are indicators of local activity and corrective action can be taken to insure the stability of the system.

ASTM test standards related to pitting corrosion include the following:

- G 46–94 Standard Guide for Examination and Evaluation of Pitting Corrosion
- G 48–00 Standard Test Methods for Pitting and Crevice Corrosion Resistance of Stainless Steels and Related Alloys by Use of Ferric Chloride Solution

- G 150–99 Standard Test Methods for Pitting and Crevice Corrosion Resistance of Stainless Steels and Related Alloys by Use of Ferric Chloride Solution

31.5 Biocorrosion

Biocorrosion, or microbiologically influenced corrosion (MIC), refers to generally corrosion affected by the excretions and activities of microbes in biofilms on metals and alloys. Microbes are capable of altering corrosion processes by changing the chemical environment at a metal surface and thereby influencing electrochemical events occurring there. Microbiologically mediated processes can alter both rates and types of electrochemical reactions and produce a broad range of outcomes, from severe localised corrosion to significant reductions in corrosion rate.[1] Pitting, crevice corrosion, differential aeration cells, metal concentration cells, selective dealloying, accelerated erosion, and enhanced galvanic corrosion can result from biocorrosion. Single-cell organisms such as bacteria, fungi, and yeast have all been shown to influence corrosion processes. Micro organisms are generally ubiquitous and can be attracted to both anodic[2] and cathodic[3] sites making spatial relationships between micro organisms and corrosion products insufficient, on their own, to be interpreted as biocorrosion. Accurate interpretation of biocorrosion generally requires a combination of surface analytical, microbiological, and electrochemical techniques.

Cells attach to wet surfaces and produce a biofilm consisting of a matrix of polymeric substances and a diverse population of cells in microcolonies, as shown schematically in Figure 31.4[4,5]. Biofilm formation consists of a sequence of steps that begin with adsorption of macromolecules (proteins, polysaccharides, and humic acids) and smaller molecules (fatty acids, lipids) at the metal surface. The adsorbed molecules form conditioning films that alter the physical and chemical characteristics of the water-surface interface, including such properties as hydrophobicity and electrical charge. This conditioning of the surface allows for the additional adsorption of microbial cells from the bulk liquid. The adsorbed micro organisms then initiate production of adhesive substances (predominantly exopolysaccharides[6]), collectively known as extracellular polymeric substances (EPS). These substances provide the biopolymer matrix (slime) that holds bacteria together and permits the formation of microcolonies and ultimately a mature biofilm. EPS are also implicated in the increased resistance of biofilm cells to biocides and other antimicrobial compounds.[6]

The most detrimental biocorrosion takes place in the presence of microbial consortia where many species of micro organisms are present within the structure of the biofilm. Participants in biocorrosion include sulphate-reducing bacteria (SRB), metal-oxidising and metal-reducing bacteria,

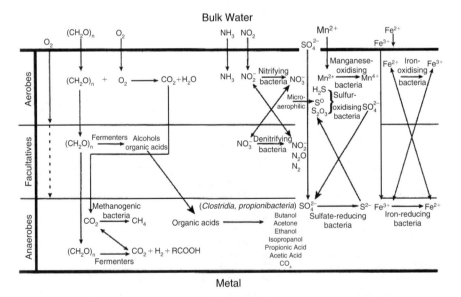

Figure 31.4 *Possible microbial reactions within a mixed-species biofilm. (Courtesy of Dr. Brenda Little, Naval Research Laboratory)*

acid-producing and slime-producing bacteria, hydrogen-consuming and hydrogen-producing bacteria, and metal-depositing bacteria, plus other forms of aerobic and anaerobic micro organisms (see Figure 31.4). Among these types, SRB are the best understood in terms of their role in biocorrosion.[1]

SRB consist of a diverse group of anaerobic bacterial species found in both natural and industrial environments that vary in their ability to accelerate corrosion[7]. In the absence of oxygen, SRB use sulphate as the terminal electron acceptor during respiration and produce hydrogen sulphide. When the respiration rate of aerobic bacteria within a biofilm is greater than the oxygen diffusion rate, the metal/biofilm interface can become anaerobic and provide a niche environment for SRB. The corrosion rate of iron and copper alloys in the presence of hydrogen sulphide is accelerated by the formation of iron sulphide minerals that stimulate the cathodic reaction. Typically, the distribution of the sulphides produced is not uniform and, therefore, leads to localised pitting of the metal surface.

Some microbes utilise a combination of manganese and iron oxidation and reduction reactions to facilitate cell growth and metabolism of organic carbon. Microbial deposition of manganese oxide on stainless and mild steel can significantly alter the corrosion potential (E_{corr}) resulting in an increased corrosion rate.[8–10] Iron-oxidising bacteria produce tubercles of iron oxides and hydroxides creating regions of relatively low oxygen concentration within the tubercle that can induce localised corrosion events. Other bacterial strains promote corrosion of iron alloys through metal reduction reactions that lead to the dissolution of protective oxide/hydroxide films on the metal surface. Passive layers are either removed or replaced by less stable films that are ineffective against further corrosion.[11]

Fungi as well as bacteria can produce a wide variety of either inorganic or organic acids (e.g. acetic, formic, lactic, succinic) as by-products of metabolism that are capable of initiating or accelerating corrosion.[12,13] Mineral acids produced by bacteria can be very corrosive. In addition, a growing body of research points to EPS/metal-ion interactions as a source of biocorrosion.[14,15] These and other microbially-mediated reactions within a biofilm are shown schematically in Figure 31.4.

Laboratory experiments[16–21] and field tests in service water[22] have demonstrated a reduction in corrosion in the presence of certain aerobic bacteria. One theory for this phenomenon is the mitigation of corrosive agents. For example, reduction of oxygen by aerobic bacteria in a biofilm[17,20] has been proposed. In this process, oxygen diffusing through a biofilm from the electrolyte is prevented from reaching the metal surface by microbial respiration. This deaeration should reduce corrosion rates as long as the concentration of oxygen is roughly uniform. Otherwise, aeration cells could develop that would result in enhanced localised corrosion. Although bacterial respiration of oxygen probably assists the observed reduction in corrosion, it is unlikely that this mechanism alone produces the substantial decreases in corrosion rate that have been observed. Furthermore, E_{corr} is often ennobled in the presence of protective aerobic bacteria.[17,22] However, simple depletion of oxygen should push E_{corr} in the opposite direction. Thus, it appears that a microbiological process other than oxygen depletion, such as the excretion of a metabolic product, is inhibiting the corrosion. Although this inhibition is not entirely understood, it has been shown that live bacteria are required in the biofilm to maintain the observed corrosion inhibition. For example, corrosion rates rise substantially, and E_{corr} becomes more negative, to levels typical of abiotic conditions when an antibiotic is introduced that kills the bacteria in the biofilm.[21]

Whether or not corrosion takes place, and at what rate, typically depends on a complex set of both abiotic and biotic environmental conditions. Consequently, serious difficulties generally arise when interpreting data from any single type of corrosion measurement or observation. Thus, it is important to incorporate a number of investigative techniques in any study involving biocorrosion. Three broad categories of techniques used to detect and monitor biocorrosion are: 1) surface inspection, 2) microbial sampling and identification, and 3) corrosion rate determination using mass loss and electrochemical measurements.[1] Surface inspection ranging in resolution from macroscopic visual, Fourier transform infrared spectroscopy (FTIR), scanning electron microscopy (SEM), to atomic force microscopy (AFM) have been found to be useful. Microbial sampling can be made from either the liquid or biofilm on the metal surface. However, it is generally agreed that biofilm sampling is more representative of the microbial population that influences corrosion processes. Identification of bacteria ranges from relatively indiscriminant techniques such as microbial activity or biomass measurements, to species-specific techniques that employ DNA sequence comparisons or gene probes. Field kits have also been developed that detect certain types of corrosion promoting bacteria such as SRB.[1] Useful electrochemical techniques include those that are non-destructive (such as redox and open circuit potential measurement, electrochemical noise analysis, linear polarisation resistance, and electrochemical impedance spectroscopy), as well as those that are destructive (such as polarisation curve and pitting scan methods).

Several approaches for preventing biocorrosion have been tried with limited success. For example, cathodic protection has been found to retard the growth of aerobic bacteria but also favour the growth of deleterious anaerobic species. Pitting has been observed in stainless steel under a biofilm containing SRB even at high cathodic potentials.[23] The success of biocides for controlling biocorrosion has also been quite limited, primarily because the bacteria in biofilms are protected by the EPS that seem to mitigate biocide efficiency or act as a barrier.[1] Biocides are also relatively expensive, often dangerous to work with, and present a potential environmental hazard if used in large quantities. Antifouling paints that contain biocides have also been developed.[24] While these coatings can be effective when fresh biocide is continually exposed by rapid flow (e.g. ship hull), they tend not to perform well in the presence of moderate flow rate or stagnant media. Another recommendation for preventing biocorrosion is to keep the industrial system extremely clean to avoid the formation of a biofilm altogether. Although this may seem like an effective measure, it can be extremely costly to maintain over long periods and may not prevent the short-term colonisation of metal surfaces that lead to biocorrosion. If this approach is taken, it must be implemented from the design of the system onward, including hydrostatic testing prior to service. Finally, it should be noted that a novel method of controlling biocorrosion has be proposed which involves the use of protective strains of bacteria.[25] Some of these strains seem to be capable of secreting both antimicrobial and corrosion inhibiting compounds. This probiotic approach, called corrosion control using regenerative biofilms (CCURB), has been successfully implemented so far in both laboratory experiments and field tests resulting in two-fold or greater reductions in corrosion rate.

BIBLIOGRAPHY

1. D. Anderson, 'Handbook of Corrosion Data', 2nd Edition, ASM International, Materials Park, OH. 1995.
2. 'ASM Handbook, vol. 13, Corrosion', ASM International, Materials Park, OH. 1987.
3. S. A. Bradford, 'Corrosion Control', 2nd Edition, ASM International, Materials Park, OH. 2001.
4. G. T. Burstein, L. L. Sheir, and R. A. Jarman, 'Corrosion', 3rd Edition, Butterworth–Heinemann, Oxford. 1994.
5. M. G. Fontana, 'Corrosion Engineering', McGraw-Hill, New York. 1986.
6. R. Francis, 'Galvanic Corrosion: A Practical Guide', NACE International, Houston, TX. 2000.
7. R. Gibala and R. F. Hehemann, (Editors), 'Hydrogen Embrittlement and Stress Corrosion Cracking', ASM International, Materials Park, OH. 1984.
8. D. L. Graves, 'Corrosion Data Survey: Metals Section', 6th Edition, NACE International, Houston, TX. 1985.
9. Jones, D. A., 'Principles and Prevention of Corrosion', MacMillan, New York. 1992.
10. R. H. Jones (Editor), 'Stress-Corrosion Cracking: Materials Performance and Evaluation', ASM International, Materials Park, OH. 1992.
11. P. Marcus and J. Oudar (Editors), 'Corrosion Mechanisms in Theory and Practice', Marcel Dekker, New York. 1995.
12. P. R. Roberge, 'Handbook of Corrosion Engineering', McGraw-Hill, New York. 1999.
13. P. A. Schweitzer and A Schweitzer, 'Corrosion Resistance Tables: Metals, Nonmetals, Coatings, Mortars, Plastics, Elastomers and Linings, and Fabrics', Marcel Dekker, New York. 1995.
14. H. -H. Strehblow, 'Mechanisms of Pitting Corrosion,' in 'Corrosion Mechanisms in Theory and Practice' (Editors: P. Marcus and J. Oudar), Marcel Dekker, New York. 1995.
15. Z. Szlarska-Smialowska, 'Pitting Corrosion of Metals', NACE International, Houston, TX. 1986.
16. Uhlig, H. H., and Revie, R. W., 'Corrosion and Corrosion Control, An Introduction to Corrosion Science and Engineering', 3rd Edition, Wiley, New York. 1985.

REFERENCES

1. H. A. Videla, 'Manual of Biocorrosion', CRC Press, Boca Raton, FL. 1996.
2. M. J. Franklin, D. C. White, B. J. Little, R. I. Ray and R. K. Pope, Biofouling, 1999, 15 (1–3), 13.
3. B. J. Little and P. A. Wagner, pp. 433–446 of 'Electrochemical Methods in Corrosion Research' (5th International Symposium, Sesimbra, Portugal, 5–8 Sept. 1994) Trans Tech Publications, Uetikon-Zurich, Switzerland, 1995.
4. Handbook of Bacterial Adhesion Principles, Methods, and Applications, Y. H. An and R. J. Friedman (Editors), Humana Press, Totowa, N.J. 2000.
5. Biofilms II: Process Analysis and Applications, J. D. Bryers (Editor), John Wiley & Sons, Inc., New York. 2000.
6. J. W. Costerton, Z. Lewandowski, D. E. Calwell, D. R. Korber and H. M. Lappin-Scott, Microbial Biofilms, *Ann. Rev. Microbiol.*, 1995, **49**, 711.
7. I. B. Beech, C. W. S. Cheung, C. S. P. Chan, M. A. Hill, R. Franco and A. R. Lino, *Mar. Biofouling Corrosion*, 1994, **34**, 4.
8. B. H. Olesen, R. Avci and Z. Lewandowski, Manganese Dioxide as a Potential Cathodic Reactant in Corrosion of Stainless Steels, *Corrosion Sci.*, 2000, **42**, 211.
9. B. H. Olesen, P. H. Nielsen and Z. Lewandowski, Effect of Biomineralised Manganese on the Corrosion Behaviour of C1008 Mild Steel, *Corrosion*, 2000, **56**, 80.
10. P. Linhardt, CORROSION/2000, Paper No. 398, NACE International, Houston, TX. 2000.

11. B. J. Little, P. A. Wagner, K. R. Hart, R. I. Ray, D. M. Lavoie, K. Nealson and C. Aguilar, The Role of Metal Reducing Bacteria in Microbiologically Influenced Corrosion, CORROSION/97, Paper No. 215, NACE International, Houston, TX. 1997.
12. R. J. Soracco, D. H. Pope, J. M. Eggers and T. N. Effinger, Microbiologically Influenced Corrosion Investigations in Electric Power Generating Stations, *Corrosion*/88, Paper No. 83, NACE International, Houston, TX, 1988.
13. B. Little, P. Wagner and D. Duquette, Microbiologically Induced Increase in Corrosion Current Density of Stainless Steel Under Cathodic Protection, *Corrosion*, 1988, **44**, 270.
14. G. G. Geesey, and M. W. Mittelman, The Role of High-Affinity, Metal-Binding Exopolymers of Adherent Bacteria in Microbial-Enhanced Corrosion, CORROSION/85, Paper No. 85, NACE International, Houston, TX. 1985.
15. G. G. Geesey, M. W. Mittelman, T. Iwaoka and P. R. Griffiths, Role of Bacterial Exopolymers in the Deterioration of Metallic Copper Surfaces, *Mat. Perform.*, 1986, **25**, 37.
16. B. W. Samuels, K. Sotoudeh and R. T. Foley, *Corrosion*, 1981, **37**, 92.
17. G. Hernandez, V. Kucera, D. Thierry, A. Pedersen and M. Hermansson, *Corrosion*, 1994, **50**, 603.
18. H. A. Videla, in *Biofouling and Biocorrosion in Industrial Water Systems,* G. G. Geesey, Z. Lewandowski and H. C. Flemmings (Editors), Lewis Publishers, Boca Raton, FL, 1994, pp. 231–241.
19. A. Jayaraman, J. C. Earthman, T. K. Wood, *Applied Microbiology and Biotechnology*, 1997, **47**, 62.
20. D. Ornek, T. K. Wood, B. C. Syrett, C. H. Hsu and F. Mansfeld, 'Corrosion Control Using Regenerative Biofilms (CCURB) on Al 2024 and 26000 Brass'. CORROSION/2001, Paper No. 1270, NACE International, Houston, TX. 2001.
21. K. Ismail, T. Gehrig, K. Trandem, P. J. Arps, A. Jayaraman, T. K. Wood and J. C. Earthman, *Corrosion*, 2002, in press.
22. K. Trandem, Z. S. Farhangrazi, T. K. Wood, P. J. Arps, and J. C. Earthman, Field Sidestream Investigations of Corrosion Control Using Regenerative Biofilms (CCURB), Corrosion/2001, Paper No. 1271, NACE International, Houston, TX. 2001.
23. M. F. L. de Mele, S. G. Gómez de Saravia and H. A. Videla, pp. 50–51 in 'Proc. of the 1995 international Conference on MIC', P. Angell et al (Editors), NACE International, Houston, TX. 1995.
24. C. A. Giudice and J. C. Benitez, *Corrosion Reviews*, 1995, **13**, 81.
25. B. C. Syrett, P. J. Arps, J. C. Earthman, F. Mansfeld and T. K. Wood, 'Biofilms that Prevent Corrosion', Corrosion/2002, Paper No. 02RTS8, NACE International, Houston, TX. 2002.

32 Electroplating and metal finishing

The processes and solutions described in this section are intended to give a general guide to surface finishing procedures. To operate these systems on an industrial scale would normally require recourse to one of the Chemical Supply Houses which retail properietary solutions. This particularly applies to electroplating baths containing brighteners.

32.1 Polishing compositions

The following abrasive powders are used for polishing metal.

ALOXITE

Aluminium oxide made by fusing bauxite. Used for cutting down in the same way as emery.

ALUMINA

Certain grades of alumina are used for polishing stainless steel and chromium. The material is generally used in the form of a composition in which the powder is mixed with stearines or other fats.

EMERY POWDER

Used principally in cutting down and for preliminary operations. It is applied to the mop by means of an adhesive, usually glue. Emery powder is an impure aluminium oxide containing about 50–60% Al_2O_3, 30–40% magnetite and small amounts of ferric oxide, silica, chromium, etc. Emery powder should never be used on magnesium or aluminium components because of the adverse effect on corrosion resistance.

TRIPOLI

A calcined diatomaceous earth used for polishing brass, steel and aluminium. It is used generally in the intermediate stages, and is usually compounded with stearines and paraffin wax to make a polishing composition which can be used directly on a mop.

CROCUS POWDER

A polishing composition consisting essentially of ferric oxide, of coarser grade than rouge, used for polishing iron and steel, and also, tin. Usually compounded with stearine and used with a mop or fibre brush.

ROUGE

A high-grade ferric oxide supplied in various degrees of fineness. It can be used in the form of a paste directly on to a soft mop or can be made into a composition with stearine. It is used essentially for finishing to obtain a very high polish on gold, silver, brass, aluminium, etc.

BLACK ROUGE

This consists of black oxide of iron and is sometimes used for finishing operations.

GREEN ROUGE

Chromic oxide used for polishing chromium and stainless steel and can be used either in the form of a composition mixed with stearine or as a paste applied directly to the mop.

VIENNA LIME

Used for making the white finish for polishing nickel, etc. It consists of a calcined dolomite and contains about 60% calcium oxide and 40% magnesia.

CARBORUNDUM

Silicon carbide used for low tensile strength materials, e.g. brass, copper, aluminium, etc. and also brittle metals, such as hard alloys and cast irons.

32.2   Cleaning and pickling processes

VAPOUR DEGREASING

Used to remove excess oil and grease. Components are suspended in a solvent vapour, such as tri- or tetrachloroethylene.

Note: Both vapours are toxic and care should be taken to ensure efficient condensation or extraction of vapours.

EMULSION CLEANING

An emulsion cleaner suitable for most metals can be prepared by diluting the mixture given below with a mixture of equal parts of white spirit and solvent naphtha.

Pine oil	62 g
Oleic acid	10.8 g
Triethanolamine	7.2 g
Ethylene glycol–monobutyl ether	20 g

This is used at room temperature and should be followed by thorough swilling.

Table 32.1   ALKALINE CLEANING SOLUTIONS

Metal to be cleaned	Composition of solution			Temperature		Remarks
		oz gal^{-1}	g l^{-1}	°F	°C	
All common metals other than aluminium and zinc, but including magnesium	Sodium hydroxide (NaOH)	6	37.5	180–200	80–90	For heavy duty
	Sodium carbonate (Na_2CO_3)	4	25.0			
	Tribasic sodium phosphate ($Na_3PO_4.12H_2O$)	1	6.2			
	Wetting agent	$\frac{1}{4}$	1.5			

(*continued*)

Table 32.1 ALKALINE CLEANING SOLUTIONS—*continued*

Metal to be cleaned	Composition of solution			Temperature		Remarks
		oz gal^{-1}	g l^{-1}	°F	°C	
	Sodium hydroxide	2	12.5	180–200	80–90	For medium duty
	Sodium carbonate	4	25.0			
	Tribasic sodium phosphate	2	12.5			
	Sodium metasilicate (Na$_2$SiO$_3$.5H$_2$O)	2	12.5			
	Wetting agent	$\frac{1}{8}$	0.75			
	Tribasic sodium phosphate	4	25.0	180–200	80–90	For light duty
	Sodium metasilicate	4	25.0			
	Wetting agent	$\frac{1}{8}$	0.75			
Aluminium and zinc	Tribasic sodium phosphate	2	12.5	180–200	80–90	—
	Sodium metasilicate	4	25.0			
	Wetting agent	$\frac{1}{8}$	0.75			
Most common metals	Sodium carbonate	2	12.5	180–200	80–90	Electrolytic cleaner, 6 V Current density 100/A ft^{-2} (10/A dm^{-2})
	Tribasic sodium phosphate	4	25.0			
	Wetting agent	$\frac{1}{4}$	1.5			Article to be cleaned may be made cathode or anode or both alternately
Most common metals	Sodium carbonate	6	37.5	Room	Room	May be used electrolytically
	Sodium hydroxide	1	6.25			
	Tribasic sodium phosphate	2	12.5			
	Sodium cyanide (NaCN)	2	12.5			
	Sodium metasilicate	1	6.25			
	Wetting agent	$\frac{1}{8}$	0.75			

Table 32.2 PICKLING SOLUTIONS

Metal to be pickled	Composition of solution			Temperature		Remarks
		oz gal^{-1}	g l^{-1}	°F	°C	
Aluminium (wrought)	*For etching* Sodium hydroxide (NaOH)	8	56	104–176	40–80	Articles dipped until they gas freely, then swilled, and dipped in nitric acid 1 part by vol. to 1 of water (room temperature)
Aluminium (cast and wrought)	Nitric acid, s.g. 1.42	1 gal	11	Room	Room	Articles first cleaned in solvent degreaser. Use polythene or PVC tanks
	Hydrofluoric acid (52%)	1 gal	11			
	Water	8 gal	81			
	Bright dip					
	Chromic acid	0.84 oz	5.2 g	195	90	Immerse for $1\frac{1}{2}$ min. Solution has limited life. AR chemicals and deionised or distilled water should be used
	Ammonium bifluoride	0.72 oz	4.5 g			
	Cane syrup	0.68 oz	4.2 g			
	Copper nitrate	0.04 oz	0.25 g			
	Nitric acid (s.g. 1.4)	4.8 oz	30 ml			
	Water (distilled) to	1 gal	11			
Aluminium and other non-ferrous metals	*Bright dip* Phosphoric acid (s.g. 1.69)	8.4 gal	9.41	195	90	Immerse for several min. Agitate work and solution. Good ventilation necessary. Addition of acetic acid useful with some alloys
	Nitric acid (s.g. 1.37)	0.6 gal	0.61			

(*continued*)

Table 32.2 PICKLING SOLUTIONS—*continued*

Metal to be pickled	Composition of solution			Temperature		Remarks
		oz gal^{-1}	g l^{-1}	°F	°C	
Copper and copper alloys	*To remove scale* Sulphuric acid* Water	1 gal 4 gal	11 41	150–170	65–75	After pickling articles can be dipped in sodium cyanide: 4 oz gal^{-1} (25 g l^{-1}) to remove tarnish
	Or Sulphuric acid* Sodium dichromate (Na$_2$Cr$_2$O$_7$.2H$_2$O) Water	1 gal 12 oz 4 gal	11 75 g 41	70–175	20–75	This solution leaves a slight passive film which helps to prevent tarnish
	Bright dip Sulphuric acid* Nitric acid Water Hydrochloric acid	2 gal 1 gal 1 gal 0.5 oz	21 11 11 25 ml	Room	Room	If any scale first dip in spent bright dip. Remove stains by dipping in sodium cyanide 4 oz gal^{-1} (25 g l^{-1})
	Matt dip Sulphuric acid* Nitric acid (s.g. 1.42) Zinc oxide (ZnO)	1 gal 1 gal 2 lb	11 11 200 g	160–180	70–80	If the finish is too fine add nitric acid. If too coarse add sulphuric acid
	Semi-matt dip Sodium dichromate Sulphuric acid* Water	3 oz 18 oz 1 gal	19 g 114 g 11	Room	Room	
Iron and steel	*Slow pickle to loosen heavy scale* Sulphuric acid* Glue	2% 0.25%	— —	Room	Room	Leave for several hours or overnight
	To remove scale Sulphuric acid*	10%	—	120–180	50–80	Or hydrochloric acid 10–20%
	Bright dip Oxalic acid crystals Hydrogen peroxide (100 vol.) Sulphuric acid (10%) Water to	4 oz 2 oz 0.02 oz 1 gal	25 g 13 g 0.1 g 11	Room	Room	This solution has so far only been used on an experimental basis
	Anode etching Sulphuric acid* Water	1 gal 2 gal	11 21	Not above 75	Not above 25	Current density: 200 A ft^{-2} (20 A dm^{-2})
	For polished work Sulphuric acid*	—	—	Not above 75	Not above 25	Density must not fall below 1.61 g cm^{-3} or work will be etched
Magnesium and magnesium alloys	*General cleaner* Chromic acid	16–32	100–200	Up to b.p.	Up to b.p.	For removal of oxide films, corrosion products, etc. Should not be used on oily or painted material

(*continued*)

Table 32.2 PICKLING SOLUTIONS—*continued*

Metal to be pickled		Composition of solution		Temperature		Remarks
		oz gal^{-1}	g l^{-1}	°F	°C	
	Sulphuric acid pickle					
	Sulphuric acid*	3%	—	Room	Room	Should be used on rough castings or heavy sheet only. Removes approx. 0.002 in. in 20–30 s
	Nitro–sulphuric pickle					
	Nitric acid	8%	—	Room	Room	
	Sulphuric acid*	2%	—			
	Bright pickle for wrought products					
	Chromic acid	23	150	Room	Room	Lustrous appearance. Involves metal removal
	Sodium nitrate	4	25			
	Calcium or magnesium fluoride	$\frac{1}{8}$	$\frac{3}{4}$			
	Bright pickle for castings					
	Chromic acid	$37\frac{1}{2}$	235	Room	Room	
	Concentrated nitric acid (70%)	$3\frac{1}{4}$	20			
	Hydrofluoric acid (50%)	1	6.2			
	Acetic acid	8 approx.	50 approx.	Room	Room	Special purpose pickles
	Citric acid	8 approx.	50 approx.	Room	Room	Special purpose pickles
Stainless steel	*To loosen scale*					
	Sulphuric acid*	13–30	80–180	130–160	60–70	Use prior to scale removal treatment, for heavy scales
	Hydrochloric acid (s.g. 1.16)	6–20	40–120			
	To remove scale					
	Nitric acid (s.g. 1.4)	32	200	130–150	55–65	
	Hydrofluoric acid (52% HF)	6	40			
	Or					
	Sulphuric acid*	10	60	Room	Room	
	Hydrofluoric acid	10	60			
	Chromic acid (CrO$_3$)	10	60			
	Bright pickle					
	Hydrochloric acid (s.g. 1.16)	40	250	140–160	60–70	
	Nitric acid (s.g. 1.4)	3	22			
	White matt finish					
	Ferric sulphate [Fe$_2$(SO$_4$)$_3$]	13	80	160–180	70–80	5–15 min
	Hydrofluoric acid (52% HF)	6	40			
Zinc and zinc alloys	*Bright dip*					
	Chromic acid (CrO$_3$)	40	250	Room	Room	5–30 s. If yellow film persists after rinsing dip in sulphuric acid: 1 fl oz per gal (6 ml l^{-1}) and rinse again
	Sodium sulphate (Na$_2$SO$_4$)	3	19			

* Sulphuric acid, pure comcl. grade, s.g. 1.84.

Note: It is almost universal practice to use an inhibitor in the pickling bath. This ensures dissolution of the scale with practically no attack on the metal. Inhibitors are usually of the long chain amine type and often proprietary materials. Examples are Galvene and Stannine made by ICI.

32.3   Anodising and plating processes

Table 32.3   ANODISING PROCESSES FOR ALUMINIUM

Good ventilation above the bath and agitation of the bath is advisable in all cases.

Composition of solution		Temperature		Current density amp ft⁻² (A dm⁻²)	Time and voltage	Cathodes	Vat	Hangers	Remarks
	oz gal⁻¹	g l⁻¹	°F	°C					
Chromic acid (CrO₃), chloride content must not exceed 0.2 g l⁻¹, sulphate less than 0.5 g l⁻¹ (After Bengough-Stuart)	5–16	30–100	103–108	38–42	Current controlled by voltage. Average 3–4 (0.3–0.4) d.c.	†1–10 min 0–40 V increased in steps of 5 V 5–35 min Maintain at 40 V 3–5 min Increase gradually to 50 V 4–5 min Maintain at 50 V	Tank or stainless steel	Steel (exhausted)	Pure aluminium or titanium	Slight agitation is required. This process cannot be used with alloys containing more than 5% copper
Sulphuric acid (s.g. 1.84)	32	200	60–75	15–24	10–20 (1–2) d.c.	12–18 V 20–40 min	Aluminium or lead plates (tank if lead lined)	Lead lined steel	Pure aluminium or titanium	The current must not exceed 0.2 A l⁻¹ of electrolyte
Hard anodising Hardas process Sulphuric acid	32	200	23–41	−5–+5	25–400 (2.5–40) d.c.	40–120 V	Lead	Lead lined steel	Aluminium or titanium	Agitation required. Gives coating 1–3 thou. thick
Eloxal GX process Oxalic acid (COOH)₂.2H₂O	12.8	80	70	20	10–20 (1–2) d.c.	50 V 30–60 min	Vat lining	Lead lined steel	Aluminium or titanium	Oxalic acid processes are more expensive than sulphuric acid anodising; but coatings are thicker and are coloured.
Eloxal WX process Oxalic acid	12.8	80	75–95	25–35	20–30 (2–3) a.c.	20–60 V 40–60 min	Vat lining	Lead lined steel	Aluminium or titanium	
Integral colour Anodising Kalcolor process Sulphuric acid Sulphosalicylic acid	0.8 16	5 100	72	22	30 (3) d.c.	25–60 V 20–45 min	Lead	Lead lined steel	Aluminium or titanium	Aluminium level solution must be maintained between 1.5 and 3 g l⁻¹

† Period according to degree of protection. Complete cycle normally 40 min.

Table 32.4   ANODISING PROCESSES FOR MAGNESIUM ALLOYS

	Composition of solution		Temperature		Current density		Time and voltage	Cathodes	Vat	Hangers	Remarks
	oz gal^{-1}	g l^{-1}	°F	°C	A ft^{-2}	(A dm^{-2})					
HAE process			<95	<35	12–15	(1.2–1.5)	90 min at 85 V approx. a.c. preferred	Mg alloy for a.c. Mg or steel if d.c. used	Mild steel or rubber lined	Mg alloy	Matt hard, brittle corrosion resistant, dark brown 25–50 μm thick, abrasion resistant
Potassium hydroxide	19.2	120									
Aluminium	1.7	10.4									
Potassium fluoride	5.5	34									
Trisodium phosphate	5.5	34									
Potassium manganate	3.2	20									
Dow 17 process			160–180	70–85	5–50	(0.5–5)	10–100 min up to 110 V a.c. or d.c.	Mg alloy for a.c. Mg or steel for d.c.	Mild steel or rubber lined	Mg alloy	Matt dark green, corrosion resistant, 25 μm thick approx., abrasion resistant
Ammonium bifluoride	39	232									
Sodium dichromate	16	100									
Phosphoric acid 85% H_3PO_4	14	88									
Cr-22 process			165–205	75–95	15	(1.5)	12 min 380 V a.c.	—	Mild steel	Mg alloy	Matt dark green, corrosion resistant, 25 μm thick approx.
Chromic acid	4	25									
Hydrofluoric acid (50%)	4	25									
Phosphoric acid H_3PO_4 (85%)	13.5	84									
Ammonia solution	25–30	160–180									
MEL process Fluoride anodise			<86	<30	5–100	(0.5–10)	30 min 120 V a.c. preferred	Mg alloy for a.c. Mg or steel for d.c.	Rubber lined	Mg alloy	Principally a cleaning process to improve corrosive resistance by dissolving or ejecting cathodic particles from the surface
Ammonium bifluoride	16	100									

32.4   Electroplating process

Table 32.5   PLATING PROCESS

Metal	Type and composition	oz gal⁻¹	g l⁻¹	Temperature °F	Temperature °C	Current density A ft⁻³	Current density A dm⁻²	Current efficiency %	Voltage	pH	Anodes	Vat	Remarks
Aluminium	Aluminium chloride	65	400	60–140	15–60	20	2	100	—	—	Aluminium	Glass sealed	Operation must be in an atmosphere of nitrogen and the work introduced through a lock. Connections must not spark
	Lithium aluminium hydride	2	13										
	Diethyl ether	Solvent											
Antimony	Antimony oxide (Sb_2O_3)	7	45	130	55	25	2.5	—	—	3.6	Antimony or carbon	Hard rubber or rubber lined	—
	Potassium citrate	20	130										
	Citric acid	24	150										
Brass	Sodium cyanide (NaCN)	7.5	45	75–100	24–40	3–5	0.3–0.5	60–70 (cathode)	2–3	10.5–11.5	Brass (80/20) cast or rolled	Steel or rubber lined steel	*Brightener:* 2 lb caustic soda in ½ gal of water to which is added 1 lb white arsenic. Use 2–4 fl oz gal⁻¹ solution (15–30 ml per 100 l). Free cyanide by analysis
	Copper cyanide (CuCN)	4	26										
	Zinc cyanide ($Zn(CN)_2$)	1.25	7.7										
	Sodium carbonate (Na_2Co_3)	4	26										
	Ammonia (s.g. 0.88)	0.2*	1.5*										
	Free cyanide	2.6	17										
Bronze imitation	*Zinc*												
	Sodium cyanide	5	33	Room for light colour, warm for red colour		2–4	0.2–0.4	—	2–3	—	Copper 92% Zinc 8%	Steel	—
	Copper cyanide	4	26										
	Zinc cyanide	0.3	2										
	Rochelle salt ($KNaC_4H_4O_6.4H_2O$)	2	13										
	Free cyanide	0.3	2										

Process	Constituents	oz gal⁻¹	g l⁻¹	Temp (°F)						Anode spacing	Anode	Tank lining	Remarks
Cadmium	Sodium cyanide	4.5	29	Room	2–5	0.2–0.5	—	2–3	—	—	Copper	Steel	Cadmium content maintained by addition of small quantities of cadmium oxide dissolved in sodium cyanide
	Copper cyanide	3	20										
	Cadmium oxide	0.25	1.5										
	Sodium carbonate (Na₂CO₃)	2	13										
	Free cyanide	1	6.5										
Bronze	Potassium cyanide	9	60	150	65	Up to 100	Up to 10	40–50	—	12.5	Copper	Steel	Must be kept free of bivalent tin. Maintain tin content by additions of potassium stannate
	Copper cyanide	4	26										
	Potassium stannate (K₂SnO₃)	5	33										
	Potassium hydroxide	1.5	10										
	Rochelle salt	6	40										
	Free cyanide	3	20										
Cadmium	Sodium cyanide	12–15	75–100	75–90	24–30	10–20	1–2	90	2–3	13	Cast cadmium or cadmium balls in a steel cage	—	*Brightener:* Organics (such as dextrin) or metallic (such as nickel salts)
	Cadmium oxide	3–5	20–33										
	Free cyanide	8–10	52–66										
	Addition agents	0.015–2.4	0.1–15										
Chromium	*Bright chrome*												
	Chromic acid (CrO₃)	72	450	95–110	35–45	70–150	7–15	12–15	4–5	—	Tin (7%) lead	Steel lined with antimonial lead 7% or PVC	This solution requires reducing: either boil with citric acid 2 oz gal⁻¹ (12.5 g l⁻¹) or with tartaric acid 3 oz gal⁻¹ (18 g l⁻¹) or with oxalic acid 4 oz gal⁻¹ (25 g l⁻¹)
	Sulphuric acid	0.72	4.5										
	Hard chrome												
	Chromic acid	40	250	120–140	50–60	200–700	20–70	12	5–7	—	Tin (7%) lead	Steel lined with antimonial lead 7% or PVC	Reduction as above: citric acid 1 oz gal⁻¹ (6.25 g l⁻¹) or tartaric acid 1½ oz gal⁻¹ (9 g l⁻¹) or oxalic acid 2 oz gal⁻¹ (12.5 g l⁻¹)
	Sulphuric acid	0.4	2.5										

* 0.2 fl oz gal⁻¹ or 1.5 ml l⁻¹.

(*continued*)

Table 32.5   PLATING PROCESS—*continued*

Metal	Type and composition	oz gal⁻¹	g l⁻¹	Temperature °F	Temperature °C	Current density A ft⁻³	Current density A dm⁻²	Current efficiency %	Voltage	pH	Anodes	Vat	Remarks
Chromium —*continued*	*Black chrome* Chromic acid Fluosilicic acid	33 0.033	220 0.22	90	30	150–450	15–45	—	—	—	Tin (7%) lead	Steel lined with antimonial lead (7%) or PVC	Before use, work solution on scrap plates until 100–150 A h per gal, has been passed
Copper	*Acid* Copper sulphate (CuSo₄.5H₂O) Sulphuric acid Phenol	32 8.0 0.16	200 50 1	60–120	16–50	10–200	1–20	95–97	1–3	—	Pure copper	Lead or rubber lined wood or steel	The phenol is sulphonated by heating with its own weight of sulphuric acid to 120°C for 1 h before use. Agitation is necessary for high current density. Constant filtration is advisable
	Through-hole plating Copper sulphate Sulphuric acid Chloride	14–17 27–30 >15 ppm	88–110 170–190	75–90	24–30	10–45	1–45	—	—	—	Copper	Lead or rubber lined steel	Chloride content serves as a deposit modifier
	Cyanide (strike) Sodium cyanide (NaCN) Copper cyanide (CuCN) Sodium carbonate (Na₂CO₃)	3 2 2	19 13 13	110–140	45–60	10–30	1–3	10–60	6	11–12	Pure copper rolled or extruded	Steel	Used to deposit thin undercoats for other metals
	Cyanide (high efficiency) Sodium cyanide Copper cyanide Sodium hydroxide	13 10 4	82 60 26	140–180	60–80	10–100	1–10	100	2–4	—	Oxide-free copper sheet	Steel	For rapid plating
	Cyanide (Rochelle) Sodium cyanide Copper cyanide Rochelle salt (KNaC₄H₄O₆.4H₂O) Sodium carbonate Free cyanide	6 4 8 5 0.5–1	37.5 26 50 30 3–6	125–160	50–70	20–60	2–6	50–60	6	12.2–12.8	Copper, rolled and annealed	Steel	—

	oz/gal	g/l							pH	Anode		Remarks
Pyrophosphate												
Copper pyrophosphate (Cu$_2$P$_2$O$_7$.3H$_2$O)	11	66	125–140	50–60	10–80	1–8	100	—	8–8.8	Copper	Steel	Commonly used for plating printed circuit boards. Use vigorous agitation
Potassium pyrophosphate	45	300										
Ammonium nitrate	1	6										
Ammonia	0.1	0.6										
Gold												
Hard												
Potassium gold cyanide (KAu(CN)$_2$)	2	12	97	35	5–15	0.5–1.5	—	—	3–4.5	Insoluble	—	—
Citric acid	16	105										
Phosphoric acid	2	12.5 ml										
Cobalt (as CoK$_2$EDTA)	0.16	1										
Alkaline cyanide												
Potassium cyanide	5	30	120–150	50–65	1–5	0.1–0.5	100	1.5–2	11	Fine gold (24-carat) or insoluble: stainless steel, platinum or graphite	Enamelled iron	If insoluble anodes are used, solution must be renewed periodically
Potassium gold cyanide	2	12										
Potassium carbonate	5	30*										
Dipotassium phosphate (K$_2$HPO$_4$)	5	30										
Indium												
(1) Indium fluoborate	38	230	70–90	20–30	50–100	5–10	75	—	1	Part indium, part insoluble	—	Use fluoboric acid to adjust pH
Boric acid	4.8	30										
Ammonium fluoborate	7.5	47										
(2) Indium (as hydroxide)	2.5–5	15–30	70–90	20–30	15–30	1.5–3	50	—	11–12	Steel	—	—
Potassium cyanide	22–25	140–160										
Potassium hydroxide	5–6	30–40										
D-glucose	3–5	20–30										
Iron												
Ferrous chloride (FeCl$_2$.4H$_2$O)	48	300	195	90	Up to 120	Up to 12	—	—	1.2–1.8	Pure iron	Lead or rubber lined	Agitation is desirable for high current densities
Calcium chloride	50	335										

(*continued*)

Table 32.5 PLATING PROCESS—*continued*

Metal	Type and composition	oz gal⁻¹	g l⁻¹	Temperature °F	°C	Current density A ft⁻³	A dm⁻²	Current efficiency %	Voltage	pH	Anodes	Vat	Remarks
Lead	Lead fluoborate (Pb(BF₄)₂)	40	240	77–100	25–40	5–70	0.5–7	100	—	—	Pure lead free from antimony	Rubber lined steel	—
	Fluoboric acid	10	60										
	Boric acid (H₃BO₃) free	4.5	27										
	Glue	0.03	0.2										
Nickel	*Watts bath*												
	Nickel sulphate (NiSO₄.6H₂O)	50	350	110–150	45–65	50	5	95	—	3–4	Cast or rolled Ni (99–100%) bagged	Lead or rubber lined	Agitation desirable for high current densities. Constant filtration desirable. *Wetting agent.* Sod. lauryl sulphate
	Nickel chloride (NiCl₂.6H₂O)	7	45										
	Boric acid	6	37										
	Wetting agent	0.015–0.075	0.1–0.5										
	For plating zinc and zinc-base alloys												
	Nickel sulphate	12–17	75–112	70–90	20–32	10–30	1–3	—	3–4	5.3–5.8	Nickel (99–100%)	Lead or rubber lined	Agitation can be used
	Sodium sulphate (anhydrous)	12–17	75–112										
	Ammonium chloride	2.4–6	15–37.5										
	Boric acid	2.4	15										
	Sulphamate bath												
	Nickel sulphamate (Ni(NH₂SO₃)₂)	48	300	80–140	25–60	20–250	2–25	—	—	3.5–4.2	Nickel (99–100%)	Lead or rubber lined	Air or mechanical agitation
	Boric acid	4.8	30										
	Nickel chloride	1	6										
	Hard nickel												
	Nickel sulphate	28	180	110–140	43–60	20–100	2–10	—	6–8	5.6–5.9	Nickel (99–100%)	—	For building up worn parts
	Ammonium chloride	4	25										
	Boric acid	4.8	30										
	(1) *Bright*												
	Nickel sulphate	50	330	110–150	45–65	25–100	2.5–10	95	—	3–4	Nickel (99–100%)	Rubber lined steel	Bright nickel plating baths are basically Watts solutions containing brighteners
	Nickel chloride	7	45										
	Boric acid	6	38										
	Sodium naphthalene trisulphonate (C₁₀H₇(SO₃)₃Na)	5.6	35										

Process	Constituent										Anode	Tank lining	Remarks
	(2) *Bright* (low metal)												
	Nickel sulphate	9.6	60										
	Nickel chloride	18	110	97–140	35–60	25–100	2.5–10	95	—	3.5–4.2	Nickel	Rubber lined steel	Agitation constant filtration necessary
	Boric acid	8	50										
	Woods nickel strike bath												
	Nickel chloride	38	240	70–80	21–27	30	3	—	—	—	Nickel	Rubber lined or plastic	Used to strike onto metals such as stainless steel or nickel. Pre-etch in bath at 3 A dm^{-2} anodic for 2 min before strike plating for 6 min
	Hydrochloric acid	13	80										
	Black or grey												
	Nickel chloride	12	75										
	Ammonium chloride	4.8	30	Room temperature		1.5	0.15	—	—	5	Nickel or insoluble	Rubber lined	—
	Zinc chloride	4.8	30										
	Sodium thiocyanate (NaCNS)	2.4	15										
Palladium	Palladium (as Pd(NH$_3$)$_4$Br$_2$)	4.8	30	120	50	40	4	—	—	9.2	Insoluble	Glass or rubber lined	Can be used to produce thick deposits for electro forming
	Ammonium bromide	7	45										
Platinum	Platinum (as dinitrodiamino platinum)	1.6	10										
	Ammonium nitrate	16	100	203	95	70	7	10	2–4	—	Platinum or insoluble	Glass or rubber lined	Solution maintained by addition of platinum salt
	Sodium nitrite	1.6	10										
	Ammonia (s.g. 0.88)	7*	44†										
Rhodium	Rhodium (as sulphate concentrate)	0.32	2	104	40	10–40	1–4	—	3–6	—	Platinum or insoluble	Glass or rubber lined	During plating remove bubbles by cathode agitation
	Sulphuric acid	3.2†	20‡										
Silver	Potassium cyanide	8–12	50–78										
	Silver cyanide (80% Ag)	5–9	31–56	70–80	20–27	5–15	0.5–1.5	99–100	<1	—	Fine silver rolled	Lead or rubber lined	Cathode bar may be rocked. *Brightener*: Carbon bisulphide dissolved in silver solution
	Potassium carbonate	2.5–14	15–90										
	Free cyanide	5.5–8	35–50										

(continued)

* 7 fl oz gal^{-1} or 44 ml l^{-1}.   †3.2 fl oz gal^{-1} or 20 ml l^{-1}.

Table 32.5   PLATING PROCESS—*continued*

Metal	Type and composition	oz gal⁻¹	g l⁻¹	Temperature °F	°C	Current density A ft⁻³	A dm⁻²	Current efficiency %	Voltage	pH	Anodes	Vat	Remarks
Silver continued	*High speed*			100–120	38–50	5–100	0.5–10	—	—	12	Pure silver (bagged)	Enamelled iron or rubber lined	Agitation is necessary and is usually effected by solution pumping or cathode movement. Constant filtration advisable
	Silver cyanide (80%)	7–25	44–150										
	Potassium cyanide (92%)	11–38	70–240										
	Caustic soda	0.6–4.8	4–30										
	Potassium carbonate	2.5–14	15–90										
	Potassium nitrate	6.4–9.6	40–60										
	Strike solution for non-ferrous metals			Room temperature		15–20	1.5–2.0	—	4–6	—	Silver or steel	Steel or earthenware	—
	Silver cyanide	0.7	4.5										
	Potassium cyanide	13	80										
Tin	*Acid*			68	20	10–100	1–10	~100	9.4–0.8	—	Pure tin bagged in terylene	Lead or rubber lined	Constant filtration advantageous. Periodic filtration essential. Use agitation at higher current densities
	Sulphuric acid	8	50										
	Cresol sulphonic acid (CH₃.C₆H₃OH.SO₃H)	6.4	40										
	Stannous sulphate (90% SnSO₄)	10	65										
	Gelatin	0.3	2										
	Beta-naphthol	0.16	1										
	Alkaline			167	75	5–30	0.5–3	85	4–6	13	Pure tin (high speed 1% Al) or insoluble	Steel	Anode must be filmed for uniform dissolution
	Caustic soda	1.6	10										
	Sodium stannate (48% SnO₂)	16	100										
	Immersion (on steel)			200–232	90–100	—	—	—	—	—	—	—	Immersion time 5–20 min, work immersed in Monel or stainless steel baskets
	Stannous sulphate	0.16–0.32	1–2										
	Sulphuric acid	0.8–2.5	5–15										

Process	Solution constituents	oz gal⁻¹	g l⁻¹	Temp °F	Temp °C	A ft⁻²	A dm⁻²	%	Volts	pH	Anodes	Tank lining	Remarks
Tin–nickel	Stannous chloride (SnCl$_2$)	8	50	154	68	10–30	1–3	100	2–3	2–2.5	Nickel	Rubber lined	Tin content maintained by regular additions of anhydrous stannous chloride
	Nickel chloride	48	300										
	Ammonium bifluoride (NH$_4$F.HF)	9	56										
Tin–lead	Tin (as fluoborate)	4	25	68–86	20–30	5–35	0.5–3.5	95	6–12	—	60/40 tin/lead alloy	Rubber lined	This process gives a 60/40 tin/lead deposit. Proprietary grain refiners are available to replace peptone
	Lead (as fluoborate)	2	12										
	Fluoboric acid	16	100										
	Peptone	0.8	4.5										
	Boric acid	5	30										
Zinc	*Acid*												
	Zinc chloride (ZnCl$_2$)	10	60	68–100	20–40	5–50	0.5–5	95	3–8	4.5–5.5	Zinc (99.9%)	Lead or rubber lined	Agitation and constant filtration necessary for high current densities
	Potassium chloride	25	150										
	Boric acid	3.7	23										
	Proprietary organic additives	3.2–8*	20–50*										
	Cyanide (decorative)												
	Sodium cyanide	8–22	50–140	68–120	20–50	25–150	2.5–15	~85	—	—	Pure zinc	Steel	—
	Caustic soda	10–20	60–120										
	Zinc oxide (ZnO)	4–9	25–55										
	Sodium carbonate	3.2–20	20–120										
	Cyanide (protective)												
	Sodium cyanide	15–25	90–150	68–120	20–50	25–150	2.5–15	—	—	—	Pure zinc free from lead	Steel	—
	Caustic soda	15–22	90–140										
	Zinc oxide	9–12	55–75										
	Sodium carbonate	5–12	30–75										
	Zincating (on aluminium)												
	Caustic soda	80	500	77	25	—	—	—	—	—	—	—	Improved adhesion by zincating, stripping in 40% HNO$_3$ and rezincating
	Zinc oxide	16	100										

* 3.2–8 fl oz gal⁻¹ or 20–50 ml⁻¹.

32.5   Plating processes for magnesium alloys

DOW PROCESS (H. K. DELONG)

This process depends on the formation of a zinc immersion coat in a bath of the following composition:

	Concentration	
Component	oz gal^{-1}	g l^{-1}
Tetrasodium pyrophosphate	16	120
Zinc sulphate	5.3	40
Potassium fluoride	1.0	7

The treatment time is 3–5 min at a temperature of 175–185°F (80–85°C) with mild agitation. The pH of the bath should be 10–10.4.

The steps of the complete process are:

1. Solvent or vapour degreasing.
2. Hot caustic soda clean or cathodic cleaning in alkaline cleaner.
3. Pickle $\frac{1}{4}$–$1\frac{1}{2}$ min in 1% hydrochloric acid and rinse.
4. Zinc immersion bath as above without drying off from the rinse.
5. Cold rinse and immediately apply copper strike as under.

	Concentration	
Component	oz gal^{-1}	g l^{-1}
Copper cyanide	4.2	26
Potassium cyanide	7.4	46
Potassium carbonate	2.4	15
Potassium hydroxide	1.2	7.5
Potassium fluoride	4.8	30
Free cyanide	1.2	7.5
pH	12.8–13.2	—
Temperature	140°F (60°C)	—

CONDITIONS

30–40 A ft^{-2} (3–4 A dm^{-2}) for $\frac{1}{2}$–1 min, reducing to 15–20 A ft^{-2} (1.5–2 A dm^{-2}) for 5 min or longer.

If required, the copper thickness from the above strike can be built up in the usual alkaline or proprietary bright plating baths. Following the above steps, further plating may be carried out in conventional electroplating baths.

ELECTROLESS PLATING ON MAGNESIUM

Deposits of a compound of nickel and phosphorus can be obtained on magnesium alloy components by direct immersion in baths of suitable compositions. Details of the process may be obtained from the inventors, The Dow Chemical Co. Inc., Midland, Michigan, USA.

'GAS PLATING' OF MAGNESIUM (VAPOUR PLATING)

Deposits of various metals on magnesium components (as on other metals) can be produced by heating the article in an atmosphere of a carbonyl or hydride of the metal in question.

32.6   Electroplating process parameters

Table 32.6   AVERAGE CURRENT EFFICIENCIES OF PLATING SOLUTIONS

The figures given below are approximate

	%
Cadmium (oxide)	85–95
Chromium	12–16
Copper (acid)	95–99
Copper (cyanide)	30–60
Copper (Rochelle)	40–65
Gold	70–85
Indium (cyanide)	30–50
Indium (sulphate)	70–90
Iron	90–95
Lead	90–100
Nickel	94–98
Silver	100
Tin (acid)	90–95
Tin (stannate)	70–85
Rhodium	35–40
Zinc (acid)	97–99
Zinc (cyanide)	85–90

Thickness of metal deposited per hour is given in mils (1 mil = 0.001 in.) by

$$\frac{CD \times W \times CE}{237 \times \Delta}$$

where

$CD =$ current density in A ft^{-2}
$W\ =$ g A^{-1} h (Table 32.7)
$CE =$ current efficiency (Table 32.6)
$\Delta\ =$ density of metal deposited

Table 32.7   THEORETICAL RELATIONS OF METAL AND CURRENT

Metal	*Metal deposited*		*Current required*	
	g A^{-1} h	oz A^{-1} h	A h lb^{-1}	A h kg^{-1}
Aluminium	0.335	0.011 8	1 356	2 989
Antimony (antimonious)	1.515	0.053 4	299	659
Cadmium	2.096	0.073 9	216	476
Chromium (hexavalent)	0.323 5	0.011 4	701	1 545
Cobalt	1.099	0.038 8	413	911
Copper (cuprous)	2.372	0.083 7	191	421
Copper (cupric)	1.186	0.041 8	383	844
Gold (auric)	2.452	0.086 5	185	408
Indium	1.428	0.050 3	318	701
Iron (ferrous)	1.042	0.036 8	435	959
Lead	3.866	0.136 3	117	258
Nickel	1.095	0.038 6	414	913
Palladium	1.990	0.070 2	228	503
Platinum	3.642	0.128 4	125	276
Rhodium	1.920	0.067 7	236	520
Silver	4.025	0.129 4 (Troy)	113 (Avoir.)	350
Tin (stannous)	2.215	0.078 1	205	451
Tin (stannic)	1.108	0.039 1	409	902
Zinc	1.220	0.043 0	372	820

32.7   Miscellaneous coating processes

(1) AUTOCATALYTIC PLATING

Autocatalytic plating is a form of electroless plating in which metal is deposited via a chemical reduction process (as opposed to immersion plating in which thin coatings are formed by electrochemical displacement of the coating metal).

Processes exist for the autocatalytic deposition of a large number of metals, particularly nickel, gold, silver and copper. Basically, the solutions contain a salt of the metal to be deposited and a suitable reducing agent (most commonly hypophosphite, but also hydrazine and boranes etc.). When a metal substrate, which is catalytic to the solution, is introduced into the bath, it becomes covered with a layer of the coating metal which is itself catalytic and thus the process can continue. This mechanism results in an extremely even distribution of deposit on the substrate, i.e. these solutions have a high 'throwing power'.

The most widely used autocatalytic process is nickel–phosphorus; a typical acid bath is as follows:

nickel chloride	4.8 oz gal^{-1}	(30 g l^{-1})
sodium hypophosphite	1.6 oz gal^{-1}	(10 g l^{-1})
sodium glycollate	8.0 oz gal^{-1}	(50 g l^{-1})

The solution is operated at temperatures between 75 and 100°C, at pH 4–6, giving deposition rates up to 0.6 thou. per hour (15 μm h^{-1}). The deposit is an alloy of nickel and phosphorus, containing about 7–10% phosphorus. A useful property of this material is that it can be hardened (typically by heat treating for 1 hour at 400°C) so as to increase the as-deposited hardness from 400 HV to almost 1 000 HV. Thus, autocatalytic nickel coatings find engineering applications, often as a replacement for hard chromium electrodeposits.

(2) COMPOSITE COATINGS

Composite coatings can be electroplated or electroless plated deposits into which a uniform distribution of a second phase material is dispersed. The incorporated material can be hard, ceramic particles for increased wear resistance, as in the Tribomet process carried out by BAJ Ltd. Alternatively, PTFE particles can be used to increase surface lubricity as in the Niflor process from Norman Hay Engineered Surfaces Ltd.

(3) ELECTROSTATIC AND ELECTROPHORETIC METHODS OF PAINT APPLICATION

These are methods which have been developed for the economic application of paint to articles of complicated shapes (and often skeleton structure) in large numbers.

When an article like a metal chair is sprayed by a conventional spray gun procedure, much of the spray overshoots the surfaces and is wasted. In the electrostatic method a very high electrical potential is developed between the gun and the article being painted. The droplets of paint assume a charge of opposite sign to the workpiece and are attracted to it. This ensures more uniform coverage, less overspray, and thus greater economy in operation.

Electrophoresis has been utilised in a somewhat similar way. When a direct current is applied to an aqueous emulsion, large dispersed molecules and even oily particles are caused to move towards one of the electrodes. In the electrophoretic process a paint is provided as an aqueous emulsion and current is applied in such a manner that the globules of paint move towards and attach themselves to the object to be painted. The process can be made automatic and continuous and results in very uniform build-up even on points and sharp edges. It is only suitable for large-scale operations but is very economical in paint.

(4) COATING WITH CERAMIC MATERIALS

Just as metals can be sprayed, certain refractory oxides and silicates and the like can be applied by flame gun. Coating thickness and uniformity can be controlled by suitable means and hard, dense coatings can be built up. The coatings resemble biscuit-ware rather than a vitreous glaze, that is, they are absorbent: their chief use is to provide abrasion resistance and to delay heat transfer.

(5) MECHANICAL PLATING

Mechanical plating is a method of plating which utilises mechanical energy to deposit metal coatings on to metal parts. Parts, glass beads, water, chemicals and metal powder are tumbled together in a barrel at around room temperature to obtain the desired coating.

The process is used primarily to provide ferrous-based parts with sacrificial coatings of zinc, cadmium, and co-deposits with tin. Parts treated by this method are most often fasteners, springs, clips and sintered iron components which are typically handled in bulk.

Parts which have been degreased, descaled, and copper-flashed are tumbled in rubber-lined barrels with water, glass bead impact media, promoter chemicals, and a finely divided powder of the metal to be plated. The promoter chemical serves to clean the metal powder and controls the size of the metal powder agglomerates that are formed. The mechanical energy generated from the barrel's rotation is transmitted through the glass impact media and causes the clean metal powder to be cold welded to the clean metal parts, thereby providing an adherent, metallic coating.

Due to the absence of an impressed current during coating, the process does not produce hydrogen diffusion into the steel substrate. Thus, a post-electroplating bake, in order to preclude hydrogen embrittlement of high strength steel components, is not required for mechanically plated deposits.

32.8   Plating formulae for non-conducting surfaces

(1) METAL POWDERS

The article to be treated may be coated with a metal powder. The best powder for this purpose is a finely ground copper, which is generally sold under the commercial title of bronze powder. This may be applied by mixing it with a cellulose lacquer to which has been added five parts by volume of thinner and spraying it on the object concerned. Alternatively, the object may first be sprayed with lacquer and before it is quite dry may be brushed over with the bronze powder using a soft camel-hair brush. After this treatment the article can be struck over in an acid copper solution.

Waxes (for gramophone records, etc.) may be coated directly by brushing with bronze powder and a soft brush. After brushing with bronze powder they may be treated in the following manner to improve the conductivity and reduce the time of covering in the plating bath:

1. Brush with a soft brush and a 50% mixture of methylated spirit and distilled water.
2. Immerse in a solution containing $30\,g\,l^{-1}$ of sodium cyanide and $6\,g\,l^{-1}$ of silver nitrate.
3. Make up two solutions as follows: (a) Pyrogallic acid $7\,g\,l^{-1}$, citric acid $4\,g\,l^{-1}$ and (b) silver nitrate, $40\,g\,l^{-1}$. Take four parts of solution (a) and one part of solution (b) and mix together and immerse article in this solution for about 10 min.
4. Swill and place in plating bath.

(2) SILVER REDUCTION

A number of articles can be treated by directly reducing silver on the surface. This process is particularly applicable to plastics and glass. Any process which will form a good silver mirror may be used, but the following will be found satisfactory for most purposes.

1. The surface of the article must be very thoroughly cleaned and completely free from grease. Glass and porcelain may be cleaned by using concentrated acid and alkali alternately. Plastics may be treated by brushing or barrelling with a mixture of Vienna lime and pumice, by treating with a suitable solvent or by immersing in a solution of chromic acid.
2. Priming. After cleaning the article is immersed in a 10% stannous chloride solution. Alternately, the solution may be swabbed on to the surface with cotton wool. The article is then thoroughly swilled.
3. The article is then silvered by immersing in a silvering solution, the formula for which is given below. The silvering operation generally takes about 20 min and the temperature must be carefully controlled during this period; usually about 21°C will be found the most satisfactory. The solution should be slightly agitated during the process.
4. The articles are thoroughly swilled and struck over in a copper tartrate bath. After being coated over with copper any desired plating can be made upon it.

The silvering solution is prepared from the following:

> Solution (a) 100 g l^{-1} Rochelle salt.
> Solution (b) 10 g l^{-1} silver nitrate.
> Solution (c) 200 cm^3 per litre ammonia (sp. gr. 0.880).

Take 100 cm^3 of solution (b) and add solution (c) carefully a little at a time until the precipitate which first forms just redissolves. If too much is added and the solution becomes quite clear, add a few drops of solution (b) until a very faint turbidity is produced. Then add 20 cm^3 of solution (a), thoroughly mix and use immediately.

[N.B. The brown precipitate formed by adding ammonia to silver nitrate is explosive if allowed to dry. Care should be taken therefore to see that this does not happen.]

(3) 'VACUUM METALLISING'

This process is carried out at less than 0.000 7 mmHg pressure and can only be applied to objects which are stable under these conditions. The metal to be deposited is heated until it evaporates and there are several ways in which this is achieved. The vapour recondenses on the first cool surface it encounters, and can be made to form a thick dense coating. The 'throwing power' is very poor since the evaporated metal travels in straight lines and steps must be taken to rotate or manipulate objects exposed to it in order to achieve uniform coating. By controlling the temperature of the work piece the crystal structure of the deposit can be varied.

32.9 Methods of stripping electroplated coatings

CADMIUM OR TIN FROM STEEL

Coatings may be stripped from steel by immersing the article in a solution containing 1 gal (4.5 litres) of hydrochloride acid, 2 oz (57 g) of antimony trioxide and $\frac{1}{2}$ pint (280 ml) of water. After stripping and rinsing the article will probably require wiping to remove smuts.

CHROMIUM

Chromium may be stripped from non-ferrous metals by dissolving it in dilute hydrochloric acid. From steel it is best stripped by making it the anode in a solution of sodium hydroxide. If the base metal is zinc or zinc base diecastings, it is best to strip the chromium by making it the anode in sodium carbonate solution as this will not attack the exposed zinc. The conditions of operation are not critical.

COPPER FROM STEEL

Copper can be stripped from steel by immersing in a solution containing 5 lb gal^{-1} (500 g l^{-1}) of chromic acid and 8 oz gal^{-1} (50 g l^{-1}) of sulphuric acid. This solution will work at room temperature but strips the copper very quickly if heated.

Alternatively, the article can be made anodic (2–6 V) in a solution containing 14 oz gal^{-1} (90 g l^{-1}) sodium cyanide and 2.4 oz gal^{-1} (15 g l^{-1}) sodium hydroxide.

COPPER FROM ZINC AND ZINC BASE DIECASTINGS

Copper may be stripped from these materials by immersing in a solution prepared by dissolving 18 g of sulphur and 250 g of sodium sulphide ($Na_2S.9H_2O$) in a litre of solution. The solution works very rapidly if warmed. The sludge formed on the surface of the object will require removing from time to time by brushing or immersion in a 120 g l^{-1} sodium cyanide solution.

COPPER, NICKEL, ETC., FROM MAGNESIUM

Most metal deposits can be removed from plated magnesium by submitting the component to the fluoride anodising process. Alternating current is used in a bath of 10% ammonium bifluoride, in which the plating gradually dissolves without affecting the magnesium. About 4–10 V are used until most of the plating has disappeared. Finally, the voltage is raised to 120 V to complete the process and cleanse the magnesium from remaining traces of foreign metal. Magnesium hangers must be used with firm connections. The bath should be operated cold. Direct current may be used if the work piece is made the anode using mild steel or magnesium cathodes.

LEAD FROM STEEL, COPPER AND BRASS

Immersion of article in a solution of 95 vol. % glacial acetic acid, 5 vol. % hydrogen peroxide (30 wt %). Dilute solutions may be used although this can lead to pitting of steel.

NICKEL FROM COPPER AND BRASS

Nickel may be stripped from copper by an anodic treatment in either 60 vol. % sulphuric acid or 15 g l^{-1} hydrochloric acid. Care must be taken with the concentration of the acids as this can affect the pitting of the substrates.

NICKEL FROM STEEL

Nickel can be anodically stripped in sulphuric acid, as for nickel from copper. Copper sulphate (30 g l^{-1}) or glycerine (30 g l^{-1}) can be added to reduce pitting of the steel. An immersion process involves the use of fuming nitric acid (85–95% HNO$_3$, sp. gr. 1.50) from which water is excluded.

SILVER FROM BRASS

Silver may be stripped from brass by immersing the object in a mixture of 1 vol. conc. nitric acid and 19 vols. conc. sulphuric acid heated to 175°F (80°C). The silver is dissolved in a few minutes. The articles should be immediately removed and swilled.

ZINC FROM STEEL

The reagent used for stripping cadmium may be used. Alternatively zinc may be stripped from steel in either warm dilute hydrochloric or sulphuric acid, or 10–15% ammonium nitrate solution, or hot sodium hydroxide solution.

32.10   Conversion coating processes

(1) PHOSPHATING

Phosphating solutions are used to produce corrosion-resistant coatings on ferrous metals and also zinc, cadmium and aluminium. Probably the most important application for these coatings is to act as bases for subsequent painting operations.

Basically, phosphate solutions comprise metal phosphates dissolved in carefully balanced solutions of phosphoric acid. When a clean metal surface is dipped into the solution, the free acid present reacts with the metal, liberating hydrogen and causing the pH of the solution, adjacent to the metal, to rise. This unbalances the solution, resulting in the precipitation of metal phosphates which form a film, chemically bonded to the substrate.

Due to the complexity of modern phosphate solutions, these processes are normally proprietary, examples of which are listed in section 32.11.

There are four main types of phosphate solution: iron, zinc, heavy zinc and manganese, and these produce increasing weights of coating from 30–90 mg ft^{-2} for iron phosphate solutions to 1 000–4 000 mg ft^{-2} for manganese phosphate solutions.

(2) CHROMATING

Chromating solutions contain hexavalent chromium ions and a mineral acid and are used to increase the corrosion resistance of metals, in particular zinc, cadmium, aluminium and magnesium, by forming a surface layer containing chromium compounds.

The process is usually performed by immersion, although spraying or brushing processes are also used. A wide variety of proprietary solutions are available which produce coatings of different thicknesses. These coatings are often distinguishable by their colour which can vary from clear, to blue, to iridescent yellow and finally to black.

(3) COLOURING OF METALS

The following solutions and operating conditions will produce coloured conversion coatings, as detailed:

(i) Copper and Brass
Black

Copper carbonate	1 lb (454 g)	
Ammonia	2 pt (950 ml)	
Water	5 pt (2.4 l)	

The copper carbonate and ammonia are mixed before adding the water. The solution is operated at 175°F (80°C).

The blue black colour may be fixed by dipping in $2\frac{1}{2}$% sodium hydroxide solution.

Green

Water	1 gal (USA)	(3.8 l)
Sodium thiosulphate	8 oz	(227 g)
Nickel ammonium sulphate	8 oz	(227 g)
or Ferric nitrate	1 oz	(28 g)
Temperature	160°F	(71°C)

Brown

Potassium chlorate	$5\frac{1}{2}$ oz	(154 g)
Nickel sulphate	$2\frac{3}{4}$ oz	(77 g)
Copper sulphate	24 oz	(680 g)
Water	1 gal (USA)	(3.8 l)
Temperature	195–212°F	(90–100°C)

(ii) Iron and steel
Black

Sodium hydroxide	8 lb	(3.6 kg)
Sodium nitrate	$1\frac{1}{2}$ oz	(42 g)
Sodium dichromate	$1\frac{1}{2}$ oz.	(42 g)
Water	1 gal.	(3.8 l)
Temperature	295°F	(146°C)

Blue

Ferric chloride	2 oz	(56 g)
Mercuric nitrate	2 oz	(56 g)
Hydrochloric acid	2 oz	(56 g)
Alcohol	8 oz	(227 g)
Water	8 oz	(227 g)

Room temperature. Parts are immersed for 20 minutes, removed and allowed to stand in air for 12 hours. Repeat and then boil in water for 1 hour. Dry, scratch-brush and oil.

Brown

Copper sulphate	3 oz gal^{-1}	(20 g l^{-1})
Mercuric chloride	0.8 oz gal^{-1}	(5 g l^{-1})
Ferric chloride	5 oz gal^{-1}	(30 g l^{-1})
Nitric acid	25 oz gal^{-1}	(150 g l^{-1})
Alcohol	93 fl oz gal^{-1}	(700 ml l^{-1})

Dip in solution, place in hot box at 175°F (80°C) for 30 minutes, stand in steam box at 150°F (65°C) until coated in red rust, immerse in boiling water to form black oxide, dry and scratch brush. Repeat this operation three times before oiling with linseed oil.

(iii) Stainless steel
Black

Sulphuric acid	180 parts
Water	200 parts
Potassium dichromate	50 parts
Temperature	210°F (99°C)

(iv) Zinc
Black

Ammonium molybdate	4 oz gal^{-1}	(24 g l^{-1})
Ammonia	6 fl oz gal^{-1}	(45 ml l^{-1})

Heat solution to obtain a deep black. Rinse in cold and then hot water; allow to dry and harden.

(v) Aluminium
Black

Potassium permanganate	1.6 oz gal^{-1}	(10 g l^{-1})
Nitric acid	0.5 fl oz gal^{-1}	(4 ml l^{-1})
Copper nitrate	4 oz gal^{-1}	(25 g l^{-1})

Operate at 70°F (24°C) for 10 minutes.
Blue

Ferric chloride	60 oz gal^{-1}	(360 g l^{-1})
Potassium ferricyanide	60 oz gal^{-1}	(360 g l^{-1})
Temperature	150°F	(66°C)

32.11   Glossary of trade names for coating processes

32.11.1   Wet processes

(1) PHOSPHATE PROCESSES

Processes by which a coating of phosphate is produced on the surface of steel or zinc base alloys by treatment in or with a solution of acid phosphates. For rustproofing, the metal must receive a finishing treatment with paint, varnish, lacquer or oil; examples of typical finishing treatments are given under *Parkerising* but it should be understood that firms using or marketing other proprietary phosphate processes may apply different designations or use different media for the necessary finishing treatment.

Bonderising   A proprietary phosphate process applied to steel and zinc, marketed by Ardrox Pyrene Ltd. (similar to Parkerising for steel but produces a thinner and less protective coating; synonymous with Parkerising for zinc alloys).

Coslettising   The original phosphate process for steel, introduced in 1903.

Electro-granodising   A proprietary phosphate process applied to steel, marketed by ICI Ltd. (Paints Division). The chemical action of the solution is assisted by electrolysis.

Granodising   Proprietary phosphate processes applied to steel and zinc marketed by ICI Ltd. (Paints Division).

Lithoform   A proprietary phosphate solution applied to zinc, marketed by ICI Ltd. (Paints Division).

Parkerising   A proprietary phosphate process applied to steel and zinc, marketed by Ardrox Pyrene Ltd. Examples of subsequent finishing treatments are:

P20	Dewatering black finish
P41	Black shellac finish
P75	Oiled finish
P96	Oiled finish
SP55	Mineral oil finish

Walterising   A proprietary phosphate process applied to steel and zinc base alloys, marketed by the Walterisation Co., Ltd.

(2) ALKALINE OXIDATION PROCESSES

Processes by which a black oxide film is formed on steel by treatment in a strongly alkaline solution containing an oxidising agent. For rustproofing, the metal must receive a finishing treatment which is usually carried out with oil.

Black Magic   Blackening processes marketed by M&T Chemicals Ltd.
Blakodising   Black chemical finishes on steel, marketed by Tool Treatments Ltd.
Ebonol   Range of conversion coating processes for both ferrous and non-ferrous metals, marketed by OMI-Imasa (UK) Ltd.
Jetal   An American process marketed by Technic Inc.

(3) CHROMATE PROCESSES

Alocrom   Chromating solutions marketed by ICI Paints Ltd.
Enthox   Chromating solutions marketed by OMI-Imasa (UK) Ltd.
Kenvert   Chromating solutions marketed by the 3M Company Ltd.

(4) ANODIC OXIDATION OF ALUMINIUM AND ITS ALLOYS (ANODISING)

For protection against corrosion and wear, for decoration, for aiding heat emission and for miscellaneous uses based on the absorptive properties of the oxide film when freshly made; used in conjunction with electrolytic 'polishing' for producing reflectors. Processes involving electrolytic treatment in solution, generally of chromic, sulphuric or oxalic acid with the production of a relatively thick film of oxide.

Anobrite   Bright anodising process operated by Anobrite Ltd.
Bengough–Stuart process (chromic acid)   The first anodising process patented in 1923.
Brytal process   A process of producing reflectors introduced by British Alcan Aluminium.
Eloxal   A generic term used in Germany for anodic oxidation.
Phosbrite process   Chemical polish for bright anodising marketed by Albright & Wilson Ltd.

(5) IMMERSION PROCESSES FOR THE TREATMENT OF ALUMINIUM ALLOYS

Alocrom process   A process of priming a thin greenish yellow film in a cold acid solution containing chromates. Marketed by ICI Ltd. (Paints Division).
Alumon   Zincating process marketed by Enthone/Imasa Ltd.
Bondal   Zincating process marketed by W. Canning Ltd.
Decoral   Oxidising process giving electrically conducting coatings capable of being coloured by dyeing. Marketed by Lea Manufacturing Co. Ltd.
MBV process   (Modified Bauer Vogel)   A process of forming a thin oxide film by immersion in an alkaline solution containing chromates.
Pylumin process   A similar process marketed by Ardrox Pyrene Ltd.

(6) NON-ELECTROLYTIC PROCESSES

Enplate   Electroless nickel plating solutions, marketed by Enthone/Imasa Ltd.
Niklad   Electroless nickel plating solutions, marketed by Lea Manufacturing Co. Ltd.
Sylek   Electroless nickel–boron plating solutions, marketed by Imasa Ltd.
Transiflo   Mechanical plating process, marketed by the 3M Company Ltd.

(7) ELECTROPLATING AND ELECTRODEPOSITION PROCESSES

Achrolyte   Tin–cobalt alloy plating process marketed by Udylite/Oxy Metals Ltd.
Alecra 3000   Trivalent chromium plating process marketed by Albright and Wilson Ltd.
Brylanising   Zinc coating of wire by the 'Bethanising' electroplating process of the Bethlehem Steel Co.
Chromonyx   Black chromium plating process, often used for coating solar panels. Marketed by Harshaw Chemicals Ltd.

Fescolising   A term applied by Fescol Ltd., to any electrodeposition process carried out by them (incomplete without mention of the metal referred to, e.g. Fescolising in chromium, etc.). Now carried out by British Metal Treatments Ltd.

Listard process–Van der Horst process   Patented processes of hard chromium plating for protecting the cylinders of internal combustion and other engines from wear. These processes give an oil-retaining surface to the chromium. The processes are operated by British Metal Treatments Ltd.

Niron   Bright nickel–iron plating process, marketed by OMI-Imasa (UK) Ltd.

Rovalising   A term applied by International Corrodeless Ltd., to any protective coating process used or marketed by them. Incomplete without description of the process referred to (e.g. Roval cadmium, etc., *see also* phosphate processes).

Tryposit   A name used by Thomas Try Ltd., to indicate their special process of electrodepositing heavy nickel or hard chromium for engineering purposes.

Zartan   Alloy deposit, used as an alternative to decorative chromium. Marketed by M&T Chemicals Ltd.

32.11.2   Dry processes

(1) THERMAL PROCESSES

The processes described below involve heating of the object to be coated in contact with the coating metal (in the form of powder or as a coating to secure inter-penetration) or with a compound of the coating element. Used for coating steel except where otherwise stated.

Aluminising   A process involving spraying with aluminium and heating to cause alloying.

Bower Barff   A process of coating steel with a black oxide by heating in contact with steam.

Calorising   A process of coating with aluminium, similar to sherardising (*see* below) but using aluminium in place of zinc.

Chromising   A process of coating with chromium involving heating to a high temperature in contact either with a vapour of chromous chloride or with metallic chromium.

Galvanising   A process of coating with zinc by dipping pickled steel into molten zinc (electrogalvanising is sometimes used to mean electroplating with zinc).

Ihrigising   A process of coating with silicon by heating to a high temperature in contact with the vapour of silicon tetrachloride.

Nitriding   A process of forming a hard layer on steel involving heating in a suitable atmosphere (usually ammonia vapour) to form a surface layer rich in nitride.

Sherardising   A process of coating with zinc involving heating in contact with zinc powder in revolving drums (introduced by Sherard Cowper Coles).

(2) METAL SPRAYING PROCESSES

A method of coating consisting of projecting a stream of molten metallic particles at high velocity against the surface to be coated. Mainly for protection against corrosion but also used for restoring the dimensions of undersized parts.

Wire or Schoop process   A process marketed by Metallisation Ltd., and Metallising Equipment Co. Ltd., in which the coating metal in the form of wire is melted and atomised.

Schori or powder process   A process marketed by Schori Metallizing Process Ltd., in which a stream of the powdered coating metal is fed into a flame and blown on to the surface to be coated.

Mellosing or molten metal process   A process in which a molten metal is fed into a jet of heated compressed air which serves to atomise it and to project it against the surface to be coated.

Arc spray process   A process in which the metal feed material is melted by an electric arc and then atomised by a stream of compressed air.

Plasma spray process   A process in which a very high temperature plasma is produced by blowing gas through an electric arc. Metal wire or powder is melted by passage through the plasma and is projected by the gas on to the surface to be coated.

Detonation or D-gun process   Oxygen, acetylene and the material to be plated, are introduced into a detonation chamber where a spark ignites the mixture. A detonation wave travelling at supersonic speed, forces the powdered material heated to $>3\,500°C$, on to the substrate. A special building is required for sound insulation. Very high density coatings of refractory materials like tungsten carbide, can be plated by the process.

33 Welding

33.1 Introduction and cross-reference

Materials joining technologies cover a very wide range of manufacturing processes to assemble products by connecting and uniting component parts. Materials joining processes are generally divided into the following groups: mechanical fastening, adhesive bonding, solid-state welding, brazing and soldering, and fusion welding. Chapters 33 and 34 are concerned only with those processes in which materials are joined through the formation of primary chemical bonds under the combined action of heat and pressure, i.e. welding, brazing and soldering.

In theory, two ideal metallic surfaces that are both perfectly clean and atomically flat will bond together if brought into intimate contact: they will be drawn together spontaneously by the inter-atomic forces until the distance separating them corresponds to the equilibrium interatomic spacing. In practice, surface oxides or other contaminants and surface roughness act to effectively prevent spontaneous bonding. Heat, pressure and/or other engineering measures are therefore required to encourage the formation of a metallurgical joint. There are three major types of materials joining mechanisms: solid-state welding, brazing/soldering, and fusion welding. Solid-state welding is achieved by deformation where no melting occurs; therefore, by definition, solid-state welding occurs at temperatures lower than the melting point of the metals to be joined. Fusion welding and brazing/soldering are achieved by local melting and epitaxial solidification. In brazing/soldering, only the melting of filler metals occurs (in initial stages at least) whereas fusion welding entails transient melting of some base metal in addition to any added filler metal.

A partial list of welding, brazing and soldering processes is given in Table 33.1. Resistance welding, a group of processes utilising localised electric resistance heating, is generally treated separately, although brazing and soldering, as well as the usual fusion welding, are all feasible processing results of local resistance heating. Space limitations preclude the consideration of all welding processes in this chapter, only the major processes being treated, with information on the welding of common materials included in each section.

For techniques related to brazing and soldering, *see* Chapter 34.

33.2 Glossary of welding terms

The 'Glossary of welding terms' is intended to explain the meaning of terms common in welding technology and used throughout this section, and is not necessarily confined to terminology defined in standards. A list of major standards relating to welding is given in Section 33.7.

Arc blow In arc welding, the deflection of the arc by magnetic forces induced by the welding current or residual magnetism in the workpiece.
Arc welding A fusion welding process wherein the source of heat is an electric arc.
Argon arc welding A fusion welding process wherein the source of heat is an electric arc struck between an argon shielded non-consumable tungsten electrode and the work. The filler wire, if required, is added separately.
Atomic hydrogen welding A fusion welding process wherein hydrogen is dissociated in an arc struck between two tungsten electrodes afterwards recombining to supply the welding heat. The filler wire is added separately.
Autogenous welding Fusion welding without a filler metal addition, in which the weld metal is provided by melting of the parent material.

Table 33.1   MAJOR WELDING, BRAZING AND SOLDERING PROCESSES

Solid-state group
  1. Pressure welding
     a. Cold (pressure) welding
     b. Hot pressure welding
  2. Friction welding
     a. Linear friction welding
     b. Radial friction welding
     c. Rotary friction welding
     d. Friction stir welding
  3. Ultrasonic welding
  4. Explosive welding
  5. Diffusion welding

Liquid-state group
  1. Fusion welding
     a. Oxyfuel gas welding
     b. Electroslag welding
     c. Thermit welding
     d. Arc welding
        – Manual metal arc welding
        – Metal inert gas welding (MIG)
        – Metal active gas welding (MAG)
        – Tungsten inert gas welding (TIG)
        – Submerged arc welding (SAW)
        – Plasma arc welding
     e. High-energy-density beam welding
        – Laser beam welding
        – Electron beam welding
  2. Resistance welding
     a. Resistance spot welding
     b. Resistance seam welding
     c. Projection welding
  3. Brazing/soldering
     a. Furnace brazing/soldering
     b. Resistance brazing/soldering
     c. Laser brazing/soldering
     d. Diffusion brazing/soldering
     e. Wave soldering

Backward welding   A fusion welding technique in which the source of heat is directed towards the already deposited weld.

Backing bar   A bar of material used for backing up a joint during welding to control penetration at the root and not contiguous with the weld.

Backing strip   A strip of metal used for backing up a joint during welding and which may or may not be left on after welding.

Backing run   See *Sealing run.*

Backstep welding   Welding in which increments of a run are deposited in a direction opposite to the general direction of welding.

Bead-on-plate weld   A single run of weld metal deposited on an unbroken surface.

Braze welding   A joining process whereby a brazing type of filler is deposited in a prepared joint using a *gas* welding technique.

Bronze welding   A joining process whereby a brazing type of filler is deposited in a prepared joint using an *arc* welding technique.

Butt weld   A weld between two members lying approximately in the same plane and not overlapping.

Carbon arc welding   A fusion welding process wherein the source of heat is an arc struck between the work and a non-consumable carbon electrode. A filler wire may or may not be used. A shielding gas may also be employed.

Carbon dioxide welding   A carbon dioxide shielded metal arc welding process using a continuous consumable bare wire electrode. (See also *MAG.*)

Chain intermittent fillet welding   Two lines of intermittent fillet welding on either side of a joint wherein the fillet welds on one side are opposite those on the other side.

Cold welding   Pressure welding at room temperature.

Constant potential (or voltage) *power source*   A power source, the output voltage/ampere characteristic of which is substantially parallel to the current axis. Ideal for self-adjusting arc
welding.

Controlled arc welding   Metal arc welding in which the arc length is kept constant by controlling,
from the arc voltage, the rate of feed of the consumable electrode.

Controlled tungsten arc welding   Tungsten arc welding in which the arc length is controlled from
the arc voltage.

Cored electrode (or cored wire)   A consumable electrode in tubular form having a core of flux
or other materials.

Corner weld   An outside weld between two members approximately at right angles forming an L.

Cover glass   A clear glass, sometimes gelatin coated, used in welders' helmets and hand shields
to protect the filter glass from spatter and fumes.

Crater   A depression left in the weld metal at the termination of a run.

Diffusion bonding   See *Diffusion brazing/soldering and diffusion welding.*

Diffusion brazing/soldering   A brazing or soldering process in which a thin layer of liquid forms
by melting of a filler metal or formation of an in-situ liquid phase. The liquid film wets the
faying surfaces and then solidifies isothermally at the bonding temperature. It is also known
as liquid phase diffusion bonding or transient liquid phase (TLP) bonding.

Diffusion welding   A solid-state welding process wherein the component parts are held together
under a pressure too low to cause significant macroscopic plastic deformation (localised
plastic deformation does occur at asperities on the faying surfaces), generally at an elevated
temperature. The resultant atomic diffusion (and other sintering type mechanisms) causes
bonding.

Dip transfer   A mode of particle transfer in gas-shielded metal arc welding wherein use is made
of controlled short circuits between electrode and pool.

Downhand weld   See *Flat position weld.*

Drooping characteristic power source   A power source, the output voltage of which falls as current demand increases. Necessary for manual metal arc welding and controlled-arc welding.

Edge weld   A weld between the edges of two or more parallel and faying members.

Electrogas welding   A vertical butt welding process using a gas-shielded consumable electrode
to deposit metal into a molten pool held in place by moving dams which move upwards as
the joint is made.

Electron beam welding   A fusion welding process employing a high voltage focussed electron
beam to supply energy.

Electroslag welding   A vertical butt welding process in which the filler metal electrode is melted
in a rising bath of conducting slag held in position by moving dams.

Explosive welding   A pressure welding process employing the energy from the controlled
explosion of sheet explosive to effect joining.

Face of weld   The exposed surface of a fusion weld on the side from which welding was
carried out.

Faying surface   That surface of a member which is in contact with another member to which it
is to be joined.

Filler wire   Metal to be added in making a weld, usually in the form of a bare or flux-coated wire.

Fillet weld   A weld of approximately triangular cross-section between two members approximately at right angles to each other.

Filter glass   A dark coloured glass used to cut down the radiation from a fusion welding process
to assist vision and protect the eyes of the welding operator.

Flash butt welding   A resistance welding process wherein coalescence is produced by the heat
obtained from resistance to the flow of a heavy electric current between two lightly abutting
surfaces and by the application of pressure upon attainment of welding temperature.

Flat position weld   A fusion weld in which the weld face is approximately horizontal and
uppermost.

Flux cored arc welding (*FCAW*)   An arc welding process that uses an arc between a continuous
consumable *cored electrode* and the weld pool.

Forge welding   A group of solid-state welding processes wherein coalescence is produced by
applying pressure or blows to material rendered plastic by heating.

Forward welding   A fusion welding technique in which the source of heat is directed ahead of
the already deposited weld.

Friction stir welding   A variation of friction welding in which a weld is made between two butting
workpieces by the frictional heating and plastic material displacement caused by a high speed
rotating tool that is slowly plunged into and then traversed along the joint line.

Friction welding   A hot pressure welding process in which frictional heat is produced by rotating one component of the joint against a stationary mating surface under slight pressure. An upset force is applied when rotation is stopped. Orbital or oscillating movement may be employed rather than rotation. (See also *Friction stir welding*.)

Fusion boundary   The boundary between weld metal and parent metal in a fusion weld.

Fusion welding   All welding processes in which the weld is made by fusion of the parent metal by means of externally applied heat but without hammering or pressure. Filler wire may or may not be added.

Gap   The minimum distance at any cross-section between edges, ends or surfaces to be joined.

Gas metal arc welding (GMAW)   An arc welding process employing an arc between a continuous consumable electrode and the weld pool shielded by an externally supplied gas. The usual North American terminology for *MIG* and *MAG* welding.

Gas tungsten arc welding (GTAW)   The usual North American terminology for *TIG* welding.

Gas welding   A fusion welding process in which the source of heat is the combustion of gas.

Hammer welding   A forge welding process in which the welding pressure is obtained by means of hammer blows.

Heat affected zone   The parent material alongside a weld which is heated during the welding operation.

High frequency induction welding   A resistance welding process wherein coalescence is produced by the application of pressure to edges heated by the skin effect of high frequency alternating current.

High frequency injection   The superimposing of a high frequency voltage on the alternating current used in argon-arc welding to ensure that the arc will restrike at the beginning of electrode positive half-cycles.

Horizontal vertical welding   A weld made in such a position that the longitudinal axis of the weld is approximately horizontal and the plane of the weld face is approximately vertical.

Inert gas metal arc welding   A group of fusion welding processes in which the source of heat is an electric arc struck between the work and an inert gas shielded, continuous, consumable electrode which acts as filler metal.

Interpass temperature   In a multiple pass weld, the lowest temperature of the deposited weld metal before the next pass is started.

Lap weld   A fillet weld between two overlapping members.

Laser welding   A fusion welding process employing the energy from the excitation of an optical laser.

Leg of a fillet weld   The distance from the root of the joint to the toe of a fillet weld.

MAG (metal active gas) welding   Gas-shielded metal-arc welding with a consumable wire electrode, shrouded by active or non-inert gas.

Manual metal arc welding   Metal arc welding using a flux-coated wire or rod electrode under manual control.

Metal arc welding   An arc welding process in which the metal electrode forms the filler metal.

MIG (metal inert gas) welding   See *Inert gas metal arc welding*.

Nugget   The fusion zone of a spot, seam or projection weld.

Overhead position weld   A weld made on a surface lying horizontally or at an angle not more than 45° to the horizontal, the weld being made from the underside of the parts joined. Alternatively a weld carried out on the underside of a joint.

Overlap   Protrusion of weld metal beyond the bond at the toe of the weld.

Penetration   The maximum depth a weld extends from its face into a joint, exclusive of reinforcement. In resistance welding, the distance from the interface to the edge of the weld nugget measured on a cross-section through the centre of the weld and normal to the surface.

Plasma arc   An arc which is constricted mechanically or magnetically to produce a high heat concentration over a small area.

Plug weld   A fusion weld made in a hole formed in one of the parts of a lapped joint to attach the other part.

Positional welding   Welding carried out in all positions other than the normal flat position.

Pressure welding   A process of joining metals in which the surfaces to be joined are brought into close contact by pressure either cold or heated below the melting point.

Projection welding   A resistance welding process in which current and pressure concentration for making a weld is achieved by means of small projections usually raised on one of the workpieces.

Pulsed welding   Arc welding, usually gas shielded, in which one or more welding variables, but most commonly current, are periodically varied between two levels.

Reinforcement   Weld metal in a butt weld lying outside the plane joining the toes of the weld.

Resistance butt welding   A resistance welding process in which the parts to be joined are butted together under pressure while current is allowed to flow until a predetermined temperature is attained and metal at the interface is upset and a weld produced.

Resistance welding   A generic term covering those welding processes in which the welding heat is produced by the electrical resistance of the weldment and interfacial contact resistance during passage of the welding current.

Reverse polarity   In direct current welding, the arrangement of current supply wherein the work is made the negative pole and the electrode the positive pole of the welding circuit. The term is misleading, and it is preferable to refer to polarity.

Roller spot welding   A spot welding process using a machine similar to that used in seam welding. Pressure is applied continuously and current intermittently.

Root   (a) The zone in the preparation for V, U, J and bevel butt welds, in the neighbourhood of and including the gap.

(b) The zone between the prepared edges adjacent to a backing strip in an open square butt weld.

(c) The zone, in parts to be fillet welded, in the neighbourhood of the actual or projected intersection of the fusion faces.

Root face   That portion of the groove face adjacent to the root and normal to the face of the weld.

Root gap   The separation between the members to be joined at the root of a joint.

Run-on and run-off tags   Small pieces of metal tacked to a weldment at the beginning and end of welds to facilitate avoidance of undercutting at free edges and to remove the start and stop positions from the components being welded.

Sealing run   A weld bead laid along the back of a groove weld.

Seam welding   A resistance welding process in which a continuous weld is produced in over-lapped sheets by means of two electrode wheels or between an electrode wheel and an electrode bar. The electrode wheels provide continuous pressure and current flow is intermittent with accurately timed periods.

Self-adjusting arc welding   Metal arc welding in which the consumable electrode is fed into the arc at a constant speed with a high current density, any alteration in arc length being corrected by naturally occurring changes in the burn-off rate.

Series welding   The making of two spot or seam welds or two or more projection welds simultaneously with electrodes forming a series circuit.

Shielded metal arc welding (*SMAW*)   North American terminology for *Manual metal arc welding*.

Size of weld   (a) Butt weld—The joint penetration.

(b) Fillet weld—The leg length of the largest isosceles right-angled triangle which can be inscribed within the fillet weld cross-section.

Spot welding   A resistance welding process in which overlapping parts are welded at one or more spots by means of shaped electrodes which give a high current density at the welding point and maintain mechanical pressure on the weld during and after current flow.

Spray transfer   A transfer mode in gas-shielded metal arc welding, wherein metal transfers from the electrode to the weld pool as a stream of droplets.

Staggered intermittent fillet welding   Two lines of intermittent fillet welding on either side of a joint wherein the fillet welds on one side are staggered relative to the fillet welds on the other side.

Stitch welding   A spot welding process in which the welds overlap.

Straight polarity   In direct current welding, the arrangement of current supply wherein the work is made the positive pole and the electrode the negative pole of the welding circuit. The term is misleading, and it is better to refer to actual polarity.

Stringer bead   A weld bead deposited without appreciable transverse oscillation of the electrode.

Strip cladding   A method of surfacing components by automatic submerged arc or gas-shielded arc welding, using a consumable electrode in the form of a strip rather than wire.

Submerged arc welding   Arc welding in which a bare wire consumable electrode is used, the arc being enveloped in a powdered flux, some of which fuses to form a protective slag covering the weld.

Surge injection   A means of arc re-ignition whereby a timed uni-directional medium voltage surge is applied across the arc gap. Used in a.c. argon-arc welding.

Synergic welding   MIG or MAG welding system in which arc parameters are automatically set and controlled, normally for a specific wire feed speed.

Tack weld   A short weld used in assembly of parts and for preventing distortion during welding.

Thermit welding   A fusion welding process where the heat for fusion is obtained from liquid steel resulting from a thermit reaction, the steel so produced being used as the added metal.

Throat of a fillet weld (a) Theoretical. The distance from the beginning of the root of the joint perpendicular to the hypotenuse of the largest right-angled triangle that can be inscribed within the fillet weld cross-section.

(b) Actual. The shortest distance from the root of a fillet weld to its face.

TIG (tungsten inert gas) welding Arc welding using a non-consumable tungsten electrode, surrounded by an inert gas shield. (See *Argon arc welding.*)

Toe of weld The junction between the face of a weld and the parent metal.

Touch welding A metal arc welding technique employing flux coated electrodes of special type, whereby the tip of the rod is rested on the parent metal during welding.

Undercut A groove melted into the parent metal at the toe of a weld and left unfilled by weld metal.

Vertical position weld A weld made in such a position that the longitudinal axis of the weld is approximately vertical.

Weaving The deposition of a weld bead with transverse oscillation of the electrode.

Weldment An assembly whose component parts are joined by welding.

Weld metal That portion of metal which has been molten during welding.

33.3 Resistance welding

There are four main types of resistance welding process: spot welding, projection welding, resistance butt welding and flash butt welding. Seam, stitch and roller spot welding are closely related to spot welding.

Spot welding finds application in the lap joining of sheet material in the range of thickness from 2×0.3 mm to 2×4 mm and occasionally thicker materials. Stitch and seam welding are used for the manufacture of pressure-tight seams, stitch welding being more suitable for irregularly shaped components and seam welding for long, straight runs or curves with regular or generous radii. Multiple pressure and heating cycles are possible in stitch welding and roller spot welding, but not in seam welding. Roller spot welding is used for producing long, straight rows of spots at higher production rates than are possible in spot welding.

Projection welding may also be used for lap joining of sheets provided the material has sufficient ductility for the production of suitable projections and sufficient strength in the projections to withstand the high loads employed. This process is also used for the production of T-joints and the attachment of studs, nuts, collars and discs to sheet materials. Cross-wire welding is regarded as a form of projection welding.

Resistance butt welding is mainly used for the butt joining of wire and light gauge rods. Heating is more widespread, power consumption is higher and the condition of the end faces is more important than in *flash butt welding.* The latter process is used for butt joining of heavier and more complex sections and bevel joining.

33.3.1 The influence of metallurgical properties on resistance weldability

The properties of most importance which influence weldability, and therefore the selection of suitable welding conditions, are conductivity, expansion characteristics, nature and condition of the surface film, high temperature strength and the structure of the fusion and or heat affected zone.

High conductivity materials require more current than mild steels to give an equivalent heat input for two reasons, first, due to the increased electrical conductivity and secondly, due to the increased conduction of heat away from the weld area. In practice, high conductivity materials such as aluminium are welded for short times with exceptionally high currents. There is, therefore, a need for more complex and expensive equipment for welding these materials than would be required for mild steel. Splashing or boiling of the weld metal may occur in low conductivity materials, and short times and high currents are used for such cases also.

Coefficient of thermal expansion and shrinkage contraction on solidification are important factors in governing the type of mechanical system used in welding machines. In materials of high thermal expansion and of high thermal conductivity, where shrinkage takes place relatively quickly, it is necessary to provide a high electrode force immediately following the current pulse, in order to avoid shrinkage cavities and cracks.

Consistency of surface condition is extremely important in most resistance welding processes (flash butt welding is an exception), particularly in materials with thick natural oxide films, such as aluminium and magnesium alloys. Cleaning procedures are invariably recommended in such cases.

Thin and uniform oxide films, such as those obtained on pickled stainless steels and Nimonic alloys, give consistent welding behaviour, particularly with high electrode forces. Rust, paint and grease influence consistency, and should be removed. Metallic coatings may interfere with weldability seriously, although consistency may suffer slightly and electrode tip life in spot welding may be reduced.

Materials which have high strength when hot require high electrode forces in order to maintain a seal around the molten weld metal.

It is important that the metallurgical structure of both weld and heat affected zone should be adequate to withstand the demands made upon the joint in service. The heat affected zone in many alloys is softened, and it is therefore important to limit its extent by the use of short times. Ferritic steels suffer from hardening, which increases with carbon and alloy content. In serious cases this may necessitate some form of post-weld heat treatment to overcome the resulting brittleness.

33.3.2   The resistance welding of various metals and alloys

STEELS

Low carbon steels may be readily resistance welded by all processes, clean deep drawing steel being commonly regarded as excellent in this respect (Table 33.2). A guide to the maximum carbon content which can be tolerated in spot and projection welding without excessive hardening is given by the formula

$$C_{\max} = 0.1 + 0.012t$$

where $t =$ thickness in millimetres of the thinnest sheet in the combination.

Spot and projection welds in medium carbon and low alloy steels may be made, and it is desirable that a high electrode force be applied after the welding current pulse to prevent cracking followed by a tempering treatment which is most conveniently carried out in the welding machine. The electrode force may be maintained at a high level during tempering or reduced to its original value. Hardening is not such a serious problem in resistance butt or flash welding, since cooling rates are lower; however, butt welding is usually confined to steels that do not form refractory oxides.

Table 33.2   THE RESISTANCE WELDING OF METALS AND ALLOYS—SUITABILITY OF PROCESSES

	Process			
Material	*Spot*	*Projection*	*Resistance butt*	*Flash butt*
Low carbon steel	S	S	S	S
Low alloy steel / Medium carbon steel	P	P	S	S
Austenitic stainless steel	S	S	N	S
Aluminium and low-strength alloys	P	N	S	S
Medium and high-strength aluminium alloys	S	N	P	S
Copper	N	N	S	S
Beryllium–copper	S	—	S	S
Gilding metals	N	N	S	N
70/30 and 60/40 brass	S	—	S	N
Tin bronze	S	S	S	S
Aluminium bronze	P	P	P	P
Silicon bronze	S	—	S	—
Cupronickel	S	—	S	—
Nickel–silver	S	—	S	—
Magnesium alloys	P	N	N	S
Nickel and its alloys	S	S	N	S
Titanium	S	P	N	S
Zirconium	S	S	N	S
Tantalum	S	—	—	—
Niobium	S	—	—	—
Molybdenum	P	—	N	S
Tungsten	P	—	N	S

S = Suitable.   N = Not recommended.   P = Possible under certain conditions.   — = Information not available.

Table 33.3   RELATIVE CONDITIONS FOR SPOT WELDING MILD STEEL AND OTHER MATERIALS

Material	Electrode force	Weld time	Welding current	Remarks
Mild steel	A-70 MPa on tip area	Short B	High C	
Low alloy steels	A × 1	B × 1	C × 1	Post-heating needed
18/8 stainless steels	A × 3	B × 1	C × 0.9	Surface oxides may cause
Nimonic alloys	A × 4	B × 1	C × 0.9	trouble in rare cases
Nickel	A × 2.5	B × 1	C × 1.5	
Monel, Inconel	A × 4	B × 1	C × 0.9	
Aluminium alloys	A × 2	B × 0.5	C × 4	Surface oxide must be removed
Cupronickel, silicon bronze, 70/30 brass, nickel–silver, etc.	A × 0.8	B × 0.8	C × 1.2–2.0	Current depends on conductivity of alloy
Titanium and its alloys	A × 2	B × 1	C × 1	Avoid excessive penetration

Taken from 'Resistance Welding', published by the former British Welding Research Association.

Austenitic stainless steels require somewhat lower currents and higher electrode forces in spot and projection welding. In flash butt welding these materials require similar currents to mild steel, but with higher open-circuit voltages and upset forces.

Relative conditions for spot welding mild steel and other materials are summarised in Table 33.3.

ALUMINIUM AND ALUMINIUM ALLOYS

The resistance weldability of aluminium alloys is governed mainly by their relatively high conductivity and the presence of a tenacious high resistance oxide film. Care is therefore required in surface preparation to produce a consistent contact resistance, and this is normally done by controlled scratch brushing or chemical dipping. The use of paste fluxes is not recommended, due to inconsistent quality. Degreasing alone is occasionally practised in applications where consistency is less important. Cleaning of outer surfaces improves tip life in spot welding.

The need for high currents and short weld times has already been mentioned, and comparative figures are given in Table 33.3.

In this group, pure aluminium is the most difficult material to spot weld, and probably the easiest to resistance butt weld. Strong aluminium alloys clad with pure aluminium do not give the same trouble in spot welding due to their lower conductivity and higher resistance to indentation. Generally, spot welding difficulties increase with decreasing parent metal strength and increasing conductivity. All the common sheet materials may be spot welded.

There is little experience with projection welding, the main difficulty being the relatively low strength in compression of the projection. Resistance butt welding is used only for pure aluminium and low strength alloys in small sections, and care with surface preparation is again necessary. Flash butt welding can be used for most aluminium alloys, provided rapid heating and careful control in the time and speed of application of the upset force is applied.

COPPER AND COPPER ALLOYS

Due to its high conductivity, copper is not normally regarded as weldable by the spot and projection welding processes, although very thin copper may be spot welded with molybdenum tipped electrodes using very high currents and short times. The alternative of interposing shim of relatively high resistance brazing filler metal is often employed, giving a resistance brazed joint rather than a weld. Pure copper may be satisfactorily resistance butt and flash butt welded, and both processes are used in the wire industry.

Cadmium–copper is difficult to resistance weld, and resistance brazing is preferred, but beryllium–copper may be readily spot or butt welded, provided the normal precautions for high conductivity materials are taken.

Brasses increase in weldability with increasing zinc content, and it is difficult to spot weld gilding metals, though 70/30 and particularly 60/40 brass may be welded satisfactorily by this process. Resistance butt welding is frequently used for brass wire, but flash butt welding is not suitable, due to zinc volatilisation. Information on current, weld time and upset force compared with mild steel is given in Table 33.3. Certain brasses should be rendered immune from season cracking by a low temperature annealing treatment after welding.

Tin bronzes have relatively low conductivity, and may be readily joined by practically all resistance welding methods. Phosphor bronzes may, however, tend to stick to copper alloy electrodes in spot welding, and plating of the electrodes may be required.

Information on aluminium bronze is meagre, but satisfactory resistance welds of all types have been produced in single phase alloys. Flash butt welding has been used for complex alloys.

Silicon bronzes have low conductivity, and may be satisfactorily spot and resistance butt welded. Weld times and electrode forces are rather lower than for mild steel and currents somewhat higher. It is important that rapid follow up of the electrodes during heating should take place.

Cupronickels and nickel-silvers behave in a similar manner to silicon bronze.

MAGNESIUM AND MAGNESIUM ALLOYS

Magnesium and its alloys may be satisfactorily spot welded for low duty service using equipment similar to that used for aluminium alloys. Reference has been made to the satisfactory weldability of Mg–Mn, Mg–Al–Zn, Mg–Zn–Zr and Mg–Th–Zr alloys. Cleaning is important, chemical methods being preferred to mechanical methods, although good results can be obtained by brushing with wire wool. Electrode cleanliness is important if copper contamination of the work-piece is to be avoided. High currents and short weld times are used together with a rapid follow up of the electrodes to give a force somewhat lower than that used for aluminium. Radiused tip electrodes help to form a pressure seal around the weld.

Flash butt welding may also be used for magnesium alloys.

NICKEL AND NICKEL ALLOYS

Nickel, Monel,* Inconel* and the Nimonic* alloys may be satisfactorily spot, projection and flash butt welded.

Conditions for spot welding Nimonic alloys are similar to those for austenitic stainless steel but with somewhat higher electrode forces. Nickel requires higher currents and lower forces although forces are still much higher than for mild steel. Monel and Inconel require currents and forces similar to those recommended for Nimonic alloys. Comparative figures are given in Table 33.3. Cleanliness is important, sulphur and lead contamination being particularly dangerous. Sticking which may occur when spot welding annealed nickel, may be overcome by silver plating of the electrodes. Cracking may be encountered with some precipitation hardening alloys if electrode pressure is too low. Increased weld time and pressure are beneficial.

Flash butt welding is satisfactory if clamping is firm to prevent arcing, clamping distances are kept short because of the low conductivity and upset forces are high. Open-circuit voltages require to be higher than used for mild steel.

REFRACTORY METALS

Among this group of materials are included titanium, zirconium, tantalum, niobium, molybdenum and tungsten. The problem of contamination by reaction with the atmosphere is not as serious in spot welding as in fusion welding, but the quality of projection welds may be improved by the use of an inert atmosphere.

Titanium may be readily spot or flash butt welded. Conditions for spot welding are similar to mild steel but with higher electrode forces. The high electrical resistance tends to give a large weld with high penetration, and seam welding may need to be done under water to prevent contamination after welding. Flash welding conditions are similar to those used for aluminium alloys. Projection welds have also been made in titanium.

Zirconium in thin gauges may be spot welded using conditions similar to those used for stainless steel. No special cleaning other than degreasing is normally necessary. Projection welding and flash butt welding are also possible.

Tantalum may be spot welded in thin gauges, and welding may be carried out in air if the time is less than one cycle. Larger times require water cooling of the weld area. Surfaces should be degreased and pickled in a sulphuric/chromic acid mixture. Niobium should be treated in a similar way.

---

* Henry Wiggin & Co. Ltd.

Resistance welds in molybdenum and tungsten are often inherently brittle due to the high brittle/ductile transition temperature of the recrystallised and as cast materials. Other problems are contamination of surfaces with electrode materials and contamination of the electrodes themselves. These troubles are minimised by welding under water. Surfaces should be thoroughly cleaned, if possible by grit blasting. Both materials may be resistance brazed readily, using shims of tantalum, zirconium or nickel. Both materials have been satisfactorily flash butt welded.

DISSIMILAR METALS

It is normally easier to make satisfactory joints between dissimilar metals by resistance welding or resistance brazing than by fusion welding or normal brazing techniques, since the problem of fluxing does not arise and techniques may be chosen to minimise the danger of brittle intermetallic phases within the joint. Copper and aluminium, for example, form a series of brittle phases when melted together, but flash butt welding of copper to aluminium is widely practised, since these phases are forced out of the joint when the upset force is applied.

When spot welding dissimilar metals, it may be necessary to use electrodes of differing conductivity against the different parts, i.e. a high conductivity electrode against the lower conductivity material.

Some indication of the spot welding characteristics of dissimilar metals is given in Table 33.4. In some cases where spot welding is not possible due to excessive formation of intermetallics, it is possible to interpose a layer of a third material compatible with both the parts requiring to be joined.

33.4   Solid-state welding

33.4.1   Friction welding

The most common method of generating frictional heat for welding is to rotate one component relative to the other under an axial load. When the metal temperature at the interface is sufficiently high, the rotation is stopped, and the workpieces are forced together under the original or higher load. Inertia welding is a variant of the process, using the kinetic energy of a flywheel to provide rotation and thermal energy. In friction welding, one component usually has a circular cross-section, but this is not always essential, particularly when orbital rather than rotational differential movement of the workpieces is used.

In radial friction welding, two lengths of tube are aligned axially with a small gap, and a collar of tapered cross-section is rotated between them to generate heat at both interfaces simultaneously. The tubes do not rotate but are forced together during the forging stage of the operation. For linear friction welding, the parts to be joined are oscillated along one axis parallel to the faying surfaces, and are aligned as required immediately prior to forging.

The principal welding variables are the welding speed, applied heating pressure, forge pressure, and heating duration. For any particular joint configuration, selection of optimum welding conditions is primarily dependent on material strength and thermal conductivity characteristics, although with dissimilar metal joints, it may be necessary to use conditions minimising the formation of brittle phases at the interface.

Welding speed is not generally critical, peripheral velocities of between 0.3 and 7 m s^{-1} being common for a range of materials. Higher welding speeds increase the width of the heat-affected zone (HAZ) and welding time.

Both heating and forge pressures must be selected to achieve uniform heating and maintain the faying surfaces in intimate contact. Typical values for mild steel are 45 MPa and 75 MPa for heating and forging, respectively. With mild steel, doubling the heating pressure increases the power required by 50%. Excessive pressure is to be avoided with alloy steels, since the steep temperature gradient may lead to the formation of hard structures, with reduced ductility. Higher pressures are required for materials of greater hot strength. The heating pressure is particularly important in determining both the torque required to make the weld, and the temperature gradient at the interface. With high conductivity materials, such as copper, it is advantageous to use a low heating pressure to obtain rapid local heating, followed by a high forge pressure. Duration of heating is selected to obtain suitable thermal conditions for the production of a sound joint on forging, without overheating of surrounding material. For mild steel, times of 0.5–5 s are general depending on the joint geometry, lower times being applicable to smaller sections, or higher conductivity materials.

Since it is to some extent self-cleaning, the method is tolerant with respect to surface preparation. The atmosphere is excluded from the weld area, and reactive metals can be joined with no special

Table 33.4 SPOT WELDING OF DISSIMILAR METALS

	Nimonic	Monel, Inconel	Nickel	Phosphor bronze	Silicon bronze	Nickel–silver	25–40% zinc brass	10–25% zinc brass	Copper	Aluminium and alloys	18/8 stainless steel	Mild steel			
												Other coatings	Tin or zinc coated	Scaly	Clean
Mild steel — Clean	G 5 7	P 8	P 8	P 8	P 8	P 8	P 6 8	U 6 8	U 6 8	P 6 8	G 5	P 3 4	G 3 4	P 2	E 1
Mild steel — Scaly				U 2 8	U 2 8	U 2 8				U 2 8 6	U 2	U 2 4 3	U 2 4 3	P 2	
Mild steel — Tin or zinc coated		P 3 8 4	P 3 4	P 3 4	P 3 4	P 3 4	P 3 6 4	U	U	U	G 3 4	P 3 4	G 3 4		
Mild steel — Other coatings		U 3 8 4	U 3 4	U	U	U	U	U	U	U	U	P 3 4			
18/8 stainless steel	G 5	P	P	P	P	P	U	U	U	U	E 5				
Aluminium and alloys	U	U	U	P 5 7	P 5 7		P 5 7	U	U	G 5 7					
Copper	U	U	U	P 6 8 7	P 6 8 7	P 6 8 7	P 6 8 7	U	U						
10–25% zinc brass	U	U	U	U	U	U	U	P							
25–40% zinc brass		P 5 7	P 5 7	P 5 7	P 5 7	P 5 7	G 5 7								
Nickel–silver		G	G	P	P	G									
Silicon bronze		P	P	P	G										
Phosphor bronze		P	P	G											
Nickel		G	G												
Monel, Inconel		G													
Nimonic	G 5 7														

Key: Spot weldability: E Excellent; G Good; P Poor; U Unsatisfactory.
Taken from 'Resistance Welding', published by the former British Welding Research Association.

Notes:

(1) Wide range of welding conditions.
(2) Inconsistent welds of poor strength. Shot blasting or pickling recommended.
(3) Coating thickness should be uniform.
(4) Electrodes should be cleaned frequently to prevent sticking.
(5) High currents and short times preferred.
(6) Thin gauges may be welded with special conditions.
(7) Welding conditions must be accurately controlled.
(8) Low weld strength.

Table 33.5   WELDABILITY OF DIFFERENT MATERIALS BY FRICTION WELDING*

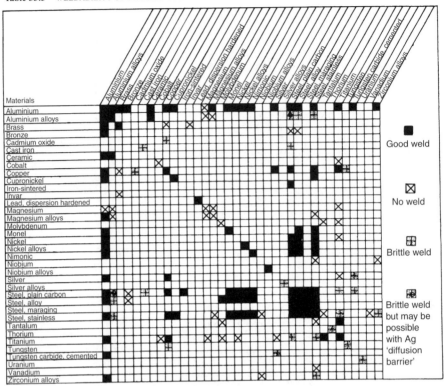

Materials (rows): Aluminium; Aluminium alloys; Brass; Bronze; Cadmium oxide; Cast iron; Ceramic; Cobalt; Copper; Cupronickel; Iron-sintered; Invar; Lead, dispersion hardened; Magnesium; Magnesium alloys; Molybdenum; Monel; Nickel; Nickel alloys; Nimonic; Niobium; Niobium alloys; Silver; Silver alloys; Steel, plain carbon; Steel, alloy; Steel, maraging; Steel, stainless; Tantalum; Thorium; Titanium; Tungsten; Tungsten carbide, cemented; Uranium; Vanadium; Zirconium alloys

Legend:
- ■ Good weld
- ☒ No weld
- ▦ Brittle weld
- ▦ Brittle weld but may be possible with Ag 'diffusion barrier'

* For further information, *see* C. R. G. Ellis, *Met. Const. & Brit. Weld. J.*, 1970, **2**, 185–188.

precautions, while the absence of a liquid phase during welding may avoid problems of porosity, cracking, etc., associated with fusion processes. In fact, if a low melting point liquid area is formed at the interface, it may act as a lubricant, and restrict the development of frictional heat essential for welding. This can arise when welding on to galvanised steel, unless the zinc is removed from the abutting region. Sulphur-bearing free machining steels cause similar difficulties.

Most common alloys are weldable to themselves, while a number of dissimilar metal combinations can be joined, that are normally weldable by other techniques. Table 33.5 illustrates the relative weldability of a range of materials.

FRICTION STIR WELDING

Friction stir welding is a variant of friction welding, in which a solid-state weld is made between two butting workpieces by the frictional heating and plastic material flow caused by a high speed rotating tool that is slowly plunged into and then traverses along the weld line. The workpieces have to be clamped onto a backing bar in a manner that prevents the abutting joint faces from being forced apart. Frictional heat is generated between the wear resistant welding tool and the material of the workpieces. This heat causes the latter to soften without reaching the melting point and allows traversing of the tool along the weld line. The plasticised material is transferred from the leading edge of the tool to the trailing edge of the tool probe and is forged by the intimate contact of the tool shoulder, which leaves a solid phase bond between the two workpieces.

Developed and patented in early 1990s at the Welding Institute (TWI) in England, friction stir welding has already been used in production, particularly for joining aluminium alloys in aerospace structures because of the advantages of this process over conventional arc processes, such as the

Table 33.6 DIFFUSION WELDING COMBINATIONS OF METALS AND ALLOYS WITHOUT AN INTERLAYER (DIRECT) AND WITH AN INTERLAYER (INDIRECT)*

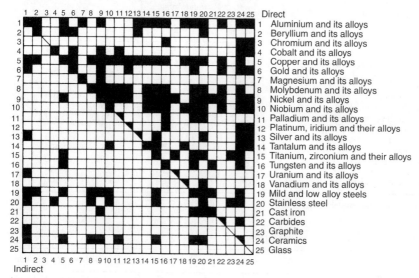

Direct
1 Aluminium and its alloys
2 Beryllium and its alloys
3 Chromium and its alloys
4 Cobalt and its alloys
5 Copper and its alloys
6 Gold and its alloys
7 Magnesium and its alloys
8 Molybdenum and its alloys
9 Nickel and its alloys
10 Niobium and its alloys
11 Palladium and its alloys
12 Platinum, iridium and their alloys
13 Silver and its alloys
14 Tantalum and its alloys
15 Titanium, zirconium and their alloys
16 Tungsten and its alloys
17 Uranium and its alloys
18 Vanadium and its alloys
19 Mild and low alloy steels
20 Stainless steel
21 Cast iron
22 Carbides
23 Graphite
24 Ceramics
25 Glass

Indirect

* For further information, *see* P. M. Bartle, *Welding J.*, 1975, **54**, 799–804.

absence of a melt zone, reduced distortion, freedom from porosity, and ease of automation. The process is well proven for many aerospace aluminium alloys up to 75 mm thick, and for carbon manganese steels, magnesium alloys and copper alloys. Work is being performed to extend the process to titanium and nickel based alloys.

33.4.2 Ultrasonic welding

Ultrasonic welding may be considered as a variation of friction welding, in which a solid-state weld is formed by applying high-frequency vibratory energy at the faying surfaces while the workpieces are clamped together under light pressure. The process is widely used in the semiconductor, microcircuit, electrical, aluminium fabricating and chemical industries. When the process is used in semiconductor and microcircuit industries, it is called (ultrasonic) wire bonding, used to install wire jumpers to connect components to conducting tracks. There are two basic types of wire bonds, ball/wedge using gold wires and wedge/wedge using aluminium wires.

33.4.3 Diffusion welding

In diffusion welding, contacting surfaces are joined by the simultaneous application of temperature and pressure, using conditions adequate for diffusional processes to occur, but insufficient for macroscopic deformation. As most commonly practised, it is a solid-state process. Interlayers can be applied by vacuum spraying, galvanising or as powder or foils, and are employed to accelerate diffusion or to prevent the formation of brittle intermetallic phases. Welding is normally carried out in vacuum, although protective gas shielding can be used. The range of metals and alloys which have been bonded either directly or indirectly via an interlayer is illustrated by Table 33.6, with butt joint properties in Table 33.7. Bonds between metallic and non-metallic components such as glass or ceramics have been made.

The process is governed by four main interrelated variables: pressure, temperature, time and surface condition. A summary of reported welding conditions is given in Table 33.8.

Table 33.7   EXAMPLES OF DIFFUSION BUTT JOINT PROPERTIES*

	Joint strength			Tensile joint efficiency, %
Material combination	tonf in^{-2}	ksi	MPa	
Mild steel/mild steel	30	66	465	100
High strength steel/high strength steel	75	165	1 160	90
Mild steel/cast iron	28	61	430	100
Stainless steel/cast iron	34	75	525	100
Cast iron/cast iron	56	124	865	100
Mild steel/aluminium alloy	10	22	155	50
Aluminium alloy/aluminium alloy	10	22	155	50
Aluminium alloy/copper	9	20	140	70
Copper/nickel	14	31	215	100
Copper/copper	14	31	215	100
Nimonic/nimonic	59	130	910	90
Stellite/stellite	60	132	925	90
Titanium alloy/titanium alloy	69	152	1 065	100

* For further information, *see* Table 33.6.

Table 33.8   DIFFUSION WELDING PARAMETERS*

Material I	*Material II*	*Interlay*	*t*, °C	*P*, MPa	*t*, min
Copper	Molybdenum	—	900	7.4	10
Copper	Steel	—	900	4.9	10
Copper	Nickel	—	900	14.7	20
Copper	Copper	—	800–850	5–7	15–20
Titanium	Nickel	—	800	9.8	10
Titanium	Copper	Molybdenum	950	4.9	30
Titanium	Copper	Niobium	950	4.9	30
Titanium	Copper	—	800	4.9	30
Molybdenum	Steel	—	1 200	4.9	10
Tantalum	Tantalum	Zirconium	870	—	—
Niobium	Niobium	Zirconium	870	—	—
Zircaloy-2	Zircaloy-2	Copper	1 040	20.6	30–120
Kovar	Kovar	—	1 000–1 110	20–25	20–25
Steel	Cast iron	Nickel	850–950	14.7	5.7
Steel	Aluminium	—	500	7.4	30
Steel	Aluminium	Silver	550	4.9	10
Steel	Steel	—	900–1 100	5–15	10–60
Titanium	Titanium	—	850–920	3–10	10–60
Al alloys	Al alloys	Silver or copper	550	1–2	15

* For further information *see* P. Wiesner, *Met. Con. & Brit. Weld. J.*, 1971. **3**, 91–93.

PRESSURE

It is necessary to obtain intimate contact between the faying surfaces, and hence the pressure must be high enough to cause plastic deformation at the tips of abutting surface asperities and to induce creep so that contact over the entire surface area can be achieved. This implies pressures below, but often close to, the material yield stress at temperature and hence welding pressures for a range of metals and alloys are typically 5–15 MPa, although higher pressures are required for refractory metals.

TEMPERATURE

To achieve sufficiently rapid creep for practical purposes, a minimum temperature of $0.7 T_m$ is normal, T_m being the material melting point in degrees kelvin. Higher temperatures may be used to accelerate welding.

TIME

The diffusion time is strongly dependent on the temperature. Welding can occur after only a few minutes, but times of 0.5–2.0 hours are preferred to give complete interface void removal by diffusion, and improved joint ductility.

SURFACE PREPARATION

The rate of welding depends on the flatness and roughness of the faying surfaces, and a finish of ~0.4 μm is preferred. Removal of thick oxide layers and degreasing prior to welding are required, although to some extent contaminants and oxides can dissolve in the parent material, especially with iron- and copper-base systems.

33.5 Fusion welding

While gas welding using an oxy-acetylene torch is still frequently used for mild steel and some non-ferrous metals in sheet gauges, arc welding processes account for the greater proportion of welding carried out at the present time.

Metal arc welding using a coated electrode is the most popular for the manual welding of mild and alloy steels, and electrodes are available for welding cast iron, nickel and its alloys, certain copper alloys and a limited range of aluminium alloys.

A greater number of processes are available for the mechanised welding of mild steel plates. The submerged arc process employs a bare wire electrode and a flux cover continuously supplied via a hopper ahead of the arc. In alternative processes, the flux is applied as a wrapping on the continuously fed electrode or in the core of an electrode formed from folded strip. Other processes using both flux and shielding gas (carbon dioxide) have been developed for mechanised fillet welding.

The gas-shielded processes may be applied to almost all metallurgical materials. They fall into two principal groups, namely those employing a non-consumable electrode, and those wherein the electrode is in the form of a continuous consumable wire.

The argon arc, or TIG, process is the best known of the former type and has been successfully employed for practically all the materials mentioned below, with the possible exception of cast iron. A.c. argon arc welding is normally recommended for aluminium and magnesium alloys and materials forming refractory films, such as aluminium bronze and beryllium copper, while d.c. electrode negative is used for steels, copper and nickel alloys. D.c. nitrogen arc welding is used for copper, and d.c. helium arc welding may be used in circumstances where the high gas cost can be justified in terms of higher welding speeds or increased penetration.

The gas-shielded metal arc processes have been applied to almost as wide a range of materials as the TIG process. The shielding gases used vary with the metals being welded, argon being used for aluminium, magnesium, nickel alloys and refractory metals, argon or nitrogen for pure copper, argon for copper alloys, and argon, argon–oxygen mixtures or carbon dioxide for all types of steel, including stainless steels. D.c. electrode positive is invariably used in present commercial equipment. CO_2 welding is particularly valuable as a low hydrogen process for welding low alloy steels.

Atomic hydrogen and carbon-arc welding are only occasionally used, mainly for mild steel sheet metal welding.

Fusion welding can also be achieved by the electron beam and laser processes. Both processes are applicable to a range of ferrous and non-ferrous alloys, and because of the high energy density, they give a rapid thermal cycle and little distortion. The former method is normally (although not solely) carried out in vacuum and hence is especially appropriate for reactive metals. Laser welding commonly employs helium shielding, this gas having a high ionisation potential with reduced energy loss from the laser beam and, for the same reason, helium is also used for out-of-vacuum electron welding.

33.5.1 The fusion welding of metals and alloys—ferrous metals

MILD STEELS

The welding of mild steel may be accomplished by the majority of processes, the choice of method being dependent upon thickness, type and position of joint and to some extent the type of steel employed. The processes most applicable to thin sheet are gas welding, TIG welding, CO_2 shielded metal arc welding, and less commonly, atomic hydrogen welding. The TIG and atomic hydrogen processes may be manual or mechanised.

Joints in thicker sheet and plate will generally employ the manual metal arc, or, if circumstances permit, one of the mechanised processes such as submerged arc or CO_2 welding. Cored wire welding, employing a continuous tubular consumable with a flux or metal power core, also finds application. Plate of around 100 mm thickness and above may be welded in a single pass using the electroslag process.

The composition of the steel to be welded may limit the choice of process. Rimming steels can give porous welds in gas-shielded welding, and if a mechanised welding process is required, atomic hydrogen, providing a reducing atmosphere, would seem to be the most suitable. Killed steels give better results with argon arc and gas-shielded metal arc welding. High sulphur-free cutting steels are not recommended for welding, since the sulphur gives rise to hot cracking. If such steel has to be welded, basic coated metal arc electrodes should be used.

Gas welding

A neutral flame and no flux is used. The filler material is preferably copper coated to ensure freedom from rust. For general purposes, filler metal to EN 12536:2000 A1 should be used. This is a low carbon mild steel giving a tensile strength of 330 MPa minimum. For somewhat higher strength A2 filler, containing manganese and silicon, is available; *see* Table 33.9.

Metal arc welding

The choice of electrode depends upon the application and the type of welding power source available. All electrodes supplied to EN 499:1994 are coded according to the following system:

1. *Compulsory part:*
   - (a) The letter E for a covered manual metal-arc electrode.
   - (b) Tensile and yield strength of deposited weld metal.
   - (c) Elongation and Charpy impact values of deposited weld metal.
   - (d) Type of flux coating.
2. *Optional part:*
   - (e) Nominal electrode efficiency.
   - (f) Welding positions in which the electrode can be used.
   - (g) Recommended current and voltage conditions.
   - (h) Whether or not the electrode is hydrogen controlled.

For full details of this system, EN 499:1994 should be consulted. The salient coating characteristics are summarised in Table 33.10.

Table 33.9 FILLER RODS FOR THE GAS WELDING OF FERRITIC STEELS (EN 12536:2000)

	Element, wt %						
Type	C	Si	Mn	Ni	Cr	Mo	*Typical application*
A1	0.10 max	—	0.60 max	0.25 max	—	—	General purpose mild steel
A2	0.10–0.20	0.10–0.35	1.00–1.60	—	—	—	Mild steel, 420 MPa
A3	0.25–0.30	0.30–0.50	1.30–1.60	0.25 max*	0.25 max*	—	Medium tensile, 480 MPa
A4	0.25–0.35	0.10–0.35	0.35–0.75	2.75–3.25	−0.30	—	Heat treatable deposit
A5	0.35–0.45	0.40–0.70	0.90–1.10	—	0.90–1.20	—	Wearing surfaces
A6	0.15 max	0.25–0.50	0.60–1.50	0.20 max*	0.20 max*	0.45–0.65	Welding $\frac{1}{2}$%Mo steels
A7	0.08–0.15	0.10–0.35	0.80–1.10	—	—	—	Similar to A2 but lower alloy
A32	0.12 max	0.20–0.90	0.40–1.60	—	1.10–1.50	0.45–0.65	Welding $1\frac{1}{4}$%Cr/$\frac{1}{2}$%Mo steels
A33	0.12 max	0.20–0.90	0.40–1.60	—	2.00–2.70	0.90–1.10	Welding $2\frac{1}{2}$%Cr/1%Mo steels

* If present as a residual element.
Both S and P to be 0.040% max, for A1 to A7, and 0.030% max for A32 and A33.

Table 33.10 TYPES OF COATINGS FOR CARBON AND CARBON MANGANESE STEEL ELECTRODES IN EN 499:1994

Coding	Type of covering	Resultant slag	Special features
B	Basic lime-fluospar	Dense	Low weld metal hydrogen contents are possible, with high resistance to solidification cracking. Generally suitable for all welding positions
BB	Basic lime-fluospar with metallic material	Dense	As **B**, but higher deposition efficiency. Not normally suitable for vertical or overhead welding
C	Cellulosic	Thin and friable	High fusion rate and penetration. Can normally be used in all welding positions. High hydrogen level
R	Rutile	Fluid	Smooth arc with little spatter, and usable in all welding positions
RR	Rutile	Fluid, but more viscous than **R**	Similar to **R**, but with heavier coating thickness. Good weld surface finish
S	Various	Various	Includes acid or oxide fluxes. Consult manufacturer

Automatic processes employing flux

As with manual metal arc welding, the flux type involved is of considerable importance in the submerged arc process and in techniques employing flux-cored consumables. The choice of flux obtainable with these processes is more limited than is the case with manual metal arc welding but both rutile and basic fluxes (equivalent to Classes **R** and **B** in Table 33.10) are available for particular applications, together with more neutral fluxes.

Submerged arc welding fluxes may be in a fused or agglomerated form, depending on the method of manufacture, and have in the past normally been of a neutral or acidic character with high silicate content. More basic fluxes have been developed, which may offer advantages in the mechanical properties of deposited weld metal, but usually with more difficult slag detachment and a more irregular weld surface. Although not giving a full description of reactivity, fluxes are commonly categorised in terms of the 'basicity index' (BI), wt% of constituents:

$$BI = \frac{CaO + CaF_2 + MgO + \frac{1}{2}(MnO + FeO)}{SiO_2 + \frac{1}{2}(Al_2O_3 + TiO_2 + ZrO_2)}$$

Values of BI range from about 0.7 for an acid flux to 3.5 for a basic system. For details of BI see S. S. Tuliani *et al.*, *Welding Metal Fabrication*, August 1969, 327–339.

Low carbon wire consumables are used, generally with added manganese and silicon as deoxidants.

Gas-shielded welding processes

Consumables used for gas-shielded welding of steels contain added deoxidants (usually silicon and manganese). Examples for mild steel are given in Table 33.11, A15 and A17. TIG welding is usually carried out using argon shielding for optimum weld pool control, although helium may be used on thicker sections to increase penetration of the weld. With gas-shielded metal arc welding of mild steel, the surrounding gas is generally CO_2 or argon/CO_2 mixtures.

LOW ALLOY STEELS

In general, the application of the various welding processes is similar for both mild and low alloy steels, the principal difference being that the choice of consumables for low alloy steels may be more limited. Manual metal arc and submerged arc welding are most widely used for low alloy steels in industry, consumables for the former process being given in Table 33.12.

It must be appreciated that in the heat-affected zone (HAZ) around a weld, very high cooling rates may be experienced. Particularly with alloy steels, hard HAZ microstructures can be produced, having a high susceptibility to cracking under the influence of residual welding stresses, and hydrogen picked up during the welding operation. The necessity to avoid hydrogen induced cracking is a major factor in the selection of welding conditions for low alloy steels.

Table 33.11   FILLER RODS AND WIRES FOR GAS-SHIELDED ARC WELDING OF FERRITIC STEELS (EN 440:1994, EN 1668:1997)

	Element, wt %							
Type	C	Si	Mn	Cr	Mo	Al	S	P
A15*	0.12 max	0.30–0.90	0.90–1.60	—	—	0.04–0.40	0.040 max	0.040 max
A16	0.25–0.30	0.30–0.50	1.30–1.60	—	—	—	0.040 max	0.040 max
A17	0.12 max	0.20–0.50	0.85–1.40	—	—	—	0.040 max	0.040 max
A18	0.12 max	0.70–1.20	0.90–1.60	—	—	—	0.040 max	0.040 max
A19	0.08–0.12	0.30–0.50	1.00–1.30	—	—	0.35–0.75	0.040 max	0.040 max
A30	0.12 max	0.02–0.90	0.40–1.60	—	0.45–0.65	—	0.030 max	0.030 max
A31	0.14 max	0.50–0.90	1.60–2.10	—	0.40–0.60	—	0.030 max	0.030 max
A32	0.12 max	0.20–0.90	0.40–1.60	1.10–1.50	0.45–0.65	—	0.030 max	0.030 max
A33	0.12 max	0.20–0.90	0.40–1.60	2.00–2.70	0.90–1.10	—	0.030 max	0.030 max
A34	0.12 max	0.20–0.90	0.40–1.60	5.00–6.00	0.45–0.65	—	0.030 max	0.030 max
A35[†]	0.10 max	0.50 max	0.60 max	8.00–10.50	0.8–1.20	—	0.030 max	0.040 max

[*] Titanium or zirconium may be present up to 0.15% max each.   [†] 0.5% nickel maximum.
Note: All grades have 0.4% copper maximum.

Table 33.12   MANUAL METAL ARC ELECTRODES FOR LOW ALLOY STEELS (EN 1599:1997, EN 757:1997)

	Element, wt %							
Composition code	C	Si	Mn	Ni	Cr	Mo	S	P
Mo-steel								
MoB*	0.10 max	0.60 max	0.75–1.20	—	—	0.40–0.70	0.030 max	0.035 max
MoC ⎱ MoR ⎰	0.10 max	0.80 max	0.35 min	—	—	0.40–0.70	0.030 max	0.035 max
Cr–Mo steel								
1CrMoLB*	0.05 max	0.50 max	0.50–1.20	—	1.0–1.8	0.40–0.70	0.025 max	0.025 max
1CrMoB*	0.10 max	0.50 max	0.75–1.20	—	1.0–1.5	0.40–0.70	0.030 max	0.035 max
1CrMoR	0.10 max	0.30 max	0.35 min	—	1.0–1.5	0.40–0.70	0.030 max	0.035 max
2CrMoLB*	0.05 max	0.50 max	0.50–1.20	—	2.0–2.5	0.90–1.20	0.025 max	0.025 max
2CrMoB*	0.10 max	0.50 max	0.75–1.20	—	2.0–2.5	0.90–1.20	0.030 max	0.035 max
2CrMoR	0.10 max	0.30 max	0.35 min	—	2.0–2.5	0.90–1.20	0.030 max	0.035 max
5CrMoB	0.10 max	0.50 max	0.50–1.00	—	4.0–6.0	0.40–0.70	0.030 max	0.035 max
7CrMoB	0.10 max	0.60 max	0.50–1.00	—	6.0–8.0	0.40–0.70	0.030 max	0.035 max
9CrMoB	0.10 max	0.60 max	0.50–1.00	—	8.0–10.0	0.90–1.20	0.030 max	0.035 max
12CrMoB[a]	0.23 max	0.80 max	0.30–1.00	0.3–0.8	11.0–12.5	0.80–1.20	0.025 max	0.025 max
12CrMoVB[b]	0.23 max	0.80 max	0.30–1.00	—	11.0–13.0	0.80–1.20	0.030 max	0.030 max
12CrMoWVB[c]	0.28 max	0.80 max	0.50–1.00	0.3–0.8	11.0–13.0	0.80–1.20	0.030 max	0.030 max
Ni-steel								
1NiLB	0.07 max	0.60 max	0.30–1.10	0.80–1.20	—	—	0.030 max	0.030 max
2NiLB	0.07 max	0.60 max	0.30–1.10	2.00–2.75	—	—	0.030 max	0.030 max
3NiLB	0.07 max	0.60 max	0.50–1.10	2.80–3.75	—	—	0.030 max	0.030 max
1NiB	0.10 max	0.80 max	0.50–1.20	0.80–1.10	—	—	0.030 max	0.035 max
2NiB	0.10 max	0.80 max	0.50–1.20	2.00–2.75	—	—	0.030 max	0.035 max
3NiB	0.10 max	0.80 max	0.50–1.20	2.80–3.50	—	—	0.030 max	0.035 max
1NiC	0.15 max	0.80 max	0.50–1.20	0.80–1.20	—	—	0.030 max	0.035 max
2NiC	0.15 max	0.80 max	0.50–1.20	2.00–2.75	—	—	0.030 max	0.035 max
Mn–Mo steel								
MnMoB	0.10 max	0.80 max	1.20–1.80	—	—	0.25–0.45	0.025 max	0.025 max
2MnMoB	0.10 max	0.80 max	1.60–2.00	—	—	0.25–0.45	0.025 max	0.025 max
High strength steel								
MnNiB	0.10 max	0.80 max	0.60–1.20	1.00–1.80	—	—	0.030 max	0.035 max
NiMoB	0.10 max	0.80 max	0.80–1.60	1.20–1.90	—	0.20–0.50	0.030 max	0.035 max
1NiMoC	0.18 max	0.80 max	0.50–1.20	1.20–1.90	—	0.20–0.50	0.030 max	0.035 max
2NiCrMoB	0.10 max	0.80 max	1.30–2.20	1.50–2.50	0.70–1.50	0.20–0.50	0.030 max	0.035 max

[*] Mn : Si at least 2 : 1.   [a] 0.3%V max.   [b] 0.5%V max.   [c] 0.5%V max: 0.7–1.0%W.

HAZ cracking in ferritic steels is influenced by the following inter-related variables:

1. Steel composition and transformation behaviour.
2. Welding process and type of consumable.
3. Energy input of the welding process, and preheating temperature.
4. Joint restraint.

In general, more highly alloyed steels have a greater tendency to cracking, due to their lower transformation temperature and the formation of harder transformation products. In practice, the important aspects of transformation behaviour of a steel are the hardenability of the steel, and the susceptibility of the hardened structure to cracking.

Both hardenability and susceptibility generally increase with increases in carbon and alloying element content. The dominant effect of carbon has been recognised by the empirical derivation of carbon equivalent (CE) formulae which are intended to describe the weldability of a steel. A widely used formula for carbon–manganese structural steels to BS 4360:1986, is as follows:

$$CE = C + \frac{Mn}{6} + \frac{Cu + Ni}{15} + \frac{Cr + Mo + V}{5}$$

This relationship is recognised by the International Institute of Welding and ISO. The following may be also considered:

$$Pcm = C + \frac{Si}{30} + \frac{Mn + Cu + Cr}{20} + \frac{Ni}{60} + \frac{Mo}{15} + \frac{V}{10} + 5B$$

$$CEN = C + A(C)\left[\frac{Si}{24} + \frac{Mn}{6} + \frac{Cu}{15} + \frac{Ni}{20} + \frac{Cr + Mo + V + Nb}{5} + 5B\right]$$

where

$$A(C) = 0.75 + 0.25\tanh[20(C - 0.12)]$$

The Pcm formula was derived primarily for higher strength low alloy steels, while the CEN relationship recognises a non-linear contribution of carbon to hardenability and cracking risk at higher carbon levels. With increased CE, more stringent precautions are necessary to avoid cracking. It should be appreciated that such formulae apply only to the particular compositions used in determining them, and are not universally applicable.

The risk of HAZ cracking is highly dependent upon the hydrogen content of the freshly deposited weld metal. With respect to manual metal arc welding, the risk of cracking in a particular steel is lower when low hydrogen basic electrodes are used rather than the common rutile type, particularly if the former electrodes are dried at temperatures above about 300°C. The submerged-arc and gas-shielded metal-arc processes can produce weld metal deposits having hydrogen contents comparable to or lower than those of basic electrodes, and some relaxation of welding conditions may be possible with these processes, provided that clean dry consumables are used. Typical weld metal hydrogen levels determined following the procedures in BS 6693 are given in Table 33.13.

Table 33.13 TYPICAL HYDROGEN CONTENTS FOR DIFFERENT STEEL WELD METALS (EN 499:1994)

Weld metal type	Range of hydrogen contents (ml per 100 g of deposited metal)
Classification C	70–100
Classification R	20–35
Classification B (dried 100–150°C)	10–15
Classification B (baked 350–450°C)[†]	3–10
Submerged arc	5*–25
Gas-shielded metal arc (Ar or CO_2)	3*–10
Cored CO_2: basic flux	3–5
rutile flux	5–15

* Lower values for clean, dry consumables.
[†] Consumable manufacturer's advice on drying conditions should be sought.

Thermal conditions prevailing during welding are controlled for one or both of two reasons. First, by control of the cooling rate, it is often possible to determine the transformation product after welding. Secondly, cracking may be avoided by holding the weld area at about 250°C (when the embrittling effect of hydrogen is negligible) for a sufficient length of time for hydrogen to diffuse away from susceptible regions.

For a given weld geometry, increased heat input during welding will result in a lower cooling rate. In many materials, such reduction in cooling rate will be sufficient to avoid transformation to hard, susceptible microstructures. Increased heat input may be achieved by increasing the welding current or voltage, or by reducing the speed of travel. If cracking cannot be controlled by increasing energy input, preheat may be applied to reduce the cooling rate after welding. The preheat temperature required in a particular case will depend upon the material composition, the joint restraint, and the welding process used, but it will usually be in the range 100–250°C. In multipass welds, the preheat temperature must be maintained between weld runs as an interpass temperature.

Joint restraint may be defined as resistance to deformation which would relieve contractural welding stresses. Restraint will, in general, increase with increasing thickness of the component parts being joined. Increase in material size will also increase the cooling rate of the weld by affording a larger heat sink. Joint type is significant in determining the number of paths along which heat may be conducted away from the weld, and the effect of joint geometry on cooling conditions can be described by a 'combined thickness' parameter. The combined thickness is the total thickness of material in mm through which heat can flow away from the weld, as indicated in Table 33.14.

When doubt exists regarding the risk of cracking in a given situation, it must be recommended that a procedural trial be undertaken.

It has been found possible with carbon–manganese structural steels to avoid cracking by controlling welding conditions so that the microstructure in the HAZ is not harder than a critical value. When using manual metal arc welding consumables of relatively high hydrogen potential, the HAZ hardness should not exceed 350 HV. With low hydrogen consumables or CO_2 welding, a hardness of 400 HV or even above is appropriate. Figure 33.1 is a nomogram for material of CE determined as above, showing welding conditions such that the critical hardness is not exceeded, and cracking can be avoided. Allowance is made for the use of processes with different hydrogen potentials, and for the combined thickness of the weldment. From the intended welding process, the relevant CE scale is selected. A vertical line is drawn from the CE of the material being considered, to intersect the intended preheat temperature. From this intersection, a line is taken horizontally to the appropriate

Table 33.14   DETERMINATION OF COMBINED THICKNESS

Type of joint and heat flow	Combined thickness (mm)
	$2t$
	$t_1 + t_2$
t_1 = average thickness over 75 mm	$t_1 + t_2$
	$t_1 + t_2 + t_3$
Twin fillet welds made simultaneously	$\frac{1}{2}(t_1 + t_2 + t_3)$

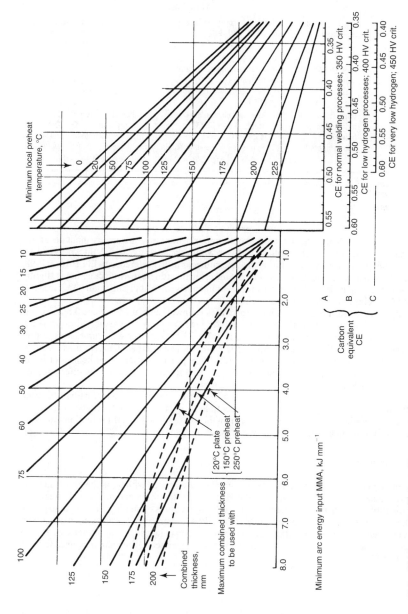

Figure 33.1 *Welding conditions for mild and C–Mn steels.*

For further information, *see* N. Bailey, *Met. Constr. & Brit. Weld J.*, 1970, **2**, 442.
CE Scale A: Normal welding processes; HAZ hardness restricted to below 350 HV, CE Scale B: Low hydrogen welding processes; HAZ hardness restricted to below 400 HV,
CE Scale C: Very low hydrogen welding processes; HAZ hardness restricted to below 450 HV.

Table 33.15 CLASSIFICATION OF LOW ALLOY STEELS

Class	Hardenability	Susceptibility to HAZ cracking	Examples	Precautions in welding
1	Low	None	Mild steel	None
2	Low	Low	(1) Mild steel with C >0.15% but <0.25%, and Mn <1.0%. (2) C–Mn steels with C <0.2% and Mn <1.4%	Thin sections, none; with thicker sections, lower H_2 processes, high heat input and some preheat are all advantageous
3	Low	High (twinned martensite)	Medium carbon steels, e.g. 080M30, 080M40*	Production of non-martensitic HAZ more important than low hydrogen. Use high heat inputs, and 250–350°C preheat
4	High	Low (low carbon martensite)	Low carbon, low alloy high strength steels, e.g. 9%Ni HY80, C–Mn steels in thick sections	Low hydrogen processes, increased heat input and some preheat generally necessary
5	High	High (twinned martensite or bainitic constituents)	(1) Medium carbon alloy steels, e.g. 530H32, 835M30* (2) Some creep resistant steels, e.g. 5Cr $\frac{1}{2}$ Mo, $2\frac{1}{4}$ Cr 1Mo	Preheat and interpass temperature in the range 200–350°C. (May be possible to use 150°C in thin sections.) Post-weld heat treatment advisable

For further information, *see* T. Boniszewski and R. G. Baker, Proc. Second Commonwealth Welding Conference, Institute of Welding, London, 1965, Paper M5.
* BS 970:Part 1:1996.

combined thickness, and then vertically down to determine the minimum arc energy input necessary during welding to avoid exceeding the critical hardness. Similarly, given a material, thickness and arc energy, the minimum preheat temperature can be determined. Further reference to EN 1011-1:1998, EN 1011-2:2001 should be made. One caveat is that the critical hardness to avoid cracking is slightly dependent on carbon equivalent. Conservatism is recommended when applying Figure 33.1 to steels of CE <0.40 for welds to be made without preheat.

The CE approach for formulating welding procedures is unworkable at CEs above 0.6. With low alloy steels, it is possible to classify susceptibility to HAZ cracking according to their transformation behaviour. Five general classifications are shown in Table 33.15, together with summaries of precautions necessary for avoiding cracking. Appropriate preheat and interpass temperatures for steels in Classes 4 and 5 can be obtained from Figure 33.2. More precise definition of behaviour is not possible since cracking susceptibility is not uniquely related to hardness, while the effect of joint geometry cannot be adequately quantified for these materials.

In general, weld metals used for mild and low alloy steels contain rather less carbon than the parent material, and in consequence are less susceptible to hydrogen-induced cracking than is a HAZ. The factors outlined above for HAZs are nonetheless applicable, and precautions to avoid hydrogen-induced weld metal cracking may be necessary in some cases.

HAZ liquation cracking

Another problem in the welding of ferritic steels is cracking associated with liquation of nonmetallic phases at prior austenite grain boundaries in the HAZ, under the influence of thermal stresses in this area. Susceptibility is dependent mainly on the relative amounts of sulphur, carbon, manganese and phosphorus in the steel and, in plain iron–carbon alloys, the problem becomes significant with carbon contents above about 0.2%. At this carbon content, a manganese to sulphur ratio of at least 20 : 1 is required to avoid the problem, and with higher carbon contents, this ratio may need to be as high as 30 : 1 or 50 : 1.

Lamellar tearing

Lamellar tearing may arise in joints in which the fusion boundary of the weld is parallel to the plate surface, and tensile residual stresses act across the plate thickness. Cracking is generally

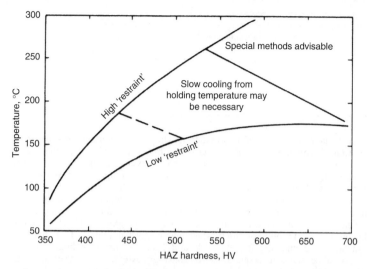

Figure 33.2 *Suggested preheat conditions for procedural tests for steels in Classes 4 and 5 of Table 33.15.*

step-like in character, and is associated with inclusions which are rolled into planar form during plate manufacture. Such inclusions result in reduced ductility in the through-thickness direction of the plate. Lamellar tearing will not normally arise in plate with a through-thickness ductility above 20% reduction in area (RA). Between 10 and 20% RA, it may be encountered in highly restrained cases, such as fully penetrating nozzles. If the ductility is below 10% RA, cracks may form even in relatively lightly restrained T-joints. Lamellar tears have been found in plate of 10 to 175 mm thickness, but are not common below 25 mm thickness. Where experience with a particular material and joint configuration indicates a risk of lamellar tearing, consideration must be given to joint design and welding procedure to reduce the effect of restraint, and to ensure that the fusion boundary of the weld runs across possible planes of weakness in the material.

Stress relief cracking

A number of ferritic steels may be subject to stress relief, or reheat, cracking. Cracking is inter-granular, and occurs in the HAZ when restrained joints are given a post-weld heat treatment. The problem is analogous to that in austenitic steels considered below, and arises primarily in alloys showing secondary hardening. Cracking behaviour cannot be predicted at the present time with any precision. Cracking has not been reported in alloys such as $2\frac{1}{4}$Cr/1Mo or $\frac{1}{2}$Mo/B steels at below 75 mm thickness, or in $\frac{1}{2}$Cr/$\frac{1}{2}$Mo/$\frac{1}{4}$V at below 18 mm thickness. The risk of cracking can be reduced by weld dressing to reduce local stress concentrations, and by controlled heat treatment procedures.

HIGH ALLOY HEAT AND CORROSION-RESISTANT STEELS

These fall into three basic categories: the martensitic chromium-containing steels, the higher chromium ferritic steels and the chromium–nickel austenitic steels. With different chromium and nickel contents, materials containing mixtures of the three main microstructural constituents may be produced. The welding behaviour of such materials is generally dependent on the major phase present.

MARTENSITIC STAINLESS STEELS

The principal problem with the martensitic alloys is that of hydrogen-induced cold cracking, as with low alloy steels. Preheating is required, generally to over 200°C, followed by slow cooling and post-weld heat treatment at 650–750°C. The likelihood of cracking depends on a number of factors, and procedural trials are normally required to determine the optimum welding conditions in any

Table 33.16   SUGGESTED WELDING CONDITIONS FOR 13%Cr STEELS

Material	Temperature, °C		
	Preheat	*Interpass*	*Post-weld heat treatment*
13%Cr/0.1%C	250	250–350	700–750
13%Cr/0.2%C	250	250–350	700–750
13%Cr/4%Ni/0.06%C	150	150–250	600–650

given instance. Carbon content is of particular importance, and thin plate material of less than 0.1% carbon can often be welded without preheat or post-weld heat treatment; with thicker material or carbon content above 0.2%, both preheat and heat treatment are necessary. At higher carbon levels, it is essential that the welds do not cool to room temperature between completion of welding and application of heat treatment (although the joint should cool sufficiently for complete transformation to martensite to occur in the HAZ, prior to heat treatment), unless sufficient time is given to permit hydrogen diffusion out of the weld area. The martensite transformation finish temperature (M_f) may not be well defined, and in such cases, a suitable cool-out temperature should be defined by procedural tests. Suggested welding conditions for typical alloys are given in Table 33.16. If matching composition consumables are required (e.g. Table 33.17), careful consideration to hydrogen-induced weld metal cracking should be paid, and adequate electrode baking is essential. The use of austenitic consumables is preferred. Although these are of lower strength than the parent material, hydrogen is more soluble in austenite than in martensite, and thus hydrogen tends to remain in the weld metal, reducing the risk of cracking. Manual metal arc welding is generally employed for martensitic stainless steels, with TIG welding finding application for thin gauge material.

FERRITIC STAINLESS STEELS

Ferritic stainless steels suffer embrittlement on welding due to grain growth and grain boundary, martensite formation in the heat affected zone that is heated above 1100°C. These materials are notch sensitive, and the grain growth raises the ductile–brittle transition temperature to above room temperature, resulting in a brittle joint when the weldment cools out. Preheat to 200°C and post-heat treatment at 750–850°C are recommended. These measures do not avoid the deleterious microstructure, but serve to reduce residual welding stresses as far as possible and obtain some HAZ softening. The toughness of matching composition weld metal (e.g. Table 33.17) may be lower than that of the heat affected zone. Thus, austenitic consumables are preferred unless differences in coefficient of expansion between the weld metal and parent material are likely to cause thermal fatigue in service. Manual metal arc and TIG welding are most commonly used for ferritic grades of stainless steel.

Ferritic stainless steels are also sensitive to intercrystalline corrosive attack in the region around a weld that is heated to above 1 100°C. The problem may be avoided by post-weld heat treatment at between 700 and 850°C. In some circumstances sigma phase may form during heat treatment, causing brittleness below 200°C. This can be eliminated by annealing at above 850°C.

The problems of loss of toughness and corrosion resistance in the weld area restrict the industrial use of fusion welded ferritic stainless steel assemblies. These materials can, however, offer good general corrosion resistance, and are considerably more resistant to chloride-induced stress corrosion than are austenitic grades. To avoid the welding problems of conventional ferritic stainless steels, alloys have been developed based either on reduction in interstitial elements, or on balanced chromium–nickel contents so that mixed ferrite–martensite or ferrite–austenite microstructures are obtained. Such materials have been successfully welded in thin gauges for service conditions where a risk of stress corrosion has precluded the use of austenitic steels.

AUSTENITIC STAINLESS STEELS

For practical purposes, austenitic stainless steels can be regarded as among the most weldable materials, and virtually any process can be used to make a joint. The manual metal arc and TIG processes are most common, with MIG and submerged arc welding being used when high deposition rates are required. Manual metal arc and gas-shielded welding consumables are given in Tables 33.17 and 33.18. Oxyacetylene welding is readily carried out with consumables as in Table 33.19, but is not preferred since excessive heating and a carburising flame can cause weld decay, as considered below. A flux is desirable, and a neutral flame is recommended, oxidising conditions causing porosity.

Table 33.17 MANUAL METAL ARC WELDING ELECTRODES FOR Cr AND Cr/Ni STEELS (EN 1600:1997)

Composition code	Element, wt %						
	C	Si	Mn	Cr	Ni	Mo	Nb
13	0.08 max	1.0 max	1.0 max	11.0–13.5	0.60 max	0.5 max	0.5 max
13.4Mo	0.06 max	1.0 max	1.0 max	11.0–13.5	4.0–5.0	0.4–0.7	
17	0.10 max	1.0 max	1.0 max	15.0–18.0	0.60 max	0.5 max	
28	0.10 max	1.0 max	1.0 max	26.0–32.0	0.60 max	0.60 max	0.5 max
19.9	0.08 max	1.0 max	0.5–2.5	18.0–21.0	9.0–11.0	0.5 max	
19.9L	0.04 max	1.0 max	0.5–2.5	18.0–21.0	9.0–11.0	0.5 max	
19.9.3	0.10 max	1.0 max	0.5–3.0	18.5–21.0	8.0–10.0	2.0–4.0	
19.9Nb	0.10 max	1.0 max	0.5–2.5	18.0–21.0	9.0–11.0	0.5 max	10 × C − 1.10
17.8.2	0.06–0.10	0.8 max	0.5–2.5	16.5–18.5	8.0–9.5	1.5–2.5	
19.12.2L	0.04 max	1.0 max	0.5–2.5	17.0–20.0	11.0–14.0	2.0–2.5	
19.12.3	0.08 max	1.0 max	0.5–2.5	17.0–20.0	10.0–14.0	2.5–3.5	
19.12.3L	0.04 max	1.0 max	0.5–2.5	17.0–20.0	10.0–14.0	2.5–3.5	
19.12.3Nb	0.10 max	1.0 max	0.5–2.5	17.0–20.0	10.0–14.0	2.5–3.5	10 × C − 1.10
19.13.4	0.08 max	1.0 max	0.5–2.5	17.0–20.0	11.0–15.0	3.5–5.5	
19.13.4L	0.04 max	1.0 max	0.5–2.5	17.0–20.0	11.0–15.0	3.5–5.5	
19.13.4Nb	0.10 max	1.0 max	0.5–2.5	17.0–20.0	11.0–15.0	3.5–5.5	10 × C − 1.10
18.15.3LNb*	0.04 max	1.0 max	0.5–7.0	16.0–20.0	13.0–18.0	2.0–3.5	
23.12	0.15 max	1.0 max	0.5–2.5	22.0–25.0	11.0–14.0	0.5 max	
23.12L	0.04 max	1.0 max	0.5–2.5	22.0–25.0	11.0–14.0	0.5 max	
23.12Nb	0.10 max	1.0 max	0.5–2.5	22.0–25.0	11.0–14.0		10 × C − 1.10
23.12.2	0.10 max	1.0 max	0.5–2.5	22.0–25.0	11.0–14.0	2.0–3.0	
23.12W[a]	0.20 max	1.0 max	0.5–2.5	22.0–25.0	11.0–14.0		
25.6.2*	0.06 max	1.0 max	0.5–2.5	24.0–28.0	4.0–8.0	1.5–4.0	
25.6.2Cu*[b]	0.06 max	1.0 max	0.5–2.5	24.0–28.0	4.0–8.0	1.5–4.0	
25.20	0.20 max	0.7 max	0.5–6.0	24.0–28.0	18.0–22.0	0.5 max	
25.20H	0.35–0.45	0.7 max	0.5–2.0	24.0–28.0	18.0–22.0	0.5 max	
25.20Nb	0.12 max	0.7 max	0.5–6.0	24.0–28.0	18.0–22.0	0.5 max	10 × C − 1.20
25.20.1	0.12 max	0.7 max	0.5–2.5	25.0–28.0	18.0–22.0	0.5–1.20	
25.20.1Nb	0.12 max	0.7 max	0.5–6.0	24.0–28.0	18.0–22.0	0.1–1.20	10 × C − 1.20
25.20.2	0.12 max	0.7 max	0.2–2.5	25.0–28.0	18.0–22.0	2.0–3.0	
25.21.2LMn*	0.04 max	1.0 max	3.0–7.0	23.0–27.0	19.0–23.0	2.0–3.0	
15.35H	0.25–0.50	1.0 max	1.0–2.5	14.0–20.0	33.0–40.0	0.5 max	
20.25.5LCuNb[c]	0.04 max	1.0 max	0.5–4.0	19.0–22.0	24.0–28.0	4.0–5.5	8 × C − 0.5
20.34.2CuNb[d]	0.07 max	0.6 max	0.5–2.5	19.0–21.0	32.0–36.0	2.0–3.0	8 × C − 1.0
25.35H	0.35–0.45	1.0 max	0.5–2.0	23.0–27.0	32.0–36.0	0.5 max	
25.35HNb	0.35–0.45	1.0 max	0.5–2.0	23.0–27.0	32.0–36.0	0.5 max	0.5–1.8
29.9	0.15 max	1.2 max	0.5–2.5	28.0–32.0	8.0–12.0	0.5 max	
29.9.1	0.15 max	1.0 max	0.5–2.5	28.0–32.0	8.0–10.5	0.5–1.20	
29.9.2	0.15 max	1.0 max	0.5–2.5	28.0–32.0	8.0–10.5	2.0–3.0	

Notes: All grades 0.030%S max and 0.040%P max.
* 0.25%N max.   [a] 2.0–4.0%W.   [b] 1.0–4.0%Cu.   [c] 1.0–2.5%Cu.   [d] 3.0–4.0%Cu.

The main metallurgical problems with austenitic stainless steels are the avoidance of weld metal and heat affected zone cracking, and maintenance of full corrosion resistance in the weld area.

Fully austenitic weld metals are susceptible to cracking both during solidification and in underlying reheated runs. The problem may be entirely avoided by the use of consumables with composition such that the deposited weld metal contains more than about 3–5% ferrite. A guide to the relationship between composition and structure is given by the Schaeffler diagram, reproduced in Figure 33.3. Dilution from parent material must be taken into account when using the Schaeffler diagram.

The presence of ferrite in the weld metal may promote transformation to the embrittling sigma phase during heat treatment or in service at temperatures between 550 and 850°C. The compositional range likely to suffer rapid embrittlement is indicated in Figure 33.3. Particularly if sigma formation is a hazard, the preferred weld metal compositions lie in the shaded area.

In certain corrosion-resistant applications, notably urea plant, the presence of ferrite may be undesirable since it can lead to preferential attack taking place on the weld metal. In such cases,

Table 33.18   FILLER RODS AND WIRES FOR GAS-SHIELDED ARC WELDING OF AUSTENITIC STAINLESS STEELS (EN 12072:1999)

Type	*Element*, wt %						
	C	Si	Mn	Ni	Cr	Mo	Nb
308S92	0.03 max	0.25–0.65	1.0–2.5	9.0–11.0	19.5–22.0	0.5 max	—
308S96	0.08 max	0.25–0.65	1.0–2.5	9.0–11.0	19.5–22.0	0.5 max	—
308S93	0.03 max	0.65–1.00	1.5–2.5	9.5–10.5	20.0–21.0	0.5 max	—
347S96	0.08 max	0.25–0.65	1.0–2.5	9.0–11.0	19.0–21.5	0.5 max	$10 \times C - 1.0$
309S94	0.12 max	0.25–0.65	1.0–2.5	12.0–14.0	23.0–25.0	0.5 max	—
309S92	0.03 max	0.25–0.65	1.0–1.5	12.0–14.0	23.0–25.0	0.5 max	—
311S94	0.12 max	0.25–0.65	1.0–2.5	12.0–14.0	23.0–25.0	0.5 max	$10 \times C - 1.3$
310S94	0.08–0.15	0.25–0.65	1.0–2.5	20.0–22.5	25.0–28.0	0.5 max	—
310S98	0.35–0.45	0.80–1.30	1.0–2.5	20.0–22.5	25.0–28.0	0.5 max	—
312S94	0.15 max	0.25–0.65	1.0–2.5	8.0–10.5	28.0–32.0	0.5 max	—
313S94	0.06–0.13	0.25–0.65	1.0–2.5	20.0–22.5	25.0–28.0	—	$10 \times C - 1.3$
316S92	0.03 max	0.25–0.65	1.0–2.5	11.0–14.0	18.0–20.0	2.0–3.0	—
316S96	0.08 max	0.25–0.65	1.0–2.5	11.0–14.0	18.0–20.0	2.0–3.0	—
316S93	0.03 max	0.65–1.00	1.5–2.5	10.0–13.5	18.0–20.0	2.5–3.0	—
317S96	0.08 max	0.25–0.65	1.0–2.5	13.0–15.0	18.5–20.5	3.0–4.0	—
318S96	0.08 max	0.25–0.65	1.0–2.5	11.0–14.0	18.0–20.0	2.0–3.0	$10 \times C - 1.0$

S and P each to be 0.030% max, Cu 0.5% max.

Table 33.19   FILLER RODS FOR THE GAS WELDING OF AUSTENITIC STAINLESS STEELS (EN 12536:2000)

Type	*Element*, wt %						
	C	Si	Mn	Ni	Cr	Mo	Nb
347S96	0.08 max	0.25–0.60	1.0–2.5	9.0–11.0	19.0–21.5	—	$10 \times C - 1.0$
309S94	0.12 max	0.25–0.60	1.0–2.5	12.0–14.0	23.0–25.0	—	—
311S94	0.12 max	0.25–0.60	1.0–2.5	12.0–14.0	23.0–25.0	—	$10 \times C - 1.3$
310S94	0.08–0.15	0.25–0.60	1.0–2.5	20.0–22.5	25.0–28.0	—	—
313S94	0.06–0.13	0.25–0.60	1.0–2.5	20.0–22.5	25.0–28.0	—	$10 \times C - 1.3$
316S96	0.08 max	0.25–0.60	1.0–2.5	11.0–14.0	18.0–20.0	2.0–3.0	—
318S96	0.08 max	0.25–0.60	1.0–2.5	11.0–14.0	18.0–20.0	2.0–3.0	$10 \times C - 1.0$

S and P each to be 0.030% max.

low or zero ferrite consumables should be used, and the risk of cracking controlled by minimising joint restraint and welding with minimum heat input. Increased manganese contents of 3–5% are beneficial.

During the welding cycle, intergranular precipitation of chromium-rich carbides can occur in areas heated to within 500–900°C. This precipitation causes local loss of chromium from the matrix, with a consequent reduction in corrosion resistance and susceptibility to intercrystalline attack, or 'weld decay'. The problem is normally avoided by the use of material either with carbon contents below 0.03%, or containing strong carbide forming elements such as niobium or titanium. Such 'stabilised' steels are not immune to intercrystalline attack under all circumstances, and heat treatment within the sensitising range 500–900°C can induce sensitisation to intercrystalline corrosion. If post-weld heat treatment is carried out, the temperature should be above 900°C. For practical purposes, intercrystalline attack due to arc welding is unlikely to be encountered in unstabilised molybdenum-free material of 0.06% carbon and below, provided that the arc energy per unit length of weld metal is below 2 kJ mm^{-1}, although service in highly oxidising media should be regarded with caution.

Austenitic steels may suffer HAZ liquation cracking during welding. The problem is minimised by the use of low arc energy welding conditions and with wrought material, by avoiding grain sizes coarser than about ASTM 3–4. In castings, cracking can often be suppressed by using material containing above 5% ferrite. The liquation cracks are of the order of 0.5 mm long, and are not generally

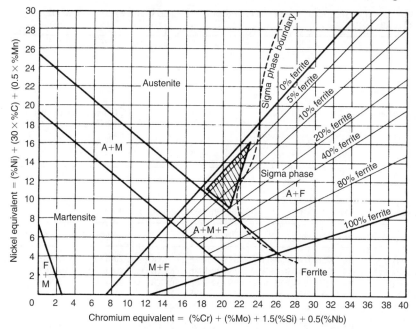

Figure 33.3 *Modified Schaeffler diagram for constitution of stainless steel weld metal*

significant in service. In welds of high restraint, however, they can form initiation points for 'reheat cracking' during elevated temperature service or post-weld heat treatment. At elevated temperature, intragranular strain-induced precipitation occurs in the HAZ. This causes a loss of creep ductility, and if joint restraint is high enough, intergranular cracking results. All common grades of austenitic stainless steel are susceptible to reheat cracking, with the exception of the 18%Cr/12%Ni/3%Mo types, provided that these do not contain residual carbide forming elements such as niobium or titanium. The risk of reheat cracking is reduced by dressing weld toes to remove liquation cracks, by the use of low hot strength weld metal, and by stress relief at above 950°C.

High proof stress variants of the common austenitic stainless steels have been developed, based either on solid solution hardening by nitrogen or on 'warm working' by rolling at down to 850°C to obtain a work hardening effect. These materials can generally be welded with normal consumables with no loss of strength in the weld area. With the nitrogen-bearing steels, excessive dilution of the weld pool by parent material causes a fully austenitic weld metal, with a risk of solidification cracking. High dilution situations, such as the root pass of a butt weld in thick plate, should therefore be regarded with caution. Joint preparation should be such that at least 50% of the molten weld pool is filler material.

CLAD STEELS

These consist of mild or low alloy steel clad with an overlay by rolling, explosively or by weld deposition. A number of overlays such as nickel, Monel, Inconel or stainless steel are available, to fulfil different requirements.

Various welding processes may be used for joining clad plate, although manual metal arc welding is normally employed in view of the range of electrodes available, and the facility of control. The recommended procedure is to prepare the ferritic side of the joint, and weld this conventionally with suitable electrodes, taking care that no cladding material is picked up by the weld metal. The clad side is then chipped out to sound metal to below the depth of the cladding, and welded using suitable filler metal. The choice of consumable for this weld is determined primarily by the necessity to accept dilution from the ferritic substrate material, and give the desired final weld metal composition without the formation of undesirable microstructures. A two-pass technique is usually specified, the first pass

Table 33.20   MANUAL METAL ARC WELDING ELECTRODE COMPOSITIONS FOR WELDING CLAD STEELS ON THE CLADDING SIDE

Cladding	Electrode	
	First pass to cover steel	*Remaining passes*
Austenitic stainless steel	25Cr/12Ni	Matching composition to cladding
Chromium stainless steel	25/12	25/12 or matching composition to cladding
Nickel	Nickel 141	Nickel 141
Monel 400	Monel 190	Monel 190
Inconel 600	Inconel 182	Inconel 182

Table 33.21   MANUAL METAL ARC WELDING PROCEDURES FOR A SELECTION OF CAST STEELS

Steel type	Specification	Electrode	Preheat, °C	Post-heat, °C
0.25%C max	BS 592:1967—Grade A	BS 638:1986 E43XXR[‡]	20–150[§]	600–650[*]
0.35%C max	BS 592:1967—Grade B	BS 639:1986 E51XXB	20–150[§]	600–650
0.45%C max	BS 592:1967—Grade C	BS 639:1986 E51XXB	20–150[§]	600–650
C/0.5% Mo	BS 1398:1967—Grade A	BS 2493:1985—MoB	150	630–680
2.25%Cr/0.5%Mo	BS 1504—622	BS 2493:1985—2CrMoB	275	640–690[†]
9%Cr/1%Mo	BS 1463:1967	BS 2926:1984—23.12	250	650–720[†]
13%Cr	BS 1630:1967—Grade A	BS 2926:1984—23.12	250	680–750[†]
18%Cr/8%Ni/Nb	BS 1631:1967—Grade B	BS 2926:1984—19.9Nb	None	None
18%Cr/8%Ni/Mo	BS 1632:1967—Grade B	BS 2926:1984—19.12.3	None	None

[*] Desirable, but not essential.
[†] Immediate post-weld heat treatment and special care essential.
[‡] Select electrode to match casting properties.
[§] *See* Figure 33.1.

employing a consumable tolerant of dilution, and the second pass being intended to deposit weld metal of matching composition to the cladding. Typical consumable compositions are given in Table 33.20.

CAST STEELS

The welding of cast steels presents no special problems additional to those encountered in wrought metal of similar composition. Silicon and manganese contents are usually high, and this has an influence on weldability. Repairs to steel castings are usually subject to the same conditions of restraint as repairs to cast iron, and preheating is often desirable for this reason alone. Plain carbon steels with less than 0.25% C may otherwise be welded without preheat following the procedure in Figure 33.1. For steels containing 0.25–0.50% C, preheat temperatures up to 300°C may be used, while for steels of carbon content greater than 0.50%, a preheat of 300°C and a post-weld stress relief at 650°C should be employed. It may be necessary to use nickel-bearing electrodes when carbon contents are very high. Alternatively, bronze welding may be used.

Preheating, when recommended, should preferably be applied to the whole casting and the figures in Table 33.21 should be regarded as minima. Manual metal arc welding is normally employed, although other methods are possible.

CAST IRONS

Malleable irons, grey iron, spheroidal graphite cast irons and austenitic cast irons may be welded, provided suitable precautions are taken.

Difficulties are due to lack of ductility in the parent material to accommodate weld shrinkage stresses, the transfer of carbon to the weld metal, resulting in hard, brittle deposits, and hardening in the heat affected zone. In addition, high sulphur contents may result in hot shortness in the weld and subsequent cracking.

Table 33.22 FILLER RODS FOR GAS WELDING CAST IRON (EN 12536:2000)

	Element, wt %						
Type	C	Si	Mn	Ni	S	P	*Applications*
B1	3.0–3.6	2.8–3.5	0.5–1.0	—	0.15 max	1.5 max	Easy machining
B2	3.0–3.6	2.0–2.5	0.5–1.0	—	0.15 max	1.5 max	Hard (valve seats)
B3	3.0–3.5	2.0–2.5	0.5–1.0	1.25–1.75	0.10 max	0.50 max	Ni cast iron

Table 33.23 ELECTRODES FOR METAL ARC AND BRONZE WELDING OF CAST IRON

Electrode type	*Applications and remarks*
High nickel	Minimum preheat, easily machined. Not for high sulphur irons
60% nickel, 40% iron	Suitable for spheroidal graphite irons. Moderate machineability
Cast iron (soft iron)	General purpose, preheat essential
Austenitic stainless	Austenitic castings
Phosphor bronze	Not affected by sulphur, poor machineability
Aluminium bronze	Good strength, wear resistance and machineability

Spheroidal graphite irons and other alloy cast irons have increased ductility and impact resistance over normal grey iron, and so may be welded with rather less difficulty. White cast irons can seldom be welded satisfactorily.

The choice of process is influenced by the type of component and its composition, gas welding and braze welding being suitable for light components, and metal arc and bronze welding for the heavier types of construction. Filler rods for gas welding are given in Table 33.22. For braze welding, consumables C4, C5, or C6 in Table 33.29 may be used. Types of electrodes for manual metal arc and bronze welding are given in Table 33.23.

To accommodate shrinkage stresses and to minimise hardening in the heat affected zone, preheating to 550°C and slow cooling is essential unless minimum penetration techniques are employed. In the latter case, repairs in thin sections may be made using high nickel, nickel–iron or bronze electrodes; and in heavy sections, buttering of the edges of the joint with nickel–iron alloy should be followed by welding with soft iron electrodes. If preheating is not employed, minimum heat input is essential, by the use of short weld runs and small diameter electrodes.

33.5.2 Non-ferrous metals

ALUMINIUM AND ALUMINIUM ALLOYS

The main processes for fusion welding this group of materials are the TIG and MIG systems. Manual metal arc and oxyacetylene welding find very limited application, and then only when alternative processes are not available.

The gas-shielded MIG and TIG processes may be used for all the weldable alloys. Sound joints with good mechanical properties can be obtained, as long as weld cleaning is carefully carried out. TIG is suitable for sheet metal work, butt welds up to 6 mm thick and fillet welds where runs are short. It is also valuable in cases where the edge preparation permits autogenous welding to be used. MIG welding is particularly suitable for fillet welding and for the butt welding of material 5 mm thick and above. MIG welding normally employs commercial purity argon as a shielding gas. This gas is also general for TIG welding, although in sections above 6 mm thickness, helium may offer advantages in increased penetration and travel speed.

The majority of aluminium alloys may be welded without difficulty, provided the correct filler wire is used. Recommendations for all processes are given in Table 33.24, and consumable compositions are in Table 33.25. Certain wrought (and cast) alloys, notably aluminium–$2\frac{1}{4}$% magnesium, aluminium–magnesium silicide and duralumin types, suffer from hot cracking when welded with parent metal fillers and such fillers should be used only under closely controlled conditions of low restraint. The fusion welding of the latter types of alloy is not recommended.

Table 33.24   FILLER WIRES AND ELECTRODES FOR WELDING WROUGHT Al ALLOYS

Parent material		Filler wire or electrode[1]		
Designation	Type	Gas welding[2]	MIG or TIG[3]	Metal arc[4]
1080A	99.8%Al	1050A	1080A (1050A)	99.5%Al
1050A	99.5%Al	1050A	1050A	99.5%Al
1200	99.0%Al	1260	1050A	99.5%Al
3103	Al–Mn	3103	3103	Al–Mn
3105	Al–Mn–Mg	3103	3103	—
5251 5154A[5]	Al–Mg	5356	5356 (5056A, 5183, 5556A)	—
5454	Al–Mg	5554	5554	—
5083	Al–Mg–Mn	5356	5556A	—
6063 6061 6082 6060 6463[6]	Al–Mg–Si	4043A (4047A)	4043A (4047A, 5356)	Al–5%Si, Al–10%Si
2014A 2024 2618A 2031	Al–Cu–Mg–Si	NR[7] (4047A)	NR[7] (4047A)	NR[7] (Al–10%Si)

1 Recommended fillers given first, and alternatives in parentheses.
2 Fillers to EN 12536:2000.
3 Filler wires to EN 1668:1997, EN 440:1994.
4 These are not covered by a British Standard.
5 These Al–Mg alloys may be susceptible to hot cracking.
6 5356 may be used with care to weld these alloys, especially when anodising is to be carried out, to give a better colour match.
7 These are not recommended as weldable alloys but 4047A gives the best chance of success.

Table 33.25   FILLER RODS AND WIRES FOR THE GAS-SHIELDED WELDING OF ALUMINIUM ALLOYS (AFTER BS 2901:PART 4:1990)

Type	Element, wt %									Notes
	Al	Cu	Mg	Si	Fe	Mn	Zn	Cr	Ti	
1080A	99.8 min	0.03 max	0.02 max	0.15 max	0.15 max	0.02 max	0.06 max		0.02 max	Cu + Si + Fe + Mn + Zn: 0.2% max
1050	99.5 min	0.05 max	0.05 max	0.25 max	0.40 max	0.05 max	0.07 max		0.05 max	Cu + Si + Fe + Mn + Zn: 0.5% max Si content should be less than that of Fe
3103*	Remainder	0.1 max	0.30 max	0.50 max	0.7 max	0.9–1.5	0.20 max	0.10 max			Cr + Ti*: 0.2% max
4043A	Remainder	0.30 max	0.20 max	4.5–6.0	0.6 max	0.15 max	0.10 max		0.15 max	
4047A	Remainder	0.30 max	0.10 max	11.0–13.0	0.6 max	0.15 max	0.20 max		0.15 max	
5154A	Remainder	0.10 max	3.1–3.9	0.5 max	0.50 max	0.1–0.5	0.20 max	0.25 max	0.2 max	Mn + Cr: 0.5% max
5554	Remainder	0.10 max	2.4–3.0	0.25 max	0.40 max	0.50–1.0	0.25 max	0.05–0.20	0.05–2.00	Si + Fe: 0.40% max
5056A	Remainder	0.10 max	4.5–5.6	0.40 max	0.50 max	0.10–0.6	0.20 max	0.20 max	0.2 max	Mn + Cr: 0.1–0.5%
5556A	Remainder	0.10 max	5.0–5.5	0.25 max	0.40 max	0.6–1.0	0.2 max	0.05–0.20	0.05–0.20	Si + Fe: 0.40% max
5356	Remainder	0.10 max	4.5–5.5	0.25 max	0.40 max	0.05–0.20	0.10 max	0.05–0.20	0.06–0.20	
5183	Remainder	0.10 max	4.3–5.2	0.40 max	0.40 max	0.5–1.0	0.25 max	0.05–0.25	0.15 max	

* Ti content can include other grain refining elements.

ALUMINIUM CASTINGS

For welding heat treatable castings, the choice of filler wire should be based upon the composition of the casting itself if post-weld heat treatment is to be employed. If cracking is encountered, a higher alloy content in the filler wire may be required. Alloys containing zinc are generally difficult to weld. For welding LM6, LM9 and LM20, as in EN 1559-1:1997, EN 1559-4:1999, 4047A wire may be used: 4043A is recommended for LM18 and LM25, while 5356 and 5556A can be used for LM5 and LM10. Parent material is suggested for the remainder of the weldable materials.

DISSIMILAR ALUMINIUM ALLOYS

Welding different aluminium alloys together frequently involves a compromise between mechanical or corrosion properties and joint soundness. Recommendations are given in Table 33.26a, b and c.

33.5.3   Copper and copper alloys

Copper is produced in several grades, which vary in weldability according to the nature and quantity of the residual elements present. The material has a high thermal conductivity, and heat conduction away from the weld area may restrict the size of molten pool that can be obtained. Preheat is applied to counteract this, particularly with thicker material. Table 33.27 gives an indication of the preheat temperatures for copper and various copper alloys.

Welding is normally carried out using the gas shielded processes, and the choice of gas influences the thickness above which preheat is desirable as in Table 33.28. In general, the welding speed and penetration increase with change in shielding gas in the order Ar, He, N_2, and the level of preheat decreases with the same order of gases.

Manual metal arc welding of copper and its alloys is possible, but the gas-shielded processes are preferred. Manual metal arc welding is used mainly when other methods or suitable gas-shielded consumables are not available.

Filler wires for gas welding and gas shielded arc welding of copper and copper alloys are given in Tables 33.29 and 33.30.

Tough pitch copper, which contains residual oxygen, is available in several degrees of purity, and only the high conductivity grades of tough pitch copper should be used for welding. The inert gas-shielded processes are suitable using boron deoxidised copper filler (Table 33.30, C21) where electrical conductivity is important, or silicon–manganese deoxidised filler (Table 33.30, C7). Argon, helium, or mixtures of these gases should be used for shielding. Tough pitch copper may also be bronze welded with silicon bronze or aluminium bronze electrodes, or braze welded with filler to

Table 33.26a   PARENT METAL GROUPS USED IN TABLE 33.26c

Parent metal	Alloys
1XXX series	1080A, 1050A, 1200, 1350, LMO*
2XXX series	2014A, 2024, 2618A
3XXX series	3103, 3105
5XXX series	5251, 5454, 5154A
6XXX series	6063, 6061, 6082, 6101A
Al–Si castings	LM6, LM9, LM18, LM20, LM25
Al–Mg castings	LM5, LM10

* Included in this group for simplicity.

Table 33.26b   FILLER METAL GROUPS USED IN TABLE 33.26c

Filler metal group	Alloys
Pure Al	1080A, 1050A
Al–Si	4043A, 4047A*
Al–Mg	5056A, 5356, 5556A, 5183

* 4047A is used to prevent weld metal cracking in joints of high dilution and restraint. In most other cases, 4043A is preferable.

Table 33.26c   SELECTION OF FILLER ALLOYS FOR GAS SHIELDED ARC WELDING MATCHING AND DISSIMILAR ALUMINIUM ALLOYS

Parent metal combination	Al–Si castings	Al–Mg castings	3XXX	2XXX	1XXX	7020	6XXX	5005	5XXX	5083
5083	NR[1]	Al–Mg	Al–Mg	NR[2]	Al–Mg	556A	Al–Mg	Al–Mg	Al–Mg	5556A
5XXX	NR[1]	Al–Mg	Al–Mg	NR[2]	Al–Mg	Al–Mg	Al–Mg	Al–Mg[3]	Al–Mg[3]	
5005	Al–Si	Al–Mg	Al–Si	NR[2]	Al–Si	Al–Mg	Al–Si	Al–Mg[3]		
6XXX	Al–Si	Al–Mg	Al–Si	NR[2]			Al–Si or Al–Mg[4]			
7020	NR[1]	Al–Mg	Al–Mg	NR[2]	Al–Si	Al–Mg				
1XXX	Al–Si	Al–Mg	Al–Si	NR[2]	Al–Mg	5556A				
2XXX	NR[2]	NR[2]	NR[2]	NR[2]	Pure Al[5]					
3XXX	Al–Si	Al–Mg	3103[4]							
Al–Mg castings	NR[1]	Al–Mg								
Al–Si castings	Al–Si									

NR = not recommended.

[1] The welding of alloys containing approximately 2% or more of Mg with Al–Si filler metal (and vice-versa) is not recommended because sufficient Mg_2Si precipitate is formed at the fusion boundary to embrittle the joint.

[2] 2XXX alloys covered by British Standards are not regarded as weldable alloys, but 4047A gives the best chance of success.

[3] The corrosion behaviour of weld metal is likely to be better if its alloy content is close to that of the parent metal and not markedly higher. For service in potentially corrosive environments it is preferable to weld 5154A with 5154A filler metal or 5454 with 5554 filler metal. This may only be possible at the expense of weld soundness.

[4] Al–Si gives better crack resistance; Al–Mg gives higher weld metal ductility.

[5] For welding 1080A to itself, 1080A filler metal should be used.

Table 33.27   SUGGESTED PREHEATING CONDITIONS FOR VARIOUS COPPER ALLOYS USING ARGON SHIELDING

Material type	Preheating temperature, °C	
	Minimum	Maximum
Copper	300*	530
Silicon bronze	20	65
Phosphor bronze	175	290
70/30 Cu/Ni	20	110
Aluminium bronze	20	150

* 350°C required for TIG welding.
For further information *see* P. G. F. du Pré, *Philips Welding Reporter*, 1972, **8**, 14–26.

Table 33.28   THICKNESS ABOVE WHICH PREHEAT MAY BE REQUIRED—COPPER AND COPPER ALLOYS

Process	Shielding gas		
	Argon	Helium	Nitrogen
TIG	3 mm	6 mm	9 mm
MIG	6 mm	9 mm	12 mm

See Table 33.27 for further information.

Table 33.29   FILLER WIRES FOR GAS WELDING COPPER AND COPPER ALLOYS (EN 12536:2000)

Type	Element, wt %							
	Cu	Zn	Pb	Al	Fe	Ni	Mn	Si**
C1	99.85 min[†]		0.010 max		0.030 max	0.10 max		
C2	57.0–63.0	rem.	0.03 max	0.03 max				0.2–0.5
C2B	56.0–60.0	rem.	0.05 max	0.01 max	0.25–1.2	0.2–0.8	0.01–0.50	0.04–0.15
C2C	56.0–60.0	rem.	0.05 max	0.01 max	0.25–1.2		0.01–0.50	0.04–0.15
C3	59.0–61.0	rem.	0.03 max	0.03 max			0.05–0.25	0.15–0.3
C4	57.0–63.0	rem.	0.03 max	0.03 max	0.1–0.5		0.5 max	0.15–0.5
C5	45.0–53.0	rem.	0.03 max	0.03 max	0.5 max	8.0–11.0	0.5 max	0.15–0.5
C6	41.0–45.0	rem.	0.03 max	0.03 max	0.3 max	14.0–16.0	0.2 max	0.2–0.5

** For other elements *see* next page.   (*continued*)

Table 33.29   FILLER WIRES FOR GAS WELDING COPPER AND COPPER ALLOYS (EN 12536:2000)—*continued*

Type	Sn	As	Sb	Bi	P	Tl	Total impurities excluding Ag, Ni, As, P
			Element**, wt %				
C1	0.01 max	0.05 max	0.005 max	0.0030 max	0.015–0.08	0.010 max*	0.060 max
C2	0.5 max						
C2B	0.8–1.1						0.50 max incl. Pb and Al
C2C	0.8–1.1						0.50 max incl. Pb and Al
C3							0.50 max
C4	0.5 max						
C5	0.5 max						
C6	1.0 max						

* Se + Tl: 0.020% max.   † Includes Ag 0.5–1.2%.   ** For other elements *see* previous page.

Table 33.30   FILLER RODS AND WIRES FOR THE GAS-SHIELDED WELDING OF COPPER AND COPPER ALLOYS (BS 2901:PART 3:1990)

Type	Cu	Al	Ti	Fe	Ni
		Element, wt %			
C7*	98.5 min	0.03 max		0.03 max	0.10 max
C8†	99.4 min	0.1–0.3 Al + Ti: 0.25–0.5	0.1–0.3	0.30 max	0.10 max
C9	Remainder	0.03 max		0.10 max	0.10 max
C10	93.8 min	0.03 max			
C11	92.3 min	0.03 max			
C12	90.0 min	6.0–7.5		(Fe + Ni + Mn)‡: 1.0–2.5	
C12Fe	89.0 min	6.5–8.5		2.3–3.5	
C13	86.0 min	9.0–11.0		0.75–1.5	1.0 max
C16	Remainder	0.03 max	0.20–0.50	1.5–1.8	10.0–12.0
C18	66.5 min	0.03 max	0.20–0.50	0.4–1.0	30.0–32.0
C20	80.5–85.0	8.0–9.5		1.5–3.5	3.5–5.0
C22	Remainder	7.0–8.5		2.0–4.0	1.5–3.0
C23	Remainder	6.0–6.4		0.5–0.7	0.1 max
C24	Remainder	0.01 max		0.10 max	
C25	Remainder	0.05 max	0.05 max	0.05 max	1.0–1.7
C26	Remainder	8.5–9.5		3.0–5.0	4.0–5.5

Type	Mn	Si	Zn	Sn	P
		Element, wt %			
C7*	0.15–0.35	0.20–0.35		1.0 max	0.015 max
C8†					0.015 max
C9	0.75–1.25	2.75–3.25			0.020 max
C10			0.5 max	4.5–6.0	0.02–0.40
C11				6.0–7.5	0.02–0.40
C12		0.10	0.2 max		
C12Fe		0.10 max	0.2 max		
C13	1.0 max	0.10 max	0.2 max		0.01 max
C16	0.5–1.0	0.01 max			0.01 max
C18	0.5–1.5	0.01 max			
C20	0.5–2.0	0.10 max	0.2 max		
C22	11.0–14.0	0.10 max	0.15 max		
C23§	0.5 max	2.0–2.4	0.4 max	0.1 max	
C24	1.5–2.5		0.2 max		
C25	0.15–0.4	0.4–0.7		4.5–6.0	
C26	0.6–3.5	0.10 max	0.10 max		

* These rods and wires are intended for welding Cu using Ar or He as the shielding gas.
† These rods and wires are intended for welding Cu using N_2 as the shielding gas: Ar or He may be used.
‡ Optional elements.
§ 0.05%Mg max.

Table 33.29, C2 (silicon brass). The presence of arsenic does not affect weldability. Oxyacetylene welding is not recommended, due to the risk of 'gassing'.

Phosphorus-deoxidised (PDO) copper may be welded by the oxy-acetylene, TIG or MIG processes. Argon, helium or nitrogen may be used for gas-shielded methods, either separately or mixed. The use of nitrogen produces a hotter arc with increased penetration, but with MIG welding, less satisfactory metal transfer may result. Phosphorus content is important, and should be as low as possible to minimise porosity. Phosphorus does not act as an efficient deoxidant in gas-shielded welding, and for this reason filler wires containing additional deoxidants should be used. Autogenous welding is not possible without the risk of porosity, and if welding without filler wire is required, zinc deoxidised (cap) copper can be employed, using the TIG process. Zinc content is relatively unimportant within the range 0.5–3.0% zinc. Oxy-acetylene welds are made using a copper–silver–phosphorus filler rod (Table 33.29, Cl) and a flux. Such welds are hot hammered during welding to remove porosity and frequently cold hammered to improve mechanical properties.

Oxygen-free high conductivity copper may be oxy-acetylene welded without gassing, but if TIG or MIG welding is employed, there is a risk of porosity formation which may be overcome by using boron–copper filler wire.

All grades of copper may be bronze welded using the manual metal arc, TIG or MIG processes and fillers of the aluminium, silicon or tin bronze types. The technique involves the use of a wide edge preparation, preferably a fillet and the use of soft arcs to minimise penetration. Deoxidised and oxygen-free copper may also be braze welded using silicon brass (Table 33.29, C2) or manganese bronze (Table 33.29, C4) fillers.

COPPER–ALUMINIUM ALLOYS

Gas welding is not recommended but carbon arc welding using cryolite flux, manual metal arc welding, or preferably the argon or helium TIG welding processes may be used. The iron-bearing single phase alloy Cu/7 Al/3 Fe is normally welded with a nickel-bearing duplex alloy (Table 33.30, C20). Most duplex and complex bronzes are welded with fillers of matching composition, as are the manganese–aluminium bronzes. An important consideration is corrosion resistance, and care should be taken to avoid a combination of manganese–aluminium bronze and the normal single phase and duplex bronzes. The nickel-bearing filler C20 is resistant to dealuminification in all but the most severe environments and is frequently used as a facing deposit in welds in BS 1400: AB2C castings, which are normally made with 10% aluminium fillers (Table 33.30, C13). If a single phase deposit is required, C12 Fe filler may be used for a corrosion resistant layer on top of a more crack resistant filler. If stress corrosion is a problem, a small tin addition may be made to both parent and filler metals.

The welding of the aluminium bronzes should be carried out with as low a heat input as possible. Thus, preheating and high inter-run temperatures should be avoided, as should weaving when depositing filler metal. It is often necessary to give a post-weld thermal treatment to eliminate the risk of stress corrosion cracking.

COPPER–NICKEL ALLOYS

The problems of embrittlement and porosity in welding cupro-nickel may be overcome by using filler wires containing manganese as a desulphuriser and titanium as a deoxidant. Filler wires are available for 90/10 and 70/30 cupro-nickels (Table 33.30, C16 and C18). Either argon arc or inert-gas metal arc welding is normally employed, but flux-coated manual metal arc electrodes are available for some alloys.

COPPER–SILICON ALLOYS

Silicon bronzes and 'Everdur' are readily weldable, the inert-gas processes being preferred. Parent metal filler is employed (Table 33.30, C9). Bronze welding is also possible.

COPPER–TIN ALLOYS

Tin bronzes usually contain phosphorus, and their welding behaviour is somewhat similar to phosphorus deoxidised copper, both oxy-acetylene and TIG welds tending to be porous. Filler rods employed are given in Table 33.30 (C10, C11). Bronze welding is preferable.

The weldability of the gunmetals depends upon the lead content. Those containing 0.1%Pb are weldable by the TIG, MIG, manual metal arc and gas welding processes. At 0.1–0.5%Pb, manual metal arc with phosphorus bronze electrodes or bronze welding offers a reasonable chance of success. The leaded gunmetals can be considered to be unweldable although single layer cosmetic repairs may sometimes be made successfully.

Both classes of material have long freezing ranges and are consequently hot short. They are also subject to coring and shrinkage porosity if large molten pools are employed.

COPPER–ZINC ALLOYS

The welding of brass is not easy, due to excessive fume formation, and fume extraction plant may prove necessary. Oxy-acetylene welding using an oxidising flame and parent metal filler or silicon or manganese–brass or nickel–silver fillers is possible. TIG and MIG welding both employ zinc-free filler alloys in preference to brass, although brass fillers are available for the former process (Table 33.30). For manual metal arc and MIG welding, silicon bronze or aluminium bronze electrodes are employed. The susceptibility of brasses to season cracking should be borne in mind, and stress relief of brass components is desirable.

BERYLLIUM–COPPER

Alternating current TIG welding is preferred. In order to obtain the maximum joint efficiency, welding speed should be high, and post-weld heat treatment is essential. Both the high Be/low Co and high Co/low Be alloys should be welded in the solution heat-treated condition using high Be filler rods. Post-weld ageing will restore most of the strength and hardness. Although possible, consumable electrode methods are not recommended due to the toxicity problem.

CADMIUM–COPPER

Experience is limited, but it is suggested that for gas welding, filler rods to BS 1453 C1 (Table 33.29) should be used. Parent metal filler may be suitable for inert-gas welding. Cadmium is extremely toxic and fusion welding should be attempted only under stringent ventilation control.

CHROMIUM–COPPER

Chromium–copper is heat-treatable and suffers from a loss in mechanical properties in the heat affected zone. It also exhibits a tendency to hot shortness. Bronze welding with aluminium bronze filler has been successfully employed, but joint efficiencies are low after post-weld heat treatment. Joints of low restraint may be made using a d.c. electrode negative TIG technique with helium shielding. The components should be solution treated prior to welding, and subsequently aged.

OTHER COPPER-RICH ALLOYS

Silver–copper may be welded by the inert-gas processes using boron–copper filler wire. Tellurium– and selenium–copper suffer from excessive weld porosity. Leaded copper is hot short, and welding is difficult.

33.5.4 Lead and lead alloys

Lead and its alloys may be welded by the oxy-acetylene process using no flux and a neutral flame. Other fuel gases, such as hydrogen, butane or coal gas, may be used. The edges should be cleaned before welding, and parent metal filler is used. Other processes are little used.

33.5.5 Magnesium alloys

Magnesium alloys are preferably welded by the a.c. TIG process with argon shielding. Normal inert-gas metal arc welding is not suitable for magnesium because of the high deposition rates required.

Table 33.31   FILLER RODS AND WIRES FOR GAS-SHIELDED WELDING OF MAGNESIUM ALLOYS (AFTER BS 2901:PART 4:1990)

Type	*Element*, wt %					
	Al	Zn	Mn	Zr	Rare earths	Th
MAG 1	7.5–9.0	0.3–1.0	0.15–0.4			
MAG 3	9.0–10.5	0.3–1.0				
MAG 5		3.5–5.0		0.4–1.0	0.75–1.75	
MAG 6		0.8–3.0		0.4–1.0	2.5–4.0	
MAG 8		1.7–2.5	0.15 max	0.4–1.0	0.10 max	2.5–4.0
MAG 9		5.0–6.0	0.15 max	0.4–1.0	0.20 max	1.5–2.3
MAG 111	2.5–3.5	0.6–1.4	0.15–0.40			
MAG 141	0.02 max	0.75–1.5	0.15 max	0.4–0.8		

However, the pulsed version of this process has found application. The use of gas welding is confined to the wrought Mg–Mn and Mg–Al–Zn alloys, and should never be used for fillet or lap welds or for welds in zirconium-bearing alloys due to the corrosion hazard from flux residues. Proprietary fluxes are available. Flux removal and chemical cleaning should be followed by chromating.

The argon arc process may be used for welding all the commercially available alloys, with the exception of the wrought Mg–3%Zn–0.7%Zr and Mg–5.5%Zn–0.7%Zr alloys and the cast Mg–4.5%Zn–0.7%Zr alloy.

British Standard filler alloys are listed in Table 33.31. For castings parent metal filler is invariably used. A guide to filler metal selection is given in Table 33.32 for wrought and cast products to BS 3370, 3372, 3373 and 2970.

33.5.6   Nickel and nickel alloys

These materials fall into two groups, namely solid solution and precipitation hardening alloys. Nickel alloys are highly sensitive to contamination during welding and to the presence of various minor impurities. To avoid weld cracking and porosity, all surfaces to be welded and filler rods must be grease-free and scrupulously clean. The presence of sulphur, lead and zinc are particularly detrimental, the effects of sulphur being most marked with the chromium-free alloys. Most nickel alloy weld metals are quite fluid and, if possible, components should be welded in the flat position.

It is normally recommended that nickel alloys be stress-free prior to welding. The solid solution hardened grades should be annealed, although a certain amount of cold work is permissible. Post-weld heat treatment is not usually necessary. Incoloy DS* containing silicon is sensitive to hot cracking and restraint must be kept to a minimum. Precipitation hardened materials should be solution treated before welding, and aged afterwards. Some materials based on the Ni–Al–Ti precipitation hardening system are susceptible to post-weld heat treatment cracking, depending on the joint restraint, as are austenitic stainless steels. In such cases, a full heat treatment after welding is recommended. HAZ liquation cracking may arise and in certain precipitation hardened alloys, such as Nimonic* 80A, 90 and PK33, restricts the thickness that can be welded to below 5 mm.

TIG welding, both manual and mechanised, is the most widely used process for sheet material. A.c. or d.c. may be employed, although the latter is preferred. Most alloys may be welded and suitable fillers are listed in Tables 33.33 and 33.34. Chromium-free grades are susceptible to porosity during autogenous welding, and the addition of filler is recommended so that at least 50% of the weld pool volume is constituted by filler metal. Commercial purity argon is generally used for shielding with additions of up to 10% hydrogen helping to reduce porosity and increase welding speeds. Helium or argon–helium mixtures are applicable and may also be advantageous in increasing welding speed. Nitrogen in the gas or from the atmosphere can cause porosity. Complete coverage of the weld area with shielding gas is essential. The largest gas nozzle possible should be used with the minimum practicable distance from the nozzle to the workpiece, while adequate gas backing must be applied. Many nickel alloys form tenacious oxide films, which tend to be patchy and interfere with weld uniformity. Argon–5% hydrogen mixtures can be used as backing and shielding gases to overcome such problems. Nickel–molybdenum alloys may give a very fluid weld pool, when the TIG process is preferred for positional welding, in view of the facility of control.

---

* Henry Wiggin & Co. Ltd.

Table 33.32 SELECTION OF FILLER ALLOYS FOR GAS-SHIELDED WELDING MATCHING AND DISSIMILAR MAGNESIUM ALLOYS (AFTER EN 1011-4:2000)

Parent metal combination	Mg-3Al-1Zn-Mn	Mg-6Al-1Zn-Mn	Mg-3Zn-0.5Zr	Mg-8Al-0.5Zn-Mn	Mg-10Al-0.5Zn-Mn	Mg-4Zn-1RE-0.5Zr	Mg-2Zn-3RE-0.5Zr	Mg-3Th-2Zn-0.5Zr	Mg-5.5Zn-2Th-0.5Zr
Mg-3Al-1Zn-Mn	MAG 111	MAG 1	NR	MAG 1	MAG 3	MAG 6 [P]	MAG 6 [P]	NR	NR
Mg-6Al-1Zn-Mn		MAG 121	NR	MAG 1	MAG 3	MAG 6	MAG 6 [P]	NR	NR
Mg-3Zn-0.5Zr			MAG 151	NR	NR	MAG 141 [P]	MAG 141 [P]	MAG 141 [P]	MAG 141 [P]
Mg-1Al-0.5Zn-Mn				MAG 1 / MAG 2 or MAG 7	MAG 3	MAG 6 [P]	MAG 6 [P]	MAG 8 [P]	MAG 8 [P]
Mg-10Al-0.5Zn-Mn					MAG 3	MAG 6 [P]	MAG 6 [P]	MAG 8 [P]	MAG 8 [P]
Mg-4Zn-1RE-0.5Zr						MAG 5	MAG 6	MAG 8 [P]	MAG 8 [P]
Mg-2Zn-3RE-0.5Zr							MAG 6	MAG 8 [P]	MAG 8 [P]
Mg-3Th-2Zn-0.5Zr								MAG 8	MAG 8
Mg-5.5Zn-2Th-0.5Zr									MAG 9

Key: NR = not recommended, P = sensitive to cracking.

Table 33.33   FILLER RODS AND WIRES FOR GAS SHIELDED WELDING OF NICKEL ALLOYS (BS 2901:PART 5:1990)

Type	Ni	Cr	Co	Fe	Mo	Ti	Al
				Element, wt %			
NA32	93.0 min			1.0 max		2.0–3.50	1.50 max
NA33	62.0–69.0			2.5 max		1.5–3.0	1.25 max
NA34	Remainder	18.0–21.0		0.5 max			
NA35	67.0 min	18.0–22.0	0.12 max	3.0 max		0.75 max	
NA36	Remainder	18.0–21.0	15.0–21.0	3.0 max		1.8–3.0	0.9–2.0
NA37	Remainder	16.0–20.0	12.0–16.0	1.0 max	5.0–9.0	1.5–3.0	1.7–2.5
NA38	Remainder	19.0–21.0	19.0–21.0	0.7 max	5.6–6.1	1.9–2.4	0.3–0.60
NA39	67.0 min	14.0–17.0		0.80 max		2.5–5.0	
NA40	Remainder	20.5–23.0	0.5–2.5	17.0–20.0	8.0–10.0		
NA41	33.0–46.0	19.5–23.5	1.0 max	Remainder	2.5–3.40	0.6–2.20	0.20 max
NA42	42.0–45.0	15.0–18.0	2.0 max	Remainder	2.5–4.0	0.9–1.5	0.9–1.5
NA43	58.0 min	20.0–23.0	1.0 max	5.0 max	8.0–10.0	0.40 max	0.40 max
NA44	Remainder	1.0 max	1.0 max	2.0 max	26.0–30.0		
NA45	Remainder	14.0–18.0	2.0 max	3.0 max	14.0–17.0	0.70 max	

Type	Mn	C	Si	Cu	Notes
			Element, wt %		
NA32	1.00 max	0.15 max	0.75 max	0.25 max	Other elements 0.5% max
NA33	3.0–4.0	0.15 max	1.25 max	Remainder	
NA34	1.2 max	0.26 max	0.50 max	0.20 max	
NA35	2.5–3.50	0.10 max	0.50 max	0.50 max	Ta 0.30% max: (Nb + Ta) 2.0–3.00%
NA36	1.0 max	0.13 max	1.5 max	0.2 max	
NA37	0.5 max	0.07 max	0.5 max	0.2 max	B 0.005% max: Zr 0.06% max
NA38	0.2–0.60	0.04–0.08	0.1–0.40	0.2 max	(Ti + Al) 2.4–2.8%
NA39	2.0–2.75	0.08 max	0.35 max	0.50 max	
NA40	1.0 max	0.15 max	1.0 max	0.50 max	W 0.2–1.0%
NA41	1.0 max	0.05 max	0.50 max	1.5–3.00	Other elements 0.5% max
NA42	0.2 max	0.10 max	0.3 max	0.5 max	B 0.005% max: Zr 0.05% max
NA43	0.5 max	0.10 max	0.50 max	0.50 max	(Nb + Ta) 3.15–4.15%
NA44	1.0 max	0.02 max	0.50 max	0.50 max	
NA45	1.0 max	0.015 max	0.50 max	0.50 max	

Table 33.34   TYPICAL CONSUMABLES FOR WELDING NICKEL ALLOYS

Parent material	Welding process		
	Manual metal arc	Gas	Inert gas shielded
Nickel	Nickel electrode 141	Nickel 41	NA32 (Nickel 61)
Monel[†] 400	Monel electrode 190	Monel 40	NA33 (Monel 60)
Monel K-500	Monel electrode 134	Monel 64	Monel 64
Inconel[†] 600	Inconel electrodes 132 or 182	NA34 (NC80/20)[†]	NA35 (Inconel 82)
Incoloy[†] 800	Incoweld[†] A	NA34 (NC80/20)	NA35 (Inconel 82)
Incoloy DS	Incoweld A	NA34 (NC80/20)	NA34 (NC80/20)
Nimonic[†]	Inconel electrodes 132 or 182	NA34 (NC80/20)	NA34 (NC80/20)
Nimonic 80A			NA36 (Nimonic 90)
Nimonic 90			NA36 (Nimonic 90)
Hastelloy* X			NA40

See Table 33.33 for NA32, 33, 34, 35, 36 and 40.
* Union Carbide U.K. Ltd.
[†] Henry Wiggin & Co. Ltd.

For welding nickel and its alloys in heavier gauges, manual metal arc and MIG welding are generally suitable, the welding technique differing little from that practised with austenitic stainless steels. With manual metal arc welding, it is advisable to use small gauge electrodes, and to allow cooling between weld runs. Complete slag removal is imperative, particularly for high temperature service where reactions between the metal and residual slag may cause severe corrosion. Table 33.34 gives typical consumables for some of the more common commercial alloys: d.c. current is essential with positive electrodes. There is no British Standard for manual metal arc electrodes for nickel alloys, but a number of proprietary makes are obtainable.

Argon shielding is preferred with MIG welding, although the addition of up to 20% helium may be of benefit in spreading the weld pool and reducing the incidence of cold laps. Spray, globular dip, or dip transfer conditions have been employed for most nickel alloys, with consumables shown in Tables 33.33 and 33.34. Spray transfer gives the highest deposition rates but may cause weld metal cracking, particularly if the welding conditions are such that a weld bead with a concave surface is produced. Pulsed MIG welding has also been found applicable for nickel alloys, especially for positional welding.

Oxy-acetylene welding is little used for nickel alloys. It is suitable for most solid solution hardened alloys and Monel K-500, although carbon pick-up may reduce corrosion resistance. Nickel and Incoloy DS can be welded without flux while fluxes are available for the Monel, Incoloy, Inconel, Brightray and Hastelloy D alloys. Fluxes used with the chromium-containing alloys must be free from boron, since the presence of boron compounds can cause weld metal cracking. Complete flux removal after welding is essential. The flame should be slightly reducing for nickel and Monel 400, more reducing with the chromium-containing alloys, and highly reducing for Monel K-500 and Hastelloy D. Filler alloys are given in Table 33.34.

Submerged-arc welding is applicable to various nickel alloys, generally when heavy section material is being joined for corrosion resistant applications. Only a limited range of flux types is available. D.c. is used, with electrode positive being preferred to obtain deeper penetration and reduced risk of slag entrapment.

Electron beam welding can be used for most nickel alloys, and is a common fabrication process for high temperature items such as gas turbine components.

33.5.7 Noble metals

SILVER

Silver is difficult to weld due to its property of oxygen absorption. A slightly reducing flame is used for oxy-acetylene welding with a borax–boric acid flux. Silver–0.5% aluminium rods reduce the tendency to porosity formation. Argon arc welding using d.c. is possible, though slight porosity is usually present. A.c. argon arc welding using aluminium bearing fillers or traces of aluminium powder in the joint preparation is also possible.

GOLD

Gold and gold–copper alloys are readily welded by the gas welding process using parent metal filler and borax fluxes. A reducing flame is recommended for the alloys.

PLATINUM

An oxidising oxy-hydrogen flame is used without flux. The platinum–rhodium and platinum–iridium alloys may also be welded using this technique.

33.5.8 Refractory metals

The problem common to the welding of this group of materials is their strong affinity for atmospheric and other gases, and the consequent need to avoid contamination, both of the molten pool and the surfaces of the cooling solidified weld bead and heat affected zones. For arc welding this problem is overcome in two ways, either by provision of a complete argon-filled chamber in which the workpiece and welding head is placed, or by the provision of extended argon shrouds and argon backing.

These materials may all be welded by the tungsten inert gas arc process, and titanium has been successfully welded by the inert-gas metal arc process. The use of this process may eventually be extended to cover other materials in this group. Electron beam welding is also generally applicable, while, given appropriate gas shielding, laser welding may be suitable in some cases. Solid state joining by friction welding and diffusion welding is summarised in Tables 33.5 and 33.6, respectively.

TITANIUM AND TITANIUM ALLOYS

The α titanium alloys (e.g. commercially pure Ti, Ti–5%Al–2$\frac{1}{2}$%Sn) are fully weldable, as are alloys with only small amounts of eutectoid formers or β stabilisers (e.g. Ti–2$\frac{1}{2}$%Cu, Ti–8%Al–1%Mo–1%V, Ti–7%Al–2%Nb–1%Ta). As the β stabiliser content rises, the weld zone becomes less ductile and eventually brittle. The Ti–6%Al–4%V alloy represents about the limiting composition for $\alpha\beta$ alloys for good weldability.

In the weldable alloys, TIG welds (d.c., electrode negative) may be made without filler or with parent metal filler depending on thickness. For $\alpha\beta$ alloys, the ductility of the weld metal can be increased by using commercially pure Ti filler, but this will not improve the ductility of the heat affected zone. MIG welding is feasible, but experience is limited. All metal reaching higher than 600–650°C should be protected by argon.

NIOBIUM AND TANTALUM AND THEIR ALLOYS

The same technique as for titanium should be used, except that argon protection will be required at and above 400°C.

MOLYBDENUM

Electron beam and TIG (d.c., electrode negative) welds can be made, but the weld zone will be brittle below 300–500°C. The techniques used for titanium should be applied. The metal produced by sintering gives grossly porous welds, and that made by arc casting should be used.

ZIRCONIUM ALLOYS

Experience is limited to the zirconium–tin and zirconium–niobium alloys, which are used in nuclear engineering. These materials are particularly susceptible to nitrogen contamination, which is harmful to the corrosion resistance. Superficial contamination during welding is removed by pickling. D.c. electrode negative is used, and gas cover should extend to areas in excess of 400°C.

URANIUM

Uranium may be welded without difficulty by the TIG process using d.c. electrode negative. It is less sensitive to contamination than titanium.

33.5.9 Zinc and zinc alloys

Zinc and zinc-base castings are difficult to weld. The oxy-acetylene process using a slightly reducing flame and a zinc-ammonium chloride flux is most suitable. Parent metal filler should be used.

33.5.10 Dissimilar metals

Direct fusion welding of dissimilar metals is possible in certain cases, and special techniques have been developed for difficult combinations.

Steels may be joined to nickel alloys, but the joint dilution and selection of filler materials must be made to give either a high nickel-alloy weld or a weld composition falling within the shaded area of the Schaeffler diagram (Figure 33.3).

Combinations of copper and chromium should be avoided, as when welding Monel to chromium bearing steels, and in such cases, nickel-base filler should be interposed.

Steels may be joined to copper and copper alloys by bronze welding techniques, but joining to aluminium involves precoating of the steel with aluminium–silicon brazing alloy, and joint ductility is low.

Copper may be welded direct to nickel, provided a deoxidant (titanium) is present either in the nickel or the filler wire.

Copper to aluminium joints are made by coating the copper with a layer of silver solder (BS 1845: Type AG1 or AG2) and welding by argon arc or inert gas metal arc processes, using aluminium 10% silicon filler.

33.6 Major standards relating to welding, brazing and soldering

It is impossible to list all the standards relating to welding, brazing and soldering, considering the quantity of standards issued by many different nations and organizations and the growth of this body of documents on a daily basis. Therefore, internationally-agreed standards are becoming more and more important, especially considering the increasing economic globalisation in the past few decades. It is a recently accelerating trend that individual European national standards, such as British Standards (BS), are being gradually replaced by European standards, and/or by the International Organization for Standardization (ISO) standards. European standards issued by the European Committee for Standardization (CEN) are increasingly similar in structure and scope to the ISO. The following tables cover some of the major ISO, European, British, AWS, and Canadian standards relating to welding, brazing and soldering, in which the ISO and European standards are listed according to the International Classification for Standards (ICS). European Standards that are identical to ISO are not included. Instead, the listing for the ISO standards (Table 1) includes cross-references. Similarly, British Standards that are identical to ISO or EN are not included. Instead, the listing for the ISO standards (Table 33.35) or European Standards (Table 33.36) includes a cross-reference.

Table 33.35 ISO STANDARDS RELATING TO WELDING, BRAZING AND SOLDERING

ICS: 25.160.01 Welding, brazing and soldering in general

ISO

581:1980	Weldability—Definition
3834:	Quality requirements for welding—Fusion welding of metallic materials Part 1:1994 Guidelines for selection and use Part 2:1994 Comprehensive quality requirements Part 3:1994 Standards quality requirements Part 4:1994 Elementary quality requirements
4063:1998	Welding and allied processes—Nomenclature of processes and reference numbers (Incorporated in EN ISO 4063:2000, BS EN ISO 4063:2000)
9606:	Approval testing of welders—Fusion welding Part 1:1994 Steels Part 1:1994/Amd 1:1998 Part 2:1994 Aluminium and aluminium alloys Part 2:1994/Amd 1:1998 Part 3:1999 Copper and copper alloys (Incorporated in EN ISO 9606-3:1999, BS EN ISO 9606-3:1999) Part 4:1999 Nickel and nickel alloys (Incorporated in EN ISO 9606-4:1999, BS EN ISO 9606-4:1999) Part 5:2000 Titanium and titanium alloys, zirconium and zirconium alloys (Incorporated in EN ISO 9606-5:2000, BS EN ISO 9606-5:2000)
10882:	Health and safety in welding and allied processes—Sampling of airborne particles and gases in the operator's breathing zone Part 1:2001 Sampling of airborne particles (Incorporated in EN ISO 10882-1:2000, BS EN ISO 10882-1:2001) Part 2:2000 Sampling of gases (Incorporated in EN ISO 10882-2:2000, BS EN ISO 10882-2:2000)
14554:	Quality requirements for welding – Resistance welding of metallic materials Part 1:2000 Comprehensive quality requirements (Incorporated in EN ISO 14554-1: 2000, BS EN ISO 14554-1:2001) Part 2:2000 Elementary quality requirements (Incorporated in EN ISO 14554-2: 2000, BS EN ISO 14554-2:2001)

(continued)

Table 33.35   ISO STANDARDS RELATING TO WELDING, BRAZING AND SOLDERING—*continued*

14731:1997	Welding coordination—Tasks and responsibilities
14732:1998	Welding personnel—Approval testing of welding operators for fusion welding and of resistance weld setters for fully mechanised and automatic welding of metallic materials
15608:2000	Welding—Guidelines for a metallic materials grouping system (Incorporated in PD CR ISO 15608:2000)
17663:2001	Welding—Guidelines for quality requirements for heat treatment in connection with welding and allied processes (ISO/TR 17663:2000) (Incorporated in PD CR IS 17663:2001)

ICS: 25.160.10 Welding processes

857:	Welding and allied processes—Vocabulary Part 1:1998 Metal welding processes
9013:1992	Welding and allied processes—Quality classification and dimensional tolerances of thermally cut (oxygen/fuel gas flame) surfaces (Incorporated in EN ISO 9013:1995, BS EN ISO 9013:1995)
9692:	Welding and allied processes—Joint preparation Part 2:1998 Submerged arc welding of steels (Incorporated in EN ISO 9692-2:1998, BS EN ISO 9692-2:1998) Part 3:2000 Metal inert gas welding and tungsten inert gas welding of aluminium and its alloys (Incorporated in EN ISO 9692-3:2001)
9956:	Specification and approval of welding procedures for metallic materials Part 1:1995 General rules for fusion welding Part 1:1995/Amd 1:1998 Part 2:1995 Welding procedure specification for arc welding Part 2:1995/Amd 1:1998 Part 3:1995 Welding procedure tests for arc welding of steels Part 3:1995/Amd 1:1998 Part 4:1995 Welding procedure tests for the arc welding of aluminium and its alloys Part 4:1995/Amd 1:1998 Part 5:1995 Approval by using approved welding consumables for arc welding Part 6:1995 Approval related to previous experience Part 7:1995 Approval by a standard welding procedure for arc welding Part 8:1995 Approval by a pre-production welding test Part 10:1996 Welding procedure specification for electron beam welding (Incorporated in EN ISO 9956-10:1996, BS EN ISO 9956-10:1997) Part 11:1996 Welding procedure specification for laser beam welding (Incorporated in EN ISO 9956-11:1996, BS EN ISO 9956-11:1997)
11970:2001	Specification and approval of welding procedures for production welding of steel castings
13916:1996	Welding—Guidance on the measurement of preheating temperature, interpass temperature and preheat maintenance temperature (Incorporated in EN ISO 13916:1996, BS EN ISO 13916:1997)
13920:1996	Welding—General tolerances for welded constructions—Dimensions for lengths and angles—Shape and position (Incorporated in EN ISO 13920:1996, BS EN ISO 13920:1997)
14555:1998	Welding—Arc stud welding of metallic materials (Incorporated in EN ISO 14555:1998, BS EN ISO 14555:1998)
15609:	Specification and qualification of welding procedures for metallic materials—Welding procedure specification Part 2:2001 Gas welding (Incorporated in EN ISO 15609-2:2001, BS EN ISO 15609-2:2001)
15620:2000	Welding—Friction welding of metallic materials (Incorporated in BS EN ISO 15620:2000)

ICS: 25.160.20 Welding consumables (including electrodes, filler metals, gases, etc.)

544:1989	Filler materials for manual welding—Size requirements
636:1989	Bare solid filler rods for oxy-acetylene and tungsten inert gas arc (TIG) welding, depositing an unalloyed or low alloyed steel—Codification

(*continued*)

Table 33.35 ISO STANDARDS RELATING TO WELDING, BRAZING AND SOLDERING—*continued*

864:1988	Arc welding—Solid and tubular cored wires which deposit carbon and carbon manganese steel—Dimensions of wires, spools, rims and coils
1071:1983	Covered electrodes for manual arc welding of cast iron—Symbolisation
2401:1972	Covered electrodes—Determination of the efficiency, metal recovery and deposition coefficient (Incorporated in EN 22401:1994, BS EN 22401:1994)
2560:1973	Covered electrodes for manual arc welding of mild steel and low alloy steel—Code of symbols for identification ISO 2560:1973/Add 1:1974
3580:1975	Covered electrodes for manual arc welding of creep-resisting steels—Code of symbols for identification
3581:1976	Covered electrodes for manual arc welding of stainless and other similar high alloy steels—Code of symbols for identification
5182:1991	Welding—Materials for resistance welding electrodes and ancillary equipment
5184:1979	Straight resistance spot welding electrodes (Incorporated in EN 25184:1994, BS EN 25184:1995)
6847:2000	Welding consumables—Deposition of a weld metal pad for chemical analysis (Incorporated in EN ISO 6847:2001, BS EN 26847:1994)
6848:1984	Tungsten electrodes for inert gas shielded arc welding, and for plasma cutting and welding—Codification (Incorporated in EN 26848:1991, BS EN 26848:1991)
10446:1990	Welding—All-weld metal test assembly for the classification of corrosion-resisting chromium and chromium–nickel steel covered arc welding electrodes
14175:1997	Welding consumables—Shielding gases for arc welding and cutting
14372:2000	Welding consumables—Determination of moisture resistance of manual metal arc welding electrodes by measurement of diffusible hydrogen
15792:	Welding consumables—Test methods Part 1:2000 Test methods for all-weld metal test specimens in steel, nickel and nickel alloys Part 2:2000 Preparation of single-run and two-run technique test specimens in steel Part 3:2000 Classification testing of positional capacity and root penetration of welding consumables in a fillet weld

ICS: 25.160.30 Welding equipment (including thermal cutting equipment)

669:2000	Resistance welding—Resistance welding equipment—Mechanical and electrical requirements (Incorporated in BS 3065:2001)
693:1982	Dimensions of seam welding wheel blanks (Incorporated in EN ISO 20693:1991, BS EN 20693:1992)
865:1981	Slots in platens for projection welding machines (Incorporated in EN 20865:1991, BS EN 20865:1992)
1089:1980	Electrode taper fits for spot welding equipment—Dimensions (Incorporated in EN 21089:1991, BS EN 21089:1992)
2503:1998	Gas welding equipment—Pressure regulators for gas cylinders used in welding, cutting and allied processes up to 300 bar
3253:1998	Gas welding equipment—Hose connections for equipment for welding, cutting and allied processes
3821:1998	Gas welding equipment—Rubber hoses for welding, cutting and allied processes
5171:1995	Pressure gauges used in welding, cutting and allied processes
5172:1995	Manual blowpipes for welding, cutting and heating—Specifications and tests (Incorporated in EN ISO 5172:1996, BS EN ISO 5172:1997) ISO 5172:1995/Amd 1:1995
5175:1987	Equipment used in gas welding, cutting and allied processes—Safety devices for fuel gases and oxygen or compressed air—General specifications, requirements and tests
5183:	Resistance welding equipment—Electrode adaptors, male taper 1:10 Part 1:1998 Conical fixing, taper 1:10 (Incorporated in EN ISO 5183-1:2000, BS EN ISO 5183-1:2001) Part 2:2000 Parallel shank fixing for end-thrust electrodes (Incorporated in EN 25183-2:1991, BS EN 25183-2:1992)

(*continued*)

Table 33.35   ISO STANDARDS RELATING TO WELDING, BRAZING AND SOLDERING—*continued*

5186:1995	Oxygen/fuel gas blowpipes (cutting machine type) with cylindrical barrels—General specifications and test methods
5821:1979	Resistance spot welding electrode caps (Incorporated in EN 25821:1991, BS EN 25821:1992)
5822:1988	Spot welding equipment—Taper plug gauges and taper ring gauges (Incorporated in EN 25822:1991, BS EN 25822:1992)
5826:1999	Electric resistance welding—Transformers—General specifications applicable to all transformers (Incorporated in BS 7125:1989)
5827:1983	Spot welding—Electrode back-ups and clamps (Incorporated in EN 25827:1992, BS EN 25827:1992)
5828:2001	Resistance welding equipment—Secondary connecting cables with terminals connected to water-cooled lugs—Dimensions and characteristics (Incorporated in EN ISO 5828: 2001, BS EN 5828:2001)
5829:1984	Resistance spot welding—Electrode adaptors, female taper 1 : 10
5830:1984	Resistance spot welding—Male electrode caps
6210:	Cylinders for robot resistance welding guns Part 1:1991 General requirements
7284:1993	Resistance welding equipment—Particular specifications applicable to transformers with two separate secondary windings for multi-spot welding, as used in the automobile industry (Incorporated in EN ISO 7284:1996, BS EN ISO 7284:1996)
7285:1995	Pneumatic cylinders for mechanised multiple spot welding
7286:1986	Graphical symbols for resistance welding equipment (Incorporated in EN 27286:1991, BS EN 27286:1995)
7287:1992	Graphical symbols for thermal cutting equipment (Incorporated in EN ISO 7287:1995, BS EN ISO 7287:1995)
7289:1996	Quick-action couplings with shut-off valves for gas welding, cutting and allied processes
7291:1999	Gas welding equipment—Pressure regulators for manifold systems used in welding, cutting and allied processes up to 300 bar (Incorporated in BS EN ISO 7291:2001)
7292:1997	Flowmeter regulators used on cylinders for welding, cutting and allied processes—Classification and specifications
7931:1985	Insulation caps and bushes for resistance welding equipment (Incorporated in EN 27931:1992)
8167:1989	Projections for resistance welding (Incorporated in BS EN 28167:1992)
8205:	Water-cooled secondary connection cables for resistance welding Part 1:1993 Dimensions and requirements for double-conductor connection cables (Incorporated in EN ISO 8205-1:1996, BS EN ISO 8205-1:1997) Part 2:1993 Dimensions and requirements for single-conductor connection cables (Incorporated in EN ISO 8205-2:1996, BS EN ISO 8205-2:1997) Part 3:1993 Test requirements (Incorporated in EN ISO 8205-3:1996, BS EN ISO 8205-3:1997)
8206:1991	Acceptance tests for oxygen cutting machines—Reproducible accuracy—Operational characteristics (Incorporated in EN 28206:1992, BS EN 28206:1992)
8207:1996	Gas welding equipment—Specification for hose assemblies for equipment for welding, cutting and allied processes
8430:	Resistance spot welding—Electrode holders Part 1:1988 Taper fixing 1:10 (Incorporated in EN 28430-1:1992, BS EN 28430-1:1992) Part 2:1988 Morse taper fixing (Incorporated in EN 28430-2:1992, BS EN 28430-2:1992) Part 3:1988 Parallel shank fixing for end thrust (Incorporated in EN 28430-3:1992, BS EN 28430-3:1992) Part 3:1988/Cor 1:1990
9012:1998	Gas welding equipment—Air-aspirated hand blowpipes—Specifications and tests
9090:1989	Gas tightness of equipment for gas welding and allied processes (Incorporated in EN 29090:1992, BS EN 29090:1992)
9312:1990	Resistance welding equipment—Insulated pins for use in electrode back-ups (Incorporated in EN ISO 9312:1994, BS EN ISO 9312:1995)

(*continued*)

Table 33.35   ISO STANDARDS RELATING TO WELDING, BRAZING AND SOLDERING—*continued*

9313:1989	Resistance spot welding equipment—Cooling tubes (Incorporated in EN 29313:1992, BS EN 29313:1992)
9539:1988	Materials for equipment used in gas welding, cutting and allied processes (Incorporated in EN 29539:1992, BS EN 29539:1992)
10656:1996	Electric resistance welding—Integrated transformers for welding guns ISO 10656:1996/Cor 1:2000
11032:1994	Manipulating industrial robots—Application oriented test—Spot welding
12145:1998	Resistance welding equipment—Angles for mounting spot welding electrodes
12166:1997	Resistance welding equipment—Particular specifications applicable to transformers with one secondary winding for multi-spot welding, as used in the automobile industry
12170:1996	Gas welding equipment—Thermoplastic hoses for welding and allied processes
14112:1996	Gas welding equipment—Small kits for gas brazing and welding
14113:1997	Gas welding equipment—Rubber and plastic hoses assembled for compressed or liquefied gases up to a maximum design pressure of 450 bar (Incorporated in 14113:1997)
14114:1999	Gas welding equipment—Acetylene manifold systems for welding, cutting and allied processes—General requirements (Incorporated in EN ISO 14114:1999, BS EN ISO 14114:1999)
14744:	Welding—Acceptance inspection of electron beam welding machines Part 1:2000 Principles and acceptance conditions (Incorporated in BS EN ISO 14744-1:2001) Part 2:2000 Measurement of accelerating voltage characteristics (Incorporated in BS EN ISO 14744-2:2001) Part 3:2000 Measurement of beam current characteristics (Incorporated in BS EN ISO 14744-3:2001) Part 4:2000 Measurement of welding speed (Incorporated in BS EN ISO 14744-4:2001) Part 5:2000 Measurement of run-out accuracy (Incorporated in BS EN ISO 14744-5:2001) Part 6:2000 Measurement of stability of spot position (Incorporated in BS EN ISO 14744-6:2001)

ICS: 25.160.40 Welded joints (including welding position and mechanical and non-destructive testing of welded joints)

1106:	Recommended practice for radiographic examination of fusion welded joints Part 1:1984 Fusion welded butt joints in steel plates up to 50 mm thick Part 2:1985 Fusion welded butt joints in steel plates thicker than 50 mm and up to and including 200 mm in thickness Part 3:1984 Fusion welded circumferential joints in steel pipes of up to 50 mm wall thickness
2400:1972	Welds in steel—Reference block for the calibration of equipment for ultrasonic examination
2437:1972	Recommended practice for the X-ray inspection of fusion welded butt joints for aluminium and its alloys and magnesium and its alloys 5 to 50 mm thick
2504:1973	Radiography of welds and viewing conditions for films—Utilisation of recommended patterns of image quality indicators (I.Q.I.)
2553:1992	Welded, brazed and soldered joints—Symbolic representation on drawings (Incorporated in EN 22553:1994, BS EN 22553:1995)
3088:1975	Welding requirements—Factors to be considered in specifying requirements for fusion welded joints in steel (technical influencing factors)
3690:2000	Welding and allied processes—Determination of hydrogen content in ferritic steel arc weld metal (Incorporated in EN ISO 3690:2000, BS EN ISO 3690:2001)
4136:2001	Destructive tests on welds in metallic materials—Transverse tensile test
5173:2000	Destructive tests on welds in metallic materials—Bend tests
5178:2001	Destructive tests on welds in metallic materials—Longitudinal tensile test on weld metal in fusion welded joints
5817:1992	Arc-welded joints in steel—Guidance on quality levels for imperfections (Incorporated in EN 25817:1992, BS EN 25817:1992)

(continued)

Table 33.35   ISO STANDARDS RELATING TO WELDING, BRAZING AND SOLDERING—*continued*

6520:	Welding and allied processes—Classification of geometric imperfections in metallic materials Part 1:1998 Fusion welding (Incorporated in EN ISO 6520-1:1998, BS EN ISO 6520-1:1998)
6947:1990	Welds—Working positions—Definitions of angles of slope and rotation (Incorporated in EN ISO 6947:1997, BS EN ISO 6947:1997)
7963:1985	Welds in steel—Calibration block No. 2 for ultrasonic examination of welds (Incorporated in EN 27963:1992, BS EN 2763:1992)
8249:2000	Welding—Determination of Ferrite Number (FN) in austenitic and duplex ferritic-austenitic Cr–Ni stainless steel weld metals (Incorporated in EN ISO 8248:2000, BS EN ISO 8249:2000)
9015:	Destructive tests on welds in metallic materials—Hardness testing Part 1:2001 Hardness test on arc welded joints
9016:2001	Destructive tests on welds in metallic materials—Impact tests—Test specimen location, notch orientation and examination
9017:2001	Destructive tests on welds in metallic materials—Fracture test
9692:1992	Metal-arc welding with covered electrode, gas-shielded metal-arc welding and gas welding—Joint preparations for steel (Incorporated in BS EN ISO 29692-3:2001)
9764:1989	Electric resistance and induction welded steel tubes for pressure purposes—Ultrasonic testing of the weld seam for the detection of longitudinal imperfections
9765:1990	Submerged arc-welded steel tubes for pressure purposes—Ultrasonic testing of the weld seam for the detection of longitudinal and/or transverse imperfections
10042:1992	Arc-welded joints in aluminium and its weldable alloys—Guidance on quality levels for imperfections (Incorporated in EN 30042:1994, BS EN 30042:1994)
10447:1991	Welding—Peel and chisel testing of resistance spot, projection and seam welds
12096:1996	Submerged arc-welded steel tubes for pressure purposes—Radiographic testing of the weld seam for the detection of imperfections
13663:1995	Welded steel tubes for pressure purposes—Ultrasonic testing of the area adjacent to the weld seam for the detection of laminar imperfections
13919:	Welding—Electron and laser-beam welded joints—Guidance on quality levels for imperfections Part 1:1996 Steel (Incorporated in EN ISO 13919-1:1996, BS EN ISO 13919-1:1997) Part 2:2001 Aluminium and its weldable alloys
14270:2000	Specimen dimensions and procedure for mechanised peel testing resistance spot, seam and embossed projection welds
14271:2000	Vickers hardness testing of resistance spot, projection and seam welds (low load and microhardness)
14272:2000	Specimen dimensions and procedure for cross tension testing resistance spot and embossed projection welds
14273:2000	Specimen dimensions and procedure for shear testing resistance spot, seam and embossed projection welds

ICS: 25.160.50 Brazing and soldering (including brazing and soldering alloys and equipment)

857:	Welding and allied processes—Vocabulary Part 1:1998 Metal welding processes
3677:1992	Filler metal for soft soldering, brazing and braze welding—Designation (Incorporated in EN ISO 3677:1995, BS EN ISO 3677:1995)
5179:1983	Investigation of brazeability using a varying gap test piece
5187:1985	Welding and allied processes—Assemblies made with soft solders and brazing filler metals—Mechanical test methods
9453:1990	Soft solder alloys—Chemical compositions and forms (Incorporated in EN 29453:1993, BS EN 29453:1994)

(continued)

Table 33.35   ISO STANDARDS RELATING TO WELDING, BRAZING AND SOLDERING—*continued*

9454:	Soft soldering fluxes—Classification and requirements Part 1:1990 Classification, labelling and packaging (Incorporated in EN 29454-1:1993, BS EN 29454-1:1994) Part 2:1998 Performance requirements (Incorporated in EN ISO 9454-2:2000, BS EN ISO 9454-2:2001)
9455:	Soft soldering fluxes—Test methods Part 1:1990 Determination of non-volatile matter, gravimetric method (Incorporated in EN 29455-1:1993, BS EN 29455-1:1994) Part 2:1993 Determination of non-volatile matter, ebulliometric method (Incorporated in EN ISO 9455-2:1995, BS EN ISO 9455-2:1996) Part 3:1992 Determination of acid value, potentiometric and visual titration methods (Incorporated in EN ISO 9455-3:1994, BS EN ISO 9455-3:1995) Part 5:1992 Copper mirror test (Incorporated in EN 29455-5:1993, BS EN 29455-5:1993) Part 6:1995 Determination and detection of halide (excluding fluoride) content (Incorporated in EN ISO 9455-6:1997, BS EN ISO 9455-6:1997) Part 8:1991 Determination of zinc content (Incorporated in EN 29455-8:1993, BS EN 29455-8:1993) Part 9:1993 Determination of ammonia content (Incorporated in EN ISO 9455-9:1995, BS EN ISO 9455-9:1996) Part 10:1998 Flux efficacy tests, solder spread method (Incorporated in EN ISO 9455-10:2000, BS EN ISO 9455-10:2001) Part 11:1991 Solubility of flux residues (Incorporated in EN 29455-11:1993, BS EN 29455-11:1993) Part 12:1992 Steel tube corrosion test (Incorporated in EN ISO 9455-12:1994, BS EN ISO 9455-12:1994) Part 13:1996 Determination of flux spattering (Incorporated in EN ISO 9455-13:1999, BS EN ISO 9455-13:2000) Part 14:1991 Assessment of tackiness of flux residues (Incorporated in EN 29455-14:1993, BS EN 29455-14:1993) Part 15:1996 Copper corrosion test (Incorporated in EN ISO 9455-15:1999, BS EN ISO 9455-15:2000) Part 16:1998 Flux efficacy tests, wetting balance method (Incorporated in EN ISO 9455-16:2001)
10564:1993	Soldering and brazing materials—Methods for the sampling of soft solders for analysis (Incorporated in EN ISO 10564:1997, BS EN ISO 10564:1997)
12224:	Solder wire, solid and flux cored—Specification and test methods Part 1:1997 Classification and performance requirements (Incorporated in EN ISO 12224-1:1998, BS EN ISO 12224-1:1998) Part 2:1997 Determination of flux content (Incorporated in EN ISO 12224-2:1999, BS EN ISO 12224-2:1999)
14112:1996	Gas welding equipment—Small kits for gas brazing and welding

Table 33.36   EUROPEAN STANDARDS RELATING TO WELDING, BRAZING AND SOLDERING

ICS: 25.160.01 Welding, brazing and soldering in general

EN

657:1994	Thermal spraying—Terminology, classification

ICS: 25.160.10 Welding processes

287:	Approval testing of welders—Fusion welding Part 1:1992 Steels (Incorporated in BS EN 287-1:1992) Part 2:1992 Aluminium and aluminium alloys (Incorporated in BS EN 278-2:1992)
288:	Specification and qualification of welding procedures for metallic materials Part 1:1992 General rules for fusion welding (Incorporated in BS EN 288-1:1992) Part 2:1992 Welding procedure specification for arc welding (Incorporated in BS EN 288-2:1992) Part 3:1992 Welding procedure tests for the arc welding of steels (Incorporated in BS EN 288-3:1992)

(*continued*)

Table 33.36 EUROPEAN STANDARDS RELATING TO WELDING, BRAZING AND SOLDERING—*continued*

	Part 4:1992 Welding procedure tests for the arc welding of aluminium and its alloys (Incorporated in BS EN 288-4:1992)
	Part 6:1994 Approval related to previous experience (Incorporated in BS EN 288-6:1995)
	Part 7:1995 Approval by a standard welding procedure for arc welding (Incorporated in BS EN 288-7:1995)
	Part 8:1995 Approval by a pre-production welding test (Incorporated in BS EN 288-8:1995)
	Part 9:1999 Welding procedure test for pipeline welding on land and offshore site butt welding of transmission pipelines (Incorporated in BS EN 288-9:1999)
719:1994	Welding coordination—Tasks and responsibilities (Incorporated in BS EN 719:1994)
729:	Quality requirements for welding—Fusion welding of metallic materials
	Part 1:1994 Guidelines for selection and use (Incorporated in BS EN 729-1:1995)
	Part 2:1994 Comprehensive quality requirements (Incorporated in BS EN 729-2:1995)
	Part 3:1994 Standard quality requirements (Incorporated in BS EN 729-3:1995)
	Part 4:1994 Elementary quality requirements (Incorporated in BS EN 729-4:1995)
1011:	Welding—Recommendations for welding of metallic materials
	Part 1:1998 General guidance for arc welding (Incorporated in BS EN 1011-1:1998)
	Part 2:2001 Arc welding of ferritic steels (Incorporated in BS EN 1011-2:2001)
	Part 3:2000 Arc welding of stainless steels (Incorporated in BS EN 1011-3:2000)
	Part 4:2000 Arc welding of aluminium and aluminium alloys (Incorporated in BS EN 1011-4:2000)
1418:1997	Welding personnel—Approval testing of welding operators for fusion welding and resistance weld setters for fully mechanised and automatic welding of metallic materials (Incorporated in BS EN 1418:1998)
1598:1997	Health and safety in welding and allied processes. Transparent welding curtains, strips and screens for arc welding processes (Incorporated in BS EN 1598:1998)
1792:1997	Welding—Multilingual list of terms for welding and related processes (Incorporated in BS EN 1792:1998)
12584:1999	Imperfections in oxyfuel flame cuts, laser beams cuts and plasma cuts—Terminology (Incorporated in BS EN 12584:1999)
13918:1998	Welding—Studs and ceramic ferrules for arc stud welding (Incorporated in EN ISO 13918:1998, BS EN ISO 13918:1998)

ICS: 25.160.20 Welding consumables (including electrodes, filler metals, gases, etc.)

288:	Specification and approval of welding procedures for metallic materials
	Part 5:1994 Approval by using approved welding consumables for arc welding (Incorporated in BS EN 288-5:1995)
439:1994	Welding consumables—Shielding gases for arc welding and cutting (Incorporated in BS EN 439:1994)
440:1994	Welding consumables—Wire electrodes and deposits for gas shielded metal arc welding of non alloy and fine grain steels—Classification (Incorporated in BS EN 440:1995)
499:1994	Welding consumables—Covered electrodes for manual metal arc welding of non alloy and fine grain steels—Classification (Incorporated in BS EN 499:1995)
60974:	Arc welding equipment
	Part 11:1995 Electrode holders (Incorporated in BS EN 60974: Part 11:1996)
756:1995	Welding consumables—Wire electrodes and wire-flux combinations for submerged arc welding of non alloy and fine grain steels—Classification (Incorporated in BS EN 756:1996)
757:1997	Welding consumables—Covered electrodes for manual metal arc welding of high strength steels—Classification (Incorporated in BS EN 757:1997)
758:1997	Welding consumables—Tubular cored electrodes for metal arc welding with and without a gas shield of non alloy and fine grain steels—Classification (Incorporated in BS EN 758:1997)
759:1997	Welding consumables—Technical delivery conditions for welding filler metals—Type of product, dimensions, tolerances and marking (Incorporated in BS EN 759:1997)
760:1996	Welding consumables—Fluxes for submerged arc welding—Classification (Incorporated in BS EN 760:1996)

(*continued*)

Table 33.36   EUROPEAN STANDARDS RELATING TO WELDING, BRAZING AND SOLDERING—*continued*

1597:	Welding consumables—Test methods Part 1:1997 Test piece for all-weld metal test specimens in steel, nickel and nickel alloys (Incorporated in BS EN 597:1997) Part 2:1997 Preparation of test piece for single-run and two-run technique test specimens in steel (Incorporated in BS EN 1597-2:1997) Part 3:1997 Testing of positional capability of welding consumables in a fillet weld (Incorporated in BS EN 1597-3:1997)
1599:1997	Welding consumables—Covered electrodes for manual arc welding of creep-resisting steels—Classification (Incorporated in BS EN 1599:1997)
1600:1997	Welding consumables—Covered electrodes for manual arc welding of stainless and heat resisting steels—Classification (Incorporated in BS EN 1600:1997)
1668:1997	Welding consumables—Rods, wires and deposits for tungsten inert gas welding of non alloy and fine grain steels—Classification (Incorporated in BS EN 1668:1997)
12070:1999	Welding consumables—Wire electrodes, wires and rods for arc welding of creep-resisting steels—Classification (Incorporated in BS EN 12070:2000)
12071:1999	Welding consumables—Tubular cored electrodes for gas shielded metal arc welding of creep-resisting steels—Classification (Incorporated in BS EN 12071:2000)
12072:1999	Welding consumables—Wire electrodes, wires and rods for arc welding of stainless and heat-resisting steels—Classification (Incorporated in BS EN 12072:2000)
12073:1999	Welding consumables—Tubular cored electrodes for metal arc welding with or without a gas shield of stainless and heat-resisting steels—Classification (Incorporated in BS EN 12073:2000)
12074:2000	Welding consumables—Quality requirements for manufacture, supply and distribution of consumables for welding and allied processes (Incorporated in BS EN 12074:2000)
12534:1999	Welding consumables—Wire electrodes, wires, rods and deposits for gas shielded metal arc welding of high strength steels—Classification (Incorporated in BS EN 12534:1999)
12535:2000	Welding consumables—Tubular cored electrodes for gas shielded metal arc welding of high strength steels—Classification (Incorporated in BS EN 12535:2000)
12536:2000	Welding consumables—Rods for gas welding of non alloy and creep-resisting steels—Classification (Incorporated in BS EN 12536:2000)
12943:1999	Filler materials for thermoplastics—Scope, designation, requirements, tests (Incorporated in BS EN 12943:2000)
28167:1992	Projections for resistance welding (8167:1989) (Incorporated in BS EN 28167:1992)

ICS: 25.160.30 Welding equipment (including thermal cutting equipment)

559:1994	Gas welding equipment. Rubber hoses for welding, cutting and allied processes (Incorporated in BS EN 559:1994)
560:1994	Gas welding equipment. Hose connections for welding, cutting and allied processes (Incorporated in BS EN 560:1995)
561:1994	Gas welding equipment. Quick-action couplings with shut-off valves for welding, cutting and allied processes (Incorporated in BS EN 561:1995)
562:1994	Gas welding equipment—Pressure gauges used in welding, cutting and allied processes
730:1995	Gas welding equipment—Equipment used in gas welding, cutting and allied processes, safety devices for fuel gases and oxygen or compressed air—General specifications, requirements and tests (Incorporated in BS EN 730:1995)
731:1995	Gas welding equipment—Air-aspirated hand blowpipes—Specifications and tests (Incorporated in BS EN 731:1996)
874:1995	Gas welding equipment—Oxygen/fuel gas blowpipes (cutting machine type) of cylindrical barrel—Type of construction, general specifications, test methods (Incorporated in BS EN 874:1995)
961:1995	Gas welding equipment. Manifold regulators used in welding, cutting and allied processes up to 200 bar (Incorporated in BS EN 961:1996)
1256:1996	Gas welding equipment. Specification for hose assemblies for equipment for welding, cutting and allied processes (Incorporated in BS EN 1256:1996)

(continued)

Table 33.36   EUROPEAN STANDARDS RELATING TO WELDING, BRAZING AND SOLDERING—*continued*

1327:1996	Gas welding equipment. Thermoplastic hoses for welding and allied processes (Incorporated in BS EN 1327:1996)
50063:1989	Specification for safety requirements for the construction and the installation of equipment for resistance welding and allied processes (Incorporated in BS 5924:1989)
50199:1995	Electromagnetic compatibility (EMC). Product standard for arc welding equipment (Incorporated in BS EN 50199:1996)
60974:	Arc welding equipment. Part 1:1998, IEC 60974-1:1998 Welding power sources (Incorporated in BS EN 60974:Part 1:1998) Part 7:2000, IEC 60974-7:2000 Arc welding equipment. Torches (Incorporated in BS EN 60974:Part 11:2000) Part 12:1995 Arc welding equipment. Coupling devices for welding cables (Incorporated in BS EN 60974:Part 12:1996)

ICS: 25.160.40 Welded joints (including welding position and mechanical and non-destructive testing of welded joints)

875:1995	Destructive tests on welds in metallic materials—Impact tests—Test specimen location, notch orientation and examination (Incorporated in BS EN 875:1995)
876:1995	Destructive tests on welds in metallic materials—Longitudinal tensile test on weld metal in fusion welded joints (Incorporated in BS EN 876:1995)
895:1995	Destructive tests on welds in metallic materials—Transverse tensile test (Incorporated in BS EN 895:1995)
910:1996	Destructive tests on welds in metallic materials—Bend tests (Incorporated in BS EN 910:1996)
970:1997	Non-destructive examination of fusion welds—Visual examination (Incorporated in BS EN 970:1997)
1043:	Destructive tests on welds in metallic materials—Hardness testing Part 1:1995 Hardness test on arc welded joints (Incorporated in BS EN 1043-1:1996) Part 2:1996 Micro hardness testing on welded joints (Incorporated in BS EN 1043-2:1997)
1289:1998	Non-destructive examination of welds—Penetrant testing of welds—Acceptance levels (Incorporated in BS EN 1289:1998)
1290:1998	Non-destructive examination of welds—Magnetic particle examination of welds (Incorporated in BS EN 1290:1998)
1291:1998	Non-destructive examination of welds—Magnetic particle testing of welds—Acceptance levels (Incorporated in BS EN 1291:1998)
1320:1996	Destructive tests on welds in metallic materials—Fracture test (Incorporated in BS EN 1320:1997)
1321:1996	Destructive tests on welds in metallic materials—Macroscopic and microscopic examination of welds (Incorporated in BS EN 1321:1997)
1435:1997	Non-destructive examination of welds—Radiographic examination of welded joints (Incorporated in BS EN 1435:1997)
1708:	Welding—Basic weld joint details in steel Part 1:1999 Pressurised components (Incorporated in BS EN 1708-1:1999) Part 2:2000 Non internal pressurised components (Incorporated in BS EN 1708-2:2000)
1711:2000	Non-destructive examination of welds—Eddy current examination of welds by complex plane analysis (Incorporated in BS EN 1711:2000)
1712:1997	Non-destructive examination of welds—Ultrasonic examination of welded joints—Acceptance levels (Incorporated in BS EN 1712:1997)
1713:1998	Non-destructive examination of welds—Ultrasonic examination—Characterisation of indications in welds (Incorporated in BS EN 1713:1998)
1714:1997	Non-destructive examination of welds—Ultrasonic examination of welded joints (Incorporated in BS EN 1714:1998)
2497:1989	Specification for dry abrasive blasting of titanium and titanium alloys (Incorporated in BS EN 2497:1990)

(*continued*)

Table 33.36   EUROPEAN STANDARDS RELATING TO WELDING, BRAZING AND SOLDERING—*continued*

2574:1990	Aerospace series—Welds—Information on drawings
10246:	Non-destructive testing of steel tubes Part 10:2000 Radiographic testing of the weld seam of automatic fusion arc welded steel tubes for the detection of imperfections (Incorporated in BS EN 10246-10:2000) Part 16:2000 Automatic ultrasonic testing of the area adjacent to the weld seam of welded steel tubes for the detection of laminar imperfections (Incorporated in BS EN 10246-16:2000) Part 8:1999 Automatic ultrasonic testing of the weld seam of electric welded steel tubes for the detection of longitudinal imperfections (Incorporated in BS EN 10246-8:2000) Part 9:2000 Automatic ultrasonic testing of the weld seam of submerged arc welded steel tubes for the detection of longitudinal and/or transverse imperfections (Incorporated in BS EN 10246-9:2000)
12062:1997	Non-destructive examination of welds—General rules for metallic materials (Incorporated in BS EN 12062:1998)
12345:1998	Welding—Multilingual terms for welded joints with illustrations (Incorporated in BS EN 12345:1999)
12517:1998	Non-destructive examination of welds—Radiographic examination of welded joints—Acceptance levels (Incorporated in BS EN 12517:1998)
12732:2000	Gas supply systems—Welding steel pipework—Functional requirements (Incorporated in BS EN 12732:2000)
12814:	Testing of welded joints of thermoplastic semi-finished products Part 1:1999 Bend test (Incorporated in BS EN 12814-1:2000) Part 2:2000 Tensile test (Incorporated in BS EN 12814-2:2000) Part 3:2000 Tensile creep test (Incorporated in BS EN 12814-3:2000) Part 4:2001 Peel test Part 5:2000 Macroscopic examination (Incorporated in BS EN 12814-5:2000) Part 6:2000 Low temperature tensile test (Incorporated in BS EN 12814-6:2000) Part 8:2001 Requirements
13100:	Non destructive testing of welded joints of thermoplastics semi-finished products Part 1:1999 Visual examination (Incorporated in BS EN 13100-1:2000)

ICS: 25.160.50 Brazing and soldering (including brazing and soldering alloys and equipment)

1044:1999	Brazing—Filler metals (Incorporated in BS EN 1044:1999)
1045:1997	Brazing—Fluxes for brazing—Classification and technical delivery conditions (Incorporated in BS EN 1045:1997)
12797:2000	Brazing—Destructive tests of brazed joints (Incorporated in BS EN 12797:2000)
12799:2000	Brazing—Non-destructive examination of brazed joints (Incorporated in BS EN 12799:2000)
13133:2000	Brazing—Brazer approval (Incorporated in BS EN 13133:2000)
13134:2000	Brazing—Procedure approval (Incorporated in BS EN 13134:2000)
1326:1996	Gas welding equipment—Small kits for gas brazing and welding (Incorporated in BS EN 1326:1996)

Note: European Standards that are identical to ISO are not included in this table. Instead, the listing for the ISO standards (Table 33.35) includes cross-references.

Table 33.37   BRITISH STANDARDS RELATING TO WELDING, BRAZING AND SOLDERING

ICS: 25.160.01 Welding, brazing and soldering in general

BS

499:	Welding terms and symbols Part 1:1991 Glossary for welding, brazing and thermal cutting Part 2C:1999 European arc welding symbols in chart form
4515:	Specification for welding of steel pipelines on land and offshore Part 1:2000 Carbon and carbon manganese steel pipelines Part 2:1999 Duplex stainless steel pipelines
4870:	Specification for approval testing of welding procedures Part 3:1985 Arc welding of tube to tube-plate joints in metallic materials

(*continued*)

Table 33.37   BRITISH STANDARDS RELATING TO WELDING, BRAZING AND SOLDERING—*continued*

4871:	Specification for approval testing of welders working to approved welding procedures Part 3:1985 Arc welding of tube to tube-plate joints in metallic materials
4872:	Specification for approval testing of welders when welding procedure approval is not required Part 1:1982 Fusion welding of steel Part 2:1976 TIG or MIG welding of aluminium and its alloys
7670:	Steel nuts and bolts for resistance projection welding Part 1:1993 Specification for dimensions and properties

ICS: 25.160.10 Welding processes

1140:1993	Specification for resistance spot welding of uncoated and coated low carbon steel
1724:1990	Specification for bronze welding by gas
1821:1982	Specification for class I oxy-acetylene welding of ferritic steel pipework for carrying fluids
2630:1982	Specification for resistance projection welding of uncoated low carbon steel sheet and strip using embossed projections
2633:1987	Specification for Class I arc welding of ferritic steel pipework for carrying fluids
2640:1982	Specification for Class II oxy-acetylene welding of carbon steel pipework for carrying fluids
2971:1991	Specification for class II arc welding of carbon steel pipework for carrying fluids
4204:1989	Specification for flash welding of steel tubes for pressure applications
4570:1985	Specification for fusion welding of steel castings
4677:1984	Specification for arc welding of austenitic stainless steel pipework for carrying fluids
6265:1982	Specification for resistance seam welding of uncoated and coated low carbon steel
6944:1988	Specification for flash welding of butt joints in ferrous metals (excluding pressure piping applications)
6990:1989	Code of practice for welding on steel pipes containing process fluids or their residuals
7123:1989	Specification for metal arc welding of steel for concrete reinforcement
7670:	Steel nuts and bolts for resistance projection welding Part 2:1997 Specification for welding of weld nuts and weld bolts

ICS: 25.160.20 Welding consumables (including electrodes, filler metals, gases, etc.)

807:1955	Specification for spot welding electrodes
1453:1972	Specification for filler materials for gas welding
2901:	Filler rods and wires for gas-shielded arc welding Part 3:1990 Specification for copper and copper alloys Part 4:1990 Specification for aluminium and aluminium alloys and magnesium alloys Part 5:1990 Specification for nickel and nickel alloys
4129:1990	Specification for welding primers and weld-through sealants, adhesives and waxes for resistance welding of sheet steel
4215:	Resistance spot welding electrodes, electrode holders and ancillary equipment Part 2:1987 Specification for straight resistance spot welding electrodes Part 6:1987 Specification for electrode adaptors, female taper 1 : 10 Part 7:1987 Specification for male electrode caps
4577:1970	Specification for materials for resistance welding electrodes and ancillary equipment
7384:1991	Guide to laboratory methods for sampling and analysis of particulate matter generated by arc welding consumables

ICS: 25.160.30 Welding equipment (including thermal cutting equipment)

499:	Welding terms and symbols Part 1:Supplement:1992 Glossary for welding, brazing and thermal cutting. Supplement Definitions for electric welding equipment
638:	Arc welding power sources, equipment and accessories Part 4:1996 Specification for welding cables Part 5:1988 Specification for accessories Part 7:1984 Specification for safety requirements for installation and use Part 9:1990, EN 50060:1989 Arc welding power sources, equipment and accessories Specification for power sources for manual arc welding with limited duty

(*continued*)

Table 33.37   BRITISH STANDARDS RELATING TO WELDING, BRAZING AND SOLDERING—*continued*

4215:	Resistance spot welding electrodes, electrode holders and ancillary equipment Part 15:1990 Specification for insulated pins for use in electrode back-ups
4819:	Resistance welding water-cooled transformers of the press-package and portable types Part 2:1989 Specification for portable transformers
6942:	Design and construction of small kits for oxy-fuel gas welding and allied processes Part 2:1989 Specification for kits using refillable gas containers for oxygen and fuel gas
7570:2000	Code of practice for validation of arc welding equipment

ICS: 25.160.40 Welded joints (including welding position and mechanical and non-destructive testing of welded joints)

709:1983	Methods of destructive testing fusion welded joints and weld metal in steel
3451:1973	Methods of testing fusion welds in aluminium and aluminium alloys
3923:	Methods for ultrasonic examination of welds Part 2:1972 Automatic examination of fusion welded butt joints in ferritic steels
4206:1967	Methods of testing fusion welds in copper and copper alloys
6084:	Methods for the evaluation of prefabrication primers Part 1:1997 Rating and weldability tests
7009:1988	Guide to application of real-time radiography to weld inspection
7363:1990	Methods for controlled thermal severity (CTS) test and bead-on-plate (BOP) test for welds
7448:	Fracture mechanics toughness tests Part 2:1997 Method for determination of KIc, critical CTOD and critical J values of welds in metallic materials
7706:1993	Guide to calibration and setting-up of the ultrasonic time of flight diffraction (TOFD) technique for the detection, location and sizing of flaws
7910:1999	Guide on methods for assessing the acceptability of flaws in metallic structures

ICS: 25.160.50 Brazing and soldering (including brazing and soldering alloys and equipment)

1723:	Brazing Part 1:1986 Specification for brazing Part 2:1986 Guide to brazing
5245:1975	Specification for phosphoric acid based flux for soft soldered joints in stainless steel

Note: British Standards that are identical to ISO or EN are not included in this table. Instead, the listing for the ISO standards (Table 33.35) or European Standards (Table 33.36) includes a cross-reference.

Table 33.38   AWS STANDARDS RELATING TO WELDING, BRAZING AND SOLDERING

Definitions and Symbols

AISI/AWS

A2.4:1998	Standard Symbols for Welding, Brazing, and Nondestructive Examination
A3.0:1994	Standard Welding Terms and Definitions

Arc and Gas Welding and Cutting

C4.2:1990	Operator's Manual Oxyfuel Gas Cutting
C5.1:1973	Recommended Practices for Plasma Arc Welding
C5.2:2001	Recommended Practices for Plasma Arc Cutting and Gouging
C5.3:2000	Recommended Practices for Air Carbon Arc Gouging and Cutting
C5.4:1993	Recommended Practices for Stud Welding
C5.5:1980R	Recommended Practices for Gas Tungsten Arc Welding
C5.6:1989R	Recommended Practices for Gas Metal Arc Welding
C5.7:2000	Recommended Practices for Electrogas Welding
C5.10:1994	Recommended Practices for Shielding Gases for Welding and Plasma Arc Cutting

(*continued*)

Table 33.38   AWS STANDARDS RELATING TO WELDING, BRAZING AND SOLDERING—*continued*

Brazing

A5.8:1992	Specification for Filler Metals for Brazing and Braze Welding
A5.31:1992	Specifications for Fluxes for Brazing and Braze Welding
C3.2:1982R	Standard Method for Evaluating the Strength of Brazed Joints in Shear
C3.3:1980R	Recommended Practices for Design, Manufacture, and Inspection of Critical Brazed Components
C3.4:1999	Specification for Torch Brazing
C3.5:1999	Specification for Induction Brazing
C3.6:1999	Specification for Furnace Brazing
C3.7:1999	Specification for Aluminium Brazing
C3.8:1990R	Recommended Practices for Ultrasonic Inspection of Brazed Joints

Resistance Welding

C1.1/C1.1M:2000	Recommended Practices for Resistance Welding
C1.4/C1.4M:1999	Specification for Resistance Welding of Carbon and Low-Alloy Steels

Surfacing Processes

C2.18:1993	Guide for the Protection of Steel with Thermal Sprayed Coatings of Aluminium and Zinc and Their Alloys and Composites

Automotive Applications

D8.6:1977	Standard for Automotive Resistance Spot Welding Electrodes
D8.7:1988R	Automotive Weld Quality—Recommended Practices for Resistance Spot Welding
D8.8:1997	Specification for Automotive and Light Truck Weld Quality—Arc Welding
D8.9:1997R	Recommended Practices for Test Methods for Evaluating the Resistance Spot Welding Behaviour of Automotive Sheet Steel Materials
D8.14/D8.14M:2000	Specification for Automotive and Light Truck Components Weld Quality—Aluminium Arc Welding

Machinery and Equipment

D14.1:1997	Specification for Welding Industrial and Mill Cranes and Other Material Handling Equipment
D14.2:1993	Specification for Metal Cutting Machine Tool Elements
D14.3/D14.3M:2000	Specification for Welding Earthmoving and Construction Equipment
D14.4:1997	Specification for Welded Joints in Machinery and Equipment
D14.5:1997	Specification for Welding of Presses and Press Components
D14.6:1996	Specification for Welding of Rotating Elements of Equipment
D15.1:1993	Railroad Welding Specification—Cars and Locomotives
D15.2:1994	Recommended Practices for the Welding of Rails and Related Rail Components for Use by Rail Vehicles

Marine Applications

D3.5:1993R	Guide for Steel Hull Welding
D3.6M:1999	Specification for Underwater Welding
D3.7:1990	Guide for Aluminium Hull Welding

Piping and Tubing

D10.4:1986R	Recommended Practices for Welding Austenitic Chromium–Nickel Stainless Steel Piping and Tubing
D10.6/D10.6M:2000	Recommended Practices for Gas Tungsten Arc Welding of Titanium Piping and Tubing
D10.7/D10.7M:2000	Guide for the Gas Shielded Arc Welding of Aluminium and Aluminium Alloy Pipe

(*continued*)

Table 33.38   AWS STANDARDS RELATING TO WELDING, BRAZING AND SOLDERING—*continued*

D10.8:1996	Recommended Practices for Welding of Chromium–Molybdenum Steel Piping and Tubing
D10.11:1987R	Recommended Practices for Root Pass Welding of Pipe Without Backing
D10.12/D10.12M:2000	Guide for Welding Mild Steel Pipe

Structural Applications

D1.1:2001	Structural Welding Code—Steel
D1.2:1997	Structural Welding Code—Aluminium
D1.3:1998	Structural Welding Code—Sheet Steel
D1.4:1998	Structural Welding Code—Reinforcing Steel
D1.5:1996	Bridge Welding Code
D1.6:1999	Structural Welding Code—Stainless Steel

Filler Metals

A5.01:1993R	Filler Metal Procurement Guidelines
A5.1:1991	Specification for Carbon Steel Electrodes for Shielded Metal Arc Welding
A5.2:1992R	Specification for Carbon and Low Alloy Steel Rods for Oxyfuel Gas Welding
A5.3/5.3M:1999	Specification for Aluminium and Aluminium Alloy Electrodes for Shielded Metal Arc Welding
A5.4:1992	Specification for Stainless Steel Electrodes for Shielded Metal Arc Welding
A5.5:1996	Specification for Low Alloy Steel Electrodes for Shielded Metal Arc Welding
A5.6:1984R	Specification for Covered Copper and Copper Alloy Arc Welding Electrodes
A5.7:1984R	Specification for Copper and Copper Alloy Bare Welding Rods and Electrodes
A5.8:1992	Specification for Filler Metals for Brazing and Braze Welding
A5.9:1993	Specification for Bare Stainless Steel Welding Electrodes and Rods
A5.10/A5.10M:1999	Specification for Bare Aluminium and Aluminium Alloy Welding Electrodes and Rods
A5.11/A5.11M:1997	Specification for Nickel and Nickel Alloy Welding Electrodes for Shielded Metal Arc Welding
A5.12/A5.12M:1998	Specification for Tungsten and Tungsten Alloy Electrodes for Arc Welding and Cutting
A5.13:1980R	Specification for Solid Surfacing Welding Rods and Electrodes
A5.14/A5.14M:1997	Specification for Nickel and Nickel Alloy Bare Welding Electrodes and Rods
A5.15:1990	Specification for Welding Electrodes and Rods for Cast Iron
A5.16:1990R	Specification for Titanium and Titanium Alloy Welding Electrodes and Rods
A.5.17/A5.17M:1997	Specification for Carbon Steel Electrodes and Fluxes for Submerged Arc Welding
A5.18:1993	Specification for Carbon Steel Filler Metals for Gas Shielded Arc Welding
A5.19:1992R	Specification for Magnesium Alloy Welding Electrodes and Rods
A5.20:1995	Specification for Carbon Steel Electrodes for Flux Cored Arc Welding
A5.21:1980R	Specification for Composite Surfacing Welding Rods and Electrodes
A5.22:1995	Specification for Stainless Steel Electrodes for Flux Cored Arc Welding and Stainless Steel Flux Cored Rods for Gas Tungsten Arc
A5.23/A5.23M:1997	Specification for Low Alloy Steel Electrodes and Fluxes for Submerged Arc Welding
A5.24:1990	Specification for Zirconium and Zirconium Alloy Welding Electrodes and Rods
A5.25/A5.25M:1997	Specification for Carbon and Low Alloy Steel Electrodes and Fluxes for Electroslag Welding
A5.26/A5.26M:1997	Specification for Carbon and Low Alloy Steel Electrodes for Electrogas Welding
A5.28:1996	Specification for Low Alloy Steel Electrodes for Gas Shielded Metal Arc Welding
A5.29:1998	Specification for Low Alloy Steel Electrodes for Flux Cored Arc Welding
A5.30:1997	Specification for Consumable Inserts
A5.32:1997	Specification for Welding Shielding Gases

(*continued*)

Table 33.38 AWS STANDARDS RELATING TO WELDING, BRAZING AND SOLDERING—*continued*

Qualification and Certification

C2.16:1992	Guide for Thermal Spray Operator Qualification
B2.2:1991	Standard for Brazing Procedure and Performance Qualification
B2.1:2000	Specification for Welding Procedure and Performance Qualification
C3.2:1982R	Standard Method for Evaluating the Strength of Brazed Joints in Shear

Inspection and Testing

B1.10:1999	Guide for the Nondestructive Examination of Welds
B4.0:1998	Standard Methods for Mechanical Testing of Welds—U.S. Customary
B4.0M:2000	Standard Methods for Mechanical Testing of Welds—Metric Only

Safety and Health

F1.1:1999	Methods for Sampling Airborne Particulates Generated by Welding and Allied Processes
F1.2:1999	Laboratory Method for Measuring Fume Generation Rates and Total Fume Emission of Welding and Allied Processes
F1.3:1999	Sampling Strategy Guide for Evaluating Contaminants in the Welding Environment
F1.4:1997	Methods for Analysis of Airborne Particulates Generated by Welding and Allied Processes
F1.5:1996	Methods for Sampling and Analysing Gases for Welding and Allied Processes
F4.1:1999	Recommended Safe Practices for Preparation for Welding and Cutting of Containers and Piping
Z49.1:1999	Safety in Welding, Cutting and Allied Processes

Table 33.39 CANADIAN STANDARDS RELATING TO WELDING, BRAZING AND SOLDERING

CAN/CSA

W47.1:1992	Certification of Companies for Fusion Welding of Steel Structures
W47.1S1:M1989	Supplement No 1 to W47.1:1983 Steel Fixed Offshore Structures
W47.2-M:1987	Certification of Companies for Fusion Welding of Aluminium
W48:2001	Filler Metals and Allied Materials for Metal Arc Welding
W55.3:1965	Resistance Welding Qualification Code for Fabricators of Structural Members Used in Buildings
W59-M:1989	Welded Steel Construction (Metal Arc Welding) (Imperial version)
W59S1:1989	Supplement to W59-M:1989 No.1-M:1989, Steel Fixed Offshore Structures-Welded Steel Construction (Metal Arc Welding)
W59.2-M:1991	Welded Aluminium Construction
W117.2:1994	Safety in Welding, Cutting and Allied Processes
W178.1:1996	Certification of Welding Inspection Organizations
W178.2:1996	Certification of Welding Inspectors
W186-M:1990	Welding of Reinforcing Bars in Reinforced Concrete Construction

33.7 Bibliography and sources of information

Welding processes

Welding processes—general

'Welding Handbook', 8th edn, Vol. 1, American Welding Society, Miami, FL, 1987.
'Welding Handbook', 8th edn, Vol. 2, American Welding Society, Miami, FL, 1991.
'ASM Handbook', Vol. 6, Welding, Brazing and Soldering, ASM International, Materials Park, Ohio, 1993.

P. T. Houldcroft, 'Welding Processes', Cambridge University Press, Cambridge, UK, 1967.
A. C. Davis, 'Science and Practice of Welding', Cambridge University Press, Cambridge, UK, 1971.
R. W. Messler, Jr., 'Principles of Welding: Processes, Physics, Chemistry and Metallurgy', John Wiley and Sons, New York, NY, 1999.

Resistance welding

'Resistance Welding Manual', 4th edn, RWMA, Philadelphia, PA, 1989.

Friction welding

'Friction welding', Special Feature, *Met. Constr. Br. Weld. J.*, 1970, **2**, 181.
C. J. Dawes and W. M. Thomas, *Weld. J.*, March 1996, 41.

Metals and alloys

Steels

G. A. Phipps, 'Projection Welding of Low Carbon Mild Steel', *Br. Weld. J.*, 1958, **5**, 549.
K. S. Irvine, 'High Strength Weldable Steels', *Metallurgia*, 1958, **58**, 13.
D. Séférian, 'The Metallurgy of Welding', Chapman and Hall, London, 1962.
G. E. Linnert, Welding metallurgy, 'Carbon and Low Alloy Steels', 3rd edn, American Welding Society, 1965, and 'Volume 2, Technology', 1967.
K. G. Richards, 'The Weldability of Steel', The Welding Institute, 1972.
N. Bailey, 'Welding Carbon: Manganese Steels', *Met. Constr. Br. Weld. J.*, 1970, **2**, 442.
R. G. Baker, F. Watkinson and R. P. Newman, 'The Metallurgical Implications of Welding Practice as Related to Low Alloy Steels', Proc. Second Commonwealth Welding Conference, Institute of Welding, London, 1965.
S. S. Tuliani *et al.*, 'Notch Toughness of Commercial Submerged Arc Weld Metal', *Weld. Metal Fabric.*, 1967, **37**, 327–329.
F. R. Coe, 'Welding Steels without Hydrogen Cracking', The Welding Institute, 1973.
N. Yurioka, S. Oshita and H. Tamehiro, 'Study on Carbon Equivalents to Assess Cold Cracking Tendency and Hardness in Steel Welding', Symposium on Pipeline Welding in the '80s, AWRA, Australia, 1981.
J. C. M. Farrar and R. E. Dolby, 'Lamellar Tearing in Welded Steel Fabrication', The Welding Institute, 1972.
T. G. Gooch and D. C. Willingham, 'Weld Decay in Austenitic Stainless Steels', The Welding Institute, Abington, Cambridge, 1975.
R. Castro and J. de Cadenet, 'Welding Metallurgy of Stainless and Heat-Resisting Steels', Cambridge University Press, Cambridge, 1974.

Aluminium and aluminium alloys

'The Gas Welding of Aluminium', Information Bulletin No. 5, Aluminium Development Association, 1967.
'Resistance Welding of Wrought Aluminium Alloys', Information Bulletin No. 6, Aluminium Development Association.
'Welding Kaiser Aluminium', Kaiser Aluminium & Chemical Sales Inc., 1967.
'Manual MIG Welding of Aluminium', Alcan Service Bulletin, 1964.
'Mechanised MIG Welding of Aluminium', Alcan Service Bulletin, 1964.

Copper and copper alloys

'Gas Shielded Arc Welding of Copper and Copper Alloys', Technical Note TN2, Copper Development Association.
'The Bronze Welding Process', Technical Note TN5, Copper Development Association.
P. G. F. du Pré, 'The Gas-shielded Arc Welding of Copper and Copper Alloys', *Philips Welding Reporter*, 1972, **8**, 14.

Magnesium alloys

'Joining', Pamphlet, Magnesium Elektron Ltd.
E. F. Emley, 'The Metallurgical Background to Magnesium Alloy Welding', *Br. Weld J.*, 1957, **4**, 321.
P. Klain, 'The Welding of Magnesium Alloys', *Weld. J.*, 1957, **36**, 321.
'Joining Magnesium', Dow Chemical Co., 1956.

Nickel and nickel alloys

'Welding, Brazing and Soldering of Wiggin Nickel Alloys', Henry Wiggin & Co., Ltd., 1971.
'The Joining of Some Nickel Alloys', Leaflet, International Nickel Co., Ltd.
J. Hinde, 'Welding of Nickel and High Nickel Alloys', *Br. Weld. J.*, 1958, **5**, 311.

Refractory metals

C. A. Terry and E. A. Taylor, 'Welding of Titanium', *Weld. Metal Fabric.*, 1958, 26 (June).
J. G. Purchas, D. R. Harris and H. Cobb, 'The Welding of Zircalloy-2', *Br. Weld J.*, 1957, **4**, 412.
G. L. Miller, 'Zirconium', Butterworths, London, 1957.
F. G. Cox, 'Tantalum', *Weld. Metal Fabric.*, 1957, **25**, 416.
F. G. Cox, 'Welding and Brazing Refractory Metals—Molybdenum, Niobium, Tantalum, Zirconium', *Murex Review*, 1956, **1**, 429.
L. Northcott, 'Molybdenum', Butterworths, London, 1956.
T. R. C. Gough and D. Roberts, 'The Welding of Uranium', *Br., Weld. J.*, 1957, **4**, 393.
'I.M.I. Titanium Fabrication', Imperial Metals Industry (Kynoch) Ltd., 1966.
M. H. Scott, 'The Joining of the Rarer Metals' edited by G. Isserlis, Chap 4, Columbine Press, 1962.
E. G. Thompson, 'Welding of Reactive and Refractory Metals', Welding Research Council Bulletin, No. 85, 1963.
M. H. Scott and P. M. Knowlson, 'The Welding and Brazing of the Refractory Metals, Niobium, Tantalum, Molybdenum and Tungsten—a review', *J. Less. Common Met.*, 1963, **5**, 205.

Non-ferrous metals—general

E. A. Taylor, 'Inert Gas Welding of Non-Ferrous Metals', *Metall, Rev.*, No. 116, 1967.

Dissimilar metals

M. C. T. Bystram, 'Welding Dissimilar Alloy Steels', *Br. Weld. J.*, 1958, **5**, 475.
J. G. Young and A. A. Smith, 'Joining Dissimilar Metals', *Weld. Metal Fabric.*, 1959, **27**, 278, 331.
'Dissimilar Metals', *Met. Constr. Br. Weld. J.*, 1969, **1**, 12.
'Dissimilar—Metal Joint', International Nickel Co. Ltd.

34 Soldering and brazing

34.1 Introduction and cross-reference

Soldering and brazing are useful, fairly simple joining processes in which metals are wetted and joined together by a dissimilar metal of lower melting temperature. When compared with mechanical joints and adhesives, the join is permanent unless remelted, has good electrical and thermal conductivity and is unaffected by organic solvents. It is also resistant to failures at temperatures under those of the original joining technique. There are many filler metals available for a variety of applications with melting temperatures from just above room temperature upwards, the majority of conventional solders being molten below 300°C. Brazing is basically similar to soldering but takes place at a higher temperature, above 450°C according to the definitions of BS 499 but, unlike welding, always below the melting temperature of the parent metals. Brazing is also distinguished from welding by the fact that the brazed joint is formed mainly by capillary action between adjacent surfaces whereas welding starts by local melting of the base metals and continues with the addition of larger fillets of welding filler metal.

The majority of soldering and brazing alloys are based on binary or ternary eutectic systems, sometimes with small amounts of other addition elements. The compositions are chosen to have good wetting and capillarity at the joining temperatures, good corrosion resistance and reasonable strength. Brazed joints are generally significantly stronger than soldered ones and are usually intended to be permanent. The higher the joining temperature, the greater will be the heat-affected zone in the parent metals, with consequential possibilities of loss of temper due to annealing effects.

For joining to be successful the metal surfaces must be clean initially and are normally protected from oxidation during heating by the presence of a suitable flux. The wetting of a parent metal by a filler creates a metallurgical bond, the depth of which is diffusion controlled with the usual time- and temperature-dependent relationships. Overheating a joint, or repeated remaking of a joint usually has a deleterious effect on properties due to the formation of brittle intermetallic compounds.

Tin is present in most common soft solders and rapidly forms intermetallics with most base metals during soldering processes. These intermetallics may then continue to grow by solid-state reactions during storage and operational life of a joint. Under some time- and temperature-controlled conditions the tin can be totally converted to an intermetallic compound. The higher-temperature brazing operations form much deeper reaction zones of modified composition. Properly chosen brazing alloys will not form brittle intermetallics but the modified composition and microstructure will elevate the melting temperature, possibly to a value as high as the parent metal which will preclude the possibility of reworking joints.

For discussion of fusion welding and other joining processes, *see* Chapter 33. Chapter 33 also includes standards related to brazing and soldering.

34.2 Quality assurance

While soldering and brazing techniques have been employed satisfactorily for many years by skilled and semi-skilled personnel, modern manufacturing and quality control systems frequently require that process instructions be fully documented and refer to standards where relevant. Typical organizations preparing product standards covering the fluxes and filler metals include British Standards Institute, American Society for Testing and Materials (ASTM), Deutsches Normenausschuss (DIN) and many others, some specifically aimed at particular industries or defence applications. The European standards are issued by the European Committee for Standardization (CEN) and are very similar in structure and scope to the ISO (International Organization for Standardization). It is a general trend for individual European national standards, such as British Standards (BS), to be gradually replaced by European standards or by ISO standards. Some major ISO, European, British and AWS standards relating to soldering and brazing are listed in Section 33.6

The requirements included in these specifications are typical of those needed for initiating a full quality control specification. For specific industries there are many other documents published both by national standards institutions and by organizations such as the European Space Agency and NASA. For military requirements for processing, cleanliness, protection and packaging there are additional defence standards such as the American MIL and QQ series of specifications.

34.3   Soldering

Soldering is commonly used for structural applications, for the assembly of pipes and fittings for gas, water or other liquid services and is vital for electrical connections, especially on printed circuit boards. The relatively low melting temperatures of solders are beneficial in having little effect on the properties of the parent metals and adjacent items.

34.3.1   General considerations

Since the solder is weaker than the parts being fastened, the design of the joint must allow for this fact. Significant overlapping areas should be allowed and tensile and peeling forces should be avoided, leaving only shear stresses at most on the solder which is mainly intended to act as filler and seal. Clearances of 0.07 to 0.25 mm are permissible between overlapping parts with approximately 0.1 mm being optimum for capillarity and joint strength. During soldering and solidification there should be no relative movement between joint faces and the use of jigs, self-locating designs or temporary solder-tagging is therefore recommended. The jigs and fixtures should not, however, act as local heat sinks that prevent efficient soldering.

Soldering itself should be carried out as quickly as possible in order to avoid the effects of overheating. The choice of heating method may be either direct, which is fast, or indirect, which reduces surface oxidation effects. For all joints it is essential to start with the components clean and for most it is also essential to protect surfaces from oxidation during soldering with a suitable flux.

34.3.2   Choice of flux

Fluxes have several purposes. They conduct heat to the component, modify or remove surface oxides and other contaminants and eliminate air from the solder wetting front.

The earliest and most active of fluxes are those water-based solutions such as the original 'killed spirits', zinc dissolved in hydrochloric acid, latterly with additions of ammonia and perhaps alcohol and/or a detergent. While very effective on the common metals such as steel, tinplate, copper and brass they are also very corrosive if not washed off thoroughly immediately soldering is completed. Aqueous solutions of orthophosphoric acid are also used for some applications, especially for the low melting point solders and for special steels. Subsequently organic-based fluxes have been developed, now forming four main groups, rosin, synthetic, synthetic resin and water-based organic fluxes.

Dependent on the application and the extent to which subsequent corrosion must be avoided, rosins may be mildly activated, with a halide content of less than 0.2%, activated with 0.2–0.5% halide or super-activated with even higher halide levels. Synthetic activated fluxes are based on formulations giving good wetting and excellent residue removal properties when cleaned with CFCs or other suitable solvents. Synthetic resin fluxes are now made with a low solids content which minimises the need for cleaning and the latest water-soluble organic fluxes combine the best of soldering results with ease of washing clean. For critical assemblies such as wave-soldered printed circuit boards, it is essential to take expert opinion from manufacturers or other specialists in order to assure reproducible success.

The choice of flux is influenced by many factors including type and mass of metal to be soldered, cleanliness, heating techniques, type of solder and after-cleaning requirements. For example, the use of torch soldering techniques will require the use of a more active flux to deal with the extra tarnish caused by the flame.

The notes in Table 34.1 cover the common, and some of the less common, commercial metals (arranged in order of electrochemical potential, *see* Table 34.2).

Table 34.1 RECOMMENDED SOLDER FLUXES FOR ENGINEERING MATERIALS

Group No.	Category	Flux type and possible protective finish with precautions
1	Gold, solid or plated; gold–platinum alloys; wrought platinum	Both gold and platinum have excellent solderability. Soldering to gold and its alloys should be avoided because gold–tin intermetallics embrittle joints. Gold platings can be removed by solder-dipping to dissolve the gold followed by pre-tinning in a second, uncontaminated, solder pot. Rosin, non-activated flux is adequate. If gold cannot be removed, use indium–lead solder. Bright, hard gold platings may be difficult to wet due to the presence of certain alloying elements or organic additives derived from plating solutions: activated rosin flux is then required.
2	Rhodium	Not easy—inorganic flux.
3	Silver, solid or plated on copper; high silver alloys	Easily soldered with mildly activated flux or, when free of surface sulphide, non-activated rosin is preferable. If chloride-contaminated (from plating bath), may need abrasion. Use silver-loaded solder when joining to thin silver plate. The silver-saturated liquid reduces danger of scavenging. Silver-plated parts are not recommended for electrical circuits due to problem of silver migration and subsequent short circuits.
4(a)	Nickel, solid or plated; Monel	Difficult to solder. Use inorganic or organic acid for pre-tinning with solder and non-activated rosin after pre-tinning.
4(b)	Titanium	Impossible to solder without prior copper plating.
5(a)	Copper, solid or plated, tin-bronzes, gunmetals	With red oxide tarnish, can be soldered with mildly activated rosin. Black oxide only removable with activated rosin, organic acids or zinc ammonium chloride solutions (e.g. for radiator plates).
5(b)	Copper–nickel alloys	Not difficult to solder if clean and well fluxed but soldering is not commonly suitable for the severe conditions for which these alloys are specified.
5(c)	Nichrome alloys, austenitic, high corrosion-resistant steels; Nilo-K, Kovar, Monel, etc.	Very difficult to solder with inorganic acid fluxes, zinc chloride and ammonium chloride solutions. Some proprietary brands are available. Can be nickel-plated, but deposit must be non-porous or substrate will become oxidised by plating salts. All ionic matter, including handling contamination, must be thoroughly and immediately removed because this group is susceptible to stress corrosion cracking.
5(d)	Copper–aluminum alloys (aluminium bronzes)	Due to the tenacious alumina film which forms so rapidly, these alloys can only be soldered after copper plating or with techniques used for aluminium alloys.
6	Gilding metals (CuZn10 and CuZn20)	If clean, quite easy to solder with mildly activated resin. May be copper or silver-plated, but ensure good plating adhesion.
7	Commercial brasses (CuZn30 and CuZn40)	Use activated rosin if tarnish is thin. Impossible to solder—even with inorganic flux—if significant oxidation is visible, due to surface film of zinc oxide. Barrier plating of more than 3 μm nickel or copper is recommended for preserving solderability during shelf-life (should prevent zinc diffusion to surface). Barrier of 5 μm necessary if brass is leaded free-machining grade.
8	18% chromium-type stainless steel	*See* Group 5(b).
9	Tin-plated metals	New coatings are easily soldered with non-activated rosin flux. Activation of rosin depends on extent of tin oxide. Fused tin is preferred as it is less porous. Pure tin coatings not recommended for electronic applications due to risk of whisker growth. Should exceed 1 μm thickness as otherwise will react with copper and completely convert to intermetallic, which is extremely difficult to solder.
10	Tin–lead coatings, solid, plated or fused	Most suitable finish for easy soldering with non-activated rosin flux. Porous platings need activated flux. Should exceed 1 μm thickness, but lead slows down intermetallic formation. Good shelf-life for fused coatings on copper wire and printed circuits.

(*continued*)

Table 34.1   RECOMMENDED SOLDER FLUXES FOR ENGINEERING MATERIALS—*continued*

Group No.	Category	Flux type and possible protective finish with precautions
11	Lead, solid or plated; high lead alloys	Mildly activated flux required to penetrate surface oxides. High dissolution of lead in tin–lead solder produces joints with extremely low shear strength.
12	Aluminium–copper alloys (e.g. Duralumin and most of AA 2xxx series)	Impossible with tin–lead alloys. Generally, welding or dip-brazing are more suitable. Plate with zincate and copper. The low melting solders are preferred to avoid thermal stressing platings.
13	Iron, wrought, grey or Armco; plain carbon and low alloy steels	Solderability depends on oxide thickness. Clean, pickled surfaces can be easily soldered with mild or activated rosin flux. Passivated or phosphated steels require special fluxes.
14	Aluminium and most alloys other than in Group 12 (e.g. AA 1xxx, 3xxx and 5xxx series)	Can be soldered, but with special alloys (e.g. tin–zinc or cadmium–zinc alloys). May be friction-soldered without flux or, when aluminium oxide is removed, by ultra-sound. Some proprietary fluxes are available. Corrosion in aluminium-soldered joints is particularly troublesome; a water-proof coating is essential.
15(a)	Cast aluminium alloys other than aluminium–silicon alloys	As for Group 14 but choice of solder and flux requires specialist advice.
15(b)	Cadmium plating	Easy to solder with mildly or fully activated rosin fluxes. If passivated by chromate film, they are very difficult and require ammonium chloride-type flux. Fumes are toxic.
16	Hot dipped zinc plate	As for Group 15(b).
17	Zinc, wrought; zinc-based casting alloys; zinc plating	Generally, very difficult to solder due to oxidation. Use inorganic acid or special proprietary flux. Water-tight protection necessary to avoid corrosion.
18	Magnesium and magnesium-based alloys, cast or wrought	Not recommended because of poor strength and corrosion.

Table 34.2   ELECTROCHEMICAL POTENTIAL OF COMMON ENGINEERING MATERIALS

Group No.	Category	Electrochemical potential V*
1	Gold, solid or plated; gold–platinum alloys; wrought platinum	+0.15
2	Rhodium	+0.05
3	Silver, solid or plated on copper; high silver alloys	0
4(a)	Nickel, solid or plated; Monel	−0.15
4(b)	Titanium	−0.15
5(a)	Copper, solid or plated, tin–bronzes, gunmetals	−0.20
5(b)	Copper–nickel alloys	−0.20
5(c)	Nichrome alloys, austenitic, high corrosion-resistant steels; Nilo-K, Kovar, Monel, etc.	−0.20
5(d)	Copper–aluminium alloys (aluminium bronzes)	−0.20
6	Gilding metals (CuZn10 and CuZn20)	−0.25
7	Commercial brasses (CuZn30 and CuZn40)	−0.30
8	18% chromium-type stainless steel	−0.35
9	Tin-plated metals	−0.45
10	Tin–lead, solid, plated or fused	−0.50
11	Lead, solid or plated; high lead alloys	−0.55
12	Aluminium–copper alloys (e.g. Duralumin and most of AA 2xxx series)	−0.60
13	Iron, wrought, grey or Armco; plain carbon and low alloy steels	−0.70
14	Aluminium and most alloys other than in Group 12 (e.g. AA 1xxx, 3xxx and 5xxx series)	−0.75
15(a)	Cast aluminium alloys other than aluminium–silicon type	−0.80
15(b)	Cadmium plating	−0.80
16	Hot dipped zinc plate	−1.05
17	Zinc, wrought; zinc-based casting alloys; zinc plating	−1.10
18	Magnesium and magnesium-based alloys, cast or wrought	−1.60

* Calomel electrode/sea water. For reasonable galvanic compatibility, the maximum potential difference between soldered metals should be less than 0.5 V.

34.3.3   Control of corrosion

If long storage periods are expected between manufacture of components and solder assembly, the surfaces of the common industrial metals can be plated initially with copper or nickel and subsequently with a pre-tinning eutectic composition of tin–lead solder which can be fused in hot oil to form a 4–12 μm thick pore-free protective layer possessing excellent solderability even after many years in storage. Less active fluxes may then be employed which will reduce the need for post-soldering cleaning.

Subsequently to joining, corrosion may occur near a joint if flux residues are present. These residues are frequently hygroscopic and can cause severe discoloration or worse problems in the joint locality given only the presence of atmospheric humidity. It is therefore essential to clean off flux residues, preferably as soon as possible after soldering. The choice of cleaning agent will depend on the type of joint, the flux used and the extent to which cleanliness is critical.

If aqueous fluxes are used, the simplest wash is to quench the joint in water prior to inspection and possible further cleaning. Pickling in dilute acid is sometimes used but the active surface left by this aggressive process must be subsequently passivated. For cleaning organic residues a variety of solvents are possible and the recommendations of the flux manufacturer should be followed in the absence of any other process standard.

In service it is possible to suffer galvanic corrosion if dissimilar metals are exposed in damp or wet service or storage environments. In any bi-metallic couple the more noble metal will not be corroded but the less noble will be attacked at a rate dependent on the environment and the electrical potential difference between the metals. Its effect may be reduced if the area of the less noble, or sacrificial metal is significantly greater than the noble metal. Fortunately tin–lead, the most common soldering alloy, is generally compatible with most solderable metals, except gold. The galvanic series is useful as a first approximation in selecting materials for both solderability and corrosion control, but for many applications it may be too simplistic because it does not provide information about corrosion rates or changes in surface chemistry which may pacify or accelerate corrosion at bi-metallic interfaces.

34.3.4   Solder formulations

Generally, solder alloys are based on the metals tin, lead, cadmium, zinc and indium. They are available in a variety of physical forms to facilitate different means of application. Solder ingots are used to replenish large baths for dip or wave-soldering. Solder creams, containing a gel of solder powder, flux and wetting agent, can be painted or screen-printed for microelectronic applications. Preforms of solder are used for furnace and torch-soldering and include formed wires and parts punched from flat sheet.

Solder alloys can be divided into four groups according to their melting ranges. Melting temperatures are not recommended soldering temperatures. Typically, 20–70°C above the liquidus temperatures ensures good alloy fluidity and wetting characteristics. Compositions, melting ranges, and typical uses for these types are given in Tables 34.3 to 34.6.

Many worldwide initiatives have been taken to find lead-free alternatives to currently used lead containing solders because of the toxicity of lead. Eutectic tin–lead solder (63Sn/37Pb) is currently very widely used in electrical and electronic assembly, particularly in surface mount printed circuit boards. The National Center for Manufacturing Sciences (NCMS) in USA has completed a project in which alternatives to eutectic tin/lead solder have been identified and evaluated from an initial list of more than 70 candidate alloys. Although the project concluded that no drop-in replacement is available for eutectic Sn/Pb solder, it has indicated that promising alternatives exist. More detailed information is needed however before implementing any wholesale changes to present industrial practice. The project has also recommended three alloys that might perform substantially better than eutectic Sn/Pb in certain applications (Table 34.7). Lead-containing solders, usually Pb/Sn or Pb/Sn/Sb, have also been very widely used in the past for assembly of copper pipework in domestic plumbing. In many parts of the world, lead-containing solder alloys are no longer approved for use on systems handling potable water. Where such prohibitions exist, the current trend is to use Sn/Sb or Sn/Ag solder alloys (i.e. similar to alloys 95A or 96S in Table 34.7).

34.3.5   Cleaning

Soldered joints should be cleaned after completion in order to prevent surface corrosion and stress corrosion cracking. Cleaning also facilitates inspection. Immersion cleaning is improved by ultrasonic agitation.

Table 34.3   VERY LOW MELTING POINT SOLDERS (BELOW 183°C)

Uses include joining heat-sensitive components or adding components to existing circuits (one solder operation on another where second soldering does not remelt initial joint). Contamination with certain other solder compositions may cause alloying and even lower melting temperatures.

Sn	Pb	Sb	Bi	Cd	In	Ag	Melting range °C	Name
49	33	—	—	18	—	—	145 m.p.	ISO 9453 Grade 'T' (DIN 1707, L-SnPb Cd 18)
50	—	—	—	—	50	—	117–125	Glass-to-metal sealing (L-SnIn 50)
14.5	28.5	9	48	—	—	—	103–227	Matrix alloy
22	28	—	50	—	—	—	96–110	Rose's
13	27	—	50	10	—	—	70–73	Lipowitz's
12.5	25	—	50	12.5	—	—	70–72	Bending (Wood's)
8.3	22.6	—	44.7	5.3	19.1	—	47 m.p.	—
12.8	25.6	—	48	9.6	4	—	62 m.p.	—
—	—	—	—	27	73	—	123 m.p.	—
37.5	37.5	—	—	—	25.0	—	134–174	—
—	15	—	—	—	80	5	149 m.p.	—
13	27	—	50	10	—	—	70 m.p.	Quaternary eutectics
26	—	—	54	20	—	—	103 m.p.	Ternary eutectics
—	—	—	60	40	—	—	144 m.p.	Binary eutectics

Table 34.4   COMMON SOLDER ALLOYS (183–270°C)

Sn	Pb	Other	Melting range °C	Specification ISO 9453:1990 grade	Typical uses (similar alternative standard)
Tin solders					
63	37	0.6Sb max	183 m.p.	A	Soldering of electrical connections to copper, brass and zinc. Capillary joints in copper and stainless steel (QQ-S-571d, type Sn63*)
63	37	0.2Sb max	183 m.p.	AP	As 'A', but for higher reliability on printed circuit boards (DIN 1707 L-Sn63Pb) (ESA QRM-08, Sn63)*
60	40	0.5Sb max	183–188	K	As 'A', but also for pre-tinning of electrical components (DIN 1707 L-Sn60PbSb) (QQ-S-571d, type Sn60)
60	40	0.2Sb max	183–188	KP	Hand and machine-soldering of electronics and pre-tinning (DIN 1707 L-Sn60Pb) (ESA QRM-08, Sn60) (ASTM-B-32-66T-60A)
50	50	0.5Sb max	183–212	F	General engineering work on copper,
45	55	0.4Sb max	183–224	R	brass and zinc. Can-soldering (DIN 1707
40	60	0.4Sb max	183–234	G	L-Sn50Pb(Sb), etc.)
35	65	0.3Sb max	183–244	H	Joining of electrical cable sheaths
30	70	0.3Sb max	183–255	J	(DIN 1707 L-PbSn35(Sb))
20	80	0.2Sb max	183–276	V	Lamp solder. Dip-soldering. For low
15	85	0.2Sb max	227–288	W	temperature service (also ISO 12224-2)
100	—	—	232 m.p.	(BS 2352)	Food-handling equipment and cans
Tin–lead–antimony					
50	50	2.5–3.0Sb	185–204	B	Hop dip coating and soldering of ferrous
45	55	2.2–2.7Sb	185–215	M	metals; jointing of copper conductors
40	60	2.0–2.4Sb	185–277	C	General engineering. Heat exchanges. General dip-soldering
32	68	1.6–1.9Sb	185–243	L	Plumbing, wiping of lead and lead alloy
30	70	1.5–1.8Sb	185–248	D	cable-sheathing. Dip-soldering
18	82	0.9–1.1Sb	185–275	N	Dip-soldering (also BS AU90-19A)
Tin–antimony					
95	—	4.75–5.25Sb	236–243	95A	High service temperatures (more than 100°C) and refrigeration equipment. Step-soldering

(*continued*)

Table 34.4 COMMON SOLDER ALLOYS (183–270°C)—*continued*

Nominal composition			Melting range °C	Specification ISO 9453:1990 grade	Typical uses (similar alternative standard)
Sn	Pb	Other			
Tin–silver					
96	0.10 max	3.5–3.7Ag	221 m.p.	96S	High service temperatures (more than 100°C)
97	0.10 max	3.0–3.5Ag	221–223	97S	Intended for making capillary joints in copper plumbing systems for potable water applications where the lead content of the solder is restricted
98	0.10 max	1.8–2.2Ag	221–230	98S	
Tin–copper					
99	—	0.45–0.85Cu	227–228	99C	
Tin–lead–silver					
5	Rest	1.4–1.6Ag	296–301	5S	For service at high (>100°C) and low (<−60°C) temperature
62.5	Rest	1.8–2.2Ag	178 m.p.	62S	Soldering of silver-coated substrates (ESA QRM-08, 62Sn) (approx. DIN 1707 L-Sn60PbAg) (QQ-S-571d, Sn62)
Solder for aluminium					
90	—	10Zn	200–210	—	Soldering aluminium by ultrasonics (DIN 8512 L-SnZn10)
60	—	40Zn	200–300	—	Frictional soldering of aluminium (DIN 8512 L-SnZn40)
Gold–tin alloy					
20	—	80Au	280 m.p.	—	Micro-electronics manufacturing
Lead–indium alloys					
—	50	50In	215–230	—	Hybrid micro-electronics (thick-thin film substrates). Prevents scavenging of gold
—	75	25In	173–190	—	

* Eutectic type solders with single discrete melting point; they solidify with smooth bright fillets which can be readily inspected—recommended for highly reliable electrical connections. These are used for hand or machine soldering of printed circuit boards but care should be taken to avoid trace impurities, e.g. Zn or Al <0.005% to prevent oxide skin; Fe or Au or Cu or As or Sb <0.2% to prevent hard or embrittled joints.

Table 34.5 INTERMEDIATE TEMPERATURE SOLDERS (270–370°C)

Elevated temperature applications where some creep strength is required. The lead-based alloys, which contain additions of up to 8% tin and/or silver, may be used to about 130°C. Higher service temperatures may be satisfied by selecting tin-free lead alloys, but these have poor wetting properties and require the use of special organic fluxes which will not decompose at the high soldering temperatures. Alloys containing in excess of 90% lead are selected for cryogenic applications because they retain ductility and can take up mismatches in thermal expansion at the low temperatures. The final inspection of cleaned connections made with these alloys may reveal high contact angles and dull surface finishes, but these should not be construed as a sign of cold solder joints.

Nominal composition						Melting range °C	Name
Sn	Pb	Sb	Ag	Cd	Zn		
2	98	—	—	—	—	320–325	DIN 1707, L-PbSn2
5	93.5	—	1.5	—	—	296–301	ISO 9453, Grade 5S and BS AU90-5S
8	91.7	0.3	—	—	—	280–305	DIN 1707, L-PbSn8(Sb)
—	—	—	2.0	82	16	270–280	DIN 1707, L-CdZnZg2
—	—	—	10	68	22	270–380	DIN 1707, L-CdZnAg10
—	97.5	—	2.5	—	—	304–305	DIN 1707, L-PbAg3
2	96.2	—	1.8	—	—	304–310	DIN 1707, L-PbAg2Sn2
70	—	—	—	—	30	200–300	Solder for aluminium (by friction when liquid)
—	—	—	—	40	60	265–350	DIN 8512, L-ZnCd40 (frictional soldering of aluminium)
3	97	5	—	—	—	245–284	BS AU90-3AX (automobile body-filling)

Table 34.6 HIGH TEMPERATURE SOLDERS (370–430°C)*

Nominal composition			wt %			
Sn	Zn	Al	Cd	Ag	Melting range °C	Name
73 (87)	8 (15)	5 (12)	—	—	193–510, depending on alloy composition	General purpose solder for aluminium alloys (MIL-S-12204, Comp. C)
—	95	5	—	—	380 m.p.	High strength solder for aluminium
—	100	—	—	—	418 m.p.	High strength solder for aluminium
—	—	—	95	5	337–393	General purpose (DIN 1707, L-CdAg5)
19	81	—	—	—	195–385	General purpose (DIN 1707, L-ZnSn20)

* Rarely used because of high dissolution of many base metals into the solder. Rapid tarnishing of adjacent areas to the joint unless special protection by flux or inert gas is provided. Only high temperature stable inorganic fluxes useful.

Table 34.7 PROMISING LEAD-FREE ALLOYS FOR PARTICULAR ELECTRONIC SMT APPLICATIONS*

Alloys	Liquidus and solidus temperature (°C)	Industrial sectors	Evidence for recommendation
Sn/58Bi	139 (eutectic)	Consumer Electronics Telecommunications	Simple two-component eutectic alloy but low eutectic temperature restricts maximum use temperature.
			SMT: Better fatigue life than eutectic tin/lead for both thermal cycling ranges; less fatigue.
			THT: Mixed results with fatigue life for CPGA-84 better than eutectic tin/lead; for CDIP-20 worse than eutectic tin/lead.
Sn/3.5Ag/ 4.8Bi	210–205	Consumer Electronics Telecommunications Aerospace Automotive	SMT: Longer fatigue life than eutectic tin/lead at 0° to 100°C; no failures in 1 206 resistors up to 6 673 cycles; fatigue life equivalent to eutectic tin/lead at −55° to 125°C; less fatigue damage than eutectic tin/lead seen in surface mount cross sections.
			THT: Most joints show failure by fillet lifting.
Sn/3.5Ag	221 (eutectic)	Consumer Electronics Telecommunications Aerospace Automotive	Simple two-component eutectic alloy.
			SMT: Fatigue life equivalent to eutectic tin/lead at 0° to 100°C; worse than eutectic tin/lead at −55° to 125°C.
			THT: Less susceptible than other high-tin solders to fillet lifting but results with Sn/2.6Ag/ 0.8Cu/0.5Sb indicate through-hole reliability still may be compromised.

* For further information, *see* NCMS Lead-free Solder Project Final Report, National Center For Manufacturing Science, Ann Arbor, Michigan, USA, August 1997. The data in this table are reproduced, with permission, from the NCMS report.

Subject to manufacturers' recommendations or other process control specifications, the following solvents are acceptable for the cleaning of electronic equipment provided cleaning takes place immediately after the soldering operation (e.g. before the flux and residues have had time to polymerise or age):

1. Ethyl alcohol, 99.5 or 95% pure by volume.
2. Isopropyl alcohol, 99% pure.
3. Trichlorotrifluorethane, clear, 99.8% pure.
4. Any mixture of the above.

5. De-ionised water at 40°C maximum may be used for certain fluxes. Items shall be thoroughly dried directly after the use of de-ionised water.

34.3.6   Product assurance

Acceptable solder connections are generally characterised as:

1. Clean, smooth, undisturbed surfaces.
2. Concave fillets between solder and joined surfaces.
3. Complete wetting as evidenced by a low contact angle between the solder and the joined surfaces.

Unacceptable solder conditions, which may be cause for rejection, are:

1. Damaged, crushed, cracked, melted, corroded, etc. surfaces.
2. Improper tinning.
3. Flux residues or other contamination.
4. Cold joints, as shown by the contact angle.
5. Fractured joints.
6. Pits, holes or voids in the joint which are not attributable to liquid-to-solid solder shrinkage.
7. Excessive or insufficient solder.
8. Splattering of flux or solder on adjacent areas.
9. De-wetting, etc.

34.4   Brazing

34.4.1   General design consideration

Brazing is a very useful technique for making permanent, strong joints between metals at moderately elevated temperatures without fusing the parent metals themselves. It should be considered when one or more of the following design requirements must be met:

1. Joints will be loaded in shear.
2. For the joining of thin sections to heavy sections.
3. For joining of sections too thin for welding.
4. When hermetically sealed assemblies are required.
5. For joints with low distortion and minimal residual stress distribution.
6. For high volume production.
7. For joining dissimilar metal combinations which cannot be welded.
8. For producing complex, permanent assemblies with multiple joints which would be inaccessible to welding.

34.4.2   Joint design

Normally only small or medium sized assemblies can be brazed as the heat required to melt the braze alloy must be applied to a broad, or usually the entire area of the component parts being joined.

Both lap and butt joints can be made with flat or tubular parts. *The strength of the join depends on the amount of bonding surface.* The simple butt joint will have a small bonding area that will not exceed the cross-sectional area of the thinner member and this type of joint is therefore not recommended. The preferred type of joint loading is in shear and lap or lap-butt joints are the most reliable. The joint should be designed to prevent stress from being concentrated at a single point. Most of the specialised textbooks and codes of good practice include diagrams showing the optimum ways in which joints such as lap-butt, tees, corners, tube-to-tube, tube caps and tube-through-plate can be designed considering the process to be used. For joints in metals such as aluminium and its alloys, tapering clearances are required to enable fluxes to float clear rather than be entrapped.

During the brazing operation provisions must be made to hold the various components to be joined. This is best done by using an interlocking design where alignment is obtained by fitting parts into each other under gravity but if this is not possible, jigs or fixtures will be needed. Parts may also be positioned by means of spot or tack welding, crimping and pinning.

Table 34.8   COMPARISONS OF MATERIALS: COEFFICIENT OF THERMAL EXPANSION[a]

Material	10^{-5} K^{-1} High	Low	Material	10^{-5} K^{-1} High	Low
Aluminium and its alloys[c]	2.5	2.1	Rhodium[b]	0.8	—
Aluminium bronzes (cast)[c]	1.7	1.6	Ruthenium[b]	0.9	—
Alumina ceramics[e]	0.7	0.6	Silicon carbide[e]	0.4	0.39
Alumina cermets[d]	0.9	0.8	Silver[b]	2.0	—
Beryllia[e]	0.9	—	Steatite[c]	0.7	0.6
Beryllium[b]	1.1	—	Steels—alloy[d]	1.5	1.1
Beryllium carbide[d]	1.0	—	—alloy (cast)[d]	1.5	1.4
Beryllium copper[c]	1.7	—	—carbon, free cutting[d]	1.5	1.5
Boron carbide[c]	0.3	—	—high temperature[d]	1.4	1.1
Boron nitride[d]	0.8	—	—nitriding[d]	1.2	—
Brasses—plain and leaded[c]	2.1	1.8	—stainless (cast)[d]	1.9	1.1
Bronzes (cast)[c]	1.8	1.8	—stainless (austenitic)[c]	1.8	1.6
Carbon[c]	0.3	0.2	—stainless (ferritic)[c]	1.1	1.0
Chromium carbide cermet[c]	1.1	1.0	—stainless (martensitic)[c]	1.2	1.0
Cobalt[d]	1.2	—	—ultra high strength[d]	1.4	1.0
Coppers[c]	1.8	—	Superalloys—Cr–Ni–Fe[d]	1.9	1.7
Copper–nickel alloys[c]	1.7	1.6	—Cr–Ni–Co–Fe[d]	1.6	1.4
Graphite[c]	0.3	0.2	—cobalt based[d]	1.7	1.2
Gold[c]	1.4	—	—nickel base	1.8	1.4
Hafnium[b]	0.6	—	Tantalum[b]	0.6	—
Heat resistant alloys (cast)[d]	1.9	1.1	Tantalum carbide[d]	0.8	—
Iridium[b]	0.7	—	Thoria[e]	0.9	—
Iron—cast (grey)[c]	1.1	—	Thorium[b]	1.1	—
—cast, nodular or ductile[d]	1.9	1.2	Tin and its alloys[b]	2.3	—
—malleable[c]	1.3	1.1	Titanium and its alloys[d]	1.3	0.9
—wrought[c]	1.3	—	Titanium carbide[d]	0.7	—
Lead and its alloys[c]	2.9	2.6	Titanium carbide cermet[d]	1.3	0.8
Magnesium alloys[b]	2.8	2.5	Tungsten[b]	0.4	—
Molybdenum and its alloys	0.6	0.5	Tungsten carbide cermet[e]	0.7	0.4
Molybdenum disilicide[e]	0.9	—	Vanadium[b]	0.9	—
Nickel and its alloys[d]	1.7	1.2	Zinc and its alloys[c]	3.5	1.9
Nickel alloys—low expansion[c]	1.0	0.3	Zircon[c]	0.3	0.2
Nickel silvers[c]	1.7	1.6	Zirconia[e]	0.6	—
Niobium and its alloys	0.7	0.68	Zirconium and its alloys[b]	0.6	0.55
Osmium[b]	0.6	—	Zirconium carbide[d]	0.7	—
Palladium[c]	1.2	—			
Phosphor bronzes[c]	1.8	1.7			
Platinum[c]	0.9	—			

[a] Values represent high and low sides of a range of typical values.
[b] Value at room temperature only.
[c] Value for a temperature range between room temperature and 100–390°C.
[d] Value for a temperature range between room temperature and 540–980°C.
[e] Value for a temperature range between room temperature and 1 205—1 508°C.
Based on data from 'Materials Selector', Reinhold Publishing Co., Penton/IPC; *see* also Chapter 14—General physical properties.

The joint clearance, which is the dimension between the interfaces of the completed brazed joint, has a direct bearing on the mechanical strength of any brazed joint. Normally the highest strengths are obtained with the smallest possible thickness of filler and this is particularly true where there is little or no solubility of the parent material in the braze filler (e.g. stainless steels brazed with silver- or copper-based alloys). More clearance is required if material solubility exists, particularly in furnace brazing when long heating cycles are required (e.g. austenitic stainless steels brazed with nickel fillers).

It is extremely important to plan a joint clearance based on the brazing temperatures and not room temperature. The coefficients of expansion of the metals being joined must be taken into account, especially in tubular assemblies in which dissimilar metals are being joined. If the metal with the

Table 34.9   RECOMMENDED JOINT CLEARANCES (mm)

Brazing material	Copper	Copper-base alloys	Ferrous metals	Aluminium and its alloys
Noble metal alloys[a]	0.025–0.10	0.025–0.10	0.025–0.1	—
Copper	—	—	Nil–0.075	—
Brasses	0.075–0.375	0.075–0.375	0.05–0.25	—
Copper–phosphorus	0.075–0.375	0.075–0.375	—	—
Silver–copper–phosphorus	0.05–0.30	0.05–0.30	—	—
Silver brazing alloys	0.03–0.25	0.03–0.25	0.025–0.15	—
Nickel-base alloys	—	—	0.075–0.375[b]	—
Aluminium brazing alloys	—	—	—	0.125–0.6

[a] Ensure that the solidus value of the parent metal is at least 100°C above the liquidus of the selected brazing material.
[b] Special formulations are available which will permit wider gaps than 0.375 mm to be brazed in a satisfactory manner.
Reprinted from Brooker and Beatson, 'Industrial Brazing', 2nd edn, Newnes-Butterworth, 1975.

greater expansion coefficient is on the inside there may be no clearance at the brazing temperature and under reverse conditions a room temperature gap can become too large for adequate capillary flow at the brazing temperature.

As a design rule the filler metal should remain in slight compression once the assembly returns to room temperature. The metal having the greater coefficient of expansion is therefore normally used as the female member of the joint. Ideally, parts to be brazed should have a similar coefficient of expansion in order to facilitate joint clearance calculation and prevent the generation of large residual stresses caused by badly matched materials. The optimum joint clearance will depend on the nature and sizes of the parts being joined, the heating methods and the configuration of the joint itself. As there are many variables involved in assuming the actual clearance at brazing temperatures it is recommended that simple calculations are made utilising the coefficients of expansion presented in Table 34.8 and these are followed up by trial operation using representative test samples. For most brazing alloys effective capillary action occurs when the braze path gap at the brazing temperature is maintained within the range 0.025–0.200 mm. Table 34.9 may also be used as a guide.

34.4.3   Precleaning and surface preparation

Uniform capillary action may only occur when all grease, oil, oxides and dirt have been removed from both the base metals and braze alloy. Fluxes cannot be relied upon to clean the area to be brazed—they are mainly intended to prevent the formation of oxides during brazing and reduce the surface tension of the filler metal. Cleaning may be performed by chemical, mechanical or vacuum methods. Chemical cleaning is most frequently carried out utilising commercial solvents, vapour degreasing, alkaline mixtures (detergents, soaps, carbonate, hydroxides), electrolytic methods (anodic and cathodic), acidic solutions and salt bath pickling. Mechanical cleaning methods include grinding, filing, machining, shot blasting and ultrasonics. In all cases care must be taken to avoid the embedment of oxides or ferrous materials into the surfaces to be brazed as these will interfere with the proper wetting of the base metal by the liquid braze alloy. Very smooth, polished surfaces are less satisfactory for brazing than slightly roughened surfaces.

34.4.4   Positioning of filler metal

Dependent on the design of the joint and the type of heating technique used, the braze filler metal may be pre-placed prior to heating or applied when the joint is at temperature. It may be in the form of wire, ring, shim, clad sheet, powder, paste or slurry, depending upon the specific application and process. These forms of filler metals can be positioned by grooves, shoulders and/or recesses which serve as reservoirs for filling the braze paths. Designs of joint suitable for hand feeding will be different from those used when pre-formed inserts of brazing filler are used. In the latter case, the manufacturers of the fillers are generally glad to advise on optimum joint design and for the positioning of the components to allow for movement as the pre-form melts.

All braze assemblies should be designed so that the parent metals and the filler metals are uniformly heated when component parts have different thermal capacities. Two-stage heating with long dwell times may be necessary to ensure that liquation and flow of the filler metal occur simultaneously in all locations. Gravity flow of the filler metal through the braze path should be provided when possible. Machined grooves should be provided when large ($>50\,mm^2$) flat surfaces are united by brazing. Unless full penetration by the brazing alloy is achieved, an internal crevice will remain as a potential site for flux and residue. In certain configurations the presence of visible filler metal fillets will aid inspection.

34.4.5 Heating methods

TORCH BRAZING

Parts should be pre-heated with a neutral or slightly reducing flame so that uniform heating or the surfaces to be joined can be achieved. Localised overheating must be avoided. Flux is usually applied (some proprietary fluxes indicate the part's surface temperature by a colour change) and the braze alloy is introduced at one of the mating surfaces. Torch brazing is frequently used for the attachment of small parts on to large surfaces, one-off jobs and repairs. Automation, generally on a rotary table, is not difficult when long production runs are planned.

FURNACE BRAZING

This process can frequently be employed with the use of little or no flux, dependent on the materials being brazed and the extent to which the furnace atmosphere is kept reducing in character. Where possible, the furnace atmosphere should be strictly controlled, particularly with respect to the hydrogen potential and the dew point which should be instrumented for continuous control. Braze alloy pre-forms are located between pre-assembled parts which are usually fluxed and self-jigging.

Vacuum furnaces may be used, employing pressures of 1.0 to 1.0×10^{-9} torr and being heated either by radiation from resistance elements or by induction coils. Metal–metal oxide equilibrium diagrams are useful to show the relationship which exists between the furnace operating temperature and the reducing or oxidising behaviour of the atmosphere. Figure 34.1 indicates the effect of dew point of hydrogen and partial pressure of water on the brazeability of pure metals.

INDUCTION BRAZING

Heating is localised, fast, and generally accomplished within one or two minutes. Coils may be shaped to follow the contours of the assembly and adjustment to the coil separation distances can cause heavier sections to heat up at the same rate as thin sections. The mating surfaces can be coated with flux, or pre-assembled units with braze alloy pre-forms may be heated in a suitable atmosphere. It must be remembered that high-resistance metals such as steel heat up rapidly whereas low-resistance metals take longer to reach the brazing temperature. Amongst other factors, the efficiency of heating is controlled by the frequency of the equipment which should be suitable for the size of assemblies normally expected.

ELECTRIC RESISTANCE BRAZING

This technique is suitable only for relatively small joint areas because it is difficult to maintain uniform current distributions. Localised heating does not usually modify the mechanical properties of the parent metals. Success will depend upon the thermal capacities and thermal conductivities of the metals being joined. The presence of an intermediate layer of flux can be detrimental if direct heating is employed (i.e. when heating current passes through two work pieces sandwiching the brazing alloy). Once the optimum brazing schedule has been established this method can become semi-automated and requires minimal operator skills.

SALT BATH DIP BRAZING

The assembled joint, with the filler metal(s) preplaced, is dipped into a bath of molten salt which acts as flux. Occasionally the parts may be dipped into a bath of molten brazing alloy covered with a layer of flux. Dip brazing is widely used for joining aluminium components. Preheating of parts is recommended as it prevents excessive freezing of the salt on a cold surface, expels water entrapped

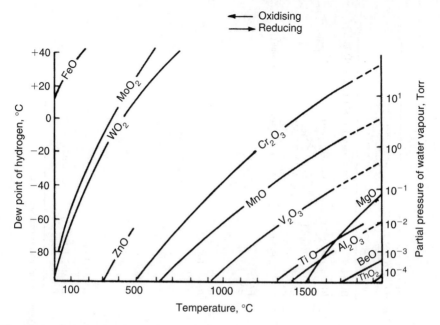

Figure 34.1 *Metal–metal oxide equilibria in hydrogen atmospheres. Oxidising conditions exist at the life of each curve: reducing conditions which facilitate brazing exist on the right of each curve. Note Au, Pt, Ag, Pd, Ir, Cu, Co, Ni and Os oxides are easier to reduce than those plotted.*

within the assembly thereby minimising the risk of explosions from immersed parts, and markedly increasing the speed of production. The specific gravity of the molten salt used is close to that of aluminium, reducing by half the weight of submerged pieces. This facilitates the dip-brazing of complex aluminium assemblies that are not strong enough to be self-supporting when brazed by other techniques. Personnel must of course wear protective clothing and face masks.

34.4.6   Brazeability of materials and braze alloy compositions

It is not possible to detail the optimum heating methods and filler metals for the wide range of base metals and combinations of base metals. Expert advice should be sought or literature searches and experimentation must be performed prior to the brazing of uncommon materials. It is strongly recommended that the referenced Standards are consulted for additional information. The brazing alloy systems in common use are listed in Table 34.10 with compositions detailed in Table 34.11.

A useful guide to the suitability of parent materials and filler metals for torch brazing, furnace brazing, induction brazing, electrical resistance brazing, and vacuum brazing is presented in Tables 34.12 to 34.16.

34.5   Diffusion soldering or brazing

Diffusion soldering or brazing is a variation of soldering or brazing processes in which a thin layer of liquid forms, by melting of a filler metal or formation of an in-situ liquid phase by reaction, and wets the faying surfaces, then solidifies isothermally at the bonding temperature as the elements responsible for the depression of melting temperature diffuse away into the surrounding solid. The process offers advantages such as joint microstructures and properties that are very similar to those of the base metal, and the need for only very low contact pressure during the process. This process is also known as liquid phase diffusion bonding or transient liquid phase (TLP) bonding.

A simple example to illustrate the process is the diffusion brazing of silver to silver using a thin copper or eutectic silver–copper interlayer. When the silver-interlayer-silver assembly is heated to

Table 34.10   COMMON BRAZING-ALLOY SYSTEMS

Alloy system	Melting range (°C)	Refer to EN 1044 table no.	AWS-ASTM classification (non-exhaustive)	DIN designations and Werkstoff numbers (in brackets) (non-exhaustive)
Copper	1083	4	BCu-1	L–Cu (2.0081), L–SFCU (2.0091)
Silver	961			
Copper–Zinc	860–890	5	RBCuZn–A	L–CuZn46 (2.0413)
Bronze welding type	920–980	5		
Copper–phosphorus	705–800	3	BCuP–1, BCuP–2	L–CuP8 (2.1465)
Silver–copper	779	2	BAg–8	L–Ag72 (2.5151)
Copper–silver–zinc	690–775	2	BAg–5	L–Ag44 (2.5147), L–Ag60
Copper–silver–phosphorus	644–780	3	BCuP–3, BCuP–5	L–Ag15P (2.1210), L–Ag5P (2.1466)
Copper–silver–tin	602–718	2		
Copper–silver–zinc–cadmium–nickel	630–660		BAg–3	L–Ag50CdNi (2.5160)
Copper–silver–zinc–cadmium	620–640	2	BAg–1, BAg–2	L–Ag50Cd (2.5143)
Cobalt–chromium–boron	984–1 240			
Nickel–chromium–boron	1 030–1 175			
Nickel–chromium–silicon–boron	950–1 100	6	BNi–2	
Nickel–silicon–boron	950–1 070		BNi–3, BNi–4	
Nickel–phosphorus	890–1 023		BNi–6	
Silver–palladium–manganese	1 000–1 200	7		(2.5154, 2.5164)
Nickel–palladium–manganese	1 120	7		(2.4304)
Copper–palladium–nickel–manganese	1 060–1 105	7		
Palladium–nickel	1 237			
Copper–nickel	1 100–1 150	7		
Copper–palladium	1 080–1 090	7		
Silver–palladium	970–1 010	7		
Silver–copper–palladium	807–950	7		
Gold–nickel	950	8	BAu–4	
Gold–copper	910	8	BAu–1, BAu–2	
Aluminium–silver–nickel (or manganese)	558–680			
Aluminium–silicon	565–625	1	BAlSi–2, BAlSi–4	L–AlSi12 (3.2285)
Auminium–silicon–copper	550–570	1	BAlSi–3	

Note: Some alloys are available in only one specific form and this should be ascertained before joint-design is considered. The available forms are wire, foil, powder, clad sheet, deposited coating and, in many cases, paste.

Table 34.11 FILLER METALS FOR BRAZING* *SEE* (BS 1845 FOR FULL COMPOSITION AND REQUIREMENTS)

Designation		Nominal composition (%) (excluding impurities)											Melting range (°C)
EN1044	ISO	Cu	Ag	Zn	P	Cd	Sn	Mn	Ni	Sb	Bo	Si	
Group AG: silver brazing filler metals													
AG1		15	50	15	—	19	—	—	—	—	—	—	620–640
AG2		17	42	15	—	25	—	—	—	—	—	—	610–620
AG3		20	38	22	—	20	—	—	—	—	—	—	605–650
AG11		25	34	20	—	21	—	—	—	—	—	—	610–670
AG12		28	30	21	—	21	—	—	—	—	—	—	600–690
AG14		21	55	22	—	—	2	—	—	—	—	—	630–660
AG20		30	40	28	—	—	2	—	—	—	—	—	650–710
AG21		36	30	32	—	—	3	—	—	—	—	—	665–755
AG5		37	43	20	—	—	—	—	—	—	—	—	690–770
AG7		28	72	—	—	—	—	—	—	—	—	—	780
AG9		15	50	15	—	16	—	—	—	—	—	—	635–655
AG13		26	60	14	—	—	—	7	3	—	—	—	695–730
AG18		16	49	23	—	—	—	15	4	—	—	—	680–705
AG19		—	85	—	—	—	—	—	—	—	—	—	960–970
Group CP: copper–phosphorus brazing filler metals													
CP1	CuAg14P5	Rem.	14.5	—	4.5	—	—	—	—	—	—	—	645–800
CP2	CuP6Ag2	Rem.	2	—	6.5	—	—	—	—	—	—	—	645–825
CP3	CuP7	Rem.	—	—	7.5	—	—	—	—	—	—	—	710–810
CP4	CuP6Ag5	Rem.	5	—	6	—	—	—	—	—	—	—	645–815
CP5	CuP6Sb	Rem.	—	—	6	—	—	—	—	2	—	—	690–825
CP6	CuP6	Rem.	—	—	6	—	—	—	—	—	—	—	710–890
Group CU: copper brazing filler metals													
CU2 (C102)	Cu-FRTP	99.90	—	—	—	—	—	—	—	—	—	—	1085
CU3 (C103)	Cu-OF	99.95	—	—	—	—	—	—	—	—	—	—	1085
CU5		99.00	—	—	—	—	—	—	—	—	—	—	1085
CU6 (C106)	Cu-DHP	99.85	—	—	0.02	—	—	—	—	—	—	—	1085
CU7	CuNi3	Rem.	—	—	—	—	—	—	3	—	—	—	1085–1100
CU8	CuMn2	Rem.	—	—	—	—	—	2	—	—	0.03	—	1045–1060
Group CZ: copper–zinc brazing filler metals													
CZ6	CuZn40Si	60	—	Rem.	—	—	—	—	—	—	—	0.3	875–895
CZ6A	CuZn40SiSn	60	—	Rem.	—	—	0.3	—	—	—	—	0.3	875–895
CZ7	CuZn40SiMn	60	—	Rem.	—	—	—	0.1	—	—	—	0.3	870–900
CZ7A	CuZn40SiSnMn	60	—	Rem.	—	—	0.3	0.1	—	—	—	0.3	870–900
CZ8	CuZn43Ni9Si	48	—	Rem.	—	—	—	—	9	—	—	0.3	920–980

* Applications for these materials are described in CDA publication TN 25 'Joining of Copper and Copper Alloys'.

Table 34.12　COMBINATIONS OF PARENT MATERIAL AND FILLER METAL FOR FLAME (TORCH) BRAZING*

Filler metal

Parent material	Copper	Silver	Copper–zinc	Bronze welding type	Copper–phosphorus	Silver–copper	Copper–silver–zinc	Copper–silver–phosphorus	Copper–silver–tin	Copper–silver–zinc–cadmium–nickel	Copper–silver–zinc–cadmium	Cobalt–chromium–boron	Nickel–chromium–boron	Nickel–chromium–silicon–boron	Nickel–silicon–boron	Nickel–phosphorus	Silver–palladium–manganese	Nickel–palladium–manganese	Copper–palladium–nickel–manganese	Palladium–nickel	Copper–nickel	Copper–palladium	Silver–palladium	Silver–copper–palladium	Gold–nickel	Gold–copper	Aluminium–silver–nickel (or manganese)	Aluminium–silicon	Aluminium–silicon–copper
Copper	N	R	R	R	R	R	R	R	P	R	R	N	N	N	N	N	N	N	N	N	N	N	N	R	P	R	N	N	N
Copper-base alloys	N	R	R	R	R	R	R	R	P	R	R	N	N	N	N	N	N	N	N	N	N	N	R	R	P	R	N	N	N
Mild steel	N	P	R	P	N	P	R	N	P	R	R	P	R	R	R	N	R	R	P	P	N	R	R	R	P	P	N	N	N
Carbon steels	N	P	R	R	N	P	R	N	N	R	R	P	R	R	R	N	R	R	P	P	N	R	R	R	P	P	N	N	N
Alloy steels	N	N	R	R	N	R	R	N	N	R	R	P	R	R	R	N	R	R	R	P	N	R	P	R	P	P	N	N	N
Stainless steels and irons	N	N	R	R	N	N	R	N	P	R	R	R	R	R	R	N	R	R	R	P	N	R	P	R	P	P	N	N	N
Malleable iron	N	N	R	R	N	N	P	N	N	R	R	P	N	R	R	N	P	R	P	P	N	R	P	P	P	P	N	N	N
Wrought iron	N	N	R	R	N	N	P	N	N	R	R	N	N	N	N	N	P	P	P	P	N	P	N	P	P	P	N	N	N
Cemented carbides	N	N	R	R	N	N	R	N	N	R	R	N	R	P	P	N	P	P	P	P	N	R	P	R	P	P	N	N	N
Nickel-base alloys	N	N	P	R	N	N	R	N	N	R	R	N	P	R	R	N	R	R	R	P	N	P	P	R	P	P	N	N	N
Cobalt-base alloys	N	N	N	R	N	N	P	N	N	P	P	P	P	P	P	N	P	P	P	P	P	R	P	R	P	P	N	N	N
Aluminium and certain aluminium alloys	N	N	N	N	N	N	N	N	N	N	N	P	N	N	N	N	N	N	N	N	N	N	N	N	N	N	N	R	R
Tungsten	N	N	P	P	P	P	R	P	P	R	R	N	P	P	P	N	P	P	P	N	N	N	P	P	P	P	N	N	N
Molybdenum	N	N	P	P	P	P	R	P	P	R	R	N	P	P	P	N	P	P	P	N	N	N	P	P	P	P	N	N	N
Titanium	N	N	N	N	N	N	N	N	N	N	N	N	N	N	N	N	N	N	N	N	N	N	N	N	N	N	N	P	P
Zirconium	N	N	N	N	N	N	N	N	N	N	N	N	N	N	N	N	N	N	N	N	N	N	N	N	N	N	N	N	N
Tantalum	N	N	N	N	N	N	N	N	N	N	N	N	N	N	N	N	N	N	N	N	N	N	N	N	N	N	N	N	N
Beryllium	N	N	N	N	N	N	N	N	N	N	N	N	N	N	N	N	N	N	N	N	N	N	N	N	N	N	N	N	N
Niobium	N	N	N	N	N	N	N	N	N	N	N	N	N	N	N	N	N	N	N	N	N	N	N	N	N	N	N	N	N

R—Recommended and known to be in general use. It should be recognised, however, that in many instances jointing by brazing results in a weakening of the parent material due to intercrystalline penetration, grain growth and other causes.
P—Possible but not known to be in general use.
N—Not recommended.
* From EN 1723.

Table 34.13 COMBINATIONS OF PARENT MATERIAL AND FILLER METAL FOR FURNACE BRAZING*

Filler metal

Parent material	Copper	Silver	Copper–zinc	Bronze welding type	Copper–phosphorus	Silver–copper	Copper–silver–zinc	Copper–silver–phosphorus	Copper–silver–tin	Copper–silver–zinc–cadmium–nickel	Copper–silver–zinc–cadmium	Cobalt–chromium–boron	Nickel–chromium–boron	Nickel–chromium–silicon–boron	Nickel–silicon–boron	Nickel–phosphorus	Silver–palladium–manganese	Nickel–palladium–manganese	Copper–palladium–nickel–manganese	Palladium–nickel	Copper–nickel	Copper–palladium	Silver–palladium	Silver–copper–palladium	Gold–nickel	Gold–copper	Aluminium–silver–nickel (or manganese)	Aluminium–silicon	Aluminium–silicon–copper
Copper	N	R	R	N	R	R	R	R	R	R	R	N	N	N	N	N	N	N	N	N	N	N	N	R	R	R	N	N	N
Copper-base alloys	R	R	P	N	R	R	R	R	R	R	R	N	N	N	N	N	N	N	N	N	N	N	N	R	R	R	N	N	N
Mild steel	R	P	R	R	N	R	R	N	R	R	R	P	R	R	R	N	R	R	R	R	R	R	R	R	R	R	N	N	N
Carbon steels	R	P	R	R	N	R	R	N	R	R	R	P	R	R	R	N	R	R	R	R	R	R	R	R	R	R	N	N	N
Alloy steels	R	P	P	R	N	R	R	N	P	R	R	P	R	R	R	N	R	R	R	P	R	R	R	R	P	R	N	N	N
Stainless steels and irons	R	P	P	N	N	R	R	N	P	P	R	P	R	R	R	N	R	R	P	P	R	R	R	R	R	R	N	N	N
Malleable iron	R	P	R	N	N	R	R	N	P	P	P	P	R	P	P	R	R	P	P	P	R	R	R	R	P	P	N	N	N
Wrought iron	R	P	R	R	N	R	R	N	P	P	P	P	R	P	P	N	R	R	R	P	R	R	R	R	P	P	N	N	N
Cemented carbides	R	P	P	R	N	P	P	N	P	P	P	N	P	P	R	N	P	P	P	R	P	R	P	R	P	P	N	N	N
Nickel-base alloys	R	P	P	P	N	R	P	N	P	P	P	R	R	R	R	N	R	R	R	P	R	R	R	R	R	R	N	N	N
Cobalt-base alloys	P	P	N	N	N	P	P	N	P	P	P	R	R	P	R	R	R	R	P	N	P	P	P	P	P	P	N	N	N
Aluminium and certain aluminium alloys	N	N	N	N	N	N	N	N	N	N	N	N	N	N	N	N	N	N	N	N	N	N	N	N	N	N	R	R	R
Tungsten	N	N	N	N	N	N	P	P	P	P	P	N	R	R	R	N	R	R	R	R	P	P	P	R	R	P	N	N	N
Molybdenum	N	N	N	N	N	N	P	P	P	N	N	N	R	R	R	N	R	R	R	R	P	R	N	R	R	P	N	N	N
Titanium	N	P	N	N	N	P	N	N	P	N	N	N	N	P	P	N	N	N	N	N	P	P	N	P	P	P	R	P	P
Zirconium	N	P	N	N	N	P	N	N	P	N	N	N	N	N	N	N	P	P	P	P	P	P	P	P	P	P	N	N	P
Tantalum	P	P	N	N	N	P	N	N	N	N	N	N	N	P	P	N	P	N	N	R	N	N	N	P	R	P	N	N	N
Beryllium	N	N	N	N	N	P	N	N	N	N	N	N	N	N	N	N	P	P	P	P	P	P	P	P	P	P	N	N	N
Niobium	N	N	N	N	N	N	N	N	N	N	N	N	N	P	P	N	N	N	N	N	N	N	N	N	P	P	N	N	N

R—Recommended and known to be in general use. It should be recognised, however, that in many instances jointing by brazing results in a weakening of the parent material due to intercrystalline penetration, grain growth and other causes.
P—Possible but not known to be in general use.
N—Not recommended.
* From EN 1723.
Note: Alloys containing zinc or cadmium require a flux.

Table 34.14   COMBINATIONS OF PARENT MATERIAL AND FILLER METAL FOR ELECTRIC INDUCTION BRAZING*

Filler metal

Parent material	Copper	Silver	Copper–zinc	Bronze welding type	Copper–phosphorus	Silver–copper	Copper–silver–zinc	Copper–silver–phosphorus	Copper–silver–tin	Copper–silver–zinc–cadmium–nickel	Copper–silver–zinc–cadmium	Cobalt–chromium–boron	Nickel–chromium–boron	Nickel–chromium–silicon–boron	Nickel–silicon–boron	Nickel–phosphorus	Silver–palladium–manganese	Nickel–palladium–manganese	Copper–palladium–nickel–manganese	Palladium–nickel	Copper–nickel	Copper–palladium	Silver–palladium	Silver–copper–palladium	Gold–nickel	Gold–copper	Aluminium–silver–nickel (or manganese)	Aluminium–silicon	Aluminium–silicon–copper
Copper	N	R	P	P	P	R	R	P	P	R	R	Z	Z	Z	Z	Z	Z	Z	Z	Z	Z	Z	Z	R	R	R	Z	Z	Z
Copper-base alloys	N	R	P	P	P	R	R	P	P	R	R	Z	Z	Z	Z	Z	Z	Z	Z	Z	Z	Z	Z	R	R	R	Z	Z	Z
Mild steel	P	Z	R	P	Z	R	R	Z	P	R	R	P	R	R	R	Z	R	R	P	P	P	R	R	R	P	R	Z	Z	Z
Carbon steels	P	Z	R	R	Z	R	R	Z	P	R	R	P	R	R	R	Z	R	R	P	P	P	R	R	R	P	P	Z	Z	Z
Alloy steels	P	Z	R	R	Z	R	R	Z	P	R	P	P	R	R	R	Z	R	R	P	P	P	R	P	R	P	P	Z	Z	Z
Stainless steels and irons	P	Z	R	R	Z	R	R	Z	P	P	P	P	R	R	R	P	R	R	P	R	P	R	P	R	R	R	Z	Z	Z
Malleable iron	P	Z	R	R	Z	P	P	Z	P	P	P	P	P	P	P	Z	R	P	P	Z	P	R	P	R	P	P	Z	Z	Z
Wrought iron	P	Z	R	R	Z	R	R	Z	P	P	P	Z	P	P	P	Z	R	P	P	Z	P	R	P	R	P	P	Z	Z	Z
Cemented carbides	R	P	P	Z	Z	P	R	Z	P	R	R	P	P	P	P	Z	P	P	R	P	P	R	R	R	P	P	Z	Z	Z
Nickel-base alloys	R	P	Z	Z	Z	P	P	Z	P	P	P	P	R	R	R	P	R	P	P	R	P	R	R	P	R	R	Z	Z	Z
Cobalt-base alloys	Z	Z	Z	Z	Z	Z	P	P	P	P	P	P	P	P	P	Z	P	P	P	P	P	P	P	R	P	P	Z	Z	Z
Aluminium and certain aluminium alloys	Z	Z	Z	Z	Z	Z	Z	Z	Z	Z	Z	Z	Z	Z	Z	Z	Z	Z	Z	Z	Z	Z	Z	Z	Z	Z	Z	P	P
Tungsten	N	Z	Z	P	P	Z	R	P	P	R	R	Z	P	R	R	Z	P	P	P	R	Z	P	P	R	R	P	Z	Z	Z
Molybdenum	N	Z	Z	P	P	Z	R	P	P	R	R	Z	P	R	R	Z	R	P	P	R	Z	R	P	R	R	P	Z	Z	Z
Titanium	N	P	Z	Z	Z	P	P	Z	Z	P	P	Z	Z	P	Z	Z	Z	Z	Z	Z	Z	Z	Z	R	P	P	P	P	P
Zirconium	N	Z	Z	Z	Z	P	P	Z	P	Z	Z	Z	Z	Z	Z	P	P	P	P	P	P	P	P	P	P	P	Z	Z	Z
Tantalum	N	Z	Z	Z	Z	Z	Z	Z	Z	Z	Z	Z	Z	Z	Z	Z	P	P	P	R	P	P	P	P	P	P	Z	Z	Z
Beryllium	N	P	Z	Z	Z	P	Z	Z	Z	Z	Z	Z	Z	Z	Z	Z	P	Z	Z	Z	Z	P	P	P	P	P	P	Z	Z
Niobium	N	Z	Z	Z	Z	Z	Z	Z	Z	Z	Z	Z	Z	Z	Z	Z	P	P	P	P	Z	P	P	P	P	P	Z	Z	Z

R—Recommended and known to be in general use. It should be recognised, however, that in many instances jointing by brazing results in a weakening of the parent material due to intercrystalline penetration, grain growth and other causes.
P—Possible but not known to be in general use.
N—Not recommended.
* From EN 1723.

Table 34.15 COMBINATIONS OF PARENT MATERIAL AND FILLER METAL FOR ELECTRIC RESISTANCE BRAZING*

Filler metal

Parent material	Copper	Silver	Copper–zinc	Bronze welding type	Copper–phosphorus	Silver–copper	Copper–silver–zinc	Copper–silver–phosphorus	Copper–silver–tin	Copper–silver–zinc–cadmium–nickel	Copper–silver–zinc–cadmium	Cobalt–chromium–boron	Nickel–chromium–boron	Nickel–chromium–silicon–boron	Nickel–silicon–boron	Nickel–phosphorus	Silver–palladium–manganese	Nickel–palladium–manganese	Copper–palladium–nickel–manganese	Palladium–nickel	Copper–nickel	Copper–palladium	Silver–palladium	Silver–copper–palladium	Gold–nickel	Gold–copper	Aluminium–silver–nickel (or manganese)	Aluminium–silicon	Aluminium–silicon–copper
Copper	N	R	R	N	P	R	R	R	P	R	R	N	N	N	N	N	N	N	N	N	N	N	N	R	R	R	N	N	N
Copper-base alloys	N	R	R	N	P	R	R	R	P	R	R	N	N	N	N	N	N	N	N	N	N	N	N	R	R	R	N	N	N
Mild steel	P	P	P	N	N	P	R	N	P	R	R	P	P	P	P	N	R	R	P	P	P	R	R	R	R	R	N	N	N
Carbon steels	P	P	P	N	N	P	R	N	P	R	R	P	P	P	P	N	R	R	P	P	P	R	R	R	P	P	N	N	N
Alloy steels	P	P	P	N	N	P	R	N	P	R	R	P	P	P	P	N	R	R	P	P	P	R	P	R	P	P	N	N	N
Stainless steels and irons	P	P	P	N	N	P	R	N	P	R	R	P	P	P	P	P	R	R	R	P	P	R	R	R	R	R	N	N	N
Malleable iron	P	N	P	N	N	P	R	N	P	R	R	N	P	P	P	N	R	R	P	P	P	P	R	P	P	P	N	N	N
Wrought iron	P	N	P	N	N	P	R	N	P	R	R	P	P	P	P	N	R	R	R	P	P	P	R	P	P	P	N	N	N
Cemented carbides	P	P	P	N	N	P	R	N	P	R	R	P	P	P	P	N	R	P	P	P	P	R	P	R	P	P	N	N	N
Nickel-base alloys	P	P	P	N	N	P	R	N	P	R	R	P	P	P	P	P	R	P	P	R	P	P	R	R	R	R	N	N	N
Cobalt-base alloys	P	P	P	N	N	P	P	N	P	P	R	P	P	P	P	N	P	R	R	P	P	R	R	R	P	P	N	N	N
Aluminium and certain aluminium alloys	N	N	N	N	N	N	N	N	N	N	N	N	N	N	N	N	N	N	N	N	N	N	N	N	N	N	P	P	P
Tungsten	N	N	P	P	P	N	R	P	P	R	R	N	N	N	N	N	P	P	P	R	P	P	P	R	R	R	N	N	N
Molybdenum	N	N	P	P	P	N	R	P	P	R	R	N	N	N	N	N	R	P	P	R	P	R	P	R	R	P	N	N	N
Titanium	N	P	N	N	N	N	P	P	P	P	P	N	N	N	N	P	N	N	N	P	N	P	P	P	P	R	P	P	P
Zirconium	N	N	N	N	N	N	P	N	N	P	P	N	N	N	N	N	P	P	P	P	P	P	P	P	P	P	N	P	P
Tantalum	N	N	N	N	N	N	N	N	N	N	N	N	N	N	N	N	P	P	R	R	N	P	P	R	P	P	N	N	N
Beryllium	N	N	N	N	N	N	N	N	N	N	N	N	N	N	N	N	N	N	P	P	N	P	P	R	P	P	N	N	N
Niobium	N	N	N	N	N	N	N	N	N	N	N	P	N	N	N	N	N	N	P	P	N	P	P	P	P	P	N	N	N

R—Recommended and known to be in general use. It should be recognised, however, that in many instances jointing by brazing results in a weakening of the parent material due to intercrystalline penetration, grain growth and other causes.
P—Possible but not known to be in general use.
N—Not recommended.
* From EN 1723.

Table 34.16   COMBINATIONS OF PARENT MATERIAL AND FILLER METAL FOR VACUUM BRAZING*

Parent material	Copper	Silver	Copper–zinc	Bronze welding type	Copper–phosphorus	Silver[a]–copper	Copper–silver–zinc	Copper–silver[a]–phosphorus	Copper–silver[a]–tin	Copper–silver–zinc–cadmium–nickel	Copper–silver[a]–zinc–cadmium	Cobalt–chromium–boron	Nickel–chromium–boron	Nickel–chromium–silicon–boron	Nickel–silicon–boron	Nickel–phosphorus	Silver–palladium–manganese	Nickel–palladium–manganese	Copper–palladium–nickel–manganese	Palladium–nickel	Copper–nickel	Copper–palladium	Silver–palladium	Silver–copper–palladium	Gold–nickel	Gold–copper	Aluminium–silver[a]–nickel (or manganese)	Aluminium–silicon	Aluminium–silicon–copper
Copper	N	P	N	N	N	R	N	N	R	N	N	N	N	N	N	N	N	N	N	N	N	N	N	R	P	R	N	N	N
Copper-base alloys	R	P	N	N	N	R	N	N	R	N	N	N	N	N	N	N	N	N	N	N	N	N	N	R	P	P	N	N	N
Mild steel	R	P	N	N	N	R	N	N	R	N	N	P	R	R	R	N	R	R	R	P	P	P	P	R	P	P	N	N	N
Carbon steels	R	P	N	N	N	P	N	N	P	N	N	P	R	R	R	N	R	R	R	P	P	P	P	R	P	P	N	N	N
Alloy steels	R	P	N	N	N	P	N	N	P	N	N	R	R	R	R	N	R	R	R	R	R	R	R	R	R	R	N	N	N
Stainless steels and irons	R	P	N	N	N	P	N	N	R	N	N	R	R	R	R	R	R	R	R	R	R	R	R	R	R	P	N	N	N
Malleable iron	R	P	N	N	N	P	N	N	P	N	N	P	P	P	P	N	P	P	P	P	R	P	P	P	P	P	N	N	N
Wrought iron	R	P	N	N	N	P	N	N	P	N	N	N	P	P	P	N	P	P	P	P	R	P	P	P	P	P	N	N	N
Cemented carbides	R	P	N	N	N	R	N	N	P	N	N	P	R	R	R	N	R	R	R	R	R	R	R	R	R	R	N	N	N
Nickel-base alloys	R	P	N	N	N	P	N	N	P	N	N	R	R	R	R	P	R	R	R	R	R	R	R	R	R	P	N	N	N
Cobalt-base alloys	P	P	N	N	N	N	N	N	N	N	N	R	P	P	P	N	P	P	P	P	P	P	P	P	P	P	N	P	P
Aluminium and certain aluminium alloys	N	N	N	N	N	N	N	N	N	N	N	N	N	N	N	N	N	N	N	N	N	N	N	N	N	N	N	P	P
Tungsten	N	N	N	N	N	N	N	N	P	N	N	N	N	R	R	N	P	P	P	R	P	R	R	R	R	P	N	N	N
Molybdenum	N	N	N	N	N	P	N	N	P	N	N	N	N	R	R	N	P	P	P	R	N	R	R	R	R	P	N	N	N
Titanium	N	P	N	N	N	P	N	N	P	N	N	N	N	N	N	N	N	N	P	N	P	N	N	R	N	P	R	N	P
Zirconium	R	P	N	N	N	P	N	N	N	N	N	N	N	P	P	P	P	P	P	P	P	P	P	P	P	R	N	N	N
Tantalum	N	P	N	N	N	P	N	N	P	N	N	N	N	P	P	N	P	P	P	R	P	P	P	P	R	R	N	N	N
Beryllium	P	P	N	N	N	P	N	N	N	N	N	N	N	P	P	N	P	P	N	N	P	P	P	R	N	P	N	N	N
Niobium	P	P	N	N	N	P	N	N	N	N	N	N	N	P	P	N	P	P	P	P	P	P	P	P	P	P	N	N	N

Filler metal

[a] Some alloys contain elements which are volatile at the temperatures and pressures involved in the brazing process and caution is required under prolonged heating.

R—Recommended and known to be in general use. It should be recognised, however, that in many instances jointing by brazing results in a weakening of the parent material due to intercrystalline penetration, grain growth and other causes.

P—Possible but not known to be in general use.

N—Not recommended.

Note: This table is intended only as a guide and any further information should be obtained from the manufacturer of the filler metal.

* From EN 1723

Table 34.17   MATERIALS COMBINATIONS USED FOR DIFFUSION BRAZING/SOLDERING

Substrate	*Filler metal (wt. %)*	*Bonding temperature (°C)*	*Remelting temperature (°C)*
Aluminium-based alloys and composites	Cu	550	600
Cobalt alloys	Ni-3.7B	1 175	1 495
Nickel alloys	Ni-15.5Cr-3.7B	1 100	1 455
	Ni-11P	900	
Silver alloys	Cu or Ag-28Cu	820	960
Steel	Fe-12Cr-4B	1 050	1 500
Titanium-base alloys and composites	Cu-50Ni	975	1 700
	Ag-15Cu-15Zn	700	700
	Ti-50Cu-25Zr	880	1 700
Copper metallisation	Sn	500	>900
	In	300	>600
Silver metallisation	Sn	250	>480
	In	300	>660
Gold metallisation	In	250	>450

a bonding temperature slightly above the silver–copper eutectic temperature (780°C), a liquid film is formed between the silver workpieces by melting of the eutectic silver–copper interlayer or by in-situ formation of eutectic silver–copper liquid by alloying between a copper interlayer and the base metal. This liquid phase wets the joint surfaces and fills the gap. If the assembly were then cooled with the liquid phase solidified during cooling, the joint would be a conventional brazed joint with a remelting temperature equivalent to the eutectic temperature. However, by maintaining the assembly at the bonding temperature, the liquid phase solidifies isothermally as the copper in the liquid phase diffuses away from the joint line into the silver workpieces. The resultant diffusion-brazed joint has a remelting temperature eventually equivalent to the melting point of silver!

Diffusion soldering/brazing is not limited only to material combinations that form solid solutions (such in the silver–copper system). Systems that form intermetallics or new phases at the joint interfaces may also be used to produce diffusion-brazed or -soldered joints because the same advantage of raised remelting temperature may also be realised. For example, during diffusion soldering of silver using tin, the tin will melt and react with the silver to form silver–tin intermetallics until no liquid tin remains. The remelt temperature of the diffusion-soldered joint is then determined by the new phase(s) formed in the process. Table 34.17 lists examples of material combinations used for diffusion brazing/soldering.

34.6   Bibliography and sources of information

G. Humpston and D. M. Jacobson, 'Principles of Soldering and Brazing', ASM International, Materials Park, OH, 1993.

P. T. Vianco, 'Soldering Handbook', 3rd edn, American Welding Society, Miami, FL, 1999.

Metals information: Institute of Metals, 1 Carlton House Terrace, London SW1Y 5DB.

British Standards Institution, Sales Office, Information Department and Technical Help to Exporters, Linford Wood, Milton Keynes MK14 6LE.

American specifications: American Technical Publications, 68a Wilbury Way, Hitchin, Herts SG4 0TP.

International Tin Research Institute, Kingston Lane, Uxbridge, Middx U88 3PJ.

British Association for Brazing and Soldering, The Welding Institute, Abington Hall, Abington, Cambridge CB1 6AL.

Copper Development Association, Orchard House, Mutton Lane, Potters Bar, Herts EN6 3AP.

Printed Circuit Group, Institute of Metal Finishing, Exeter House, 48 Holloway Head, Birmingham B1 1NQ.

DTI Soldering Club, National Physical Laboratory, Teddington, Middx.

Lead Development Association, 42 Weymouth Street, London W1N 3LQ.

Low melting point solders: Mining and Chemical Products Ltd, 38 Rosemont Road, Alperton, Wembley, Middx HA0 4PE.

Solders: Fry's Metals Ltd, Tandem Works, Christchurch Road, Wimbledon, London SW19 2PD and United Alloys Ltd, Crosslands Lane North Cave, Brough HU15 2PG.

Hard solders: Johnson Matthey Metals Ltd, Metal Joining, Orchard Road, Royston, Herts SG8 5HE.

Brazing: Solder Products Manufacturing Ltd, 8–10 Stanley Street, Sheffield S3 8HJ and Thessco Ltd, Royds Mills, Windsor Street, Sheffield S4 7WB.

C. J. Thwaites and B. T. K. Barry, 'Soldering', Engineering Design Guides, Oxford University Press, 1975.

'Properties and uses of MCP low melting point alloys', Mining and Chemical Products Ltd, 1987.

G. Leonida, 'Handbook of Printed Circuit Design, Manufacture, Components and Assembly', Electrochemical Publications, Ayr, 1981.

F. R. Thorns, Choosing a flux (for wave soldering), *Electronic Production*, 1989, June, 35–37.

H. R. Brooker and E. V. Beatson (revised by P. M. Roberts), 'Industrial Brazing', Newnes-Butterworth, 1975.

Technical Information Sheets B1–B15, British Association for Brazing and Soldering.

'Joining of Copper and Copper Alloys', Copper Development Association, 1980.

'Brazing Materials and Applications', Johnson Matthey Ltd, 1987.

'Brazing Manual', American Welding Society, New York, 1963.

'Metals Handbook', 9th edn, vol. 6 'Welding, Brazing and Soldering', American Society for Metals, Ohio, 1983.

'The Welding, Brazing and Soldering of Copper and its Alloys', Copper Development Association, 1958.

D. R. Andrews, 'Soldering, Brazing, Welding and Adhesives', Institute of Production Engineers, London, 1978.

P. M. Roberts, 'Brazing', Engineering Design Guides, Oxford University Press, 1975.

35 Vapour deposited coatings and thermal spraying

Vapour deposition processes are of two kinds: physical vapour deposition (PVD) and chemical vapour deposition (CVD). Process details and references for elements are given in Tables 35.1 and 35.2, for oxides in Tables 35.3 and 35.4, for nitrides in Tables 35.5 and 35.6, and for carbides in Tables 35.7 and 35.8. Section 35.3 discusses thermal spraying (*see* Chapter 33 for relevant standards).

35.1 Physical vapour deposition

Physical vapour deposition processes use a physical effect such as evaporation or sputtering to transport material, usually a metal, from a source to the substrate to be coated. If the material is transformed (e.g. into a carbide) during transport, then the process is described as reactive. All PVD processes are carried out in a relatively high vacuum (i.e. pressure $<10^{-4}$ Torr).

35.1.1 Evaporation (E)

The substrate to be coated is placed in a vacuum chamber with a line-of-sight to the source which is a pool of molten material. The pool is heated either by an electron beam or an arc or by resistance heating. The electron beam method is best for source materials with a high melting point.

REACTIVE EVAPORATION (RE)

The same process as evaporation except that the zone through which the evaporated species is passing contains a very small concentration of a reactive gas, usually a hydrocarbon, N_2, NH_3 or O_2. This results in the deposition of the carbide, nitride or oxide of the source material.

ACTIVATED REACTIVE EVAPORATION (ARE)

The same process as reactive evaporation except that an electrically excited plasma is established in the region where the evaporated species encounters the reactive gas.

ION PLATING (IP)

The same process as evaporation except that substrate is biassed negatively with respect to the source. This usually results in a plasma region around the substrate which can be enhanced by r.f. coupling. The ionised species actually represents only a small fraction of the total material flux.

REACTIVE ION PLATING (RIP)

The same process as ion plating except that a reactive gas (usually a hydrocarbon, N_2, NH_3 or O_2) is introduced into the flux travelling towards the substrate so that the carbide, oxide or nitride of the source material is formed.

35.1.2 Sputtering (S)

A process in which material is transferred from a target and deposited on a substrate by means of ionic bombardment of the target. Usually the ion bombardment is achieved by establishing a d.c. or r.f. plasma at the surface of the target, although ion guns can also be used.

SPUTTER ION PLATING (SIP)

The same process as sputter plating except that the substrate has an electrical negative bias, d.c. and r.f. possible (for non conductive materials), so that it is bombarded with positive ions as well as the neutral flux created by the sputtering.

REACTIVE SPUTTER PLATING (RSP)

The same process as sputter plating except that a reactive gas is introduced into the plasma region so that the transported target material reacts before it arrives at the substrate. For example, the conversion of Ti into a deposit of TiN by the introduction of nitrogen.

MAGNETRON SPUTTERING (MS)

The same process as sputter plating except that a magnetic field is used to direct the ion flux at the surface of the target.

35.1.3 Ion cleaning

An argon plasma may be used to bombard a surface with excited argon atoms. These transfer their momentum to atoms in the surface and some of these are ejected from the surface. This has the effect of gradually removing material from a surface, hence cleaning the surface. Ion cleaning is often used as a pretreatment in PVD coating.

35.2 Chemical vapour deposition

Chemical vapour deposition processes use the vapour phase to transport reactive material to the surface of a substrate where a chemical reaction occurs to form a coating, e.g. tantalum by the reaction: $TaCl_5 + H_2 \rightarrow Ta + 5HCl$. Normally the substrate is heated to activate the reaction. In many CVD processes the substrate takes no part in the formation of the coating, but in *diffusion coatings* the substrate takes part in the formation of the coating (e.g. aluminising).

CVD processes can be divided into two groups, hot wall and cold wall, as follows:

A hot-wall reactor is a chamber which is heated from the outside so that both the chamber and the work pieces reach the reaction temperature and so become coated. Consequently there is a certain amount of redundant plating which is undesirable with expensive materials.

A cold-wall reactor is a chamber in which only the work pieces are heated, usually by induction. Only the work pieces become coated.

CVD processes are a rather diffuse group because the physical conditions for each process differ widely so that there are no standard items of equipment as there are for PVD. Many CVD processes use a reduced pressure to improve throwing power (cf. PVD).

In the references *CVD* stands for the International Chemical Vapor Depositions Conference Proceedings published by The Electrochemical Society, Vapour Deposition in the refs. refers to that title by Powell, Oxley and Blocher, J. Wiley, N.Y. 1965. Process details Tables 35.2, 35.4, 35.6, 35.8.

Plasma Activated or Enhanced Chemical Vapour Deposition (PACVD or PECVD)

The same process as CVD except that a d.c. or r.f. plasma is established around the substrate, usually to permit a low coating temperature.

Metal Organic Chemical Vapour Deposition (MOCVD)

The same process as CVD except that organometallic chemicals (e.g. trimethylgallium) are used as the feed reagents. Many organometallics have a high vapour pressure and a low decomposition temperature so that coatings can be formed at low substrate temperatures. It is particularly used for the growth of epitaxial layers.

Chemical Vapour Infiltration (CVI)

The same process as CVD except that the substrate is a porous and usually fibrous material into which the deposited material is infiltrated. It is an extremely slow process, but it leads to composites of low final porosity.

Table 35.1 PHYSICAL VAPOUR DEPOSITION OF ELEMENTS AND METALLIC ALLOYS

Element	Process*	Coating thickness (μm)	Typical applications	Ref.
Ag	S RIP	0.02–0.35	Window glass Photographic materials Foil bearings	2 6 31
Al**	E S MS	~1 ~1	IC metallisation Mirror coating	1 3 8
Au**	IP E S	0.2 0.02–0.1 1.4	 Window glass, electron microscopy Coating micro-spheres for laser fusion	4 7 5
Bi	S E	 ~1	Undercoat for photographic materials	6
C	TS Arc	0.05 350	Conductive layers for electron microscopy samples	32 19
(Co (Cr/Al))***	E		Turbine blades	14, 15
Cr	3	0.1–0.6	Ceramics Plastics Ni/Cr for window glass Switch contacts	11 12 16
Cu	IP S MS	0.2 0.02 0.01–1.0	Window glass coating Metallisation of plastics prior to electroplating	4 11 13
Mo	IBS S	 0.15	Multi-level metallisation for LSI	17 18
Nb	S	0.4	Superconductors	22
Ni	E, S	—	—	25
Pb	E (EG)	0.4	Pb/oxide/Pb Josephson junctions	21
Pd	E E (EG)	0.06 0.1–0.15	Thermoelectric power Contact layer	21 23, 24
Pt	MS E	1.0	Schottky barrier solar cells	10 —
Si	IBS E	0.4	Integrated circuits	—
Ta	S, E	6–20	Corrosion resistant layer	26, 27
Ti	S	—	—	12
Ti–W	MS	500–5 000 Å	Resistance films	—
V	E	0.086	—	30
W	S	500–5 000 Å	Barrier layer for ICs	28, 29

* ARE — Activated reactive evaporation.
** also used for compact disc metallising.
*** Typically MCrAly *alloys* M = Co, Ni.

E	— Evaporaiton.	RIS	— Reactive ion sputtering.
AE	— Arc evaporation.	RIP	— Reactive ion plating.
EG	— Electron gun.	RE	— Reactive evaporation.
IBS	— Ion beam sputtering.	RAE	— Reactive arc evaporation.
IP	— Ion Plating.	RSP	— Reactive sputtering.
MS	— Magnetron sputtering.	S	— Sputtering.
		SIP	— Sputter ion plating.
		TS	— Triode sputtering.

REFERENCES TO TABLE 35.1

1. R. J. Hill, 'Physical Vap. Dep.,' Airco Temescal, Berkeley, California, 1976, p. 114.
2. T. Abe and T. Yamashina, *Thin Solid Films*, 1975, **30**, 19.
3. C. R. Fuller and P. B. Ghate, *Thin Solid Films*, 1979, **64**, 25.
4. T. Spalvins, *Thin Solid Films*, 1979, **64**, 143.
5. L. Buene *et al.*, *Thin Solid Films*, 1980, **65**, 247.
6. S. K. Sharma and J. Spitz, *Thin Solid Films*, 1980, **65**, 339.

7. A. T. Lowe and C. D. Hosford, *J. Vac. Sci. Tech.*, 1979, **16**, 197.
8. A. J. Learn, *J. Electrochem. Soc.*, 1976, **123**, 894.
9. A. Aronson and S. Weinig, *Vacuum*, 1977, **27**, 151.
10. S. Schiller *et al.*, *J. Vac. Sci. Tech.*, 1977, **14**, 813; 1975, **12**, 858.
11. S. Hurwitt, Trans. Conf. Sputter, Mats. Res. Corp., Orangeberg, NY, 1974, p. 31.
12. D. W. Hoffmann and J. A. Thornton, *Thin Solid Films*, 1977, **40**, 355.
13. T. Tsutada and N. Hosokawa, *J. Vac. Sci. Tech.*, 1979, **16**, 348.
14. D. H. Boone *et al.*, *Thin Solid Films*, 1979, **64**, 299.
15. C. J. Spengler and S. Y. Lee, *Thin Solid Films*, 1979, **64**, 263.
16. Matsushita Electric Works, JP 5 476 972.
17. P. H. Schmidt *et al.*, *J. Appl. Phys.*, 1973, **44**, 1833.
18. J. Nagano, *Thin Solid Films*, 1980, **67**, 1.
19. Y. Sakai *et al.*, *J. Nucl. Mater.*, 1988, 155.
20. M. Murahami, *Thin Solid Films*, 1980, **69**, 253.
21. G. Wedler and R. Chander, *Thin Solid Films*, 1980, **65**, 53.
22. C. T. Wu, *Thin Solid Films*, 1979, **64**, 103.
23. D. J. Sharp, *J. Vac. Sci. Tech.*, 1979, **16**, 204.
24. V. Köster *et al.*, *Thin Solid Films*, 1980, **67**, 35.
25. S. Schiller *et al.*, *J. Vac. Sci. Tech.*, 1975, **12**, 858.
26. S. Kashu *et al.*, *J. Vac. Sci. Tech.*, 1972, **9**, 1399.
27. J. Spitzel and J. Chevallier, *CVD*, 1975, **V**, 204.
28. Kossowsky, *Electrochem. Soc. Ext. Abst.*, 1977, **77-2**, 429.
29. J. B. Bindell and T. C. Tisone, *Thin Solid Films*, 1974, **23**, 31.
30. A. Borodziuk-kulpa *et al.*, *Thin Solid Films,* 1980, **67**, 21.
31. C. Dellacorte *et al.*, Nasa Lewis Research Center, Nasa Tech. Memo. NASA TM-100783, 1988.
32. A. G. Fitzgerald *et al.*, *Carbon*, 1988, **26** (2), 229.

Table 35.2   CHEMICAL VAPOUR DEPOSITION OF ELEMENTS

Elements	Process	Temp (°C)	Thickness (μm)	Typical applications	Ref.
Al	Al R or Al HR$_2$ thermal decomp. (R usually isobutyl)	200–600	50	Impervious layer on plastics	6, 9
	Al + AlX^3 (X = halide)	700–1 000	20–100	Aluminising steel and	10
	AlCl$_3$ + H$_2$	900–1 000		super alloys	15
As	AsH$_3$ thermal decomp.	230–300	—	GaAlAs lasers	7
	AsH$_3$ plasma assisted				1
B	B$_2$H$_6$ thermal decomp.	400–700	—	B fibres	13
	BBr$_3$ thermal decomp.	1 000–1 300			
	BBr$_3$ + H$_2$	600–1 100		High purity boron	5
	BCl$_3$ + H$_2$ (also with laser excitation)	1 000–1 500	5–1 000	Jewelled bearings, fibres	2, 5, 12, 72
C (graphite)	CH$_4$, C$_2$H$_6$ thermal decomp.	2 000–2 500	500–3 000	Free-standing components	14
C (carbon)	CH$_4$, C$_2$H$_6$ thermal decomp.	900–1 200	1–50	Barrier layers, C–C composites	3, 4, 11
C (diamond and diamond-like carbon)	CH$_4$, C$_3$H$_8$ microwave plasma assisted	200–800	0.1–20	Anit-reflectance coating	70, 20
Co	Co(acac)$_2$* thermal decomp.	300–500	1.5	—	17
Cr	CrI$_2$ thermal decomp.	900–1 400	—	High purity chromium	19
	Cr dicumene	300–400	25–50	Coating nuclear fuel elements	14
	Cr + NH$_4$Cl pack	950–1 050	20–50	Coating turbine blades	16
Cu	Cu(acac)$_2$* thermal decomp.	260–450	—	—	18
Ge	GeCl$_4$ + H$_2$	600	—	—	22
	GeH$_4$ thermal decomp.	400–900		Epitaxial semiconductor layer	21
Hf	HfX$_4$ thermal decomp. (X = I or Br)	—	—	—	25

(continued)

Table 35.2 CHEMICAL VAPOUR DEPOSITION OF ELEMENTS—*continued*

Elements	Process	Temp (°C)	Thickness (μm)	Typical applications	Ref.
Ir	$Ir(CO)_2Cl_2$ thermal decomp.	600	—	—	28
Mo	$MoX_5 + H_2$ (X = Cl, F)	650–1 400	8	Thin film resistors	26, 30
	$(Mo(CO)_6 + H_2$	300–600	0.2	Coating nuclear fuels	29
	$Mo(CO)_6$ plasma			Anti-reflection on surfaces for photochem. conversion	27, 24, 23
Nb	$NbCl_5 + H_2$	1 000–1 400	5–50	Nuclear fuel cladding	32
Ni	$Ni(CO)_4$ thermal decomp.	180–200	1–1 000	Vapour forming	33
	$Ni(acac)_2$* thermal decomp.	250–450		Coating of uranium	34, 38
	$Ni(CO)_4$ plasma assisted				24
	$Ni(CO)_4$ laser assisted		0.055	Localised deposit	31
Pb	$PbEt_4$ thermal decomp.	300–500	—	—	35
Pd	$Pd(acac)_2$* thermal decomp.	350–450	—	—	39
Pt	$Pt(CO)_2Cl_2$ thermal decomp.	600	—	—	46
	$Pt(acac)_2$* thermal decomp.	350–450			40
Re	$ReF_5 + H_2$	850–1 100	—		—
	$ReCl_5 + H_2$	600–1 200			42
	$ReOCl_4$ thermal decomp.	1 250–1 500			37
	$ReCl_5$ thermal decomp.	1 100–1 300		Thermionic emitters	41
Rh	$Rh(tfa)_3$† $+ H_2$	250	—	—	38
	$Rh(CO)_2Cl_2$ thermal decomp.	600			46
Ru	$Ru(CO)_5$ thermal decomp.	200	—	—	47
	$Ru(CO)_2Cl_2$ thermal decomp.	600			46
Sb	$SbCl_3$ thermal decomp.	500–600	—	—	49
	SbH_3 thermal decomp.	~150		Prod. of high purity Sb	48
Si	SiH_4 thermal decomp.	300–1 200	—	—	51, 71
	$SiCl_4 + H_2$	900–1 800		Prod. of polycryst. Si	
	$SiHCl_3 + H_2$	950–1 250	30	Solar cells	50
	SiH_4 plasma decomp.	300–1 150		Semiconductor layers	44
	Siliconising Ni alloys			Corrosion-resistant layers on gas turbine parts	45, 43
Sn	$Sn + NH_4Cl/MgO$ pack diffusion	450–750	10	Corrosion and wear resistant coating	53
Ta	$TaCl_5 + H_2$	900–1 100	5–20	Corrosion-resistant layers	54
				Thin film resistors	58
				Impregnation of C fibres	59
Ti	$TiBr_4$ thermal decomp.	1 100–1 400	—	Metal production	55
	TiI_4 thermal decomp.	1 200–1 500			56
U	UI_6 thermal decomp.	~1 500	—	—	57
V	$VCl_4 + H_2$	800–1 000	—	—	60
	VI_2	1 000–1 200	10–50		69, 61
W	$WF_6 + H_2$	400–700	1–1 000	Thin wall components	62
	$WCl_6 + H_2$	600–700	1–1 000	Erosion-resistant coatings on rocket motors	66, 63, 64
	$W(CO)_6$ thermal decomp.	350–600		Electron emission surfaces	67, 68
Zr	ZrI_4 thermal decomp.	1 000–1 500	5–50	Nuclear fuel cladding	65

* acac = acetylacetonate.   † tfa = trifluoro-acetylacetonate.

REFERENCES TO TABLE 35.2

1. J. C. Knights and J. E. Matan, *Solid State Corrosion*, 1977, **21**, 983.
2. R. M. Mehalso and R. J. Diefendorf, *CVD*, 1975, **V**, 84.
3. R. P. Gower and J. Hill, *CVD*, 1975, **V**, 114.
4. W. F. Knippenberg *et al., Philips Tech. Rev.*, 1977, **37**, 189.
5. D. R. Stern and L. Lynds, *J. Electrochem Soc.*, 1958, **105**, 676.
6. H. O. Pierson, *Thin Solid Films*, 1977, **45**, 257.
7. K. Tamaru, *J. Phys. Chem.*, 1955, **59**, 777.
8. 'Vapor Deposition', p. 283.
9. E. R. Breining *et al.*, Ger. 1 235 106, 1967.
10. 'Vapor Deposition', p. 277.
11. J. Chin *et al., CVD*, 1975, **VI**, 364.
12. H. E. Hintermann *et al., Ext. Abs. Electrochem. Soc.*, 1973, **73/2**, 218.
13. R. B. Reeves and J. J. Gelhardt, *SAMPE*, 1979, **10** (1), 13.
14. J. H. Oxley *et al., Ind. and Eng. Chem.: Prod. Res. and Dev.*, 1962, **1**, 102.
15. L. H. Marshall, US 1 893 782, 1933.
16. H. M. J. Mazille, *Thin Solid Films*, 1980, **65**, 67.
17. E. J. Jablonwski, *Cobalt*, 1962, **14**, 28.
18. P. Palovlyk, US 2 704 728, 1955.
19. Powell *et al.*, 'Vapor Deposition', p. 290.
20. A. Sawabe *et al., Appl. Phys. Lett.*, 1987, **50**, 728.
21. D. J. Dumin *et al., RCA Rev.*, 1970, **31**, 620.
22. E. C. Cave and B. B. Czorny, *RCA Rev.*, 1963, **24**, 523.
23. B. O. Seraphin, *J. Vac. Sci. Tech.*, 1979, **16**, 193.
24. H. F. Sterling *et al., Vide*, 1966, **21**, 80.
25. F. B. Litton, *J. Electrochem. Soc.*, 1951, **98**, 488.
26. J. E. Cline and J. Wulff, *J. Electrochem. Soc.*, 1951, **98**, 385.
27. J. J. Lander and L. H. Germer, Am. Inst. Min. Met. Eng.: Inst. Met. Div. Met. Tech., **14**, 6. Tech. Prod. 2259, 1947.
28. J. A. M. van Liempt, *Metallwerkschaft*, 1932, **11**, 357.
29. K. Hieber and M. Stolz, *CVD*, 1975, **V**, 436.
30. R. R. Jaeger and S. T. Cohen, *CVD*, 1972, **III**, 500.
31. S. D. Allen and M. Bass, *J. Vac. Sci. Tech.*, 1979, **16**, 431.
32. W. A. Jenkins and H. W. Jacobson, US 3 020 148, 1962.
33. L. W. Owen, *J. Less-Common Met.*, 1962, **4**, 35.
34. E. C. Marboe, US 2 430 520, 1947.
35. R. N. Meinert, *J. Am. Chem. Soc.*, 1933, **55**, 979.
36. E. H. Reerin, *Z. f. Anorg. Chem.*, 1928, **173**, 45.
37. A. N. Zelikman *et al., Russian Metall.*, 1963, **4**, 120.
38. R. L. Van Hemert *et al., J. Electrochem. Soc.*, 1965, **112**, 1123.
39. C. F. Powell, J. H. Oxley and J. M. Blocher, in 'Vapor Deposition', Wiley, NY, 1968.
40. 'Vapor Deposition', p. 314.
41. L. Yang *et al., CVD*, 1972, **III**, 253.
42. P. J. Sherwood, *CVD*, 191073, **III**, 728.
43. A. R. Nicoll *et al., Thin Solid Films*, 1974, **64**, 321.
44. H. F. Sterling and R. C. G. Swan, *Solid State Electronics*, 1965, **8**, 653.
45. W. G. Townsend and M. E. Uddin, *Solid State Electronics*, 1973, **16**, 39.
46. E. H. Reerin, US 1 818 909, 1931.
47. Smelin, 'Handbuch der Anor. Chem.', 8th edn, Verlag Chemie Berlin, p. 63, 1938.
48. K. Tamaru, *J. Phys. Chem.*, 1955, **59**, 1084.
49. 'Vapor Deposition', p. 283.
50. T. L. Chu *et al., CVD*, 1975, **V**, 653.
51. CVD of Si, Semiconductor Silicon 1977, Electrochem. Soc., Princeton, **NJ**, 1977.
52. A. K. Praturi, *CVD*, 1977, **VI**, 20.
53. S. Andisio, *J. Electrochem. Soc.*, 1980, **127**, 2299.
54. C. F. Powell *et al., J. Electrochem. Soc.*, 1948, **93**, 258.
55. I. E. Campbell *et al., J. Electrochem. Soc.*, 1948, **93**, 271.
56. B. W. Gosner, Titanium Rep. Symp., ONR, Washington, DC, 1948.
57. G. Derge and G. P. Monet, US 2 743 173, 1956.
58. J. Spitz and J. Chevallier, *CVD*, 1975, **V**, 204.
59. C. M. Hollabaugh *et al., CVD*, 1977, **VI**, 559.
60. A. E. van Arkel, 'Reine Metalle', Springer, Berlin, 1939.
61. A. E. van Arkel *et al.*, US 1 891 124, 1932.
62. A. Bremer and W. E. Reid, US 3 072 983, 1963.
63. C. F. Powell *et al., J. Electrochem. Soc.*, 1948, **93**, 258.
64. J. J. Lander, US 2 516 058, 1930.
65. Z. M. Shapiro, 'Metallurgy of Zirconium', McGraw-Hill, NY, 1955.
66. P. J. Sherwood *et al., CVD*, 1975, **V**, 801.

67. A. M. Shroff *et al., CVD*, 1975, **V**, 351.
68. J. A. Papke and R. D. Stevenson, *CVD*, 1969, **I**, 193.
69. K. J. Miller *et al., J. Electrochem. Soc.*, 1966, **113**, 902.
70. A. R. Badzian and T. Badzian, *Surf. Coat. Technol.*, 1988, **36**, 283.
71. S. Furuno *et al.*, JP 6,381,810, April 1988.
72. V. Hoffe and A. Tehel, *Wiss. Tag. Tech. Univ. Karl-Marx-Stadt*, 1987, 201.
73. A. Sawabe *et al., Appl. Phys. Lett.*, 1987, **50**, 728.

Table 35.3   PHYSICAL VAPOUR DEPOSITION OF OXIDES

Compound	Process*	Coating thickness (μm)	Typical applications	Ref.
Al_2O_3	S RSP IP	1–6	Passive sealing layers Wear-resistant layers	— 1, 2
CuO/Cu_2O	r.f. S Cu + O_2	—	—	3
Cr_2O_3	r.f. S RE	0.1 0.01	Wear resistant layer Anti-reflectance coating	4, 5
NiO	S	1	—	6
SiO_2	MS r.f. S	0.006 13	Mask layer on IC Multilayer laser fusion test	— 7, 8
SiO–Cr	r.f. S	—	Resistor film	9
SnO_2	RSP	—	Transparent electrodes for electro-optics	10
Ta_2O_5	r.f. S Ta + O_2	—	Optical wave guides	11
TiO_2	RSP Ti + O_2	1–2	Anti-reflectance coatings on Al mirror	12
ZnO	r.f. S	—	Surface acoustic wave (SAW) devices	13
ZrO_2 (Y-stabilised)	d.c./r.f. S EB	25–100	Thermal barrier coatings for gas-turbine parts	14 15

* *See* Table 35.1 for the key

REFERENCES TO TABLE 35.3

1. R. S. Nowicki, *J. Vac. Sci. Tech.*, 1977, **14**, 127.
2. R. F. Bunshah and R. J. Schramm, *Thin Solid Films*, 1977, **40**, 211.
3. M. Samirant *et al., Thin Solid Films*, 1971, **8**, 293.
4. B. Bhushan, *Thin Solid Films*, 1979, **64**, 231.
5. E. Ritter *et al.*, US 4 172 156, 1976.
6. P. V. Plunkett *et al., Thin Solid Films*, 1979, **64**, 121.
7. K. Urbanek, *Solid State Tech.*, 1977, **20**, 87.
8. S. F. Meyer and E. J. Hsieh, *Thin Solid Films*, 1979, **64**, 383.
9. V. Fronz *et al., Thin Solid Films*, 1980, **65**, 33.
10. E. Leja *et al., Thin Solid Films*, 1980, **67**, 45.
11. W. M. Paulson *et al., J. Vac. Sci. Tech.*, 1979, **16**, 307.
12. L. D. Hartsough and P. S. McLeod, *J. Vac. Sci. Tech.*, 1977, **14**, 123.
13. K. Ohji *et al., J. Vac. Sci. Tech.*, 1978, **14**, 123.
14. J. W. Patten, *Thin Solid Films*, 1979, **64**, 337.
15. A. S. James *et al., Surf. Coat. Technol.*, 1987, **32**, 377.

Table 35.4 CHEMICAL VAPOUR DEPOSITION OF OXIDES

Material	Process	Temp. (°C)	Coating thickness (μm)	Typical applications	Ref.
Al_2O_3	$AlCl_3 + CO_2 + H_2$	850–1 800	2–8	Hard coating for tool tips. Coating for nuclear fuels	1–5
	$AlCl_3 + O_2$ plasma assisted	230–350		Passive sealing layer for integrated circuits. Single crystal growth	
	$Al(CH_3)_3$ or $Al(OC_3H_7)_3 + O_2$	275–500			
B_2O_3–SiO_2	$B_2H_6 + SiH_4 + O_2$	350	0.4–0.6	Diffusion source for doping electronic materials. Cladding for optical fibres	23
Cr_2O_3	$Cr(CO)_6 + O_2$	400–600	<25	Oxidation-resistant coating	6
Fe_2O_3	$Fe(acac)_3 + O_2$ plasma assisted	400–600	—	—	24
SiO_2	$SiH_4 + O_2 (+h\nu)$	300–430	5–25	Passive layer-encapsulant for electronics	7–12
	$Si(OEt)_4$ thermal decomp.	800–1 000		Anti-reflectance coating	25
	$Si(OEt)_4 + H_2O$	500–1 000		Corrosion-resistant coating for chemical plant. AGR plant	
	$SiH_4 + N_2O$	300–1 500			
	$SiCl_4 + O_2$ plasma assisted	>1 000	20–50	Prep. of optical fibres	
SnO_2	$SnCl_4 + O_2$	600–1 000	0.1	Semiconductor assemblies. Optoelectronic applications.	13–15
	$SnCl_4 + H_2O$	>430		Transparent electrically conducting coatings	
	$Sn(CH_3)_4 + O_2$	400–600	0.1		
TiO_2	$TiCl_4 + O_2 +$ hydrocarbon (in flame)	450–800	—	Pigment production	16–19
	$TiCl_4 + CO$ plasma assisted or $TiCl_4 + O_2$ plasma assisted			Dielectric films in MOS structure	
	$Ti(OC_3H_7)_4 + O_2$	450			26
$YBa_2Cu_3O_{7-x}$	$Y/Ba/Cu(acac) + O_2$	400–600	1–20	Superconductor with high T_c	27, 28
ZrO_2	$Zr(acac)_4 + O_2$	450–700	—	MOS devices	20–22
	$ZrCl_4 + CO_2 + CO + H_2$	1 000	5–10	Hard coatings on sintered carbides. Oxidation-resistant coatings	
	$Zr(OC_3H_7)_4$	500–600	12–75		

REFERENCES TO TABLE 35.4

1. R. Funk *et al., J. Electrochem. Soc.*, 1976, **123**, 285.
2. J. N. Lindström and R. T. Johannesson, *J. Electrochem. Soc.*, 1976, **123**, 555.
3. P. S. Schaffer, *J. Am. Ceram. Soc.*, 1965, **48**, 508.
4. H. Katto and Y. Koga, *J. Electrochem. Soc.*, 1971, **118**, 1619.
5. M. T. Duffy and W. Kern, *RCA Rev.*, 1970, **31**, 754.
6. J. J. Lander, US 2 671 739, 1954; 'Vapor Deposition', p. 390.
7. J. Graham, *High Temperatures—High Pressures*, 1974, **6**, 577.
8. J. Middelhoek and A. J. Klinkhamer, *CVD*, 1975, **V**, 19.
9. 'Vapor Deposition', p. 392.
10. J. Irven and A. Robinson, *Phys. Chem. of Glass*, 1980, **21**, 47.
11. D. Küppers and H. Lydtin, *CVD*, 1977, **VI**, 461.
12. J. Irven and A. Robinson, *Electronic Letters*, 1979, **15**, 252.
13. O. Tabata, *CVD*, 1975, **V**, 681.
14. 'Vapor Deposition', Wiley, NY, p. 398, 1966.

15. R. N. Ghostagore, *J. Electrochem. Soc.*, 1971, **118**, 1619.
16. A. Pechukas and G. Atkinson, US 2 394 633, 1946.
17. H. F. Sterling *et al.*, *Vide*, 1966, **21**, 80.
18. J. H. Alexander *et al.*, 'Thin Film Dielectrics', F. Vratny, Electrochem. Soc., NY, 1969.
19. C. C. Wang *et al.*, *RCA Rev.*, 1970, **31**, 728.
20. B. Balog *et al.*, *J. Cryst. Growth*, 1972, **17**, 298.
21. J. N. Lindström *et al.*, US 3 837 896, 1974.
22. K. S. Kagdiyasin and C. T. Lynch, USAF Rept. ASD-DR-322, 1963.
23. J. Wong, *J. Electrochem. Soc.*, 1980, **127**, 62.
24. A. Fuji *et al.*, JP 6 394 312, May 1988.
25. K. Ito *et al.*, JP 6 314 872, January 1988.
26. Y. Hokari, JP 6 372 883, April 1988.
27. H. Yamane *et al.*, *Chem. Lett.*, 1988, 939.
28. A. D. Berry *et al.*, *Appl. Phys. Lett.*, 1988, **52**, 1743.

Table 35.5   PHYSICAL VAPOUR DEPOSITION OF NITRIDES

Compound	Process*	Coating thickness (μm)	Typical applications	Ref.
AlN	RIS Al + N$_2$	—	Electronic material	1–3
	RIP Al + N$_2$ or NH$_3$—substrate 1 000°C	2	Refractory dielectric	
			Acoustic transducers	
	RE Al + NH$_3$	0.6–1.5	SAW diodes	
CrN	RIP Cr + N	5	Die casting dies	13
HfN	RSP	—	Thin film dielectrics	4
	RIP Hf + N$_2$	0.2–0.3		
NbN	RSP	—	Superconducting layers	5
TaN	RSP	115 Å min^{-1}	Resistor film	6
Si$_3$N$_4$	RIS Si + N$_2$	0.25–0.08	Encapsulant for GaAs	1,
	r.f. RSP Si + N$_2$		Optical wave guides	7–9
	RSP Si$_3$N$_4$ + N$_2$			
TiN	RSP Ti + N$_2$	—	Hard coatings on steel tools	10–11
	ARE Ti + N$_2$	5–15		
	RAE Ti + N$_2$			
W$_2$N	RSP	100 Å min^{-1}	Superconductivity	6
ZrN	RIP Zr + N$_2$	3–5	Wear resistant coatings	12

See Table 35.1 for key.

REFERENCES TO TABLE 35.5

1. H. J. Erler *et al.*, *Thin Solid Films*, 1980, **65**, 233.
2. Y. Murayama *et al.*, *J. Vac. Sci. Tech.*, 1980, **17**, 796.
3. S. Yoshida *et al.*, *Appl. Phys. Lett.*, 1975, **26**, 461.
4. F. T. J. Smith, *J. Appl. Phys.*, 1970, **41**, 4227.
5. J. Spitz and A. Aubert, *CVD*, 1975, **V**, 258.
6. F. M. Kilbane and P. S. Habig, *J. Vac. Sci. Tech.*, 1975, **12**, 107.
7. L. E. Bradley and J. S. Sites, *J. Vac. Sci. Tech.*, 1979, **16**, 189.
8. F. H. Eisen *et al.*, *Solid State Electronics*, 1977, **20**, 219.
9. W. M. Paulson *et al.*, *J. Vac. Sci. Tech.*, 1979, **16**, 307.
10. T. Abe and T. Yamashiua, *Thin Solid Films*, 1975, **30**, 19.
11. K. Nakamura *et al.*, *Thin Solid Films*, 1977, **40**, 155.
12. O. Knofek *et al.*, *Surf. and Coat. Techn.*, 1994, **68/69**, 489.
13. O. Knofek *et al.*, *Surf. and Coat. Techn.*, 1993, **62**, 630.

Table 35.6   CHEMICAL VAPOUR DEPOSITION OF NITRIDES

Material	Process	Temp. (°C)	Coating thickness (μm)	Typical applications	Ref.
AlN	$AlCl_3 + NH_3 + H_2$	800–1 000	0.5	Electronic applications Refractory dielectrics SAW devices	1–5
	$Al + N_2$	1 800–2 000	Whiskers	High strength refractories III/V semiconductors	
	$AlCl_3 + N_2 + H_2$ plasma assisted	800–1 200			
	$(CH_3)_3Al + NH_3 + H_2$	1 200	Epitaxial		
BN	$BCl_3 + NH_3$	1 000–2 000	1–2 000	Formation of free-standing crucibles, tubes and plates	6–12
	$BF_3 + NH_3$	1 000–2 000		Travelling wave tube isolators Insulating layers in semi-cond. manufacture	
	$B_2H_6 + NH_3$ plasma assisted	400–700	0.2–0.6	Diffusion doping of Si	
	$B_2H_6 + N_2$ microwave plasma assisted	800	4	Cubic BN for wear resistance	24
HfN	$HfCl_n + N_2 + H_2$ $(n = 2, 3, 4)$	900–1 300	5–10	Wear-resistant coatings Diffusion barriers	13
TaN	$TaCl_5/TaBr_5 + N_2$	800–1 500	—	Resistor film	14–15
Si_3N_4	$SiH_4 + N_2/NH_3$ plasma assisted	350	0.15	Encapsulation of III/V semiconductor devices	
	$SiF_4 + NH_3$	1 000–1 500		Anti-reflection coatings on solar cells	12, 16–20
	$SiCl_4 + NH_3$ $SiHCl_3 + NH_3$ $SiH_4 + NH_3 + N_2H_4$ chem. decomp.	1 000–1 500 1 100–1 400	10–2 000	Gas turbine parts	
TiN	$TiCl_4 + N_2 + H_2$ $Ti(NMe_2)_4$ thermal decomp. $TiCl_4 + N_2 + H_2$ plasma assisted	650–1 700 300–500 <550	2–8	Hard coatings for hard-metal cutting tools. Scratch-resistant decorative surfaces. Diffusion barrier layers	21–23, 25
ZrN	$ZrCl_4 + N_2 + H_2$	900–1 200	2–8	Hard coatings for cutting tools	13

REFERENCES TO TABLE 35.6

1. A. J. Noreika and D. W. Ing, *J. Appl. Phys.*, 1968, **39**, 5578.
2. T. L. Chu and R. W. Kelm, Jr., *J. Electrochem. Soc.*, 1975, **122**, 995.
3. K. M. Taylor and C. Lenie, *J. Electrochem. Soc.*, 1960, **107**, 308.
4. J. Bauer *et al., Physica Status Solid* (a), 1977, 39, 173.
5. H. M. Manasevit *et al., J. Electrochem. Soc.*, 1971, **118**, 1864.
6. G. Clerc and P. Gerlach, *CVD*, 1975, **V**, 777.
7. D. Morin and M. Le Clercq, French Pat. 2 232 613, 1975.
8. N. J. Archer, High Temp. Chem. of Inorg. and Ceram. Mats., Chem. Soc. Pub. 30, 1977.
9. H. O. Pierson, *J. Compos. Mat.*, 1975, **9**, 228.
10. R. Francis and E. P. Flint, US Army Report, WAL-766, 41/1, 1961.
11. M. Hirayama and K. Shohno, *J. Electrochem. Soc.*, 1975, **122**, 1671.
12. K. Shohno et al., *J. Electrochem. Soc.*, 1980, **127**, 1546.
13. M. J. Hakim, *CVD*, 1975, **V**, 634.
14. R. Kieffer *et al., Powder Metall. Inst.*, 1973, **5**, 188.
15. K. Hieber, *Thin Solid Films*, 1974, **5**, 188.
16. Dietrich and Reid, *Ext. Abst., Electrochem. Soc.*, 1977, **77-2**, 510.
17. M. Ružička, *Czech. J. Phys.*, 1974, **B24**, 465.
18. J. Gebhart *et al., CVD*, 1975, **V**, **786**.
19. J. Bühler *et al., CVD*, 1977, **VI**, 493.
20. Hughes Aircraft US Pub. 4 181 751, 1980.
21. W. Schintlmeister *et al., CVD*, 1975, **V**, 523.
22. K. Sugiyama and S. Motojima, *CVD*, 1975, **V**, 147.
23. R. Warren and M. Carlsson, *CVD*, 1975, **V**, 611.
24. K. Fukushima and M. Tobioka, JP 6 340 800, February 1988.
25. T. Arai *et al., Thin Solid Films*, 1988, **165**, 139.

Table 35.7 PHYSICAL VAPOUR DEPOSITION OF CARBIDES

Compound	Process*	Coating thickness (μm)	Typical applications	Ref.
Cr_3C_2	RSP IP	0.3–2.5	Solar energy collectors Wear-resistant coatings	1–2
Cr_7C_3	ARE			
HfC	ARE/RE $Ar + C_2H_2 + Hf$	$2.5\,\mu m\,min^{-1}$	—	2
Mo_2C	RSP IP $Ar + CH_4$	0.18	Solar energy absorption	1
NbC	ARE/RE $Ar + CH_4$ or $C_2H_4 + Nb$	$2.5\,\mu m\,min^{-1}$	Optical/semiconductor	2
SiC	RSP $Ar + C_2H_2$	1	Electronic barrier coating	3–4
Ta_2C	RSP	0.8	Solar energy absorption	1, 2
TaC	$Ar + CH_4, C_2H_2$ ARE/RE $Ta + C_2H_2$	— $1.5\,\mu m\,min^{-1}$ at 590°C	—	5
TiC	RSP $A + CH4$ ARE ARE/RE $Ti + C_2H_4$	$5.0\,\mu m$ at 700/ 900°C $4.0\,\mu m\,min^{-1}$ at 450°C	Cutting tools Wear-resistant surfaces	2 6,7,8
VC	ARE $V + C_2H_2$	$3.0\,\mu m\,min^{-1}$ at 555°C	—	2
W_2C	RSP SIP W/C	0.8	Solar energy absorption Drill tips, pump components	1 10
WC	$W + Ar + CH_4$	—	—	—
ZrC	ARE/RE $Zr + C_2H_2$	$5.0\,\mu m\,min^{-1}$ at 540°C	Diffusion barrier	2, 9

* *See* Table 35.1 for key.

REFERENCES TO TABLE 35.7

1. G. L. Harding, *J. Vac. Sci. Tech.*, 1976, **13**, 1070.
2. R. F. Bunshah and A. C. Raghuram, *J. Vac. Sci. Tech.*, 1972, **9**, 1385.
3. K. E. Haq, *Appl. Phys. Lett.*, 1975, **26**, 255.
4. Y. Murayama and T. Takao, *Thin Solid Films*, 1977, **40**, 309.
5. W. Grossklaus and R. F. Bunshah, *J. Vac. Sci. Tech.*, 1975, **12**, 811.
6. F. Shinoki and A. Itoh, *Jap. J. Appl. Phys. Suppl.*, 1974, **2** (1), 505.
7. K. Nakamura *et al., Thin Solid Films*, 1977, **40**, 155.
8. W. R. Stowell, *Thin Solid Films*, 1974, **22**, 111.
9. K. D. Kennedy and Scheuermann, US 3 900 592, 1975.
10. J. P. Coad *et al.*, BP 2 174 678, March 1987.

Table 35.8 CHEMICAL VAPOUR DEPOSITION OF CARBIDES

Material	Process	Temp. (°C)	Coating thickness (μm)	Typical applications	Ref.
B_4C	$BCl_3 + CO + H_2$ $B_{10}C_2H_{12}$ thermal decomp. $B_2H_6 + CH_4$ BMe_3 thermal decomp. $BCl_3 + CCl_4$ thermal decomp. $BCl_3 + CH_4 + H_2$	1 200–1 800 1 000–1 300 ~550 1 200–2 000 1 300–1 900	 10–1 000	Nuclear industry Armament Abrasive blasting nozzles Rocket nozzles Fibres	1–6
Cr_3C_2 Cr_7C_3	$Cr(CO)_6 + H_2$ $CrCl_2 + CH_4 + H_2$ $CrCl_2 + H_2$ Cr dicumene–thermal decomp.	300–650 ~1,000 1 000 450–650	10–40	Wear-resistant coatings HV 2250 Chromising of steel	7 8 9 10
HfC	$HfCl_4 + H2 + CH_4$	1 000–1 500	5	Diffusion barrier Wear resistant coating	11 12
Mo_2C	$Mo(CO)_6$ thermal decomp. $MoF_6 + C_6H_6 + H_2$	350–475 400–1 000	—	Wear-resistant layer	13 14

(*continued*)

Table 35.8 CHEMICAL VAPOUR DEPOSITION OF CARBIDES—*continued*

Material	Process	Temp. (°C)	Coating thickness (μm)	Typical applications	Ref.
NbC	$NbCl_5 + CCl_4 + H_2$	1 500–1 900	—	{ Wear-resistant coating { Diffusion barrier	15
SiC	$SiCl_4 + CH_4$	1 000–2 000	10–1 000	Oxidation-resistant coatings Single fibres of high strength	16
	$CH_3SiCl_3 + H_2$	1 000–1 600		Heating elements Electrical heat stylus	17
	$SiH_4 + C_2H_4$ or CH_4 plasma assisted		~1	Electrochemical machining tool	18
Ta$_2$C TaC	$TaCl_5 + CH_4$	1 100		Coating W filaments for incorporation into superalloy matrices	19
TiC	$TiCl_4 + H_2 + CH_4$ or C_6H_6 $TiCl_4 + CCl_4 + H_2$ Ti organic	800–1 400 >1 000 150–500	2–12	Wear-resistant coatings on steel and carbides— particularly cutting tools	20–24 35 36
VC	$VCl_2 + CH_4 + H_2$ Diffusion coating VC_x $(x = 0.84$–$0.89)$	1 050–1 130	—	Wear-resistant coating	25–26
WC W$_2$C	$WF_6 + C_6H_6 + H_2$ $W(CO)_6$ thermal decomp. $WF_6 + CO + H_2$ $WCl_6 + CH_4 + H_2$	400–900 300–500 600–1 000 900–1 150	— 5–50	Wear-resistant coatings	27–31
ZrC	$ZrCl_4 + CH_4 + H_2$	1 050–1 500	50–100	Wear-resistant coating Diffusion barrier layer	32–33
	$ZrCl_4 + C_3H_6 + H_2$	1 150–1 300	50	High temperature insulation Nuclear fuel coating	34

REFERENCES TO TABLE 35.8

1. M. Formstecher and E. Ryskevic, *Compt. rend.*, 1945, **221**, 558.
2. R. L. Hough, *SAMPE*, 1979, **10** (1), 25.
3. G. R. Martin, US 2 484 519, 1949.
4. As 3.
5. S. Mierzejewska and T. Niemyski, *J. Less-Common Met.*, 1965, **8**, 368.
6. R. G. Bourdeau, US 3 334 967, 1967.
7. 'Vapor Deposition', p. 360.
8. B. B. Owen and R. T. Weber, Am. Inst. Min. Met. Tech. Pub. 2306, 1948.
9. S. Csch, Proc. 3rd Int. Symp. Met. and Heat Treatment Met., Warsaw, 1967.
10. J. E. Gates *et al.*, US 3 951 612, 1980.
11. M. J. Hakim, *CVD*, 1975, **V**, 634.
12. D. Hertz *et al., High Temp.–High Pressure*, 1974, **6**, 423.
13. 'Vapor Deposition', p. 362.
14. R. H. Lewin and C. Hayman, BP 1 326 769.
15. T. A. Lyndvinskaya *et al.*, 'Refractory Carbides', N. Y. Consultants Bur., 1974.
16. E. Fitzer *et al., CVD*, 1975, **V**, 523.
17. H. Beutler *et al., CVD*, 1975, **V**, 749.
18. G. Verspui, *CVD*, 1977, **VI**, 366.
19. W. J. Heffernan *et al., CVD*, 1973, **IV**, 498.
20. W. Schintlemeister *et al., CVD*, 1979, **V**, 523.
21. M. Maillat *et al., Thin Solid Films*, 1979, **64**, 243.
22. Many papers, *CVD*, 1972, **III**.
23. V. K. Sarin and J. N. Lindstrom, *CVD*, 1977, **VI**, 389.
24. J. J. Nickl *et al., J. Less-Common Met.*, 1972, **26**, 335.
25. G. Ebersbach *et al., Die Technik*, 1974, **29**, 273.
26. E. Horvath and A. J. Perry, *Thin Solid Films*, 1980, **65**, 309.
27. N. J. Archer, *CVD*, 1975, **V**, 556.

28. N. J. Archer and K. K. Yee, *Wear*, 1978, **48**, 237.
29. 'Vapor Deposition', p. 372.
30. D. A. Tarver, US 3 574 672, 1971.
31. H. Mantle *et al.*, *CVD*, 1975,**V**, 540.
32. T. C. Wallace, *CVD*, 1973, **IV**, 91.
33. C. Hollabaugh *et al.*, *CVD*, 1977, **VI**, 419.
34. A. R. Driesner *et al.*, *CVD*, 1973, **IV**, 473.
35. A. Inzenhofer, *Jud.-Anz.*, 1988, **11**, 28.
36. A. E. Kaloyeros *et al.*, *Rev. Sci. Instrum.*, 1988, **59**, 1209.

35.3   Thermal spraying

Thermal spraying is a process in which a material is heated to a molten, partly-molten or highly plastic solid state by a heat source and is accelerated to the surface on a substrate by a gas jet. On impacting the substrate, the spray material droplets flatten, solidify and cool down on the surface or overlap on the fore-flattened droplets. In this way, the spray material is deposited to form a coating. Figure 35.1 shows the principle of thermal spraying schematically. A wide range of spray materials from ceramics to metal alloys can be deposited to protect a component against wear, corrosion, oxidation or heat or to generate special electric, magnetic or biomedical properties. Table 35.9 gives some of spray materials used commonly. The significant advantages of thermal spraying are:

1. Coatings of nearly all materials can be deposited on various substrates
2. Low heat load on the substrate material
3. Local coating is possible
4. Some facilities are mobile
5. High deposition rate.

Thermal spraying processes are usually distinguished by the heat sources used. The most widely used techniques are conventional combustion spraying, high velocity oxy-fuel spraying, wire arc spraying and plasma spraying. There are different processes variants under each one of the above mentioned heat sources.[1]

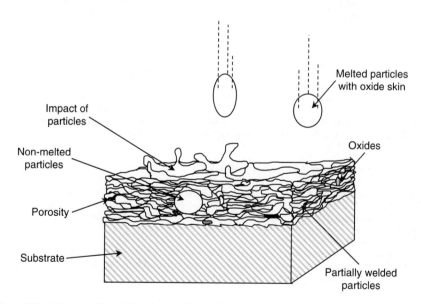

Figure 35.1   *Principle of deposition of a thermal sprayed coating*

Table 35.9   EXAMPLES FOR COATING MATERIALS FOR THERMAL SPRAYING[1,4]

Material group	Example
Pure metals	Al, Co, Mo, Ni, Ti, Zn,...
Alloys	Ni–Al, Ni–Cr, Ni–Cr–Al, Steel, ...
Pseudo-Alloys	Cu–W, Bronze-Steel
Oxide-Ceramics	Al_2O_3, Cr_2O_3, TiO_2, ...
Hard materials	Cr_3C_2/Ni, WC/Co, ...
Synthetics	EMAA, ETFE,...

For most thermal spray processes, the particles are heated up very fast and can react with the gases of the surrounding atmosphere. Because of this, the material of the particles can change its characteristics during the spraying process. The different effects are selective evaporation of components, reaction of metal compounds and formation of non-volatile metal compounds like oxides, nitrides, and hydrides.[5]

35.3.1   Combustion wire spraying

In the combustion wire spray process the spray material in wire form is fed continually into a flame where it is melted. The spike of the wire is atomised by the surrounding compressed air. The flame reaches temperatures up to 3 160°C. The accelerated particles deposit on the workpiece surface. Typical fuel gases are acetylene, propane and hydrogen.[1]

35.3.2   Combustion powder spraying

The powder flame spray process is similar to the combustion wire spray process with the difference, that powder and not a wire is injected into the flame. This offers a much wider range of coating material options than the wire flame spray process. The powdered material is fed continually into the fuel gas-oxygen flame where it is typically melted or semi-melted by the heat of combustion. A powder feed carrier gas transports the powder particles into the combustion flame, and the gases transport the material towards the prepared workpiece surface. Typical choices for fuel gases are acetylene or hydrogen.[1]

35.3.3   Electric wire arc spraying

Two identical or different metallic wires are moved into a spray gun with a controlled speed. The two wires are electrically charged with opposing polarity. When the wires are brought together at the nozzle the arising arc melts the material. Compressed air atomises and accelerates the molten material onto the workpiece surface. The temperatures in the electric arc are approximately 6 500°C. This process can also be driven with three or four wires in the same way.[1]

35.3.4   High velocity oxy-fuel spraying (HVOF)

High velocity oxygen fuel spraying is a further development of the combustion powder spray process. HVOF has reached a good market position in a relatively short time. HVOF is characterised by very high gas and particle velocities. A reaction of oxygen and fuel gas in the combustion chamber generates the heat required for melting the powder. The expanding gas accelerates the particles to velocities up to 300–600 m/s.

Low particle temperature and high particle speed are the advantages of the HVOF process for a lot of materials and applications. The coatings are characterised by a low porosity, high adhesion, high hardness and a good resistance against wear.[1]

35.3.5   Plasma spraying

Plasma spraying is a very flexible thermal spray process. It can provide energy in the form of a plasma jet with temperatures higher than 20 000°C to melt any material. The number of coating materials that can be deposited with the plasma spray process is almost unlimited. For the plasma spray process the plasma is generated by a arc between two water cooled electrodes in a torch (Figure 35.2). The atoms of the plasma gas are ionised when passing the arc. For a diatomic gas molecular dissociation will also happen. As plasma gas argon, helium, hydrogen, nitrogen and any mixture of these gases are used. During the ion/electron recombination process heat is released. The coating material is injected into the gas plume, melted, accelerated and deposited on a surface.

The sensitivity of the plasma spraying process to operating conditions and the great number of adjustable parameters makes the optimisation of the process for each application necessary.[2]

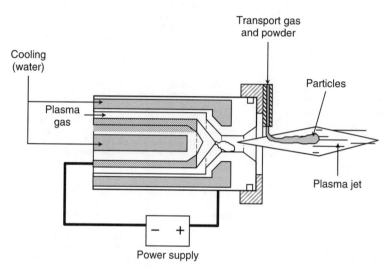

Figure 35.2　*Plasma spraying process*

Table 35.10　THERMAL SPRAY PROCESSES, CHARACTERISATION AND TYPICAL APPLICATIONS

Process	*Process temperature/ particle velocity/ deposition rate*	*Typical applications*
Combustion powder spray	<3 160°C/<50 m/s/ 1–6 kg/h	Repair operations Resistance to impact Abrasion and wear resistance Resistance to chemical attack Resistance to atmospheric and heat corrosion
Combustion wire spray	<3 160°C/<200 m/s/ 6–8 kg/h	Resistance to atmospheric and heat corrosion Repair operations
Electric wire arc spray	≈4 000°C/≈150 m/s/ 8–20 kg/h	Resistance to atmospheric and heat corrosion Repair operations Bond coats Electrically conductive and solderable coatings Anti-skid and traction coatings
High velocity oxy-fuel thermal spray (HVOF)	<3 160°C/<550 m/s/ 2–8 kg/h	Repair operations Abrasion, wear or erosion resistance Sliding wear resistance Resistance against cavitation Resistance to chemical attack Control of oxidation and sulphidation Resistance to atmospheric and heat corrosion

(*continued*)

Table 35.10 THERMAL SPRAY PROCESSES, CHARACTERISATION AND TYPICAL APPLICATIONS—*continued*

Process	Process temperature/ particle velocity/ deposition rate	Typical applications
Plasma spray	>20 000°C/<450 m/s/ 4–8 kg/h	Repair operations Abrasion or erosion resistance Sliding wear resistance Resistance to fretting, galling or adhesive wear Resistance to impact and cavitation Resistance to chemical attack Control of oxidation and sulphidation Resistance to atmospheric and heat corrosion Galvanic corrosion control Thermal or electrical insulation Abradables
Laser spraying	>10 000°C/>1 m/s	Partial coating of tolls
Detonation spraying	<3 160°C/ca. 600 m/s/ 2–8 kg/h	Abrasion or erosion resistance Sliding wear resistance Resistance to fretting, galling or adhesive wear Resistance to cavitation Resistance to chemical attack Control of oxidation and sulphidation

Table 35.11 FUNCTIONS AND PROPERTIES OF SELECTED THERMAL SPRAYED COATINGS[3]

Coating	Resistance against…				Slide capability	Hardness [HV]	Maximum application temperature [°C]
	Corrosion	Oxidation	Wear abrasive	Adhesive			
Aluminium	++[1]						
Zinc	++[1]						
Molybdenum	+	+	++	++	++	1 000	350
Steel	+					250–500 [2]	
MCrAlY	+++	+++					1 000
Hard alloy	+		++			600	900
Tribaloy	++		++	++	++	600	1 000
WC/Co			++	+++	++	1 000	540
CrC/NiCr			+++	+++	++	800	1 000
Alumina			++	+++	+	2 000	[2]
Chromium oxide			++	+++	++	2 300	[2]
Zirconia (stabilised)			++	+++		1 500	
Co-alloy/ Chromium oxide			+++	+++		600	1 000
Nickel/graphite					4	3	500

[1] cathodic protection,   [2] depends on alloy,   [3] depends on porosity of the coating on steel,   [4] abradable.
+: low,   ++: medium,   +++: high.

REFERENCES TO SECTION 35.3

1. Fr. W. Bach, T. Duda, Moderne Beschichtungsverfahren, 2000, WILEY-VCH Verlag, ISBN 3-527-30117-8.
2. E. Lugscheider, I. Kvernes, Thermal Barrier Coatings, Powder Spray Process and Coating Technology, Intermetallic And Ceramic Coatings, 1999, Marcel Dekker, Inc., ISBN 0-8247-9913-5.
3. H. Simon, M. Thoma, Angewandte Oberflaechentechnik für metallische Werkstoffe, Carl Hanser Verlag, 1985.
4. M. Allen, C. C. Berndt, J. A. Brogan, D. Ottersond, Thermal Sprayed Polymer Coatings for Corrosion Protection In A Biochemical Treatment Process, Proc. Of the 15th ITSC, 25–29 May 1998, Nice, France.
5. R. A. Haefer, Oberflaechen- und Duennschichttechnologie, Springer Verlag Berlin, 1987.

36 Superplasticity

Superplasticity is the ability of polycrystalline materials to sustain large plastic deformation in tension without rupture, and is phenomenologically delineated by tensile fracture elongations (e_f) in excess of $\sim$300%. Originally discovered in 1912 by Bengough[1], superplasticity is now employed industrially in metal shape forming operations around the world, with primary applications in the aerospace field (*see*, e.g. Ref. [2]).

Superplasticity is a viscous deformation process, and is frequently characterised by a simple flow law relating uniaxial stress σ and strain rate $\dot{\varepsilon}$:

$$\sigma = K \cdot \dot{\varepsilon}^m \tag{1}$$

where K is a temperature- and microstructure-dependent parameter, and m is referred to as the strain rate sensitivity; its reciprocal value, the stress-sensitivity, n, is also commonly reported. Whereas values of m are usually less than 0.25 for typical creep mechanisms in polycrystals, superplasticity is distinguished by values of $m \geq 0.3$. Higher values of m directly promote tensile ductility, by increasing the stability of necks, and by suppressing the tendency for grain-boundary cavitation. In the Newtonian limit of $m = 1$, there is no theoretical limitation to the achievable ductility, although inhomogeneities in temperature or microstructure can substantially impact superplastic extension.

There are two main classes of superplasticity, broadly referred to as *microstructural superplasticity* and *internal-stress superplasticity*, distinguished by deformation mechanisms at the microstructural level.

Microstructural superplasticity is by far the more studied mode of superplastic deformation, and relies on the increase in strain-rate sensitivity, m, observed in fine-grained materials (grain size, d, less than about 10 μm in most metals, and less than about 1 μm in ceramics). At these grain sizes, bulk deformation mechanisms occurring within each grain (i.e. dislocation creep) become secondary to the process of *grain boundary sliding*, involving the relative motion and reorientation of grains with respect to one another. Some general features of microstructural superplasticity are listed below; the reader is referred to Ref. [2] for details:

- The power-law form of Eq. (1), with a constant value of m, does not generally hold over a very broad range of stresses and strain rates in superplastic materials. Rather, a characteristic sigmoidal $\sigma - \dot{\varepsilon}$ relationship is observed, with a peak value of $m \geq 0.3$ occurring over a relatively narrow range of strain rate. This transition is associated with changes in the mechanism for accommodation of grain boundary sliding, from diffusional accommodation at low strain rates to dislocation-motion accommodation at high rates. In superplastic forming operations, it is of primary importance to identify this optimum condition for superplastic elongation.
- The parameter K incorporates an Arrhenius temperature dependence and a power-law grain-size dependence. In general, a finer grain size leads to improved superplastic flow, and the optimum temperature is high enough to promote rapid deformation, but not so high as to encourage rapid grain growth.
- In two-phase superplastic alloys, the optimum superplastic conditions are usually associated with equal proportions of the two phases.
- The optimum strain rate for superplasticity commonly falls in the range 10^{-5}–$10^{-3}\,\text{s}^{-1}$. Recent research has focused on the possibility of high strain rate superplasticity, most commonly observed in aluminium-matrix composite and mechanically alloyed materials at very high temperatures. In such materials, the optimum strain rate may be as high as $10\,\text{s}^{-1}$.
- The fine grain structure required for microstructural superplasticity is usually attained by a schedule of thermomechanical processing, such as rolling or forging, although many more exotic methods such as severe plastic deformation have been developed to refine metallic microstructures.

Table 36.1 summarises 184 investigations of superplasticity in metals and alloys, with an emphasis on the more technologically-relevant aluminium, copper, iron and titanium systems which form the basis of industrial superplastic forming operations. Tabulated are the range of experimental conditions (temperatures, strain rates, and grain sizes), as well as the strain rate sensitivity m, the tensile elongation to failure, e_f, and the optimum conditions for superplasticity. Experimental research in superplasticity has recently focused on emerging structural materials, such as intermetallics, metal-matrix composites, and ceramics. Tables 36.2 and 36.3 summarise recent work on microstructural superplasticity in intermetallics and metal-matrix composites, respectively.

Internal-Stress Superplasticity does not require a fine microstructure, but rather depends upon a renewable source of internal stresses within the material. Internal stresses are generated by, e.g. thermal expansion mismatch between phases or anisotropic grains, or by a polymorphic phase transformation. When an external stress is applied concurrently, the internal mismatch strains are biased to develop preferentially in the direction of the applied external stress, leading to macroscopic plastic deformation. In general, both an internal and a superimposed external stress are required for this deformation mechanism to be active, and the internal stresses must be renewed many times to accumulate large deformations. Thermal cycling is a common means of renewing internal stresses, especially in systems with a thermal expansion mismatch or a thermally-driven phase transformation. Upon each thermal cycle, an increment of plastic strain is developed, and many cycles are required to achieve superplastic strains.

Unlike microstructural superplasticity, which can exhibit a range of strain-rate sensitivities, internal-stress plasticity occurs under ideal Newtonian conditions, with $m = 1$. Unfortunately, this mechanism operates only at low strain rates, in the range 10^{-7}–10^{-4} s^{-1}, but because it does not require a fine microstructure it can be applied to hard-to-form engineering materials such as metal-matrix composites or alloys with coarse microstructures. Internal stress superplasticity has primarily been studied during thermal cycling for the case of thermal expansion-mismatch in aluminium-matrix composites, as well as during allotropic phase transformation (transformation superplasticity) in iron and titanium alloys. Table 36.4 summarises research on internal stress superplasticity.

Table 36.1   INVESTIGATIONS ON SUPERPLASTICITY OF METAL ALLOYS

Material [wt %]	Grain size [μm]	Temp. range [°C]	Strain rate range [s⁻¹]	m	e_f [%]	Optimum condition [s⁻¹]/[°C]	Ref.
Binary Aluminium Alloys							
Al–6Cu	—	375–455	$7 \cdot 10^{-5}$–$3 \cdot 10^{-2}$	0.21	137	—/425	3
Al–33Cu	1.5–18	380–470	10^{-7}–10^{-2}	0.5	1 300	10^{-7}/380 $d < 13\,\mu m$	4
Al–33Cu	3.4–8.7	450	$7 \cdot 10^{-7}$–$3 \cdot 10^{-2}$	0.7	1 475	$1.3 \cdot 10^{-5}$/—	5
Al–33.23Cu	2–9	360–520	$2 \cdot 10^{-5}$–$3 \cdot 10^{-2}$	0.75	>500	$8 \cdot 10^{-4}$/520	6
Al–2.8Mg	30	200–500	10^{-4}–10^{-1}	0.33	325	10^{-4}/400	7
Al–4Mg	0.5–2	400–538	$8 \cdot 10^{-5}$–$4 \cdot 10^{-2}$	0.3	210	$2 \cdot 10^{-2}$/538	8
Al–5.5Mg	250	200–500	10^{-4}–10^{-1}	0.33	284	$2 \cdot 10^{-2}$/500	7
Al–10Mg	1.6	300	$2 \cdot 10^{-4}$–$8 \cdot 10^{-2}$	—	360	10^{-3}/—	9
Al–0.54Sc	0.5–2	400–538	$8 \cdot 10^{-5}$–$4 \cdot 10^{-2}$	0.25	157	$2 \cdot 10^{-2}$/538	8
Al–13Si	18	490–558	$5 \cdot 10^{-6}$–$2 \cdot 10^{-1}$	0.55	300	$2 \cdot 10^{-1}$/558	10
Al–4Ti	2.6	425–605	$^{b}10^{-4}$–$3 \cdot 10^{0}$	0.45	—	—	11
Al–10Ti	0.9	600–650	$2 \cdot 10^{-4}$–10^{0}	0.5	493	10^{-1}/650	12
Al–0.5Zr	—	375–455	$7 \cdot 10^{-5}$–$3 \cdot 10^{-2}$	0.36	104	—/423	3
Multicomponent Aluminium Alloys							
Al–4.87Ca–4.64Zn–0.13Fe	1	275–550	10^{-3}–10^{-2}	0.35	305	10^{-3}/550	13
Al–5Ca–4.8Zn	1–5	450–550	$2 \cdot 10^{-3}$–$2 \cdot 10^{-1}$	0.4	900	—	14
Al–5Ca–5Zn	1.2	550–600	10^{-3}–10^{-1}	0.63	900	10^{-1}/600	15
Al–2Cu–0.5Zr	—	375–455	$7 \cdot 10^{-5}$–$3 \cdot 10^{-2}$	0.41	304	—/420	3
Al–2.3Cu–1.7Mg–1.2Ni–1Fe–0.19Si	3	470–550	10^{-3}	1	240	$3 \cdot 10^{-1}$/530	16
Al–3Cu–2Li–1Mg–0.15Zr	2–4	400–500	10^{-4}–$3 \cdot 10^{-2}$	0.4	810	10^{-3}/500	17
	1–7	460–540	10^{-3}–10^{-1}	0.57	1 200	$2 \cdot 10^{-2}$/530	18

(continued)

Table 36.1   INVESTIGATIONS ON SUPERPLASTICITY OF METAL ALLOYS—*continued*

Material [wt %]	Grain size [μm]	Temp. range [°C]	Strain rate range [s⁻¹]	m	e_f [%]	Optimum condition [s⁻¹]/[°C]	Ref.
Al–3Cu–2Li–1Mg–0.2Zr	2–4	400–500	$3 \cdot 10^{-5}$–$3 \cdot 10^{-2}$	0.4	660	10^{-3}/500	17, 19
Al–4Cu–2Li–0.5Zr	2.3	400–500	$3 \cdot 10^{-4}$–$3 \cdot 10^{-2}$	0.4	>1 000	$5 \cdot 10^{-3}$/450	20
Al–2.4Li–2.6Cu–0.2Zr	0.8–4	450–510	10^{-4}–$2 \cdot 10^{-2}$	0.6	758	10^{-3}/500	21
Al–4Cu–3Li–0.5Zr	2–8	450	10^{-4}–$3 \cdot 10^{-2}$	0.45	850	$7 \cdot 10^{-2}$/—	22
Al–4Cu–1.5Mg–1.1C–0.8O	—	400–500	$8 \cdot 10^{-4}$–10^{0}	0.3	300	$7 \cdot 10^{-1}$/425	23
	—	300–550	10^{-3}–$2 \cdot 10^{2}$	0.5	1 250	$5 \cdot 10^{1}$/550	24
Al–4Cu–0.5Zr	—	375–455	$7 \cdot 10^{-5}$–$3 \cdot 10^{-2}$	0.43	418	—/420	3
	0.3–8	220–500	$3 \cdot 10$–$4 \cdot 10^{0}$	0.5	800	—/500, d = 8 μm	25
Al–6Cu–0.2Zr	—	375–455	$7 \cdot 10^{-5}$–$3 \cdot 10^{-2}$	0.45	191	—/400	3
Al–6Cu–0.26Zr	—	375–455	$7 \cdot 10^{-5}$–$3 \cdot 10^{-2}$	0.41	233	—/425	3
Al–6Cu–0.33Zr	—	375–455	$7 \cdot 10^{-5}$–$3 \cdot 10^{-2}$	0.44	554	—/450	3
Al–6Cu–0.41Zr	—	375–455	$7 \cdot 10^{-5}$–$3 \cdot 10^{-2}$	0.42	612	—/450	3
Al–6Cu–0.5Zr	—	375–455	$7 \cdot 10^{-5}$–$3 \cdot 10^{-2}$	0.39	612	—/455	3
	—	375–500	$2 \cdot 10^{-4}$–$2 \cdot 10^{-2}$	0.4	600	$2 \cdot 10^{-2}$/600	26
	1.5–10	300–505	$1 \cdot 10^{-4}$–$2 \cdot 10^{-3}$	—	508	—	27
Al–12Cu–0.5Zr	—	375–455	$7 \cdot 10^{-5}$–$3 \cdot 10^{-2}$	0.35	390	—/455	3
Al–25Cu–11Mg	5–6	420–480	$4 \cdot 10^{-6}$–$3 \cdot 10^{-3}$	0.7	>600	a>10^{-3}/450–80	28
Al–33Cu–7Mg	5–6	420–480	10^{-6}–10^{-2}	0.72	>600	a>10^{-3}/450–80	28
Al–3Li–0.5Zr	—	450	10^{-4}–$3 \cdot 10^{-2}$	0.45	1 035	$3 \cdot 10^{-2}$/—	22
Al–3.2Li–1Mg–0.3Cu–0.18Zr	<4	490–580	10^{-3}–10^{0}	0.5	250	$1 \cdot 10^{-1}$/570	29
Al–2.7Mg–2.04Li–0.5Cu–0.13Zr	0.5	350–500	$5 \cdot 10^{-4}$–10^{-1}	0.5	670	$8 \cdot 10^{-4}$/450	30
Al–4Mg–1.5Li–1.2C–0.4O	0.4	425–575	10^{-3}–$2 \cdot 10^{2}$	0.5	180	$2 \cdot 10^{1}$/575	31
Al–4Mg–0.56Sc	0.5–2	400–538	$8 \cdot 10^{-5}$–$4 \cdot 10^{-2}$	0.4	>1 020	$2 \cdot 10^{-2}$/538	8
Al–4Mg–1.5O–0.8C	0.3	25–325	$5 \cdot 10^{-4}$–$5 \cdot 10^{-2}$	—	130	$5 \cdot 10^{-2}$/300	32
Al–4.3Mg–0.2Zr	10	475	$4 \cdot 10^{-5}$–$4 \cdot 10^{0}$	0.15	130	$4 \cdot 10^{0}$/—	33
Al–4.3Mg–0.3Zr	1	475	10^{-4}–$4 \cdot 10^{0}$	0.3	220	10^{-2}/—	33
Al–4.3Mg–0.8Zr	1	475	10^{-4}–$4 \cdot 10^{0}$	0.3	450	10^{-2}–10^{-1}/—	33
Al–5Mg–1.2Cr	3	372–550	10^{-4}–10^{0}	0.5	1 000	$2 \cdot 10^{-2}$/550	34
Al–5Mg–2Li–0.1Zr	0.1	250–350	10^{-1}	—	350	—/250	35
	0.3–10	180–400	$5 \cdot 10^{-4}$	0.42	320	—/400, d = 10 μm	25
Al–5Mg–1.5Mn	5	550	10^{-4}–$2 \cdot 10^{0}$	0.45	500	10^{-2}/—	33
Al–5Mg–2.2Mn	3	550	10^{-4}–$2 \cdot 10^{0}$	0.45	400	$2 \cdot 10^{-3}$/—	33
Al–5Mg–0.2Zr	8	500	$2 \cdot 10^{-5}$–$4 \cdot 10^{0}$	0.15	150	$3 \cdot 10^{-4}$/—	33
Al–5Mg–0.5Zr	1	500	$1 \cdot 10^{-4}$–$4 \cdot 10^{0}$	0.3	400	10^{-2}–10^{-1}/—	33
Al–5Mg–0.8Zr	1	500	$1 \cdot 10^{-4}$–10^{0}	0.3	575	10^{-1}/—	33
Al–5.5Mg–2.2Li–0.12Zr	1.2	350	10^{-2}–10^{0}	—	>1 180	10^{-2}/—	36
Al–5.76Mg–0.32Sc–0.3Mn–0.1Fe–0.2Si–0.1Zn	3	450–560	10^{-4}–10^{0}	0.6	1 130	$1 \cdot 10^{-2}$/475	37
Al–6Mg–0.54Sc	0.5–2	316–538	$8 \cdot 10^{-5}$–$4 \cdot 10^{-2}$	—	>1 020	$2 \cdot 10^{-2}$/538	8
Al–10Mg–0.5Mn	0.5	200–425	$7 \cdot 10^{-5}$–$1 \cdot 10^{-1}$	0.45	700	$6 \cdot 10^{-3}$/300	38
Al–9.89Mg–0.09Zr	2–5	300	$7 \cdot 10^{-5}$–$2 \cdot 10^{-2}$	0.45	600	$2 \cdot 10^{-3}$/—	39
Al–10Mg–0.1Zr	2	150–400	$7 \cdot 10^{-5}$–$7 \cdot 10^{-2}$	0.5	1 100	10^{-3}/300	40
	1.5	300	$2 \cdot 10^{-4}$–$8 \cdot 10^{-2}$	—	420	10^{-3}/—	9
	2	200–425	$7 \cdot 10^{-5}$–$1 \cdot 10^{-1}$	0.5	1 100	$7 \cdot 10^{-3}$/300	38
Al–10Mg–0.4Zr	0.9	300	$2 \cdot 10^{-4}$–$8 \cdot 10^{-2}$	—	640	10^{-3}/—	9
Al–10Mg–0.6Zr	0.7	300	$2 \cdot 10^{-4}$–$8 \cdot 10^{-2}$	—	600	10^{-2}/—	9
Al–10.2Mg–0.52Mn	5	25–425	10^{-4}–10^{-1}	0.45	580	$5 \cdot 10^{-3}$/300	41
Al–9Zn–1Mg	9.8	520	$5 \cdot 10^{-4}$–$5 \cdot 10^{-1}$	0.65	>200	$5 \cdot 10^{-3}$/—	42

(*continued*)

Table 36.1   INVESTIGATIONS ON SUPERPLASTICITY OF METAL ALLOYS—*continued*

Material [wt %]	Grain size [μm]	Temp. range [°C]	Strain rate range [s^{-1}]	m	e_f [%]	Optimum condition [s^{-1}]/[°C]	Ref.
Engineering Aluminium Alloys							
Al-2004 (Supral 100): Al–6Cu–0.4Zr–0.25Mg–0.1Fe–0.05Zn–<0.1Mn							
	0.5	300–350	10^{-2}–10^{-1}	—	970	10^{-2}/300	36
Supral 220: Al–6Cu–0.4Zr–<0.2 Si, Ge, Zn, Fe							
	4	460	$5 \cdot 10^{-5}$–$5 \cdot 10^{-2}$	—	1 100	—	43
	—	480	10^{-3}–10^{-1}	0.451	—	—	44
Al-2024: Al–4.4Cu–1.5Mg–0.6Mn (nominal)							
	5–8	430–485	$2 \cdot 10^{-4}$–$2 \cdot 10^{-2}$	0.6	485	$4 \cdot 10^{-4}$/480	45
+0.7Mg	6.6	475–490	10^{-5}–$4 \cdot 10^{-2}$	0.6	400	$5 \cdot 10^{-4}$/—	46
Al-2090: Al–2.8Cu–2.2Li–0.25Mg–0.15Zr (nominal)							
	10	518	$4 \cdot 10^{-4}$–10^{-3}	—	1 373	$4 \cdot 10^{-4}$/—	47
	—	485	$5 \cdot 10^{-5}$–10^{-2}	0.42	850	$3 \cdot 10^{-4}$/—	48
	10–15	493–527	$2 \cdot 10^{-4}$–10^{-2}	0.5	1 690	$2 \cdot 10^{-4}$/510	49
	20	510–530	$2 \cdot 10^{-5}$–10^{-2}	0.6	900	$2 \cdot 10^{-3}$/510	50
Al-2195 (Weldalite): Al–4.8Cu–1.3Li–0.4Mg–0.4Ag–0.14Zr (nominal)							
	<20	463–496	$5 \cdot 10^{-5}$–10^{-3}	0.64	700	$5 \cdot 10^{-4}$/485	51
Al-5056: Al–5Mg–0.12Mn–0.12Cr (nominal)							
	0.3	200–275	10^{-5}–10^{-4}	0.3	185	10^{-5}/275	52
Al-5083: Al–4.4Mg–0.7Mn–0.12Cr (nominal)							
	6–14	500–550	$5 \cdot 10^{-4}$–10^{-3}	0.65	450	$5 \cdot 10^{-4}$/550	53
	6.7	450–550	10^{-4}–10^{-1}	0.58	425	$4 \cdot 10^{-4}$/550	54
	6–10	510	10^{-5}–10^{-2}	0.48	320	$4 \cdot 10^{-4}$/—	55
	8.5–9.5	510–550	$2 \cdot 10^{-4}$–10^{-3}	0.57	450	$2 \cdot 10^{-4}$/510	56
	7–14	450–570	$3 \cdot 10^{-5}$–$3 \cdot 10^{-3}$	0.75	700	$2 \cdot 10^{-4}$/525	57
	9–39	500–565	10^{-5}–10^{-2}	0.65	630	10^{-2}/550	58
+0.19Zr	7.5	450–550	10^{-4}–10^{-1}	0.55	350	$4 \cdot 10^{-4}$/550	54
+1.60Mn	5.9	450–550	10^{-4}–10^{-1}	0.6	425	$2 \cdot 10^{-3}$/550	54
+1.57Mn, 0.2Zr	4.8	450–550	10^{-4}–10^{-1}	0.65	425	$2 \cdot 10^{-3}$, 550	54
Al-5454: Al–2.69Mg–0.74Mn–0.12Si (nominal)							
	10–18	500–550	$5 \cdot 10^{-4}$–10^{-3}	0.44	239	$5 \cdot 10^{-4}$, 500	53
Al-5456: Al–5.1Mg–0.8Mn–0.12Cr (nominal)							
	20	25–400	10^{-5}–10^{-2}	0.42	150	$2 \cdot 10^{-5}$, 275	59
Al-5754: Al–3.15Mg–0.25Mn–0.18Fe (nominal)							
	21–30	500	10^{-3}	0.32	140	—	53
Al-6061: Al–1.0Mg–0.6Si–0.28Mn–0.20Cr (nominal)							
	200	500–610	10^{-5}–$2 \cdot 10^{-2}$	0.9	350	10^{-4}, 590	60
	10	475–600	$3 \cdot 10^{-4}$	—	1 250	—/590	60
Al-7075: Al–5.6Zn–2.5Mg–1.6Cu–0.23Cr (nominal)							
	4	440–520	10^{-5}–10^{-2}	0.63	500	10^{-3}–10^{-2}/—	9
	10	480–520	10^{-4}–$3 \cdot 10^{-3}$	0.75	1 200	—	61
	10–17	510	$6 \cdot 10^{-4}$–$9 \cdot 10^{-3}$	—	2 100	$8 \cdot 10^{-4}$/—	62
Al-7091: Al–6.5Cu–2.5Mg–1.5Cu–0.4Co (nominal)							
	3–5	300–500	$8 \cdot 10^{-5}$–$2 \cdot 10^{-2}$	0.38	445	$8 \cdot 10^{-5}$, 300	63
Al-7475: Al–5.7Zn–2.2Mg–1.6Cu–0.22Cr (nominal)							
	2–7	515	10^{-3}–10^{-1}	0.4	1 750	10^{-1}/—	64
	4–7	515	$2 \cdot 10^{-5}$–$2 \cdot 10^{-2}$	0.9	1 020	$8 \cdot 10^{-4}$/—	65
	5.5–13	516	$2 \cdot 10^{-4}$–$2 \cdot 10^{-2}$	0.67	2 000	$2 \cdot 10^{-3}$/— $d = 5.5\,\mu m$	66

(continued)

Table 36.1   INVESTIGATIONS ON SUPERPLASTICITY OF METAL ALLOYS—*continued*

Material [wt %]	Grain size [μm]	Temp. range [°C]	Strain rate range [s⁻¹]	m	e_f [%]	Optimum condition [s⁻¹]/[°C]	Ref.
	6.8	505	10^{-5}–$6\cdot10^{-3}$	0.5	>400	—	67
	9	437–516	$^e10^{-4}$–10^{-2}	0.58	550	10^{-4}/516	68
	10	516	$2\cdot10^{-4}$–$5\cdot10^{-3}$	0.4	500	—	69
	11	400–516	$6\cdot10^{-6}$–10^{-2}	0.8	780	$2\cdot10^{-4}$/516	70
	11	400–516	$6\cdot10^{-5}$–10^{-2}	0.67	780	10^{-4}/516	71
	12.3	371–516	$2\cdot10^{-5}$–10^{-2}	0.8	1 400	$2\cdot10^{-4}$/516	72
	16.3	371–516	$2\cdot10^{-5}$–10^{-2}	0.9	850	$2\cdot10^{-4}$/516	72
	15–20	516	$6\cdot10^{-5}$–$2\cdot10^{-3}$	0.5	500	—	73
	22	371–516	$2\cdot10^{-5}$–10^{-2}	0.9	730	$2\cdot10^{-4}$/516	72
	69.4	371–516	$2\cdot10^{-5}$–10^{-2}	0.75	210	$2\cdot10^{-4}$/516	72
	156.3	371–516	$2\cdot10^{-5}$–10^{-2}	0.5	170	$2\cdot10^{-4}$/516	72
+0.72Zr	1–2.8	520	$5\cdot10^{-3}$–$3\cdot10^{-1}$	0.62	900	$5\cdot10^{-2}$/—	74, 75
+1.2Fe, 0.1Si	—	516	$5\cdot10^{-4}$–$5\cdot10^{-2}$	0.6	—	—	44
Al-8090: Al–2.5Li–1.3Cu–0.8Mg–0.12Zr–0.15Fe–0.04Si (nominal)							
	1	300–525	$8\cdot10^{-5}$–$5\cdot10^{-3}$	0.5	645	$8\cdot10^{-4}$/350	76
	2	450–550	$4\cdot10^{-5}$–$6\cdot10^{-3}$	0.5	760	$6\cdot10^{-4}$/500	77
	3	300–540	$\geq2\cdot10^{-4}$	0.6	550	$8\cdot10^{-4}$/525	78
	2–4	530	$5\cdot10^{-4}$–$5\cdot10^{-2}$	0.45	—	—	44
	1.6–300	460–540	$3\cdot10^{-5}$–$3\cdot10^{-2}$	—	475	$3\cdot10^{-4}$/540	79
	14	518	$2\cdot10^{-4}$–10^{-3}	—	963	$4\cdot10^{-4}$/—	47
Al-9021: Al–4Cu–1.5Mg–1.1C–0.8O (nominal)							
	0.5	550	10^{-3}–$3\cdot10^{2}$	0.5	1 100	10^{2}/—	80
	0.2–3	424–475	10^{-4}–$3\cdot10^{-2}$	0.25	525	$2\cdot10^{0}$/475	81
Al-9052: Al–4Mg–1.1C–0.8O (nominal)							
	0.5	500–590	10^{-3}–10^{2}	0.5	330	10^{1}/590	82
LFC-X1: Al–(5–7)Mg–(1–2)Zn–(0.5–1)Mn–(0.1–0.2)Cr–(0.3–0.5)Zr							
	<10	510–550	$4\cdot10^{-5}$–$8\cdot10^{-4}$	0.6	800	$3\cdot10^{-4}$/—	83
Formall 570: Al–4.5Mg–3.5Zn–0.35Mn–0.25Cr–0.1Si–0.1Fe							
	7	485–530	$6\cdot10^{-5}$–$2\cdot10^{-2}$	0.6	600	$2\cdot10^{-4}$/515	84
Al-PM64: Al–7.15Zn–2.39Mg–2.02Cu–0.09Fe–0.19Zr–0.12Cr–0.20Co							
	6	500–516	10^{-5}–$2\cdot10^{-2}$	0.85	800	$2\cdot10^{-4}$/—	85
Copper Alloys							
Cu–2.8Al–1.8Si–0.4Co	1.25	550	$5\cdot10^{-4}$–$2\cdot10^{-2}$	0.42	500	$2\cdot10^{-4}$/—	86
	2.8	550	10^{-6}–10^{0}	0.5	2 550	$7\cdot10^{-2}$/—	87
Cu–8.2Al–4.4Fe–5.2Ni–1.5Mn	—	800	$2\cdot10^{-5}$–$6\cdot10^{-1}$	0.52	>1 300	$2\cdot10^{-3}$/—	88
Cu–8.9Al–4.1Fe–3.9Ni–0.9Mn	2	750–850	$2\cdot10^{-5}$–$6\cdot10^{-1}$	0.68	650	10^{-3}/—	88
Cu–9.1Al–5.9Fe–5.1Ni–1.6Mn	—	800	$2\cdot10^{-5}$–$6\cdot10^{-1}$	0.56	>1 300	$8\cdot10^{-3}$/—	88
Cu–10Al–4.5Fe–6.0Ni–1.7Mn	2	800	$2\cdot10^{-5}$–$6\cdot10^{-1}$	0.57	>1 300	$6\cdot10^{-3}$/—	88
Cu–10.3Al–5.2Fe–5.0Ni–1.5Mn	2–4	750–850	$2\cdot10^{-5}$–$6\cdot10^{-1}$	0.68	>5 500	$6\cdot10^{-3}$/750	88, 89
Cu–13.5Al–3Ni–(X)Cr (X = 0–1)	3–6	680	$2\cdot10^{-3}$	—	620	X = 0.5	90
Cu–14Al–3Ni–(X)Cr (X = 0–1)	3–6	680	$2\cdot10^{-3}$	—	1 000	X = 0.5	90
Cu–1.9Be–0.2Ni	—	550	$2\cdot10^{-3}$	—	1 300	—	91
Cu–21.25Zn–6.58Al–3.13Fe–2.32Mn	70	550–750	10^{-3}–$4\cdot10^{-2}$	—	377	$3\cdot10^{-2}$/700	92

(continued)

Table 36.1 INVESTIGATIONS ON SUPERPLASTICITY OF METAL ALLOYS—*continued*

Material [wt %]	Grain size [μm]	Temp. range [°C]	Strain rate range [s⁻¹]	m	e_f [%]	Optimum condition [s⁻¹]/[°C]	Ref.
Cu–28.1Zn–15.0Ni–13.3Mn	1–3	570	10^{-5}–$2\cdot10^{-3}$	0.5	620	$2\cdot10^{-4}$/—	93, 94
Cu–38.4Zn–15.7Ni–0.13Mn	1–3	462–652	10^{-5}–$2\cdot10^{-3}$	0.5	750	$6\cdot10^{-4}$/580	93, 94
Cu–38.5Zn–3Fe	10–80	500–800	$2\cdot10^{-5}$–10^{-2}	0.5	350	$2\cdot10^{-4}$/700	95
Cu–40Zn	3	560–640	$3\cdot10^{-5}$–$6\cdot10^{-3}$	0.63	410	—	96
Cu–40Zn–1.5Pb	2 000–3 000	670–720	$4\cdot10^{-4}$–$2\cdot10^{-3}$	0.56	192	$6\cdot10^{-4}$/700	97
Iron Alloys							
Fe–0.03C–4Ni–3Mo–1.6Ti	6	900–960	$4\cdot10^{-5}$–$4\cdot10^{-3}$	0.53	>500	$8\cdot10^{-4}$/800	98
Fe–0.03C–4Ni–3Mo–1.6Ti + 600 ppm B	6	900–960	$4\cdot10^{-5}$–$4\cdot10^{-3}$	0.53	634	$2\cdot10^{-4}$/900	98
Fe–(0.13–0.34)C	2–12	700–950	$3\cdot10^{-5}$–$7\cdot10^{-3}$	0.58	376	$3\cdot10^{-4}$/900	99
Fe–0.42C–1.9Mn	1.4–2.6	718–782	$2\cdot10^{-6}$–$2\cdot10^{-2}$	0.65	460	$3\cdot10^{-4}$/727	100
Fe–0.88C–1.15Mn–0.22V–0.5W–0.48Cr–0.31Si	<3	650	$4\cdot10^{-5}$–10^{2}	0.5	1 200	$2\cdot10^{-4}$/—	101
Fe–0.91C–0.45Mn–0.12Si	1–5	716–927	10^{-3}–10^{2}	0.42	133	$4\cdot10^{-5}$/716	102
Fe–1.11C–0.48Mn–1.45Cr–9.30Al	3–4	700–900	$7\cdot10^{-6}$–10^{-2}	0.74	—	—	103
Fe–1.17C–5.12Al	1–2	750–850	10^{-4}–$8\cdot10^{-2}$	0.64	490	$2\cdot10^{-3}$/850	104
Fe–1.25C–0.65Mn–0.1Si	1.5–3	605–800	10^{-4}–$2\cdot10^{-2}$	0.5	700	10^{-4}/630	105
Fe–1.25C–10Al–1.5Cr–0.54Mn	10	750–920	10^{-5}–$2\cdot10^{-2}$	0.74	>700	$5\cdot10^{0}$/920	106
Fe–1.32C–0.6Mn–1.68Cr–1.9Al	1–2	700–780	$7\cdot10^{-6}$–10^{-2}	0.43	—	—	103
Fe–1.34C–3.05Al	1–2	750–850	10^{-4}–$8\cdot10^{-2}$	0.5	583	$8\cdot10^{-3}$/750	104
Fe–1.51C–3.38Al	1–2	750–850	10^{-4}–$8\cdot10^{-2}$	0.49	783	$8\cdot10^{-3}$/750	104
Fe–1.55C–5.22Al	1–2	750–850	10^{-4}–$8\cdot10^{-2}$	0.63	450	$2\cdot10^{-3}$/850	104
Fe–1.6C–0.73Mn–0.28Si	1.5–3	600–800	$7\cdot10^{-5}$–$3\cdot10^{-2}$	0.5	760	10^{-4}/630	105
Fe–1.92C–0.82Mn–0.3Si	1.5–3	600–800	$2\cdot10^{-4}$–$2\cdot10^{-3}$	0.5	380	$2\cdot10^{-4}$/650	105
Fe–2.4C	2	630–725	$2\cdot10^{-5}$–$5\cdot10^{-3}$	0.5	500	$2\cdot10^{-4}$/700	107
Fe–3C	2	650–700	$2\cdot10^{-5}$–$5\cdot10^{-4}$	0.5	940	$2\cdot10^{-4}$/700	107
Fe–3C–1.5Cr	1	630–725	$2\cdot10^{-5}$–$5\cdot10^{-3}$	0.5	1 410	$2\cdot10^{-4}$/700	107
Fe–3.26C–4.1N–2.1Cr–0.38Si–0.45Mn	1–2	650–750	$7\cdot10^{-5}$–$2\cdot10^{-3}$	0.46	210	$2\cdot10^{-4}$/675	108
Fe–4.3C–0.18Si	1–2	650–750	$9\cdot10^{-5}$–$3\cdot10^{-3}$	0.41	150	$2\cdot10^{-4}$/675	108
Stainless Steels							
Fe–20Cr–10Ni–0.7Ni	—	700–1 000	$3\cdot10^{-4}$–$2\cdot10^{-2}$	0.65	527	—/800	109
Fe–21.3Cr–4.2Mn–2.1Ni–0.1Si–0.74Cu	10–20	900–1 050	10^{-4}–$6\cdot10^{-2}$	0.56	—	—	110
Fe–23Cr–4Ni–0.1N	7.5	880–1 000	$5\cdot10^{-6}$–$3\cdot10^{-2}$	0.8	—	—	111
Fe–24.66Cr–6.82Ni–2.79Mo–0.85Mn–0.48Si–0.28W–0.14N	<20	900–1 050	$4\cdot10^{-4}$–$4\cdot10^{-2}$	0.45	2 000	$4\cdot10^{-3}$/900	112
Fe–25Cr–7Ni–3Mo	5	900–1 050	10^{-4}–$6\cdot10^{-2}$	0.89	—	—	110
	1–2	1 000	10^{-4}–$2\cdot10^{0}$	0.66	>1 600	10^{-2}/—	113

(*continued*)

Table 36.1 INVESTIGATIONS ON SUPERPLASTICITY OF METAL ALLOYS—*continued*

Material [wt %]	Grain size [μm]	Temp. range [°C]	Strain rate range [s^{-1}]	m	e_f [%]	Optimum condition [s^{-1}]/[°C]	Ref.
Fe–25Cr–7Ni–3Mo–0.14N (SuperDux 64)	2–3	1 050–1 350	10^{-3}–10^{0}	0.6	1 000	$2 \cdot 10^{-2}$/1350	114
Fe–26Cr–6.5Ni–0.4Ti–0.5Mn–0.5Si	3–4	700–1 020	$2 \cdot 10^{-6}$–$2 \cdot 10^{-2}$	0.54	1 050	$3 \cdot 10^{-3}$/960	115
Magnesium Alloys							
Mg–2.83Al–0.09Zr (AK20)	73.1	300–500	$8 \cdot 10^{-5}$–$8 \cdot 10^{-2}$	—	330	$8 \cdot 10^{-4}$/450	116
Mg–5.14Al–0.18Zr (AK50)	43.9	300–500	$8 \cdot 10^{-5}$–$8 \cdot 10^{-2}$	—	350	$8 \cdot 10^{-4}$/450	116
Mg–6.19Al	66.8	300–500	$8 \cdot 10^{-5}$–$8 \cdot 10^{-2}$	—	260	$8 \cdot 10^{-4}$/450	116
Mg–8.3Al–8.1Ga	0.2	250–350	10^{-4}–10^{0}	0.5	1 080	10^{-2}/250	117
Mg–9Al–0.7Zr–0.15Mn (AK91)	1–20	250–300	$4 \cdot 10^{-4}$–10^{-2}	0.62	1 480	$4 \cdot 10^{-1}$/300, d = 1 μm	118
	1	175–200	$2 \cdot 10^{-5}$–10^{-3}	0.33	661	$6 \cdot 10^{-5}$/200	119
	0.5–3	150–250	$2 \cdot 10^{-5}$–$3 \cdot 10^{-4}$	0.5	950	$7 \cdot 10^{-5}$/200	120
	<5	270–400	$2 \cdot 10^{-5}$–10^{0}	0.5	425	$3 \cdot 10^{-4}$/250	121
Mg–11.28Al–0.42Zr	19.7	300–500	$8 \cdot 10^{-5}$–$8 \cdot 10^{-2}$	—	110	$8 \cdot 10^{-4}$/450	116
Mg–33.6Al	1–2	250–400	$3 \cdot 10^{-7}$–$3 \cdot 10^{-2}$	0.85	2 100	$3 \cdot 10^{-2}$/400	122
Mg–6.5Li	10.5	125–300	10^{-5}–$8 \cdot 10^{-3}$	0.33	—	—	123
Mg–9Li	2.7	25–175	$7 \cdot 10^{-6}$–$5 \cdot 10^{-3}$	0.5	450	$6 \cdot 10^{-4}$/100	124
	6–34	150–250	$4 \cdot 10^{-5}$–$3 \cdot 10^{-2}$	0.5	460	$3 \cdot 10^{-4}$/180, d = 6 μm	125
Mg–5.8Zn–0.65Zr (ZK60)	0.8–14	150–450	10^{-5}–10^{-1}	0.38	700	10^{-4}/250	126
	<2	250–350	10^{-3}–10	0.5	425	10^{-1}/400	121
Mg–6Zn–0.5Zr (ZK60)	3.4	150–400	10^{-5}–10^{-2}	0.5	540	10^{-2}/325	127
	0.8–10	270	$3 \cdot 10^{-6}$–$6 \cdot 10^{-2}$	0.52	—	$^a10^{-4}$/—, d = 4 μm	128
	6.5	200–250	$3 \cdot 10^{-6}$–$3 \cdot 10^{-5}$	0.5	450	$<1 \cdot 10^{-5}$/—	129
	<7	250–450	$2 \cdot 10^{-4}$–10^{-1}	0.5	725	$4 \cdot 10^{-3}$/300	121
Mg–6Zn–0.65Zr	5–12	250–500	10^{-4}–$4 \cdot 10^{-3}$	0.38	—	—	130
Nickel Alloys							
Ni–12.1Cr–5.96Al–0.78Ti–4.61Mo–0.13Zr	1.3–3	1 050–1 100	$^b10^{-5}$–10^{0}	0.5	—	—	131
Ni–12.38Cr–3.53Mo–18.52Co–4.26Ti–4.90Al	3	1 050–1 100	$^b8 \cdot 10^{-5}$–$4 \cdot 10^{-3}$	0.6	—	$^a6 \cdot 10^{-4}$/1075	132, 133
Ni–12.5Cr–18.3Co–4.4Ti–3.2Mo–4.9Al–0.76V	2.1	900–1 040	$5 \cdot 10^{-5}$–$6 \cdot 10^{-1}$	0.66	>850	$<10^{-1}$/—	134
Ni–13.1Cr–10.7Co–4.6Mo–2.8W–3.2Al–3.4Nb–2.6Ti	0.25–5.5	800–1 075	$7 \cdot 10^{-5}$–10^{-2}	0.6	>550	10^{-3}/1075, d = 5.5 μm	135
Ni–15Cr–4.5Al–2.5Ti–2Mo–4W–2Ta/1.1 vol% Y$_2$O$_3$^{e}	0.26	900–1 100	$3 \cdot 10^{-8}$–$6 \cdot 10^{0}$	0.47	308	$5 \cdot 10^{-1}$/1 000	136
Ni–20Cr–0.5Ti–0.3Al/ 0.6 vol% Y$_2$O$_3$^{e}	0.7	900–1 100	$2 \cdot 10^{-4}$–$3 \cdot 10^{0}$	0.3	200	10^{-1}/1 000	136
Ni–20Cr–18Fe–5Nb–3Mo–1Co (Inconel 718)	<6	982	10^{-4}	0.75	500	—	137, 138
	1–10	900–980	$2 \cdot 10^{-4}$–$5 \cdot 10^{-3}$	0.77	1 050	10^{-3}/920	139
	5–12	927–982	$4 \cdot 10^{-5}$–10^{-2}	0.9	760	$3 \cdot 10^{-5}$/954	140
Ni–29.97Cr–5.27Al	0.6–1.5	852–1 002	$3 \cdot 10^{-4}$–$2 \cdot 10^{-2}$	0.5	500	$2 \cdot 10^{-3}$/852	141
Ni–39Cr–9Fe–2Ti–1Al	<5	816–982	$2 \cdot 10^{-5}$–$2 \cdot 10^{-1}$	0.5	>880	$2 \cdot 10^{-3}$–$2 \cdot 10^{-2}$/ 927–982	142

(continued)

Table 36.1   INVESTIGATIONS ON SUPERPLASTICITY OF METAL ALLOYS—*continued*

Material [wt %]	Grain size [µm]	Temp. range [°C]	Strain rate range [s^{-1}]	m	e$_f$ [%]	Optimum condition [s^{-1}]/[°C]	Ref.
Tin Alloys							
Sn–1Bi	2–20	25	$2\cdot10^{-8}$–$2\cdot10^{-2}$	0.45	300	$2\cdot10^{-4}$/—	143
Sn–32Pb–5.2Bi	1.5	25	$5\cdot10^{-6}$–10^{-2}	0.48	1 275	10^{-2}/—	144
Sn–35.4Pb–6.9Ag	1.7	25	$5\cdot10^{-6}$–10^{-2}	0.4	350	$3\cdot10^{-5}$/—	144
Sn–35.4Pb–7.4Sb	11	25	$5\cdot10^{-6}$–10^{-2}	0.4	490	10^{-3}/—	144
Sn–36Pb–5.6Ag	2–3	25	$4\cdot10^{-5}$–$4\cdot10^{-3}$	0.4	362	$8\cdot10^{-4}$/—	145
Sn–36.6Pb–4.1Ag	1.9	25	$5\cdot10^{-6}$–10^{-2}	0.42	550	10^{-3}/—	144
Sn–36.6Pb–4.3Cu	2.3	25	$5\cdot10^{-6}$–10^{-2}	0.43	220	10^{-3}–10^{-4}/—	144
Sn–36.6Pb–4.1Sb	1	25	$5\cdot10^{-6}$–10^{-2}	0.41	675	$3\cdot10^{-4}$/—	144
Sn–38.1Pb	2	25	$4\cdot10^{-5}$–$4\cdot10^{-3}$	0.53	1 812	$8\cdot10^{-4}$/—	145
	2.2	25	$5\cdot10^{-6}$–10^{-2}	0.45	—	—	144
	6.9	140	10^{-4}	—	>4 850	—	146
Sn–9.8Zn	1–2	121–177	$3\cdot10^{-6}$–10^{-3}	0.5	570	$2\cdot10^{-4}$/163	147
Titanium Alloys							
Ti–2.5Al–0.5Mn–0.3Zr–<0.3Fe	15–20	800–900	$4\cdot10^{-5}$–$5\cdot10^{-3}$	0.3	340	10^{-3}/850	77
Ti–4Al–4Mo–2Sn–0.5Si	2–5	805–915	10^{-5}–10^{-2}	0.8	1 230	-/915	148
Ti–4Al–0.25O	23.5	800–1 025	$4\cdot10^{-5}$–$4\cdot10^{-4}$	0.6	—	—	149
Ti–4.5Al–3V–2Fe–2Mo	4	700–850	$2\cdot10^{-5}$–$2\cdot10^{-3}$	0.5	300	$2\cdot10^{-5}$–10^{-3}/—	150
Ti–5Al–2.5Sn	18.5	800–1 055	$3\cdot10^{-5}$–10^{-2}	0.72	>400	—	149
Ti–5.8Al–4.02Sn–3.49Zr–0.71Nb–0.52Mo–0.33Si–0.05C	8–20	990	$2\cdot10^{-5}$–$2\cdot10^{-3}$	0.7	>300	10^{-4}/—	151
Ti–6Al–3.2Mo	0.06	25–600	$3\cdot10^{-8}$–$2\cdot10^{-4}$	0.35	600	$5\cdot10^{-4}$/600	152
	0.1	25–600	$3\cdot10^{-8}$–$2\cdot10^{-4}$	0.33	480	$5\cdot10^{-4}$/600	152
	0.4	25–600	$3\cdot10^{-8}$–$2\cdot10^{-4}$	0.3	225	$5\cdot10^{-4}$/600	152
	5	25–600	$3\cdot10^{-8}$–$2\cdot10^{-4}$	0.1	35	$5\cdot10^{-4}$/600	152
Ti–6Al–4V	3	750–925	$1\cdot10^{-5}$–10^{-2}	0.71	490	10^{-4}/750	153
	4.5	760–940	$2\cdot10^{-4}$–$2\cdot10^{-3}$	0.75	1 050	$2\cdot10^{-4}$/880	154
	6	800–1 000	$3\cdot10^{-5}$–$2\cdot10^{-2}$	0.86	>1 000	—	149
	6.4	927	$2\cdot10^{-5}$–$2\cdot10^{-2}$	0.9	—	[a]$3\cdot10^{-4}$/—	155
	6.8	880	$2\cdot10^{-5}$–$2\cdot10^{-2}$	0.67	—	—	11
	7.7	927	$3\cdot10^{-5}$–$2\cdot10^{-3}$	0.73	—	$7\cdot10^{-4}$/—	156
	6–12	927	$2\cdot10^{-5}$–$8\cdot10^{-3}$	0.94	—	10^{-4}/—, d = 8 µm	157
	5–16	815–927	$2\cdot10^{-5}$–$2\cdot10^{-2}$	0.7	600	$2\cdot10^{-4}$/927	158
	9	927	$2\cdot10^{-5}$–$2\cdot10^{-2}$	0.8	—	[a]10^{-4}/—	155
	11.5	927	$2\cdot10^{-5}$–$2\cdot10^{-2}$	0.75	—	[a]$8\cdot10^{-5}$/—	155
	20	927	$2\cdot10^{-5}$–$2\cdot10^{-2}$	0.7	—	[a]10^{-5}/—	155
Ti–6Al–4V–1Co–1Ni	3.4–22	815–871	$2\cdot10^{-5}$–$2\cdot10^{-2}$	0.82	550	$2\cdot10^{-4}$/815	158
Ti–6Al–4V–(X)Co, X = 0–2	6–12	750–950	10^{-5}–10^{-2}	1.0	—	[a]$3\cdot10^{-5}$/950, X = 1	159
Ti–6Al–4V–2Co	4–63	815–871	$2\cdot10^{-5}$–$2\cdot10^{-2}$	0.53	670	$2\cdot10^{-4}$/815	158
Ti–6Al–4V–1Fe–1Co	2.7–10	815–871	$2\cdot10^{-5}$–$2\cdot10^{-2}$	0.48	525	$2\cdot10^{-4}$/815	158
Ti–6Al–4V–1Fe–1Ni	3.4–20	815–871	$2\cdot10^{-5}$–$2\cdot10^{-2}$	0.68	550	$2\cdot10^{-4}$/815	158
Ti–6Al–4V–2Fe	3.5–8.6	815–871	$2\cdot10^{-5}$–$2\cdot10^{-2}$	0.54	650	$2\cdot10^{-4}$/815	158
Ti–6Al–4V–0.05H	2	750–870	10^{-5}–10^{-2}	0.61	610	10^{-4}/750	153
Ti–6Al–4V–0.07H	6.8	820	$2\cdot10^{-5}$–$2\cdot10^{-2}$	0.69	—	—	11
Ti–6Al–4V–0.1H	2	700–850	$5\cdot10^{-6}$–10^{-2}	0.6	545	10^{-3}/750	153
Ti–6Al–4V–0.14H	6.8	780	$2\cdot10^{-5}$–$2\cdot10^{-2}$	0.6	—	—	11
Ti–6Al–4V–0.17H	6.8	765	$2\cdot10^{-5}$–$2\cdot10^{-2}$	0.55	—	—	11

(continued)

Table 36.1 INVESTIGATIONS ON SUPERPLASTICITY OF METAL ALLOYS—*continued*

Material [wt %]	Grain size [μm]	Temp. range [°C]	Strain rate range [s⁻¹]	m	e_f [%]	Optimum condition [s⁻¹]/[°C]	Ref.
Ti–6Al–4V–0.29H	6.8	735	$2 \cdot 10^{-5}$–$2 \cdot 10^{-2}$	0.28	—	—	11
Ti–6Al–4V–0.3H	—	700–750	10^{-5}–$9 \cdot 10^{-2}$	0.5	180	$5 \cdot 10^{-4}$/750	153
Ti–6Al–4V–0.40H	6.8	720	$2 \cdot 10^{-5}$–$2 \cdot 10^{-2}$	0.29	—	—	11
Ti–6Al–4V–0.53H	6.8	715	$2 \cdot 10^{-5}$–$2 \cdot 10^{-2}$	0.35	—	—	11
Ti–6Al–4V–(X)Ni, X = 0–2	6–12	750–950	10^{-5}–10^{-2}	0.98	—	[a]$3 \cdot 10^{-5}$/950, X = 1	159
Ti–6Al–4V–2Ni	3.5–35	815–871	$2 \cdot 10^{-5}$–$2 \cdot 10^{-2}$	0.85	720	$2 \cdot 10^{-4}$/815	158
Ti–6Al–4Zr–2Sn–2Mo	4.5	820–970	$2 \cdot 10^{-4}$–$2 \cdot 10^{-3}$	0.7	775	$2 \cdot 10^{-4}$/940	154
Ti–6.7Al–4.7Mo	0.06	500–600	10^{-6}–$3 \cdot 10^{-2}$	0.45	1 200	$2 \cdot 10^{-4}$/575	160
Ti–5.95Co–6.15Ni–5.05Al	<2	700–750	10^{-3}–10^{-1}	0.6	>1 500	$3 \cdot 10^{-3}$/700	161
Ti–10Co–4Al	0.5	550–750	$9 \cdot 10^{-6}$–$7 \cdot 10^{-2}$	0.48	>1 100	$2 \cdot 10^{-3}$/725	162
Ti–12.55Co–4.95Al	<2	700–750	10^{-3}–10^{-1}	0.6	>2 000	10^{-2}/700	161
Ti–11Mo–5.5Sn–4.2Zr	10	900	$3 \cdot 10^{-5}$–$3 \cdot 10^{-2}$	0.4	>300	$6 \cdot 10^{-5}$/—	163
Ti–10V–3Al–2Fe	<10	625–850	$4 \cdot 10^{-4}$–$4 \cdot 10^{-3}$	0.57	910	10^{-3}/700	164
Uranium Alloys							
U–1.82Mo	40	590–735	$3 \cdot 10^{-4}$–$5 \cdot 10^{-2}$	0.44	692	$3 \cdot 10^{-3}$/695	165
U–5.8Nb	1.2	600–750	$3 \cdot 10^{-4}$–$5 \cdot 10^{-3}$	0.85	658	$3 \cdot 10^{-4}$/685	166
Zinc Alloys							
Zn–1.1Al	1.1	20–220	10^{-5}–$4 \cdot 10^{-2}$	0.5	700	10^{-2}/102	27
Zn–18Al	0.6–2.2	250–350	$2 \cdot 10^{-6}$–$2 \cdot 10^{-2}$	0.53	—	—	167
Zn–20Al	0.5–1.5	20–250	$7 \cdot 10^{-5}$–$2 \cdot 10^{-1}$	0.55	>1 000	[a]$2 \cdot 10^{-1}$/250	168
Zn–22Al	1	60–200	$3 \cdot 10^{-7}$–$3 \cdot 10^{-1}$	0.5	—	—	169
	0.4–2	200–250	$3 \cdot 10^{-5}$–$2 \cdot 10^{-2}$	0.5	500	10^{-1}/250	170
	0.6–2.2	200–250	$2 \cdot 10^{-6}$–$2 \cdot 10^{-1}$	0.6	900	$2 \cdot 10^{-2}$/250	167
	2	220–300	$8 \cdot 10^{-4}$–$8 \cdot 10^{-1}$	0.44	1 200	—	171
	2.1	200	10^{-5}–10^{0}	0.41	—	—	172
	1.3–3.7	177–252	[c]10^{-7}–10^{-1}	0.45	—	—	173
	2.4	200	$6 \cdot 10^{-6}$–$4 \cdot 10^{-1}$	—	1 800	—	174
	2.5	200	10^{-5}–10^{0}	—	2 900	10^{-2}/—	175
	2.6	230	10^{-5}–10^{-1}	0.5	2 550	$7 \cdot 10^{-2}$/—	176
	1–5	175–250	[c]$3 \cdot 10^{-9}$–$2 \cdot 10^{-2}$	0.5	—	[a]$2 \cdot 10^{-5}$–$2 \cdot 10^{-3}$	177
	2.3–4.6	136–230	10^{-7}–$2 \cdot 10^{-2}$	0.5	—	[a]10^{-4}–10^{-2}	178
	2.5–4.2	150–230	10^{-5}–10^{0}	0.5	2 900	10^{-2}/230, d = 2.5 μm	179
	4–8	25–250	$3 \cdot 10^{-4}$–$3 \cdot 10^{-1}$	—	770	$3 \cdot 10^{-2}$/250	180
	—	25–350	$2 \cdot 10^{-5}$–$5 \cdot 10^{-2}$	0.7	900	$2 \cdot 10^{-2}$/250	181
Zn–22.5Al	0.5–1.8	250	[b]$2 \cdot 10^{-6}$–2	0.5	—	—	182
Zn–23.55Al–0.43Cu	1	180	10^{-6}–10^{-1}	0.4	—	—	183
Zn–36Al	0.6–2.2	22–325	$2 \cdot 10^{-6}$–$2 \cdot 10^{-2}$	0.5	—	—	167
Zn–0.6Cu–0.1Ti	2.5	150–190	$8 \cdot 10^{-4}$	0.54	290	—/190	184
Zn–22Cu–20Al	2	20	$2 \cdot 10^{-4}$–$4 \cdot 10^{-2}$	0.34	180	—	185
Zn–0.2Mn	2–4	275	10^{-5}–1	0.17	102	—	186
Zn–0.7Mn	2–4	275	10^{-5}–1	0.39	620	—	186
Zn–1.2Mn	2–4	275	10^{-5}–1	0.36	500	—	186
Zn–1.5Mn	2–4	275	10^{-5}–1	0.27	132	—	186
Zn–2Mn	2–4	275	10^{-5}–1	0.22	150	—	186

[a] Optimum condition based on strain rate sensitivity, m, instead of elongation.
[b] Testing performed in compression.
[c] Testing performed in double-shear configuration.
[d] Dispersion-strengthened alloy.
[e] Tested under a 2 kV/cm applied electric field.

Table 36.2 INVESTIGATIONS ON INTERMETALLIC ALLOYS EXHIBITING SUPERPLASTICITY

Note that compositions are given in atomic %.

Material [at %]	Grain size [μm]	Temp. range [°C]	Strain rate range [s⁻¹]	m	e_f [%]	Optimum condition [s⁻¹]/[°C]	Ref.
Iron Aluminide Alloys							
Fe₃ Al Alloys							
Fe–27Al	100–800	600–800	10^{-4}–10^{-2}	0.33	180	10^{-4}/800	187
Fe–28Al–2Ti	100	800–900	$2 \cdot 10^{-4}$–$4 \cdot 10^{-3}$	0.42	330	10^{-3}/850	188
	100	850	10^{-4}–$2 \cdot 10^{-2}$	0.42	333	10^{-3}/—	189
FeAl Alloys							
Fe–36.5Al	>100	875–1 000	10^{-4}–$3 \cdot 10^{-1}$	0.33	—	—	190
Fe–36.5Al–1Ti	>100	875–1 000	10^{-4}–$3 \cdot 10^{-1}$	0.35	—	—	190
Fe–36.5Al–2Ti	350	900–1 000	10^{-4}–$3 \cdot 10^{-2}$	0.42	208	10^{-2}/1 000	191
	>100	875–1 000	10^{-4}–$3 \cdot 10^{-1}$	0.38	—	—	190
Iron Silicide (Fe₃Si) Alloys							
Fe–14Si–0.25B	72	800–950	$3 \cdot 10^{-5}$–$2 \cdot 10^{-3}$	0.33	200	$3 \cdot 10^{-5}$/800	192
Fe–18Si–0.25B	72	800–950	$3 \cdot 10^{-5}$–$2 \cdot 10^{-3}$	0.33	250	$3 \cdot 10^{-5}$/800	192
Nickel Aluminide Alloys							
NiAl Alloys							
NiAl	200	1 000–1 100	$2 \cdot 10^{-4}$–$2 \cdot 10^{-2}$	0.34	211	$2 \cdot 10^{-3}$/1100	193
Ni–28.5Al–20.4Fe	30–50	850–980	10^{-4}–10^{-2}	—	233	$5 \cdot 10^{-4}$/850	194
Ni₃ Al Alloys							
Ni–15.5Al–7.4Cr–	0.05	650–725	10^{-3}	—	375	10^{-3}/650	35
0.4Zr–0.09B	6	1 000–1 100	10^{-4}–$2 \cdot 10^{-2}$	0.8	641	$9 \cdot 10^{-4}$/1100	195
	13–20	25–1100	$8 \cdot 10^{-4}$–$2 \cdot 10^{-2}$	0.6	638	$8 \cdot 10^{-4}$/1100	196
Ni–24Al–0.24B	1.6	700	$5 \cdot 10^{-5}$–$2 \cdot 10^{-3}$	0.45	155	$5 \cdot 10^{-5}$/—	197
Nickel Silicide (Ni₃Si) Alloys							
Ni–17.2Si–3.3V–	15	1 050–1 100	10^{-3}–$3 \cdot 10^{-2}$	0.5	500	10^{-3}/1 050	198
1.1Mo							
Ni–17.3Si–3.3V–	8–20	1 040–1 090	$9 \cdot 10^{-5}$–$2 \cdot 10^{-2}$	0.6	500	10^{-3}/1 070	199
2.3Mo	14	1 025–1 090	$6 \cdot 10^{-5}$–10^{-1}	0.6	665	$6 \cdot 10^{-4}$/1 090	200
Ni₃(Si,Ti)	4–14	800–900	$6 \cdot 10^{-5}$–10^{-3}	0.43	190	$6 \cdot 10^{-5}$/960	200
Titanium Aluminide Alloys							
TiAl Alloys							
Ti–40Al	0.5	900	$7 \cdot 10^{-4}$–$3 \cdot 10^{-3}$	0.56	—	—	201
Ti–43Al	5	1 000–1 100	10^{-5}–$2 \cdot 10^{-2}$	0.53	275	$3 \cdot 10^{-4}$/1 050	202
Ti–43.8Al–12.1V	<20	797–1147	$3 \cdot 10^{-4}$–10^{-1}	—	580	$3 \cdot 10^{-4}$/1147	203
Ti–45Al	0.5	900	$7 \cdot 10^{-4}$–$3 \cdot 10^{-3}$	0.52	—	—	201
Ti–45.5Al–2Cr–2Nb	3–12	900–1 200	10^{-3}	0.63	483	—/1 200	204
Ti–46Al	0.2–2.5	650–950	10^{-4}–10^{-1}	0.52	700	10^{-3}/900	205
Ti–46.1Al–3.1Cr	18	27–1 200	$5 \cdot 10^{-2}$–10^{-4}	0.58	383	$5 \cdot 10^{-4}$/1 200	175
Ti–47Al–2Cr–2Nb	0.5–5	650–1 000	10^{-6}–10^{-4}	0.5	310	$2 \cdot 10^{-5}$/800	206
Ti–47.3Al–1.9Nb–	20	1 180–1 310	$2 \cdot 10^{-5}$–$2 \cdot 10^{-3}$	0.7	470	$8 \cdot 10^{-5}$/1280	207
1.6Cr–0.5Si–							
0.4Mn							
Ti–48Al–2Cr–2Nb	10–22	1 100–1 300	[b]$2 \cdot 10^{-4}$–$2 \cdot 10^{-1}$	0.33	—	—	208
Ti–48.1Al–0.8Mo	2–5	25–1100	$2 \cdot 10^{-5}$–$2 \cdot 10^{-3}$	0.8	228	$4 \cdot 10^{-5}$/228	209
Ti–49.7Al	0.4	600–900	$2 \cdot 10^{-4}$–$2 \cdot 10^{-3}$	0.47	260	$8 \cdot 10^{-4}$/850	210
	0.4–22	20–850	$8 \cdot 10^{-4}$	—	260	—/850, d = 0.4 μm	211
	2–13	600–1 050	$2 \cdot 10^{-4}$–$8 \cdot 10^{-3}$	0.43	250	$8 \cdot 10^{-4}$/1 025	212
	2–15	900–1 050	10^{-4}–10^{-2}	0.43	250	$9 \cdot 10^{-4}$/1 025	213
Ti–50Al	8	1 025	10^{-4}–10^{-2}	0.43	250	10^{-3}/1 025	163
	0.3	900	$7 \cdot 10^{-4}$–$3 \cdot 10^{-3}$	0.37	—	—	201

(continued)

Table 36.2 INVESTIGATIONS ON INTERMETALLIC ALLOYS EXHIBITING SUPERPLASTICITY—*continued*

Material [at %]	*Grain size* [μm]	*Temp. range* [°C]	*Strain rate range* [s⁻¹]	*m*	e_f [%]	*Optimum condition* [s⁻¹]/[°C]	*Ref.*
Ti₃Al Alloys							
Ti–24Al–11Nb	3	920–1 040	$2 \cdot 10^{-5}$–$5 \cdot 10^{-3}$	0.7	810	$5 \cdot 10^{-5}$/980	214
	4	960–1 020	$3 \cdot 10^{-6}$–10^{-2}	0.6	520	$3 \cdot 10^{-4}$/980	215
Ti–24.4–10.6Nb–3V–1Mo	3.4	850–1100	$5 \cdot 10^{-5}$–$5 \cdot 10^{-3}$	0.7	—	—	216
Ti–25Al–10Nb–3V–1Mo	1–10	950–1 040	$8 \cdot 10^{-6}$–$7 \cdot 10^{-2}$	0.55	800	$8 \cdot 10^{-5}$/950	217
	2–12	900–1 000	$2 \cdot 10^{-5}$–10^{-2}	0.75	1 500	$2 \cdot 10^{-4}$/960	218
	<20	950–1 010	$2 \cdot 10^{-5}$–$5 \cdot 10^{-3}$	0.62	570	$2 \cdot 10^{-4}$/980	219
Ti–30Al	1.5	900	$7 \cdot 10^{-4}$–$3 \cdot 10^{-3}$	0.41	—	—	201

[b] Testing performed in compression.

Table 36.3 SUMMARY OF RESEARCH ON SUPERPLASTIC METAL MATRIX COMPOSITES

Grain size, d, refers to the matrix material, and V_f denotes the volume fraction of the reinforcing phase.

Matrix	*d* [μm]	*Reinforcement*	V_f [%]	*m*	e_f [%]	*Optimum condition* T [°C]	$\dot{\varepsilon}$ [s⁻¹]	*Ref.*
Al	2	AlN$_P$, 1.8 μm	—	0.3	200	640[f]	10^{-1}	220
Al–2014	2.5	Al$_2$O$_{3P}$	20	0.45	210	500[f]	$2 \cdot 10^{-2}$	221
	—	SiC$_P$, 6–10 μm	15	0.4	442	480	10^{-4}	222
Al–2024	7	SiC$_P$, 10 μm	10	0.44	685	515	$5 \cdot 10^{-4}$	223
	3–13	SiC$_W$	20	0.5	150	550	10^{0}	224
Al–2124	<10	SiC$_P$, 3 μm	17.8	0.41	425	490	$8 \cdot 10^{-2}$	225
	—	SiC$_W$	20	0.33	300	525[f]	$3 \cdot 10^{-1}$	226
	2	β- Si$_3$N$_{4W}$, 0.5–1.5 by 10–20 μm	20	0.5	225	525[f]	$2 \cdot 10^{-1}$	227
	—	Si$_3$N$_{4W}$, 0.1–15 by 10–30 μm	20	>0.3	280	545	$4 \cdot 10^{-2}$	228
	—	Si$_3$N$_{4P}$, <1 μm	20	>0.3	830	515	$4 \cdot 10^{-2}$	228
Al–2324	1	SiC$_W$	20	0.55	520	520	$5 \cdot 10^{-2}$	229
Al–5052	—	Si$_3$N$_{4P}$, <0.2 μm	20	>0.3	700	545	10^{0}	228
Al–6061	—	AlN$_P$, 1–10 μm	~20	0.5	509	600[f]	$9 \cdot 10^{-1}$	230
	—	SiC$_W$	20	0.32	300	550	10^{-1}	231
	1.9	Si$_3$N$_{4P}$, 0.5 μm	20	0.5	500	560[f]	$5 \cdot 10^{-4}$	232
	<3	Si$_3$N$_{4P}$, 1 μm	20	0.45	450	545[f]	10^{-1}	233
	—	β-Si$_3$N$_{4P}$	25	0.46	600	545[f]	$2 \cdot 10^{-1}$	234
	—	Si$_3$N$_{4W}$, 0.1–15 by 10–30 μm	20	>0.3	285	560	10^{-1}	228
	—	Si$_3$N$_{4P}$, <0.2 μm	20	>0.3	620	560	$2 \cdot 10^{0}$	228
	—	Si$_3$N$_{4P}$, <1 μm	20	>0.3	615	545	$2 \cdot 10^{-1}$	228
Al–7064	—	α-Si$_3$N$_{4W}$, 0.1–15 by 5–200 μm	20	0.42	150	525	$3 \cdot 10^{-1}$	235
	—	β-Si$_3$N$_{4W}$, 0.5–15 by 10–20 μm	15	0.22	150	545	$2 \cdot 10^{-1}$	235
	—	β-Si$_3$N$_{4W}$, 0.5–15 by 10–20 μm	20	0.35	150	525	$2 \cdot 10^{-1}$	235
	—	β-Si$_3$N$_{4W}$, 0.5–15 by 10–20 μm	25	0.24	70	605	10^{0}	235
	—	Si$_3$N$_{4W}$, 0.1–15 by 10–30 μm	20	>0.3	380	560	10^{-1}	228
	—	Si$_3$N$_{4P}$, <1 μm	20	>0.3	330	545	10^{0}	228

(*continued*)

Table 36.3   SUMMARY OF RESEARCH ON SUPERPLASTIC METAL MATRIX COMPOSITES—*continued*

Matrix	d [μm]	Reinforcement	V_f [%]	m	e_f [%]	Optimum condition T [°C]	Optimum condition $\dot{\varepsilon}$ [s^{-1}]	Ref.
Al–7475	—	SiC$_P$, 0.6 μm	20	0.33	200	580^f	10^{-1}	236
	7–22	SiC$_P$, 6–10 μm	15	0.43	395	480	$8 \cdot 10^{-4}$	222
	5–10	SiC$_P$, 10 μm	16	0.67	315	515	$2 \cdot 10^{-4}$	237
Al–IN9021	0.45	SiC$_P$, 2 μm	15	0.5	610	550^f	$5 \cdot 10^{0}$	80
Mg–ZK60	1.8	SiC$_P$, 0.5 μm	17	0.5	450	350	10^{-1}	238
	2	SiC$_P$, 2 μm	17	0.5	450	350	10^{-1}	239
	0.3	SiC$_P$, 2 μm	17	0.33	360	450	10^{0}	240
	0.5	SiC$_P$, 2 μm	17	>0.5	320	400	10^{0}	241
Mg–5Zn	—	TiC$_P$, 2–5 μm	10	0.33	200	470	$2 \cdot 10^{-2}$	242
	—	TiC$_P$, 2–5 μm	20	0.43	300	470	$7 \cdot 10^{-2}$	242
Mg–10.6Si–4Al	1.4	Mg$_2$Si$_P$, 1 μm	20	0.3	346	500^f	10^{0}	243
Ti–6Al–4V	<5	TiC$_P$, 5 μm	10	0.64	270	925^f	$2 \cdot 10^{-4}$	244
	<5	TiC$_P$, 5 μm	20	0.56	270	925^f	$2 \cdot 10^{-4}$	244
	<5	TiN$_P$, 5 μm	10	0.57	270	925^f	$2 \cdot 10^{-4}$	244
	<5	TiN$_P$, 5 μm	20	0.62	270	925^f	$2 \cdot 10^{-4}$	244

^{f} only one temperature was investigated.

Table 36.4   SUMMARY OF RESEARCH ON INTERNAL STRESS SUPERPLASTICITY

The source of internal mismatch stress is denoted by C = thermal expansion mismatch between coexisting phases in a composite, A = thermal expansion mismatch between adjacent grains in an anisotropic material, and T = phase transformation volume mismatch.

Material	Mismatch	Temp. range [°C]	Frequency [hr^{-1}]	e_f [%]	Ref.
Al–0.8Be	C	300–400	36–180	200	245
Al–0.8Be	C	250–350	90	—	246
		300–400			
		350–450			
		400–500			
Al–12Si	C	165–300	15	220	247
		125–235			
		141–281			
		164–334			
Al–Al$_3$Ni	C	400–500	90	>120	248
Al/5 vol% SiC$_P$	C	130–450	8.5	—	249
Al/10 vol% SiC$_P$		130–350			
Al/20 vol% SiC$_P$		80–200			
Al/30 vol% SiC$_P$					
Al/40 vol% SiC$_P$					
Al/20 vol% SiC$_P$	C	40–150	8.5	>90	250
		110–350			
Al/20 vol% SiC$_P$	C	80–200	8.5	>150	251
Al/30 vol% SiC$_P$		130–350			
		130–450			
Al/20 vol% SiC$_P$	C	100–300	6	—	252
		115–350			
Al–2024/10 vol% SiC$_W$	C	100–450	18	—	253
Al–2024/20 vol% SiC$_W$					
Al–2024/20 vol% SiC$_W$	C	100–450	18	300	254
Al–2024/20 vol% SiC$_W$	C	100–450	20–22.5	300	255
Al–2124/20 vol% SiC$_W$	C	100–350	7	220	256
Al–6061/20 vol% SiC$_W$		Compressibility mismatch during pressure cycling at 420°C		96	257

(*continued*)

Table 36.4   SUMMARY OF RESEARCH ON INTERNAL STRESS
SUPERPLASTICITY—*continued*

Material	Mismatch	Temp. range [°C]	Frequency [hr^{-1}]	e_f [%]	Ref.
Al–6061/20 vol% SiC$_W$	C	100–450	18	1 300	254
Al–6061/20 vol% SiC$_W$	C	100–450	18	1 400	258
Al–7090/25 vol% SiC$_P$	C	135–200	6	—	259
		100–200			
		70–200			
Co	T	367–467	—	—	260
Co	T	Single heating		—	261
		excursions, 1–4 K/min			
Cu–Zn Alloys	T	−180–20	—	—	262
Fe	T	750–850	—	—	263
		800–1 000			
Fe	T	860–930	1–6	—	264
Fe	T	770–960	50	—	265
Fe	T	860–960	—	—	260
Fe	T	700–900	6–10	>454	266
Fe/10 vol% TiC$_P$	T	Various,	6–30	231	266
		700–950			
Fe/20 vol% TiC$_P$	T	700–900	6–10	—	266
Fe–0.2C	T	600–900	36	—	267
Fe–0.2C	T	740–800	72	—	264
Fe–0.39C	T	673–850	—	—	260
Fe–AISI 1018	T	540–815	50	510	268
Fe–AISI 1045	T	540–815	50	500	268
Fe–AISI 1095	T	540–815	50	580	268
Fe–AISI 52100	T	540–815	50	695	268
Fe–0.91C–0.45Mn–0.12Si	T	538–815	~20	>490	102
Fe–2 at% Cr	T	Single cooling		—	269
		excursions, 1–65 K/s			
Mg–AZ31	A	275–375	90	—	270
		300–400			
		325–425			
Ni–Mar–M247	C	1 100–1 200	180	>100	271
NiAl–Cr	C	950–1 150	90	—	272
		975–1 175			
		1 000–1 200			
NiAl/10–20 vol% ZrO$_2$	T	700–1 200	7.5–15	>30	273
		850–1 250			
		750–1 150			
Ti	T	Single excursions,		—	274
		110–420 K/h			
Ti	T	800–900	0.18–90	—	275, 276
Ti	T	650–925	120	—	277
Ti	T	832–932	—	—	260
Ti	T	830–1030	4.2	—	278
Ti	T	Cycles of chemical		—	279
		composition through			
		cyclic hydrogen			
		introduction			
Ti/10 vol% TiC	T	830–1 010	4.2	135	278, 280
Ti–6Al–4V	T	760–981	120	—	277
Ti–6Al–4V	T	840–1 030	7.5	260–398	281, 282
Ti–6Al–4V	T	Various,	7.5	—	283
		840–990			
Ti–6Al–4V/5 vol% TiB$_W$	T	840–1 030	7.5	390	284
Ti–6Al–4V/10 vol% TiB$_W$	T	840–1 030	7.5	260	282
Ti$_3$Al, Super α_2	T	950–1 150	4–15	610	285

(*continued*)

Table 36.4   SUMMARY OF RESEARCH ON INTERNAL STRESS
SUPERPLASTICITY—*continued*

Material	Mismatch	Temp. range [°C]	Frequency [hr^{-1}]	e_f [%]	Ref.
U	T	Various, 200–450	50–170	—	286
U	T	613–713 720–820	—	—	260
Zn	A	60–150 110–330	14.4 8.5	—	251
Zn	A	50–250 100–300 150–350	12–180	>200	287
Zn	A	200–300 230–330 250–350 270–370 290–390	90	—	288
Zn–30 vol% Al$_2$O$_3$P	A,C	100–300 150–350	120	—	289
Zr	T	750–950	—	—	263
Zr	T	813–913	—	—	260
Zr	T	810–910 810–940	6–15	>270	290
Zr/2 vol% ZrH$_2$	T	23–425	0.036	—	291

REFERENCES

1. G. D. Bengough, *J. Inst. Metals*, 1912, **7**, 123.
2. T. G. Nieh, J. Wadsworth and O. D. Sherby, 'Superplasticity in Metals and Ceramics', Cambridge University Press, New York. 1997.
3. B. M. Watts, M. J. Stowell, B. L. Baikie and D. G. E. Owen, *Metal Sci.*, 1976, **10**, 189.
4. G. Rai and N. J. Grant, *Metall. Trans.*, 1975, **6A**, 385.
5. A. H. Chokshi and T. G. Langdon, *J. Mater. Sci.*, 1989, **24**, 143.
6. D. L. Holt and W. A. Backofen, *Trans. ASM*, 1966, **59**, 755.
7. E. Taleff, G. A. Henshall, D. R. Lesuer, T. G. Nieh and J. Wadsworth, p. 3 in 'Superplasticity and Superplastic Forming 1995' (eds A. K. Ghosh and T. R. Bieler), The Minerals, Metals & Materials Society, Warrendale, PA. 1995.
8. R. R. Sawtell and C. L. Jensen, *Metall. Trans.*, 1990, **21A**, 421.
9. H. Aiko and N. Furushiro, p. 423 in 'International Conference on Superplasticity in Advanced Materials (ICSAM-91)' (eds S. Hori, M. Tokizane and N. Furushiro), The Japan Society for Research on Superplasticity, Osaka, Japan. 1991.
10. J. W. Chung and J. R. Cahoon, *Metal Sci.*, 1979, **13**, 635.
11. L. R. Zhao, S. Q. Zhang and M. G. Yan, p. 459 in 'Superplasticity and Superplastic Forming' (eds C. H. Hamilton and N. E. Paton), TMS, Warrendale, PA. 1988.
12. D. Kum, *Mater. Sci. Forum*, 1997, 243–5, 287.
13. K. Swaminathan and K. A. Padmanabhan, p. 687 in 'Superplasticity in Advanced Materials' (eds S. Hori, M. Tokizane and N. Furushiro), The Japan Society of Research on Superplasticity, Osaka, Japan. 1991.
14. W. Qingling, M. Hongsen and M. Lung-Xiang, p. 168 in 'Superplasticity in Aerospace Aluminium' (eds R. Pearce and L. Kelly), Ashford Press, Culridge, UK. 1985.
15. I. I. Novikov, V. K. Portnoy, V. M. Iljenko and V. S. Levchenko, p. 121 in 'Superplasticity in Advanced Materials' (eds S. Hori, M. Tokizane and N. Furushiro), The Japan Society of Research on Superplasticity, Osaka, Japan. 1991.
16. Z. Cui, W. Zhong, J. Bao and L. Yong, *Scr. Metall. Mater.*, 1994, **31**, 1311.
17. J. Wadsworth, A. R. Pelton and R. E. Lewis, *Metall. Trans.*, 1985, **16A**, 2319.
18. I. G. Moon, J. W. Park and J. E. Yoo, *Mater. Sci. Forum*, 1994, 170–2, 255.
19. J. Wadsworth and A. R. Pelton, *Scr. Metall.*, 1984, **18**, 387.
20. J. Wadsworth, C. A. Henshall, A. R. Pelton and B. Ward, *J. Mater. Sci. Lett.*, 1985, **4**, 674.
21. B. Ash and C. H. Hamilton, p. 239 in 'Superplasticity and Superplastic Forming' (eds C. H. Hamilton and N. E. Paton), TMS, Warrendale, PA. 1988.
22. J. Wadsworth, I. G. Palmer and D. D. Crooks, *Scr. Metall.*, 1983, **17**, 347.
23. T. G. Nieh, P. S. Gilman and J. Wadsworth, *Scr. Metall.*, 1985, **19**, 1375.
24. K. Higashi, T. Okada, T. Muka and S. Tanimura, *Scr. Metall. Mater.*, 1991, **25**, 2053.

25. R. Z. Valiev, N. A. Krasilnikov and N. K. Tsenev, *Mater. Sci. Eng.*, 1991, **A137**, 35.
26. T. Endo, M. Hirano and K. Yoshida, p. 157 in 'Superplasticity in Advanced Materials' (eds S. Hori, M. Tokizane and N. Furushiro), The Japan Society of Research on Superplasticity, Osaka, Japan. 1991.
27. P. Lukac, p. 109 in 'Superplasticity and Superplastic Forming' (eds C. H. Hamilton and N. E. Paton), TMS, Warrendale, PA. 1988.
28. R. Horiuchi, A. B. El-Sebai and M. Otsuka, *Scr. Metall.*, 1973, **7**, 1101.
29. Z. Cui, W. Zhong and Q. Wei, *Scr. Metall. Mater.*, 1994, **30**, 123.
30. E. W. Lee and T. R. McNelley, p. 223 in 'Superplasticity in Aerospace II' (eds T. R. McNelly and H. C. Heikkenen), TMS, Warrendale, PA. 1990.
31. K. Higashi, T. Okada, T. Mukai and S. Tanimura, *Scr. Metall. Mater.*, 1992, **26**, 761.
32. Y. W. Kim and L. R. Bidwell, *Scr. Metall.*, 1982, **16**, 799.
33. K. Higashi, S. Itsumi, M. Hoshikawa, Y. Matsumara, T. Ito, S. Tanimura, and H. Yoshida, p. 575 in 'International Conference on Superplasticity in Advanced Materials (ICSAM-91)' (eds S. Hori, M. Tokizane and N. Furushiro), The Japan Society for Research on Superplasticity, Osaka, Japan. 1991.
34. O. D. Sherby and J. Wadsworth, *Mater. Sci. Forum*, 1996, 233–34, 125.
35. S. X. McFadden, R. S. Mishra, R. Z. Valiev, A. P. Zhilyaev and A. K. Mukherjee, Nature, 1999, 398, 684.
36. P. B. Berbon, N. K. Tsenev, R. S. Valiev, M. Furukawa, Z. Horita, M. Nemoto and T. G. Langdon, p. 127 in 'Superplasticity and Superplastic Forming' (eds A. K. Ghosh and T. R. Biehler), TMS, Warrendale, PA. 1998.
37. T. G. Nieh, L. M. Hsiung, J. Wadsworth and R. Kaibyshev, *Acta Mater.*, 1998, **46**, 2789.
38. T. R. McNelley and S. J. Hales, p. 207 in 'Superplasticity in Aerospace II' (eds T. R. McNelly and H. C. Heikkenen), TMS, Warrendale, PA. 1990.
39. R. Crooks, S. J. Hales and T. R. McNelley, p. 389 in 'Superplasticity and Superplastic Forming' (eds C. H. Hamilton and N. E. Paton), TMS, Warrendale, PA. 1988.
40. T. R. McNelly and P.N. Kalu, p. 413 in 'International Conference on Superplasticity in Advanced Materials (ICSAM-91)' (eds S. Hori, M. Tokizane and N. Furushiro), The Japan Society for Research on Superplasticity, Osaka, Japan, Osaka, Japan. 1991.
41. E. W. Lee, T. R. McNelley and A. F. Stengel, *Metall. Trans.*, 1986, **17A**, 1043.
42. K. Matsuki, H. Morita, M. Yamada and Y. Murakami, *Metal Sci.*, 1977, **11**, 156.
43. B. Geary, J. Pilling and N. Ridley, p. 127 in 'Superplasticity in Aerospace-Aluminium' (eds R. Pearce and L. Kelly), Ashford Press, Curdridge, Southampton, Hampshire, Cranfield, UK. 1985.
44. P. G. Partridge, A. W. Bowen and D. S. McDarmaid, p. 215 in 'Superplasticity in Aerospace Aluminium' (eds R. Pearce and L. Kelly), Ashford Press, Culridge, UK. 1985.
45. H. Huang, Q. Wu and J. Hua, p. 465 in 'Superplasticity and Superplastic Forming' (eds C.H. Hamilton and N.E. Paton), TMS, Warrendale, PA. 1988.
46. V. S. Levchenko, O. V. Solovjeva, V. K. Portnoy and Y. V. Shevnuk, *Mater. Sci. Forum*, 1994, 170–2, 261.
47. R. E. Goforth, M. N. Srinivasan and N. Chandra, p. 145 in 'Superplasticity in Advanced Materials' (eds S. Hori, M. Tokizane and N. Furushiro), The Japan Society of Research on Superplasticity, Osaka, Japan. 1991.
48. H.-S. Lee and A. K. Mukherjee, p. 121 in 'Superplasticity in Aerospace II' (eds T. R. McNelly and H. C. Heikkenen), TMS, Warrendale, PA. 1990.
49. R. E. Goforth, M. Srinivasan, N. Chandra and L. Douskos, p. 285 in 'Superplasticity in Aerospace II' (eds T. R. McNelly and H. C. Heikkenen), TMS, Warrendale, PA. 1990.
50. C. C. Bampton, B. A. Cheney, A. Cho, A. K. Ghosh, and C. Gandhi, p. 247 in 'Superplasticity in Aerospace' (eds H. C. Heikkenen and T. R. McNelly), TMS, Warrendale, PA. 1988.
51. G. T. Kridli, A. S. El-Gizawy and R. Lederich, *Materials Science and Engineering*, 1998, **A244**, 224.
52. M. Kawazoe, T. Shibata, T. Mukai and K. Higashi, *Scr. Mater.*, 1997, **36**, 699.
53. M. T. Smith, J. S. Vetrano, E. A. Nyberg and D. R. Herling, p. 99 in 'Superplasticity and Superplastic Forming' (eds A. K. Ghosh and T. R. Biehler), TMS, Warrendale, PA. 1998.
54. C. A. Lavender, J. S. Vetrano, M. T. Smith, S. M. Bruemmer and C. H. Hamilton, *Mater. Sci. Forum*, 1994, 170–2, 279.
55. J. S. Vetrano, C. A. Lavender, C. H. Hamilton, M. T. Smith and S. M. Bruemmer, *Scr. Metall. Mater.*, 1994, **30**, 565.
56. H. Iwasaki, K. Higashi, S. Tanimura, T. Komatubara and S. Hayami, p. 447 in 'International Conference on Superplasticity in Advanced Materials (ICSAM-91)' (eds S. Hori, M. Tokizane and N. Furushiro), The Japan Society for Research on Superplasticity, Osaka, Japan. 1991.
57. H. Imamura and N. Ridley, p. 453 in 'International Conference on Superplasticity in Advanced Materials (ICSAM-91)' (eds S. Hori, M. Tokizane and N. Furushiro), The Japan Society for Research on Superplasticity, Osaka, Japan. 1991.
58. R. Verma, A. K. Ghosh, S. Kim and C. Kim, *Mater. Sci. Eng.*, 1995, **A191**, 143.
59. A. Tavassoli, S. E. Razavi and N. M. Fallah, *Metall. Trans.*, 1975, **6A**, 591.
60. T. G. Nieh, R. Kaibyshev, F. Musin and D. R. Lesuer, p. 137 in 'Superplasticity and Superplastic Forming–1998' (eds A. K. Ghosh and T. R. Bieler), The Minerals, Metals, and Materials Society, Warrendale, PA. 1998.
61. K. Xia and L.-X. Ma, p. 160 in 'Superplasticity in Aerospace Aluminium' (eds R. Pearce and L. Kelly), Ashford Press, Culridge, UK. 1985.
62. J. Xinggang, W. Qingling, C. Jianzhong and Longxiang, p. 263 in 'Superplasticity in Aerospace II' (eds T. R. McNelly and H. C. Heikkenen), TMS, Warrendale, PA. 1990.
63. H. N. Azari, G. S. Murty and G.S. Upadhyaya, *Metall. Mater. Trans. A*, 1994, **25**, 2153.

64. S. Matsuda, Y. Okubo and H. Yoshida, p. 441 in 'Superplasticity in Advanced Materials' (eds S. Hori, M. Tokizane and N. Furushiro), The Japan Society of Research on Superplasticity, Osaka, Japan. 1991.

65. S. A. McCoy, J. White and N. Ridley, p. 435 in 'Superplasticity in Advanced Materials' (eds S. Hori, M. Tokizane and N. Furushiro), The Japan Society of Research on Superplasticity, Osaka, Japan. 1991.

66. D. H. Shin, C.S. Lee and W.-J. Kim, *Acta Mater.*, 1997, **45**, 5195.

67. S. Zhou, L. Wang and C. Liu, p. 39 in 'Superplasticity and Superplastic Forming' (eds C. H. Hamilton and N. E. Paton), TMS, Warrendale, PA. 1988.

68. W. Cao, X.-P. Ling, A. F. Sprecher and H. Conrad, p. 269 in 'Superplasticity in Aerospace II' (eds T. R. McNelly and H. C. Heikkenen), TMS, Warrendale, PA. 1990.

69. J. M. Story, J. I. Petit, D. J. Lege and B. L. Hazard, p. 67 in 'Superplasticity in Aerospace Aluminium' (eds R. Pearce and L. Kelly), Ashford Press, Culridge, UK. 1985.

70. P. Comely, p. 353 in 'Superplasticity in Aerospace Aluminium' (eds R. Pearce and L. Kelly), Ashford Press, Culridge, UK. 1985.

71. T. Eto, M. Hirano, M. Hino and Y. Mayagi, p. 199 in 'Superplasticity in Aerospace' (eds H. C. Heikkenen and T. R. McNelly), TMS, Warrendale, PA. 1988.

72. C. H. Hamilton, C. C. Bampton and N. E. Paton, p. 173 in 'Superplastic Forming of Structural Alloys' (eds N. E. Paton and C. H. Hamilton), TMS-AIME, Warrendale, PA, Warrendale, PA. 1982.

73. A. Varloteaux and M. Suery, p. 55 in 'Superplasticity in Aerospace Aluminium' (eds R. Pearce and L. Kelly), Ashford Press, Culridge, UK. 1985.

74. K. Matsuki, H. Matsumoto, M. Tokizawa and Y. Murakami, p. 551 in 'International Conference on Superplasticity in Advanced Materials (ICSAM-91)' (eds S. Hori, M. Tokizane and N. Furushiro), The Japan Society for Research on Superplasticity, Osaka, Japan. 1991.

75. K. Matsuki, M. Tokizawa and G. Staniek, p. 395 in 'Superplastic and Superplastic Forming' (eds C. H. Hamilton and N. E. Paton), The Minerals, Metals & Materials Society, Warrendale, PA. 1988.

76. H. P. Pu and J. C. Huang, *Scr. Metall. Mater.*, 1993, **28**, 1125.

77. S. S. Bhattacharya and K.A. Padmanabhan, p. 459 in 'Superplasticity in Advanced Materials' (eds S. Hori, M. Tokizane and N. Furushiro), The Japan Society of Research on Superplasticity, Osaka, Japan. 1991.

78. H. P. Pu, F. C. Liu and J.C. Huang, *Metall. Mater. Trans.*, 1995, **26A**, 1153.

79. Y. Ma and T. G. Langdon, p. 173 in 'Superplasticity and Superplastic Forming' (eds C. H. Hamilton and N. E. Paton), TMS, Warrendale, PA. 1988.

80. K. Higashi, T. G. Nieh and J. Wadsworth, *Mater. Sci. Eng.*, 1994, **188A**, 167.

81. T. R. Bieler and A. K. Mukherjee, *Mater. Sci Eng.*, 1990, **A128**, 171.

82. K. Higashi, T. Okada, T. Mukai and S. Tanimura, *Mater. Sci. Eng.*, 1992, **159A**, L1.

83. Z. Tiecheng, Z. Zhimin and Z. Yanhui, p. 383 in 'Superplasticity and Superplastic Forming' (eds C. H. Hamilton and N. E. Paton), TMS, Warrendale, PA. 1988.

84. P. Fernandez, p. 675 in 'International Conference on Superplasticity in Advanced Materials (ICSAM-91)' (eds S. Hori, M. Tokizane and N. Furushiro), The Japan Society for Research on Superplasticity, Osaka, Japan. 1991.

85. M. W. Mahoney and A. K. Ghosh, *Metall. Trans.*, 1987, **18A**, 653.

86. C. H. Caceres and D. S. Wilkinson, *Acta Metall.*, 1984, **32**, 423.

87. A. H. Chokshi and T. G. Langdon, *Acta Metall. Mater.*, 1990, **38**, 867.

88. K. Higashi and N. Ridley, p. 447 in 'Superplasticity and Superplastic Forming' (eds C.H. Hamilton and N.E. Paton), The Minerals, Metals, and Materials Society, Warrendale, PA, Blaine, Washington. 1988.

89. K. Higashi, T. Ohnishi and Y. Nakatani, *Scr. Metall.*, 1985, **19**, 821.

90. M. Miki and Y. Ogino, p. 527 in 'Superplasticity in Advanced Materials' (eds S. Hori, M. Tokizane and N. Furushiro), The Japan Society of Research on Superplasticity, Osaka, Japan. 1991.

91. J. Tao, Z. Min and C. Puquan, p. 63 in 'Superplasticity and Superplastic Forming' (eds C. H. Hamilton and N. E. Paton), TMS, Warrendale, PA. 1988.

92. S. Shenggui, S. Huanxiang and S. Shengzhe, p. 441 in 'Superplasticity and Superplastic Forming' (eds C.H. Hamilton and N.E. Paton), TMS, Warrendale, PA. 1988.

93. D. W. Livesey and N. Ridley, *Metall Trans.*, 1978, **9A**, 519.

94. D. W. Livesey and N. Ridley, *Metall. Trans. A*, 1982, **13**, 1619.

95. S. Sagat and D. M. R. Taplin, *Acta Metall.*, 1976, **24**, 307.

96. C. W. Humphries and N. Ridley, *J. Mater. Sci.*, 1978, **13**, 2477.

97. C. Liu, D. Chang, C. Liang and J. Liu, p. 81 in 'Superplasticity in Advanced Materials' (eds S. Hori, M. Tokizane and N. Furushiro), The Japan Society of Research on Superplasticity, Osaka, Japan. 1991.

98. C. W. Humphries and N. Ridley, *J. Mater. Sci.*, 1974, **9**, 1429.

99. W. B. Morrison, *Trans. ASM*, 1968, **61**, 423.

100. H. W. Schadler, *Trans. AIME*, 1968, 1281.

101. J. Wadsworth, J. H. Lin and O. D. Sherby, *Metals Technol.*, 1981, **8**, 190.

102. G. R. Yoder and V. Weiss, *Metall. Trans.*, 1972, **3A**, 675.

103. J. P. Wittenauer, P. Schepp and B. Walser, p. 507 in 'Superplasticity and Superplastic Forming' (eds C. H. Hamilton and N.E. Paton), The Minerals, Metals & Materials Science Society, Warrendale, PA, Blaine, WA. 1988.

104. D.-W. Kum, H. Kang and S. H. Hong, p. 503 in 'Superplasticity in Advanced Materials' (eds S. Hori, M. Tokizane and N. Furushiro), The Japan Society of Research on Superplasticity, Osaka, Japan. 1991.

105. B. Walser and O. D. Sherby, *Metall. Trans.*, 1979, **10A**, 1461.

106. H. Fukuyo, H. C. Tsai, T. Oyama and O. D. Sherby, *ISIJ International*, 1991, **31**, 76.

107. O. A. Ruano, L. E. Eiselstein and O. D. Sherby, *Metall. Trans.*, 1982, **13A**, 1785.

108. D. W. Kum, G. Frommeyer, N. J. Grant and O. D. Sherby, *Metall. Trans.*, 1987, **18A**, 1703.
109. K. Mineura and K. Tanaka, *J. Mater. Sci.*, 1989, **24**, 2967.
110. D. R. Lesuer, T. G. Nieh, C. K. Syn and E. M. Taleff, *Mater. Sci. Forum*, 1997, **243–5**, 469.
111. J. Pilling, Z. C. Wang and N. Ridley, p. 297 in 'Superplasticity and Superplastic Forming' (eds A. K. Ghosh and T. R. Biehler), TMS, Warrendale, PA. 1998.
112. Y. Maehara and Y. Ohmori, *Metall. Trans.*, 1987, **18A**, 663.
113. K. Tsuzaki, H. Matsuyama, M. Nagao and T. Maki, *J. JIM*, 1990, **54**, 878.
114. Y. Maehara, *Metall. Trans.*, 1991, **22A**, 1083.
115. C. I. Smith, B. Norgate and N. Ridley, *Metal Sci.*, 1976, **10**, 182.
116. H. Takuda, S. Kikuchi, H. Fujimoto and N. Hatta, p. 497 in 'Superplasticity in Advanced Materials' (eds S. Hori, M. Tokizane and N. Furushiro), The Japan Society of Research on Superplasticity, Osaka, Japan. 1991.
117. A. Uoya, T. Shibata, K. Higashi, A. Inoue and T. Masumoto, *J. Mater. Res.*, 1996, **11**, 2731.
118. J. K. Solberg, J. Torklep, O. Bauger and H. Gjestland, *Mater. Sci. Eng.*, 1991, **A134**, 1201.
119. M. Mabuchi, H. Iwasaki, K. Yanase and K. Higashi, *Scr. Mater.*, 1997, **36**, 681.
120. M. Mabuchi, K. Ameyama, H. Iwasaki and K. Higashi, *Acta Mater.*, 1999, **47**, 2047.
121. M. Mabuchi, T. Asahina, H. Iwasaki and K. Higashi, *Mater. Sci. Technol.*, 1997, **13**, 825.
122. D. Lee, *Acta Metall.*, 1969, **17**, 1057.
123. E. M. Taleff and O. D. Sherby, *J. Mater. Res.*, 1994, **9**, 1392.
124. E. M. Taleff, O. A. Ruano, J. Wolfenstine and O.D. Sherby, *J. Mater. Res.*, 1992, **7**, 2131.
125. P. Metenier, G. Gonzalez-Doncel, O.R. Ruano, J. Wolfenstine and O. D. Sherby, *Mater. Sci. Eng.*, 1990, **A125**, 195.
126. R. Kaibyshev and A. Galiyev, *Mater. Sci. Forum*, 1997, 243–5, 131.
127. H. Watanabe, T. Mukai and K. Higashi, p. 179 in 'Superplasticity and Superplastic Forming' (eds A. K. Ghosh and T. R. Biehler), TMS, Warrendale, PA. 1998.
128. A. Karim and W. A. Backofen, *Mater. Sci. Eng.*, 1968–69, **3**, 306.
129. H. Watanabe, T. Mukai, M. Kohzu, S. Tanabe and K. Higashi, *Mater. Trans. JIM*, 1999, **40**, 806.
130. N. G. Zaripov and R. O. Kaibyshev, p. 91 in 'Superplasticity and Superplastic Forming' (eds C. H. Hamilton and N. E. Paton), TMS, Warrendale, PA. 1988.
131. J.-P. A. Immarigeon and P. H. Floyd, *Metall. Trans.*, 1981, **12A**, 1177.
132. S. Kikuchi, S. Ando, S. Futami, T. Kitamura and M. Koiwa, *J. Mater. Sci.*, 1990, **25**, 4712.
133. S. Kikuchi, p. 485 in 'International Conference on Superplasticity in Advanced Materials (ICSAM-91)' (eds S. Hori, M. Tokizane and N. Furushiro), The Japan Society for Research on Superplasticity, Osaka, Japan. 1991.
134. R. G. Menzies, J. W. Edington and G. J. Davies, *Metal Sci.*, 1981, **15**, 210.
135. V. A. Valitov, G. A. Salishchev and S. K. Mukhtarov, *Mater. Sci. Forum*, 1997, **243–5**, 557.
136. J. K. Gregory, J. C. Gibeling and W.D. Nix, *Metall. Trans.*, 1985, **16A**, 777.
137. M. W. Mahoney and R. Crooks, p. 73 in 'Superplasticity and Superplastic Forming' (eds C.H. Hamilton and N.E. Paton), The Minerals, Metals, and Materials Society, Warrendale, PA, Blaine, Washington. 1988.
138. M. W. Mahoney and R. Crooks, p. 331 in 'Superplasticity in Aerospace' (eds C. Heikkenen and T. R. McNelly), TMS-AIME, Warrendale, PA. 1988.
139. L. Ceschini, G. P. Cammarota, G. L. Garagnani, F. Persiani and A. Afrikantov, p. 351 in 'Superplasticity in Advanced Materials - ICSAM-94' (eds T. G. Langdon), Trans Tech Publications Ltd, Switzerland, Moscow, Russia. 1994.
140. G. D. Smith and H. L. Flower, p. in press in 'Superplasticity and Superplastic Forming 1995' (eds A. K. Ghosh and T. Bieler), The Minerals, Metals & Materials Science Society, Warrendale, PA. 1995.
141. I. Kuboki, Y. Motohashi and M. Imabayashi, p. 413 in 'Superplasticity and Superplastic Forming' (eds C. H. Hamilton and N. E. Paton), TMS-AIME, Warrendale, PA. 1988.
142. H. W. Hayden, R. C. Gibson, H. P. Merrick and J. H. Brophy, *Trans. ASM*, 1967, **60**, 3.
143. M. A. Clark and T. H. Alden, *Acta Metall.*, 1973, **21**, 1195.
144. D. W. Livesey and N. Ridley, *J. Mater. Sci.*, 1978, **13**, 825.
145. C. W. Humphries and N. Ridley, *J. Mater. Sci.*, 1977, **12**, 851.
146. M. M. I. Ahmed and T. G. Langdon, *Metall. Trans.*, 1977, **8A**, 1832.
147. R. J. Prematta, P. S. Venkatesan and A. Pense, *Metall. Trans.*, 1976, **7A**, 1235.
148. J. Ma, R. Kent and C. Hammond, *J. Mater. Sci.*, 1986, **21**, 475.
149. D. Lee and W. A. Backofen, *Trans. AIME*, 1967, **239**, 1034.
150. A. Wisbey, B. Geary, D. P. Davies, and C. M. Ward-Close, *Mater. Sci. Forum*, 1994, 170–2, 293.
151. A. Wisbey and P. G. Partrige, p. 465 in 'International Conference on Superplasticity in Advanced Materials (ICSAM-91)' (eds S. Hori, M. Tokizane and N. Furushiro), The Japan Society for Research on Superplasticity, Osaka, Japan. 1991.
152. G. A. Salishchev, O. R. Valiakhmetov, V. A. Valitov and S. K. Mukhtarov, *Mater. Sci. Forum*, 1994, 170–2, 121.
153. J. V. Sirina, I. L. Fedotov, V. K. Portnoy, A. A. Ilyin and A. M. Mamonov, *Mater. Sci. Forum*, 1994, 170–2, 299.
154. M. T. Cope, D. R. Evetts and N. Ridley, *J. Mater. Sci.*, 1986, **21**, 4003.
155. A. K. Ghosh and C. H. Hamilton, *Metall. Trans.*, 1979, **10A**, 699.
156. C. H. Hamilton and A. K. Ghosh, *Metall. Trans.*, 1980, **11A**, 1494.
157. N. E. Paton and C. H. Hamilton, *Metall. Trans.*, 1979, **10A**, 241.
158. J. A. Wert and N. E. Paton, *Metall. Trans.*, 1983, **14A**, 2535.

159. J. R. Leader, D. F. Neal and C. Hammond, *Metall. Trans.*, 1986, **17A**, 93.
160. G. A. Salishchev, R. M. Galeyev, S. P. Malisheva and O. R. Valiakhmetov, *Mater. Sci. Forum*, 1997, 243–5, 585.
161. Q. Liu, W. Yang and G. Chen, *Acta Metall. Mater.*, 1995, **43**, 3571.
162. G. Frommeyer, H. Homann and W. Herzog, *Mater. Sci. Forum*, 1994, 170–2, 483.
163. G. A. Salishchev, R. M. Galeyev and R. M. Imaev, p. 163 in 'Superplasticity in Advanced Materials' (eds S. Hori, M. Tokizane and N. Furushiro), The Japan Society of Research on Superplasticity, Osaka, Japan. 1991.
164. P. Y. Qin, L. Weimin and S. Zuozhou, p. 263 in 'Superplasticity and Superplastic Forming' (eds C. H. Hamilton and N. E. Paton), TMS, Warrendale, PA. 1988.
165. G. M. Ludtka, R. E. Oakes, R. L. Bridges and J. L. Griffith, *Metall. Trans. A*, 1993, **24A**, 369.
166. G. M. Ludtka, R. E. Oakes, R. L. Bridges and J. L. Griffith, *Metall. Trans. A*, 1993, **24A**, 379.
167. T. H. Alden and H. W. Schadler, *TMS-AIME*, 1968, **242**, 825.
168. H. Naziri, R. Pearce, M. Brown and K. F. Hale, *Acta Metall.*, 1975, **23**, 489.
169. R. I. Todd, p. 13 in 'Superplasticity and Superplastic Forming' (eds A. K. Ghosh and T. R. Biehler), TMS, Warrendale, PA. 1998.
170. O. N. Senkov and M. M. Myshlyaev, *Acta Metall.*, 1986, **34**, 97.
171. J. S. Kim, J. Kaneko and M. Sugamata, p. 391 in 'International Conference on Superplasticity in Advanced Materials (ICSAM-91)' (eds S. Hori, M. Tokizane and N. Furushiro), The Japan Society for Research on Superplasticity, Osaka, Japan. 1991.
172. N. Furushiro and T. G. Langdon, p. 197 in 'Superplasticity and Superplastic Forming' (eds C. H. Hamilton and N. E. Paton), TMS, Warrendale, PA. 1988.
173. A. Arieli, A. K. S. Yu and A. K. Mukherjee, *Metall. Trans.*, 1980, **11A**, 181.
174. P. Shariat and T. G. Langdon, p. 227 in 'Superplasticity and Superplastic Forming' (eds C. H. Hamilton and N. E. Paton), TMS, Warrendale, PA. 1988.
175. H. Ishikawa, D.G. Bhat, F.A. Mohamed, and T.G. Langdon, *Metall. Trans.*, 1977, **8A**, 523.
176. A. H. Chokshi and T. G. Langdon, *Acta Metall. Mater.*, 1989, **37**, 715.
177. M. L. Vaidya, K. L. Murty and J. E. Dorn, *Acta Metall.*, 1973, **21**, 1615.
178. F. A. Mohamed, S. Shei and T. G. Langdon, *Acta Metall.*, 1975, **23**, 1443.
179. F. A. Mohamed, M. J. Ahmed and T. G. Langdon, *Metall. Trans*, 1977, **8A**, 933.
180. R. Kossowsky and J. H. Bechtold, Trans. *TMS-AIME*, 1968, **242**, 716.
181. W. A. Backofen, I. R. Turner and D. H. Avery, *Trans. ASM*, 1964, **57**, 980.
182. D. L. Holt, Trans. *TMS-AIME*, 1968, **242**, 25.
183. G. S. Murty, p. 45 in 'Superplasticity in Advanced Materials' (eds S. Hori, M. Tokizane and N. Furushiro), The Japan Society of Research on Superplasticity, Osaka, Japan. 1991.
184. C. Liu, P. Si, D. Chang, and J. Wong, p. 75 in 'Superplasticity in Advanced Materials' (eds S. Hori, M. Tokizane and N. Furushiro), The Japan Society of Research on Superplasticity, Osaka, Japan. 1991.
185. G. T. Villasenor and J. Negrete, p. 51 in 'Superplasticity and Superplastic Forming' (eds C. H. Hamilton and N. E. Paton), TMS, Warrendale, PA. 1988.
186. N. Dyulgerov, A. Istatkov, N. Mitev, and I. Spirov, p. 419 in 'Superplasticity and Superplastic Forming' (eds C. H. Hamilton and N. E. Paton), TMS, Warrendale, PA. 1988.
187. J. P. Chu, I. M. Liu, J. H. Wu, W. Kai, J. Y. Wang, and K. Inoue, *Mater. Sci. Eng.*, 1998, **A258**, 236.
188. A. Shan, M. Chen, D. Lin, and D. Li, *Mater. Sci. Forum*, 1994, **170–2**, 489.
189. D. Lin, A. Shan and D. Li, *Scr. Metall. Mater.*, 1994, **31**, 1455.
190. D. Li and D. Lin, *Scr. Metall. Mater.*, 1997, **36**, 1289.
191. D. Li, A. Shan, Y. Liu, and D. Lin, *Scr. Metall. Mater.*, 1995, **33**, 681.
192. W.-Y. Kim, S. Hanada and T. Sakai, p. 279 in 'Towards Innovation in Superplasticity I' (eds T. Sakuma, T. Aizawa and K. Higashi), Trans Tech Publications, Switzerland. 1997.
193. X. H. Du, J. T. Guo and B. D. Zhou, *Scripta Mater.*, 2001, **45**, 69.
194. W. L. Zhou, J. T. Guo, R. S. Chen, G. S. Li and J. Y. Zhou, *Mater. Let.*, 2001, **47**, 30.
195. J. Mukhopadhyay, G. Kaschner and A. K. Mukherjee, *Scr. Metall.*, 1990, **24**, 857.
196. A. Choudhury, A. K. Mukherjee and V. K. Sikka, *J. Mater. Sci.*, 1990, **25**, 3142.
197. M. S. Kim, S. Hanada, S. Wantanabe and O. Izumi, *Mater. Trans. JIM*, 1989, **30**, 77.
198. T. G. Nieh and W. C. Oliver, *Scr. Metall.*, 1989, **23**, 851.
199. S. L. Stoner and A. K. Mukherjee, p. 323 in 'International Conference on Superplasticity in Advanced Materials (ICSAM-91)' (eds S. Hori, M. Tokizane and N. Furushiro), The Japan Society for Research on Superplasticity, Osaka, Japan. 1991.
200. S. L. Stoner and A. K. Mukherjee, *Mater. Sci. Eng.*, 1992, **A153**, 465.
201. K. Ameyama, A. Miyazaki and M. Tokizane, p. 317 in 'Superplasticity in Advanced Materials' (eds S. Hori, M. Tokizane and N. Furushiro), The Japan Society of Research on Superplasticity, Osaka, Japan. 1991.
202. S. C. Cheng, J. Wolfenstine and O. D. Sherby, *Metall. Trans.*, 1992, **23A**, 1509.
203. D. Vanderschueren, M. Nobuki and M. Nakamura, *Scr. Metall. Mater.*, 1993, **28**, 605.
204. C. M. Lombard, A. K. Ghosh and S. L. Semiatin, p. 267 in 'Superplasticity and Superplastic Forming' (eds A. K. Ghosh and T. R. Biehler), TMS, Warrendale, PA. 1998.
205. R. M. Imaev, M. Shagiev, G. Salishchev, V. M. Imayev and V. Valitov, *Scr. Mater.*, 1996, **34**, 985.
206. T. G. Nieh, J. N. Wang, L. M. Hsiung, J. Wadsworth and V. Sikka, *Scr. Mater.*, 1997, **37**, 733.
207. W. B. Lee, H. S. Yang, Y.-W. Kim and A. K. Mukherjee, *Scr. Metall. Mater.*, 1993, **29**, 1403.
208. G. E. Fuchs, p. 277 in 'Superplasticity and Superplastic Forming' (eds A. K. Ghosh and T. R. Biehler), TMS, Warrendale, PA. 1998.

209. T. Maeda, M. Okada and Y. Shida, p. 311 in 'International Conference on Superplasticity in Advanced Materials (ICSAM-91)' (eds S. Hori, M. Tokizane and N. Furushiro), The Japan Society for Research on Superplasticity, Osaka, Japan. 1991.
210. R. M. Imayev and V. M. Imayev, *Scr. Metall. Mater.*, 1991, **25**, 2041.
211. R. Imayev, V. Imayev and G. Salishchev, *Scr. Metall. Mater.*, 1993, **29**, 713.
212. R. M. Imayev, V. M. Imayev and G. A. Salishchev, *J. Mater. Sci.*, 1992, **27**, 4465.
213. R. M. Imayev, O. A. Kaibyshev and G. A. Salishchev, *Acta Metall. Mater.*, 1992, **40**, 581.
214. H. S. Yang, P. Jin and A. K. Mukherjee, *Mater. Sci. Eng.*, 1992, **A153**, 457.
215. A. Dutta and D. Banerjee, *Scr. Metall. Mater.*, 1990, **24**, 1319.
216. D. Jobart and J. J. Blandin, *J. Mater. Sci.*, 1996, **31**, 881.
217. A. K. Ghosh and C.-H. Cheng, p. 299 in 'International Conference on Superplasticity in Advanced Materials (ICSAM-91)' (eds S. Hori, M. Tokizane and N. Furushiro), The Japan Society for Research on Superplasticity, Osaka, Japan. 1991.
218. H. C. Fu, J. C. Huang, T. D. Wang and C.C. Bampton, *Acta Mater.*, 1998, **46**, 465.
219. H. S. Yang, P. Jin, E. Dalder and A. K. Mukherjee, *Scr. Metall. Mater.*, 1991, **25**, 1223.
220. T. Imai, G. L'Esperance, B. D. Hong, and S. Kojima, *Scr. Metall. Mater.*, 1995, **33**, 1333.
221. R. Kaibyshev, V. Kazykhanov, V. Astanin and E. Evangelista, *Mater. Sci. Forum*, 1994, 170–2, 531.
222. J. Pilling, *Scripta Metall.*, 1989, **23**, 1375.
223. W. Zheng and B. Zhang, *Mater. Sci. Lett.*, 1994, **13**, 1806.
224. H. Y. Kim and S. H. Hong, *Scr. Metall. Mater.*, 1994, **30**, 297.
225. G. H. Zahid, R. I. Todd and P. B. Prangnell, p. 227 in 'Superplasticity and Superplastic Forming' (eds A. K. Ghosh and T. R. Biehler), TMS, Warrendale, PA. 1998.
226. T. G. Nieh, C. A. Henshall and J. Wadsworth, *Scr. Metall.*, 1984, **18**, 1405.
227. G. L'Esperance and T. Imai, p. 379 in 'International Conference on Superplasticity in Advanced Materials (ICSAM-91)' (eds S. Hori, M. Tokizane and N. Furushiro), The Japan Society for Research on Superplasticity, Osaka, Japan. 1991.
228. M. Mabuchi, T. Imai and K. Higashi, *J. Mater. Sci.*, 1993, **28**, 6582.
229. M. Kon, M. Sugamata and J. Kaneko, *Mater. Sci. Forum*, 1994, 170–2, 513.
230. T. Imai, G. L'Esperance and B. Hong, *Scr. Metall. Mater.*, 1994, **31**, 321.
231. X. Huang, Q. Liu, C. Yao and M. Yao, *J. Mater. Sci. Lett.*, 1991, **10**, 964.
232. H. Iwasaki, M. Taceuchi, T. Mori, M. Mabuchi and K. Higashi, *Scr. Metall. Mater.*, 1994, **31**, 255.
233. M. Mabuchi, K. Higashi, Y. Okada, S. Tanimura, T. Imai and K. Kubo, *Scr. Metall. Mater.*, 1991, **25**, 2003.
234. I. Tochigi, T. Imai and K. Ai, *Scr. Metall. Mater.*, 1995, **32**, 1801.
235. T. Imai, M. Mabuchi and Y. Tozawa, p. 373 in 'International Conference on Superplasticity in Advanced Materials (ICSAM-91)' (eds S. Hori, M. Tokizane and N. Furushiro), The Japan Society for Research on Superplasticity, Osaka, Japan. 1991.
236. T. Hikosaka, T. Imai, T. G. Nieh and J. Wadsworth, *Scr. Metall. Mater.*, 1994, **31**, 1181.
237. P. Virro and J. Pilling, p. 47 in 'Superplasticity in Aerospace II' (eds T. R. McNelly and H. C. Heikkenen), TMS, Warrendale, PA. 1990.
238. T. Mukai, H. Iwasaki, K. Higashi and T. G. Nieh, *Mater. Sci. Technol.*, 1998, **14**, 32.
239. T. Mukai, T. G. Nieh and K. Higashi, p. 313 in 'Superplasticity and Superplastic Forming' (eds A. K. Ghosh and T. R. Biehler), TMS, Warrendale, PA. 1998.
240. T. G. Nieh and J. Wadsworth, *Scr. Metall. Mater.*, 1995, **32**, 1133.
241. T. G. Nieh, A. J. Schwartz and J. Wadsworth, *Mater. Sci. Eng.*, 1996, **A208**, 30.
242. S.-W. Lim, T. Imai, Y. Nishida and I. Chou, *Scr. Metall. Mater.*, 1995, **32**, 1713.
243. M. Mabuchi, K. Kubota and K. Higashi, *Scr. Metall. Mater.*, 1995, **33**, 331.
244. M. Kobayashi, S. Ochiai, K. Funami, C. Ouchi and S. Suzuki, *Mater. Sci. Forum*, 1994, 170–2, 549.
245. K. Kitazono and E. Sato, *Acta Mater.*, 1998, **46**, 207.
246. K. Kitazono, E. Sato and K. Kuribayashi, *Acta Mater.*, 1999, **47**, 1653.
247. Y. C. Chen and G. S. Daehn, *Metall. Trans.*, 1991, **22A**, 1113.
248. K. Kitazono and E. Sato, *Acta Mater.*, 1998, **47**, 135.
249. S. M. Pickard and B. Derby, *Acta Metall. Mater.*, 1990, **38**, 2537.
250. S. M. Pickard and B. Derby, p. 447 in '9th Risø International Symposium on Metallurgy and Materials Science' (eds S.I. Andersen, H. Lilholt and O. B. Pedersen), Risø National Laboratory, Roskilde, Denmark. 1988.
251. S. M. Pickard and B. Derby, *Mater. Sci. Eng.*, 1991, **A135**, 213.
252. J. K. Lee, M. Taya and D. J. LLoyd, p. 29 in 'The Johannes Weertman Symposium' (eds R. J. Arsenault, D. Cole, T. Gross, G. Kostorz, P. K. Liaw, S. Parameswaran and H. Sizek), TMS, Warrendale PA. 1996.
253. S. H. Hong, O. D. Sherby, A. P. Divecha, S. D. Karmarkar and B. A. MacDonald, *J. Comp. Mater.*, 1988, **22**, 102.
254. G. Gonzalez-Doncel and O. D. Sherby, *Metall. Mater. Trans.*, 1996, **27A**, 2837.
255. M. Y. Wu and O. D. Sherby, *Scr. Metall.*, 1984, **18**, 773.
256. Y. C. Chen, G. S. Daehn and R. H. Wagoner, *Scr. Metall. Mater.*, 1990, **24**, 2157.
257. C. Y. Huang and G. S. Daehn, p. 135 in 'Superplasticity and Superplastic Forming 1995' (eds A. K. Ghosh and T. R. Bieler), TMS, Warrendale PA. 1996.
258. G. Gonzales-Doncel, S. D. Karmarkar, A. P. Divecha and O. D. Sherby, *Comp. Sci. Technol.*, 1989, **35**, 105.
259. J. C. LeFlour and R. Locicero, *Scripta Metall.*, 1987, **21**, 1071.
260. G. W. Greenwood and R. H. Johnson, *Proc. Roy. Soc.-London*, 1965, **283A**, 403.
261. M. Zamora and J. P. Poirier, *Mech. Mater.*, 1983, **2**, 193.

262. V. E. Hornbogen and G. Wassermann, *Z. Metallk.*, 1956, **47**, 427.
263. M. G. Lozinsky and I. S. Simeonova, *Acta Metall.*, 1961, **9**, 689.
264. M. de Jong and G. W. Rathenau, *Acta Metall.*, 1959, **7**, 246.
265. F. Clinard and O. D. Sherby, *Acta Metall.*, 1964, **12**, 911.
266. P. Zwigl and D. C. Dunand, *Metall. Mater. Trans.*, 1998, **29A**, 565.
267. M. de Jong and G. W. Rathenau, *Acta Metall.*, 1961, **9**, 714.
268. D. Oelschlagel and V. Weiss, *Trans. Am. Soc. Metals*, 1966, **59**, 143.
269. T. B. Massalski, S. K. Bhattacharyya and J. H. Perepezko, *Metall. Trans.*, 1978, **9A**, 53.
270. K. Kitazono, E. Sato and K. Kuribayashi, *Scripta Mater.*, 2001, **44**, 2695.
271. K. Kitazono, E. Sato and K. Kuribayashi, *Scr. Mater.*, 1999, **41**, 263.
272. R. S. Sundar, K. Kitazono, E. Sato and K. Kuribayashi, *Intermetallics*, 2001, **9**, 279.
273. P. Zwigl and D. C. Dunand, *Materials Science and Engineering*, 2001, **298**, 63.
274. C. Chaix and A. Lasalmonie, *Res Mech.*, 1981, **2**, 241.
275. N. Furushiro, H. Kuramoto, Y. Takayama and S. Hori, *Trans. ISIJ*, 1987, **27**, 725.
276. Y. Takayama, N. Furushiro and S. Hori, p. 753 in 'Titanium Science and Technology' (eds G. Lutjering, U. Zwicker and W. Bunk), Deutsche Gesellschaft fur Metallkunde, Munich. 1985.
277. R. Kot, G. Krause and V. Weiss, p. 597 in 'The Science, Technology and Applications of Titanium' (eds R. I. Jaffe and N.E. Promisel), Pergamon, Oxford. 1970.
278. C. Schuh and D. C. Dunand, *Scripta Mater.*, 1999, **40**, 1305.
279. D. C. Dunand and P. Zwigl, *Metall. Mater. Trans.*, 2001, **32A**, 841.
280. D. C. Dunand and C. M. Bedell, *Acta Metall. Mater.*, 1996, **44**, 1063.
281. C. Schuh and D. C. Dunand, *J. Mater. Res.*, 2001, **16**, 865.
282. C. Schuh and D. C. Dunand, *Int. J. Plastic.*, 2001, **17**, 317.
283. C. Schuh and D. C. Dunand, *Acta Mater.*, 2001, **49**, 199.
284. C. Schuh and D. C. Dunand, *Scripta Mater.*, 2001, **45**, 631.
285. C. Schuh and D. C. Dunand, *Acta Mater.*, 1998, **46**, 5663.
286. A. G. Young, K. M. Gardner and W. B. Rotsey, *Journal of Nuclear Materials*, 1960, **2**, 234.
287. M. Y. Wu, J. Wadsworth and O. D. Sherby, *Metall. Trans.*, 1987, **18A**, 451.
288. K. Kitazono, R. Hirasaka, E. Sato, K. Kuribayashi and T. Motegi, *Acta Mater.*, 2001, **49**, 473.
289. M. Y. Wu, J. Wadsworth and O. D. Sherby, *Scr. Metall.*, 1987, **21**, 1159.
290. P. Zwigl and D. C. Dunand, *Metall. Mater. Trans.*, 1998, **29A**, 2571.
291. K. Nuttall and D. P. McCooeye, p. 129 in 'Mechanical Behavior of Materials', The Society of Materials Science, Japan. 1974.

37   Metal-matrix composites

Metal-matrix composites are engineered materials typically comprising reinforceants of high elastic modulus and high strength in a matrix of a more ductile and tougher metal of lower elastic modulus and strength. The metal-matrix composite has a better combination of properties than can be achieved by either component material by itself. Normally, the objective of adding the reinforceant is to transfer the load from the matrix to the reinforceant so that the strength and elastic modulus of the composite are increased in proportion to the strength, modulus and volume fraction of the added material. Improvements in non-mechanical properties can also be achieved by adding a second phase, for example electrical conductivity, as compared with conventional alloying for strength.

The reinforcement can take one of several forms. The least expensive and most readily available on the market are the particulates. These can be round but are usually irregular particles of ceramics, of which SiC and Al_2O_3 are most frequently used. Composites reinforced by particulates are isotropic in properties but do not make best use of the reinforceant. Fine fibres are much more effective though usually more costly to use. Most effective in load transfer are long parallel continuous fibres. Somewhat less effective are short parallel fibres. Long fibres give high axial strength and stiffness, low coefficients of thermal expansion and, in appropriate matrices, high creep strength. These properties are very anisotropic and the composites can be weak and brittle in directions normal to that of the fibres. Where high two-dimensional properties are needed, cross-ply or interwoven fibres can be used. Short or long randomly oriented fibres provide lower efficiencies in strengthening (but are still more effective than particulates). These are most frequently available as SiC whiskers or as short random alumina ('Saffil') fibres or alumino-silicate matts.

Long continuous fibres include drawn metallic wires, mono-filaments deposited by CVD or multifilaments made by pyrolysis of polymers. The properties of some typical fibres are compared in Table 37.1. The relative prices are given as a very approximate guide. Because most composites are engineered materials, the matrix and the reinforcement are not in thermodynamic equilibrium and so at a high enough temperature, reaction will occur between them which can degrade the properties of the fibre in particular and reduce strength and more especially fatigue resistance. As many composites are manufactured by infiltration of the liquid metal matrix into the pack of fibres, reaction may occur at this stage. Some typical examples of interaction are listed in Table 37.2.

In order to obtain load transfer in service, it is essential to ensure that the reinforceant is fully wetted by the matrix during manufacture. In many cases, this requires that the fibre is coated with a thin interlayer which is compatible with both fibre and matrix. Commonly, this also has the advantage of preventing deleterious inter-diffusion between the two component materials. The data on most coatings are proprietary knowledge. However, it is well known that silicon carbide is used as an interlayer on boron and on carbon fibres to aid wetting by aluminium alloys.

The routes for manufacturing composites are still being developed but the most successful and lowest cost so far is by mixing particulates in molten metal and casting to either foundry ingot or as billets for extrusion or rolling. This is applied commercially to aluminium alloy composites. Another practicable route is co-spraying in which SiC particles are injected into an atomised stream of aluminium alloy and both are collected on a substrate as a co-deposited billet which can then be processed conventionally. This is a development of the Osprey process and can be applied more widely to aluminium and other alloys. Other routes involve the infiltration of molten metal into fibre pre-forms of the required shape often contained within a mould to ensure the correct final shape. This can be done by squeeze casting or by infiltrating semi-solid alloys to minimise interaction between the fibre and metal. Fibres can also be drawn through a melt to coat them and then be consolidated by hot-pressing.

To reduce interaction, solid state methods can be used, e.g. cold isostatic pressing (CIP) and sintering or hot isostatic pressing (HIP) of metal powders mixed with short fibres. Diffusion-bonding

37–1

Table 37.1   PROPERTIES OF REINFORCING FIBRES AT ROOM TEMPERATURE (BASED ON REFERENCES 9 and 16)

Fibre	Form	Preparation route	Diameter μm	Density g cm^{-3}	Weibull modulus	Fracture stress MPa	Elastic modulus GPa	Coefficient of thermal expansion K^{-1} × 10^6	Price relative to glass fibre
Tungsten	Wire	Drawn	10–500	19.2		2 500	400	5.0	1 000–50
Steel	Wire	Drawn	10–250	7.8		2 500	210	15.0	30–1
Boron	Cont. mono-filament	Chemical vapour depos.	150	2.6	7–19	3 500	400	8.0	250
SiC	Cont. mono-filament	Chemical vapour depos.	150	3.4	8	3 800	450	4.5	500
SiC	Cont. multi-filament	Polymer fibre pyrolysis	10–15	2.6		2 500	200	4.5	100
SiC	Whisker (random, short)	Polymer fibre pyrolysis	0.1–2.0	3.2	1.8	10 000	700	4.5	150
~Al$_2$O$_3$	Multi-filament	Oxide/salt fibre	15–25	3.9		1 500	380	7.0	100
~Al$_2$O$_3$	Random short fibres	pyrolysis	2–4	3.5	6	2 000	300	7.0	25
C (high modulus)	Cont. multi-filament	Polymer fibre pyrolysis	10	2.0	3–8	3 000	600	0	1 000
C (med. strength)	Cont. multi-filament	Polymer fibre pyrolysis	8	1.9	5–8	4 200	300	0	100
Alumino-silicate	Random short fibres	Polymer fibre pyrolysis	3	3.0		850	150	—	20
Alumino-silicate Glass (27% SiO$_2$)	Random short fibres	Drawn from melt	3	2.7		1 750	105	—	1
Glass (47% SiO$_2$) S-Glass	Cont. multi-filament	Drawn from melt	3–20	2.5		4 000	90	3.0	2.5

Table 37.2 TYPICAL INTERACTIONS IN SOME FIBRE-MATRIX SYSTEMS

System	Potential interaction	Temperature of significant interaction °C	References
Al-C	Formation of Al_4C_3 at interface. Degradation of C fibre properties.	550 ~495	9
Al-Al_2O_3	No significant reaction at normal fabrication temperatures		9
Al-oxide (Al_2O_3-SiO_2-B_2O_3)	B_2O_3 reacts with Al to form borides.	770	10
Al-B	Boride formation; interlayer of SiC needed.	500	9
Al/Li-Al_2O_3	Interfacial layer of $LiAl_5O_8$ on liquid infiltration.	~650	13
Al-SiC	No significant reaction below melting point. Al_3C_3 and Si can form in liquid Al.	m.p. 660 >700	9 12
Al-steel	Formation of iron aluminides.	500	9
Co-TaC	Directionally solidified eutectic; dissolution and re-precipitation.	1 200	9
Cu-W	No interaction up to melting point.	1 083	9
Fe-W	Formation of Fe_7W_6: dissolution of fibre.	1 000	9
Mg(AZ91)-C	No significant reaction at m.p. of alloy provided O and N avoided during infiltration.		11, 14
Ni-Al_2O_3	Slight reaction to pit fibre. $NiAl_2O_4$ spinel forms in air.	1 100 1 100	9
Ni-C(I)	Ni activates recrystallisation of fibre.	1 150–1 300	9
Ni-(II) Ni-SiC	Ni activates recrystallisation of fibre. Formation of nickel silicides.	800 800	9 9
Ni-W	Recrystallisation of fibre. Degradation of creep properties.	1 000 900	9
Ti-B	Formation of TiB_2.	750	9
Ti-SiC	Formation of TiC, $TiSi_2$ and Ti_5Si_3.	700	9

laminates of layers of fibres and metal foil is also effective. Another successful method of fabricating continuous fibre composites is to pre-coat the individual fibres with metal-matrix by PVD prior to consolidation by hot pressing. Very fine composite microstructures can be made by co-deforming a dispersed metal in a metal matrix, for example, Cu-Cr, Cu-Nb, in these cases very high strengths can be achieved for large drawing strains.

The properties which can be achieved in composites include higher strength, higher stiffness, improved high-temperature properties, lower or matched coefficients of thermal expansion and improved wear resistance. This is usually at the expense of conventional ductility but in most cases improved fracture toughness can be obtained. The following tables illustrate what has been achieved in aluminium, magnesium, titanium, zinc and copper alloys. The improvement in strength is often larger in the weaker alloys with less benefit being realised in the stronger alloys, e.g. see Table 37.3. All the alloys show an improvement in elastic modulus when reinforced.

Table 37.3   MECHANICAL PROPERTIES OF ALUMINIUM ALLOY COMPOSITES AT ROOM TEMPERATURE

Base alloy	Nominal composition	Form	Heat treatment	% particulate	0.2% proof stress MPa	Ultimate tensile stress MPa	Elongation %	Elastic modulus GPa	Fracture toughness MPa m$^{-1/2}$	Density g cm^{-3}	References
6061	Mg 1.0 Si 0.6 Cu 0.2 Cr 0.25	Extrusion	T6	Nil	276	310	20.0	69.9	29.7	2.71	1, 2
				10% Al$_2$O$_3$	297	338	7.6	81.4	24.1	2.81	1, 2
				15% Al$_2$O$_3$	317	359	5.4	87.6	22	2.86	1, 2
				20% Al$_2$O$_3$	359	379	2.1	98.6	21.5	2.94	1, 2
				13% SiC	317	356	4.9	89.5	17.9	—	1
				20% SiC	440	585	4.0	120.0	—	—	3
				30% SiC	570	795	2.0	140.0	—	—	3
2014A	Cu 4.4 Mg 0.7 Si 0.8 Mn 0.75	Extrusion	T6	Nil	414	483	13.0	73.1	25.3	2.80	1, 2
				10% Al$_2$O$_3$	483	517	3.3	84.1	18.0	2.92	1, 2
				15% Al$_2$O$_3$	476	503	2.3	91.7	18.8	2.97	1, 2
				20% Al$_2$O$_3$	483	503	0.9	101.4	—	2.98	1, 2
				10% SiC	457	508	1.8	91.2	17.7	—	1
		Sheet	T6	8.2% SiC	448	516	4.5	82.5	—	—	1
2219	Cu 6.0 Mn 0.3 V 0.1	Extrusion	T6	Nil	290	414	10.0	73.1	—	—	1, 2
				15% Al$_2$O$_3$	359	428	3.8	88.3	—	—	1, 2
				20% Al$_2$O$_3$	359	421	3.1	91.7	—	—	1, 2
2618	Cu 2.0 Mg 1.5 Si 0.9 Fe 0.9 Ni 1.0	Sheet	T6	~ 10% SiC	396	468	3.3	93.6	—	—	1, 6
		Extrusion	T6	Nil	320	400	—	75.0	—	—	1, 6
		Extrusion	T6	13% SiC	333	450	6.0	89.0	28.9	—	1, 6
7075	Zn 5.6 Mg 2.2 Cu 1.5 Cr 0.2	Extrusion	T6	Nil	617	659	11.3	71.1	—	—	6
				12% SiC	597	646	2.6	92.2	—	—	6
8090	Li 2.5 Cu 1.3 Mg 0.95 Zr 0.1	Extrusion (18 mm)	T6	Nil	480	550	5.0	79.5	—	—	1, 6
			T6	12% SiC	486	529	2.6	100.1	—	—	1, 6

Table 37.4 MECHANICAL PROPERTIES OF ALUMINIUM ALLOY COMPOSITES AT ELEVATED TEMPERATURES

Base alloy	Nominal composition		Form	Heat treatment	% particulate	Temperature °C	0.2% proof stress MPa	Tensile strength MPa	References
6061	Mg	1.0	Extrusion	T6	15% Al_2O_3	22	317	359	1
	Si	0.6			15% Al_2O_3	93	290	331	1
	Cu	0.2			15% Al_2O_3	150	269	303	1
	Cr	0.25			15% Al_2O_3	204	241	262	1
					15% Al_2O_3	260	172	179	1
					15% Al_2O_3	316	110	117	1
					15% Al_2O_3	371	62	69	1
2014A	Cu	4.4	Extrusion	T6	15% Al_2O_3	22	476 (413)	503 (483)	1
	Mg	0.7			15% Al_2O_3	93	455 (393)	490 (434)	1
	Si	0.8			15% Al_2O_3	150	407 (352)	434 (379)	1
	Mn	0.75			15% Al_2O_3	204	317 (283)	338 (310)	1
					15% Al_2O_3	260	200 (159)	214 (172)	1
					15% Al_2O_3	316	103 (62)	110 (76)	1
					15% Al_2O_3	371	55 (35)	55 (41)	1

Figures in parentheses are for basic alloy without particulate.

Table 37.5 MECHANICAL PROPERTIES OF MAGNESIUM ALLOY COMPOSITES AT ROOM TEMPERATURE

Base alloy	Nominal composition	Form	% reinforcement	0.2% proof stress MPa	Tensile strength MPa	Elongation %	Elastic modulus GPa	References
ZK60A	Mg 5.5	Extruded rod	Nil	260	325	15.0	44	5
	Zn 0.5Zr		15% SiC (partic.)	330	420	4.7	68	5
			20% SiC (partic.)	370	455	3.9	74	5
			15% SiC (whisker)	450	570	2.0	83	5
			20% B_4C (partic.)	405	490	2.0	83	5
	MG 12Li	Squeeze infiltration	Nil	—	80	8.0	—	7
		Squeeze infiltration	12% Al_2O_3 (fibre)	—	200	3.5	—	7
		Squeeze infiltration	24% Al_2O_3 (fibre)	—	280	2.0	—	7
		Extruded rod	Nil	—	75	10.0	45	7
		Extruded rod	20% SiC(whisker)	—	338	0.8	112	7

Table 37.6 MECHANICAL PROPERTIES OF TITANIUM ALLOY COMPOSITES

Base alloy	Form	% reinforceant	Temperature °C	0.2% proof stress MPa	Tensile strength MPa	Elongation %	Elastic modulus GPa	References
Ti-6Al-4V	Particulate	10% TiC	21	800	806	1.13	106–120	4
	Particulate	10% TiC	427	476 (393)	524 (510)	1.70 (11.6)	—	4
	Particulate	10% TiC	538	414 (359)	455 (441)	2.40 (8.5)	—	4
	Particulate	10% TiC	649	369 (221)	317 (310)	2.90 (4.2)	—	4
	Particulate	10% B_4C	21	—	1 055 (890)	—	205	5
Ti-6Al-4V	SCS6 fibre	35% SiC	20	—	1 634 (1 024)	1.1	213	15
Ti-15Al-3V	SCS6 fibre	35% SiC	20	—	1 572 (772)	0.7	210	15

Figures in parentheses are for basic alloy without reinforcement.

Table 37.7 MECHANICAL PROPERTIES OF ZINC ALLOY COMPOSITES AT ROOM TEMPERATURE

Base alloy		Nominal composition	Form	Tensile strength MPa	Elongation %	Elastic modulus GPa	References
ZA-12	Zn-12% Al	As cast	Nil	300	5	87	8
		As rolled	Nil	350	22	115	8
			10% SiC	323	0	119	8
			20% SiC	373	0	129	8
		Extruded	Nil	313	—	102	8
			20% Al_2O_3	532	—	120	8
ZA-27	Zn-27% Al	As cast	Nil	410	2	73	8
			10% SiC	396	0	92	8
			20% SiC	330	0	110	8
			50% SiC	310	0	220	8
		As rolled	Nil	393	16	85	8
			10% SiC	370	3	89	8
			20% SiC	349	0	102	8
		Extruded	Nil	340	—	83	8
			12% SiC	382	—	105	8
			16% SiC	466	—	130	8
			Nil	314	—	78	8
			20% Al_2O_3	382	—	100	8

Table 37.8 MECHANICAL PROPERTIES OF CO-DEFORMED COPPER COMPOSITES AT ROOM TEMPERATURE

System	Ultimate strength MPa	Drawing strain	References
Cu-15wt% Cr	906	7.0	17
Cu-18vol% V	1 850	11.0	18
Cu-30vol% Fe	1 500	8.5	18
Cu-12vol% Nb	400	3.1	19
Cu-12vol% Nb	810	8.2	19
Cu-12vol% Nb	1 180	10.3	19
Cu-12vol% Nb	1 520	11.9	19
Cu-20vol% Nb	430	3.1	19
Cu-20vol% Nb	970	8.2	19
Cu-20vol% Nb	1 410	10.3	19
Cu-20vol% Nb	1 840	11.9	19

REFERENCES

1. Private communication, *Alcan Aerospace*, 1989.
2. C. Baker, *Alcan International*, 1990.
3. F. H. Froes, *Materials Edge*, May/June, 1989, 17.
4. S. Abkowitz and P. Weihrauch, *Adv. Mat. and Proc.*, 1989 (7), 31.
5. F. H. Froes and J. Wadsworth, BNF 7th Int. Conf., 1990, Paper 1.
6. J. White *et al.*, in 'Adv. Mat. Tech. Int.', ed. G. B. Brook, publ. Sterling Publ. Ltd, 1990, 96.
7. J. F. Mason *et al.*, *J. Mat. Sci.*, 1989, **24**, 3934.
8. J. A. Cornie *et al.*, ASM Conf. on Cast Metal Matrix Composites, Chicago, Sept. 1988, 155.
9. R. Warren, in 'Adv. Mat. Tech. Int.', ed. G. B. Brook, Publ. Sterling Publ. Ltd, 1990, 100.
10. J. Yang and D. D. L. Chung, *J. Mat. Sci.*, 1989, **24**, 3605.
11. T. Iseki, T. Kameda and T. Maruyama, *J. Mat. Sci.*, 1984, **19**, 1692.
12. S. P. Rawal, L. F. Allard and M. S. Misra, in 'Interfaces in Metal-Matrix Composites', Proc. Symp. AIME, New Orleans, March, 1986 (ed. A. K. Dhinga and S. G. Fishman), 211.
13. A. R. Champion *et al.*, Proc. Int. Conf. on Composite Mat., AIME, ICCM, 1978, 883.
14. M. H. Richman, A. P. Levitt and E. S. DiCesare, *Metallography*, 1973, **6**, 497.
15. S. M. Jeng, J. M. Yang and C. J. Yang, *Mat. Sci. Eng. A*, 1991, A138, 169.
16. A. Parvizi-Majidi, in Mat. Sci. & Tech.—A Comprehensive Treatment, (Ed. R. W. Cahn, Phaasen, E. J. Kramer), VCH, Weinheim, **13**, 1994, pp 27–88.
17. K. Adachi, T. Sumiyuki, T. Takeuchi and H. G. Suzuki, *J. Jap. Inst. Metals*, 1997, **61**, 397.
18. A. M. Russell, S. Chumbley and Y. Tian, *Adv. Eng. Mat.*, 2000, **2**, 11.
19. W. A. Spitzig and P. D. Krotz, Scripta Metal, 1987, **21**, 1143.

38 Non-conventional and emerging metallic materials

38.1 Introduction

Metallic materials that are outside the mainstream of conventional metals and alloys, with respect to composition, structure and/or processing, are discussed in this chapter. The materials considered are as follows:

a. structural intermetallic compounds;
b. metallic foams;
c. metallic glasses;
d. micro- and nanoscale materials;

These materials are in different stages of development. Metallic glasses are technologically mature materials, with long-established commercial applications, although there continue to be exciting developments in the area of bulk metallic glasses. Metallic foams are the subject of a quite intense research effort at the time of writing, but have already found commercial success in a number of markets. Structural intermetallic compounds have been the subject of major world-wide research activities dating back over many years. However, significant barriers remain to the widespread commercial implementation of these materials. Metallic nanomaterials feature in much contemporary research and the details of these are still very much subject to change. However, work on the fundamentals of these materials has reached a point where this can usefully be summarised.

38.2 Cross references

The following contain information that is relevant to the contents of this Chapter:

a. crystallographic data on intermetallics—Chapter 6;
b. shape memory alloys—Chapter 15;
c. thermoelectric materials—Chapter 16;
d. superconducting materials—Chapter 19;
e. magnetic materials—Chapter 20;
f. metal matrix composites—Chapter 37.

38.3 Structural intermetallic compounds

Intermetallic compounds are employed commonly to strengthen conventional alloys. Two examples are:

• Age-hardenable aluminium alloys, strengthened by a variety of intermetallic phases, as discussed in Section 29.3.
• Nickel-base superalloys (for which an overview can be found in reference 1), with a γ/γ' microstructure. In this case γ represents a nickel-base solid solution and γ' is the intermetallic phase $Ni_3(Al,Ti)$.

More recently, intermetallic compounds have attracted great interest as monolithic (stand-alone) structural materials, for high-temperature applications. Table 38.1 summarises some of the

Table 38.1   SOME EXAMPLES OF INTERMETALLIC COMPOUNDS
OF INTEREST FOR HIGH-TEMPERATURE APPLICATIONS

Phases listed in brackets, within a given system, are relatively
technologically immature. This table does not represent an
exhaustive list, especially with respect to the 'exotic intermetallics'.
For example, numerous ternary compounds are not included.

System	Compounds of interest
Aluminides	
Ni	Ni_3Al, $NiAl$
Ti	Ti_3Al, $TiAl$, $(TiAl_3)$
Fe	Fe_3Al, $FeAl$
Refractory metal	$NbAl_3$ etc.
Silicides	$MoSi_2$, $(NbSi_2)$, $(TiSi_2)$, $(CrSi_2)$, (Co_2Si), (Ti_5Si_3) etc.

intermetallics that have been considered as structural materials (see [2] for discussion of the screening criteria needed for selecting structural intermetallic compounds).

Much of the interest in structural intermetallics has centred on the aluminides of nickel (Ni_3Al and $NiAl$) and titanium (Ti_3Al and $TiAl$). However, other aluminides feature prominently in the literature on intermetallics. In particular, interest has focused on the iron aluminides Fe_3Al and $FeAl$ (see some of the general references in the nickel aluminide section). However, other aluminide systems, for example refractory metal aluminides (e.g. [3–7]), have also been investigated. There is also a body of work on non-aluminide phases, such as silicides, in particular $MoSi_2$ (see e.g. [8, 9] for overviews). Nonetheless, in the interests of brevity, the discussion in this section is confined to the nickel and titanium aluminides.

38.3.1   Sources of information on structural intermetallic compounds

General overviews of intermetallic compounds, may be found in [10–14], including a historical perspective in the case of the first of these references. Other useful sources of collected work on structural intermetallics are the proceedings of the following:

- The Materials Research Society (MRS) has sponsored a long-running symposium series, 'High-Temperature Ordered Intermetallic Alloys'.[15]
- The Minerals, Metals and Materials Society (TMS), in some cases in conjunction with ASM International, has organised a number of conferences and symposia on structural intermetallics. An example of these is the 'International Symposium on Structural Intermetallics' series.[16]

A useful compilation of mechanical and other property data, collected up to the early to mid 1990s, for aluminides and other intermetallics, may be found in [17].

38.3.2   Focus of the section

At the time of writing, there has been nearly 20 years of intense research effort in the US, Europe, Japan and elsewhere on structural intermetallics and sporadic research dating back far longer. Following this work, the general features of the leading structural intermetallic systems are now relatively mature, for good or ill. In contrast, the detailed compositions and processing paths for these materials are still evolving. Hence, this section concentrates on the general features of these materials, and a large body of alloy-specific numerical data are not included (since the latter are still subject to change).

38.3.3   The nature of ordered intermetallics

In a solid-solution (A–B), the solute atoms (B) sit on random sites in the solvent (A) lattice. Small solute atoms (e.g. C in Fe) become interstitials and larger solute atoms (e.g. Cr in Fe) substitute for the solvent atoms. In contrast to a solid-solution, in an intermetallic compound (A_xB_y) the constituent (A and B) atoms occupy well defined sites within the structure of this phase. Hence, intermetallic

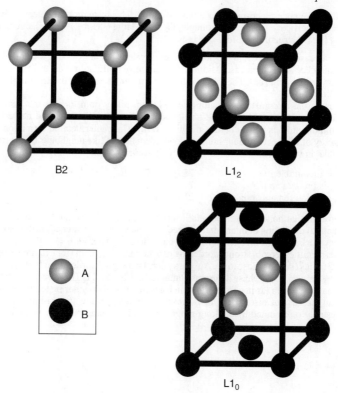

Figure 38.1 *The crystal structures of* NiAl (B2), Ni_3Al ($L1_2$) *and* TiAl ($L1_0$). *Crystallographic data for these phases (and many other intermetallics) may be found in* [18]. *A significant body of crystal structure data is available in Chapter 6*

compounds are referred to as 'ordered' and random solid-solutions as 'disordered'.* For example, Al atoms occupy random substitutional sites in the face centred cubic (FCC) Ni–Al solid solution. In contrast, ordering of this solid solution, to form the intermetallic compound Ni_3Al, involves placing the aluminium atoms at the cube corners and the nickel atoms at the face centre sites of the unit cell (Figure 38.1). Note, this ordering process results in a change in Bravais lattice and Ni_3Al is primitive cubic, not FCC.[18] Ordering is often described in terms of the production of a 'superlattice' that is superimposed on the disordered phase. Not all ordered intermetallic phases have crystal structures that can be related to those of their constituent metals. Consider an example. The Ni–Al phase diagram (which can be found in Chapter 11), contains several intermetallic compounds, including the phase NiAl. NiAl has the CsCl structure (Figure 38.1), whereas both nickel and aluminium are FCC.

Given the free electron model of metallic bonding, it might seem at first sight that metals of the same crystal structure should be fully mutually soluble (and this can be the case, as in the Ni–Cu system, for example). However, the Hume-Rothery rules (see e.g. [19, 20]) describe the circumstances under which solid-solubility is limited. These circumstances and the resulting phases are as follows:

a. Size misfit between the solvent and solute atoms. This leads to a 'size factor' compound. Examples include Laves phases, such as $MgZn_2$ and MX type interstitial compounds, for example Ti(C,N).

b. Large differences in electronegativity between the solvent and solute. In this case, there will be a tendency towards ionic bonding. Thus ionically bonded compounds, such as Mg_2Sn can form.

---

* The term 'disordered' in the present context refers to a random solid solution. Confusingly, however, the same term is sometimes applied to amorphous metals (see Section 38.5), even though a disordered solid solution possesses a well defined Bravais lattice and an amorphous metal does not.

c. Unsuitable valence electron (e) to atom (a) ratios. The formation of 'electron compounds' occurs at specific e/a values. For example, there is a large class of 1.5 e/a electron compounds, sharing the B2 structure of ordered β brass.

Note, only in case 'b' above will the normal rules of valency be followed in the chemical formula of the intermetallic. A detailed discussion of the nature of solid solutions and the breakdown in solubility may be found in [21]. The same volume also includes a useful discussion of the structure of intermetallics.[22]

Regardless of the circumstances leading to ordering, a key point is that the formation of an intermetallic will result in unlike nearest neighbour atoms (e.g. in NiAl, the nearest neighbour of each aluminium atom is a nickel atom and vice versa). Thus, intermetallic formation is favoured when the energy of unlike neighbour bonds is lower than that of like neighbour bonds (the opposite is also possible, in which case cluster formation will occur), so that:

ordering is favourable if: $\qquad 2E_{AB} < E_{AA} + E_{BB}$
clustering is favourable if: $\qquad 2E_{AB} > E_{AA} + E_{BB}$
the solid-solution remains random if: $\quad 2E_{AB} = E_{AA} + E_{BB}$

Note, however, that bonding in intermetallics can be quite complex and the reader is referred to work on structural modelling of intermetallics (see [23] and specific references in the text; Chapter 39 of the present work contains an overview of the modelling of materials).

Some intermetallics remain strongly ordered up to their melting point (for example NiAl), whereas other intermetallics (such as those in the Au–Cu system) undergo disordering at high temperatures, as can be seen by perusal of the phase diagrams contained in Chapter 11. The transition from ordered to disordered may be sudden (as in Cu_3Au, for which a well defined disordering temperature is seen) or gradual (as in the case of disordering of β-brass), depending on the system involved. The degree of order is usually represented by the long-range order (LRO) parameter (S), which is defined[24] as follows:

$$S = \frac{p - r}{1 - r}$$

where

p = fraction of species X on X sites
r = atom fraction of species X in the alloy.

Thus, for a fully ordered alloy, S equals unity and the value of S falls as the degree of order decreases. In cases where a solid-solution is not completely random, but does not possess a well-defined superlattice, the short-range order (SRO) parameter (σ) can be employed. The SRO parameter (see [25] for the original discussion), at a given temperature, is defined as:

$$\sigma = \frac{q - q_r}{q_m - q_r}$$

where

q $\ $ = fraction of unlike nearest neighbour atoms observed
q_r = fraction of unlike nearest neighbour atoms that would be present if the solid-solution were completely random
q_m = fraction of unlike nearest neighbour atoms that would be present under the condition of maximum possible order.

Two important features of ordered intermetallic compounds are antisite defects and antiphase (domain) boundaries (APBs or APDBs). Consider a notional intermetallic AB. In this case, an antisite defect would be an A atom on a B site, or vice versa. An APB would be formed, as illustrated in Figure 38.2 (for early work on the observations of antiphase domains, see [26, 27]).

The domains at the top and bottom of the figure are both perfectly ordered. However, there is an antiphase displacement between the ordering of these two domains. As a result, AA and BB bonds are formed at the boundary between the domains (i.e. the APB).

In the event that the intermetallic is strongly ordered, the creation of an antisite defect will carry a large energy penalty and the APB energy (APBE) will be high, in some cases to the point where APBs can not be produced. Note, however, in intermetallics with a wide compositional range, the presence of *structural* antisite defects provides a possible mechanism for accommodating deviations

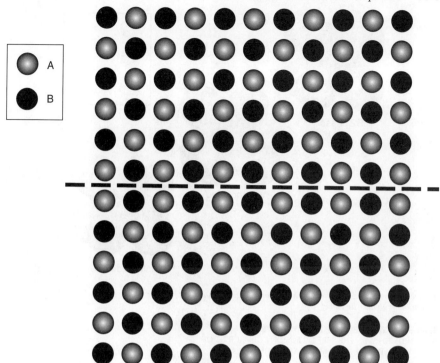

Figure 38.2 *Schematic representation of an APB (the plane of the APB corresponds to the dashed line at the centre of the figure)*

from stoichiometry, as an alternative to the formation of structural vacancies. Such structural antisite defects can be a normal part of the structure of a non-stoichiometric intermetallic and can form even in strongly ordered alloys (this is discussed further, with respect to NiAl, in Section 38.3.6 below).

In long period superlattice structures, such as those that form in the Au–Cu system, regularly spaced APBs form an integral part of the structure of the intermetallic (a concise discussion of these and other superlattice structures is contained in [28]). However, in the case of intermetallics of interest as structural materials, APBs are of the greatest importance with respect to their role in the behaviour of dislocations. Issues related to both antisite defects and APBE values are discussed further in the next section and in more detail (again with respect to NiAl) in Section 38.3.6.

38.3.4 The effect of ordering on the properties of intermetallics

In cases where there is a strong affinity between the constituent species, the melting point of an intermetallic can be significantly higher than that of any of its constituent elements. For example, as can be seem from the Ni–Al phase diagram in Chapter 11, the melting point of stoichiometric NiAl is nearly 200°C higher than that of nickel and far exceeds the melting temperature of aluminium. In a very strongly ordered intermetallic:

- Diffusion of the constituents of the intermetallic will be impeded, when vacancy—atom interchanges lead to the creation of antisite defects. The simplest example of this phenomenon* is B2 compounds (as shown in Figure 38.3), such as NiAl. In NiAl, strong ordering forces diffusion to occur via cyclic mechanisms requiring a series of co-ordinated vacancy—atom interchanges.[29] Diffusion is a random walk process and so coordinated cyclic jumps have a much lower probability than a single vacancy—atom interchange event. Thus, ordering will significantly reduce

---

* The effect of ordering on diffusion can be seen most clearly by looking at the effect of temperature on diffusion coefficients in reversibly ordered intermetallics.

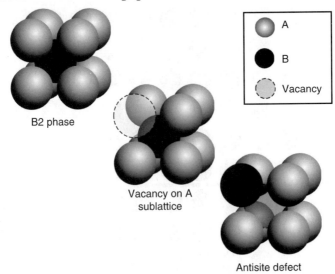

Figure 38.3   *The B2 structure, showing that a single vacancy—atom interchange event will result in the formation of an antisite defect*

Table 38.2   THE EFFECT OF DEVIATIONS FROM STOICHIOMETRY ON THE SELF DIFFUSION COEFFICIENT OF NICKEL (D_{Ni}) IN NiAl, AT A TEMPERATURE OF 1 000°C [33]

Ni (at. %)	D_{Ni} (m² s⁻¹)
50	$\sim 1 \times 10^{-16}$
48.5	7×10^{-16}
58	5×10^{-14}

the observed diffusion coefficients* Non-stoichiometric intermetallics can have quite different diffusion coefficients than stoichiometric compounds (see Table 38.2). This follows because deviations from stoichiometry are accommodated by constitutional point defects, These constitutional point defects can be either antisite atoms (as in nickel-rich NiAl), or structural vacancies (as in aluminium-rich NiAl—see references in the discussion of NiAl).

- The possible Burger's vectors may be restricted, to those which do not involve the formation of an APB (see Figure 38.4). Hence the number of slip systems may be severely restricted. The operation of the remaining available slip systems may be changed significantly (in some cases such that these are difficult to activate).
- Other properties may be affected, for example the rate of work hardening can increase on ordering and recrystallisation may become sluggish (for an overview of the latter, see [34]).

The formation of a high melting temperature intermetallic, in which diffusion (and possibly dislocation motion) may be difficult, raises the possibility of employing such materials under conditions

---

* Diffusion of ternary additions in B2 compounds is influenced by the extent to which the alloying addition partitions to one of the two sublattices (i.e. Ni sublattice and Al sublattice in the case of NiAl). An example of recent work on diffusion in B2 compounds with ternary additions can be found in [30]. See [31] for recent work on site occupancy in NiAl and FeAl with ternary additions.

The exact character of the intermetallic is very important in determining the extent to which ordering influences diffusion. For example, in the case of L1₂ type Ni₃Al, diffusion on the Al sublattice might be expected to be much more difficult than diffusion on the Ni sublattices, since only the former would disrupt local order. However, in recent work on Ni₃Al, the intrinsic diffusion coefficients of Ni and Al were found to be almost the same,[32] which suggests that Al can diffuse readily by an antisite mechanism.

In discussing diffusion in an intermetallic $A_x B_y$, it is especially important to distinguish between the interdiffusion coefficient ($\bar{D}$) and the self diffusion coefficients (D_A and D_B). These are related to the atom fractions of A and B (X_A and X_B) by

$$\bar{D} = (X_B D_A + X_A D_B) F$$

The value of the parameter F depends on the character of the intermetallic and in some cases can actually be larger than unity.[33]

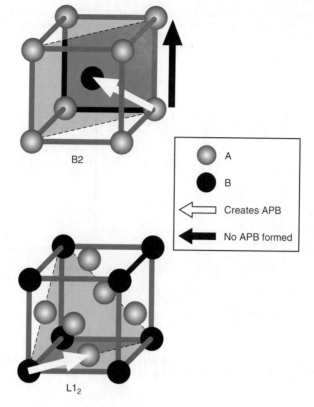

Figure 38.4 *Some possible Burgers vectors in the B2 and L1$_2$ systems. Notice that 1/2 <111>$_{B2}$ and 1/2 <110>$_{L1_2}$ lead to the creation of an APB, whereas <001>$_{B2}$ does not*

where creep resistance is required (see [35] for an overview of creep in structural intermetallic compounds). However, it must be cautioned that in cases where dislocation motion is difficult and/or an inadequate number of slip systems operate, problems are experienced with loss of low-temperature ductility. As will be discussed below, the lack of room-temperature ductility is perhaps the largest single obstacle impeding the widespread commercial application of structural intermetallics. An overview of plastic deformation in intermetallic compounds is available in [36].

38.3.5 Overview of aluminide intermetallics

Much of the work on monolithic intermetallic compounds has aimed at enabling the use of these materials in high-temperature structural applications, typically as an alternative to nickel-base superalloys. The driving force behind this work has largely been the relatively low densities of nickel and titanium aluminide intermetallics, when compared with those of typical nickel-base superalloys (Table 38.3). Given the low density of these materials, some specific properties of aluminide intermetallics, such as specific stiffness, are attractive. However, it should be cautioned that the creep and environmental resistance of these materials do not necessarily rival those of single crystal γ/γ' superalloys, as can be seen in Table 38.4 (an overview of oxidation and corrosion of intermetallics can be found in [37]). Note also the wide variation in fracture toughness between these different alloys. These various points will be discussed further in subsequent sections.

38.3.6 Nickel aluminides

The Ni–Al binary system contains several intermetallic phases. However engineering interest has focused on Ni$_3$Al and NiAl. The aluminium-rich phases in this system have relatively low melting

Table 38.3   THE DENSITY (ρ), ROOM-TEMPERATURE YOUNG'S MODULUS (E) AND SPECIFIC STIFFNESS (E/ρ) OF SOME COMMON INTERMETALLIC COMPOUNDS, IN COMPARISON WITH THOSE OF NICKEL-BASE SUPERALLOYS AND TITANIUM ALLOYS

Material	ρ (Mg m^{-3})	E (GPa)	E/ρ (kPa g^{-1} m^3)	References
Ni$_3$Al	7.5	180	24	38, 39
NiAl				
SC	5.9	95–270*	16–46	
PC		170		
Ti$_3$Al				38–41
S	4.3	145	31–35	
A	4.1–4.7			
TiAl				
S	3.8	176	45–48	
A	3.7–3.9			
Ni-Base superalloys				
SC	~8.0–9.0	~ 125$^{\#}$	14–25	For density data see Chapter 14
PC		~ 200		For modulus data see Chapter 22 and [39]
Ti alloys	~4.5–4.9	105–125	21–28	For density data see Chapter 14
				For modulus data see Chapter 22

These data are for polycrystalline alloys unless denoted otherwise (SC = single crystal, PC = polycrystalline, S = stoichiometric binary phase, A = typical ternary alloys, * = depending on orientation, $^{\#}$ = <001> oriented).

points (and are very brittle) and so are not candidates for structural applications. The structure and properties of Ni$_3$Al and NiAl diverge considerably.

Ni$_3$Al

The primitive cubic L1$_2$ type structure of γ'–Ni$_3$Al phase is the same as that of the similar phase in γ/γ' superalloys, although the latter has a composition that is usually closer to Ni$_3$(Al,Ti), as has already been remarked. In the binary Ni–Al system, Ni$_3$Al has quite a narrow range of composition. However, ternary alloying of this phase is possible (for an overview of partitioning behaviour and phase stabilisation in the $\gamma/\gamma'/\beta$ system see [42]).

Dislocations in γ' travel in the form of pairs.[13] Each dislocation in such a pair is of the 1/2 <110> type (but see discussion below on further dissociation). This Burgers vector is the same as that observed in disordered FCC metals and alloys—see e.g. [43] for a concise overview of dislocations in FCC metals). A dislocation with a Burger's vector of 1/2 <110> is a perfect dislocation in the disordered γ phase. However, with respect to the superlattice of γ', 1/2 <110> is a partial dislocation, since this displacement places Al onto Ni sites (Figure 38.4). The terminology describing such dislocations is far from standardised, but the term 'superlattice partial' dislocation will be used here (superpartial is another common term).

Given that 1/2 <110> is a superlattice partial, the first dislocation in the pair creates an APB, which is then removed by the second dislocation. Hence, the two 1/2 <110> dislocations in a pair are separated by a ribbon of APB, the formation of which is opposed by the APBE. The spacing between the superlattice partials is determined by equilibrium of the tension, induced by the APB, and the elastic repulsion of the like 1/2 <110> dislocations. Such pairs of dislocations, bound together by APBs, are often referred to as 'superdislocations'.

The APBE in γ' is relatively low. Measured and calculated values of the APBE in γ' have been summarised in [46]. These lie in the range of 90–196 mJ m^{-2} for APBs on {100} and 110–283 mJ m^{-2} for APBs on {111} (compared with stacking fault energies in the range of 5–40 mJ m^{-2} in the same system). Given the relatively low APBE, the spacing between the 1/2 <110> dislocations can be quite large. Just as in disordered FCC metals and alloys, the 1/2 <110> dislocations dissociate into pairs of 1/6 <112> partial dislocations (these are partials with respect to the fundamental FCC structure and not just to the primitive cubic superlattice). Each pair of 1/6 <112> partials is separated by a ribbon of stacking fault (SF), so that the spacing of these partials is influenced by the SF energy (SFE) in the same fashion as in an FCC metal or alloy. Note, in the case of ordered γ', the result

Table 38.4   STRUCTURES AND REPRESENTATIVE PROPERTIES OF NICKEL AND TITANIUM ALUMINIDES, IN COMPARISON WITH NICKEL-BASE SUPERALLOYS

See the footnotes to the table for the sources of these data. It should be noted that, within these general classes of material, there are very significant variations in properties depending on the alloy and processing route chosen. Also, some information on the current generation of alloys remains proprietary at the time of writing. Thus, these data should be treated only as a general guide.

Type of alloy	Usual notation for major phase	Structure of major phase(s)[1]	Likely maximum operational temperature of alloys based on this phase (°C)		Type of alloy (see text for details)	Representative room temperature properties (polycrystals unless otherwise stated)			
			Due to creep	Due to oxidation[2]		Yield stress (MPa)	Ultimate tensile strength (MPa)	Tensile ductility (%)	Fracture toughness (MPa $\sqrt{m}$)
Ni_3Al	γ'	$L1_2$	1 000	>1 100	Microalloyed with B[5] / Single crystal[5]	$\sim$240–$\sim$700 / <100–$\sim$300	$\sim$700–$\sim$1 200 / $\sim$200–$\sim$900	20–30 / 10–50	$\geq$60 (alloyed with Cr)
$NiAl$[3]	β	$B2$	>1 100	$\gg$1 100	Stoichiometric[5,6] / Stoichiometric single crystal[5,6]	180 / <150–>1 200	260 / 125–183	$\sim$0 ($\sim$20 for $\beta-(\gamma)/(\gamma')$ ternary alloys)[8]	5–6 (>15 for eutectics)[8]
Ti_3Al[4]	α_2	$D0_{19}$	760	650	$\alpha_2+\beta$[9]	700–990	800–1 140	2–10	13–30
$TiAl$[4]	γ	$L1_0$	<1 000	900	γ/α_2[10]	400–650	450–800	1–4	10–20
Ni-Base[3]	γ/γ'	$A1/L1_2$	$\sim$1 100[7]	>1 100[7]	Single crystal[7]	>1 100	<1 200	$\sim$10	30–35

[1] Data from [18].
[2] For more detailed comments on oxidation see [44].
[3] Data from [40] except where indicated otherwise.
[4] Data from [41] except were indicated otherwise.
[5] Data from [17].
[6] Data from [45].
[7] Additional data from [39].
[8] See references to multiphase NiAl alloys in text.
[9] In this alloy, β indicates a B2 phase, typically stabilised by Nb additions.
[10] The best ductility is produced by duplex alloys (equiaxed γ + lamellar γ/α_2), whereas fully lamellar (γ/α_2) materials offer the maximum fracture toughness and creep resistance.

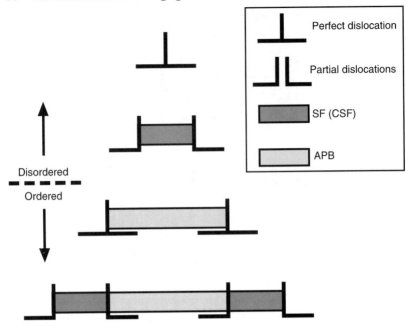

Figure 38.5 *Schematic representation of the dislocation configuration in γ (disordered) and γ' (ordered)*

Table 38.5 THE EFFECT OF TEMPERATURE (T) ON THE YIELD
STRESS (σ_y) OF Ni_3Al (PEAK INDICATES THE HIGHEST FLOW
STRESS OBSERVED)

These values are from evaluated data in [17] for <001> oriented
Ni–24.9 at. % Al. Both the yield stress at any given temperature
and the temperature at which the yield stress reaches a
maximum are sensitive to both composition and orientation.

T (K)	σ_y (MPa)
300	151
700 (~ peak)	534
1 000	407

is a complex stacking fault (CSF). Thus, dislocation motion in γ' involves a train of four partial dislocations, arranged in two pairs, with the partials in each pair bound together by CSFs and the pairs held together by an APB, as shown in Figure 38.5.

Unalloyed single crystal Ni_3Al is quite ductile, even at room temperature. In contrast, poly-crystalline binary Ni_3Al suffers from brittle grain boundaries. Fortunately, however, microalloying with boron[47] overcomes this problem for hypostoichiometric, i.e. Ni-rich, Ni_3Al. Modelling work suggests that the grain boundaries in Ni_3Al are intrinsically weak and so the boron may be acting as a grain boundary strengthener. However, it has also been found that boron-free Ni_3Al suffers from grain boundary environmental embrittlement under normal conditions and that good ductility can be produced in very dry environments even without the addition of boron.[48,49] Environmental embrittlement mechanisms for Ni_3Al and other intermetallics are discussed in [50].

Ni_3Al exhibits an anomalous increase in yield and flow stresses with increasing temperature, as can be seen in Table 38.5. The yield stress anomaly in γ' is the result of thermally activated cross-slip from the usual {111} slip plane (the same as in disordered FCC metals and alloys) to {100}. Once screw dislocations cross slip to {100} their mobility is reduced and so the flow stress increases. The driving force for cross-slip from {111} to {100} may come from anisotropy in the APBE; as can be seen in the data above, the APBE may be somewhat lower for APBs on {100} than on {111}. Anisotropic elastic properties may also produce a driving force for cross-slip. A detailed discussion of the mechanisms underlying the yield stress anomaly is beyond the scope of the present work. However, such a discussion can be found in [51], along with numerous references to the original work in this area.

Although high creep resistance γ/γ' superalloys[52] contain at least 60 vol. % γ' (in some cases considerably more than this), this does not imply that a 100 vol. % γ' material will have superior creep resistance. There are a number of mechanisms that contribute to the high creep resistance of superalloys (see for example [1] for a useful summary). Unfortunately, many of these are not applicable to single phase γ'. Thus, in comparison with a high γ' volume fraction single crystal γ/γ' superalloy, the creep resistance of single phase γ' is quite poor. An elegant demonstration of this problem may be found in [53] which compares the creep behaviour of a superalloy with single phase γ' of the same composition as that in the superalloy.

In summary, the density of Ni_3Al is only a little lower than that of a superalloy and the creep resistance is markedly inferior to that of the best superalloys. Hence, there is little hope that Ni_3Al will find application in aerospace, or other cases where high specific creep resistance is a key requirement. However, suitably alloyed Ni_3Al is much more 'forgiving' than more brittle intermetallics. Also, this material does possess some intriguing properties in terms of strong environmental resistance. For example, Ni_3Al has as excellent resistance to carburising atmospheres.[54] Hence, there has been interest[54] in using this material in applications such as belts, pans, rollers and other heat treatment furnace furniture, for high volume production in the automotive and other industries. Likewise, the diesel engine industry has expressed interest in the selective application of Ni_3Al.[55]

NiAl

Unlike Ni_3Al, the B2 compound β-NiAl occupies a significant range of composition in the Ni–Al binary system. As has already been remarked, nickel-rich NiAl forms Ni antisite defects. When aluminium-rich, this phase has nickel vacancies. A discussion of site occupancy and point defects in non-stoichiometric and alloyed NiAl may be found in [56–61]. As has already been remarked, stoichiometric NiAl has quite a high melting point (around 300°C above that of current superalloys). Also, the relatively low density of NiAl (approximately 2/3 that of modern superalloys) offers the possibility of good specific properties. Overviews of NiAl may be found in [62–67] and comparison with other B2 Compounds in [68–71].

NiAl has a high thermal conductivity (at around 80 W m^{-1} K^{-1} some 4–8 times those of typical superalloys).[62] This property is important with respect to the potential application of NiAl as an aero gas turbine blade material. Replacement of superalloy blades by NiAl would result in a highly significant 50°C reduction in blade temperatures.[62] Furthermore, NiAl offers a good match with superalloy thermal expansion coefficients.[62]

Given the high aluminium content of this phase, NiAl is an unequivocal Al_2O_3 former. NiAl therefore provides excellent oxidation resistance.[72–75] Indeed NiAl offers an enhancement in oxidation resistance over that available with alloyed Ni_3Al (see [76] for a an example of work on the oxidation of Ni_3Al alloys).

NiAl has markedly better specific creep resistance than Ni_3Al (see [77] for a discussion of the creep resistance of (Fe, Ni) Al alloys). Furthermore, the creep resistance of NiAl can be markedly enhanced through alloying, to produce a suitable two phase microstructure (see e.g. [53, 78–88] for a selection of the literature relevant to this topic). This approach to creep resistance has some similarities with that employed in γ/γ' superalloys. However, in the case of creep resistant NiAl alloys, both the matrix (which retains the B2 structure) and the second phase, are ordered. A common choice of second phase, in the references listed above, is an L2$_1$ type Heusler phase, with a composition Ni_2AlX, where X is titanium, hafnium or a range of other elements. In NiAl alloys, phases of this type are commonly known* as β'. The use of other ternary intermetallic phases to enhance creep resistance is also possible. Figure 38.6 provides an overview of the various microstructural modifications to NiAl alloys discussed in the text.

NiAl also offers good chemical compatibility with common ceramic phases. NiAl is therefore suitable for use as the matrix of intermetallic matrix composites and can also be dispersion strengthened (these topics are discussed further in Section 38.3.8).

There is currently a major problem with NiAl. The usual slip systems in this phase are <001>{110} or {001} and NiAl shows great reluctance to undergo non <001> slip at room-temperature, except under unusual circumstances. With only <001> Burgers vectors available at room-temperature, there are insufficient slip systems for polycrystalline NiAl to satisfy the Von Mises yielding criterion. Hence, polycrystalline NiAl is exceedingly brittle (although textured materials can show some

---

* β' is also used sometimes to indicate the L1$_0$ type martensite that forms from nickel-rich NiAl. However, in the present work, this symbol will be used exclusively to represent the Heusler phase.

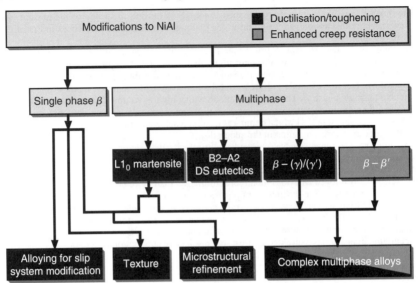

Figure 38.6   *An overview of* NiAl *alloys. For details of specific microstructures and references, see text*

ductility).[89] Likewise, the ductility of single-crystal NiAl becomes very sensitive to the loading axis (and can be quite poor even with soft orientations).

A key problem with NiAl is that ordering of this phase is very strong and consequently there is a large energy penalty associated with the formation of APBs. Calculated APBEs for NiAl are in the range of 240 to 1 000 mJ m^{-2} (see references below). The upper end of this predicted APBE range is very high (comparable in magnitude, for example, to high angle grain boundary energies—see [90] for a compilation of interfacial energy data).

In cases where non <001> slip is produced, e.g. because a single crystal is tested on the hard <001> axis, <111> dislocations are observed, but these remain perfect with respect to the superlattice* and dissociation into 1/2 <111> superlattice partials is not observed.[91] There has been quite an intense effort to model bonding and its effects for example on cleavage fracture, APBE, grain boundary properties and dislocation core structures (see e.g. [92–99]) in NiAl.

NiAl also appears to suffer from problems with embrittlement as a result of strain ageing.[100–103] In recent work, it has been suggested that phosphorus-induced softening of polycrystalline NiAl is the result of gettering or trapping of interstitials by this element.[104]

NiAl also has unusual fracture behaviour,[105,106] involving cleavage on {115} and {117}. In this case, the fracture path involves alternating {115} or {117} planes so that the fracture surface has a 'rooftop' appearance. Cleavage on these high index planes is very unusual. Instead, as a strongly ordered B2 compound, NiAl might be expected to cleave on {110}, thus preserving the Ni:Al ratio at the fracture surface. More weakly ordered B2 compounds cleave on {100}, as for disordered body centred cubic metals.

Various approaches have been investigated in an attempt to enhance the low-temperature ductility and/or fracture toughness of NiAl (see for example [45, 89, 107–109] and Figure 38.6). These approaches are discussed below.

Although mixed results have been claimed, solid-solution alloying (to reduce the APBE) does not seem to offer a clear solution to the problem of enhancing the low-temperature ductility of NiAl. A few examples (covering a wide time span) of the extensive body of literature on the topic of slip and slip system modification in NiAl may be found in [110–117] (for some early work on slip in B2 compounds, see [118]).

A possible solution to the poor fracture toughness of NiAl is composite type toughening using *in-situ* composites (e.g. [119–127]) consisting of, usually directionally solidified, eutectics of NiAl–Z. In this case, Z is typically a disordered body centred cubic solid solution, based for example

---

* Within the capabilities of weak-beam dark field transmission electron microscopy, no dissociation of <111> dislocations is observed and this restricts the maximum dissociation (if any) to ~1.5 nm. See the overviews of B2 compounds mentioned elsewhere in this section for comparison of NiAl with less strongly ordered FeAl or (Fe, Ni) Al.

on Cr, Mo or W, although other additions have been investigated. In this fashion, significant increases in room temperature fracture toughness can be produced (>10 MPa $\sqrt{m}$, compared to about 5–6 MPa $\sqrt{m}$ for binary NiAl).

The poor room temperature ductility of NiAl can be addressed by the addition of a significant volume fraction of a ductile second phase, for example γ' or a number of disordered FCC solid solution γ phases (see e.g. [128–139] for a representative selection of the research that has been conducted on this topic). In this case, the second phase serves two functions. Firstly, the ductile phase serves as an 'adhesive' that bonds the NiAl together, so that the limited number of slip systems in the latter ceases to be a problem. Secondly, glissile dislocations are 'injected' into the β-phase, so that the available slip systems in the β-phase work more efficiently (additional slip systems are not necessarily activated). In these circumstances, it is very important to have a morphology that removes the high angle β–β grain boundaries, that would initiate cracks in the absence of non $<001>$ slip in the β-phase. Also the occurrence of a suitable orientation relationship (i.e. Kurdjumov–Sachs or Nishiyama–Wasserman between β and γ or γ') facilitates slip transfer to the β-phase. Two phase β–γ' alloys can of course be formed in the binary Ni–Al binary system. However, the formation of a β–γ alloy requires ternary alloying, for example with iron.

Although useful toughening or ductilisation of NiAl can be achieved by the use of multiphase microstructures of the two types discussed above, there is a price to be paid. Increases in density result from the use of alloys that are nickel-rich and/or have extensive additions of heavy alloying elements. Thus, there is reason to question the extent to which these multiphase alloys would be useful in density critical applications (e.g. aerospace). Furthermore, creep resistance may be compromised as a result of:

- the presence of significant volume fractions of low creep resistance second phases, in the form of contiguous regions.
- moving the β-phase away from stoichiometry.

An additional problem (see e.g. [140–146]) is that the microstructures developed in complex multiphase NiAl alloys can be quite complex and in some cases show limited stability. Those alloys that attempt to combine ductilisation and/or toughening (through the addition of γ or γ' or an eutectic involving a body centred cubic metal), with creep resistance (via the formation of β'), can be especially problematical with regard to the stability of the desired microstructure.

There have been a number of studies (e.g. [147–151]) on micro- and nanostructured NiAl alloys. The intent of these studies was to enhance the fracture toughness of this alloy. In some cases, this work avoided sacrificing the low density of stoichiometric NiAl, whereas a significant portion of this effort focused on nickel-rich materials. In the latter case, there are overlaps with work on γ' and martensite-containing NiAl-derived materials, as discussed elsewhere in this section. Although these efforts have met with at least some success, the microstructural stability of these materials at high-temperature and their likely creep resistance remain concerns. For a general overview of grain size effects in intermetallics see [152].

As with much of the work on structural intermetallics, the intended applications of NiAl as a structural material are mostly in aerospace (e.g. gas turbine stators and rotors). However, unless the low temperature ductility/toughness of this phase can be improved, without compromising its low density, it is difficult to envisage large scale structural application of this material.

Nickel-rich NiAl will undergo a martensitic transformation,* by twinning, to an $L1_0$ type martensite. This transformation is thermoelastic and is capable of showing shape memory. The martensite start (M_s) temperature of NiAl is strongly dependent on composition and ranges from 30 K ($-243°$C) for a 60 at. % Ni alloy to 1 146 K ($873°$C) for 69 at. % Ni material.[161] Thus NiAl offers M_s temperatures that are far higher than for a commercial shape memory alloy (SMA), such as NiTi (Nitinol). Hence, there has been some interest in using NiAl as a high-temperature SMA.[162–166] However, the maximum recoverable strain from the shape memory effect (SME) in NiAl is very low ($\ll 1$ %). Also both the diffusional and diffusionless phase transformations in nickel-rich NiAl are quite complex (see [167–200] for some of the literature on this topic) and many of the possible product phases do not show the SME. The $L1_0$ martensite does have somewhat better fracture toughness ($K_{IC} > 10$ MPa $\sqrt{m}$ versus around 5–6 MPa $\sqrt{m}$) than stoichiometric NiAl. Indeed, the formation of

---

* In common with many other phases based on the B2 type β CuZn structure (see e.g. [153, 154,]), even stoichiometric NiAl possesses a modulated 'tweed' microstructure. This microstructure involves small displacements from the nominal lattice of the β-phase (not to be confused with spinodal decomposition, which can occur in some NiAl-derived alloys). An important cause of this modulation is that the β phase is intrinsically unstable with respect to $\{110\} <1\bar{1}0>$ shear. In NiAl (and some other β phases), this instability is manifested in the formation of shear displacement waves, with propagation vectors of $<110>$ and polarisation vectors of $<1\bar{1}0>$. Notwithstanding the shear displacements, there have been suggestions that the tweed microstructure is not necessarily a precursor of the martensitic transformation of β. Some of the literature on the tweed microstructure can be found in [155–160]. Information on the elastic properties of NiAl is available in [160].

martensite been considered as a possible toughening route for NiAl (see references in the discussion of micro- and nanostructured NiAl alloys). Nonetheless, the addition of a significant volume fraction of non-shape-memory ductilising second phase is still likely to be needed. This will further dilute the already low recoverable strain obtainable with NiAl.

38.3.7   Titanium aluminides

The intermetallic phases in the Ti–Al system that have attracted major interest as structural materials are $D0_{19}$ type α_2–Ti_3Al and $L1_0$ type γ–TiAl, both of which have a significant range of composition in the Ti–Al binary system (see Chapter 11). $TiAl_3$ is $D0_{22}$ in the Ti–Al system, but an $L1_2$ structure can be stabilised by ternary alloying (see e.g. [201]). The use of $TiAl_3$ would be attractive from the standpoint of both density ($3.37\,\mathrm{Mg\,m^{-3}}$) and oxidation resistance (since this phase contains enough aluminium to be a reliable alumina former). However, alloys based on $TiAl_3$ are very brittle (and seem likely to remain brittle for the foreseeable future) and so will not be discussed here. See for example [202] for work on dislocations in $L1_2$ stabilised $TiAl_3$.

The interest in the titanium aluminides has primarily been driven by potential applications in aerospace, both as engine components and airframe materials. Unlike the nickel aluminides, the titanium aluminides do not offer a sufficiently high operating temperature to compete with nickel-base superalloys in applications such as aero gas turbine blades and guide vanes. Instead, the titanium aluminides are largely intended to serve in applications at intermediate temperatures ($<900°C$ even for TiAl alloys). These applications exceed the operating-temperature limits of conventional titanium alloys, thus forcing currently the use of superalloys. Section 38.3.10 provides further discussion of the potential applications of structural intermetallics.

Ti_3Al

The density of Ti_3Al and its alloys is somewhat higher than for TiAl-based materials, as can be seen from the data given in Table 38.3. However, Ti_3Al alloys offer a significantly lower density than for stoichiometric NiAl. Indeed, the density of Ti_3Al alloys (in the low to mid $4\,000\,\mathrm{kg\,m^{-3}}$ range), is comparable to those of typical titanium alloys (data on the latter may be found in Section 7 of Chapter 14).

Ti_3Al alloys of potential commercial interest invariably contain ternary alloying additions. Most commonly, significant additions of niobium (as for example in the alloy Ti–24 at. % Al–11 at. % Nb) are made to stabilise the B2 type β-phase (see e.g. [203, 204]). Note, the high-temperature β phase in the Ti–Al binary system is disordered and of the A2 type (i.e. body centred cubic), as can be seen from the relevant phase diagram in Chapter 11. In some Ti_3Al alloys with β stabilised by ternary additions, the β-phase may not be fully ordered.

The orthorhombic 'O' phase forms in high niobium materials.[205] Detailed discussion of this phase is beyond the scope of the present work. Instead, for structure–property relationships in alloys containing the O phase, see [206, 207]. Comparison of β and O alloys can be found in [208]. Phases of the ω type can also form in some cases (e.g. [209]; for discussion of ω formation in conventional titanium alloys see [210–212]).

The relationship between the processing and the microstructure of these α_2–β alloys is quite complex (although these do form a logical extension of the physical metallurgy of conventional titanium alloys) and will not be discussed here. Instead, readers interested in composition/processing–structure–property relationships are referred to an overview of Ti_3Al alloys.[203]

The ductility and fracture toughness of two phase Ti_3Al alloys are far superior to those of stoichiometric NiAl, for example, and the use of Ti_3Al alloys would be attractive if it were not for the relatively poor high-temperature creep and oxidation resistance of these alloys. Low oxidation resistance is only to be expected in a material of low aluminium content, although additions of niobium are helpful in this regard. With any intermetallic, the extent to which ordering impedes diffusion is a key consideration in determining creep resistance (see Section 38.3.4) and recent work on diffusion in Ti_3Al and TiAl can be found in [213–218].

In recent years, interest has tended to switch from Ti_3Al to TiAl-based materials, given the superior creep and oxidation resistance of the latter.

TiAl alloys

Of the materials discussed in this section, alloys derived from TiAl perhaps offer the best hope of seeing commercial application within a reasonable timeframe. Although both the oxidation and creep resistance of TiAl alloys (see e.g. [219] for an example of recent work on creep in TiAl alloys) lag noticeably behind those of NiAl, the former offers considerably lower density and much higher

Table 38.6   THE CHARACTERISTICS ASSOCIATED WITH VARIOUS MICROSTRUCTURES OF
TiAl ALLOYS

For an overview of the effects of microstructure on deformation and fracture of TiAl alloys see
[261]. Numerical data in this table is from [262], except for data marked *, which is from [263].

Microstructure	Typical characteristics (all properties shown are at room-temperature)
Near single phase and single phase (γ)	Can be fine grained ($<20\,\mu$m) Ductility depends strongly on composition (0% elongation for high Al alloys*) Worst fracture toughness ($\sim$12 MPa $\sqrt{m}$)
Fully lamellar (γ/α_2)	Coarse grained ($\sim$300 μm, often coarser) Poor tensile ductility ($<$1% elongation*) Best fracture toughness ($>$20 MPa $\sqrt{m}$) Best creep resistance
Duplex ($\gamma + \gamma/\alpha_2$)	Can be fine grained ($\sim$20 μm) Best ductility ($\sim$4% elongation*) Poor fracture toughness ($\sim$15 MPa $\sqrt{m}$) Poor creep resistance

room-temperature fracture toughness (see data in Table 38.4). Indeed, the fracture behaviour of TiAl alloys can be regarded as somewhere in between brittle and ductile[220] (although the properties of TiAl alloys have been claimed to be amenable to analysis by Weibull statistics;[221] for a recent example of work on the fracture of TiAl alloys see [222]).

Single phase TiAl is of no potential value as a structural material, as this material lacks both ductility and fracture toughness, especially when aluminium-rich (which is unfortunate, given the benefits this would have for oxidation resistance). Instead, interest focuses on materials that are titanium-rich and consequently have a two phase lamellar microstructure comprised of γ and α_2 (useful overviews of TiAl alloys can be found in [223–228] and [229] provides a comparison with Ti$_3$Al alloys). A noticeable feature of the lamellar γ/α_2 microstructure is that γ forms a number of twin variants when the α_2 second phase is present. This microstructure is described as being 'polysynthetically twinned' (PST). Details of the PST γ/α_2 microstructure are available elsewhere (microstructural features are described in references [230–237]; for discussions of mechanical twinning in TiAl alloys and the stability of the PST microstructure during creep deformation see [238] and [239] respectively). Recent work suggests that interstitials can have an effect on the lamellar spacing (and hence the properties) of PST alloys.[240]

Two microstructures are of particular interest. The first of these is fully-lamellar (FL), or in some cases near-fully-lamellar (NFL), γ/α_2. The second is a duplex (DP) microstructure containing γ grains plus lamellar γ/α_2. Note, the formation of these microstructures does not necessarily occur in the fashion that might be expected from perusal of the Ti–Al phase diagram. In any case this phase diagram has been subject to revision and controversy. A discussion of the various transformation sequences involved in the formation of lamellar and duplex microstructures is beyond the scope of the present work. However, such a discussion may be found, for example, in [241] or [242] (there has been recent interest in the occurrence of massive transformations in these alloys, as can be seen in [243, 244]).

The key features of various TiAl-alloy microstructures and their properties are summarised in Table 38.6. Notice that it is necessary to make a compromise between fracture toughness and creep resistance on the one hand and ductility on the other (for an example of recent work on an integrated study of tensile, creep and fatigue behaviour in these materials see [245]). A discussion of the toughening mechanism in multiphase intermetallics, including γ/α_2 can be found in [246]. As with other intermetallics, TiAl alloys can suffer from strain ageing and it has been suggested recently that antisite defects play a role in this problem.[247]

TiAl alloys of potential commercial interest invariably contain ternary alloying additions. A large number of alloying additions have been investigated and so only some examples will be included here (a more detailed summary of the effects of ternary alloying can be found in [248]). The titanium-rich TiAl alloys (i.e. the alloys that are of potential commercial interest), greatly benefit from alloying to promote oxidation resistance, since these alloys are at best only borderline alumina formers. For some examples of work, and overviews, on oxidation of TiAl alloys see [249–257]. A common example of alloying to promote oxidation resistance is the addition of Nb. However, it must be cautioned that the effect on oxidation resistance of alloying additions can be strongly influenced by concentration. For example, chromium is added in small quantities (as in the well established alloy Ti–48 at. % Al–2 at. % Cr–2 at. % Nb) for control of mechanical properties. In contrast, much larger chromium levels (~8 at. % [248]) than those used typically in TiAl alloys would be needed for this element to have a beneficial effect on oxidation resistance.

Although most attention on TiAl alloys has focused on γ/α_2 microstructures, there has also been interest in the addition of elements such as W to stabilise the formation of B2 type β phases. In some alloys (e.g. those containing Mn), disordered body centred cubic β can play an important role in their processing.[258] There has also been some work on the formation of ω and other phases—see e.g. [259, 260]. As with other intermetallics, considerable interest has been shown in dispersion strengthened TiAl alloys and TiAl matrix composites (see next section).

Although the creep and oxidation resistance of TiAl alloys are noticeably better than those of Ti$_3$Al alloys, they can not compete with NiAl, for example. Thus, as has already been discussed, likely applications of TiAl alloys are expected to involve relatively low temperatures (below 900°C).

38.3.8 Dispersion strengthened intermetallics and intermetallic matrix composites (IMCs)

There has been considerable interest in dispersion strengthening of structural intermetallics.[264] The driving force behind work on dispersion strengthening has been the potential for enhanced creep resistance. An overview of dispersion strengthened intermetallics may be found in [264]. With conventional alloys, interest has focused principally on oxide dispersion strengthening (e.g. Y$_2$O$_3$ for nickel-base and ferritic ODS alloys; see [265] for an overview of these materials). However, with intermetallics there has been consideration of a wider range of dispersoids, for example phases such as AlN, HfC and TiB$_2$ have been considered (see references below), in addition to oxide dispersions, such as Y$_2$O$_3$.[266]

There has been a fairly large body of work on intermetallic matrix composites (IMCs). A significant portion of the work conducted on IMCs has focused on *in-situ* composites (as discussed above for NiAl), primarily for enhanced fracture toughness. However, there has also been considerable interest in IMCs with a variety of ceramic reinforcing phases, such as TiB$_2$, AlN and sapphire. An overview of fibres and fibre coatings for IMCs can be found in [267].

Some examples of work on IMCs and dispersion strengthened intermetallics can be found in [268–278]. A good source of overview information on the topic of IMCs is the proceedings of the MRS symposium series 'Intermetallic Matrix Composites'.[279] In particular, for a concise overview of application-related issues for IMCs, see [280].

For conventional alloys, there is a clear distinction between ODS alloys, with low volume fraction of fine dispersoids, and particulate metal matrix composites (MMCs), with a much higher volume fraction of coarse reinforcements. In ODS alloys, the dispersion serves to strengthen the matrix (by controlling the motion of grain boundaries and/or impeding Orowan bypass by dislocations), whereas MMCs show true composite properties. In contrast, in some of the work on intermetallic matrix materials, this distinction is less well defined, due to the use of intermediate volume fractions and particle sizes.

38.3.9 Processing and fabrication of structural intermetallics

Structural intermetallics can be (and often are) processed via a cast, or wrought route, much as in the case of conventional alloys. There has also been extensive interest in the powder metallurgical processing and mechanical alloying of intermetallics (see [281] for an overview of processing of intermetallics). However, some special features of intermetallic compounds do need to be borne in mind when processing these materials:

- The melting temperature of structural intermetallics, such as NiAl for example, is considerably higher than for nickel base superalloys. Thus existing processing technologies (e.g. mould cores for casting[67]) may need modification for use with intermetallics.
- Properties such as fracture toughness can be very sensitive to even very minor changes in the composition and/or microstructure of the intermetallic. Therefore, compositional and process tolerances may have to be considerably tighter than for more 'forgiving' materials. Furthermore, balancing ease of processing and optimum properties in service can be an issue.[282] For a review of phase stability during processing of intermetallics see [283].
- The poor low temperature ductility and fracture toughness of structural intermetallics must be allowed for in selecting a processing route.

- The potential for environmental embrittlement of some intermetallics (as discussed above) must be allowed for, as this can be a problem during processing.[284]
- With the aluminide intermetallics, evaporation of aluminium can be a major problem during processing and needs to be controlled.[285,286]

In some cases, the intrinsic characteristics of intermetallics provide opportunities for processing, as for example in the self-heating synthesis (SHS) of a number of intermetallic materials ([287–291] provide some recent examples). Some intermetallics show superplastic behaviour (see e.g. [292]) and so offer opportunities for superplastic forming (and also hybrid process routes such as diffusion bonding—superplastic forming).

With regard to both primary fabrication and pre/post-service repair, a number of joining processes have at least some applicability to intermetallics. There has been a body of research on the fusion welding of intermetallics (e.g. [293, 294]). This work has shown that fusion welding by electron beam or gas-tungsten-arc processes is quite attractive for alloys such as Ni_3Al with good intrinsic ductility and reasonably practicable even with materials such as TiAl. In contrast, with materials such as NiAl that possess poor intrinsic ductility/toughness, fusion welding would be far less suitable. Furthermore, the fusion welding of aluminide intermetallics to nickel-base superalloys would present a major challenge. Friction welding offers a means of fabricating a variety of difficult to join materials. Indeed, friction welding of TiAl appears promising.[294] However, with intermetallics such as NiAl that are extremely brittle at low-temperatures, initial attempts at friction welding[295] have not met with success.

Diffusion bonding provides a means of joining a wide range of materials and (in cases where an interlayer is not required) offers the possibility of a relatively low degree of microstructural disruption. In the case of TiAl, diffusion bonded components can maintain the strength of the parent material.[296] The major limitation of diffusion bonding is that the presence of continuous oxide layers on the faying surfaces has a strongly detrimental effect on the ability to produce sound bonds. Thus, the diffusion bonding of strongly alumina forming intermetallics (such as NiAl) is not attractive.

In common with brazing, the transient liquid phase (TLP) bonding process is tolerant of the presence of stable oxide layers on the faying surfaces. TLP bonding is therefore well suited to the joining of strongly alumina forming intermetallics (such as NiAl), unlike diffusion bonding. The TLP bonding process is conducted under isothermal conditions and requires (at most) only a small applied load for fixturing. Thus, TLP bonding is well suited to the joining of non-weldable brittle materials, such as many intermetallics. An overview of TLP bonding, including details of the issues related to the application of this technique to intermetallic compounds (and numerous references) can be found in [297].

TLP bonding of intermetallics can involve non-conventional processes. For example, in the case of NiAl substrates,[298] a high-temperature vacuum treatment of the faying surfaces produces a nickel-rich layer, due to evaporation of aluminium. This nickel-rich layer has a lower melting-point than stoichiometric NiAl and can be used as a liquid-former. In some cases, TLP bonding starts to resemble self-heating synthesis (SHS). There has also been investigation of the joining of intermetallics alloys by other techniques, such as reactive casting.[299]

38.3.10 Current and potential applications of intermetallics

The current applications of intermetallic compounds are largely non-structural. For example, pack aluminide diffusion coatings (in some cases with platinum modification[300,301]) have been employed for many years to enhance the oxidation resistance of nickel-base superalloys. Details of aluminide coatings can be found in [302–305]. Many of the microstructural features observed in aluminide coatings (see e.g. [306–311]) are similar to those of multiphase structural NiAl alloys. However, it should be cautioned that (especially due to incorporation of alloying elements from the superalloy) aluminide diffusion coatings have very complex microstructures. These contain numerous phases of varying stability and are influenced strongly by thermal/environmental exposure.

Overlay coatings of the MCrAlY type (where M is usually nickel and/or cobalt, although iron is used in some cases), find extensive industrial application. The coatings can be deposited, for example by low pressure plasma spraying (see Chapter 35 for details of thermal spray technologies). The MCrAlY system is used both to form oxidation resistant coatings and as the bond-coat[312] for thermal barrier coatings (TBCs). For details of MCrAlY coatings, see [313, 314], whilst information on TBCs can be found in [315]. Overviews of intermetallic-based protective coatings for superalloys can be found in [316–318]. There has also been some interest in employing powder eutectic coating, involving $TiAl_3$, FeAl and other phases, to protect steels.[319]

Bulk intermetallic compounds have found successful commercial application as functional (as opposed to structural) alloys. A major commercial application (see e.g. [320]) of intermetallic compounds is in the form of shape memory alloys (SMAs), most commonly based on NiTi ('Nitinol'). Applications of SMAs can involve the shape memory effect (SME), superelasticity, or vibration damping. The commercial uses of SMAs are widespread. However, some examples include:

- medical (stents to reinforce blood vessels, surgical guide wires etc.) and consumer applications (spectacle frames, underwires for brassieres);
- self clamping pipe couplings, especially for aerospace;
- a wide variety of actuators (e.g. on the NASA Mars lander) and thermal control systems (thermostats, thermal cut-outs, anti-scald valves etc.).

Further details of SMAs may be found in Chapter 15. Overviews of SMAs are available in [321–323] and the following provide details of NiTi.[324–327]

A further major commercial application of intermetallics is the use of $MoSi_2$ in heating elements for a range of commercial applications. Intermetallic (A15 type) superconducting materials continue to be widely used, notwithstanding the development of high-temperature ceramic superconductors (see Chapter 19 for a discussion of superconducting materials). Intermetallic compounds are also important as hard magnetic materials (as discussed in Chapter 20).

A major structural application of intermetallic compounds, although clearly not involving their use at high-temperatures, is in the form of dental alloys (e.g. [328, 329]. In contrast, high-temperature structural applications of intermetallic compounds are typically still at the experimental stage. Interest in these materials focuses mostly on aerospace applications in gas turbines and airframe components for advanced air and spacecraft,[40,330–332] although use in terrestrial transportation (e.g. turbocharger rotors) has also attracted interest.[333,334] An overview of the high-temperature applications of structural intermetallics can be found in [335] and other papers in the same volume include information on non-structural applications of intermetallics. Comments on the application status of individual intermetallic systems may be found in the relevant sections above. For an overview of the requirements imposed on structural intermetallics by potential aerospace applications, see [38].

Work continues on materials development for a wide range of intermetallic systems. In some cases, research on intermetallics has advanced to the point of examining end-application specific issues, such as damage due to particle impact in gas turbine engines[336] and shock loading, as would be experienced in a bird strike.[337] Fatigue resistance of intermetallics has not been discussed here, but a review of this topic may be found in [338].

38.4 Metallic foams

Metallic foams have attracted considerable interest in recent years and have a variety of commercial applications (some current and potential uses of metallic foams are summarised in Table 38.7). Foams have been produced in a variety of other materials (Table 38.8 provides some commercial examples), such as magnesium, titanium, austenitic stainless steels and nickel-base superalloys, see e.g. [339, 340]. However, there has been particular interest in foamed aluminium alloys. For representative properties of aluminium alloy foams, see Table 38.9. As can be seen from this table, metallic foams offer a very low density and can provide high specific properties.

References [341, 342] provide a useful introduction to the properties of foamed materials, whilst an overview of both biological and engineered foams can be found in [343]. Extensive information

Table 38.7 SOME APPLICATIONS OF METALLIC FOAMS AND FEATURES OF THESE MATERIALS THAT ARE RELEVANT TO END APPLICATIONS

This information is abstracted from a much more detailed table in [344].

Application	Relevant features of metallic foams
Lightweight structures (sandwich panels etc.)	High specific stiffness and specific yield stress.
Mechanical damping, vibration control and acoustic absorption	Damping capacity is an order of magnitude higher than for solid metals. Vibration control and acoustic absorption characteristics are attractive on a specific basis.
Impact absorption	High energy absorption capability.
Thermal management and flame arrest	High surface area for heat transfer, combined with high thermal conductivity cell edges.
Filters	Controlled pores sizes can be produced in open cell foams.
Electrodes and catalyst supports	High surface area per unit volume in open cell foams.

on metallic foams is available in [344] and the papers in [345, 346]. For an analysis of the economics of metallic foams, see [347].

Both open and closed-cell foams are available. The choice of which type of foam to employ depends on the intended application. For applications where fluid access to the walls of the foam is required (electrodes or filters for example) an open cell foam is needed. In contrast, many foams for load-bearing applications have closed cells.

Metal foams can be produced by a wide variety of routes, as can be seen from the examples shown in Figure 38.7 and an extensive discussion of foam production methods is beyond the scope of the present work (a summary of foam processing methods can be found in [349]). However, the following are representative examples of process routes for foamed products:

- Inco offers an open cell nickel foam, produced by a chemical vapour deposition route,[350] for applications in batteries, fuel cells, catalysis and filtration.
- A commercial example of a closed cell aluminium foam is produced Shinko Wire,[351] using a batch casting process. This process employs 1.5 wt. % Ca as a thickening agent (which operates through the formation of metal oxides). This addition is followed by 1.6 wt. % TiH_2 which decomposes to liberate hydrogen gas, thus producing a foaming effect. The thickening agent is needed to ensure a viscous enough melt to prevent gas escape.

In contrast to the Shinko process, Al–SiC particulate metal matrix composites (MMCs) can be produced with a foamed matrix,[352] since the ceramic particulate makes the melt sufficiently viscous to prevent escape of the foaming gas phase. Regardless of the processing route, it is important that gas release occurs at the appropriate moment during processing and this can be challenging to achieve (see e.g. [353]). There is continuing work on production techniques (for an example of recent work on open cell foams see [354]).

Table 38.8 SOME COMMERCIALLY AVAILABLE FOAM MATERIALS AND A COMPARISON OF THEIR DENSITY, WHEN IN THE FORM OF FOAMS PRODUCED NOMINALLY AT 5% OF THE FULL DENSITY OF THE MATERIAL CONCERNED

The materials shown are selected from the list available in [348].

Material	ρ_f (kg m^{-3})
MCrAlY (M = Fe or Ni–Co)	360–420
Cu/60–40 Brass	450/420
Steel (Eutectoid and 316 Austenitic Stainless)	390
Nickel	450
Silver	530
Titanium	230

Table 38.9 REPRESENTATIVE CHARACTERISTICS AND PROPERTIES OF SOME COMMERCIALLY AVAILABLE ALUMINIUM FOAMS

This information is abstracted from a much more detailed table in [344].

Property (units)	*Type or value*
Material	Pure Al or Al–SiC$_{particulate}$ composite
Available morphologies	Open and closed cell
ρ_f (kg m^{-3})	70–1 000
ρ_f/ρ_s	0.02–0.35
E (MPa)	20–12 000
σ_c (kPa)	40–14 000
σ_y (kPa)	40–20 000
σ_t (kPa)	50–30 000
MOR (kPa)	40–25 000
K_{IC} (kPa $\sqrt{m}$)	30–1 600
C_p (J kg^{-1} K^{-1})	830–950
λ (mW m^{-1} K^{-1})	300–35 000

ρ_f = density of foam, ρ_s = density of equivalent fully dense material, E = Young's modulus, σ_c = compressive strength, σ_y = tensile yield stress, σ_t = tensile strength, MOR = modulus of rupture, K_{IC} = mode I fracture toughness, C_p = specific heat capacity, λ = thermal conductivity.

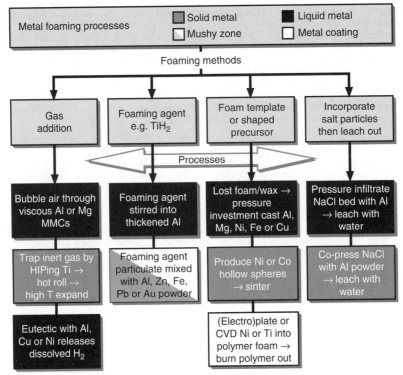

Figure 38.7 *Some methods for the production of metal foams. The methods are applicable to the metals shown and certain of their alloys. Some of the methods listed in this figure are appropriate only for laboratory batch production and are unsuitable for commercial use. For more detail on these methods, see* [344, 349].

Given the interest in using foams for structural applications (and hybrid applications with both structural and functional aspects) there is quite an extensive body of literature on structure—mechanical property relationships for metal foams. Detailed discussion of these is beyond the scope of the present work, but some representative examples of research in this area can be found in [355–365]. An overview of creep in foams and other cellular materials can be found in [366].

When foams are used in sandwich structures, bonding of the face sheets to the foam is required (in some applications, foam—foam joining may also be needed). Foams can be joined by a variety of methods, including brazing (and the braze itself can be made foaming),[367] spot welding,[368] laser welding[369] and diffusion bonding.[370]

Foams are not the only type of non-fully-dense metallic structures. Other examples include:

- honeycombs, as employed widely in aerospace;
- wire mesh products ranging from the macroscopic scale down to sintered microfibrous materials for catalyst supports and filters;
- felt metals for abradable seals in gas turbine engines.

38.5 Metallic glasses

Metallic materials are crystalline in their equilibrium state and solidification is a crystallisation process. with a distinct transition from liquid to solid. Most metals and alloys are able to crystallise, even when cooled very rapidly. However, in some alloy systems, when a sufficiently high cooling rate is employed the solidification phase transformation does not occur.[371] Instead, the liquid congeals, resulting in a metallic glass. The amorphous structure and hence the properties of metallic glasses are quite different from those of crystalline metallic solids. Nonetheless, it should be noted that metallic glasses are still metallically bonded. Indeed, metallic glasses are not transparent and have a reflective appearance that is similar to a crystalline metallic solid! The underlying science of amorphous alloys

Table 38.10   SOME EXAMPLES OF COMMERCIALLY AVAILABLE METALLIC GLASS FOILS (DATA FROM [376])

Composition (wt. %, except for compositions marked #, which are at. %)											
Cr	Fe	Si	C	B	P	Mo	W	Ni	Pd	Co	*Applications*
13.0	4.2	4.5	0.06	2.8	—	—		Bal.	—	—	
7.0	3.0	4.5	0.06	3.2	—	—		Bal.	—	—	
—	—	4.5	0.06	3.2	—	—		Bal.	—	—	
19.0	—	7.3	0.08	1.5	—	—		Bal.	—	—	Brazing/TLP
15.0	—	7.25	0.06	1.4	—	5.0		Bal.	—	—	Bonding
—	—	—	0.1	—	11	—		Bal.	—	—	
21.0	—	1.6	—	2.15	—	—	4.5	—	—	Bal.	
—	—	6.1	—	—	—	—	—	Bal.	46.7	—	
—	Bal.	9#	—	11#	—	—	—	—	—	—	Magnetic

Table 38.11   TYPICAL DIMENSIONAL RANGES (w = WIDTH, h = THICKNESS) FOR COMMERCIAL METALLIC GLASS BRAZING FOILS (DATA FROM [376])

w (mm)		h (μm)	
Minimum	*Maximum*	*Minimum*	*Maximum*
75	215	25	75

is beyond the scope of the present work, which concentrates on applications. Instead, readers are referred to a detailed discussion in [372].

Metals and alloys generally crystallise much more readily than many ceramic materials and only a subset of metallic alloys are glass forming. Even then, high cooling rates (commonly of the order of 10^5–10^6 K s^{-1}),[371,372] may be needed to suppress nucleation of solid grains. Thus, rapid solidification processing (RSP), for example melt-spinning, has traditionally been required for metallic glasses.* An important exception to this situation is what are known as bulk metallic glass (BMG) materials. BMGs will amorphatise when cooled at moderate rates ($<$100 K s^{-1} in some cases far lower), which are within the capabilities of conventional metals processing routes.

38.5.1   Metallic glasses requiring rapid quenching

A wide range of metallic-glass-forming alloys exist.[373–375] However, of particular commercial interest are those involving late transition metals and metalloids, such as Ni–(Cr)–(Si)–B, Ni–(Cr)–P and Fe–(Si)–B. These materials are commercially available in the form of melt-spun foils, as in the examples shown in Table 38.10 and Table 38.11.

An important application of metallic glasses is as a soft magnetic material. Since amorphous materials lack microstructural features, such as dislocations and grain boundaries, they allow easy motion of magnetic domain walls (see e.g. [377] for an overview of magnetic metallic glasses and their applications). The characteristics of metallic glasses as magnetic materials are discussed in Chapter 20 and so will not be repeated here. Applications of these materials include: transformers for electric power distribution and electronics, security tags for retailing and components for pulsed power systems.

A further major commercial application of metallic glasses is as interlayer foils for brazing or transient liquid phase (TLP) bonding of nickel-base and other alloys. Many of the alloys in the American Welding Society's BNi series of nickel-base brazes contain boron or phosphorus.[378] These materials are too brittle to produce in the form of rolled foils (due to the formation of phases such as Ni_3B or Ni_3P) and so conventional brazing alloys are powder-based. Use of brazes in powder form both complicates handling and compromises wettability. Thus, the ability to produce a boride-free (and hence flexible) foil by rapid solidification is very attractive (see [379] for a recent detailed overview of metallic glass brazing alloys; some additional literature on amorphous brazing foils and their applications can be found in [380–384]). Amorphous brazing foils do crystallise during heating

---

* It is important to note that RSP can be used both to produce metallic glasses and micro/nanomaterials (the latter are discussed in the next section). In the event that solidification is not suppressed entirely by RSP (due either to the use of a non-glass forming alloy or an insufficient cooling rate) then the result will be solidification at a very high undercooling. Solidification at a high undercooling implies very rapid nucleation, but slow diffusional growth and hence an extremely fine (sometimes nanoscale) grain size.

to the brazing temperature, but this does not present a problem, as by that time the foil has already been placed into the desired location.

38.5.2   Bulk metallic glasses

The requirement that conventional metallic glass forming alloys undergo very rapid cooling limits the production of these materials to thin foils. In contrast, BMG materials can be produced, for example, by centrifugal casting[385] or water quenching.[386] BMGs form in a variety of systems based on Pd,[387-389] Au,[390] La,[391] Mg,[392] Zr,[393,394] Ti,[395] Fe,[396] Ni,[397] Co,[398] and Cu.[399-401] Typically, these systems contain three to five constituents, with a large atomic size mismatch and a composition close to a deep eutectic. For an overview of BMG materials, see [402].

In some cases, BMGs can be generated[403] by mechanical alloying (MA) at room-temperature. Examples of recent work in this area can be found in [404–407]. However, the duration of the MA process is crucial and as this is increased, the metastable glass can be replaced by nanocrystalline phases, that can also be metastable (e.g. [408]).

BMGs offer the possibility of producing thick sections and have therefore attracted much recent interest. Some representative examples of the large body of recent work in this area can be found in [409–413]. Unlike conventional metallic glasses, which are mature materials with long-established commercial applications, the field of BMGs is evolving rapidly.

A key requirement for a BMG is that the material shall have a high glass forming ability (GFA). A variety of work has been conducted recently on the fundamentals of BMGs. Recent work on the theory of bulk metallic glass formation,[414,415] involves both electron concentration and size factor rules, as was the case for earlier studies of metallic glasses.[416-418] Indeed, it is interesting to compare the rules (*anti-Hume-Rothery rules*) for a lack of crystallisation with the Hume-Rothery rules for intermetallic formation,[19,20] as discussed in Section 38.3.3. An example of recent research on the GFA of BMGs can be found in [419]. In addition to the GFA, the thermal conductivity of bulk metallic glass formers is clearly a key practical issue in achieving amorphatisation on cooling (and is also important in the deformation of these materials). For an example of recent work on thermal conduction in BMG alloys, see [420].

There has been interest recently in employing amorphous bulk Mg–Ni based alloys as an electrode material for Ni-metal hydride batteries (see e.g. [421–425]). The use of amorphous and quasicrystalline Ti–Zr–Ni has also been considered for hydrogen storage.[426] For recent work on diffusible hydrogen in amorphous alloys, see [427]. Soft magnetic applications of iron-based bulk metallic glasses have also attracted attention (see e.g. [428–430]), as has the production of corrosion-resistant BMGs.[431]

The mechanical properties of BMGs (see Table 38.12) have attracted very considerable interest, since these properties are important both in structural and functional applications. Metallic glasses do not have a lattice and so can not form dislocations. Instead, plastic deformation of metallic glasses[433] involves the rapid propagation of intense shear bands (for an example of recent work on this topic, see [434]). However, it should be cautioned that BMGs behave differently in tension and compression. In cases where plasticity is observed, this may only occur in compression.[435] The plastic flow characteristics of these materials are also strain rate sensitive.[436]

There has been much recent interest in enhancing the mechanical properties of bulk metallic glasses through the formation of nanocrystals, produced by partial devitrification, so that an amorphous–nanocrystalline composite results. In some of these composites, plasticity is observed.

Table 38.12   REPRESENTATIVE MECHANICAL PROPERTIES OF BULK METALLIC GLASSES

Based on data provided in [401, 432] and various references given in the text.

Property (units)	Value
E (GPa)	~100–140
σ_{yc} (GPa)	~1.8–2
σ_c (GPa)	>2
σ_t (GPa)	~2–2.1
ε_e(%)	>2
ε_{pc}(%)	~0.8–2

E = Young's modulus, σ_{yc} = compressive yield stress, σ_c = compressive strength, σ_t = tensile strength, ε_e = elastic elongation to failure, ε_{pc} = compressive plastic strain to failure.

These materials can also have unusual magnetic properties.[437] For some recent examples of work in the area of amorphous–nanocrsystalline composites, see.[438–443] Research has also been conducted[444] on composites with a BMG matrix, plus high volume fractions (up to ~50 vol. %) of coarse crystalline metallic particles (with particle sizes up to 200 μm). There has also been interest in metal fibre reinforced BMG matrix composites.[445]

As with any metallic glass, BMGs are metastable and will crystallise if heated to a suitable temperature. Depending on the intended application, this can be a problem (if unwanted devitrification occurs during processing or service), or a benefit (as in controlled devitrification to produce the composite materials discussed above). Some recent research on crystallisation and the thermal stability of BMGs can be found in [446–452]. In addition to thermally-induced devitrification, it is also possible to precipitate metastable nanocrsystalline phases under electron irradiation.[453]

It should be cautioned that low-temperature annealing, below the glass transition temperature (T_g) can lead to embrittlement in some cases.[454] For work on the effect of heat treatment on the mechanical properties of BMGs, see for example [455].

38.6   Mechanical behaviour of micro and nanoscale materials

Until recently, the field of mechanics of materials has dealt mainly with how solids behave when acted upon by static and dynamic forces. The resulting interplay of stress and strain in controlling how materials deform is organised into a continuum describing material behaviour. However, with the advent of microelectronics and micro-electro-mechanical systems (MEMS), traditional continuum mechanics has been unable to explain the deleterious effects that stresses cause in thin film devices. Consequently, considerable work has been carried out towards investigating and understanding the origin of distinctive thin film behaviour as well as determining means to minimise and control its effects. In addition to stresses in thin films, recent advances in nanotechnology and the arrival of carbon nanotubes have again challenged traditional theories describing material behaviour. In these systems, typically one or more of the structure's dimensions is in the order of 1 to 100 nm in size, and so the mechanisms controlling deformation revolve around interatomic (or molecular) forces rather than defect generation, density and mobility. The study of these issues no longer lies only in the field of mechanics but has become cross-disciplinary and must borrow from other fields, such as physics and materials science. This section is focused on introducing the reader to recent advances in the micro and nanomechanics field and providing a basic understanding of the issues involved therein.

38.6.1   Mechanics of scale

It has been known for quite some time that materials and structures with small-scale dimensions do not behave mechanically in the same manner as their bulk counterparts. This aspect was first observed in thin films where stresses were found to have deleterious effects on the film's structural integrity and reliability. This occurs as one of a specimen's physical dimensions begins to approach that of the microstructural features. The material's mechanical properties then begin to exhibit a dependence on the specimen size. In metallic thin films, this translates to plastic yielding occurring at increased stresses, over their bulk counterparts. Although this phenomenon was observed as early as 1959,[456] no consensus or common basic understanding of it has yet been arrived upon. Besides plastic behaviour, other mechanical properties, such as fracture and fatigue also exhibit size effects. Each of these properties operates on a characteristic length scale that can be compared with the characteristic dimensions of microelectronics, microdevices or nanodevices. This is shown schematically in Figure 38.8, which utilises a logarithmic length scale map beginning at the atomic scale and onwards up to the macro world.

On the left of Figure 38.8 are four categories of devices and the regime where their dimensions fit on the length scale. On the right are regimes indicating where dimensional size-effects begin to affect the material's mechanical properties and theories used to predict behaviour. Elastic properties are dependent on the nature of bonding of the material and only exhibit size effects at the atomic scale. In contrast, plastic, fatigue and fracture properties all exhibit size effects in the micrometre and sub-micrometre regime. These properties all depend on defect generation and evolution, the mechanisms of which operate on characteristic length scales.[457] In terms of devices, these fundamental changes in mechanical behaviour occur in the size scale of the MEMS and microelectronics regimes, and thus, a better understanding of inelastic mechanisms is required to better predict their limits of strength.

The right side of Figure 38.8 lists the theories used to predict behaviour and the regimes on the length scale where they are applicable. These include classical plasticity at the upper end of the scale and dislocation/interatomic mechanics at the lower end. Classical plasticity is described as traditional

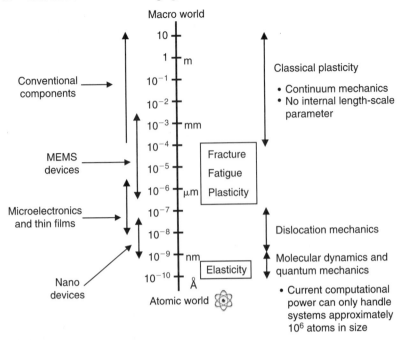

Figure 38.8   *Illustration of length-scale effects on the mechanical properties of materials*

continuum mechanics, which uses traditional relationships describing the interplay between stress and strain and is applicable for predicting behaviour from a range of approximately 100 μm and up. Dislocation mechanics (or molecular mechanics) is at the other end of the scale and involves the generation, mobility, and interaction between individual dislocations. The latter is applicable only at the lower end of the scale since it is based on numerical simulation and is consequently limited by current computational power, i.e. systems approximately one million atoms in size. Also, at the lower end is intermolecular mechanics that deals with the interatomic potentials between bonded atoms. In the region between classical and dislocation/interatomic mechanics is a grey area that continues to elude prediction of behaviour. This is a critical region where many size effects fit on the length scale. There has been solid work in the field by a number of researchers who have developed theories and models to explain thin film behaviour.[458–463] Unfortunately, these tend to address the subject within a particular range of length scale and some with only particular geometrically induced effects. An additional issue also stems from a lack of experimentally resolved real-world phenomena in the literature. Moreover, no consensus has yet been reached on exactly how the length scale affects the mechanical properties of thin films, e.g. failure mode and fracture toughness, yield strength, etc. The higher yield point of metallic thin films mentioned above is likely the result of a combined interaction of strain hardening and constraint of dislocation motion due to specimen size[464] and/or an effect of grain boundary dislocation sources[465] and geometrically necessary dislocations.[466] Clearly, a better understanding of inelastic mechanisms is required to better predict limits of strength in this region.

Several pioneering studies have identified experimentally the existence of size effects on the plasticity of metals.[467–471] These include: nanoindentation data showing a strong size effect in finding that hardness decreased as indentation depth increased;[468,472–479] bending metal strips of varying thickness around a rigid rod of scaling diameter whereby when strained to the same degree the thinner strips required a larger bending moment[471] and applying a torque load to copper rods of varying diameter whereby an identical applied twist resulted in an increase in strength by a factor of three for the smallest wire over the largest.[470] In these pioneering studies, the size dependence of the mechanical properties has been considered to be a result of non-uniform straining.[470,475,480] It was shown that classical continuum plasticity cannot predict the size dependence under these conditions. A new theory termed *strain gradient plasticity* has been proposed to describe this behaviour, but is only applicable under particular geometrically induced effects.[461–463,470,480–483] However, a comprehensive theory that predicts behaviour in this intermediate region continues to elude researchers.

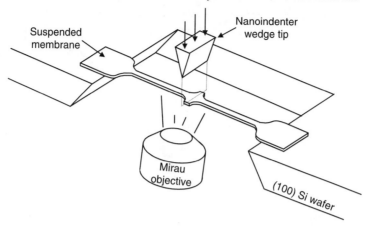

Figure 38.9   3-D *schematic view of the membrane deflection experiment* (*MDE*)

38.6.2   Thin films

Although thin films have been employed widely in protective coatings, microelectronics and micro-electro-mechanical systems, little is understood about their mechanical behaviour. This deficiency stems from difficulties in performing actual mechanical tests on these microscale structures. Since the physical dimensions of thin films are of a very small magnitude (from a few hundred μm down to as small as 1 nm), conventional mechanical testing methods have not been successful at measuring properties. Specimens in this size range are easily damaged through handling and they are difficult to position to ensure uniform loading along specimen axes. They are also difficult to attach to the testing instrument's grips, in a consistent fashion. Testing has been shown to suffer from inadequate load resolution as well as having analytical solutions that are highly sensitive to precise dimensional measurements. Several small-scale testing techniques have been employed to investigate size effects on mechanical properties. A number of reviews detail the particulars of these methods.[459,484–486]

One of the more noteworthy micro-scale tensile tests, called the membrane deflection experiment (MDE), was developed and involves the stretching of freestanding, thin-film membranes in a fixed-fixed configuration with sub-micrometre thickness.[487–490] In this technique, the membrane is attached at both ends and spans a micromachined window located beneath the membrane (see Figure 38.9). A nanoindenter applies a line-load at the centre of the span to achieve a deflection. Simultaneously, an interferometer focused on the lower side of the membrane records the deflection. The geometry of the membranes is such that they contain tapered regions to eliminate boundary-bending effects. The result is direct tension, in the absence of strain gradients, of the gauged regions of the membrane, with load and deflection being measured independently.

The MDE test has certain advantages, for instance, the simplicity of sample microfabrication and ease of handling lend confidence to repeatability. The loading procedure is straightforward and accomplished in a highly sensitive manner, while preserving the independent measurement of stress and strain. It can also test specimens of widely varying geometry, with thicknesses from sub-micrometre to several micrometres and width from one micrometre to tens of micrometres.

Recent work, employing the MDE technique, on tensile testing of thin gold films of sub-micrometre thickness has shown that size effects do indeed exist in the absence of strain gradients.[487–490] In these studies, grain size was held constant at approximately 250 nm while specimen thickness and width were varied systematically. Figure 38.10(a) consists of diagrams that show the side view of the three studied membranes with different thicknesses. Each thickness has a characteristic number of grains composing the thickness. Stress–strain plots for these films are shown in Figure 38.10(b) and show that the yield stress more than doubled when film thickness was decreased, with the thinner specimens exhibiting brittle-like failure and the thicker samples a strain softening behaviour. It is believed that these size effects stem from the limited number of grains through the thickness, which limits the number of dislocation sources and active slip systems; hence, other deformation modes such as grain rotation and grain boundary shearing accompanied by diffusion become dominant.

These data show great promise of an ability to test ever-smaller specimens, generating information that is expected to have a great impact on the development of micro and nanoscale devices. When

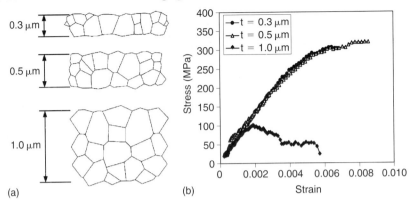

Figure 38.10   *Thin film size effects (a) Schematic representation of the number of grains within the film thickness (b) Stress–strain plot for gold membranes, 0.3, 0.5 and 1.0 μm thick*[488]

coupled with finite element modelling, this information should be able to provide an accurate description of nanoscale structures and features, in order to predict behaviour. These sorts of data are important in exploiting micro and nanoscale properties in the design of more reliable devices with increased functionality. A compilation of mechanical behaviour for various thin film materials is given in [491].

38.6.3 Nanomaterials

Nanomaterials have been receiving increased attention in recent years due to the promising and unique properties that result from their distinctive structure. They are described physically as materials possessing a nominal grain size in the range of 1 to 100 nm, a condition where the interface to volume ratio, or in other words the grain boundary area, greatly increases over traditionally coarse-grained systems. These characteristics are expected to influence strongly a material's chemical and physical properties. Thus, the major focus of study in the nanomaterials field is how properties change as grain size is reduced to the nanometre scale. Recent reviews on nanomaterials can be found in the literature.[492–495]

By controlling the grain size, properties such as the hardness, wear and fatigue resistance, ductility, electric resistivity, magnetic properties, etc. can be optimised.[492–498] An example of these improvements is the yield strength increase observed with decreasing grain size in the *Hall-Petch* relationship.[499,500] This relationship which predicts realistically increases in material strength as grain size is reduced, is as follows:

$$\sigma_y = \sigma_0 + \frac{k}{\sqrt{d}} \tag{1}$$

where σ_y is the yield strength, σ_0 is a friction stress below which dislocations will not move, d is the grain diameter and k is an experimental constant. A similar relationship exists for material hardness. This expression predicts, that as the grain size is reduced, the yield strength will increase. This behaviour has indeed been demonstrated experimentally in many material systems. Extrapolation of the *Hall-Petch* relationship to the nanometre scale predicts an enormous increase in strength and hardness. However, experimental measurements of these properties actually fall well below the extrapolation. The understanding of relationships between nanostructure and mechanical properties of nanocrystalline materials is in its infancy however, and much research remains to be accomplished.

Several models have been developed to account for *Hall-Petch* behaviour;[501–504] graphical representations of three leading models are given in Figure 38.11. The first is a traditional approach and involves the pile-up of dislocations at the boundary of a grain.[501] A grain boundary is considered the ultimate obstacle for a dislocation to overcome. Thus dislocations generated in a grain, in response to an applied stress, will travel along slip planes until a boundary is met, whereby, the dislocations then arrest and begin piling-up. After a critical number of dislocations pile-up, a number dependent on the misorientation of the neighbouring grain sharing the boundary, stress in the boundary causes dislocations to arise in the neighbouring grain. This allows yielding to propagate across the boundary. This theory is applicable only when the grain size is large enough to support the generation of at least two dislocations. Further grain refinement would then result in a plateau of strength or

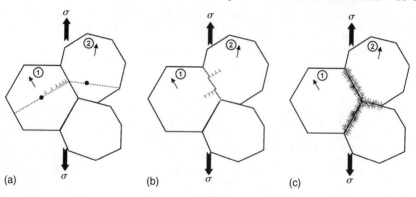

Figure 38.11 *Schematic representations of models proposed to explain* Hall-Petch *behaviour,*
(a) Cottrell,[501] *(b)* Li[502,503] *and (c) Meyers*[504]

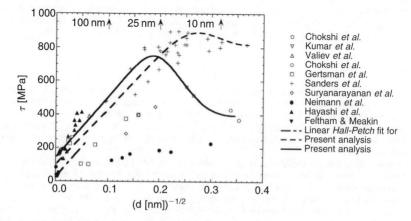

Figure 38.12 *Compilation of yield stress data on nanocrystalline copper on a* Hall-Petch *plot.*
Note that the references given in the plot are from [506]

even a reduction,[505] the so-called *negative Hall-Petch behaviour*. Other theories include dislocation
generation at grain boundary ledges [502,503] or as a consequence of stress concentrations at grain
boundaries resulting from elastic anisotropy.[504]

It has not yet been agreed upon as to whether dislocation based deformation processes are active
in the nanosized grain size regime or are merely suppressed because of the difficulty in activating
sources. In moving from coarse-grained to nano-grained materials three types of deformation, based
upon the *Hall-Petch gradient* $d\sigma/d(d^{-1/2})$, have been documented experimentally. These include:
coarse-grained materials exhibiting a constant positive *Hall-Petch gradient*, a reduction in the positive gradient when a critical grain size is reached and finally the so-called *negative Hall-Petch*
behaviour with a negative gradient.[492,494,495] Masumura et al.[506] have collected data on the yield
stress of nanocrystalline copper from various sources and summarised them on a *Hall-Petch* plot
(see Figure 38.12). The plot shows a continued increase in strength, with decrease in grain size down
to a critical size of approximately 15–25 nm. Here, the transition to *negative Hall-Petch behaviour*
is distinct where the difference in strength between the grain sizes can readily be seen. Recent work
in molecular dynamics (MD) simulations have also predicted that, at very small grain sizes, further
grain refinement would lead to a decrease in strength.[507–509]

Beyond hardness and strength, other additional aspects of nanocrystalline materials are affected
by their distinctive structure. These include elasticity, ductility and toughness, and superplastic
behaviour. In terms of material elastic response, the first documented evidence showed that the
elastic moduli of nanocrystalline materials can sometimes be only a fraction of their coarse-grained
counterparts.[510–514] The agreed upon explanation of this is rooted in extrinsic defects such as pores

and cracks that result from the compaction of powders. Experimental results on pore-free Ni–P, produced by electroplating, have verified that modulus does return to coarse-grained levels with the removal of pores,[515] also confirmed by other studies.[513,516]

Ductility and toughness of coarse-grained materials has been shown to depend on dislocations and cracks, both their interplay or competition, the effect of which are determined by grain size, whereby a reduction in grain size limits the propagation of each and results in a corresponding increase in fracture toughness.[517,518] In nanocrystalline materials, one would then expect further grain refinement to be a simple extrapolation of coarse-grained behaviour. However, experimental results have been inconclusive in this respect, a consequence of specimen inconsistencies such as porosity, flaws, and surface conditions as well as the testing methodology.[519] This may be the case for metallic materials, however, ductility has been observed in nanocrystalline ceramic materials such as CaF_2 and TiO_2 at temperatures as low as 80 to 180°C.[520,521]

Superplasticity is the ability of a material to withstand elongations in excess of 1000% without the occurrence of necking or fracture. Traditionally this occurs in materials at elevated temperatures, $T > 0.5\,T_m$, containing a second phase that inhibits grain growth, and possessing a fine grain size, on the order of 10 μm.[522,523] Microscopic examination of these materials has concluded that grain boundary sliding accompanied by considerable grain rotation is the incipient mechanism. As grain size is further refined, so is the temperature where superplasticity occurs. However, superplasticity in nanocrystalline materials has not been observed much below $0.5\,T_m$, most likely a result of nanocrystalline creep being comparable to coarse grained creep as exhibited by creep measurements in this regime.[524] Nevertheless, several experimental investigations have found enhancements in superplastic behaviour of nanocrystalline materials above $0.5\,T_m$.[525–527]

38.6.4   Nanostructures

As structures move beyond the sub-micrometre to the nanometre scale, description of mechanical behaviour will be focused on concerns other than the traditional ensemble of defects. For instance, the length scale of a typical dislocation and the volume of material required for it to have significant influence on deformation is large compared with the typical nanovolume. Therefore, it can be argued that beyond a certain point, other types of defects, surface forces, and intermolecular processes will control behaviour.

Nanostructures can be described as either a grain structure with a nominal size in the range of 1 to 100 nm or structures with one or more dimensions on the same order. An example of a nanostructure is a carbon nanotube, which is a molecular scale fibrous structure made of carbon. Carbon nanotubes, discovered by Sumio Iijima in 1991, are a subset of the family of fullerene structures.[528] Note that fullerenes were originally discovered by Curl, Kroto and Smalley, for which they were awarded the 1996 Nobel Prize in Chemistry. The simplest description of their structure would be to imagine a flat plane of carbon graphite rolled into a tube, much the same as a piece of paper (see Figure 38.13).

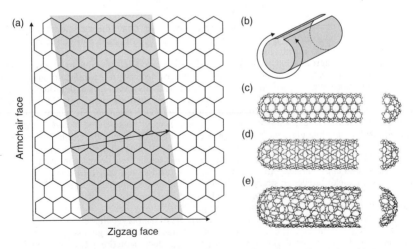

Figure 38.13   *Schematic drawings of a 2-D graphene sheet (a) rolled-up sheet (b) armchair (c) zigzag (d) chiral (e) nanotubes [(c), (d), and (e) from reference [536]]*

Like the paper, the graphene plane can be rolled in several directions to achieve varying structures. Since their discovery, many scientists have been fascinated by their ensemble of unique properties.

A dominant characteristic of nanostructures is that they possess a rather large surface area to volume ratio. As this ratio increases, interfaces and interfacial energy as well as surface topography, are expected to acquire a commanding role in deformation and failure processes. The picture of nanoscale behaviour can be viewed as the following. At the larger end of the length scale, i.e. 50 to 100 nm, dislocation generation and motion will continue to dictate behaviour. As the grain size or structural dimensions proceed below this, behaviour control will transition to surface and intermolecular mechanisms.

A few experimental studies detail the tensile testing of carbon nanotubes.[529–535] The main difficulties that have to be overcome in the testing of these nanoscale structures are similar to those for thin films; specimen handling, position, gripping and alignment. Young's modulus was found to be in the TPa regime and the tensile strength was found to be in the 11–63 GPa range, depending on the technique. These rather large numbers are reflective of the nanotubes' very low defect density.

Clearly there are many things still to be learned about materials at small scales. Understanding the mechanics of these materials and structures and the competition and interplay between their deformation mechanisms will be essential to predicting their behaviour. Such knowledge is essential to their application in nanoscale electronics and devices. These types of concerns compose the field of nanomechanics, the foundation of which is currently being laid out.

REFERENCES

1. C. T. Sims, N. S. Stoloff and W. C. Hagel, Superalloys II High—Temperature Materials for Aerospace and Industrial Power, John Wiley and Sons, New York, NY, 1987.
2. R. L. Fleischer, C. L. Briant and R. D. Field, *MRS Symp. Proc.*, 1991, **213**, 463–474.
3. S. Naka, M. Marty, M. Thomas and T. Khan, *Mater. Sci. Engin.*, 1995, **A192/193**, 69–76.
4. D. L. Anton and D. M. Shah, *MRS Symp. Proc.*, 1993, **288**, 141–150.
5. S. Sircar, K. Chattopadhyay and J. Mazumder, *Metall. Trans. A*, 1992, **23A**, 2419–2429.
6. M. G. Hebsur, I. E. Locci, S. V. Raj and M. V. Nathal, *J. Mater. Res.*, 1992, **7**, 1696–1706.
7. S. V. Raj, M. Hebsur, I. E. Locci and J. Doychak, *ibid.*, 1992, **7**, 3219–3234.
8. J. J. Petrovic, *Mater. Sci. Engin.*, 1995, **A192/193**, 31–37.
9. *idem, MRS Bull.*, 1993, **18**, 35–40.
10. J. H. Westbrook, pp 1–15 in Structural Intermetallics (Editors: R. Darolia, J. J. Lewandowski, C. T. Liu, P. L. Martin, D. B. Miracle and M. V. Nathal), TMS, Warrendale, PA, 1993.
11. O. Izumi, *Mater. Trans. JIM*, 1987, **30**, 627.
12. J. R. Stephens, *MRS Symp. Proc.*, 1985, **39**, 381–395.
13. N. S. Stoloff, *ibid.*, 1985, **39**, 3–27.
14. S. Naka, M. Thomas and T. Khan, pp 645–662 in Ordered Intermetallics—Physical Metallurgy and Mechanical Behaviour (Editors: C. T. Liu, R. W. Cahn and G. Sauthoff), Kluwer Academic Publishers, Dordrecht, The Netherlands, 1992.
15. High-Temperature Ordered Intermetallic Alloys, various volumes and dates in the MRS Symp. Proc. series (Materials Research Society, Pittsburgh, PA).
16. Structural Intermetallics, various volumes and dates in the International Symposium on Structural Intermetallics (ISSI) series, (TMS, Warrendale, PA).
17. Various editors, Properties of Intermetallic Alloys, Metals Information and Analysis Center, West Lafayette, IN, 1994 onwards.
18. P. Villars and L. D. Calvert, Pearson's Handbook of Crystallographic Data for Intermetallic Phases, 2nd Edition, ASM International, Materials Park, OH, 1991.
19. W. Hume-Rothery, Elements of Structural Metallurgy, Institute of Metals Monograph and Report Series, Volume 26, The Institute of Metals, London, UK, 1961.
20. W. Hume-Rothery, R. E. Smallman and C. W. Haworth, The Structure of Metals and Alloys, 5th Edition, The Institute of Metals, London, UK, 1969.
21. T. B. Massalski, pp 135–204 in Physical Metallurgy (Editors: R. W. Cahn and P. Haasen), 4th Edition, North Holland, Amsterdam, The Netherlands, 1996.
22. R. Ferro and A. Saccone, *ibid.*, pp 205–369.
23. D. G. Pettifor, *Mater. Sci. Technol.*, 1992, **8**, 345–349.
24. W. L. Bragg and E. J. Williams, *Proc. Roy. Soc.*, 1934, **A145**, Issue 855, 699–730.
25. H. A. Bethe, *ibid.*, 1935, **A150**, Issue 871, 552–575.
26. A. B. Glossop and D. W. Pashley, *ibid.*, 1959, **A250**, Issue 1260, 132–146.
27. H. Sato and R. H. Toth, *Phys. Rev.*, 1961, **124**, 1833–1847.
28. C. Barrett and T.B. Massalski, Structure of Metals, 3rd Edition, Pergamon Press, Oxford, UK, 1980.
29. B. S. De Bas and D. Farkas, *Acta Mater.*, 2003, **51**, 1437–1446.
30. E. Rabkin, V. N. Semenov and A. Winkler, *ibid.*, 2002, **50**, 3227–3237.
31. R. Banerjee, S. Amancherla, S. Banerjee and H.L. Fraser, *ibid.*, 2002, **50**, 633–641.
32. K. Fujiwara and Z. Horita, *ibid.*, 2002, **50**, 1571–1579.
33. H. Wever, J. Hünecke, and G. Frohberg, *Z. Metallkd.*, 1989, **80**, 389–397.

34. F. J. Humphreys and M. Hatherly, Recrystallisation and Related Phenomena, Pergamon Press, Oxford, UK, 2002.
35. G. Sauthoff, *Mater. Sci. Technol.*, 1992, **8**, 363–366.
36. M. Yamaguchi and Y. Umakoshi, *Prog. Mater. Sci.*, 1990, **34**, 1–148.
37. M. P. Brady, B. A. Pint, P. F. Tortorelli, I. G. Wright and R. J. Hanrahan, Jr., pp 229–336 in Corrosion and Environmental Degradation, Vol. II (Editors: M. Schütze, R. W. Cahn, P. Haasen and E. J. Kramer), Wiley-VCH Verlag GmbH, Weinheim, Germany, 2002.
38. P. K. Wright, pp 885–893 in Structural Intermetallics (Editors: R. Darolia, J. J. Lewandowski, C. T. Liu, P. L. Martin, D. B. Miracle and M. V. Nathal), TMS, Warrendale, PA, 1993.
39. J. R. Davis, Heat Resistant Materials, ASM International, Materials Park, OH, 1997.
40. D. M. Dimiduk, D. B. Miracle and C. H. Ward, *Mater. Sci. Technol.*, 1992, **8**, 367–375.
41. Y-W Kim and F. H. Froes, pp 465–492 in High Temperature Aluminides and Intermetallics (Editors: S. H. Whang, C. T. Liu, D. P. Pope and J. O. Stiegler), TMS, Warrendale, PA, 1990.
42. C. C. Jia, K. Ishida and T. Nishizawa, *Metall. Mater. Trans. A*, 1994, **25A**, 473–484.
43. D. Hull and D. J. Bacon, Introduction to Dislocations, 4th Edition, Butterworth-Heinemann, Oxford, UK, 2001.
44. J. L. Smialek, J. A. Nesbitt, W. J. Brindley, M. P. Brady, J. Doychak, R. M. Dickerson and D. R. Hull, *MRS Symp. Proc.*, 1995, **364**, 1273–1284.
45. K. S. Kumar, S. K. Mannan, and R. K. Viswanadham, *Acta Metall. Mater.*, 1992, **40**, 1201–1222.
46. C. T. Liu, pp 365–377 in Structural Intermetallics (Editors: R. Darolia, J. J. Lewandowski, C. T. Liu, P. L. Martin, D. B. Miracle and M. V. Nathal), TMS, Warrendale, PA, 1993.
47. O. Izumi and K. Aoki, *J. Japan Inst. Met.*, 1979, **43**, 1190–1196.
48. E. P. George, C. T. Liu, H. Lin and D. P. Pope, *Mater. Sci. Engin.*, 1995, **A192/193**, 277–288.
49. C. T. Liu and C. G. McKamey, pp 133–151 in High Temperature Aluminides and Intermetallics (Editors: S. H. Whang, C. T. Liu, D. P. Pope and J. O. Stiegler), TMS, Warrendale, PA, 1990.
50. E. P. George and C. T. Liu, *MRS Symp. Proc.*, 1995, **364**, 1131–1145.
51. D. M. Dimiduk, *J. Phys. III*, 1991, **1**, 1025–1053.
52. M. J. Donachie and S. J. Donachie, Superalloys: A Technical Guide, 2nd Edition, ASM International, Materials Park, OH, 2002.
53. S. Naka, M. Thomas and T. Khan, *Mater. Sci. Technol.*, 1992, **8**, 291–299.
54. M. Santella, Presentation at 6th International Conference in Trends in Welding Research, Pine Mountain, GA, 2002.
55. J. W. Patten, pp 493–501 in High Temperature Aluminides and Intermetallics (Editors: S. H. Whang, C. T. Liu, D. P. Pope and J. O. Stiegler), TMS, Warrendale, PA, 1990.
56. P. Georgopoulos and J. B. Cohen, *Scripta Metall.*, 1977, **11**, 147–150.
57. C. L. Fu and J. Zou, *MRS Symp. Proc.*, 1995, **364**, 91–96.
58. H. Hasoda, K. Inoue and Y. Mishima, *ibid.*, 1995, **364**, 437–442.
59. M. Kogachi, S. Minamigawa and K. Nakahigashi, *Acta Metall. Mater.*, 1992, **40**, 1113–1120.
60. M. Kogachi, T. Tanahashi, Y. Shirai and M. Yamaguchi, *Scripta Mater.*, 1996, **34**, 243–248.
61. H. C. Liu, E. Chang and T. E. Mitchell, pp 58–59 in Proceedings 39th Annual Meeting, Electron Microscopy Society of America (Editor: G. W. Bailey), Claitors Publishing Division, Baton Rouge, LA, 1981.
62. R. Darolia, D. F. Lahrman, R. D. Field, J. R. Dobbs, K. M. Chang, E. H. Goldman and D. G. Konitzer, pp 679–698 in Ordered Intermetallics—Physical Metallurgy and Mechanical Behaviour (Editors: C. T. Liu, R. W. Cahn and G. Sauthoff), Kluwer Academic Publishers, Dordrecht, The Netherlands, 1992.
63. D. B. Miracle, *Acta Metall. Mater.*, 1993, **41**, 649–684.
64. R. D. Noebe, , R. R. Bowman and M. V. Nathal, *Int. Mater. Rev.*, 1993, **38**, 193–232.
65. K. Vedula, V. Pathare, I. Aslandis and R. H. Titran, *MRS Symp. Proc.*, 1985, **39**, 411–421.
66. R. Darolia, *JOM*, 1991, **43**(3), 44–49.
67. R. Darolia, W. S. Walston and M. V. Nathal, pp 561–570 in Superalloys 1996 (Editors: R. D. Kissinger, D. J. Deye, D. L. Anton, A. D. Cetel, M. V. Nathal, T. M. Pollock and D. A. Woodford), TMS, Warrendale, PA, 1996.
68. I. Baker, and P. R. Munroe, pp 425–452 in High Temperature Aluminides and Intermetallics (Editors: S. H. Whang, C. T. Liu, D. P. Pope and J. O. Stiegler), TMS, Warrendale, PA, 1990.
69. J. R. Stephens, *MRS Symp. Proc.*, 1984, **39**, 381–395.
70. K. Vedula and J. R. Stephens, *ibid.*, 1987, **81**, 381–391.
71. I. Baker, *Mater. Sci. Engin.*, 1995, **A192/193**, 1–13.
72. J. V. Cathcart, *MRS Symp. Proc.*, 445–459, **39**, 1984.
73. J. K. Doychak and T. E. Mitchell, *ibid.*, 1984, **39**, 475–484.
74. H. J. Grabke, M. Brumm and M. Steinhorst, *Mater. Sci. Technol.*, 1992, **8**, 339–344.
75. G. H. Meier and F. S. Pettit, *ibid.*, 1992, **8**, 331–338.
76. G. Jiangting, S. Chao, L. Hui and G. Hengrong, *MRS Symp. Proc.*, 1989, **133**, 591–596.
77. M. Rudy and G. Sauthoff, *ibid.*, 1985, **33**, 327–333.
78. N. C. Tso and J. M. Sanchez, *ibid.*, 1989, **133**, 63–68.
79. M. Enomoto and T. Kumeta, *Intermetallics*, 1997, **5**, 103–109.
80. R. D. Field, R. Darolia and D. F. Lahrman, *Scripta Metall.*, 1989, **23**, 1469–1474.
81. A. Garg, R. D. Noebe and R. Darolia, *Acta Mater.*, 1996, **44**, 2809–2820.
82. P. R. Strutt and B. H. Kear, *MRS Symp. Proc.*, 1985, **39**, 279–292.
83. R. Yang, J. A. Leake and R. W. Cahn, *J. Mater. Res.*, 1991, **6**, 343–354.
84. R. Yang, J. A. Leake, R. W. Cahn and C. Small, *Scripta Metall. Mater.*, 1992, **26**, 1169–1174.

85. R. Yang, J. A. Leake, R. W. Cahn, A. Couret, D. Caillard and G. Molenat, *ibid.*, 1991, **25**, 2463–2468.
86. R. Yang, N. Saunders, J. A. Leake and R. W. Cahn, *Acta Metall. Mater.*, 1992, **40**, 1553–1562.
87. L. C. Hsiung and H. K. D. H. Bhadeshia, *Metall. Trans. A*, 1995, **26A**, 1895–1903.
88. I. E. Locci, R. M. Dickerson, A. Garg, R. D. Noebe, J. D. Whittenberger, M. V. Nathal and R. Darolia, *J. Mater. Res.*, 1996, **11**, 3024–3038.
89. K. Vedula, K. H. Hahn and B. Boulogne, *MRS Symp. Proc.*, 1989, **133**, 299–304.
90. L. E. Murr, Interfacial Phenomena in Metals and Alloys, TechBooks, Herndon, VA (originally published by Addison-Wesley, 1975).
91. P. Veyssière and R. Noebe, *Philos. Mag. A*, 1992, **65**, 1–13.
92. A. G. Fox and M. A. Tabbernor, *Acta Metall. Mater.*, 1991, **39**, 669–678.
93. C. L. Fu and M. H. Yoo, *ibid.*, 1992, **40**, 703–711.
94. P. A. Schultz and J. W. Davenport, *Scripta Metall. Mater.*, 1992, **27**, 629–634.
95. T. A. Parthasarathy, S. I. Rao and D. M. Dimiduk, *Philos. Mag. A*, 1993, **67**, 643–662.
96. M. H. Yoo and C. L. Fu, *Scripta Metall. Mater.*, 1991, **25**, 2345–2350.
97. T. Hong and A. J. Freeman, *MRS Symp. Proc.* 1989, **133**, 75–80.
98. G. Petton and D. Farkas, *Scripta Metall. Mater.*, 1991, **25**, 55–60.
99. X. Xie and Y. Mishin, *Acta Mater.*, 2002, **50**, 4303–4313.
100. J. M. Brzeski, J. E. Hack, R. Darolia and R.D. Field, *Mater. Sci. Engin.*, 1993, **A170**, 11–18.
101. M. L. Weaver, M. J. Kaufman and R. D. Noebe, *Intermetallics*, 1996, **4**, 121–129.
102. M. L. Weaver, R. D. Noebe, J. J. Lewandowski, B. F. Oliver and M. J. Kaufman, *Mater. Sci. Engin.*, 1995, **A192/193**, 179–185.
103. M. L. Weaver, R. D. Noebe and M. J. Kaufman, *Scripta Mater.*, 1996, **34**, 941–948.
104. J. T. Guo and J. Zhou, *J. Alloys Compd.*, 2003, **352**, 255–259.
105. J. H. Schneibel, R. Darolia, D. F. Lahrman and S. Schmauder, *Metall. Trans. A*, 1993, **24A**, 1363–1371.
106. K.-M. Chang, R. Darolia and H. A. Lipsitt, *Acta Metall. Mater.*, 1992, **40**, 2727–2737.
107. D. R. Pank, M. V. Nathal and D. A. Koss, *MRS Symp. Proc.*, 1989, **133**, 561–566.
108. K. Vedula and P. S. Khadkikar, pp 197–217 in High Temperature Aluminides and Intermetallics (Editors: S. H. Whang, C. T. Liu, D. P. Pope and J. O. Stiegler), TMS, Warrendale, PA, 1990.
109. C. T. Liu, E. H. Lee, E. P George and A. J. Duncan, *Scripta Metall. Mater.*, 1994, **30**, 387–392.
110. R. D. Field, D. F. Lahrman and R. Darolia, *Acta Metall. Mater.*, 1991, **39**, 2961–2969.
111. *idem, Acta. Metall. Mater.*, **39**, 2951–2959.
112. V. Glatzel, K. R. Forbes and W. D. Nix, *Philos. Mag. A*, 1993, **67**, 307–323.
113. D. B. Miracle, *Acta Metall. Mater.*, 1991, **39**, 1457–1468.
114. D. B. Miracle, S. Russell and C. C. Law, *MRS Symp. Proc.*, 1989, **133**, 225–231.
115. R. J. Wasilewski, S. R. Butler and J. E. Hanlon, *Trans. Metall. Soc. AIME*, 1967, **239**, 1357–1364
116. W. R. Kanne Jr., P. R. Strutt and R. A. Dodd, *Trans. Metall. Soc. AIME*, 1969, **245**, 1259–1267.
117. R. Darolia, D. F. Lahrman, R. D. Field and A. J. Freeman, *MRS Symp. Proc.*, 1989, **133**, 113–118.
118. W. A. Rachinger and A. H. Cottrell, *Acta Metall.*, 1956, **4**, 109–113.
119. F. E. Heredia and J. J. Valencia, *MRS Symp. Proc.*, 1992, **273**, 197–204.
120. F. E. Heredia, M. Y. He, G. E. Lucas, A. G. Evans, H. E. Dēve and D. Konitzer, *Acta Metall. Mater.*, 1993, **41**, 505–511.
121. D. R. Johnson, B. F. Oliver, R. D. Noebe and J. D. Whittenberger, *Intermetallics*, 1995, **3**, 493–503.
122. M. T. Kush, J. W. Holmes and R. Gibala, *MRS Symp. Proc.*, 1999, **552**, KK9.3.1–KK9.3.6.
123. D. P. Mason, D. C. Van Aken and J. G. Webber, *ibid.*, 1990, **194**, 341–348.
124. D. P. Mason, D. C. Van Aken, R. D. Noebe, I. E. Locci and K. L. King, *ibid.*, 1991, **213**, 1033–1038.
125. S. V. Raj, I. E. Locci, J. A. Salem and R. J. Pawlik, *Metall. Mater. Trans. A*, 2002, **33A**, 597–612.
126. J. D. Whittenberger, S. V. Raj, I. E. Locci and J. A. Salem, *ibid.*, 2002, **33A**, 1385–1397.
127. C. Y. Cui, J. T. Guo, Y. H. Qi and H. Q. Ye, *Intermetallics*, 2002, **10**, 1001–1009.
128. K. Ishida, R. Kainuma, N. Ueno and T. Nishizawa, *Metall. Trans. A*, 1991, **22A**, 441–446.
129. R. Yang, J. A. Leake and R. W. Cahn, *MRS Symp. Proc.*, 1993, **288**, 489–494.
130. I. Baker, S. Guha and J. A. Horton, *Philos. Mag. A*, 1993, **67**, 663–674.
131. R. D. Field, D. D. Krueger and S. C. Huang, *MRS Symp. Proc.*, 1989, **133**, 567–572.
132. S. Guha, I. Baker and P. R. Munroe, *Mater. Charact.*, 1995, **34**, 181–188.
133. M. Larsen, A. Misra, S. Hartfield-Wunsch, R. D. Noebe and R. Gibala, *MRS Symp. Proc.*, 1990, **194**, 191–198.
134. A. Misra, R. D. Noebe and R. Gibala, *ibid.*, 1992, **273**, 205–210.
135. *idem, MRS Symp. Proc.*, 1994, **350**, 243–248.
136. R. Kainuma, S. Imano, H. Ohtani and K. Ishida, *Intermetallics*, 1996, **4**, 37–45.
137. I. Baker and S. Guha, *Scripta Mater.*, 1996, **4**, 557–559.
138. P. S. Khadkikar, K. Vedula and B. S. Shabel, *MRS Symp. Proc.*, 1987, **81**, 157–165.
139. T. Sakata, H. Y. Yasuda and Y. Umakoshi, *Scripta Mater.*, 2003, **48**, 749–753.
140. R. Yang, J. A. Leake and R. W. Cahn, *Philos. Mag. A*, 1992, **65**, 961–980.
141. W. F. Gale and Z. A. M. Abdo, *J. Mater. Sci.*, 1999, **34**, 4425–4437.
142. *idem, J. Mater. Sci.*, 1998, **33**, 2299–2304.
143. W. F. Gale, Z. A. M. Abdo and R. V. Nemani, *ibid.*, 1999, **34**, 407–416.
144. W. F. Gale, R. V. Nemani and J. A. Horton, *ibid.*, 1996, **31**, 1681–1688.
145. W. F. Gale and R. V. Nemani, *Mater. Sci. Engin.*, 1995, **A192/193**, 868–872.
146. J. T. Guo, C. Y. Cui, Y. H. Qi and H. Q. Ye, *J. Alloys Compd.*, 2002, **343**, 142–150.
147. G. B. Schaffer, *Scripta Metall. Mater.*, 1992, **27**, 1–5.

148. T. Cheng, *Nanostruct. Mater.*, 1992, **1**, 19–27.
149. *idem, J. Mater. Sci.*, 1993, **28**, 5909–5916.
150. T. Cheng, H. M. Flower and M. McClean, *Scripta Metall. Mater.*, 1992, **26**, 1913–1918.
151. I. E. Locci, R. D. Noebe , J. A. Moser, D. S. Lee and M. Nathal, *MRS Symp. Proc.*, 1989, **133**, 639–646.
152. E. M. Schulson, *ibid.*, 1985, **39**, 193–204.
153. L. Delaey, A. J. Perkins and T. B. Massalski, *J. Mater. Sci.*, 1972, **7**, 1197–1215.
154. C. M. Wayman, *MRS Symp. Proc.*, 1985, **39**, 77–91.
155. I. M. Robertson and C. M. Wayman, *Philos. Mag. A*, 1983, **48**, 629–647.
156. *idem, Philos. Mag. A*, 1983, **48**, 443–467.
157. *idem, Philos. Mag. A*, 1983, **48**, 421–442.
158. *idem, Metall. Trans. A*, 1984, **15A**, 1353–1357.
159. G. Van Tendeloo and S. Amelinckx, *Scripta Metall.*, 1986, **20**, 335–339.
160. N. Ruscovic and H. Warlimont, *Phys. Stat. Sol. (a)*, 1977, **44**, 609–619.
161. J. L. Smialek and R. H. Hehemann, *Metall. Trans.*, 1973, **4**, 1571–1575.
162. K. K. Jee, P. L. Potapov, S. Y. Song and M. C. Shin, *Scripta Mater.*, 1997, **36**, 201–212.
163. J. A. Horton, C. T. Liu and E. P. George, *Mater. Sci. Engin.*, 1995, **A192/193**, 873–880.
164. K. Enami, and S. Nenno, *Metall. Trans. A*, 1971, **2A**, 1487–1490.
165. R. Kainuma, K. Ishida and T. Nishizawa, *ibid.*, 1992, **23A**, 1147–1153.
166. Y. D. Kim and C. M. Wayman, *Scripta Metall. Mater.*, 1991, **25**, 1863–1868.
167. Y. Yamada, Y. Noda and K. Fuchizaki, *Phys. Rev. B*, 1990, **42**, 10405–10414.
168. J. H. Yang and C. M. Wayman, *Intermetallics*, 1994, **2**, 121–126.
169. *idem, Intermetallics*, 1994, **2**, 111–119.
170. *idem, Mater. Sci. Engin.*, 1993, **A160**, 241–249.
171. A. S. Murthy and E. Goo, *Acta Metall. Mater.*, 1993, **41**, 2135–2142.
172. *idem, Acta Metall. Mater.*, 1993, **41**, 3435–3443.
173. *idem, Metall. Mater. Trans. A*, 1994, **25A**, 57–61.
174. S. Muto, N. Merk, D. Schryvers and L.E. Tanner, *Philos. Mag. B*, 1993, **67**, 673–689.
175. S. Muto, D. Schryvers, N. Merk and L. E. Tanner, *Acta Metall. Mater.*, 1993, **41**, 2377–2383.
176. D. Schryvers and L. E. Tanner, *Ultramicroscopy*, 1990, **32**, 241–254.
177. D. Schryvers, Y. Ma, L. Toth and L.E. Tanner, *Acta Metall. Mater.*, 1995, **43**, 4057–4065.
178. *idem, Acta Metall. Mater.*, 1995, **43**, 4045–4056.
179. D. Schryvers, *Philos. Mag. A*, 1993, **68**, 1017–1032.
180. I. M. Robertson and C. M. Wayman, *Metallog.*, 1984, **17**, 43–55.
181. S. Rosen and J. A. Goebel, *Trans. Metall. Soc. AIME*, 1968, **242**, 722–725.
182. S. M. Shapiro, J. Z. Larese, Y. Noda, S. C. Moss and L. E. Tanner, *Phys. Rev. Lett.*, 1986, **57**, 3199–3202.
183. S. M. Shapiro, B. X. Yang, G. Shirane, Y. Noda and L.E. Tanner, *ibid.*, 1985, **62**, 1298–1301.
184. D. Schryvers and Y. Ma, *Mater. Lett.*, 1995, **23**, 105–111.
185. *idem, J. Alloys Compd.*, 1995, **221**, 227–234.
186. W. F. Gale, *Intermetallics*, 1996, **4**, 585–587.
187. D. Schryvers, B. De Saegher and J. Van Landuyt, *Mater. Res. Bull*, 1991, **26**, 57–66.
188. K. Enami, , S. Nenno and K. Shimizu, *Trans. JIM*, 1973, **14**, 161–165.
189. R. Kainuma, H. Ohtani and K. Ishida, *Metall. Mater. Trans. A*, 1996, **27A**, 2445–2453.
190. S.-H. Kang, S.-J. Jeon, and H. C. Lee, *MRS Symp. Proc.*, 1991, **213**, 385–390.
191. A. G. Khachaturyan, S. M. Shapiro and S. Semenovskaya, *Phys. Rev. B*, 1991, **43**, 10832–10843.
192. D. Kim, P. C. Clapp and J. A. Rifkin, *MRS Symp. Proc.*, 1991, **213**, 249–254.
193. V. S. Litvinov and A. A. Arkhangelskaya, *Phys. Met. Metall.*, 1978, **44**, 131–137.
194. Y. Noda, S. M. Shapiro, G. Shirane, Y. Yamada and L. E. Tanner, *Phys. Rev. B*, 1990, **42**, 10397–10404.
195. M. Liu, T. R. Findlayson, T. F. Smith and L. E. Tanner, *Mater. Sci. Engin.*, 1992, **A157**, 225–232.
196. K. Otsuka, T. Ohba, M. Tokonami, and C. M. Wayman, *Scripta Metall. Mater.*, 1993, **29**, 1359–1364.
197. L. E. Tanner, A. R. Pelton, G. Van Tendeloo, D. Schryvers and M.E. Wall, *ibid.*, 1990, **24**, 1731–1736.
198. T. Cheng, *J. Mater. Sci. Lett.*, 1996, **15**, 285–289.
199. V. V. Martynov, K. Enami, L. G. Khandros, A. V. Tkachenko and S. Nenno, *Scripta Metall.*, 1983, **17**, 1167–1171.
200. S. H. Kim, M. C. Kim, M. H. Oh, T. Hirano and D. M. Wee, *Scripta Mater.*, 2003, **48**, 443–448.
201. K. Kita, G. Itoh and M. Kanno, *MRS Symp. Proc.*, 1995, **364**, 1241–1246.
202. J. Douin, K. Shavrin Kumar and P. Veyssière, *Mater. Sci. Engin.*, 1995, **A192/193**, 92–96.
203. D. Banerjee, A. K. Gogia, T. K. Nandy, K. Muraleedharan and R. S. Mishra, pp 19–33 in Structural Inter-metallics (Editors: R. Darolia, J. J. Lewandowski, C. T. Liu, P. L. Martin, D. B. Miracle and M. V. Nathal), TMS, Warrendale, PA, 1993.
204. L. A. Bendersky and W. J. Boettinger, *MRS Symp. Proc.*, 1989, **133**, 45–50.
205. L. H. Hsiung and H. N. G. Wadley, *Mater. Sci. Engin.*, 1995, **A192/193**, 908–913.
206. B. S. Majumdar, C. Boehlert, A. K. Rai and D. B. Miracle, *MRS Symp. Proc.*, 1995, **364**, 1259–1264.
207. S. Emura, A. Araoka and M. Hagiwara, *Scripta Mater.*, 2003, **48**, 629–634.
208. H. A. Lipsitt, *MRS Symp. Proc.*, 1993, **288**, 119–130.
209. R. Xu, Y. Y Cui, D. M. Xu, D. Li and Z. Q. Hu, *Metall. Mater. Trans. A*, 1996, **27A**, 2221–2228.
210. A. T. Balcerzak and S. L. Sass, *Metall. Trans*, 1972, **3**, 1601–1605.
211. S. L. Sass, *Acta Metall.*, 1969, **17**, 813–820.
212. K. K. McCabe and S. L. Sass, *Philos. Mag.*, 1971, **23**, 957–970.
213. Y. Mishin and Chr. Herzig, *Acta Mater.*, 2000, **48**, 589–623.

214. Chr. Herzig, T. Wilger, T. Przeorski, F. Hisker and S. Divinski, *Intermetallics*, 2001, **9**, 431–442.
215. Chr. Herzig, T. Przeorski, M. Friesel, F. Hisker and S. Divinski, *ibid.*, 2001, **9**, 461–472.
216. J. Breuer, T. Wilger, M. Friesel and Chr. Herzig, *ibid.*, 1999, **7**, 381–388.
217. J Rüsing and Chr. Herzig, *ibid.*, 1996, **4**, 647–657.
218. J Rüsing and Chr. Herzig, *Scripta Metall. Mater.*, 1995, **33**, 561–566.
219. G. B. Viswanathan, S. Karthikeyan, R. W. Hayes and M. J. Mills, *Metall. Mater. Trans. A*, 2002, **33A**, 329–336.
220. J. E. Milke, J. L. Beuth, N. E. Biery and H. Tang, *ibid.*, 2002, **33A**, 417–426.
221. N. Biery, M. De Graef, J. Beuth, R. Raban, A. Elliott, C. Austin and T.M. Pollock, *ibid.*, 2002, **33A**, 3127–3136.
222. Y.-W. Kim and K. V. Jata, *ibid.*, 2002, **33A**, 2847–2857.
223. Y.-W. Kim and D. M. Dimiduk, JOM, 1991, **43**(8), 40–47.
224. F. Appel and R. Wagner, *Mater. Sci. Engin.*, 1998, **R22**, 187–268.
225. Y. Yamaguchi and H. Inui, pp 127–142 in Structural Intermetallics (Editors: R. Darolia, J. J. Lewandowski, C. T. Liu, P. L. Martin, D. B. Miracle and M. V. Nathal), TMS, Warrendale, PA, 1993.
226. M. Yamaguchi, *Mater. Sci. Technol.*, 1992, **8**, 299–308.
227. Y.-W. Kim, *JOM*, 1989, **41**(7), 24–30.
228. *idem*, *JOM*, 1994, **46**(7), 30–39.
229. F. H. Froes, C. Suryanarayana and D. Eliezer, *J. Mater. Sci.*, 1992, **27**, 5113–5140.
230. C. R. Feng, D. J. Michel and C. R. Crowe, *Scripta Metall.*, 1988, **22**, 1481–1486
231. *idem*, *Scripta Metall.*, 1989, **23**, 1135–1140.
232. *idem*, *Philos. Mag. Lett.*, 1990, **3**, 95–100.
233. H. Inui, A. Nakamura, M. H. Oh and M. Yamaguchi, *Ultramicroscopy*, 1991, **39**, 268–278.
234. H. Inui, M. H. Oh, A. Nakamura and M. Yamaguchi, *Philos. Mag. A*, 1992, **66**, 539–555.
235. D. S. Schwartz and S. M. L. Sastry, *Scripta Metall.*, 1989, **23**, 1621–1626.
236. Y. S. Yang and S. K. Wu, *Philos. Mag. A*, 1992, **65**, 15–28.
237. S. H. Chen, G. Schumacher, D. Mukherji, G. Frohberg and R. P. Wahi, *Scripta Mater.*, 2002, **47**, 757–762.
238. F. D. Fischer, F. Appel and H. Clemens, *Acta Mater.*, 2003, **51**, 1249–1260.
239. H. Y. Kim and K. Maruyama, *ibid.*, 2003, **51**, 2191–2204.
240. C. Y. Nam, M. H. Oh, K. S. Kumar and D. M. Wee, *Scripta Mater.*, 2002, **46**, 441–446.
241. Y.-W. Kim, *Acta Metall. Mater.*, 1992, **40**, 1121–1134.
242. H. M. Flower and J. Christodoulou, *Mater. Sci. Technol.*, 1989, **15**, 45–52.
243. J. M. Howe, W. T. Reynolds Jr. and V. K. Vasudevan, *Metall. Mater. Trans. A*, 2002, **33A**, 2391–2411.
244. J. F. Nie and B. C. Muddle, *ibid.*, 2002, **33A**, 2381–2389.
245. V. Recina, D. Lundström and B. Karlsson, *ibid.*, 2002, **33A**, 2869–2881.
246. K.S. Chan, *MRS Symp. Proc.*, 1995, **364**, 469–480.
247. U. Fröbel and F. Appel, *Acta Mater.*, 2002, **50**, 3693–3707.
248. S.-C. Huang, pp 299–307 in Structural Intermetallics (Editors: R. Darolia, J. J. Lewandowski, C. T. Liu, P. L. Martin, D. B. Miracle and M. V. Nathal), TMS, Warrendale, PA, 1993.
249. A. Takaski, K. Ojima, Y. Taneda, T. Hoshiya and A. Mitsuhashi, *Scripta Metall. Mater*, 1992, **27**, 401–405.
250. J. W. Fergus, *Mater. Sci. Engin.*, 2002, **A338**, 108–125.
251. J. W. Fergus, N. L. Harris, C. J. Long, V. L. Salazar, T. Zhou and W. F. Gale, in Gamma Titanium Aluminides 2003 (Editors: Y.-W. Kim, H. Clemens and A. H. Rosenberger), TMS Warrendale, PA, In Press 2003.
252. B. Dang, J. W. Fergus and W. F. Gale, *Oxid. Met.*, 2001, **56**, 15–32.
253. A. Rahmel, W. J. Quadakkers and M. Schütze, *Mater. Corros.*, 1995, **46**, 271–285.
254. V. A. C. Haanappel and M. F. Stroosnijder, *Surf. Eng.*, 1999, **15**, 119–125.
255. V. A. C. Haanappel, J. D. Sunderkötter and M. F. Stroosnijder, *Intermetallics*, 1999, **7**, 529–541.
256. M. P. Brady, W. J. Brindley, J. L. Smialek and I. E. Locci, *JOM*, 1996, **48**(11), 46–50.
257. V. A. C. Haanappel, H. Clemens and M. F. Stroosnijder, *Intermetallics*, 2002, **10**, 293–305.
258. T. Tetsui, K. Shindo, S. Kobayashi and M. Takeyama, *Scripta Mater.*, 2002, **47**, 399–403.
259. R. Ducher, B. Viguier and J. Lacaze, *ibid.*, 2002, **47**, 307–313.
260. Z. W. Huang, W. Voice and P. Bowen, *Scripta Mater.*, 2003, **48**, 79–84.
261. Y.-W. Kim, *Mater. Sci. Engin.*, 1995, **A192/193**, 519–533.
262. H. Clemens, W. Glatz, N. Eberhardt, H.-P. Martinz and W. Knabl, *MRS Symp. Proc.*, 1997, **460**, 29–43.
263. M. Yamaguchi, H. Inui, K. Kishida, M. Matsumuro and Y. Shirai, *ibid.*, 1995, **364**, 3–16.
264. C. C. Koch, *ibid.*, 1987, **81**, 369–380.
265. H. K. D. H. Bhadeshia, *Mater. Sci. Engin.*, 1997, **A223**, 65–77.
266. S.-C. Ur and P. Nash, *Scripta Mater.*, 2002, **47**, 405–409.
267. A. K. Misra, *MRS Symp. Proc.*, 1994, **350**, 73–88.
268. J. D. Whittenberger, R. Ray and S. C. Jha, *Mater. Sci. Engin.*, 1992, **A151**, 137–146.
269. K. S. Kumar and J. D. Whittenberger, *Mater. Sci. Technol.*, 1992, **8**, 317–330.
270. J. D. Whittenberger, R. Ray, S. C. Jha and S. Draper, *Mater. Sci. Eng.*, 1991, **A138**, 83–93.
271. S. N. Tewari, R. Asthana and R. D. Noebe, *Metall. Trans. A*, 1993, **24A**, 2119–2125.
272. J. D. Whittenberger, T. J. Moore and D. L. Kuruzar, *J. Mater. Sci. Lett.*, 1987, **6**, 1016–1018.
273. J. D. Whittenberger and M. J. Luton, *J. Mater. Res.*, 1992, **7**, 2724–2732.
274. *idem*, *J. Mater. Sci.*, 1995, **10**, 1171–1185.
275. J. D. Whittenberger, E. Arzt and M. J. Luton, *Scripta Metall. Mater.*, 1992, **26**, 1925–1930.
276. *idem*, *J. Mater. Res.*, 1990, **5**, 271–277.
277. *idem*, *MRS Symp. Proc.*, 1990, **194**, 211–218.

278. R. R. Bowman, *ibid.*, 1992, **273**, 145–156.
279. Intermetallic Matrix Composites, various volumes and dates in the MRS Symp. Proc. series (Materials Research Society, Pittsburgh, PA).
280. C. H. Ward and A. S. Culbertson, *MRS Symp. Proc.*, 1994, **350**, 3–12.
281. S. L. Semiatin, J. C. Chesnutt, C. Austin and V. Seetharaman, pp 263–276 in Structural Intermetallics 1997 (Editors: M. V. Nathal, R. Darolia, C. T. Liu, P. L. Martin, D. B. Miracle, R. Wagner and M. Yamaguchi), TMS, Warrendale, PA, 1997.
282. V. Seetharaman and S. L. Semiatin, *Metall. Mater. Trans. A*, 2002, **33A**, 3817–3830.
283. J. H. Perepezko, C. A. Nuñes, S.-H. Yi and D. J. Thoma, *MRS Symp. Proc.*, 1997, **460**, 3–14.
284. T. Tsuyumu, Y. Kaneno, H. Inoue and T. Takasugi, *Metall. Mater. Trans. A*, 2003, **34A**, 645–655.
285. J. Guo, G. Liu, Y. Su, H. Ding, J. Jia and H. Fu, *ibid.*, 2002, **33A**, 3249–3253.
286. S. Yanqing , G. Jingjie, J. Jun, L. Guizhong and L. Yuan, *J. Alloys Compd.*, 2002, **334**, 261–266.
287. J. Oh, W. C. Lee, S. Gyu Pyo, W. Park, S. Lee and N. J. Kim, *Metall. Mater. Trans. A*, 2002, **33A**, 3649–3659.
288. A. Biswas, S. K. Roy, K. R. Gurumurthy, N. Prabhu and S. Banerjee, *Acta Mater.*, 2002, **50**, 757–773.
289. D. Horvitz and I. Gotman, *ibid.*, 2002, **50**, 1961–1971.
290. S. Dong, P. Houb, H. Yanga and G. Zoua, *Intermetallics*, 2002, **10**, 217–223.
291. Q. Fan, , H. Chai and Z. Jin, *ibid.*, 2002, **10**, 541–554.
292. Y. Rosenberg and A. K. Mukherjee, *Mater. Sci. Engin.*, 1995, **A192/193**, 788–792.
293. M. C. Maguire, G. R. Edwards and S. A. David, *Weld. J. (Res. Suppl.)*, 1992, **71**, 231s–242s.
294. P. L. Threadgill, *Mater. Sci. Engin.*, 1995, **A192/193**, 640–646.
295. J. D. Whittenberger, T. J. Moore and D. L. Kurzar, *J. Mater. Sci. Lett.*, 1987, **6**, 1016–1018.
296. P. Yan and E. R. Wallach, *Intermetallics*, 1993, **1**, 83–98.
297. W. F. Gale and D. A. Butts, *Sci. Technol. Weld. Join.*, In Press, 2003.
298. T. J. Moore and J. M. Kalinowski, *MRS Symp. Proc.*, 1993, **288**, 1173–1178.
299. K. Matsuura, M. Kudoh, H. Kinoshita and H. Takahashi, *Metall. Mater. Trans. A*, 2002, **33A**, 2073–2080.
300. P. C. Patnaik, R. Thamburaj and T. S. Sudarshan, pp 759–777 in Surface Modification Technologies III (Editors: T. S. Sudarshan and D. G. Bhad), TMS, Warrendale, PA, 1990.
301. H. M. Tawancy, N. M. Abbas and T. N. Rhys-Jones, *Surf. Coat. Technol.*, 1992, **49**, 1–7.
302. J. T. Bowker, Ph.D. Thesis, University of Sheffield, Sheffield, UK, 1978.
303. M. J. Fleetwood, *J. Inst. Metals*, 1970, **98**, 1–7.
304. G. W. Goward, D. H. Boone and C. S. Giggins, *Trans. ASM*, 1967, **60**, 228–241.
305. S. R. Levine and R. M. Caves, *J. Electrochem. Soc.*, 1974, **8**, 1051–1065.
306. P. Shen, D. Gan, and C. C. Lin, *Mater. Sci. Engin.*, 1986, **78**, 171–178.
307. *idem*, *Mater. Sci. Engin.*, 1986, **78**, 163–170.
308. W. F. Gale, T. C. Totemeier and J. E. King, *Metall. Mater. Trans. A*, 1995, **26A**, 949–956.
309. W. F. Gale and J. E. King, *Metall. Trans. A*, 1992, **23A**, 2657–2665.
310. *idem*, *J. Mater. Sci.*, 1993, **28**, 4347–4354.
311. J. L. Smialek, *Metall. Trans.*, 1971, **2**, 913–915.
312. R. Panat, S. Zhang and J. Hsia, *Acta Mater.*, 2003, **51**, 239–249.
313. F. J. Pennisi and D. K. Gupta, *Thin Solid Films*, 1981, **84**, 49–58.
314. R. W. Smith, *ibid.*, 1981, **84**, 59–72.
315. T. A. Taylor, pp 53–57 in Metallurgical Coatings and Thin Films, Volume 1 (Editors: B. D. Sartwell, G. E. McGuire and S. Hofmann), Elsevier, Amsterdam, The Netherlands, 1992.
316. T. N. Rhys-Jones, pp 218–223 in Materials Development in Turbo-Machinery Design (Editors: D. M. R. Taplin, J. F. Knott and M. H. Lewis), Institute of Metals, London, UK and Parsons Press, Dublin, Ireland, 1989.
317. T. N. Rhys-Jones, *Mater. Sci. Technol.*, 1988, **4**, 421–430.
318. T. N Rhys-Jones and D. F. Bettridge, pp 129–158 in Advanced Materials and Processing Techniques for Structural Applications (Editors: T. Khan, and A. Lasalmonie), Office National d'Etudes et de Recherches Aerospatiales (ONERA), Chatillon, France, 1988.
319. H. Kafuku, Y. Tomota, M. Isaka and T. Suzuki, *Metall. Mater. Trans. A*, 2002, **33**, 3235–3240.
320. C. M. Wayman, *JOM*, 1980, **32**(6), 129–137.
321. E. Hornbogen, *Metall*, 1987, **41**, 488–493.
322. C. M. Wayman and J. D. Harrison, *JOM*, 1989, **9**, 26–28.
323. C. M. Wayman, *MRS Bull.*, 1993, **18**(4), 49–56.
324. J. A. Shaw and S. Kyriakides, *J. Mech. Phys. Solids*, 1995, **43**, 1243–1281.
325. H. C. Ling and R. Kaplow, *Metall. Trans. A*, 1980, **11A**, 77–83.
326. *idem*, *Metall. Trans. A*, 1981, **12A**, 2101–2111.
327. H. A. Mohamed and J. Washburn, *ibid.*, 1976, **7A**, 1041–1043.
328. M. P. Dariel, U. Admon, D. S. Lashmore, M. Ratzker, A. Giuseppetti and F. C. Eichmiller, *J. Mater. Res.*, 1995, **10**, 505–511.
329. T. K. Hooghan, R. F. Pinizzotto, J. H. Watkins and T. Okabe, *ibid.*, 1996, **11**, 2474–2485.
330. J. M. Larsen, K. A. Williams, S. J. Balsone and M. A. Stucke, pp 521–556 in High Temperature Aluminides and Intermetallics (Editors: S. H. Whang, C. T. Liu, D. P. Pope and J. O. Stiegler), TMS, Warrendale, PA, 1990.
331. R. LeHolm, B. Norris, and A. Gurney, *Adv. Mater. Process.*, 2001, **159**(5), 27–31.
332. C. M. Austin and T. J. Kelly, pp 143–150 in Structural Intermetallics (Editors: R. Darolia, J. J. Lewandowski, C. T. Liu, P. L. Martin, D. B. Miracle and M. V. Nathal), TMS, Warrendale, PA, 1993.

333. V. K. Sikka, pp 505–520 in High Temperature Aluminides and Intermetallics (Editors: S. H. Whang, C. T. Liu, D. P. Pope and J. O. Stiegler), TMS, Warrendale, PA, 1990.
334. Y. Nishiyama, T. Miyashita, S. Isobe and T. Noda, *ibid.*, pp 557–584.
335. D. P. Pope and R. Darolia, *MRS Bull*, 1996, **21**(5), 30–36.
336. V. T. McKenna, M. P. Rubal, P. S. Steif, J. M. Pereira and G. T. Gray III, *Metall. Mater. Trans. A*, 2002, **33**, 581–589.
337. J. C. F. Millett, N. K. Bourne, G. T. Gray III and I. P. Jones, *Acta Mater.*, 2002, **50**, 4801–4811.
338. N. S. Stoloff, pp 33–42 in Structural Intermetallics 1997 (Editors: M. V. Nathal, R. Darolia, C. T. Liu, P. L. Martin, D. B. Miracle, R. Wagner and M. Yamaguchi), TMS, Warrendale, PA, 1997.
339. M. Bram and C. Stiller, *Adv. Eng. Mater.*, 2000, **2**(4), 196–199.
340. Y. Yamada, K. Shimojima, Y. Sakaguchi, M. Mabuchi, M. Nakamura, T. Asahina, T. Mukai, H. Kanahashi and K. Higashi, *ibid.*, 2000, **2**(4), 184–187.
341. L. J. Gibson and M. F. Ashby, Cellular Solids–Structure and Properties, 2nd Edition, Cambridge University Press, Cambridge, UK, 1998.
342. *Several Papers in MRS Bull.*, 2003, **28**(4).
343. D. Weaire and S. Hutzler, The Physics of Foams, Oxford University Press, Oxford, UK, 1999.
344. M. F. Ashby, A. Evans, N. A. Fleck, L. J. Gibson, J. W. Hutchinson and H. N. G. Wadley, Metal Foams—A Design Guide, Butterworth Heinemann, Boston, MA, 2000.
345. J. Banhart, M. F. Ashby and N. A. Fleck (Editors), Cellular Metals and Metal Foaming Technology, Verlag MIT Publishing, Bremen, Germany, 2001.
346. *Various papers in Adv. Eng. Mater.*, 2000, **2**(4).
347. E. Maine and M. F. Ashby, *ibid.*, 2000, **2**(4), 205–209.
348. Porvair Product Literature, Porvair Advanced Materials, Hendersonville, NC.
349. C. Körner and R. F. Singer, *Adv. Eng. Mater.*, 2000, **2**(4), 159–165.
350. Incofoam Product Literature, Inco Special Products, Toronto, ON, Canada, 2003.
351. T. Miyoshi, M. Itoh, S. Akiyama and A. Kitahara, *Adv. Eng. Mater.*, 2000, **2**(4), 179–183.
352. V. Gergely and T. W. Clyne, *ibid*, 2000, **2**(4), 175–178.
353. A. R. Kennedy, *Scripta Mater.*, 2002, **47**, 63–767.
354. K.-S. Chou and M.-A. Song, *Scripta Mater.*, 2002, **46**, 379–382.
355. C. Motz and R. Pippan, *Acta Mater.*, 2002, **50**, 2013–2033.
356. I. W. Hall, M. Guden and T. D. Claar, *Scripta Mater.*, 2002, **46**, 513–518.
357. T.-J. Lim, B. Smith and D. L. McDowell, *Acta Mater.*, 2002, **50**, 2867–2879.
358. K. C. Chan and L. S. Xie, *Scripta Mater.*, 2003, **48**, 1147–1152
359. O. Kesler and L.J. Gibson, *Mater. Sci. Engin.*, 2002, **A326**, 228–234.
360. A. E. Markaki and T. W. Clyne, *Acta Mater.*, 2001, **49**, 1677–1686.
361. P. S. Liu, K. M. Liang, S. W. Tu, S. R. Gu, Q. Yu, T. F. Li and C. Fu, *Mater. Sci. Technol.*, 2001, **17**, 1069–1072.
362. J. Zhou, C. Mercer and W. O. Soboyejo, *Metall. Mater. Trans. A*, 2002, **33A**, 1413–1427.
363. T. Bernard, J. Burzer and H. W. Bergmann, *J. Mater. Process Technol.*, 2001, **115**, 20–24.
364. X. Badiche, S. Forest, T. Guibert, Y. Bienvenu, J.-D. Bartout, P. Ienny, M. Croset and H. Bernet, *Mater. Sci. Engin.*, 2000, **A289**, 276–288.
365. O. Kesler, L. K. Crews and L. J. Gibson, *Mater. Sci. Engin.*, In Press, 2003.
366. E. W. Andrews, L. J. Gibson and M. F. Ashby, *Acta Mater.*, 1999, **47**, 2853–2863.
367. K.-J. Matthes and H. Lang, pp 501–504 in Cellular Metals and Metal Foaming Technology (Editors: J. Banhart, M. F. Ashby and N. A Fleck), Verlag MIT Publishing, Bremen, Germany, 2001.
368. Ch. Born, H. Kuckert, G. Wagner and D. Eifler, *ibid.*, pp 485–488.
369. J. Burzer, T. Bernard and H. W. Bergmann, *MRS Symp. Proc.*, 1998, **521**, 159–164.
370. K. Kitazono, A. Kitajima, E. Sato, J. Matsushita and K. Kuribayashi, *Mater. Sci. Engin.*, 2002, **A327**, 128–132.
371. K. Klement, R. H. Willens and P. Duwez, *Nature*, 1960, **187**, 869–870.
372. R. W. Cahn and A. L. Greer, pp 1723–1830 in Physical Metallurgy (Editors: R. W. Cahn and P. Haasen), 4th Edition, North Holland, Amsterdam, The Netherlands, 1996.
373. S. Takayama, *J. Mater. Sci.*, 1976, **11**, 164–185.
374. C. Suryanarayana, Rapidly Quenched Metals—A Bibliography, 1973–1979, Plenum Press, New York, NY, 1980.
375. W. L. Johnson, *Prog. Mater. Sci.*, 1986, **30**, 81–134.
376. Honeywell Metglas Solutions, Product Data, Conway, SC, 2003.
377. N. DeCristofaro, *MRS Bull.*, 1998, **23**(5), 50–56.
378. G. E. Sheward, High Temperature Brazing in Controlled Atmospheres, Pergamon Press, Oxford, UK, 1985.
379. A. Rabinkin, *Sci. Technol. Weld. Join.*, In Press, 2003.
380. Author Not Specified, *Weld. J.*, 1983, **62**(10), 57–58.
381. W. Ozgowicz, J. Tyrlik-Held, G. Thomas, A. Zahara, and J. Le Coze, *Scripta Metall.*, 1983, **17**, 295–298.
382. A. Rabinkin, *Weld. J.*, 1989, **68**(10), 39–45.
383. T. R. Tucker and J. D. Ayers, pp 206–211 in Rapid Solidification Processing—Principles and Technologies II (Editors: R. Mehrabian, B. H. Kear, and M. Cohen), Claitor's Publishing Division, Baton Rouge, LA, 1980.
384. *idem*, *Metall. Trans. A*, 1981, **12A**, 1801–1807.
385. Q. S. Zhang, , D. Y. Guo, A. M. Wang, H. F. Zhang, B. Z. Ding and Z. Q. Hu, *Intermetallics*, 2002, **10**, 1197–1201.
386. Y. Zhang, D. Q. Zhao, R. J. Wang and W. H. Wang, *Acta Mater.*, 2003, **51**, 1971–1979.
387. H. S. Chen, *Acta Metall.*, 1974, **22**, 1505–1511.

388. A. J. Drehman, A. L. Greer and D. Turnbull, *Appl. Phys. Lett.*, 1982, **41**, 716–717.
389. A. Inoue, N. Nishiyama and T. Matsuda, *Mater. Trans JIM*, 1996, **37**, 181–184.
390. M. C. Lee, J. M. Kendall and W. L. Johnson, *Appl. Phys. Lett.*, 1982, **40**, 382–384.
391. A. Inoue, T. Zhang and T. Masumoto, *Mater. Trans JIM*, 1989, **30**, 965–972.
392. A. Inoue, A. Kato, T. Zhang, S.G. Kim and T. Masumoto, *ibid.*, 1991, **32**, 609–616.
393. A. Inoue, T. Zhang and T. Masumoto, *ibid.*, 1990, **31**, 177–183.
394. A. Peker and W.L. Johnson, *Appl. Phys. Lett.*, 1993, **63**, 2342–2344.
395. A. Inoue, N. Nishiyama, K. Amiya, T. Zhang and T. Masumoto, *Mater. Lett.*, 1994, **19**, 131–135.
396. A. Inoue and J. S. Gook, *Mater. Trans. JIM*, 1995, **36**, 1180–1183.
397. X. Wang, I. Yoshii and A. Inoue, *ibid*, 2000, **41**, 539–542.
398. T. Itoi and A. Inoue, *ibid.*, 2000, **41**, 1256–1262.
399. A. Inoue, W. Zhang, T. Zhang and K. Kurosaka, *ibid.*, 2001, **42**, 1149–1151.
400. *idem*, *Acta Mater.*, 2001, **49**, 2645–2652.
401. J. Z. Jiang, B. Yang, K. Saksl, H. Franz and N. Pryds, *J. Mater. Res.*, 2003, **18**, 895–898.
402. R. Busch, *JOM*, 2000, **52**(7), 39–42.
403. C. Suryanarayana, *Prog. Mater. Sci.*, 2001, **46**, 1–184.
404. F. Delogu and G. Cocco, *J. Alloys Compd.*, 2003, **352**, 92–98.
405. M. S. El-Eskandarany, W. Zhang and A. Inoue, *J. Alloys Compd.*, 2003, **350**, 232–245.
406. *idem*, *J. Alloys Compd.*, 2003, **350**, 222–231.
407. M. S. El-Eskandarany, and A. Inoue, *Metall. Mater. Trans A*, 2002, **33A**, 2145–2153.
408. M. S. El-Eskandarany, J. Saida and A. Inoue, *ibid.*, 2003, **34A**, 893–898.
409. D. H. Bae, H. K. Lim, S. H. Kim, D. H. Kim and W. T. Kim, *Acta Mater.*, 2002, **50**, 1749–1759.
410. G. He, Z. F. Zhang, W. Löser, J. Eckert and L. Schultz, *ibid.*, 2003, **51**, 2383–2395.
411. G. He, J. Eckert and W. Löser, *ibid.*, 2003, **51**, 1621–1631.
412. H. J. Jin, X. J. Gu, F. Zhou and K. Lu, *Scripta Mater.*, 2002, **47**, 787–791.
413. H. Men, Z.Q. Hu and J. Xu, *ibid.*, 2002, **46**, 699–703.
414. W. Chen, Y. Wang, J. Qiang and C. Dong, *Acta Mater.*, 2003, **51**, 1899–1907.
415. Y. M. Wang, C. H. Shek, J. B. Qiang, C. H. Wong, W. R. Chen and C. Dong, *Scripta Mater.*, 2003, **48**, 1525–1529.
416. S. R. Nagel and J. Taue, *Phys. Rev. Lett.*, 1975, **35**, 380–383.
417. P. Haussler, *Z. Phys. B*, 1983, **53**, 15–26.
418. G. J. Van der Kolk, A. R. Miedema and A, Niessen, *J. Less Common Metals*, 1988, **145**, 1–17.
419. Z. P. Lu and C. T. Liu, *Acta Mater.*, 2002, **50**, 3501–3512.
420. U. Harms, T. D. Shen and R. B. Schwarz, *Scripta Mater.*, 2002, **47**, 411–414.
421. T. Abe , S. Inoue, D. Mu, Y. Hatano and K. Watanabe, *J. Alloys Compd.*, 2003, **349**, 279–283.
422. W. Liu , H. Wu , Y. Lei and Q. Wang, *ibid.*, 2002, **346**, 244–249.
423. S. Ruggeri , C. Lenain , L. Roue , G. Liang , J. Huot and R. Schulz, *ibid.*, 2002, **339**, 195–201.
424. S.-I. Yamaura , H.-Y. Kim , H. Kimura, A. Inoue and Y. Arata, *ibid.*, 2002, **339**, 230–235.
425. *idem*, *J. Alloys Compd.*, 2002, **347**, 239–243.
426. A. Takasaki and K.F. Kelton, *ibid.*, 2002, **347**, 295–300.
427. K. Tompa, Bánki, M. Bokor, G. Lasanda and L. Vasáros, *ibid.*, 2003, **350**, 52–55.
428. Z. H. Gan, H. Y. Yi, J . Pu, J. F. Wang and J. Z. Xiao, *Scripta Mater.*, 2003, **48**, 1543–1547.
429. E. Jartych, K. Pekala, P. Jéskiewicz, J. Latuch, M. Pekala and J. Grabski, *J. Alloys Compd.*, 2002, **343**, 211–216.
430. M. Xu , M. X. Quan, Z. Q. Hu, L. Z. Cheng and K. Y. He, *ibid.*, 2002, **334**, 238–242.
431. S. J. Pang, T. Zhang, K. Asami and A. Inoue, *Acta Mater.*, 2002, **50**, 489–497.
432. M. Calin, J. Eckert and L. Schultz, *Scripta Mater.*, 2003, **48**, 653–658.
433. C. A. Pampillo, *J. Mater. Sci.*, 1975, **10**, 1194–1227.
434. C. A. Schuh and T. G. Nieh, *Acta Mater.*, 2003, **51**, 87–99.
435. Z. F. Zhang, J. Eckert and L. Schultz, *ibid.*, 2003, **51**, 1167–1179.
436. W. H. Jiang and M. Atzmon, *J. Mater. Res.*, 2003, **18**, 755–757.
437. G. Kumar, J. Eckert, S. Roth, W. Löser, L. Schultz and S. Ram, *Acta Mater.*, 2003, **51**, 229–238.
438. Z. Bian, G. He and G.L. Chen, *Scripta Mater.*, 2002, **46**, 407–412.
439. M. Galano and G. H. Rubiolo, *ibid.*, 2003, **48**, 617–622.
440. W. H. Jiang, F. E. Pinkerton and M. Atzmon, *ibid.*, 2003, **48**, 1195–1200.
441. H. S. Kim, *ibid.*, 2003, **48**, 43–49.
442. L. Wang, L. Ma and A. Inoue, *J. Alloys Compd.*, 2003, **352**, 265–269.
443. P. Wesseling, B. C. Ko and J. J. Lewandowski, *Scripta Mater.*, 2003, **48**, 1537–1541.
444. H. Choi-Yim, R. D. Conner, F. Szuecs and W. L. Johnson, *Acta Mater.*, 2002, **50**, 2737–2745.
445. K. Q. Qiu, Wang, H. F. Zhang, B. Z. Ding and Z. Q. Hu, *Intermetallics*, 2002, **10**, 1283–1288.
446. T. W. Kempen, F. Sommer and E. J. Mittemeijer, *Acta Mater.*, 2002, **50**, 1319–1329.
447. L. Liu , Z. F. Wu and J. Zhang, *J. Alloys Compd.*, 2002, **339**, 90–95.
448. D. S. dos Santos and R. S. de Biasi, *ibid.*, 2002, **335**, 266–269.
449. T. Spassov, P. Solsona, S. Suriñach and M.D. Baró, *ibid.*, 2002, **345**, 123–129.
450. H.-R. Wang, Y.-L. Gao, Y.-F. Ye, G.-H. Min, Y. Chen and X.-Y. Teng, *ibid.*, 2003, **353**, 200–206.
451. B. C. Wei, W. Löser, L. Xia, S. Roth, M. X. Pan, W. H. Wang and J. Eckert, *Acta Mater.*, 2002, **50**, 357–4367.
452. L. Zhang, E. Brück, O. Tegus, K. H. J. Buschow and F. R. de Boer, *J. Alloys Compd.*, 2003, **352**, 99–102.
453. T. Nagase and Y. Umakoshi, *Scripta Mater.*, 2003, **48**, 1237–1242.
454. U. Ramamurty, M. L. Lee, J. Basu and Y . Li, *ibid.*, 2002, **47**, 107–111.

455. M. L. Vaillant, V. Keryvin, T. Rouxel and Y. Kawamura, *ibid.*, 2002, **47**, 19–23.
456. C. A. Neugebauer, J. B. Newkirk and D. A. Vermilyea, Structure and Properties of Thin Films, John Wiley and Sons, New York, NY, 1959.
457. E. Arzt, *Acta Mater.*, 1998, **46**, 5611–5626.
458. P. Chaudhari, *Philos. Mag. A*, 1979, **39**, 507–516.
459. W. D. Nix, *Metall. Trans. A*, 1989, **20A**, 2217–2245.
460. C. V. Thompson, *J. Mater. Res.*, 1993, **8**, 237–238.
461. H. Gao, Y. Huang, W. D. Nix and J. W. Hutchinson, *J. Mech. Phys. Solids*, 1999, **47**, 1239–1263.
462. Y. Huang, H. Gao, W. D. Nix and J. W. Hutchinson, *ibid.*, 2000, **48**, 99–128.
463. N. A. Fleck and J. W. Hutchinson, *Adv. Appl. Mech.*, 1997, **33**, 295–361.
464. R. M. Keller, S. P. Baker and E. Arzt, *J. Mater. Res.*, 1998, **13**, 1307–1317.
465. L. E. Murr and S. S. Hecker, *Scripta Metall.*, 1979, **13**, 167–171.
466. M. F. Ashby, *Philos. Mag.*, 1970, **21**, 399–424.
467. M. S. De Guzman, G. Neubauer, P. Flinn and W. D. Nix, *MRS Symp. Proc.*, 1993, **308**, 613–618.
468. Q. Ma and D. R. Clarke, *J. Mater. Res.*, 1995, **10**, 853–863.
469. N. A. Stelmashenko, M. G. Walls, L. M. Brown and Y. V. Milman, *Acta Metall. Mater.*, 1993, **41**, 2855–2865.
470. N. A. Fleck, G. M. Muller, M. F. Ashby and J. W. Hutchinson, *ibid.*, 1994, **42**, 475–487.
471. J. S. Stolken and A. G. Evans, *Acta Mater.*, 1998, **46**, 5109–5115.
472. M. Atkinson, *J. Mater Res.*, 1995, **10**, 2908–2915.
473. W. J. Poole, M. F. Ashby and N. A. Fleck, *Scripta Mater.*, 1996, **34**, 559–564.
474. W. D. Nix, *Mater. Sci. Eng.*, 1997, **A234**, 37–44.
475. W. D. Nix and H. J. Gao, *J. Mech. Phys. Solids*, 1998, **46**, 411–425.
476. K. W. McElhaney, J. J. Vlassak and W. D. Nix, *J. Mater. Res.*, 1998, **13**, 1300–1306.
477. M. Goken and M. Kempf, *Acta Mater.*, 1999, **47**, 1043–1052.
478. S. Suresh, T. G. Nieh and B. W. Choi, *Scripta Mater.*, 1999, **41**, 951–957.
479. M. R. Begley and J. W. Hutchinson, *J. Mech. Phys. Solids*, 1998, **46**, 2049–2068.
480. N. A. Fleck and J. W. Hutchinson, *ibid.*, 1993, **41**, 1825–1857.
481. E. C. Aifantis, *Int. J. Eng. Sci.*, 1992, **30**, 1279–1299.
482. H. Gao, Y. Huang and W. D. Nix, *Naturwissenschaften*, 1999, **86**, 507–515.
483. J. W. Hutchinson, *Int. J. Solids Struct.*, 2000, **37**, 225–238.
484. O. Kraft and C. A. Volkert, *Advanced Engineering Materials*, 2001, **3**, 99–110.
485. F. R. Brotzen, *Int. Mater. Rev.*, 1994, **39**, 24–45.
486. R. P. Vinci and J. J. Vlassak, *Annu. Rev. Mater. Rev.*, 1996, **26**, 431–462.
487. B. C. Prorok and H. D. Espinosa, *J. Nanoscience Nanotechnology*, 2002, **2**, 427–433.
488. H. D. Espinosa, B. C. Prorok and M. Fischer, *J. Mech. Phys. Solids*, 2003, **51**, 47–67.
489. H. D. Espinosa, B. C. Prorok and B. Peng, *ibid.*, in press, 2003.
490. H. D. Espinosa and B. C. Prorok, *MRS Symp. Proc.*, 2002, **695**, 349–354.
491. W. N. Sharpe, Jr., pp 3-1-3-33 in The MEMS Handbook (Editor: M. Gad-el-Hak), CRC Press, New York, 2002.
492. H. W. Song, S. R. Guo and Z. Q. Hu, *Nanostruct. Mater.*, 1999, **11**, 203–210.
493. J. R. Weertman, pp 397–422 in Nanostructured Materials—Processing, Properties, and Potential Applications (Editors: C.C. Koch), William Andrew Publishing, Norwich, NY, 2002.
494. R. W. Siegel and G. E. Fougere, *Nanostruct. Mater.*, 1996, **6**, 205–216.
495. H. Hahn and K. A. Padmanabhan, *ibid.*, 1996, **6**, 191–200.
496. H. Gleiter, *Acta Mater.*, 2000, **48**, 1–29.
497. G.-M. Chow, I. Ovid'ko and T. Tsakalakos(Editors), Nanostructured Films and Coatings, Kluwer Academic Publishers, Dordrecht, 2000.
498. M. C. Roco, R. S. Williams and P. Alivisatos (Editors), Nanotechnology Research Directions: IWGN Workshop Report, Kluwer Academic Publishers, Dordrecht, 2000.
499. E. O. Hall, *P. Phys. Soc, Lond. B*, 1951, **64**, 747–753.
500. N. J. Petch, *J. Iron Steel I.*, 1953, **174**, 25–28.
501. A. H. Cottrell, *T. Am. I. Min. Met. Eng.*, 1958, **212**, 192–203.
502. J. C. M. Li, *Trans. Metall. Soc. AIME*, 1963, **227**, 239–247.
503. J. C. M. Li and Y. T. Chou, *Metall. Trans.*, 1970, **1**, 1145–1149.
504. M. A. Meyers and E. Ashworth, *Philos. Mag. A*, 1982, **46**, 737–759.
505. T. G. Nieh and J. Wadsworth, *Scripta Metall. Mater.*, 1991, **25**, 955–958.
506. R. A. Masumura, P. M. Hazzledine and C. S. Pande, *Acta Mater.*, 1998, **46**, 4527–4534.
507. H. Van Swygenhoven, M. Spaczer and A. Caro, *ibid.*, 1999, **47**, 3117–3126.
508. J. Schiotz, T. Vegge, F. D. Di Tolla and K. W. Jacobsen, *Phys. Rev. B*, 1999, **60**, 11971–11983.
509. J. Schiotz, F. D. Di Tolla and K. W. Jacobsen, *Nature*, 1998, **391**, 561–563.
510. G. W. Nieman, J. R. Weertman and R. W. Siegel, *J. Mater. Res.*, 1991, **6**, 1012–1027.
511. H. Gleiter, *Prog. Mater. Sci.*, 1989, **33**, 223–315.
512. V. Krstic, U. Erb and G. Palumbo, *Scripta Metall. Mater.*, 1993, **29**, 1501–1504.
513. P. G. Sanders, J. A. Eastman and J. R. Weertman, pp 379–386 in Processing and Properties of Nanocrystalline Materials, (Editors: C. Suryanarayana, J. Singh and F. H. Froes), TMS,: Warrendale, PA, 1996.
514. T. D. Shen, C. C. Koch, T. Y. Tsui and G. M. Pharr, *J. Mater. Res.*, 1995, **10**, 2892–2896.
515. L. Wong, D. Ostrander, U. Erb, G. Palumbo and K. Aust, pp 85–93 in Nanophases and Nanostructured Materials (Editors: R. D. Shull and J. M. Sanchez), TMS, Warrendale, PA, 1993.

516. M. N. Rittner, J. R. Weertman, J. A. Eastman, K. B. Yoder and D. S. Stone, pp 399–406 in Processing and Properties of Nanocrystalline Materials (Editors: C. Suryanarayana, J. Singh and F. H. Froes), TMS: Warrendale, PA, 1996.
517. R. Thomson, pp 2207–2291 in Physical Metallurgy (Editors: R. W. Cahn and P. Haasen), 4th Edition, Elsevier Science, Amsterdam: 1996.
518. P. Nagpal and I. Baker, *Scripta Metall. Mater.*, 1990, **24**, 2381–2384.
519. C. Koch, pp. 93–111 in Nanostructure Science and Technology, R & D Status and Trends in Nanoparticles, Nanostructured Materials and Nanodevices, (Editors: R. W. Siegel, E. Hu and M. C. Roco), Kluwer Academic Publishers, Dordrecht, The Netherlands, 1999.
520. J. Karch, R. Birringer and H. Gleiter, *Nature*, 1987, **330**, 556–558.
521. R. Bohn, T. Haubold, R. Birringer and H. Gleiter, *Scripta Metall. Mater.*, 1991, **25**, 811–816.
522. J. W. Edington, K. N. Melton and C. P. Cutler, *Prog. Mater. Sci.*, 1976, **21**, 63–170.
523. M. M. I. Ahmed and T. G. Langdon, *Metall. Trans. A*, 1977, **8A**, 1832–1833.
524. P. G. Sanders, M. Rittner, E. Kiedaisch, J. R. Weertman, H. Kung and Y. C. Lu, *Nanostructured Materials*, 1997, **9**, 433–440.
525. R. S. Mishra and A. K. Mukherjee, *Mater. Sci. Forum*, 1997, **243**, 315–320.
526. R. S. Mishra, R. Z. Valiev and A. K. Mukherjee, *Nanostructured Materials*, 1997, **9**, 473–476.
527. V. A. Valitov, G. A. Salishchev and S. K. Mukhtarov, *Russ. Metall.*, 1994, 109–112.
528. S. Iijima, *Nature*, 1991, **354**, 56–58.
529. F. Li, H. M. Cheng, S. Bai, G. Su and M. S. Dresselhaus, *Appl. Phys. Lett.*, 2000, **77**, 3161–3163.
530. M. F. Yu, B. S. Files, S. Arepalli and R. S. Ruoff, *Phys. Rev. Lett.*, 2000, **84**, 5552–5555.
531. M. F. Yu, O. Lourie, M. J. Dyer, K. Moloni, T. F. Kelly and R. S. Ruoff, *Science*, 2000, **287**, 637–640.
532. D. A. Walters, L. M. Ericson, M. J. Casavant, J. Liu, D. T. Colbert, K. A. Smith and R. E. Smalley, *Appl. Phys. Lett.*, 1999, **74**, 3803–3805.
533. Z. W. Pan, S. S. Xie, L. Lu, B. H. Chang, L. F. Sun, W. Y. Zhou, G. Wang and D. L. Zhang, *ibid.*, 1999, **74**, 3152–3154.
534. M. B. Nardelli, B. I. Yakobson and J. Bernholc, *Phys. Rev. Lett.*, 1998, **81**, 4656–4659.
535. H. D. Wagner, O. Lourie, Y. Feldman and R. Tenne, *Appl. Phys. Lett.*, 1998, **72**, 188–190.
536. M. S. Dresselhaus, G. Dresselhaus and R. Saito, *Carbon*, 1995, **33**, 883–891.

39 Modelling and simulation

39.1 Introduction

The design of useful materials generally involves the simultaneous optimisation of large numbers of parameters, often in circumstances where the interactions between the parameters are ill-defined. This chapter describes some of the methods by which the task can be enhanced using quantitative models which cover a range of disciplines and which include both rigour and empiricism. From the scientific point of view, modelling can lead to the creation of new theory capable of dealing with complexity; the practical goal is to accelerate the design process and to minimise the use of resources. Each of the following sections is intended to communicate the concepts associated with a particular method; appropriate references are provided where the details can be sought.

39.2 Electron theory

A metal is created when the atoms are brought sufficiently close together so that the electrostatic repulsion in transferring a valency electron between adjacent atoms is offset by the gain due to delocalisation of the electron.[1] This enables the valency electrons to move within the metal.

The essential characteristics of the metallic state can be understood in terms of a single-electron wave function which in one-dimension (x) and for the description of stationary states takes the time-independent form:

$$\psi\{x\} = C \exp\{ikx\}$$

where k is the wave number, C is a constant. All such wave functions must become zero at the boundaries of the metal so that allowed values of k are discrete, each of which defines a quantum state. The Pauli exclusion principle permits only two electrons with opposite spins to occupy each state. The energy of each state scales with k^2 and the highest occupied state at $0\,\mathrm{K}$ is said to have the Fermi energy.

The delocalised electrons feel only a weak electrostatic field ('pseudopotential'), from the positively charged atomic cores. This is because the attraction of the valence electrons to the positive cores is greatly reduced by the repulsion from the core electrons. The delocalised electrons are also partly screened from each other by small exclusion zones (positive holes).

It is useful to understand why the electrons are able to move within the potential of the screened positive ion cores without being scattered. Bloch showed that the effect of the periodic potential $u\{x\}$ is simply to modulate the free-electron wave function by a term $u\{x\}$:

$$\psi\{x\} = u\{x\} \exp\{ikx\}$$

Difficulties only arise when the wave vector **k** is such that the wave satisfies the Bragg condition; the electron cannot then, on average, move forward or backward through the reflecting planes. In wave theory this is represented by using two functions corresponding to $\pm k$, one of which places the maximum electron density where the positive potential is lowest, giving a solution with an energy that is lower than that of a completely free electron. Similarly the other solution has an energy higher than that of a free electron. This introduces *band gaps* in the distribution of electron energies. Electrons are able to move, and hence the metallic state is able to exist, if the valency bands are only partly filled.

We have so far considered the metallic state in terms of a single-electron wave function; this fails to account for the vast numbers of electrons that are interacting within the metal. There will be Coulomb correlations between the electrons. Their spins may also correlate. The density functional method which is the basis of much of the modern theory of metals, provides an expression of these effects whilst at the same time exploiting the one-electron equation. The energy E of an electron gas is written as a function of a function (i.e. a functional) in terms of the electron density n and a radial coordinate r:

$$E\{n\} = \int V\{r\}n\{r\}\,\mathrm{d}r + \frac{1}{2}\int\int \frac{e^2}{r_{12}}n\{r_1\}n\{r_2\}\,\mathrm{d}r_1\,\mathrm{d}r_2 + T\{n\} + E_{xc}\{n\}$$

where the first term is the pseudopotential due to the ionic cores, the second term is the classical Coulomb interaction energy (e is the electronic charge), the third term is the kinetic energy (i.e. Fermi energy) and the fourth term accounts for the exchange and correlation effects. The use of this equation, and the adaptations necessary to account for various complications (e.g. to take account of partly covalent characteristics), represents the skill in using electron theory. The most rigorous application of the theory requires the largest amount of computing power and hence is limited to circumstances where there is strict periodicity in the atomic structure so that only small numbers of atoms need to be considered with periodic boundary conditions. It is then possible to calculate, for example, the cohesive energy of arbitrary crystal structures, the elastic properties and surface energies.

Some calculations for iron are illustrated in Figure 39.1, which shows the cohesive energy as a function of the density and crystal structure. Of all the test structures, hexagonal close-packed iron (h.c.p.) is found to show the highest cohesion and therefore should represent the most stable form. This contradicts experience, because body-centred cubic iron (b.c.c.) is the equilibrium form at low temperatures. This discrepancy arises because the model ignores magnetic effects—it is ferromagnetism which stabilises b.c.c. iron over the h.c.p. form.

The example illustrates how all models begin with intentional or unintentional simplifying assumptions and yet can be useful. The electron theory not only highlights the importance of magnetic terms, but in addition makes it possible to study crystal structures of iron which have yet to be achieved in practice. The diamond form of iron would have a density of only 5 g cm^{-3}. Unfortunately, the calculations show that the difference in energy between the diamond and b.c.c. forms is so large that it is improbable for the b.c.c. → diamond transformation to be induced by alloying. The methodology has been used to see whether certain intermetallic compounds can be forced to occur into desirable crystal structures.[3]

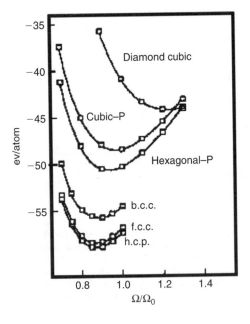

Figure 39.1   *Plot of cohesive energy* (0 K) *versus the normalised volume per atom for a variety of crystal structures of iron. Hexagonal–P and Cubic–P are primitive structures*[2]

One application of the theory is to derive empirical potentials representing pairwise or higher order interactions between neighbouring atoms. Such potentials are used in molecular dynamics simulations in which the motion of hundreds of thousands of atoms can be followed simultaneously.[4]

39.3   Thermodynamics and equilibrium phase diagrams

Electron theory is not yet able to give sufficiently accurate calculations of phase diagrams; thermodynamic models which exploit experimental databases are much more suited for this purpose.[5–7] The basic information to be gained from a binary phase diagram in which temperature is plotted against concentration, is the chemical composition and proportion of each phase at a specified temperature and average composition. Plotting such diagrams becomes tedious for ternary alloys, and ceases to be useful for higher order alloys. This is not a problem in practice because the information to be gained from a multicomponent, multiphase diagram is identical to that expressed for a binary system. The volume fractions and chemical compositions can readily be communicated numerically instead of using complicated diagrams, as in Table 39.1.

Given thermodynamic data, calculations such as those presented in Table 39.1 are based on the simple requirement that the chemical potential must be uniform at equilibrium:

$$\mu_i^\alpha = \mu_i^\beta = \dots \quad \text{for } i = 1, 2, 3, \dots \quad \text{and} \quad \text{phase} = \alpha, \beta \dots \tag{1}$$

where μ_i^j represents the chemical potential of component i. There are many thermodynamic models for expressing the chemical potential as a function of the mixing the solutes. Each model will have a contribution from the entropy of mixing (ΔS_M) and enthalpy of mixing (ΔH_M) to derive a free energy of mixing (ΔG_M). The equations defining these quantities depend on the assumptions: in ideal mixing the solutes are dispersed at random and $\Delta H_M = 0$; the regular solution model has $\Delta H_M \neq 0$ but it is still assumed that ΔS_M can be calculated assuming random mixing; quasichemical theory also has a finite enthalpy of mixing but does a better estimation of ΔS_M. The method adopted for generalised computation of phase diagrams recognises that in practice it is necessary to implement calculations over a wide range of concentrations, parameters and phases. The molar free energy of mixing in a binary solution is often written as:

$$\Delta G_M = \Delta_e G + N_a kT[(1 - x)\ln\{1 - x\} + x\ln\{x\}] \tag{2}$$

where N_a is Avogadro's number, k is the Boltzmann constant, x is the mole fraction of solute, T is the absolute temperature. The last term in equation 2 comes from the ideal entropy of mixing. The excess Gibbs free energy $\Delta_e G$ expresses the deviation from an ideal solution and is written empirically for a binary solution with components A and B as:

$$\Delta_e G_{AB} = x_A x_B \sum_i L_{AB,i}(x_A - x_B)^i \tag{3}$$

The excess free energy of a ternary solution can as a first approximation be expressed in terms of purely binary interactions:

$$\begin{aligned} \Delta_e G_{ABC} &= x_A x_B \sum_i L_{AB,i}(x_A - x_B)^i \\ &+ x_B x_C \sum_i L_{BC,i}(x_B - x_C)^i \\ &+ x_C x_A \sum_i L_{CA,i}(x_C - x_A)^i \end{aligned} \tag{4}$$

Table 39.1   EQUILIBRIUM PHASE MIXTURE FOR Fe–0.1C–0.2Si–0.5Mn–9Cr–1W wt%, AT 873 K

Phase fractions and concentrations are as mole fractions

Phase	Fraction	C	Si	Mn	Cr	W
Ferrite	0.976 9	0.000 001 9	0.006 067 5	0.005 124 6	0.083 947 3	0.001 998 7
$M_{23}C_6$	0.022 3	0.206 896 6	0.000 000 0	0.001 977 6	0.624 445 9	0.036 302 9
Laves	0.000 8	0.000 000 0	0.000 000 0	0.000 000 0	0.134 653 0	0.333 333 3

The advantage of the representation embodied in equation 4 is that for the ternary case, the relation reduces to the binary problem when one of the components is set to be identical to another, e.g. $B \equiv C$.

There might exist ternary interactions, in which case a term $x_A x_B x_C L_{ABC,0}$ is added to the excess free energy. If this does not adequately represent the deviation from the binary summation, then it can be converted into a series which properly reduces to a binary formulation when there are only two components:

$$x_A x_B x_C \left[L_{ABC,0} + \frac{1}{3}(1 + 2x_A - x_B - x_C)L_{ABC,1} \right.$$

$$+ \frac{1}{3}(1 + 2x_B - x_C - x_A)L_{BCA,1}$$

$$\left. + \frac{1}{3}(1 + 2x_C - x_A - x_B)L_{CAB,1} \right]$$

The method can clearly be extended to deal with any number of components, with the great advantage that few coefficients have to be changed when the data due to one component are improved. The experimental thermodynamic data necessary to derive the coefficients may not be available for systems higher than ternary so high order interactions are often set to zero.

39.4   Thermodynamics of irreversible processes

Equilibrium as described above represents a state in which there is no perceptible change no matter how long one observes the state. A process of change (i.e. an irreversible process) is by contrast associated with the dissipation of free energy and is best described using kinetic theory. Before dealing with full-blown kinetic theory, it is possible to study a compromise model in which thermodynamics is used to represent systems in which there is *apparently* no change but free energy is nevertheless being dissipated. This is the case of the *steady state*, for example, diffusion across a constant gradient; neither the flux nor the concentration at any point changes with time, and yet the free energy of the system is decreasing since diffusion occurs to minimise free energy. The rate at which energy is dissipated is the product of the temperature and the rate of entropy production (i.e. $T\sigma$) with:[8,9]

$$T\sigma = JX \tag{5}$$

where J is a generalised flux of some kind, and X a generalised force. In the case of an electrical current, the heat dissipation is the product of the current (J) and the electromotive force (X).

Provided the flux–force sets can be expressed as in equation 5, the flux must naturally depend in some way on the force. It may then be written as a function $J\{X\}$ of the force X. It is found that for small deviations from equilibrium, the flux is proportional to force, Table 39.2.

There are many circumstances where a number of irreversible processes occur together. In a ternary Fe–Mn–C alloy, the diffusion flux of carbon depends not only on the gradient of carbon, but also on that of manganese. Thus, a uniform distribution of carbon will tend to become inhomogeneous in the presence of a manganese concentration gradient. Similarly, the flux of heat may not depend on the temperature gradient alone; heat can be driven also by an electromotive force. Electromigration

Table 39.2   EXAMPLES OF FORCES AND THEIR CONJUGATE FLUXES

z is distance, ϕ is the electrical potential in volts, and μ is a chemical potential

Force	Flux
Electromotive force (e.m.f.) $= \dfrac{\partial \phi}{\partial z}$	Electrical current
$-\dfrac{1}{T}\dfrac{\partial T}{\partial z}$	Heat flux
$-\dfrac{\partial \mu_i}{\partial z}$	Diffusion flux

involves diffusion driven by an electromotive force. When there is more than one dissipative process, the total energy dissipation rate can still be written.

$$T\sigma = \sum_i J_i X_i.$$

(6)

For multiple irreversible processes, it is found experimentally that each flow J_i is related not only to its conjugate force X_i, but also is related linearly to all other forces present. Thus,

$$J_i = M_{ij} X_j$$

(7)

with $i, j = 1, 2, 3, \ldots$. Therefore, a given flux depends on all the forces causing the dissipation of energy. Onsager has shown that to maintain dynamic equilibrium, and provided that the forces and fluxes are chosen from the dissipation equation and are independent, $M_{ij} = M_{ji}$. An exception occurs with magnetic fields in which case there is a sign difference $M_{ij} = -M_{ji}$.

39.5   Kinetics

Many kinetic processes are not consistent with the steady-state. The vast majority of such processes in metals involve aspects of nucleation and growth; the theory for nucleation and growth is standard and is assumed here, in order to focus on the estimation of the volume fraction, which requires *impingement* between particles to be taken into account.

This is done using the extended volume concept of Kolmogorov, Johnson, Mehl and Avrami.[10] Suppose that two particles (Fig. 39.2) exist at time t; a small interval δt later, new regions marked a, b, c & d are formed assuming that they are able to grow unrestricted in extended space whether or not the region into which they grow is already transformed. However, only those components of a, b, c & d which lie in previously untransformed matrix can contribute to a change in the real volume of the product phase (α):

$$dV^\alpha = \left(1 - \frac{V^\alpha}{V}\right) dV_e^\alpha$$

(8)

where it is assumed that the microstructure develops at random. The subscript e refers to extended volume, V^α is the volume of α and V is the total volume. Multiplying the change in extended volume by the probability of finding untransformed regions has the effect of excluding regions such as b, which clearly cannot contribute to the real change in volume of the product. For a random distribution of precipitated particles, this equation can easily be integrated to obtain the real volume fraction,

$$\frac{V^\alpha}{V} = 1 - \exp\left\{-\frac{V_e^\alpha}{V}\right\}$$

(9)

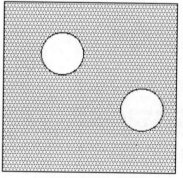
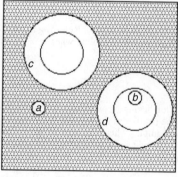

Time = t Time = $t + \Delta t$

Figure 39.2   *An illustration of the concept of extended volume. Two precipitate particles have nucleated together and grown to a finite size in the time t. New regions c and d are formed as the original particles grow, but a & b are new particles, of which b has formed in a region which is already transformed*

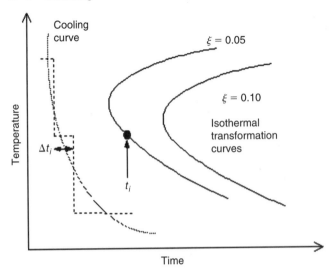

Figure 39.3   *The Scheil method for converting between isothermal and anisothermal transformation data*

The extended volume V_e^{α} is straightforward to calculate using nucleation and growth models and neglecting any impingement effects. Solutions typically take the form:

$$\xi = 1 - \exp\{-k_A t^n\} \tag{10}$$

where k_A and n characterise the reaction as a function of mechanism of transformation, time, temperature and other variables.

Reactions sometimes do not happen in isolation. For example, a steel designed to serve at 600°C over a period of 30 years may contain more than six different kinds of precipitates so that it can sustain a load without creeping. A simple modification for the simultaneous formation of two precipitates (α and β) is that the relation between extended and real space becomes a coupled set of two equations,[11,12]

$$\mathrm{d}V^{\alpha} = \left(1 - \frac{V^{\alpha} + V^{\beta}}{V}\right)\mathrm{d}V_e^{\alpha} \quad \text{and} \quad \mathrm{d}V^{\beta} = \left(1 - \frac{V^{\alpha} + V^{\beta}}{V}\right)\mathrm{d}V_e^{\beta} \tag{11}$$

This can be done for any number of reactions happening together. The resulting set of equations must in general be solved numerically.

A popular method of converting between isothermal and anisothermal transformation data is the *additive reaction rule* of Scheil.[10] A cooling curve is treated as a combination of a sufficiently large number of isothermal reaction steps. Referring to Figure 39.3, a fraction $\xi = 0.05$ of transformation is achieved during continuous cooling when

$$\sum_i \frac{\Delta t_i}{t_i} = 1 \tag{12}$$

with the summation beginning as soon as the parent phase cools below the equilibrium temperature.

The rule can be justified if the reaction rate depends solely on ξ and T. Although this is unlikely, there are many examples where the rule has been empirically applied with success. Reactions for which the additivity rule is justified are called isokinetic, implying that the fraction transformed at any temperature depends only on time and a single function of temperature.

39.6   Monte Carlo simulations

The complete description of microstructure requires both field and feature parameters; the former describes averaged quantities such as volume fraction and the amount of surface per unit volume;

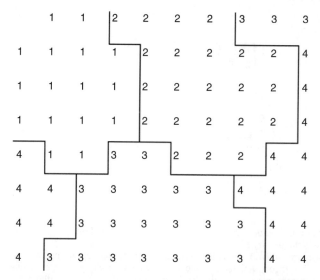

Figure 39.4 *Mapping of variable defining crystallographic orientation; the orientation domains are separated by grain boundaries. In this example, the elements within each grain are arranged in a square pattern*

the latter includes for example, size distributions. One disadvantage of the Avrami method is that it yields only the field parameters since information about individual particles is lost during the conversion from extended to real space. Methods such as Monte Carlo and phase field simulations permit the visualisation of microstructural development and can therefore reveal feature parameters.

The simulation of microstructure using the Monte Carlo method begins with the definition of domains using an appropriate variable which can take on discrete values out of a set of possible values.[4] A case where the variable is used to represent crystallographic orientation is illustrated in Figure 39.4, with grain boundaries located wherever the variable changes value.

We shall consider isothermal grain coarsening to illustrate the method. Having defined the starting grain structure as in Figure 39.4, the elements located at the grain boundaries are identified and are the only ones considered in further calculations. Each of these grain boundary variables is then randomly perturbed. If the perturbation leads to an overall reduction in energy, it is accepted. If it leads to an increase ΔE in energy, then it is accepted only if the term $\exp\{-\Delta E/kT\} \leq p$ where p is a random, computer-generated number between 0 and 1.

The accepted perturbations lead to a new grain structure once all the elements have been sampled. The process of sampling of all of the elements defined in the model corresponds to one Monte Carlo step, equivalent to the progress of time. Since each of the elements illustrated in Figure 39.4 is much coarser than an atom, care must be taken to investigate the effect of this 'coarse-graining'; indeed, the symmetry of the pattern in which the elements are arranged within each grain is known to influence the outcome.[4]

39.7 Phase field method

Consider the growth of a precipitate which is separated from the matrix by an interface. There are then three distinct quantities: the precipitate, matrix and interface. The interface is described as an evolving surface whose motion is controlled according to the boundary conditions imposed to describe the physical mechanisms which lead to the growth of one phase at the expense of the other. This mathematical description categorises the boundary as a two-dimensional surface with no width or structure, a *sharp interface*.

By contrast, the phase-field method[13] begins with a description of the entire microstructure, including the interface, in terms of a single variable known as the *order parameter*. The precipitate and matrix each have a particular value of the order parameter. Likewise, the interface is located by the position where the order parameter changes from its precipitate-value to its matrix-value. The order parameter is thus continuous as one traverses the precipitate and enters the matrix (Fig. 39.5).

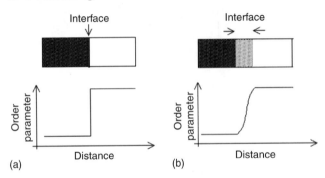

Figure 39.5   *(a) Sharp interface. (b) Diffuse interface*

The range over which it changes is the *width* of the interface. The set of values of the order parameter over the whole microstructure is the *phase field*.

The order parameter changes smoothly through the interfacial region from one limiting value to the other, in a phase-field model. A theory which tracks the dynamics of the order parameter across the entire phase field, would allow the evolution of the microstructure to be calculated without the need to track the interface. Such a theory would require the free energy as a function of the order parameter so that the field can follow a path which leads to a maximisation of entropy production.

Two examples of phase-field modelling are as follows: the first is that of grain growth, where the order parameter is not conserved since the amount of grain surface per unit volume decreases with grain coarsening. The second conserves the order parameter. When composition fluctuations evolve into precipitates, it is the average chemical composition which is conserved. We now consider this second example in a little more detail.

In solutions that tend to exhibit the clustering of atoms, it is possible for a homogeneous phase to become unstable to infinitesimal perturbations of chemical composition. The free energy of a solid solution which is chemically heterogeneous can be factorised into three components.[14] First, there is the free energy of a small region of the solution in isolation, given by the usual plot of the free energy of a homogeneous solution as a function of chemical composition. The second term comes about because the small region is surrounded by others which have different chemical compositions. This *gradient term* is an additional free energy in a heterogeneous system, and is regarded as an interfacial energy describing a 'soft interface' of the type illustrated in Figure 39.5b. In this example, the soft-interface is due to chemical composition variations, but it could equally well represent a structural change. The third term arises because a variation in chemical composition also causes lattice strains in the solid-state. We shall neglect these coherency strains.

The free energy per atom of an inhomogeneous solution is then given by:

$$g_{ih} = \int [g\{c_0\} + v^3 \kappa (\nabla c)^2] \mathrm{d}V \tag{13}$$

where $g\{c\}$ is the free energy per atom in a homogeneous solution of concentration c_0, v is the volume per atom and κ is called the *gradient energy coefficient*. g_{ih} is often referred to as a free energy *functional* since as in density functional theory, it is a function of a function.

Equilibrium in a heterogeneous system is obtained by minimising the functional, subject to the requirement that the average concentration is maintained constant:

$$\int (c - c_0) \mathrm{d}V = 0$$

where c_0 is the average concentration. Spinodal decomposition can therefore be simulated using the functional defined in equation 13. The system would initially be set to be homogeneous but with some compositional noise. It would then be perturbed, allowing those perturbations which reduce free energy to survive. In this way, the whole decomposition process can be modelled without explicitly introducing an interface. The interface is instead represented by the gradient energy coefficient.

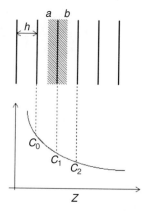

Figure 39.6 *Finite difference representation of diffusion*

39.8 Finite difference method

The finite difference method is useful in the numerical solution of second order boundary value problems.[15] Consider one-dimensional diffusion in a concentration gradient along a coordinate z (Fig. 39.6). The concentration profile is divided into slices, each of thickness h. The matter entering a unit area of the face at a in a time increment τ is given approximately by $J_a = -D\tau(C_1 - C_0)/h$. That leaving the face at b is $J_b = -D\tau(C_2 - C_1)/h$. If C_1' is the new concentration in slice 1, then the net gain in solute is $(C_1' - C_1)h$ so that

$$C_1' - C_1 = \frac{D\tau}{h^2}(C_0 - 2C_1 + C_2) \tag{14}$$

This allows the concentration at a point to be calculated as a function of that at the two neighbouring points. By successively applying this relation at each slice, and advancing the time τ, the entire concentration profile can be estimated as a function of time.

The approximation here is that we have applied Fick's first law to each slice, i.e. assumed that the concentration gradient within each slice is constant. Since the profile is in fact a curve, the approximation will be better for smaller values of h, but the computation times will be correspondingly longer. Considerations of numerical accuracy are vital in such methods; the accuracy can be assessed by changing h and seeing whether it makes a significant difference to the calculated profile.

39.9 Finite element method

In finite element analysis, continuous functions are replaced by piecewise approximations. Thus, a finite element representation of a circle would be a circumscribed polygon, with each edge being a finite element. We shall discuss this in terms of the forces involved in the deflection of springs.[16]

It is assumed that for any spring, the force varies linearly with displacement: $F = k\delta$ where k is the stiffness of the spring. Consider the forces at the nodes of a spring in a system of springs at equilibrium, as illustrated in Figure 39.7a. Since $F_1 = -F_2$,

$$\begin{bmatrix} F_1 \\ F_2 \end{bmatrix} = \begin{pmatrix} k & -k \\ -k & k \end{pmatrix} \begin{bmatrix} \delta_1 \\ \delta_2 \end{bmatrix}$$

The forces at the nodes of the springs illustrated in Figure 39.7b are therefore

$$\begin{bmatrix} F_1 \\ F_2 \\ 0 \end{bmatrix} = \begin{pmatrix} k_1 & -k_1 & 0 \\ -k_1 & k_1 & 0 \\ 0 & 0 & 0 \end{pmatrix} \begin{bmatrix} \delta_1 \\ \delta_2 \\ \delta_3 \end{bmatrix} \qquad \begin{bmatrix} 0 \\ F_2 \\ F_3 \end{bmatrix} = \begin{pmatrix} 0 & 0 & 0 \\ 0 & k_2 & -k_2 \\ 0 & -k_2 & k_2 \end{pmatrix} \begin{bmatrix} \delta_1 \\ \delta_2 \\ \delta_3 \end{bmatrix}$$

$$\begin{bmatrix} F_1 \\ F_2 \\ F_3 \end{bmatrix} = \underbrace{\begin{pmatrix} k_1 & -k_1 & 0 \\ -k_1 & k_1 & 0 \\ 0 & 0 & 0 \end{pmatrix} + \begin{pmatrix} 0 & 0 & 0 \\ 0 & k_2 & -k_2 \\ 0 & -k_2 & k_2 \end{pmatrix}}_{\text{component stiffnesses}} \begin{bmatrix} \delta_1 \\ \delta_2 \\ \delta_3 \end{bmatrix} \equiv \underbrace{\begin{pmatrix} k_1 & -k_1 & 0 \\ -k_1 & k_1 + k_2 & -k_1 \\ 0 & -k_2 & k_2 \end{pmatrix}}_{\text{overall stiffness}} \begin{bmatrix} \delta_1 \\ \delta_2 \\ \delta_3 \end{bmatrix}$$

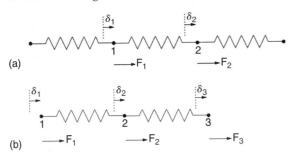

(a)

(b)

Figure 39.7   *Forces on springs*

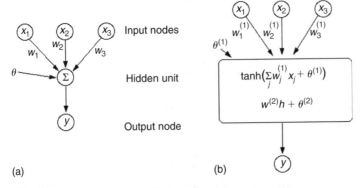

(a)   (b)

Figure 39.8   *(a) A neural network representation of linear regression. (b) A non-linear network representation*

This simple case illustrates how the properties of the elements can be combined to yield an overall response function.

39.10   Empirical modelling: neural networks

There are problems in materials science where the concepts might be understood but which are not amenable to thorough scientific treatment. Empirical regression then becomes useful. By fitting data to a specified relationship, an equation is obtained which relates the inputs x_j via weights w_j and a constant θ to obtain an estimate of the output $y = \sum_j w_j x_j + \theta$. Because the variables are assumed to be independent, this equation can be stated to apply over a certain range of the inputs.

This linear regression method can be represented as a neural network (Fig. 39.8a). The inputs x_i define the *input nodes*, and there is an *output node*. Each input is multiplied by a random weight w_i and the products are summed together with a constant θ to give the output $y = \sum_i w_i x_i + \theta$. The summation is an operation which is hidden at the *hidden node*. Since the weights and the constant θ were chosen at random, the value of the output will not match with experimental data. The weights are changed systematically until a best-fit description of the output is obtained as a function of the inputs; this operation is known as *training* the network.

The network can be made non-linear as shown in Figure 39.8b. As before, the input data x_j are multiplied by weights ($w_j^{(1)}$), but the sum of all these products forms the argument of a hyperbolic tangent:

$$h = \tanh\left(\sum_j w_j^{(1)} x_j + \theta\right) \quad \text{with} \quad y = w^{(2)} h + \theta^{(2)} \tag{15}$$

where $w^{(2)}$ is a weight and $\theta^{(2)}$ another constant. The strength of the hyperbolic tangent *transfer function* is determined by the weight w_j. The output y is therefore a non-linear function of x_j, the

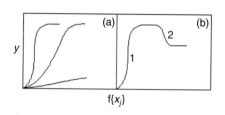

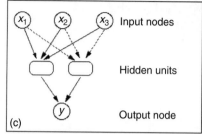

Figure 39.9 *(a) Three different hyperbolic tangent functions; the 'strength' of each depends on the weights. (b) A combination of two hyperbolic tangents to produce a more complex model. (c) A two hidden-unit network*

function usually chosen being the hyperbolic tangent because of its flexibility. The exact shape of the hyperbolic tangent can be varied by altering the weights (Fig. 39.9a).

A one hidden-unit model may not be sufficiently flexible. Further degrees of non-linearity can be introduced by combining several of the hyperbolic tangents (Fig. 39.9b), permitting the neural network method to capture almost arbitrarily non-linear relationships. The number of tanh functions per input is the number of hidden units; the structure of a two hidden unit network is shown in Figure 39.9c.

The function for a network with i hidden units is given by

$$y = \sum_i w_i^{(2)} h_i + \theta^{(2)} \quad \text{where} \quad h_i = \tanh\left(\sum_j w_{ij}^{(1)} x_j + \theta_i^{(1)}\right) \tag{16}$$

Notice that the complexity of the function is related to the number of hidden units. A neural network like this can capture interactions between the inputs because the hidden units are nonlinear. Appropriate measures must be taken to avoid overfitting. With complex networks it is not possible to easily specify a range of applicability. MacKay has developed a particularly useful treatment of neural networks in a Bayesian framework;[18] instead of calculating a unique set of weights, a probability distribution of sets of weights is used to define the fitting uncertainty. The error bars therefore become large when data are sparse or locally noisy.

REFERENCES

1. A. H. Cottrell, Introduction to the Modern Electron Theory of Alloys, Institute of Materials, London, U.K., 1989.
2. A. T. Paxton, M. Methfessel and H. M. Polatoglou, *Physical Review B*, 1990, **41**, 8127.
3. D. G. Pettifor, *Materials Science and Technology*, 1992, **8**, 345.
4. D. Rabbe, Computational Materials Science, Wiley–VCH, Germany, 1998.
5. L. Kaufman, *Prog. in Materials Science*, 1969, **14**, 57.
6. M. Hillert, Hardenability Concepts with Applications to Steels, eds D. V. Doane and J. S. Kirkaldy, TMS–AIME, Warrendale, Pennsylvania, U.S.A., 1977, 5.
7. K. Hack, SGTE Casebook: Thermodynamics at Work, Institute of Materials, London, U.K., 1996.
8. D. G. Miller, *Chemical Reviews*, 1960, **60**, 15.
9. K. G. Denbigh, Thermodynamics of the Steady State, John Wiley and Sons, Inc., New York, U.S.A., 1955.
10. J. W. Christian, Theory of Transformations in Metals and Alloys, Part 1, 2nd edition, Pergamon Press, Oxford, U.K., 1975.
11. J. D. Robson and H. K. D. H. Bhadeshia, *Materials Science and Technology*, 1997, **13**, 631.
12. S. J. Jones and H. K. D. H. Bhadeshia, *Acta Materialia*, 1997, **45**, 2911.
13. J. A. Warren, IEEE Computational Science and Engineering, Summer 1995, 38.
14. J. W. Cahn and J. E. Hilliard, *Journal of Chemical Physics*, 1959, **31**, 688.
15. J. Crank, The Mathematics of Diffusion, 2nd edition, Clarendon Press, Oxford, 1975.
16. K. M. Entwistle, Basic Principles of Finite Element Analysis, Institute of Materials, London, U.K., 1999.
17. D. J. C. MacKay, *Neural Computation*, 1992, **4**, 415.
18. H. K. D. H. Bhadeshia, *ISIJ International*, 1999, **39**, 966.

40 Supporting technologies for the processing of metals and alloys

40.1 Introduction and cross-references

This chapter brings together three topics that support the processing of metals and alloys, namely: the design of metallurgical furnaces, vacuum technology and the control of metallurgical processes. Material in this chapter relates to the following:

- The statistical basis of process control is discussed in Chapter 2.
- Thermocouple-based temperature measurement and information on emissivity relevant to pyrometry are outlined respectively in Chapters 16 and 17.
- Foundry practice and heat-treatment are considered respectively in Chapters 26 and 29.
- Refractory materials form the topic of Chapter 27.
- For a discussion of fuels, *see* Chapter 28.

40.2 Furnace design

40.2.1 Introduction

There are many considerations to be made when designing a metallurgical processing furnace. A furnace is not just simply a box of bricks; it is a unit which encompasses many engineering aspects including:

- Furnace safety
- Energy input
- Temperature measurement
- Temperature control
- Temperature uniformity
- Temperature recording
- Mechanical handling
- Door mechanisms
- Heat loss
- Batch or continuous operation
- Pre-cleaning
- Post-cleaning
- Method of quenching
- Quench media
- Quench agitation
- Heat transfer from quench medium
- Furnace atmosphere
- Generation of furnace atmosphere
- Control of furnace atmosphere
- Mass (load weight) of work pieces
- Heat transfer to load.

The above are just a few of the considerations that are necessary in the design of the metallurgical processing furnace. Furnace design requires the engineer to have, or to have access to, multiple disciplines which range from:

- Mechanical engineering
- Electrical engineering
- Metallurgical engineering
- Physics
- Chemistry.

These disciplines will assist the engineer in designing a cost-effective, functioning, metallurgical processing furnace that will process the necessary work in an efficient and safe manner to the benefit of both the furnace operator and the resulting product's end user.

Metallurgical processing furnaces can be categorised into two main groups, which relate to the types of materials to be processed.

- Ferrous materials
- Non-ferrous materials

Within the above two groups, further subdivisions can be made in relation to heating methods. These subdivisions are:

Electrical heating

The electrical heating methods and techniques can be further subdivided into elemental type heating systems and other electrical methods of heating such as laser, radio frequency, electron beam, infrared. The latter four types of heating will not be considered in this section.

Gas heating

Furnaces can be further categorised into three distinctive groupings which relate to the general types of application. These distinctive groupings are:

Ovens: An oven is considered to be a low operating temperature system. This means that the unit will operate from room temperature up to approximately 300°C (600°F).

Heat treatment furnaces: A heat treatment furnace will operate over a temperature range from 150°C (300°F) up to (as in the case of vacuum furnaces) 1 650°C (3 000°F), although specialist furnaces can reach far higher temperatures.

Melting and holding furnaces: These furnaces will melt both ferrous and non-ferrous materials ranging from lead to aluminium, brasses, cast irons and steels.

The focus of the discussion in this section will be on the design of heat treatment furnaces only, i.e. this means furnaces that are capable of heat treating aluminium alloys and various types of steels, whose only function is to accomplish phase changes within the selected metal.

40.2.2   Types of furnaces

There are three simple furnace configurations, which are as follows:

- Horizontal furnaces
- Pit furnaces
- Bottom load elevator hearth furnaces.

The first group of furnaces to be considered will be the horizontal furnace configuration.

40.2.2.1   *Box furnaces*

The box furnace is perhaps the most simple of all of the heat treatment furnaces. This is usually of a rugged and simple construction. The furnace will literally be a box which will contain insulation material (refractory insulation or thermal blanket insulation). The furnace usually has a range of operating temperatures from 150°C (300°F) up to 1 200°C (2 200°F). Some high-temperature box furnaces that are heated by silicon carbide elements are capable of reaching temperatures above 1 300°C (2 400°F). Heating methods, such as the use of $MoSi_2$ elements, for higher temperature vertical furnaces are not discussed here. The furnace can have either a single access door or an entry and exit door at each end of the furnace. The furnace heating system can be:

Direct heating

In this instance, the products of combustion (if gas-fired) can provide an atmosphere. This atmosphere can be either oxidising, carburising or decarburising. If the furnace is heated electrically, then heat transfer will be either by convection, conduction or both, depending on the furnace's maximum operating temperature, with radiant heat transfer becoming important at high temperature.

Indirect heating

A furnace heated by indirect methods can have its heat source located in an external heating chamber, and the heated air driven by an air blower which will distribute the heated air into the work processing chamber through a series of dampers, which are set to distribute the heated air and ensure good temperature uniformity.

40.2.2.2 *Integral quench furnaces*

The integral quench furnace is perhaps the 'workhorse' of the heat treatment furnaces. The design of the integral quench furnace dates back to the late 1940s. The original aim of this type of furnace was to transform a box type (batch) furnace into a continuous/semi continuous type furnace. The design of the furnace integrates the use of two separate process chambers, the first chamber being the entry vestibule which contains the quench tank and is separated by an intermediate insulated door. Behind the intermediate door is the high heat chamber in which the heat treatment procedure could be carried out. The high heat chamber contains the furnace heating system and the furnace atmosphere system, as well as the mechanical handling equipment which enables the workload to be transferred into the furnace from an external load table, through the entry vestibule and on into the heating chamber. Upon completion of the heat treatment procedure, the workload is transferred back into the front entry vestibule and on to an elevator table which will lower the workload into the quench medium (if required). If the process is, for example, an annealing process, then the workload can sit on the elevator table without being immersed into the quench medium. This will allow a slow cool.

The integral quench furnace is used for many different heat treatment processes, such as:

– Annealing
– Carburising
– Carbonitriding
– Ferritic nitrocarburising
– Solutionising
– Neutral hardening.

The limiting factors of the integral quench furnaces can be both the volume of quench medium in the quench tank and size limitations of the heating chamber. Generally, these types of furnaces will have a temperature limitation of 950°C (1 750°F).

Many advances have been made since the late 1940s in terms of heating efficiency, operating costs, work load handling and most importantly, process control in terms of atmosphere control, temperature control, temperature uniformity and minimal heat losses.

40.2.2.3 *Pit furnaces*

As the name implies, this type of furnace can be installed into a pit or be mounted on the shop floor. It will be the size of the furnace, height clearance above the furnace, and foundation/floor conditions that will determine the positioning of the furnace.

These furnaces are also known as vertical air circulation furnaces. This means that the heated air/atmosphere within the furnace is circulated throughout the heating chamber, which provides good temperature uniformity. The air circulation is accomplished by the use of a fan, located either in the base of the furnace or in the furnace lid. This type of furnace is used typically for the following heat treatment processes:

– Tempering
– Stress relieving
– Normalising
– Annealing

- Nitriding
- Ferritic nitrocarburising
- Carbonitriding
- Carburising.

The furnace construction is very simple, once again, and offers excellent uniformity of both temperature and atmosphere (if used). Improved temperature uniformity is a result of the shape of the furnace chamber, which is circular.

40.2.2.4 *Horizontal car bottom furnaces*

The car bottom furnace is designed usually to process large products such as fabrications, pressure vessels, ingots, forgings and castings. The heat treatment procedures usually carried out in these furnaces are:

- Annealing
- Normalising
- Stress relieving
- Hardening
- Tempering.

The furnace consists of a static steel framework into which is installed the heating system (gas or electric), insulation (refractory brick or low thermal mass fibre), entry door (which can be a single entry door, or two doors i.e. one front, one rear) and finally the car bottom. If the furnace is of the single door entry type, then it is supplied usually with one insulated refractory brick/castable transportable base on wheels. The drive mechanism of the hearth can be an external steel rope and pulley system, or a motorised drive system located on the underside of the hearth. If the furnace has a two-door system, then a single hearth can still be used on a straight through basis or two load transportation cars can be utilised i.e. one for processing and one for loading/unloading.

If the furnace is to be used for hardening followed by quenching, then a lifting or tilting mechanism will be necessary to discharge the load into the quench tank.

Another variation of the car bottom type of furnace is the lift off bell furnace. The bell can be circular and cylindrical, or a box type furnace. The circular/cylindrical bell furnace can be heated by electrical methods or by gas firing. The heating system is generally on the external side of the bell. The box type furnace is usually fired within the insulated box and can be fired by either electrical or gaseous heating methods.

40.2.2.5 *Continuous furnaces*

The continuous type of furnace is used normally for high-volume production of manufactured parts. The mechanical handling of the work piece within this type of furnace can be designed for the particular product that the furnace is to handle, such as:

Ingots, slabs bars: These products can be handled within the furnace grouping known as walking beam furnaces. This work handling method is managed by two sets of beams within the furnace heating chamber. The first group of beams are static beams, which are fixed to the hearth of the furnace. A second group of beams is able to move. This group of beams manipulates the products through the furnace. This type of furnace is generally fired internally by gas and air mixtures.

Smaller components (e.g. small billets):   These can be heated inside a rotary hearth furnace. This type of furnace is round in shape and is of a static type. The only moving parts of this furnace being the entry/exit doors and the rotary hearth upon which the work is placed. The furnace can be heated either by gaseous means or electrically. The furnace atmosphere can either be generated from the products of combustion, from direct gas firing, or by any generated atmosphere such as an endothermic atmosphere or a nitrogen/methanol atmosphere. If these types of atmospheres are used, then the firing method will be encapsulated in the burner combustion tubes.

Mass production: The continuous type of furnace for this application can be a roller hearth furnace. In this type of furnace, the heating method can be either direct or indirect heating, using gas or electrical heating. The work is transported through the furnace on specially designed work load trays that are placed (with the work) on to driven rollers. Each of these rollers are driven from a master drive system and each roller is supported on either side of the roller with water cooled bearings. The furnaces are usually long and have an entry and exit door at each end of the furnace.

Where the integrity of the atmosphere is vital, the choice of furnace would be a pusher type furnace. This type of furnace is usually heated indirectly and the heating method (gas or electrical) is encapsulated in the heating tubes. Generally, the furnaces are designed for atmosphere conditions using either endothermic or nitrogen/methanol atmospheres. Because of these types of atmospheres, a front and rear vestibule is generally fitted to the furnace. Generally these furnaces are fitted with an atmosphere circulation system compromising of multiple fans located either in the roof or side walls of the furnace. This type of furnace can maintain a very accurate atmosphere for a variety of processes.

When small components are required to be heat-treated, the furnace of choice would be a shaker hearth furnace or a continuous mesh belt furnace. A shaker hearth furnace (a.k.a. a shuffle hearth furnace) comprised of a long horizontal tetragonal furnace with a single hearth that moves rapidly forwards and backwards over a distance of approximately 150 mm (6 in). This causes the work load to move forward through the heating chamber. A variety of processes can be accomplished in this type of furnace, such as hardening, tempering, carbonitriding and carburising. Usually, at the end of the table, the work is allowed to free fall into a vertical chute which is immediately above the quench tank. The disadvantage of this type of furnace is that there is no control of the presentation of the work to the quench medium, therefore the problem of distortion becomes quite high. A mesh belt furnace is a long horizontal tetragonal type of furnace that is capable of holding the furnace atmosphere as it is with the shaker hearth furnace. The difference between the two furnaces is simply in the method of transportation of the work. In the mesh belt furnace, the mesh belt is a woven belt of heat resistant stainless-steel which transports the work through the heating chamber and discharges it either into a quench tank or into a controlled cooling area. The mesh belt can be returned to the driven end of the furnace either externally or internally. If the belt is returned externally, it means that the belt has to be reheated. These furnaces can be heated either electrically or by gaseous methods. The heat source is usually encapsulated, irrespective of which method is chosen.

Rotary barrel furnaces contain a heat resisting process barrel which is fitted internally with a screw drive mechanism. The work is loaded, on a continuous basis, through an external sealed feeding system. The barrel is rotated by an external drive system which can be either a direct chain drive or a roller drive system. Sometimes, the whole furnace will tilt to discharge the workload into the quench medium, or the parts will be screw driven out of the furnace through a sealed discharge aperture and into the quench medium.

The single most important problem with any of the above furnaces is that of the potential for distortion to occur, simply because the work is allowed to free fall into the quench medium, without any control of how the component is presented to the quench medium.

40.2.3 Heat calculations

Irrespective of the type of furnace selected, it is necessary to calculate the amount of energy that is required to raise the temperature of the workload from ambient up to the metallurgical processing temperature. Such calculations can range from simple 'back of the envelope' methods (*e.g.* energy requirement = specific heat capacity × gross load mass × [target temperature − ambient temperature]) to sophisticated heat transfer models. The former type of calculation will be familiar to readers, whilst the latter forms an entire branch of the technical literature and hence cannot be discussed here.

Once the heating requirement calculations have been accomplished, then it is necessary to calculate the hearth length.

The furnace length will be determined by considering the following parameters and factors which are in relation to the operation process parameters. Note that load support fixtures as well as furnace support and carrier trays must be considered in the heating calculation. The process parameters to be considered are listed below:

- Total work throughput (mass per hour)
- Number of trays per hour
- Trays and support mass per hour
- Number of trays per hour required for the desired rate of production
- Method of heating
- Process time at temperature.

It is necessary to consider the heating method to determine if there is space to equip the furnace with either heating elements or burners and burner tubes.

40.2.4   Refractory design

The choice of insulation materials are wide and varied (*see* Chapter 27 for more detail).

REFRACTORY BRICKS VERSUS CERAMIC FIBRE BLANKETS

The maximum operating temperature of the furnace will determine the primary choice of insulation material, in addition to which air velocity and atmosphere movement within the process chamber must be considered. On this basis, a choice must be made between refractory brick and a low mass ceramic fibre material. It cannot be stated that one type of material is universally better or worse than the other. Instead, the choice of insulating material is application specific and depends upon the heat up and cool down rate required, maximum furnace operating temperature and the amount of wind abrasion that will occur within the furnace process chamber.

Another consideration in regard to temperature, will be the cold face temperature. That is the temperature that will occur on the external steel walls of the furnace. The accepted safe outside wall temperature is 65°C (150°F) maximum. However, if the furnace maximum operating temperatures is say, 1 000°C (1 800°F), then the external wall temperature will be determined by:

The hot face insulation of the furnace wall is made up of refractory material (in the form of bricks or a fibre blanket as discussed above). Behind the hot face insulation material will be another course of lower temperature heat resistance material, followed by possibly a third or fourth layer, to reduce the transfer of the heat that is generated at the hot face, in order to maintain a safe external wall temperature.

The disadvantage of using refractory brick is that the insulating wall thickness is generally quite high, which means additional weight within the furnace construction. A further disadvantage of all refractory bricks (which is probably the primary disadvantage) is the thermal energy necessary to heat the furnace chamber up to its operating temperature. The major advantage of refractory brick however, is that it has good heat retention and storage capacity. Generally (excluding fire brick) the brick is very porous. The degree of porosity will determine the bricks' heat storage capacity. The high porosity bricks can work against the heat treatment practitioner, in so much as, if the work load requires a controlled carbon atmosphere, or a carburising atmosphere, then the brick will begin to soak up the carbon atmosphere (depending of course on the carbon potential of the furnace atmosphere). This will mean a frequent burnout of the process chamber. Another consideration when using refractory brick, is the iron content. Generally when firing the furnace electrically, one should choose a brick with a low iron content. Excess iron can lead to 'tracking' or small arcs occurring at the point of mounting the element onto the wall. If the furnace is gas fired, then the iron content is not as critical to the refractory selection process.

Low mass ceramic fibre blanket is not very capable of heat storage. However, it has the ability to heat up to the operating temperature very quickly when compared with refractory brick. The blanket is very easy to install, and in order to reduce the effects of wind erosion, due to air movement within the process chamber, a surface hardener can be applied to the surface of the blanket. Periodic inspections must be made to assess the condition of the surface hardener. The surface can also be enhanced for longer life by the application of wire mesh with a hardener painted over the mesh. The mesh is generally made of stainless steel.

The low mass ceramic fibre blanket is not very good for use with controlled carbon atmosphere potentials. The blanket has a far greater porosity than the brick. Thus the blanket will also take carbon away from the furnace atmosphere, making it difficult to control the atmosphere accurately.

MODULES

When contemplating the use of low mass fibre, consideration can be given to the use of preformed square blanket modules. This will reduce the installation time dramatically as a result of increased ease of application.

PREFORMS

Another insulation material to consider, particularly when designing and building repeat types of furnaces and sometimes with complex designs, are pre-formed/vacuum formed insulation modules.

MORTAR

Careful consideration should be given to the selection of mortar used to install and cement the brick together. The mortar should be applied:

- Uniformly
- Thinly
- With a small joint clearance.

The mortar should have similar insulating characteristics to those of the refractory brick. When mixing the mortar, the consistency of the mortar will be determined by the amount of water added. The mixed mortar should have a firm consistency.

DRYOUT

Any new or re-bricked furnace must be dried out before commencing heat treatment operations. This is not as critical when the furnace has been relined with, for example, a low mass thermal insulation blanket, simply because the fibre is flexible and can accommodate any moisture that might be present within the blanket.

A furnace that is insulated with refractory brick will absorb atmospheric moisture as well as moisture contained in the mortar. If the furnace is raised from room temperature to operating temperature immediately after bricking or re-bricking, the residual moisture present in the brick will expand very rapidly and will have the potential to crack the brick. Therefore the furnace should be raised to a temperature of say, 105°C (220°F), and held at this temperature for up to 12 hours. This should be done with the furnace doors open, to allow the moisture bearing air to disperse from the furnace process chamber.

The next step in the dry out procedure will be to raise the furnace temperature up to say 260°C (500°F) and hold it for approximately 12 hours, again with the doors open. After this segment is completed, the furnace can be heated up to perhaps 535°C (1 000°F) and held for approximately six hours. At this point the furnace doors can be closed.

The furnace temperature can then be ramped up to say 750°C (1 400°F) then the furnace can be heated up to its process temperature with a ramp rate of no more than typically 35°C per hour (100°F per hour). This is to reduce the risk of thermal shock to the refractory brick, thus minimising the potential to crack the refractory, which might occur if the temperature is raised too quickly.

In the case of a new furnace, then it is most important to adhere to the manufacturer's dry out procedure. The above description of dry out, gives both the furnace engineer and the service technician an idea of the potential for cracking the refractory, if the furnace is raised to temperature too quickly.

40.2.5 Vacuum furnaces (see also section 40.3 on vacuum technology)

The dry out procedure of a vacuum furnace is somewhat different to that of a refractory lined furnace. Vacuum furnaces can be both heated and insulated in two different fashions:

All metal This means refractory metals and stainless steels for both heat shield and elements.
Graphite Graphite or carbon reinforced carbon can be used as the heating elements. The insulation can be manufactured from pure graphite or graphite fibres with a sealed reflective surface on the hot face of the insulation material.

Both oxygen and moisture can be detrimental to each of the above insulating materials, insomuch as oxygen contamination of a metallic heat shield will disturb the reflective surface of the insulating metal by oxidising the metal surface.

If graphite or fibrous graphitic material is used as an insulator or a heating element, and is left exposed to atmosphere, it will absorb moisture. The presence of moisture within the vacuum chamber will make it increasingly difficult to pump down to the appropriate pressure level, due to the evaporation of moisture. Moisture can be a technician's enemy in terms of vacuum operation, due to its ability to migrate anywhere within the vacuum process chamber. Once moisture is in the chamber, and in particular within the fibrous insulation materials, the moisture will 'stick' to the insulating fibres and will be difficult to remove on pump down.

Therefore, when a new vacuum furnace is shipped to its end-user, it is usually shipped with the door clamped down and the interior under vacuum conditions. Once the furnace is in position at the installation site, the vacuum will be broken, and the door opened. Once the installation is complete, the door is closed and the vacuum pump started to take the furnace down to its ultimate vacuum level. The temperature of the furnace is raised to its maximum operating temperature and allowed to hold at this temperature for a short period of time. After the outgassing period of time has expired, the furnace is allowed to cool down to ambient temperature. The vacuum pump is then turned off and the leak up rate of the process chamber is noted. The normal leak up rate of any type of commercial-scale vacuum furnace that is cold, clean and empty should be in the range of 0.7 Pa (5 mtorr) per hour maximum.

The act of pumping down the vessel and raising the furnace up to its maximum operating temperature is designed to ensure that not only the insulating material, but also all other materials of construction within the heating chamber, have completely outgassed. If the vessel then leaks up at a rate greater than 0.7 Pa (5 mtorr) per hour, it means that there is a potential leak within the process chamber. It is then necessary to begin to establish the nature of the problem.

There are two types of problems that produce the appearance of leaks, namely:

– Real leak (actual physical leak)
– Apparent leak (internal or virtual leak or outgassing)

REAL LEAK

A real leak is a leak through a hole in the furnace wall. This means that somewhere within the construction of the furnace vessel there is a passage which will allow air, or moisture to leak into the process chamber, thus steadily raising the furnace chamber vacuum pressure. The leak can be at any vacuum sealing point, such as the main door, power feedthroughs, the thermocouple feedthroughs, recirculation fan seal, or any other aperture into the furnace.

APPARENT LEAK

An apparent leak is a situation where there is a rate of rise in pressure, but after leak detection procedures are completed, no leak is visible, yet the pressure continues to rise. This is the most difficult type of vacuum problem to handle, simply because it has no external cause. The apparent leak up rate is being caused by something within the process chamber that is outgassing. This can be caused, of course, by a pin hole leak within a weldment on the inner wall of the chamber. This allows water from the water jacket to migrate into and vaporise in the process chamber. Another cause of this condition can be oxides on furnace support fixtures that are outgassing, or the components being processed outgassing. This last type of apparent 'leak' is the most difficult to locate. Poorly designed vacuum systems can include constrictions that make pump down of some regions difficult, thus producing the appearance of a leak (i.e. a virtual leak).

Assuming that the vacuum furnace is tight and that there are no apparent or real leaks in the furnace system, then the vacuum furnace should be kept closed and under partial vacuum when not in use.

40.2.6   Cooling

Once the work has been processed, it needs to be cooled. The cooling rate of the work piece will be determined by the desired metallurgy necessary to impart the required mechanical properties to the material. Cooling can be accomplished (see Chapter 29 for more detail) as follows:

Rapidly   (by a variety of quenching processes)

Slowly   The word 'slowly' in this instance, is relative. The work piece can be cooled down in still air or forced air (fan cooling). It can also be cooled down in air with a specially designed cooling chamber which would be surrounded by a water jacket. The work piece could be cooled down in a specially designed cooling chamber but insulated with a low mass ceramic fibre blanket. The work piece could further be cooled down by quenching into a liquid salt bath at a predetermined temperature. The physical act of cooling can be accomplished in either the batch type operation or a continuous operation.

40.3   Vacuum technology

40.3.1   Introduction

In recent years, the field of vacuum technology has expanded greatly in terms of volume production of equipment and in the diversity of equipment available. To a large extent, this has been driven by the semiconductor industry with its requirements for many large scale wafer fabs operating at high vacuum levels, although other industrial uses of vacuum have also burgeoned. These cover such diverse applications as coating window glass with reflective films, plasma processing of baby's disposable diapers, freeze drying, etc. Figure 40.1 shows some typical industrial processes using vacuum and the range of pressures over which they tend to operate. Despite this growth, the underlying technologies of pumping and measurement of vacuum have changed little. Such advances as there have been have tended to be better implementations of existing technologies.

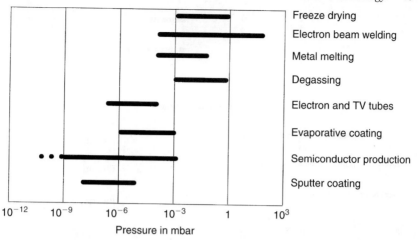

Figure 40.1 *Some industrial processes using vacuum*

Table 40.1 UNITS OF PRESSURE

Unit	Common area of usage	Equivalent in mbar	Comment
Inches of water	Industry where a 'negative' pressure is required e.g. vacuum forming	2.49	Widely used, but definitely 'non-standard'
Torr	Very widely used in industry	0.752	Non-standard but refuses to die!
mbar	Widely used in laboratories and often in industry	1.0	An alternative unit in the SI system
Pascal	Largely confined to standards laboratories	0.01	The SI unit

Table 40.2 VACUUM RANGES WITH APPROPRIATE PRESSURE MEASUREMENT TECHNIQUES

Pressure region	Pressure range (mbar)	Pressure measurement technique
Low (rough) vacuum	1 000–1	Direct acting gauges—e.g. Bourdon tube
Medium (fine) vacuum	$1–10^{-4}$	Gas bulk properties—e.g. Pirani gauge
High vacuum	$10^{-4}–10^{-9}$	Molecule 'counting'—e.g. Penning gauge
Ultra high vacuum (UHV)	$<10^{-9}$	Molecule 'counting'—e.g. Bayard Alpert gauge

In this section, a short review will be given which should enable the practicing technologist to understand some of the aspects to be considered when choosing appropriate vacuum technology for the job in hand. Space limitations force this review to be highly selective.

40.3.2 Pressure units and vacuum regions

Table 40.1 lists the most common units in which the pressure in a vacuum system is measured and their relationship. Table 40.2 lists a common set of terminology for different pressure regimes, although it should be noted that there is no universally agreed way of doing this.

40.3.3 Pressure measurement

There is no single way of measuring sub atmospheric pressure over the entire range. Broadly, the measurement techniques used fall into three categories: direct; those which use bulk properties of

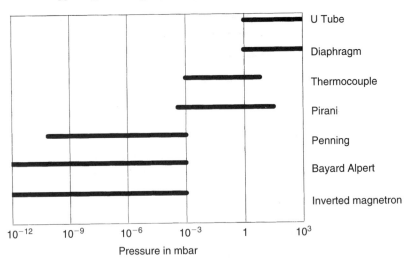

Figure 40.2   *Operating ranges of some vacuum gauges*

a gas; and those which essentially 'count' gas molecules. Here we only discuss types of pressure measurement devices (i.e. gauges) which are widely available and are commonly used in industrial applications. More specialist gauges are described in the references to be found in the bibliography. Figure 40.2 shows the operating pressure ranges of some gauges.

40.3.3.1   Direct measurement gauges

These use the difference in forces exerted by the gas in the vacuum system at a given pressure and a reference pressure—usually atmosphere. In a sense, they are exactly analogous to gauges measuring pressures above atmosphere. Such gauges are insensitive to the gas species being measured.

Occasionally, one will still see a *water column* being used to measure 'negative' pressures of an inch or two of water. This is usually a simple U-tube with one side open to atmosphere and the other connected to the vacuum and the U-bend filled with water. The pressure difference is then the difference in height between the surfaces of the water in the two arms of the U.

More common is a *diaphragm gauge*, where the deflection under the differential force of a thin diaphragm or a capsule like an aneroid barometer is measured. Often a system of levers and gears causes a pointer to move over a dial to give an indication of the pressure. Modern gauges may use a strain gauge to measure the deflection and display it electrically. The *capacitance manometer* is a sophisticated implementation of such a device. An equivalent type of gauge uses a *Bourdon tube*—a thin metal or glass spiral which curls or uncurls as the pressure inside changes. Similar display techniques to diaphragm gauges are used.

40.3.3.2   Gas bulk property gauges

These gauges measure some physical property of the gas which is pressure dependent. The most common property used is heat conduction, although some specialist gauges measure viscosity changes (e.g. *spinning rotor gauges*). *Thermocouple gauges* measure the temperature of a filament which is heated by a constant electrical current. *Pirani gauges* measure the electrical resistance of a filament wire whose resistance varies with temperature. In each case as the pressure changes, heat conduction through the gas changes and so the temperature of the filament changes. Some forms of Pirani gauge keep the filament temperature (and hence resistance) constant by varying the current flowing through the wire.

Gauges of these types are clearly only useful when the property actually varies with pressure. Those relying on thermal conductivity are therefore limited to measuring pressures between about 1 mbar and 10^{-3} mbar. The upper range may be extended to above atmospheric pressures by taking convection into account.

There are two main drawbacks to bulk property gauges. In general, they are gas species sensitive—so the gas composition must be known if accurate pressure measurement is required. This is very important when safety issues need to be considered. Thermal conductivity gauges are also dependent on the state of the filament—surface contamination changes its thermal transfer properties. The hot filament itself may interact with the gas which may be explosive, or may 'crack' to leave e.g. carbon deposits on the filament.

40.3.3.3 Molecular counting techniques

Below about 10^{-3} mbar, physical properties of gases vary little with pressure, so the only readily accessible technique to measure pressure is to count gas molecules in some way. Such techniques of course do not actually measure the real physical pressure of the gas, but provide a measure of the number density of gas molecules. Calibration is therefore required to convert this to an equivalent pressure and the gauge sensitivity will be gas species dependent.

A simple way to count gas molecules is to ionise the gas and collect the resultant ions—this ion current is a good measure over many decades of pressure. Two ionisation techniques are commonly employed. *Cold cathode gauges* generate a plasma which is confined by electrical and magnetic fields and ions are extracted to a cathode plate of some sort. *Hot cathode gauges* use a beam of electrons extracted from a hot filament to directly ionise the gas. Classical implementations of these techniques are *Penning gauges* and *Bayard-Alpert gauges* respectively. Both are useful and relatively stable, although the sensitivity of each type of gauge is dependent on its history. It can vary somewhat unpredictably by factors of 50% or more in use. Both types of gauge are prone to contamination changing sensitivity, the Penning gauge being somewhat less so. However, this gauge is subject to random fluctuations in the plasma discharge conditions which change the sensitivity. The Penning gauge requires a strong magnetic field at the gauge head, which can be a problem in some applications and the gauge head generates some particulates. The Bayard-Alpert gauge has a hot filament with similar drawbacks to those noted above for thermal conductivity gauges. It is also susceptible to external magnetic fields which can change the indicated pressure dramatically. At low pressures, it can release a surprising quantity of gas from the walls of its housing.

Penning gauges are reliable down to about 10^{-8} mbar and Bayard-Alpert gauges to 10^{-11} mbar, in both cases the lower limit being dependent on detailed gauge design. A variant on the cold cathode gauge is the *Inverted Magnetron gauge* which is somewhat more stable in sensitivity and can work down to 10^{-11} mbar. It is also more expensive. All of these gauges can measure pressures up to 10^{-3} mbar.

40.3.3.4 Control and measurement units

All gauges which use indirect methods to indicate pressure—and indeed some direct measurement gauges also—require electronic control and measurement units to drive the gauge head, to measure currents, to display the pressure, etc. Nowadays, all of these can readily interface to computer systems to provide gauge control, plant interlocking and data logging for process control or monitoring.

A recent trend in control units is the development of so called compact gauges where the electronics to drive the head and to measure the appropriate analogue of pressure is contained in a miniaturised unit fixed directly to the gauge head. This unit usually requires only a dc electrical supply (often $+24$ V) and a computer connection (e.g. RS232 or Ethernet). This is an economical solution where many gauges have to be integrated into a process control system, or where long distances are involved. However, where vacuum systems operate at high temperatures or generate radiation, such close-coupled units may not be reliable.

40.3.3.5 Can gauge readings be believed?

All vacuum gauges in common use in industry require calibration. As discussed above, the indicated pressure often depends on the gas species being measured, so the effects of this must be understood. In some cases, such differences can be calculated from the kinetic theory of gases, etc., but more often gauges must be specifically calibrated for the gas species of interest.

Most of the vacuum gauges described above—certainly all the indirect gauges—exhibit variations in sensitivity dependent on their history—even a simple exposure to atmosphere during a system vent can change gauge sensitivity in an unpredictable way. Such variations may be of the order of a few percent, although in the worst cases much larger factors are not unknown. Thus, unless a gauge is calibrated in situ against a standard of some sort, then it is wise to treat pressure indications as a guide rather than a precise measurement.

When accurate and repeatable measurements are required, specially built gauges whose repeatability is much better are available at correspondingly increased cost. Manufacturers should be consulted for further information.

40.3.4 Pumping technologies

40.3.4.1 *Low and medium vacuum pumps*

The classic workhorse pump in this range is the oil sealed *rotary vane pump* often supplemented by a *Roots blower* to achieve lower pressures or higher gas throughput. Newer technologies in this area have sought to eliminate the problems of contamination of the work system with pump oil by using clean, 'oil free' techniques. In some processes, pumping of particulate contaminated gas streams is required—in others generation of particulates which can get back to the process must be minimised. In any pumping situation, care must be taken to ensure that the gases being pumped do not attack the pump structure or react adversely with oils and greases in the pump. A classic example is pumping oxygen-rich gases with a mineral oil sealed pump. This might well result in an explosion. Chemically inert pumps with passivated surfaces and using inert synthetic oils and greases are available. Pump manufacturers should always be consulted about appropriate pumps and fluids for particular situations. When pumping large quantities of water vapour or other condensable gases, pumps may need to be heated to prevent the gas condensing in the pump, possibly diluting the lubricant.

Table 40.3 summarises some of the more widely available low vacuum pumping techniques.

40.3.4.2 *Medium and high vacuum pumps*

At one time, the *diffusion pump* reigned supreme in this region. It still does when either very high pumping speeds or extreme economy is required. It can also handle particulate laden gas streams. For many applications, the diffusion pump has been largely superseded by the *turbomolecular pump*. Although initially more expensive to purchase, the turbomolecular pump is much cleaner, is more reliable, more suited to unattended running since it (and the process) can be protected against many failure modes. It requires less down time for servicing with modern pump bearings being lubricated for life and having lifetimes in excess of 15 000 hrs dependent on application. Versions where a molecular drag pump is combined on the same drive shaft as the turbine stage are available, so that backing pumps which are capable of reaching only a few mbar may be used, resulting in cost savings. Versions of turbomolecular pumps are available which use magnetic bearings and are therefore completely clean. Most modern lubricated pumps use only very small quantities of bearing lubricant, often in the form of a grease, so the possibility of contamination is much reduced. Indeed most contamination is likely to come from the backing pump unless a clean or dry pump is used.

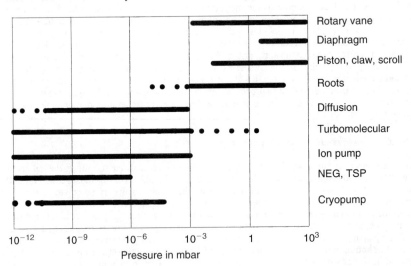

Figure 40.3 *Operating ranges of some vacuum pumps*

Turbomolecular pumps can handle modest quantities of small particulates. They must be protected (by means of a simple mesh in the pump throat) against larger items falling into them. Particulate generation is low.

Modern synthetic diffusion pump oils are available which are much less prone to cracking when exposed to atmosphere and therefore down time to service the pumps can be reduced. These oils are relatively inert, so the pumps can be made to be chemically inert. The safety problems associated with having a hot boiler remain however and they are difficult to protect adequately.

Pumps in this pressure range do not exhaust directly to atmosphere and require a backing pump with a matched throughput.

40.3.4.3 *Ultra high vacuum pumps*

Pumps for this pressure range are more specialised. In addition to *turbomolecular pumps* where pressures below 10^{-9} mbar are reasonably easily achieved, most pumps in this range work by trapping gas molecules in some way. Further details on their operation can be found in the references of the bibliography, but Table 40.5 lists some pumps and their main characteristics. Rather than using backing pumps, gas trapping pumps rely on the system being pumped down to a reasonable pressure

Table 40.3 LOW VACUUM PUMPS

(Note $1\,\mathrm{m^3 hr^{-1}} = 0.59\,\mathrm{cu.\ ft\,min^{-1}}$)

Pumping technology	Typical lowest pressure (mbar)	Typical throughputs available ($\mathrm{m^3\ hr^{-1}}$)	Comments
Diaphragm pump	10	5	Clean. Pumping and generation of particulates can be a problem.
Piston pump	10^{-2}	50	Modern versions clean (oil free). Low particulate
Claw pump	10^{-2}	500	generation possible. Possible problems when pumping particulates. Can be chemically 'inert'.
Scroll pump	10^{-2}	35	Modern versions clean (oil free). Possible problems with pumping particulates and particulate generation.
Rotary vane (single stage)	10^{-3}	1 200	Oil sealed so difficult to remove backstreaming
Rotary vane (two stage)	10^{-4}	350	contamination. Can use inert synthetic oils.
Roots blower	10^{-5}	15 000	Can generate particulates. Can be reasonably clean with care. Only operates below about 100 mbar.

Table 40.4 MEDIUM–HIGH VACUUM PUMPS

Pumping technology	Typical pressure range (mbar)	Backing pump max pressure (mbar)	Typical throughput (pumping speed) available ($1\ \mathrm{sec^{-1}}$)	Comments
Diffusion pump	$10^{-2}-10^{-8}$	0.1	50 000	Can have very high throughputs. Can be chemically inert. Can handle particulates. Relatively cheap.
Turbomolecular pump	$10^{-3}-10^{-9}$	10^{-3}	2 000	Good throughput. Reliable. Low particulate generation. Can handle modest particulate load. Can be chemically inert.
Wide-range turbomolecular pump	$10-10^{-10}$	10	1 500	Throughput at high pressure restricted. More sensitive to particulates.
Magnetic bearing turbomolecular pump	$10^{-3}-10^{-9}$	10^{-3}	2 000	Inherently clean. Quite expensive.

Table 40.5   UHV PUMPS

Pumping technology	Starting pressure (mbar)	Ultimate pressure (mbar)	Typical max pumping speed available (1 sec^{-1})	Comment
Sputter ion pump	10^{-3}	10^{-10}	500	Pump lifetime depends on amount of gas pumped. The pump can be unstable against air leaks. Speed depends on gas species being pumped. Can generate particulates.
Evaporable getter pumps	10^{-5}	10^{-12}	Limited by area of evaporated film	Typical example uses titanium. Only pumps active gases. Can generate particulates.
Non evaporable getter pump	10^{-5}	10^{-14}	Limited by surface area of carrier	Requires high temperature (200–450°C) activation. Does not pump inert gases.
Cryogenic pump	10^{-4}	10^{-14}	Limited by area of cold surface	Requires provision of liquid cryogens (helium or nitrogen) or a refrigerator system.

(listed here as starting pressure) by an auxiliary pump. This is then usually valved off from the system being pumped. For UHV pumps, pumping speed is usually limited by the rate at which gas molecules arrive at the pump.

40.3.4.4   *Control units and power supplies*

Vacuum pumps in all pressure ranges can be provided with controllers or power supplies which may be interfaced directly to computer control systems for monitoring and control. As with gauges, there is a trend towards integrating these controllers with the pump body, attached to, or even inside, the pump housing.

40.3.5   Vacuum systems

The design of vacuum systems is of course a very wide topic, since a vacuum system will in general be tailored to meet the requirements of the job in hand. Nevertheless a few general points can be made.

Good engineering practice will be required in order to ensure that the system will be able to work satisfactorily. For example, deflections of walls when the system is evacuated may be important.

In most cases, the vacuum performance of a vacuum system will be defined in terms of the base pressure required and the time taken to pump down to that pressure. In designing a vacuum system to meet such requirements, the designer will have to take into account the gas load to be removed. This will consist of three main parts:

- the gas inside the system from its exposure to atmosphere
- the gas load generated from the things inside the system (evaporation from any liquids present, offgassing from work pieces, outgassing from walls, wires and any contamination present)
- the removal of gas flowing into the system from process gases or leaks (some of which may be internal leaks from trapped volumes).

Many of these will not be known to any great accuracy but reasonable estimates can be made by an experienced vacuum designer. Some can be eliminated or minimised by good design and careful operation. In general, it is easier to minimise gas load than to increase pump throughput or speed. For the low to high vacuum regions, there are computer packages commercially available from the major vacuum equipment manufacturers that can be used to calculate necessary pump sizes and pump down times so that an optimum economic result can be achieved.

As the required pressure becomes lower, design becomes more complicated and specialist techniques may be required to calculate system performance.

Design of systems to work in the UHV region is an involved subject, requiring much attention to detail and the advice of an experienced specialist designer should normally be sought. Most of the vacuum equipment manufacturers who supply equipment to work in this range will be able to provide this as a service.

Table 40.6 SOME IMPORTANT PARAMETERS FOR A RESIDUAL GAS ANALYZER

Parameter	What it means	Typical value	Comment
Mass range	Maximum and minimum masses which will pass through the filter	1–100 amu	Some process gases or oils may have higher masses
Resolution	How well species with masses close to each other can be separated	1 amu at 10% peak height (equivalent to 20% valley)	Most instruments are adjusted so that peak widths are the same across the mass range. This means that gauge sensitivity will vary across the mass range
Mass discrimination	The variation in transmission of the mass filter for different masses of ion	Rarely specified!	Important if quantitative or semi-quantitative analysis is required. Software packages should automatically compensate for this
Sensitivity	Usually specified as the minimum detectable partial pressure of nitrogen (often at a signal-noise ratio of 2)	Without electron multiplier detector: 10^{-9} mbar. With electron multiplier detector: 10^{-14} mbar	Sensitivity for different gas species depends not only on mass discrimination (above) but also on ionisation cross sections. Again, software packages should automatically compensate for this
Abundance sensitivity or tailing contribution	The smallest peak which can be detected as a 'bump' on the shoulder of a peak at an adjacent mass. Usually expressed as a percentage of the large peak	0.01%	In practice, only important for trace analysis or isotopic analysis. It will be different 'above' and 'below' the large peak—usually only the low mass side value (which will be worse than the high side value) is quoted

40.3.6 Residual gas analysis

It is probably fair to say that most processes which involve vacuum require one of two things—a simple sub atmospheric pressure so that something can be moved or removed (vacuum forming or freeze drying for instance) or the production of an inert atmosphere. The latter may be for something as simple as vacuum packing where removing the oxygen delays the onset of decay processes in food or as 'high tech' as controlling surface processes in a semiconductor fabrication facility.

There are many occasions when the vacuum system user will want to know what is left in the vacuum system—and the last example above may well be one such case. This is relatively easily achieved using a *residual gas analyzer*. These devices, the most common of which are based on a simple form of mass spectrometer known as a radio frequency quadrupole analyzer, are readily available at modest cost. Such systems comprise an ion source (usually a hot filament source akin to a Bayard-Alpert gauge), a mass filter and an ion detector.

Modern residual gas analyzers are computer driven by dedicated software packages and are simple and easy to operate. Results can be displayed in terms of a spectrum, a tabulated set of masses present, trends with time of selected masses and so on. Process control can be achieved and instruments can be set to generate alarms if gas compositions go outside preset limits. Unknown peaks in a spectrum can be compared to built-in libraries of 'fingerprint' spectra of many molecular species.

Important points to note in selecting and using such analysers include those shown in Table 40.6. In this table are listed some of the key parameters and their definition with typical values for a modestly priced class of instrument. Such an instrument will suffice for most routine vacuum applications and semi-quantitative analysis will be possible. A more highly specified (and more expensive) system will be required for accurate quantitative analysis and such a device will need frequent calibration. Modestly priced equipment can provide the trained user with a wealth of information about the 'health' of a vacuum system or a process.

One very important point to note in using residual gas analyzers is that there is usually a hot filament in the ion source of the instrument so that especially at low pressures the measured residual gas spectrum may differ considerably from the actual spectrum of gases present. Experienced users will be able to assess the effect of this on the results.

40.3.7   Safety

Safety is a major consideration in the design and operation of vacuum systems. An understanding of how the vacuum system is to be used is important in this. For example in the above discussion of vacuum pumps, some emphasis has been placed on the importance of understanding the possible chemical interaction between process gases and lubricants and working fluids in pumps. Failure to take this into account may result not simply in seizure of the pump but possibly in explosions. A turbomolecular pump contains a turbine rotating at high speed (perhaps 50 000 rpm) and pump seizure may result in a catastrophic disintegration of the pump housing. Objects falling into such pumps will result in blades being stripped from the shaft.

Great care must be taken to ensure that gases being exhausted from vacuum pumps are dealt with safely. If they are in any way hazardous, they must not be exhausted directly to atmosphere but must be passed into properly designed gas handling plant. Where a range of different gas species is being pumped, consideration must be given to the effects of mixing these in the exhaust system. When particulates are being pumped, exhaust plumes must use filtration to remove these. Systems for safely removing and disposing of contaminated pump oils must be put in place.

Ion pumps operate at high voltages (3–7 kV) so adequate shielding and safety precautions against disconnection of leads is required. They also contain powerful magnets.

Cold cathode gauges operate at about 3 kV and hot cathode gauges get hot! Where hot filaments are present, there is the possibility of explosion when pumping flammable gases if pressures get too high. As has been mentioned, many types of gauge give pressure readings which are dependent on the gas species being measured, so when used as safety indicators must be calibrated for the appropriate gas.

Clearly, any vacuum vessel must be designed to withstand an external differential pressure of 1 bar without collapsing. However, it should also be capable of withstanding small internal overpressures if it is backfilled with piped or bottled gas before, for example, opening to atmosphere. Fragile components like viewing windows and feedthroughs should be protected against breakage.

Careful design will result in a vacuum system which meets its specifications and is safe to use.

40.3.8   Selective bibliography

The best source of information on vacuum pumps and gauges is to be obtained from the manufacturers of such equipment. A useful source of information on manufacturers is the Buyers' Guide on the AVS website http://www.avs.org

A reasonably extensive Bibliography on Vacuum is located on the website of the International Union of Vacuum Science, Technique and its Applications (IUVSTA) at http://www.iuvsta.org/vsd/biblio.htm

Some general references on vacuum are:

Basic vacuum technology. A. Chambers, R.K. Fitch, and B.S. Halliday: 2nd ed. Bristol: Institute of Physics 1998. ISBN: 0750304952

High-vacuum technology, a practical guide. Marsbled H. Hablanian: 2nd ed. New York: M. Dekker 1997. ISBN: 0824798341

Handbook of vacuum science and technology. Edited by Dorothy M. Hoffman, Bawa Singh, John H. Thomas, III. San Diego, Calif., London. Academic Press 1998. ISBN: 0123520657.

Vacuum technology and applications. David J. Hucknall. Oxford: Butterworth-Heinemann 1991. ISBN: 0750611456

Foundations of vacuum science and technology. Edited by James M. Lafferty. New York, Chichester: Wiley 1998. ISBN: 0471175935.

A user's guide to vacuum technology. John F O'Hanlon: 2nd ed. New York: Wiley 1980. ISBN: 0471016241

The physical basis of ultrahigh vacuum. P.A. Redhead *et al.* London: Chapman and Hall 1968. (Reissued AVS Classics Series, New York 1993) ISBN: 1563961229

40.4   Metallurgical process control

40.4.1   Metals production and processing

Metals are produced through a long chain of processes. For primary resources, metals production goes through mining, mineral processing, ore smelting (oxidation or reduction) in high temperature

Products
Society

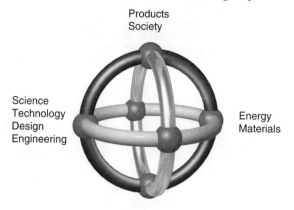

Science
Technology
Design
Engineering

Energy
Materials

Figure 40.4   *Achieving sustainability*

furnaces or via low temperature hydrometallurgical processes, or a combination of both pyro- and hydrometallurgical operations. Taking ironmaking and steelmaking as an example, it normally follows a number of consecutive operations from blast furnace ironmaking, BOF or EAF steelmaking, steel refining e.g. in a ladle furnace, steel casting, to cold or hot rolling. For the secondary resources, metals are separated and recovered from metal scrap, which play an increasingly important role in the metal supply of the world. After being produced, either from primary or secondary resources, metals have to be further refined or manufactured into various types of alloys and finally shaped into different products suited to the end-users. Since the end-user will discard products eventually at the end of products' life the contained materials find their way back into the resource cycle as scraps. Therefore, controlling the metallurgical reactors also controls in an even much broader sense, the material quality in the complete metals cycle, from the production phase, to the end-use phase, up to the recycling phase. The recycling phase includes separation of materials into saleable intermediate products, which are reintroduced into the production phase as a secondary resource. Therefore, the metallurgical processes in each unit operation in the metals production and processing chain have to be controlled, so as to achieve optimal performance of high product quality, low energy and materials consumption, and low environmental impact. Symbolically these interactions are best illustrated by Figure 40.4, depicting the links between three interconnected cycles: the life cycle—the technology cycle—the resource cycle. The optimal control of all these cycles is affected substantially by good control of metal production and refining.

40.4.2   Modelling and control of metallurgical processes[2,31,32,34,35]

In view of the above, the discussion focuses on various industrial supervisory control, optimisation and modelling examples in which the goal was to recognise the above-mentioned philosophy. This is done by discussing (i) low level control, (ii) supervisory control and optimisation of metallurgical furnaces, (iii) modelling and simulation techniques ranging from 'black box' to 'grey box', and (iv) logistic control of metal flows between furnaces and processes. The objective of these control actions are very clear, i.e. they (i) optimise product quality, (ii) minimise energy consumption, (iii) minimise environmental impact, and (iv) optimise metal flows, or in other words optimise energy flows in a plant and between plants. Therefore, control should not only maximise economic benefits, but at the same time also minimise associated ecological effects.

Implicitly it should also become evident that metallurgical control and optimisation is the reason why modelling and simulation is done. All the related experimental work in the laboratory and on the plant is performed to parameterise these metallurgical models. In other words it can be argued that control is the umbrella for performing all the tasks depicted by Figure 40.5. This document will cover a number of the aspects depicted in Figure 40.5, i.e. the importance of good process data will be mentioned and a short overview to various modelling and simulation techniques, ranging from 'black box' or empirical to 'grey box' (some call them 'white box') semi-empirical approaches will be given. These will all be discussed very briefly in the context of various industrial process control examples.

The various techniques discussed in this section will be placed in the context of modelling and controlling poorly defined industrial metallurgical reactor systems. Due to often poorly definable

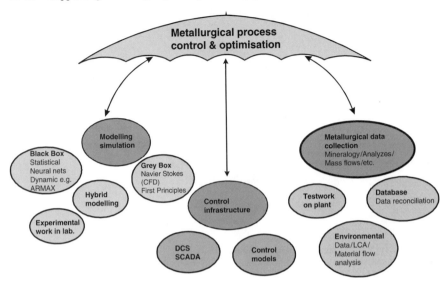

Figure 40.5   *Umbrella of process control*

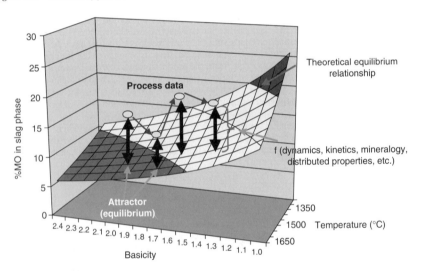

Figure 40.6   *Loci of fictitious data points for a furnace positioned above their respective equilibrium data (the attractor) to illustrate the effect of process dynamics, kinetics, etc.*

parameters, that include (a) the effects of changing mineralogy and distributed parameters such as density, viscosity (as a function of conditions and spatial position within the furnace), (b) the effects of different reductant types and their respective reactivities, ash contents, etc., (c) heterogeneous distributed conditions within the furnace such as temperature, that creates a distribution of properties of the system making pure thermodynamic models difficult to apply, and (d) synergetic effects created by the feed mix, it is often very difficult to create sufficiently accurate and practically useful predictive first principles metallurgical models. Furthermore, since furnaces are often not at equilibrium, the idea of a metallurgical equilibrium attractor is discussed, which is basically the equilibrium a furnace at any point in time would like to reach but finds difficult due to production constraints and process dynamics. Therefore a good process model should not only include the thermodynamic basis but also try to quantify the effects of process dynamics. Figure 40.6 suggests that in a real furnace the measured

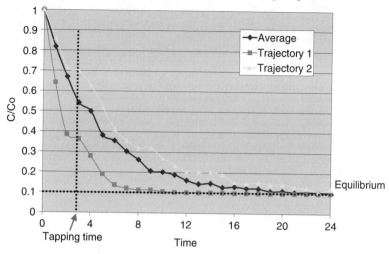

Figure 40.7   *Change of concentration in two different lumps of ore following different paths through the furnace adjacent the mean conversion $\overline{C}$ as an integral of all trajectories*

process data points mostly lie above (or below) the equilibrium, i.e. for instance the analysis of a metal oxide in the slag would lie above the equilibrium value predicted for the same slag conditions and temperature as measured during the tap. This difference is created due to heterogeneities in the furnace created by a distribution of temperatures, mineral properties, reaction paths, and histories of each lump of ore and reductant in the reactor. In other words, the difference could also be an indication of how well the furnace is being operated. It is obvious that the fewer heterogeneities are present, such as in a well-mixed steel converter, this difference becomes much smaller. This applies also to casting and reheating operations that can be quantified and modelled in a very much more precise way than a multi-phase system such as a metallurgical furnace. In order to further explain Figure 40.6, Figure 40.7 is given which depicts the remaining concentration of an unreacted species of two different lumps of ore with different mineralogies following two different paths in the furnace, which have seen different temperatures, conditions, reductants, etc. In short, each trajectory is a fingerprint of the history of the particle. In order to establish the mean conversion it is necessary to integrate over all paths and conditions to obtain a true understanding of what is happening in the furnace.

Incorporating the distribution of all histories for all particles is defined best by the equation below. This equation defines the average concentration (average curve in Figure 40.7) as a function of the residence time distribution E, the temperature distribution ψ, reaction rate distribution χ, time t, and spatial coordinates Ω, as well as production constraints that dictate when to tap. In the equation, k is reaction rate, and T is temperature.

$$\overline{C} = C_0 \cdot \int\limits_{0}^{t_1 = \text{tap time}} \int\limits_{k_1}^{k_2} \int\limits_{T_1}^{T_2} \int\limits_{\Omega} E(t) \cdot \psi(\Omega, T, t) \cdot \chi(k, T, \text{Mineralogy}) \cdot C(k, T, t) \cdot d\Omega \cdot dT \cdot dk \cdot dt$$

Figures 40.6 and 40.7 suggest that it is important to consider the following steps when modelling industrial metallurgical systems: (a) Assimilate reconciled plausible data for the reactor system in a real-time database and control software (see below). (b) Ensure that a reconciled mass balance produces a statistical useful mass balance from these data, from which also statistically significant recoveries, energy balances and statistically consistent analyzes are estimated. (c) For these data thermodynamic equilibrium models can be calibrated from which the equilibrium attractor of the process can be estimated (if this is at all possible). (d) The difference between the attractor and the measured (reconciled) analyzes may then be applied to estimate how far the process is from equilibrium. This difference is mostly a function of process dynamics, which also includes kinetic effects, or of mineralogy and other factors such as metal entrainment rendering the system inhomogeneous. (e) The difference between equilibrium and dynamic data should then be interpreted and investigated to obtain a true picture of what happens in the furnace. (f) Useful methods such as first principles

CFD (computational fluid-dynamics) models can be applied to estimate distribution relationships (equation above), incorporating kinetics and black box methods to further improve the understanding of the system (if this is possible).

It would be clear from the above that this suggested approach does not permit the forcing of dynamic process data into thermodynamic and/or kinetic metallurgical models at all costs by fitting with parameters. More often than not these parameters are distributed and are therefore not really suitable for inclusion in classical kinetic and thermodynamic modelling approaches. This classical modelling approach (*also followed without often passing through a data reconciliation step*) usually leads to erroneous results due to not incorporating process dynamics, leading to models that often have little or no practical value and do not lead to true furnace optimisation and control. Rather consideration of the distributed nature of the system is required, and methods should be applied to extract useful information from the data. Conversely, if the simulation becomes very difficult due to poor parameterisation, empirical methods should complement semi-empirical models to ensure good predictive control in hybrid architecture.

Although much of the discussion focuses on furnace control, the discussion is completed by providing examples on casting, reheating and milling operations, and by providing a general overview of the components of control systems.

40.4.3 Process control techniques[32,35]

Process control plays a critical role in metals production and treatment. Metallurgical processes, both in batch and in continuous operations, normally exhibit time delays with large time constants, uncertainties, non-linearities, unmodelled dynamics, etc. Taking this into account, different control techniques have been developed since 1960's. In the early days, the process control was based on mechanical, electrical or pneumatic analogue controllers. Later on, the broad use of digital computers came to the process control thanks to the rapid development of control theory, from single-input single-output linear system to a wide range of multivariable non-linear systems. Nowadays, the extensive application of expert systems, neural networks, fuzzy control, process monitoring and on-line diagnostics has become important parts of a modern process control system in metallurgical processes.

Figure 40.8 illustrates a general control structure for a typical metallurgical process,[35] which indicates different levels of a modern control system and the importance of database central to the whole system. For the benefit of the discussion below 5 levels are defined, however, various levels have been defined in the past by various vendors and providers. The figure explains the significance of each level. From Figure 40.8, discussed with reference to a submerged arc furnace for ferroalloy production, various levels can be discerned, each containing a mixed bag of the methodologies discussed above.[32] This section will dwell on the process control structure and will specifically

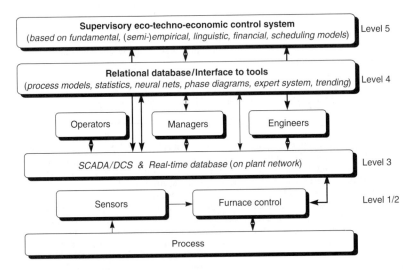

Figure 40.8 *Control structure for a metallurgical furnace*[35]

attempt to explain the structure and function of the database, which is central to the industrially implemented system.

Levels 1 & 2: Mintek's patented low-level control system (Minstral[TM]) for submerged-arc furnaces[31] has an international proven track record, having been implemented in various plants all over the world. This control system, packaged in different forms, i.e. from a stand-alone unit to a Windows based network version, has various proven features (a) control based on a proven resistance based algorithm, which de-couples the movement of electrodes, (b) optimisation of power input into the furnace as a function of MVA, resistance, MW and current set-points as well as a function of various dead bands and feedback loop settings, (c) differential tapping of transformers, (d) balancing of the electric circuit and hence minimisation of the asymmetry of the electric circuit, (e) the TCP/IP Windows based network version of the Minstral[TM] which permits plant wide maximum demand control, and (f) control of electrode baking. Level 3: The SCADA system creates the man-machine-interface (MMI) between the control system, the user and subsequently between all higher-level control and modelling applications. The networked SCADA system (Table 40.8) provides all the standard features of such state-of-the-art packages e.g. graphical interfaces, data logging, historical trending, and SQL-features, intelligent multiple symptom alarming e.g. through the command language, statistical process control (SPC), on-line documentation, checking of trends and setting of corresponding alarms, checking instrument readings for failure and setting alarms, and monitoring furnace pressures, temperatures, etc. and setting of alarms. In addition to the basic SCADA software customised for an application, various modular Windows utilities could also be accessed from the user-SCADA's user-interfaces. These model-based utilities provide the user with simulation tools with which various furnace modes of operation can be simulated hence providing very useful operator guidance. Level 4: A good database (for Mintek's Furnstar[TM] system[31]—MS Access® front end to SQL Server®) residing on the server of a network is the nucleus of a supervisory control system. It links databases containing data originating from different sources such as real-time data from a SCADA, production, planning and human resources data from e.g. SAP®, etc. It accommodates all furnaces in a plant and collates data in a metallurgically meaningful way in a metallurgically useful database. This structure ensures that all data can be considered simultaneously during decision making, be it by management, metallurgist, operator, fundamental process models, the control system or even an expert system. Since the furnace, its feeds and products, are only poorly defined, rather accurate process data sets are required in order to be able to calibrate (semi-)empirical models, create statistical models, or provide inputs to expert systems, etc. It can, therefore, be stated that: the more poorly a system is defined, the more heavily the (feed-forward) control system should make use of a well-structured and metallurgically useful database. Level 5: As is clear from Figure 40.8 various types of on-line models can be developed for this level, which could include fundamental models, expert systems, neural nets, process scheduling, ecological simulation and plant wide optimisation models. Level 6: Although not indicated, man (still) plays a critical part in supervisory control systems in extractive metallurgy. For submerged-arc furnaces man's involvement is still large. What this in fact also implies is that, if everyone on a plant does not buy into the control philosophy of a plant, it can be stated that optimal results will only be achieved with difficulty. Man is often still really the learning machine on an extractive metallurgical plant!

In Table 40.7 standard control techniques are summarised and listed for various operations of metals processing. For extensive applications in different minerals and metals processing industries, please read the paper by Jämsä-Jounela.[5]

40.4.4   Instrumentation for process control[1–6]

Instrumentation is an area that is currently under rapid changes as field communications move from analogue (4–20 mA) one-way to digital two-way protocols (e.g. PROFIBUS, Ethernet). Any control theory and technique have to be implemented to a process through hardware and lower level control software systems on a network. Therefore, one needs measurement devices e.g. sensors or transmitters, controllers, controlling devices, actuators and lower level control software on a plant. This constitutes levels 1 to 3 in Figure 40.8.

Control software: Control software is very important for any process control system, and it is closely associated with the controllers and the control techniques. Any good control technique needs control software to realise its functions. Along with the rapid development of control techniques, various control software have been developed and applied in different unit operations of metallurgical industry. Table 40.8 lists a few examples of commonly used modern control software.

Network: Ethernet Industrial Protocol (Ethernet/IP) is an open industrial networking standard that supports implicit messaging (real-time I/O messaging), explicit messaging (message exchange), or

Table 40.7 STANDARD CONTROL TECHNIQUES USED FOR METALS PRODUCTION AND PROCESSING[5]

Control technique	Application examples
Classical methods	
P, PI, PID control	Submerged-arc furnace (e.g. FeMn, FeCr, SiMn, SiCr, FeSi, Si), rotary kiln, hot and cold rolling mills
Feed forward control	Submerged-arc furnace
Model-based control and model predictive control	Thermal treatment, reheat furnace, extrusion, ferrosilicon production, hot and cold rolling
Multivariable control	Hot rolling mill, Al production cell,
Adaptive control	Al production cell, BOF, ingot/slab/continuous casting, reheat furnace, cold rolling
Artificial intelligence (AI) control	
Expert system	Blast furnace, BOF, hot/cold rolling
Fuzzy logic	Sintering machine, BOF, Zn roaster, continuous casting
Neural networks	Reheat furnace, EAF, BOF, ladle furnace, continuous casting, hot rolling
Genetic algorithms	
Diagnostics and monitoring	
Fault diagnostics and process monitoring	Zn hydrometallurgy plant, Outokumpu flash smelting, blast furnace, ferroalloy furnace, steelmaking, continuous casting, hot rolling
Quality monitoring	Steel surface inspection, continuous casting, hot rolling
Modelling techniques	
Thermodynamic (ChemSage, Metsim)	Equilibrium calculations for metallurgical reactors
Dynamic modelling	Various ARMAX type models to quantify the dynamics of metallurgical reactors
Statistical:	
Data reconciliation	Blast furnaces, converters, arc-furnaces,
Neural networks	Sn/Cu/Zn smelters
Principal component analysis	
CFD	Casting, modelling of flow in various metallurgical furnaces, off-gas systems, rotary kilns, waste incinerator

both and uses commercial off-the-shelf Ethernet communication chips and physical media. Because Ethernet technology has been used since the mid 1970s and is widely accepted throughout the world, Ethernet products serve a large community of vendors. Ethernet/IP emerged due to the high demand for using the Ethernet network for control applications. Ethernet/IP is an open network because it uses (a) IEEE 802.3 Physical and Data Link standard, (b) Ethernet TCP/IP protocol suite (Transmission Control Protocol/Internet Protocol), the Ethernet industry standard, and (c) Control and Information Protocol (CIP), the protocol that provides real-time I/O messaging and information/peer-to-peer messaging. ControlNet and DeviceNet networks also use CIP. TCP/IP is the transport and network layer protocol of the Internet and is commonly linked with Ethernet installations and the business world. TCP/IP provides a set of services that any two devices can use to share data. Because Ethernet technology and standard protocol suites such as TCP/IP have been published for public use, standardised software tools and physical media have been mass-produced and are readily available offering two benefits, established technology and accessibility. The UDP/IP (User Datagram Protocol) is also used in conjunction with the Ethernet network. UDP/IP provides the fast, efficient data transport necessary for real-time data exchange. To make Ethernet/IP successful, CIP has been added on top of TCP/UDP/IP to provide a common application layer.

Actuators: An actuator is the final driving element in a control loop, which are usually control valves, variable speed drives (pumps, etc.), and other control switches.

Sensors: Process variables in metallurgical processes are normally physical variables such as temperature, pressure, and differential pressure and flow rate. These variables are easy to measure with conventional sensors and transducers for control and monitoring purposes. For instance, thermocouples are widely used in temperature measurement and control of metallurgical processes. However, to control the chemical composition of intermediate or final products, on-line chemical analysis is also highly demanded. In reality, it is very difficult to measure directly on-line, though many on-line analyzers are used in chemical processes. Quick off-line measurements are normally taken to control the product quality, and in this case, a certain delay of the process control may happen, for instance tapping time for liquid steel in a BOF converter (2–5 minutes to sample and analyze by

Table 40.8 COMMONLY USED LOWER LEVEL (1 TO 3) CONTROL SOFTWARE FOR METALS PRODUCTION AND PROCESSING

Name of software	Functionality
PI system (OSI software) (www.osisoft.com)	PI is a universal production data acquisition system with an integrated analysis system. The data from various systems of your production plant (PCS, PLC, laboratory data, manual inputs) is stored in a long-term archive on a data server and can be visualised and evaluated on a PC within the Windows environment. In other words, a set of server and client based software programs, designed to fully automate the collection, storage and presentation of plant information, also permitting real-time data reconciliation.
G2 (Gensym) (www.gensym.com)	A graphical, object-oriented, customisable software platform for rapidly building expert manufacturing applications. G2 applications model and simulate operations, intelligently monitor and control processes and diagnose faults.
Emerson process management (www.emersonprocess.com)	Process management software.
FIX DMACS^{TM} iFIX (www.intellution.com)	SCADA (Supervisory Control and Data Acquisition) software, iFIX, the HMI/SCADA component of the Intellution Dynamics family of automationsoftware, is a Windows NT-based industrial automation solution for monitoring and controlling manufacturing operations.
Fisher-Rosemount systems (www.emersonprocess.com)	Part of Emerson Process Management. Provides software for process management; Process control, automation, and optimisation; Asset management, monitoring, maintenance, and optimisation.
Honeywell (www.honeywell.com)	Provides advanced software applications for home/building control and industrial optimisation; sensors, switches, controls systems and instruments for measuring pressure, air flow, temperature, electrical current and more.
InTouch^{TM} and FactorySuite (www.wonderware.com)	InTouch, a HMI, provides a single integrated view of all the control and information resources. InTouch enables engineers, supervisors, managers and operators to view and interact with the workings of an entire operation through graphical representations of their production processes. FactorySuite is an integrated suite of industrial automation software.
iBaan for metals (www.baan.com)	Business software for the metals industry e.g. management of inventory to optimise supply and value chain.
ABB Industrial^{IT} Solutions for Metals (www.abb.com)	ABB Industrial IT integrates diverse automation and information technologies in real time to provide better business decision support, standardisation of global processes.
Minstral^{TM} (www.mintek.ac.za)	Mintek's Submerged Arc Furnace Controller, an integrated electrode and tap-changer control system.
CSense^{TM} Crusader Systems (www.crusader.co.za)	CSense^{TM}, interfaces with SCADAs, DCSs and historians, for diagnosis of process deviations: Real-time analyzes of deviations, Alarming and messaging in HMI environment, Automatic reporting. Real-time soft sensor with a variety of artificial intelligence tools such as neural nets, fuzzy as well as classical statistics.
MDC Technology (www.mdctech.com)	MDC Technology is a major specialist provider of Production Optimisation and Performance Monitoring software solutions to the oil & gas, hydrocarbon processing, chemicals and power/utility industries. Part of Emerson Process Management.

automated robot driven XRF/XRD equipment). For some variables, which are difficult to measure, 'soft sensors' can be applied, as discussed by Reuter and Grund.[33] These use directly measurable variables to infer non-measurable data by the use of appropriate mathematical methods. The Optical Production Control (OPC) system developed by Semtech has been used for on-line production control in ferrous and non-ferrous metallurgical industry[38] e.g. copper converting and steelmaking to determine the process end point. The technology is based on the spontaneous measurement of vapour phase from molten melts by spectrometer. In addition, solid electrolytes as sensors have been used for measuring oxygen content in the molten melt of steel and copper.[37] Non-visible refractory wear and melt penetration of metallurgical furnaces has long been a difficult problem, which cause major plant damages and process shut-downs.[39] In many cases the real lining profile could only be seen when the furnace is shutdown. However, some on-line wear detecting systems are available such as SAVEWAY described by Hopf.[39] The technique has been widely used in induction furnaces for copper or copper–nickel alloys, and ISP furnaces for zinc and lead extraction.

Controllers: A controller is the heart of a control system, where different control techniques are implemented. The simplest one is the PID controller. Conventional controllers have the advantages in small control systems. For a large-scale plant-wide control system, Distribute Control System (DCS) and a centralised operating and processing centre as well as plant-wide networking are commonly used. Fieldbus is a new generation control system: field distributed control system (FDCS), with an all-digital, high performance, multi-drop communications protocol. With bi-directional communication capabilities, the field engineers will be able to access field devices, furthermore the field devices themselves will be able to accomplish tasks as broadcasting messages over the process network and feeding other controllers with ongoing information regarding process status. These programmable logic controllers (PLC) therefore form the basis of control systems, and various types are available on the market ranging from Allen-Bradley (www.ab.com), Siemens (www.siemens.com), Bailey ABB (www.ab.com), Modicon (Schneider) (www.modicon.com), and Honeywell (www.honeywell.com), to name a few.

40.4.5   Process control examples in metals production

Various process control examples in the metals production industry have been reviewed by Jämsä-Jounela[5] from resources of IFAC MMM events. In the following paragraphs, an effort is made to include information from various sources. However, the aim is not to give a complete review of the control applications in these areas, rather to use these wide application areas to show the latest control techniques and instrumentation.

Blast furnace ironmaking:[7–11]   Blast furnace (BF) is the dominating ironmaking reactor to provide hot metal for steelmaking. As Jämsä-Jounela[5] indicated, AI methods (expert system, fuzzy control, neural networks) are applied extensively to BFs for process monitoring and fault diagnostics. Expert systems for BFs are advisory systems to support the operation of the furnace. The system monitors the heat level of the furnace, the position and shape of the cohesive zone, and forecasts abnormal conditions such as slipping, channelling, sliding and scaffolding. Warren and Harvey[11] from CORUS gave a comprehensive description on the development and implementation of a generic blast furnace expert system. G2 (Table 40.8) was selected for the BF control system, comprising data acquisition software, a comprehensive relational database, a G2 database bridge, and an expert system developed with G2. Some of the key control features include: injecting flow rate, liquid level via EMF measurement, cooling water leakage monitoring, trend analysis on hearth thermocouple readings for temperature to monitor the protective skull on the carbon blocks. Brunnbauer *et al.*[8] presented a 3rd generation close-loop expert system for a fully automatic blast furnace. Kramer and Parker described the implementation of ironmaking (level 2) supervisory control system for Dofassco's blast furnace.

Oxygen steelmaking and continuous casting:[12–18]   About 60% of the crude steel is made with BOF or LD converters, where steel scraps are melted and mixed with liquid iron from blast furnace (the hot metal). An oxygen lance is used to oxidise the impurities in the melt such as carbon, silicon, phosphorus, sulphur, etc. before further refining and casting. The blowing process takes about 15–20 minutes. Most important control parameters in BOF converter are blowing control, endpoint control of temperature and chemical composition, and exhaust gas pressure control. The converter is controlled by an operator through judgment of the state of the process based on a number of measurements such as (1) a sound level measurement with a sonicmeter to get information on the thickness of the foaming slag; (2) off-gas analysis for the information on the decarburisation rate. The operator will stop the blowing process, when the carbon content reaches the preset value based on the online or semi-online (2–3 minutes or so) measurement. Commercial systems for online estimation of metal analysis, e.g. MEFCON from MEFOS are available on the market. New development is made for carbon content control by Johansen *et al.*[13] to apply nonlinear feedback to a nonlinear process model so as to create a nonlinear observer. Superior results are obtained with real data from SSAB Oxelösund AB. Yoshida *et al.*[18] described a hybrid expert system of empirical knowledge of operators combined with a mathematical model for blowing control at Nippon Steel Corporation. The control system consists of two subsystems: (1) static control model using rule-based reasoning, based on the information available before the start of blowing, (2) dynamic control model using fuzzy reasoning, based on process signals for the state of refining during blowing. Kohl *et al.*[15] presented their development of a scheduling expert system VASE (Voest Alpine Scheduling Expert) for a LD steelmaking plant. A combined approach of expert system technology, conventional programming, and database management proved to be an adequate solution. During the blowing operation, foam level control is important. Sound signals, provided by sonic meter or microphones, are sometimes used as control signals for static control (not for dynamic control). The sonic meter signal is estimated from off-gas flow rate and CO content in the gas. There the deviation of estimated signal from the measured signal is used to detect sloping, combining sonic meter and gas analysis[12]. Kobayashi *et al.*[14] discussed the control of slag slopping or excessive slag foaming by using a microwave gauge. Using the established prediction criterion, the sloping is predicted 30 seconds before the event with a reliability of over 80%. Then the oxygen flow rate, lance position and sub-materials addition are used for preventing the sloping. The frequency of the sloping has been reduced by four times. It has been used online at the Wakayama Works of Sumitomo Metal Industries. Various control algorithms have been developed for continuous casting, as for example by Siemens (www.siemens.com).

Electric arc furnace (EAF) steelmaking:[1]　Electric arc furnaces (both AC arc and DC arc furnaces) are used to produce about 40% of the world total crude steel. Instead of using hot metal from blast furnaces, EAF steelmaking uses almost 100% steel scrap as raw materials. The operation of modern high-powered arc furnaces is characterised by the conflict between a high power input and the limit of the thermal load of the water-cooled wall panels, as reported by Ameling from VDEh[1]. VDEh-BFI has developed a model-based on-line control system for the appropriate distribution of the power among the arcs of AC furnaces. The system was installed at a 150-t arc furnace with 150 MVA power and worked reliably with about 10% reduction in gunning materials consumption and about 8% increased productivity. There are various examples of artificial neural networks that have been applied to model arc furnaces to stabilise the power supply e.g. the Expert Furnace System Optimisation Process (EFSOP) (*www.stantec.com/efsop*), Siemens (*www.siemens.com*) and Hatch (*www.hatch.com*) to name a few.

Steel reheat furnace and rolling mill:[19–22]　Before a hot rolling operation, the steel slabs or billets have to be kept at a proper temperature so that rolling operation could be feasible. A large body of information is available on the process control of the reheating furnaces. Ameling[1] gives a comprehensive presentation about on-line control of rolling mill furnaces. Norberg[19] described practical control systems for the reheating and annealing process. The reheating furnace control systems have evolved from early control method of slab surface temperature control by adjusting fuel and air manually and then by computer based temperature measurements. Subsequently feedback and feed-forward control techniques as well as model-based technique were applied, in combination with radiation pyrometers. Presently model aided furnace control systems for reheating furnaces are state of the art especially in rolling mills[1], applied to minimisation of energy consumption, optimisation of throughput, improvement of temperature uniformity, heating and product quality, and reduction of environmental pollution. Shulkosky *et al.*[20] introduced a reheat furnace supervisory control system THERMOD for pusher, walking beam and rotary hearth furnaces. The THERMOD uses a matrix of set point and target temperature values prepared through off-line modelling and stored on hard disk for on-line retrieval keypad to scheduling information. An on-line model determines product temperature throughout the furnace and provides for priority heating of the product with the most stringent heating requirements. Indirect and direct biasing provides two stages of on-line correction to the off-line prepared set points. When stabilisation is present, THERMOD artificial intelligence modifies the set points to remove the bias for subsequent processing. Valentine *et al.*[22] described the optimisation of reheat furnace control level I and level II systems (their definition). The level I control system uses Micron P-100 single loop controllers set up in the 2-loop format. The controllers are configured to do PID control of the zone temperatures, air and gas flows using standard cross-limiting techniques, and ancillary loops such as combustion air, dilution air, and furnace pressure control. The single loop controllers also act as a conduit to pass the profile thermocouple temperature information on to the Level II supervisory computer. In the automatic mode, the zone temperature control is done via operator entered set points or the Level II supervisory computer download of set points. The control system also uses a Micon MCD-200 CRT base console for the convenient control of all the single loop controllers. The level II supervisory control system automatically regulates each furnace control zone to heat the billets to a desired temperature while minimising fuel consumption. The supervisory control software resides on a MicroVax 2000 computer and has communication links with two computers and a field device interface. A data link to the plant-tracking computer provides information on the identity and desired heating strategy for individual heats. The link to the level I system enables access of real time process variable information and downloading of set points for furnace control. The field device interface link provides the supervisory system with real time tracking signals. Siemens has also developed and implemented on various sites in the world intelligent controllers for rolling mills with the goal of obtaining closer tolerances, higher productivity, and longer service life (www.siemens.com).

Takahashi[21] gave an excellent review on the state of the art in hot rolling process control. As Takahashi summarised, since the 1960s the computer control systems have been installed in hot strip mills. Nowadays the areas cover reheating furnace combustion control, roughing mill setup, finishing mill setup, dimensional control of thickness, width and profile, control of finishing mill and coiling temperature. Various kinds of sensors are installed and numerous data are gathered through the plant-wide network. However, increasingly severe demands for dimensional accuracy, mechanical and surface properties require further development of advanced control. The imaging processing technology combined with charge coupled device (CCD) has been applied to in-line sensors for functions such as flatness measurement by projected-fringe-method and surface defect inspection. The interstand sensors are used to improve the dimensional accuracy, such as an interstand thickness meter and width meter, and an interstand strip velocity meter. The modern high performance actuators are increasingly used such as a hydraulic roll gap control device and a hydraulic looper; AC drive of high response and reliability; profile and shape control apparatus such as pair cross, roll shifting, variable crown roll; slab sizing press; coil box or induction heater to maintain homogeneous temperature of roughed transfer bar. The rolling process at the finishing mill is a multivariable system of the rolling force, strip gauge, interstand strip tension and lopper angle. The conventional automatic gauge control (AGC) system is single-input and single-output controller. More advanced control techniques applied include least quadratic (LQ) multivariable control, H-infinity control, inverse least quadratic (ILQ) control, adaptive control, sliding mode control, observer-based multivariable control, internal model control (IMC), and the predictive feed-forward control. In the area of temperature control, the model-based feed-forward control method is used to the finisher delivery temperature and coil temperature control. In addition, rolling schedule design is very important and becomes the multi-criteria optimising problem, and the dynamic programming or the genetic algorithm has been used to solve this problem.

Aluminium reduction cell:[23–25]   In order to produce aluminium metal, aluminium oxide (alumina) is first prepared through Bayer process, and then the pure aluminium is produced in an electrolytic cell in a molten salt at high temperatures. The reduction cell is a very complicated reactor, which involves complex electromagnetic flows, and electrolytic chemical reactions. McFadden *et al.*[25] gave a good review on the advanced process control for the aluminium reduction cell. Control of the reduction cell is carried out using single input–single output (SISO) control loops that operate independently of each other. All are discrete or sampled data feedback loops, either manual or automatic/online. They are either rule-based (e.g. alumina feed control), or conventional three-term or PID controllers. In addition, there is also considerable human activity in the areas of process supervision, and fault detection and diagnostics. Normally controlled variables include cell resistance (and its noise, rise or slope), bath and metal depth, bath temperature and chemistry (AlF_3, LiF, CaF_2, MgF_2), metal purity (Fe, Si), and individual anode current. Through application of advanced control techniques, performance of aluminium electrolytic cells can be achieved. Examples include alumina feed control by application of linear and non-linear models, alumina feed control and cell resistance control using fuzzy logic control technique, online supervision of alumina feed and cell resistance control by adaptive, fuzzy logic expert and neural networks, as well as different ways to control AlF_3 concentration or bath temperature and heat balance.

Cu-flash smelting/Aussmelt Ni-Smelting/FeMn Smelting/Sn–Zn–Pb-Smelting:[26–35]   Copper metal is produced either hydrometallurgically or pyrometallurgically from its oxidic or more from its sulphide ores. Flash smelting of copper sulphide concentrates is a modern pyrometallurgical copper making process, where the sulphide concentrates are firstly smelted to copper matte in a flash furnace, and then the matte is further converted or oxidised into metallic copper either through a PS-converter or more recently through a flash converter. The control of a flash smelter has been described by Bonekamp *et al.*[27] These authors apply data reconciliation in combination with neural nets and dynamic modelling to predict future operating conditions. Lance based copper–nickel matte converters have become an established technology in the field of base metal sulphide smelting and converting. An Ausmelt® converter is currently being constructed at Anglo Platinum in South Africa to convert a nickel–copper rich matte containing platinum group metals (PGMs). The accurate prediction of the matte chemistry is required to control the composition at the endpoint which influences segregation of PGMs during the subsequent slow cooling and magnetic separation processes. Eksteen *et al.*[29] discussed a hybrid modelling approach that incorporates both the effect of thermochemistry and system dynamics. It is shown how the model relates to fundamental conservation laws, while still providing enough flexibility to adapt to changing operating conditions. It will also show how off-line modelling can be incorporated into a robust on-line model. Based on the data from many pilot plant trials, the model prediction performance is shown to be within the accuracy requirement set by industry and therefore constitutes a reliable basis for model-based control. The final generic model structure takes on the form of a thermodynamically integrated ARMAX hybrid model. The final model has the advantages of being dynamic, adaptive, robust and it can be related to thermochemical fundamentals. Reuter and Moolman[31,32,34,35] discuss the dynamic modelling and control of ferromanganese furnaces while Aldrich and Reuter[26] discuss the use of artificial intelligence techniques to monitor the operation of a submerged arc furnace.

Process control in recycling: Pyrometallurgical reactors close all recycling chains. Dalmijn, Reuter and De Jong[28] explored the complex interrelationship between the resource cycle, recycling and metal production as a function of economics, environmental legislation, and product design. It is argued that central to optimising this resource cycle lies the control in particular of physical separation plants by the application of advanced sensor technology and sorting equipment as well suitable control and sampling algorithms supported by optimising mass balance models. The modelling and control of a furnace for the recycling of Sn/Zn/Pb containing material is discussed by Grund *et al.*[30] Furthermore, the modelling and control of a chemical waste incineration kiln by the use of computational fluid dynamics is discussed by Yang *et al.*[36]

40.4.6 Keywords in process control

Actuator, Adaptive control, Artificial Intelligence, Automation, Controller, Dynamic modelling, Expert system, Feed forward control, Feedback control, Fuzzy Logic control, Model predictive control, Model-based control, Neural networks, PLC, Process control, TCP/IP, Sensor.

BIBLIOGRAPHY

General review articles

1. D. Ameling, Examples for on-line control developments in ironmaking, steelmaking and rolling processes in Germany, *Stahl und Eisen*, 2002(1), **122**, 23–31.
2. O. A. Bascur and J. P. Kennedy, pp. 111 to 124 in '*Heft 89 Automatisierung in der Metallurgie*: Improving metallurgical performance: Data unification and measurement management' (ed. GDMB), Clausthal-Zellerfeld, Germany, 2001.
3. J. Chu, H. Su, F. Gao and J. Wu, Process control: art or practice, *Annual Review in Control*, 1998, **22**, 59–72.
4. D. Galara and J. P. Hennebicq, Process control engineering trends, *Annual Review in Control*, 1999, **23**, 1–11.
5. S.-L. Jämsä-Jounela, Current status and future trends in the automation of mineral and metal processing, *Control Engineering Practice*, 2001, **9**, 1021–1035.
6. M. V. Le Lann, M. Cabassud and G. Casmatta, Modelling, Optimisation, and control of batch chemical reactors in fine chemical production, *Annual Review in Control*, 1999, **23**, 25–34.

On blast furnace ironmaking

7. G. R. Bryer, K. J. Legeard and J. E. DiCosimo, pp. 419 to 437 in '*2001 Ironmaking Conference Proceedings*: The development of process control systems and strategies at Lake Brie company's #1 blast furnace', ISS, Baltimore, Maryland, USA, 2001.
8. G. Brunnbauer, B. Rummer, H. Nogratnig, H. Druckenthaner, B. Schürz, M. Schaler and K. Stohl, pp. 677 to 688 in '*2001 Ironmaking Conference Proceedings*: The fully automatic blast furnace—only a vision?', ISS, Baltimore, Maryland, USA, 2001.
9. M. M. Kramer and B. J. Parker, pp. 405 to 417 in '*2001 Ironmaking Conference Proceedings*: Implementation of ironmaking supervisory control systems', ISS, Baltimore, Maryland, USA, 2001.
10. B. Stackhouse, J. E. Dalsimer, T. J. Wittmer and K. Stohl, pp. 439 to 448 in '*2001 Ironmaking Conference Proceedings*: Process control changes on 'L' blast furnace', ISS, Baltimore, Maryland, USA, 2001.
11. P. Warren and S. Harvey, Development and implementation of a generic blast-furnace expert system, *Transactions of the Institution of Mining and Metallurgy—Section C—Mineral Process. Extr. Metall.*, 2001, **110**, C43–C49.

On steelmaking

12. W. Birk, I. Arvanitidis, A. Medvedev and P. Jönsson, pp. 223 to 228 in '*10th IFAC Symposium on Automation in Mining, Mineral, and Metal Processing (MMM 2001)*: Foam level control in a water model of the LD converter process' (ed. M. Araki), IFAC, Tokyo, Japan, 2001.
13. A. Johansson, A. Medvedev and D. Widlund, pp. 217 to 222 in '*10th IFAC Symposium on Automation in Mining, Mineral, and Metal Processing (MMM 2001)*: Model-based estimation of metal analysis in the steel converter: linear versus non-linear approach' (ed. M. Araki), Tokyo, Japan, 2001.
14. S. Kobayashi, A. Hatono, K. Katohgi, A. Kuriyama and K. Ichihara, pp. 297 to 301 in '*IFAC Symposium on Automation in Mining, Mineral, and Metal Processing (MMM 2001)*: Prediction and control of slag sloping in BOF using microwave gauge' Helsinki, Finland, 1983.
15. K. Stohl, W. Snopek, Th. Weigert and Th. Moritz, pp. 39 to 44 in '*Expert Systems in Mineral and Metal Processing*: Development of a scheduling expert system for a steel plant' (eds S.-L. Jämsä-Jousela and A.J. Niemi), IFAC, Espoo, Finland, 1991.

16. B. Lang, T. Poppe, T. Runkler, R. Döll, M. Jansen and K. Weinzierl, pp. 87 to 110 in '*Heft 89 Automatisierung in der Metallurgie*: Application of artificial intelligence in steel processing' (ed. GDMB), Clausthal-Zellerfeld, Germany, 2001.
17. L. Schöttler, pp. 255 to 268 in '*Proceedings of EMC 2001*: Computerised process control for stainless steel metallurgy, Volume 3' (ed. GDMB), Clausthal-Zellerfeld, Germany, 2001.
18. T. Yoshida, H. Tottori, K. Sakane, K. Arima, H. Yamane and M. Kanemoto, pp. 51 to 56 in '*Expert Systems in Mineral and Metal Processing*: A hybrid expert system combined with a mathematical model for BOF process control' (eds S.-L. Jämsä-Jousela and A.J. Niemi), IFAC, Espoo, Finland, 1992.

On rolling mill and reheating furnaces

19. P-O Norberg, Challenges in the control of the reheating and annealing process. *Scandinavian Journal of Metallurgy*, 1997, **26**, 206–214.
20. R. A. Schulkosky and K. E. Soroka, pp. 18 to 25 in '*Proceedings of Int. Symp. On Steel Reheat Furnace Technology*: THERMOD – a reheat furnace supervisory control system' (ed. F. Mucciardi), Hamilton, Canada, 1990.
21. R. Takahashi, State of the art in hot rolling process control, *Control Engineering Practice*, 2001, **9**, 987–993.
22. T. J. Valentine and M. K. Hanne, pp. 42 to 51 in '*Proceedings of Int. Symp. On Steel Reheat Furnace Technology*: Optimisation of reheat furnace control Level I and Level II system' (ed. F. Mucciardi), Hamilton, Canada, 1990.

On aluminium production

23. A. Heime and K. Esser, pp. 213 to 220 in '*Proceedings of EMC 2001, Volume 3*: Process control of aluminium reduction cell—a contribution to future developments,' (ed. GDMB), Clausthal-Zellerfeld, Germany, 2001.
24. Postlethwaite, P. A. Atsck and I. S. Robinson, The improved control for an aluminium hot reversing mill using the combination of adaptive process models and an expert system, *J. Materials Processing Technology*, 1996, **60**, 393–398.
25. F. J. Stevens McFadden, G. P. Bearne, P. C. Austin and B. J. Welch, pp. 1233–1242 in '*Proceedings Light Metals 2001*: Application of advanced process control to aluminum reduction cells—A review' (ed. J. L. Anjier), Warrendale, PA, 2001 TMS Annual Meeting, New Orleans, Louisiana, USA, 2001.

Non-ferrous, ferro-alloy production and recycling systems

26. C. Aldrich and M. A. Reuter, Monitoring of metallurgical reactors by the use of topographic mapping of process data, *Minerals Engineering*, 1999, **12**(11), 1301–1312.
27. I. H. Bonekamp, J. H. Groeneveld and M. A. Reuter, pp. 361–375 in '*Proceedings of Copper 99—Cobre 99, International Environment Conference, Vol. VI—Pyrometallurgy*: Quantification of the dynamics of the flash smelter' (eds B. A. Hancock and M. R. L. Pon), TMS, Warrendale, USA, 1999.
28. W. L. Dalmijn, M. A. Reuter and T. De Jong, pp. 8 to 18 in '*10th IFAC Symposium on Automation in Mining, Mineral, and Metal Processing (MMM 2001)*: Recycling—The role of automation in the resource cycle' (ed. M. Araki), Tokyo, Japan, 2001. (to appear in Control Engineering Practice).
29. J. J. Eksteen, G. A. Georgalli and M. A. Reuter, pp. 457 to 468 in '*Proceedings 3rd Sulfide Smelting Conference*: Online prediction of the actual melt chemistry in an Ausmelt converter using a thermodynamic—system identification hybrid modeling technique' (ed. R. Stevens and H. Y. Sohn), Warrendale, PA, TMS Annual Meeting, Seattle, USA, 2002.
30. S. C. Grund, M. A. Reuter, R. A. G. Janssen and A. Nolte, pp. 851 to 864 in '*Proceedings Extraction & Processing Division EPD Congress*: Optimisation of a tin–lead-smelting furnace with the aid of statistical modelling techniques' (ed. P.R. Taylor), Warrendale, PA, TMS Annual Meeting, San Diego, USA, 1999.
31. R. L. Moolman, M. S. Rennie, C. Pretorius, M. A. Reuter, J. Klopper and N. Dawson, pp. 179–188 in '*Proceedings 54th Electric Furnace Conference*: PC Based control technology in the ferroalloy industry', Electric Furnace Division of the Iron & Steel Society, Warrendale, PA, Dallas, Texas, 1996.
32. M. A. Reuter, C. Pretorius, C. West, P. Dixon and M. Oosthuizen, Intelligent control of submerged-arc furnaces for ferroalloys, *Journal of Metals*, December, 1996, 51–53.
33. M. A. Reuter, and S. C. Grund, pp. 629 to 642 in '*Proceedings Extraction & Processing Division EPD Congress*: The use of data reconciliation as a soft sensor in various hybrid modelling and control architectures for pyrometallurgical furnaces' (ed. P.R. Taylor), Warrendale, PA, 2001 TMS Annual Meeting, New Orleans, Louisiana, USA, 2001.
34. M. A. Reuter and D. W. Moolman, Feedforward quality control methodology for reactors using thermodynamics and system identification techniques, *Erzmetall*, 1999, **52**(9), 472–483.
35. M. A. Reuter, Some aspects of the control structure of ill-defined metallurgical furnaces, *Erzmetall*, 1998, **51**(3), 181–194.
36. Y. Yang, M. A. Reuter, J. H. L. Voncken and J. Verwoerd, Understanding of hazardous waste incineration through computational fluid-dynamics simulation. *Journal of Environmental Science and Health, PART A: Toxic/Hazardous Substance & Environmental Engineering*, 2002, **A37**(4), 693–705.

Sensors

37. D. J. Fray, The use of solid electrolytes as sensors for applications in molten metals. *Solid State Ionics*, 86–88, Part 2, July 1996, 1045–1054.

38. W. Persson and W. Wendt, pp. 169 to 183 in '*Proceedings of EMC 2001, Volume 3*: On-line production control utilising optics and spectroscopy' (ed. GDMB), Clausthal-Zellerfeld, Germany, 2001.
39. M. Hopf, pp. 199 to 211 in '*Proceedings of EMC 2001, Volume 3*: The refractory wear measurement system SAVEWAY—increase of process safety and service life of lining' (ed. GDMB), Clausthal-Zellerfeld, Germany, 2001.

Other information source on process control

International Federation of Automatic Control (IFAC) conference proceedings on mining, mineral and metal processing MMM (since 1973).

Control Engineering Practice (Elsevier Science Ltd. Pergamon, ISSN: 0967-0661).

International Journal of Control (Taylor & Francis Ltd., ISSN 1366-5820).

Annual Review in Control (IFAC, Pergamon, ISSN: 1367-5788).

IEEE Proceedings: Control Theory and Applications (ISSN: 1350-2379).

Journal of Process Control (Elsevier Science Ltd. ISSN: 0959-1524).

Automatica (Elsevier Science Ltd. Pergamon, ISSN: 0005-1098).

41 Bibliography of some sources of metallurgical information

This chapter provides a short bibliography of some major sources of metallurgical reference. The intent is to supplement, rather than replace, the detailed references provided in each chapter of Smithells. Thus, the present chapter concentrates on works of general utility. For the convenience of readers, this bibliography is split into a limited number of categories. Inevitably, there is some overlap between categories and where this occurs, works have been assigned to the most relevant-seeming category. Within some of these categories, a distinction is drawn between works that are sources of data, those with significant narrative content and works containing both narrative and data. In the case of web-based resources, a URL has not been included, as these are subject to change, but sufficient information is provided for these items to be located easily with any standard search tool.

General metallurgical and related works

Data only

Title	Author(s)/ Editor(s)	Bibliographic Details	In Print	Notes
Aerospace Structural Metals Handbook	W. F. Brown, Jr., H. Mindin and C. Y. Ho	CINDAS/USAF CRDA Handbooks, 1999.	Yes	Compilation of a variety of data for common alloys used in aerospace.
Alloy Digest (1952–2000)	D. Rahoi	ASM International, Materials Park, OH, 2001.	Yes	Wide range of data on commercial alloys.
ANSI Standards	Various	American National Standards Institute, Washington, DC, various dates.	Yes	US standards.
ASM Metals Reference Book	M. Bauccio	3rd Edition, ASM International, Materials Park, OH, 1993.	Yes	Concise source of data. No index.
ASTM Standards	Various	ASTM International, West Conshohocken, PA, various dates.	Yes	Comprehensive collection of standards for materials and their testing.
British Standards	Various	BSI, London, UK, various dates.	Yes	Comprehensive collection of standards for materials, their use and testing.
Cambridge Engineering Selector	Various	Granta Design Ltd. Cambridge, UK, continuously updated.	Yes	Materials selection tool.
Control of Substances Hazardous to Health (COSHH) data sheets	Various	Provided by materials vendors, numerous on-line collections available.	Yes	Provides safety information mandated by UK law.
CRC Handbook of Chemistry and Physics	D. R. Lide	83rd Edition, CRC Press, Boca Raton, FL, 2002.	Yes	Comprehensive source of general physical science data. More for pure metals than alloys.
CRC Materials Science and Engineering Handbook	J. F. Shackelford and W. Alexander	3rd Edition, CRC Press, Boca Raton, FL, 2001.	Yes	Wide range of data on metallic and non metallic materials.
DIN Standards	Various	International Library Service, Provo, UT, various dates.	Yes	German standards (available in English).
EN Standards	Various	CEN, Brussels, Belgium, various dates	Yes	European standards.
ISO Standards	Various	International Organization for Standardization, Geneva, Switzerland, various dates.	Yes	International standards.

Title	Author	Publication		Description
Japanese Standards	Various	Japanese Standards Association, Tokyo, Japan, various dates.	Yes	Japanese standards.
Material Safety Data Sheets (MSDS)	Various	Provided by materials vendors, numerous on-line collections available, various dates.	Yes	Provides safety information mandated by US Federal law.
Metals and Alloys in the Unified Numbering System	Multiple	9th Edition, ASM International. Materials Park, OH, 2001 (co-published with SAE).	Yes	Detailed cross-reference for alloy numbering systems.
MIL Standards	Various	Document Automation and Production Service, Philadelphia, PA, various dates.	Yes	US military standards.
NIST Standards Database	Various	NIST, Gaithersburg, MD, various dates.	Yes	Comprehensive database of US civil and military, plus international standards.
Physical and Chemical Constants and Some Mathematical Functions, Tables of	G. W. C. Kaye and T. H. Laby	16th Edition, Longman Science and Technology, 1995.	Yes	General physical property data.
Pure Metal Properties: A Scientific and Technical Handbook	A. Buch	Co-published by ASM International, Materials Park, OH and Freund Publishing House Ltd., 1999.	Yes	Physical and mechanical property data for pure metals.
SAE Standards	Various	SAE, Warrendale, PA, various dates.	Yes	Standards for alloys used in both ground transportation and aerospace.
Sax's Dangerous Properties of Industrial Materials	R. J. Lewis, Sr.	10th Edition, Wiley Interscience, New York, NY, 2000.	Yes	Provides detailed safety information. Complimentary to MSDS.
Science Data Book	R. M. Tennent	Oliver and Boyd, London, UK, 1971.	No	Highly concise work, yet with fairly comprehensive coverage.
Structural Alloys Handbook	J. M. Holt, H. Mindin and C.Y. Ho	CINDAS/Purdue University, 1996.	Yes	Compilation of a variety of data for common alloys.
Woldman's Engineering Alloys	J. Frick (Editor)	9th Edition, ASM International, Materials Park, OH, 2001.	Yes	Comprehensive, cross-referenced guide to commercial alloys.
Narrative only				
ASM Materials Engineering Dictionary	J. R. Davis	ASM International, Materials Park, OH, 1992.	Yes	Alphabetical listing of key materials terminology.
Atomic Theory for Students of Metallurgy	W. Hume-Rothery and B. R. Coles	6th Reprint, Institute of Metals, London, UK, 1988.	Yes	Authored by the originator of the Hume-Rothery rules.

(continued)

Title	Author(s)/Editor(s)	Bibliographic Details	In Print	Notes
Concise Chemical Thermodynamics	J. R. W. Warn and A. P. H. Peters	2nd Edition, Nelson Thornes Ltd., Cheltenham, UK, 1996.	Yes	Concise basic introduction to chemical thermodynamics.
Engineering Materials I and II	M. F. Ashby and D. R. H. Jones	Butterworth Heinemann, Oxford, UK, various dates.	Yes	Series of introductory texts. Numerous case studies.
Engineering Materials III	D. R. H. Jones	Pergamon Press, Oxford, UK, 1993.	Yes	
Materials Selection in Mechanical Design	M. F. Ashby	2nd Edition, Butterworth-Heinemann, Oxford, UK, 1999.	Yes	Introductory level text.
Metallurgy for the Non-Metallurgist	H. Chandler	ASM International, Materials Park, OH, 1998.	Yes	Introductory level text.
Metallurgy, An Introduction to	A. H. Cottrell	2nd Edition, Institute of Metals, London, UK, 1995.	Yes	Contains more advanced material than the title implies, but assumes relatively little prior knowledge.
Open University Materials in Action Series	Various	Butterworth-Heinemann, Oxford, UK, various dates.	Yes	Series of introductory texts. Numerous examples.
Selection and Use of Engineering Materials	J. A. Charles, F. A. A. Crane and J. A. G. Furness	3rd Edition, Butterworth-Heinemann, Oxford, UK, 2001.	Yes	Introductory level text.
Ternary Phase Diagrams in Materials Science	D. R. F. West and N. Saunders	3rd Edition, Institute of Metals, London, UK, 2002.	No	Guide to interpretation and use of ternary phase diagrams
Narrative and data				
ASM Handbook	Various	ASM International, Materials Park, OH, various dates.	Yes	Multivolume work (currently 21 volumes) covering most aspects of metallurgy.
ASM Metals Handbook Desk Edition	J. R. Davis	2nd Edition, ASM International, Materials Park, OH, 1998.	Yes	Single volume form of the ASM Materials Handbook (revised more recently than some volumes of the full handbook)
ASTM Standard Technical Publications (STP)	Various	ASTM International, West Conshohocken, PA, various dates.	Yes	Cover a wide variety of materials related topics.
Metallurgical Design, Handbook of	G. E. Totten, K. Funatani and L. Xie	Marcel Dekker, New York, NY, 2003.	In Press	Design with metallic materials.

Physical and chemical metallurgy

Data

Title	Author(s)	Publisher	Data	Description
ASM Alloy Phase Diagrams Collection	Depends on edition	Various editions published by ASM International, Materials Park, OH, various dates.	Yes	Comprehensive collection of evaluated binary and ternary phase diagrams. Binary phase diagrams supersede earlier series of phase diagrams collected by Hansen and Anderko, Elliott, Shunk and Moffatt. Available in a variety of editions, including volumes for specific alloy systems.
ASM Materials Property Atlases	Various	ASM International, Materials Park, OH, various dates.	Yes	Series includes atlases of stress-strain curves, fatigue curves, creep and stress-rupture curves and corrosion fatigue/stress corrosion curves.
Corrosion Data, Handbook of	D. Anderson and B. Craig	2nd Edition, ASM International, Materials Park, OH, 1995.	Yes	Detailed manual on corrosion in a wide range of media.
NIST-JANAF Thermochemical Tables	M. W. Chase, Jr.	4th Edition, American Chemical Society, Washington, DC, 1998.	Yes	Comprehensive collection of thermodynamic data.
Pearson's Handbook of Crystallographic Data for Intermetallic Phases	P. Villars and L. D. Calvert	2nd Edition, ASM International, Materials Park, OH, 1991.	Yes	Comprehensive collection of crystallographic data on pure metals and alloy phases (4 volumes).
Powder Diffraction File (formerly the Joint Center on Powder Diffraction Studies Database)	Numerous	International Center for Diffraction Data, Newtown Square, PA, updated annually.	Yes	Extremely comprehensive database of crystal structure and powder diffraction data.
Selected Values of the Thermodynamic Properties of Binary Alloys	R. Hultgren, O. D. Desai, D. T. Hawkins, M. Gleiser and K. K. Welley	ASM International, Materials Park, OH, 1973.	No	Thermodynamic data for binary alloys.
Ternary Alloys: A Comprehensive Compendium of Evaluated Constitutional Data and Phase Diagrams	G. Petzow and G. Effenberg (Editors)	John Wiley and Sons, New York, NY, various dates.	No	Currently incomplete series of volumes of ternary phase diagrams.
Thermochemical Properties of Inorganic Substances (2 Volumes)	O. Knacke, O. Kubaschewski and K. Hesselmann	2nd Edition, Springer-Verlag, Berlin, Germany, 1991.	No	Tables of thermodynamic data.

Narrative

Title	Author(s)	Publisher	Data	Description
CALPHAD (Calculation of Phase Diagrams): A Comprehensive Guide	N. Saunders and A. P. Miodownik	Elsevier Science Ltd, Oxford, UK, 1998.	Yes	Textbook on thermodynamics and calculation of phase diagrams.

(continued)

Title	Author(s)/ Editor(s)	Bibliographic Details	In Print	Notes
Corrosion, The Fundamentals of (International Series on Materials Science and Technology, Vol. 44)	J. C. Scully	3rd Edition, Elsevier Science Ltd., Oxford, UK, 1990.	Yes	Introductory level textbook on aqueous corrosion and oxidation.
Crystal Chemistry	R. C. Evans	TechBooks, Herndon, VA, 1980 (originally published by Cambridge University Press, 1964).	Yes	Contains descriptions of important crystal structures.
Deformation and Fracture Mechanics of Engineering Materials	R. W. Hertzberg	4th Edition, John Wiley and Sons, New York, NY, 1995.	Yes	Introductory level textbook on mechanical properties.
Diffusion in Solids	P. G. Shewmon	2nd Edition, TMS, Warrendale, PA, 1991.	Yes	A concise work, covering both diffusion phenomena and atomic mechanisms.
Dislocations, Introduction to	D. Hull and D. J. Bacon	4th Edition, Butterworth-Heinemann, London, UK, 2001.	Yes	A comprehensive guide to dislocation theory.
Fatigue of Materials	S. Suresh	2nd Edition, Cambridge University Press, Cambridge, UK, 1999.	Yes	Comprehensive coverage of fatigue. Concentrates more on mechanisms than on testing procedures.
Fracture Mechanics, Fundamentals of	J. F. Knott	Butterworth's, London, UK, 1981.	No	Detailed discussion of fracture mechanics, including fatigue.
High Temperature Corrosion	P. Kofstad	Elsevier Science, New York, NY, 1988.	Yes	Textbook on high-temperature oxidation and corrosion.
Mathematics of Diffusion	J. Crank	2nd Edition, Oxford University Press, Oxford, UK, 1979.	Yes	Comprehensive guide to phenomenological diffusion theory, including numerous different geometries. Does not cover the atomic mechanisms of diffusion theory.
Mechanical Metallurgy	G. E. Dieter (adapted by D. Bacon)	SI Edition, McGraw-Hill, New York, NY, 1988.	Yes	Introductory text covering mechanical properties, mechanical testing and metalworking.
Modern Physical Metallurgy and Materials Engineering: Science, Process, Applications	R. E. Smallman and R. J. Bishop	6th Edition, Butterworth-Heinemann, Oxford, UK, 2000.	Yes	An expanded version of Smallman's "Modern Physical Metallurgy". Covers most major physical metallurgy principles.
Phase Transformations in Metals and Alloys	D. A. Porter and K. E. Easterling	2nd Edition, Chapman and Hall, London, UK, 1992.	No	Covers a range of topics needed to understand phase transformations.
Physical Metallurgy	P. Haasen and B. L. Mordike	3rd Edition, Cambridge University Press, Cambridge, UK, 1996.	Yes	Advanced text that assumes significant familiarity with solid-state physics.

Title	Author(s)	Publication	In print	Description
Physical Metallurgy	R. W. Cahn and P. Haasen	4th Edition, North Holland, Amsterdam, The Netherlands, 1996.	Yes	Three volume multi-author work, offering comprehensive coverage.
Physical Metallurgy Principles	R. E. Reed-Hill and R. Abbaschian	3rd Edition, PWS Publishers, Boston, MA, 1991.	Yes	Covers general physical metallurgy at an introductory level.
Plastic deformation of Metals, The	R. W. K. Honeycombe	2nd Edition, Edward Arnold, London, 1984.	No	Introduction to plasticity and dislocation theory.
Principles and Prevention of Corrosion	D. A. Jones	2nd Edition, Prentice Hall, New York, NY, 1995.	Yes	General textbook on corrosion.
Solidification, Fundamentals of	W. Kurz; D. J. Fisher (Editor)	4th Edition, Trans Tech Publications Ltd., Aedermannsdorf, Switzerland, 1998.	Yes	Covers most aspects of solidification theory.
Structure of Metals, Crystallographic Methods, Principles and Data	C. S. Barrett and T. B. Massalski	3rd Revised Edition, Elsevier Science Ltd, Oxford, UK, 1980.	Yes	This work is part physical metallurgy, part crystallography.

Narrative and data

Title	Author(s)	Publication	In print	Description
Corrosion	L. L. Shreir, R. A. Jarman and G. T. Burstein (Editors)	3rd Edition, Butterworth-Heinemann, UK, 1994.	Yes	Comprehensive two volume guide to corrosion.
Interfacial Phenomena in Metals and Alloys	L. E. Murr	TechBooks, Herndon, VA (originally published by Addison-Wesley, 1975).	Yes	Primarily a narrative work, however, an extensive collection of interfacial energy data is included.
Introduction to Crystallography, An	F. C. Phillips	4th Edition, Longman, Harlow, Essex, 1971.	No	A systematic guide to point and space groups, with descriptions of their symmetry.
Materials Thermochemistry	O. Kubaschewski, C. B. Alcock and P. J. Spencer	6th Edition, Elsevier Science Ltd, Oxford, UK, 1993.	Yes	Textbook on thermochemistry with data in appendices.
Physical Chemistry of High Temperature Technology	E. T. Turkdogan	Academic Press, 1980.	Yes	Textbook on thermochemistry with table of free energy data.
Surface Wear: Analysis, Treatment and Prevention	R. Chattopadhyay	ASM International, Materials Park, OH, 2001.	Yes	Overview of tribology.
Uhlig's Corrosion Handbook	R. W. Revie (Editor)	2nd Edition, Wiley Interscience, 2000.	Yes	Comprehensive manual on corrosion.

(*continued*)

Specific alloy systems and classes of materials

Data

Title	Author(s)/Editor(s)	Bibliographic Details	In Print	Notes
Aluminium Standards and Data	Not identified	The Aluminum Association, Washington DC, 2003.	Yes	Includes alloy and temper designations, nomenclature and representative property data for commercial aluminium alloys.
ASM Specialty Handbooks	Various	ASM International, Materials Park, OH, various dates.	Yes	Handbooks on various specific classes of metals and alloys.
Atlas of Time-Temperature Diagrams	G. Vander Voort (Editor)	ASM International, Materials Park, OH, 1991.	Yes	Includes diagrams for isothermal transformation, continuous cooling, precipitation, embrittlement and ordering. Two volumes (steels and nonferrous alloys, respectively).
Properties of Aluminium Alloys	J. G. Kaufmann (Editor)	ASM International, Metals Park, OH and the Aluminum Association, Washington, DC, 1999.	Yes	Tensile, creep and fatigue data for aluminium alloys.
Properties of Intermetallic Alloys	Various	Metals Information and Analysis Center, West Lafayette, IN, 1994 onwards.	Yes	Three volume detailed collection of data on aluminide, silicide and other intermetallic compounds. Includes physical properties, mechanical properties and some information on oxidation resistance.
Steel Product Manuals and Databooks	Various	Iron and Steel Society, Warrendale PA, various dates.	Yes	Series of manuals covering major steel forms.
Worldwide Guide to Equivalent Irons and Steels	Various	4th edition, ASM International, Materials Park, OH, 2000.	Yes	Lists equivalent alloy specifications and designations.
Worldwide Guide to Equivalent Nonferrous Alloys	Various	4th edition, ASM International, Materials Park, OH, 2001.	Yes	Lists equivalent alloy specifications and designations.

Narrative

Title	Author(s)/Editor(s)	Bibliographic Details	In Print	Notes
Austenitic Stainless Steels: Microstructure and Mechanical Properties	P. Marshall (Editor)	Kluwer Academic Publishers, Dordrecht, The Netherlands, 1984.	Yes	Comprehensive discussion of structure-property relationships in austenitic stainless steels.
Light Alloys: Metallurgy of the Light Metals	I. J. Polmear	3rd Edition, John Wiley & Sons, New York, NY, 1996.	Yes	Covers aluminium, titanium and magnesium alloys.

Title	Author	Publisher	In print	Description
Making, Shaping and Treating of Steel, The	R. J. Fruehan (Editor)	11th Edition, The AISE Steel Foundation, 1998.	Yes	Comprehensive guide from steelmaking to final shape.
Metal Matrix Composites, An Introduction to	T. W. Clyne and P.J. Withers	Cambridge University Press, Cambridge, UK, 1995.	Yes	Comprehensive guide to metal matrix composites.
Metal Matrix Composites, Fundamentals of	S. Suresh, A. Mortensen and A. Needleman (Editors)	Butterworth-Heinemann, London, UK, 1993.	Yes	Comprehensive guide to metal matrix composites.
Physical Metallurgy and the Design of Steels	F. B. Pickering	Applied Science Publishers, London, UK, 1978.	No	Assuming a basic knowledge of physical metallurgy, this work explains how metallurgical principles are applied in commercial alloys.
Physical Metallurgy of Steels, The	W. C. Leslie	CBLS Publishers, 1991 (originally published by McGraw Hill, London, UK, 1983).	Yes	Discussion of ferrous physical metallurgy.
Steel Heat Treatment Handbook	G. E. Totten and M. A. H. Howes (Editors)	Marcel Dekker, Inc., New York, 1997.	Yes	Guide to heat-treatment of ferrous alloys.
Steels: Microstructure and Properties	R. W. K. Honeycombe and H. K. D. H. Bhadeshia	2nd Edition, Butterworth-Heinemann, Oxford, UK, 1996.	Yes	Discussion of ferrous physical metallurgy. Expanded in scope in the 2nd edition to include newer work, such as modelling.
Steels: Heat Treatment and Processing Principles	G. Krauss	ASM International, Materials Park, OH, 1990.	Yes	A guide to ferrous physical metallurgy with an emphasis on heat-treatment.
Narrative and data				
Aluminum Alloys and Tempers, Introduction to	J. G. Kaufmann	ASM International, Materials Park, OH, 2000.	Yes	Fairly broad coverage of aluminium alloys, with a focus on alloy and temper designations.
Aluminum: Physical Metallurgy and Processes, Handbook of	G. E. Totten and D. S. MacKenzie	Marcel Dekker, New York, NY, 2003.	In Press	Two volume guide to aluminium alloys.
Aluminum: Properties and Physical Metallurgy	J. E. Hatch	ASM International, Materials Park, OH, 1984.	Yes	Metallurgy of aluminium and its alloys.
Copper Development Association Publications	Various	Copper Development Association, New York, NY.	Yes	Wide range of publications on copper and copper-based alloys.

(continued)

Title	Author(s)/Editor(s)	Bibliographic Details	In Print	Notes
Metal Foams: A Design Guide	M. F. Ashby, A. Evans, N. A. Fleck, L. J. Gibson, J. W. Hutchinson and H. N. G. Wadley	Butterworth-Heinemann, London UK, 2000.	Yes	Introduction to metallic foams.
Nickel Development Institute Library	Various	Nickel Development Institute, Toronto, Ontario, Canada, various dates.	Yes	Wide range of publications on nickel and nickel-based alloys.
Superalloys II High-Temperature Materials for Aerospace and Industrial Power	C. T. Sims, N. S. Stoloff and W. C. Hagel	John Wiley and Sons, New York, NY, 1987.	Yes	Comprehensive coverage of processing, structure and properties of nickel-base superalloys. Includes alloy data as an appendix.
Superalloys: A Technical Guide	M. J. Donachie and S. J. Donachie	2nd Edition, ASM International, Materials Park, OH, 2002.	Yes	Superalloy metallurgy and processing.
Titanium: a Technical Guide	M. J. Donachie, Jr.	2nd Edition, ASM International, Materials Park, OH, 2000.	Yes	Metallurgy and processing of titanium and its alloys.

Process metallurgy

Data

Title	Author(s)/Editor(s)	Bibliographic Details	In Print	Notes
Coatings and Coating Processes for Metals	J. H. Lindsay (Editor)	ASM International, Materials Park, OH, 1997.	Yes	Comprehensive list of coating processes and trade information.
Foseco Ferrous Foundryman's Handbook	J. Brown	11th Edition, Butterworth-Heinemann, London, UK, 2000.	Yes	Comprehensive data on ferrous metalcasting.
Foseco Non-Ferrous Foundryman's Handbook	J. Brown	11th Edition, Butterworth-Heinemann, London, UK, 1999.	Yes	Comprehensive data on non-ferrous metalcasting.
Heat-Treater's Guide: Practices and Procedures for Irons and Steels	H. Chandler (Editor)	2nd Edition, ASM International, Materials Park, OH, 1995.	Yes	Comprehensive heat-treatment data.
Heat-Treater's Guide: Practices and Procedures for Nonferrous Alloys	H. Chandler (Editor)	2nd Edition, ASM International, Materials Park, OH, 1996.	Yes	Comprehensive heat-treatment data.
Machinery's Handbook	E. Oberg, C. J. McCauley, R. Heald, M. I. Hussain, F. D. Jones and H. H. Ryffel (Editors)	26th Edition, Industrial Press, NY, 2000.	Yes	Comprehensive data on machining.

Narrative

Title	Author	Edition/Publisher		Description
Castings	J. Campbell	2nd Edition, Butterworth Heinemann, Oxford, UK, 2003.	Yes	Introduction to metalcasting.
Extraction Metallurgy	J. D. Gilchrist	3rd Edition, Pergamon Press, Oxford, UK, 1989.	No	Fundamentals of extractive metallurgy.
Foundry Technology	P. R. Beeley	2nd Edition, Butterworth-Heinemann, London, UK, 2001.	Yes	Comprehensive information on metalcasting.
Metal Forming: Fundamentals and Applications	T. Altan, S. Oh and H. Gegel	ASM International, Materials Park, OH, 1983.	Yes	Detailed introduction to metal forming.
Metallurgy of Welding	J. F. Lancaster	6th Edition, William Andrew Publishing, Norwich, NY, 1999.	Yes	Covers fusion welding, adhesive bonding, brazing, soldering, various solid-phase welding processes as well as fusion welding.
Physical Metallurgy of Welding, Introduction to the	K. Easterling	2nd Edition, Butterworth Heinemann, Oxford, UK, 1992.	No	Processing–structure–property relationships in welding.
Steelmaking, Fundamentals of	E. T. Turkdogan	The Institute of Materials, London, 1996.		Textbook on steelmaking.
Transport Phenomena in Material Processing	D. R. Poirier and G. H. Geiger	The Minerals, Metals & Materials Society, Warrendale, PA, 1994.	Yes	Covers fluid dynamics, mass and heat-transfer.

Narrative and data

Title	Author	Edition/Publisher		Description
Principles of Extractive Metallurgy	T. Rosenqvist	2nd Edition, McGraw-Hill, Tokyo, Japan, 1983.	No	Primarily a narrative work, but includes appendices with a significant body of thermodynamic data.
Hot Working Guide: A Compendium of Processing Maps	Y. V. R. K. Prasad, S. Sasidhara, H. L. Gegel and J. C. Malas (Editors)	ASM International, Materials Park, OH, 1997.	Yes	Data on the hot working of ferrous and non-ferrous materials with supporting discussion.
Welding Handbook	Various	8th Edition, American Welding Society, Miami Florida, various dates.	Yes	Multivolume handbook providing comprehensive coverage of a wide range of materials joining technologies, plus related cutting and coating processes.

Microstructural characterisation, mechanical and non-destructive testing of metals and alloys

Data

Title	Author	Edition/Publisher		Description
Handbook of Crystallography for Electron Microscopists and Others	A. G. Jackson	Springer-Verlag, New York, NY, 1991.	Yes	Compact reference manual for crystallography.

(continued)

Title	Author(s)/Editor(s)	Bibliographic Details	In Print	Notes
International Tables for Crystallography	Various	International Union for Crystallography/Kluwer Academic Publishers, Chester, UK, various dates.	Yes	Multivolume comprehensive reference work on crystallography.
Metallographic Etching	G. Petzow	2nd Edition, ASM International, Materials Park, OH, 1999.	Yes	Contains etchant recipes and etching procedures for most types of metals and alloys.
Narrative				
Chemical Microanalysis using Electron Beams	I. P. Jones	The Institute of Materials, London, UK, 1992.	Yes	Concise yet comprehensive guide to microanalysis.
Crystallography and Diffraction, The Basics of	C. Hammond	International Union of Crystallography/Oxford University Press, Oxford, UK, 1997.	Yes	Coverage largely compliments, rather than duplicates, Cullity and Stock.
Electron Beam Analysis of Materials	M. H. Loretto	2nd Edition, Kluwer Academic Publishers, London, UK, 1993.	Yes	A concise work that concentrates on transmission electron microscopy, but also covers scanning electron microscopy and auger electron spectroscopy.
Electron Microdiffraction	J. C. H. Spence and J. M. Zuo	Kluwer Academic/Plenum Publishers, 1992.	Yes	Detailed guide to convergent beam diffraction. Compliments the more general transmission electron microscopy texts listed here.
Energy Dispersive X-ray Analysis, Fundamentals of	J. C. Russ	Butterworth-Heinemann, UK, 1984.	No	Covers energy dispersive x-ray spectroscopy in both TEM and SEM.
High-Resolution Transmission Electron Microscopy	P. Buseck, J. Cowley and L. Eyring (Editors)	Oxford University Press, Oxford UK, 1989.	Yes	Includes numerous case studies. Best suited for intermediate to advanced users.
Light Microscopy of Carbon Steels	L. E. Samuels	ASM International, Materials Park, OH, 1999.	Yes	Updated version of 'Optical Microscopy of Carbon Steels'.
Metallography: Principles and Practice	G. F. VanderVoort	ASM International, Materials Park, OH, 1984.	Yes	Covers most aspects of metallography.
Metallographic Polishing by Mechanical Methods	L. E. Samuels	4th Edition, ASM International, Materials Park, OH, 2003.	Yes	Metallurgical specimen preparation.
Royal Microscopical Society Handbooks	Various	Royal Microscopical Society, Oxford, UK, various dates.	Yes	Extensive collection of works on various aspects of microscopy.

Title	Author(s)	Publication		Description
Scanning and Transmission Electron Microscopy—An Introduction	S. L. Flegler, J. W. Heckman Jr. and K. L. Klomparens	Oxford University Press, 1993.	Yes	Introductory level textbook covering the basics of transmission and scanning electron microscopy in a concise fashion.
Scanning Electron Microscopy and X-Ray Microanalysis	J. Goldstein, D. E. Newbury, D. C. Joy, C. E. Lyman, P. Echlin, E. Lifshin, L. C. Sawyer and J. R. Michael	3rd Edition, Kluwer Academic/Plenum Publishers, 2003.	Yes	Comprehensive guide to scanning electron microscopy and associated analytical techniques.
Surface and Interface Analysis, Handbook of	J. C. Riviere and S. Myhra (Editors)	Marcel Dekker, New York, NY, 1998.	Yes	Covers both surface analysis and surface modification techniques.
Transmission Electron Microscopy of Materials	G. Thomas and M. J. Goringe	2nd Edition, TechBooks, Fairfax, VA, 1981 (originally published by John Wiley and Sons, 1979).	Yes	Concise, yet comprehensive coverage of transmission electron microscopy.
Transmission Electron Microscopy: A Textbook for Materials Science	D. B. Williams and C. B. Carter	Kluwer Academic/Plenum Publishers, New York, 1996.	Yes	Comprehensive information on transmission electron microscopy techniques and the interpretation of images, diffraction patterns and analytical data.
Narrative and data				
Encyclopedia of Materials Characterisation Surfaces, Interfaces, Thin Films	C. Evans, R. Brundle and S. Wilson	1st Edition, Butterworth-Heinemann, UK, 1992.	Yes	Comprehensive survey of analytical techniques.
Hardness Testing	H. Chandler	2nd Edition, ASM International, Materials Park, OH, 1999.	Yes	Comprehensive overview of hardness testing.
NDE Handbook: Non-destructive Examination Methods for Condition Monitoring	K. Boving (Editor)	Woodhead Publishing Ltd., Cambridge, UK, 1989.	Yes	Comprehensive overview of non-destructive testing.
Nondestructive Testing	L. Cartz	ASM International, Materials Park, OH, 1995.	Yes	Introduction to non-destructive testing, covering major techniques.
Practical Electron Microscopy in Materials Science	J. W. Edington	CBLS, 1976 (originally published as the Philips Technical Library Monographs).	Yes	Single volume compilation of the first four volumes of this five volume series. Some data provided in appendices.
Tensile Testing	P. Han (Editor)	ASM International, Materials Park, OH, 1992.	Yes	Comprehensive overview of tensile testing.
X-Ray Diffraction, Elements of	B. D. Cullity and S. R. Stock	3rd Edition, Prentice Hall College Div., Upper Saddle River, NJ, 2001.	Yes	Recently updated with expanded coverage of related topics, such as electron diffraction. Limited data provided in appendices.

(continued)

Title	Author(s)/Editor(s)	Bibliographic Details	In Print	Notes
X-Ray Diffraction. A Practical Approach	C. Suryanarayana and M. Grant Norton	Kluwer Academic/Plenum Publishers, 1998.	Yes	Introductory text, using a workbook approach. Limited data provided in appendices.
Metallurgical and related search tools				
Data				
Cambridge Scientific Abstracts Databases	Various	Cambridge Scientific Abstracts, Bethesda, Maryland, continuously updated.	Yes	Provides a gateway into a number of bibliographic databases relevant to materials.
SciFinder Scholar (incorporates the Chemical Abstracts)	Various	Chemical Abstracts Service, American Chemical Society Washington DC, continuously updated.	Yes	On-line search tool for literature on all aspects of chemistry. Includes extensive information on metallic materials.
Compendex	Various	Engineering Information Inc., Hoboken, NJ, continuously updated.	Yes	On-line search tool for literature on all aspects of engineering. Extensive coverage of metallurgy and materials.
Dictionary of Physical Metallurgy in Five Languages: English, German, French, Russian and Spanish.	R. Freiwillig	Elsevier Science Ltd, 1987.	Yes	Equivalent terms.
Metadex	Various	Chemical Abstracts Service, American Chemical Society Washington DC, continuously updated.	Yes	On-line search tool for literature on all aspects of metallurgy.
Metallurgy/Materials Education Yearbook	K. Mukherjee (Editor)	40th Edition, ASM International Foundation, Materials Park, OH, 2001.	Yes	Comprehensive list of metallurgy and materials-related programmes at US and Canadian Universities. Lists research interests and contact details for individual faculty. Provides some details of materials-related programmes at universities outside North America.
Web of Science (incorporates the Science Citation Index)	Various	Institute for Scientific Information, Philadelphia, PA, continuously updated.	Yes	Comprehensive bibliographic search tools for all aspects of science.
Science Direct	Various	Elsevier Science, Amsterdam, The Netherlands, continuously updated.	Yes	On-line search tool for literature on all aspects of science. Allows searches to easily be restricted to materials science.
UMI dissertation databases	Various	UMI (formerly University Microfilms), Ann Arbor, MI, continuously updated.	Yes	Permits bibliographic searches for dissertations.

Peer-reviewed metallurgical and related journals

There are numerous journals covering metallurgy and materials science. For a fairly comprehensive list see the Science Citation Index (SCI), Journal Citation Reports (JCR) which lists journals by subject and provides ranking metrics.

Acta Materialia	Covers both experimental and modelling work on materials, extensive metallurgical content.
Corrosion	Covers corrosion of materials.
Corrosion Science	Covers corrosion of materials.
International Materials Reviews	Consists of lengthy review papers, each with numerous references.
Journal of Alloys and Compounds	Covers materials with low-volume usage. Formerly called the 'Journal of Less-Common Metals'.
Journal of Materials Research	Covers both experimental and modelling work on materials, especially non-metallic materials.
Journal of Materials Science	Covers general materials science, with subsidiary journals covering specific topics.
Journal of Materials Science Letters	Short papers.
Journal of Nuclear Materials	Covers a wide variety of metallurgical topics for materials used in the nuclear industry.
Materials Characterisation	Covers both microstructural characterisation techniques and application of microscopy to materials.
Materials Science and Engineering	Covers general materials science. Includes four series: A consists of research papers on structural materials, B consists of research papers on solid-state materials, C consists of research papers on biomaterials and R consists of long review papers.
Materials Science and Technology	Covers general materials science.
Metallurgical and Materials Transactions	Covers both experimental and modelling work, almost all on metallic materials. Has two series: A covers physical metallurgy and materials science, B covers chemical metallurgy, with some overlap between the two.
Oxidation of Metals	Covers high-temperature reaction of materials.
Philosophical Magazine	Focuses on the pure science of materials.
Science and Technology of Welding and Joining	Covers joining of materials.
Scripta Materialia	Short papers.
Ultramicroscopy	Covers issues related to advanced microscopy techniques.
Welding Research Supplement (Welding Journal)	Covers research on joining of materials, often with a focus on more applied aspects.
Zeitschrift für Metallkunde	Covers general metallurgy.

(continued)

Professional magazines (see the professional societies list below for acronyms)

Advanced Materials and Processes	Published by ASM International.
JOM	Published by TMS.
MRS Bulletin	Published by MRS.
Materials World	Published by IoM.
Welding Journal	Published by AWS.

Metallurgical and related conference series

There are numerous metallurgy-related conferences held worldwide (many with English as the conference language or with simultaneous translation) and so the following is only a limited selection. Most conferences are sponsored by one or more professional societies (see the professional societies list below for abbreviations).

AFS Transactions	Covers metalcasting.
ASM Materials Solutions Conference	Annual conference with sessions on a wide variety of materials topics.
ASM Subject-Specific Conferences	Numerous conference series.
AWS Convention	Annual conference and trade show on all aspects of joining.
AWS Subject-Specific Conferences	Numerous conference series on joining and related topics.
DGM Conferences	Various materials conferences, a number of which have English as the conference language.
DVS Conferences	Various joining-related conferences.
Euromat Conference	Conference held annually in a different FEMS member country.
Gordon Research Conferences	Conferences devoted to a series of specific materials topics.
IoM Conferences	General materials conferences and meetings on specific materials topics.
MRS Symposia	Semi-annual conferences comprised of a large number of symposia on different aspects of materials.
MSA Conference	Annual conference on all aspects of microscopy in the materials and biological sciences.
TMS Materials Week	Annual conference with sessions on a wide variety of materials topics.
TMS Subject-Specific Conferences	Numerous conference series.

Metallurgical and related professional societies

Many countries have one or more metallurgical/materials-related professional societies. The list below provides some examples.

American Foundry Society (AFS) — Society for metalcasting.

American Welding Society (AWS) — Professional society for all forms of materials joining, plus related cutting and surface treatment processes. Develops standards and a variety of accreditation programmes for welding personnel.

ASM International — Comprehensive US-based materials society devoted primarily to publishing and the organisation of conferences.

ASM Affiliate Societies — Semi-autonomous societies devoted to specific aspects of metallurgy and materials (e.g. International Metallographic Society).

Deutsche Gesellschaft für Materialkunde (DGM) — German materials society.

Deutscher Verband für Schweißen (DVS) — German welding society.

Federation of European Materials Societies (FEMS) — Consortium of professional materials societies throughout Europe.

Institute of Materials (IoM) — UK professional society for materials engineers. Administers the award of chartered-engineer status to materials engineers.

Materials Research Society (MRS) — Comprehensive US-based materials society devoted primarily to publishing and the organisation of conferences.

Microscopy Society of America (MSA) — US society devoted primarily to publishing and the organisation of conferences on both materials and biological microscopy.

Society of Materials Science, Japan (JSMS) — Japanese Materials Society.

The Minerals Metals and Materials Society (TMS) — US professional society for materials engineers. Addresses accreditation issues for materials-related bachelor's degree programmes at US universities with ABET (the Accreditation Board for Engineering and Technology).

Index

A (arrest) temperatures, *see* Steel(s)
Abrasive:
  cutting, *see* Metal cutting
  wear, *see* Wear
Acknowledgements, xix, *see also* relevant
    location in the text
Acoustic:
  emission, **21**–22
  microscopy, **10**–82 to **10**–83
Activation energy:
  diffusion, **13**–7
Adhesive wear, *see* Wear
Ageing, natural and artificial, *see* Heat
    treatment of aluminium alloys
  aluminium alloys, **29**–74 to **29**–75
Air plasma spraying (APS), *see*
    Thermal spraying
Algebra, **2**–13, **2**–15 to **2**–19
Alloy:
  designations, **1**–1 to **1**–14
    nickel-base alloys, **22**–61 to **22**–66
  steels, *see* Stainless steel(s); Steel(s)
  thermodynamic data, *see* Thermochemistry
Alpha:
  emitters, *see* Emitters
  sources, *see* Emitters
Alnico alloys, use as magnetic material, **20**–7
Alumel, *see* Thermocouple(s)
Alumina, *see* Ceramics; Composites
Aluminide structural intermetallics,
    38–7 to **38**–16
Aluminium and aluminium alloys, *see*
    Elements
  ageing, natural and artificial, *see*
    Heat treatment
  anodising, **32**–6
  cast, composition and properties,
    26–20 to **26**–38
  characteristic X-ray emission lines and
    absorption edges, **4**–5
  composites, **37**–4 to **37**–5
  creep properties, **22**–22 to **22**–24
  damping, **15**–9
  density, **14**–16 to **14**–19
  designations, **1**–10 to **1**–11, **22**–1, **29**–66
  dislocation glide planes, **4**–39
  elastic modulus, **14**–16 to **14**–19
  electrical resistivity, **14**–16 to **14**–19,
    19–3, **19**–7
  excitation potentials for characteristic
    X-rays, **4**–4
  fatigue properties:
    elevated temperature, **22**–22 to **22**–24
    room temperature, **22**–3 to **22**–15,
      22–22 to **22**–24
  foams, **38**–19 to **38**–20
  fracture toughness:
    low temperature, **22**–20 to **22**–22
    room temperature, **22**–3 to **22**–15
  heat treatment, **29**–66 to **29**–79
  impact strength, **22**–3 to **22**–15
  metallography, **10**–25, **10**–29 to **10**–35
  ores, **7**–2
  phase diagrams, Indexed on pages
    11–1 to **11**–6
  phase transformations, **29**–67 to **29**–68
  physical properties, **14**–16 to **14**–19
  powder metallurgy alloys:
    processing, **23**–13, **23**–27
    properties, **23**–14, **23**–30
  quenching, *see* Heat treatment
  soldering, **34**–4
  solution treatment, *see* Heat treatment
  superplastic alloys, **36**–20 to **36**–5
  tensile properties:
    effect of heat treatment, *see*
      Heat treatment
    elevated temperature, **22**–16 to **22**–20
    low temperature, **22**–20 to **22**–22
    room temperature, **22**–3 to **22**–15
  texture, **4**–41 to **4**–42
  thermal conductivity, **14**–16 to **14**–19
  thermal expansion coefficient, **14**–16 to
    14–19
  welding, **33**–29 to **33**–31
American alloy designations, *see* related
    designations for alloys, tables of
Americium, *see* Elements
  phase diagrams, Indexed on pages **11**–1 to
    11–6
Amorphous metals, *see* Metallic glasses
Anelastic damping, *see* Damping
Angles, *see* Geometrical relationships
  interplanar, table of, **4**–16

Annealing and stress relieving, **29**–22, **29**–23 to **29**–28, **29**–29
Anodising, **32**–6 to **32**–7
Antimony, *see* Elements
  characteristic X-ray emission lines and absorption edges, **4**–6
  excitation potentials for characteristic X-rays, **4**–4
  metallography, **10**–35
  ores, **7**–2
  phase diagrams, Indexed on pages **11**–1 to **11**–6
Aperture size, *see* Test sieve mesh number
Applications, *see* Material or process of interest
APS (air plasma spraying), *see* Thermal spraying
Arc:
  spraying, *see* Thermal spraying
  welding, *see* Joining
Archetypes of crystal structures, **6**–30 to **6**–53
Arrest (A) temperatures, *see* Steel(s)
Argon, *see* Noble gases
Arrhenius equation, **13**–7
Arsenic, *see* Elements
  characteristic X-ray emission lines and absorption edges, **4**–5
  excitation potentials for characteristic X-rays, **4**–4
  ores, **7**–2
  phase diagrams, Indexed on pages **11**–1 to **11**–6
Astroloy, *see* Nickel and nickel alloys
Atomic:
  numbers:
    periodic table (chart), **3**–4
    table for the elements, **3**–1
  radii:
    table of, **4**–44 to **4**–48
    scattering factors, table of, **4**–27 to **4**–31
  weights:
    periodic table (chart), **3**–4
    table for the elements, **3**–1
Atomization, powder production by, **23**–2
Auger Electron:
  spectroscopy (AES), **10**–80 to **10**–81
  emission, **18**–6 to **18**–7
Austempering, **29**–40 to **29**–41
Austenite, retained, *see* Heat treatment
  determination of, **4**–17
Austenitic stainless steels, *see* Stainless steels
Austenitisation, **29**–21 to **29**–23
Authors, list of, xxii to xxiv

B type (Pt-30%Rh/Pt-6%Rh) thermocouple, **16**–6 to **16**–7
Backscattered electron imaging, *see* Electron microscopy
Barium, *see* Elements
  characteristic X-ray emission lines and absorption edges, **4**–6

excitation potentials for characteristic X-rays, **4**–4
  ores, **7**–2
  phase diagrams, Indexed on pages **11**–1 to **11**–6
Bearing metals/alloys:
  mechanical properties, **22**–161 to **22**–162
  metallography, **10**–71 to **10**–72
  oil replacing, **23**–11 to **23**–12
Becquerels:
  conversion factors for, *see* Conversion factors
  definition of, **2**–2
Belt furnaces, *see* Continuous furnaces
Beryllium and beryllium alloys, *see* Elements
  characteristic X-ray emission lines and absorption edges, **4**–6
  metallography, **10**–35 to **10**–36
  phase diagrams, Indexed on pages **11**–1 to **11**–6
Beta:
  emitters, *see* Emitters
  sources, *see* Emitters
Bibliography:
  general, **41**–1 to **41**–17
  specific topic, *see* Topic of interest
Binary alloy phase diagrams, Indexed on pages **11**–1 to **11**–6
Binders, foundry, *see* Casting
Biocorrosion, **31**–10 to **31**–12
Bismuth and bismuth alloys, *see* Elements
  excitation potentials for characteristic X-rays, **4**–5
  metallography, **10**–35
  ores, **7**–2
  phase diagrams, Indexed on pages **11**–1 to **11**–6
  use as thermoelectric materials, **16**–10, **16**–11
Blast furnaces:
  energy use data, *see* Fuels
  process control, *see* Metallurgical process control
Boiling points, table of, **8**–2 to **8**–8, **14**–1 to **14**–2, **14**–29
Boltzmann-Matano method, **13**–5
Bonding, *see* Joining
Borides, *see* Boron
Boring, **30**–1 to **30**–2
Boron, *see* Elements
  borides:
    thermodynamic data, **8**–21, **8**–43
    phase diagrams, Indexed on pages **11**–1 to **11**–6
Bourdon tube, *see* Vacuum technology
Box furnaces, **40**–2 to **40**–3
Brasses, *see* Copper and copper alloys
Bravais lattices, *see* Crystallography
Brazing, *see* Joining
Bright field, *see* Electron microscopy
Brinell hardness, *see* Hardness

British alloy designations, *see* related designations for alloys, tables of
Bromide, *see* Bromine
Bromine, *see* Elements
  characteristic X-ray emission lines and absorption edges, **4**–5
  excitation potentials for characteristic X-rays, **4**–4
  bromides:
    minerals, **7**–2
    thermodynamic data, *see* Thermochemistry
Bronzes, *see* Copper and copper alloys
Bulk:
  metallic glasses, **38**–22 to **38**–23
  modulus, *see* Elastic modulus
Burgers vector, transmission electron microscopy for identifying, **10**–78

C type (W-5%Re/W-25%Re) thermocouple, **16**–9
Cadmium, *see* Elements
  characteristic X-ray emission lines and absorption edges, **4**–6
  excitation potentials for characteristic X-rays, **4**–5
  metallography, **10**–36
  ores, **7**–3
  phase diagrams, Indexed on pages **11**–1 to **11**–6
Caesium, *see* Elements
  characteristic X-ray emission lines and absorption edges, **4**–6
  excitation potentials for characteristic X-rays
  phase diagrams, Indexed on pages **11**–1 to **11**–6
Calcium, *see* Elements
  characteristic X-ray emission lines and absorption edges, **4**–5
  excitation potentials for characteristic X-rays, **4**–4
  ores, **7**–3
  phase diagrams, Indexed on pages **11**–1 to **11**–6
Calculus, **2**–25 to **2**–28
Calorific value, *see* Fuels
Cannon-Muskegon (CM) and Cannon-Muskegon single crystal (CMSX) alloys, *see* Nickel and nickel alloys
Carbides, *see* Carbon
  cemented:
    mechanical properties, **23**–33
    metallography, **10**–72 to **10**–73
    powder metallurgy processing, **23**–31 to **23**–36
    wear rates, **25**–19
  thermodynamic data, *see* Thermochemistry

Carbon, *see* Elements; Carbides
  equivalent, **29**–21, **33**–19
  fibres, *see* Composites
  phase diagrams, Indexed on pages **11**–1 to **11**–6
  steels, *see* Steel(s)
Carbonates:
  thermodynamic data, *see* Thermochemistry
Carbonitriding, **29**–59 to **29**–66
Carburising, **29**–42 to **29**–58
Case hardening, *see* Carburising; Heat treatment
Cast iron, *see* Casting; Iron and steel
Casting(s):
  aluminium alloys, **26**–20 to **26**–38
  binders, *see* Mould and core materials in this entry
  cast irons, **26**–84 to **26**–100
  ceramics, castable, **27**–8 to **27**–13
  contraction allowances, **26**–10 to **26**–11
  copper alloys, **26**–39 to **26**–51
  cores, *see* Mould and core materials in this entry
  crucibles, **26**–15
  fluxing and inoculation, **26**–16 to **26**–19
  fuels, *see* Fuels
  grey cast irons, *see* Cast irons in this entry
  magnesium alloys, **26**–56 to **26**–67
  mechanical properties, *see* relevant material in this entry
  mould and core materials, **26**–4 to **26**–10
  nickel-alloys, **26**–52 to **26**–55
  parting powders/liquids, mould dressing and paints, **26**–13 to **26**–14
  patterns, **26**–12 to **26**–13
  steels, **26**–70 to **26**–82
  techniques:
    expendable mould and pattern, **26**–1
    expendable mould, **26**–2
    non-expendable mould, **26**–3
  texture, table of, **4**–43
  white cast irons, *see* Cast irons in this entry
  zinc alloys, **26**–68 to **26**–69
Cavitation, *see* Wear
CCD, CCT (Continuous cooling diagrams), *see* Continuous cooling diagrams (CCD)
Cells, unit, *see* Crystal structure
Cellular metals, *see* Metal foams
Celsius temperature scale, definition of, **16**–1, *see also* Temperature
Cemented carbides, *see* Carbides
Cements, **27**–7 to **27**–8
Ceramics (refractories):
  castable, **27**–8 to **27**–13
  cements, **27**–7 to **27**–8
  crucibles, foundry, **26**–15
  design, **27**–14 to **27**–15
  furnaces, use in, **40**–6 to **40**–7
  gunning, **27**–14
  mould and core materials, foundry, **26**–4 to **26**–10
  mouldable, **27**–13 to **27**–14

Ceramics (refractories) (*cont.*)
  physical and mechanical properties data,
    27–1 to 27–4, 27–15
  polishing media:
    metal finishing, 32–1 to 32–2
    metallographic, *see* Metallography
  ramming, 27–13 to 27–14
  standards, 27–15 to 27–18
  thermodynamic data, *see* Thermochemistry
  unshaped, introduction to, 27–7 to 27–8
Cerium, *see* Elements
  characteristic X-ray emission lines and
    absorption edges, 4–6
  excitation potentials for characteristic
    X-rays, 4–4
  phase diagrams, Indexed on pages 11–1 to
    11–6
Charpy impact energy, *see* Impact strength
Chemical:
  polishing:
    vapour deposition (CVD) and infiltration
      (CVI), 35–2, 35–4 to 35–5, 35–8,
      35–10 to 35–12, *see also*
      Metallography
Chinese alloy designations, *see* related
  designations for alloys, tables of
Chlorine, *see* Elements
  characteristic X-ray emission lines and
    absorption edges, 4–5
  chlorides:
    minerals, 7–3
    thermodynamic data, *see*
      Thermochemistry
  excitation potentials for characteristic
    X-rays, 4–4
Chromate coatings, *see* Conversion coatings
Chromel, *see* Thermocouples
Chromium, *see* Elements
  characteristic X-ray emission lines and
    adsorption edges, 4–5
  excitation potentials for characteristic
    X-rays, 4–4
  metallography, 10–37
  phase diagrams, Indexed on pages 11–1 to
    11–6
Cladding, *see* Coating and metal finishing
Cleaning and pickling, 32–2 to 32–5, 32–20 to
  32–21
CM and CMSX (Cannon-Muskegon and
  Cannon-Muskegon single crystal) alloys,
  see Nickel and nickel alloys
Coal and coke, 28–1 to 28–18
Coating and metal finishing:
  anodising:
    aluminium, 32–6
    magnesium, 32–7
  arc spraying, *see* Thermal spraying in this
    entry
  chemical vapour deposition (CVD) and
    infiltration (CVI), 35–2, 35–4 to 35–5,
    35–8, 35–10 to 35–12

cleaning and pickling, 32–2 to 32–5, 32–20
  to 32–21
coating, other processes, 32–18 to 32–20
combustion spraying, *see* Thermal spraying
  in this entry
conversion coatings, 32–21 to 32–23
electroplating, 32–8 to 32–17, 32–20 to
  32–21
evaporation, *see* physical vapor deposition
  (PVD) in this entry
grinding:
  metal finishing, 30–4 to 30–5
  metallographic, 10–9 to 10–10
high velocity oxy-fuel spraying (HVOF),
  see Thermal spraying in this entry
ion plating, *see* physical vapor deposition
  (PVD) in this entry
intermetallic (diffusion and overlay), 38–17
  to 38–18
physical vapour deposition (PVD), 35–1 to
  35–3, 35–7, 35–9, 35–11
plasma spraying, *see* thermal spraying in
  this entry
polishing compounds:
  metal finishing, 32–1 to 32–2
  metallographic, *see* Metallography
sputtering, *see* physical vapor deposition
  (PVD) in this entry
thermal spraying, 35–13 to 35–16
trade names, 32–23 to 32–25
Cobalt and cobalt alloys, *see* Elements
  characteristic X-ray emission lines and
    adsorption edges, 4–5
  excitation potentials for characteristic
    X-rays, 4–4
  metallography, 10–37, *see also*
    Metallography
  ores, 7–3
  phase diagrams, Indexed on pages 11–1 to
    11–6
Coke, metallurgical, *see* Coal and coke
Colour:
  oxide films versus temperature, table of,
    29–35
  temperature relationship, table of, 29–5
Columbium, *see* Niobium
Combustion spraying, *see* Thermal
  spraying
Commercial alloys, *see* Alloy; *see also* entry
  for the relevant metal and its alloys
Composites, *see* Metal matrix composites;
  Intermetallic matrix composites
Compounds, *see* Ceramics; Intermetallic
  compounds
Compressibility:
  unit conversion factors, 2–5
Conductivity:
  electrical, *see* Electrical conductivity
  thermal:
    data, *see* Physical properties
    unit conversion factors, 2–8
Constantan, *see* Thermocouples

Constants:
fundamental physical, table of, **3**–2
general, table of, **3**–2
Continuous:
casting, *see* Casting techniques
cooling diagrams (CCD), also known as time, continuous cooling transformation (CCT), **29**–8, **29**–10 to **29**–14
Contraction, *see* Thermal expansion
Contributors, list of, xxii to xxiv
Control chart, statistical, **2**–48, **2**–49
Conversion:
coatings, **32**–21 to **32**–23
factors, table of, **2**–4 to **2**–10
Coolants, *see* Lubrication
Copper and copper alloys, *see* Elements
cast, composition and properties, **26**–39 to **26**–51
characteristic X-ray emission lines and absorption edges, **4**–5
composites, **37**–6
creep properties, **22**–40 to **22**–46
damping, **15**–9
density, **14**–19 to **14**–21
dislocation glide elements, **4**–39
elastic modulus, **14**–19 to **14**–21
electrical conductivity/resistivity (absolute and relative to int. standard), **14**–19 to **14**–21, **19**–4, **19**–7
excitation potentials for characteristic X-rays, **4**–4
fatigue properties, **22**–37 to **22**–38
impact strength:
low temperature, **22**–35 to **22**–36, **22**–39
room and elevated temperature, **22**–39
metallography, **10**–37 to **10**–42
ores, **7**–4
phase diagrams, Indexed on pages **11**–1 to **11**–6
physical properties, **14**–19 to **14**–21
powder metallurgy alloys:
processing, **23**–9, **23**–25, **23**–27
properties, **23**–22, **23**–25
related designations, **1**–12 to **1**–13
tensile properties:
elevated temperature, **22**–33 to **22**–34
low temperature, **22**–35 to **22**–36, **22**–45 to **22**–46
room temperature, **22**–27 to **22**–32
texture, **4**–41 to **4**–43
thermal conductivity, **14**–19 to **14**–21
thermal expansion coefficient, **14**–19 to **14**–21
welding, **33**–31 to **33**–35
Cores and moulds, foundry, **26**–4 to **26**–10
Corrosion:
bio, **31**–10 to **31**–12
conversion factors, weight loss to penetration depth, **2**–12
crevice, **31**–4 to **31**–5
de-alloying, **31**–5 to **31**–6
erosion, cavitation and fretting, **31**–3
galvanic, **31**–2 to **31**–4
hydrogen induced, **31**–7 to **37**–8
intergranular (sensitisation, weld decay), **31**–8
introduction, **31**–1
localised, **31**–4 to **31**–10
pitting, **31**–9 to **31**–10
stress corrosion and corrosion fatigue, **31**–6 to **31**–10
uniform, **31**–1 to **31**–4
Creep:
aluminium and aluminium alloys, **22**–22 to **22**–24
carbon and low alloy steels, **22**–141 to **22**–142
cast alloys, *see* Casting(s)
copper and copper alloys, **22**–40 to **22**–46
magnesium and magnesium alloys, **22**–54 to **22**–56
nickel and nickel alloys, **22**–75 to **22**–77, **22**–79 to **22**–80
ordering, effect on rate of, **38**–6 to **38**–7
stainless steels, **22**–142 to **22**–143
test procedures and standards, **21**–19 to **21**–20
titanium and titanium alloys, **22**–89 to **22**–90
Crevice corrosion, **31**–4 to **31**–5
Critical:
diameter, *see* Hardenability
fields, *see* Superconductivity
temperatures, *see* Superconductivity; Thermochemistry
Crucibles, foundry, **26**–15, *see also* Ceramics
Cryogenic properties of metals and alloys, *see* entry for the relevant metal and its alloys
Crystal structure:
archetypes, **6**–30 to **6**–53
data for specific phases, **6**–1 to **6**–53
fundamentals of crystal structure, *see* Crystallography
intermetallic compounds:
data on, **6**–1 to **6**–30
interplanar:
angles, table of, **4**–15
spacing data on, **6**–1 to **6**–30
spacing relationships for, **4**–24 to **4**–25
metals:
data on, **6**–1 to **6**–53
notation:
comparison table with Pearson symbols, **6**–54 to **6**–56
Pearson symbols
comparison table with Strukturbericht notation, **6**–54 to **6**–56
prototypes, **6**–30 to **6**–53
Strukturbericht notation:
comparison table with Pearson symbols, **6**–54 to **6**–56

Crystal structure (*cont.*)
  unit cell:
    data for archetypes and specific phases,
      6–1 to **6**–53
    volume, relationships for, **4**–24 to **4**–25
Crystallography:
  bravais lattices
    table of, **5**–2
  data for archetypes and specific phases, **6**–1
    to **6**–53
  Hermann–Mauguin system, **5**–3
  point groups:
    introduction, **5**–2
    tables of, **5**–4 to **5**–11
  Schoenflies system, **5**–3
  space groups:
    introduction, **5**–3
    tables of, **5**–4 to **5**–11
  symmetry elements:
    description of, **5**–2
    tables of point and space groups, **5**–4
    translation groups, **5**–1 to **5**–11
Cupronickels, *see* Copper and copper alloys
Customary units, US ("English units"), SI
  conversion factors for, **2**–4 to **2**–10
Curie temperature, *see* Magnetic materials and
  properties
Cutting, *see* Metal cutting
  fluids, *see* Lubrication; Metal cutting
CVD and CVI (chemical vapour deposition
  and infiltration), **35**–2, **35**–4 to **35**–5,
  35–8, **35**–10 to **35**–12

Damping:
  aluminium alloys, **15**–9
  anelastic, **15**–8, **15**–10 to **15**–32
  cast irons, **15**–8 to **15**–9
  copper alloys, **15**–9
  definitions (anelastic and hysteric), **15**–8
  magnesium and its alloys, **14**–22 to **14**–24,
    15–9
  manganese alloys, **15**–9
  nickel alloys, **15**–9
  shape memory alloys, **15**–9, **15**–37 to **15**–44
  steels, **15**–9
  titanium alloys, **15**–9
Dark field microscopy:
  electron, *see* Electron microscopy
  light, **10**–22
De-alloying corrosion, **31**–5 to **31**–6
Deburring, **30**–5
Definitions, *see* relevant material or property
Deformation, *see* Elastic; Mechanical
  properties; Metalworking and
  Superplasticity
Density:
  aluminium and its alloys, **14**–16 to **14**–19
  copper and its alloys, **14**–19 to **14**–21
  fuels, *see* Fuels
  liquid metals, **14**–9 to **14**–11
  magnesium and its alloys, **14**–22 to **14**–24

nickel and its alloys, **14**–25 to **14**–27
pure metals, **14**–1 to **14**–2, **14**–29
salts:
  molten, **9**–1 to **9**–16
  solid, **9**–18
steels, **14**–30 to **14**–43
titanium and its alloys, **14**–28
unit conversion factors, **2**–5
zinc and its alloys, **14**–29
zirconium and its alloys, **14**–29
Deposition, *see* Coating and metal finishing
Design, *see* relevant material or process
Designations, related alloys, tables of,
  1–1 to **1**–14
  nickel-base alloys, **22**–61 to **22**–66
Diagrams, phase, Indexed on pages **11**–1 to
  11–6
Die casting, *see* Casting techniques
Differential equations, solutions to, **2**–26
Differentials, table of, **2**–25
Diffraction:
  contrast, *see* Electron microscopy
  electron:
    X-ray, *see* X-ray
Diffusion:
  activation energy, **13**–7
  Arrhenius equation, **13**–7
  Boltzmann-Matano method, **13**–5
  bonding, brazing, welding, *see* Joining
  coefficients:
    interdiffusion, **13**–2
    measurement methods, **13**–4 to **13**–8
    temperature effect on, **13**–8
    unit conversion factors, table of, **2**–5
  couples, **13**–5
  data, solid-state:
    carbon in austenite, **29**–46
    chemical diffusion, **13**–73 to **13**–119
    homogeneous alloys, **13**–43 to **13**–70
    self-diffusion, **13**–10 to **13**–14
    tracer, **13**–16 to **13**–37
  data, liquid, **13**–119 to **13**–120
  interdiffusion, **13**–2
  Matano-Boltzmann method, **13**–5
  mechanisms, **13**–8
  non-steady-state, **13**–1
  ordering, effect on, **38**–5 to **38**–6
  pumps (DP), *see* Vacuum technology
  steady-state, **13**–1
  theory, **13**–1 to **13**–8
  thin film (layer), **13**–6
Disclaimer, xxi
Dislocations:
  glide elements (planes/directions), **4**–39
  imaging:
    etchants for, **10**–27 to **10**–28
    transmission electron microscopy, **10**–78
  ordering, effect on, *see* Intermetallic
    compounds
Dispersion strengthened intermetallics, **38**–16
Dissociation pressures:
  thermodynamic data, **8**–36 to **8**–38

DP (diffusion pump), *see* Vacuum
    technology
Drawing, **30**–7
    texture, table of, **4**–42
Drilling, **30**–2
Dysprosium, *see* Elements
    characteristic X-ray emission lines and
        absorption edges, **4**–6
    excitation potentials for characteristic
        X-rays, **4**–4
    phase diagrams, Indexed on pages **11**–1 to
        11–6

E type (chrome-constantan)
    thermocouple, **16**–7
Eddy current testing, *see* Non-destructive
    testing and evaluation
EDM (electrical discharge machining), **30**–10
Elastic:
    constants, discussion and tables, **15**–4 to
        15–7
    damping, *see* Damping
    modulus:
        aluminium and its alloys, **14**–16 to
            14–19, **15**–2
        cast irons, **15**–2
        copper and its alloys, **14**–19, **15**–2
        discussion, **15**–1, **15**–4 to **15**–7
        magnesium and its alloys, **14**–22 to **14**–24
        nickel and its alloys, **14**–25 to **14**–27,
            15–2 to **15**–3
        pure metals, **15**–2 to **15**–3
        steels, **14**–30 to **14**–43, **15**–3
        titanium and its alloys, **14**–28
        zinc and its alloys, **14**–29, **15**–3
        zirconium and its alloys, **14**–29, **15**–3
Electrical:
    conductivity, *see* Electrical resistivity
    salts, molten data for, **9**–19 to **9**–39
    resistivity, *see* Electrical conductivity
        aluminium and its alloys, **14**–16 to
            14–19, **19**–3, **19**–7
        copper and its alloys (absolute and
            relative to int. standard), **14**–19 to
            14–21, **19**–4, **19**–7
        liquid metals, **14**–12 to **14**–15
        magnesium and its alloys, **14**–22 to
            14–24, **19**–5
        nickel and its alloys, **14**–25 to **14**–27,
            19–5, **19**–6
        pure metals, **14**–1 to **14**–7, **14**–30, **19**–1
            to **19**–2
        steels, **14**–30 to **14**–43, **19**–4 to **19**–5
        titanium and its alloys, **14**–28, **19**–5
        zinc and its alloys, **14**–29, **19**–5
        zirconium and its alloys, **14**–29
    discharge machining (EDM), **30**–10
Electrochemical series, *see* Galvanic corrosion
Electro(lytic)/(chemical):
    etching, *see* Metallography
    machining, **30**–11

polishing:
    for light microscopy, *see* Metallography
    for transmission electron microscopy, *see*
        Electron microscopy
Electromotive force (EMF):
    absolute electromotive power of Pt, **16**–5,
        see also Thermocouples
    binary alloys with respect to Pt, **16**–5
    elements with respect to Pt, **14**–30
Electron:
    emission, **18**–1 to **18**–10
        auger, **18**–6 to **18**–7
        field, **18**–8 to **18**–9
        photoelectric, **18**–4 to **18**–5
        secondary, **18**–5 to **18**–6
        thermionic, **18**–1 to **18**–2
    microscopy, **10**–74 to **10**–84
    theory, **39**–1 to **39**–3
        intermetallic compounds, *see*
            Intermetallic compounds
Electroplating and electrodeposition, **32**–8 to
    32–17, **32**–20 to **32**–21
    texture, table of, **4**–42
Elements, *see* entry for the individual element
    of interest
    atomic numbers, **3**–1
    atomic weights, **3**–1
    crystal structure data, metals, *see* Crystal
        structure
    physical properties, metals, *see* Physical
        properties
    work function, *see* Electron emission
    thermodynamic data, *see* Thermochemistry
Elevated temperature properties, *see* entries
    under the relevant property
EMF (electromotive force), *see* Electromotive
    force
Emissivity, *see* Emittance
Emittance:
    definitions of, **17**–1 to **17**–4
    tables of, **17**–6 to **17**–11
    use in temperature measurement, **17**–4
Emitters:
    alpha:
        useful nuclides (energies and half
            lives), **3**–10
    beta:
        table of energies and half lives, **3**–6
        useful nuclides (energies and half
            lives), **3**–10
    gamma:
        table of energies and half lives, **3**–8
    neutron:
        useful nuclides (energies and half lives),
            3–10 to **3**–11
    positron:
        useful nuclides (energies and half
            lives), **3**–5
Empirical models, methods for constructing,
    39–10 to **39**–11
Energies:
    ionising radiation, *see* Emitters

Energy:
  dispersive X-ray analysis (EDS, EDX, EDXA), *see* Electron microscopy
  surface, *see* Surface tension
  unit conversion factors, **2**–5, **2**–10
  usage in metallurgical processes, **28**–22 to **28**–23
Engineering ceramics, *see* Ceramics
"English units" (i.e. US customary units), SI conversions for, **2**–4 to **2**–10
Enthalpy:
  thermodynamic data, *see* Thermochemistry
Entropy:
  thermodynamic data, *see* Thermochemistry
  unit conversion factors, **2**–6
Equilibrium:
  phase diagrams, Indexed on pages **11**–1 to **11**–6
  thermodynamic data, *see* Thermochemistry
Erbium, *see* Elements
  characteristic X-ray emission lines and absorption edges, **4**–6
  excitation potentials for characteristic X-rays, **4**–4
  phase diagrams, Indexed on pages **11**–1 to **11**–6
Erosion, *see* Wear
Etching (etchants), *see* Metallography
European alloy designations, *see* related designations for alloys, tables of
Europium, *see* Elements
  characteristic X-ray emission lines and absorption edges, **4**–6
  excitation potentials for characteristic X-rays, **4**–4
  phase diagrams, Indexed on pages **11**–1 to **11**–6
Evapouration, *see* Physical vapour deposition; Thermochemistry
Expansion, thermal, *see* Thermal expansion
Extractive metallurgy
  mineral sources, *see* Minerals
  process control for, *see* Metallurgical process control
Extrusion, **30**–6
  texture, table of, **4**–42

Fabrication, *see* Joining
Failure analysis, applications of microscopy in, **10**–83 to **10**–84
Fatigue:
  aluminium and aluminium alloys, **22**–24 to **22**–26
  bearing alloys, **22**–162
  carbon and low alloy steels, **22**–138 to **22**–139
  cast alloys, *see* Casting(s)
  copper and copper alloys, **22**–37 to **22**–38
  corrosion, *see* Corrosion
  lead and lead alloys, **22**–47

  magnesium and magnesium alloys, **22**–57 to **22**–59
  nickel and nickel alloys, **22**–74 to **22**–75
  stainless steels, **22**–139 to **22**–140
  test procedures and standards, **21**–16 to **21**–19
  titanium and titanium alloys, **22**–90 to **22**–92
Felt metals, **38**–20
Ferrites, use as magnetic material, **20**–12
Ferritic steels, *see* Stainless steel(s); Steels
Ferroalloys, *see* Casting
Ferrous metals, *see* Iron and steel; Steel(s); Stainless steel(s)
Fluorescence, X-ray (XRF), **4**–49
Flash points, *see* Fuels
Flouride:
  minerals, **7**–4
  thermodynamic data, *see* Thermochemistry
Fibre(s):
  metal matrix composites for, **37**–1 to **37**–3
  texture, table of, **4**–42
Fick's laws of diffusion, **13**–1
Field emission, *see* Electron emission
Finishing, *see* Coating and metal finishing
Finite element method, **39**–9 to **39**–10
Fixed temperature points, *see* Temperature
Flux, *see* Diffusion
Fluxing and inoculation of castings, **26**–16 to **26**–19
Foams, *see* Metallic foams
Force:
  unit conversion factors, **2**–6, **2**–10
Forging, **30**–6 to **30**–7
Forming and formability, *see* Metalworking
Foundry, *see* Casting
Fracture toughness:
  aluminium and aluminium alloys:
    low temperature, **22**–20 to **22**–22
    room temperature, **22**–3 to **22**–15
  cast alloys, *see* Casting
  elastic-plastic (K-R, JIc), **21**–14 to **21**–15
  linear-elastic (KIc), **21**–12 to **21**–14
  test procedures and standards, **21**–12 to **21**–15
  unit conversion factors, **2**–6
Free energy, thermodynamic data, *see* Thermochemistry
French alloy designations, *see* Related designations for alloys, tables of
Fretting, *see* Wear
Friction, **25**–1 to **25**–12
  coefficient of:
    ceramics, **25**–6
    elevated temperature, **25**–5
    lubricated surfaces, **25**–10 to **25**–11
    metals on metals, **25**–3 to **25**–4
    polymers on metals, **25**–7 to **25**–8
  welding (including friction stir), *see* Joining
Fuels:
  coal and coke:
    analysis and testing, **28**–1 to **28**–5, **28**–8
    classification, **28**–5 to **28**–8

metallurgical cokes, **28**–8 to **28**–18
physical and mechanical properties, *see*
Metallurgical cokes in this entry
gaseous and liquid:
gaseous, **28**–18 to **28**–22
liquid, **28**–13 to **28**–18
thermochemistry, *see* Entries for liquid or
gaseous above
metallurgical applications, energy use in,
28–22 to **28**–23
Full mould casting, *see* Casting techniques
Fundamental physical constants, table of, **3**–2
Furnaces:
design of, **40**–1 to **40**–8
aluminum heat treatment, **29**–70 to **29**–71
box, **40**–2 to **40**–3
continuous, **40**–4 to **40**–5
heat calculations, **40**–5
horizontal car bottom, **40**–4
integral quench, **40**–3
pit, **40**–3 to **40**–4
vacuum, **40**–7 to **40**–8
systems for, *see* vacuum technology
fuels for, *see* Fuels
process control for, *see* Metallurgical
process control
refractories for, *see* Ceramics
Fusion welding, *see* Joining

Gadolinium, *see* Elements
characteristic X-ray emission lines and
absorption edges, **4**–6
excitation potentials for characteristic
X-rays, **4**–4
phase diagrams, Indexed on pages **11**–1 to
11–6
Gallium, *see* Elements
characteristic X-ray emission lines and
absorption edges, **4**–5
excitation potentials for characteristic
X-rays, **4**–4
ores, **7**–4
phase diagrams, Indexed on pages **11**–1 to
11–6
Galvanic corrosion, **31**–2 to **34**–4
Gamma:
emitters, *see* Emitters
prime alloys, *see* Nickel and nickel alloys
sources, *see* Emitters
Garnets, magentic properties of, **20**–15
Gas and gasses:
fuels, **28**–18 to **28**–22
residual, analysis, *see* Vacuum technology
solutions in metals, **12**–1 to **12**–24, **14**–30
welding, *see* Joining
Gauges:
process, *see* Metallurgical process control
vacuum, *see* Vacuum technology
General physical constants, table of, **3**–2 to
3–3
Geometrical relationships, **2**–22 to **2**–24

German alloy designations, *see*
related designations for alloys, tables of
Germanium, *see* Elements
characteristic X-ray emission lines and
absorption edges, **4**–5
excitation potentials for characteristic
X-rays, **4**–4
ores, **7**–4
phase diagrams, Indexed on pages **11**–1 to
11–6
Gibbs energy, data for, *see*
Thermochemistry
Glasses, metallic, *see* Metallic glasses
Glide elements (planes/directions),
dislocation, **4**–39
Gold, *see* Elements
characteristic X-ray emission lines and
absorption edges, **4**–6
excitation potentials for characteristic
X-rays, **4**–5
metallography, **10**–42 to **10**–44
ores, **7**–4
phase diagrams, Indexed on pages **11**–1 to
11–6
plating, *see* Coating and metal finishing
welding, **33**–39
Goldschmidt atomic radii, table of, **4**–44 to
4–48
Gravitational constant, *see* Fundamental
physical constants
Gravity die casting, *see* Casting techniques
Greases:
use as lubricants, **24**–8 to **24**–11
classification, table of, **24**–10
Grinding:
metal cutting, **30**–4 to **30**–5
metallography, **10**–9 to **10**–11
Grossmann Hardenability, **29**–20 to **29**–21
Gunmetals, *see* Copper and copper alloys
Gunning materials, **27**–14

Hafnium, *see* Elements
characteristic X-ray emission lines and
absorption edges, **4**–6
excitation potentials for characteristic
X-rays, **4**–4
metallography, *see* Metallography
ores, **7**–4
phase diagrams, Indexed on pages **11**–1 to
11–6
Halides:
thermodynamic data, *see*
Thermochemistry
Half lives:
tables of, **3**–5 to **3**–11
Hall-Petch equation, **29**–22, **38**–26
Hard:
magnets, *see* Magnetic materials
and properties
metals, *see* Carbides, cemented
Hardenability, *see* Heat treatment

Hardness:
  aluminium and aluminium alloys, **22**–3 to
    22–15
  bearing alloys, **22**–161
  Brinell, test procedures, **21**–1
  carbon and low alloy steels, **22**–99 to
    22–130
  cast alloys, *see* Casting
  copper and copper alloys, **22**–27 to **22**–32
  conversion tables, **21**–4 to **21**–7
  magnesium and magnesium alloys, **22**–49 to
    22–50
  micro-indentation, test procedures,
    21–3, **21**–4
  other metals, **22**–157 to **22**–159
  rockwell, test procedures, **21**–2, **21**–3
  stainless steels, **22**–114 to **22**–129
  tool steels, **22**–153 to **22**–156
  Vickers, test procedures, **21**–3
  zinc and zinc alloys, **22**–93 to **22**–94
  zirconium and zirconium alloys, **22**–94 to
    22–95
Hastelloy, *see* Nickel and nickel alloys
HAZ (heat affected zone), *see* Corrosion;
  Joining
H-factor (Grossmann hardenability), **29**–20 to
  29–21
Heat:
  affected zone (HAZ), *see* Corrosion;
    Joining
  calculations for furnaces, **40**–5
  capacity:
    thermodynamic data, *see*
      Thermochemistry
    unit conversion factors, **2**–8
  formation, of, thermodynamic data, *see*
    Thermochemistry
  solution, of, thermodynamic data, *see*
    Thermochemistry
  transfer:
    quenching, *see* Quenching and quenching
      media (quenchants)
  treatment of aluminium alloys:
    ageing, natural and artificial, **29**–74 to
      29–75
    defects resulting from, **29**–68, *see also*
      relevant process step in this entry
    designations, alloys, **29**–66, *see also*
      Designations
    furnaces, **29**–70 to **29**–71, *see also*
      Furnaces
    introduction, **29**–66 to **29**–67
    mechanical property data, *see* Aluminium
      and its alloys
    quenching and quenching, **29**–71 to
      29–79
    phase transformations, **29**–67 to **29**–68
    solution treatment, **29**–68 to **29**–71,
      29–73
  treatment of steels:
    annealing and stress relieving, **29**–22 to
      29–29
    arrest (A) temperatures, *see*
      transformation temperatures in this
      entry
    austempering, **29**–40 to **29**–41
    austenitisation, **29**–21 to **29**–23
    carbon equivalent, **29**–21, **33**–19
    carbonitriding, **29**–59 to **29**–66
    carburising, **29**–42 to **29**–59
    castings, **26**–70 to **26**–71
    colour–temperature relationship, table of,
      29–5
    colour of oxide films versus temperature,
      table, **29**–35
    continuous cooling diagrams (CCD), also
      known as time, continuous cooling
      transformation (CCT) diagrams,
      29–8, **29**–10 to **29**–14
    classification of steels, **29**–5 to **29**–7
    critical diameter, *see* Hardenability in this
      entry
    diffusion coefficient of carbon in
      austenite, **29**–46
    furnaces, *see* Furnaces
    Grossmann (H) hardenability, **29**–20 to
      29–21
    Hall-Petch equation, **29**–22
    hardenability, **29**–12 to **29**–21, *see also*
      Quenching and quenching media
      (quenchants) in this entry
    heat transfer, data for quenching media,
      29–30
    H-factor, *see* Grossmann hardenability in
      this entry
    Holloman parameter, **29**–29, **29**–36 to
      29–37
    hot working temperatures, **29**–4
    introduction and definitions, **29**–1
    isothermal transformation diagrams
      (ITD), also known as time,
      temperature, transformation (TTT)
      diagrams, **29**–8 to **29**–9
    iron-carbon phase diagram, discussion of,
      29–3 to **29**–5
    Jominy bar end-quench, **29**–18 to **29**–19
    kinetics of carburising and carbonitriding,
      29–43 to **29**–46, **29**–59 to **29**–60
    quenching and quenching media
      (quenchants), **29**–30 to **29**–33, *see
      also* hardenability in this entry
    martempering, **29**–41 to **29**–42
    martensite start and finish temperatures,
      see Transformation temperatures in
      this entry
    mechanical property data, *see* Iron and
      steel, steels; relevant treatment in
      this entry
    nitriding, *see* Carbonitriding
    normalising, **29**–22, **29**–28 to **29**–29
    phase transformations during, **29**–1 to
      29–7, **29**–34, **29**–55
    recommended temperatures for, **29**–22

retained Austenite, *see* Quenching and
quenching media (quenchants)
in this entry
standards, *see* relevant topic in this
entry
stress relieving, *see* Annealing and stress
relieving
tempering, **29**–33 to **29**–41, *see also*
austempering and martempering in
this entry
thermodynamics of carburising and
carbonitriding, **29**–42 to **29**–43,
29–59 to **29**–60
tool steels, **22**–153 to **22**–156
transformation temperatures, **29**–5 to
29–7
Hermann–Mauguin system, **5**–3
High:
alloy steels, *see* Stainless steel(s)
resolution electron microscopy, *see* Electron
microscopy
strength low alloy (HSLA) steels, *see*
Steel(s)
temperature properties, *see* entry for the
relevant metal and its alloys
vacuum, *see* Vacuum technology
velocity oxy-fuel spraying (HVOF), *see*
Thermal spraying
Helium, *see* Noble gases
Holloman parameter, **29**–29, **29**–36 to **29**–37
Holmium, *see* Elements
characteristic X-ray emission lines and
absorption edges, **4**–6
excitation potentials for characteristic
X-rays, **4**–4
phase diagrams, Indexed on pages **11**–1 to
11–6
Honeycombs, **38**–20, *see* Metal foams
Horizontal car bottom furnaces, **40**–4
Hot working, *see* Metalworking
steels, temperature for, **29**–4
HSLA (high strength low alloy) steels, *see*
Steel(s)
HVOF (high velocity oxy fuel) spraying, *see*
Thermal spraying
Hydrogen, *see* Elements
embrittlement, **31**–7 to **37**–8
phase diagrams, Indexed on pages **11**–1 to
11–6
solution in metals, **12**–2 to **12**–8, **12**–9 to
12–15, **12**–16
Hysteresis, magnetic, *see* Magnetic materials
and properties
Hysteric damping, *see* Damping

IACS (International Annealed Copper
Standard), *see* Electrical resistivity
Image analysis, **10**–81 to **10**–82
IMC (intermetallic matrix composites), *see*
Intermetallic matrix composites
Impact energy, *see* Impact strength

Impact strength:
aluminium and aluminium alloys, **22**–3 to
22–15
carbon and low alloy steels:
low temperature, **22**–144 to **22**–146
room temperature, **22**–99 to **22**–129
cast alloys, *see* Casting
charpy test procedures and standards, **21**–10
to **21**–12
copper and copper alloys:
elevated and room temperature, **22**–39
low temperature, **22**–35 to **22**–36, **22**–39
Izod test procedures and standards, **21**–10
magnesium and magnesium alloys, **22**–57 to
22–59
other metals, **22**–157 to **22**–159
stainless steels, **22**–114 to **22**–129, **22**–146
to **22**–147
titanium and titanium alloys, **22**–93
Imperial units, SI conversion factors for, **2**–4
to **2**–10
Inconel and Incoloy alloys, *see* Nickel and
nickel alloys
Indentation hardness, *see* Hardness
Information, sources of, *see* Bibliography
Indium, *see* Elements
characteristic X-ray emission lines and
adsorption edges, **4**–6
excitation potentials for characteristic
X-rays, **4**–4
metallography, **10**–44 to **10**–45
ores, **7**–4
phase diagrams, Indexed on pages **11**–1 to
11–6
Injection moulding, **23**–29 to **23**–31
Integral quench furnaces, **40**–3
Intermetallic:
compounds:
aluminides, **38**–7 to **38**–16
applications, **38**–17 to **38**–18
coatings, use in, **38**–17
comparison with conventional alloys,
38–8 to **38**–9
crystallographic data, *see* Crystal
structure
dispersion strengthened, **38**–16
dislocations in, *see* nickel aluminides in
this entry
nickel aluminides, **38**–7 to **38**–14
processing and fabrication, **38**–16 to
38–17
shape memory, *see* Shape memory
alloys
structural, **38**–1 to **38**–18
theory, **38**–2 to **38**–7
titanium aluminides, **38**–14 to **38**–16
thermodynamic data, *see*
Thermochemistry
matrix composites (IMC), **38**–16
International:
alloy designations (ISO), *see* related
designations for alloys, tables of

International (*cont.*)
  annealed copper standard (IACS), *see*
      Electrical resistivity
  temperature scales (ITS and ITPS), *see*
      Temperature
Interplanar angles and spacing, *see* Crystal
      structure
Inert gases, *see* Noble gases
Inertia, moments of, table, 3–3
Inoculation and fluxing of castings, 26–16 to
      26–19
Instrumentation:
  process, *see* Metallurgical process control
  vacuum, *see* Vacuum technology
Interdiffusion, *see* Diffusion
  definition, 13–2
Intergranular corrosion, 31–8 to 31–9
Interference microscopy, *see* Metallography
Integrals, table of, 2–25 to 2–26
Iodides, thermodynamic data, *see*
      Thermochemistry
Iodine, *see* Elements
  characteristic X-ray emission lines and
      adsorption edges, 4–6
  excitation potentials for characteristic
      X-rays, 4–4
  iodide minerals, 7–5
Ion:
  cleaning, 35–2
  etching, 10–24
  gauges, *see* Vacuum technology
  plating, *see* Physical vapour deposition
  pumps, *see* Vacuum technology
Ionic:
  compounds, *see* Ceramics; Minerals; Salts
  radii:
    table of, 4–44 to 4–48
Ionising radiation, sources of, *see* Emitters
Iridium, *see* Elements
  characteristic X-ray emission lines and
      absorption edges, 4–6
  excitation potentials for characteristic
      X-rays, 4–5
  ores, 7–5
  phase diagrams, Indexed on pages 11–1 to
      11–6
Iron and steel, *see* Elements; Stainless steel(s);
      Steel(s)
  casting and cast irons and steels, *see* Casting
  characteristic X-ray emission lines and
      absorption edges, 4–5
  damping:
    cast irons, 15–8 to 15–9
    steels, 15–9
  dislocation elements, 4–39
  emittance, 17–6 to 17–11
  excitation potentials for characteristic
      X-rays, 4–4
  magnetic properties, 20–1, 20–3 to 20–5,
      20–10, 20–11 to 20–13
  metallography, 10–25, 10–45 to 10–51
  ores, 7–5

  phase diagrams, Indexed on pages 11–1 to
      11–6
  iron-carbon, discussion of, 29–3 to 29–5
  texture, 4–41 to 4–43
ISO (International Standards Organisation)
  alloy designations, *see* related
      designations for alloys, tables of
Isothermal transformation diagrams (ITD),
      also known as time, temperature,
      transformation (TTT) diagrams, 29–8 to
      29–9
Isotopes, radio(active), *see* Emitters
ITD (isothermal transformation diagrams), *see*
      Isothermal transformation diagrams
IT(P)S (International temperature scale), *see*
      Temperature
Izod impact energy, *see* Impact strength

J type (iron-constantan) thermocouple, 16–7
      to 16–8
Japanese alloy designations, *see* Related
      designations for alloys, tables of
Jet polishing, *see* Electron microscopy
JIc, *see* Fracture toughness
Joining:
  bibliography:
    brazing and soldering, 34–21 to 24–22
    welding and general, 33–56 to 33–58
  bonding:
    diffusion, *see* welding, diffusion in this
        entry
    transient liquid phase (TLP), *see* brazing,
        diffusion in this entry
  brazing and soldering:
    brazing, 34–9 to 34–21
    corrosion, 34–4 to 34–5
    diffusion, 34–13
    fluxes, 34–2 to 34–4
    lead-free, 34–8
    metallic glass foils, 38–21 to 38–22
    quality assurance, 34–1 to 34–2, 34–9
    soldering, 34–2 to 34–9
    thermal expansion, role in brazing, 34–9
        to 34–10
    tin-rich solders, bulk mechanical
        properties, 22–96
  coating, spraying and surfacing, *see* Coating
      and metal finishing
  diffusion:
    bonding, *see* welding, diffusion in this
        entry
    brazing, *see* brazing and soldering,
        diffusion in this entry
    welding, *see* welding, diffusion in this
        entry
  glossary, welding, 33–1 to 33–6
  introduction:
    brazing and soldering, 34–1
    general and welding, 33–1
  soldering, *see* brazing and soldering in this
      entry

standards, **33**–41 to **33**–56
welding:
  aluminium and its alloys, **33**–29 to **33**–31
  arc, *see* fusion welding in this entry
  copper and its alloys, **33**–31 to **33**–35
  diffusion, **33**–13 to **33**–15
  dissimilar metals, **33**–40 to **33**–41
  friction and friction stir, **33**–10 to **33**–13
  fusion, **33**–15 to **33**–41
  gas, *see* fusion welding in this entry
  gold, **33**–39
  lead and its alloys, **33**–35
  magnesium and its alloys, **14**–22 to
    14–24, **33**–35 to **33**–36
  molybdenum, **33**–40
  nickel and its alloys, **33**–36 to **33**–39
  niobium, tantalum and their alloys,
    33–40
  platinum, **33**–39
  resistance, **33**–6 to **33**–10
  Schaeffler diagram, **33**–27
  sensitisation (weld decay), *see* Corrosion
  silver, **33**–39
  solid-state, **33**–10 to **33**–15
  spot, *see* resistance welding in this entry
  steels and cast irons, **33**–15 to **33**–29
  titanium and its alloys, **33**–40
  ultrasonic, **33**–13
  uranium, **33**–40
  zirconium alloys, **33**–40
Jominy bar end-quench, **29**–18 to **29**–19

K type (chromel-alumel) thermocouple, **16**–8
Kelvin temperature scale, *see* Temperature
  definition, **16**–1
KIc, *see* Fracture
Kinetics, *see* Diffusion
  carburising and carbonitriding, **29**–43 to
    29–46, **29**–59 to **29**–60
  modelling, **39**–5 to **39**–6
Krypton, *see* Elements; Noble gases

Lanthanum, *see* Elements
  characteristic X-ray emission lines and
    absorption edges, **4**–6
  excitation potentials for characteristic
    X-rays, **4**–4
  ores, **7**–4
  phase diagrams, Indexed on pages **11**–1 to
    11–6
Laser, **30**–11 to **30**–12, **33**–2, **33**–4, **33**–15,
  33–40, **33**–42, **33**–46, **33**–48
Latent heat:
  thermodynamic data, *see* Thermochemistry
Lathing, **30**–1
Lattice constants, *see* Crystal structure
Lead and lead alloys, *see* Elements
  characteristic X-ray emission lines and
    absorption edges, **4**–6
  excitation potentials for characteristic
    X-rays, **4**–5

important alloys, mechanical properties,
  22–46 to **22**–48
fatigue properties, **22**–47
free solders, **34**–8
metallography, **10**–51 to **10**–52
ores, **7**–5
phase diagrams, Indexed on pages **11**–1 to
  11–6
tensile properties, **22**–47
welding, **33**–35
LEED (low energy electron diffraction), *see*
  Electron microscopy
LEFM (linear elastic fracture mechanics), *see*
  Fracture
Light microscopy, *see* Metallography
Linear elastic fracture mechanics (LEFM), *see*
  Fracture
Liquid:
  fuels, **28**–13 to **28**–17
  metals:
    physical properties, *see* Physical
      properties
    surface tension, **14**–9 to **14**–11, **14**–30
    thermodynamic data, *see*
      Thermochemistry
    viscosity, **14**–10 to **14**–11
  phase sintering, **23**–36
Lithium, *see* Elements
  ores, **7**–5
  phase diagrams, Indexed on pages **11**–1 to
    11–6
Lorentz-polarization factors,
  table of, **4**–17
Low:
  alloy steels, *see* Steel(s)
  energy electron diffraction (LEED), *see*
    Electron microscopy
  pressure plasma spraying (LPPS),
    see Thermal spraying
  temperature properties, *see* entry for the
    relevant metal and its alloys
LPPS (low pressure plasma spraying),
  see Thermal spraying
Lubrication, **24**–1 to **24**–15, **25**–9 to **25**–12,
  see also Friction
  applications in metal cutting and working,
    30–9 to **30**–10
  greases, **24**–8 to **24**–11
  oils, **24**–2 to **24**–8, **24**–11 to **24**–13
  solid metal lubricants, **24**–13 to **23**–15
Lutetium, *see* Elements
  characteristic X-ray emission lines and
    absorption edges, **4**–6
  excitation potentials for characteristic
    X-rays, **4**–4

Machining and machinability, *see*
  Metal cutting
Magnesite, *see* Ceramics
Magnesium and magnesium alloys, *see*
  Elements

Magnesium and magnesium alloys (*cont.*)
   alloys, related designations, **1**–14
   anodising, **32**–7
   cast, composition and properties, **26**–56 to
      26–67
   characteristic X-ray emission lines and
      absorption edges, **4**–5
   composites, **37**–5
   creep properties, **22**–54 to **22**–56
   damping, **14**–22 to **14**–24, **15**–9
   density, **14**–22 to **14**–24
   dislocation glide elements, **4**–39
   elastic modulus, **14**–22 to **14**–24
   electrical resistivity, **14**–22 to **14**–24, **19**–5
   excitation potentials for characteristic
      X-rays, **4**–4
   fatigue properties, **22**–57 to **22**–59
   hardness, **22**–49 to **22**–50
   heat treatments for castings, **22**–60
   impact strength, **22**–57 to **22**–59
   metallography, **10**–53 to **10**–57
   ores, **7**–5
   phase diagrams, Indexed on pages **11**–1 to
      11–6
   physical properties, **14**–22 to **14**–24
   tensile properties:
      elevated temperature, **22**–51 to **22**–53
      room temperature, **22**–49 to **22**–50
   texture, **4**–42
   thermal conductivity, **14**–22 to **14**–24
   thermal expansion coefficient, **14**–22 to
      14–24
   welding, **14**–22 to **14**–24, **33**–35 to **33**–36
Magnetic materials and properties:
   alnico alloys, **20**–7
   bibliography, **20**–23 to **20**–24
   cast iron, **20**–5, **20**–18 to **20**–20
   elements, **20**–3
   ferrites, **20**–12
   iron-silicon alloys, **20**–10 to **20**–11, **20**–13
   metallic glass, **20**–16, **38**–21
   nickel-iron alloys, **20**–15
   non-magnetic steels and cast irons, **20**–18 to
      20–20
   permanent magnet materials, **20**–2 to **20**–3,
      20–6, **20**–10
   powder metallurgy production of, **23**–37
   pure metals and non-metallic elements, **20**–3
   rare earth alloys, **20**–8
   steels, **20**–4 to **20**–5, **20**–18 to **20**–20
   units, conversion factors and terminology,
      2–3, **20**–21 to **20**–23
Manganese and manganese alloys, *see*
      Elements
   characteristic X-ray emission lines and
      absorption edges, **4**–5
   damping, **15**–9
   excitation potentials for characteristic
      X-rays, **4**–4
   ores, **7**–5
   phase diagrams, Indexed on pages **11**–1 to
      11–6

Martempering, **29**–41 to **29**–42
Martensite, *see* Heat treatment
Martin-Marietta (Mar-M) alloys, *see* Nickel
      and nickel alloys
Matano-Boltzmann method, **13**–5
Mathematical formulae, **2**–13 to **2**–49
Macrography, *see* Metallography
Malleable cast iron, *see* Iron and steel
Mechanical:
   alloying, *see* Powder metallurgy
   properties:
      cast alloys, *see* Casting
      ceramics, *see* Ceramics
      coal, *see* Fuels
      creep, *see* Creep
      data tables, metals and alloys, *see* Entry
         for metal and its alloys of interest
      fatigue, *see* Fatigue
      fracture toughness, *see* Fracture
         toughness
      hardness, *see* Hardness
      heat treatment, effect on, *see* Heat
         treatment
      impact strength, *see* Impact strength
      superplasticity, **36**–1 to **36**–14
      tensile, *see* Tensile properties and testing
   testing:
      creep, **21**–19 to **21**–20
      fatigue, **21**–16 to **21**–18
      fracture toughness, **21**–12 to **21**–15
      hardness, **21**–1 to **21**–7
      impact, **21**–9 to **21**–12
      non-destructive, **21**–20 to **21**–23
      tensile, **21**–8 to **21**–9
Mechanics, Nanoscale, *see* Nanomaterials
Melting, *see* Casting(s)
   point, data tables of, **8**–2 to **8**–8, **14**–1 to
      14–2
   volume change on, data tables, **8**–2 to **8**–8
Memory alloys, shape, **15**–9, **15**–37 to **15**–44,
      38–17 to **38**–18
Mensuration, **2**–22 to **2**–25
Mercury, *see* Elements
   characteristic X-ray emission lines and
      absorption edges, **4**–6
   excitation potentials for characteristic
      X-rays, **4**–5
   ores, **7**–6
   phase diagrams, Indexed on pages **11**–1 to
      11–6
Mesh:
   metals, **38**–20
   size, *see* Test sieve mesh number
Metal:
   casting, *see* Casting
   cutting, **30**–1 to **30**–5, **30**–9 to **30**–12
      boring, **30**–1 to **30**–2
      deburring, **30**–5
      drilling, **30**–2
      electrical discharge machining
         (EDM), **30**–10
      electrochemical, **30**–11

grinding, **30**–4 to **30**–5, *see also* Metallography
  laser, **30**–11 to **30**–12
  machinability, **30**–8
  milling, **30**–2 to **30**–4
  photochemical, **30**–11
  plasma, **30**–10 to **30**–11
  rapid prototyping and freeforming, **30**–11 to **30**–12
  reaming, **30**–2
  turning, **30**–1
  ultrasonic, **30**–11
  waterjet and abrasive waterjet, **30**–10
fatigue, *see* Fatigue
finishing, *see* Coating and metal finishing
foams:
  aluminium, **38**–19 to **38**–20
  applications, **38**–18 to **38**–19
  materials, **38**–19 to **38**–20
  production (open and closed cells), **38**–19 to **38**–20
injection moulding (MIM), **23**–29 to **23**–31
matrix composites (MMC), **37**–1 to **33**–6
  aluminium alloy, **37**–4 to **37**–5
  copper alloy, **37**–6
  intermetallic matrix, *see* Intermetallic matrix composites
  magnesium alloy, **37**–5
  reinforcements (fibres), **37**–1 to **37**–3
  titanium alloy, **37**–5
  zinc alloy, **37**–6
working, **30**–5 to **30**–12
  drawing, **30**–7
  extrusion, **30**–6
  forging, **30**–6 to **30**–7
  formability, **30**–8 to **30**–9
  rapid prototyping and freeforming, **30**–11 to **30**–12
  rolling, **30**–7
  superplastic, **30**–7 to **30**–8, **36**–1 to **36**–14
Metallic:
  elements, *see* Elements
  foams, *see* Metal foams
  glasses, **38**–20 to **38**–23
    applications, **38**–21 to **38**–22
    brazing foils, **38**–21 to **38**–22
    bulk, **38**–22 to **38**–23
    soft magnets, **20**–16, **38**–21
Metallising, *see* Coating and surface finishing
Metalloids, *see* Element of interest
Metallography:
  acoustic microscopy, **10**–82 to **10**–83
  aluminium and its alloys, **10**–25, **10**–29 to **10**–35
  antimony, **10**–35
  auger electron spectroscopy (AES), *see* Spectroscopy and surface characterisation, see below
  bearing and type metals, solders, **10**–71 to **10**–72
  beryllium, **10**–35 to **10**–36

bismuth, **10**–35
brass, *see* Copper and its alloys, see below
bronze, *see* Copper and its alloys, see below
cadmium, **10**–36
cast iron, *see* Iron and steel, see below
cemented carbides and other hard alloys, **10**–72 to **10**–73
chemical:
  etching, *see* Etching, see below
  polishing, *see* Polishing, see below
chromium, **10**–37
cobalt, **10**–37
colour etching, *see* Etching, see below
copper and its alloys, **10**–37 to **10**–42
dark field (light optics), **10**–22
definition, **10**–1
dislocations, etching for, *see* Etching, see below
drying, **10**–26
electrochemical etching, *see* Etching, see below
electroetching, *see* Etching, see below
electrolytic etching, *see* Etching, see below
electron:
  microscopy and diffraction, **10**–74 to **10**–84
  spectroscopy:
    auger electron (AES), *see* Spectroscopy and surface characterisation, see below
    X-ray photo electron (XPS, ESCA), *see* Spectroscopy and surface characterisation, see below
electrochemical polishing, *see* Polishing, see below
electrolytic polishing, *see* Polishing, see below
electropolishing, *see* Polishing, see below
energy dispersive X-ray analysis (EDS, EDX, EDXA), *see* Spectroscopy and surface characterisation, see below
etching, *see* Also listed by specific metals and alloys
  chemical, **10**–24, **10**–26
  colour, **10**–24 to **10**–25
  data tables and discussion, **10**–24 to **10**–74
  dislocations, **10**–27 to **10**–28
  electro(lytic)/(chemical), **10**–24 to **10**–74
  ion, **10**–24
  macroscopic, **10**–2 to **10**–6
  microscopic, **10**–21 to **10**–74
etchants, *see* Etching, see above
failure analysis, applications in, **10**–83 to **10**–84
gold and its alloys, **10**–42 to **10**–44
grinding, **10**–9 to **10**–11
hafnium, **10**–70 to **10**–71
ion etching, **10**–24
image analysis, **10**–81 to **10**–82
indium, **10**–44 to **10**–45

Metallography (*cont.*)
  interference (light optics), **10**–22
  iron and steel, **10**–25, **10**–45 to **10**–51
  lead and its alloys, **10**–51 to **10**–52
  light, *see* Optical, see below
  low energy electron diffraction (LEED), *see*
      Spectroscopy and surface
      characterisation, see below
  macroscopic, **10**–2
  magnesium and its alloys, **10**–53 to **10**–57
  microscopic, **10**–7
  molybdenum, **10**–57
  mounting:
    description, **10**–7 to **10**–9
    resins, **10**–8
  nickel and its alloys, **10**–57 to **10**–59
  niobium and its alloys, **10**–60
  optical:
    colour, **10**–24
    dark field, **10**–22
    description, **10**–22
    interference microscopy, **10**–22
    polarised light, **10**–22 to **10**–24
  platinum and its alloys, **10**–60 to **10**–61
  palladium and its alloys, **10**–60 to **10**–61
  polarised light:
    microscopy, **10**–22 to **10**–24
  polishing:
    attack, **10**–13 to **10**–14
    chemical, **10**–15 to **10**–21
    electro(lytic)/(chemical), **10**–14 to
        10–19, Electron microscopy, see
        above
    jet, *see* Electron microscopy, see above
    mechanical, **10**–11 to **10**–14
  powder metallurgical samples, **10**–73 to
      10–74
  refractory metals, *see* Entries for the
      individual metals
  resins, mounting, **10**–8
  rhodium, **10**–60 to **10**–61
  Rutherford backscattering (RBS), *see*
      Spectroscopy and surface
      characterisation, see below
  scanning:
    acoustic microscopy, **10**–82 to **10**–83
    auger spectroscopy, *see* Spectroscopy and
        surface characterisation, see below
    electron microscopy (SEM), **10**–80
  sectioning, taper, **10**–28 to **10**–29
  silicon, **10**–61 to **10**–62
  silver and its alloys, **10**–62 to **10**–63
  sintered metals and alloys, **10**–73
  spectroscopy and surface characterisation:
    auger electron (AES), **10**–80 to **10**–81
    energy dispersive X-ray (EDS, EDX,
        EDXA), **10**–80 to **10**–81
    low energy electron diffraction (LEED),
        10–80 to **10**–81
    Rutherford backscattering (RBS), **10**–80
        to **10**–81

  wavelength dispersive X-ray
      (WDS, WDX), **10**–80 to **10**–81
  X-ray fluorescence (XRF), **4**–49
  X-ray photoelectron (XPS, ESCA),
      10–80 to **10**–81
  steels, *see* Iron and steel, see above
  taper sectioning, **10**–28 to **10**–29
  tin and its alloys, **10**–63 to **10**–64
  titanium and its alloys, **10**–64 to **10**–67
  transmission electron microscopy (TEM),
      10–74 to **10**–80
  tungsten and its alloys, **10**–66 to **10**–67
  uranium and its alloys, **10**–67 to **10**–68
  washing, **10**–26
  wavelength dispersive X-ray spectroscopy
      (EDS, EDX, EDXA), *see* Spectroscopy
      and surface characterisation, see above
  X-ray spectroscopy/analysis, *see*
      Spectroscopy and surface
      characterisation, see above
  zinc, **10**–68 to **10**–70
  zirconium and its alloys, **10**–70 to **10**–71
Metallurgical process control, **40**–16 to **40**–29
  applications, examples of, **40**–24 to **40**–27
  bibliography, **40**–27 to **4**–29
  introduction, **40**–16 to **40**–21
  instrumentation for, **40**–21 to **40**–24
  keywords, **40**–27
  statistical basis of, **2**–46 to **2**–49
Microscopy, *see* Metallography
Micro-scale materials, *see* Nanomaterials
Milling:
  ion, *see* Electron microscopy
  metalcutting, **30**–2 to **30**–4
MIM (metal injection moulding), **23**–29 to
    23–31
Minerals:
  sources, Table of, **7**–2 to **7**–9
  uses, Table of, **7**–2 to **7**–9
MM (Martin-Marietta) alloys, *see* Nickel and
    nickel alloys
MMC (metal matrix composites), *see* Metal
    matrix composites
Modelling and simulation, **39**–1 to **39**–11
  electron theory, **39**–1 to **39**–3
  finite difference and element methods, **39**–9
      to **39**–10
  kinetics, **39**–5 to **39**–6
  monte Carlo methods, **39**–6 to **39**–7
  neural networks (empirical), **39**–10 to **39**–11
  phase field method, **39**–7 to **39**–9
  thermodynamics and phase diagrams, **39**–3
      to **39**–5
Modulus, *see* Elastic modulus
Molar properties, thermodynamic data, *see*
    Thermochemistry
Molybdenum, *see* Elements
  characteristic X-ray emission lines and
      adsorption edges, **4**–5
  excitation potentials for characteristic
      X-rays, **4**–5
  metallography, *see* Metallography

ores, **7**–6
phase diagrams, Indexed on pages **11**–1 to
    11–6
welding, **33**–40
Molten:
  metals, *see* Liquid metals
  salts, *see* Salts, Molten
Moments of inertia:
  table of, **3**–3
  unit conversion factors, **2**–7
Monels, *see* Nickel and nickel alloys
Monte Carlo methods, **39**–6 to **39**–7
Moulds and cores, foundry, **26**–4 to **26**–10
Mounting, metallographic specimens, **10**–7 to
    10–9

N type (nicrosil-nisil) thermocouple, **16**–9
Nanomaterials, **38**–23 to **38**–29
  bulk, **38**–26 to **38**–28
  structures, **38**–28 to **38**–29
  thin films, **38**–25 to **38**–26
NDE/NDT (non destructive
    evaluation/testing), *see* Non-destructive
    testing and evaluation
Neel point, *see* Magnetic materials and
    properties
Neodymium, *see* Elements
  characteristic X-ray emission lines and
    absorption edges, **4**–6
  excitation potentials for characteristic
    X-rays, **4**–4
  use as magnetic material, **20**–8
Neural networks:
  modelling of metallic materials, use in,
    39–10 to **39**–11
  process control, use in, *see* Metallurgical
    process control
Neutron:
  emitters, *see* Emitters
  sources, see Emitters
Nickel and nickel alloys, *see* Elements
  aluminide structural intermetallics and
    comparison with conventional
    nickel-base alloys, **38**–7 to **38**–16
  cast, composition and properties, **26**–52 to
    26–55
  characteristic X-ray emission lines and
    absorption edges, **4**–5
  creep properties, **22**–75 to **22**–77, **22**–79 to
    22–80
  damping, *see* Damping
  density, **14**–25 to **14**–27
  directionally solidified alloys, **22**–80 to
    22–81
  dislocation glide elements, **4**–39
  elastic modulus, **14**–25 to **14**–27
  excitation potentials for characteristic
    X-rays, **4**–4
  electrical resistivity, **14**–25 to **14**–27
  emittance, **17**–6 to **17**–11
  fatigue properties, **22**–74 to **22**–75

metallography, **10**–57 to **10**–59,
    see also Metallography
ores, **7**–6
oxide dispersion strengthened (ODS),
    see Powder metallurgy, see below
phase diagrams, Indexed on pages **11**–1 to
    11–6
physical properties, **14**–25 to **14**–27
powder metallurgy, **23**–25 to **23**–27
related designations and specifications,
    22–61 to **22**–66
single crystal, **22**–80 to **22**–81
tensile properties:
  elevated temperatures, **22**–70 to **22**–73,
    22–78 to **22**–79
  low temperatures, **22**–74
  room temperature, **22**–67 to **22**–70,
    22–78 to **22**–79
thermal conductivity, **14**–25 to **14**–27
thermal expansion coefficient, **14**–25 to
    14–27
welding, **33**–36 to **33**–39
Nimonic alloys, *see* Nickel and nickel alloys
Niobium and niobium alloys, *see* Elements
  characteristic X-ray emission lines and
    absorption edges, **4**–5
  excitation potentials for characteristic
    X-rays, **4**–5
  metallography, **10**–60
  phase diagrams, Indexed on pages **11**–1
    to **11**–6
  ores, **7**–6
  welding, **33**–40
Nitinol, *see* Shape memory alloys
Nitrides, *see* Nitrogen
Nitriding, *see* Carbonitriding
Nitrogen, *see* Elements
  solution in metals, **12**–8, **12**–17 to **12**–20
  nitrides, thermodynamic data, *see*
    Thermochemistry
  phase diagrams, Indexed on pages **11**–1 to
    11–6
Noble gases, *see* Elements
  solution in metals, **12**–15, **12**–24
Nodular cast iron, *see* Iron and steel
Non-destructive testing (NDT) and
    evaluation (NDE), **21**–20 to **21**–22
Normalising, **29**–22, **29**–28 to **29**–29
Notation, crystallographic, *see*
    Crystal structure
Nuclides, *see* Emitters
Numbering system, unified, *see* Unified
    numbering system

ODS (oxide dispersion strengthened) alloys,
    see Oxide dispersion strengthened
    alloys
Oils:
  use as lubricants, **24**–2 to **24**–8
  use in metal cutting, **30**–9
Optical microscopy, *see* Metallography

Ordered intermetallics and ordering, *see*
  Intermetallic compounds
Ores:
  sources, Table of, **7**–2 to **7**–9
  uses, Table of, **7**–2 to **7**–9
Osmium, *see* Elements
  characteristic X-ray emission lines and
    absorption edges, **4**–6
  excitation potentials for characteristic
    X-rays, **4**–5
  ores, **7**–6
  phase diagrams, Indexed on pages **11**–1 to
    11–6
Oxygen, *see* Elements
  oxides:
    thermodynamic data, *see*
      Thermochemistry
    phase diagrams, Indexed on pages **11**–1 to
      11–6
    solution in metals, **12**–8, **12**–21 to **12**–23
Oxide dispersion strengthened (ODS) alloys,
  23–3, **23**–27, **23**–28, **38**–16, *see also*
    Powder metallurgy
Oxidation, *see* Corrosion
Oxides, *see* Oxygen

Painting, *see* Coating and metal finishing
Palladium and palladium alloys, *see* Elements
  characteristic X-ray emission lines and
    absorption edges, **4**–6
  excitation potentials for characteristic
    X-rays, **4**–5
  metallography, **10**–60 to **10**–61
  phase diagrams, Indexed on pages **11**–1 to
    11–6
Particulate reinforcements, *see* Metal matrix
  composites
Parting powders/liquids, mould dressing and
  paints, **26**–13 to **26**–14
Patterns, foundry, **26**–12 to **26**–13
Pearson symbols, **6**–54 to **6**–56
Peltier effect, **16**–10
Penning gauges, *see* Vacuum technology
Periodic table (chart) of the elements, **3**–4
Permanent:
  magnets, *see* Magnetic materials and
    properties
  mould casting, *see* Casting techniques
Permeability, magnetic, *see* Magnetic
  materials and properties
Phase:
  contrast, *see* Electron microscopy
  diagrams, Indexed on pages **11**–1 to **11**–6
  simulation of, **39**–7 to **39**–9
  field method, **39**–7 to **39**–9
  identification, *see* Metallography; X-ray
  transformations:
    aluminium alloys, **29**–67 to **29**–68
    shape memory alloys, *see* Shape memory
      alloys
    steels, **29**–1 to **29**–7, **29**–34, **29**–55

thermodynamic data on, *see*
  Thermochemistry
Phosphating, *see* Coating and metal finishing
Phosphide, *see* Phosphorus
Phosphorus, *see* Elements
  characteristic X-ray emission lines and
    absorption edges, **4**–5
  excitation potentials for characteristic
    X-rays, **4**–4
  phase diagrams, Indexed on pages **11**–1 to
    11–6
  phosphide:
    minerals, **7**–6
    thermodynamic data, *see*
      Thermochemistry
Photochemical machining, **30**–11
Photoelectric emission, **18**–4 to **18**–5
Physical:
  constants, fundamental, **3**–2 to **3**–3
  properties:
    aluminium and its alloys, **14**–16 to **14**–19
    boiling points of pure metals, **14**–1 to
      14–2, **14**–29
    ceramics, *see* Ceramics
    composites, *see* Metal matrix composites
    copper and its alloys, **14**–19 to **14**–21
    damping capacity of magnesium and its
      alloys, **14**–22 to **14**–24
    density of aluminium and its alloys,
      14–16 to **14**–19
    density of copper and its alloys, **14**–19 to
      14–21
    density of liquid metals, **14**–9 to **14**–11
    density of magnesium and its alloys,
      14–22 to **14**–24
    density of nickel and its alloys, **14**–25 to
      14–27
    density of pure metals, **14**–1 to **14**–2,
      14–30
    density of steels, **14**–30 to **14**–43
    density of titanium and its alloys, **14**–28
    density of zinc and its alloys, **14**–29
    density of zirconium and its alloys, **14**–29
    elastic moduli of aluminium and its
      alloys, **14**–16 to **14**–19
    electrical resistivity of aluminium and its
      alloys, **14**–16 to **14**–19, **19**–3, **19**–7
    electrical resistivity of copper and its
      alloys, **14**–19 to **14**–21, **19**–4, **19**–7
    electrical resistivity of liquid metals,
      14–12 to **14**–15
    electrical resistivity of magnesium and its
      alloys, **14**–22 to **14**–24, **19**–5
    electrical resistivity of nickel and its
      alloys, **14**–25 to **14**–27, **19**–5, **19**–6
    electrical resistivity of pure metals, **14**–1
      to **14**–7, **14**–30, **19**–1 to **19**–2
    electrical resistivity of steels, **14**–30 to
      14–43, **19**–4 to **19**–5
    electrical resistivity of titanium and its
      alloys, **14**–28, **19**–5

electrical resistivity of zinc and its alloys, **14**–29, **19**–5

electrical resistivity of zirconium and its alloys, **14**–29

elevated temperature, pure metals, **14**–3 to **14**–7

liquid metals, **14**–9 to **14**–16

magnesium and its alloys, **14**–22 to **14**–24

magnetic, *see* Magnetic materials and properties

melting points of pure metals, **14**–1 to **14**–2

metal matrix composites, *see* Metal matrix composites

molten Salts, *see* Salts, molten

nickel and its alloys, **14**–25 to **14**–27

pure metals, **14**–1 to **14**–7

specific heat capacity of liquid metals, **14**–12 to **14**–15

specific heat capacity of pure metals, **14**–1 to **14**–7

steels, **14**–30 to **14**–43, **19**–4 to **19**–5

surface tension of liquid metals, **14**–9 to **14**–11, **14**–30

thermal conductivity of aluminium and its alloys, **14**–16 to **14**–19

thermal conductivity of copper and its alloys, **14**–19 to **14**–21

thermal conductivity of liquid metals, **14**–12 to **14**–16

thermal conductivity of magnesium and its alloys, **14**–22 to **14**–24

thermal conductivity of nickel and its alloys, **14**–25 to **14**–27

thermal conductivity of pure metals, **14**–1 to **14**–7

thermal conductivity of steels, **14**–30 to **14**–44

thermal conductivity of titanium and its alloys, **14**–28

thermal conductivity of zinc and its alloys, **14**–29

thermal conductivity of zirconium and its alloys, **14**–29

thermal expansion coefficient of aluminium and its alloys, **14**–16 to **14**–19

thermal expansion coefficient of copper and its alloys, **14**–19 to **14**–21

thermal expansion coefficient of liquid metals, **14**–9 to **14**–11

thermal expansion coefficient of magnesium and its alloys, **14**–22 to **14**–24

thermal expansion coefficient of nickel and its alloys, **14**–25 to **14**–27

thermal expansion coefficient of pure, **14**–1 to **14**–7

thermal expansion coefficient of steels, **14**–30 to **14**–44

thermal expansion coefficient of titanium and its alloys, **14**–28

thermal expansion coefficient of zinc and its alloys, **14**–29

thermal expansion coefficient of zirconium and its alloys, **14**–29

tin (pure), **14**–29 to **14**–30

titanium and its alloys, **14**–28

viscosity of liquid metals, **14**–10 to **14**–11

zinc and its alloys, **14**–29

zirconium and its alloys, **14**–29

vapour deposition (PVD), **35**–1 to **35**–3, **35**–7, **35**–9, **35**–11

texture of evaporative films, table of, **4**–43

Pickling, *see* Cleaning and pickling

Pig iron, *see* Iron and steel

PIM (powder injection moulding), *see* Powder metallurgy

Pirani gauges, *see* Vacuum technology

Pit furnaces, **40**–3 to **40**–4

Pitting corrosion, **31**–9 to **31**–10

Plain carbon steels, *see* Steel(s)

Plasma:

cutting, **30**–10 to **30**–11

spraying, *see* Thermal spraying

Plasticity, *see* Mechanical properties; Superplasticity

Plastic working, *see* Metalworking

Plating, *see* Coating and metal finishing

Platinum and its alloys, *see* Elements

characteristic X-ray emission lines and absorption edges, **4**–6

electromotive force and power, **16**–4 to **16**–5

excitation potentials for characteristic X-rays, **4**–5

metallography, **10**–60 to **10**–61

ores, **7**–6 to **7**–7

phase diagrams, Indexed on pages **11**–1 to **11**–6

thermocouples, use in, **16**–4 to **16**–7, *see also* thermocouples

welding, **33**–39

Plutonium, *see* Elements

phase diagrams, Indexed on pages **11**–1 to **11**–6

PM (powder metallurgy), *see* Powder metallurgy

Point groups, *see* Crystallography

Poisson ratio, data for metals and alloys, **15**–2 to **15**–3

Polarised light microscopy, *see* Metallography

Polishing:

metallographic, *see* Metallography

surface finishing, **32**–01 to **32**–02

Positron:

emitters, *see* Emitters

sources, *see* Emitters

Potassium, *see* Elements

characteristic X-ray emission lines and absorption edges, **4**–5

excitation potentials for characteristic X-rays, **4**–4

ores, **7**–7

Potassium (*cont.*)
  phase diagrams, Indexed on pages **11**–1 to
    11–6
Powder metallurgy (PM):
  aluminium and aluminium alloys:
    description, **23**–13, **23**–27
    mechanical properties, **23**–14, **23**–30
  copper and copper alloys:
    description, **23**–9, **23**–25, **23**–27
    mechanical properties, **23**–22, **23**–25
  injection moulding, **23**–29, **23**–31
  iron and steels:
    description, **23**–8 to **23**–9, **23**–15, **23**–27
    mechanical properties, **23**–10, **23**–15,
      23–16 to **23**–21, **23**–30 to **23**–31
  metallography, **10**–73
  nickel and nickel alloys, **23**–25 to **23**–27
  powder characterization, **23**–4 to **23**–5
  powder production, **23**–2 to **23**–3
  process description, **23**–1 to **23**–8
  refractory metals, **23**–24, **23**–36
  thermal management materials, **23**–37
Preferred orientation, *see* Texture
Pressing, *see* Metal working
Producer gas, *see* Fuels
Properties, *see* Entry on the relevant property
Praseodymium, *see* Elements
  characteristic X-ray emission lines and
    absorption edges, **4**–6
  excitation potentials for characteristic
    X-rays, **4**–4
  phase diagrams, Indexed on pages **11**–1 to
    11–6
Pratt and Whitney alloys (PWA), *see* Nickel
  and nickel alloys
Precipitate extraction, techniques for, **4**–23
Preferred orientation, *see* Texture
Prefixes, SI, **2**–4
Pressure:
  data, *see* Entry for relevant property
  die casting, *see* Casting techniques
  instrumentation, *see* Vacuum technology
  units, **2**–2, **40**–9, *see also* Stress
Probability, *see* Statistics
Process control:
  metallurgical, *see* Metallurgical process
    control
  statistical basis for, *see* Statistical process
    control
Promethium, *see* Elements
  characteristic X-ray emission lines and
    absorption edges, **4**–6
  excitation potentials for characteristic
    X-rays, **4**–4
Prototypes, crystallographic, *see* Crystal
  structure
Pumps, vacuum, *see* Vacuum technology
Pure metals, *see* Elements
  physical properties, **14**–1 to **14**–7
PVD (Physical vapour deposition), *see*
  Physical vapour deposition

PWA (Pratt and Whitney alloys), *see* Nickel
  and nickel alloys
Pyrometry, corrections for emittance, **17**–5

Quantitative image analysis, **10**–81 to **10**–82
Quenching and quenching media
  (quenchants), **2**–30 to **29**–33, **29**–71 to
    29–79, **40**–3

R type (Pt-13%Rh/Pt) thermocouple, **16**–6
Radiation:
  ionising:
    sources of, *see* Emitters
    units for, **2**–8 to 2-9
  pyrometry, *see* Pyrometry
Radiative properties, *see* Emittance
Radii:
  atomic and ionic, table of, **4**–44 to **4**–48, *see*
    also Geometrical relationships
Radio isotopes, *see* Emitters
Radioactive isotopes, *see* Emitters
Radioactivity, *see* Radiation, ionising
Radiography, **21**–21 to **21**–22
Radium, *see* Elements
Ramming, ceramics, **27**–13 to **27**–14
Rapid:
  prototyping and freeforming, **30**–11 to
    30–12
  solidification, *see* Solidification
Rare earth materials, magnetic properties of,
  see Magnetic materials and applications
Reaming, **30**–2
References, *see* Bibliography; Entry for the
  topic of interest
Refinery gas, *see* Fuels
Refractories, *see* Ceramics
Refractory metals, *see* Entries for individual
  metals
Reinforcements for metal matrix composites,
  37–1 to **37**–3
Related designations for alloys, tables of, **1**–1
  to **1**–14
  nickel-base alloys, **22**–61 to **22**–66
Rene alloys, *see* Nickel and nickel alloys
Residual:
  gas analysis, **40**–15
  stress measurement, **4**–18 to **4**–22
Resistance and resistivity:
  welding, *see* Joining; Electrical
    conductivity; Electrical resistivity
Retained austenite, *see* Heat treatment
  determination of, **4**–17
Rhenium, *see* Elements
  characteristic X-ray emission lines and
    absorption edges, **4**–6
  excitation potentials for characteristic
    X-rays, **4**–4
  ores, **7**–7
  phase diagrams, Indexed on pages **11**–1 to
    11–6
Rhodium *see* Elements

characteristic X-ray emission lines and
absorption edges, **4**–6
excitation potentials for characteristic
X-rays, **4**–5
metallography, **10**–60 to **10**–61
phase diagrams, Indexed on pages **11**–1 to
11–6
Rockwell hardness, *see* Hardness
Rolling, **30**–7
process control, *see* Metallurgical process
control
texture, table of, **4**–41
Rolls-Royce (RR) alloys, *see* Nickel
and nickel alloys
Rotary pumps (RP), *see* Vacuum
technology
RP (Rotary pumps), *see* Vacuum technology
RR (Rolls-Royce) alloys, *see* Nickel
and nickel alloys
Rubidium, *see* Elements
characteristic X-ray emission lines and
absorption edges, **4**–5
excitation potentials for characteristic
X-rays, **4**–4
phase diagrams, Indexed on pages **11**–1 to
11–6
Russian alloy designations, *see* Related
designations for alloys, tables of
Rusting, *see* Corrosion
Ruthenium, *see* Elements
characteristic X-ray emission lines and
absorption edges, **4**–5
excitation potentials for characteristic
X-rays, **4**–5
phase diagrams, Indexed on pages **11**–1 to
11–6
Rutherford backscattering spectroscopy
(RBS), *see* Metallography

S type (Pt-10%Rh/Pt) thermocouple, **16**–6
Safety, *see* Disclaimer; Relevant
technique
Salts:
molten:
density, **9**–1 to **9**–16
electrical conductivity, **9**–19 to **9**–39
surface tension, **9**–41 to **9**–48
viscosity, **9**–50 to **9**–54
solid:
density, **9**–18
Samarium, *see* Elements
characteristic X-ray emission lines and
absorption edges, **4**–6
excitation potentials for characteristic
X-rays, **4**–4
phase diagrams, Indexed on pages **11**–1 to
11–6
Sand casting, *see* Casting(s)
Scale, *see* Elements
oxide, *see* Corrosion; Oxygen
microstructural, *see* Nano materials

Scandium, *see* Elements
characteristic X-ray emission lines and
absorption edges, **4**–5
excitation potentials for characteristic
X-rays, **4**–4
phase diagrams, Indexed on pages **11**–1 to
11–6
Scanning:
acoustic microscopy, **10**–82 to **10**–83
auger analysis, *see* Metallography
electron microscopy, **10**–80
Scattering factors, atomic, table of:
atomic, table of, **4**–27 to **4**–31
Schaeffler diagram, **33**–27
Schoenflies system, **5**–3
Scroll pumps, *see* Vacuum technology
Secondary electron:
imaging, *see* Electron microscopy
emission, **18**–5 to **18**–6
Sectioning, taper for metallography, **10**–28 to
10–29
Seebeck effect/coefficient, **16**–10 to **16**–11
Selenides, *see* Selenium
Selenium, *see* Elements
characteristic X-ray emission lines and
absorption edges, **4**–5
excitation potentials for characteristic
X-rays, **4**–4
ores, **7**–7
phase diagrams, Indexed on pages **11**–1 to
11–6
selenides, thermodynamic data, *see*
Thermochemistry
Self diffusion, data on, **13**–10 to **13**–14
SEM (scanning electron microscopy), **10**–80
Sensitisation, *see* Corrosion
Shape memory alloys, **15**–9, **15**–37 to **15**–44,
38–17 to **38**–18
Shear modulus, *see* Elastic modulus
Shewhart control charts, **2**–48, **2**–49
SI Units and Prefixes, **2**–2 to **2**–4
Sieve, *see* Test sieve mesh number
Sieverts, conversion factors for, *see*
Conversion factors
Silicates, *see* Silicon
Silicides, *see* Silicon
Silicon, *see* Elements
characteristic X-ray emission lines and
absorption edges, **4**–5
excitation potentials for characteristic
X-rays, **4**–4
metallography, *see* Metallography
ores, **7**–7
phase diagrams, Indexed on pages **11**–1 to
11–6
silicates, thermodynamic data, *see*
Thermochemistry
silicides, thermodynamic data, *see*
Thermochemistry
Silver, *see* Elements
characteristic X-ray emission lines and
absorption edges, **4**–6

Silver (*cont.*)
  excitation potentials for characteristic
        X-rays, **4**–5
  metallography, *see* Metallography
  ores, **7**–7
  phase diagrams, Indexed on pages **11**–1 to
        11–6
  welding, **33**–39
Simulation of metals and alloys, *see*
        Modelling and simulation
Single crystal, *see* Nickel and nickel alloys
Sintered metals, *see* Powder metallurgy
Size effects, *see* Nanomaterials
Sodium, *see* Elements
  characteristic X-ray emission lines and
        adsorption edges, **4**–5
  excitation potentials for characteristic
        X-rays, **4**–4
  ores, **7**–7
  phase diagrams, Indexed on pages **11**–1 to
        11–6
Soft magnets, *see* Magnetic materials
        and properties
Solders and soldering, *see* Joining
Solidification, *see* Casting(s)
  rapid, *see* Metallic glasses
        powder metallurgy applications, **23**–2
Solid:
  solutions, thermodynamic data, *see*
        Thermochemistry
  state welding, *see* Joining
Solubility, *see* Solutions; Thermochemistry
Solutions:
  solid, thermodynamic data, **8**–14 to **8**–15,
        8–18 to **8**–20, **8**–42
  gases in metals, **12**–1 to **12**–24, **14**–30
Solution treatment, aluminium alloys, **29**–68
        to **29**–71, **29**–73
Sources:
  information, *see* Bibliography
  metals, of, *see* Minerals
  radioactive, *see* Emitters
Space groups, *see* Crystallography
Specific heat capacities, data, *see*
        Thermochemistry
Specifications, alloy, *see* Designations
Spectroscopy, **10**–80 to **10**–81
Spot welding, *see* Joining
Sputtering, *see* Physical vapour deposition
Spray forming, **23**–28
Spraying, *see* Coating and metal finishing
Stainless steel(s), *see* Iron and steel; Steels
  creep properties, **22**–142 to **22**–143
  fatigue properties, **22**–139 to **22**–140
  hardness, **22**–114 to **22**–129
  impact strength, **22**–114 to **22**–129, **22**–146
        to **22**–147
  metallography, *see* Steel(s)
  related designations, **1**–5 to **1**–8
  Schaeffler diagram, **33**–27
  tensile properties:
        elevated temperature, **22**–135 to **22**–137

        low temperature, **22**–146 to **22**–147
        room temperature, **22**–114 to **22**–129
Standards, *see* Entry for the relevant topic
Statistical process control, **2**–46 to **2**–49
  applications in metallurgical processes, *see*
        Metallurgical process control
  Shewhart control charts, **2**–48, **2**–49
Statistics, introduction to, **2**–28 to **2**–49
Steel(s), *see* Iron and steel; Stainless steel(s)
  alloy, *see* Stainless steel(s); Tool in this entry
        creep properties, **22**–141 to **22**–142
        fatigue properties, **22**–138 to **22**–139
        hardness, **22**–103 to **22**–114, **22**–130 to
            22–132
        impact strength, **22**–103 to **22**–114,
            22–130 to **22**–132, **22**–145 to
            22–146
        related designations, **1**–3 to **1**–5
        tensile properties (elevated temperature),
            22–133 to **22**–135
        tensile properties (low temperature),
            22–145 to **22**–146
        tensile properties (room temperature),
            22–103 to **22**–114, **22**–130 to
            22–132
  annealing and stress relieving, **29**–22 to
        29–29
  arrest (A) temperatures, *see* Transformation
        temperatures in this entry
  austempering, **29**–40 to **29**–41
  austenite, retained, determination of, **4**–17
  austenitisation, **29**–21 to **29**–23
  carbon:
        creep properties, **22**–141
        equivalent, **29**–21, **33**–19
        fatigue properties, **22**–138
        hardness, **22**–99 to **22**–103
        impact strength, **22**–99 to **22**–103
        related designations, **1**–2
        tensile properties (elevated
            temperature), **22**–133
        tensile properties (low
            temperature), **22**–144
        tensile properties (room temperature),
            22–99 to **22**–103
  carbonitriding, **29**–59 to **29**–66
  carburising, **29**–42 to **29**–58
  cast, composition and properties, **26**–70 to
        26–82
  castings, heat treatment of, **26**–70 to **26**–71
  continuous cooling diagrams (CCD), also
        known as time, continuous cooling
        transformation (CCT) diagrams, **29**–8,
        29–10 to **29**–14
  corrosion, *see* Corrosion
  damping, **15**–9
  density, **14**–30 to **14**–43
  discussion (general), **29**–5 to **29**–7
  elastic modulus, **14**–30 to **14**–43
  electrical resistivity, **14**–30 to **14**–43
  emittance, **17**–6 to **17**–11
  hardenability, *see* Heat treatment

heat treatment, **29**–1 to **29**–66, *see also* Heat treatment for detailed index
high:
    alloy, *see* Stainless steel(s)
    strength low alloy (HSLA), *see* Alloy in this entry
hot working, **29**–4
HSLA (high strength low alloy), *see* Alloy in this entry
iron-carbon phase diagram, *see* Phase diagrams in this entry
isothermal transformation diagrams (ITD), also known as time, temperature, transformation (TTT) diagrams, **29**–8 to **29**–9
low alloy, *see* Alloy in this entry
making, process control for, *see* Metallurgical process control
martempering, **29**–41 to **29**–42
martensite start and finish temperatures, *see* Transformation temperatures in this entry
mechanical properties, *see* Heat treatment; Relevant class of steel
metallography, **10**–25, **10**–45 to **10**–51
nitriding, *see* Carbonitriding in this entry
normalising, **29**–22, **29**–28 to **29**–29
phase diagrams, Indexed on pages **11**–1 to **11**–6
    iron-carbon, discussion of, **29**–3 to **29**–5
phase transformations, **29**–1 to **29**–7, **29**–34, **29**–55
physical properties, **14**–30 to **14**–45
plain carbon, *see* Carbon in this entry
powder metallurgy alloys:
    description, **23**–8 to **23**–9, **23**–15, **23**–27
    processing and properties, **23**–10, **23**–15, **23**–16 to **23**–21, **23**–30 to **23**–31
retained austenite determination, **4**–17
stainless, *see* Stainless steel(s)
thermal conductivity, **14**–30 to **14**–44
tempering, **29**–33 to **29**–41
tensile properties, *see* Relevant class of steel
thermal expansion coefficient, **14**–30 to **14**–44
tool, **1**–8 to **1**–9, **22**–148 to **22**–156
    hardness, **22**–153 to **22**–156
    heat treatment, **22**–153 to **22**–156
    related designations, **1**–8 to **1**–9
    uses and compositions, **22**–148 to **22**–152
transformation temperatures, **29**–5 to **29**–7
welding, **33**–15 to **33**–29
Stress:
    corrosion and corrosion fatigue, **31**–6 to **31**–10
    relieving, *see* Annealing and stress relieving
    rupture, *see* Creep
    ultimate tensile, *see* Entry for the metal and its alloys of interest
    unit conversion factors, **2**–8, **2**–10
    yield, *see* Entry for the metal and its alloys of interest

Strontium, *see* Elements
    characteristic X-ray emission lines and adsorption edges, **4**–5
    excitation potentials for characteristic X-rays, **4**–5
    ores, **7**–8
    phase diagrams, Indexed on pages **11**–1 to **11**–6
Structural intermetallic compounds, **38**–1 to **38**–18
Structure, metals and intermetallic compounds, *see* Crystal structure
Strukturbericht notation, **6**–54 to **6**–56
Surface coating, finishing and treatment, *see* Coating and metal finishing
Sulphides, *see* Sulphur
Sulphur, *see* Elements
    characteristic X-ray emission lines and adsorption edges, **4**–5
    excitation potentials for characteristic X-rays, **4**–4
    mineral, native, **7**–8
    phase diagrams, Indexed on pages **11**–1 to **11**–6
    sulphide:
        minerals, **7**–8
        thermodynamic data, *see* Thermochemistry
Superalloys, *see* Cobalt and cobalt alloys; nickel and nickel alloys
Superconductivity, **14**–30, **19**–7 to **19**–9
Superplastic forming, **30**–7 to **30**–8
Superplasticity, **36**–1 to **36**–14
Susceptibility, magnetic, *see* Magnetic properties
    elements, data for, **20**–3
Surface:
    energy, *see* Surface tension
    hardness, modification of, **25**–20, *see also* Carburising and carbonitriding
    tension:
        liquid metals, data for, **14**–9 to **14**–11, **14**–30
        molten salts, data for, **9**–41 to **9**–48
        unit conversion factors, **2**–8, **2**–10
Swedish alloy designations, *see* Related designations for alloys, tables of
Symmetry elements:
    description of, **5**–2
    specific phases, *see* Crystal structure
Systems, vacuum, *see* Vacuum technology

T type (copper-constantan) thermocouple, **16**–8
Tantalum and tantalum alloys, *see* Elements
    characteristic X-ray emission lines and absorption edges, **4**–6
    excitation potentials for characteristic X-rays, **4**–4
    ores, **7**–8

Tantalum and tantalum alloys (*cont.*)
  phase diagrams, Indexed on pages **11**–1 to
    11–6
  welding, **33**–40
Taper sectioning, **10**–28 to **10**–29
Techniques, *see* entry for the relevant process
Tellurides, *see* Tellurium
Tellurium, *see* Elements
  characteristic X-ray emission lines and
    absorption edges, **4**–6
  excitation potentials for characteristic
    X-rays, **4**–4
  ores, **7**–8
  phase diagrams, Indexed on pages **11**–1 to
    11–6
  tellurides, thermodynamic data, *see*
    Thermochemistry
Ternary alloys, *see* Alloy
TEM, *see* Electron microscopy
Temperature:
  international temperature scale (ITS and
    ITPS), defining points, **16**–3
  ITPS correction factors, table of, **2**–11
  measurement standards, **16**–1 to **16**–4
  transformation, of, data for, *see*
    Thermochemistry
  unit conversion factors, **2**–8, **2**–13 to **2**–14
Tempering, **29**–33 to **29**–41, *see also*
    Austempering and martempering
Tensile properties and testing:
  aluminium and aluminium alloys:
    elevated temperature, **22**–16 to **22**–20
    low temperature, **22**–20 to **22**–22
    room temperature, **22**–3 to **22**–15
  bearing alloys:
  carbon and low alloy steels:
    elevated temperature, **22**–133 to **22**–135
    low temperature, **22**–144 to **22**–146
    room temperature, **22**–99 to **22**–114,
      22–130 to **22**–132
  cast metals and alloys, *see* Casting
  copper and its alloys:
    elevated temperature, **22**–33 to **22**–34
    low temperature, **22**–35 to **22**–36, **22**–45
      to **22**–46
    room temperature, **22**–27 to **22**–32
  lead and lead alloys, **22**–47
  magnesium and its alloys:
    elevated temperature, **22**–51 to **22**–53
    room temperature, **22**–49 to **22**–50
  nickel and its alloys:
    elevated temperatures, **22**–70 to **22**–73,
      22–78 to **22**–79
    low temperatures, **22**–74
    room temperature, **22**–67 to **22**–70,
      22–78 to **22**–79
  other metals, **22**–157 to **22**–159
  stainless steels:
    elevated temperature, **22**–135 to **22**–137
    low temperature, **22**–146 to **22**–147
    room temperature, **22**–114 to **22**–129
  test procedures and standards, **21**–8 to **21**–9

titanium and its alloys:
  elevated and low temperature, **22**–85 to
    22–88
  room temperature, **22**–83 to **22**–84
Tension, surface, *see* Surface tension
Terbium, *see* Elements
  characteristic X-ray emission lines and
    absorption edges, **4**–6
  excitation potentials for characteristic
    X-rays, **4**–4
  phase diagrams, Indexed on pages **11**–1 to
    11–6
Test sieve mesh number:
  conversion to aperture size, table of, **2**–12
  related standards and aperture sizes, **23**–5
Texture:
  cast metals, table of, **4**–43
  electrodeposits, table of, **4**–42
  evaporative films, table of, **4**–43
  fibre (Drawn and extruded wire),
    table of, **4**–42
  measurement techniques, **4**–22 to **4**–23
  rolling, table of, **4**–41
Thallium, *see* Elements
  characteristic X-ray emission lines and
    absorption edges, **4**–6
  excitation potentials for characteristic
    X-rays, **4**–5
  ores, **7**–8
  phase diagrams, Indexed on pages **11**–1 to
    11–6
Thermal:
  conductivity:
    aluminium and its alloys, **14**–16 to **14**–19
    copper and its alloys, **14**–19 to **14**–21
    liquid metals, **14**–12 to **14**–16
    magnesium and its alloys, **14**–22 to **14**–24
    nickel and its alloys, **14**–25 to **14**–27
    pure metals, **14**–1 to **14**–7
    steels, **14**–30 to **14**–44
    titanium and its alloys, **14**–28
    thermal conductivity of zinc and its
      alloys, **14**–29
    thermal conductivity of zirconium and its
      alloys, **14**–29
    unit conversion factors, **2**–8
  expansion:
    coefficient of aluminium and its alloys,
      14–16 to **14**–19
    coefficient of copper and its alloys,
      14–19 to **14**–21
    coefficient of liquid metals, **14**–9 to
      14–11
    coefficient of magnesium and its alloys,
      14–22 to **14**–24
    coefficient of nickel and its alloys, **14**–25
      to **14**–27
    coefficient of pure metals, **14**–1 to **14**–7
    coefficient of steels, **14**–30 to **14**–44
    coefficient of titanium and its alloys,
      14–28
    coefficient of zinc and its alloys, **14**–29

coefficient of zirconium and its alloys, **14**–29

contraction allowances in casting, **26**–10 to **26**–11

spraying, **35**–13 to **35**–16

management materials, powder metallurgy production of, **23**–37

Thermionic electron emission, **18**–1 to **18**–4

Theory, *see* relevant topic

modelling and simulation, **39**–1 to **39**–11

Thermochemistry:

binary alloys, see entries below for: Intermetallic compounds; Liquid metals; Solid solutions

data, tables of:

boiling points, **8**–2 to **8**–8

borides, **8**–21, **8**–43

bromides, see entry below for halides

calorific value of fuels, *see* Fuels

carbides, **8**–5, **8**–22, **8**–44

carbonates, **8**–33 to **8**–34

chlorides, see entry below for halides

compounds, **8**–5 to **8**–8, *see also* entries under Thermochemistry data for: Borides; Carbides; Carbonates; Halides; Intermetallic compounds; Nitrides; Oxides; Phosphides; Selenides; Silicates; Silicides; Sulphide; Tellurides

dissociation pressures, **8**–36 to **8**–38

elements, **8**–2 to **8**–3, **8**–8, **8**–39 to **8**–41, **8**–51 to **8**–53

enthalpy (of formation), **8**–8 to **8**–36

entropy (standard), **8**–8 to **8**–36

fluorides, see entry below for halides

free energy of formation, **8**–14 to **8**–15, **8**–21 to **8**–36

fuels, *see* Fuels

halides, **8**–5 to **8**–7, **8**–28 to **8**–33, **8**–45, **8**–49 to **8**–51, **8**–53 to **8**–58

heat capacities, **8**–39 to **8**–51, **14**–1 to **14**–7, **14**–12 to **14**–15

heat of formation, **8**–8 to **8**–36

intermetallic compounds, **8**–3 to **8**–4, **8**–9 to **8**–12, **8**–14 to **8**–15, **8**–41 to **8**–43

iodides, see entry above for halides

latent heat, **8**–2 to **8**–8, **14**–30

liquid metals, **8**–16 to **8**–20, **14**–12 to **14**–15

melting points, **8**–2 to **8**–8, **14**–1 to **14**–2, **22**–97

molar properties, see entry for the relevant property

nitrides, **8**–7, **8**–22 to **8**–23, **8**–44 to **8**–45

oxides, **8**–7, **8**–24 to **8**–26, **8**–34 to **8**–35, **8**–46 to **8**–47, **8**–53 to **8**–58

phosphides, **8**–36 to **8**–37

pressure, see entries for vapour and dissociation pressures

selenides, **8**–12 to **8**–13, **8**–48

silicates, **8**–33 to **8**–34

silicides, **8**–23 to **8**–24, **8**–45

specific heat capacities, see entry above for heat capacities

solid solutions, **8**–14 to 8-15, **8**–18 to **8**–20, **8**–42

steels, **29**–6 to **29**–7

sulphides, **8**–7, **8**–26 to 8-28, **8**–38, **8**–48

tellurides, **8**–12 to 8-13, **8**–48

transformation temperatures, **8**–2 to 8-8, **14**–29, **15**–38 to **15**–44, **29**–6 to **29**–7

vapour pressures, **8**–51 to **8**–58

volume change on melting, **8**–2 to **8**–8

Thermocouple(s), *see* Electromotive force gauges, vacuum, *see* Vacuum technology

reference tables, **16**–4 to **16**–9

type B (Pt-30%Rh/Pt-6%Rh), **16**–6 to **16**–7

type C (W-5%Re/W-25%Re), **16**–9

type E (chrome-constantan), **16**–7

type J (iron-constantan), **16**–7 to **16**–8

type K (chromel-alumel), **16**–8

type N (nicrosil-nisil), **16**–9

type R (Pt-13%Rh/Pt), **16**–6

type S (Pt-10%Rh/Pt), **16**–6

type T (copper-constantan), **16**–8

Thermodynamic(s):

carburising and carbonitriding, basis of, **29**–42 to **29**–43, **29**–59 to **29**–60

data, *see* Thermochemistry

modelling of, **39**–3 to **39**–5

Thermoelectric materials, **16**–10 to **16**–12, *see also* Thermocouple(s)

Thin film(s):

deposition, *see* Coating and metal finishing

diffusion, **13**–6 to **13**–7

nanoscale, **38**–25 to **38**–26

Thorium, *see* Elements

characteristic X-ray emission lines and adsorption edges, **4**–6

excitation potentials for characteristic X-rays, **4**–5

ores, **7**–8

phase diagrams, Indexed on pages **11**–1 to **11**–6

Thulium, *see* Elements

characteristic X-ray emission lines and adsorption edges, **4**–6

excitation potentials for characteristic X-rays, **4**–4

Time temperature transformation (TTT) diagrams, *see* Isothermal transformation diagrams

Tin and tin alloys, *see* Elements

characteristic X-ray emission lines and absorption edges, **4**–6

creep properties, **22**–96

density, **22**–97

excitation potentials for characteristic X-rays, **4**–4

hardness, **22**–96 to **22**–97

impact strength, **22**–96, *see also* Metallography

ores, **7**–8

Tin and tin alloys (*cont.*)
   physical properties, **14**–29 to **14**–30,
      22–97
   phase diagrams, Indexed on pages **11**–1 to
      11–6
   tensile properties, **22**–96 to **22**–97
Titanium and titanium alloys, *see* Elements
   aluminide structural intermetallics, **38**–14 to
      38–16
   characteristic X-ray emission lines and
      absorption edges, **4**–5
   composites, **37**–5
   creep properties, **22**–89 to **22**–90
   damping, **15**–9
   density, **14**–28
   dislocation glide elements, **4**–39
   elastic modulus, **14**–28
   electrical resistivity, **14**–28
   excitation potentials for characteristic
      X-rays, **4**–4
   fatigue properties, **22**–90 to **22**–92
   impact strength, **22**–93
   metallography, **10**–64 to **10**–67
   ores, **7**–9
   phase diagrams, Indexed on pages **11**–1 to
      11–6
   physical properties, **14**–28
   tensile properties:
      elevated and low temperature, **22**–85 to
         22–88
      room temperature, **22**–83 to **22**–84
   texture, **4**–42
   thermal:
      conductivity, **14**–28
      expansion coefficient, **14**–28
   welding, **33**–40
TMP (turbomolecular pump), *see* Vacuum
   technology
Tool steels, *see* Steel(s)
Town gas, *see* Fuels
Transformation:
   phase:
      thermodynamic data on, *see*
         Thermochemistry
   temperature:
      thermodynamic data on, *see*
         Thermochemistry
Translation groups, *see* Crystallography
Transmission electron microscopy, *see*
   Electron microscopy
Trigonometry, **2**–19 to **2**–21
TTT diagrams, *see* Isothermal transformation
   diagrams
Tungsten, *see* Elements
   carbide, *see* Carbides
   characteristic X-ray emission lines and
      absorption edges, **4**–6
   excitation potentials for characteristic
      X-rays, **4**–4
   field (electron) emission, **18**–8 to **18**–9
   metallography, **10**–66 to **10**–67
   phase diagrams, Indexed on pages **11**–1 to
      11–6

powder metallurgy alloys, **23**–24, **23**–36
   ores
Turbomolecular pumps (TMP), *see* Vacuum
   technology
Turning, **30**–1
Twin(ning) elements (planes/directions),
   table of, **4**–40
Type, thermocouple, *see* Thermocouples

Udimet alloys, *see* Nickel and nickel alloys
UHV (ultrahigh vacuum), *see*
   Vacuum technology
Ultrahigh vacuum (UHV), *see*
   Vacuum technology
Ultrasonic, *see* Non-destructive testing and
   evaluation
   machining, **30**–11
Unified numbering system (UNS), *see* related
   designations for alloys, tables of
   description of, **1**–1
Unit(s):
   cells, *see* Crystal structure
   SI, table of, **2**–2 to **2**–4, *see also* Conversion
      factors
United States:
   alloy designations, *see* Related designations
      for alloys, tables of
   customary units ("English units"), SI
      conversion factors for, **2**–4 to **2**–10
UNS (unified numbering system), *see* Unified
   numbering system
Uranium and uranium alloys, *see* Elements
   characteristic X-ray emission lines and
      absorption edges, **4**–6
   excitation potentials for characteristic
      X-rays, **4**–5
   metallography, **10**–67 to **10**–68
   ores, **7**–9
   phase diagrams, Indexed on pages **11**–1 to
      11–6
   welding, **33**–40
US:
   alloy designations, *see* Related designations
      for alloys, tables of
   customary units ("English units"), SI
      conversion factors for, **2**–4 to **2**–10
Unshaped ceramics, introduction to, **27**–7

Vanadium, *see* Elements
   characteristic X-ray emission lines and
      absorption edges, **4**–5
   excitation potentials for characteristic
      X-rays, **4**–4
   ores, **7**–9
   phase diagrams, Indexed on pages **11**–1 to
      11–6
Vacuum:
   furnaces, **40**–7 to **40**–8
   plasma spraying (VPS), *see* Thermal
      spraying
   technology, **40**–8 to **40**–16
      bibliography, **40**–16
      instrumentation, **40**–9 to **40**–12

pumps and systems, **40**–12 to **40**–15
residual gas analysis, **40**–15
safety, **40**–16
units of pressure, **40**–9
Vapour:
  deposition, *see* Chemical vapour deposition;
    Physical vapour deposition
  pressures, data on, *see* Thermochemistry
Vickers hardness, *see* Hardness
Viscosity:
  liquid metals, data for, **14**–10 to **14**–11
  molten salts, data for, *see* Salts
  unit conversion factors, **2**–9
Volume change on melting, data on, **8**–2 to **8**–8
VPS (vacuum plasma spraying), *see* Thermal
  spraying

Waspaloy, *see* Nickel and nickel alloys
Waterjet and abrasive jet cutting, **30**–10
Wavelength:
  dispersive X-ray spectroscopy (EDS, EDX,
    EDXA), *see* Electron microscopy
  X-ray emission and absorption, *see* Element
    of interest; X-ray
Wear:
  abrasive, **25**–12 to **25**–13
    rates, tables of, **25**–14 to **25**–17
  adhesive, **25**–13 to **25**–14
    rates, tables of, **25**–21
  cavitation, **25**–14
    rates, tables of, **25**–24 to **25**–25
  corrosive, **25**–15, *see also* Corrosion
  erosive, **25**–14
    rates, tables of, **25**–23 to **25**–25
  fretting, **25**–15
  materials for wear resistance, **25**–14 to
    25–16
Weights, atomic, *see* Atomic weights
Weld decay, *see* Corrosion
Welding, *see* Joining
Work function, *see* Electron emission
Working, metal, *see* Metalworking

Xenon, *see* Elements; Noble gases
X-ray:
  crystals to produce monochromatic
    X-rays, **4**–10
  diffraction techniques:
    general, **4**–11 to **4**–13
    quantitative analysis, **4**–14
    phase identification, **4**–14
    preferred orientation (texture)
      measurement, **4**–22 to **4**–23
    residual stress measurement, **4**–18 to
      4–22
    retained austenite determination, **4**–17
  emission lines and absorption edges, table
    of, **4**–5 to **4**–9
  excitation potentials, table of, **4**–4
  fluorescence (XRF), **4**–49
  Lorentz-polarization factors, table of, **4**–17
  mass adsorption coefficients, **4**–32 to **4**–38
  multiplicity factors, table of, **4**–16

scattering factors, atomic, **4**–27 to **4**–31
sources and beta-filters for crystallographic
  work, **4**–3
spectroscopy/analysis, *see* Metallography
XRF (X-ray fluorescence), **4**–49

Young's modulus, *see* Elastic modulus
Ytterbium, *see* Elements
  characteristic X-ray emission lines and
    absorption edges, **4**–6
  excitation potentials for characteristic
    X-rays, **4**–4
  ores, **7**–9
  phase diagrams, Indexed on pages **11**–1 to
    11–6
Yttrium, *see* Elements
  characteristic X-ray emission lines and
    absorption edges, **4**–5
  excitation potentials for characteristic
    X-rays, **4**–5
  ores, **7**–9
  phase diagrams, Indexed on pages **11**–1 to
    11–6

Zinc and zinc alloys, *see* Elements
  cast, composition and properties, **26**–68 to
    26–69
  characteristic X-ray emission lines and
    absorption edges, **4**–5
  composites, **37**–6
  density, **14**–29
  elastic modulus, **14**–29
  electrical resistivity, **14**–29
  excitation potentials for characteristic
    X-rays, **4**–4
  hardness, **22**–93 to **22**–94
  metallography, **10**–68 to **10**–70
  ores, **7**–9
  phase diagrams, Indexed on pages **11**–1 to
    11–6
  physical properties, **14**–29
  tensile properties, **22**–93 to **22**–94
  thermal conductivity, **14**–29
  thermal expansion coefficient, **14**–29
Zirconium and zirconium alloys, *see* Elements
  characteristic X-ray emission lines and
    absorption edges, **4**–5
  density, **14**–29
  elastic modulus, **14**–29
  electrical resistivity, **14**–29
  excitation potentials for characteristic
    X-rays, **4**–5
  hardness, **22**–94 to **22**–95
  metallography, **10**–70 to **10**–71
  ores, **7**–9
  phase diagrams, Indexed on pages **11**–1 to
    11–6
  physical properties, **14**–29
  tensile properties, **22**–94 to **22**–95
  thermal conductivity, **14**–29
  thermal expansion coefficient, **14**–29
  welding, **33**–40